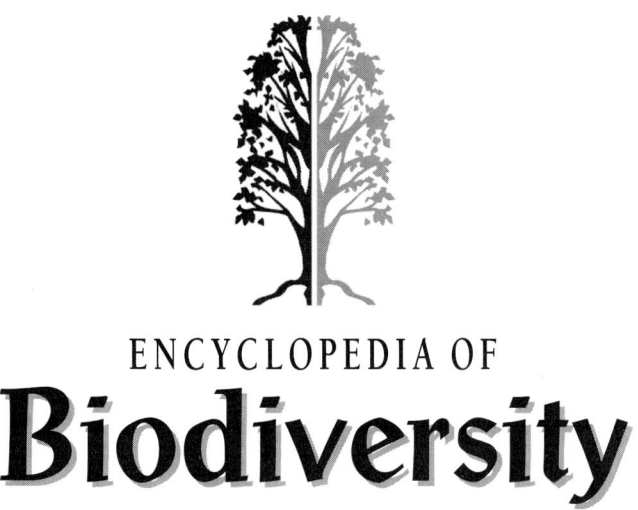

ENCYCLOPEDIA OF
Biodiversity

VOLUME 5

R–Z

EDITORIAL BOARD

Editor-in-Chief
Simon Asher Levin

Associate Editors

Robert Colwell
University of Connecticut
Storrs, Connecticut, USA

Gretchen Daily
Stanford University
Stanford, California, USA

Jane Lubchenco
Oregon State University
Corvallis, Oregon, USA

Harold A. Mooney
Stanford University
Stanford, California, USA

Ernst-Detlef Schulze
Universität Bayreuth
Bayreuth, Germany

G. David Tilman
University of Minnesota
St. Paul, Minnesota, USA

International Editorial Advisors

Dan Cohen
Hebrew University of Jerusalem
Jerusalem, Israel

Rita R. Colwell
National Science Foundation
Arlington, Virginia, USA

Francesco di Castri
National Research Center of France
Montpellier, France

Paul R. Ehrlich
Stanford University
Stanford, California, USA

Thomas Eisner
Cornell University
Ithaca, New York, USA

Niles Eldredge
American Museum of Natural History
New York, New York, USA

Paul Falkowski
Rutgers University
New Brunswick, New Jersey, USA

Tom Fenchel
University of Copenhagen
Helsingoer, Denmark

Diana H. Wall
Colorado State University
Fort Collins, Colorado, USA

Madhav Gadgil
Indian Institute of Science
Bangalore, India

Stephen Jay Gould
Harvard University
Cambridge, Massachusetts, USA

Francesca Grifo
American Museum of Natural History
New York, New York, USA

Masahiko Higashi
Kyoto University (deceased)
Kyoto, Japan

Yoh Iwasa
Kyushu University
Fukuoka, Japan

John H. Lawton
Imperial College at Silwood Park
Ascot, Berks, United Kingdom

Sir Robert May
University of Oxford
Oxford, United Kingdom

Ortwin Meyer
Universität Bayreuth
Bayreuth, Germany

Norman Myers
Consultant in Environment and Development
Headington, Oxford, United Kingdom

Michael J. Novacek
American Museum of Natural History
New York, New York, USA

Sir Ghillean Prance
Royal Botanic Gardens
Richmond, Surrey, United Kingdom

Michael Rosenzweig
University of Arizona
Tucson, Arizona, USA

Nigel Stork
Research Center for Tropical Rainforest
Ecology and Management
Cairns, Queensland, Australia

Monica G. Turner
University of Wisconsin
Madison, Wisconsin, USA

Marvalee H. Wake
University of California, Berkeley
Berkeley, California, USA

Brian H. Walker
Commonwealth Scientific and Industrial
Research Organization
Lyneham, Australia

Edward O. Wilson
Museum of Comparative Zoology
Harvard University
Cambridge, Massachusetts, USA

*Dedicated to the memory of three encyclopedia authors,
Takuya Abe, Masahiko Higashi, and Gary Polis, and
their colleagues Shigeru Nakano and Michael Rose,
who perished March 27, 2000 in a tragic boating
accident while on a research trip in Baja California.
Masahiko Higashi was also a member of the
Board of International Editorial Advisors.*

ENCYCLOPEDIA OF
Biodiversity

VOLUME 5
R–Z

Editor-in-Chief

Simon Asher Levin

Moffett Professor of Biology
Princeton University
Princeton, New Jersey, USA

ACADEMIC PRESS

A Harcourt Science and Technology Company

SAN DIEGO   SAN FRANCISCO   BOSTON   NEW YORK   LONDON   SYDNEY   TOKYO

STAFFORD LIBRARY
COLUMBIA COLLEGE
1001 ROGERS STREET
COLUMBIA, MO 65216

This book is printed on acid-free paper. ∞

Copyright © 2001 by ACADEMIC PRESS

All Rights Reserved.
No part of this publication may be reproduced or transmitted in any form or by any means, electronic or mechanical, including photocopy, recording, or any information storage and retrieval system, without permission in writing from the publisher.

Requests for permission to make copies of any part of the work should be mailed to:
Permissions Department, Harcourt Inc., 6277 Sea Harbor Drive,
Orlando, Florida 32887-6777

Academic Press
A Harcourt Science and Technology Company
525 B Street, Suite 1900, San Diego, California 92101-4495, USA
http://www.academicpress.com

Academic Press
Harcourt Place, 32 Jamestown Road, London NW1 7BY, UK
http://www.academicpress.com

Library of Congress Catalog Card Number: 00-105903

International Standard Book Number: 0-12-226865-2 set
International Standard Book Number: 0-12-226866-0 Volume 1
International Standard Book Number: 0-12-226867-9 Volume 2
International Standard Book Number: 0-12-226868-7 Volume 3
International Standard Book Number: 0-12-226869-5 Volume 4
International Standard Book Number: 0-12-226864-4 Volume 5

PRINTED IN THE UNITED STATES OF AMERICA
00 01 02 03 04 05 MM 9 8 7 6 5 4 3 2 1

Contents

CONTENTS OF OTHER VOLUMES   xi
CONTENTS BY SUBJECT AREA   xix
FOREWORD   xxv
PREFACE   xxvii
GUIDE TO THE ENCYCLOPEDIA   xxix

RAINFOREST ECOSYSTEMS, ANIMAL DIVERSITY   1
Gregory H. Adler

RAINFOREST ECOSYSTEMS, PLANT DIVERSITY   13
Ian Turner

RAINFOREST LOSS AND CHANGE   25
K. D. Singh

RANGE ECOLOGY, GLOBAL LIVESTOCK INFLUENCES   33
J. Boone Kauffman and David Pyke

RECOMBINATION   53
Abraham Korol

REEF ECOSYSTEMS: THREATS TO THEIR BIODIVERSITY   73
James Porter and Jennifer Tougas

REFORESTATION   97
David Lamb

RELIGIOUS TRADITIONS AND BIODIVERSITY   109
Fikret Berkes

REMOTE SENSING AND IMAGE PROCESSING   121
Ronen Kadmon

REPTILES, BIODIVERSITY OF   145
F. Harvey Pough

RESOURCE EXPLOITATION, FISHERIES   161
John Beddington

RESOURCE PARTITIONING   173
Fakhri A. Bazzaz and Sabastain Catovsky

RESTORATION OF ANIMAL, PLANT, AND MICROBIAL DIVERSITY   185
Edith Allen, Michael Allen, and Joel S. Brown

RESTORATION OF BIODIVERSITY, OVERVIEW   203
Joy B. Zedler, Roberto Lindig-Cisneros, Cristina Bonilla-Warford, and Isa Woo

RIVER ECOSYSTEMS   213
Karin Limburg, Dennis P. Swaney, and David L. Strayer

SALMON   233
Michael Schiewe and Peter Kareiva

SCALE, CONCEPT AND EFFECTS OF   245
David Clayton Schneider

SEAGRASSES   255
Carlos Duarte

SLASH AND BURN AGRICULTURE, EFFECTS OF   269
Stefan Hauser and Lindsey Norgrove

SOCIAL AND CULTURAL FACTORS   285
Jeffrey McNeely

SOCIAL BEHAVIOR   295
Daniel I. Rubenstein

SOIL BIOTA, SOIL SYSTEMS, AND PROCESSES   305
David C. Coleman

SOIL CONSERVATION   315
Dorota L. Porazinska and Diana Wall

SOUTH AMERICA, ECOSYSTEMS OF   327
Luis A. Solórzano C.

SOUTH AMERICAN NATURAL ECOSYSTEMS, STATUS OF   345
Philip Fearnside

SOUTHERN (AUSTRAL) ECOSYSTEMS   361
Robert Hill and Peter Weston

SPECIATION, PROCESS OF   371
Guy L. Bush

SPECIATION, THEORIES OF   383
Hope Hollocher

SPECIES-AREA RELATIONSHIPS   397
Edward Connor and Earl D. McCoy

SPECIES COEXISTENCE   413
Robert Holt

SPECIES, CONCEPTS OF   427
James Mallet

SPECIES DIVERSITY, OVERVIEW   441
A. Ross Kiester

SPECIES INTERACTIONS   453
Jessica Hellmann

STABILITY, CONCEPT OF   467
Clarence Lehman

STEWARDSHIP, CONCEPT OF   481
Peter Alpert

STORAGE, ECOLOGY OF   495
Caroline Pond

STRESS, ENVIRONMENTAL   515
John Cairns, Jr.

SUBSPECIES, SEMISPECIES AND SUPERSPECIES   523
James Mallet

SUBTERRANEAN ECOSYSTEMS   527
David Culver

SUCCESSION, PHENOMENON OF   541
H. H. Shugart

SUSTAINABILITY, CONCEPT AND PRACTICE OF   553
Kai N. Lee

SYSTEMATICS, OVERVIEW   569
Quentin Wheeler

TAXONOMY, METHODS OF   589
R. I. Vane-Wright, Ian Kitching, and David Williams

TEMPERATE FORESTS   607
John A. Silander, Jr.

TEMPERATE GRASSLAND AND SHRUBLAND ECOSYSTEMS   627
Osvaldo Sala, Amy T. Austin, and Lucía Vivanco

TERRESTRIAL ECOSYSTEMS   637
Ian Noble and Stephen Roxburgh

THERMOPHILES, ORIGIN OF   647
Anna-Louise Reysenbach and Margaret L. Rising

TIMBER INDUSTRY   655
Seppo Kellomaki, Jari Kouki, Pekka Niemelä, and Heli Peltola

TOURISM, ROLE OF   667
Richard W. Braithwaite

TRADITIONAL CONSERVATION PRACTICES   681
Carl Folke and Johan Colding

TROPHIC LEVELS 695
Peter Yodzis

TROPICAL FOREST ECOSYSTEMS 701
Gary Hartshorn

TRUE BUGS AND THEIR RELATIVES 711
Carl Schaefer

ULTRAVIOLET RADIATION 723
Andrew Blaustein

URBAN/SUBURBAN ECOLOGY 733
Ann P. Kinzig and J. Morgan Grove

VENTS 747
Cindy Lee Van Dover

VERTEBRATES, OVERVIEW 755
Carl Gans and Christopher A. Bell

VICARIANCE BIOGEOGRAPHY 767
Christopher J. Humphries

WETLANDS ECOSYSTEMS 781
Barbara L. Bedford, Donald Leopold, and James Gibbs

WETLANDS RESTORATION 805
Philip Benstead and Paul Jose

WILDLIFE MANAGEMENT 823
David Saltz

WORMS, ANNELIDA 831
Kristian Fauchald

WORMS, NEMATODA 843
Scott L. Gardner and Tom Powers

WORMS, PLATYHELMINTHES 863
Janine N. Caira and Timothy J. Littlewood

ZOOS AND ZOOLOGICAL PARKS 901
Anna Marie Lyles

CONTRIBUTORS 913
GLOSSARY 931
INDEX 1021

Contents of Other Volumes

CONTENTS OF VOLUME 1

ACID RAIN AND DEPOSITION   1
George Hendrey

ADAPTATION   17
Michael R. Rose

ADAPTIVE RADIATION   25
Rosemary G. Gillespie, Francis G. Howarth, and George K. Roderick

AESTHETIC FACTORS   45
Gordon H. Orians

AFRICA, ECOSYSTEMS OF   55
J. Michael Lock

AGRICULTURAL INVASIONS   71
David Pimentel

AGRICULTURE, INDUSTRIALIZED   85
Phrabhu Pingali and Melinda Smale

AGRICULTURE, SUSTAINABLE   99
G. Philip Robertson and Richard R. Harwood

AGRICULTURE, TRADITIONAL   109
Miguel A. Altieri

AIR POLLUTION   119
Michael Ashmore

ALPINE ECOSYSTEMS   133
Christian Körner

AMAZON ECOSYSTEMS   145
Ghillean Prance

AMPHIBIANS, BIODIVERSITY OF   159
Ross A. Alford, Stephen J. Richards, and Keith R. McDonald

ANTARCTIC ECOSYSTEMS   171
Peter Convey

AQUACULTURE   185
Nils Kautsky, Carl Folke, Patrik Rönnbäck, Max Troell, Malcolm Beveridge, and Jurgenne Primavera

ARACHNIDS   199
Jonathan A. Coddington and Robert K. Colwell

ARCHAEA, ORIGIN OF   219
Constantino Vetriani

ARCTIC ECOSYSTEMS   231
Terry V. Callaghan, Nadya Matveyeva, Yuri Chernov, and Rob Brooker

ARTHROPODS (TERRESTRIAL), AMAZONIAN   249
Joachim Adis

ASIA, ECOSYSTEMS OF   261
Elgene O. Box and Kazue Fujiwara

ATMOSPHERIC GASES   293
Donald J. Wuebbles

AUSTRALIA, ECOSYSTEMS OF   307
Raymond L. Specht and Alison Specht

BACTERIAL BIODIVERSITY   325
Erko Stackebrandt

BACTERIAL GENETICS   339
Michael Travisano

BEETLES   351
Henry Hespenheide

BIODIVERSITY AS A COMMODITY   359
Geoffrey Heal

BIODIVERSITY, DEFINITION OF   377
Ian R. Swingland

BIODIVERSITY, EVOLUTION AND   393
Gregg Hartvigsen

BIODIVERSITY GENERATION, OVERVIEW   403
Paul H. Harvey

BIODIVERSITY, ORIGIN OF   411
Mark E. J. Newman and G. J. Eble

BIODIVERSITY-RICH COUNTRIES   419
José A. Sarukhán and Rodolfo Dirzo

BIOGEOCHEMICAL CYCLES   437
Paul G. Falkowski

BIOGEOGRAPHY, OVERVIEW   455
Mark Lomolino

BIOPROSPECTING   471
Nicolás Mateo, Werner Nader, and Giselle Tamayo

BIRDS, BIODIVERSITY OF   489
Jeremy J. D. Greenwood

BOREAL FOREST ECOSYSTEMS   521
Roy Turkington

BREEDING OF ANIMALS   533
David R. Notter

BREEDING OF PLANTS   547
Donald Duvick

BUTTERFLIES   559
Philip J. DeVries

C_4 PLANTS   575
Rowan F. Sage

CAPTIVE BREEDING AND REINTRODUCTION   599
Katherine Ralls and Robin Meadows

CARBON CYCLE   609
John Grace

CARNIVORES   629
Hans Kruuk

CARRYING CAPACITY, CONCEPT OF   641
Gregg Hartvigsen

CATTLE, SHEEP, AND GOATS, ECOLOGICAL ROLE OF   651
Andreas Troumbis

CENTRAL AMERICA, ECOSYSTEMS OF   665
Rodolfo Dirzo

CLADISTICS   677
Ian J. Kitching, Peter L. Forey, and David M. Williams

CLADOGENESIS   693
Christopher J. Humphries

CLIMATE CHANGE AND ECOLOGY, SYNERGISM OF   709
Stephen H. Schneider and Terry L. Root

CLIMATE, EFFECTS OF   727
F. Ian Woodward

COASTAL BEACH ECOSYSTEMS   741
Anton McLachlan

COEVOLUTION   753
Douglas J. Futuyma and André Levy

COMMONS, CONCEPT AND THEORY OF   769
Colin W. Clark

COMMONS, INSTITUTIONAL DIVERSITY OF   777
Elinor Ostrom

COMPETITION, INTERSPECIFIC   793
Bryan Shorrocks

COMPLEMENTARITY   813
Paul Williams

COMPLEXITY VERSUS DIVERSITY   831
Shahid Naeem

COMPUTER SYSTEMS AND MODELS, USE OF   845
Louis J. Gross

CONSERVATION BIOLOGY, DISCIPLINE OF   855
Andrew P. Dobson and Jon Paul Rodriguez

CONSERVATION EFFORTS, CONTEMPORARY   865
Kristiina Vogt, Oswald J. Schmitz, Karen H. Beard, Jennifer L. O'Hara, and Michael G. Booth

CONSERVATION MOVEMENT, HISTORICAL   883
Curt Meine

CROP IMPROVEMENT AND BIODIVERSITY   897
Giorgini Augusto Venturieri

CRUSTACEANS   915
Marjorie L. Reaka-Kudla

CONTENTS OF VOLUME 2

DARWIN, CHARLES   1
Michael J. Ghiselin

DEFENSES, ECOLOGY OF   11
Phyllis D. Coley and John A. Barone

DEFORESTATION AND LAND CLEARING   23
Jaboury Ghazoul and Julian Evans

DESERT ECOSYSTEMS   37
James A. MacMahon

DESERTIFICATION   61
James F. Reynolds

DIAPAUSE AND DORMANCY   79
Nelson G. Hairston, Jr.

DIFFERENTIATION   85
Nicholas H. Barton

DINOSAURS, EXTINCTION THEORIES FOR   95
J. David Archibald

DISEASES, CONSERVATION AND   109
Sonia Altizer, Johannes Foufopoulos, and Andrea Gager

DISPERSAL BIOGEOGRAPHY   127
Ran Nathan

DISTURBANCE, MECHANISMS OF   153
Frank Davis and Max Moritz

DIVERSITY, COMMUNITY/ REGIONAL LEVEL   161
Howard V. Cornell

DIVERSITY, MOLECULAR LEVEL   179
Carlos Machado and Marcos Antezana

DIVERSITY, ORGANISM LEVEL   191
Daniel R. Brooks

DIVERSITY, TAXONOMIC VERSUS FUNCTIONAL   205
John C. Moore

DOMESTICATION OF CROP PLANTS   217
Daniel Zohary

ECOLOGICAL FOOTPRINT, CONCEPT OF   229
William Rees

ECOLOGICAL GENETICS   245
Beate Nürnberger

ECOLOGY, CONCEPTS AND THEORIES IN   259
Peter Kareiva and Michelle Marvier

ECOLOGY OF AGRICULTURE   269
Alison G. Power

ECONOMIC GROWTH AND THE ENVIRONMENT   277
Karl-Göran Mäler

ECONOMIC VALUE OF BIODIVERSITY, MEASUREMENTS OF   285
Robert Mendelsohn

ECONOMIC VALUE OF BIODIVERSITY, OVERVIEW   291
Partha Dasgupta

ECOSYSTEM, CONCEPT OF   305
Eugene P. Odum

ECOSYSTEM FUNCTION MEASUREMENT, AQUATIC AND MARINE COMMUNITIES   311
John Lehman

ECOSYSTEM FUNCTION MEASUREMENT, TERRESTRIAL COMMUNITIES   321
Sandra Diaz

ECOSYSTEM FUNCTION, PRINCIPLES OF   345
Ross A. Virginia and Diana Wall

ECOSYSTEM SERVICES, CONCEPT OF   353
Gretchen Daily and Shamik Dasgupta

ECOTOXICOLOGY   363
J. M. Lynch, A. Wiseman, and F. A. A. M. De Leij

EDIBLE PLANTS   375
Eduardo H. Rapoport and Barbara S. Drausal

EDUCATION AND BIODIVERSITY   383
Shirley Malcom

ENDANGERED BIRDS   395
Nigel Collar

ENDANGERED ECOSYSTEMS   407
Raymond C. Nias

ENDANGERED FRESHWATER INVERTEBRATES   425
David L. Strayer

ENDANGERED MAMMALS 441
Peter Zahler

ENDANGERED MARINE INVERTEBRATES 455
James T. Carlton

ENDANGERED PLANTS 465
Thomas J. Stohlgren

ENDANGERED REPTILES AND AMPHIBIANS 479
Tim Halliday

ENDANGERED TERRESTRIAL
INVERTEBRATES 487
Mark Deyrup

ENDEMISM 497
R. M. Cowling

ENERGY FLOW AND ECOSYSTEMS 509
Alan P. Covich

ENERGY USE, HUMAN 525
Patrick Gonzalez

ENVIRONMENTAL ETHICS 545
Richard Primack and Philip Cafaro

ENVIRONMENTAL IMPACT, CONCEPT AND
MEASUREMENT OF 557
Ellen W. Chu and James R. Karr

ESTUARINE ECOSYSTEMS 579
G. Carleton Ray

ETHICAL ISSUES IN BIODIVERSITY
PROTECTION 593
Philip Cafaro and Richard Primack

ETHNOBIOLOGY AND ETHNOECOLOGY 609
Gary Martin

EUKARYOTES, ORIGIN OF 623
Dorion Sagan and Lynn Margulis

EUROPE, ECOSYSTEMS OF 635
Ladislav Mucina

EUTROPHICATION AND
OLIGOTROPHICATION 649
JoAnn M. Burkholder

EVOLUTION, THEORY OF 671
Catherine Craig

EX SITU, IN SITU CONSERVATION 683
Nigel Maxted

EXTINCTION, CAUSES OF 697
Richard Primack

EXTINCTION, RATES OF 715
Jeffrey S. Levinton

EXTINCTIONS, MODERN EXAMPLES OF 731
Gábor Lövei

FIRES, ECOLOGICAL EFFECTS OF 745
William Bond

FISH, BIODIVERSITY OF 755
Gene S. Helfman

FISH CONSERVATION 783
Carl Safina

FISH STOCKS 801
Daniel Pauly and Rainer Froese

FLIES, GNATS, AND MOSQUITOES 815
Brian Brown

CONTENTS OF VOLUME 3

FOOD WEBS 1
Gary Huxel and Gary Polis

FOREST CANOPIES, ANIMAL DIVERSITY 19
Terry L. Erwin

FOREST CANOPIES, PLANT DIVERSITY 27
Nalini Nadkarni, Mark Merwin, and Jurgen Nieder

FOREST ECOLOGY 41
Timothy J. Fahey

FOSSIL RECORD 53
Sean Connolly

FRAMEWORK FOR ASSESSMENT AND MONITORING
OF BIODIVERSITY 63
James A. Comiskey, Francisco Dallmeier, and
Alfonso Alonso

FRESHWATER ECOSYSTEMS 75
Robert G. Wetzel

FRESHWATER ECOSYSTEMS, HUMAN
IMPACT ON 89
Kaj Sand-Jensen

FUNCTIONAL DIVERSITY   109
David Tilman

FUNCTIONAL GROUPS   121
Robert Steneck

FUNGI   141
Thomas J. Volk

GENE BANKS   165
Simon Linington and Hugh Pritchard

GENES, DESCRIPTION OF   183
Michael Antolin and William C. Black III

GENETIC DIVERSITY   195
Eviatar Nevo

GEOLOGIC TIME, HISTORY OF
BIODIVERSITY IN   215
James W. Valentine

GOVERNMENT LEGISLATION AND
REGULATION   233
Kathryn Saterson

GRASSHOPPERS AND THEIR RELATIVES   247
Piotr Naskrecki

GRAZING, EFFECTS OF   265
Mark Hay and Cynthia Kicklighter

GREENHOUSE EFFECT   277
Jennifer Dunne and John Harte

GUILDS   295
Richard B. Root

HABITAT AND NICHE, CONCEPT OF   303
Kenneth Petren

HEMIPARASITISM   317
David L. Smith, Todd J. Barkman, and
Claude W. dePamphilis

HERBACEOUS VEGETATION, SPECIES
RICHNESS IN   329
J. P. Grime

HERBICIDES   339
Jodie S. Holt

HIGH-TEMPERATURE ECOSYSTEMS   349
Richard Weigert

HISTORICAL AWARENESS OF
BIODIVERSITY   363
David Takacs

HOTSPOTS   371
Norman Myers

HUMAN EFFECTS ON ECOSYSTEMS,
OVERVIEW   383
Paul Ehrlich and Claire Kremen

HUMAN IMPACT ON BIODIVERSITY,
OVERVIEW   395
Leslie Sponsel

HUNTER-GATHERER SOCIETIES, ECOLOGICAL
IMPACT OF   411
Kathleen Galvin

HYMENOPTERA   417
Norman F. Johnson

INBREEDING AND OUTBREEDING   427
Katherine S. Ralls, Richard Frankham, and
Jonathon Ballou

INDICATOR SPECIES   437
John H. Lawton and Kevin Gaston

INDIGENOUS PEOPLES, BIODIVERSITY AND   451
Victor M. Toledo

INSECTICIDE RESISTANCE   465
Ian Denholm and Greg Devine

INSECTS, OVERVIEW   479
Brian V. Brown

INTERTIDAL ECOSYSTEMS   485
Antony Underwood and M. G. Chapman

INTRODUCED PLANTS, NEGATIVE
EFFECTS OF   501
William G. Lee

INTRODUCED SPECIES, EFFECT AND
DISTRIBUTION   517
Daniel Simberloff

INVERTEBRATES, FRESHWATER,
OVERVIEW   531
Margaret Palmer and P. Sam Lake

INVERTEBRATES, MARINE, OVERVIEW   543
John Lambshead and Peter Schalk

INVERTEBRATES, TERRESTRIAL,
OVERVIEW   561
Olof Andrén

ISLAND BIOGEOGRAPHY 565
Dieter Mueller-Dombois

ISOPTERA 581
Takuya Abe and Masahiko Higashi

KEYSTONE SPECIES 613
Bruce Menge and Tess Freidenburg

LAKE AND POND ECOSYSTEMS 633
Christian Leveque

LANDSCAPE DIVERSITY 645
Debra P. Coffin and Sarah C. Golslee

LAND-USE ISSUES 659
John Marzluff and Nathalie Hamel

LAND-USE PATTERNS, HISTORIC 675
Oliver Rackham

LATENT EXTINCTIONS—THE LIVING DEAD 689
Daniel H. Janzen

LATITUDE, COMMON TRENDS WITHIN 701
Michael Willig

LIFE HISTORY, EVOLUTION OF 715
Derek Roff

LIMITS TO BIODIVERSITY (SPECIES PACKING) 729
Larry Slobodkin

LITERARY PERSPECTIVES ON BIODIVERSITY 739
William Howarth

LOGGED FORESTS 747
Reinmar Seidler and Kamaljit Bawa

LOSS OF BIODIVERSITY, OVERVIEW 761
Robert Barbault

MAMMALS, BIODIVERSITY OF 777
Joshua Ginsberg

MAMMALS, CONSERVATION EFFORTS FOR 811
E. J. Milner-Gulland and R. Woodroffe

MAMMALS (LATE QUATERNARY), EXTINCTIONS OF 825
Paul S. Martin

MAMMALS (PRE-QUATERNARY), EXTINCTIONS OF 841
William Clemens

MANGROVE ECOSYSTEMS 853
Peter Hogarth

CONTENTS OF VOLUME 4

MARINE AND AQUATIC COMMUNITIES, STRESS FROM EUTROPHICATION 1
Jonathan Sharp

MARINE ECOSYSTEMS 13
J. Frederick Grassle

MARINE ECOSYSTEMS, HUMAN IMPACTS ON 27
Juan C. Castilla

MARINE MAMMALS, EXTINCTIONS OF 37
Glenn VanBlaricom, Leah Gerber, and Robert Brownell

MARINE SEDIMENTS 71
Paul Snelgrove

MARKET ECONOMY AND BIODIVERSITY 85
R. David Simpson and Pamela Jagger

MASS EXTINCTIONS, CONCEPT OF 97
John Sepkoski, Jr.

MASS EXTINCTIONS, NOTABLE EXAMPLES OF 111
Douglas Erwin

MEASUREMENT AND ANALYSIS OF BIODIVERSITY 123
Wade Leitner and Will Turner

MEDITERRANEAN-CLIMATE ECOSYSTEMS 145
Philip W. Rundel

METAPOPULATIONS 161
Peter Chesson

MICROBIAL BIODIVERSITY, MEASUREMENT OF 177
Kate M. Scow, Egbert Schwartz, Mara J. Johnson, and Jennifer L. Macalady

MICROBIAL DIVERSITY 191
Paul V. Dunlap

MICROORGANISMS, ROLE OF 201
Tom Fenchel

MIGRATION 221
Mace A. Hack and Daniel Rubenstein

MOLLUSCS 235
David R. Lindberg

MOTHS 249
David Wagner

MUSEUMS AND INSTITUTIONS 271
Paul Henderson and Neil Chalmers

MUTUALISM, EVOLUTION OF 281
Egbert Giles Leigh, Jr.

MYRIAPODS 291
Alessandro Minelli and Sergei I. Golovatch

NATURAL EXTINCTIONS (NOT HUMAN-INFLUENCED) 305
Christopher Johnson

NATURAL RESERVES AND PRESERVES 317
Alexander Glazer

NEAR EAST ECOSYSTEMS, ANIMAL DIVERSITY 329
Joseph Heller

NEAR EAST ECOSYSTEMS, PLANT DIVERSITY 353
Avinoam Danin

NEST PARASITISM 365
Scott Robinson and Stephen Rothstein

NITROGEN, NITROGEN CYCLE 377
Robert W. Howarth and Sandy Tartowski

NOMENCLATURE, SYSTEMS OF 389
David Hawksworth

NORTH AMERICA, PATTERNS OF BIODIVERSITY IN 403
Martin Lechowicz

NUCLEIC ACID BIODIVERSITY 415
Tamara L. Horton and Laura F. Landweber

OCEAN ECOSYSTEMS 427
Richard T. Barber

ORIGIN OF LIFE, THEORIES OF 439
Susanne Brakmann

PALEOECOLOGY 451
Thompson Webb III

PARASITISM 463
Klaus Rohde

PARASITOIDS 485
Charles Godfray

PELAGIC ECOSYSTEMS 497
Andrea Belgrano, Sonia D. Batten, and Philip C. Reid

PESTICIDES, USE AND EFFECTS OF 509
Paul C. Jepson

PHARMACOLOGY, BIODIVERSITY AND 523
Paul Alan Cox

PHENOTYPE, A HISTORICAL PERSPECTIVE 537
R. J. Berry

PHOTOSYNTHESIS, MECHANISMS OF 549
John A. Raven

PHYLOGENY 559
Kevin Nixon

PLANKTON, STATUS AND ROLE OF 569
Colin S. Reynolds

PLANT–ANIMAL INTERACTIONS 601
Ellen Simms

PLANT BIODIVERSITY, OVERVIEW 621
Jeannette Whitton and Nishanta Rajakaruna

PLANT COMMUNITIES, EVOLUTION OF 631
Karl J. Niklas, Bruce Tiffney, Brian Enquist, and John Haskell

PLANT CONSERVATION, OVERVIEW 645
Mike Maunder

PLANT HYBRIDS 659
Robert S. Fritz

PLANT INVASIONS 677
David M. Richardson

PLANT–SOIL INTERACTIONS 689
Joan Ehrenfeld

PLANT SOURCES OF DRUGS AND CHEMICALS 711
William H. Gerwick, Brian Marquez, Ken Milligan, Lik Tong Tan, and Thomas Williamson

POLLINATORS, ROLE OF 723
David Inouye

POLLUTION, OVERVIEW 731
William H. Smith

POPULATION DENSITY 745
Brian McArdle

POPULATION DIVERSITY, OVERVIEW   759
Jennifer B. Hughes

POPULATION DYNAMICS   769
Alan M. Hastings

POPULATION GENETICS   777
Brian Charlesworth

POPULATIONS, SPECIES, AND CONSERVATION GENETICS   799
David S. Woodruff

POPULATION STABILIZATION, HUMAN   819
Alene Gelbard

POPULATION VIABILITY ANALYSIS   831
Hugh Possingham, David B. Lindenmayer, and Michael A. McCarthy

POVERTY AND BIODIVERSITY   845
Madhav Gadgil

PREDATORS, ECOLOGICAL ROLE OF   857
James Estes, Kevin Crooks, and Robert Holt

PRIMATE POPULATIONS, CONSERVATION OF   879
Russell A. Mittermeier and William Konstant

PROPERTY RIGHTS AND BIODIVERSITY   891
Susan Hanna

PROTOZOA   901
Bland J. Finlay

PSYCHROPHILES, ORIGIN OF   917
Richard Morita and Craig L. Moyer

Contents by Subject Area

AGRICULTURE

AGRICULTURAL INVASIONS
AGRICULTURE, INDUSTRIALIZED
AGRICULTURE, SUSTAINABLE
AGRICULTURE, TRADITIONAL
AQUACULTURE
BREEDING OF ANIMALS
BREEDING OF PLANTS
CATTLE, SHEEP, AND GOATS, ECOLOGICAL ROLE OF
CROP IMPROVEMENT AND BIODIVERSITY
DOMESTICATION OF CROP PLANTS
ECOLOGY OF AGRICULTURE
HERBICIDES
PESTICIDES, USE AND EFFECTS OF
RANGE ECOLOGY, GLOBAL LIVESTOCK INFLUENCES
RESOURCE EXPLOITATION, FISHERIES
SLASH AND BURN AGRICULTURE, EFFECTS OF
TIMBER INDUSTRY

CONSERVATION AND RESTORATION

CAPTIVE BREEDING AND REINTRODUCTION
CONSERVATION BIOLOGY, DISCIPLINE OF
CONSERVATION EFFORTS, CONTEMPORARY
CONSERVATION MOVEMENT, HISTORICAL
DISEASES, CONSERVATION AND
EX SITU, IN SITU CONSERVATION
FISH CONSERVATION
MAMMALS, CONSERVATION EFFORTS FOR
PLANT CONSERVATION, OVERVIEW
PRIMATE POPULATIONS, CONSERVATION OF
REFORESTATION
RESTORATION OF ANIMAL, PLANT, AND MICROBIAL DIVERSITY
RESTORATION OF BIODIVERSITY, OVERVIEW
SOIL CONSERVATION
SUSTAINABILITY, CONCEPT AND PRACTICE OF
TRADITIONAL CONSERVATION PRACTICES
WETLANDS RESTORATION
ZOOS AND ZOOLOGICAL PARKS

ECONOMICS OF BIODIVERSITY

BIODIVERSITY AS A COMMODITY
ECONOMIC GROWTH AND THE ENVIRONMENT
ECONOMIC VALUE OF BIODIVERSITY, MEASUREMENTS OF
ECONOMIC VALUE OF BIODIVERSITY, OVERVIEW
ECOSYSTEM SERVICES, CONCEPT OF
LAND-USE ISSUES
MARKET ECONOMY AND BIODIVERSITY
PHARMACOLOGY, BIODIVERSITY AND
PLANT SOURCES OF DRUGS AND CHEMICALS
POVERTY AND BIODIVERSITY
PROPERTY RIGHTS AND BIODIVERSITY
TIMBER INDUSTRY
TOURISM, ROLE OF

ENVIRONMENTAL CONDITIONS AND EFFECTS

ACID RAIN AND DEPOSITION
AIR POLLUTION
ATMOSPHERIC GASES
CARBON CYCLE
CLIMATE CHANGE AND ECOLOGY, SYNERGISM OF
CLIMATE, EFFECTS OF
DESERTIFICATION
DISTURBANCE, MECHANISMS OF
ECONOMIC GROWTH AND THE ENVIRONMENT
ECOTOXICOLOGY
ENERGY FLOW AND ECOSYSTEMS
ENERGY USE, HUMAN
ENVIRONMENTAL ETHICS
EUTROPHICATION AND OLIGOTROPHICATION
FIRES, ECOLOGICAL EFFECTS OF
GRAZING, EFFECTS OF
GREENHOUSE EFFECT
INSECTICIDE RESISTANCE
MARINE AND AQUATIC COMMUNITIES, STRESS FROM EUTROPHICATION
NITROGEN, NITROGEN CYCLE
PESTICIDES, USE AND EFFECTS OF
POLLUTION, OVERVIEW
SOIL BIOTA, SOIL SYSTEMS, AND PROCESSES
STRESS, ENVIRONMENTAL
ULTRAVIOLET RADIATION

EVOLUTION

ADAPTATION
ADAPTIVE RADIATION
BIODIVERSITY, EVOLUTION AND
BIODIVERSITY GENERATION, OVERVIEW
BIODIVERSITY, ORIGIN OF
COEVOLUTION
COMPLEMENTARITY
DARWIN, CHARLES
DEFENSES, ECOLOGY OF
DIAPAUSE AND DORMANCY
EUKARYOTES, ORIGIN OF
EVOLUTION, THEORY OF
FOSSIL RECORD
GEOLOGIC TIME, HISTORY OF BIODIVERSITY IN
LIFE HISTORY, EVOLUTION OF
MUTUALISM, EVOLUTION OF
PHYLOGENY
PLANT COMMUNITIES, EVOLUTION OF

EXTINCTIONS

DINOSAURS, EXTINCTION THEORIES FOR
ENDANGERED BIRDS
ENDANGERED ECOSYSTEMS
ENDANGERED FRESHWATER INVERTEBRATES
ENDANGERED MAMMALS
ENDANGERED MARINE INVERTEBRATES
ENDANGERED PLANTS
ENDANGERED REPTILES AND AMPHIBIANS
ENDANGERED TERRESTRIAL INVERTEBRATES
EXTINCTION, CAUSES OF
EXTINCTION, RATES OF
EXTINCTIONS, MODERN EXAMPLES OF
HUMAN EFFECTS ON ECOSYSTEMS, OVERVIEW
LATENT EXTINCTIONS: THE LIVING DEAD
LOSS OF BIODIVERSITY, OVERVIEW
MAMMALS (LATE QUATERNARY), EXTINCTIONS OF
MAMMALS (PRE-QUATERNARY), EXTINCTIONS OF
MARINE MAMMALS, EXTINCTIONS OF
MASS EXTINCTIONS, CONCEPT OF
MASS EXTINCTIONS, NOTABLE EXAMPLES OF
NATURAL EXTINCTIONS (NOT HUMAN-INFLUENCED)

GENETICS

BACTERIAL GENETICS
ECOLOGICAL GENETICS
GENE BANKS
GENES, DESCRIPTION OF

GENETIC DIVERSITY
INBREEDING AND OUTBREEDING
NUCLEIC ACID BIODIVERSITY
PHENOTYPE, A HISTORICAL PERSPECTIVE
POPULATION GENETICS
RECOMBINATION

GEOGRAPHIC AND GLOBAL ISSUES

BIODIVERSITY-RICH COUNTRIES
BIOGEOCHEMICAL CYCLES
BIOGEOGRAPHY, OVERVIEW
CLIMATE CHANGE AND ECOLOGY, SYNERGISM OF
DISPERSAL BIOGEOGRAPHY
DIVERSITY, COMMUNITY/REGIONAL LEVEL
ENDEMISM
ENERGY USE, HUMAN
HOTSPOTS
INTRODUCED PLANTS, NEGATIVE EFFECTS OF
ISLAND BIOGEOGRAPHY
LATITUDE, COMMON TRENDS WITHIN
MIGRATION
RAINFOREST LOSS AND CHANGE
VICARIANCE BIOGEOGRAPHY

HABITATS AND ECOSYSTEMS

AFRICA, ECOSYSTEMS OF
ALPINE ECOSYSTEMS
AMAZON ECOSYSTEMS
ANTARCTIC ECOSYSTEMS
ARCTIC ECOSYSTEMS
ASIA, ECOSYSTEMS OF
AUSTRALIA, ECOSYSTEMS OF
BOREAL FOREST ECOSYSTEMS
CENTRAL AMERICA, ECOSYSTEMS OF
COASTAL BEACH ECOSYSTEMS
DESERT ECOSYSTEMS
ECOSYSTEM, CONCEPT OF

ECOSYSTEM FUNCTION, PRINCIPLES OF
ENDANGERED ECOSYSTEMS
ESTUARINE ECOSYSTEMS
EUROPE, ECOSYSTEMS OF
FRESHWATER ECOSYSTEMS
FRESHWATER ECOSYSTEMS, HUMAN IMPACT ON
HABITAT AND NICHE, CONCEPT OF
HIGH-TEMPERATURE ECOSYSTEMS
HUMAN EFFECTS ON ECOSYSTEMS, OVERVIEW
INTERTIDAL ECOSYSTEMS
LAKE AND POND ECOSYSTEMS
LOGGED FORESTS
MANGROVE ECOSYSTEMS
MARINE ECOSYSTEMS
MARINE ECOSYSTEMS, HUMAN IMPACTS ON
MARINE SEDIMENTS
MEDITERRANEAN-CLIMATE ECOSYSTEMS
NEAR EAST ECOSYSTEMS, ANIMAL DIVERSITY
NEAR EAST ECOSYSTEMS, PLANT DIVERSITY
NORTH AMERICA, PATTERNS OF BIODIVERSITY IN
OCEAN ECOSYSTEMS
PELAGIC ECOSYSTEMS
RAINFOREST ECOSYSTEMS, ANIMAL DIVERSITY
RAINFOREST ECOSYSTEMS, PLANT DIVERSITY
REEF ECOSYSTEMS
RIVER ECOSYSTEMS
SEAGRASSES
SOUTH AMERICA, ECOSYSTEMS OF
SOUTHERN (AUSTRAL) ECOSYSTEMS
SUBTERRANEAN ECOSYSTEMS
TEMPERATE FORESTS
TEMPERATE GRASSLAND AND SHRUBLAND ECOSYSTEMS
TERRESTRIAL ECOSYSTEMS
TROPICAL FOREST ECOSYSTEMS
URBAN/SUBURBAN ECOLOGY
VENTS
WETLANDS ECOSYSTEMS

HUMAN EFFECTS AND INTERVENTIONS

ACID RAIN AND DEPOSITION
DEFORESTATION AND LAND CLEARING
ECOLOGICAL FOOTPRINT, CONCEPT OF
ENERGY USE, HUMAN
ETHNOBIOLOGY AND ETHNOECOLOGY
EXTINCTIONS, MODERN EXAMPLES OF
FRESHWATER ECOSYSTEMS, HUMAN IMPACT ON
GREENHOUSE EFFECT
HUMAN IMPACT ON BIODIVERSITY, OVERVIEW
HUNTER-GATHERER SOCIETIES, ECOLOGICAL IMPACT OF
INDIGENOUS PEOPLES, BIODIVERSITY AND
LAND-USE PATTERNS, HISTORIC
MARINE ECOSYSTEMS, HUMAN IMPACT ON
NATURAL RESERVES AND PRESERVES
POLLUTION, OVERVIEW
POPULATION STABILIZATION, HUMAN
SLASH AND BURN FARMING, EFFECTS OF
SOUTH AMERICAN NATURAL ECOSYSTEMS, STATUS OF
WILDLIFE MANAGEMENT

INVERTEBRATES

ARACHNIDS
ARTHROPODS (TERRESTRIAL), AMAZONIAN
BEETLES
BUTTERFLIES
CRUSTACEANS
ENDANGERED FRESHWATER INVERTEBRATES
ENDANGERED MARINE INVERTEBRATES
ENDANGERED TERRESTRIAL INVERTEBRATES
FLIES, GNATS, AND MOSQUITOES
GRASSHOPPERS AND THEIR RELATIVES
HEMIPARASITISM
HYMENOPTERA
INSECTICIDE RESISTANCE
INSECTS, OVERVIEW
INVERTEBRATES, FRESHWATER, OVERVIEW
INVERTEBRATES, MARINE, OVERVIEW
INVERTEBRATES, TERRESTRIAL, OVERVIEW
ISOPTERA
MOLLUSCS
MOTHS
MYRIAPODS
PARASITOIDS
TRUE BUGS AND THEIR RELATIVES
WORMS, ANNELIDA
WORMS, NEMATODA
WORMS, PLATYHELMINTHES

MICROBIAL BIODIVERSITY

ARCHAEA, ORIGIN OF
BACTERIAL BIODIVERSITY
BACTERIAL GENETICS
FUNGI
HIGH-TEMPERATURE ECOSYSTEMS
MICROBIAL BIODIVERSITY, MEASUREMENT OF
MICROBIAL DIVERSITY
MICROORGANISMS, ROLE OF
PLANKTON, STATUS AND ROLE OF
PROTOZOA
PSYCHROPHILES, ORIGIN OF
THERMOPHILES, ORIGIN OF

PLANT BIODIVERSITY

BREEDING OF PLANTS
C_4 PLANTS
DEFORESTATION AND LAND CLEARING
ECOLOGY OF AGRICULTURE
EDIBLE PLANTS
ENDANGERED PLANTS
FOREST CANOPIES, ANIMAL DIVERSITY
FOREST CANOPIES, PLANT DIVERSITY
FOREST ECOLOGY
HEMIPARASITISM
HERBACEOUS VEGETATION, SPECIES RICHNESS IN
HERBICIDES

INTRODUCED PLANTS, NEGATIVE EFFECTS OF
LANDSCAPE DIVERSITY
PHOTOSYNTHESIS, MECHANISMS OF
PLANT-ANIMAL INTERACTIONS
PLANT BIODIVERSITY, OVERVIEW
PLANT COMMUNITIES, EVOLUTION OF
PLANT CONSERVATION, OVERVIEW
PLANT HYBRIDS
PLANT INVASIONS
PLANT-SOIL INTERACTIONS
PLANT SOURCES OF DRUGS AND CHEMICALS
POLLINATORS, ROLE OF
TEMPERATE FORESTS

POPULATION ISSUES

INTRODUCED SPECIES, EFFECT AND DISTRIBUTION
METAPOPULATIONS
POPULATION DENSITY
POPULATION DIVERSITY, OVERVIEW
POPULATION DYNAMICS
POPULATION GENETICS
POPULATION STABILIZATION, HUMAN
POPULATION VIABILITY ANALYSIS (PVA)
PREDATORS, ECOLOGICAL ROLE OF
STABILITY, CONCEPT OF
SUCCESSION, PHENOMENON OF

PUBLIC POLICIES AND ATTITUDES

AESTHETIC FACTORS
BIODIVERSITY-RICH COUNTRIES
EDUCATION AND BIODIVERSITY
ETHICAL ISSUES IN BIODIVERSITY PROTECTION
GOVERNMENT LEGISLATION AND REGULATION
HISTORICAL AWARENESS OF BIODIVERSITY
HUMAN IMPACT ON BIODIVERSITY, OVERVIEW
LITERARY PERSPECTIVES ON BIODIVERSITY
EUMS AND INSTITUTIONS
RELIGIOUS TRADITIONS AND BIODIVERSITY

SOCIAL AND CULTURAL FACTORS
STEWARDSHIP, CONCEPT OF
TOURISM, ROLE OF
ZOOS AND ZOOLOGICAL PARKS

SPECIES INTERACTIONS AND INTERRELATIONSHIPS

COEVOLUTION
COMPETITION, INTERSPECIFIC
FOOD WEBS
INDICATOR SPECIES
KEYSTONE SPECIES
INTRODUCED SPECIES, EFFECT AND DISTRIBUTION
LIMITS TO BIODIVERSITY (SPECIES PACKING)
MUTUALISM, EVOLUTION OF
NEST PARASITISM
PARASITISM
PLANT-ANIMAL INTERACTIONS
POLLINATORS, ROLE OF
PREDATORS, ECOLOGICAL ROLE OF
SOCIAL BEHAVIOR
SPECIES-AREA RELATIONSHIPS
SPECIES COEXISTENCE
SPECIES INTERACTIONS
TROPHIC LEVELS

SYSTEMATICS AND SPECIES CONCEPT

CLADISTICS
CLADOGENESIS
DIFFERENTIATION
DIVERSITY, MOLECULAR LEVEL
DIVERSITY, ORGANISM LEVEL
DIVERSITY, TAXONOMIC VERSUS FUNCTIONAL
FUNCTIONAL GROUPS
NOMENCLATURE, SYSTEMS OF
PHYLOGENY
SPECIATION, PROCESS OF
SPECIATION, THEORIES OF
SPECIES, CONCEPTS OF

SPECIES DIVERSITY, OVERVIEW
SUBSPECIES, SEMISPECIES AND SUPERSPECIES
SYSTEMATICS, OVERVIEW
TAXONOMY, METHODS OF

TECHNIQUES AND MEASUREMENTS

BIOPROSPECTING
COMPUTER SYSTEMS AND MODELS, USE OF
ECONOMIC VALUE OF BIODIVERSITY, MEASUREMENTS OF
ECOSYSTEM FUNCTION MEASUREMENT, AQUATIC AND MARINE COMMUNITIES
ECOSYSTEM FUNCTION MEASUREMENT, TERRESTRIAL COMMUNITIES
ENVIRONMENTAL IMPACT, CONCEPT AND MEASUREMENT OF
FRAMEWORK FOR ASSESSMENT AND MONITORING OF BIODIVERSITY
GENE BANKS
MEASUREMENT AND ANALYSIS OF BIODIVERSITY
MICROBIAL BIODIVERSITY, MEASUREMENT OF
PALEOECOLOGY
REMOTE SENSING AND IMAGE PROCESSING

THEORIES AND CONCEPTS OF BIODIVERSITY

BIODIVERSITY, DEFINITION OF
CARRYING CAPACITY, CONCEPT OF
COMMONS, CONCEPT AND THEORY OF
COMMONS, INSTITUTIONAL DIVERSITY OF
COMPLEXITY VERSUS DIVERSITY
ECOLOGICAL FOOTPRINT, CONCEPT OF
ECOLOGY, CONCEPTS AND THEORIES IN
ECOSYSTEM, CONCEPT OF
ECOSYSTEM FUNCTION, PRINCIPLES OF
ECOSYSTEM SERVICES, CONCEPT OF
FUNCTIONAL DIVERSITY
GUILDS
HABITAT AND NICHE, CONCEPT OF
METAPOPULATIONS
ORIGIN OF LIFE, THEORIES OF
RESOURCE PARTITIONING
SCALE, CONCEPT AND EFFECTS OF
STABILITY, CONCEPT OF
STEWARDSHIP, CONCEPT OF
STORAGE, ECOLOGY OF
SUSTAINABILITY, CONCEPT AND PRACTICE OF

VERTEBRATES

AMPHIBIANS, BIODIVERSITY OF
BIRDS, BIODIVERSITY OF
CARNIVORES
DINOSAURS, EXTINCTION THEORIES FOR
ENDANGERED BIRDS
ENDANGERED MAMMALS
ENDANGERED REPTILES AND AMPHIBIANS
FISH, BIODIVERSITY OF
FISH CONSERVATION
FISH STOCKS
MAMMALS, BIODIVERSITY OF
MAMMALS, CONSERVATION EFFORTS FOR
MAMMALS, LATE QUATERNARY, EXTINCTIONS OF
MAMMALS, PRE-QUATERNARY, EXTINCTIONS OF
MARINE MAMMALS, EXTINCTIONS OF
PRIMATE POPULATIONS, CONSERVATION OF
REPTILES, BIODIVERSITY OF
SALMON
VERTEBRATES, OVERVIEW

Foreword

Biodiversity is the totality of the inherited variation of all forms of life across all levels of variation, from ecosystem to species to gene. Soon after the term was introduced at the first National Forum on Biodiversity in 1986, and after it began its rapid spread around the world, there occurred a reconfiguration in the way much of the science of biology is conceived. Where previously comparative biology had been almost entirely focused on the fundamentals of classification, evolution, behavior, and ecology, now it was augmented by a wide range of analyses from the social sciences. Where taxonomy and biogeography had been marginalized through the middle half of the twentieth century, now they moved back toward center stage. And where extinction had been little more than a phenomenon recognized and lamented, now it became a major concern of science. Much of ecology shifted toward the study of biodiversity's role in the assembly and maintenance of ecosystems. A growing number of economists, political scientists, and bioethicists took up the issue as part of their scholarly agenda. From this mix the discipline of conservation biology was born, and the Society of Conservation Biology became one of the fastest growing organizations in modern science. In 1992 the Rio Summit catapulted biodiversity to global prominence, from which most of the nations of the world endorsed the Convention on Biodiversity and have since used it as a guideline for conservation programs.

The new biodiversity initiative gave organismic and evolutionary biologists a global mission worthy of their science. It confirmed for those who labored in the vineyards that, as medicine is to molecular and cellular biology, the environment is to organismic and evolutionary biology. The first is responsible for personal health, and the second for planetary health. The additional evidence adduced moreover made clear that we are in the midst of an episode of massive extinction, unprecedented since that closing of the Mesozoic Era—and that scientists must lead the attempt to save the Creation.

The articles in the *Encyclopedia of Biodiversity* are unusually eclectic, yet organized by a set of easily articulated goals. They are the following: to carry the systematics and biogeography of the world fauna and flora toward completion; map the hot spots where conservation will save the most biodiversity; orient studies of natural history to understand and save threatened species; advance ecosystems studies and biogeography to create the needed principles of community assembly and maintenance; acquire the knowledge of resource use, economics, and polity to advance conservation programs based on sustainability; and enrich the ethic of global conservation in terms persuasive to all.

The road ahead, down which we must urgently travel, will be smoothed by the exponential growth of information and a growing public awareness and support. The *Encyclopedia of Biodiversity* will serve as an important knowledge base to guide this supremely important effort.

Edward O. Wilson
Museum of Comparative Zoology
Harvard University

Preface

The science of biodiversity has become the science of our future. Our awareness of the disappearance of biodiversity has brought with it a long-overdue appreciation of the magnitude of our loss, and a determination to develop the tools to protect our future. This encyclopedia brings together, for the first time in its completeness, study of the dimensions of diversity with examination of the services that biodiversity provides, and measures to protect it.

The entries in the encyclopedia have been arranged alphabetically, but the coverage is designed functionally. At the core is a comprehensive survey of biodiversity, across taxonomic groups and ecological regions. The emergence of biodiversity is then placed in an evolutionary perspective, as background for an understanding of current trends. Particular attention is given to the loss of services—for example, in fisheries, forestry and climate mediation—that are derived from natural systems. These are placed in an economic framework through a comprehensive set of papers that address problems of valuation, costs, and benefits, and develop a framework for prioritizing actions. Finally, a review is given of institutions and other mechanisms that exist and are needed for the preservation of biodiversity and, with it, the services that humans derive from nature.

The background for understanding biodiversity is to be found in the fossil record, and in the evolutionary patterns and trends that it reveals. The encyclopedia hence discusses these patterns, the origins of biodiversity, the effects of geological events, the mechanisms of evolution, and the uniqueness of the evolutionary process, with implications for conservation and restoration. The essential processes in macroevolution are those of speciation and extinction, which together govern the dynamics of diversity at higher levels of organization. These are given extensive coverage, both from a mechanistic and from a historical perspective, and provide an essential context for understanding the rest of the contributions.

The classification of organisms into species and higher taxa, and the elucidation of the mechanisms of natural selection, were the essential intellectual advances that allowed the development of the science of biodiversity. Carl Linnaeus introduced a systematic framework for understanding phylogenies, which continues to provide the foundation for evolutionary studies today; and Charles Darwin's great legacy—the theory of evolution by natural selection—is the essential organizing principle for understanding the processes that gave rise to the patterns Linnaeus recognized. The encyclopedia provides unmatched taxonomic coverage of the organization of diversity into taxonomic groups and complements that with an extensive examination of ecosystems by biogeographic region and by functional type. These chapters elucidate latitudinal trends, life zones, species–area relationships and the distribution of diversity within and among ecological communities.

Throughout the core chapters, there is a healthy balance between empirical facts and conceptual theories. Such theories help to illuminate principles that cross systems and levels of organization, and transform the study of biodiversity into a science. Basic ecological constructs, such as the habitat and the niche, are given extensive treatment, as are key ecological mechanisms such as competition, predation, herbivory, parasitism and mutualism. These treatments are complemented by exploration of fundamental evolutionary mechanisms related to local differentiation, aspect diversity, sex, and recombination, and especially theories of extinction.

With these foundational chapters in hand, one can turn to the contemporary problems in biodiversity and compare today's rapid rates of change to the historical patterns. Key chapters examine agriculture, fisheries, and forests, their importance to human needs, and their status and trends in response to changing land-use patterns, population growth, overexploitation, and climate change. Threatened and endangered species are discussed in detail, with relation to the consequences of the spread of invading species.

The utilization of nature's bounty for food, fiber and fuel provides some of the most obvious benefits of biodiversity to humanity. Equally important, however, are the things that are less well appreciated: the potential for the discovery of new pharmaceuticals that can improve human health, the role of biodiversity in pollinating crops and wild species, and the importance of natural systems in regulating climate, mediating nutrient fluxes, and sequestering carbon as well as toxic materials. Each of these services provides humans with direct and indirect benefits, and somehow we must find ways to weigh these benefits, along with the ethical and aesthetic values we place on natural systems and biodiversity, to provide priorities for action. Only recently have economists recognized the importance of such issues as intellectual challenges essential to our survival on the planet. Much of biodiversity is exploited by humanity as part of a global commons, in which one does not pay in fair measure for extracting parts, or affecting the commons otherwise through land use or pollution. Economists have come to realize, along with ecologists and others, the magnitude of the externalities involved. When such externalities are involved, the market does not function as it must to maintain the resource, and new measures are needed if the sustainability and resilience of these resources are to be preserved. There is as yet, however, no ecological equivalent to the power of financial institutions, such as the Federal Reserve Board in the United States, to modify individual incentives sufficiently to maintain regional or global stability in the system of interest. In this encyclopedia, some of the most enlightened and thoughtful economists turn their attention to the economic challenges, and discuss the mechanisms and institutions that might be needed.

Together, the state-of-the-art entries in this encyclopedia tell an exciting story of how biodiversity arose, continues to arise, and is maintained. It is a story of a complex, self-organizing system—the biosphere—whose pieces can be examined individually, but cannot be understood outside the context of the whole. It is also a story of the coevolution of the biosphere and *Homo sapiens*, the first species whose own activities can feed back to influence the evolution of the biosphere on time scales that could lead to its own demise. The articles in the encyclopedia can be used as material for a wide spectrum of courses, tracing the history of the emergence of biodiversity from its origins to the challenges we face today.

This has been a massive effort, but one of the most rewarding I have ever undertaken. So many people have played a role that it is difficult to know where to begin. The project began through the initiative of Scott Bentley at Academic Press, and then was managed flawlessly by Chris Morris at AP. I cannot recall ever having dealt with an editor who operated more professionally than Chris, who combined a true vision and enthusiasm for the project with a sense of economic realities and the energy and insight to make the whole project work. In this he was ably assisted by outstanding Academic Press staff, especially Naomi Henning, Nick Panissidi, and Ann Marie Martin. At every step, it has been a pleasure to work with Academic Press, and I especially single out Chris for his fantastic and scholarly efforts.

At the next stage, the Editorial Boards were terrific in generating and commenting on ideas, suggesting authors, and critiquing contributions. More than 400 authors then adopted our view of the importance of the project, accepted the task of writing, and produced timely and comprehensive articles that make this Encyclopedia like no other source available today. To all of these, I extend my thanks and congratulations.

And finally, special acknowledgment and gratitude are due my wife, Carole, and my assistant, Amy Bordvik. Carole put up with the late nights and obsessiveness that were essential to the process, and Amy worked tirelessly and without complaint through the whole long process. To them, any expression of thanks is insufficient.

Simon Levin
Princeton University

References

Simon A. Levin (1999). *Fragile Dominion*, Perseus Books. Reading, Massachusetts

Edward O. Wilson (1992). *The Diversity of Life*. Norton. New York, New York.

Guide to the Encyclopedia

The *Encyclopedia of Biodiversity* is a comprehensive study of the topic of diversity in the natural world, contained within the covers of a single unified work. It consists of five volumes and includes 313 separate full-length articles by leading international authors.

Each article in the encyclopedia provides a comprehensive overview of the selected topic to inform a broad spectrum of readers, from research professionals to students to the interested general public. In order that you, the reader, will derive the greatest possible benefit from the *Encyclopedia of Biodiversity*, we have provided this Guide. It explains how the encyclopedia was developed, how it is organized, and how the information within it can be located.

ENTRY SELECTION

This encyclopedia was conceived with the goal of providing a complete description of all the issues contained within, or impacting upon, the field of biodiversity. To that end, a thorough and systematic method of entry selection was devised for the work.

To begin the selection process, the project's chief editor, Simon Levin, prepared a bibliography of leading source materials in the field, including books, journal articles, conference proceedings, Websites, and so on. Then the reference staff of Academic Press combed through these materials to develop a list of potential article topics for the encyclopedia. This preliminary list was refined and approved by Prof. Levin; at this point the number of possible entries was approximately twice as large as the eventual total in the published encyclopedia.

The entry list was then provided to all the associate editors and the international editorial board for their evaluation. Their mandate was to read through the list and rate each topic on a numerical scale according to how important they deemed it to be for inclusion in the encyclopedia. The editors were also encouraged to recommend new topics not on the existing list, and to make other comments on the list as appropriate. A number of additions to the entry list emerged from this process.

The editors' ratings and comments were returned to Academic Press for scoring, and an overall tabulation was created that indicated the consensus of the group as to the priority of each topic. Then the list was sent to Prof. Levin for a final evaluation in which he made "tie-breaker" decisions for certain topics on which the editors' vote was split, and also other adjustments based on his expert judgment. The result was a working entry list of about 325 topics which, after some attrition and the combining of related topics, resulted in the final table of contents of 313 articles.

ORGANIZATION

The *Encyclopedia of Biodiversity* is organized in a single alphabetical sequence by title. Articles whose titles begin with the letters A to C are in Volume 1, articles with titles from D through Fl are in Volume 2, then Fo through Man in Volume 3, Mar through Q in Volume 4, and R to Z in Volume 5.

Volume 5 also includes a complete subject index for the entire work, an alphabetical list of the authors who contributed to the encyclopedia, and a glossary of key terms used in the articles.

TABLE OF CONTENTS

A complete table of contents for the *Encyclopedia of Biodiversity* appears at the front of each volume. This alphabetical list of article titles (see p. vii) is followed by a second contents list (p. xix) in which the titles are listed according to their subject area within the overall field of biodiversity.

Articles are classified in 20 different subject areas, including not only core disciplines of biodiversity such as evolution, speciation, populations, extinction, and ecosystems, but also areas that link biodiversity to other disciplines, such as environmental science, agriculture, public policy, and economics.

ARTICLE TITLES

Article titles generally begin with the key term describing the topic, and have inverted word order if necessary to begin the title with this term. For example, "Archaea, Origin of" is the article title rather than "Origin of Archaea," "Grazing, Effects of" is the title rather than "Effects of Grazing," and so on with other titles such as "Species, Concepts of," "Mammals, Biodiversity of," "Pollinators, Role of," and so on. This is done so that the reader can more easily locate a desired topic. For example, eight different articles on endangered groups (e.g., endangered birds) appear in succession in the "En-" section of the encyclopedia.

INDEX

The index appears as the last element of Volume 5. Subjects are listed alphabetically and indicate the volume and page number where information on this topic can be found. In addition, the table of contents by subject area also functions as an index, since it lists all the topics covered in a given area; e.g., the encyclopedia has 26 different articles dealing with invertebrates.

ARTICLE FORMAT

Articles in the *Encyclopedia of Biodiversity* are arranged in a standard format, as follows:

- Title and Author
- Outline
- Glossary
- Defining Statement
- Main Body of the Article
- Cross-References
- Bibliography

OUTLINE

Entries in the encyclopedia begin with a topical outline that indicates the general content of the article. This outline serves two functions. First, it provides a preview of the article, so that the reader can get a sense of what is contained there without having to leaf through the pages. Second, it serves to highlight important subtopics that are discussed within the article. For example, the article "Greenhouse Effect" includes subtopics such as "Climatic Consequences: Global Warming" and "Climate Change and Biodiversity."

The outline is intended as an overview and thus it lists only the major headings of the article. In addition, extensive second-level and third-level headings will be found within the article.

GLOSSARY

The Glossary section contains terms that are important to an understanding of the article and that may be unfamiliar to the reader. Each term is defined in the context of the article in which it is used. The same term may appear as a glossary entry in different articles, with the details of the definition varying slightly from one article to another. The encyclopedia includes approximately 2,500 glossary entries. For example, the article "Mangrove Ecosystems" has the following entry:

aerenchyma A spongy plant tissue composed largely of air spaces enabling gas exchange to take place by diffusion in underground mangrove roots.

In addition, Volume 5 has a comprehensive glossary that presents the core vocabulary of biodiversity in one A-Z list. This section can be consulted for definitions of unfamiliar terms not found in the individual glossary for a given article.

DEFINING STATEMENT

The text of each article in the encyclopedia begins with a single introductory paragraph that defines the topic under discussion and summarizes the content of the article. For example, the article "Agriculture, Sustainable" begins with the following statement:

> Sustainable agriculture describes a food and fiber production system that is economically viable, environmentally safe, and socially acceptable over long periods.

CROSS REFERENCES

The entry list for *Encyclopedia of Biodiversity* has been constructed so that each entry is supported by one or more other entries that provide additional information. Therefore all articles in the encyclopedia have references to other articles. These cross references appear at the conclusion of the article text. They indicate articles that can be consulted for further information on the same issue, or for pertinent information on a related issue. The encyclopedia includes a total of about 1,750 cross references to other articles. For example, the article "Biodiversity-Rich Countries" contains the following list of references:

> Biodiversity as a Commodity • Deforestation • Economic Growth and the Environment • Indigenous Peoples, Biodiversity and • Social and Cultural Factors • Tropical Ecosystems

BIBLIOGRAPHY

The Bibliography section appears as the last element in an article. Entries in this section include not only print sources but relevant Websites as well.

The bibliography entries in this encyclopedia are for the benefit of the reader and do not represent a complete list of all the materials consulted by the author in preparing the article. Rather, the sources listed are the author's recommendations of the most appropriate materials for further research on the given topic. For example, the article "Fires, Ecological Effects of" lists as references (among others) the works *Fire and Plants, Fire in the Environment, Fire in the Tropical Biota,* and *The Role of Fire in Mediterranean Ecosystems.*

COMPANION WORKS

Encyclopedia of Biodiversity is part of a continuing program of multivolume reference works published by Academic Press. This program encompasses many different areas of science, ranging from organismal biology (e.g., *Encyclopedia of Dinosaurs, Encyclopedia of Microbiology*) to biomedical topics (*Encyclopedia of Reproduction, Encyclopedia of Stress*), to physical science (*Encyclopedia of the Solar System, Encyclopedia of Volcanoes*) to social and political issues (*Encyclopedia of Applied Ethics, Encyclopedia of Creativity, Encyclopedia of Nationalism, Encyclopedia of Violence, Peace, and Conflict*).

For information on these and other Academic Press reference titles, please see the Website at:

www.academicpress.com/reference/

RAINFOREST ECOSYSTEMS, ANIMAL DIVERSITY

Gregory H. Adler
University of Wisconsin at Oshkosh

I. Definition and Geographical Context
II. Types of Rainforests
III. Overview of Animal Diversity
IV. Diversity of Beetles
V. Diversity of Butterflies
VI. Diversity of Frogs
VII. Diversity of Birds
VIII. Diversity of Mammals
IX. Geographical Patterns of Animal Diversity
X. Hypotheses of High Species Richness
XI. The Future of Rainforest Animal Diversity

GLOSSARY

allopatric speciation The evolutionary development of new species in the presence of a geographical barrier which reduces gene flow and promotes genetic divergence.
neotropics A biogeographic region that includes the New World tropics, extending from southern Mexico south through the Southern Cone of South America to Tierra del Fuego. Many different ecosystems are found here, including tropical rainforest.
species diversity This has two connotations. In a broad sense, it simply refers to the number of species of a particular taxonomic group living within a given area and is used synonymously with species richness. In a narrow sense, it refers to the number of species within a given area while simultaneously taking into account their relative abundances. In this article, species diversity is used in the broad sense and is used interchangeably with species richness.
species richness The number of species of a particular taxonomic group living within a given area.

TROPICAL RAINFORESTS contain more species of animals than any other ecosystem on Earth. This article reviews the distributions and types of rainforests and the diversity of selected groups of animals in tropical rainforests. The article concludes with a discussion of global and regional patterns of rainforest animal diversity, some hypotheses that have been developed to explain the high diversity, and the future of rainforest animal diversity.

It is indeed ironic that Linneaus, cataloging specimens collected by early naturalists for the museums of Europe, considered the tropics to be of low diversity. However, the first naturalists collected solely in villages that contained second-growth vegetation and where hunting was intense. It was not until naturalists penetrated the dense border of vegetation between villages and undisturbed forest that the high diversity became evident. Single collecting expeditions routinely returned to Europe with thousands of undescribed species of plants and animals. Discoveries of new species of animals continue into the twenty-first century. It has been estimated that several million species of insects alone are yet to be described from tropical rainforests. Even larger and presumably more conspicuous animals, such as birds and mammals, are being discovered every year in the world's tropical rainforests. For instance, in the 1990s new species of monkeys were found in Brazil and two new species of deer and a relative of the goat were discovered in Laos and Vietnam.

I. DEFINITION AND GEOGRAPHICAL CONTEXT

Tropical forests cover much of the land area that lies between the Tropics of Cancer and Capricorn (23°30′N and S latitude). Tropical forests also extend outside of the two tropics in southeastern Brazil and northeastern Australia. The lowland tropics have a relatively high constant temperature, where the mean temperature of the coldest month is at least 18°C. However, not all warm areas in the tropics support tropical forest. Rainfall, in concert with high stable temperature, is a crucial factor determining the development of forest. In many areas, closed-canopy forest begins to develop where there is at least 800–1000 mm of rain per year. Although the term rainforest has been applied loosely and sometimes inappropriately to virtually any type of tropical forest, a more precise definition of rainforest has been developed. Tropical rainforests occur in areas of high, relatively constant temperature (below approximately 1000 m in elevation) and where rainfall exceeds 2000 mm per year. A dry season may occur but generally does not exceed 4 months in duration. Forests growing under such conditions are composed primarily of evergreen trees (<15% of the trees are deciduous) and have a closed canopy of at least 25 m in height. Larger trees, called emergents, often protrude above the canopy and may reach heights of up to 50 m. Thus, rainforests are defined based on the temperature and precipitation regimes under which tropical forests develop and also on the structural characteristics of these forests.

An understanding of the geographical context of rainforests is necessary to a discussion of animal diversity in these forests. For historical reasons (i.e., evolutionary origins of taxa in specific geographical regions and subsequent inability to disperse to other regions), many groups of animals are geographically restricted to single tropical regions. Vertebrates illustrate this point. The species-rich frog family Leptodactylidae is restricted to the New World and reaches its greatest diversity in rainforests. Iguanas (Iguanidae) and anoles (Polychrotidae), two species-rich families of lizards with greatest diversity in rainforests, are restricted to the New World, except for iguanas in Fiji that stem from New World colonizers. Families of birds such as toucans (Ramphastidae), motmots (Momotidae), jacamars (Galbulidae), puffbirds (Bucconidae), woodcreepers (Dendrocolaptidae), ovenbirds (Furnariidae), antbirds (Thamnophilidae), antthrushes (Formicariidae), and cotingas (Cotingidae) are restricted to the New World tropics, with many species occurring in rainforests. Birds such as hornbills (Bucerotidae), honeyguides (Indicatoridae), broadbills (Eurylaimidae), sunbirds (Nectariniidae), white-eyes (Zosteropidae), and pittas (Pittidae) are restricted largely to the tropics of the Old World (with a few species found also in adjacent subtropical regions). Birds of paradise (Paradisaeidae) and bowerbirds (Ptilonorhynchidae) are restricted to the tropics of Australia and New Guinea. Marsupials, which reach their greatest diversity in rainforests and surrounding areas, are restricted to the New World and the Australopapuan region. Among primates, two families of monkeys (Callitrichidae and Cebidae) are restricted to the New World tropics, whereas Cercopithecidae is restricted to the Old World and Hylobatidae is restricted to Southeast Asia. The lemurs, a group of primates composed of five families, are found solely in Madagascar.

Rainforests are found in three major regions of the world. The rainforests of the neotropics cover the greatest area (approximately 4 million km^2) and extend from southern Mexico to southeastern Brazil. The neotropics contain four fairly distinct regions that are isolated from each other by mountain ranges, savannas, or scrub vegetation. Trans-Andean rainforests are distributed from Chiapas and Veracruz, Mexico, to the Pacific slope of Colombia and Ecuador. Venezuelan coastal rainforests are found in northern Colombia, Venezuela, and the Guianas. Amazonian rainforests occupy most of the Amazon basin and constitute the largest contiguous area of tropical rainforest in the world. Atlantic rainforests formerly covered a long narrow swath of coastal southeastern Brazil but have been reduced to remnant patches by human activity. Only about 10% of the Atlantic rainforest remains. The Asia-Pacific tropics, also known as Malesia, extend from China's Yunnan Province southward to Australia. Included in this region are large parts of Southeast Asia (Burma, Thailand, Malaysia, Indonesia, Singapore, Cambodia, Laos, and Vietnam) and the Australopapuan region (New Guinea, northeastern Australia, and tropical Pacific islands such as New Caledonia, the Bismarck Archipelago, Solomon Islands, Fiji, and Vanuatu). Asia-Pacific tropical rainforest also occurs far to the west in the Western Ghats of India. This discontinuous and largely insular block of rainforest is second in areal extent only to the neotropics, covering approximately 2.5 million km^2. The Australopapuan region harbors faunal elements that are sufficiently distinct to sometimes warrant separate treatment. For instance, monotremes (egg-laying mammals), marsupials, and several families of birds are found there but not in the remainder of the Asia-Pacific tropics. Conversely, many families of birds and mammals are not found in the Australopapuan region but occur in the remainder of the Asia-Pacific tropics. Africa contains a large block of rainforest in the Congo River basin in central Africa and a smaller block of rainforest in West Africa. An outlying area of rainforest occurs in eastern Madagascar. The African rainforests cover approximately 1.8 million km^2.

II. TYPES OF RAINFORESTS

Researchers frequently delimit lowland tropical forests based on annual amounts of rainfall such forests receive and the temporal distribution of rainfall. These delimitations are somewhat arbitrary, and it is important to recognize that contiguous forest often occurs along a rainfall gradient where one type of forest grades into another with little obvious demarcation. Tropical scrub forest and tropical dry forest typically receive less than 2000 mm of rain per year, most of which falls in a few months. Severe dry seasons of up to 8 or 10 months punctuate the short rainy seasons, and many trees are deciduous and shed their leaves during the dry season to conserve water. Such forests, particularly scrub forests, often do not have a closed canopy. If we adhere to our definition of rainforests given previously, then these tropical forests do not constitute true rainforests (although they frequently abut rainforests and share many species of animals), and this article will not discuss their animal diversity. The remaining forest types that will be discussed are sometimes referred to as semievergreen and evergreen forests and conform to the definition of tropical rainforests. Tropical moist forests receive from about 2000 to 4000 mm of rain per year. These forests frequently have a pronounced dry season of up to 4 months, but heavy rains may occur during the dry season. Deciduous trees may occur in such forests, but they are much less frequent than in the preceding two types of forests. Tropical moist forests occupy the greatest aereal extent of rainforests and occur throughout much of the three major geographical rainforest regions. Tropical wet forests typically receive more than 4000 mm of rain per year and frequently have no extended dry season. The wettest of these forests (those forests receiving more than 8000 mm of rain per year) sometimes are designated as pluvial forests and are essentially aseasonal. The "driest" months in these forests typically receive more rain than the wettest months in tropical moist forests.

Tropical forests may also be defined based on the elevations at which they occur. In mountainous areas, contiguous forest may be found from near sea level to the tree line. However, forest characteristics change dramatically with elevation. The upper elevational limit of lowland rainforests is frequently considered to be 1000 m, where premontane forest often occurs, although this limit is arbitrarily chosen. Lower montane and upper montane forests, which do not fall within the operational definition of lowland rainforests, occur at increasingly higher elevations. In some areas, montane forests may occur at elevations lower than 1000 m, depending on latitude and stature of mountain ranges. For instance, small mountain ranges and lower spurs from larger ranges often support montane forests at lower elevations, which is termed the Massenerhebung effect. This article will focus on animal diversity in lowland rainforests (i.e., tropical moist and wet forests lower than 1000 m).

III. OVERVIEW OF ANIMAL DIVERSITY

Tropical–temperate comparisons of diversity are commonly made to highlight the high diversity of the tropics, and this latitudinal diversity gradient has attracted considerable attention from ecologists and evolutionary biologists. These comparisons show that many major groups of animals reach their greatest diversity in tropical rainforests. Conspicuous among groups of animals that reach their greatest diversity in such forests are beetles, butterflies, frogs, birds, and mammals. This article will focus on these groups because they best exemplify the diversity of rainforest animals, and tropical–temperate comparisons will be included to highlight rainforest diversity.

It should be noted, however, that other major groups of animals, particularly among invertebrates, have been poorly sampled in tropical rainforests. It is evident, however, that other such groups also exhibit high rainforest diversity. For example, 498 species of spiders in 33 families were collected during 6 weeks at Pakitza in Manu National Park, Peru. Many of these specimens represented new species. Several weeks of sampling at the same site yielded 224 species of caddisflies (order Trichoptera), 66% of which could not be identified confidently. Sampling of dragonflies and damselflies (order Odonata) in the same area yielded 117 lowland rainforest species. At Tambopata Reserve in Peru, 135 species of ants were found, and one individual tree harbored 43 species, which is approximately the total number of ant species found in the entire British Isles or in all of Canada. The 1500-ha La Selva reserve in Costa Rica contains an estimated 4000 species of moths. Thus, when invertebrates have been sampled in tropical rainforests, diversity is usually very high, and new species invariably are found.

IV. DIVERSITY OF BEETLES

Approximately 1.4 million species of organisms have been described, about 400,000 of which are beetles. Beetles therefore constitute one-fourth of all living species that have been cataloged by taxonomists. Beetles have been poorly sampled in rainforests, and any discussion of their diversity will be incomplete. However, it is clear from even preliminary sampling that beetle diversity is extremely high and that beetles no doubt are the most species-rich group of rainforest animals, comprising an estimated 40% of all arthropods. Indeed, the majority of rainforest animal species may be still undescribed beetles.

There has been considerable conjecture over how many species of beetles occur in rainforests. Rainforest beetles reach their greatest richness in the canopy, in which standardized sampling efforts have been concentrated. Such sampling involves the deployment of insecticidal sprays into the canopy, a process known among coleopterists (scientists who study beetles) as fogging. Sampling of beetles (excluding weevils) by fogging 19 individuals of a single species of rainforest tree (*Lueehea seemanni*) in Panama produced 955 species, 163 species of which were found only on this 1 species of tree. This preliminary sampling was used to estimate the total number of arthropod species in tropical rainforests (Erwin, 1982). The estimate was based on the number of beetle species restricted to a single species of tree, the approximate number of tropical tree species, the proportion of arthropods that are beetles, and the contribution of canopy species to the overall arthropod species pool. Based on this reasoning, estimates of 30–50 million species of insects and >7 million species of beetles have been derived, which indicate that insects may constitute about 97% of all species of living organisms on Earth and that tropical rainforest beetles contribute the greatest proportion of species. These estimates of global biodiversity based on rainforest beetles stimulated biologists to reassess their more traditional estimates of biodiversity.

Sampling of ground beetles (Carabidae) at Pakitza in Manu National Park, Peru (not all species of which are confined to the ground), has revealed this 4000-ha site to have the greatest species richness of this family ever recorded. Indeed, nearly as many species have been found in this small area as have been found in all of New Guinea. More than 600 species have been recorded from Pakitza, and not only many new species but also new genera have been discovered. It should be noted that sampling was incomplete and that dozens of additional species are expected.

Comparison of beetle species richness among sites is difficult because of differences in sampling methodology and intensity and the paucity of sufficiently rigorous studies. However, several studies using the standardized fogging methodology allow researchers to calculate species richness on equal footing as the number of species/m^3/m. Among canopy and subcanopy beetles, estimates range from 0.02 species/m^3/m in an Australian rainforest tree and in rainforest in Sulawesi to 1.5 species/m^3/m in a Peruvian rainforest. Other estimates for rainforest sites are 0.29 species/m^3/m in New Guinea, 0.32 species/m^3/m in Brunei, and 1.17 species/m^3/m in Panama. The estimate for Panama was derived from the same data set from which global biodiversity estimates were made.

V. DIVERSITY OF BUTTERFLIES

Unlike most groups of insects, butterflies have been studied so extensively that about 90% of all living species have been described. An estimated 13,750 species of butterflies, excluding skippers (family Hesperiidae), occur in the world. With skippers included, the number is approximately 17,500 species. Because of our thorough understanding of butterfly systematics, they are perhaps the most appropriate insects for examining rainforest diversity.

A tropical–temperate comparison of butterfly diversity highlights the great richness of tropical rainforest butterflies. Table I shows the numbers of butterfly species in four families in temperate forests of West Virginia and tropical rainforests of Costa Rica. These two areas are of comparable size, and the butterfly faunas in the four families are well-known. Butterflies in two additional families (Lycaenidae and Hesperiidae) are too poorly known to make a valid comparison with those in West Virginia. Included in the species tally for West Virginia are those species known to occur in all forest types below 1000 m, whereas the species tally for

TABLE I
Approximate Numbers of Butterfly Species in Four Families in Lowland Rain Forests of Costa Rica Compared with Lowland Temperate Forests of West Virginia

Family	West Virginia	Costa Rica
Papilionidae	5	25
Pieridae	4	31
Riodinidae	1	208
Nymphalidae	16	234
Total	26	498

Costa Rica includes only those species that are known to occur in lowland rainforests. Thus, butterflies occurring in both dry and wet temperate deciduous forests are included in the tally for West Virginia, but habitats such as tropical scrub and dry forests and mangrove forests are excluded from the tally for Costa Rica.

It is important to note that species tallies will likely vary depending on definitions of forest types and the compilation of results from additional sampling and natural history studies. Many species of butterflies live in disturbed areas adjacent to rainforest but are not included in the species tally because they are not true denizens of the rainforest.

Regardless of such qualifications, it is clear that rainforest butterfly diversity is much higher than in temperate forests. Costa Rican rainforests harbor nearly 20 times as many species of butterflies in the four families. Of particular note is the extremely high relative diversity in two Costa Rican families, Riodinidae and Nymphalidae, which together constitute nearly 90% of Costa Rica's rainforest butterfly fauna. Although the majority of West Virginia forest species are nymphalids, only one species of riodinid is found in West Virginia's lowland forests, and the two families together constitute only 65% of the total butterfly fauna.

Sampling at single sites further reveals the high species richness of Costa Rica's butterfly fauna. At the 1500-ha La Selva reserve in Costa Rica, 204 species have been recorded. However, Costa Rica's rainforest butterfly fauna is not particularly rich relative to that of other rainforest sites, especially in the neotropics. On a nineteenth-century expedition to the Amazon, Bates found 550 species of butterflies at a single site. This number is small compared with the total butterfly list for sites in the Peruvian Amazon. Within a 200-ha study site at the Explorer's Inn Reserve in southern Peru, 1234 species have been recorded, and approximately 1300 species have been recorded from Pakitza in Manu National Park. Thus, these single sites contain approximately 10% of all butterfly species in the world. The total numbers of species at these sites and other sites in the neotropics are not final; virtually every collecting expedition to these sites finds additional species. Indeed, the 1300 species from Pakitza were collected during only five sampling trips that averaged <3 weeks in duration. Nonetheless, the butterfly lists for these single sites exceed the lists for virtually all countries in the African and Asia-Pacific tropics.

VI. DIVERSITY OF FROGS

Frogs are conspicuously species rich in tropical rainforests, and their richness is particularly evident when breeding. At a single site in the Amazon rainforest in Santa Cecilia, Ecuador, 81 species of frogs have been recorded, which to date is the highest tally of frog species for any single site in the world. The number of species at this site equals the total number of frog species in the entire United States. High frog species richness has also been recorded from other tropical rainforest sites. In Panama, 59 species of frogs have been recorded on the 1500-ha Barro Colorado Island. More than 40 species have been recorded from La Selva, Costa Rica, 38 species from Makoukou, Gabon, 25 species from Pasoh, Malaysia, and 23 species from Gogol, Papua New Guinea. On the island of Borneo, 33, 51, and 46 species of frogs have been recorded at three different lowland rainforest sites, respectively, and 29 species of larval frogs have been recorded from rainforest streams at Nanga Tekalit in Sarawak. At Sakaerat, northern Thailand, 19 species of frogs have been recorded within rainforest, with an additional 5 species found in adjacent dry forest.

VII. DIVERSITY OF BIRDS

Tropical rainforests typically harbor the greatest diversity of birds compared with other ecosystems. Costa Rica's bird fauna is well-known, and a comparison of the number of resident bird species in that country's lowland rainforests with resident birds in all forest types in West Virginia illustrates this point. Table II gives the approximate number of bird species that are resident in Costa Rica's lowland rainforests compared with the number that are resident in lowland forests of all types in West Virginia. Excluded from this list are migratory

TABLE II

Approximate Numbers of Resident Breeding Bird Species in Lowland Rain Forests of Costa Rica Compared with Lowland Temperate Forests of West Virginia

Order and Family	West Virginia	Costa Rica	Order and Family	West Virginia	Costa Rica
Tinamiformes			Momotidae	0	6
Tinamidae	0	3	Piciformes		
Ciconiiformes			Galbulidae	0	2
Ardeidae	0	3	Bucconidae	0	5
Threskiornithidae	0	1	Capitonidae	0	2
Falconiformes			Ramphastidae	0	6
Cathartidae	2	3	Picidae	6	11
Accipitridae	4	21	Passeriformes		
Falconidae	0	5	Dendrocolaptidae	0	13
Galliformes			Furnariidae	0	13
Cracidae	0	2	Thamnophilidae	0	19
Phasianidae	3	4	Formicariidae	0	8
Gruiformes			Cotingidae	0	8
Rallidae	0	1	Tyrannidae	1	44
Eurypygidae	0	1	Pipridae	0	9
Columbiformes			Corvidae	3	2
Columbidae	1	12	Paridae	3	0
Psittaciformes			Sittidae	2	0
Psittacidae	0	11	Certhiidae	1	0
Cuculiformes			Regulidae	1	0
Cuculidae	0	3	Troglodytidae	2	14
Strigiformes			Turdidae	1	5
Strigidae	4	9	Mimidae	1	0
Caprimulgiformes			Bombycillidae	1	0
Nyctibiidae	0	2	Sylviidae	0	3
Caprimulgidae	0	2	Vireonidae	0	4
Apodiformes			Parulidae	1	6
Apodidae	0	4	Icteridae	0	3
Trochilidae	0	28	Thraupidae	0	40
Trogoniformes			Emberizidae	5	11
Trogonidae	0	8	Fringillidae	1	0
Coraciiformes			Total	43	362
Alcedinidae	0	5			

species that breed in West Virginia or that overwinter in Costa Rica. Also excluded are species that are found in disturbed areas adjacent to rainforests, in clearings within rainforests, or along major watercourses that may pass through rainforests. Although the same cautionary note applies to this comparison as it did with butterflies, it is again evident that Costa Rica's resident rainforest avifauna is much more diverse than that of West Virginia's lowland forests.

Costa Rican rainforests contain more than eight times as many species of birds as do all lowland forest types in West Virginia. Much of this greater diversity stems from two taxonomic sources. First, Costa Rican rainforests harbor more families of birds with resident species (42 compared with 18 in West Virginia). There are 29 families of birds that have representatives that are resident in Costa Rican rainforests but not in West Virginia's forests, and these families contribute 218 species to the total rainforest avifauna. In contrast, only 7 families have resident species in West Virginia's forests but not in Costa Rican rainforests, and these families contribute only 10 species to West Virginia's resident forest avifauna. Second, several families that have resident representatives in both areas are much more diverse in Costa Rican rainforests. For example, approximately 21 species of hawks are rainforest residents, compared with only 4 resident forest species in West Virginia; 12 species of pigeons and doves are rainforest residents, but only 1 species is resident in West Virginia's forests; 44 species of tyrant flycatchers are rainforest residents but only 1 is a resident in West Virginia (although many species are migratory); and 14 species of wrens are rainforest residents but only 2 species are resident in West Virginia. This comparison shows that bird diversity is higher not only at the species level but also at higher taxonomic levels and highlights a common theme in tropical–temperate comparisons of diversity: There is often greater representation at higher taxonomic levels such as genus and family in tropical rainforests, and there are generally several higher level groups that are particularly rich in species in rainforests.

High bird species richness is typical of tropical rainforests, and Costa Rica's rainforests may be considered about average with respect to number of species. Approximately 410 species of birds have been found at La Selva in Costa Rica, an area of approximately 1500 ha. However, not all of these species are truly rainforest birds. If only rainforest birds are included (i.e., birds frequenting open water, marshes, cleared areas such as pastures, etc. are excluded), the total list is approximately 338 species. Approximately 314 rainforest species have been found on Barro Colorado Island and adjacent mainland areas, with a list of all species nearing 400. By contrast, about 554 species of birds have been recorded from Cocha Cashu, Peru, with nearly 400 truly rainforest species. To date, 575 species of birds have been found adjacent to the Explorer's Inn Reserve, an area of 5500 ha in the Amazon rainforest of southern Peru. The majority of these birds are rainforest species.

Other rainforest regions also have high bird diversity. Khao Yai National Park, Thailand, harbors 318 species, approximately 216 of which are true rainforest species. A single plot of <50 ha in southern Vietnam contains 164 species of forest birds. Makoukou, Gabon (2000 km^2), has 342 species of birds; 212 species have been recorded from Pasoh, Malaysia (800 ha), and 162 species from Gogol, Papua New Guinea (1000 ha). Although not all species recorded from these last three sites are true rainforest residents, the majority can be considered rainforest species.

VIII. DIVERSITY OF MAMMALS

Table III highlights the high diversity of rainforest mammals by comparing the approximate number of species that are found in Costa Rica's lowland rainforests with that of lowland forests of West Virginia (including species that were extirpated following European colonization). Costa Rican rainforests contain nearly three times as many species of mammals as do forests in West Virginia. A comparison of families shows the same trend with birds. There are 31 families represented in Costa Rica's rainforest mammal fauna and only 17 represented in West Virginia forests. The 20 families that have rainforest representatives in Costa Rica but not in West Virginia forests contribute 97 species. In contrast, 6 families contributing only 8 species have lowland forest species in West Virginia but not in Costa Rican forests. Particularly noteworthy are bats (order Chiroptera). Only 11 species of bats, all from a single family, are found in West Virginia forests, but approximately 89 species of bats in 9 families are found in Costa Rican rainforests. A single family, Phyllostomidae, contains 56 rainforest species. However, not all higher level taxonomic groups are more species rich in rainforests. For instance, the order Insectivora contains 9 lowland forest species in West Virginia but only 2 in Costa Rican rainforests. These 2 species are confined to the upper elevational limits of rainforest in Costa Rica and reach greater richness in montane forests. Similarly, there are 18 species of carnivores in West Virginia forests (4 of which have been extirpated from the state) and only 16 in Costa Rican rainforests.

Mammal diversity at single sites within rainforests is concomitantly high. However, species lists for most sites are certainly low because of inadequate sampling. In La Selva, Costa Rica, the mammal list comprises 117 species, with as many as 138 species expected to occur there based on geographical distributions of species that overlap the area but have not been recorded. As with birds, Costa Rica's rainforest mammal diversity can be considered average. Barro Colorado Island's list comprises 113 species, with as many as 144 species expected. Several study sites located in Amazonian rainforests have species lists of >130 species, with nearly 200 species expected from these sites. Makoukou's list

TABLE III
Approximate Numbers of Mammal Species in Lowland Rain Forests of Costa Rica Compared with Lowland Temperate Forests of West Virginia

Order and Family	West Virginia	Costa Rica	Order and Family	West Virginia	Costa Rica
Didelphimorphia			Castoridae	1	0
Didelphidae	1	8	Heteromyidae	0	1
Xenarthra			Zapodidae	1	0
Bradypodidae	0	1	Muridae	6	13
Megalonychidae	0	1	Erethizontidae	0	1
Dasypodidae	0	2	Dasyproctidae	0	1
Myrmecophagidae	0	3	Agoutidae	0	1
Insectivora			Echimyidae	0	2
Soricidae	7	2	Lagomorpha		
Talpidae	2	0	Leporidae	3	1
Chiroptera			Carnivora		
Emballonuridae	0	10	Canidae	5	1
Noctilionidae	0	2	Ursidae	1	0
Mormoopidae	0	4	Procyonidae	1	6
Phyllostomidae	0	56	Mustelidae	6	4
Natalidae	0	1	Mephitidae	2	0
Furipteridae	0	1	Felidae	3	5
Thyropteridae	0	1	Perissodactyla		
Vespertilionidae	11	9	Tapiridae	0	1
Molossidae	0	5	Artiodactyla		
Primates			Tayassuidae	0	2
Cebidae	0	4	Bovidae	1	0
Rodentia			Cervidae	2	2
Sciuridae	5	3	Total	58	154

of mammals is 199 species, Pasoh's is 89 species, and Gogol's is 27 species.

By far the most species-rich order of mammals in these rainforests is bats (Chiroptera). Even in well-studied sites, though, bat communities are incompletely known because of difficulties in sampling species such as those that forage above the canopy. New techniques that allow researchers to identify bats based on ultrasonic recordings are adding substantially to bat inventories at sites that have been sampled for decades by more traditional means such as mist netting.

IX. GEOGRAPHICAL PATTERNS OF ANIMAL DIVERSITY

Researchers have searched for both global and regional patterns of animal diversity. In particular, researchers have attempted to identify the major rainforest region that harbors the greatest diversity. The answer depends to some extent on the taxonomic group under consideration. At higher taxonomic levels (e.g., class and order), however, it is clear that the neotropics are generally richest in species. As noted for beetle species richness, Peru and Panama have the greatest numbers of recorded species per volume of canopy and subcanopy foliage, with the Asia-Pacific tropics, particularly Australia, having the lowest richness. Among butterflies, the neotropics again exceed other tropical regions with respect to numbers of species. Comparing species lists of countries or regions of approximately comparable size that occur in the tropics and that have substantial rainforest highlights this point. Panama in the neotropics has approximately 1550 species of butterflies, Liberia in the African tropics has about 720 species, and the Malay Peninsula in the Asia-Pacific tropics has 1031 species. Although not all species in these lists are true rainforest species, the trend of higher butterfly species richness in the neotropics is clear.

Among frogs, a similar trend is evident, with the Neotropical rainforests having the highest species richness. The greatest single-site frog species count is in Ecuador, with other Neotropical sites in Peru and Central America having nearly as many species. Sites in the African and Asia-Pacific tropics have considerably lower frog species richness, with the lowest being in insular sites in the Australopapuan region and rainforests of mainland Asia such as Thailand. However, some sites on the island of Borneo have nearly the same levels of frog diversity seen in some parts of the neotropics.

Ornithologists have long referred to South America as the bird continent because of its high bird diversity; approximately one-third of all species of birds occur on this continent. Within single rainforest sites, the highest bird diversity clearly occurs in the neotropics of Amazonia in Peru and surrounding countries on the eastern slope of the Andes. Neotropical rainforests are followed by the African and finally the Asia-Pacific rainforests. Lowest rainforest bird diversity on a per site basis is found in the Australopapuan region, particularly on smaller islands.

Panglobal comparisons of mammal diversity provide a somewhat different view. On a per site basis, mammal diversity is comparable in the neotropics and African tropics. However, bats are better represented in the neotropics, whereas primates are considerably more species rich in the African tropics.

On a regional scale (i.e., within one of the three major rainforest regions), researchers have noted substantial variability in species richness. For instance, with respect to butterflies, a belt of high species richness apparently extends across the neotropics from southern Colombia and the border of Peru and Bolivia at the base of the Andes eastward into the Brazilian states of Rondonia and Acre. However, even within this belt of high butterfly diversity lie areas with apparently lower diversity.

Within the Asia-Pacific tropics, lowest diversity of most taxa is found in the Australopapuan region, particularly in smaller and more isolated islands such as New Caledonia and those of the Bismarck Archipelago, Solomon Islands, Fiji, and Vanuatu. Many of these islands, particularly those in Fiji, New Caledonia, and Vanuatu, lack entire groups such as native rodents and amphibians that have been unable to disperse long distances over open ocean. Other taxa such as butterflies, lizards, and birds, although represented, are depauperate because of island isolation and small island area and a subsequent lack of opportunity for substantial allopatric speciation within an island. Only New Guinea, by far the largest island in the region, harbors taxa that have undergone substantial allopatric speciation.

Soils have a major effect on forest characteristics and productivity and consequently on animal diversity. Although soils of tropical rainforests are notoriously poor in nutrients, there is substantial variability. Where soils change abruptly, differences in forest structure and animal species richness are often evident. For example, the most species-rich sites in the world, those at the eastern base of the Andes, are situated on relatively richer soils. In contrast, the soils of the Guyana Shield in northern South America are particularly poor, and faunas are notably depauperate. Similarly, the sandy soils in parts of the Orinoco drainage of Venezuela are also very poor in nutrients, and animal diversity is often lower than that in surrounding areas with more fertile soils.

X. HYPOTHESES OF HIGH SPECIES RICHNESS

Explaining the high species richness of tropical rainforests has preoccupied ecologists and evolutionary biologists since naturalists first recognized the tropical–temperate latitudinal richness gradient and began to appreciate such richness. Many hypotheses have been advanced, but it is important to realize that none of these hypotheses has been adequately tested. It is also important to note that the great richness may be due to a combination of factors, and invoking a single-factor explanation ignores the complexities and intricacies of evolutionary and ecological processes and the long history of geological and climatalogical changes that certainly influenced diversity. Although the high species richness of animal consumers in tropical rainforests no doubt is linked with the high species richness of the plants that act as primary producers, invoking high plant diversity to explain high animal diversity is largely inadequate because it does not explain the high plant diversity. In other words, the root cause of higher tropical diversity is left unanswered by such circular reasoning.

Researchers have sought explanations that would account for high diversity of all taxa. Several of the more important hypotheses are briefly reviewed here and may be divided into biotic and abiotic hypotheses. Biotic hypotheses emphasize interactions among species to account for high species richness in the tropics relative to temperate and polar regions. Abiotic theories of species richness rely on climatological and geological processes that may promote greater speciation in the tropics. Several hypotheses are specific to terrestrial ecosystems, but tropical–temperate richness gradients are also evident in marine organisms. Ecologists and evolutionary biologists must therefore grapple simultaneously with the phenomena of high species richness in both terrestrial and marine systems to develop explanations that will apply to both systems.

The first of the biotic theories is the spatial heterogeneity theory, which states that the tropics are more heterogeneous in space and structurally more complex, thereby providing more niches for animals to exploit and thus a greater number of species. This theory relies on the development of greater vegetational complexity in tropical forests but does not explain adequately how this vegetational complexity arose.

The competition theory states that because of a more benign climate in the tropics, organisms live closer to the carrying capacity. Because organisms are closer to the carrying capacity, interspecific competition for limiting resources is more intense, and such competition promotes morphological and ecological divergence and specialization on a narrower range of resources. Specialization then allows more species to coexist, thereby promoting greater diversity. This theory is based on the presumption of a more benign climate, but quantifying how benign a climate is from each organism's perspective is not possible. It is also not possible to measure whether each organism in the tropics lives closer to its respective carrying capacity.

The predation theory is contrary to the competition theory and states that predation and parasitism are more intense in the tropics, which keep organisms below their carrying capacities. Because organisms are kept below their carrying capacities, more resources are available to support a greater number of species. It has not been determined whether or not predation rates are indeed higher in the tropics (at least not for the vast majority of animal species), and again it is not possible to determine whether or not most animals are below their carrying capacities.

The animal pollinators theory states that there is less wind in the tropics, and therefore plants have evolved a greater reliance on animals to pollinate their flowers. This greater reliance promotes a closer relationship with animals and greater specialization to enhance the mutualistic relationship, which therefore promotes diversity. Establishing a definitive link between a greater reliance on animal pollinators to less wind in the tropics is not possible.

The first of the abiotic theories is the ecological time theory, which states that the tropics are older because they have not been subjected to the devastating effects of glaciation that occurred in north temperate regions and that periodically covered entire ecosystems under the great ice sheets. A corollary of this theory is that tropical forests underwent a series of contractions and expansions with cyclic changes in precipitation as water was tied up in the expanding ice sheets and subsequently was released with global warming and contraction of the ice sheets. As the tropical forests contracted, they became isolated as refugia within a matrix of drier savanna and scrub ecosystems. Isolation into refugia promoted allopatric speciation and greater species richness. This corollary was an attractive explanation for pockets of endemism found in organisms such as birds in Amazonia but has become controversial.

The area theory states that the tropics occupy a greater area and that more species occur there because of the species–area relationship. This theory also states that there are more geographical barriers, such as mountain ranges and large rivers, in the tropics, which isolate populations, thereby reducing gene flow and promoting genetic divergence and allopatric speciation.

The energy theory states that there is greater incident solar radiation striking the tropics and that this solar radiation is more equitably distributed throughout the year. Primary production is consequently greater, which provides more resources for consumers, and more species can consequently coexist. This theory does not address the origin of the greater diversity of species.

The evolutionary speed theory states that generation times are shorter in the tropics and more mutations occur. More mutations give natural selection a greater genetic base on which to operate, and speciation rates are consequently higher, which leads to an accumulation of species. This theory and most of the preceding abiotic theories do not explain how more species can coexist in the tropics and consequently do not explain the maintenance of high species richness in the tropics.

XI. THE FUTURE OF RAINFOREST ANIMAL DIVERSITY

Much has been written about the rapid rate of tropical rainforest clearance by humans and subsequent impacts on animal diversity. Indeed, virtually every volume that has been written in the past two decades on rainforests addresses the potential for a catastrophic decline in global biodiversity. Several biologists have argued that the earth is on the verge of another mass extinction event of the magnitude of the Cretaceous event that eradicated the dinosaurs. Although such arguments are persuasive, it is not possible to predict accurately the magnitude of an impending extinction event, particularly because current knowledge of rainforest biodiversity is very limited and no precise estimate of numbers of species is available. However, with approximately 10% of the earth's entire butterfly fauna and untold thousands of species of other animals found within a single 200-ha tract of tropical rainforest, it is not difficult to surmise that the clearing of large tracts of such rainforests indeed will lead to a major decline in global animal diversity.

See Also the Following Articles

AMAZON ECOSYSTEMS • BIODIVERSITY-RICH COUNTRIES • FOREST CANOPIES, ANIMAL DIVERSITY • RAINFOREST ECOSYSTEMS, PLANT DIVERSITY • RAINFOREST LOSS AND CHANGE • TROPICAL ECOSYSTEMS

Bibliography

DeVries, P. J. (1987). *The Butterflies of Costa Rica and Their Natural History. Volume I: Papilionidae, Pieridae, Nymphalidae.* Princeton Univ. Press, Princeton, NJ.

DeVries, P. J. (1997). *The Butterflies of Costa Rica and Their Natural History. Volume II: Riodinidae.* Princeton Univ. Press, Princeton, NJ.

Dobson, A. P. (1996). *Conservation and Biodiversity.* Freeman, New York.

Erwin, T. L. (1982). Tropical forests: Their richness in Coleoptera and other arthropod species. *Coleopterist's Bull.* 36, 74–75.

Flannery, T. F. (1995). *Mammals of New Guinea.* Cornell Univ. Press, Ithaca, NY.

Gentry, A. H. (1990). *Four Neotropical Rainforests.* Yale Univ. Press, New Haven, CT.

Inger, R. F. (1980). Relative abundances of frogs and lizards in forests of Southeast Asia. *Biotropica* 12, 14–22.

Kricher, J. (1997). *A Neotropical Companion.* Princeton Univ. Press, Princeton, NJ.

Leigh, E. G., Jr. (1999). *Tropical Forest Ecology: A View from Barro Colorado Island.* Oxford Univ. Press, Oxford.

Mabberley, D. J. (1992). *Tropical Rain Forest Ecology,* 2nd ed. Blackie, Glasgow.

Reaka-Kudla, M. L., Wilson, D. E., and Wilson, E. O. (1997). *Biodiversity II: Understanding and Protecting Our Biological Resources.* Joseph Henry, Washington, D.C.

Stotz, D. F., Fitzpatrick, J. W., Parker, T. A., III, and Moskovits, D. K. (1996). *Neotropical Birds: Ecology and Conservation.* Univ. of Chicago Press, Chicago.

Voss, R. S., and Emmons, L. H. (1996). Mammalian diversity in Neotropical lowland rainforests: A preliminary assessment. *Bull. Am. Mus. Nat. History* 230, 1–15.

Whitmore, T. C. (1998). *An Introduction to Tropical Rain Forests,* 2nd ed. Oxford Univ. Press, New York.

Wilson, D. E., and Sandoval, A. (1996). *Manu: The Biodiversity of Southeastern Peru.* Smithsonian Institute, Washington, D.C.

Wilson, E. O., and Peter, F. M. (1988). *Biodiversity.* Academic Press, New York.

RAINFOREST ECOSYSTEMS, PLANT DIVERSITY

I. M. Turner
Singapore Botanic Gardens

I. Introduction
II. Overview of Tropical Rainforest Biodiversity
III. The Rainforest Community
IV. Why Are Tropical Rainforests So Diverse?
V. The Value of Tropical Rainforest Plant Diversity
VI. The Fate of the Tropical Rainforest

VEGETATION STRUCTURE, PHYSIOGNOMY, FLORISTIC COMPOSITION, AND CLIMATIC CONDITIONS have all been employed as means of defining tropical rain forest—one of the world's major biomes. There is no universally agreed definition, but nevertheless a strong consensus position emphasizes features such as a closed evergreen canopy; trees 25 m tall or higher; abundant epiphytes; the presence of large, thick-stemmed woody climbers (lianas); and warm, wet, and relatively aseasonal climate. Substantial spatial variation in the structure, physiognomy, and composition of tropical rainforest due to factors such as altitude, edaphic conditions, and local climate means that any circumscription tends either to be somewhat arbitrary or to lack general applicability.

I. INTRODUCTION

The tropical rainforests grow in conditions of high rainfall (usually more than 2000 mm per year) with few, if any, dry months, where a month would be considered dry if less than 100 mm of rain fell. The temperatures in the lowlands generally average approximately 27°C, with little variation throughout the year. The day course of temperature with peaks in the afternoon and lows before dawn is frequently the major signal in long-term records—the so-called tropical diurnal climate. Tropical rainforest will grow under climatic conditions of relatively short dry seasons, or even quite long ones if groundwater is available to the trees. As the length and severity of the seasonal drought increases, the proportion of deciduous species in the canopy of the forest tends to increase. When more than about one-third of the canopy trees are drought-deciduous, the forest is no longer tropical rainforest by many definitions.

The plant species of the lowland tropical rainforest are generally intolerant of freezing and frequently sensitive to chilling temperatures. Many major tropical plant families have few members outside the tropics (e.g., the nutmegs, Myristicaceae), possibly because they have not developed cold tolerance.

The tropical rainforests constitute a belt of evergreen vegetation that until relatively recently almost provided a continuous link across the landmasses near the equator. Only in East Africa is rainforest naturally rare. The largest contiguous areas of rainforest were the basins of two major rivers, the Amazon in South America and the Congo in Africa. These essentially continental rainforests can be contrasted with the largely insular forests of the Asia-Pacific region.

II. OVERVIEW OF TROPICAL RAINFOREST BIODIVERSITY

Biodiversity can be assessed at many different levels ranging from the genetic diversity within populations to landscape or regional heterogeneity. The scale of assessment most widely studied and best understood (though still poorly known) is that of species diversity, which is the focus of this review. The meaning of the term species remains a contentious issue in biology, but in practice in the tropics species are defined by the judgment of taxonomists based largely on the gross morphology of specimens.

There can be little doubt that tropical rainforests are the most species rich of terrestrial ecosystems. Naturally they would cover about 6% of the earth's land surface. Therefore, the high species diversity is not merely a reflection of the tropical rainforests covering an enormous area. The high diversity, relative inaccessibility, and remoteness from major centers of taxonomic research (mostly situated in Europe and North America) have compounded to slow progress in completing the basic inventory of the vascular plants of the humid tropics. Most tropical areas are studied by ongoing projects to provide floras, but progress on most of these is very slow. The *Flora of Tropical West Africa* is the only major tropical flora to have been completed. Others, such as *Flora Neotropica* and *Flora Malesiana*, have published accounts for relatively small proportions of the floras they cover despite decades of work. At this pace, it may take centuries to complete the flora production. However, the state of affairs may not be as bad as it seems because much information is available in monographs, local floras, and checklists. Nevertheless, some tropical areas, notably Central Amazonia, are still poorly known and estimates of plant diversity in terms of named species must be considered as provisional and tentative. Of extant living organisms, vascular plants are among the most completely described, but there remain many undescribed taxa. Exactly how many species remain unnamed cannot be estimated readily, but I suspect that the floristic inventory of the tropical forests has reached a stage at which the number of recognized species will change relatively little because descriptions of new species are heavily compensated for by synonymization among overdescribed taxa.

In Table I, I have attempted to estimate the total plant diversity (in terms of species richness) of the tropical rainforest regions of the world. It is difficult to assess the accuracy of this estimate, and until a species list is compiled it will remain uncertain. However, the

TABLE I

Estimates of the Number of Vascular Plant Species in the Tropical Rainforests of the World[a]

Area	No. of species	Notes
Neotropics	93,500	This includes some areas of vegetation other than tropical rainforest, but these are less species rich than the predominant tropical rainforest, so the overestimate is not too large.
African tropics	20,000	
Mainland Africa	16,000	Guineo–Congolian and Afromontane phytochoria totals combined. The overlap in species may be compensated by the failure to include species from the small rainforest areas of East Africa.
Madagascar	4,000	Assuming 5000 of the approximately 9500 species are from the island's tropical rainforest and that 80% are endemic (i.e., no overlap with mainland Africa).
Asia-Pacific tropics	61,700	
Malesia	45,000	I suspect this to be an overestimate, but New Guinea is still poorly known botanically and recent studies indicate high levels of diversity and endemism.
Indo-China and adjacent areas	10,000	This is an educated guess on my part. There have been few attempts to consider the flora of this region as a single entity.
Southern India	4,000	Largely the flora of the Western Ghats rainforests.
Sri Lanka	1,000	The endemic rainforest element.
Australia	700	The endemics of the Queensland rainforests that are largely Malesian floristically.
Pacific Islands	1,000	There is true rainforest on some of the Pacific Islands but because of their isolation they are relatively poor in species.
Total	175,200	

[a] Data mostly from WWF/IUCN (1994–1997) and Johns (1995).

derived data provide some interesting results. First, the vascular plant species richness of the tropical rainforests amounts to more than half of the estimated global total of 240,000. This clearly indicates the important contribution of tropical rainforests to global biodiversity. Second, on the broad geographic scale the data allow comparison of the three main rainforest blocks: the neotropics, tropical Africa (including Madagasacar), and the Far Eastern tropics from India into the Pacific. Africa evidently has far fewer species than the other two areas, and the neotropics are the most diverse. To a first approximation, a 3 : 2 : 1 ratio of species richness between tropical America, tropical Asia-Pacific, and tropical Africa can be considered, with an approximately equal proportion of diversity between the Old and New Worlds. At higher taxonomic levels the Neotropical dominance tends to disappear. For instance, it has been estimated that 292 higher plant families are represented in the neotropics, but Malesia, with fewer than half the number of species, has 310 families in its flora (Prance, 1994). Southern Indo-China is the region of the world most diverse in angiosperm families (Williams et al., 1994).

III. THE RAINFOREST COMMUNITY

Any visitor to a tropical rainforest can readily appreciate the botanical diversity they contain because of the many forms of plants to be observed ranging from tiny herbs to gigantic trees. There have been numerous attempts to classify plant species by their life-forms. Although there is no universally accepted or universally applicable system, most of these classifications use a similar set of criteria for determining groups within the structure. This set, at its simplest, generally includes plant mature stature and degree of woodiness, need for mechanical support from other plants, and information on trophic requirements. A classification of plant life-forms found in the tropical rainforest, based on these criteria, is given in Table II.

An area of lowland forest in Brazil would have relatively few plant species in common with a site in Costa Rica and very few indeed in common with sites in Congo or Indonesia, but all would be classified as lowland tropical rainforest. Despite this strong geographic dissimilarity in species composition, there are floristic affinities throughout the world. These occur at higher taxonomic levels, particularly at the rank of family, but also include some strong generic similarities in makeup. Tropical floras tend to be more similar to each other, in terms of species abundance in common families, than to subtropical and temperate floras (Turner, 1997). Table III summarizes floristic data by life-form and geographic area. Gentry (1988, 1993), in particular, emphasized the consistent patterns of family abundances, in terms of numbers of species represented in inventories of forest areas, across the tropics. Among woody stems, families such as Leguminosae, Rubiaceae, Euphorbiaceae, and Myristicaceae are nearly always among the most speciose at a lowland tropical rainforest site. Pteridophytes,[1] Araceae, and Zingiberaceae are frequently the most common ground herbs, and Orchidaceae and Pteridophytes almost invariably dominate the epiphytes. Table III lists 17 very large pantropical genera that are well represented in all regions, again emphasizing the floristic similarity of diverse tropical areas. However, there are also differences between regions.

The "dipterocarp substitution" has been noted as a major anomaly in the Asia-Pacific region (Gentry, 1988). That is, in this region the Dipterocarpaceae substitute for Leguminosae as the dominant woody family. The Dipterocarpaceae are a pantropical family, but the vast majority of its species occur in the region from India east to Wallace's line. In this area, dipterocarps form a majority of the large trees in lowland forests. However, this does not translate into dominance in terms of species abundance in the complete flora, or even the tree flora, unless a very large stem diameter limit is chosen. Legumes are not as numerically important as in American or African floras, but nevertheless they are still well represented, so the dipterocarp substitution is not as important a phenomenon as might be thought. Additionally, East Malesian rainforests have only a small representation of Dipterocarpaceae. Table III indicates other regional variations in representation of more important families. Perhaps the most notable of which are the virtual restriction of the important ground herb and epiphyte family Bromeliaceae to the neotropics; the importance of Bignoniaceae and Sapindaceae as climbers in the same region; the paucity of palms, particularly understory species, in Mainland Africa (although not Madagascar); and the abundance of Dichapetalaceae. There are many other smaller families with restricted ranges in the tropics.

Because of the range of spatial scales that can be considered and the relative inclusivity of the sampling, many approaches are available for assessing plant species richness of tropical forests. These include plot, transect, and plotless sampling at small spatial scales

---

[1] Pteridophyta are frequently accorded "family status" in these analyses.

TABLE II
Classification of Vascular Plant Life-Forms in the Tropical Rainforest and a Summary of Their Taxonomic Composition

Life-form group	Definition	Examples from tropical rainforest	Subdivisions of the group
Trees	Autotrophic, woody, and mechanically independent (self-supporting) plants	Many taxa within the dicotyledons, palms among the monocots, a few conifers and gnetums, some cycads and tree ferns.	Mature size and form have often been used to subdivide the tree category. Unfortunately, no consistent definition has been employed.
Herbs	Autotrophic, mechanically independent plants with little or no lignification and secondary thickening of stems or roots	Many dicot and monocot taxa, many ferns; some characteristically epiphytic species are facultatively terrestrial when they fall from the trees.	The giant herbs, largely from the Zingiberales, can be distinguished by size, and in ecological terms may have more in common with trees than the other herbs.
Climbers	Autotrophic, woody, or herbaceous plants, rooted in the soil and dependent on other plants for support, at least in later stages	A wide range of taxa including dicots, monocots (e.g. palms and aroids), gymnosperms (*Gnetum* spp.), and ferns.	Woody climbers (lianas) are sometimes distinguished from herbaceous ones; means of climbing can also be employed
Epiphytes	Autotrophic, woody, or herbaceous species habitually growing on other plants and not rooting in the soil	Epiphytes are dominated by three groups—orchids, pteridophytes, and aroids; bromeliads are an almost exclusively Neotropical epiphyte group.	Division into herbaceous and woody forms might be possible. Often, a distinction is made between species on their relative preference for crown and bole locations on their host trees.
Hemiepiphytes	Plant, mostly woody, that begin life as epiphytes but may eventually grow roots down to the ground	Strangling figs (*Ficus* spp.) are the largest group of hemiepiphytes.	
Plant parasites	Plants that directly parasitise other plants	Direct plant parasitism is confined to the dicots, and with the exception of *Cassytha* (Lauraceae) is restricted to probably advanced groups. Mistletoes (Loranthaceae and Viscaceae) and related root-parasitic groups (Santalaceae and Olacaceae) and the mostly herbaceous Scrophulariaceae form the two largest groups.	The parasites can be divided into hemiparasites that are chlorophyllous but possess haustorial connection to host xylem (e.g., mistletoes), holoparasites that are achlorophyllous and entirely heterotrophic (e.g., dodders, broomrapes, and Balanophoraceae), and endoparasites that live inside the host and are only visible externally as reproductive structures (Rafflesiaceae).
Mycotrophs	Heterophic plants that derive carbohydrate from a fungus; the fungus may derive its carbon from dead material or from another living plant	All orchids are mycotrophic as seedlings, but most become autotrophic as they develop. The small proportion that remain entirely heterotrophic represent the largest group of mycotrophic species. Other monocot mycotrophs include members of the families Petrosaviaceae, Triuridaceae, Iridaceae, Burmanniaceae, and Corsiaceae. Ericaceae, Gentianaceae, and Polygalaceae are the only dicot families with mycotrophic species.	Detailed investigations can distinguish the "parasitic" from the "saprophytic" species but these have been conducted for relatively few of the tropical species.

that may include all plant taxa present, larger plot sampling that considers only trees and climbers exceeding a certain lower stem diameter limit, and local or regional floristic accounts based on intensive collection within the area concerned. Relatively few efforts have been made to identify all the individual plants present within an area of rainforest largely because few people possess the botanical expertise to identify the enormous amount

TABLE III

Summary of the Floristic Composition of Tropical Rainforest, with a Breakdown by the Major Geographic Regions

		Regional breakdown[a]		
	All tropics	Neotropics	Africa (including Madagascar)	Asia-Pacific
Important tree families	Leguminosae	**Vochysiaceae**	Dichapetalaceae	**Dipterocarpaceae**
	Lauraceae	Bignoniaceae	Olacaceae	Myrtaceae
	Annonaceae	**Cyclanthaceae**		
	Rubiaceae	Lecythidaceae	<u>Moraceae</u>	
	Moraceae		<u>Palmae</u>	
	Myristicaceae	<u>Ebenaceae</u>	<u>Fagaceae</u>	
	Sapotaceae	**<u>Pandanaceae</u>**	<u>Bombacaceae</u>	
	Meliaceae			
	Palmae			
	Euphorbiaceae			
Important herb families	Pteridophyta	Bromeliaceae		
	Araceae	Gramineae		
	Zingiberaceae	Heliconiaceae		
	Cyperaceae			
	Rubiaceae			
	Gesneriaceae			
	Orchidaceae			
Important climber families	Asclepiadaceae	Compositae	Annonaceae	Annonaceae
	Convolvulaceae	Bignoniaceae	Dichapetalaceae	Palmae
	Leguminosae	Sapindaceae		**Nepenthaceae**
	Araceae	Malpighiaceae		
	Apocynaceae	Passifloraceae		
	Cucurbitaceae			
	Rubiaceae			
Important epiphyte families	Orchidaceae	**Bromeliaceae**		Asclepiadaceae
	Pteridophyta	**Cyclanthaceae**		Rubiaceae
	Araceae	**Cactaceae**		
	Piperaceae			
	Melastomataceae			
	Gesneriaceae			
Important genera (>500 spp.)	*Asplenium*	*Anthurium*		*Dendrobium*
	Begonia	*Epidendrum*		*Eria*
	Bulbophyllum	*Eugenia*		*Rhododendron*
	Clerodendrum	*Lepanthes*		*Syzygium*
	Croton	*Maxillaria*		
	Cyathea	*Miconia*		
	Dioscorea	*Oncidium*		
	Ficus	*Peperomia*		
	Habenaria	*Pleurothallis*		
	Impatiens	*Stelis*		
	Justicia			
	Phyllanthus			
	Piper			
	Psychotria			
	Schefflera			
	Selaginella			
	Solanum			

[a] Names in bold indicate families entirely restricted to a particular geographic region, or nearly so. Names underlined indicate families notable for their scarcity or absence from a particular region. Names in bold and underlined are entirely absent from a particular region, or almost so.

of juvenile and sterile plant material that such total inventories generate. Recently, an attempt was made to enumerate all the plant species occurring in 1 ha of forest in lowland Amazonian Ecuador. A total of 942 species were encountered in the 100 × 100-m plot (Balslev et al., 1998). Lower montane forest at 1500 m in Java was estimated to contain 333 species in 1 ha (Meijer, 1959). A single plot of 10 × 10 m in Costa Rica contained 233 species (Whitmore et al., 1985), more than any among eight plots of the same size recorded in Gabon (74–130 spp.; Reitsma, 1988). Duivenvoorden (1994) enumerated all the vascular plant species in a series of ten 0.1-ha plots in Colombia. The most species-rich plot contained 313 species. Gentry and Dodson (1987a) recorded 365 species from 0.1 ha at Rio Palenque, Ecuador, but this was from 10 noncontiguous subplots. Two other sites in Ecuador, sampled with the same methods, were considerably less diverse, with 169 and 173 species recorded (Gentry and Dodson 1987b).

Most estimates of plant diversity in tropical forests involve the enumeration of stems above a certain diameter (or girth) at breast height (1.3 m) or above obvious outgrowths, such as buttresses or stilt roots. Many different sampling methods and size limits have been employed in such investigations, but fortunately a few approaches have been applied relatively consistently and allow intersite comparisons of diversity. Three commonly employed methods of assessing plant diversity in tropical forests produce scales of species-richness values that are comparable, at least in the broad view (Table IV). I used these data to define categories of

TABLE IV

Categories of Plant Diversity in Lowland Tropical Rainforests

Species-richness class	No. of species[a]	Examples from different geographic regions[a]		
		Neotropics	Africa	Asia-Pacific
Low	Less than 100	Barro Colorado Island, Panama;[f] San Carlos, Venezuela;[e] Belém, Brazil[k]	Kibale, Uganda;[e] Kade, Ghana;[e] Lopé, Gabon[f]	Mulu (limestone), Malaysia[o]
Intermediate	100–199	Many, including La Selva, Costa Rica[e]	Makoukou, Gabon;[h] Korup, Cameroon;[h] Oveng, Gabon[f]	Davies River, Queensland;[j] Xishuangbanna, China;[l] Danum Valley, Malaysia[m]
High	200–249	Coca, Ecuador;[g] Jatun Sacha, Ecuador;[g] Mishana, Peru[i]	Nosy Mangabe, Madagascar[h]	Pasoh, Malaysia;[j] Lambir, Malaysia;[e] CMBRS, Papua New Guinea[n]
Very high	250 or more	Cuyabeno, Ecuador;[b] Yanamono, Peru;[e] Bajo Calima, Colombia;[c] Tutunendo, Colombia;[c] Manaus, Brazil[d]	None recorded	None recorded

[a] Species richness can be measured in one of three ways: (x) Species represented among the trees and lianas of 10 cm dbh (or 30 cm gbh) or greater on a 100 × 100-m (1 ha) plot; (y) species represented among the first 500 tree and liana individuals of 10 cm dbh (or 30 cm gbh) or greater on a contiguous plot; and, (z) species represented among the trees and lianas of 2.5 cm dbh or greater on a 0.1-ha sample consisting of ten 50 × 2-m transects.
[b] Valencia et al. (1994) (x).
[c] Faber-Langendoen and Gentry (1991) (x), Gentry (1986) (z).
[d] Ferreira and Rankin-de-Mérona 1998 (x).
[e] Phillips et al. (1994) (y).
[f] Condit et al. (1996) (x).
[g] Valencia et al. (1998) (x).
[h] Gentry (1993) (z).
[i] Gentry and Emmons (1987) (z).
[j] Gentry (1988) (z).
[k] Black et al. (1950) (x).
[l] Cao et al. (1996) (x).
[m] Newbery et al. (1992) (x).
[n] Wright et al. (1997) (x).
[o] Proctor et al. (1983) (x).
[p] Reitsma (1988) (x).

relative diversity for lowland tropical forests. Forests of very high diversity have thus far only been reported from the New World. They are known from the lowland forests of the Pacific coast of Colombia in Chocó Province and the Upper Amazon Basin stretching in a belt below the eastern slopes of the Andes from Ecuador to Peru. The occurrence of 250 species in a 1-ha plot near Manaus in the center of the Brazilian Amazon (Ferreira and Rankin-de-Mérona, 1998) countered the impression that the forests of the middle and lower Amazon were of uniformly lower diversity than the uppermost reaches. High-diversity forests are well represented in the South American tropics and are of similar species richness as the lowland dipterocarp forests of West Malesia and at least some of the lowland forests of New Guinea. To date, only Madagascan rainforests have been shown to attain high-diversity status among those from Africa. The most diverse mainland African rainforests occur in Cameroon and Gabon in West Africa, but these are substantially less rich than many forest sites in the neotropics or the Asia-Pacific region and on a global scale could only be classed as of intermediate diversity. Many sites where detailed ecological research has been conducted, such as Barro Colorado Island, Panama, San Carlos de Rio Negro, Venezuela, and Kibale, Uganda, are low-diversity rainforests. These geographic patterns in local species richness appear to mirror quite closely the tropical regions of the map of plant species diversity per 10,000 km^2 (Barthlott *et al.*, 1996; see *http://www.botanik.uni-bonn.de/biodiv/phytodiv.htm*).

An important point is that the range of values for diversity within a locale, i.e., an area of approximately identical climate, can cover much of the general range exhibited across wide geographic and climatic gradients. For example, at Caquetá in Colombia total vascular plant diversity in ten 0.1-ha plots was 40–313 species (Duivenvoorden, 1994), at Tambopata in Peru five 1-ha plots had 60–173 species for stems of 10 cm dbh or larger (Phillips *et al.*, 1994), and at Gunung Mulu National Park in Sarawak four plots varied between 73 and 223 species for the same plot and tree sizes as those of the previous example (Proctor *et al.*, 1983). In all these cases the lowest diversity plots were ones on predictably harsher substrates, such as permanently waterlogged soils (Caquetá and Tambopata) or limestone karst (Gunung Mulu). Soil physicochemical properties are highly influential on the composition of the forest community and clearly affect species richness. However, there are also other environmental factors that can lead to gradients in plant diversity (Table V). Rainfall appears to be the major determinant of diversity when old-growth sites on freely draining, nonextreme soils are compared. A relatively strong positive correlation

TABLE V

Environmental Trends in Tree Species Richness per Unit Area in Tropical Rainforest[a]

Number of species increases with total precipitation.
Number of species decreases with increasing seasonality of rainfall.
Number of species increases with soil fertility, although diversity may level off, or even decline, on the most fertile sites.
Number of species declines with altitude.

[a] After Givnish (1999).

between average annual precipitation and diversity has been observed in many tropical data sets (Givnish, 1999), although good data for total plant diversity is lacking so that only estimates for large woody stems can be used. This correlation is evident over the range of true lowland tropical rainforest and is not merely a comparison of diversity between drier forest formations and proper rainforest. Rainfall seasonality (the presence and severity of a dry season) is of key importance, but it is difficult to quantify this from standard meteoriologial data, and total annual rainfall is a good predictor of diversity in most cases. It is not immediately evident why further increasing the wetness of a wet climate should continue to increase the local species diversity, although highly ombrogenous sites may favor epiphytes. As can be seen from the data presented in Table VI, the proportion of the flora that epiphytes represent increases quite sharply with average annual rainfall.

IV. WHY ARE TROPICAL RAINFORESTS SO DIVERSE?

Lowland tropical rainforests undoubtedly represent the zenith of terrestrial biodiversity, probably with more than half the world's biota on less than a twentieth of the land surface. High plant diversity may well be a major contributory factor to the overall abundance of species in the rainforest through the specialization of a myriad of invertebrate species to each plant. However, why are there so many plant species in the tropics? In evolutionary terms, high species diversity must imply many speciation events and relatively few extinctions of species because diversity is the balance between these two processes. Therefore, the tropics may be richer in species than other areas of the world because there has been more rapid speciation here and/or a slower rate of extinction. It is possible that the benign and equable climate of the tropics makes it less likely that a species

TABLE VI

Representation of Major Life-Form Groups in the Floras of Different Tropical Rainforest Sites[a]

Site	Mean annual rainfall (mm)	Total No. spp.	Percentage of species in major life-form groups			
			Trees	Herbs	Climbers	Epiphytes
La Selva, Costa Rica[b]	3950	1450	39	27	7	27
Rio Palenque, Ecuador[c]	3000	1055	25	37	16	23
Barro Colorado Island, Panama[d]	2650	966	43	21	16	20
Manu floodplain, Peru[e]	2050	1217	57	13	18	12
Singapore (forest flora)[f]	2000	1673	56	13	18	13
Jauneche, Ecuador[c]	1850	608	28	38	22	12

[a] Sites are listed in descending order of mean total annual rainfall.
[b] Hammel (1990).
[c] Gentry and Dodson (1987b).
[d] Foster and Hubbell (1990).
[e] Foster (1990).
[f] Turner (1994).

will become extinct than in the harsher climates elsewhere. The positive correlation between forest diversity and annual precipitation within the tropics also indicates a link between environmental adversity and reduced diversity.

Switching from an evolutionary to an ecological perspective, tropical rainforest diversity maintains a perplexing fascination. A commonly held view of natural communities is that they consist of populations of species in which intraspecific competition is stronger than interspecific competition. In other words, species occupy separate niches within the community and have reached a competetive equilibrium. An equilibrium model of the structure of a lowland tropical rainforest plant community requires that all the species present, including the many species of similar life-form and basic ecology such as canopy trees, or understory treelets, occupy separate niches. Given the simple requirements of plants—light, water, and nutrients from the soil—it is not immediately obvious where so many niches could come from to support the observed diversity. Differences in requirements for reproduction and regeneration may allow more axes from which to partition niches. These niches could be extremely narrow because of the constancy of the tropical environment allowing persistence of small or sparse populations. I suspect, however, that few tropical forest ecologists believe that all the plant species in a forest are ecologically distinct. A more plausible hypothesis is that most species have ecologically equivalent or similar species with which they cooccur. It is worth noting that a high proportion of the species present in a rainforest exist as populations intermingled with those of congenerics. For instance, of the 814 tree species recorded from 50 ha at Pasoh, Malaysia, 82% had a congeneric present in the plot and 70% had a congeneric present in the same broad stature class (Manokaran et al., 1992). For descriptive convenience, ecologists may group species into guilds of ecologically similar entities such as pioneer trees or bole climbers, but it must be remembered that these are not necessarily discrete groupings in terms of niche space in the community. They may represent arbitrary divisions of whatever axes of differentiation are being considered.

If competetive equilibrium has not been obtained then something must be preventing competitive exclusion within the community from occurring. One possible process by which this could happen is through frequency-dependent mortality operating within the community setting an upper limit on the abundance of a species and thus making it unlikely that it can totally exclude other species from the landscape. This compensatory mortality, particularly among tropical trees, is often referred to as the Janzen–Connell effect after the two researchers who independently put forward the hypothesis that species-specific predators could produce the density-dependent mortality required to promote the coexistence of many plant species in a forest. Invertebrate seed and seedling predators, and plant pathogens such as fungi, have frequently been observed to cause severe mortality of germinants and juveniles occurring in high densities near parent trees. Detailed analyses of spatial patterns of sapling mortality on Barro Colorado Island, Panama, showed evidence of compensatory mortality in recruitment for 67 of 84 of the most common tree species (Wills *et al.*, 1997). Rare species, which make up the bulk of the diversity in the forest, may not reach sufficient adult densities to suffer compensatory mortality. However, they may just be too rare to provide statistically significant results with the available data sets. Givnish (1999) hypothesized that compensatory mortality may increase in intensity with rainfall because pest and pathogen populations are less likely to be checked by a dry season in an ever-wet forest. This may explain, at least partially, the observed increase in forest diversity with precipitation.

An alternative nonequilibrium view of the community is that chance is the major determinant of community composition and relative abundance. The majority of species are so infrequent that they very rarely come into direct competition with other ecologically equivalent, similarly sparse species. Chance, and not competitive ability, is an important determinant of which species establishes an adult successfully at a given site because most species are unable to produce enough seeds, disperse them well enough, or survive the early onslaughts of pests and pathogens to ensure that they have a candidate for any vacancy in the community (Hubbell *et al.*, 1999). Many potential episodes of interspecific competition may be won by default, and by competitively inferior species, because no adversaries were present at the right time and place.

At present we have no simple answer to the question of why there are so many species in the rainforests and how they manage to coexist. Nor does it seem likely that there is a simple answer. Probably all the factors mentioned previously, and possibly others not yet considered, are operating within any rainforest but probably not affecting all members of the community simultaneously.

V. THE VALUE OF TROPICAL RAINFOREST PLANT DIVERSITY

If tropical rainforests are to be conserved, conservationists need to convince a wide range of people that tropical rainforests have sufficient value to merit conservation. In attempting to do so, conservationists face a series of dilemmas. The first is whether to consider value as an economic parameter or not. The global economy hardly has a good record for preventing extinction or supporting sustainable exploitation, but the reality is that people in the tropics exploit the rainforest so that they can improve their standard of living, even if only in the short term. Conservationists can promote the development of a truly sustainable global economy, but in the meantime they must address the immediate problem of impending extinctions. Purely economic analyses generally favor rapid liquidation of forest resources, although a notable exception was a comparison of exploitation of minor forest products, such as fruits and latexes, with logging and timber tree plantations in Peru (Peters *et al.*, 1989) which showed that the relatively sustainable nontimber production generated greater revenues and thus put the highest net present value on the resource. However, the study has been criticized as being atypical in having a forest with a very high density of fruit and latex trees relatively close to a major population center providing a ready market for the products.

Are there any economic benefits to be gained from leaving rainforest undisturbed? A group of people for which the answer to this question is generally in the affirmative are traditional forest dwellers. Few of the tropical rainforest regions are without people who live in the forest, either making their living by hunting and gathering in the forest or by farming on a small scale, usually as some form of traditional shifting agriculture. Large-scale forest clearance dispossesses the traditional forest dwellers of their entire resource base and often much of their social fabric.

Tourism potential is perhaps the most economically lucrative opportunity offered by virgin tropical rainforest, but fortunately we are not yet in a situation in which every remaining area of forest can attract enough paying visitors to make conservation profitable. However, at a local level, ecotourism probably offers the best hope of generating an income from unexploited forest. Plant biodiversity, in addition to providing the glorious backcloth, is rarely likely to be the focus of ecotourism, although *Rafflesia* may be a notable exception.

The value of the tropical rainforest as a storehouse of potentially useful genetic diversity is probably the best claim for a high intrinsic value to be placed on biodiversity. For vascular plants the potential uses are largely of three main types: medicinal, novel crops, and novel genes for improvement of existing crop varieties. A new drug that is superior to other available medicines can be worth millions of dollars, not to mention that it can have tremendous humanitarian value in the reduction in suffering. New or improved crops also have substantial economic value. Given the massive concentration of biodiversity in the tropics, it is almost certain that superior and economically rewarding new products await discovery in the rainforest. However, searching the haystack for the few golden needles is a difficult, expensive, and often fruitless procedure. Island biogeography theory predicts that a 90% loss of habitat equates to a 10% reduction in diversity. Therefore, the potential economic returns from undiscovered rainforest products seem unlikely to provide strictly economic arguments for preventing the destruction of tropical forests until the area left is very small.

Forests are important for their so-called "ecosystem services"—the amelioration of atmospheric composition, water quality, and local climatic regime among a variety of beneficial emergent properties of the rainforest community. Despite considerable speculation and some very simplistic experimentation, there is no solid evidence that species richness per se has a major influence on these community-level processes. Most species in the rainforest are very rare, so it seems unlikely that the random removal of a certain proportion of the flora would consistently influence oxygen production, carbon sequestration, nutrient cycling, and transpiration.

In the prevailing international financial framework it is difficult to find a sound, strictly economic, argument for conserving tropical rainforest. When environmental and ecological sustainability become the key goals of the global economy, conservation will make economic sense. The biodiversity of the tropics has a unique and irreplaceable intrinsic value as well as cultural, spiritual, educational, and humanitarian worth to millions of people, many of whom have never been in a tropical rainforest.

VI. THE FATE OF THE TROPICAL RAINFOREST

The human population of the tropics has already destroyed the rainforest occupying half the area it would be expected to cover in today's climatic regime. Much of the remaining forest is fragmented into small remnant patches, damaged, burnt, or defaunated. Extensive, relatively pristine forests are largely confined to Amazonia, the Congo Basin, and New Guinea. Most tropical rainforest nations have little remaining forest, and although they generally have designated national parks, forest reserves, or other protected area status, the actual protection that these sites receive is often inadequate to prevent deterioration. There are of course exceptions, and excellent tropical rainforest nature reserves do exist.

There is no sign that the global rates of deforestation in the tropics are declining. Tropical forests are cleared for a variety of purposes and the relative importance of these varies between regions (Bawa and Dayanandan, 1997). In Africa and parts of South America, the demand for land by a large population of poor people who need to subsist on what they can grow leads to deforestation. Clearance for cattle ranching is also common in Latin America. Establishment of plantations for commercial tree crops, such as oil palm (*Elaeis guineensis*) and rubber (*Hevea brasiliensis*), is a more destructive force in the Asian tropics. Human populations continue to grow rapidly throughout the tropics, making it unlikely that the pressure for land for food production or cash crops will decline.

There is increasing evidence that the tropical rainforest community, particularly its botanical component, is more resistant to human disturbance and fragmentation than has been feared. Studies have shown that selective logging does not reduce tree diversity very much (Cannon *et al.*, 1998) and long-isolated forest fragments maintain relatively high proportions of their original floras (Turner and Corlett, 1996). This information is met with some ambivalence by conservationists, who have tended to stress the fragility of tropical forest ecosystems. It is positive and hopeful in that it provides some scientific support to the aspirations for sustainable exploitation of tropical forests and the persistence of the native biota in nations in which only fragments remain. However, there is a real concern that knowledge of this relative robustness will be used as an excuse by those with political power to allow greater levels of exploitation and deforestation, possibly including inside currently protected areas.

From a pragmatic viewpoint, it is obvious that a multiplicity of approaches will be required to conserve the tropical rainforest biota. Where there are still large areas of relatively undisturbed forest with low human population densities, efforts to establish large inviolate parks are warranted, but in areas with a landscape dominated by human activity the remnant patches of the original forest have an important conservation value in their own context. These fragments may not contain all the original species, but they still support a rich diversity that otherwise might be completely lost. In highly degraded tropical landscapes, which are very common, forest fragments may also play a vital role in the reforestation of these derelict lands by acting as foci for tree establishment and sources of planting material.

See Also the Following Articles

AMAZON ECOSYSTEMS • BIODIVERSITY-RICH COUNTRIES • DEFORESTATION AND LAND CLEARING • ECONOMIC VALUE OF BIODIVERSITY, OVERVIEW • FOREST CANOPIES, PLANT DIVERSITY • PLANT BIODIVERSITY, OVERVIEW • RAINFOREST ECOSYSTEMS,

ANIMAL DIVERSITY • RAINFOREST LOSS AND
CHANGE • TROPICAL ECOSYSTEMS

Bibliography

Balslev, H., Valencia, R., Paz y Miño, G., Christensen, H., and Nielsen, I. (1998). Species count of vascular plants in one hectare of humid lowland forest in Amazonian Ecuador. In *Forest Biodiversity in North, Central and South America, and the Caribbean* (F. Dallmeier and J. A. Comiskey, Eds.), pp. 585–594. UNESCO, Paris.

Barthlott, W., Lauer, W., and Placke, A. (1996). Global distribution of species diversity in vascular plants: Towards a world map of phytodiversity. *Erdkunde* 50, 317–327.

Bawa, K. S., and Dayanandan, S. (1997). Socioeconomic factors and tropical deforestation. *Nature* 386, 562–563.

Black, G. A., Dobzhansky, T., and Pavan, C. (1950). Some attempts to estimate species diversity and population density of trees in Amazonian forests. *Bot. Gaz.* 111, 413–425.

Cannon, C. H., Peart, D. R., and Leighton, M. (1998). Tree species diversity in commercially logged Bornean rainforest. *Science* 281, 1366–1368.

Cao, M., Zhang, J., Feng, Z., Deng, J., and Deng, X. (1996). Tree species composition of a seasonal rain forest in Xishuangbanna, Southwest China. *Trop. Ecol.* 37, 183–192.

Condit, R., Hubbell, S. P., LaFrankie, J. V., Sukumar, R., Manokaran, N., Foster, R. B., and Ashton, P. S. (1996). Species–area and species individual relationships for tropical trees: A comparison of three 50-ha plots. *J. Ecol.* 84, 549–562.

Duivenvoorden, J. F. (1994). Vascular plant species counts in rain forests of the middle Caquetá area, Colombian Amazonia. *Biodivers. Conserv.* 3, 685–715.

Faber-Langendoen, D., and Gentry, A. H. (1991). The structure and diversity of rain forests at Bajo Calima, Chocó Region, Western Colombia. *Biotropica* 23, 2–11.

Ferreira, L. V., and Rankin-de-Mérona, J. M. (1998). Floristic composition and structure of a one-hectare plot in terra firme forest in Central Amazonia. In *Forest Biodiversity in North, Central and South America, and the Caribbean* (F. Dallmeier and J. A. Comiskey, Eds.), pp. 649–662. UNESCO, Paris.

Foster, R. B. (1990). The floristic composition of the Rio Manu floodplain forest. In *Four Neotropical Forests* (A. H. Gentry, Ed.), pp. 99–111. Yale Univ. Press, New Haven, CT.

Foster, R. B., and Hubbell, S. P. (1990). The floristic composition of the Barro Colorado Island forest. In *Four Neotropical Forests* (A. H. Gentry, Ed.), pp. 85–98. Yale Univ. Press, New Haven, CT.

Gentry, A. H. (1986). Species richness and floristic composition of Choco Region plant communities. *Caldasia* 15, 71–91.

Gentry, A. H. (1988). Changes in plant community diversity and floristic composition on environmental and geographical gradients. *Ann. Missouri Bot. Gardens* 75, 1–34.

Gentry, A. H. (1993). Diversity and floristic composition of lowland tropical forest in Africa and South America. In *Biological Relationships between Africa and South America* (P. Goldblatt, Ed.), pp. 500–547. Yale Univ. Press, New Haven, CT.

Gentry, A. H., and Dodson, C. (1987a). Contribution of nontrees to species richness of a tropical rain forest. *Biotropica* 19, 149–156.

Gentry, A. H., and Dodson, C. H. (1987b). Diversity and biogeography of neotropical vascular epiphytes. *Ann. Mis. Bot. Gardens* 74, 205–233.

Givnish, T. J. (1999). On the causes of gradients in tropical tree diversity. *J. Ecol.* 87, 193–210.

Hammel, B. (1990). The distribution of diversity among families, genera, and habit types in the La Selva flora. In *Four Neotropical Forests* (A. H. Gentry, Ed.), pp. 75–84. Yale Univ. Press, New Haven, CT.

Hubbell, S. P., Foster, R. B., O'Brien, S. T., Harms, K. E., Condit, R., Wechsler, B., Wright, S. J., and Loo de Lao, S. (1999). Light-gap disturbances, recruitment limitation, and tree diversity in a neotropical forest. *Science* 283, 554–557.

Johns, R. J. (1995). Endemism in the Malesian flora. *Curtis's Bot. Mag.* 12, 95–110.

Monokaran, N., LaFrankie, J. V., Kochummen, K. M., Quah, E. S., Klahn, J., Ashton, P. S., and Hubbell, S. P. (1992). Stand table and distribution of species in the 50-ha research plot at Pasoh Forest Reserve. Forest Research Institute of Malaysia, Research Data 1, pp. 1–454.

Meijer, W. (1959). Phytosociological analysis of montane rainforest near Tjibodas, West Java. *Acta Bot. Neerl.* 8, 277–291.

Newbery, D. M., Campbell, E. J. F., Lee, Y. F., Ridsdale, C. E., and Still, M. J. (1992). Primary lowland dipterocarp forest at Danum Valley, Sabah, Malaysia: Structure, relative abundance and family composition. *Philos. Trans. R. Soc. London B* 335, 341–356.

Peters, C. M., Gentry, A. H., and Mendelsohn, R. O. (1989). Valuation of an Amazonian rain forest. *Nature* 339, 655–656.

Phillips, O. L., Hall, P., Gentry, A. H., Sawyer, S. A., and Vázquez, R. (1994). Dynamics and species richness of tropical rain forests. *Proc. Natl. Acad. Sci. USA* 91, 2805–2809.

Prance, G. T. (1994). A comparison of the efficacy of higher taxa and species numbers in the assessment of biodiversity in the neotropics. *Philos. Trans. R. Soc. London B* 345, 89–99.

Proctor, J., Anderson, J. M., Chai, P., and Vallack, H. W. (1983). Ecological studies in four contrasting lowland rain forests in Gunung Mulu National Park, Sarawak I. Forest environment, structure and floristics. *J. Ecol.* 71, 237–260.

Reitsma, J. M. (1988). Forest vegetation of Gabon. *Tropenbos Technol. Ser.* 1, 1–103.

Turner, I. M. (1994). The taxonomy and ecology of the vascular plant flora of Singapore: A statistical analysis. *Bot. J. Linnean Soc.* 114, 215–227.

Turner, I. M. (1997). A tropical flora summarized—A statistical analysis of the vascular plant diversity of Malaya. *Flora* 192, 157–163.

Turner, I. M., and Corlett, R. T. (1996). The conservation value of small, isolated fragments of lowland tropical rain forest. *TREE* 11, 330–333.

Valencia, R., Balslev, H., and Paz y Miño, G. (1994). High tree alpha diversity in Amazonian Ecuador. *Biodiversity Conserv.* 3, 21–28.

Valencia, R., Balslev, H., Palacios, W., Neill, D., Josse, C., Tirado, M., and Skov, F. (1998). Diversity and family composition of trees in different regions of Ecuador: A sample of 18 one-hectare plots. In *Forest Biodiversity in North, Central and South America, and the Caribbean* (F. Dallmeier and J. A. Comiskey, Eds.), pp. 569–584. UNESCO, Paris.

Whitmore, T. C., Peralta, R., and Brown, K. (1985). Total species count in a Costa Rican tropical rain forest. *J. Trop. Ecol.* 1, 375–378.

Wills, C., Condit, R., Foster, R. B., and Hubbell, S. P. (1997). Strong density- and diversity-related effects help to maintain species diversity in a neotropical forest. *Proc. Natl. Acad. Sci. USA* 94, 1252–1257.

World Wildlife Fund/International Union for the Conservation of Nature (WWF/IUCN) (1994–1997). *Centres of Plant Diversity*, 3 vols. IUCN, Cambridge, UK.

Wright, D. D., Jessen, J. H., Burke, P., and Gomez de Silva Garza, H. (1997). Tree and liana enumeration and diversity on a one-hectare plot in Papua New Guinea. *Biotropica* 29, 250–260.

RAINFOREST LOSS AND CHANGE

K. D. Singh
Harvard University

I. Assessment Techniques
II. The Historic Rainforest Area
III. The Current Rainforest Area and the Rate of Loss
IV. Types of Changes within the Forest Land
V. Implications for Biological Diversity

GLOSSARY

change matrices A way to relate area of land use classes at the beginning with that at the end of a period. From such matrices, it is possible to follow the exact path of change viz. how much land was transferred from a given use to other uses, how much came into that use, and what remained unchanged during the period.

deforestation The change from forest to other land uses such as agriculture or ranching; or depletion of forest crown cover to less than 10% density. Because shifting cultivation involves a change of land use, it is considered deforestation.

forest change processes A collective term for all activities on the forest land which affect the stand or site and, in particular, the biological diversity. These processes include forest exploitation, forest fragmentation, and establishment of new plantations.

historic rainforest All land areas with a potential to support the tropical rainforest as determined by the climatic and physiographic conditions whether currently forested or not. The concept is interesting because it provides a baseline to calculate the rainforest loss in a country or region over a historic period.

rainforest In this context, the "tropical rainforest formation" which occurs approximately within a latitudinal belt of 10° on either side of the equator. It may be noted that temperate and sub-tropical zones have forest formations structurally similar, but biologically not as diverse as the tropical rainforest.

THERE HAS BEEN GROWING WORLDWIDE CONCERN ABOUT THE FATE OF THE TROPICAL FORESTS and, in particular, the rainforest. Incomplete and often contradictory information has added to the uneasiness of the international community. Widely differing estimates of loss have appeared in the media and the issue has also taken a political overtone. This article presents the reasons for differing estimates and discusses how recent international initiatives have improved the information base. It then presents the most reliable data on the historic and current area of the tropical rainforest, the current rate of loss, and the process of ongoing changes. Finally, some implications for biological diversity are presented at the conclusion.

I. ASSESSMENT TECHNIQUES

A detailed review of the problems associated with assessment of deforestation has been published by the

Food and Agriculture Organization of the United Nations (FAO) in Forestry Papers 112 and 130 released 1993 and 1996, respectively. According to these reports, the assessment of global deforestation or that within a country is complicated due to several reasons:

- There is no globally accepted definition of forest or deforestation. Some include only primary forest in the term "forest," whereas others include all forests whether primary or disturbed, closed or open.
- Assessing deforestation requires a minimum set of two consistent observations over time. In many countries or regions, there are no observations due to a weak capacity for conducting forest inventories.
- Rainforests are located in inaccessible terrain costly to survey. Perpetual clouds in some equatorial zones prevent acquisition of cloud-free images or the taking of aerial photographs. Radar images, due to their cloud-penetrating capabilities, hold promise but have only limited information to offer.
- Deforestation has very uneven and patchy spatial distribution. Special statistical techniques are required for a cost-effective assessment.
- Sometimes, inappropriate technology has been employed and data have been reported for a country without the reporting of the associated error. Fearnside (1993) presents an account of varying estimates of deforestation reported for the Brazilian Amazonia.
- Even in countries with a tradition of forest inventory, the techniques used have not been very appropriate for monitoring changes and do not enable a statistically sound comparison of estimates on two dates.

FAO, as a part of Forest Resources Assessment (FRA) 1990 project, organized a comprehensive study on tropical deforestation as a collaborative effort of a team of more than 100 national and international experts. Most of the statistics presented in this article originate from this project. A study by FAO for the Year 2000 is currently in progress and reports are expected to be published by the end of 2000 or beginning of 2001.

The FAO FRA 1990 project made use of the following data sources and procedures to estimate tropical deforestation:

1. The existing reliable information on tropical forests since 1960 was compiled at a subnational level in the form of a database called FORIS (Forest Resources Information System). All repeated observations for the same area were recorded in the form of a time series. The database includes information relating to forest area, volume/biomass, forest management status, and forest utilization.

2. Maps were made showing the current and historic distribution of tropical forests (the latter map is also termed the ecological zone map) using forest-type maps of countries, available remote sensing data, and ecological information related to forest-type distribution such as climate, physiography, and soil. This major work was accomplished in cooperation with the Laboratoire d'ecologie Terrestre in Toulouse, France. Subnational boundaries for all tropical countries were georeferenced and corresponding time series of demographic data since 1960 compiled in the form of a database.

3. A model was developed relating the time series data on forest cover with the time series data on demography by subnational-level unit stratified by ecological zone.

4. A pantropical sample survey was conducted using pairs of high-resolution satellite images, one taken in approximately 1980 and the other in approximately 1990, with a goal to produce change matrices at the sample, regional, and global tropical levels. The survey divided the tropical zone into 10 geographic and 3 forest cover strata using the forest cover and ecological zone maps. All 117 satellite images, each showing approximately 3 million ha, were randomly selected and distributed to various strata. The study, covering nearly 10% of the area of the tropics, was conducted under a very controlled environment. The images of each sample site were of comparable quality and taken during the same season and interpreted by the same person with a centralized quality control and training. The (interdependent) image-to-image interpretation procedure, specifically developed to capture deforestation, enabled assessment of changes with a much lower standard error compared to that of the existing method of independent assessments at two dates.

II. THE HISTORIC RAINFOREST AREA

Rainforest is the natural vegetation of the equatorial zone. It is associated with a very uniform temperature throughout the year (26–28°C) and wet conditions with precipitation exceeding evaporation at least for 10 months. The historic distribution of rainforests, as described in Singh (1974), is as follows:

South and Central America: The Amazon basin extending west to the lower slopes of the Andes and east to the coast of Guyanas, south to the Gran

Chaco, and north along the eastern side of Central America.

Africa: The Congo basin (mainly Zaire and Congo), Gabon, and Cameroon. It extends westward into Nigeria, Ghana, Ivory Coast, Liberia, and Guinea. On the eastern side it extends to Uganda.

Asia: mainly Southeast Asia, including the Malay Peninsula, the islands of Sumatra, Borneo, Sulavesi, and New Guinea, part of India, Sri Lanka, Burma, Thailand, and the Philippines.

Although climate in the rainforest regions is relatively uniform, there are local variations caused by changes in soil, topography, and geology. These site variations give rise to a corresponding change in vegetation forms. Some typical forest formations in the equatorial zone are the following: The rainforest proper (also called climax rainforest) usually occurs on well-drained soils and below elevations of 500–800 m. This is generally what the common person means when he or she talks of the rainforest. Other types are mangrove forest, swamp forest, periodically flooded forest, riparian forest, hill forest (usually at altitudes higher than 800–1000 m), forest on tropical podsols (a soil type), and forest on coastal sand.

Ecological studies show evidence of striking similarities among the rainforest formations across continents even though they occur in widely separated areas. According to Richards (1952), they are composed of plant species with comparable characteristics and great similarity in their spatial arrangement, life form, and physiognomy. The fundamental pattern of structure is thus the same through the whole extent of the rainforest.

According to the FAO ecological zone map, the potential area of the tropical lowland rainforest worldwide was 935 million ha, with 127 million ha (13.5%) in Africa, 308 million ha (33.0%) in Asia, and 500 million ha (53.5%) in South America. South China and Australia also have small areas of tropical rainforests that were not included in the previous total.

III. THE CURRENT RAINFOREST AREA AND THE RATE OF LOSS

Based on FAO FRA 1990, Table I gives the regional distribution of the lowland rainforest in 1990 and the rate of loss during 1980–1990. An approximate estimation of the historic rainforest area, based on the FAO ecological zone map, is also given to make comparisons with the current area. Accordingly, the worldwide area was 715 million ha, with 87 million ha in Africa, 178 million ha in Asia, and 451 million ha in America. A comparison with the historic areas shows a global decline of 220 million ha, with Africa contributing nearly 40 million ha (32%), Asia 130 million ha (42%), and South America 50 million ha (~10%). These data are very approximate but still interesting. The area losses by continent, as discussed in Section V, have varying significance for biological diversity.

The time series data of FORIS permit a study of trends in the rainforest loss (Singh *et al.*, 1995). The curve shows an acceleration of loss from 1960 to 1980 followed by some stabilization from 1980 to 1990. Assuming this trend will continue, the rate of loss during 1990–2000 is projected to be approximately 50 million ha. The worldwide loss during 1960–2000 is approximately 180 million ha. Compared to the historic loss of 220 million ha, this suggests that most of the rainforest loss has happened during 1960–2000.

Some discussion of the reliability of FRA 1990 data sources may be useful in interpreting the previous estimates. For the purposes of global assessments, FAO continuously searches all published forest inventory

TABLE I

The Current Rainforest Area and the Rate of Loss[a]

Continent	Potential rainforest area (ha)	Current forest area (1990)		Loss rate 1980–1990 (ha)
		ha	(%) of potential	
Africa	126,561	86,604	68	467
Asia	307,979	177,837	58	2235
America	500,060	450,891	90	1900
Total	934,600	715,332	76	4602

[a] Source: FAO (1993).

reports and maps. The published reports on country inventories are reviewed, relevant statistics are archived in the database, and the survey year and quality of estimates are noted. In general, the reference date and definition of data vary from country to country. However, data quality in FORIS is usually quite good and mostly derived from the interpretation of high-resolution satellite images by local professional staff. The FRA 1990 staff apply an adjustment for the definition and date to make all country estimates comparable on a global basis. These comparable estimates are termed standard data in contrast to source data.

A study was done by FAO (1996) on the differences between estimates derived from FORIS and the remote sensing-based sample survey. Means of forest cover and rate of deforestation were calculated by the two methods for each of the 10 forest cover strata. The correlation coefficient (R^2) between the forest cover area was 0.94 and that between deforestation rates 0.64. A lower value of correlation for the deforestation is expected because the estimation is subject to many sources of errors as explained previously. Skole and Tucker (1993) and INPE (1992) studied the rate of deforestation in the same area (legal Amazonia), during the same period (1978–1988), and for comparable data sets. The first reported deforestation rate was 1.5 million ha and the second was 2.1 million ha, a difference of 30% in a geographic area of 5000 million ha. This highlights the complexity associated with the assessment of deforestation.

IV. TYPES OF CHANGES WITHIN THE FOREST LAND

The statistics presented in Section III mainly relate to the transfer of land from forest to nonforest. Changes within forest land are of equal if not greater interest from an ecological as well as an economic perspective. However, representative data on these changes, on a global basis, are very scarce or not available. The only source of regionally consistent information is the FRA 1990 Survey of the Tropics, which provided information on change processes in a matrix form (Table II). The matrices are ideal for tracking changes but not very easy to read. To give a bird's-eye view of the change information, FRA 1990 developed a pictorial presentation called a flux diagram which shows area changes on the x-axis and associated average biomass changes on the y-axis. Thus, the figure as a whole shows the total biomass change in the form of a rectangle. Such a presentation is easy to read and informative for studies on climate change.

The FAO change matrices have nine rows and nine columns that represent the full range of land use and forest cover changes important for a study of biological diversity and climate change. The transition among the first four classes relates to changes within the forest domain and represents various grades of density or forest cover changes (closed and open), spatial disturbance (fragmented), and temporal disturbance (long fallow). The transition among the next four classes relates to transfers from forest to nonforest domain including shrub, short fallow, other land cover, and water. The last class, plantation, could be in the forest or nonforest domain (e.g., forest or agricultural plantation). These domains are grouped as one because it is not easy to separate them with certainty on satellite images.

A distinction needs to be made between the long fallow and short fallow categories, both of which are affected by shifting cultivation. The distinction between long and short has been made with the purpose of dividing the total area under shifting cultivation into "predominantly forested," no matter how much degradation has occurred, and "predominantly nonforested" (but woody) on the basis of estimated intensity of cultivation or, specifically, of the rotation cycle. This distinction brings some clarity to the controversy about "shifting cultivation" which, without further subdivision, is sometimes considered nonforest (i.e., classified as "other wooded land" according to the FAO–FORIS definition) and sometimes forest (i.e., according to the definition used by some countries).

The information appearing in the flux diagram (Fig. 1) clearly shows that most of the changes result in the loss of the rainforest. Of the total rainforest lost, about 31% goes to permanent agriculture, 45% to short and long fallow (viz. shifting cultivation), 7% to fragmented forest, 3% to open forest, 11% to plantations, and 3% to other land uses. Most of these changes have far-reaching implications for biological diversity because they involve the introduction of fire and an external gene pool in the forest ecosystem. Logging is also a major cause of changes within forest land (Table III). Asia is the leader in terms of the area logged and the timber extracted, followed by Africa and South America.

V. IMPLICATIONS FOR BIOLOGICAL DIVERSITY

What is the implication of the previously mentioned changes for biological diversity? As expected, forest area lost and the condition of the remaining forests

TABLE II

Area Transition Matrix for the Period 1980–1990 for the Wet and Very Moist Regions of the Tropical World[a]

	Classes at year 1990 (million ha)								Total 1980	
Classes at year 1980	Closed forest	Open forest	Long fallow	Fragmented forest	Shrub	Short fallow	Other land cover	Water	Plantations	Million ha	%
Closed forest	685.36	0.96	5.35	2.32	0.42	9.16	10.11	0.36	3.70	717.74	80.5
Open forest	0.10	11.54	0.02	0.08	0.01	0.36	0.20	0.01	0.02	12.32	1.4
Long fallow	0.37	0.06	21.63	0.06	0.13	1.31	0.47	0.01	ε	24.02	2.7
Fragmented forest	0.07	0.03	0.33	7.78	0.19	1.22	1.44	0.05	0.14	11.24	1.3
Shrubs	ε		0.04	0.01	3.52	0.03	0.20	ε	0.01	3.81	0.4
Short fallow	0.27	0.04	0.27	0.22	0.02	55.78	5.04	0.13	0.14	61.92	6.9
Other land	0.13	0.01	0.03	0.02	0.64	0.62	47.32	0.09	0.05	48.93	5.5
Water	0.11	ε	0.01	0.03		0.02	0.13	(Water)	0.01	0.29	ε
Plantation	ε	ε	ε			0.01	0.02		11.45	11.48	1.3
Total 1990											
Million ha	686.41	12.65	27.68	10.52	4.93	68.50	64.92	0.66	15.50	891.76	
%	77.0	1.4	3.1	1.2	0.6	7.7	7.3	0.1	1.7		100

[a] Source: FAO (1996).
The number of sampling units was 26; land area studied was 45.54 million ha. Sampling results have been expanded to the total land area covering 891.76 million ha.

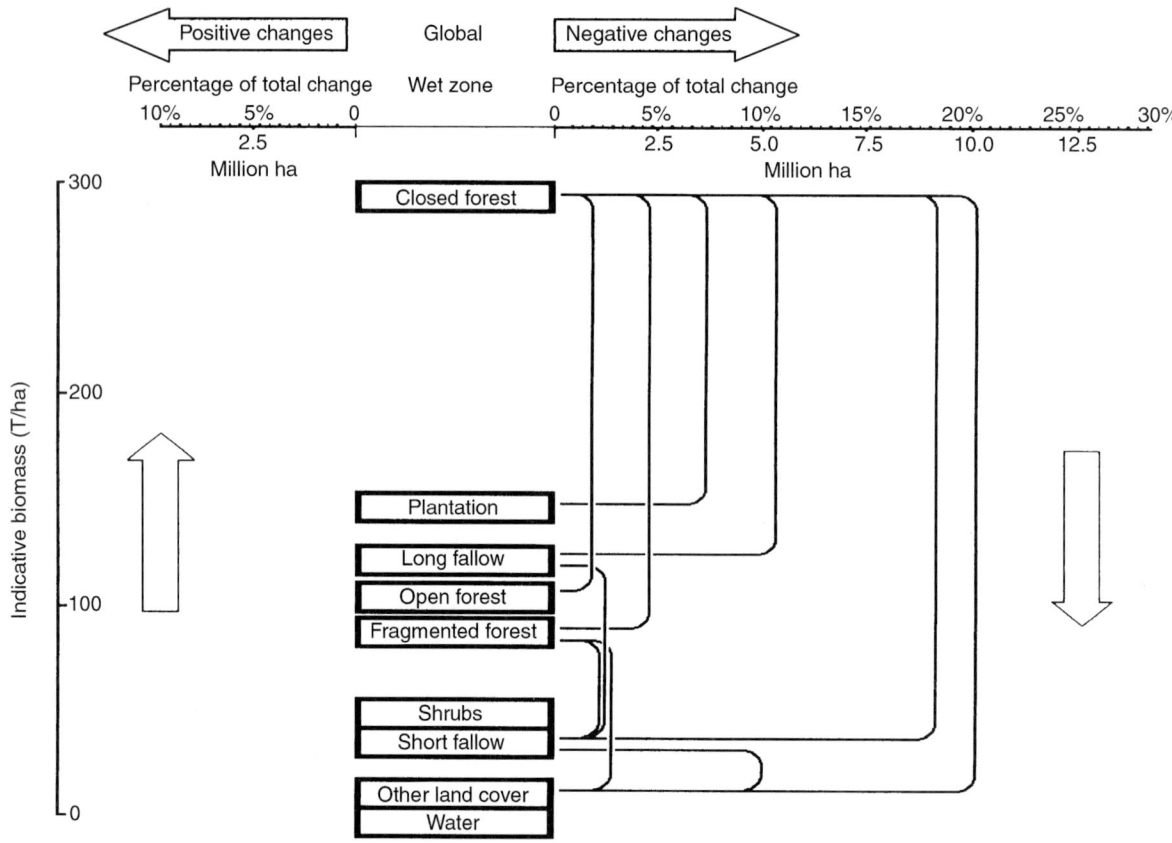

FIGURE 1 Flux diagram showing area transition on the x-axis and biomass transition on the y-axis. Each rectangle represents total biomass change associated with the transition (reproduced with permission from FAO, 1996).

(including degradation and fragmentation) jointly determine the loss of biological diversity. Overall, Asia has lost a major part of its historic rainforest area (about 42%). The current rate of deforestation is the highest among the three regions in absolute area terms. The area logged annually and the intensity of logging are

TABLE III
Intensity of Logging in Different Regions[a]

Continent	Logged area (ha)	Logging intensity (m³/ha)	Log production (million m³)
Africa	91	13	13
Asia	215	33	33
America	258	8	8
Total	564	19	54

[a] Source: FAO (1993).

also the highest compared to those of the other two regions. In Asia, only Indonesia still has some large areas of tropical rainforests left; the rainforests in the other countries are in a highly fragmented form, with species diversity under a high risk of loss (FAO, 1993).

Africa initially had a small share of the world's rainforest and had lost nearly 32% of the total by 1990. In particular, the west African region has undergone a very high loss of the historic rainforest area, with the remaining forests occurring in a highly fragmented manner like islands in a sea of forest fallow (FAO, 1993). The ongoing changes in the central African region were observed to be relatively slow during 1980–1990. However, the recent flux of population at the boarder of Zaire and Rwanda must have affected the rate of loss and biological diversity, but this effect is not yet known. Second, the ongoing forest area changes in Africa are from closed forest to short fallow, which is an extensive and destructive form of land use change

TABLE IV

Transfer of Land from Forestry to Nonforestry Uses[a]

Region	Total area transfer		Transfer to nonwooded areas (mainly agriculture)		Transfer to other wooded areas (mainly short fallow)	
	ha	%	ha	%	ha	%
Africa	2,036	100	52	3	1,984	97
S. America	11,510	100	7,775	68	3,736	32
Asia	9,941	100	3,824	38	6,117	62
Total	23,487	100	11,651	49	11,837	51

[a] Source: FAO (1996).

from the biological diversity perspective. The pattern of transfer is different from that of South America or Asia, where the land use change is more planned and intensive (Table IV). This implies that more land would be needed in Africa to feed the same number of people as are fed in other regions. With a high rate of population growth, such land use changes are a cause of concern from a forestry and a biological diversity perspective.

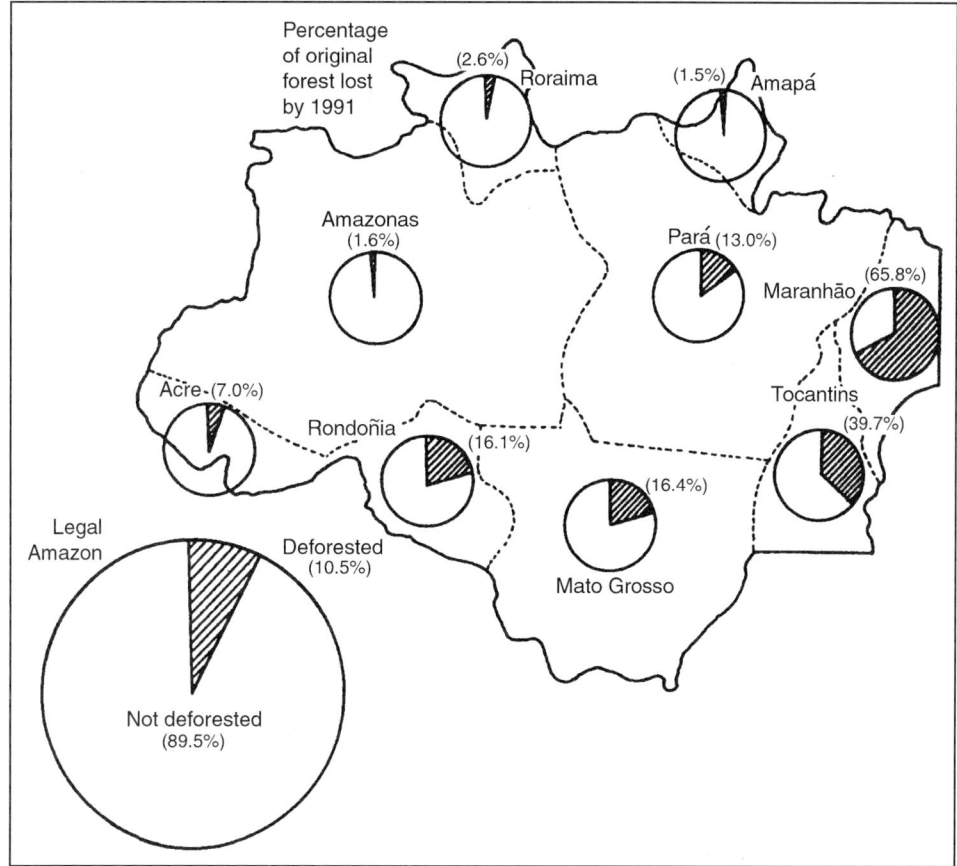

FIGURE 2 Percentage of original forest lost in the Brazilian Amazonia (reproduced with permission from Fernside, 1993).

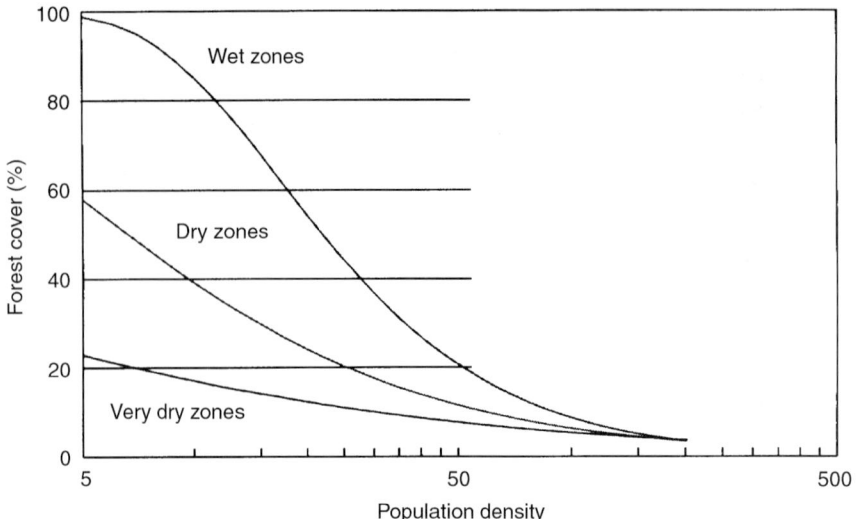

FIGURE 3 Illustration of Model Curves for Different Ecological Zones (reproduced with permission from FAO, 1993).

The land use changes in South America and the Amazon, in particular, are of more recent origin compared to those in Asia and Africa. The total area of the rainforest is still very high and the total area lost, therefore, is relatively low. However, a state-wise distribution of the remaining forest cover and the rate of deforestation provides a different view (Fig. 2). Only the states of Amazonas, Roraima, and Ampa have large tracts of inaccessible forests; other states, such as Maranhao, Tocantins, Para, Mato Grosso, Rondonia, and Acre, are undergoing larger than average forest area changes per capita compared to other regions of the world. Cattle ranching seems to be the main cause of deforestation, which is quite different from the causes of deforestation in the other regions of the tropics.

The rainforests are fragile ecosystems. The time series data on forest cover and population changes, collected by the FRA 1990 project from all regions of the tropics, show a more rapid forest area loss with an increase of the population pressure in the rainforest region compared to other ecological zones (Fig. 3). The rainforest, per unit area, is also the most biodiverse. The two facts imply a larger loss of diversity by human settlement and hence the need for a long-term perspective and care in planning of land use and forestry changes in the tropical rainforest zone.

See Also the Following Articles

DEFORESTATION AND LAND CLEARING • RAINFOREST ECOSYSTEMS, ANIMAL DIVERSITY • RAINFOREST ECOSYSTEMS, PLANT DIVERSITY • TROPICAL ECOSYSTEMS

Bibliography

Fearnside, P. M. (1993). Deforestation in Brazilian Amazonia: The effect of population and land tenure. *Ambio* 22(8), 537–545.
Food and Agriculture Organization (FAO) (1993). Forest resources assessment 1990: Tropical countries. Forestry Paper No. 112. FAO, Rome.
Food and Agriculture Organization (FAO) (1996). Forest resources assessment 1990: Survey of tropical forest cover and study of change processes, Forestry Paper No. 130. FAO, Rome.
Instituto Nacional De Pesquisaa Espacias (INPE) (1992). Deforestation in Brazilian Amazonia. INPE.
Richards, P. W. (1952). *The Tropical Rain Forest*. Cambridge Univ. Press, Cambridge, UK.
Singh, K. D. (1974). Spatial variation patterns in the tropical rain forest. *Unasylva* 26(106).
Singh, K. D., and Marzoli, A. (1995). Deforestation trends in the tropics: A time series analysis. World Wildlife Fund/International Institute for Tropical Forestry Conference on "The potential impact of climate change on tropical forest ecosystem," Puerto Rico.
Skole, D., and Tucker, C. (1993). Tropical deforestation habitat fragmentation in the Amazon: Satellite data from 1978 to 1988. *Science* 260, 1905–1910.

RANGE ECOLOGY, GLOBAL LIVESTOCK INFLUENCES

J. Boone Kauffman* and David A. Pyke[†]
*Oregon State University and [†]USGS Forest and Rangeland Ecosystems Science Center

I. Rangelands of the World
II. Biological Impoverishment Associated with Livestock Grazing
III. Conclusions: Human Institutions and Livestock-Caused Degradation

GLOSSARY

animal unit month (AUM) Quantity of forage needed to sustain one 1000-lb cow or five sheep for one month.

biological soil crusts Collective term referring to the combination of nonvascular and non-seed-forming plants that commonly occur on the soil surface of many arid and semiarid ecosystems. These crusts may consist of mosses, lichens, liverworts, green algae, and cyanobacteria. Synonyms: cryptogams, microfloral, cryptobiotic, cryptogamic, microbiotic, organogenic, biogenic, biotic, and microphytic soil crusts.

browse That part of leaf, twig, and reproductive growth of shrubs, woody vines, and trees available for animal consumption. The act of consuming leaves, twigs, or reproductive parts of shrubs, woody vines, and trees.

degradation Decrease in ecosystem productivity or structure, and/or declines in native species diversity (sometimes with concomitant increases in exotic species) due to land use practices. Directly related to declines in biodiversity, ecosystem degradation encompasses soil impoverishment (e.g., compaction, erosion, salinization, loss of biological soil crusts) and hydrological alterations that diminish water availability. Similar to desertification.

desertification Process by which an area becomes more arid through the decline in primary productivity, elimination of biological cover, shifts in plant diversity, or degradation of soils due to human uses whose impacts exceed the sustainability of the landscape. Similar to degradation.

forbs Non-graminoid herbaceous plants.

graminoids Grasses (family Poaceae) and grass-like plants, mostly sedges (family Cyperaceae) and rushes (family Juncaceae).

graze Act of consuming herbaceous plants, including grass, grass-like plants, and forbs.

overgraze Consumption of forage to the extent that declines in productivity or desirable species composition are probable (i.e., degradation). Overgrazing is excessive use in terms of the amount, duration, and frequency of plant utilization.

pasture Lands in which the native vegetation has been removed in favor of cultivated grasses or other forage plants.

range condition Current productivity and composition of plants on a rangeland relative to its ecological potential. Often range condition is classified into arbitrary classes (i.e., excellent, good, fair, and poor).

rangeland Land on which the potential native vegetation is predominately grasses, grass-like plants, forbs, or shrubs. Rangelands include prairies, marshes, tundra, wet meadows, savannas, shrubland steppe, chaparral, desert grasslands, and woodlands.

riparian zone Interface between streams and terrestrial uplands. These zones are three-dimensional areas of direct interaction between the terrestrial and aquatic ecosystems. They often contain unique species and high levels of biological diversity.

ruminant Grazing mammal with a complex four-chambered stomach (including a rumen). Common ruminants include cattle, sheep, goats, deer, antelopes, and giraffes.

savanna Ecosystem with a more or less continuous herbaceous layer dominated by graminoids and broad-leaved herbs with an overstory of trees or shrubs that covers less than 10% of the area. In contrast, grasslands are typified by a pure graminoid or herbaceous layer with no (or very few) tree or shrub elements that rise above the grass layer.

woodland Ecosystem in which the mature overstory tree or tall shrub cover makes up 10 to 60% of the area. Typically, a well-developed herbaceous understory is present. Note: forests typically have a mature tree cover >60%.

LIVESTOCK HAVE PLAYED A PIVOTAL AND POSITIVE ROLE IN HUMAN DEVELOPMENT. Yet this has come at a heavy cost to the biological diversity of the world's rangelands and forests. Livestock grazing is the most widespread land use occurring on Earth. In many regions of the world, grazing has reduced the density and biomass of many plant and animal species, reduced biodiversity, aided in the spread of exotic species and disease, altered ecological succession and landscape heterogeneity, altered nutrient cycles and distribution, accelerated erosion, and diminished both the productivity and land use options for future generations. Rangelands, a general term for grasslands, savannas, semi-arid shrublands, and woodlands, are a dominant terrestrial land cover type in all continents (except Antarctica). They contain a significant portion of the world's biodiversity. Ecosystem degradation or desertification associated with livestock grazing is a significant problem throughout these rangelands. In addition, the majority of tropical forests in South and Central America are being deforested at alarming rates primarily for conversion to cattle pasture. In effect, the most biologically diverse ecosystems on Earth are being converted to simple pastures dominated by exotic grasses. To halt this degradation or to restore native ecosystems, innovative approaches to land use and livestock management need to be implemented. Ecosystems should be managed such that natural ecological and physical processes are allowed to occur with minimal disruption. Governmental institutions and policies are among the most significant barriers to land use change.

I. RANGELANDS OF THE WORLD

Rangelands are defined as those lands where the native vegetation is predominantly grasses, shrubs, or open woodlands. Rangelands include grasslands, shrublands, savannas, open woodlands, and most desert, tundra (arctic and alpine), meadow, wetland, and riparian ecosystems. Occurring from the tropics to polar regions, rangelands cover more land area than any other type on Earth. Because of climatic or edaphic limitations, they tend to be lands that are incapable of growing marketable timber or agronomic crops without irrigation or some other modification (i.e., they are over poor-quality soils and/or are too wet, too dry, or too cold for cultivation). The composition, structure, productivity, and diversity of these ecosystems are governed by a combination of climate, geography, topography, and geology, including soil development. In addition, rangelands are used by a large number of vertebrate and invertebrate herbivores, including a diverse combination of domestic or native ungulates.

A. Tundra

Tundra ecosystems are dominated by perennial grasses, forbs, shrubs, and biological soil crusts consisting of cyanobacteria, lichens, and mosses (Fig. 1). These species can tolerate the climatic conditions of high latitudes or altitudes. Tundra ecosystems are generally set apart from forested ecosystems by a climatic tree line defined by the 10°C isotherm for the mean temperature of the warmest month. Tundra plants are low in stature with growth buds near the soil surface, where temperature is less variable. The morphology of herbaceous plants are often cushion or rosette growth forms and lichens are usually a foliose form. Shrubs are low-growing with short internodes between leaves. Temperate alpine tundra communities have short growing seasons similar to those in arctic tundra. In contrast, tropical alpine tundra communities of Africa, South America, and Oceania exist in conditions of freezing or near-freezing temperatures each night, with daytime temperatures allowing for active plant growth throughout the year.

The important herbivores of the Arctic are rodents and ungulates. Lemmings (*Dicrostonyx* spp. and *Lem-*

FIGURE 1 The arctic tundra landscape of the Brooks Range, Arctic National Wildlife Refuge, Alaska. (Photo by Boone Kauffman.)

mus spp.) can have great impacts on arctic ecosystems during years of high population densities. Lemmings exhibit different seasonal patterns of use within various parts of their habitats. During the winter, they use wetlands and graze the soil-level parts of plants while discarding the upper portions of the plants. This grazing pattern results in piles of graminoid litter known as lemming hay. During spring, lemmings move to south-facing slopes, where solar insolation results in the earliest available green biomass. During peak population cycles when densities are high, lemmings can consume between 20 and 70% of the available plant mass.

The dominant native ungulates in the Arctic include caribou or reindeer (*Rangifer tarandus*) and musk ox (*Ovibos moschatos*). Domestication of reindeer and musk ox has led to their introduction throughout the arctic tundra and to reindeer introduction into northern Antarctica. Alpine tundra provides seasonal habitat for wild ungulates, including elk, deer, red deer, mountain sheep, and mountain goats. Domestic livestock breeds adapted to cooler environments are also present in various arctic or alpine tundra communities of the world (e.g., cattle, sheep, goats, horses, llamas, and yak).

B. Grasslands and Savannas

Natural grasslands, also known as campos, llanos, pampas, plains, prairie, steppe, and veld, are frequently associated with semi-permanent high air pressure systems and high amounts of solar radiation. Annual precipitation typically ranges from 160 to 1700 mm with distinct wet and dry seasons. Grasses or graminoids dominate the vegetation composition, with forbs being a secondary component and short-stature shrubs contributing a lesser amount. The grass (or graminoid) dominance of grasslands is the result of current climatic conditions and/or natural disturbance processes such as fire. With human degradation such as overgrazing or fire suppression, shrubs often increase in abundance. In the tropics and in temperate locations with summer precipitation, C_4 grasses (warm-season grasses) will predominate because of their optimal growth at warmer temperatures. Northern temperate locations, where moisture is most available when temperatures are cooler, favor C_3 grasses (cool-season grasses). Biological soil crusts frequently occupy the soil surface below or between the plant cover, particularly in drier desert grasslands.

In equatorial regions, tropical savannas are a dominant rangeland ecosystem. They form a transitional zone along a precipitation gradient from tropical rain forests to arid deserts. Dense grasslands with occasional trees and shrubs characterize tropical savannas. As in grasslands, distinct wet and dry seasons typify the climate of tropical savannas. Disturbances are a prevalent feature of savannas; during the dry season, lightning-caused fires may occur. Human-ignited fires are very common throughout the world and maintain the open nature of many tropical savannas. Fire-return intervals in savannas are among the most frequent of any landscape on Earth. It is estimated that 50–75% of the humid savannas of Africa and South America burn each year.

Tropical savannas are highly productive. Grass biomass in humid savannas (>700 mm precipitation yr^{-1}) averages about 6.6 Mg ha^{-1} in Africa and South America, and 4.9 Mg ha^{-1} in tropical Asia. In savanna-woodlands, aboveground biomass may be as high as 25–61 Mg ha^{-1}. Roots and other subterranean tissues account for the majority of the plant biomass in savannas and grasslands (de Castro and Kauffman, 1998).

Tropical savannas are also very species-rich. For example, the largest savanna type in South America is the Brazilian cerrado, which often contains over 400 species ha^{-1} (Fig. 2). Except for tropical rain forest, this is among the richest vascular plant assemblages on Earth. For example, about 3500 plant species and 400 bird species have been identified in the cerrado of the 5800-km^2 Federal District surrounding Brasilia, the capital of Brazil.

Savannas can experience rapid changes in species composition and structure following the introduction of an anthropogenic perturbation. Changes in the patterns of fire are one example; fire suppression results in an increase in the dominance of tree and shrub vegetation. Conversely, increasing the occurrence of fire enhances grass abundance. Linkages between fire, grazing animals, disease, and vegetation structure were described by Sinclair (1979). He related how the elimination of a ruminant disease, rinderpest, was a perturbation that changed the structure and composition of landscapes within the Serengeti of Africa. The immediate effect of rinderpest elimination was an increase in wildebeest numbers. On the plains, increased grazing pressures reduced grasses and increased forbs. Grant's gazelle increased along with the forbs, thus leading to population increases in cheetah. In woodlands, grazing reduced the grasses, which in turn reduced the fuel to support fires. The reduction in grasses resulted in a reduction in buffalo. Tree numbers then began to increase, which favored an increase in the giraffe populations.

Plants within natural temperate grasslands and tropical savannas vary in their tolerance to large numbers of grazing animals. The images of grasslands and savannas for many people include large herds of ungulates, such as bison (*Bison bison*) in the Great Plains of North America and wildebeest in East Africa. In these ecosystems, large herds of ungulates were a keystone feature that strongly influenced composition and structure. Plants are well adapted to herbivory in these grass-dominated ecosystems. However, large herds of ungu-

FIGURE 2 Overview of the Brazilian cerrado. This is a mosaic of grasslands, savannas, and woodlands that have a high level of biotic richness. Note the fire in the background, which is a common disturbance in this ecosystem. Land conversion to cattle pastures, soybeans, and other croplands is occurring at alarming rates in this ecosystem. (Photo by Boone Kauffman.)

lates are not a natural feature in all grasslands and savannas (e.g., the Intermountain and Palouse grasslands of the United States and the Brazilian cerrado). In these grasslands, rodents and insects are the dominant native herbivores.

The most widely distributed wild ungulates in North American grasslands and woodlands are pronghorn (*Antilocapra americana*), elk (*Cervus elaphus*), and deer (*Odocoileus* spp.). However, only bison comprised large herds within the Great Plains until they were decimated in the late nineteenth century by hunters. Bison also occurred in other North American grasslands, but not in large herds. Rodents and lagomorphs are also common grazers in North America. Population eruptions can result in years when the vegetation is heavily grazed by these animals in local areas, but within a few years disease, predators, competition, or other density-dependent factors reduce their numbers.

The common ungulates in South American grasslands and savannas include tapirs (*Tapirus* spp.), peccaries (*Catagonus*, *Dicotyles*, and *Tayassu* spp.), camelids (*Lama* spp.), and deer (*Odocoileus* and *Ozotoceros* spp.). Some of these are primarily forest or gallery forest (riparian) dwellers, but they also use savannas or grasslands. Only one native lagomorph (*Sylvilagus brasiliensis*) occurs in South America. The dominant vertebrate grazers in South American savannas are a diverse group of rodents. The largest rodent, the capybara (*Hydrochoerus hydrochaeris*), is widely distributed, but prefers wetland savannas such as the Brazilian Pantanal. Capybara graze in herds as large as 1500 animals.

The African grasslands and savannas maintain the greatest species richness of wild ungulates of any rangeland ecosystem in the world (Fig. 3). Rodents comprise a lower herbivore biomass than ungulates in most regions of Africa. The African ungulates consist of three taxonomic orders: Proboscidea (1 species; elephant); Perissodactyla (6 species; rhinoceros and zebras); and Artiodactyla, with five families. The families include Suidae (3 species; pigs), Hippopotamidae (2 species; hippopotamus), Tragulidae (1 species; water chevrotain), Giraffidae (2 species; giraffes), and Bovidae (79 species; buffalo and antelope). Although a few of the African ungulates are forest-specific species, most have distributions that include grasslands, savannas, and woodlands. Riverine and wetland locations with high rainfall in East Africa maintain populations of hippopotamus, waterbuck, and buffalo, whereas midheight grass regions with intermediate moisture support zebra, giraffe, and hartebeest. Native ungulates have been extirpated or driven to extinction in many regions of Africa. In the Fété Olé savanna of the Sahel, 11 species of native ungulates are now extinct.

Northern Asia has few native ungulates within its grassland/savanna ecosystems. Common ungulates include gazelle (*Procarpa* or *Gazella* spp.), wild horse (*Equus przewalskii*), and the Bactrian camel (*Camelus bactrianus*), and none of these form large herds. Rather,

FIGURE 3 An African woodland/savanna in South Africa. (Photo by Ron Shea.)

rodents are the primary vertebrate grazers within the northern Asian grasslands. They influence these ecosystems by creating local areas of high disturbance through digging and grazing. Voles, marmots, and ground squirrels often undergo population cycles that may result in the consumption of large amounts of vegetation when populations are high, thus reducing forage for local livestock herds. Such situations can exacerbate overgrazing since livestock densities are not typically reduced in years of high rodent numbers. Southern Asia is more diverse in numbers of native ungulate species with the inclusion of several species of wild ass, gazelle, and antelope. Asian elephants and rhinoceros were also present, but are now endangered and extirpated from practically all of their former range.

In the Australian grasslands, large grazing animals are represented by marsupials (euros and kangaroos, *Macropus* spp.; wallabies, *Petrogale* spp.). The red kangaroo (*Macropus rufus*) grazes in the arid and semi-arid zones, whereas the antilopine kangaroo is restricted to tropical grasslands of the north. Introduced herbivores, such as the European rabbit (*Oryctolagus cunculus*) and the Timor water buffalo (*Bubalus bubalis*), compete for herbage with native grazers in southern Australia. In Australia as throughout the world, the high productivity of grasslands and savannas has led to their widespread use by domestic cattle, sheep, and goats.

C. Temperate or Cold-Desert Shrublands and Semi-deserts

Deserts or desert shrublands can be broadly classified into two different types—hot deserts of subtropical and tropical latitudes and cold deserts of temperate latitudes. Temperate deserts and semi-deserts occur at higher latitudes than hot deserts (between 30° and 50°), which allows for moisture from oceans to move into these regions by way of prevailing circulation patterns. Most of the temperate deserts and semi-deserts are characterized by sparse vegetation that is usually dominated by low shrubs and herbaceous plants. Total vascular plant cover rarely exceeds 50%, and biological soil crusts commonly occupy interspaces between vascular plants.

Cold deserts are characterized by cold winters and hot summers. Most occur in North and South America (Great Basin, Intermountain West, and Patagonia) and Eurasia (Kazakho-Dzungarian deserts and semi-deserts). In North America and western Eurasia, most of the annual precipitation arrives as snow in the winter, while summers are dry. However, eastern Eurasia receives most of its moisture during the summer from frontal or cyclonal storms that move west from East Asia, while the winter remains dry. Patagonian semi-deserts receive low amounts of precipitation throughout the year, with strong winds exacerbating arid conditions during the summer from elevated evaporation.

The importance of native ungulates in cold deserts varies widely among these geographic regions. Large herds of ungulates, largely saiga antelope (*Saiga tatarica*), zheiran gazelle (*Gazella subgutturosa*), and wild ass (*Equus hemionus*), once roamed Eurasian ecosystems (until 100–200 years ago), but are now largely restricted to reserves. Wild sheep are common to both Northern Hemisphere ecosystems, with *Ovis ammon cyclocero* present in Afghanistan and Iran and *Ovis canadensis* in North America. Common native ungulates in North American cold deserts include pronghorn, deer, and elk. Feral horses and burros are now locally common in areas. Although bison were once found in this region, they never attained large numbers as found on the Great Plains and were largely absent by the time European exploration began. Only one native Patagonian ungulate exists, the guanaco (*Lama guanicoe*). Today domestic herbivores are predominate throughout all of these regions.

D. Hot Deserts

Hot deserts (or hot desert shrublands) are the most arid form of rangeland in the world. The hot deserts generally occur between latitudes of 15° and 30°N and S, the belt of the subtropical anticyclones that create near permanent high-pressure systems. Most hot deserts receive less than 120 mm of rain annually. The position within the high-pressure belts also results in a wide variation in the amount, location, and timing of rainfall. A wide spacing of drought- and heat-adapted shrubs is characteristic of hot desert shrubland vegetation. Perennial vascular plants normally cover less than 10% of the land area within hot deserts. Hot desert vegetation must be adapted to the uncertainty in the timing of adequate growing conditions that are characteristic of this ecosystem. The vascular plants persist in the variable and unpredictable growing conditions by using a host of adaptations to avoid dry seasons while quickly responding to moisture when it becomes available. Perennial plants often have persistent belowground organs that facilitate survival during the dry season. When moisture is available, some perennial plants can respond with a rapid production of adventitious roots and leaves. Many desert species (the Cactaceae, some Euphorbiaceae, and others) have photosynthetic stems that may allow growth during sporadic storm events without the cost of leaf production. Plant reproduction is limited to years with sufficient moisture. Ephemeral annuals survive in deserts by germinating and completing their life cycles during wet periods and then persisting in the soil as a dormant seed during dry periods. In locations where fog is the major source of water (coastal deserts of Chile, Peru, and Namibia), fog desert vegetation forms, with biological soil crust species often being the dominant ground cover.

Native ungulates are virtually missing from most hot deserts except along major rivers and wetlands. They may occupy the fringes of the hot deserts and migrate into the deserts during favorable conditions. An example is the desert big-horn sheep in the North American Mojave Desert. Though populations are found in adjacent semi-deserts where greater food and water are available, use of the Mojave Desert by bighorn only occurs during wet periods. Australian marsupials follow a similar pattern, with most only using the arid desert areas during favorable times. A few wallabies are found in rocky arid areas, but never in large numbers.

Herbivores are found in low densities in the North African (Sahel and Sahara) and Middle Eastern deserts. Native ungulates of these deserts include gazelle, ibex, oryx, wild asses, and wild boar. The exception to a low density of ungulates in hot deserts is found in the Karoo and Kalahari Deserts of southern Africa. Here, springbok, gemsbok, and hartebeest disperse in small herds or treks during dry periods. Once moisture produces aboveground growth of grass, these animals congregate into larger herds. Populations of these animals 200 years ago were larger than those today. Although forage is scarce and unpredictable, domestic livestock (cattle, sheep, and goats) grazing remains common in hot deserts of the world.

II. BIOLOGICAL IMPOVERISHMENT ASSOCIATED WITH LIVESTOCK GRAZING

Domesticated animals (livestock) have played prominent and largely beneficial roles in human society for thousands of years, providing food, fuel, fertilizer, transport, and clothing. Yet livestock have had a dramatic negative impact on global biodiversity. The influence of livestock on global biodiversity is of concern simply because cattle and all other ruminant livestock graze about one-third to one-half of the planet's total land area. Along with pigs and poultry, they eat feed and fodder raised on one-fourth of the cropland. Livestock grazing is the most ubiquitous human activity on Earth and occurs on more area than any other land use. The impacts of livestock are far-reaching and often not immediately apparent (Table I). In many regions of the world, grazing has reduced the density and biomass of many plant and animal species, reduced biodiversity, aided in the spread of exotic species and disease, altered ecological succession and landscape heterogeneity, altered nutrient cycles and distribution, accelerated erosion, and decreased both the productivity and land use options for future generations.

Given the widespread degradation caused by livestock on the diversity and structure of native rangelands, one would assume that rangelands are the source of a large proportion of the feed utilized by the grazing animals on Earth. Yet it is estimated that three-quarters of the world's 3 billion domestic ruminants are raised in conjunction with farming and are fed crop residues, hay, and other forages such as alfalfa. Roughly 38% of the world's grain, especially corn, barley, sorghum, and oats, is fed to livestock. In the United States, livestock accounts for 70% of the domestic grain consumption; in contrast, India and sub-Saharan Africa provide about 2% of their cereal harvest to livestock. A 1983 U.S. Department of Agriculture report stated that $\approx$103 million ha of fertilized pastures and forage crops provided 84% of the U.S. livestock forage. The $\approx$312–319 million ha of U.S. rangelands provided only 16% of the forage. Two-thirds of these rangelands are privately owned, and the rest are publicly owned. In 1984, permits issued by the U.S. Bureau of Land Management accounted for less than 4% of the forage consumed by the U.S. herds.

Rangelands of the world are important reservoirs of biological diversity as well as water, natural beauty, and other plant, animal, and mineral resources. For example, native rangelands provide essential habitat for a wide variety of wildlife and plant species. An estimated 84% of U.S. mammal species and 74% of bird species are associated with rangeland ecosystems (particularly using riparian zones within these semi-arid landscapes). Public rangelands in the United States are home to more than 3000 wildlife species, including about one-third of the nation's threatened or endangered species. Compared to forests, rangelands have received comparatively less attention with respect to environmental degradation in general, and losses in biodiversity in particular.

Because of the great differences in composition, soils, climate, and historical and current land uses among rangelands of the world, grazing effects will vary (see Table I). Degradation occurs when the frequency, intensity, or season of use exceeds the biological thresholds

TABLE I
Some Direct and Indirect Influences of Domestic Livestock on Biological Diversity

Desertification or degradation—the impoverishment of native plant and animal ecosystems

Deforestation/forest conversion to pasture

Contribution of greenhouse gases to the atmosphere—directly from methane emissions and carbon losses due to forest conversion

Conversion of native rangeland to cropland

Dispersal of exotic species and/or the creation of conditions suitable for their increase

Spread of disease to native herbivores

Geomorphic influences on stream channels (widening and incision) that disrupt floodplain/stream interactions

Water diversions for livestock forage production

Development of permanent water sources in areas where they traditionally did not occur

Erosion/sedimentation or fecal inputs into aquatic ecosystems

Influences on biogeochemical cycling

Control of wildlife species perceived as competitors or predators

Applications of herbicides and pesticides to control undesirable plants and insects

Altering natural fire regimes—elimination of fires in natural glasslands and shrublands or frequent introduction in pastures converted from tropical forests

Social causations—overpopulation, government incentives, culture, and customs

---

for ecosystem persistence or recovery. Plant communities that evolved with large herds of ungulates may have adaptations that allow persistence with high levels of defoliation by domestic livestock. Conversely, those rangelands that evolved with low densities of large herbivores may be susceptible to alterations following the introduction of even low levels of domestic livestock impacts.

Unsustainable grazing and ranching practices result in the loss of forests, grasslands, shrublands, savannas, and the indigenous species that inhabit them. As the most pervasive land use, livestock grazing has had widespread and dramatic ecological impacts, including loss of native species, changes in species composition, soil deterioration, degradation of fish and wildlife habitat, and changes in ecosystem structure and function. The loss of biological diversity and productivity of rangelands due to overgrazing is a concern because it diminishes both human and ecological welfare. However, livestock influences on biodiversity are not limited to rangelands. The range of environmental effects of livestock encompasses such varied land uses as conversion of tropical rain forest to pasture in the Amazon Basin, water pollution from manure associated with large livestock factory farms in Europe and the United States, and the desertification of semiarid rangelands in the African Sahel, western United States, and interior Australia.

A. Effects of Livestock on Terrestrial and Aquatic Ecosystems

Figure 4 presents a simple model depicting how livestock influence biological diversity and ecosystem integrity. The direct or primary livestock influences are immediate and readily observable in the field. These direct influences include: (1) removal of vegetation through grazing; (2) the trampling of soils, vegetation, and biological soil crusts; (3) the redistribution of nutrients via forage removal, defecation, urination, gaseous loss, and animal gain; and (4) the dispersal of exotic plant species and pathogens. The cumulative effects of these four primary influences lead to a suite of physical and biotic responses or adjustments within the ecosystem or landscape. These are the secondary influences and they include alteration in disturbance regimes (fire cycles), accelerated erosion, altered hydrology (e.g., runoff, infiltration rates, and water-holding capacity), altered competitive relationships among organisms, and changes in plant and animal reproductive success and/or establishment patterns of plant seedlings. Tertiary

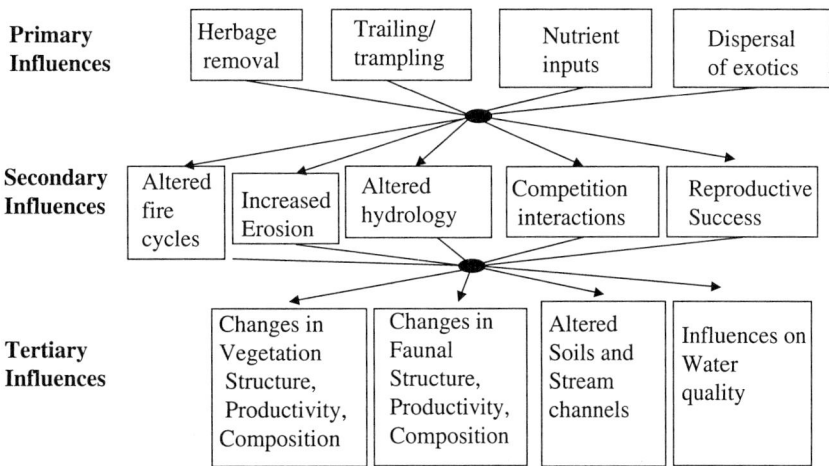

FIGURE 4 A conceptual model of the influences of livestock on ecosystem structure and function. Direct influences are those that are immediate and readily observable in the field. The synergistic effects of these influences lead to the secondary influences, the physical and biological adjustments in ecosystems. The tertiary influences are the long-term alterations in landscapes due to the introduction of livestock. Tertiary influences are the cumulative effects of the primary and secondary influences of livestock and are reflective of varying degrees of desertification or degradation. The degree of alteration would be affected by unique characteristics of the ecosystem (i.e., soils, topography, climate), as well as the intensity and frequency of livestock use.

influences are the long-term cumulative effects of domestic livestock on rangelands. These are the changes in structure, composition, and productivity of plant and animal communities that occur at community, ecosystem, and landscape scales. Tertiary influences are the overall declines in the biotic richness or diversity of affected aquatic and terrestrial areas as a result of long-term unsustainable grazing practices.

Alterations in aquatic and riparian ecosystems may include structural changes in stream channels, changes in water quality (temperature, chemistry, and sediments), and the loss of native aquatic biota. Tertiary influences are the manifestations of desertification and landscape degradation.

1. Direct or Primary Influences

Forage removal by livestock can decrease the quality and quantity of forage for wild grazers as well as decrease nesting or hiding cover for other wildlife species. Vegetation removal by livestock can also reduce the protective litter layer, thus decreasing the quantity of carbon and nutrients incorporated into the soil as organic matter. Along streams, vegetation removal may affect stream shading as well as the amount of vegetation that enters into the stream. Vegetation inputs into streams are important because they are often the primary energy and nutrient source for headwater aquatic ecosystems in forests and rangelands.

The effects of herbage removal depend on a number of environmental, animal, and land management factors. The quantity of plant mass consumed and season, duration, and frequency of grazing are important determinants of how herbivory affects ecosystems. Overgrazing and degradation of ecosystems are a combination of excesses in all of these factors. The season of use refers to the phenological stage during which the plants are grazed. Typically, grazing when plants are dormant is less detrimental than grazing during active growth. Season of use may also influence the reproductive success of plants if flowers and seed heads are not allowed to form or are eaten. In addition, the degree of harm is dependent on the duration and amount of grazing. Declines in plant vigor will occur if grazing is of such duration or intensity that plants are not allowed to regrow and replenish carbon and nutrient reserves. For shrubs and trees, grazing more than the current year's growth will reduce the size of the plant and, depending on the plant's morphology, can result in the death of the plant. Finally, the frequency of grazing is an important determinant. Plants, other organisms, or soils in ecosystems in which grazing is interspersed with periods of rest have a greater opportunity for recovery than in ecosystems that are grazed continuously or annually.

Trampling damage occurs via the mechanical compression and compaction of soils or the physical destruction of biological soil crusts or vegetation as ani-

> **Box 1**
>
> ### Riparian Zones: Hot Spots of Biodiversity on Rangelands
>
> Riparian zones are the unique environments adjacent to rivers and streams and comprise assemblages of plant and animal communities whose presence can be either directly or indirectly attributed to factors that are stream-induced or stream-related (Kauffman and Krueger, 1984). Functionally, they are three-dimensional zones of direct interaction between terrestrial and aquatic ecosystems. Boundaries of riparian zones extend outward to the limits of flooding and upward into the canopy of streamside vegetation (Gregory et al., 1991). These zones perform a variety of ecosystem functions, including influences on water quality, flood attenuation, and provision of valuable fish and wildlife habitats.
>
> Riparian areas are hot spots of biodiversity. This is particularly true in arid and semiarid environments, where riparian zones may be the only tree-dominated ecosystem in the landscape. The presence of water, increased productivity, favorable microclimate, and unique disturbance (flooding) regimes combine to create a disproportionately higher biological diversity than the surrounding uplands. In the Intermountain West and Great Basin of the United States, about 85% of the wildlife species are dependent on riparian zones for all or part of their life cycles. In these same riparian zones, well over 100 plant species commonly can be found on a single gravel bar (<50 m in length along the streambank).
>
> Livestock grazing is among the most significant land uses affecting the structure and productivity of riparian zones. Because of the high plant productivity, close proximity to water, favorable microclimate, and level ground, these areas are preferred habitat for livestock, and overgrazing commonly occurs. Grazing changes the composition and structure of the riparian zones, which lead to dramatic effects on the large numbers of wildlife that use the areas for feeding, breeding, or hiding cover. Of particular importance are the effects of livestock grazing on streamside forests of cottonwood, aspen, and willows. Long-term overgrazing can eliminate the woody vegetation that provides essential breeding bird habitat. In addition, trampling in riparian zones decreases infiltration and breaks down streambanks, resulting in channel incision or widening. These influences on streambank structure sever the linkages between the riparian and aquatic zones through the elimination of overbank flooding. The removal or loss of riparian vegetation results in the decline of organic inputs into the stream and the loss of overstory vegetation shade, which can dramatically affect the aquatic biota, such as salmon and trout.
>
> Because of the high level of biodiversity and other ecosystem values of riparian zones, particularly in semiarid and arid environments, their restoration yields many positive benefits. Riparian zones are naturally an environment with frequent disturbances (floods), and because of this, the adapted biota often exhibit considerable resilience following the cessation of land uses that cause degradation or that prevent recovery. In rangelands, this most often includes abusive grazing practices. Thus, the cessation of overgrazing is the first important step in riparian restoration and should be implemented prior to determining if other active restoration measures need to be taken. For lasting riparian restoration, a larger effort should be aimed at the recovery of the entire watershed (Kauffman et al., 1997).
>
> Sources: Gregory, S. V., F. J. Swanson, W. A. McKee, and K. W. Cummins. (1991). An ecosystem perspective of riparian zones. *BioScience* **41**, 540–550.
> Kauffman, J. B., and W. C. Krueger. (1984). Livestock impacts on riparian ecosystems and streamside management implications . . . A review. *J. Range Management* **5**, 430–437.
> Kauffman, J. B., R. L. Beschta, N. Otting, and D. Lytjen. (1997). An ecological perspective of riparian and stream restoration in the western United States. *Fisheries* **22**, 12–24.

mals move across the land. The break-up of soil crusts (both physical and biological) can increase soil loss via wind and water erosion. The susceptibility of soils to compaction is related to soil texture (sand, silt, and clay content), as well as the season of grazing. Clay soils are more susceptible to compaction than sandy soils and wet soils are more susceptible than drier soils. Soil compaction results in reductions in soil porosity, water-holding capacity, and rates of infiltration. All of these result in less water available for plant growth and more water lost via surface runoff.

Another direct impact of livestock is the influence on nutrient inputs and redistribution. The effects of herbivory on litter reduction, trampling, and soil compaction include direct influences on nutrient cycles.

Trampling of biological soil crusts can reduce nitrogen inputs while increasing nitrogen losses via erosion. Urination and defecation are forms of nutrient redistribution. These inputs can enhance soil fertility, but can also have serious effects on water quality and aquatic organisms. Concentrated nutrient inputs can increase bacterial and protozoan pathogens, promote algal growth, and alter water chemistry—particularly the depletion of dissolved oxygen.

The final direct influence of livestock is the dispersal of undesirable exotic organisms (as well as native weeds). Seeds of exotic species may be transported on the fur or hooves of animals or deposited in the feces. As livestock move through an area, they may deposit exotic plant seeds while simultaneously preparing a seedbed via trampling. A grazing preference for desirable native species may further improve the competitive advantage of the less desired weeds, thus increasing their probability of establishment and dominance. Animals do not graze indiscriminately. Different animals have preferences and select some plant species over others. This will confer a selective advantage on less palatable species over those preferred by livestock. Often exotic species are less palatable and more grazing tolerant than native species. Livestock grazing in rangelands where native species are not well adapted to large herbivores will confer a selective advantage to exotic plant species.

Livestock also spread exotic diseases to native herbivores. Numerous herds of bighorn sheep in the United States have been eliminated following disease transmission from domestic sheep.

2. Indirect Effects: Secondary and Tertiary Influences

The direct or primary influences of livestock elicit a number of feedback responses that affect additional species, functions, and processes. The outcomes of these far-reaching effects are described as secondary influences and include changes in landscape disturbance cycles (e.g., fire regimes), accelerated rates of erosion, alterations in hydrology and plant available water, and alterations in successional patterns due to changes in competition and reproduction of both the native and exotic species (see Fig. 4).

The cumulative impacts of livestock grazing are apparent in many desertified or degraded ecosystems throughout the world. The tertiary influences of livestock overgrazing on ecosystems are the characteristic endpoint of desertification or land degradation (Fig. 4). The tertiary effects include dramatic alterations to the biotic structure, composition, and productivity of ecosystems. These are closely linked to simultaneous degradation of hydrological and soil properties. Declines in the cover of native perennial plants and biological soil crusts and concomitant increases in bare ground, unpalatable shrubs, or noxious weeds and annuals are examples of the long-term degradation of ecosystems. Tertiary effects of livestock are often difficult to separate from other significant influences on ecosystems. Degraded landscapes are often the cumulative result of inappropriate livestock management in concert with other poor land management practices, including logging, road building, inappropriate agronomy, fuel-wood harvesting, mining, water diversions, and over-harvesting of plant and/or animal resources (i.e., hunting, poaching).

B. Livestock Grazing and the Desertification of Native Rangelands

Desertification of rangelands has had strong negative effects on Earth's biodiversity and the continued rate of desertification represents a major global threat. Desertification is manifested in a combination of declines in native species diversity, ecosystem productivity, loss of topsoil, changes in nutrient cycles, and alterations in hydrology and microclimate. For example, Flather *et al.* (1994) reported that 40% of all U.S. federally listed Threatened and Endangered (T&E) species were associated with rangelands. More T&E plant species occur on rangelands in the United States than any other land cover type. Trends in biodiversity loss are likely similar in other rangelands of the world. This is particularly true for large native herbivores, which are now largely restricted to small preserves or parks. Nearly all of Earth's rangelands have been degraded by human activities, including livestock grazing, the introduction of exotic species, fuelwood harvesting, alteration of natural fire cycles, wildlife depredation, and conversion to cropland or urbanization.

Estimates by the United Nations Environmental Programme (UNEP, 1990) indicated that 73% of the world's 3.3 billion ha of dry land is at least moderately desertified, having lost more than 25% of its productive capacity. It has been estimated that ≈84% of the world's rangelands are at least moderately desertified (i.e., conditions in which plant productivity has declined by ≥25%; or indications that accelerated wind or water erosion has occurred due to land use; or soil salinity has reduced crop yields 10–50%). Of the 3.1 billion ha of rangeland in the world, Mabbutt (1984) estimated that as many as 1.3 billion ha are severely degraded (i.e., productivity loss exceeds 50%).

The UNEP study on soil degradation reported that over the past 45 years about 11% (1.2 billion ha) of Earth's vegetated soils have become degraded to the point that their original biotic functions were damaged; reclamation would be costly or in some cases impossible. Most of these degraded soils are suffering from "moderate" degradation, that is, productivity is greatly reduced but can still be used for agriculture. The soil's original biotic functions (its capacity to process nutrients into a form usable by plants) have been partially destroyed and only with major improvements can productivity be restored. A smaller proportion ($\approx$3% of Earth's surface) showed severe degradation, where original biotic functions are largely destroyed and may be reclaimable only with major financial and technical assistance. Nine million hectares of land were classified as extremely degraded, that is, unreclaimable and beyond restoration. On this land the original biotic functions are fully destroyed. At continental scales, land areas with soil degradation classified as moderate to severe ranged from 4.4% in North America, to 14.4% in Africa, to 24.1% in Central America and Mexico.

Throughout the world, accelerated wind and water erosion is the process that causes the majority of soil degradation, and overgrazing is the most ubiquitous cause of the acceleration in the erosion process. Globally 35% of the soil degradation that has occurred since 1945 was attributed to overgrazing (UNEP, 1990). Thus overgrazing is the most pervasive cause of soil degradation, affecting 679 million ha. In Africa and North America, grazing was the principal cause for 49 and 30% of the soil degradation, respectively. Soil erosion resulting from the decrease in plant cover and trampling may require thousands of years for recovery, particularly in semiarid and arid ecosystems.

Perhaps the largest data set showing trends in the ecological condition and biological impoverishment of rangelands at large scales (country or continental scales) are those of public lands in the western United States. Determining the pristine state of rangelands is difficult, and current approaches are being challenged as to their usefulness for truly ascertaining ecological condition. This is because few areas of undisturbed or pristine rangeland exist (i.e., areas with no history of livestock grazing), unlike the case with old growth or primary forest (i.e., never deforested). In addition, comparison of the degree of departure from a single hypothetical climax composition is specious at best. Keeping in mind these uncertainties associated with ecological interpretations of rangeland condition, only about 4% of nonarctic rangeland in the United States is considered to be in excellent condition (i.e., covered with native vegetation and suffering no loss in productivity). Most U.S. rangeland is in fair or good condition (i.e., slight to moderate degrees of degradation), but a significant portion (15%) is categorized as poor or severely degraded. In 1990, the U.S. Bureau of Land Management reported that only 33% of its holdings were in good or excellent condition.

Range trends, whether the rangelands are improving, degrading, or static, are a measure of how well rangelands are being managed. The results vary depending on the parameters examined. The U.S. land-managing agencies suggest that about 86% of rangelands are either improving or static (i.e., exhibit no discernable trend), and that 14% are degrading. Yet portions considered static or stable may be in such a degraded ecological state that further decline would be difficult to discern. For example, many western rangelands are completely dominated by exotic annuals or weeds, and others have experienced extensive shrub encroachments during the latter half of the twentieth century. These patterns make it difficult to interpret landscape-scale changes in ecological condition.

Though many scientists believe that many U.S. rangelands are currently in the best ecological condition of the last 60 years, others point out that more than 50% of the rangelands contain less than half of their potential natural plant composition after more than 50 years of "modern range management" (i.e., directed toward recovery and sustained use). Because of the loss of soil and native species, coupled with ongoing invasions of exotic species, many suggest that desertified North American rangelands will not likely regain their former diversity in time frames of less than a century. Research and innovative approaches for the restoration of biological diversity of rangelands are needed.

C. Conversion of Tropical Forests to Livestock Pasture

Livestock impacts are not limited to Earth's rangelands. Rain forests are among the most biologically diverse ecosystems in the world. Tropical forests cover about 720 million ha and contain 40–90% of the world's species. The dramatic rate of conversion of tropical forests to cattle pastures is one of the most deleterious environmental impacts on global biodiversity in the latter half of the twentieth century. Conversion of tropical forest to pasture not only results in the extirpation of native rain forest plants and animals, but also alters the microclimate, hydrology, and fire cycles, and is a significant source of greenhouse gases.

> **Box 2**
>
> ### Biological Soil Crusts: Their Role and Susceptibility in Rangeland Ecosystems
>
> Biological or microphytic soil crusts consist of a combination of nonvascular plants (also known as cryptogams; these include cyanobacteria, mosses, lichens, liverworts, and green algae). Biological soil crusts serve a number of ecosystem functions that make them an important component of the biological diversity of many of the semiarid and arid environments of the world. These soil crusts exist in nearly all ecosystems where vascular plant material (live or dead) covers less than 100% of the ground surface; they occupy the interspaces between vascular plants. Spatial coverage within some ecosystems may be greater than that of vascular plants, ranging from 10 to 100% coverage in some undisturbed temperate and arctic ecosystems. Biological soil crusts are adapted to extreme environmental conditions of extended drought and extreme high and low temperatures. These species are metabolically active only after hydration. Some are able to hydrate under conditions of saturated air, such as fog, dew point, or high vapor pressure; others require liquid water.
>
> The functional role of biological soil crusts in semiarid and arid ecosystems is multifaceted. Some species are involved in the breakdown of humus and in the release of nutrients. However, a more significant role in some ecosystems may be their direct contribution of increasing nitrogen availability to other plants and organisms within the ecosystem through biological nitrogen fixation. Nitrogen is often an important limiting factor for ecosystem productivity. Biological soil crusts (cyanobacteria and cyanolichens) can be important sources of nitrogen for many semiarid ecosystems, such as the North American Great Basin and Colorado Plateau. Fixed nitrogen is immediately released by these organisms into the surrounding soils; however, the amount taken up by vascular plants has not been widely studied.
>
> Biological soil crusts can diminish the rate of wind and water erosion from rangelands. Soil stability is enhanced by these soil crusts through the formation of exudates of polysaccharides that bind soil particles into aggregates. They also protect the soil from raindrop erosion (the dislodging of soil particles by individual raindrops, also known as splash erosion) by absorbing the kinetic energy of raindrops. The potential to diminish soil erosion varies with biological crust composition, increasing from algal, to lichen, to moss-covered crusts. Crusts create a surface roughness that slows the surface runoff of water and may trap soil particles arriving via wind deposition. The phototropic nature of the crust species is an adaptation that contributes to their upward growth and soil entrapment when soil buries the crusts. They are important contributors to dune stability when dunes consist of sufficient amounts of fine soil particles (silts and clays).
>
> Complex direct and indirect mutualistic relationships exist between biological soil crust and vascular plants. In addition to nitrogen fixation and soil stability, the presence of soil crusts is positively correlated with floristic diversity in several ecosystems around the world. This positive correlation may be related to the increased seedling establishment and survival of vascular plants when grown in association with biological soil crusts. The surface variation that crust species provide may also contribute to greater variation in suitable sites for successful seedling germination and establishment.
>
> Throughout the rangelands of the world, the introduction of large numbers of domestic ungulates has damaged biological soil crusts via herbivory and trampling. Herbivory by ungulates is generally restricted to locations where domestic reindeer have overgrazed lichen-dominated winter habitat in arctic tundra ecosystems. Globally, the most prevalent ungulate impacts are related to trampling, which breaks the dry, brittle lichens into small pieces that are then blown from the site. In many rangelands, intense grazing by livestock around watering places results in the elimination of lichens near the water; as distance from water increases, livestock impacts on lichen abundance decrease. Biological soil crusts growing on soils with low aggregate stability (i.e., sandy soils) are more susceptible to trampling damage during dry periods. The disruption of the dry soil surface breaks the bonds between the biological soil crust and the soil. Both the soil particles and the crust species are then susceptible to wind and water erosion.
>
> Following disturbance or loss, biological crust recovery time depends on the severity of the disturbance and the environmental characteristics of the ecosystem. Primary recovery begins with cyanobacteria within one year after disturbance.

Belnap (1993) estimated that some ecosystems in the U.S. Southwest may require 30–40 years for the replacement of sheath material in the soil, 45–85 years for lichen diversity to recover, and over 250 years for moss cover to return if left undisturbed. Recovery of crust species is more rapid in locations with higher effective precipitation and finer-textured soils, and if inoculating material is present. Clearly, ecosystems in which biological soil crusts are a major component need appropriate grazing management or protection to maintain or restore biodiversity and to prevent degradation, erosion, and desertification.

Source: Belnap, J. (1993). Recovery rates of cryptobiotic crusts: Inoculate use and assessment methods. *Great Basin Naturalist* 53, 89–95.

Most of the deforestation in the tropical rain forests, moist forests, and dry forests of Latin America is for conversion to cattle pasture. In addition to soil, climatic, and vegetative characteristics, socioeconomic status of the landowner is a determinant of the patterns of land use. For example, on large ranches in the Brazilian Amazon, tropical forests are converted directly to cattle pasture. In contrast, small-scale subsistence farmers may use the land for one or two shifting cultivation cycles before ultimately converting to cattle pasture.

Slashing and burning tropical forest for cattle production converts some of the most structurally and biologically diverse ecosystems on Earth to simple pastures dominated by exotic grasses (Fig. 5). Rain forests frequently have over 100 canopy tree species per hectare and millions of highly interdependent organisms living within the forest and soil. In contrast, the only trees, if any, in pastures are cultivated fruit and nut trees or highly invasive second-growth trees. Rather than the diverse array of birds, mammals, insects, reptiles, amphibians, and other native tropical fauna, the pastures are dominated by the cattle and the few other animals that thrive in human-modified environments.

Deforestation affects biological diversity in three ways: (1) destruction of habitat; (2) fragmentation of formerly contiguous forest habitat; and (3) edge effects within the forest that is immediately adjacent to pastures. In fragments and edge forests, the microclimate is altered such that temperatures and wind speeds are higher, and relative humidity is lower. Higher wind speeds and drier conditions result in increased tree losses due to windthrow and maybe increased water stress. The drier conditions, in concert with increased quantities of dead wood, increase the susceptibility of fragmented and edge forests to wildfires. Ignition

FIGURE 5 Tropical rain forests are among the most biologically diverse ecosystems on Earth. The principal cause of deforestation in the Neotropics (and elsewhere) is for cattle pasture conversion. In this photo, the biologically diverse rain forest has been replaced by a pasture dominated by a few exotic grasses and invasive plants. (Photo by Boone Kauffman.)

sources for wildfire are widespread in a tropical landscape mosaic of fragmented forest and pastures because fire is frequently used to maintain the pastures. Because native flora and fauna are poorly adapted to fires in tropical rain forests, even light fires result in high rates of mortality. Low-intensity fires that spread from cattle pastures to adjacent edge forest have been found to completely kill the overstory canopy, while only $\approx 40\%$ of the species had the capacity to sprout after burning (Kauffman, 1991).

Data from the Brazilian space agency (INPE) and the U.S. Space Agency (NASA), as well as that from other Brazilian and U.S. scientists, indicate that the Amazon Basin is experiencing the world's highest absolute rate of forest destruction. The rate of deforestation averaged 15,000 km^2 yr^{-1} from 1978 to 1988, while the rate of habitat degradation (i.e., the sum of cleared and fragmented lands) was 38,000 km^2 yr^{-1}. The rate of deforestation has accelerated in recent years from about 11,000 km^2 yr^{-1} in 1991 to just over 20,000 km^2 yr^{-1} from 1995 to 1997. At the rates experienced in the late 1990s, 5479 ha of forest are lost every day; this is equivalent to about 3.8 ha destroyed every minute.

It is difficult to determine the total area of tropical rain forest that has been converted to cattle pasture. Fearnside (1993) reported that by 1991 the area of forest cleared in the Amazon Basin had reached 426,000 km^2. This estimate did not take into account forest areas affected by fragmentation and edge effects. Cumulatively, areas in which biodiversity has been affected by forest clearing (primarily for cattle pasture) may be over twice that of deforestation alone (Skole and Tucker, 1993). Because about 13% of the region has now been deforested (INPE, 1996,1998), the total area affected by deforestation and fragmentation could comprise a third of the entire Amazon Basin.

Rain forests tend to receive the most attention, but tropical dry forests have been deforested to an even greater degree. These are also the most abundant type of tropical forest. Few areas of intact (i.e., uncut or primary) tropical dry forest still exist in Latin America today (e.g., the Brazilian caatinga and tropical deciduous forests of Central America and Mexico).

Although scientists and economists have questioned both the sustainability and economic viability of cattle pastures in areas of former tropical forest, the deleterious consequences to biological diversity are evident. If cattle pasture production is neither sustainable nor economically viable, then why are rates of deforestation so high? Contrary to popular belief, deforestation in the Americas has not been a response to international demands for beef nor a response to increasing human populations (hoof-and-mouth disease is prevalent in Brazilian cattle, which precludes importation into the United States, and the Amazon Basin supports only $\approx 5\%$ of all cattle production in Brazil). Deforestation largely occurs because it is a way to obtain title to the land, or to profit from land speculation, government financial incentives, and subsidies. In addition, labor costs on ranches are quite low compared to intensive agriculture or agroforestry systems. The environmental limitations in many tropical landscapes, coupled with governmental and social forces, combine to promote a land use that produces little food and little direct monetary return, but tremendous environmental degradation (Hecht, 1983).

D. Livestock Grazing and the Greenhouse Effect

Livestock production and related land uses are significant sources of many greenhouse gases. In addition, changes in climate due to the greenhouse effect are predicted to have strong feedbacks on the environments that provide feed and water for livestock. Among the most significant influences of the livestock industry in relation to the greenhouse effect is slash burning, which occurs during the conversion of tropical rain forest to pasture. When slashed tropical rain forest burns, as much as one-half of the aboveground carbon (76–112 Mg ha^{-1}) may be released as CO_2, CO, and other radiatively active gases. As much as 800–1600 kg nitrogen ha^{-1} are also released by burning forest (Guild et al., 1998). When tropical dry forest is slashed and burned, the proportion of biomass consumed is even greater—as much as 95% of the aboveground carbon pool may be released as greenhouse gases during slash fires. The burning of the African savannas, most often initiated by herders for forage enhancement, is another important source of CO_2. Savanna fires contribute as much as 16% of the total annual CO_2 arising from agricultural sources. However, much of this will be absorbed the following year by regrowing grasses. The same is not true for forest that is converted to pasture; the biomass of pasture grasses will be about 5% of the forests that it replaced.

Methane is a powerful greenhouse gas with a radiative absorption capacity that is 21 times that of CO_2. The global increase in methane results from human activities such as livestock production, manure management, rice production, landfills, and the production and use of oil, gas, and coal. Livestock and manure management contribute about 16% of the total annual methane production (about 87 million Mg out of 550 million Mg produced each year; de Haan et al., 1996). Methane emissions from ruminants (cattle, sheep, and goats) result from their capacity to utilize large amounts of fibrous grasses. Low productivity and poor feed quality are characteristic for most land-based ruminant production systems in arid regions. Poor feed quality also typifies pastures in the tropics and subtropics. In these scenarios, emissions per animal are higher than in scenarios where animals are on higher-quality feeds. With the conversion of tropical forest to cattle pasture, not only are large carbon sinks shifted to significant global atmospheric carbon sources, but the resulting land use becomes a significant global methane source.

E. Other Factors

1. Conversion of Native Rangelands to Croplands or Pasture

In addition to the effects previously mentioned, there are many other livestock-related land use activities and industries that affect global biological diversity (see Table I). Grazing on native grasslands and shrublands supports a dwindling share of the livestock forage supply of the world. As intensive livestock production has expanded and the importance of grazing native rangeland has declined, there have been large conversions of rangelands to seemingly more lucrative land uses. The highly diverse Brazilian cerrado, a mosaic of savanna, grasslands, and open evergreen woodlands, covers about 2 million km^2 in Brazil. The cerrado is undergoing a rapid conversion from species-rich savanna/woodland to planted pastures, soybeans, and other annual crops. This region has experienced more land cover change than the Amazon forests (i.e., about 600,000 km^2 of the cerrado had been cleared as of 1991, compared to about 400,000 km^2 for the Brazilian Amazon; Klink et al., 1994).

Many rangelands of the U.S. Southwest and southern Great Plains were plowed into croplands when government subsidies made it cost-effective to utilize groundwater for irrigation (i.e., the Ogalala aquifer). Often, these water sources were rapidly depleted and groundwater became prohibitively costly to pump. This resulted in the abandonment of lands that are now in a desertified state—a depauperate cover of native vegetation coupled with high rates of soil erosion.

2. Exotic Species and Pathogens

Invasions of exotic species are among the greatest threats to global biodiversity and livestock play significant direct and indirect roles in facilitating their establishment. Livestock influence exotic plant distribution through seed dispersal in fur and dung. Establishment of exotics is enhanced via creation of seedbeds through trampling and forage removal. In those regions where large herbivores were not a strong selective pressure on the native flora (i.e., much of the Pacific Northwest, Intermountain West, and California, in the United States), exotic annuals now dominate areas formerly dominated by perennial bunchgrasses. Indirectly, the livestock industries also facilitated introductions and spread of exotic plant species through cropland plantings (fodder and feed grains) that contained weed seeds.

In contrast to unanticipated shifts of native ecosystems to exotic species dominance, many native rangelands and forests have been converted to exotic grass dominance through purposeful seeding. The goals of these so-called "range improvements" were to increase forage for livestock, decrease soil erosion, and provide a desired alternative to weeds. However, there was scant consideration for influences on biological diversity. For example, in the United States millions of hectares formerly occupied by shrub/bunchgrass ecosystems have been seeded with grasses of exotic origin (e.g., crested wheatgrass, *Agropyron cristatum*, and Lehmann lovegrass, *Eragrostis lehmanniana*).

3. Livestock, Water Developments, and Aquatic Ecosystems

Livestock and related management activities have strong effects on the biodiversity of aquatic ecosystems. These not only include direct effects such as consumption of streamside or wetland vegetation, trampling of streambanks, and fecal inputs, but also indirect effects such as irrigation withdrawals, water developments, and wetland draining. These activities can alter aquatic diversity through degradation of aquatic habitats as well as water quality (temperature, chemistry, and microbial composition). Numerous studies have documented accelerated streambank erosion and losses of streambank structure due to livestock grazing. The changes and the accrual of sediment in the channel degrade spawning and reproductive habitats for fishes and aquatic insects. Overgrazing results in the simplification of stream channels, which may include loss of channel sinuosity, increases in channel widths, increased channel incision, and decreases in deep pools. Incised and simplified channels result in the elimination of important linkages between the floodplain and stream channel that positively influence biodiversity.

Water diversions for pasture or forage crop irrigation are a widespread influence on the biodiversity of rivers streams and their associated riparian zones. Diversions range from the de-watering of small streams for the irrigation of a single pasture or farm field to huge dams that span major rivers. Regardless of the scale, when water is diverted or dammed, the natural hydroperiod of the river or stream is altered (i.e., changes in the timing and magnitude of peak and low flows throughout the year). This can have dramatic influences on channel-forming processes and riparian community development. Because many aquatic and riparian organisms are adapted to cycles of floods and flood effects, these hydrological alterations can completely alter their environment, leading to species losses. Throughout the western United States, livestock-related agriculture is a dominant user of irrigation water stored in federally-constructed dams. Significant detrimental effects on native fishes, as well as on biologically diverse riparian plant and animal communities, occur both upstream and downstream of the dams or diversions.

Land rehabilitation and forage enhancement efforts have been attempted throughout degraded rangelands of the world. However, these have often had disappointing results. Because of misinterpretations of ecosystem needs or ignorance of ecological consequences, rehabilitation efforts have sometimes exacerbated the deterioration and degradation that they were intended to reverse. A striking example is the establishment of permanent water sources in dry season grazing areas in Africa (similar developments have also been implemented in the U.S. West). Water sources were developed in areas so distant from surface water that they were not traditionally used by livestock during the dry seasons. Following developments such as the drilling of wells or construction of small catchment basins, lands around these African water sources were often severely overgrazed and trampled. In addition to losses in biodiversity, these water sources have fostered the growth of herds beyond the carrying capacity of the land, resulting in range deterioration and livestock deaths during droughts. During droughts, livestock did not die from lack of water but from starvation. The ultimate consequences include degradation of the native ecosystem coupled with increased human suffering and hardship.

Water pollution is a common problem associated with intensive livestock production in relatively confined spaces. Manure is a valuable organic fertilizer and soil builder in modest amounts, but it is a dangerous environmental hazard when waste production exceeds the absorptive capacity of land and water. In fecal-polluted waterways, pathogens detrimental to both the native biota and humans are present. In addition, increased nitrogen and phosphorus concentrations can result in the eutrophication of waterways, in which algae blooms rapidly consume oxygen to the point where fish kills occur.

4. Biogeochemical Cycling

Livestock grazing can affect nutrient cycling in numerous ways. The most obvious would arise from the dramatic effects associated with accelerated erosion, trampling of soils, and the replacement of perennial plants (and often biological soil crusts) with annuals or shrubs and bare ground. The direct consumption of vegetation changes the patterns and distribution of litter, which affects decomposition and nutrient cycling. Consumption of herbage resulting in less litter also increases the proportion of bare ground, thereby increasing susceptibility to erosion. Accelerated erosion results in the loss of nutrients, soil organic matter, and the capacity of the soil to store water and retain nutrients. In riparian zones, diminished litterfall can also affect the aquatic biota. For many headwater streams, the nutrients and energy that drive in-stream aquatic ecosystems are derived from terrestrial inputs. It is particularly deleterious to aquatic ecosystems when long-term grazing results in the elimination of shrub and tree overstories in streamside environments.

The synergistic effects of forage removal and trampling damage by livestock have been described in Oregon riparian meadows that were grazed for over a century. Grazed meadows had lower rates of nitrogen mineralization and lower soil organic matter than areas where grazing had been halted for 9–12 years. These indicators of decreased productivity were associated with dramatically lower water infiltration rates, higher soil bulk density, lower levels of residual litter, and lower root biomass in the grazed sites.

Desertification often results in the alteration of natural desert grasslands and savannas into shrublands with bare soil in the intershrub spaces. Under these scenarios, nutrient and water resources become concentrated under shrubs. Livestock grazing can also concentrate nutrients in areas where livestock may congregate through the concentration of urine and feces (i.e., resting places, water holes, or streamsides).

5. Attempts to Control Undesirable Species

Other activities associated with livestock management that affect biodiversity include predator control (e.g., eliminating carnivores via trapping, poisoning, or shooting) and other forms of wildlife control for species that are perceived as competitors with the livestock industry. These activities, called "animal damage control," are often carried out without consideration of their effects on biodiversity, ecological processes, or ecosystem functions. Techniques of animal damage control are often not specific to the "target species" and other "nontarget" species can be eliminated as well. Poison bait traps and many pesticides often kill nontarget animals (and insects). The black-footed ferret (*Mustela nigripes*) was almost driven to extinction because of the decline of the prairie dog (*Cynomys* spp.), its major prey. Prairie dog populations have precipitously fallen due to habitat loss by agricultural development and from purposeful elimination because they eat the same forage as livestock. Many large carnivores have been eliminated in regions because of their real or perceived threats to livestock. The livestock industry interest groups provide the most vehement opposition to the reintroductions of large carnivores to extirpated ranges (e.g., wolves to Yellowstone National Park).

As with predator control, the use of petrochemicals (herbicides and pesticides) to control unwanted species of plants and insects has sometimes resulted in undesirable effects on non-target species. In animal, plant, and insect control programs, scant attention has been paid to the functional role that these organisms may play in ecosystems and possible feedback responses to their removal. In addition, questions regarding the underlying reasons for the increase in pest or weed damage frequently were not asked or left unanswered. In the western United States, practices such as mechanical and chemical removal of shrubs and trees were implemented for the benefit of increased cattle forage, but at the expense of other human and ecological values. For example, widespread rangeland herbicide applications eliminated traditional food and medicinal plants used by indigenous peoples on many western U.S. rangelands. This is because herbicides applied to control unwanted plant species also kill a suite of other native broad-leaved species. Species with limited capacities for reinvasion may be locally extirpated from sites after a single herbicide application. Similarly, insecticides may lack selectivity, such that a suite of species performing a multitude of ecosystem functions are eliminated. Many of these chemicals act as environmental estrogens and their elevated concentrations within secondary consumers (predators and scavengers) have resulted in reproductive and physiological maladies.

6. Altering Fire Regimes

In many grasslands and savannas of the world, fire is a dominant disturbance feature of the landscape. The high productivity of flammable grasses, climatic conditions with long dry periods, and a prevalent ignition source (lightning or humans) result in the frequent occurrence of fires. Fires in grasslands and savannas are important in cycling nutrients and in influencing biotic structure and composition. Both the flora and fauna of grasslands and savannas are adapted to, and often dependent on, fire for their continued existence. Altering the fire regime can lead to a vastly different structure and composition. Livestock alter fire regimes by removing the fuels (i.e., the grasses) that carry the fire in rangelands and savannas. For example, throughout the U.S. West, livestock grazing has resulted in the diminution of fires in rangelands with a concomitant increase in shrub species that are not as fire-adapted. Increases in the abundance of juniper (*Juniperus* spp.) and mesquite (*Prosopis* spp.), as well as declines in aspen (*Populus tremuloides*), are related to the synergism of livestock and the decline of fire occurrence. Long-term changes in ecosystem structure due to grazing and fire suppression have resulted in a scenario in which simple livestock removal would not completely remedy the ecosystem decline; vegetation manipulations coupled with the reintroduction of fire would also be necessary components of ecological restoration.

III. CONCLUSIONS: HUMAN INSTITUTIONS AND LIVESTOCK-CAUSED DEGRADATION

Livestock create an array of environmental problems, not because cows, sheep, goats, and other grazers are hazards in themselves, but because human institutions have forced animal farming out of alignment with the ecosystems in which they are practiced. The root causes of land degradation, including deforestation and desertification, lay not in the animals, but in the social and political institutions and population pressures placed upon the environment. Sociopolitical causations must be addressed if degradation of biodiversity by livestock is to be halted and ecosystems restored.

In many countries there persists government-based subsidies and economic incentives that result in the expansion of livestock production that accelerates environmental degradation. Government policies on pricing, taxing, and land titling incentives affect resource use by influencing decisions about type of use, inputs, technology adoption, and investments in development. Policies have misguided livestock development through the subsidized pricing of inputs and products that induced the non-sustainable use and degradation of natural resources. Resources have been priced too low to reflect their true environmental costs, thus passing the debt of environmental degradation along to future generations. In the United States, subsidized grazing on public lands and federal dams that provide water for irrigated pasture and fodder production are prominent examples. In the Brazilian Amazon, land speculation, titling procedures, and government financial incentives have been the main reasons for the conversion of tropical rain forests into cattle pastures. In many less-developed countries, strong human population growth is fueling demand for livestock products while at the same time limiting the traditional sources for livestock production. Increases in per capita income and urbanization are raising the demand for livestock products. The growing social inequalities among the rich and the poor result in different motivations for degradation: the greed of the rich and the desperation of the poor.

Desertification is the result of inappropriate land use activities, including overgrazing, overcultivation, salinization due to irrigation practices, and deforestation. Although differences exist among different rangelands, cultures, and countries of the developing world, the underlying forces of degradation include population densities that are greater than the land can sustain and, more fundamentally, social and economic inequities that push people into marginal environments and vulnerable livelihoods. Particularly important in the developing world is the inequitable distribution of land among the rich and poor. When a disproportionate share of the land is kept in the hands of a few, the ability of the impoverished majority to sustainably manage the marginal lands that they control is severely compromised.

The first step in the ecological restoration of degraded rangelands is the cessation of those activities that are causing degradation or preventing recovery (Kauffman et al., 1997). Changing the social and institutional incentives that result in the degradation of ecosystems may be more difficult to achieve than the biological and physical aspects of ecosystem restoration. It must be recognized that sustained productivity and livelihoods will most likely occur when lands are managed in a manner that closely approximates their evolutionary development and where ecosystem processes are allowed to occur with minimal disruptions or alterations. In rangelands, management that mimics the natural grazing patterns of large wild herbivores may be the most sustainable and successful in maintaining biological diversity. Practices that maintain functional ecosystems rather than attempt to maximize short-term meat production are an integral feature of sustained land management. Similarly, if functional features and biodiversity are to be maintained in forests, the utilization of standing rain forest should be encouraged rather than replacing them with cattle pastures.

See Also the Following Articles

AGRICULTURE, SUSTAINABLE • AGRICULTURE, TRADITIONAL • BIOGEOCHEMICAL CYCLES • CATTLE, SHEEP, AND GOATS, ECOLOGICAL ROLE OF • DEFORESTATION • DESERTIFICATION • FIRES, ECOLOGICAL EFFECTS OF • PLANT-ANIMAL INTERACTIONS • SAVANNAH AND GRASSLAND ECOSYSTEMS, NONTROPICAL

Bibliography

de Castro, E. A., and J. B. Kauffman. (1998). Ecosystem structure in the Brazilian Cerrado: A vegetation gradient of aboveground biomass, root mass and consumption by fire. *J. Tropical Ecol.* 14, 263–283.

de Haan, C., H. Henning, and H. Blackburn. (1996). *Livestock and the Environment. Finding a Balance*. European Commission Directorate-General for Development.

Dodd, J. L. (1994). Desertification and degradation in sub-Saharan Africa: The role of livestock. *Bioscience* 44, 28–34.

Fearnside, P. M. (1993). Deforestation in Brazilian Amazonia: The effect of population and land tenure. *Ambio* 22, 537–545.

Flather, C. H., L. A. Joyce, and C. A. Bloomgarden. (1994). *Species Endangerment Patterns in the United States*. USDA Forest Service General Technical Report RM 241. Rocky Mountain Forest and Range Experiment Station, Ft. Collins, Colorado.

Guild, L. S., J. B. Kauffman, L. J. Ellingson, D. L. Cummings, E. A. Castro, R. E. Babbitt, and D. E. Ward. (1998). Dynamics associated with total aboveground biomass, C, nutrient pools and biomass burning of primary forest and pasture in Rondonia during SCAR-B. *J. Geophys. Res.* 103, 32,091–32,100.

Hecht, S. B. (1993). The logic of livestock and deforestation in Amazonia. *Bioscience* 43, 687–695.

INPE. (1996). *Deforestation Estimates in the Brazilian Amazon, 1992–1994*. Instituto Nacional de Pesquisas Espaciais, Diario Official da Uniao, Brasilia, Brasil.

INPE. (1998). *Deforestation Estimates in the Brazilian Amazon, 1995–1997*. Instituto Nacional de Pesquisas Espaciais, Diario Official da Uniao, Brasilia, Brasil.

Kauffman, J. B., R. L. Beschta, N. Otting, and D. Lytjen. (1997). An ecological perspective of riparian and stream restoration in the western United States. *Fisheries* 22, 12–24.

Klink, C. A., R. H. Macedo, and C. C. Mueller. (1994). *Cerrado: Processo de Ocupacao e Implicacoes para a Conservacao e Utilizacao Sustenavel de Sua Diversidade Biologica*. World Wildlife Fund, Brasil Report. 104.

Mabbutt, J. A. (1984). A new global assessment of the status and trends of desertification. *Environ. Conservation* 11, 103–113.

Oldeman, L. R., R. T. A. Hakkeling, and W. G. Sombroek. (1990). *World Map of the Status of Human-Induced Soil Degradation: An Explanatory Note*, revised 2nd ed. International Soil Reference and Information Center/United Nations Environmental Programme, Wageningen, Netherlands.

Sinclair, A. R. E. (1979). The eruption of the ruminants. *Serengeti: Dynamics of an Ecosystem* (A. R. E. Sinclair and M. Norton-Griffiths, eds.), pp. 82–103. University of Chicago Press, Chicago.

Skole, D., and C. Tucker. (1993). Tropical deforestation and habitat fragmentation in the Amazon: Satellite data from 1978 to 1988. *Science* **260**, 1905–1910.

Steinfeld, H., C. de Haan, and H. Blackburn. (1996). *Livestock–Environment Interactions—Issues and Options*. European Commission Directorate-General for Development.

United Nations Environmental Programme. (1990). *Global Assessment of Soil Degradation*. UNEP, Nairobi, Kenya.

Wolf, E. C. (1986). Managing rangelands. In *State of the World 1986*. Worldwatch Institute, Washington, D.C.

World Resources Institute. (1994). *World Resources 1994–1995*. Oxford University Press, New York.

RECOMBINATION

Abraham B. Korol
University of Haifa

I. Types of Recombination
II. Basic Features of Reciprocal Homologous Recombination (Crossing-Over)
III. Genetic Control and Mechanisms of Recombination
IV. Biodiversity of Recombination Systems
V. Evolutionary Effects of Recombination
VI. Evolution of Recombination
VII. Applied Aspects

GLOSSARY

crossing-over Complex interaction of homologs within bivalents at pachytene resulting in reciprocal exchange of genetic material between non-sister chromatids.
chiasmata χ-like configurations.
gene conversion A process of nonreciprocal (unidirectional) transfer of genetic information.
synaptonemal complex A three-part proteinaceous structure that affects the number and distribution of crossovers and converts crossovers into functional chiasmata.

THE GENERAL NOTION "RECOMBINATION" includes a range of genetic phenomena. The common dominator of various recombination processes is reassociation of pieces of genetic information resulting in the formation of new combinations that differ from the parental ones. Systematic recombination studies initiated by T. Morgan at the beginning of the twentieth century developed into one of the most fruitful branches of classical and modern genetics. The main types of genetic recombination include recombination of whole chromosomes (the basis of Mendel's law of independent assortment of genetic factors), reciprocal exchange between homologous chromosomes (crossing-over), nonreciprocal exchange (gene conversion), site-specific recombination occurring between homologous sequences, illegitimate recombination that involves nonhomologous sequences, transformation, and transduction. Genetics has accumulated abundant evidence on common features and peculiarities of these processes in various organisms, their genetic control, and molecular mechanisms. In fact, the achieved understanding is a product of combined efforts of genetics, cytology, molecular biology, biochemistry, population genetics, and evolutionary biology. It appears that recombination plays a key role in a remarkably broad spectrum of genetic processes related to DNA repair, sexual reproduction, immunity, genetic diversity, adaptation and speciation, evolution of genes, gene families, chromosomes, and the entire genome. Recombination is the basis for traditional breeding, genetic mapping, recombinant DNA technology and genetic engineering, gene targeting and gene therapy, developmental biology, and cancer genetics. However, despite a long history and thousands of studies, recombination continues to be a puzzle with respect to its mechanisms, diversity of genetic and evolutionary effects, and factors determining its evolution. The importance of recombination as a crucial factor determining the dynamic balance between the stability and flexibility of genetic organization as well as the interplay between heredity and environment justifies the keen interest in this process within the "biodiversity" paradigm. As such, in demonstrating recombination-related phenomena I use examples from a wide range of species in addition to traditional *Drosophila*, fungi, and humans.

I. TYPES OF RECOMBINATION

There are various ways to classify different types of recombination. No matter how logical a classification system may seem, the simplest way is to compare alternative aspects (modes) of the general phenomenon of recombination. These may be based on types of participating cells (meiotic or mitotic and male or female), cell compartments (nuclear or mitochondrial), chromosomes and DNA sequences (homologous or nonhomologous), and types of exchanges (reciprocal or nonreciprocal).

A. Crossing-Over and Chiasmata

Two types of cell division are known for eukaryotic organisms: mitotic and meiotic. Recombination takes place in both. In cells that enter meiosis, the chromosomes have already undergone DNA replication so that each chromosome consists of two sister chromatids connected by a shared centromere. During the meiotic prophase, pairing of homologs brings about bivalent formation; each bivalent combines four chromatids. Complex interaction of homologs within bivalents at pachytene results in reciprocal exchange of genetic material between non-sister chromatids, a process referred to as crossing-over (Fig. 1). Observation of corresponding meiotic stages (diplotene–diakinesis) reveals characteristic χ-like configurations called chiasmata. These configurations are interpreted as cytological manifestation of crossing-over, although their precise 1:1 correspondence is still debated (Nilsson *et al.*, 1993; Tease and Jones, 1995). Chiasmata were observed and interpreted as a result of an unknown process of rejoining genetic material at the end of nineteenth century, a decade before the discovery of partial linkage between

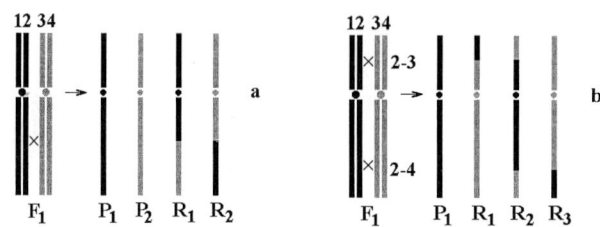

FIGURE 1 Crossing-over at the four-chromatid stage: (a) single exchange resulting in two parental (P_1 and P_2) and two recombinant (R_1 and R_2) chromosomes and (b) double exchange involving non-sister chromatids 2 and 3 at the crossover point in the short arm and 2–4 in the long arm—the result is one nonrecombinant (P_1), two single-recombinant (R_1 and R_2), and one double-recombinant (R_3) chromosomes. Note that if the same two nonsister chromatids participate in the exchange at both positions (e.g., 1–3 and 1–3) then two parental and two double-recombinant products will appear. Meiotic configurations involving all four chromatids in the double exchange (e.g., 1–4 at the first site and 2–3 at the second) result in four single-recombination products.

genes in plants and two decades before Morgan's studies on crossing-over on *Drosophila*.

B. Nonhomologous Chromosome Segregation

After crossing-over occurs, chromosomes of each bivalent segregate independently relative to chromosomes of other bivalents in two consequent meiotic divisions. These two processes, crossing-over and recombination of nonhomologs, are characteristic of all sexual eukaryotes. With independent segregation, the ratio of parental and nonparental combinations of corresponding alleles is 1:1, although a departure from independent assortment of unlinked genes, termed quasi-linkage, was observed in both plants and animals, especially in interspecific hybrids (Korol *et al.*, 1994).

C. Mitotic Recombination

Exchange of genetic information can also occur during mitotic cell division, but it does so with a frequency of a few orders of magnitude lower than that of meiotic ones. The major exclusion from this rule is mitotic recombination resulting in hyper-mutation of somatic cells of the vertebrate immune system. Mitotic recombination may occur in the germline cells, e.g., in *Drosophila* males in which meiotic recombination is normally suppressed. Some genetic factors increase the rate of mitotic recombination manyfold. For example, homozygosity for an inserted mobile element may result in

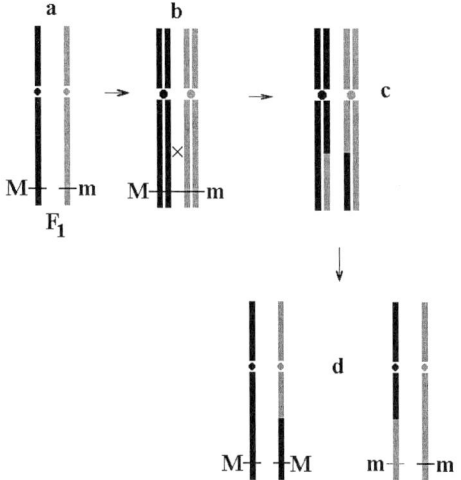

FIGURE 2 Segmental homozygosity resulting from mitotic crossing-over. (a) A diploid heterozygote for a marker locus (M/m); (b) after DNA reduplication each homolog is represented by two sister chromatids (the position of mitotic crossing-over between nonsister chromatids is marked by ×); (c) the structure of the homologs after crossing-over; and (d) the daughter cells resulting from mitosis are homozygous for the entire segment distal to the point of exchange.

male (premeiotic) recombination rates in Drosophila comparable to that in female meiosis (Sved et al., 1991). Mitotic crossing-over between the centromere and any locus results in two daughter cells homozygous for the entire segment distal to the site of exchange (Fig. 2). This may have a strong negative impact if the initial cell was heterozygous for some detrimental alleles (see Section VII).

D. Nonreciprocal Exchange

Normally, meiotic and mitotic crossovers are based on reciprocal physical exchange of genetic material between the parental chromosomes. Studies of recombination in fungi (Sordaria, Neurospora, and Ascobolus) have revealed a process of nonreciprocal (unidirectional) transfer of genetic information called gene conversion that accompanies reciprocal exchange (crossing-over). Gene conversion is based on DNA repair of mismatched strands along the "recombination tract." The demonstration of this process was due to the possibility of classifying marker segregation (spore color) of ordered spores in asci resulting from individual meiotic cells in hybrids between colored and wild-type strains. In addition to the expected regular 4:4 (colored:noncolored) octads, aberrant ratios 2:6, 6:2, 3:5, and 5:3 are also observed, reflecting nonreciprocal exchange (see Section III).

E. Ectopic Recombination

Crossing-over occurring between homologous sequences of homologous chromosomes generates new genetic variation without changing genome structure. However, crossing-over may also involve homologous sequences of nonhomologous parts of the genome. This process is called ectopic recombination. In particular, numerous families of repeated elements that are spread in eukaryotic genomes may participate in ectopic recombination. Especially active can be retrotransposons, either from nonhomologous chromosomes or from different sites of the same chromosome (giving rise to inter- and intrachomosomal ectopic recombination), resulting in chromosome rearrangements. In fact, the frequency of such events is very low and it is unclear which mechanisms reduce this danger. One possibility may be a strong bias of recombination interaction in favor of gene conversion if such ectopic contacts occur. Ectopic recombination may also occur between such partners as chromosome and a plasmid, such as a two-micron plasmid in yeast harboring the gene for site-specific recombinase Flp and the tumor-inducing T-DNA segment of the Ti plasmid of Agrobacterim tumefacience interacting with the host plant chromosome.

F. Illegitimate Recombination and Horizontal Transfer

Unlike homologous and site-specific ectopic recombination, illegitimate recombination is a process of joining DNA molecules that have only a small sequence homology. Illegitimate recombination is common in nature, especially in interactions between retroelements with host genome DNA and in some host–parasite systems. One of the most common examples is the previously mentioned interaction between bacteria, such as A. tumefasciens, and their host plants mediated by the bacterial plasmid transformation. Although this process can be considered as a "physiological" rather than a genetic transfer, it may also have evolutionary consequences as one of mediators of horizontal gene transfer. Illegitimate recombination is involved in many situations of foreign DNA integration in direct transformation experiments aimed at obtaining genetically engineered (transgenic) organisms.

G. Extranuclear Recombination, Exchanges between Cell Compartments

Recombination was proven to affect the organization of mitochondrial DNA. Convincing evidence of widespread recombination in mitochondrial genomes has been obtained for plants. In fact, a specific substrate for frequent mitochondrial recombination was discovered and sequenced in a few plant species. Especially interesting from the evolutionary viewpoint are putative exchanges between different cell compartments: nuclear–mitochondrial, nuclear–chloroplast, and mitochondrial–chloroplast. Many such sequences demonstrating "chloroplast → mitochondrial" transfer were obtained for cereals.

II. BASIC FEATURES OF RECIPROCAL HOMOLOGOUS RECOMBINATION (CROSSING-OVER)

A. Recombination Rate and Map Distance

The average number of crossovers in a given segment represents its genetic length. Correspondingly, genetic distance between two loci can be defined as the average number of recombination events occurring in the segment. Therefore, for loci flanking a unit length segment one exchange per meiosis is expected on average, resulting in 50% recombinant gametes. This notion of genetic distance serves as a basis for gene mapping. Two fundamental facts make possible genetic mapping: (i) linear organization of genetic material in chromosomes and (ii) increasing recombination rate between loci with their physical distance. Map distance is of additive nature: For any subdivision of a segment, the segment length is the sum of lengths of its subintervals. A limitation of map distance as a measure is that it cannot be estimated directly: An odd number of exchanges in a segment marked by two loci leads to the appearance, in the progeny, of the same recombinant genotypes as those resulting from one exchange, whereas an even number results in no observable recombinants. Recombination frequencies (RFs) estimated from the observed proportions of recombinant and nonrecombinant individuals coincide with map distances only for short segments when multiple exchanges can be ignored. Therefore, transformation of experimentally estimated RF values into map distances (x) is needed. Relations of the form RF = RF(x) are referred to as mapping functions.

B. Genetic Interference and Mapping Functions

Multiple exchanges are relatively rare among eukaryotes. These vary between 1 and 3 for the majority of higher organisms, with 5 or 6 occurring seldomly, although values up to 12 have been reported (Korol *et al.*, 1994). In *Drosophila* the frequency of tetrads with four or more crossovers is extremely low (usually much less than 1%).

1. Interference: Its Measurement and Basic Properties

Consider two adjacent marked segments m_1–m_2 and m_2–m_3 with recombination rates RF_1 and RF_2. If exchanges in adjacent segments occur independently, the probability of a double exchange within m_1–m_3 can be determined as $RF_{12} = RF_1 RF_2$. Under the same assumption, recombination frequency RF_3 between loci m_1 and m_3 is given by

$$RF_3 = RF_1(1 - RF_2) + RF_2(1 - RF_1)$$
$$= RF_1 + RF_2 - 2RF_{12}$$

Much evidence indicates that the observed frequency of double crossovers usually differs from the expected one, a phenomenon termed genetic interference. The degree of interference is measured by the coefficient of coincidence (c):

$$c = RF_{12}(\text{observed})/RF_{12}(\text{expected})$$
$$= RF_{12}(\text{observed})/(RF_1 RF_2)$$

The existence of interference was shown both genetically and cytologically. It is ubiquitous in eukaryotes. To date, only in three species (*Shizosaccharomyces pombe*, *Aspergillus nidulans*, and *Ascobolus immersus*) has recombination been found to proceed with no interference (Egel-Mitani *et al.*, 1982). If the occurrence of crossing-over in one segment reduces the probability of exchange in an adjacent segment ($c < 1$), then it is positive interference; when one exchange increases the frequency of another ($c > 1$), it is negative interference.

Two types of interference have been distinguished: chiasma interference, in which an exchange occurring in a segment affects the probability of another one occurring in an adjacent segment, and chromatid interference, which determines the pattern of chromatid configurations in successive pairs of chiasmata in a bivalent (i.e., the proportions of two-, three-, and four-strand double exchanges) (Fig. 1). In the absence of chromatid interference, the ratio should be 1:2:1. Evidence of chromatid interference is scarce. Chiasma interference determines the pattern of distribution between successive exchanges along a chromosome. The amount of interference within a segment depends on its distance from the centromere. Near the centromere, interference between exchanges in the adjacent segments of the same chromosome arm is maximal. It is generally believed that no interference exists across the centromere, although many exclusions have been registered.

2. Mapping Functions

Positive interference ($c < 1$) is a widespread phenomenon. It was found that in many cases the approximation $c(RF) = 2RF$ (Kosambi interference) fits the data well (i.e., $c \approx 0$ at small distances and $c \approx 1$ at large distances). The corresponding mapping function

$$RF = 0.5 \tanh(2x) \text{ or}$$
$$x = 0.25 \ln((1 + 2RF)/(1 - 2RF))$$

appeared to be a good approximation for Drosophila, rice, mouse, and other eukaryotes. The formula for RF values in adjacent intervals, corresponding to Kosambi's function, takes the following form:

$$RF_3 = (RF_1 + RF_2)/(1 + 4RF_1RF_2)$$

Situations with no interference can be approximated by the mapping function proposed by J. Haldane:

$$RF(x) = \tfrac{1}{2}[1 - \exp(-2x)]$$

Some results indicate that negative interference (i.e., an excess of double exchanges over the level expected with independence assumption), may also occur. Such data were observed for Drosophila and some plants, fungi, and viruses. One should take into account that values of $c > 1$ can also result from data pooling, owing to heterogeneity of individual RF values.

C. Distribution of Crossovers within Eukaryotic Chromosomes

The efficiency of crossing-over strongly depends on between- and within-chromosome patterns of exchange distribution in the nucleus. Here, two different (albeit related) problems are generally distinguished: (1) crossover distribution in relation to major components of the chromosome, i.e., the centromere, telomeres, heterochromatin (intensively stained cytological segments of chromosomes—gene-poor regions), and euchromatin (faintly stained gene-rich regions), and (ii) relative positions of crossovers with respect to one another. These questions were thoroughly studied based on scoring recombination between multiple markers complemented by cytological observations. A wide range of spatial variation in exchange frequency was revealed— from long stretches of heterochromatin in which crossing-over is almost blocked to short segments containing recombination hot spots (see Section III.E). Heterochromatic blocks affect the distribution of crossovers in the entire genome. Recombination nonuniformity of the chromosomes was first established in Drosophila and then confirmed in a wide range of organisms. Thus, in the B genome of wheat the physical length of DNA per exchange varies more than 150-fold along the chromosome (Lukaszewsky and Curtis, 1993). In the centromeric region of the human X chromosome the recombination rate is at least eight-fold lower than the average rate of female recombination on the X chromosome. The centromere also seems to inhibit gene conversion.

1. Chiasmata and Recombination

Of critical concern is the interpretation of chiasma frequency and distribution records as objective characteristics of the recombination process. The recent development of numerous molecular markers has resulted in a fast production of detailed genetic maps. This has allowed for new comparisons of cytogenetic and genetic maps. Skepticism was displayed by some authors regarding the compatibility of these two types of maps for certain organisms (e.g., cereals) (Nilsson et al., 1993) because there is a presumably higher proportion of multiple exchanges and reduced interference distance (e.g., in wheat) than previously thought (Gill et al., 1996). Consequently, cytological inspection of chiasmata may underestimate the recombination rate if close chiasmata are indistinguishable. Although these concerns require further analysis, convincing cytogenetic evidence indicates a one-to-one correspondence between recombination and chiasmata as its microscopic phenotype in organisms with favorable cytology (Tease and Jones, 1995).

D. The Effects of Sex

By the Haldane–Huxley rule, if crossing-over is strongly suppressed or lacking in meiosis of one of the sexes, then that sex is invariably the heterogametic one. The effect of sex on RF has been established in a variety of animal species, with this rule holding in all cases of alternative (all-or-none type) sex differences. Provided crossing-over occurs normally during the formation of both male and female gametes, then RF in the heterogametic sex may be either lower or higher than that in the homogametic one, with the sign and magnitude of sex difference in RF being segment specific. Sex differences in recombination have also been revealed by cytological analysis. Recent years have seen rapid accumulation of data on human recombination owing to the use of molecular markers. The total length of female maps is nearly twice as long as that from male maps; in some regions (mainly subtelomeric ones), the opposite tendency is observed. Most of the maize genomic regions examined exhibit higher male recombination or no difference; the extent of the difference (if it exists) may strongly depend on the genotype. A substantially higher crossing-over rate in male compared to female meiosis has been found in *Arabidopsis* (Korol *et al.*, 1994). This is also true for some animals exhibiting a higher degree of linkage and/or more localized pattern of exchange distribution in the homogametic sex than in the heterogametic one, such as marsupials (Trivers, 1988).

The mechanisms responsible for differences in RF between male and female meiosis are unknown. In most animals, sex is determined by a certain chromosome. Thus, it can be suggested that the difference between male and female meiosis is controlled by genes located on this chromosome. Alternatively, sex chromosome can canalize the development of general physiological (biochemical) sex differences (e.g., in the hormonal status) in a toggle-like manner, with specific recombination features of male and female meiosis being a secondary effect of the main regulator. The latter suggestion is supported by data from the fish *Oryzias latipes* in which the heterogametic sex is represented by XY males (Yamamoto, 1961). When males XY were transformed into functional XY females by a hormonal treatment, RF increased by a factor of 5 compared to the level in normal males. In humans, it was found that the distribution of sex difference in RF is related to genomic imprinting manifested in region-specific differential modification (methylation) of parental alleles.

E. The Effects of Environmental Conditions

The environment can affect recombination in two ways. First, it can affect recombination indirectly when changes in the frequency of alleles at recombination controlling loci depend, through a feedback mechanism, on changes in the frequency of the selected (fitness-related) loci (see Section VI). Second, it can affect recombination directly when in response to environmental factors the organism modifies (within genetically determined limits) the value of RF, i.e., when not RF but the norm of reaction with respect to RF is genetically determined. The fact of direct influence of the environment on recombination has long been known in genetics. As early as 1917, H. Plough observed in *Drosophila* a U-shaped dependence of RF on temperature: Recombination was minimum at optimum temperature and it increased with the departure of temperature from the optimum. Similar data indicating an increase in recombination level with deteriorating environment have been reported in other works (Korol *et al.*, 1994). Extreme conditions also tend to reduce interference. Moreover, variation of recombination parameters in response to stressful factors can be modulated by the organism's fitness: Higher fitness provides a better preservation of gene complexes formed in previous generations. Experimental evidence for the last principle has been obtained for *Drosophila* and tomato (Korol *et al.*, 1994). In *Drosophila*, the increase in male recombination that is induced by a heat treatment is negatively correlated with flies' resistance to high temperatures. Results reported by Laurencon and co-authors (1997) indicate that this may also apply to DNA repair and transposition of mobile elements. Therefore, recombination is not absolutely random relative to an individual's adaptation to environmental stresses affecting recombination rate and distribution. The foregoing feedback mechanism may assist the selection process in preserving adaptive gene combinations, thereby reducing the recombination load.

III. GENETIC CONTROL AND MECHANISMS OF RECOMBINATION

A. Selection for Changed Recombination, Genetic Modifiers of Recombination

Selection for altered recombination proved an important tool in analyzing genetic control of recombination. Such experiments have been conducted on dozens of organisms, including *Drosophila*, flour beetle *Tribolium castaneum*, grasshopper *Schistocerca gregaria*, silkworm *Bombix mori*, lima bean, and fungi *Neurospora* and *Schizophyllum* (Korol *et al.*, 1994). These experiments complemented the numerous findings on the existence of a considerable amount of genetic variation in RF in natural and laboratory populations (see Section IV.B) available for selection to act on. The effectiveness of directional selection for altered RF has been shown to depend on the segment under study; the size, structure, and origin of the start population; the breeding system; the estimation procedure; and the intensity and duration of selection. Genetic analysis of the accumulated differences between selected lines enabled the question of the genetic basis of variation in recombination to be addressed. In as few as 10 generations, divergent selection for RF in the p–Y region of silkworm chromosome 2 succeeded to produce lines with RF = 37–39 and 5–7%, starting from 25.6%. The obtained evidence suggests that recombination (frequency and distribution) in eukaryotic genomes is under complex control of polymorphic modifiers with small to moderate effect. The rate of recombination in a given region may depend on both linked modifiers and genes located on other chromosomes.

B. Meiotic Mutants as an Analytical Tool in Recombination Studies: Overlapping of DNA Repair, Recombination, and Segregation Systems

The first point mutation affecting recombination was *Drosophila* gene *c(3)G* (discovered in 1922), which almost completely blocks crossing-over in females. During subsequent years, many other genes with strong effect on different meiotic steps and recombination were isolated in eukaryotes (mainly in fungi, *Drosophila*, maize, and recently *Arabidopsis* and the nematode *Caernohabtidis elegans*). The availability of diverse mutants affecting the coordinated steps of meiosis and recombination allowed a detailed cytogenetic characterization of meiosis. This includes commitment to meiosis, regulation of major steps in chromosome conjugation and formation of the synaptonemal complex (see Section III.C), DNA breaks formation and strand exchange, chiasma maintenance, and chromosome disjunction. The broad network of recombination–meiosis mutants identified in yeast and other fungi and in *Drosophila* serves as a basis for cloning recombination genes and studying the molecular mechanisms of recombination. Numerous results were obtained on this rich material showing a deep overlapping in pathways related to DNA recombination and DNA repair. Thus, mutations for loci controlling meiosis and recombination also tend to exhibit disturbed repair functions. Such a duality seems to be an ancient phenomenon. In prokaryotes, the best example is the RecA protein of *Escherichia coli* that is involved in early stages of recombination and, simultaneously, in induced SOS repair. Recombination is also strongly affected by mutations in genes controlling other enzymatic steps in DNA metabolism, e.g., replication. Genes for DNA repair and recombination exhibit a high degree of sequence, structural, or functional similarity throughout life. This conservation was established using a combination of genetic and cytological methods with molecular cloning, sequencing, and genetic transformation. Thus, defects in the recombination–repair system of one species may be complemented by a gene transferred from another species, sometimes a distant one (e.g., from a prokaryote to eukaryote). However, this does not exclude high interspecific differences in many important features of recombination (Korol *et al.*, 1994).

Comparison of recombination in meiotic mutants and in normal genotypes provides an interesting conclusion. It appears that genomic distribution of exchanges in normal meiosis is more restricted, less random, than in meiosis altered by mutations, indicating that these restrictions are largely a result of evolutionary adjustment of recombination. Meiotic mutants tend to lessen the restrictions, and regions normally excluded from exchanges become involved in crossing-over. Simultaneously, a reduction of interference is also observed despite a general tightening of linkage. In meiotic mutants, the correspondence between physical and recombination length of chromosome segments becomes closer. In many cases, *mei* mutations referred to as *hyper-rec* can significantly enhance crossing-over (and not only in specific regions) or increase the rate of multiple exchanges, intragenic recombination, and/or mitotic crossing-over many-fold.

The meiotic system manifests a close connection between homologous conjugation, crossing-over, and regulation of chromosome segregation. This association is invariable for normal meiosis and served as a basis for the view that regulation of chromosome segregation is one of the main functions of crossing-over. Meiotic mutants were isolated where the two phenomena, segregation and recombination, are not coupled. For instance, crossing-over is normally almost completely suppressed in chromosome 4 of *D. melanogaster*. Some *mei* mutants with reduced recombination and/or altered exchange distribution within large chromosomes exhibit crossing-over within chromosome 4 and disturbed segregation of this chromosome (Sandler and Szauter, 1978).

C. Ultrastructure of Recombination: Synaptonemal Complex and Recombination Nodules

Electron microscopy reveals meiosis-specific organization of chromosome associations at the meiotic prophase I. A three-part proteinaceous structure, called synaptonemal complex (SC), is initiated at the zygotene stage, maintains the paired homologs together during pachytene, and disappears at the diplotene. Short stretches of DNA of the homologs are involved in the central part of SC where intimate molecular pairing takes place. SCs are considered as structures that affect the number and distribution of crossovers and convert crossovers into functional chiasmata. Small electron-dense bodies, recombination nodules (RNs), were detected within the central part of the SC. Early RNs that appear before pachytene are more or less evenly distributed along the bivalent and are supposed to participate in synapsis. Late, or pachytene, RNs seem to represent recombination enzymes associated with exchanging DNA molecules. The distribution of late RNs corresponds well to chiasma distribution and is believed to represent the sites for recombination. Numerous meiotic mutants were tested with respect to their ultrastructural and molecular phenotype—the morphology and biochemistry of SC. An interesting fact is that only in three organisms was meiotic recombination normally found without SC (*S. pombe, A. nidulans*, and *A. immersus*). However, in many cases SC was found in achiamatic meiosis (with no recombination). In the three fungi with recombination occurring without SC, the frequency of exchanges per bivalent is high, and especially interesting is the absence of interference (Egel-Mitani *et al.*, 1982). Thus, the absence of SC may allow for more reshuffling of the homologs. Analysis of SC appeared to be an important tool in determining the causes of altered fertility in farm mammals.

D. Molecular Mechanisms of Recombination

When considering molecular mechanisms of recombination in eukaryotes, a broader spectrum of basic processes involved may also be addressed, including chromosome conjugation, recombination, and segregation. This is partly due to the prevailing experimental methodology based on analyzing cytogenetic and molecular effects of numerous mutants defective for meiosis–recombination and DNA repair. Historically, the studies of molecular mechanisms of recombination in prokaryotes served as the main source of models to be tested in eukaryotes. The basic enzymatic activities involved in prokaryote recombination include those of endo- and exonucleases, strand transfer, DNA unwinding, ligation, and resolving Holliday's heteroduplex structure. Eukaryotic analogs were found and (are to be) characterized in the best studied eukaryotes, such as yeast, *Drosophila*, lily, and mammals. Clearly, eukaryotic recombination systems are supposed to be more complex due to additional problems related to meiosis, such as homologous pairing and segregation.

The classical concept of molecular recombination included the following steps: (i) intimate pairing of homologs; (ii) formation of single-stand DNA nicks at homologous sites and local denaturation of double-stranded DNA; (iii) formation of recombination intermediate (Holliday junction) by reciprocal strand transfer and renaturation, its migration from the initiation site, and heteroduplex formation (with local mismatched DNA due to heterozygosity of the parental chromosomes); and (iv) resolution of the Holliday structure, either after or before isomerization. In the first case, the content from both sides of the considered regions remains unchanged, whereas in the second case it results in reciprocal exchange. In both cases, the mismatched positions should undergo correction. Depending on the strand used as a template in the repair process, two alternatives will be observed: recovering the initial state and nonreciprocal information transfer (gene conversion).

A new understanding of eukaryotic recombination mechanisms resulted from the discovery of the leading role of double-strand DNA breaks (DSBs) in yeast *Saccharomyces cerevisiae*. It was demonstrated that meiotic recombination is closely associated with the repair of DSBs. Details on different pathways related to this process are being studied in yeast, mammalian cells, *Drosophila*, and plants based mainly on recombination–repair mutants. Another strategy is to study specially engineered genetic models with a modified recombination system (Shalev *et al.*, 1999). Homology of recombination genes in one species to genes controlling DSB-dependent recombination in yeast may be indicative for a shared DSB mechanism of recombination initiation, as exemplified by the *Drosophila mei- W68* gene and yeast *Spo11* (McKim and Hayashi-Hagihara, 1998).

E. Recombination Signals, Hot Spots of Recombination, and Transcription

Although recombination occurs all over the genome, the accumulated data points to the existence of hot spots and cold spots of recombination. In certain cases, highly specific interactions were found between recombination–repair mutations and "signal" DNA sequences resulting in a sharp increase in RF within defined genomic segments. The best known recombination signal is the octamere 5′GCTGGTGG-3′ of *E. coli* referred to as χ element. An example of a short signal acting as a meiotic recombination hot spot is the *M26* heptamer 5′ATGACGT-3′ of *S. pombe*. Likewise, in eukaryotes, mobile elements may act as recombination enhancers. Thus, with identical locations on homologous chromosomes, a pair of P elements of *D. melanogaster* can induce recombination at a frequency of 20% in males in which recombination is normally absent (Sved *et al.*, 1991). This effect serves as a basis for an efficient method of fine genetic mapping, in which P elements are used to promote recombination at their insertion sites. Hot spots may represent specific signals such as χ or *M26*, certain genes or target sequences within these genes, or more extended chromosomal segments such as gene-rich regions. In the maize *bronze* locus, recombination rate per physical length is 100-fold higher than the average level in the genome (Dooner and Martinez-Ferez, 1997). A few recombination hot spots have been characterized in mammals, with the major histocompatibility complex being one of the best known examples.

In the 1960s and 1970s, it was speculated that initiation of recombination is attached to transcription promoters (Whitehouse, 1966). In moving from lower to higher organisms, the recombination density over the genome is reduced by three or four orders of magnitude with no marked trend in structural genes. Thuriaux (1977) attributed this paradox to preferential localization of recombination in structural genes. It was suggested that the requirement for common specific changes in chromatin organization may cause an overlapping in control of transcription, recombination, repair and mutation (Whitehouse, 1966; Thuriaux, 1977; Korol *et al.*, 1994; Laurencon *et al.*, 1997). The evidence from many eukaryotes indicates that regions with high RF correspond to gene-rich regions (Lichten and Goldman, 1995; Gill *et al.*, 1996), although an opposite trend was found in *Caenorhabditis elegans*. The results obtained during the past few years mainly for yeast (*S. cerevisiae* and *S. pombe*) strongly support the idea of a possible recombination–transcription relationship (Nicolas, 1998). In general, the available information shows high intragenomic heterogeneity and nonrandomness of recombination (Korol *et al.*, 1994). These are caused by a hierarchical control system that involves interaction of a complex web of recombination–repair genes with peculiarities of macro- and micro-organization of the genetic material (Table I). The important role of gene distribution in this regulation indicates its adaptive role in the evolution of genetic systems.

IV. BIODIVERSITY OF RECOMBINATION SYSTEMS

A. Macroevolutionary Trends in Genome-Wise Recombination Rate

A considerable (by several orders of magnitude) decrease in RF per unit physical length of DNA is observed when moving from prokaryotes and lower eukaryotes to higher organisms (Thuriaux, 1977). Even among eukaryotes, the density of recombination events within the genome varies widely (Table II). The genetic length of the genome varies less and strongly depends on haploid chromosome number. Higher "regularity" of recombination due to frequent outcrossing, the obligatory occurrence of at least one exchange per bivalent, and increasing genome size are considered to determine the evolutionary importance of recombination in higher sexual eukaryotes. In lower organisms, recombination was supposed to play a less important role due to a slight chance of sexual contacts between individuals (Maynard Smith *et al.*, 1991). Consequently, natural populations of such species should manifest a clonal structure, with strong correlation [linkage disequilibrium (LD)] between alleles at different loci. However, this may not be the case in some protozoan species, such as malaria *Plasmodium falciparum* (Conway *et al.*, 1999). The revealed population pattern of *P. falciparum* includes high polymorphism and a rapid decay of LD with distance (reaching independence between polymorphic sites of 0.3–1.0 kb) that looks like that of higher sexually reproducing eukaryotes.

TABLE I
Hierarchical Control of Recombination Events[a]

System effects
- Control of the nucleus as a whole, manifested in the classical *interchromosomal effect* of rearrangements, as well as effects of eu- and aneuploidy and supernumerary (*B*) chromosomes
- Effect of chromosome size
- Regulation based on cytonuclear interactions
- Sex differences in recombination frequency and distribution
- Effect of environmental conditions (stress) and age
- Position of euchromatin (as main target for recombination) relative to the centromere, telomeres, and heterochromatic blocks

DNA sequence organization of the target region
- Heterozygosity for micro- and macroinversions, deletions, and translocations
- Microsite organization of DNA sequences including distribution of specific regulatory sequences such as recombination hot spots and micro- and minisatellites
- Distribution of epigenetically modified (e.g., methylated) DNA stretches
- Effects of mobile genetic elements
- Distribution of gene-rich islands

Genes controlling DNA methabolism
- Genome-, chromosome-, and segment-specific effects of major *rec* genes of the "*coarse control*" system affecting the basic steps in recombination mechanics (pairing, DNA strand exchanges, and recombination repair)
- Segment-specific regulation of crossover frequency by genes of the "*fine control*" system with relatively small effects of individual components
- Genes encoding active transposases

[a] Interaction of factors both within and between the groups plays an important role in observed patterns of recombination frequency and distribution within the genome and its variation within and between species.

TABLE II
Recombination Characteristics of Eukaryotic Genomes

Species	Genome length (cM)	Haploid number (n)	Chromosome length (cM)	kb:cM ratio
Magnaporthe grisea (Pyricularia oryzae)	600	7	85	80
Loblolly pine (Pinus taeda)	1700	12	140	
Arabidopsis	600	5	120	230
Brassica rapa	1900	10	190	280
Common bean (Fhaseolus vulgaris)	1200	11	110	530
Barley (Hordeum spontaneum)	1200	7	170	4000
Tomato (Lycopersicum esculentum)	1400	12	120	900
Fruitfly (Drosophila melanogaster)	300	4	75	570
Silkworm (Bombix mori)	900	28	30	590
Flour beetle (Tribolium castaneum)	600	10	60	350
Honeybee (Apis mellifera)	3400	26	130	50
Zebrafish (Danio rerio)	2900	25	120	590
Chicken	3000	28	110	
Mouse	1600	20	80	2000
Cat	3300	19	170	

B. Intraspecific Variation for the Rate of Recombination

The analysis of genetic variation in recombination parameters allows a deeper insight into patterns, mechanisms, and the evolutionary role of recombination. The genetic differences in crossing-over rate seem to have been originally established by Sturtevant in 1913, who discovered modifiers sharply reducing crossing-over during oogenesis in heterozygous females of Drosophila. Later, he suggested these to be inversions, a prediction that was subsequently confirmed by cytological analysis of salivary gland chromosomes. Among point mutations affecting recombination, the first to be discovered was $c(3)G$, which almost completely blocks crossing-over in Drosophila females. Additional studies demonstrated high genetic variation in RF in D. melanogaster and other Drosophila species (Korol et al., 1994; see Section III.B). Recombination studies in D. ananassae males revealed high variation in crossing-over rate in the second and third chromosomes; individual RF values in the ml-ru segment of chromosome 3 varied between 2.2 and 42.8% (i.e., 20-fold). Differences between maize plants in RF in the sh-su-wx segment increased as the degree of relatedness decreased and there was correspondence between micro- and macrosporogenesis. Polymorphism for B chromosomes and heterochromatic knobs and their interaction play a significant role in the variation of RF in maize. B chromosomes affect pairing and crossing-over in many plants and animals.

Abundant evidence for intraspecific genetic variation in crossing-over rate and/or chiasma frequency is available for many eukaryotes, both lower (Coprinus, Schizophyllum, Neurospora, Saccharomyces, and Sordaria) and higher (pearl millet, pea, jute, ryegrass, barley, wheat, tomato, snails, silkworm, grasshoppers, locusts, flour beetle, lizards, mouse, rat, cattle, and human) (Korol et al., 1994). This variation may be of primary importance in better understanding the genetic basis of biodiversity in natural systems, optimizing genetic conservation strategies, designing experiments on genetic mapping of complex traits, and map-based cloning.

C. Recombination and Domestication Evolution

Two questions are of major interest when discussing the interface domestication–recombination: (i) How have recombination features of the progenitors affected the pattern of selection response and selection advance? and (ii) How has selection for domestication affected the recombination system? Despite many theoretical studies on the interaction between selection and recombination, the very limited available evidence concerns only a few organisms out of a few hundreds domesticated plant and animal species. Burt and Bell (1987) studied the relationship between male chiasma frequency and features of reproductive strategy, longevity, body size, etc. of more than 30 mammalian species. They found that domesticated species exhibit higher rates of exchanges than might be expected based on a general relationship among these parameters. The authors attributed this difference to the advantage of increased recombination that facilitated to overcome unwanted correlations between characters in the course of selection. At the stage of primitive selection, recombination was a major factor of genetic variation allowing for a deep and very fast (on the evolutionary scale) reorganization of domesticated organisms. Consequently, this selection pressure could become a driving force of evolution of increased recombination, at least during the first phase of domestication (thousands of generations). However, at the second phase, after regular agriculture had become established, with more stable environmental conditions, stabilizing selection had to play an ever-increasing role. This might reverse the direction of recombination evolution, at least in some species. For example, Williams et al. (1995) established a higher chiasma frequency in primitive forms of maize compared to modern industrial lines.

Polyploid formation is considered a unique evolutionary pathway exploited by both animal and plant organisms. A few of the most important crops, such as wheat, cotton, or potato, are allopolyploids. The recombination–domestication interface may have very interesting aspects when related to polyploids. These include the evolution of diploid-like chromosome behavior in meiotic pairing, recombination and segregation, the effect of various forms of recombination on the coevolution and sequence polymorphism of genomes sharing one nucleus, and the effect of selection on the foregoing processes. Detailed mapping and DNA sequencing efforts recently initiated on the foregoing crops combined with comparative evidence on their diploid relatives and/or ancestors may be considered an important way of highlighting the role of recombination in evolution and domestication.

V. EVOLUTIONARY EFFECTS OF RECOMBINATION

A. Recombination and Genetic Variation for Adaptation to Varying Environment

Producing variation is considered one of the major evolutionary functions of recombination that is supposed to assist in adaptation to both spatial and temporal environmental changes. Abundant genetic variation manifested by natural population is a result of a complex interaction between a few factors: mutation, selection, population subdivision, breeding system, and recombination in its various forms. All other things being equal, genetic variance for a sexual population with free recombination may be several times that for a population with no recombination (Charlesworth, 1993). Classical estimates from *Drosophila* species indicated that up to 25–40% of variation in fitness in natural populations is regenerated by crossing-over in one or two generations from a randomly drawn chromosome pair. Thus, a temporary arrest of the mutation process would not result in a significant decrease in variation over many generations. Such a view corroborates recent studies showing the presence of polymorphism in small populations in nature and recovery of variation after severe bottlenecks (Nevo et al., 1997).

An important question is how recombination affects genetic variation. In a population at linkage equilibrium (LD = 0), changes in recombination rate do not affect the amount of variation. However, this does not mean that the level of variation is independent of RF. Indeed, the stability of polymorphism varies with RF: The general trend is a reduction of polymorphism stability with increased recombination. This will occur when one models the dynamics of a diploid population subjected to stabilizing or directional selection with all selected loci linked within a block. Stability of polymorphism for these loci inversely depends on recombination within the block. However, the effect of linkage may be opposite for configurations with multiple blocks of selected loci (Nevo et al., 1997). Another example is selection (either diploid or haploid) in a fluctuating environment (Korol et al., 1996). The effect of recombination on the amount of variation may also result from a dependence of the steady-state LD on RF: LD $\neq 0$ is achievable (if at all) only at tight linkage and/or strong selection.

B. Recombination as a Source of Evolutionary Novelties

Through gene duplications (by unequal exchanges), recombination creates prerequisites for increasing variability. In fact, it is a general belief that the evolutionary formation of a new gene by mutation–selection interaction is preceded by the duplication of the genetic material. Local duplications may result from unequal crossing-over, unequal sister chromatid exchange, polymerase slippage, or replicative transposition. There is abundant evidence (from both pro- and eukaryotes) suggesting that structurally (and in many cases functionally) related genes are linked to one another. These may include tandemly repeated gene families, gene clusters (with arbitrary spacing and orientation of the individual genes), and dispersed distribution of the family members across the genome. Different mechanisms were proposed for allele diversification after duplication. For example, a new allele may be produced by intrachromosomal homologous recombination mediated by DNA sequences within the duplicated elements. The creative role of different forms of recombination (unequal crossing-over, gene conversion, reciprocal exchange, and ectopic and illegitimate recombination) is well documented for different organisms.

Recombination was recognized to play an important role in the evolution of gene families that are usually organized in the genome as arrays of tandem repeats residing in one or a few chromosomes. Coordinated change of individual elements of such families is referred to as concerted evolution. The diversification of the elements by mutations is opposed to homogenization processes caused by unequal crossing-over and gene conversion. This process is opposite the scenario of evolution of new functions through "recombination (unequal crossing over) → duplication → diversification (via mutation)." In the case of multigene families, selection presumably favors the most homogeneous arrays. Different variants of homogenized arrays may coexist within a population, as shown on the rDNA cluster in *Drosophila*. This implies a low recombination between different arrays combined with high intrachromosomal recombination within arrays.

C. Recombination and Sequence Polymorphism

During the past decade, new evidence became available on the association between recombination and DNA sequence polymorphism in natural populations. The first group of data came from *Drosophila*: Begun and Aquadro (1992) compared variation in RF per physical DNA length in different regions of *D. melanogaster* genome with polymorphism of DNA sequences of these regions. Positive correlation between recombination and sequence polymorphism was established. Similar results were obtained in different *Drosophila* species, humans, and some plants. Recombination seems to play an important role in sequence variation of natural populations for the mitochondrial genome.

Two opposite though not mutually exclusive hypotheses were proposed as explanations for the association between recombination and sequence polymorphism. The first, referred to as selective sweeping, is based on selection of new favorable mutations. In genomic regions with a low recombination rate, the process of fixation of positively selected mutations will result in fixation of closely linked neutral or even slightly deleterious sequence variations. The second explanation is based on the theory of background selection (Nordborg *et al.*, 1996) that considers the consequences of purifying selection against deleterious mutations. This process can also reduce sequence variation at closely linked sequences. Combination of these two mechanisms seems to explain the main pattern of polymorphism–recombination association, although the remaining difficulties may point to the importance of additional factors.

D. Horizontal Gene Transfer

Numerous mechanisms exist that normally prevent the exchange of genetic material among distant species. Nevertheless, if this process is possible even at a low frequency, it may have important consequences on the long-term evolutionary scale. Only 20 years ago such a possibility was considered an unrealistic one. Now, rich evidence is available showing that the opposite is likely to be true. Illegitimate recombination seems to be the main responsible mechanism. Extensive similarities have been revealed between house-keeping genes of *Archaea* and *Eubacteria*, although for the majority of informational genes (involved in transcription and translation) the difference is very large. An interesting fact is that some of the foregoing genes showing high similarity between the kingdoms are clustered in the genome that may result from horizontal transfer events.

If horizontal transfer is indeed an evolutionary significant form of genetic recombination, one could consider ecological interaction between species a replacement of sexual interaction (vertical transfer). In other words, ecologically interacting species are the most probable exchanging partners. Viral DNA integration in the host nuclear genome is considered a relatively frequent event (on the evolutionary scale) for animal hosts and recently also for plants. Likewise, there is evidence of an opposite, host → parasite flow of genetic information: Many transfers of signaling domains of pathogenesis-related plant genes were detected in bacterial genomes.

VI. EVOLUTION OF RECOMBINATION

A. Theoretical Models of Recombination Evolution

Artificial selection experiments suggest that almost every population has enough stored genetic variability to ensure response to direct selection for changed recombination frequency (see Sections III.A and IV.B). The observed polymorphism at loci affecting recombination could be either balanced (selected) or transient. Theoretical analysis shows that under a stable environment a panmictic population should evolve toward a minimum possible level of recombination. This can be formulated in terms of the fate of a selectively neutral modifier locus affecting recombination.

Understanding the forces maintaining sex and recombination is considered one of the most challenging problems in evolutionary theory (Michod and Levin, 1988). Shared genetic control and molecular mechanisms of DNA recombination and DNA repair across life (see Sections III.B and III.D) indicate their common evolutionary origin and functional overlap in extant organisms. It could probably be supposed that repair functions played the leading role at early evolutionary stages, having provided opportunities for a large increase in the genome size and the transition from haploidy to diploidy. The latter offered the possibility of recombination repair of two-strand DNA lesions that is impossible in haploid systems. Some authors hypothesize that the subsequent stages in the evolution of recombination and sex were associated with repair alone. These explanations of repair functions were called "physiological" (Maynard Smith, 1978). Another physiological explanation of recombination in sexuals is its association with chromosome segregation (see Section III.B), although numerous examples are known, such as male *Drosophila* and female silkworm, in which normal segregation is associated with achiasmatic (without crossing-over) meiosis.

Combinative, or generative, hypotheses consider the main function of sex and recombination in shuffling genes. This removes negative correlation between favorable alleles at different loci, thereby increasing the efficiency of natural selection. Generative models can be classified according to the source of linkage disequilibria between selected loci: stochastic or deterministic, caused by new mutations or variation of external conditions (Kondrashov, 1993). In the 1930s, Fisher and Muller proposed that sex may be advantageous by combining beneficial mutations randomly occurring in different individuals (Otto and Barton, 1997). A complementary version of the stochastic–mutation explanation considers the role of recombination in selection against deleterious mutations (Muller, 1964). According to Muller, deleterious mutations tend to be fixed in a finite asexual population due to random drift despite purifying selection (Muller's ratchet), whereas recombination helps to stop this process. In the 1980s, a few deterministic models of selection against deleterious alleles were proposed, with the evolutionary advantage of recombination being dependent on linkage disequilibria resulted from synergistic interaction between harmful mutations produced at a high rate (Kondrashov, 1988). Other models of deterministic sources of linkage disequilibria include (i) adaptation to temporarily or spatially varying environment, (ii) antagonistic species interaction (the Red Queen hypothesis), and (iii) instraspecific competition (sib competition or tangled bank hypothesis) (Hamilton *et al.*, 1990; Kondrashov, 1993; Feldman *et al.*, 1997; Barton and Charlesworth, 1998; Otto and Michalakis, 1998). In most of these models, the conditions favoring increased recombination are associated with negative linkage disequilibria among selected loci owing to stringent conditions for epistasis. Some mechanisms (e.g., selection of beneficial mutations) may favor recombination in the absence of epistasis, with linkage disequilibirum caused by finite population size (Otto and Barton, 1997).

B. Testing the Theoretical Assumptions

Despite voluminous literature, the problem has been studied little experimentally or by observations in natural populations. In the classical period of genetics it was proposed that sex and recombination may facilitate selection for advantageous and against deleterious mutations and assist in adaptation to changing environment. The major question is whether these processes in turn are able to promote evolution of the recombination system. Are the resulting changed patterns of recombination detectable experimentally? Is it possible to find corresponding changes in nature? Theoretical models answer positively to these questions. However, do these models (and/or their parameter values) fit the real world? Unfortunately, there is not enough evidence to answer this question. Two different and complementary approaches can be considered (Barton and Charlesworth, 1998; Korol, 1999).

1. Testing the Initial Assumptions

The aim is to test whether the real estimates of the main parameters underlying the proposed mechanism(s) fit the expectations. These include mutation rate, effective population size, mode of interaction of deleterious or beneficial mutations, relative role of stabilizing and directional selection, and cost of sex and meiosis. The major advantage of this approach is that it allows the accumulation of data on basic parameters affecting the evolution of recombination. The drawback is that it is difficult to believe that even the main factors were taken into account so that the predicted direction of recombination evolution is determined correctly. As a recent example of a successful application of this approach, competition experiments with yeast *S. cerevisiae* (Greig *et al.*, 1998) are worth mentioning. Using sexual and asexual strains it was found, contrary to the expectations following from the concept of costs of sex or meiosis, that sex can confer significant selective advantage to its carriers. Direct tests have also been conducted with the nematode *C. elegans* demonstrating the possible role of recombination in purifying selection (Zetka *et al.*, 1987). It was shown that increased RF owing to enhancer *Rec-1* results in a higher fitness of genotypes carrying mutator gene *mut-6* compared to analogous genotypes with a normal recombination rate. Therefore, the presence of a gene for increased recombination, which has no effect on mutation, reduces the genetic load caused by the mutator system. However, recent studies of normal (wild-type) genotypes of *C. elegans* give very low estimates of mutation rate, which is inconsistent with the hypothesis of selection against deleterious mutations.

2. Searching for Changed Recombination

The second approach is to test the "final effect" by comparing rates and genomic distribution of recombination in populations subjected to contrasted selection regimes during many generations. The advantage of such an analysis is that the target effect (changed pattern of recombination) is estimated directly (see Sections VI.D and VI.E), although one cannot ensure that it is causally connected only with the presumed process.

C. Recombination System and Life History Traits

Rasmusson (1927) was possibly the first to discuss the genetic basis of putative differences in recombination systems of outbreeding and inbreeding species. On the basis of extremely high variation in the recombination rate in peas that is usually not found in *Drosophila*, he advanced a hypothesis of possible differences in the system of genetic control of recombination between outbreeders and selfers. The species origin and life conditions determine the strategy for adaptation and hence the "optimum" amount of recombination. Evidence of variation of chiasma frequency in nature and its correlation with ecological and life historical traits is available for different organisms and may be employed to compare alternative theoretical models (Burt and Bell, 1987; Koella, 1993). In mammals with low reproduction rates, longer life cycles, and small progeny sizes, recombination is higher than in species with short generation times and large progeny sizes (Burt and Bell, 1987), fitting the Red Queen model. Koella (1993) found that in plants with animal-dispersed seeds recombination is higher than in other plants, and perennials show a higher recombination than annuals when compared among genera but an opposite tendency is displayed on the species level.

D. Recombination System and Species Ecology

Attempts to correlate life history traits and the level of recombination to justify theoretical concepts have occasionally encountered serious criticism. The fact that interspecific variation usually includes a broad complex of traits does not allow clear conclusions to be reached when testing theoretical predictions. One may take advantage of intraspecific comparisons, especially when testing hypotheses of recombination maintenance due to its role in adaptation to adverse environments. In particular, it is of great interest to compare, within the same species, the rate of recombination in populations inhabiting stressful versus mild conditions. Recombination rate was evaluated in *D. melanogaster* males from the opposing slopes of "Evolution Canyon" on Mount Carmel (Israel) manifesting strong microclimatic differentiation, mainly in temperature and humidity (Korol, 1999). A significant difference in RF was found in chromosome 3 in flies derived from the opposing slopes. In particular, higher (four-fold) RF was manifested by the sub-population of the more stressful south-facing slope compared to the north-facing slope. Parallel patterns have been found in Evolution Canyon in the soil fungus *Sordaria fimicola* in the rates of mutation, crossing-over, and gene conversion, all being higher on the south-facing slope (Lamb et al., 1998).

E. Experimental Evolution of Recombination

1. Changes in Recombination Caused by Selection for Adaptive Traits

Theoretical models indicate that directional or variable selection for multilocus traits may promote evolution toward increased recombination if the selected population is polymorphic for recombination-modifying loci (Charlesworth, 1993; Korol, et al., 1994; Otto and Barton, 1997). Only a few experiments seem to have a direct bearing on this question. The following are examples with regard to Drosophila (Korol 1999). First, Clegg and coworkers (1979) attempted to test Fisher's (1930) prediction that stable conditions may favor reduced recombination. The experiment started with a large cage population of D. melanogaster with maximum linkage disequilibrium between allozyme loci of chromosome 3. No general trend in RF for 20 generations was revealed, which was explained as an indication of a low epistatic component of selection within the marked interval. Second, Flexon and Rodell (1982) established a synthetic population of D. melanogaster and subjected it to selection for DDT resistance for 300 days. By the time of estimation, resistance increased 10- to 100-fold relative to the control level. The control and treatment populations were compared with respect to recombination rate across the genome. Increased RF between marker loci on chromosomes 2 and 3 but not X was found in the selected population. Third, most experimental studies on recombination evolution have employed one-way selection for fitness traits. The limitation of this scheme is the danger of a change in RF caused by initial linkage disequilibria between selected genes and recombination modifiers. Therefore, Korol and Iliadi (1994), in an attempt to assess the effect of selection for geotaxis on recombination in D. melanogaster, employed two-way selection. Forty generations of selection for altered geotaxis resulted in an increment in recombination for the genome of 78 cM for geo$^+$ and 66 cM for geo$^-$ compared to non-selected control. Selection for negative geotaxis did not affect RF in chromosome 2, whereas selection in the opposite direction caused a 4 fold increase in RF in the b-cn segment spanning the centromere of chromosome 2. The observed change in RF, caused by two-way selection for geotaxis, was attributed to the advantage conferred by selection on recombinants.

2. Experimental Evolution of Recombination Caused by Adaptation to Stress

I review a series of studies with D. melanogaster conducted in my laboratory and aimed to demonstrate that population adaptation to adverse conditions may promote evolution for increased recombination (Korol et al., 1994, Korol, 1999). The experiments were performed on four large heterogeneous populations with the objective of simulating the process of population adaptation to stress and evaluating its effect on recombination. At the start, each population was subdivided randomly into two parts, control (C) and treatment (T); T variants were then subjected to daily fluctuating temperature with amplitude increasing with generations (from 23–27°C at the beginning to 12–32°C at the end of the experiment). C variants were maintained at 25°C.

a. Increased Recombination Resulting from Adaptation to Stressful Conditions

The control and treatment populations appeared to diverge genetically with respect to thermotolerance. Changes in recombination were evaluated using multiple marked lines. All four populations showed one pattern—a segment-specific increase in RF, resulting from selection for thermoadaptation. Those that reacted most were the near-centromeric regions of chromosomes 2 and 3 and the left arm of chromosome 3 (Fig. 3). The T and C variants were also compared with respect to the frequency of double exchanges (interference). Significant differences in the values of coefficients of coincidence have been found for adjacent and non-adjacent intervals of large autosomes. The general tendency was a reduced interference in T variants, corroborating the results for recombination rate.

b. Reduction of Recombination Rate under Optimal Stable Conditions

The described divergence over time with respect to recombination rate resulted from two independent trends. An increase in RF was observed in populations adapting to adverse conditions, and simultaneously tightening of linkage in control populations that were maintained under optimal temperature. The proportion of these two effects varied in different "treatment–control" pairs corresponding to the structure of the starting population. These results can be considered as the first experimental verification of Fisher's (1930) hypothesis that in a constant environment selection favors tighter linkage. Indirectly, these results also indi-

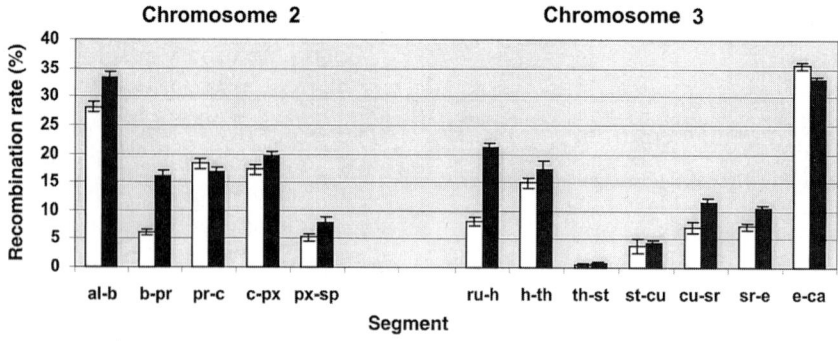

FIGURE 3 Adaptation to daily fluctuating temperature results in a segment-specific increase in recombination rate in large autosomes of *D. melanogaster* (based on data from Korol *et al.*, 1994).

cate that selection against harmful mutations alone seems to be insufficient for maintenance of the normal level of recombination, at least in the genetic material that was tested.

VII. APPLIED ASPECTS

A. Recombination as a Tool of Genetic Mapping

Recombination-based mapping as a method of studying the genetic material topography has been a major analytical approach in genetics for nearly 80 years. However, until recently this technique was of little applied interest because of a lack of easily scorable markers. The situation changed dramatically with the appearance of DNA markers. Establishing high-quality maps is considered one of the most important objectives in the Human Genome Project, which also includes a few model organisms. Mapping efforts allow a better understanding of the organization of genetic material and also allow one to conduct far-reaching comparative mapping. This approach is a power tool of ever-increasing importance for studies of genome structure, functioning, and evolution, and for diverse medical and commercial applications of modern genomics. Fine genetic mapping is a basis of positional cloning, an approach proved successful in many organisms. Its efficiency strongly depends on the kb:cM ratio in the region harboring the target gene. If the target region is very rarely involved in recombination, then map-based cloning is impractical. Induced recombination may be a nontraditional approach for such situations. The availability of dense maps facilitates mapping of genetic disease genes in humans with subsequent cloning, sequencing, and medical applications. An important application of mapping is genetic dissection of complex traits, or quantitative trait loci mapping, especially in the framework of new marker-assisted strategies in plant and animal breeding.

B. Recombination as a Major Source of Genetic Variation in Breeding

When viewed at the level of phenotypic traits, genetic innovations produced by recombination upon artificial hybridization can arbitrarily be divided into three main types: (i) transgression for individual traits with the range of trait values in the segregating progeny exceeding the respective parental ranges; (ii) formation of new trait combinations of the crossed components; and (iii) the appearance of new traits ("anomalous" variation) as a result of recombination in genetic complexes with strong nonallelic interactions. These forms of recombination variability play a significant role in breeding programs, providing the raw material for selection. The importance of controlling recombination increases with the necessity of involving new genetic resources for breeding purposes, especially in light of ever-decreasing homeostasis and tolerance to abiotic and biotic stresses of elite animal breeds and plant cultivars. New genes are supposed to be introgressed from exotic germplasm based on recombination. This problem is complicated by reduced recombination on distant crosses and linkage drag (close linkage between the target gene and undesirable genes). Hence, manipulating recombination may become an efficient tool of modern breeding.

C. Recombination and Cancer

There are many aspects of anomalous chromosome and cell behavior and cancer related to DNA metabolism and recombination–repair. A few are discussed here.

1. Mitotic Recombination

A surprisingly large proportion of human neoplasms appear to result from loss of function of tumor-suppressing genes, for example, due to loss of heterozygosity of tumor suppressors. In retinoblastomas caused by loss of heterozygosity, about half of the homozygosity cases appear to be the result of mitotic recombination.

2. Specificity of Chromosome Translocations

Many types of human cancers are associated with nonhomologous recombination events producing specific translocations. This results in an enhanced expression of oncogenes residing in the vicinity of the breakpoint. Some of these oncogenes are transcription factors participating in normal development. It seems reasonable to assume that interaction between recombination and transcription (see Section III.E) may also involve ectopic recombination resulting in the specificity of some tumorigenic translocations (Barr, 1998).

3. Defects in DNA Repair–Recombination

A few cancer-related genetic disorders of this type have been characterized, including Bloom syndrome, which manifests mutator and hyperrecombination phenotypes; Ataxia telangiectasia mutation, which increase V(D)J-mediated chromosome rearrangement in T lymphocytes; and Omenn syndrome, a disorder caused by mutations of recombination–activation genes *Rag1* and *Rag2* [related to V(D)J recombination] resulting in severe immune deficiency combined with a complete lack of circulating T and B cells. One of the most important tumor-suppressing genes controlling cell apoptosis, *p53*, is involved in repair–recombination processes.

D. Recombination Shuffling of Genes: A New Technology of Artificial Sequence Optimization

Recombination combined with selection has recently become a basis for a new technology of "molecular breeding" aimed at improving protein functions via directed evolution *in vitro*. It includes reshuffling of a few copies of the coding DNA by homologous or site-specific recombination with subsequent screening of the resulting sequences for functional (enzymatic) activity. The reshuffling is provided by template switching in self-priming polymerase chain reaction (sexual PCR). The efficiency of such artificial evolution may be enhanced by increasing the initial genetic variation involved in recombination, e.g., by mutagenesis of the target sequences based on error-prone PCR. Another possibility is to include a wider spectrum of parental sequences using the natural diversity of homologous genes of the same or related species or even distant taxa (Crameri *et al.*, 1998). This approach is especially promising for the improvement of molecules that provide resistance against toxic agents or increase survival on minimal media. Experiments including transformation of the engineered sequences into single-cell organisms (*E. coli*) and selection of the transformants showed dozens-to-hundreds- and even thousands-fold increases in efficiency of the target molecules obtained after recombination–selection cycles. DNA shuffling can improve the targeted pathway by complex mechanisms, even though it may be impossible to predict the optimized sequence by a rational design.

E. Targeted Gene Replacement

Gene targeting is an experimental approach aimed at producing site-specific changes in the genome based on genetic recombination. Predetermined changes can be introduced in the target site, in both somatic and germ-line cells. Usually, the targeting system involves a gene mediating site-specific recombination and a specific target locus. The best known examples are *Cre* and *Flp* recombinases (of bacteriophage P_1 and yeast) with *loxP* and *FRT* targets, respectively. The basic idea is to first transform the target sites flanking a selectable marker into the recipient genome. Subsequent introduction of the recombinase activity leads to the excision of the marker. Many important applications have been proposed and many others are expected using the powerful strategies of targeted gene replacement. These include controlled expression of transformed genes, disruption of a chosen gene *in vivo* for a better understanding of its role in physiological and pathological processes, disruption of genes to rechannel the biosynthesis and achieve overproduction of certain molecules, site-specific targeting of DNA into embryonic stem cells for production of gene-modified model organisms, targeted delivery of specific constructs aimed at producing anti-tumor effects or complementing a mutated gene, and testing of gene expression in donor versus recipient tissues in biomedical transplantation studies.

Acknowledgments

I thank E. Nevo, A. Kondrashov, A. Levy, and three anonymous referees for useful comments and suggestions.

See Also the Following Articles

ADAPTATION • COEVOLUTION • DIVERSITY, MOLECULAR LEVEL • GENETIC DIVERSITY • LIFE HISTORY, EVOLUTION OF

Bibliography

Barr, F. G. (1998). Translocations, cancer and the puzzle of specificity. *Nature Genet.* **19**, 121–124.

Barton, N. H., and Charlesworth, B. (1998). Why sex and recombination? *Science* **281**, 1986–1990.

Begun, D. J., and Aquadro, C. F. (1992). Levels of naturally occurring DNA polymorphism correlate with recombination rates in *D. melanogaster*. *Nature* **356**, 519–520.

Burt, A., and Bell, B. (1987). Mammalian chiasma frequencies as a test of two theories of recombination. *Nature* **326**, 803–805.

Charlesworth, B. (1993). Directional selection and evolution of sex and recombination. *Genet. Res.* **61**, 205–224.

Clegg, M. T., Horch, C. R., and Kidwell, J. F. (1979). Dynamics of correlated genetic systems. VI. Variation in recombination rates in experimental population of *Drosophila melanogaster*. *J. Heredity* **70**, 297–300.

Conway, D. J., Roper, C., Oduola, A. M., Arnot, D. E., Kremsner, P. G., Grobusch, M. P., Curtis, C. F., and Greenwood, B. M. (1999). High recombination rate in natural populations of *Plasmodium falciparum*. *Proc. Natl. Acad. Sci. USA* **96**, 4506–4511.

Crameri, A., Raillard, S. A., Bermudez, E., and Stemmer, W. P. (1998). DNA shuffling of a family of genes from diverse species accelerates directed evolution. *Nature* **391**, 288–291.

Dooner, H. K., and Martinez-Ferez, I. M. (1997). Recombination occurs uniformly within the *bronze* gene, a meiotic recombination hotspot in the maize genome. *Plant Cell* **9**, 1633–1646.

Egel-Mitani, M., Olson, L. W., and Egel, R. (1982). Meiosis in *Aspergillus nidulans*: Another example for lacking synaptonemal complexes in the absence of crossover interference. *Hereditas* **97**, 179–187.

Feldman, M. W., Otto, S. P., and Christiansen, F. B. (1997). Population genetic perspectives on the evolution of recombination. *Annu. Rev. Genet.* **30**, 261–295.

Fisher, R. A. (1930). *The Genetical Theory of Natural Selection.* Clarendon, Oxford.

Flexon, P. B., and Rodell, C. F. (1982). Genetic recombination and directional selection for DDT resistance in *Drosophila melanogaster*. *Nature* **298**, 672–675.

Gill, K. S., Gill, B. S., Endo, T. R., and Boyko, E. V. (1996). Identification and high-density mapping of gene-rich regions in chromosome group 5 of wheat. *Genetics* **143**, 1001–1012.

Greig, D., Borts, R. H., and Louis, E. J. (1998). The effect of sex on adaptation to high temperature in heterozygous and homozygous yeast. *Proc. R. Soc. London B* **265**, 1017–1023.

Hamilton, W. D., Axelrod, R. A., and Tanese, R. (1990). Sexual reproduction as an adaptation to resist parasites (a review). *Proc. Natl. Acad. Sci. USA* **87**, 3566–3573.

Koella, J. C. (1993). Ecological correlates of chiasma frequency and recombination index of plants. *Biol. J. Linnean Soc.* **48**, 227–238.

Kondrashov, A. S. (1988). Deleterious mutations and the evolution of sexual reproduction. *Nature* **336**, 435–440.

Kondrashov, A. S. (1993). Classification of hypotheses on the advantage of amphimixis. *J. Heredity* **84**, 372–387.

Korol, A. B. (1999). Selection for adaptive traits as a factor of recombination evolution: Evidence from natural and experimental populations. In *Evolutionary Theory and Processes: Modern Perspectives* (S. P. Vasser, Ed.), pp. 31–53. Kluwer, Dordrecht.

Korol, A. B., and Iliadi, K. G. (1994). Recombination increase resulting from directional selection for geotaxis in *Drosophila*. *Heredity* **72**, 64–68.

Korol, A. B., Preygel, I. A., and Preygel, S. I. (1994). *Recombination Variability and Evolution.* Chapman & Hall, London.

Korol, A. B., Kirzhner, V. M., Ronin, Y. I., and Nevo, E. (1996). Cyclical environmental changes as a factor maintaining genetic polymorphism. II. Two-locus diploid selection. *Evolution* **50**, 1432–1441.

Lamb, B. C., Saleem, M., Scott, N., Thapa, N., and Nevo, E. (1998). Inherited and environmentally-induced differences in mutation frequencies between wild strains of *Sordaria fimicola* from "Evolution Canyon." *Genetics* **149**, 87–99.

Laurencon, A., Gay, F., Ducau, J., and Bregliano, J. C. (1997). Evidence for an inducible repair-recombination system in the female germ line of *Drosophila melanogaster*. III. Correlation between reactivity levels, crossover frequency and repair efficiency. *Genetics* **146**, 1333–1344.

Lichten, M., and Goldman, A. S. H. (1995). Meiotic recombination hotspots. *Annu. Rev. Genet.* **29**, 423–444.

Lukaszewski, A. J., and Curtis, C. A. (1993). Physical distribution of recombination in B-genome chromosomes of tetraploid wheat. *Theor. Appl. Genet.* **86**, 121–127.

Maynard Smith, J. (1978). *The Evolution of Sex.* Cambridge Univ. Press, Cambridge, UK.

Maynard Smith, J., Dowson, C. G., and Spratt, B. G. (1991). Localized sex in bacteria. *Nature* **349**, 29–31.

McKim, K. S., and Hayashi-Hagihara, A. (1998). mei-W68 in *Drosophila melanogaster* encodes a Spo11 homolog: Evidence that the mechanism for initiating meiotic recombination is conserved. *Genes Dev.* **12**, 2932–2942.

Michod, R. E., and Levin, B. R. (Eds.) (1988). *The Evolution of Sex: An Examination of Current Ideas.* Sinauer, Sunderland, MA.

Muller, H. J. (1964). The relation of recombination to mutational advance. *Mutat. Res.* **1**, 2–9.

Nevo, E., Kirzhner, V., Beiles, A., and Korol, A. (1997). Selection versus random drift: Long-term polymorphism persistence in small populations (evidence and modelling). *Philos. Trans. R. Soc. B* **352**, 381–389.

Nicolas, A. (1998). Relationship between transcription and initiation of meiotic recombination: Toward chromatin accessibility. *Proc. Natl. Acad. Sci. USA* **95**, 87–89.

Nilsson, N. O., Säll, T., and Bengtsson, B. O. (1993). Chiasma and recombination data in plants: Are they compatible? *Trends Genet.* **9**, 344–348.

Nordborg, M., Charlesworth, B., and Charlesworth, D. (1996). The effect of recombination on background selection. *Genet. Res.* **67**, 159–174.

Otto, S. P., and Barton, N. H. (1997). The evolution of recombination: Removing the limits to natural selection. *Genetics* **147**, 879–906.

Otto, S. P., and Michalakis, Y. (1998). The evolution of recombination in changing environments. *Trends Ecol. Evol.* **13**, 145–151.

Rasmusson, J. (1927). Genetically changed linkage values in *Pisum*. *Hereditas* **10**, 1–52.

Sandler, L., and Szauter, P. (1978). The effect of recombination-defective meiotic mutants on fourth-chromosome crossing-over in *Drosophila melanogaster*. *Genetics* **90**, 699–712.

Shalev, G., Sitrit, Y., Avivi-Ragolski, N., Lichtenstein, C., Levy, A. A. (1999). Stimulation of homologous recombination in plants by expression of the bacterial resolvase ruvC. *Proc. Natl. Acad. Sci. USA* **96**, 7398–7402.

Sved, J. A., Blackman, L. M., Gilchrist, A. S., and Engels, W. R. (1991). High level of recombination induced by homologous P elements in *Drosophila melanogaster*. *Mol. Gen. Genet.* **225**, 443–447.

Tease, C., and Jones, G. H. (1995). Do chiasmata disappear? An examination of whether closely spaced chiasmata are liable to reduction or loss. *Chromosome Res.* **3**, 162–168.

Thuriaux, P. (1977). Is recombination confined to structural genes of the eukaryotic chromosome? *Nature* **268**, 460–462.

Trivers, R. (1988). Sex differences in rates of recombination and sexual selection. In *The Evolution of Sex: An Examination of Current Ideas* (R. E. Michod and B. R. Levin, Eds.), pp. 270–286. Sinauer, Sunderland MA.

Whitehouse, H. L. K. (1966). An operator model of crossing over. *Nature* **211**, 708–713.

Williams, C. G., Goodman, M. M., and Stuber, C. W. (1995). Comparative recombination distances among *Zea mays* L. inbreds, wide crosses and interspecific hybrids. *Genetics* **141**, 1573–1581.

Yamamoto, T. (1961). Progenies of sex-reversal females mated with sex-reversal males in the Medaka, *Oryzias latipes*. *J. Exp. Zool.* **146**, 163–179.

Zetka, M. C., Rose, A. M., and Baillie, D. L. (1987). A test of the Muller Ratchet Hypothesis using the *REC*-1 strain of *Caenorhabditis elegans*. *Genetics* **116** (Suppl.), 30.

REEF ECOSYSTEMS: THREATS TO THEIR BIODIVERSITY

James W. Porter and Jennifer I. Tougas
University of Georgia

I. Coral Reef Biodiversity
II. Coral Biology
III. Anthropogenic Causes of Coral Decline
IV. Coral Disease
V. Coral Reefs and Global Climate Change

GLOSSARY

biodiversity Refers to the diversity of life, including genetic biodiversity (diversity within a species), species biodiversity (diversity among species), and ecosystem biodiversity (diversity among ecosystems).

bleaching The loss of symbiotic zooxanthellae from corals. Bleaching is usually caused by elevated sea surface temperatures, but it can also be caused by sedimentation, salinity variation, or bacterial infection.

calcification The deposition of calcium carbonate skeletons by aquatic plants or animals. In reef-building corals, calcium is deposited in its aragonitic mineral form.

Cnidaria The marine invertebrate phylum containing the reef-building corals.

disease Any impairment of the normal physiological functions of an organism. While disease normally refers to infection by bacterial, fungal, protozoan, or viral pathogens, technically bleaching could also be classified as a disease based on its physiological effect.

epizootic Disease outbreaks among animal populations (as distinguished from an epidemic in human populations).

eutrophication Nutrient enrichment, typically in the form of nitrates or phosphates, most often from human sources such as agriculture, sewage, or urban runoff from land.

extinction Extinction is said to occur when a species is not definitely located in the wild during the past 50 years.

global climate change Refers to a suite of changes in the Earth's climate, including phenomena such as global warming, severe storm frequency and intensity, and glacial melting. Increasingly, scientists believe that global climate change is being accelerated by anthropogenic inputs of CO_2.

gonochoric A mode of reproduction in which individuals of the species are either male or female and produce either eggs or sperm within a single colony.

hermaphroditic A mode of reproduction in which individuals of the species produce both eggs and sperm within a single colony, sometimes within the same polyp.

hermatypic Reef-building; more recently, this term has been replaced by the term zooxanthellate to refer to those coral species with symbiotic algae.

nematocysts Harpoon-like stinging cells found in the tentacles of all cnidarians. They are used to pierce, immobilize, and capture zooplankton food.

oligotrophic Low in nutrients and low in primary production. Coral reefs grow in oligotrophic water.

planula A coral larva. This ciliated planktonic stage

rarely lasts for more than 1 or 2 weeks prior to settlement.

P/R The ratio between photosynthetic and respiratory rates of the combined coral host and zooxanthellate symbiont. A ratio greater than 1 ($P/R > 1$) indicates a net gain of energy that is then available for growth and reproduction.

Scleractinia The taxonomic order of cnidarians that includes the reef-building corals.

sedimentation Particulate material falling out of the water column onto the seafloor.

trophic efficiency The percentage of material or energy that moves, without loss, from one trophic level to the next. Most food chains have trophic efficiencies around 10%. Through tight internal recycling, corals routinely achieve trophic efficiencies in excess of 90%.

trophic level Position within the food chain, e.g., primary producer, herbivore, and carnivore. Corals, however, with their symbiotic algae and their ability to feed on zooplankton, exist at all three trophic levels simultaneously.

turbidity Particulate material suspended in the water column that reduces water clarity, light penetration, and hence photosynthesis.

zooxanthellae Symbiotic dinoflagellate algae in corals and other tropical marine invertebrates.

‰ The oceanographic symbol for salinity, or the salt content of seawater in parts per thousand.

CORAL REEFS ARE the oldest and most diverse communities on Earth. With 32 of the 34 presently known animal phyla, reef ecosystems are vastly more diverse than tropical rain forests, which support only 9 free-living phyla. There are many close analogies between coral reefs and tropical rain forests. Both exhibit high species diversity, both have high topographic complexity (trees in the rain forest, corals on the reef), and both have a high proportion of their organic material resident in the living biota rather than in organic-rich soils or sediments. However, it is probable that no other ecosystem on Earth has, or ever had, as many higher level taxa as are present on modern-day coral reefs.

To a certain extent, coral reefs are an enigma: on the one hand, they are the most luxuriant ecosystems on Earth, supporting high diversity and high biomass, and yet on the other, they achieve this status in the least fertile waters on Earth. Corals solve this problem by tight recycling and high efficiency. The flesh of corals is a symbiotic association between algae, called zooxanthellae, and cnidarians (10% plant, 90% animal). Corals are primary producers, herbivores, and carnivores all at the same time. This tightly knit symbiosis produces trophic efficiencies as high as 90%. Furthermore, filter-feeding invertebrates, which create and cover the topographically complex three-dimensional structure of the reef, capture and retain a high proportion of the material that moves over them.

From a geological perspective, reefs may be defined as masses of carbonate limestone, built up from the seafloor by the accumulation of the skeletal material of many coral reef plants and animals. For every gram of carbon dioxide fixed into organic (living) material by coral photosynthesis, an equal amount of carbon dioxide is deposited into inorganic material (limestone) by calcification. Reef growth has shaped the face of the Earth by creating limestone structures over 1.3 km thick (Enewetak Atoll) to over 2000 km long (Great Barrier Reef). Depending on their proximity to land, coral reefs are classified as either fringing reefs (paralleling the coast line at a distance of <1 km from shore), barrier reefs (paralleling the coast line >5 km from shore), or atoll reefs (midoceanic reefs without any relationship to continental or island land masses). Reefs can be further subdivided into back-reef, patch-reef, or offshore-reef habitats.

Coral reefs flourish on stable substrates within a very narrow range of physical parameters. These requirements include shallow depths (0–50 m), normal oceanic salinities (32–38 parts per thousand), warm sea surface temperatures (mean annual values of 22–26°C), high ambient light levels (100–2000 $\mu E\ m^{-2}\ s^{-1}$ at solar noon), high water clarity (transmittance values above 90%), high oxygen concentrations (near 90% full saturation), and extremely low nutrient concentrations (<1.0 μM dissolved inorganic nitrogen; <0.1 μM soluble reactive phosphorus). Although some coral reefs can exist under conditions slightly suboptimal to these, such reefs are never the richest, fastest growing, or most diverse. As a result of these requirements, coral reefs are restricted to the tropics, generally between 25° north and south latitude, and predominantly on the western boundaries of the world's oceans in the Caribbean and the Indo-Pacific. Tropical coastal zones cover 9.8×10^6 km^2, or 1.9% of the Earth's surface; coral reefs are thought to occupy only 0.6×10^6 km^2, or slightly less than 0.1% of the planet. Humans have a special responsibility and a special challenge to preserve these environments as they house the fullest expression of the evolution of life on Earth.

I. CORAL REEF BIODIVERSITY

A. Phyletic Diversity

Coral reefs harbor extraordinary biodiversity. At the phyletic level, a level that more accurately tallies the diversity of evolved life forms in an ecosystem, 32 of the 34 described phyla are found on coral reefs. In contrast, only 9 are found free-living in the tropical rain forest (Table I). Even if freshwater and parasitic forms are included in the count, the rain forest total rises to 17 phyla, approximately half of the phyletic diversity of coral reefs.

This observation raises important concerns in the conservation of biodiversity. Whereas most biologists focus on issues pertaining to species loss, geologists frequently examine extinction patters in higher level taxa (Veron, 1995). Ninety percent of the 83 described animal classes are marine. Almost all of these are found on coral reefs, and some, such as the class Sclerospongiae, are exclusively tropical. If coral reef habitats worldwide become significantly degraded, then it might be reasonable for ecologists (as well as geologists) to contemplate the loss, over the next century, of some of the Earth's higher taxa.

B. Species Diversity

The species diversity of coral reefs greatly exceeds that of any other marine environment. Of the roughly 1.86 million plant and animal species described, 274,000 are thought to be marine and more than half of these are tropical (Table II). At present, there are thought to be 93,000 described species of coral reef plants and animals. Almost 66,000 of these are macroscopic invertebrates. Specific examples of this extraordinary diversity exist in the disparate coral reef literature; a few of these remarkable numbers are listed in Table III. At present, no fully comprehensive all-taxa biodiversity inventory has ever been conducted on a coral reef (Ormond et al., 1997), but it is obvious that were this to be done, the total biodiversity would be extremely high.

As in the rain forest, estimates of coral reef species diversity based on the number of described species are considered to be a gross underestimate of the actual number of species there. Also, as in the rain forest, the tiniest members of the community (insects in the rain forest and microinvertebrates on the coral reef) are thought to be the most diverse, and least well described, component of the fauna. Reaka-Kudla (1997) has pointed out that most of the diversity and most of the biomass of coral reefs reside within the cryptofauna,

TABLE I

The Phyletic Diversity of Coral Reefs Vastly Exceeds That of Any Other Habitat on Earth[a]

Phylum	Tropical coral reef	Tropical fresh-water	Tropical rain forest
Placozoa	X		
Porifera	X	X	
Cnidaria	X	X	
Ctenophora	X		
Mesozoa	X		
Platyhelminthes	X	X	X
Nemertina	X	X	X
Gnathostomulida	X		
Gastrotricha	X	X	
Rotifera	X	X	
Kinorhyncha	X		
Loricifera	X		
Acanthocephala	X	X[b]	
Entoprocta	X		
Cycliophora	X		
Nematoda	X	X	X
Nematomorpha	X	X[b]	
Ectoprocta	X	X	
Phoronida	X		
Brachiopoda	X		
Mollusca	X	X	X
Priapulida	X		
Sipuncula	X		
Echiura	X		
Annelida	X	X	X
Tardigrada	X	X	X
Pentastoma	X	X[b]	
Onychophora			X
Arthropoda	X	X	X
Pogonophora			
Echinodermata	X		
Chaetognatha	X		
Hemichordata	X		
Chordata	X	X	X
Total	32	16	9

[a] Of the 34 animal phyla, 32 are found on coral reefs. Only the phylum Onychophora is found exclusively in moist forests; all other rain forest phyla are also found on coral reefs. The deep-sea phylum Pogonophora is the only phylum found neither on coral reefs nor in tropical rain forests.

[b] Found in terrestrial organisms as internal parasites only.

TABLE II

Biodiversity Patterns Suggest That, as with Terrestrial Organisms, Species Diversity among Marine Organisms Is Higher in the Tropics Than in the Temperate or Arctic Zones[a]

Group	Number of described species (to nearest 1,000)	Percentage of total described species (1.87 million)
Observed species diversity		
Total described global biodiversity	1,868,000	—
Total marine species, all taxa	274,000	14.7
Total macroscopic marine species	200,000	10.7
Total animals	193,000	10.3
Macroinvertebrates	180,000	9.6
Total algae	4,000–8,000	0.2–0.4
Total described tropical coastal species	195,000	10.4
Total described coral reef species	93,000	5.0
Total macroscopic coral reef species	68,000	3.6
Animals	66,000	3.5
Algae	2,000–3,000	0.1–0.2
Expected species diversity		
Total expected coral reef species		
Most conservative estimate	618,000	34.3
Intermediate estimate	948,000	—
Least conservative estimate	9,477,000	—

[a] Despite the paucity of data on marine biodiversity, it also appears (1) that most of the biodiversity of coral reefs has not been described and (2) that many species may already have gone extinct. The data are summarized from Reaka-Kudla (1997).

not the large spectacular corals and fishes that sit on or swim over the reef. This generality is reflected in the proportion of undescribed species observed in samples taken in a systematic fashion from some of the world's richest coral reef habitats (Table IV). In general, this table reveals that the smaller the body size of the organisms, the greater the proportion of undescribed species in the sample. Since most of the species on a coral reef are small and cryptic, it follows that most are also undescribed.

Reaka-Kudla (1997) has attempted to estimate the actual number of species on a coral reef based on the comparative species richness of coral reefs versus tropical rain forests and their relative surface areas. Depending on which assumptions are accepted, her formulae result in a low estimate of 618,000 species and a high estimate of 9,477,000. The most reasonable intermediate estimate puts the biodiversity estimate at slightly less than 1 million species (Table II). Briggs (1999) has argued against such extrapolations, pointing out that statistical errors are compounded unrealistically when small sample sizes are increased by several orders of magnitude, e.g., from 93,000 observed species to 9,477,000 expected species (Table II). After an exhaustive review of the literature and advice from marine systematists, Poore and Wilson (1993) argue that only 1 in 20 marine species have been described, producing a conservative estimate for tropical marine biodiversity of 1,870,000 species. This 1.87 million estimate suggests that the number of species to be found on a coral reef equals all of the currently described life forms on our planet (Table II).

If we accept as fact that tropical marine biota is almost certainly more poorly described than temperate biota, that is, that the ratio of undescribed to described species is greater than 20 to 1, then there appear to be somewhere between 1.86 and 9.47 million species on coral reefs. Regardless of the estimating technique used,

TABLE III

Examples of the Extraordinary Biodiversity of Coral Reefs

Group	Number of species	Sampling unit	Location	Source[a]
Organisms > 0.2 mm (all groups)	309	Colonies of the coral *Oculina arbuscula*	Florida	McCloskey, 1970
Infaunal invertebrates	800	10 m^2	Australia	Poore and Wilson, 1993
	350	10 m^2	Aldabra Atoll	Hughes and Gamble, 1977
Polycheates	158	6 liters of sediment	Oahu, HI	Butman and Carlton, 1993
	103	One colony of living coral	Heron Island, Australia	Grassle, 1973
Motile cryptofauna	776	One reef flat	Moorea	Peyrot-Clausade, 1983
Mollusks	637	Milne Bay, Papua	New Guinea	Werner and Allen, 1998
Boring cryptofauna	220	Dead coral	Solomon Islands	Gibbs, 1971
Cheilostome bryozoans	46	Hard substrates	Jamaica	Jackson, 1984
Hermatypic corals	362	Milne Bay, Papua	New Guinea	Werner and Allen, 1998
	350	Great Barrier Reef	Australia	Veron, 1985
	242	Ishigaki Island	Indo-Pacific	Veron, 1985
	53	Discovery Bay	Jamaica	Wells, 1973
Fishes (all groups)	1500	Great Barrier Reef	Australia	Sale, 1977
	1039	Milne Bay, Papua	New Guinea	Werner and Allen, 1998
	496	Bahamas	Caribbean	Bohlke and Chapin, 1968
	442	Dry Tortugas	Florida	Longley and Hildegrand, 1941
	517	Alligator Reef	Florida	Starck, 1968
	23	Single coral head, Big Pine Key	Florida	Bohnsack, 1979

[a] Bohlke, J., and Chapin, C. (1968). *Fishes of the Bahamas and Adjacent Tropical Waters.* Livingston, Wynnewood, PA. Bohnsack, J. A. (1979). "The Ecology of Reef Fishes on Isolated Coral Heads: An Experimental Approach with an Emphasis on Island Biogeographic Theory," Ph.D. Dissertation, University of Miami, Coral Gables, FL. Butman, C. A., and Carlton, J. T. (1993). *Biological Diversity in Marine Systems.* National Science Foundation, Washington, D.C. Gibbs, P. E. (1971). *Bull. Br. Mus. (Nat. Hist.) Zool.* **21**, 99–211. Grassle, J. F. (1973). In *Biology and Geology of Coral Reefs* (O. A. Jones and R. Endean, Eds.), pp. 247–270. Academic Press, New York. Hughes, R., and Gamble, J. (1977). *Philos. Trans. R. Soc. London, B* **279**, 324–355. Jackson, J. B. C. (1984). *J. Exp. Mar. Biol. Ecol.* **75**, 37–57. Longley, W., and Hildegrand, S. (1941). *Pap. Tortugas Lab.* **34**, 1–331. McCloskey, L. (1970). *Int. Rev. Ges. Hydrobiol.* **55**, 13–81. Peyrot-Clausade, M. (1983). *Thalassgraphica* **6**, 27–48. Poore, G. B. C., and Wilson, G. D. F. (1993). *Nature* **361**, 597–598. Sale, P. F. (1977). *Am. Nat.* **111**, 337–359. Starck, W. (1968). *Undersea Biol.* **1**, 1–40. Veron, J. (1985). *Proc. Fifth Int. Coral Reef Cong.* **4**, 83–88. Wells, J. (1973). *Bull. Mar. Sci.* **23**, 16–58. Werner, T., and Allen, G. (Eds.) (1998). *A Rapid Biodiversity Assessment of the Coral Reefs of Milne Bay Province, Papua New Guinea.* Conservation International, Washington, D.C.

and regardless of how fully we accept Brigg's caveat, the gathering impression is that, with the exception of species in a few showy classes and orders, the vast majority of coral reef species are as yet undescribed.

C. Control of Scleractinian Coral Biodiversity

Geography, age, and temperature appear to control biodiversity patterns in reef-building corals. The Indo-Pacific region, with its vastly greater age and geographic extent is richer by far than the Caribbean (Veron, 1995). For instance, whereas 362 species of coral are found on the eastern end of Papua New Guinea, only 53 are found in Jamaica (Table III). The Indo-Pacific has a "species-generating" topography: tens of thousands of isolated islands scattered across vast spaces. Despite the recent discovery of sibling species of corals even within the relatively well known Caribbean genus *Montastrea* (Knowlton et al., 1992), the numerical disparity between the two regions will persist as the list of described species from both oceans lengthens.

Figure 1 shows the widespread distribution of coral genera throughout the Caribbean. The dense packing of generic diversity isopleths along the eastern coast of

TABLE IV

Examples of Undescribed Biodiversity among Several Tropical Marine Invertebrate Faunae from Familiar and Easily Accessible Marine Habitats (Merrell, 1995)[a]

Site	Taxon	Number of undescribed species out of the total collected in the taxon	Source[b]
New Guinea	Corals	14 of 362	Werner and Allen, 1998
New Guinea	Fish	3 of 1039	Werner and Allen, 1998
New Guinea	Snails, sea slugs	310 of 564	Gosliner, 1993
Phillipines (one island, multiple sites)	Snails, sea slugs	135 of 320	Gosliner, 1993
Hawaii (one island, 6 liters of coral reef sediment)	Marine polycheate worms	112 of 158	Dutch, 1988
Great Barrier Reef (two islands)	Marine flatworms (Polyclads)	123 of 134	Newman and Cannon, 1994
Gulf of Mexico	Copepods (Harpacticoids)	19–27 out of 29	Merrell, 1995

[a] This table has been arranged from larger to smaller body size and suggests that, as with fauna everywhere, especially in the tropics, the smaller the body size, the higher the percentage of undescribed species in the sample.

[b] Sampling effort and number of samples varied among studies. Dutch, M. (1988). "A Characterization of Polycheate Assemblages on a Hawaiian Fringing Reef" Master's Thesis, Zoology Department. University of Hawaii, Honolulu, HI. Gosliner, T. M. (1993). *Proc. Seventh Int. Coral Reef Symp.* **2**, 702–709. Merrell, W. J. (1995). *Understanding Marine Biodiversity: A Research Agenda for the Nation.* National Academy Press, Washington, D.C. Newman, L. J., and Cannon, L. R. G. (1994). *Mem. Queensl. Mus.* **36**, 159–163. Werner, T., and Allen, G. (Eds.) (1998). *A Rapid Biodiversity Assessment of the Coral Reefs of Milne Bay Province, Papua New Guinea.* Conservation International, Washington, D.C.

Florida correlates with the frequency and intensity of cold water disturbances (Birkeland, 1996). Cold temperature limits the distribution of coral reefs northward in the Northern Hemisphere and southward in the Southern Hemisphere.

From a biodiversity perspective, Bermuda (with 14 hermatypic scleractinian coral genera) may have more in common with the coral reefs of the Florida Middle Grounds (12 genera) and the Flower Garden Banks (13 genera) than with coral reefs of the Florida Keys (24 genera; Fig. 1). The absence of the family Acroporidae from all three of these northern localities is probably more a function of winter cold kills than of limits to dispersal due to geographic isolation. These three reef localities demonstrate that temperature tolerances of individual Caribbean species are probably more important than distance in determining which species are present at a given location.

Local environmental conditions can also influence coral diversity. The loss of coral diversity along the northern and eastern coast of South America is probably due to sedimentation, not temperature. The presence of extensive terrigenous beaches and sediments transported from tropical rivers, such as the Orinoco and the Amazon, diminishes coral reef survival in these locations. It would be interesting to see if the species and genera waning as one travels south along the coast of South America (where low temperature is not a factor) are the same as those disappearing as one travels north along the coast of Florida (where low temperature is the controlling factor).

D. Species Loss

Both IUCN (International Union for Conservation of Nature and Natural Resources) and CITES (Convention on the International Trade in Endangered Species) define extinction as occurring when a species is not definitely located in the wild during the past 50 years. With this strict definition, and in the complete absence of monitoring efforts at the appropriate temporal and spatial scales, extinctions in the marine environment in general, and on coral reefs in particular, are almost impossible to prove. An example of this kind of difficulty can be seen in the announcement of the extinction of an eastern Pacific coral species due to severe El Niño conditions, and an almost immediate retraction when it was subsequently rediscovered alive (Glynn and Feingold, 1992). Nevertheless, Carlton (1993), in his review of modern marine invertebrate extinctions, includes one tropical species in his list, the Indo-Pacific mangrove periwinkle, *Littoraria flammea*, which was last seen in the mid-1800s.

There are several methodological reasons why marine extinctions might be especially difficult to detect. As pointed out by Ray (1988), "The last fallen mahogany

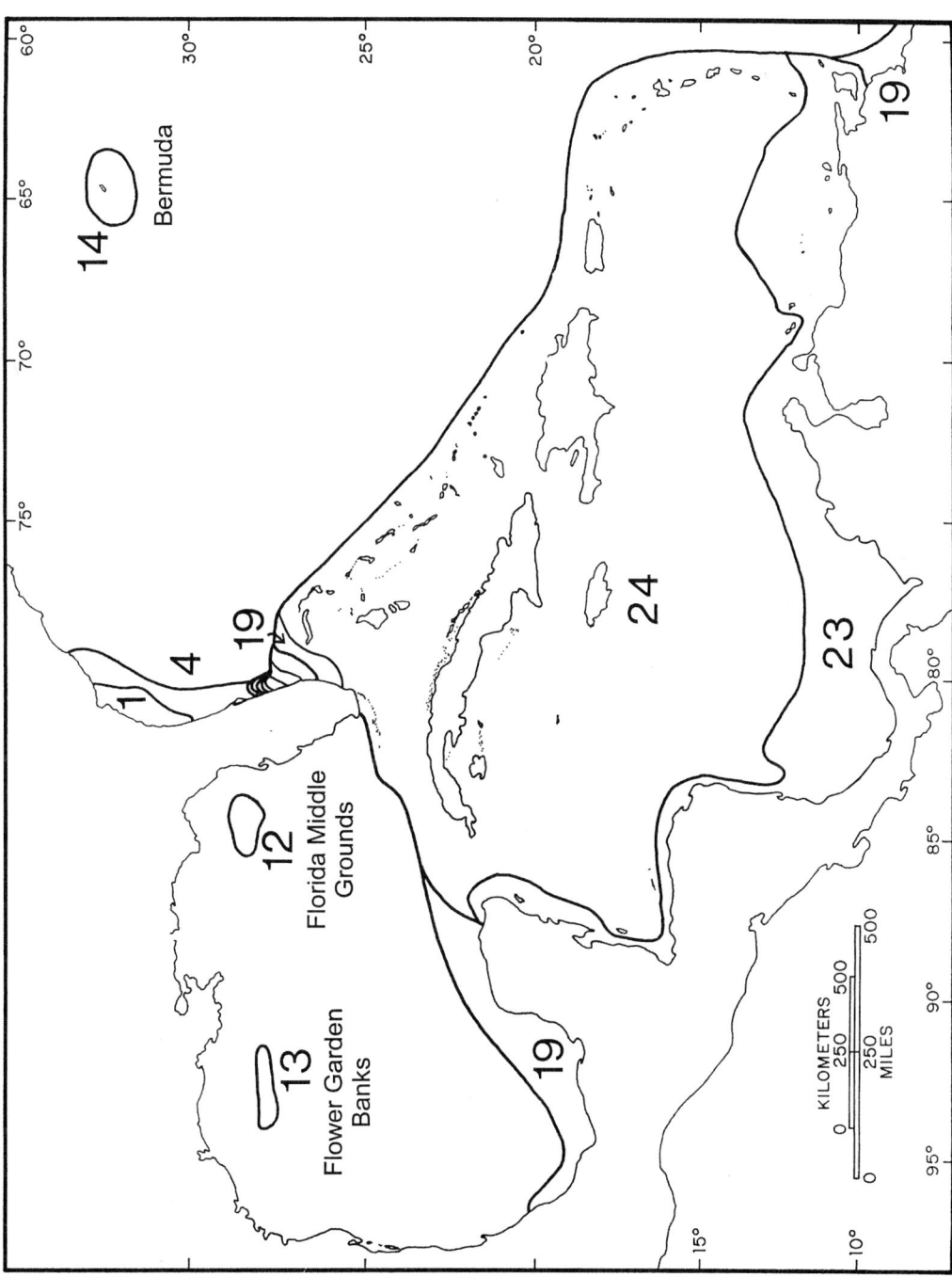

FIGURE 1 Patterns of generic scleractinian coral diversity in the Caribbean reveal no endemism within the region, but rather broad-scale distribution followed by rapid faunal diminution north and south. Coral loss northward along the coast of Florida is due to cold temperature limitations; faunal loss southward along the coast of South America is probably due to the influence of river sediments pouring into the coastal zone. While the northward distribution of corals in the Caribbean is due to historical patterns of global temperature regimes, the sediment load of costal environments in South America is increasingly influenced by anthropogenic upland management practices in the coastal zone.

would lie perceptibly on the landscape, and the last black rhino would be obvious in its loneliness, but a marine species may disappear beneath the waves unobserved, and the sea would seem to roll on the same as always." In addition, there is the perception that marine species are somehow less susceptible to extinctions. Lamark, in his 1809 *Philosophie Zoologique*, states that "Animals living in the waters, especially the sea waters . . . are protected from the destruction of their species by man. Their multiplication is so rapid and their means of evading pursuit or traps are so great, that there is no likelihood of his being able to destroy the entire species of any of these animals."

While this argument may carry some validity, Reaka-Kudla (1997) points out that most species on coral reefs are small and that these smaller species also have much smaller geographic ranges. This leads to the conclusion that most species on coral reefs may be much more vulnerable to extinction than has been widely assumed. The few clear examples of marine extinctions have in common a vulnerable, extinguishable habitat. Coral reefs, especially those located near population centers, fall into this category. The most interesting perspective on the complex, worrisome, but poorly researched topic of marine extinctions may belong to Carlton (1993), "At the end of the 20th Century, one of the major crises in global marine invertebrate conservation is not so much that invertebrates are becoming extinct at a rapid rate (although they may be)—the crisis is that we do not know."

II. CORAL BIOLOGY

A. Anatomy

Corals are benthic marine invertebrates belonging to the phylum Cnidaria, which is characterized by two distinct tissue layers, the inner endoderm and outer ectoderm, separated by an amorphous collection of cells called the mesoglea. A single coral polyp has a central mouth cavity surrounded by tentacles armed with stinging cells called nematocysts. Corals can be solitary, consisting of a single large polyp, or colonial, consisting of thousands of interconnected polyps. Colonies form through budding—one polyp produces a daughter polyp that is genetically identical to the original.

B. Reproduction and Recruitment

Corals can reproduce asexually through fragmentation or self-generation of brooded larvae. This form of reproduction restricts genetic diversity of coral populations. In contrast, sexual reproduction through fertilization of gametes originating from genetically distinct colonies increases the genetic diversity of coral populations.

Sexual reproduction in corals occurs in one of two ways: either through mass spawning, in which thousand of gametes (eggs and sperm) are released simultaneously into the water column where fertilization takes place, or by brooding, in which sperm are released into the water column and are taken inside the maternal coral polyp to fertilize the eggs stored there. Depending on the species, a given colony may be hermaphroditic, producing both eggs and sperm, or gonochoric, producing either eggs or sperm. In both instances, ciliated planulae larvae are produced (Birkeland, 1996).

The coral larvae spend between 3 days and 3 weeks in the water column, during which time they disperse. They may travel only a few meters away on the same reef or to entirely different reefs kilometers away. Dispersion maintains gene flow in coral populations. After dispersion, larvae settle onto relatively clean, hard surfaces on the reefs, metamorphose into polyps, and begin to form new colonies through asexual budding (Birkeland, 1996).

Coral recruitment is favored by nutrient-poor conditions with high light availability, low sedimentation rates (Rogers, 1990), limited competition for space by algae, and decreased predation by fish, sea urchins, and starfish. The patterns of settlement, survival, and growth of coral recruits directly influence the structure and function of coral communities and associated reefs.

C. Calcification

Common to all scleractinian (stony) corals is the ability to secrete calcium carbonate. The shape of the resulting skeleton is species specific at the polyp level, but the overall shape of the colony is influenced, within limits, by environmental conditions. Colony morphologies aid in the removal of trapped sediments (Rogers, 1990) and the capture of food and influence both zooxanthellate and host physiology (Sebens, 1994).

As a chemical process, deposition of $CaCO_3$ is influenced by the ambient concentration of CO_2, which is directly related to temperature, pressure, and concentrations of other dissolved materials. As a biological process, calcification is driven by photosynthesis and is closely controlled by temperature (Dubinsky, 1990). Under optimal conditions, growth rates of branching corals, such as the Caribbean coral *Acropora cervicornis*, can exceed 10 cm per year. However, local variables such as nutrient concentrations and sedimentation rates reduce realized growth rates (Birkeland, 1996).

Calcification by thousands of colonies over hundreds of thousands of years produces the complex, three-dimensional structure of modern reefs essential to the maintenance of reef biodiversity. For example, topographical features are important for the distribution, survival, and resulting abundance of many reef fishes and invertebrates (Sebens, 1994).

D. Photosynthesis

While corals can capture prey with their tentacles, many sclerectinian corals rely on endosymbiotic algae for nourishment. Known as zooxanthellae, the algae are located within the ectoderm of the coral. Depending on the species, corals may host a variety of zooxanthellae within a colony through space and time (Rowan et al., 1997). Photosynthesis by the zooxanthellae provides nutrients required by the coral for growth and reproduction and drives calcification and subsequent reef formation. As a result, the bathymetric distribution of reef-building corals is largely restricted to high light environments, typically less than 50-m depth, which can sustain this symbiotic relationship (Dubinsky, 1990).

Depending on the clarity of the water, ultraviolet light penetrates the ocean to about 5 m. Ultraviolet radiation inhibits photosynthesis and is damaging to many organisms, including corals and zooxanthellae. However, some coral species have developed protective pigments that allow the transmission of visible light while blocking ultraviolet radiation (Dubinsky, 1990). As not all corals have this ability, the distribution of corals is also influenced by the presence of ultraviolet radiation.

If the relationship between the coral and its symbiotic zooxanthellae is disturbed through increased temperatures or exposure to elevated UV light, bleaching may occur. The term "bleaching" describes the condition in which the zooxanthellae exit, or are expelled from, the coral, thus showing the stark white skeleton beneath the coral tissue. Without the symbiotic algae, corals lose their vital source of nutrition, slow their growth rates, stop reproducing, and sometimes die (Birkeland, 1996). When environmental conditions return to normal, zooxanthellae repopulate the coral. Susceptibility to bleaching is influenced by the species of coral in question and the species of zooxanthellae it hosts (Rowan et al., 1997). Consequently, two colonies of the same species may have dramatically different bleaching responses to the same stresses.

E. Physiological Limitations

While availability of light limits the depth distribution of corals, temperature limits the latitudinal and longitudinal distribution and is one of the best predictors of coral diversity (Veron, 1995). Optimal temperature for coral growth and reproduction ranges from 22 to 26°C, depending on geographic location and species in question. Corals generally do not grow in waters in which minimum temperatures drop below 18°C, and such a thermal barrier also limits dispersal of larvae. A few corals survive in temperatures above 30°C, such as those found in some locations in the Middle East. To some extent, corals are able to adapt to ambient conditions; consequently, upper lethal temperatures for a species in the tropical zone will be higher than those of the same species in a subtropical zone (Dubinsky, 1990).

Salinities can also influence the distribution of corals. Corals grow well in water that has a constant salinity of 32–36‰. Low salinity (<20‰), due to increased freshwater flow from localized flooding or exposure to heavy rainfall during low tides, limits coral distribution and reduces diversity. High salinities (>38‰) can also inhibit coral growth, particularly in the Persian Gulf (Dubinsky, 1990).

III. ANTHROPOGENIC CAUSES OF CORAL DECLINE

A. Benefits from Coral Reefs

Humans benefit from both the resources and recreation that coral reefs provide. Coral is used for building materials in areas where there is no viable alternative. In fact, many inhabited tropical islands around the world were, at one point, coral reefs themselves. In addition, coral reefs reduce coastal erosion by protecting coastlines from severe storms (Hoegh-Guldberg, 1999). This is particularly important in tropical waters where hurricanes and tropical storms occur frequently (Richmond, 1993). Indeed, entire islands have been washed into the sea when their surrounding living coral reefs were removed.

Coral reefs are important for the development of local economies (Birkeland, 1996). The reefs support valuable fisheries for local consumption and for the aquarium trade. Throughout the Caribbean and Indo-Pacific, local diets derive nearly 60% of their intake of protein from these reefs. The life cycles of many commercially important fish and shellfish are dependent on the presence of healthy mangrove swamps, coral reefs, sea grass beds, and coastal lagoons.

A multibillion dollar tourism industry is supported by tropical coral reefs (Hoegh-Guldberg, 1999), and is a critical source of income, particularly for small island nations with few alternative resources to exploit. For the 1992–1993 fiscal year, attendance at the Coral Reef State Park on Key Largo, Florida, had the highest visitation of any state park in Florida that year (Fig. 2). Tourism can only be a viable option for economic development if reefs are healthy.

Finally, pharmaceutical companies have discovered naturally occurring bioactive compounds among the organisms found on coral reefs (Birkeland, 1996): antitumor compounds have been found in the mucus of corals, anti-inflammatory agents have been isolated from soft corals, and coral has successfully been used as a bone substitute in reconstructive surgeries.

B. Coastal Urbanization

Despite their importance, coral reefs around the world have been declining at an alarming rate. At the core of this decline are human activities (Fig. 3), spurred by population growth (Table V). Nearly 15% of the human population lives within 100 km of coral reefs (Hoegh-Guldberg, 1999). The geographical locations of highest coral diversity also coincide with large human populations. More than 100 countries have coral reefs within their territorial boundaries (Birkeland, 1996). Most of these are developing nations and have, by far, the fastest growing populations due to advances in medicine, technology, and public health services. Coastal cities are growing rapidly by attracting immigrants from country interiors to these bustling centers of trade and commerce. Influxes of tourists can substantially increase the effective population of an area and place additional demands on potable fresh water, power, and sewage systems. For example, in 1990, the population of the Florida Keys was 80,500, but there were over 2,000,000 visitors per year to this tourist destination.

As populations continue to expand, human pressures on coral reefs will increase. There is a direct correlation between reef degradation and proximity to urban centers. Activities associated with urbanization include waste disposal and power and desalination plant operation. Rapid urbanization has outpaced sewage treatment capacities in several regions and has caused eutrophication in coastal zones as raw sewage is often discharged directly into nearshore waters (Richmond, 1993; Sebens, 1994). The effluents from operating power and desalination plants are up to 5 or 6°C warmer than ambient temperatures.

Industrialization often accompanies urbanization and is encouraged by economic demands for hard currency and international commerce. Effluents from some

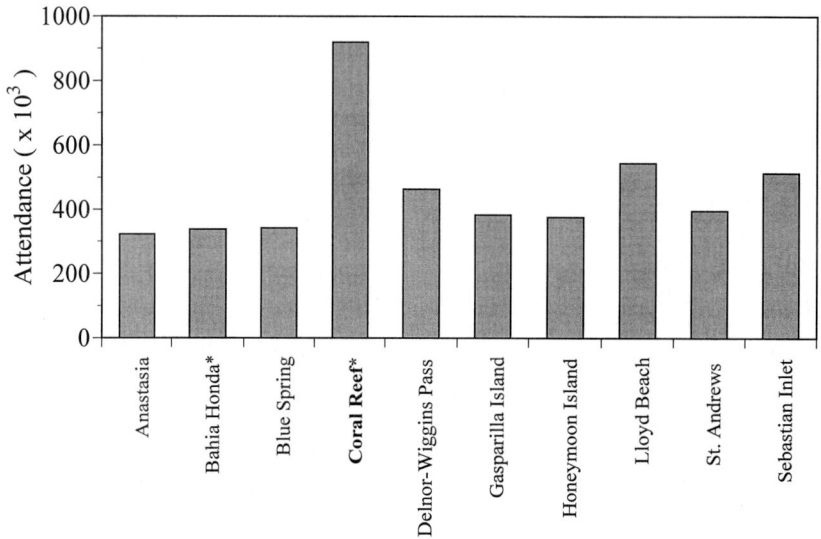

FIGURE 2 Coral reefs are popular tourist destinations. For fiscal year 1992–1993, the most frequently visited state park in Florida was Coral Reef State Park. Attendance at this park was nearly twice that of Lloyd Beach, the second most visited park. Asterisks indicate parks with coral reefs. Tourism provides an important source of economic development for tropical island countries with coral reefs.

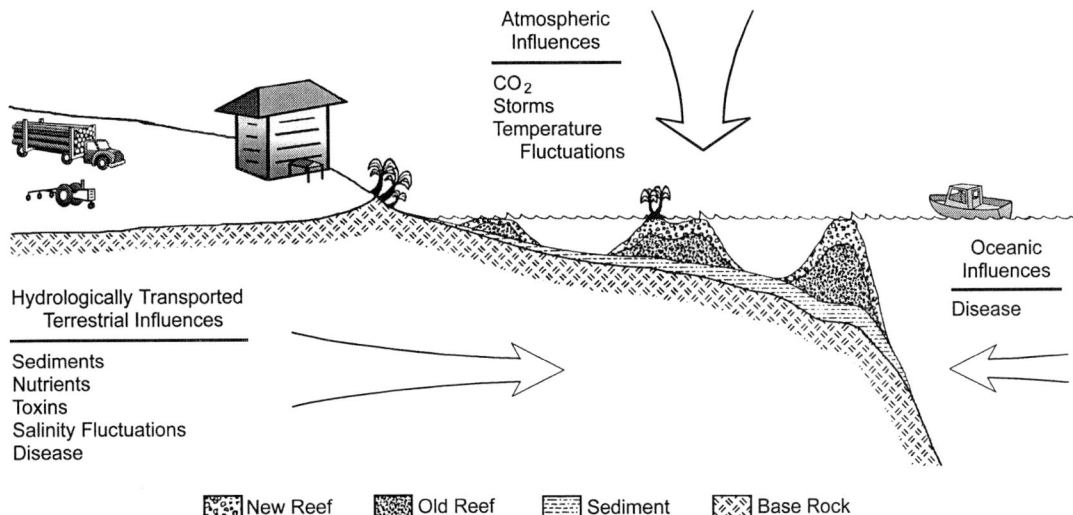

FIGURE 3 Coral reefs are subject to terrestrial, atmospheric, and oceanic influences. Sediments, nutrients, and toxins, released from activities such as deforestation, agriculture, and industry, are hydrologically transported to coral reefs through local rivers. CO_2 buildup in the atmosphere increases CO_2 concentrations in the ocean and alters climate patterns. Finally, diseases are circulated by ocean currents. Reefs located near human population centers are subjected to multiple stresses simultaneously and so suffer losses in diversity and cover [adapted from Wilkinson, C. R., and Buddemeier, R. W. (1994). *Global Climate Change and Coral Reefs: Implications for People and Reefs. Report of the UNEP–IOC–ASPEI–IUCN Global Task Team on the Implications of Climate Change on Coral Reefs*. IUCN, Gland, Switzerland].

industries, such as rum distilleries and fertilizer plants, contribute to coastal eutrophication and heavy metal contamination. Ores are one of the few available resources for economic development in the tropics and mining activities can be a significant source of sediments. Furthermore, heavy metals readily bind to clays and are transported by terrigenous sediments to the reefs. Mining activities throughout the Indo-Pacific and Caribbean deliver thousands of tons of toxin-laden sediments to coral reefs each year.

Another land-based activity that affects coral reefs is deforestation. Nearly 70% of all tropical hardwood products originate from Southeast Asia. In the Philippines, forests have been reduced to 25% of their original cover. Upland areas of French Polynesia are cleared for residential and commercial construction as well as for agriculture and hydroelectricity. Deforestation, particularly of coastal mangroves (Rogers, 1990), increases erosion and the amount of soils transported from the land to the reefs. Erosion can be severe when heavy rains fall on logged areas (Birkeland, 1996).

Agricultural activities often take over land cleared for timber production. Millions of hectares of mangrove forests have been reclaimed for aquaculture and agriculture. In Southeast Asia, farming has become universally dependent on the use of agrochemicals. As a result of agricultural activities, nearshore waters are subjected to increased nutrients, sediments, and agrochemicals (Richmond, 1993). Heavy metals found in corals from Panama and Costa Rica were common components of agricultural pesticides (Guzmán and Jiménez, 1992).

Upland and coastal ecosystems on land are intimately linked with coral reefs in the sea (Porter *et al.*, 1999). At the organismal level, corals subjected to land-based pollution undergo metabolic changes that lead to bleaching, reduced growth and reproduction rates, and, on occasion, death (Richmond, 1993). Coral recovery after natural disturbances is inhibited by pollution. At the ecosystem level, these effects lead to losses in coral diversity, coral cover (Edinger *et al.*, 1998), and shifts in dominant benthic organisms (Lapointe, 1999). The causal agents include sedimentation, eutrophication, altered temperatures, and altered salinities.

C. Sedimentation

Sedimentation influences coral communities through lethal and sublethal mechanisms, depending on the sediment load and the life cycle of the marine organisms. While increased sedimentation causes direct mortality of corals by smothering them, most effects are sublethal. Corals remove sediments by secreting copious amounts of mucous that trap the sediments. These mucous sheets, which are moved off of the coral through ciliary

TABLE V
Disturbances to Reef Ecosystems: Their Sources and Consequences

Disturbance	Effect of disturbance	Source of disturbance	Cause of source
Sedimentation	Lethal effects	Deforestation	Human population growth
	Smothering	Infrastructure construction	Migration to cities
	Reduced coral cover	Road construction	Tourism
	Reduced coral diversity	Logging in the watershed	Economic demands
	Sublethal effects	Clearing for agriculture	
	Decreased water clarity	Clearing mangroves	
	Shift toward shallower community	Industry	Introduction of new technologies
	Decreased photosynthesis	Mining	Economic demands
	Increased respiration		Introduction of new technologies
	Increased mucus production		
	Reduced coral recruitment		
Eutrophication	Lethal effects	Waste management	Human population growth
	Overgrowth by macroalgae		Migration to cities
	Reduced coral cover		Tourism
	Reduced coral diversity	Agriculture	Introduction of new technologies
	Sublethal effects	Fertilizer application	Food needs
	Decreased water clarity	Ranching (raising pigs)	
	Shift toward shallower community	Industry	Introduction of new technologies
	Decreased photosynthesis	Fertilizer plant operation	Economic demands
	Reduced coral reproduction	Rum distillery operation	
	Reduced coral recruitment		
	Increased activity of boring algae		
Toxic contamination	Lethal effects	Agriculture	Economic demands
Heavy metals	Death	Pesticide application	Food needs
Pesticides	Increased bacterial infections	Herbicide application	Introduction of new technologies
Herbicides	Sublethal effects	Industry	Human population growth
	Increased mucus production	Mining	Migration to cities
	Increased respiration rates	Fertilizer plant operation	Economic demands
	Decreased photosynthesis	Power plant operation	Introduction of new technologies
	Decreased growth	Desalination plant operation	
	Decreased reproduction		
	Bleaching		
Altered temperatures	Lethal effects	Industry	Human population growth
Increases	Bleaching, leading to death	Power plant operation	Migration to cities
Decreases	Decreased coral cover	Desalination plant operation	Tourism
	Decreased coral diversity	Altered hydrology	
	Sublethal effects	Global climate change	Fossil fuel consumption
	Bleaching and recovery		Human population growth
	Increased respiration		Introduction of new technologies
	Decreased photosynthesis		Urban development
	Reduced reproduction		

continues

Continued

Disturbance	Effect of disturbance	Source of disturbance	Cause of source
Altered salinity	Lethal effects	High salinities	Human population growth
	Bleaching, leading to death	Desalination plant operations	Migration to cities
	Decreased coral cover	Reduction of freshwater input	Tourism
	Decreased coral diversity		
	Sublethal effects	Low salinities	Introduction of new technologies
	Bleaching and recovery	Increased freshwater runoff from deforestation of watersheds	Urban development
	Increased mucus production		
	Decreased photosynthesis		
	Decreased respiration		
	Reduced fertilization		
Disease	Lethal effects	Largely unidentified pathogens	Increased susceptibility to disease by multiple stressors
	Tissue death		
	Sublethal effects		
	Decreased photosynthesis		
	Decreased growth		
	Decreased reproduction		
Storms	Lethal effects	Global climate change	Fossil fuel consumption
Increased frequency	Scouring		Human population growth
Increased intensity	Fragmentation		Introduction of new technologies
	Sublethal effects		Urban development
	Fragmentation		
	Increased sedimentation		
	Increased turbidity		
	Increased nutrients		

action, constitute a tremendous energy drain for the corals and cause a decrease in the P/R by increasing respiration (Rogers, 1990). Despite this removal process, sediments tend to accumulate in depressions on large, massive colonies and cause death to those patches. Consequently, there is a positive correlation between the amount of terrigenous sediments and the amount of coral injury.

Water turbidity, which increases when sediments are suspended in the water column, decreases the amount of light available for photosynthesis. As photosynthetic rates decrease, so do growth and reproduction rates. Because of the reduced availability of light, the maximum depth at which corals can grow decreases and the coral community compresses into shallower environments (Dubinsky, 1990).

Adult corals are more tolerant to sedimentation stresses than juveniles. Coral larvae are not able to settle on loose sediments (Rogers, 1990). Consequently, if a fine layer of sediments covers the reef benthos, then coral settlement patterns shift toward vertical surfaces (Rogers, 1990) and successful recruitment drops dramatically (Richmond, 1993).

D. Eutrophication

The effect of eutrophication varies according to the quantity and quality of the nutrient source, as well as the hydrographic regime in the area, and becomes especially apparent when high nutrients are present for an extended period of time. On naturally oligotrophic reefs, tight nutrient cycling between the coral host and zooxanthellate symbionts affords a competitive advantage to the coral: corals are able to flourish and outcompete many other primary producers on the reef. When nutrients are added to the system, the competitive edge shifts to faster growing macroalgae (Lapointe, 1999; Richmond, 1993) and filter feeders. The algae proceed to overgrow the corals and effectively shade them until the corals die. Coral recruitment is reduced because algae occupy space on the reef and prevent coral larvae from settling (Lapointe, 1999). Finally, the growth of boring organisms is promoted, which weakens the reef structure itself and increases the probability of storm damage (Richmond, 1993).

Eutrophication is also associated with increased turbidity and a concomitant decrease in light availability (Richmond, 1993), largely due to an increase in phytoplankton densities. When photosynthesis decreases, growth rates and reproduction diminish. This leads to decreases in coral diversity and coral cover (Birkeland, 1996). Furthermore, vertical zonation becomes truncated under decreased light availability (Dubinsky, 1990).

E. Heavy Metals and Toxins

Howard and Brown (1984) reviewed the effects heavy metals have on corals. Corals are able to directly absorb soluble metals from seawater. Alternatively, they may ingest metals directly by catching particulate matter in mucous nets or indirectly as a result of feeding on copepods, which accumulate metals in their chitinous skeletons. Some metals may be deposited directly into the skeleton and become immobilized. Others remain in the coral tissues and cause dramatic physiological responses. These include excessive mucous production, increased bacterial infections, bleaching, decreased skeletal deposition, which decreases vertical growth rates, reduced reproduction, and death (Howard and Brown, 1984). Similarly, when corals are exposed to agrochemicals, responses include increased respiration, decreased photosynthesis, increased mucous production, increased planulae release (a common stress response for brooding species), and decreased larval settlement (Birkeland, 1996).

F. Altered Temperatures

Although the warmth of tropical water may seem benign to humans, reef-building corals live much closer to their upper lethal temperature (the temperature that will kill or disable them) than to their lower lethal temperature. In fact, a rise in the water temperature of only 2–3°C above the normal summertime average is much more stressful physiologically than a drop of 2–3°C below this value. There are two ways in which elevated temperatures affect coral: increased respiration and decreased photosynthesis. Under severe thermal stress, bleaching occurs. Anything that acts to increase temperatures has the potential to adversely affect the health and survival of coral reefs.

Coral respiration increases with increasing temperatures (Porter et al., 1999). Bleaching under increased temperature is correlated with increased respiratory rates and a decrease in photosynthesis (Porter et al., 1999). Under both of these conditions, the P/R ratio decreases for the coral, and growth and reproduction decrease. If exposed for an extended period of time to temperatures above the average maximum temperature they are accustomed to, bleaching can occur and the coral colonies can die (Birkeland, 1996).

Reproductive success decreases with increased temperatures and is far more sensitive to temperature fluctuations than growth rates. Consequently, healthy adult corals could live in environments unsuitable for reproduction. Nonetheless, as temperature-sensitive species die, or fail to reproduce, the composition of coral communities will change (Hoegh-Guldberg, 1999).

G. Altered Salinity

Hyposalinity results from increased discharge or runoff associated with deforestation, particularly of mangrove forests, and from urban development. On the other hand, hypersalinity is associated with power and desalination plant effluents as well as large-scale reductions of freshwater flow from land.

Responses of corals to altered salinities vary according to species and region. In Florida, the coral *Siderastrea siderea* can grow in areas where salinity fluctuations are prevalent. Changes in salinity of up to 10‰ away from the mean produces little response in the coral. Beyond 10‰ above the mean, respiration and photosynthesis decreased and some bleaching was observed. In contrast, *Porites* species have demonstrated a narrower tolerance to salinity fluctuations: an increase of 10‰ causes corals to contract their polyps, shed copious amounts of mucus, and bleach [as cited in Porter et al. (1999)].

In Kaneohe Bay, Hawaii, widespread coral death has accompanied increased freshwater runoff. Low salinities also inhibit fertilization and larval survival (Richmond, 1993). Fertilization of mass-spawning species takes place at the water surface, where eggs and sperm mix. Once the eggs have been fertilized, the resultant larvae float near the water surface for several days. Freshwater also floats on seawater. Therefore, the gametes and larvae could be exposed to lowered salinities if mass spawning occurs during heavy rainfall. In one study, fertilization rates and larval survivorship dropped by more than 50% when the salinity dropped to 28‰. These results demonstrated that terrestrial runoff can have a major influence on reproductive success (Richmond, 1993).

Water from Florida Bay naturally flows through channels between the keys and out over the reef tract. Extensive channelization of water for use in Miami and agricultural areas and for flood control has decreased the amount of freshwater entering the bay. Consequently, salinities in Florida Bay rose dramatically in the 1980s, particularly during drought years. Because Florida Bay is shallow, temperatures fluctuate with the seasons. Warm, hypersaline waters originating from Florida Bay have been recorded at depth on reefs along the Florida Keys (Porter *et al.*, 1999). In a study of six sites along the Florida Keys reef tract, between 1984 and 1991, all six sites lost coral diversity, and five out of six sites lost coral cover. Looe Reef, the southernmost and hardest hit reef, lost 43.9% of its coral cover (Porter *et al.*, 1999). Porter *et al.* (1999) suggested that the reef degradation observed could result from the influence of poor water quality originating from Florida Bay. Another potential source of stress from Florida Bay water is eutrophication (Lapointe, 1999).

IV. CORAL DISEASE

A. Identification of Diseases

Coral reefs are no exception to the truism that, even in healthy ecosystems, disease is part of the natural environment. Diseases in the ocean, however, are poorly understood because of the conceptual and methodological challenges in studying ephemeral phenomena in an alien environment. This explains why most coral reef pathogens are unidentified (Table VI). For instance, of the twelve scleractinian coral diseases easily recognized by their symptoms, only two have been positively identified (Table VI). Nonetheless, an increase in either the frequency or severity of disease epidemics, called epizootics in animal populations, can be cause for legitimate concern.

While there is a perception that the incidence of coral disease has increased (Harvell *et al.*, 1999), it is easy to dismiss these accounts as either biased by heightened environmental concern or unfounded in the

TABLE VI

Coral Disease Conditions Commonly Observed in the Florida Keys

General disease category[a]	Common name	Pathogen	Reference[b]
Black line disease	Black band	*Phormidium corallyticum*	Rutzler and Santavy, 1983
White line diseases	White pox	Unknown	Porter *et al.*, in press
			Holden, 1996
	White band, Type I	Unknown	Williams and Bunkley-Williams, 1990
	White band, Type II	Unknown	Ritchie and Smith, 1998
	White plague, Type I	*Sphingomonas* sp. nov.	Richardson *et al.*, 1998b
	White plague, Type II	Unknown	Richardson *et al.*, 1998a
Other diseases	Yellow blotch	Unknown	Santavy *et al.*, 1999
	Dark spot	Unknown	Goreau *et al.*, 1998
	Ridge mortality	Unknown	Goreau *et al.*, 1998
	Red band	*Oscillatoria*?	Goreau *et al.*, 1998
	Rapid wasting	Fungal/predation	Cervino *et al.*, 1998
	Neoplasia	Cancer?	Goreau *et al.*, 1998

[a] Criteria for disease designation: Active tissue mortality, tissue necrosis, bared skeleton, mucus production, bisected or partial polyps.
[b] Cervino, J., Goreau, T., Hayes, R., Kaufman, L., Nagelkerken, I., Patterson, K., Porter, J., Smith, G., and Quirolo, C. (1998). *Science* **199**, 1302–1310. Goreau, T., Cervino, J., Goreau, M., Hayes, R., Richardson, L., Smith, G., DeMeyer, K., Nagelkerken, I., Garzon-Ferra, J., Gil, D., Garrison, G., Williams, E., Bunkley-Williams, L., Quirolo, C., Patterson, K., Porter, J., and Porter, K. (1998). *Rev. Biol. Trop.* **46** (Suppl. 5), 157–172. Holden, C. (1996). *Science* **274**, 2017. Porter, J., Patterson, K., Porter, K., Peters, E., Mueller, E., Santavy, D., and Quirolo, C. *Coral Reefs* (in press). Richardson, L., Goldberg, W., Carlton, R., and Halas, J. (1998a). *Rev. Biol. Trop.* **46** (Suppl. 5), 117–198. Richardson, L., Goldberg, W. M., Kuta, K. G., Aronson, R. B., Smith, G. W., Ritchie, K. B., Halas, J. C., Feingold, J. S., and Miller, S. L. (1998b). *Nature* **392**, 557–558. Ritchie, K., and Smith, W. (1998). *Rev. Biol. Trop.* **46** (Suppl. 5), 199–203. Rutzler, K., and Santavy, D. (1983). *Mar. Ecol* **4**, 301–319. Santavy, D., Peters, E., Quiolor, C., Porter, J., and Bianchi, N. (1999). *Coral Reefs* **18**, 97. Williams, E., and Bunkley-Williams, L. (1990). *Atoll Res. Bull.* **335**, 1–71.

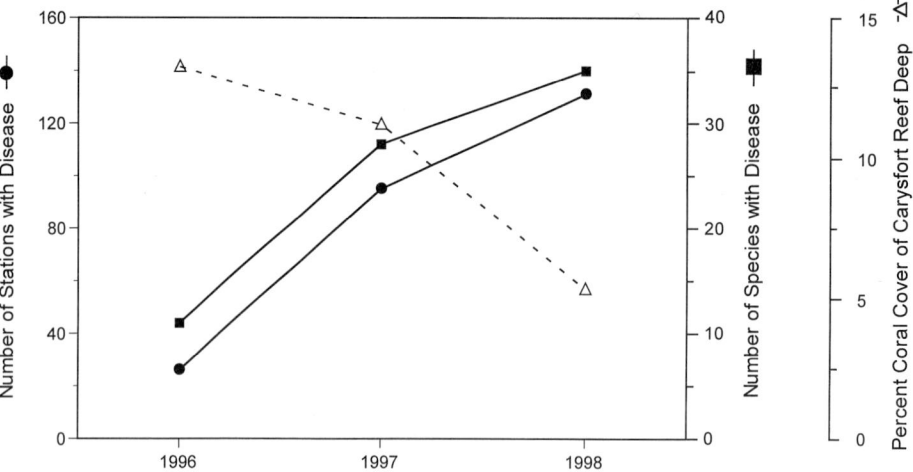

FIGURE 4 Coral disease and coral mortality in the Florida Keys, 1996–1998. The number of stations exhibiting disease (out of a possible total of 160 stations) and number of coral species exhibiting disease (out of a total of 41 species in the survey) are plotted against percent coral cover data from one of the hardest hit reef sites (out of a total of 40 reef sites in the survey), Carysfort Reef Deep. The data show that between 1996 and 1998, coral disease became more widespread in the Keys, affected more species, and had a devastating effect on the live coral cover of at least one reef in the Florida Keys (Porter et al., in press).

absence of baseline data. However, evidence to the contrary is mounting. White band disease in the Caribbean inflicted heavy losses in both St. Croix and Belize (Aronson and Precht, 1997). Paleontological evidence has demonstrated that disease outbreaks in Belize have no historical precedence over the past 5000 years (Aronson and Precht, 1997), lending credence to the idea that disease outbreaks on the present scale are a recent phenomenon.

To date, most of the well-documented epizootics are from the Caribbean, but it is not clear if this represents a real difference between the Caribbean and the Indo-Pacific or merely a difference in observational coverage. The need to know more is urgent. Only multidisciplinary teams will be able to provide the ecological information necessary to devise appropriate management strategies.

B. Effects of Diseases on Diversity

Coral reef scientists are coming late to the realization that disease may exert a major control on diversity. In his review of factors explaining the biological diversity of coral reefs, Connell (1978) does not mention disease. The Environmental Protection Agency's Coral Reef Monitoring Project in the Florida Keys has been collecting information on coral disease since 1996 (Fig. 4). Because these data have been collected systematically, they allow one to resolve whether coral diseases are more widespread now than in the past. For the 3-year period covered by the survey, these data show significant increases in all disease parameters measured, including the number of stations and the number of species with diseases present. Of the 160 stations surveyed from Key Largo (in the Upper Florida Keys) to Key West (in the Lower Keys), the number of stations with diseased corals rose from 26 in 1996 to 131 in 1998, an increase of 404%. Over the same period, the number of species affected by disease rose from 11 to 35, an increase of 218%. Many of the rarest corals disappeared from the study sites due to disease.

C. Ecosystem Effects of Disease

When diseases dramatically affect populations of a single species, the effects can influence whole ecosystems. The first documented coral reef epizootic occurred between 1982 and 1983 when almost all of the black-spined sea urchins, *Diadema antillarum*, in the Caribbean died from an unknown pathogen. From its point of origin near the Atlantic terminus of the Panama Canal, this disease spread throughout the Caribbean as a waterborne agent moving at the same speed and in the same direction as well-mapped Caribbean oceanic currents (Lessios et al., 1984). *Diadema* is a major herbivore on Caribbean coral reefs, and its loss led to an increase in algal abundance, especially on reefs with reduced herbivorous fish populations due to overfishing

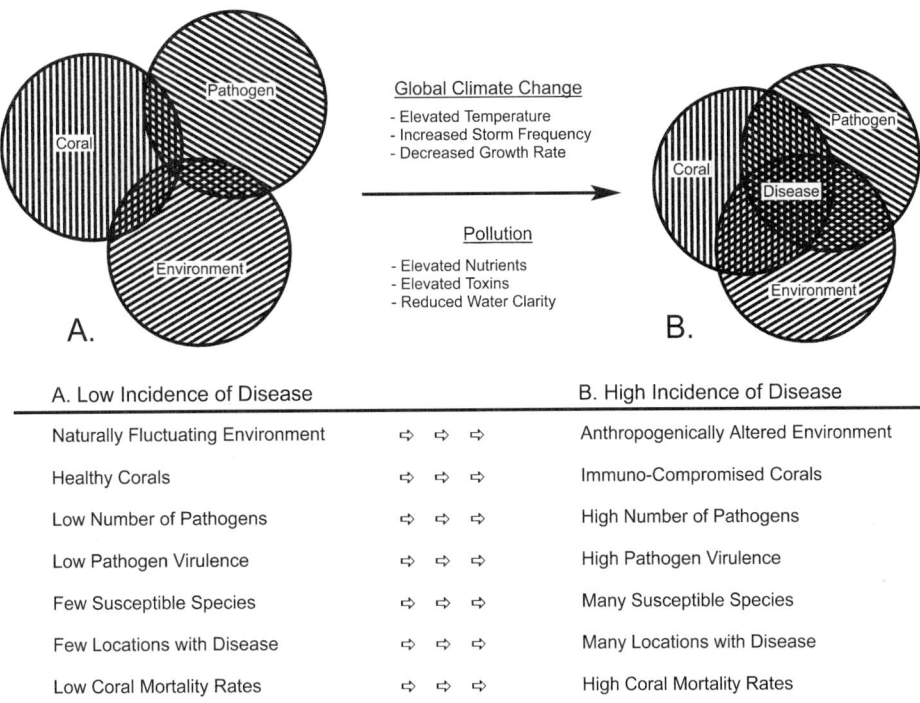

FIGURE 5 Coral disease stress model. While natural background levels of disease are expected even in healthy ecosystems, a variety of stresses could lead to the suppression of the immune and disease defense systems in coral. The consequence of reduced health would be an increase in the number of pathogenic organisms, susceptible species, locations, and mortality rates. All of these results have been observed in the Florida Keys, and while they do not prove the accuracy of this model, this hypothesis is at present the only one that explains all of the observations.

(Hughes and Connell, 1999) or on reefs with elevated nutrient levels from coastal eutrophication (Lapointe, 1999).

Corals themselves have experienced mass mortalities due to epizootics. A new disease, white pox, has inflicted high mortality among *Acropora palmata* stands on some Key West coral reefs (Table VI). For some white pox and white-band outbreaks, coral mortality rates were as high as coral losses during the worst crown-of-thorns starfish "plagues" in the Indo-Pacific (Birkeland, 1996). In the Florida Keys, the most dramatic change linked to coral disease can be seen in the loss of living coral exhibited at the deep site (18 m) on Carysfort Reef in the Upper Keys (Fig. 4). Sixty percent of the living coral there died in 2 years, mostly due to disease. Clearly, Floridian coral reefs cannot survive if these mortality rates continue.

Because corals grow slowly and live for decades or centuries, epizootics will have far-reaching impact on coral reefs on geological time scales. When deadly diseases decimate coral populations to this extent, geological rates of carbonate deposition in the Caribbean may actually be affected (Aronson and Precht, 1997). It is clear that disease epidemics can have a real impact on coral reefs.

D. The Coral Disease Model

We propose a coral reef disease model (Fig. 5) that depicts how changes in environmental conditions alter the interactions between hosts and pathogens and subsequently enable disease outbreaks. Stress factors (Porter *et al.*, 1999) are considered relevant, even for corals whose immune systems are not well known, because the ability to resist infection is a function of the host's overall health. Compromised immune systems result in increased susceptibility to disease. One of the most striking aspects of the disease patterns seen in the Florida Keys is the simultaneous increase in all disease parameters measured (Fig. 4). Only a hypothesis that addresses environmental quality will explain the simultaneous increase in the number of diseases, the number of species affected, and the rates of coral mortality throughout such a large geographic area. If this model is correct, then the incidence of disease would be expected to be higher near polluted population centers,

FIGURE 6 In 1997 and 1998, severe coral bleaching episodes were caused by dramatically elevated sea surface temperatures worldwide (see also Fig. 8). A vast majority of elk horn corals (*Acropora palmata*) on Looe Key, in the Florida Keys, bleached stark white. It is not known whether these colonies would have recovered because on August 27, 1998, Hurricane Georges removed nearly all branching corals from this reef, including the bleached colonies shown here (photograph by James W. Porter).

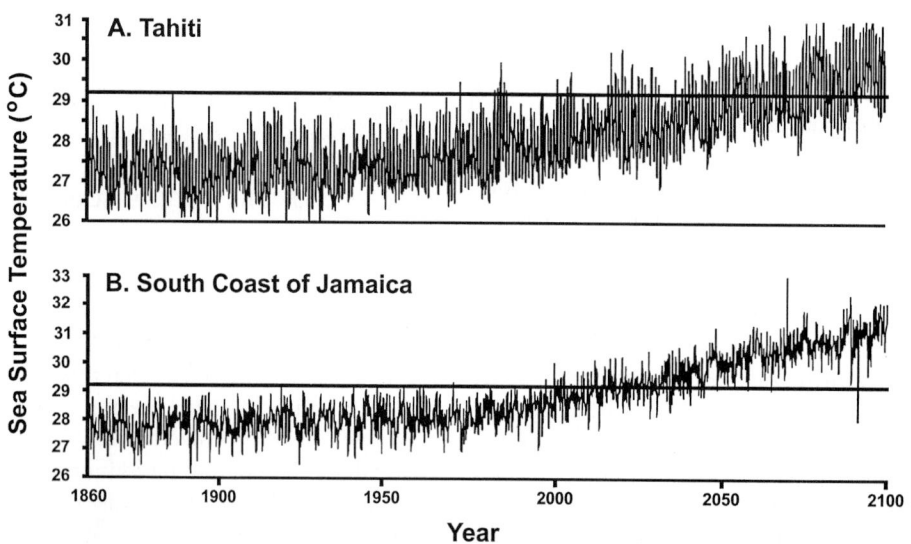

FIGURE 7 A model of sea surface temperatures based on greenhouse gas concentrations and El Niño Southern Oscillation events predicts temperatures will exceed normal thresholds for many reefs in the very near future. The horizontal lines indicate the temperature thresholds at which corals begin to bleach. As the twenty-first century proceeds, a higher percentage of time is spent above this line [Hoegh-Guldberg, O. (1999). *Mar. Freshwater Res.* **50**, 839–866].

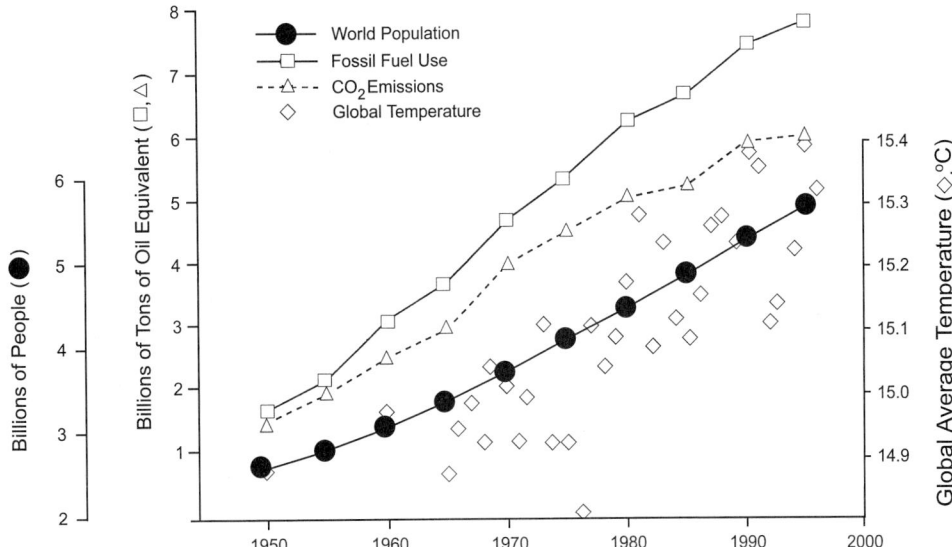

FIGURE 8 The burning of fossil fuels (coal, oil, and natural gas) and the destruction of a majority of the Earth's forests by an increasing human population have caused an increase in atmospheric CO_2 concentrations over the past half century. It is highly likely that these elevated atmospheric CO_2 concentrations have also caused the incontrovertible increase in the Earth's temperature over the same time period. Even if population growth shows some signs of slowing down as we enter the twenty-first century, energy consumption shows no such sign of declining. An increased reliance on coal, especially toward the latter half of the twenty-first century, could exacerbate rising CO_2 levels considerably [Houghton, J. T., Meira Filho, L. G., Callander, B. A., Harris, N., Kattenberg, A., and Maskell, K. (1996). *Climate Change 1995. The Science of Climate Change.* Cambridge Univ. Press, Cambridge, UK]. The low temperature value in 1976 resulted in coral death in the Florida Keys [Porter, J., Battey, J., and Smith, G. (1981). *Proc. Natl. Acad. Sci. USA* 79, 1678–1681].

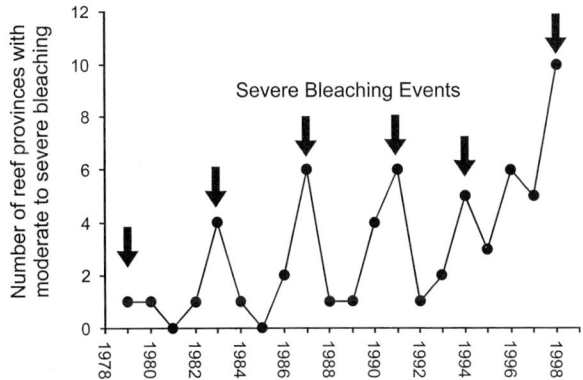

FIGURE 9 Recently, the frequency, intensity, and geographical extent of bleaching episodes have increased. During the strongest bleaching event to date, 1998, bleached corals were recorded for the first time in many provinces [Hoegh-Guldberg, O. (1999). *Mar. Freshwater Res.* 50, 839–866].

for example, or following bleaching events (Fig. 6), both of which might be expected to compromise the coral's immune system.

E. The Human Connection

Oceanic diseases (Harvell *et al.*, 1999), and wildlife diseases in general (McCallum and Dobson, 1995), appear to have increased. It is not premature to ask whether or not these disease outbreaks are caused by, or influenced by, humans. At present, the historical novelty of the outbreaks is a suggestive, but not a definitive, answer to this question (Aronson and Precht, 1997). Recently, however, the disease link to human activities has been strengthened by an examination of a fungal pathogen, *Aspergillus sydowii*, of sea fans (Harvell *et al.*, 1999). These authors have proposed that this marine pathogen is a terrestrial fungus that has secondarily invaded the marine environment via sediment runoff from land.

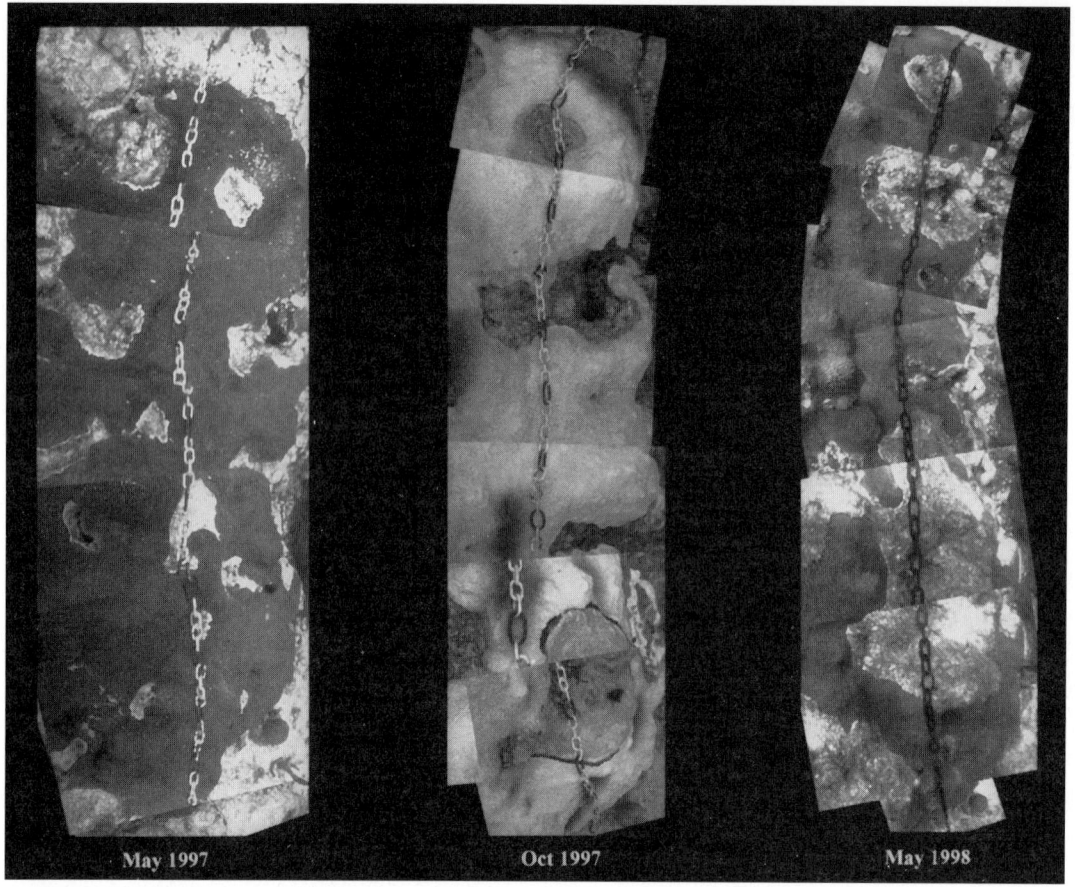

FIGURE 10 The potential interaction between coral bleaching and disease can be seen in this montage of images from 1997 to 1998. Healthy colonies of *Montastrea annularis* [left (May, 1997)] on Looe Key, in the Florida Keys, bleached in late summer due to elevated sea surface temperatures [middle (October, 1997)]. This colony also contracted black band disease (middle, lower part of the image). By May, 1998 (left), most of the colony had recovered, but the black band damaged tissue did not.

V. CORAL REEFS AND GLOBAL CLIMATE CHANGE

A. CO_2, Temperature, and Human Population Growth

The Earth is warming. Data from analyses of tree rings, sea ice extent, and ice cores, as well as direct measures of air and sea surface temperatures in both the Atlantic and the Pacific (Hoegh-Guldberg (1999), Fig. 7) demonstrate that the Earth is warmer now than a half century ago. The temperature rise closely parallels human population growth and the growth of atmospheric CO_2 inputs from the burning of fossil fuels (Fig. 8). CO_2 is one of the greenhouse gases, and the general consensus is that the buildup of this gas in the Earth's atmosphere is causally related to the measured temperature increases.

The frequency and intensity of major storms, such as hurricanes, are expected to increase with increasing temperatures. These storms cause direct physical destruction of corals by increased wave action and scouring (Birkeland, 1996). Indirect effects include increased sedimentation and turbidity and release of nutrients from dying tissues. Some species are more resistant to storm damage than others, so the frequency with which storms strike could influence the diversity of corals present on a reef (Birkeland, 1996). There is an ongoing debate as to whether storms increase coral diversity (Connell, 1978), and there is substantial evidence for both sides of the argument (Sebens, 1994). Ultimately, the effects of storms will depend on the ability of the reef corals to recover from this disturbance.

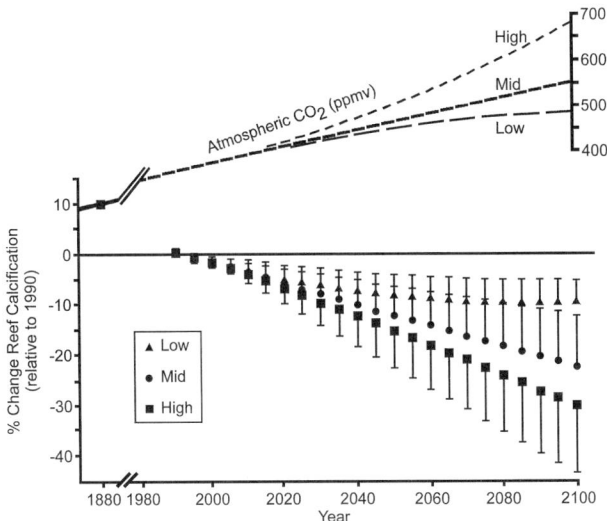

FIGURE 11 The percent change in coral reef calcification through time (1880–2100) is plotted as a function of atmospheric CO_2 concentration [Gattuso, J.-P., Allemand, D., and Frankignoulle, M. (1999). *Am. Zool.* 39, 160–183]. This graph demonstrates the linkage between anthropogenic carbon dioxide production from the burning of fossil fuels and decling coral reef growth rates.

Their recovery ability may be severely compromised in areas subjected to strong anthropogenic influences (Sebens, 1994).

B. Coral Bleaching and Elevated Sea Surface Temperatures

Coral bleaching is the loss of the symbiotic algae and is caused by elevated temperature. All marine organisms harboring zooxanthellae loose their symbiotic algae when exposed to high temperatures. Temperature-induced bleaching occurs in one of two ways, either by brief exposure to moderately increased temperature (1.5–2.0°C above average summertime temperatures for several days) or by prolonged exposure to slightly elevated temperature (only 1.0–1.5°C above normal for 3–4 weeks beyond the end of the typical summer warm season). Bleached corals appear white and lifeless (Fig. 6). The ghostlike appearance is deceptive. The chalky coloration is not due to the coral's death but instead due to the fact that, in the absence of pigmentation conferred by the symbiotic algae, the flesh of the coral is transparent. The white limestone skeleton of the coral is visible underneath its tissue. If very high temperatures persist for a few weeks, or even if moderately high temperatures persist for more than a month, the coral will die.

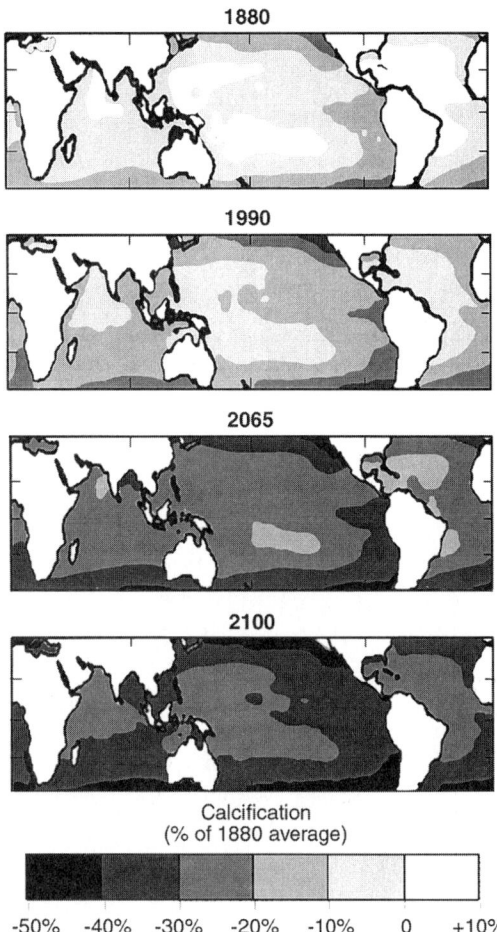

FIGURE 12 Projected changes in reef calcification rates are depicted as a percent of conditions from 1880 [reprinted from Kleypas, J. A., Buddemeier, R. W., Archer, D., Gattuso, J.-P., Langdon, C., and Opdyke, B. N. (1999). *Science* 284, 118–120 © 1999 American Association for the Advancement of Science]. This model suggests that oceanic conditions in the year 2100 will be substantially less optimal for coral growth than in the nineteenth century.

The evidence suggests that coral reefs are at serious risk from high temperatures. Over the past 20 years, there has been a dramatic increase in the number of reef provinces bleaching (Fig. 9) and in the severity of these bleaching episodes. During the 1982–1983 bleaching event, Glynn and Feingold (1992) documented up to 95% loss of corals in the Galapagos Islands. Mass mortalities have also been reported recently for Australia and the Indian Ocean (Hoegh-Guldberg, 1999). Unfortunately, arguments over the cause of high temperatures have clouded the unambiguous connection between rising temperatures and increased coral mortality. As the earth warms, more corals will die. It remains to be seen whether corals can evolve genetic

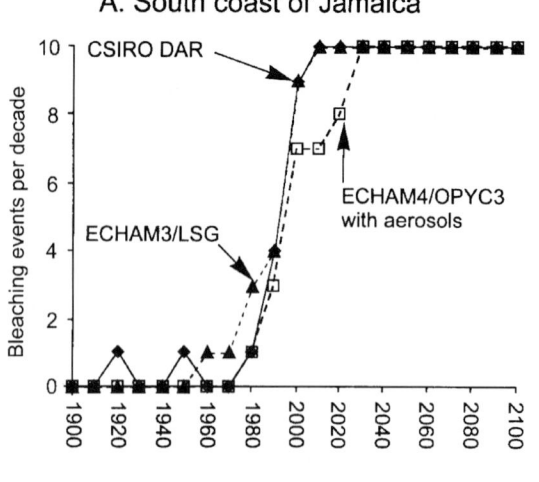

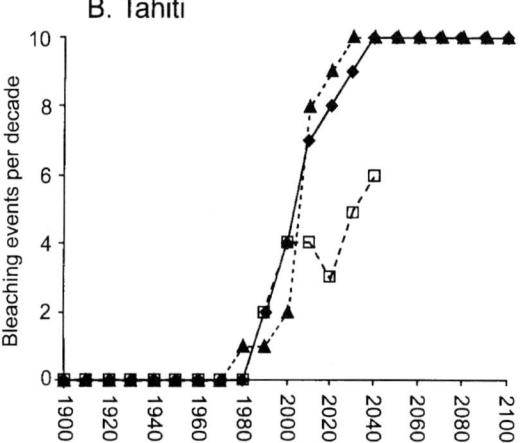

FIGURE 13 Based on predicted increases in sea surface temperatures, models indicate bleaching events per decade will increase [Hoegh-Guldberg, O. (1999). *Mar. Freshwater Res.* **50**, 839–866].

resistance fast enough to survive this coming thermal stress, or alternatively, whether human energy policy can evolve fast enough to prevent corals from the necessity of having to do so.

Bleaching represents a life-threatening stress to corals. The disease model presented in Fig. 5 suggests that bleaching should promote increased susceptibility to disease. This hypothesis has not been tested experimentally, but anecdotal observations from the Florida Keys suggest it may be correct (Fig. 10).

C. Coral Calcification and Elevated CO_2

Since the ocean is in equilibrium with the atmosphere, rising CO_2 concentrations will cause an immediate increase in the amount of carbon dioxide dissolved in seawater. While this increase is not expected to modify the highly buffered pH of the ocean, it will alter the ocean's chemistry (Fig. 11). Tropical surface waters are supersaturated with dissolved calcium carbonate. Corals exploit this supersaturation to manufacture their calcium carbonate skeletons at a substantially reduced metabolic cost. Over the next century, grossly elevated atmospheric CO_2 concentrations are expected to reduce this supersaturation and reduce coral growth (Fig. 11). Kleypas *et al.* (1999) argue convincingly that this reduction in coral reef calcification has already begun (Fig. 12). The end point of this global experiment is not known, but it is extremely worrisome.

D. Global Climate Change and Coral Reef Survival

Climate change models predict that tropical sea surface temperatures will continue to rise (Figs. 7 and 13). If these scenarios are correct, then bleaching will be (a) more frequent, (b) more prolonged, and (c) more lethal (Hoegh-Guldberg, 1999). These predictions are not for the distant future, but for the near future, only a few decades away. It is also becoming clear that although coastal zone management practices are critical in protecting the well-being of some coral reefs, especially those near population centers, over the next century, global climate change, and how humans mitigate this anthropogenic stress, will determine the long-term survival of the most diverse environment on Earth.

Acknowledgments

We thank G. P. Schmahl (Sanctuary Program, NOAA) for coral biodiversity information about the Flower Gardens Bank, Kathryn L. Patterson (University of Georgia, Athens), Stephanie Borrett (University of Georgia, Athens), DeeVon and Craig Quirolo (Reef Relief, Key West) for editorial suggestions.

See Also the Following Articles

CRUSTACEANS • EUTROPHICATION AND OLIGOTROPHICATION • GRAZING, EFFECTS OF • INVERTEBRATES, MARINE, OVERVIEW • MARINE ECOSYSTEMS • PLANKTON, STATUS AND ROLE OF

Bibliography

Aronson, R., and Precht, W. (1997). Stasis, biological disturbance and community structure of a Holocene coral reef. *Paleobiology* **23**, 326–346.
Birkeland, C. (1996). *Life and Death of Coral Reefs*. Chapman & Hall, New York.

Briggs, J. C. (1999). Marine species diversity. *BioScience* **49**, 351.

Carlton, J. T. (1993). Neoextinctions of marine invertebrates. *Am. Zool.* **33**, 499–509.

Connell, J. H. (1978). Diversity in tropical rain forests and coral reefs. *Science* **199**, 1302–1310.

Dubinsky, Z. (Ed.) (1990). *Coral Reefs*. Elsevier, Amsterdam.

Edinger, E., Jompa, J., Limmon, G., Widjatmoko, W., and Risk, M. (1998). Reef degradation and coral biodiversity in Indonesia: Effects of land-based pollution, destructive fishing practices and changes over time. *Mar. Pollut. Bull.* **36**, 617–630.

Glynn, P. W., and Feingold, J. S. (1992). Hydrocoral species not extinct. *Science* **257**, 1845.

Guzmán, H. M., and Jiménez, C. E. (1992). Contamination of coral reefs by heavy metals along the Caribbean coast of Central America (Costa Rica and Panama). *Mar. Pollut. Bull.* **24**, 554–561.

Harvell, C. D., Kim, K., Burkholder, J. M., Colwell, R. R., Epstein, P. R., Grimes, J., Hofmann, E. E., Lipp, E., Osterhaus, A. D. M. E., Overstreet, R., Porter, J. W., Smith, G. G. W., and Vasta, G. (1999). Emerging marine diseases—Climate links and anthropogenic factors. *Science* **285**, 1505–1510.

Hoegh-Guldberg, O. (1999). Climate change, coral bleaching and the future of the world's coral reefs. *Mar. Freshwater Res.* **50**, 839–866.

Howard, L. S., and Brown, B. E. (1984). Heavy metals and reef corals. *Oceanogr. Mar. Biol. Annu. Rev.* **22**, 195–210.

Hughes, T. P., and Connell, J. H. (1999). Multiple stressors on coral reefs: A long-term perspective. *Limnol. Oceanogr.* **44**, 932–940.

Knowlton, N., Weil, W., Weight, A., and Guzman, H. M. (1992). Sibling species in *Montastrea annularis*, coral bleaching and the coral climate record. *Science* **255**, 330–333.

Lapointe, B. E. (1999). Simultaneous top-down and bottom-up forces control macroalgal blooms on coral reefs (reply to the comment by Hughes *et al.*). *Limnol. Oceanogr.* **44**, 1586–1592.

Lessios, H. A., Robertson, D. R., and Cubit, J. D. (1984). Spread of *Diadema* mass mortality through the Caribbean. *Science* **226**, 335–337.

McCallum, H., and Dobson, A. (1995). Detecting disease and parasite threats to endangered species and ecosystems. *Trends Ecol. Evol.* **10**, 190–191.

Ormond, R., Gage, J., and Angel, M. (Eds.) (1997). *Marine Biodiversity: Patterns and Processes*. Cambridge Univ. Press, Cambridge, UK.

Poore, G. B. C., and Wilson, G. D. F. (1993). Marine species richness. *Nature* **361**, 597–598.

Porter, J. W., Lewis, S. K., and Porter, K. G. (1999). The effect of multiple stressors on the Florida Keys coral reef ecosystem: A landscape hypothesis and a physiological test. *Limnol. Oceanogr.* **44**, 941–949.

Ray, G. C. (1988). Ecological diversity in coastal zones and oceans. In *Biodiversity* (E. O. Wilson and F. M. Peter, Eds.), pp. 36–50. National Academy Press, Washington, D.C.

Reaka-Kudla, M., Wilson, D., and Wilson, E. (Eds.) (1997). *Biodiversity II. Understanding and Protecting Our Biological Resources*. Joseph Henry Press, Washington, D.C.

Richmond, R. H. (1993). Coral reefs: Present problems and future concerns resulting from anthropogenic disturbance. *Am. Zool.* **33**, 524–536.

Rogers, C. S. (1990). Responses of coral reefs and reef organisms to sedimentation. *Mar. Ecol. Prog. Ser.* **62**, 185–202.

Rowan, R., Knowlton, N., Baker, A., and Jara, J. (1997). Landscape ecology of algal symbionts creates variation in episodes of coral bleaching. *Nature* **388**, 265–269.

Sebens, K. P. (1994). Biodiversity of coral reefs: What are we losing and why? *Am. Zool.* **34**, 115–133.

Veron, J. (1995). *Corals in Space and Time. The Biogeography and Evolution of the Scleractinia*. Comstock–Cornell Press, Ithaca, NY.

REFORESTATION

David Lamb
University of Queensland

I. Need for Reforestation
II. Methods of Overcoming Degradation
III. Can Restoration Ever Be Achieved?
IV. Methods of Reforestation
V. Ecosystem Recovery after Reforestation
VI. Choosing Areas to Reforest
VII. Specialized Conditions or Sites with Particular Problems
VIII. Socioeconomic Issues

GLOSSARY

degradation A loss of forest structure, productivity, and native species diversity. A degraded site might still contain trees (i.e., a degraded site is not necessarily deforested) but it has lost at least some of its former ecological integrity.
reclamation To recover productivity at a degraded site using mostly exotic tree species. The original biodiversity is not recovered although the protective function and many of the original ecological services may be reestablished.
reforestation The reestablishment of trees and understory plants at a site previously occupied by forest cover.
rehabilitation To reestablish the productivity and some, but not necessarily all, of the plant and animal species thought to be originally present at a site. For ecological or economic reasons, the new forest might also include species not originally present at the site. The protective function and many of the ecological services of the original forest may be reestablished.
restoration To reestablish the presumed structure, productivity, and species diversity of the forest originally present at a site. The ecological processes and functions of the restored forest will closely match those of the original forest (see text for alternative definition).

BIODIVERSITY AND PRODUCTIVITY are being lost as natural ecosystems are fragmented and degraded. Reforestation is used to reverse this process but this process can take a variety of forms. In some cases, reforestation can restore significant amounts of plant and animal biodiversity to degraded sites. In other cases, only productivity is restored. True ecosystem restoration is very difficult and more modest goals are often all that is possible, particularly over large areas. Successful reforestation usually requires solving social and economic problems as well as just ecological problems. These socioeconomic issues are often the most difficult to resolve.

I. NEED FOR REFORESTATION

Large areas of the world's forested land have been cleared to enable agricultural development. Estimates by FAO (1995) suggest 163 million ha of forest were lost between 1980 and 1990, of which more than 90% was in the tropics.

Even if the rate of the world's population growth is reduced and human populations are stabilized, it is likely that much more of the world's remaining natural ecosystems will be cleared. Many of these cleared lands are now used for food or fiber production but many have been degraded through misuse and have since been abandoned. In the face of this process of biological simplification, much effort has gone into developing a comprehensive, representative, and adequate series of nature reserves to protect the biodiversity remaining in the intact ecosystems still unaffected by these changes. But in many places there are doubts about whether this network can be sufficient since it is based on a residual series of plant and animal communities left behind in the landscape after the original ecosystems have been fragmented. This problem might be diminished if some of these degraded lands could be reforested and recover some of the biodiversity they once contained. Then they, too, might also contribute to the conservation of biodiversity in this developing reserve network.

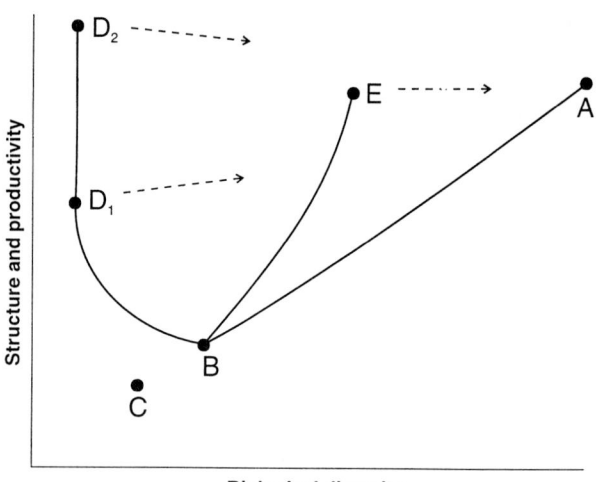

FIGURE 1 Various methods of reforestation after degradation. Degradation has moved the system from its original state at point A to a degraded state at point B. Over time, the site can degrade further to C or recover naturally back to A. Reclamation via monocultures can lead the system to D1 or D2, depending on the productivity achieved. Rehabilitation may recover most of the structure and productivity but only part of the former biodiversity (E). Over time, some of the original species may recolonize the forests at D and E from intact forest, causing them to drift back toward point A. (After Bradshaw, 1995. Reprinted with permission from Cairns, J. (Ed.) (1995). *Rehabilitating Damaged Ecosystems*. Lewis Publishers. Copyright 1995, CRC Press, Boca Raton, FL.)

II. METHODS OF OVERCOMING DEGRADATION

Degradation is a subjective term. A newly cleared area of forest might be regarded as prime agricultural land by a farmer but as degraded wildlife habitat by a bird enthusiast. To some extent, degradation is in the eye of the beholder. On the other hand, many would agree that once-forested land that has lost both its structure and diversity and is not used in any productive way is now, in some sense, "degraded." The two main components of degradation are shown in Fig. 1. One component is the proportion of the original biodiversity still present and the other is the proportion of the original structure or productivity remaining. The original undisturbed ecosystem is at point A and point B represents the same site in a degraded state. The degraded area now has less biodiversity and is at a lower level of productivity or structure. The degraded site might degrade even further, perhaps because of recurrent fires (point C) or, if not disturbed further, might gradually recover, unaided, back to point A. The circumstances under which this recovery could occur are described further below.

The process of recovery can be accelerated by reforestation but there are several ways in which this might be done. Each leads to a different outcome. *Restoration* describes the process in which the aim of reforestation is to restore the ecosystem back to its former condition (point A) containing the original complement of plant and animal species. At this point the ecosystem will have regained its original structure and productivity and the ecological processes that sustained the system will have been reestablished. (This definition may be too exclusive in some situations and the notion of "restoration" is discussed further below.) *Reclamation*, on the other hand, attempts to restore structure and productivity but not the original biodiversity (point D). Reforestation may be carried out using a single species and that species might be an exotic tree species planted for commercial purposes. The structure and productivity may be partially regained (point D1) or even exceed that of the original ecosystem because of the use of fertilizers or other management inputs (point D2).

Between these two approaches lies a midway position that might best be described as *rehabilitation*. This may recover the original structure and productivity but may not recover all of the original biodiversity (point E). The new ecosystem may contain a mix of native and exotic species. Over time, the new systems at D and E may gradually drift toward point A as some of the original species recolonize from intact forest nearby or they may remain in a partially restored state if no recolonization takes place.

These three forms of reforestation are a necessary simplification of the variety of methods of reforestation that might take place on degraded lands. They differ in the degree to which biodiversity is recovered but they also share certain common attributes. These include the fact that a new stable and productive land use is achieved and that at least some of the ecological services and protective functions of the original forest have been recovered.

III. CAN RESTORATION EVER BE ACHIEVED?

Some have questioned if reforestation can ever achieve complete ecological restoration and recover all the biodiversity once present at a degraded site. For example, restoration implies that the original species complement at a site is known and that this composition was static. But the original state of many long-degraded sites is often poorly known and successional development is more common than any static "climax" condition. Even without degradation, changes might have occurred to the original condition in both space and time. And what of places such as the Mediterranean Basin, where some of the original species are no longer present in the landscape following a long process of deforestation and fragmentation. Or where other exotic species have become naturalized over a long period, because of particular cultural practices, and are now impossible to eradicate? These difficulties may mean that "restoration" is sometimes an uncertain target that might be difficult to achieve in practice. For these and other reasons, the Society for Ecological Restoration defines ecological restoration as "the process of assisting the recovery and management of ecological integrity. Ecological integrity includes a critical range of variability in biodiversity, ecological processes and structures, regional and historical context, and sustainable cultural practices."

Ecological restoration may also be difficult for other reasons. In some severely degraded sites the numbers of species to be restored may be too high or the magnitude of the changes such as exposure, topsoil erosion, or increases in salinity may be so great that restoration is too difficult to achieve even if the technical means were available. The costs of attempting full restoration of the original system might simply be too high. In some situations social constraints may also apply. Some traditional landowners or managers may be unwilling to agree to restoration of degraded sites they are not currently using because it is not a goal they share or because they believe restoration might somehow lessen their rights to its future use. In such cases outside intervention to achieve restoration is unlikely to be successful.

None of these constraints mean that reforestation to improve the biodiversity across degraded landscapes is not worth attempting. Rather, they simply point to the fact that different ecological and social situations will require different approaches.

IV. METHODS OF REFORESTATION

In some situations degraded forest ecosystems can recover unaided. A well-known example is the reforestation that has occurred over the past 100 years on much of the northeastern United States following the abandoning of some of the land previously cleared for agriculture. The former farm sites are now occupied by deciduous woodlands with a structure and diversity probably similar to that which was originally present. This same natural recovery process can be found in many other temperate and tropical forests. The rate at which such successions take place varies widely but most are usually slow. Natural recovery after degradation is not an invariable process and it only takes place under certain conditions. In all cases the disturbing agents that were the causes of degradation must have been removed and the original topsoils must have remained more or less intact. Further, remnant communities of the original forest species must have persisted on the landscape to act as a source of propagules and colonists for the new succession.

A particular problem with such slow recovery processes is that it increases the risk that sites might be damaged again by disturbing agents such as fire or grazing that degrade the site once more (and take it back to point B or C on Fig. 1). Alternatively, new land users may appear and think a site has been abandoned because they are unaware it is being reforested. More active reforestation programs may be needed to accelerate the process and overcome these risks. These programs can take a variety of forms but all require that the causes of degradation (fire, grazing, firewood collection, etc.) be removed and that sites be actively protected. The subsequent reforestation process can then be carried out by direct seeding or by planting seedlings.

FIGURE 2 Seedling establishment after direct sowing of seed from the air at a bauxite mine in northern Australia. Note the lack of competing weeds at this early stage of recovery. Areas with poor seedling establishment can be filled by direct planting if necessary.

A. Direct Seeding

The advantages of direct seeding are that a costly nursery stage is avoided and seed can be broadcast in the field comparatively cheaply. Seed can be sown by hand or vehicle or can be distributed from the air and large areas can be covered relatively cheaply (Fig. 2). The disadvantage of direct seeding is that only a small proportion of the seed distributed is likely to be established as seedlings as many are removed by insects or fail to germinate under field conditions or are outcompeted by weeds. This means the method may require large volumes of seed to achieve a desired tree density and it may be impossible to acquire such volumes for many of the less common species being reestablished at a particular site. Situations where direct seeding has been most useful have been mine sites where topsoil has been respread after mining has ceased and where competing weeds have not yet reestablished. Direct seeding is also possible in areas where competing weeds can be removed by plowing or weedicide. In such circumstances a cocktail of species might be reintroduced as seed although the diversity of species in such a mix can be limited by the availability of seed for some less common species and these might need to be added to the site as seedlings.

B. Planting Seedlings

Planting of seedlings raised from seed (or cuttings) in a nursery is a more expensive operation but it may be necessary for species with very small seeds or particular germination or mycorrhizal requirements. It also necessitates other management inputs such as site preparation (ripping or plowing), weed control, or fertilizing which improve survival and foster more rapid growth of the planted seedlings. In plantation monocultures being established for commercial reasons (i.e., reclamation), it is common to plant trees in regularly spaced rows. This also permits the use of planting machines pulled behind tractors which can lead to very rapid planting rates. In a restoration operation a more irregular planting design might be used with a random distribution of species.

C. The Choice of Species

A key issue in such restoration plantings is the number of species to use. It is rarely the case that all of the original species can be planted and usually some subset must be chosen. The choice of which species to use will be guided initially by the objectives of the reforestation program. If the goal of reforestation is commercial production, then a fast-growing, exotic species might be chosen. Such species usually have readily available seed, well-established nursery and planting methods, and defined silvicultural prescriptions to govern their management. If, on the other hand, ecological restoration of a former ecosystem is the goal, then it is likely that a larger number of native species will be used. In these cases the seed of many species may not be as readily available and much less might be known about how to raise large numbers of seedlings of these species or establish these in the field. At some heavily degraded areas it might be necessary to adopt a two-stage operation and use a tolerant exotic species to first alter the environment (e.g., reduce exposure, improve soil fertility, or lower water tables) before reestablishing the original native species that are not tolerant of these changed site conditions.

Complete restoration of plant diversity by sowing seeds or planting seedlings is rarely possible except

TABLE I

Plant Species That Might Be Chosen for Reforestation Programs

Type of species	Reason
Native species	To restore the original communities once present at the site; where possible the seed of these species should be from local sources
Exotic species	Since only exotics can tolerate the conditions now present at the degraded site or because they are fast growing and provide a commercial return
Commercially useful species	For example, providing timber, fruit, oils, and nuts; financial return needed to justify funding of reforestation
Traditionally or culturally useful species	For example, those providing fuel, food, or medicines used by local communities
Tolerant species	Able to tolerate the adverse conditions likely to be found at degraded sites such as low soil fertility, exposed windy conditions, soil salinity, or occasional fires; in some cases, only exotic species may have the required attributes
Species with fleshy fruit	To attract seed-dispersing frugivores to the site
Poorly dispersed species	Species with large fruit or those with high specialized animal dispersers unable to tolerate the conditions now present at the degraded site
Nitrogen-fixing species	To improve soil nitrogen levels
Fast-growing species	To rapidly occupy the site, shade out weeds, and improve microclimatic conditions
Understory species	To provide the structural complexity and habitat conditions required by some wildlife species
Multipurpose species	Trees that fulfill more than one requirement such as providing timber and food
"Special" species	Species required for some special purpose; e.g., rare or endangered trees requiring enlarged populations, "keystone" species producing food required by wildlife at certain times of the year, etc.

perhaps in the most simple of ecosystems. In most situations plantings can only initiate a new succession and further ecosystem development is dependent on the remaining species colonizing the site from intact forests remaining elsewhere on the landscape (Case Study 1). In these cases particular attention might be paid to facilitating conditions suitable for the return of wildlife since many of these are mutualists and able to foster the seed dispersal needed for future successional development.

Attributes of some species that might make them attractive to establish early in any reforestation program (by direct seeding or as seedlings) are shown in Table I. The actual numbers of species to use are discussed further below.

D. Planting Densities and Species Richness

The choice of planting density and species will be determined by the objectives of reforestation and the resources available. Commercial tree growing in well-watered temperate or tropical locations usually requires moderately close initial spacings (perhaps 1100 trees per hectare) to exclude weeds and ensure rapid site occupancy. Close spacings also reduce side branching and promote rapid height growth. In most commercial plantations a single species is commonly planted. Where ecological restoration is the objective, close spacings might also be used. Again, the dense planting excludes weeds and ensures rapid site occupancy. But high planting densities are costly and many of these seedlings may die as competition develops. In drier areas it may be preferable to use lower density plantings to ensure adequate water is available for each tree as it develops.

1. Plant Density

Low-density plantings may also be useful if large areas of degraded land must be restored but resources are limited. Observations of isolated trees in vacant fields suggest these trees often attract seed-dispersing birds. The birds frequently deposit seeds around the base of the tree that germinate and grow into a cluster of seedlings around each tree. Planting isolated trees or tree clusters to act as bird perches could simulate this process and foster successional development over large areas, albeit at the expense of much slower rates of recovery.

2. Species Richness

The numbers of species that should be included in these plantings can vary. In landscapes where large numbers of forest remnants remain close by (i.e., the dispersal distances are short), only a modest number of species might need to be planted to initiate the succession because natural colonization by other species will take place over time and enrich the planted area. In these circumstances a "dense" planting of a short-lived early colonizer may be sufficient to accelerate the successional process and only a small number of species might need to be used. Likewise, if the number of canopy species in the natural forest is low, it may be possible to replant most of these species and rely on successional development to restore understory diversity. In both cases the recovery process is initiated by planting but is completed by natural successional processes.

A different approach is needed where remnant vegetation is not close by or seed dispersal might be limited. In such cases a much higher proportion of the total plant diversity may need to be brought to the site to achieve any degree of biodiversity within a useful time period. Under these circumstances true restoration (as defined earlier) might be difficult to achieve.

A common difficulty in all of these circumstances is that seed dispersers such as birds may distribute weed species as well as native species. All restoration programs, but particularly those relying on successional development to achieve ecosystem diversity, may therefore need substantial weed control before full site occupancy by the original species is achieved. Even then, tending may be needed for some years until canopy cover is achieved and the new community is less prone to colonization by weeds.

V. ECOSYSTEM RECOVERY AFTER REFORESTATION

Random plantings of a variety of trees will not necessarily restore the ecosystem that was once present at a site. These species must be able to reproduce and regenerate in the formerly degraded site and the animal community must be introduced as well. Some plant species may require particular soil or microclimatic conditions to germinate and grow and will not become established from seed dispersed into a newly forested area until these conditions are met. Other species may need particular regeneration niches to reproduce after they have been planted at a site. For Species A to successfully colonize or reproduce at a site, Species B may need to be established first. In most ecosystems such "rules of assembly" are only poorly understood. What is clear, however, is that many animal species (including insects) are part of this development process because of the mutualistic partnerships they often form with plants that enable processes such as pollination and seed dispersal. Species A (a plant) may require Species B (a bird) for pollination that, in turn, also needs Species C (another plant flowering and supplying food at a different time of the year). Ecosystem restoration is dependent on the formation of these complex trophic processes and interactions.

It is commonly assumed that animals will colonize new plant communities once successional processes have created appropriate habitat conditions. Like plants, animals may be classified as early colonists, tolerant of a broad range of conditions, or later colonists that have more specialized habitat requirements and are only able to enter the succession when it has reached a relatively mature phase. The extent to which animals actually join the new forest community will depend, of course, on the extent to which the appropriate habitat conditions develop. A commercial plantation established using a single exotic tree species is likely to be much less attractive to many wildlife than a spatially and structurally complex new forest of native tree and undergrowth species. Likewise, the structural features of an old-growth forest such as thick litter layers, hollow trees, or rotting logs on the forest floor may take years to develop following restoration. These might need to be artificially simulated to foster the reentry of particular wildlife (e.g., by dragging logs into a restored area or by using nest boxes).

Wildlife from early successional stages are often able to recolonize restored forests relatively quickly. But two factors can limit the extent to which recolonization occurs more generally. Some wildlife, especially those with habitats in mature or old-growth forests, are unable to cross degraded landscapes lacking trees. Restored forests not linked with natural forest remnants may therefore be permanently deprived of such animal species unless some means can be found to overcome the blockage. This topic will be discussed further below. A second limiting factor is that small forest remnants in degraded landscapes often contain only a subset of the original species formerly present. Vertebrates appear to disappear more rapidly from fragments than do plants while top-carnivores and large-bodied animal species in particular have large area requirements, making them especially sensitive to fragmentation. This means that recolonization may be impossible unless these species are brought in from more distant geographic locations. Wildlife recolonization may also be limited by predatory behavior or competitive relations between various animal species. Naturalized exotic species such as foxes or cats represent a particular problem.

Threatened or endangered wildlife species may require rather special habitat conditions. Where the threat to these species is from naturalized exotic predators, these new habitats might be most easily restored on offshore islands, which ensures the exotics are excluded (that is, the "isolation" effect noted above becomes an advantage). However, a dilemma can then emerge over whether the specialized habits required by the endangered species means other species are discriminated against. Is "ecological restoration" the goal or is a more specialized and restrictive objective such as maximizing the population of a particular species more appropriate? Answers to such questions obviously depend on local circumstances. This topic is discussed further below.

All natural ecosystems are subject to disturbances and the well-known intermediate disturbance hypothesis suggests ecosystem diversity reaches an optima when the frequency of disturbances reaches some "intermediate" level. This means that the initial protection offered to reforestation projects to prevent further degradation must be relaxed at some stage to enable normal ecosystem processes to occur. There are few guidelines to describe how this should be carried out. The disturbance regime that produced degradation is different from that fostering diversity but the nature of these differences is rarely understood. For example, at what point can fire be allowed to occur? Indeed, when should fire be actively encouraged? And how intense should this fire be? These issues will not be important in reclamation projects where enhanced productivity is the objective but might be crucial to the ultimate success of many restoration projects.

VI. CHOOSING AREAS TO REFOREST

Reforestation is expensive and takes time. It is rarely possible to quickly reforest large areas at once and strategic planning is needed to identify priority areas where the early benefits from reforestation can be maximized. When reforestation is primarily a reclamation process and is carried out for commercial reasons, the location of a reforestation area will be dictated by markets and soils. The best locations will be those where transport conditions are shortest and where productivity is highest.

When the purpose of reforestation is to restore biodiversity to a degraded site, the choice is more complex. Possible locations are outlined in Table II. Common targets are degraded areas within forested nature reserves or the buffer strips around such reserves. In these areas the task is to restore the ecological integrity of a particular site and hence that of the larger reserve area as well. Since colonization distances are likely to be short, there is a high probability of success if disturbing agents can be removed and weeds eradicated.

Another common target is to establish corridors linking forest fragments still remaining in a formerly forested landscape. The purpose of this is to foster connectivity between forest fragments as well as to increase the area of habitats for species in these landscapes. Some have argued that narrow corridors are more likely to be areas of increased predation rather than a means by which wildlife species movement across the landscape and genetic exchange are enhanced. But others have argued that any increase in spatial heterogeneity should

TABLE II

Priority Areas for Reforestation

Location	Reason
Degraded areas in or around nature reserves	To remove weeds and restore ecological integrity to the reserve
Corridors between forest fragments or reserves	To increase the overall habitat area and to foster spatial heterogeneity across a landscape
Habitats of endangered or threatened species	To increase the populations of these species
Offshore islands	Because islands may be areas free from naturalized weeds or pests that would otherwise preclude restoration on a mainland
Riparian areas	To protect streamsides and because riparian areas are often centers of wildlife activity in many drier landscapes
Upper watersheds	To stabilize catchments, prevent erosion, and improve downstream water quality
Lower watersheds	To lower water tables or reduce salinity
Mine sites or other severely degraded areas	To overcome the extreme degradation that may have occurred and to prevent these areas from becoming erosion sources or sources of toxic leachates
General landscape matrix between forest fragments	To restore biodiversity or productivity across the general landscape wherever it was formerly forested

be beneficial and the advantages would seem to outweigh any disadvantages provided corridor widths are not too narrow.

Key targets for many restoration projects are the habitats of particular species such as those whose populations have declined or are endangered by other predatory species. This presumes, of course, that these habitat requirements are known. In the latter case, offshore islands have been especially useful areas to use because they are of sufficient size to contemplate eradicating such predators and to be assured that recolonization will not occur. Encouraging results have been reported from New Zealand (Case Study 2).

Many reforestation programs are carried out because of the improvements in ecological services rather than because they restore biodiversity. For example, streamsides are often degraded by grazing cattle that cause bank erosion and slumping. Reforestation can stabilize these areas and the vegetation can act as a filter by trapping topsoil erosion. Such reforestation can also have conservation benefits, of course, since these areas are often the focal points within drier landscapes for much wildlife activity. Reforestation of upper or lower watershed areas may be necessary to improve waterlogging or salinity and, again, indirectly improve landscape biodiversity. Some of the most innovative and impressive restoration has been carried out at mine sites. These areas are intensively degraded and can cause serious pollution problems (e.g., heavy metals, acid drainage, salinity) in large areas of the catchment below the mine site even though the size of the actual mine itself may be small. Mine site restoration usually requires careful stockpiling and respreading of topsoils but the large expenses of doing this are commonly only a small percentage of total mine expenses.

VII. SPECIALIZED CONDITIONS OR SITES WITH PARTICULAR PROBLEMS

Some sites have particular problems. This might be because of the degree of degradation suffered (compacted or infertile soils, saline areas, waterlogged sites) or because of the stressful environmental conditions present (e.g., unstable soils, windswept or cold mountain areas, desert areas). A variety of techniques have been developed to deal with many of these conditions, including ripping and plowing, fertilizing, using drains or mounds to allow more tolerant species to become established at waterlogged or saline sites, etc. In some cases mycorrhiza and soil fauna such as earthworms must be reintroduced as well as plants to promote nutrient cycling. Sites in extreme environments such as mountainous or exposed areas are particularly difficult. Sometimes tolerant exotic species can be planted to facilitate the later establishment of native species. But the survival rates of planted seedlings in especially severe environments can be very low. This raises the question of whether it is worth even attempting to restore severely degraded ecosystems or whether a better return on the resources to be invested might come from restoring only "moderately" degraded landscapes? In such cases the balance between what is biologically possible is confronted by what is economically feasible.

VIII. SOCIOECONOMIC ISSUES

Ecological restoration can be difficult because it requires detailed knowledge of the autecology of a number of species and of the processes of successional development. But it is frequently less complex than the socioeconomic issues surrounding many reforestation projects. Degradation is commonly caused by some kind of socioeconomic problem and it can only be overcome if these problems are also dealt with. That is, successful reforestation projects invariably involve resolving problems concerning ecological, social, and economic systems. Our understanding of the interactions between these three systems is usually poor.

Many reforestation projects directly or indirectly involve a number of stakeholders. These might be landowners, leaseholders, or people who have some tradition claim to the land although that claim is not recognized by the state. Others may be land users downstream from the land or simply the general public. As noted earlier, these various groups may view degradation in quite different ways and many may not want change because they perceive themselves being disadvantaged by the change (e.g., they lose access to grazing land or food-producing areas). The planning stage of reforestation projects must resolve these differences if a project is to succeed. Experience in many places suggests this is best done by involving all stakeholders in the planning and management process, including those in a weak bargaining position and those with little political power. In this way, all interests can be protected from the beginning. Some form of educational program may also be necessary to allow the various stakeholders to see the reasons for the project and the benefits it may have for themselves (e.g., perhaps in employment opportunities) as well as for the community as a whole. If these collaborative interactions do not take place or

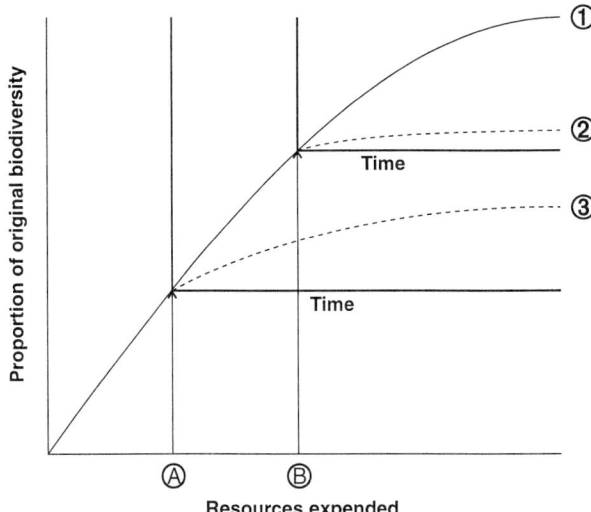

FIGURE 3 The degree to which biodiversity is recovered can vary with the resources invested. Unlimited resources may regain most of the original biodiversity (1), while a small resource expenditure (A) is likely to restore a smaller proportion of the original biodiversity than a larger expenditure (B). Further recovery may take place over time (i.e., time substitutes for resources invested on the x axis). Further recovery may be small (2) or larger (3), depending on the extent of subsequent recolonization from nearby intact forest.

the concerns of stakeholders are not resolved, there may be a high risk of continued disturbances and further degradation (Case Study 3).

A particular problem is cost. In most situations there is a resources–response relationship of the kind shown in Fig. 3. Complete restoration usually requires substantial financial and other resources and the more resources that can be utilized, the greater the degree of restoration that might be achieved and the faster this is likely to take place. For example, restoration is obviously faster if more species are planted at the commencement rather than waiting for recolonization to occur. Likewise, planted seedlings are more likely to survive if weed control is carried out than if it is not. However, these large resources are rarely available, except, perhaps, in some postmining restoration projects. In such cases species recovery becomes dependent on time and this replaces resources on the x axis of Fig. 3 (at point A if limited resources are available or point B if more are available). But time may be an incomplete substitute because nearby remnants do not contain the species needed or because colonization is impeded. In those cases a much lower level of "restoration" is achieved (with little further recruitment producing outcome 2 or only modest levels producing outcome 3 rather than the complete restoration of outcome 1). These several factors mean that complete ecological restoration is a difficult task to achieve and that sometimes more modest goals might be all that is possible, at least in the short term.

All of this raises the question of how to finance the restoration of diversity over large areas. There is no doubt that promising techniques are now available to accelerate restoration over small areas. How can these methods be expanded to cover the large areas of degraded land now being created each year? One way of doing this might be to modify other commercially attractive forms of land use such as timber plantations or agroforestry and use these to enhance biological diversity. Most commercial timber plantations, for example, are single-species monocultures that contribute little to landscape biodiversity. But these might be broken up by a network of species-rich buffer strips (natural or restored). Such strips would have benefits as firebreaks or as streamside filters and they could substantially increase landscape heterogeneity. Or the plantation might use more than one species planted in a mosaic of monocultures across the landscape, with each species being targeted to fit a particular site type (e.g., plantings would vary with soil type, slope position, or aspect). Or species mixtures might be planted instead of monocultures. Or, finally, advantage might be taken of the fact that many plantations acquire a diverse understory over time, especially if intact forest is nearby (Fig. 4). The value of such plantations as wildlife habitats is likely to be substantially enhanced by this understory.

In the same way, agroforestry also has the capacity to improve regional biodiversity while providing eco-

FIGURE 4 Native tree seedling colonization in the understory of a tropical hardwood plantation monoculture. The plantation is within 200 m of natural forest and its presence has catalysed the recovery of substantial plant biodiversity at the site.

nomic benefits to land users. The traditional home gardens of Indonesia and Sri Lanka, for example, are well-known agricultural systems utilizing a very rich mix of species. Perhaps even more striking are the man-made forests in parts of Indonesia based around trees that produce fruit, gums, spices, and timbers. These are managed for the tree products plus a variety of agricultural goods but they also contain a substantial degree of indigenous plant and animal biodiversity.

Neither timber plantations nor agroforestry will achieve complete ecological restoration. In the case of the timber plantations, this biodiversity is temporarily lost when each plantation unit is harvested. But both situations can potentially recover some of the former biodiversity across large areas of landscape and reverse the trend toward biological simplification that forest plantations and agriculture usually generate. And they do this while improving the socioeconomic conditions of the land users.

Much more work needs to be done to explore new forms of reforestation and to redesign timber plantations and agroforestry systems to suit the various ecological and socioeconomic situations present in the increasing areas of degraded forest landscape likely to be found in the future.

A. Case Study 1: Reforestation after Mining Tropical Forest Lands in Brazil

Mining causes complete degradation of an ecosystem and perhaps represents one of the greatest challenges to those interested in restoration, particularly if the site was previously occupied by a biologically rich tropical rain forest. The Trombetas bauxite mine in Amazonia has an annual reforestation program of 100 ha. Following mining, the sites are leveled and stockpiled topsoil is spread back over the area to a depth of 15 cm. The site is then ripped and replanted using seed, stumps, and seedlings. About 70 species are planted at a density of around 2500 trees per hectare.

Studies carried out in an area that had been replanted 10 years earlier found forest development had been rapid. A number of plant species had colonized the site from the adjoining rain forest. After 10 years these represented 60% of the total number of woody species present and the site contained around 50% of the species present in the adjoining, undisturbed forest. Some areas had rather sparser forest cover than others and the colonization process was less vigorous in these, most probably because of the presence of grasses in the understory. The density and composition of the understory developing beneath the planted trees were related to the overstory basal area and topsoil depth but negatively related to the distance from the undisturbed forest. Primary forest species were able to colonize up to 640 m although most colonists had smaller diaspores. Most of the colonists in the more open sites with grass were early pioneer species.

Surveys of wildlife found low densities of birds but high densities of bats. Most of the bird species were representative of secondary forests and are unlikely to have carried many seeds from the undisturbed forest nearby. On the other hand, the bats included several important seed dispersers which feed on a variety of tree species and are able to carry seed over large distances. A number of other mammal species (e.g., opossums, deer, agouti, and armadillo) were also observed within the study site.

It appears that a dynamic and sustainable successional environment has been established in a much shorter time than it would otherwise take to occur. This was due to the respreading of topsoil which contained a number of important tree species and the choice of trees planted in the original planting program. The program was less successful where tree survival or growth allowed grasses to become established and at these sites the successional process may be rather longer.

Further reading: Parrotta *et al.* (1997)

B. Case Study 2: Restoration of Habitats of Endangered Biota in New Zealand

The island (or minicontinent) of New Zealand has a unique biota but many of these have become endangered since the arrival of humans. In the case of birds, for example, 48% of species have been lost in the past 200 years because of the introduction of predatory and browsing mammals. Even on protected areas, attempts to maintain populations of some presently endangered species are difficult because of the difficulty in controlling predatory species over large areas. New Zealand has many offshore islands that were once part of the mainland (and consequently contain many mainland species). Although a number have been deforested or degraded, they are potential safe havens for endangered species since it may be feasible to eradicate predators and browsers over these smaller and more isolated locations. This has fostered a shift in attitude from trying to simply preserve species to restoring their habitats.

The first step has been to mount a vigorous campaign to eradicate exotic species such as rodents, cats, goats, or cattle from the islands. While the larger grazing animals are comparatively easy to remove, it has only been in more recent years that techniques for eradicating rodents have become successful. In many situations this alone has led to improvements in habitat quality although the responses of many resident species is often unpredictable. On islands where deforestation has been severe, it has been necessary to replant trees and shrubs to reestablish appropriate habitat conditions. Similarly, translocations have been necessary in the case of some animals to establish new breeding populations.

The objective of this program has been quite specific; it is not to necessarily restore the original ecosystems (because of difficulties in deciding which previous time period to target) but rather to seek to reintroduce sufficient elements of the original ecosystem to enable survival of nominated species, communities, or ecosystems. This goal acknowledges that restoration may never be complete because some exotic organism may always be present or because knowledge of the system may always be incomplete. In the New Zealand context, however, the goal is a way of ensuring the protection of a significant proportion of the country's biological heritage.

Further Reading: Towns and Ballantine (1993)

C. Case Study 3: Reforestation of Degraded Lands in Nepal

Forests are an integral part of the farming system in Nepal, with leaves and grass being used for fodder and animal bedding. As well, such material is also used in compost to maintain productivity of agricultural lands. Timber is regularly harvested for fuel.

Some parts of the middle hills of Nepal have been highly degraded over time because of intensive harvesting. However, externally sponsored attempts to encourage land users to reforest have not always been successful. If there is forest near villages, there is commonly little interest in forest protection or tree planting although indigenous management systems (primarily to define user access rights) for these forests may exist. By contrast, where there is a severe shortage of forest products and any remaining forests are more distant, there is likely to be genuine interest in forest development activities and well-developed indigenous management systems for the remaining forests. These systems are likely to have biological objectives (e.g., rotational cutting) as well as simply defining user access rights. That is, interest in reforestation depends on the "perceived need" of land users and not on an external perception that forest degradation has occurred.

Indigenous management systems can foster reforestation under certain conditions and some areas of private farmland in Nepal now have more forest cover than in earlier years. However, this does not invariably occur. For example, larger trees may be protected but grazing beneath these may lead to a failure of regeneration. Likewise cutting of smaller trees for firewood may be tolerated while the larger trees are saved. A particularly difficult situation can be where a cash market for forest products develops. In such cases poor people can succumb to the temptation to cut firewood, often at night, to sell the next day. Indigenous management systems must be very strong to resist such pressures.

Tree planting has been carried out at particularly degraded sites. The most successful reforestation has been with pines, although broad-leaved species are more valued by villagers as leaf fodder. Manipulative trials with young pine plantations have found that many broad-leaved species can also then regenerate from suppressed stumps and root stocks or seed once the sites are protected. This natural successional development increases the silvicultural options available, giving local communities greater flexibility in determining their preferred forest management strategy. It is also possible to thin the young pine stand and enrich it using other broad-leaved species that might otherwise be difficult to establish. This, too, increases the options available.

Reforestation of degraded sites and the development of indigenous management systems may not always take place without some outside intervention. Once benefits start to accrue to land users, however, neighbors are often quick to pick up the ideas and follow the example. At this point it is important that government agencies who might legally "own" the land are cautious in imposing rigid bureaucratic controls within which the new forests must fit. For example, allowing someone from outside the user group to harvest resources from the new forest could easily unleash a splurge of unregulated harvesting, leading to renewed degradation.

Further reading: Gilmour (1990), Gilmour et al. (1990)

See Also the Following Articles

DEFORESTATION AND LAND CLEARING • FOREST ECOLOGY • LAND-USE ISSUES • RESTORATION OF BIODIVERSITY, OVERVIEW • TIMBER INDUSTRY • TROPICAL ECOSYSTEMS

Bibliography

Banerjee, A. K. (1995). *Rehabilitation of Degraded Forests in Asia*, World Bank Technical Paper No. 270. World Bank, Washington, D.C.

Bradshaw, A. D., and Chadwick, M. J. (1980). *The Restoration of*

Land: The Ecology and Reclamation of Derelict and Degraded Land. Blackwell Sci., Oxford.

FAO (1995). *Forest Resources Assessment 1990: Global Synthesis*. Food and Agriculture Organization of the United Nations.

Gilmour, D. A. (1990). Resource availability and indigenous management systems in Nepal. *Soc. Nat. Res.* 3, 145–158.

Gilmour, D. A., King, G. C., Applegate, G. B., and Mohns, B. (1990). Silviculture of plantation forest in central Nepal to maximise community benefits. *For. Ecol. Manage.* 32, 173–186.

Jordan, W. R., Gilpin, M. E., and Aber, J. D. (Eds.) (1987). *Restoration Ecology: A Synthetic Approach to Ecological Research*. Cambridge Univ. Press, Cambridge, UK.

Lamb, D. (1998). Large-scale ecological restoration of degraded tropical forested lands: The potential role of timber plantations. *Restor. Ecol.* 6, 271–279.

Luken, J. O. (1990). *Directing Ecological Successions*. Chapman & Hall, London.

Majer, J. D. (Ed.) (1989). *Animals in Primary Succession: The Role of Fauna in Reclaimed Lands*. Cambridge Univ. Press, Cambridge, UK.

Munshower, F. F. (1994). *Practical Handbook of Disturbed Land Revegetation*. CRC Press, Boca Raton, FL.

Parrotta, J. A., Knowles, O. H., and Wunderlie, J. M. (1997). Development of floristic diversity in 10 year old restoration forests on a bauxite mined site in Amazonia. *For. Ecol. Manage.* 99, 21–42.

Saunders, D. A., Hobbs, R. J., and Ehrlich, P. R. (Eds.) (1993). *Reconstruction of Fragmented Ecosystems: Global and Regional Perspectives*. Surrey Beatty, Chipping Norton, Sydney.

Society for Ecological Restoration Home page at http://nabalu.flas.ufl.edu/ser/SERhome.html.

Towns, D. R., and Ballantine, W. J. (1993). Conservation and restoration of New Zealand island ecosystems. *Trends Ecol. Evol.* 8, 452–457.

RELIGIOUS TRADITIONS AND BIODIVERSITY

Fikret Berkes
University of Manitoba

I. The Context: Religion for Encoding Ethics
II. Monotheistic Traditions
III. Asian Traditions
IV. Pantheistic Traditions
V. Conclusions

GLOSSARY

animism Belief in spiritual beings. The term is associated with the anthropology of E. B. Tylor, who described the origin of religion and primitive beliefs in terms of animism in *Primitive Culture* (1871). Tylor considered animism a minimum definition of religion and asserted that all religions, from the simplest to the most complex, involve some form of animism. From Latin *anima*, "breath" or "soul."

ethics Codes that exert a palpable influence on human behavior. Embedded in worldviews, ethics provide models to emulate, goals to strive for, and norms by which to evaluate actual behavior.

monotheism Belief in the unity of the Godhead, or in one God, as opposed to pantheism and polytheism. Monotheism is a firm tenet of Islam and Judaism. Christianity, with its concept of Trinity, alone among the three monotheistic religions, dilutes monotheism. From Greek *mono*, "one," and Greek *theos*, "god."

pantheism The doctrine that identifies the universe with God. In Western thought, the term is associated with the Dutch philosopher Baruch Spinoza. His view represents an important criticism of the "orthodox" view of a god whose reality is somehow external to the reality of the world. From Greek *pan*, "all," and Greek *theos*, "god."

religion Human recognition of superhuman controlling power, and especially of a personal God or gods entitled to obedience and worship; the effect of such recognition on conduct or mental attitude.

traditional ecological knowledge A cumulative body of knowledge, practice, and belief, evolving by adaptive processes and handed down through generations by cultural transmission, about the relationship of living beings (including humans) with one another and with their environment. A subset of indigenous knowledge, which is local knowledge held by indigenous peoples or local knowledge unique to a given culture or society.

traditional societies Groups in which knowledge, practice, and belief are handed down through generations largely by cultural transmission. Tradition itself evolves by adaptive processes, but not all tradition is necessarily adaptive.

worldview The larger conceptual complex in which ethics are embedded. A. N. Whitehead called it the conceptual order, or one's general way of conceiving the universe, which supplies the concepts by which one's observations of nature are invariably interpreted. In general, world-views limit and inspire human behavior, shape observations, and perceptions. A. Toynbee's *Weltanschauung*.

THIS ARTICLE IS about religions and attitudes toward the natural environment as relevant to biodiversity conservation. Religious traditions have little to say specifically about biodiversity, but they provide the values, worldviews, or environmental ethics that shape the way in which different societies interact with biological diversity and nature in general. In this sense, religion can be part of the problem or part of the solution. Roy Rappoport suggests a general theory of religion for encoding information and for involving human emotions. Thus, religion can encode adaptive strategies for resource management and biodiversity use, and supply emotionally powerful beliefs to put these strategies into practice. The anthropologist Eugene Anderson observes that all traditional societies that have succeeded in keeping their resources productive over time have done so in part through religious or ritual representation of resource management. The key point, he says, is not religion per se, but the use of emotionally powerful cultural symbols to help maintain a sense of sacred respect.

I. THE CONTEXT: RELIGION FOR ENCODING ETHICS

Different groups of people in different parts of the world perceive and value sacredness differently. Religion is a general term that has become established in the Western world for a whole class of cultural codes and rituals. In the West, there often is a distinction between a person's religion and environmental ethics. In many other societies, religion is part of a way of life as well as a worldview. For example, in Chinese there is no one world that translates as "religion" in the Western sense. In some traditional societies, including many Amerindian groups, religion, worldview, and environmental ethics and practice are inseparable. Therefore, it is necessary to define religion in its broader sense of "human recognition of superhuman controlling power," inclusive of both monotheistic and pantheistic traditions.

Religions provide a central organizing myth and include cultural symbols for a moral code. Conceived that way, religions can be thought to include a wider variety of beliefs, within the definition of a superhuman controlling power. For example, Rodney Dobell refers to the "religion of the market." Peter Timmerman uses the term "econotheism" as the adoption of market economics "to bring the rest of the earth under new management." Econotheism, argues Timmerman, "contains within it all the elements of a religion: a priesthood, scriptures, catechisms (Econ 100 textbooks), an explanation of happiness and misery, a way to get to heaven, and a core model of the human."

Anderson holds that religion is best regarded as something providing an emotionally powerful way to "sell" a moral code. The content of the moral code is negotiable and variable; it is not in itself a part of the religion. One can have a moral code without a religion, as in secular humanism, but at the risk of losing the emotional content supplied by religion.

Religion is not a particularly popular topic among those involved in biodiversity conservation. The index of the 1140-page *Global Biodiversity Assessment* contains a grand total of only two minor entries on religion. One problem with religion is that it has often been bent to serve all possible ends, including those destructive of biodiversity and cultural diversity. Thus, the role of religion in biodiversity conservation, and more generally in environmental ethics, is controversial.

There are several cautions in interpreting the discussion on religion and biodiversity. First, generalizations are always risky. As Donald Worster put it in *The Ends of the Earth*, "not a few scholars have fallen into the trap of speaking of 'the Buddhist view of nature' or 'the Christian view' or 'the American Indian view,' as though people in those cultures were all simple-minded, uncomplicated, unanimous, and totally lacking in ambivalence. Every culture, we should assume, has within it a range of perceptions and values, and no culture has ever really wanted to live in total harmony with its surroundings."

Second, there is invariably a gap between "the ideal" and "the actual" in making sense of how societies deal with biodiversity. Ethics do not describe how people actually behave but rather set out how people *ought* to behave. There often is a discrepancy between belief and practice. Some scholars argue that one could spend too much time on values and beliefs, neglecting to examine behavior or the actual effects of these values on human practices regarding the environment. The field of environmental history studies the historical record of changes in the landscape as a way of evaluating the actual effects of the belief-and-practice complex of a culture.

A third caution is the intangible nature of traditions and their record in the literature. Not only do traditions change and adapt all the time, but the literature is based largely on outsiders' accounts of various religious traditions, especially those that do not have a written record. Since traditional societies are, by definition, those cultures in which much knowledge is transmitted orally, available documentation on such groups is almost always suspect. This is because insiders do not normally record their teachings in writing, and outsiders do not always interpret these teachings correctly and tend to add their own interpretations, as in *Black Elk Speaks.*

II. MONOTHEISTIC TRADITIONS

A. Debates over the Judeo-Christian Tradition

The controversy about environmental values in Christianity and Judaism has centered on the appropriate interpretation of the relationship between God, "man," and nature as set out in the Book of Genesis 1:26–28:

> 26 And God said, Let us make man in our image, after our likeness: and let them have dominion over the fish of the sea, and over the fowl of the air, and over the cattle, and over all the earth, and over every creeping thing that creepeth upon the earth.
>
> 27 So God created man in His own image, in the image of God created He him; male and female created He them.
>
> 28 And God blessed them, and God said unto them, Be fruitful, and multiply, and replenish the earth, and subdue it: and have dominion over the fish of the sea, and over the fowl of the air, and over every living thing that moveth upon the earth.

There are two widely held interpretations of these relationships. The "mastery" interpretation is that Genesis gives humans a unique status among the species by virtue of creation in the image of God, and awards humans a God-given right to exploit nature without moral restraint, except where it affects human welfare. God, according to this view, seems to have intended humans to be his viceroy on earth. Humans are to the rest of earth's species as God is to humans. Humans are given dominion over the earth, and are expressly asked to subdue (Hebrew *kabas*, stamp down) the earth, as if nature needed humans to put it in order.

The "stewardship" interpretation also relies on the verses above, as well others before and after it, and emphasizes responsibility. Since humans are created in the image of God, they have not only rights and privileges but also special duties and responsibilties. Foremost among these is the human duty to rule the dominion of nature wisely. To degrade nature and destroy God's other species would violate the trust God has placed in his viceroy. Thus, far from giving a free hand to exploit and degrade nature, according to this view, God expects humans to exercise stewardship in the wise use of nature.

The five Declarations of Assisi are the addresses by religious leaders to their own faithful in the Buddhist, Christian, Hindu, Jewish, and Moslem worlds. The Declarations were the highlight of an interfaith ceremoney at the World Wildlife Fund's 25th anniversary celebrations (*WWF News*, Nov/Dec 1986). The three monotheistic traditions represented in Boxes 1, 2, and 3 (Christian, Jewish, and Moslem) make it clear that all three strongly support the stewardship interpretation of Genesis. It should be noted that Islam's holy book, the Qur'an, is similar in content to the Old Testament, and gives humans dominion over nature, but at the same

---

Box 1

Excerpts from the Five Declarations of Assisi: The Christian Declaration by Father Lanfranco Serrini

Because of the responsibilities which flow from his dual citizenship, man's dominion cannot be understood as license to abuse, spoil, squander or destroy what God has made to manifest his glory. That dominion cannot be anything else than a stewardship in symbiosis with all creatures. On the other hand, his self-mastery in symbiosis with creation must manifest the Lord's exclusive and absolute dominion over everything, over man and over his stewardship. At the risk of destroying himself, man may not reduce to chaos or disorder, or, worse still, destroy God's bountiful treasures.

For St. Francis, work was a God-given grace to be exercised in that spirit of faith and devotion to which every temporal consideration must be subordinate: uncontrolled use of technology for immediate economic growth, with little or no consideration for the planet's resources and their possible renewal; disregard for just and peaceful relations among peoples; destruction of cultures and environments during war; ill-considered exploitation of natural resources by consumer-oriented societies; unmastered and unregulated urbanization; and, the exclusive preoccupation with the present without any regard for the future quality of life.

> **Box 2**
>
> **Excerpts from the Five Declarations of Assisi: The Jewish Declaration by Rabbi Arthur Hertzberg**
>
> The encounter of God and man in nature is conceived in Judaism as a seamless web, with man as the leader, and custodian, of the natural world. Even in the many centuries when Jews were most involved in their own immediate dangers and destiny, this universalist concern has never withered. . . Now, when the whole world is in peril, when the environment is in danger of being poisoned, and various species, both plant and animal, are becoming extinct, it is our Jewish responsibility to put the defence of the whole of nature at the very centre of our concern.
>
> . . . Man was given dominion over nature, but he was commanded to behave towards the rest of creation with justice and compassion. Man lives, always, in tension between his power and the limits set by conscience.
>
> Our ancestor Abraham inherited his passion for nature from Adam. The later rabbis never forgot it. Some 20 centuries ago they told they story of two men who were out on the water in a rowboat. Suddenly, one of them started to saw under his feet. He maintained that it was his right to do whatever he wished with the place which belonged to him. The other answered him that they were in the rowboat together—the hole that he was making would sink both of them (Vayikra Rabbah 4:6).

> **Box 3**
>
> **Excerpts from the Five Declarations of Assisi: The Moslem Declaration by Dr. Abdullah Omar Nasseef**
>
> Unity, trusteeship and accountability, that is *tawheed, khalifa and akhrah*, the three central concepts of Islam, are also the pillars of the environmental ethics of Islam. They constitute the basic values taught by the Qur'an. It is these values which led Muhamad, the Prophet of Islam to say: "Whoever plants a tree and diligently looks after it until it matures and bears fruit is rewarded," and "If a Moslem plants a tree or sows a field and men and beasts and birds eat from it, all of it is charity on his part," and again. "The world is green and beautiful and God has appointed you his stewards over it." Environmental consciousness is born when such values are adopted and become an intrinsic part of our mental and physical makeup.
>
> Moslems need to return to this nexus of values, this way of understanding themselves and their environment. The notions of unity, trusteeship and accountability should not be reduced to matters of personal piety; they must guide all aspects of their life and work. Shariah should not be relegated just to issues of crime and punishment, it must also become the vanguard for environmental legislation. We often say that Islam is a complete way of life, by which it is meant that our ethical systems provides the bearings for all our actions.

time, as the declaration notes, vests in them trusteeship and accountability.

B. Searching for a New Religious Base for Environmental Ethics

The debates over the interpretations have been going on since the 1970s. Many environmentalists do not see monotheistic stewardship as a sufficient solution. The critics include the historian Lynn White, Jr., who started the debate over Judeo-Christian environmental ethics with his 1967 paper in *Science*, "The historical roots of our ecologic crisis." According to White, Christianity is the most anthropocentric (human-centered) religion the world has ever seen, especially in its Western form. In pre-Christian times

> every tree, every spring, every stream, every hill had its own *genius loci*, its guardian spirit. These spirits were accessible to men, but were very unlike men; centaurs, fauns, and mermaids show their ambivalence. Before one cut a tree, mined a mountain, or dammed a brook, it was important to placate the spirit in charge of that particular situation, and to keep it placated. By destroying pagan animism, Christianity made it possible to exploit nature in a mood of indifference to the feelings of natural objects.

Since the roots of the enviromental crisis "are so largely religious." White continues, "the remedy must also be essentially religious." He briefly entertains and then rejects Zen Buddhism as a solution: "Zen is as deeply conditioned by Asian history as Christianity is by the experience of the west, and I am dubious of its viability among us." Seeking instead an alternative Christian view, White proposes Saint Francis of Assisi as a patron saint for ecologists. "The key to an understanding of Francis is his belief in the virtue of humility—not merely for the individual but for man as a species. Francis tried to depose man from his monarchy over creation and set up a democracy of all of God's creatures."

White may be taken to recommend a Franciscan theology as the basis for a Christian environmental ethic. Others have provided other Christian theologies. For example, Rene Dubos in *A God Within* recommends St. Benedict, as the Benedictines were the original environmental managers of Europe who drained swamps and made the countryside both productive and beautiful. The theologian John B. Cobb, Jr., casts his net wider, beyond European cultures, in looking for a religious solution in *Is It Too Late? A Theology for Ecology*.

Accepting White's view that Christianity is largely responsible for the environmental crisis, Cobb examines in particular American Indian and Chinese worldviews. He rejects the American Indian view because, he thinks, Indians did not respect human life sufficiently and because the Indian way of life cannot support the existing North American population. Turning to the Chinese tradition (represented by Taoism), Cobb argues that it was unable to prevent deforestation and other ecological ills. He concludes that it is more prudent for the West to fix the Western tradition than to find a non-Western alternative. Rejecting Francis as too radical (Saint Francis had also preached poverty to the Catholic Church!), Cobb finally opts for Albert Schweitzer's Christianity and his reverence-for-life ethic.

The historian Arnold Toynbee differs from the above-mentioned commentators in that he sees no fundamental problem in developing a new environmental ethic that can be integrated with Asian traditions. As a classically educated scholar, he sees a historical continuity from pantheistic traditions ("the original religion of all mankind") to the Graeco-Roman religion and to Asian religions.

> At the western end of the Old World, this nature-worship, which is the original religion of all mankind, has been overlaid by an opaque veneer of Christianity and Islam; but, when a native of the monotheistic portion of the present-day World travels eastward beyond the easternmost limits of Islam, he finds himself in a living premonotheistic World, and, if he has had a Greek and Latin education which he has taken seriously, the religion of present-day East Asia will be more familiar to him.

Monotheism is exceptional among the religions of the world in its doctrine about subduing nature, says Toynbee. As enunciated in the Book of Genesis, monotheism has removed the sense of awe and, with it, the age-old restraint that was once an effective check on human greed. His analysis does not so much negate the stewardship interpretation as makes it irrelevant. The real solution to the crisis, in his thinking, has to do with restoring a sense of sacred respect for the earth and its creatures. For Toynbee, the remedy lies in reverting from a worldview of monotheism to a worldview of pantheism.

Of course, the historical roots of the ecological crisis are not only found in the area of religion. A complex of social changes related to the Age of Enlightenment, the Industrial Era, and the commodification of nature has been implicated, as part of what Karl Polanyi calls the *Great Transformation*. A major factor, according to C. J. Glacken (*Traces on the Rhodian Shore*) and many others, is the Age of Enlightenment concept of an external environment analytically separate from humans. Many scholars, including Gregory Bateson (*Mind and Nature*), hold that this separation is the basis of the Cartesian dualism of mind versus matter, and hence humans versus the environment.

There is no agreement among experts regarding the relative importance of the emergence of monotheistic traditions in the West vs that of the Enlightenment philosophy and the great transformation many centuries later. The change in values accompanying the former was no doubt complemented and reinforced by the latter. Together, these two waves of change left the Western world with a fundamentally altered relationship with nature. The sociologist Max Weber sees modern society as having become "disenchanted" with the world, literally losing spiritual enchantment with life, whereas for pre-monotheistic people, nature was not just a treasure trove of natural resources; nature was a goddess and the whole of the environment was sacred.

III. ASIAN TRADITIONS

The religious traditions of South Asia and East Asia, with their many gods and non-dominant relationship to nature, are similar in many ways to the premonotheistic traditions of Europe, and fundamentally different from the monotheistic tradition, as Toynbee points out. "Confucianism and Taoism and Shinto, like the pre-Christian Greek cults of the corn-goddess Demeter (Ceres in Latin) and the wine-god Dionysus, counsel man to respect nature even when he is applying his human science to coax nature into bestowing her bounty on man." Elements of this thinking can be seen in the Buddhist and Hindu declarations in Assisi. The statements, "we should be wary of justifying the right of any species to survive solely on the basis of its usefulness to human beings" (Box 4) and "the divine is not exterior to creation, but expresses itself through natural phenomena" (Box 5) signal a worldview that is fundamentally different from the monotheistic one.

Teachings of Hinduism can be traced in the environmental literature on species preservation and animal rights. The philosopher Arne Naess adopted the concepts of "identification" and "self-realization" from Hindu thought and used them as central ideas in Deep Ecology. Zen Buddhism (a form of Buddhism) is one of the few major Asian religions to put firm roots in contemporary Western culture through the writings of D. T. Suzuki, Alan Watts, Peter Timmerman, and the poet Gary Snyder. However, "pop Zen" or "West Coast Zen," as Callicott and Ames call it, has been used to

---

Box 4

Excerpts from the Five Declarations of Assisi: The Buddhist Declaration by The Venerable Lungrig Nmgyal

Buddhism is a religion of love, understanding and compassion and is committed towards the ideal of non-violence. As such it also attaches great importance towards wildlife and the protection of the environment on which every being in this world depends for survival. The underlying reason why beings other than humans need to be taken into account is that, like human beings, they too are sensitive to happiness and suffering. We should therefore be wary of justifying the right of any species to survive solely on the basis of its usefulness to human beings.

We are told that history is a record of human society in the past. From existing sources there is evidence to suggest that for all their limitations, people in the past were aware of this need for harmony between human beings and nature. They loved the environment. They revered it as the source of life and well-being in the world.

We regard our survival as an undeniable right; as coinhabitants of this planet, other species too have this right for survival. And since human beings as well as other non-human sentient beings depend upon the environment as the ultimate source of life and well-being, let us share the conviction that the conservation of the environment, the restoration of the imbalance caused by our negligence in the past, be implemented with courage and determination.

Box 5

Excerpts from the Five Declarations of Assisi: The Hindu Declaration by Dr. Karan Singh

Not only in the Vedas, but in later scriptures such as the Upanishads, the Puranas and subsequent texts, the Hindu view-point on nature has been clearly enunciated. It is permeated by a reverence for life, and an awareness that the great forces of nature—the Earth, the sky, the air, the water and fire—as well as various orders of life including plants and trees, forests and animals, are all bound to each other within the great rhythms of nature. The divine is not exterior to creation, but expresses itself through natural phenomena.

The Yajurveda lays down that "no person should kill animals helpful to all. Rather, by serving them, one should attain happiness." This view was later developed by the great Jain Tirthankara, Lord Mahmavira, who regenerated the ancient Jain faith that lives down to the present day. For the Jains, Ahimsa, or non-violence, is the greatest good, and on no account should life be taken. This philosophy was emphasized more recently by Mahatma Gandhi, who always spoke of the importance of Ahimsa and looked upon the cow as a symbol of the benign element in animal life. All this strengthens the attitude of reverence for all life including animals and insects. The Hindu tradition of reverence for nature and all forms of life, vegetable or animal, represents a powerful tradition which needs to be renurtured and re-applied in our contemporary context. India, the population of which is over 80% Hindu, has in recent years taken a special interest in conservation.

reinforce selfish individualism already endemic in Western culture, instead of the self-discipline, simplicity, and enlightenment of true Zen.

Buddhism has also influenced some environmental writing. The most radical chapter in E. F. Schumacher's *Small Is Beautiful* is entitled "Buddhist economics." West is rich in technological and economic means, Schumacher points out, but it has little clue about the *ends* that are most appropriate for the use of those means. Buddhist economics proposes the solution that appropriate means should be used to achieve appropriate ends. The discovery of these appropriate ends, in turn, can be guided through our place as caretakers of the world around us, and the appropriate means will then become obvious.

Taoism, one of the Chinese religious traditions, considers nature through the concept of *tao*, the way, and emphasizes living in harmony with nature of "flowing with nature." Taoism has influenced Western culture and environmental thinking through such books as *The Tao of Physics* by Fritjof Capra. In a Taoist story told by Timmerman, a philosopher falls asleep and dreams that he is a butterfly. From that moment on, he is never sure if he is a philosopher dreaming that he is a butterfly, or a butterfly dreaming that he is a philosopher. In Taoist thinking, the crucial point is that there is no right answer: the moment an answer is attempted, the original symmetry is broken and one crystallizes out as a philosopher or a butterfly—but not both. Capra, a physicist, argues that quantum physics works in much the same way as electrons move out of a probability sphere into determined existence.

Romantic notions of Asian religions are debunked by the geographer Yi-Fu Tuan, who pointed out the discrepancies between attitude and behavior. "Western humanists commonly show bias in favor of Taoist and Buddhist traditions," says Tuan. But the people of China, through their long history, have transformed and degraded the landscape in a major way. Specifically, Taoism, with its vaguely anti-urban and anti-humanistic stance, according to Tuan, represents little more than a hermit's point of view in the larger scheme of environmental destruction. Tuan's exposition of episode after episode of land abuse by the ancient Chinese, long before they were influenced by the West, leads to the conclusion that Chinese religions are unlikely to provide a solution for all. However, this is not to deny that there existed a tradition of natural philosophy in China consistent with contemporary ecological ideals. Rather, Tuan's critique points out that there often exist "glaring contradictions of professed ideal and actual practice."

IV. PANTHEISTIC TRADITIONS

A. Traditional Peoples, Indigenous Knowledge

Some traditional societies retain elements of pantheism, the "original religion" of all humankind. In these societies, religion, worldview, and environmental practice are often intertwined. Religious sanctions may be invoked in two ways in direct support of biodiversity conservation: through the prohibition of areas or of species. "Sacred groves" or sacred forests occur throughout the world, especially in India, Indonesia, South America, and parts of Africa (Box 6). Even small sacred groves may be surprisingly effective in conserving biodiversity. A botanical survey in a Nigerian sacred grove yielded 330 plant species as compared to only 23 in surrounding non-protected areas (Warren and Pinkston in *Linking Social and Ecological Systems*). In the case of species taboos (the word is borrowed from the Polynesian *tabu*), a study by Colding and Folke showed that about 30% of taboos identified worldwide prohibited the use of species that also happened to be listed as threatened by the IUCN *Red Data Book*.

Traditional views may also help conserve biodiversity indirectly through the emotional involvement of people with their land and living things. This may entail, as in some American Indian traditions, a community-of-beings worldview in which animals and plants are considered persons—not human persons but persons nonetheless. Land may also provide a strong sense of place or sense of identity for a social group, as seen with American Indians and Inuit, Southeast Asian indigenous peoples, and cultures in Oceania.

The report *Our Common Future* identified indigenous peoples as "the repositories of vast accumulations of traditional knowledge and experience that link humanity with its ancient origins." Agenda 21 of the 1992 Rio Conference encouraged the use of this indigenous knowledge to develop new adaptive strategies for conservation. Who are these traditional societies and indigenous peoples? According to international criteria, four characteristics distinguish indigenous peoples from others. They are descendants of groups inhabiting an area prior to the arrival of other populations; they are politically not dominant; they are culturally different from the dominant population; and they identify themselves as indigenous.

Indigenous peoples are found in many parts of the world. Callicott's book on environmental ethics and religious traditions provides a representative survey of

> **Box 6**
>
> ### Sacred Groves and Traditional Forest Conservation in Nigeria
>
> Among the Yoruba of southwestern Nigeria, there were a number of categories of traditional sacred forests, although these have fallen into disuse in recent years. According to the Nigerian environmental study *Nigeria's Threatened Environment* (1991), land was regularly set aside as hunting forests, religious groves, and isolation and quarantine forests, and to serve as the abode of fairies and spirits.
>
> *Igbo ode* (hunting forest): Lands located at some distance from settlements and devoted to game. *Igbo egan* (high forest), abandoned secondary forest, when put to use for game-hunting activities becomes *Igbo ode*. Lands are often named according to their wildlife inhabitants, e.g., *Igbo erin* (elephant forest) *and Igbo efon* (buffalo forest). Only brave hunters dare use such specialized forests.
>
> *Igbo oro* (religious groves). Places set aside for religious worship of many of the elements of the physical environment. They are not extensive (usually less than a quarter of a hectare) and are uncultivated forests located on the borders of a settlement and in as many separate locations as there are families of the deities.
>
> *Igbo egbee* (religious groves "land of sorrows"): Reserved forests for the burial of people whose deaths are considered mysterious. Such lands, isolated further away from settlements, were never put under cultivation in the past when diseases were rampant and sudden deaths were attributed to the anger of the gods. Only brave hunters dare enter such lands.

indigenous ethics from various parts of the world, with major sections entitled "Polynesian Paganism," "American Indian Land Wisdom" (with major cases of Lakota shamanism and Ojibwa totemism), "South American Eco-eroticism" (cases of Tukano systems theory and Kayapo agro-ecology), "African Biocommunitarianism" (cases of Yoruba anthropo-theology and San etiquette of freedom), and "Australian Aboriginal Conservators."

The globalization of Western culture has meant, among other things, the globalization of Western modes of production (e.g., monocultures) and resource conservation (expert-knows-best positivist science). Has ancient knowledge become irrelevant, or perhaps simply swamped by Western science and practice? Conversely, are there useful lessons that can be learned from indigenous knowledge and practice? These are some of the central questions in a growing body of literature on traditional ecological knowledge.

B. The Context of Traditional Ecological Knowledge

Berkes and colleagues define traditional ecological knowledge as a cumulative body of knowledge, practice, and belief, evolving by adaptive processes and handed down through generations by cultural transmission, about the relationship of living beings (including humans) with one another and with their environment. As a knowledge–practice–belief complex, traditional ecological knowledge includes the religious traditions of a society. It is both cumulative and dynamic, building on experience and adapting to changes. It is an attribute of societies with historical continuity in resource use on a particular land. By and large, these are non-industrial or less technologically oriented societies, many of them indigenous or tribal, but not exclusively so. Traditional ecological knowledge is a subset of indigenous knowledge, generally defined as local knowledge held by indigenous peoples or local knowledge unique to a given culture or society.

Any discussion of traditional ecological knowledge and indigenous conservation needs to be qualified. The question of the relevance of traditional societies for biodiversity conservation is confounded by what has been called the "noble savage" syndrome. On the one hand, there is the romantic, Rousseauian view of "the primitive," intrinsically attuned to nature and somehow living "in balance" with the environment. Such noble savages tend to become "fallen angels" when they come into contact with the dominant society. Lest they become a threat to the very ecosystem in which they live, they should continue to live as primitives. On the other hand, there is the view that primitive peoples are not noble savages but tend to be ignorant and superstitious. Historically, they lived as biological populations at the mercy of natural forces and supernatural beliefs, not as organized communities with their own knowledge systems and ecologically adaptive practices. They had a tendency, even as primitive hunters, to cause massive species extinctions, as in New Zealand, Polynesia, Madagascar, and perhaps the Americas and Australia.

As with all myths, there are elements of truth at

the basis of these views, and many people, including conservationists and indigenous peoples themselves, no doubt believe parts of one or more of these views. An alternative view, however, is to start with the recognition that traditional ecological knowledge is not mere tradition but a set adaptive responses that have evolved over time. All societies, pre-scientific and scientific, strive to make sense of how the world behaves, and to apply this knowledge to guide practice. Because people were dependent for their survival on resources in their immediate environments, there were strong incentives for them to use resources sustainably. They could not mask this dependence with fossil fuel subsidies or capital markets in a globalized economy. Thus, the ability to nurture and sustain biodiversity was a selective pressure on these societies. That is, there was a survival value in conserving and augmenting local biodiversity.

Pre-scientific, traditional systems of management have been the main ways by which societies have managed their natural resources for millennia. In many cases, the main reason we still have any biodiversity to speak about is because of these systems of management. Within this context, biodiversity conservation is the indirect outcome, rather than the objective, of traditional practices, but the practices themselves had adaptive value—biodiversity conservation was often a matter of survival.

Such a probabilistic and evolutionary view clearly rejects the notion that all traditional peoples always manage resources well. What the literature suggests is that groups can make mistakes but can also learn from their mistakes. They are capable of responding to resource collapses and to environmental change, along the lines suggested by the books *Linking Social and Ecological Systems* and *Sacred Ecology*. Detailed accounts of traditional ecological knowledge do not support the noble savage. What they do is something very different: they provide a documentation of extremely detailed, pragmatic, empirical knowledge, represented in religious traditions and rituals, bound up with an emotional tie to the land and living beings.

C. Conserving Biodiversity: Traditional "Rules of Thumb"

How indigenous knowledge of nature and human social behavior is translated into resource use practices that tend to promote sustainable use of the biota and conservation of biodiversity is little known. Better known, many traditional peoples exhibit resource use restraints that promote conservation. But knowledge, belief, and practice leading to restraint tend to be intermingled, making it difficult to trace linkages among them. Conservation researchers often ignore social restraints on grounds that they are not considered rational in our worldview, often involving beliefs in supernatural forces, for example, as in the case of taboos. To arrive at an appropriate set of social restraints is an order of magnitude more complicated than to employ knowledge of habitat preference and behavior for efficient hunting strategies. To implement a set of social restraints is complicated also because it requires continued cooperation of a large number of individuals.

The dealings of traditional societies with nature are often hedged by prescriptions as to what, when, and how much is to be left undisturbed. These prescriptions become part of a culture and are mediated by religious traditions. Madhav Gadgil and colleagues identified four kinds of widely used "rules of thumb" as social restraints leading to indigenous biodiversity conservation practice:

1. *Provide total protection to some biological communities or habitat patches*   These may include pools along river courses, sacred ponds, sacred mountains, meadows, and forests. For example, sacred groves were once widely protected from Africa to China, and in fact, throughout the Old World. In the tribal state of Mizoram in northeastern India, they continue to be protected even after conversion to Christianity. It is now called a "safety forest," while the village woodlot from which regulated harvests are made is called the "supply forest." Ecological theory suggests that providing such absolute protection in "refugia" can be a very effective way of ensuring persistence of biological populations.

2. *Provide total protection to certain K-selected species*   Trees of all species of the genus *Ficus* are protected in many parts of the Old World. It is notable that *Ficus* is considered a keystone genus significant to the conservation of overall biodiversity. Local people seem to be often aware of the importance of *Ficus* as affording food and shelter for a wide range of birds, bats, and primates, and it is not difficult to imagine that such understanding was converted into widespread protection of *Ficus* trees at some point in the distant past. Taboos with apparent functional significance may also be placed on some less obvious species within the ecological community. For example, some Amazon fish species considered important for folk medicine are taboo and are avoided as food.

3. *Protect critical life history stages*   In south India, fruit bats may be hunted when foraging, but not at daytime roosts on trees that may be in the midst of villages. Many waders are hunted outside the breeding season, but not at heronaries, which may again be on

trees lining village streets. Cree Indians of James Bay in the subarctic hunt Canada goose, a major subsistence resource, but rarely kill or even disturb nesting geese. The danger of overharvest and depletion of a population is clearly far greater if these vulnerable stages are hunted and the protection afforded to them seems to be a clear case of ecological prudence.

4. *Mandate local stewards to supervise resource use* Traditional resource harvesting systems in diverse parts of the world rely on the guidance of a traditional expert to organize the harvest, control access, supervise local rules, and generally act as a "steward." This practice also ensures the proper use and transmission of knowledge. Further, in some societies, major events of resource harvest are carried out as a short-term, prescribed group effort. Thus, many tribal groups engage once a year in a large-scale communal hunt. Such a group exercise may also serve the purpose of assessing the status of prey populations and their habitat and may help to adjust resource harvest practices to sustain yields and conserve biodiversity.

Considering that contemporary scientific prescriptions for biodiversity conservation are little more than "rules of thumb," these traditional prescriptions add up to a reasonably good set of practices. Thus, traditional knowledge, with its time depth of observations, can help us arrive at practical, locally tailored prescriptions for resource use relevant to conservation needs.

D. Ecosystem People, Local Knowledge, and Sacred Respect

Many practices used by indigenous peoples contribute to the conservation of biodiversity through the use of more varieties, species, and landscape patches than do modern agricultural food production systems (which tend to rely heavily on the monoculture of a few varieties). The chapter by Berkes, Folke, and Gadgil in *Biodiversity Conservation* documents many cases. Gary Nabhan's work in Arizona and Mexico provides an illustration of biodiversity enhancement due to traditional multi-species agriculture, in this case, as compared to a protected area. Investigating two similar oases in the Sonoran Desert, Nabhan found 32 species of birds in the U.S. Organ Pipe Cactus National Monument on the Arizona side of the desert. The other oasis, 50 km away on the Mexico side of the border and still being farmed in the traditional Papago style by a group of Indian villagers, supported 65 species. A possible explanation was that the farmed oasis provided greater diversity and amounts of food available for the birds, as compared to the protected oasis.

There is evidence that indigenous practices based on good local observations and natural history can provide the capability to respond to feedbacks from the environment and to ecosystem change. The book *Linking Social and Ecological Systems* documents examples of multiple-species management, resource rotation, succession management, landscape patchiness management, and other ways of responding to and managing pulses and ecological surprises. Even assuming that biodiversity conservation is the *indirect outcome* of these practices, for the most part, the amount of ecological understanding suggested by these practices is considerable. Social mechanisms behind these practices include adaptations for the generation, accumulation, and transmission of knowledge and the use of local institutions and rules for social regulation. Religious traditions are important for the cultural internalization of traditional practices and for the development of worldviews and cultural values appropriate for them.

Some traditional societies have ecosystem-like concepts. Ancient conceptualizations of ecosystems exist in several American Indian, Asia-Pacific, European, and African cultures. Among some indigenous peoples of the North American Subarctic, *land* is more than a physical landscape; it encompasses the living environment, including humans. For example, the term used by the Dene groups of the Western Subarctic, *ndé* (*ndeh*), is usually translated as *land* but its meaning is closer to *ecosystem* because it conveys a sense of relations of living and nonliving things on the land. However, it differs from the scientific concept of ecosystem in that *ndé* is based on the idea that everything in the environment has life and spirit.

Table I provides a summary of traditional ecosystem-like concepts in which the unit of nature is often defined in terms of a physical boundary (such as a watershed) and living and non-living elements are considered interlinked. For example, among the Gitksan Indians of the Pacific Northwest, tribal chiefs describe their land boundaries as "from mountain top to mountain top" and orient themselves by two directional axes within this watershed framework: vertically up and down from valley bottom to mountain top, and horizontally, upstream and downstream. Detailed land use maps of the kinship-based *house* groups (*wilps*) of the Gitksan show that there is a close correspondence between watershed areas and *wilps* or clusters of *wilps*. Clearly, these are not merely political boundaries but watershed-ecosystems-as-territories.

In Oceania, there was a wealth of ecosystem-like

TABLE I

Examples of Traditional Applications of the Ecosystem View

System	Country/region
Watershed management of salmon rivers and associated hunting and gathering areas by tribal groups	American Indians of the Pacific Northwest
Delta and lagoon management for fish culture (*tambak* in Java), and the integrated cultivation of rice and fish	South and Southeast Asia
Vanua (in Fiji), a named area of land and sea, seen as an integrated whole with its human occupants	Oceania, including Fiji, Solomon Islands, and ancient Hawaii
Family groups claiming individual watersheads (*iworu*) as their domain for hunting, fishing, and gathering	The Ainu of northern Japan
Integrated floodplain management (*dina*) in which resource areas are shared by social groups through reciprocal access arrangements	Mali, Africa

Source: Berkes, 1999.

concepts. Examples include the ancient Hawaiian *ahupua'a*, wedge-shaped land units controlled by local chiefs, encompassing entire valleys stretching from the top of mountains to the coast and shallow waters. The variations of the Hawaiian system may be found in the Yap *tabinau* the Fijian *vanua*, and the Solomon Islands *puava*. The common point in each is that the term refers to an intimate association of a group of people with land, reef, and lagoon and all that grows on or in them. It is the "personal ecosystem" of a specific group of people. In the Solomons, for example, a *puava* is a named territory consisting of land and sea, and it includes all lands and resources associated with a kinship group.

These examples provide insights into what Raymond Dasmann (in *The Ends of the Earth*) calls ecosystem people—people who depend heavily on natural resources of their own localities, and hence develop a very detailed understanding of their environment, as well as a spiritual intimacy with the land. Such indigenous ecological understanding is different from that of the scientist. These traditional views tend to depict ecosystems not as lifeless, mechanical, and distinct from people, but as fully alive and encompassing humans. In some cases, traditional concepts of land also incorporate spirits of animals and other natural objects (as among the Dene of Northern Canada and Alaska), and spirits of human ancestors, as among some African groups and among the Australian aborigines (the concept of Dreamtime).

V. CONCLUSIONS

In monotheistic traditions, humans have dominion over the earth, they have a God-given right to subdue the earth, and they are stewards of God. By contrast, pantheistic traditions do not have a dominant relationship with nature. Asian religious traditions and pre-Christian European traditions also have a non-dominant view of nature. The record of different religious traditions in conserving biodiversity is mixed. This is because human behavior is conditioned not only by religious beliefs but also by many other forces.

Traditional knowledge—practice—belief systems, tapping the wisdom of many traditional cultures with pantheistic traditions, offers a number of lessons. These systems are characterized by a similarity of concepts of nature, in which humans are part of nature. As well, these knowledge systems are characterized by similarity in design. An example is shifting cultivation, developed apparently independently, by tropical forest peoples in Africa, South America, South Asia, and New Guinea. At the same time, traditional knowledge systems are characterized by a remarkable diversity in practice, even in adjacent areas. For example, as Robert Johannes and others have shown, taboos used in the social control of resources in Oceania can vary greatly from one island to another, and even from one part of an island to another. Such diversity in resource use practice contrasts with the relatively uniform conservation prescriptions of government agencies, and highlights the need for conceptual pluralism (Richard Norgaard).

One important lesson from traditional ecological knowledge is that values and beliefs are important in encoding ethics, including the ethics of conservation. As Rappoport and Anderson point out, the use of emotionally powerful cultural symbols is important to implement a moral code. If this is true, the incorporation of values and beliefs into biodiversity conservation efforts is more likely to succeed than the use of purely scientific arguments or purely economic incentives.

A number of contemporary concepts appear to be exploring the combination of emotions and ideas that may help restore a sense of sacred respect. They include Naess' deep ecology, E. O. Wilson and Stephen Kellert's biophilia hypothesis (love of living beings), bioregionalism (with its combination of local self-reliance and sense of belonging), the related notion of "sense of place," and the Gaia hypothesis of James Lovelock, which is the contemporary version of the Mother Earth idea. Each provides an approach to the understanding of reciprocal ties that bind humans with the natural world, and these ties invariably have a spiritual or religious aspect.

See Also the Following Articles

CONSERVATION MOVEMENT, HISTORICAL • ECOSYSTEM, CONCEPT OF • ENVIRONMENTAL ETHICS • HISTORICAL AWARENESS OF BIODIVERSITY • INDIGENOUS PEOPLES, BIODIVERSITY AND • LITERARY PERSPECTIVES ON BIODIVERSITY • STEWARDSHIP

Bibliography

Anderson, E. N. (1996). *Ecologies of the Heart. Emotion, Belief and the Environment.* Oxford Univ. Press, New York and Oxford.

Berkes, F. (1999). *Sacred Ecology. Traditional Ecological Knowledge and Resource Management.* Taylor & Francis, Philadelphia and London.

Berkes, F., and Folke, C. (Eds.) (1998). *Linking Social and Ecological Systems. Management Practices and Social Mechanisms for Building Resilience.* Cambridge Univ. Press, Cambridge. UK.

Callicott, J. B. (1994). *Earth's Insights. A Multicultural Survey of Ecological Ethics from the Mediterranean Basin to the Australian Outback.* Univ. of California Press, Berkeley.

Callicott, J. B., and Ames, R. T. (Eds.) (1989). *Nature in Asian Traditions of Thought: Essays in Environmental Philosophy.* State Univ. of New York Press, Albany.

Capra, F. (1996). *The Web of Life.* Anchor Books/Doubleday, New York.

Engel, J. R., and Engel, J. G. (Eds.) (1990). *Ethics of Environment and Development.* Belhaven Press, London.

Kellert, S. R., and Wilson, E. O. (Eds.) (1993). *The Biophilia Hypothesis.* Island Press, Washington, D.C.

Levine, M. P. (1994). *Pantheism.* Routledge, London.

Marshall, P. (1992). *Nature's Web. An Exploration of Ecological Thinking.* Simon & Schuster, London.

Naess, A., with Rothenberg, D. (1989). *Ecology, Community and Lifestyle: Outline of an Ecosophy.* Cambridge Univ. Press, Cambridge, UK.

Norgaard, R. B. (1994). *Development Betrayed: The End of Progress and a Coevolutionary Revisioning of the Future.* Routledge, London.

Perrings, C., Maler, K.-G., Folke, C., Holling, C. S., and Jansson, B.-O. (Eds.) (1995). *Biodiversity Conservation.* Kluwer, Dordrecht.

Rappaport, R. A. (1984). *Pigs for the Ancestors.* 2nd ed. Yale Univ. Press, New Haven, CT.

Worster, D. (Ed.) (1988). *The Ends of the Earth: Perspectives on Modern Environmental History.* Cambridge Univ. Press, Cambridge, UK.

REMOTE SENSING AND IMAGE PROCESSING

Ronen Kadmon
The Hebrew University of Jerusalem

I. Introduction
II. Fundamentals of Remote Sensing
III. Remote Sensing Applications
IV. Summary

GLOSSARY

classification The process of assigning the pixels of an image to discrete categories.
electromagnetic radiation The energy transmitted through space in the form of electric and magnetic waves. Can be detected by various types of sensors.
geographic information systems (GIS) Computer systems that store, enhance, analyze, and display layers of geographic data and connect these to alphanumeric databases.
image processing The manipulation of digital image data, including operations such as image display, restoration, enhancement, and classification.
pixel Abbreviated from "picture element"; the smallest part of a picture (image).
sensor A device that gathers energy, converts it to a digital value, and presents it in a form suitable for obtaining information about the environment

REMOTE SENSING IS one of the main tools available for mapping and monitoring patterns of biodiversity across large spatial scales. Data derived from remote sensing provide information on landscape characteristics that influence biodiversity, structural and functional properties of ecosystems, spatial distribution of different components of biodiversity, patterns of natural and human-induced vegetation changes, and impacts of various disturbances and ecological interactions. Remotely sensed data can be integrated with information on the physical and human environment for analysis and modeling purposes. When combined with field surveys, the potential contribution of remote sensing to studies of biodiversity is literally infinite.

I. INTRODUCTION

It is now widely recognized that biodiversity is a multidimensional and multiscale phenomenon encompassing different organization levels (populations, species, functional groups, communities, ecosystems, and landscapes) and a wide range of spatial scales (from microhabitat heterogeneity to global-scale patterns). Identifying patterns of biodiversity and their causal factors is therefore an enormous task that requires (1) mapping and monitoring of biological patterns across different spatial and temporal scales and (2) analysis of such patterns with respect to diverse aspects of the physical and human environment. Remote sensing is one of the best tools available for coping with this challenge. Information obtained from remote sensing is intrinsically multidimensional (horizontally, vertically,

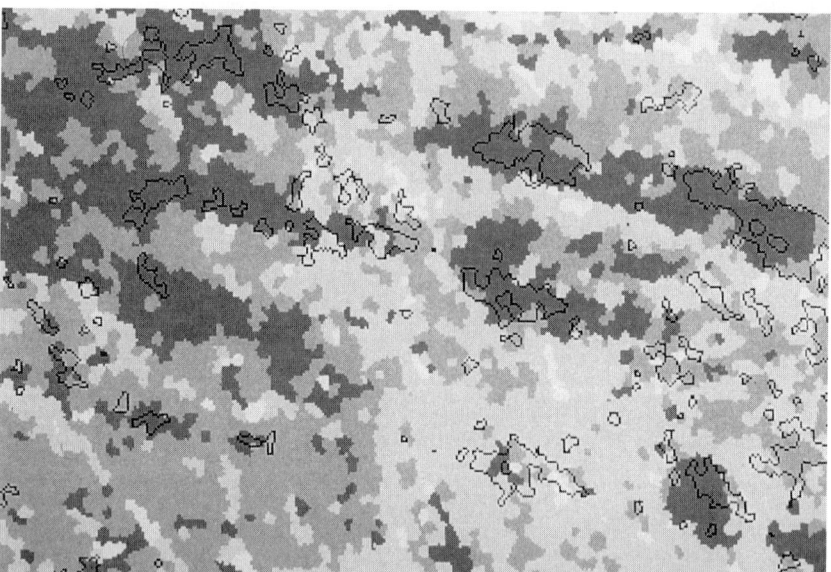

FIGURE 1 A vegetation map produced for a grassland area of 50 × 31 m using image processing of a color infrared aerial photograph. Dark gray, bunchgrasses; gray, dense annuals; light gray, sparse annuals; white, rocks and stones. Black outlines indicate perimeters of gopher disturbances. Pixel size is 13.5 cm. Reproduced from Lobo *et al.* (1998), with permission from Kluwer Academic Publishers.

temporally, and spectrally) and may cover spatial scales ranging from a few centimeters (Fig. 1) to entire continents (Fig. 2). Data derived from remote sensing provide information on all levels of biological organization, including species, functional groups, ecosystems, and biomes. These capabilities make remote sensing an extremely valuable tool for studies of biodiversity.

The aim of this article is to draw attention to the potential and actual contribution of remote sensing to studies of biodiversity. The article consists of two main sections. The first section introduces fundamental concepts of remote sensing. The second section provides selected examples of how remote sensing is being used for mapping, monitoring, and modeling patterns of biodiversity. The examples were chosen to represent different aspects of biodiversity as well as a wide range of data sources and analytical methodologies. The article is concluded by a brief summary that points to some current and expected future developments in this field.

II. FUNDAMENTALS OF REMOTE SENSING

A. An Overview of the Remote Sensing Process

Remote sensing can be broadly defined as "the technique of obtaining information about objects through the analysis of data collected by special instruments that are not in physical contact with the objects of investigation" (Avery and Berlin, 1992). A refined definition of remote sensing that better fits the scope of this article is: *The practice of obtaining information about the Earth's surface, using images acquired from airborne or spaceborne vehicles by measuring reflected or emitted electromagnetic radiation.*

This narrower definition restricts our discussion to applications in which the general target of the observation is the Earth's surface, the measured energy is electromagnetic radiation, the sensors are positioned on airborne or spaceborne platforms, and the recorded data are available as two-dimensional images. While some techniques of remote sensing do not fit this formalization (e.g., laser, acoustical or sonar technologies), the above definition covers most of the methods that are currently applied for the documentation, monitoring, and mapping of biodiversity.

Figure 3 schematically illustrates the basic elements of the remote sensing process. Any remote sensing application consists of two distinct processes: data acquisition (detection and recording of electromagnetic radiation), and data analysis (extraction of information from the recorded data). The first process can be performed either photographically (i.e., by a photographic camera) or electronically (by electronic sensors). A photographic camera uses optical lenses and a light-sensitive film to detect electromagnetic radiation. The film acts

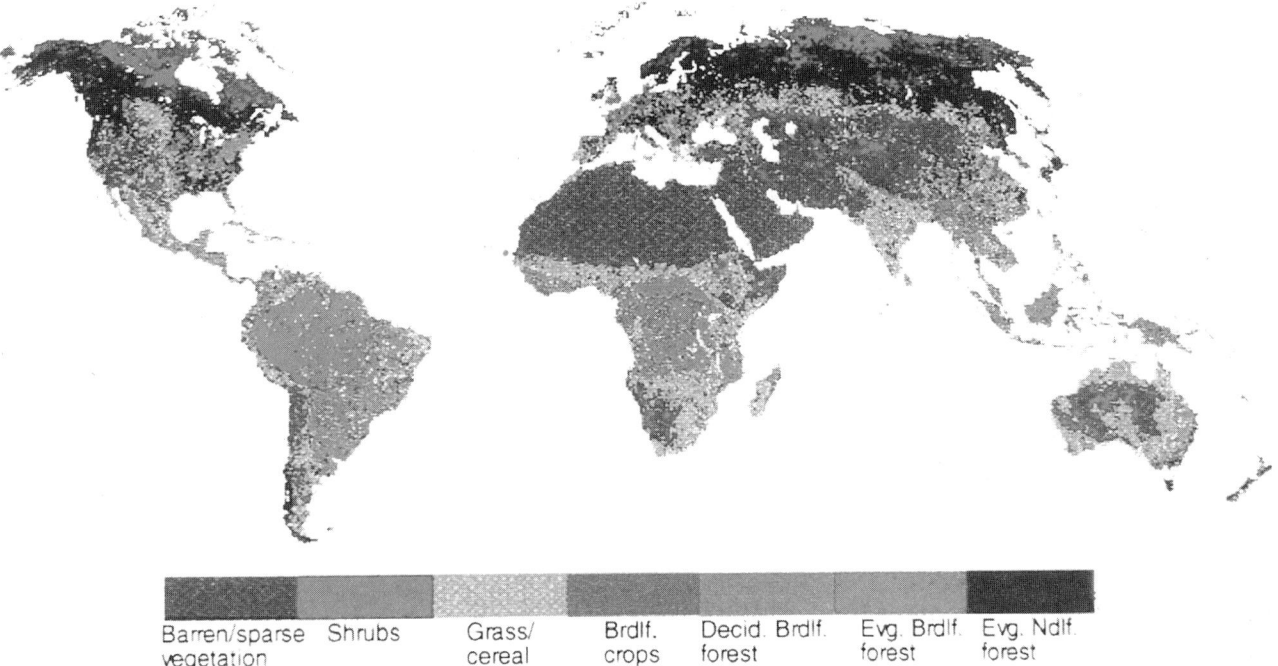

FIGURE 2 A global land cover map derived from satellite imagery. Reproduced from Nemani and Running (1997), with permission from the Ecological Society of America.

as both the detector of the radiation and the recording medium. Electronic sensors convert electromagnetic radiation into electronic signals that can be stored as digital images. The end product of this stage of remote sensing is therefore either a photograph or a digital image.

The second step of the remote sensing process is data analysis (Fig. 3). Remotely sensed data, either photographs or digital images, must be analyzed in order to provide useful information about the observed features. Analytical techniques available for the analysis of remotely sensed data are diverse, ranging from traditional methods of visual interpretation to sophisticated computer-based image processing. The final product of both visual interpretation and computer-based image processing is usually a map showing the spatial distribution of the variables of interest (Figs. 1–3). Maps generated from remotely sensed data can be used directly, or in combination with other data sources, for a variety of applications, including research, decision making, planning, and management.

B. Electromagnetic Radiation and Its Measurement

As already defined, remote sensing is concerned with the measurement of electromagnetic radiation reflected and emitted from objects and features on the Earth's surface. The main source of radiation reaching the Earth is the sun. Some of the solar radiation is reflected from the Earth's surface and can therefore be detected by cameras or electronic sensors. Other components of the solar radiation are absorbed at the Earth's surface and are reradiated as thermal energy. This form of electromagnetic radiation can also be detected and recorded by appropriate sensors. As different objects emit and reflect different types and amounts of radiation, images

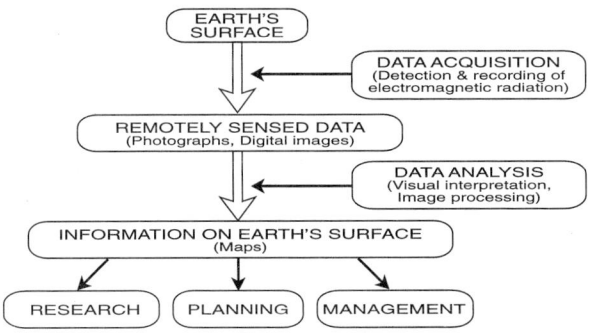

FIGURE 3 Basic elements of the remote sensing process.

capturing such differences contain much information on the nature of the Earth's surface.

1. The Electromagnetic Spectrum

A basic characteristic of electromagnetic radiation is its wavelength. Solar radiation consists of a wide and continuous range of wavelengths, stretching from less than 10^{-6} to more than 10^7 μm (1 μm equals 10^{-6} m). This continuum is subdivided into several divisions called spectral bands, which share similar characteristics (Fig. 4a). The main bands used in remote sensing are ultraviolet, visible, infrared, and microwave.

The ultraviolet (UV) band lies between the X-rays and visible light with wavelength limits of 0.01 and 0.4 μm (Fig. 4a). Wavelengths shorter than 0.3 μm are unable to pass through the atmosphere and therefore only the 0.3- to 0.4-μm wavelength interval, or near-UV, is available for remote sensing (Fig. 4b).

The visible band stretches from 0.4 to 0.7 μm (Fig. 4a). This spectral range constitutes an extremely small portion of the full electromagnetic spectrum and its boundaries are defined by the wavelength limits of human vision (Fig. 4b). The visible spectrum is further subdivided into three equal segments: blue (0.4–0.5 μm), green (0.5–0.6 μm), and red (0.6–0.7 μm).

The infrared (IR) band extends from 0.7 to 1000 μm (Fig. 4a). In remote sensing this wide band is divided into two major components: the reflected IR band (from 0.7 to 3.0 μm) and the emitted, or thermal, IR band (3.0–1000 μm). The first component is essentially solar radiation reflected from the Earth's surface. This component is further subdivided into near-IR (<1.3 μm) and mid-IR (>1.3 μm). Radiation in the near-IR behaves, with respect to optical systems, in a manner analogous to radiation in the visible spectrum. It can therefore be detected and recorded by films and cameras similar to those used for the visible spectrum. The thermal component of the IR spectrum represents heat energy that is continuously emitted by all objects on the Earth's surface. It ranges from 3 to 1000 μm, but wavelengths beyond 14 μm are largely absorbed by the atmosphere and are therefore not available for remote sensing (Fig. 4b).

The microwave band falls between the infrared and the radio bands (Fig. 4a). It ranges from about 1 mm to 1 m and contains the longest wavelengths used in remote sensing. An important feature of microwave radiation is its ability to pass through clouds, precipitation, and tree canopies.

2. Types of Remote Sensors

The sensors used in remote sensing can be divided into four main groups: (1) photographic camera, (2) electro-optical, (3) passive microwave, and (4) radar.

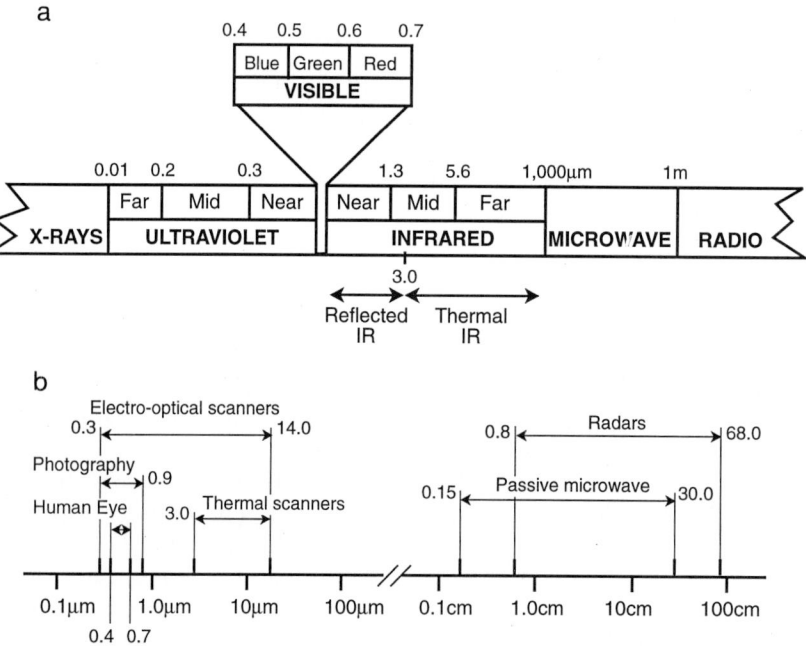

FIGURE 4 Characteristics of electromagnetic radiation. (a) Major divisions of the electromagnetic spectrum (not to scale). (b) Spectral ranges of common remote sensing systems.

Photographic cameras can detect wavelengths ranging from about 0.3 to 0.9 μm (Fig. 4b). This includes the near-ultraviolet band (0.3–0.4 μm), all the visible light (0.4–0.7 μm), and the relatively short wavelengths of the near-IR (0.7–0.9 μm). This spectral range, which is nearly twice as wide as the human vision, can be recorded directly into film and is therefore referred to as the photographic spectrum.

Electro-optical sensors measure radiation in narrow wavelength ranges that can be located at various points along the near-UV, visible, reflected IR, and thermal IR spectral bands (Fig. 4b). This spectral range (0.3–14 μm) is termed the optical spectrum because wavelengths in this range can be reflected and refracted using optical devices such as lenses and mirrors. An important characteristic of electro-optical sensors operating at the thermal IR band is their ability to collect data day and night without being sensitive to dust and haze.

Passive microwave sensors are designed to measure microwave radiation that is naturally emitted from the Earth's surface. These sensors operate at a spectral range of 0.15–30 cm (Fig. 4b). Radar systems also operate at the microwave region. However, in contrast to passive microwave sensors that measure naturally emitted radiation, radars are "active" sensors that transmit artificially produced energy to the Earth's surface and record the reflected component of this radiation.

Three forms of remote sensing can therefore be distinguished: the detection of solar radiation reflected from the Earth's surface, the detection of radiation emitted from the Earth's surface, and the detection of radiation transmitted from the sensor itself.

C. Sources of Remotely Sensed Data

1. Aerial Photography

Aerial photography is the oldest form of remote sensing. Despite the increasing availability of more sophisticated imaging systems, aerial photographs remain one of the most reliable and widely used sources of remotely sensed data. In addition, aerial photographs provide some advantages that cannot be achieved using alternative sources of image data, particularly high spatial resolution and long periods of documentation. High spatial resolution enables one to detect and analyze patterns at very small spatial scales (Fig. 1, Case Study 4). The availability of historical aerial photographs makes it possible to measure and analyze long-term ecological processes (Case Study 8).

Both black-and-white and color photographs are used in remote sensing. Black-and-white photographs can be taken with either panchromatic film or infrared-sensitive film. Panchromatic film is characterized by an aggregate spectral sensitivity that includes the near-UV, blue, green, and red wavelengths (0.25–0.7 μm). It provides a tonal rendition that closely approximates the brightness of the scene being photographed and can be used to distinguish between objects of truly different color. Infrared-sensitive film has spectral sensitivity that extends from 0.25 to about 0.9 μm. This range encompasses the near-IR band in addition to the near-UV and visible wavelengths. Thus, infrared-sensitive film operates beyond the confines of the visible spectrum and can be used to detect information that is unavailable for human vision. Infrared-sensitive films are especially valuable for vegetation classification because differences between vegetation classes are often more distinct in the near-infrared than in the visible range (Fig. 5).

While the information contained in black-and-white photographs is expressed by differences in gray levels, color films offer the additional qualities of hue (dominant wavelength), chroma (color strength), and value (color intensity). This added information greatly improves interpretation ability. Two kinds of color films used in aerial photography are normal color films and color infrared films. Normal color films are sensitive primarily to wavelengths of the visible spectrum and provide a color rendition that approximates the original scene as it would be viewed by the human eye. This facilitates photo interpretation because objects appear in their natural colors. Color infrared films are sensitive to green, red, and near-IR radiation (0.5–0.9 μm). Since maximum reflectance from vegetation occurs in the near-IR spectral region, such films are highly valuable for studies of vegetated landscapes.

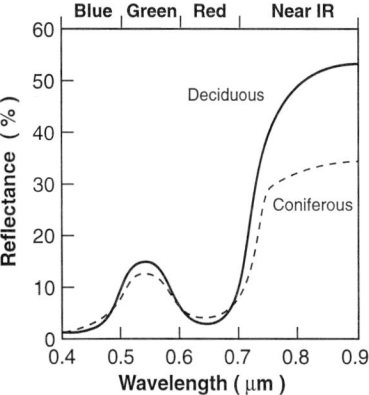

FIGURE 5 Generalized spectral reflectance curves for deciduous (broad-leaved) and coniferous (needle-bearing) trees. Note that both reflectance intensity and differentiation ability are highest in the near-IR region.

Remote sensing applications based on aerial photography are presented in Case Studies 4 and 8.

2. Satellite Imagery

Data acquired by spaceborne imaging systems are usually recorded in digital form and are therefore available as digital images. A digital image is composed of a two-dimensional array of picture elements (pixels), with each pixel corresponding to a particular area on the ground. The radiance measured over the ground area represented by each pixel is translated into a digital number (DN). This number (also referred to as brightness) is a positive integer expressing the average intensity of radiance measured from the particular ground area. Two important characteristics of digital images are spatial resolution and radiometric resolution. Spatial resolution refers to the size of the ground area represented by each pixel in the image. For example, a spatial resolution of 10×10 m means that each pixel in the image corresponds to a ground unit of 10×10 m. The term radiometric resolution refers to the number of DN values, or brightness levels, used for recording differences in radiation intensity. Thus, a higher radiometric resolution permits the detection of finer differences in radiation intensity. Both spatial and radiometric resolution have important consequences for the amount of information available in remotely sensed data.

Satellite images suitable for mapping, monitoring, and modeling patterns of biodiversity have usually been derived from multispectral sensors measuring reflected or emitted radiation in the optical spectrum (0.3–14 μm). Imaging systems operating at this spectral range have been used in a variety of national and international Earth observation programs. Three of these programs, namely, Landsat, SPOT, and NOAA AVHRR, are of particular importance for studies of biodiversity. The main characteristics of these programs are outlined below.

a. Landsat Program

The Landsat program is the longest running project for acquisition of repetitive multispectral data of the Earth's surface. The first Landsat satellite was launched in 1972 and the most recent one (Landsat 7) in 1999. Landsats 1–3 covered the Earth every 18 days. Landsats 4, 5, and 7 have a coverage cycle of 16 days (Landsat 6 has failed to achieve orbit). Currently (1999), the only active satellites in this program are Landsats 5 and 7. However, data recorded by earlier Landsat missions are still being used in many remote sensing applications.

Table I compares the characteristics of data acquired by the two main sensors used in the Landsat program, the Multispectral Scanner (MSS) and the Thematic Mapper (TM). The MSS collects reflected solar radiation in four contiguous spectral bands, two in the visible spectrum at 0.5–0.6 μm (green) and 0.6–0.7 μm (red), and two in the near-IR at 0.7–0.8 and 0.8–1.1 μm. This imaging system was used by Landsats 1–5 (note that the same spectral bands had different numbering systems in Landsats 1–3 and Landsats 4 and 5). Data collected by the MSS are framed into individual scenes that cover a nominal ground area of 185×185 km at a spatial resolution of 79 m in Landsats 1–3 and 82 m in Landsats 4 and 5 (Table I).

The Thematic Mapper (TM) was designed to improve the imaging capabilities of Landsat satellites. A major difference between the TM and the MSS is the acquisition of data in seven bands instead of four, with new bands in the visible (blue-green), mid-infrared, and thermal portions of the spectrum (Table I). Also, based on experience with MSS data and extensive research, the wavelength range and location of the TM scanner have been modified to facilitate the spectral differentiability of major Earth surface features. The radiometric resolution of the TM is higher than that of the MSS (256 vs 64 digital numbers) and its spatial resolution is also higher (30 m for all bands except the thermal band, as compared to about 80 m in the case of the MSS). These overall capabilities significantly improved the spectral coverage, radiometric resolution, and spatial resolution of Landsat images.

TABLE I
Characteristics of Landsat MSS and TM Data

Band	Wavelength range (μm)	Nominal spectral location	Spatial resolution (m)
MSS[a]			
4 (1)	0.5–0.6	Green	79 (82)
5 (2)	0.6–0.7	Red	79 (82)
6 (3)	0.7–0.8	Near-IR	79 (82)
7 (4)	0.8–1.1	Near-IR	79 (82)
TM			
1	0.45–0.52	Blue-green	30
2	0.52–0.60	Green	30
3	0.63–0.69	Red	30
4	0.76–0.90	Near-IR	30
5	1.55–1.75	Mid-IR	30
6	10.4–12.5	Thermal IR	120
7	2.08–2.35	Mid-IR	30

[a] MSS bands of Landsats 1, 2, and 3 are numbered 4–7. The same bands are numbered 1–4 in Landsats 4 and 5. Spatial resolution is 82 m in Landsats 1–3 and 79 m in Landsats 4 and 5.

Applications based Landsat data are presented in Case Studies 1, 6, and 9.

b. SPOT Program

SPOT (Systeme Pour l'Observation de la Terre) is an international Earth observation program initiated by the French government in the late 1970s. The first SPOT satellite was launched in 1986, and the most recent one (SPOT 4) in 1998. All satellites have an Earth coverage cycle of 26 days. The sensors of SPOT 1, 2, and 3 consist of two identical high-resolution visible (HRV) imaging systems, each designed to operate in either of two modes: a black-and-white panchromatic mode over the range 0.51–0.73 μm (green-red) and a multispectral mode over the ranges 0.50–0.59 μm (green), 0.61–0.68 μm (red), and 0.79–0.89 μm (near-IR). Spatial resolution of the panchromatic mode is 10 m while that of the multispectral mode is 20 m. The data are encoded over a range of 256 digital numbers, and individual images are formatted to cover a 60 $\times$ 60 km area. An important advantage of the HRV imaging system is the ability to adjust its viewing angle by ground commands and thus to observe the same location several times within the normal 26-day coverage cycle. This "revisit" capability facilitates the documentation and monitoring of short-term dynamic processes (e.g., vegetation responses to episodic rainfall or drought events). It also increases the chance of obtaining cloud-free images in areas where cloud cover is a problem. Applications based on SPOT data are presented in Case Studies 3 and 7.

c. The NOAA Program

The National Oceanic and Atmospheric Administration (NOAA) of the United States operates national weather programs that rely to a large extent on data collected by meteorological satellites. Imaging systems of such satellites are usually designed to have a relatively low spatial resolution, but very high temporal resolution. Satellites of the NOAA series, one of the longest weather satellite programs, observe the complete Earth twice a day at a ground resolution of 1.1 $\times$ 1.1 km at nadir. These satellites carry a multispectral scanner known as the Advanced Very High Resolution Radiometer (AVHRR). The AVHRR was designed primarily for meteorological applications, but it has been applied successfully for many land-oriented applications. Details of the specific spectral coverage vary with the specific mission, but in general AVHRR sensors collect data in four or five bands falling in the visible, near-IR, and thermal IR spectral regions. Images acquired by AVHRR are available at a full 1.1-km resolution (local area coverage, or LAC) and at a subsampled resolution of 4 km (global area coverage, or GAC). Applications based on AVHRR data are presented in Case Studies 2, 5, and 10.

D. Analysis of Remotely Sensed Data

Data obtained from remote sensing must be interpreted in order to become usable information. The outcome of the interpretation process is usually a map showing the spatial distribution of the objects or variables of interest. Such maps can be analyzed with respect to other sources of information using geographical information systems (see Section D.2).

1. Interpretation of Remotely Sensed Data

In the case of aerial photographs, the interpretation process is usually based on visual examination: the interpreter systematically examines the image (and possibly supporting materials such as maps or field reports) and tries to identify the nature of the objects and phenomena seen in the image. This form of image interpretation relies on the ability of the human mind to qualitatively evaluate spatial patterns in an image. In most cases, it involves the identification and delineation of discrete areal units throughout the image. For example, in studies of vegetation variation, the interpreter outlines the boundaries between areas covered by different vegetation types.

Digital images are usually interpreted using computer-based image processing. Recall that the brightness of each pixel in a digital image is expressed by a numerical integer (its DN value); the general objective of image processing is to "translate" the original image into a more informative image by mathematically manipulating the recorded DN values. Practically this is done by subjecting the original image, pixel by pixel, to appropriate mathematical operations and storing the results of the computation as a new image. The mathematical algorithms used in image processing are literally infinite, but most kinds of algorithms can be categorized into one of three main operations: image restoration, image enhancement, and thematic classification.

Image restoration involves various operations of spectral and geometric corrections. Digital images are noisy and suffer from different kinds of spectral and geometric distortions that reduce their quality and limit their use. For example, the spectral characteristics of objects detected in a digital image are distorted by differences in sun angle, topographic variation, atmospheric scattering, and sensor characteristics. All of these factors may influence the intensity of electromagnetic radiation measured by the sensor. Other factors may introduce positional errors to digital images. Radiometric corrections are mathematical operations that compensate for various sources of spectral distortion in the data (Case Study 6). Geometric corrections attempt to correct for positional errors and to transform the original image into a new image that has the geometric characteristics of a map.

Image enhancement operations attempt to improve the detectability of objects or patterns in the image. For example, contrast stretching algorithms increase the contrast between features of interest and thus improve interpretation ability. Edge enhancement algorithms are designed to automatically detect boundaries between different kinds of features (Case Study 3). Other algorithms operate on multispectral data. For example, vegetation indices (VI) represent a group of spectral operations that attempt to enhance vegetation mapping and monitoring. The most widely used VI is the normalized difference vegetation index (NDVI) defined by the general equation

$$\text{NDVI} = \frac{\text{near-IR band} - \text{red band}}{\text{near-IR band} + \text{red band}}$$

This index can be computed using data acquired by a variety of imaging systems (e.g., Landsat, SPOT, and NOAA AVHRR) and has proved to be a sensitive indicator of the presence and condition of green vegetation (Case Studies 5 and 10).

Thematic classification is a process of assigning all pixels of an image into land cover classes (themes) based on their DN values. In applications such as land cover or vegetation mapping, thematic classification is the main object of the analysis. The outcome of the classification process is a new image where each pixel represents a particular land cover type. Such an image can be interpreted as a map depicting the spatial distribution of the selected land cover types within the scene (Figs. 1, 2, and 6).

Two standard approaches for image classification are unsupervised classification and supervised classification. In unsupervised classification the pixels of the image are numerically classified into a number of spectral classes based on the likeness of their DN values. The land cover identity of each spectral class is then determined using reference information such as field data, aerial photographs, or published maps (Case Study 2). In supervised classification, the analyst first selects areas of known cover type in the image. These "training sites" are analyzed statistically to identify the spectral characteristics of each cover type. With appropriate algorithms, the DN values of each unknown pixel

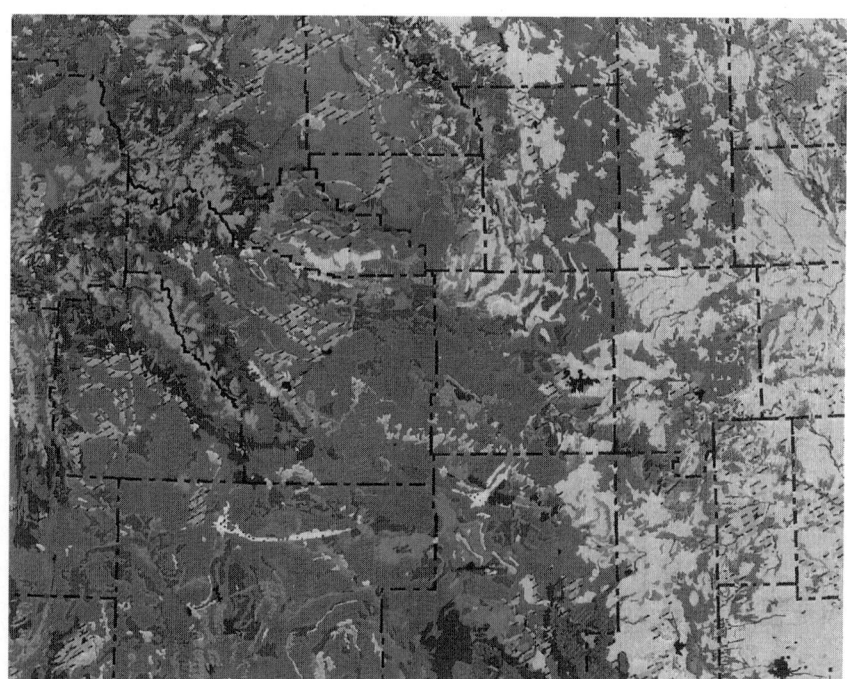

FIGURE 6 A land cover map produced for Wyoming using visual interpretation of Landsat TM data. The most common cover types in Wyoming are Wyoming big sagebrush shrubland (the pink areas that dominate the left side of the map) and mixed grass prairie (the creme-colored areas on the right). Reproduced from Driese et al. (1997), with permission from OPULUSPRESS. See also color insert, Volume 1.

in the image are compared with those of the training sites, and the pixel is assigned to the land cover type to which it is most similar (Case Study 6). It is important to emphasize that classification of digital images is never perfect. The success of image classification depends on many factors and may vary considerably from one case to another. Therefore, any classification of remotely sensed data must be followed by some form of accuracy assessment.

2. Integration of Digital Images within a GIS Environment

Geographical information systems (GIS) are computer-based systems capable of storing, manipulating, and visualizing geographically referenced data. Basically, any kind of information that can be presented by standard maps can be converted into digital form and handled by GIS. An important benefit of GIS is the ability to spatially integrate multiple types of data stemming from different sources of information. For example, in Case Study 9, GIS was used to integrate field-based data on the distribution of buzzard nests in northwestern Scotland with remotely sensed data of vegetation conditions and a set of topographic maps derived from a digital terrain model. By integrating these data in the GIS, it was possible to identify factors affecting the spatial distribution of buzzard nesting sites.

The foregoing example illustrates a typical GIS operation called overlay analysis. Many other kinds of spatial operations can be performed with GIS. For example, aggregation is an operation by which detailed map categories are aggregated into less detailed categories (e.g., aggregation of detailed vegetation units into a few broad vegetation types). Buffering is an operation by which a zone of user-defined width is created around map objects. Case Study 7 exemplifies the use of buffering as a tool for testing the effect of distance to road on deforestation rates in the Philippines. Data merging is an operation by which different grid maps of the same area are coregistered into a uniform, multidimensional database. This operation is commonly applied to combine digital images representing different wavelengths, different times of observation, or different sensors. For example, monitoring seasonal variation in NDVI requires the coregistration of images taken at different times (Case Study 5). Low-resolution images (e.g., AVHRR) can be merged with images of higher resolution (e.g., Landsat TM) for calibration or interpretation purposes (Case Study 2). Using the merging capability of GIS, one is able to coregister digital images with "ancillary" data (e.g., elevation and slope) that correspond to each pixel in the image. Once such data are spatially registered, information available in the GIS can be used to improve image classification. Thus, the interaction between remote sensing and GIS is two way in nature; remote sensing is used to generate digital maps that can be fed into GIS, while GIS data can be used to interpret and classify remotely sensed data.

GIS can also be used for modeling purposes. For example, by overlaying vegetation maps derived from remote sensing with corresponding maps of climatic and topographic variables, one is able to develop regression models that predict vegetation characteristics from information on climate and topography (e.g., Case Study 1).

III. REMOTE SENSING APPLICATIONS

The previous section has focused on general concepts and elements of remote sensing. Its main purpose was to provide the scientific and technical background required to understand how remote sensing technology can be used for collecting, analyzing, and visualizing images of Earth's surface. This section presents 10 case studies that were chosen to illustrate the wide spectrum of biodiversity issues that can be studied with remote sensing.

Three main kinds of applications are exemplified: (1) mapping of land cover and vegetation patterns, (2) measurement of temporal changes in land cover and vegetation characteristics, and (3) modeling of species distribution patterns. Each case study is described with respect to its scientific questions, data sources, and methodologies. The case studies also differ from each other considerably in their spatial and temporal scales. For example, Case Study 5 focuses on continental patterns of vegetation changes (spatial resolution 4 km), while Case Study 4 focuses on individual trees within a small forest stand (spatial resolution 10 cm). This trade-off between spatial scale and spatial resolution is a common characteristic of remote sensing applications. Temporal intervals between successive images of the same system range from 1 month in Case Study 5 to four decades in Case Study 8. These differences point to the diversity of spatial and temporal scales that can be investigated using remotely sensed data.

A. Mapping Vegetation and Land Cover Patterns

Mapping vegetation and land cover patterns has been the most common application of remote sensing. It may contribute to studies of biodiversity both directly (by documenting the distribution of different components of biodiversity) and indirectly (by providing knowledge on landscape and habitat characteristics that influence patterns of biodiversity). The case studies described here were chosen to illustrate four different levels of land cover and vegetation mapping. The first study was designed to provide a general map of land cover types. The second study attempted to quantify variation in the percentage cover of one particular vegetation type (conifer forests). The aim of the third study was to develop a method of counting and mapping tree densities, while the fourth study was designed to develop methods for mapping individual tree crowns. The four studies also differ from each other in their data sources: Landsat TM in Case Study 1, AVHRR in Case Study 2, SPOT panchromatic in Case Study 3, and color IR aerial photographs in Case Study 4.

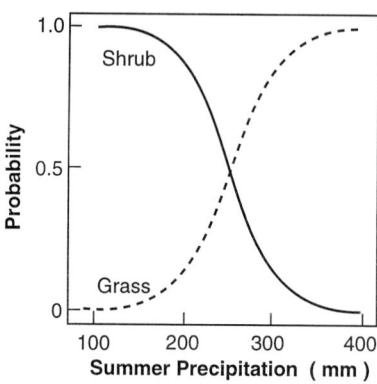

FIGURE 7 Effect of summer precipitation on the probability of a site being dominated by either mixed grass prairie or Wyoming big sagebrush. The vegetation data used for the analysis were derived using visual interpretation of Landsat TM images. Adapted from Driese *et al.* (1997), with permission from OPULUSPRESS.

1. Case Study 1: Producing a Land Cover Map of Wyoming

The objective of this project was to produce a detailed land cover map of Wyoming for purposes of biodiversity conservation and management (Driese *et al.*, 1997). The map was based on Landsat TM data from 1988 to 1993. Classification was performed using visual interpretation of color composites produced from TM bands 3 (red), 4 (infrared), and 5 (infrared). The color composites were displayed on a computer screen and boundaries between areas interpreted as different land cover types were delineated using on-screen digitizing. A total of 14,690 polygons were digitized, and each polygon was classified into a particular land cover type based on image interpretation and supporting information from existing maps, literature, and extensive field surveys. Boundaries between polygons classified into the same land cover type were dissolved using tools available in the GIS. The final map included 6167 polygons representing 41 land cover types (Fig. 6).

With standard GIS tools, the total area occupied by each land cover type was calculated. It was found that the two dominant land cover types in Wyoming are Wyoming big sagebrush (*Artemisia tridentata* ssp. *wyomingensis*) shrubland (33.4% of the area) and mixed grass prairie (17.5% of the area). Which of these types is more likely to occur in a particular site was found to be a function of the amount of summer rainfall (Fig. 7).

By merging the Wyoming land cover map with digital topographic data, it was possible to test the hypothesis that treeline elevation decreases from north to south. The results supported this hypothesis and indicated that upper treeline decreases at an average rate of 0.5 m/km from north to south. Lower treeline showed a corresponding decrease of 1.3 m/km. Treeline also decreased in elevation by 0.35 m for every degree of aspect away from south.

This case study shows two aspects of remote sensing that are relevant for studies of biodiversity: (1) the ability to produce detailed land cover maps and (2) the ability to test ecological hypotheses by integrating such maps with other types of data (e.g., climate and topography) within a GIS environment. Land cover maps such as the one produced for Wyoming are often used for making predictions on patterns of species distribution and species richness (see Case Study 10).

2. Case Study 2: Estimating Percentage Cover of Closed Coniferous Forests in Oregon

This project was designed to test the hypothesis that AVHRR data can be used to estimate the percentage cover of closed conifer forests in the Pacific Northwest (Ripple, 1994). The motivation for this project was legislation, approved by the U.S. Congress, that such forests should be protected in order to facilitate the long-term survival of the northern spotted owl, an endangered species that needs large stands of old-growth forests to survive extinction.

An area of about 2600 km² in the western Cascade Mountains of Oregon was chosen for the study. A sub-scene of NOAA-9 AVHRR data (from July 19, 1988) covering this area was resampled to 1×1 km resolution and was registered to UTM coordinates. Each pixel in the scene was characterized by its visible (0.58–0.68 μm) and near-IR (0.725–1.1 μm) band values. Landsat MSS images of the same area (from August 31, 1988) were resampled from the original 79-m resolution to a

50-m pixel size and were merged to the AVHRR scene such that each pixel in the AVHRR image corresponded to 20 × 20 pixels of the MSS image. Using unsupervised classification, the MSS data were classified into 100 spectral classes, and these classes were further classified into two broad categories: "closed-canopy conifer forest" and "other areas." By counting the number of MSS pixels classified as closed canopy conifer forest within each AVHRR pixel, it was possible to evaluate the relationships between the spectral characteristics of the AVHRR data and the proportion of closed conifer cover. Such analyses indicated that both visible and near-IR values of the AVHRR data decrease linearly with the proportion of closed conifer cover within a pixel (Fig. 8).

A regression model calibrated from these results was applied to a larger set of AVHRR data in an attempt to create a statewide map of the percentage cover of closed conifer forests. The resulting map (Fig. 9) was used to identify areas characterized by highly fragmented vs spatially continuous patches of closed conifer forests as well as to locate forest corridors that may link different regions within Oregon.

3. Case Study 3: Estimating Tree Density in Oak Woodland of Southern Spain

This project was designed to test the hypothesis that SPOT panchromatic data can be used to estimate tree density in savanna-like vegetation systems (Joffre and Lacaze, 1993). The spatial resolution of SPOT panchromatic data (10 m) is not much larger than the canopy of mature trees. Since trees differ in their reflectance in the visible spectrum from herbaceous plants, it was expected that SPOT panchromatic data would discriminate between pixels occupied by trees and pixels representing herbaceous vegetation. It was also expected that edge enhancement algorithms would facilitate image classification by amplifying the contrast between trees and their surrounding herbaceous vegetation.

These hypotheses were tested using data obtained for a landscape characterized by oak woodland in southwestern Spain. Figure 10 shows the relationships between actual tree density in a set of representative test sites and predictions of tree density based on two analytical methods. The first method was based on the original DN values of the panchromatic data. The second method was based on the thresholded Laplacian index, a measure that was calculated by integrating two types of edge enhancement algorithms. Both methods showed reasonable fit between predicted and observed densities. However, as expected, the application of edge enhancement algorithms significantly improved classification accuracy. Figure 11 presents a map of tree density classes derived from the enhanced data. Maps of this sort can be used to predict spatial patterns of tree density over large geographic areas.

4. Case Study 4: Delineation of Individual Tree Crowns in a Temperate Forest

In this study, a multiple-scale approach based on a series of image processing operations (image smoothing, edge detection, curvature estimation, and segmentation) was developed for automatically delineating individual tree crowns in high-resolution images (Brandtberg and Walter, 1998). The method was developed and tested using color IR aerial photographs of a temperate forest dominated by mixed and pure stands of Scotch pine (*Pinus silvestris*), Norway spruce (*Picea abies*), and Aspen (*Populus tremula*) trees. The photographs were scanned and the resulting images were resampled to a pixel size corresponding to 10 cm on the ground. Figure 12 shows a section of an original image and the result of the

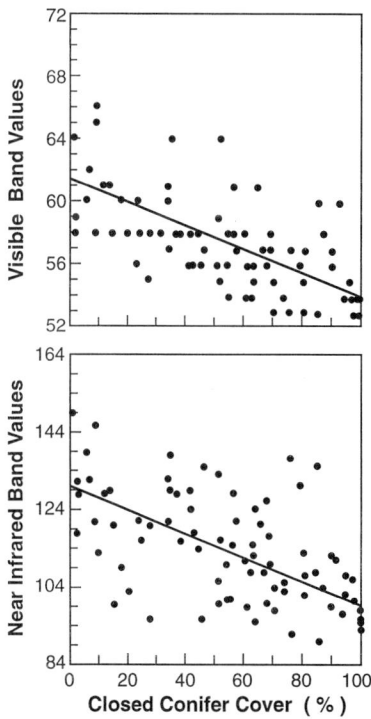

FIGURE 8 Relationships between spectral characteristics of AVHRR data (above, visible band values; below, near-IR band values) and the percentage cover of closed conifer forest in Oregon. Adapted with permission, the American Society for Photogrammetry and Remote Sensing. Ripple, W. J., Determining coniferous forest cover and forest fragmentation with NOAA-9 advanced very high resolution radiometer data. *Photogramm. Eng. Remote Sens.* 60, 533–540, 1994.

FIGURE 9 A map showing the distribution of four classes of closed conifer cover in Oregon, derived from AVHRR data. Reproduced with permission, the American Society for Photogrammetry and Remote Sensing. Ripple, W. J., Determining coniferous forest cover and forest fragmentation with NOAA-9 advanced very high resolution radiometer data. *Photogramm. Eng. Remote Sens.* **60**, 533–540, 1994.

delineation algorithm. Accuracy assessment based on 43 ground truthing plots indicated that, on average, 70% of the trees were identified correctly. This was almost equivalent to visual interpretation, suggesting that this method can be used for mapping large forest areas at the individual tree level.

B. Measuring Vegetation Changes

Identifying patterns, scales, and trends of vegetation changes is important for understanding the dynamics and structure of both plant and animal communities. As with mapping applications, measurements of vegetation changes based on remotely sensed data may provide direct evidence for changes in biodiversity (e.g., changes in the relative frequency of different vegetation types) as well as information on environmental changes that may influence biodiversity patterns (e.g., deforestation and forest fragmentation). The case studies presented here illustrate four different aspects of vegetation changes: interannual fluctuations in vegetation cover (Case Study 5), forest succession (Case Study 6), deforestation (Case Study 7), and tree demography (Case Study 8). The respective data sources are AVHRR, Landsat MSS, SPOT multispectral, and black-and-white aerial photographs. Time spans analyzed in these case studies range from 10 to 54 years.

1. Case Study 5: Interannual Vegetation Changes in Sub-Saharan Africa

This study was designed to evaluate the magnitude of interannual changes in the vegetation of sub-Saharan Africa and to identify the spatial and temporal characteristics of these changes (Lambin and Ehrlich, 1997). It was based on daily global area coverage (GAC) data acquired by NOAA AVHRR from 1982 to 1991. Two measures were calculated from the AVHRR data, the normalized difference vegetation index (NDVI) and an index of surface temperature (T_s) calculated from the thermal bands of the AVHRR. Theoretical considera-

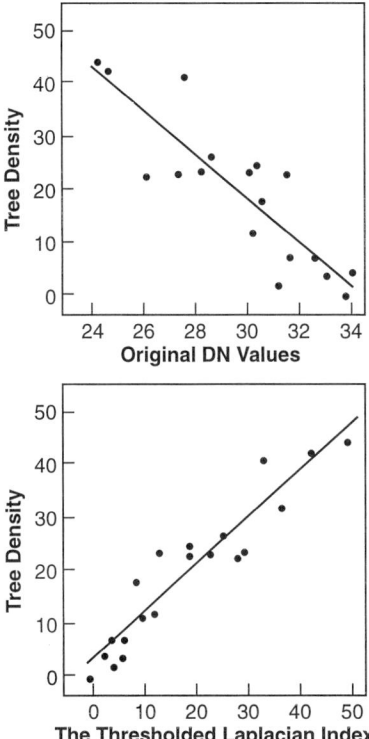

FIGURE 10 Relationships between actual tree density in representative sites of oak woodland and predictions of density based on two kinds of measures derived from SPOT panchromatic data: the original DN values (above), and the thresholded Laplacian index, a measure that integrates two types of edge enhancement algorithms (below). Predictions based on edge enhancement were more accurate than those based on the original data. Adapted from Joffre and Lacaze (1993). Estimating tree density in oak savanna-like 'dehesa' of southern Spain from SPOT data. *Int. J. Remote Sens.* 14, 685–697. Copyright 1993, with permission from Taylor & Francis (http://www.tandf.co.uk/journal).

FIGURE 11 A map showing the distribution of five classes of tree density, produced for a savanna-like ecosystem in southwestern Spain using SPOT panchromatic data. Reproduced from Joffre and Lacaze (1993). Estimating tree density in oak savanna-like 'dehesa' of southern Spain from SPOT data. *Int. J. Remote Sens.* 14, 685–697. Copyright 1993, with permission from Taylor & Francis (http://www.tandf.co.uk/journal).

tions and empirical data have indicated that the arctangent of the ratio between T_s and NDVI ($\arctan(T_s/\text{NDVI})$) is negatively and linearly correlated with the amount of vegetation cover. This index was therefore computed for each monthly period for the 10 years of observation. The status of the vegetation in a given pixel in a given year was determined by subtracting the vector of the monthly values of $\arctan(T_s/\text{NDVI})$ recorded for the relevant pixel during the year of observation from a corresponding vector representing the "best" conditions recorded for that particular pixel during the study period. Images of this standardized vegetation index were constructed for every year from 1982 to 1991.

Figure 13a presents an image of the standardized vegetation index produced for 1983. The brightness of a given pixel in this image indicates how much vegetation conditions in 1983 were different from the best possible conditions recorded for this particular pixel during the whole study period. A comparison of this image with a map of 1983 drought events produced by the U.S. Department of Commerce Climate Impact Assessment (the areas outlined by dashed lines) indicates that areas that experienced droughts in 1983 were characterized by relatively high brightness values. Analysis of similar images constructed for other years confirmed the conclusion that the standardized vegetation index can be used effectively to identify patterns of vegetation changes caused by rainfall fluctuations. Interestingly, while differences between images of successive years revealed a high degree of interannual variability in the vegetation, only small parts of the continent exhibited continuous trends of vegetation changes (Fig. 13b). This result contrasts the "marching desert" prospect and similar views concerning the extent of deforestation and land degradation in Africa.

The overall results of this project indicate that spectral indices derived from AVHRR data can be used effectively for monitoring and interpreting continental-scale patterns of changes in vegetation cover.

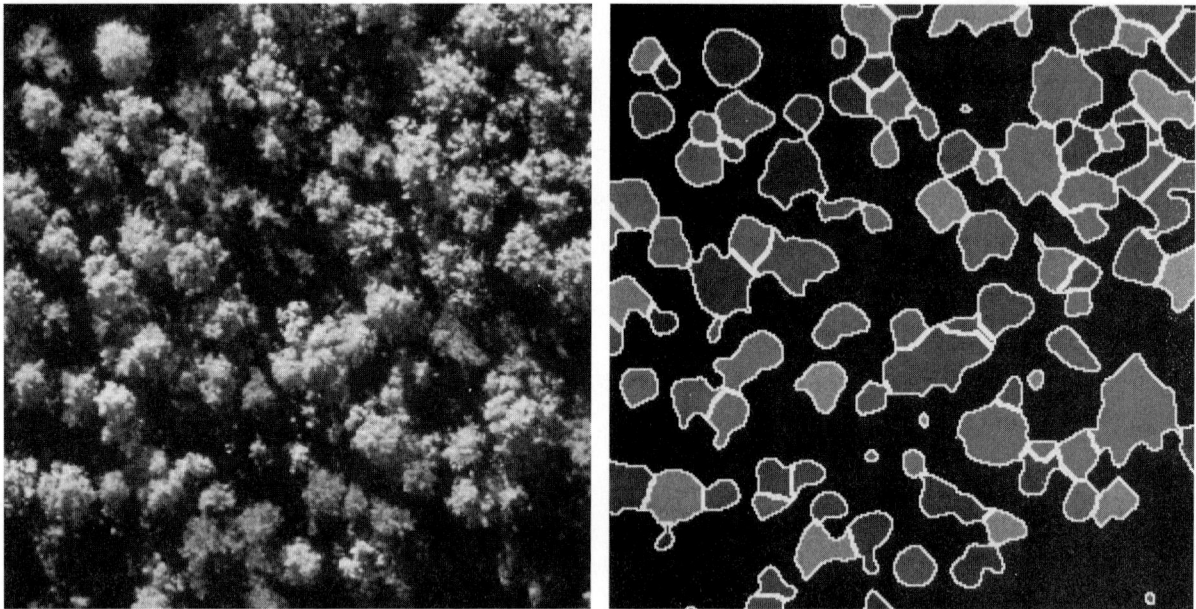

FIGURE 12 A section of an aerial photograph of a temperate forest (left) and the result of a canopy delineation algorithm applied to this image (right). Pixel size is 10 cm. Reproduced from *Machine Vision and Applications*, Automated delineation of individual tree crowns in high spatial resolution aerial images by multiple scale analysis, Brandtberg, T., and Walter, F., Vol. 11, pp. 64–73, Figs. 2 and 11. Copyright 1998, with permission from Springer-Verlag.

2. Case Study 6: Forest Succession in Northeastern Minnesota

In this study, Landsat MSS data were used to investigate succession processes in boreal forests of northeastern Minnesota (Hall *et al.*, 1991). The analysis was based on a comparison of two images: one from July 3, 1973 (acquired by Landsat 1), and the second from August 18, 1983 (acquired by Landsat 4). Training sites located on the 1983 image were used to define the spectral characteristics of five major states of forest succession: clearing, regeneration, broadleaf, mixed, and conifer. Figure 14 shows the spectral ranges of these states in a two-dimensional space defined by the green and the near-IR MSS bands. The spectral boundaries defined by the training sites were used to classify all unknown pixels in the two images. The 1973 image was subjected to radiometric corrections prior to the classification process in order to compensate for differences in growing season, atmospheric conditions, and sensor response characteristics between the two acquisition dates.

To quantify succession rates, the two images were coregistered and the frequencies of all possible state transitions (clearing to clearing, clearing to regeneration, clearing to broadleaf, etc.) were summarized in the form of a transition probability matrix (Table II). Each element in such a matrix indicates a probability of transition from one state to another within a certain time period (10 years in this study). Examination of the values obtained for the diagonal elements of the matrix indicates that late successional states were more stable during the study period than earlier states (i.e., had higher probabilities of staying in the same state). It can also be seen that the primary succession pathway (defined by the largest transition probability in each row) was from clearings to regeneration, to broadleaf, to mixed, to conifer. This result was consistent with ecological expectations.

A transition matrix like the one computed in this study can be used to project future changes in the vegetation, to estimate the steady-state composition of the system, and to calculate dynamic characteristics such as the recurrence time of various successional states. Remote sensing is the only tool available for studying such aspects of succession over large spatial and temporal scales.

3. Case Study 7: Deforestation in the Philippines

This project was designed to investigate factors leading to deforestation in the Philippines (Liu *et al.*, 1993). Previous studies have suggested that road building may increase deforestation rates by improving accessibility to undisturbed forest stands. Other studies proposed

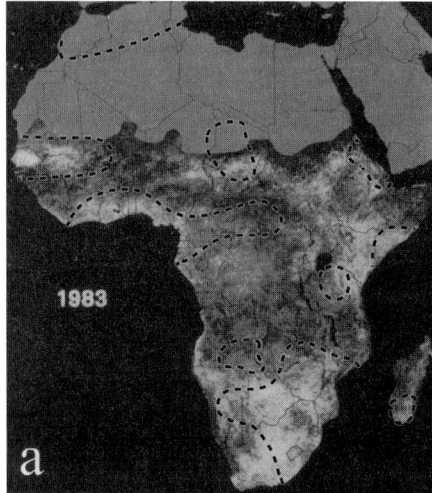

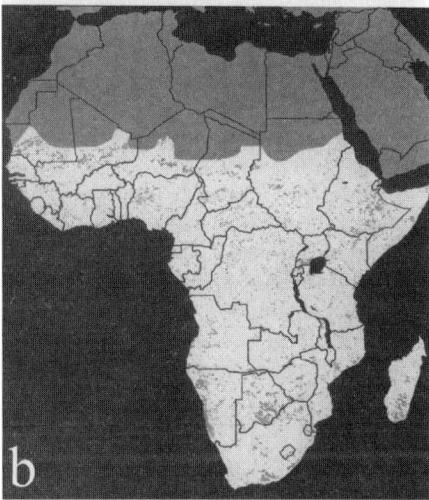

FIGURE 13 Vegetation changes in Africa as obtained from the analysis of AVHRR data. (a) An image of a standardized vegetation index based on monthly NDVI and T_s values produced for 1983. The brightness of each pixel in the image indicates how much vegetation conditions in 1983 were different from the best conditions recorded for this particular pixel between 1982 and 1991. Thus, bright areas indicate relatively unfavorable vegetation conditions. Dashed lines indicate areas that experienced droughts in 1983. Note that such areas are characterized by relatively high values of the standardized vegetation index. (b) Areas showing a continuous gain (gray spots) and continuous loss (black spots) in vegetation cover over a period of at least 6 years between 1982 and 1991. Only small parts of sub-Saharan Africa experienced a continuous change in vegetation cover during this period. Adapted from *Remote Sensing of Environment* **61**, Lambin and Ehrlich, Land cover changes in Sub-Saharan Africa (1982–1991): Application of a change index based on remotely-sensed surface temperature and vegetation indices at a continental scale, pp. 191–200. Copyright 1997, with permission from Elsevier Science.

that small forest patches and patches with long perimeter per unit area (i.e., more edge exposed to humans) are more likely to be disturbed than large patches. It was therefore expected that (1) deforestation rates in

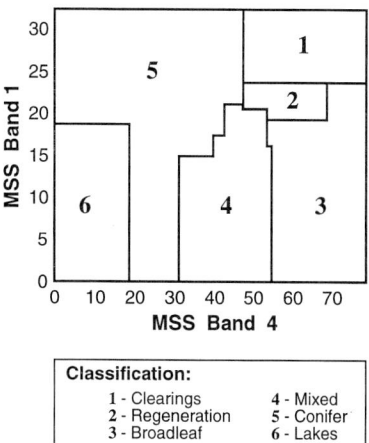

FIGURE 14 Spectral ranges of different states of boreal forest succession in a two-dimensional space defined by the green and the near-IR MSS bands. Boundaries between successional states were determined by analyzing the spectral characteristics of data obtained from a set of representative training sites. Adapted from Hall *et al.* (1991), with permission from the Ecological Society of America.

the Philippines would decrease with increasing distance to the closest road and (2) forest patches characterized by a high perimeter-to-area (P/A) ratio would show higher rates of forest loss than patches characterized by a low P/A ratio.

These hypotheses were tested using GIS tools by integrating three types of data: a land-use map of the Philippines derived from 187 SPOT multispectral images recorded during 1987–1988, a land-use map from 1934 based on survey data, and a road map of the Philippines from 1941. Due to differences in classification systems between the two land-use maps, all land cover types in both maps were aggregated into two main groups: forest and nonforest.

The effect of roads on deforestation rates was analyzed using a buffering approach. Buffer zones were created at 1.5-km intervals around the 1941 road network. The smallest buffer included areas at a maximum distance of 1.5 km from the closest road, the second buffer covered areas up to 3 km from the closest road, etc. The largest buffer (22.5 km) covered almost the whole land area of the Philippines. These buffers were overlaid in the GIS on the forest/nonforest maps, and the percentage change in forest area between 1934 and 1988 was determined for each buffer ring (0–1.5 km, 1.5–3 km, 3–4.5 km, etc.). The results (Fig. 15) supported the hypothesis that road building accelerate deforestation. The percentage of forest loss was highest (78%) for areas within 1.5 km of roads and decreased linearly to about 40% at areas located 15 km from the closest road. Further analyses of the data indicated that forest loss was concentrated in patches characterized

TABLE II

Transition Matrix Showing Probabilities of Transition between Different Successional States in a Temperate Northern Minnesota Forest As Calculated from Landsat MSS Data[a]

From	To				
	Clearings	Regeneration	Broadleaf	Mixed	Conifer
Clearings	0.17	0.46	0.17	0.15	0.05
Regeneration	0.05	0.31	0.17	0.37	0.10
Broadleaf	0.01	0.20	0.47	0.28	0.04
Mixed	0.01	0.07	0.11	0.58	0.23
Conifer	0.01	0.04	0.02	0.31	0.58

[a] Probabilities are calculated for a time step of 10 years (1973–1983). Adapted from Hall et al. (1991), with permission from the Ecological Society of America.

by a high P/A ratio (Fig. 16). These results are consistent with the hypothesis that distance from road and the P/A ratio of forest patches are important determinants of deforestation processes in the Philippines.

4. Case Study 8: Demography of *Acacia* Trees in the Negev Desert

This study was designed to investigate the long-term demography of *Acacia* trees in the Negev desert, Israel (Lahav-Ginott and Kadmon, unpublished data). *Acacia* trees function as keystone species in this ecosystem by providing food and hospitable habitat conditions for many organisms, including insects, reptiles, birds, and mammals. Recent field studies have documented large-scale mortality of *Acacia* trees in the Negev and pointed to the importance of evaluating the long-term viability of these populations. This study applied a demographic approach based on the analysis of historical aerial photographs, to evaluate the viability of *Acacia* populations in the Negev.

Long-term changes in the density of *Acacia* populations were studied using image processing of black-and-white aerial photographs from 1956 and 1996. The photographs were scanned and the resulting images were corrected for geometric distortions and resampled into a spatial resolution of 30 cm. Since canopies of trees growing in this desert environment rarely overlap, it was possible to derive maps of individual tree distribution from the classified images (Fig. 17a). These maps were analyzed using a GIS in an attempt to (1) estimate changes in density of *Acacia* populations in two different sites, (2) calculate corresponding rates of mortality and recruitment, and (3) test for differences in these demographic rates between habitats characterized by high vs low input of runoff water.

The results (Table III) revealed a general trend of increase, rather than a decrease, in tree density. The magnitude of this increase varied between the two sites as well as between the two habitat types. Mortality and recruitment rates were determined by overlaying the 1956 and 1996 images in the GIS and distinguishing between three states of trees: trees that were present in 1956 but absent in 1996, trees that were present in both years, and trees that were present only in 1996 (Fig. 17b). The results of these GIS operations revealed relatively high rates of mortality and recruitment (Table III) and pointed to some patchiness in the distribution of mortality and recruitment events (Fig. 17b). The overall results of this project indicate that image processing of historical aerial photographs may provide important insights into the long-term dynamics of desert tree populations.

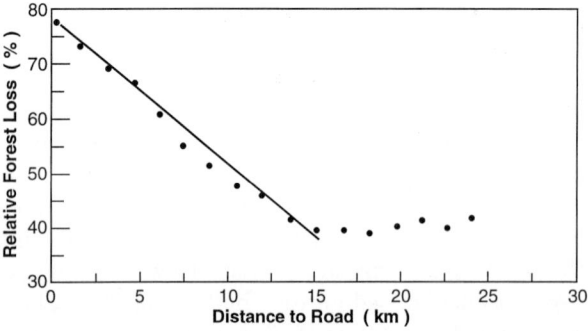

FIGURE 15 The relationship between relative forest loss and distance to the closest road in the Philippines. Adapted from *Forest Ecology and Management* 57, Liu et al., Rates and patterns of deforestation in the Philippines: Application of geographic information system analysis, pp. 1–16, Copyright 1993, with permission from Elsevier Science.

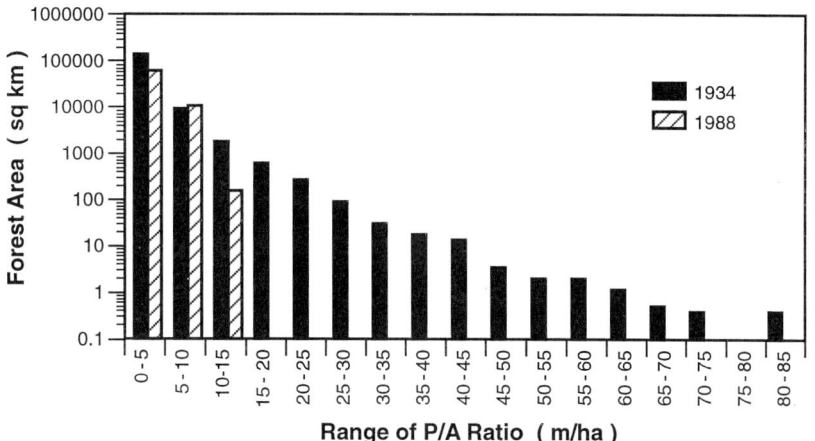

FIGURE 16 The area of forest patches by perimeter-to-area (P/A) ratio classes in 1934 and 1988 for the Philippines. Adapted from *Forest Ecology and Management* 57, Liu et al., Rates and patterns of deforestation in the Philippines: Application of geographic information system analysis, pp. 1–16, Copyright 1993, with permission from Elsevier Science.

C. Modeling Species Distribution and Species Richness

Obviously, most plant and animal species cannot be detected directly by remote sensing. However, the integration of data obtained from remote sensing with field-based observations of plants and animals enables one to develop statistical models that predict distribution patterns of individual species from information on the environment. A similar approach can be applied to model and predict patterns of species diversity. Examples for these two types of applications are presented in the following case studies.

1. Case Study 9: Spatial Distribution of Buzzard Nesting Sites in Scotland

This study used vegetation cover data derived from satellite imagery and topographic variables derived from a digital terrain model to identify factors affecting the distribution of buzzard (*Buteo buteo*) nesting sites in upland Argyll, Scotland (Austin et al., 1996).

A vegetation map of the study area was produced using supervised classification of a Landsat 5 image (TM bands 3, 4, and 5) recorded in May 1990. A grid at a resolution of 500 m was superimposed on the vegetation map, and each grid cell was classified based on whether it was known to contain a buzzard nest or not. With GIS tools, the relative area of different vegetation categories was calculated for circles of radii 500, 1000, and 1500 m about the center of each grid cell. The length of boundaries between different vegetation categories was also determined for each radius, to give an index of habitat heterogeneity. Finally, a set of topographic indices was extracted from a digital terrain model to describe the topography within, and around, each grid cell. Overall, 165 habitat variables were derived from the satellite and topographic data as potential predictors of nesting site locations.

Analysis of this huge data set indicated that buzzard nesting sites could be predicted with an accuracy level of about 90% using regression models based on only four variables: (1) area of heathland within 500 m, (2) length of borders between prethicket forestry and open land within 500 m, (3) total length of borders between major land cover categories within 1500 m, and (4) median altitude within 500 m. The ecological significance of these results was interpreted in terms of food availability, food preference, territoriality, and foraging and hunting behavior of the buzzard. A probability map of buzzard nesting sites based on model predictions showed a close agreement with observed patterns of nesting site distribution (Fig. 18). These results indicated that satellite imagery may help to predict the location of buzzard nesting sites over large geographic areas.

2. Case Study 10: Vertebrate Species Richness in Wyoming

This study used data derived from AVHRR and Landsat TM imagery to investigate factors affecting vertebrate species richness in Wyoming (Fraser, 1998). Previous studies of North American vertebrates have pointed to

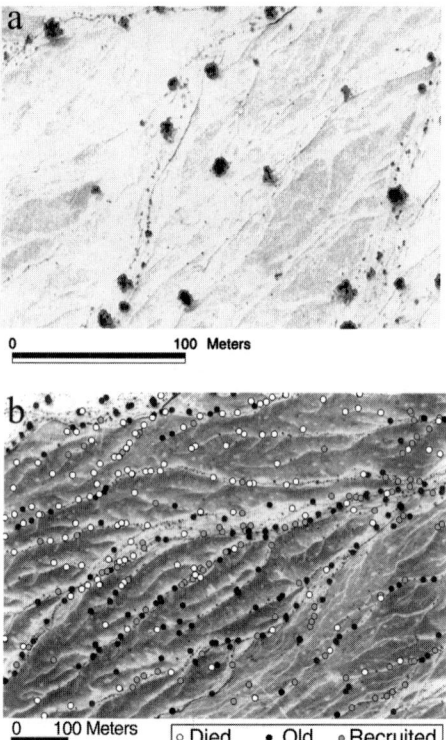

FIGURE 17 Analysis of *Acacia* tree populations in the Negev desert using image processing of aerial photographs. (a) A section of an aerial photograph from 1996 with a corresponding map of individual tree distribution derived by computerized delineation of tree canopies. Only trees with a canopy >6 m² were delineated. (b) A section of an aerial photograph from 1956 with data on *Acacia* tree distribution. Trees larger than 6 m² were classified into three states: trees that were present in 1956 but absent in 1996 (died, open circle), trees that were present in both years (old, black circle), and trees that were present only in 1996 (recruited, gray circle). Note the magnitude of turnover in tree density and the patchiness in mortality and recruitment events. S. Lahav-Ginott and R. Kadmon, unpublished data.

TABLE III

Demographic Parameters Calculated for *Acacia* Populations in the Negev Desert Using Image Processing of Aerial Photographs from 1956 and 1996[a]

	Density (number/ha)		Mortality (number/ha)	Recruitment (number/ha)
	1956	1996		
Site a				
Low water input	5.5	5.7	2.6	2.8
High water input	3.9	5.0	0.9	2.0
Site b				
Low water input	5.2	7.2	1.3	3.3
High water input	2.3	4.7	0.3	2.7

[a] Data are mean values for two sites and for two types of habitats differing from each other in the contribution of runoff water (Lahav-Ginott, S., and Kadmon, R., unpublished data).

and the number of land cover types as derived from Landsat TM data). Figure 19 shows the relationships between species richness and the most strongly related energy (maximum July NDVI) and heterogeneity (land cover variety) parameters. These results indicate that environmental heterogeneity may be more important than energy availability in determining mesoscale patterns of species richness.

It is interesting to compare these results with a study focusing on bird species diversity in Senegal (Nohr and Jorgensen, 1997). As with the Wyoming study, patterns

the importance of energy as a factor controlling continental-scale patterns of species richness (the "species-energy" hypothesis). In contrast, data from local-scale studies have usually emphasized the importance of habitat heterogeneity as a determinant of species richness. This study focused on intermediate spatial scales, and its main purpose was to evaluate the relative importance of energy availability vs environmental heterogeneity as factors affecting mesoscale vertebrate species richness.

In general, parameters of energy availability (e.g., maximum July NDVI and forest biomass, both derived from AVHRR data) were less correlated with species richness than parameters of environmental heterogeneity (e.g., NDVI variability, forest biomass variability,

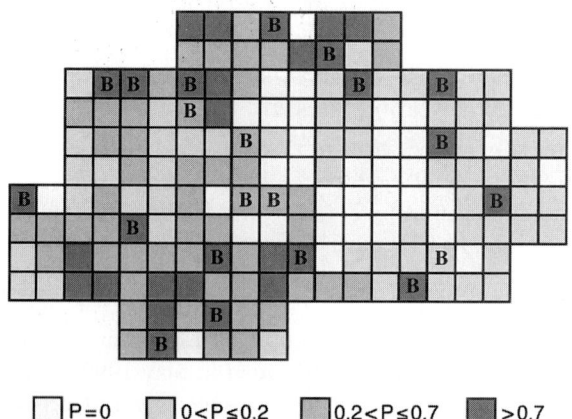

FIGURE 18 A probability map of buzzard nesting sites. Probabilities were estimated using a regression model based on vegetation variables derived from Landsat TM data. Observed nesting sites are indicated as B. Grid cells are 500 × 500 m. Adapted from Austin *et al.* (1996), with permission from Blackwell Science Ltd. See also color insert, Volume 1.

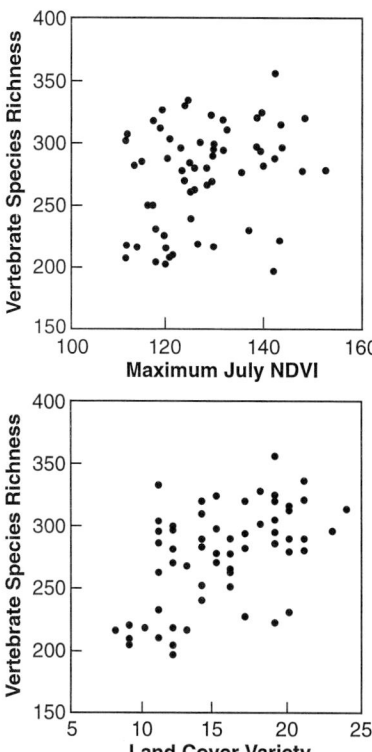

FIGURE 19 Relationships between vertebrate species richness in Wyoming and two environmental indices based on satellite imagery: maximum July NDVI (derived from AVHRR data) and land cover variety (derived from Landsat TM data). Adapted from Fraser (1998), with permission from Blackwell Science Ltd.

of species richness were analyzed with respect to parameters of vegetation biomass (derived from AVHRR data) and landscape heterogeneity (derived from Landsat data). A regression model integrating the effects of biomass and landscape diversity (Fig. 20) accounted for 36% of the observed variation in bird species diversity (as expressed by Simpson's diversity index). Based on this model, a predictive map of bird species diversity was produced for an area of about 30,000 km² in northern Senegal (Fig. 21).

These studies demonstrate that information derived from satellite imagery can be used to investigate factors affecting large-scale patterns of species diversity.

D. Other Applications

The case studies presented above exemplify various aspects of biodiversity that can be investigated using remote sensing. Three additional aspects, not covered by these case studies, are mentioned here to provide a more complete picture of the subject. These are fire, grazing, and canopy gap dynamics.

Fire is a major source of disturbance in many kinds of ecosystems. Studies of fire distribution in space and time may therefore contribute to our understanding of the dynamics and structure of ecological communities. Remotely sensed data have been used to reconstruct fire histories, to estimate spatial and temporal characteristics of fire disturbances (e.g., patch size and frequency), to assess fire severity, to evaluate the susceptibility of different vegetation types to fire, and to measure rates of vegetation recovery following fires. Patterns of biomass burning were mapped using remotely sensed data for spatial scales ranging from local sites to the entire Earth. As one example, Fig. 22 presents a fire map constructed for western Africa by integrating data acquired by two imaging systems: the Defense Meteorological Satellite Program (DMSP) administrated by the U.S. Air Force and the NOAA AVHRR radiometer. This map was used to evaluate the role of fire as a determinant of regional vegetation changes in Africa (Ehrlich et al., 1997).

Another aspect of biodiversity that has been studied extensively with remote sensing is grazing. Data obtained from remote sensing were used to investigate vegetation responses to both natural and domestic grazing. Some studies have focused on spatial patterns (e.g., differences between fenced and nonfenced areas), while other studies were designed to quantify temporal vegetation changes caused by grazing (e.g., biomass removal, replacement of herbaceous vegetation by woody species, or shrub encroachment). A study of vegetation responses to grazing is exemplified in Fig. 23. This study investigated the effect of cattle and goat grazing on long-term vegetation dynamics in a disturbed Mediterranean ecosystem. By integrating data on changes in vegetation cover derived from image processing of historical aerial photographs (Fig. 23), with corresponding maps of grazing type (cattle vs goats) and intensity (high, moderate, or low), it was possible to evaluate the effect of the grazing regime on long-term changes in the vegetation. Such analyses indicated that both cattle grazing and goat grazing reduce the rate of tree regeneration in this ecosystem and that the negative impact of goats on tree regeneration is stronger than that of the cattle.

Canopy gaps caused by the death of one or more trees are a major disturbance mode in many temperate and tropical ecosystems. Such disturbances may have considerable effects on the diversity of plant and animal communities. Traditionally, studies of canopy gap dynamics have focused on simple "static" parameters such

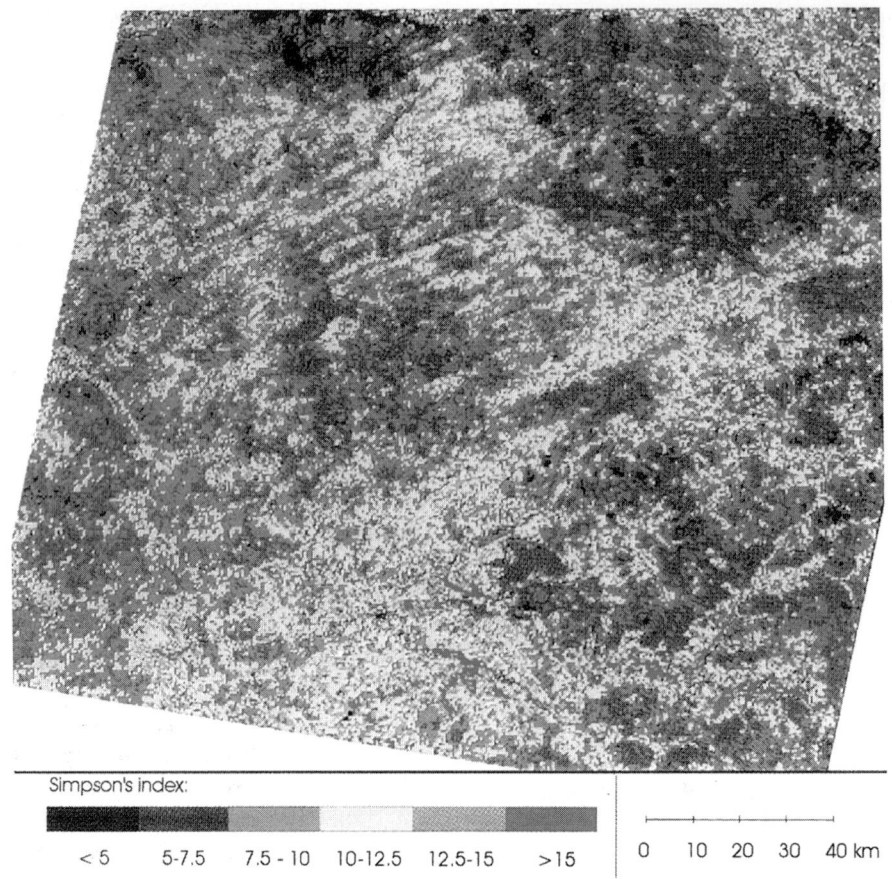

FIGURE 20 A predictive map of bird species diversity (expressed by Simpson's index) constructed for an area of about 30,000 km² in Senegal using the regression model presented in Fig. 20. Reproduced from Nohr and Jorgensen (1997), with permission from Kluwer Academic Publishers.

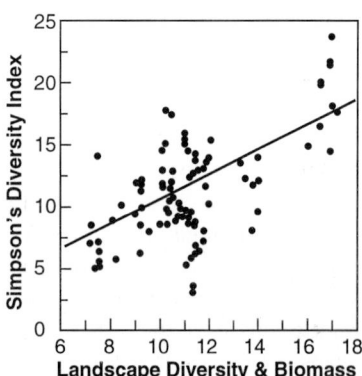

FIGURE 21 Bird species diversity in Senegal plotted against an index combining information on landscape diversity (derived from Landsat TM data) and vegetation biomass (derived from AVHRR data). Adapted from Nohr and Jorgensen (1997), with permission from Kluwer Academic Publishers.

as gap size distribution and total gap area. An alternative (and more direct) approach for studying gap dynamics is to analyze time sequences of remotely sensed data. Such an approach is exemplified in Fig. 24. This figure shows a map of canopy gaps generated for a temperate forest in Japan using digital analysis of aerial photographs from 1981 and 1986 (Tanaka and Nakashizuka, 1997). Data derived from such maps can be used to parameterize models of gap dynamics and to predict the equilibrium structure of the forest canopy.

IV. SUMMARY

Remote sensing in general, and Earth observation satellites in particular, have brought a new dimension to our understanding of the Earth's surface. Since the launch of the first Landsat satellite in 1972, millions of images of the Earth have been acquired from spaceborne imaging

FIGURE 22 A fire map constructed for western Africa using satellite images from the period 1984 to 1988. Areas for which a fire was detected once are represented in red. Areas for which two or more fires were detected during this period are identified by yellow and green, respectively. Reproduced from *Remote Sensing of Environment* **61**, Ehrlich et al., Biomass burning and broad scale land cover changes in Western Africa, pp. 201–209. Copyright 1997, with permission from Elsevier Science. See also color insert, Volume 1.

systems. Earth observation satellites have provided images over spatial scales ranging from local regions to global cover, at spatial resolution of a few meters to tenths of a kilometer, and at spectral wavelengths ranging from near-ultraviolet to microwave radiation. Some of these data have been acquired systematically and repetitively over many years, with temporal resolution ranging from hours to a few weeks. In addition, most of the Earth observation satellites were designed to provide multispectral images of the Earth's surface. This enormous amount of information has been utilized extensively for studies of biodiversity. As demonstrated in this article, data obtained from remote sensing have been used for mapping various components of biodiversity, measuring different ecosystem characteristics (e.g., standing biomass and primary productivity), analyzing factors affecting the distribution of plant and animal species, monitoring natural and human-induced vegetation changes (e.g., succession and deforestation), and evaluating the impact of various disturbances (e.g., fire) and ecological interactions (e.g., grazing). The case studies presented in the previous section exemplify some of these capabilities.

It should be noted, however, that this review focuses on the two most common forms of remote sensing, namely, aerial photography and electro-optical satellite imagery. Two other forms of remote sensing that should be mentioned here are hyperspectral scanners and radars. Hyperspectral scanners are imaging systems that acquire multispectral images in many, very narrow, contiguous spectral bands throughout the visible, near-IR, and mid-IR portions of the spectrum. This spectral capability permits discrimination among objects that have fine diagnostic reflection characteristics that are "lost" within the bands of conventional electro-optical scanners. For example, AVIRIS (Airborne Visible/Infrared Imaging Spectrometer), one of the main hyperspectral scanners now in use, collects data in 224 contiguous bands between 0.4 and 2.45 μm.

Imaging radars have been used for remote sensing since the late seventies. However, most of these applications were of experimental nature. For example, the Shuttle Imaging Radar (SIR) carried by the space shuttle Columbia in its second flight was used to collect data from different continents in an attempt to construct a library of images representing a wide variety of environments differing with respect to climate, geology, vegetation, and other qualities. More recent studies have evaluated the potential of radar data for vegetation mapping, detection of deforestation, determination of vegetation biomass, and estimation of various forest parameters. Such applications have led to a better understanding of the capabilities of radar imagery and demonstrated that this form of remote sensing may complement many characteristics of images based on the optical spectrum.

There are many reasons to believe that in the near future much more information on biodiversity will be obtained from remote sensing. First, recent imaging systems are equipped with more advanced sensors. For example, the imaging system of SPOT 4 (launched on March 24, 1998) has been improved by adding a 20-m resolution band in the mid-IR portion of the spectrum

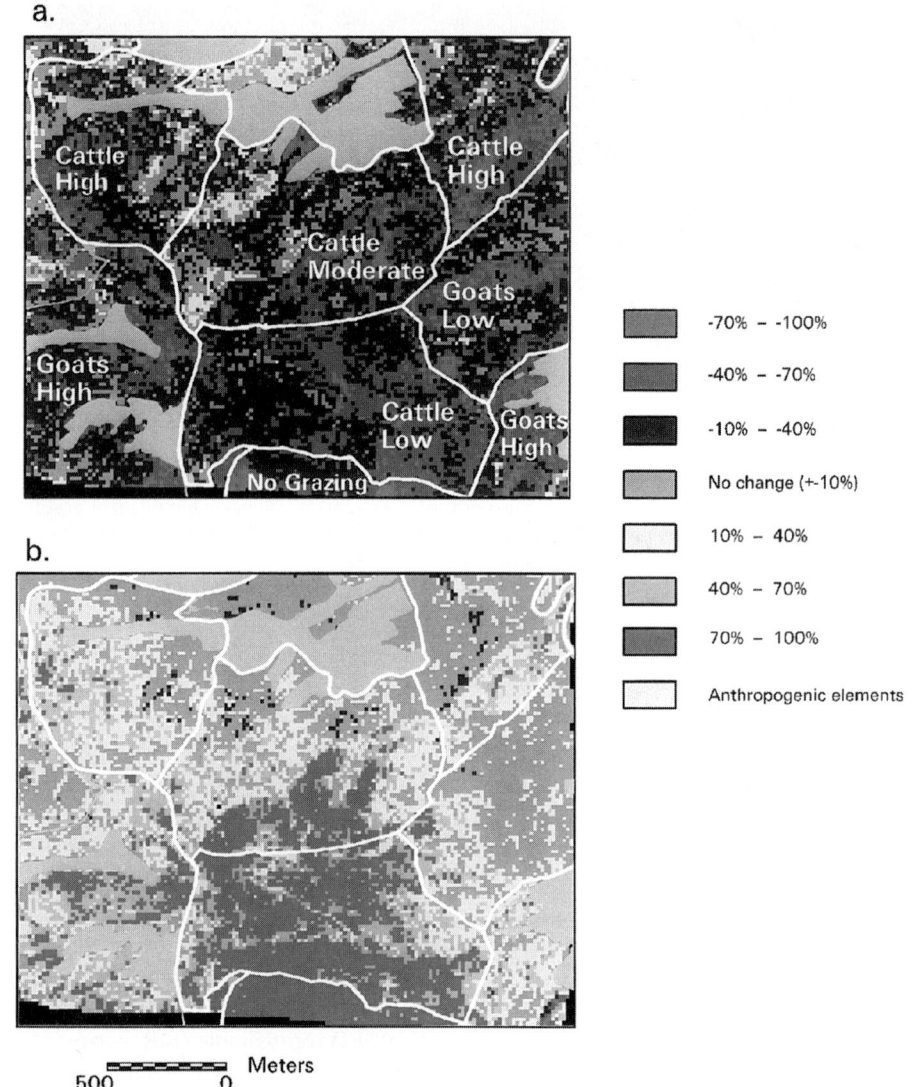

FIGURE 23 Maps of changes in the percentage cover of herbaceous vegetation (a) and trees (b) between 1964 and 1992 in a Mediterranean ecosystem. Data on vegetation cover were obtained using image analysis of black-and-white aerial photographs. Changes in vegetation cover were calculated as (1992 cover) − (1964 cover). Boundaries of plots subjected to different intensities of cattle and goat grazing are superimposed on the maps. Y. Carmel and R. Kadmon, unpublished data.

(1.58–1.75 μm). SPOT 4 also carries a new imaging instrument called VEGETATION, which was designed to provide daily, global data of the continental biosphere. Landsat 7, the most recent satellite of the Landsat program (launched on April 15, 1999) carries a new, eight-band sensor called the Enhanced Thematic Mapper Plus (ETM+) that includes an additional panchromatic band (0.52–0.90 μm) with 15-m spatial resolution. Another advantage of the ETM+ is an increase in the spatial resolution of the thermal IR channel from 120 to 60 m. During the writing of this article (September 24, 1999), an imaging satellite called IKONOS was launched and opened a new generation of imaging capabilities. This satellite will provide panchromatic images at a spatial resolution of 1 m and multispectral images at a resolution of 4 m. These levels of spatial resolution are much higher than those of any other satellite data available today and approximate the resolution obtained by standard aerial photography. In parallel to these advances in space technology, new analytical

FIGURE 24 Spatial distribution of canopy gaps in a temperate forest, based on the analysis of aerial photographs from 1981 and 1986. Gaps existing during 1981 are shaded gray and gaps created between 1981 and 1986 are red. Note that newly created gaps tend to occur at the edges of old gaps. Adapted from Tanaka and Nakashizuka (1997), with permission from the Ecological Society of America. See also color insert, Volume 1.

methodologies are being developed for the analysis and interpretation of high-resolution and hyperspectral images. We can therefore expect a considerable growth in the contribution of remote sensing to studies of biodiversity during the next decade.

See Also the Following Articles

COMPUTER SYSTEMS AND MODELS, USE OF • MEASUREMENT AND ANALYSIS OF BIODIVERSITY • NATURAL RESERVES AMD PRESERVES

Bibliography

Austin, G. E., Thomas, C. J., Houston, D. C., and Thompson, D. B. A. (1996). Predicting the spatial distribution of buzzard *Buteo buteo* nesting areas using a geographical information system and remote sensing. *J. Appl. Ecol.* 33, 1541–1550.

Avery, T. E., and Berlin, G. L. (1992). *Fundamentals of Remote Sensing and Airphoto Interpretation,* 5th ed. Prentice Hall, Englewood Cliffs, NJ.

Brandtberg, T., and Walter, F. (1998). Automated delineation of individual tree crowns in high spatial resolution aerial images by multiple-scale analysis. *Mach. Vis. Appl.* 11, 64–73.

Campbell, J. B. (1996). *Introduction to Remote Sensing,* 2nd ed. Taylor & Francis, London.

Driese, K. L., Reiners, W. A., Merrill, E. H., and Gerow, K. G. (1997). A digital land cover map of Wyoming, USA: A tool for vegetation analysis. *J. Veg. Sci.* 8, 133–146.

Ehrlich, D., Lambin, E. F., and Malingreau, J. P. (1997). Biomass burning and broad-scale land-cover changes in Western Africa. *Remote Sens. Environ.* 61, 201–209.

Fraser, R. H. (1998). Vertebrate species richness at the mesoscale: Relative roles of energy and heterogeneity. *Glob. Ecol. Biogeogr. Lett.* 7, 215–220.

Hall, F. G., Botkin, D. B., Strebel, D. E., Woods, K. D., and Goetz, S. J. (1991). Large-scale patterns of forest succession as determined by remote sensing. *Ecology* 72, 628–640.

Joffre, R., and Lacaze, B. (1993). Estimating tree density in oak savanna-like 'dehesa' of southern Spain from SPOT data. *Int. J. Remote Sens.* 14, 685–697.

Lambin, E. F., and Ehrlich, D. (1997). Land-cover changes in Sub-Saharan Africa (1982–1991): Application of a change index based on remotely-sensed surface temperature and vegetation indices at a continental scale. *Remote Sens. Environ.* 61, 181–200.

Lillesand, T. M., and Kiefer, R. W. (1994). *Remote Sensing and Image Interpretation,* 3rd ed. Wiley, New York.

Liu, D. S., Iverson, L. R., and Brown, S. (1993). Rates and patterns of deforestation in the Philippines: Application of geographic information system analysis. *For. Ecol. Manage.* 57, 1–16.

Lobo, A., Moloney, K., Chic, O., and Chiariello, N. (1998). Analysis of fine-scale pattern of a grassland from remotely-sensed imagery and field collected data. *Landscape Ecol.* 13, 111–131.

Nemani, R., and Running, S. (1997). Land cover characterization using multitemporal red, near IR, and thermal-IR data from NOAA/AVHRR. *Ecol. Appl.* 7, 79–90.

Nohr, H., and Jorgensen, A. F. (1997). Mapping of biological diversity in Sahel by means of satellite image analyses and ornithological surveys. *Biodivers. Conserv.* 6, 545–566.

Ripple, W. J. (1994). Determining coniferous forest cover and forest fragmentation with NOAA-9 advanced very high resolution radiometer data. *Photogramm. Eng. Remote Sens.* 60, 533–540.

Tanaka, H., and Nakashizuka, T. (1997). Fifteen years of canopy dynamics analyzed by aerial photographs in a temperate deciduous forest, Japan. *Ecology* 78, 612–620.

REPTILES, BIODIVERSITY OF

F. Harvey Pough
Arizona State University West

I. Extant Evolutionary Lineages
II. Ancestral and Derived Characters
III. Characteristics of Groups
IV. Conservation Issues

GLOSSARY

ancestral Describes a character or character state of the organism being considered that retains the primitive condition for its evolutionary lineage.
derived Describes a character or character state of the organism being considered that has changed from the primitive condition for its evolutionary lineage.
ectothermy Deriving the energy needed to raise body temperature from sources outside the body.
endothermy Deriving the energy needed to raise body temperature from within the body—i.e., from metabolic heat production.
paraphyletic A taxonomic grouping of animals that does not meet the cladistic criterion of including the most recent common ancestor and all its descendants.
sister group The evolutionary lineage most closely related to the one being discussed.

THE WIDESPREAD ADOPTION of phylogenetic systematics (cladistics) as the basis for taxonomic designations has had more impact on our understanding of reptiles than of any other group of terrestrial vertebrates. As our knowledge of ancestral and derived characters of reptiles has grown and been applied over the past decade, our perception of evolutionary relationships of lineages within the Reptilia has changed, and some lineages have even moved in and out of the Reptilia. Our current understanding of the relationships of extant reptiles recognizes three major monophyletic lineages (Pough *et al.*, 2001): Turtles, crocodilians and birds, and lepidosaurs (with tuatara, lizards, and snakes as the extant representatives). Birds are derived from the lineage that includes crocodilians as well as dinosaurs, pterosaurs, and several other groups of extinct reptiles (Dingus and Rowe, 1997), but the anatomical and physiological specializations seen at the avian grade (most notably endothermy, insulation, and flight) make birds functionally different from other reptiles. As a result, the pre-cladistic allocation of nonavian reptiles to the field of herpetology and avian reptiles to ornithology is useful, and will be followed here. Nonetheless, the evolutionary origins of the characters of birds provide a useful context for considering the present and past diversity of other reptiles.

I. EXTANT EVOLUTIONARY LINEAGES

Reptiles, exclusive of birds, include about 7150 living species (Pough *et al.*, 1999). The vast majority of these (about 6850 species) are lepidosaurs (tuatara, lizards, and snakes). There are approximately 260 living species of turtles and 22 crocodilians. Species diversity is greatest in the tropics, but reptiles are not limited to warm regions—temperate deserts in the Holarctic, Africa, and Australia have rich lizard faunas, and temperate aquatic habitats in North America and Asia are home to many species of turtles. Reptiles occur on every continent except Antarctica, occupy habitats ranging from freshwater swamps to desert salt flats, and exploit ecological zones from subterranean to arboreal.

A. Turtles (Testudines or Chelonia)

The shell that makes turtles instantly recognizable is formed by bone overlain by horny epidermal scales called scutes. The ribs, vertebrae, and parts of the pectoral girdle are fused to the dorsal shell (carapace), which is connected to the lower shell (plastron) by a bony bridge. The scutes and/or bone are secondarily reduced in some lineages of turtles, most notably the freshwater soft-shell turtles (Trionychidae) and the marine leatherback turtle (*Dermochelys coriacea*).

The earliest fossils of turtles are from the late Triassic, and two major lineages, Pleurodira and Cryptodira, can be distinguished by the early Jurassic. Pleurodires have the common name side-necked turtles because they bend the neck horizontally when they retract their heads, whereas cryptodires bend their necks vertically. The geographic occurrence of pleurodires is currently restricted to the Southern Hemisphere (although fossil pleurodires are known from the Northern Hemisphere), and all of the approximately 65 species are aquatic. Pleurodires are the only turtles native to Australia and New Guinea and the only aquatic turtles in sub-Saharan Africa.

The nearly 200 species of cryptodires have a worldwide distribution and include specialized marine and terrestrial forms as well as aquatic and semiaquatic species. Tortoises (Testudinidae, about 40 species with a worldwide distribution in temperate and tropical regions) are the most terrestrial turtles. Most tortoises have domed carapaces, and many have forelimbs that are modified for digging. Semiaquatic and aquatic turtles generally have low, streamlined shells and webbed feet. Most of the 35 species of emydid turtles fit this description. Emydids are found primarily in the New World, with one European species, the pond turtle *Emys*. The paraphyletic group "Bataguridae" is the Old World equivalent of the emydids, with about 50 species in Europe and Asia and one genus (*Rhinoclemys*, 9 species) in the Americas. The sea turtles (Cheloniidae [7 species] and Dermochelyidae [1 species]) are still more specialized for swimming, with forelimbs that are modified as flippers and short necks that cannot be retracted. Sea turtles have a worldwide distribution, primarily in tropical and subtropical regions although the range of the leatherback sea turtle extends well into the North Atlantic and South Pacific.

B. Crocodilians (Crocodylia)

Crocodilians are part of the archosaur lineage that includes dinosaurs and birds as well as a number of other groups. Most crocodilians are large and have bony plates in the skin. These characteristics have contributed to an extensive fossil record that extends back to the middle of the Triassic. The history of crocodilians includes terrestrial species, a marine group (metriorhynchids) with the forelimbs modified as paddles and a lobed tail, and species as large as tyrannosaur dinosaurs.

Crocodilians are the smallest group of reptiles, with only 22 living species. Crocodilians are primarily inhabitants of tropical and subtropical regions, but the American and Chinese alligators (*Alligator missippiensis* and *A. sinensis*) are found in temperate areas. All living crocodilians are aquatic; they swim with lateral undulations of the tail, which is somewhat flattened laterally. The nostrils of crocodilians are at the tip of the snout and are slightly elevated, as are the eyes. Alone among nonavian reptiles, crocodilians have developed a secondary palate that separates the passageway for air from the oral cavity. This combination of characters allows crocodilians to float with only the eyes and nostrils exposed.

Three lineages of extant crocodilians are recognized: The Alligatoridae includes seven species of freshwater crocodilians. All except the Chinese alligator occur in the New World. *Paleosuchus* (dwarf caimans) are among the smallest crocodilians at 1.7 m total length, and *Caiman niger* (6 m) is one of the largest. Alligatorids have relatively broad snouts and the teeth of the lower jaw fit into pits in the upper jaw and cannot be seen when the mouth is closed. Crocodylids have narrower snouts than alligators, and the fourth tooth of the lower jaw of a crocodile is visible when the mouth is closed. The 13 species of crocodiles have a worldwide distribution, primarily in tropical and subtropical regions. Several species of crocodiles, especially the American crocodile (*Crocodylus acutus*) and the saltwater crocodile (*C. porosus*), regularly make long sea journeys, and the saltwater crocodile has colonized islands 1000 km from the nearest mainland. The saltwater crocodile is the largest species, growing to a length of more than 7 m. The African dwarf crocodile, *Osteolaemus tetraspis* (2 m), is the smallest member of the family. The Gavialidae includes two species from Asia, the gharial (*Gavialis gangeticus*) and the false gharial (*Tomistoma schlegeli*). Both are specialized fish eaters with long, extremely narrow snouts.

C. Lepidosaurs (Lepidosauria)

More than 95% of living reptiles are lepidosaurs, and unlike turtles and crocodilians, this lineage displays enormous diversity in body form and ecology. Lepidosaurs are covered by scales (the name means "scaly reptile") and shed the skin at regular intervals, they have a transverse cloacal slit (rather than the longitudinal slit seen in crocodilians and turtles), and they exhibit caudal autotomy that is facilitated by fracture planes within the vertebrae of the tail. Lepidosaurs include two lineages of extant forms, the Rhynchocephalia (the tuatara of New Zealand) and the Squamata (snakes and lizards). Like their sister group the Archosauria, lepidosaurs are diapsid (i.e., primitively they have two fenestrae defined by bony arches on the sides of the skull). Modifications of the diapsid condition help to define subgroups within lepidosaurs and contribute to functional differences in prey handling.

1. Rhynchocephalia

The Rhynchocephalia, as it is currently understood (Gauthier *et al.*, 1988), includes several extinct forms dating back to the Triassic and a single extant lineage, the Sphenodontidae, represented by two species of tuatara (*Sphenodon punctatus* and *S. guntheri*) found only on islands off the coast of New Zealand. (Tuatara is a Maori word meaning "spines on the back" and does not add an *s* to form the plural.) Tuatara are lizardlike in body form and reach a total length of about 50 cm.

Unlike their sister group the Squamata, tuatara retain the fully diapsid skull condition, with a lateral opening that is bounded dorsally by the postorbital and supratemporal bones and ventrally by the supratemporal and jugal. These bony connections impart a rigidity to the skull that allows tuatara to deliver a forceful bite. The single row of teeth on the lower jaw fits between two rows of teeth in the upper jaw—an outer row on the maxilla and an inner row on the palatines—and prey is chewed to a pulp before it is swallowed. Tuatara lack the paired hemipenes (intromittent organs) that characterize squamates.

2. Squamata

A large number of anatomical features characterize squamates. Prominent characteristics are reductions in the bony connections forming the arches in the skull, allowing a degree of kinesis unknown in other nonavian reptiles, and the presence in male squamates of paired hemipenes that develop as outgrowths of the posterior wall of the cloaca and are housed in the base of the tail. The hemipenes, which are everted in use, are held in the female's cloaca by spines, and sperm passes along a lateral groove.

Reduction or loss of limbs, usually accompanied by elongation of the trunk and tail, is a recurrent theme among squamates that reaches its zenith in snakes. Phylogenetically snakes are included within lizards, but the abundance and diversity of snakes combined with their morphological, ecological, and behavioral specializations make it helpful to discuss lizards and snakes separately.

a. Lizards

Fossils of lizards are known from the middle Jurassic, and the major lineages were distinct by the end of the Mesozoic. Lizards are found on every continent except Antarctica and are well represented on oceanic islands because large species, such as iguanas, can survive long journeys from the mainland on floating vegetation and small species (most notably skinks and geckos) often travel as stowaways with humans.

The more than 3000 species of extant lizards can be assigned to 19 families, three of which are very large: Scincidae (more than 1000 species), Iguanidae (more than 900 species), and Gekkonidae (870 species). Skinks and geckos have worldwide distributions, whereas only three genera of iguanids are found outside the New World (*Brachylophus* on Fiji, and *Oplurus* and *Chalarodon* on Madagascar).

Most species of iguanids are diurnal, visually oriented lizards whose activities and social behaviors can readily be observed, and they occur in areas accessible to biologists. As a result, iguanids figure prominently in ecological and behavioral studies of lizards. The family as it has traditionally been defined included several genera of large herbivorous lizards (the Iguaninae) and a variety of smaller, primarily insectivorous species that were placed in other subfamilies (Etheridge and de Queiroz, 1988). A review of anatomical characters elevated the subfamilies of Iguanidae to family status, leaving only the large herbivores in the Iguanidae (Frost and Etheridge, 1989). That reclassification had a broad impact in the biological literature, because the newly elevated families include species that have formed the basis for much of the research on lizards. More recently, an analysis based on molecular characters has supported the traditional interpretation of Iguanidae (Macey *et al.*, 1997), and that classification is used here.

b. Snakes

Phylogenetically snakes are included within lizards, and the most primitive snakes, aniliids and uropeltids, lack many of the derived characters that are responsible for the distinctive ecology and behavior of advanced snakes. Even primitive snakes have greatly increased the number of precloacal vertebrae, however, and have lost the bony connection between the postorbital and supratemporal bones, freeing the supratemporal to rotate on its flat articulation with the parietal. This anatomical change allows the supratemporal and quadrate bones to act as part of the lower jaw, and these bones show varying patterns of elongation in advanced lineages of snakes. The brain of snakes is enclosed in a rigid box formed by ventral extension of the frontal and parietal bones and their articulation with the sphenoid bone.

Snakes are clearly derived lizards, but the sister group of snakes and the conditions that led to their evolution are still controversial topics. Snakes focus images on the retina by moving the lens, whereas lizards focus images by changing the shape of the lens. This difference, as well as differences in the photosensitive cells in the retinae of lizards and snakes, has long been interpreted as indicating that snakes are derived from a subterranean lineage of lizards with degenerate eyes. Other hypotheses of the origins of snakes are plausible, although they do not specifically account for differences in the eye. For example, legs are not particularly useful to small squamates that live in dense vegetation such as clumps of grass, and this is the habitat occupied by many legless lizards. This epigean adaptive zone might provide an intermediate stage in the evolution of snakes. Alternatively, aquatic habits could have been an intermediate stage in the evolution of snakes. *Pachyrachis problematicus*, a recently described elongate aquatic lepidosaur from Cretaceous marine deposits, was over a meter long and had more than 100 presacral vertebrae and a skull that shows many of the derived features of snakes, including envelopment of the brain by bone (Caldwell and Lee, 1997).

Snakes were apparently widespread by the late Cretaceous; fossils of that age are known from India, Africa, and South America. Most fossils consist of isolated vertebrae, but vertebral characters play a minor role in classification of living snakes and it is difficult to associate these fossils with extant lineages. Snakes radiated extensively in the Cenozoic, and about a dozen families are usually recognized, containing more than 2500 species. Primitive snakes, including boas and pythons, can be compared to more advanced forms, the Colubroidea, which include elapids (cobras, coral snakes, sea snakes, and their relatives), viperids (pit vipers and true vipers), and colubrids (the largest family, containing about 1700 species, which is 70% of the total number of extant species of snakes).

II. ANCESTRAL AND DERIVED CHARACTERS

The functional ecology of reptiles is shaped by a combination of ancestral and derived characters, and an appreciation of these features and the ways in which they interact is central to understanding how reptiles function as organisms. Important derived characters include reproduction via a shelled egg (with the secondary development of viviparity in many lineages), excretion of nitrogenous wastes as urate salts, and a dry skin that is relatively impermeable to the passage of water. Retained ancestral characters include ectothermal thermoregulation (with the accompanying low metabolic rate and the absence of an insulating layer of hair or feathers) and an anatomically three-chambered heart (except in crocodilians) that permits intracardiac blood shunting.

A. Temperature and Water Relations

With the exception of a few very large species, nonavian reptiles are ectotherms. That is, they obtain the energy they need to raise body temperatures to activity levels from external sources, either directly from the sun or via contact with surfaces that have been heated by the sun. Reptiles employ a mixture of behavioral and physiological mechanisms of thermoregulation. Moving between sun and shade, orienting the body to maximize or minimize interception of solar radiation, and concentrating or dispersing melanin to change reflectivity adjust heat exchange in relation to body temperature. These mechanisms are supplemented by intracardiac shunting of blood and adjustments of the peripheral circulation to accelerate or retard exchange of heat with the environment. In contrast, birds and mammals are endotherms, changing the insulation provided by hair or feathers and adjusting metabolic rate to match rates of heat production and heat loss.

Ectothermy imposes some limits on reptiles. When

heat sources are not available—at night for example, in water, or beneath the closed canopy of a tropical forest—reptiles must either accommodate their temperature requirements to the ambient temperature or cease activity. In the absence of environmental constraints, however, ectothermal thermoregulation can produce body temperatures that are as high and as stable as those of birds or mammals. Lizards and snakes, especially species that live in open habitats, generally control their body temperatures within activity temperature ranges of 2 or 3°C. Terrestrial turtles, too, control their body temperatures by moving between sun and shade during periods of activity. The large body sizes of some turtles and crocodilians stabilize their body temperatures. Leatherback sea turtles, which weigh up to 1000 kg, are found off the coasts of New England and Nova Scotia in the summer in water temperatures as low as 8°C. A countercurrent exchange mechanism in the flippers conserves the heat produced by muscular activity as the turtles swim, and body temperatures can exceed water temperature by at least 18°C (Spotila and Standora, 1985).

The dry skin of reptiles is an essential component of their ectothermal thermoregulation. In contrast, the high rate of evaporation from the moist skins of extant amphibians limits their capacity to maintain body temperatures above ambient levels as well as their ability to be active during the day when evaporative stress is high. Reptilian water balance is additionally facilitated by enzymatic pathways that convert nitrogenous waste to uric acid, which is excreted as a complex mixture of urate salts. The shelled eggs of reptiles absorb water from the substrate during development, and the availability of water in the nest environment influences the body size and, probably, the viability of newly hatched reptiles.

B. Metabolism and Energetics

Because ectotherms do not depend on metabolic heat production for thermoregulation, their resting metabolic rates are low—one-seventh to one-tenth those of endotherms of the same body size. The influence of low resting metabolic rates can be seen in the energy use, body size, and the exercise physiology and foraging behavior of reptiles.

1. Energy Requirements and Conversion Efficiency

The daily energy budget for a lizard is only 3% of the energy requirement for a mammal of the same body size (Bennett and Nagy, 1977). Low energy use by the lizard results partly from its low resting metabolic rate during the day when it is active and partly from a reduction in body temperature (with a consequent further reduction in metabolic rate) at night when it is inactive. In contrast, a mammal maintains a higher metabolic rate during the day and may increase its metabolic rate above resting levels at night to counteract the effect of falling ambient temperature.

Low metabolic rates translate to low energy requirements for reptiles. As a result, reptiles can cope with environments in which food is limited, either episodically or permanently. Areas of low primary production, such as deserts, and even regions where there is no primary production, such as desert salt flats and mobile sand dunes, are home to a variety of lizards.

Efficient conversion of food to biomass is a second consequence of the low metabolic rates of reptiles. Endotherms use most of the energy they ingest to stay warm and be active—19 species of birds and mammals have an average net conversion efficiency of 1.4% (range 0.5–3%), whereas 9 species of reptiles have an average net conversion of 46.6% (range 18–86%) (data from Pough, 1980).

2. Body Size

Mass-specific metabolic rates ($W \cdot g^{-1}$) are related to body mass by an exponent that approximates -0.25, although statistically significant variation from that value is associated with both phylogeny and ecology (Hayssen and Lacey, 1985; Andrews and Pough, 1985). The consequence of that negative allometry is a mass-specific energy requirement that increases steeply as body size decreases (Fig. 1).

The low metabolic rate of reptiles makes extremely small body size energetically feasible, and lizards, in particular, include very small species. The modal body mass of lizards falls in the range 1–10 g, and nearly 20% of lizards have adult body masses less than 1 g. In contrast, there are no mammals with adult body masses less than 1 g, and fewer than 5% of mammals have body masses below 10 g (see Pough et al., 1999, p. 483). More than half the living species of lizards occupy a zone of body sizes in which there are practically no mammals, and the smallest lizards have their major competitive and predatory interactions with invertebrates. In contrast to lizards, snakes are relatively large reptiles—nearly 80% of snakes are larger than 20 g. Thus, snakes and lizards illustrate two different ways to succeed as a squamate in a world that is largely dominated by endotherms: Small body size is a morphological specialization of lizards, just as elongation is a morphological specialization of snakes.

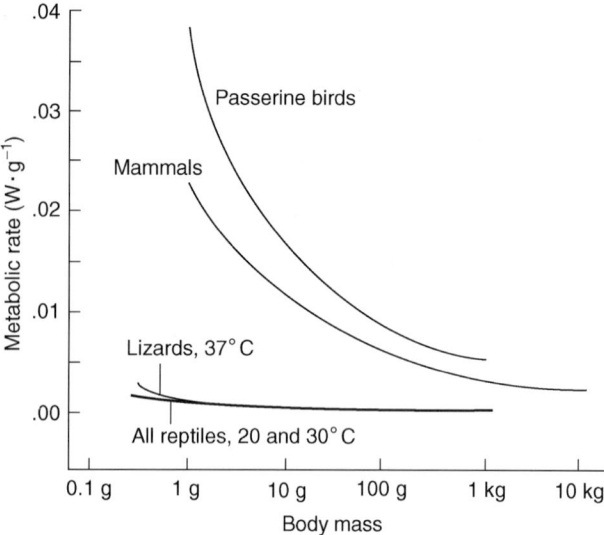

FIGURE 1 The mass-specific energy requirements of animals increase at small body sizes. The metabolic rates of reptiles are one-seventh to one-tenth those of birds or mammals of the same body size, and the smallest reptiles have adult body masses less than 1 g, whereas very few birds and mammals are smaller than 10 g as adults. (Modified from Pough, 1980. © 1980 by The University of Chicago.)

3. Exercise Physiology and Behavior

The low resting metabolic rates of reptiles limit their aerobic metabolic capacity—i.e., the maximum rates of oxygen consumption they can achieve during strenuous activity. Both reptiles and mammals increase resting rates of oxygen consumption about 10-fold when they run at maximum speed, so the differential in metabolic rates of ectotherms and endotherms persists during activity. Most reptiles have limited aerobic metabolic capacity, and they derive most of the ATP used during high levels of activity from anaerobic pathways.

Glycogen, the metabolic substrate for anaerobic metabolism, is stored in the muscle cells and a brief period of activity can exhaust the muscles' glycogen supply. As a result, anaerobic metabolism is an effective way to support bouts of intense muscular activity, but it is not suitable for sustained behaviors such as long-distance locomotion. The activity patterns of many reptiles are built on this relationship—they employ sit-and-wait modes of foraging, they escape predators by fleeing to a nearby shelter, and their social behaviors do not include sustained high levels of activity.

C. Active Reptiles and the Evolution of Endothermy

In an ecological context, the generalization that reptiles are ectotherms with low metabolic rates and low levels of activity applies to extant species, although some modifications of that characterization will be addressed in subsequent sections. In an evolutionary context, however, the generalization is patently false because birds are reptiles and they are endotherms with high metabolic rates. Clearly the reptilian lineage has a capacity for endothermy that is barely expressed in nonavian reptiles. An examination of the evolution of endothermy explains that dichotomy and emphasizes how tightly anatomical and physiological characteristics are linked to thermal ecology.

Ectothermy is the ancestral condition for vertebrates, and the derived condition of whole-body endothermy has evolved at least twice, in mammals and in birds. In a broader view, regional endothermy has evolved independently in sharks, tunas, and billfishes, and whole-body endothermy may have been characteristic of pterosaurs (flying archosaurian reptiles of the Mesozoic) and some lineages of dinosaurs.

Ectotherms and endotherms have very different relationships to their physical environments: Ectotherms rely primarily on behavioral thermoregulation to raise their body temperatures because they have low metabolic rates, and the absence of insulation facilitates uptake of heat from the environment. In contrast, endotherms use internal heat production from high metabolic rates to regulate body temperature, and they require insulation to retain metabolic heat in their bodies.

The evolutionary transition from ectothermy to endothermy is impeded by a Catch-22—adding insulation to an ectotherm impedes its behavioral thermoregulation, but in the absence of insulation any heat produced by increasing its resting metabolic rate is lost to the environment. The solution to this paradox lies in finding a basis for the evolution of insulation or the evolution of an increased metabolic rate that does not depend on the preexisting occurrence of the other character.

Mammals are the sister group of reptiles (including birds) and the common ancestor of mammals and reptiles was ectothermal. Thus, the evolution of endothermy in the mammalian lineage may provide a model for the evolution of endothermy among reptiles. Anatomical changes seen in the fossil record of predatory synapsids, the sister group of mammals, strongly support the hypothesis that the initial step in the evolution of mammalian endothermy was selection for increased locomotor capacity. These changes include the evolution of a cursorial body form, changes in the rib cage that suggest the presence of a diaphragm, and increased surface area in the nasal passages to warm and humidify large volumes of air. Increasing levels of locomotor activity require an increase in metabolic rate, and internal heat production would create a selective value for insulation (see Pough *et al.*, 1999, Chapters 4 and 19, for details and references).

Some features of the fossil record of birds suggest that a similar scenario can be applied to the evolution of avian endothermy, but others appear to contradict that interpretation. Like the predatory synapsids, the small maniraptoran dinosaurs that form the sister group of birds appear to have been fleet-footed predators that pursued their prey. That interpretation suggests that these dinosaurs may have evolved the metabolic capacity for endothermy just as synapsids did, and if the recent report of feathers in fossil dinosaurs is correct, it would support that interpretation. However, two lines of evidence cast doubt on the hypothesis that the dinosaurian precursors of birds had high metabolic rates. Examination of an excellently preserved specimen of the small dinosaur *Sinosauropteryx* suggests that it had simple septate lungs that were ventilated by a pistonlike movement of the liver like those of living crocodilians. Lungs of this sort would not support high rates of oxygen consumption. That interpretation is supported by CAT scans of the nasal passages of dinosaurs that reveal no trace of modifications of the nasal passages to warm and humidify large volumes of air (see Pough *et al.*, 1999, Chapter 13, for details and references).

This evolutionary perspective emphasizes intricate interconnections among the anatomical and physiological characters of extant reptiles and their ecology and behavior, as well as the evolutionary capacity for breaking those links. An examination of living reptiles reveals additional connections among anatomy, physiology, ecology, and behavior.

III. CHARACTERISTICS OF GROUPS

Evolution proceeds in a mosaic fashion; all descendant lineages inherit ancestral characteristics, and each lineage develops a set of derived characters that interact with ancestral characters and with each other to shape the way animals live. The following sections focus on combinations of characters found in specific groups of reptiles that are important in shaping their ecology and behavior. Temperature-dependent sex determination is widespread among turtles, for example, and crocodilians show extensive parental care. To avoid repetition, examples from other lineages are included when a topic is discussed. Additional details, examples, and references can be found in Pough *et al.* (1999), Pough *et al.* (2001), and the sources cited.

A. Turtles

1. The Nest Environment and the Size and Sex of Hatchlings

All turtles lay eggs in holes they excavate with their hind legs. Physical conditions within the nest determine both the size and the sex of hatchlings of many species of turtles (Packard and Packard, 1988). In general, embryos in nests constructed in moist substrates metabolize most of their yolk before they hatch and grow larger than embryos in drier nests. Hatchlings of snapping turtles (*Chelydra serpentina*) from moist nests run and swim faster than the smaller hatchlings that emerge from dry nests. This difference in locomotor performance, which persists for at least a year after hatching, might result in higher first-year survival of hatchlings from moist nests.

Temperature-dependent sex determination is widespread among both pleurodires and cryptodires. For most species of turtles, low temperatures during incubation produce males (Pattern Ia), but in a few species males are produced at high temperatures (Pattern Ib), and some species show a biphasic response in which females develop at both low and high temperatures and males are produced at intermediate temperatures (Pattern II). Environmental sex determination can produce both male and female hatchlings from the same nest, because the shift between sexes typically occurs within a range of only 1 or 2°C. Thus, eggs from the top of a nest can develop into females while eggs deeper in the same nest produce males, or clutches laid early in the season when nest sites are exposed to the sun can produce females, whereas clutches laid in the same sites later in the summer produce males because the nests are now shaded by vegetation.

Temperature-dependent sex determination is universal among crocodilians and tuatara and is known for some iguanid, agamid, gekkonid, and lacertid lizards. No snake has been shown to have temperature-dependent sex determination.

2. Gigantism

Giant tortoises are among the most spectacular reptiles. The largest species of living tortoise are found on the Galápagos and Aldabra Islands and reach carapace lengths of more than a meter and body weights exceeding 100 kg, but large tortoises are not limited to islands. Extant members of the genus *Geochelone* have a pantropical distribution, with mainland representatives in Africa, India, Asia, and South America as well as additional island species on Madagascar, Sri Lanka, and Indonesia (Ernst and Barbour, 1989; Pritchard, 1996). The largest living mainland species, the African spur-thighed tortoise *G. sulcata*, reaches a carapace length of 76 cm, and fossil tortoises with carapace lengths from 60 cm to 1.5 m are known from Pleistocene deposits on all of the continents except Australia and Antarctica. Giant tortoises were present in North America when humans arrived, and a subfossil carapace from Florida contains a spear point and is charred as if it had been placed upside down on a fire and cooked.

B. Crocodilians

1. Reproduction and Parental Care

All crocodilians are oviparous, laying eggs either in nests that the female excavates in the ground or in mounds of vegetation that she builds. Crocodilians in some areas are reported to lay eggs on floating mats of vegetation. Parental care is probably universal among extant crocodilians, although the reproductive behavior of some species is unknown and parental care appears to have been reduced in populations that have been hunted extensively (Lang, 1989). Females are the primary caregivers, but males often participate. A female crocodilian remains near her nest while the eggs are incubating. As the young approach hatching, they start to vocalize within the eggshells, and these vocalizations signal the female to begin opening the nest. The male may participate in opening the nest, and both parents pick up hatchlings in their mouths and transport them to water. Adult crocodilians show remarkable dexterity in this process, sometimes picking up an intact egg and cracking the shell to release the hatchling.

Parental care is virtually unknown among turtles, but about 100 species of squamates exhibit some degree of parental care (Shine, 1988; Gans, 1996). These behaviors range from defense of the nest site to brooding the eggs. In an exception to the generalization that reptiles are ectotherms, females of some species of python gather their eggs into a clump encircled by coils of the body and use muscular contractions to keep the temperature of the eggs close to 32°C. The frequency of contractions, and thus the python's rates of oxygen consumption and heat production, increases as ambient temperature falls. A female Indian python (*Python molurus*) can maintain her egg mass at 31–32°C at ambient temperatures between 23 and 32°C, and during this period she is functioning as an endotherm.

C. Lepidosaurs

1. Tuatara

Two species of tuatara are known, *Sphenodon punctatus* and *S. guntheri*, but the existence of *S. guntheri* was largely forgotten after it was described in 1877 until it was rediscovered by a molecular analysis more than a century later (Daugherty *et al.*, 1990). The entire population of *S. guntheri* consists of about 300 individuals living on 1.7 hectares of scrub on the top of North Brother Island; *S. punctatus* is represented by about 30 populations on small islands off the coast of New Zealand.

Tuatara have a number of unusual ecological and physiological characteristics. They are crepuscular and forage on cool, foggy nights, with body temperatures as low as 6°C. During the day tuatara bask in the sun and raise their body temperatures to 28°C. Tuatara feed on invertebrates and nestling shearwaters. Prey is crushed as the lower jaw closes with an initial vertical movement and is then sheared by the interdigitating rows of triangular teeth as the lower jaw slides forward.

Tuatara, unlike other lepidosaurs, lack hemipenes. Sperm is transferred by apposition of the cloacas of the male and female. In late spring, female tuatara excavate shallow depressions in which they deposit 6–15 eggs. Embryonic development requires 12–15 months (Saint Girons, 1985).

Tuatara are the sister group of squamates, and their anatomy shows characters that are probably ancestral for squamates, including the connection between the jugal and supratemporal bone forming the lower border of the lateral arch of the skull. Some authors have used the phylogenetic position and primitive anatomical characters of tuatara as a basis to suggest that their ecology and physiology also reveal ancestral conditions for squamates, but that is probably not the case. New Zealand is a cold place, and the native geckos and skinks are active at body temperatures well below those of geckos and skinks in warmer parts of the world. It seems most likely that the low activity temperatures of tuatara and the slow development of their eggs are derived characters that reflect current ecological conditions.

2. Squamata

Lizards have proven to be ideal study animals. Many lizards are diurnal and have limited home ranges, so their activities can readily be observed. As a result of their ease of study, species of *Anolis*, *Sceloporus*, *Pogona*, *Urosaurus*, and *Uta* have played seminal roles in the fields of animal behavior, ecology, and evolutionary morphology and physiology (Greer, 1989; Vitt and Pianka, 1994; Wainwright and Reilly, 1994). Most snakes, in contrast, are secretive in their behavior and direct observation is much more difficult. Nonetheless, the combination of perseverance with radiotelemetry has recently produced several important field studies (Shine, 1991; Seigel and Collins, 1993; Greene, 1997).

a. Lizards

Classifying species of lizards in terms of the foraging modes they employ provides a framework that organizes diverse aspects of their biology into a series of useful generalizations (Table I). Like other mobile animals, lizards display a spectrum of foraging behaviors that extends from species that perch motionless on lookouts and make short dashes to capture prey to species that are in nearly continuous motion as they forage, searching for hidden prey under leaf litter and in holes and cavities. Species at the inactive end of this spectrum are called sit-and-wait predators, those in the middle are cruising foragers, and the species that move continuously are called active foragers.

TABLE I
Ecological and Behavioral Characteristics Associated with the Foraging Modes of Lizards[a]

Character	Sit-and-wait	Foraging mode cruising forager	Widely foraging
Foraging behavior			
Movements/hour	Few	Intermediate	Many
Distance traveled/hour	Small	Intermediate	Large
Primary sensory mode	Vision	Vision and olfaction	Olfaction
Exploratory behavior	Low	Intermediate	High
Predators and prey			
Types of prey	Mobile, large	Intermediate	Sedentary, often small
Types of predators	Widely foraging	?	Sit-and-wait and widely foraging
Risk of predation	Low	?	High
Mode of escape	Cryptic, flee a short distance to shelter when attacked	?	Conspicuous, may flee a long distance when attacked
Physiological characteristics			
Sprint speed	High	?	Low
Endurance	Low	?	High
Aerobic metabolic capacity	Low	?	High
Anaerobic metabolic capacity	High	?	Low
Heart mass	Low	?	High
Hematocrit	Low	?	High
Energetics			
Daily energy expenditure	Low	?	High
Daily energy intake	Low	?	High
Social behavior			
Social system	Territorial	?	Not territorial
Home range	Small	Intermediate	Large
Body form			
Trunk	Stocky	Intermediate	Slim, elongate
Tail	Often short	?	Often long
Color and pattern	Often blotched, may match background, rarely colorful	?	Often solid or striped, sometimes colorful
Reproduction			
Mass of clutch relative to mass of female	High	?	Low
Viviparity	Often	Sometimes	Rare

[a] Foraging modes are presented as a continuum from sit-and-wait to widely foraging species. In most cases, data are available only for species at the extremes of the spectrum. (Modified from Pough et al., 1999.)

Sit-and-wait lizards rely on vision to detect moving prey and may be selective in deciding which prey to attack. Widely foraging lizards examine places that prey may be hiding; olfaction is an important sensory mode for these species, and they usually take any prey items they encounter.

Sit-and-wait lizards are usually cryptic and are most likely to be detected by widely foraging predators, whereas widely foraging lizards are conspicuous because of their movement and are consequently vulnerable to sit-and-wait predators. Thus, there may be an alternation of predatory modes at successive levels in a food chain—widely foraging predators find sedentary prey and are eaten by sit-and-wait predators. In the American Southwest, for example, whiptail lizards (*Cnemidophorus*) forage widely and locate sedentary insects such as termites. Leopard lizards (*Gambelia wislizenii*) wait motionless beneath bushes and capture whiptail lizards when they come within range. In turn, leopard lizards are captured by coachwhip snakes (*Masticophis flagellum*), which move from bush to bush in search of prey.

The quick dash that a sit-and-wait lizard makes to capture prey requires high sprint speed, whereas the continuous movement of a widely foraging predator requires endurance. These characteristics are reflected in the physiology of lizards—sit-and-wait lizards can sprint rapidly but they rely primarily on anaerobic metabolism and have limited endurance. Widely foraging lizards cannot sprint so fast, but they have relatively larger hearts and higher hematocrits than sit-and-wait species, so they have greater aerobic metabolic capacities and greater endurance.

The colors of sit-and-wait lizards often match the backgrounds on which they rest and their patterns obscure the outlines of the body, making them cryptic when they are motionless. Widely foraging lizards move across a variety of backgrounds and do not necessarily match any one. They are often solid colored or striped, and these patterns may confuse predators when a lizard is moving. Sit-and-wait lizards are exhausted by brief periods of rapid locomotion, and when they are attacked, they sprint to shelter. Widely foraging lizards often run from bush to bush, changing direction when they are out of sight and traveling a long distance before they stop. The generally long tails of widely foraging lizards may facilitate caudal autotomy as a last-ditch means of escape.

Sit-and-wait lizards can survey their surroundings from their perches whereas widely foraging lizards can see little beyond their immediate horizon. Sit-and-wait species have small home ranges and are often territorial, whereas widely foraging species have larger home ranges and are rarely territorial. The bulk of eggs within the oviducts is probably a handicap for a widely foraging lizard, and widely foraging species have smaller ratios of clutch mass to maternal mass than do sit-and-wait species. Viviparity is more common among sit-and-wait lizards than among widely foraging species.

These generalizations organize a large number of empirical observations into a coherent whole, but assigning evolutionary cause-and-effect interpretations to the correlations may confound phylogeny with ecology. Most of the sit-and-wait species that have been studied are in the iguanian lineage, whereas the widely foraging lizards are scincomorphs. Thus, we cannot separate the effects of phylogeny (iguanian versus scincomorph) from those of ecology (sit-and-wait versus widely foraging). However, a study of several closely related species of lacertid lizards identified sit-and-wait and widely foraging species and detected many of the same behavioral and physiological correlations that are found in the broader comparison (Huey et al., 1984).

Because many species of lizards use behavioral mechanisms to regulate their body temperatures within narrow limits while they are active, the thermal environment plays a central role in defining the niche of a species. The thermoregulatory characteristics of habitats can be defined as high cost or low cost, depending on the behavioral and ecological trade-offs a lizard must make to maintain a body temperature that is different from ambient temperatures. Lizards that live in habitats in which basking sites, perches, and escape sites are abundant and close to each other have low ecological and behavioral costs of thermoregulation and are likely to maintain body temperatures within a narrow range during activity. In contrast, lizards that live in habitats where basking sites are scarce or ephemeral, such as the understory of a tropical forest, have high costs of thermoregulation. These lizards often have body temperatures indistinguishable from ambient temperature.

The costs of thermoregulation in a particular habitat are not the same for all lizards because body size profoundly affects the thermoregulatory options available to a lizard. An example is provided by three species of *Ameiva* that live in microsympatry on the Osa Peninsula of Costa Rica. The three species partition the structural habitat on the basis of the radiant environment. Adults of the smallest species (*A. quadrilineata*) weigh about 10 g and forage in short vegetation in full sun, whereas the middle species (*A. festiva*) weighs about 32 g and forages in areas where taller vegetation provides broken shade. The largest species (*A. leptophrys*) weighs

83 g and forages in deep shade beneath the forest canopy.

Ameiva are widely foraging lizards, and the three species display the same alternation of thermoregulatory and foraging behaviors: a lizard basks in the sun until its body temperature reaches 39 or 40°C and then forages until its body temperature cools to 35°C, when it ceases foraging and basks again. The rate of cooling is inversely proportional to body size, and relative rates of cooling appear to explain the differences in foraging sites of the three species. *Ameiva quadrilineata* cools from 40 to 35°C in 4 min, compared to 7 min for *A. festiva* and 12 min for *A. leptophrys*. The slow cooling of *A. leptophrys* allows it to move deep under the canopy before it has to return to the forest edge to bask, whereas the two smaller species cool too rapidly to move far from basking sites.

A different physical relationship apparently prevents the two larger species from foraging in the sun-drenched habitat occupied by *A. quadrilineata*. The equilibrium temperature reached by an object in the sun is sensitive to its size. The equilibrium temperatures of small objects are strongly affected by convective heat exchange, whereas the equilibrium temperatures of larger objects are closely tied to the radiant environment. As a result of this physical relationship, under the conditions on the Osa Peninsula, an *A. quadrilineata* in full sun reaches equilibrium at 39°C, which is within its activity temperature range. In contrast, if an *A. festiva* or *A. leptophrys* remained in the sun, its body temperature would rise to a lethally high level.

Thus, the effectiveness with which each species can use a specific thermal environment appears to contribute to habitat partitioning by these species—the forest floor is a high-cost environment for the small and medium species and only the largest species can forage effectively in that habitat. Conversely, the absence of shade creates high costs of thermoregulation for the large and medium species in the open habitat, and only the smallest species can exploit this area.

b. Snakes

Elongation of the body is the immediately distinctive feature of snakes, and specializations associated with feeding form the core of their diversity. The body form of snakes results primarily from an increase in the number of vertebrae in the neck and trunk, and the longest species have more than 400 precaudal vertebrae. Even generalized snakes—rat snakes (*Elaphe*), gopher snakes (*Pituophis*), Australian whipsnakes (*Demansia*), and blacksnakes (*Pseudechis*)—have head-plus-trunk lengths that are 10–14 times the maximum trunk circumference. In contrast, elongate lizards (*Cnemidophorus*, *Egernia*, *Lacerta*, and *Eumeces*) have length/circumference ratios of 2.0–2.5.

In a functional sense, repackaging the mass of a lizard into a long, thin body form means that a snake must feed a large body through a small mouth. Lizards and snakes almost always swallow their prey whole; many lizards manipulate prey items in their mouths until they have been softened by repeated bites, whereas nearly all snakes swallow prey without any oral processing. Most legless lizards eat large numbers of small prey items; they retain the connection between the postorbital and supratemporal bones on the side of the skull, and this structure gives the skull sufficient rigidity to crush the exoskeletons of insects. In contrast, snakes have lost the postorbital–supratemporal connection and have replaced the body symphysis at the front of the lower jaw with a flexible ligament. These changes, which increase skull kinesis and allow the lower jaws to move independently, also reduce the ability of snakes to exert pressure on food in the mouth. A few snakes have secondarily evolved modifications of the skull and jaws that allow them to eat hard-bodied prey, but most snakes feed on soft prey.

Lizards normally consume several small prey items on a daily basis, whereas most snakes eat larger prey items less frequently. The prey of snakes is large relative to the predator, and many snakes have specializations that allow them to immobilize prey before it can escape or injure the snake. Snakes often swallow prey items with diameters larger than the snake's head, and these snakes are capable of large gapes.

Prey size must be evaluated relative to the size of the snake in two respects—mass and diameter. That is, a prey item can be light or heavy compared to the snake that eats it and it can also be slim or bulky. The first comparison can be called relative mass and the second relative girth. These two variables create four categories of prey (Greene, 1997).

- Type I prey have both small mass and small girth ratios. Insects are Type I prey, and this is the prey type that is characteristic of most lizards and probably of the earliest snakes.
- Type II prey are elongate and combine a large mass ratio with a small girth ratio. Earthworms, caecilians (elongate, burrowing amphibians), eels, and other snakes are examples of Type II prey. These food items require specializations to immobilize the prey (constriction or venom), but do not require large gapes. Most primitive snakes (e.g., the pipe

snakes, *Cylindrophis* and *Anilius*) are limited to Type II prey.
- Type III prey are heavy and bulky and have large mass and girth ratios. Frogs, mammals, and some lizards are Type III prey. These prey require both immobilization and large gape.
- Type IV prey are bulky and light, with small mass ratios and high girth ratios. Birds (which are bulky because of their feathers) are Type IV prey and require gape specializations but not immobilization. Boas and pythons take Type III and IV prey, as do most of the advanced snakes (elapids, vipers, and colubrids).

The foraging behavior of snakes extends from active searchers that use visual and/or olfactory cues to find prey to mobile ambushers that move to places they are likely to encounter prey. Rattlesnakes (*Crotalus*), for example, coil beside trails followed by rodents and wait for prey, and tropical racers (*Mastigodryas*) wait beside sunspots on the forest floor to capture lizards (*Ameiva*) that pause to bask.

The exercise physiology of snakes appears to parallel differences in their foraging behavior. Widely foraging species, such as North American whipsnakes (*Masticophis*), have high aerobic metabolic capacities and high endurance, whereas species that rely on ambush (*Lichanura*) have lower aerobic capacities and endurance. Widely foraging species probably eat more often than ambushing species and take smaller prey, but even a frequent feeder such as the coachwhip snake (*Masticophis flagellum*) eats at 10-day intervals (Secor and Diamond, 1998), whereas most lizards probably eat daily.

Type III prey (heavy and bulky) can pose a risk of injury to a snake, especially those species that swallow prey while it is still struggling. Constriction and venom are specializations that allow snakes to subdue prey before they are swallowed. Constriction is characteristic of primitive snake lineages and of boas and pythons. Apparently it evolved as a method of controlling elongate prey and subsequently proved effective with bulky prey as well. Constriction requires enclosing a prey item with coils of the snake's body. Little pressure is exerted on the prey, and friction between adjacent coils holds them in place. Death may result from asphyxiation (because the snake takes up the slack each time the prey exhales) or because increased internal pressure interferes with and ultimately stops the action of the heart. The short-radius curves required for constriction require short vertebrae and trunk muscles that span only a few vertebrae. This morphology is not compatible with rapid locomotion, which requires large-radius curves formed by long vertebrae and muscles that span many vertebrae. The evolutionary appearance of advanced snakes in the Miocene coincides with the spread of grasslands. Constrictors (largely small boas) dominated the snake fauna at the beginning of the Miocene, but by the end of that epoch the snake fauna was composed largely of colubroids. These were fast-moving snakes, but their locomotor specializations precluded constriction. Although some lineages of colubroids secondarily evolved constriction, venom is the distinctively colubroid method of immobilizing prey with minimum risk to the snake.

Venom evolved early in the history of colubroids as part of the feeding apparatus. The defensive use of venom is a secondary development, enhanced in some cases by morphological specializations. The venom channel in the fangs of spitting cobras, for example, makes an abrupt right-angle turn near the tip of the fang that enables the cobra to emit a spray of venom that travels a meter or more. The snake aims for the predator's eyes, and accounts by people who have been sprayed confirm the powerful deterrent effect of this defense. Attacking a venomous snake can have severe consequences for a predator, and the warning mechanisms of venomous snakes (such as the rattle of rattlesnakes, the hood of cobras, and the distinctive colors of coral snakes) foster both learned and innate avoidance behavior by potential predators (Greene, 1988; Pough, 1988).

Duvernoy's gland, which lies at the rear of the maxilla and produces venom that immobilizes prey, is a derived character of colubroids. The embryonic development of Duvernoy's gland is linked to development of the posterior pair of maxillary teeth. These rearmost teeth are enlarged and grooved in many colubrids. Although colubrids are often called "the nonvenomous snakes" in popular literature, many colubrids use venom to subdue prey, and a few species known as rear-fanged snakes (notably the African boomslang, *Dispholidus typus*) can deliver defensive bites that are lethal to animals as large as humans. In the two groups popularly known as venomous snakes, elapids (cobras, coral snakes, sea snakes, and their relatives) and viperids (vipers and pit vipers), the maxilla has shortened, thereby moving the enlarged fang on the maxilla to the front of the mouth. Elapids have short fangs in a relatively immobile maxilla, whereas viperids have long fangs and the maxilla rotates so the fang lies horizontally when the mouth is closed.

The venoms of colubrids and elapids attack the nervous system and rapidly immobilize prey. In contrast, the venoms of vipers and pit vipers have high proteolytic activity and cause tissue destruction. The proteolytic activity of venom speeds digestion because the venom hydrolyzes the body tissues of the prey from the inside while the snake's digestive enzymes attack from the outside. Some vipers, especially species of the African genus *Bitis,* appear to combine proteolytic venom, injected deep into the prey by very long fangs, with modifications of the body form to feed on very bulky prey. Large species of *Bitis* are remarkably heavy-bodied—length/circumference ratios of the puff adder (*B. arietans*) and Gaboon viper (*B. gabonica*) are 4–5. The thick bodies of these snakes accommodate bulky prey without disrupting locomotion or interfering with the function of internal organs.

The physiology of the digestive system of snakes is as specialized as their trophic anatomy (Secor and Diamond, 1998). Large meals taken at infrequent intervals pose a challenge—the physiological cost of digesting a meal is high, but occasional periods of high activity are separated by long periods during which the digestive system is inactive. Within 24 h after feeding, snakes display remarkable factorial increases in rates of oxygen consumption; the masses of organs in the cardiovascular, respiratory, digestive, and excretory systems; and activities of digestive enzymes. Between feedings, the sizes of organs and activities of enzymes regress to a resting state. This cyclic hypertrophy and regression, which is interpreted as a mechanism that conserves energy compared to the alternative of maintaining the physiological systems continuously at high levels, emphasizes the extent to which snakes have developed suites of trophic specializations not seen among lizards.

IV. CONSERVATION ISSUES

Many of the causes of declining populations of reptiles are depressingly familiar and affect other kinds of animals and plants as well. Habitat loss as a result of human activities is probably the major reason for shrinking populations, although mortality on roads may be more significant than is generally recognized. Road kills are considered the major form of human-induced mortality for Australian reptiles and amphibians, exceeding the effect of habitat loss by nearly 70% (Ehmann and Cogger, 1985). Some European countries have built drift fences and constructed tunnels beneath roads to reduce mortality of reptiles and amphibians. Introductions of exotic species (especially on islands) and pollution (potentially including global warming [Dunham, 1993]) are additional factors that affect animals and plants on a large scale. Other threats are limited to particular groups and result from combinations of human activities and specific biological characteristics.

A. Turtles

Many species of turtles are long-lived and display a suite of coevolved traits, including delayed sexual maturity and high annual survivorship of adults (Congdon *et al.*, 1994). Adult turtles are often nearly immune from nonhuman predators, and a turtle that survives to adulthood can expect a long reproductive period. When adult mortality is increased by human activity, as is the case for sea turtles which are drowned at sea in the nets of shrimp trawlers and killed for food when they come ashore to nest, a crucial component of the life history strategy is destroyed. Sea turtles are the most visible examples of excessive adult mortality resulting from human activities, but automobiles kill adult turtles and tortoises all over the world.

Well-intentioned efforts at management can founder on aspects of the biology of a species. Sea turtles, for example, exhibit temperature-dependent sex determination—males are produced at low temperatures and females at higher temperatures. Conservation activities at the nesting beaches of sea turtles usually include moving the eggs from natural nests to a protected site where they are buried in the sand to complete incubation. Under these circumstances all of the eggs are exposed to similar temperatures, and as a result the hatchlings will be predominantly one sex. Even worse, in some cases the incubation conditions used by management programs produced high proportions of male hatchlings.

B. Crocodilians

The skins of crocodilians have long been used for expensive leather goods, and hunting has depleted many populations. The recovery of the American alligator, *Alligator missippiensis*, following passage of protective legislation shows the potential for managing and restoring crocodilians. Success requires enforcement of laws and control of commerce, and these are difficult goals in many areas where crocodilians occur.

Crocodilians may be especially susceptible to some forms of pollution as a result of their position near the top of aquatic food chains. Pollution of Lake Apopka by the pesticides dicofol and DDT and their metabolites has reduced the hatching success and hatchling viability of alligators in the lake and has interfered with normal sexual development (Guillette *et al.*, 1994). Dicofol and

DDT have estrogen-analogue properties and have affected the development of both sexes. Six-month-old female alligators from Lake Apopka have abnormal ovarian morphology and higher plasma levels of estradiol-17β than females from the control site. Male hatchlings from Lake Apopka had abnormally small penises and plasma testosterone levels only one-quarter those of males from the control site.

C. Lepidosaurs

Most extant species of lepidosaurs are relatively small. Of course, small size does not protect a species, but it does tend to make its problems less conspicuous—and hence less appreciated—than those of animals like sea turtles and crocodilians. With a few exceptions, which have earned special status by being unique (e.g., tuatara, *Sphenodon*) or colorful (the San Francisco garter snake, *Thamnophis sirtalis tetrataenia*), the status of populations of lepidosaurs is unknown.

Larger species of snakes and lizards are killed for food and for their skins. Professional hunters in Central and South America ship iguanas (*Iguana*) to food markets in the cities and collect tegu lizards (*Tupinambis*) for their skins. Monitor lizards (*Varanus*) are hunted for their skins in Africa and Asia. During 1986 (the most recent year for which figures are available), more than 8 million skins of snakes and lizards were traded legally. The major countries listed as sources of the skins were Indonesia, Singapore, Thailand, and Argentina, and the major importing countries were Singapore, the United States, Italy, and Spain. The lizards and snakes hunted for the leather trade are large animals with long adult life spans and are high in the food chain. We have little information about the status of wild populations, but it seems unlikely that animals with those biological characteristics can withstand high rates of human predation indefinitely.

See Also the Following Articles

AMPHIBIANS, BIODIVERSITY OF • BIRDS, BIODIVERSITY OF • ENDANGERED REPTILES AND AMPHIBIANS • FISH, BIODIVERSITY OF • MAMMALS, BIODIVERSITY OF • VERTEBRATES, OVERVIEW

Bibliography

Andrews, R. M., and Pough, F. H. (1985). Metabolism of squamate reptiles: Allometric and ecological relationships. *Physiol. Zool.* 58, 214.

Bennett, A. F., and Nagy, K. A. (1977). Energy expenditure in free-ranging lizards. *Ecology* 58, 697.

Caldwell, M. W., and Lee, M. S. Y. (1997). A snake with legs from the marine Cretaceous of the Middle East. *Nature* 386, 705.

Congdon, J. D., Dunham, A. E., and van Loben Sels, R. C. (1994). Demographics of common snapping turtles (*Chelydra serpentina*): Implications for conservation and management of long-lived organisms. *Am. Zool.* 34, 397.

Daugherty, C. H., Cree, A., Hay, J. M., and Thompson, M. B. (1990). Neglected taxonomy and continuing extinctions of tuatara (Sphenodon). *Nature* 347, 177.

Dingus, L., and Rowe, T. (1997). *The Mistaken Extinction: Dinosaur Evolution and the Origin of Birds*. Freeman, New York.

Dunham, A. E. (1993). Population responses to environmental change. In *Biotic Interactions and Global Changes* (P. M. Kareiva, J. G. Kingsolver, and R. B. Huey, Eds.), pp. 95–119. Sinauer, Sunderland, MA.

Ehmann, H., and Cogger, H. (1985). Australia's endangered herpetofauna: A review of criteria and policies. In *The Biology of Australasian Frogs and Reptiles* (G. Grigg, R. Shine, and H. Ehmann, Eds.), pp. 435–461. Surrey Beatty, Chipping Norton, NSW, Australia.

Ernst, C. H., and Barbour, R. W. (1989). *Turtles of the World*. Smithsonian Institution Press, Washington, D.C.

Etheridge, R., and de Queiroz, K. (1988). A phylogeny of the Iguanidae. In *Phylogenetic Relationships of the Lizard Families, Essays Commemorating Charles L. Camp* (R. Estes and G. Pregill, Eds.), pp. 283–367. Stanford Univ. Press, Stanford, CA.

Frost, D. R., and Etheridge, R. (1989). A phylogenetic analysis and taxonomy of iguanian lizards (Reptilia: Squamata). *Misc. Pub. Mus. Nat. Hist., Univ. Kansas* 81, 1.

Gans, C. (1996). An overview of parental care among the Reptilia. *Adv. Study Behav.* 25, 145.

Gauthier, J. A., Kluge, A. G., and Padian, K. (1988). Amniote phylogeny and the importance of fossils. *Cladistics* 4, 105.

Greene, H. W. (1988). Antipredator mechanisms in reptiles. In *Biology of the Reptilia* (C. Gans and R. B. Huey, Eds.), Vol. 16, pp. 1–152. A. R. Liss, New York.

Greene, H. W. (1997). *Snakes, The Evolution of Mystery in Nature*. Univ. of California Press, Berkeley.

Greer, A. E. (1989). *The Biology and Evolution of Australian Lizards*. Surrey Beatty, Chipping Norton, NSW, Australia.

Guillette, L. J., Jr., Gross, T. S., Mason, G. R., Matter, J. M., Percival, H. F., and Woodward, A. R. (1994). Developmental abnormalities of the gonad and abnormal sex hormone concentrations in juvenile alligators from contaminated and control lakes in Florida. *Environ. Health Persp.* 102, 680.

Hayssen, V., and Lacey, R. C. (1985). Basal metabolic rate in mammals: Taxonomic differences in the allometry of BMR and body mass. *Comp. Biochem. Physiol.* 81A, 741.

Huey, R. B., John-Alder, H., and Nagy, K. A. (1984). Locomotor capacity and foraging behavior of Kalahari lizards. *Anim. Behav.* 32, 41.

Lang, J. W. (1989). *Social Behavior in Crocodiles and Alligators* (C. A. Ross, Ed.), pp. 102–117. Facts on File, New York.

Macey, J. R., Larson, A., Ananjeva, N. B., and Papenfuss, T. J. (1997). Evolutionary shifts in three major structural features of the mitochondrial genome among iguanian lizards. *J. Mol. Evol.* 44, 660.

Packard, G. C., and Packard, M. J. (1988). The physiological ecology of reptilian eggs and embryos. In *Biology of the Reptilia* (C. Gans and R. B. Huey, Eds.), Vol. 16, pp. 523–605. A. R. Liss, New York.

Pough, F. H. (1980). The advantages of ectothermy for tetrapods. *Am. Nat.* **115**, 92.

Pough, F. H. (1988). Mimicry and related phenomena. In *Biology of the Reptilia* (C. Gans and R. B. Huey, Eds.), Vol. **16**, pp. 153–234. A. R. Liss, New York.

Pough, F. H., Andrews, R. M., Cadle, J. E., Crump, M. L., Savitzky, A. H., and Wells, K. D. (2001). *Herpetology*, 2nd Ed. Prentice Hall, Saddle River, NJ.

Pough, F. H., Janis, C. M., and Heiser, J. B. (1999). *Vertebrate Life*, 5th ed. Prentice Hall, Saddle River, NJ.

Pritchard, P. C. H. (1996). The Galápagos tortoises: Nomenclatural and survival status. *Chelonian Research Monographs No. 1.* Chelonia Research Foundation, Lunenburg, MA.

Saint Girons, H. (1985). Comparative data on lepidosaurian reproduction and some time tables. In *Biology of the Reptilia* (C. Gans and F. Billett, Eds.), Vol. **15**, pp. 59–148. Wiley, New York.

Secor, S. M., and Diamond, J. (1998). A vertebrate model of extreme physiological regulation. *Nature* **395**, 659.

Seigel, R. A., and Collins, J. T. (1993). *Snakes, Ecology and Behavior.* McGraw-Hill, New York.

Shine, R. (1985). The evolution of viviparity in reptiles: An ecological analysis. In *Biology of the Reptilia* (C. Gans and F. Billett, Eds.), Vol. **15**, pp. 59–148. Wiley, New York.

Shine, R. (1988). Parental care in reptiles. In *Biology of the Reptilia* (C. Gans and R. B. Huey, Eds.), Vol. **16**, pp. 275–329. A. R. Liss, New York.

Shine, R. (1991). *Australian Snakes, A Natural History.* Reed Books, NSW, Balgowlah, Australia.

Spotila, J. R., and Standora, E. A. (1985). Environmental constraints of the thermal energetics of sea turtles. *Copeia* 1985, 694.

Vitt, L. J., and Pianka, E. R. (1994). *Lizard Ecology, Historical and Experimental Perspectives.* Princeton Univ. Press, Princeton, NJ.

Wainwright, P. C., and Reilly, S. M. (1994). *Ecological Morphology, Integrative Organismal Biology.* Univ. of Chicago Press, Chicago, IL.

RESOURCE EXPLOITATION, FISHERIES

John Beddington
Imperial College of Science, Technology, and Medicine

I. History
II. The Current Situation
III. Fisheries Science
IV. The Future of Commercial Fisheries

GLOSSARY

fishing effort The level of fishing activity quantified in terms of the number and power of vessels and duration of fishing.
maximum sustainable yield The maximum level of catch that can be removed continually from a fish population without depleting the population.
mesh size The size of the holes in a fishing net.
recruits The young fish that are first caught by a fishery.
stock recruitment relationship The relationship between the adult breeding stock and the number of recruits produced by that stock.
total allowable catch The level of permitted catch established by fishery regulations.

FISHERIES INVOLVE THE active removal of individual organisms from aquatic ecosystems using a variety of different techniques. Commercial fisheries occur in all major aquatic ecosystems. This article briefly reviews the history of fisheries and the current global situation, sets out in simple terms the scientific basis for fisheries management, and examines the possibilities for the future.

I. HISTORY

Fishing as a commercial activity has been conducted since the very earliest of times. Early records indicate that it was occurring in both Egyptian and Indian societies as long ago as 5000 B.C., and fisheries in the Mediterranean were already significant as long ago as 1000 B.C.

Since the early years, fishing has increased largely as a result of the changes that have occurred in the human population and its activities. In Fig. 1a, the UN estimates for world population from 1800 are shown. This shows an increase from around 1 billion in 1800 to approximately 6 billion at present. The major increase in population has occurred in the past 100 years or so, and particularly since 1950 the rate of increase has been very high.

In Fig. 1b, estimates of fish catch since around 1800 are shown. The figure indicates a catch of around 2 million tonnes in 1800, increasing to around 20 million tonnes in 1950. Since that time, catches have increased to a current level of around 120 million tonnes. The striking similarity between Fig. 1a, which shows the world population, and Fig. 1b, which shows the global fish catch, results from the fact that fisheries to a large extent have mimicked the change in population and

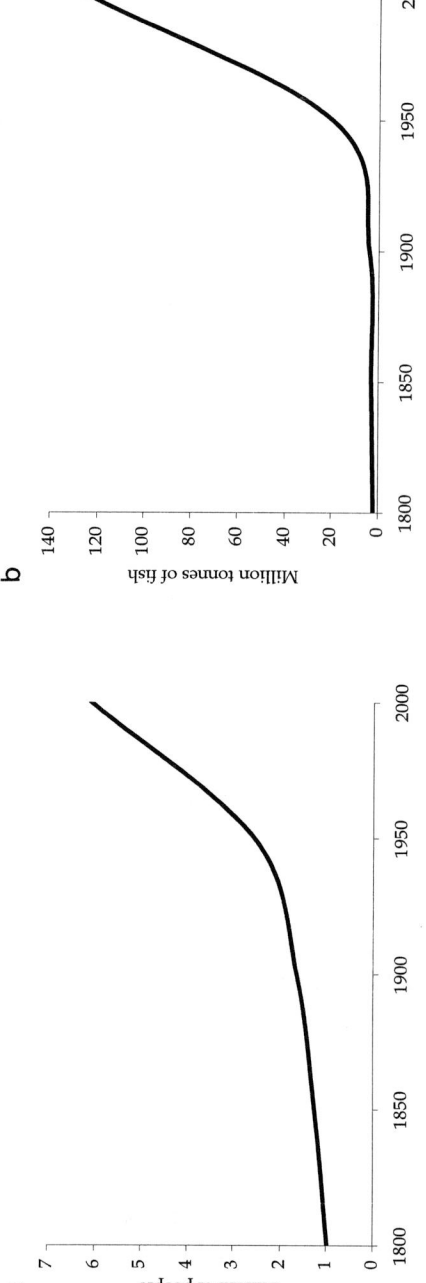

FIGURE 1 (a) World population (source: *UN Population Trends: The 1998 Revision*). (b) World fish catch, 1800–1997, including catches from inland fisheries and aquaculture. Capture fisheries alone were around 95 million tonnes in 1997.

the changes in human activity reflected by increasing industrialization.

Fishing as a commercial activity can usefully be divided into two elements: that involved with catching and that involved with preserving and processing the catch.

A. Methods of Catching Fish

There are four main methods:

Hooks and lines: This is perhaps the simplest method of fishing, from the earliest times handlines have been used, but current commercial longlines can be up to 100 km in length.

Gill nets: In this operation, nets are hung from vessels or floating buoys and are largely invisible to the fish, which are caught as they try to pass through the net. Such fishing methods can catch many species other than the main target species and large high seas gill nets are now banned under international treaties.

Trawls: This type of fishing involves the vessel dragging a large net (or trawl). This can be done either through the water, close to or along the bottom of the sea (demersal trawling), or in the water column (pelagic trawling).

Seines: The principle of this operation is that a school of fish is encircled by a net which is then tightened, reducing the circumference of the net until the school is caught. In modern tuna purse seining, a fast motorboat is used to encircle a shoal of tuna and the net is then pulled in from the main vessel.

Although the basic methods of fishing are little changed since earliest times, the development of motor vessels to succeed sailing or manpower and the increasing power of engines and gear to locate fish has dramatically increased the fishing power of vessels over the years.

B. Processing the Catch

Original methods for processing or preserving fish involved sun drying, smoking, salting, and pickling. Subsequent developments as a result of industrialization have involved preserving fish on ice, freezing and processing on board the vessel, and reduction to meal. Modern factory vessels are highly industrialized operations that catch, process, and package fish ready for delivery directly to markets.

C. Changes Since 1945

The increased industrialization prior to World War II increased the pressure on fish stocks, but the war provided a brief respite as commercial fishing activity was reduced, for obvious reasons. Analyses of this period indicate significant recoveries in major fish stocks that had been heavily exploited prior to the start of the war.

Since then, total fish catch has increased by almost a factor of 5. Much of the increase has come from the coastal regions, where fishing activity has expanded with increases in population, creating more demand for fish products. However, in this period an activity of global significance has been the development of distant-water fishing by fleets primarily, but not exclusively, of the USSR and Japan. These fleets consisted of large vessels capable of operating for many months away from the home port and with substantial industrial freezing capacities. In the maritime regimes applying in the 1950s and 1960s, these vessels fished the rich continental shelf waters off the coast of many countries. In this period, the catch of distant-water fleets increased from around 500,000 tonnes in 1950 to a peak of 8.5 million tonnes in 1972. The change in the maritime regime brought about by the UN Conference on the Law of the Sea and the consequent UN Convention on the Law of the Sea has altered this in a dramatic way. Coastal states have set up Exclusive Economic Zones (EEZs) and restricted fishing on their continental shelf waters. This has resulted in a decline in the proportion of the world fish catch coming from distant-water fleets from 15% in 1972 to 5% currently.

D. Fisheries Management

The need to manage fisheries as an activity has been recognized since early times, and as long ago as the fifteenth-century, Dutch vessels operating in the North Atlantic had a closed season and restricted gear types in order to reduce pressure on fish stocks. Probably with an analogy to farming, it was also recognized that in addition to restrictions on season and types of gear, there was merit in protecting young fish that had yet to breed or fish that were currently breeding. However, regulations were comparatively few and were largely unenforceable except via peer pressure at a local level. The need for management was recognized at an international level after World War II, and a number of international fishery commissions were set up. These varied from the International Whaling Commission, with its focus on whaling worldwide, to commissions specifically aimed at dealing with fisheries in a particular area, which were the most typical. These commissions were set up via international conventions agreed between the coastal states of the region with fishing interests and states with distant-water fishing fleets operating in the area.

It is fair to say that with rare exceptions these international fishery commissions were largely ineffective in directly regulating the level of fishing pressure on fish stocks either via restricting the amount of fishing (effort) or by regulating the catch. The main difficulty was that the commissions typically found the necessary degree of consensus difficult to achieve among states with differing economic imperatives. Most agreements reached concerned readily enforceable regulations such as restrictions on gear types and on fishing seasons. Despite these difficulties, the fisheries management in these bodies was dependent on proper scientific assessment (discussed below) and routinely collected statistics on the catches of species under exploitation. Such data have proved invaluable in assessing the history of exploitation of particular fish stocks and are still of use today.

E. Post-UN Convention on the Law of the Sea

Since the Law of the Sea convention came into force, the need for international regulation of fisheries by international commissions has been reduced. Most states now have an Exclusive Economic Zone which extends 200 miles from their coast. The reason for 200 miles is that this is an approximation to the extent of the continental shelf where, with very rare exceptions, the vast majority of fish are caught, the nutrient-rich environment providing food for plankton and the fish communities dependent on it. Management in these EEZs has had mixed success, but they do in principle provide the opportunity for a single entity, the coastal state, to directly regulate the fisheries in the area.

II. THE CURRENT SITUATION

The UN Food and Agriculture Organization (FAO) has been collecting comprehensive data on fisheries since around 1950. Their figures indicate that the total world catch of fish is dominated by a small group of very large states, or those states that have particularly rich fish resources or a high level of distant-water fleet activities.

TABLE I
Fishery Catches in 1996 by Country

Country	Catch ($\times 10^6$ tonnes)
China	14.2
Peru	9.5
Chile	6.7
Japan	6.0
United States	5.0
Russian Federation	4.7
Indonesia	3.7
India	3.5
Thailand	3.1
Norway	2.6
Korea Republic	2.4
Iceland	2.0

In fact, 12 countries take around 70% of the total world catch (Table I).

In the context of biodiversity, the world fish catch is dominated by a relatively small number of species: some 30% of the total world catch consists of only 11 species (Table II) and the many hundreds of other species caught only form a very modest proportion of this total catch. Nevertheless, fisheries for certain highly valued species such as squid, shrimp, and tuna are extremely important in global terms despite the volume of catch being small.

The development of commercial fisheries in the last three decades of the twentieth century has led to such pressure on fish resources that the paramount question

TABLE II
Fishery Catches in 1996 by Main Species

Species	Catch ($\times 10^6$ tonnes)
Anchoveta	8.9
Alaska pollack	4.5
Chilean jack mackerel	4.4
Atlantic herring	2.3
Chub mackerel	2.2
Capelin	1.5
South American pilchard	1.5
Skipjack tuna	1.5
Atlantic cod	1.3
Largeheaded hairtail	1.3
Japanese anchovy	1.25

is whether the stocks can continue to sustain these levels of exploitation. Such questions have been posed for a number of decades and it has been recognized for several centuries that overexploitation can occur unless proper management action is taken. However, the current situation at a global level is clearly of serious concern. In a recent study by FAO, it has been calculated that of those important commercial fish stocks where sufficient information was available to allow judgment to be made, 44% are heavily or fully exploited, 16% are overfished, 6% are depleted with no evidence of any recovery, and 3% are depleted with some evidence of recovery. In other words, almost 70% of the key commercial fish stocks in the world are fully exploited or overexploited and 25% have been depleted to an extent that their capacity to deliver a sustainable yield has been seriously eroded.

In another study, FAO characterized three different phases in the fishery development process. This analysis has been based on a detailed historical record of the changes in fishery catches over the history of some 200 key fish stocks. The summary of these ideas is set out in Fig. 2, which shows the four phases of a fishery. These comprise an initial underdeveloped phase where exploitation is low and catches are low, a developing phase where the fishery is growing and catches increase, a mature phase where catches reach their peak, and finally a senescent phase where catches decline as the fishery is overexploited and the capacity of the resource to provide a sustainable yield has been eroded. Of the stocks analyzed in this way, 35% are senescent, 25% are mature, 40% are developing, and 0% have low exploitation. Clearly, the 0% is not a representative sample of all fish stocks, but is indicative of a low figure.

That these two sets of results are similar is of significant importance as they have been derived in rather different ways. The first explored the underlying scientific assessment of stocks and the second used a simple analysis of catch trends. Both show that there is a major problem of overexploitation and that some limit appears to be being reached to the overall global level of catches. The total catch of commercial fish stocks appears to be stabilizing around some 95 million tonnes. It is important to recognize that this apparent stabilization is not an intrinsic characteristic of the biological productivity of the oceans but rather a combination of that productivity and the plethora of individual fisheries that exploit it. Indeed, productivity varies quite significantly from area to area in the world.

A. Economic Issues

The historical changes that have led to the current level of overexploitation of the world's fisheries have been paralleled by a somewhat unsatisfactory economic situation. Studies on changes in fishing capacity are relatively few as data are often difficult to come by and changes in overall levels of fishing are difficult to quantify, particularly where diverse regimes and types of fishing vessel operate. However, a recent attempt at synthesis has been made by FAO by referring to the number of decked vessels which are reported as being licensed by states. The distinction between a decked and an undecked vessel is a reasonable one between

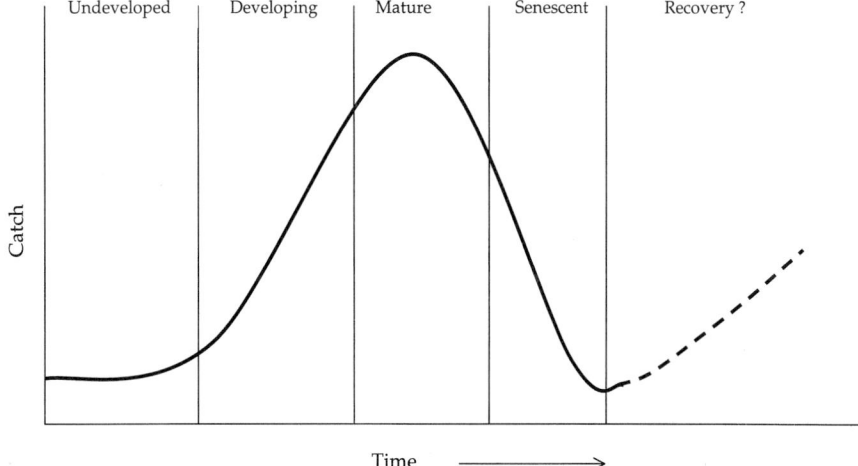

FIGURE 2 Schematic representation of fisheries development (after *FAO: World Fisheries Resources 1997*).

primarily commercial vessels, which have a reasonable level of construction, and simple undecked vessels, canoes, etc. Such data that are available indicate that the numbers of undecked vessels primarily operating in subsistence fisheries in Africa and Asia have remained constant over the past two decades. By contrast, since 1970, the number of decked vessels has doubled. It is worth noting that the vast majority of these increases in vessel numbers have come from one state alone: China. In the period between 1980 and 1997, FAO reported that the number of decked vessels in China increased from 60,000 to 460,000.

This significant increase in the capacity of fleets as measured by the number of vessels also conceals the phenomenon that, on the whole, vessels are getting larger and more technologically sophisticated. Their costs are also correspondingly higher and thus the value and operating costs of the vessels of the world fleet in recent times are likely to be significantly more than those of a few decades ago—certainly more than the simple doubling indicated by the statistics.

In a recent study, Garcia and Newton analyzed the behavior of the world fleet in a very simple but informative way. They calculated that the total revenue obtained by the world's fishing fleets (essentially the total first-sale value of the world catch) was less than the costs of operating the fleets by a very large amount. They quote a figure of some $54 billion for the operating loss in the year 1989.

How such operating losses are being sustained is an obvious question, to which there is an equally simple answer. These losses have got to be sustainable by either public, i.e., state, or private subsidies of the fishing operations. Given that most fishery operators are small, owning at most a few vessels, it is quite clear that such losses could not be sustained by the private sector and hence it is state subsidies that are underwriting the continuing losses of the global fleet.

That this is occurring has two unfortunate consequences. First, it means that substantial world resources are being devoted to the continuation of a failing economic activity. Second, vessels are being subsidized to continue in a fishery when normal economic forces would have driven them out much earlier. Hence the fishing pressure on already heavily exploited resources has been continued artificially.

This issue of subsidy in world fisheries has provoked major concern since the work of Garcia and Newton. A recent examination by Milazzo for the World Bank indicated the scale of the problem and showed how substantial and ubiquitous these subsidies are. Clearly, an immediate removal of subsidies from the fishing fleets around the world would lead to quite unacceptable levels of social deprivation and loss of food supplies. However, the overcapacity of fishing fleets has serious consequences for the continuation of sustainable fishing. States have recognized this and in some areas a reverse process is happening. For example, within the European Union there is a program, the Multi Annual Guidance Programme, which is aimed at ensuring, by fiscal means, a reduction in the European fleet. This and other mechanisms for reducing fishing effort are going to be essential if fisheries are to be managed sustainably in the future.

B. Ecosystem Considerations

1. Fishing Down the Trophic Web

Much of the discussion in the review so far has focused on the general characteristics of fisheries and it has treated individual fish resources as if they were in isolation from the ecosystem in which they are embedded. However, the level of exploitation that has been occurring has been accompanied, as might be expected from simple ecological principles, by a change in the species composition of the ecosystems that are being exploited. In a recent review, Pauly and coauthors have shown that the average trophic level of catches has been declining over the past five decades. In other words, fish catches are coming from lower in the food chain: fewer predators and more prey. It is too simplistic to consider this as a simple fishing down of the trophic web in which first the predators are removed followed by an increase in abundance of their prey. Fish species interact within an ecosystem in complex ways, often at different trophic levels and hence such a simplification is inappropriate. Nevertheless, it indicates that there is a secondary effect where levels of exploitation are high.

2. The Issue of Bycatch

It is well known that certain commercial fisheries operate inefficiently, in that the species targeted are not the only ones to be removed, damaged, or killed. Recently, Alverson and co-workers estimated that the annual level of bycatch in commercial fisheries is on average some 27 million tonnes (a very significant portion of the overall total world catch). This bycatch occurs primarily in fisheries where the gear, almost by definition, is unselective. The elimination, of such bycatch by gear change and regulation and/or the utilization of the catch for human consumption offers a significant potential for increasing overall productivity.

III. FISHERIES SCIENCE

Although work in the late nineteenth century anticipated some developments, the main pioneers of fishery science worked in the first two decades after World War II. Of the many scientists involved in the development of fisheries as a science, five individuals stand out: Schaefer, an American working at the Inter American Tropical Tuna Commission; Ricker, a Canadian working at the Nanaimo Laboratory in Canada; and three British scientists, Beverton, Gulland, and Holt, all working at the Fisheries Laboratory in Lowestoft. It is their work that set the scene for fisheries science as it is today. All of this work was developed prior to the availability of modern computing power and the scale of their achievements, set in this context, is impressive.

A. Maximum Sustainable Yield (MSY)

The basic question of fisheries science is how exploitation, the removal of individuals from a population, affects the dynamics of that population: the processes of birth, growth, and death. A central concept in this exploration is sustainable yield. The idea is simple but it requires an explanation in the context of population ecology. In its unexploited state, a population will on average be in balance with its environment, the processes of birth and growth will be offset by death, and the population will be at equilibrium. A simple differential equation, commonly used in other branches of ecology, expresses this idea well. The rate of change of the biomass B of a population over time is given by

$$dB/dt = rB(1 - B/B_0). \quad (1)$$

When the biomass is equal to B_0, the rate of change is zero and the population stays at its equilibrium (unexploited) level B_0. When B is very small compared to B_0, the population will grow exponentially at a rate r. At intermediate levels, the rate of change will decline from its maximum at $B = 0$ as B increases to zero when $B = B_0$. It is possible to plot the right-hand side of Eq. (1) to show how the increase in biomass of a population is affected by its size.

This is done in Fig. 3a. The figure shows that there is a maximum increase in biomass which occurs at a level where the biomass is equal to $B_0/2$. Some simple calculus can be used to show that this maximum is given by the expression $rB_0/4$. If exploitation is removing biomass from the population, then as long as the removals are less than the maximum, the removals will be replenished by the population. These removals (the yield) are thus sustainable as long as they do not exceed this maximum, which is called the maximum sustainable yield (MSY). The MSY will vary with the abundance of the fish stock concerned (B_0) and with its capacity to increase (r).

The exploitation of the population can be modeled by a simple modification of Eq. (1). Let p be the proportion of the population that is removed by exploitation. The equation for exploitation is then

$$dB/dt = rB(1 - B/B_0) - pB. \quad (2)$$

This is shown by a simple modification of Fig. 3a in Fig. 3b.

The expression for the removals pB is a simple straight line with slope p. The larger the proportion removed, the steeper the line. Where the line intersects the curve showing the increase in biomass, the removals are exactly replenished by the increase in biomass. Simple calculations can show that where $p = r/2$, the maximum sustainable yield is reached. Where p is greater than this level, there will be a sustainable yield, but it will be less than the maximum and the biomass of the population will be reduced below the level where it produces the maximum sustainable yield. Increasing the intensity of harvesting then will actually reduce rather than increase the yield. By contrast, if p is less than $r/2$, increasing the intensity of harvesting will increase the yield.

In the practical situation, the proportion of the fish stock removed will be determined by the level of fishing effort: the number of boats, their size, the type of fishing gear, the duration of fishing, etc. If the level of fishing effort is less than that required to take the maximum sustainable yield, increasing the effort will result in an increased yield. If, however, the level of fishing effort is higher than that required to take the maximum sustainable yield, further increases in that effort will result in a decrease in sustainable yield. Such a situation is often termed overcapacity as removing fishing effort will actually increase yield in the longer term. It should be emphasized that the idea of MSY developed above is an oversimplification and that environmental variability and other factors render its simple application problematic.

The simple models used to illustrate the ideas of sustainable yield and fishing capacity do not allow a proper exploration of some of the other key areas of fisheries science. To explore these requires a deeper understanding of the way in which fish populations behave. The idea can be developed simply using as a

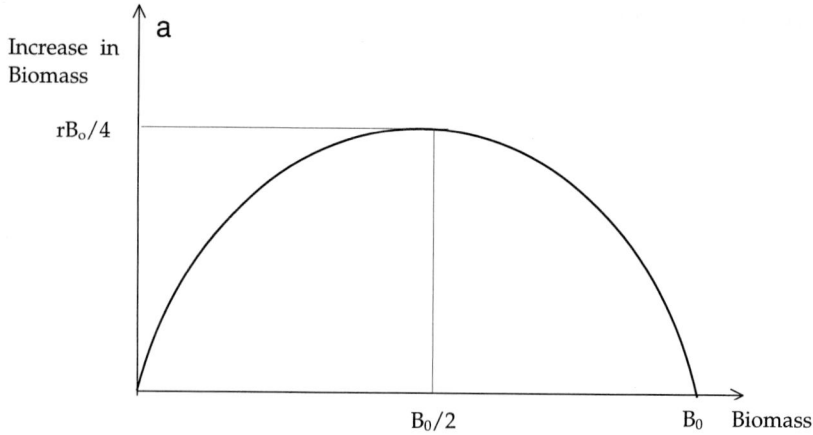

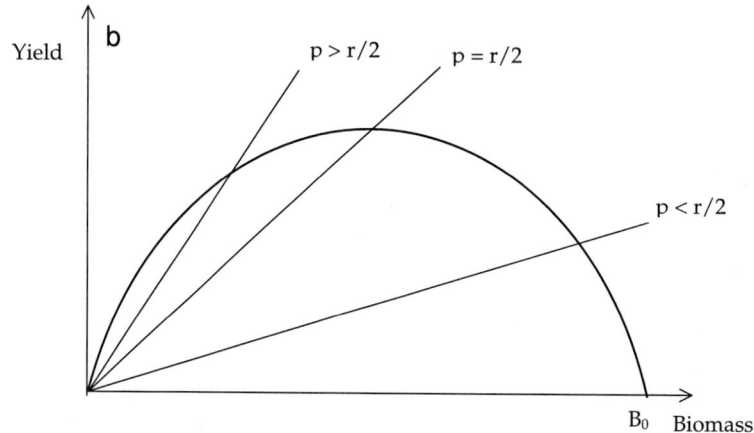

FIGURE 3 (a) Rate of change (increase) of biomass related to biomass: The position of stock size when MSY is taken is $B_0/2$ and the MSY level is $rB_0/4$. (b) Catch (yield) levels and the rate of change of biomass. Where intersection occurs, an equilibrium yield is being taken.

subject an idealized fish population; here young fish develop from their larval form and become part of the population at a particular time of the year. A year later they will have grown and some of the cohort will have died, either through predation or other natural causes. The process continues and a typical temperate fish population will consist of a number of age groups, with fewer fish surviving to greater and greater ages. This is illustrated in Fig. 4a.

Exploitation of this population can in principle start at any age, but in practice will be determined by fishing practices. For example, the size of the net will determine what age groups are caught; smaller and younger fish can escape through the mesh. Figure 4b shows how this will affect the age structure of the population. In the exploited population, fewer fish survive to greater ages and the heavier the exploitation, the more extreme will be the reduction in older fish. However, the biomass of the population is affected not just by natural deaths and fishing but also by growth. As fish get older, they increase in size according to some relationship. Figure 5 illustrates this for a typical fish.

In a comprehensive work first published in 1957, Beverton and Holt explored how the yield could be affected by changing the age at which fishing commenced and the intensity of fishing. Their results show how the interplay between these two parameters, which can be changed by fishing practice, can alter the level of sustainable yield. A centrally important result is that in many cases delaying the age of which fishing starts to operate can improve yields. This was in part the basis for many national regulations and international agreements to regulate the mesh sizes of nets and the type of fishing gear permitted.

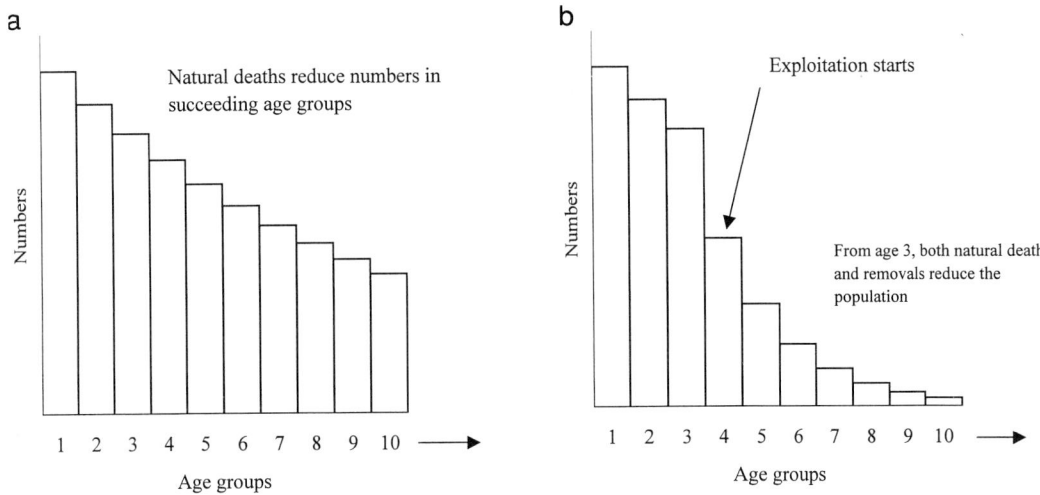

FIGURE 4 (a) Numbers of fish in each age group in an unexploited population. (b) Numbers of fish in each age group where fishing starts from age 3.

B. Stock and Recruitment

One further element is missing from the description of the dynamics of an exploited fish population: reproduction. Typical fish are enormously fecund, laying as many as several hundred thousand eggs at a time. The eggs and the larvae into which they develop become part of the plankton, where they are subject to high levels of mortality from predators and from starvation. To illustrate the enormous mortality involved, a male and female fish might produce 100,000 eggs. For the population to replenish itself, neither grow nor decline, two eggs need to survive to adulthood. There are two issues here. The first is conservation, for clearly sufficient adult fish must be permitted to survive to breed to allow population renewal. The second is variability, for it is hard to conceive of a natural process that can provide such control: if on average less than two individuals survive, the population is failing to replenish itself; if more than two survive, it will increase. Unsurprisingly, observations on fish recruitment, as it is termed, show high variation.

C. Management

The management implications of this analysis are important. If fishing starts on age groups that are not yet mature, then fishing intensity must be controlled to ensure sufficient adults survive to breed so that declines in recruitment do not occur. If fishing does not start on age groups until after breeding has occurred, the conservation issues become less intense, although care is still needed. For example, larger, older fish produce more eggs.

An overexploited fish population will thus exhibit characteristic symptoms. First, there will be few age classes in the population and hence the average age/size of the catch will be lower than in less intensively exploited times. Second, the recruitment of young fish will be in decline, reflecting overexploitation of the adult stock. In such a situation, the management action needed is straightforward, a reduction in fishing intensity and an increase in the age of first fishing. However, depending on the severity of overexploitation, fishing may need to cease altogether.

Although the management action needed to address overexploitation is well understood, once severe overexploitation has occurred, there is no guarantee that

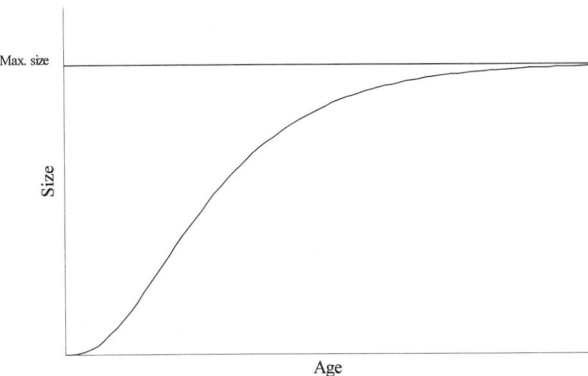

FIGURE 5 Schematic growth curve for typical fish growth.

reducing or ceasing fishing will reverse the process and that the stock will recover. There are plausible mechanisms that can prevent this, of which the most important are likely to be increased competition or predation by other species. Accordingly, the best fisheries management practice is aimed at preventing overexploitation from occurring.

D. Monitoring

Clearly, a key to managing fish stocks is the ability to monitor what is happening to the stock under exploitation. There are direct methods of surveying and monitoring stocks using research vessels which can work well in many cases. However, in other situations such methods are either inefficient or economically unjustifiable. Regular monitoring of information from the catches of the commercial fleets permits a number of different methods to be used. These methods use this statistical information to reconstruct the behavior of the stock. One common method involves returning to the model of Eq. (2). The yield or catch from the stock was given by the equation

$$\text{Catch} = pB. \qquad (3)$$

As noted earlier, p depends on the level of fishing effort. The definition of fishing effort (F) is dependent on circumstances, hence for a particular definition of fishing effort, e.g., hours fished:

$$p = qF, \qquad (4)$$

where q is a scaling factor (known in the literature as the catchability coefficient)

$$\text{Catch} = qFB. \qquad (5)$$

Statistics can provide information on catch for known effort and hence catch per unit of effort (CPUE) is proportional to stock biomass, with the constant of proportionality being the catchability coefficient. A variety of statistical methods exist to estimate the catchability coefficient and as a consequence the stock size. Time series of information on catch and effort thus provide direct insight into the way in which the stock has responded to exploitation.

A second widely used method uses information not just on the catch but on the age composition of that catch. This extra information is used in a method known as virtual population analysis (initially developed by Gulland) to reconstruct the changes in the population under exploitation. The method is simple in conception, although computationally quite laborious. In essence, a simple accounting procedure is used: a catch of 1000 eight-year-olds this year implies that at least 1000 seven-year-olds were alive in the previous year. Adjustments for natural deaths and fish catches then permit the changes in population to be estimated.

E. Fisheries Economics

It is reasonable to enquire why fisheries are in such a poor state when the science of fisheries is reasonably well developed, methods of monitoring stocks exist, and management actions needed to address overexploitation are understood. The answer lies primarily in the domain of social science.

Fisheries as an economic activity has one fundamental difference from other industrial activities. In manufacturing industry, increases in supply develop in response to demand by increasing inputs of labor, raw materials, capital, etc. In fisheries (and in other industries that exploit renewable resources) there is a limit to supply which is determined by the maximum sustainable yield. Increasing the inputs to a fishery—more fishermen, better gear, more vessels, etc.—actually reduces the yield if the stock is already being exploited at or above the level of effort that produces the maximum sustainable yield. In this situation, increases in demand reflected in higher prices can lead to increases in fishing effort, lower catches, and hence higher prices. Depending on the situation, vessels may still be profitable or covering costs even though the effort being expended is well above that required to take the catch and the price paid by consumers well above that which would prevail if effort were lower and the catch higher. In a situation where there is no control of the fishery, this process can continue until resources are severely overexploited and individuals in the industry barely cover their operating costs. The issue is one of ownership; many fisheries are still (and most were) common property resources. In such cases, access is unrestricted and additional effort enters the fishery as long as the return is better than in other activities. Because fishing is a specialized activity (fishing gear and vessels have few alternative uses), other alternative activities offer very low returns. Hence most fisheries without management tend to be overcapitalized and economic returns are poor.

Most government action in the past few decades has been the reverse of what would be economically efficient. Fishing operators have been subsidized and have stayed with the industry rather than leave, thus accentuating the overexploitation problem. Clearly, social issues play an important role here; it is obviously problematic to let fishing communities face economic collapse, but the scale of subsidy involved is substantial (see above).

F. Fishery Management in Practice

The management of fisheries involves some simple principles based on sustainable yield ideas yet its practice is often fraught with difficulty. Management measures typically control fishing by regulations that impose limits on catch (quotas, or total allowable catches (TACs)) or limits on access or effort (licensing, closed areas, and seasons) or by indirect controls such as minimum size of catch or restrictions on types of gear.

It is fair to say that such methods have not been entirely successful, for a variety of reasons. Control and policing of regulations are difficult and can be extremely costly. Some regulations, TACs in particular, provide an economic motivation for misreporting of catches and hence undermine the scientific basis of the process. Political issues, particularly when there is significant overcapacity, can override fisheries management concerns. Often in the face of uncertainty, regulations are set with an optimistic view of the stock, which subsequent scientific information reveals to be unfounded. Even when regulation has succeeded in its conservation goals, the fishing industry remains overcapitalized and continues to need assistance.

One development where there appear grounds for some optimism is the construction by certain states of regimes based on the allocation of property rights to operators: Australia, Iceland, and New Zealand have been particularly active. In these regimes, operators are allocated the rights to a particular level of catch (or proportion of some fixed catch). These rights, known as Individual Transferable Quotas, can be bought and sold. In this way the inefficient operations can be bought out by the more efficient ones, which can adjust their capacity to the appropriate level for exploiting the fish stock. The attraction of such a regime is that it is in the interests of the operators (owners) to ensure that the stock is exploited sustainably.

G. New International Measures

Although most states now operate 200-mile zones, in which they control fishing, in certain areas the continental shelf extends beyond 200 miles and exploitation of the same stock can be possible both within the 200-mile zone and beyond it on the high seas. A recent international agreement, the UN Agreement on Straddling Stocks, offers a new international framework for addressing problems of this sort. Its aim is to facilitate the creation of regional bodies composed of both coastal states and those with distant-water fleets operating in the area. It is too early to assess whether this will be successful.

Certain types of species, tuna in particular, are highly migratory and are exploited over very large areas of ocean. Management regimes exist under International Treaty to regulate these fisheries. The latest commission of this sort to be set up is the Indian Ocean Tuna Commission, which held its first meeting in 1997. These regimes rely on considerable political goodwill. A recent development (August 1999) involved Australia and New Zealand taking Japan to the International Tribunal for the Law of the Sea to seek judgment for Japan to reduce fishing on southern bluefin tuna. The three states are in the Commission for the Conservation of Southern Bluefin Tuna but have proved unable to agree on management action within the Commission. The Tribunal made an interim judgment effectively asking Japan to reduce catches pending a full arbitration on the merits of the case. Whether similar cases will be brought and the Tribunal becomes the international arbitrator of fishery management regimes on the high seas remains to be seen.

H. Ecosystem Management

Fish stocks exist in aquatic ecosystems, forming the prey of some species and being predators of others, yet most fisheries science and almost all fisheries management regimes tend to focus on individual species. The regime that was set up to deal with the Southern Ocean, the Commission for the Conservation of Antarctic Marine Living Resources (CCAMLR), had a different focus. This regime was specifically orientated to the Southern Ocean ecosystem and regulations were aimed at ensuring that each element of the ecosystem was protected from the effects of exploitation. The primary reason for this markedly different approach was that the large baleen whales of the Southern Ocean had been grossly overexploited by distant-water fleets this century. A significant development of a fishery for their common prey, krill, could hinder, or indeed stop, the recovery of the populations following the cessation of whaling. To date, CCAMLR has been reasonably successful in its operation, setting precautionary levels of catch for krill as well as regulating other fisheries.

However, better scientific understanding of the implications of exploitation of groups of interacting species is needed. Current understanding is really little more than educated commonsense—e.g., a predator cannot be exploited at its MSY level if its prey is simultaneously harvested at a high level.

IV. THE FUTURE OF COMMERCIAL FISHERIES

The increase in the world population over the next two decades will clearly produce an increase in basic demand for food. This is likely to be enhanced in the case of fish as it is a preferred protein source and increasing wealth tends to be associated with increasing fish consumption. Additionally, much of the population increase is estimated to be occurring in Asia, where fish is a preferred source of animal protein.

Statistical analyses of trends in capture fisheries point to a maximum harvest level in the region of 95 million tonnes, a value close to recent average landings. However, more recent analysis by FAO indicates that there is a potential for increasing these catch levels if the full potential of underexploited areas and species were to be realized. Similarly, there is scope, as discussed earlier, for significant improvement in the reduction of unwanted bycatch (around 27 million tonnes is estimated to be discarded each year).

Improved fishery management could in principle add significantly to the global production—recall the high proportion of overexploited stocks. However, the extent to which species interactions may mean that the recovery of one stock may lead to lower catches from another is not well understood.

FAO has concluded that a substantial increase in catch levels of around 20 million tonnes could be achieved if (a) degraded resources are rehabilitated, (b) underexploited resources are successfully managed to increase yields, (c) fully exploited resources are not degraded, and (d) discarding and wastage are reduced. This is the challenge for fisheries science and management in the new century.

See Also the Following Articles

AQUACULTURE • FISH BIODIVERSITY, OVERVIEW • FISH CONSERVATION • FISH STOCKS

Bibliography

Alverson, D. L., Freeberg, M. H., Pope, J. G., and Murawski, S. A. (1994). A global assessment of fisheries bycatch and discards. *FAO Fisheries Tech. Pap. 339.*

Anon. (1997). Review of the state of world fishery resources: Marine fisheries. *FAO Fisheries Circ. No. 920.*

Anon. (1997). Review of the state of world aquaculture. *FAO Fisheries Circ. No. 886, Rev. I.*

Garcia, S. M., and Newton, C. (1997). Current situation, trends and prospects in world capture fisheries. In *Global Trends in Fisheries Management* (E. Pikitch, D. D. Huppert, and M. Sisseswine, Eds.), pp. 3–27. (Am. Fisheries Soc. Symp. 20, Bethesda, MD).

Hilborn, R., and Walters, C. J. (1992). Quantitative fisheries stock assessment. Choice, dynamics and uncertainty. Chapman & Hall, New York.

Matteo, M. (1998). Subsidies in world fisheries, a reexamination. *World Bank Tech. Pap. No. 406,* Washington, D.C.

Pauly, D., Christensen, V., Dalsgaar, J., Froese, R., and Torres, F. (1998). Fishing down marine food webs. *Science* **279**, 860–863.

RESOURCE PARTITIONING

F. A. Bazzaz and S. Catovsky
Harvard University

I. Introduction
II. Assessment of Diversity and Its Functional Significance
III. Resources and Niche Differentiation: Theoretical Considerations
IV. Resources and Niche Differentiation: Empirical Evidence
V. Additional Mechanisms for Maintenance of Species Diversity
VI. Conclusion

GLOSSARY

coexistence Ability of species to persist together in an ecological community indefinitely in the absence of any major environmental change.
ecological community An assemblage of species occurring together in a given space and time.
heterogeneity Variation in environmental conditions through space (spatial) or time (temporal).
niche Pattern of response of an individual, a population, or a species to the physical and biological gradients of its environment.
niche differentiation Differential resource use or response that results from long-term, consistent reciprocal selection.
reciprocal selection Evolutionary change as a result of species acting as selective agents on each other through interspecific interactions.
resource A consumable or depletable substance, such as a nutrient, water, or light, that is required by plants for maintenance, growth, and reproduction.
trade-off Constraint on expression of a species' trait/behavior as a result of expression of a correlated trait/behavior.

RESOURCES ARE ESSENTIAL to the life of every organism. Species constantly compete with one another for these resources but may be able to coexist in the long-term if their resource requirements are sufficiently different, known as "resource partitioning." Maintenance of species diversity within a community is contingent on this ability of species to coexist together. In this article, we discuss the evidence that differences in plant species' resource requirements are sufficient to permit their coexistence. We emphasize the importance of considering species' responses to multiple, interacting resource axes and to variation in resource availability in both space and time. Species do partition the resource environment in multiple ways, but in plant communities of high diversity, e.g., tropical rain forest, there is considerable overlap in species' resource requirements. In such systems, other diversity mechanisms might play a more significant role.

I. INTRODUCTION

The scientific community and the public at large are expressing great concerns about the global loss of biological diversity. Currently, biological diversity is threatened worldwide by the widespread destruction of forests and by impacts of global environmental change, such as nitrogen deposition and global warming. In addition, ease of travel, the dramatic increase in the number of people who travel, and deliberate introductions of organisms into various habitats can all lead to the modification of ecosystems and the loss of biological diversity. Invasions are now rampant and may lead to a great deal of homogenization of the Earth's biota. This situation has led the renowned biologist Gordon Orians to call this epoch the "Homogecene."

If we are to understand and predict the consequences of loss of species diversity for the functioning of the biosphere, we need to ascertain what factors control patterns of species richness both within and across ecological communities. The notion of community diversity has its roots in the writings of Henry Gleason and Frederic Clements. These American ecologists differed sharply in their interpretation of the plant community. The Clementsian conception of the community assumed it to be a cohesive active unit. The notions of reciprocal selection and niche differentiation are both imbedded in this idea, although not explicitly. In contrast, Gleason thought of the community as a chance assemblage of species with overlapping physiological requirements. This view may assume that there is no interaction between species or that these interactions are incidental. More recent discussion of why there are so many species was rekindled by G. E. Hutchinson in 1959 when he asked this question in his now very famous "Homage to Santa Rosalia" paper.

Currently, there are many hypotheses about the generation and maintenance of plant diversity (see discussions in both Tilman and Pacala (1993) and Huston (1994). Much like the Clements–Gleason debate, discussion of the maintenance of species diversity often focuses on the distinction between theories that assume evolution has driven species to partition resources so as to prevent competitive exclusion (equilibrium) and those that explain diversity through the action of demographic dynamics and do not assume a past evolutionary history between species (nonequilbrium). Ecologists commonly favor one mechanism for maintenance of diversity over others, but it is more likely that the strength of any cause differs under different locations and circumstances. It is now critical that we ascertain what characteristics of species and ecosystems lead to the dominance of any one diversity mechanism, so that we can develop a predictive framework for determining causes for maintenance of species diversity across all ecological communities.

To move the science of biodiversity from a descriptive to a predictive discipline, we need to ascertain what mechanisms underlie global patterns of species diversity. This requires clearly defining what we mean by diversity and recognizing its importance (Section II). We must then determine what biological processes lead to the development of diversity within ecological communities. In this paper, we take a predominantly resource-based approach to understanding the causes of biological diversity. We highlight why resources play a central role in the maintenance of species diversity (Sections III and IV) and then discuss how this resource-focused view relates to other hypotheses that have been proposed to explain global patterns of species diversity (Section V).

II. ASSESSMENT OF DIVERSITY AND ITS FUNCTIONAL SIGNIFICANCE

Presently, there is no agreed quantitative way to assess diversity, except to use the taxonomic species concept. However, the use of species as the primary unit of biological diversity is not always appropriate for the situation. We have previously argued that we should focus on different components of ecological diversity depending on the patterns and processes being studied. Indeed, there is now an increasing awareness that biological diversity should encompass the full range of organizational units in ecology, from genes to ecosystems. If we wish to incorporate this broader definition of diversity into our study and stewardship of ecological communities, we must present a clear framework for the accurate assessment of diversity. Developing such a framework involves explicitly outlining the different kinds of biological diversity that should be recognized and how the level of diversity can be quantified.

In many cases, the confusion arises from not recognizing the appropriate scale for the ecological phenomenon of interest. The need for clarification of scale was first recognized by Robert H. Whittaker in 1972, when he addressed the importance of distinguishing within- and between-patch diversity (known as α-diversity and β-diversity, respectively). The temporal scale of ecological dynamics may be as relevant to our discussion of biodiversity assessment as the spatial scale. This notion is well illustrated by the need in certain systems to distinguish between apparent diversity and total diversity, which may be quite different from one other. Many ecological communities depend on an intermittent availability of resources. For example, in deserts, when water occasionally becomes available in large quantities, the number of plant species (and perhaps microbes and the animals that depend on these plants) dramatically increases. In such systems, apparent diversity, which usually involves the enumeration of the species present at a given time, can vastly underestimate total diversity. Other kinds of diversity measurement may be appropriate, depending on the ultimate goals of the study. For example, aesthetic diversity for humans may depend on the number of flower colors present in a community rather than species diversity assessed by a professional (Fig. 1).

Specifying what components of ecological communities should be measured has become increasingly important since expanding our definition of biodiversity. As we begin to recognize the central role that ecosystems play in the global cycling of elements, functional diversity has been identified as an important characteristic in ecological assessment. Functional diversity measures the range of energy and matter processing that a particular community can undertake, e.g., carbon uptake and nitrogen fixation. Explicit in this concept is the idea of functional redundancy, which states that often more than one species in a community can play a role in a particular ecosystem process (see chapters in Schulze and Mooney, 1993). Functional diversity may be very important in the discussion of biological diversity because the removal of a given species may be much more critical to a particular system function than the removal of another species, even one with more biomass.

Although plant communities are only theoretical constructs (see Bazzaz, 1996), much discussion has gone into the notion of biological diversity and its impact on the functioning of ecosystems. A number of recent, high-profile experiments have addressed how many species must be present for a given ecosystem to function appropriately, e.g., the Ecotron experiment (Shahid Naeem and others), field studies on California serpentine grassland communities (David Hooper and Peter Vitousek) and on midwestern tallgrass prairie communities (David Tilman and others), and the BIO-DEPTH experiment carried out at eight grassland sites across Europe. Current data in these grassland ecosystems seem to suggest that the number of species for greatest ecosystem productivity hovers close to nine. Unfortunately, there is still considerable uncertainty about what mechanisms are driving these diversity–productivity patterns (see discussion in recent BIO-DEPTH paper: Hector et al., 1999). Some ecologists currently believe that complementarity in species' resource use could make a multispecies community more effective at utilizing available resources (to produce higher net productivity), while others argue that more diverse communities are merely more likely to contain more productive species (known as the sampling effect). As we currently lack a complete mechanistic understanding of the relationship between species diversity and ecosystem function and as relevant data exist for few other ecosystem types, we cannot yet know how general these results will turn out to be. In these particular experiments, the species are all somewhat similar in their behavior and no "keystone" species exists. In communities containing a species that performs a special function (such as nitrogen fixation), the removal or addition of such a species may severely impact ecosystem function, e.g., *Myrica faya* invasion in Hawaii. We must also pay close attention to species that supply unique goods and services to humans.

III. RESOURCES AND NICHE DIFFERENTIATION: THEORETICAL CONSIDERATIONS

Maintenance of species diversity is often thought to be a result of niche differentiation. This idea was implicit in much early writing on community structure that recognized that species could not coexist with one another unless they had sufficiently different patterns of resource use. As plants are predominantly sessile organisms, it is assumed that plant species that are compo-

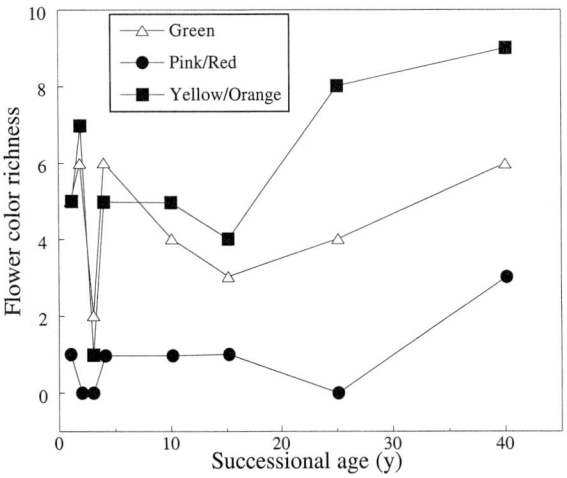

FIGURE 1 Changes in floral color diversity in midwestern old-field successions. Modified from Bazzaz (1996), *Plants in changing environments*.

nents of a community have coexisted and competed for a long time. In this way, they are assumed to have shaped each other's responses to the multitude of critical physical and biological resources in their local environment through reciprocal selection, which could result in niche differentiation in the long term (Connell, 1980). This idea is similar to character displacement in animals. However, requirements for niche differentiation are stringent and may not always be met. Long-term species co-occurrence is needed and the frequency of encounter between plant species will depend on dispersal patterns (see Section V.C). In addition, reciprocal shaping of species' responses to their environment is not easy to accomplish, unless initially traits related to fitness are somewhat independent of each other. To avoid assuming that niche differentiation has taken place in the past, Bazzaz (1996) suggested previously that it is more tractable to consider copresence rather than coexistence, especially in communities with a poorly known history. Studies on the postglacial migrations of temperate tree species during the Holocene in North America have clearly demonstrated how species have moved individualistically across the landscape and have not necessarily coexisted in the same community for long periods of time (see papers by Margaret Davis). Here, we consider communities where species are found together without regard to whether or not they co-occurred over evolutionary time and therefore have shaped each other's responses to the environment.

While niche differentiation has been touted as a major cause of the generation and maintenance of biodiversity, particularly in late successional stable ecosystems, it was unclear for many years how the concept applied to plant communities. The notion of resource partitioning was originally developed for animal communities, where species can potentially utilize a wide range of food resources and thus can easily partition up their resource environment. This idea was harder to apply to plant communities, as plants all have relatively similar resource requirements (Connell, 1978). As a result, much ecological research has focused on trying to determine how species partition themselves along resource axes. Progress on understanding factors controlling plant community diversity has been achieved by considering species' responses to multiple resources in the environment. David Tilman (1982) tackled this issue by using a mathematical model to show that species coexistence was feasible if there were trade-offs in species' requirements for different ratios of resources. This approach clearly demonstrated the potential for a resource-based approach to contribute to explaining plant species diversity. Often, however, species' responses to

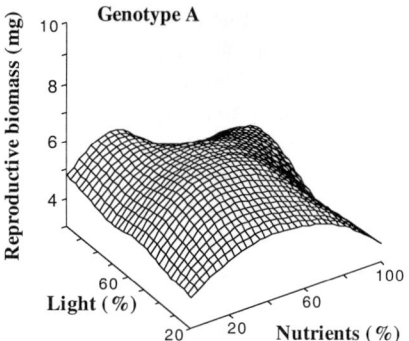

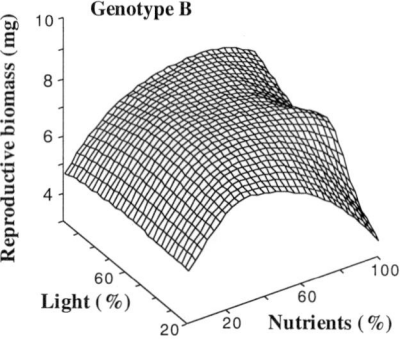

FIGURE 2 Differential responses of two genotypes of *Polygonum* to varying light and nutrient gradients. Modified from Bazzaz (1996), *Plants in changing environments*.

environmental factors are not independent of each other. Thus, it may be more informative to develop a multidimensional response surface that approximates the response of the species to its environment (Fig. 2) (Bazzaz, 1996). Gradient techniques from vector calculus now allow us to quantify these response surfaces.

In this multidimensional view of the plant niche, maintenance of diversity depends on species separating themselves along multiple resource axes, such that they can coexist for a long time. It is assumed that there must be a minimal overlap between members of a community. In 1973, Sir Robert May attempted to quantify the minimum allowable overlap of the species for their stable coexistence and concluded that, if the difference between the means of their responses is more than twice that of the spread of each individual curve, stable coexistence should be possible. If species are highly specialized (narrow response curves), they do not need to position themselves very far apart on one particular resource axis to prevent competitive exclusion, whereas species that are more generalist in their resource use (wide curves) must be further separated in niche space. This kind of analysis has to be extended to multidimensional response surfaces. We still do not know, even theoretically, what the minimum overlap to allow coexistence should be when we consider species' responses to more than one critical environmental gradient.

These minimum overlap conditions for coexistence may differ for contrasting community types. For example, it is reasonably well established that plants of early successional communities have broader responses to resource gradients than do plants of late successional communities. This pattern alters the length and intensity with which species interact with one another, and thus the degree to which species partition the environment. However, it has not been determined whether this pattern results from or leads to the lower species diversity of early successional communities relative to late successional ones. Early successional plants have a near-geometric type of distribution (each species takes approximately the same proportion of resources to produce the same amount of biomass). As succession proceeds, there is a shift toward a more log-normal distribution (Fig. 3) and then finally, without major disturbance, a return to a near-geometric distribution. This variation in patterns of resource use between species of different successional position creates a hump-backed relationship between diversity and successional time, as predicted by the intermediate disturbance hypothesis (Connell, 1978).

We have seen that mathematical models of community structure predict that diversity is largely a function of how species partition themselves along multiple environmental resource axes. Theoretical considerations suggest that trade-offs in plant species' responses along multiple environmental gradients could permit multispecies coexistence. But, do we see such trade-offs in nature and are they sufficient to explain long-term species coexistence? In the next section, we discuss the evidence for niche differentiation in plant communities and consider whether or not the degree of resource partitioning in these communities is adequate to explain species diversity in different plant communities. Resources may be partitioning both spatially and temporally and we present evidence for both kinds of niche differentiation among plant species.

IV. RESOURCES AND NICHE DIFFERENTIATION: EMPIRICAL EVIDENCE

A. Spatial Heterogeneity

Most resource partitioning arguments proposed to explain species diversity assume that the environment is heterogeneous. Coexistence is possible because species have different resource requirements and are specialized to succeed on particular patch types. Diversity is maintained through the presence of an array of patch types. As our ability to measure and quantify environmental variation has improved, we have developed clear evidence that the environment is heterogeneous with regard to resources that are critical to plants. To date, however, we have not critically assessed if the degree of heterogeneity recorded by our instruments is representative of the resource environment perceived by the plants themselves (Bazzaz, 1996). Despite this uncertainty, there is now clear evidence that, at least on the broad scale, species do partition themselves along resource axes. Studies that relate species' distributions to multiple environmental factors using either direct or indirect gradient analysis have identified numerous soil-related axes of specialization, e.g., R. H. Whittaker's 1956 work on the vegetation of the Great Smoky Mountains. Research on experimentally produced plant communities also supports the notion that species may be differentiated along soil resource axes (see studies discussed in Bazzaz, 1996).

Although plants themselves can influence their local environment and thus produce intrinsic heterogeneity within the community, much spatial resource heterogeneity is generated by extrinsic factors, the most common of which is disturbance. In fact, disturbance is judged both by change in availability of resources and by per-

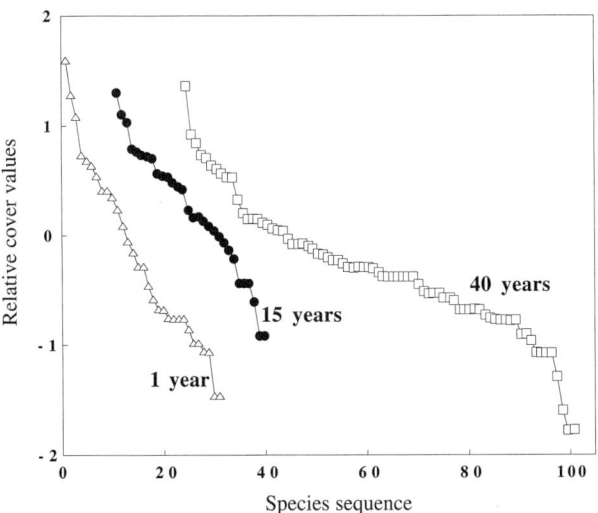

FIGURE 3 Changes in community dominance–diversity relationships during 40 years of old-field succession. Modified from Bazzaz (1975), *Ecology* 56, 485–488.

ception of that change. Disturbance events create heterogeneity in resource availability at multiple spatial and temporal scales, and these different scales can allow species to partition the environment in more ways. Awareness of the importance of disturbance events for understanding the maintenance of plant species diversity was heightened when Peter Grubb (1977) proposed that regeneration traits of species should provide a critical axis for differentiation, the so-called regeneration niche. Much of the research on partitioning along this regeneration axis has focused on the broad-scale effects of canopy disturbance, addressing differences in conditions and plant responses between gaps and closed forest. Gap creation generates variability in the light environment and has been advanced as a major factor in increasing diversity at the stand level. Broad guilds of plant species that respond similarly to gap formation (light demanding vs shade tolerant) have been identified in numerous forest communities where vertical canopy stratification creates a wide range of light microenvironments (Whitmore, 1989). More recently, sophisticated mathematical techniques have been used to classify both physiological and demographic responses of tree seedlings to light availability more precisely (see SORTIE papers of Steve Pacala, Richard Kobe, and others). Much of this research into species' responses to total light quantity has found considerable overlap in response, and species' partitioning of the light microenvironment is not usually sufficient to explain maintenance of species diversity, particular in communities with high numbers of co-occurring species, such as lowland tropical rain forest.

An additional mechanism for the maintenance of diversity arose from the gap partitioning hypothesis, originally developed by Ricklefs (1977) and applied specifically to the tropical rain forest by Julie Denslow (1980). Here, finer-scale variation in resource availability across the gap–understory continuum was suggested as a possible mechanism for coexistence in multispecies communities. Unfortunately, this hypothesis has yet not been validated in tropical rain forests, where we lack a good understanding of the maintenance of species diversity. Experimental work by Tim Sipe and F. A. Bazzaz in temperate forests in eastern North America, in contrast, provides some support for the gap partitioning hypothesis. In experimentally created gaps at Harvard Forest in central Massachusetts, we found clear microenvironmental differences between gaps of different sizes (especially with regard to light). We compared morphological and physiological responses of three co-occurring maple (*Acer*) species to this light variation and found clear evidence that the species differed in

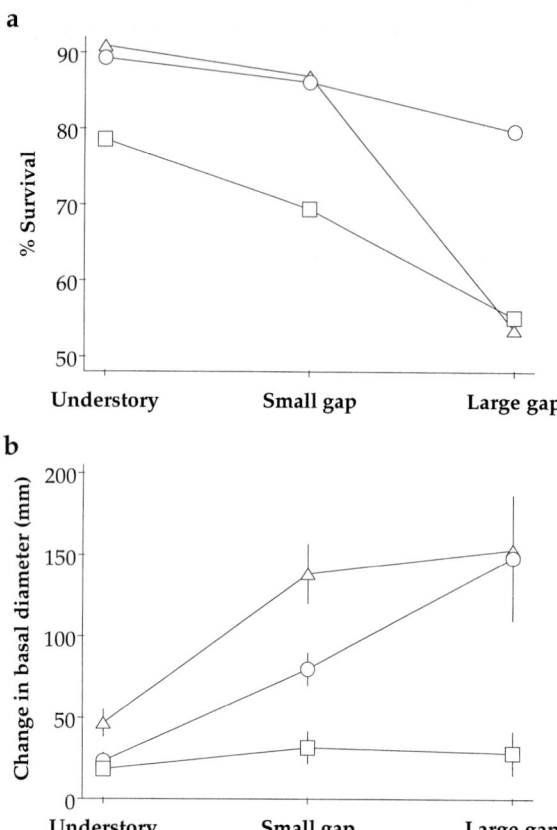

FIGURE 4 Survival (a) and growth (b) responses of striped maple (*Acer pensylvanicum*, triangles), red maple (*Acer rubrum*, circles), and sugar maple (*Acer saccharum*, squares) seedlings to canopy gap size. Data redrawn from Sipe and Bazzaz (1994), *Ecology* 75, 2318–2332, and Sipe and Bazzaz (1995), *Ecology* 76, 1587–1602.

their response to gap size. Many response variables showed significant differences between large gaps, small gaps, and understory plots, and these differences often varied between species, creating distinct species' preferences for canopy gap environment (Fig. 4). We also observed substantial variation in light availability between different parts of canopy gaps, due to seasonal and diurnal trends in solar patterns. This variation, however, was rarely reflected in differences in species' responses to position within the gap.

B. Temporal Heterogeneity

As resources vary in both space and time, species could partition their resource environment temporally as well as spatially. Temporal heterogeneity in resource availability is often not considered in discussions of maintenance of diversity, but could prove central to our under-

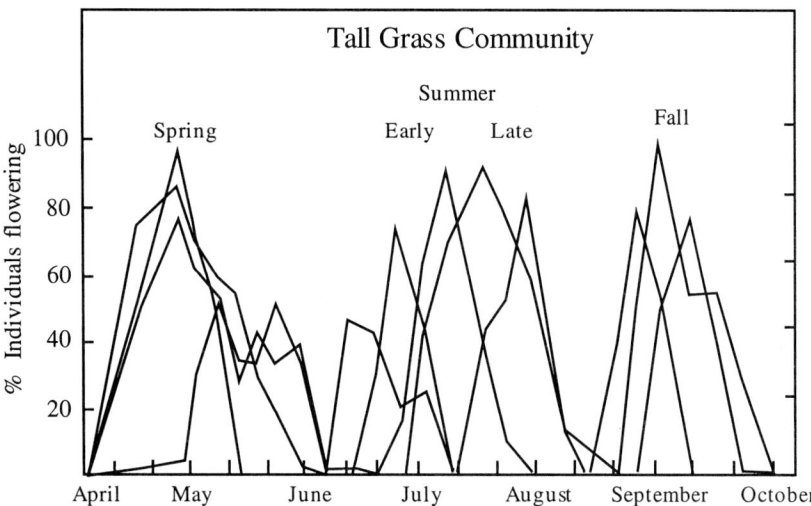

FIGURE 5 Seasonal flowering patterns of a midwestern prairie community with the major species shown as separate lines. Redrawn from Parrish and Bazzaz (1979), *Ecology* 60, 597–610.

standing of species coexistence. Differences in the timing of plant developmental events, such as germination, flowering, and fruit drop, could play a critical role in mechanisms of species coexistence. In some communities that have existed for a long time, there is evidence for clear separation in flowering between groups, e.g., Peter Ashton's work on staggered mast-flowering in the Dipterocarpaceae in the tropical rain forests of Southeast Asia. In the tallgrass prairie of the American Midwest (communities that have a long evolutionary history and great species diversity), we have demonstrated clearly how these species separate their life history events along the entire growing season, i.e., phenological separation. One can recognize three fairly distinct groups of flowering species (Fig. 5). Also, in these grasslands one can observe the separation of the phenology of the introduced species, *Poa pratensis*. Although it is now considered by many investigators to be part of the tallgrass prairie, it is an introduced species (from European grasslands) that grows well in the cool portion of the season. Thus, it is able to grow at two distinct times of the year: before and after the main growing season of its competitors, the native, warm-season grasses of the community (Fig. 6). In this way, this species separates its growth activity from that of the native species and reduces competition for critical resources.

This kind of phenological separation is also evident in some early successional communities where species generally have broad niches (see earlier discussion in Section IV.A). In old-field communities in the American Midwest, *Chenopodium album* delays its growth and reproduction much later than all other species in the community (Bazzaz, 1996), thereby reducing competition during the main portion of the growing season. In this case, it is not clear that competition has shaped these phenological responses, as species in this early successional community have not necessarily been present together over evolutionary time. Many arguments about coexistence assume that competition has shaped the niches of co-occurring species in the past (see earlier discussion). However, in communities of unknown history, the question remains whether a species has modified its phenology by competition with the native species or was preadapted to function in these communities. An example of such preadaptation is found in the shifted daily flowering behavior of the native *Erigeron*

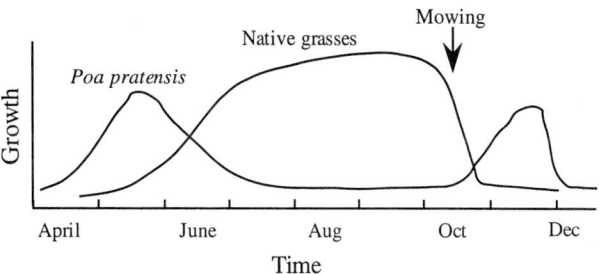

FIGURE 6 Separation of growth phenology of the introduced grass, *Poa pratensis* (bluegrass), from that of the predominant native prairie grasslands. Redrawn from Bazzaz and Parrish (1982), in *Grasses and grasslands*.

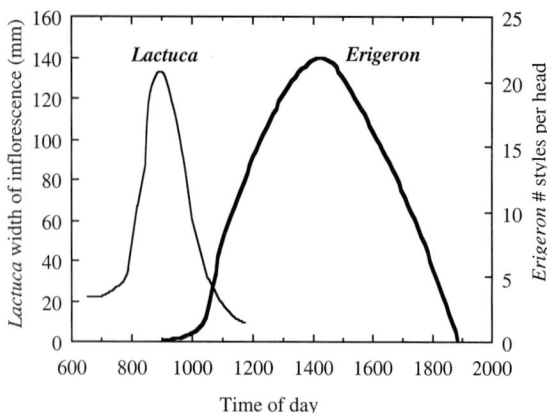

FIGURE 7 Importance of the daily timing of flowering for two members (*Erigeron annuus* and *Lactuca scariola*) of a midwestern winter annual plant community. Redrawn from Parrish and Bazzaz (1978), *Oecologia* (Berl.) 35, 133–140.

annuus and the introduced *Lactuca scariola*. In winter annual communities of the midwestern United States, *Lactuca* opens its flowers in the morning hours 7:00 to 10:00 a.m. and requires a large number of pollinators. After 10:00 a.m., however, *Erigeron* starts to open and *Lactuca* closes its flowers (Fig. 7). As a result, the pollinators shift their foraging to *Erigeron*. Thus, while both species use the same pollinators, the daily separation in the opening of the flowers ensures that both are pollinated. At first glance, this may appear to be a convincing example of niche differentiation. However, when examined carefully, one finds that the introduced *Lactuca* developed this flowering behavior in its native habitat in Europe and not by competition for pollinators in the successional fields of the Midwest.

Temporal differences in species' traits may extend to very early stages of the life cycle. The formation of seed banks that can persist in the soil for a long time may also lead to maintenance of species diversity. A persistent seed bank may buffer species' responses to short-term variation in environmental conditions and prevent irreversible population crashes. Weedy species are known to form persistent seed banks. In good years for a particular species, many seeds are produced and enter the soil seed bank. In bad years, in contrast, few seeds are produced. The species can, therefore, partition the time axis to contribute the most to future generations in years when conditions are favorable for growth and reproduction. The development of this sort of persistent seed bank has been mathematically modeled and is often called the storage effect (see work by Peter Chesson). Coexistence is promoted if species have differing sensitivities to temporal variation in environmental conditions.

Further niche axes may potentially be created through the combination of temporal heterogeneity in multiple resources (see discussion in Section III). Because plant responses to a particular resource are contingent on prevailing local environmental conditions, e.g., temperature and relative humidity, the time course of resource availability patterns could critically determine a plant's performance. If resource availability is high when other conditions are favorable (i.e., resources are congruent), then plants may be able to make good use of that resource. If, however, conditions are unfavorable and plant activity is inhibited to some extent, then plants may not be able to take full advantage of the resource in ample supply. Species might have different levels of tolerance for the extent of resource congruency in a given plant community (Bazzaz, 1996). Additional research by our laboratory at Harvard Forest has demonstrated that, for seedlings in a canopy gap where environmental conditions show distinct temporal patterns, resource congruency could play an important role in permitting species coexistence (see papers by Peter Wayne and Gary Carlton).

C. Resource Partitioning: Is This Sufficient to Understand Species Diversity?

There is now good empirical evidence for substantial resource partitioning between species in the same community. Underlying variation in soil resources and disturbance-mediated changes in the aboveground microenvironment provide primary axes for species' specialization. Further axes may be produced by responses to multiple environmental gradients, e.g., resource congruency, and by substantial temporal separation of developmental events, such as germination and flowering. It is not clear, however, that these differences are adequate to prevent competitive exclusion in the long term, as the theoretical considerations for community-level consequences of niche differentiation in multidimensional space have not been adequately explored (see Section III). In transient, early successional communities, this discussion may not be relevant, but in communities with a long-term evolutionary history, we certainly need to consider whether resource partitioning plays a central role in community diversity. Communities characterized by long-term species' interactions, such as the tropical rain forest and the tallgrass prairie, typically contain a high number of species, and so it is critical that we unequivocally establish the role of niche differentiation in maintenance of species diversity. To date, experimental evidence suggests that resource partitioning is not sufficient to fully explain the high species richness of some such communities, e.g., tropical rain forest. Work by Hubbell et al. (1999) on Barro Colorado Island in Panama demonstrated that gaps did not explain variation in species richness. Gap formation did increase seedling establishment, but this effect was nonspecific and unpredictable.

V. ADDITIONAL MECHANISMS FOR MAINTENANCE OF SPECIES DIVERSITY

Resource partitioning is clearly not the only diversity mechanism operating in plant communities containing a large number of species. Species' specialization on different portions of multidimensional resource space prevents competitive exclusion by reducing the degree of interspecific interactions. However, other processes within plant communities might also reduce the extent or slow the rate of competitive exclusion. In this section, we discuss three additional mechanisms that have been suggested as causes of maintaining plant diversity.

A. Frequency-Dependent Effects

Mathematical models of competition clearly demonstrate that stable coexistence is possible if intraspecific interactions are stronger than interspecific ones. Weak interspecific interactions commonly arise from reciprocal selection acting over evolutionary time to produce niche differentiation. However, other intraspecific density-dependent effects could act to effectively reduce interspecific interactions. A mechanism for such density-dependent effects in species-rich plant communities like tropical rain forest was first suggested by Joseph Connell and Daniel Janzen. They both argued that the most common and dominant species in the community would be most commonly attacked by herbivores and pathogens, resulting in community-level compensatory mortality. In this way, no single species would be able to dominate the community and reduce overall species diversity. Host-specific herbivores or pathogens could act in a similar compensatory manner, increasing the average distance between parent plants and effective offspring recruitment. Currently, however, empirical support for frequency-dependent mortality in tropical rain forests has been variable. Broad-scale examination of size-specific population dynamics of tropical trees has demonstrated negative density-dependent effects in some cases, e.g., the work of Cam Webb and David Peart in Borneo, but not others (Hubbell and Foster, 1986).

B. Chance

The strongest alternative theory to the hypothesis that resource partitioning determines community diversity proposes that plant communities are dynamic entities containing a random assemblage of species not organized by reciprocal selection (see Clements vs Gleason dichotomy earlier). According to this view, higher diversity communities result from larger regional species pools rather than from species dividing up the resource environment more finely. Under these so-called lottery model scenarios, coexistence is possible as all individuals have an equal chance of reaching a particular location and whichever individual reaches the spot first will successfully establish there.

Finding little evidence for compensatory mortality in tropical rain forests (see above), Hubbell and Foster (1986) have argued that these lottery models may well represent tropical rain forest dynamics and explain their high species diversity. In these communities, beyond the basic early vs late successional dichotomy, many tropical forest species are essentially ecologically identical. Fluctuations in numbers result simply from chance events, and these fluctuations lead to a random walk in population densities with no stabilizing tendencies at all. Species coexistence is possible because the time for competitive elimination is even longer than the rate of new species formation.

Can we reconcile the niche differentiation vs chance perspectives on tropical rain forest diversity? It might be that the relative importance of deterministic and stochastic processes in community structure depends on the characteristics of the particular community. Chance could play a more important role in community diversity when communities are dynamic and have a fast rate of change, e.g., the tropical forest on Barro Colorado Island. In contrast, in older communities, where the rate of change is much slower and where there is likely to have been considerable sympatric speciation, e.g., the forests of Southeast Asia, niche differentiation could be the more dominant process maintaining species diversity. Until we develop a broad conceptual model of plant communities that incorporates both deterministic and stochastic processes, it will be difficult to test these ideas explicitly. Currently, the relative importance of niche differentiation vs chance effects in structuring plant communities is not well resolved.

C. Spatial Structure

Implicit in the notion of the regeneration niche is the idea that trade-offs between different components of a species' life cycle may contribute to species diversity. For plant species, different strategies can be employed during the recruitment and establishment phases of the life cycle, and these strategies could be subject to selection pressure. The importance of life cycle trade-offs for maintenance of species diversity was effectively demonstrated following awareness of the importance of spatial structure in population and community dynamics. Incorporation of spatial structure into our conceptual models of species diversity reduces the requirements for long-term stable coexistence of species, as dominance of a community by one species is commonly prevented by restrictions on a species' dispersal capabilities. By assuming that there is indeed a trade-off between a species' competitive and colonization ability, i.e., no "superspecies" exists, Tilman (1994) was able to demonstrate that coexistence of an unlimited number of species on a single resource was possible. Although we have much anecdotal evidence that trade-offs in species' competitive and colonization abilities do exist, e.g., during old-field succession (Bazzaz, 1996), we currently do not know if the exact nature of the trade-offs is sufficient to explain multispecies coexistence. It is clear, however, that recruitment limitation is a major factor controlling the structure and diversity of both temperate and tropical forests.

As plants only interact with close neighbors, spatial structure (position of individuals within the community) is a critical consideration for plant diversity as it determines the extent and intensity of intra- and interspecific interactions. Spatial heterogeneity generated by the plant community itself, through differential species' effects on their local environment, could reduce interspecific competition between dominant species if contrasting patch types favor different species. For example, it is now becoming clear that, in many multistory plant communities, the understory herb/shrub layer may play a role in the spatial structure and diversity of the community. In a number of forests worldwide, the understory is dominated by a specific species, such as bamboo in Japan and ferns in low-nutrient forests in the eastern United States. These dense understory layers act as filters for seedlings trying to ascend to the canopy. As the density of the understory is spatially variable and dependent on resource conditions (soil moisture, light availability), the strength of this filtering capability will also be spatially heterogeneous. In experimental work at Harvard Forest, Lisa George and F. A. Bazzaz have found that ferns substantially influence conditions at the forest floor, in terms of both microenvironmental characteristics and the activity of seed/seedling predators. Examination of seedling demography patterns revealed the strong species selectivity of the fern layer (Fig. 8). Red maple seedlings dominated in fern areas, while the contribution of yellow birch and red oak to seedling bank composition increased in fern-free areas. These fern-induced seedling spatial dynamics have important implications for neighborhood competitive interactions between co-occurring tree species and ultimately for species coexistence.

Spatial structure in forest dynamics may also be produced by the distribution of tree species themselves. Canopy trees can alter light and soil characteristics in the understory and thus influence patterns of seedling regeneration. When a species influences conditions at a site to promote recruitment of conspecifics, this dynamic will set up a stand-level positive feedback with a result that this species may dominate a stand until the next major disturbance. The outcome of positive feedbacks for diversity will depend on the characteristics of the particular community. In relatively species poor temperate forests, these positive feedback effects may promote species coexistence by creating stands dominated by different species and thus reducing interspecific interactions. Sugar maple and hemlock are able to coexist in eastern United States forests as a result of each stand type favoring regeneration of its own species. In contrast, in species-rich tropical forests, these spatially structured positive feedback effects may act to encourage development of single-species stands and thus reduce overall forest species diversity.

VI. CONCLUSION

Biological diversity and its maintenance remain central to the science of ecology. As we lose more and more species as a result of human activities, the need to document, understand, and evaluate biological diversity becomes increasingly pressing. In this article, we have attempted to clarify our understanding of resource partitioning, as it may influence local or regional diversity. Precisely defining what we mean by diversity is the first step in any scientific endeavor on the subject, and the way that biodiversity is interpreted will depend on the goals of the study. Understanding the functional significance of biological diversity for ecosystem pro-

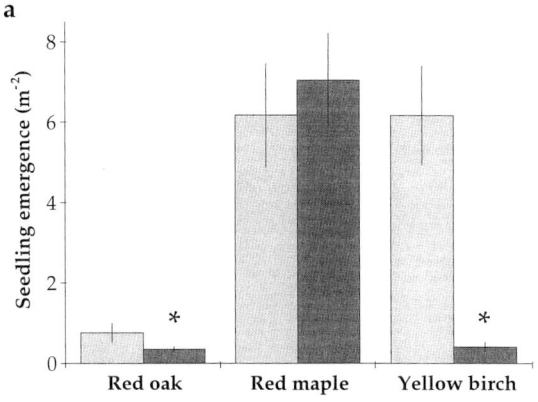
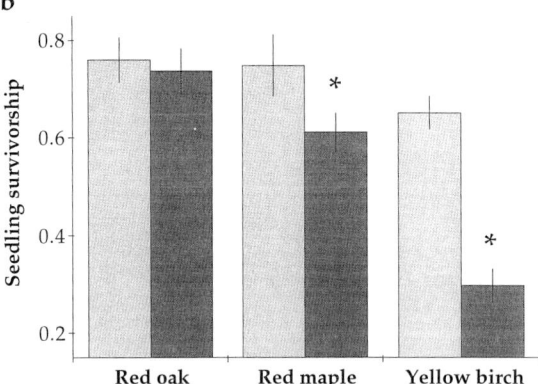
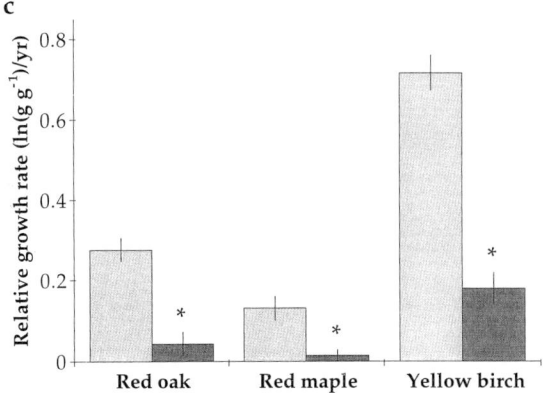

FIGURE 8 Variation in seedling demography in experimental fern plots: ferns removed (light shading) and ferns present (dark shading). Redrawn from George and Bazzaz (1999), *Ecology*, 80, 833–845, 846–856.

cesses has become particularly relevant for assessing the full impact of species loss on the functioning of the biosphere.

Any attempts to determine the underlying causes for global diversity patterns rely on a mechanistic understanding of the processes driving species diversity in ecological communities. We have evaluated the extent to which we can understand species diversity from a resource-based approach. Theory suggests that long-term species coexistence is possible if species' responses to their environment are sufficiently separated in multi-dimensional resource space. Empirical evidence from a wide range of plant communities supports the notion that species separate themselves along spatial and temporal resource axes. As yet, however, we have not been able to combine this information with theoretical models to establish if the degree of separation is sufficient to favor long-term coexistence and maintenance of high species diversity. Certainly, in communities with very high species diversity such as tropical rain forest, it becomes more challenging to understand coexistence in terms of resource partitioning. Other mechanisms for maintenance of diversity such as frequency-dependent mortality, chance, and spatial structure may play more significant roles in these systems. Now that we have established likely contenders for mechanisms of species coexistence, it is important that we develop a comprehensive, integrative conceptual model that can incorporate these distinct factors into a cohesive framework of biological diversity.

See Also the Following Articles

DIVERSITY, TAXONOMIC VERSUS FUNCTIONAL • HABITAT AND NICHE, CONCEPT OF • SPECIES COEXISTENCE • SPECIES DIVERSITY, OVERVIEW • STABILITY, CONCEPT OF

Bibliography

Bazzaz, F. A. (1996). *Plants in Changing Environments: Linking Physiological, Population, and Community Ecology.* Cambridge Univ. Press, Cambridge, UK.

Connell, J. H. (1978). Diversity in tropical rain forests and coral reefs. *Science* 199, 1302–1310.

Connell, J. H. (1980). Diversity and the coevolution of competitors, or the ghost of competition past. *Oikos* 35, 131–138.

Denslow, J. S. (1980). Gap partitioning among tropical rainforest trees. *Biotropica* 12, S47–S55.

Grubb, P. J. (1977). The maintenance of species-richness in plant communities: The importance of the regeneration niche. *Biol. Rev.* 52, 107–145.

Hector, A., Schmid, B., Beierkuhnlein, C., Caldeira, M. C., Diemer, M., Dimitrakopoulos, P. G., Finn, J. A., Freitas, H., *et al.* (1999). Plant diversity and productivity experiments in European grasslands. *Science* 286, 1123–1127.

Hubbell, S. P., and Foster, R. B. (1986). Biology, chance, and history and the structure of tropical rain forest tree communities. In *Community Ecology* (J. Diamond and T. J. Case, Eds.), pp. 314–329. Harper & Row, New York.

Hubbell, S. P., Foster, R. B., O'Brien, S. T., Harms, K. E., Condit, R., Wechsler, B., Wright, S. J., and Loo de Lao, S. (1999). Light-gap disturbances, recruitment limitation, and tree diversity in a neotropical forest. *Science* 283, 554–557.

Huston, M. A. (1994). *Biological Diversity*. Cambridge Univ. Press, Cambridge, UK.

Ricklefs, R. E. (1977). Environmental heterogeneity and plant species diversity: An hypothesis. *Am. Nat.* **111**, 376–381.

Schulze, E.-D., and Mooney, H. A. (Eds.) (1993). *Biodiversity and Ecosystem Function*. Springer-Verlag, Berlin.

Tilman, D. (1982). *Resource Competition and Community Structure*. Princeton Univ. Press, Princeton, NJ.

Tilman, D. (1994). Competition and biodiversity in spatially structured habitats. *Ecology* **75**, 2–16.

Tilman, D., and Pacala, S. (1993). The maintenance of species richness in plant communities. In *Species diversity in ecological communities* (R. E. Ricklefs and D. Schluter, Eds.), pp. 13–25. Univ. of Chicago Press, Chicago.

Whitmore, T. C. (1989). Canopy gaps and the major groups of forest trees. *Ecology* **70**, 536–538.

RESTORATION OF ANIMAL, PLANT, AND MICROBIAL DIVERSITY

Edith B. Allen,* Joel S. Brown,† and Michael F. Allen‡
*University of California, †University of Illinois, and ‡Center for Conservation Biology, University of California

I. Restoration Goals: Conservation and Biodiversity
II. Restoring a Diverse Plant Community
III. Restoring Animal Diversity
IV. Promoting Microbial Diversity
V. Conclusions

GLOSSARY

active restoration Requiring manipulation by humans for successful colonization and/or establishment of organisms and ecosystem functioning.
functional diversity Having all of the functions required for maintenance of ecosystem processes, but not necessarily all of the species richness.
functional redundancy, functional similarity Having species that may be substituted because their contributions to ecosystem processes are similar or overlapping.
passive restoration Relying on natural successional processes for restoration after the stresses that caused the disturbance have been removed.
reclamation A revegetation or land management goal that includes a lower diversity of species and may include substitutions by introduced species.
reference area An undisturbed or natural area chosen to compare with a restored site to determine the success of restoration.
rehabilitation Creation of an alternative ecosystem following a disturbance, different from the original and having utilitarian rather than conservation values.
restoration The manipulation of organisms and ecological processes to create self-organizing ecosystems that resemble predisturbance structure and functioning and promote conservation of biodiversity.

CONSERVATION OF BIODIVERSITY is the central goal of most restoration efforts and ranges from reintroductions of individual species of rare plants and animals to efforts to reintroduce a high diversity of species. Restoration may be defined as the manipulation of organisms and ecological processes to create self-organizing, sustainable, native ecosystems as integral parts of the landscape, as much as possible as they existed before disruptive human disturbances. In this article, we will examine the possibilities and the limits to restoration of biodiversity.

I. RESTORATION GOALS: CONSERVATION AND BIODIVERSITY

Conservation of biodiversity is the central goal of most restoration efforts and ranges from reintroductions of individual species of rare plants and animals to efforts to reintroduce a high diversity of species (Jordan et al., 1987, 1988; Bowles and Whelan, 1994; Falk et al., 1996). Restoration may be defined as the manipulation of organisms and ecological processes to create self-organizing, sustainable, native ecosystems as integral parts of the landscape, as much as possible as they existed before disruptive human disturbances. Where propagules of native organisms are remnant, restoration may require reintroducing ecosystem functions such as fire or hydrologic regime to enable natural recolonization and recovery processes. Re-creation of a close replica of a previously existing ecosystem type is increasingly difficult in a world with a growing human population, considering fragmentation and limits to dispersal of organisms; global invasions of exotic species that cause large-scale replacement and even extinction of native species; air, water, and soil eutrophication; and habitat loss by land conversion to urbanization and agriculture. Even nature reserves and parks suffer from visitor overutilization and other impacts, and restoration coupled with more careful management is required to preserve the original flora and fauna.

There are limits to restoration whether biodiversity or functioning of the ecosystem is concerned, but restoring all of the species richness that was originally present on a site is usually more difficult than restoring similar functioning. The reasons are that many rare species are difficult to propagate, their basic biology has often not been studied, they have lost genetic diversity, and their propagules are limited due to habitat loss. They are also less likely to be chosen for many kinds of restoration because they are more expensive to reintroduce and they contribute relatively little as individual species to ecosystem functioning. On the other hand, abundant species are generally better known ecologically, and in some cases, individual abundant species will regulate ecosystem functioning. Restorationists more often focus on reintroducing the abundant plant species or the "matrix" species to initiate succession and recovery of disturbed lands. Reintroducing dominant species enables a recovery of major functions and most of the vegetation vertical and horizontal structure. For instance, John Ewel showed that low-diversity replanted (but not restored) tropical forest in Costa Rica may have similar levels of soil nutrients, organic matter, and nutrient loss, compared to natural forest, but will not have the same conservation value for species preservation.

Additional limits to restoration of diversity are economic, political, and social. Restoration for the intrinsic value of nature, with its complement of biodiversity, often includes participation by laypersons who wish to see nature returned to a state they remember from years ago or that was documented in historical texts or anecdotally. Lesser goals than restoration include reclamation or rehabilitation (National Academy of Sciences, 1974), which have utilitarian values and less emphasis on conservation and biodiversity values. Reclamation may include a less diverse mix of species and exotic substitutions, while rehabilitation is simply to make the land useful again and may produce an alternative ecosystem, such as turning a forest into a pasture.

In this article, we will examine the possibilities and the limits to restoration of biodiversity. Different species have different limits and require different approaches for restoration. We divide the discussion into restoration of plants, animals, and soil microorganisms. This is admittedly an overly simplistic approach, as they all interact. However, points of overlap among the groups are brought out numerous times within each section. The division is logical in that the three groups require different degrees of active or passive restoration, depending on the level of disturbance of a site (Fig. 1). The kinds of disturbances that require restoration are varied. They range from drastic alteration of the ecosystem, such as surface mining or other construction projects that remove the topsoil; to abandonment from agriculture that leaves soil nutrients and a largely weedy seed bank; to invasion of natural vegetation by exotic weeds that leaves the soil intact but depletes native biodiversity; to alteration of certain ecosystem functions such as fire cycle or hydrologic regime; to extirpation of individual species with no other associated impacts to the physical habitat. Plant introduction and management are central to restoration projects where disturbance has caused vegetation removal or weed invasion. Where plant propagules remain or are readily dispersed, as after abandonment from agriculture in the northeastern United States, passive restoration may suffice. For animals, habitat may be created by manipulating the structure of the plant community, with the hope that the animals will recolonize. This has come to be called the "build it and they will come" hypothesis (see article by Palmer et al. in Restoration Ecology, Vol. 5, 1997). However, recolonization may be restricted by fragmentation, lack of corridors, or lack of propagules, so reintroduction of the animal into restored or intact vegetation may be needed.

For microorganisms, passive restoration is generally the rule in contrast to higher organisms that may be purposefully reintroduced. Very few species of microorganisms that are members of natural ecosystems are

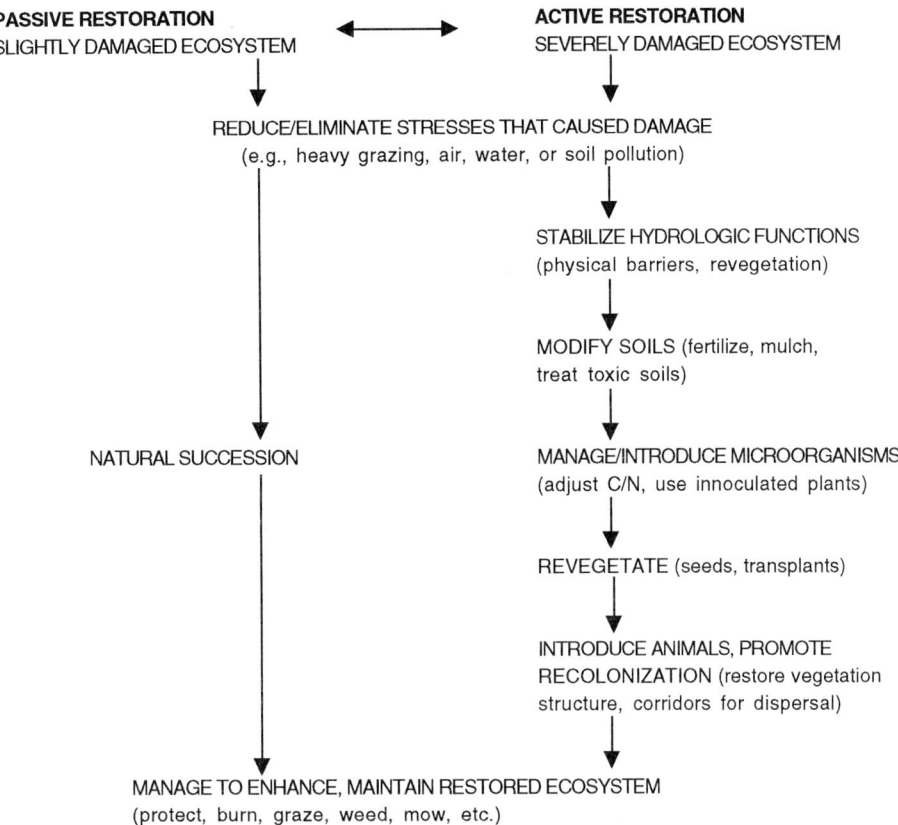

FIGURE 1   A comparison of active and passive restoration. Active restoration requires various degrees of human intervention, depending on the degree of disturbance, while passive restoration only requires removal of the stresses that caused the initial disturbance, such as grazing or pollution, followed by natural recovery processes.

cultured, so they are not even available for restoration purposes. The exceptions are symbionts (N-fixing bacteria, mycorrhizae) that are used routinely in agriculture and reforestation, but less often in restoration. All the other taxa that contribute to belowground functioning of native ecosystems are never or rarely cultured, including soil saprotrophic fungi, bacteria, nematodes, and other soil micro- and mesofauna. For all of these groups, we must learn how to manage the soil to promote their recolonization and to understand the distances from which they may be able to recolonize. In addition, there may be 1000s of species of microorganisms even in 1 cm^3 of soil. So the argument must be made whether all these species can or should be reintroduced. Many ecologists argue that there is functional redundancy among so many microbial species and that a minimum number of functional groups rather than a minimum number of species are required for restoration of ecosystem functioning.

The goals for restoration may vary for these three groups of organisms. For plants and animals, species diversity concerns are high, but for soil microorganisms, functional diversity is more often expressed as the concern. The limits and possibilities for restoration of diversity vary for the three groups, and the discussion below will expand upon these topics.

II. RESTORING A DIVERSE PLANT COMMUNITY

To improve plant species diversity, the kind of revegetation or vegetation management is determined by the degree of disturbance. These form a gradient of disturbance types and recovery possibilities that require the entire range of active to passive restoration. Heavily disturbed sites such as surface mines need to have all species reintroduced, whereas slightly disturbed sites may be missing certain species that are not adapted to the new postdisturbance conditions, such as alteration of fire regime. Other sites may require weed control to increase native species diversity. There are no examples of severely disturbed sites that have been restored to their original complement of species. This may be related to the high costs involved, the changed environmental conditions, and the sheer impossibility of artificially reintroducing a large number of species. As richness increases in more productive habitats, the probability of reintroducing even a majority of the species diminishes. One of the best examples of attempts to restore a diversity of species comes from bauxite mining in southwestern Australia, where 80 species or more may be included in the seed mix. However, the adjacent undisturbed jarrah (*Eucalyptus marginata*) forest may contain more than 200 species. In Wyoming sagebrush–grasslands, mining reclamation regulations are not as strict, and typically only 5–10 native species are planted in an area that may have 50 or more naturally occurring (Fig. 2). The dominant rather than the rare species are typically chosen for revegetation, and the long "tail" of rare species that typifies a dominance–diversity curve for natural areas is missing (Fig. 2). Reclamation and restoration both generally create com-

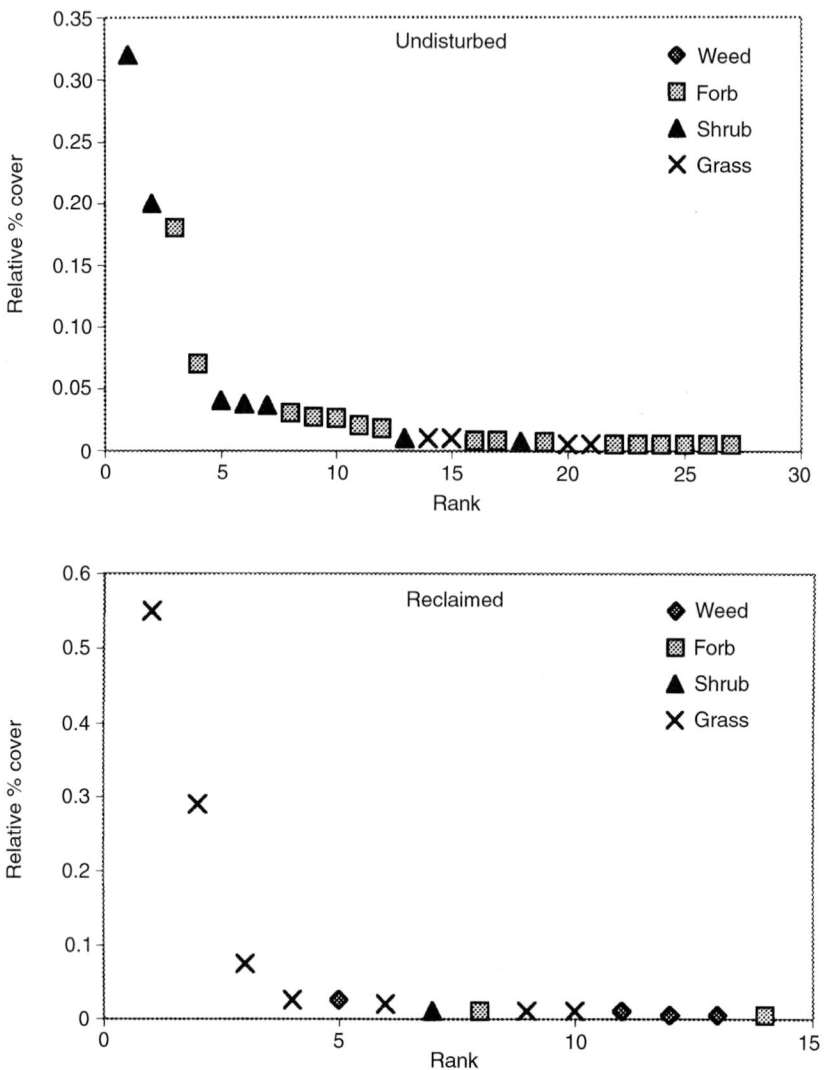

FIGURE 2 Dominance–diversity curves for reclaimed and natural vegetation in Wyoming sagebrush–grassland. The long tail of inabundant species found in natural vegetation is seldom reestablished in revegetated areas.

munities of a few abundant species but few rare species, whereas undisturbed communities may have the same few abundant species, but in addition will have many rare species.

A. Limits to Restoring a Diverse Plant Community

The reasons for omitting the many species that form the rare species tail in Fig. 2 are many. Seeds or other propagules must be collected locally for the restoration to reflect the local genetic populations, but these are typically not available unless the project is planned well in advance, usually two growing seasons, and the seeds are collected specifically for that project. Collecting native seed is becoming a large industry but it is unusual to have all the species available on the open market at the time they are needed, and even rarer to have the seed collections from local populations (see chapters in Falk et al., 1996).

Loss of genetic diversity and lack of local ecotypes are also a limitation to restoration. Where local extirpations have occurred, the nearest populations may no longer have all the genetic diversity of the original, and restoration in the true sense of restoring genetic as well as species diversity is no longer possible. Locally selected ecotypes have the "home team advantage" of being better adapted to local conditions and also avoid problems of outcrossing and hybrid depression with remnant native individuals (see article by Montalvo et al. in *Restoration Ecology*, Vol. 5, 1997). Additional discussion of genetic issues is in the section on animal restoration.

Once the appropriate local seeds or propagules have been collected, there is little information on seed dormancy and propagation of most species. The emphasis on conservation biology in the past decade has resulted in increased concern for research on rare species, and information on propagation, microenvironment requirements, reasons for disappearance, interactions with pollinators and other species, and so forth have enabled restorationists to include rare species in revegetation plans. More often, restorationists work with relatively unknown species and must begin research anew for each species, as, for example, the work on restoring the endangered *Amsinckia grandiflora* (see chapter by Pavlik in Falk et al., 1996). Lack of biological information on species translates into practical economic limitations. Most often, only the species that are best understood with the most available propagules are used.

Once germination and propagation requirements are understood, the plants must go into a field setting that presents a whole new set of problems. Different species germinate at different times and have different growth rates, so some will never emerge from certain seed mixes. The northern Great Plains of the United States are dominated by *Bouteloua gracilis* in the coal mining regions, but this has proven to be one of the most difficult species to reestablish for mining reclamation. It is a slow-growing, late-germinating, warm-season grass, but when it is seeded as part of a mix of native species, the cool-season grasses germinate first, grow quickly, and dominate. Reclamationists have devised numerous methods to reestablish shortgrass prairie, such as alternating seed drill rows or planting *Bouteloua* seeds a year earlier. However, the most important consideration for companies that have spent millions of dollars on earthmoving is to stabilize the soil, rather than to establish high levels of diversity that are not required by law in any case. Preventing soil erosion is the first goal, and the mines of this region use a fast-growing native plant mix that reduces the establishment of slow-growing species. The goals of soil stabilization using productive plant species and establishing a diverse mixture are often at odds (Fig. 3).

Reestablishing the full complement of species may require reintroduction in stages, as shown in examples from the tallgrass prairie. When Robert Betz began prairie restoration in the late 1960s at the Fermi Lab prairie in Illinois, he quickly learned that the dominant native grasses could be readily established from seed. He used the grasses to form the matrix of vegetation, followed by later introductions of the less common forbs. These forbs could not colonize naturally into the restored grassland because of the high density of grasses, but could be hand planted. Thus Betz was able to restore 116 plant species to the Fermi Lab prairie, but only by expensive hand labor. Similarly, the Curtis Prairie in Wisconsin, the oldest restored prairie, was initially planted during the 1930s, but plants have been continually introduced since then (see chapter by Cottam in Jordan et al., 1987). Their survival has been monitored, and they now contribute to the most diverse tallgrass prairie anywhere. The Curtis Prairie has over 300 species in 10 acres, more than any remnant natural prairie.

Competition from invasive plant species is another important limit to restoration of diversity. The invasives may be weeds that are part of the successional process, but a more difficult problem than early seral weeds are those invasive plant species that persist and cause vegetation type conversions. In this case, the major restoration activities involve mowing, fire, or selective weeding to remove the offensive species, planting or managing for regrowth of natives, and then continual management to keep the invasives from becoming dom-

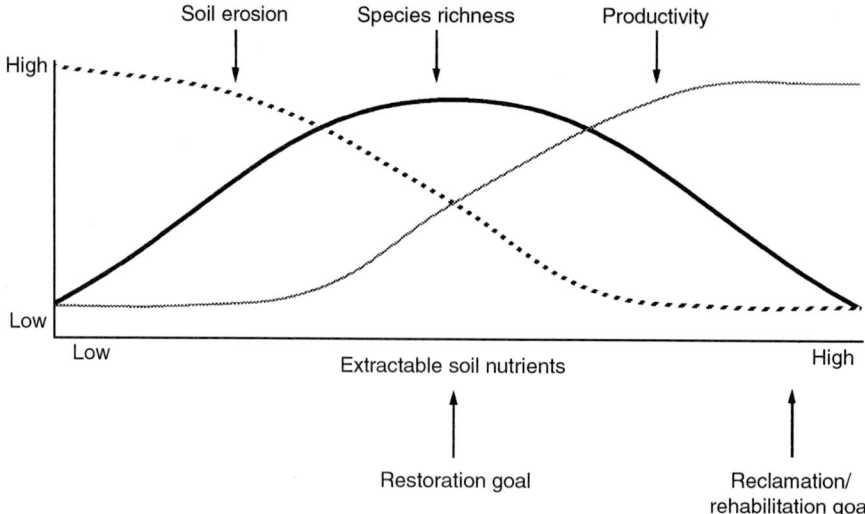

FIGURE 3 The relationships among soil fertilizer level, plant species richness, plant productivity, and soil erosion. The trade-off for using fertilizer to promote high initial rates of plant productivity to reduce soil erosion is a loss in diversity.

inant once again (see chapters in Allen, 1988; Falk et al., 1996). Such activities have been undertaken in Florida, where *Schinus terebinthifolium* and *Melaleuca quinquenervia* dominate wetlands, Hawaii, where exotic perennial grasses and *Myrica faya* have replaced native vegetation, California, where Mediterranean annual grasses have replaced native perennial grasslands, and in many other sites worldwide.

Poor, toxic, eutrophied, or erosive soils also impair the conservation value of restored communities. Any practice that changes the soil from its original physical or chemical state will also change the plant species composition. An example comes from the china clay mines in England, where the mining refuse is high in sand. Local plants are poorly adapted to the droughty and nutrient-poor sand, so strand plants are imported for revegetation (Bradshaw and Chadwick, 1980). As expected, these artificial strand communities are poor in diversity. However, Bradshaw has documented some relatively high species diversity following natural colonization onto abandoned waste sites, although not as high as surrounding native areas. There is a tendency during revegetation to overfertilize soils to agricultural levels when fertilizer is applied, with expected poor results for native plant diversity. The reason for overfertilizing is usually economic, to promote vegetation establishment and therefore prevent soil erosion, but may also be aesthetic, to please the paying public as rapidly as possible. As shown by the experiments of Tilman, overfertilization, especially by nitrogen, will promote the productivity of a few species to the detriment of stand richness (Fig. 3). For restoration to promote the natural levels of species richness, the soils must be in their natural fertility condition as much as possible. If erosion is a problem, then other means to control erosion, such as mulch or temporary mechanical barriers, could be used to stabilize the soil and allow the slow-growing plant species to establish within a mixture of fast-growing species.

B. Improving Plant Species Diversity

Restorationists have used many techniques to increase richness of plant communities when the seed and propagule sources are limiting. Taking advantage of the existing seedbank is a primary one, where the seedbank still exists. Mined land reclamation laws in developed nations often require retopsoiling with fresh topsoil, and many studies document the importance of this source. Certain species still need to be supplemented, as in the case of late seral species absent from the seedbank of jarrah forest in southwestern Australia (see chapter by Bell in Allen, 1988). Jarrah seed is never found in the seedbank, and it reproduces vegetatively after fire in natural communities. It must be reintroduced as nursery transplants for revegetation to be successful. While the seedbank is an important source of diversity in many kinds of restorations, it is also notorious for harboring a large complement of early successional and weedy species. Bradshaw and Chadwick

(1980) have recommended against using topsoil just for this reason, if appropriate weed-free subsoil can be found and amended to support plant growth.

Dispersal of propagules into the restored site is another way to increase diversity that depends upon landscape structure such as proximity of source populations or existence of corridors. In a study of 2- to 18-year-old revegetated, untopsoiled roadsides in San Diego County, dispersal from adjacent mediterranean shrubland more than doubled the richness from 12 planted species to a maximum of 16 colonizing native species. However, dispersal of native species only occurred when the shrublands were adjacent to the highway, and occurred rarely in urban areas of the highway. Reclaimed surface mines in Wyoming that are surrounded by native sagebrush–grassland have a higher richness of colonizing species than, for instance, a revegetated landfill in Staten Island that is surrounded by suburbia (see articles by Bell and by Ehrenfeld in *Restoration Ecology*, Vol. 5, 1997). While dispersal is often limited by vectors in terrestrial habitats, aquatic and riparian habitats typically are recolonized rapidly (National Academy of Sciences, 1992). Water is a very effective medium for propagule dispersal, and these habitats are often not even revegetated, unless the soil needs to be stabilized rapidly in riparian edges or the dispersing propagules include unwanted exotic species.

Vegetation may be replanted to attract animals that are seed dispersers or pollinators and thus may create a positive feedback on the future reproduction and diversity of a site. When a limited suite of species is chosen, the selected species become especially critical. Synthetic grasslands that simulated tallgrass prairie had sufficient diversity and especially structure to attract birds and small mammals (Howe and Brown, 1999). Shrub islands were planted on a surface mine in Wyoming to increase the movement of animals and microorganisms onto a site that was otherwise dominated by grasses (Allen, 1988). The feedback of these shrub species and patterns determined both their mortality and recolonization patterns of additional plant species. In Costa Rica, Karen Holl and Daniel Janzen have shown that tree and shrub "islands" within pastures are critical for attracting animals that disperse seeds and then increase the diversity of the plant community.

C. Single-Species Plant Introductions

Individual plant species have been reintroduced where they were extirpated, largely because of laws protecting endangered species. Mitigation laws often require transplantation of a rare species from a site that is about to be destroyed to a safe site. For instance, Howald (see chapter in Falk *et al.*, 1996) documents 40 instances where rare species were transplanted for habitat mitigation purposes in California. Of these, only 5 were considered successful, 7 had limited success, 13 were not successful, and others were unknown or ongoing and too recent to evaluate. The reasons for lack of success were varied but included moving plants to sites where they did not exist previously, having different environmental conditions in the transplant site, or using poor horticultural techniques. Primack (chapter in Falk *et al.*, 1996) planted seeds of 41 species into sites in Massachusetts where they were once known to occur. Of these, only 10 produced seedlings, and only three produced a second generation. In this study, the reasons for failure may have been due to exotic species, changes in natural disturbance regime, and other changes such as anthropogenic nitrogen deposition that causes inappropriately high levels of soil fertility. The relatively few successful transplantations indicate that mitigation transplantation in general is poor policy because the kind of planning and postintroduction management that are required is seldom done in practice. Often little consideration is given to the quality of the habitat into which organisms are being reintroduced. Overall, the reintroduction of one species does little to improve the diversity of a site, but it may be the only option to avoid extinction, provided restorationists and land managers use their best practices.

While many single-species introductions focus on rare species, they may alternatively focus on abundant species. The reintroduction of a keystone species or ecosystem engineer (see Palmer *et al.* in *Restoration Ecology*, Vol. 5, 1997) is one way to "jump start" natural successional processes in a community. Shrubs and trees were removed from rangelands for many years in an effort to increase forage production for domestic animals, but this was short-lived and in the long run promoted degradation of the community because the deep-rooted shrubs changed the nutrient and moisture balance of the entire stand. The reintroductions of *Pistacia lentiscus* in Israel, oak trees in California, or *Artemisia tridentata* in mined lands in Wyoming enable the shrub or tree island effect to reinitiate, building up soil and biotic resources around the transplanted shrub (Allen, 1988). This could potentially result in further increases in diversity as the shrubs enable recolonization of additional species.

D. Succession on Restored Lands

Restoration depends upon succession to complete the process that was started by revegetation, but succession alone does not always bring us to the goal we wish to achieve. Revegetated lands are not static and can be expected to undergo change in species density, richness, and relative composition. The degree to which they undergo succession varies, depending upon the amount of colonization, initial species complement, and site conditions. Little change in species composition occurred after bauxite mining in the Australian jarrah forest. The site typified Egler's initial floristics model of succession, where late seral dominants colonize in the earliest stages of succession, because very few species that did not occur in the original seed mix became dominant later on. Sites that were 16 years old appeared structurally different from sites that were 1–2 years old in that woody plants were larger and more evident in the older sites, but most of the same species were present in sites of all age classes. A few surprising species did colonize the older sites, such as a native orchid. In Anthony Bradshaw's experiments on plants colonizing industrial waste heaps in England, most plants that did not establish appeared to be limited by dispersal and establishment, rather than poor substrate. The sites were too distant from propagule sources or had dense stands of initially colonizing species that prevented establishment of a diversity of species. The initial floristics model can also be applied to Wyoming, Pennsylvania, and Czech surface mined lands, where the native grasses that dominated early on allowed little colonization of native trees, shrubs, or forbs after 10–30 years.

Even when soil conditions are optimal and sources of propagules for local colonization are available, the initial revegetation treatments will affect long-term successional processes, as shown by recent studies in the journal *Restoration Ecology*. A 30-year-old planted forest that had been severely disturbed by hydroelectric development in New Zealand showed little change over time when 46 native species were planted. The species composition was chosen to resemble the dominants during secondary succession after fire in natural forest, but had only a fraction of the total diversity. However, an even less diverse stand developed when no vegetation was planted, consisting of a stand of exotic *Citisus scoparius* and European grasses. A comparison of planted forest stands on bauxite mines in Brazilian tropical forest showed that planting a low-diversity commercial mix resulted in a low-diversity, but high-productivity forest, while seeding and planting mixes of 70 or more species promoted a diverse, but less productive forest. In all cases, colonization of certain late successional forest species that were not part of the original mix was poor. An interesting contrast comes from the Arctic on the North Slope in Alaska, where more than 100 species of native plants colonized 20 years after planting two native and one exotic grass species over a dozen sites.

Thus the productivity and species richness of the initial planted stand will make a large difference in the ability of other species to colonize later. From these examples, and many others in the literature, it is apparent that succession may promote an increase in richness if the planted stand is not so aggressive that it precludes establishment of colonizers. Most often, additional interventions are needed if the desired diversity is to be achieved.

III. RESTORING ANIMAL DIVERSITY

A. An Introduction to Animal Restoration

Like plant restoration, animal restoration involves reintroducing or encouraging the return of native species to an area or region from which they have been lost. An animal restoration project may involve just a single species, as in the reintroduction of wolves into Yellowstone National Park, or an entire community, as in Augrubies Falls National Park in South Africa, where over 10 species of large mammals have been reintroduced (including black rhinoceros, eland, giraffe, and springbok).

Animal and plant restorations will generally involve the same ecological considerations, and most differences between the two concern a matter of degree and emphasis (Jordan *et al.*, 1988). Save for insects (which represent a challenging form of restoration more in line with promoting microbial diversity), plant restorations generally call for a larger number of species. In terms of ecosystem function, animals have significant roles in restoration as pollinators and herbivores of the plants. Animal roles in ecosystem function also occur when herbivores alter nutrient cycles and where animals may be "ecological engineers" of the physical environment, such as the contributions to soil function from the digging and foraging activities of earthworms or burrowing mammals (Jordan *et al.*, 1987). In general, animals are more mobile than plants and while plants may disperse widely and unexpectedly as seeds and pollen, a single animal may range over scales that are much larger than that of the restoration project and site. Some of John Laundre's radiocollared mountain lions from south central Idaho have appeared as far away as Yellowstone National Park in Wyoming and mountain ranges in central Nevada. As a consequence of mobility, restoration sites from the perspective of animals will be open rather than closed systems. In animal restoration, the environs of the restoration site may deserve equal attention. Much attention has been given by all stakeholders to the possible fates and consequences of wolves wandering in and out of their Yellowstone restoration site.

The ecology of animal restoration draws less heavily from theories of ecological succession and more from the ecology of invasions and the ecology of small population sizes (Major, 1989). Plant restorations, as succession, start with an inoculum of plants that initiate a process of changes in species composition that eventually leads to the desired natural community in terms of structure, function, and appearance. This aspect of succession is absent from most animal restorations except in the case of the mostly passive development of an insect and soil invertebrate community that fits and complements the plant community. With most animal restorations, each species enters the community at a very small population size either as an invader or as a reintroduction. Small and alien describes the starting population. Most random introductions of animals to a community fail. Naive animals fall to predators more easily, compete less successfully for resources, often fail to develop or find suitable denning or nesting sites, and sometimes attempt to emigrate from the site. In the absence of preexisting burrows, reintroduced prairie dogs simply roam far and wide in search of colonies. Small animal populations face twin genetic and demographic threats. Small, sexually reproducing populations may suffer inbreeding depression and produce a higher proportion of young with genetic defects. Demographically, small populations may teeter on the edge of extinction. An accidental death, a missed breeding opportunity, or a chance skewing of sex ratios or age distributions may compromise the population irrecoverably.

Animal restoration begins with knowledge of why the species or animal community is currently absent or threatened at the site. Next comes an assessment of whether the site is currently suitable for the animals and an evaluation of what site preparations are necessary to promote success. Then the current status of the animals receives attention to determine the best and likeliest sources of individuals. The animal(s) may be present on site but at low numbers, they may be present near the site or far from the site, or they may only exist as captive populations. Finally, the passive or active restoration of the animals begins with appropriate considerations of how to manage and monitor the project's success. We will briefly examine each of these steps.

B. Reasons for the Absence of an Animal or Animal Community from a Site or Region

Habitat fragmentation or changes in land use may reduce or eliminate particular animal species, subsets of species, or entire taxonomic groups of animals from an area. For instance, most animals associate with particular types and structures of vegetation. When altered, by livestock grazing or agriculture for instance, some of the original animals disappear as they cannot find food, nesting, or denning sites or they may be excluded by changing intensities of competition or predation from other animals more suited to the new circumstances. In rivers and lakes, pollution, erosion, and sedimentation drastically change species composition. Everything from invertebrate larvae to fish may die off or be replaced by other species more tolerant of the polluted or modified waterways.

Hunting, poaching, and commercial harvesting may be so intense as to extirpate an animal or group of animals from an area or region. The African gray parrot and South American macaws face threats from those supplying the demand for pets. Incidental mortality in fishing nets threatens many of the world's populations and species of sea turtle. Overharvest of commercially valuable species has created a litany of crashes such as California sardine, North Sea herring, blue whales, kaluga sturgeon, and southern bluefin tuna. As a valuable source of meat and a denizen of potential rangeland and farmland, the pronghorn antelope of the western half of the United States, like the American bison, faced extirpation by the 1900s.

Exotic species, those species accidentally or intentionally transported by humans into new places, often eliminate native animals via predation, competition, or keystone effects on structure and function of the ecosystem. The introduction of mosquitoes into Hawaii brought bird malaria for which the birds were no more prepared for than the peoples of the Americas were for the introduction of smallpox from Europe. An inoculum of larvae in ship ballast introduced zebra mussels from Europe into the Great Lakes of North America. In the early 1990s, the mussel's spread was spectacular, and with it has come dramatic declines in the abundances of phytoplankton and the fish and invertebrates relying on the phytoplankton. Competition from introduced North American gray squirrels has eliminated the native European red squirrel from much of its former range in England and Wales.

With respect to restoring their wildlife, Australia and Israel provide interesting contrasts. Both have seen their native mammals ravaged. In Israel this has gone on dramatically for millennia whereas it is more recent in Australia. In Israel, overexploitation and habitat modification have been the bane of most animals like lions and crocodiles, both extirpated long ago. Deforestation, reaching its nadir under the Turks just prior to World War I, nearly extirpated the subspecies of European red squirrel and jay from Israel's wooded habitats. In Australia, wildlife habitat is much more available and overexploitation generally more benign than in the Middle East. However, introduced species (red foxes, feral house cats, and European rabbit) of animals and plants have dictated the decline in numbers and range of comparable Australian mammals. Having contributed to their extirpation on the mainland, cats now occupy Kangaroo Island off of southern Australia and threaten the last remaining population of Tamar wallaby. Feral goats compete with and threaten the dwindling population of yellow-footed rock wallabies in the Flinder's Range. Red foxes prey heavily upon Australia's native small mammals and consume the eggs of ground-nesting birds.

The reasons for the absence of animal species from their former range are manifold but fall roughly into the categories of habitat change, overexploitation, or exclusion by animal species nonnative to the particular location. Understanding the reason for the absence of an animal or community of animals is the essential starting point for animal restoration.

C. Site Assessment and Habitat Restoration

All species have the capacity to grow exponentially under ideal conditions, but no population can grow exponentially forever. There are limits to growth, and population interactions such as competition and predation from other species can exclude a species from a community. The goal of site assessment and habitat restoration is to ensure near ideal conditions for the single species or the community of animals under restoration. Ideal conditions means ample food, space, shelter, and safety from predation. Limits to growth invites a consideration of specific factors in the environment that might currently or ultimately limit the growth and size of the animal's population. Considerations of competition and predation focus on which current species at the site, possibly exotic species, might preclude a successful introduction of the desired species to the site.

Site assessment evaluates whether the area offers an appropriate combination of food, safety, shelter, and space for a small population of animals to grow exponentially and prosper. If conditions are not appropriate, then habitat restoration must precede animal restoration. If the vegetation of a site has been heavily degraded, animal restoration may await plant restoration. At Midewin National Grassland south of Chicago, long-term hopes include first restoring the tallgrass prairie and then reintroducing elk and bison. The site is currently a mix of abandoned munitions facilities, farm fields, pastures and assorted oldfields, woodlots, and groves. Wetland restorations often require establishing the aquatic and surrounding vegetation and then passively or actively reintroducing the associated aquatic invertebrates (usually all passive), vertebrates (usually passive but nest boxes or nesting platforms may be used to encourage the return of waterfowl, egrets, or herons), and fish (sometimes passive or active). Different species have different needs, and reconstructing those habitats that fill an animal's needs is central to attracting wildlife or to introducing wildlife.

The possible reasons for the animals' absence from a site focuses site assessment. Often several factors combine to explain an animal's absence. When habitat alteration or fragmentation changes and reduces the diversity of animals, the first step is to determine whether the site has been, or needs to be, revegetated to its former state. This condition of the site does not have to be exact with respect to its former state, but rather it need only include the salient environmental factors that favor the animal's ecological requirements and aptitudes. For instance, many bird species are most responsive to the structure rather than the exact composition of the vegetation. Hence, bird community restoration can commence with promoting the appropriate vegetation structure rather than being particular to its composition (which may be the goal of a corresponding plant restoration). Habitat fragmentation offers unique challenges because the remaining area may simply be too small to successfully support the desired animal or animal communities. The site assessment may examine the need for more space or wildlife easements, the need for corridors to connect habitat fragments, or the need to enhance the quality of a habitat fragment beyond its natural state to permit the successful persistence of animals in a smaller, more confined space. At Jackson Hole, Wyoming, a wintering elk population is maintained on supplemental hay within a fenced pasture surrounded by the extensive human developments that have robbed the elk of most of their wintering habitat. In spring, the elk return to their natural mountain pastures. In Israel and elsewhere, feeding stations stocked with animal carcasses or offal have been used to facilitate the maintenance and restoration of such animals as wolves, griffon vultures, and eagles.

The California condor provides an illustration of a complex site assessment. Reasons for the decline of condors that culminated in their removal from the wild in the late 1980s included too small a population (demographic threats), habitat fragmentation, lack of carcasses, and lead poisoning from carcasses containing lead shot or bullets. The restoration plan called for captive breeding, gazetting of habitat and habitat corridors, restrictions on development, and for reasons beyond just the condor, the banning of lead shot for hunting. The actual restoration took several approaches to evaluating the complexities of habitat suitability and the ability to introduce a small population of naive animals. First, less rare Andean condors were introduced as a surrogate species. They permitted tests of the reintroduction techniques and indicators of ranging patterns and habitat suitability. The actual reintroductions have used the original site north of Los Angeles but also additional sites, including one near the Grand Canyon. The California release site has more recently had condors but suffers much greater degrees of habitat fragmentation and development. The alternative sites have seen condors less recently, but have the advantage of offering extensive tracts of original and natural habitat.

In response to habitat degradation or fragmentation, habitat restoration may be as simple as adding nest boxes, as in the case of bluebirds at some sites in northeastern Illinois, or as involved as adding proper soil, microbes, and restoring an entire vegetation community, as in the case of toxic waste sites or former mining operations. Ceasing the use of pesticides or release of pollutants may be all that is necessary to restore a habitat. Peregrine falcons in North America have benefited from the ban on DDT. Portions of fish communities and whole invertebrate communities can recover just from preventing sewage and pollution discharges into waterways. The structure of the environment with respect to shrub cover in arid lands, the mixture of ages and types of trees in forests, and the availability of salt licks and waterholes for wildlife all become considerations for particular habitat restorations. Koala in Australia require particular species of eucalyptus trees as food, whereas gravel beds at particular depths provide spawning grounds for trout of the Great Lakes.

When a species absence is the consequence of overexploitation, success often depends less on the availability of suitable habitat and more on the cessation of hunting, poisoning (including mortality from hazardous chemicals and pollutants), harvesting, or poaching. Wolves were exterminated throughout the western half of the United States through individual hunting and trapping and by explicit eradication programs. Their reintroductions into Yellowstone National Park, Wyoming, and the White Mountains of Arizona involve presumably moderate to high quality habitat. The success of the programs probably depends most on the cessation of poaching, which at present is a minor problem for Yellowstone and a major threat to the Arizona reintroduction. In 1989, it was discovered that stock assessments of cod within the Canadian North Atlantic were wildly optimistic. The stock had crashed. Subsequent quotas follow from the premise that habitat quality remains high and that reduced harvest will be sufficient to promote recovery. In Southeast Asia, edible-nest swiftlets face decline and extirpation. The birds nest in caves, use saliva for nest building, and face true "nest predation." Humans gather the nests even as the nests may have eggs or nestlings. Even with complete bans on nest harvesting, places like Sarawak, Malaysia, still face intense nest destruction from local people, pirates from countries such as Indonesia, and organized collecting groups. Policing thousands of cave entrances in often remote places throughout the country is impractical. But, until poaching relaxes or ceases, restoration and recovery cannot be achieved. Having reduced poaching of black rhinoceros throughout the 1980s, the Kenyan Wildlife Service has begun the recovery and restoration of current and former populations.

Habitat restoration for overexploited species requires the first step of controlling the harvesting. However, the absence of wildlife from an area may coincide with other changes to the habitat, some which at first may seem subtle. "Nature abhors a vacuum" is a saying recognizing that unused food or opportunities in an environment often become filled by alternative species or by exotics. Habitat restoration may require controlling the abundance of exotic species or those species that compete or prey. Factors contributing to the decline of red squirrels in England include an increase in oaks and a decline in conifers. Both factors assist the exotic gray squirrel in outcompeting the red squirrel. Habitat restoration may require both changes in forest composition and active control measures of gray squirrels. The draining of reservoirs, netting, or poisoning of lakes infested with exotic fish such as carp has preceded several fish restorations in North America.

D. Sources of Animals for Restorations

Small remnant populations, dispersal from other areas, or active reintroductions provide the sources for animal restorations. Ideally, animal restoration should begin while a remnant population still occupies the site. Such a population, while often small, has the advantages of already being established and acclimated to the site. This avoids problems associated with naive animals unfamiliar with the site. In general, once habitat restoration has assured a quality site, resident populations recover faster than immigrants from other areas, which recover faster than reintroduced populations from captive stock. When remnant populations exist, restoration activities can immediately focus on improving vegetation, improving food availability, improving removal of exotics or competitor species, and/or ceasing harvesting and pollution. Success is measured by the recovery of the species or the expansion of the species into previously unoccupied parts of the site.

When a restoration relies on immigrants, there are the twin concerns of: "Is the site suitable for the establishment of immigrants?" and "Will the immigrants find the site?" When a site such as an oldfield, mine tailings, or forest clearcut is restored, there will usually be successful invasions of insects, soil invertebrates, aquatic invertebrates, mammals, and birds from the environs. For instance, in an oldfield subject to small experimental prairie restorations (Howe and Brown, 1999), there was an invasion of two small mammals (prairie vole and white-footed mouse) and some red-wing blackbirds have shifted their breeding from a nearby pond to the prairie plots. Mourning doves preferentially feed on rather than off of the prairies. Invasions of insects, birds, and mammals ease the restoration effort but sacrifice complete control of the resulting composition of animals. Species that are already well established and abundant in the region of a restoration are the most likely invaders. Rare and remote species are the least likely. Target species may require coaxing in the form of habitat corridors connecting the restoration site with existing populations. Temporary augmentation with food or nesting boxes can encourage reestablishment. In Tsavo West National Park, Kenya, rhinos have been corralled into a sanctuary (ca. 70 km^2) that is much smaller than the whole park (ca. 9000 km^2). The sanctuary is fenced to keep the rhinos in, patrolled to protect the animals from poaching, and supplied with piped water to maintain permanent water holes.

Animal restoration may require the active reintroduction of animals (Bowles et al., 1994). This is best done by translocating individuals from wild populations, but may require captive breeding. Wild populations provide experienced individuals, whereas captive animals may be particularly naive and unfamiliar with the wild. The reintroduction population should have the appropriate balance of sexes and age classes and sufficient genetic diversity to preclude serious consequences of inbreeding. When the reintroduction is very expensive, is politically sensitive, or is from a very small source population, great care goes into ensuring that the site offers close to ideal conditions and that the animals are prepared for the new environment. When animals are relatively cheap and available, the reintroductions can be more numerous and there is room for experimentation. Wild asses (onagers) in the Negev Desert of Israel, pronghorn antelope in Arizona, lynx in Switzerland, white rhinoceros in Lake Nakuru of Kenya, and peregrine falcons in various midwestern cities all provide examples of reintroductions from captive or wild source populations. In the case of the Guam rail, this bird had been driven extinct on Guam by the introduction of the brown tree snake. Rather than attempting the impossible task of eradicating the snakes on Guam, the rail has been reintroduced on a neighboring, snake-free island.

IV. PROMOTING MICROBIAL DIVERSITY

A. Microbial Diversity and Functional Redundancy

Compared to plants and animals, microorganisms are highly diverse and offer a special challenge to understanding biodiversity and to assuring successful restoration. Microbes should be the underpinnings of any discussion of biodiversity as they constitute the vast majority of the diversity of any ecosystem at any location, yet are rarely even mentioned in terms of maintaining diversity. Microbes should be a focus, not an afterthought, for restoring disturbed lands. Without animals and most species of plants, ecosystems would stabilize and most ecosystem functions would be performed. Without microbes, the ecosystem would cease to function.

Estimates of microbial diversity range up to 10^5 prokaryotic taxa per gram of soil, although 10^4 is generally a more accepted value. In a study of restoration of a Wyoming coal surface mine, 57 different fungal taxa were found out of 135 colonies randomly chosen from approximately 12,500 colony sources (spores, hyphal fragments, bits of organic matter). These taxa were evaluated within a 4-cm^2 area (using 5 g of soil). Given the numbers, it is not surprising that a large fraction of these organisms remain undescribed. Recent estimates suggest that less than 10% of soil organisms ranging from bacteria to spiders have been described. And, a few efforts that extracted DNA directly from soils have even found new kingdoms of prokaryotes. Despite the difficulty in estimating diversity, every study published has shown a loss in the richness of taxa with severe disturbance. This ranges from soils in the Pumice Plain of Mount St. Helens, which was sterilized, to burned areas in which only the aboveground material and soil surface were directly affected. Because of the extremely high microbial diversity, the concept of microbial functional redundancy has been raised. Among those thousands of species, how many need to be restored to maintain ecosystem functions? This is the single largest challenge for studying soil microbes and restoration.

There are three critical issues for evaluating microbial diversity and restoration. The first issue is defining the spatial and temporal distribution of species and functional groups and their relationship to ecosystem processes. The second concern is assessing the richness of organisms within the different functional groups. The third is the system-level capacity for dispersal and natural reestablishment versus the need for artificial introduction of microbial inoculum.

B. Spatial and Temporal Arrangement

Just as important as the richness of organisms are the changes in their spatial and temporal distributions. Unfortunately, few studies have evaluated microbial communities using species increment curves or overlap estimates. Several types of analyses are critical to understanding biodiversity and restoration. These include species × area, species × time, and dispersion relative to ecosystem processes. Unfortunately, few data sets are available to evaluate microbial recolonization on this basis.

Microbial richness estimates tend to be taken on a per-sample basis. However, plants and animals tend to be analyzed on an area basis. This makes comparative studies difficult but opens an important area of research. Nevertheless, evaluating the spatial array is absolutely critical. Fungi, for example, exist as a network of hyphae (a mycelium) extending from a few millimeters (such as a *Trichoderma* colony occupying a single fern petiole) to tens of meters in diameter (fairy rings, or the giant Wisconsin *Armillaria*, for example). For the same Wyoming data set, the species increment rate was the same up to the size of a 400-cm² patch of disturbed as well as reference area (Fig. 4). However, in the disturbed site, as one expanded outward, the species increment rate declined whereas it continued to increase in the reference area. In the reference area, new species were added as the habitat changed. In the disturbed area, the habitat was rather uniform across the site (in this case, mixed, respread topsoil on a surface mine). Thus, one conclusion is that microbial activity and composition become more diffuse and repetitive across scales in severely disturbed areas, and overall landscape diversity is lower than in native undisturbed areas.

Frequency of sampling over time is also crucial for

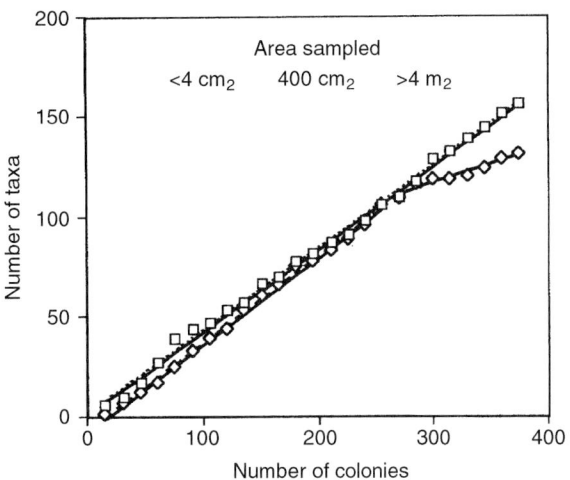

FIGURE 4 Species increment curve for microfungi at a Wyoming surface mined site. Shown are the taxa in the reference area (squares) versus the disturbed area (diamonds). The increment curve for the reference area is linear through the 375 colonies across the entire area ($y = 1.88 + 0.413x$, $r^2 = 0.997$). The curve for the disturbed area for a small area (4 m²) is not significantly different from the reference area ($y = -6.743 + 0.435x$, $r^2 = 0.998$) but the slope (0.187) is significantly lower ($r^2 = 0.94$) across the larger areas. These results indicate that at small scales, microbial diversity remains very high. However, at the landscape level, microbial diversity is lost in disturbed communities.

describing microbial biodiversity. Many microorganisms are only identified based on sporulating structures. However, these organisms may be present continuously but can only be found periodically. For example, macrofungi are spread widely in the mycelial stage and live for many years. Several continuous years of observations are needed for the right conditions to occur before a sporocarp forms. In many cases, fruiting times may occur over successional time. For example, on Mount St. Helens, establishing the first ectomycorrhiza found on the Pumice Plain took 5 years. We never saw a sporocarp to identify the fungus that formed on the ectomycorrhizal root tips of conifers. Development of techniques for DNA fingerprinting will eventually allow us to identify more of these organisms even when they do not sporulate.

Samples are often taken at the wrong time, leading to erroneous conclusions about microbial diversity. For example, soil animals migrate vertically in response to soil moisture conditions. Soil samples are normally taken from surface soils. Thus, the mesofauna may only be detected when they return to the surface. Usually soil organisms are sampled at the convenience of the investigator, but microbial populations are often event driven (e.g., precipitation). Thus, frequent or a deterministic sampling regime is needed to detect their presence.

Spatial relationships are just as crucial as temporal ones. Ecosystem processes (e.g., decomposition, mineralization, and immobilization) are not uniform across an undisturbed area. Microbial-regulated processes tend to be highly patchy and organized to optimize production. However, possibly the greatest impact of disturbance on ecosystem dynamics is spatially mixing soils and creating relatively uniform conditions across a site. In fact, this led to an oldfield view of succession that still largely dominates restoration practices, where a relatively uniform aboveground community is planted. However, succession may be a patchwork of starts and stops, with a few initial colonists acting as islands that become the nuclei for future colonists. Succession and microbial composition and activity are tightly coupled to the developing patchwork. In restored or recovering ecosystems, these patch recovery patterns are evident. In many abandoned, disturbed sites, no spatial recovery is detectable.

C. Diversity and Functional Groups

The biodiversity of types of microbes in ecosystems is daunting. However, to a certain degree, maintaining or recovering the functional groups of microbes is the first critical task in restoration efforts. Microbes play every ecosystem role at every site. In fact, microbes alone can, and do, form fully functioning ecosystems without higher plants or larger animals. In the most extreme environments of the Sahara desert and the uplands of the Antarctic Dry Valleys, microbes are the only living organisms, existing on aeolian-deposited or ancient carbon inputs. In many extreme environments, microbes make up the entire ecosystem. These range from the simple endolithic (inside rocks) communities of the Dry Valleys of Antarctica to the thermal pools of New Zealand and Yellowstone geysers. As one proceeds to more favorable environments, more and more types of microbes emerge, subdividing the processes of primary production and decomposition. At all sites, microbes undertake primary production (bacteria, cyanobacteria, algae). The relative contribution of the microbes to the overall proportion of net primary production tends to range from high in more extreme environments (such as deserts and tundra) or situations with dispersed nutrients at low concentrations (open oceans) to low in conditions highly favorable such as tropical rain forests.

In addition to directly fixing C, microbes also catalyze the nutrient cycling processes that transfer elements directly to plants or convert unavailable nutrients into forms that can be taken up and utilized. Thus, they are indirectly linked to carbon fixation by providing limiting resources. Although these organisms are generally modeled as "microbial mass," they often live symbiotically with plants. Mycorrhizal fungi probably have the largest biomass within this group. These fungi form mutualisms with plants and transfer from soil to plant a range of soil resources, from water to N to P. Importantly, they can also make unavailable soil resources, such as bound P, available, by producing organic acids and phosphatases, and Fe with siderophores. Other prokaryotes fix atmospheric N_2, ranging from free-living forms such as cyanobacteria and *Arthrobacter*, to symbionts such as *Frankia* and *Rhizobium*. Other microbes catalyze almost every other nutrient transformation that is biotically important to the sustainability of ecosystems, from N and S transformations, to Fe state transitions, to immobilization of heavy metals and bioremediation of toxic organics. In the case of mutualistic symbionts, many studies have demonstrated that an increasing diversity of species and genotypes can be critical to establishing and maintaining a diversity of plants.

Microbes are the dominant decomposers. Higher animals only take a small fraction (1–10%) of the NPP; the remainder of the energy goes to microbes. The animals themselves constitute a source of a slightly different C source from plant material, making a new type of C resource. Microbes then utilize almost all of the remaining plant material, thereby releasing the nutrients immobilized in plant tissues. Only a small fraction of C remains, as highly complex plant constituents or recalcitrant microbial compounds. These are critical in that this forms the organic matter essential to recovery of all sites.

1. Free-Living Saprobes

In every study, microbial diversity even of disturbed lands continues to increase with increasing sampling. While the actual slope of species increment curves may be lower than for undisturbed areas, it still remains very high. It is not clear if the reduced diversity of microbes is a factor in these detrimental responses. However, it is clear that if the environmental conditions for free-living microorganisms are present, a high diversity of species and a high density of individual cells will reestablish. Thus, restoration of free-living microbes is largely a matter of management of the soils, rarely by inoculation with bags of "beneficial" microorganisms. To date, there is no evidence that biodiversity per se of free-living microorganisms limits saprobic microbial activity in restored lands.

Free-living saprobes form the bulk of the microbial diversity in both functional pathways and the diversity of taxa. As we look at the known studies, those processes catalyzed by free-living microbes always occur, sometimes in detrimental levels. For example, *Thiobacillus ferrooxidans* uses Fe^{2+} in pyrite, which results in the release of sulfuric acid, detrimentally reducing the pH of streams. Immediately following disturbance, there is a rapid mineralization of N, resulting in N leaching and denitrification. *Reduction* in some of these microbial-catalyzed processes often is an important restoration task.

2. Soil Animals and Food Webs

Microbes consist of prokaryotes and fungi. These are capable of immobilizing nutrients such as N in the presence of excess C. Soil formation is also dependent on mixing of surface organic matter down through the horizon. These activities are undertaken by a food web of enormous complexity. Soil animals generally invade rapidly, either dispersing directly or by moving with soils or other materials. Soil food webs are generally characterized using functional groups as the richness of species is simply too high to characterize in detail. Food web analyses indicate that there are distinct channels (Fig. 5) that can be affected by the soil conditions and the composition of the microbes. Undertaking detailed studies of the role of biodiversity in these food webs is a critical future task. We currently do not know if species changes really matter to the recovery process.

3. Symbionts

Symbiotic microorganisms are much less diverse and clearly play critical roles in the establishment and persistence of vascular plants and plant composition. These roles are basically of two types, pathogens that inhibit plant growth, and mutualists that extract resources and exchange those resources with plants for energy or provide protection in some form, again in exchange for energy.

Plant pathogens are of two basic types for our purposes, specialists and generalists. Specialist pathogens are those that are associated with only a single species or group of host species. They tend to be highly diverse. Generalist pathogens tend to be widely spread across plant groups. Specialist pathogens are known to be devastating in agricultural ecosystems. However, they tend to be much less of a problem in restoration efforts. This probably results from the efforts made to restore a diversity of plants, making it more difficult for a pathogen to find a host and build up adequate inoculum densities. Exceptions exist when there is a high prevalence by a single species coupled with an exotic introduction.

Generalist pathogens may be another matter. These are highly diverse organisms that often live as saprobes except when conditions prove favorable to a parasitic lifestyle. For example, there are a wide variety of fusaria and rhizoctonia fungi found in virtually all soils. These can destroy a wide variety of plants under appropriate conditions. *Phytophthora cinnamomi* is responsible for loss of plants ranging such as the eucalyptus in the Jarrah forest in Australia. Often, these are almost undetectable except for very short times. In Wyoming, snow mould reduced sagebrush densities up to 60% and reduced growth in the survivors. This "mould" was a complex mix of fungi, not *Typhula sp.* found in the snow mould diseases of wheat. The disease was opportunistic and only found during El Niño years of high autumn rains and locations of high snowfall accumulation. It was found only one year and only in locations of high snowfall. Plant parasitic nematodes are always found in soils. They are responsible for high levels grazing, but, remarkably, rarely can nematode damage be observed in a restoration project.

Thus, despite the examples where disease was present, there are remarkably few demonstrations where diseases were highly diverse or markedly changed the outcomes of a restoration effort. Even under rather

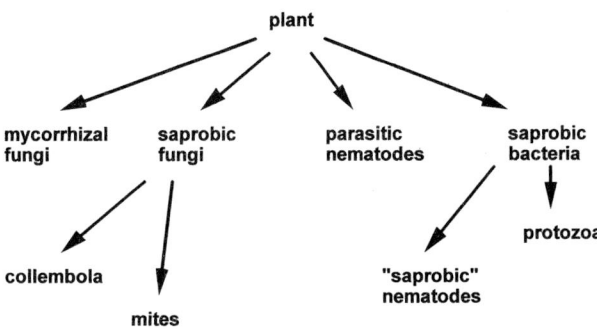

FIGURE 5 Food web channels. Carbon is allocated from plants initially to four critical microbial functional groups. Parasites, especially nematodes, take a large fraction of the net primary production and largely respire it away. This provides little benefit to soil organic matter and should be discouraged, probably largely by weed reduction and initiating a diverse plant mix. Mycorrhizal fungi provide resources in exchange for their carbon and also appear to be critical for producing stable soil organic matter. Soil bacteria and fungi are two separate webs that are grazed by different animals. While the fungi and bacteria mineralize C, they can simultaneously immobilize critical nutrients. The animals are important to retain the nutrient mineralization process as well as mixing in soil organic matter.

optimum conditions, such as tropical seasonal forests, we have observed few instances of root or shoot disease and then, it tended to be single root tips or individual leaves, but not widespread across a site. This supports the need to establish a diverse plant community. Clearly, it is generally not desirable to restore pathogens to a restoration site.

Mutualistic symbionts are relatively diverse but that diversity may play unique roles in restoration. Probably the best known are N-fixing prokaryotes. Legumes tend to be important early colonizers as N often limits primary production. In croplands, nodulation tends to be highly specific. This led to the all too common practice of using commercial inoculum. However, at sites ranging from glacial outwash in Alaska to Mount St. Helens to a seasonal tropical forest, we have planted or observed invading legumes. In no case were legumes present and functioning nodules absent. We know little about dispersal of rhizobia, but they do appear to be dispersed readily. Further, we now know that host specificity, at least for *Rhizobium*, appears to be largely associated with plasmids and not a nuclear genomic component. Thus, limitation in nodulation is likely not a function of the presence of rhizobia, but of the conditions of the site.

The effectiveness of the nodulation, however, could be an important question. In well-established hot deserts, *Bradyrhizobium* was an efficient bacterium stimulating high rates of N fixation. However, it was slow-growing and deep in the soil profile. *Rhizobium* was fast-growing and found near the surface. It rapidly colonized plants but was not an effective fixer. In pasture soils, different rhizobia are distributed in patches scattered across a site. Thus, while the presence of rhizobia is likely not limiting to restoration, having a diversity of populations capable of acting with a range of plants under a range of conditions may be important.

We know far less about *Frankia*, although there is a wide diversity of associated plants. In bioassays of respread cold-desert soils, we found that the plants failed to become nodulated. However, the soils had high N concentrations, which may have restricted activity. Alternatively, invading species such as Russian Olive has nodules even in areas where it has previously not been found. Unfortunately, beyond just a few observations of groups and N fixation rates, there are no studies of the diversity of N-fixing species or genotypes in restored areas of these critical groups.

Mycorrhizal fungi have been studied in much greater detail. Their diversity is highly variable. In desert sites, we have found as few as two or three species. Alternatively, in forests, there can be hundreds of species and thousands of genotypes. These fungi are often eliminated by the disturbance event. However, even when they are not completely eliminated, the diversity of species is often radically altered.

Moreover, many species depend on a mycelial network that can extend up to many meters across. This spatial structure is always broken up, providing opportunities for new taxa to invade. The resulting pattern is an increase in the intraspecies diversity with more, smaller clones. As these clones expand, some die and disappear while others continue expanding into the open habitats. Thus, intraspecific diversity initially increases as many propagules arrive and then declines as fewer colonies come to predominate.

Recovery of symbionts is a critical limiting step in restoration. There are two limiting steps: first, invasion of propagules, and second, establishment on site. Invasion is by physical or biological vectors. The most notable physical vector is wind. Wind has been shown to move organisms as large as mycorrhizal fungal spores up to 2 km. However, there are important limitations. Spores larger than 70–100 μm in diameter are rarely wind-dispersed. In those cases, animals are the vectors for microbes. Many animals feed on microbial spores. This can occur directly. For example, the diet of many rodents can be predominantly fungal sporocarps. This was the major means for mycorrhizal recolonization on Mount St. Helens following the eruption. Other propagules are transported unintentionally. Ungulates and rabbits feed on forbs and grasses, but in doing so, they tear plants from the soil, bringing fungal hyphae and internal spores and vesicles. Animals such as gophers preferentially feed on the nodules of legumes in addition to mycorrhizal fungi. The microbes are adapted to pass through the guts and are deposited across restored sites. Thus, just as for plants and animals, a key factor in restoration is the proximity of the source areas (Fig. 6).

D. Microbial Establishment

Different microbial species have different abilities to reestablish on a disturbed site. Because of their remarkable diversity, we are unable to artificially return even a small fraction of the microbes necessary for successful restoration. Thus, dispersal from surrounding areas is critical. In all of these invasions, two factors emerge as critical: distance and directionality for the appropriate vectors, and a suitable site. Adjacent source areas are important for reinvasion. At Mount St. Helens, for example, disturbed areas within or adjacent to surviving patches were rapidly recolonized. It took several years for sites at a greater distance to recover (Fig. 6). This pattern can be found in many other areas. Both physical and biotic vectors travel along specific pathways. These

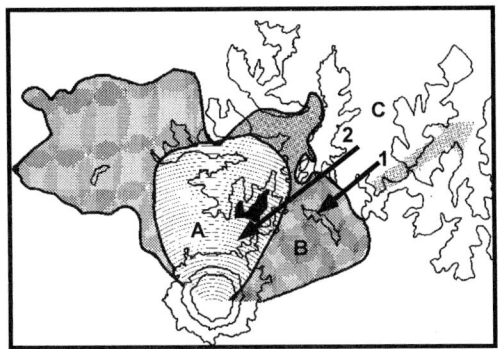

FIGURE 6 Invasion pathways and source areas. Recovery of microbes on Mount St. Helens can serve as a model of reestablishment. There were three critical types of disturbance. In area A, virtually everything was sterilized. Area B was the blast zone, where everything aboveground was eliminated and ash was deposited up to a meter in depth. Area C was the area of high ashfall. Area C recovered rapidly as most organisms survived in protected areas, establishing on the ash within the first year. This included mycorrhizae, nematodes, nitrogen-fixing bacteria, basidiomycetous fungi, mites, and collembola. Some small patches, where pocket gophers survived and emerged, served as islands. Invasion via animal dispersal also occurred from area C (pathway 1). Recovery across the site was well underway within 2–5 years. Area A was largely invaded from vectors moving from area C (pathway 2). This process was quite slow. Windborne saprophytic fungi (primarily imperfect fungi and bacteria) came in almost immediately. Symbiotic nitrogen-fixing bacteria established in 2 years, from legumes probably dispersed by birds. Other symbiotic microbes took 5–10 years to reestablish.

pathways must be present. In areas without appropriate connections allowing vectors (wind flows or animal immigration), the microbes will not invade.

Just because a microbe reaches a site does not mean it establishes. An appropriate host must be present and the soil conditions must be favorable. This can be facilitated in the planting regimes. For example, planting in patches provides windbreaks, increasing harvest of wind-dispersed microbes in addition to organic matter and other seeds. These patches also are lures to animals that are dispersing microbes. Thus, restoration of the microbes on a site often rests on reconstructing a spatially complex structure that provides protection from wind and for animals.

Another mechanism for restoring microorganisms is simply to respread the original, preferably fresh, topsoil. When topsoil is limiting, it can be stored for a short while and respread. However, mining studies have repeatedly demonstrated that as a mine moves, the newly stripped topsoil can be immediately replaced into the newly restored area. This facilitates microbial recovery rapidly. Salvaged transplants would have the same effect of introducing microbes into a disturbed soil.

In circumstances where soil microorganisms are completely lost, it may be necessary to inoculate. We know of no cases where inoculation of bacteria or saprobes has been done directly to facilitate microbial processes, although use of fresh topsoil or salvaged transplants is commonly done. However, mycorrhizal fungal inoculation can improve establishment, particularly of trees and shrubs, in areas where dispersal is limited due to long distances to source areas or where toxic conditions eliminate the native microbes. A few ectomycorrhizal fungi can be cultivated and, in some cases, limited endomycorrhizal inoculum is available through a few companies.

There are no standards for microbes and restoration. Some protocols require the addition of symbiotic mutualists, but all others assume that microbes recover just fine and will "do their jobs." In fact, soil microbes probably never "stabilize." Their short individual life spans, coupled with the ability of some members of each functional group to invade and establish, makes assessment of composition and activity difficult. Inoculation can be an important practice in conditions where little or no inoculum for an entire functional group remains and has little chance for reinvasion. However, restoration requires management of soils, plants, and animals to encourage natural migration, patch structure for concentration of resources, and a complex structure that facilitates spatial and temporal diversity in ecosystem processes. If these conditions are met, it is likely that microbes will be capable of taking care of themselves quite well.

V. CONCLUSIONS

A. Assessing Restoration Success

These examples of restoration of plants, animals, and soil microorganisms all show the difficulties and limitations of restoration of the entire richness of a prior existing community. While dominant plants and animals may be reintroduced, microorganisms are all expected to recolonize naturally. The resultant lower diversity restored communities indicated that if preservation of biodiversity is the goal, then conservation prior to disturbance is the preferred alternative, rather than restoration after disturbance. In focusing on species richness, we have placed little emphasis on ecosystem functioning, even though restoration of functioning is one of the major goals of restoration. Natural ecosystems provide ecosystem services, such as water supply, oxygen, soil stability, natural products, and so forth for free. Reclamation or rehabilitation is usually sufficient to provide these basic services, without the necessity of reintroducing all of the original biodiversity.

Measurements of both structure and functioning are used to assess restoration success. Restoration success is usually assessed by comparing the restoration site to a reference area, a native site with structure and functioning that are predetermined as the restoration goal. Measurements are compared between the reference and restoration sites. Structural measurements, such as the richness, density, and relative composition of species, are easier to measure than functional measures such as decomposition, nutrient cycling, erosion rate, or biological functions such as species reproduction and mortality or food web energy throughput. Yet it is the functional measurements that we need to determine whether the restored land has really stabilized, not simply the relatively easy measurements that require species counts. Measurements of restoration success are often not legally required, so many restoration/reclamation efforts receive no assessment at all. When they are, a species count, density, or percent cover is often all that is required to declare success.

B. Designer Ecosystems

Preservation of certain rare species may require manipulation of the ecosystem to stabilize their populations, possibly to the detriment of associated species. Such actions are already taken in numerous situations. For instance, wetland parks for shore birds have been diked, dredged, and dammed to create aquatic habitat for bird species with different water depth requirements. Pastures in Europe receiving high anthropogenic nitrogen deposition are mowed at critical times of the year to reduce the growth of nitrophilous-dominant plants and promote survival of rare plant species. James MacMahon has termed these "designer ecosystems" because they are highly managed ecosystems that have a specific conservation goal, compared to the ecosystem where the species in question may occur naturally. Biodiversity has become highly manipulated in many areas where human populations are dense and where the remnant landscape is managed to promote as high a diversity of species as possible. Virtually all restored communities are missing species, so in one sense restoration may be considered an unintended experiment to determine the impacts of rare or other missing species on community and ecosystem functioning. This will require more research and monitoring than has been done in the past. One aspect of ecological restoration that has not been emphasized in this article is the general lack of data. Many sites are restored that have never received any kind of monitoring or research, or the data are simply not available. The generalization that restoration will not return the original diversity holds for the limited number of sites that have been studied, but as more data become available, we will understand more about how to manipulate ecosystems to maximize diversity.

Acknowledgments

We thank Karen Holl and Zev Naveh for reviewing the manuscript. The research reported here was supported by grants from the Division of Environmental Biology, National Science Foundation, and the National Research Initiative, U.S. Department of Agriculture.

See Also the Following Articles

CAPTIVE BREEDING AND REINTRODUCTION • CONSERVATION BIOLOGY, DISCIPLINE OF • CONSERVATION EFFORTS, CONTEMPORARY • MICROBIAL BIODIVERSITY • PLANT CONSERVATION, OVERVIEW • REFORESTATION • RESTORATION OF BIODIVERSITY, OVERVIEW

Bibliography

Allen, E. B. (Ed.) (1988). *The Reconstruction of Disturbed Arid Lands*. Westview Press, Boulder, CO.

Allen, E. B. (Ed.) (1997). *Restor. Ecol.* **5**, 275–354.

Allen, M. F. (1988). Re-establishment of VA mycorrhizas following severe disturbance: Comparative patch dynamics of a shrub desert and a subalpine volcano. *Proc. R. Soc. Edinburgh, Sect. B: Biol. Sci.* **94**, 63–72.

Allen, M. F. (1991). *The Ecology of Mycorrhizae*. Cambridge Univ. Press, Cambridge, UK.

Bowles, M. L., and Whelan, C. J. (Eds.) (1994). *Restoration of Endangered Species: Conceptual Issues, Planning and Implementation*. Cambridge Univ. Press, Cambridge, England, UK.

Bradshaw, A. D., and Chadwick, M. J. (1980). *The Restoration of Land*. Univ. of California Press, Berkeley.

Buckley, G. P. (Ed.) (1989). *Biological Habitat Reconstruction*. Belhaven Press, London.

Collins, H. P., Robertson, G. P., and Klug, M. J. (Eds.) (1995). *The Significance and Regulation of Soil Biodiversity*. Kluwer Academic, Dordrecht/Norwell, MA.

Falk, D. A., Millar, C. I., and Olwell, M. (Eds.) (1996). *Restoring Diversity: Strategies for Reintroduction of Endangered Plants*. Island Press, Washington, D.C.

Howe, H. F., and Brown, J. S. (1999). Effects of birds and rodents on synthetic tallgrass communities. *Ecology* **80**, 1776–1781.

Jordan, W. R., Gilpin, M. E., and Aber, J. D. (Eds.) (1987). *Restoration Ecology: A Synthetic Approach to Ecological Research*. Cambridge Univ. Press, Cambridge, UK.

Jordan, W. R., Peters, R. L., and Allen, E. B. (1988). Ecological restoration as a strategy for conserving biological diversity. *Environ. Manage.* **12**, 55–72.

Major, J. (1989). *Animals in Primary Succession*. Cambridge Univ. Press, Cambridge, UK.

National Academy of Sciences. (1974). *Rehabilitation of Western Coal Lands*. Ballinger, Cambridge, MA.

National Academy of Sciences. (1992). *Restoration of Aquatic Ecosystems*. National Academy Press, Washington, D.C.

RESTORATION OF BIODIVERSITY, OVERVIEW

Joy B. Zedler, Roberto Lindig-Cisneros, Cristina Bonilla-Warford, and Isa Woo
University of Wisconsin

I. Introduction
II. Multispecies Approaches
III. Single-Species Approaches
IV. Novel Opportunities for Restoration of Biodiversity
V. Conclusion

GLOSSARY

assembly rules Constraints that one community imposes on subsequent configurations of that community.
habitat creation Construction of one habitat type from another, often a disturbed upland excavated to make a wetland.
matrix species Dominant plant with broad coverage.
nonindigenous species A species occurring beyond its natural range or potential natural dispersal range. Synonyms used here are exotic and alien species.
nurse plant A plant that shelters and facilitates the growth of others.
propagules Reproductive units (spore, seed, bulb, cyst, egg, bud, larva, etc.) that give rise to new individuals.
restoration The return of an ecosystem to its condition prior to disturbance, or to that of a nearby reference system (representative, little-disturbed ecosystem).

HABITATS THAT HAVE LOST populations of native species have potential for biodiversity restoration, that is, the return of species-rich conditions. Attempts to restore biodiversity involve multispecies efforts (e.g., sowing seed mixtures or using restoration tools such as fire to encourage colonization or spread of native species) and single-species reintroductions. Genetic issues in biodiversity restoration involve the potential to reduce intraspecific diversity, especially in reintroduced populations of clonal plants and captive-reared animals. The deliberate introduction of nonindigenous species or species not known to occur naturally at a site does not constitute biodiversity restoration; however, the area formerly occupied by a taxon is rarely known in detail, making it difficult to distinguish reintroductions from deliberate range expansions. Spatial scale is also an issue, as a taxon may be restored to a region without being restored to its exact historical location. Several assumptions underlie biodiversity restoration efforts, namely, that ecologists know what biodiversity elements historically occurred at the restoration site, that the site can support more taxa, that dispersal and environmental constraints can be overcome, and that suitable plant and animal propagules can be found on site or nearby for expansion or reintroduction. Tests of these assumptions are limited and rarely consider more than a single species or single site. In this article we review examples of multispecies and single-species approaches, reiterate concern that community-level restoration efforts often lead to "generic assemblages" rather than replicas of natural systems, and describe novel opportunities for biodiversity restoration.

I. INTRODUCTION

The desire to restore biodiversity stems from a long history of biodiversity loss as described elsewhere in this encyclopedia. The result of habitat destruction and degradation in the United States is that 1175 native plant and animal species are considered threatened or endangered and 371 native plant communities are rare (Grossman et al., 1994). Restoring biodiversity is thus a huge task—a part of the growing effort to protect and restore ecosystem structure and functioning that is being undertaken by governmental and nongovernmental organizations around the globe.

Ecologists are still struggling to understand what leads to biologically diverse communities at various spatial scales. Because our knowledge of how natural ecosystems support high species richness is inadequate, it is no surprise that there are few guidebooks for restoring diversity (although see Packard and Mutel, 1997). Yet the hypotheses that predict causes of richness are useful in identifying constraints at restoration sites. Natural communities that have high species richness share several features: great age (they have had time for speciation and time for many species to arrive by dispersal); heterogeneous habitats that can support more habitat specialists; large area (cf. species–area curves, island biogeography models); low-stress environments (e.g., tropical climates); and/or intermediate disturbance regimes. Few restoration sites have any or all of these attributes. Additional constraints on species richness in a natural community concern the sequence of species introductions, called assembly rules. For example, in greenhouse experiments several grassland plant species increased their germination rates when sown after another species (Eriksson and Eriksson, 1998), indicating the need for a facilitator. Many species require or benefit from the presence of facilitators (nurse plants, host plants, or mycorrhizae). Once established, species may be inhibited or extirpated by competitors or predators. The concept of assembly rules is particularly attractive for restoration, but the development of rules has been limited to specific cases.

Ecologists are also unsure what dictates the sequence in which species are lost as habitats become degraded. In prairie remnants, legumes are readily eliminated as prairies become fragmented. Because small prairies rarely burn (and less nitrogen is volatilized), the soil can accumulate nitrogen, reducing the competitive advantage of nitrogen-fixing legumes. As indicated below, nutrient-rich sites often support fewer species of plants than nutrient-poor sites.

In every natural community, some species are common and others are more rare. There is little guidance on which species that drop out of a degraded habitat will be restored most easily. Rarity derives from several situations, including a requirement for rare microsites or dependence on other species that are rare. Species with high habitat specificity (e.g., host-specific insects that rely on rare plants) are more likely to drop out as habitats are lost. Panzer et al. (1995) identified 256 insects from the Chicago area that were dependent on habitat remnants and absent in more disturbed habitats. An exhaustive study of this same region's vegetation showed that over two-thirds of the native plant species were remnant-dependent, with one-third having broader distributions. Thus, a minority of species occupies a majority of the landscape. Plant and animals species that are not remnant-dependent are considered adaptable and disturbance-tolerant.

There are concerns that the adaptable species will dominate restored habitats and landscapes and that restoration efforts will result in "generic communities" (Zedler, 1999). Habitat restoration involves disturbance, such as contouring and grading that creates topography lacking in heterogeneity. The species attracted to restored habitats are likely to be adaptable, disturbance-tolerant generalists. In prairie restoration efforts, the sowing of species-rich mixtures of grasses and forbs often results in dominance by a few species of grasses. Such low-quality restoration efforts have a floristic quality index of 2–5, compared to that of the original prairie's value of 20 or higher (Packard and Mutel, 1997). [Floristic quality indices consider both the number of species and their fidelity to high-quality habitats.]

The paucity of habitat specialists in restoration sites has many explanations. Some forbs may require specific germination sites, such as open patches disturbed by small mammals. Recently, the absence of diverse microflora (mycorrhizae) has been indicated as a cause of the low diversity of old field sites (van der Heijden et al., 1998). The scarcity of specific microsites and rarity of facilitators both limit reestablishment of rare species.

Efforts to restore biodiversity take place in a broad range of contexts and involve minimal or major effort. In a recent overview of international case studies in wetland restoration, Zedler (1999) described the diversity of contexts and efforts as comprising an ecological

Degree of degradation → / Degree of effort ↓	Minor (small area, minor damage)	Minor-Moderate	Moderate	Major-Moderate	Major (large area, intensive damage)
>1 action, all components ($H_{ijk} + S_{ijk} + V_{ijk} + F_{ijk}$)	Examples unlikely in this sector				Numerous actions needed
>1 action, >1 component (e.g., $S_{ijk} + V_{ijk}$)				C: Marsh excavated from fill, connected to tidal source (H_{ij}), fertilized (S_i), planted and weeded (V_{ij})	
1 action, >1 component (e.g., $H_i + S_i$)					
>1 action in 1 component (e.g., H_{ijk})				B: Wetland and channels excavated from fill, connected to tidal source (H_{ij}), to attract fish	
1 action, (H_i, S_i, V_i, or F_i)	Less action needed in this sector	A: Endangered plant reintroduced (V_i), to existing wetland			Low effort not likely to succeed

FIGURE 1 The ecological restoration spectrum (redrawn from Zedler, 1999): Actual restoration projects include many restoration situations (arranged here by degree of degradation of the landscape and the site). Sites have been treated with different actions (e.g., i, j, k) involving one or more ecosystem components: H, hydrology; S, substrate, including microbes; V, vegetation, including planting or weeding; F, faunal introduction or removal. For example, restoration of tidal wetlands at San Diego Bay for three purposes (A–C from Zedler, 1998) involved different starting conditions (A occurred in an area that was much less degraded than B and C) and different degrees of effort (C >> B > A).

restoration spectrum (Fig. 1). The effectiveness of restoration efforts was hypothesized to relate to two major variables, the degree of degradation of the site and the degree of effort expended in restoration. Where biodiversity is relatively intact, less restoration effort is necessary, and outcomes may more likely match expectations. Where massive damages have occurred, however, the challenge is daunting, and many actions are required, including restoration of the hydrology and soils and reintroduction of the microbes, vegetation, and fauna. Regrettably, comprehensive reviews of case studies have not been done; hence, we focus on various constraints to restoring biodiversity that we identified from case studies. An emphasis on constraints is appropriate because biodiversity restoration involves the removal of conditions that limit species establishment and persistence. We describe multispecies approaches, single-species approaches, and concerns about intraspecific genetic diversity in biodiversity restoration.

II. MULTISPECIES APPROACHES

A. Restored Habitats That Became Species Rich

Establishing a community that has a known composition of native species from local reference sites constitutes biodiversity restoration. Four "success stories," one for plants and three for animals, show that species richness can be restored, although the factors responsible are not entirely clear. All four cases occur in habitats that were severely modified, but only the first two had deliberate introduction of the propagules of the species assessed. For the others, highly dispersible species were evaluated. A feature that is common to all four case studies is that propagules were readily available from plantings or dispersal from nearby natural habitats, allowing colonization and development of species-rich ecosystems.

1. Pine forests: An area of pygmy pine forest in New Jersey (an Air National Guard Weapons Range) was severely damaged by military exercises, which eliminated the woody vegetation. Planting of different species mixtures and the use of soil fertility amendments (fertilizer, sewage sludge, or shredded bark mulch) produced diversities comparable to those of reference sites after two growing seasons. Although species richness was similar, a shrub component was lacking. The restoration sites were near or adjacent to nondisturbed areas (Fimbel and Kuser, 1993).

2. Dipterans: Streever *et al.* (1996) found comparable numbers of dipteran species in created and natural wetlands, and the 20 most common species had similar densities. The wetlands were constructed following phosphate mining by contouring depressions, importing stockpiled sediments, and mulching with substrate from nearby wetlands. The mulch would likely have contained insect propagules. Both created and natural wetlands were within a 33-km radius of central Florida, where agricultural land uses dominated the landscape.

3. Epibenthic invertebrates: Salt marsh invertebrates readily colonized restoration sites in San Diego Bay (Scatolini and Zedler, 1996) and the North Carolina coast (Sacco *et al.*, 1994) but in both cases, densities were lower than in reference sites. Lower soil organic matter may have limited densities without precluding the presence of many species. An aggressive exotic mussel (*Musculista senhausii*) was more abundant in the constructed than natural channels of San Diego Bay (Scatolini and Zedler, 1996). The native and exotic species were all present in the adjacent natural wetlands. Tidal channels connected the sites and allowed larval dispersal, as no transplants were done (Scatolini and Zedler, 1996).

4. Fishes: Fish species composition in the San Diego Bay wetlands (discussed above) was similar to reference sites. Fish densities were higher in constructed channels, which were deeper and broader than reference sites; however, the occurrence of an exotic fish and an exotic shrimp caused concern. As with the invertebrates, fish dispersal occurred via tidal channels that connected the natural and constructed habitats.

Without many more examples of biodiversity equivalence between restored and natural ecosystems, it is difficult to generalize the features that are essential. Questions remain about the natural diversity levels of the communities under concern (are they naturally species-poor or dominated by habitat generalists?) and the taxa assessed (are they the best dispersers and do they have the broadest ranges of tolerances?).

B. Restoration Cases That Identified Abiotic and Biotic Constraints on Biodiversity Restoration

Most assessments of biodiversity restoration identify one or more shortcomings of ecosystem structure or function. We list potential constraints from case studies that document outcomes with less than natural biodiversity.

1. Excess nutrients: In a series of experiments to establish species-rich grasslands in a reclaimed opencast coal site in Northumberland, UK, Chapman and Younger (1995) used a commercial seed mix but did not compare outcomes with natural grasslands. They found that fertilizers decreased diversity while other treatments (e.g., grazing) increased diversity (Chapman and Younger, 1995). Several strategies were examined to control the exotic grasses *Agropyron cristatum* and *Bromus inermis* in an old field in Saskatchewan, Canada. Just introducing seeds of native species did not restore grasslands. Low nutrient levels, especially for nitrogen, favored native species restoration. In addition, spraying herbicide on nonindigenous grasses prior to planting increased the number of natives that established (Wilson and Gerry, 1995).

2. Insufficient nutrients: Soils that are very low in nutrients can support too few species to satisfy biodiversity restoration goals. In the Appalachians, open-pit coal mine reclamation is constrained by low nitrogen and by the precipitation of phosphorus as these iron-rich mineral soils weather, oxidize, and form complexes that remove phosphorus from solution (Daniels and Zipper, 1995).

3. Lack of seed banks or propagules and limitations to dispersal: Prairie pothole wetlands in North America that were restored by eliminating drainage structures (tiles, ditches) had fewer plant species than natural wetlands. Three years after restoration, restored wetlands had a mean of 27 species and natural wetlands had a mean of 46 species. This is a consequence of poor representation of wetland species in the seed bank and low dispersal rates (Galatowitsch and van der Valk, 1996).

4. Improper timing of site preparation: Efforts to increase the botanical diversity following bauxite mining in *Eucalyptus marginata* forests in southern Australia often produce lower plant diversity than natural refer-

ence sites, even though overburden and topsoil are returned as part of the restoration process. Rainfall is highly seasonal and the timing of ripping, cultivation, and seed sowing were suspected as important to the outcomes. Early site preparation should maximize the number of species that establish.

5. Improper timing of propagule introduction: Some species may be difficult to reestablish because recruitment may occur only at particular times of the year or following unusual events, such as forest fires, unusually rainy periods for desert perennials, or hurricanes or other disturbances that uproot trees and expose the soil (Primack, 1996).

6. Insufficient spatial heterogeneity: Under experimental conditions microtopographic heterogeneity (microsites) increases plant species richness of experimental wetlands (Vivian-Smith, 1997). Boeken and Shachak (1994) restored desert wasteland by creating depressions and adjacent mounds that funneled water and materials into the depressions. Higher species richness occurred where water, organic matter, and nutrients collected.

7. Excessive fragmentation and insufficient connectivity: Results of experiments in which up to 54 species were added to 2 × 2 m plots of grassland communities show that dispersal and recruitment limitations have a strong influence in species abundance. Thus, processes that disrupt natural dispersal dynamics, such as habitat fragmentation, should lead to changes in species abundances. Processes that overcome natural dispersal barriers can allow novel species to invade habitats and consequently cause marked changes in community composition, diversity, and functioning.

8. Unfavorable pH: Exposed fen soils in The Netherlands became acidic upon exposure to oxygen, and calciphilic species dropped out of the community. Attempts to raise pH by liming failed. Because acidification was not reversed, native species did not reestablish, even after 5 years (van Duren et al., 1998). Also in The Netherlands, acidic heath, degraded by atmospheric nitrogen deposition, was depleted in species and several methods were employed to restore the native plant community. Both liming (to increase pH) and cutting the sod to the mineral soil layer (to reduce nutrient concentrations) were employed. The highest increase of species occurred with a combination of both treatments. (De Graaf et al., 1998).

9. Lack of nurse plants: Existing vegetation provides suitable conditions for some rare plant species. In limestone grasslands of northern Switzerland, drought and substrate-heaving limit plant establishment. *Arabis hirsuta* and *Primula veris* were unable to establish in gaps; they were found to depend on the shelter of neighboring plants for seedling recruitment (Ryser, 1993). Under experimental restoration efforts in an arid region, cluster plantings provided better conditions for establishment than more dispersed ones because they facilitated organic matter accumulation, soil water accumulation, and higher rates of formation of mycorrhizal associations (MacMahon, 1998).

10. Lack of mycorrhizae: Experimental evidence has recently been published showing that a more diverse mycorrhizal community will sustain more diverse plant communities. This work concerned grassland species in Europe and old-field assemblages in Canada (van der Heijden et al., 1998).

11. Insufficient safe sites: Many rare species require specific places/conditions for seeds to germinate and establish (Primack, 1996). In an alpine community, above timberline, large-scale removal of vegetation and mechanical disturbance to the substrate exacerbate an already severe environment. In revegetating graded areas of a downhill ski run, increasing biodiversity was dependent on two factors, immigration from nearby populations and the provision of safe sites for seedling establishment. The extant vegetation was an important source of propagules and seeding establishment was aided by use of fibrous mats (Urbanska, 1995).

12. Lack of fire: Fire is commonly reintroduced to combat exotic plants and slow the invasion of woody species (Wheeler et al., 1995). Where habitats are small or where controlled burns are a hazard to people or structures, the elimination of fire is a constraint on biodiversity restoration (Packard and Mutel, 1997).

13. Inhibition by nonindigenous species: Introduced weeds strongly impact native plant populations, both in their occurrence (Wilcove et al., 1998) and in reestablishment. Restoration sites are particularly susceptible to dominance by nonindigenous plants because of disturbances before and during restoration and because of the small size and isolation of many restoration projects. For example, many prairie and savanna restorations have a high proportion of edge, allowing ready access to propagules of nonindigenous species, just as is true in natural ecosystems. In wetland plant microcosms with varying levels of water, nutrients, and leaf litter cover, *Bidens cernua*, a native annual, dominated all 24 treatments in the first year. After 5 years, most of the microcosms were becoming dominated by the exotic purple loosestrife (*Lythrum salicaria*), with only a few of the original species remaining. Flooding regimes and soil fertility were the main factors influencing establishment and growth of *Lythrum*, with high water and low fertility limiting its dominance (Weiher et al.,

1996). In another ecosystem type, a semidesert area of Arizona, efforts to establish native grasses in mesquite (*Prosopis juliflora* var. *velutina*) woodland were impaired by the dominance of the understory by an exotic grass, *Eragrostis lehmannian*. Reintroduced natives were sparse relative to the exotic plant, depending on the interaction of several factors, such as fire (a treatment variable) and rain intensity (introductions were done in 3 successive years) (Biedenbender and Roundy, 1996).

Restoration efforts often produce assemblages with lower species richness and fewer rare species than in natural communities. Attempts to reintroduce entire animal assemblages were not found (but see single-species efforts, below). For the most part, animals are expected to disperse into restoration sites on their own. Hence, species with high mobility are favored.

III. SINGLE-SPECIES APPROACHES

A. Abiotic and Biotic Constraints

Habitat loss and competition from nonindigenous species are thought to be the most important causes of the imperilment of rare and endangered species in the United States (Wilcove *et al.*, 1998). As with multispecies efforts, attempts to reintroduce single species to a habitat it formerly occupied have suggested both abiotic (e.g., substrate type, fire, habitat size, and contiguity) and biotic (competitors, predators, disease organisms) factors that reduce population viability and limit the restoration of biodiversity. Hall (1987, in Primack, 1996) reviewed 15 plant reintroduction projects and identified five key elements of a successful project: (a) the appropriateness of planting techniques and effective execution, i.e., matching microsite conditions for each species; (b) effective site selection (safe sites); (c) complete documentation of the reintroduction project; (d) maintenance of good growing conditions, elimination of competitors, etc.; and (e) long-term monitoring, with success determined as expansion of the newly established population. Griffith *et al.* (1989) reviewed animal reintroduction attempts, focusing on efforts in the United States following passage of the Endangered Species Act. Only 7% of the 93 species of birds and mammals treated in their review were rare species; most of the rest were game species not protected under the Act. These latter species were more easily reintroduced than endangered or threatened species. Features identified as facilitating reintroduction of animal species were (a) selection of quality habitat; and (b) consideration of life history features (see item 9), introduction where competitors are lacking, use of wild-caught animals, and transplantation of large numbers of individuals.

We offer examples of these and other traits that constrain single-species efforts to restore biodiversity.

1. Inappropriate substrate: Prior to constructing habitat to attract nesting by Caspian terns (*Sterna caspia*) adjacent to Lake Ontario, substrate preferences were determined in a field experiment. The terns preferred sand substrates over other experimental substrates or the preexisting compacted ground. Experimental results were used to design a larger nesting site using the preferred substrate, which was successfully colonized by the birds (Quinn and Sirdevan, 1998).

2. Elimination of natural fire regimes: Changes in the frequency, intensity, and timing of fire have reduced rare species populations, and reintroduction of fire is beneficial to ecosystems that burned naturally. For example, fire improved the survival of the endangered annual plant *Amsinckia grandifolia*, which was introduced to habitat dominated by exotic annual grasses (Pavlik, 1996).

3. Landscape context: A key factor in restoring single species is proximity to large natural habitats that support the same species and its matrix community. For example, in New Jersey, a major utility company breached dikes that restricted tidal flows to former salt marshes. Rapid restoration of *Spartina alterniflora* occurred, in part attributed to the landscape context—some 32,000 ha of fully tidal wetlands surrounded the ~1000 ha of diked wetlands that were breached to allow tidal action and natural colonization by the target clonal grass (John Teal, Woods Hole Oceanographic Institute, personal communication).

4. Behavioral rigidity: Animal species that cannot tolerate human influences are unlikely to persist upon reintroduction to human-dominated landscapes, while those that can take advantage of modifications may flourish. For example, laboratory-reared and released Mauritius kestrel (*Falco punctatus*) changed their behavior to make use of agricultural areas and to prey on exotic species. Their numbers increased from 6 individuals in the 1970s to at least 30 mating pairs in 1992 (Crade and Jones, 1993).

5. Presence of aggressive predators: The unsuccessful reintroduction of a marsupial, the quokka (*Setonix brachyurus*), to a field station near Perth, Australia, was attributed to competition for food by rabbits and predation by foxes, both nonindigenous species. In general, invasive predators, particularly foxes and cats, challenge reintroduction efforts of many macropod spe-

cies in Australia (Short *et al.*, 1992). The *Partula* snail, a native to Moorea, French Polynesia, was driven to extinction by the introduction of the predatory snail *Euglandina rosea*. *E. rosea* was originally introduced to control *Achatina fulica*, another introduced snail raised for escargot. *Partula* snails were reintroduced in enclosures with a physical and chemical barrier to keep out the predatory snail, but in the wild, the reintroduced species is still vulnerable to predatory snails (Mace *et al.*, 1998).

6. Abundant competitors: Weedy vegetation (nonindigenous annual grasses) reduced germination, growth, and survival of an endangered annual plant, *Amsinckia grandiflora*; fire and herbicides were effective in facilitating its reintroduction. Griffith *et al.* (1989) did not highlight competitors as a factor in constraining animal reintroduction efforts.

7. Method of reintroduction: Hall (1987, in Primack, 1996) listed the use of appropriate planting techniques, effective execution, and appropriate maintenance of good growing conditions, including watering, as important to reintroduction efforts. Similar concerns were expressed by Griffith *et al.* (1989) for animal reintroductions (see item 9).

8. Insect outbreaks: Plant species that are naturally rare may not be restorable when planted in dense populations. Such was the case for a tropical woody plant that endured high mortality from insects when planted as a monotype but grew well as isolated individuals (P. Kagayama, University of Sao Paulo, personal communication).

9. Animal species life history: In their survey of reintroduction efforts for birds and mammals, Griffith *et al.* (1989) found that carnivores and omnivores were less readily translocated than herbivores; and late breeders with small clutches were slightly less likely to establish in new surroundings.

10. Scarcity of pollinators: An endangered annual plant was restored to a San Diego Bay salt marsh by sowing seeds on a small island where most of the habitat was not suitable for some of its pollinators, ground-nesting bees, that require upland, not intertidal, habitat. Insufficient pollination was shown to limit seed production in this species even when introduced to marshes next to disturbed upland.

11. Improper canopy structure: Habitats created for the endangered light-footed clapper rail (*Rallus longirostris levipes*) supported the right plant species, the cordgrass *Spartina foliosa*, but not the tall canopy that this rail species requires for nesting and camouflage. The ultimate cause of the short vegetation was the coarse soil, a sandy dredge spoil, which was much coarser than salt marsh soils and which failed to supply or retain nitrogen.

Restoration efforts begin with eliminating known constraints, and work will be most efficient if the reasons for a species' decline or extirpation are known. If limiting factors are unknown, then experimental approaches with a suite of management tools are advisable. From detailed field experiments, Pavlik (1996) determined that at least two restoration methods were needed to reestablish *Amsinckia grandiflora* to grassland in northern California, namely, fire and control of nonindigenous annual grasses. But even with his considerable understanding of the constraints, one reintroduced population spread without much treatment, while another required both fire and removal of nonindigenous grasses. His work suggests that outcomes of reintroduction may be quite site-specific.

Where experimental approaches are not feasible, knowledge of the species natural history can be employed in designing the reintroduction program. In restoring an endangered annual plant (*Cordylanthus maritimus* ssp. *maritimus*) to San Diego Bay, seeds were sown in microhabitats thought to be optimal for this hemiparasite: small openings in the canopy of a favored host plant. Additional limiting factors were later identified (canopy closure and loss of regeneration sites, drought, and hypersalinity, but none prevented the reintroduced population from persisting in the short term (8 years to date). A major decline and loss of seed banks occurred in response to a dry year, so a series of years of low rainfall could reduce the population to nonviable proportions. As in most restoration projects, it is unlikely that monitoring will continue long enough to discover the link between environmental constraints and persistence/extirpation.

B. Genetic Diversity Issues

Genetic diversity is of great concern in single-species efforts to restore biodiversity, because

1. Propagule sources may be far from the reintroduction site. Managers (e.g., M. Kenney and B. Collins, US FWS, personal communication) have recognized several alternatives: (a) importing material from distant donor populations; (b) mixing gene pools from several potential donor populations; and (c) selecting the nearest donor population. The issues have been discussed, but no standard guidelines are available (Bowles and Whelan, 1994). Tijuana Estuary was the nearest donor site for an attempt to restore an endangered salt marsh

plant (*Cordylanthus maritimus* ssp. *maritimus*) to San Diego Bay. Seeds from that one site were sown repeatedly at the recipient site in order to ensure that the annual hemiparasite would become established. Although thousands of seedlings were present and reproducing after several years of seeding, the resulting population had low genetic diversity and was considered at risk from genetic drift.

2. Reintroduction efforts involve small populations, and genetics are very important in small populations. For example, the Florida panther is experiencing inbreeding depression; males have low sperm quality and quantity and are cryptorchid (having only one descended testicle) (Lacy, 1994). While inbreeding depression can lead to the decline of population viability, there are also risks with outbreeding. Outbreeding depression can result in an intermediate form of hybrid that is adapted to neither parent's habitats. Also outbreeding can break up beneficial coadapted gene complexes. In restoration, if genes can be selected for an additive effect, then heterosis (hybridizing) may outweigh outbreeding depression. However, if populations are locally adapted through the evolution of coadapted gene complexes, then outbreeding and the mixing of alleles can possibly decrease fitness.

3. Captive breeding is employed to increase animal population sizes before reintroduction to the wild. In captivity, species can go through genetic changes that compromise their ability to survive in the wild. For example, captive-bred tamarins were released in Brazil, but these individuals moved slowly, were confused by scarce food, and were unwary of predators; preconditioning and training and some harsh experience helped other released tamarins integrate with natural populations (Lacy, 1994). Lacy (1994) recommends a population size of at least 50 for short-term captive breeding to help avoid immediate deleterious effects of inbreeding. For long-term efforts, a population size of at least 500 is desirable to maintain genetic variability. The desired number of individuals would depend on demography, sex ratio, nonrandom production of offspring, and fluctuations of population across generations. If possible, individuals from the wild should be introduced to the captive population at different breeding stages to maintain diversity.

4. Restoration of matrix species that reproduce vegetatively (e.g., *Spartina* species, *Zostera* species) are easily propagated, but when restorationists draw on small areas for propagules, the reestablished population may be primarily one genotype. Williams and Davis (1996) documented reduced genetic variability in *Zostera marina* in Mission Bay relative to more natural stands in nearby San Diego Bay. Whether or not genetic diversity is important to the long-term functioning of restored habitats is another issue. Seliskar (1995) showed that three different genotypes of *Spartina alterniflora* planted to a salt marsh restoration site in Delaware differed in many structural and functional attributes, including support of animal populations.

IV. NOVEL OPPORTUNITIES FOR RESTORATION OF BIODIVERSITY

Numerous opportunities exist for partial restoration of regional diversity. Human-influenced landscapes include many patches of habitat that currently support nonindigenous species, either intentionally or unintentionally. Highway edges and privately owned properties could support some or many native plants. Even if grown in horticultural arrangements, gene pools would be sustained by planting propagules native to the region. We found papers describing two suitable places to introduce native plants that once occurred within the region, if not on the exact site:

A. Tree Plantations

Plantations can be managed to provide a diversity of understory plants. In New Zealand, more indigenous plants and animals occur in forest plantations that are older and closer to natural habitats (Norton, 1998). In European boreal forests, biodiversity is being restored by managing fire regimes, e.g., privately owned plantations in Sweden (Angelstam, 1998). In northern Australia, plantations of native and exotic tree species span a gradient from higher to lower species richness of understory natives. For similar sites, the natives *Flindersia brayleyana* and *Araucaria cunninghamii* support more diverse understory vegetation than the exotic *Pinus caribea* (Keenan et al., 1997).

Planting a diversity of trees can also be beneficial to forest production by reducing chances of fungal (*Armillaria* spp.) infection. An experiment in Minnesota varied the mixtures of tree species (six conifers and four hardwoods) and planting densities. Species were differentially susceptible to infection, and mortality was highest when the conifers were most numerous in the plots. Seedling mortality was correlated with the species planted, the density of planting, and the proportion of conifers (Gerlach et al., 1998).

B. Utility Line Rights-of-Way

An opportunity exists for restoration of vegetation native to areas where power lines have been installed. Since tall trees cannot be allowed beneath high-voltage lines, tree growth is controlled by cutting and use of herbicide. These methods of tree control result in communities low in species diversity and wildlife. Brown (1995) suggested that the planting of competitive cover crops, including grasses, legumes and forbs, could inhibit the establishment of trees. *Dactylis glomerata* (a nonindigenous grass) reduced survival of two tree species (*Fraxinus pennsylvanica* and *Acer saccharum*) planted into experimental plots as desired, but biomass and diversity of forbs was also reduced. This alternative approach can produce communities that require less physical and chemical maintenance, as well as greater species diversity and habitat for wildlife.

V. CONCLUSION

Community-level approaches to biodiversity restoration often produce systems that are less diverse than reference systems. Prevention of natural habitat loss is thus the preferred conservation approach over allowing damages and attempting to reverse losses. More understanding of factors responsible for a species being common or rare in the region would assist restoration efforts. The relationship between a species commonness, adaptability, the breadth of its tolerance range, and its ability to be reintroduced needs to be known before the outcomes of biodiversity restoration efforts can be more predictable.

Many rare species have been the subject of reintroduction, and studies of such efforts have yielded more information about individual species requirements, especially populations that have variable reestablishment in space or over time. Comparisons of conditions associated with greater and lesser recruitment have identified many abiotic and biotic constraints on species reintroductions. Restoration efforts need to begin with an understanding of what factors were responsible for greater diversity prior to species declines.

Little work has been done on the restoration of animal assemblages, particularly terrestrial species, although wildlife ecologists have considerable experience introducing selected game species and endangered animals. In tidal wetlands, fishes readily invade sites with suitable hydrology; many marine invertebrates are ready invaders, but they may be accompanied by nonindigenous species, and their restored populations may have lower numbers of individuals.

Genetic diversity is a concern for reintroduced populations because small populations may not have sufficiently diverse gene pools and because large populations of matrix species are sometimes too homogeneous genetically due to vegetative propagation from a few clones.

Tree plantations and utility rights-of-way offer novel opportunities for biodiversity restoration. As landscapes become more fragmented and open habitat becomes more limited, sites managed for vegetation will become more important in conservation, and using such sites to establish diverse, native plant populations will aid overall biodiversity efforts.

Acknowledgments

We thank Kimberly Hamblin Hart for assistance in manuscript preparation. This review was supported in part by a grant from the National Science Foundation, Award DEB 96-19875, to J.Z., J.C., and G.S.

See Also the Following Articles

CONSERVATION EFFORTS, CONTEMPORARY • DIVERSITY, COMMUNITY/REGIONAL LEVEL • GENETIC DIVERSITY • RESTORATION, CHARACTERISTICS AND REQUIREMENTS

Bibliography

Angelstam, P. K. (1998). Maintaining and restoring biodiversity in European boreal forests by developing natural disturbance regimes. *J. Veg. Sci.* 9, 593–602.

Biedenbender, S. H., and Roundy, B. A. (1996). Establishment of semidesert grasses into existing stands of *Eragriostis lehmanniana* in southern Arizona. *Restor. Ecol.* 4, 155–162.

Boeken, B., and Shachak, M. (1994). Desert plant communities in human-made patches—Implications for management. *Ecol. Appl.* 4, 702–716.

Bowles, M. L., and Whelan, C. J. (Eds.) (1994). *Restoration of Endangered Species*. Cambridge Univ. Press, Cambridge, UK.

Brown, D. (1995). The impact of species introduced to control tree invasion on the vegetation of an electrical utility right-of-way. *Can. J. Bot.* 73, 1217–1228.

Chapman, R., and Younger, A. (1995). The establishment of a species-rich grassland on a reclaimed opencast coal site. *Restor. Ecol.* 3, 39–50.

Crade, T. J., and Jones, C. G. (1993). Progress in restoration of the Mauritius kestrel. *Conserv. Biol.* 7, 169–175.

Daniels, W., and Zipper, C. (1995). Improving coal surface mine reclamation in the Central Appalachian Region. In *Rehabilitating Damaged Ecosystems* (J. Cairns, Jr., Ed.), pp. 187–217. Lewis Publishers, Boca Raton, FL.

De Graaf, M. C. C., Verbeek, P. J. M., Bobbink, R., and Roelofs, J. G. M. (1998). Restoration of species-rich dry heaths: The importance of appropriate soil conditions. *Acta Bot. Neer.* 47, 89–111.

Eriksson, O., and Eriksson, A. (1998). Effects of arrival order and seed size on germination of grassland plants: Are there assembly rules during recruitment? *Ecol. Res.* 13, 229–239.

Fimbel, R. A., and Kuser, J. E. (1993). Restoring the pygmy pine forests of New Jersey's Pine Barrens. *Restor. Ecol.* 2, 117–129.

Galatowitsch, S. M., and van der Valk, A. G. (1996). The vegetation of restored and natural prairie wetlands. *Ecol. Appl.* 6, 102–112.

Gerlach, J. P., Reich, P. B., Puettmann, K., and Baker, T. (1998). Species, diversity, and density affect tree seedling mortality from *Armillaria* root rot. *Can. J. For. Res.* 27, 1509–1512.

Griffith, B., Scott, J. M., Carpenter, J. W., and Reed, C. (1989). Translocation as a species conservation tool: Status and strategy. *Science* 245, 477–480.

Grossman, D. H., Goodin, K., and Reuss, C. (1994). *Rare Plant Communities of the Conterminous United States: An Initial Survey.* The Nature Conservancy, Arlington, VA.

Keenan, R., Lamb, D., Woldring, O., Irvine, T., and Jensen, R. (1997). Restoration of plant biodiversity beneath tropical tree plantations in Northern Australia. *For. Ecol. Manage.* 99, 117–131.

Lacy, R. C. (1994). Managing genetic diversity in captive populations of animals. In *Restoration of Endangered Species* (M. L. Bowles and C. J. Whelan, Eds.), pp. 63–89. Cambridge Univ. Press, Cambridge, UK.

Mace, G. M., Pearce-Kelly, P., and Clarke, D. (1998). An integrated conservation programme for the tree snails (Partulidae) of Polynesia: A review of captive and wild elements. *J. Conchol. Suppl.* 2, 89–96.

MacMahon, J. A. (1998). Ecology as a basis for restoration. In *Successes, Limitations and Frontiers in Ecosystem Science* (M. L. Pace and P. M. Groffman, Eds.). Springer-Verlag, New York.

Norton, D. A. (1998). Indigenous biodiversity conservation and plantation forestry: Options for the future. *New Z. For.* 43, 34–39.

Packard, S., and Mutel, C. F. (Eds.) (1997). *The Tallgrass Restoration Handbook.* Island Press, Washington, D.C.

Panzer, R., Stillwaugh, D., Gnaedinger, R., and Derkovitz, G. (1995). Prevalence of remnant dependence among the prairie and savanna-inhabiting insects of the Chicago region. *Nat. Areas J.* 15, 101–116.

Pavlik, B. (1996). Defining and measuring success. In *Restoring Diversity: Strategies for Reintroduction of Endangered Plants* (D. A. Falk, C. I. Millar, and M. Olwell, Eds.), pp. 127–155. Island Press, Washington, D.C.

Primack, R. B. (1996). Lessons from ecological theory: Dispersal, establishment, and population structure. In *Restoring Diversity: Strategies for Reintroduction of Endangered Plants* (D. A. Falk, C. I. Millar, and M. Olwell, Eds.). Island Press, Washington, D.C.

Quinn, J. S., and Sirdevan, J. (1998). Experimental measurement of nesting substrate preference in Caspian terns, *Sterna caspia*, and the successful colonisation of human constructed islands. *Biol. Conserv.* 85, 63–68.

Ryser, P. (1993). Influences of neighbouring plants on seedling establishment in limestone grassland. *J. Veg. Sci.* 4, 195–202.

Sacco, J. N., Seneca, E. D., and Wentworth, T. R. (1994). Infaunal community development of artificially established salt marshes in North Carolina. *Estuaries* 17, 489–500.

Scatolini, S. R., and Zedler, J. B. (1996). Epibenthic invertebrates of natural and constructed marshes of San Diego Bay. *Wetlands* 16, 24–37.

Seliskar, D. (1995). Exploiting plant genotypic diversity for coastal salt marsh creation and restoration. *Biology of Salt-Tolerant Plants* (M. A. Khan and I. A. Ungar, Eds.), pp. 407–416. Department of Botany, Univ. of Karachi, Karachi, Pakistan.

Short, J., Bradshaw, S. D., Giles, J., Prince, R. I. T., and Wilson, G. R. (1992). Reintroduction of macropods (Marsupalia: Macropodoidea) in Australia: A review. *Biol. Conserv.* 62, 189–204.

Streever, W. J., Portier, K. M., and Crisman, T. L. (1996). A comparison of dipterans from ten created and ten natural wetlands. *Wetlands* 16, 416–428.

Urbanska, K. M. (1995). Biodiversity assessment in ecological restoration above the timberline. *Biodivers. Conserv.* 4, 679–695.

van der Heijden, M. G. A., Klironomos, J. N., Ursic, M., Moutoglis, P., Streitwolf-Engel, R., Boller, T., Wiemken, A., and Sanders, I. R. (1998). Mycorrhizal fungal diversity determines plant biodiversity, ecosystem variability and productivity. *Nature* 36, 69–72.

van Duren, I. C., Strykstra, R. J., Grootjans, A. P., ter Heerdt, G. N. J., and Pegtel, D. M. (1998). A multidisciplinary evaluation of restoration measures in a degraded Cirsio-molinetum fen meadow. *Appl. Veg. Sci.* 1, 115–130.

Vivian-Smith, G. (1997). Microtopographic heterogeneity and floristic diversity in experimental wetland communities. *J. Ecol.* 85, 71–82.

Weiher, E., Wisheu, I. C., Keddy, P. A., and Moore, D. R. (1996). Establishment, persistence and management implications of experimental wetland plant communities. *Wetlands* 16, 208–218.

Wheeler, B. D., Shaw, S. C., Fojt, W. J., and Robertson, R. A. (Eds.) (1995). *Restoration of Temperate Wetlands.* Wiley, Chichester.

Wilcove, D. S., Rothstein, D., Dubow, J., Phillips, A., and Losos, E. (1998). Quantifying threats to imperiled species in the United States. *BioScience* 48, 607–615.

Williams, S. L., and Davis, C. A. (1996). Population genetic analysis of transplanted eelgrass (*Zostera marina*) beds reveals reduced genetic diversity in Southern California. *Restor. Ecol.* 4, 163–180.

Wilson, S. D., and Gerry, A. K. (1995). Strategies for mixed-grass prairie restoration: Herbicide, tilling, and nitrogen manipulation. *Restor. Ecol.* 3, 290–298.

Zedler, J. B. (1999). The ecological restoration spectrum. In *An International Perspective on Wetland Rehabilitation* (W. Streever, Ed.). Kluwer Academic, Dordrecht/Norwell, MA.

RIVER ECOSYSTEMS

K. E. Limburg,* D. P. Swaney,[†] D. L. Strayer[‡]
*State University of New York, [†]Cornell University, and [‡]Institute of Ecosystem Studies

I. Introduction
II. Anatomy of Rivers
III. Energy and Matter Transformation (Biogeochemistry)
IV. Different Conceptual Models of Riverine Ecosystems
V. Riverine Biota
VI. Threats to Biodiversity
VII. Assessment and Management

GLOSSARY

allochthonous Referring to the production of organic matter outside of the ecosystem; in streams and rivers, this would be inputs from the upstream watershed.
autochthonous Referring to the production of organic matter within the ecosystem, for example, primary production by aquatic plants.
cultural eutrophication The acceleration of nutrient overenrichment caused by humans; can be caused by direct, point-source pollution (sewers) or by diffuse, non-point-source pollution (such as fertilizer runoff from farm fields).
hyporheic zone The area of saturated soils beneath a stream or river channel; interacts with the stream through the processes of hydraulic upwelling and downwelling.
riparian zone The vegetated areas on either side of a stream or river, with saturated soils usually underlying aerated soils.
river continuum concept The first of a series of conceptual, unifying models to explain and predict the structure and function of river ecosystems.
stream order A numbering system to denote stream size within a network of streams. There are several stream order systems. Stream order is also dependent on the scale of observation.
watershed (catchment) A unit of landscape in which precipitation drains to a common stream or lake; alternatively, a watershed is the divide between catchments.

RIVER ECOSYSTEMS ARE NETWORKS of streams that drain the landscape. They are thus composed of a hierarchical series of fluvial channels, beginning with small headwater streams and enlarging, ultimately, to estuaries that meet the sea. This article introduces concepts of stream flow within the catchment area, or watershed, drained by the stream network. We describe several conceptual models, which attempt to provide unifying concepts about the connections of rivers with the landscape in terms of ecosystem properties such as processing of energy and matter, habitat, biodiversity, and resilience in the face of disturbance. The main groups of riverine biota include organisms that live in the water column (such as the free-floating plankton, or fish), organisms (the benthos) that live in the sediments, organisms that use the shoreline area (the littoral zone), and organisms that dwell within adjacent areas (hyporheic and riparian zones) but which are connected to the stream in some fashion. Many human activities pose threats to the well-being of river ecosystems, including the shifting of land from forests, grasslands, and wetlands into urban or agricultural uses, construction of dams, pollutant loadings, alteration of natural drainage characteristics, introduced species, overharvesting, and climate change.

I. INTRODUCTION

Rivers are the sinuous conduits of overland flow that drain the world's continents. They displace very little of the world's land area and their standing stock of water is small relative to other hydrologic storages. Of the estimated 1,385,984,610 km^3 of global water, the average volume of rivers composes about two ten-thousandths of one percent, or 0.006% of the freshwater supply (Maidment, 1993). Nevertheless, rivers are of disproportionate importance in terms of geomorphology, ecology, and biodiversity, for they serve as transport vectors for sediments, organic matter, nutrients, and species. In contrast to the image of a river as a pipeline from source to mouth, river networks form bifurcating, fractal patterns through the landscape which are among the most complex in nature. Along any reach of the river channel, a fully three-dimensional world lies between the riverbanks; therein are found habitats in which organisms form ecological communities and process energy and matter in various ways, depending upon where the reach is located. Because of the intimate connections of rivers with the landscape, the burgeoning pressures of human-accelerated environmental change are becoming particularly acute in these systems.

II. ANATOMY OF RIVERS

To a hydrologist, the basic unit of the landscape is the *catchment* or *watershed*, in which water organizes itself into *flowpaths*. Watersheds are formed by the various forces of *continental uplift* and wearing down (*weathering*) of the bedrock by water and wind. Geomorphologically, vegetation within watersheds plays a dual role: it aids in weathering and conversion of rock to soils, and it retards surface runoff, diverting some of it to gaseous water vapor by *evapotranspiration*. Watershed boundaries, or *divides*, are those more resistant zones where water flows either one way or the other into alternative drainages. The Continental Divide in the United States separates the western drainages (which flow into the Pacific Ocean) from those (principally the Mississippi) which drain east or south. Divides at the continental scale can often limit the flow of species, but the process of species flow is affected at all scales of watersheds.

Watersheds and their accompanying flowing waters are organized hierarchically; that is, smaller watersheds usually form into larger watersheds (unless a physical boundary prevents this), which in turn may form parts of regional drainages. Hydrologists in the 1940s and 1950s recognized that streams and rivers form drainage networks and defined topological metrics to describe these. The most common metric is *stream order*. Though there are various definitions of order, Strahler (1952) considers that a first-order stream has no tributaries, a second-order stream is formed at the meeting of two first-order tributaries, a third-order stream is formed at the meeting of two second-order tributaries, etc. Shreve's link magnitude system (Shreve, 1966) defines the magnitude of a stream, downstream of the meeting of two streams, as the sum of the magnitudes of those streams. It should be noted that *order* is inevitably related to the scale of the map on which the stream and its tributaries is plotted (typically, in the United States, the smallest streams delineated at 1 : 24000 scale USGS topographic maps are considered "first-order" streams, but higher resolution maps produce higher order streams). This is a characteristic of the fractal nature of drainage networks (Rodriguez-Iturbe and Rinaldo, 1997). Another metric is the *bifurcation ratio*, or the ratio of first-order to second-order, second-order to third-order, etc. streams within a drainage network. The *drainage density*, or ratio of cumulative length of streams (L) to watershed area (L^2), is a third measure of network organization. High bifurcation ratios might indicate the presence of many small streams and, in conjunction with a high drainage density, also describe a highly dissected landscape. More complex models, which take into account differences in elevation, geology, and soils, provide even more detail for water flow and habitat creation.

Early models of watershed hydrology assumed that precipitation events saturated the soils, producing overland flow, beginning from the base of a drainage and moving uphill, much like filling a bathtub. However, in the 1970s the *variable source area* model, in which flow originates from source areas that expand and contract in different parts of the watershed in response to the dynamics of precipitation and drainage characteristics, gained acceptance. Much overland flow during a

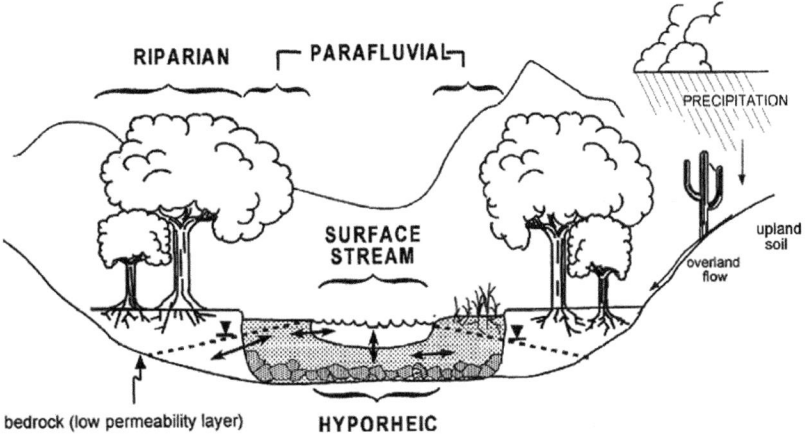

FIGURE 1 Cross-sectional view of the stream channel and zones that interact with it. From Fisher et al., 1998, with permission.

storm is actually water that had been stored in the soil, which is displaced by the new water coming in.

From an ecological perspective, the multitude of geomorphologic forms and flowpaths gives rise to a similar multitude of habitats and ways for matter to be transformed, at scales ranging from the microscopic to the entire basin. A typical cross-sectional view of a stream is given in Fig. 1.

In this figure, it is clear that the stream is in contact with several areas: a zone of saturated sediments below, called the *hyporheic zone*; an area with emergent mud or gravel bars overlying saturated soils, called the *parafluvial*; and beyond that a zone of bankside vegetation, also in contact with saturated soils, the *riparian zone*. In upland streams, where the terrain is steep and rugged, these zones may be compressed around the stream channel; but farther downstream the zones may spread out across the floodplain that was carved out over geological time by the flowing water.

Hynes (1975) was among the first stream ecologists to point out the intimacy of connection between rivers and their watersheds. As systems, rivers and streams are inherently open and serve as connectors, transporters, and processors of material. Among the biological connections of streams and watersheds are the inputs of organic matter from vegetation: leaves, insect frass, and partial or entire trees (which form debris dams). Water does much of the physical work in carving channels and carrying sediment; but organisms mediate this by regulating the rates of mineral use and transport, fixing organic matter and passing it along the food chain, and even altering landforms so that hydrology is affected (e.g., the work of beavers).

III. ENERGY AND MATTER TRANSFORMATION (BIOGEOCHEMISTRY)

From a biogeochemical standpoint, rivers differ from other ecosystems in at least three important ways: first, they are flowing systems that provide a source of kinetic energy, as well as externally produced nutrients, to resident organisms. Second, they are spatially complex and extensive systems, with a relatively large interfacial connection to contiguous terrestrial ecosystems. Such a large "buffer" zone has significant implications for processing nutrients and other fluxes into the river. Finally, as discussed above, they are transitional systems, changing in character from small, shallow, flashy first-order streams that are closely connected to the landscape to large, relatively deep and steady rivers that may be tidally influenced as they become estuarine in character near their mouths. The interplay between light, nutrients, and turbulent energy and circulation structures the biotic community, and this structure changes along the river from sources to mouth.

Riverine ecosystems are fueled by inputs of nutrients, including carbon, nitrogen, phosphorus, and silicon. Nutrients that are available to primary producers must generally exist in *inorganic* (mineral) forms. In pristine watersheds, rock weathering and decomposition of organic material in forests, grasslands, and soils provide the sources of these nutrients. In most watersheds today, atmospheric deposition of nutrients from distant industrial and automotive sources occurs via rainfall and dust. Within the watershed, fertilizer from agricultural and urban applications enters rivers in surface runoff and groundwater; discharges of sewage treatment plants and industry provide other sources. Frequently, nutrients also enter rivers in various *organic* forms (that is, as byproducts of other organisms, such as soil particles, plant and animal remains, and feces). Sources of organic nutrients associated with human activity include untreated sewage and manure runoff from farms. Organic nutrients entering a river segment from the watershed upstream augment the local primary production of biomass within it to fuel heterotrophic organisms. Such external sources are termed *allochthonous* as opposed to the *autochthonous* production occurring within the river. Regardless of whether nutrients enter rivers in organic or inorganic forms, excessive loading can result in accelerated riverine *metabolism* in which organic matter is consumed in respiration, with a consequent increased biological oxygen demand (BOD). When BOD exceeds the local oxygen supply in a reach of the river, available oxygen levels may decline precipitously (resulting in the "oxygen sag curve" downstream of nutrient sources familiar to environmental engineers), with catastrophic results to riverine organisms. Such anoxic zones typically resemble the popular image of "polluted waters": smelly masses of decaying algae, fish kills, and greatly reduced biodiversity.

Because organisms require nutrients in particular ratios (in oceanography, these are termed *Redfield ratios* after their discoverer Alfred C. Redfield (1890–1983)) to maintain their metabolism, deviations of environmental concentrations from these ratios can result in species shifts. For example, the shift of the available nitrogen:phosphorus (N:P) ratio below the Redfield ratio of 16:1 (moles of N per mole of P) in freshwaters is often associated with the onset of a bloom of cyanobacteria (blue-green algae), which are capable of fixing nitrogen from the atmosphere. Depletion of silicate below the Si:N molar ratio of 1:1 has been observed to shift the dominant species of algae from diatoms (which require silicate to synthesize their siliceous "skeletons") to species that do not utilize silicate. Such species shifts are sometimes associated with nuisance algal blooms.

It must be emphasized that the dominant physical controls of riverine ecosystems change dramatically as stream order increases. The ecosystem of the shallow, well-mixed, intermittent stream is controlled almost completely by weather and the character of the contiguous terrestrial ecosystem—geology and soils, nutrient processing in the riparian zone, light levels established by the degree of canopy closure, etc. Near its mouth, the river is deeper and broader and influenced by the sea. Relatively dense seawater meets the lighter freshwater, sometimes resulting in stratification, which, together with the effects of the tide, yields complicated patterns of circulation. The cumulative sediment load of the watershed results in decreased water clarity and increased sediment deposition (the deltas of many large rivers). The combined workings of internal circulation and turbidity structure the light and nutrient fields of the water column as well as the distribution of biological communities in ways that are not completely understood at present.

IV. DIFFERENT CONCEPTUAL MODELS OF RIVERINE ECOSYSTEMS

Beginning in the 1970s, synthetic conceptual models of rivers began to be developed as part of the ongoing movement to understand the world's ecosystems, under the auspices of the International Biological Programme (IBP). One of the fundamental insights that came to the fore was the necessity of integrating stream ecosystem dynamics over a range of spatial and temporal scales.

A. River Continuum Concept, or RCC

The RCC (Vannote *et al.*, 1980) originated as a theoretical concept to organize and predict the structure and function of riverine ecosystems from source to sink. The premise is that river ecosystems are dominated by the physical factors that create the drainage network and that ecological communities tend to be selected in response to fluvial geomorphic processes, which along with hydrology force a more or less predictable template onto the biotic components of the system. According to the RCC, the kinds of producers, consumers, magnitude of organic inputs, and degree of heterotrophy or autotrophy should be predictable along the physical gradient. Ecological communities should be in dynamic equilibrium with the stream, and the openness of streams, connecting upstream and downstream processes, is made manifest in the RCC.

Figure 2 depicts the RCC. The first thing to notice is that the stream is organized as a gradient, from an idealized headwater stream to a large river (stream order is on the left of the diagram, and it suggests a logarithmic scale). The figure also attempts to unify biological communities with functional and ecosystem-level properties.

The RCC authors divide streams into headwaters (orders 1–3), medium-sized streams (orders 4–6), and

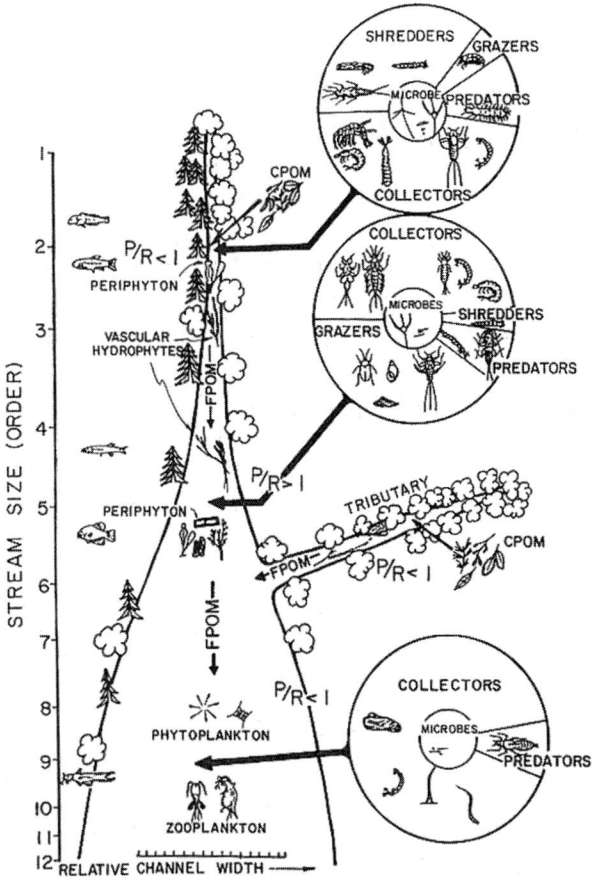

FIGURE 2 The river continuum concept. From Cummins et al., 1995.

is greater than 1, indicating this is the most productive (*in situ*) part of the continuum.

The large river end of the gradient is dominated by FPOM inputs and primary production, but the latter can be limited due to turbidity (due in part to inorganic sediment, not discussed explicitly in the RCC). Turbidity attenuates the light in the water column, so phytoplankton production may be limited and $P:R$ is predicted to decrease. Because large rivers increase in depth, they become more lakelike, with a deep water column and a muddy benthic zone, so that many of the stream insects drop out of the system.

The RCC also makes predictions about where in the system maximum diversity of organisms should occur (in the medium-sized reaches, or areas with high physical variation, because these are arguably the most diverse of habitats) and that the biotic components of the system will serve to enhance the stability of the ecosystem (referring to a long-ongoing debate as to the importance of biodiversity to ecosystem stability).

One final prediction, or observation, in the RCC is that rivers have both spatially and temporally redundant groups of organisms that tend to perform the same kinds of ecosystem functions, i.e., breaking down the organic matter inputs and remineralizing nutrients; they also argue that one should not expect to see the development of successional stages of biotic communities in rivers, ". . . because the communities in each reach have a continuous heritage rather than an isolated temporal composition within a sequence of discrete successional stages" (Vannote et al., 1980, p. 135). Later on we shall see that scientists dispute this idea.

B. Resource (Nutrient) Spiraling

"All land represents a downhill flow of nutrients from the hills to the sea. This flow has a rolling motion. Plants and animals suck nutrients out of the soil and air and pump them upward through the food chains; the gravity of death spills them back into the soil and air. Mineral nutrients, between their successive trips through this circuit, tend to be washed downhill. Without the impounding action of soils and lakes, plants and animals would have to follow their salts to the coast line." (Leopold, 1941)

The famous conservationist Aldo Leopold makes important points in his elegant prose: first, there is a tendency for elements to run downhill from the uplifted land to the sea; and second, biota play a key role in slowing down this transfer. Third, he speaks of the

large streams (orders >6). They point out that many headwater streams are dominated, in terms of the energy coming in, by the allochthonous inputs from out-of-channel, because the tree canopy tends to close off the stream from direct sunlight. In this case, the $P:R$ ratio (ratio of primary production to ecosystem respiration) is less than 1, and most of what enters the stream is coarse particulate organic matter (CPOM, matter >1 mm), which, in the case of tree boles, can be very coarse indeed. The aquatic food chain here is dominated by organisms that can deal with CPOM, mediated by microbial processing—the so-called guilds of shredder, grazers, collectors, and their predators.

Moving downstream, the relative contribution from allochthonous inputs declines, and autochthonous production (from submersed macrophytes, phytoplankton, and epiphytes) and processed inputs from upstream (now fine particulate organic matter, or FPOM, 0.005- to 1-mm grain size) become more important. Somewhere in this midzone, $P:R$ reaches a maximum and "rolling motion" of mineral transfer across the landscape and waterscape. Webster (1975) used the term *spiraling* to describe the combined downhill movement and cycling aspects of nutrients and organic matter, and Elwood et al. (1983) derive a specific model to describe the dynamics. The concept describes how an atom of a resource (a nutrient like N, P, or C) moves into a stream and how it is retained, processed, and released downstream. The atom may travel downstream, be picked up in a part of the system, cycled there many times, buried for some time, or released relatively quickly. The spiraling model provides an analytic framework to quantify the relationships between biogeochemical processing of materials and hydrodynamic transport, proposing the *spiraling length* as a measurable index of this interaction.

Spiraling of nutrients in streams is slowed by biotic retention (through sorption and ingestion) of dissolved and particulate forms. Recycling of nutrients *in situ* further slows down their downstream transport. Taking a steady-state, averaged approach for a first derivation, the *total downstream flux of a nutrient* (F_T, g/s) is related to the standing stock of nutrient available per unit length of stream (N, g/m) and the average downstream travel speed (velocity) of the nutrient (v, m/s):

$$F_T = Nv$$

As the biota take up the nutrient (postulated to happen mostly on the stream bottom), the average downstream velocity of the nutrient slows. The channel waters are assumed to be highly advective and the channel bottom retentive.

A second aspect of spiraling is *the rate at which nutrients are cycled* in the stream. The use rate of a nutrient N is defined as

$$U = kN,$$

where k (1/S) is a constant of proportionality (the fraction of the standing stock of N that is recycled per unit time, time being seconds in this case). A nutrient atom on average completes a cycle in $T = 1/k$.

By dividing the first equation by the second one, a ratio of retention to cycling is obtained:

$$Nv/Nk = v/k = vT = F_T/U = S.$$

The parameter S is the average downstream distance traveled by a nutrient atom during a cycle [between bottom and flowing water] and is called the spiraling length. If the value of k is large (high cycling rate within the stream section) and v is low ("resistance to stream transport"), then S will be short, and a greater retention of the nutrient is implied.

Figure 3 illustrates several predictions for how the spiraling length will be affected by different retention and biological processing rates. Ecosystem stability and response to nutrient loading (a disturbance) are also predicted as a consequence of spiraling.

The nutrient spiraling model can be elaborated for any number of compartments, and different nutrient spirals can be measured simultaneously as well. Carbon is one exception which is treated separately, because carbon moves so readily in and out of the dissolved and gaseous phases.

Another important point about spiraling is that it potentially identifies fast and slow components of the system. Slow components (with tight nutrient spirals and short S) are postulated to be relatively resistant to perturbation, implying that these help to stabilize a system in the face of disturbance.

C. Serial Discontinuity Concept

This heuristic model is really an early corollary to the RCC, developed by two river scientists (Ward and Stanford, 1983) from the western United States, where many rivers do not flow freely. Whereas the RCC is intended to describe more or less pristine systems, Ward and Stanford were concerned about rivers that contain dams and impoundments. They argued that such regulating structures "reset" the river continuum, although not always in the low-order to high-order direction. The key feature of serial discontinuity is that it can cause a stream reach to "behave" or function like reaches that the RCC would predict should occur in a different stream order. Figure 4 illustrates the general concept.

The serial discontinuity concept proposes that in regulated rivers, two measurable parameters of interest are the *discontinuity distance* (i.e., how far upstream or downstream a given stream parameter becomes displaced) and the *parameter intensity* (how much has the maximum value, duration, etc. of the parameter been changed?).

D. Flood Pulse Concept

In contrast to the RCC, which predicts a downstream decreasing influence of the riparian zone, the flood pulse concept (Junk *et al.*, 1989) deals with rivers that interact strongly with the floodplain by rising out of the channel bed. This theory describes the effect of floods on river channel and floodplain in large, unregu-

Mechanism		Effect on Nutrient Cycling		Ecosystem Response to Nutrient Addition	Ecosystem Stability	Categorization of study Streams
	Retention	Biological Activity	Rate of Recycling	Distance Between Spiral Loops			

	Retention	Biological Activity				
A.	HIGH	HIGH	FAST / SHORT 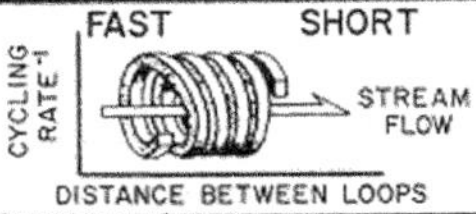	CONSERVATIVE (I>E)	HIGH	MI 2,3 PA 1,2,3
B.	HIGH	LOW	SLOW / SHORT 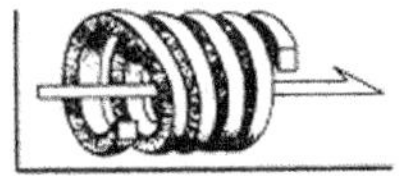	STORING (I>E)	HIGH	OR 1,2 ID 1 MI 1
C.	LOW	HIGH	FAST / LONG 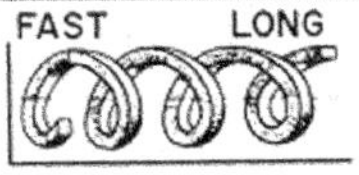	INTERMEDIATELY CONSERVATIVE < A but > D	LOW	ID 3 MI 4 PA 4
D.	LOW	LOW	SLOW / LONG 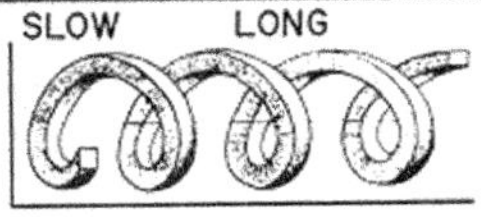	EXPORTING (I=E)	LOW	OR 3,4 ID 2,4

FIGURE 3 Predicted effects of interactions of downstream transport and measures of biological activity, and postulated responses of the river ecosystem to disturbance, here in the form of nutrient addition. I, import; E, export. The last column refers to streams in a study (by state) and their approximate stream order. From Minshall, 1983, reproduced in Cummins *et al.*, 1995.

lated rivers of this type and thereby incorporates a lateral dimension in the consideration of ecosystem structure and function.

During nonflood periods (see Fig. 5), the floodplain (which is functionally similar to a large wetland) develops and maintains its own nutrient cycles, and depending on the geomorphic configuration of the floodplain, matter may accumulate in place or enter the stream as described in other models. The pulse of a seasonal flood causes the stream to push out, sending nutrients and river biota over the floodplain. If such a pulse is predictable, then communities of organisms

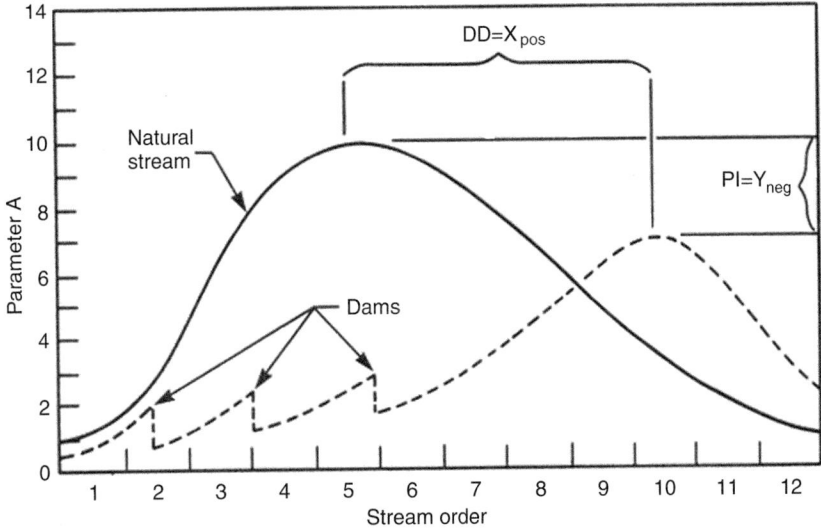

FIGURE 4 Main features of the serial discontinuity concept, describing the effect of an impoundment on a river ecosystem. DD, discontinuity distance is the distance that a parameter is displaced; PI, = parameter intensity, or degree to which a parameter is altered.

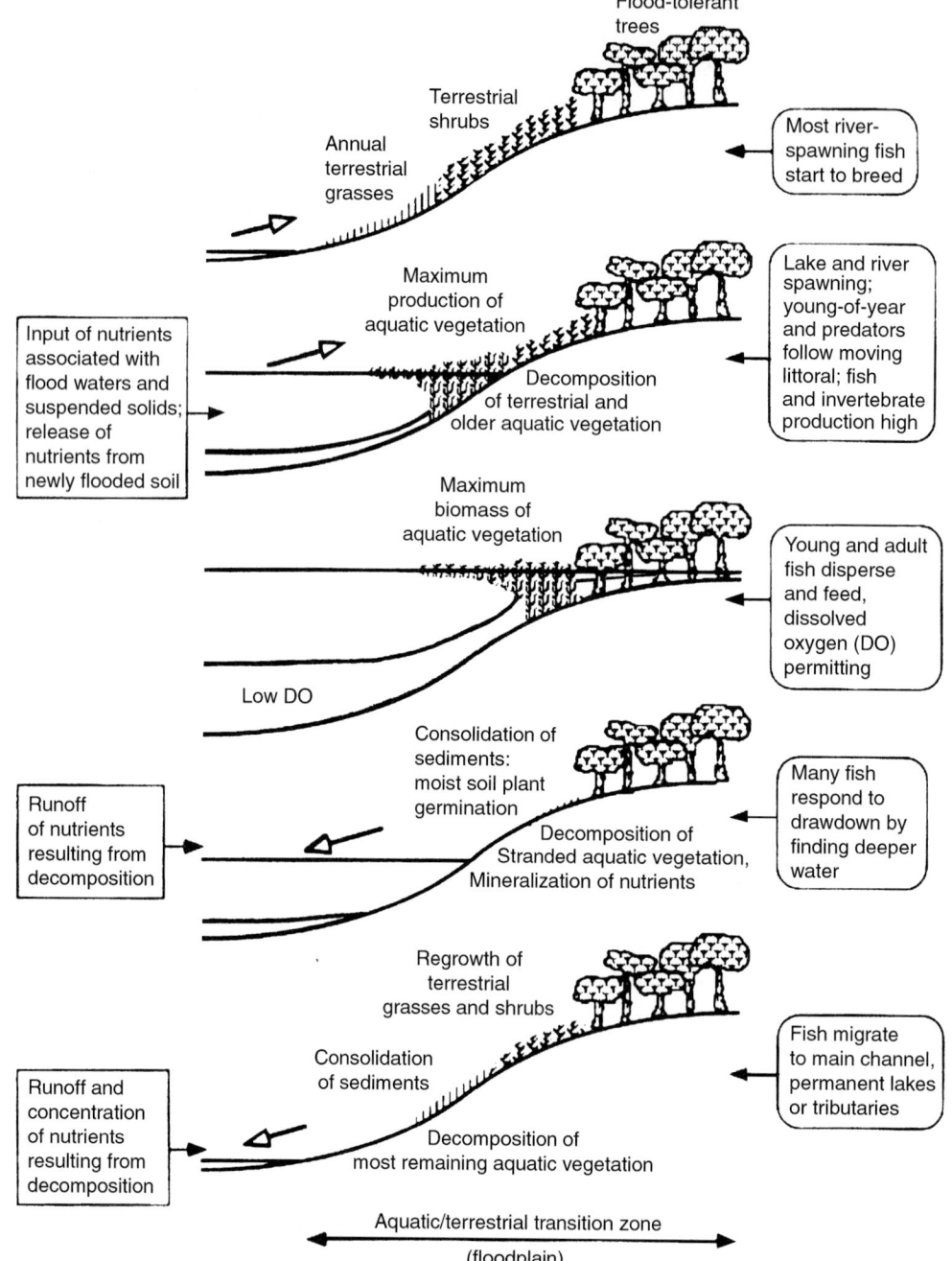

FIGURE 5 Schematic of the flood pulse concept. From Bayley, 1995.

will evolve to take advantage of the pulse because it amplifies resource availability. The flood pulse is postulated to enhance diversity and productivity by structuring the dynamics of plants, nutrients, detritus, and sediments.

During floods, matter that was remineralized in the floodplain is quickly dissolved and made available to aquatic primary producers, and production in many of these systems (including the Amazon; Bayley, 1995) exceeds decomposition. As the floodwaters stop rising, decomposition in deeper, slower-moving parts of the stream increases, often becoming anoxic. Terrestrial plants, in the meantime, recruit into the sites of retreating waters and begin to build up terrestrial biomass again.

The "moving littoral zone" is used by many fish as a nursery for young and by aquatic invertebrates that can utilize shallow environments. Many fish have adapted to the pulse by spawning just before the floods, so their offspring can be on hand when the pulse triggers off the high primary production with associated high secondary production.

E. Telescoping Ecosystem Model (TEM)

A recent paper by Fisher et al. (1998) presents another way of looking at the holistic, biogeochemical functioning of river systems and their environs, and bears some discussion. In contrast to the RCC that visualizes riverine systems as continua, and in contrast with models such as nutrient spiraling, which assumes rivers are simple, uniform chemostats, these authors emphasize heterogeneities in space and time:

> "The landscape is a patchwork of many patch sizes and shapes. Materials follow a large number of routes from uplands to oceans. The path is stochastic and often downhill because the predominant transporting vehicle is water. Because these hydrologic linkages are driven by weather, the path is episodic and jerky. Within patches, characteristic rates of uptake, transformation, and release occur: thus, as the vehicle (water) travels, its load (elements) is adjusted in response to patch-specific material dynamics." (p. 20)

This perspective corresponds to the understanding of variable source flow in watershed hydrology and takes into account the patchiness and position of patches in the landscape.

As connectors of landforms with lakes and oceans, rivers with relatively fast throughput link land and water units with considerably larger residence times; thus, the connection of land–river–lake or land–river–ocean is (to some extent) slow–fast–slow in terms of biogeochemical movements or high–low–high in terms of retention. Because of the difference in time constants among these units of the landscape, Fisher et al. (1998) focus on what landscape factors influence the ability of rivers to retain matter.

River systems are composed not only of the surface flow channel but of several adjacent parts (see Fig. 1). The stream is embedded in a saturated zone called the *hyporheic* ("underneath the flow") zone, and around that is a zone often seen as gravel bars at the surface, with underlying saturated sediments termed the *parafluvial*. The *riparian* (bank side) zone beyond this forms a larger buffer between the stream and upland areas and also overlies saturated soils. These zones are often readily distinguishable, particularly in arid landscapes. Fisher et al. (1998) term the spatially explicit arrangement of these units as the configuration of the system, and this can affect the rates of transfer, uptake, processing, and release of materials.

Similar to the spiraling length index developed by Webster (1975) and elaborated upon by Elwood et al. (1983), Fisher et al. define a parameter they call the *processing length*, or the length of subsystem required for biogeochemical transformation of some substance (note this is left intentionally vague). The key element that is different about processing length vs spiraling length is that the former applies to all these zones (stream channel, hyporheic, parafluvial, and riparian) and that these characteristic processing lengths are expected to be quite different. For example, NO_3 concentration in the authors' study stream, Sycamore Creek in Arizona, is controlled by algal uptake rate (the predominating N-transforming process), with a processing length of 42 m. In the parafluvial zone, nitrate is controlled by nitrification (due to microbial decomposition) and has a shorter processing length of 13 m.

The RCC posits that classical ecological succession does not occur in river systems, because the physical constraints dominate the development of ecological communities. Fisher and colleagues, who work in desert ecosystems where rains are infrequent and episodic, view river systems as going through succession with processing lengths becoming shorter over time, until an episodic disturbance (flood, fire, dumping toxicants) resets the system.

Fisher et al. (1998) liken their observation of stream system zonation, and dynamic differences in processing lengths within each, to the cylinders of a telescope (hence the model name). In particular, with respect to disturbance, they hypothesize that

1. When a stream system is disturbed, the different subsystems will alter their processing lengths, but differently. The *resistance* of the system, or how little it changes when perturbed, is inversely proportional to processing length change and will be greatest away from the center of the "telescope."

2. The *resilience* of the stream telescope (how fast

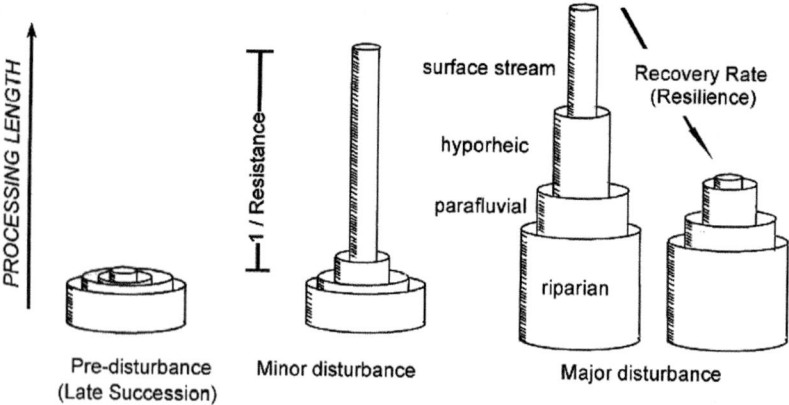

FIGURE 6 Diagram of the telescoping ecosystem model (from Fisher et al., 1998). The cylinders represent different zones of the system. See text for explanation.

the system returns to the previous state) decreases away from the center.

These two emergent properties of the TEM are shown in Fig. 6. However, *cross-links* (interactions between the subsystems) are proposed to enhance resilience of the system. Thus, the TEM is a model that can predict the rates of system change, but also maintains a hierarchical structure.

F. Geomorphic–Trophic Hypothesis and Other Geomorphological Controls on Ecology

Recent work in relatively undisturbed ecosystems has demonstrated how topography and landscape position can influence not only the biogeochemistry of waters within a catchment but even the invasibility of aquatic habitats by organisms, which sets the stage for subsequent ecological interactions. At a Long-Term Ecological Research (LTER) site in northern Alaska, researchers have found that the fish communities of lakes, connected by streams, are determined by such factors as basin steepness (which determines waterfalls that may not be passable by certain fish species), lake depth and size, etc. (Hershey et al., 1999). This results in a limited number of fish communities, and the species present appear to regulate the presence and abundance of other organisms (invertebrates and phytoplankton). Within this simple, Arctic ecosystem, the "geomorphic–trophic" hypothesis provides insight into how ecological communities develop and, in particular, how basin geomorphology can constrain the mix of species. This would be much more difficult, if not impossible, to observe in temperate systems subject to human alteration.

V. RIVERINE BIOTA

A. Microbiota

Microbiota are widely recognized as being the most abundant and diverse organisms on Earth. One cubic centimeter of water can hold over 10^9 bacteria. Microbes are responsible for most of the biogeochemical processing of nutrients and organic matter in rivers. Some important functional groups of bacteria are nitrifiers (which convert organic N into nitrate and nitrite), the denitrifiers (which convert nitrate and nitrite back into atmospheric N), sulfate reducers, and methanogens (which convert carbohydrates into methane). Microbes can be free-swimming or attached to particles; they are found in the hyporheic and riparian zones as well as in the water column itself. Microbial diversity, although recognized as vast, is difficult to assess, but new techniques involving DNA are improving the process of identification and classification.

B. Plants

Riverine plants fall into two general groups, the *littoral*, mostly rooted vegetation (*macrophytes*, or, literally, large plants), and the microscopic, free-living *phytoplankton*. Macrophytes can be further divided into those that send their leaves up above the water surface (*emergent macrophytes*) and those that photosynthesize within the water column (*submersed macrophytes*). A number of macrophytes are also free-floating, such as the water hyacinth (*Eichornia crassipes*), which has enlarged, buoyant leaf stems. Some algal species are also macrophytic.

Aquatic vascular plants have several special adaptations for dealing with the conditions imposed by the aquatic environment. Many plants have roots that extend into the water-saturated sediments, and as a result, are faced with an anoxic environment. To avoid root anoxia, many species have evolved air spaces (*aerenchyma*) in their roots and shoots that conduct oxygen down from the leaves. An example is the American water celery (*Vallisneria americana*), which translocates so much oxygen to its roots that it actually oxidizes the surrounding sediments, which in turn causes iron to precipitate and form concretions of iron oxides on the roots.

Macrophytes that grow in the brackish or salty parts of rivers have more freshwater within their cells than in the surrounding water, so the osmotic concentration gradient tends to force water out of the plant into the water. To avoid dehydration in this manner, vascular plants have a number of adaptations for maintaining their osmotic balance, including specialized cells that either serve as barriers to salt intrusion or glands that actively excrete salt. In mangroves (*Rhizophora* spp., *Avicennia* spp.), for example, the root endodermis is highly salt-resistant and serves as a salt filter, and leaf cell membranes also serve as filters by excluding sodium and selectively transporting potassium. The emergent salt-marsh plant, *Spartina*, excretes salt crystals from specialized glands in its leaves.

Phytoplankton, the microscopic plants that float in the water column, rely on turbulent flow to remain high enough in the water column to photosynthesize. In this *photic zone*, phytoplankton can fix carbon and reproduce, but below this zone, they can only respire away their carbon. In shallow rivers, this is usually not a problem, as the entire water column may be photic, but larger rivers usually become turbid due to the downstream transport of sediments, organic detritus, etc.

The main groups of phytoplankton are the *green algae* (*colonial greens, diatoms, desmids*, and others), blue-green algae, or *cyanobacteria*, and *dinoflagellates* (which are actually both autotrophic and heterotrophic). Diatoms have hard exteriors, called frustules, that are composed of silicate, so they cannot persist if this mineral is in short supply. They are an important taxonomic group in rivers, and often the riverine populations are maintained by inputs from tributaries. Algae in small streams often grows directly on rocks (*epilithic* algae) and can be an important source of organic matter in open (not covered by tree canopy) river segments. Blue-green algae, such as *Anabaena* or *Oscillatoria*, can fix nitrogen from the atmosphere by means of specialized cells called *heterocysts*; they will do this if there is an excess of available phosphorus, such as from cultural eutrophication.

Macrophytic algae are attached to hard substrates and are common in streams or along the littoral zones of larger rivers. Some, such as *Cladophora*, grow as filaments, while others (*Ulothrix, Oedogonium*) grow in tufts; *Chara* and *Nitella* are branched and their cell walls are reinforced with calcium carbonate. The filamentous green alga *Cladophora glomerata* is a cosmopolitan species that does particularly well in the presence of elevated nutrients, and therefore large colonies serve as bellwethers of cultural eutrophication.

In a review of phytoplankton diversity from 67 rivers around the world, Rojo *et al.* (1994) found that average species richness was 126, and about half of the species were reported as sporadic. Average species richness was higher in temperate rivers (130 species) than in the Tropics (75 species). The total numbers of species enumerated, by family and zone, are given in Table I.

Although the tropical rivers may have been underrepresented in the survey, nevertheless there are some clear trends. Diatoms are the most diverse taxon and predominate in temperate systems, whereas desmids are of greater importance in tropical rivers.

TABLE I
Species Richness of Riverine Phytoplankton[a]

Family	Total number of species	Temperate (%)	Tropical (%)
Chlorophyceae	385	28	27
Chrysophyceae	43	3	3
Cryptophyceae	25	—	—
Cyanobacteria	141	10	9
Diatomophyceae	408	40	25
Dinophyceae	27	—	—
Euglenophyceae	112	7	4
Xanthophyceae	11	—	—
Zygophyceae (desmids)	221	9	29

[a] Temperate and tropical percentages indicate the taxon representation in temperate and tropical planktonic assemblages, respectively (source: Rojo *et al.*, 1994).

C. Invertebrates

The invertebrates that live in rivers can be divided into two groups, the *zooplankton* (animals that live suspended in the water) and the *zoobenthos* (animals that live in or on surfaces such as the river bottom, plants, etc.). River currents usually are much faster than the swimming speeds of the zooplankton, so washout downstream is a major problem for river zooplankton. Consequently, zooplankton are common in rivers only where their growth rates (or inputs from nearby lakes and wetlands) exceed loss rates from washout and other losses (e.g., predation). Thus, river zooplankton usually is dominated by rotifers and small cladoceran crustaceans, which have high growth rates. Zooplankton densities often are highest when the water is warm and low, when growth rates are high and washout rates are low. The biomass of zooplankton in the main channel of rivers usually is much lower than that in lakes, but off-channel habitats such as oxbow lakes may support a rich zooplankton. River zooplankton often includes a few dozen species of rotifers, cladocerans, and copepods, fewer than 10 of which are abundant. At least some of these species probably are sustained by inputs from nearby habitats that are more hospitable to zooplankton, such as backwaters, impoundments, or lakes. Most riverine zooplankton are neither specialized nor endemic: the same species and genera of zooplankton that live in rivers also are common in lakes and occur in rivers worldwide. Despite its low density and richness, river zooplankton plays an important role as food for young of many species of riverine fish as well as adults of a few specialized zooplanktivorous fish such as paddlefish, *Polyodon spathula*.

By comparison with the zooplankton, the riverine zoobenthos often is rich and dense. A typical large river contains several hundred species of benthic invertebrates representing a biomass of $1-100$ g/m^2 (ash-free dry matter). Dominant groups include insects (chironomid midges and many others), bivalves, prosobranch (gill-breathing) snails, tubificid oligochaetes, and (in tropical rivers) crabs and prawns. As is the case with zooplankton, most of these species are not associated with the open waters of the main channel, but with shorelines, backwaters, vegetation, and other structurally complex habitats along the river margin (Table II). Unlike the zooplankton, many zoobenthic species are found only in rivers and are endemic to particular river systems. Thus, the rivers of southeastern North America contain several hundred species of mollusks and crayfish that are found neither in American lakes nor in rivers elsewhere in the world. Local density and species composition of riverine zoobenthos are controlled by the character of the bottom (particle size, organic content, stability), flow conditions (current speed, frequency and size of floods and droughts), the availability of food supplied by phytoplankton, rooted vegetation, attached algae, and the watershed, and biological interactions such as predation and competition. According to the river continuum concept, large rivers should be dominated by filter-feeding animals (both benthic and planktonic) and burrowing deposit-feeders. Data to test this idea in large rivers have not been collected, but it appears that large rivers actually contain a rich variety of feeding types, including filter-feeders, deposit-feeders, predators, algal scrapers, and herbivores that eat rooted plants.

Like the zooplankton, the zoobenthos is an impor-

TABLE II

Number of Species of Zoobenthos Living in the Main Channel and Backwaters of the River Danube in Austria

Taxonomic group	Main channel	Dammed backwaters	Forested backwaters	Total
Porifera (sponges)	0	1	2	2
Hydrozao (hydras)	0	0	2	2
Turbellaria (flatworks)	9	8	15	23
Mollusca (snails and mussels)	32	42	53	62
Annelida (worms)	28	57	40	69
Crustacea	11	10	27	28
Insecta	224	234	536	703
Other	2	2	8	8
Total	306	354	683	897

Source: Humpesch, U. (1996). *Arch. Hydrobiol. Suppl.* **113**, 239–266.

tant source of food to riverine fishes. The insects and crustaceans appear to be especially important as fish food, perhaps because many species are active and live above the sediment–water interface, where they are accessible to fish. Benthic filter-feeders (especially bivalves) are sometimes so abundant that they control the number and kind of phytoplankton and other suspended particles in rivers. Through their burrowing and feeding activities, benthic animals may also affect sediment mixing and the exchange of substances (e.g., nutrients, toxins) across the sediment–water interface. Also, humans use riverine invertebrates as food (e.g., prawns, crabs, crayfish, mussels, oysters), bait (crayfish, large insects), or ornament (e.g., pearly mussels, whose shells and pearls have been collected for thousands of years).

Human alteration of the hydrology of large rivers, introduction of toxins and exotic species, and destruction and isolation of the lateral habitats that are important to both zooplankton and zoobenthos have had strong effects on the biodiversity and function of riverine invertebrates. Because of their much greater degree of habitat specialization and endemism, the large-river zoobenthos has been affected more severely than the zooplankton, with hundreds to thousands of species now extinct or endangered.

D. Fish

Fish compose over half the known species of vertebrates. At present around 25,000 valid species of fish have been described, but the actual number may be closer to 28,500 (Nelson, 1994). Of these, approximately 9970 species live in freshwater, and another 500 species use freshwater habitats over some part of the life cycle (recall that freshwater composes only 1% of the world's water!). Many of the freshwater resident species, and all of the occasionals, use rivers for part or all of their lives.

Riverine fish faunas can be very depauperate, as in small, oligotrophic headwater streams, or they can be highly diverse, as in many large, tropical rivers. Fish occupy virtually every heterotrophic ecological niche, from parasite to detritus-feeder to herbivore, planktivore, insectivore, and piscivore (fish-eater). Large, predatory fish can also consume other organisms, such as small mammals, herps, and birds. In the Tropics, where seasonal floods cause rivers to spread out over the floodplain (see Section IV.D), many fish species time their reproduction just prior to the floods, so that their offspring use the flooded forests as nursery habitat. Many tropical species also directly consume leaves, fruits, and seeds while foraging in the flooded forest.

Riverine fish species are adapted to flow regimes that are considerably more turbulent than lakes or oceans. In the riffles and rapids of lower order streams, many fish spend much of their time in "dead zones," or locally quiet waters, such as under rocks, behind boulders or tree roots, or along the riparian banks, emerging briefly to feed. The larvae of freshwater lamprey (ammocoetes) actually live under the stream, in sediments, for up to 6 years. In higher order streams, current velocities are slower, and fish may swim more freely in the water column. An exception is in tidal estuaries, where currents at peak flood and ebb may be as high as a meter per second or more. Some fish apparently "ride the tides" selectively, entering a tidal current to move either up- or downstream, then moving out of the flow to remain in place.

A fairly small, but important, group of fish undertakes large-scale migrations between freshwater and marine water. These so-called *diadromous* fish number about 200 species, most of which are *anadromous*, or freshwater spawning. This group includes many of the salmons, the Southern Hemisphere galaxiids, and many herrings. These species use the freshwater environment to spawn; the young rear in the relatively protected riverine or estuarine environment and subsequently migrate to sea to feed, grow, and mature. A few species, notably the freshwater eels (Anguillidae), have the opposite strategy (*catadromy*), spawning at sea and then moving into freshwater to mature. Diadromous fish are important in many fisheries, as their spawning migrations tend to concentrate their numbers and make them relatively easy to catch.

Studies of riverine fish diversity on a global scale reveal that species richness increases with the size of the drainage basin and with river discharge (note: discharge and drainage size are closely correlated in unimpounded rivers). Tropical rivers also tend to be more speciose, and this is in part due to warmer water temperatures, which stimulates primary productivity (Fig. 7). The Amazon River, being both the largest river and tropical, supports close to 4000 species. Some of the high-diversity groups in tropical rivers include the catfishes (Siluriformes), the minnows (Cypriniformes), and characins (Characiformes).

E. Other Large Animals

Although the numbers of other animal species that use rivers is far smaller, they can be important components of these ecosystems. Among the amphibians and reptiles

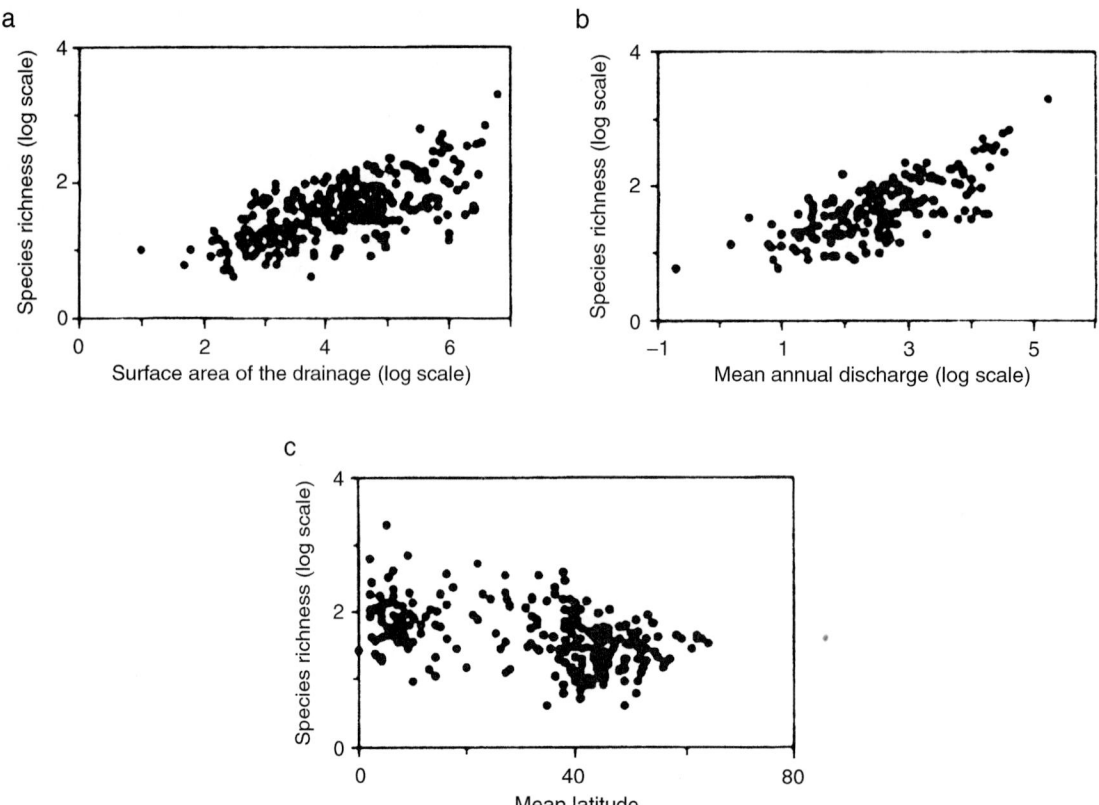

FIGURE 7 Patterns of riverine fish species richness as functions of (a) drainage basin size, (b) mean annual river discharge, and (c) latitude. Note the logarithmic scales. From Oberdorff et al., 1995.

are frogs and water snakes, many turtles, and the Crocodylia (alligators, crocodiles, and allies). Shoreline bird species are numerous, including herons, ibises, cranes, and top predators such as eagles and ospreys. Mammals, such as beaver (*Castor* spp.), otter (*Lutra* spp.), and hippopotamus (*Hippopotamus amphibius*) function as keystone species in many systems, either by virtue of their physical alteration of the habitat (beavers, hippos) or trophic influence (otters). Beavers build dams, which create ponds that change the flow characteristics, and hence the processing of organic matter, in small- to medium-sized (first to fifth order) streams in North America and Europe. Hippos, by creating "aquatic highways" between their daytime resting areas and nighttime foraging grounds, can cause drainage alterations that can be seen in satellite imagery (McCarthy et al., 1998). Some tropical Asian rivers support freshwater dolphins and sharks. Further, many large mammals, including many of the threatened or endangered "charismatic megafauna," occur in riparian zones of rivers; for example, flooded tropical grasslands are an important riparian habitat for such animals.

VI. THREATS TO BIODIVERSITY

Humans have always modified rivers and their watersheds, both purposely and inadvertently. Human settlement and development transform the landscape from its natural patterns of vegetative cover to networks of roads, drainage works, farms, and urban areas aimed at efficient transportation, food production, and sanitation. In the last century, technological advances have enabled transformation of the landscape on an unprecedented scale. Now, the effects of human activities even threaten to modify climatic factors. Effects of these transformations on the resident species of the watershed have been underestimated in the past, largely due to ignorance. Table III reviews some examples of effects of human activities on riverine ecosystems.

A. Dams

In many areas, dams are constructed for flood control, water supply, and hydroelectricity. In order to keep up with world demand for water and energy, dams are

TABLE III
Threats of Biodiversity in Rivers

Human impact	Proximate effects	Consequent short-term change in		
		Productivity	Dominant species	Species richness
Damming	Reduction and change of temporal patterns of flow, sediment flux, and turbidity	Decreases downstream associated with nutrient reduction	Shifts associated with changes in nutrient ratios (e.g., N:P, Si:N), flow reduction, and barriers to migration	Probable decline, particularly in migratory species
Stream channelization	Increase in flow velocity; disruption of benthos, hyporheic zone	?	Species shifts associated with altered flow regime	?
Nutrient loading	Fertilization of productivity; fueling of increased respiration	Increase in GPP, respiration; BOD	Species shifts associated with changes in nutrient ratios and oxygen levels	Declines associated with anoxia; potential collapse of some communities
Toxic substance loading (heavy metals, pesticides, toxic organics)	Increased mortality of resident species; reproductive failure	Probable reduction in productivity in response to toxicity	Shifts to tolerant species	Probable short-term decline, but possible long-term increase
Introductions of alien species	Colonization of ecosystem by new competitors	?	Potentially dramatic shifts if the introduction is successful	Potentially dramatic decreases as the new species become established
Land use change (logging, agriculture, urbanization)	Typically, increased sediment and nutrient load; hydrological alterations; possible increases in loading of toxics	Probable shift toward heterotrophy in response to nutrient additions and light reduction	Shifts associated with the first four impacts above	Potential decreases associated with the first four impacts above
Overharvesting of species	Depletion of target species; perturbations in predator and prey populations	?	Shifts to nontarget species; other food-chain-dependent shifts	?
Climate change	Changes in temperature, precipitation, evaporation, and atmospheric CO_2	Increases or decreases, depending upon location	?	?

being constructed at an unprecedented rate. Approximately 40,000 large dams have been constructed in the past 50 years, and the total area of land flooded by large dams exceeds 400,000 km²—an area the size of California. In the United States alone, there are nearly 2,000,000 small dams, and 68,000 dams two stories (6 m) or higher.

The most obvious ecological effect of a dam is to act as a barrier for water flowing downstream. Sediments are no longer transported downstream, but rather accumulate behind the dam. For example, prior to the construction of the Aswan Dam in the 1960s, the Nile River in Egypt transported over 120 million tons of sediment to the Mediterranean annually, and 10 million tons were dispersed in the floodplain and delta, providing fertile soils which were farmed for millennia. The dam now retains 98% of the sediments, cutting off the vital mechanism of soil renewal, causing a decline in floodplain agriculture, and also resulting in coastal erosion.

In ecological terms, a dam is a most effective barrier for migrating fish and other species, and the presence of dams is known to eliminate or drastically reduce populations of these species unless mitigating structures (e.g., fish ladders) are put in place. Even these are only partially successful: in many rivers with hydroelectric dams, the passage of fish to upstream spawning grounds is a "one-way ride," for most of these fish are killed as they attempt to move downstream and pass through the turbines.

Less obvious is the effect of dams on flow regime and biogeochemistry. Dams alter the temporal pattern of river discharge as well as the flow pattern of groundwater. Nutrient and sediment fluxes are typically diminished downstream because of settling or biological uptake in the impoundment. Resulting changes in elemental ratios (for example, the decrease in available silicate compared to nitrogen and phosphorus; Humborg *et al.*, 1997) downstream have been associated with species shifts, reductions in productivity, and (it is claimed in some cases) decreased yields in the downstream fishery.

B. Stream Channelization

Stream channelization is designed to improve the navigability of rivers or to reduce flooding potential in streams. In the former case, the river channel is deepened and widened by dredging, which destroys benthic habitat. In the latter, the streambed is sometimes straightened and "paved" to increase the capacity of the stream to transport water downstream. These processes change the flow regime of the stream, favoring species that tolerate faster, turbulent currents, and excluding others. Changes to the streambed can affect the conductivity of water through the hyporheic zone, which affects nutrient processing (see Section IV).

C. Nutrient Loading

Nutrient loading, as discussed above, fuels the ecosystem. Excessive loading results in catastrophic collapse when the resulting increase in metabolism exceeds the available oxygen supply of the river. Cultural eutrophication of rivers is severe to dire in some parts of the developing world, in particular in India, Southeast Asia, and Africa, as well as around many urban centers worldwide.

D. Toxic Substances

Associated with pesticide application in agriculture (or suburban gardens), industrial effluents, and chemical spills, toxic substances have a spectrum of effects, all deleterious to organisms. Some herbicides dramatically reduce productivity by direct mortality of aquatic plants. Other chemicals kill sensitive species higher in the food chain. Subtler effects include behavioral changes (due to endocrine-disrupting compounds) in predators or prey species, which can shift the structure of food chains, increase susceptibility to disease, and cause infertility. What complicates this story is that different organisms do not respond uniformly to exposure to a particular toxic substance, so that the effect on a particular riverine system is determined largely on the mix of species present. Chronic exposure to a stream of toxic material will gradually shift the species present to those that can tolerate it, generally resulting in a decline in species richness.

E. Overharvesting Species

In addition to humanity's indirect impacts on biota, humans threaten biodiversity directly through the overharvesting of species. In the United States and Canada, a number of river-associated fisheries are currently in decline, including those for several species of Pacific salmon, Atlantic salmon, American shad, sturgeons, and oysters. Worldwide, fishing pressures in rivers are enormous. Many river fisheries in developing countries are *artisanal*, providing food for a local community, and little is known about harvesting pressures. However, responses of fish communities to overharvesting are becoming understood. For instance, in one West African river, catch per unit of fishing effort remained relatively constant from the 1950s through the 1990s, but the trophic structure of the harvested fish went from large piscivores to small planktivores, with loss of food web complexity (Fig. 8; Welcomme, 1999). This phenomenon is called "fishing down food webs" and was first noted in marine fisheries (Pauly *et al.*, 1998). At the population level, overharvesting often selectively removes older age classes, pushing the age of first reproduction down, and often skews the sex ratios in the population (for example, female sturgeon are selectively sought for their roe from which caviar is produced).

F. Introduced Species

Humans are also introducing new species into ecosystems at an unprecedented rate. Some of these introductions are intentional: for example, salmon, trout, and bass are stocked into rivers and lakes all around the world for sport fishing, and grass carp are introduced to control aquatic weeds. Many introductions are unintentional, arising from the pumping out of ballast water in large tankers, overland transfer of boats with attached organisms from one drainage to another, or accidental spills. Although most organisms that are introduced into a novel environment perish, the increase in commercial boat traffic particularly increases the likelihood of a successful introduction.

Rates of exotic species introductions are difficult to quantify, in part because many of the species are too small to notice. However, Mills *et al.* (1996) documented the rate of species introductions into the Hud-

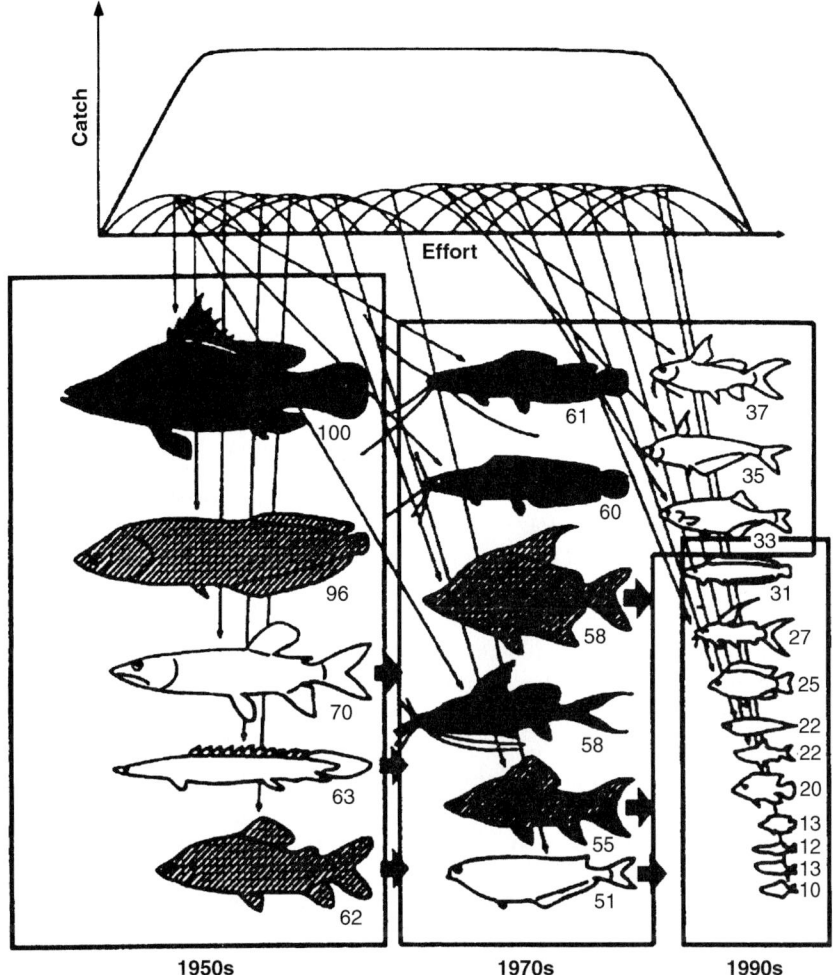

FIGURE 8 Example of a "fished-down" riverine food web, Oueme River, West Africa. Diagram shows how fish catches changed from larger fish to smaller and smaller target species over time. Fish colored in black disappeared before 1965; fish in gray are fish whose numbers were seriously reduced. Arrows indicate species which reduced the age at first reproduction. Note that overall catch per unit of effort remained largely constant. From Welcomme, 1999, with permission.

son River, in New York State; since 1840, the rate has been about one successful introduction per year—much higher than natural rates of species flow.

Direct perturbations of the mix of species, either by introduction of new species or by removal of extant species by overharvesting, may cause dramatic changes in relations of the food web, thereby also affecting predators and prey. Some species additions affect their physical and chemical environment as well: in North America, zebra mussels, which were introduced into the Great Lakes in 1988 and which spread rapidly through the Mississippi and eastern drainages, have been documented to increase the clarity of the water column by virtue of their filter feeding and also carpet large areas of the benthos, where they have successfully colonized.

VII. ASSESSMENT AND MANAGEMENT

River ecosystems are clearly important and at risk. Their direct economic importance to societies includes their use in transportation, water supply, energy, and provision of harvestable products. Their indirect importance, sometimes termed *essential ecosystem services*, includes their fundamental role as biogeochemical transformers of energy and matter, physical transformers of the landscape (shaping the land through fluvial processes), and the provision of a wide variety of ecological habitats along the river continuum.

Historically, the assessment of river ecosystems was hindered by two factors. First, on a practical basis, it is difficult to adopt a single methodology by which to assess the state of a large river system from headwaters to mouth. Many of the methods that were developed for small-stream study are not easily adapted to the higher order parts of the system, and at the other end of the continuum, oceanographic methodologies may be designed for scales of study that are larger than rivers. Second, until the development of scale-spanning river paradigms such as the river continuum concept, ecological studies of rivers tended to consider a river in isolation from its watershed; as is now widely recognized, upstream processes in the watershed (including human-accelerated land use change) have major downstream effects on rivers. (A third difficulty may be due to the somewhat arbitrary tendency of researchers to specialize on broadly different parts of the system: there are many more small stream and estuarine scientists than there are "river scientists.")

Recent revival of the concept of "ecosystem health" has proven to be a useful point of departure for current study and management of river ecosystems (see, for example, Karr and Chu, 1999; and Scow et al., 2000). Metrics of ecological communities, such as those that compose *indices of biotic integrity* (IBI; see Karr and Chu, 1999), can be combined with in-stream biogeochemical measurements (e.g., processing lengths of critical chemical cycles, ecosystem metabolism, or food web relationships as identified by tracers), in-stream habitat assessments, and whole-watershed metrics (primarily land use, soils, geology, topography, etc.) to provide an integrated assessment of the status and threats to rivers. An emerging consensus is that assessment and management no longer fall solely in the domain of scientists and managers, but must involve those who share a stake in the management process. This is evidenced in the growth of watershed management councils, which struggle to deal fairly with the trade-offs between economic prosperity and sustainability of rivers and lakes in watersheds. The difficulty of finding equitable solutions rises as the scales of political boundaries increase: rivers that share or cross international jurisdictions require a higher level of political cooperation. Maintaining river ecosystems as healthy components of the landscape will be one of the great challenges of the twenty-first century.

See Also the Following Articles

ENDANGERED FRESHWATER INVERTEBRATES • ESTUARINE ECOSYSTEMS • FISH, BIODIVERSITY OF • LAKE AND POND ECOSYSTEMS • INVERTEBRATES, FRESHWATER, OVERVIEW • MARINE ECOSYSTEMS • WETLANDS ECOSYSTEMS

Bibliography

Bayley, P. B. (1995). Understanding large river–floodplain ecosystems. *BioScience* 45, 153–158.

Cummins, K. W., Cushing, C. E., and Minshall, G. W. (1995). Introduction: An overview of stream communities. In *Ecosystems of the World: River and Stream Ecosystems* (C. E. Cushing, K. W. Cummins, and G. W. Minshall, Eds.), Vol. 22, pp. 1–8. Elsevier, Amsterdam.

Elwood, J. W., Newbold, J. D., O'Neill, R. V., and Van Winkle, W. (1983). Resource spiraling: An operational paradigm for analyzing lotic ecosystems. In *Dynamics of Lotic Ecosystems* (T. D. Fontaine and S. M. Bartell, Eds.), pp. 3–27. Ann Arbor Science, Ann Arbor, MI.

Fisher, S. G., Grimm, N. B., Marti, E., Holmes, R. M., and Jones, J. B., Jr. (1998). Material spiraling in stream corridors: A telescoping ecosystem model. *Ecosystems* 1, 19–34.

Humborg, C., Ittekkot, V., Cociasu, A., and von Bodungen, B. (1997). Effect of Danube River dam on Black Sea biogeochemistry and ecosystem structure. *Nature* 386, 385–388.

Humpesch, U. (1996). Case study: The River Danube in Austria. *Arch. Hydrobiol. Suppl.* 113, 239–266.

Hershey, A. E., Gettel, G. M., McDonald, M. E., Miller, M. C., Mooers, H., O'Brien, W. J., Pastor, J., Richards, C., and Schuldt, J. A. (1999). A geomorphic–trophic model for landscape control of Arctic lake food webs. *BioScience* 49, 887–897.

Hynes, H. B. N. (1975). The stream and its valley. *Verh. Int. Ver. Theor. Angew. Limnol.* 19, 1–15.

Junk, W. J., Bayley, P. B., and Sparks, R. B. (1989). The flood pulse concept in river–floodplain systems. International Large Rivers Symposium (D. P. Dodge, Ed.). *Can. Spec. Publ. Fisheries Aquat. Sci.* 106, 110–127.

Karr, J. R., and Chu, E. W. (1999). *Restoring Life in Running Waters: Better Biological Monitoring.* Island Press, Washington, D.C.

Leopold, A. (1941). Lakes in relation to terrestrial life patterns. In *A Symposium on Hydrobiology*, pp. 17–22. Univ. of Wisconsin Press, Madison WI.

Maidment, D. R. (Ed.). (1993). *Handbook of Hydrology.* McGraw-Hill, New York.

McCarthy, T. S., Ellery, W. N., and Bloem, A. (1998). Some observations on the geomorphological impact of hippopotamus (Hippopotamus amphibius L.) in the Okavango Delta, Botswana. *Afr. J. Ecol.* 36, 44–56.

Minshall, G. W., Petersen, R. W., Cummins, K. W., Bott, T. L., Sedell, J. R., Cushing, C. E., and Vannote, R. L. (1983). Interbiome comparison of stream dynamics. *Ecol. Monogr.* 53, 1–25.

Nelson, J. S. (1994). *Fishes of the World*, 3rd ed. Wiley, New York.

Oberdorff, T., Guégan, J.-F., and Hugueny, B. (1995). Global scale patterns of fish species richness in rivers. *Ecography* 18, 345–352.

Pauly, D., Christensen, V., Dalsgaard, J., Froese, R., and Torres, F. (1998). Fishing down marine food webs. *Science* 279, 860–863.

Rodriguez-Iturbe, I., and Rinaldo, A. (1997). *Fractal River Basins: Chance and Self-Organization.* Cambridge Univ. Press, Cambridge, UK.

Rojo, C., Alvarez Cobelas, M., and Arauzo, M. (1994). An elementary, structural analysis of river phytoplankton. *Hydrobiologia* **289**, 43–55.

Scow, K. M., Fogg, G. E., Hinton, D. E., and Johnston, M. L. (Eds.). (2000). Integrated Assessment of Ecosystem Health. Lewis, Boca Raton, FL.

Shreve, R. L. (1966). Statistical law of stream numbers. *J. Geol.* **74**, 17–37.

Strahler, A. N. (1952). Dynamic basis of geomorphology. *Geol. Soc. Am. Bull.* **63**, 923–938.

Vannote, R. L., Minshall, G. W., Cummins, K. W., Sedell, J. R., and Cushing, C. E. (1980). The river continuum concept. *Can. J. Fisheries Aquat. Sci.* **37**, 130–137.

Vörösmarty, C. J., Sharma, K. P., Fekete, B. M., Copeland, A. H., Holden, J., Marble, J., and Lough, J. A. (1997). The storage and aging of continental runoff in large reservoir systems of the world. *Ambio* **26**, 210–219.

Ward, J. V., and Stanford, J. A. (1983). The serial discontinuity concept of lotic ecosystems. In *Dynamics of Lotic Ecosystems* (T. D. Fontaine and S. M. Bartell, Eds.), pp. 29–42. Ann Arbor Science, Ann Arbor, MI.

Webster, J. R. (1975). "Analysis of Potassium and Calcium Dynamics in Stream Ecosystems on Three Southern Appalachian Watersheds of Contrasting Vegetation," Ph.D. Dissertation, Univ. of Georgia, Athens, GA.

Welcomme, R. L. (1999). A review of a model for qualitative evaluation of exploitation levels in multi-species fisheries. *Fisheries Manage. Ecol.* **6**, 1–19.

SALMON

Michael H. Schiewe and Peter Kareiva
National Marine Fisheries Service

I. Classification
II. Life Histories
III. Special Adaptations
IV. Historical Status and Overview of Decline
V. Habitat, Harvest, Hatcheries, and Hydropower as Major Threats to Salmonid Diversity
VI. What Is the Value of Salmonid Biodiversity and Can We Rescue Salmon from Extinction?

GLOSSARY

alvin Newly hatched salmon with unabsorbed yolk sac.
anadromous Aquatic organisms that spawn and undergo early development in freshwater, migrate to sea to grow and mature, and return to freshwater to reproduce.
fry Juvenile salmon that have absorbed their yolk sac and emerged from the gravel.
grilse Atlantic salmon that spends only one year at sea before returning to spawn.
iteroparous Aquatic species that can spawn more than one time.
jack A sexually precocious male salmon that spends one winter or less at sea before returning to its natal stream to spawn.
kelt A spawned out Atlantic salmon or steelhead that is returning to the ocean to recuperate, sexually mature, and spawn again.
parr Older juvenile salmon that are distinguished by parr marks—prominent oval spots on their sides.
semelparous Aquatic species that die after reproduction.
smolt A downstream migrating salmon that has begun the physiological transition to seawater.

THE PACIFIC AND ATLANTIC SALMON are a diverse group of fish distributed throughout the Northern Pacific and Northern Atlantic Oceans and their adjacent freshwater coastal habitats. The Pacific Salmon include seven species in the genus *Oncorhynchus*: pink (*O. gorbuscha*), chum (*O. keta*), sockeye (*O. nerka*), coho (*O. kisutch*), and chinook (*O. tshawytscha*) salmon are found in North America and throughout the Pacific rim; masu (*O. masou*) and amago (*O. amago*) salmon are found only in Asia. In addition, there are two trout members of the genus *Oncorhynchus* that have anadromous forms: steelhead (*O. mykiss*, the anadromous form of rainbow trout) and the sea-run cutthroat trout (*O. clarkii*). There is a single species of species of Atlantic salmon (*Salmo salar*). This chapter focuses on the anadromous salmonids of North America.

I. CLASSIFICATION

Over the past two centuries, fish have been the subject of extensive systematic research and countless taxonomic revisions—and among the some 24,000+ named species of fishes, the salmonids have been no exception.

Currently, the salmonids are members of the class *Osteichthyes* (the bony fishes), the subclass *Actinopterygii* (the ray-finned fishes), the division *Teleostei*, and the super order *Protacanthopterygii*. The *Protacanthopterygii*, which contains some 310-plus species, is divided into three orders: the *Salmoniformes* (the salmonids), the *Esociformes* (the pike and mudminnow), and the *Osmeriformes* (the smelts). There are about 70 species in the family *Salmonidae* (order *Salmoniformes*), which can be divided into three subfamilies: *Salmoninae* (salmon and trout), *Coregoninae* (whitefishes), and *Thymallinae* (graylings).

According to Behnke (1992), the earliest fossil record of the salmonids is *Eosalmo driftwoodensis*, which was discovered in Eocene deposits some 40 to 50 million years ago in British Columbia. The separation of the Atlantic Ocean group (*Salmo*) and the Pacific Ocean group (*Oncorhynchus*) likely occurred by the mid-Miocene about 15 million years ago. By the end of the Miocene (about 5 million years ago), the ancestral form of *Oncorhynchus* in North America evolved into two distinct lineages, one leading to the Pacific salmon and the other to the Western trout. The evolution of Pacific salmon is still the subject of scientific debate. Particularly intriguing is the question of whether salmon evolved first in freshwater, only later developing the ability to migrate to the ocean, or whether they evolved first as marine species that later developed the ability move into freshwater refugia to spawn and rear. No matter which the case, the currently recognized species are generally believed to have differentiated only 500,000 to 1 million years ago, which represents an extremely short time period for such extensive speciation to occur. Rapid speciation was probably favored by the geological and climatic history of North American landscape—constantly being shaped and reshaped by volcanism, alternating periods of warm and cold climate, alternating arid and pluvial periods, and rearrangement of major drainage basins.

The classification of North American salmonids does not stop with the traditional Linnean system of classification for plants and animals. In particular, salmon have recently been classified into evolutionary significant units or ESUs. An ESU is any group of populations that is distinct according to two criteria: (a) It must infrequently exchange genes with other conspecific population units, and (b) It must represent an important component in the evolutionary legacy of the species (Waples, 1995). The concept of the ESU as a fundamental unit of conservation under the Endangered Species Act has been reviewed and endorsed by the National Research Council (1995) and is an important landmark in evolutionary thinking applied to conservation. This is because the ESU emphasizes genetic structure and life history adaptation as a focus for protection as opposed to more typological definitions of species or subspecies based on appearance.

In Western North America, 56 distinct salmonid ESUs have been identified, with 25 of them currently listed as threatened or endangered under the U.S. Endangered Species Act.

II. LIFE HISTORIES

Salmon and anadromous trout display an amazing variety of life history patterns, both among and within species. However, many features are held in common. The life cycle begins with females depositing eggs in nests or redds in gravel bottoms of rivers and lakes. The young emerge from the gravel as alevins or fry to rear in fresh water for periods varying from a few days to several years. As juveniles or parr approach the time for their seaward migration, they undergo physiological metamorphosis or smoltification in which, among other changes, they develop the ability to osmoregulate in the hypertonic marine environment. In the ocean, they often undertake extensive migrations, covering many thousands of miles, while they grow and mature. After periods that range anywhere from a few months to 4 or more years, adult salmon return with remarkable fidelity to their natal river and tributary to spawn and continue the cycle.

The key characteristic that distinguishes the Atlantic salmon and anadromous trouts from the five species of Pacific salmon is the fact that Pacific salmon species are semelparous and die after spawning only once, whereas anadromous trout and Atlantic salmon are iteroparous and may spawn multiple times. As is evident in the ensuing sketches of species-specific life history traits, diversification among salmon primarily has involved permutations that alter the schedule and timing of freshwater versus ocean habitat use (Busby *et al.*, 1997; Groot and Margolis, 1991; Shearer, 1992; Trotter 1992).

A. Pink Salmon (Also Known as Humpback Salmon)

Pink salmon, which are distributed throughout the North Pacific and Bering Sea north of 40°N, are the most abundant of the Pacific salmon. In North America, pink salmon spawn from August to November and generally return at smaller sizes than other salmon—

ranging between 1.0–2.5 kg. Returning adults undergo only modest migrations, often spawning in intertidal areas. After emerging from the gravel, they swim directly to sea, where they migrate and feed extensively in the North Pacific. After 18 months at sea, they return to their river of origin to spawn and die. This fixed 2-year cycle leads to the genetic isolation of what are referred to as odd- or even-year runs. Typically, different geographic regions are dominated by odd- or even-year runs, but not both. The tendency of having only one type of run (odd or even) in each river may exist because the rivers are not productive enough to support spawning by pink salmon every year.

B. Chum Salmon (Also Known as Dog Salmon)

Chum salmon are second only to pink salmon in abundance, but because of their large size they comprise up to 50% of the marine biomass of salmon in the North Pacific. Historically ranging from the San Lorenzo River in California to the Mackenzie River on the Beaufort Sea to the north, returning adults can weigh as much as 20 kg. Like pink salmon, chum salmon tend to spawn low in river systems and migrate quickly to estuarine and marine waters, often spending extended periods rearing in estuarine habitats. Once offshore, chum salmon are widely distributed throughout the North Pacific, typically spending 2 to 5 years at sea, but occasionally as long as 7 years. There are two generally recognized races of chum salmon, which are distinguished by their time of return to fresh water: early and late stocks, or summer- and fall-runs, with the fall-run populations far more prevalent.

C. Sockeye Salmon (Also Known as Red Salmon)

Sockeye salmon are the third most abundant of Pacific salmon, comprising 17% by weight and 14% by number of the salmon total. They display a wide variety of life history patterns, including a subspecies that spends its entire life in fresh water—the kokanee. The vast majority of sockeye salmon populations spawn either in tributaries associated with lakes or directly in lakes on gravel shoals. After emerging from the gravel, the juvenile sockeye salmon spend from 1 to 3 years rearing in lakes. There are, however, rare populations that spawn in the rivers not associated with lakes and populations that migrate directly to sea shortly after emergence from the gravel. Like the other salmon species with extended freshwater rearing, smolts tend to move quickly downstream, spending little time in the estuarine and nearshore habitats and are found widely distributed throughout the North Pacific Ocean. Adults typically spend 1 to 4 years at sea, before returning to spawn in their natal lakes and tributaries. Maturing adults enter fresh water from late spring to early summer, with spawning occurring from late July through January.

D. Coho Salmon (Also Known as Silver Salmon)

Coho salmon are the most widely distributed of the Pacific salmon, but they are the least abundant. The vast majority of the populations have a 3-year fixed cycle—spending about 18 months in fresh water and 18 months at sea. In North America, they range from Monterey Bay, California, in the south to the Kukpuk River in Alaska. Coho salmon typically enter fresh water from August to January and quickly move upstream to spawn in coastal streams and the smaller tributaries of large river systems. After emergence from gravel, juveniles rear in the smaller tributaries, and overwinter an additional year before beginning their seaward migration during the spring of their second year of life. At sea, coho salmon tend to stay on the continental shelf, feeding and growing in marine waters at roughly the same latitude as their stream of origin.

E. Chinook Salmon (Also Known as King Salmon)

Chinook salmon are the largest of the Pacific salmon—returning at weights up to 45 kg. In North America they range from central California to Kotzebue Sound, Alaska, and spawn in diverse habitats ranging from tidewater to small tributaries located as far as 3200 km inland. Locally, runs tend to be identified by the time at which adults enter fresh water and begin their spawning migrations, hence the colloquial names spring-, summer-, fall-, late fall-, and winter-run chinook salmon. However, these designations do not capture the true biological diversity of this species.

Along with sockeye salmon, chinook salmon display the greatest within-species variation in life histories of all the Pacific salmon, including variation in age at seaward migration, variation in duration of freshwater, estuarine, oceanic residency, variation in ocean distribution and oceanic migratory patterns, and variation in age and season of spawning migration. This variation is derived from two sources, one racial and the other environmental. The racial differences result from geographic and hence genetic isolation that divides chinook salmon populations into one of two behavior forms, stream-type and ocean-type. Stream-type populations spend one or more years in fresh water before seaward migration; undertake extensive offshore migrations, and return to their natal streams during the spring or summer, often holding in fresh water for several months before spawning in late summer. In contrast, populations of ocean-type chinook salmon migrate to sea prior to age 1, utilize estuarine and nearshore habitats for extended periods, often spending their entire ocean residence on the continental shelf, and return to their natal stream in the fall immediately before spawning.

The second major source of life history variation in chinook salmon is expressed as run-timing and age-at-spawning differences within races, within populations, and sometimes even within cohorts. The adaptive value of such variation is well known in evolutionary theory—by spreading the timing of spawning among years, these fish are effectively "spreading the risk" and reducing the likelihood of total failure if any single year or even week is disastrous (the argument works at a variety of temporal and spatial scales). Uncertainty in juvenile survival and marine productivity is enormous for salmon, and this is reflected by the fact that year-to-year variability in spawner counts for salmon (see Fig. 1) have coefficients of variation typical of pest insects and other notoriously fluctuating species.

F. Atlantic Salmon

Atlantic salmon are the only anadromous salmonid native to the Atlantic coast of North America. Historically found in nearly every major coastal river north of the Hudson River, the extinction of several U.S. populations by the end of the 19th century has shifted the southern extent of the species distribution over 200 kilometers northward.

The majority of adult Atlantic salmon enter fresh water to spawn during the spring (late May to early June primarily), spending most of the summer holding in fresh water and then spawning in fall. In general, the more northern populations arrive earlier, and spawn earlier. However, a few populations do not enter freshwater until late summer or early fall, bypassing the summer holding period and consequently spawning directly. Juvenile Atlantic salmon emerge from the gravel in the spring and typically spend from 2 to 4 years rearing in fresh water (although some populations spend as little as 1 year and others as much as 8 years in fresh water). Outmigration of juveniles usually occurs from April to mid-June, with the smolts spending little time in the estuarine and nearshore zone. Atlantic salmon are widely distributed in the North Atlantic, with American and European stocks often concentrating and mixing in the waters off Greenland—a distribution which enabled the development of highly efficient

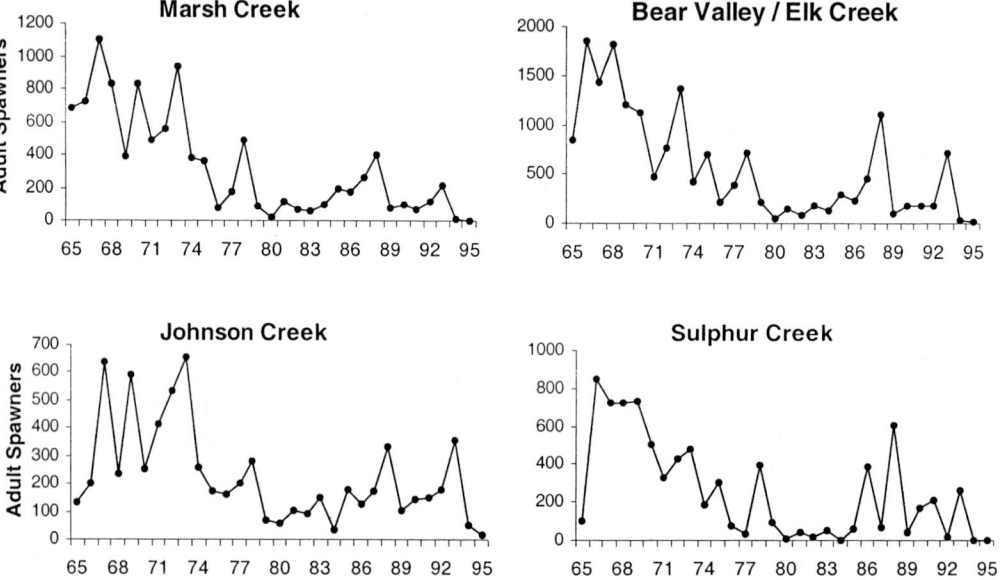

FIGURE 1 Counts of spring chinook salmon returning to spawning grounds at four Snake River tributaries. Note that the populations appear to be clearly at risk, yet population fluctuations are so large that occasionally large "returns" interrupt what appears to be a severe decline.

commercial fisheries that nearly drove the species to extinction in the 1960s. The majority of Atlantic salmon spend two winters at sea before returning to spawn.

G. Steelhead

There are two major genetic groups of *O. mykiss* presently recognized in North America: the *inland group* and *coastal group*, with the distinguishing character being their use of freshwater habitat. As the name implies, the coastal form populates the wet coastal regions of western North America, whereas the inland form is found in the more arid interior regions of western North America. Both forms occur in British Columbia, Washington, and Oregon; Idaho has only the inland form, and California has only the coastal form. Both forms display a wealth of life history variation, including an anadromous form that is known as steelhead. Although steelhead can be observed entering fresh water to spawn throughout the year, they usually display striking seasonal peaks in spawning activity. This has resulted in the use of seasonal run timings to distinguish different populations (fall-run, summer-run, etc.). Steelhead can be divided into two basic reproductive ecotypes based on their state of sexual maturity at the time of river entry and the duration of spawning migration. The stream-maturing type (commonly known as fall steelhead in Alaska, summer steelhead in the Pacific Northwest and northern California) enters fresh water in a sexually immature state and spends several months maturing before they spawn. In contrast, the ocean-maturing type (spring steelhead in Alaska and winter-run elsewhere) enters fresh water in a sexually mature state and spawns shortly thereafter. While many rivers contain populations of both ecotypes, coastal streams tend to be dominated by winter steelhead, whereas inland reaches tend to be predominately the summer steelhead.

Despite the major differences in migration timing of the adults, both ecotypes of steelhead generally spawn from December through March. After emerging from the gravel in late spring, juveniles spend up to 7 years in fresh water, with 1 to 3 years being the norm. Most populations in Alaska and British Columbia tend to smolt and migrate to sea at 3 years of age, whereas the more southern stocks tend to smolt at 1 or 2 years of age. Most North American steelhead spend 2 years at sea; however, some of the more southern stocks characteristically spend only 1 year at sea. An unusual variant in life history found among selected steelhead populations in Southern Oregon and Northern California is that of the half-pounder. Half-pounders are juvenile fish that migrate to sea, only to return to fresh water in an immature state after only 2 to 4 months. The biological basis for this unusual behavior is unknown. As an iteroparous species steelhead can spawn multiple times, with the frequency varying both within and among populations. The more northern populations tend to have a lower proportion of repeat spawners than those further south.

H. Sea-Run Cutthroat Trout

Cutthroat trout display a complex suite of life histories—some exclusively fresh water and others involving a period of marine residency. While it is the latter life history strategy that justifies their inclusion in this chapter, the reader should be aware that the interrelationships between the different life history variants are not fully understood, and sea-run forms probably cannot be treated as isolated from the freshwater forms (Trotter, 1989).

In North America, sea-run cutthroat trout are distributed along the Pacific Coast from the Eel River in California to Prince William Sound in Alaska. Adults typically spawn in the small tributaries of low gradient streams and the lower-gradient downstream reaches of large river systems from late winter to late spring. Juveniles rear in fresh water for 2 to 4 years, generally utilizing slower moving waters and pools. Those populations that migrate into the protected waters of large estuaries and sheltered bays tend to smolt at age 2, whereas those populations migrating into the nearshore ocean often delay smoltification until age 3 or 4. Sea-run cutthroat trout seldom spend more than a few months in marine waters and stay well within the nearshore zone, typically returning to fresh water in late summer, fall, or winter of the same year they migrate to sea. Although fish returning to fresh water to spawn return to their natal stream with high fidelity, nonmaturing fish are known to leave the ocean and overwinter in nonnatal streams. Like steelhead and Atlantic salmon, sea-run cutthroat are iteroparous, and multiple spawning is quite common.

III. SPECIAL SALMON ADAPTATIONS

A. Anadromy

Anadromous fishes are those that spawn in fresh water, migrate to the ocean to forage and mature, and return to fresh water to spawn and begin the cycle again. As anadromous species, the salmonids pass through multiple habitat transitions and face major physiological challenges—the most critical adaptation for this lifestyle is the ability to osmoregulate in both freshwater and marine environments, and to do so at precisely prescribed times. The evolution of anadromy has intrigued biologists for as long as salmon have been studied. Although many theories have been put forth to explain the evolution of anadromy, such a major life history character must, on balance, provide an overall advantage to individual fish. The potential advantages of anadromy include (a) a mechanism for dispersal and rapid recolonization of regions suddenly made available to fish (such as the retreat of a glacier), and (b) being able to get the "best of two worlds" (freshwater and marine). The "best" of the freshwater world involves reduced predation on eggs and juvenile fish because the embryos develop in the protected environment of gravel until they emerge at a large enough size to actively avoid stream-dwelling invertebrate predators. The "best" of the marine world involves higher growth rates for larger maturing fish in the oceans, made possible because in upwelling areas and years of high productivity the regions of the ocean frequented by salmon provide an abundance of larger prey items that far exceeds the prey base of most freshwater ecosystems.

B. Homing

The legendary ability of Pacific and Atlantic salmon to home with high fidelity to their stream of origin is one of the more spectacular migrations in nature. Beginning with the pioneering work of Hasler, Quinn, and others, it is growing increasingly clear that this ability involves a combination of magnetic detection, solar navigation, and olfactory cues. The salmon's ability to navigate the high seas appears to be based on the ability to detect inclination and declination of the earth's magnetic field, combined with celestial orientation and an "endogenous circannual rhythm, synchronized by day length or rate of daylight change to sense latitude" (Quinn, 1982). Once in the vicinity of their home stream, it appears that salmon homing is facilitated by following unique gradients of odors that are learned during the downstream migration of juveniles.

Despite uncertainty about the exact mechanism of homing, the fact that salmon return to their natal streams is ecologically quite significant. From a very practical perspective, it assures that salmon return to an area with favorable spawning and rearing conditions. In addition, although often considered a failure in homing, the small proportion of salmon in a population that stray to neighboring streams play an important role in the colonization of vacant habitat. Such a nearest neighbor strategy is more likely to be successful in matching salmon with habitat than a random pattern of straying, and hence it is a more efficient means of reestablishing sustainable populations.

IV. HISTORICAL STATUS AND OVERVIEW OF DECLINE

Historically, spawning populations of Atlantic salmon were abundant in most rivers and their tributaries throughout Europe, the British Isles, and the Baltic North, and in all the major North America rivers north of the Hudson River along the Atlantic seaboard. Likewise, the thousands of rivers and streams dissecting the coastal lands surrounding the North Pacific Ocean supported major populations of Pacific salmon and anadromous trout. Today, however, these once plentiful species are greatly reduced in both abundance and distribution. Due largely to the actions of humans and the encroachment of "civilization," populations of Atlantic salmon throughout much of Europe and in New England have all but disappeared from the landscape. Similarly, in the Pacific Northwest, only a fraction of the streams and rivers that once supported robust salmon populations continue to do so. Although there are some obvious differences, there are also some striking similarities in the causes and patterns of decline of Pacific and Atlantic salmon—in Europe the onset of the Industrial Revolution, in New England the colonization by European settlers, and in the Pacific Northwest the arrival of the Euro-American settlers. The 1991 review of Nehlsen *et al.* perhaps best captures the status of salmon in the Pacific Northwest by noting that of 214 individual populations of native naturally spawning populations of salmon in California, Oregon, Idaho, and Washington, 101 were at high risk of extinction, 58 were at moderate risk of extinction, and 54 were of special concern. The situation has become so dire that the 25 currently endangered and threatened salmonid ESUs span almost every major freshwater ecosystem from the Sacramento River northward, with vast political and social consequences for the western United States (if the Endangered Species Act is enforced with vigor).

V. HABITAT, HARVEST, HATCHERIES, AND HYDROPOWER AS MAJOR THREATS TO SALMONID DIVERSITY

Many factors, both natural and human caused, are major sources of mortality among populations North

American salmon. While there is a handful of examples of a single factor causing populations to go extinct (e.g., a dam completely blocking access to spawning and rearing habitat), more often than not there is a constellation of factors that place salmon at risk of going extinct. An extremely thorough review of the current status of Pacific Northwest salmon and the factors contributing to their decline was recently published by Lichatowich (1999). In *Salmon without Rivers*, Lichatowich chronicles the methodical extirpation of one salmon population after another as the Euro-American settlers and their industrial economy replaced what he refers to as the "gift economy" of the native Indian tribes. Given the widespread destruction of habitat and a "modern" harvest philosophy built on the false premise of unlimited productivity, it is indeed amazing we still have viable populations of salmon in the Pacific Northwest.

The human-caused threats to salmon are often described in a framework of the four H's: habitat, harvest, hatcheries, and hydropower. This convenient framework is followed here, but it is also a little misleading in how it categorizes risk factors in tidy compartments. For example, while it has been traditional to view habitat threats as fresh water and affected only by human destruction, this view grossly neglects the importance of estuarine and marine habitats and the role they play in salmon productivity. Further, while there is a tendency to dismiss the ocean as a homogeneous "black box" that is unaffected by human activities, there is likely to be a strong interaction between some of the H's such as hatcheries and ocean conditions. For example, the industrial-scale production of hatchery fish in large river systems such as the Columbia River Basin may not be problematic except during periods of low primary and secondary productivity in the nearshore ocean. Hatchery and naturally produced smolts may compete only during periods of low ocean productivity. In addition, the ill effects of individual fitness compromised by gene flow from hatchery fish may be felt only if wild fish are already stressed by poor ocean conditions.

A. Habitat

The destruction of freshwater habitat is perhaps the single most significant factor affecting Pacific and Atlantic salmon in North America. Logging, mining, and agriculture (including grazing, irrigation withdrawals, soil erosion, and chemical contamination) have been major causes of salmon decline. While many of the impacts of these activities are similar, some are unique. For example, the logging practices of the past two centuries has caused widespread destruction of riparian vegetation along stream corridors (producing increased instream temperatures) and has caused landslides due to poor road-building practices (resulting in sedimentation and destruction of spawning and rearing habitat). Other impacts, such as the creation of thousands of acres of impermeable surfaces, are unique to urbanization. While it is beyond the scope of this chapter to describe in detail the nature of all of the different forms of habitat destruction, it is important to note that most are clearly the product of human actions and hence controllable.

B. Harvest

Pacific and Atlantic salmon have been the target of harvest by the indigenous peoples of North America for many thousands of years. However, it was not until the early- to mid-1800s that the Euro-American settlers (with their advanced methods of capture and preservation) began to so overfish salmon that extinction became a possibility. Particularly destructive have been offshore fisheries than indiscriminately harvest mixtures of plentiful hatchery fish and dramatically less abundant wild fish. Although harvest is currently regulated to some extent, it still continues at substantial rates, even on endangered and threatened ESUs. For example, fall chinook salmon from the Snake River are currently harvested at an annual rate that exceeds 20%, and some listed ESUs in the Puget Sound area are harvested at annual rates that exceed 40%. Fish taken by harvest represent individuals with maximum reproductive value and the impact of their mortality prior to spawning is unarguably large. It may be that the models guiding our harvest policy are themselves flawed, as suggested by recent theoretical papers which suggest that harvest models built with the concept of an optimum percentage take and maximum sustained yield often fail in highly variable environments (Lande *et al.*, 1997; Ludwig, 1998). The good news is that simple models suggest that threatened ESUs such as Snake River fall chinook salmon may be readily rebuilt by temporarily curtailing harvest—indeed with only a 10-year moratorium populations may recovery sufficiently to then sustain a vigorous harvest that annually yields double the number of fish currently removed from the river (Fig. 2). Certainly, the issue of harvest for these threatened stocks needs to be reevaluated—for perspective, it is hard to imagine anyone tolerating the hunting of threatened whales at a rate leading to 20% adult mortality per year, given the status of whales as listed under the Endangered Species Act.

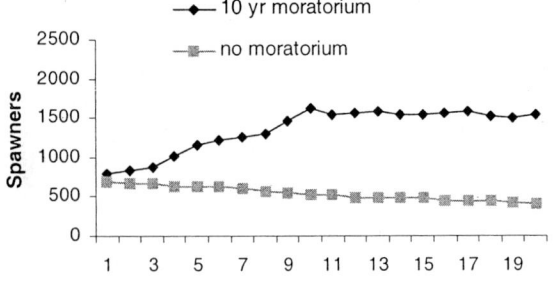
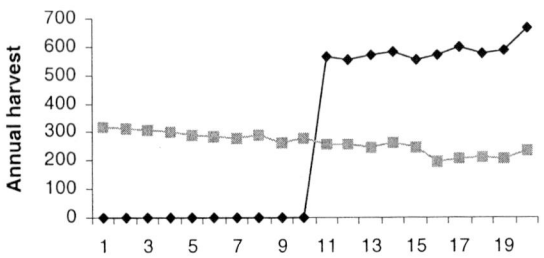
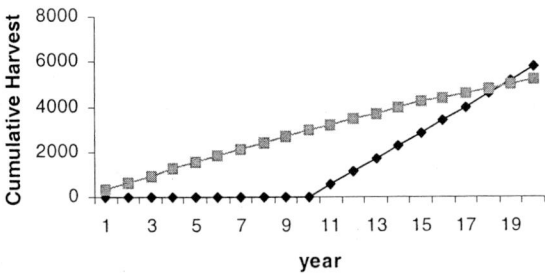

FIGURE 2 Analysis of the benefits of a temporary moratorium on harvest for Snake River fall chinook salmon. A simple age-structured model iterates salmon populations forward 20 years, using estimates of variability and recruitment from 1980 onward. Two alternative simulations are run: simply continuing to harvest at the current rate versus halting harvest for 10 years and then reinitiating harvest for years 11 through 20, but at 75% of the current rate. Note that if there is a 10-year moratorium followed by harvest at 75% of the current rate, the population rebuilds and total cumulative harvest after 20 years (even including 10 years without any harvest) ends up being higher than if current practices are continued. One may be able to have fish and harvest, if harvest is temporarily halted.

C. Hatcheries

The building of hatcheries and the production and release of millions of hatchery salmonids in the Pacific Northwest closely tracked the wholesale destruction of habitat and overfishing. Indeed, hatcheries were widely viewed as the solution—in fact, a solution that allowed the extraction of natural resources to proceed unchecked. It was not until the past several decades that biologists began recognizing that such use of hatcheries were, by themselves, a potential problem. Beyond the obvious creation of harvest opportunities and destructive mixed-stock fishing, there is now much greater recognition that the release of large numbers of hatchery fish can have detrimental genetic and ecological impacts on natural populations. Interbreeding of hatchery and wild fish can reduce the fitness of wild fish through the disruption of local adaptation; and increased interactions between hatchery and wild fish can lead to competition for food and space, as well as the spread of infectious diseases.

D. Hydropower

Dams block access to spawning habitats, create reservoirs that flood spawning habitats, and may stress fish migrating through them to such a large extent that females cannot replace themselves due to depressed physiological fitness. Dams affect both upstream adult spawning migrations and downstream juvenile migrations. Primary mechanisms for the negative impacts of dams involve increased mortality of juveniles associated with passage through turbines, atmospheric gas supersaturation, and decreased flow and turbidity. On the other hand, engineering improvements in dams can mitigate some of these effects, although certainly not to the extent that rivers return to the equivalent of their free-flowing natural state. Because dams are the most visible impediment to salmon survival, getting fish around, over, or through dams has been the focus of much research. Unfortunately, the understandable focus on hydropower as a risk factor for salmon has left the "other H's" less well studied. Recovery of salmon is likely to often require more than simply removing dams—we may also need improvements in spawning habitats and in ocean and estuarine environments, further reductions in harvest, and perhaps major modifications of hatchery programs. Finally, it is worth pointing out that although removal of dams is often championed in terms of "returning to a normative river" (ISG, 1999), most proposed dam removals still leave many dams untouched and our rivers so altered by other factors that they remain a long way away from the ideal of a free-flowing "wild river."

E. Natural Environmental Variability and Climate Change

Although the tendency in recent years has been for resource managers to focus on the human-caused factors that have contributed to the decline of salmon, natural variation in environmental conditions, both freshwater and marine, are also major factors affecting the abundance and distribution of salmon. Operating on seasonal, annual, and decade-long timescales, events such as regime shifts, El Ninos, and changes in seasonal patterns of coastal circulation (particularly upwelling) all can have a profound effect on oceanic survival of salmon. Particularly noteworthy has been the research documenting correlations among broad patterns of productivity, and major physical features of the ocean-atmosphere interaction. For example, Hare *et al.* (1999) found that harvest (and presumably abundance) of Pacific salmon in Alaska has varied inversely with catches along the U.S. West Coast over the past 70 years. In exploring various explanations for this observation, they found a striking correlation between switching pattern of north-south abundance and the Pacific Decadal Oscillation (PDO; a recurring pattern of pan-Pacific atmosphere-ocean variability). While there are several ocean-atmosphere indices such as the PDO, they all, in concept, attempt to capture the key physical forces that drive decadal-scale basin-wide climate variability. For example, since the last major climate shift of 1976–1977 in the North Pacific Ocean, oceanographers have documented decreased mixed-layer depths in the Subarctic Gyre and increased sea surface temperatures along the west coast. The biological response to these physical changes were profound, including major changes in primary and secondary production, and micro- and mesoscale changes in zooplankton biomass and species composition. In addition to salmon, a large number of higher trophic level fish, birds, and mammals show similarly dramatic shifts in abundance and distribution coincident with these changes in ocean conditions.

Operating on an even grander scale, the impacts of long-term climate change will undoubtedly have profound consequences for all animal and plant species. Salmon, because of their anadromous nature, will suffer the dubious distinction of being subject to impacts in both the marine and freshwater habitats—a distinction that may well challenge their ability to adapt and survive over much of their current range. Of the many studies designed to predict the consequences of climate change on aquatic species, one of the more dramatic is the recent work of Welch *et al.* (1998). To explore potential marine impacts on salmon, Welch and colleagues compiled an extensive data set on distribution of salmon at sea and found that, for sockeye salmon, marine distribution was limited by sharp thermal boundaries. Using a climate model, and assuming a doubling of greenhouse gases over the next 50 years (which is consistent with the current rate of input), it was predicted that sockeye salmon would be excluded from most of the Pacific Ocean and restricted to a relatively small area of the Subarctic Pacific for much of the year. Similar analyses revealed marked reduced ocean habitat for other species of Pacific salmon as well. Climate change is also expected to depress oceanic productivity by disrupting upwelling patterns in a way that reduces the delivery of nutrients (Bakun, 1990). To the extent that salmonid population fluctuations are determined by bottom-up forces, the possibility of reduced nutrient upwelling could have dire consequences for these and many other fish species.

VI. WHAT IS THE VALUE OF SALMONID BIODIVERSITY AND CAN WE RESCUE SALMON FROM EXTINCTION?

Two arguments dominate the mainstream ecological literature regarding the "value of biodiversity": (a) diversity fosters enhanced productivity by allowing a more complete exploitation of the environment via a variety of different forms, each adapted to different habitats or resources, and (b) diversity fosters enhanced stability by spreading the risk and providing redundancy in the face of unpredictable catastrophes. Both hypotheses regarding the value of biodiversity apply well to salmon. First, different salmonid ESUs spawn in different portions of the freshwater environment (inland versus coastal, mainstream rivers versus streams, lakes versus rivers, and so forth) and migrate to different regions of the ocean. Any loss of salmonid diversity would mean that the productivity of these ecosystems would be reduced, with minimal likelihood that any other species could take the place of salmon (since the anadromous lifestyle is so unique). Second, due to climatic fluctuations, different months of the year will be favorable for migration; in fact, recent data indicate that even a shift of a few days with respect to the time at which juvenile chinook salmon arrive at the ocean can alter survival rates by a factor of two. If salmon diversity is reduced, then there will be less variety in run timing and life history schedule and consequently much less insurance against the vagaries of climatic fluctuations. The precipitous decline of Pacific and Atlantic salmon throughout much of their range raises a number of important questions: Can we save the salmon? Should we save the salmon? And what do we need to do to save the salmon?

A. Can We Save the Salmon?

The answer to this question is an unequivocal yes. A species complex that has survived and flourished under the wide range of conditions from which salmon have evolved has almost certainly weathered numerous survival bottlenecks in the past, caused by environmental perturbations ranging from volcanic eruptions to the total loss of huge blocks of habitat to glaciers, fire, and floods. Perhaps the key attribute that has allowed the salmonids to survive these events is the tremendous diversity of life histories and patterns of habitat use. The salmonids represent a plastic species complex with a robust evolutionary potential that should facilitate survival under diverse conditions.

That said, many populations in the Pacific Northwest and Northeast North America are already extinct, and many others have never been at lower levels of abundance. Saving Pacific and Atlantic salmon will clearly require immediate action that will include economic sacrifice, societal discipline, and a commitment to a science-based recovery strategy—all at a very significant cost to the people of the affected regions.

B. Should We Save the Salmon?

Because of the societal implications, this question perhaps extends well beyond the boundaries of what should be the subject of this volume. However, from strictly an ecological perspective, the salmonids are a vital component of the North Pacific and North Atlantic ecosystems and the associated coastal habitats. The ecological importance of salmonids as predators, prey, and nutrient recyclers is obvious to even the most casual observers of nature. While as scientists we can marvel at the complexity of an ecosystem and strive to understand species and environmental interactions in the broadest sense, we must also acknowledge that such complex and dynamic systems will never be fully understood. Nonetheless, the loss of such a key species complex as salmonids over such a broad geographic area would no doubt be a biological event without precedent.

C. What Do We Need to Do to Save the Salmon?

As noted earlier, the keys to solving the salmon problem are societal commitment, discipline, and a science-based recovery strategy. While societal commitment and discipline are not scientific issues, they certainly are not devoid of science. To make a decision on whether and how to proceed, *society* must have knowledge—knowledge of the complexity of the problem and the certainty (or lack thereof) associated with various courses of action. This knowledge is vital to the economic decision-making process.

In contrast, the design of recovery strategies is largely a science issue. At the core of every science-based recovery strategy should be an analytical framework that allows (a) the methodical determination of survivorship in each of the life history stanzas of salmon, (b) the identification of actions and opportunities to increase survivorship, (c) the assessment of biological (and societal) feasibility of each action or suite of actions, and (d) a rigorous monitoring and evaluation plan. Whatever models are used should be clear and understandable to the average citizen, and the data used as model inputs must be available to all.

Clearly, society cannot continue the extraction-based economy of the past 200 years and expect even remnant populations of salmon to exist over even a fraction of their former range. A major shift in environmental decision making will be required. In particular, we must abandon the habits of the past two centuries, which have involved almost universally approaching conservation through a "reduction from the status quo" perspective. For example, rather than arguing about whether we should reduce harvest by 10, 20, or 50% from current levels, any harvest above 0% should be viewed with great skepticism. Likewise, the issue should be whether *any* timber harvest is allowed in salmon watersheds, not whether the current acreage of timber harvest should be reduced. When populations face a high probability of extinction, modest adjustments of the status quo are not enough. Unless society seriously acts to rebuild salmon populations, the future of salmon is not assured, and moderate actions ought to be recognized as optimistic gambles (as opposed to reasonable moderation).

See Also the Following Articles

AQUACULTURE • ENVIRONMENTAL IMPACT, CONCEPT AND MEASUREMENT OF • FISH, BIODIVERSITY OF • MIGRATION • RESOURCE EXPLOITATION, FISHERIES

Bibliography

Bakun, A. (1990). Global climate change and intensification of coastal ocean upwelling. *Science* **247**, 198–201.

Behnke, R. J. (1992). *Native Trout of Western North America.* American Fisheries Society Monograph 6. American Fisheries Society, Bethesda, MD.

Groot, C., and Margolis, L. (Eds.) (1991). *Pacific Salmon Life Histories.* University of British Columbia Press, Vancouver, British Columbia.

Groot, C, Margolis, L., and Clarke, W. C. (1995). *Physiological Ecology of Salmon*. University of British Columbia Press, Vancouver, British Columbia.

Hare, S. R., Mantua, N. J., and Francis, R. C. (1999). Inverse production regimes: Alaska and West Coast Pacific salmon. *Fisheries* 24, 6–14.

Independent Science Group (ISG). (1999). Scientific issues in the restoration of salmonid fishes in the Columbia River. *Fisheries* 24, 10–19.

Lande, R., Saether, B., and Engen, S. (1997). Threshold harvesting for sustainability of fluctuating resources. *Ecology* 78, 1341–1350.

Lichatowich, J. (1999) *Salmon without Rivers*. Island Press, Washington, D.C.

Ludwig, D. Management of stocks that may collapse. (1998). *Oikos* 83, 397–402.

Moyle, P. B., and Cech, J. J.,Jr. (1996). *Fishes: An Introduction to Ichthyology*. Prentice-Hall, Englewood Cliffs, NJ.

National Research Council. (1995). *Science and the Endangered Species Act*. National Academy Press, Washington, D.C.

National Research Council. (1996). *Upstream: Salmon and Society in the Pacific Northwest*. National Academy Press, Washington, D.C.

Netboy, A. (1974) *The Salmon: Their Fight for Survival*. Houghton Mifflin, Boston.

Quinn, T. P. (1982). A model for salmon navigation on the high seas. In *Proceedings of the Salmon and Trout Migratory Behavior Symposium* (E. L. Brannon and E. O. Salo, Eds.), pp 229–237. School of Fisheries, University of Washington, Seattle, WA.

Shearer, W. M. (1992). *The Atlantic Salmon*. Fishing News Books. Blackwell Scientific Publications, Oxford, England.

Trotter, P. C. (1989). Coastal cutthroat trout: A life history compendium. *Transactions of the American Fisheries Society* 118, 463–473.

Waples, R. (1995). Evolutionary significant units and the conservation of biological diversity under the Endangered Species Act. *Amer. Fish. Soc. Symp.* 17, 8–27.

Welch, D. W., Ishida, Y., and Nagasawa, K. (1998). Thermal limits and ocean migrations of sockeye salmon (*Oncorhynchus nerka*): Long-term consequences of global warming. *Canadian Journal of Fisheries and Aquatic Sciences* 55, 937–948.

SCALE, CONCEPT AND EFFECTS OF

David Clayton Schneider
Memorial University of Newfoundland

I. Biodiversity as a Measurable Quantity
II. Concept of Scale
III. Effects of Scale

GLOSSARY

biodiversity Includes taxonomic diversity along with habitat and genetic diversity.
dimension Set of measurement units that are completely similar. Commonly used dimensions are length (inches, meters, etc.), time (seconds, years, etc.), and mass (grams, pound-mass, etc.).
diversity distribution A frequency distribution showing the number of classes (usually species) that have one, two, three, etc. individuals per class (per species).
diversity index A single number used to characterize a diversity distribution. The most common is s, the number of classes (species) into which a collection has been sorted.
fractal Length dimension with exponent other than L^1 (length), or L^2 (area), or L^3 (volume).
scale Unit measure (resolution, inner scale) relative to largest multiple (range, outer scale).
scope Ratio of largest multiple to unit measure.

IT IS A FAMILIAR experience that taxonomic diversity increases as more area is examined, as more time is spent watching an area, as more organisms are collected, and as a wider range of body sizes are accumulated. These increases in diversity do not scale directly with area, time, collection size, or body size. More than 50 years ago, C. B. Williams found that, on average, a 10-fold increase in area will increase the number of plant species by a factor of 2 rather than 10. Williams ascribed divergence from a 1:1 scaling to the vagaries of sampling (at the scale of plots), to habitat diversity (at the scale of biogeographic regions), and to evolutionary history (at the scale of continents). Subsequent work has confirmed this empirical scaling, developed a wealth of empirical scalings by area, developed alternatives to scaling by area, and tested theoretical explanations for these scaling relations. Biodiversity now includes genetic and habitat diversity in addition to taxonomic diversity. This article presents the quantitative basis for scaling of biodiversity based on rapid development of scaling theory in ecology in the past decade.

I. BIODIVERSITY AS A MEASURABLE QUANTITY

A. Scalable Quantities

Biodiversity, no matter how it is measured, is a quantity with units on one of several types of measurement scale.

Consequently, some care is needed in defining the measure of diversity at hand to obtain accurate scaling relations with area or time.

A well-defined measure of biodiversity has five parts: (i) a name, (ii) a symbol, (iii) a procedural statement that sets forth the method and conditions for measurement, (iv) a set of numbers generated by that procedural statement, and (v) units on one of several types of measurement scales. A scaled quantity is conveniently represented as a symbol equated to a set of numbers arranged inside brackets, multiplied by the unit of measurement. An example is the number of cichlid fish species found in each of six African lakes, as reported by Ricklefs and Schluter (1993):

Procedural statement	Name	Symbol	Numbers	× Units
Ricklefs and Schluter (1993), p. 358	Species richness	$S_j =$	$\begin{bmatrix} 200 \\ 136 \\ 200 \\ 7 \\ 9 \\ 40 \end{bmatrix}$	× # lake^{-1}

In this example the lakes are listed in order from large (Lake Victoria = 69,484 km^2) to small (Lake Edward = 2150 km^2).

Measurements occur on one of one types of scale: nominal (yes/no), ordinal (ranks), interval (arbitrary zero point such as longitude in degrees), and ratio (true zero such as distance from the equator in kilometers). The mathematical rules for working with a ratio scale quantity (such as number of species per island) differ from those for working with quantities on other types of scales (such as ranking of lake area from large to small in the previous example). If our interest is in scale and its effects, then we need to work with the species-abundance distribution (on a ratio type of scale) rather than the rank-abundance distribution (on a rank type of scale).

B. Taxonomic Diversity Distributions

Diversity studies typically employ three basic quantities: N is the number of organisms, A is the area, and T is time or duration of measurement. A collection of N organisms from a unit area A_o during a unit period T_o is sorted into s groups (typically species) as shown in Fig. 1. Any taxonomic level can be used in sorting, but for the sake of clarity species will be used to illustrate taxonomic diversity distributions. The number of organisms n_i is recorded for each species. A familiar characterization of this taxonomic diversity is the curve of abundance across species, ranked from high (largest n_i) to low (smallest n_i, usually one organism per species). The information in the rank-abundance curve can be re-expressed (see Fig. 1) as a frequency distribution $s(n_i = n_k)$. The symbol $s(n_i = n_k)$ represents the number of species at abundances ranging from $n_k = 1$ upward to the abundance n_k of the most common species in the collection. This frequency distribution shows, in full detail, the information on species diversity in a single collection of size $\Sigma n_i = N$, taken from unit area A_o in unit time T_o. If our interest is in the effects of scaling, then we will need to work with the frequency distribution $s(n_i = n_k)$, which is on a ratio type of scale, rather than with the rank abundance curve (Fig. 1), which is on an ordinal type of scale.

Following practice in the literature on diversity, we can reduce the information in the species abundance distribution $s(n_i = n_k)$ to an index with a single value. The most common index is the total number of species s, obtained by counting classes or equivalently by summing the distribution $s(n_i = n_k)$ across abundance classes n_k. We might also compute Shannon Weaver diversity or the Simpson index for the collection. Both go beyond gross measures of diversity such as species number s, but fall short of describing all the information in the species diversity distribution $s(n_i = n_k)$ for the collection N.

If we obtain a second collection, again from an area A_o during a period T_o, we expect a somewhat different number of organisms $n_j = 2$, a new set of species $s_{j=2}$ (many the same as in the first collection), and a new diversity distribution $s_2(n_i = n_k)$. With several collections $s_j(n)$, now labeled $j = 1$ through jt, we have a fuller characterization of diversity. We can construct the combined distribution $S(n_i = n_k)$ across some number of collections, jt. The total number of species S across the combined collection will exceed the number of species s_j in any one collection.

If the collections were always drawn by the same procedural statement, they can be viewed as samples from a set of all possible collections having unit area A_o and duration T_o. The diversity distribution for this hypothetical set is $S(n_i = n_k)$. This is the species abundance distribution for the community, provided all collection sites have an equal (or at least known) chance of appearing in the sample. The total number of species S and total number of organisms N are computed respectively as $S = \Sigma s(n_i = n_k)$, $N_{tot} = \Sigma n_i s(n_i)$. Integral signs ($\int$) replace the summation signs for distributions based on continuous values of population size n_k (May,

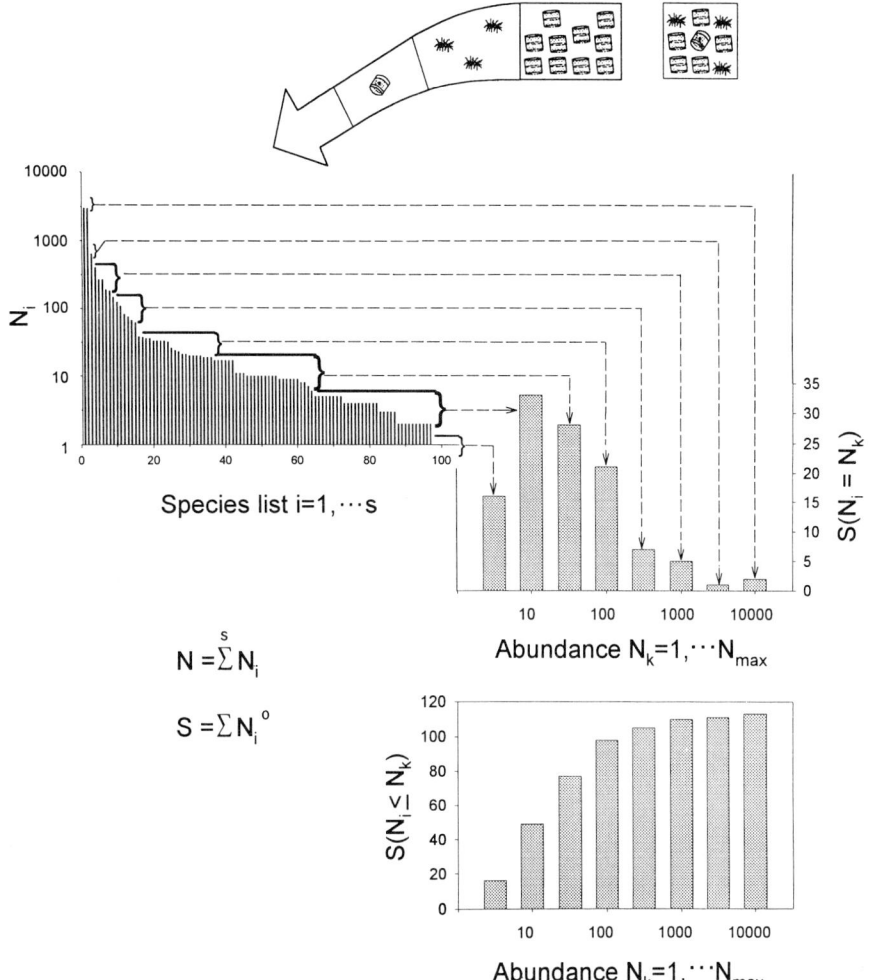

FIGURE 1 Collection of organisms sorted into a rank-abundance curve. Arrows show how the taxonomic diversity distribution $S(N_i = N_k)$ is constructed from the rank-abundance curve. The cumulative diversity distribution $S(N_i \leq N_k)$ is then constructed from the diversity distribution. N is the number of organisms in the collection, and S is the number of species.

1975). Following May (1975) we can compare an observed distribution $s(n_i = n_k)$ to a mathematical function, allowing us to compute species numbers at unobserved values of abundance n_k. An impressive number of methods (Bunge and Fitzpatrick, 1993) have been devised to estimate S from several collections $s_j(n_i = n_k)$, taken to be samples.

When making statistical estimates from taxonomic distributions, it often proves convenient to use the cumulative distribution $s(n_i \leq n_k)$ rather than the distribution $s(n_i = n_k)$. The cumulative distribution rises monotonically from left to right, as it records the number of species with n_k or fewer organisms per species, rather than recording the number of species at each value of n_k. Both distributions express exactly the same information. The cumulative distribution is less lumpy and therefore easier to use.

C. Other Diversity Distributions

Genetic variability within a species can be characterized in the same fashion as taxonomic diversity. For a collection of N individuals the usual single-locus measure is the proportional presence of each type of allele $q_i = n_i/2N$. This information can be expressed as a rank-abundance distribution of allelic proportions from common (q_i large) to rare (q_i small). For convenience, the information in this distribution is reduced to an index such as the number of classes (alleles) or expected heterozygosity ($H_e = 1 - \Sigma q_i^2$). The information in several collections can be expressed as a rank-abundance curve for each collection and a rank-abundance curve for the combined collection. This information can in turn be reduced to an index, such as F_{st}, which compares expected heterozygosity for the entire collection H_e to the average over k collections ($\bar{H}_e = k^{-1} \Sigma H_e$). This approach, applied to genetic diversity at the single-locus level, can be extended to coarser classifications, such as recognizable genotypes in a population (Mallet as cited in Gaston, 1996, p. 16). If we are interested in the effects of scale on genetic diversity, then the logic of working with ratio scale quantities will compel us to take the (highly unusual) step of re-expressing the rank-abundance curve as the genetic diversity distribution $q(n_i = n_k)$, which is on a ratio type of scale. This is the frequency of genotypic classes for which there was one individual ($n_k = 1$), two individuals ($n_k = 2$), etc. This can be converted to a cumulative frequency distribution $q(n_i \leq n_k)$ to again smooth away the inconvenient lumpiness of the distribution $q(n_i = n_k)$.

We can construct a diversity distribution to summarize habitat diversity. This requires defining a collection area A divided into N units of size A_o, which are sorted into habitat classes. From this we can construct the usual rank-abundance curve, running from the most common habitat (n_i large) to rarest (n_i small). The same information can be plotted on a ratio scale as a habitat diversity distribution $h(n_i = n_k)$, expressing the frequency of habitats found in one unit ($n_k = 1$), in two units ($n_k = 2$), etc.

D. Alpha, Beta, and Gamma Diversity

In 1960, R. H. Whittaker distinguished taxonomic diversity within a habitat (alpha diversity) from diversity within larger geographic units composed of several habitats (gamma diversity). Examples of these larger units include islands, lakes, landscapes, and coastal embayments. Whittaker defined beta diversity as the among-habitat diversity, or diversity change going from a single habitat to the larger geographic unit. Later (Box 1) Whittaker added two more spatial scales or "levels"—that of point diversity within a habitat and that of regional (epsilon) diversity for broader geographic units containing smaller units, each characterized by its own gamma diversity. Whittaker presented these as convenient labels, recognizing that point, alpha, gamma, and epsilon diversity were arbitrary points on a continuum, and that the size of these units might be defined differently in any given study. Starting with Whittaker, the

---

Box 1

Alpha, Beta, and Gamma Diversity

In an evolutionary context, Whittaker (1977) defined inventory diversity at four spatial scales or levels: the point sample, the habitat, the landscape, and the region. Differentiation diversity is defined as the change in diversity between these four levels.

Point diversity The number of species for a small or microhabitat sample within a community regarded as homogeneous. Also called internal alpha or subsample diversity.

Pattern diversity Change going from one point to another within a habitat. Also called internal beta diversity.

Within-habitat or alpha (α) diversity The number of species in a sample representing a community regarded as homogeneous (despite its internal pattern). It measures the number of potentially interacting species.

Between-habitat or beta (β) diversity Change along an environmental gradient or among the different communities of a landscape. Developed by Whittaker (1960) as a measure of packing of competing species along a gradient, it was defined as the ratio of regional (gamma) diversity to average within-habitat (alpha) diversity. Wilson and Shmida (1984) review other measures of beta diversity.

Landscape or gamma (γ) diversity The number of species in a set of samples including more than one kind of community.

Geographic differentiation or delta (δ) diversity Change along a climatic gradient or between geographic areas.

Regional or epsilon (ε) diversity The number of species in a broad geographic area including differing landscapes.

---

most effective use of these labels has been in distinguishing diversity patterns at the scale of habitats from patterns at larger scale. The labels are also useful in discussing the very different processes responsible for changes in diversity at habitat, regional, and global scales. Alpha, beta, and gamma diversity have remained in wide use during the past 40 years; point and epsilon diversity did not come in to wide use. In practice, gamma diversity is sometimes dropped in favor of two

Box 2

	Evolutionary biology	Terrestrial ecology		Marine ecology	
	Whittaker (1977, Fig. 14)	Delcourt and Delcourt (1988)		Haury et al. (1978)	Steele (1991)
	Area	Area	Time	Area	Time
-10^{15}					
-10^{12}		Mega	$10^6 +$ years	Mega	
		Macro	10^4–10^6 years	Macro Meso	1–10 years
-10^9	—ε—				
	δ	Meso	500–10,000 years	Coarse	Days–years
-10^6	—γ—				
-10^3	β				
	—α—	Micro	1–500 years	Fine	1–10 days
$-1\ m^2$	pattern				
	—point—				
-10^{-3}				Micro	

labels—within-habitat (alpha) and among-habitat (beta) diversity.

It is of note that similar sets of arbitrary labels to distinguish pattern and process over a range of spatial scales have been developed in both marine (Haury et al., 1978; Box 2) and terrestrial ecology (Delcourt and Delcourt, 1988; Box 2). Habitat features persist longer in terrestrial than marine systems, and hence the time-scales associated with any given spatial scale are greater for terrestrial than marine systems (Steele, 1991; Box 2).

II. CONCEPT OF SCALE

A. Multiscale Analysis

Multiscale questions naturally arise for biodiversity: How does diversity change with collection size? How does it change with area? How does it change with duration of collection? These questions can be addressed quantitatively by examining the change in a biotic diversity distribution with change in collection size N, in collection area A, and in collection duration T. For habitat diversity similar questions of scale arise: How does the habitat diversity distribution $h(n_i = n_k)$ change when the collection area A is doubled? As with taxonomic diversity, we expect genetic and habitat diversity to be greater in large collections than in small collections, greater in widely separated samples than in closely placed samples, and greater in large areas than in small areas. We do not expect that a doubling in area will simply double the number of habitats or the number of alleles.

There are several ways to compare diversity at large and small scales. One of the more venerable is to rate (or plot) diversity relative to island size (Fig. 2). Rating diversity against area is equivalent to rating against collection size N if organism density is the same on large and small islands. Another venerable method is to make a series of collections and then plot a measure of diversity against effort, measured as the accumulated number of collections (Fig. 2). These "collector's curves" are equivalent to plots against accumulated collection size N if organism catch per unit effort remains constant across collections. Figure 3 shows the increase in diatom species number with increase in effort (number of slides). A third method, currently popular but less venerable, is to fix the unit size, measure the separation between units (in time or space), and then plot some measure of difference in diversity between units as a function of separation or lag (Fig. 2). An additional method is to construct a diversity distribution from a large area and then take the average diversity distribution for each half of the area, each quarter, etc. This method of coarse graining (Fig. 2) is better known in

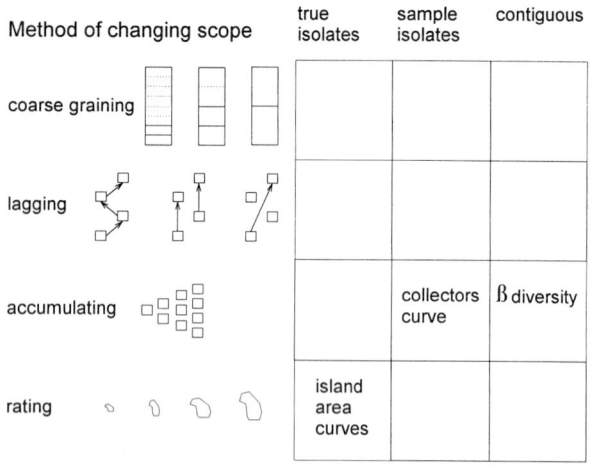

FIGURE 2 Scaling methods change the spatial scope by one of several different methods, explained in the text. For biodiversity, the traditional applications are collector's curves (increasing the number of samples), island area curves (islands rated from small to large), and beta (β) diversity (increase in number with the number of contiguous samples along a transect).

geophysics (Rodriguez-Iturbe and Rinaldo, 1997; Barenblatt, 1996) than in biology. Coarse graining expresses the same information as lagging, an equivalence that goes unrecognized. Both lagging and coarse graining assume that, on average, organism number N does not depend on distance separating samples, and hence diversity differences are due to spatial scale (separation) rather than spatial gradients in organism number N.

These various methods of multiscale comparison express the same information in principle but not in practice. Lagging and coarse graining (in space) express the same information if the spatial layout is the same. In practice, somewhat different information is obtained because coarse graining is applied to contiguous units, whereas lagging is applied to separated units. Collector's curves introduce time-scales. In a seasonal environment, collections made monthly for a year are expected to grow more rapidly in species number than collections made annually for a decade. Rating relative to area introduces another factor—the dynamics of population exchange within species. A plot or rating of diversity against area will differ in scope for evolutionary isolates (continents), ecological isolates (islands), partial isolates (islands in an archipelago), and quadrat samples from a larger area (Rosenzweig, 1995).

B. Scope

In comparing diversity within or among studies we would like to have a measure that does not depend on choice of units. To accomplish this, the scope of a quantity is defined as the ratio of its maximum to minimum value. For the cichlid example, the scope in species number is (200 species lake^{-1})/(7 species lake^{-1}) = 29. For non-cichlid species the scope in number per lake is different: (111 species lake^{-1})/(17 species lake^{-1}) = 6. The scope of a quantity has units that cancel and hence are expressed as dimensionless numbers, independent of units of mass, length, or time. However, the units that form the dimensionless ratio should be kept in view to avoid drawing conclusions outside the scope of measurement.

Care is required in comparing quantities with scopes that differ procedurally. One can increase the scope of collection size by increasing area from unit area A_o to total area A_{tot}. The scope is $\Sigma A/A_o$. This generates accumulation curves relative to area: accumulation of organisms $N(A)$, of species $S(A)$, of genotypes $Q(A)$, and of habitats $H(A)$. One can increase the scope of collection size by repetitive collection at one location, resulting in accumulation curves relative to time: $N(T)$, $S(T)$, $Q(T)$, and $H(T)$. There is no reason to expect that change in diversity relative to change in spatial scope will match change in diversity relative to change in temporal scope.

Spatial scope can be expanded in several ways. Searching out a wide range of isolated units—islands in an ocean and lakes on a landscape—will increase the scope. With this method the scope is the ratio of the largest to smallest spatial unit, $A_{max}A_{min}^{-1}$. For the lake example, the spatial scope is (69,484 km^2 lake^{-1})/(2150 km^2 lake^{-1}) = 32. Accruing units will increase spatial scope if these are distributed over an area. Accrual can be by adjacency (along a transect or concentric rings around a point) or by random sampling within a larger area (in statistical jargon, within the frame of all possible units). With accrual, the scope is $A_o^{-1} \Sigma A$ = number of units. Lagging can apply to contiguous units; the scope is $(L_{max} - L_{min})/L_{min}$, where L_{max} is the distance across the block or along the transect, and L_{min} is the distance across the unit block. Fine graining increases scope via an increase in the frequency of measurement. With fine graining, the scope is A_{tot}/A_{min}, where A_{min} is the smallest resolvable area within a large block A_{tot}.

Temporal scope can be varied as well. In principle, one can compare short time periods with long periods, but there are no examples of this. Temporal scope increases as units of equal duration T_o accrue; the scope is the ratio of the time series to the minimum unit: T_o^{-1}

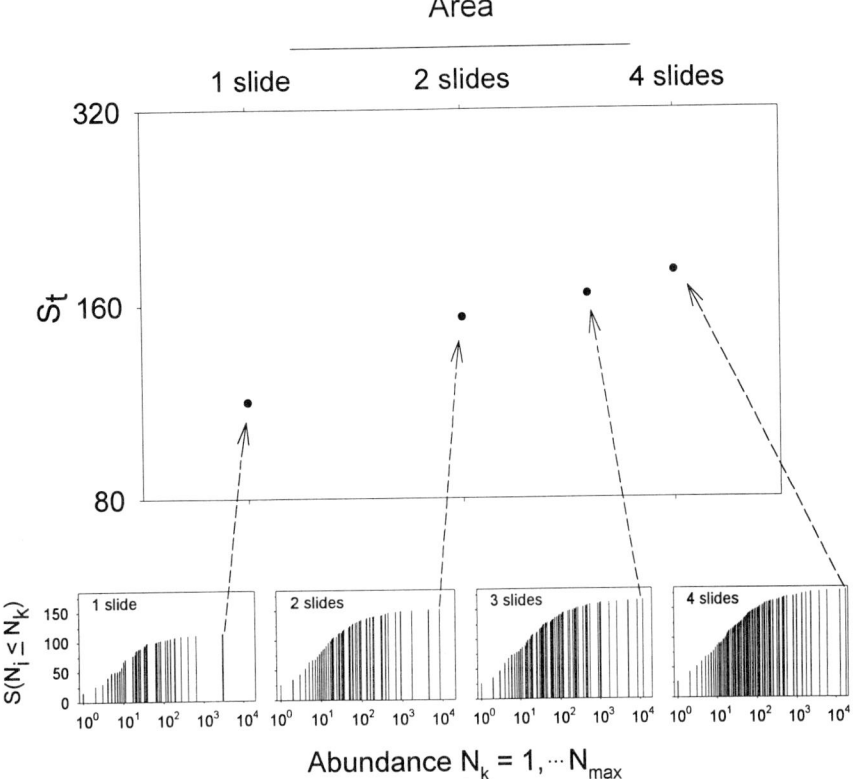

FIGURE 3 Cumulative diversity distributions from areas of increasing size (one, two, three, and four slides). Note the linear relation (on a logarithmic scale) between species number S and slide area. However, species number S represents only part of the information in the cumulative diversity distributions from areas of increasing size. Data from Patrick (1968).

ΣT = number of units. Temporal scope also increases by fine graining down to the smallest time units T_o within a continuous period T. The scope is T/T_o.

C. Complete and Incomplete Similarity

The classical solution to the problem of change in scale has been to define two systems (such as a model boat and full-scale prototype), define groups of similar measurement units (typically mass, length, and time), and then work out the scaling from model to prototype with reference to groups of similar units called dimensions mass $[M]$, time $[T]$, Euclidean lengths $[L]$, areas $[L^2]$, and volumes $[L^3]$. If the model and prototype are completely similar (Barenblatt, 1996) then we expect that a 10-fold increase in length of prototype relative to model will result in a 10^3-fold increase in volume.

$$\frac{\text{Volume}_{\text{prototype}}}{\text{Volume}_{\text{model}}} = \left(\frac{\text{Length}_{\text{prototype}}}{\text{Length}_{\text{model}}}\right)^3 \quad (1)$$

We also expect a 1000-fold increase in mass if model and prototype are constructed of the same material and have the same density. The scaling relation is mass $\cong$ volume $\cong$ length^3. Dimensional analysis, based on the principle of complete similarity, has been notably successful in engineering and several areas of physics, including geophysical fluid dynamics. It has been notably unsuccessful with natural rather than engineered objects, or with dynamics that "lurch" rather than ticking along in a clock-like fashion.

In ecology, the failure of complete similarity was addressed by comparative methods of "hierarchy theory," which recognizes that large-scale spatial structure or dynamics cannot be computed directly from local scale structure or rates. Using Herbert Simon's concept of "nearly decomposable levels," the hierarchical method describes structure or computes dynamics at two (or more) levels and then compares the results. For example, in the biodiversity literature it is customary to distinguish local or within-habitat diversity from regional (among-habitat) diversity. One widely recognized finding (Holt as cited in Ricklefs and Schluter, 1993, p. 77) is that the relation of species number to area at the habitat level (i.e., within a habitat) does not match that at the ecosystem level (i.e., among larger areas having multiple habitats).

When completely similar scaling fails, a property (e.g., coastline length) at a large scale can often be computed from the average property at a smaller scale by a power law expressing incomplete similarity (Barenblatt, 1996). The most familiar example of incomplete similarity is that of a fractal relation between coastline length and Euclidean length L. If we take large steps along a coastline and then count these as if on a straight line, we obtain a coastline perimeter of length $P(L_{large})$. If we take small steps and then count these as if in a line, we obtain a length $P(L_{small})$ that exceeds $P(L_{large})$. The relation of the two measures of coastline length is

$$\frac{P(L_{large})}{P(L_{small})} = \left(\frac{L_{large}}{L_{small}}\right)^\alpha \quad (2)$$

In briefer form, $P(L) = c \times L^\alpha$. In even briefer form, $P(L) \cong L^\alpha$. The exponent α is not an integer, as it would be for completely similar Euclidean objects. Mandelbrot (1977) coined the term "fractal" to describe non-Euclidean objects where α is a non-integral constant.

More generally, incomplete similarity relates some property $S(x)$ to some measure x according to a power law:

$$S(x) = cx^\alpha \quad (3)$$

This relation states that when x is rescaled (e.g., by a factor of 2), the property $S(x)$ is still proportional to x^α. In the example of African lakes, how does species number scale with area? A rough scaling is obtained by applying the principle of homogeneity of scope (Schneider, 1994), which requires that all terms in an equation have the same scope. An exponent is applied to equate the scope of diversity with that of lake area:

$$\frac{s(A_{big})}{s(A_{small})} = \left(\frac{A_{big}}{A_{small}}\right)^\alpha \quad (4a)$$

$$\frac{200}{40} = \left(\frac{69,484}{2150}\right)^\alpha \quad (4b)$$

Solving for the exponent, we obtain the scaling for incomplete similarity

$$\alpha = \log(200/40)/\log(69484/2150) = 0.463.$$

This is a rough estimate that could be improved by using regression to include more information in the scaling. Based on the rough scaling, a doubling in lake area from, for example, 2150 km² to 4300 km² is expected to amplify cichlid species number by $2^{0.463} = 1.4$ rather than by a factor of 2.

Incomplete similarity applies to cumulative frequency distributions (Fig. 3) as well as to single indices such as species numbers. The scaling based on incomplete similarity is

$$P[X \geq x] = c \cdot x^\alpha \quad (5)$$

where x is again some measure of interest, such as length, area, or time. For some property (such as species per unit area) the probability distribution relative to the measure is

$$S(x) = c \cdot x^{-(\alpha+1)} \quad (6)$$

The exponent that quantifies incomplete similarity can be estimated by selecting isolated units, as in the cichlid example. The exponent can be estimated via autocorrelation techniques, which measure correlation of a quantity with itself at increasingly large spatial separations. Scaling exponents can be estimated by fine graining (Milne, 1997) or equivalently by coarse graining (Fig. 2). Estimates via fine or coarse graining are equivalent to those from lagging (Fig. 2), a fact that is well recognized for time series analysis but largely overlooked in the literature on spatial structure. Fine graining (or equivalently, coarse graining) has been used to quantify the fractal scaling exponent for a variety of landscape features (Rodriguez-Iturbe and Rinaldo, 1997). Fine graining (under the name of box counting or the dividers method) has been used to quantify the scaling exponent for habitats as diverse as bare soil, leaf perimeter, seaweed, and coral heads; it has been used to quantify coastline nesting habitat of eagles, pine nesting habitat of woodpeckers, and foraging habitat of rabbits (Milne, 1997).

The dynamics of systems that are incompletely similar at small and large scales are also examined via renormalization and coarse graining (Barenblatt, 1996). Renormalization has been used to investigate the dynamics of landscapes (Rodriguez-Iturbe and Rinaldo, 1997) and population interaction (Levin and Pacala as cited in Tilman and Kareiva, 1997, p. 271). Renormalization is an appropriate technique for examining the dynamics of biodiversity, which do not proceed by evenly paced transitions, but instead jump and lurch by episodes of invasion, local extinction, speciation, and anthropogenic extinction.

Incomplete similarity provides an alternative to hierarchical (usually two-level) analysis of structure and dynamics in ecology. Currently, the hierarchical approach is more familiar and remains the dominant approach when analyzing ecological interactions (Tilman and Kareiva, 1997) or biodiversity patterns at multiple scales (Ricklefs and Schluter, 1993; Rosenzweig, 1995).

D. Scaling Relations for Biodiversity

The development of scaling relations for biodiversity has been vigorous for taxonomic diversity. It has just begun for habitat diversity and does not yet exist for genetic diversity. For taxonomic diversity, the history of theoretical development can be summarized briefly as a sequence of scaling relations.

1. $S \cong A^\beta$

Species number scales incompletely with area. This is the oldest and best known scaling relation in ecology, with a rich history (Ricklefs and Schluter, 1993; Rosenzweig, 1995). The scaling coefficient was recognized to be less than 1 for most situations, but values were completely empirical and could not be generalized beyond the data used to estimate the coefficient.

2. $S \cong N^{1/4}$

Species number scales incompletely with numbers according to a ¼ power law. This relation, due to Preston, was discussed in detail by May (1975), who provided a list of models. It is worth noting that the canonical value (¼) depends on the assumption that the taxonomic diversity distribution has a lognormal form.

3. $S \cong N^{1/4} \cong A^{1/4}$

Species number scales with total numbers and hence with area according to a ¼ power law. Preston's canonical theory and MacArthur and Wilson's theory for isolated communities attribute the scaling to equilibrium conditions (May, 1975), but it is now known that scaling laws such as this can arise in non-equilibrium conditions (Barenblatt, 1996). This scaling assumes that organism number N scales in a 1:1 fashion with area.

4. $N \cong A^1$

Numbers scale completely with area. This scaling will always hold for coarse graining—we can always compute the average numbers in smaller areas from numbers in larger areas. This scaling will hold on average for accumulation if we take multiple sequences of organism counts. However, this scaling is usually untrue for any one sequence of accruals because of patchiness. Often, the rate of accrual will appear to accelerate because often the densest site will be sampled late rather than early during accrual. Complete scaling of numbers N with area A^1 is easier to assume than demonstrate for islands or other isolated units.

5. $S \cong A^\beta$, where $\beta_{island} > \beta_{block}$

The scaling exponent is steeper for islands ($0.25 < \beta_{island} < 0.33$) than for isolated blocks of land on continents ($0.13 < \beta_{block} < 0.18$). Rescue effects, where small populations are maintained by frequent migration from surrounding areas on continents, readily explain the shallow scaling for continental communities (Rosenzweig, 1995). Milne (1997) describes a correction factor to account for scaling effects in comparing the area of Euclidean blocks to the area of islands with fractal shapes.

6. $S \cong A^\beta$, where $\beta_{province} > \beta_{block}$

Biogeographic provinces, isolated at evolutionary timescales (Rosenzweig, 1995), have scaling exponents that exceed blocks within a province. Estimates of $\beta_{province}$ fall closer to unity (Rosenzweig, 1995) than to typical values for β_{block}.

7. $S(n_i \leq n_k) \cong S(A_i \leq A_k) \cong H(A_i \leq A_k)$

Species diversity, as a distribution, scales with habitat diversity, again as a distribution. Substantial qualitative support exists as correlations of species richness with several habitat variables (Wright *et al.* as cited in Ricklefs and Schluter, 1993, p. 73; Rosenzweig, 1995), including fractal measures of habitat complexity (Milne, 1997). An analytic review, similar to that of May (1975), is needed to integrate the taxonomic diversity literature with the literature on the structure and dynamics of landscapes (Rodriguez-Iturbe and Rinaldo, 1997).

III. EFFECTS OF SCALE

A. On Estimates of Diversity

Estimates of diversity and of rate of extinction are required at both local and global scales, but measurements are restricted to small scales. The principal effect of incomplete similarity on estimates of diversity and extinction rate is that it precludes intuitive scalings or extrapolations. If species number, or any other index of taxonomic diversity, scales completely with area or with number of organisms, then scaling could be handled with intuition-based 1:1 scalings. A survey could be used to estimate diversity in a larger area by knowing only the area of interest relative to the area surveyed to make the extrapolation. Extinction rates could be estimated over large areas by knowing only what percentage of the area had been examined directly. Complete similarity of taxonomic diversity with any other quantity has yet to appear in the literature, which is vast and growing (Gaston, 1996; Ricklefs and Schluter, 1993).

Incomplete similarity poses little problem if the scaling exponent is used to interpolate within the scope at which it was estimated. Extrapolation to larger scales poses a problem because of uncertainty in choice of exponents. Extrapolation becomes possible if the scaling exponent increases in a predictable way with increase in the degree of isolation (Rosenzweig, 1995). Choosing an incomplete scaling exponent, however, remains fraught with uncertainty.

Problems of non-intuitive (incomplete) scaling and uncertainty in choice of scaling exponent attend any effort to estimate change in taxonomic diversity. For habitat diversity, the problems are less acute because remote sensing can provide maps at high resolution over large areas. Consequently, scaling exponents can be estimated over a scope wide enough to be useful, from large areas of interest down to the small scale of plots and quadrats amenable to experiment and direct measurement of population and community dynamics. The number of variables that can be sensed remotely is limited. Hence, there is need for scaling relations that connect remotely sensed variables to population variables, such as density, biomass, production, movement, recruitment, and mortality. There is also a need to connect remotely sensed variables to community variables such as extinction rate.

B. On Conservation Planning

Conservation planning is carried out at far larger scales than measurements of the underlying dynamics. The problem is inescapable but not solvable by intuitive modes of thinking based on complete similarity. One cannot execute a field study, and apply it with certainty to conservation issues at larger scales, using the intuitive notion that rates (e.g., per capita mortality) are independent of spatial scale (i.e., scale as A^0). Nor can one proceed on the intuitive notion that a variable will scale completely with area on a 1:1 relation. The idea of incomplete similarity has already started to take its place in conservation planning. An example is the "single large vs several small" debate over refuge size—a debate that recognizes that doubling area does not leave rates unchanged nor does it double quantities such as the number of species. Several other examples are provided by Milne (1997). The next step will be to develop incomplete scalings that are reliable and based on biologically based theory rather than empirical coefficients. One particularly exciting topic, which may soon yield to theoretical development, is the scaling of taxonomic diversity to habitat diversity. The basis in biological theory already exists. Species lists increase by invasion or speciation, for which adaptation to habitat via natural selection is important. Species lists shrink by local or global extinction, for which change in habitat is important. There is thus the biological basis for quantitative scaling of taxonomic diversity with habitat diversity. It is a promising topic for theoretical development, for which reliable computations are needed to address urgent questions of local and planet scale change in biodiversity.

See Also the Following Article

MEASUREMENT AND ANALYSIS OF BIODIVERSITY

Bibliography

Barenblatt, G. I. (1996). *Scaling, Self-Similarity, and Intermediate Asymptotics*. Cambridge Univ. Press, Cambridge, UK.

Bunge, J., and Fitzpatrick, M. (1993). Estimating the number of species: A review. *J. Am. Stat. Assoc.* **88**, 364–373.

Delcourt and Delcourt (1988). Quaternary landscape ecology. Relevant scales in space and time. *Landscape Ecol.* **2**, 45–61.

Gaston, K. J. (Ed.) (1996). *Biodiversity*. Blackwell, Oxford.

Haury et al. (1978). Patterns and processes in the time-space scales of plankton distributions. In *Spatial Pattern in Plankton Communities* (pp. 227–327). Plenum, New York.

Mandelbrot, B. (1977). Fractals. Form, Chance, and Dimension. Freeman, San Francisco.

May, R. M. (1975). Patterns of species abundance and diversity. In *Ecology and Evolution of Communities* (M. L. Cody and J. M. Diamond, Eds.), pp. 81–120. Belknap, Cambridge, MA.

Milne, B. T. (1997). Applications of fractal geometry in wildlife biology. In *Wildlife and Landscape Ecology* (J. A. Bissonnette, Ed.), pp. 32–69. Springer-Verlag, New York.

Patrick, R. (1968). The structure of diatom communities in similar ecological conditions. *Am. Nat.* **102**, 173–183.

Ricklefs, R. E., and Schluter, D. (Eds.) (1993). *Species Diversity in Ecological Communities*. Univ. of Chicago Press, Chicago.

Rodriguez-Iturbe, I., and Rinaldo, A. (1997). *Fractal River Basins. Chance and Self-Organization*. Cambridge Univ. Press, Cambridge, UK.

Rosenzweig, M. L. (1995). *Species Diversity in Space and Time*. Cambridge Univ. Press, Cambridge, UK.

Schneider, D. C. (1994). *Quantitative Ecology. Spatial and Temporal Scaling*. Academic Press, San Diego.

Steele (1991). Can ecological theory cross the land-sea boundary? *J. Theor. Biol.* **153**, 426–436.

Tilman, D., and Kareiva, P. (1997). *Spatial Ecology*. Princeton Univ. Press, Princeton, NJ.

Whittaker (1960). Vegetation of the Siskiyou Mountains, Oregon and California. *Ecol. Monogr.* **30**, 279–338.

Whittaker (1977). Evolution of species diversity in land communities. *Evol. Biol.* **10**, 1–68.

Wilson and Shmida (1984). Measuring beta diversity with presence-absence data. *J. Ecol.* **72**, 1055–1064.

SEAGRASSES

Carlos M. Duarte
IMEDEA, CSIC-Universitat de les illes Balears

I. The Origin and Evolution of Seagrasses
II. The Seagrass Flora: Species Richness
III. Distribution and Controls of Seagrass Species Diversity
IV. Seagrass Genetic Diversity
V. Biological Diversity in the Seagrass Habitat
VI. The Function and Management of Seagrass Diversity

GLOSSARY

hydrophily Underwater pollination developed by most seagrasses and many freshwater angiosperms.
seagrasses Marine angiosperms, all monocots, able to grow submersed in marine waters, to which they are restricted.

SEAGRASSES ARE ANGIOSPERMS restricted to marine life. They encompass a limited (approximately 50; Table I) set of closely related species which have successfully colonized the coastal areas across the coastal ocean from intertidal areas to depths in excess of 50 m in the clearest waters (Duarte, 1991), except for polar seas, from which they are absent (den Hartog, 1970). Seagrasses often develop vast meadows (Fig. 1), which are important seascapes covering approximately 0.6×10^6 km² in the ocean (Duarte and Cebrián, 1996) and are responsible for a primary production of approximately 0.60 Gt C year^{-1} (Duarte and Chiscano, 1999) and 15% of the total net CO_2 uptake by oceanic biota (Duarte and Chiscano, 1999). In addition, seagrasses provide habitat for animals, many of which are economically important or endangered, and also play an important role by estabilizing sediments in coastal areas (Hemminga and Duarte, 2000). Hence, the elucidation of their diversity, its maintenance, and its influence on the functions of seagrasses in the ecosystem are important goals for the sustainable management of these key ecosystems. This article provides a comprehensive view of seagrass diversity by examining the origin and current distribution of seagrasses, the controls on seagrass species and genetic diversity and their functions in the ecosystem, and threats to the sustainability of seagrass biodiversity.

I. THE ORIGIN AND EVOLUTION OF SEAGRASSES

Seagrass originated early in the evolution of angiosperms, approximately 100 million years ago, from ancestors that have been hypothesized to be freshwater plants or mangrove- or salt-marsh-like plants. The origin of seagrasses appears to be polyphyletic, and recent analyses based on molecular phylogeny have supported the development of seagrass species from three clades of primitive plants (Les *et al.*, 1997). There are two main

TABLE I

Seagrass Genera and Accepted Species Indicating Their Membership to the Different Seagrass Floras Identified in Fig. 2[a]

Species	Biogeographic membership
Amphibolis antarctica	S. Australian flora
Amphibolis griffithii	S. Australian flora
Cymodocea angustata	Indo-Pacific flora
Cymodocea nodosa	Mediterranean flora
Cymodocea rotundata	Indo-Pacific flora
Cymodocea serrulata	Indo-Pacific flora
Enhalus acoroides	Indo-Pacific flora
Halodule pinnifolia	Indo-Pacific flora
Halodule uninervis	Indo-Pacific flora
Halodule wrightii	Caribbean flora
Halophila baillonis	Caribbean flora
Halophila beccarii	Indo-Pacific flora
Halophila capricornii	Indo-Pacific flora
Halophila decipiens	Caribbean and Indo-Pacific floras
Halophila engelmannii	Caribbean flora
Halophila minor	Indo-Pacific flora
Halophila ovalis	Indo-Pacific flora
Halophila spinulosa	Indo-Pacific flora
Halophila stipulacea	Indo-Pacific flora
Heterozostera tasmanica	S. Australian flora
Phyllospadix iwatensis	Temperate W. Pacific flora
Phyllospadix japonicus	Temperate W. Pacific flora
Phyllospadix scouleri	Temperate E. Pacific flora
Phyllospadix serrulatus	Temperate E. Pacific flora
Phyllospadix torreyi	Temperate E. Pacific flora
Posidonia australis	S. Australian flora
Posidonia oceanica	Mediterranean flora
Posidonia ostenfeldii	S. Australian flora
Posidonia sinuosa	S. Australian flora
Posidonia angustifolia	S. Australian flora
Posidonia coriacea	S. Australian flora
Posidonia denhartogii	S. Australian flora
Posidonia kirkmanii	S. Australian flora
Posidonia robertsoniae	S. Australian flora
Syringodium filiforme	Caribbean flora
Syringodium isoetifolium	Indo-Pacific flora
Thalassia hemprichii	Indo-Pacific flora
Thalassia testudinum	Caribbean flora
Thalassodendron ciliatum	Indo-Pacific flora
Thalassodendron pachyrhizum	S. Australian flora
Zostera asiatica	Temperate W. Pacific flora
Zostera capensis	S. Atlantic flora
Zostera capricorni	S. Australian flora
Zostera caulescens	Temperate W. Pacific flora
Zostera japonica	Temperate W. Pacific flora
Zostera marina	N. Atlantic, Mediterranean, W. and E. Pacific floras
Zostera mucronata	S. Australian flora
Zostera mulleri	S. Australian flora
Zostera noltii	N. Atlantic and Mediterranean floras
Zostera novazelandica	New Zealand flora

[a] After Phillips and Meñez (1998) and Kirkman and Walker (1989).

FIGURE 1 A mixed (seven species) seagrass meadow in Bolinao (Pangasinan, the Philippines). See also color insert, Volume 1.

hypotheses on the origin of seagrasses. One identifies marsh- or mangrove-like plants as their ancestors (den Hartog 1970), which is supported by the observations that some seagrass species present lignified rhizomes and that two genera present viviparous reproduction, similar to some mangrove taxa. The second hypothesis identifies, on the basis of molecular phylogeny, freshwater plants to be the seagrass ancestors (Les *et al.*, 1997). A freshwater ancestor appears clearer for the genus *Enhalus* (*Enhalus acoroides*), the only seagrass species without submarine pollination, which has been associated with related freshwater plants (*Vallisneria* sp.), that it resembles in form and pollination mode.

The fossil record of seagrass is very scarce, and has not been very helpful in testing hypotheses on the origin or evolution of seagrasses. The oldest specimens in the record consist of, at most, three genera: *Archeozostera*, believed to be the ancestor to the extant genera *Heterozostera*, *Zostera*, and *Phyllospadix*; *Thalassocharis*, ancestor to the extant genera *Amphibolis*, *Thalassodendron*, *Halodule*, *Cymodocea*, and *Syringodium*; and *Posidonia*, the only extant genera among the three from the Cretaceous.

A. Evolution of Seagrasses

Although the path of seagrass evolution is not very clear due to the small size of the fossil record, there is evidence that most modern seagrass genera were already established in the Eocene, but the genera *Phyllospadix* and *Enhalus* appear to have evolved in the Palaeocene. The evolution of seagrasses has been postulated to be linked to the development of four major capacities essential for life in the marine environment: (i) the capacity to grow underwater, (ii) the development of mechanisms to tolerate the high salinity of the sea, (iii) the development of substantial anchoring capacity so as to enable seagrasses to remain attached to the substrate in the high-energy marine environment, and (iv) the capacity for underwater pollination (hydrophily). The net result of the action of these constraints is that seagrasses represent a rather homogenous plant group: All seagrasses are rhizomatous plants, with basal meristems, generally blade-like leaves, and mostly hydrophilous pollination. These adaptations may have all been present in some of the freshwater plants, particularly those resistant to high salinity, that may have acted as ancestors to the seagrasses. In contrast, the hypothesis of possible marsh or mangrove ancestors would have required the evolution of most of these adaptations. In particular, the capacity to grow underwater relies critically on the development of an extensive lacunar system able to transport gases between the different organs of the plants, thereby supplying photosynthetic oxygen to the roots and, possibly, CO_2 derived from respiration to enhance leaf photosynthesis. This lacunar system may have already been present in possible freshwater ancestors, such as that for *E. acoroides*. Hydrophily, which is believed to have played an important role in seagrass evolution, affecting their speciation, was already present in possible freshwater ancestors but must have been evolved independently under the hypothesis of mangrove- or marsh-like ancestors. Although its importance is undisputed, the capacity for underwater pollination is not an essential adaptation to life in the marine environment. For example, *Enhalus*, the most recent seagrass genus, pollinates over the water surface as did its ancestor freshwater plants, and yet it is clearly a successful plant in the Indo-Pacific region. A freshwater ancestor for seagrasses is further coherent with new evidence suggesting that the first angiosperms may have also originated in freshwater environments.

II. THE SEAGRASS FLORA: SPECIES RICHNESS

A. Seagrass Taxonomy: Status and Uncertainties

The seagrass flora is composed of approximately 50 species, distributed in 12 genera, belonging to only two families from two closely related orders (Table I). Despite the small number of species, the taxonomy of seagrasses is far from settled, and there is considerable uncertainty not only at the level of species but also at the level of genera. Indeed, the total number of species described could, if uncritically accepted, exceed 60. Although species descriptions are often based on reproductive structures, these characters are, in practice, not always useful as diagnostic criteria to separate them. In practice, the identification of species in field studies is often dependent on vegetative characters to discriminate between genera, whereas the discrimination of species within genera is often based on the biogeographical segregation between species because congeneric species are often present in different floras. The diagnostic characters most commonly used to classify seagrasses into genera and species include differences in the venation, the presence of tannin cells, denticulation, and the tip shape of leaves (den Hartog, 1970). Although differences between reproductive structures are likely to be robust indicators of the broad genetic differences expected from species, differences in leaf anatomy are not as reliable. Leaf characteristics are remarkably plastic within seagrass species, even within adjacent shoots, rendering their diagnostic value questionable. Classification systems based on differences in leaf characteristics are therefore conducive to an ever-increasing number of species, particularly in the small and very plastic genus *Halophila*.

It is clear that a major revision of the taxonomy of seagrass species is needed, and that this revision cannot be based on anatomical or morphological characters alone. Experiments to test for phenotypic plasticity and the more parsimonic methods of molecular taxonomy must play a critical role in this revision. The use of biogeographic separation as diagnostic criteria for species and genera is also questionable because the current distribution of seagrasses may result from local extinctions isolating previously continuous distributions. Indeed, populations of some species with a wide distribution are genetically isolated across biogeographical regions (e.g., *Zostera marina*), although there is no evidence that such isolation may have resulted in sufficient genetic changes as to render them different species. In fact, the description of some seagrasses as separate species on the basis of their biogeographic separation (e.g., *Zostera* and *Phyllospadix* species at either side of the Pacific) has yet to be validated by robust genetic analyses. Indeed, examination of seagrass phylogeny using molecular markers has led to the postulation that there is no basis to separate *Zostera* and *Heterozostera* species into different genera (Les *et al.*, 1997) and has generated a new, more parsimonic, tree of genetic relationships

TABLE II

Classification of Seagrass Families, Subfamilies, and Genera Derived from Analyses of Morphological and Reproductive Structures (den Hartog, 1970) and the More Robust Analysis of Parsimonic Molecular Phylogeny (Les et al., 1997)[a]

Morphological phylogeny	Molecular phylogeny
Potamogetonaceae	Zosteraceae
Subfamily Zosteroideae	*Heterozostera*
Heterozostera	*Phyllospadix*
Phyllospadix	*Zostera*
Zostera	Cymodoceaceae
Subfamily Posidonoideae	*Amphibolis*
Posidonia	*Cymodocea*
Subfamily Cymodoceoideae	*Halodule*
Amphibolis	*Syringodium*
Cymodocea	*Thalassodendeon*
Halodule	Posidonieaceae
Syringodium	*Posidonia*
Thalassodendron	Hydrocharitaceae
Hydrocharitaceae	*Enhalus*
Subfamily Halophiloudeae	*Halophila*
Halophila	*Thalassia*
Subfamily Hydrocharitoideae	
Enhalus	
Subfamily Thalassioideae	
Thalassia	

[a] Ruppiaceae were included in the original analysis of molecular phylogeny. Although closely related to the seagrass taxa, Ruppiaceae are not restricted to life in the sea; therefore they are often excluded from the seagrasses as an ecological category of obligated marine plants.

between the species (Table II) that represents a useful departure point for the needed reassessment.

Independently of current uncertainties regarding the exact size of the seagrass flora, it is clear that seagrasses are minor contributors to the richness of plant species in the sea, where they represent <0.5% of the oxygen-evolving photosynthetic species. This minor contribution cannot be accounted for by the comparatively much longer evolutionary history of marine algae because seagrasses also represent <0.02% of the angiosperm flora, despite an early origin in the angiosperm evolution. Although extinctions of some genera are documented in the fossil record, it is unlikely that past seagrass floras ever comprised a substantially greater number of species than does the extant flora.

The reason for the paucity of speciation in seagrasses remains a subject of intense debate, involving many possible constraints which operate collectively to result in the remarkably low membership of the seagrass flora.

The evolution of the land angiosperm flora is closely associated to a coevolution with insect pollinators. The fact that there are no insects in the sea has been used to argue that the very low extent of speciation in the seagrass flora is attributable to the inefficiency of hydrophilous pollination (Van der Hage, 1996). The stress associated with submerged life in a highly saline, high-energy environment may also further constrain the number of species which have been capable of undergoing the needed physiological, anatomical, and architectural (all seagrass species are rhizomatous) adjustments. No single constraint can be invoked in isolation to account for the low number of marine angiosperm species. The constraints imposed by hydrophilous pollination are not likely to differ much from those associated with the fertilization of red algae, the most species-rich phylum of marine algae. The adaptation of angiosperms to underwater life has led to approximately 500 species in fresh waters, and hundreds of angiosperms have adapted to life in saline environments. In addition to these factors, the greater possibilities for exchange of genetic material between distant populations in the sea, which reduces allopatric isolation, may have played an important role in accounting for the small number of marine angiosperms when compared with the apparent shorter-range exchange between angiosperms on land or in fresh water (Ackerman, 1998).

B. Seagrass Biogeography and Species Richness

The continuity of the marine environment across large spatial scales is best portrayed by the vast distributional ranges of many seagrass species, as exemplified by *Zostera marina*, which extends from subtropical to subartic waters. Basically, only nine floras, including some areas of overlap, can be identified (Table I and Fig. 2) in the shallow, coastal areas inhabited by seagrasses in between polar waters (Hemminga and Duarte, 2000). Six of these species assemblages are composed of temperate species, and only two are represented by exclusively tropical (or subtropical) species.

The biogeographic patterns of seagrass floras are parallel to those followed by hermatypic corals, tropical fish, and mangrove species, all of which reach their highest diversity in the Indo-Pacific region, centered at Malaysia, which is believed to represent the center of origin of these organisms. Indeed, the number of seagrass species declines sharply with increasing distance from the region of highest diversity along the major oceanic currents in the region (Mukai, 1993).

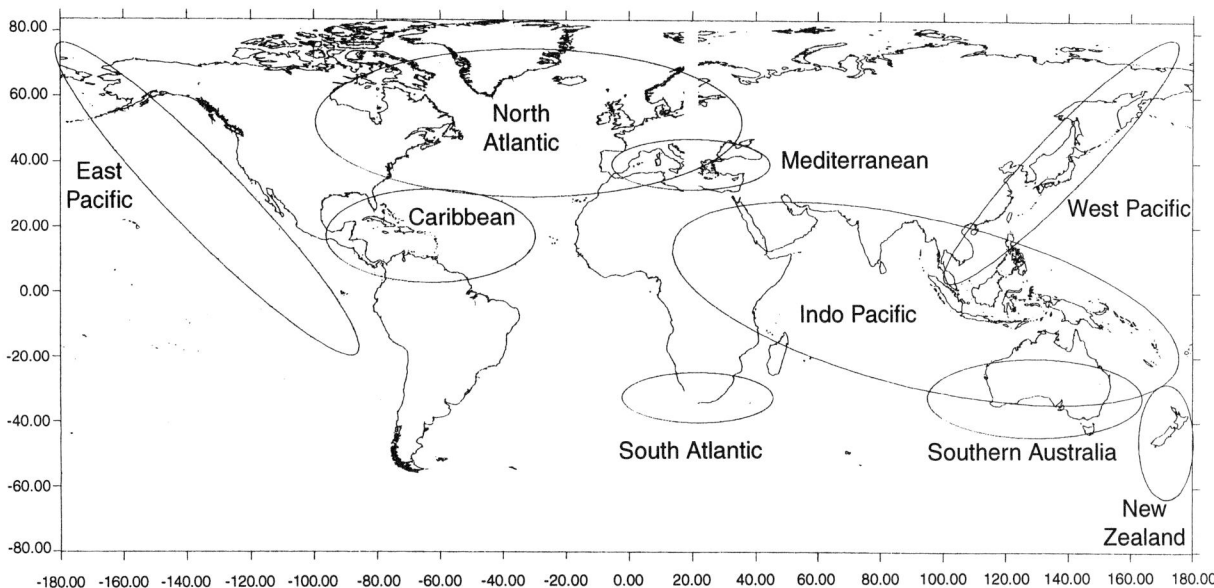

FIGURE 2 The distribution of seagrass floras. The species present in each flora are listed in Table I.

The boundaries between seagrass floras are often sharp, corresponding to frontal areas separating different water masses, such as the Almería-Oran density front separating Atlantic from Mediterranean surface waters. Oceanographic fronts act as physical barriers for the dispersal of organisms, preventing the dispersal of seagrass propagules. The importance of major oceanographic fronts as boundaries between different seagrass floras indirectly validates the perception that the low number of seagrass floras and species result from the continuity of the marine environment over large scales.

Geographically isolated floras, however, have many common characteristics, such as the presence of congeneric species assemblages in distant floras such as the twin assemblage of *Thalassia–Halodule–Syrigingodium* in the Caribbean and the Indo-Pacific regions and *Zostera–Phyllospadix* species in both sides of the temperate Pacific zone (Fig. 2). Single genera are also common to different floras, such as *Posidonia* present in the Mediterranean and Southern Australian floras (Table I and Fig. 2). The occurrence of twin species suggests an even wider early distribution of the species, separated by the rearrangements of continental mass and local extinctions. The high species richness in the Indo-Pacific region (Fig. 2) suggests that this may have represented an area of particularly high speciation acting as a focus for radiative dispersal of the species or that, alternatively, extinctions have been fewer there.

III. DISTRIBUTION AND CONTROLS OF SEAGRASS SPECIES DIVERSITY

A. Distribution and Controls at the Global Scale

The biogeographic factor constraining the distribution of floras is the major determinant of species richness at the global scale because it sets the maximum species richness possible at any one site. This maximum, however, may never be realized. For example, in Western Australia, the flora contains many species which often grow segregated, forming mostly monospecific stands. The maximum species richness reported at any one stand is 12 species, but the majority of meadows studied are monospecific, with an exponential decline in the number of meadows with increasing number of species (Fig. 3). The mean richness of 596 descriptions of seagrass meadows encountered in a thorough revision of the published literature is approximately two seagrass species per meadow (mean $\pm$ SE = 1.99 $\pm$ 0.07; Hemminga and Duarte, 2000). The resulting diversity, as calculated from the biomass distribution of the seagrasses using the Shanon–Weaver index, is closely linked to the species richness. The majority of seagrass meadows are monospecific and therefore have the minimum diversity index of 0. The highest Shanon–Weaver diversity index for a seagrass meadow was 1.56 for a

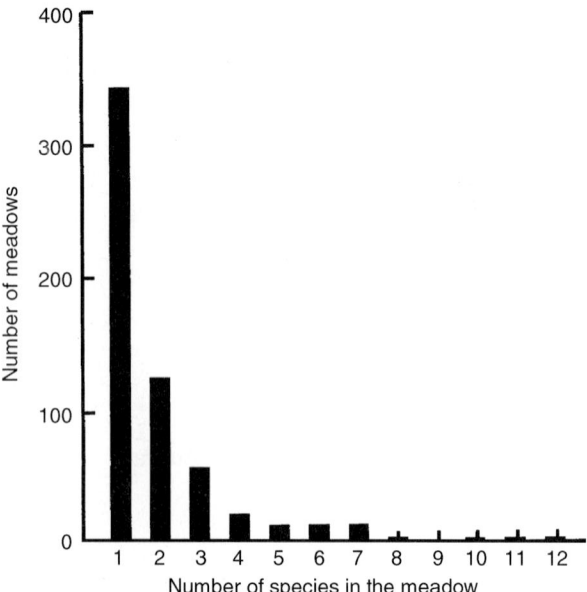

FIGURE 3 The number of species present in 596 seagrass meadows reported in the literature.

seagrass meadow containing seven seagrass species in Bolinao (Pangasinan Province, the Philippines), and indices >1 have only been reported for meadows in the Indo-Pacific region, although the available estimates for such comparisons are few. These values compare poorly with diversity indices calculated for most other marine taxa, which generally oscillate between 1 and 4 for marine algae, invertebrates, and fish assemblages (Margalef, 1980).

The low species richness of seagrass stands, with most meadows being monospecific or containing two species, extends to areas occupied by floras with high numbers of species, such as Western Australia, in which species tend to be segregated. Mixed meadows, however, are common in the Red Sea and the Indo-Pacific region (Fig. 4), with the highest species richness found in Philippines waters. As a result of these patterns, seagrass species richness shows a clear latitudinal pattern, with the average number of seagrass species being greatest in meadows near the equator and least in meadows in cold, temperate areas (Fig. 5). The seagrass species richness in the subtropics in the Southern Hemisphere is somewhat greater than that for similar latitudes in the Northern Hemisphere (Fig. 5), reflecting the high species richness in the Australian continent and the Indian Ocean. The pattern toward species richness declining from the tropics to the cold temperate zone is comparable to those described for terrestrial taxa.

B. Distribution and Controls at the Regional Scale

The species richness encountered at any one location is constrained by the size of the flora present in the particular biogeographic region and the suitability of growth conditions in the area to support the different species. The study of the habitat requirements of seagrasses has received considerable attention and is relatively well delineated. Seagrass life is restricted to marine waters, extending to estuarine areas with low salinity, such as those encountered in the Baltic Sea. Seagrasses are further constrained by exposure to desiccation and physical disturbance by waves or, in cold seas, ice scouring in shallow waters, and the downslope limit is imposed by light, with seagrasses penetrating to a depth receiving, on average, 11% of the surface irradiance (Duarte, 1991). Within these limits, seagrasses require appropriate substratum which, for most species—except *Phyllospadix*, *Amphibolis*, and *Posidonia*, which are able to grow attached to rocks—is sandy to silty sediments. Highly silted sediments, which are also associated with reduced sediment conditions and accumulation of phytotoxins, are not appropriate to support seagrass life, which is restricted to sediments with low organic contents, redox potentials above -200 mV, and sulfide concentrations <100 μM. Although these growth requirements define habitats appropriate to support seagrass life, different species are expected to have different tolerances to these stress factors, which should lead to differences in species richness among locations at the regional scale.

Species-specific growth requirements are poorly documented for seagrasses, but some regional patterns are emerging. Seagrass species richness declines along gradients of siltation in Southeast Asian coastal waters, with a steep reduction in species richness as the silt content of the sediments exceeds 15%, and only monospecific (*E. acoroides*) meadows occur at highly silted locations (Fig. 6). That these patterns derive from species-specific differences in the resistance to disturbance has been predicted by theory and verified experimentally (Duarte et al., 1997).

C. The Maintenance of Local Species Diversity

Seagrass species are closely related and therefore often engage in competitive interactions in the resource-poor environments which they inhabit. Competitive interactions, if strong, could lead to species replacement and

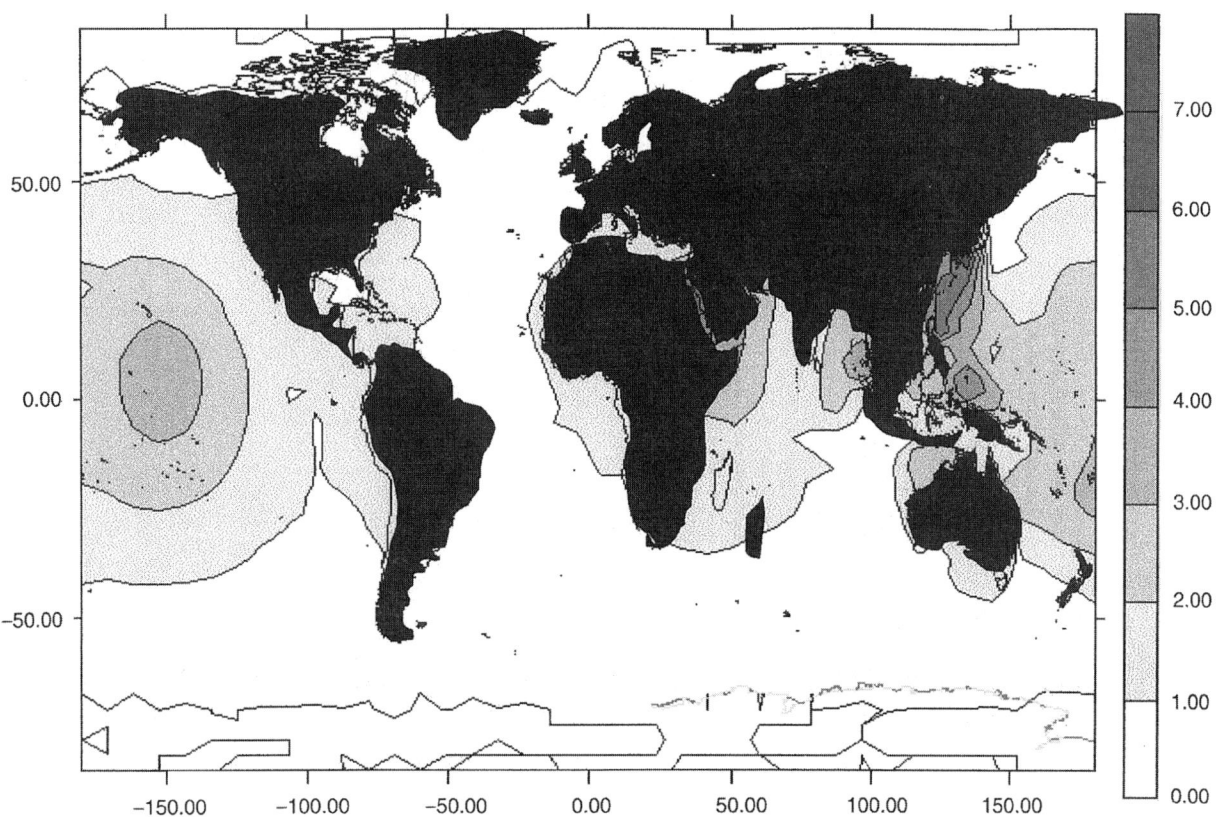

FIGURE 4 Contour isolines of seagrass species richness generated from data on 596 seagrass meadows reported in the literature.

the dominance of one species following a period of coexistence. Indeed, disturbance experiments have documented how a period of high diversity during intermediate stages of colonization is followed by the dominance by the climax species, which is typically slow growing. That the occurrence of such monospecific stands of climax species may results from strong resource limitation has been demonstrated by long-term fertilization experiments of a monospecific *Thalassia testudinum* meadow, which was colonized by the pioneer species *Halodule wrightii* following experimental nutrient additions (Fourqurean et al., 1995).

High disturbance is also expected to lead to species-poor meadows which contain only small, pioneer species capable of fast colonization and growth, often arrested in a permanent stage of colonization. This is best illustrated by the occurrence of meadows containing only *Halophila* species in areas visited by dugongs in Southeast Asia. The grazing mode of the dugong uproots the seagrass, creating small gaps (feeding tracks about 2 m long and 30 cm wide) that must be recolonized in between successive visits. Only the small *Halophila* species, able to grow 9 m annually, are sufficiently fast colonizers to withstand this pressure. Because *Halophila* leaves are also tender, low in fibers, and nutrient rich, dugongs have been postulated to be actively farming these seagrass species (Preen, 1995).

Hence, seagrass species diversity is expected to comply with the intermediate disturbance hypothesis, in which the highest diversity is expected at moderate levels of disturbance. Multispecific seagrass meadows would therefore develop in the presence of moderate disturbance levels. The maintenance of the high species diversity in the most diverse meadow reported to date, located in the Philippines, has been attributed to the disturbance induced by the activity of burrowing shrimps, which maintain small-scale ($<m^{-2}$) patchiness by generating gaps that allow the proliferation of pioneer species in meadows otherwise dominated by large, climax species (Duarte et al., 1997).

Although the preceding discussion highlights the role of competitive interactions, there is also evidence of positive interactions whereby the presence of some species may facilitate the occurrence of other species,

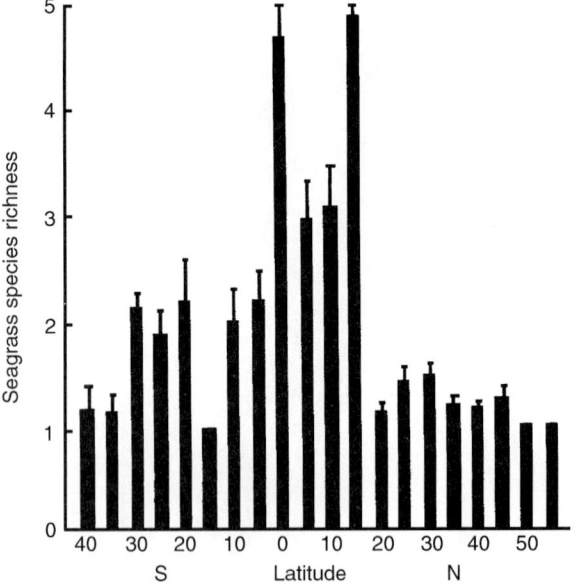

FIGURE 5 The mean (±SE) species richness in seagrass meadows growing in different latitudinal ranges. Compiled from data on 596 seagrass meadows encountered in a thorough revision of the literature.

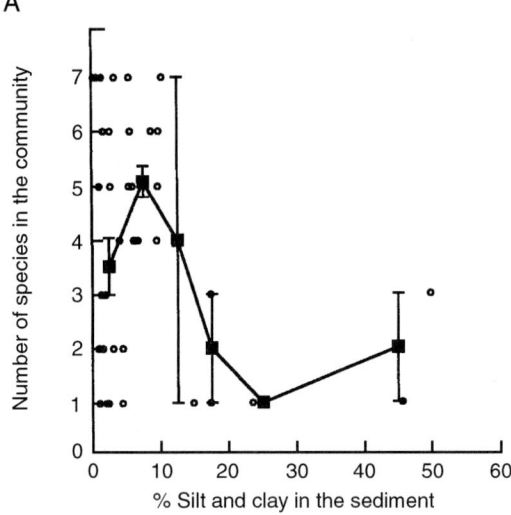

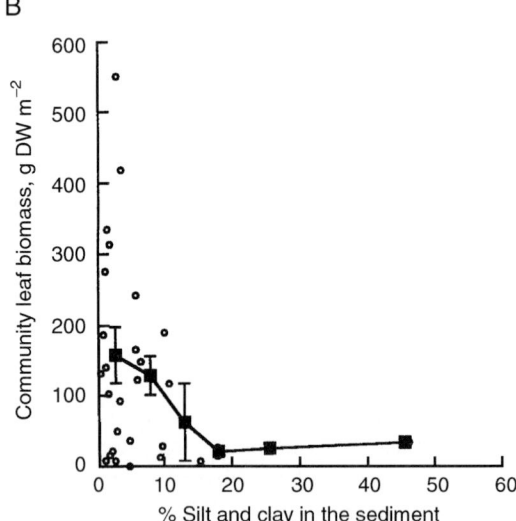

FIGURE 6 The reduction in the average species richness and community biomass of Southeast Asian seagrass meadows with increasing sediment silt content (reproduced with permission from Terrados et al., 1997).

thereby maintaining high seagrass diversity. This role has been assigned to "engineering species" (Duarte et al., 1997) able to modify the growth conditions to improve their suitability to maintain plant growth. The occurrence of some seagrass species in Southeast Asia has been reported to be associated with the presence of the dominant species, *Thalassia hemprichii*, which is also linked to high biomass development in the meadows (Terrados et al., 1997). Seagrasses are able to alter the sedimentary conditions in which they grow, injecting oxygen into the sediments through their roots, thereby avoiding the accumulation of phytotoxins produced under anoxic conditions. Through this effect, seagrass species with extensive root systems, such as *T. hemprichii*, may maintain sediment conditions tolerable for seagrass growth where the organic loading derived from the seagrass growth would otherwise render sediments highly reduced and unsuitable for seagrass growth.

IV. SEAGRASS GENETIC DIVERSITY

Seagrasses are clonal organisms, so the genetic diversity present in meadows may be low. Indeed, vegetative reproduction is by far the dominant mode of propagation of seagrasses; therefore, the hypothesis that entire meadows may be composed of one or a few organisms cannot be rejected without serious consideration. Some aspects of seagrass biology, such as the occurrence of mass mortality across large spatial scales, have been ascribed to the dominance of vegetative reproduction in seagrasses and an assumed low genetic diversity to resist diseases. However, examinations of seagrass genetic diversity are still few and have lagged well behind studies of genetic diversity in other angiosperm groups.

The introduction in the mid-1990s of molecular techniques to quantify the extent of genetic diversity of seagrass populations is now providing the needed information to test the assumed low genetic diversity of the meadows. Examination of the genetic structure of seagrass populations has shown considerable variability in the extent of genetic diversity within and among species. Genetic diversity can be important in some of the species investigated (e.g., *Posidonia australis*, *T. testudinum*, *Halodule uninervis*, and *Z. marina*), whereas very low levels of genetic variability have been demonstrated for other species (e.g., *Amphibolis antarctica* and *Posidonia oceanica*). Even in the genetically variable species, most of the variation is found within local populations, and there is reduced gene flow at scales of only a few kilometers or even meters along the depth gradients.

The level of genetic uniformity of seagrasses appears to be greater than that typical of terrestrial plants, supporting the view that the low levels of speciation in seagrasses could derive from inefficient sexual reproduction. The significant within-meadow variability in some species falsifies the assumption that seagrasses are primarily clonal and challenges the explanation of high clonality as the factor responsible for large-scale decline in *Z. marina*, the populations of which show considerable genetic variability. In contrast, the widespread decline of *P. oceanica* in the Mediterranean could be associated with its high clonality. Although some genetic variation has been reported for populations along depth gradients, sizable variation has only been found near *P. oceanica*'s biogeographical limit in the Mediterranean–Atlantic transition zone. Indeed, a seagrass (*Z. marina*) clone growing in the Baltic has been identified as the largest and oldest marine organism found to date (Reusch *et al.*, 1999). The results to date provide no patterns of possible regulatory factors in seagrass genetic diversity because the differences observed among species do not appear to correspond to differences in reproductive mode or dispersal properties. The study of the genetic structure of the seagrass population is an active area of research; therefore, more robust conclusions than those currently available are expected to be derived in the near future.

V. BIOLOGICAL DIVERSITY IN THE SEAGRASS HABITAT

As indicated previously, seagrass ecosystems are important reservoirs of biological diversity in the sea. Seagrass ecosystems tend to contain more species than the adjacent unvegetated bottoms, and the species common to both seagrass meadows and unvegetated areas tend to be more abundant within the meadows. The enhanced biological diversity and richness within seagrass meadows result from their importance as habitat for (i) species that use seagrasses, and the rich detritus they produce, as food sources; (ii) species that use seagrass surfaces as substratum; and (iii) species that seek refuge within seagrass meadows. These species encompass a broad taxonomic spectrum from fungi to mammals.

A. Microbial Diversity

The abundance of microbial communities appears to be enhanced within seagrass meadows compared to adjacent bare sediments. These microbial communities benefit from the large amounts of detritus seagrass produces as well as the efficient retention of carbon deposited from the water column in vegetated sediments. As a result, there seems to be clear links between bacterial activity and seagrass production and nutritional status, and exploratory analyses have shown that seagrass meadows host highly diverse bacterial communities. The association between seagrass meadows and fungi is even clearer because seagrass rhizomes and roots are heavily colonized by fungi, although mychorrizal-type associations do not appear to have been developed in the sea. In fact, some of these fungal species may be pathogenic; in particular, the pathogenic *Laberynthula* species have been found to be widespread in seagrass stands. *Laberynthula* species were associated with large-scale seagrass (*Z. marina*) declines in the Atlantic in the 1930s, although the role of these pathogens on the decline remains hypothetical. Recent analyses have also reported a rich ciliate community to be associated with seagrass leaves. This ciliate community encompasses a wide diversity of taxa, and their role in the seagrass ecosystems has yet to be elucidated. These microbial communities may rely directly on detritus produced by seagrasses or benefit from the trapping of particles within the quiescent waters within seagrass canopies.

The best studied microbial community associated with seagrass meadows is the epiphytic microalgal community that develops on the seagrass leaves. This community is largely represented by large benthic diatoms, which colonize newly produced seagrass leaves and are replaced, as the age of the leaf substratum increases, by macroalgae and sessile invertebrate communities. Hence, there is a gradation in epiphytic community structure from the younger leaf base toward the leaf apex, particularly in species with long leaf life spans such as the Mediterranean seagrass *P. oceanica*, the leaves of which last for about a year. The ephiphytic community is an important component of the ecosystem metabolism within seagrass meadows, often supporting a primary production comparable to that of the host plants; it also makes an important contribution to heterotrophic processes. In addition, the epiphytic community is responsible for high deposition rates of carbonates on seagrass surfaces, particularly in warm waters, and is therefore responsible for the high carbonate deposition in seagrass sediments.

B. Macroalgae

Macroalgae often live in association with seagrasses, prominently as epiphytes on the leaf apices of species with long leaf life spans. The number of epiphytic macroalgae species exceeds 350, including all major taxonomic divisions, in which blue-green algae are relatively rare. Competitive interactions and other negative effects are particularly intense in eutrophic environments because the abundance of opportunistic algae (*Ulva*, *Enteromorpha*, and *Chaetomorpha*) is promoted under high nutrient loading. The growth of these opportunistic algae can be so extreme that they can develop thick blankets over the seagrass meadows, which may suffocate or uplift (from the buoyancy due to gaseous release) the plants. In addition, competitive interactions may also be established with the siphonales, notably species of the genus *Caulerpa*, the rhizoids of which confer them the ability to extend horizontally over soft sediments and exploit the sediment nutrient pools. Seagrass meadows in Southeast Asia often contain many *Caulerpa* species. Introduction of *Caulerpa* species (*C. taxifolia* and, recently, *C. racemosa*) to the Mediterranean poses an important threat, where invasive expansion occurs, to the seagrass meadows (e.g., *P. oceanica*).

C. Invertebrates

The invertebrate fauna associated with seagrasses is very rich, including a multitude of organisms living epiphytically on their leaves and on their rhizomes and associated with seagrasses on the sediments. Epiphytic fauna comprise more than 170 species. Meiofauna include nematods, rotifers, tardigrades, copepods (planktonic and epibenthic harpacticoid species), and ostracods. Copepods and other free-living zooplankters often swarm within seagrass beds. The leaves support a rich community of epiphytic filter-feeding sessile organisms, including hydrozans, ascideans, and bryozoans. Spongi grow attached to the seagrass rhizomes exposed aboveground and sometimes epiphytically on their leaves (e.g., *E. acoroides*). Polychaetes are abundant within seagrass sediments and also reach high densities on exposed seagrass rhizomes. Borer polychaetes have been reported to cause damage to seagrass leaves and rhizomes.

Gastropods rank among the most abundant organisms on seagrass leaves, on which they graze the epiphytic algae and sometimes erode the leaf epidermis, damaging the seagrasses. The large gastropod *Strombus gigas*, the queen conch, is an important consumer in Caribbean *T. testudinum* meadows, but heavy fishing pressure has decimated its populations. Seagrass beds are important habitat for bivalves including scallops, which show behavioral adaptations toward a preferential recruitment to seagrass beds. Bivalves may experience an increased growth rate within seagrass beds, as reported for Mercenaria clams and mussels, due to the particle trapping by seagrasses and may also be less prone to predation there. These bivalve populations engage in mutualistic relationships with the plants, such as those reported for blue mussels (*Mytilus edulis*) and temperate seagrasses, because their enhanced activity within seagrass beds provides inorganic nutrients to the surrounding waters and sediments which promote seagrass growth. An outstanding association between seagrasses and bivalves is that by *Pinna nobilis*, a very large (up to 80 cm tall) bivalve that is largely restricted to *P. oceanica* beds. *Pinna nobilis* is a protected Mediterranean species and is particularly endangered because, in addition to direct threats from collection and predation, its populations are declining as a result of the widespread decline of *P. oceanica* across the western Mediterranean. Furthermore, *P. nobilis* is the host of a crustacean species (*Pontonia pinnophylax*) which lives in mutualistic association inside its large shell and is also an endangered species because of the decimated populations of the host bivalve. This nested association between *P. oceanica*, *P. nobilis*, and *P. pinnophylax* has been termed a "biodiversity Russian doll" (Richardson et al., 1997) and used as a paradigm of how the threat to landscape components, such as seagrass meadows, involves a threat to the many species therein contained. Although nudibranchs are not particularly prominent components of seagrass beds, sea hares (*Aplysia*) have been observed to graze on seagrass leaves, particularly when gathering in large numbers during the reproductive season.

Isopods and amphipods are major consumers of both seagrasses and their epiphytes. In particular, the isopod *Idotea* plays an important role as a link between seagrasses and higher trophic levels in food webs. Amphipods rank among the most abundant organisms within seagrass meadows. Thalassinid shrimps are important components of seagrass meadows, particularly in tropical waters in which their burrowing activity generates gaps within the seagrass and affects the growth of the adjacent shoots. Thalassinid shrimps can also be major grazers in seagrass meadows, removing up to 30% of their production. Thalassinid shrimps clip leaves of seagrass species adjacent to their burrows and transport them into the burrows to decompose prior to consumption. In addition, their burrowing activity and the active and passive ventilation of their burrows appear to have a major effect on the sediment conditions, attenuating the reduced stage of the sediment while increasing nutrient availability, which may promote seagrass growth. Hence, Thalassinid shrimps are important engineering species within (particularly tropical) seagrass beds, in which their disturbance effect has been equated to that of typhoons and is responsible for generating small-scale, highly dynamics gaps that are believed to be essential for maintaining multispecific seagrass beds (see Section III). Seagrass beds are also important nursery areas for prawns such as tiger prawns (e.g., *Penaeus semisulcatus* and *P. esculentus* in Australian waters), which have been found to feed on seagrass seeds, and caridean shrimps are abundant in subtropical and tropical seagrass meadows. Decapod crustaceans such as *Callinectes* spp. can also be important predators of seagrass seeds. Fiddler crabs (*Uca* spp.) are also very abundant in tropical and subtropical intertidal beds, in which they use their developed claw to remove the epiphytes of the seagrass leaves for consumption.

Detritivore echinoderms, such as holthurids and ophiurids, and herbivores such as sea urchins are abundant within seagrass meadows and are among the dominant invertebrate components therein. Sea urchins can graze heavily on seagrass leaves—to the point that sudden population outbreaks may defoliate seagrass meadows over vast areas.

D. Vertebrates

The fish communities within seagrass beds typically comprise between 12 and 50 species in any one meadow. Fish communities tend to be more abundant within seagrass meadows than in adjacent bare bottoms, and they seem to seek refuge from predation there or benefit from the higher prey densities often encountered within the seagrass bed. Only approximately 20 species worldwide graze on seagrass. Some species, however, graze on seagrass and/or the epiphytes present on their leaves, such as *Boops salpa* (a herbivorous species in the Mediterranean), parrot fish, and rabbit fish (*Syganus* sp.). Cryptic species such as stonefish are often encountered within seagrass meadows, and gobids are also abundant in seagrass meadows, sometimes in association with Thalassinid shrimps. Active predators are often observed visiting seagrass meadows to prey on the abundant animal resources present therein. Other species, such as rays, can act as important disturbance agents by uprooting the plants when hiding under the sediment. Demersal fish production has been found to correlate with seagrass biomass, likely reflecting the higher prey production within these ecosystems.

The main reptile users of seagrass beds are adult sea turtles, which are important consumers of seagrass meadows. The most studied species is the green turtle, *Chelonia mydas*, which grazes on the seagrass. The association of turtles with seagrass meadows is so strong as to be reflected in some vernacular names (e.g., turtlegrass for the Caribbean species *T. testudinum*). Their importance as grazers in seagrass beds, however, has decreased due to the decline of the populations. In Southeast Asia, sea snakes are often observed foraging on seagrass beds.

In temperate climates, intertidal seagrass beds (mostly *Zostera* sp.) often support migrant goose and swan populations. They are the most important vertebrate grazers in cold, temperate waters, in which low water temperature would hamper the digestibility of seagrasses by poikilotherm grazers. Goose and swans feed on both leaves and rhizomes, and they may play an important role in the dispersal of the plant seeds. Many bird species, including waders, shore birds, gulls, pelicans, and diving birds, feed on seagrass meadows, particularly on intertidal ones.

The syrenidae or marine cows are large marine herbivores that feed exclusively on seagrasses and, in the case of manatees (Caribbean waters), also on freshwater plants growing in rivers. Dungongs (*Dugong dugong*) prefer fast-growing, small seagrasses, especially *Halophila*, which in Southeast Asia develop large stands in intertidal flats (Fig. 7), have low lignin and cellulose contents, and regenerate fast. Dugongs feed by absorbing the sediments through their snouts and filtering out the seagrass while dropping the finer sediment particles. Their feeding activity leaves behind distinct feeding trails, usually approximately 20–30 cm wide and 1 or 2 m long, that are bare of seagrasses. In contrast, manatees graze only on the leaves and do not demonstrate a digging behavior. Dungongs and manatees are threatened and protected throughout their ranges. Al-

FIGURE 7 A *Halophila ovalis* population in a large intertidal flat visited by dugongs in Con Dao Island (South Vietnam). See also color insert, Volume 1.

though the manatee population is still significant, dugong (Southeast Asia to Australia) populations have disappeared from most of their original range, and sizable populations have only been reported from the Great Barrier Reef. Although active hunting of sea cows has been abandoned, their loss to the point of extinction of local populations continues due to mortality induced by fishing gear, dynamite and cyanide fishing, and accidental collisions. As a result, the dugong is now among the most endangered marine mammals at great risk of extinction.

VI. THE FUNCTION AND MANAGEMENT OF SEAGRASS DIVERSITY

Seagrass meadows are one of the ecosystems that provide most valuable services for the biosphere. They produce approximately 0.60 Gt C year^{-1} and are responsible for one-sixth of the total net CO_2 uptake by oceanic biota (Duarte and Chiscano, 1999). Seagrasses provide habitat for animals, many of which are economically important or endangered, and also play an important role by stabilizing sediments in coastal areas (Hemminga and Duarte, 2000). These functions are closely linked to seagrass diversity because there is evidence indicating that mixed meadows can support considerably more biomass and be much more productive than less diverse meadows within regions (Terrados et al., 1997). High seagrass diversity may also be linked to a more effective retention of sediments because the more complex canopies of mixed meadows will achieve a greater dissipation of energy by acting at different scales of the energy spectrum than the simpler canopies of monospecific meadows. They are also expected to host a larger number of animal taxa than monospecific ones because of their diversity of substrate and food types. However, the possible functional roles of seagrass diversity remain largely untested, as is unfortunately the case for most other marine taxa.

The importance of seagrass meadows as components of marine diversity does not stem primarily from the diversity of the angiosperm but rather, as indicated previously, from the high diversity that these ecosystems harbor. The role of seagrass meadows as sites of enhanced marine diversity was first addressed in the study of commercially important species, which often use seagrass meadows as nurseries. These studies were later extended to the study of endangered marine organisms, such as marine turtles and dugongs, which rely on seagrasses for food. Indeed, marine biodiversity presents a nested structure within seagrass meadows, which contain endangered species that often act as hosts for species that are necessarily rare, such as the *P. oceanica*—*P. nobilis* (mollusca)—*P. pinnophyla* (decapoda) association described in Mediterranean waters (Richardson et al., 1997). The pattern toward high biological diversity within seagrass meadows has been further strengthened by the demonstration of high microbial diversity, with the leaves of seagrass species containing complex microbial assemblages (fungi, bacteria, microalgae, ciliates, and flagellates) which benefit from the substratum offered by the plants as well as their release of detrital carbon to support these active microbial food webs. The plants also benefit from the activity of epiphyte grazers that prevent an excessive accretion on seagrass leaves, alleviating the negative effects of moderate levels of eutrophication for the seagrasses. Seagrass ecosystems also support food webs distant from the meadows; exported seagrass material has been documented as the carbon basis for some deep-sea communities, and seagrass material accumulating on shore, which may form mounds up to 3 m in height, is used by insects and other terrestrial animals.

A. Managing Seagrass Diversity: Threats

The limited number of seagrass species in the ocean implies that threats to one or a few species may have devastating consequences for the overall diversity of these plants. Unfortunately, catastrophic seagrass decline has been reported for many seagrass species (Short and Wyllie-Echevarria, 1996), both locally and across entire basins, such as the wasting disease that decimated *Z. marina* in the Atlantic Ocean in the 1930s and the widespread decline of *P. oceanica* in the Mediterranean Sea (Marbá *et al.*, 1996). These losses often result from direct human-induced disturbance, primarily reduced water clarity derived from increased nutrient loading, but they also involve diffuse sources likely associated with changes in climate (Marbá *et al.*, 1996) which may increase the occurrence of diseases (Short and Wyllie-Echevarria, 1996). Human disturbance leading to seagrass loss includes many factors in addition to eutrophication, such as siltation derived from deforestation (Terrados *et al.*, 1997), mechanical damage from boating, dredging, filling, dynamite fishing and trawling, and altered sediment dynamics after coastal constructions (Short and Wyllie-Echevarria, 1996). Remarkably, seagrasses have considerable resistance to the release of toxic compounds (Short and Wyllie-Echevarria, 1996).

The seagrass area lost cannot be quantified due to lack of a reliable baseline information, but it is certainly large since even large meadows may be lost within a few years (Marbá *et al.*, 1996). Concern about the loss of seagrasses, and the associated loss of ecosystem functions, biological diversity, and resources, has been translated into legislation aimed at protecting seagrasses in many countries (e.g., USA, Spain and France), and particular emphasis has been placed on the importance of preserving seagrass meadows at the International Convention of Biological Diversity (i.e., the Rio Convention). The effectiveness of these measures to promote seagrass recovery, however, is uncertain because, although disturbed seagrass meadows can recover, the current capacity to predict this process is meager. Models of seagrass recovery timescales predict vast differences in recovery time among species, with small, fast-growing species being able to recover within a few years provided that adjacent reproductive populations exist, and the recovery of slow-growing, long-lived species such as *P. oceanica* involving timescales of centuries (Duarte, 1995). Therefore, restoration efforts have been designed to speed up recovery, but these efforts have a high cost and can only be implemented in the countries with the strongest economies.

Hence, the management of seagrass biodiversity must be based on prevention measures aimed at achieving sustainability targets. In particular, management practices should be based on the known habitat requirements of seagrasses to maintain water quality targets (particularly water transparency) and sedimentary fluxes (particularly to prevent erosion and reduce burial rates) compatible with seagrass life. Deterioration of these properties will lead to reduced seagrass diversity, species loss, and the loss of the associated seagrass functions and biological diversity in the meadows they form.

See Also the Following Articles

GRAZING, EFFECTS OF • MARINE ECOSYSTEMS

Bibliography

Ackerman, J. D. (1998). Is the limited diversity of higher plants in marine systems the result of biophysical limitations for reproduction or evolutionary constraints? *Functional Ecol.* **12**, 975.

den Hartog, C. (1970). *The Seagrasses of the World*. North Holland, Amsterdam.

Duarte, C. M. (1991). Seagrass depth limits. *Aquat. Bot.* **40**, 363–77.

Duarte, C. M. (1995). Submerged aquatic vegetation in relation to different nutrient regimes. *Ophelia* **41**, 87–112.

Duarte, C. M., and Cebrián, J. (1996). The fate of marine autotrophic production. *Limnol. Oceanogr.* **41**, 1758–1766.

Duarte, C. M., and Chiscano, C. L. (1999). Seagrass biomass and production: A reassessment. *Aquat. Bot.* **65**, 159–174.

Duarte, C. M., Terrados, J., Agawin, N. S. W., Fortes, M. D., Bach, S., and Kenworthy, W. J. (1997). Response of a mixed Philippine seagrass meadow to experimental burial. *Mar. Ecol. Prog. Ser.* **147**, 285–294.

Fourqurean, J. W., Powell, G. V. N., Kenworthy, W. J., and Zieman, J. C. (1995). The effects of long-term manipulation of nutrient supply on competition between the seagrasses *Thalassia testudinum* and *Halodule wrightii* in Florida Bay. *Oikos* **72**, 349–358.

Hemminga, M., and Duarte, C. M. (2000). *Seagrass Ecology*. Cambridge Univ. Press., London.

Jones, C. G., Lawton, J. H., and Shachak, M. (1997). Positive and negative effects of organisms as physical ecosystem engineers. *Ecology* **78**, 1946–1957.

Kirkman, H., and Walker, D. I. (1986). Regional studies—Western Australian seagrass. In *Biology of Seagrasses* (A. W. D. Larkum, A. J. McComb, and S. A. Shepherd, Eds.). Elsevier, New York.

Les, D. H., Cleland, M. A., and Waycott, M. A. (1997). Phylogenetic studies in Alismatidae. II. Evolution of marine angiosperms (seagrasses) and hydrophily. *Syst. Bot.* **22**, 443.

Marbá, N., Duarte, C. M., Cebrián, J., Enríquez, S., Gallegos, M. E., Olesen, B., and Sand-Jensen, K. (1996). Growth and population dynamics of *Posidonia oceanica* on the Spanish Mediterranean coast: Elucidating seagrass decline. *Mar. Ecol. Prog. Ser.* **137**, 203–213.

Margalef, R. (1980). *Ecología*. Omega, Barcelona.

Mukai, H. (1993). Biogeography of the tropical seagrasses in the western Pacific. *Aust. J. Mar. Freshwater Res.* **44**, 1.

Pettitt, J. M. (1984). Aspects of flowering and pollination in marine angiosperms. *Oceanogr. Mar. Biol. Ann Rev.* **22**, 315–342.

Phillips, R. C., and Meñez, E. G. (1988). Seagrasses. Smithsonian Contributions to Marine Science No. 34. Smithsonian Institute, Washington, DC.

Preen, A. (1995). Impacts of dugong foraging on seagrass habitats: Observational and experimental evidence for cultivation grazing. *Mar. Ecol. Prog. Ser.* **124**, 201–213.

Reusch, T. B. H., Borström, C., Stam, W. T., and Olsen, J. L. (1999). An ancient eelgrass clone in the Baltic. *Mar. Ecol. Prog. Ser.* **183**, 301–304.

Richardson, C. A., Kennedy, H., Duarte, C. M., and Proud, S. V. (1997). The occurrence of *Pontona pinnophylax* (Decapoda: Natanti: Pontoniinae) in the Fan mussel *Pinna nobilis* (Mollusca: Bivalvia: Pinnidae) from the Mediterranean. *J. Mar. Biol. Assoc. U.K.* **77**, 1227–1230.

Short, F. T., and Wyllie-Echevarria, S. (1996). Natural and human-induced disturbance of seagrasses. *Environ. Conserv.* **23**, 17–27.

Terrados, J., Duarte, C. M., Fortes, M. D., Borum, J., Agawin, N. S. R., Bach, S., Thampanya, U., Kamp-Nielsen, L., Kenworthy, W. J., Geertz-Hansen, O., and Vermaat, J. (1997). Changes in community structure and biomass of seagrass communities along gradients of siltation in SE Asia. *Est. Coastal Shelf Sci.* **46**, 757–768.

Van der Hage, J. C. H. (1996). Why are there no insects and so few higher plants in the sea? New thoughts on an old problem. *Functional Ecol.* **10**, 546.

SLASH-AND-BURN AGRICULTURE, EFFECTS OF

Stefan Hauser* and Lindsey Norgrove[†]
*International Institute of Tropical Agriculture and [†]King's College London

I. Reasons for Slash-and-Burn Agriculture
II. The Importance of Fallow Length: Long and Short Fallow Slash-and-Burn Systems
III. Geographic Distribution and Characteristics of Slash-and-Burn Agriculture
IV. Effects of Clearing at Field Establishment
V. Consequences of Burning
VI. Effects during the Cultivation Phase
VII. Effects during the Fallow Phase
VIII. The Scale Dependence of the Effects Caused by Slash-and-Burn Agriculture
IX. Alternative Systems to Slash-and-Burn Agriculture

GLOSSARY

bulk density The mass of the undisturbed dry soil per unit volume.
cation exchange capacity The capacity of a soil to adsorb cations such as calcium, magnesium, and potassium on the surfaces of soil particles and organic matter.
dormant/dormancy Condition under which a seed will not germinate, even if all environmental requirements for germination are met.
fallow A phase in which no crop is on the land and the volunteer vegetation regrows.
leaf area index Single-sided total leaf area of the vegetation per unit area of land.
mulch Retained slash covering soil surface; this can be *in situ* slashed or imported biomass.
pegging The penetration of the fertilized groundnut flower into the soil.
penetrometer resistance Energy required to push a probe of defined size and shape into the soil.
transpiration Amount of water taken up by the vegetation and released through the leaves to the atmosphere.

SLASH-AND-BURN AGRICULTURE is a generic term for agricultural systems in which the fallow vegetation is manually slashed, left to dry and cleared from the field by burning before crop cultivation. "Swidden" is an English dialect word for a burned clearing, thus swidden agriculture is a synonym for slash-and-burn agriculture. After a cropping phase, the land is abandoned to a fallow phase. Later, the cycle is repeated. Only systems that alternate between crop and fallow phases are included in this definition. Multistory tree gardens, home gardens, and cocoa plantations, where crops are permanently cultivated, are excluded. With the exception of labor, slash-and-burn farmers use few or no external inputs. Implements such as machetes and hoes are most commonly used. Systems in which machinery is used for clearance and irrigated systems are excluded from this definition. Not considered in this context are the systems such as the *ankara* of the western Highlands in Cameroon, the *nkule* of the Tanzanian grasslands, and the gy of Ethiopia, in which vegetation is slashed, gathered, covered with soil, and then burned inside the soil mounds.

I. REASONS FOR SLASH-AND-BURN AGRICULTURE

Most slash-and-burn farmers are poor (section III.B). Unlike many people in developed countries, they have little choice in how to live their lives. Often the only resource available to them is land. Thus, farming, whether subsistence or market oriented, might be their only option. However, even cash-crop production may not provide sufficient income for farmers to change to other economic pursuits. Thus, farmers they remain. Farming methods used reflect the economic constraints of the farmers and tend to minimize labor requirements (section III, C).

A. Removal of Biomass

The amounts of biomass debris on the soil are extremely variable ranging from less than 10 Mg ha^{-1} to more than 500 Mg ha^{-1} dry matter. Large amounts of biomass make planting difficult and would require too much labor to remove manually. Thus, farmers use fire to remove the biomass, giving easier access to the soil for planting.

B. Release and Addition of Nutrients from Biomass

Where the fallow vegetation has a high proportion of aboveground biomass as wood, the nutrients bound in it are not easily available to annual crops, as wood does not decompose rapidly. Burning the biomass releases the nutrients.

C. Weed Avoidance and Weed Suppression

Slash-and-burn agriculture farmers have three principal methods to suppress weed infestation: site selection, the timing of the burn, and tree retention. In forested areas, farmers might choose older secondary or primary forest for clearing because the arable weed seed bank is depleted or absent because the fallow phase has exceeded the viable period of arable weed seeds. If fields get too weedy, farmers may abandon them and clear a new field as the labor requirement for a new clearing may be less than weeding an old field. In short fallow rotations, where the fallow length is insufficient to deplete the weed seed bank, some farmers delay the burn until a flush of weeds has germinated. These weeds are destroyed in the burn.

D. Pest and Disease Avoidance

As pesticides are not available to most slash-and-burn farmers, they attempt to avoid crop pests and diseases by appropriate site selection. This option is limited to regions where long fallow phases would withdraw the hosts of pests and diseases for a sufficiently long time to eliminate them. However, pests such as rodents and monkeys would not be affected.

E. Declining Crop Yields

The combination of nutrient depletion, increasing weed, and pest and disease pressures leads to declining crop yields. After 1 to 4 years the field is abandoned to a fallow phase usually because of crop yield declines, or high weed biomass decreasing labor productivity to the point where clearing a field requires less labor than maintaining an old one. In South America, farmers often abandon their fields when they cannot expect the yield of the subsequent crop to be more than half that of the first crop. However, superimposed on this is the social custom of caring less for crops that are expected to yield less, which accentuates and accelerates any decline. When a cash crop is grown, reasons for abandonment can be external. Farmers in Indonesia reportedly abandoned their slash-and-burn pepper farms before crop harvest, when the market price for pepper fell.

II. THE IMPORTANCE OF FALLOW LENGTH: LONG AND SHORT FALLOW SLASH-AND-BURN SYSTEMS

Two main types of slash-and-burn agriculture are distinguished, which differ in their effects on the environment: long fallow systems (shifting cultivation) and short fallow systems.

The fallow is the successional vegetation that follows the cropping phase. It may be dominated by trees, shrubs, or grasses, depending on the climax vegetation type, management history, and successional stage. From a utilitarian perspective, its purpose is to reverse the degradative processes of cropping. Here, "degradative" refers to soil fertility decline, weed buildup, crop pest or disease buildup, or a combination of these. The fallow can act as a weed-, pest-, or disease break and fallow vegetation accumulates nutrients in the aboveground biomass, improves soil physical properties through root penetration, and usually permits the recovery of soil macrofauna. Thus, the sustainability of slash-and-burn agriculture depends primarily on the rate of degradation during the cropping phase, regeneration during the fallow phase, and their relative time allocation.

A model of this is detailed in Fig. 1. However, the diagram is a simplification and the rates of change are unknown. At low human population density, even on nutrient-poor soils, fallow phases may exceed 20 years; thus the system is a case 1 type "long-fallow system." Labor is the main factor limiting production. For a cropping phase of time a, fallow length ($b + c$) may exceed the optimum. The optimum fallow length, b, is defined as the minimum fallow length required to maintain and maximize crop yields per unit area in the long term and thus is just when the system has recovered enough to permit this.

When population densities increase, more land is required and fallow phases are shortened, up to the minimum length required to restore soil fertility (case 2). However, where human population densities increase to the extent that more land needs to be cropped than in case 2, fallows are further shortened and system recovery is not possible. If no additional inputs are made, this can cause soil fertility, weed, pest, and disease problems and lead to lower crop yields (case 3 type short fallow system).

III. GEOGRAPHIC DISTRIBUTION AND CHARACTERISTICS OF SLASH-AND-BURN AGRICULTURE

A. History

Slash-and-burn agriculture is probably the oldest method of land preparation for planting crops. It was used as a technique in China for establishing rice fields as early as the late Stone Age, 8000 to 10,000 years ago, and in Mexico at least 5500 years ago, concurrent with the domestication of maize. It was used in central

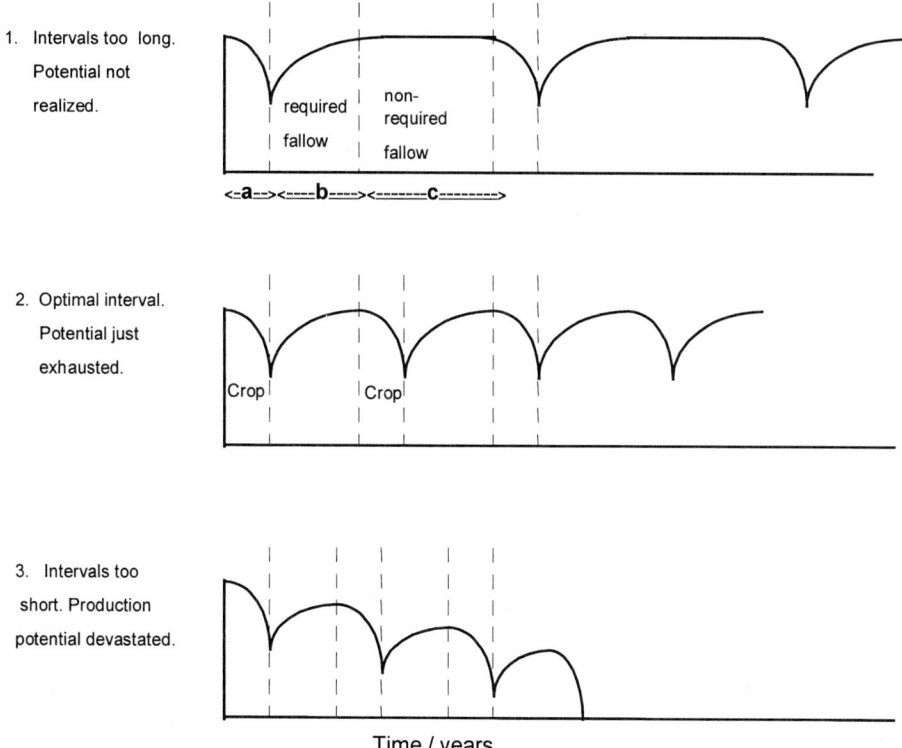

FIGURE 1 Degradation and recovery in slash-and-burn systems as related to the cropping interval and when no agronomic adjustments are made. Modified after Guillemin (1956).

Europe, along the Danube River, to penetrate the postglacial European forests, for cereal production and, later, in the boreal forests of Scandinavia.

B. Geographic Distribution

Today, slash-and-burn agriculture is practiced in the moist savanna, transitional, and forest zones of the tropics and subtropics of Central and South America, Africa, Asia, and the islands of Australasia. The majority of countries in these regions have weak economies. Slash-and-burn agriculture is also practiced in other nonindustrial areas, such as Bhutan. An estimated 36 million km^2 of land are under slash-and-burn agriculture, approximately 30% of the global soil resource. Human population densities in these areas can vary tremendously. In Africa, rural population densities in some parts of the Congo basin are less than three people per square kilometer, yet exceed 400 people per square kilometer in parts of southeast Nigeria.

C. Labor Requirements in Slash-and-Burn Agriculture

In areas with low population density, long-fallow systems predominate. Long fallow systems in forested areas have large land, yet low-capital and low-labor requirements. Soil fertility, weed, pests, and disease problems are avoided, rather than managed, by shifting to a new field. In the Amazon basin, only 8% of the human energy input to cultivate a cassava field, including postharvest processing, is used for slashing and burning. The energy or labor efficiency of long fallow slash-and-burn systems is the highest among all agricultural systems in terms of energy invested versus the amount of energy gained with the crop yield. This is largely due to the absence of fossil fuel use and chemical inputs. However, the destruction of the biomass and the release of the accumulated energy and carbon (section V) in the burn is not considered in such calculations.

In areas with higher population densities and consequently shortened fallow periods, economic conditions become more suitable for market-oriented, commercial farming as the higher population densities naturally create markets. While labor is available, small-scale farmers have scarce financial resources to purchase agricultural inputs. Due to infrastructural, economic, and soil-related problems of pesticide and fertilizer use, high-input, intensive agriculture is rarely practiced. With reduced fallow length, the labor requirement increases as additional field work, such as weeding and tillage, become necessary (section VI). In short fallow systems of southern Cameroon, land preparation, including slashing, burning, and cleaning the soil surface of unburned debris and weed stumps, was 10 to 13% of the total labor required to establish and harvest a groundnut/maize/cassava intercrop.

While population densities are increasing in many tropical areas and thus fallows shorten, in some situations the reverse has occurred. There were decreases in the population density of the Mayan lowlands and this was followed by a transition from intensive agriculture to long-fallow slash-and burn-systems.

D. Typical Fallow/Crop Sequences in Slash-and-Burn Agriculture

In West Kalimantan in the mid-1990s, fallow lengths averaged 17 years after a 1-year cropping phase, thus less than 10% of land was cultivated at any time. Such areas are characterized by low availability of labor and a lack of infrastructure, leading to very limited exchange of products and thus predominantly subsistence farming. In 1996, in southern Cameroon, 32% of fields cultivated had a previous fallow length of 8 or more years or had not been previously cultivated. In the early 1980s, in north Sierra Leone in the forest-savanna transition zone, most fields had been established from fallows of 30 years or more. Most farmers planted upland rice (*Oryza sativa*) for 1 year before abandoning the land. A minority of farmers followed rice by a groundnut then millet (*Setaria* spp.) sequence before abandonment.

In long fallow systems, crop diversity within one field can be high. A tuber species is often the dominant crop, which is intercropped with other roots and tubers, grains, vegetables, and herbs and spices. In the early 1980s, researchers counted an average of 10 crop species per 25 m^2 in 3-month-old fields of the Maring people of Papua New Guinea. The dominant crop was taro (*Colocasia esculenta*). Other crops included yams (*Dioscorea* spp.), sweet potato (*Ipomoea batatas*), maize (*Zea mays*), beans (*Phaseolus* spp.), sugarcane (*Saccharum officinarum*), and summer squash (*Cucurbita pepo*). Taro monocrops were also common. In the Columbian and Peruvian Amazon, fields may be dominated by cassava, constituting 80% of crop numbers. In Pará, in the Brazilian Amazon, monocrop maize (*Zea mays*) fields are common and farmers may use fallow from 3 to 30 years old or even primary forest.

In southern Cameroon, four of the most common slash-and-burn field types are as follows (Fig. 2):

- The *essep* or long fallow or primary forest conversion field, which is used to gain title to the land. This is approximately 0.25 ha and a few larger forest trees are retained. Relatively shade-tolerant crops, particularly tannia (*Xanthosoma sagittifolium*), egusi melon (*Cucumeropsis mannii*), and plantain (*Musa* spp. AAB), are grown with many minor crops. After the plantain harvest, the fields are either abandoned for 20 or more years or are burned and recultivated after 1 to 4 years as a groundnut/maize/cassava field.
- The groundnut/maize/cassava field (*Arachis hypogaea, Zea mays, Manihot esculenta*) with plantains (*Musa* spp. AAB) and leafy vegetables as minor components. Fallow length is 8 to 10 years near the border of Equatorial Guinea yet 4 years near the capital city, Yaoundé.
- Monocrop horticultural systems, particularly tomatoes (*Lycopersicum lycopersicon*) or maize for fresh consumption. These fields are most common around Yaoundé and external inputs, both inorganic and organic, are used.
- *Musa* spp. (banana and plantain) monocrops. These are usually established after long-fallow forest clearing.

Clearly, slash-and-burn systems may have high or low crop diversity. This does not depend on ecoregion or population density, but rather on the preferences of the farmers.

IV. EFFECTS OF CLEARING AT FIELD ESTABLISHMENT

Farmers in different parts of the world use slash-and-burn techniques yet their goals may vary. These range from completely and irreversibly removing the initial vegetation to retaining a rich repository of indigenous and useful plant species while introducing crops. A number of elaborate clearing and tree-felling techniques were developed by different indigenous cultures to facilitate the drying and burning of biomass. These techniques are summarized by Peters and Neunschwander (1988).

The types and numbers of trees retained vary considerably, depending on the tradition of the farmers and their intentions. However, usually very large high-canopy trees are not felled as: it is not worth the considerable effort involved; it would require paying for the labor of a skilled chainsaw operator; and the felled tree would take up a considerable part of the plot. Other trees retained include those producing fruits, nuts, or medicine, or those believed to maintain or enhance soil fertility or have a commercial value. Such trees are a source of seed, although such species cannot always establish in an environment imposed by slash and burn. Furthermore, these trees can harbor a large number of species of animals and possibly host microsymbionts. The retention of trees in slash-and-burn fields can facilitate the return of the initial vegetation, depending on the stand density of trees retained.

Reductions in tree densities, through slash-and-burn or other forms of clearance, may affect densities and species numbers of arboreal and litter-dwelling ants. Yet these species may also have a role in pest control. In areas cleared of forest in the Amazon, it was found that isolated trees can retain considerable ant diversity, although the species composition was not identical to that of the nearby forest.

When slash-and-burn farmers in the forests of West and Central Africa clear for a short-fallow groundnut/maize/cassava field, they completely clear the fallow vegetation and remove all trees so that there is abundant light for these crops, which do not tolerate shade. Additionally, as any mulch layer would impede groundnut growth, the farmer attempts a complete burn. The soil is tilled by hoeing after burning and weeding is frequent to reduce competition. In contrast, the root, tuber, and plantain crops of the "essep" field can be planted even if trees are retained, and if the slash is not completely burned, the soil is not tilled and weeding is neither as intensive nor as frequent as in the short fallow fields.

Partial or complete removal of vegetation has differential effects on the microclimatic and hydrological conditions after clearing. Soil temperatures are usually higher in cleared areas and even higher when the slash is burned. The absence of tall vegetation leads to a different redistribution of rain, as less or no rain will be intercepted in the canopy. Raindrops reach the soil surface with their full impact, potentially contributing to compaction and crust formation at the surface. Surface compaction can lead to temporarily ponding water at the surface and, if on sloping land, can cause erosion. Furthermore, crust formation can impede the gas exchange between soil and atmosphere resulting in deteriorating conditions for the soil fauna and flora.

Slash-and-burn fields have lower biomass than forests and crops may not cover the land for the entire year. The leaf area index may not be as high as in a

FIGURE 2 Typical crop association in slash-and-burn agriculture in southern Cameroon. Note the variety of crops including groundnuts, maize, sugarcane, cassava, tannia and plantain. Photo courtesy of U. Buettner.

forest, thus transpiration is lower. With reduced transpiration, water flux through the soil is increased, which can increase mineral losses due to leaching. The root systems of crops have to grow for each cropping cycle while in the forest the root system of the trees remains in place. Access to and uptake of water at the start of the rains is thus immediate in forest. Rooting depths of crops are restricted to the upper layers of the soil, while forest vegetation roots may penetrate deeper layers drawing on water during dry seasons, not accessible and available to food crops. The more permanent canopy and root system of forest contributes to less strong fluctuations of most climatic and hydrological factors.

V. CONSEQUENCES OF BURNING

A. Heat Evolved during the Burn

Few studies have measured the heat evolving from the burning of slashed biomass. The amount of heat evolved depends on the amount of biomass and its water content. Climatic conditions during the burn, such as wind speed, topographic features such as aspect, and slope of the field, have an impact on the intensity of the burn and the maximum temperatures attained. Slash-and-burn agriculture in savanna regions or in short fallow systems, where small amounts of biomass are burned, does not cause significant heat-related effects on the soil. Temperatures at the soil surface are relatively low and heat penetration into the soil is limited to 1 to 2 cm depth and the exposure time to the heat is short. However, even fast burns, which do not attain high temperatures, can kill seeds lying on the soil surface. There are no reports of light burns affecting seeds buried in the soil.

Burning of cleared forest biomass, depending on its water content, can reach 800°C near the soil surface, and these temperatures will be maintained for a longer period. If temperatures reach 400°C, soil organic matter may be lost by combustion, resulting in decreases in soil organic C and N, and clay particles may fuse, altering the soil texture and its cation exchange capacity. The spatial variability of temperatures is higher in a cleared forest than in short fallow or savanna systems. Tree trunks may burn for more than a week, evolving large amounts of heat and depositing large amounts of ash in a small area. In southern Cameroon, on an Ultisol, the maximum temperatures attained during burning of 3000 Mg ha^{-1} of wood, equivalent to a trunk of approximately 0.65 m diameter at a wood density of 0.6 Mg m^{-3}, were 788°C at the soil surface, 225°C at 5 cm, and 172°C at 10 cm and 105°C at 20 cm depths. The burn lasted for approximately 24 hr. Such severe burns usually kill plant seeds even in deeper layers and reduce living microbial biomass in the soil, including mycorrhizae, rhizobia, and other microsymbionts. Further, all soil meso- and macrofauna are affected, depending on the temperature reached and the soil depth to which it penetrated.

B. Soil, Soil Nutrients, and Soil Physical Properties

Depending on the level of disturbance, local consequences range from nutrient inputs from the atmosphere such as sulfur dioxide and flying ash to changes in microclimate and increased variation in rainfall pattern, and influxes of pests, diseases, and invasive weeds.

Many soils in forested areas of the tropics have a low nutrient status, aluminium toxicity problems, high levels of phosphorus fixation by iron oxides, and a low cation exchange capacity due to the dominant kaolinitic clays, which are also susceptible to structural collapse under mechanization. Where the fallow vegetation has a high proportion of aboveground biomass as wood, the nutrients bound in it are not easily available to annual crops, as wood does not decompose rapidly. Burning the biomass releases the nutrients, and with the exception of nitrogen and sulfur, a large proportion will remain in the ash on the soil surface. While the nitrogen is almost entirely lost to the atmosphere, the sulfur, as sulfur dioxide, can be redeposited as sulfuric acid near the burn, if the humidity of the air is high. When the fuel (biomass debris) is very dry, the burn is fast and hot and strong upward air currents result, which carry away the ash particles. Windy conditions during the burn will increase such losses. In the Amazon, element transfer to the atmosphere due to ash particle transport and volatilization have been reported as up to 98% C, 98% N, 33% P, 31% K, 24% Ca, and 43% Mg of the initial amounts in the fuel. Sulfur losses to the atmosphere were estimated between 69 and 76% of that in the original biomass.

Wood ash contains calcium, magnesium, and potassium in the form of phosphates, carbonates, and silicates and also other elements. The nutrients in the ash are partially water soluble and will be released into the soil solution with the rain. Thus, slash-and-burn agriculture makes nutrients available to crops. While farmers realize the fertilizing effect of the ash, they are more concerned to achieve complete burns to remove the biomass. The longevity of increases in calcium, magnesium, and potassium concentrations in the topsoil depends partially on the cation exchange capacity. While the increased calcium concentrations are likely to be long term because of low calcium mobility and solubility, magnesium and potassium may be prone to leaching. Burning usually reduces soil organic matter content. On kaolinitic soils, much of the cation exchange capacity (CEC) is on organic matter thus a reduction in organic matter will reduce CEC and increase the risk of leaching. In southern Cameroon, it could be shown that increases in exch. K were measurable down to 50 cm within 6 months of burning forest biomass.

However, with the rain dissolving the ash, soil pH usually increases rapidly to above pH 8 and remains high for an extended period. Topsoil pH usually increases after burning. In soils where aluminum toxicity problems occur, burning reduces extractable aluminum concentrations and increases phosphorus availability.

It has been reported in the literature that the burn loosens the soil and facilitates tillage and planting by reducing bulk density and penetrometer resistance. Direct measurements in southern Cameroon could not confirm these reports. Farmers experience the forest floor to be difficult to penetrate with tools before the burn, probably because of the dense network of small but woody roots between the litter layer and the soil surface. The litter layer and the roots will be destroyed in the burn and will no longer be an obstacle. Furthermore, attempts to work the soil before burning are on dry soil, which is harder, while planting after the burn usually happens after some rains, softening the soil. However, where burns are hot and lengthy, such as under a log, the resultant ash is white. Soil color changes to bright red and compaction and a collapse of its soil structure can be observed. This is probably due to a relative excess of monovalent cations, which leads to a breakdown of cation bridges between clay minerals, the combustion of soil organic carbon and thus substances binding soil particles, and the elimination of soil macro and microfauna mixing soil particles and organic materials to form stable aggregates. Slash-and-burn farmers recognize such areas in their fields and usually do not plant crops there.

In savanna regions with small amounts of biomass, the effects of burning upon soil are not discernible. Ash inputs do not cause a measurable increase in pH or cation concentrations. Neither soil organic carbon nor total soil nitrogen is affected by such light burns. In grass fallows, the amounts of biomass may not be an obstacle to planting. However, the quality of the slashed biomass is low. Biomass retention as mulch would have adverse effects on crop growth, because soil nitrogen might be immobilized in the decomposition process, causing nitrogen deficiency for the crop.

Most slash-and-burn agriculture systems use only the *in situ* slashed vegetation. There are, however, systems such as the *chitemene* system in southern Africa in which wood from a larger cleared area is collected and concentrated in a smaller area where it is burned. Farmers seek nutrient augmentation or concentration in the topsoil by piling large amounts of wood on the soil surface. The wood is transported to the field from the surrounding miombo woodland. The area cleared may be up to 20 times larger than the field in which the wood is burned, adding large quantities of nutrients. Although this system appears to be designed to concentrate nutrients, its efficiency in retaining these nutrients remains low. In an experiment conducted in Zambia, 84% of the phosphorus contained initially in the vegetation was accounted for by increases in the top 50 cm of soil at 40 days after burning. However, 57% of this phosphorus was at 20 to 50 cm depth, in a region already inaccessible to any growing crop. For potassium, only about 10% of the original input was retained by the top 50 cm soil at 40 days after burning. The remainder presumably had leached below 50 cm depth.

In other systems, only part of the biomass is burned while a certain portion is used for other purposes such as fencing the field against animals, as fuelwood or for construction.

C. Soil Bacteria and Fungi

Burning can reduce living microbial biomass in the topsoil and potentially their decomposing activities. However, there are reports that burning increased microbial biomass and activity in tropical savanna and in a tropical plantation forest. It has been shown in Kenyan soils that total bacteria counts dropped after burning. Nitrogen-fixing bacteria are very susceptible to pH changes after burning. Such effects of the burn on microsymbionts can subsequently affect the composition and productivity of the vegetation as certain species will depend on symbionts.

Many crops have mycorrhizae and they are able to access organic phosphorus using the enzyme phosphatase in a hydrolysis reaction. Thus, although burning can increase phosphate availability, if burning reduces mycorrhizae, overall effects could be negated, depending on whether the crops grown are mycorrhizal and whether the planting material is infected. Studies in India comparing vesicular-arbuscular mycorrhizal (VAM) infection of plants between burned and non-burned areas of a tropical forest found that infection was reduced in burned areas. However, studies from wet tropical forests in Costa Rica found that mycorrhizae survived burning.

D. Invertebrates: Soil Fauna

The soil fauna are only directly affected by the burn if their active stages are present in the topsoil during the burn. As fields are usually burned toward the end of the dry season, groups such as earthworms are inactive and thus unaffected. After burning large amounts of biomass, the ash is dissolved in the first rains. This solution is very high in pH (up to 10). Such solution infiltrating the soil might kill certain species that used to live under rather acidic conditions under the previous forest cover (pH of around 4.0 to 4.5 in most tropical forests).

E. Weeds

Severe burns usually kill plant seeds even in deeper layers and can reduce the weed seed bank. Although burning is believed to reduce weed pressure, it can facilitate weed invasion. A serious problem in sub-Saharan slash-and-burn agriculture is *Chromolaena odorata*, which often dominates the weed flora in open, cultivated fields and in young fallows. *C. odorata* seeds have higher germination rates when they are exposed to light on the soil surface. Thus, airborne seeds can colonize newly burned fields. On the contrary, where slash is retained, seeds fall onto the slash, which dries out quickly preventing germination. Once established, *C. odorata* plants can survive burning as their woody stumps may remain unaffected in the upper soil layers. Burning destroys seedlings of other species and thus can reduce competition from other weeds on *C. odorata*.

F. Other Plant Species

Slash-and-burn agriculture in a tropical dry deciduous forest in Mexico eliminated 29% of resprouting species, particularly those that were present at low abundance. This reduction was exacerbated by repeated burning.

G. Vertebrates

The effect at field establishment and specifically at burning depends on the species and its mobility. While mammals may escape the slash and the burn by moving into surrounding unaffected vegetation, some reptiles and amphibians are killed when forest is cleared and burned, as they cannot escape quickly. Depending on the season and the temporal pattern of activity, animals that hibernate through the dry season may escape the fire if burning happens before larger rains have fallen. Where vertebrate densities decrease after slash-and-burn agriculture, the potential for fallow recovery may be reduced where vertebrates are responsible for seed dispersal.

H. Global Effects

The destruction of natural vegetation and its replacement by crops in slash-and-burn agriculture systems is accompanied by the release to the atmosphere of carbon as carbon dioxide during the burn. Decomposition rates of soil organic matter may increase during the cropping phase due to tillage and soil chemical changes, releasing additional amounts of carbon. Furthermore, food crops will assimilate less carbon from the atmosphere than the initial vegetation would have assimilated, thus reducing the C-fixation in biomass and contributing to a larger net release of carbon.

VI. EFFECTS DURING THE CULTIVATION PHASE

In traditional shifting cultivation with long fallows, the cropping phase lasts 1 to 4 years. As the length of cultivation increases, the potential of land to recover is reduced. Operations such as tillage and weeding frequently disturb the soil and the viability of seeds of the initial vegetation is reduced. Regrowth from stumps is slashed more frequently, increasing the risk of dieback. There are changes in microclimatic and edaphic conditions such that species of the initial vegetation cannot survive.

A. Tillage

The traditional shifting cultivator does not use many tools to work the soil. Most crops are planted with a simple planting stick or by opening small planting holes with a machete or hoe. Most crops, such as cassava, tannia, taro, plantain, melon, and maize, do not require tillage to succeed. If slash and burn is practiced in short fallow systems in degraded land, crops such as melon and plantain are not grown as they are considered not to produce well, while other crops can perform well but require tillage. The soil is tilled either to mix the ash with the soil, to bring the seed to the required planting depth, or to concentrate nutrient-rich topsoil around the crops. Most intensive soil disturbance is caused by tillage for crops such as yams (*Dioscorea* spp.), which are mounded, groundnuts (*Arachis hypogaea*) for which the soil is tilled at planting and a second time before pegging, potatoes (*Solanum tuberosum*), and sweet potato (*Ipomea batatas*) for which ridges or small mounds are made. Farmers throughout have made their own observations and know which crops require tillage. However, tillage is a labor-intensive operation and is more frequently used where labor availability is high.

Tillage disrupts the natural layering of the soil and breaks up soil aggregates. Naturally formed pores (earthworm channels, termite and ant galleries, root channels) are interrupted and often are filled with loose fine material. Soil fauna dwelling in such pores will be negatively affected by tillage, as their habitats are destroyed or the number of potential niches is reduced. Tillage changes the water regime in the topsoil and might lead to a more rapid drying out of the top layer, causing stronger moisture and temperature fluctuations. Soil mites and springtails respond to such fluctuations by withdrawing into deeper layers, which may be less rich in substrate and thus do not permit the same level of activity and population density.

When land is cleared, slashed, burned, and tilled, earthworm density, diversity, and activity are reduced. This has been shown in southwest Nigeria (Fig. 3), where the earthworm fauna is dominated by *Hyperiodrilus africanus* and the epigeic *Eudrilus eugeniae*, in the Peruvian Amazon, in southern Cameroon and in southeast Mexico (Figs. 4 and 5). Earthworms are classified functionally into epigeic, anecic, and endogeic categories. Epigeics live in the litter layer. Anecics feed on litter and soil, dragging litter into their vertical burrows. Endogeics live in the soil feeding on soil organic matter and dead roots. Anecic and epigeic are the groups most immediately affected by slash, burn, and tillage due to the loss of current and future substrates and physical habitat disruption. These groups will take longer than endogeics to recolonize the recovering fallow.

Tillage affects the seed bank in the soil. This is less important in long fallow slash-and-burn systems as seeds will normally be on the soil surface where they are most likely to be killed by the burn. In younger fallow, with light burns, not all seeds on the soil surface may be killed. Where the land has been previously cultivated with tillage or any soil-moving operation (harvest of root and tuber crops) has been conducted, a seed bank, protecting the buried portion of seeds from the effect of subsequent burns, may already exist in the soil. Buried seeds can become dormant and survive for several years. Tillage after the burn will move such seeds closer to the soil surface, increasing their chance of germination.

B. Weeding

Most weeding operations are conducted with the same or similar tools as tillage. Therefore, in addition to the tillage, weeding affects the soil, soil inhabitants, and the weed seed bank. It is difficult to separate the terms "seed bank" and "weed seed bank" as the same species may be both: a weed in the cropping phase and a desired plant in the fallow phase of a slash-and-burn rotation (cycle). To simplify, all nonplanted species will be treated as weeds in the cropping phase, although in some slash-and-burn systems farmers tend certain "weeds" as they produce useful items, spices, or food. The weeding of a field, established after forest clearing, will eliminate seedlings and reduce stump regrowth of forest species. Depending on the cutting height and the frequency of weedings, these species may eventually die. When the land reverts to fallow, these species will

FIGURE 3 Burning savanna grassland in central Nigeria. Photo by S. Hauser.

be unable to reestablish immediately. The fallow may thus consist only of weeds from the cropping phase.

C. Pest and Disease Suppression

Certain pests and diseases of crops may be avoided by choosing land that was under fallow for a long time. The efficiency of pest and disease avoidance depends on the distances between fields, the mobility of pests, and the vectors by which pests and diseases are spread. Soil-borne diseases might be affected by the dramatic soil chemical changes after the burn and the associated increase in soil pH and the changed microclimate. One important, yet not directly "slash-and-burn" related mechanism in pest and disease suppression is the complex mixture of crop species in slash-and-burn fields. Up to 40 species might be planted or tended within a single plot. Pests and disease buildup is not possible or is strongly delayed if the organisms are confined to a few isolated plants. The large number of pests and diseases with their different ecological requirements does not permit a general statement of the effects of slash-and-burn agriculture on their presence and severity in crops.

Radopholus similis is a plant-parasitic nematode, cosmopolitan in the tropics. It is the greatest cause of yield loss in bananas and plantains worldwide. Groundnut and maize are also hosts, as are *Commelina benghalensis* and *Fleurya aestuans*, minor components of weed and fallow communities in West and Central Africa. *Radopholus similis* does not survive in the soil for more than 6 months when host roots and corm pieces are absent. Thus, even a single year fallow can prevent infection of the following crop, if *C. benghalensis* and *F. aestuans* are uprooted and the new planting material is pest-free. The Maring people of Papua New Guinea plant banana suckers into their field during the burn. Heat would penetrate the outer layers of the banana sucker and would be likely to kill nematodes infecting the planting material. Other peoples roll banana suckers in ash before planting, believed both to kill nematodes and have a fertilizing effect.

Certain crop diseases are affected by the degree of shade in the plot. For example, *Mycosphaerella fijiensis* fungus is the causal agent of black sigatoka disease of plantains and bananas. *Mycosphaerella musicola* is the less virulent yellow sigatoka, dominant at higher altitude. Workers in the Caribbean, Central America, and Africa have noted that when trees are eliminated from slash-and-burn fields, these diseases destroy larger areas of the leaves and yields are reduced by up to 40%.

Nowadays, short fallow slash-and-burn farmers have

FIGURE 4 Forest clearing in southern Cameroon after burning. Note the large tree stumps and remaining trunks on the field. The wood may be piled and burned again or allowed to rot, depending on the crop to be planted. Photo by S. Hauser.

a limited choice of land, but in some areas they have access to pesticides. With a broad range of crops and more recently intensive vegetable production, the use of pesticides has become more common in many developing countries. Slash-and-burn farmers do resort to pesticide use. While any pesticide may be targeted against a specific pest or disease, it usually affects other organisms as well and indirectly organisms relying on the affected ones. With often little knowledge of the effects of pesticides on slash-and-burn agriculture, cultivators may destroy organisms that are important to the ecosystem's functioning. Thurston (1992) provides a very comprehensive description of pest and disease control in such low-input cropping systems.

D. Management of Crop and Weed Residues

Some effects of slash-and-burn land preparation are aggravated by crop management. In the natural vegetation, in both long and short fallows, litterfall will accumulate on the soil surface, contributing to reduced soil temperature and providing nutrient and substrate inputs. During the cropping phase, some crops and weeds do produce litter. Yet mulching or simple retention of such residues is not a common practice amongst slash-and-burn farmers. In most cases, litter and weeds are removed, thrown outside the field, or piled on tree stumps. Retaining residues may create niches in which some of the fauna may survive the cropping phase. Yet slash-and-burn farmers are aware that mulch can also provide potential niches for pests.

E. Nutient Export through Crop Harvesting

In addition to unproductive losses (combustion, wind, leaching, erosion, transfer to unavailable forms), the harvested crops contain nutrients that are exported from the field. As slash-and-burn crop yields are usually low, crop export losses are considered marginal compared to other losses, especially in situations in which old fallows are cleared. However, in short fallow systems, thus with more frequent cropping or prolonged cropping phases, the crop export losses will become more important in the overall nutrient balance of a field.

F. Effects of Prolonged Cropping Phases

Prolonged cropping phases permit the invasion of weeds suited for conditions in a clearing. Such weed species die out if the field is abandoned after a short

FIGURE 5 Bush fallow clearing in southern Cameroon after burning. Note the absence of large tree stumps. Only small tree trunks are remaining, which will be burned in piles. Photo by S. Hauser.

cropping phase. However, if the cropping phases are prolonged, competition from these invasive weeds further reduces the chances of recovery of the initial vegetation, particularly if the plot is burned repeatedly. Repeated burns kill tree stumps and seeds and can select for grasses (Poaceae), which establish from seeds but often propagate by rhizomes in the soil and thus are less affected than other species by the burn. The result is that the plot may revert to "arrested succession" grassland and be rendered unsuitable for cropping in the medium term. This may be a goal of a farmer who wishes to create pasture or may be an unintended consequence. In southeast Asia, prolonged cropping after forest clearing has facilitated the invasion of the land by *Imperata cylindrica*.

Generally, climax forest species have seeds that remain viable for a short time. Thus, extended cropping phases will compromise the regeneration potential of the fallow. However, where the intended purpose is to recrop after short fallow phases, the absence of viable seeds is academic, as they would have no possibility to establish on that site.

VII. EFFECTS DURING THE FALLOW PHASE

The effects of slash and burn on biodiversity in the fallow phase are strongly related to the conditions prevailing during the cropping phase. In long fallow systems with short cropping phases, the transition into a fallow dominated by pioneer trees is fast, without a phase in which arable weeds dominate. In the highlands of Papua New Guinea, the Maring people tend the resprouting trunks of the pioneer tree *rokunt* (*Saurauia* sp.) in their fields near the end of the cultivation phase, even though these compete with their crops. Seedlings of *Casuarina oligodon*, a nitrogen-fixing tree, are transplanted to the field at this time to enhance soil fertility recovery during the fallow. Such reestablishment of the fallow vegetation will probably ensure the reestablishment of the soil micro-, meso-, and macrofauna. With an increased length of the cropping phase, fields tend to weediness as species not existing in the previous fallow might have established in the cultivation phase. These are likely to form at least part of the first fallow succession community. In such a situation, the establishment of a fallow vegetation as before the slash-and-burn cycle is delayed, in some cases impossible (Fig. 6).

Biomass production and nutrient uptake and cycling are the primary processes of soil fertility restoration. The amounts of biomass and nutrients accumulated are a function of the fallow length. The minimum fallow length to completely restore soil fertility (Fig. 1, case 2) will vary depending on soil type, the types of weeds, crop pests and diseases present, and the growth rates, composition, and succession of the fallow species.

Aboveground biomass accumulation rates of forest fallows vary between sites and may vary between fallow ages. Most data from 4-year-old fallows range from 24 to 56 Mg ha^{-1}, for 8-year-old fallows ranging from 40 to 150 Mg ha^{-1}, and with older fallows up to 500 Mg ha^{-1}. Biomass accumulates for at least the first 8 years. Some authors have reported that half the biomass in a 20-year-old fallow had already accumulated by year 4. In contrast, others showed that accumulation rates of approximately 5.5 Mg ha^{-1}yr^{-1} remained constant from the 4th to the 15th year.

FIGURE 6 Yam field, slashed and burned and tilled into mounds awaiting the start of the rains in west central Nigeria. Photo by S. Hauser.

Generally soil organic matter levels increase with time under fallow. During 4 years of bush fallow in southwest Nigeria, soil organic carbon content increased by 46% and total nitrogen by 59%. A 15- to 20-year-old fallow had soil organic matter levels of 3.8% compared with 1.1% in cultivated plots. In the drier region of northeast Brazil, a 10-year-old fallow was sufficient to restore soil carbon and nitrogen levels to those under undisturbed vegetation. Many authors have reported that soil organic matter and soil nitrogen concentrations do not increase substantially after 8 to 10 years of fallow.

Occasionally farmers use indicators to judge when to recrop a fallow. In southwest Cameroon, farmers stated that there is a definite minimum fallow time to regain fertility, expressed in years but judged by the type and maturity of the vegetation growing there. The Turumbu of Yangambi in eastern Zaire judged when to break fallow by the girth of umbrella trees (*Musanga cecropioides*) and the biomass of the undergrowth. They cleared when they could walk easily beneath the trees. These secondary considerations coincided with soil fertility regeneration (Jurion and Henry, 1969). In southwest Nigeria, farmers claim to use the appearance of the vegetation cover, the presence of particular plant species, and the presence of earthworm casts to decide whether a fallow is ready for recropping.

In Papua New Guinea, bird and reptile species diversities were low in young fallows but increased as succession proceeded. Reptiles were less affected than birds. The bird communities in young, grassy fallows were dominated by obligate granivores, whereas the forest supported specialist feeders (frugivores, nectarivores, and branch gleaners). Butterfly species richness showed a similar pattern.

In southern Cameroon, species richness of birds in *Chromolaena odorata* fallow was only a fifth of that in secondary or primary forest. Butterflies and termites had similar magnitudes of declines. Canopy ants and canopy beetles were completely absent in such fallows, given the near absence of trees. The species richness of flying beetles, leaf-litter ants, and soil nematodes was not different between forest and young fallow. Presumably, the young fallow, located within a mosaic of recovering fallows and secondary forest, provided sufficient food and conditions for their continued survival. For more details, see Lawton *et al.* (1998).

In southern Cameroon, earthworm activity, indexed by surface cast production declined rapidly after forest clearance, burning, and cropping and did not recover during 2 years of fallow. At the same site, cast production decreased after clearing, burning, and cropping a 4-year-old *Chromolaena odorata* bush fallow, but did recover again during the fallow phase. The change in living conditions from forest to cropland was too drastic for the existing earthworm community to maintain its activity level. A new community structure comprising species capable to cope with the new conditions could not establish within 2 years of fallow. On the contrary, the community in bush fallow had probably already undergone some shifts thus recovering as soon as bush conditions were reestablished. The response of termites was different. In both forest and bush clearings, termite densities dropped to about 1% of densities in the retained forest. There was no recovery during the fallow phase. The number of species in the cleared area was only half as high as in the forest.

VIII. THE SCALE DEPENDENCE OF THE EFFECTS CAUSED BY SLASH-AND-BURN AGRICULTURE

A. Topographic Preferences and Fragmentation

Slash-and-burn farmers have topographic preferences. Long-fallow farmers often prefer gently sloping land as this facilitates the burn. They avoid slopes that are too steep as this increases the risk of soil erosion. Yet their choice is limited by the type of terrain available. In other parts of the tropics, farmers prefer the plateaux and avoid cropping even on the gentle slopes. Valley fringes, although difficult to access and waterlogged during the rainy season, are under greater pressure from farmers for dry season cropping of high value market crops. In contrast, riverine forests and swamps are avoided and therefore remain undisturbed.

Such preferences result in fragmentation. Ultimately, in areas of high-population density, fragments of the previously contiguous vegetation will only remain in agriculturally unsuitable areas. Not all species of plants and animals are represented in such fragments of vegetation and thus a recovery of the surroundings to the initial condition is not possible. However, if slash and burn in particular topographic positions occurs in low population areas where systems have sufficiently long fallow to restore the forest, it does not have major effects on biodiversity.

B. Temporal and Spatial Scales

Traditional systems at low population densities, with small fields (0.25–3 ha) in 1-year cropping phases and long fallows of 20 years, use 5% of the land at any given time (Fig. 7). Fields are surrounded by fallow land at various stages of recovery. Even if the fields are cropped for up to 4 years, not more than 20% of the land will be cultivated. Recolonization of a cleared plot by species from the surrounding fallow occurs rapidly after farmers abandon the land. The rate of recolonization by animals depends on the species' mobility. Plant species' invasion or recolonization of the site depends on the species' type of seed and the method of seed dispersion. Unless species are extremely immobile, or in the case of plants, if the fallow length is too short for species to reach the reproductive phase, this system can maintain the initial biodiversity.

With increasing human population densities, slash-and-burn agriculture is practiced in shortened bush or grass fallow cycles, with prolonged cropping phases. Both processes lead inevitably to a larger proportion of land being under cultivation. This occurs rapidly when fallows are shorter than 10 years in systems cropping for more than 1 year and when fallows are shorter than 5 years in systems cropping for 1 year (Fig. 7).

Topographic preferences and an increasing proportion of land cropped at any given time lead to fragmentation, with areas of the initial or recovering vegetation isolated by cropped land and early stages of fallow. Spatial fragmentation can compromise (a) the ability of the remaining undisturbed or recovering habitat fragments to maintain diversity, which determines the potential to recolonize the disturbed habitats, and (b) the ability of species to be dispersed to and establish in disturbed areas. Both aspects depend on the species

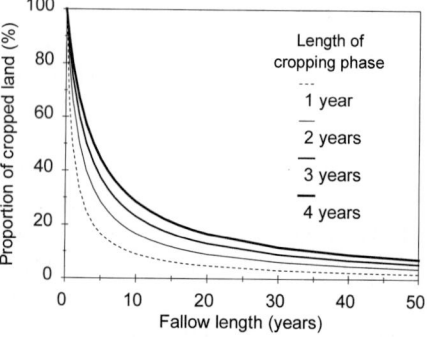

FIGURE 7 Proportion of cultivated land at 1 to 4 years of cropping phases, as a function of fallow length.

investigated. Generally, with increasing size of animals, spatial requirements for foraging and hunting increase and with increasing fragmentation their chances of survival diminish. A number of additional factors, such as mating habits and the ability to cross disturbed areas to reach other undisturbed fragments, will modify the likeliness of the species' maintenance. Smaller animals, largely invertebrates, might not be affected as severely by fragmentation as larger animals. On isolated trees retained in pastures in the Amazon, a considerable diversity of ants was found, with a large proportion of species usually found in the forest. Ant diversity was positively related to the epiphyte load of trees.

Fragmentation of forest by pastureland such as in the Amazon basin negatively affects certain tree species retained in the forest fragments. Some species need cross pollination to produce seeds. If fragments become too small or too distant from each other, the tree density might fall below a threshold required for pollen to reach other trees. Such a process, observed in mahogany (*Swietenia macrophylla*), reduces or eliminates seed production and may lead to local extinction.

The potential to regenerate degraded land to the initial vegetation and habitat depends on seed dispersion and mobility of animals. Seed dispersion by birds, bats, and monkeys has been shown to be important in disturbed and undisturbed areas. However, in Uganda, it was shown that bats and birds did not considerably contribute to the dispersal of forest species' seeds to degraded grassland areas. The vast majority of seeds moved by these animals were of species dominantly found in disturbed areas or of species unable to establish in the grassland. Similarly, seed dispersion by gibbons confined to small forest fragments after large-scale slash-and-burn agriculture in Kalimantan was found to be uncertain as individuals were extremely hesitant or unable to cross cleared areas to reach other fragments.

The effects of slash-and-burn agriculture on biodiversity are various. However, generally, the shorter the cropping phase and the longer the fallow phase, and the smaller the proportion of cropped land, the greater the possibility of recovery of the previous vegetation. However, these criteria limit the productivity of slash-and-burn systems. To maintain their livelihoods, the choice for many farmers whose land allocation is fixed is either to prolong the cropping phase or to shorten the fallow phase. Clearly research should also compare and contrast the effects of these options on biodiversity, rather than focusing exclusively on comparisons with undisturbed ecosystems, which are increasingly unrepresentative.

IX. ALTERNATIVE SYSTEMS TO SLASH-AND-BURN AGRICULTURE

A. Plantations of Perennial Crops and Trees

Plantation systems include monocrop oil-palm, fast-growing timber species, coconut, rubber, tea, coffee, and cocoa. Most plantation systems are established by slash and burn. Timber plantations are often established in forests by selective felling or poisoning of undesired trees, underbrushing, and sometimes light burning of the understory. In such systems, the forest habitat is only slightly altered or reestablishes quickly. In southern Cameroon, timber plantations had similar levels of termite species richness to secondary forests and more species of leaf-litter ants than both secondary and near-primary forest (Lawton *et al.*, 1998). However, the dominance of one tree species may affect the underbrush regrowth and the conditions for soil fauna through a narrower range of substrate. Many plantations, such as oil palm, coconut, rubber, tea, and other species that demand full sun light, are planted after clear-cut felling. In contrast, cocoa plantations in southern Bahia, Brazil, are established in old secondary or primary forest. Cocoa is shade tolerant, so while the lower-canopy trees and herbaceous components are slashed, many upper-canopy trees are retained. These include many valuable timber species, rosewood (*Dalbergia nigra*), pau Brasil (*Caeselapinia esplinata*), and cedro (*Cedrela odorata*), that have been nearly eliminated elsewhere by loggers. Consequently associated fauna have been maintained, including rare primate species, for example, the sagui (*Callathrix kulhii*) and the lion tamarin (*Leontopithicus rasalia*).

B. Pastures

In large parts of the Amazon basin and in Central America, forest is clearcut to establish pastures. While in tree and palm plantation systems a forest-like microclimate will be reestablished after some years, pastures will remain unshaded. Any forest regrowth is regularly slashed and in some situations pastures are regularly burned to reduce forest regrowth and promote grasses.

C. Improved Short Fallow Systems

Slash-and-burn agriculture is demanding on the natural resource base, irrespective of the fallow type (forest or short fallow). It requires large areas for cultivation as yields are low. In many situations the natural regrowth in short fallow systems is not capable to restore soil fertility because of the shortened period of time available in which the volunteer regrowth cannot produce sufficient biomass and accumulate sufficient amounts of nutrients. Improved fallow systems seek to replace the, usually undesired, volunteer regrowth by species that produce more biomass faster, contain more nutrients, or produce a type of biomass that is of a different physical structure than the natural regrowth. Such species are expected to outcompete the weed flora quickly and not to contribute to propagation or maintenance of crop pests and diseases. Often legumes fixing atmospheric nitrogen are used to augment the nitrogen resources of the soil. Depending on the plant type used for improved fallows (trees, shrubs, or herbaceous), the biomass is slashed and burned or only slashed and retained as mulch. Systems using herbaceous legumes, such as *Pueraria phaseoloides*, reduce the number of plant species surviving in the fallow rather drastically. However, depending on the crop planted after the fallow, the slashed mulch can be retained as a mulch layer, with minimal disturbance to soil invertebrates. The tree- and shrub-based systems are more likely to be managed in a similar way as any other slash-and-burn system, because the woody proportion of the biomass is more easily cleared by burning. However, these systems are more productive per unit area or restore soil fertility faster than systems relying on natural regrowth. Therefore less land is used for agricultural production and consequently more land remains undisturbed, retaining biodiversity.

See Also the Following Articles

AGRICULTURE, TRADITIONAL • CARBON CYCLE • DEFORESTATION AND LAND CLEARING • FIRES, ECOLOGICAL EFFECTS OF • INDIGENOUS PEOPLES, BIODIVERSITY AND • POVERTY AND BIODIVERSITY • SOIL BIOTA, SOIL SYSTEMS, AND PROCESSES

Bibliography

Beets, W. C. (1990). *Raising and Sustaining Productivity of Smallholder Farming Systems in the Tropics*. AgBé Publishing, Alkmaar, The Netherlands.

Guillemin, R. (1956). Evolution de l'agriculture autochtone dans les savanes de l'Oubangui. *Agronomie Tropicale* 11, 143–176.

Jurion, F., and Henry, J. (1969). *Can Primitive Agriculture Be Modernised?* Agra Europe, London (translation from French).

Lawton, J. H., Bignell, D. E., Bolton, B., Bloemers, G. F., Eggleton, P., Hammond, P., Hodda, M., Holt, R. D., Larsen, T. B., Mawdsley, N. A., Stork, N. E., Srivastava, D. S., and Watts, A. D. (1998). Biodiversity inventories, indicator taxa and effects of habitat modification in tropical forest. *Nature* 391, 72–76.

Peters, W. J., and Neuenschwander, L. F. (1988). *Slash and Burn: Farming in the Third World Forest*. University of Idaho Press, Moscow, ID.

Thurston, H. D. (1992) *Sustainable Practices for Plant Disease Management in Traditional Farming Systems*. Westview Press, Boulder, CO.

SOCIAL AND CULTURAL FACTORS

Jeffrey A. McNeely
IUCN—The World Conservation Union

I. Religion and Biodiversity
II. Language and Biodiversity
III. Indigenous Peoples, Resource Management, and Biodiversity
IV. Urban Versus Rural Perceptions
V. Conclusions

GLOSSARY

adaptation Any behavioral pattern that makes a human or a cultural group more fit to survive and reproduce in comparison with other individuals or cultures.

carrying capacity The maximum population that can be sustained indefinitely in a given area without changing the ecosystem in ways that will eventually reduce the sustainable population. This balance between population and resources is a dynamic one that involves changes in technology and other factors.

community A social unit consisting of members who are in direct interaction with each other and who have a collective identity, however defined. Relationships in such a group are principally primary rather than secondary in nature, and conformity to group norms is achieved mainly by peer pressure.

culture All capabilities and habits acquired by people as members of society.

indigenous peoples Those groups of human beings who share and preserve direct, everyday connections to their distinguishable distinctive cultural roots (even though they may have willingly migrated or been forcefully moved from their homelands, including moves to cities). These indigenous peoples consciously or unconsciously draw on specific knowledge and strategies developed and tested by past generations to address current problems. Cultural traditions emerge and are maintained in a dynamic process of creative invention and reinvention, as well as borrowing and adaptation from other subgroups and cultures.

language A system of conventional spoken or written symbols that enable human beings to communicate as members of a social group and participants in its culture.

management The efforts of humans to select, plan, organize, and implement programs designed to achieve specified goals; activities can range from protective measures to ensure that human influences on natural resources are minimized to greater interventions required to maintain diversity, install facilities, control populations, and eliminate unwanted elements.

myth A story, usually of unspecified origin and at least partially traditional, that ostensibly relates actual events to explain some practice, belief, institution, or natural phenomenon and that is frequently associated with religious beliefs; myths provide models for human behavior and often emphasize behaviors that support conservation of biodiversity.

religion The relationship between people and that which they regard as holy, often in supernatural terms. Religions include a wide range of ceremonies and ritual practices aimed at supporting the moral and ethical values of a society and maintaining natural or built sacred spaces where such ceremonies and practices can take place.

BIODIVERSITY IS OFTEN CONCEIVED in primarily biological terms, but the term has substantial cultural elements as well. Culture at its most basic includes all capabilities and habits acquired by people as members of society. It consists of language, ideas, beliefs, customs, institutions, technologies, works of art, religions, and ceremonies. Every human society has its own particular culture, often with its own language. Cultures change over time based on historical factors, physical habitats, and the availability of local resources. Thus culture and biodiversity are closely related, with culture carrying the attitudes, values, ideals, and beliefs that enable individuals to live within the particular constraints of their local ecosystem and to relate to other ecosystems. Particularly important components of culture in the context of biodiversity include religion, language, and traditional knowledge about resource management.

I. RELIGION AND BIODIVERSITY

Religions order the spiritual and physical relationships of people with other humans and with their environment. Approaches to conserving biodiversity that are based on cultural and religious values are often much more sustainable than those based only on legislation or regulation. While religions offer important guidance on the relationship between people and the rest of biodiversity, all religions are characterized by a wide gap between their philosophy and the practices of the people who have accepted the philosophy.

Religion frequently has had the effect of preventing excess human demands from outstripping the environmental resources that are required to support them. For example, in Roman times, high priests dictated that the well-being of humanity depended on thousands of animals being sacrificed every year, thereby keeping the size of herds within the carrying capacity of the grazing lands. More fundamentally, religions provide a complex holistic view of how to use natural resources, supporting this knowledge with an ethical perspective that is built on the implicit social contract that enables communities to function.

All of the world's major religions are today sensitive to the importance of biodiversity, though of course their historical writings do not use today's conservation vocabulary. The following is a brief summary of some major religious belief systems and how they relate to modern biodiversity concerns.

Animism is a term applied to a wide variety of beliefs that focus on numerous spiritual beings concerned with human affairs and capable of either helping or harming people. Animism is a view of the world found among many traditional peoples who conceptualize an intrinsic spiritual connection between humans and nature. This connection has been described by Harvard University biologist Edward O. Wilson as "biophilia," an innate human need for contact with a diversity of life-forms. For example, in the North American desert, the Yaqui Indians believe that close communication exists among the plants, animals, birds, fish, rocks, and springs that inhabit the Sonoran Desert. Many traditional approaches to conservation are supported by religious beliefs, often based on various kinds of animism that have the effect of fostering respect for plants and animals. In many parts of the world, people have established sacred sites on the basis of inherent spiritual or religious significance. Such sacred sites, based ultimately on animistic beliefs, are often sanctuaries for biodiversity. They may well survive substantial cultural changes; for example, the Parliament of Kenya (a very modern institution) voted in 1990 to protect all of the country's remaining sacred forests, known as *kaya*. Many animistic systems of belief are accompanied by the idea of *taboo*, that which is forbidden; breaking a taboo can bring sanctions such as illness, social ostracism, or even death. Taboo often applies to certain sets of natural resources that are particularly vulnerable to overexploitation. Animism is also often associated with *totemism*, a complex of ideas and practices based on the belief in a mystical relationship (often kinship) between people and certain animals or plants; these relationships often include reverential and genealogical relationships between social groups or individuals and the totems. Totems normally are associated with taboos of avoidance, so an Amazonian Indian hunter within a social group that has the peccary as a totem may be forbidden from hunting peccaries. Thus totemism also helps to restrict exploitation of harvestable resources.

Buddhism, with a total of about 300 million practitioners found in many Asian countries, teaches that a behavior has a natural relationship to its resulting consequences in the physical world. All Buddhist teachings and practice come under the heading of *Dharma*, which means truth and the path to truth. Based on the teachings of the Buddha, who lived in Nepal more than 2500 years ago, Buddhism teaches that people are responsible for their actions and go through a cycle of rebirths before finally reaching *nirvana*. Right actions lead to progress toward *nirvana* while negative actions, such as killing animals, lead to regression from that goal. Committed to the ideal of nonviolence, Buddhism also attaches great importance to wildlife and the protection of biodiversity (though of course not using such modern terms in its ancient teachings). Respect for life in the natural world is essential, and by living simply one can be in harmony with other creatures and learn to appreciate the interconnectedness of all that lives. This is not to claim that nature is unchanging; on the contrary, Buddhism recognizes change as the very essence of nature. The Buddha taught that all things are interrelated and do not have an autonomous existence, so the health of the whole is inseparably linked with the health of the parts and the health of the parts is inseparably linked with the whole.

Christianity, with some 1.6 billion members, is the dominant religion in Europe, sub-Saharan Africa, the Pacific, and the western hemisphere. It has numerous denominations, all based on the belief in a single all-powerful God who created nothing unnecessarily and omitted nothing that is necessary. Christians believe that Jesus Christ was the son of God and a great teacher whose message was that all of creation is the loving action of God, who continues to care for all aspects of existence. The very nature of biodiversity considered in itself, without regard to humanity's convenience or inconvenience, is considered to give glory to the Christian God. Christianity teaches in effect that humanity may not disorder biodiversity or destroy God's creations, at the risk of destroying itself; in the Holy Bible, Ecclesiastes 3:19 says: "For that which befalleth the sons of men befalleth beasts . . . as the one dieth, so dieth the other . . . a man hath no preeminence above a beast." Human acts of irresponsibility toward creatures are contrary to the divine wisdom, which sustains and gives purpose to the interdependent harmony of the universe. Modern Christian teachings are that humanity, both individually and collectively, should perceive the natural order as a sign and sacrament of God, recognizing that all creatures and objects have a unique place in God's creation. On the other hand, Christianity also recognizes a special role of humanity within creation. The Holy Bible contains Christianity's main teachings. It provides strong support to the protection of natural resources, including passages mandating the preservation of fruit trees (Deuteronomy 20:19, Genesis 19:23–25), agricultural lands (Leviticus 25:2–4), and wildlife (Deuteronomy 22:6–7, Genesis 9). The Bible often refers to the impressive intelligence of wild creatures, such as in Jeremiah 8:7–8, Proverbs 6:6–8 and 30:24–28, Numbers 22:22–35, and Isaiah 1:3.

Hinduism, the dominant religion in India with about 700 million followers, teaches the all-encompassing sovereignty of the divine, manifesting itself in a graded scale of evolution. While the human race is currently at the top of the evolutionary pyramid, it is not seen as something apart from the earth and its biodiversity. Hinduism is permeated by a reverence for life and an awareness that the great forces of nature—earth, sky, air, water, and fire—as well as biodiversity are all bound together within the great rhythms of nature. Hindus recognize that all lives play their fixed roles and function together so that no link in the chain of life is lost. The divine is not exterior to creation but expresses itself through natural phenomena. Hindus consider at least some forests and groves as sacred and associate various plants and animals with gods and goddesses in the Hindu pantheon. The most important teachings of Hinduism are contained in the Bhagavad Gita, a dialogue between Sri Krishna and Arjuna, which has a clear description of ecology as a cycle of life dependent on everything from bacteria to birds. Like animists, Hindus believe that all plants and animals have souls, and that people must do penance even for killing plants and animals for food. Hinduism involves sacrifice, forms of ritual worship that are designed to protect life through reinvigorating the powers that sustain the world by securing cosmic stability and social order. Hindus recognize certain rivers and mountains as sacred, because they give and sustain life.

Islam, with about a billion adherents, is the dominant religion in North Africa, the Middle East, and many Asian countries. Its teachings are very similar to those of Christianity in relation to biodiversity. The entire universe is God's creation; Allah makes the waters flow upon the earth, upholds the heaven, makes the rain fall, keeps the boundaries between day and night, creates all of biodiversity, and gives it the means to multiply. *Tawheed* is the principle of Oneness of Allah, teaching that "there is no God but God and Mohammed is His messenger." This testifies to the unity of all creation and the interlocking grid of the natural order of which people intrinsically are parts. The *Holy Koran* provides a set of principles that define the relationship of man to God and of God to the environment in its totality. However, humans are considered a very special creation because they alone have the power to reason and think, giving them the potential to do great good, or great harm. For Islam, the role of people on earth is that of a *khalifa*, or trustee of God, so humans are entrusted with the safekeeping of earth and its biodiversity. People are answerable for their actions, including maintaining the unity of Allah's creation, the integrity of the earth, and its biodiversity. The Islamic relationship to biodiversity is that each generation uses and makes the best use of biological resources, without compromising the interests of future generations. The Koran says, "With it we have produced diverse pairs of plants each separate from the others. Eat (for yourselves) and pasture your cattle; verily, in this are signs for men endowed with understanding" (Sura 20 Aya 53). While humans have the right to utilize and subjugate natural resources, this involves a commitment to conserve them both quantitatively and qualitatively. In the Koran, God says, "There is not an animal (that lives) on the Earth, nor a being that flies on its wings, but (forms part of) communities like you" (Sura 13 Aya 15). Thus Islam looks upon biodiversity as an expression of God's wisdom and omnipotence and as support for human development. This leads to the necessity of conserving and developing biodiversity for its own sake as well as for the benefit of humans. And the Prophet Mohammed said, "There is a reward in doing good to every living thing." Such beliefs can lead to biodiversity being conserved. For example, in Pakistan, representatives of original tree species still persist in old Moslem graveyards, because of a taboo against cutting these trees. And the only surviving population of a freshwater turtle, *Trionyx nigricans*, survives in a sacred pond dedicated to a Moslem saint in Bangladesh.

Jainism is one of the oldest living religions, beginning in India at least 2800 years ago. Jains believe that all living beings have an individual soul (*Jiva*), which occupies the body until it dies, then leaves the body and immediately takes birth in another. Attaining nirvana and thereby terminating this cycle of birth and death is the goal of Jain practice, much like Buddhism. Jainism is based on the principle of *ahimsa* (nonviolence) toward human beings and all of nature. Jains teach that no human quality is more subtle than nonviolence and no virtue greater than reverence for life. While biodiversity often is affected negatively by people, the intention to harm is what makes an action violent, and without violent thought no violent action is recognized. Jain cosmology recognizes the fundamental natural phenomenon of symbiosis or mutual dependence, with all aspects of nature belonging together and bound in a physical as well as a metaphysical relationship—an ancient perspective that is reflected in modern ideas about biodiversity.

Judaism, which originated in the Middle East but whose 18 million practitioners today are thinly spread around the world, teaches that God created the world, making order out of primal chaos. The sun, the moon, the stars, plants, animals, and ultimately humanity were each created with a rightful and necessary place in the universe. The encounter of God and man in nature is conceived in Judaism as a seamless web with man as the leader and custodian of the natural world. Judaism maintains that the earth is the arena that God created for man, half beast and half angel, to prove that he could behave as a moral being. Judaism views God as the divine "Giver of the Torah," which sets out a series of ethical obligations including many relevant to biodiversity. It teaches that the relationship between man and nature is one of ownership, though limited by good sense; man was commanded to behave toward the rest of biodiversity with justice and compassion. But humanity inevitably lives in tension between its power and the limits set by conscience. The Bible contains much environmental wisdom, for example allowing agricultural fields to go fallow one year out of seven (Leviticus 25:1–5). The Bible also warns that even in time of war, it is forbidden to destroy fruit-bearing trees of one's enemies (Deuteronomy 20:19).

Shinto is the system of indigenous religious beliefs and practices of Japan, first appearing in written form around 1400 years ago, after Buddhism had been introduced into Japan. Shinto is based on beliefs concerning the nature and attributes of *Kami*, sacred power, which is found in all individual things. Shinto temples are often established on sites that have particular spiritual integrity and force, often with large groves of trees (totaling nearly 120,000 ha in Japan). Shinto is based on numerous sacred texts, collectively known as Shinten. Shinto is strongly based in rural agricultural practices, involving various ceremonies and festivals that guide the relationship between people and nature.

Sikhism, a modern religion that began in India the late fifteenth century, builds on the message of the oneness of a universe created by an Almighty God, who is master of all forms in the universe and the source of the birth, life, and death of all beings. God has created the natural beauty, which exists and can be found in all of biodiversity. Without his *hukum* (order) nothing exists, changes, or develops, but having brought the world into being, God sustains, nourishes, and protects it. Biodiversity exists under God's command and with God's grace. Sikhism teaches against a life of conspicuous consumption, emphasizing mastery over the self rather than mastery over nature. A major value of Sikhs is to be in harmony with the earth and all creation. Obviously, people are moving farther in the opposite direction, but Sikhs argue that the current instability of the natural system of the earth is a reflection of the instability and pain within humans.

Taoism has a history of more than 2500 years and has been one of the main components of Chinese traditional culture. *Tao* means simply "the way" and is the origin of everything. The Tao took form in the being of the Grandmother Goddess, who came to earth to enlighten humanity. In Taoism, everything is composed of two opposite forces known as *yin* and *yang*. When the two forces reach harmony, the energy of life is created. Taoism judges affluence by the number of different species; thus a society with healthy levels of biodiversity is affluent, while societies with declining biodiversity are themselves in decline. Chinese Taoism, as reflected in the Tao Te Ching (c. 600 B.C.), calls for people to have a noninterventionist role in the natural world.

These short descriptions indicate that virtually all religions are very sensitive to biodiversity concerns, essentially asserting the interdependence of all life-forms and the need to treat all life with respect. But this religious perspective has not always been sufficient to conserve biodiversity. The special role of humanity within Christianity has often resulted in overexploitation of natural resources and alienation of land from people; the recognition of the Ganges as a river sacred to the Hindus has not prevented significant pollution and biodiversity loss at the hands of people; the unity maintained by balance advocated by Islam is today being replaced by environmental pressures that lead to disunity and imbalance; and the Tao noninterventionist perspective did not prevent the ancient Chinese from making profound changes to the land and water around them. Numerous indigenous cultures have been destroyed or fundamentally modified by missionaries from the major religions, who often are accompanied by technological innovations. But religions constantly evolve, and it is possible that a stronger religious response to biodiversity loss will evolve as the concerns about such loss begin to affect people more profoundly.

II. LANGUAGE AND BIODIVERSITY[1]

As an essential part of human cultures, languages provide the categories to conceive the natural and social world. A species of plant or animal that is important to a community gets named, and by learning that name, the people within the culture also learn what is vital to know in their natural environment. Thus language helps to organize the world and frees energy for other tasks, using words like pegs on which to hang the meanings stored in the storehouse of the human mind and providing a framework that binds together the details into a meaningful whole. Words for objects and phenomena help us learn the connotations, associations, emotions, and value judgments of a culture. Thus knowledge, beliefs, and values are linguistically encoded, thereby helping to promote a diversity of adaptive ideas and support various forms of biodiversity.

Many scientists argue that the preservation of the world's linguistic diversity (totaling more than 6000 languages) and the distinct forms of local knowledge that local languages encode must be incorporated as an essential goal in bioculturally oriented diversity conservation programs. Indeed, the diversity of human languages may be the best available indicator of human cultural diversity.

Many indigenous peoples believe that the external reality or environment is no different than the description of this reality or environment in a linguistic ecology. Variation in humans is adaptive, and language helps to support cultural diversity in humans. Human culture is a powerful adaptation tool and language both enables and conveys much cultural behavior.

Many countries with high biodiversity also have high cultural diversity, as indicated by their large number of languages. Languages have multiplied and thrived in places where natural selection has produced a rich variety of landscapes, animals, and plants. Researchers have found that linguistic diversity in sixteenth century North America was greatest in areas with the greatest diversity of habitats, irrespective of the latitude. Ten of the 12 "megadiversity countries" are among the top 25 countries for endemic languages. This correlation between biodiversity and linguistic diversity may have been fostered by a process of coevolution of small-scale human groups with their local ecosystems, in which humans interacted closely with the environment, modified it as they adapted to it, and acquired intimate knowledge of it. This knowledge was encoded and transmitted through language, supporting the contention of some linguists that life in a particular human environment depends on the ability of people to talk about it.

In many parts of the world, anthropologists have found a strong correlation between biodiversity, linguistic diversity, and ethnobiological knowledge. More generally, landscapes are anthropogenic not only in the sense that they are physically modified by human intervention, but also because they are symbolically

---
[1] This section draws from a contribution by Luisa Maffi.

brought into the sphere of human communication by language—by the words, expressions, stories, legends, and songs that encode and convey human relationships with the environment and that inscribe the history of those relationships onto the land. When people name places, they are identifying where things happen within the local environment, thus providing an entry in a mental encyclopedia that helps to describe the ecological niche occupied by the local peoples.

For example, certain American Indian groups in the Pacific Northwest have an extensive vocabulary for talking about salmon and other fish, vegetation, streams, and trails, along with extensive development of grammatical devices that express the direction, distances, and relative positionings of these valuable features. Thus language helps to enable these seminomadic hunter-gatherers to be very clear about where they were and where they needed to go to find food and other necessities. Many cultures in the far north have numerous names for various types of snow, and pastoral peoples have detailed terms for various types of livestock. The richness of language thereby reflects the many ways people relate to their environment.

III. INDIGENOUS PEOPLES, RESOURCE MANAGEMENT, AND BIODIVERSITY

Humans earn a living from the earth and its biodiversity in four main ways: hunting and gathering, pastoralism, shifting cultivation, and permanent agriculture. Only the latter is sufficiently productive to enable large cities to flourish, and the very high productivity enabled by irrigation may have been needed for civilization. Historical evidence indicates that traditional activities related to agriculture, fishing, and livestock husbandry sometimes have led to sustainable systems that were in the self-interest of the people involved, at least at their current level of population and technology. But all four forms of using biological resources carry with them the dangers of overexploitation. Broadly speaking, the combination of technology and population growth potential requires human cultures to develop effective ways of living in a sort of balance with the resources available with a given technology. The extinction of numerous cultures over time indicates how frequently the limits have been exceeded. On the other hand, many cultures have developed belief systems and practices that served to limit overexploitation and to foster greater biodiversity (even if these benefits sometimes were unintended).

For most of our history as a species, humans were hunters and gatherers, living from harvesting a very wide range of plants and animals; a few such groups still survive today, usually where agriculture is impossible or impractical. Hunter-gatherers were usually nomadic and fit ecologically into their habitat in much the same way as other omnivorous species. An important discovery that interrupted this presumed harmony and may have had major ecological impacts on biodiversity was the deliberate use of fire for driving game and clearing forest. This cultural activity produced major modification in vegetation patterns and fauna distribution in at least some regions. The invention and use of spears, bows and arrows, harpoons, and nets also must have affected the interrelationships between human beings and populations of other animal species. The effects of culture on humans themselves in the hunter-gatherer phase included impacts on techniques of food gathering, especially hunting. Nevertheless, with the exception of the use of fire in some situations, it is difficult to accurately determine the impacts on biodiversity of these changes.

Given the very great variability in approaches to resource management taken by various groups of hunting peoples, it is not surprising that various hypotheses have been developed to explain why some groups of indigenous peoples appear to have been able to live in better balance with their resources than other groups. A common observation is that ethnic groups that have lived in an area for extensive periods of time and appear most dependent on their locally available resources are most likely to have developed sustainable forms of resource use. Many of the population processes that link human hunters and their prey occur over timescales that elude both ethnographic and archaeological fieldwork. The critical factor is the distinctive features of the foraging economy and their outcomes for resource population ecology, as a matter of behavior with practical consequences.

Fishing is a specialized form of hunting and gathering. People who depend primarily on fish for their protein typically have developed both technology and management techniques to ensure sustainable yields. For example, Amazonian Indians may avoid certain parts of the river during breeding seasons; eastern Indonesian fishermen have complex systems of taboos, known as *sasi*, which control how and when the fishing grounds are to be used; and Polynesian fishermen had various taboos that served the function of managing their fisheries.

The development of domestication of plants and animals about 10,000 years ago marked the beginning of an entirely new era in the interplay between human society and biological systems. The controlled breeding of plants and animals, using a small sample of the world's biodiversity, established a more reliable and expandable resource base than subsistence hunting and gathering, enabling humans to reduce substantially the space required for sustaining each individual and allowing the human population to increase. All early farming activities consisted of redistributing plant and animal species in a given area by deliberately increasing the local concentrations of the species humans valued and decreasing concentrations of species that competed with the favored ones. In terms of biodiversity, the selective breeding of species also expanded the range of human impact from habitats and species to genes. The 40 or so species of animals that have been domesticated have often greatly increased their genetic diversity, range, and populations due to human management, as have the 100 or so main domesticated plant species and the thousands of other plants selected for use as food, spice, medicine, decoration, and construction. But the gains for these species typically have been at the expense of other species which people found less easy to mold to their vision of the desirable.

Domestication of livestock apparently arose after domestication of plants and remains the most significant use of arid and semiarid lands by people. The dangers of overgrazing have led to many beliefs about human relationships, livestock management, and range management. For example, many Middle Eastern pastoralists had complex ways of sharing grazing lands, including the idea of *hema*, a traditional grazing reserve. At least some species of trees typically are considered sacred by pastoral peoples and are not destroyed except in times of dire need; conservation of such vegetation thereby provides an emergency resource reserve (as well as conserving biodiversity).

Fallowing was the agricultural technique originally followed in most places; it remains a component of the shifting cultivation systems in many parts of Asia, Africa, and Latin America today. Such systems often use fire as a means of clearing the land, so they are sometimes called "slash-and-burn" systems. Such a system involves cultivating a plot of land—known as a "swidden"—for a few years and leaving it to develop into successional stages for a much longer period. This process creates a mosaic of vegetation, much of which is useful for foods, medicines, building materials, and other useful products (such as dyes, colorings, ceremonial objects, etc.). Many species of wildlife, including birds and terrestrial mammals, are attracted to the fallow swiddens because of their high productivity of nutritious vegetation. Furthermore, part of the land was often left out of cultivation. Thus in northeast Indian states like Manipur, as much as 10 to 30% of the land was permanently maintained under natural mature vegetation in the form of sacred groves. Historically, this would have ensured the persistence of almost all the natural elements of biodiversity, coupled with stimulation of overall productivity by favoring faster-growing early successional species in the patchwork of successional stages covering 70 to 90% of the land. Selection for adaptation to highly heterogeneous local environments also promoted considerable genetic variation in the cultivated species.

As an example of how shifting cultivation affects biodiversity in the Ecuadorian portion of the Amazon forest, Runa Indian swiddens resemble agroforestry systems rather than the slash and burn that merely results in temporary clearings in the forest canopy. Compared to unmanaged fallows, management actually increases species diversity in 5-year-old fallows. Between 14 and 35% of this enhanced species diversity is attributed to direct planting and production of secondary species. Thus Runa agroforestry can be seen as a low-intensity succession management system that alters forest composition and structure in the long term.

Among the most diverse of agricultural systems known are the home gardens in the humid tropics of tropical Asia, the result of long historical development of technology designed to meet the needs of local agricultural communities. In West Java, the typical home garden appears as a crowded assemblage of trees, shrubs, climbers, herbs, and creeping plants that are used for fruit, vegetables, starchy food crops, spices, ornamentals, medicines, fodder, fuel, and building material, and involve over a hundred species useful to people.

The Kantu of Kalimantan, Indonesia, plant at least 44 varieties in one area with an average of 17 per household. A gene pool of potatoes of some 3000 varieties representing 8 species is traditionally under cultivation in the Andes. In Papua New Guinea, as many as 5000 varieties of sweet potato are under cultivation, with as many as 20 varieties being planted in a single garden. Thus indigenous peoples often maintain high biodiversity, at least among those species useful to them.

A. Traditional Belief Systems and Resource Management

The way of life of the world's tribal or indigenous peoples depends closely on biodiversity, often using cultural and religious beliefs to avoid disastrous overexploitation. Notions of sustainability of use are inherent in the value systems of many traditional societies, usually manifested in some notion of intergenerational equity. For example, the Iroquois tribe in North America would plan for the seventh generation when making their decisions, the life span, incidentally, of the dominant tree in their region. And the Koyukon Indians believe that future events will depend on the way people behave today and that the world can be nurtured by prudent use or harmed by unrestrained abuse; but equally important, the natural world will respond to gestures of respect given by those who recognize its sensitivity and awareness and humble themselves to its power. Plains Indians generally believed that their relations with other species were regulated by expectations and obligations similar to those that governed relations between kin or allies; such relations could vary from beneficial to harmful and could be mediated by ritual specialists.

In Colombia, the forests of the northwest Amazon basin harbor the world's most diverse array of plants and animals. This region is often considered part of the world's greatest remaining tropical wilderness. But the Tukano people who live there perceive their "wilderness" environment to be anthropogenic, transformed and structured in the past by the symbolic meaning their ancestors gave to resources and the knowledge they obtained about the plants and animals that enabled people to survive. Their forest is a system of resources in which the energy produced is directly proportional to the amount of energy it receives, a very modern perspective. They know that they cannot harvest more than the forest can produce, and they apply sophisticated knowledge of individual species and their uses. Their myths tell of animal species that were punished by the spirits for indulging in gluttony, boastfulness, improvidence, and aggressiveness. These myths serve as object lessons to human society, in which animals are metaphors for survival. By analyzing animal behavior, the Tukano find an order in the physical world within which human activities can be adjusted.

The restraints on harvesting may involve protection to keystone species that may support the persistence of a range of other species. Thus fig trees belonging to the genus *Ficus* are recognized as important sources of fleshy fruits available in seasons when no other species are producing such fruits in the tropical forest communities. *Ficus* thereby promote persistence of a number of insect, bird, bat, squirrel, and primate species for which they serve as a critical resource in a period of fruit shortage. All species of *Ficus* are even today to some extent protected as sacred trees through much of tropical Asia and Africa, where the local communities are aware of this ecological role. In other parts of the world, other species of useful trees may also be sacred, such as date palms in desert areas, baobab trees in Africa, palmyra in Brazil, and kapok among the Mayas.

Harvesting restraints also include protection of critical stages in the life history of species that are especially vulnerable to overharvest. Thus a nomadic hunting tribe of Western India has the tradition of releasing any pregnant does or fawns of antelope or deer caught in their snares, and egrets, storks, herons, pelicans, ibises, and cormorants at their colonial nesting colonies are given immunity from hunting over most of India, although these birds are hunted in the nonbreeding season. In many Asian villages, fruit bats are not hunted at their daytime roosts but may be killed at a distance from the roost during the night. Numerous such examples can be found in all parts of the world

Some traditional communities have developed detailed resource regulation systems. For example, many traditional communities—whether "indigenous" or "tribal" like the Tara'n Dayaks of West Kalimantan or the nontribal, such as the ribereños of Amazonian Peru—establish community reserves and formulate rules about how the species therein can be exploited, with the expressed intention of preserving these resources for the existing community and for future generations. Therefore, in some cases reserves may be protected by religious sanctions and in others by a more overtly "social" contract and enforced by strong economic and social sanctions; combinations of approaches are also found.

Biological diversity may well have been higher in rural areas in former times, when large numbers of farmers of many different cultures had long-term stakes in the land they farmed and had control over their own technology. These historical systems of land management were highly variable, following a range of different rules to take into account specific attributes of the physical systems within which they were found, cultural views of the world, and the economic and political relationships that existed in the setting. Despite their great diversity, such systems often had characteristics such as clearly defined boundaries, specific rules on the harvesting of different products, involvement of the affected people in these collective choices, a system of monitoring the use of resources, cultural sanctions for those who violated the operational rules, inexpensive local mechanisms for resolving conflict, and ways of organizing these activities so that different types of decisions were taken at different levels.

IV. URBAN VERSUS RURAL PERCEPTIONS

About half the world's population today lives in urban settings far removed from the agricultural systems on which their lives depend. While rural peoples typically have a very good practical understanding of biodiversity, those living in cities increasingly have no more than a theoretical understanding, perhaps reinforced by an occasional visit to a national park or by a television program about nature. The growing numbers of nature-based films, protected areas, and forms of tourism based on idealized visions of nature indicate that urban people still feel a need to connect to biodiversity. Of course, different cultures have different perceptions, and the diversity of human societies is reflected in the diversity of perceptions urban people have of their environment. Any classification of such cultural values will necessarily mask some of this diversity, but rural people who live in close contact with the realities of nature typically have very different attitudes about biodiversity from urban people who have only a distant relationship with nature. Examples of these contrasting attitudes are presented in Table I.

These attitudes are ideal types, and will not necessarily be found in their pure forms in any rural or urban community. Rather, societies both over time and over space can be placed on a continuum between these opposites, as human attitudes change due to both internal social dynamics and external influences. However, it is clear that such attitudes greatly influence how humans use natural resources and therefore the impacts they have on biodiversity.

In marked contrast to the rural areas where people still live in a close relationship with biodiversity, urban dwellers are able to gain the benefits from the biodiversity of the entire globe. Their connections through trade and communications enable urban dwellers to have rapid and unprecedented access to biodiversity. Ironically, the expanding access to more biodiversity has tended to weaken the feedback between human welfare and the way biological resources are managed; the urban dwellers typically have only the very faintest notion of how their consumption might have impacts on biodiversity in the countryside of a distant country.

V. CONCLUSIONS

Establishing a connection between specific cultural practices and conservation or enhancement of biological diversity is by no means a simple matter, for the overtly declared purpose of a practice that seems to help conserve biological diversity may in fact be quite different. Thus, in South Asia many sacred ponds have helped conserve the indigenous fish fauna. But people may leave these ponds alone out of respect for some deities, not with an expressly declared purpose of conserving fish diversity. It is then quite possible that many cultural or religious practices that seem to promote conservation may have originated from different motivations, while others declared as promoting conservation may in reality achieve something very different.

Traditional societies have often protected parts of the natural landscape they live in, or left untouched some of its elements. Most such societies, for instance, have considered certain sites as sacred, where most or all human activities are prohibited. Most societies have also considered certain species as sacred, with elaborate myths and folk tales about how humans originated from such species, or how these species are incarnations of gods and deities, or in some way associated with them, or how they obtained magical powers. These could be wild species that are left undisturbed (and therefore have survived in even the most densely populated and radically altered human landscapes, such as rhesus and bonnet macaques in Indian cities) or domesticated plants and animals whose utilitarian value is intertwined with spiritual values (such as the cow in Hinduism). Such belief systems can have the effect (although perhaps unintentional) of regulating resource use.

TABLE I
Rural and Urban Attitudes about Biodiversity

Rural attitudes	Urban attitudes
Humans as one part of nature's biodiversity	Humans not part of biodiversity
All living creatures considered equal	Humans superior to all creation
Culture/nature as a continuum (no such thing as wilderness)	Culture/civilization human, and nature considered wild
Natural way is right, and human activities should be molded along nature's rhythms	Human technology is superior, needs to mold nature to suit human needs
Biodiversity has an integral set of multiple values (cultural, spiritual, material)	Biodiversity is predominantly a material (economic) resource

A rich literature of traditional conservation practices indicates that they have been a common feature for many cultures over many years. This literature indicates that such practices serve group interests of communities and remain viable only so long as (a) local communities continue substantial levels of dependence on resources harvested from their immediate vicinity, (b) local communities have full control over the local resource base, and (c) local communities retain a sufficiently high level of internal cohesion. In most cases, these conditions are no longer fulfilled when outside state or corporate bodies establish control over natural resources.

When local people are part of a local ecosystem, their behavior directly affects their own survival. But cultural mechanisms that have been developed as adaptations to the environment over tens or hundreds of generations are quickly cast aside when trade or new technology frees people from traditional ecological constraints, changing them from "ecosystem people," who are adapted to their local ecosystem, into "biosphere people," who often live in cities and can draw from the resources of the entire world.

While changes in traditional attitudes toward biodiversity can be a result of internal dynamics (e.g., population increase), it is perhaps more often a result of outside influences: interaction with a modern culture, intrusion of the market, a technical innovation imported from another culture, and so on. Some have argued that traditional resource use patterns may be sustainable only under conditions of low population density, abundant land, simple technology, and limited involvement with a market economy. When confronted with market pressures, higher population densities, new technologies, and increased opportunities, few indigenous peoples appear able to maintain the integrity of their traditional methods, even when these are reinforced by religious beliefs.

See Also the Following Articles

AESTHETIC FACTORS • AGRICULTURE, TRADITIONAL • COMMONS, THEORY AND CONCEPT OF • ETHNOBIOLOGY AND ETHNOECOLOGY • HISTORICAL AWARENESS OF BIODIVERSITY • HUNTER-GATHERER SOCIETIES, ECOLOGICAL IMPACT OF • INDIGENOUS PEOPLES, BIODIVERSITY AND • LITERARY PERSPECTIVES ON BIODIVERSITY

Bibliography

Bennett, J. W. (1976). *The Ecological Transition: Cultural Anthropology and Human Adaptation*. Pergamon Press, New York.

Boyden, S. (1992). *Biohistory: The Interplay Between Human Society and the Biosphere, Past and Present*. Parthenon, London.

Burger, Julian. (1990). *The Gaia Atlas of First People*. Doubleday, New York.

Crosby, A. W. (1987). *Ecological Imperialism: The Biological Expansion of Europe 900–1900*. Cambridge University Press, Cambridge, UK.

Gadgil, M., and R. Guha. (1992). *This Fissured Land: An Ecological History of India*. Oxford University Press, Delhi.

Hamilton, L. S. (1993). *Ethics, Religion and Biodiversity*. The Whitehorse Press, Cambridge.

Maybury-Lewis, D. (1992). *Millennium: Tribal Wisdom and the Modern World*. Viking Press, New York.

Posey, Darrell A. (compiler and editor). (1999). *Cultural and Spiritual Values of Biodiversity*. United Nations Environment Programme, Nairobi, Kenya.

Ruddle, K. E., and R. E. Johannes. (1985). *The Traditional Knowledge and Management of Coastal Systems in Asia and the Pacific*. UNESCO Regional Office for Science and Technology for Southeast Asia, Jakarta, Indonesia.

Suzuki, D., and P. Knudtson. (1992). *Wisdom of the Elders*. Bantam Books, New York.

Turner, B. L., Clark, W., Kates, R., Richards, J., Mathews, J. T., and Meyer, W. B. (1990). *The Earth as Transformed by Human Action*. Cambridge University Press, Cambridge, UK.

SOCIAL BEHAVIOR

Daniel I. Rubenstein
Princeton University

I. Introduction
II. From Social Behavior to Social Organization: Patterns and Mechanisms of Formation
III. From Social Organization to Ecological Impacts

GLOSSARY

affiliative behavior Behavior that is supportive and brings individuals together.
agonistic behavior Behavior that is aggressive, threatening, combative, and submissive.
altruism Behavior that enhances the recipient's reproduction while reducing the actor's.
musth Annual reproductive period in male elephants characterized by extreme aggressiveness and secretions from temporal glands.
polyandry Mating system in which a female mates with more than one male.
polygyny Mating system in which a male mates with more than one female.
protandry Breeding system in which individuals change sex from male to female.
protogyny Breeding system in which individuals change sex from female to male.

SOCIAL BEHAVIOR characterizes the interactions that occur among individuals. These can be aggressive, mutualistic, cooperative, altruistic, and parental. When individuals interact repeatedly, social relationships develop and these can form among strangers, relatives, members of the same or opposite sex, and members of the same or different generations. Sets of consistent social relationships produce social systems or social organizations that can be variations on monogamous or polygamous themes and involve various types of helpers. The nature of any social system is ultimately determined by ecological and social circumstances, demography, and kinship.

I. INTRODUCTION

No animals are completely asocial. All must come together at some point to breed. Few, however, are highly social coming together and interacting repeatedly. For those that do, social interactions—both affiliative and agonistic—occur sometimes with relatives, sometimes with strangers, sometimes with members of the same sex, sometimes with members of the other sex; and sometimes with members of the same generation while at other times with members of other generations. The interactions themselves can be aggressive, cooperative, or even altruistic and can develop into strong relationships among particular individuals. Depending on the nature of these relationships and with whom they form, a variety of social systems can develop. Some may be made up mostly of kin or mostly of nonkin, some may be based on territorial separation or on the aggregation of competitors, some will exhibit monogamous as opposed to polygamous relationships between the sexes, and some will rely on the help of nonmates in the rearing of offspring. This seemingly bewildering pattern of social diversity is shaped ultimately by ecological circumstances and patterns of demography. But the very nature of a population's social structure itself will in turn affect the population's demography and its place in the biological community. It is these dual connections between social activities and the environment and the resulting feedbacks that mandates understanding both the causes and consequences of animal sociality so that effective and efficient management or conservation policies can be developed for protecting endangered species and for preserving biodiversity and ecosystem integrity.

II. FROM SOCIAL BEHAVIOR TO SOCIAL ORGANIZATION: PATTERNS AND MECHANISMS OF FORMATION

Social animals form groups. While some are temporary, others are more permanent. Given that animals in groups incur automatic costs of increased disease and parasite transmission as well as intensified competition, groups will only form when there are sufficiently large benefits to offset these costs. Benefits largely come in three forms. First, animals can develop forms of social behavior specific to stable groups that compensate them for the costs of group living. One such example is forming mutual grooming partnerships as do olive baboons (*Papio anubis*) to lower disease and parasite transmission. Second, by forming groups animals can enhance foraging by being better able to find, acquire, or defend food. Examples include colonial cliff swallows (*Hirundo pyrrhonota*) that transfer information about the locations of rich but ephemeral feeding sites or troops of monkeys who drive smaller troops away from feeding trees. And third, animals in groups can reduce their risk of being preyed upon by either increasing the likelihood of detecting predators, diluting their personal risks, or by decreasing the likelihood that predators can make a kill; confusion and cooperative defense are mechanisms that provide such antipredator benefits. Examples include the scattering of fish in schools, or the gathering of young inside a ring of adult musk oxen (*Ovibos moschatus*) facing outwards toward approaching predators with upturned horns.

Depending on the nature of the social relationships that develop, groups take on particular organizational forms. As Fig. 1 illustrates, these relationships are shaped by the features of an individual's physical and social environment. Particular distributions of food, water (bottom-up factors) and predators (top-down factors), in conjunction with the physiological demands of individuals differing in body sizes or reproductive states, will determine the frequency and magnitude of competitive and cooperative interactions that occur. The outcome of these repeated interactions will shape overall time budgets and activity patterns. Because females maximize their reproductive success relative to other females by their ability to raise offspring to the age of independence, females are forced to efficiently solve the three ecological problems listed earlier. What particular associations and distributions develop will depend on the particular ecological and physiological circumstances that females experience. These associations and distributions in turn will shape male associations and distributions, because male fitness is mostly determined by their ability to acquire mating access to disproportionate numbers of females. The mating

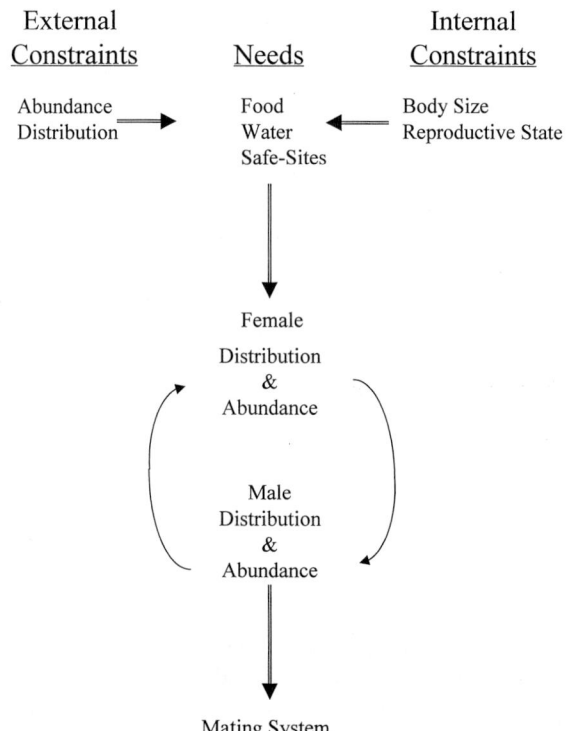

FIGURE 1 Model of ecological and physiological pressures influencing both intra- and intersexual relationships and showing how they shape social systems. Arrows depict social relationships. What type of society develops depends on details and perceptions of environmental conditions.

system that develops becomes the core of a species' social system.

A. Ecological Challenges and Phenotypic Constraints

Typically, in environments in which resources are abundant, especially when they are evenly distributed, competition is low enough to permit females to aggregate. If sufficiently large foraging or antipredator benefits can be derived by females that aggregate, and the groups that form are not too large, then these groups can be defended by single males and so-called harem defense polygyny results. If resources, however, are more patchily distributed so that competition among females intensifies periodically, then female group sizes will vary and female associations become more transitory. Rather then defending unstable groups of females, males instead attempt to defend resource patches sought by females. Typically, in these systems of "resource defense polygyny" the most able males defend the best patches and thus gain access to the largest number of females for the longest periods of time. If resources are not only patchily distributed, but the patches are large, widely separated, and fluctuate seasonally in abundance, then competition among females becomes so low that the formation of large groups is even more likely, provided that females can range widely and follow the shifting locations of peaks in food abundance. Males will thus be forced either to follow these large groups competing for, and then tending, one reproductive female at a time ("wandering") or to position themselves at the intersection of female migratory routes and wait for females to visit them ("lekking"). In either case, intense male-male competition generates a mating system based on "male dominance polygyny," and in the latter case females are afforded the exquisite opportunity to simultaneously compare many males before choosing with which one to mate! Whenever resources are sparsely but somewhat evenly distributed, high levels of competition prevent females from forming groups. As a result, individual females defend territories thus ensuring a regular supply of a renewing resource. Since solitary individuals searching for members of the opposite sex will face increased predation risk, pairs often share territories and monogamy results.

This model accounts for the diversity of mating systems for many different groups of animals including insects, fish, reptiles, rodents, and many varieties of birds, ungulates, carnivores, marine mammals, and primates. One of the best illustrations showing how environmental forces interact with physiological constraints to shape a species' social system emerges from Peter Jarman's classic study of African antelopes. By showing how body size affected the ways in which different species perceived, and then responded to, the distribution and abundance of forage and predators of grasslands, Jarman showed why particular social systems increased survival and reproductive prospects for particular species. He argued that the smallest bodied species, such as did-dik (*Madoqua kirkii*), duikers (*Cephalophus* spp., *Sylvicapra* spp.), suni (*Neotragus moschatus*), and klipspringers (*Oreotragus oreotragus*) require limited amounts of high-quality vegetation. But given their small size, such food items often appear as if they are widely scattered. Faced with high levels of competition and intensified risks of predation, territoriality and monogamy appear to be the best strategies. Pairs generally live in wooded or shrub-rich areas where moisture enables vegetation to grow and renew itself well into the dry season. By signaling territorial ownership via scent rather than by means of sound or visual display these small-bodied species reduce the chances that any of a large number of carnivores will prey on them.

As species increase in body size, both physiologically determined dietary needs and the way acceptable forage becomes distributed on the landscape changes. Since "crypsis" becomes an untenable antipredator strategy for larger and more widely ranging species, forming groups becomes the best strategy for larger species to lower predation risk. Fortunately, with larger size also comes an ability to subsist on more abundant, lower-quality vegetation. When it is patchily distributed, as it is for impala (*Aepyceros melampus*), reedbuck (*Redunca* spp.), and some gazelles (*Gazella* spp.), males defend the best patches that females prefer. When the vegetation is more evenly distributed, which often results simply from the fact that larger species such as eland (*Taurotragus oryx*) and Cape buffalo (*Syncerus caffer*) can utilize even the lowest quality items, larger groups form. Because the largest species view large continuous swards of a landscape as acceptable, competition is virtually eliminated and many males associate with many females. With such high levels of male-male competition, defense of a small subgroup of females becomes impossible and dominance defense systems develop.

The same sorts of connections between changing ecological circumstances shape the types of sociality exhibited in other taxa. A brief survey of mammals illustrates some of the more general patterns. For the equids the close association between food and water enable horse (*Equus caballus*) and plains zebra (*Equus burchelli*) females of different reproductive states to associate permanently. Thus males are able to defend such groups and so-called harems form. When these two resources are widely dispersed, as for Grevy's zebra (*Equus grevyi*) and the Asiatic wild ass (*Equus hemionous*), females of different states are precluded by metabolic constraints from foraging together. As a result, males compete for territories along traveling routes that take females from feeding areas to watering points.

In felids, females remain separated when food is scarce and the habitats are densely wooded because in such circumstances individual prey can be caught and consumed before competitors can intervene. In more open habitats and where both prey and competitors are large and much more numerous, coalitions of females form to help hold on to kills until they are completely devoured. If these female coalitions are themselves large, then there is pressure for males to aggregate to control reproductive access to females. Thus the only highly social felid is the savanna-living lion (*Panthera leo*), yet even in this plentiful landscape the leopard (*Panthera pardus*) remains solitary because it can safely cache its large prey in trees.

For canids monogamy is the rule. But variations do occur with small-bodied foxes (*Vulpes* spp.) sometimes exhibiting polygyny and large-bodied hunting dogs (*Lycaon pictus*) and timber wolves (*Canis lupus*) developing polyandry. Typically canids need help from nonbreeding foragers to help nourish lactating mothers. For the mid-sized jackals, such as the silverback jackal (*Canis mesomelas*), the helpers are young from previous litters that cannot themselves find successful breeding locations. For the smallest species, however, prepartum investment in young is relatively small for a female and thus her mate provides all the help that is needed. In years when prey resources are very high, competition between mothers and their soon-to-be-fecund daughters is low and their daughters are not forced to disperse. As a result they provide neighboring males with additional mating opportunities. For the largest bodied species, however, female prepartum investment is very high and they need all the help they can get. To enlist the support of other adult males that must hunt cooperatively to capture large prey, a dominant female not only kills the offspring of other females in the group, she mates with their mates so that these males behave as if they are the sires of the dominant female's young. Hence, depending on size-determined metabolic investments and needs, social systems of canids can vary from polygyny to polyandry and sometimes use the services of juvenile or adult nonbreeders to help rear offspring.

B. Kinship and Demography

Although the abundance and distribution of key ecological resources are important in determining the particular pattern of sociality exhibited by a population, other factors such as kinship and demography also play significant roles. Ever since W. D. Hamilton formulated the concepts of "inclusive fitness" and "kin selection," the magnitude of the costs and benefits associated with particular social behaviors, and that ultimately determine which strategy is best, had to be adjusted by degree of relatedness between social partners. Altruistic or cooperative behavior between relatives should be favored whenever Benefits $>$ Costs $* r$ where r is the degree of relatedness. For both parents and offspring and among full siblings, $r = \frac{1}{2}$; for grandparents and grand-offspring and among half-siblings, $r = \frac{1}{4}$; and among first cousins, $r = \frac{1}{8}$. Hence, the stronger the degree of relatedness among relatives, the more likely altruists are to enhance the reproductive opportunities of kin while incurring costs associated with diminished personal reproduction. Thus in the examples presented here, it is not surprising to find that the coalitions that form among lionesses when protecting kills and among male lions when protecting mating opportunities with females are formed most often among full siblings; nor is it surprising that the helpers jackals recruit to rear additional offspring are themselves the full siblings of the offspring being raised. In general, strong kinship lowers the threshold for the appearance of altruistic and cooperative behavior and may substantially affect the costs of living in groups.

But as both these cases illustrate, high population density and the intensified competition it engenders for finding suitable territories with sufficient food and habitable burrows is the factor ultimately responsible for favoring the establishment of coalitions or the recruitment of helpers. Thus demographic factors, such as population density as well as sex ratio and age structure, that result from differences in phenotype-specific vital rates also shape patterns of social organization. And since these features of populations are often altered by human activities, understanding how they shape, and are shaped by, patterns of sociality is essential if conservation assessment and planning is to be effective.

Hypothetically, if mortality, for example, were age specific and higher for prereproductive females than males, then high breeding sex ratios (males/females) would result and mating relationships would generally become more polygynous. If the sex-specific patterns of mortality were reversed, however, then polyandry would become more common. Since some of these mortality concerns were responsible for the variation in canid social structure described earlier, these mortality schedules can have important consequences. Similarly, if mortality rates were size specific and happened to be greater for larger males rather than smaller ones, then discrete size polymorphisms could arise in populations. This appears to be the case regarding the maintenance of so-called alternative male mating strategies. Often the typical strategy adopted by older and larger males of defending harems or resources yields the most reproductive gains but it often incurs the highest costs as well. Because displaying, fighting, attracting the attention of predators, and delaying reproduction while growing are all costly activities, males adopting less successful but also less costly tactics can also flourish.

The maintenance of such alternative mating patterns is common among many species of insects, fish, amphibians, birds, and mammals. For example, in bluegill sunfish (*Lepomis macrochirus*) males typically defend nest sites where they display, attract females with which to mate, and then fertilize and guard the eggs they lay. Since only the largest males have the ability to defend such nest sites, they must delay breeding in this fashion often for more than seven years. As a result, smaller and younger males have evolved various cuckolding strategies that may in fact be equally successful evolutionarily. In one, the so-called sneak begins breeding at two years of age. Although very small, such sneaks are virtually all testes and because of their inconspicuousness can dive from the surface just as a mating pair release sperm and eggs. By exuding large volumes of sperm, some make it to the eggs and fertilize a few. In the other, males delay breeding for as long as do females. By being the same size and color as females, these males join the mating pair and apparently fool the displaying male into thinking that he is courting two females. Then as the original pair releases their gametes, he does too and thus fertilizes some of the females' eggs. For these two strategies to be equally successful alternatives, the costs and benefits of each must vary inversely with frequency. Although they sometimes do, as in the case of the bluegill sunfish, in other species they do not.

A variant on this theme is associated with sex change. For many species, especially among the fishes, individuals begin life as one sex and end life as another. In bluehead wrasse (*Thalassoma bifasciatum*), for example, most individuals begin reproducing as females while residing with one male. If he disappears, then the largest female changes sex and becomes the harem tending male. Alternatively, in anemonefish (*Amphiprion* spp.), some individuals change from male to female. In either protandrous (male first) or protogynous (female first) species, the sex that is last is the one whose reproductive success is most influenced by body size, and this ultimately depends on the nature of the social system.

Clearly, a variety of environmental features influence the patterns of sociality that populations of animals develop. In many cases these patterns are flexible and species can vary in the system of social organization they exhibit depending on environmental conditions. Thus although female burros and asses typically live in transitory groups whose membership changes, when populations move from arid to mesic areas, social relationships can change. As was found on an island off the coast of Georgia, with food and water both more abundant and less separated, females with differing needs can and do coalesce into permanent groups. Males, which in arid areas are forced to establish territories along routes to and from water, respond by defending these groups much like males of horses, their close kin, and harem-like societies emerge.

III. FROM SOCIAL ORGANIZATION TO ECOLOGICAL IMPACTS

Despite the fact that some species exhibit social flexibility and that there exists significant "environmental determinism" in the development of patterns of sociality, knowledge is only now emerging about how changes in social systems impact important vital rates, alter interspecific competitive abilities, and shape a population's growth rate and genetic structure. Never before have the social systems of so many species been forced to respond to human activities, most of which lead to reductions in population size, through overharvesting, fragmentation of landscapes or habitats, and changes in the earth's climate. How are we able to predict which species will respond and how will a species' response impact its long-term growth and the health of the ecosystem in which it resides? When we intervene will we do so with a sufficient understanding of what the consequences of changing social behavior and social systems are likely to be? For example, will we know when a species switches from a monogamous to a polygynous social system and whether its growth rate or genetic diversity will increase or decrease? Will we know whether it will be more or less able to withstand selective harvesting or habitat fragmentation? Such consequences of differing patterns of sociality must be understood before endangered species or their habitats can be protected.

A. Consequences of Social Change

Many examples are now appearing of human induced changes in species' social systems. One population of feral horses inhabiting a barrier island off the east coast of the United States has been impacted twice by human activities and both times major changes to social behavior and social organization resulted in large-scale changes in the population's vital rates. The first instance occurred when dredging of surrounding ship channels altered currents and sediment deposition around the island. Before the emergence of hundreds of hectares of a new continuous sward of grassland, the island's horses exhibited two types of social organization. Where the vegetation was abundant and evenly distributed, females formed permanent membership groups and associated with a single male. Such harem groups were stable and males protected their females from harassment by intruders and rarely herded their own females. Where the preferred high-quality vegetation was broken up into patches of variable size by extensive ridges of sand dunes, females were not able to aggregate into permanent membership groups. As a result, males wandered in search of reproductively receptive females and, when doing so, harassed females and limited their ability to forage. Not surprisingly, the body condition and reproductive success of the harassed females was much lower than that of females bonded to males. After the emergence of the new sward of grassland in the region where females lived in the fission-fusion type of society, females were able to aggregate and males were able to establish harems and the growth rate of the entire population increased markedly. Thus when the abundance and distribution of key ecological resources alter social systems so that they more closely resemble a species' typical system, vital rates and the overall health of the population improve.

When the perturbation moves a population away from its normal pattern, as happened during the feral horse population's second brush with human activity, vital rates decline. As the horse population approached its carrying capacity, population control measures were instituted. During the initial round of management actions, over 85% of the harem stallions were removed from the island. The resulting social disruptions were massive; many females were separated from their males and young, inexperienced bachelor males took over. Extremely high levels of harassment led to low reproductive rates despite the fact that overall density had also been reduced and food supplies were expanding. As these two disruptions illustrate, human activities that alter existing patterns of social relationships dramatically alter reproductive rates.

Similar outcomes occurred when large-scale El Niño induced climatic changes resulted in major adjustments to time and activity budgets of Alaskan red foxes (*Canis aureus*). Before the 1982–1984 El Niño, many of the foxes on an island in the Bering Sea bred polygynously and produced litters of sizes equivalent to, and often greater than, those of monogamous pairs. After the El Niño and the reproductive failure of the large seabirds, the foxes shifted their diet to smaller, less abundant, and harder to catch prey. Dietary shortages were common, fewer foxes bred and polygyny disappeared. Moreover, the litter sizes of these monogamous pairs were smaller as well. Clearly, changes in social behavior in response to altered environmental conditions enabled some individuals to make the best of a bad situation. Overall, however, the health of the population suffered.

Major changes in the breeding patterns of elephants (*Loxodonta africana*) could result if climate changes alter the patterning of grasslands in East Africa. Ordinarily group sizes and compositions change seasonally in response to changing abundances of vegetation. After the rains, when grasses grow rapidly on the extensive upland plains, small family groups of elephants aggregate into large matrilineal assemblages. With the flush of new vegetation many come into estrus and become reproductively active. At this time only the most dominant bulls come into musth, a heightened sexual and aggressive state. By aggregating, females incite male-male competition thus ensuring that they mate only with the best males. During the dry season, however, after the grasses on the plains stop growing and have been consumed by elephants and other ungulates, the elephants retreat to the swamps where other grasses and browse remain. The patchy nature of the habitat, however, prevents large groups of females from remaining together and the huge herds fission into smaller family units. In addition, few females remain to be mated and the dominant bulls go out of musth. Subordinates become sexually active at this time, but reproductive opportunities are limited. Thus, under a normal environmental regime two evolutionary consequences result: most males do not mate, but those that do are the most fit; and because genetic effective population size is affected by the number of males relative to the number of females mating, the effective population size of elephants is much smaller than the census population size. If global warming increases the duration and intensity of dry seasons, then selection for the best males will be relaxed but the effective population size of the population will increase, provided that elephants can sustain their high metabolic levels and maintain their high fertility and survival rates. With global changes in climate appearing in many different habitats, other examples of social readjustments that either change a species' evolutionary potential or that act as demographic shock absorbers in the short run, but are unlikely to provide compensatory relief in the long run, are likely to become more common.

B. Consequences of Demographic Change

Many demographic changes leave their mark by altering critical aspects of social behavior or by disrupting the development of important social relationships. As the elephant example illustrates, changing environmental conditions can change operational sex ratios thus altering a population's ability to maintain genetic diversity. But by changing sex ratios in other ways, humans can put populations at risk by making it difficult for them to grow in size. In lions, for example, trophy hunting, if it removed pride defending males and nomads in proportion to their abundances, would not severely affect population growth and effective population sizes would be maintained at normal levels. But if such hunting focused on pride males, then increased turnover would foster increased infanticide by take over males. While such diversification of the gene pool might increase effective population sizes, it would certainly lower population growth rates because infant survival, which is already low, would plummet. Much clearly depends on whether or not the population is polygynous or monogamous and whether the perturbation will accentuate or reverse the pattern. Large-scale climate changes could reverse and ameliorate the normal pattern as is likely to be the case in elephants, but selective hunting or poaching, as in the case of lions, could exacerbate an already critical situation.

Selective poaching could even make it difficult for populations to recover once the poaching has been eliminated. Since hunting elephants for ivory meant that poachers sought the individuals with the largest tusks first, male numbers were reduced to dangerously low levels before large tusked females and smaller males were hunted. Because females prefer to mate with older males, lowered sex ratios and distorted age structures could make it difficult for all females to mate. As a result, population growth rates could remain low for historically heavily poached populations.

Harvesting in fish might be accelerating already rapid population declines because of disruptions to normal social interactions. When harvesting is size selective and reduces the number of large females in a population, fecundity might be disproportionately reduced since ovary volume scales allometrically with linear body size dimensions; larger females have disproportionately larger ovaries and high fecundity. And as sex ratios increase, male harassment of the remaining females may further reduce recruitment and population growth rate. When such size selective harvesting removes more males than females, sex ratios are reduced and females may become less choosy and more aggressive when selecting mates. While such changes in the intensity of sexual selection might not alter population growth rates, they might alter the genetic diversity of the population. Yet if males provide the majority of parental care and large males are removed from the population as if often the case in sports fisheries, then recruitment will be markedly reduced as unguarded eggs are cannibalized.

The implications of overzealous size-selective harvesting in sex-changing fish or in fish that exhibit alternative male mating behaviors appear to be equally severe. For species such as anemonefish in which the largest males change to females, removal of the largest females will force males to change sex at ever smaller sizes, thus producing fewer eggs. Population growth rates will be curtailed and the species will find itself at risk all because of overzealous aquarium traders.

Migratory Atlantic salmon (*Salmo salar*) provide perhaps the clearest example of where human activities are changing the balance of male mating strategies within a population and hence the patterns of sociality. Atlantic salmon typically are born in fresh water and develop for 1 to 2 years before smolting and heading to sea to grow and fatten by feasting on a rich supply of marine invertebrates. Once they attain a certain size, they become sexually mature and return to rivers. There they travel long distances upstream until they reach clear and cool breeding grounds. Some individuals, however, never head to the sea. Instead, they remain in their natal streams and mature sexually at young ages and at small sizes. Such individuals are called parr and they never go through the smolting process that adapts them to a marine lifestyle. Under pristine environmental conditions, the fraction of the population that becomes parr is small since the reproductive gains of such a strategy are low. When competing for mates with larger more aggressive males, parr fare poorly. What matings they obtain are derived by "sneaking" among a mating pair and releasing milt at just the right time. Such events are rare and the mixing of milt is poor so such "sneak" matings result in few young being sired. But the survival prospects of parr are high since they don't incur the risks of going to sea and, moreover, they begin breeding at a very early age. Thus over a lifetime, reproductive success is moderate. But as human fishing increases and the netting of the older larger salmon intensifies, the *relative* lifetime reproductive success of parr is improving. Since the costs of migrating to sea and back again have increased dramatically, the long life span of parr give them a relative advantage. And given that competition with the larger males for mates is also being reduced, the chances of parr securing matings are also improving. Thus it is not surprising that the composition of the population is changing as parr increase markedly in abundance. The impact on the long-term stability of the population is unclear. But the long-term impact on the balancing of mating morphs and sexual relationships is.

C. Consequences of Inappropriate Interventions

When humans intervene and develop management and conservation plans they sometimes do so without accounting for important aspects of social behavior. As a result some disastrous consequences have ensued.

Ignorance about the mating system of sperm whales (*Physeter macrocephalus*) has led to the implementation of harvesting plans that have almost led to their extinction. By using data from previous harvests in which the counts of males were much fewer than those of females, sperm whales were thought to exhibit a harem breeding system. Such a conclusion could only have been derived by assuming that sperm whales lived in permanent, closed membership groups. By making this assumption, the International Whaling Commission concluded that most males sighted would be superfluous thus disproportionate hunting of males would not limit the growth rate of the population; most of the males caught would not be necessary for fertilizing females.

Unfortunately, the social relationships of sperm whales are very different from those assumed. Females are the core of the society and do appear to live in permanent membership groups. But these groups often merge when diving for fish and divide labor between fishing and tending the young that cannot dive deeply. Males are forced to leave their natal group when reaching sexual maturity, but they cannot accompany the breeding associations to the tropics. Instead they remain at the higher latitudes foraging in the cooler more productive waters. Thus when males are hunted on the breeding grounds it is breeders and prereproductives that are taken. With breeding males severely reduced in abundance and the younger, subadult upstarts located thousands of kilometers away, many females are going unmated. As a result, the population is failing to recover and it will be years before the next generation of males is ready to start the population, already handicapped by such a low recruitment potential, on the road to recovery. Knowledge about sperm whale social relationships could have prevented the mistaken overharvesting of mature males.

Another species in which ignorance about important features of social behavior is problematic is the black rhinoceros (*Diceros bicornis*). During the past 25 years, numbers have declined from 65,000 to fewer than 2500 because demand for their horns is so great. In an attempt at making rhinos less desirable, some nations have implemented a strategy of dehorning. Other nations have opted to translocate rhinos to extremely well protected areas. Unfortunately, horns, and thus the value of rhinos, regrow quickly. But what is more troubling about the dehorning treatment is that it appears to disrupt effective maternal antipredator behavior. Although dehorned rhinos were no more likely to flee from predators than intact rhinos, the disproportionate disappearance of offspring being reared by dehorned mothers as opposed to intact mothers in areas with large numbers of lions and spotted hyenas suggests that horns play a vital parental and protective function. Dehorning, although initially thought of a cure for the disappearance of adults via poaching, became a contributing factor to the disappearance of young via predation. Although dehorning might enhance adult survival in the short run, it would limit a population's growth potential in the long run.

A third example where ignorance hindered, but did not harm, a conservation program involved the reintroduction of Asiatic wild asses (*Equus hemionus*) to habitats within its historical range. A goal of the Israeli government is to repopulate Judea and Samaria with biblical animals. Since the Palestinian race of the wild ass was extirpated by the Ottaman Turks at the beginning of the twentieth century, the onager was a prime candidate for translocation. The first reintroduction took place in 1982 with subsequent additions throughout the 1980s and increased the number of breeding females to 14. By the mid-1990s, however, the population had hardly grown and the number of breeding females totaled 16. What was not known at the time was that breeding success is bolstered by being reared in the wild and that onagers could facutatively adjust the sex ratio of the offspring they produced. In the Negev Desert, free-ranging onagers gave birth disproportionately to sons.

Trivers and Willard were the first to suggest that differences in the ability to invest in the successful rearing of offspring should lead to individual differences in the primary sex ratios of offspring. They argued that females with sufficient resources should invest in offspring of the sex with the higher variance in reproductive success. By producing well-endowed offspring of this sex, mothers would increase the number of grand-offspring they were likely to have. In polygynous species like the onager, males are usually the sex with the greatest variance. Since onager females compete more on the basis of resource exploitation, or utilization efficiency, rather than via direct conflicts settled by social status, sex biases in offspring production were not expected. However, because age and experience affect utilization efficiency, differences in bodily condition and ability to invest in offspring were expected to exist among females in this onager population. Thus it was heartening to find that middle-aged females—those in the best physical condition—gave birth to more sons than daughters. Conversely, those breeding for the first or second time gave birth to daughters as did older females nearing the end of their reproductive lives.

The strategy the Israelis employed of translocating middle-aged females to the desert was sensible in economic terms since high risks were attached to releasing very young females and low future expectations were associated with releasing older females. But had the Israelis released old females with yearling daughters they would have quickly accelerated the growth of the population. At the time, an understanding of how environments shape patterns of onager social behavior and the impacts that these behaviors would have on population dynamics was unknown. By understanding the relationships between environments and social behavior and between social organization and population processes, we should be in a better position to understand the problems endangered species are likely to face and what interventions to protect them are likely to be effective.

See Also the Following Articles

MUTUALISM, EVOLUTION OF • POPULATION DENSITY • SALMON • SPECIES COEXISTENCE

Bibliography

Caro, T. (Ed.) (1998). *Behavioral Ecology and Conservation Biology*. Oxford University Press, New York.
Caughley, G., and Gunn, A. (1996). *Conservation Biology in Theory and Practice*. Cambridge University Press, Cambridge.
Jarman, P. J., and Rossiter, A. (Eds.) (1994). *Animal Societies: Individuals, Interactions and Organisation*. Kyoto University Press, Kyoto, Japan.
Lott, D. F. (1991). *Intraspecific Variation in Social Systems of Wild Vertebrates*. Cambridge University Press, Cambridge.
Peters, R. L., and Lovejoy, T. E. (Eds.) (1992). *Global Warming and Biological Diversity*. Yale University Press, New Haven, CT.
Rubenstein, D. I., and Wrangham, R. W. (Eds.) (1986). *Ecological Aspects of Social Evolution*. Princeton University Press, Princeton, NJ.

SOIL BIOTA, SOIL SYSTEMS, AND PROCESSES

David C. Coleman
University of Georgia

I. Soils as Components of Ecosystems
II. Soils as Organizing Centers in Ecosystems
III. Major Soil Processes
IV. Biodiversity in Soils
V. Conclusions

GLOSSARY

domains The major divisions of the biota on earth, namely: Bacteria, Archaea, and the Eucarya.
ecosystem engineers The concept whereby members of the macrofauna (e.g., termites and earthworms) are actually moving parts of the soil volume for their own uses (e.g., making macropores, which permit flow of large amounts of water rapidly through the soil).
immobilization The process wherein nutrients are taken up or immobilized in litter and other organic detritus until later (usually weeks to months) in the decomposition process.
microfauna Animals that have high turnover rates, and live in water films in soils. Small mesofauna, such as nematodes, are also water-film dwellers.
larger Mesofauna and macrofauna live in pores in portions of soil profiles.
mineralization The availability of inorganic nutrients in the decomposition of organic detritus, occurring after the immobilization phase; see above.
organizing centers Soils are centers of history and activity in terrestrial ecosystems. See the legacy concept in forest ecology.

SOIL BIODIVERSITY is an intriguing, largely unappreciated facet of global biodiversity. There are many phyla, even "domains," within soils, which are largely unseen, making use of the uniquely diverse physicochemical complexity of soils, which is an intersection of mineral, organic, aquatic, and aerial habitats. Organisms have evolved in soils literally since pre-Cambrian times (more than 600 million years ago). They are still largely undescribed, and this is particularly true for the prokaryotes, which have awaited the development of new techniques to characterize them. By linking several organismal groups to major processes in global biogeochemistry, it is proving possible to appreciate the wide array and diverse nature of soil organism functions in the biosphere.

I. SOILS AS COMPONENTS OF ECOSYSTEMS

A. Soil-Forming Factors

Soils are an intriguing, relatively thin (often <1 m. depth) zone of physical-chemical and biological weathering of the earth's land surface. Soils are formed by an array of factors, namely climate, organisms, parent material, the extent of slope, and aspect (relief) operating over time (Fig. 1). These factors affect major ecosystem processes, such as primary production, decomposition, and nutrient cycling, which lead to the development of ecosystem properties unique to that soil type, as a result of its previous history. For example, a deep loess soil in Iowa, with a very fertile and deep surface or "A" horizon, containing considerable amounts of organic matter, will be very different from an "A" horizon developed in the Nebraska sandhills, with much greater porosity and lower water retention due to the nature of the sandy surface material. As noted in the soil-forming factors diagram (Fig. 1), the array of biota—namely microbiota, vegetation, and consumers (herbivores, carnivores, detritivores)—is influenced by soil processes and in turn has an impact on the soil system.

B. Poly-Phasic Nature of Soils, Influence on the Biota

Soils are perhaps the ultimate in interface media, located at the intersection of four principal entities: the atmo-

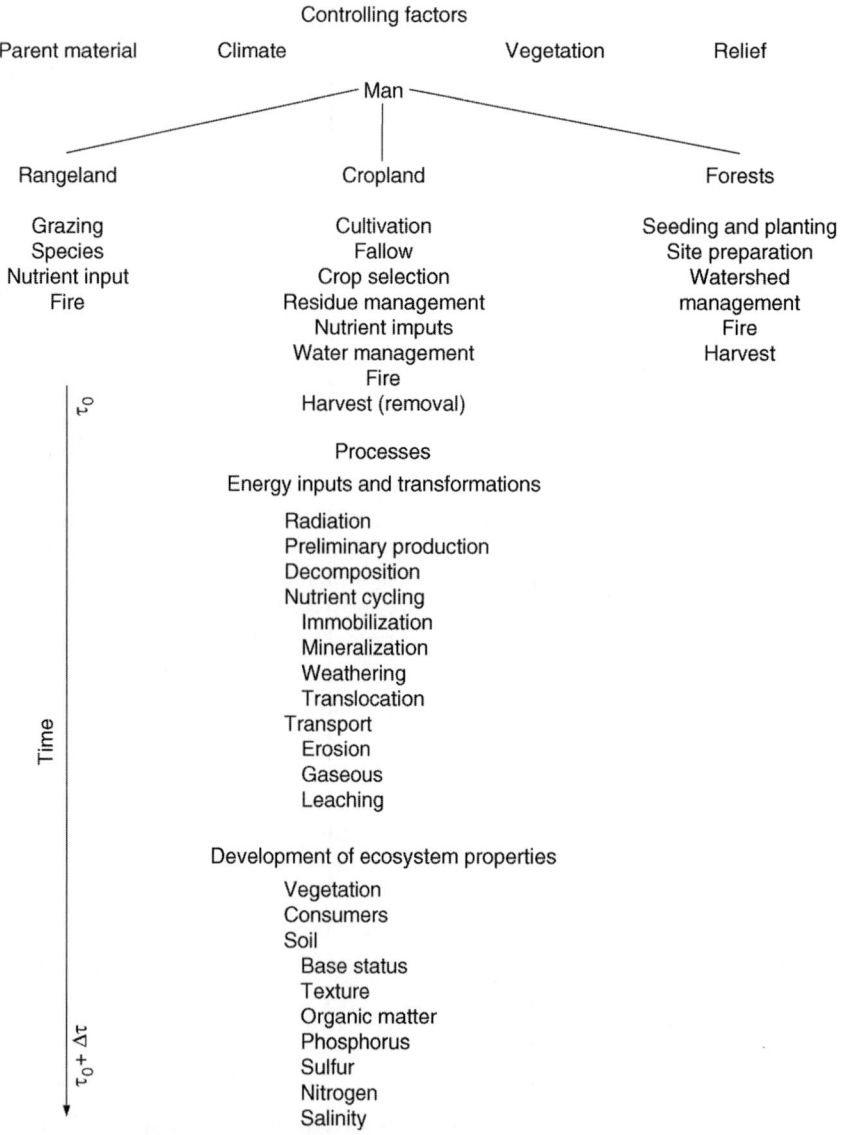

FIGURE 1 Soil-forming factors and processes, and the interaction over time. From Coleman *et al.* (1983) and Coleman and Crossley (1996).

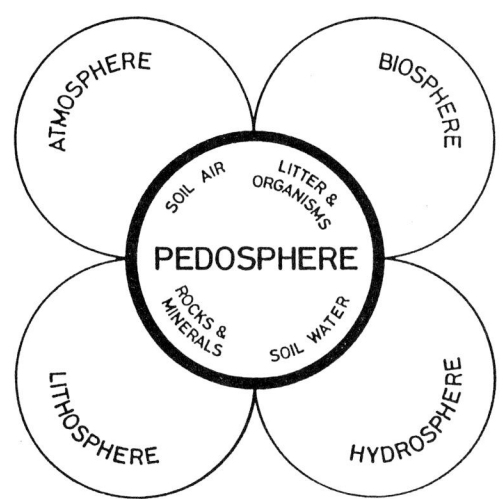

FIGURE 2 The pedosphere showing interactions of abiotic and biotic entities in the soil matrix. From Fitzpatrick (1984) and Coleman and Crossley (1996).

sphere, biosphere, lithosphere, and pedosphere (Fig. 2). Soils provide a wide range and variety of microhabitats, thus accommodating a very diverse biota. The microbes (bacteria and fungi) are found in numerous microsites, well-aerated or not; bacteria may thus respire either aerobically or anaerobically. There is an enormous amount of surface area (hundreds of m^2 per gram of soil) on the soil particles, which range in size classes from clays (0.1–2 μm in diameter), to silts (2–50 μm in diameter), and sands (0.05–2 mm diameter). Numerous microbes and micro- and meso-fauna (protozoa and nematodes) exist in water films on these particles and in or on the surfaces of microaggregates formed from the primary particles. In turn, the more mobile fauna, from collembola and mites (larger meso-fauna) to the macrofauna (earthworms, millipedes, ants, termites, and fossorial or earth-dwelling vertebrates), move through macro- and micro-pores in the soil. The macrofauna play a role in moving parts of the soil profile around and form many sorts of burrows and pores; they are often termed "ecological engineers."

II. SOILS AS ORGANIZING CENTERS IN ECOSYSTEMS

Soils may be viewed as the organizing centers for terrestrial ecosystems. Major functions such as ecosystem production, respiration, and nutrient recycling are controlled by the rates at which nutrients are released by decomposition in the soil and litter horizons and transported to the photosynthetic layers of the ecosystem. This is particularly true for less heavily managed, near-natural ecosystems, many of which occur on soils of relatively poor nutrient status. In these systems, mycorrhizas are often obligate partners in the obtaining of adequate nutrients for the growing plants. Mycorrhizas are known to be efficient at extracting nutrients from both mineral and organic sources, enabling plants to thrive in habitats that are considered poor in nutrients. We need to be aware of these and other mutualistic associations between microorganisms and roots, such as rhizobia and actinorrhizas, various root-associated symbiotic bacteria that facilitate the nitrogen fixation process on which the entire ecosystem often depends. These associations have arisen in soils over evolutionary time and are key to an understanding of ecosystem function.

III. MAJOR SOIL PROCESSES

A. Decomposition: Immobilization and Mineralization

A very large proportion (greater than 90%) of the terrestrial net primary production is returned to the soil as dead organic litter. This litter, consisting of leaves, roots, and wood from trees and organic residues from agricultural fields, is decomposed on or in the soil, and the nutrients contained within it recycled for further use. The decomposition process drives complex food webs in the soil, with numerous interactions between the initial agents of decomposition, the bacteria and fungi, and the fauna which in turn feed on them.

Decomposition is the catabolism of organic compounds in plant litter and other organic detritus. Decomposition is principally the result of microbial activities; few soil animals have cellulases in their guts, which allows them to hydrolyze the celluloses in plant residues. The decomposition of organic residues involves the activities of a variety of soil biota, including both microbes and fauna, which interact conjointly in the process. For example, the initial breaking up of plant litter usually is conducted by the chewing and macerating action of both large and small animals. This initial breaking into smaller pieces, or "comminution," is a process that benefits the fauna, which derive nutritional benefit from the litter or microbes initially colonizing the plant material. The increased surface area and further inoculation of the smaller pieces enhances the microbial access to, and breakdown of, these tissues.

B. Nitrogen Cycle: Major Processes

Nitrogen enters the ecosystem via nitrogen fixation, in which the dinitrogen molecule (N_2) is separated into two nitrogen atoms, with considerable expenditure of energy and the assistance of the nitrogenase enzyme, to break the triple covalent bond. The atoms are ammonified and then used in the production of amino acids and proteins in the plant. Another avenue for nitrogen entry into soils is by lightning fixation, in which the extensive high-voltage energy in the lightning charge ruptures the dinitrogen molecule, hydrogens are attached, and then the ammonium is brought in by rainfall. As shown in Fig. 3, nitrogen is lost from the system via harvest and erosion of organic forms of N, it can be ammonified in decomposition, and then undergoes nitrification to nitrate (NO^{3-}), whereupon it can be taken up by biota, either plant roots or into microbial tissues. If there is adequate energy and low amounts of oxygen present, there can be denitrification, in which the nitrogen is lost as either nitrogen gas (N_2) or N_2O, nitrous oxide. For further details, consult textbooks on ecology or ecosystem studies.

The nitrogen cycle is of critical importance to biodiversity considerations, because key points in the cycle are dependent on relatively species-poor assemblages of microbes, including the nitrogen fixation and nitrification steps. There are only a few species of nitrogen fixing rhizobia, in the genera *Rhizobium* and *Bradyrhizobium*. The other principal nitrogen-fixing symbiont, the bacterium *Frankia* (Actinomycetaceae) forming the actinorrhiza (literally actinomycete-root), contains only a few species in the genus. However, approximately 194 plant species in eight families and four different subclasses of flowering plants have been identified as hosts. These plants share the general tendency to grow in marginal soils and play an important role as pioneer species in early successional habitats.

In the nitrification steps, noted earlier, there are only a few genera and species of nitrifiers. Most of them are autotrophic and quite sensitive to changes in soil pH. This means that these organisms may be unusually prone to being diminished or eliminated in regions where there is considerable acid rain.

IV. BIODIVERSITY IN SOILS

A. Evolutionary History

Soils, as we know them, with well-differentiated profiles, probably developed concurrently with the origin of a land flora in the early Devonian era, about 425 million years ago. The microorganisms that inhabit the soils, particularly the prokaryotic microbes such as the cyanobacteria, originated perhaps 3 billion years ago.

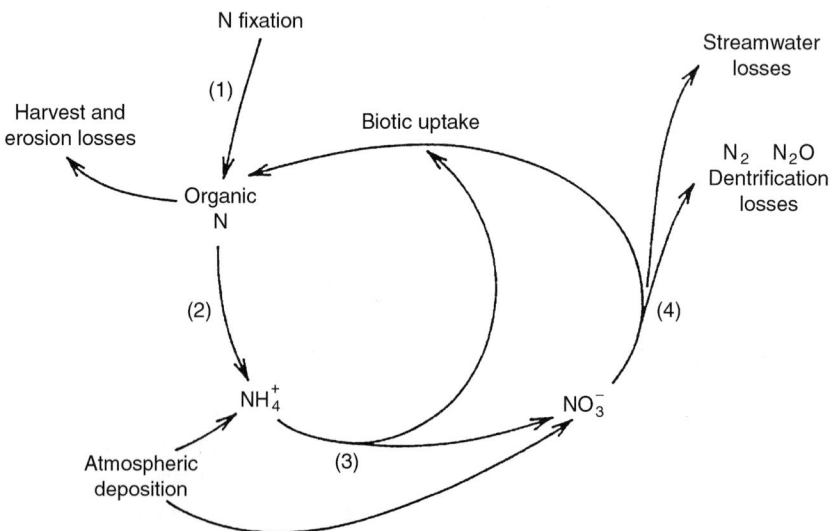

FIGURE 3 Inputs and outputs of N, which make up the intersystem transfers in an ecosystem. Numbers indicate groups of organisms active at a given stage of the N cycle. (1) Rhizobia, *Frankia*, (2) ammonifiers, (3) nitrifying bacteria, and (4) denitrifying bacteria. Denitrifying bacteria must compete with biotic uptake of NO_3 by other microbes and plants; thus the rate of nitrification sets an upper limit on denitrification losses. Streamwater losses represent the excess of available N over that taken up by biotic processes. Modified from Waring and Schlesinger (1985).

B. Diversity of Biota

Biodiversity is an inclusive concept, including a wide range of functional attributes in ecosystems in addition to being concerned with numbers of species present in the system. This differentiates it from the concept of species diversity, which is concerned with the identity and distribution of species in a given habitat or region.

Soil biodiversity is best considered by focusing on the groups of soil organisms that play key roles in ecosystem functioning. Spheres of influence (SOI) of soil biota are recognized, such as the root biota, the shredders of organic matter, and the soil bioturbators. These organisms influence or control ecosystem processes and have further influence via their interactions with key soil biota (e.g., plants). What is the extent of redundancy within functional groups within these SOI? Some soil organisms, such as the fungus and litter-consuming microarthropods, are very speciose. For example, there are up to 170 species in one Order of mites, Oribatida (members of the Arachnida, eight-legged arthropods), in the forest floor of one watershed in western North Carolina. The soil biota considered at present to be most at risk are some of the species-poor functional groups, such as specialized bacteria, that is, nitrifiers and nitrogen fixers (see diagram). Others include fungi forming mycorrhiza (literally fungus-root), a symbiotic association that benefits both plant and fungus, with the plant supplying high-quality carbon to the fungus, and the fungal hyphae exploring a greater volume of the soil, obtaining scarce mineral nutrients, particularly phosphorus. Other species-poor functional groups include macrofaunal shredders of organic matter (e.g., millipedes) and bioturbators of soils, which includes various types of earthworms and termites.

C. Three Great "Domains" of Organisms on Earth

All of life exists in three great "urkingdoms," or domains. These domains are (a) the *Bacteria* (eubacteria), which are the bacteria as generally considered; (b) *Archaea* (archaebacteria), which include the methanogens (methane-producers), most extreme halophiles (ones living in hypersaline environments), and hyperthermophiles (ones living in volcanic hot springs, and in mid-sea ocean hot-water vents); and (c) *Eucarya* (eukaryotes) (Fig. 4). The first two domains are prokaryotes, which are unicellular organisms, lacking a unit membrane-bound nucleus and other organelles, usually having their DNA in a single circular molecule. Eukaryotes, in comparison, consist of all of the organisms that have a unit membrane-bound nucleus and other organelles, such as mitochondria. Eukaryotic organisms are often multicellular. This scheme is based on an increasing body of evidence from ribosomal RNA (rRNA) phylogenies, that the archaebacteria are worthy of the same taxonomic status as eukaryotes and bacteria. As shown in Fig. 4, the universal rRNA tree develops from a postulated "cenancestor," leading to the relative positions of the three great domains.

1. Number of Species of Prokaryotes

Recent estimates of the number of prokaryotic species range from 100,000 to 10 million. Interestingly, the number of described species of bacteria in soil amount only to about 4000. This discrepancy is due largely to the fact that only a small proportion, usually less than 1%, of the bacteria present in soil or any other medium are amenable to culturing and subsequent microscopic observation.

It should be noted that, on the basis of the accepted criterion for separating taxa in microbial studies, which is a greater than 70% DNA homology, a mouse and a human would be considered as being in the same species. This leads to complications, as we shall see, in discussing the total amount of genetic diversity of all organisms, including the as-yet largely unknown diversity of Archaea and Eubacteria. The latter now are estimated to have an array of 36 kingdoms, which are genetically as diverse as the Kingdoms Animalia, Plantae, and Fungi in older classification systems.

2. Biomass and Numbers of Bacterial Species on Earth

This figure is vastly underestimated. We are just now delineating the overall genetic makeup of isolates taken from soils, which are determined by the use of molecular probes. The total numbers of bacteria on earth in all habitats is truly staggering: $4-6 \times 10^{30}$ cells, or 350 to 550 petagrams of Carbon. One petagram is 10^{15} g, or one billion metric tonnes. The amount of the total that is calculated to exist in soils is approximately 2.6×10^{29} cells, or about 5% of the total on earth. A majority of bacteria exist in oceanic and terrestrial subsurfaces, especially in the deep mantle regions, extending several kilometers below the earth's surface.

3. Viruses as Quasi-Organisms

Viruses are quasi-organisms, not included in the three domains. Viruses are RNA or DNA molecules contained within protein envelopes. Viral particles are metabolically inert, carrying out neither biosynthetic nor respiratory functions. They multiply only within host cells, by inducing a living host cell to produce the necessary

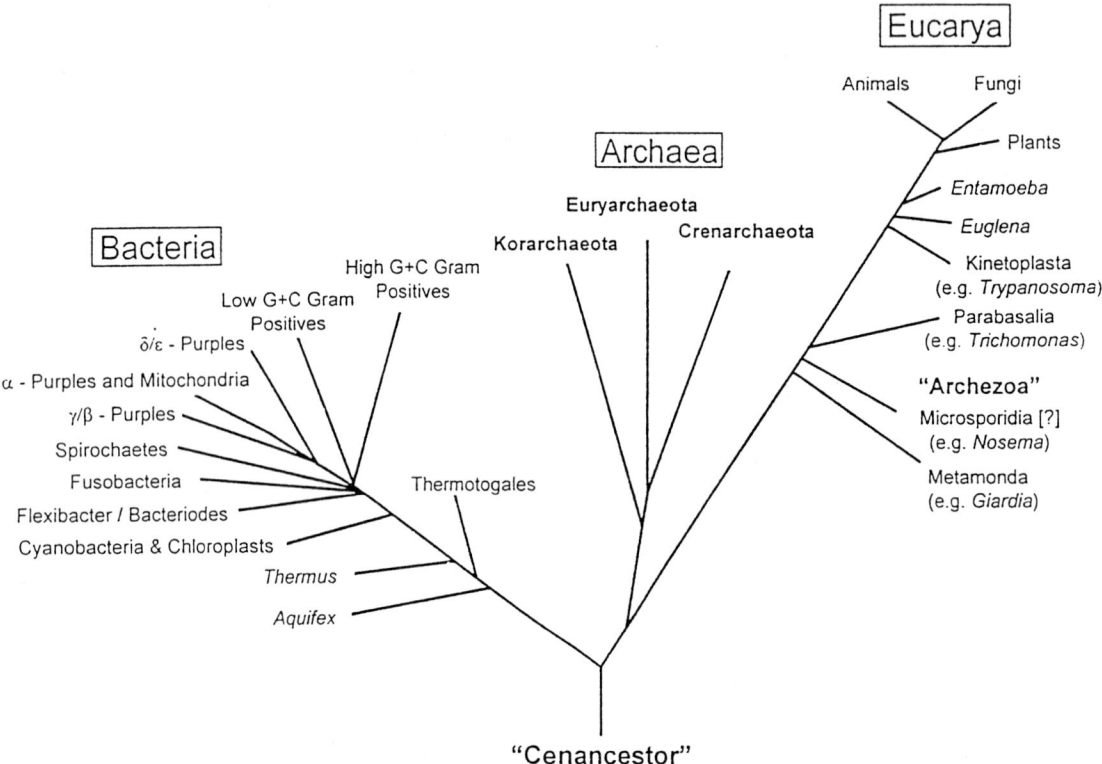

FIGURE 4 Schematic drawing of a universal rRNA tree showing the relative positions of evolutionary pivotal groups in the domains *Bacteria, Archaea,* and *Eucarya*. The location of the root (the cenancestor) corresponds to that proposed by reciprocally rooted gene phylogenies. The question mark beside the Archezoa group Microsporidia denotes recent suggestions that it might branch higher in the eukaryotic portion of the tree. (Branch lengths have no meaning in this tree.) From Brown and Doolittle (1997).

viral components. Once assembled, the replicated viruses escape from the cells. Viruses infect all sorts of animals, plants, and microbes. Viruses parasitizing bacterial cells are commonly called bacteriophages, or simply phages. Although little is known about the ecology of viruses, they can persist in soils for many years and decades. Some research on viruses in deserts showed that they were inactivated in soils at acid pH levels between 4.5 and 6. There is little information on the overall species diversity of viruses in soils. Current estimates are 5000 species known and perhaps 130,000 in existence.

4. Numbers and Biodiversity of Eukaryotes

a. Fungal Diversity

Fungi are multicellular eukaryotes that are found in many habitats worldwide. They have long, ramifying strands (hyphae), which can grow into and explore many microhabitats, and are used for obtaining water and nutrients. The hyphae secrete a considerable array of enzymes, such as cellulases, and even lignases in some specialized forms, decomposing substrates *in situ,* imbibing the decomposed subunits and translocating them back through the hyphal network. Fungi are very abundant, particularly in undisturbed forest floors in which literally thousands of kilometers of hyphal filaments will occur per gram of leaf litter.

Fungi are still little-described, with possibly less than 5% of them known to Science (69,000 described; perhaps 1,500,000 in existence (Table I)). This is largely because of the fact that so many fungi are associated with tropical plants and animals, and these in turn have not been described.

As noted earlier, the roles of mycorrhizas in soil systems are being increasingly viewed as central to much of terrestrial ecosystem function. The total number of mycorrhizal species may be just 1000 or 2000, but they are essential to the growth and reproduction of numerous families of plants. Recent experimental studies have noted that species richness, namely with large versus small numbers of species of Arbuscular

TABLE I
Comparison of the Numbers of Known and Estimated Total Species Globally of Selected Groups or Organisms

Group	Known species	Estimated total species	Percentage known
Vascular plants	220,000	270,000	81
Bryophytes	17,000	25,000	68
Algae	40,000	60,000	67
Fungi	69,000	1,500,000	5
Bacteria	3,000	30,000	10
Viruses	5,000	130,000	4

Source: Hawksworth (1991).

mycorrhiza, has a positive impact on plant primary production in macrocosms of North American old fields (fields undergoing succession and not intensively managed).

b. Microfauna

The unicellular eukaryotes, or Protoctista, include a wide range of organisms, which are more often called protozoans. These include the flagellates, naked amoebae, testacea, and ciliates (Fig. 5). These organisms range in size from a few cubic micrometers in volume to larger ciliates, which may be up to 500 micrometers in length and 20 to 30 micrometers in width. Protozoa are quite numerous, reaching densities of from 100,000 to 200,000 per gram of soil. Bacteria, their principal prey, often exist in numbers up to 1 billion per gram of soil. All of these organisms are true water-film dwellers and become dormant or inactive during episodes of drying in the soil. They can exist in inactive or resting stages for literally decades at a time in very exeric environments.

About 40,000 extant protozoan species have been described, but many more undoubtedly are awaiting scientific discovery. Foissner (1997) notes that about 360 protozoan species per year are being discovered. In an extensive survey of soils from Africa, Australia, and Antarctica, in some cases nearly half of the total species described were new to science. This was particularly true in Africa, where of 507 species identified, 240 of them, or 47%, were previously undescribed. Even in a more extensively investigated region, Australia, 43% of the total of 361 species were new to science. In Antarctica, 95 species were described, with only 14, or 15%, being unknown.

Because many habitats have been uninvestigated yet, and the isolation procedures are still imperfect, from 70 to 80% of all soil ciliates may yet be unknown. This high proportion may hold true for the other protozoan groups as well.

c. Mesofauna

i. Nematodes Nematodes feed on a wide range of foods. A general trophic grouping is bacterial feeders, fungal feeders, plant feeders, and predators and omnivores. Anterior (stomal or mouth) structures can be used to differentiate general feeding or trophic groups. The feeding categories are a good introduction, but feeding habits of many genera are complex or poorly known. For example, some genera in immature phases will feed on bacteria and then become predators on other fauna once they have matured. Because of the wide range of feeding types and the fact that nematodes seem to reflect ages of the systems in which they occur (e.g., annual versus perennial crops, or old fields and pastures and more mature forests), they have been used as indicators of overall ecosystem condition. This is a growing area of research in soil ecology, and one in which the intersection of community analysis and ecosystem function could prove very fruitful. Current species described total some 5000, and upward of 20,000 may exist.

ii. Collembola Collembolans, or "springtails," are primitive Apterygote (wingless) insects. They are called "springtails" because many of them have a spring-like lever, or furcula, which enables them to move many body lengths away from predators by use of it, in a springing fashion. Collembolans are ubiquitous members of the soil fauna, often reaching abundances on 100,000 or more per square meter. They occur throughout the soil profile, where their major diet is decaying vegetation and associated microbes (usually fungi). However, like many members of the soil fauna, collembolans defy placement in exact trophic groups. Many collembolan species will eat nematodes when those are abundant. Some feed on live plants or their roots. One family (Onychiuridae) may feed in the rhizosphere and ingest mycorrhizae or even plant pathogenic fungi.

Eight families of collembolans occur in soils. Many collembolans are opportunistic species, capable of rapid population growth under suitable conditions. Eggs are laid in groups. Collembolans become sexually mature with the fifth or sixth instar, but they continue to molt throughout life. Although many species are bisexual, some of the common species are parthenogenic, consisting of females only. Collembolan "blooms" are a phenomenon of late winter or early spring, when some species may appear in large numbers on the surface of snow banks, on the surface ice of pond water, or on

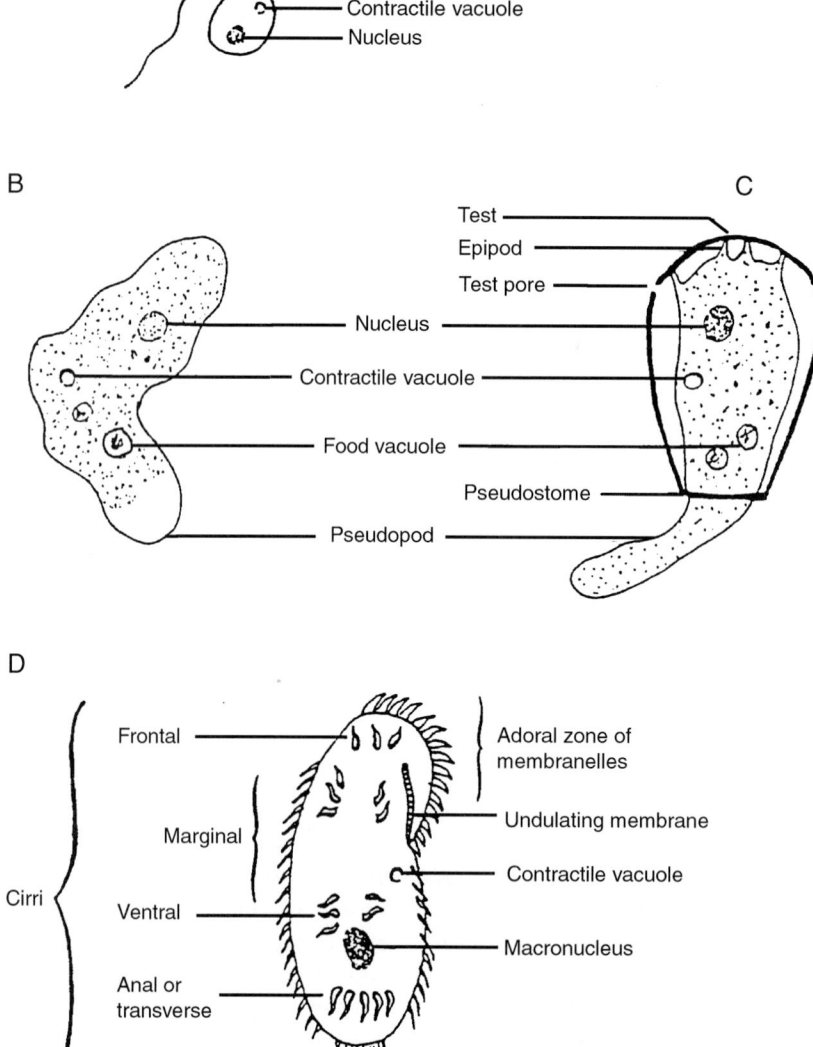

FIGURE 5 Morphology of four types of soil Protozoa: (a) flagellate (*Bodo*), (b) naked amoeba (*Naegleria*), (c) testacean (*Hyalosphenia*), and (d) ciliate (*Oxytricha*). From Coleman and Crossley (1996).

lichen-covered granite outcrops. There are some 6500 described species and possibly more than 10,000 in existence.

iii. Mites (Acari) The soil mites, Acari, are chelicerate arthropods related to the spiders. They are often the most abundant microarthropods in many types of soils. A 100-g sample may contain as many as 500 mites representing nearly 100 genera. This diverse array includes participants in three or more trophic levels, with varied strategies for feeding, reproduction, and dispersal.

Four suborders of mites occur frequently in soils: the Oribatei, Prostigmata, Mesostigmata, and Astigmata. Occasionally, mites from other habitats are extracted from soil samples. These include, for example, plant mites (also called spider mites), predaceous mites normally found on green vegetation, and parasites of vertebrates or invertebrates. The most numerous ones are the true soil mites. The oribatid mites (Oribatei) are the characteristic mites of the soil and are usually fungivorous or detritivorous. Mesostigmatid mites are nearly all predators on other small fauna, although a few species are fungivores and may become numerous at times. Astigmatid mites are associated with rich, decomposing nitrogen sources and are rare except in agricultural soils. The Prostigmata contains a broad diversity of mites with several feeding habits. Very little is known of the niches or ecological requirements of most soil mite species, but some interesting information is emerging. For further details on the life-history characteristics of these interesting animals, refer to Coleman and Crossley (1996). About 20,000 species have been described and possibly in excess of 80,000 exist.

d. Macrofauna

i. Termites Termites (Isoptera) are one of the major ecosystem "engineers" particularly in tropical regions. Termites are social insects with a well-developed caste system. By their ability to digest wood, they have become economic pests of major importance in some regions of the world. Termites are arranged in five different families. The termites in a more primitive family, the Kalotermitidae, possess a gut flora of protozoans, which enables them to digest cellulose. Their normal food is wood that has come into contact with soil. Many species of termites construct runways of soil, or along root channels, and some are builders of large, spectacular mounds. Members of the phylogenetically advanced family Termitidae possess a formidable array of microbial symbionts (bacteria and fungi, but not protozoa), which enable them to process and digest the humified organic matter in tropical soils and to grow and thrive on such a diet.

Although termites are mainly tropical in distribution, they occur in temperate zones and deserts as well. Termites are often considered the tropical analogs of earthworms since they reach large abundances in the tropics and process large amounts of litter. Termites parallel earthworms in ingestive and soil turnover functions. The principal difference is that earthworms egest much of what they ingest in altered form (that enriches microbial action), whereas termites can transfer large amounts of soil/organic material into building nests and mounds (carbon sinks). More than 2000 species of termites have been described, and probably up to 10,000 exist.

ii. Earthworms The earthworm fauna of North America is surprisingly poorly known, given the importance of these animals to soil processes and soil structure. Much of the evidence for earthworm effects on soil processes comes from agroecosystems and involves a small group of European lumbricids (family Lumbricidae in the order Oligochaeta). In North America, south of the southern limit of the Wisconsinan glaciation, several native genera exist. However, exotic (often peregrine European lumbricids) earthworm species have been introduced into much of this area following human population changes and colonizations. Impacts of exotic earthworms on native species are not well understood, although there is evidence that when native habitat is destroyed and native earthworm species extirpated, exotic earthworms colonize the newly empty habitat. As more extensive studies are carried out, it is becoming clear that earthworms are present in a wide variety of tropical as well as temperate ecosystems.

Earthworms have important roles in the fragmentation, breakdown, and incorporation of soil organic matter (SOM). This affects the distribution of SOM and also its chemical and physical characteristics. Changes in any of these soil parameters may have significant effects on other soil biota, by changing their resource base (e.g., distribution and quality of SOM, microbes, or microarthropods) or by changing the physical structure of the soil. Recent evidence indicates that earthworm activities impact the communities of other soil biota through their effects on the chemical and physical characteristics of SOM, causing changes in oribatid species richness and microarthropod abundances. It is probable that earthworm-induced changes in the microbial and microarthropod communities will also have impacts both higher and lower in the soil food web. Some 3650 species of earthworms have been described and possibly as many as 8000 exist.

V. CONCLUSIONS

It is apparent that a large proportion of the biota associated with soils are as yet undescribed, with the most extreme cases being the bacteria and fungi. However, of even somewhat more extensively studied groups, such as Oribatid mites, more than half remain unknown to science. Therefore, it is premature to give even a rough estimate of the total numbers of species that occur in many of these taxa, as such large percentages of the total number of organisms are skill unknown. It is incumbent on the rising generation of ecologists and biologists to develop more innovative ways to describe, catalog, and understand the myriad patterns and processes in the biosphere, which are due in large part to the actions of the biota. It is hoped that some of the observations in this chapter, plus the insights offered by the references cited in the bibliography, will encourage this effort.

See Also the Following Articles

ARCHAEA, ORIGIN OF • BACTERIAL BIODIVERSITY • EUKARYOTES, ORIGIN OF • FOOD WEBS • FUNGI • NITROGEN AND NITROGEN CYCLE • SOIL CONSERVATION

Bibliography

Behan-Pelletier, V. M., and Bissett, B. (1993). Biodiversity of nearctic soil arthropods. *Canadian Biodiversity* 2, 5–14.

Brown, J. R., and Doolittle, W. F. (1997). Archaea and the prokaryote-to-eukaryote transition. *Microbiology and Molecular Biology Reviews* 61, 456–502.

Brussaard, L., Behan-Pelletier, V. M., Bignell, D. E., Brown, V. K., Didden, W., Folgarait, P., Fragoso, C., Freckman, D. W., Gupta, V. V. S. R., Hattori, T., Hawksworth, D. L., Klopatek, C., Lavelle, P., Malloch, D. W., Rusek, J., Soderstrom, B., Tiedje, J. M., and Virginia, R. A. (1997). Biodiversity and ecosystem functioning in soil. *Ambio* 26, 563–570.

Coleman, D. C., and Crossley, D. A., Jr. (1996). *Fundamentals of Soil Ecology*. Academic Press, San Diego.

Coleman, D. C., Dighton, J., Ritz, K., and Giller, K. E. (1994). In *Perspectives on the compositional and functional analysis of soil communities* (K. Ritz, J. Dighton, and K. E. Giller, Eds.), pp. 261–271. Wiley-Sayce, Chichester, England.

Coleman, D. C., Odum, E. P., and Crossley, D. A., Jr. (1992). Soil Biology, soil ecology, and global change. *Biology and Fertility of Soils* 14, 104–111.

Coleman, D. C., Reid, C. P. P., and Cole, C. V. (1983). Biological strategies of nutrient cycling in soil systems. *Advances in Ecological Research* 13, 1–55.

FitzPatrick, E. A. (1984). *Micromorphology of Soils*. Chapman and Hall, London.

Foissner, W. (1997). Global soil ciliate (Protozoa, ciliophora) diversity: A probability-based approach using large sample collections from Africa, Australia and Antarctica. *Biodiversity and Conservation* 6, 1627–1638.

Hammond, P. M. (1994). Described and estimated species numbers: An objective assessment of current knowledge. In *Microbial Diversity and Ecosystem Function* (D. Allsopp, R. R. Colwell, and D. L. Hawksworth, Eds.), pp. 29–71. CAB International, Wallingford, UK.

Hansen, R. A., and Coleman, D. C. (1998). Litter complexity and composition are determinants of the diversity and species composition of oribatid mites (Acari: Oribatida) in litterbags. *Applied Soil Ecology* 9, 17–23.

Harrison, M. J. (1997). The arbuscular mycorrhizal symbiosis: An underground association. *Trends in Plant Science* 2, 54–60.

Hawksworth, D. L. (1991). The fungal dimension of biodiversity: Magnitude, significance and conservation. *Mycological Research* 95, 641–655.

Jones, C. G., Lawton, J. H., and Shachak, M. (1994). Organisms as ecosystem engineers. *Oikos* 69, 373–386.

Read, D. J. (1991). Mycorrhizas in ecosystems. *Experientia* 47, 376–391.

Swensen, S. M. (1996). The evolution of actinorhizal symbioses: Evidence for multiple origins of the symbiotic association. *American Journal of Botany* 83, 1503–1512.

van der Heijden, M. G. A., Klironomos, J. N., Ursic, M., Moutoglis, P., Streitwolf-Engel, R., Boller, T., Wiemken, A., and Sanders, I. R. (1998). Mycorrhizal fungal diversity determines plant biodiversity, ecosystem variability and productivity. *Nature* 396, 69–72.

Wall, D. H., and Moore, J. C. (1999). Interactions Underground: Soil biodiversity, mutualism, and ecosystem processes. *BioScience* 49, 109–117.

Waring, R. H., and Schlesinger, W. H. (1985). *Forest Ecosystems*. Academic Press, Orlando, FL.

Whitman, W. B., Coleman, D. C., and Wiebe, W. J. (1998). Prokaryotes: The unseen majority. *Proceedings of the National Academy of Sciences* 95, 6578–6583.

Woese, C. R., Kandler, O., and Wheelis, M. L. (1990). Towards a natural system of organisms: Proposal for the domains *Archaea*, *Bacteria*, and *Eucarya*. *Proceedings of the National Academy of Sciences* 87, 4576–4579.

SOIL CONSERVATION

Dorota L. Porazinska and Diana H. Wall
Colorado State University

I. Definition of Soil
II. Soil Quality
III. Conservation and Restoration
IV. Summary

GLOSSARY

agroecosystem Agricultural ecosystem (e.g., crop field and grazing pasture).
biocontrol Control of agricultural pests with the use of predators and other beneficial organisms (e.g., control of turf grass crickets with insect parasitic nematodes).
eutrophication The process by which a body of water becomes enriched in dissolved nutrients (nitrogen, phosphorous) that stimulate growth of aquatic plant life (e.g., algae), usually resulting in the depletion of oxygen from water. This frequently creates unfavorable conditions for fish and other biota.
herbivory Animals feeding on plants.
mineralization Process of transforming from organic to inorganic form.
soil food web Representation of all feeding interactions among organisms in the soil (who eats whom).
soil structure Spatial arrangement of soil particles.
soil texture Percent composition of clay, silt, and sand in soil.
trophic Describing feeding habits or the kind of nutrition used by a group of organisms.
water-holding capacity Capacity of soil to hold water (e.g., sandy soils have very low water-holding capacity).

SOIL IS AN IMPORTANT natural resource supporting plant, animal, and human populations and is a habitat for a diversity of species. Productive soils throughout the world are being degraded rapidly due to human activities. Declining soil quality over the past century has manifested itself in loss of agriculturally productive lands, forests, and wetlands. Soil conservation efforts aim for preserving, sustaining, or improving the quality of soils. This chapter discusses soil quality and the many methods for conserving, sustaining, and restoring soil ecosystems.

I. DEFINITION OF SOIL

A. A Complex and Dynamic System

Soil is composed of living and nonliving components organized vertically, in a profile of horizontal layers or horizons. Soil is the habitat for a great abundance and diversity of living organisms, many of which are microscopic (e.g., bacteria, fungi, protozoa, nematodes, mi-

TABLE I
Examples of Some Groups of Soil Fauna Ordered by Body Width and the Number of Described Species in Soil

Soil fauna	Estimated species number
Microfauna (1–130 μm body width)	
Protozoans	1,500
Nematodes	5,000
Mesofauna (80 μm–4 mm body width)	
Mites	20,000–30,000
Springtails	6,500
Diplurans	660–800
Pot worms	600
Termites	1,600
Macrofauna (1–35 mm of body width)	
Isopods	5,000
Centipedes	2,500
Millipedes	10,000
Earthworms	3,600
Ants	8,800
Flies (larvae)	60,000

Only about 1 to 5% of soil biota have been described to species level. Modified from Wall and Virginia (1999).

croarthropods) (Table I). The nonliving component of soil consists of the solid phase, or weathered parent geological material, which contributes to physical (e.g., soil texture, structure, density, porosity [the space within and between aggregates]) and chemical (e.g., fertility, moisture, acidity) properties. The formation of soil is a function of climate, parent material, biota, and topography, which through geologic time has formed a multitude of natural soil types of varying properties across geographic landscapes (Jenny, 1980). Hundreds to thousands of years may be required to form just a centimeter of soil.

B. Role of Soil

Fertile, productive soils provide a basis for the economic wealth of a nation by providing food, fiber, and fuel. Soil supports human civilization by supplying nutrients for growth of plants, including agricultural crops, by regulating the water flow from rainfall to groundwater, and by filtering and transferring damaging substances that might enter the atmosphere or groundwater. Water quality is largely dependent on the filtering capacity of the living and nonliving components of soil. Soil as a natural resource, therefore, contributes to the foundation of our social and industrial infrastructure.

Belowground systems, as fundamental constituents of all terrestrial ecosystems, influence and are influenced by the functioning (the performance) of the aboveground systems. For example, soils affect aboveground biodiversity, ecosystem nutrient and energy cycles and fluxes, and even certain atmospheric components (e.g., global cycles of C and N). Soils are one of the largest reservoirs of global carbon and are the major terrestrial reservoir of dead organisms—plants, animals, and microorganisms. The quality (chemical composition) and quantity of dead material (organic matter) received from both aboveground and within the soil determine the primary nutrient and energy base for the soil biota. Plant litter (dead leaves, twigs, roots) and other dead organic matter is decayed by soil microbial and faunal assemblages (organized in detritus food webs) during decomposition, a process of transforming organic matter to inorganic compounds (e.g., nitrate, ammonium, and phosphate), which supplies nutrients for plant growth. Thus, factors affecting the quality and quantity of soil organic matter (SOM), such as climate, cultivation, invasive species, and atmospheric nitrogen (N) inputs (e.g., acid rain), can alter the diversity of above- and belowground organisms, the rate of decomposition, the structure of soil, and the availability of nutrients needed for plant growth.

II. SOIL QUALITY

A. Attributes of High-Quality Soils

In the broader sense, the quality of soil refers to an ability to sustain biological productivity and the diversity of plant, animal, and human populations, and to maintain quality of water and air. High soil quality also implies an ability of soils to maintain high fertility, productivity, and resist erosion. Natural differences in the quality of soils indicate different capacities of soils to resist stress whether of natural (e.g., wind, fire, and rainstorm) or anthropogenic (e.g., plow, invasive species, and pesticides) origin. In the narrower sense, soil quality relates to the inherent combination of biotic, physical, and chemical properties that allows soils to have long-term productivity.

A number of factors are involved in defining the quality and productivity of soils: parent material (the geologic weathering of rock), climate, soil organic matter, soil structure (aggregation of soil particles), soil stabilization, and soil biota (including roots) (Coleman and Crossley, 1996). One of the most important factors is soil organic matter. The amount and type of soil organic matter and the products of decomposition affect several soil properties. A higher content of SOM results in higher cation exchange capacity (CEC), higher water-holding capacity, higher infiltration rates, better soil aeration, and increased soil particle aggregation, all leading to improved moisture infiltration and retention, reduced runoff of nutrients, and less soil erosion. Cation exchange capacity refers to the ability of soil to store nutrients (e.g., calcium, magnesium, and potassium) for future plant uptake. Organic matter, soil biota abundance and diversity, plant roots, and water and air movement affect soil structure formation and aggregation (Hartel, 1998; Paul and Clark, 1996). Soil stabilization, which is the ability of soil to maintain its structural integrity when subjected to natural or anthropogenic stress, occurs when there is some degree of aggregation of soil particles. The soils are then more resistant to soil degradation (e.g., erosion, loss of fertility, reduced filtering, and buffering capacity).

B. Erosion

Erosion is a general term for the removal of the surface of the earth by abrasive actions of wind, water, waves, and glaciers. There are two types of erosion: geological and accelerated. The Grand Canyon in the United States is an example of the effects of geological erosion that occurred over millions of years. Geologic erosion in natural ecosystems is a slow process, typically occurring at a rate slower than the rate of soil formation. This is because the soil is protected by vegetation. For example, aboveground vegetation can reduce wind speed and roots help anchor the soil. Accelerated erosion happens when the rate of soil loss is higher than the rate of soil formation. This type of erosion occurs when the lighter individual particles of soil aggregates are detached and transported by wind or water. These particles can be blown for great distances as dust (the Dust Bowl of the 1930s in the United States) and be deposited to form new soils or washed into streams, rivers, and oceans. The risk of soil erosion depends on the natural conditions (climate, slope, vegetation cover, and soil) and land use (e.g., removal of the protective cover of vegetation). Erosion results in a deterioration of soil quality and imposes hazards to humans and organisms in terrestrial and aquatic ecosystems (e.g., loss of soil fertility, reduced plant productivity, loss of water and air quality, flooding, mudslides, and sedimentation).

Erosion is the primary factor degrading soils on a global scale (Fig. 1). In many areas of the world eroded land is no longer productive and is often abandoned. A recent estimate indicates the amount of arable land on earth is about 1.44 billion hectares ("Conserving Land") (Fig. 2). In many areas, continuous intensive agricultural cultivation leads to soil erosion, salinization, and desertification and often to loss of the land from agricultural production. It is estimated that globally 25 billion metric tons of soil erodes each year from agricultural land (17 tons per cultivated hectare, or 4.5 tons per person) (Food and Agriculture Organization, 1992). To accommodate the rapidly increasing human population, agriculture will have to increase globally the quantity or quality of food produced on a given area or increase cropland area into marginal areas of lower soil quality. Thus, global efforts to conserve soils are of immediate importance for human well-being.

Water is estimated to be responsible for more than one-half the global soil degradation, followed by wind, chemical (e.g., salinization, acidification, pollutants, atmospheric nitrogen deposition, excessive fertilizers, pesticides and manures), and physical (soil compaction, water logging, subsidence) degradation. However, whether a region is affected more by wind or water is dependent to a large degree on its climate. Wind erosion is a more serious problem for agricultural lands in the arid and semiarid regions of the world (e.g., North Africa, the Near East, parts of Asia, Australia, northwest China, southern South America, and North America). Salinization, the increased concentration of salts in the topsoil, can occur in irrigated lands of the world due to increased evaporation, susceptibility of soils to salty groundwater, and invasion of seawater. These forms of soil degradation continue to decrease the availability of productive land for future food, fiber, and fuel production.

As high-quality and agriculturally productive land diminishes, other ecosystems (e.g., forests, wetlands, and meadows) become vulnerable. Within the past 3 centuries, 2.2 billion ha of forests (down from 4–6.2 billion ha worldwide) have been converted to agriculture. Forests generally have a high biodiversity and are a reservoir of hundreds of billions of tons of carbon stored in trees and soils. On a global scale, conversion of forests to croplands releases CO_2 into the atmosphere, contributing to global warming. Conservation of soil is

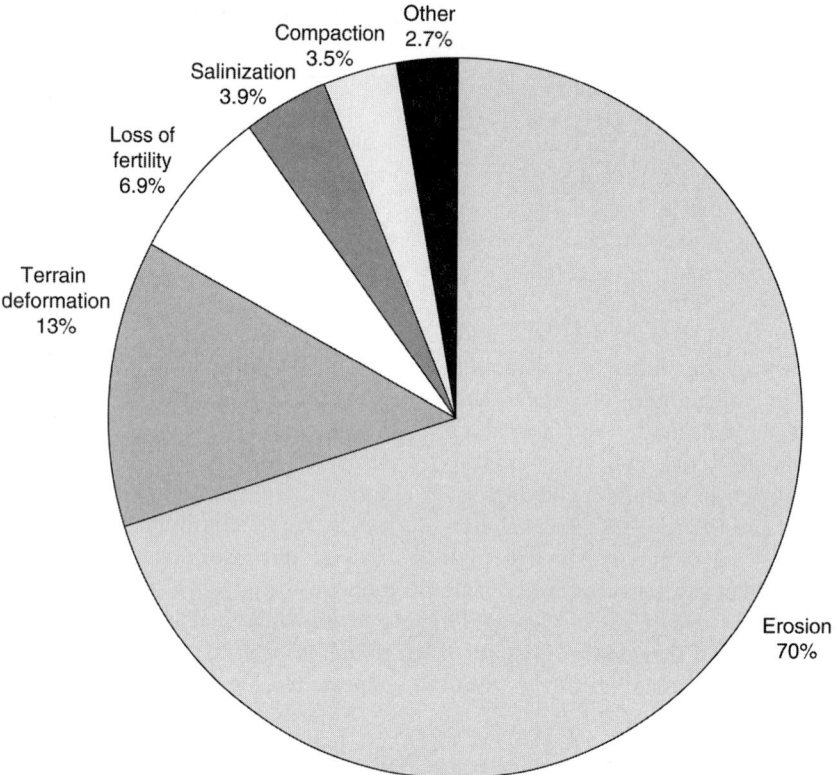

FIGURE 1 Causes of degradation of the world's land surface. 1995. The percentage indicates the respective impact of the various forms of degradation. The "Other" section includes pollution (1.1%), overblowing (0.6%), waterlogging (0.5%), acidification (0.3%), and subsidence (0.2%).

critical to the maintenance of sustainable ecosystems for the future.

C. Effects of Conversion of Natural Systems on Soil Quality

In natural systems, the amount of SOM is maintained by high (generally greater than 90% of primary production) litter inputs, but in agricultural and grassland grazed systems due to harvesting the yield or herbivory, only about 50% of primary production contributes to SOM. In agriculture, the lower amounts of organic matter input to SOM results in less substrate (energy and nutrients) for the soil food web and fewer nutrients available to plants over time. In intensive agriculture, nutrients removed in crops or lost with erosion of topsoil have to be replaced with synthetic fertilizers in order to maintain high productivity.

Human activity changes soil chemical and physical properties, soil biodiversity, and soil quality. Agriculture (including forestry and grazing), urbanization, and industrialization are the most significant modifiers of soil quality. Tillage (plowing), one of the most successful agricultural practices, has been used for centuries to reduce weeds, aerate soils, and break up compacted soils. However, tillage along with other intensive agricultural practices, such as application of fertilizers and pesticides, use of heavy machinery, irrigation, and monocropping have resulted in degradation of soils in many regions of the world. Intensive tillage destroys soil aggregates, redistributes and enhances SOM turnover by increasing rates of decomposition, decreases the quality and quantity of organic matter, changes soil climatic conditions, and decreases soil moisture, soil food web complexity, soil fertility, and soil stability (Paustian et al., 1997). These changes increase microbial activity, organic matter oxidation, soil compaction, and water runoff. The timescale of agricultural soil degradation varies from decades to a few years (in tropical soils) to a single rainstorm. The loss of topsoil depends on the original status of soil quality, farming practices, climatic events, and the amount of topsoil (Jenny, 1980).

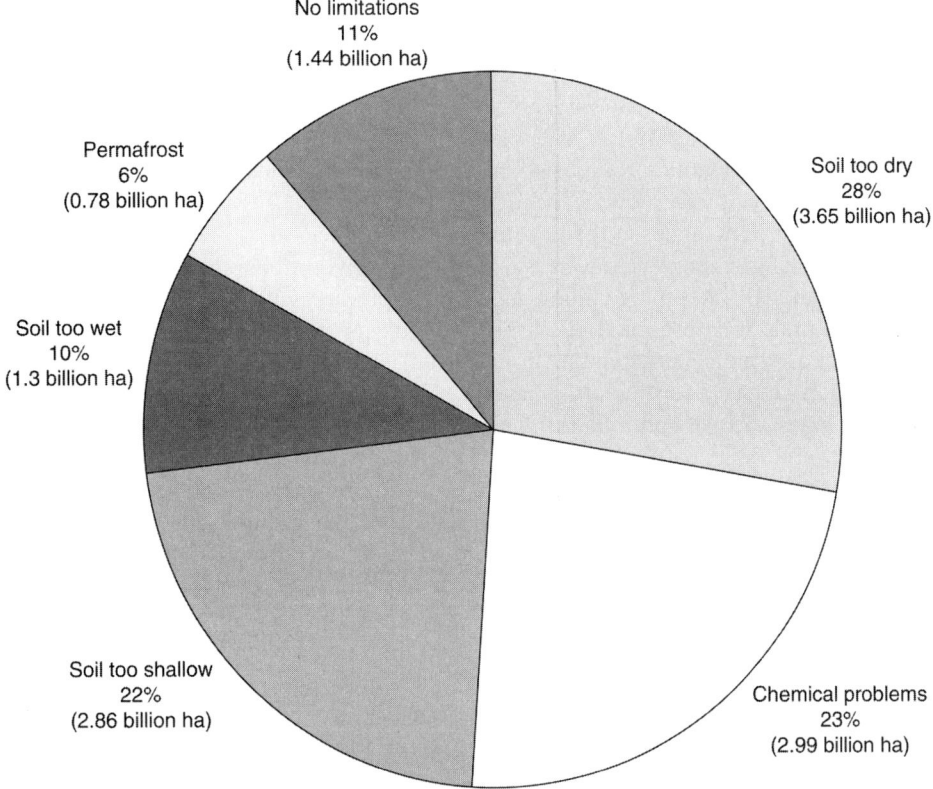

FIGURE 2 Total world's land surface and reasons why the entire land on earth cannot be used for agricultural activity. The percentage indicates fraction of total land, and numbers in parenthesis indicate the actual area in hectares (ha).

Construction (e.g., building of dams, urban development) and deforestation also contribute to an increased rate of erosion and loss of soil quality. Local effects of soil degradation are manifested on a landscape scale (watersheds, rivers, wetlands, oceans) through surface and groundwater pollution and soil sedimentation, with resulting impacts on the quality of water and air, and economic impacts on tourism and the fishing industry.

III. CONSERVATION AND RESTORATION

A. Methods of Soil Conservation in Agriculture

The rate of soil degradation presently exceeds the rate of soil formation in systems impacted by humans (Jenny, 1980; Jenny, 1984). All vegetative productivity depends to a great extent on the nutrient availability in the soil, but in managed systems, such as intensive agricultural systems with continuous monocultures, a disproportionate amount of nutrients are removed from the system as crop yield. Although synthetic fertilizers renew the nutrient pool necessary for plant growth, they do not regenerate the quantity or the heterogeneity in the quality of soil organic matter. Thus, slower decomposing organic materials (e.g., fertilizers such as manures) are used increasingly to replenish soil organic matter. Crop harvest also leaves the soil bare and exposed to erosion by rain and wind. Thus, the diminishing quality of soil resources in natural and managed ecosystems, and the increasing demand for food, fiber, and fuel supply, has resulted in a reevaluation and selection of land management strategies that enhance longevity and quality of soil ecosystems. These alternatives focus on maintenance of soil organic matter, soil fertility, soil biodiversity, soil structure, and soil stabilization, and a reduction of soil erosion as a means of sustaining ecosystems for the long term.

Methods to reduce soil erosion are directed at protecting the surface of the soil (topsoil) and include tillage practices (reduced tillage, conservation tillage, no till, minimum till or zero tillage—techniques that use specially designed machines and herbicides for minimal impact on the soil); diversified farming or cover crops; polycultures; adding organic composts or mulches; crop rotations; plowing techniques, such as contour farming (plowing at right angles to the land to create ridges to hold water); terracing (leveling areas on a slope to prevent water runoff); and timing of plowing.

Methods of agricultural soil conservation have proven valuable whether used alone or in combination. In the United States, soil conservation methods have decreased soil erosion rates so that only one-third of the agricultural lands are eroding faster than the average rate of soil formation. Although this is a considerable improvement over the Dust Bowl days of the 1930s in the United States, soil conservation still appears to be necessary. A recent summary of long-term experiments has evaluated the methods of conservation tillage intensity and crop management used alone or in combination on soil carbon storage. The combination of reduced tillage, bare fallow, increased inputs of crop residue, and crop rotations (with use of perennial vegetation) was more efficient at restoring soil organic matter and soil carbon and reducing levels of soil erosion than any of these management methods used separately. Rates of erosion decreased in Creek Basin, Wisconsin, over the past 140-year period (historical and current data) because of improvements in local agricultural land management in 1975 through 1993. These studies provide evidence supporting the efficacy of soil conservation practices.

1. Reduced Tillage

One of the most important factors decreasing soil quality is tillage. Reduced tillage practices that incorporate crop residues into the soil are among the best alternatives to conventional tillage and have increased globally over the past 10 years. Although reduced tillage practices increase SOM and soil moisture and contribute in many cases to an improved soil food web, they may require a greater use of herbicides due to the establishment and spread of weeds and soil pathogens. In some situations, crop yields may decrease due to the presence of the weeds and pathogen diseases.

Conventional and reduced tillage systems appear to be economically comparable, particularly if crop rotation is used in the reduced tillage system. The reduced tillage system, however, promotes greater long-term benefits for ecosystems locally and globally. Soil biotic complexity is generally positively affected by the retention of crop residues, and the rate of decomposition generally changes from a microbial to a fungal pathway resulting in a slower pulsed release of nutrients for plant uptake. The retention of soil residues also results in a reduction in CO_2 emissions to the atmosphere compared to conventional tillage, and a long-term conservation of beneficial soil chemical and physical properties.

In conventional high-input agroecosystems, high crop productivity is achieved by application of synthetic fertilizers (mainly nitrogen, potassium, and phosphorus), not by decomposition of soil organic matter. As SOM diminishes in conventionally tilled systems, N and P retention is reduced and significant amounts of added fertilizers (nitrogen and phosphorus) are lost from the system. Fertilizer runoff and nutrient leaching reduces surface and groundwater quality. At a regional and global scale, nutrients from agricultural fields are lost to the atmosphere or lead to eutrophication of coastal waters with effects on coral reefs, estuaries, and fisheries. For example, in the Mississippi River, the nutrient levels have doubled or tripled since the 1950s, resulting in large algal blooms that deplete oxygen from the water, killing fish and shrimp, as well as other organisms.

In contrast, in most reduced tillage systems, the quantity of nitrogen applied as fertilizer is reduced, and there is an increase in the amount of applied nitrogen that becomes immobilized in the soil's organic matter. The nitrate leaching potential is reduced and subsequently the risk of water pollution is lower. However, increased earthworm populations co-occurring with reduced tillage practices can result in increased soil porosity and higher nitrate leaching.

2. Diversified Cropping

Organic matter can also be conserved in soils by growing cover crops, termed diversified or multiple cropping, during fallowing. Advantages of cover crops are numerous. Cover crops not only reduce soil nitrogen leaching (nitrates become immobilized in plant biomass), but they also provide a relatively effective means for weed control. They also decrease erosion since the plants and not the soil intercept the raindrops. Cover crops are selected based on several attributes, including plant species that improve soil (e.g., depth of roots, type of organic matter produced, legumes that fix atmospheric nitrogen), their effect on plant pathogens and their predators, or for their economic benefit. Although multiple cropping requires an extensive knowledge on the type of crop, and in what sequence to grow cover crops, the benefits have been recognized for centuries. Multiple cropping systems are increasingly being used as a tool in modern agriculture. Today's research provides scientific support for diversified agricultural systems as they maintain soil quality (SOM, soil fertility), low insect and disease occurrence, and high plant productivity.

3. Organic Amendments

Organic materials or mulches offer in many instances, a long-term option for maintaining soils with higher soil organic matter for long-term agricultural production. Composts provide sources of nutrients that promote plant productivity and soil quality. Other benefits include improved soil aggregation, soil aeration, water-holding capacity, and cation exchange capacity. Examples of organic amendments include green manures (herbaceous crops plowed under while green), chicken and cow manure, pig slurry, urban grass cuttings, and homeowner garden composts of leaves, grass, and food remains. When conformed to quality standards set up by national environmental agencies, composts such as those derived from municipal and industrial wastes have the potential to improve soil quality and at the same time reduce the conversion of land to landfills (land conservation). Currently, however, the supply of compost does not meet demands, and with the higher cost of compost as compared to synthetic fertilizers, the use of organic amendments is not an economical option for many growers in industrialized nations. An increase of consumer interest in the purchase of organically grown food products may increase future use of composted organic materials.

4. Bioremediation of Pollutants

Soil organic matter, soil structure, fertility, and stability—and thus soil quality—can be affected by pollutants. Chemicals may enter the soil system purposefully (e.g., fertilizers and pesticides) or accidentally (via spills or failures of technological processes). Depending on the final concentration, the presence of many hazardous and toxic chemicals can have an immediate or long-term effect on soil quality. Restoration of polluted soils through bioremediation not only avoids the risk of human and animal health hazards, but restores, to some degree, otherwise degraded land. Bioremediation, unlike other methods, takes advantage of soil biota or plants to detoxify contaminants in the soil.

Soil is a natural habitat to a diversity of microorganisms and other soil biota that assist in the reduction of soil contaminants. Microorganisms, for example, may use the pollutant as a substrate or energy source for their metabolism and change the composition of the compound to a less harmful chemical. Pollutants can be changed from harmful to environmentally safe compounds through natural bioremediation (use of indigenous microflora), biostimulation (addition of nutrients to soil to stimulate activity of indigenous microflora), bioventing (addition of gases such as oxygen or methane to stimulate activity of microflora), bioaugmentation (inoculation of soil with exogenous microorganisms), landfarming (mixing healthy and toxic soil), and phytoremediation (use of plants) (Skipper, 1998).

Bioremediation offers an interesting and economical alternative to the conventional way of cleaning soil contamination. The entire treatment takes place at the polluted site as opposed to more traditional method of moving large amounts of soil to treatment facilities and then back to the field. Bioremediation is often cheaper, resulting in a savings on energy used to remove and transport soil, and typically has a lower environmental impact, as it depends on natural processes with no hazardous byproducts (Skipper, 1998). The major constraints of bioremediation apply to the nature of the pollutants, as microorganisms cannot access many new synthetic compounds. Bioremediation can be further complicated by soil contaminants that are mixtures of pollutants, degradation of which may require the presence of several types of organisms each capable of transforming a different chemical in the mixture.

5. Manipulating the Biotic Community

Another aspect of soil quality relates to the diversity of biotic communities. Native land transformation, agricultural management, and pollution significantly affect, in most cases negatively, the abundance and diversity of organisms (Foissner, 1999; Wall and Virginia, in press). The array of multiple plant species (trees, shrubs, grasses) seen in native systems contribute to soil stabilization because the heterogeneity of rooting distribution, rooting depth, and plant chemical composition help to maintain soil structure and a diverse biotic community. In conventional agricultural systems, monocropping or cropping of one plant species (if not one variety) provides homogeneous root morphology, root depth, and litter quality across the landscape. The differences between the belowground heterogeneity of natural systems and the belowground homogeneity imposed by monoculture agriculture must be a consideration when developing long-term soil conservation plans. Using polycultures or intercropping is an agricultural practice that mimics, to some degree, the belowground heterogeneity of natural systems. An unresolved question is whether creating or restoring the structure of soil by this practice will actually recreate and restore the function provided by the soil biota.

The abundance and diversity of microflora and fauna in soil provides beneficial services to humans and ecosystems, including, but not limited to, nutrient mineralization, biological control, nitrogen fixation, soil aggregation, and soil stabilization. In natural systems, the decomposition of organic matter (plant litter, dead roots, animals) to inorganic chemicals necessary for plant growth involves many different groups of soil organisms. Microflora (bacteria and fungi) influence the amount of C stored in the soil and are responsible for the global cycling of many minerals (e.g., N, C, S, P) (Gregorich et al., 1997). Microflora play a critical role by aggregating soil particles, which helps to prevent soil erosion. For example, they produce chemicals (e.g., polysaccharides) that bind soil particles, or have morphological structures (e.g., fungal hyphae) that connect soil particles. Microfauna (e.g., protoza, nematodes) affect soil fertility by trophic interactions with microflora that increase available N and P for plants. Nematodes, for instance, by grazing on bacteria can alter bacterial abundance and activity and thus significantly affect the rate of organic matter turnover and nutrient availability. Invertebrates transport microflora throughout the soil. In addition to functions similar to microfauna, mesofauna (e.g. mites and springtails) feed on fungi and other small fauna (Brussaard et al., 1997). Macrofauna (e.g., earthworms, termites, ants, and snails) comminute and redistribute organic matter within the soil profile and by burrowing affect soil physical properties (Lee and Foster, 1991). They have been termed "soil engineers" for their role in mixing soils and microflora.

Tillage, monocropping, and additions of fertilizers and pesticides often reduce the diversity of many components of soil biotic communities, limiting at the same time their potential positive effects on many ecosystem processes. Improvement of soil quality will depend on integrated management practices that reduce the artificial energy inputs (fertilizers) and exploit the food webs of soil organisms. Maintenance of or exploitation of soil biodiversity for beneficial human services may require a significant change from intensive conventional tillage. Reduced tillage, for instance, promotes more diverse decomposer communities (often with a higher fungal proportion) and lower mineralization rates (reduction of N losses). A long-term addition of organic matter (compost or no till) promotes microbial and faunal activities through which restoration of nutrients and soil organic matter is possible. However, considerable research is needed to determine how these practices will affect the food web. Just as all soils differ in their composition, food webs vary with soil, climate, and the quality and quantity of organic matter. Therefore, different organic composts cannot be expected to create the same food web in all soils.

Another way to maintain soil biodiversity and belowground heterogeneity is development of agricultural mosaics. Traditional agriculture (still common in the tropics and many European countries) promotes a patchy landscape with the use of strip weed margins, hedgerows, and shelterbelts. These latter two types create vertical structures to reduce wind, and also provide microhabitats and refugia for many beneficial soil organisms. A huge area with a monocropping system increases the risk of pest outbreaks. Among many characters of agroecosystems with low pest potential are diversified crops in time and space, crop rotations, a structural mosaic of cultivated and uncultivated lands, the presence of perennial crops, and high crop genetic diversity. All these elements not only stimulate development of favorable habitats for biological control agents, but also they provide a reservoir for recolonization of organisms involved in decomposition, detoxification, and other renewal processes.

6. Conservation of Urban Soils

Human population growth has been followed by urbanization. Conversion of native land for housing, parking, roads, industry, landfills, mining, an so on not only reduces potentially arable land but also affects global nutrient cycles (C and N). Additional consequences of urbanization include an increased use of fertilizers and pesticides in home lawns and gardens. These contribute to soil and water pollution, decreased soil biotic complexity, and increased nitrogen runoff. With urban development it is important to adopt urbanization strategies that are safe environmentally. Proposed methods of environmental urbanization, including vertical rather than horizontal sprawl, smaller house acreage, and preservation of open spaces between developed areas, are already being socially accepted and implemented. On a smaller scale, planning home gardens based on the local climate and natural vegetation is becoming more popular.

B. Assessment and Monitoring of Soil Quality

The goal of providing future generations with an opportunity for a high quality of life cannot be detached from the preservation or improvement of environmental quality. In order to evaluate the status or change of the environment, assessment, monitoring, and regulation programs are being developed. On the basis of monitoring data, farming strategies, urban development techniques, and industrial technologies are being modified to prevent further degradation of a particular system. Typically, to evaluate the quality of the soil system, or its state of sustainability, the system attributes are compared against native or minimally disturbed reference sites. This procedure can provide an initial estimate of human impact (e.g., agriculture, urbanization, and pollution) and through monitoring (an assessment of a soil ecosystem attributes through time) gives us a perspective on the positive or negative aspects of a management strategy. To be effective, monitoring has to be connected with the process of management decisions; therefore adjustments or modifications of management strategies should accompany these decisions (a process termed "adaptive management").

1. Indicators

Indicators are measures that tell us about the status of the environment over time. Monitoring of, for example, air and water quality, as well as CO_2 concentrations, are officially regulated by many countries. There have been many indicators proposed for monitoring soil quality status, but it has been difficult to select a single indicator that would aid policy makers.

The quality of soil can be defined in terms of many variables or combinations of variables. These variables include items such as biodiversity, levels of specific mineral elements and pollutants, levels of primary productivity, or profitability. Some aspects of high-quality soils may contradict each other. For instance, higher levels of nitrogen in the soil may stimulate primary production but may suppress biodiversity. The choice of an indicator is further complicated by the spatial heterogeneity of soils. Different soil systems may require a different suite of indicators. In addition, soil has many functions and maintaining those functions may have different impacts at different scales. For example, continuous cultivation may increase local crop productivity, but may deplete carbon stored in the soil, contributing to global elevated atmospheric CO_2. Theoretically, some optimal condition involving a combination of different aspects of soil quality can be achieved.

a. Physical Attributes of Soil as Indicators

The ability of soil to provide mechanical support for plant growth, diversity of life forms, and agricultural activities is another measure of soil quality. The most important physical attributes of potentially indicative functions of soil quality are soil porosity, soil stability, water storage capacity, aeration, and soil structure. Any of these measures would indicate a change in the soil quality, but might not indicate other changes that are important to ecosystem function.

Soil porosity determines how well liquids, gases, and heat can be stored and transmitted within the soil matrix. Greater porosity often indicates greater storage and transmission ability. Soils have greater stability and resistance to natural and anthropogenic stress when the structural integrity of the soil is intact. Soil stability decreases with tillage, the use of heavy equipment, and heavy rainfall or irrigation, because soil aggregates are destroyed and greater compaction and erosion occurs. Soil water storage capacity is particularly important for the biotic component (plants, microbes, fauna) of soil ecosystems. The best plant performance occurs in soils almost saturated with water and declines as soil dries. Many of the soil fauna are aquatic organisms requiring a film of water around soil particles for reproduction and mobility. The abundance and diversity of soil organisms can also be negatively affected by extremes in soil moisture.

Soil porosity and soil water content are directly related to soil aeration. Aeration is considered as an ability of the soil to store and transmit gases, particularly oxygen and carbon dioxide, as these are the components of roots, microbes, and fauna activity. Low porosity and waterlogged soils might induce inadequate aeration and negatively affect plant growth, root activity, soil biodiversity, and trace gas flux.

These physical soil characteristics are inherently related to soil structure. The extent, size, and shape of aggregates can strongly affect physical, chemical, and biological properties of soil. Typical indicators of good soil structure are well-established aggregates with a desirable porosity, water-holding capacity, and aeration, resulting in environments more favorable for biodiversity and plant productivity as well as resistance to degradation.

b. Chemical Attributes of Soil as Indicators

In terms of chemical attributes, the major functions of soil are to store and supply sufficient amounts of nutrients to sustain an ecosystem's primary productivity and immobilize or detoxify hazardous compounds that are toxic to plants and animals. Among the chemical soil properties most useful as indicators of soil quality are mineralogy, organic matter content, cation exchange capacity, salinity, and pH (acidity or alkalinity). Soil mineralogy determines bioavailability (adsorption and precipitation) and mobility of nutrients in the soil solution and is dependent on the content and type of silicate clay minerals and the content of Fe and Al oxide and hydrous oxide minerals.

Soil organic matter (or soil carbon) is usually the most important factor affecting soil quality and productivity. SOM has a direct influence on physical and chemical soil characteristics and an indirect effect on plant production, and it is the major energy source for the abundance and diversity of soil organisms. Soil organic matter increases water retention and decreases runoff by preventing the sealing of the soil surface and promoting infiltration (and less erosion). When bound to soil particles as aggregates, it improves soil structure. The upper part of the soil typically has the most soil organic matter, generally about 1 to 10 percent, compared to 1 percent or less at greater depths. Mineral soils with high SOM are generally the most productive, and the carbon contained in the soil organic matter is an important component of the global carbon cycle. Because of the positive correlation between high levels of organic matter and desirable attributes of mineral soil, it has been suggested as a reliable measure of soil quality. Even though changes in SOM are slow and vary locally and regionally, SOM is reliable as a long-term, rather than short-term, indicator of change in soils.

Cation exchange capacity is another useful indicator of changes in soil quality. In general, a higher content of soil organic matter and clay particles has a positive effect on CEC. Soil pH can also significantly influence nutrient availability. For instance, Ca, Mg, and K deficiencies are often present in acidic soils, with Fe and Zn deficiencies appearing in alkaline soils. Trace elements and heavy metals may become more bioavailable in acidic soils. Plant productivity not only depends on the storage and supply of nutrients, but also on the ability of soil to detoxify pollutants. Phytotoxicity and bioaccumulation may occur in areas under strong influence of human activities (e.g., fuel burning, mining, and industry) particularly in soils deficient in organic matter and low in CEC and pH.

c. Biological Attributes of Soil as Indicators

The environmental impact of human activities can be indicated, as outlined earlier, by changes in the physical and chemical properties of the soil ecosystem. Despite their usefulness in characterization of the general status and quality of the soil, physical and chemical indicators have limitations. Do we know the biologically important thresholds of soil chemicals? Do we know how different sets of physical soil factors affect the bioavailability of soil chemicals? Are they stable enough to permit detection of environmental change? Are they easily interpretable? How useful are the indicators across regional and global scales? These and other questions have led many scientists to examine other indicators.

As biological attributes of the soil system directly relate and respond to physical and chemical attributes, soil biota have been investigated as indicators for soil quality. Many organisms are sensitive to changes in the physical and chemical properties of the soil-air or soil-water interfaces. Research shows that disturbances affect the soil biota differently depending on their physiology and life histories. Specific groups of organisms are associated with the soil at spatial (vertical versus horizontal, rhizosphere versus bulk soil) and temporal scales. Moreover, biota vary in their geographic distribution, therefore those groups of organisms that occur in all soils (nematodes, protozoa, microbes) are promising as indicators on a global scale, while those geographically limited groups (e.g., earthworms) may be more valuable on a local scale. The use of soil invertebrates as bioindicators might be problematic because the taxonomic and life history diversity of soil fauna requires extensive knowledge and training. As with soil physical and chemical indicators, no single bioindicator has been yet proposed.

Microbes, protozoa, nematodes, mites, and earthworms are the most noted indicators for soil quality monitoring. Changes in community structure of soil biota can be based on a direct analysis of the dynamics of all identified taxa, or on data derived from taxonomic or ecological indices. Measures for comparisons of soil communities include species, genus, family, or functional group level. Because identification of soil organisms to species or genus level is difficult, a higher level of taxonomic hierarchy or functional group level is often preferred. To relate the status of soil quality to ecosystem change, the ecosystem processes driven by soil biota (e.g., decomposition, mineralization, and respiration) can be measured.

Many of the smaller organisms in soil share a number of attributes that are necessary for useful indicators. These include high abundance for ease of monitoring, occurrence in all soils (microbes, fungi, protozoa, nematodes), sensitivity to changes in soil chemistry and pollutants (microbes, fungi, protozoa, nematodes, mites, earthworms), sensitivity to soil physical changes (most taxa), and representation of a wide range of groups in the soil food web (nematodes, mites). Both protozoa and nematodes have been used to monitor the effects of pesticides and heavy metals. Nematode community structure has also been used to illustrate the effects of natural ecosystem succession, environmental disturbance, and land management practices. Limitations to both protozoa and nematodes as indicators include the difficulties in enumeration and identification of species and their applicability at larger geographic scales. This pattern of local scale specificity seems to apply to most soil biota. Although common indices for abundance, diversity, richness, or rate of a particular process for different ecosystems may not exist, the applicability of bioindicators at a local scale might provide a powerful long-term tool for soil conservation efforts.

2. Policies and Regulations

Adoption of soil conservation measures by farmers may not always be an easy choice. Surveys of farmers reveal that a conservation ethic is a less effective motivation for adoption of soil conservation practices than demonstration of economic benefits. However, farmers knowledgeable about erosion and soil degradation are more likely to adopt soil conservation practices than are uninformed farmers. Thus, education and access to information are important aspects of any conservation effort, as are regulations and policies that conserve or sustain the land.

When private incentives differ from societal incentives, the influence of government policies and regulations can have a major impact on managing the natural resources. In the United States, 25 percent of arable land is regulated with beneficial effects for soil conservation. Conservation tillage in the United States was part of the 1985 Food Security Act (FSA), which encouraged farmers to take erosion-control measures by providing farm subsidy payments. The U.S. Food, Agriculture, Conservation, and Trade Act of 1990 (amendment to FSA) established financial penalties and ineligibility for most farmer program subsidies to farmers who produced agricultural crops on wetlands that were converted after enactment. This act also established the Conservation Reserve Program, providing an opportunity for farmers to take highly erodible lands out of production by receiving annual rental payments from their 10-year contracts with the Department of Agriculture. The Federal Agriculture Improvement and Reform Act of 1996 introduced "planting" flexibility, giving farmers entering commodity programs freedom to choose crops on the contracted acreage. Under the same act, soil erosion control and wetland restoration regulations were improved. These few examples illustrate that government policies can slow land degradation.

The effects of land cultivation and soil erosion expand beyond the border of an agricultural field, state, or even country. Separate agricultural ecosystems are connected via a network of groundwater, streams, and rivers. Silt, sediments, nutrients, and other agricultural pollutants that are transported to streams, rivers, and ultimately marine systems can restrict possibilities for navigation, irrigation, food production, and fisheries, and can affect water and air quality. Thus, the local ecological impacts of agricultural practices have been recognized globally. This has resulted in national and international policy frameworks including such organizations as: the Food and Agriculture Organization (FAO), the International Geosphere and Biosphere Program—Global Change in Terrestrial Ecosystems (IGBP—GCTE), the International Union of Soil Science (IUSS), Global Assessment of Soil Degradation (GLASOD), and Global Change International Panel on Climate Change (IPCC), all of whom have been active in either research, communication, or assistance to countries requiring immediate soil conservation measures and agricultural improvement.

There are also many international agreements whose policies include or are directed at ensuring soil sustainability. Agenda 21 is the action plan signed at the 1992 United Nations Conference on Environment and Development in Rio de Janeiro by leaders from 169 countries. Agenda 21 devotes an entire chapter to sustainable agriculture and rural development both at national and international levels. Among many others, it aims for policies on land reform, less environmentally destructive developments, and conservation and rehabilitation of soil resources. Another example is the Convention on Biodiversity, which encourages agricultural practices that sustain biodiversity and ecosystem functioning. As our knowledge increases about the components of soil and their complexity, we realize that soil is a rare natural resource that should be used wisely.

IV. SUMMARY

Soils provide the basis for the world's food, fiber, and fuel production, as well as for the functioning of global ecosystems. They provide a habitat for a diversity of species comparable to the diversity of life above ground. Human activity changes the soil chemical and physical properties and biodiversity of soil. Agriculture, urbanization, and industrial development are the primary activities that are accelerating the loss of soil quality. Because ecosystems are so interconnected, degradation of soils has detrimental effects on all life. Practices exist that can be selected to conserve and allow the sustainable use of soils.

The most effective methods improving the quality of agricultural soils are reduced tillage, cropping mosaics, multiple cropping, crop rotations, and application of organic amendments. Not only do they prevent further loss of soil organic matter and soil carbon (an important property of soil environments), but also loss of other favorable soil physical and chemical soil characteristics. To determine the status of soil environment, assessment and monitoring programs are being developed. Soil quality can be estimated with the use of various indicators reflecting physical, chemical, or biological soil attributes. The collected information serves as a basis for evaluation and modification of current management methods. Different soil ecosystems may require different suites of indicators.

When private incentives differ from societal, government policies and regulations can have a pronounced impact on managing soil resources. Since the problem of soil degradation crosses spatial scales (local, regional, and global), development and implementation of national or international policies is essential.

See Also the Following Articles

AGRICULTURE, SUSTAINABLE • AGRICULTURE, TRADITIONAL • ECOLOGY OF AGRICULTURE • GREENHOUSE EFFECT • POLLUTION, OVERVIEW • SOIL BIOTA, SOIL SYSTEMS, AND PROCESSES

Bibliography

Brussaard, L., Behan-Pelletier, V. M., Bignell, D. E., Brown, V. K., Didden, W., Folgarait, P., Fragoso, C., Wall Freckman, D., Gupta, V. V. S. R., Hattori, T., Hawksworth, D. L., Klopatek, C., Lavelle, P., Malloch, D. W., Rusek, J., Soderstrom, B., Tiedje, J. M., and Virginia, R. A. (1997). Biodiversity and ecosystem functioning in the soil. *Ambio* 26, 563–570.

Coleman, D. C., and Crossley, D. A., Jr. (1996). *Fundamentals of Soil Ecology*. Academic Press, San Diego, CA.

Foissner, W. (1999). Soil protozoa as bioindicators: Pros and cons, methods, diversity, representative examples. *Agriculture, Ecosystems and Environment* 74, 95–112.

Food and Agriculture Organization (FIOA). (1992). *Protect and Produce: Putting the Pieces Together*. FAO, Rome.

Gregorich, E. G., Carter, M. R., Doran, J. W., Pankhurst, C. E., and Dwyer, L. M. (1997). Biological attributes of soil quality. In *Soil Quality for Crop Production and Ecosystem Health* (E. C. Gregorich and M. R. Carter, Eds.), pp. 81–113. Elsevier, New York

Hartel, P. G. (1998). The soil habitat. In *Principles and Applications of Soil Microbiology* (D. M Sylvia, J. J. Furman, P. G. Hartel, and D. A. Zuberer, Eds.), pp. 21–44. Prince Hall, Upper Saddle River, NY. Conserving land: Population and sustainable food production. Limits. http://www.cnie.org/pop.consering/landuse2.htm.

Jenny, H. (1980). The soil resource: Origin and behavior. *Ecological Studies* 37, Springer-Verlag, New York.

Jenny, H. (1984). The making and unmaking of fertile soil. In *Meeting the Expectation of the Land* (W. Jackson, W. Berry, and B. Coleman, Eds.), pp. 44–55. North Point Press, San Francisco, CA.

Lee, K. E., and Foster, R. C. (1991). Soil fauna and soil structure. *Australian Journal of Soil Resources* 29, 745–775.

Paul, E. A., and Clark, F. E. (1996). *Soil Microbiology and Biochemistry*. Academic Press, San Diego, CA.

Paustian, K., Collins, H. P., and Paul, E. A. (1997). Management controls on soil carbon. In *Soil Organic Matter in Temperate Agroecosystems: Long Term Experiments in North America* (E. A. Paul, Ed.), pp. 15–49. CRC Press, Boca Raton, FL.

Skipper, H. D. (1998). Bioremediation of contaminated soils. In *Principles and Applications of Soil Microbiology* (D. M Sylvia, J. J. Furman, P. G. Hartel, and D. A. Zuberer, Eds.), pp. 469–481. Prince Hall, Upper Saddle River, NY.

Wall, D. H., and R. A. Virginia. (in press). The world beneath our feet: Soil biodiversity and ecosystem functioning. In *Nature and Human Society: The Quest for a Sustainable World* (P. R. Raven and T. Williams, Eds.). National Academy Press, Washington, D.C.

SOUTH AMERICA, ECOSYSTEMS OF

Luis A. Solórzano C.
The Woods Hole Research Center

I. Physical Environment
II. Biogeography
III. Major Ecosystems of South America

GLOSSARY

caatinga Collective name assigned to the semiarid ecosystems of eastern South America. Some authors use this term as a general name for a type of thin Amazonian forests. In this chapter however, the term is reserved for the xeric caatinga region of northeastern Brazil.

cerrado Name assigned by phytogeographers to the tropical savannas of the central Brazilian shield. Cerrado vegetation is further divided in four categories: (a) campo limpio (i.e., clean field), (b) campo sujo (i.e., dirty field or grasslands with scattered shrubs), (c) campo cerrado (i.e., closed fields or grasslands with numerous trees and shrubs), and (d) cerradao (i.e., when the vegetation is dominated by a closed canopy of trees).

chaco Region located between the Paraná basin, the central Brazilian shield, and the Andes. The ecosystems present in the Chaco region include temperate grasslands, savannas, and arid and semiarid environments.

igapo Floodplain of the blackwater rivers of the Amazon basin.

mangal/mangle Name assigned to the vegetation of mangrove ecosystems.

morichal Plant community characteristic of tropical savannas, it is seasonally flooded and the presence of the palm *Mauritia flexuosa* is conspicuous.

llanos Region of tropical South America, in the east side of the Andean range of Colombia and Venezuela. The llanos landscapes occupy lower elevations (2–300 m), and the landforms vary from flat to rolling terrain. The dominant vegetation is savanna with areas of dry forest, gallery forest, and morichales.

pampa(s) Region of temperate South America, west of the Andes in Argentina. The Pampa region contains several ecosystem types, including temperate grasslands, dry forest, and xeric shrub lands.

pantanal Largest wetland of South America drained by the Paraguay River and its tributaries.

páramo Ecosystem of the high Neotropical mountains, found in South America above tree line in the northern Andean mountains of Colombia, Venezuela, Ecuador, Peru, and Bolivia. The dominant vegetation is similar to an alpine meadow or grassland; cacti and giant rosette plants are conspicuous.

puna Ecosystem of the high Andean mountains found above tree line from Bolivia to southern Peru. The Puna comprises four distinct ecological regions: wet, dry, thorny, and desert Puna. The dominant vegetation types vary from grassland/shrub lands to desert.

varzea Floodplain of the whitewater rivers of the Amazon basin.

SOUTH AMERICA CONTAINS a diversity of ecosystems from tropical, subtropical, alpine, and temperate environments. The geology, climate, and biogeographical history of the continent have important consequences in shaping the current geographic distribution of ecosystem types and their structural attributes and functioning. The stability of the old Precambrian shields and the recent orogenic evolution of the Andean range have been the principal factors that determine geomorphologic, edaphic, and continental climatic patterns that correlate with the geographic distribution of major ecosystem types. The Andean range causes an altitudinal zoning of different environments and dramatically increases the diversity of ecosystems that occur in the continent. Most of the area of South America lies in tropical regions and therefore tropical ecosystems dominate the landscapes. The main tropical ecosystems include tropical rain forest, dry forest, cloud forest, savannas, shrub lands, xeric formations, and high Andean ecosystems. Ecosystems of the southern temperate regions include temperate grasslands, the Mediterranean Matorral, and temperate forest and deserts. There are extensive wetland ecosystems across the continent occupying areas along the floodplains of major rivers, lakes, estuarine areas, and seasonally flooded savannas. Mangrove ecosystems occur along the tropical coastlines of both the Atlantic and Pacific oceans.

I. PHYSICAL ENVIRONMENT

South America occupies an area of approximately 18 million square kilometers and extends from about 11°N to 56°S, and from about 35°W to 81°W. The continent has been connected to Central and North America by the land bridge of the Panamanian Isthmus for approximately 3.5 million years and is bounded on the east by the Pacific Ocean, on the north by the Caribbean Sea, and on the west by the Atlantic Ocean.

A. Geology and Geomorphology

The geological history of South America cannot be fully covered in this summary as it is very complex and reflects many tectonic and orogenic processes operating over millions of years. However, the structural characteristics of South American ecosystems—in particular the development, properties, and geography of their soils—are significantly associated with the geophysical environment. Likewise, the major landforms of the continent create distinctive regional climatic patterns, which ultimately affect the functioning of ecosystems. A general overview of the geology and physiography of the continent will help to explain the distribution and general properties of its major ecosystems.

The basic geomorphic features of South America are (a) the high Andean Mountain chain on the western border of the continent spreading from Venezuela to Chile, (b) the low plains of the piedmont that occupy the eastern side of the Andes, (c) the very large sedimentary lowlands of the Amazon valley, and (d) the continental shields outcropping on the east (Fig. 1). These major geotectonic categories of the continent have important differences in the origin, thickness, and stage of weathering and composition of the underlying materials, which are of essential importance to determine the type, depth, and physical composition of soils. Geomorphologic processes such as soil formation and erosion have produced characteristic surface features, which identify discrete areas that developed under different climatic conditions.

The central and eastern part of the continent is made of three cratonic areas: the Guyana Shield, the central Brazilian Shield, and the Coastal Brazilian Shield (Fig. 1). These cratonic areas have not been geologically disturbed since the Silurian ($\sim 408.5 \times 10^6$ years ago). Large plateaus and rolling highlands occupy the most eastern part of the continent covering most of the Guyana and Brazilian shields. From remote geological times, in the east side the stable Precambrian shields of the continent have held positions above sea level and have been affected by simple deformations. Landscapes in these areas, therefore, have evolved continuously under conditions of direct exposure to the atmosphere, and this has resulted in the widespread development of old erosion surfaces. The shields consist primarily of crystalline basement complexes from the Precambrian period and the dominant rocks are granites, genisses, and mica schists. The waste mantle may be very thick on the old erosion surfaces of the shields. The materials that form these rocks are at an advanced stage of *in situ* weathering and can be millions of years old. A shallower layer of waste material covers the landscapes of residual relief. The valley bottoms and lowlands are covered with drift material that can attain considerable depths.

Along the Andean belt (Fig. 1) there are Precambrian rocks masked by different younger deposits. Since the Paleozoic, the ocean has invaded pre-Andean areas several times, and extensive marine deposits mixed with metamorphic and intrusive rocks occur all along the range from Venezuela to Chile. One important consequence of marine transgressions (i.e., ingressions and regressions) is the inland deposition of rich marine sediments. Marine regressions disrupt the deposition

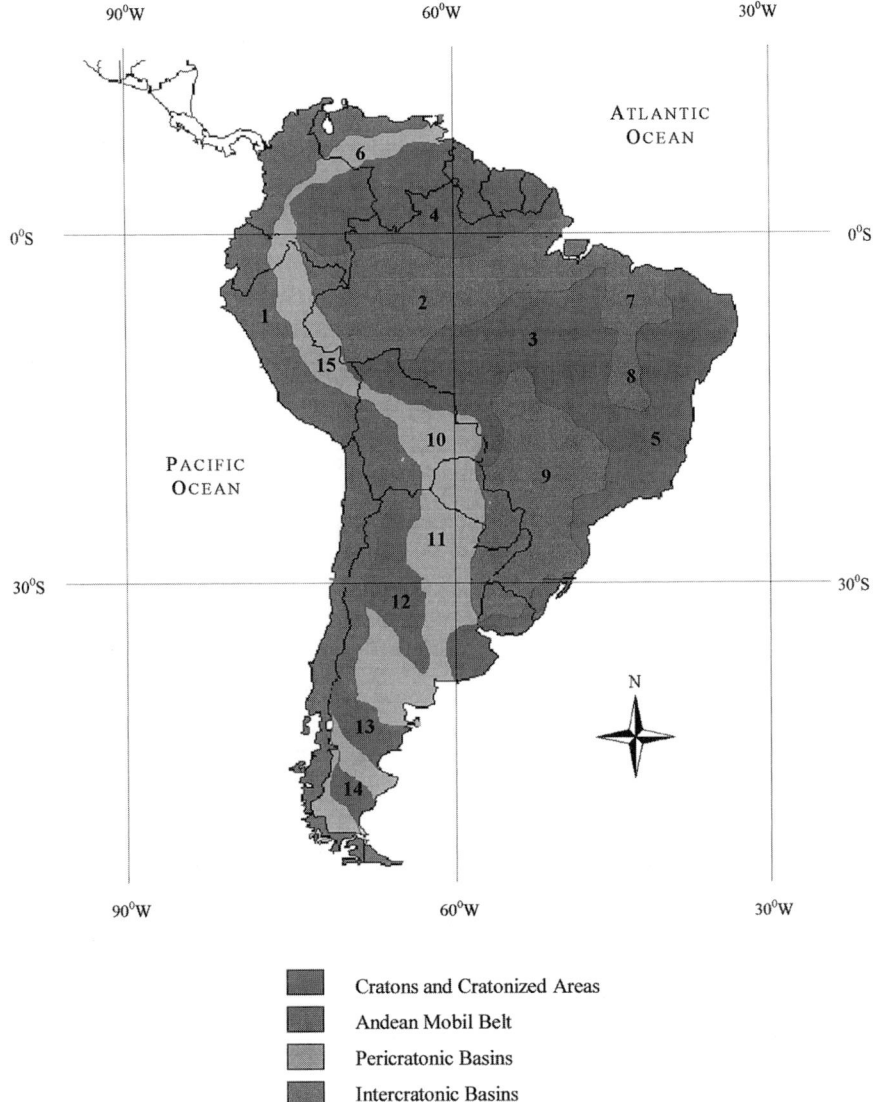

FIGURE 1 Major geotectonic components of South America. Adapted from Harrington (1973). 1 = Andean Belt; 2 = Amazon Basin; 3 = Central Brazilian Shield; 4 = Guyana Shield; 5 = Coastal Brazilian Shield; 6 = Llanos; 7 = Parnaiba Basin; 8 = Sao Francisco Basin; 9 = Paraná Basin; 10 = Chaco; 11 = Pampas; 12 = Pampean Massif; 13 = Patagonian Massif; 14 = Deseado Massif; 15 = Lowlands of Iquitos, Acre, and El Beni.

of marine sediments on land and further erosion may increase the loss of previously deposited layers. In the early Devonian (408–362 × 10^6 years ago), the largest of all marine transgressions covered the Amazon and Parnaiba basins (Fig. 1), which remained submerged until the middle Devonian. In the upper Paleozoic (290–245 × 10^6 years ago) only the Parnaiba basin received mixed marine and continental sediments. In the early Mesozoic (245~ × 10^6, years ago), eolian and fluviatile sediments were deposited in the Parnaiba and Amazon basins. The Triassic sediments (208~ × 10^6, years ago) came from continental areas and were deposited in an arid environment. During the Jurassic (208–145 × 10^6, years ago), the products of major volcanic activity in central and western Argentina were deposited in the Parana, Sao Francisco, Parnaiba, and Amazon basins (Fig. 1). During the late Cretaceous, marine depositions also occurred over large areas of eastern South America.

Most of the current physical features of South

America—topography, coastline, and river systems—developed toward the end of the Tertiary period during the Pliocene ($2-8 \times 10^6$, years ago) when the major uplift of the Andes took place. During the Tertiary, several tectonic movements accompanied by acid intrusions and volcanic activity lifted the western part of the continent forming what is now the Andean Cordillera (Fig. 1). During the uplifting, continuous sedimentation persisted in the inter-Andean valleys and along the eastern piedmont, forming deposits that were later involved in Andean orogenic movements that resulted in complex folded geological sequences. During the Miocene-Pliocene, most of the northern Andean ranges were strongly lifted while the southern Pampas were broken into faulted blocks. These movements were accompanied by intense volcanic activity in Colombia, Ecuador, Peru, Bolivia, Argentina and Chile, which also cause the uplifting of the Brazilian and Guianan highlands in the eastern part of the continent. These orogenic and volcanic activities continue today, particularly in Colombia, Ecuador, and Chile, and the areas affected have very fertile soils. At present, the Andean mountains have fluctuating altitudes from 3000 to 5000 m with higher summits between Ecuador and Central Chile and the highest peak of the Western Hemisphere—Cerro Aconcagua—reaching an altitude of 7005 m in Argentina. Due to the position of the Andes, most of the hydrological networks of the continent drain to the Atlantic Ocean through three main river systems: the Amazon River, the Paraná-Paraguay Rivers, and the Orinoco River.

A wide belt of low sedimentary plains extends along the eastern piedmont of the Andes, from the Llanos of Colombia and Venezuela to the Argentinean Pampas, and eastward along the valley of the Amazon River (Fig. 1). The majority of these plains reach altitudes from 200 to 500 m. These plains are composed of Tertiary and Quaternary sediments deposited from the South American and Brazilian shields and more recently from the Andes. The intercratonic lowlands of the Amazon basin are made of heavy sedimentary clays originated from erosion of uplifted deposits washed from the Andes (Fig. 1). Those deep clay sediments were deposited on the flat bottom of an ancient inland sea that covered most of the basin during the late Tertiary–early Pleistocene when the ocean level was higher.

B. Climate

Climate is an important factor that determines in some degree the type of soils and vegetation of any given region and, therefore, influences the structure and functioning of ecosystems. Figure 2 illustrates the major South American climate types according to the classification of Köppen. The shape of South America is of paramount importance in determining its large-scale climatic patterns, which in turn influence the geographical distribution of ecosystems. South America has the shape of an acute triangle with most of its area situated in tropical latitudes and its southern portion extending well into high latitudes. Although the southern portion South America reaches more than 20° farther south than the tip of Africa, the continent is so narrow at the southern end that the landmass of this region rather constitutes a continental peninsula. Because most of South America occupies tropical regions and the continental area decreases at higher latitudes, the property of climatic continentality is absent. Accordingly, the more important factors that determine the diversity of climatic regimes observed across South America are the major global atmospheric circulation patterns, the proximity to oceans, the Andean relief, and other coastal and inland topographic variations.

In general, tropical temperatures dominate the northern part of the continent, declining smoothly toward the south. Major variations in temperature are due to the Andean Mountains and ocean currents. The presence of the Andes causes climatic diversity over the continent due to altitudinal variations that change wind circulation patterns and to adiabatic responses of air masses along the altitudinal gradients of the mountain chain. South American environments can range from hot deserts and warm rainy climates in the lowlands to cold deserts, temperate, and iced polar climates in the Andean highlands (Fig. 2). Two ocean currents are particularly important to the regional temperatures: the cold Humboldt current that brings lower temperatures from the south up north along the West Coast and the warm Brazilian current that brings warmer temperatures down south along the eastern part of the continent.

Due to the lack of continentality, the distinct climates of South America are often characterized by differences in precipitation. Rainfall across most of the northern and central parts of the continent depends on the intertropical convergence (ITC). The ITC occupies its most northerly position from June to September, when is located between 7° and 9°N. Its advance and retreat migration does not take place parallel to the equatorial line. When the ITC drifts south, a prevalent High Pressure Cell in the Atlantic Ocean restrains its movement in the eastern part of the continent causing an arch of precipitation that reaches southeast Brazil in February-March, while dry conditions persist in the northeast.

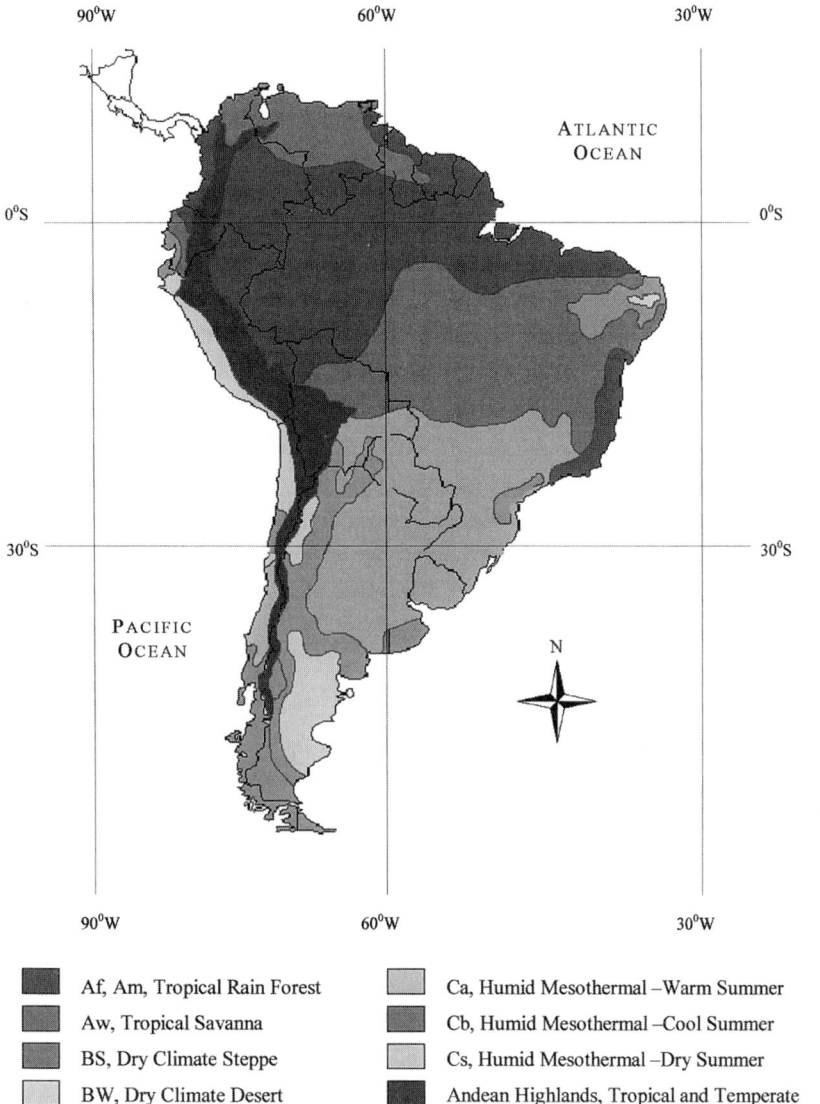

FIGURE 2 Broad climatic regions of South America defined according to the Köppen classification system.

This Atlantic high cell also causes higher precipitation in most of the East Coast north of Patagonia.

The Andean chain also has a significant effect on the continental patterns of precipitation. Due to adiabatic cooling, the Andean windward slopes remove moisture from the air masses moving from the Pacific causing a large rain shadow that extends from east of the Andes from Bolivia to Tierra del Fuego (Fig. 2). Conversely, in the windward slopes of the Pacific coasts of Colombia and Chile the Andes cause exceptional orographic rainfall levels, which in the Choco region of Colombia can surpass 10,000 mm/year (Fig. 2). A similar effect is observed in the eastern Andean slopes of central South America where along the western edge of the Amazon basin moist winds rise and deliver higher precipitation levels than those register in lower elevations of the basin. In the north of Venezuela, the northern Andean ranges block the easterly trade winds and create a tropical savanna climate that extends from northern Colombia to Surinam (Fig. 2). In the western coast of the continent, between 27° and 10°S, the edge of the western Pacific high pressure cell combined with the cold Humboldt current produce an arid zone called in Chile the Atacama Desert (Fig. 2). The structural complexity of the Andean relief includes inter-Andean cordilleras and valleys (e.g., the Eastern, Central, and Western Colombian cordilleras) and high plateaus (e.g., the Altiplanos of Peru and Bolivia) that further create many local climates according to variations in wind exposure and altitude.

Other extra-Andean topographic features that produce orographic climatic modifications of rain shadows or rainfalls are the highlands of the Brazilian shield that block onshore winds between Porto Alegre and Bahia; the uplands of northeast Brazil that cause a dry rain shadow in the interior northeast region; the Guyana Highlands that block the trade winds and cause a dry winter effect in their leeward slopes; and the Sierras de Cordoba and Patagonian Plateau in Argentina, which affect air temperature and can collect moisture from humid air masses crossing Patagonia (Fig. 2).

II. BIOGEOGRAPHY

A. Early History and Associations

Biogeographers include central and northern South America in the Neotropical Kingdom, which also contains the Antilles and most of tropical Central America. Until approximately the early Cretaceous, South America, Africa, Australia, and Antarctica were joined together composing the supercontinent Gondwanaland, thus allowing for a continuous interchange of their ancient biotas. Africa and South America share many plant families like Annonaceae, Myristicaceae, Cecropiaceae, Sterculiaceae, and Bombacaceae. Some plant genera are very diverse in one continent and barely represented on the other, for instance, *Mayaca* (eight species in America and one in Africa), *Duvernoya* (35 species in Africa and 3 in America), and *Hyptis* (400 species in America and 2 in Africa)

The southern cone of South America is included within the Holantartic Kingdom, which includes the temperate mesic-adapted floras of southern South America, New Zealand, and southwestern Australia. The genera of trees *Araucaria*, *Podocarpus*, and *Nothofagus* are typical although not exclusive to the southern cone. Fossil records have also shown strong ancient faunal associations between South America, Africa, Antarctica, and Australia. For instance, some of the existing species of fishes of South America, as well as their parasites, show affinities with African groups. There are also close phylogenetic relations among groups of South American, African, and New Zealand invertebrates including, among others, arachnids and mollusks.

After the separation of South America from the rest of Gondwanaland in the early Cretaceous, the biota of the continent was isolated for millions of years. Because of their isolation, South American flora and fauna are extraordinarily rich in endemic groups and species. Among the endemic plant families are the Bromeliaceae, Marcgraviaceae, Nolanaceae, Cactaceae, Tropeolaceae, Quiinaceae, Lacistemataceae, Bixaceae, Brunelliaceae, Krameriaceae, Cyclanthaceae, and palm tree genera such as *Jubaea* and *Mauritia*. It is estimated that about 7% of the total world's superior plants are endemic species from Brazil, 6.8% from Colombia, 3.2% from Venezuela, 2.1% from Peru, and 2% from Ecuador.

Brazil is the country with more mammal species reported in the world (524 of which 131 are endemic), Colombia the fourth (456), and Peru the ninth (344 of which 46 are endemic). Endemic faunal families include the armadillos Dasypodidae, the anteaters Myrmecophagidae, and the monkeys Cebidae. Particularly important are the marsupials that in South America are represented by 2 genera, 2 families, and 87 species. There are no fewer than 3000 bird species with 2 orders and 30 endemic families. The hummingbirds, which are endemic of the New World, have more than 250 species in South America and populate ecosystems from the Amazonian lowlands to the high Andean Mountains. The diversity of endemic reptiles is exceptional; there are 520 endemic species reported for Colombia, 468 in Brazil, 374 in Ecuador, 298 in Peru, and 293 in Venezuela, including species of snakes, turtles, and iguanas—which other than in South America only occur in the Fiji and Tonga Islands of the Pacific. Brazil is also considered the country with highest diversity of freshwater fish species in the world with > 3000 species reported, Colombia the second (>1.500), Venezuela the fourth (>1.200), and Peru the seventh (> 850). The diversity of invertebrates is also outstanding; for example, there are 350 endemic species of butterflies reported for Peru, 300 for Colombia, 200 for Brazil, and 200 for Bolivia.

B. Late Tertiary

Although it is likely that some exchange occurred between North and South America during the Eocene and Miocene, the land bridge of the Panama Isthmus connected permanently South, Central, and North America about 3.5 million years ago. The completion of the Panamanian land bridge in the early Pliocene was perhaps one of the most influential biogeographical events that have occurred in the continent and dramatically changed the ancient biota of South America. The significant exchange of South and North American biotas is call by biogeographers "the great American Interchange". Among the many invaders that entered South America through the isthmus are rattlesnakes, rodents, tapirs, deer, peccaries, felids (including jaguars and pumas), canids, and humans. About 40% of the mammal fauna of South America are species that arrived since the Pliocene. Among the South American forms that migrated to North America are the ground sloth, which disappeared about 10,000 years ago, and opossums and armadillos, which are still expanding their range. The dynamic interchange of plant and animal forms between South and North American continues at present.

C. Quaternary

During the past 2 million years, climatic fluctuations of dry glacial and wet interglacial periods affected both the extents of temperate and tropical areas. Palynological sequences indicate that at least during the Pleistocene major climatic changes affected speciation and biogeographic patterns in South America. The modern geographical distributions of many endemic taxa resulted from Quaternary speciation and redistribution of species overlaid with previously existing patterns. The explicit mechanisms that generated current biogeographical patterns are not fully understood, in part due to their inherently historical nature that is difficult to reconstruct.

Paleoecological studies initially assumed that the richness and complexity of plant communities was a consequence of continuous favorable growing conditions persisting over long time periods. Many areas have been identified as centers of endemism and proposed as tropical forest refugia during the period spanning 18,000 to 13,000 years ago (Haffer, 1969). According to this hypothesis, during the last four glaciating cycles of the Pleistocene, tropical forest contracted during dry glacial periods causing most of the Amazon Basin, as well as other forested areas, to be occupied by savanna vegetation. Continuous but isolated islands of forest (refugia) might have contained populations of plant and animal species that underwent speciation while they remained isolated. During mesic interglacial periods similar to currently existing conditions, the forest expanded again and new geographic patterns appeared including the signal of speciation processes that occurred within the refugia themselves. Paleoecological data have also shown that in the Andes the tree line has migrated down and up-slope by as much as 1,500 m, matching the frequency of each northern hemispheric ice age. The estimated cooling experienced during the glacial maxima at the Sabana de Bogota is 7 to 9°C. A consequence of the lowering of the tree line would be the connection of some of the high Andean Páramo ecosystems, which are currently isolated on top of mountains. The modern distributions of several plant and animal species appear to agree with the refugial hypothesis.

An alternative explanation for the distribution of centers of endemism proposes changes in temperature rather than precipitation as the driving force of vegetation re-arrangements during glacial periods. It has been proposed that the areas of refugia were not "isolated islands of stability" but rather of "maximal disturbance" due to cooler glacial temperatures, reduced atmospheric CO_2, and moderate reductions of precipitation (Bush, 1994; Colinvaux, 1989). Far from being static places of refuge, under this hypothesis the regions of endemism would be the dynamic edges between the forest below and the cool-tolerant vegetation above.

The uplifting of the Andes during the Quaternary has also contributed to the isolation or dispersal of some taxa. While the mountain range has served as a dispersal corridor from north to south, it also represents a dispersal barrier (like the Atacama Desert) to some plants and animals. For instance, the floristic individuality of the High Andean deserts is attributed to the short history of these environments that appeared after the recent and fast uplifting of the cordillera. The age of these ecosystems might have not allowed enough evolutionary time for many species to adapt and disperse under those new conditions. There is evidence also that during interglacial dry periods the southern Andes cone experienced a cold and arid climate that served as an effective dispersal obstacle for rodents and plant species from the northern arid ecosystems.

Not only long-term events like Pleistocene glaciations or orogenic processes such as the uplifting of the Andes have had significant effects on the ecosystems of South America, but short-term climatic episodes like El Niño cause important impacts as well. For example, changes in water temperatures of the western Pacific during El Niño years trigger profound modifications in rainfall patterns across all of South America. Major large-scale ecosystem disturbances have been associated with El Niño: fires in savannas and forests of Amazonia; droughts in the llanos of northern South America, in the Chaco region and in northeastern Brazil; and floods in Peru, Ecuador, and Colombia. There is evidence that these perturbations affect the dynamics of populations and ecosystems, which over evolutionary time could modify the composition and geographic patterns of their biota.

III. MAJOR ECOSYSTEMS OF SOUTH AMERICA

Although it is possible to make some generalizations about the nature of the biomes of the continent and their major ecosystem types, it is important to bear in mind that across several spatial scales they are in fact complex mosaics of microhabitats. The categories of ecosystem types applied in this summary are very broad and based on the classification system developed by FAO-Unesco (1971), the satellite-based map of Stone *et al.* (1994), and the ecoregions study of the World Bank/World Wildlife Fund (Dinerstein *et al.*, 1995). In this review, very general categories are used and ecosystems are geographically delimited based on broad vegetation formations. Only some important features of these major ecosystem types are briefly discussed.

A. Tropical Forest

Tropical forests are among the most complex ecological communities of the world and offer a variety of niches that support an immense diversity of plants and animals. South America contains over 40% of the tropical forest of the world.

1. Distribution and Structure

a. Tropical Rain Forest

Tropical wet forests occupy the drainage of the Amazon and southern Orinoco basin, the piedmont of the Andes in northern and western South America, the Choco-Darien region between the Andes and the Pacific Ocean in western Colombia, and the southwest and Atlantic coast of Brazil (Fig. 3). Among other types of tropical rain forests are *Babassu palm forests*, which occupy the southeastern border of Amazonia, and the *liana forests*, which are found in southern Amazonia between the rivers Xingu, Tapajos, and Tocantins. The liana forest is an open forest with well-spaced trees, rich in woody climbers, and often completely entwined by lianas. It occurs on all soil types, but there are no detailed studies of this vegetation, nor explanation of the factors that cause its wide distribution.

b. Tropical Semideciduous and Dry Forests

These ecosystems occupy part of the eastern edge of the Andes in Bolivia and Brazil, the inter-Andean valleys of Colombia, the northern edge of the Venezuelan Llanos, and the western regions of Ecuador and Peru (Fig. 3). Most of the South American regions originally covered by dry forests are today populated areas, and much of the primary vegetation has been cleared. Tropical semideciduous forests usually occur on alluvial lowlands, submontane, and mountain terrains up to 1,000 m.

Between the Amazonian rain forest and the Cerrado (discussed later) of central Brazil, there is a so-called *transition forest*, which comprises a large number of plant communities. Transition forests are generally less species rich than moister forest in terms of plant and vertebrate species; however, plant physiognomy diversity seems to be greater in dry than in wet forests. Both transition dry forest and savannas can be found under similar climatic conditions. In eastern South America, transition forests run along most of the boundary between Amazonia and the Guyana and Brazilian shields, and they vary in width from a few kilometers to over 100 km in some areas. The transition forest is taller and with a more closed canopy than the closed Cerrado, *cerradao* (discussed later), but is lower and with a more open canopy than the typical Amazonian rain forests.

c. Mountain Cloud Forests

These ecosystems occur in the humid tropical slopes of the Andes from Venezuela to northwest Argentina and in the Santa Marta Massif in Colombia. They occupy defined altitudinal belts with the lower belts intermixing with lowland forests and the upper belts limiting with the high Andean grasslands at the tree line about 3,000 m above see level. Fog is persistent in these ecosystems and the vegetation is luxuriant with an extraordinary variety and abundance of epiphytic forms and mosses.

d. General Patterns

Structural variations between tropical forests are mainly governed by differences in the amount and seasonality of precipitation and the temperature regimes. In general, the proportion of deciduous woody components increases along a gradient of rainfall, as the amount of annual precipitation decreases below 2,000 mm. Although exact boundaries between ecosystems are somewhat arbitrary—because vegetation is not always so sharply delimited—the transition from arid types to rain forest seems to be primarily determined by continental climatic patterns, especially rainfall intensity and seasonality. When nonevergreen trees are present in wet tropical forests, they never shed all their foliage at the same time. A large number of woody species, of which most have evergreen foliage, characterize these

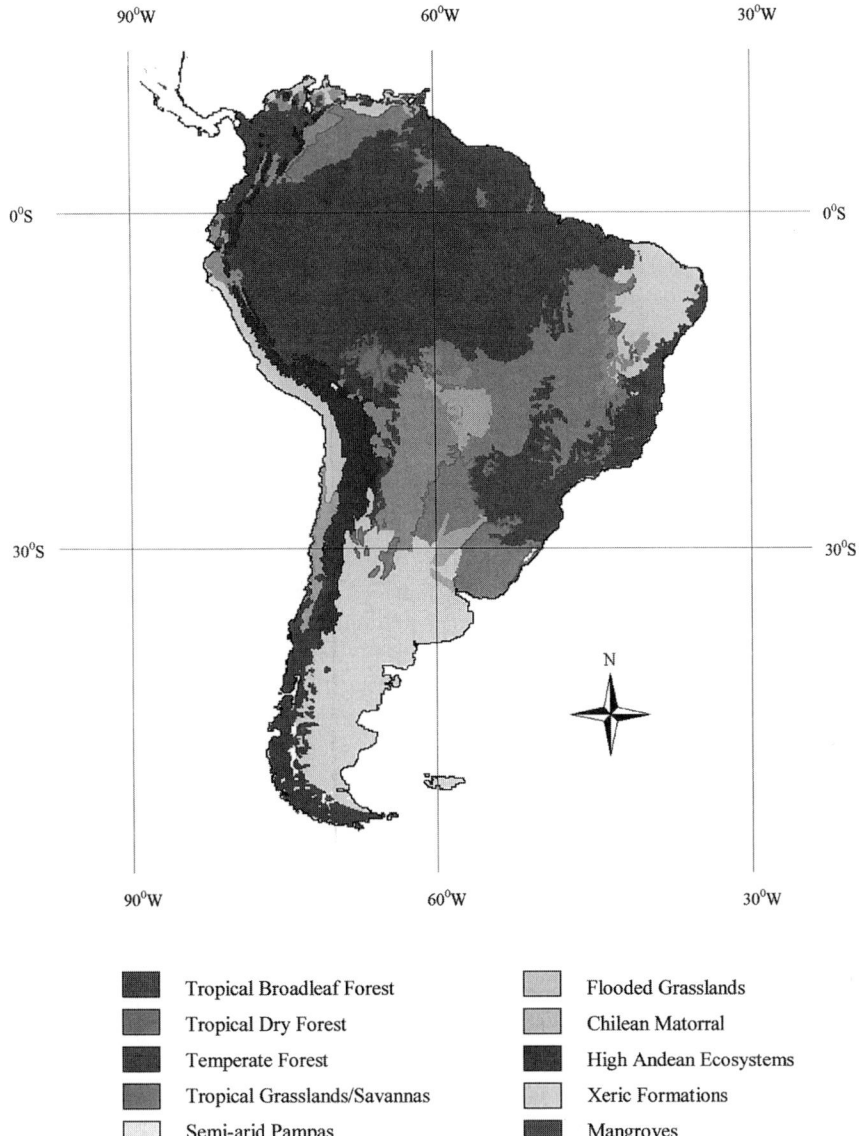

FIGURE 3 Simplified map of the major ecosystem types of South America geographically defined on the basis of dominant vegetation. Adapted from Stone et al. (1994) and Dinerstein et al. (1995).

ecosystems and the most frequent tree species rarely represent more than 15% of all species present.

In general, tropical forest have a vertical structure with several strata where the tallest trees form a discontinuous layer of emergent individuals, which in South America can reach heights of 30 to 40 m. A denser canopy grows below and consists of two or more layers of shorter trees, small palms, tall herbs, ferns, and shrubs. Lowland forest trees are generally shallow rooted (but see the discussion presented later on deep rooting) and many of them develop buttresses at the base of their trunks, an adaptation thought to provide mechanical support. The epiphytes and vines are conspicuous in tropical forests, with the epiphytic Orchidaceae and Bromeliaceae are usually prominent. The barks of lowland tropical rain forest trees are usually thin and smooth, and this morphological characteristic makes difficult the anchorage of epiphytic forms. In the Andean cloud forests, the heights of trees decrease at increased altitudes and their shapes become irregular (e.g., elfin forest). In cloud forests, the bark of trees and the forest floor are cover with mosses, lichens, and ferns, and the diversity of fungi, algae, bryophytes, and seedless vascular plants is remarkable.

2. Functional Aspects

a. Tropical Rain Forests

A mean monthly temperature of >26 and mean annual rainfall usually exceeding 1,800 mm characterizes the Af climate that coincides with the geographic distribution of tropical rain forest (Figs. 2 and 3). Some areas of tropical rain forests are the result of orographic precipitation. Isolated mountains (e.g., Sierra Nevada de Santa Marta in Colombia) or coastal escarpments (e.g., the Atlantic Brazilian Shield and the lower slopes of the Andes in the Pacific coast of Colombia and Ecuador) act as barriers to airflow, forcing moist air to ascend upslopes where adiabatic cooling generates clouds and precipitation. These geophysical conditions conduce to forested landscapes with a physiognomy similar to the Amazon wet evergreen forest. In areas of lowland tropical forest (e.g., Amazonia), minimum changes in temperature between summer and winter and continuous heating coupled with the high evapotranspiration rates of vegetation create a predictable pattern of daily convectional rain that falls during the middle afternoon. It is estimated that up to 80% of the precipitation falling in Amazonia is retained within the hydrological cycle of the basin.

The geological material underlying the hydrographic network also influences the development of forest ecosystems. In the Amazon lowlands, large areas of forest are subject to inundation (varzea) and these ecosystems differ from those that occupy high terrain (terra firme). Rivers running from the Andes carry large amounts of silt and have a light brown color (whitewater). These large loads of silt are deposited in areas of low relief and form broad floodplains of high fertility. During the flooding season, these areas remain under several meters of water. In contrast, rivers running though the old Precambrian Shield are poor in nutrients and minerals and their waters are clear. The Shields have been exposed to millions of years of leaching and in order to conserve nutrients plants protect their foliage with tannins. The tannins leached into the rivers give a tainted coloration to the water (blackwater). The Rio Negro and other blackwater rivers also have floodplains, but they are different because of the lack of sediments and because plant communities are adapted to acidic conditions (igapo).

Because of the high temperatures and abundant rainfall of tropical forest ecosystems, there is intense chemical action on bedrock, and soils and all soluble components continuously leach, producing characteristic Ultisol and Oxisol soils. These soils are rich in iron, manganese, and aluminum, which stay behind after all other soil soluble constituents have been leached. Large amounts of these minerals arrange in stratified layers forming laterites.

Recent studies have shown that tropical forest exploit a larger volume of soil than previously thought. Soil depth is therefore quite important on water balance variability because it allows evergreen forest to keep evapotranspiring during dry periods by absorbing water from soil depths of more than 8 m (Nepstad et al., 1994). Although the most rapid changes in water availability take place at the soil surface, where it is depleted by plants, the process of hydraulic lift can take water from deep soil layers and discharge it into the dryer upper layers. During the day, the water potential (Ψ) gradient causes water to move from the ground into the roots, shoots, and through stomata to the atmosphere. During the night when stomata are closed, Ψ in the plant equals Ψ in the deep soil and a water potential gradient can move water from the deeper moist soil and plant interior into the dryer upper soil layers.

Decomposition and nutrient dynamics proceed rapidly under the warm, moist conditions of tropical forest ecosystems. High temperatures also favor an intense bacterial activity on upper soil layers and therefore there is no accumulation of humus. Many nutrients are stored in the forest biomass where they are kept from leaching. Nutrients are partitioned in different compartments of the forest and, on a weight basis, leaves contain the highest concentrations. Most of the phosphorous is stored in the leafy biomass and potassium and calcium in stem tissues. Nutrient losses in intact forests are generally low because of the high concentration of roots in the upper soil layers. The presence of mycorrhizal associations in roots enhances their nutrient uptake. Epiphytes also reduce the loss of nutrients washed away from standing biomass. Over long successional periods, the forest biomass is progressively stored in the woody components and nutrient accumulation in leaves decreases.

Invertebrates can account for up to 60% of the animal biomass of tropical forest ecosystems. About 19% of the woody fraction of the biomass is consumed by termites, beetles, and the larvae of insects. Litter is the main source of food for the animal community and leaves, flowers, and fruits are consumed mainly by insects, birds, bats, and small mammals. Chemical defenses are common among tropical forest plants and many species have developed mechanical defenses by incorporating silica, lignin, and fibers in their tissues.

b. Tropical Semideciduous and Dry Forests

Along gradients from tropical rain forest to dry forest, the structural and functional attributes of the ecosystems change. Tropical deciduous forests correlate with climates with two well-determined seasons: a rainy period followed by a dry one. More than 50% of the tree species shed their foliage during the dry period. The predominant characteristic of the habitat of these forests is the prolonged seasonal drought, causing desiccation of the topsoil and lowering atmospheric humidity. The length of the dry season determines the degree of divergence in physiognomy and structure of the seasonal forest. From wet evergreen to dry deciduous forests, prolonged drought correlates with a reduction in physiognomy and floristic composition and the lowering of canopy height. At the same time, the degree of deciduousness of the trees of the overstory increases. The phenology of these ecosystems is determined by a seasonal climate with a period of intense summer rainfalls followed by a dry winter. In the southern dry deciduous forests, the winter season brings also changes in mean temperature down to 15°C. Although between 20 to 50% of the woody species found in these ecosystems are deciduous, these forests are still dominated by Amazonian genera widely distributed across South America, such as *Parapiptadenia, Peltophorum, Cariniana, Lecythis, Tabebuia, Astronium*, and others of lesser physiognomic importance.

Although tropical rain forests contain the highest diversity of species measured in terrestrial ecosystems, life-form diversity is higher in dry forest probably because of habitat heterogeneity. Woody plants in tropical rain forest tend to converge to a small number of life-forms. In drier more seasonal environments, the proportion of deciduous trees and shrubs increases, the presence of epiphytes decreases, and vines become more frequent. The presence of epiphytes is associated with high air humidity and the presence of dew. Along seasonality gradients succulent plants conspicuously increase, including those with Crassulacean acid metabolism (CAM). Succulent-stemmed plants have very stable water relations and are drought resistant. Evergreen woody plants dominate at both extremes of the gradient, and although they all belong to the C_3 photosynthesis type, there are differences in their leaf structure and drought resistance, which allow them to function in both mesic and arid environments.

B. Savannas, Grasslands, and Shrub Lands

1. Distribution and Structure

Savannas and grasslands occupy nearly 25% of South America in both tropical and subtropical regions.

a. Tropical Savannas

These ecosystems are characteristic of the warm lowland tropics and occur in areas with a strongly seasonal rainfall regime and a dry period lasting from 4 to 7 or 8 months. Although a herbaceous cover consisting mostly of bunch grasses and sedges is dominant in savannas, the term "savanna" embraces a variety of vegetation types found in tropical latitudes. An important characteristic of tropical savannas is a clear seasonality in their phenology and a period of lower activity associated with times of water stress. According to Sarmiento (1984), tropical savannas can be divided into four functionally distinct types: (a) semiseasonal savannas, with weak water stress conditions; (b) seasonal savannas, with a distinct rainfall seasonality and common fires during the dry season; (c) hyperseasonal savannas, where in addition to a marked dry season there is also a period of water flooding; and (d) "esteros," which are areas without a clear dry season, but an excess of soil water for most of the year.

The largest extension of savanna in South America occupies the old Precambrian Brazilian shield and has been dubbed by phytogeographers "Cerrados." Savanna ecosystems are also found in the Orinoco Llanos of Colombia and Venezuela, in Suriname, Guyana, in some Amazonian regions of Brazil, and in Paraguay, Bolivia, Argentina, and the northern Chaco region (Fig. 3). The vegetation of the Guyanan savannas is more related floristically to the llanos than the cerrado.

Perennial grasses herbs and shrubs are the dominant life-forms in savanna ecosystems. They are well adapted to the seasonal climate, soils poor in nutrients and the fire disturbance regime that dominate in these environments. At least five phenological groups of grasses have been recognized in savanna ecosystems: perennial with a seasonal semidormant period, annual ephemeral with a short cycle, annual with long cycle, perennial with a seasonal dormant period, and of continuous flowering and growth. Tree species found in savannas have adapted to these conditions by increased allocation to underground biomass. Trees in woody savannas can reach 25 to 30 m in height, but their diameter rarely exceeds 50 cm. Some savanna tree species are described as subterranean trees because their roots can penetrate to depths of 18 m. Other species like *Curatella ameri-*

cana have shallow roots that can grow more than 20 m horizontally and secondary roots that can reach depths of 6 m. Other conspicuous components of savanna landscapes are the gallery forest along rivers and smaller streams and the seasonally inundated palm communities (*morichales*) dominated by the palm *Mauritia flexuosa*.

In the Brazilian Cerrado, there is a series of plant communities from open grasslands to dense woodlands and more or less recognizable stages in this continuum receive vernacular names. Dry grasslands without shrubs or trees are called "campo limpio" (i.e., clean field); grasslands with shrubs scattered are called "campo sujo" (i.e., dirty field); grasslands with numerous trees and shrubs are called "campo cerrado" (i.e., closed field); when the vegetation is dominated by a closed canopy of trees it is called "cerradao" (i.e., the vegetation is closed). The latter is a woodland composed of trees often 8 to 12 m or even taller, with ground vegetation reduced because of the shade. Many factors probably determine which of these forms of Cerrado vegetation occurs in a given locality, among them topography and soil texture seem to have the main causal effect. Cerrados have a markedly seasonal climate and posses a large characteristic flora of fire-resistant plants, including about 800 species of trees and large shrubs, and many times that number of herbs and subshrubs. The vast majority of these species are endemic to the Cerrado, an ancient vegetation formation dating back perhaps 50 million years.

b. Temperate Grasslands

These ecosystems occupy the eastern part of southern South America in Argentina, Uruguay, Chile, and southern Brazil. They expand across the pampas of Rio de la Plata in Argentina, the Chaco region, the semiarid region west of the humid pampa (*Monte*), and reach the edges of desert and semidesert areas of the Patagonian region (Fig. 3). The dominant landforms are large plains with a few table-shaped outcroppings no more than 500 m above the plains. The dominant climate across the large grassland region is characterized by a mean temperature of 10 to 20 C° with an annual precipitation ranging from 400 to 1,600 mm. The dominant soil types are young and include Mollisols, Alfisols, Vertisols, and Entisols.

Vegetation in the Pampean temperate grasslands is composed of about 1,000 species and vary from grass-dominated communities in the east, to xeric woodlands and semidesert communities in the west. In the Argentinean Pampas, plant communities are dominated by species of *Agrostis, Bouteloa, Elyonurus, Festuca, Panicum, Paspalum*, and *Stipa*. In Chile and northward in the Andes, the ecosystems of grasslands are dominated by species *of Andropogon, Bromus, Calamagrostis, Poa, Sporobolus, Eragrostis*, and *Distichlis*. It is difficult to determine the proportion of Patagonia that can be classified as grasslands, because the sparse vegetation is mostly dwarf shrubs and grasses do not form a continuous cover, therefore many areas are treated rather as deserts.

c. High Andean Grasslands

These ecosystems are found above the tree line (3,000–4,500 m) of the Andean range in Colombia, Venezuela, Ecuador (*Páramo*), Peru, Bolivia, and Chile (*Puna*). The actual lower limit of páramo and puna ecosystems depends on exposure, the climatic regime, and in particular precipitation. The dominant life-forms of the Andean grasslands are giant-rosette plants (known locally as "*frailejon*"), tussocks of grasses or sedges, acaulescent rosette plants, cushion plants, cacti, and sclerophyllous shrubs. The giant-rosette plants can reach heights of several meters and have developed a number of morphological and anatomical adaptations including the retention of dead leaves that cover the shoot and provide insulation. At higher elevations, the vegetation cover becomes sparser and shrubs and herbs replace bunch grasses.

2. Functional Aspects

a. Tropical Savannas

Relatively homogeneous temperature and a well-defined wet season characterize the *Aw* tropical savanna climate or dry-wet tropical climate (Fig. 2). The alternation of wet and dry periods coincides with the presence of tropical savanna vegetation, characterized by open spaces covered by grasses and spaced trees. Soils are mainly latosols of low fertility, formed by similar processes but not as deep as those found in the *Af* climate.

Savanna ecosystems are characterized by seasonal growth cycles that follow the annual precipitation cycles. However, the phenology patterns of savanna plants are very asynchronous resulting in different patterns of productivity, reproduction, and decomposition. Most grasses usually start growing at the beginning of the rainy season and reach reproductive stages early in the wet season or a few months later. Some vegetative growth continues throughout the year but at the end of the dry season all aboveground biomass is dead. In perennial grasses, all the aboveground biomass dies by the end of the dry season and in the following year is replaced by vegetative growth of subterranean perennial organs. In evergreen species, growth of leaf and shoot biomass and reproduction generally occur during the dry season. Leaf development stops at the beginning of the wet season, and before the dry season begins leaves are shaded. In most evergreen species, however, growth of stems and other woody parts only occurs during the wet season.

Usually there are layers of laterites under savanna soils that were originated during the Cretaceous and Tertiary periods. Continuous weathering has produced soils poor in nutrients and with high acidity and concentration of aluminum that immobilizes phosphorous as iron-aluminium phosphate. In the soil environment, marked annual differences occur particularly in the upper soil layers where moisture levels can exceed field capacity during the wet season and diminish to wilting point during the dry period.

The stock of nutrients in savannas varies over the year as they move from different compartments of the ecosystem. The largest fraction of nutrients is stored in the aboveground biomass. Savanna trees tend to accumulate fewer nutrients that similar dry forest species and some nutrients reallocate from leaves to other organs before shedding occurs. Some nutrients accumulated in grasses return to the soil after fires but large fractions of nitrogen, phosphorous, and potassium are lost to the atmosphere when biomass burns. Decomposition of leaf litter by microorganisms occurs during the wet season and termites consume most of the woody material during the dry period.

The flora of the Brazilian cerrado and other South American savannas has typical features of pyrophytic vegetation. Trees are low, of contorted form, with thick, corky, fire-resistant barks; sclerophylly is common, and many leaves have thick cuticles, silicified tissues, and are often of considerable longevity. The underground organs of perennial grass species are protected from fire and their seeds as well as those of annual species are adapted to fire; in fact, the germination of many grass species is stimulated by fire events. Because savanna vegetation is adapted to resist burning, occasional natural fires must have been an environmental factor throughout its history. Frequent burning favors the herbaceous, as opposed to the woody, component of the vegetation and when this occurs savanna woodlands are transformed into grasslands. Conversely, protection from fire often has the opposite effect and a type of dense savanna woodland with considerable shade and only a very sparse grass layer is established. Some authors have correlated increasing production of woody vegetation with an increasing soil fertility gradient. Soil moisture is another factor that controls many of the most obvious differences in vegetation in the savanna landscape. In general, woody vegetation occurs only on soils well drained throughout the year. Swampy gallery forest is found where the water table is permanently high.

b. Temperate Grasslands

There are different hypotheses that attempt to explain the physiognomic origin of the temperate grasslands. An initial hypothesis proposed that pre-Columbian peoples burned an original scrub forest. Another hypothesis points to an adaptive dominance of grasses due to a climatic effect that maintains a negative water balance during part of the year. Although there are significant regional differences, edaphic factors such as the fine texture of soils are proposed as major causes for the dominance of grasses. A lack of air spaces in the soil combined with summer dry periods would favor the competitive ability of grasses over trees. A variety of root growth habits is observed among both grasses and shrubs and the stratification of root biomass along the soil profile minimizes competition for water and nutrients. The dominant species of grasses are perennial, live for 3 to 4 years, and are short and leafy.

The soils of the southern grasslands usually present bad drainage conditions, relatively acid pH, and the humus layer can be 30 to 40 cm deep. Deeper soil layers are less organic, more alkaline, and composed mostly of clay or calcareous sediments. From west to east, soil depth and organic matter content decrease because in drier areas productivity is lower. Soils are relatively poor in nitrogen and phosphorous, which frequently limit grassland productivity. Nitrogen accumulates mainly in organic forms; it is quickly absorbed and stored in leaves and live root biomass. Fungi are the most important decomposers in grassland ecosystems, and nematodes are the principal soil invertebrates. In the flooded Pampean grasslands, phenological studies have shown that there is not a well-defined growing season. These grasslands are subject to periodic flooding and drought almost every year and floristically different plant communities occur according to various degrees of perturbation.

South American grasslands have been greatly modified by humans through burning, grazing livestock, and agriculture. Before these intensive activities took place, there was a vast number of ungulates like the Guanaco (*Lama guanicoe*) and the Pampas deer (*Ozotoceros bezoarticus*). Birds are still abundant with more than 390 species described, of which the most conspicuous is the Ñandu (*Rhea americana*). Reptiles and amphibians are also abundant with poisonous snakes (e.g., *Bohtrops alternata*), caimans (e.g., *Caiman latirostris*), lizards (e.g., *Teius teyou*), toads (e.g., *Bufo refus*), and frogs (e.g., *Hyla pulchella*).

c. High Andean Grasslands

The most important functional attributes of these ecosystems are associated with the extraordinary amplitude of daily fluctuations in temperature and humidity, which are more pronounced than the seasonal ones. In these environments, high daytime temperatures follow nocturnal frost causing significant physiological stress in both plants and animals. Plants have evolved mechanisms that allow them to grow during the whole year. The frailejon giant-rosettes, for instance, at night protect their buds in the center of the rosette by adjusting the position of their leaves. In addition, these plants have large intercellular air spaces and excrete mucilage that contributes to their heat-storage capacity. The discontinuous areas of páramo ecosystems are isolated by montane wet (cloud) forests that grow at lower altitudes and as a result there is a large number of endemic species. Nutrient dynamics in high Andean ecosystems are not fully documented. In general, low temperatures and moist acid soils slow decomposition rates, and a thick layer of organic humus usually accumulates in sites where the relief allows deposition.

C. Xeric Formations

In South America there is a variety of xerophytic ecosystems, all marked by an intensive and prolonged period of seasonal drought during which the vegetation suffers water deficit.

1. Distribution and Structure

A mean annual precipitation of 400 to 1,200 mm and more extreme daily changes of temperature characterize the *BS* climate. Soils are brown and yellowish, reflecting a higher content of humus. The *BS* climate coincides with the geographic distribution of semiarid environments and generally represents transitions between desert environments of the type *BW* and savanna environments of the type *Aw* (Figs. 2 and 3). In South America, xeric ecosystems are represented by the xeric scrubs of La Guajira peninsula in Colombia and Venezuela, the Paraguaná xeric scrub of Venezuela, the dry forest of west Ecuador, the Sechura and Atacama deserts in Chile, the Gran Chaco region in southern Bolivia, west Paraguay and north Argentina, and the Caatinga xeric scrub of northeastern Brazil (Fig. 3). The dry highland plateaus of the Puna region and the thorn woodlands of Patagonia are usually included within the dry formations of the continent. There is also a xeric formation called "agreste," which forms a narrow strip between the Caatinga and the seasonal forest of eastern Brazil. Although not a strictly xeric ecosystem, the Matorral of Chile is also usually included as a semiarid type of vegetation (Fig. 3).

Deciduous forest trees that severely restrict their transpiration during the middle of the day—even during the rainy period—characterize the thorny woodlands of Colombia, Venezuela, and northeastern Brazil. Significant areas of the Guajira region consist of sand, moving dunes. Only a few trees are evergreen, but along the periodically dry channels of streams, riparian vegetation retain their leaves and reflect the presence of groundwater. The mean canopy height of the woody components is 5 m, sometimes reaching 7 m. The most important genera in these plant communities are *Acacia, Bulnesia, Bursera, Caesalpina, Capparis, Croton, Jacquina, Opuntia,* and *Prosopis*.

The caatinga ecosystems in the larger "Sertao" depression of northeastern Brazil are dominated by the genera *Canavillesia, Chorisia,* and *Schinopsis*. Communities on the dry tablelands are dominated by the genera *Spondias, Commiphora, Cnidoscolus, Aspidosperma,* and *Mimosa*. The driest parts of the Sertao are dominated by the genera *Mimosa, Auxema, Combretum,* and *Aristida*. The Agreste formation between the Caatinga and seasonal forest of Brazil has an open canopy of xerophytic trees (mostly Leguminoseae and Myrtaceae) with a very scanty understory of palms (e.g., *Copernicia cerifera*) and cactus.

The ecosystems of the Gran Chaco region are structurally diverse reflecting the varied environmental changes observed across the Chaco range. Precipitation increases from the center of the region, both to the east and west. The landforms are flat in the west and north and during the rainy season, flooding is common. The thorn woodlands of Patagonia occupy the west margin of the Pampa region, where annual precipitation is lower (300–500 mm) and the woody components of the vegetation (Acacias and palms) are characteristically undersized.

Due to the effect of the Andes and climatic circulation patterns (see section I.B), an arid zone occupies the zone from the coasts of Peru and Chile across the mountains into the Patagonian steppes (Fig. 3). The Pacific coastal area is a desolated desert where rain can be absent for years. The Atacama Desert is the driest climatic region of South America and occupies the eastern Pacific coast of the continent from about 5°S to 26°S (Figs. 2 and 3). In many areas there is no fog to provide moisture for vegetation and the cover is dominated by lichens and algae that depend on water condensed as dew. In those areas where fog and clouds develop, bringing a few millimeters of annual precipitation, there are some short-cycled annual plants from the genera *Aristida, Loasa, Malesherbia, Nolana,* and *Tropelum*. On the west-facing slopes, between 100 and 150 m, moisture from fog carried by the Humboldt Current allows the presence of some vegetation. Some of the common species in these ecosystems are *Prosopis juliflora, Distichilis spicata, Capparis angulata, Sporobolus virginicus, Acacia macracanta,* and the giant cactus *Neoraimondia gigantea*. In the high Andean slopes, the lowland forests and the grassland Puna limit the dry ecosystems of the Andean Peruvian valleys. The dominant genera there are *Acacia, Bombax, Bursera, Caesalpinia, Cercidicum, Prosopis, Puya,* and several cacti.

The Mediterranean Matorral occupies the coastal west-facing slopes of the Andes in central Chile (Fig. 3). Plant communities in these ecosystems are composed of evergreen shrub species that reach 3 to 5 m in height and have characteristic sclerophyllous leaves (e.g., *Lithaea caustica* and *Fluorencia thurifera*), while succulents and herbs cover the ground. Some Acacias and *Nothofagus* are present in communities of higher areas and those found further inland. Despite its latitudinal position, the flora of the Chilean Matorral is very rich, with more than 2,000 species reported.

2. Functional Aspects

There are only a few studies on the physiology and phenology of South American xeric ecosystems. Most of these ecosystems are characterized by at most 3 to 5 months of rainy season with no more than 400 to 800 mm/year. In tropical lowland areas, temperatures are 23 to 27°C year round. Relative humidity is stable around 50% and the potential evapotranspiration is usually 1,500 to 2,000 mm/year. Plant communities are affected by water deficit, which varies greatly from region to region, and so do the physiognomic and physiologic responses of the vegetation. Some annual plants germinate, grow, and reproduce only during periods of available water (e.g., most herbaceous). Other plants evade stress conditions by restricting growth to periods of available water, while the shedding of leaves is a common adaptation to cope with the long dry periods. Since leaves are only present during the rainy season, many species have not developed mechanisms to control evapotranspiration. In addition, many species from these ecosystems produce leaves with chemicals that inhibit the germination of other plants. Crassulacean acid metabolism (CAM) is common and succulent perennials use this type of photosynthesis to endure the scarcity of water.

Annual precipitation of about 300 to 900 mm and a dry summer climate characterize Mediterranean ecosystems. The dry, hot weather favors the occurrence of fires, which are the main disturbance in these ecosystems. Topographical conditions are also important in Mediterranean ecosystems, where erosion and continuous rearrangement of surface materials are common. In Chile, the duration and intensity of the drought period increases southward and Matorral communities become dominated by deciduous evergreen shrubs and succulent species. The climatic conditions allow all shrub species to use C_3 photosynthesis to fix carbon and there are not species with C_4 metabolism. As already pointed out, the availability of soil resources is important in determining the investment that plants make on belowground biomass. In the Chilean Matorral it has been estimated that shrubs have an average root to shoot ratio of 1.5 and the roots spread in a radius at least two times the area of their crowns. Most of the fine absorbing roots, however, are found under the crown where they absorb nutrients from decomposing leaves and debris. Water availability appears to be the most important controlling factor of productivity in Matorral ecosystems. Annual productivity in these ecosystems range from 2 to 6.5 kg m^2.

D. Coastal Ecosystems and Wetlands

1. Mangroves

In South America, mangroves occur along fragments of the Atlantic coastlines of Colombia, Venezuela, Guyana, Surinam, French Guyana, and Brazil, and the Pacific coastlines of Colombia, Ecuador, and Peru (Fig. 3). Mangroves develop well in areas where the ocean temperature remains above 24 to 27°C. Because of the effect of ocean currents (see section I.B.), the southern limits of mangroves in South America occur at about 25°S in the Atlantic coast of Brazil and 4°S in the Pacific coasts of Ecuador and Peru.

These woody communities can grow up to 30 m. The mangle is best developed in coastal areas where high rates of sedimentation are enhanced by the mangroves themselves. Local topography, runoff channels, and sediment stability determine distinct zonal patterns. The vegetation of mangrove ecosystems (mangal) is adapted to salinity and is dominated by *Rhizophora mangle, Avicenia* sp. and in high ground by *Laguncularia racemosa*. Plant communities directly influenced by seawater but occupying high grounds are characterized by genera typical of seashore environments: *Remirea, Salicorna, Canavalia, Acicarpa,* and *Vignia*.

Mangrove areas are sometimes associated with river deltas and coastal swamps that have habitats in which the soils are inundated for at least part of the year. In the mouth of the Amazon, however, mangroves do not grow well. According to the quality of the water, it is possible to distinguish freshwater swamps and brackish water swamps. To the first type belong the freshwater swamp forests of the Atlantic coast between the Orinoco delta and the mouth of the Parnaiba River. Most of the other mangrove swamp forests belong to the second type. Other conspicuous coastal ecosystems are the Restingas of the Brazilian coast. These ecosystems are characterized by sand dunes where low scrubby vegetation grows sparsely. Their flora is rich in endemic species.

2. Wetlands

In South America, there are many dispersed wetland ecosystems. Some savanna areas of the Llanos region in the Orinoco watershed and the Rio Branco in Brazil have bad drainage and are seasonally inundated. In Ecuador, west of the Daule River, there is also a region of wetlands. The complex of grassland and woodland savanna vegetation occupies flat plains with tropical seasonal climate. The different types of vegetation are closely associated with the duration of the period of inundation. Areas with a long inundation period maintain communities dominated by the genera *Euterpe* and *Mauritia*. The genera *Typha, Cyperus,* and *Juncus* dominate swampy areas. In well-drained floodplains, the dominant genera are *Panicum, Paspalum,* and *Thalia*.

In western and eastern Amazonia there are also some areas of flooded grasslands but larger wetland ecosystems usually occupy the floodplain of large rivers (e.g., varzea and igapo in Amazonia) and the structure of the plant communities depends on the level and duration of the flooding period (see section III.A.2). The larger wetlands of the continent are in the Gran Pantanal region in Brazil Bolivia and Paraguay, and the Paraná flooded savannas in Argentina (Fig. 3). The Pantanal occupies the upper watershed of the Paraguay basin and is one of the world's largest wetlands, which is flooded in more than 80% of its area during the rainy season (from May to December). The geomorphology is essentially flat and small high-ground islands separate hundreds of small lagoons. Cerrado tree species and palms occupy the high-ground islands of the Pantanal. The region accommodates large seasonal migrations of wildlife including aquatic birds and caimans.

E. Temperate Forests

In South America, temperate forest have a limited distribution and are restricted to relatively small areas. These ecosystems occur in the southern Andean areas of Chile and Argentina (Fig. 3). Evergreen temperate forest occupies areas of frost-free oceanic climate. These ecosystems are dominated by tall trees, some reaching heights up to 50 m, and composed of species such as *Eucryphia cordifloria* and *Laurelia philippiana* and some conifers like *Araucaria araucana*.

In the western humid slopes of the Andes, temperate rain forests are divided into the Valdivian and Magellanic forests. The Valdivian rain forest has an upper limit of 500 m and occurs from 40°S to 47 to 49°S where it becomes the Magellan forest. A lush vegetation rich in evergreen species, abundant epiphytes, and lianas characterizes Valdivian ecosystems. The Valdivian forests can support an extraordinary biomass, in part because the climate is very humid and the mean annual temperature ranges from 10 to 12°C. In areas with excessive humidity, coniferous trees (*Fitzroya patagonica*)—whose stands can reach heights of 50 to 60 meters—replace common species of the rain forests. Coniferous forests also grow well at higher altitudes up to the tree line, which in the southern cone occur at about 1,600 to 1,900 m above sea level. Between the Valdivian forests and the sclerophyllous vegetation of central Chile there are limited extensions of deciduous forests composed of *Nothofagus obliqua*, which are found up to 1,200 m above sea level.

The Magellanic forests occupy the western slopes of the Cordillera Patagonica in southern Chile. In this region, the coastline is very rocky, curved by fjords and with an abrupt slope. Abundant rainfall, low evapotranspiration, and strong winds characterize the climate. The Magellanic forests are poor in species. As precipitation decreases rapidly to the east, in the leeward slopes deciduous forests occur and are composed almost of pure stands of *Nothofagus* with an understory of small evergreen trees. These forests grow to heights of about 30 m, but they become shorter as altitude increases. The tree line falls from 1,200 to 1,400 m in the north to about 600 m in the region of Ushuaia.

See Also the Following Articles

AFRICA, ECOSYSTEMS OF • CENTRAL AMERICA, ECOSYSTEMS OF • EUROPE ECOSYSTEMS OF • NEAR EAST ECOSYSTEMS • NORTH AMERICA, PATTERNS OF BIODIVERSITY IN

Bibliography

Archibold, O. W. (1995). *Ecology of World Vegetation*. Chapman & Hall, London.

Beard, S. J. (1955). The classification of tropical American vegetation-types. *Ecology* 36(1) 89–100.

Bush, M. B. (1994). Amazonian speciation: A necessarily complex model. *Journal of Biogeography* 21, 5–17.

Colinvaux, P. A. (1989). Ice-age Amazon revisited. *Nature* 340, 188–189.

Coupland, R. T. (1992). Overview of South American Grasslands. In *Natural Grasslands: Introduction and Western Hemisphere* (R. T. Coupland, Ed.), pp. 363–365. Elsevier Scientific Publishing Company, Amsterdam.

Dinerstein, E., Olson, D. M., Graham, D., Webster, A., Primm, S., Bookbinder, M. (1995). *A Conservation Assessment of the Terrestrial Ecoregions of Latin America and the Caribbean.* The World Bank & World Wildlife Fund, Washington, D.C.

Edit, R. C. (1968). The climatology of South America. In *Biogeography and Ecology in South America* (E. Fittaku, J. Illes, G. Klinge, G. Schwabe, and H. Sioli, Eds.). Junk, The Hague.

Eiten, G. (1972). The cerrado vegetation of Brazil. *The Botanical Review* 38(2), 201–341.

FAO-Unesco (1971). *Soil map of the world, 1:5.000.000.* Food and Agriculture Organization of the United Nations and the United Nations Educational, Scientific and Cultural Organization, Paris.

Gentry, A. H. (1995). Diversity and floristic composition of Neotropical dry forests. In *Seasonally Dry Tropical Forests* (S. H. Bullock, H. A. Mooney, and E. Medina, Eds.), pp. 146–194. Cambridge University Press, New York.

Haffer, J. (1969). Speciation in Amazonian forest birds. *Science* 165(3889), 131–137.

Harrington, H. (1973). Geology of South America. In *The Encyclopedia of Earth Sciences* (R. W. Fairbridge, Ed.), vol. VIII, pp. 456–465. Stroudsburg, Dowden, Hutchinson & Ross.

Leigh, E. G. (1975). Structure and climate in tropical rain forest. *Ann. Rev. Ecol. and Syst.* 6, 67–86.

Murphy, P. G., and A. E. Lugo (1986). Ecology of tropical dry forest. *Ann. Rev. Ecol. Syst.* 17, 67–88.

Nepstad, D. C., Carvalho, C. R., Davidson, E., Jipp, P., Lefevbre, P., Negrelros, G., daSilva, E., Stone, T., Trumbore, S., Vieira, S., (1994). The role of deep roots in the hydrological and carbon cycles of Amazonian forests and pastures. *Nature* 372, 666–669.

Sarmiento, G. (1984). *The Ecology of Tropical Savannas.* Harvard University Press, Cambridge.

Stone, T. A., Schlesinger, P., Houghton, R., Woodwell, G. (1994). A map of the vegetation of South America based on satellite imagery. *Photogrammetric Engineering & Remote Sensing* 60(5), 541–551.

Terborgh, J. (1992). *Diversity and the Tropical Rain Forest.* New York.

Whitmore, T. C., and G. T. Prance (1987). *Biogeography and Quaternary History in Tropical America.* Cambridge University Press, Cambridge.

SOUTH AMERICAN NATURAL ECOSYSTEMS, STATUS OF

Philip M. Fearnside
National Institute for Research in the Amazon (INPA)

I. Original Extent of Terrestrial Ecosystems
II. Present Extent of Terrestrial Ecosystems
III. Human Use of Converted Areas
IV. Human Use of Remaining Natural Habitats
V. Threats to Remaining Natural Habitats
VI. Status of Protected Areas
VII. Priorities for Conservation

GLOSSARY

bioregion One of six biogeographic divisions of South America consisting of contiguous ecoregions. Bioregions are delimited to better address the biogeographic distinctiveness of ecoregions.
ecoregion A geographically distinct assemblage of natural communities that share a large majority of their species and ecological dynamics, share similar environmental conditions, and interact ecologically in ways that are critical for their long-term persistence.
ecosystem A set of interacting living and nonliving components in a defined geographic space. Ecosystems include both plant and animal communities and the soil, water, and other physical elements of their environment.
major ecosystem type Groups of ecoregions that share minimum area requirements for conservation, response characteristics to major disturbance, and similar levels of β diversity (i.e., the rate of species turnover with distance).
major habitat type Groups of ecoregions that have similar general structure, climatic regimes, major ecological processes, β diversity, and flora and fauna with similar guild structures and life histories.

THE TERM ECOREGION, as used in this article, refers to "natural" ecological systems, or terrestrial and aquatic areas as they were when Europeans first arrived in the New World. The original extent of natural ecoregions is presented, grouped by bioregion, major habitat type, and major ecosystem type. The definitions of these terms, given in the Glossary, are taken from Dinerstein *et al.* (1995); the rating codes are given in the footnotes to the table. Indications of the extent of remaining natural ecosystems, the threats to their continued existence, and the status of protected areas are discussed, together with priorities for conservation.

I. ORIGINAL EXTENT OF TERRESTRIAL ECOSYSTEMS

Ecosystems can be classified in many ways, making the number of categories vary widely depending on the use intended. Here, the system adopted by Dinerstein *et al.* (1995) is used. This divides the continent into 95 terrestrial "ecoregions," exclusive of mangroves. These are grouped into four "major ecosystem types:" tropical broadleaf forests, conifer/temperate broadleaf forests, grasslands/savannas/shrublands, and xeric formations.

Within each of these categories are varying numbers of "major habitat types," such as tropical moist broadleaf forests. These are further divided into nine "bioregions." Amazonian tropical moist forests, for example, is a bioregion.

The 95 ecoregions, with their hierarchical groupings, are presented in Table I. Also included are the ratings for conservation status, biological distinctiveness, and biodiversity priority derived by Dinerstein et al. (1995). This study made a systematic survey of the status of natural ecosystems in Latin America and the Caribbean (LAC) and applied a uniform methodology to assigning priorities to these ecosystems for conservation efforts. The work was done for the United States Agency for International Development (USAID) by the WWF–US Biodiversity Support Program (BSP). The document is based on three workshops, plus consultations with relevant organizations and individual experts (the list of contributors contains 178 names).

The classification system is hierarchical, starting with four "major ecosystem types" (e.g., Tropical Broadleaf Forests), which are divided into 10 "major habitat types" (e.g., Tropical Moist Broadleaf Forests). These are crossed with 6 bioregions (e.g., Amazonia) and divided into 95 ecoregions (e.g., Rondônia/Mato Grosso moist forests). The system allows the priority of some ecoregions to be promoted upward based on uniqueness and regional representation, even if indicators of diversity and vulnerability are not so high.

The effort was unusual in emphasizing protection of areas with high β diversity (a measure of the turnover of species along ecological gradients), as well as the more commonly used α diversity (species diversity within a habitat). In the case of mangroves, the diversity assessed is ecosystem diversity, including aquatic animal life. This avoids mangroves receiving the unjustly low diversity ratings that tend to result when assessments are restrained to terrestrial organisms, especially trees.

Although the ecoregions identified in Table I refer to "natural" (pre-Columbian) ecosystems, it should be emphasized that these had already been subject to millennia of influence by indigenous peoples prior to the arrival of Europeans. This influence continues today, together with much more rapid alterations from such activities as deforestation and logging done by nonindigenous residents. "South America" is taken to include the three Guianas (different from usage by the Food and Agriculture Organization of the United Nations (FAO)) and to exclude Panama (however, in the case of ecoregions that extend into Panama, the area estimates in Table I include the Panamanian portions).

The ecoregions are mapped in Fig. 1. The ecoregion numbering corresponds to Table I and also to the report by Dinerstein et al. (1995); the numbering presented here is not continuous, since the report also includes ecoregions in Mexico, Central America, and the Caribbean. Extensive bibliographic material on the delimitation of the ecoregions and on the state of knowledge about them can be found in Dinerstein et al. (1995).

Mangroves occur along the coasts of Brazil, the three Guianas, Venezuela, Colombia, Ecuador, and northern Peru. Dinerstein et al. (1995) divide them into five complexes: Pacific South America, Continental Caribbean, Amazon–Orinoco–Maranhão, Northeast Brazil, and Southeast Brazil. Each complex is further subdivided into 2–5 units, corresponding to distinct segments of coastline. Mangroves are essential to maintaining populations and ecological processes in surrounding marine, freshwater, and terrestrial ecosystems.

II. PRESENT EXTENT OF TERRESTRIAL ECOSYSTEMS

Unfortunately, information is not available on the present extent of each of the 95 ecoregions listed in Table I. Information on the extent of tropical forests in approximately 1990 is available from the FAO Tropical Forest Resources Survey (FAO, 1993). These data are tabulated by country in Table II. Nontropical areas are covered by a variety of national surveys (Harcourt and Sayer, 1996). National data are important because decisions regarding land-use policies and conservation are taken at the national level—not at the levels of bioregions or ecosystem types. Over half of the South American continent is represented by a single country: Brazil (Fig. 2).

An idea of the extent of existing ecosystems can be gained from measurements of land cover in 1988 made using 1×1 km resolution data from the AVHRR sensor on the NOAA satellite series (Stone et al., 1994). These are tabulated in Table III.

It should be emphasized that many ecosystems can be heavily disturbed by logging and other activities without the change being evident on satellite imagery. This is true for Landsat TM imagery (30×30 m resolution) used for deforestation estimates in Brazil, and the limitations are much greater for 1×1 km AVHRR data.

Brazil is the country with the most extensive satellite information on forest cover and its loss. Unfortunately, information on nonforest vegetation types such as cerrado is much less complete. Considerable confusion arises between the FAO (1993) classification and others such as the one adopted here because FAO classifies cerrado, caatinga, and chaco as "forests."

Brazil's Legal Amazon region originally had 4 million km^2 of forests, the rest being cerrado and other types of savannas. Agricultural advance was slow until recent decades because of human diseases (especially yellow fever and malaria), infertile soil, and vast distances to markets. These barriers have progressively crumbled, although a range of limiting factors restricts the extent and the duration over which many uses of deforested areas can be maintained (Fearnside, 1997a). Deforestation in the region has been predominantly for cattle pasture, with critical contributions to the motivations for the transformation coming from the role of clearing as a means of establishing land tenure and in allowing land to be held and sold for speculative purposes (Fearnside, 1993).

The Atlantic forests of Brazil (ecoregions 54 and 55) have been almost completely (>95%) destroyed, mainly for agriculture, silviculture, and real estate development. Most of what remains of this extraordinarily rich ecosystem is in protected areas, but unprotected areas continue in rapid retreat. These forests are recognized as major "hot spots" of biodiversity (Heywood and Watson, 1995; Stotz et al., 1996).

In Andean countries, clearing by small farmers has predominated in driving deforestation, in contrast to the predominant role of medium and large cattle ranchers in Brazil. Migration from densely populated areas in the Andean highlands (altiplano) has led to settlement in lowland forests areas, with consequent upsurges in clearing (e.g., Rudel and Horowitz, 1993).

Savanna ecosystems have suffered heavy human pressure. The pampas of Argentina and the Uruguayan savannas of Uruguay and southern Brazil (ecoregions 120 and 121) have largely been converted to agriculture. The Brazilian cerrado, originally covering 2 million km^2, is the largest ecoregion in South America, as well as holding the largest number of species of any of the world's savannas. The cerrado was largely intact until the mid-1970s. Clearing, especially for soybeans and planted pasture, reduced the cerrado to 65% of its original area by 1993 according to Landsat imagery interpreted by Brazil's National Institute for Space Research (INPE). The advance of clearing has proceeded at an accelerating pace, speeded by infrastructure projects and an array of government subsidies.

The temperate and coniferous forests of the Southern Cone have been under severe pressure from logging. These forests are usually logged by clear-cutting in a manner similar to their counterparts in the North American temperate zone. This contrasts with the "selective" logging (highgrading for a few species) that characterizes timber extraction from the diverse forests of the tropical region.

III. HUMAN USE OF CONVERTED AREAS

Conversion of natural ecosystems to agroecosystems and secondary forests creates landscapes that maintain biodiversity to varying degrees. "Shifting cultivation" as practiced by indigenous peoples and by traditional nonindigenous residents (caboclos) in Amazonian forests maintains a substantial part of the original biodiversity. This contrasts with the effect of the vast expanses of cattle pasture that have replaced this, either directly or following a phase of use in pioneer agriculture by small farmers who have recently arrived from other places.

In densely settled areas along the coast of Brazil and in the southern portions of the country, agricultural use has gone through a series of "cycles," such as sugarcane and coffee. The productivity of many areas has been damaged by soil erosion and other forms of degradation. Cattle pasture is often the land use replacing these crops. Since the 1970s, plantation silviculture (which now covers over 70,000 km^2) and soybeans (130,000 km^2) have made large advances.

In Argentina and Uruguay, cattle ranching and wheat and rice farming are major land uses. Natural vegetation is better represented in areas with little agricultural potential, such as mountain and polar areas and arid and semiarid zones.

IV. HUMAN USE OF REMAINING NATURAL HABITATS

Areas that remain under natural vegetation cover, rather than being converted to other land uses through clearing, are also subject to human use and alteration. Selective logging in tropical forests, for example, leaves much of the basic structure of the ecosystem intact, but also can lead to significant changes that can set in motion a sequence of events leading to complete destruction of the ecosystem. Logging leaves a substantial amount of dead biomass in the forest, including the crowns and stumps of harvested trees and all of the biomass of the many additional trees that are killed by damage

TABLE I
Terrestrial Ecoregions of South America[a]

Major ecosystem type	Major habitat type	Bioregion	Ecoregion name	Ecoregion number	Countries	Original area (km²)	Conservation status[b]	Biological distinctiveness[c]	Biodiversity priority[d]
Tropical broadleaf forests	Tropical moist broadleaf forests	Orinoco tropical moist forests	Cordillera La Costa montane forests	17	Venezuela	13,481	3	2	I
			Orinoco Delta swamp forests	18	Venezuela, Guyana	31,698	4	3	III
			Guianan Highlands moist forests	20	Venezuela, Brazil, Guyana	248,018	5	2	III
			Tepuis	21	Venezuela, Brazil, Guyana, Suriname, Colombia	49,157	5	1	II
			Napo moist forests	22	Peru, Ecuador, Colombia	369,847	4	1	I
		Amazonian tropical moist forests	Macarena montane forests	23	Colombia	2,366	3	2	I
			Japurá/Negro moist forests	24	Colombia, Venezuela, Brazil	718,551	5	1	II
			Uatumã moist forests	25	Brazil, Venezuela, Guyana	288,128	4	3	III
			Amapá moist forests	26	Brazil, Suriname	195,120	4	3	III
			Guianan moist forests	27	Venezuela, Guyana, Suriname, Brazil, French Guiana	457,017	4	3	III
			Paramaribo swamp forests	28	Suriname	7,760	3	3	III
			Ucayali moist forests	29	Brazil, Peru	173,527	2	1	I
			Western Amazonian swamp forests	30	Peru, Colombia	8,315	4	1	I
			Southwestern Amazonian moist forests	31	Brazil, Peru, Bolivia	534,316	4	1	I
			Juruá moist forests	32	Brazil	361,055	5	2	III
			Várzea forests	33	Brazil, Peru, Colombia	193,129	3	1	I
			Purús/Madeira moist forests	34	Brazil	561,765	4	4	IV
			Rondônia/Mato Grosso moist forests	35	Brazil, Bolivia	645,089	3	2	II

348

	Beni swamp and gallery forests	36	Bolivia	31,329	4	4	IV
	Tapajós/Xingu moist forests	37	Brazil	630,905	3	4	IV
	Tocantins moist forests	38	Brazil	279,419	2	4	III
Northern Andean tropical moist forests	Chocó/Darién moist forests	39	Colombia, Panama, Ecuador	82,079	3	1	I
	Eastern Panamanian montane forests	40	Panama, Colombia	2,905	2	1	I
	Northwestern Andean montane forests	41	Colombia, Ecuador	52,937	2	1	I
	Western Ecuador moist forests	42	Ecuador, Colombia	40,218	1	2	I
	Cauca Valley montane forests	43	Colombia	32,412	1	1	I
	Magdalena Valley montane forests	44	Colombia	49,322	1	1	I
	Magdalena/Urabá moist forests	45	Colombia	73,660	2	3	II
	Cordillera Oriental montane forests	46	Colombia	66,712	3	1	I
	Eastern Cordillera Real montane forests	47	Ecuador, Colombia, Peru	84,442	3	1	I
	Santa Marta montane forests	48	Colombia	4,707	3	2	I
	Venezuelan Andes montane forests	49	Venezuela, Colombia	16,638	2	1	I
	Catatumbo moist forests	50	Venezuela, Colombia	21,813	1	4	III
Central Andean tropical moist forests	Peruvian Yungas	51	Peru	188,735	2	1	I
	Bolivian Yungas	52	Bolivia, Argentina	72,517	2	2	I
	Andean Yungas	53	Argentina, Bolivia	55,457	3	3	III
Eastern South American tropical moist forests	Brazilian Coastal Atlantic forests	54	Brazil	233,266	1	1	I
	Brazilian interior Atlantic forests	55	Brazil	803,908	2	2	I
Orinoco tropical dry forests	Llanos dry forests	74	Venezuela	44,177	2	4	III
Amazonian tropical dry forests	Bolivian Lowland dry forests	76	Bolivia, Brazil	156,814	1	1	I
Tropical dry broadleaf forests							

continues

Continued

Major ecosystem type	Major habitat type	Bioregion	Ecoregion name	Ecoregion number	Countries	Original area (km²)	Conservation status[b]	Biological distinctiveness[c]	Biodiversity priority[d]
		Northern Andean tropical dry forests	Cauca Valley dry forests	77	Colombia	5,130	1	4	III
			Magdalena Valley dry forests	78	Colombia	13,837	1	4	II
			Patía Valley dry forests	79	Colombia	1,291	1	4	III
			Sinú Valley dry forests	80	Colombia	55,473	1	4	III
			Ecuadorian dry forests	81	Ecuador	22,271	1	1	I
			Tumbes/Piura dry forests	82	Ecuador, Peru	64,588	2	1	I
			Marañon dry forests	83	Peru	14,921	2	3	II
			Maracaibo dry forests	84	Venezuela	31,471	2	4	III
			Lara/Falcón dry forests	85	Venezuela	16,178	2	4	III
		Central Andean tropical dry forests	Bolivian montane dry forests	86	Bolivia	39,368	1	3	II
Conifer/temperate broadleaf forests	Temperate forests	Southern South American temperate forests	Chilean winter rain forests	87	Chile	24,937	2	2	I
			Valdivian temperate forests	88	Chile, Argentina	166,248	3	1	I
			Subpolar *Nothofagus* forests	89	Chile, Argentina	141,120	3	3	III
	Tropical and subtropical coniferous forests	Eastern South American tropical and subtropical coniferous forests	Brazilian *Araucaria* forests	105	Brazil, Argentina	206,459	1	3	II
Grasslands/savannas/shrublands	Grasslands, savannas, and shrublands	Orinoco grasslands, savannas, and shrublands	Llanos	110	Venezuela, Colombia	355,112	4	3	III
		Amazonian grasslands, savannas, and shrublands	Guianan savannas	111	Suriname, Guyana, Brazil, Venezuela	128,375	4	3	III
			Amazonian savannas	112	Brazil, Colombia, Venezuela	120,124	4	3	III
			Beni savannas	113	Bolivia	165,445	2	3	II
		Eastern South American grasslands, savannas, and shrublands	Cerrado	114	Brazil, Paraguay, Bolivia	1,982,249	3	1	I
			Chaco savannas	115	Argentina, Paraguay, Bolivia, Brazil	611,053	3	2	I
			Humid Chaco	116	Argentina, Paraguay, Uruguay, Brazil	474,340	3	4	IV

			#					
	Southern South American grasslands, savannas, and shrublands	Córdoba montane savannas	117	Argentina	55,798	3	4	IV
		Argentine Monte	118	Argentina	197,710	4	3	III
		Argentine Espinal	119	Argentina	207,054	4	3	III
		Pampas	120	Argentina	426,577	2	3	III
		Uruguayan savannas	121	Uruguay, Brazil, Argentina	336,846	3	3	III
Flooded grasslands	Orinoco flooded grasslands	Orinoco wetlands	128	Venezuela	6,403	4	3	III
	Amazonian flooded grasslands	Western Amazonian flooded grasslands	129	Peru, Bolivia	10,111	4	3	III
		Eastern Amazonian flooded grasslands	130	Brazil	69,533	3	3	III
		São Luis flooded grasslands	131	Brazil	1,681	2	4	III
	Northern Andean flooded grasslands	Guayaquil flooded grassland	132	Ecuador	3,617	2	3	II
	Eastern South American flooded grasslands	Pantanal	133	Brazil, Bolivia, Paraguay	140,927	3	1	I
		Paraná flooded savannas	134	Argentina	36,452	2	3	II
Montane grasslands	Northern Andean montane grasslands	Santa Marta paramo	137	Colombia	1,329	3	1	I
		Cordillera de Mérida paramo	138	Venezuela	3,518	4	1	I
		Northern Andean paramo	139	Ecuador	58,806	3	1	I
	Central Andean montane grasslands	Cordillera Central paramo	140	Peru, Ecuador	14,128	3	1	I
		Central Andean puna	141	Bolivia, Argentina, Peru, Chile	183,868	3	2	I
		Central Andean wet puna	142	Chile	188,911	3	2	I
		Central Andean dry puna	143	Argentina, Bolivia, Chile	232,958	3	2	I
	Southern South American montane grasslands	Southern Andean steppe	144	Argentina, Chile	198,643	4	4	IV
		Patagonian steppe	145	Argentina, Chile	474,757	3	2	I
		Patagonian grasslands	146	Argentina, Chile	59,585	3	3	III
Xeric formations	Mediterranean scrub	Central Andean Mediterranean scrub	148	Chile	141,643	2	1	I

continues

Continued

Major ecosystem type	Major habitat type	Bioregion	Ecoregion name	Ecoregion number	Countries	Original area (km^2)	Conservation status[b]	Biological distinctiveness[c]	Biodiversity priority[d]
	Deserts and xeric shrublands	Orinoco deserts and xeric shrublands	La Costa xeric shrublands	168	Venezuela	64,379	2	4	III
			Arayua and Paría xeric scrub	169	Venezuela	5,424	2	3	II
		Northern Andean deserts and xeric shrublands	Galapagos Islands xeric scrub	170	Ecuador	9,122	3	1	I
			Guajira/Barranquilla xeric scrub	171	Colombia, Venezuela	32,404	2	3	II
			Paraguaná xeric scrub	172	Venezuela	15,987	2	3	II
		Central Andean deserts and xeric shrublands	Sechura Desert	173	Peru, Chile	189,928	3	3	III
			Atacama Desert	174	Chile	103,841	3	3	III
		Eastern South American deserts and xeric shrublands	Caatinga	175	Brazil	752,606	3	3	III
	Restingas	Northern Andean restingas	Paranaguá restingas	176	Venezuela	15,987	2	3	II
		Amazonian restingas	Northeastern Brazil restingas	177	Brazil	10,248	1	1	I
		Eastern South American restingas	Brazilian Atlantic coast restinga	178	Brazil	8,740	1	1	I

[a] Data source: Dinerstein et al., 1995.
[b] Conservation status codes: 1, critical; 2, endangered; 3, vulnerable; 4, relatively stable; 5, relatively intact.
[c] Biological distinctiveness codes: 1, globally outstanding; 2, regionally outstanding; 3, bioregionally outstanding; 4, locally important.
[d] Biodiversity priority codes: I, highest priority at regional scale; II, high priority at regional scale; III, moderate priority at regional scale; IV, important at national scale.

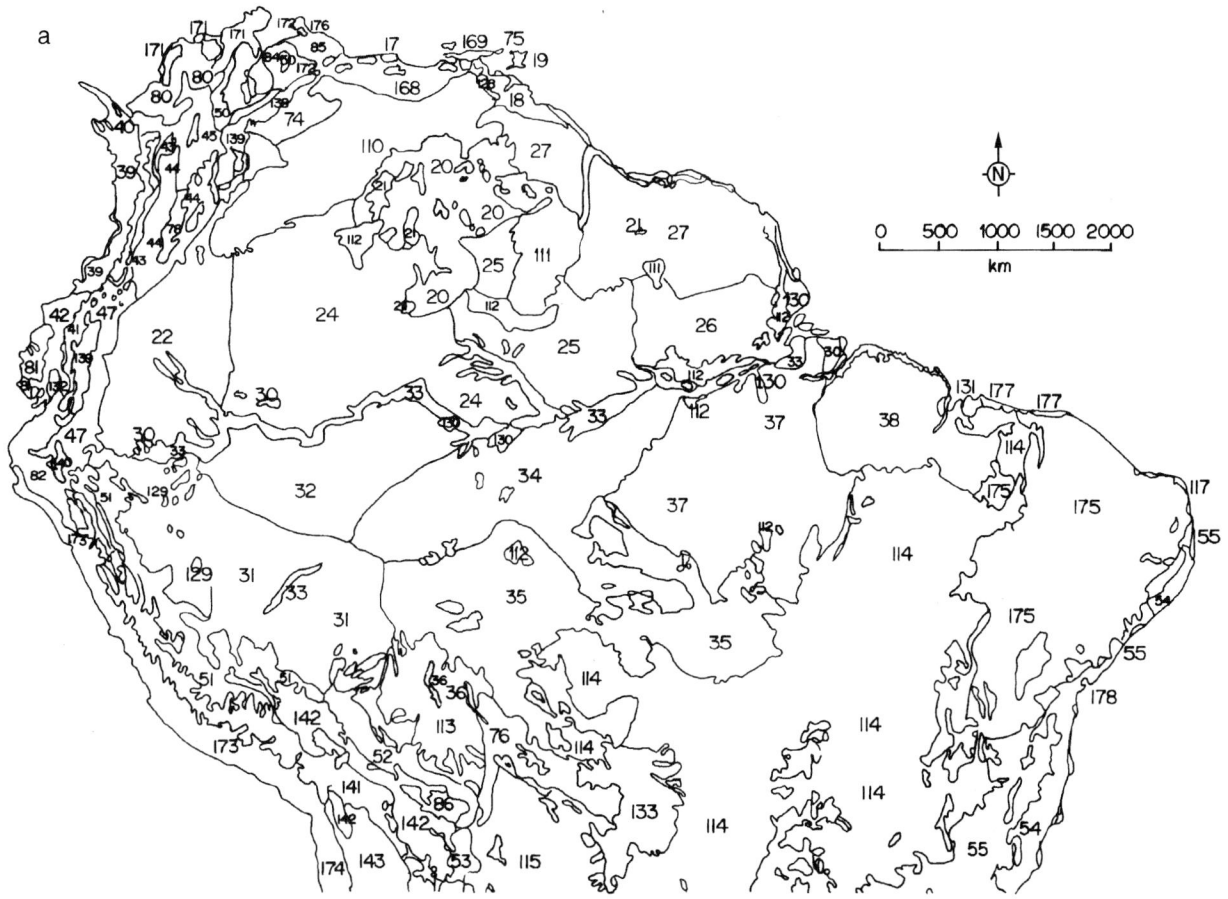

FIGURE 1 Ecoregions for pre-Columbian vegetation of South America. Numbers correspond to Table I. (Adapted from Dinerstein et al., 1995).

sustained during the logging process. Openings created in the canopy allow sunlight and heat to penetrate to the forest floor, drying out the fuel bed more quickly than in unlogged forests. Climatic variations such as those provoked by the El Niño phenomenon make logged forests especially susceptible to entry of fires. Ample opportunities for fires are provided as fields are burned to prepare land for planting and as cattle pastures are burned to control invading weeds. The fires burn slowly through the understory, charring the bases of trees as they go. Many of these trees then die, leading to a positive-feedback process whereby more dead biomass and canopy openings are provided and subsequent fires begin with greater ease, killing still more trees. This can degrade the entire forest within a few years (Nepstad et al., 1999).

Tropical forests are also used for "extractivism," or the collection of nontimber forest products (NTFPs) such as rubber and Brazil nuts. This does relatively little damage to the forest, although extractivists do have an impact through hunting and through clearing for subsistence crops. The extractivist population can also play a protective role in defending the forest against encroachment by more aggressive actors such as ranchers and loggers. This is the basis of the extractive reserve system in Brazil (see Anderson, 1990).

Savannas are often grazed by cattle without cutting trees. Cerrado (ecoregion 114), "lavrado," or Guianan savannas (ecoregion 111), the Pantanal wetlands (ecoregion 133), and the llanos of Venezuela (ecoregion 110) are among the savannas often used in this way. Increasing fire frequency, virtually all a result of human-initiated burning, can lead to shifts in species composition and to a drain of nutrients.

Aquatic ecosystems are traditionally exploited by fisheries. This alters the relative abundance of the species present. Use of watercourses as recipients for sewage and other pollutants also affects aquatic life in many ways.

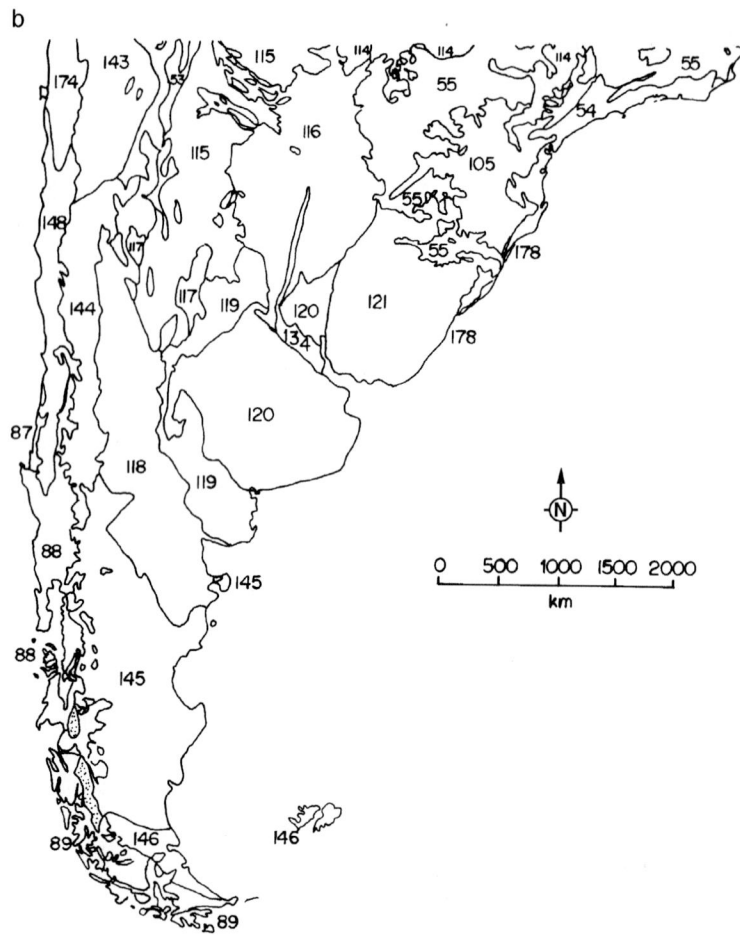

FIGURE 1 (continued)

TABLE II

Area of Tropical Forest Present in 1990 (km^2)[a]

Country	Tropical rain forests	Moist deciduous forest	Dry deciduous forest[b]	Very dry forest	Desert	Hill and montane forest	All forests[b]
Bolivia	0	355,820	73,460	0	40	63,850	493,170
Brazil	2,915,970	1,970,820	288,630	0	0	435,650	5,611,070
Colombia	474,550	41,010	180	0	0	24,900	540,640
Ecuador	71,500	16,690	440	0	0	31,000	119,620
French Guiana	79,930	30	0	0	0	0	79,970
Guyana	133,370	31,670	0	0	0	19,120	184,160
Paraguay	0	60,370	67,940	0	0	270	128,590
Peru	403,580	122,990	190	2,690	1,840	147,770	679,060
Suriname	114,400	33,280	0	0	0	0	147,680
Venezuela	196,020	154,650	2,220	1	0	103,900	456,910
Total	4,389,320	2,787,330	433,060	2,691	1,880	826,460	8,440,870

[a] Data source: FAO, 1993.
[b] Includes cerrado, caatinga, and chaco.

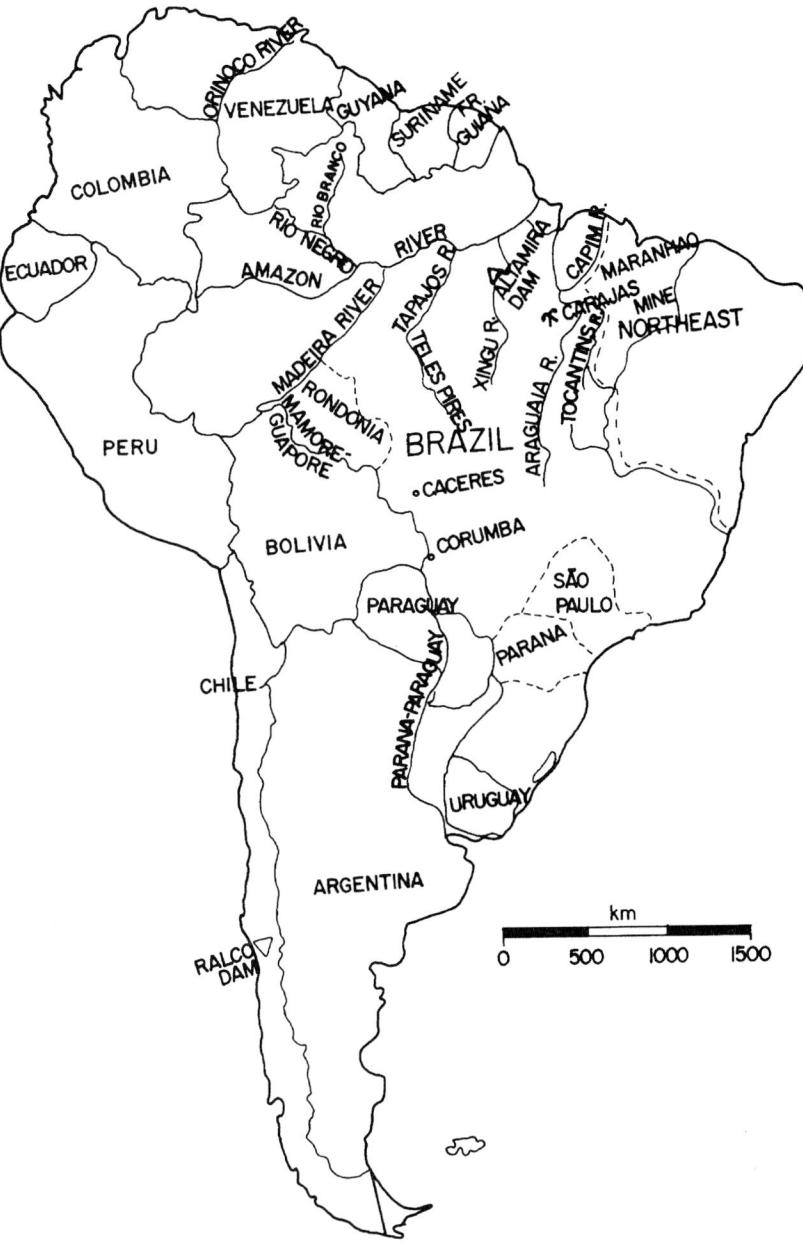

FIGURE 2 Locations mentioned in the text.

TABLE III

Land Cover in South America in 1988

Country	Closed tropical moist forest	Recently degraded TMF	Closed forest	Degraded closed forest	Woodlands	Degraded woodlands	Savanna, grasslands	Degraded savanna, grasslands	Scrublands, Shrublands	Desert, bare soil	Water	Snow, rock, ice	Other	Total
Argentina	1.2	0.0	96.8	0.6	645.4	15.7	755.4	232.8	894.8	37.9	34.0	31.4	35.7	2779.8
Bolivia	323.5	12.7	409.2	24.6	345.1	102.2	87.7	86.2	4.8	16.5	11.9	0.1	1.1	1089.4
Brazil	3522.3	519.7	3686.0	1692.2	1555.9	330.0	740.0	179.4	0.0	0.0	80.9	0.0	124.0	8388.5
Chile	0.0	0.0	134.1	29.1	75.2	29.8	101.1	14.0	86.9	186.8	7.0	16.6	3.8	684.5
Colombia	581.6	5.4	622.5	11.4	116.3	14.5	255.5	64.0	0.0	0.0	3.1	0.0	22.8	1110.1
Ecuador	115.5	1.7	121.0	1.7	33.7	4.3	41.9	13.3	3.2	2.5	0.6	0.0	0.8	223.1
French Guiana	78.8	0.0	79.8	2.4	0.6	0.0	0.2	0.0	0.0	0.0	0.1	0.0	1.0	84.1
Guyana	159.4	2.0	171.6	2.4	5.4	0.3	18.4	1.5	0.0	0.0	1.2	0.0	3.7	204.3
Paraguay	0.3	0.0	8.9	0.2	209.1	50.7	104.0	26.5	0.0	0.0	0.6	0.0	1.1	401.1
Peru	620.8	19.1	654.7	19.1	88.0	78.8	139.0	97.4	64.3	88.0	8.3	0.7	5.6	1244.1
Suriname	126.0	2.5	128.5	10.0	0.5	0.3	1.2	0.4	0.0	0.0	1.1	0.0	3.3	145.2
Uruguay	1.4	0.0	2.1	0.0	0.9	0.0	154.1	11.0	0.0	0.0	3.0	0.0	5.9	177.0
Venezuela	379.1	0.2	415.5	9.9	33.9	40.2	243.3	82.0	27.2	0.0	11.4	0.0	8.4	871.8
Unclassified														313.0
Total	5909.9	563.4	6530.7	1803.7	3109.8	666.9	2642.0	808.5	1080.6	331.7	163.2	48.9	217.2	17,716.1
Continent (%)	33.4	3.2	36.9	10.2	17.6	3.8	14.9	4.6	6.1	1.9	0.9	0.3	1.2	
Category (%)	8.7		21.6		17.7		23.4							100.0

Note. All values in thousands of km² or percent. "TMF" includes tropical moist, semideciduous, and gallery forests. "Closed Forest" includes TMF, montane forests, cool and temperate deciduous forests, and tropical seasonal forests. "Degraded Grasslands" includes agriculture. "Desert, Bare Soil" includes inland salt marsh communities, and "Other" includes wet vegetation and mangroves. Source: Stone et al., 1994. Reproduced with permission, the American Society for Photogrammetry and Remote Sensing. Stone, T. A., Schlesinger, P., Houghton, R. A., and Woodwell, G. M. (1994). A map of the vegetation of South America based on satellite imagery. *Photogramm. Eng. Remote Sens.* **60**(5), 541–551.

V. THREATS TO REMAINING NATURAL HABITATS

A. Terrestrial Ecosystems

1. Deforestation

Deforestation is the dominant transformation of forested ecosystems that threatens biodiversity. In Brazil, which holds most of the continent's remaining forests, ranching is the dominant use for land once deforested. In the 1990s, soybeans began to enter forested regions, representing a new force in this process (they had already been a major factor in transformation of the cerrado since the 1970s). The most important effect of soybeans is not loss of forest directly planted to the crop, but the extensive infrastructure of waterways, railways, and highways that are built to transport soybeans and the inputs needed to grow them. The cycle of deforestation that has repeatedly occurred along Amazonian highways can be expected to accompany these new access routes.

Population growth is a fundamental contributor to deforestation and other forms of natural habitat loss. In recent years, however, the redistribution of population through migration has overshadowed the impact of absolute growth in population size. These include migrations from the semiarid Northeast of Brazil to Amazonia, from Paraná to Rondônia, from the highlands of Bolivia, Peru, and Ecuador to the Amazonian lowlands and, in the case of Ecuador, to the Pacific lowlands as well.

2. Logging and Charcoal Manufacture

Logging is an increasingly important factor in Amazonia, and the catalytic role of this activity in increasing the flammability of the logged forest gives it potential impact far beyond its direct damage. So far, logging in Brazil has been dominated by domestic demand for sawn wood, plywood, and particleboard, which is almost entirely supplied from tropical forests rather than from silvicultural plantations (which produce wood for pulp and, to a lesser extent, charcoal). However, global markets for tropical timber are presently dependent on supplies from Asian forests that will soon come to an end if current rates of exploitation continue. In the 1990s, Asian logging companies began buying land and/or obtaining concessions in such countries as Brazil, Guyana, and Suriname, and pressure from global timber markets can be expected to increase in the future. Asian loggers are also the principal forces in clear-cutting the Valdivian and *Nothofagus* forests of Chile (ecoregions 88 and 89).

In eastern Amazonia, demand for charcoal for pig-iron smelting in the Carajás area is a potential threat to forests. Carajás, with the world's largest deposit of high-grade iron ore, is expected to be mined for 400 years at the present rate of exploitation. Wood from native forests is inherently cheaper as a source of biomass for charcoal production as compared to plantation-grown sources. Charcoal manufacture has an impact on the forest both through direct removal (including officially sanctioned forestry management systems) and by increasing the profitability of logging and deforestation (see Anderson, 1990).

Deforestation impacts are magnified by fragmentation and edge effects (Laurance and Bierregaard, 1997). This division of the remaining natural habitat into many small islands surrounded by cattle pastures or other highly modified land uses, together with forming edges with increased entry of light, wind, and foreign organisms, results in many changes in the remaining natural ecosystems. Most of these changes are forms of degradation, such as greatly increased mortality in the trees that provide the dominant component of forest structure. Vine loads on trees near edges also increase, leading to further increase in mortality and susceptibility to windthrow.

3. Other Threats

Climate change represents a major long-term threat to many South American ecosystems. The Intergovernmental Panel on Climate Change (IPCC) has prepared detailed reviews of potential climatic impacts on South America in its 1998 Special Report on Regional Impacts (Chapter 6) and its 2000 Third Assessment Report (Working Group II, Chapter 14).

Removal of fauna through hunting is a virtually universal consequence of proximity of human settlements to natural habitats. The removal of fauna can affect seed dispersal, pollination, and other processes needed for maintaining plant and animal communities. Introduction of exotic species also represents a threat to natural ecosystems. Exotic species are a particularly severe problem in the Valdivian and *Nothofagus* forests of Chile (ecoregions 88 and 89).

Mangrove ecosystems are subject to some unique threats. Shrimp culture in mangrove areas has had severe impacts on the coast of Ecuador. Mangroves in Maranhão have been subject to pressure for charcoal manufacture. In São Paulo state mangroves have often suffered from oil spills and are also losing ground to real estate development. This has also affected restingas (ecoregions 176–178).

B. Aquatic Ecosystems

1. Dams

Hydroelectric dams have major impacts on river ecosystems by blocking fish migration, by eliminating rapids and replacing well-oxygenated running water with reservoirs that usually have anoxic water in their lower layers. The composition of fish present changes radically and undergoes a succession of changes as reservoirs age. Anoxic water released through the turbines severely reduces fish and freshwater shrimp productivity in the rivers downstream of the dams.

In Brazil, the 2010 Plan, released in 1987, listed over 300 dams for eventual construction in Brazil, independent of the expected date of completion. Of these, 65 dams were in the Amazon region. Economic difficulties have caused projected construction dates to be successively postponed, but the ultimate number of dams has not changed. Most contentious is the Babaquara Dam on the Xingu River, which would flood over 6000 km^2 of forest, much of it in indigenous areas. This has been renamed the "Altamira Dam" and appears in the current decennial plan for construction by 2013.

In Chile, the dams planned and under construction on the Bio-Bio River are expected to have major environmental impacts. The Ralco Dam is particularly contentious. In Uruguay, at least five major dams are planned for construction in the next few years.

2. Waterways

Industrial waterways, known as *hidrovias* in Brazil, greatly alter aquatic habitats. No less than seven waterways are under construction or planned for soybean transport on barges: the Paraguay–Paraná (*Hidrovia do Pantanal*), the Madeira River waterway, the Tocantins–Araguaia waterway, the Teles Pires–Tapajós waterway, the Capim River waterway, the Mamoré–Guaporé waterway, and the Rio Branco and Rio Negro–Orinoco waterways. Waterway construction involves blasting rock obstructions, cutting sharp curves, and dredging sediment from the river beds. The Corumbá–Cáceres stretch of the *Hidrovia do Pantanal*, if built, would lower the water level in the Pantanal wetlands (ecoregion 133), threatening one of the world's most renowned concentrations of wildlife.

3. Other Threats

Other threats to aquatic habitats include sedimentation from soil erosion and landslides. This is severe, for example, in rivers draining steep areas of former Atlantic forest in the coastal mountains of Brazil. Mining for gold, tin, and diamonds in Amazonia can also inject large amounts of sediment into streams and rivers.

Destruction of varzea forest (ecoregion 33) in Amazonia can affect aquatic life through loss of important fish breeding areas and food sources for fruit- and seed-eating fish. Destruction of varzea lakes and overfishing represent additional threats.

VI. STATUS OF PROTECTED AREAS

The choice and design of reserves depend on the financial costs and biodiversity benefits of different strategies. In Brazil, rapid creation of lightly protected "paper parks" has been a means of keeping ahead of the advance of barriers to establishment of new conservation units, but emphasis must eventually shift to better protection of existing reserves (Fearnside, 1999).

Creating reserves that include human occupants has a variety of pros and cons (Kramer *et al.*, 1997). Although the effect of humans is not always benign, much larger areas can be brought under protection regimes if human occupants are included. Additional considerations apply to buffer zones around protected areas. A "fortress approach," whereby uninhabited reserves are guarded against encroachment by a hostile population in the surrounding area, is believed to be unworkable as a means of protecting biodiversity, in addition to causing injustices for many of the human populations involved.

VII. PRIORITIES FOR CONSERVATION

Indigenous peoples have the best record of maintaining forest, but negotiation with these peoples is essential in order to ensure maintenance of the large areas of forest they inhabit (Fearnside and Ferraz, 1995). The benefits of environmental services provided by the forest must accrue to those who maintain these forests. Development of mechanisms to capture the value of these services will be a key factor affecting the long-term prospects of natural ecosystems.

In the case of deforestation in Amazonia, a variety of measures could be taken immediately through government action, including changing land tenure establishment procedures so as not to reward deforestation, revoking remaining incentives, restricting road building and improvement, strengthening requirements for environmental impact statements for proposed development projects, creating employment alternatives, and, in the case of Brazil, levying and collecting taxes that discourage land speculation. A key need is for a better informed process of making decisions on building roads and other infrastructure such that the full array of impacts is taken into account.

Environmental services represent a major value of natural ecosystems, and mechanisms that convert the value of these services into monetary flows that benefit the people who maintain natural habitats could significantly influence future events in the region (Fearnside, 1997b). Environmental services of tropical forests include maintenance of biodiversity, carbon stocks, and water cycling. The water cycling function, although very important for countries in the region, does not affect other continents as the first two services do. At present, avoiding global warming by keeping carbon out of the atmosphere represents a service for which monetary flows are much more likely to result from international negotiations. Activities under the United Nations Framework Convention on Climate Change (UN-FCCC) are at a much more advanced stage of negotiation than is the case either for the Biodiversity Convention or for the "Non-Binding Statement of Principles" and possible future convention on forests.

In the case of carbon, major decisions regarding credits for tropical forest maintenance are likely to be taken at the sixth Conference of the Parties (COP-6) to the Kyoto Protocol, at the end of 2000, considering the IPCC Special Report on Land Use, Land-Use Change and Forestry (SR-LUCF), released in May 2000. Regardless of what is decided at COP-6, global warming is a permanent consideration that can be expected to receive increasing weight in decision making. The threats to natural ecosystems in South America are many, and recognition of the multiple environmental services provided by them is a key factor in ensuring that substantial areas of each of these ecosystems continue to exist, thereby maintaining their biodiversity.

Acknowledgments

I thank Eric Dinerstein and the World Bank for permission to publish Fig. 1 and Table I, and Tom Stone and the American Society for Photogrammetry and Remote Sensing for permission to publish Table III. Brazil's National Council of Scientific and Technological Development (CNPq AI 523980/96-5) and National Institute for Research in the Amazon (INPA PPI 1-3160) provided financial support. S. V. Wilson and two anonymous reviewers made helpful comments on the manuscript.

See Also the Following Articles

AMAZON, ECOSYSTEMS OF • DEFORESTATION AND LAND CLEARING • FIRES, ECOLOGICAL EFFECTS OF • GRAZING, EFFECTS OF • INDIGENOUS PEOPLES, BIODIVERSITY AND • LOGGED FORESTS • RAINFOREST LOSS AND CHANGE

Bibliography

Anderson, A. B. (Ed.) (1990). *Alternatives to Deforestation: Towards Sustainable Use of the Amazon Rain Forest*. Columbia Univ. Press, New York.

Dinerstein, E., Olson, D. M., Graham, D. J., Webster, A. L., Primm, S. A., Bookbinder, M. P., and Ledec, G. (1995). *A Conservation Assessment of the Terrestrial Ecoregions of Latin America and the Caribbean*. The World Bank, Washington, D.C.

FAO (Food and Agriculture Organization of the United Nations). (1993). *Forest Resources Assessment 1990: Tropical Countries* (FAO Forestry Paper 112). FAO, Rome, Italy.

Fearnside, P. M. (1993). Deforestation in Brazilian Amazonia: The effect of population and land tenure. *Ambio* 22(8), 537–545.

Fearnside, P. M. (1997a). Limiting factors for development of agriculture and ranching in Brazilian Amazonia. *Rev. Brasil. Biol.* 57(4), 531–549.

Fearnside, P. M. (1997b). Environmental services as a strategy for sustainable development in rural Amazonia. *Ecol. Econ.* 20(1), 53–70.

Fearnside, P. M. (1999). Biodiversity as an environmental service in Brazil's Amazonian forests: Risks, value and conservation. *Environ. Conserv.* 26(4), 305–321.

Fearnside, P. M., and Ferraz, J. (1995). A conservation gap analysis of Brazil's Amazonian vegetation. *Conserv. Biol.* 9(5), 1134–1147.

Harcourt, C. S., and Sayer, J. A. (Eds.) (1996). *The Conservation Atlas of Tropical Forests: The Americas*. Simon & Schuster, New York.

Heywood, V. H., and Watson, R. T. (Eds.). (1995). *Global Biodiversity Assessment*. Cambridge Univ. Press, Cambridge, UK.

Kramer, R., van Schaik, C., and Johnson, J. (Eds.) (1997). *Last Stand: Protected Areas and the Defense of Tropical Biodiversity*. Oxford Univ. Press, Oxford.

Laurance, W. F., and Bierregaard, R. O. (Eds.) (1997). *Tropical Forest Remnants: Ecology, Management, and Conservation of Fragmented Communities*. Univ. of Chicago Press, Chicago, IL.

Nepstad, D. C., Moreira, A. G., and Alencar, A. A. (1999). *Flames in the Rain Forest: Origins, Impacts and Alternatives to Amazonian Fire*. Pilot Program to Conserve the Brazilian Rain Forest, Brasilia, Brazil.

Rudel, T. K., and Horowitz, B. (1993). *Tropical Deforestation: Small Farmers and Land Clearing in the Ecuadorian Amazon*. Columbia Univ. Press, New York.

Stone, T. A., Schlesinger, P., Houghton, R. A., and Woodwell, G. M. (1994). A map of the vegetation of South America based on satellite imagery. *Photogramm. Eng. Remote Sens.* 60(5), 541–551.

Stotz, D. F., Fitzpatrick, J. W., Parker, T. A., III, and Moskovitz, D. K. (1996). *Neotropical Birds: Ecology and Conservation*. Univ. of Chicago Press, Chicago, IL.

SOUTHERN (AUSTRAL) ECOSYSTEMS

Robert S. Hill* and Peter H. Weston[†]
*University of Adelaide and [†]Royal Botanic Gardens, Sydney

I. Introduction
II. History
III. Patterns
IV. Processes
V. Conclusion

GLOSSARY

continental drift The movement of continents as a result of plate tectonic processes. This was especially important for Southern Hemisphere continents.
disturbance A set of processes that lead to an ecological point crisis, that is, an event or events that are not part of the normal environment for the organisms that are influenced by it and that can be catastrophic in its impact.
Eucalyptus The dominant tree genus in Australia today.
Gondwana The southern supercontinent, named after an ancient kingdom in India. Gondwana rifted apart over many millions of years, giving rise to the current Southern Hemisphere landmasses.
Nothofagus A genus of Southern Hemisphere trees with a classic "Gondwanic" distribution in southern South America, southeastern Australia, New Zealand, New Caledonia, and New Guinea.

IN THIS ARTICLE we describe the biogeographic patterns on which the concept of southern ecosystems is based. We outline how several different research traditions have interpreted those patterns, briefly discussing the strengths and weaknesses of each approach. Finally, we discuss the evolutionary, geological, and climatic processes that might have caused or constrained the patterns. It is easy to show that the living biotas of these landmasses are the result of their post-Gondwanic history, and by comparing them, we demonstrate processes that influenced the entire hemisphere during this phase leading to the modern world. We concentrate mostly on Australia, where both the extant biota and the fossil record are relatively well documented, comparing the other landmasses to it. The impact of European settlement in the Southern Hemisphere is not considered, since despite its magnitude it is well documented elsewhere.

I. INTRODUCTION

A large number of plant taxa are shared by New Zealand, southern South America, and southeastern Australia (including Tasmania) but are found nowhere else. Joseph Hooker was so impressed by this pattern that in 1853 he wrote in his *Introductory Essay* to the *Flora Novae-Zelandiae*:

"... many of the peculiarities of each of the three great areas of land in the southern latitudes are representative ones, effecting a botanical relationship as strong as that which prevails throughout the lands within the Arctic and Northern hemisphere zones, and which is not to be accounted for by any theory of transport or variation, but which is agreeable to the hypothesis of all being members of a once more extensive flora, which has been broken up by geological and climatic causes."

Thus was born the concept of "southern (austral) ecosystems." Hooker knew nothing of the geological basis for the patterns that he observed but he thought that they demanded explanation. We now know that the "once more extensive flora" about which he speculated did exist and that the "geological and climatic causes" that he hypothesized were the consequences of continental drift. "Southern (austral) ecosystems" thus comprise an historical entity rather than an ecological class as one finds in a volume of *Ecosystems of the World*.

The Southern Hemisphere is dominated by oceans, and the development of these oceans is critical in understanding the evolution of terrestrial biodiversity of this part of the world. All major southern landmasses, along with some that are now mostly in the Northern Hemisphere, were once part of the supercontinent Gondwana. The breakup of Gondwana, and the consequent movement of the continents that were once part of it, produced vicariant distribution patterns, as observed by Hooker, but it also had other profound effects on the Gondwanic biotas.

II. HISTORY

In 1859, Joseph Hooker further emphasized the close relationship between the floras of the southern landmasses in his *Introductory Essay* to the *Flora of Tasmania* and persisted in maintaining that this must be the relic of a continuous, ancestral flora. His friends and colleagues soon disagreed. In 1859, Charles Darwin in his *Origin of Species* cited Hooker's austral floristic pattern but not his geological speculations, explaining the pattern primarily as the result of repeated, independent episodes of long-distance dispersal across the Southern Ocean. In 1880, Alfred Russell Wallace in his *Island Life*, while noting that a number of animal taxa, such as the ratite birds, also showed an austral distributional pattern, chose to explain this as the result of waves of southward dispersal from the Northern Hemisphere, followed by extinction in the north. He agreed with Darwin that the austral floristic relationship must be the result of long-distance dispersal. Neither had any time for geological instability on the scale that Hooker had proposed.

These pioneering biogeographers spawned different, competing research traditions that remain alive and distinguishable today. Hooker inspired an intellectual lineage of phytogeographers to search for repeated distributional patterns and to postulate general explanations for them, invoking dispersal and vicariance of entire floras. This approach was applied in extraordinary breadth by Croizat in his "panbiogeographic" analyses of numerous plant and animal distributions. Croizat's work produced not only a general distributional pattern ("generalized track") crossing the Southern Ocean, consistent with Hooker's analysis, but plenty of other general patterns too, including tracks spanning all ocean basins. After 100 years, the Hookerian tradition was still looking for an explanatory geological theory.

A competing (and for long, dominant) research tradition, inspired by Darwin and Wallace, flourished among zoogeographers, paleontologists, and many phytogeographers. The primary aim of this approach was to construct plausible scenarios describing the centers of origin and dispersal histories for particular taxa, based chiefly on their patterns of occurrence in the fossil record, knowledge of their phylogeny, the ecology of extant species, and knowledge of geological history. An example of this tradition is the work of Darlington (1965).

Corroboration of the theory of continental drift through the discovery of plate tectonics in the 1960s vindicated Hooker's approach by providing the former land connections that he had predicted, in the form of the ancient supercontinent Gondwana. The transformation from "austral flora" to "Gondwanan biota" did, however, involve a substantial expansion in scope. Its potential boundaries shifted northward to include the whole of South America and Africa, Madagascar, India, all of Australia plus southern New Guinea, and New Caledonia. Refinements in geological reconstruction progressively added more fragments to Gondwana, such as eastern Sulawesi, the "Outer Melanesian Arc" (Solomons to Vanuatu and Fiji), and, eventually, large parts of eastern Asia.

The immediate effect of this geological revolution on biology was to flip the conventional null hypothesis in historical biogeography from long-distance dispersal to dispersal over land. "Darwinian" biogeographers changed the paleogeographic maps on which they plotted dispersal routes but continued to use the same methods as before. A new question was added to their task list: "To what extent can extant distributions be explained by continental drift?"

Different problems exercised the minds of "Hookerian" biogeographers. While a number of Croizat's tracks were corroborated in surprising detail as vicariant patterns, others, such as his trans-Pacific tracks, remained geologically anomalous. It could be argued that all approaches to historical biogeography had been shown by continental drift to be inadequate in some way or other.

The heuristic success of Croizat's track method and the emergence of new cladistic techniques for reconstructing phylogeny inspired the development of a new approach, cladistic (or vicariance) biogeography, in the 1970s and 1980s (see Humphries, this volume). This is based on the idea that historical biogeographic relationships can be summarized in the same way as phylogenetic relationships: in the form of tree diagrams. However, cladistic biogeographic trees connect areas of endemism occupied by taxa, rather than the taxa themselves. Methods of analysis have been developed that reconcile "area cladograms" for different taxa to produce "general-area cladograms," which summarize episodes of biotic dispersal and fragmentation. The optimistic expectation of this approach is that general patterns will dominate such analyses, producing a small number of general-area cladograms. Several cladistic biogeographic analyses of austral biotas have been published, most of which reproduce expanded, cladistically resolved versions of Hooker's austral biota.

Another new approach requiring detailed phylogenetic knowledge is based on the assumption of "molecular clocks." If a set of homologous DNA sequences, sampled from a clade of biogeographic interest, has accumulated site mutations at a constant rate, then it can be used to calculate the relative age of speciation events. Such a clock can be "calibrated" by reference to the minimum age of one or more clades, as indicated by the fossil record. This method requires copious amounts of sequence data and few applications relevant to austral biogeography have been published. These have suggested that overland dispersal, vicariance, and long-distance dispersal have all contributed to Gondwanic patterns.

III. PATTERNS

In 1853, Hooker counted 77 plant species and "upwards of 100 genera, subgenera, or other well-marked groups of plants" entirely or nearly confined to New Zealand, Australia, and extratropical South America. His analysis has not been replicated but subsequent authors have added additional taxa without attempting to be comprehensive. For example, Crisci et al. (1991) added 10 vascular plants as well as 41 invertebrate taxa, one vertebrate family, and one genus of fungi. Some of these additions were taxa found in New Caledonia, New Guinea, and southern Africa, and hence ignored by Hooker. Some include numerous subclades showing overlapping Gondwanic distributions and therefore are more informative than other taxa. The Diamesine and Podonomine Chironomid midges are classic examples of such groups (Brundin, 1966), as is the plant family Proteaceae, for which Weston and Crisp (1996) list 12 higher taxa showing separate, overlapping Gondwanic distributions.

Clearly, the pattern on which Hooker originally based his concept of an austral flora has not been weakened by further research, but has been substantially strengthened by the addition of more botanical examples as well as a large number of zoological ones. What other patterns are relevant to the reconstruction of Gondwanic biogeographic history? The relationships between distributional patterns and other variables such as means of dispersal, area, and climatic factors have been discussed in great detail by Darwinian biogeographers. General-area cladograms that specify the putative splitting sequence of Gondwanic biotas are potentially useful as are estimates of the age of particular taxa derived from molecular clocks. Finally, biogeographers of all persuasions have sought to explain patterns of occurrence of austral taxa in the fossil record.

A. Distributional Patterns in Relation to Means of Dispersal, Climate, and Area

Darwinian biogeographers emphasized a number of patterns that they thought outweighed Hooker's austral pattern and these were reviewed by Darlington (1965). He pointed to the low number of vertebrate taxa that belong to the austral biota: marsupials, chelyid turtles, leptodactylid and hylid frogs, and galaxiid fishes, and of these, only the galaxiids occur in New Zealand. If these dispersed over land between Australia and South America, then why did no other vertebrate groups follow them? Why did hardly any mammals, reptiles, or amphibians walk to New Zealand? Darlington considered plants and insects, which comprise the great majority of austral taxa, to have more effective means of dispersal than most terrestrial vertebrates.

Darlington also argued that climatic patterns were highly significant in explaining austral distributions. The southern parts of South America, Tasmania, and

New Zealand are cold and their western sides very wet because of the strong prevailing westerly winds. The austral biota (as circumscribed by Hooker) is largely restricted to these areas. Moreover, the strong westerly winds blow directly over parts of Tasmania, New Zealand, and southern South America. The area of land in the Southern Hemisphere tapers off markedly to the south, greatly restricting the amount of land available to the austral biota. This suggested to Darlington that the "southern end of the world" would be a poor evolutionary center. Biological diversity also tapers off to the south. The extreme southern parts of Australia, New Zealand, and South America all have lower biological diversity per unit area than the less climatically extreme areas further north.

B. General-Area Cladograms

Do Gondwanic landmasses share a common ancestral biota to the exclusion of Laurasian landmasses? What are the cladistic interrelationships between different Gondwanic biotas? These are the kinds of questions that cladistic biogeographers are inclined to ask about the Southern Hemisphere. Among the first studies to focus primarily on these questions was that of Humphries (1981), which was based on cladograms for *Nothofagus* and 24 other taxa. Most of these cladograms were derived from precladistic taxonomic treatments and all relied entirely on morphological evidence. Nevertheless, he discerned two general cladistic patterns, summarized as:

1. (((((Australia, New Guinea, Tasmania, New Zealand, New Caledonia) South America) Africa) (North America, Europe))
2. ((Australia, New Guinea) (South America, North America, Europe))

He concluded that "the two positions for South America ... are probably due to the fact that it is a huge composite area and shouldn't be treated as a single area of endemism" (Humphries, 1981, p. 205).

More recent attempts to resolve a general-area cladogram for Gondwanic landmasses have tended to subdivide South America, and sometimes other continents and islands such as Australia, New Zealand, and New Caledonia, into two or more areas for purposes of analysis. They have also been able to use more rigorously produced cladograms, some of which have incorporated molecular data.

For example, Crisci *et al.* (1991) analyzed 17 cladograms (8 insect, 8 angiosperm, 1 fungal) and found that southern South America, New Zealand, Tasmania, Australia, New Caledonia, and New Guinea consistently grouped together as an austral biota, to the exclusion of northern South America, North America, and Africa. Interrelationships within the austral group, however, remained ambiguous.

Muona (1991), on the other hand, intuitively examining cladograms for 84 genera of Eucnemid beetles, concluded that "fifty-seven genera shared a pattern coinciding with the traditional model of Laurasia–Gondwana breakup" (Muona, 1991, p. 165). He did add, however, that "twenty five groups showed an anomalous feature within the Gondwanan pattern having an Indomalesian clade as the sister group of the Australia–New Guinea clade" (Muona, 1991, p. 174).

A feature common to all of these general-area cladograms is the clustering of most, but usually not all, Gondwanic landmasses. Africa, Madagascar, and northern and eastern South America often group with Laurasian areas such as Europe and North America. Within the austral cluster, relationships tend to be unstable. These oddities probably reflect some of the limitations of cladistic biogeographic methods, which handle hierarchical vicariant patterns elegantly, but which require ad hoc adjustment to cope with reticulate patterns. The latter may be due to long-distance dispersal or secondary mixing of biotas caused by geological processes such as continental collision.

C. Molecular Clocks

The obvious limitation of the molecular clock is its reliance on an assumption that is known to be false: a constant rate of molecular evolution. Molecular rates differ within and between genes and among lineages but the idea of a biological dating technique analogous to radioactive decay is so attractive that many are prepared to tinker with the assumption or with their sampling in order to apply it to even a limited extent.

Waddell *et al.* (1999), for example, applied relative rate tests to mitochondrial DNA sequences for mammals and birds in order to select the largest subset of sequences consistent with the assumption of rate equality. Unfortunately, this procedure consigned their only marsupial sequence to the dustbin. However, they did manage to get two ratite sequences into their bird analysis and concluded that the Rhea and Ostrich are too similar for their most recent common ancestor to have walked between Africa and South America.

Bremer and Gustafsson (1997) took a different approach. They relaxed the assumption from rate homogeneity to minimal heterogeneity in their analysis of rbcL sequences for the plant order Asterales. They concluded that this clade is of mid to late Cretaceous age. This implies, for instance, that the austral distribution pattern of the family Carpodetaceae may be of Gondwanic origin while those of the austral daisy genera are all due to long-distance dispersal.

Manos (1997) was more cautious in interpreting relative levels of molecular divergence in *Nothofagus*, finding that rate heterogeneity was significant within and between subgenera. He concluded (p. 1148) "a combined gene approach using multiple substitution rates and various calibration points taken from the fossil and geologic record may be able to statistically support or refute long-distance dispersal as a biologically realistic event in the biogeographic history of *Nothofagus*."

Molecular clocks look as though they will be useful in rejecting vicariance as an explanation for very recently dispersed, disjunct taxa but much less decisive in detecting ancient dispersal events.

D. Fossil Record

The patterns of occurrence of taxa in the fossil record provide evidence of the minimum ages of taxa and often indicate that they previously had more extensive distributional ranges than today. In most cases, absence of fossils of a taxon from deposits of a particular age or geographic location cannot reasonably be construed as evidence that the taxon really did not exist at that time or place. Absence of fossil evidence can also be due to incompleteness of the fossil record. However, a few taxa fossilize so readily and abundantly and are so reliably identified as fossils that their absence from the record has to be taken seriously as an indication of either true absence or extreme rarity. The angiosperm genus *Nothofagus* falls into this class (see below).

The fossil record has other limitations, the most restrictive of which is the relatively low proportion of taxonomic characters that are preserved. Vertebrates, for example, are usually only represented by skeletal remains. Their "soft parts" and DNA are rarely preserved. Similarly, many plant taxa are represented only by their fossilized pollen grains. While these sometimes display several distinctive synapomorphies diagnostic of particular clades, most often they do not. Moreover, pollen grains are such simple structures that the probability of convergent evolution of a distinctive pollen type in the ancestor of an extant clade and in a different, extinct taxon, while not being high, is not negligible.

Despite these limitations, the fossil record is a rich source of historical information. Rather than inadequately reviewing the paleontological literature pertaining to all Gondwanic taxa, we concentrate on examples of a couple of key angiosperms.

Probably no taxon has been as strongly associated with the concept of the austral biota as *Nothofagus*. In 1853, Hooker included two subgroups (as *Fagus*) in his list of Antarctic plant groups, Darlington (1965) came close to defining his "southern end of the world" on the basis of its distribution, and van Steenis considered it a key genus for plant geography. *Nothofagus* has an exceptionally detailed fossil record, with both pollen and macrofossils relatively common. At least five distinct types of fossil *Nothofagus* pollen are recorded and they include all the extant subgenera, *Lophozonia, Fuscospora, Nothofagus,* and *Brassospora,* respectively.

Nothofagus pollen is produced in abundance and fossil pollen is extremely common. If *Nothofagus* pollen is not present in sediments then the genus probably was not present in the immediate vicinity. Therefore, the only limitation to determining the early biogeographic history of the genus is the lack of sediments at critical times and places. To a certain extent, dating the earliest appearances of *Nothofagus* in early Campanian sediments of southern Gondwana (Dettmann et al., 1990) by pollen assemblages alone could involve some circularity, since they are sometimes indicator species for the pollen stratigraphy. However, in both southern Australia and the Antarctic Peninsula, where the earliest *Nothofagus* pollen has been found, the grains are preserved in marine sediments that are independently dated using dinoflagellate and invertebrate fossil assemblages. Earliest records of *Nothofagus* pollen date from the early Campanian, making *Nothofagus* one of the oldest extant angiosperm genera in the fossil record. Records of this pollen are from (in the order they appear) Australia, West Antarctica, New Zealand, and South America (Dettmann et al., 1990). By the early Maastrichtian, the *Nothofagus* clade had diversified into all four modern subgenera, first appearing in West Antarctica, shortly thereafter in South America, and subsequently in Australia some 12–15 million years later (Dettmann et al., 1990). According to the pollen records, the last of the modern subgenera to appear in Australia was either the most basal subgenus *Lophozonia* or the advanced *Brassospora*. New Caledonia and New Guinea are devoid of fossils other than subgenus *Brassospora*, the current earliest record being from the Miocene in New Guinea (Dettmann et al., 1990).

In contrast to *Nothofagus*, the eucalypts (Myrtaceae:

Eucalyptus alliance) have only been proposed as a Gondwanic clade relatively recently. Like many other plant taxa, the pollen grains of most eucalypts are not particularly distinctive and may be confused with those of other myrtaceous groups such as *Syncarpia* and *Metrosideros* (Hill, 1994). They are pollinated by insects, birds, and mammals and produce fewer, less widely dispersed pollen grains than wind-pollinated *Nothofagus*.

Phylogenetic studies (e.g., Udovicic *et al.*, 1995) have resulted in the recognition of seven genera of eucalypts: *Eucalyptus*, with more than 500 species, nearly all of which are endemic to Australia (several species occur in New Guinea, one of which extends west and northwest to Sulawesi and Mindanao, several occur in the lesser Sunda Islands as far west as Flores); *Corymbia*, with 113 species, restricted to Australia and New Guinea; *Angophora*, with about 20 species restricted to eastern Australia; *Arillastrum* with one species endemic in New Caledonia; *Eucalyptopsis* with one species endemic in New Guinea; *Allosyncarpia*, with one species restricted to Arnhem Land in northern Australia; an unnamed species most closely related to the previous two genera and surviving as a single population in the Queensland wet tropics. Unfortunately, fossil names for eucalypts have not yet been changed to match this new classification, and so all fossils mentioned here are referred to *Eucalyptus*.

In form the species range from low shrubs to the tallest flowering plants in the world. Eucalypts have been estimated to contribute 75% of Australian vegetation, and at the wetter margins of the continent, they dominate nearly all vegetation except rain forest and allied mesic types. Only in the arid Australian interior are eucalypts generally lacking in dominance.

Given their diversity and current dominance, the eucalypts would be expected to yield a complex and challenging fossil record. That this is not the case has been well documented (Hill, 1994). The oldest reliably dated and described Australian macrofossils associated with *Eucalyptus* include tree stumps, leaves and umbels about 20–22 million years old from several sites in the southeast of Australia (Hill, 1994).

The early pollen record of *Eucalyptus* is enigmatic mostly because of the difficulty in identifying eucalypt pollen. However, probable eucalypt pollen is now recorded from the late Paleocene of the Lake Eyre Basin in inland Australia (Hill, 1994).

An intriguing aspect of the fossil record of *Eucalyptus* is its apparent presence in both South America and New Zealand. A group of three fruits from probable Eocene sediments in Patagonia have been described as *Eucalyptus patagonica*, although there is some doubt as to whether they belong to the *Eucalyptus* alliance (Hill, 1994). Leaves from early Miocene sediments in the South Island of New Zealand are very similar to many *Eucalyptus* species and occur with eucalypt-like fructifications. This record is far more reliable at present than that from South America, especially when it is considered in conjunction with the pollen record, since *Eucalyptus* pollen in New Zealand occurs from the Miocene–early Pleistocene, suggesting a relatively recent extinction.

IV. PROCESSES

The complex patterns described here for the Southern Hemisphere biota suggest complex processes, and we still have much to learn about the way in which the Southern Hemisphere biota was molded into its living form. In the following section we briefly mention some of the more likely processes that must have had a significant impact.

A. Continental Drift

Although the precise detail of the breakup of Gondwana is still being refined, our broad understanding is robust and very important for the history of the biota. When continent-sized landmasses rift apart there must be profound direct and/or indirect consequences for the biota. The most obvious direct consequence is the physical separation of once continuous populations of species, which must have often led to allopatric speciation as conditions on the resultant landmasses changed relative to one another. Indirect consequences include changes in climate as the various land masses move into and out of different climatic zones and the altered relative position of the continents can also influence climate, for example by altering ocean currents on a massive scale.

B. Long-Distance Dispersal

The Southern Hemisphere is dominated by oceans, which provide formidable barriers to dispersal. However, it seems likely that many organisms have successfully crossed these ocean barriers and formed successful populations at the other end. This is likely to have been in the direction of prevailing winds. While long-distance dispersal is a fact of life, it is very difficult to provide clear evidence for it and so it is largely recorded on a narrative basis. However, the methodological inadequacy of long-distance dispersal as a general explanation should not be taken to imply its unreality as a process.

Another example of long-distance dispersal that may have occurred predominantly overland involves north to south dispersal in more recent times. As the Gondwanic landmasses moved close to their present positions, the opportunity for relatively simple, predominantly land-based dispersal opened in three major regions: from Southeast Asia into Australia, from North America to South America, and from Europe to Africa. There are many compelling examples of such dispersals.

C. Climate

The climate history of the Southern Hemisphere is relatively well known and has clearly undergone major change during the rifting of Gondwana. This has been due to a combination of effects, including altered ocean currents, changing atmospheric CO_2 levels and shifting continental positions. The most obvious place where this has had an impact is Antarctica, and this makes a useful case study to demonstrate the potential impact of climate change.

Antarctica has a long history as a continent close to the South Pole. Throughout the rifting of Gondwana, Antarctica has been one of the more stable landmasses, and although its history is complex, it can be considered as a long-standing high-latitude region. The idea that Antarctica was a source of plants and animals for surrounding high-latitude landmasses such as southern South America, southern Australia, and New Zealand is central to Southern Hemisphere biogeography. However, Antarctica today not only is geographically isolated from the other southern landmasses, but is also far too climatically hostile to support the biota concerned.

Antarctica now has only a very sparse covering of plants near the coastline, only two of which are vascular. Most of the continent is without plant cover, and where terrestrial plants occur, they are usually lichens and mosses. The Antarctic vertebrate fauna is sea-based, using the land only temporarily. In contrast, there is abundant evidence from Antarctica and adjoining landmasses to demonstrate the presence of complex biotas near the South Pole at times in the past. The almost total lack of vascular plants on Antarctica today testifies to major extinctions on this continent. This rates as one of the most complete regional extinction events in the Earth's history, and it is important for us to understand it now more than ever, since it was indisputably a climate-based extinction.

There is now abundant fossil evidence for complex Cretaceous and Cenozoic forests on Antarctica that contain taxa that often still occur (or at least their descendants do) in high southern latitudes in South America, Australia, and New Zealand (Hill and Scriven, 1995). A combination of plate tectonics and other less well defined factors can be used to explain a very different climate at high latitudes in the past, and physiological research on the extant flora has demonstrated that diverse plant life was possible in Antarctica in the past without other physical changes.

During the Cretaceous at high southern latitudes massive speciation occurred, and there is no doubt that lineages were produced that today occupy niches from the equator down to the furthest southern latitudes at which the biota can survive.

The Cenozoic Antarctic vegetation is known from only limited data, but it appears that it continued to show basic similarities to Paleogene floras from Australia and South America. The separation of Australia from Antarctica began in the late Cretaceous, although opportunities for floristic interchange across restricted water gaps probably persisted well into the Cenozoic. Opening of the Drake Passage began in the late Oligocene, but details are lacking concerning the extent of dry land, probably as islands, between the Antarctic Peninsula and South America prior to that date.

The question of when the Cenozoic vegetation of Antarctica was eliminated by increasing cold and ice cover has not yet been answered by the fossil record. This is largely because of the confusion introduced into the spore and pollen record by the recycling of plant microfossils once processes of glacial erosion and sedimentation had begun. However, there is reasonable evidence for a cover of temperate forest in the Eocene, perhaps over much of Antarctica. Truswell (1991) noted that sea-surface temperatures about Antarctica were close to 0°C after the end of the Eocene, allowing the formation of sea ice; temperatures on land were probably lower, and it is unlikely that anything but a highly specialized, restricted flora would have persisted for long in such conditions. However, sparse Oligocene palynofloras containing relatively abundant *Nothofagus* (at least three subgenera) and a low diversity of other vascular plants have been described from Antarctica (Hill and Scriven, 1995). The data indicate a coastal *Nothofagus* forest that was diverse enough to suggest an extension of temperate forest in the region beyond the Eocene.

Milder sea-surface temperatures adjacent to the lower latitude regions of the northern Antarctic Peninsula probably allowed the persistence there of some form of woody vegetation for even longer, and there are good fossil data to suggest vegetation cover into the late Oligocene at least.

The probable middle to late Pliocene Sirius Formation sediments contain leaves, wood, and pollen in an excellent state of preservation, and there can be no doubt that the plants which produced these fossils were growing *in situ* on the site at the time the fossils were deposited. The main fossil evidence is for a single species of *Nothofagus*, which had a low growth habit similar to woody plants at the extreme of their growing range in the Arctic today (Hill and Scriven, 1995). Therefore the evidence available suggests progressive extinction from an earlier vegetation in which *Nothofagus* was very diverse. It is important to note that the Sirius Formation fossils are well inland in Antarctica and were deposited at some altitude (although this is difficult to quantify). In that case it can be speculated that there may have been other, more diverse vegetation present at lower sites closer to the coast as recently as the late Pliocene.

D. Disturbance

The angiosperms (flowering plants) play a critical role in the current Southern Hemisphere biota, providing a substantial component of the biodiversity and also acting as an important part of the food chain and niche requirements for many other organisms. Angiosperms probably evolved in the early Cretaceous and radiated worldwide. Early angiosperms probably had generalized pollination and dispersal strategies which made them ideal for long-range dispersal by pioneering the fresh sedimentary surfaces of coastal deltas, lagoons, and tidal flats. During the early Cretaceous, these pioneering early angiosperms or angiosperm ancestors dispersed along coastlines out of the rift valley of West Gondwana (between Africa and South America).

Angiosperms were established generally in Gondwana by Barremian–Aptian times (Dettmann, 1989), and those lineages that were widespread in southern Gondwana by the close of the Cretaceous formed the foundation for the living austral floras. Dettmann (1989) has suggested that rift valley systems at high southern latitudes acted as migrational pathways for early angiosperms, and this may be because of the high disturbance level in these environments and the early successional nature of these angiosperms.

Some prominent angiosperm taxa from the early high southern latitude flora still exist today, and their ecology is well understood. A good example is *Nothofagus*, which has been studied extensively. South American *Nothofagus* does not usually regenerate continuously at low altitudes, especially at lower latitudes, where climatic conditions favor more complex forests and *Nothofagus*s seedlings cannot establish. Instead, *Nothofagus* relies on catastrophic disturbance, a relatively common phenomenon in the highly unstable Andes mountain chain and the Southern Alps of New Zealand, to provide a fresh, cleared substrate which is readily colonized by seedlings. This regeneration behavior is probably very similar to that which occurred early in *Nothofagus* history, when it either occupied dense forests on unstable sites and was unable to regenerate in the absence of disturbance, or was in more sparse forest on less equable sites where its seedlings could survive in the understory. Hill (1994) hypothesized that the large reduction in *Nothofagus* diversity at high latitudes today, especially away from the Andes, may be due to the relative modern stability of the landscape, which, coupled with the density of the forest cover, does not provide the right environment for the continued presence of many *Nothofagus* species.

Thus, at high southern latitudes during the Cretaceous–Paleogene, unstable habitats provided both a pathway for the migration of early angiosperms and also the potential for peripheral isolation of populations of widespread species, which may have been a critical factor for evolution in the region (Hill and Scriven, 1995).

A more recent form of disturbance has been the widespread impact of fire. In Australia, fire history as a major impact on the biota can be traced back to the mid Cenozoic at least, but its influence has been particularly profound only in the past 100,000 years or so. Fire is now very significant in areas of the Southern Hemisphere with hot, dry summers.

E. Plant Growth at High Southern Latitudes

Woody plants no longer grow at very high southern latitudes (e.g., Antarctica), and yet they were a key component of Cretaceous biotas very close to the South Pole. In the past, debate has centered on whether this was possible, since the photoperiods at such high latitudes require long, dark winter periods. The issue is more one of suitable temperatures than suitable photoperiods, but nevertheless, the question remains: How did plants grow and what form did the communities take in such an unusual photoperiod? It is a real limitation of the fossil record in general that behavioral traits such as dormancy rarely leave a clear imprint (Truswell, 1991). Nevertheless, information on physiological adaptations to life at high latitudes can be obtained, especially from fossil wood. Cretaceous gymnospermous wood from paleolatitudes probably higher than 70°S shows a consistent growth pattern, with increments, probably annual in origin, clearly delineated, and indicating a pronounced seasonal influence (Hill and Scriven, 1995). The amount of wood added annually was usually large, even when compared with modern low-latitude trees, but there is considerable variation in ring widths from year to year, suggesting that the trees were highly sensitive to environmental fluctuations (Truswell, 1991). A fossil forest at one early Cretaceous site had trees spaced 3–5 m apart and the tallest preserved trunk height is 7 m. Hill (1994) used these data to explain the reduction in wind pollination in understory plants during the very early Paleogene in southern Australia. With the sun at a low angle in the sky during summer and tracking an almost circular path around the horizon during the day, he concluded that it was likely that the forest structure was quite different to that observed in heavily forested regions at lower latitudes today. Relatively widely spaced, conical trees probably dominated and this may have continued until at least the late Cretaceous, and thus wind pollination was a viable strategy for understory plants below the open canopy. As today's forest-bearing landmasses moved to lower latitudes and the sun angle increased, a closed forest structure became possible, due to the increase in incoming solar radiation. This may have been critical for the loss of wind pollination as a viable strategy in the associated understory plants.

The presence of a relatively dense plant cover, including large and rapidly growing trees, in regions south of the Antarctic circle, was for a long time viewed as difficult to explain, given that today growth at high latitudes is considered to be constrained by characteristic day-length patterns and by the temperatures associated with a regime of long rigorous winters and short cool summers (Truswell, 1991). However, it is now generally accepted that the light energy at high latitudes was sufficient, at present values of axial obliquity, to maintain forest growth in the past provided the temperatures, particularly in winter, were higher.

Enhanced atmospheric CO_2 levels may have been influential in producing the high annual increments of growth observed in Cretaceous trees. The overall effect of the high predicted levels of CO_2 in the Cretaceous on plant growth are difficult to quantify, since experiments carried out on living plants examine the instantaneous effects of large changes in CO_2 concentration on plants which have evolved to a particular ambient level. When plants have evolved to accommodate higher levels (as in the Cretaceous) the result may have been even more spectacular (Hill and Scriven, 1995).

V. CONCLUSION

The biota of the Southern Hemisphere is extremely diverse and is the result of a complex history. However, clear biotic linkages occur among the major landmasses. These linkages have historical significance, with vicariance and long-distance dispersal contributing to the geographic spread of organisms and climate change and disturbance being among the major processes that have led to the living species that occur there. Our knowledge of the history of the Southern Hemisphere is incomplete, but we have enough information from interrelationships among living organisms and the fossil record to be able to reconstruct at least part of this history with reasonable accuracy. The most important event in the history of the Southern Hemisphere as it influences the living biota was the rifting of Gondwana. Not only did this separate the biota on to smaller landmasses that subsequently had very different histories, but it also influenced other processes such as climate change and disturbance regime. This distinguishes the Southern Hemisphere biota very clearly from that of the Northern Hemisphere, and it is important not to presume that the processes in the two hemispheres were at all similar.

See Also the Following Articles

ANTARCTIC ECOSYSTEMS • AUSTRALIA, ECOSYSTEMS OF • SOUTH AMERICA, ECOSYSTEMS OF

Bibliography

Bremer, K., and Gustafsson, M. H. G. (1997). East Gondwana ancestry of the sunflower alliance of families. *Proc. Natl. Acad. Sci. USA* **94**, 9188–9190.

Brundin, L. Z. (1966). Transantarctic relationships and their significance as evidenced by chironomid midges. *K. Sven. Vetenskapsakad. Handl., Ser. 4*, **11**, 1–472.

Crisci, J. V., Cigliano, M. M., Morrone, J. J., and Roig-Junent, S. (1991). Historical biogeography of southern South America. *Syst. Zool.* **40**, 152–171.

Darlington, P. J. (1965). *Biogeography of the Southern End of the World*. Harvard Univ. Press, Cambridge, MA.

Dettmann, M. E. (1989). Antarctica: Cretaceous cradle of austral temperate rainforests? In *Origins and Evolution of the Antarctic Biota*. (J. A. Crame, Ed.). *Geol. Soc. Spec. Publ.* **47**, 89–105.

Dettmann, M. E., Pocknall, D. T., Romero, E. J., and Zamaloa, M. de C. (1990). *Nothofagidites* Erdtman ex Potonié 1960—A catalogue of species with notes on the palaeogeographic distribution of *Nothofagus* Bl. (southern beech). *NZ Geol. Surv. Palaeontol. Bull.*, No. 60.

Hill, R. S. (1994). The history of selected Australian taxa. In *History of the Australian Vegetation: Cretaceous to Recent* (R. S. Hill, Ed.), pp. 390–420. Cambridge Univ. Press, Cambridge, MA.

Hill, R. S., and Scriven, L. J. (1995). The angiosperm-dominated woody vegetation of Antarctica: A review. *Rev. Palaeobot. Palynol.* 86, 175–198.

Humphries, C. J. (1981). Biogeographical methods and the Southern Beeches (Fagaceae: *Nothofagus*). In *Advances in Cladistics* (V. A. Funk and D. R. Brooks, Eds.), pp. 177–207. The New York Botanical Garden, New York.

Manos, P. S. (1997). Systematics of *Nothofagus* (Nothofagaceae) based on rDNA spacer sequences (ITS): Taxonomic congruence with morphology and plastid sequences. *Am. J. Bot.* 84, 1137–1155.

Muona, J. (1991). The Eucnemidae of Southeast Asia and the western Pacific—A biogeographical study. *Aust. Syst. Bot.* 4, 165–182.

Truswell, E. M. (1991). Antarctica—A history of terrestrial vegetation. In *The Geology of Antarctica* (R. J. Tingey, Ed.), pp. 499–537. Clarendon, Oxford.

Udovicic, F., Ladiges, P. Y., and Drinnan, A. N. (1995). Eucalypt phylogeny–Molecules and morphology. *Aust. Syst. Bot.* 8, 483–497.

Waddell, P. J., Cao, Y., Hasegawa, M., and Mindell, D. P. (1999). Assessing the Cretaceous superordinal divergence times within birds and placental mammals by using whole mitochondrial protein sequences and an extended statistical framework. *Syst. Biol.* 48, 119–137.

Weston, P. H., and Crisp, M. D. (1996). Trans-Pacific biogeographic patterns in the Proteaceae. In *The Origin and Evolution of Pacific Island Biotas, New Guinea to Eastern Polynesia: Patterns and Processes* (A. Keast and S. E. Miller, Eds.), pp. 215–232. SPB Academic Publishing BV, Amsterdam.

SPECIATION, PROCESS OF

Guy L. Bush
Michigan State University

I. Speciation and Species
II. Reproductive Barriers to Gene Flow and the Evolution of Mate Recognition Systems
III. Two Modes of Speciation
IV. Summary

GLOSSARY

allopatric Populations, species, or taxa occupying different and disjunct geographical areas.
ecotone The boundary or transitional zone between adjacent ecological communities or biomes.
epistasis The interaction of non-allelic genes in which one gene (epistatic gene) masks the expression of another gene at a different locus.
genetic drift The occurrence of random changes in the gene frequencies of small isolated populations not due to selection, mutation, or immigration.
isolating mechanisms Any intrinsic or extrinsic mechanism or barrier that inhibits the free exchange of genes between populations.
pleiotropic gene A gene that has more than one independent phenotypic effect.
polyploid Having more than one set of homologous chromosomes.
sister species A pair of species that have arisen from a single speciation event; each is the other's closest relative.
sympatric Populations, species, or taxa that occur together in the same geographical area within the dispersal range of one another.

SPECIATION is the process by which new species are formed. Understanding the origin of species and their contribution to the origin of biodiversity requires an understanding of the process of speciation and an appreciation of the problem of how to define a species.

I. SPECIATION AND SPECIES

A. Speciation

The ultimate source of the earth's biodiversity is speciation, a process that occurs when gene flow is reduced sufficiently between sister populations to allow each to become irrevocably committed to different evolutionary lineages. Speciation results in the splitting of a lineage into two or more species (cladogenesis). However, when a species is transformed over time (anagenesis) by the acquisition of phenotypic and genetic modification, there is no increase in the number of species and thus no speciation event (Fig. 1a).

B. The Species Problem

Although the end product of speciation seems clear, defining when and under what conditions two popula-

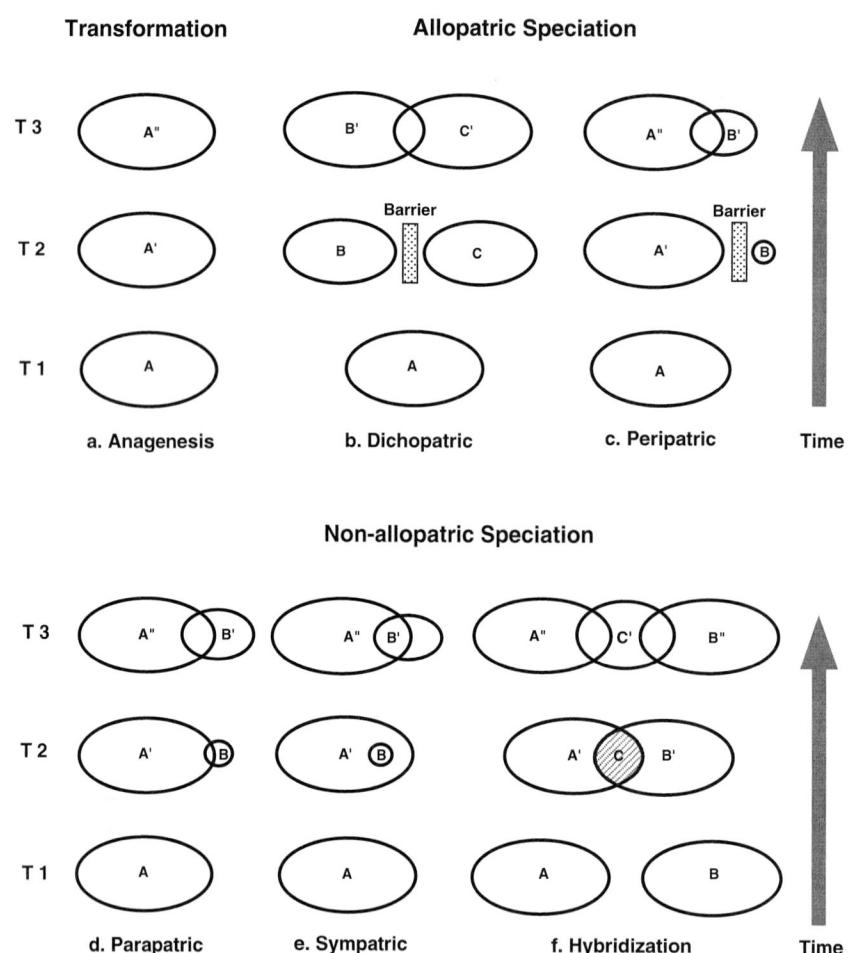

FIGURE 1 Major modes of speciation. See text for details on the processes involved in divergence and speciation.

tions become irrevocably committed to different evolutionary lineages has proved to be difficult and controversial. A frequently invoked species definition is the biological species concept (BSC) of Mayr (1942), who proposed that "species are groups of actually or potentially interbreeding natural populations that are reproductively isolated from other such groups" (p. 120).

However, the BSC and its later modifications have several limitations. It can be applied objectively only to sexually reproducing taxa and to natural populations that are sympatric (i.e., those that occur in the same geographic area) where they have the opportunity to interbreed. Asexually reproducing taxa and fossils must be treated subjectively. A more serious limitation of the BSC is that sympatric taxa that maintain distinct phenotypic and genotypic differences, but sustain some gene flow through interbreeding with close relatives, are generally not recognized as species. Recent molecular studies on closely related sympatric animal and plant taxa recognized as distinct, closely related sister species (the products of one speciation event) reveal that gene flow between them may still occur (Feder et al., 1995; Taylor et al., 1997). Such neospecies have acquired sufficient genetically based differences to maintain independent genotypes and distinct phenotypes but are still capable of sustaining some gene flow as they diverge. Several other species concepts have been proposed (Futuyma, 1998), but all suffer from problems of circularity, inconsistencies, and untestable qualities.

For this reason, the genetic-based species concept of Mallet (1995) provides a useful, less assumption-laden, operational species definition for identifying sister species. Mallet proposed that sister species be recognized only when they maintain distinctly different sympatric genotypic clusters. Defining a species based on genotypic clusters represents the level of differentiation most interesting to those studying speciation. Generally, sympatric sister taxa that maintain distinct genotypic clusters exploit different habitats or hosts. The wide use of modern DNA analysis in evolutionary studies makes this a useful approach particularly when coupled with sound biological studies. However, as with the BSC, using genotypic clusters to determine the biological status of geographically isolated sister populations without a "test of sympatry" is still a subjective although somewhat more quantitative call.

II. REPRODUCTIVE BARRIERS TO GENE FLOW AND THE EVOLUTION OF MATE RECOGNITION SYSTEMS

A. Reproductive Barriers

The free exchange of genes between populations adapting to different niches inhibits the evolution of independent genetic systems. As soon as adaptive gene combinations are formed, they are broken up by recombination through interbreeding. There are several biologically based reproductive barriers or "isolating mechanisms" that evolve during the course of speciation which limit gene flow between sister species (Table I). These biological barriers constitute components of a mate recognition system that promotes assortative mating between individuals adapted or adapting to shared environmental and reproductive conditions.

III. TWO MODES OF SPECIATION

There are two primary modes of speciation recognized by most evolutionary biologists (Table II). Allopatric (geographic) speciation occurs when sister populations are isolated for a period by a physical barrier such as a mountain range or an expanse of uninhabitable terrain or water. In the absence of gene flow, such isolated populations inevitably accumulate unique mutations and, over time, may diverge genetically by genetic drift and in response to divergent selection pressures. If sufficient genetic divergence occurs during isolation to establish and maintain distinct genotypic clusters, particularly when their ranges later overlap, then the taxa are regarded as species.

Non-allopatric (ecological) speciation occurs in the absence of physical barriers to gene flow when sister populations diverge genetically and become ecologically and reproductively isolated as they adapt to different habitats. New sister species thus evolve within the dispersal range of the offspring from a single deme. Several different patterns of geographic and ecological speciation are recognized based on the factors involved in promoting their divergence, such as population

TABLE I

Classification of Reproductive Isolating Mechanisms

Prezygotic isolating mechanisms: mechanisms that prevent interspecific mating
    Temporal isolation (potential mates have overlapping ranges but reproduction occurs at different times)
    Habitat isolation (potential mates have overlapping ranges but reproduction occurs in different habitats)
    Ethological isolation (potential mates meet but do not mate)
    Mechanical isolation (potential mates attempt to copulate but no sperm is transferred)
Mechanism that prevents fertilization
    Gametic mortality (sperm is transferred but egg is not fertilized)
    Gametic incompatibility (gametes meet but fertilization is not completed)
Postzygotic isolating mechanisms: mechanisms that prevent the development of interspecific hybrids
    Zygote mortality (eggs fertilized but zygote dies)
    Hybrid inviability (zygote produces F_1 hybrid of reduced viability)
    Hybrid sterility (F_1 hybrid zygote is fully viable but partially or completely sterile or produces deficient F_2 hybrid)
    Coevolutionary or cytoplasmic interactions (individuals from a population infected by an endoparasite or with a particular cytoplasmic element are fertile with each other, but fertility and/or viability break down when matings occur between infected and uninfected individuals)

TABLE II
Classification of Modes of Speciation

Allopatric speciation
   Dichopatric speciation (vicariant division of wide-ranging species by an extrinsic barrier or extinction of intervening populations)
   Peripatric speciation (by evolution in a very small isolated colony)
Nonallopatric speciation
   Parapatric speciation
   Sympatric speciation
   Autopolyploid speciation
   Spontaneous thelytokous parthenogenetic speciation
   Hybrid speciation
     Homoploid hybrid speciation
       Introgressive speciation
       Recombinational speciation
     Polyploid speciation
       Allopolyploid speciation
     Direct and reticulate allogenous speciation
   Symbiont-induced speciation

structure, habitat specialization, hybridization, and polyploidy.

A. Geography, Geographic Races, and Allopatric Speciation

The process of geographic speciation (often referred to as allopatric speciation) may happen in two ways. Dichopatric or vicariant speciation (Fig. 1b) occurs when a widespread species becomes geographically divided into two or more large subpopulations by an insurmountable or impassable barrier. A second mode of geographic speciation, usually called peripatric or founder event speciation (Fig. 1c), occurs when a geographically isolated population is established by a single fertilized female or a very small number of founding individuals.

However, determining the actual biological status of sister taxa that have arisen in isolation following either a vicariant or a founder event is difficult and controversial. Species can be recognized when artificial hybridization of allopatric taxa in the laboratory reveals post-mating reproductive isolation. If no post-mating incompatibility is noted and mating occurs, then it is necessary to determine if there are factors important in pre-mating isolation not provided in the laboratory environment of the tests. Whether or not such populations are species or geographic races may require an unequivocal "test of sympatry" in which the two taxa coexist without fusing. Only when previously isolated taxa have the opportunity to interbreed under natural field conditions can the species status of closely related taxa lacking post-mating reproductive isolation be confirmed. If they maintain distinct sympatric genotypic clusters, they should be recognized as distinct species. Because such tests cannot often be performed, deciding on whether isolated taxa represent species or geographic races (i.e., subspecies) is a subjective decision. However, it should be stressed that when two taxa overlap naturally (i.e., are parapatric) and do not interbreed, this does not necessarily mean that they speciated sympatrically. This is because it is often impossible to establish if the overlap represents a case of secondary contact following allopatric speciation or if the taxa evolved as the result of non-allopatric divergence.

1. Dichopatric Speciation

When populations of geographically isolated sister taxa remain large during and following geographic subdivision, they will slowly diverge genetically by genetic drift and as they adapt to local conditions in isolation. Over time, they may accumulate sufficient genetic differences to cause negative pleiotropic gene interaction and hybrid incompatibility among genes responsible for proper mate recognition and genome integration. Reproductive isolation therefore occurs by chance as the by-product of genetic divergence in isolation rather than the outcome of natural selection acting directly to promote reproductive isolation.

a. Dichopatric Race Formation and Speciation in Salamanders

The long-term study of evolution in the plethodontid salamander genus *Ensatina* by David Wake and colleagues (Wake and Schneider, 1998; Wake, 1997) provides an excellent example of dichopatric speciation and incipient species formation. The *Ensatina eschscholtzii* complex represents an ancient lineage that has undergone several instances of range contraction, isolation, and expansion and divergence of populations along broad ecological gradients. It is composed of seven contiguous subspecies wrapped in ring-like fashion around the Central San Joaquin Valley of California (Fig. 2). Historically, populations of *Ensatina* slowly differentiated into highly distinct forms from the northern subspecies *E. e. oregonensis* in color and pattern as they expanded southward. Blotched forms occur in the Sierra Nevada and mountains of southern California, whereas unblotched forms are found in the coastal and northern region of their range. The animals, which are long-lived (10–15 years) and take 4 years to mature, are

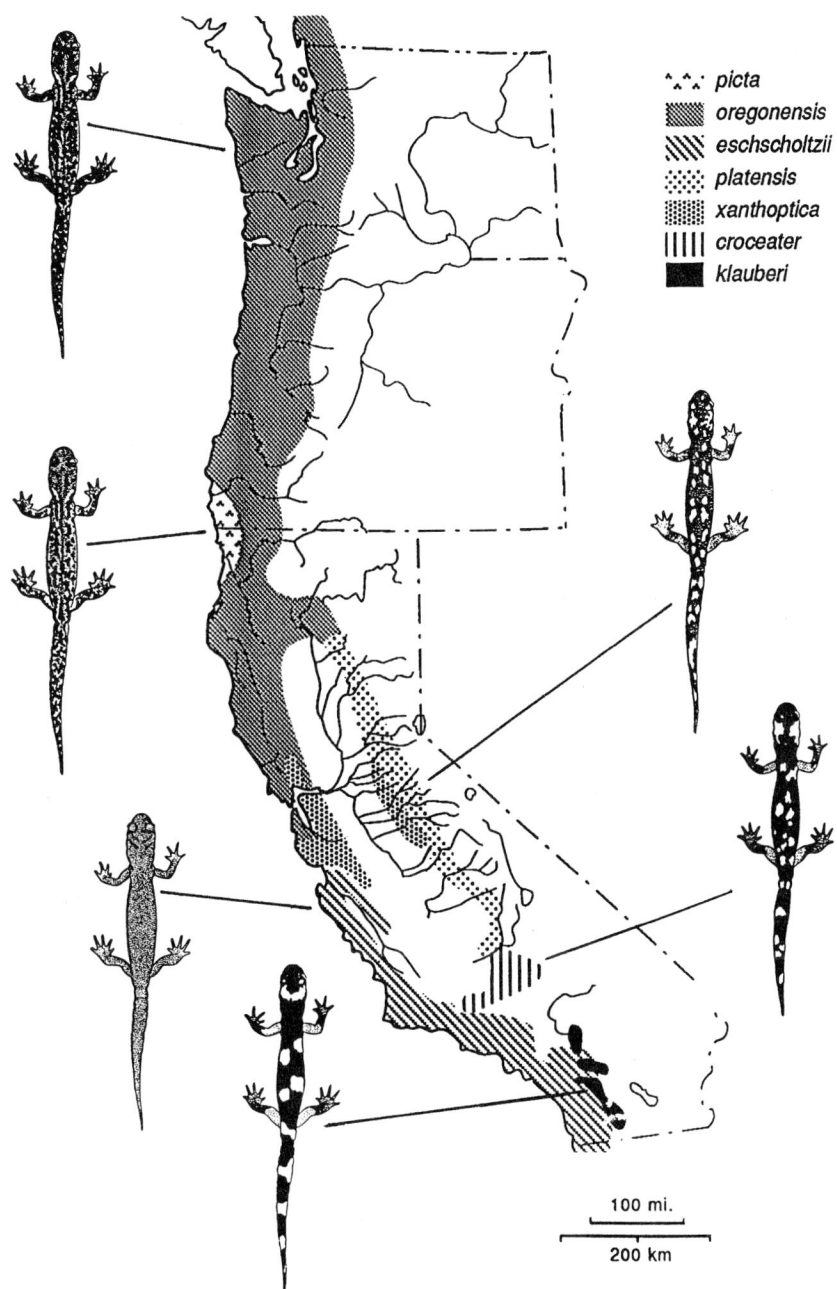

FIGURE 2 Geographic distribution of taxa in the salamander *Ensatina* complex based on molecular markers rather than morphological traits. Modified from Wake, 1997 (Incipient species formation in salamanders of the *Ensatina* complex. *Proc. Natl. Acad. Sci. USA* **94**, 7761–7767. Copyright 1997 National Academy of Sciences, U.S.A.) and Futuyuma, 1998.

sedentary and disperse only short distances (home range 10–20 m).

Hybridization and intergradation occur between adjacent subspecies to varying degrees in all except one contact zone. The exception occurs in the Cuyamaca Mountains of San Diego County, in which *E. e. eschscholtzii* and *E. e. klauberi* overlap but rarely interbreed in sympatry and are regarded by some biologists as distinct species. In other areas, the complex includes many geographically and genetically distinct entities, some of which are near the species level of differentiation.

A limited narrow secondary hybrid zone also occurs in Calaveras County between the upland *E. e. platensis* and a coastal, lowland subspecies, *E. e. xanthoptica* that established an outpost in the Sierran foothills sometime in the past. *Ensatina e. platensis* lives in the cool and moist, closed canopy forest at higher elevation than those of *E. e. xanthoptica,* which prefers warmer and drier lower elevations. Hybridization between the pure subspecies is rare, with most hybrids representing progeny from backcrosses or from crosses between hybrid individuals. Hybrid individuals appear to be at a selective disadvantage and thus subject to strong negative selection. They have reduced success in competing for preferred habitats and mates and seem to be more prone to predation. Because little or no gene flow occurs across the narrow hybrid zone, the coastal and Sierra taxons behave as incipient species or taxa in the final stages of speciation. Four other hybrid zones in central California involving *E. e. xanthoptica* have been characterized in detail. The *Ensatina* complex presents a full array of differentiation from geographically variable populations or races to well-marked species. It provides an excellent example of how speciation has progressed slowly by adaptive divergence and stochastic processes during periods of geographic isolation.

2. Peripatric Speciation and the Founder Effect Principle

There are several examples in which species appear to have evolved in small, isolated populations at the periphery of the range of a sister species. Unlike the slow rate of evolution in dichopatric speciation, peripatric speciation is postulated to take a relatively short time. How and how often this divergence occurs is controversial (Barton, 1996; Hollocher, 1996).

Three models, each based on a founder event in which a population is established by only a few individuals, have been proposed to account for the rapid evolution of species observed on islands and elsewhere (Fig. 1c). The founder effect principle was developed by Mayr (1954). It is based on the assumption that reproductive isolation from the parent species can evolve rapidly in a population established by a very small number of founding individuals (i.e., 2–10). He postulated that in such populations a genetic revolution could take place as a by-product of inbreeding, selection, drift, and genome reorganization. While the population is small, these genetic changes may promote substantial morphological and ecological shifts.

A modification of the founder effect principle was proposed by Carson (1975). In his founder-flush model of speciation, isolated populations undergo a series of population expansions and drastic contractions to a very small number of individuals. Carson believes that founder-flush speciation can occur only in certain cross-fertilizing diploid organisms with "open" genetic systems. The genome in such organisms represents a clique of harmoniously collaborating or coadapted genes united by strong epistatic interactions. Their genomes also have abundant pleiotropic interacting genetic polymorphisms and share a high recombination index. Carson hypothesized that these attributes provide great genetic flexibility that predisposes such organisms to speciate by the founder-flush process.

In Carson's view, the drastic events required to reorganize the original genetic system and restore new balances that are incompatible with ancestors are accomplished during cycles of the founder-flush process. Selection, which is relaxed during the flush phase of population expansion, is greatly intensified as the population crashes. Repeated disorganization and reconstitution of the genome results in the rapid evolution of reproductive isolation.

Templeton (1980) proposed a third modification of founder-induced speciation that involves a genetic transilience (Carson and Templeton, 1984). It is similar to Carson's founder-flush speciation but requires changes in only one or a few segregating units, commonly with epistatic modifiers responsible for reproductive isolation that occurs when a population rapidly passes through an extremely unstable intermediate genetic state. As in the case of the founder-flush model, a genetic transilience involves strong inbreeding and large variance in population size. The critical trigger that initiates a transilience requires a reweighting of fitness components owing to drift-induced shifts in allele frequencies at one or more major loci that have pleiotropic effects on reproductive isolation.

It seems clear that speciation may occur rapidly following the geographic isolation of a small population. It is not clear, however, whether speciation results from a founder-flush or transilience process or solely from natural selection in response to factors such as runaway sexual selection or rapid adaptation to divergent ecological and reproductive conditions. Because conditions required for stochastic transitions are severe and well-documented cases are lacking in which drastic genetic changes caused by founder effects result in speciation, it appears that the process of speciation in small, peripheral populations is the same as that which occurs in dichopatric speciation. Although certain kinds of epistasis can promote strong reproductive isolation, divergence occurs by selection and not random genetic drift (Barton, 1996).

B. Ecology, Ecological Races, and Nonallopatric (Ecological) Speciation

Reproductive isolation between sister populations in cases of allopatric speciation arises as a by-product of genetic changes that originate and accumulate independently in each population during periods of geographic isolation. Speciation may or may not be accompanied by an ecological transformation. In contrast, nonallopatric or ecological speciation inevitably involves a shift to a new niche at the time of speciation, and adaptation to different habitats is the driving force initiating and sustaining genetic divergence (Figs. 1d–1f). Divergent selection on existing genetic variation results in the evolution of reproductive isolation as different suites of genetic variants are favored in each habitat. The emergence of reproductive isolation is also facilitated when mating occurs within a preferred habitat or if mate preference that is based on a phenotypic variant of a sexually selected trait becomes associated with a preferred habitat (Johnson et al., 1996).

A misconception concerning nonallopatric speciation is that it requires a process called reinforcement, which involves the divergence of the mate recognition system between populations during the speciation process as an outcome of selection against hybrids (Butlin, 1987). Because theoretical models suggest that recombination will not allow reproductive reinforcement to develop, critics have argued that parapatric and sympatric speciation are unlikely to occur in nature. These models, however, fail to factor in the role of habitat preference and specialization in speciation and the effects of ecological divergence, which always accompanies sympatric speciation. When habitat choice is taken into consideration, conditions for sympatric speciation are greatly relaxed and its occurrence is probable (Johnson et al., 1996).

Three modes of nonallopatric speciation are generally recognized in sexually reproducing organisms. Parapatric speciation (Fig. 1d) takes place when sister species evolve while adapting to contiguous but spatially segregated habitats or ecotones across a narrow contact zone (Bush, 1994). Sympatric speciation (Fig. 1e) occurs in the absence of geographic segregation when sister species evolve within the dispersal range of the offspring of a single deme. During the course of sympatric speciation, the probability of mating between two individuals depends on their genotypes, and divergence occurs between populations adapting to alternate habitats within the "cruising range" of each other. Parapatric and sympatric speciation actually represent extremes of a continuum in the pattern and extent of habitat and geographic-imposed spatial segregation and gene flow reduction that occurs during nonallopatric divergence. Hybridization (Fig. 1f) is a third mode of nonallopatric speciation. New species rapidly evolve from a mating between individuals of two closely related species by a variety of different processes. It is a frequent mode of speciation in plants but appears to be rare in animals.

1. Parapatric Speciation

This mode of speciation occurs when sister species evolve while adapting to contiguous habitats (or ecotones) along a zone of contact. Examples from nature involve situations in which individuals from a species adapted to one habitat invade and colonize an adjacent habitat. Individuals bearing novel genetic recombinants capable of exploiting and reproducing in the new habitat are the first to invade and colonize the narrow zone where the two habitats meet. As adaptation proceeds, the new colony expands throughout the range of the new habitat. Only along the adjoining borders between the original and new habitat is gene exchange possible between the two populations. As the new population adapts to conditions imposed by the new habitat, divergent selection promotes the evolution of reproductive isolation and eventually parapatric species.

a. Parapatric Speciation in *Mimulus*

An example of rapid parapatric adaptation and speciation of a plant to a newly established vacant niche is that of the monkey flower *Mimulus cupriphilis* in California (Macnair and Gardner, 1998). This species grows only on relatively dry and toxic mine tailings of two small copper mines in California. It grows in close proximity and is recently derived from the widespread hydrophylic *M. guttatus*.

The two species differ in many ways. *Mimulus cupriphilis*, an obligate annual, flowers earlier and produces many small flowers that differ in shape and color from those of *M. guttatus*. *Mimulus cupriphilis* also has higher fitness on the dry tailings of the copper mines and flowers early when pollinators are rare. Because it is self-fertilizing, it is reproductively isolated from the outcrosser *M. guttatus*, which blooms later when its larger flowers are fertilized by bumblebees. Genetic studies have revealed that the species differ in a few major genes controlling flowering time, flower size, corolla spot number, and general size. Because all the genetic systems are recessive with the exception of flowering time, recessive alleles would be spread by natural selection in the original outbreeding parent species. In inbreeders, there is no difference between dominant and recessive alleles.

Mimulus cupriphilis has evolved recently since the copper mines are less than 150 years old. The shift to the new soils created by the mines involved the development of a copper-tolerant ecotype with a primary semidominant adaptive mutation that shifted flowering time earlier in the plants growing in dry habitats. In the absence of pollinators early in the season, selection favored alleles for self-fertilization. This was accomplished by flower size reduction that brought the stigma and anthers in close proximity. Reduction in size of corolla and associated structures may have also freed up resources for seed production. The result is a new selfing species that has evolved locally in a very short time and that is well adapted to the unique conditions of the mine tailing fields.

There is no evidence of a genetic revolution or reduction in genetic diversity in the new species and F_2 progeny. Nor is there evidence of significant breakdown in so-called gene complexes or major epistatic interactions. Therefore, it is concluded that speciation has been achieved by new selection pressures on normal *M. guttatus* as it colonized a new and unusual habitat. Major genes of large effect that play important roles in reproductive isolation have also been demonstrated in other sister species of *Mimulus* species (Bradshaw *et al.*, 1995). In such cases, new species have the potential to appear nearly full-blown in relatively few evolutionary steps because of changes in a few essential genes.

2. Sympatric Speciation

It is now apparent that sympatric speciation, once thought to be rare in animals and plants, occurs more frequently than previously realized. Because two closely related sympatric sister species cannot share the same resources, nonallopatric speciation is inevitably accompanied by the shift and exploitation of a new habitat by the daughter species. Such habitat or resource shifts reduce competition between sister species. It is for this reason that sympatric speciation is often referred to as ecological speciation.

a. Sympatric Speciation in Cichlid Fish

The important role of ecology in sympatric speciation is exemplified by the study of Schliewen *et al.* (1994). In a molecular phylogenetic study they discovered that two endemic cichlid fish species flocks (9 and 11 species) in two small, ecologically monotonous volcanic lakes (4.15 and 0.6 km²) in Cameroon speciated sympatrically. Because each crater is isolated from rivers and lacks internal structure, past lake levels are not responsible for physically isolating sister populations during speciation. Field observations, stomach content analyses, and the presence of specialized morphological features related to feeding confirm that the species within each lake have different benthic or pelagic trophic and reproductive ecologies. Although all species nest on the bottom and are capable of encountering one another at moderately high frequencies along ecotones, no hybrids are found because the species mate assortatively and do not interbreed.

Phylogenetic trees were constructed based on an analyses of a 340-base pair fragment of mtDNA cytochrome b and an additional 350 base pairs from the rapidly evolving mitochondrial control region for all 20 species and all tilapine species in neighboring river systems and lakes. This analysis confirmed that the lake species are monophyletic with respect to the river species, i.e., each flock evolved within each lake after a single colonization event.

b. Sympatric Host Race Formation and Speciation in *Rhagoletis* Fruit Flies

The importance of habitat and resource shifts in the nonallopatric speciation process is exemplified in the case of recent host race formation and speciation in the tephritid fruit fly genus *Rhagoletis* (Bush and Smith, 1998; Feder, 1998). As in other phytophagous and parasitic insects, mating in these flies occurs on the host plant, primarily on the host fruit in which eggs are deposited and larvae develop. *Rhagoletis pomonella*, whose native hosts are fruits of several hawthorn (*Crataegus*) species, colonized apples in approximately 1860, more than 200 years after this fruit tree was introduced to North America by Europeans. Early infestations on apples were restricted to a small area of the Hudson River valley in southern New York where hawthorn is common. The apple flies later spread over the entire northeastern United States and southeastern Canada, where it became a major pest. In a brief span of 150 years, the apple and hawthorn populations diverged in several genetically based traits, such as eclosion time of the adult in summer, and host recognition and acceptance. In addition, strong frequency differences in alleles at several loci coding for proteins are maintained by strong selection and greatly reduced gene flow between the races. The apple and hawthorn races now behave as semispecies or species. All seven members of the *R. pomonella* species group are sympatric in eastern North America and speciation in this species group has always been accompanied by a shift to a new host plant family or genus.

C. Other Modes of Nonallopatric Speciation

1. Spontaneous Thelytokous Speciation

This mode of sympatric speciation occurs when a unisexually reproducing taxa arises spontaneously from an unfertilized egg of a diploid bisexual species. Subsequent reproduction in taxa originating in this way produces only females from unfertilized eggs. Several good examples are provided by White (1978).

2. Autopolyploid Speciation

Occasionally, polyploidization of a diploid species may occur spontaneously in one or more individuals. Because autopolyploid individuals have three or more chromosome sets, each chromosome has more than one homologous pairing partner. During meiosis, multivalents are produced leading to unbalanced gametes and zygotes, sterility, and other problems. Only rarely does autopolyploidy result in the origin of new species, such as in the common potato (*Solanum tuberosum*) and its relatives (Grant, 1981). These usually originate from crosses between races whose chromosomes differ only slightly.

3. Speciation by Interspecific Hybridization

A new species can arise two ways by interspecific hybridization (Fig. 1f). Homoploid hybrid speciation results in a diploid-derived species, whereas polyploid hybrid speciation produces a species that combines a complete set of chromosomes from each hybridizing parental species. Hybrid species occupy habitats different from those of the parental species, thus reducing competition and the level of gene flow between them. Hybrid speciation, which is far more common in plants than in animals, can occur in at least four ways.

a. Introgressive Hybrid Speciation

Individual gene exchange among closely related species provides recombinant offspring that shift to and exploit a new habitat not utilized by either parental species. The hybrid species may be interfertile with one or both parental species, but it is reproductively isolated from them by premating barriers to gene flow. This mode of speciation has been reported in several plant species (Grant, 1981), but confirmation of this mode of speciation remains controversial and requires definitive experimental and analytical studies.

b. Recombinational Hybrid Speciation

A far more common mode of hybrid speciation involves the formation and establishment in the progeny of a chromosomally sterile or semisterile species hybrid of a new, structurally homozygous recombination type. Individuals are fertile within the line but isolated from other lines and from the parental species by a chromosomal sterility barrier. It is most likely to occur when the hybrid interface is long and the organisms involved are predominantly selfing, relatively fertile, and possess few structural chromosome differences between the parental species.

i. Hybrid Speciation in Wild Sunflowers A molecular study of hybrid speciation in the wild sunflowers *Helianthus* by Rieseberg et al. (1995) revealed that F_1 hybrids of *H. annus* and *H. petiolaris* are semisterile with pollen viabilities less than 10% and seed set less than 1%. F_2 pollen viability is highly variable, ranging from 13 to 97%. The two species are distinguished by several morphological and chromosomal features, and based on chloroplast DNA and nuclear ribosomal DNA variation they occur in divergent clades. Although the species are sympatric throughout much of the western United States, they have different ecological requirements. *Helianthus annus* is restricted to heavy, clay soils, whereas *H. petiolaris* predominantly inhabits dry, sandy soils.

Helianthus anomalus is a rare endemic to xeric habitats in northern Arizona and southern Utah. It is well within range of parental species and is a recombinational hybrid resulting from a cross between *H. annus* and *H. petiolaris*. The F_1 hybrids with parental species are partially sterile because chromosomal structural differences enhance reproductive isolation. A preliminary survey of 126 loci in natural populations of the parental species indicated that *H. anomalus* has loci derived from both *H. annus* and *H. petiolaris*. Some blocks of markers, possibly protected from recombination, are transmitted intact.

Helianthus anomalus combines rDNA repeat units and allozymes of *H. annus* and *H. petiolaris* as predicted for diploid hybrid species, although individuals possess chlDNA haplotypes of *H. annus* and *H. petiolaris* rather than a unique haplotype. Genetic linkage maps generated for all three species using random amplified polymorphic DNA markers reveal loci distributed onto 17 linkage groups corresponding to the haploid chromosome number of the three species. Although levels of polymorphisms vary from 212 in *H. annus* to 400 in *H. petiolaris*, map density is similar among species. By comparing genomic location and linear order of homologous markers, chromosomal structural relationships were inferred among the three species.

Even though 6 linkage groups showed no changes

in all three species, the remaining 11 linkages were not conserved in gene order. The parental species differ from *H. anomalus* by at least 10 separate structural rearrangements, 3 inversions and a minimum of 7 interchromosomal translocations. The genome of *H. anomalus* is thus extensively rearranged relative to its parents. All 7 novel rearrangements in *H. anomalus* involve linkage groups that are structurally divergent in parental species, suggesting that structural differences may induce additional chromosomal rearrangements upon recombination.

c. Allopolyploid (Amphiploid) Speciation

Interspecific hybridization can also result in combining two or more complete chromosome sets. F_1 hybrids produced between two established related species are often sterile because chromosomes lack sufficient homology to pair well at meiosis. Fertility is restored if hybrids persist long enough by asexual reproduction until somatic doubling of the chromosomes can occur in a flower, or until there is a rare union between two unreduced gametes. A new sexually reproducing species is then established that is "instantaneously" isolated from both parental species (Grant, 1981).

i. Allopolyploid Speciation in Spartina anglica The recent natural rapid evolution of the amphiploid perennial salt marsh grass, *Spartina anglica*, provides an example of allopolyploid speciation (Raybould et al., 1991). This species originated on the south coast of England at the end of the nineteenth century. It arose as a result of chromosome doubling in *S. x townsendii*, a hybrid between the native British *S. maritima* and the North American *S. alterniflora*, introduced by shipping (Fig. 3). *Spartina anglica* is now widespread along the English coast and is highly successful.

Although more than half of all plant species are directly or indirectly the by-products of allopolyploid speciation, allopolyploid speciation is relatively rare in animals (White, 1978).

d. Direct and Reticulate Allogenous Speciation

There are two modes of allogenous speciation (i.e., combining the genomes of two distinct species). In the case of direct allogenous speciation, hybridization between two bisexual, closely related species combines the genomes of two distinct parental species giving rise to a new, unisexual species (Bullini, 1994). The hybridization event produces either an allodiploid or an allopolyploid unisexual species that acquires clonal (parthenogenesis) or hemiclonal (hybridogenesis; i.e., it must mate with males of a bisexual parental species)

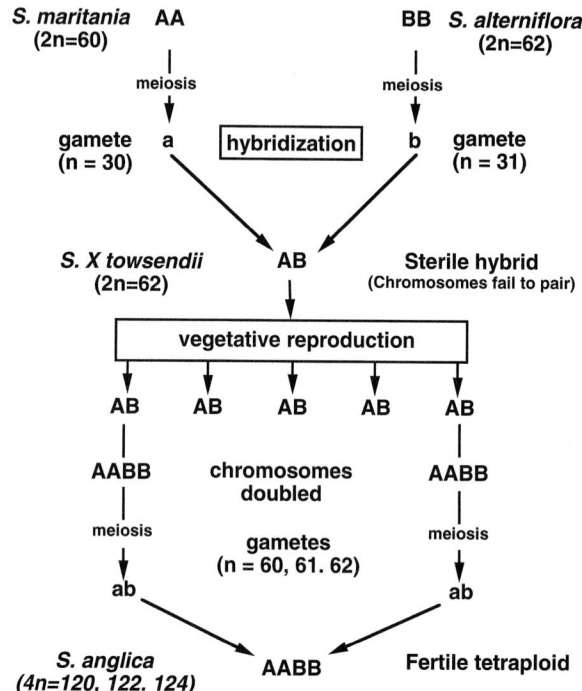

FIGURE 3 The origin of the allopolyploid marsh grass species *Spartina anglica* (AABB) that resulted from hybridization between *S. maritima* (AA) and *S. alternifolia* (BB). Each letter in parentheses represents a haploid genome. The sterile hybrid, *S. x townsendii* (AB), reproduced vegetatively until the genomes of individual plants doubled by autopolyploidy. This results in the establishment of a new fertile sexually reproducing tetraploid species, *S. anglica*, with a full complement of chromosomes from *S. maritima* and *S. alternifolia* (AABB) that can produce normal gametes (ab).

modes of reproduction. In most such clonal and hemiclonal organisms, the heterozygous genetic structure of the parental species is retained.

In the case of reticulate allogenous speciation, individuals of unisexual hybrid taxon hybridize with a bisexual relative giving rise to new, unisexual species, often with a higher ploidy level than that of the parental species (Bullini, 1994). Most have highly heterozygous genomes that display heterosis. Clones produced by direct and indirect allogenous speciation exhibit heterosis and demographic advantage over both parental species and individuals.

IV. SUMMARY

After 140 years of speciation research, it is now clear that animal and plant species can originate in a variety of ways. Rapid autopolyploid speciation is common in plants and accounts for many plant species. In animals, this mode of speciation is relatively rare. Dichopatric speciation gives rise to many, possibly the majority, of the species in some animal groups such as the land vertebrates. In fish, both allopatric and nonallopatric speciation have been reported. The majority of living organisms, however, are insects. Some authorities estimate there are as many as 30–40 million species, and all agree that there are more than 10 million. Approximately 70–75% are highly specialized parasites that feed in or on plant and animal tissue. Although many of these insects probably originated by nonallopatric speciation, it is not clear what percentage have done so. The same is true for the many mites and nematode species that also may number in the millions. Because the mode of speciation in only an extremely small number of animal and plant species has been established, it is impossible to estimate how often each mode of speciation occurs in any particular group of organisms. Such estimates must await research on a great many more taxa.

See Also the Following Articles

CLADOGENESIS • SPECIATION, THEORIES OF • SPECIES, CONCEPT(S) OF • SPECIES DIVERSITY, OVERVIEW

Bibliography

Barton, N. H. (1996). Natural selection and random genetic drift as causes of evolution on islands. *Philos. Trans. R. Soc. London B* 353, 785–794.

Bradshaw, H. D., Wilbert, S. M., Otto, K. G., and Schemske, D. W. (1995). Genetic mapping of floral traits associated with reproductive isolation in monkeyflowers (*Mimulus*). *Nature* 376, 762–765.

Bullini, L. (1994). Origin and evolution of animal hybrid species. *Trends Ecol. Evol.* 9, 422–426.

Bush, G. L. (1994). Sympatric speciation in animals: New wine in old bottles. *Trends Ecol. Evol.* 9, 285–288.

Bush, G. L., and Smith, J. J. (1998). The genetics and ecology of sympatric speciation: A case study. *Res. Population Ecol.* 40, 175–187.

Butlin, R. (1987). Speciation by reinforcement. *Trends Ecol. Evol.* 2, 8–13.

Carson, H. L. (1975). The genetics of speciation at the diploid level. *Am. Nat.* 109, 83–92.

Carson, H. L., and Templeton, A. R. (1984). Genetic revolutions in relation to speciation phenomena: The founding of new populations. *Annu. Rev. Ecol. Syst.* 15, 97–131.

Feder, J. L. (1998). The apple maggot fly, *Rhagoletis pomonella*, flies in the face of conventional wisdom about speciation. In *Endless Forms: Species and Speciation* (D. J. Howard and S. H. Berlocher, Eds.), pp. 130–144. Oxford Univ. Press, Oxford.

Feder, J. L., Reynolds, K., Go, W., and Wang, E. C. (1995). Intra- and interspecific competition and host race formation in the apple maggot fly, *Rhagoletis pomonella* (Diptera: Tephritidae). *Oecologia* 101, 416–425.

Futuyma, D. J. (1998). *Evolutionary Biology*. Sinauer, Sunderland, MA.

Grant, V. (1981). *Plant Speciation*. Columbia Univ. Press, New York.

Hollocher, H. (1996). Island hopping in *Drosophila*: Patterns and processes. *Philos. Trans. R. Soc. London B* 351, 735–743.

Johnson, P., Hoppensteadt, F., Smith, J., and Bush, G. L. (1996). Conditions for sympatric speciation: A diploid model incorporating habitat fidelity and nonhabitat assortative mating. *Evol. Ecol.* 10, 187–205.

Macnair, M. R., and Gardner, M. (1998). The evolution of edaphic endemics. In *Endless Forms* (D. J. Howard and S. H. Berlocher, Eds.), pp. 157–171. Oxford Univ. Press, Oxford.

Mallet, J. (1995). A species definition for the new synthesis. *Trends Ecol. Evol.* 10, 294–299.

Mayr, E. (1942). *Systematics and the Origin of Species*. Columbia Univ. Press, New York.

Mayr, E. (1954). Change of genetic environment and evolution. In *Evolution as a Process* (J. Huxley, A. C. Hardy, and E. B. Ford, Eds.), pp. 157–180. Allen and Unwin, London.

Raybould, A. F., Gray, A. J., Lawrence, M. J., and Marshall, D. F. (1991). The evolution of *Spartina anglica*. *Biol. J. Linnean Soc.* 44, 369–380.

Rieseberg, L. H., Fossen, C. V., and Desrochers, A. M. (1995). Hybrid speciation accompanied by genomic reorganization in wild sunflowers. *Nature* 375, 313–316.

Schliewen, U., Tautz, D., and Pääbo, S. (1994). Sympatric speciation suggested by monophyly of crater lake cichlids. *Nature* 368, 629–632.

Taylor, E. B., McPhail, J. D., and Schluter, D. (1997). History of ecological selection in sticklebacks: Uniting experimental and phylogenetic approaches. In *Molecular Evolution and Adaptive Radiation* (T. J. Givnish and K. J. Sytsma, Eds.), pp. 513–534. Cambridge Univ. Press, Cambridge, UK.

Templeton, A. R. (1980). The theory of speciation via the founder principle. *Genetics* 94, 1011–1038.

Wake, D. B. (1997). Incipient species formation in salamanders of the *Ensatina* complex. *Proc. Natl. Acad. Sci. USA* 94, 7761–7767.

Wake, D. B., and Schneider, C. J. (1998). Taxonomy of the Plenthodontid salamander genus *Ensatina*. *Herpetologia* 54, 279–298.

White, M. J. D. (1978). *Modes of Speciation*. Freeman, San Francisco.

SPECIATION, THEORIES OF

Hope Hollocher
Princeton University

I. Theoretical Considerations
II. Defining Species
III. Evolutionary Forces Involved in Speciation
IV. Common Modes of Speciation
V. Genetic Patterns of Species Differentiation
VI. New Directions for Studying Speciation
VII. Conclusions

GLOSSARY

assortative mating The tendency of like to mate with like whether it is based on similarities in genotypes or similarities in phenotypes.

coadapted gene complexes Genes that have been selected to work in a coordinated fashion to confer high fitness.

epistasis An interaction between different alleles at two or more loci such that the phenotype differs from what would be expected if each locus acted independently.

evolution Change in allele frequencies over time.

fitness The average number of offspring produced by individuals with a certain genotype relative to the number produced by individuals with other genotypes as a result of differences in survival and reproductive success.

gametogenesis The developmental process for the formation of functional male and female haploid reproductive cells that combine at fertilization to produce the zygote.

gene pool The collective set of genes in a population at a particular time.

genetic architecture Characterization of the number and types of genes and their interactions that underlie a particular trait.

genetic recombination The physical exchange of genetic material between a pair of chromosomes during meiosis.

genotype The state of an individual with respect to a specific genetic locus or set of loci.

heterogamety Having two different sex chromosomes.

hybrid inviability The phenomenon whereby one or both sexes of progeny produced in crosses between two different species are unable to survive.

hybrid sterility The phenomenon whereby one or both sexes of progeny produced in crosses between two different species are able to survive, yet are unable to reproduce.

monogamy One male mates with one female exclusively.

morphometrics A statistical procedure by which morphological traits are quantified and analyzed.

phenotype The state of an individual with respect to a specific trait.

pleiotropy One gene affects more than one trait.

polygamy Males and females both have several different mates.

reproductive isolation Intrinsic barriers to the production of offspring.

SPECIATION IS THE PROCESS by which new species are formed. Thus, its central importance to the study of evolution and issues of biodiversity cannot be overstated. The seemingly simple definition of speciation as the origin of species belies the fact that the actual process of speciation is almost as complex and varied as the diversity of life itself. No one mechanism of speciation is sufficient to describe the origination of all the diversity we see in the world and more often than not several speciation mechanisms come into play simultaneously in the formation of a single new species. The study of speciation attempts to analyze these competing mechanisms to determine the relative importance of different evolutionary forces promoting species divergence under particular circumstances. By looking for recurring patterns across a wide spectrum of individual case studies, we are devising models that help explain speciation more generally. This knowledge then serves as a framework for making policy decisions regarding the preservation of biodiversity, for the ultimate goal of conservation biology is not just to preserve current species, but also to preserve the natural processes that help generate new ones.

I. THEORETICAL CONSIDERATIONS

Although speciation has been studied intensively for more than 140 years, surprisingly very little is understood about the patterns of genetic change that occur during the process. This lack of knowledge is a true stumbling block for describing speciation using a population genetic framework. A key aspect of all population genetic models of speciation is the assumptions made about the underlying genetic architecture—the number and types of genes and their interactions—of traits that have changed during speciation. The underlying genetic architecture influences directly how traits change during speciation even when the same evolutionary forces are known to be in operation (reviewed in Hollocher, 1998). Therefore, the examination of the genetic basis of species differences reveals information about the specific evolutionary processes that were involved in creating them.

Equally important to theories of speciation is determining which traits are most influential for promoting species divergence. We recognize species because they are different from one another, be it morphologically, behaviorally, or reproductively. In all cases, a sequence of events has led to the eventual fixation of these observed differences. Knowing whether certain traits are more prone to change than others during speciation or whether certain traits have a larger impact on future patterns of divergence directly informs our models of speciation. This knowledge improves our ability to target groups of organisms at the nascent stage of divergence and to focus our attention on specific traits for future studies of speciation that will bring a better understanding of the process.

II. DEFINING SPECIES

A. Why Is There More Than One Definition of Species?

One would think that by now evolutionary biologists would agree on what constitutes a species. However, it turns out that species concepts are almost as varied as the types of researchers that work on the question. To date there is no single, universally accepted definition of a species, although certain frameworks for thinking about species have come in and out of favor over the years (see "Species, Concepts of," by Mallet in this volume; also see Ereshefsky, 1992; Ridley, 1996; Howard and Berlocher, 1998). Most definitions of species serve a useful research purpose and are a function of the types of questions that are being addressed (e.g., whether species designations should serve a strictly taxonomic purpose as in systematics or represent the active evolutionary unit of diversification as in population genetics). This is not to imply that the definition of species is purely a semantic issue, but rather that the diversity of definitions reflects very different perspectives as to what factors should be emphasized in describing species. Beyond these purely theoretical issues, species concepts have an additional problem of not always being easy to apply in real-life situations.

As a result of these conflicting purposes of species concepts, every subdiscipline within speciation research has formulated its own particular species concept. However, the problems surrounding species definitions are more fundamental than simple differences in perspective. There is an inherent problem of trying to categorize groups of organisms that are undergoing a continual process of change. Speciation is a process, not an instantaneous event. It is also not caused by a single unified mechanism, but by a collection of different mechanisms of divergence affecting different traits at potentially different rates. The formation of natural groups does eventually result from the ongoing action of each of these processes of divergence and at some point in time these natural groups can and should be delineated as separate species. However, because of the fluidity of the process, the boundaries that circumscribe these natural groups are often at best blurry given the fact that each underlying mechanism of divergence can potentially delineate a separate natural group. In some instances, this creates no problem and these different boundaries will actually coincide with each other. In these cases, several different species concepts will correspondingly give the same species designations. In other instances, these boundaries may not correspond well at all. Hence, depending on which process is being emphasized, the actual definition of a particular species may also vary in these circumstances. In most cases, conflicting species boundaries are not so much a problem of the competing species definitions as they are a reflection of the complicated nature of the underlying biological process that governs speciation. For a population geneticist interested in the process of speciation, it is those cases where the boundaries of divergence based on different criteria do not correspond well that often prove to be most interesting to study.

B. Most Commonly Used Species Concepts

Because of space constraints, a thorough critique of competing species concepts is beyond the scope of this article (for a more complete review, see "Species, Concepts of," by Mallet in this volume; also see Ereshefsky, 1992; Ridley, 1996; and the chapters by De Queiroz, Harrison, Shaw, and Templeton in Howard and Berlocher, 1998). However, a general overview of the most commonly used species concepts will be presented here, paying particular attention to how each concept relates to the elucidation of underlying speciation processes.

1. Morphological and Phenetic Species Concepts

Intuitively, using morphological criteria to get a first approximation of species boundaries works rather well. Different groups of organisms look different from each other. This simple "morphological species concept" was used by the early naturalists of the eighteenth and nineteenth centuries and is still often used today by field biologists and paleontologists. This concept is also used by most evolutionary biologists as a starting point from which other criteria are then later added (the fact that species appear morphologically distinct is usually taken for granted by most practicing evolutionary biologists). The "phenetic species concept" embellishes on this first approximation by incorporating specific morphometric analyses to describe explicitly the phenotypic groupings. From a functional point of view, groups of organisms that are morphologically distinct most likely represent different groups genetically as well.

Using morphological criteria alone, however, carries with it the risk of categorizing groups of organisms as the same species when indeed there are reproductive barriers between them (as is the case for cryptic species, which are morphologically identical but nevertheless represent distinctly different gene pools) or categorizing groups of organisms as separate species when in fact the difference is not genetically based, but solely a phenotypic response to varying ecological conditions (such is the case for some plants that can change the morphology of their leaves in response to varying light levels). In addition, the categorization of species based solely on morphological criteria without the incorporation of an evolutionary framework is generally unsatisfactory to the majority of people working in the field of speciation.

2. The Genotypic Species Cluster Definition

An improvement over most morphological approaches to defining species has recently been derived by Mallet (1995) using genetic criteria to cluster groups of organisms rather than strictly morphological traits (termed the "genotypic species cluster" definition). In this case species are defined as genetically distinguishable groups of individuals that exhibit few or no intermediate genotypes (both for single and multiple loci) when in contact. This is clearly an advantage over strict morphological criteria because it gets right to the heart of speciation—genetic differentiation. It is also very satisfying as an operational definition of species groups, because genetic clustering is easily diagnosable for all taxa, whether sexual or asexual. However, this definition is only easily applicable to the situation where species are in direct contact with each other and cannot be generally applied. More importantly, it is still a phenetic approach to species definitions and lacks any inherent evolutionary framework for the interpretation of species groups from a mechanistic point of view.

3. The Biological and Recognition Species Concepts

Until recently, the most enduring species concept has been the "biological species concept," first outlined in detail by Dobzhansky (1937) and subsequently popularized by Mayr (reviewed in Mayr, 1982). The concept defines species as interbreeding groups of organisms that are reproductively isolated from other groups. In other words, species are groups of organisms that share a common gene pool. Speciation, under this concept, is the development of a biological barrier to gene flow and the subsequent differentiation of the two resulting gene pools. Biological barriers to gene flow were termed isolating mechanisms and covered the full gamut from prezygotic barriers (including ecological or habitat isolation, seasonal or temporal isolation, sexual or ethological isolation, mechanical or physiological barriers to mating, isolation by different pollinators, and gametic isolation blocking fertilization) to postzygotic barriers (including hybrid sterility, hybrid inviability, and other forms of reduced fitness in progeny resulting from crosses between differentiated populations). The biological species concept is firmly rooted in speciation processes and encompasses a number of different mechanisms that can serve to reduce the exchange of genes between different populations and thus allow them to evolve independently.

The "recognition species concept" is very similar to the biological species concept in that species are defined by interactions among its members (Paterson, 1985). In addition, both concepts try to pinpoint traits that are biologically relevant in species divergence. In the case of the recognition concept, species are defined as groups of organisms that share a common system of fertilization or specific mate recognition system; i.e., the emphasis is placed on what holds species together as a cohesive group rather than on what may separate them. The same general concept is embedded in the biological species concept, but the emphasis on mate recognition systems versus isolating barriers does influence how questions about speciation mechanisms are formulated and places more importance on evolutionary forces that positively influence the specific mating interactions between members of a species rather than on the more negative effects associated with preventing matings from occurring between members of different species. In either case, both of these concepts are mechanistically based and represent a prospective view of species distinctions by giving some indication of how isolated groups are now and will continue to be in the future.

4. Phylogenetic, Evolutionary, and Genealogical Species Concepts

Other more recently derived species concepts rely more heavily on a retrospective view by defining a species in a strictly historical sense as a separate evolutionary lineage that is internally connected through time (i.e., the "phylogenetic species concept" (Cracraft article in Otte and Endler, 1989; Nixon and Wheeler, 1990), the "evolutionary species concept" (Wiley, 1978), and the "genealogical species concept" (Baum and Shaw, 1995); for a good overview, see Harrison article in Howard and Berlocher, 1998). These species concepts are less oriented toward process and identifying specific biological traits that maintain cohesion within species or promote divergence between species and instead are more oriented toward the final evolutionary result—lineage divergence. That is not to say that process cannot be usefully inferred from looking at historical patterns of lineage sorting and splitting (see The Cohesion Species Concept below and Section VI.A), but the emphasis in these particular species concepts is clearly on pattern rather than on trying to incorporate a variety of biological processes into the species definitions themselves.

5. The Cohesion Species Concept

Perhaps the most inclusive species concept that has been proposed to date is the "cohesion species concept" (see Templeton in Otte and Endler, 1989; see also Templeton in Howard and Berlocher, 1998). This singlespecies concept combines key elements of the biological, recognition, phylogenetic, evolutionary, and genealogical species concepts in order to form a comprehensive framework for defining species and studying speciation mechanisms. Like the recognition species concept, the emphasis in this case is on defining species based on mechanisms that provide cohesion among members of a population. Unlike the recognition concept, cohesion involves not just shared fertilization systems promoting gene flow (termed genetic exchangeability in this model) but also shared historical fates (i.e., belonging to a common evolutionary lineage) as well as shared evolutionary tendencies (i.e., experiencing common selective regimes), both of which will maintain species cohesion (through demographic exchangeability) in the absence of gene flow. With this definition, the repertoire of traits important to speciation is broadened beyond traits determined primarily by their effects on reproduction and includes any traits involved in ecological adaptation, regardless of whether or not they directly affect the probability of genetic exchange. Although some would argue this more inclusive, pluralist species concept obfuscates the ability to clearly define species (if anything goes, then what should be the object of study?), in reality, the concept serves as a better starting point for speciation studies precisely because it is not narrowly focused on any one particular speciation mechanism to the exclusion of all others. The functionality of the cohesion species concept has been improved of late by incorporation of historical patterns of genetic variation to use as an explicit framework for both testing hypotheses of species boundaries and investigating speciation processes by modeling how different evolutionary forces pattern this genetic variation under different geographical scenarios (Templeton in Howard and Berlocher, 1998; see also Section VI.A).

III. EVOLUTIONARY FORCES INVOLVED IN SPECIATION

Because genetic change lies at the heart of speciation, the formation of new species involves the action of the same evolutionary forces known to cause genetic changes within populations (see Charlesworth in this volume for an in-depth description of population genetics). The principal evolutionary forces most commonly involved in species divergence are natural selection, genetic drift, sexual selection, and mutation. All of these forces act to change allele frequencies within populations over time and there is no question that they operate during speciation as well. The challenge of speciation studies comes from understanding how these evolutionary forces operate alone and in conjunction with each other within very specific geographical, ecological, behavioral, and genetic contexts. Change any one of these contexts and the same evolutionary force can give rise to very different evolutionary outcomes, creating very different patterns of speciation.

A. Natural Selection

Darwin (1872) was the first to popularize the idea that speciation could readily occur through the prolonged action of natural selection, the process by which genetic variants that are better suited to the natural environment increase in frequency while variants that are not well suited decrease. In Darwin's model of speciation, competition to survive was thought to be extremely intense. Therefore, the modification of descendants through natural selection served to increase the adaptation of any one population to the environment while also acting to decrease the intense competition between populations by promoting diversification. Over time, speciation would result automatically under these conditions.

Because of the general acceptance of Darwin's idea that natural selection is a potent driving force in species divergence, it is often erroneously considered the only evolutionary force important in species formation. However, natural selection can act to cause change (e.g., through directional or disruptive selection as imagined originally by Darwin) as well as to prevent change (e.g., through balancing selection—see Charlesworth's article on population genetics in this volume, Section III.E). Because of the dual nature of selection, speciation studies need to distinguish between those circumstances under which natural selection is promoting speciation versus those where natural selection is a force that actually helps populations remain similar to each other and needs to be overcome in order for change to occur.

Natural selection's role as a powerful agent for stasis rather than change comes into play in speciation studies in many instances. One particular instance that has been the focus of a prolonged debate in speciation studies surrounds the issue of coadapted gene complexes. Organisms are required not only to adapt to the external environment but also to their own internal genetic environments. Organisms develop and survive as a result of the coordinated control of a complex network of genes selected to function together properly. All possible genetic changes that may help an organism survive in a particular external environment must also function within the constraints posed by this internal genetic environment in order for these changes to be propagated through the population. Genes that have been selected to work particularly well together are called coadapted gene complexes. Once coadapted gene complexes have evolved, it can be very difficult for natural selection to deviate from the established optimum (termed a fitness peak) in order to reach a new optimum that may ultimately be better for the population, but would require the population to experience a temporary loss of fitness (through suboptimal gene interactions) during the transition.

Sewall Wright (1932) coined the phrase fitness landscape or adaptive landscape to describe this complicated behavior of fitness as a function of gene interactions. Different genotypic combinations plotted in two dimensions create a complex fitness landscape in the third dimension, much like how a mountainscape appears from an airplane. Mountain peaks correspond to regions of high fitness (for particularly good combinations of genes) separated by valleys of low fitness (for not so good combinations of genes). Natural selection, being somewhat myopic, would work to drive populations up a local peak, but would not be able to take populations across valleys to neighboring peaks of fitness even if those peaks represented greater overall fitness.

Because the true nature of adaptive landscapes is a point of contention for competing models of divergence (for a review, see Hollocher article in Grant, 1998; see also Coyne et al., 1997; and Wade and Goodnight, 1998), much work in population genetics has been geared toward understanding what the fitness landscape really looks like under a variety of circumstances (for a review, see Whitlock et al., 1995). In spite of its importance, it has been relatively difficult to get a firm handle on the topology of adaptive landscapes partly because at any given moment natural populations tend to occupy a single peak. Advances have been made more recently by studying microbial and viral populations, which can be followed for tens of hundreds of generations (Lenski and Travisano, 1994; Burch and Chao, 1999), and through in vitro studies of the evolution of new functions for specific classes of nucleic acids (for example, see Huynen, 1996), both of which provide more direct information about the actual shape of adaptive landscapes. Different species themselves represent the successful transition from one optimal state to another that has been frozen in time. Characterizing the genetic transitions that distinguish closely related species gives us clues to the actual shape of the fitness landscape from a complementary perspective. It also allows us to generate genetic models of speciation that describe how different species are able to make this shift from one fitness peak to another.

B. Genetic Drift

Given the possible constraints posed by selection acting on coadapted gene complexes, an additional mechanism that would facilitate the movement of a population from one fitness peak to another is often incorporated into models of speciation. Genetic drift is the random change in allele frequencies caused by sampling across generations in a finite population (see Charlesworth's article on population genetics in this volume, Section IV). The actual change in allele frequency caused by genetic drift is random for any given generation; however, the effects of drift will accumulate over time. In the absence of the introduction of new variation through mutation or immigration from another source, genetic drift will serve to decrease the amount of genetic variation present in a population. Because this process is entirely random, how one population responds can differ from how another population responds just by chance, and drift can actually serve to differentiate two populations in the complete absence of natural selection.

In Wright's original model describing adaptive landscapes, genetic drift was one mechanism by which populations could move from one adaptive peak to another. For small population sizes, drift effects can be so strong as to overcome the effects of natural selection. If genetic drift could free the population from the bonds of natural selection holding it on one particular fitness peak, then the population could randomly explore the adaptive landscape and possibly find another fitness peak better than the original one. Individually, genetic drift and natural selection can be very effective for bringing about change. Together they may be even more effective. Although speciation is not specifically addressed in Wright's original model, it does form the theoretical basis for Founder Effect Speciation, which will be discussed in more detail later in this article (see Section IV.B).

C. Sexual Selection

Adapting to the natural environment is only one manner in which organisms can increase their ability to leave more offspring for future generations. Another method is to improve their ability to secure mates through the action of sexual selection. Sexual selection arises from differences in reproductive success caused by competition over mates (Darwin, 1872; Andersson, 1994). Because variance in reproductive success can often be quite high (especially for males), adaptation to the mating environment can sometimes be even more effective at promoting change than adapting to the natural environment and has resulted in the rapid divergence of traits related to reproduction, such as the morphology of male genitalia, sperm composition and abundance, and male secondary sexual characteristics (Eberhard, 1985; Grant and Grant, 1997).

The pressure for sexually reproducing organisms to have compatible reproductive systems ensures that male and female reproductive systems will coevolve. Reproductive systems include behavioral traits related to mate recognition, courtship, and mating as well as physiological traits related to gametogenesis, fertilization, and the production of offspring. Reproductive systems are generally subject to several counteracting evolutionary forces that drive the system in different directions; therefore, mating systems within any particular population tend to reach a stable equilibrium (Lande, 1981; Iwasa and Pomiankowski, 1995). Although the population may be evolving directionally through time, individuals who deviate too much from the norm at any given point along the evolutionary trajectory will generally be disfavored either because of mate discrimination or because of natural selection pressures. Although stabilized with respect to any one population, reproductive systems can be very fluid with respect to the number of stable equilibria that are theoretically possible. Therefore, across several populations, traits related to reproduction are particularly susceptible to changes that can eventually lead to speciation (Lande, 1981; Kaneshiro, 1989).

The pressure to maintain sexual compatibility (positive male–female coevolution) within sexually reproducing populations is only one important evolutionary consideration for the role of sexual selection in speciation. In addition, there is a second evolutionary force operating that involves antagonistic male–female coevolution which is also very effective at bringing about change, especially in the case of internal fertilization (see Rice in Howard and Berlocher, 1998). The reproductive interests of males and females are not always identical, especially when there is any deviation from strict monogamy. In these cases, evolutionary conflicts can easily arise. For example, females may mate with multiple males to maximize their own reproductive success, but that creates a battleground for sperm competition among males. To combat the intensified selection for sperm to perform better under these circumstances, the chemical composition of male seminal fluid may evolve to include proteins that act to kill other sperm. These proteins in turn may also have negative pleiotropic effects on the female reproductive system itself, which is then selected to combat these negative effects. In this scenario, polygamy sets the stage for continual antagonistic selection to occur between male and female reproductive systems. As seen earlier, within any one particular population, antagonistic coevolution will force the male and female reproductive systems to track each other very closely. However, the effect is very localized and populations that are not in contact with each other have the potential to diverge rapidly with respect to their reproductive physiology (Wu et al., 1996; Rice article in Howard and Berlocher, 1998).

D. Mutation

Neither selection nor drift can operate in the absence of genetic variability. Although quite a bit of variability can be generated just through genetic recombination and the random assortment of chromosomes during reproduction, ultimately spontaneous mutation is the primary source of genetic variation (see Charlesworth's article on population genetics in this volume, Section III.B). Mutation is considered random with respect to the environmental challenges faced by organisms, although that is not to say it is without bias. Because mutations are important in the context of their effects on the phenotype of organisms, certain phenotypic classes of mutations can be more common than other classes of mutations. In this respect, most spontaneous mutations tend to be deleterious to the fitness of an organism. In addition, mutations are constrained by the chemical nature of DNA or particular properties of the DNA sequence itself, resulting in certain types of changes being more common than others. This can lead to certain classes of base substitutions being more common and to recurring mutation producing the same allele several times independently as is seen for mutations involving slippage during DNA replication of tandemly repeated sequences.

An important consideration for speciation studies is whether new mutations tend to have large or small phenotypic effects (i.e., does divergence occur in leaps and bounds or through the accumulation of numerous small changes?). In order to model speciation, it is also important to know whether mutations tend to have dominant or recessive effects with respect to fitness—the dynamics of how allele frequencies change over time are dependent on whether or not alleles at low frequencies (as is always the case for new mutations) are readily subject to natural selection or are masked by the effects of other alleles.

IV. COMMON MODES OF SPECIATION

The crux of speciation studies involves understanding the processes by which a single population splits to form two diverging lineages. Speciation involves the weakening of forces that hold populations together (such as the presence of gene flow, shared mating systems, recurring mutation, and natural selection in common environments) in favor of forces that drive them to diverge (such as the action of genetic drift, mutation, and natural and sexual selection in different environments in the absence of gene flow). Most modes of speciation incorporate the action of all the evolutionary forces discussed earlier (natural selection, genetic drift, sexual selection, and mutation), but emphasize the operation of certain forces over that of others. Several of these modes of speciation are discussed in another article in this volume (see "Speciation, Processes of," by Bush in this volume; see also Ridley, 1996); therefore I will focus my discussion here on a subset of all possible modes of speciation, concentrating on the specific interaction of different evolutionary forces operating within the context of different geographical and ecological situations.

A. Classical Allopatric Speciation

Perhaps the simplest conceptual framework for speciation is the allopatric model. Intuitively, it is easy to recognize that if a physical barrier prevents populations from exchanging genes, they will be free to follow independent evolutionary trajectories. Extensive geographical variation exists (reviewed in Mayr, 1963); therefore, any complete barrier to gene flow most likely means the operation of natural selection will differ to some extent on either side of the barrier. Even in the absence of varying environmental pressures, the continual action of mutation, genetic drift, and sexual selection (see Section III) can change the populations relative to each other over time, if they are completely isolated from each other and no longer actively exchanging genes.

Given enough time, speciation will most likely occur under this scenario of no gene flow between isolated populations. Therefore, the interesting issues to explore in allopatric speciation involve examining the relative rates of divergence caused by the action of natural selection, sexual selection, mutation, and drift as well as determining which traits are most easily affected. There are no general rules that can be presented for what happens during allopatric speciation; examples for all possible scenarios (where each of the evolutionary forces has been shown to play a role to a varying degree) can be cited. The specific outcome is very much dependent on the circumstances surrounding the isolation. It has also been shown well that changes in the reproductive system (both prezygotic behavioral changes in mating propensity and postzygotic physiological changes in the ability to generate offspring) occur rather commonly when populations are separated (Rice and Hostert, 1993; Ridley, 1996).

B. Peripatric or Founder Effect Speciation

Peripatric speciation represents a variation on allopatric speciation. In this case, a small population forms at the periphery of a larger population. This type of geographical isolation can happen anywhere, but is most easy to visualize in the situation of founders colonizing oceanic islands or, more generally, isolated pockets of habitat. Because the founding of peripheral populations can sometimes involve the movement of only a few individuals (or even a single gravid female), this type of speciation has also become known as founder effect speciation. The emphasis here is on the interaction between genetic drift and natural or sexual selection that occurs during the early stages of speciation.

If a new population is founded by a small number of individuals, just by chance the genetic composition of the founding population may differ significantly from that of the original source population because of genetic drift (see Section III.B). The population need not remain small for very long in order for this sampling effect to influence the future evolutionary trajectory of the population. In addition to this immediate genetic change, oftentimes small populations founded in peripheral habitats or on islands also experience changed environments (both the physical environment, including such things as the quality of the habitat, the distribution of resources, or the presence of competitors or predators, as well as the mating environment, represented by a shift in the distribution of available mating types and preferences), creating new selection regimes. Even in the complete absence of new selective environments, the shift in allele frequencies alone can potentially have a profound effect on how the population will respond to selection because of the changed internal genetic environment that results from drift. The combination of shifting gene frequencies by drift and the presence of potentially new selection regimes under this scenario has led some researchers to propose that this type of speciation can occur more rapidly than the more standard form of allopatric speciation which generally involves populations of larger size and less drastic changes in the physical and mating environment upon isolation (for a review, see Hollocher article in Grant, 1998; see also Ridley, 1996).

The theoretical framework used to justify the conclusion that speciation would be accelerated during founder effect speciation stems directly from Wright's model of an adaptive landscape (see Sections III.A and III.B). A major underlying genetic assumption that enters into the idea that the random sampling of alleles during the founder event can have a profound effect on the evolutionary trajectory of a population is that epistasis (where interactions between alleles at different loci produce phenotypic effects that are not predicted by the action of the individual allelic effects considered alone) and pleiotropy (where a single locus can directly influence more than one phenotypic trait) are quite common. It is under the assumptions of this type of genetic architecture that fitness peaks of varying heights will exist in the adaptive landscape and where random shifts in allele frequencies can have profound effects (for example, see Gavrilets and Hastings, 1996; for a review, see Hollocher in Grant, 1998). If allelic effects are more additive (where interactions between alleles at different loci are minimal), then allele frequency changes do not greatly affect the action of natural selection.

Much of the debate surrounding the likelihood of founder effects accelerating the process of speciation has focused on the specific influence drift alone would have on the probability of shifting from one fitness peak to another (for a review, see the Barton and Hollocher articles in Grant, 1998). What has emerged from these theoretical studies has been the idea that the actual size of the founding population does not play as crucial a role in determining the probability of shifting from one fitness peak to another as does the underlying genetic architecture of fitness. On the basis of these theoretical results, researchers have begun to shift their focus to evaluating the genetic architecture underlying traits that change during speciation to see how often epistasis is an important component (see Section V). Although it has been shown that drift alone is not likely to cause populations to shift from one fitness equilibrium to another (see Barton article in Grant, 1998), it is becoming clear that genetic drift acting in concert with natural selection can facilitate speciation when divergence involves traits characterized by extensive epistasis (see Hollocher article in Grant, 1998).

In addition to the genetic architecture influencing rates of change, the actual nature of the trait itself can affect the type of response that is expected under founder effect speciation. Reproductive isolation (both prezygotic and postzygotic) can be particularly susceptible to rapid change under this scenario because of the tight coevolution of male and female traits that normally occurs via sexual selection (see Section III.C). The random sampling of individuals during a founder event can easily move the population away from the stable equilibrium that characterizes the reproductive system in the original population. Reestablishment of a new equilibrium can often involve a radical shift in the mating system of the new population relative to the ancestral one. For sexually selected traits, random genetic drift coupled with sexual selection can act as a particularly powerful mechanism for driving speciation (Lande, 1981; Kaneshiro, 1989; Grant and Grant, 1997).

C. Parapatry and Speciation through Hybridization

Parapatric speciation occurs when new species evolve in contiguous, yet spatially segregated habitats. Unlike allopatric speciation, the populations that are diverging during parapatric speciation maintain a zone of contact and do not cease the exchange of genes completely. In this case, a balance is achieved between continual gene flow and strong natural selection to maintain divergent populations at the two ends of the contiguous habitats. The zone of contact between the two diverging populations, where hybridization between the differently adapted types takes place, is called a hybrid zone (for more general information on hybrid zones, see Hewitt article in Otte and Endler, 1989; Harrison, 1993; Butlin article in Howard and Berlocher, 1998). There can be a single zone of contact along a linear environmental gradient or several points of contact between habitats that are more patchily distributed. In both cases, the basic dynamics of hybrid zones are similar. In practice, it is generally impossible to distinguish between the situation in which the two populations continually maintained contact during the process of divergence (in which case the hybrid zone would then be considered a primary zone of contact) and the scenario in which the populations were actually allopatric at some point during divergence and then more recently came back into contact (in which case the hybrid zone would be considered a secondary zone of contact).

In either event, hybrid zones are particularly compelling to population geneticists because of the opportunity they present for examining the rate of exchange of genes that may be under different selection pressures in the two habitats. For most stable hybrid zones, the point of contact between the two species represents a semipermeable barrier to gene flow. Examining the distribution of different alleles sampled along a transect through the two habitat types can provide insights into the nature of the selective forces operating during divergence. The distribution of alleles at loci that have absolutely no effect on fitness will be governed entirely by migration distances and rates of exchange between the two species. Genes that are important for maintaining adaptive differences will show entirely different dynamics relative to these more neutral genes, revealing different patterns depending on the number of genes involved in the adaptations and the strength of selection operating in the two different habitats. Examination of the fates of alleles that show differential movement across the hybrid zone is also a powerful method for actually mapping the genes responsible for adaptation (see Barton article in Harrison, 1993).

In addition, hybrid zones represent natural laboratories for determining whether there is speciation by reinforcement. If two species coming into contact exhibit some degree of postzygotic isolation, it is then theoretically possible for selection to act on mating behavior (increasing assortative mating) to eliminate the production of hybrids, thus reinforcing the divergence that has already occurred and completing the speciation process (Dobzhansky, 1940; Butlin article in Otte and Endler, 1989; Butlin article in Howard and Berlocher, 1998). Reinforcement is a very appealing concept because it allows postzygotic isolation to play an active role in driving speciation rather than simply being a pleiotropic consequence of divergence. Although undoubtedly reinforcement is a selection pressure that does occur in nature, many theoretical arguments have narrowly limited the range of circumstances under which it is likely to occur and the general consensus is that reinforcement is possible, but probably not prevalent, in hybrid zones.

As is clear from the above discussion, hybridization generally creates zones of tension between the operation of natural selection and gene flow. Speciation is thought to be proceeding through the constant action of natural selection serving to increase adaptation at the two extremes of the zone in the face of a continuous influx of genes that hamper adaptation. Intermediate types that form at the zone of contact are often inferior in fitness and disfavored by natural selection. The actual situation is far more complicated than this simple scenario suggests, even for hybrid zones that generally operate in this fashion (see Hewitt article in Otte and Endler, 1989; Harrison, 1993). More importantly, it has become increasingly clear that hybridization does not always play such a negative role in speciation. Instead, it has been shown repeatedly that hybridization can actually provide an important arena for evolutionary innovation (reviewed in Arnold article in Howard and Berlocher, 1998). In this case, hybridization presents the opportunity for the formation of unique genetic combinations through the mixing of different gene pools. In certain circumstances, these unique gene combinations end up performing better than either parental type in particular habitats and can result in the establishment of novel independent evolutionary lineages and the formation of new species.

D. Sympatric Speciation

Sympatric speciation represents the extreme opposite of allopatric speciation by being wholly independent of geographical context. New species form well within the dispersal range of the ancestral species through divergent natural selection to adapt to alternative habitats. Although ecological adaptation is thought to be the most important process in most cases, sympatric speciation can also occur through the sole operation of sexual selection (Higashi *et al.*, 1999). Because gene flow is known to have a powerful homogenizing effect on populations, it was generally thought that sympatric speciation would be an impossibility. More recently, modeling and several important case studies have revealed this skepticism to be unwarranted (see Bush in this volume, "Speciation, Processes of"; see also the article by Johnson and Gullberg as well as that by Feder in Howard and Berlocher, 1998). Most well-studied examples of sympatric speciation involve shifts from one host to another that occur in the same geographical region. The genetic adaptations involved in host shifts do not appear to be simple and often involve several different loci that interact to influence a wide variety of traits associated with fitness on a particular host. Often adaptation to one host will preclude the ability to do well on the other, especially if changes in life history are needed in order to track the host species more closely. Host-specific fitness trade-offs will then result, creating the opportunity for divergent selection to cause different host-adapted genotypes to segregate in the population. This type of divergent selection alone will probably not result in speciation, but if it is combined with a certain level of host fidelity in which individuals who are reared on a particular host and then return to feed and mate on that same host as adults, then there is a much greater probability that different populations specifically adapted to one host or another will evolve and eventually give rise to new species.

V. GENETIC PATTERNS OF SPECIES DIFFERENTIATION

It is clear from the above discussion that the tempo and mode of speciation not only are affected by geographical and ecological considerations but also can be directly influenced by the nature and genetic architecture of the traits that are diverging. Whether genetic drift can act as a facilitator of speciation during founder effect speciation, or whether specific adaptations generally result in fitness trade-offs that lead to divergence in sympatric speciation, is partially dependent on how commonly epistasis and pleiotropy underlie traits subject to change during speciation. In addition, specific models to explain genetic patterns of postzygotic reproductive isolation (see Section V.A) predict very different outcomes depending on whether isolation involves a few or many genes and whether they act recessively or not. To begin to sort through these different possibilities, it is necessary to evaluate the genetic basis of traits that have diverged during speciation to see if any common genetic patterns begin to emerge.

Ultimately, assaying the genetic basis of traits having diverged between species could be helpful in determining whether genetic patterns of change inform us about the evolutionary forces themselves that played a major role during divergence. If this is true, then genetic patterns of change alone could serve as genetic road maps for navigating the process of speciation. Although we are just in the beginning stages of research that examines speciation from a strictly genetic point of view, population genetic models investigating what the properties and distribution of these genetic changes would look like under different evolutionary scenarios have begun to be formulated (articles by Hey and Templeton in Schierwater et al., 1994; Templeton article in Howard and Berlocher, 1998; Orr, 1998).

A. Postzygotic Reproductive Isolation and Haldane's Rule

A remarkable and repeated pattern in speciation is the tendency for interspecific hybrids to be either sterile or inviable (collectively known as postzygotic reproductive isolation). This phenomenon is so striking and common that it forms the central premise of the biological species concept (see Section II.B) and has been the focus of intense genetic study over the past several decades. A central organizing principle for the study of the genetics of postzygotic reproductive isolation has been Haldane's rule. Haldane (1922) observed that "When in the F1 offspring of two different animal races one sex is absent, rare or sterile, that sex is the heterogametic sex." As Haldane's rule is obeyed when males are the heterogametic sex as well as when females are the heterogametic sex, the genetic mechanisms that explain Haldane's rule are thought to be fundamental to speciation in all taxa.

Several explanations for Haldane's rule have been posited and an almost equal number have been rejected (reviewed in Laurie, 1997; Orr, 1997; Hollocher, 1998). Part of this controversy stems from the difficulties inherent in performing genetic analyses on such traits as hybrid sterility and hybrid inviability. Part also stems from the fact that most researchers naturally sought a *single*, universally applicable genetic mechanism to account for all cases conforming to Haldane's rule. The general consensus today is that Haldane's rule requires separate explanations depending on whether or not hybrid inviability or hybrid sterility is being considered.

For hybrid inviability, Haldane's rule results because genetic incompatibilities causing inviability that evolve between species tend to act recessively. Given this recessivity, genes causing hybrid inviability will be expressed in the heterogametic sex while remaining masked in the homogametic sex, thus generating Haldane's rule. For hybrid sterility, the explanation is more complicated and involves the joint action of several different processes. As is the case for hybrid inviability, Haldane's rule for hybrid sterility results because genetic incompatibilities causing sterility also tend to act recessively. In addition, however, when males are the heterogametic sex, the evolution of hybrid male sterility is accelerated, most likely due to the additional action of sexual selection driving the rapid divergence of male sexual traits (see Section III.C; Wu et al., 1996; Rice article in Howard and Berlocher, 1998). Interestingly, much detailed work on the characterization of the genetic basis of postzygotic reproductive isolation (namely male hybrid sterility) has revealed that many genes of small effect are involved and that epistasis is of primary importance for this trait (reviewed in Wu and Palopoli, 1994; see Wu and Hollocher article in Howard and Berlocher, 1998; Hollocher, 1998); therefore, it is conceivable that the general evolution of postzygotic reproductive isolation may occur so rapidly simply because it involves epistasis, regardless of whether or not drift or selection is the primary force driving the change.

B. Patterns of Genetic Divergence for Other Species Traits

Postzygotic reproductive isolation is only one trait that has generally diverged between species. In order to formulate a broader picture of the genetic patterns of divergence, it is necessary to compare the genetic architecture of this trait with what is seen for other traits that diverge between species (reviewed in Hollocher, 1998). The traits that have been looked at in some genetic detail include interspecific mate discrimination and interspecific differences in secondary sexual characteristics. Overall, the general pattern that emerges is that these traits, too, are governed by many genes of small effect. However, in contrast to what was found for the genetic basis of postzygotic reproductive isolation, epistasis does not play a dominant role in governing the evolution of these traits.

Comparison of patterns of genetic variation within species versus patterns of genetic variation between species for the same traits can be useful for gaining insights into the evolutionary mechanisms that may have played a role during species divergence (reviewed in Hollocher, 1998). If these within-species versus between-species comparisons reveal strong similarities, then generally it can be concluded that speciation proceeded through the same general action of evolutionary forces (in terms of type, direction, and strength) normally operating on these traits within species. In contrast, strikingly different patterns of within-species versus between-species comparisons could reveal the operation of a different set of evolutionary forces operating during divergence of species than what normally occurs within species.

Interestingly, within- and between-species patterns

of genetic variation for mate discrimination and secondary sexual characteristics are very similar, indicating that divergence of these traits probably reflects the direct extension of the same evolutionary forces (most likely directional sexual selection in this case) that operate on these traits within species. In contrast, within- and between-species genetic patterns of sterility and inviability do not show similar patterns at all. Not only is the role of epistasis drastically different between the two comparisons, the relative frequency of genes that affect sterility versus inviability is completely reversed depending on whether within- or between-species patterns are considered. This disjunction between the genetic patterns observed within species versus those observed between species suggests that the evolution of postzygotic reproductive incompatibilities may result from the accumulated action of relatively rare evolutionary events happening over long periods of time. Such rare events may include periodic episodes of random genetic drift happening alone or in combination with natural and sexual selection working on these traits with varying intensity or changes in direction over the course of evolution.

VI. NEW DIRECTIONS FOR STUDYING SPECIATION

A. Incorporation of Genealogical Models

As discussed above, the classic approach for determining patterns of genetic change between species has been to cross species and analyze the genetic basis of their differences. Most detailed analyses have focused on the genetic bases of postzygotic and prezygotic reproductive isolation. The emphasis of this approach is on understanding how specific biological properties of species change as they diverge from one another. Therefore, the method involves starting with groups of organisms that show interesting differences in evolutionarily important traits and working down to the genetic level to gain insights into speciation mechanisms that may involve these traits specifically.

A very different approach is to use gene genealogies to trace the evolutionary history of genetic variation without regard to the phenotypic effects (in fact, assuming that the genes being investigated are neutral). Population genetic theory involving gene genealogies was originally developed to study the patterns of genetic variation within and among populations of a species to investigate microevolutionary processes such as gene flow, genetic drift, and natural selection. This theory is now being extended from populations up to the species level to investigate how speciation affects patterns of variation within and between species (Hey and Templeton articles in Schierwater et al., 1994; Templeton article in Howard and Berlocher, 1998; Wang et al., 1997). This approach examines the genetic structuring of neutral variation within and between species to evaluate the role such things as random genetic drift, continual gene flow, and different geographical patterns play in speciation.

The two approaches described above supply very different information about speciation. Each approach in isolation has contributed substantially to formulating hypotheses about speciation. Using a combined approach could potentially be even more powerful for investigating important issues in speciation. Future studies of speciation would benefit from continuing to use traditional genetic methods to identify candidate genes for phenotypic traits that may have been critically important in promoting divergence and then analyzing the genealogical structure of these gene regions in the context of the structuring of genes presumed to be neutral to test specific hypotheses about the role the candidate gene regions may have played during speciation. The evolutionary history of genes that contribute directly to phenotypic differences between species is embedded in the evolutionary history of all genes. A combined approach would allow us to tease apart the population dynamics of genes directly involved in speciation from the dynamics of neutral genes. In other words, the genealogical structuring of neutral genes can serve as a baseline measure of population history (supplying information on such things as the geographical patterns of divergence) to compare to the genealogical structuring of genes thought to contribute more directly to the speciation process itself (which would supply information on selective forces important in divergence). In taking this dual approach to understanding the genetics of speciation, we will be able to move one step closer to inferring the evolutionary process of speciation based on the examination of the genetic patterns of divergence.

B. Incorporation of Developmental Genetics

For a complete understanding of the process of speciation, it is not enough simply to understand in broad terms the number of genes involved in species divergence and the general distribution of their effects. As this article has revealed, this approach has been very helpful so far to give us new insights into genetic mechanisms of speciation. However, ultimately it will be necessary to identify the specific products of genes that have changed during speciation to try to understand how the changes in these gene products have specifically led to the changes in the phenotypes that we observe between species. To achieve understanding at all these different levels, it will be necessary to incorporate developmental biology into speciation theories in order to model how specific genetic changes are translated through the developmental program of an organism to give rise to new species. Although the very roots of evolutionary biology are firmly anchored in the field of development, it is only recently that attempts have been made to meld these two disciplines (for example, see Dickinson et al., 1993; Stern, 1998). The union of these two fields is important not only from a mechanistic point of view for understanding how genes are translated into phenotypes but also for investigating how the developmental systems themselves can serve as direct conduits for the action of natural selection in generating phenotypic diversity.

VII. CONCLUSIONS

The study of speciation is a rich and exciting field. Despite being the object of intense study for the past 140 years, speciation studies continue to capture the imagination of evolutionary biologists and offer a wide range of important questions ripe for investigation. The continued vibrancy of the field comes from its ability to track technological and theoretical advances being made in disciplines such as population genetics and molecular and developmental biology, and to synthesize these distinct approaches to provide fresh perspectives on important ongoing issues in speciation, which due to their innate complexity will never offer simple explanations.

See Also the Following Articles

BIODIVERSITY, EVOLUTION AND • PHENOTYPE, A HISTORICAL PERSPECTIVE • POPULATION GENETICS • SPECIATION, PROCESS OF • SPECIES, CONCEPTS OF • SPECIES DIVERSITY, OVERVIEW

Bibliography

Andersson, M. (1994). *Sexual Selection*. Princeton Univ. Press, Princeton, NJ.
Baum, D. A., and Shaw, K. L. (1995). Genealogical perspectives on the species problem. In *Experimental and Molecular Approaches to Plant Biosystematics*. (P. C. Hoch and A. G. Stevenson, Eds.). Missouri Botanical Garden, St. Louis, MO.
Burch, C. L., and Chao, L. (1999). Evolution by small steps and rugged landscapes in the RNA virus phi6. *Genetics* 151, 921–927.
Coyne, J., Barton, N. H., and Turelli, M. (1997) Perspective: A critique of Sewall Wright's shifting balance theory of evolution. *Evolution* 51, 643–671.
Darwin, C. (1872). *The Origin of Species*, 6th ed. Reprinted in 1976 by Macmillan, New York.
Dickinson, W. J., Yang, Y., Schuske, K., and Akam, M. (1993). Conservation of molecular prepatterns during the evolution of cuticle morphology in *Drosophila* larvae. *Evolution* 47, 1396–1406.
Dobzhansky, T. (1937). *Genetics and the Origin of Species*. Columbia Univ. Press, New York.
Dobzhansky, T. (1940). Speciation as a stage of evolutionary divergence. *Am. Nat.* 74, 312–321.
Eberhard, W. G. (1985). *Sexual Selection and Animal Genitalia*. Harvard Univ. Press, Cambridge, MA.
Ereshefsky, M. (Ed.) (1992). *The Units of Evolution: Essays on the Nature of Species*. MIT Press, Cambridge, MA.
Gavrilets, S., and Hastings, A. (1996). Founder effect speciation: A theoretical reassessment. *Am. Nat.* 147, 466–491.
Grant, P. R. (Ed.) (1998). *Evolution on Islands*. Oxford Univ. Press, Oxford.
Grant, P. R., and Grant, B. R. (1997). Genetics and the origin of bird species. *Proc. Natl. Acad. Sci. USA* 94, 7768–7775.
Haldane, J. B. S. (1922). Sex-ratio and unisexual sterility in hybrid animals. *J. Genet.* 12, 101–109.
Harrison, R. G. (Ed.) (1993). *Hybrid Zones and the Evolutionary Process*. Oxford Univ. Press, Oxford.
Higashi, M., Takimoto, G., and Yamamura, N. (1999). Sympatric speciation by sexual selection. *Nature* 402, 523–526.
Hollocher, H. (1998). Reproductive isolation in *Drosophila*: How close are we to untangling the genetics of speciation? *Curr. Opin. Genet. Dev.* 8, 709–714.
Howard, D. J., and Berlocher, S. H. (Eds.) (1998). *Endless Forms: Species and Speciation*. Oxford Univ. Press, Oxford.
Huynen, M. A. (1996). Exploring phenotype space through neutral evolution. *J. Mol. Evo.* 43, 165–169.
Iwasa, Y., and Pomiankowski, A. (1995). Continual change in mate preferences. *Nature* 377, 420–422.
Kaneshiro, K. Y. (1989). The dynamics of sexual selection and founder effects in species formation. In *Genetics, Speciation, and the Founder Principle* (L. V. Giddings, K. Y. Kaneshiro, and W. W. Anderson, Eds.). Oxford Univ. Press, Oxford.
Lande, R. (1981). Models of speciation by sexual selection on polygenic traits. *Proc. Natl. Acad. Sci. USA* 78, 3721–3725.
Laurie, C. C. (1997). The weaker sex is heterogametic: 75 years of Haldane's rule. *Genetics* 147, 937–951.
Lenski, R. E., and Travisano, M. (1994). Dynamics of adaptation and diversification: A 10,000-generation experiment with bacterial populations. *Proc. Natl. Acad. Sci. USA* 91, 6808–6814.
Mallet, J. (1995). A species definition for the modern synthesis. *Trends Ecol. Evol.* 10, 294–299.
Mayr, E. (1963). *Animal Species and Evolution*. Harvard Univ. Press, Cambridge, MA.
Mayr, E. (1982). *The Growth of Biological Thought: Diversity, Evolution, and Inheritance*. Harvard Univ. Press, Cambridge, MA.
Nixon, K. C., and Wheeler, Q. D. (1990). An amplification of the phylogenetics species concept. *Cladistics* 6, 211–223.
Orr, H. A. (1997). Haldane's rule. *Annu. Rev. Ecol. Syst.* 28, 195–218.
Orr, H. A. (1998). The population genetics of adaptation: The distri-

bution of factors fixed during adaptive evolution. *Evolution* 52, 935–949.
Otte, D., and Endler, J. A. (1989). *Speciation and Its Consequences.* Sinauer, Sunderland, MA.
Paterson, H. E. H. (1985). The recognition concept of species. In *Species and Speciation* (E. S. Vrba, Ed.). Transvaal Museum, Pretoria.
Rice, W. R., and Hostert, E. E. (1993). Laboratory experiments on speciation: What have we learned in 40 years? *Evolution* 47, 1637–1653.
Ridley, M. (1996). *Evolution,* 2nd ed. Blackwell Sci., Oxford.
Schierwater, B., Streit, B., Wagner, G. P., and DeSalle, R. (Eds.) (1994). *Molecular Ecology and Evolution: Approaches and Applications.* Birkhauser Verlag, Basel.
Stern, D. L. (1998). A role of *Ultrabithorax* in morphological differences between *Drosophila* species. *Nature* 396, 463–466.
Wade, M. J., and Goodnight, C. J. (1998). Perspective: The theories of Fisher and Wright in the context of metapopulations: when nature does many small experiments. *Evolution* 52, 1537–1553.
Wang, R. L., Wakeley, J., and Hey, J. (1997). Gene flow and natural selection in the origin of *Drosophila pseudoobscura* and close relatives. *Genetics* 147, 1091–1106.
Whitlock, M. C., Phillips, P. C., Moore, F. B.-G., and Tonsor, S. J. (1995). Multiple fitness peaks and epistasis. *Annu. Rev. Ecol. Syst.* 26, 601–629.
Wiley, E. O. (1978). The evolutionary species concept reconsidered. *Syst. Zool.* 27, 17–26.
Wright, S. (1932). The roles of inbreeding, crossbreeding and selection in *evolution. Proc. Sixth Int. Cong. Genet.* 2, 356–366.
Wu, C.-I., and Palopoli, M. F. (1994) Genetics of postmating reproductive isolation in animals. *Annu. Rev. Genet.* 28, 283–308.
Wu, C.-I., Johnson, N. A., and Palopoli, M. F. (1996). Haldane's rule and its legacy: Why are there so many sterile males. *Trends Ecol. Evol.* 11, 281–284.

SPECIES–AREA RELATIONSHIPS

Edward F. Connor* and Earl D. McCoy[†]
*San Francisco State University and [†]University of South Florida

I. Introduction and Underlying Mechanisms
II. Sampling and Statistical Practice in Description of Species–Area Relationships
III. Functional Form of the Species–Area Relationship
IV. Interpretation of the Parameters of Species–Area Models
V. Use of Species–Area Curves in Conservation Biology

GLOSSARY

relative abundance distribution The frequency distribution depicting the number of species in a community as a function of the number of individuals comprising each species.
species–area curve A graphical depiction of the dependence of species richness on area.
species–area model A function used to describe species–area curves.
species–area relationship The dependence of the number of species in a sample region on the area or size of the region.

A SPECIES–AREA RELATIONSHIP is simply the observation that the number of biological species found in a region is a positive function of the area of the region. Species–area relationships are depicted graphically as a bivariate plot of species richness on the ordinate and area on the abscissa, a species–area curve (Fig. 1). Species–area relationships appear to be ubiquitous, having been observed for a wide array of taxa ranging from diatoms to fish, insects, birds, vascular plants, and mammals and for geographical entities such as islands, political entities, woodland, grassland, and cropland habitat patches, lakes, river drainages, and artificial substrates from microscope slides to synthetic sponges and slates.

I. INTRODUCTION AND UNDERLYING MECHANISMS

H. G. Watson first described the species–area relationship in 1835 by remarking that as the area of a county in England increases by a factor of 10 the number of plant species found in that county increases by a factor of 2. The long history of discourse on species–area relationships has evolved from a focus prior to the 1960s on its empirical utility in determining optimal sample size and sample number for community description, for determining the minimum area of a community, and in extrapolating predictions of species richness to areas larger than those sampled to a focus in the 1960s and 1970s on the mechanisms underlying the species–area relationship, in finding the best mathematical

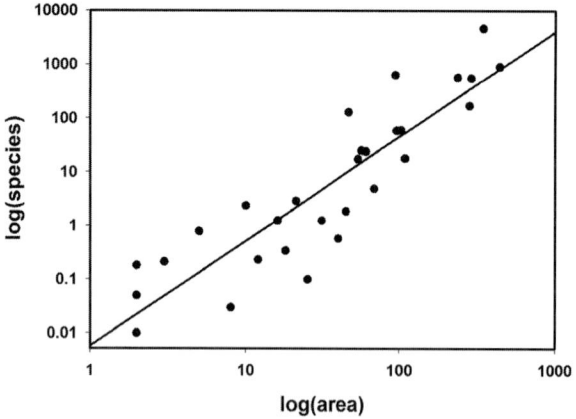

FIGURE 1 Species–area curve for the vascular plants of the Galapagos Archipelago. The logarithm of species richness is commonly plotted as a function of the logarithm of area, but species richness may also be plotted as a function of area or log(area).

model for species–area relationships, and in explaining species–area relationships in the context of the equilibrium theory of island biogeography (Connor and McCoy, 1979). Since 1980, the primary focus of discussion involving species–area relationships has been their use and application in conservation biology to determine the optimal design of nature reserves and to project the expected loss of species richness from a region undergoing specified levels of area reduction (habitat loss). We review the discourse on species–area relationships beginning in the 1960s and focus our synthesis on the application of species–area relationships in conservation biology.

Three biological mechanisms have been proposed to account for species–area relationships: (1) the habitat diversity hypothesis, (2) the area per se hypothesis, and (3) the passive sampling hypothesis.

A. The Habitat Diversity Hypothesis

The habitat diversity hypothesis (Williams, 1964) proposes that the increase in species richness in large areas relative to small areas arises because large areas have a greater variety of habitats than small areas. This greater variety of habitats permits species that are only found in specific habitats to occur in large areas and permits species that require multiple habitats to persist in large areas, resulting in higher species richness in large areas than in small areas and the existence of species–area relationships. The habitat diversity hypothesis views area as affecting species richness indirectly because of its association with habitat diversity rather than any direct effect of area on the ability of species to colonize or persist in larger areas. Studies that demonstrate positive correlations between the number of species and the number of habitats on an island have been viewed as supporting the operation of the habitat diversity hypothesis.

B. The Area Per Se Hypothesis

The area per se hypothesis (Simberloff, 1976) is based on the assumptions that the abundance of each species in a sample region varies as a positive function of that region's area and that the probability of each species going stochastically extinct in that area is a negative function of abundance, and therefore of area. Given these assumptions, large areas would have more species than small areas because more species would persist (not go locally extinct) in large areas. The area per se hypothesis suggests that even in a group of patches consisting of a single type of habitat, one would observe a species–area relationship. Correlations between species richness and area in studies that purport to examine a single habitat type and experimental studies that examine the consequences of reducing patch area are viewed as evidence supporting the operation of the area per se hypothesis.

C. The Passive Sampling Hypothesis

The passive sampling hypothesis (Connor and McCoy, 1979) conjectures that larger areas are more likely to receive more colonists than small areas and that these colonists are likely to represent a wider array of species than the pool of colonists arriving on small areas. Therefore, purely as a result of the higher abundance of colonists expected for large areas and independent of any increase in habitat diversity or reduction in extinction probabilities, one would expect more species to arrive on large areas, leading to species–area relationships. Several authors have examined the number of colonists arriving at their study sites and conclude that passive sampling accounts in part for the species richness of invertebrates on intertidal boulders, fish in stream pools and riffles, and birds on forest habitat islands. Each of these studies shows that for habitats that are colonized seasonally, those patches receiving more colonists have higher species richness.

Two other area-dependent factors that affect species richness have more recently entered discussions concerning species–area relationships, particularly for habitat patches: the resource concentration hypothesis and edge effects.

D. The Resource Concentration Hypothesis

The resource concentration hypothesis attempts to explain the phenomenon that habitat patches with large amounts of resources (e.g., monocultures, areas of high plant density, or large patches) have higher densities of insects. Therefore, the resource concentration hypothesis conjectures that population density should be positively correlated with patch area. In 1973, R. B. Root conjectured that the higher density of animals in larger patches might be solely a consequence of movement behavior (the movement hypothesis); herbivores are more likely to find, and remain in, large, monospecific stands of their host plant than in small or heterogeneous patches. If many species have higher population densities in large patches because of the resource concentration hypothesis, then extinction probabilities should be even lower than expected from the area per se hypothesis, and hence resource concentration could contribute to observed species–area relationships. E. F. Connor and co-workers have recently reviewed the existing literature on the relationships between animal population density and patch area and found for insects and birds that positive area–density relationships are common, but not so for mammals.

E. Edge Effects

Edge effects, or habitat edge dependent changes in abundance or risk of mortality, have been reported for species in a variety of taxa in habitat patches. Given that the proportion of a patch that occurs within any fixed distance from its edge is inversely related to area, edge effects could lead to species–area relationships even within a single habitat type. Such an edge effect on species richness would be mediated by a reduction in the abundance of a species on small patches because of a larger amount of "edge habitat," leading to higher probabilities of local extinction. Therefore, part of the dependence of species richness on area that has previously been attributed to area per se may actually be caused by edge effects.

F. Multiple Causes of Species–Area Relationships

Habitat diversity, area per se, passive sampling, edge effects, and resource concentration are not mutually exclusive mechanisms and may operate individually or in combination to cause species–area relationships (Connor and McCoy, 1979). Experiments on invertebrate colonization on artificial substrates of varying size and habitat diversity clearly demonstrate the joint contribution of habitat diversity and either the area per se or passive sampling hypothesis, or both, to the generation of species–area relationships. Studies that subsampled islands using a constant-sized quadrat on each island in an attempt to eliminate the effect of habitat diversity found strong effects of habitat diversity, but species richness remained a positive function of island area even for data derived from constant-sized quadrats. Using path analysis, D. D. Kohn and D. M. Walsh concluded that area had both a direct effect on species richness (area per se and/or passive sampling) and an indirect effect mediated by the effect of area on habitat diversity.

The evidence that would allow one to clearly partition the causes of a specific species–area relationship among these underlying mechanisms is exacting. Habitat diversity is difficult to measure, and edge effects might make it impossible to separate the effects of increasing area from the effects of area on habitat diversity. The area per se, resource concentration, and passive sampling hypotheses act by affecting the abundances of species on an island. Area per se and resource concentration do so by reducing extinction probabilities via allowing larger populations (or more dense populations) to persist on large islands, and passive sampling does so by proposing that greater numbers of colonists arrive on large islands. To demonstrate an effect of area per se or resource concentration, one must ultimately show that species on average have larger population sizes (or in the case of resource concentration, higher population densities) and lower extinction probabilities on large islands or habitat patches. To demonstrate an effect of passive sampling, one must show that the arrival rate of colonists on large islands or habitat patches is greater than for small patches and that the colonizing individuals arriving on large patches comprise a greater number of species than those arriving on small patches.

To demonstrate that edge effects contribute to species–area relationships would require evidence that species absences from small patches could be uniquely attributed to edge effects and not to the alternative mechanisms discussed above.

II. SAMPLING AND STATISTICAL PRACTICE IN DESCRIPTION OF SPECIES–AREA RELATIONSHIPS

A. Sampling Practice

Two main sampling schemes have been used to generate data on the relationship between species richness and area. The most widely used approach has been to sample physically separated areas such as islands or habitat patches or to sample adjacent or abutting areas of continuous habitat as independent, nonoverlapping replicates (Fig. 2). For physically separated areas, sampling from the observed natural range of areas generates a range of sample areas. For continuous habitat, the region is divided arbitrarily into a series of nonoverlapping subregions to generate a range of sample areas. The alternative approach, which has been used widely by plant ecologists, is to enumerate the locations of species within a larger region and generate a sample of areas by subsampling a range of areas within the larger region. These subsamples may be overlapping or not, depending on the choice of the researcher (Fig. 2).

1. Independent Areas

Physically independent sample areas have been widely used because species–area data could be readily generated for natural geographical units, such as islands, by combining published monographic species lists with published data on island areas. The majority of species–area curves were published in the 1960s and 1970s by scientists gleaning existing data from the monographic literature and combining data on species richness with published estimates of area. Species–area curves based on gleaned data could be produced relatively quickly via a trip to the library. Another advantage of using physically independent sample areas is that they are

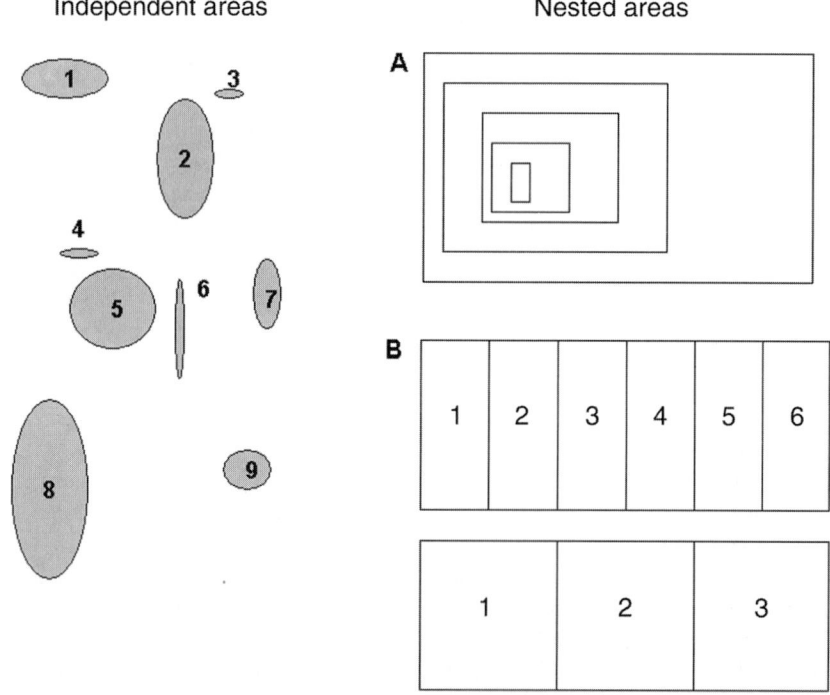

FIGURE 2 Independent and nested sampling designs used in species–area studies. Independent areas are physically separate. Nested areas may be serially self-contained as in A or abutting as in B. If abutting areas are used, then the sampled region is partitioned into a series of equal-sized, adjacent areas, and this process is repeated for the range of different sized areas. Only two areas are depicted in B, one with six replicates and one with three.

arguably also statistically independent. The statistical independence of separate, nonoverlapping areas allows use of the methods of statistical inference and hypothesis testing associated with ordinary least-squares (OLS) regression. The application of OLS regression permitted the estimation of the parameters (and their standard errors) for any of the proposed functional forms of the species–area relationship that could be made sufficiently linear by transformation of the data and an assessment of the statistical significance of the fit of a particular model. Furthermore, the use of statistically independent areas allowed the rigorous comparison of parameter estimates between studies via analysis of covariance or other simpler techniques.

2. Nested Areas

Plant ecologists have generated data for many years on the relationship between species richness and plot or quadrat area for a variety of plant communities. Much of this work was done to determine the optimal plot size to use when describing plant communities. Presumably because of the saving in sampling effort, plant ecologists used serially self-contained or nested quadrats rather than physically independent quadrats to determine optimal quadrat size. However, nested quadrats are not statistically independent so it is inappropriate to use OLS regression to fit a linear model to such data and assess the fit of the model or estimate its parameters. Therefore, use of nested quadrats to examine species–area relationships in plant communities was largely discontinued by the end of the 1970s. Recent research programs aimed at understanding the structure and dynamics of tropical forests have once again led to studies of species–area relationships using nonindependent data. However, Condit *et al.* (1996) partition their largest quadrat into as many equal-sized quadrats as possible and use the mean for all nonoverlapping quadrats to estimate species richness for that quadrat size. They repeat this process of partitioning their large quadrat (50 ha) into the maximum number possible of nonoverlapping subquadrats for a variety of quadrat sizes. Condit *et al.* (1996) then plot the mean and standard error of species richness for each quadrat size as a function of quadrat area to produce a species–area curve. This approach leads to appropriate estimates of species richness and its standard error for each individual quadrat size, since within a size category the quadrats used to estimate the mean and standard error of species richness are independent. However, since the same sampled area is used to estimate species richness for each quadrat size, these estimates are not statistically independent and Condit *et al.* (1996) appropriately refrain from using regression techniques that require independence to fit statistical models to species–area data.

B. Statistical Practice

OLS regression has been used to fit models to species–area data. By applying OLS regression, estimates of the model's parameters and their associated standard errors can be obtained readily. If one makes the additional assumptions that the model's errors are independent, homoscedastic, and normally distributed and that species richness or the logarithm of species richness is a linear function of area or its logarithm, then rigorous statistical inferences about the parameters may be made. In recent years, nonlinear regression procedures have become widely available so that inherently curvilinear species–area models may be directly fit to data. Nevertheless, OLS regression continues to be used widely.

P. J. Vincent, J. M. Haworth, and M. R. Williams point out that the assumptions of normality and homogeneity of error variances are unlikely to be met by species–area data. First, species richness is unlikely to be normally or log-normally distributed and probably should be treated as a discrete, rather than a continuous, variable. Second, the variance in species richness is likely to be a function of the mean species richness. These properties arise in part because on very small areas both the mean and variance of species richness must be nearly zero. Vincent and Haworth suggest analyzing species–area data using a generalized linear model and treating species richness as a Poisson-distributed variable. Williams also suggests analyzing species–area data using a generalized linear model but recommends that species richness be treated as a binomially distributed variable (see discussion of Williams' extreme value model below). In both cases, parameters and their standard errors can be estimated and rigorous statistical inferences can be made. Unfortunately, few authors have explored the treatment of species-richness data as Poisson or binomial variables in studies of species–area relationships.

III. FUNCTIONAL FORM OF THE SPECIES–AREA RELATIONSHIP

When species richness is plotted as a function of area, the resulting plot is a curve that may be linear, concave downward, concave upward, or sigmoid (Fig. 3). The shape of species–area curves appears to be a function of the particular range of areas studied, and this may,

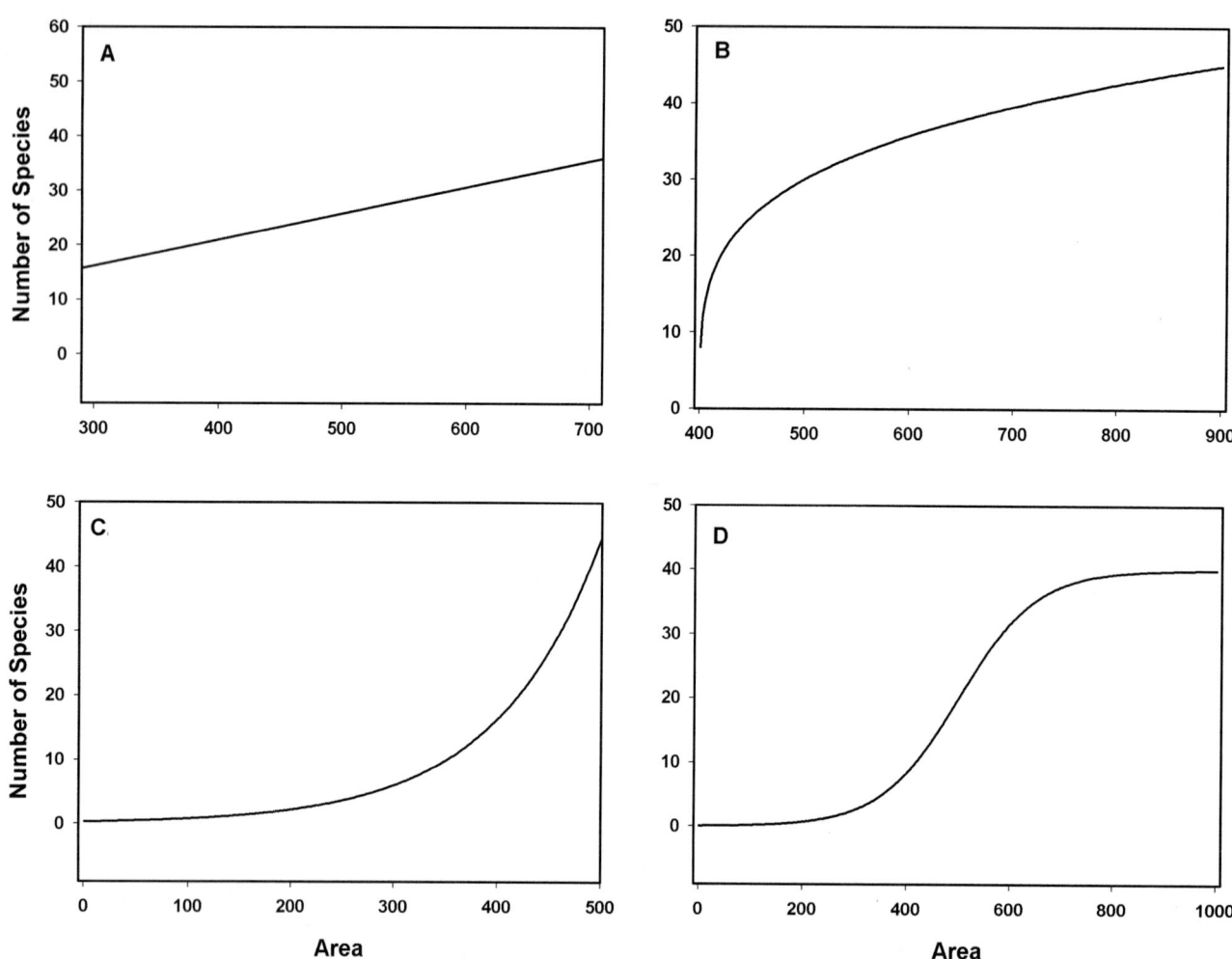

FIGURE 3 Shapes of species–area curves encountered in empirical studies: (A) linear species–area curve; (B) concave-downward species–area curve (linearized by the log–log transformation); (C) concave-upward species–area curve (linearized by a semilogarithmic transformation, area log-transformed); (D) sigmoid species–area curve (usually not transformed). Linear curves are encountered at intermediate spatial scales, concave-downward curves at larger spatial scales, concave-upward curves at small spatial scales, and sigmoid curves when a wide range of spatial scales is studied.

in part, explain the variety of transformations used to linearize species–area curves.

Historically, much of the discussion of species–area relationships focused on the specific functional form of the relationship between species richness and area. In 1921, O. Arrhenius proposed that species–area relationships followed a power function (Fig. 3B):

$$S = cA^z,$$

which, for statistical convenience, has commonly been approximated by the log–log transformation:

$$\log S = \log c + z \log A.$$

with c and z as constants. The double-logarithmic transformation linearizes the power-function model, so that species–area curves are linear on a double-log plot (Fig. 1). In 1922, H. A. Gleason championed an exponential model of the species–area relationship (Fig. 3C):

$$S = \log c + z \log A,$$

because he observed the power-function model to predict impossibly large numbers of species for large areas. The exponential model produces species–area curves that are linear on a semilogarithmic plot. Both of these models were supported because they seemed to fit specific data sets using independent areas reasonably well and because it was argued that they could be derived as a consequence of assuming that the distribution of individuals among species (the relative abundance distribution) was either log-normal or log-series, respectively. Other ecologists reported that when a wide range of areas were sampled, neither the power-function model nor the exponential model fit their data since species–area curves appeared sigmoid in shape (Fig. 3D).

Connor and McCoy (1979) examined the fit of the power-function model, the exponential model, and two other models to data from 100 species–area studies. They found that while the power-function model often fit the data well, it was not found to be the best fit model substantially more often than other models. Based on this analysis, the observation that the log–log transformation can linearize a wide range of curves, and because of its widespread use, Connor and McCoy (1979) recommended continued use of the power-function model of species–area relationships.

Recent discussions of the functional form of the species–area relationship have continued to promote use of the power-function model but have continued to justify its use because of its hypothesized connection to log-normal relative abundance distributions (Rosenzweig, 1995). Given a common log-normal relative abundance distribution from which the abundances of species on all sites (islands or habitat patches) arise, it is possible to derive species–area data that are reasonably well fit by a power-function model. However, simply because a particular set of species–area data is reasonably well fit by a power-function model does not imply that this fit derives as a consequence of sampling from a common log-normal relative abundance distribution (Hanski and Gyllenberg, 1997). Many island archipelagos are colonized from multiple source regions rather than from a single source pool, and individual islands within such archipelagos may receive differing proportions of their colonists from each source. On the other hand, habitat patches may be more accurately viewed as units whose species composition is determined by sampling from a source pool with a single relative abundance distribution.

Continued debate about the functional form of the species–area relationship assumes that underlying the variability we observe in empirically estimated species–area curves lies a single true function that will characterize this relationship. However, most data on species–area relationships using independent areas have small sample sizes, cover a narrow range of areas, and may be confounded by area-dependent sampling effort, among other problems. Given the quality of the data available, it will be devilishly difficult to discern the functional form of the species–area relationship, even if a single form exists. Furthermore, it is possible that differences among taxa or spatial scales could generate species–area relationships of different functional forms.

A. Self-Similarity and the Power-Function Model of Species–Area Relationships

J. Harte and co-workers have recently proposed an alternative derivation of the power-function model of the species–area relationship. This derivation comes from examining nested areas and making the assumption that successive partitions of a continuous area have the fractal property of "self-similarity." That is, the distribution of species in successive partitions is independent of spatial scale. Beginning with a large continuous area A_0 with S_0 species, bisections of A_0 such that the rectangles comprising the two halves of A_0 have area $A_i = A_0/2^i$ each will have on average S_i species, where i indicates the ith bisection of A_0. The assumption of self-similarity requires that a species known to be in A_i will have probability a of being found in at least a specific one of the two rectangles of area A_{i+1} created by bisection, and therefore that the fraction of species found in A_i that are found in a specific one of the A_{i+1} rectangles equals the same constant a for all i bisections. The constancy of a for all i bisections follows from the assumption of self-similarity since a, the probability of occurrence in a half-patch under bisection, must be independent of spatial scale. Harte et al. (1999) show that the assumption of self-similarity leads directly to the power-function model of the species–area relationship: $S_i = cA_i^z$, with the self-similarity parameter $a = 2^{-z}$. Furthermore, since under successive bisections of an initial continuous area the value of a must lie between 0.5 and 1, the relationship between a and z dictates that the slope of the power-function model, z, must lie between 0 and 1. Hence, the power-function model of the species–area relationship is derivable without recourse to any assumptions about the underlying relative abundance distribution of species.

Harte and colleagues have extended the assumption of self-similarity to generate an expectation for the relationship between the number of endemic species and area and the expected spatial turnover between two patches isolated by a known distance, d. These extensions of the assumption of self-similarity provide a basis for estimating the expected loss of species from areas undergoing habitat reduction, for calculating z from spatially separated sites, and for estimating species richness at spatial scales larger than commonly possible in previous studies of species–area relationships. However, as Harte and colleagues point out, the assumption of self-similarity is not likely to hold for a wide range of spatial scales, so care must be taken to test the assumption of self-similarity for the taxa and spatial scales of inference. While Harte and colleagues have published a few examples where the assumption of self-similarity seems reasonable over a specific range of spatial scales, Plotkin and colleagues show that tropical forest trees are not self-similar in distribution over spatial scales from 1 to 10^5 m^2.

B. Extreme Value Model and Random Placement

An alternative model of species–area relationships based on "random placement" was first proposed by B. D. Coleman in 1981 (Coleman et al., 1982) and extended by Williams (1995). The random placement model of species–area relationships derives the expected number of species on a site, $\bar{s}(\alpha_k)$, of area A_k as a consequence of placing the n_i individuals of the ith species on sites independently and at random with probability $\alpha_k = A_k/A_t$, where A_k is the area of the kth site, A_t is the combined area of all sites, and S is the total number of species among all sites:

$$\bar{s}(\alpha_k) = S - \sum_{i=1}^{S} (1 - \alpha_k)^{n_i}.$$

As presented by Coleman et al. (1982), data on the total abundance of each species combined among all sites were required to estimate the expected species–area curve under the random placement model. Because the random placement model required data that ecologists seldom have, censuses of the abundances of each species at each site, and because this model does not yield fitted parameters comparable to other species–area models, few authors attempted to fit the random placement model to their data.

Williams (1995) extended and adapted the random placement model to be approximated by an extreme value function which permits model fitting within the context of the generalized linear model and requires data only on species richness and area, not the abundances on the individual species. The extreme value function model of species richness, $\bar{s}$, in log A is then

$$\bar{s} = P[1 - \exp(-\exp(y \log A + \log d))],$$

with P being the number of species in the species pool, and y and log d the slope and intercept of the model, respectively. Williams (1995) outlines a method for estimating P if the species composition of the biota is unknown but suggests that the best estimate of P is the total number of species found. The extreme value function model has been fit to a limited number of data sets but appears sigmoid when species richness is plotted as a function of log(area). The random placement/extreme value function model of species–area relationships is appealing because it is derived under a hypothesis of independence within and between species. However, as Williams (1995) points out, discriminating between the power function and the extreme value function models with most existing data sets derived from sampling independent areas will be difficult.

IV. INTERPRETATION OF THE PARAMETERS OF SPECIES–AREA MODELS

The widespread use of the power-function model of the species–area relationship coupled with specific numerical expectations for its slope parameter, proposed by F. W. Preston in 1960, led to numerous attempts to infer biological significance to values of this parameter. The slope parameter measures the rate at which species are added as area increases, and the intercept parameter has been considered a function of taxon-specific attributes and environmental variation. Only a limited attempt has been made to offer biological interpretations of the intercept parameter, partly because in the power-function model the absolute value of c depends on the units in which area is measured (Connor and McCoy, 1979; Rosenzweig, 1995). The parameters of other models of the species–area relationship have generally been treated as fitted statistical constants, with no attempt to interpret them biologically.

The slope of the power-function model of the species–area relationship and patterns of variation in this parameter have been subjected to considerable analysis and interpretation. Connor and McCoy critiqued many of these interpretations in their 1979 review. We briefly examine a few of these interpretations and touch on those proposed more recently.

A. Canonical Slope Values

F. W. Preston proposed that isolated islands in equilibrium that sample colonists from a common log-normal relative abundance distribution with parameter $\gamma = 1$ (Preston's canonical hypothesis) will have a slope of

0.262 in the power-function model. He subsequently broadened the range of slope values that he expected from isolates to values between 0.17 and 0.33. R. M. May showed in 1975 that using a wide range of lognormal parameter values leads to power-function slope values in the 0.15–0.39 range. Many authors have generated slopes from the power-function model of the species–area relationship and interpreted values in the range of 0.2–0.4 to be consistent with Preston's idea that species–area curves arise because the species richness of islands results from sampling from an underlying lognormal relative abundance distribution. However, Connor and McCoy (1979) challenged Preston's idea of a canonical range of slope values by showing that slopes in the 0.2–0.4 range are expected purely as a consequence of the tendency to publish studies that show a high correlation between species richness and area and because the variance in species richness will always be less than the variance in area. Connor and McCoy (1979) concluded that slope values from the power-function model of species–area relationship in the 0.2–0.4 range could not be used as evidence for the existence of an underlying log-normal relative abundance distribution.

B. Island–Mainland Differences in Slope Values

Preston extended his idea that islands or isolates should have power-function slopes in the 0.2–0.4 range to project that nonisolated or mainland areas should have lower slope values than isolated areas. His rationale derived from his belief that nonisolated areas would sample from a truncated relative abundance distribution with a higher ratio of species to individuals. In their monograph "The Theory of Island Biogeography," MacArthur and Wilson (1967) modified Preston's idea, suggesting a specific range to be expected for slopes derived from nonisolated areas, 0.12–0.19, and explaining the lower slopes as arising because of the "transient hypothesis." The transient hypothesis suggests that more transient individuals will be encountered in small, nonisolated areas than in small, isolated areas, which, in turn, will lead to more species being encountered on small, nonisolated areas than on small, isolated areas. The greater number of species found on small, nonisolated areas would depress the slope of the species–area curve for mainland areas relative to that expected for islands. The available evidence, while limited, is consistent with the idea that power-function slopes for mainland areas are lower than those for islands.

Hanski and Gyllenberg (1997) develop dynamical models of species incidence that generate predictions about the slopes of species–area curves as a function of the moments of the relative abundance distribution and the ratio of species' colonization and extinction rates. Hanski and Gyllenberg (1997) develop two models, one in which sites are colonized from an external source, their "island–mainland model," and one in which the sources of colonists are internal to the system of sites, their "metapopulation model." They find that species–area slopes are lower for their metapopulation model and claim that such a model is analogous to mainland areas, while their island–mainland model is analogous to truly insular situations. Hanski and Gyllenberg (1997) suggest that their models imply that the observation of lower slope values on mainland sites is a result of metapopulation dynamics rather than the transient hypothesis.

C. The Effect of Isolation on Slope Values

MacArthur and Wilson extended their transient hypothesis to predict that the slopes of species–area curves should be lower for distant archipelagos of islands than for islands located close to mainland areas. However, T. W. Schoener illustrated with species–area curves for birds in 23 archipelagos that exactly the opposite pattern occurs. Schoener's result suggests that the slope of species–area curves depends on the size of the source pool of colonizing species, which would be smaller for distant than near island groups. In other words, the low slopes for species–area curves observed for isolated archipelagos tell us no more than that isolated biotas are depauperate. Hanski and Gyllenberg (1997) suggest that Schoener's observation of lower slope values for isolated archipelagos arises because isolated archipelagos behave according to their metapopulation model, not as an island–mainland system.

D. Other Interpretations of Slope Values

Many other efforts have been made to explain variation in the slope of the power-function model of the species–area relationship. These efforts include (1) attempts to equate the slope of the species–area curve with β, or between-habitat diversity, and the intercept with α, or within-habitat diversity, (2) attempts to predict patterns in the latitudinal dependence of the species–area relationship (both slope and intercept), and (3) attempts to identify taxonomic and trophic group differences in species–area curves, among others. Connor and McCoy (1979) critiqued many of the attempts to interpret the parameters of the power-function model of the species–area relationship and recommended that these parameters be viewed as fitted constants, with no specific biological interpretation.

E. Conclusions

Species–area relationships represent a pattern expected in nature that may arise from the colonization and development of quasi-independent biotas on islands or habitat patches (island-colonization model) or from the loss, reduction, and fragmentation of a previously continuous or widespread biota into remnant habitat patches (mainland-vicariance model). Historically, inquiry into the mechanisms underlying the species–area relationship and its functional form has been biased toward the island-colonization model. However, models based on self-similarity more closely represent the mainland-vicariance model and provide a basis to unify species–area relationships with other patterns in the geographical distribution of species (e.g., distribution and abundance relationships, compositional similarity among sites, and relative abundance distributions). Continued empirical study of the species–area relationships and inquiry into the functional forms expected under specific biological and sampling models will continue to improve our understanding of spatial patterns of species richness.

V. USE OF SPECIES–AREA CURVES IN CONSERVATION BIOLOGY

What can an understanding of the species–area relationship contribute to the preservation of biodiversity? The question has interested ecologists for more than 30 years since the pioneering work of N. W. Moore, E. Maarel, and others and has led, almost from the beginning, to a bewildering confusion of missteps and dead ends (see Shafer (1990) for an overview). Implicit in this question are the important assumptions that species richness is the primary object of preservation and that area is the primary influence on species richness. Here, we shall explore two separate, but similar, aspects of the question. We shall see what ecologists have been able to conclude about the loss of species accompanying area reduction and what they have been able to conclude about the best way to slow the loss.

A. Loss of Species from Area Reduction

As we have illustrated, the species–area relationship is an extraordinarily common pattern in nature, and it has been documented numerous times. It would seem, therefore, that calculating the loss of species accompanying a certain amount of area reduction would be a rather straightforward exercise. All that one would need to know to perform the calculation would be the original number of species ($S_{original}$), the amount of area reduction ($A_{reduced}/A_{original}$), and the slope of the species–area relationship (z). For example, if one assumes a power-function model of the species–area relationship ($S = cA^z$), then $S_{original} = c(A_{original})^z$, $S_{reduced} = c(A_{reduced})^z$, and, therefore, $S_{reduced}/S_{original} = (A_{reduced}/A_{original})^z$ (Fig. 4). Although calculations of this sort are reasonably com-

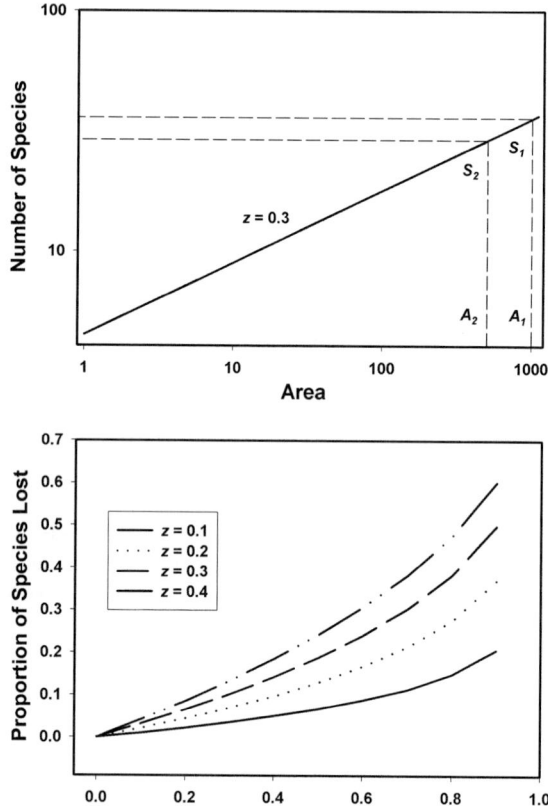

FIGURE 4 Loss of species richness under specified levels of habitat loss expected using the power-function model of the species–area curve. (A) Illustrates the SLOSS controversy calculation of the number of species expected for a single large reserve of 1000 units of area (34.5 species) versus the number of species expected for a reserve half that size (500 area units), 29 species. If the species lists from the two half-sized reserves do not overlap extensively, then the two half reserves should contain more species than the single large reserve. For this example, $S = 4.5A^{0.3}$. (B) The expected proportion of species lost when area is reduced by a fixed proportion as calculated using the power-function model of the species–area relationship for various values of z and $S = 4.5A^z$.

mon, they require at least five assumptions. First, area reduction completely eliminates species that were originally present (but some species may survive on relict fragments or disturbed lands). Second, all of the species that were originally present were distributed homogeneously (but most species are unlikely to be distributed in this manner). Third, an appropriate model has been selected to describe the species–area relationship. Fourth, the slope of the species–area relationship is accurate and a constant (but it is not likely to be, for a number of reasons). Fifth, the loss of species is a direct consequence only of area reduction (but it is not likely to be, for a number of reasons). The last two assumptions have received the most attention from ecologists. We shall discuss both of them further.

Correct choice of the slope of the species–area relationship is critical. Clearly, if one knew the original number of species and area and the reduced number of species and area, then the slope would be determined precisely. The point of the calculations presented above, however, is to estimate potential species loss, before it happens, so the slope of the species–area relationship also must be estimated, from theory and empirical study. A great deal of uncertainty accompanies such estimation. First, one must decide which model of species loss is appropriate. For example, if the relationship between species and area were assumed to be linear (Fig. 3A), then a much different rate of loss of species would be predicted to occur than if the relationship were assumed to be a power function (Fig. 3B). Second, one must assume that the slope of the species–area curve is constant between the spatial scales over which species loss is to be estimated. A nonconstant slope value would suggest that an appropriate species–area model has not been completely specified and could lead to substantial misestimation of species loss. Third, one must decide if the reduced area better represents an isolate (a "true" island) or simply a subsample of the original area. Typically, slopes of species–area relationships are greater in the first case (approximately 0.20–0.40, on a log–log scale) than in the second (approximately 0.12–0.19, on a log–log scale). Furthermore, it may be that the reduced area could come to function less like a subsample and more like a true island over time, so that the range of potential slopes could expand dramatically over time. Fourth, even if one managed to choose the more accurate of the two ranges of slope values, an impressive amount of variation would still remain. For example, with the equations presented above, if area is reduced by 80% and $z = 0.20$, then species loss is estimated at 27%, whereas if area is reduced by the same percentage but $z = 0.30$, then species loss is estimated at 38%. Fifth, even if one knew precisely the correct slope of the species–area relationship in one location at one time, Connor and McCoy (1979) and others have shown that slope is not likely to be amenable to translation across locations, latitudes, or time.

One empirical means for focusing the choice of slope values is the body of literature documenting decline in species richness accompanying area reduction. J. M. Diamond termed this process "relaxation." A number of studies, for example, have examined the relaxation in the species of mammals that may have occurred on nature reserves, both in North America and in Africa. The results of these studies clearly illustrate the difficulty in choosing a slope of the species–area relationship. For African savanna reserves, R. East has shown that estimates of species loss accompanying a 100-fold reduction in area vary from about 40% to more than 90%, depending on the required minimum viable population size (i.e., the population size that ensures persistence with a certain probability for a certain length of time). W. D. Newmark has shown that similar high variability exists for western North American national parks.

At least three potential reasons underlie the poor predictability generally seen in studies of relaxation. The first two reasons, the presence of statistical or sampling errors and the lack of sufficient time since area reduction took place to allow full relaxation, could be addressed by careful experimentation. The protocol is to document the species richness of a relatively large area, reduce the area, and then document the species richness of the reduced area. Unfortunately, most of these kinds of experiments have been carried out at such a small scale that their general value at the scale of typical nature conservation efforts is questionable. Slope values derived from these experiments are, therefore, not likely to be of much practical value. A few reasonably large scale experiments in area reduction have been undertaken, most notably The Minimum Critical Size of Ecosystem project of D. H. Janzen, T. E. Lovejoy, R. O. Bierregaard, and others. These relatively large scale experiments suggest that the species–area relationship alone does not predict species loss particularly well. Slope values derived from these experiments, therefore, also are not likely to be of much practical value. The third reason, lack of the autecological (i.e., species-specific ecological) information needed to infer that the loss of species is attributable directly to relaxation, cannot be addressed as readily by experimentation.

Recall that one of the assumptions necessary to predict loss of species accompanying area reduction was that the loss of species is a direct consequence only of area reduction. Loss of species actually may occur only as an indirect consequence of area, because of some factor that happens to be correlated with area reduction, such as level of disturbance, complexity of habitat structure, or—probably most importantly—fragmentation. It seems clear that in virtually all real-world cases, area reduction is likely to be accompanied by fragmentation. That is, one relatively large area is not likely to be reduced to a single relatively small area, but rather to an archipelago of relatively small fragments. For example, Simberloff (1992) has shown that fragmentation effects, as well as size-dependent habitat changes, are at least as important as area reduction in predicting species loss in the relatively large scale experiments mentioned previously. It has been known for quite some time, at least since the pioneering work of N. W. Moore and others, that fragmentation can have adverse effects on organisms, independently of the effects of area reduction. The species–area relationship predicts that each of the fragments will support fewer species than the original area, but it alone cannot predict the cumulative number of species in the entire archipelago of fragments. As D. Simberloff and L. G. Abele have clearly shown, we also must know the degree of overlap in the species compositions of the fragments. If, for example, the compositions of the fragments were identical, then the archipelago would support fewer species than the original area. If, on the other hand, the species compositions of the fragments did not overlap at all, then the archipelago could support as many—or, potentially, even more—species than the original area. Assuming, realistically, that compositions overlap to an intermediate degree, calculating the loss of species by the method we have just illustrated could yield either an underestimate, if many species were confined to but a small subset of fragments, or an overestimate, if they were not. Fragmentation, therefore, adds more uncertainty—the degree of overlap in species compositions—to any prediction of the loss of species accompanying area reduction. We could predict the degree of overlap in the species compositions of fragments better if we had detailed information on the habitat requirements and minimum viable population sizes of individual species and on spatial variation in species richness, but as Simberloff and others have shown, such information is notoriously difficult to obtain and, therefore, is in very short supply.

Recently, a method has been suggested by Harte and co-workers whereby the slope value for species–area relationships among fragments might be obtained more easily. This method is based on the idea of self-similarity that we discussed previously. Essentially, the method requires only measurement of the degree to which species lists in separated, small, censused patches overlap as a function of the distance between patches and the area of the patches. Self-similarity, then, allows translation of the resulting slope value across scales, so that censusing of prohibitively large areas or obtaining extensive autecological information becomes, in theory, unnecessary. A. P. Kinzig and J. Harte have also developed a procedure for estimating species loss from habitat reduction that derives from the assumption of self-similarity and from examining the relationship between the number of "endemic species" and area, an endemics–area relationship (EAR). In the examples examined thus far, the estimated species loss from area reduction is less using Kinzig and Harte's EAR approach than from the traditional approach based on the species–area relationship outlined above. Whether these approaches to estimating species–area slopes and calculating species loss will lead to yet another dead end remains to be determined but will depend on how reasonable it is to assume self-similarity over a range of spatial scales.

B. Slowing the Loss of Species

In the previous section, we explored what the species–area relationship was able to tell us about species loss accompanying area reduction. In this section, we shall explore what the relationship can tell us about slowing that loss. In doing so, we shall treat area reduction with concomitant loss of species as an accomplished fact and concern ourselves solely with maximizing the remaining species richness. We shall, in turn, focus on maximizing species richness by the judicious choice, or design, of nature reserves.

The serious study of the relationship between the species–area relationship and the design of nature reserves began in the mid-1970s. At that time, several papers by J. Terborgh, J. M. Diamond, E. O Wilson, E. Willis, and others came out more or less simultaneously touting the overarching importance of largeness of nature reserves. Some of these papers maintained further that the species–area relationship indicates that nature reserves should be subdivided as little as possible, that is, that a single intact reserve of a certain area will support more species than a series of individually smaller reserves that are cumulatively of the same area. Diamond, for example, said "many species that would have a good chance of surviving in a single large reserve would have their survival chances reduced if the same area were apportioned among several smaller reserves." The basic problem with these ostensibly reasonable proposals—call them largeness and singularity—is that they may be too simplistic for real-world conservation challenges. They may be too simplistic because of the substantial set of assumptions that accompanies them. We shall address perhaps the most basic of these assumptions in detail. This assumption is that the importance of largeness and singularity follows directly from the species–area relationship.

Most of the early papers based their arguments for largeness and singularity on R. H. MacArthur and E. O. Wilson's theory of island biogeography. This theory of island biogeography takes a dynamic view of the maintenance of species richness on "islands," as a balance between colonization and extinction. The resulting species richness, therefore, is considered an "equilibrium number of species." If relatively large areas support relatively large populations, then relatively large areas also should have relatively low extinction rates and, consequently, support relatively high equilibrium numbers of species. Accordingly, any subdivision of a relatively large area essentially creates smaller areas, each with higher extinction rates, so that the equilibrium number of species on each of the areas necessarily will fall ("relaxation"). The early papers also made a variety of proposals to enhance colonization, reflecting the potential importance of clustering and corridors and of shape, but we shall not address these proposals.

Almost immediately, the arguments for the overarching importance of largeness and singularity were challenged. Simberloff and Abele showed that the species–area relationship did not unambiguously favor a single intact reserve over an archipelago of smaller reserves. For example, if one assumes a power-function model of the species–area relationship ($S = cA^z$) and that $A_{small} = A_{large}/2$, then the number of species in the large reserve is $S_{large} = c(A_{large})^z = c(2A_{small})^z = 1.2 S_{small}$ (Fig. 4). This equation could yield $S_{large} < 2 S_{small}$ in a variety of realistic circumstances. Subsequently, R. W. Rafe elaborated the mathematical underpinnings of the comparison of the relative abilities of a single large reserve (SLR) and two reserves of half the area (THR) to support species. R. W. Rafe, A. J. Higgs, M. B. Usher, and others clearly illustrated that the proportional overlap of species between the smaller reserves is a critical issue. In essence, the greater the overlap, the more a single large reserve is favored, and vice versa. The relevant, and critical, information necessary to predict the amount of overlap mostly is lacking, a fact mentioned early on by Simberloff and Abele in response to criticisms of their analysis.

By the early 1980s, the controversy over choice of nature reserves had evolved into the "single-large-or-several-small (SLOSS) debate." O. Järvinen, D. Simberloff, L. G. Abele, and others emphasized, once again, that the species–area relationship—or the underlying island biogeographic theory, for that matter—did not unambiguously favor a single intact reserve over an archipelago of smaller reserves. That some species might need relatively large areas for their continued well-being, while others could thrive in, or actually required, certain relatively small areas, had become well established in the ecological literature, beginning with the early work of J. Terborgh, R. T. T. Foreman, and others, so their proposal should not have been very controversial. Such was not the case, however, and the debate continued. In 1986, M. E. Soulé and D. Simberloff gave the SLOSS debate last rites: "The SLOSS debate is no longer an issue in the discussion about the optimal size of nature reserves." Yet, rumors of the death of the SLOSS debate were greatly exaggerated, for it continued beyond the middle of the 1980s—briefly mutating into the SLOPP ("single-large-or-plentifully-patchy") debate of M. E. Gilpin in the late 1980s—right up to today.

M. E. Soulé and D. Simberloff suggested that the truly important remaining question was "the dynamics of species extinction after the reserves are set up and surrounded by habitat modified by human activities," as had J. A. Kushlan, H. Picton, and others nearly a decade earlier. Clearly, this statement implies that too much emphasis had been placed on area (the "extinction" side of island biogeography) and too little on isolation (the "colonization" side of island biogeography). It also implies that relatively small fragments may have "worth" despite their small sizes. Relatively small fragments often are not simply random samples of larger habitat units, but rather may represent "special" places that have been left either inadvertently or purposely. For example, R. T. T. Foreman, I. Hanski, E. D. McCoy, and others have shown that relatively small fragments could harbor unusually high population densities of species or disproportionate representations of rare species. Finally, this statement implies that fragmentation cannot proceed indefinitely without severe consequences. Simberloff, for example, has pointed out that "thousands of small fragments, with large aggregate area, will not be expected to allow conservation of many species." If this question—about the dynamics of species extinction—is indeed the only important one remaining, then the information needed to slow the loss of species accompanying area reduction largely is the same information that is needed to predict the loss of species in the first place (refer to the previous section). For example, J. G. Dony, M. E. Soulé, D. Simberloff, and others have recommended identifying target species, determining their minimum viable population sizes (MVPs), and then using known densities to estimate needed area. Unfortunately, even if sound MVPs could be calculated for several target species in a location, the areas required to support those population sizes are likely to be poorly known, if known at all. This lack of information also plagued earlier proposals for estimating needed area, such as E. D. McCoy's "minimum refuge area," T. E. Lovejoy and D. C. Oren's "minimum critical size," and S. T. A. Pickett and J. N. Thompson's "minimum dynamic area."

C. How Incidence Functions and Nestedness Fit In

Recall that the SLOSS debate was generated largely by one of the early nature reserve design principles: a single intact reserve of a certain area will support more species than a series of individually smaller reserves that are cumulatively of the same area. For this principle to apply universally, the minimal assumptions are that the habitats included on areas of different sizes are more or less uniform, that the population densities of species on areas of different sizes are similar, and that the process of relaxation largely is deterministic. Subsequent ecological research has made it clear that none of these assumptions always holds. However, do these assumptions hold in enough instances or to such a degree that it is possible to know the species composition of various-sized fragments with a high degree of certainty? We shall address two, related, methods developed to address this question.

J. M. Diamond developed "incidence functions" as a tool for determining minimum nature reserve area. The incidence function, J, equals the number of fragments ("islands") of a certain size harboring a certain species divided by the total number of fragments of the same size. Supposedly, the higher the value of J at a particular size, the higher the probability that the species can persist in fragments of that size. E. F. Connor, D. Simberloff, M. Williamson, and others questioned the validity of these conclusions, but, if they do have any validity, even in a general way, then certain patterns of incidence functions could indicate that species disappear from increasingly fragmented habitats in a predictable manner.

B. D. Patterson and W. Atmar developed a sophisticated extension of incidence functions, the "nestedness" of species' geographical distribution. Species' distributions are nested when the species on the most species-poor (and, likely, smallest) fragment comprise a subset of those on the next most species-poor fragment, which, in turn, comprise a subset of those on the next most species-poor fragment, and so on. K. B. Jones and co-workers supplied the first explicit evidence for nested distributions in the mid-1980s, although implicit evidence can be found in papers by N. W. Moore, M. D. Hooper, R. T. T. Foreman, and others a decade earlier. Jones and co-workers attributed the pattern to selective extinctions in the absence of colonizations, as have most subsequent authors, although D. T. Bolger, M. V. Lomolino, and others have shown that the pattern actually can result from a number of causes. Nestedness could have important implications for nature reserve design. If the species on an archipelago of habitat fragments were a perfect subset of the species in a single large unit of habitat, then the archipelago could never contain more species than the single large unit. Furthermore, if fragments of similar size were to harbor similar suites of species, because species loss with area reduction is deterministic, then the archipelago must contain fewer species than the single large unit. The temptation is great, therefore, to assume that simultaneously significant species–area and nested subset relationships must indicate that smaller fragments have the fewest species, but recent evidence supplied by D. Doak, E. D. McCoy, W. J. Boecklen, and others suggests that this assumption is not a good one.

D. Conclusions

Regardless of the reason for it, poor predictability of the effects of area reduction can stall effective decision making. A good example is the controversy over the amount of species loss accompanying tropical deforestation. For more than 30 years, ecologists have sought "shortcut methods" to substitute for the autecological information that is needed to improve predictability but is so difficult to obtain. To date, they have not been particularly successful. An interesting additional example that involves species–area relationships is the controversy over the amount of species loss accompanying global warming discussed by K. A. McDonald, J. H. Brown, R. W. Skaggs, W. J. Boecklen, T. E. Lawlor, and others. So, we are saddled with a seemingly unresolvable dilemma: we need more information, but we cannot afford to wait to get it. Faced with this dilemma, some biologists, for example R. East and M. Kent, have suggested that we view the species–area relationship as perhaps the best tool available for making conservation decisions and have promoted the idea of retaining large units of habitat as a matter of general course. If one is focusing solely on predicting loss of species with area reduction and refrains from becoming overly specific, then this may be a reasonable strategy. Others, for example C. F. Mason and E. D. McCoy, have suggested that reliance on species–area relationships may lead to undesirable conservation decisions and have promoted a "save-all-the-pieces" strategy. If one is focusing on forestalling loss of species, then this may be a reasonable strategy (which has been termed SLATS ("several little all too small") by E. D. McCoy and H. R. Mushinsky) for habitats suffering from extreme area reduction and fragmentation. The SLATS strategy is likely to become increasingly relevant for most habitats, because the amount of habitat needed to allow species to persist, let alone to flourish or to evolve, appears to be much larger than humans are willing to grant.

See Also the Following Articles

DIVERSITY, COMMUNITY/REGIONAL LEVEL • HABITAT AND NICHE, CONCEPT OF • ISLAND BIOGEOGRAPHY • METAPOPULATIONS • POPULATION DENSITY • POPULATION DIVERSITY, OVERVIEW

Bibliography

Coleman, B. D., Mares, M. A., Willig, M. R., and Hseih, Y. (1982). Randomness, area, and species-richness. *Ecology* **63**, 1121–1133.

Condit, R., Hubbell, S. P., LaFrankie, J. V., Sukumar, R., Manokaran, N., Foster, R., and Ashton, P. S. (1996). Species–area and species–individual relationships for tropical trees: A comparison of three 50-ha plots. *J. Ecol.* **84**, 549–562.

Connor, E. F., and McCoy, E. D. (1979). The statistics and biology of the species–area relationship. *Am. Nat.* **113**, 791–833.

Hanski, I., and Gyllenberg, M. (1997). Uniting two general patterns in the distribution of species. *Science* **275**, 397–400.

Harte, J., Kinzig, A., and Green, J. (1999). Self-similarity in the distribution and abundance of species. *Science* **284**, 334–336.

MacArthur, R. H., and Wilson, E. O. (1967). *The Theory of Island Biogeography*. Princeton Univ. Press, Princeton, NJ.

Rosenzweig, M. L. (1995). *Species Diversity in Space and Time*. Cambridge Univ. Press, Cambridge, UK.

Shafer, C. L. (1990). *Nature Reserves: Island Theory and Conservation Practice*. Smithsonian Institution Press, Washington, D.C.

Simberloff, D. (1976). Experimental zoogeography of islands: Effects of island size. *Ecology* **57**, 629–648.

Simberloff, D. (1992). Do species–area curves predict extinction in fragmented forests? In *Tropical Deforestation and Species Extinction* (T. C. Whitmore and J. A. Sayer, Eds.), pp. 75–89. Chapman & Hall, London.

Williams, C. B. (1964). *Patterns in the Balance of Nature*. Academic Press, London.

Williams, M. R. (1995). An extreme-value function model of the species incidence and species–area relations. *Ecology* **76**, 2607–2616.

SPECIES COEXISTENCE

Robert D. Holt
Museum of Natural History and Center for Biodiversity Research, University of Kansas

I. General Issues
II. Constraints on Coexistence: Examples from Introduced Species
III. Coexistence and Exclusion: Messages from Bottles
IV. Constraints on Coexistence: Rules of Dominance
V. Mechanisms of Coexistence
VI. Traditional Approaches to Coexistence
VII. Mechanisms for Local Coexistence: Current Approaches
VIII. Temporal Niche Partitioning
IX. Coexistence in Open Communities
X. Future Directions
XI. Conclusions

GLOSSARY

coexistence The state of two or more species being found in the same place at the same time.
community The assemblage of species found in a defined area, in which these species can interact.
competition The most widely used definition: use or defense of a resource by an individual which reduces resource availability to other individuals. Alternative definition: a reduction in one species' growth rate because of the effects of another species.
food web The pattern of feeding relationships among organisms.
Lotka–Volterra competition model A model describing the per-capita growth rates of two competing species as linear, declining functions of the abundances of each species.
mechanistic models of population dynamics Models which are explicit about resource consumption, the relationship of consumption to demography, and mortality factors.
microcosm studies Laboratory experiments involving population dynamics of small organisms with short generation lengths.
natural enemy Any species which consumes or parasitizes another species; a general term that includes predators, herbivores, parasites, pathogens, and parasitoids.
niche (Grinnellian/Hutchinsonian) The range of resources and conditions within which populations of a species are expected to persist. (Eltonian) The role of a species in a community.
open communities Communities which receive immigrants from external sources and may export emigrants as well.
permanence A property of communities that ensures long-term coexistence of species because community trajectories (variation in numbers through time) have no species approaching very low numbers.
zero-growth isocline In a graph with axes describing factors important to population growth (e.g., resource abundance), a line along which a population's growth rate is zero.

THE OXFORD ENGLISH DICTIONARY states that to "coexist" is "to live together in the same place, at the same time, with another." The topic of "coexistence" focuses on how biological species are organized into communities, in space and through time. Identifying the factors that influence coexistence is fundamental to understanding biodiversity. Important factors that influence coexistence include interspecific interactions, spatial and temporal scales, and historical contingencies. Insights into coexistence come from laboratory and field experiments, historical reconstructions and observational studies, and theoretical explorations. The topic of coexistence has had a long history in community ecology, some appreciation of which is necessary to understand contemporary perspectives. After a few general remarks about how scale, interactions, and the notion of permanence are related to coexistence, I provide a survey of the "phenomenology" of coexistence and exclusion, and lessons derived from field and classic lab studies. I then discuss key ingredients in both historical and current theoretical interpretations of species coexistence, emphasizing conceptual generalities rather than model details, and conclude with suggestions regarding significant unanswered questions about coexistence.

I. GENERAL ISSUES

A. The Importance of Scale

In some sense, all the species in the teeming plenitude of life present on Earth coexist because they are found in one place (Earth) at one time (now). However, from the perspective that matters most to an individual organism such as a vampire bat, a tulip, or a flatfish—the perspective defined by the space that an individual occupies or moves within over its life span—populations of any given species co-occur with only a tiny fraction of all living things. The same holds for community samples, for instance, those gathered in the quadrats beloved of plant ecologists or the towed plankton nets of limnologists; the number of species recorded is always much less than the potential number. That is, what counts as "coexistence" varies with the scale of ecological inquiry. Viewed at small spatial scales, tailored to the ecology of individual organisms or seen through the lenses of conventional field methods, most species in the global flora and fauna do not coexist. However, even at this scale many (and sometimes very many) species can coexist. Species coexistence is also temporally bounded. In the long term most species face extinction, and even in the short term coexistence may be transient rather than permanent.

B. The Importance of Interactions

Species (like ships passing in the night) may either coexist or fail to do so for reasons having nothing to do with each other. A null model of community organization is that communities arise from the independent responses of species to the environment. However, coexistence often reflects the impact of interactions among species. Familiar categories of interactions include competition for resources, natural enemy–victim interactions (e.g., predator–prey and host–pathogen interactions), commensalisms and mutualisms (e.g., plant–pollinator relationships), allelopathy (poisoning), and a broad range of environmental modifications (collectively called "ecological engineering"). In the study of coexistence, competition has received by far the most attention, but there has also been considerable work on how natural enemy–victim interactions influence coexistence. Interactions are not fixed properties of species, or pairs of species, but can vary in their strength and pattern as a function of the physical environment and many contingent details of community structure. If species A affects species B, and species B in turn influences species C, then one says there is an indirect interaction between A and C. A complete understanding of coexistence requires one to consider indirect and direct interactions. Interspecific interactions may be strongly asymmetrical, and indeed, interactions frequently cause species exclusion (constraints on coexistence). Analyzing mechanisms of coexistence and exclusion forms the conceptual core of the discipline of community ecology.

C. Coexistence and Permanence

Historically, coexistence is an outcome of the assembly of communities by colonization from species pools at larger spatial scales. If rules describe the historical process of community assembly, these rules must in large measure reflect the importance of species' interactions in determining coexistence rather than extinction as assembly occurs.

The paleontological record shows that species' distributions are not stable but move in concert with climate change, fluctuations in sea level, and the stately movement of the continents. Moreover, for many species, chance vicissitudes of dispersal provide a small trickle of potential colonists "testing the waters" outside their normal range. Local coexistence at the very least requires species' geographical ranges to overlap. Many species do not coexist together today in local communities because of historical factors preventing them from ever encountering one another.

Within a species' geographical range, for that species to be a resident member of a local community, at one time it must have colonized, increasing from very low numbers. Once present, given the vagaries of climate and fluctuations in the abundance of other species, it is likely that resident species experience times of low abundance and thus extinction risk. To persist, a species must be able to rebound from these dangerous troughs of low population size. These observations suggest that a robust form of coexistence is ensured if each species can increase when rare. If all trajectories describing fluctuations in abundance in a community are bounded away from zero, all species in the community will coexist. This criterion for coexistence is called "permanence" in the jargon of mathematical ecology. Examining conditions for increase when rare provides a natural protocol for experimental studies of coexistence and analysis of mathematical models of interacting species. Permanence is always assessed relative to a certain spatial and temporal scale. A given pair of species may not coexist indefinitely in any single site but nonetheless coexist at larger scales in ensembles of sites.

II. CONSTRAINTS ON COEXISTENCE: EXAMPLES FROM INTRODUCED SPECIES

In recent times, humans have greatly accelerated movements of species into novel habitats, both via deliberate introductions (e.g., most of the lowland birds found in Hawaii) and as incidental "hitchhikers" tracking human transport (e.g., most benthic invertebrates in San Francisco Bay are exotics released as by-products of ship ballast water dumps). Analyses of introductions suggest species interactions can constrain coexistence.

A. Competition

Classical biological control involves deliberate introductions of species that one hopes will control a pest species. Between 1947 and 1952 the Hawaii Agriculture Department released parasitoid species to control the oriental fruit fly, an economically significant pest. (Parasitoids are insects, e.g., braconid wasps, whose larvae live within and ultimately kill their host, such as caterpillars.) Three wasps in the genus *Opius* were considered to be potential control agents. The first species established, *Opius longicaudatus*, parasitized approximately 20% of the fruit fly hosts. The second species introduced, *O. vandenboschi*, was more effective, parasitizing about 30% of the hosts; as *O. vandenboschi* increased, *O. longicaudatus* decreased toward extinction. The third species, *O. oophilus*, was even more effective, parasitizing up to 80% of the hosts and replacing in turn *O. vandenboschi*. This system provides an example of competitive exclusion in exploitative competition, where species dependent on a single limiting resource (here, host insects) cannot coexist. Increasing parasitism reduces the availability of unattacked hosts. The competitive dominant in this pattern of competitive displacement is the parasitoid species that persists at the lowest host abundance—an example of a simple rule of dominance in resource competition. Understanding how multiple species can persist on a limited resource base is a perennial theme in the study of coexistence.

B. Natural Enemies

Very effective control agents can, after introduction, eliminate the target species (and thus themselves) over broad areas. The floating fern *Salvinia molesta* from Brazil escaped from a garden in Sri Lanka and became a serious aquatic pest in much of the Old World wet tropics. The beetle *Cyrtobagous salviniae* was introduced in Australia and proved highly successful at limiting the fern, which (with the beetle) is now found in only a few scattered populations. This example illustrates another species coexistence problem; that is, how effective natural enemies manage to coexist with their prey without overexploiting them and thus driving themselves to extinction.

C. Community and Spatial Contexts

Many game departments routinely introduce game species. Failed introductions reveal constraints on species coexistence. Caribou were once common in the Maritime Provinces but declined during the 1800s to extinction by 1915. Attempted reintroductions have failed because the caribou pick up a widespread nematode parasite (carried by snails and incidentally consumed as the caribou forage on vegetation). The nematode lodges in the caribou brain and is fatal within a few months. The nematode population is sustained by another host, white-tailed deer, which tolerate parasitism. Deer are abundant because land practices by European settlers in the nineteenth century increased the area of the second-growth habitats that deer favor; this may explain the disappearance of caribou. Caribou herds wander very widely, which makes it likely that they will encounter nematode-laden snails. Another ungulate species, moose, coexists with deer, even though infected moose also quickly die. Moose and deer overlap in diet and can compete; in contrast, caribou and deer do not overlap in diet. Nonetheless, it is caribou, not moose, which have been indirectly excluded by deer. Moose are relatively sedentary (for their body size) and can occupy patches of highland forests with deep winter snows, where deer do not penetrate. At the coarse scale of North America, caribou and white-tailed deer coexist but with little spatial overlap among populations. Moose and deer, in contrast, overlap geographically, coexisting at the level of landscapes, but at a finer scale they have considerable spatial segregation in habitat use.

This example illustrates several points. First, natural enemies (a term that refers to any species which consumes a species of interest) can prevent coexistence. Second, coexistence or exclusion may arise from complex webs of multispecies interactions. The nematode directly causes exclusion of caribou, but viewed more expansively exclusion is caused by the entire vegetation–snail–deer ensemble, which collectively governs nematode abundance and hence caribou infection rates. Third, movement patterns can influence interspecific interactions. Moose and white-tailed deer coexist at the landscape scale because moose have spatial refuges from infection, whereas caribou do not and are thus excluded.

III. COEXISTENCE AND EXCLUSION: MESSAGES FROM BOTTLES

The previous examples strongly suggest the importance of species interactions in determining coexistence but are not conclusive. Without detailed observational studies tied to parameterized models or rigorous manipulative field experiments, it is difficult to persuade skeptics that other hypotheses are not also plausible. As an alternative, one simple but illuminating approach to the study of coexistence is to put together a few species in a confined setting and see what happens.

The study of "nature in a bottle" simplifies the world in important respects (e.g., a few species in closed and temporally constant environments), and such experiments are viewed skeptically by some ecologists. Despite these limitations, a wealth of important messages about species coexistence emerged from classic bottle experiments—messages which have proven robust when applied to natural communities. Microcosm studies of ecological processes are now enjoying a renaissance in community ecology.

A. Competition Experiments

The term "competition" usually refers to an interaction among individuals (within or among species) that arises because they seek a resource in short supply. If this involves direct harm, the interaction is referred to as "interference competition." If instead competition involves depletion of a resource, one refers to "exploitative competition."

1. Protozoa

Most ecology textbooks (Hutchinson, 1978) recount experiments by the Russian ecologist G. F. Gause, who as a young man in the 1930s put mixed cultures of protozoa into vials full of liquid media to study species coexistence. Gause's famous competition experiments compared populations of the ciliate protozoans *Paramecium aurelia* and *P. caudatum* grown separately and together on a nutritive medium containing their essential resource (bacterial food). Both species thrived when alone, but *P. aurelia* usually displaced its congener in joint cultures within 30–50 generations. This outcome was reversed if the medium was completely replenished with fresh nutrient on a regular basis. Gause argued that metabolic by-products were building up in the experiments, and that part of the dominance of *P. aurelia* may involve its resistance to the chemical by-products of metabolic activity as well as its superior ability to exploit the food base. Here, competition combines environmental modification and exploitation of a limiting food resource.

In other experiments Gause found that *P. aurelia* could coexist with another species, *P. bursaria*, even in the confines of a closed culture. *Paramecium bursaria* contains symbiotic algae, which release oxygen in photosynthesis. In incompletely mixed cultures, bacteria accumulate on the bottom, creating a zone slightly depleted of oxygen. The protozoan with the algae in effect carries its own oxygen supply into this anoxic habitat and so can use a food source unavailable to the other, competitively superior species. Here, coexistence depends on both the availability of different habitats and differential species' abilities to utilize those habitats.

2. Beetles

Many ecologists have examined the dynamics of mixtures of beetle species competing for grain stored in containers. L. C. Birch examined several species combinations that fed on wheat or maize under different conditions of temperature and moisture. Usually, one species won. The criterion for dominance was simple: The winner always had the highest intrinsic growth rate (the maximal rate of population growth when a species is rare and growing alone). However, the identity of the winner depended on both resource type and physical conditions. Moreover, competitive exclusion sometimes required many generations; in one experiment, the apparent loser survived until the experiment had to be terminated.

Experiments by T. Park with a different set of stored grain pests (*Tribolium castaneum* and *T. confusum*) also showed that the species identity of the winner varied with temperature and humidity. However, for some abiotic conditions the outcome was indeterminate; either species could win, with the winner tending to be the species initially most abundant. The net interaction between these beetles combines several mechanisms, including exploitation of a shared resource and cannibalism (both within and between species). In one set of experiments, Park intriguingly observed that when a coccidean parasite, *Adelina* sp., was present, the normal dominant *T. castaneum* became a weak competitor, thus permitting the persistence of *T. confusum* (and even its dominance).

Sometimes, coexistence in microcosms reflects very subtle biological differences between species. Crombie raised two grain beetles (*Rhizopertha dominica* and *Oryzaephilus surinamensis*) in vats of cracked wheat and found indefinite persistence, despite the fact that there seemed to be just a single resource. The main difference in the two species seemed to be in larval feeding habits; larvae of *Oryzaephilus* lived and fed from outside the wheat grains, whereas larvae of *Rhizopertha* were sufficiently small to live and feed from within the grains. This slight difference sufficed for the competitors to coexist.

3. Key Insights

Several insights emerged from "bottle" studies of competition. First, species competing in a closed homogeneous medium for a single limiting resource typically do not coexist. Second, the winner depends on environmental conditions, and the outcome can change if the environment is altered—there is no universally superior competitor. Third, exclusion is not instantaneous and may require many generations; this leads to the possibility of transient coexistence. Fourth, indefinite coexistence occurs but requires heterogeneity in the environment as well as differences in species' responses to this heterogeneity. However, the species' differences permitting coexistence can be quite subtle. Fifth, pairs of species can interact in many distinct mechanistic ways (e.g., via impacts on resource levels or on levels of pollution from metabolic waste). It is difficult to generalize among studies if one has not clearly identified and characterized specific mechanisms of interaction. Finally, the outcome of competition can vary due to effects of other species, including parasites or mutualists.

B. Predation

In classic literature, compared to studies of competition, somewhat less attention was given to predator–prey and host–parasite interactions (and essentially none to mutualism). Gause did carry out experiments with the predatory protozoan *Didinium nasutum* and a prey protozoan, *P. caudatum*. The predator quickly overexploited its prey and went extinct. To achieve coexistence, Gause concluded that there needed to be a spatial refuge for prey, inaccessible to the predator, or recurrent immigration. A qualitatively similar message emerged from experiments by C. Huffaker and colleagues, who studied an intrinsically unstable interaction between a voracious predatory mite (*Typhlodromus*) and a prey mite (*Eotetranychus*) living on oranges. In single oranges, the predator rapidly drove its prey to extinction. In contrast, in a large "universe" of oranges laid out in a grid, separated by barriers to dispersal, the interaction persisted over many generations, with complex patterns of spatial occupancy for the two species. Luckinbill reexamined the unstable predator–prey pair studied by Gause but reduced the encounter rate between predators and prey by adding methyl cellulose (a thickener) to the medium so as to slow predator movement. Reducing movement rates in effect expands the size of the spatial arena of the interaction (scaled by the distance a predator moves per unit time). This led to stable coexistence. The qualitative message of these microcosm experiments on predation is that patchiness, localized dispersal, and spatial heterogeneity (e.g., refuges) may facilitate the coexistence of effective specialist predators and their vulnerable prey.

IV. CONSTRAINTS ON COEXISTENCE: RULES OF DOMINANCE

A. The Competitive Exclusion Principle

The experimental observation that in homogeneous well-mixed environments it was often difficult to achieve coexistence between similar species became enshrined in ecology as Gause's principle, or the "competitive exclusion principle." Another way to state this principle is to note that, to coexist, different species must have distinct ecologies.

Assume we are examining a local community defined by a spatial scale in which all individuals can reach all sites over a single generation. The basic logic of the competitive exclusion principle is impeccable (Levin, 1970): Consider two species with continuous generations, where both species respond to the same environmental factors, denoted by E. The quantity E could be many things (e.g., resource availability, predator abundance, or a weighted sum of the competitors' own numbers). The growth rate of species i is $dN_i/dt = N_i f_i(E)$, where N_i is the density of species i. We make three assumptions: (i) There is a single limiting factor, (ii) the species interact in a closed habitat (i.e., no immigration), and (iii) each species when alone settles down to an equilibrium at constant densities (i.e., the environment is temporally constant).

For species i, there will be some value of the environmental factor, E^*, at which that species equilibrates. It is a biological truism that any two species will almost surely differ in some way. Hence, it is very improbable that they will have exactly the same value of E at which they reach equilibrium. In other words, there should be no long-term persistence of two species limited by a single factor in a constant, closed environment. For this argument to work, there should be some effect of the species themselves on the magnitude of E, leading to either direct or indirect density dependence in demographic parameters such as birth or death rates. For example, if two species are consuming a single resource, then consumption should depress resource levels.

1. Exploitative Competition: The R^* Rule

What counts as a "limiting factor" needs to be interpreted quite broadly and is often quite difficult to identify in practice. Nonetheless, sometimes one can observe a single, simply characterized limiting factor, which then defines a rule of species' dominance. For instance, the Hawaiian parasitoids fit the "R^* rule" proposed by David Tilman; that is, the dominant species, given exploitative competition for a single resource, is the species persisting at the lowest resource level. (The asterisk denotes equilibrium, and R^* is the equilibrial resource level at which a given consumer species is in demographic equilibrium.) A virtue of this dominance rule is that one can measure R^* of each species when alone and then predict the outcome of competition. A species may have a low R^* and be competitively superior either because it is efficient at resource consumption or because it has low mortality (e.g., it can escape predation).

The R^* rule successfully predicts the outcome of competition among phytoplankton competing for nutrients in microcosms and also characterizes competition in some natural systems. For example, in the nitrogen-poor soils of Cedar Creek, Minnesota, plant species compete for nitrogen. Wedin and Tilman showed that species with low R^* for nitrogen won in pairwise competition experiments. If two species had similar values for R^*, the rate of competitive displacement was greatly reduced, as expected by theory.

2. Apparent Competition: The P^* Rule

Analogous "rules of thumb" arise in other situations. For instance, Sharon Lawler carried out a microcosm experiment in which a predatory protozoan (*Euplotes patella*) coexisted with either of two prey protozoans (*Tetrahymena pyriformis* and *Chilomonas paramecium*) grown alone. However, when all three species were together, *Chilomonas* was driven extinct. In the single-prey cultures, *Tetrahymena* sustained four times the number of predators as did *Chilomonas* cultures. In the mixed cultures, the latter species suffered higher predation than it could sustain. The limiting factor is the abundance of a shared predator, which can respond numerically to its prey. Let P^* be the abundance of the predator sustained (and tolerated) by a prey species. A "P^* rule" now describes dominance: The winning prey species is the one with greater P^*. This form of indirect competition between species arising from shared natural enemies (including parasites) is called apparent competition (Holt and Lawton, 1994); the word "apparent" is used because the interaction has the same consequences for coexistence as does classical exploitative competition for resources but may occur even between species with totally different resource requirements. Dominance in apparent competition may occur because of different vulnerabilities to a natural enemy (as in the caribou example) or because one prey is highly productive and sustains an abundant enemy population (as in Lawler's experiment).

V. MECHANISMS OF COEXISTENCE

The previous discussion emphasized constraints on species coexistence arising from interspecific interactions. Rules of dominance are important conceptual tools which quantify these constraints and help identify biological traits leading to dominance. However, even in simple microcosms coexistence can occur, and most natural communities are rich in species. To understand coexistence, the competitive exclusion principle is reconsidered (Crawley, 1997; Grover, 1997; Tokeshi, 1999). Species may coexist when any of the assumptions leading to the competitive exclusion principle are violated. This suggests three classes of mechanisms promoting species coexistence of potentially competing species in a local community:

1. Species may coexist in a closed, temporally constant world if they experience different limiting factors at the spatial scale of the local community; this includes classical niche partitioning of resources, as well as mechanisms involving predation and parasitism, and direct interference.

2. Species may coexist, even though they experience the same limiting factor, if the environment is temporally variable and species respond differently to this temporal variation (temporal niche partitioning).

3. Species may coexist if the environment is spatially open; this includes spatial niche partitioning at scales broader than the local community and mechanisms such as colonization–competition tradeoffs in metapopulations.

From the 1950s to the mid-1970s, stimulated largely by G. E. Hutchinson and his brilliant student Robert MacArthur, most community ecologists emphasized classical niche partitioning in studies of species coexistence. In recent years, the balance of attention has shifted markedly to a broader range of coexistence mechanisms. Ecologists now believe that maintenance of diversity—coexistence writ large—often depends on spatial dynamics in open communities, food web interactions (including predation and parasitism), and nonequilibrial dynamics reflecting either extrinsic temporal variation or the endogenous instability of complex ecological system.

VI. TRADITIONAL APPROACHES TO COEXISTENCE

A. Classical Niche Partitioning

Competitive exclusion is expected if growth rates of two (or more) species are determined by a single limiting factor. Species may coexist, even in an unvarying and spatially confined bottle, given multiple limiting factors, such that each species is limited more strongly by its own distinct set of factors. When species coexist, one sensible approach to begin to understand this coexistence is to map their niche requirements against the spectrum of limiting factors present in the environment. This basic methodology has been a source of significant insights for much of the history of ecology and continues to be fruitful. The classic paper in this genre is the 1958 study by Robert MacArthur of wood warblers in a New England boreal forest. Five warbler species in the genus *Dendroica* occurred in the same tract of forest; all five species have similar body sizes and eat the same range of insect taxa. MacArthur found that the species segregated with respect to microhabitat (with one species feeding at treetop, another in low branches, and so on). He argued that because of this microhabitat partitioning, each species consumed an independent pool of insect prey, thereby reducing the potential for competitive exclusion. Following publication of this study, there was a proliferation of field studies of niche partitioning patterns.

B. Lotka–Volterra Competition Model

MacArthur's study was motivated in part by Gause's experiments on competitive exclusion, which in turn were stimulated by theoretical models of interacting species explored by the Italian mathematician Vito Volterra. In the usual textbook formulation (the famous Lotka–Volterra equations), the competitive effect of one species on another is expressed as a direct, density-dependent reduction in abundance:

$$\frac{dN_1}{dt} = r_1 N_1 (K_1 - N_1 - \alpha_{12} N_2)/K_2$$

where N_1 and N_2 are the abundance of competing species 1 and 2, respectively, K_1 is the carrying capacity of species 1, r_1 is its intrinsic rate of increase, and α_{12} is a competition coefficient (the equation for species 2 is the same, with the subscripts 1 and 2 switched). The quantity α_{12} measures the effect an individual of species 2 has on reducing the per capita growth rate of species 1 compared to the effect of an individual of 1 on its own species. Both species increase when rare and hence coexist if

$$\frac{1}{\alpha_{21}} > \frac{K_1}{K_2} > \alpha_{12}$$

The outer inequalities imply $1 > \alpha_{12}\alpha_{21}$. This necessary condition for coexistence states that one (or both) of the competing species experiences stronger intraspecific than interspecific competition. If the inequalities are reversed, either species can exclude the other if it is initially sufficiently abundant (as in Park's experiments with *Tribolium*). If the competition coefficients are near unity (i.e., density dependence occurs uniformly within and between species), the species with higher carrying capacity wins.

The Lotka–Volterra model usefully describes competition, and multispecies extensions to it have been the focus of a rich body of theoretical work. However, the model is difficult to use predictively because competition coefficients and direct estimates of density dependence can only be measured during competitive trials rather than a priori. Moreover, taken literally, the model best describes systems in which competition is due to direct interference. Recognizing these limitations led to the development of a wide range of mechanistic models of resource–consumer and other interspecific interactions beginning in the late 1960s.

C. Whither Limiting Similarity?

This development was concordant with interest in the important concept of "limiting similarity," which is the notion that there might be a quantifiable limit to how similar species could be in their utilization of resources and still coexist. [To gain an understanding of the idea of limiting similarity, note that in Eq. (2) similar species should have competition coefficients near 1, so coexistence will not occur unless carrying capacities are finely balanced.] For a period in the history of the discipline, the goal of quantifying limiting similarity seemed to represent a "holy grail" for community ecology. Early theoretical explorations suggested simple rules might permit predictions of the maximal number of species which could persist on a defined resource base, providing a basic tool for understanding biodiversity. These studies were based on the Lotka–Volterra model coupled with assumptions about resource use. In some limiting cases, the Lotka–Volterra model emerges as a reasonable approximation of more complex resource–consumer interactions, and one can directly map niche overlap onto measures of competition. With simplifying assumptions about symmetry in resource use, the role of environmental variation, and other factors, a limiting similarity of competing species could be calculated and then compared against observed similarities. For instance, using MacArthur's warbler data, it was assumed that spatial overlap in foraging was directly related to competition coefficients. Substitution into the model showed that observed overlap was consistent with long-term coexistence. More broadly, if niche overlap were proved to be closely related to the strength of competition, observational studies of overlap and niche partitioning would provide a powerful link between descriptive analyses of community patterns and the dynamical forces of species interactions.

Further exploration has tempered the initial flush of enthusiasm for this approach to coexistence. After considerable grinding of mathematical gears, theoretical ecologists have concluded that limiting similarity, if it occurs at all, can only be characterized for models tailored to specific situations. Instead of a single general theory of limiting similarity and coexistence, there are many special theories. Moreover, there is a real sense in which increasing species similarity can facilitate, rather than hamper, coexistence. Equation (2) describes bounds on the permissible differences in the two species' carrying capacities. If the α_{ij} are less than unity, coexistence is more likely if carrying capacities are nearly equal rather than very different. The more similar species are in their demographic responses to the environment, the more similar their carrying capacities are likely to be. All else being equal, this kind of similarity promotes the coexistence of competing species.

D. Manipulative Field Experiments

Recognizing the limitation of observational studies of niche overlap and partitioning as evidence for competition, ecologists have turned to manipulative field experiments, typically removing one species and monitoring changes in the abundance of others. Reviews of such experiments show that when species are suspected to compete (e.g., because of overlap in resource requirements and habitat use), they often in fact do compete. For instance, Hairston noted that in the Great Smoky Mountains two species of *Plethodon* salamanders had altitudinal ranges which were nearly mutually exclusive, with only a narrow range of overlap. In contrast, in the Balsam Mountains altitudinal overlap was extensive. Hairston hypothesized that this was due to stronger competition in the Smokies, and using reciprocal transplants and removals he demonstrated that competition (due to behavioral aggression) was indeed much stronger in the narrow overlap zone in the Smokies.

Unfortunately, relatively few field experiments have been directly tied to mechanistic models of competition, which makes it difficult to generalize among studies. Moreover, such experiments tend to focus on species which already coexist, at least to a degree; removals then assess the magnitude of competition, given coexistence, rather than coexistence or exclusion per se. One study which directly addressed coexistence was performed by Bengtsson, who studied three species in the zooplankton genus *Daphnia* (which compete exploitatively for algal food) on the coast of the Baltic Sea.

Observational studies suggested that usually just one, or more infrequently two, species was present in any given pool. Bengtsson added all possible combinations of the three species to artificial pools and found that there were no extinctions in the single-species pools. Yet high extinction rates occurred in the two- and three-species sets. This directly demonstrates the importance of competition in constraining coexistence in natural conditions.

VII. MECHANISMS FOR LOCAL COEXISTENCE: CURRENT APPROACHES

A. Mechanistic Models of Multiple Limiting Factors

Most theoretical studies of competition today focus on models in which the mechanisms of interaction are clearly described. Detailed analyses of mechanistic models of competition are often mathematically challenging, but important insights can often be gleaned from simple graphical analyses. A generalization of the competitive exclusion principle is that "coexistence of n species requires n limiting factors." Mathematically, one expresses the growth rate of each species as a function of a vector of limiting factors (e.g., $dN_i/dt = N_i f_i(E_1, E_2, \ldots)$). For illustrative purposes, I discuss one example in detail (Tilman, 1982; Grover, 1997; Pacala as cited in Crawley, 1997).

Consider two species competing for two limiting resources. Assume exploitation depresses resources and that resource–consumer dynamics tend toward a stable equilibrium. For each species, there will be some combination of resources that allows equilibrium (with births matching deaths). On a graph with axes of resource abundance, assume this combination can be portrayed as a straight line, as in the line marked 1 in Fig. 1a (for species 1) (MacArthur, 1972). This line is the "zero-growth isocline" of species 1. The linear form of the isocline implies that resources are qualitatively substitutable so that a sufficient supply of one compensates for low abundances in the other. [Tilman (1982) examines exploitative competition for a broader range of resource types.] At equilibrium, resource abundances should lie along this line; if resources lie outside this line, species 1 should increase, depressing resource levels (with reverse dynamics inside the line). An equilibrium with coexistence requires resource levels at which both consumers have zero growth; graphically, the isoclines

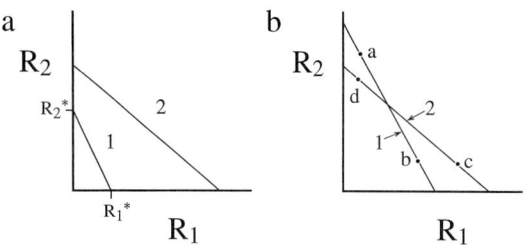

FIGURE 1 Graphical model of resource competition. The isocline of species i ($=1,2$) is a combination of resource abundances where it has zero growth. At lower resources (toward the origin) a species declines. (a) Competitive exclusion of species 2 by species 1; (b) Potential coexistence. (See text for details.)

must cross. The isoclines do not intersect (Fig. 1a) species 1 has a lower R^* for each resource. If species 1 is resident and at equilibrium, and a few individuals of species 2 are introduced, the introduction will fail due to competitive exclusion. If isoclines cross (Fig. 1b) species 1 has a lower R^* for resource 1 than does species 2, and the converse is true for resource 2. Coexistence is now possible.

Crossing isoclines reflect differences in species' ecologies and the existence of two distinct limiting factors (here, linear combinations of resources). However, such niche differences do not suffice for coexistence. As in the Lotka–Volterra model, there must be broad similarities in how the two species respond to the environment. In Fig. 1a, exclusion occurs because the species are too different in their overall requirements for resources. Coexistence, in contrast, is permitted because overall resource requirements are approximately the same for the two species (Fig. 1b).

Having isoclines that cross does not guarantee coexistence. If species 1 is alone, and resources equilibrate at point a, species 2 invades. However, if resources instead equilibrate at point b, species 2 is excluded. If species 2 is resident, species 1 invades if resources are at point c but not at point d. Comparing these points to each species' R^* (for each resource) suggests a necessary requirement for coexistence: Given two resources, each competitor must have the greater impact on that resource for which it has the lower R^*. In effect, each species must limit itself (via resource consumption) more strongly than it limits the other species. Whether or not this occurs depends on both the intrinsic renewal rates of the resources and the rate of consumption of each resource. (A complete analysis of the conditions for coexistence requires a model with equations for dynamics of both consumers and each resource.)

This graphical model illustrates the important insight that coexistence depends on a balancing of overall similarities and differences in species' niche requirements, as well as differences in species' impacts on their environments. The former involves species' intrinsic properties, whereas the latter depends on the system in which the interaction is embedded, including ecosystem processes (e.g., resource renewal rates).

B. Food Web Effects on Coexistence

This general approach can be extended in many ways, for instance, by including interactions among species at multiple trophic levels. With predation, herbivory, or parasitism inflicted on competing species, one greatly increases the number of potential limiting factors. This is a large and complex topic, and I briefly discuss some highlights here.

Specialist predators and parasites typically reduce the abundance of their favored species, freeing up resources for other nontarget species. This can facilitate coexistence if dominant competitors tend to attract more specialist predators or parasites than do subordinates. Generalist predators can have the same effect via differential fixed preferences for dominant competitors or if they reduce attacks on whichever species is temporarily rarest. However, predators or pathogens which attack two or more species can also hamper coexistence (as in the examples discussed previously). This is particularly likely if the prey species under attack do not strongly compete, and if some prey species are sufficiently productive to sustain high abundances of the predator or parasite (Holt and Lawton, 1994).

Whenever species interact via multiple mechanisms (providing different limiting factors), coexistence may occur. For instance, if two species compete for a single limiting resource, the species with higher R^* may nevertheless persist if it can also consume the superior competitor. This mixture of predation and competition is called intraguild predation.

C. Local Habitat Heterogeneity

Traditional ecological models of interacting species such as Eq. (1) assume that populations are spatially well mixed and average over local environmental variation. Relaxing this assumption often promotes coexistence. As shown in Gause's experiments, even in the confines of a microcosm there can be spatial heterogeneity in abiotic conditions that influences coexistence. Assume that within the microcosm each species has a set of conditions in which it is superior. If competitive interactions are sufficiently localized (e.g., because of limited movement or habitat selection), and each species spends more time in the microhabitat in which it is superior, one can readily generate coexistence. This simple mechanism for coexistence via local habitat partitioning is very important. A review by Schoener (1974) of resource partitioning studies following in the footsteps of MacArthur's warbler study revealed the ubiquity of habitat differences among potential competitors; subsequent years have not altered this basic message.

Competition for light is clearly important in plant communities, but it is not well understood. There is an asymmetry in that tall plants shade small plants but not the reverse. Because light absorption is imperfect, some light penetrates any canopy. Theoretical studies of light competition suggest that two (or more) plant species can coexist, given appropriate tradeoffs between size and the ability to use light at various levels. The physical structure of plants absorbing a directional flux of light almost inevitably leads to local gradients in light availability and quality (due to differential absorption of different wavelengths), which provides the opportunity for plant niche differentiation and coexistence.

In animals, habitat selection can be an important mechanism promoting coexistence with local spatial heterogeneity. If individuals of a species (when rare) can discriminate among local microhabitats and spend more time in those which provide the greatest fitness rewards, the rate of increase at low N will be increased, relative to that of a species which utilizes habitats at random. If local habitat heterogeneity permits each species to have a habitat in which it is competitively dominant, habitat selection will sharpen habitat partitioning and thereby make coexistence more likely. Habitat selection can permit subordinate species to withstand superior competitors. Recent studies in east Africa suggest that the fleetness of the cheetah helps it persist in the face of direct aggression by lions and hyenas (who kill cheetah cubs) by allowing it to seek out areas with low lion and hyena densities.

D. Consequences of Individual Discreteness

Traditional ecological models also assume that populations are sufficiently large that abundances can be treated as continuous variables. The growth in computing power in recent decades has stimulated ecologists to analyze population dynamics by tracking the fates of individuals. This reveals novel mechanisms of coexistence. For many organisms (particularly terrestrial plants), an individual occupies a small site and interacts with only a few neighbors during its lifetime. Probabilistic events of individual life histories (birth, death, and dispersal) lead to spatial variance in competition, even among sites with homogeneous physical conditions. Moreover, dispersal typically occurs over small spatial scales. Combining these two general facts together in models leads to nonuniform spatial patterns in which competition is typically stronger within species (because they tend to occur in clumps) than between species (which tend to become segregated spatially) (Pacala as cited in Crawley, 1997; Pacala and Levin as cited in Tilman and Kareiva, 1997). These effects are particularly strong for species that are relatively similar to each other. This mechanism may go a long way toward explaining the puzzling coexistence of large numbers of similar coexisting species in many plant communities.

VIII. TEMPORAL NICHE PARTITIONING

Outside the controlled confines of the lab, environments are rarely constant. Temporal variation occurs on scales ranging from diurnal cycles to climatic changes over millennia. Many ecologists have intuitively argued that temporal variability by itself weakens negative interspecific interactions and thereby facilitates coexistence. Theoretical studies (reviewed in Chesson and Huntly, 1997) have conclusively shown that this is not the case. Instead, temporal variation can provide a rich arena for differentiation among species in responses to the environments, which can promote coexistence. The basic idea is that if species A does better when the local environment is in state A' and species B does better when the local environment is in state B', and the environment alternates between A' and B', coexistence is possible. Two plausible scenarios are presented in the following sections.

A. Nonlinear Consumption

Assume two consumer species exploit a single limiting resource and have saturating, nonlinear relationships between feeding rates and resource availability. Species A has a higher rate at low resource levels, whereas species B enjoys a higher feeding rate at high levels. In a constant environment the resource level will equilibrate at a constant level, and the species with lower R^* will win. However, because of large-scale seasonal variation in resource supply rates, resource levels may fluctuate between high and low levels. Such variation can permit coexistence of these two competitors on a single resource. This mechanism hinges crucially on nonlinearity in species' responses to the shared limiting factor; if feeding rate were to increase linearly with resource availability in both species, then even in a variable environment one would observe competitive exclusion.

B. The Storage Effect

In a desert, rain falls sporadically and at different times in different years, sometimes early in the spring when the air is cool and at other times later in the hot summer. At each rain, a mixture of plant species appear in abundance from the resting seed bank and compete for water and soil nutrients. These plants are short-lived as adults, often completing their entire aboveground life cycle in just a few weeks. Some species do disproportionately better following warm rains, whereas others do better following cold rains. The seeds each species produces enter the seed bank and germinate gradually during the following years (like a time-release capsule of past recruitment). The "storage" of good years of recruitment in a long-lived seed bank can facilitate coexistence of competitors. Again, it is not variability alone which allows coexistence. One also needs (i) niche partitioning, with different species being superior at different times; (ii) nonlinear effects of the environment on responses to competition; and (iii) correlations between the varying environmental factors and the strength of competition (Chesson and Huntly, 1997). The existence of distinct life stages (e.g., long-lived seeds) can induce the needed nonlinearity in responses to temporal variation that promotes the coexistence of competing species.

IX. COEXISTENCE IN OPEN COMMUNITIES

In contrast to laboratory microcosms, many natural communities are open, coupled to the external environment via dispersal. Such coupling influences local coexistence of species in many distinct ways.

A. Autecology and Population Size Effects

A consumer species which relies on a sparse or sporadic resource, or extracts different essential resources in different habitats, may need to be mobile to persist. If absolute population size is small, extinction is risked even in favorable environments. Specialist consumers are likely to go extinct if their resources are rare. All these problems are aggravated when there are barriers to dispersal and habitable area is small. These considerations may help explain why food chains are often short on oceanic islands or isolated habitat patches. Moreover, as one increases the number of species which coexist deterministically on a fixed resource base, the abundance per species typically declines. Small population sizes have an increased risk of extinction; this can put a loose limit on coexistence.

B. Colonization–Extinction Dynamics

It was previously mentioned that strong specialist predator–prey interactions are prone to local extinctions. Coexistence may require dispersal among habitats, with prey dispersing sufficiently fast so that they can find and reproduce within empty habitat patches before being discovered by predators. Similarly, if a guild of competitors utilizes a single resource, species with successively lower values of R^* should eventually colonize and displace species with higher R^*. If there are extinctions, asynchronous among patches, inferior competitors may have a temporary window of opportunity during which they can occupy habitat patches and reproduce sufficiently to colonize other patches. This metapopulation mechanism for coexistence can be promoted by a tradeoff between competitive ability and colonization ability, and in principle it can promote the coexistence of many competing species (Tilman and Lehmann as cited in Tilman and Kareiva, 1997). Even if the inferior competitor is not a superior colonizer, it may be able to persist if it has a lower basal extinction rate when it occurs in patches alone compared to that of the superior competitor.

C. Landscape Heterogeneity

1. Spillover Effects

If a species can persist in one local community, then with emigration it can also be found in other nearby communities in which otherwise it would be excluded by local interactions. To model this "spillover" effect, assume the species when rare declines at rate $r < 0$ because of competitive exclusion but immigrates from an external source community at rate I. Its local dynamics are thus described by $dN/dt = I + rN$, which implies $N^* = I/|r|$ at equilibrium. A species which should be absent, considering only local interactions, may not just persist but even be abundant if (i) there is a large rate of immigration (e.g., from a productive source habitat into an unproductive habitat) and (ii) the rate of exclusion is slow. In this scenario, the answer to the species coexistence problem ultimately requires analyzing population dynamics at the appropriate spatial scale, larger than the local community; spatial niche partitioning at this larger scale is responsible for coexistence.

2. Landscape Mechanisms of Exclusion

Spatial coupling can also generate novel mechanisms for exclusion. For instance, a predator sustained by a prey species in a productive habitat may move through another habitat and inflict mortality at a sufficiently high rate to exclude a prey species there. Likewise, a species which is a superior competitor in one habitat may be excluded by a species which is inferior there but sufficiently abundant elsewhere so that via dispersal it can "swamp" the local habitat. For instance, Ted Case and associates recently described impacts of an invasive ant species, the Argentine ant, on an entire community of ants in coastal southern California. In contrast to many ants, the Argentine ant shows little intercolonial hostility and so has a high carrying capacity; it competes exploitatively for food but also preys on the juvenile life stages of some ant species. The Argentine ant readily becomes established in disturbed habitats. It can then spread into fragments of natural habitat resulting from intense development. Case suggests that in these fragments many other ant species have declined to the point of extinction because of the sheer force of numbers. These effects are particularly dramatic in small fragments, and at the edges of large fragments, because of incursions from surrounding human-disturbed habitats in which the Argentine ant is abundant.

I now return to the lab and field case studies of competition discussed previously. In light of what we now know about mechanisms of coexistence, it is easy to see why competitive exclusion was often observed in the classic bottle experiments with protists and grain beetles. The design of these experiments included most of the elements assumed in the syllogism leading to the competitive exclusion principle. The lab environments were climatically controlled, and the culture bottles were spatially closed, precluding two of the major classes of coexistence mechanisms. The culture media and conditions (e.g., stirring) were set up to have a restricted number of limiting resources (ideally, one) and little within-culture spatial heterogeneity in limiting food resources; the communities are very simple, so there is little opportunity for complex food web interactions. The most interesting result of these experiments may be that coexistence is observed at all. In the field study of introduced parasitoids, these species were specialized to the same host species, and the environment was relatively constant in time and homogeneous in space (the lowlands of tropical islands such as Hawaii have very stable climates, and the habitat is a deliberately homogenized landscape—plantations with regularly spaced fruit trees). It is likely that the communities are simple (islands tend to be low in species richness, particularly in heavily disturbed habitats dominated by introduced species). Most natural systems, in contrast, are temporally variable and spatially heterogeneous, and they harbor rich, complex communities; therefore, it should not be surprising that it is often difficult to document competitive exclusion in the wild.

X. FUTURE DIRECTIONS

There are many important themes relevant to species coexistence which have not been addressed in-depth by ecologists but are likely to receive considerable attention in the near future.

A. Transient Coexistence

Ecological theory has traditionally emphasized equilibrial community states and the development of criteria for exclusion and indefinite persistence. However, exclusion takes place over some timescale. If this is long (e.g., relative to climate change), communities may exhibit transient dynamics and be far from equilibrium. This is particularly likely when considering interactions among species which are very similar in their niche requirements and environmental effects; their abundances can vary through time in an essentially random fashion (a process known as community "drift," championed by Steve Hubbell). An important task for future ecological theory and empirical work is to derive a deeper understanding of transient dynamics and the drift hypothesis.

B. Allee Effects and Coexistence

Most ecological theory applies literally only to clonal, asexual organisms. However, most species of concern to ecologists are sexual. At very low population sizes, outcrossed sexual species should have depressed growth rates simply because the two sexes have to get together to reproduce—an "Allee" effect (positive density dependence in growth rates at low densities). Allee effects can influence coexistence. For instance, consider two sexual species with identical ecologies in a landscape with many habitat patches. If colonization is infrequent, whichever species arrives first in a patch should be able to dominate there because the first colonist increases until births just match deaths. When the second species appears, its potential birth rate will then also just match its death rate (because its ecology is the same as that of the resident), but its realized birth rate will be depressed because it has difficulties in pair formation at low numbers. Hence, in this patch the second species to arrive will be excluded. In the landscape as a whole, the two species may coexist if by chance each species is the first to occupy a subset of patches. In some patches, both species may by chance appear initially in sufficient numbers so as not to experience the Allee effect; in these patches, abundances should drift through time. Over the entire landscape, patterns in abundance will not be correlated with any discernible factors in the external environment. The magnitude of the Allee effect varies greatly among species, depending on details of the mating system and mate-finding strategies (e.g., selfing in plants should greatly weaken Allee effects). This influence of mating ecology on coexistence may provide an as yet poorly understood source of variation among taxa in community structure.

C. Speciation Mode and Coexistence

Over long timescales, all species originate from other species. Understanding the mode of speciation may provide insight into coexistence. For instance, sympatric speciation occurs when a lineage diverges within a single community. Most models of sympatric speciation depend on substantial ecological differentiation being present from the beginning; as speciation unfolds, ecological differences between the two emerging species must be sufficient to withstand cross-mating. In other words, with this mode of speciation, mechanisms of coexistence (e.g., substantial habitat or resource segregation) are built into the very branching pattern of the phylogenetic tree. In contrast, allopatric speciation requires geographical isolation. This can occur without any ecological difference arising between the daughter lineages (although speciation should occur more rapidly if correlated with ecological divergence). Dispersal can later bring the daughter species together, with essentially any degree of ecological difference being possible at this stage. Speciation may also often involve changes in sexual selection and mating systems, which can occur with no change in ecological requirements. After speciation, when these species' ranges begin to overlap, relative abundances should be particularly prone to community drift.

XI. CONCLUSIONS

In conclusion, understanding the factors which promote or constrain the coexistence of species is an ongoing enterprise at the intellectual core of the study of biodiversity. Community ecologists have a rich smorgasbord of hypotheses to explain both species coexistence and exclusion. In most natural communities it is likely that many mechanisms operate simultaneously. A challenging problem is to ascertain the relative contribution of these ecological mechanisms of coexistence to explain major patterns in biodiversity in space and time. There is much work to be done in analyzing mechanisms of coexistence in the context of food web dynamics and metapopulations. Moreover, very little mention was made in this article about commensalism, mutualism, and nontrophic interactions, all of which can be crucial to coexistence. An improved understanding of the factors governing species coexistence is needed to address many applied problems, particularly the conservation of natural communities. Given the importance of spatial openness for coexistence, the anthropogenic alteration and simplification of landscapes pose particular dangers to the continued coexistence of many species. Indeed, the basic unstated problem of conservation is how to structure our activities and modify our impacts so that most of the world's biota can manage to coexist (at some spatial scale and, it is hoped, over long timescales) with just a single species, namely, ourselves.

Acknowledgments

I thank many friends for conversations during the years about species coexistence, particularly Mark McPeek, Peter Chesson, John Lawton, David Tilman, Peter Abrams, Peter Morin, and James Grover.

See Also the Following Articles

COEVOLUTION • FOOD WEBS • MUTUALISM, EVOLUTION OF • INTRODUCED SPECIES, EFFECT AND DISTRIBUTION • LANDSCAPE DIVERSITY • PARASITISM • PREDATORS, ECOLOGICAL ROLE OF • POPULATION DYNAMICS • SCALE, CONCEPT AND EFFECTS OF

Bibliography

Chesson, P., and Huntly, N. (1997). The roles of harsh and fluctuating conditions in the dynamics of ecological communities. *Am. Nat.* 150, 519–553.

Crawley, M. J. (1997). *Plant Ecology*, 2nd ed. Blackwell, Oxford. [See Chapters 8 (Tilman), 14 (Crawley), 15 (Pacala), and 16 (Grover and Holt)].

Grover, J. P. (1997). *Resource Competition*. Chapman & Hall, London.

Holt, R. D., and Lawton, J. H. (1994). The ecological consequences of shared natural enemies. *Annu. Rev. Ecol. Syst.* 25, 495–520.

Hutchinson, G. E. (1978). *An Introduction to Population Ecology*. Yale Univ. Press, New Haven, CT.

Levin, S. A. (1970). Community equilibrium and stability, and an extension of the competitive exclusion principle. *Am. Nat.* 104, 413–423.

Morin, P. J. (1999). *Community Ecology*. Blackwell, Oxford.

Schoener, T. (1974). Resource partitioning in ecological communities. *Science* 185, 27–39.

Tilman, D. (1982). *Resource Competition and Community Structure*. Princeton Univ. Press, Princeton, NJ.

Tilman, D., and Kareiva, P. (Eds.) (1997). *Spatial Ecology: The Role of Space in Population Dynamics and Interspecific Interactions*. Princeton Univ. Press, Princeton, NJ.

Tokeshi, M. (1999). *Species Coexistence: Ecological and Evolutionary Perspectives*. Blackwell, Oxford.

Whittaker, R. H., and Levin, S. A. (1975). *Niche: Theory and Application* (A collection of classic papers), Benchmark Papers in Ecology No. 3. Dowden, Hutchinson, & Ross, Stroudberg, PA.

SPECIES, CONCEPTS OF

James Mallet
University College London

I. What Are Species Concepts For?
II. Statement of Bias
III. Darwinian Species Criteria
IV. The Philosophization of Species: The Interbreeding Species Concept
V. Alternative Species Concepts
VI. Species Concepts Based on History
VII. Combined Species Concepts
VIII. Dissent: Maybe Species Are Not Real
IX. The Importance of Species Concepts for Biodiversity and Conservation

GLOSSARY

cladistic A classification based entirely on monophyletic taxonomic groupings within a phylogeny; taxonomic units that are paraphyletic or polyphyletic are rejected. A cladist is one who practices cladistics, usually in the sense of using parsimony to adjudicate between data from multiple characters in the construction of a cladogram, which is an estimate of the true phylogeny.

cohesion The sum total of forces or systems that hold a species together. The term is used especially in the interbreeding and cohesion species concepts. Cohesion mechanisms include isolating mechanisms in sexual species as well as stabilizing ecological selection, which may cause cohesion even within asexual lineages.

disruptive selection Selection acting to preserve extreme phenotypes in a population. Speciation usually involves disruptive selection, because intermediates (hybrids between incipient species) are disfavored (see also stabilizing selection).

gene flow Movement of genes between populations, usually via immigration and mating of whole genotypes, but sometimes single genes may undergo horizontal gene transfer via transfection by microorganisms.

gene pool The sum total of the genetic variation within a reproductively isolated species population; this term is mostly used by supporters of the interbreeding species concept.

isolating mechanisms The sum total of all types of factors that prevent gene flow between species, including premating mechanisms (mate choice) and postmating mechanisms (hybrid sterility and inviability). Modern authors deny that these "mechanisms" have necessarily evolved to preserve the species' integrity as originally assumed, though this may sometimes be the case in reinforcement of premating isolation. Isolating mechanisms are a subset of the factors that cause cohesion of species under the interbreeding and cohesion species concepts.

monophyletic A grouping that contains all of the descendants of a particular node in a phylogeny. Monophyly is the state of such groupings. Compare paraphyletic and polyphyletic. Butterflies (Rhopalocera) and birds (Aves) are examples of two groups thought to be monophyletic.

paraphyletic A grouping that contains some, but not all, of the descendants of a particular node in a phylogeny. Paraphyly is the state of such groupings. Compare monophyletic and polyphyletic. Moths (Lepidoptera, excluding butterflies) and reptiles (amniotes, excluding birds and mammals) are examples of two groups thought to be paraphyletic.

phenetic A classification or grouping based purely on overall similarity. Pheneticists use matrices of overall similarity rather than parsimony to construct a phenogram as an estimate of the phylogeny. Examples of phenetic methods of estimation include unweighted pair group analysis (UPGMA) and neighbor joining. Cladists reject phenetic classifications on the grounds that they may result in paraphyletic or polyphyletic groupings.

phylogenetic Pertaining to the true (i.e., evolutionary) pattern of relationship, usually expressed in the form of a binary branching tree, or phylogeny. If hybridization produces new lineages, as is common in many plants and some animals, the phylogeny is said to be "reticulate." Phylogenies may be estimated using phenetics, parsimony (cladistics), or methods based on statistical likelihood.

polyphyletic Groupings contain taxa with more than one ancestor. Polyphyly is the state of such groupings. Compare paraphyletic and monophyletic. "Winged vertebrates" (including birds and bats) give an example of a polyphyletic group.

real, reality Two tricky words found frequently in the species concept debate. "Reality" is typically used to support one's own species concept, as in: "The conclusions set forth above . . . lead to a belief in the reality of species" (Poulton, 1904); similar examples can by found in the writings of Dobzhansky, Mayr, and especially phylogeneticists. The term reality in this sense is similar to an Aristotelian "essence," a hypothetical pure, albeit obscure, truth that underlies the messy actuality; unfortunately, in everyday language "real" also means "actual" (curiously, a "reality" in the first sense may be "unreal" under the second!). By rejecting the "reality" of species, one can therefore send very mixed messages: some readers will understand the author to be a nominalist who merely believes useful terms require little theoretical underpinning; others assume the author is nonsensically using some definition that does not apply to actual organisms. Here, when I discuss the reality underlying a species concept, I mean it in the sense of a hypothesized truth. Many authors of species concepts and some philosophers of science argue that definitions must be underpinned by a theoretical justification or reality. Other philosophers such as Wittgenstein and Popper agree that terms need no such definition to be useful.

speciation The evolutionary process of the origin of a new species.

specific mate recognition systems (SMRS) Fertilization and mate recognition systems in the recognition concept of species, the factors leading to premating compatibility within a species. See also cohesion, which is similar to SMRS, but includes postmating compatibility as well.

sibling species A pair of closely related, morphologically similar species (usually sister species).

stabilizing selection Selection that favors intermediate phenotypes.

SPECIES ARE CRUCIAL in many biodiversity issues: much of conservation, biodiversity studies, ecology, and legislation concerns this taxonomic level. It may therefore seem rather surprising that biologists have failed to agree on a single species concept. The disagreement means that species counts could easily differ by an order of magnitude or more when the same data are examined by different taxonomies. This article explores the controversy on species concepts and its implications for evolution and conservation.

I. WHAT ARE SPECIES CONCEPTS FOR?

Individual organisms can usually be recognized, but the larger units we use to describe the diversity of life, such as populations, subspecies, or species are not so easily identifiable. Taxonomists further group species into genera, families, orders, kingdoms, etc., while ecologists group species into higher structures such as communities and ecosystems. The justification for these group terms is utility, rather than intrinsic naturalness, but as far as possible we attempt to delimit groups of organisms along natural fault lines, so that approximately the same groupings can be recovered by independent observers. However there will be a virtually infinite number of different, albeit nested, ways of classifying the same organisms, given that life has evolved hierarchically.

Darwin (1859) felt that species were similar in kind to groupings at lower and higher taxonomic levels; in contrast, most recent authors suggest that species are more objectively identifiable, and thus more "real" than, say, populations or genera. Much of ecology and biodiversity today appears to depend on the idea that the species is the fundamental taxon, and these fields could be undermined if, say, genera, or subspecies had the same logical status.

Species concepts originate in taxonomy, where the species is "the basic rank of classification" according to the International Commission of Zoological Nomenclature. The main use of species in taxonomy and derivative sciences is to order and retrieve information on individual specimens in collections or data banks. In evolution, we would like to delimit a particular kind of evolution, speciation, which produces a result qualitatively different from within-population evolution, although it may of course involve the same processes. In ecology, the species is a group of individuals within which variation can often be ignored for the purposes of studying local populations or communities, so that species can compete, for example, while subspecies or genera are not usually considered in this light. In biodiversity and conservation studies, and in environmental legislation, species are important as units which we would like to be able to count both regionally and globally.

It would be enormously helpful if a single definition of species could satisfy all these uses, but a generally accepted definition has yet to be found, and indeed is believed by some to be an impossibility. A unitary definition should be possible, however, if species are "real," objectively definable, and fundamental biological units. Conversely, even if species have no greater objectivity than other taxa, unitary nominalistic guidelines for a definition could still be found, perhaps after much diplomacy, via international agreement among biologists; after all, if we can adopt meters and kilograms, we should be able to agree on units of biodiversity in a similar way. In either case, knowledge of the full gamut of today's competing solutions to the species concept problem will probably be necessary for a universal species definition to be found. This article reviews the proposals currently on the table and their usefulness in ecology, evolution, and conservation.

II. STATEMENT OF BIAS

I am of the opinion that the "reality" of species in evolution and in ecological and biodiversity studies over large areas has been overestimated. In contrast, it is clear to any naturalist that species are usually somewhat objectively definable in local communities. It is my belief that confusion over species concepts has been caused by scientists not only attempting to extend this local objectivity of species over space and evolutionary time but also arguing fruitlessly among themselves as to the nature of the important reality that underlies this illusory spatiotemporal objectivity. To me, agreement on a unified species-level taxonomy is possible, but will be forthcoming only if we accept that species lack a single, interpretable biological reality over their geographic range and across geological time.

Just as Marxist theory may be wrong, yet remains a convenient tool for studying political history, I hope that my own views can provide, even for the skeptic, a useful framework on which the history of proposals for species concepts can be compared. A variety of other outlooks can be found in Mayr (1982), Cracraft (1989), Ridley (1996), Claridge *et al.* (1997), and Howard and Berlocher (1998).

III. DARWINIAN SPECIES CRITERIA

A. Darwin's Morphological Species Criterion

Before Darwin, it was often assumed that each species had an Aristotelian "form" or "essence" and that variation within a species was due to imperfections in the actualization of this form. Each individual species was defined by its essence, which itself was unvarying and inherently different from all other species essences. This mode of thought of course precluded transformation of one species into another and was associated with belief that each form was separately created by God. Darwin's extensive travels and knowledge of taxonomy led to a realization that the distinction between intraspecific and interspecific variation was false. His abandonment of the essentialist philosophy and its species concept went hand in hand with his appreciation that variation itself was among the most important characteristics of living organisms, because it was this variation which allowed species to evolve.

Darwin guessed (correctly) that essentialist species would be hard to give up: "... we shall have to treat species in the same manner as those naturalists treat genera, who admit that genera are merely artificial combinations made for convenience. This may not be a cheering prospect; but we shall at least be freed from the vain search for the undiscovered and undiscoverable essence of the term species" (Darwin, 1859, p. 485). He argued that species were little more than varieties which acquired their claim to a greater reality only when intermediates died out leaving a morphological gap: "... I believe that species come to be tolerably well-defined objects, and do not at any one period present an inextricable chaos of varying and intermediate links" (Darwin, 1859, p. 177). This morphological gap criterion, which seems to have been accepted by most early evolutionists (e.g., Wallace, 1865; Robson, 1928), has been called a "morphological species concept" because Darwin used the gaps in morphology to delimit species; however, it would be easy to extend his species criterion to ecology, behavior, or genetics (see Section VIII.D).

B. Polytypic Species

A major revolution in zoological taxonomy occurred around 1900. As the great museum collections became more complete, it became obvious that apparently distinct "species" found in different areas frequently intergraded where they overlapped. These replacement series were usually combined as subspecies within a "polytypic" species (see *Subspecies, Semispecies, Superspecies*), an idea suggested for "geographical varieties" by early systematists and Darwinists such as Wallace (1865). The taxonomic clarification that followed, which allowed identifiable geographic varieties to be named below the species level as subspecies, was conceptually more or less complete by the 1920s and 1930s. At the same time, other infraspecific animal taxa such as local varieties or forms were deemed unnameable in the Linnean taxonomy. These changes are now incorporated into the International Code of Zoological Nomenclature (see also *Subspecies, Semispecies, Superspecies*). Similar ideas were promoted in botany by G. L. Stebbins (see Mayr, 1982), although local varieties and polymorphic forms remain valid and nameable taxa in the International Code of Botanical Nomenclature.

IV. THE PHILOSOPHIZATION OF SPECIES: THE INTERBREEDING SPECIES CONCEPT

In January 1904, E. B. Poulton read his famous presidential address—"What is a species?"—to the Entomological Society in London. Following up some ideas raised (but immediately dismissed) by Wallace (1865), Poulton proposed "syngamy" (i.e., interbreeding) as the true meaning of species. Poulton and Wallace were both particularly knowledgeable about swallowtail butterflies (Papilionidae). In swallowtails, there are strong sexual dimorphisms: the female color pattern often mimics unrelated unpalatable butterflies while the male is nonmimetic. The females themselves are often polymorphic, each female form mimicking a different distasteful model. Under a morphological criterion each form could be designated as a different species, whereas mating observations in the wild showed that they were part of the same interbreeding group. Similar ideas were promoted by the botanist J. P. Lotsy, who termed the interbreeding species a "syngameon." In the 1930s, T. Dobzhansky studied morphologically indistinguishable sibling species of *Drosophila* fruit flies and concluded that Lotsy's approach had some value. A species will rarely if ever interbreed with its sibling; each chooses mates from within its own species. Dobzhansky proposed his own interbreeding species concept, later popularized by Mayr as the "biological species concept," so named because interbreeding was considered the single true biological meaning or reality of the term species (reviewed by Mayr, 1970, 1982).

A short definition of the biological species concept is: "*Species are groups of interbreeding natural populations that are reproductively isolated from other such groups*" (Mayr, 1970). This concept was not so much new as a clarification of two distinct threads: (1) a local component, the Poulton–Dobzhansky interbreeding concept, and (2) a global component which extended the interbreeding concept to cover geographical replacement series of actually or potentially interbreeding subspecies (Mayr, 1970), as in the preexisting idea of polytypic species (see also *Subspecies, Semispecies, Superspecies*).

This extended interbreeding concept was, until recently, almost universally adopted by evolutionists. The species concept problem appeared to have been solved; species were interbreeding communities, each of which formed a gene pool reproductively incompatible with other such communities. The new concept answered both perceived problems of Darwin's morphological approach: (1) that under morphological criteria, distinct mutants and polymorphic variants within populations might be considered separate species, and (2) that sibling species might be misclassified morphologically as members of the same species. The new approach was promoted in a long series of books and articles by Dobzhansky, Mayr, and their followers. Mayr in particular was highly influential by justifying the taxonomic application of the polytypic species criteria in terms of the new concept of gene flow.

In order to adopt this change, it was necessary to see species in a new post-Darwinian light. Instead of species being defined simply, using manmade criteria based on demonstrable characters such as morphology, species became defined by characteristics important in their own maintenance, that is, by means of their biological function (Mayr, 1982). Significantly, the philosophical term "concept" came into vogue along with these ideas about species, and the term "species problem," which hitherto referred to the problem of how species arose (Robson, 1928), became instead the problem of defining what species were. The important features of species defined by the "biological concept" were that they were protected from gene flow by what Dobzhansky termed "isolating mechanisms," including prezygotic factors (ecological, mate choice, and fertilization incompatibilities) and postzygotic factors (hybrid inviability and sterility caused by genomic incompatibilities). By going beyond simple character-based identification of species, the concept could not apply to all living things; for example, Dobzhansky simply concluded that asexuals (between which no interbreeding is possible) could not have species.

Poulton, Mayr, and Dobzhansky emphasized that their new concept was based on the reality that underlay species, rather than being merely a criterion useful in taxonomy; in this new philosophical approach, taxonomic criteria and conceptual issues of species became separate, with taxonomic criteria taking a more minor role. The concept was true from first principles, and was therefore untestable: difficulties such as hybridization, intermediates, or inapplicability to many plants and asexuals caused taxonomic problems, but did not disprove or even challenge the underlying truth of the concept. These imperfect actualizations of species' true reality were expected in nature. Mayr claimed that the biological concept would do away with "typology" (his term for species definitions based on a fixed, unvarying type or Aristotelian essence), but in many ways it can be seen that the biological concept reverts to a new kind of essentialism, where evolutionary maintenance via interbreeding is the underlying reality of species.

V. ALTERNATIVE SPECIES CONCEPTS

It is interesting that exactly this kind of search for the essence of species had been criticized by Darwin (1859). In his chapter "Hybridism," he specifically argued against using hybrid sterility and zygote inviability as a cut-and-dried characteristic of species (although he made little mention of premating factors). Oddly, Mayr (1982, p. 269) claimed both that Darwin treated species "purely typologically [i.e., as an essentialist] as characterized by degree of difference" and also that Darwin "had strong, even though perhaps unconscious, motivation ... to demonstrate that species lack the constancy and distinctiveness claimed for them by the creationists." Whether it is reasonable to criticize Darwin in such a contradictory way can be debated, but it is clear that Mayr's proposition that interbreeding is the true essence or reality of species immediately laid itself open to debate. Although the interbreeding concept had a long run (and still does), proposals for different kinds of biological reality of species were eventually forthcoming. By proposing a single reality for species, Poulton, Dobzhansky, and Mayr opened the Pandora's box of alternative essences deemed more important by other biologists.

A. Ecological Species Concept

Asexual organisms such as the bdelloid rotifers can clearly be clustered into groups recognizable as taxonomic species (Hutchinson, 1968). On the other hand, distinct forms such as oaks (*Quercus*), between which there are high rates of hybridization, can remain recognizably distinct even where they co-occur. This suggested to L. van Valen and others that the true meaning of species was occupancy of an ecological niche rather than interbreeding. This ecological idea became known as the "ecological species concept." It became clear to Mayr during the 1970s also (see Mayr, 1982) that gene flow could not unite every population in a polytypic, biological species' range and that stabilization of phenotype might be effected by ecologically mediated stabilizing selection (see also Sections VII.B and VIII.B) rather than purely because of gene flow.

B. Recognition Concept of Species

An important attack on the biological species concept came from H. E. H. Paterson in the early 1980s. His claims were twofold: first, the Dobzhansky–Mayr term "isolating mechanisms" implied reproductive isolation was adaptive, which Paterson felt was unlikely; second, the true reality underlying species was proposed to be prezygotic compatibility, consisting of mating signals and fertilization signals. According to Paterson (1985), this compatibility is strongly conserved by stabilizing selection, whereas isolating mechanisms such as hybrid sterility or inviability are nonadaptive and can be argued to be a result rather than a cause of species separateness. To Paterson, the true reality of species must be adaptive. He termed his idea of species the "recognition concept" versus Mayr's "isolation concept," and its important characteristics "specific mate recognition systems"

(SMRS) instead of isolating mechanisms. Species were defined as *"that most inclusive population of individual biparental organisms which share a common fertilization system"* (Paterson, 1985).

The idea is generally recognized as a useful critique and has gained strong currency in some circles. However, it has been pointed out that SMRS are more or less the inverse of prezygotic isolating mechanisms and that the recognition concept therefore differs from the biological species concept mainly by focussing on a subset of isolating mechanisms used within the interbreeding concept. The interbreeding concept had always stressed a common gene pool and compatibility within a species, as well as isolation between species.

VI. SPECIES CONCEPTS BASED ON HISTORY

A. Monophyly

The rise of cladistic methods revolutionized systematics by proposing that all classification should be based on the idea of monophyly. This new systematics formalized the principle that paraphyletic and polyphyletic taxa were unnatural groupings which should not be used in taxonomy. It was logical to attempt to apply this idea throughout systematics, all the way down to the species level, leading to a monophyly criterion of species, a type of "phylogenetic species concept" (Hennig, 1968; see also the diagnostic definition below). Species were seen as forming when a single interbreeding population split into two branches or lineages that did not exchange genetic material. In a somewhat different formulation, the "cladistic species concept," species are branch segments in the phylogeny, with every branching event leading to a new pair of species (Ridley, 1996). Otherwise, if only one of the two branches were recognized as new, the other branch would become paraphyletic.

Perhaps the main criticism of this idea is that it could, if applied in taxonomy, cause great nomenclatural instability. Monophyly exhibits fractal self-similarity and can exist at very high or very low levels of the phylogeny, so the precise level at which species taxa exist becomes unclear. Supposing a new monophyletic form is discovered overlapping with, but remaining distinct from, a closely related local form in the terminal branches of an existing species. Recognition of this taxon as a species would leave the remaining branches within the original species paraphyletic. Many other branch segments would then need to be recognized at the species level, even if they interbreed and have reticulate, intermingling phylogenies. Many phylogenetic systematists therefore adopt a different phylogenetic concept, the diagnostic concept (see VI.C below), which can allow paraphyly at the species level.

B. Genealogy

Another problem with a monophyly concept is that a single, true phylogeny of taxa may rarely exist: an organismal phylogeny is in fact an abstraction of the actual genetic history, consisting of multiple gene genealogies, some of which may undergo genetic exchange with other taxa. There is now good evidence that occasional horizontal gene transfer and hybridization may selectively transfer genetic material between unrelated forms. Furthermore there are multiple gene lineages within any population, so that, if such a population were to become geographically or genetically split into two distinct forms, it would be some time before each branch became fixed for different, reciprocally monophyletic gene lineages at any single gene. The idea of monophyly for whole genomes then becomes hard to define, especially near the species boundary. However annoying, phylogenetic methods and evolutionary theory must face up to these facts (Avise and Ball, 1990; Maddison, 1997). It has therefore been suggested that species should be defined when a consensus between multiple gene genealogies indicates reciprocal monophyly. This is called the "genealogical species concept" (Baum and Shaw, 1995).

Critics argue that this idea has many problems in common with other monophyly concepts of species (Davis, 1997). Geographic forms that have become isolated in small populations or on islands, say, could rapidly become fixed for gene lineages and become viewed as separate species without any biologically important evolution taking place. On the other hand, clearly distinct sister taxa such as humans and chimpanzees still share gene genealogy polymorphisms at some genes such as the HLA complex involved in immunological defense and might therefore be classified as the same species under genealogical considerations.

C. Diagnostic Species Concept

The motivation for the diagnostic concept, usually called the "phylogenetic species concept" by its adherents, was again to incorporate phylogenetic Hennigian thinking into species-level taxonomy. There are many cases of hybridization between taxa on very different branches of species level phylogenies, which suggest that interbreeding and phylogenetic realities conflict.

Cracraft (1989) also noted that many bird taxa normally thought of as subspecies were far more recognizable and stable nomenclaturally than the polytypic species to which they supposedly belonged (see also Section VIII.D). Cracraft therefore argued that the polytypic/interbreeding species concept should be rejected, and in its place we should use a diagnostic criterion in the form of at least one fixed difference at some inherited character. "*A phylogenetic species is an irreducible (basal) cluster of organisms, diagnosably distinct from other such clusters, and within which there is a parental pattern of ancestry and descent*" (Cracraft, 1989). According to Cracraft, species defined in this way are the proper basal, real taxa suitable for phylogenetic analysis and evolutionary studies.

Of course, if diagnostic criteria are applied strictly, rather small groups of individuals, or even single specimens might be defined as separate species. Cracraft recognized this and argued that such diagnosable groups have no "parental pattern of ancestry and descent;" i.e., they are not proper populations. However, this qualification appears similar to an interbreeding criterion of species, whereas the whole approach of using diagnostic characters was an attempt to get away from interbreeding.

Most evolutionary biologists balk at the idea of speciation being merely the acquisition of a new geographic diagnostic character, a DNA base pair or color pattern change perhaps (Harrison, 1998). "Speciation" is only a different, or special kind of evolution if the new "species" is a distinct population which can coexist locally with its sibling or parent population without losing its integrity.

Characters used to diagnose phylogenetic species may not be shared derived characters; they may be primitive (plesiomorphic) characters, or they may have evolved several times. Therefore, phylogenetic species need not be monophyletic, and could presumably be paraphyletic and perhaps polyphyletic. Cracraft appears confused on this matter: on the one hand, he claims that phylogenetic species "will never be nonmonophyletic, except through error" (Cracraft, 1989, p. 35), but on the other, recognizes that "their historical status may [sometimes] be unresolved because relative to their sister species they are primitive in all respects. Whether they ... [are] truly paraphyletic ... is probably unresolvable" (Cracraft, 1989, p. 35). It seems odd to allow a "phylogenetic" species to be even paraphyletic (let alone polyphyletic) because paraphyly contravenes one of the basic tenets of phylogenetic systematics and because one of the main justifications for a phylogenetic species concept is that species defined via other concepts might sometimes be paraphyletic: "The biological species concept cannot be applied to the *Thomomys umbrinus* complex unless one is willing to accept paraphyletic species, and to do so would be a *de facto* admission that biological species are not units of evolution" (Cracraft, 1989, p. 46; see also Davis, 1997, p. 374). The phylogeneticists' resolution of that problem, using diagnostic characters, leads to the same difficulty all over again! This rather glaring logical inconsistency considerably undercuts the argument for a diagnostic species concept.

In spite of these problems, Cracraft highlighted some genuine practical problems with the polytypic application of the interbreeding concept, and as a result this phylogenetic species concept has been influential. Ornithologists in particular have used diagnostic characters to reassign many taxa long thought of as subspecies to the level of full species. Recently, many molecular systematists, including botanists (Davis, 1997) have taken up Cracraft's suggestion and used diagnostic differences between populations, in some cases at single DNA base pairs, as evidence that two forms are separate species even if they intergrade freely at the boundaries of their distribution.

VII. COMBINED SPECIES CONCEPTS

A. Evolutionary and Lineage Concepts

Faced with the problem of studying the evolution of species through time, the paleontologist G. G. Simpson (1951) proposed his "evolutionary species concept," in which a species is "*a lineage (an ancestral-descendant sequence of populations) evolving separately from others and with its own unitary evolutionary role and tendencies.*" In other words, Simpson combined the idea that species were historical lineages with the concept of their evolutionary and ecological role. The key essence here appears to be "evolutionary independence." This concept appeals to phylogenetic systematists and paleontologists alike, because of its historical dimension, and to neontologists because of its acknowledgment that biological mechanisms are what make the species real. De Queiroz (1998) is perhaps the most recent reviewer to propose that a single concept, which he calls "the general lineage concept," under which "*species are segments of population-level lineages,*" underlies all other species concepts. According to De Queiroz, apparently competing species concepts merely emphasize different characters or criteria for species definition, but all acknowledge implicitly or explicitly that evolutionary separateness of lineage is the primary concept. This is a nice ideal, but evolutionary independence has little logical force in its application to actual forms that hybridize or undergo genetic exchange.

B. Cohesion Concept of Species

In similar vein, Templeton's (1998) "cohesion concept" fuses a number of competing ideas of species. Templeton argues that a combination of ecological and reproductive cohesion are important for maintaining a species' evolutionary unity and integrity, thereby incorporating components of the evolutionary, ecological, recognition, and interbreeding concepts. As well as applying to asexual taxa ("too little sex"), Templeton's idea also applied to species like oaks that undergo frequent hybridization and gene flow ("too much sex" for the interbreeding concepts). He further argues that separateness of genealogy is another important characteristic of species.

We are perhaps nearing the apogee of the species debate with these combined concepts. By incorporating evolutionary and phylogenetic origins together with every possible biological means by which species are currently maintained, these combined concepts "cover all the bases." One can acknowledge that species evolve and are maintained as cohesive wholes by all of these multifarious processes, yet at the same time argue that species can, and perhaps should be seen as separate from their histories of origin and from current reasons for their integrity. If groups with very different and conflicting biological and evolutionary characteristics are all considered species, there should exist a simpler criterion that unites them. It can also be argued that to conflate the origin and evolutionary role of a taxon with the definition of that taxon itself may lead to circularity, particularly in conservation or ecological studies, or when investigating speciation.

VIII. DISSENT: MAYBE SPECIES ARE NOT REAL

Throughout the history of the species debate, starting with Darwin, there have been some who argue that species are not individual "real" objects, but should instead be considered merely as manmade constructs merely useful in understanding of biodiversity and its evolution. These people are not necessarily nihilists, who deny that species exist: they simply argue that actual morphological and genetic gaps between populations would be more useful for delimiting species than inferred processes underlying evolution or maintenance of these gaps. By their refusal to unite these ideas under a single named concept, this biologist "silent majority" have rarely found a common voice.

A. Taxonomic Practice

Taxonomists are on the front line of the species battle, because it is they who ultimately decide whether to lump or split taxa, and at what level to name them as species. If the objectivity and individuality of species as the primary taxon exists, taxonomists' activities have not been made any easier; and many taxonomists have simply ignored or denied belief in the evolutionary reality of species. In general, it is probably true to say that at least 10% of taxonomic species are subject to revision because of these practical difficulties in delimitation.

For this reason, since the rise of the polytypic/interbreeding species concept, there has been little impact of the postwar species concepts on practicing taxonomists, even while the debate raged around them. Procedure, at least in zoology, was more or less as follows: geographic variants which blended (or were thought to be able to blend) together at their boundaries were united within a single, polytypic species, unless morphological or genetic differences were so great that it seemed necessary to recognize two species. On the other hand, whenever two divergent forms differing at several unrelated traits, overlapped spatially, they were recognized as separate species even if a few intermediates suggested some hybridization or gene flow. Some taxonomists regarded subspecies as artificial taxa to be avoided, and may either have ignored geographic variation, or elevated subspecies of polytypic species to the rank of full species. But good taxonomic practice on species remained broadly similar across most branches of systematics, and involved careful analysis of multiple, chiefly morphological character sets tested in large samples of specimens collected from as many geographic regions as possible.

This view on species and subspecies has led in zoology to a steady reduction in the numbers of recognized species, as more and more dubiously separated taxa, previously ranked as species, became inserted as subspecies into larger and larger polytypic species. Recently, however, the diagnostic version of the phylogenetic species concept (Section VI.C) has been making strong inroads into zoological nomenclature, with the result is that counts of species on continents are again climbing as former subspecies are reelevated to the species level, in spite of intergradation at their boundaries. However, the situation could get much worse; many *Heliconius* butterflies, for example, have over 30 geographic subspecies per species, all of which can be diagnosed easily. The numbers of bird and butterfly species could easily increase 2–10 times in some groups if the diagnostic criterion were generally adopted. The one reality that is clear in species-level taxonomy is that the species is not real enough to remain at the same phylogenetic level while taxonomic fashions change. This is good evidence that actual species have been and still are purely manmade taxonomic units lacking in any objectively determinable underlying essence, even if such an essence exists.

B. Populations Are Evolutionary Units, Not Species

Botanists deal with geographically variable organisms with low powers of dispersal, and have therefore never been very happy with the polytypic/interbreeding concept applied with such apparent success in zoology. Meanwhile, the strong surge in experimental population genetics and evolutionary studies that followed the books by Dobzhansky, Mayr, and Stebbins has led to a greatly improved understanding of gene flow in natural populations. Gene flow in quite mobile animals, such as birds or butterflies, may not unite even adjacent local populations into a common gene pool. If local populations only rarely exchange genes, then gene flow across the range of a continental species is clearly insufficient to explain species integrity, because it would be outweighed easily by weak local patterns of adaptation or genetic drift (Endler, 1977).

This increasing input of population biology into systematics and evolution led to the proposal by Ehrlich and Raven (1969), Levin (1979), and others that species are not real biological units at all; instead, local populations are the only real groupings united by gene flow within a common gene pool, and which adapt to local conditions, compete, and so on. Any homogeneity of ecological niche or genetics over the range of a species might be due either to simple evolutionary inertia or to similar stabilizing selection everywhere. To these authors species exist and are real in local communities; but it is fallacious to treat distant populations in the same way (see also Section VIII.D).

This viewpoint is generally understood and respected by population biologists, but curiously has not been incorporated explicitly into current thinking on species in systematics and evolution. Perhaps there is a sneaking suspicion that even very weak levels of gene flow may explain species integrity over wide areas.

C. Phenetic Species Concept

In the 1960s and 1970s, a major systematics movement proposed numerical methods in taxonomy now usually referred to as phenetics. Pheneticists, as they were called, argued that taxonomy and systematics should be based on multivariate statistical analysis of characters rather than on underlying evolutionary or biological process information. If taxa were defined by nonevolutionary criteria, studies of evolution would be freed from the tautology of testing hypotheses about processes, when those same processes are used as assumptions in the definitions of taxa under study. Species, like other taxa, would be defined in numerical taxonomy on the basis of multivariate statistics, as clusters in phenotypic space (Sokal and Crovello, 1970).

Phenetics is reviled by those who believe that classifications should be phylogenetic. However, the approach is closely similar to the intuitive methods adopted by most actual taxonomists, who use multiple morphological or genetic characteristics to sort individual specimens into discrete groups between which there are few intermediates (see Section V.A). Some large areas of practical taxonomy are based purely on this phenetic approach. Bacterial systematists, for instance, use multiple biochemical tests to assign microbes to species taxa. The usefulness of this taxonomic method is attested by its success in hospitals for predicting pathogeneticity and antibiotic sensitivity.

Phenetic classifications based on morphology introduce the danger that, if convergent characters are used as data, one may group unrelated forms into paraphyletic or even polyphyletic taxa. In addition, single gene polymorphisms and sexual dimorphism can affect multiple morphological characters. This could lead to recognition of multiple species within polymorphic populations. Sibling species, on the other hand, could be lumped into the same species using a phenetic approach, unless a set of highly diagnostic characters could be found.

D. Genotypic Cluster Criterion

For morphological or genetic gaps to exist between species, gene flow between species (if any) must be balanced by an opposing force of disruptive selection. In my own work, I had studied hybrid zones between geographic forms of butterflies, and I attempted to show that a practical statistical definition of species could be constructed using morphological and genetic gaps alone, rather than using phylogetic or evolutionary processes that caused the gaps to exist.

However, to define species by means of the gaps between them requires consideration of the nature of the gaps to avoid falling into the trap of defining polymorphic forms as separate species or of lumping sibling species. Rather than merely using external morphology, in difficult cases I proposed that we could consider the genetics as well. DNA has a digital, rather than analog code, so there is a genetic gap between virtually any pair of individuals. Clearly, then, we cannot use just any discreteness at the genetic level to define species. Separate sexes and polymorphic female forms of mimetic *Papilio* butterflies also have gaps between them in exactly this way. A genetic element, which may be a single base pair, an allele at a gene, the entire mitochondrial genome, a chromosomal rearrangement, or perhaps a sex chromosome, may determine the genetic or morphological differences between such polymorphic forms.

To be considered part of a single local population, and therefore part of the same local species, we expect that polymorphic genetic elements like mimicry genes and sex chromosomes will be mixed in with polymorphisms at genetic elements found on other chromosomes or extrachromosomal DNA. Each individual may be a distinct multilocus genotype, but we recognize groupings of these genotypes because a polymorphism at one genetic element is more or less independent of polymorphisms at another. Conversely, if alleles at one locus are strongly associated with alleles at other, unlinked elements, we have evidence for two separate populations or species.

I therefore proposed a "genotypic cluster criterion" of species. Species are recognized by morphological and genetic gaps between populations in a local area rather than by means of the phylogeny, cohesion, or reproductive isolation that are responsible for these gaps (Mallet, 1995). In a local area, a single species (the null hypothesis) is recognized if there is but a single cluster in the frequency distribution of multilocus phenotypes and genotypes. Separate species are recognized if there are several clusters separated by multilocus phenotypic or genotypic gaps. These gaps may be entirely vacant, or they may contain low frequencies of intermediate genotypes, or hybrids (Fig. 1). The definition is useful because one avoids tautological thinking: hypotheses about speciation or phylogeny become independent of assumptions about the nature of reproductive isolation or phylogeny underlying the taxa studied.

Genotypic clusters are neither profound nor original;

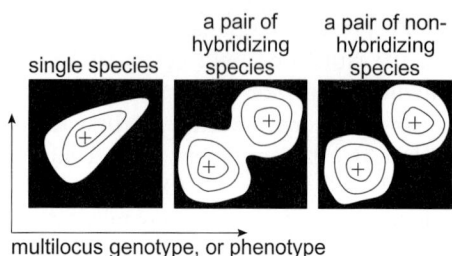

FIGURE 1 Genotypic cluster criterion for species. A sample of individuals is made at a single place and time. Numbers of individuals are represented by the contours in multidimensional genotypic space. Peaks in the abundance are represented by "+". Two species are detected if there are two peaks in these distributions; otherwise the null hypothesis of a single species is not rejected. Note: the axes represent multidimensional morphological/genotypic space, not geographic space.

I trace the idea to Darwin (1859, see Section III.A), although it undoubtedly goes further back since its use does not require acceptance of evolution. Many similar proposals have been made (Simpson, 1937; Hutchinson, 1968; Sokal and Crovello, 1970; Avise and Ball, 1990; Cohan, 1994; Smith, 1994); in fact, the approach is similar to the practice used on morphology by most taxonomists (see Section VIII.A), and indeed by those who attempt a practical application of the biological species concept (see below). This widespread use of direct genotypic or morphological data, as opposed to reproductive or phylogenetic inferences made from such data, has apparently lacked support because its practitioners were loath to label it as a "concept." I intended merely to justify the Darwinian and practical taxonomists' species definition statistically and in terms of genetics, rather than to enforce the use of genotypes instead of morphology to define species. Most genotypic cluster species can be recognized morphologically; for example, minor pattern elements in *Papilio* can be used to unite the various polymorphic forms; however, with abundant molecular marker data we could easily use the criterion to sort actual specimens.

There is also nothing wrong with concluding after seeing a male butterfly mating with an unlike female that they belong to the same species, but, because hybridization does occur occasionally between forms normally thought of as different species, one is not so much using the mating behavior itself to define species as inferring that the mating behavior is a common event, so that, if we were to analyze their genomes, the two forms would have similar genetic characteristics apart from those determining sexual dimorphism; i.e., they would belong to the same genotypic cluster. Instead of reproductive compatibility being the primary criterion of species, we can turn the argument on its head and infer from limited data on reproductive compatibility that a single genotypic cluster may result.

Asexual forms, unclassifiable under the interbreeding concept, and arbitrarily definable at any level under concepts depending on phylogeny, can be clustered and classified as genotypic clusters in exactly the same way as sexual species. The precise taxonomic level of species clustering is potentially somewhat arbitrary, as in the phylogenetic concepts, but at least the method acknowledges this arbitrariness rather than pretending that some higher evolutionary principle is being used. However, many asexual forms such as bdelloid rotifers have easily distinguishable species taxa (Hutchinson, 1968). In bacteria, competition is thought to structure promiscuous, but largely asexual, populations into recognizable genetic clusters (Cohan, 1994; see also Section V.A).

Critics have argued that the genotypic cluster criterion in sexual species is nothing other than a gene flow concept of species under a different guise. This is true for one specialized interpretation of "gene flow." If we define gene flow as the successful or effective as opposed to actual input of genes, then it is easy to see a "gene flow criterion" becomes similar to the "genotypic cluster" criterion: to find whether a hybridization or gene flow event is successful, we must either follow every gene through all possible descendants for all time, or we may examine the genotypic state of a population and determine if genes from one form are mixed randomly with genes from another form. It seems clear that looking for random association of genes within genotypes in the genotypic cluster approach will be methodologically the same as genotypic state analysis to determine whether a population is interbreeding, but the latter requires additional assumptions. The genotypic cluster criterion in sexual species could be looked upon simply as a practical application of the biological species concept. However, I distinguish the genotypic cluster criterion from the interbreeding concept, if only because its name emphasizes that the definition is character-based, rather than actually based on interbreeding, and is thus applicable to asexuals as well as sexual species.

If a single geographic race, which previously intergraded at all its boundaries with other geographic races, were to split into two forms that coexist as separate genotypic clusters, we could have the situation that the original polytypic species became paraphyletic. The new species has been derived from only one of the component subspecies. Thus paraphyly of species must be recognized as a possibility under this definition, as with both interbreeding and diagnostic concepts.

E. The Unreality of Species in Space and Time

Geographic races often form clusters differing at multiple loci from other races in the same species. The interbreeding concept or genotypic cluster criterion can be used to justify a classical polytypic species if the various geographic races are separated by zones which contain abundant intermediates (hybrids). We sample multilocus genotypes or phenotypes in local areas of overlap and determine whether a single peak (one species, i.e., abundant hybrids), or two peaks (two different species, i.e., rare hybrids) are evident in the local genotypic distribution (Fig. 1). Hybridization may occur, but if it is rare so that character and genotypic distributions remain distinctly bimodal in zones of overlap, we usually classify them as separate species, even under the interbreeding concept.

Although this spatial extension of the local species is practical to apply to any pair of forms in contact, it is unlikely to lead to general agreement. The problem is that hybrid zones can be very narrow and may separate forms that are highly distinct at multiple characters or loci, in spite of complete unimodal blending in local areas of overlap. Even adherents of the interbreeding concept are reluctant to lump such geographic forms within the same species. Examples include North American swallowtail butterflies (*Papilio glaucus/P. canadensis*; see Hagen *et al.*, 1991) and European toads (*Bombina bombina/B. variegata*; see Szymura, 1993).

An even worse problem is found in "ring species," which form a continuous band of intergrading subspecies, but whose terminal taxa may be incompatible, and overlap without intergrading. A commonly cited example are the herring gulls and lesser black-backed gulls (*Larus argentatus* complex; Mayr, 1970). Similarly, while most hybrid zones between European *Bombina* are unimodal, the same pair of taxa may have bimodal genotypic distributions in other zones of overlap (Szymura, 1993). Thus, geographic forms may be apparently conspecific in some areas, but overlap as separate species in other areas. Finally, if distinct populations are geographically isolated and there is no area of overlap, one cannot disprove the null hypothesis of "same species" under interbreeding or genotypic cluster criteria, but biologists are reluctant to unite such populations if they are very divergent. Laboratory hybridization could be tried, but many overlapping species are known to hybridize freely in captivity, while remaining separate in nature. There are good examples in the great apes, for instance, the bonobo (*Pan paniscus*) versus the chimpanzee (*P. troglodytes*) and the gorillas (Uchida, 1996), but similar decisions must be made in almost any animal and plant group.

The problem for spatial extensions of species criteria is in fact due to the evolutionary divergence of spatially separated lineages: time and space are correlated. Paleontologists face a similar temporal problem when classifying fossils in different strata. Evolutionary rates may vary, but all lineages must ultimately be continuous, so there is no very logical place to put a species boundary in time any more than there is in space. Paleontologists typically use operational species on the basis of morphological gaps between taxa from the same and different time periods (Simpson, 1937; Smith, 1994).

These difficulties show that there is no easy way to tell whether related geographic or temporal forms belong to the same or different species. Species gaps can be verified only locally and at a single time. One is forced to admit that Darwin's insight is correct: any local reality or integrity of species is greatly reduced over large geographic ranges and time periods (see also Mayr, 1970, and Section IX.D).

IX. THE IMPORTANCE OF SPECIES CONCEPTS FOR BIODIVERSITY AND CONSERVATION

A. Traditional: Species as Real Entities

Different species concepts seek to define species in mutually incompatible ways. Thus, a monophyletic species concept seems not very useful to evolutionary biologists because of difficulties with multiple gene genealogies and paraphyletic remnants. In contrast, the interbreeding concept and other combined concepts incorporating biological processes of species maintenance suffer in the eyes of phylogenetic systematists because they lack phylogenetic coherence and produce paraphyletic taxa, or worse. If we were to allow the basal unit of our taxonomy to incorporate paraphyly, it would be harder to justify a strict adherence to monophyly at other taxonomic levels. It is beyond the scope of this article to resolve these difficult issues, but these conceptual conflicts fuel the continuation of the debate, and also highlight the fact that if species are indeed real, objective biological units, their unifying reality has been extremely difficult to verify.

Many ecological and biodiversity studies of actual organisms ignore these difficulties, and assume that species are objectively real basal units. Thus in ecology, we have theories of global species diversity. In conservation, we have the Endangered Species Act in the United States, which prescribes the conservation of threatened taxa we call species. Populations not viewed as species, particularly putative hybrid taxa (like the red wolf, *Canis rufus*, of the southeastern United States), are seen as less valuable, even if rare. How do we recognize that a taxon is hybridized? Obviously, to be a hybrid, it must be a mere intergrade between two, real, objectively identifiable entities. The Endangered Species Act views species as important real conservation units and hybrids as unimportant. It did this because it incorporated the species concept in vogue at the time of its enactment, i.e., the biological species concept, in which hybridization is seen as a "breakdown in isolating mechanisms" (Mayr, 1970).

B. Alternatives: Genetic Differences More Valuable Than Species Status

If this Act were to be rewritten today, what would it say? There is undoubtedly a greater realization today that other levels in the taxonomic hierarchy are important elements of biodiversity. The diagnostic concept of species, while claiming to support the basal, objective nature of species, can at least have the beneficial effect of allowing its basal unit of biodiversity to be recognized at a lower level, in this case as subspecies within polytypic species. Some molecular geneticists have advocated conservation of "evolutionary significant units," "management units," or "stocks" (a fisheries term) defined on criteria of continuous genetic differentiation at molecular markers (Moritz, 1994) as being more important than worrying about the species level at all. The true reality of spatiotemporally extended species eludes us, if it exists. If this is so, then it seems best to adopt some other measure of conservation value that relies purely on the degree of genetic differentiation, for instance, at molecular genetic markers.

C. Species Differences as Ecologically Important Markers

However, there are many who oppose this view. Species within a local area such as a nature reserve are, for the most part, easily and objectively identifiable using morphology, behavior, genetics, or phylogeny. A pair of similar species must usually be ecologically different to coexist. To remain distinct, sexual species will need some prezygotic isolation, so their mating behavior must also be different. Thus, counting species in a local area makes ecological sense, and conserving species diversity in a local area would conserve actual ecological and behavioral diversity. Behavioral and morphological differences that cause speciesness are evolutionarily more valuable than potentially neutral genetic differences at molecular markers.

D. Biodiversity in Space and Time

As we have seen, this local view breaks down when we try to apply the term "species" over large areas or geological time scales. In some cases, there is excellent homogeneity over large areas; for example, the painted lady butterfly (*Vanessa cardui*) and the barn owl (*Tyto alba*) have a virtually worldwide distribution and look nearly identical everywhere. Other species are not so homogeneous: the familiar mallard group of ducks (*Anas platyrhynchos*) has become highly differentiated into some 18 or so forms in far-flung outposts of the world, but exactly how many are good species, and how many are races, or indeed, how many races there are in total, is a matter of taste. Current authorities recognize about 10 species, but there might easily be five or 15 in alternative treatments. One of the forms, the Mexican duck, *A. platyrhynchos diazi*, is threatened with hybridization by its mallard relative, *Anas platyrhynchos platyrhynchos*, which has been expanding from the north, and the American black duck (*Anas rubripes*) also hybridizes with the mallard but appears to resist hybridization somewhat better than the Mexican form—hence its species status.

Faced with these difficulties, should we worry about the species level when conserving endangered taxa over large areas? Whatever the answer to this question, it does not seem very sensible to rely too much on the spatiotemporal reality of species for an answer. We might rename the Mexican duck as a separate species instead of a subspecies, but ideally this should not affect its conservation value since there has been no actual change in the knowledge of biological characteristics that affect conservation value. Most conservationists now agree that the former fetish for species-level legislation was a mistake: conservation and legislation should now recognize that living, evolving populations form fractal continua over time and space, rather than attempting a division into spurious "fundamental" units.

Species are certainly fundamental units of *local* biodiversity, but they have this clarity only in a small zone of time and space, and so species counts become less and less meaningful as larger and larger areas are covered. Taxonomists might come to nominalistic agreements on a case-by-case basis, but even this shows little sign of happening yet. Ecological theory, as well as conservation and biodiversity studies must recognize that species counts over large expanses of space and time represent only a sketchy measure of biodiversity, a measure which owes more to taxonomic and metaphysical fashion than to science.

See Also the Following Articles

BIODIVERSITY, DEFINITION OF • CLADISTICS • DARWIN, CHARLES • EVOLUTION, THEORY OF • PHYLOGENY • SPECIATION, PROCESS OF • SPECIES DIVERSITY, OVERVIEW • SUBSPECIES, SEMISPECIES • TAXONOMY, METHODS OF

Bibliography

Avise, J. C., and Ball, R. M. (1990). Principles of genealogical concordance in species concepts and biological taxonomy. In *Oxford Surveys in Evolutionary Biology* (D. J. Futuyma and J. Antonovics, Eds.), Vol. 7, pp. 45–67. Oxford Univ. Press, Oxford.

Baum, D. A., and Shaw, K. L. (1995). Genealogical perspectives on the species problem. In *Experimental and Molecular Approaches to Plant Biosystematics* (P. C. Hoch and A. G. Stephenson, Eds.), pp. 289–303. Missouri Botanical Garden, St. Louis, MO. (*Monographs in Systematic Botany from the Missouri Botanical Garden*, No. 53).

Claridge, M. F., Dawah, H. A., and Wilson, M. R. (Eds). (1997). *Species: The Units of Biodiversity*. Chapman & Hall, London.

Cohan, F. M. (1994). The effects of rare but promiscuous genetic exchange on evolutionary divergence in prokaryotes. *Am. Nat.* **143**, 965–986.

Cracraft, J. (1989). Speciation and its ontology: The empirical consequences of alternative species concepts for understanding patterns and differentiation. In *Speciation and Its Consequences* (D. Otte and J. A. Endler, Eds.), pp. 28–59. Sinauer Associates, Sunderland, MA.

Darwin, C. (1859). *On the Origin of Species by Means of Natural Selection, or the Preservation of Favoured Races in the Struggle for Life*, 1st ed. John Murray, London.

Davis, J. I. (1997). Evolution, evidence, and the role of species concepts in systematics. *Syst. Bot.* **22**, 373–403.

de Queiroz, K. (1998). The general lineage concept of species, species criteria, and the process of speciation. A conceptual unification and terminological recommendations. In *Endless Forms. Species and Speciation* (D. J. Howard and S. H. Berlocher, Eds.), pp. 57–75. Oxford Univ. Press, New York.

Ehrlich, P. R., and Raven, P. H. (1969). Differentiation of populations. *Science* **165**, 1228–1232.

Endler, J. A. (1977). *Geographic Variation, Speciation, and Clines*. Princeton Univ. Press, Princeton, NJ.

Hagen, R. H., Lederhouse, R. C., Bossart, J. L., and Scriber, J. M. (1991). *Papilio canadensis* and *P. glaucus* (Papilionidae) are distinct species. *J. Lepid. Soc.* **45**, 245–258.

Harrison, R. G. (1998). Linking evolutionary pattern and process. The relevance of species concepts for the study of speciation. In *Endless Forms. Species and Speciation* (D. J. Howard and S. H. Berlocher, Eds.), pp. 19–31. Oxford Univ. Press, New York.

Hennig, W. (1966). *Phylogenetic Systematics*. Univ. of Illinois Press, Urbana, IL.

Howard, D. J., and Berlocher, S. H. (Eds). (1998). *Endless Forms. Species and Speciation*. Oxford Univ. Press, New York.

Hutchinson, G. E. (1968). When are species necessary? In *Population Biology and Evolution* (R. C. Lewontin, Ed.), pp. 177–186. Syracuse Univ. Press, Syracuse, NY. Proceedings of the International Symposium Sponsored by Syracuse University and the New York State Science and Technology Foundation, June 7–9, 1967, Syracuse, NY.

Levin, D. A. (1979). The nature of plant species. *Science* 204, 381–384.

Maddison, W. P. (1997). Gene trees in species trees. *Syst. Biol.* 46, 523–536.

Mallet, J. (1995). A species definition for the Modern Synthesis. *Trends Ecol. Evol.* 10, 294–299.

Mayr, E. (1970). *Populations, Species, and Evolution*. Harvard Univ. Press, Cambridge, MA.

Mayr, E. (1982). *The Growth of Biological Thought. Diversity, Evolution, and Inheritance*. Belknap, Cambridge, MA.

Moritz, C. (1994). Defining 'Evolutionarily Significant Units' for conservation. *Trends Ecol. Evol.* 9, 373–375.

Paterson, H. E. H. (1985). The recognition concept of species. *Species and Speciation* (E. S. Vrba, Ed.), pp. 21–29. Transvaal Museum, Pretoria. (*Transvaal Museum Monograph No. 4*).

Poulton, E. B. (1904). What is a species? *Proc. Entomol. Soc. London* 1903, lxxvii–cxvi.

Ridley, M. (1996). *Evolution*. Blackwell Sci., Oxford.

Robson, G. C. (1928). *The Species Problem. An Introduction to the Study of Evolutionary Divergence in Natural Populations*. Oliver and Boyd, Edinburgh.

Simpson, G. G. (1937). Patterns of phyletic evolution. *Bull. Geol. Soc. Am.* 48, 303–314.

Simpson, G. G. (1951). The species concept. *Evolution* 5, 285–298.

Smith, A. B. (1994). *Systematics and the Fossil Record. Documenting Evolutionary Patterns*. Blackwell Sci., Oxford.

Sokal, R. R., and Crovello, T. J. (1970). The biological species concept: A critical evaluation. *Am. Nat.* 104, 107–123.

Szymura, J. M. (1993). Analysis of hybrid zones with *Bombina*. In *Hybrid Zones and the Evolutionary Process* (R. G. Harrison, Ed.), pp. 261–289. Oxford Univ. Press, New York.

Templeton, A. R. (1998). Species and speciation. Geography, population structure, ecology, and gene trees. In *Endless Forms. Species and Speciation* (D. J. Howard and S. H. Berlocher, Eds.), pp. 32–43. Oxford Univ. Press, New York.

Uchida, A. (1996). What we don't know about great ape variation. *Trends Ecol. Evol.* 11, 163–167.

Wallace, A. R. (1865). On the phenomena of variation and geographical distribution as illustrated by the Papilionidae of the Malayan region. *Trans. Linn. Soc. London* 25, 1–71.

SPECIES DIVERSITY, OVERVIEW

A. Ross Kiester
U.S. Department of Agriculture Forest Service

I. Introduction
II. Species Richness
III. Patterns of Species Richness
IV. Determinants of Species Richness
V. A Diversity of Species Diversities
VI. Summary: Structure of Taxa and Ecological Assemblages
VII. Species Diversity and Biodiversity Policy

GLOSSARY

clique A group of organisms in a food web in which any pair of species in the clique shares at least one prey species. Thus, it is a group of species that have similar diets and are different from other such groups.

gap analysis A biodiversity policy tool which compares the actual distribution of species and vegetation classes to areas managed for the long-term protection of biodiversity and other classes of land management. Species and classes that are poorly represented in protected areas are gaps in the protection of all biodiversity. Such species and classes are candidates for proactive protection.

lognormal distribution A statistical distribution which is normal or Gaussian ("bell curve") when the logarithm of the original data is used. Species abundance data from many communities fit this distribution well. The lognormal distribution often occurs as the result of the law of large numbers in which many small factors interact multiplicatively to produce a result such as species diversity.

Simpson diversity index An index of diversity based on species abundance data. If p_i is the frequency of species i in a community of n different species, then

$$\text{Simpson diversity} = \left(1 - \sum_{i=1}^{n} p_i^2\right)$$

and is equal to the probability that two randomly chosen individuals from a given community are of the same species. This index of diversity is statistically unbiased, making it especially useful in practice.

species diversity The variability of a group of species classified in some way. For example, individual species can be classified by size and then a species size diversity measure can be developed which expresses the variability with respect to size. The most common classification is by abundance or population numbers. Then species diversity measures the variability of the number of individuals of each species. The Simpson diversity is an example of such an index. Many such indices have been developed for species diversity.

species richness The number of species occupying a particular area (such as an island) or biological entity (such as a branch of a tree) without regard to any other properties of the species. Species richness may also be expressed as the list of species that generates that number.

taxonomic rank The level in the Linnean hierarchy to which a particular taxon or group of species is assigned. For example, mammals are at the level of the class (Mammalia) and hippopotamuses are at the level of family (Hippopotamidae).

SPECIES DIVERSITY is one of the most fundamental aspects of biodiversity. However, there are many ways to think about it and ever more ways to measure it. The most basic idea is that of species richness, which is simply a count of the number of species inhabiting a given area or habitat. Species richness may also be thought of as the list of species that were counted. Patterns of species richness in space and time are some of the most foundational data in ecology, biogeography, and paleontology. These patterns both suggest and test theories of species richness dynamics. In addition to species richness, there are many concepts of species diversity. Usually species diversity is formed from species richness by further classifying the species by some attribute, such as abundance, size, or ecological role. When species are classified by abundance or population size, species diversity is the variability in the distribution of individuals into species. This species abundance relation and other measures of species diversity link species richness to many ecological processes such as population dynamics and competition. They also link to evolutionary processes such as adaptive radiation and the evolution of phenotypic plasticity. Species diversity is the building block for the diversity of higher taxa and for the diversity of ecological associations such as communities and biomes. Species diversity is an important policy tool because species diversity is relatively easy to measure and is understandable by many people. Conservation programs and goals are often cast in terms of the number of species that might be protected by a given action.

I. INTRODUCTION

The study of biodiversity often begins with species diversity because it is the most familiar aspect of biodiversity as a whole. Popular interest in biodiversity such as bird watching usually begins with learning to distinguish species, and this leads to a direct appreciation of species diversity. Later, other aspects of biodiversity, such as morphological variants or subspecies, the phylogenetic grouping of species, or the habitats and landscapes associated with particular species, are added to the initial view. Species diversity also lies at the heart of much of evolution and ecology, the sciences most concerned with biodiversity. Each discipline, in its own ways, seeks to explain as much as possible about species diversity and its patterns in time and space. Both sciences use the patterns of species diversity found in nature as tests for ideas about various ecological and evolutionary processes. The relation of species diversity to all other elements of the biodiversity hierarchies of taxa and ecological assemblages (such as communities and ecosystems) is also an important focus of these sciences.

In efforts to conserve the world's biodiversity, species diversity is a key policy concept. It is often the easiest to measure, and it is possible to frame management goals in terms of species diversity. As a practical matter, it is frequently necessary to use species diversity as a surrogate for other levels of biodiversity that may be much more difficult to measure.

Here, we initially focus on species richness (the number of species in a given area) and then discuss other approaches to species diversity. Finally, we discuss the roles of species diversity in biodiversity policy.

II. SPECIES RICHNESS

The simplest form of species diversity is species richness, which is the number of species which occur in given area. For example, the species richness of hippopotamuses in Africa is two. However, even within this simple definition there are many important issues. For example, species richness is thought of as a number, but corresponding to each number is also a list of species. The list [*Hippopotamus amphibius, Choeropsis liberiensis*] corresponds to the richness number of 2 for living African hippos.

Computing species richness requires choosing a concept of species and a higher taxon for which a list of species exists. There are many species concepts available, but most are grounded in the theory of evolution and have a theoretical basis. That is, they are entities for which there is a theory that provides methods of determining similarity and difference. The theory is what is used to resolve differences between various workers classifications. Species are thus in contrast to many other biodiversity entities such as higher taxonomic categories or ecological associations such as communities and ecosystems. These entities are more creations of humanity and their classifications are more artificial. Thus, a vegetation classification is created by ecologists arriving at a consensus for the purpose of accomplishing some work. Since diversity of any entities is a function of the method of classification which creates the units of diversity, theoretical kinds such as species (and genes) offer the possibility of more objective measures of diversity. As a practical matter, calculating species richness depends on the taxonomy available and hence on the taxonomic philosophy of the practitioners for a given taxon. Furthermore, species richness is a function of the rank of the taxon chosen: Species richness will depend on whether one chooses the family Hippopotamidae, the order Artiodactyla, the class Mammalia, and so on.

Species richness is most commonly expressed in terms of area. It may be expressed in terms of a political subdivision of the earth or an ecological subdivision of the earth, or it may be calculated by reference to an equal area grid placed arbitrarily with regard to political boundaries. In all cases, a process of discretization occurs. A species is counted simply as present or absent within a given area. Attributes such as the abundance of the species in the area or the percentage of the area occupied are not taken into account. For equal area grids we obtain an approximation of the underlying continuous variation in the species richness in such a way that we are able to quantify richness. Figure 1 shows a map of species richness for the reptiles (crocodilians, turtles, lizards, and snakes) of the continental United States. The grid cells are 100 miles on a side or 10,000 square miles in area. Figure 2 shows hand-drawn isolines of species richness which give a sense of the underlying continuous distribution.

However, species richness may also be expressed in terms of other kinds of aggregates of species. We may determine the species richness of the parasite load of a single animal or we may determine how many species of epiphytes live on a given tree.

As with any process of discretization, the patterns of species richness revealed are very much a function of the underlying scale of the discretization. Figures 3 and 4 show species richness of terrestrial vertebrates for the state of Oregon at two scales of discretization. It is clear that the pattern of species richness is scale dependent. Any inferences made from a map of species richness must clearly take the effect of scale into account. It is worth noting that finer resolution maps are not necessarily better for all purposes. There is an uncertainty tradeoff between accuracy and comparability in the choice of grid size. A smaller grid size is more accurate, but a larger grid size may produce cells that are more comparable since the effect of data of varying accuracy is smoothed out. For example, a coarser resolution map may provide better insights into large-scale biogeographic patterns because ecological noise may be averaged out. Furthermore, a minimum scale of discretization must be chosen that is appropriate given the accuracy of the individual range maps from which species richness is calculated. The data for Fig. 1 were compiled from the *Peterson Field Guides to Amphibians and Reptiles* (Conant and Stebbins). The range maps in these books show a great deal of cartographic generalization, so a relatively large-scale grid is appropriate. In general, it is a good idea to make the grid cell size several times larger than the spatial scale of the assumed errors in the individual maps.

Most work has been done with species richness as a number because the manipulation and analysis of large lists have only become possible with relatively recent advances in computer science. However, there is clearly more information in the list of species richness than the simple number. Two areas may easily have the same number of species but quite different lists. In general, the further apart two areas are, the fewer species they will have in common. Species richness thus has two components: the absolute number of species in an area and a relative measure of the uniqueness of one area compared to another. This measure is frequently computed by comparing the lists of species from the two areas and noting the number of species that are unique to each area and the number that are common to both areas.

The maps shown in Figs. 1, 3, and 4 are typical of species richness maps derived from the mapped ranges of individual species on which they depend. Calculation of species richness from a set of individual maps assumes that the maps are relatively consistent with regard to accuracy and underlying scale. This in turn requires that the taxonomy of the group under consideration be relatively complete and stable. This requirement is true only for certain taxa and for some parts of the world. In the United States and Europe, vertebrates, butterflies, and trees are generally sufficiently well-known to allow the construction of species richness maps. For the entire planet, only a few groups, such as birds, turtles, and milkweed butterflies, are sufficiently well-known to allow global species richness maps to be constructed. It is difficult or impossible to calculate species richness for poorly known taxa that have many undescribed species or whose taxonomy is controversial. Even for very well-known groups, it is not clear how up-to-date the range maps are since they are often necessarily based on collections made over a long period of time and almost never on a recent specieswide sam-

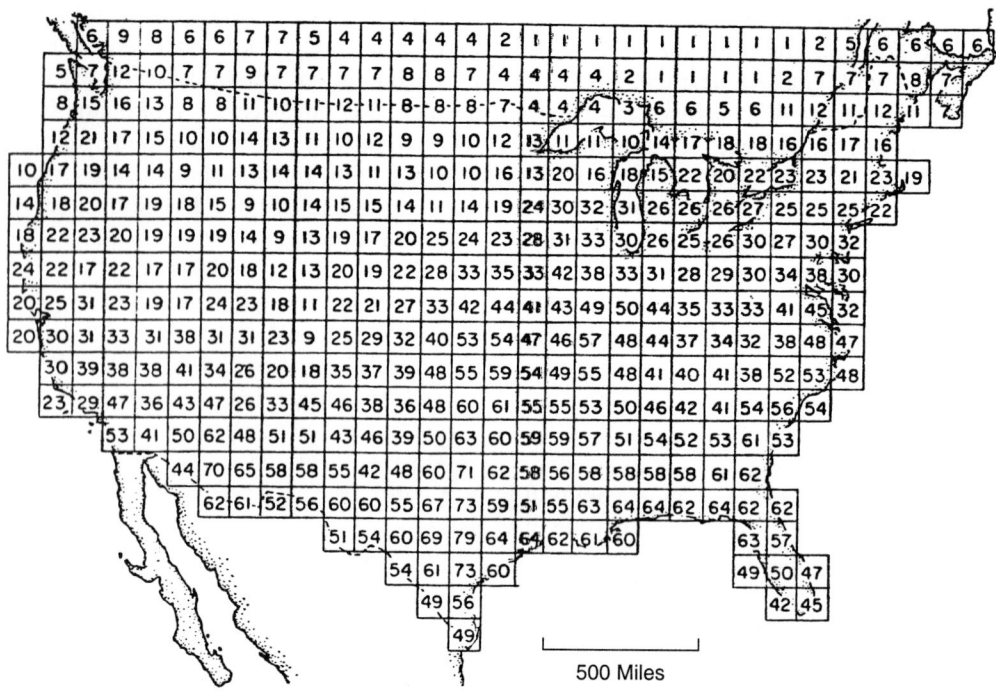

FIGURE 1 Species richness of North American reptiles; raw richness on a grid scale of 10,000 square miles per cell (reproduced with permission from Kiester, 1971).

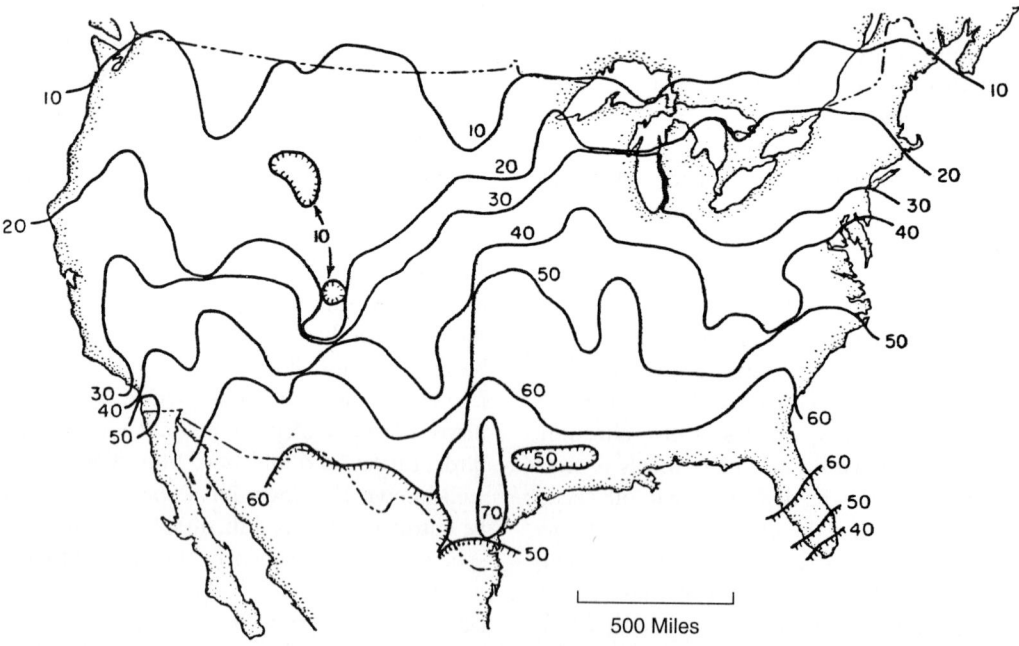

FIGURE 2 Isolines of North American reptile species richness. The lines were hand interpolated from the data of Fig. 1 (reproduced with permission from Kiester, 1971).

SPECIES DIVERSITY, OVERVIEW 445

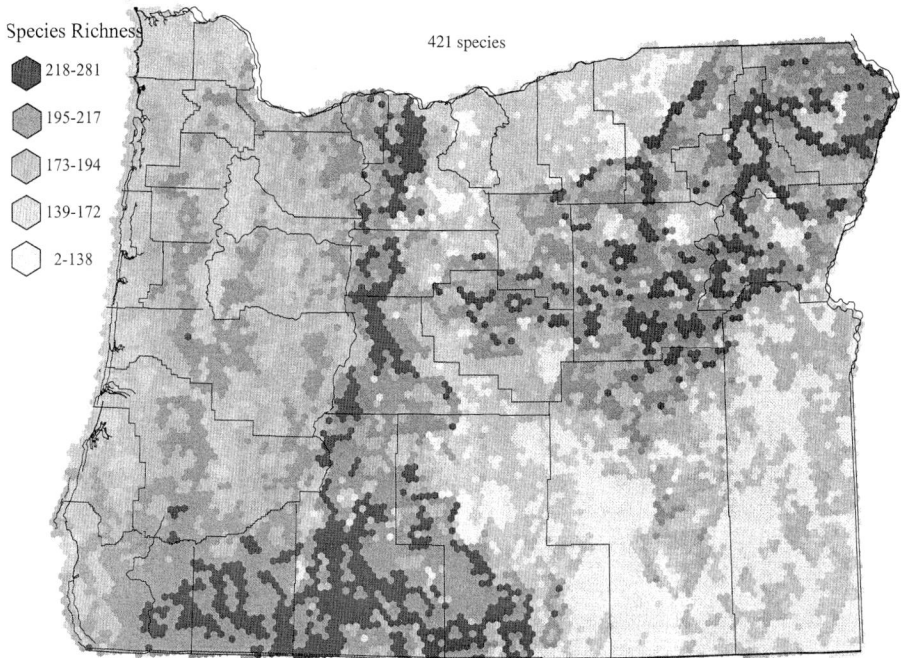

FIGURE 3 Terrestrial vertebrate (amphibians, reptiles, breeding birds, and mammals) richness for Oregon; raw richness on a grid scale of 71 km^2 (reproduced with permission from Kennelly, 1998).

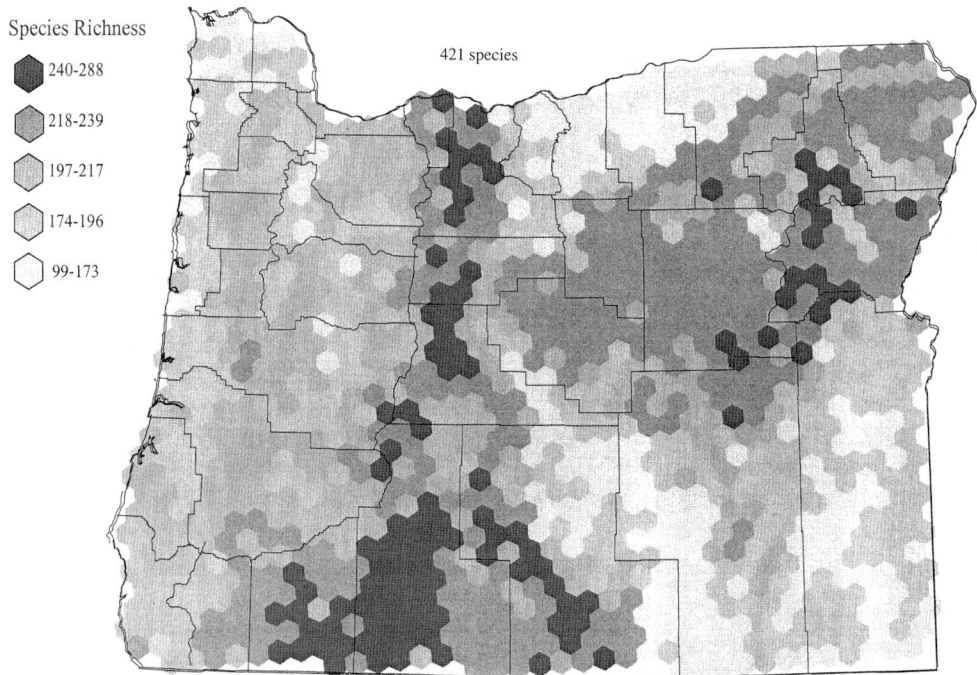

FIGURE 4 Terrestrial vertebrate (amphibians, reptiles, breeding birds, and mammals) richness for Oregon; raw richness on a grid scale of 213 km^2 (reproduced with permission from Kennelly, 1998).

pling program. For example, the data used to make the maps of Oregon in Figs. 3 and 4 were based on locality records during a 30-year period in order to have relatively complete coverage. Therefore, data for species whose ranges have changed drastically in the past two decades will be out-of-date and consequently the richness maps are in some sense always out-of-date.

There are two ways in which species richness data are accumulated. The first is the slow process of accumulation of the description of new species and the second is a more rapid method in which species richness of a given area is estimated from a statistically designed sampling effort.

As science progresses, our knowledge of species richness increases. This pattern, which is usually an increasing function of time, is called a species accumulation curve. We can think of species as being accumulated on several scales of effort and geography. For example, at the largest scale, that of all scientific efforts taken together for the entire world, we can see a historical accumulation of species. In 1758, Linnaeus recognized 11 species of turtles, in 1889 Boulenger listed 201, and in 1992 Iverson listed 257. New species of turtles are still being described even though they are among the best known animals. Turtle taxonomists estimate that by 2005 the number will be more than 300. The number of turtle species has not increased due to any organized or statistically designed effort. The work has simply proceeded as a haphazard attempt at exhaustive enumeration.

In contrast, several attempts have been made to statistically estimate species diversity at a particular location. Usually these efforts determine both the number of species and the number of individuals. In such efforts the number of species caught is a biased estimator since rare species are often necessarily missed. The contrast between haphazard enumeration and statistical estimation is that between more complete knowledge without error estimates and less complete knowledge with error estimates. The best method of synthesizing these two approaches via meta-analysis is an outstanding research problem.

For many taxa and most localities throughout the world, species richness is known only for certain restricted localities such as refuges and research field stations. In some cases, the exact area of knowledge is known, but in others the site may be only generally described. These point locality data are sometimes used to extrapolate species richness over larger areas. This is a new area of research reviewed by Colwell and Coddington (1994), who found many unanswered questions. However, extrapolation may be the best hope of obtaining an estimate of global diversity in the foreseeable future.

III. PATTERNS OF SPECIES RICHNESS

The most basic pattern of species richness is that between the number of species and the size of the area considered. In general, the larger the area, the more species that will be found in it. This species–area relationship frequently takes the form

$$S = cA^z$$

where S is the number of species, A is the area considered, and c and z are estimated constants. By taking logarithms of both sides of this equation, we find that the logarithm of species richness is a linear increasing function of area. However, many empirical studies have found that the data scatter about a wide band of values of z and that other mathematical forms of the relationship may be equally likely. The analysis of the mechanisms by which area determines species richness is a major line of enquiry for ecologists and biogeographers.

Species richness is very unevenly distributed over the entire Earth. The most important general pattern is that richness increases as one goes from the poles to the equator. Figure 2 shows that this latitudinal gradient in species richness can be quite striking. The average number of species of reptiles increases from 10 at the Canadian border to more than 60 at the Mexican border (Fig. 2). Of course, many taxa are exceptions (penguin species diversity is highest closer to the south pole and there is only one species at the equator), but overall the pattern is strong. Again, studies of the latitudinal gradient in species richness have been a major line of enquiry for ecologists and biogeographers. Another pattern is the peninsula effect whereby species richness decreases toward the end of a peninsula. For example, the number of species of mammals and reptiles decreases in Florida and Baja California as one moves toward the end of these peninsulas. This peninsular pattern can be seen for reptiles on the Florida peninsula in Fig. 2.

Species richness varies strikingly in time. The paleontological record provides good evidence that overall species richness increases with time but that there may be massive extinction events that drastically reduce species richness in a relatively short time. At ecological timescales species turnover is common so that the list of species is often changing. The total number of species usually changes less and usually in response to a change in the environment.

IV. DETERMINANTS OF SPECIES RICHNESS

Over the continuum of scales of time and space, the sciences of ecology, biogeography, and evolution all strive to understand the processes which determine species richness. Indeed, concern for species richness unites these fields (as well as bridging fields such as macroecology and paleoecology). One can easily imagine that they are all part of a single science concerned with species richness. It was recently proposed that this field be called the study of biodiversity dynamics (McKinney and Drake, 1998) and that it should include all kinds of diversity, not just species diversity.

The first question to be addressed is the role of history and the question of equilibrium versus nonequilibrium dynamics. Are the patterns of diversity that we see a result of the vicissitudes of history or of the action of relatively general mechanisms? For example, is the latitudinal gradient in species richness in North America fully recovered from the end of the last ice age. For certain taxa with good dispersal abilities, such as birds or butterflies, it is very likely that they have finished responding to that episode of climate change and are near equilibrium. Other groups, such as salamanders, with very poor dispersal capabilities may still be in the process of moving into areas opened up at the end of the glaciation and hence their latitudinal gradient may not yet be at equilibrium. Another way of viewing this question is to ask to what extent are there general mechanisms that determine species richness in particular cases and to what extent are there only particular cases. Some ecologists believe that there are only case studies because the details of any particular case overwhelm any instance of a potential general pattern. Others are much more sanguine that general rules and even laws will be able to account for species richness. In any event, for all of the biodiversity sciences the relation between the general and the particular is a difficult problem.

One useful way of dealing with the issue of the general and the particular was developed by E. E. Williams (1969), who introduced the distinction between the distant and close views in island biogeography. Island biogeography traditionally has received special attention from students of diversity because the patterns and apparent processes seem to be simpler and easier to understand than those on continents. In Williams' dichotomy the distant view is typified by MacArthur and Wilson's (1963) theory of island biogeography that accounts for the patterns of species richness seen on islands that are of different size and different distances from a source mainland. This model predicts an equilibrium number of species as a function of rates of colonization and extinction. These rates are statistical composites formed from an ensemble of many species and are the mechanisms of the distant view. Here, only the number view of species richness is used. The equilibrium number of species on an island is predicted, but the actual list of species is not. This is not viewed as a problem because the theory predicts that species are constantly turning over and so the list is always changing. The close view is taken by Williams (1969) in his work on the lizards of the genus *Anolis* of the Caribbean or by David Lack (1976) in his study of the land birds of Jamaica. In this approach, the focus is on a list of species and the details of the biology of each species examined are considered important. Mechanisms such as colonizing ability, competition, ecological plasticity, and the details of the interactions of species and available habitats are studied on a species by species basis. Adaptive radiation is often considered an important mechanism for generating diversity on a given island. The distant view has the strength of numbers and interprets broad patterns well. It does not account for the actual biology of any particular species. The close view gives detailed accounts of the biology of particular species but generally is not able to predict the number of species. Thus, the distant view focuses on the number version of species richness, whereas the close view is concerned with the actual list of particular species. The close view is not contained within the distant view. Rather it is complementary to it.

V. A DIVERSITY OF SPECIES DIVERSITIES

Beyond species richness are an enormous number of species diversities. Generally, these take the form of a species richness of species classified by some property of the species. Species may be classified by abundance (population size), body size, trophic position, taxa, membership in various ecological assemblages, and in many other ways. As with species richness, the actual number of species in a given class or the list of species in a given class may be analyzed. Because there may be several classes to which individual species are assigned, it is possible to develop an enormous variety of species diversity indices which summarize in various ways the pattern of distribution of species into the classes of a particular classification. The study of these diversity indices, their sampling properties, and the inferences about ecology and evolution that can be drawn from them have been the subject of major scientific activity in the past few decades.

The most common extension of species richness is to the species abundance relation. Here, a set of species is classified by the abundance or population size of each species. Some species are rare and others common; the distribution of all species abundances is called a species abundance curve. Figure 5 shows the species abundance curve for a classic data set of macro-Lepidoptera (mostly large moths) collected by a light trap at Rothamsted in England in 1935. A total of 6814 individuals of 197 species were collected. Thirty-seven species were represented by a single individual, whereas the most common species was represented by 1799 individuals. In this example, the number of classes of the classification is just the number of different integer valued numbers of abundance. All abundances greater that 65 were represented by single individuals. As with many data sets of species abundance, this one fits a lognormal distribution well. The many attempts to understand these data both as empirical distributions and as data to test predictions made by theories of community structure have been reviewed by May (1975) and Magurran (1988). Many of the indices fall on a continuum from emphasizing rare species to emphasizing common species. Simple species richness is at one end of this continuum since it weights all species equally. At the other end is the Simpson index, which gives the most weight to the species with the most individuals. Of the many empirical indices of species diversity that combine species abundance data, Lande (1996) found that only the Simpson index is statistically unbiased and leads naturally to measures of similarity between multiple communities. Using both species richness and the Simpson index seems to provide a useful characterization of the species abundance properties of a community.

Living organisms vary enormously in size and it is commonplace that there are fewer large species than small species. Hutchinson and MacArthur (1959) pioneered an ecological approach to understanding the species diversity of body sizes. Figure 6 shows the patterns of species diversity by body size for mammals over three scales in North America. At the scale of the continent, the distribution is unimodal centered on medium and small sizes. At the finest scale (ponderosa pine community of the Oregon Cascade Mountains) the distribution is near flat and shows distinct clumps. This area of research has many competing hypotheses which have been reviewed by Kelt and Brown (1998). Among these hypotheses, a basic idea is that there is more environmental variation at smaller size scales: The world looks more complex to a small beetle than to a larger grazing mammal. This environmental complexity in turn provides a great number of potential niches at smaller scales, allowing coexistence of more species. A provocative analysis by Holling (1993) presented evidence that species sizes of birds and mammals in individual ecological communities are clumped into a small number of discrete, separate size groups. He interpreted this pattern to be the result of a small number of structuring processes ranging from vegetative growth to disturbance dynamics and geomorphological processes which form a discontinuous temporal hierarchy and generate the discontinuous size distributions.

The familiar concept of a food web leads to a classification of species by their position in the web. Here, the basic pattern is that there are fewer species higher in the food web: There are fewer species of top predators than herbivores. Beyond these simple trophic level gen-

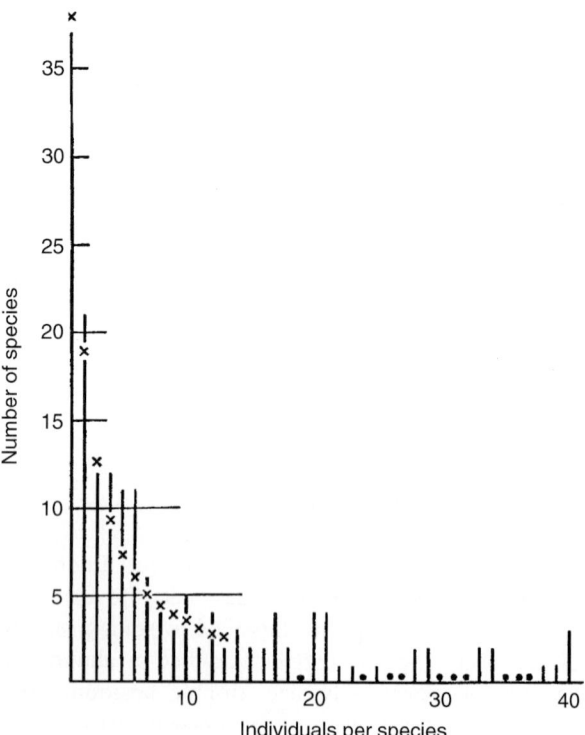

FIGURE 5 Species abundance curve for macro-Lepidoptera captured by light trap at Rothamsted, England, in 1935. The x's represent the fit of a lognormal distribution. Beyond an abundance of 40 are 26 values ranging from 42 to 1799, each representing 1 or 2 species (reproduced with permission from Williams, 1964).

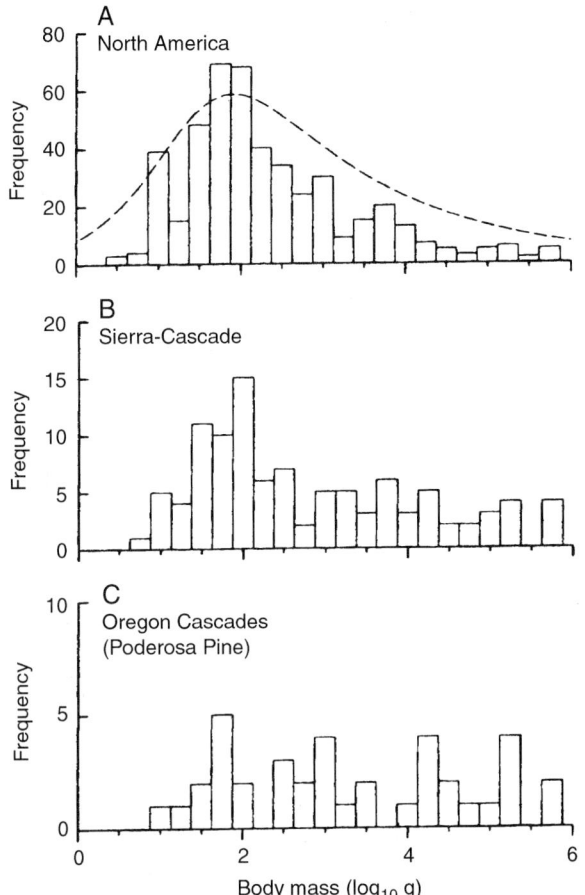

FIGURE 6 Species diversity of body size for North America mammals at three scales: (A) continental North America, (B) Sierra Nevada and Cascade ranges, and (C) Oregon Cascades ponderosa pine habitat (reproduced with permission from Kelt and Brown, 1998).

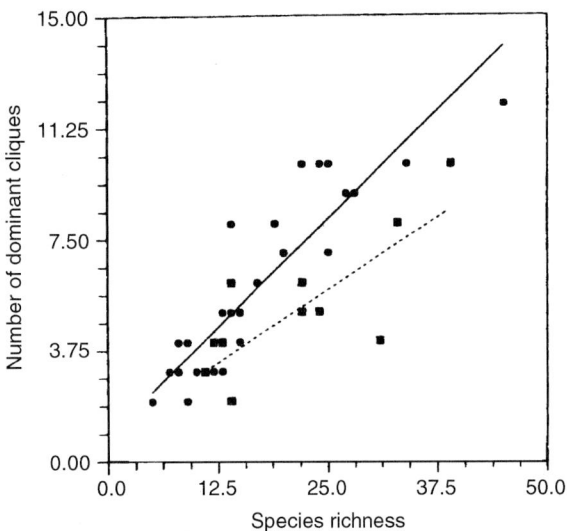

FIGURE 7 Species richness versus number of dominant cliques for 40 food webs. Circles represent data from food webs from fluctuating environments, and the solid line is a linear regression for these points. Squares represent data from food webs in constant environments, and the dashed line is a linear regression for these points (reproduced with permission from Yodzis, 1993).

eralizations, the field is controversial (Yodzis, 1993). One intriguing result concerns the relationship between the species diversity of a web and the number of dominant cliques in the web. A dominant clique of a food web is a group of species in which the predator species share at least one prey species in common and in which the groups are isolated from other such groups and do not share any of their prey species. Figure 7 shows this relatively strong relationship. This pattern may be interpreted to mean that food webs are organized into functional units (groups of interacting predators and prey) and that species diversity increases with the number of such groups.

The species diversity of higher taxa ranging from genera to phyla varies a great deal: The phylum Uniramia, which contains the myriapods and insects, has well over 1 million described species "but many millions more probably await scientific discovery" (Barnes, 1998, p. 247), whereas the phylum Placozoa apparently has only a single species—*Trichoplax adhaerens* (p. 179). Other ranks in the taxonomic hierarchy vary similarly in their diversity. We are just beginning to understand the patterns of species diversity of higher taxa and clearly this requires a synthesis of evolutionary and ecological theories. Why orchids and beetles are more diverse than many comparable groups is a function of evolutionary plasticity and ecological opportunity. Evolutionary plasticity is a result of both genetic diversity and developmental constraints and opportunities, whereas ecological opportunity is influenced by structural diversity and energy availability. Currently, we lack a detailed understanding of both of these concepts. However, recent paleontological work has demonstrated some interesting patterns of the temporal structure of species diversity of higher taxa. In particular, as origin and extinction probabilities of higher taxa are followed from the Paleozoic to the present there appears to be an accumulation of higher taxa with lower extinction probabilities. Certain groups seem to have greater staying power than others and so they are more prevalent today. These patterns offer new possibilities for testing ideas about the evolution of higher taxa species diversity.

At the other end of the taxonomic hierarchy, species may be classified by their internal variability. This variability has something of a hierarchical structure of its own: subspecies, geographic structure, population structure, and genetic structure. Subspecies represent the largest scale of variation within species and are usually thought of as being composed of a major portion of a species range within which several characters are concordant. Geographic variation may also be continuous as with clines or nonconcordant with different characters varying with different geographic patterns. Population structure ranges from continuous to highly fragmented and island-like with many varieties of metapopulation structure in between. Genetic structure follows in part from population structure and provides variability beyond that. The degree of genetic polymorphism shown by local populations provides the most basic variation shown by species. There have been few attempts to quantify species diversity of within-species diversity, although such analyses should lead to further integration of ecological and evolutionary approaches to species diversity.

VI. SUMMARY: STRUCTURE OF TAXA AND ECOLOGICAL ASSEMBLAGES

Almost any classification of species results in the fact that species richness of the classes is unequal and often shows interesting patterns. The description of these patterns and their interpretation in terms of evolutionary and ecological processes is a cornerstone of biodiversity science. Two of the most important kinds of classification are those that classify species by membership in higher taxonomic categories (which, it is hoped, represent phylogenetic groups as well) and into various ecological assemblages such as food webs, communities, landscapes, and biomes. These patterns of species diversity function both to generate and to test hypotheses about the phylogenetic and ecological processes. A great deal of effort has been devoted to analyzing some classifications such as abundance. However, it is clear that attempting to explain the pattern produced by a single classification will inevitably be incomplete. Abundance, body size, and trophic position are strongly related—large top predators are uncommon—and rather than try to explain these characteristics separately, work now focuses on attempting to explain the whole syndrome. The methodology of progressive synthesis outlined by Ford (2000) offers one way to counteract the usual pattern of fragmenting questions. The ultimate challenge to students of biodiversity is to devise better ways of understanding how the phylogenetic and ecological classifications of species diversity interact.

VII. SPECIES DIVERSITY AND BIODIVERSITY POLICY

Of the many levels of the hierarchy of biological diversity, from genes to the world, species diversity historically has played the dominant role in attempts to maintain and conserve biodiversity. Generally, this has led to a species by species approach such as is exemplified by the Endangered Species Act. This approach has usually been reactive and has been analogized to emergency room medicine. However, recent efforts such as the Gap Analysis Program have focused on sets of species and attempted to lay out proactive approaches. These approaches have in turn been analogized to public health medicine. Generally, it is less expensive to protect a species from extinction while it is relatively common rather than to wait until is it endangered. Both of these approaches use species diversity as a key level of biological diversity (gap analysis also uses vegetation communities as an important level). This focus on species diversity is due to several factors. First, the idea of conserving species is well understood by many people who have a good intuitive understanding of the concept. Second, as mentioned previously, species are scientifically well-defined entities. Finally, species diversity is apparently a good surrogate for many other levels in the hierarchy, such as genes, which are much more difficult to measure in nature. Therefore, it is likely that species diversity will remain a central focus of conservation biology even as other levels of biodiversity become better understood.

See Also the Following Articles

DIVERSITY, COMMUNITY/REGIONAL LEVEL • DIVERSITY, MOLECULAR LEVEL • DIVERSITY, ORGANISM LEVEL • DIVERSITY, TAXONOMIC VERSUS FUNCTIONAL • LANDSCAPE DIVERSITY • SPECIES-AREA RELATIONSHIPS • SPECIES, CONCEPTS OF SUBSPECIES, SEMISPECIES

Bibliography

Barnes, R. S. K. (Ed.) (1998). *The Diversity of Living Organisms.* Blackwell, Oxford.

Colwell, R. K., and Coddington, J. A. (1994). Estimating terrestrial biodiversity through extrapolation. *Philos. Trans. R. Soc. London B* **345**, 101–118.

Ford, E. D. (2000). *Scientific Method for Ecological Research.* Cambridge Univ. Press, Cambridge, UK.

Holling, C. S. (1993). Cross-scale morphology, geometry, and dynamics of ecosystems. *Ecol. Monogr.* **62**, 447–502.

Hutchinson, G. E., and MacArthur, R. H. (1959). A theoretical ecological model of size distribution among species. *Am. Nat.* 93, 117–123.

Kelt, D. A., and Brown, J. H. (1998). Diversification of body sizes: Patterns and processes in the assembly of terrestrial mammal faunas. In *Biodiversity Dynamics* (M. L. McKinney and J. A. Drake, Eds.), pp. 109–131. Columbia Univ. Press, New York.

Lack, D. (1976). *Island Biology.* Univ. of California Press, Berkeley.

Lande, R. (1996). Statistics and partitioning of species diversity, and similarity among multiple communities. *Oikos* 76, 5–13.

Magurran, A. E. (1988). *Ecological Diversity and Its Measurement.* Princeton Univ. Press, Princeton, NJ.

May, R. M. (1975). Patterns of species abundance and diversity. In *Ecology and Evolution of Communities* (M. L. Cody and J. M. Diamond, Eds.), pp. 81–120. Harvard Univ. Press, Cambridge, MA.

McKinney, M. L., and Drake, J. A. (Eds.) (1998). *Biodiversity Dynamics.* Columbia Univ. Press, New York.

Ricklefs, R. E., and Schluter, D. (Eds.) (1993). *Species Diversity in Ecological Communities.* Univ. of Chicago Press, Chicago.

Tokeshi, M. (1993). Species abundance patterns and community structure. *Adv. Ecol. Res.* 24, 111–186.

Williams, E. E. (1969). The ecology of colonization as seen in the zoogeography of anoline lizards on small islands. *Quart. Rev. Biol.* 44, 345–389.

Yodzis, P. (1993). Environment and trophodiversity. In *Species Diversity in Ecological Communities* (R. E. Ricklefs and D. Schluter, Eds.), pp. 26–38. Univ. of Chicago Press, Chicago.

SPECIES INTERACTIONS

Jessica J. Hellmann
Stanford University

I. Introduction
II. Pairwise Interactions
III. Interactions and Evolution
IV. Multispecies Interactions
V. Human Impacts on Interactions
VI. Summary

GLOSSARY

community The collection of all species which occur together in space and time.
intrinsic rate of growth A parameter, often designated as the variable r, reflecting the rate of change in population size due to births and deaths.
population A collection of individuals of the same species in a defined geographic area.
population dynamics The change in the number of individuals in a population over time.
resource Any necessity that is used or consumed for organismal growth, reproduction, or survival, including food, space, shelter, and nutrients.
species A group of individuals that is capable of interbreeding; a species is composed of one or more populations.
trait A quality possessed by an individual organism that affects its ability to survive and reproduce.

SPECIES INTERACTIONS, including competition, predation, mutualism, and parasitism, are associations among individuals of two or more different species such that each species' presence has an influence on the population dynamics of the other.

I. INTRODUCTION

The world can be divided into two parts—the living and the nonliving. Scientists refer to the nonliving component of the earth as the abiotic environment, and all living organisms are labeled biotic. Organisms survive and reproduce under many abiotic and biotic influences. Abiotic nutrients and elements are needed for survival and reproduction, and nutrient limitation is a common factor regulating limits to growth. Biotic factors also influence how an organism lives and grows. When organisms come into contact, they can facilitate, control, harm, use, or even help to propagate each other. Biotic associations that involve members of two or more different species are called species interactions.

Because the world contains a tremendous diversity of species, interactions among the biota are common. Many of these interactions are conspicuous and have a strong influence on the organization of biological communities. Studies of species interactions have produced a large body of data in a great variety of ecosystems, and these studies comprise an entire discipline called community ecology. The treatment of species interactions given here will briefly discuss the types of interactions found in nature and the simple mathematical representations of their effects, discuss the role evolution plays in shaping interactions, and provide some empirical examples. Because an increasing number of community ecologists are examining the effect of human influences on species interactions, the intersection of community ecology and conservation biology is also briefly discussed.

II. PAIRWISE INTERACTIONS

Species interact in a tremendous variety of ways, and associations can involve two, several, or many species. First, the various kinds of interactions that include only two species are discussed. (A discussion of multiple species interactions is provided in Section IV.) When individuals of two species interact, this association is called a pairwise interaction. Pairwise interactions, like all interactions, are described by the direction and the strength of their effects. Direction refers to whether the impact on each associating organism is positive or negative, and interaction types are classified according to the direction of their effect (Table I). Interaction strength refers to how much an interaction controls species dynamics.

More specifically, the strength of an interaction is a measure of how much one species affects the population size and growth rate of another species. Strength is often detected as the effect on the population of one species by removing the second species. Imagine two interacting species, species A and B. If we would like to test the effect of B on A, we could remove B and measure the resulting change in population size of A. If species A responds to the removal of species B with a large change (increase or decrease) in abundance or population growth rate, we say that B has a strong effect on A. If little effect on the population of species A is seen by removing species B, the interaction plays a small role in the population dynamics of species A, and the influence of B on A is weak. In turn, the effect of A on B can be tested by removing A and observing how the population size or growth rate of B responds.

Many factors determine how strongly two species interact. For example, the population size of either species, the behavior and age of the interacting individuals, the abundance and availability of abiotic resources, and the patchy nature of the environment all influence how strongly two species affect one another. In other words, understanding the nature of an interaction is more complicated than simply identifying that two species come into contact. In fact, even a simple experiment of interaction strength such as that described previously may not provide enough information on exactly how organisms interact with one another. Furthermore, the strength of an interaction can change at different times or in different places in the species' habitat. In the long term the strength of pairwise interactions can change due to the forces of evolution, but first the types and effects of pairwise interactions in ecological time are discussed.

A. Competition

Competition is characterized by the common use of resources by two or more species, and the effect of competition on all participating species is negative (Table I). Members of the same or different species can compete for resources. Within-species competition is known as intraspecific, and between-species competition is called interspecific. Of these, interspecific competition meets the definition of a species interaction, and intraspecific competition will not be discussed.

Interspecific competition can be further divided into two categories: interference and exploitative. Interference competition is the more direct form of competition, in which individuals of one species actively dominate a resource, preventing or decreasing the access of another species to those resources. Exploitative competition, on the other hand, is less combative. It is the diminishment of a limited resource by a species so that the other species is not able to utilize as much of the

TABLE I

Pairwise Interaction Types

Direction of effect on species 1	Direction of effect on species 2		
	+	−	None
+	Mutualism	Predation/parasitism	Commensalism
−	Predation/parasitism	Competition	Amensalism
None	Commensalism	Amensalism	—

resource as it would in the absence of the exploitative competitor.

We can predict how the population dynamics of two competing species will be affected by their interaction with the use of a mathematical model. A mathematical theory of competition began with the work of Lotka and Volterra in the late 1920s and early 1930s. The Lotka–Volterra model quantifies the influence of competition by expressing the population dynamics of each species in an equation that includes both growth in the absence of competition and the influence of competition itself. In other words, Lotka–Volterra represents the change in population size through time as a function of the growth rate of each species and the detrimental effect of competition on each population. Remember that the presence of competitors causes harm. This harm reduces the total number of individuals in each population that would be possible in the absence of competitors.

Briefly, I discuss the components of the Lotka–Volterra model. Two equations are needed to represent the population dynamics of competition (one equation for each species). The pair of equations describes the rate of change in population size with time for the two species. Subscripts represent a variable's species designation; hence, N_1 refers to the population size of species 1, and N_2 refers to the population size of species 2. The relative effect of each species on the population size of the other is determined by the value of the competition parameter, α. This competition parameter can also be interpreted as the amount of overlap between the two species in the resources they require.

The effect of one member of species 2 on one member of species 1 is represented by the parameter α_{12}, and the effect of species 1 on species 2 is the value α_{21}. The values of α_{12} and α_{21} can range from 0 to 1. If they take on a value close to 1, then the effect of one species on the other is very strong. If they take on a value close to 0, then the effect of one species on the other is weak. The values of α_{12} and α_{21} can be different from each other because it is not necessary that the two species affect each other equally. If the values of α_{12} and α_{21} are the same, this special case is called symmetric competition.

If both α_{12} and α_{21} are zero, then the two species do not compete at all and grow according to a simple expression known as logistic growth. In logistic growth, a population grows from zero to some carrying capacity, K, according to an intrinsic growth parameter, r, as shown in Fig. 1. The hallmark of logistic growth is its density dependence. In density-dependent growth, the expansion of a population slows as the number of indi-

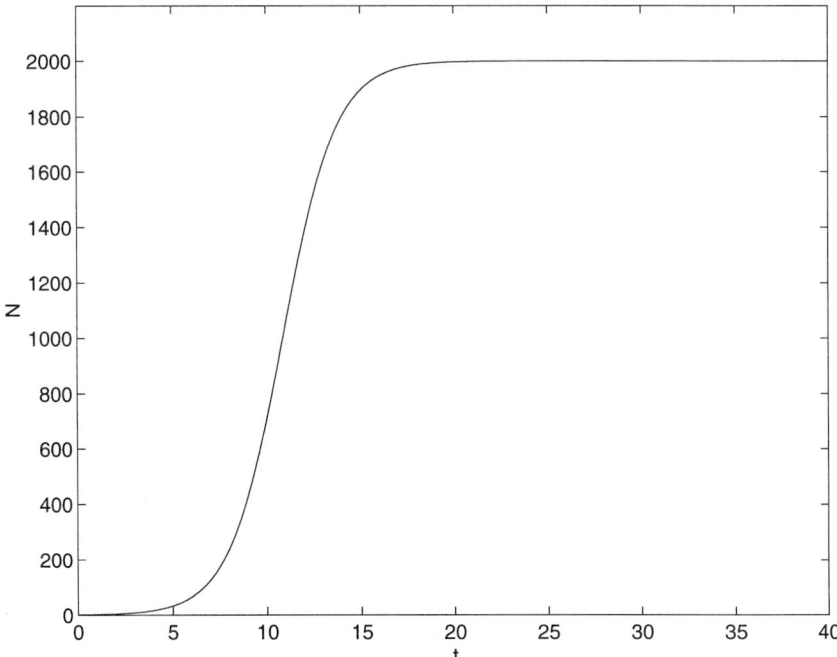

FIGURE 1 Sample plot of logistic growth. N is the number of individuals in a population of a single species, and t is time. In logistic growth, the number of individuals increases quickly when the population size is small, but the rate of growth decreases to zero as N approaches the carrying capacity. Here, the carrying capacity is 2000 individuals.

viduals in the population approaches the carrying capacity or the maximum number of individuals that a habitat can support. All of the terms in the following equation for species 1, except the component $\alpha_{12}N_2$, represent the logistic growth of species 1. The remaining $\alpha_{12}N_2$ term reflects the slowing or reduction in population growth of species 1 that results from competition with species 2. The same is true for the equation of species 2; the term $\alpha_{12}N_1$ represents the negative effect of competition with species 1 on population 2's otherwise logistic growth.

$$\text{Species 1:} \quad \frac{dN_1}{dt} = \frac{r_1 N_1 (K_1 - N_1 - \alpha_{12}N_2)}{K_1}$$

$$\text{Species 2:} \quad \frac{dN_2}{dt} = \frac{r_2 N_2 (K_2 - N_2 - \alpha_{21}N_1)}{K_2}$$

These equations can predict the impact on N_1 and N_2 for different values of r_1, K_1, α_{12}, r_2, K_2, and α_{21}. For some values of r, K, and α, one species will drive the other to extinction. In other cases, however, each species can maintain a positive population size in the presence of each other through time ($N_1 > 0$ and $N_2 > 0$). One such case is shown in Fig. 2, in which the population size of both species is plotted against time. To produce this graph, one begins with population sizes just larger than zero for both species and uses the competition equations to predict the change in population sizes (N_1 and N_2) as time passes. Note that in Fig. 2, both species eventually reach a stable and constant population size.

What conditions produce an outcome in which two competing species can coexist? It can be shown that coexistence depends on the relative value of the carrying capacities of the two species in comparison to the values of the competition coefficients, α. In short, coexistence is possible whenever the value of the competition coefficients is less than the ratio of the carrying capacities (if $\alpha_{12} < K_1/K_2$ and $\alpha_{21} < K_2/K_1$). This makes intuitive sense: Two species are able to persist on the same resource if they have a relatively small effect on

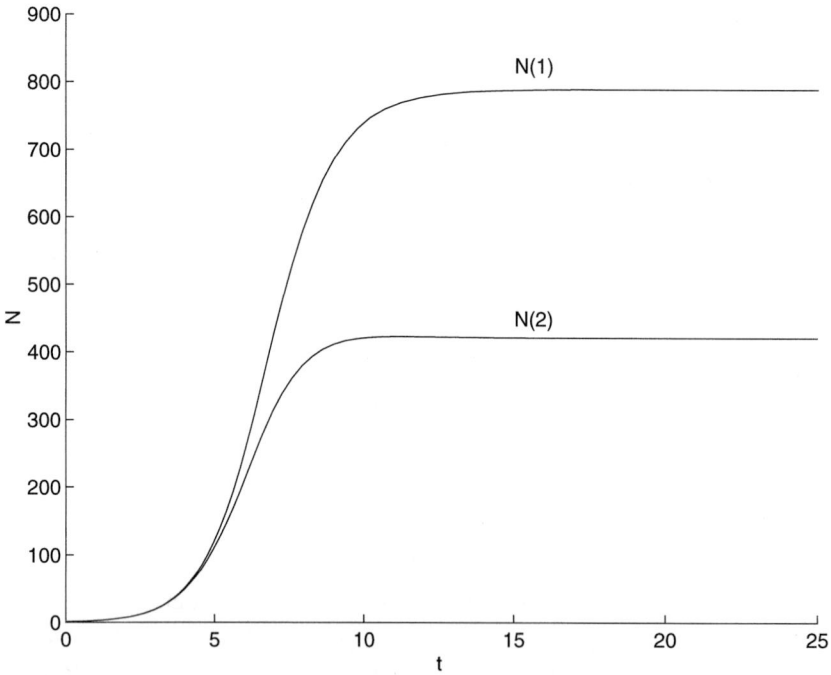

FIGURE 2 Sample plot of the population growth of two competing species according to the Lotka–Volterra competition equations. N is the number of individuals; $N(1)$ refers to the number of individuals in one species, and $N(2)$ refers to the number of individuals in the other species. In this diagram, species 1 and species 2 coexist because the abundance of both populations is positive through time. The growth of each population appears logistic (see Fig. 1), but because of competition the level at which each curve stabilizes is less than the carrying capacity for each species. In other words, the number of individuals in a population in a competitive environment is smaller than the total number of individuals that is possible in the absence of a competitor.

one another. In other words, there is a constraint on the amount of resource overlap (α_{12} and α_{21}) between two competing species that is given by ratios of their carrying capacities.

The idea that two species, based on the Lotka–Volterra equations, cannot significantly overlap in resource use is known as the competitive exclusion principle. Generally stated, no two species can occupy the same niche. (A niche can be imprecisely defined as a species' role in a community or the entire set of resources that a species requires.) If the same niche were filled by two species, or if two species used exactly the same resources, the species would compete strongly, their α values would approach 1, and one of the two would be driven to extinction.

Additional factors not captured in the Lotka–Volterra equations influence coexistence, however, and species may coexist in situations in which classical theory would predict otherwise. The role that these factors play in coexistence reduces the general applicability of the competitive exclusion principle. Spatial and temporal heterogeneity and variability of resources, for example, may enable a greater diversity of species to persist in an area than simple theory would predict. In other words, the patchy nature of resources in a habitat may enable two species to reduce their overall competitive overlap. Alternatively, species may compete only when resources are in short supply.

Controversy regarding the role of competition in natural communities has resulted in a tumultuous debate in the ecological literature. Some authors argue that competition is a critically important factor regulating community composition. Others contend that competition is overemphasized at the expense of other types of interactions that are equally (if not more) important in determining the number and kinds of species that can persist in a community. This debate as to the importance of competition in controlling community structure is ongoing, but there is little argument that competitive interactions can be found and quantified.

Gause was the first to study competition in 1934 using experiments with two *Paramecium* species that shared a single resource. Gause found that one species always went extinct when both species were put together, suggesting that, as the Lotka–Volterra theory predicted, coexistence of two competitors is possible only in special cases. In 1946, however, Cromie placed two beetle competitors in a mixed environment of flour (i.e., resources) and small pieces of fine glass tubing (i.e., environmental complexity) and found that the species coexisted. Interestingly, in the absence of tubing coexistence was not possible, and one species always drove the other to extinction. Cromie's experiments suggest that some factors or influences that are not found in the simple Lotka–Volterra model may be important in facilitating coexistence. Since the time of Gause and Cromie, many other ecologists have studied the role of competition in natural ecosystems. In the early 1980s, Schoener and Connell each reviewed the literature for the prevalence of competition and found that a majority of published field studies successfully detected the impact of competition on the population dynamics of species.

B. Predation

Predation is the consumption of one species by another. Consumption leads to a negative effect on the population of one of the two interacting species, known as the prey, and a positive effect on the second population or species, known as the predator (Table I). Also as in competition, predation has a rich history of mathematical theory, beginning with the work of Volterra. The predator–prey interaction is modeled with two equations—one for the prey (species 1) and one for the predator (species 2). Again, the components of these equations are discussed without going into great detail.

$$\text{Species 1:} \quad \frac{dN_1}{dt} = rN_1 - aN_1N_2$$

$$\text{Species 2:} \quad \frac{dN_2}{dt} = baN_1N_2 - dN_2$$

According to this model, in the absence of a predator, species 1 would grow exponentially according to the rN_1 portion of the first equation; unlike the Lotka–Volterra competition equations, no carrying capacity limits growth in this model. This type of growth is called density independent. In other words, the rate of increase in the prey population remains constant through time and is independent of the number of individuals in the two populations. (Again, r represents the intrinsic rate of growth.) The number of prey removed from the prey population by predation is represented by the term aN_1N_2, where a describes the fraction of encounters between predator and prey that ultimately results in kills. Predators consume prey for a single purpose—to produce new predators. The number of predators born (i.e., new predators added to the N_2 population in the second equation) is a function of how efficient predators are at producing offspring from consuming prey. The variable b describes this efficiency, and the term baN_1N_2

thus represents births into the predator population. In this model, prey die only from predation, but predators die from other causes, and this death rate is the variable d. The number of individuals removed from the predator population due to death is the term dN_2, and the fraction of predators removed is constant through time.

Several outcomes are predicted from this simple model. First, predators can drive prey to extinction by consuming them faster than prey can reproduce. Second, the predators can go extinct if their death rate exceeds their ability to produce new predators from the consumption of prey. Lastly, coexistence is possible. However, unlike the competition situation discussed previously, these predator–prey equations do not reach a stable and persistent number of predators and prey in cases of coexistence; instead, the abundance of each species cycles through time. These cycles are called oscillations, with the predator abundance "lagging" behind that of the prey. As the number of prey increases, it is followed by an increase in the number of predators, driving a decrease in prey, in turn driving a decrease in the number of predators. Once the predators decrease, the pressure of predation declines and prey can increase again, and the cycle repeats. The cyclic dynamics of predator–prey can be seen in Fig. 3.

To determine whether the oscillations predicted by the Volterra theory reflect the true nature of predator–prey interactions, researchers have sought evidence of population cycles in nature. The most well-known example of predator–prey oscillations was shown in lynx feeding on hare in Canada. Recent evidence, however, calls into question whether or not the oscillations observed in the lynx–hare system are in fact caused by a predator–prey interaction or by some nonpredation factor affecting hares such as seasonal food shortage.

There is agreement that, oscillating or not, predator–prey interactions are common and are often strong. For example, following the extinction of many important large forest predators, in conjunction with human alteration of the forested landscape, eastern U.S. deer populations expanded to extraordinarily high levels. This suggests that predators once played a large role in suppressing the numbers of deer. A second example is the introduction of the brown tree snake to the island

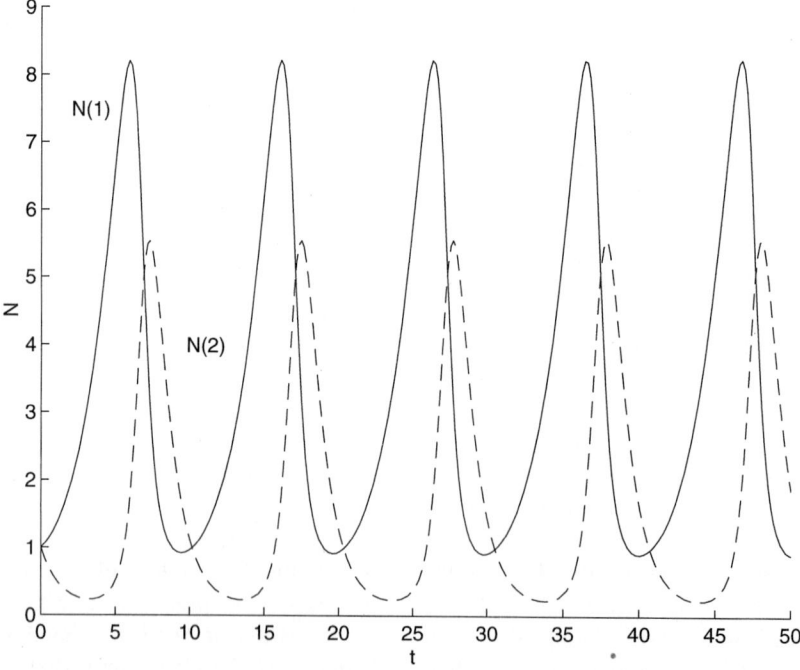

FIGURE 3 Sample diagram of a predator–prey system as predicted by the Volterra equations. N is the number of individuals $\times 10^3$ and t is time. The solid line, $N(1)$, is the number of individuals in the prey population, and the dashed line, $N(2)$, is the number of individuals in the predator population. Both predator and prey coexist in this diagram because neither goes extinct. The abundance of predator and prey is not stable through time but goes through oscillations or cycles. Notice that increases and decreases in predator population size follow those of the prey.

of Guam. In this case, the snake decimated the island's bird community, driving most of its prey to local extinction.

Several factors that are not included in the Volterra model can influence the dynamics of a predator–prey interaction, increasing the likelihood of coexistence and dampening cycles. Environmental spatial heterogeneity in abiotic and biotic factors can benefit either predators or prey. For example, in some portions of a species' habitat, it may be more difficult (or easier) to find or capture prey. It is also possible that the Volterra model does not reflect the true nature of a specific predator–prey interaction. For example, it may be unrealistic to model some prey species as density independent. Researchers have modified the Volterra model in many ways to increase its realism, including versions that incorporate density-dependent growth of prey.

C. Parasitism and Herbivory

Several other types of interactions have effects similar to those of predation, where one species benefits to the detriment of the other. Herbivory, the consumption of plants by animals, is a subtype of predation but differs in some important ways. First, most herbivores consume only a part of their "prey," and consumption infrequently results in death. Second, plants cannot escape their predators as many animal prey do, changing the nature of the predator–prey interaction. Just as predators can greatly affect prey, herbivores play an important role in the dynamics of their host plants. For example, the introduction of the gypsy moth to the eastern United States dramatically reduced the population size of many forest tree species, especially oaks and aspen. And the effects of herbivorous agricultural pests on crops are the impetus for widespread application of pesticide.

An interaction known as parasitism is also distinct from simple predation. Parasites are usually smaller than their prey (also called a "host") and often feed on their hosts without killing them. When parasitic interactions are lethal, parasites kill their host after using or feeding on it for a period of time. Many human diseases such as malaria (a protozoan), intestinal roundworms, and disease-causing trematodes such as liver flukes are parasitic. Viruses can also be considered a type of parasite. One interesting form of parasitism is a phenomenon known as brood parasitism in which one species of bird lays its eggs in the nest of another, leaving the offspring to be raised by a surrogate parent. In this case, the parasite does not consume the host but steals some of its parental care. Brood parasites, like all parasites, reduce the success and reproductive output of their hosts. An often fatal form of parasitism in many insect populations is infection by parasitic wasps or flies called parasitoids. These wasps lay eggs in the body cavity of their host; the parasitic larvae then feed on the host during development, killing it upon emergence. Some types of parasitic interactions may have dynamics similar to those predicted by Volterra predator–prey theory. Recent evidence shows, for example, that cyclic red grouse populations in Britain may be driven by the effects of a nematode parasite on host reproductive ability.

D. Mutualism and Symbiosis

When both participants in an interaction benefit from their association, the interaction is known as mutualism (Table I). Many types of mutualisms can be found in nature, and increasing evidence indicates that positive interactions may be more important than many ecologists previously considered in determining who can coexist in a community. There is no well-developed theory as to how mutualism promotes coexistence or how mutualistic species impact the population dynamics of one another. There is a large and developing empirical literature, however, that documents the presence of mutualism in nature. Many of the common interactions among organisms, such as animal pollination of plants and some animal dispersal of seeds, are in fact mutualisms. Fruit-eating birds, for example, get nutrition from a plant in exchange for dispersing the plant's seeds across a wide area.

Mutualistic interactions can be facultative or obligatory. Facultative interactions are those in which neither species requires the other to persist but each benefits from the interaction when the opportunity arises. The association between a fruit-bearing tree and its avian visitors is facultative. Many species of lycaenid butterflies have mutually beneficial, and facultative, associations with ants. The ants are thought to protect the larvae from predators while lycaenid larvae offer the ants a nutritious secretion in return.

Obligatory interactions are those in which each species requires the other for survival. These are typically highly specialized species associations. A classic pollination-based obligatory mutualism is that of the fig and fig wasp. Nearly every species of fig is pollinated exclusively by a specialized species of fig wasp, which lives inside the fig fruit. The figs rely on the wasps for pollination and the wasps rely on the fig as a habitat in which to execute their life cycle. Several plant species also have obligate mutualistic interactions with fungi. For example, many species of orchid require a fungal association for seed germination. The fungus, in turn, derives nutrients from the orchid. A third example is an obligatory interaction between some species of algae and fungi. The algae and fungus live together, creating an "organism" known as lichen.

The lichen example also illustrates a critical vocabulary point. Lichen is an example of a symbiotic mutualism. The terms "mutualism" and "symbiosis" are often used interchangeably, but they actually differ subtly. The term symbiosis stresses that two species live close to one another in prolonged association. While many symbiotic interactions can also be mutualistic, this need not be the case. For example, you and the bacteria inhabiting your stomach which aide in your digestion are symbiotic and mutualistic, but you and the flukes feeding in your liver form a parasitic symbiosis.

E. Asymmetrical Interactions

One species can harm or help another species without any benefit or detriment in return. Commensalism refers to the benefit of one species, species A, from the presence of another species, species B, while B experiences no effect from the presence of A (Table I). Conversely, amensalism refers to the detrimental effect of species B on A while B experiences no effect of A in return (Table I). Far less research has focused on commensalism and amensalism than on other types of interactions, and the strength of commensalism and amensalism is generally thought to be weak. A famous example of commensalism is an association between cattle egrets and cattle. The egrets eat insects flushed by the cattle. The presence of the egrets, however, has no measurable effect on the cattle. Amensalism often occurs as the incidental damage to one species from the presence or activity of another. For example, in the cattle–egret example, some ground-dwelling insects suffer incidental mortality from the cows that step on them.

III. INTERACTIONS AND EVOLUTION

Previously, the direction of interaction effects and differences in interaction strength were discussed. It was also mentioned that interactions can change through time and space. In ecological time, interactions between species affect population size. In the longer term, however, interactions can lead to evolutionary change in the two participating species, either reducing or increasing the strength of the association. For example, in competitive and predatory interactions, one species may evolve mechanisms to avoid strong interaction with others in response to the pressures of natural selection. In mutualistic interactions, on the contrary, specialized behaviors and morphology can be selected through time to promote efficiency, and species can grow to be more dependent on one other.

Evolution typically results from a pressure known as natural selection. Natural selection acts in the following way: Among a collection of organisms in a population, those individuals that are best suited for their environment will be the most successful and will produce a greater fraction of a population's offspring than will other individuals. Successful individuals will genetically pass their beneficial traits to their offspring if these traits are heritable. Through many generations of enhanced survival and reproduction of individuals with beneficial qualities, directional change in physiological, morphological, or behavioral traits at the population level can result. Natural selection can act in a variety of ways. For example, natural selection may favor traits that confer efficiency in food digestion or nutrient acquisition. Interactions between species can also be a force of natural selection for either or both species involved.

When natural selection results in reciprocal trait evolution in two interacting species, it is called coevolution. In other words, if morphological, physiological, or behavioral traits determined by genes in two interacting species evolve in response to an interspecific interaction, coevolution has taken place. The pressure of natural selection (called selection pressure) in coevolution stems from the positive or negative effects that species have on each other. Depending on the type of interaction driving coevolution (competition, predation, mutualism, etc.), many trait changes are possible.

A. Evolution and Competition

Remember that competition is thought to limit how similar two coexisting species can be. If, as seen in the Lotka–Volterra model of competition, resource overlap can drive one of two competitors to extinction, then competition can also be a strong selection pressure. Individuals with traits that enable them to compete less with other species will have an advantage compared to other members of their population. Over many generations of selection for minimizing overlap, populations can evolve away from interaction with their competitors.

The evolutionary change in a trait in response to competition is called character displacement. Character displacement results from selection for traits that reduce the overlap in resource use with a similar species. If two or more competing species evolve in response to one another, character displacement can be considered a type of coevolution. For example, in the Galapagos Islands, a group of coexisting finches have distinctly different body and beak sizes as well as beak shapes. Size specialization allows each finch to exploit a different food resource. The seemingly well-organized categories of finches suggests that there was once a strong influence of competition on these species and that competition drove the finches to evolve different strategies and diverge. In fact, the type of pattern seen among finches on the Galapagos is widespread. Many species of similar type in a single habitat (e.g., desert rodents or island lizards) appear divided into distinct body size categories. This division suggests that selective pressure to reduce resource overlap was once (and may still be) an important influence.

It should be noted that the pattern of body size classes often seen in nature need not necessarily evoke character displacement as an explanation. An alternative hypothesis is the idea of selective retention of invaders. This hypothesis states that as a community is being filled by new arrivals to the system, those invading species that are able to persist are ones whose niche is currently vacant. In other words, invaders who find their niche already occupied are not able to persist in the system and go extinct shortly following invasion. This hypothesis is distinct from the character displacement hypothesis in that no evolutionary adaptation is necessary to explain the morphological differences seen among similar species.

B. Evolution and Mutualism

Change in morphologies in response to competitive pressure as in the Galapagos finches is one type of interaction-driven evolution. A second type can occur in mutualistic interactions. In mutualism, natural selection may favor those individuals that benefit more from an interaction than do other members of the same species. If the partner species in a mutualism is under analogous selective pressure, reciprocal, genetic changes in both species may result. Over many generations, coevolution increases the benefit each species gains from its interaction with the other.

Many plant–pollinator systems appear to have experienced some coevolutionary change. For instance, it may be advantageous for plants to attract pollinators that visit only a few or a single plant species, and many plants attract pollinators with specialized flashy petals and nutritious rewards. Because the next flower visited by such a specialist pollinator is more likely to be a member of the same plant species, a plant that attracts specialists enjoys an increased rate of successful pollination and a decreased expense of excess pollen production. Many pollination associations seem to be coevolved in this way, including hummingbirds with long curved bills feeding on species with long curved flowers and orchid species dispersing fragrances that attract specialized bee visitors.

Other coevolved mutualisms between animals and plants are also known. For instance, some acacia trees have mutualistic associations with ants. An ant colony resides in the tree and uses tree secretions called "nectaries" for food. The acacia uses the ants for protection from other insects and plants. Ants can attack herbivores that land on the plant, and in some cases the ants even remove competing plant material from around the acacia's base. Studies have shown that mutualistic acacias perform very poorly without ant inhabitants. In effect, acacia success and survival are dependent on the ants. Similarly, several species of ants are found only on acacia, indicating that for some, acacia is an obligatory ant resource.

In both the acacia–ant and plant–pollinator examples, it is difficult to imagine how such specialized associations could evolve without multiple steps of reciprocal morphological and behavioral change. It is thought that specialized interactions evolved from generalized interactions, where survival and reproduction benefits of increased mutualism brought two select species closer together. Such increased association ended with, in some cases, obligatory dependence.

C. Evolution and Predation

Predators can place selection pressure on their prey to avoid capture, and the benefits of efficient catch and consumption of prey can drive the evolution of predators. The evolution of a trait in prey that makes predation more difficult can be followed by the reciprocal evolution of a trait in predators that reduces the prey's new advantage. Conversely, traits that confer advantages to predators can lead to increased natural selection on prey to evolve predation avoidance traits. Again, if two or more predators and prey evolve in response to one another, this is called coevolution.

Through evolutionary time, many species of prey have evolved defenses against predation, including the ability to flee, avoid detection, or be difficult to eat. Predators have also evolved techniques and physical characters to improve their efficiency or probability of successful capture of prey. Predators can become less conspicuous, faster, or more skilled at capturing and killing prey. This phenomenon was first discussed by Darwin in the late 1800s; he wrote of a constant struggle between prey to escape their enemies and predators to consume their prey. Today, we refer to the "constant struggle" as an evolutionary "arms race," with the analogy to nations that reciprocally accumulate weapons to maintain an even balance of defense.

One frequently cited example of a coevolutionary arms race exists among plants and their herbivores. Plants must defend themselves against their predators since they cannot flee, and they often do so with chemicals called secondary compounds. These compounds either have a negative impact on the physiology of herbivores or decrease the ability of herbivores to consume the plant. Studies have shown that many plants increase the tissue concentration of secondary compounds in response to herbivory, suggesting that such compounds are indeed defensive. In addition, the presence of secondary compounds has been shown to effectively reduce the number of herbivores feeding on a plant. Some herbivores, however, have overcome the negative influence of secondary compounds by adapting ways to metabolize or resist the effects of the plant's defenses. Some herbivore species even use defensive compounds as a cue for identifying the plant as a food source. The pressure to evolve chemicals to escape herbivory, as well as the pressure to overcome antiherbivory compounds in order to sequester food, may have driven the degree of specialization seen in some herbivore–plant relationships.

A second example of a predation-like coevolutionary process can occur between brood parasites and their hosts. Recall that brood parasites lay their eggs in the nests of other species; parasite offspring hatch and are raised by the surrogate parent. Brood parasites place a selective pressure on their host to identify parasitic eggs and either remove them or abandon the nest. However, as hosts become better at recognizing parasitic eggs, parasites are also selected to produce eggs that appear more similar to the host species. Parasites can become specialized on an individual host, producing eggs resembling only a single host species.

IV. MULTISPECIES INTERACTIONS

Limiting the treatment of interactions to two species helps us to classify associations and consider the effect an interaction might have on the evolution of participating species. In actuality, however, all organisms live in communities composed of more than two species. This reality is not reflected in the simple theories of competition and predation. Unfortunately, the addition of multiple species to simple models can quickly make them complex and difficult to manipulate. Despite these difficulties, a large body of empirical and theoretical work considers multispecies interactions and the role of multispecies interactions in driving evolutionary change.

A. Food Webs and Trophic Levels

One way to characterize interactions in a community is to document who eats whom. A food chain is a series of prey species and their predators (Fig. 4). Food chains begin with plants (known as primary producers), followed by the animals that consume plants (primary consumers or herbivores), the animals that eat herbivores (secondary consumers), and so on. The collections of all food chains in a community comprise a food web. One can think of a food web as a diagram that describes how all species or groups of species in a community use one another for food (Fig. 5). The positions (producer, secondary consumer, etc.) within a food chain or food web are known as trophic levels. Food webs and trophic levels are useful paradigms for thinking about species interactions; not only do they show the relationships among predators and prey but they also suggest when species might compete or interact in other ways. Because organisms live in a world of multispecies interactions, food webs enable us to count interactions and consider the relative importance of multiple interactions on a single species.

One focus of research in the study of food webs and community composition is the issue of food web

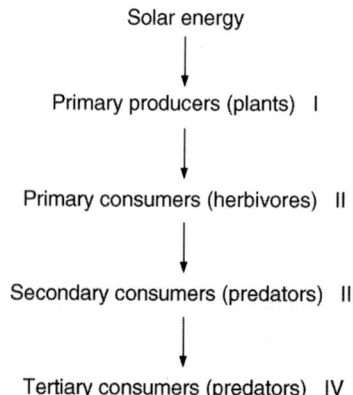

FIGURE 4 General diagram of a food chain. All food resources originate from the energy of the sun. Primary producers use this energy to create plant biomass. Herbivores consume plants to create herbivore biomass. Carnivores, in turn, consume herbivores and other predators. Roman numerals represent different trophic levels; multiple species can exist within a single trophic level.

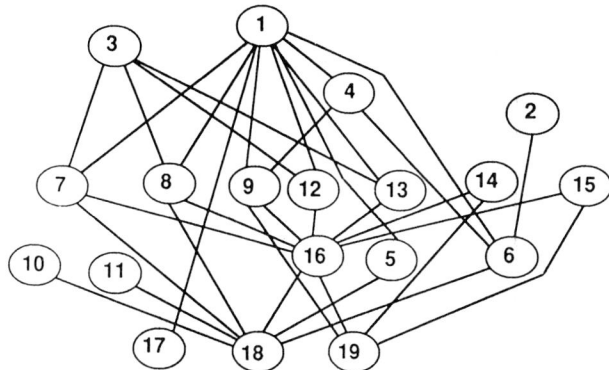

FIGURE 5 A food web of inhabitants in a pitcher plant in West Malaysia. Pitcher plants have modified leaves that hold water and small communities of insects and microorganisms. All numbers except 16 (bacteria and protozoans), 17 (live insects that have fallen in the pitcher), 18 (drowned pitcher residents), and 19 (other organic debris) represent an insect species. Predators are higher in the web than are prey. For example, Nos. 17–19 are at the bottom of the food web, and they are preyed on by those species that are connected by a line and are higher in the diagram. Top predators are among the highest in the figure (e.g., 1, 3, and 4), and they prey on all species to which they are connected. Species that prey on some species but are also prey themselves can be found in the middle of the diagram (e.g., 8, 9, 12, and 13) (reproduced with permission from Pimm et al., 1991).

stability and resistance to disturbance and change. Stability refers to how a food web will respond to the removal, addition, or a large change in the population size of one or more community members. In general, if the extinction of one community member is followed by the extinction of several other members, the web is unstable. If the web adjusts only slightly in response to extinction or large population shifts, it is stable. The issue of stability in food webs is important because the environment is very dynamic, and over long periods of time events such as large changes in resources or natural disasters are likely to occur. If a community persists through time, it must be resistant to such extreme events.

In 1972, May was the first to propose that webs with many members and many links were unstable. Using a mathematical model, May showed that the more interactions in a community, the greater the potential for a small perturbation to affect multiple web members and the structure of the web. This result, however, seems counterintuitive because many of Earth's communities are diverse, with many species and many links. Also using mathematical theory, other researchers have shown that in certain cases diversity can lead to stability. For example, food web stability may be related to the asymmetry or symmetry of its interactions or to the location or arrangement of strong interactions in the web. To date, little consensus on the issue of community and food web stability has emerged.

B. Keystone Species

A species that exerts a strong influence on the dynamics of several members of its community is called a keystone species. A keystone species' influence is one that is larger than we might expect based on its abundance alone. In other words, a keystone species must exert its influence based on the important interactions it has in the community and not simply on its dominance or high population size.

For example, the sea star, *Pisaster ochraceus*, is a classic keystone species in Oregon and Washington. It preys on several mussel species, including the mussel *Mytilus californianus*. Mussel species compete strongly for places to adhere on intertidal rocks. *Mytilus californianus* is highly successful at dominating intertidal spatial resources, blocking out space needed by other mussel species. By feeding on *M. californianus*, *P. ochraceus* opens up rock space, resulting in an increase in the diversity of the community. In the absence of *P. ochraceus*, several species of mussel would be driven to extinction by the dominant competitor *M. californianus*, but in the presence of *P. ochraceus* the number of mussel species that can persist in the community is increased. Table II shows some experimental results of removing *P. ochraceus* from the intertidal zone in Washington.

Other predatory keystones include the sea otter and the bass; however, not all keystone species are found at the tops of food chains. For example, nitrogen-fixing bacteria may be keystone members of forest communities if they increase diversity by reducing the effects of nitrogen limitation. Beavers influence species diversity in their communities by altering the habitat structure of streams and ponds. Because keystones maintain the biodiversity of their communities, identifying and preserving them is an important conservation goal. To date, however, there is no general theory that predicts which members of a community will be keystones. The only way to identify a keystone species is to study its community in detail.

C. Trophic Cascades and Indirect Interactions

Keystone species exert their strong influence on community composition by a combination of direct and indirect effects. Each of the interactions previously discussed in the context of pairwise interactions were di-

TABLE II
Species Composition in the Presence and Absence of *Pisaster ochraceus*[a]

	Survey date			
	July 1963	April 1973	August 1966	June 1971
Plot type[b]	Control	Control	Removal	Removal
% cover by species[c]				
Balanus cariosus	10	12		
Balanus glandula	~15	14		
Chthamalus tissus	~15	10		
Pollicipes polymerus	1	2	5	
Mytilus californianus	5	2	95	100
Endocladia muricata	2	3		
Corallina vancouveriensis	20	28		
Lithothamnium sp.	16	5		
Hedophyllum sessile				
Halichondria panacea	5	5		
Total % cover[d]	89	85	100	100

[a] Data from Mukkaw Bay, Washington (reproduced with permission from Paine, 1974).
[b] Plots were of two types: those in which *P. ochraceus* was removed ("removal") and those in which *P. ochraceus* remained ("control").
[c] All species are sessile invertebrates or algae that live on the rocks of the intertidal zone.
[d] Numbers do not necessarily total 100%; remaining space was either unoccupied or held by a few rarer species.

rect—two species in contact and each influences the dynamics of the other. An indirect interaction occurs when species influence one another through their interactions with other species. According to a review of indirect interactions by Wooton (1994), five types of indirect interactions have been identified. I focus on three here: apparent competition, interaction modifications, and trophic cascades.

Apparent competition is the case in which two prey species share a single predator. If one of the two prey species were to increase in abundance, an increase in predator population size might follow. An increase in predators can then lead to a decline in both prey species. For example, an increase in the number of one of two grasshopper species has been shown to increase the number of parasitoids infecting both host species. Interaction modification is the case in which one species changes the interaction of two directly interacting species. For example, the amount of phytoplankton (small green plants) in the water column of a lake may affect a predatory fish's ability to find and capture a prey species.

A trophic cascade refers to the indirect effects on many species in a food chain to a change in the population size of a single chain member. Trophic cascades were first discussed by Hairston *et al.* (1960), and the effect of cascades on population sizes at each level in a food chain was thereafter named the HSS hypothesis. HSS predicts that predators indirectly control the abundance of primary producers by reducing the numbers of herbivores through predation and releasing producers from some herbivory pressure. Predation on herbivores keeps herbivore abundance low, and a smaller number of herbivores consumes fewer producers than they would in the absence of predators. Some data exist to support the HSS claim that predation can limit the consumption of producers. As mentioned previously, following the human-caused extinction of large predators, some herbivore densities (such as that of eastern deer) have greatly increased. The increase in herbivores has in turn increased the amount of herbivory damage to forest plant species and has changed forest plant dynamics.

Trophic cascades can also be expanded to four trophic levels. For example, in lakes with piscivorous (fish-eating) fish, there is a decrease in the number of planktivores (plankton-eating fish), an increase in the number of herbivores, and a decrease in the amount of green phytoplankton in comparison to lakes without piscivores. This alternating suppression and release of predatory pressure might explain why some lakes are green and others are clear; water clarity (i.e., the amount of phytoplankton) may be determined in part by the number of trophic levels in the lake food web.

D. Multispecies and Evolution

Just as direct and pairwise interactions can influence evolution, species assemblages may influence selection pressures as well. Multiple and indirect effects, however, can select traits in complicated ways. In multispecies assemblages, many interactions can simultaneously place evolutionary pressure on a species. Where these pressures counterbalance, no change in morphology or behavior may result. When the pressures coincide, evolution by natural selection may occur. In general, a trait will be selected if an organism benefits from possessing the trait over the sum total of all its interactions. There are many examples of evolution in the balance of multiple pressures. For example, a mutualistic interaction between a lycaenid butterfly and its associated ants may have arisen in response to the pressures of parasitic infection of larvae by parasitoids.

V. HUMAN IMPACTS ON INTERACTIONS

Many important services and goods used by humans result from species interactions. Services include pollination of plants and the maintenance of soil fertility, and the products of these services provide goods such as agricultural crops, timber, and clean water. Increasing evidence suggests that human influences on the environment are impacting the very interactions that create these useful goods and services. Influences such as global warming, alteration of the earth's biogeochemical cycles, habitat loss, overharvesting, and pollution can cause disassociations among once-interacting species or change the nature of a service-providing association.

Human stresses on the environment can drive local extinction, and via species interactions single extinctions can impact multiple community members. Stresses such as global warming are changing the abiotic environment in a systematic way, forcing local extinctions in some areas and population expansion in other areas. The responses of individual species to these changes are unique, shifts in species ranges can result, and the overlap of species can break down.

For example, historical data show that vegetation responds very slowly to regional temperature change, and rates of plant species migration will be slow in response to global warming. Some insects, on the other hand, have the potential for long-range dispersal and will respond to climatic changes more quickly than will plants. If insects shift faster than plants, some plant populations may be left without pollinators. Furthermore, even if interacting species, such as plants and insects, were able to shift at the same pace, other stresses such as habitat fragmentation stand in their way. If one species is able to shift over a fragmented landscape and another species is not, disassociation can result.

Humans can alter the timing of interactions as well as the ability of species to persist in the same geographic location. For example, evidence suggests that human influences on biogeochemical cycles, including increases in anthropogenic nitrogen deposition, can impact the growth rate of plants, affecting their flowering and development timing (i.e., phenology). Changes in plant phenology may affect the temporal overlap between herbivores and their food plants, possibly leading to herbivore population decline.

Lastly, human domination of the earth has greatly increased the distribution of select species. Many organisms travel with people as they disperse throughout the world, and some of these species have established themselves in locations other than where they originally evolved. These species are often dominant competitors; they spread quickly and displace other organisms. Because they drive native species extinct or harm native populations, such introductions have significantly altered native interactions in many parts of the world. For example, the introduction of European grasses to California has excluded native plants from all but very small relic patches throughout the state. The loss of native plants has in turn affected the viability of many herbivores that are specialized to forage on native California plants.

An alarming number of Earth's species are going extinct at the hand of human civilization. Identifying and conserving keystone species, as well as predicting and preventing extinction cascades, will be critical to successful mitigation and evasion of further biodiversity loss. To preserve the interactions and communities we need, community ecologists will be asked to provide predictive and restorative information on a great diversity of community and interaction types. Ecologists also will be asked to increase the generality of their findings because species are going extinct more quickly than we can identify taxa and describe communities.

VI. SUMMARY

Species interactions are classified by the direction of their effects and are divided into the direct, pairwise categories of competition, predator–prey, mutualism, commensalism, and amensalism (Table I). The strength of an interaction, or the amount an interaction affects the population size of its participants, can be determined by experimentally removing one species and observing the population response of the second species and vice versa.

A simple mathematical theory for competition predicts that the amount of resource overlap between two coexisting, competing species must be small (Fig. 2). A simple mathematical theory for predation predicts that for coexisting predators and prey, population sizes will oscillate through time (Fig. 3). In both cases, theory may oversimplify the opportunities for species coexistence. Factors such as variation in resources, space, and time, as well as the complexity of a habitat, can minimize resource overlap among competing species or dampen predator–prey oscillations. Mutualistic or positive interactions do not have a simple theory for predicting when and where they will occur, but they are common in nature.

Pairwise interactions between species can select for changes in morphological, physiological, and behavioral traits through evolutionary time. Coevolution occurs when there is reciprocal genetic change between two interacting species. Coevolution in competitive systems can lead to character displacement. Coevolution in predator–prey systems can lead to an arms race of capture and escape. And coevolution in mutualistic systems can lead to highly specialized associations between mutually benefiting species.

Communities are composed of many species, and indirect as well as direct interactions control community composition (Fig. 5). Indirect interactions include trophic cascades, apparent competition, and the modification of a direct interaction. A keystone is a species that strongly influences the biodiversity of its community via indirect and direct effects (Table II). Keystones are often, but not always, found at the top of a trophic chain or food web. Both indirect and direct interactions can influence the evolution of community members.

Humans are causing widespread extinction of species, and the loss of species is leading to the decline of many goods and services provided by species interactions in natural systems. Human stresses on the environment, such as climate change, alteration of geochemical cycles, and destruction of habitat, can break down species associations because individual species, not communities, respond to changes in the environment.

See Also the Following Articles

COEVOLUTION • COMPETITION, INTERSPECIFIC • MUTUALISM, EVOLUTION OF • NEST PARASITISM • PARASITISM • PREDATORS, ECOLOGICAL ROLE OF • TROPHIC LEVELS

Bibliography

Connell, J. H. (1983). On the prevalence and relative importance of interspecific competition: Evidence from field experiments. *Am. Nat.* **122**, 661–696.

Diamond, J., and Case, T. J. (Eds.) (1986). *Community Ecology*. Harper & Row, New York.

Futuyma, D. J., and Slatkin, M. (Eds.) (1983). *Coevolution*. Sinauer, Sunderland, MA.

Hairston, N. G., Smith, F. E., and Slobodkin, L. B. (1960). Community structure, population control, and competition. *Am. Nat.* **94**, 421–425.

May, R. M. (1972). Will a large complex system be stable? *Nature* **238**, 413–414.

Paine, R. T. (1974). Intertidal community structure: Experimental studies on the relationship between a dominant competitor and its principal predator. *Oecologia* **15**, 93–120.

Pimm, S. L., Lawton, J. H., and Cohen, J. E. (1991). Food web patterns and their consequences. *Nature* **350**, 669–674.

Polis, G. A., and Winemiller, K. O. (Eds.) (1996). *Food Webs: Integration of Patterns and Dynamics*. Chapman & Hall, New York.

Schoener, T. W. (1983). Field experiments on interspecific competition. *Am. Nat.* **122**, 240–285.

Thompson, J. N. (1994). *The Coevolutionary Process*. Univ. of Chicago Press, Chicago.

Wooton, T. J. (1994). The nature and consequences of indirect effects in ecological communities. *Annu. Rev. Ecol. Syst.* **25**, 443–466.

STABILITY, CONCEPT OF

Clarence L. Lehman
University of Minnesota

I. Background
II. Basic Theory of Stability
III. Stability under Changing Conditions
IV. Other Kinds of Stability

GLOSSARY

community The collection of all species living together in a given area. Alternatively, some designated subset, such as the avian community or the vascular plant community.
deterministic system A dynamical system whose detailed future behavior can be predicted, in principle, for all time, assuming perfect knowledge of the system at the present.
dynamical system A set of rules defining how certain variables change with time. Ecological models are dynamical systems representing what are believed to be (vastly more complex) dynamical systems in nature.
dynamical variable A quantity in a dynamical system that changes with time according to the rules of the system (contrast with parameter).
ecological model A vastly simplified mathematical representation of an ecological system intended to capture the full system's essence. Qualitatively, ecological models are to natural ecological systems as line drawings are to full-color photographs.
equilibrium A condition of stasis in some dynamical variable.
parameter A quantity in a dynamical system that is fixed as part of the rules of the system (contrast with dynamical variable).
perturbation A temporary change in one or more dynamical variables or parameters of a dynamical system due to external factors.
population A collection of individuals of the same species, often interacting ecologically and genetically.
stochastic system A dynamical system whose detailed future behavior cannot be predicted, even in principle, due to random forces inherent in the system.

STABILITY may be defined broadly as the tendency of a system to return to its former state after some disturbance. In the natural world, the term can be applied to the capacity of an ecosystem to resist environmental disturbances. Given the fact that human activity now produces such disturbances on a vast global scale, the ability of ecosystems to remain stable has become an issue of great significance. Current investigators study the interrelationship of stability and biodiversity; i.e., the effect that the biodiversity of an ecosystem has on its stability, and the corresponding role of stability in maintaining the biodiversity of the system.

Falling down once makes a building unstable.
—GERALD WEINBERG (1975)

I. BACKGROUND

A. The Meaning of Stability

Something is said to be stable if its condition tends to remain unchanged despite external influences. The population sizes of two different species may both be relatively constant in a steady, unchanging environment, but such constancy by itself discloses nothing of the stability of the two populations. If both populations were somehow perturbed—for example, by a spring flood eliminating half of each population—and if the first population subsequently recovered to former levels while the second declined to extinction, then the first population is said to have been stable while the second was unstable. Stability is not mere constancy—it also implies an ability to recover from perturbations.

The condition that recovers from perturbations need not be as simple as a constant population value. Populations of predators and prey, for example, may have an intrinsic tendency to cycle repeatedly through high and low values, but the cycle may be stable. Suppose the populations of a predator and its prey were both maintained at fixed levels by artificial management, such as by hunting. If the cycle returned after management ceased, then the cycle could be said to be stable, even though neither population is constant during the cycle. The cycle would be an abstract condition to which the populations return after a perturbation. In fact, the condition that recovers from perturbations can be any recognizable property of the system. It could be the number of species in the system, the collective biomass of all species together, the length of the food chain, the average quantity of nutrients leached from the soil, the degree of susceptibility to disease, or many others.

In the natural world, the stability of an object is closely tied to our perception of the very existence of that object. Any property that is unstable will not retain its condition after a disturbance, by the definition of stability. Given that the world is filled with continual disturbance, any property that is unstable is therefore likely to soon change to something else and hence not be observed. For example, if food chains beyond a certain length tend to be unstable in diverse communities, then a very long food chain, should it ever appear by chance combination of species in an ecosystem, will eventually collapse to a length that is stable.

As human domination of global biogeochemical systems increases—including eutrophication of habitats, changes in atmospheric gases, fragmentation of habitats, distribution of toxic organic compounds, translocation of species, and a general reduction of biodiversity—the resulting perturbations test the stability of ecosystems. Also, as global changes take effect, induced changes in the systems alter the stability of their parts. Stability in ecosystems is therefore not just an abstract concept but something deeply connected to the persistence of the services on which all living beings, humans included, depend.

B. The Diversity-Stability Question

In the 1950s and 1960s, investigators such as Charles Elton argued that ecosystems containing more species would be more stable—less subject to fluctuations due to the myriad forces acting on them—owing to such factors as the greater number of pathways for energy to flow through them. This is the diversity–stability hypothesis. The hypothesis was generally accepted in the 1960s, although not without controversy.

Theoretical studies on the topic in the 1970s shook this general acceptance. Working with mathematical models of ecological communities, Robert May and others showed that such communities with more species were less stable: Populations of individual species returned to normal after a disturbance less rapidly as the number of species in the community increased. Although May (1974) later pointed out that certain ecosystem properties, such as total biomass of the community, can be more stable than biomasses of the constituent species, and despite objections by McNaughton (1977) and others to rejecting an accepted hypothesis based solely on theory, the acceptance of the diversity–stability hypothesis waned. During the 1970s and 1980s, ecologists generally expected that greater biodiversity would reduce stability, or at least that there was no consistent connection between biodiversity and stability.

By the 1990s, experimental evidence had accumulated that seemed to support both views simultaneously, depending on the level of focus. As the number of species increased, population densities of individual species became more variable from year to year but ecosystem parameters such as total productivity and nutrient leaching became less variable. In other words, the system seemed less stable when viewed in detail, from the perspectives of individual species, but more stable when viewed in total, from the perspective of the ecosystem (Tilman, 1999) (see Section III.A).

At the end of the twentieth century, issues concerning the effect of biodiversity on stability, and of environmental stability on biodiversity, were still being resolved, but greater biodiversity could be seen to stabilize certain important properties of the ecosystem as a whole. Many of the past controversies and apparent conflicts resulted from the application of different aspects and definitions of stability in different contexts.

C. Aspects of Stability

The seemingly simple idea of whether a system returns to its former condition after a disturbance has several aspects. Which is relevant depends on the ecological question at hand:

1. Does the system return to its former state after a disturbance, given unlimited time without further disturbance? This is the basic idea of stability, and if the system does return from all small disturbances, that state is said to be asymptotically stable or, commonly in the ecological literature, simply stable.

2. How much does the system fluctuate under variable conditions? This is called variability and it is typically measured statistically, for example, by the standard deviation of a time series, by the standard deviation divided by the mean, or by the reciprocal of that quantity. (The standard deviation being the square root of the average squared deviation from the mean.)

3. How small does the disturbance have to be in order that the system return? The set of all possible disturbances that allow the system to return to its former condition is called the domain of attraction. As the domain of attraction becomes vanishingly small, the system becomes unstable. Small domains of attraction correspond to fragile systems, whereas large domains of attraction correspond to robust systems.

4. How much is the system changed by a given disturbance? The more a given condition is changed, the lower the resistance associated with that condition. For example, suppose population values in a given area do not change much when average temperature increases above its normal but may change considerably when rainfall does so. Then the system would be said to be resistant to temperature changes but not to rainfall changes.

5. How fast does a perturbation decay? Resilience is the amount of time needed for a perturbation to be reduced to a specified fraction of its initial size. There has been variation in the use of this term in the ecological literature. Before the 1980s, resilience often simply referred to factors that reduced the chance that one or more species would become extinct (e.g., the size of the domain of attraction and the populations densities in that domain).

6. How long does the system take to return after a given disturbance? This is related to both how far the system is perturbed by the disturbance (resistance) and how fast the perturbation decays (resilience). The lower the resistance and resilience, the longer it takes to return.

7. How long can the condition be expected to last? The longer on average it lasts, the greater its persistence. The idea of persistence applies to systems subject to random variations, where the chance of encountering dangerously large variations during a time interval increases as the interval grows longer. The longer the time interval, for example, the greater the chance that one or more species will be driven extinct by random events.

8. Does the condition remain intact as the parameters of the system change slightly? If it does, the system is structurally stable with respect to those parameters. For example, if precise birth and death rates of the constituent species change slightly occasionally, does the condition remain?

9. Does the condition remain intact under evolutionary pressures? If it does, the condition is evolutionarily stable. A condition may appear stable over short time, ignoring evolution, but may change on longer timescales. Indeed, over the course of geological time, no ecosystem is invariant.

10. Does the condition depend on spatial scale? What is stable in a small area may be unstable over a larger region and vice versa.

Pimm (1984) enumerated five definitions of stability in the ecological literature at three levels of complexity and three levels of organization—potentially 45 different nuances to the meaning of stability. Hence, care is necessary in defining and applying the concept.

II. BASIC THEORY OF STABILITY

A. Equilibrium

The basic concepts of stability become precise when considered in the abstract. The many complexities in natural ecological communities, including factors that cannot be completely understood, mean that any abstraction of an ecological system is necessarily a vast reduction. To start this reduction, suppose the population increases solely by births and decreases solely by deaths. That is, processes such as immigration are not active. Suppose the rate of birth is highest and death is lowest when conditions are least crowded. As the population increases, resources become more limited—less light, reduced nutrient levels, less food, and so

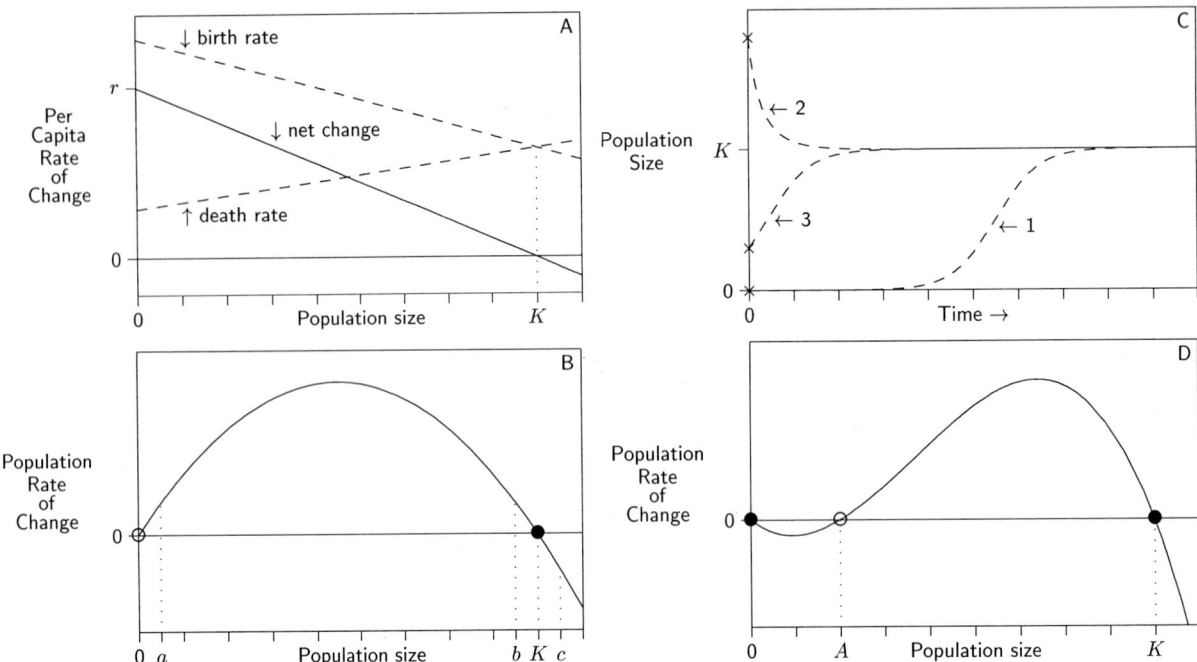

FIGURE 1 Geometric interpretation of stability. Solid and open circles represent stable and unstable equilibria, respectively. (A) Per capita birth and death rates are functions of population size; (B) growth of the entire population as a function of population size; (C) populations approaching carrying capacity with the passage of time, for the functions of A and B. The horizontal axis is time; the vertical axis is the population at the corresponding time. Cross symbols at the far left indicate the starting points. Curve 1 starts from a small population near zero. Curve 2 starts above the carrying capacity. Curve 3 starts below the carrying capacity but well above zero. (D) Growth of a population subject to the Allee effect, where 0 is a stable equilibrium.

forth; the birth rate decreases and the death rate increases. At some point, the population will increase enough so that births just balance deaths. This occurs at the carrying capacity, at which an individual, on average during its lifetime, just replaces itself with a single offspring. Above this carrying capacity, deaths exceed births, so an individual on average does not quite replace itself. In the simplest case, births decrease and deaths increase along straight lines as the population increases (Fig. 1A). The net rate of change per individual (the difference between births and deaths) is largest at the far left in Fig. 1A (r) but decreases to zero when the population size reaches the carrying capacity (K) and becomes negative when the population size is greater than the carrying capacity.

The rate of change of the entire population is simply the rate of change per individual times the number of individuals in the population. This is graphed in Fig. 1B, which shows the growth rate of the entire population (vertical axis) as a function of the population (horizontal axis). If the population is nonexistent (zero), then of course population growth is zero; with no individuals there are no births and can be no deaths. If the population is at carrying capacity, population growth is also zero; births balance deaths. These are two population equilibria—places where the population remains constant.

Between these two equilibria, the population grows. In this simplest case, in which the birth and death rates are straight lines, the population reaches the greatest rate of growth at half its carrying capacity. (In economic models, this is called the point of maximum yield.)

B. Geometric Interpretation

1. Single Species

To understand the stability of an equilibrium, the properties of the growth rate are first examined in the near neighborhood of the equilibrium. Consider the equilibrium at the carrying capacity (solid circle in Fig. 1B at position K). The graph representing population change crosses the horizontal axis precisely at the carrying capacity, which means the population neither grows nor declines there. At that point the growth curve slopes downward to the right (the slope is negative). If the population for any reason falls below the carrying capacity (e.g., point *b* in Fig. 1C), the growth function is above the axis, meaning that population growth is positive or that births exceed deaths. Thus, the population increases toward the carrying capacity. On the other hand, if the population is above the carrying capacity (e.g., point *c* in Fig. 1B), the growth function is below the axis, meaning that population growth is negative or that deaths exceeds births. Thus, the population decreases toward the carrying capacity. In both cases, an external change moving the population away from its carrying capacity induces population growth or decline in exactly the right way to counteract the external change. The population returns to the carrying capacity. That equilibrium is stable.

Now consider the equilibrium at 0 (open circle in Fig. 1B). As before, the graph representing population growth touches the horizontal axis precisely at 0. However, at this point the curve slopes upward toward the right (the slope is positive). Therefore, if the population ever increases, however slightly, above zero (e.g., point *a* in Fig. 1B)—due, for example, to the arrival of propagules from some outside source—the growth function is above the axis, meaning that population growth is positive or that births exceed deaths. Thus, the population increases away from 0. The zero equilibrium is unstable.

Thus, internal dynamics carry these populations toward their carrying capacity (which is called an attractor) and away from 0 (called a repellor). If the population starts near zero, but not precisely at zero, it may linger a long time at low values and then begin a rapid growth phase before leveling off toward its carrying capacity (Fig. 1C, curve 1). If the population starts above the carrying capacity, it rapidly decays back (Fig. 1C, curve 2). If it starts below the carrying capacity but well above zero, it can increase to the carrying capacity without the long lag period (Fig. 1C, curve 3).

2. Local versus Global Stability

In the previous example, the carrying capacity is said to be a global attractor or globally stable because almost all replicate populations eventually arrive there (all but those starting at zero in this case). Because the equilibrium at 0 is unstable, the population has a level of permanence; if driven to low values, but not completely to zero, it can recover spontaneously.

This contrasts with populations operating under an Allee effect, wherein the equilibrium at 0 is an attractor. Figure. 1D depicts the growth function in such a situation. Very low densities inhibit reproduction, for example, by reducing chances of encountering a mate. There are three equilibrium points—at 0, *A*, and *K*. Slopes both at 0 and at *K* are negative, meaning that these equilibria are stable, whereas the slope at *A* is positive, meaning that equilibrium is unstable. If such populations are driven to a low enough level (below *A*), they will spontaneously become extinct.

In this case, both 0 and *K* are local attractors (locally stable), but neither is a global attractor (neither is globally stable). If the system is pushed a small distance away, it will return to its former state. However, if pushed too far (beyond the domain of attraction of that equilibrium), the system will switch to another state. This is a bistable system. The region between 0 and *A* is the domain of attraction of the 0 equilibrium; the region between *A* and *K* and the entire region above *K* make up the domain of attraction of the carrying capacity. There is no domain of attraction for the equilibrium at *A* because this equilibrium is unstable (it is a repellor).

3. Multiple Species

The geometric example of Fig. 1 represents a single species in isolation. In this case, geometric stability arguments are direct and intuitive. The geometric arguments are similar for more than one species, but instead of curves representing growth rates, surfaces in multidimensional space are used. Each additional species requires another dimension. Unfortunately, such surfaces are difficult to visualize for more than two species.

The growth surface for a single species will in general cut through the plane of zero growth (analogous to the horizontal axis in Figs. 1B and 1D) along a curve. This curve is called the zero net growth isocline, and each species in general will have its own such curve. If the zero net growth isoclines for many species meet at a single point, then this is a multispecies equilibrium point. For this equilibrium to be stable, however, it is not sufficient that the growth surfaces of all species, considered separately, have negative slopes there. Interactions among the species must be accounted for, and this is best done algebraically.

C. Algebraic Methods

The geometric interpretation of stability, as depicted in Fig. 1B, has a direct algebraic interpretation in terms of derivatives and eigenvalues, which correspond geometrically to slopes. Where an algebraic description of the ecological system is available (i.e., in ecological models), the algebraic method is widely used and can be followed as a recipe. The algebraic method is essential for theoretical work in ecology, but the material in this section is not essential for an intuitive understanding of the concept of stability.

1. Single Species

Consider an ecological system of a single species in which the symbol N represents the population size. This may be measured in individuals, biomass, or other units. The rate of change of this population is represented by dN/dt, which tells how large an increment (dN) occurs in the population during a small increment of time (dt). Assuming that this rate of change is a function of the population size, as in the previous discussion, then the system will be described by $dN/dt = f(N)$. For example, in Fig. 1, $f(N) = rN(1 - N/K)$, which is the well-known logistic equation described in most introductory ecology texts.

Equilibria are commonly designated by symbols like $\hat{N}$ (pronounced "hat N"). They occur at such $\hat{N}$ where $f(\hat{N}) = 0$. If the slope of the function $f(N)$ at any such equilibrium is negative (i.e., if $df/dN|_{N=\hat{N}} < 0$, as shown by the solid circles in Figs. 1B and 1D), then the equilibrium is stable. If the slope is positive (as shown by the open circles in Figs. 1B and 1D), then the equilibrium is unstable. If the slope is zero, then further information is needed to determine stability. In the case of zero slope, the equilibrium may be stable, unstable, or neutrally stable, depending on the exact shape of the function near the equilibrium. In the ecological literature, neutral stability means that if the system is perturbed from equilibrium a small amount, it neither returns nor moves further away but rather maintains its new value. (Note that there is a difference between the way ecologists use the word "stable" and the way mathematicians do, as summarized in (Table I.)

2. Multiple Species

Multiple species are represented by multidimensional dynamics, in general described by m equations of the form $dN_i/dt = f_i(N_1, N_2, N_3, \ldots, N_m)$, one equation for each species. Equilibria are multidimensional points $(\hat{N}_1, \hat{N}_2, \hat{N}_3, \ldots, \hat{N}_m)$ that cause all dN_i/dt to be simultaneously zero. Some of the $\hat{N}_i$ may be zero, in which case fewer than m species coexist at that equilibrium.

The slope of a system in one dimension generalizes to eigenvalues of the Jacobian matrix in multiple dimensions. (In one dimension, the slope is a special case of an eigenvalue.) To determine stability of an equilibrium, (1) construct the Jacobian matrix of the system, $J = \{\partial f_i/\partial N_j\}$, (2) substitute the equilibrium value ($\hat{N}_1$, $\hat{N}_2$, $\hat{N}_3$, \ldots, $\hat{N}_m$) into the matrix of step 1, and (3) determine the eigenvalues of the matrix of step 2. Neuhauser (2000) or other mathematical texts may be consulted for full details.

The resulting eigenvalues are complex numbers, and as such may contain both real and imaginary parts. (That is, an eigenvalue has the form $a + bi$, where a is the real part, bi is the imaginary part, and $i = \sqrt{-1}$). The real part determines the rate of approach to or retreat from equilibrium after perturbation. In one dimension, the imaginary part is always zero, but in higher dimensions it need not be. When it is not zero, the approach to or retreat from equilibrium follows a spiral path.

Only the real parts of the eigenvalues affect stability. If the real parts of all eigenvalues are negative, that equilibrium point is stable. If one or more real parts is positive, that equilibrium point is unstable. If none is positive but one or more real parts are zero, then further information is needed to determine stability. If some eigenvalues have negative real parts but others have positive, then the equilibrium is a saddle point (so called because of its shape in the two-dimensional case). Such saddle equilibria are attracting in some directions and repelling in others; hence, they are unstable.

3. Discrete-Time Systems

Many ecological models are more naturally defined on a discrete time axis. For example, insects may emerge together at a specific time each year or seeds may be set once per year in the fall. In such models, time is better represented not as a continuum but rather as a series of integers (1, 2, 3, . . .), for example, representing a series of years. In simplest form, discrete-time models have the structure $N_i(t + 1) = f_i(N_1(t), N_2(t), N_3(t), \ldots, N_m(t))$, where $N_i(t)$ is the population level of species i at time t. To determine stability of an equilibrium in a discrete-time system, the three steps outlined previously are followed, but the resulting eigenvalues must be interpreted differently. Again, the eigenvalues are complex numbers. In the discrete-time case, it is

TABLE I

Common Ecological Usage vs Common Mathematical Usage

Term	Size of perturbation with time	
	Ecological literature	Mathematical literature
Asymptotically stable	Decreases	Decreases
Stable	Decreases	Does not increase
Unstable	Increases	Increases
Neutrally stable	Remains unchanged	(Typically not used)

the absolute values of the eigenvalues that determine stability. That is, it is the positive square root of the sum of the squares of the real and imaginary parts. (For the eigenvalue $\lambda = a + bi$, $|\lambda| = \sqrt{a^2 + b^2}$.) If the absolute value of all eigenvalues is less than 1, then the equilibrium is stable. If the absolute value of one or more eigenvalues is greater than 1, that equilibrium point is unstable. If none is greater than 1 but one or more absolute values are precisely equal to 1, then further information is needed to determine stability, as in the case of zero eigenvalues discussed previously.

4. Strength of Return

The magnitude of the eigenvalues (or, equivalently, the steepness of slope) determines the resilience, or the strength of return to equilibrium. If the eigenvalues are all very negative, the corresponding slopes will be very steep. Small deviations from equilibrium will result in large rates of growth or decline, and the system will return rapidly. On the other hand, if any of the eigenvalues is negative but close to zero, then the corresponding slopes will be shallow. Small deviations from equilibrium will not lead to rapid recovery. Of course, given a sufficiently long time, and given the absence from disturbance, the system will return regardless of the eigenvalue's magnitude, given that it is negative. However, over short times, or with repeated disturbances, the magnitude of the eigenvalues is an important part of stability.

III. STABILITY UNDER CHANGING CONDITIONS

A. Biodiversity and Temporal Stability

If a system is subject to repeated disturbances, variations in a given property of that system over time will be related to the asymptotic stability of that property. Because asymptotic stability is an abstract concept, well defined in ecological models but difficult to determine in natural systems, various measures of fluctuations are often used instead to quantify stability of natural systems.

Suppose two populations are fluctuating through time about their individual mean abundance levels due to stochastic effects in the environment and possibly due to internal dynamics as well. Suppose fluctuations in the first population are on average relatively small compared to the first population's mean abundance, whereas fluctuations in the second population are noticeably larger compared to the second population's

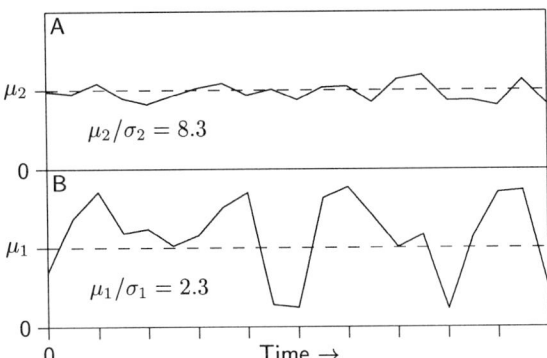

FIGURE 2 Temporal stability. Dashed lines indicate the long-term means, and solid lines indicate fluctuating variables. (A) Relatively small fluctuations about the mean, with a relatively high measure of temporal stability (8.3); (B) larger fluctuation about the mean, indicating lower stability, with a lower measure of temporal stability (2.3).

mean abundance (Fig. 2). Independent of the idea of asymptotic stability, the first system can be defined to be more stable with respect to constancy in abundance. Precisely how this fluctuational stability is related to the concept of asymptotic stability discussed in Section II is a separate theoretical issue.

The average fluctuation can be quantified in alternative ways; for example, as the average absolute value deviation from the mean or as the standard deviation of the mean. In any case, if fluctuations are large relative to the mean, the system is likely to obtain values very far from the mean. Also, if the underlying dynamics have a domain of attraction that is responsible for maintaining the mean, then large fluctuations are more likely, on rare occasions, to combine in the wrong direction and push the system out of that domain, inducing a switch to a new mode of behavior. Thus, the relative amount of fluctuation can carry information on the long-term persistence of the system.

Measures proportional to the coefficient of variation—the standard deviation over a time series divided by the mean of that time series—have been used as a measure of fluctuation. These are actually proportional to instability because larger values in the coefficient of variation correspond to lower stability. The reciprocal of the coefficient of variation (i.e., the mean divided by the standard deviation, μ/σ) has been termed temporal stability (Tilman, 1999). It carries the same information, but larger values of temporal stability correspond to greater stability of the system. In this form (μ/σ), it is reminiscent of signal-to-noise ratio in engineering. Clearly, temporal stability is greater if the mean is greater (μ, the numerator), if the standard deviation is smaller (σ, the denominator), or both. Therefore, any forces that tend to increase the mean or decrease the variation will increase temporal stability. Temporal stability is maximal (infinite) when the standard deviation is zero—when there is no variation at all.

Evidence from both experiment and theory suggests that the temporal stability of certain community characteristics of competitive plant communities tends to increase as the number of species they comprise increases (Tilman, 1999). Along a long-term nitrogen fertilization gradient, the number of species varied, changing from diverse prairie communities at low nitrogen levels to only a few agricultural grass species at high nitrogen levels. Individual species abundances varied from year to year due to many causes, both environmental and ecological. As the number of species increased, (i) the total biomass of the community tended to increase and (ii) the total biomass of the community became less variable. Simultaneously, the biomass of an individual species tended to become more variable relative to the mean biomass for that species embedded in the community. In other words, increased biodiversity appears to stabilize the community while destabilizing the individual species.

Given this pattern from experimental systems, a theoretical question immediately arises: Do standard ecological models predict similar phenomena? This is a rapidly developing area, but initial results appear to indicate that they do (Tilman, 1999). In a model of competition for a single resource, and in other more general models, temporal stability of total community biomass increased steadily and linearly as the number of species increased (Fig. 3A). Temporal stability of individual species biomass decreased sharply as biodiversity increased from one to about five species, and then it leveled off at higher diversities (Fig. 3B). In other words, simple ecological models appear to predict the effects observed along the nitrogen gradient—increased biodiversity will stabilize the community but destabilize individual species.

The importance of various possible causes of stabilization of community properties remain under discussion. When abundances of individual species are perturbed by complex and effectively random forces, some abundances may be perturbed to higher levels and others to lower levels. On statistical grounds, the variation in the total biomass will likely contain both positive and negative perturbations. Hence, the total is likely to fluctuate less, relative to its mean value. This has been called the portfolio effect (Tilman, 1999) by analogy to economics, wherein a portfolio diversified over many investment instruments will fluctuate less than one con-

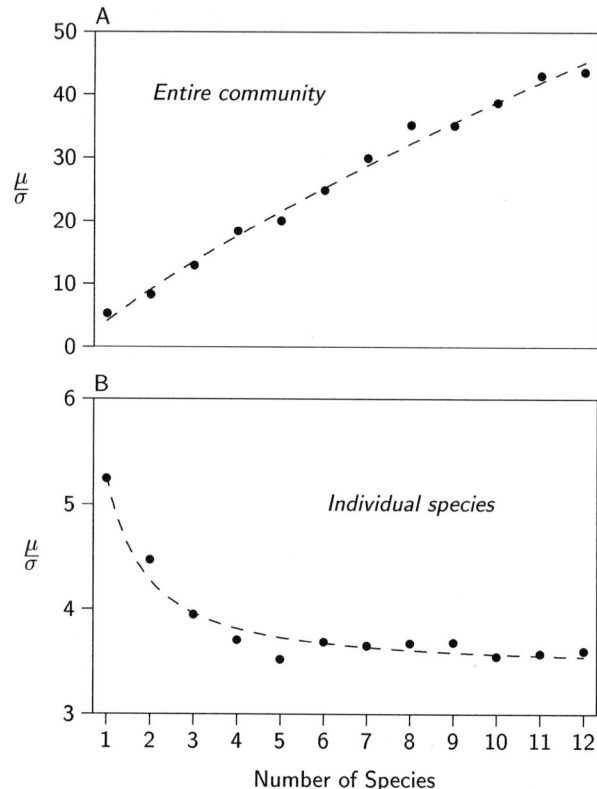

FIGURE 3 Biodiversity affects temporal stability. From a resource model reported by Tilman (1999) showing patterns similar to those observed in nature. Large dots indicate means of 100 replicates. (A) Increasing biodiversity stabilized total biomass; (B) increasing biodiversity destabilized individual species biomass.

taining only a few. It depends on species responding differently to perturbations, if only for statistical reasons. Effects of competition allow species to expand in abundance when a competing species is suppressed by external forces. This expansion, in turn, buffers any change in total biomass. This is called the covariance effect because it is evident when there is strong negative covariance between pairs of species. Finally, increased biodiversity also often leads to increased total abundance, and this increases stability by making any given level of fluctuations smaller in proportion. This is called the overyielding effect.

Regardless of the source of the stabilization, however, it appears that certain composite community properties may be stabilized by increased biodiversity. More direct experiments that will eventually be able to clarify the effects of biodiversity on stability and other ecosystem properties have been established in many countries (Hector *et al.*, 1999; Tilman, 1999). Results of this coordinated set of experiments will emerge during the first decade of the twenty-first century and bear watching for information they will provide on the diversity–stability question.

B. Temporal Niches

Temporal stability, as defined previously, certainly applies to a system continually perturbed from equilibrium. However, equilibrium was not part of the definition—only the mean and the variation about the mean participated in the measure of temporal stability. Temporal stability therefore applies to cases in which equilibrium, stable or otherwise, may not even exist.

Changing environmental conditions, such as variation in average rainfall from year to year, can favor different species at different times. This, in effect, can partition resources among species, setting up temporal niches and permitting long-term coexistence of species that greatly outnumber the resources they consume (Chesson, 1994). Without changing conditions, biodiversity would be reduced—coexistence would be limited by the number of resources available.

For a given regime of environmental fluctuations, and for a given community, each species will have some pattern of fluctuations in response to the fluctuations in the environment and those of other species in the community. Again, the issue with respect to biodiversity is how the relative size of the fluctuations—the temporal stability—of each species depends on the number of species in the community and how community-level properties such as total productivity or nutrient leaching depend on the number of species.

Coexistence among species in the absence of a multispecies equilibrium can result from stochastic variations in the environment, as explained previously. However, it is not necessary that the variations be stochastic, nor that their source be the external environment. The variations could result from some regular, well-determined periodic changes in the environment, or they could result from population cycles set up by dynamics of the ecological community (Armstrong and McGehee, 1980; Huisman and Weissing, 1999). Whether the effects are external or internal, stable coexistence at increased biodiversity can exist under fluctuating conditions.

C. Emerging Deterministic Stability

Stable deterministic characteristics such as those described in Section II can emerge from stochastic systems. Individuals may live and die, populations may fluctuate in abundance, and some species may become extinct while others appear; however, amid all this complexity, simple patterns of biodiversity can emerge.

One of the early successful theoretical explanations of biodiversity concerned island biogeography (MacArthur and Wilson, 1967). On islands, simple deterministic characteristics emerge amid complex ecological change. It is a conspicuous fact that oceanic islands have fewer species than adjacent mainlands. This effect can be attributed, to a large extent, to a stable balance between local extinction of resident species and immigration of new species.

A new individual arriving on the island may be a member of a species already resident on the island, or it may represent a novel species. If the island is devoid of life, the new individual is certain to be a member of a novel species. As the island becomes more populated with mainland species, the chance that the individual is a member of a novel species decreases. Similarly, the chance that an existing species vanishes from the island increases as the number of resident species increases.

Such considerations lead to immigration and extinction curves such as those in Fig. 4. Notice the qualitative similarity of the immigration and extinction curves in Fig. 4 to the birth and death curves, respectively, in Fig. 1. The immigration and extinction curves intersect at an equilibrium point, where resident species becoming extinct are balanced by new species arriving. Note that at this equilibrium, neither population values nor

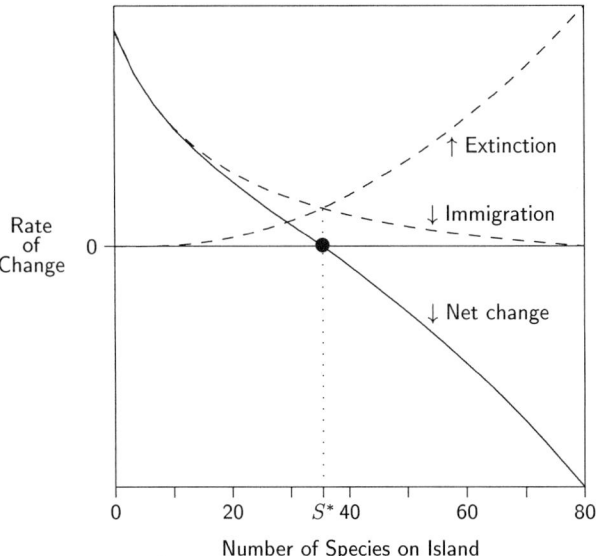

FIGURE 4 Stable island equilibrium. Island biogeographic immigration and extinction curves patterned after observed rates for real islands. Biodiversity reaches a stable equilibrium when immigration balances extinction.

community composition are constant. In fact, the individual species making up the community are constantly in flux. It is simply the biodiversity that remains constant.

Is this biodiversity equilibrium stable? The net rate of increase or decrease in the number of species is simply the difference between the immigration and extinction rates, which is the solid curve in Fig. 4. Unlike the curves of Fig. 1A, this net rate represents the entire island; it is not a per capita or per species rate. Therefore, its slope at the equilibrium directly corresponds to stability—it need not first be multiplied by the number of species on the island. Because the slope is negative, by the arguments of Section II this equilibrium is stable.

D. Stability Amid Chaos

Consider a population that changes deterministically according to the rules of Box 1—a straight-line relative of the logistic equation of Fig. 1. Populations oscillate chaotically between high and low values, but except for certain infinitely rare starting conditions they never return to a previous state (Fig. 5). This system is chaotic in the sense that slight deviations are always magnified. In almost all cases, two slightly different starting populations, no matter how nearly identical, grow increasingly different with time.

However, amid such complete chaos can be stability. After sufficient time, population values fall into a pair of disjoint intervals, shaded in Fig. 5. This pair of intervals is an attractor. The population then alternates regularly between the intervals, though its position within either interval cannot be predicted for very long from measurements of finite precision. This is a deterministic system with dynamics that appear superficially stochastic, but among all its instabilities arises another level of stability.

Box 1

$$N_{t+1} = \begin{cases} 2N_t & \text{if } N_t < a \\ 2a(1 - N_t)/(1 - a) & \text{if } N_t \geq a \end{cases}$$

Piecewise linear discrete-time population growth. If $a < 1/3$, there is a stable equilibrium at $N = 2a/(1 + a)$. If $a > 1/3$, there are no stable equilibria or cycles (see Fig. 5).

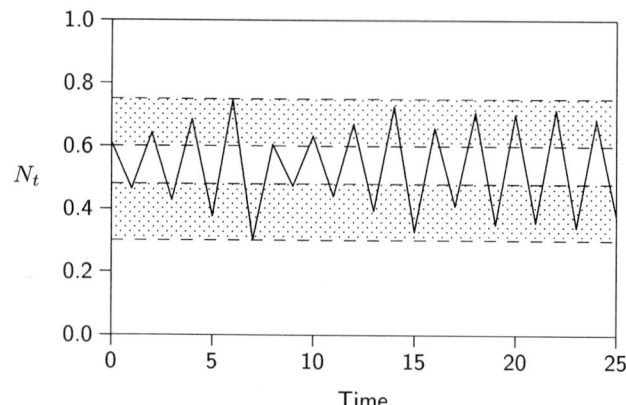

FIGURE 5 Stable attractor enclosing unstable population cycles. The system of Box 1 with $a = 3/8$ is chaotic, but population fluctuations are pulled into the shaded intervals (0.3, 0.48) and (0.6, 0.75). The horizontal axis is time; the vertical axis is the population at the corresponding time.

IV. OTHER KINDS OF STABILITY

A. Structural Stability

In the cases discussed previously, the ecological system did not change. Dynamical variables such as population levels were perturbed, but ecological parameters such as birth rates remained fixed. Ecological parameters, however, constantly change: Global temperatures increase, glaciers retreat, and spruce forests give way to pine and then hardwoods. Rainfall, soil substrates, and a host of environmental conditions alter the parameters and the structure of ecological systems. Hence, not only may the dynamical variables such as population abundances be perturbed but also the very structure of the ecological system may change.

If a given property of a system persists under small changes in the system itself, then the system is said to be structurally stable with respect to that property. The equilibrium carrying capacity in the system of Fig. 1, for example, is structurally stable. Changes in the slopes of the birth and death rates, or small changes that make the birth and death rates curves rather than straight lines, still leave a carrying capacity that all nonzero initial populations eventually reach, provided that the net per capita growth declines smoothly with increasing population. The precise size of the carrying capacity may change, but the fact that it exists and is stable does not.

1. Classical Predator–Prey Systems

Some ecological models are on a razor edge of structural instability. In the simplest form of the classical Lotka–

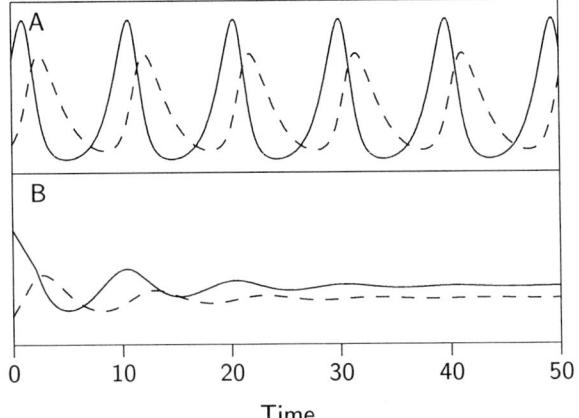

FIGURE 6 Structural instability in Lotka–Volterra predator–prey cycling. Solid curves represent population densities of the prey, and dashed curves represent that of the predator. Here, $r_N = 1$, $r_P = 1/2$, $a = 1$, $b = 1/2$, $K = 5$, $N(0) = 2$, and $P(0) = 1/2$. (A) The prey population is limited only by predation (top two equations in Box 2). The populations oscillate on neutrally stable cycles. This effect is structurally unstable. (B) The same starting conditions when the prey is also limited by its own carrying capacity in the absence of the predator (bottom two equations in Box 2). Initial variations damp out. This effect is structurally stable.

Volterra predator–prey system, included in most introductory ecology texts, the prey population is limited by predation and the predator population is limited by availability of prey (first pair of equations in Box 2). Predator and prey populations oscillate indefinitely about an equilibrium, with the predator population lagging behind that of the prey (Fig. 6A). This is a common

---

Box 2

$$\frac{dN}{dt} = r_N N - aNP$$

$$\frac{dP}{dt} = r_P NP - bP$$

$$\frac{dN}{dt} = r_N N \left(1 - \frac{N}{K}\right) - aNP$$

$$\frac{dP}{dt} = r_P NP - bP$$

Lotka–Volterra predator–prey system. Top two equations, no carrying capacity for prey; bottom two equations, carrying capacity included (see Fig. 6).

---

characteristic of predator–prey systems or, more generally, of producer–consumer systems. However, this particular system has a peculiar property for an ecological system: It possesses a "memory" of past events. If something perturbs either population, or both populations, to a new level, that new level will be revisited on each subsequent cycle. This system has no asymptotically stable behavior; instead, everything is neutrally stable.

Such neutral stability would not necessarily be pathological in an idealized physical system, such as a perfect harmonic oscillator, but here it is pathological. All abstractions are simplifications, and here one of the simplifications is the assumption that the prey population is limited only by the predators; in the absence of predators, prey can increase in numbers without bound. Inclusion of any carrying capacity for the prey, no matter how large or small (as in the second pair of equations in Box 2), changes the dynamics completely. The equilibrium becomes stable, oscillations die out, and memory of past perturbations fades with passing time (Fig. 6B). Another simplification is that predators are never satiated: They consume all the prey they encounter. Inclusion of a more realistic predator response can make the equilibrium unstable. The oscillations converge to a stable cycle of fixed amplitude, again with memory of past perturbations fading.

Thus, this simplest Lotka–Volterra formulation is structurally unstable with respect to important ecological factors. Structurally unstable ecological systems do not commonly appear in nature, so conclusions drawn from structurally unstable models must be used with caution.

2. Related Effects of Eutrophication

The change in structure from stable equilibrium to stable limit cycle, as described previously, may be induced by changes to parameters of the ecosystem. Eutrophication is enrichment by high levels of nutrients, such as increased phosphorus in a lake or nitrogen in the soil. Eutrophication is one of the principal effects of human domination of ecosystems, and it may affect both their biodiversity and their stability.

Graphical analysis by Rosenzweig and others in the 1970 (Rosenzweig, 1990) showed that stable equilibria in producer–consumer systems were favored when the carrying capacity of the consumer was relatively small. If the carrying capacity of the consumer were higher, the equilibrium would be less stable (less resilient). At a sufficiently high carrying capacity, the system would pass through a structural instability and the equilibrium would lose its stability entirely. Oscillations in producers and consumers would then occur. At even higher carrying capacities, the size of the oscillations would increase, driving both producer and consumer periodically to low population levels, thereby increasing the chance of extinction.

Now, carrying capacity is directly related to the level of resources available. Eutrophication increases these resources, thereby increasing the carrying capacity of the producer. Ironically, the act of providing more food or resource to the producer can lead to its destruction as a result of induced instability. This effect has been called the paradox of enrichment. A common result in observed fertilized terrestrial systems is a reduction in the number of species. Although there can be many reasons for this, loss of stability upon enrichment has been suggested as a contributor (Rosenzweig, 1990).

B. Spatial Stability

A property that is stable at one spatial scale need not be stable at larger or smaller scales (Levin and Segel, 1985). This was decisively demonstrated in the mid-twentieth century by the mathematician Alan Turing. Turing's simplest example was a (nonecological) dynamical system operating in two separate cells, optionally with some migration between the cells (Box 3). If the two cells are completely disconnected, then there is an asymptotically stable equilibrium that both cells approach. Thus, if the density in either cell is perturbed, it will return to its equilibrial value (Fig. 7, times 0–4).

Box 3

$$\frac{dx_1}{dt} = ax_1 - by_1 + h - m_x(x_1 - x_2)$$

$$\frac{dy_1}{dt} = bx_1 - cy_1 + h - m_y(y_1 - y_2)$$

$$\frac{dx_2}{dt} = ax_2 - by_2 + h - m_x(x_2 - x_1)$$

$$\frac{dy_2}{dt} = bx_2 - cy_2 + h - m_y(y_2 - y_1)$$

Turing's basic example of diffusive instability. Variables x_i and y_i are densities in cell i. The remaining symbols are constant coefficients, $a = 5$, $b = 6$, $c = 7$, $h = 1$, $m_x = 1/2$, and $m_y = 9/2$ (see Fig. 7).

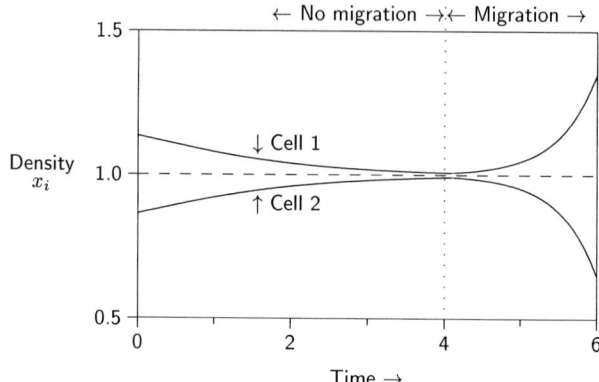

FIGURE 7 Spatial diffusive instability in a two-cell example. The vertical axis is density, or concentration, of two of the variables (x_1 and x_2) for the example shown in Box 3. The two cells are isolated until time 4, when a migration corridor between the cells is opened. Isolated, each cell approaches the same stable equilibrium (times 0–4). Connected, this equilibrium is destabilized by migration and the two cells diverge rapidly (times 4–6). Initial conditions are $x_1(0) = 1.13533$, $y_1(0) = 1.1218$, $x_2(0) = 0.864666$ and $y_2(0) = 0.8782$.

However, if there is sufficient migration between the cells, then the equilibrium, though it still exists, becomes unstable. Any random deviation that causes a difference between cells, however slight, unbalances migration and causes a sustained and accelerating net transport toward the cell with the higher concentration (Fig. 7, times 4–6). This is called a diffusive instability or a Turing instability. An equilibrium that is stable at a point of space can become unstable when its components can diffuse to neighboring parts of the space. This resulting instability can lead to variations and patterns in biodiversity over the landscape, with the resulting patterns being stable.

The opposite can also hold. A system that is unstable in a point of space can become stable, or at least persistent, when extended over a region. Predators clearly coexist with their prey and parasites coexist with their hosts for long periods in nature. However, such coexistence has been difficult to scale down to the size of experiments. Many of the early experiments on predator–prey and host–parasite systems found them to be unstable at small scales. Confined to a small area, the predators captured all the prey, and then themselves disappeared from the experiment for lack of food (or, analogously, the parasites infected all the hosts and then themselves disappeared). However, as the spatial extent of the experiment was increased, persistence of the system increased dramatically. Ecological models of such systems show a similar behavior (Hassel and

Wilson, 1997). Although each individual cell quickly runs to extinction, migration "rescues" empty cells, and the host and the parasite can have a stable average density over a large group of cells. Thus, an equilibrium that does not exist or is unstable in a point of space can be replaced by a stable equilibrium averaged over the entire region.

Something similar happens in ecological systems described by metapopulation dynamics (Hanski, 1997). Individual populations of a species may be separated by distances or barriers that inhibit movement between populations. If movement is strongly inhibited, individual populations may be driven extinct locally by stochastic effects, only to be restored later by propagules from some other population. Despite individual populations going in and out of existence, the portion of local sites occupied at any time can approach a stable equilibrium. This locally nonpersistent system can be both persistent regionally and have a stable regional equilibrium.

C. Evolutionary Stability

On the ecological timescale, ecological parameters such as birth and death rate are considered fixed. They are taken to be immutable characteristics of the species under consideration. However, ecological systems have many layers of complexity beyond our simple abstractions of them. On the evolutionary timescale, the parameters are malleable through the process of mutation and natural selection. Moreover, species formerly isolated may come into contact either as a result of natural causes or, now with great frequency, through the actions of humans. Both invading species and mutant phenotypes present new parameters that test the evolutionary stability of the system. A condition that is stable with ecological parameters fixed need not be stable when these parameters can change.

How the parameters might change can be seen by examining the growth rate of potential invaders, initially at negligible densities, entering the community at equilibrium. If the growth rate of any phenotype is positive, then the system can be invaded by that phenotype. The resident species, and hence the parameters of the system, are evolutionarily stable only if the growth rate of every potential invader is negative. Originally developed for behavioral systems, this idea is being applied to ecological communities (Geritz et al., 1998), and such work promises to shed light on the properties of invasions and ultimately on the evolution of biodiversity.

See Also the Following Articles

CARRYING CAPACITY, CONCEPT OF • DISTURBANCE, MECHANISMS OF • EUTROPHICATION AND OLIGOTROPHICATION • HABITAT AND NICHE, CONCEPT OF • ISLAND BIOGEOGRAPHY • POPULATION DYNAMICS • SPECIES, CONCEPTS OF • STRESS, ENVIRONMENTAL

Bibliography

Armstrong, R. A., and McGehee, R. (1980). Competitive exclusion. *Am. Nat.* 115, 151–170.

Chesson, P. (1994). Multispecies competition in variable environments. *Theor. Population Biol.* 45, 227–276.

Geritz, S. A. H., Metz, J. A. J., Kisdi, É., and Meszéna, G. (1998). Evolutionarily singular strategies and the adaptive growth and branching of the evolutionary tree. *Evol. Ecol.* 12, 35–37.

Hanski, I. (1997). Predictive and practical metapopulation models: The incidence function approach. In *Spatial Ecology: The Role of Space in Population Dynamics and Interspecific Interactions* (D. Tilman and P. Kareiva, Eds.), pp. 21–45. Princeton Univ. Press, Princeton, NJ.

Hassel, M. P., and Wilson, H. B. (1997). The dynamics of spatially distributed host–parasatoid systems. In *Spatial Ecology: The Role of Space in Population Dynamics and Interspecific Interactions* (D. Tilman and P. Kareiva, Eds.), pp. 75–110. Princeton Univ. Press, Princeton, NJ.

Hector, A., Schmid, B., Beierkuhnlein, C., Caldeira, M. C., Diemer, M., Dimitrakopoulos, P. G., Finn, J., Freitas, H., Giller, P. S., Good, J., Harris, R., Högberg, P., Huss-Danell, K., Joshi, J., Jumpponen, A., Körner, C., Ladley, P. W., Loreau, M., Minns, A., Mulder, C. P. H., O'Donovan, G., Otway, S. J., Pereira, J. S., Prinz, A., Read, D. J., Scherer-Lorenzen, M., Schulze, E. D., Siamantziouras, A. S. D., Spehn, E. M., Terry, A. C., Troumbis, A. Y., Woodward, F. I., Yachi, S., and Lawton, J. H. (1999). Plant diversity and productivity experiments in European grasslands. *Science* 286, 1123–1127.

Huisman, J., and Weissing, F. J. (1999). Biodiversity of plankton by species oscillation and chaos. *Nature* 402, 407–410.

Levin, S. A., and Segel, L. A. (1985). Pattern generation in space and aspect. *SIAM Rev.* 27, 45–67.

MacArthur, R. H., and Wilson, E. O. (1967). *The Theory of Island Biogeography*, Monographs in Population Biology 1. Princeton Univ. Press, Princeton, NJ.

May, R. M. (1974). *Stability and Complexity in Model Ecosystems*, Monographs in Population Biology 6. Princeton Univ. Press, Princeton, NJ.

McNaughton, S. J. (1977). Diversity and stability of ecological communities: A comment on the role of empiricism in ecology. *Am. Nat.* 111, 515–525.

Neuhauser, C. (2000). *Calculus for Biology and Medicine*. Prentice Hall, New York.

Pimm, S. L. (1991). *The Balance of Nature? Ecological Issues in the Conservation of Species and Communities*. Univ. of Chicago Press, Chicago.

Rosenzweig, M. L. (1995). *Species Diversity in Space and Time*. Cambridge Univ. Press, Cambridge, UK.

Tilman, D. (1999). The ecological consequences of changes in biodiversity: A search for general principles. *Ecology* 80, 1455–1474.

STEWARDSHIP, CONCEPT OF

Peter Alpert
University of Massachusetts at Amherst

I. Etymology
II. Rationales
III. Contexts
IV. Types of Stewards

GLOSSARY

community-based conservation Conservation-oriented management of communal or public property by local residents.

ecosystem management Management of an ecologically defined geographical area as an integrated whole for both sustainable yields of natural resources and maintenance of ecological processes and species populations.

integrated conservation and development project Project that links biological conservation with human development in a local area or set of areas.

land ethic Philosophy of proper treatment of lands and the natural communities on them.

natural resources management Management of natural systems to maintain or provide yields of wood, water, and other resources for people.

nongovernmental organization An organization that is not an agency or part of a government, typically one that is not-for-profit and not owned by individuals.

stewardship Management of a thing for someone or something else.

STEWARDSHIP is taking care of something for someone else. Originally, a steward was a person who managed household affairs for a landowner. Today, the concept of stewardship extends to the management of intangible things, such as beliefs, and to management of things legally owned by the manager, as long as he or she is acting on behalf of interests greater than just his or her own. In natural resource management, stewardship often refers to voluntary actions taken by private landowners to promote ecological goals on their own lands. Stewardship may have an instrumental rationale, such as the human need for natural resources; or an intrinsic rationale, such as the moral rights of species to exist. Stewardship of biodiversity occurs within a context of ecological, economic, cultural, and political factors that help determine the forms stewardship can take and how well it will work. For example, the large scale of some ecological processes requires forms of international stewardship. Governmental agencies, nongovernmental organizations, local communities, classes of citizens, companies, private individuals, and partnerships of these different types of groups have all undertaken stewardship of biodiversity. Their roles may be just to protect species and habitats against intrusion, or to actively maintain population sizes and ecological processes in habitats too isolated or small to retain biodiversity naturally, or even to reintroduce extirpated species and restore habitats that have lost diversity or function through previous human action.

I. ETYMOLOGY

The term "steward" has existed for approximately 1000 years. According to the *Oxford English Dictionary*, its early form "stigweard" first appeared in manuscripts in the eleventh century. The prefix "stig-" is probably the Old English word for "house or some part of a house." "Weard" is an Old English word for keeper. A steward is literally a "household-keeper." This gives the word an interesting kinship with the word "ecology," coined eight centuries later from Greek words for "house" or "household" and "study." The steward manages what the ecologist studies.

The essence of the idea of a keeper is one who manages something for someone else. For example, the Old Teutonic root "wardo," presumably related to "weard," gave rise to the word "warden." This word was used in a conservation sense during the Renaissance. A royal act under Edward I in 1543 proclaimed that "The wardeyns shall kepe and susteyne the lands without makynge dystruction of any thynge." In thirteenth-century France, wardo became "garde" and then "guardian" in fifteenth-century England, meaning "one to whom the care and preservation of any thing is committed." Like his Latinate equivalent, a "custodian," a "guardian" could manage but not rule: "From that yere . . . were al custodyes & gardeyns and no mayres" (Robert Fabyan, *The Newe Chronycles of Englande and of Fraunce*, 1516).

The "household" in "steward" may at first have meant exactly that. The first meaning of "steward" in the *Oxford English Dictionary* is "an official who controls the domestic affairs of a household, supervising the service of his master's table, directing the domestics, and regulating household expenditures." However, usage of "stewardship" soon extended beyond the kitchen and pantry. As the "royal households" in England and Ireland became more exalted, "steward" became a title of high office. In early Scotland, the Lord High Steward was the first officer of the king. One such official, Robert the Steward, claimed the throne. As the new ruler of the royal household, he was no longer its steward, and his epithet changed, founding the royal house of Stuart. On the other hand, a king threatened with the loss of his crown might find it expedient to claim to be in the employ of a still higher authority and therefore to be a steward. Shakespeare had Richard II make this case: "Show us the hand of God, That hath dismiss'd us from our stewardship." The formal usages of stewardship now include the management of any property owned by someone else.

Some contemporary common usages of "steward" adhere to these formal meanings. People today probably encounter the word steward most often in the person of an airline or ship's steward, someone in charge of food service on an airplane or ship owned by someone else. Organizations such as The Nature Conservancy employ "land" or "reserve stewards" to manage natural areas. These stewards are paid to manage part of an estate for the owner, which is the organization.

Other common usages of "steward" include the management of intangible things, such as rights or beliefs. A "shop steward" is a union official charged with protecting the rights of a group of employees; a "religious steward" may be a volunteer charged with helping to maintain a faith or church. Stewardship can also refer to the management of things owned by the steward, as long as he or she is acting on behalf of interests greater than his or her own. Currently, the most common use of "steward" in relation to the conservation of biodiversity is for private individuals or companies that agree to abide by special practices designed to protect or restore natural species or ecology on their own lands. A wide variety of national, state, and local governmental and nongovernmental agencies and organizations sponsor programs that recruit private landowners to volunteer as such "ecological," "environmental," or "habitat stewards."

Despite these extensions of the original meaning of "steward," the concept of stewardship remains distinct from the concept of belonging, on the one hand, and from the concept of domination, on the other hand. Belonging implies being an equal or subordinate part of a thing rather than managing it. The first premise of stewardship is the competence and right to act on the thing in one's charge so as to produce certain results. Domination, or proprietorship, implies controlling or managing a thing in one's own interests. The second premise of stewardship is that the desired outcomes of management are designed to satisfy interests greater than those of the steward alone. Stewardship is a three-way relationship between a manager, a thing being managed, and the principle or party in whose interests the manager acts.

II. RATIONALES

Stewardship can have a variety of rationales. These rationales can be classed as instrumental, if management is viewed as a means to an end, or as intrinsic if management is viewed as an end in itself. For example, the guardian of a child can be considered to be a steward of the child if he or she is taking care of the child for the sake of someone besides the guardian himself or herself. The guardian might be doing so for instrumental reasons, such as because his or her community needed more adults. Alternatively, a guardian might be taking care of a child for intrinsic reasons, such as because he or she believed it to be his or her duty to a deceased parent or because raising a child to realize his or her full potential is intrinsically good. The following sections summarize some of the instrumental and intrinsic arguments for the stewardship of biodiversity and some of the counterarguments to the use of stewardship models as the basis for conserving biodiversity.

A. Instrumental Arguments for Stewardship of Biodiversity

Ecological stewardship has been justified largely on instrumental or practical grounds. George Sessions (as cited in Oelschlaeger, 1992, pp. 90–130) engagingly categorized types of instrumental arguments for preserving wilderness as a "silo" for storing organisms that might have eventual use; as a "lab" for long-term, large-scale natural experiments and baseline data; as a "gym" with spectacular recreational opportunities; as a "cathedral" for the appreciation of beauty or one's natural heritage; as a "minding animals" zone needed for normal human development; as an "antitotalitarian" object lesson in how to be free and make decisions for oneself; and as a "life-support system" to furnish water, nutrients, and the right mix of gases.

Much of the history of conservation can be viewed as a succession of instrumental schools of ecological stewardship, referred to collectively by R. Edward Grumbine as "resourcism." Natural forest management has followed instrumentally motivated stewardship models for at least two centuries. Georg Ludwig Hartig, organizer of the Prussian Forest Service, wrote in 1795 that, "All wise forest management must endeavor to utilize as much as possible, but in such a way that later generations will be able to derive at least as much benefit from them as the present generation claims for itself." This view was embodied in the European concept of a "dauerwald," a permanent or perpetual forest, generally characterized by uneven-age forestry and selective cutting. Gifford Pinchot, head of the U.S. Forest Service at its inception in 1905, called conservation "the one great central problem of the use of the earth for the good of man."

The set of practices referred to as "sustainable development" is based on an instrumental argument for stewardship of ecological systems and natural resources on behalf of all people and their posterity. Herman E. Daly succinctly defines sustainable development as "development without growth." Its objective, as stated in the 1987 "Brundtland Report" of the World Commission on Environment and Development, is "meeting the needs of the present without compromising the ability of future generations to meet their own needs." On a more humble scale, this is like passing on the family farm in better shape then you found it, and one of the most resonant arguments for sustainable development is that is it essential for the welfare of one's children. Author Peter Mattheisen has appealed for stewardship of biodiversity because "indifference to the loss of species is, in effect, indifference to the future, and therefore a shameful carelessness about our children." When Margaret Jacobson and Garth Owen Smith asked Himba pastoralists in northwestern Namibia in the 1980s and 1990s why they wanted to conserve local wildlife, a common reply was so that their children could see them.

Ecosystem management focuses on natural resources management rather than on human development but makes the same argument for environmental stewardship as does sustainable development—that it is beneficial to people in the present and in the future (Johnson et al., 1999). Steven L. Yaffee of the University of Michigan noted that the term "ecosystem management" covers a range of approaches to natural resources management, from "anthropocentric" plans for multiple use of natural resources through "biocentric," whole-ecosystem management to "ecocentric," natural regionwide plans. The first approach was codified as early as 1906 in the Multiple Use Sustained Yield Act of the U.S. Congress under Theodore Roosevelt. Whole-system management was in principle implemented by the National Park Service in 1963 when it accepted the recommendation in the "Leopold Report" of its Advisory Board on Wildlife Management, chaired by Starker Leopold, to manage its lands as "biotic wholes." The statutes that govern the Oregon Board of Forest Lands prescribe a regionwide plan to manage for "greatest permanent value," defined in a 1998 administrative rule as "healthy, productive and sustainable forest ecosystems that over time and across the landscape provide a full range of social, economic, and environmental benefits to the people of Oregon."

There is evidence that people in the United States find instrumental rationales for ecological stewardship more convincing than intrinsic ones. As part of a study funded by the U.S. Forest Service Office of Communication, 30,000 on-line news stories about national forests published from 1992 to 1996 were searched by computer for references to benefits in four categories: "recreational," "commodity," "ecological," and "moral/spiritual/aesthetic." Recreational benefits accounted for about 40% of the citations, commodity benefits for 30%, ecology for 20%, and moral benefits for only about 10%. Recreational benefits tend to translate into local economic benefits and are important not only to people who recreate but also to those who sell to them or tax them. A 1999 report from the U.S. Department of Agriculture on population and economic growth in nonmetropolitan counties in the contiguous United States from 1970 to 1996 found that growth was strongly dependent on local natural amenities, as measured by an index based on weather, topographic variation, and water surface area. Populations grew by 120% on average in the counties that scored high on the amenity index and only by 1% in the counties that scored low. High-scoring counties also had a threefold higher increase in number of jobs compared to low-scoring ones.

B. Intrinsic Arguments for Stewardship of Biodiversity

Moral philosophy has provided many intrinsic arguments for the stewardship of species. Concepts of "moral status" have been held to require varying degrees of solicitude for different species depending on their ability to make decisions, be self-aware, feel desire or pleasure or pain, and strive to exist. A being has moral status when it has "moral duties directed to it" by "moral agents," those "able to deliberate and act in a responsible and answerable way" (Wetlesen, 1999). Some philosophers, such as Plato and Kant, assign moral status only to beings with free will, meaning normal humans. Thomas Regan has argued for moral status for beings that experience emotions and have a sense of their own individuality, meaning normal mammals more than 1 year old and possibly other animals such as birds. Animal rights advocate Peter Singer has proposed sentience, the ability to have conscious feelings of pleasure or pain, as the qualification for moral status.

In addition to the issue of who or what should be accorded moral status, there is the issue of how equal this status should be. One might have equal moral duties toward all things with moral status or greater duties toward some of these things than toward others. Many ethical philosophers accept that persons have equal moral status. As signatories to the Universal Declaration of Human Rights of the United Nations, many nations accept this as well: "All human beings are born free and equal in dignity and rights. They are endowed with reason and consciousness and should act towards one another in a spirit of brotherhood." Ethical philosophers generally assign nonpersons a lesser moral status. For example, Mary Anne Weaver reasoned that people have a moral responsibility to accept the equal rights of all persons, to avoid cruelty to sentient beings, and to respect all life. Wetlesen (1999) proposed "equal value for moral persons and agents and lesser values for nonpersons, in proportion to their similarity with moral persons."

Stewardship of things that have no moral status can be morally justified on the basis of their importance to things that do have moral status. For instance, it could be held to be wrong to shatter a stalactite because it will deprive other people of the enjoyment of seeing it or wrong to divert water from a pool because it will endanger a fish. In most minds, this provides the only intrinsic justification for stewardship of nonliving things. However, thinkers such as Baird Caldicott elaborate reasons for attributing intrinsic values to wholes, which could apply to ecological systems.

In addition to its instrumental schools, the history of conservation contains a tradition of intrinsic arguments for stewardship, documented in the writings of John Muir, Henry David Thoreau, David Brower, and others. In 1864, George Perkins Marsh wrote that "Man has too long forgotten that the earth was given him for usufruct alone, not for consumption." Toward the end of his career in 1949, Aldo Leopold proposed that "A thing is right when it tends to preserve the integrity, stability and beauty of the biotic community. It is wrong when it tends otherwise" and that "The last word in ignorance is the man who says of an animal or plant, 'What good is it?'" Ecologist Charles S. Elton ventured in 1958 that the real reason for conservation is that "animals have a right to exist and be left alone, or at any rate that they should not be persecuted or made extinct as species." Robert H. Nelson summed up how environmentalists who justify biological conservation on moral grounds might view things: "To argue for building a dam would be like arguing for the institution of slavery because it was economically efficient."

Intrinsic arguments for the stewardship of biodiversity have found their way into legal philosophy and animal rights legislation. Roderick Nash (1989) made a case for regarding legal rights for nonhumans as a logical continuation of the progressive enfranchisement of different economic classes, races, and genders of people. Christopher D. Stone raised the question of legal rights for nonanimals in his 1972 essay, "Should Trees Have Standing?" [Southern California Law Review 45(2), republished in 1974 by William Kaufmann,

Los Altos]. Some natural lands have been set aside for use by wild animals only. Several areas in the Sierra Nevada of California that have bighorn sheep are closed to human entry during most of the year.

Does it make a difference whether the rationale of stewardship of biodiversity is instrumental or intrinsic? Two ways that it might are if one type of rationale is more likely to withstand arguments against conservation or if the two types lead to different ways of managing. It has been widely argued that only instrumental justifications for biological conservation will lead to political action or stand up to conflicting economic pressures. The plant most often cited as an example of why it is important to conserve biodiversity is the rosy periwinkle of Madagascar. Chemicals extracted from this wild plant were discovered to cure childhood leukemia. Others argue that only a belief in the right of species to exist will motivate stewardship of biodiversity since most species have no demonstrable human use or known ecological significance. One interesting possibility is that it takes intrinsic arguments to generate instrumental ones. Officials of the United Nations Commission on Economic Development interviewed in the early 1990s by P. P. Craig and colleagues confided that although they made only instrumental arguments for environmental conservation in their professional work, they privately believed that the real arguments were intrinsic.

C. Counterarguments against Stewardship as a Model for Conserving Biodiversity

Without necessarily rejecting the idea that it is important to conserve biodiversity, many groups and individuals have rejected the concept of stewardship of biodiversity. Objections have been made to each of the two basic premises of stewardship: that management of biodiversity is needed and justified and that people should care for biodiversity on behalf of interests greater than their own.

One objection to the idea of managing biodiversity is that it is self-defeating because the essence of the biological world is its wildness. In an essay on "the etiquette of freedom," Gary Snyder (as cited in Oelschlaeger, 1992, pp. 21–39) contrasts the meanings of "natural" and "wild." He argues that managed things may be natural but not wild and that wildness is the world itself. Paul Taylor distinguished the realm of "bioculture," in which husbandry and agriculture may take place, from that of wilderness, in which "noninterference ethics" should prevail. George Sessions (as cited in Oelschlaeger, 1992, pp. 90–130) rejects the model of stewardship for wilderness and admits only that "perhaps some ecologically enlightened version of the 'stewardship' model is appropriate for the bioculture."

A second objection to the idea of managing biodiversity is that it is foolhardy because we do not know enough about wild species and ecological systems to know which actions to take. This objection is supported by past management errors. Human suppression of fires in western North American parks and national forests in which natural fires occur regularly has inhibited reproduction of plant species and reduced the diversity of some forests. Alston Chase argued that ecosystem management has altered wildlife processes in Yellowstone National Park.

These objections to the concept of stewardship of biodiversity can be viewed as cautions about what stewards should do rather than reasons to abandon stewardship altogether. Those who argue against managing wildness do not necessarily preclude protecting wilderness areas from development. Those who argue that humans need more knowledge to manage ecological systems successfully may also argue that the best course is to try to acquire this knowledge through "adaptive management," in which the results from each management prescription are analyzed and used to plan the next. Jack Ward Thomas said that "an ecosystem is not more complicated than we think, it is more complicated than we can think." However, as then head of the U.S. Forest Service, he probably did not mean by this that ecosystem management should be abolished but rather approached with humility. Richard L. Knight has likewise proposed that an appreciation of the complexity of nature and a willingness to learn from mistakes and modify techniques are essential components of good stewardship of biodiversity.

A counterobjection to the notion that human management of biodiversity is foolhardy is that humans now have no choice but to manage natural systems because we already dominate them, foolishly or not. Globally, humans are said to consume two-fifths of the earth's photosynthetic productivity and to have doubled the rate at which nitrogen is being supplied to plants (Vitousek et al., 1997). On some islands and some continents, human influences have been judged to now determine the distributions of species and the risks of extinction of animal populations more than natural factors. Many remaining areas of habitat are too small to maintain their biodiversity naturally. In southern Africa, where elephant populations have been largely restricted to movement within reserves, elephants can multiply to the point at which they deforest large areas, leading managers to kill elephants to conserve trees. In other systems, humans have eliminated key elements and may have to replace their actions artificially. Short periods of grazing by livestock have been found to promote biodiversity in interior North American grasslands, possibly because the grazing simulates the effects of the great herds of buffalo that formerly wandered the grasslands.

A second counterobjection to the charge of foolhardiness is that humans are an integral part of some ecological systems. Aboriginal burning of savanna in what is now Kakadu National Park in Australia may have been part of the fire regime for 50,000 years. This is clearly long enough for native species to have evolved in response to human-caused fire frequency.

A third objection to human management of biodiversity is that we have no right to be managers, that we stand in relation to other species, not as stewards, but as siblings. The native American writer Black Elk represented his people's belief that "We are of Earth and belong to You. Every step that we take upon You should be done in a sacred manner." In his book *The Green Fuse*, biologist John Harte explicitly rejected the idea of humans as stewards of ecological systems; it is more accurate, he said, to think of nature as the steward of humanity and to call upon humans to be stewards of their own impulses. "Self-Stewardship" is the title of the chapter that deals with stewardship in Vice President Al Gore's book *Earth in the Balance*.

The second premise for the stewardship of biodiversity is that persons should manage it for the sake of interests greater than their own. Many people reject this in favor of the premise that humans rightfully dominate nature and are the rulers and not the stewards of biodiversity. Captain Vancouver is said to have remarked upon seeing the virgin temperate rainforests of his eponymous island that here were enough masts to supply the British Navy forever. The belief that man should rule over nature figures prominently in Judeo-Christian tradition. The Book of Genesis in the Bible records that humans shall "have dominion over the fish of the sea, and over the fowl of the air, and over every living thing that moveth upon the earth." Confusingly, some recent Christian models called stewardship models still portray humans as dominant over the earth (Sessions as cited in Oelschlaeger, 1992). It should be noted that individual Judeo-Christian theologians have long argued the opposite—that humans do not have dominion over other living things. The medieval Jewish philosopher Moses Maimonides wrote, "It should not be believed that all the beings exist for humanity. On the contrary, all the other beings too have been intended for their own sake and not for the sake of something else." According to the contemporary Catholic theologian Thomas Berry, "The earth belongs to itself and to all the component members of its community." In *The Ecology of Eden*, Evan Eisenberg traces these contradictory theological viewpoints back to the biblical injunction to Adam and Eve to both work and protect the Garden of Eden, that is, to have dominion over the organisms but safeguard the divine creation.

A recent legal embodiment of the belief that people can own other life is the extension of intellectual property rights to cover species or aspects of the genetics or metabolism of a species. This principle, upheld by U.S. courts and by international governmental trade organizations, has been touted as an essential economic incentive for the development of biological and medical technology and as a way to ensure that developing countries and local residents derive benefits from biodiversity. In 1971, the first patent on a genetically engineered organism was granted in the United States to General Electric and its employee, Anand Mohan Chakravarty, for a *Pseudomonas* bacterium into which plasmids from three other bacterial strains had been introduced. The General Agreement on Trade and Tariffs that governs much of world trade contains a Trade Related Intellectual Property Rights agreement that can cover wild species and their products. Even aspects of human biodiversity can become private property. The government of Iceland has debated transferring the intellectual rights to a national database of citizens' genetic information to a corporation.

Throughout history, many societies have held that humans can own individual living things as property, including other humans. Real property rights often include the rights to plants, animals, and minerals on the property. However, it is also usual to recognize that there are multiple interests in real property, that public and private interests are commingled on almost all lands, and that property owners have responsibilities to avoid using their property in ways that harm their neighbors or society. The interests of landowners in the natural resources on their property can be legally balanced against the recreational and environmental interests of the public. On a more moral plane, Aldo Leopold wrote in "The Ecological Conscience" in 1947 that "when a farmer owns a rarity he should feel some obligation as its custodian and a community should feel some obligation to help him carry the economic cost of custodianship."

A final counterargument against stewardship as a model for the conservation of biodiversity is that the most effective basis for conservation is love, which does not involve acting in interests greater than one's own.

To explain this argument, it is useful to return to the example of a guardian and a child. Instead of taking care of a child to produce an adult or out of duty or principle, a guardian might do so out of love for the child. Love is neither an instrumental nor an intrinsic rationale; it is not rational and involves no third party or principle in whose interests one acts. Just as many will argue that love of a child leads to the best child care, Wendell Berry has proposed that "love of place" leads people to care for species and habitats. The land ethic propounded by Aldo Leopold (1966) also contains an element of love: "When we see land as a community to which we belong, we may begin to use it with love and respect" and "affection based on utility alone leads to the same pitfalls and contradictions in land as in people."

One practical implication of this view is that the best way to promote the conservation of biodiversity is to take people to see it. One cannot love without at least first sight. In southern Africa, many programs are dedicated to taking children to see wildlife. Courses for "decision makers" run by the Organization of Tropical Studies in Costa Rica discuss arguments for conservation but also take their prominent students to watch sea turtles lay their eggs in the moonlight on a tropical beach.

III. CONTEXTS

The concept of stewardship has been applied to the conservation of biodiversity in a multitude of forms and with varying success. This is partly due to the ecological, economic, cultural, political, and historic contexts in which stewardship of biodiversity has been undertaken. These contexts include the mobility of animal populations, disequilibrium in ecological systems, economic incentives and disincentives for conservation, poverty and industrialization, the scale of trade, traditional views of the place of humans in nature, legal systems of land and resource tenure, and the displacement or migration of peoples. Some of these contexts for stewardship are briefly discussed in the following sections.

A. Ecological Factors

Stewardship requires that the scale of management match the scale of the thing being managed. In the phrase of Kai Lee, there is often a "mismatch of scales" between ecological processes and human societies. For example, many species and ecological processes cross the jurisdictional boundaries that constrain managers. More than 100 species of Neotropical migratory birds breed in North America and winter in Central or South America. Declines in North American breeding populations have been associated with forest losses in Central and South America. Marine organisms with planktonic larvae often disperse widely on ocean currents and depend on long-distance dispersal to maintain local populations. The establishment of new juveniles in marine reserves may depend almost entirely on dispersal from outside reserves. Aerial nitrogen deposition, which occurs on a global scale (Vitousek et al., 1997), is thought to be responsible for replacement of heathland by grassland in The Netherlands. Due to the "openness" of ecological systems, reserves may be, as Mark H. Carr and colleagues suggest, "necessary but not sufficient for conservation."

Ecological fluxes and connectedness complicate the notions of ownership and management. Many species and the resources on which they depend fall into the category of "common pool resources," which have been defined as "those from which multiple independent users cannot be excluded and in which consumption by one user detracts from another's consumption." These include species outside national boundaries, as in international waters, and species on communally owned lands. Models for the stewardship of biodiversity in commons range from international legal agreements such as the United Nations Convention on Marine Law, which assigns states bordering oceans and seas specific rights and responsibilities for their use, to the customary land tenure systems of African pastoralists.

Land stewardship may be more difficult in habitats in which resources are relatively scarce or where natural conditions are unpredictable or rapidly changing. E. Steen observed that cultures in the relatively dry habitats around the Mediterranean have tended to exhaust soil fertility except where there are reliable pulses of water with low salt and high alluvial content, such as along the Nile. Frequent erosion and deposition of soil maintain plant species diversity in highly dynamic habitats, for example, along rivers and on coastal sand dunes, making it necessary to manage for land instability. The prevailing view of ecological systems has shifted from a notion of dynamic equilibrium, in which systems are expected to tend to reach and return to one steady state, to one of disequilibrium, in which systems are expected to change directionally over time. This makes biodiversity a moving target for management. Some ecological systems switch between alternative states: They may show a nonlinear response to human use, remaining relatively stable until use reaches a certain intensity or cumulative level and then change relatively rapidly, or they may not change back even when use discontinues. This makes it difficult for managers to learn from experience. In habitats in which rare events strongly affect the distribution and abundance of species, managers may need to be "stewards of catastrophe," preserving or mimicking the potential for major disturbance.

B. Economic Factors

Among the many economic factors that impinge on stewardship are disincentives and incentives for conservation, degree of industrialization, and trade across ecological boundaries. When land is taxed based on its most lucrative potential use instead of its current use, landholders are more likely to sell or develop their land. Many laws, such as the Williamson Act in California, have been passed to remove this economic disincentive for farmers. Often, farmers agree not to develop their land in exchange for current use valuation of the land for tax purposes. Laws have also been passed to create economic incentives for stewardship of biodiversity. One widely used instrument is the conservation easement, an agreement under which a landowner agrees not to use his or her lands in certain ways in exchange for relief from taxation. The Federal Agriculture Improvement and Reform Act of 1996 created or redirected a set of U.S. Department of Agriculture programs that provide economic encouragement to private landowners to protect wildlife habitats. The Wildlife Habitat Incentives Program provides cost-sharing for habitat improvement, the Conservation Reserve Program pays farmers to fallow cropland under a vegetation cover for a decade, the Wetlands Reserve Program provides conservation easements and cost-sharing for restoration and protection of wetlands for 30 years or more, and the Environmental Quality Incentive Program combines education and technical and financial assistance to help landowners adopt practices that reduce environmental problems.

Economic conditions in developing countries have been linked to poor stewardship of biodiversity in several ways. Poor people in rural areas are likely to depend heavily on local natural resources, and a large proportion of land area is likely to be settled by farmers. Over time and with population growth, use tends to become unsustainable. Governmental agencies are likely to lack the resources to enforce protection of species in nature reserves. Wood and charcoal are a main source of fuel in cities, leading to deforestation for hundreds of kilometers into the countryside, as around Kinshasa in Congo. On the other hand, it is by no means clear that economic development improves stewardship. In central Africa, "de facto" conservation of biodiverse areas by their remoteness and local economic poverty has probably been as effective as intentional stewardship in the region. One hypothesis is that development will tend to improve stewardship of biodiversity where subsistence resource use is the main cause of its loss, as is probably the case for forests in Madagascar, but not where commercial exploitation is the main cause, as in the forests of Cameroon.

The differences between economic conditions in developing and industrialized countries are often held to require different approaches to stewardship. Conservationists in developing countries largely argue that local rural communities must receive economic benefits from protected areas. This is the basis for the spread of integrated conservation and development projects, which generally link biological conservation within a protected area to human development in the surrounding communities (Wells and Brandon, 1997). These projects are mainly limited to developing countries.

The globalization of commerce has to some extent reversed the mismatch of scales between ecology and society. It is increasingly the case that trade occurs across ecological boundaries. Global trade is often held to threaten stewardship of biodiversity because it makes it more likely that parties using or profiting from resources will live outside the region from which the resources come, and that they will therefore be less concerned about stewardship or constrained by environmental regulations. This is one basis for "bioregionalism," advocated by Sale Kirkpatrick and others, which counsels that people use only the natural resources from within their ecological region.

C. Cultural and Political Factors

Human cultures probably differ in their attitude toward stewardship. For example, D. R. Given argued that the Maori have an environmental ethic that emphasizes guardianship and stewardship, in contrast to a European tradition of dominance over nature. He predicted that laws giving the Maori greater control over resource management would help develop a multicultural, ecocentric biodiversity ethic in New Zealand. Others contend that economic factors tend to supersede cultural traditions. Rodrigo Sierra found no recent differences in impacts between three indigenous and three nonindigenous populations in northwest Ecuador, although there had been differences in the past when economic conditions were different. He concluded that the degree of impact of different peoples on natural systems in this region was primarily associated with economic conditions.

Cultural and political factors can result in "incidental stewardship" of biodiversity, achieved as a by-product of management for other purposes. Some of the most effective stewardship of natural habitat in both the United States and developing countries has been incidental. Lands reserved by the U.S. Army for military bases have become Nature Conservancy preserves and parts of national wildlife refuges and National Park Service units. In the western United States, the military was politically able to completely exclude resource extraction over large tracts of land, some of which now constitute the most pristine examples of semiarid habitats in the region. In some areas of Africa and Asia, small groves held sacred and reserved for religious purposes are among the least disturbed forests.

Local political authority over resource and land use has been associated with good stewardship. Al Gore has proposed that "freedom is a necessary condition for an effective stewardship of the environment" and called to witness the environmental degradation in eastern Europe and hazardous waste sites in poor communities with relatively little political power. Secure land tenure is widely regarded as a precondition for effective land stewardship by individuals and communities. In African countries, governmental ownership of wildlife on private or communal land has been blamed for lack of interest in wildlife conservation.

However, a given political arrangement may have opposite effects on stewardship in different economic contexts. The establishment of wildlife management agencies or funds as "parastatals"—governmental bodies that are economically independent of the rest of the government—has promoted wildlife conservation in Kenya and Zambia. In these countries, tourism in some parks and game management areas was generating more revenues for the central treasury than the treasury was providing to the parks largely because the tourists were almost all foreign and could afford to pay large amounts relative to the local economies. Once given political authority to raise and disburse its own revenues, the Kenya Wildlife Service was able to retain gate receipts at Amboseli National Park and share them with the Maasai around the park. During the wet season, the large ungulates in the Amboseli system rely largely on the Maasai lands for grazing, making the cooperation of the Maasai essential for conservation of biodiversity in the park. In Zambia, governmental revenues from European and North American sports hunters were channeled to the park and wildlife service for management, providing hundreds of jobs for local residents in game management areas. In contrast, requiring parks to be economically self-sufficient in the United States, in which the national park service receives more governmental revenues than it generates, would probably lead to a decrease in resources for park management.

IV. TYPES OF STEWARDS

Almost every possible type of social or political group has undertaken, or purported to undertake, the management of species and habitats on behalf of interests greater than those of the group. The possible stewards of biodiversity include governmental agencies, nongovernmental organizations (NGOs), corporations, local communities, and private individuals. Governmental agencies and NGOs include international, national, and local bodies. They have worked separately and in partnerships of different types of groups. Through legal and political action, classes of citizens have also acted as stewards. The following sections provide examples of these types of stewards and discuss some of their apparent strengths and weaknesses: Are certain types of stewards more effective or trustworthy as guardians of biodiversity?

A. Governmental Stewards

International governmental agencies exercise stewardship of nature in the form of conventions, declarations, and moral and financial support for conservation efforts. Through trade agreements and support for development, international bodies have both negative and positive effects on biodiversity. International governmental bodies are uniquely able to enforce stewardship of global ecology such as upper atmospheric conditions and of common pool resources such as marine fisheries and the Antarctic continent. They can finance and lobby for the protection of local sites that are internationally prized.

The most visible international governmental steward of biodiversity has been the United Nations (UN). In 1982, its General Assembly adopted a Charter for Nature that recognized an intrinsic basis for the stewardship of biodiversity:

> Every form of life is unique, warranting respect regardless of its worth to man, and to accord other organisms such recognition, man must be guided by a moral code of action. . . . Nature shall be

respected and its essential processes shall not be disrupted.

The vote was 114 for, 17 abstaining, and 1 (the United States), opposed. UNESCO, an organization within the UN, designates areas of special cultural and conservation interest as Man and the Biosphere Reserves and World Heritage Sites. Most Biosphere Reserves center on existing protected areas, and designation as a reserve is largely symbolic. However, it may have an important educational value by affirming the preciousness of natural places and help inspire support from other sources. The National Natural Landmark program of the U.S. National Park Service is an example of a similar program on a national scale. The United Nations Environmental Programme directly funds conservation efforts and projects.

The "Montreal Protocol" is often cited as a successful example of international governmental cooperation to protect the environment. This agreement banned most release of chlorofluorocarbons into the air. These chemicals can lower the concentration of ozone in the upper atmosphere and increase the amount of cell-damaging ultraviolet radiation that reaches the earth's surface. Important scientific analysis to support this political agreement was supplied by an international nongovernmental body, the Scientific Committee on Problems of the Environment of the International Council of Scientific Unions. This is an association of professional scientific associations such as the American Association for the Advancement of Science.

International governmental trade and financial institutions such as the World Bank do not have ecological stewardship as their primary goal. The World Bank has supported projects agreed to have been environmentally harmful, notably the Aswan Dam in Egypt. However, World Bank policies formulated during the 1980s explicitly called for conservation of tropical forests. The bank set up an internal environmental review process for its projects and funded the Global Environmental Facility, which gave unusually large amounts of money to conservation projects throughout the world. Some international trade agreements, such as the Convention on International Trade in Endangered Species, have been directed at conservation of biodiversity. On the other hand, international trade agreements such as the North American Free Trade Agreement and organizations such as the World Trade Organization have been severely criticized for failing to include adequate environmental protections.

Although dedicated primarily to promoting human development, national overseas aid agencies such as the Department for International Development of the United Kingdom or the Gesellschaft für Technische Zusammenarbeiten of Germany have been funding conservation efforts for more than a decade. In the mid-1980s, the U.S. Agency for International Development (USAID) received a specific mandate from the U.S. Congress to conserve biodiversity and tropical forests. As stewards, foreign aid agencies have the advantages of considerable independence from local pressures and enormous buying power in foreign countries. Some of their disadvantages are political inability to provide long-term support for projects or to manage them directly. For example, USAID funds rather than manages projects and expects them to become self-sufficient within 5–10 years.

National governmental agencies that manage natural resources include park, wildlife, forest, fisheries, land management, environmental, and tourism ministries and services. These agencies are stewards inasmuch as they manage the resources on behalf of groups outside the agency. The U.S. Fish and Wildlife Service refers to the National Wildlife Refuges it administers as "stewardship lands." However, the primary charge of these agencies is often to maximize the economic returns from resource use rather than to conserve biodiversity. Georgia M. Mace characterized the role of natural resource managers in governmental agencies as generally aiming to maximize continuing yields and collect extensive data on a few species rather than assessing management risks to a wide diversity of species. This would tend to make them poor stewards of biodiversity. For example, maximum productivity is generally negatively associated with maximum species richness. On the other hand, governmental agencies have significant advantages as land stewards. They often command relatively extensive resources and manage a large proportion of the natural lands in a country, including most of the large natural areas. Depending on how their professional incentives are structured, managers within agencies may be largely free from personal conflicts of interest with conservation.

Many governmental resource agencies emphasize resource use over biodiversity conservation for both economic and political reasons. In many countries, including the United States, the operating budgets of national forests are tied to the fees paid by companies for logging concessions. This provides a powerful incentive to manage forests for wood production and to focus on a few economically valuable species. The charter of the U.S. National Park Service calls upon it to protect parks for the enjoyment of visitors. Parks experience pressure to manage for maximum continuing yields of visitors and therefore to develop lodging and shops and to focus on species that are appealing and easy to see, the "charismatic megafauna."

Research institutions and research departments within resource management agencies operate natural reserves throughout the world. The approximately 30 reserves of the University of California comprise the largest system of any single university. The U.S. Forest Service has designated hundreds of Natural Research Areas, often prime examples of distinctive forest types, that are open to research use only. One goal of the sites funded under the Long-Term Ecological Research program of the U.S. National Science Foundation is to provide baseline data on ecological function and community structure in diverse habitat types that are as little disturbed by humans as possible. An advantage of research reserves is that they are generally completely protected from consumptive use and largely protected from tourism. However, the reserves are mostly small and subject to experimental manipulations such as animal exclosures and vegetation removal and nutrient addition in plots.

B. Nongovernmental Organizations and Communities as Stewards

Conservation land trusts, NGOs that purchase land or the development rights to land for the express purpose of maintaining it in an undeveloped state, are likely to have the conservation of species or habitats as their primary goal. Another advantage of land trusts relative to governmental agencies seems to be that they are able to act quickly to acquire habitat in immediate threat of development. The Nature Conservancy is the largest private land trust for natural lands in the United States, with more than 10 million acres of reserves throughout the country. One of its mottos, "the last of the least and the best of the rest," epitomizes its dual objectives to prevent the extinction of rare species and to conserve whole natural communities and ecological systems. The first objective requires a large number of reserves, some of which may be very small, for example, if the species of concern is a grass. The second requires large reserves, which have to be fewer in number for reasons of cost and availability. Local land trusts have also played important stewardship roles. Starting in the early 1900s, the Save-the-Redwoods League purchased a significant proportion of the remaining old-growth forests of the coast redwood, *Sequoia sempervirens*, which the trust eventually deeded to the government for parks. A disadvantage of land trusts as stewards is that they are rarely able to purchase large areas and generally control only a small proportion of the natural lands in an area. According to the Land Trust Alliance, local land trusts owned about 0.002% of land in the United States in 1999.

A highly controversial type of steward of biodiversity is the local community, the set of human residents in a place. "Community-based conservation" is an important option in societies in which a large proportion of lands are owned or customarily used by communities as a group. It is viewed as an alternative both to the establishment of protected areas that exclude consumptive uses of natural resources by local residents and to natural resources management by governmental agencies. During the 1980s and 1990s, many projects were established to encourage local communities to assume or resume the major responsibility for local environmental stewardship, especially in the developing countries of Africa and Asia. Sets of these projects have been reported on in books edited by Western *et al.* (1994), Mark Poffenburger, and others and in networking organs such as the Forest, Trees and People Newsletter (*www-trees.slu.se*). Two of the best known examples in Africa, the CAMPFIRE project in Zimbabwe and the ADMADE project in Zambia, have given varying degrees of authority for local wildlife management to communities. Community-based forest management has become particularly widespread in India. In the late 1990s, the Van Gujjars, a forest-dwelling pastoral tribe in Uttar Pradesh, proposed to resolve their objections to the establishment of the proposed Rajaji National Park by taking responsibility for forest management in the park.

Reasons cited for turning to community-based conservation are that it is morally right, that local people have expert, "indigenous knowledge" of local species and habitats, that local people have lived in harmony with local nature for centuries, that traditional indigenous populations use a wide variety of wild species and thus have a practical incentive to conserve biodiversity, and that management by governmental agencies or in nature preserves in the area has failed to prevent habitat degradation. The Dayak groups of central Kalimantan in Borneo are reported by W. de Jong and others to provide successful examples of ethnoconservation; the Dayaks have manipulated the forest to increase productivity of species they use without reducing species diversity or simplifying structure.

One of the main reasons for doubting the efficacy of communities as stewards is the world history of environmental degradation. O. M'Hirit concluded that peoples around the Mediterranean have generally harmed their local forests and that only in the twentieth century has there been effective forest conservation in the regions. In 1957, A. Starker Leopold took a dark view:

> It is surprising that in the long history of man's conquest of the earth there is no evidence of sustained effort on the part of any people to preserve native landscape for its own sake, until our national park system began to take form late in the nineteenth century.

The consensus at the end of the 1990s was that local communities sometimes do and sometimes do not make effective stewards of biodiversity. Communities of recent immigrants in tropical rainforest tend to cause extensive deforestation because they want cleared land for crops and livestock. Even long-settled, traditional societies do not always use sustainable practices; a recent study concluded that hunters from the Piro people in the Peruvian Amazon make no attempt to conserve species vulnerable to overhunting. Practices may be sustainable only as long as human population density remains low and may change when economic and social conditions change. In a 1999 essay on community-based conservation projects in Africa, J. D. Hackel concluded that they tended to lack clear rules for the protection of local ecology and plans for what to do if conservation goals were not met. In both developing and industrialized countries, local communities may have a limited capacity to ask key questions about conservation values or to provide potential answers and limited time or will to bring issues before the public and decision makers. Others believe that community-based conservation has rarely been properly tested because governments have failed to devolve the authority over resources to communities at the same time that they have decentralized the responsibilities for resource conservation.

A different definition of a community is the set of people that have interests in a thing. In this sense, class actions by citizen groups and lobbying and public education by conservation organizations on behalf of their members are also forms of community-based conservation. The political and legal systems in the United States provide a direct route for stewardship by citizen groups through referenda and class action suits. An example of stewardship by class action is the work in the 1990s of the Mono Lake Committee in California to prevent Los Angeles County from diverting water from the streams that flow into this large desert lake. Diversion threatened, among other things, to allow coyotes to walk out to the nesting sites of birds on islands in the lake. The Sierra Club is credited with having dissuaded the U.S. federal government from creating a reservoir in the Grand Canyon in Arizona. The club's public advertising campaign asked "Would you flood the Sistine Chapel to get a better view of the ceiling?" Duck hunters are said to be largely responsible for motivating governmental protection of wetlands, which are now, under the RAMSAR convention, the first globally protected habitat type.

Such "community-of-interest-based conservation" can have the advantage of a longer economic view but the disadvantage of pitching legal and sometimes corporeal battles. The controversy over protection of old-growth Douglas fir forests in the Pacific Northwest, in the name of the northern spotted owl, pitted local logging interests against national conservation ones. According to one estimate, the protection plan that was adopted should result in economic gains of about $110 billion versus losses of only $32.5 billion. However, the losses are concentrated within the region, whereas the gains are spread throughout the country.

To avoid such battles, a broader community-based approach to conservation has been advocated based on the notion of "stakeholder rights." Stakeholders are all parties with interests of any sort in a given thing, be they economic, political, cultural, moral, or personal. One formulation of this "community-of-different-interests-based conservation" is to provide a forum for stakeholders to produce a common management plan for an area. For instance, the town library in Quincy, California, became a neutral meeting place for local environmentalists and loggers, who produced a plan for an adjacent area of national forest. To the surprise of many outside the "Quincy Library Group," the plan called for relatively intensive harvests, and its adoption by the Forest Service was challenged in court. Truman Young argued that many major conservation successes of recent decades, such as the international bans on whaling and ivory, have essentially rejected current economic stakeholder rights.

C. Private Companies and Individuals as Stewards

Private companies have also been engaged as stewards. One strategy has been to set up certification programs for commercial products or services. Companies or their taxes pay an independent body to certify that a product has been produced in an environmentally sound way. The expected benefit to the company is that environmentally concerned consumers will buy the product. A well-known certification program is run by the Forest Stewardship Council, an international NGO founded in

1993 "to support environmentally appropriate, socially beneficial, and economically viable management of the world's forests." The council accredits and monitors certification bodies for forest products. Certified products may carry the council's logo. A newer, Marine Stewardship Council aims to perform an analogous function for marine fisheries.

A second strategy has been to encourage private companies to support in-house conservation projects. The nonprofit Wildlife Habitat Council coordinates a program that enlists corporations to voluntarily manage for wildlife and biodiversity protection. Participants account for about 120,000 ha in the United States. In a survey of benefits to corporations at 164 sites, as reported by H. Cardskadden and D. J. Lober, most respondents believed that the program had improved employee morale (95% of respondents) and relations with environmental groups (72%) and the local community (64%). Half of the respondents reported that programs had resulted in annual cost savings, such as through waste reduction. In the absence of scientific monitoring, there was no indication of whether there had been any benefits to wildlife or biodiversity. This is a common problem in many nongovernmental stewardship programs.

By far the most common current use of stewardship in relation to conservation is for programs in which individual private landowners serve as stewards on their own lands, often with economic incentives, technical assistance, or just moral support provided by government or NGOs. The Saskatchewan Wetland Conservation Corporation provides land management advice and a certificate of appreciation to owners of native prairies who make a Voluntary Stewardship Agreement to protect their land from development. The Massachusetts Forest Stewardship Program, run by the state government, shares the costs of habitat improvement for wildlife and rare native fish with private owners of nonindustrial woodlands of 10–1000 acres. The Georgia Endangered Plant Stewardship Network of the State Botanical Garden of Georgia trains school classes to propagate rare and threatened wild plants. The Pesticide Environmental Stewardship Program of the U.S. Environmental Protection Agency helps pesticide users to write and implement strategies for pesticide risk reduction.

One reason for appealing to private landowners to serve as stewards of their property in the United States is that development of private land is a significant cause of species decline. In 1998, the Wilderness Society estimated that conversion of private land for commercial and residential development was at least partly responsible for population losses in 35% of the 1880 imperiled plants and animals in the United States. According to Richard L. Knight, "unchecked conversion of U.S. private, open lands to human-dominated development is causing a simplification of our native biodiversity." In some regions, such as northern New England in the United States, most of the lands with natural or seminatural vegetation are privately owned.

An objection to relying on private landowners as stewards is that private landholdings are generally small compared to public ones and cannot accommodate large-scale ecological needs. Remote imaging of land use has documented that habitat patches may tend to be smaller on private than on public lands. This objective can in some cases be partly met by cooperation between landholders. In about 1986, local landowners in southern Vermont formed the Newfane Wildlife Habitat Improvement Group to plan together for wildlife management. In 1999, the group included more than 50 properties and 7000 acres in three towns. However, the total amount of land in the United States under the protection of voluntary agreements by private landowners is also small—just 0.02% of total land according to a 1999 article by Keith Wiebe and colleagues.

As the previous examples of stewardship on private lands suggest, stewardship is often undertaken through partnerships between public agencies and private parties. This is just one sort of "cross-boundary stewardship"—coordinated land management by multiple stewards, often of different types (Knight and Landres, 1998). Many integrated conservation and development projects are organized as partnerships among governmental agencies, NGOs, and local communities (Wells and Brandon, 1997).

A major advantage of cross-boundary stewardship is that is allows for land use to be "not separate but not equal," to have a diversified portfolio of stewardship types arranged side by side within ecological regions. Arguments for zoning as an alternative to multiple use are not new (Sessions as cited in Oelschlaeger, 1992, pp. 90–130). In 1971, Eugene Odum and John Phillips proposed a landscape of urban-industrial, production, and protection zones. The Biosphere Reserve concept promoted by UNESCO since 1980 envisions protected areas surrounded by an inner zone of human use unlikely to impact the protected core and an outer zone of unrestricted use. Some species do require protected areas in which there is virtually no human impact. A recent analysis of conservation versus human land use by Kent Redford and B. Richter indicated that only extremely limited and largely nonextractive use will protect all components of biodiversity. However, as Sara

Vickerman argued, biodiversity cannot be conserved on reserves and with regulation alone (Defenders of Wildlife, 1998). A managed landscape can support important elements of biodiversity while meeting human needs. Significant components of biodiversity can and must be under the stewardship of private landowners.

The relative effectiveness of different types of stewards in conserving biodiversity seems to increase with the total amounts and individual sizes of the land areas they manage, the economic and technical resources they command, their independence from conflicting economic pressures and political constraints, the degree of authority they can exercise, and how close their primary objective is to conserving biodiversity. No one type of steward has greatest effectiveness in all these regards. Since it is also true in many landscapes that no one type of steward is responsible for enough land to ensure the conservation of biodiversity, it appears that a mix of stewardship types, arranged side by side in a landscape, is necessary and desirable.

See Also the Following Articles

COMMONS, THEORY AND CONCEPT OF • ECOSYSTEM SERVICES, CONCEPT OF • ETHICAL ISSUES IN BIODIVERSITY PROTECTION • HISTORICAL AWARENESS OF BIODIVERSITY • LITERARY PERSPECTIVES ON BIODIVERSITY • RELIGIOUS TRADITIONS AND BIODIVERSITY • SOCIAL AND CULTURAL FACTORS • SUSTAINABILITY, CONCEPT AND PRACTICE OF • WILDLIFE MANAGEMENT

Bibliography

Defenders of Wildlife (1998). *National Stewardship Incentives: Conservation Strategies for U.S. Landowners*. Defenders of Wildlife, Washington, D.C.

Johnson, N. C., Malk, A. J., Sexton, W. T., and Szaro, R. C. (Eds.) (1999). *Ecological Stewardship: A Common Reference for Ecosystem Management*. Elsevier, New York.

Knight, R., and Landres, P. (Eds.) (1998). *Stewardship Across Boundaries*. Island Press, Washington, D.C.

Leopold, A. (1966). *A Sand County Almanac with Essays on Conservation from Round River*. Oxford Univ. Press, New York.

Nash, R. F. (1989). *The Rights of Nature: A History of Environmental Ethics*. Univ. of Wisconsin Press, Madison.

Oelschlaeger, M. (Ed.) (1992). *The Wilderness Condition: Essays on Environment and Civilization*. Sierra Club Books, San Francisco.

Vitousek, P. K., Mooney, H. A., Lubchenco, J., and Melillo, J. M. (1997). Human domination of earth's ecosystems. *Science* 277, 494–499.

Wells, M., and Brandon, K. (1997). *People and Parks: Linking Protected Area Management with Local Communities*. World Bank, World Wildlife Fund, and Agency for International Development, Washington, D.C.

Western, D., Wright, M., and Strum, S. C. (Eds.) (1994). Natural connections: Perspectives in community-based conservation. Island Press, Washington, D.C.

Wetlesen, J. (1999). The moral status of beings who are not persons: A casuistic argument. *Environ. Values* 8, 287–323.

STORAGE, ECOLOGY OF

Caroline M. Pond
The Open University

I. What Can Be Stored?
II. Storage in Invertebrates
III. Plants
IV. Vertebrate Adipose Tissue
V. Adipose Tissue Function
VI. Energy Storage in the Life History

GLOSSARY

adipocytes Large cells unique to vertebrates that take up and release fatty acids, produce receptors and respond to circulating and locally produced agonists, and secrete a variety of protein and lipid informational molecules.

adipose tissue, brown Thermogenic tissue unique to mammals that consists of small adipocytes containing numerous mitochondria and many small lipid droplets plus vascular and neural tissue. Thermogenesis is controlled by the degree of uncoupling of ATP synthesis in the mitochondria and can be very high in neonates and adults emerging from hibernation.

adipose tissue, white Storage tissue unique to vertebrates and best developed in tetrapods that consists almost entirely of expandable adipocytes (q.v.) that contain a large droplet of lipid plus vascular and neural tissue.

blubber Specialized superficial adipose tissue found only in pinnipeds, cetaceans, and sirenians that serves as thermal insulation as well as providing energy storage.

fat body Storage tissue of arthropods that stores lipids and also has endocrine and immunological functions.

fatty acid Any of hundreds of different aliphatic hydrocarbons with an acid group. The main differences are the number and position of one or more double bonds and the presence of substitutions.

glycogen An insoluble polymer of glucose; the main storage carbohydrate in animals.

polyunsaturated fatty acid Fatty acid with two or more double bonds in the chain of carbon atoms. They occur in distinct families defined by the position of the double bonds, of which the most common are n-3 and n-6.

saturated fatty acid A fatty acid with the maximum complement of hydrogen atoms and no double bonds.

storage organ Any structure that sequesters storage materials (usually the carbohydrates, starch or glycogen, or triacylglycerol) for export to other tissues that utilize it.

triacylglycerol Ester of glycerol and three fatty acids; the most common lipid storage molecule in animals and green plants.

MOST ORGANISMS STORE LIPIDS and/or carbohydrates for energy production when food is unobtainable.

Carbohydrates are bulky but easily synthesized and transported. Much more energy per unit mass can be stored as lipids, and because they are less dense than water they aid buoyancy; however, long-chain fatty acids are insoluble and must be transported as lipoproteins. Vertebrates and higher arthropods have tissues specialized for storage and management of lipids for provisioning the rest of the body and for gamete formation. Adipose tissue can comprise 50% of the body mass before migration or breeding fasts with superficial depots expanding the most, especially in large animals. Up to 85% of adipocyte volume can be triacylglycerols and occur in various intra-abdominal, intermuscular, or superficial sites in all tetrapods and some fish; their histological appearance is similar but biochemical studies reveal that some depots have site-specific properties consistent with specialized local functions.

I. WHAT CAN BE STORED?

Few organisms have continuous access to nutrients derived from feeding or generated by photosynthesis; therefore, some storage is necessary. The role requires molecules that are insoluble, compact, and nontoxic to the holding cell, even when present in large quantities. Thus, oxygen, which is only slightly soluble in water, is stored in substantial quantities only when bound to special pigments in blood or muscle. Vertebrates and a few invertebrates and micro-organisms can hold oxygen bound to the iron-based pigment hemoglobin. Vertebrate cardiac and skeletal muscles contain the closely related oxygen-binding molecule myoglobin. However, even specialized diving animals such as seals and whales, whose muscles are dark red because of the high concentration of myoglobin, can store enough oxygen to support strenuous activity for at most about 1 h. Stores held by sedentary invertebrates and microorganisms may last longer, but storage capacity for oxygen is still very minor compared to that of other nutrients.

Many vitamins and essential minerals, including iron, are too toxic to be stored in more than small quantities, and then only when bound to specialized proteins, usually in the liver. In many kinds of animals, the skeleton can serve as a store of calcium and phosphate; in mammals, bone is laid down before breeding and withdrawn during lactation as the minerals passing to the offspring in the milk. However, the quantities thus stored are limited by the extra weight of the dense minerals and the need to maintain normal functioning of the mother's skeleton. The materials stored by the widest variety of organisms and in the largest quantities are those that can be broken down to provide metabolic energy.

Energy-providing molecules that animals can safely store in large quantities are the complex carbohydrates, glycogen, and triacylglycerols—esters of glycerol and three fatty acids. The complex carbohydrates generally require less energy for synthesis but they are denser and less compact than lipids because they are associated with large quantities of water. Lipid molecules have such high affinity for each other that they form almost pure droplets that are significantly less dense than other cell components, but their synthesis and reclamation are biochemically quite complex and usually slower than those of carbohydrates. Although carbohydrates (and the glycerol component of lipids) can be broken down by aerobic or anaerobic metabolism, long-chain fatty acids must be oxidized aerobically; therefore, storage lipids can only fuel prolonged, strenuous activities in animals in which the uptake and transport of oxygen are highly efficient.

Most animals can synthesize glucose from amino acids released by the breakdown of proteins but the process is slow and wasteful (because the amino groups have to be excreted). Small amounts of protein and free amino acids, particularly glutamine, may exist in the liver only as a storage material for use as precursors for the synthesis of proteins and nucleic acids as well as for glucose production, but most protein utilization entails "wasting" of lean tissue. The associated loss of function is particularly obvious when muscle proteins are broken down to make glucose, but protein depletion may also impair the immune system and digestive tract. These deleterious effects mean that, especially in animals that have dedicated storage tissues holding reserves of carbohydrate or lipid, proteins are usually only reclaimed for energy production when the other storage materials are nearly exhausted.

II. STORAGE IN INVERTEBRATES

Many invertebrates are capable of surviving long periods without food, but there is little detailed information about how these fasts are sustained. Many unicellular organisms and lower invertebrates have intracellular storage molecules and organelles that contain sufficient energy-generating material to sustain them for hours or days. In prolonged starvation, the body fabric, including structural proteins, is broken down to produce metabolic energy. The first to be reabsorbed are usually reproductive tissues, early embryos as well as gametes, but at least in less complex invertebrates, skeletal, locomotory, and digestive tissues can all be utilized if necessary. Some animals can shrink to an impressive extent while remaining capable of normal life functions and can rebuild themselves when food is again available. For example, after 4 months of starvation, the medusae of the hydrozoan jellyfish *Aurelia aurita* shrink to less than one-fourth of their original diameter but they continue to swim and can regrow if adequate feeding is restored. The ability to retain more or less normal functioning after prolonged starvation may be one of the advantages of the simple cnidarian body plan. More complex animals that have hard skeletons cannot lose as much body fabric without serious impairment of function.

Vertebrates and higher invertebrates have specialized storage tissues that sequester energy-providing molecules at much higher concentrations than is normally possible when materials are stored in cells with other primary functions. Annelids, especially oligochaetes, have variable quantities of chloragogenous tissue (so called because cell inclusions confer a green or yellow color) in the coelom around the intestinal wall. It has several functions, including that of storing lipids and glycogen, and may be ancestral to the arthropod "fat body" or "hepatopancreas" as it is known in crustaceans. The insect fat body is the most thoroughly studied invertebrate storage organ. It is present in larval and adult insects, and although originating in the abdomen it can grow to be massive, extending into the thorax in which it surrounds the major locomotory muscles. The stored lipids confer a white or yellow color when replete, as in locusts before migration and many other species before metamorphosis, but the proportion of lipid is rarely as high as it can be in adipose tissue of laboratory rodents. Storage lipids reach the muscles and other tissues as diacylglycerols or triacylglycerols in lipoproteins called lipophorins.

Several hormones involved in energy metabolism and activity were shown to be secreted from the fat body in the 1950s and 1960s, leading to the suggestion that the insect fat body had gland-like functions comparable to those of the vertebrate liver. Recent interest in invertebrate immunity has revealed that it also contributes to innate immunity: When stimulated by bacterial extracts, *Drosophila* fat body cells produce at least seven antimicrobial peptides. Like vertebrate adipose tissue, invertebrate storage organs are much more than just repositories; they play an active role in energy metabolism and defense against disease.

III. PLANTS

The aerial structures of many temperate-zone and arctic plants die back in winter when cold or lack of light make photosynthesis inefficient. Regrowth in spring is supported by storage materials distributed through roots and surviving parts of the stem, which may be enlarged to form corms or tubers in some species. These structures are not severely limited in volume and density, so the storage material is nearly always starch with much associated water. Only a few higher plants, notably sugar beet (*Beta vulgaris*), store large quantities of disaccharides or monosaccharides. Such concentrated nutrients are often protected from herbivores by stout casings or by the presence of toxic secondary compounds. Removing or inactivating these plant defenses by cooking (e.g., potatoes, carrots, and onions) or by leaching (e.g., cassava) was among the major technological advances of prehistoric humans and greatly increased the range and nutritional quality of plant foods available to them.

A. Energy Storage in Seeds

The main function of energy stores in seeds is to supply the germinating seedling until it grows its own leaves and can photosynthesize on its own. The huge range of sizes of seeds is due mainly to the chemical composition and abundance of their storage material. The storage materials of the seeds of plants native to cool or cold climates, such as oats, rye, acorns, beechnuts, and hazelnuts, are usually starches, as are those of grasses such as rice and wheat. Seed starches generally contain less water and therefore are more compact than those found in vegetative storage tissues. Like these storage tissues, seeds are often protected from herbivory by toxins. The need to compress much energy into a small space makes synthesizing triacylglycerols from primary photosynthetic products worthwhile, and the storage materials of many large, longer-lasting seeds that are dispersed by animals are lipids or mixtures of carbohydrates and lipids.

Green plants have a special biochemical mechanism, not known to be present in animals other than a few kinds of parasitic worms, that permits the incorporation of fatty acid carbon atoms into precursors for the synthesis of glucose, which is the starting point for the formation of the structural carbohydrates that build the vegetative structures that enable them to synthesize sugars in sunlight. By converting the fatty acid components of triacylglycerols into carbohydrate, plants can use seed lipids in ways that animals cannot use their lipid stores. Because they can be used in this way, instead of being oxidized in mitochondria or incorporated into phospholipids, seed lipids contain a much wider variety of fatty acids than do animal triacylglycerols.

The triacylglycerols in seeds are always dispersed into tiny compartments called oil bodies, each only a few micrometers (thousandths of a millimeter) across, that are enclosed in membranes. At least some of these membranes contain lipases that enable plant storage lipids to be mobilized quickly. Such structures contrast with the (relatively) huge droplet of triacylglycerols in vertebrate adipocytes that may be more than 100 μm (0.1 mm) in diameter or up to 1 million times the volume of oil bodies in seeds. This arrangement explains why nuts do not feel or taste as "greasy" as do vertebrate adipose tissue and purified oils even though they may be more than 50% lipid. Mobilization of seed storage materials begins at germination with the uptake of water that swells the proteins and starches. Preformed enzymes are activated and new ones synthesized, but, as in the case of animal systems, water-based enzymes cannot easily attack large droplets of lipid.

B. Chemical Composition of Plant Lipids

To take advantage of a short growing season, the seeds germinate early in the spring while the soil is still cold. The temperature at which major physiological processes occur seems to be the most important determinant of the composition of plant lipids. Most biochemical reactions occur in solution so the reactant molecules must be dissolved for metabolic processes to proceed. Unsaturated fatty acids make phospholipid membranes more fluid, so other molecules can move within and across them more readily, and also lower the freezing temperature of triacylglycerols. Complete breakdown of unsaturated fatty acids releases slightly less metabolic energy (about 1 or 2% less for each double bond) than the saturated equivalents with the same numbers of carbon atoms. Triacylglycerols containing saturated fatty acids also pack more neatly (i.e., they have higher melting points) than those containing mono- and polyunsaturates, and homogeneous assemblages solidify at higher temperatures than do mixtures. Therefore, in general, more energy can be stored in a small space by means of saturated triacylglycerols, the more homogeneous, the better.

All the plants in which saturated fatty acids are abundant as storage lipids are native to the warm, equable climates of tropical lowlands. Plants such as coconut, oil palm, nutmeg, and cacao, which live only in hot climates, do not need polyunsaturated fatty acids to keep their storage lipids fluid: They have the luxury of being able to maximize the energy obtainable from the smallest volume of storage organ because they have triacylglycerols that consist mostly of saturated fatty acids. Those of the seeds of the cinnamon tree, for example, are up to 95% lauric acid (C12:0), although the fatty acids in the lipids of the fruits of the same species and of its relative, the avocado pear, are approximately 50% oleic acid and 25% palmitic acid, with very few medium-chain saturates—a typical composition for an oily rather than a fatty fruit. Large seeds such as oil palms that contain saturated triacylglycerols provide their seedlings with supplies that last weeks or months.

Plant oils rich in polyunsaturated fatty acids are found in species adapted to temperate climates. Olive trees, with fruits and seeds containing mostly monounsaturated fatty acids, and nuts such as walnuts and pistachios, which contain both storage lipids and starch, grow in cooler climates than that required by oil palm or cacao but are still not worth cultivating for their oil outside areas that have long, hot summers. Lipids containing mixtures of different kinds of fatty acids remain liquid at lower temperatures than do those in which all the fatty acids are similar; therefore, those of nearly all temperate-zone and polar species are mixtures of many different fatty acids. However, several possible mixtures of fatty acids in triacylglycerols may have similar physical properties and each plant may be adapted to germinate under slightly different conditions, so the fatty acid composition of seed lipids may differ between species growing in similar climates. Some species can make major alterations to the composition of their seed lipids according to the climate in which they are growing. The fatty acids of the seeds of the sunflower (*Helianthus annuus*) are 44–72% linoleic acid (18:2n-6), with the remainder being mostly oleic acid (C18:1). Those growing in the coolest conditions contain the most linoleic acid, but their total yield is smaller and the crop takes longer to mature.

The seeds of the jojoba bush (*Simmondsia chinensis*) are almost unique in that the storage material is a wax and not a triacylglycerol. Jojoba oil is liquid at normal temperatures (in contrast to beeswax, which is used by bees as a solid) because both the fatty acid and the alcohol components of the wax are long-chain molecules. The fatty acids are mostly (approximately 74%) C20:1, with smaller quantities of C22:1 and oleic acid (C18:1), and the alcohols also contain 20 or 22 carbons. Although the proportion of lipid in the seeds varies widely between strains and according to the conditions under which the plant is grown, the fatty acid composition of the wax, in contrast to that of triacylglycerols, is remarkably constant. Jojoba oil, like other waxes, is indigestible to most animals but the plant is cultivated widely in California and other warm, semidesert areas of the United States, where the extracts are used as lubricants and in cosmetics. Jojoba oil would also be suitable as fuel, so the crop has been proposed as a renewable substitute for gasoline.

C. Protection from Seed Predators

The lipids and starches concentrated in grains, nuts, and seeds provide a rich source of food for animals including humans. The partitioning of seed lipids into tiny oil bodies makes them very digestible to animals because less mechanical emulsification is required. Seed-eating mammals such as rodents gnaw or crush seeds with their hard, continuously growing teeth, whereas parrots, finches, and other seed-eating birds use their powerful, finely controlled beaks to shred nuts and seeds into tiny fragments before swallowing them. Nuts are digestible, highly nutritious food that is especially suitable for flying birds, which cannot afford to carry around heavy gut contents for long periods of slow digestion.

Seeds such as walnuts, almonds, cacao (chocolate) beans, and coconuts protect their nutritious contents with a hard shell (in many species, the nutshell is much harder than the tree's wood) that only the most powerful beaks or persistent gnawing teeth can open. Other nuts, such as peanuts, form underground, avoiding seed predators that cannot dig to reach them. Small seeds such as those of wild strains of flax, sesame, and sunflower scatter as they fall, making it difficult for animals to collect them, and their coats are very tough relative to their size. Many plants have evolved chemical and mechanical means of deterring predation on their seeds.

Some seeds are rendered poisonous to seed predators by the presence of toxic amino acids, proteins, and alkaloids and/or because the storage materials are poisonous to animals. The beans of the castor oil plant, *Ricinus communis*, are protected from herbivory by the poisonous protein ricin and by the fatty acid ricinoleic acid, which can account for more than 90% of the triacylglycerol fatty acids. Animal lipases cannot easily hydrolyze triacylglycerols containing ricinoleic acid, so most pass undigested through the gut. Eaten in small quantities, castor oil acts as a lubricant and slight irritant, thus relieving constipation in people and dogs and cats. Oil from unimproved oilseed rape (and turnips, swedes, and cabbages and other Brassicaceae) contains more than 40% erucic acid (C22:1), which renders the seeds toxic to weevils and other seed-eating insects as well as to vertebrate seed predators such as voles, mice, and finches. In the West, rapeseed oil (known as canola oil) is produced from plants selectively bred, and recently genetically engineered, to form seed triacylglycerols that contain as little as 5% erucic acid. Insects as well as people find the artificially modified plants much more appetizing than the wild forms, and more pesticides are needed to protect such crops from herbivore damage.

The caterpillars of the cabbage white butterfly feed exclusively on the leaves and flowers of cabbage and related brassicas, including cauliflower, broccoli, and oilseed rape, but they have not acquired more of a taste for erucic acid than have mammals. Even when present in quite high concentrations in the diet, the unusual fatty acid is absent from phospholipids and triacylglycerols in the insects' tissues. Either it is not absorbed through the gut or not esterified into triacylglycerols or, less likely, it is all used as fuel at once, before it has a chance to become incorporated into structural or storage lipids. Erucic acid seems to be toxic even to the plants. Highly specific enzymes direct it into seed triacylglycerols only. None is found in phospholipids or any other non-storage lipid, suggesting that its presence in such molecules would disrupt membrane function.

IV. VERTEBRATE ADIPOSE TISSUE

Proper adipose tissue is unique to vertebrates. It occurs sporadically among fish, often in a greatly modified form, but it is most extensive in mammals, birds, reptiles, and amphibians. Its distinctive feature is a unique type of cell called an adipocyte, which can accommodate much larger quantities of triacylglycerols than any other kind of cell. Almost all the adipocytes' stores of lipid are not for their own use but for export to other tissues as required. Most physiological studies of adipose tissue are aimed at controlling obesity or diabetes, and the primary purpose of most studies of plant storage tissues is food production. Consequently, the basic principles governing the cellular structure, gross anatomy, and metabolism of these tissues are only slowly coming to light.

Compared to other animal cells, replete adipocytes are huge, thousands of times larger than red blood cells, most brain cells, and the cells of the immune system that protect the body from disease. In Fig. 1, only the top layer of adipocytes is in the plane of focus and because the tissue is not stained, the nucleus, cyto-

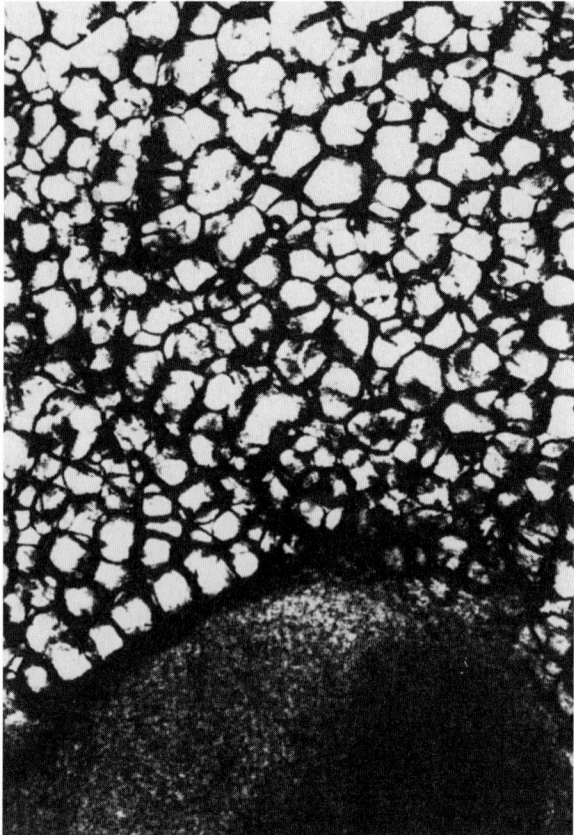

FIGURE 1 A thick, unstained section of part of the popliteal adipose depot of a young, lean rat photographed with light shining through it from below. The elongated shadows are fine blood vessels that permeate the whole tissue. The structure at the bottom is part of the popliteal lymph node that is embedded in this depot. The field of view is about 1 mm across, so these adipocytes average approximately 70 μm in diameter or about 0.2 nl in volume.

plasm, and lipid droplet cannot be distinguished. Unless severely depleted of lipid, adipocytes are spherical, or very nearly so, which is an unusual shape for functionally mature animal cells (other than eggs). Adipocytes are packed closely in a tight mesh of collagen, so together they occupy most of the volume of the adipose tissue.

The cells' most distinctive feature is that most of the cytoplasm is occupied by lipid droplets, a single large one in white adipocytes and many smaller ones in brown adipocytes. Comparative studies of angiosperm seeds and fruit, yeasts, and various animal cells show that the intracellular mechanisms of the formation of lipid droplets are essentially similar in all eukaryote cells; however, the process is greatly exaggerated in adipocytes. Lipid droplets arise from specialized regions of the endoplasmic reticulum and are controlled by specific enzymes and other associated proteins called oleosin in seeds and perilipin in adipocytes.

The compartmental arrangement of plant storage lipids is not necessary in white adipocytes because such cells do not metabolize significant amounts of their own triacylglycerols. However, brown adipose tissue oxidizes its own lipids to generate heat, although when such supplies run low it can, like most other animal cells, take up metabolites from the blood circulation. The similarity in size and general appearance between plant oil bodies and brown adipocytes may arise from the fact that in both cases storage lipids are utilized in the cells in which they are sequestered.

A. Size and Number of Adipocytes

Among mammals and birds (there are no data for other classes), the adipose tissue of larger species consists of fewer, larger adipocytes. Adipocyte number scales to (body mass)$^{0.75}$, and the adipocytes range in volume from 0.01 nl in bats and shrews to up to 4 nl in well-fed baleen whales. Carnivorous mammals and ruminants (which utilize mainly fatty acids rather than glucose) have about four times more adipocytes than do non-ruminant herbivores (whose energy metabolism is based mainly on glucose) of the same size and fatness; however, they are not fatter because the adipocytes are smaller. By coincidence, the adipocytes of rats, small non-ruminant herbivores, are about the same size (0.1–1 nl) as those of humans, who are large omnivores on a high-fat diet.

Thorough studies of many specimens of the same species in the wild reveal much interindividual variation in the total number of adipocytes in the body as a whole that cannot be attributed to age, sex, or any obvious feature of dietary history. For example, most Svalbard reindeer, which accumulate enough fat during summer and autumn to sustain them during the long winter when only low-quality dry vegetation can be reached by digging through the snow, have only two or three times more adipocytes than would be expected in temperate-zone and tropical mammals of similar size.

The number of adipocytes (relative to total body mass) proved to be even more variable in carnivores. Some arctic foxes (*Alopex lagopus*) and wolverines (*Gulo gulo*) have up to five times as many adipocytes as expected from their overall size, but a significant minority actually have fewer than the expected number. At the time the specimens were collected, there was no consistent relationship between the total number of adipocytes and fatness, measured as the relative mass of adipose tissue in the body. The number of adipocytes in the adipose tissue is not a major determinant of the capacity for fattening. If fewer adipocytes are present, each must expand to a larger volume to accommodate as much storage lipid as that in specimens that, for whatever reason, have proportionately more adipocytes.

In humans, and also in laboratory animals kept under highly artificial regimes of diet and exercise, the number of adipocytes increases with age after sexual maturity, but all these effects are slight, generating at most a doubling of the adipocyte population. They cannot be induced in naturally obese wild mammals such as dormice (*Glis glis*), and there is almost no evidence for an increase in adipocyte number after growth of lean tissue stops in any wild mammal investigated to date. Until recently, biologists believed that once fully formed and functional, adipocytes did not die during the lifetime of their owner; however, adipocytes and preadipocytes, like most other cells, have been found to disappear by apoptosis, an orderly and often energy-consuming process in which the cells synthesize enzymes that destroy their fabric. Apoptosis is rarely observed because such large cells do not die very often and death is swift, probably only a few hours from the first signs of decline to total disintegration.

In people and sedentary domestic animals, the proportion of lipid in whole adipose tissue is rarely less than 40% and it increases with fattening, reaching as much as 85% of the total mass of the tissue in middle-aged men who are 44% by weight adipose tissue. The rest of the tissue is mostly protein and water. The proportion of lipid is always lower in wild animals and does not change significantly with fatness. Dwarf hamsters (*Phodopus*) naturally become very fat during the summer, but chemical analysis shows that their adipose tissue never exceeds 46% lipid, even when it amounts to 35% of the body mass. The proportion of lipid in adipose tissue is almost constant (42–66% depending on the depot) in arctic foxes living wild in the high Arctic, whose body composition ranges from 3 to 33% by weight adipose tissue.

B. Anatomical Arrangement of Adipose Tissue

Adipose tissue in most reptiles and amphibians is concentrated in paired depots inside the abdomen or, in the case of lizards, on the tail. Adipose tissue in tortoises and turtles is more widely distributed, filling spaces between muscles and in corners of the shell.

Birds have both superficial and intra-abdominal depot, but they lack true intermuscular adipose tissue. The main superficial depots are at the base of the neck around the furcula ("wishbone"), on the anterior surface of the thigh, over the breast muscle, and on the back near the tail. Biochemical differences between depots in birds have not been as thoroughly studied as those of mammals; however, as in mammals, the superficial depots usually enlarge more than the internal ones. They expand laterally and thicken, forming an almost continuous layer that is thickest over the back, breast, and the anterior surface of the thigh in birds that fatten before migration or prolonged fasts associated with molting or breeding.

Mammalian adipose tissue is always partitioned into a few large and many small depots. Its comparative anatomy is based on measurements of the size and number of the adipocytes. Figures 2 and 3 show the patterns of distribution of larger and smaller cells for various mammals. Larger adipocytes do not always make larger depots. Some of the smallest depots, such as the popliteal behind the knee, consist of relatively large cells. Nor do adjacent depots, such as those on either side of the forelimb, necessarily consist of cells of similar size.

In the adult camel (Fig. 3A), more than one-third of all adipose tissue is in the humps, compared with 4–8% in the corresponding site of hares (Fig. 2B), up to 4% in squirrels, and a maximum of 1% in stoats (Fig. 2C). However, less than 1% of the camel's adipose tissue is in the inguinal depot, on the front of the thighs and the sides of the abdomen. This depot contains more than one-third of the total adipose tissue in stoats and squirrels, more than one-fifth of that of the lioness (Fig. 3C), and one-eighth of that of horses (Fig. 3B). There are also contrasts in the proportions of adipose tissue located on the dorsal wall of the abdomen, from nearly half in squirrels and hares (Figs. 2A and 2B) to one-eighth or less in the large animals (Fig. 3). Although the range of sizes of adipocytes differs greatly between species and between specimens, partly because some specimens were fatter than others, the anatomical pattern of relative sizes of adipocytes is similar in all terrestrial mammals. What differs between species is the relative abundance of adipocytes in each depot. The relative size of their adipocytes forms the same general pattern. Taxonomically related species have similar patterns of distribution of adipose tissue.

Although much is known about the biochemical steps by which adipocytes differentiate and mature, almost all of the information was derived from the study of cells in culture so that it cannot be explained in detail why adipose tissue forms where it does or what determines the species differences in the relative sizes of the depots. The gross anatomy must have implica-

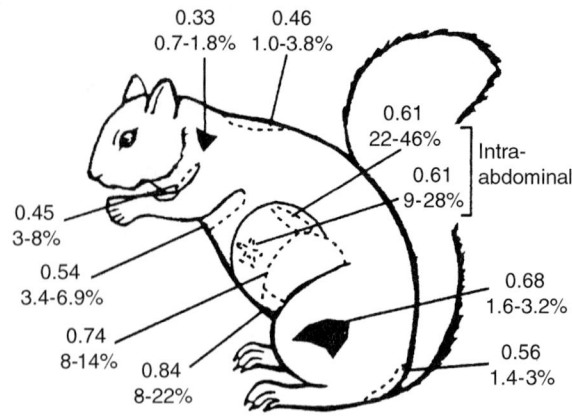
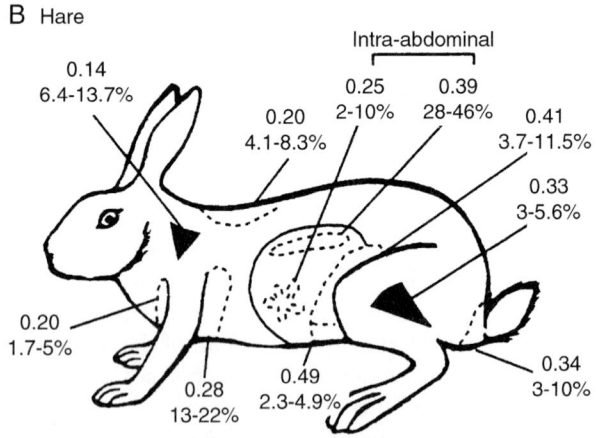
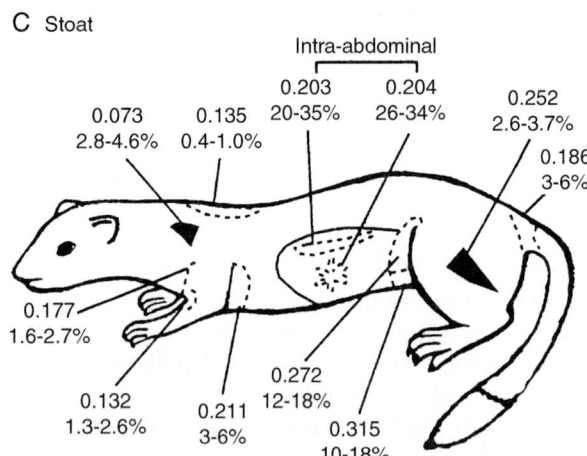

tions for the animals' habits and capabilities because there are site-specific differences between adipocytes in many physiological properties. Some depots seem to be equipped mainly for taking up lipids from the blood after a meal, whereas others seem to be equipped for quick responses to the onset of strenuous exercise. Many of the minor depots that enclose lymph nodes may be mainly or entirely devoted to supporting immune processes and contribute little to whole-body energy storage. They are among the first to form and the last to be depleted in starvation. In very lean wild animals such as rabbits, the only replete adipocytes may be those associated with lymph nodes.

Relatively minor depots may be almost undetectable in small animals, especially if the specimens are lean. Therefore, perhaps it is not surprising that larger animals (Fig. 3) appear to have more "extra" adipose depots for which there are no exact equivalents in the smaller species (Fig. 2). Some are just too large to be thus explained away. Up to one-third of all adipose tissue of horses (Fig. 3B) and donkeys forms a sheet, sometimes several centimeters thick, on the inner ventral wall of the abdomen. There is a small quantity of adipose tissue in the corresponding depot of large carnivores and camels (Figs. 3A and 3C), but it is absent in most other species.

In primates, the most conspicuous such depot is the "paunch" that arises from the midline on the outer wall of the abdomen and expands laterally as it thickens, sometimes becoming very thick at the midline and extending as far as the chest and the hip bones. In humans, the paunch adipose tissue is centered around the navel and is usually slightly thicker above it in men and below it in women. The paunch depot seems to accumulate fat selectively in primates. It may be minimal in lean (but not emaciated) specimens, but it often becomes very massive in humans and also in monkeys and apes that become obese in captivity, whereas the other depots expand more slowly with increasing fatness or, in the case of the intermuscular depots, expand hardly at all. An abdominal paunch with these properties seems to

---

FIGURE 2 The organization of adipose tissue in various wild mammals. (A) The gray squirrel (*Sciurus carolinensis*): Data are the means of four specimens; body mass, 0.38–0.67 kg; 3.1–14.4% dissectible adipose tissue. (B) European hare (*Lepus timidus scotius*): Data are the means of five specimens; body mass, 2.4–3.5 kg; 0.3–3.2% fat. (C) Stoat (*Mustela erminea*): Data are the means of five specimens; body mass, 0.19–0.4 kg; 3.2–6.3%. Top sets of data are the mean volume of adipocytes in nanoliters (10^{-9} liters); bottom sets of data are the proportions of the total dissectible adipose tissue in the depot. The intermuscular depots are shaded (data from Pond, 1998).

be a unique and distinctive feature of this group, and it is as conspicuous in lemurs as it is in apes, indicating that it appeared early in their evolutionary history. Despite having this extra depot, with the exception of modern humans, primates are generally no fatter than other mammals when living under natural conditions (although some do become obese in zoos).

There is no trace of this depot in hares, squirrels, or any related small mammals, but Carnivora have a paunch depot on the outer ventral wall of the abdomen. The depot can become thick in cats and their relatives but it is relatively small in canids, ursids, and mustelids. It does not expand disproportionately any more than any other superficial depot, even in very obese wild carnivores such as arctic foxes. Very little is known about the physiological basis for its special properties because the paunch depot is completely absent in rodents so it cannot easily be studied in the laboratory.

C. Measuring Fatness in Wild Animals

Because adipose tissue is the most variable component of the body, the simplest and most widely used measure of fatness is body mass, sometimes refined to include a term for body length. In small animals, the guts represent a large and variable fraction of the total mass that such methods are not very accurate or reproducible. If samples of adipose tissue can be obtained by biopsy or post mortem, adipocyte volume is also used to estimate fatness, but such measurements are not as reliable for most wild animals as experiments on laboratory rodents might suggest. It is essential to correct for the effects of body size and natural diet on the cellular structure of adipose tissue, for site-specific differences in adipocyte volume, and for variability in the cellular structure of the adipose tissue.

Figure 4 shows data on adipocyte volume from wild polar bears. Females (but not males) stop growing when they become sexually mature and fatten enormously during pregnancy, producing a predictable relationship between body mass and adipocyte volume. The correlation is weak for cubs and males that are still growing because adipocyte number is also increasing. However, these data illustrate some important aspects of bear ecology. Cubs still with the mother are generally fatter than those of similar size that have been abandoned to forage alone, and there is great variation in the fatness of cubs and adult males, especially among the larger specimens, probably reflecting differences in hunting success.

For children and young adult humans, measuring skinfold thickness with calipers is widely used to assess fatness, but the skin of most mammals is too thick and tough and the underlying adipose tissue too thin and patchy for skinfold thickness measurements to be of much use. As wild mammals become fatter, superficial adipose tissue expands more rapidly than that inside the abdomen. The depots shown in Figs. 2 and 3 become thicker and expand laterally to cover a greater area. In very fat specimens, including many humans, adjacent superficial depots overlap and merge so that they become difficult to distinguish, but measurements of adipocyte volume can reveal their identity. Since expanding depots spread out as well as thicken, thickness alone is only a crude measure of the total adipose tissue, even if the site from which the measurement is taken is kept strictly constant. The fatness of small birds is often assessed by inspection and/or palpation of the superficial depots around the neck or leg. This simple method takes account, at least in a qualitative way, of changes in both the thickness and the area of the adipose tissue.

A high-tech version of skinfold thickness measurements is the ultrasound scanner, which can measure fat thickness fairly accurately in large mammals, especially if, as in pigs, the hair is sparse. The method depends on the fact that sound vibrations travel at different speeds through chemically different materials and are reflected, producing an echo, at the interfaces between the two tissues. A pulse of high-frequency sound is directed into the body from a source placed on the skin, and the timing and intensity of the echo are recorded. With a resolution of only about 1 mm at best, ultrasound scanners are not accurate for animals smaller than dogs and for any animal in which the superficial adipose tissue is very thin.

Lipids are much poorer conductors of electricity than the watery components of cells and body fluids. This contrast has been exploited in the development of an apparatus that measures whole-body conductance of electric current as an indicator of the proportion of lipid present. More fat produces lower whole-body conductance. The apparatus has to be used carefully because hair, feathers, and the pockets of air they enclose also have high resistance to electric currents. However, it is a useful technique, especially for small birds and mammals, and it works equally well on living or freshly dead specimens.

Because these new techniques are impractical for most large animals, dissection is still widely used. For more than 30 years, assessment of lipid reserves in wild vertebrates *in vivo* and post mortem has depended mainly on empirically derived "indices" which are based on the dimensions of the whole body or that of the

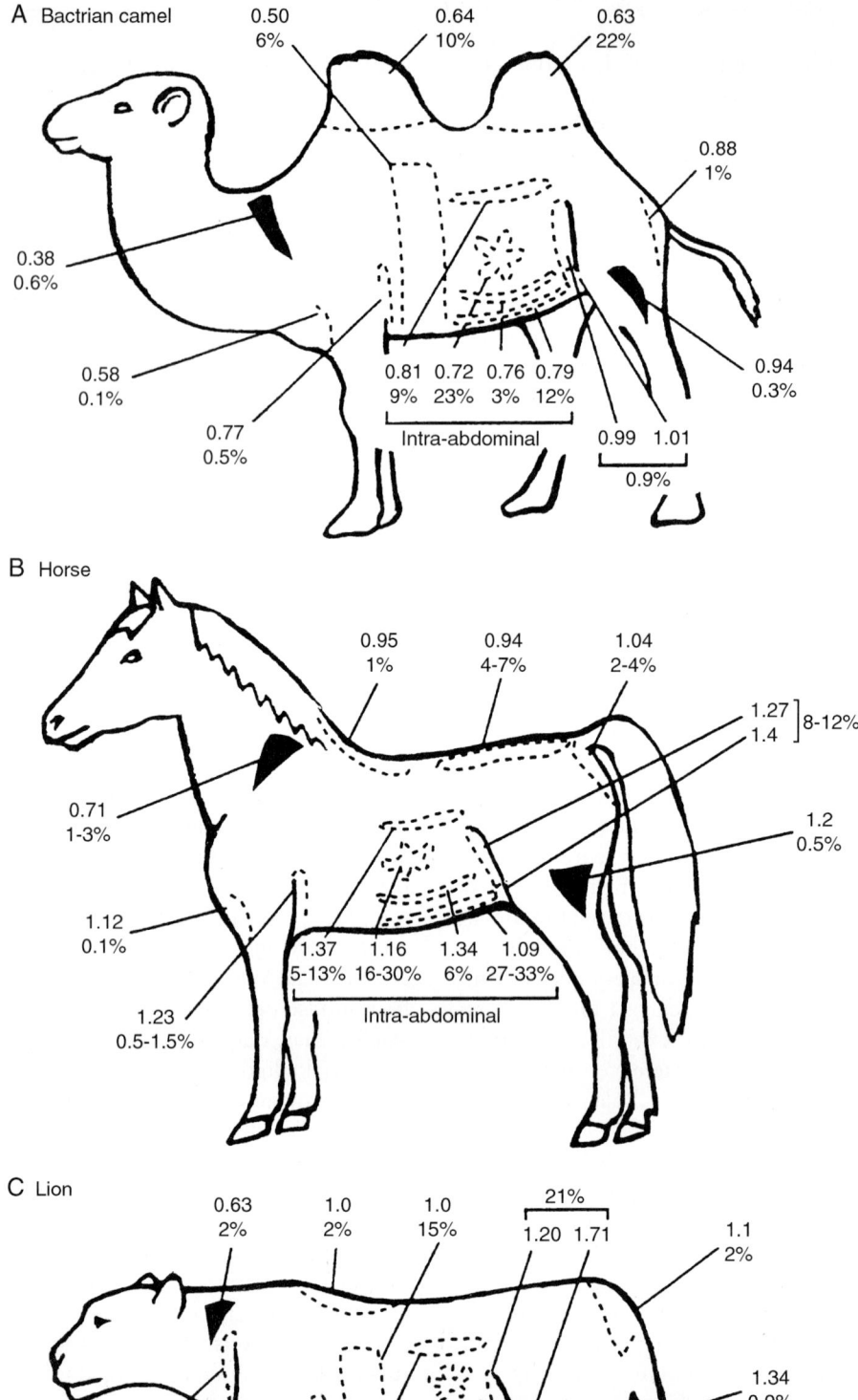

thickness or mass of one or a few adipose depots. The sample sites are chosen at random or for some technical reason such as simplicity of dissection. The enormous diversity of sample sites and indices is cumbersome and makes comparisons between studies very difficult and imprecise. One of the favorites is the kidney fat index, the mass of the perirenal adipose tissue as a percentage of that of the kidneys. This index works fairly well for lean deer, antelope, and similar ungulates, unless the mass of the kidneys changes seasonally, which can happen in some drought-adapted species. It is less accurate in carnivores (see Figs. 2 and 3), in which the perirenal depot represents only a small fraction of the total adipose tissue and is indistinguishable from that on the rest of the dorsal wall of the abdomen.

Wildlife biologists also try to estimate total fatness of mammals from the lipid content of the marrow in the long bones, thereby eliminating the need for a complete dissection. However, "marrow fat indices" only work on species such as ruminants in which the marrow adipocytes are depleted in starvation. In many other mammals, these adipocytes fail to respond to fasting, probably because they are specialized to support the immune functions of the red marrow.

An additional problem with using the mass of one or a few depots as an index of fatness is the inherent variability of the partitioning of adipose tissue between depots. In all species for which there are sufficient data, there is substantial inter-individual variation in the relative masses of depots that cannot be attributed to age, sex, or lean body mass, even in genetically homogeneous wild populations and in animals bred and maintained under carefully controlled laboratory conditions. Of the intra-abdominal depots, the dorsal wall of abdomen depot (which includes the perirenal) undergoes the largest changes, especially in deer, sheep, cattle, antelope, and other ruminant animals. It expands greatly in extreme obesity and shrinks to almost nothing after prolonged starvation. Its anatomical position means that, like the fat bodies of lower vertebrates, such large and sometimes rapid changes in size have minimal impact on the shape or size of any adjacent organs. Depots that contain lymph nodes undergo the

FIGURE 3 The organization of adipose tissue in some large zoo-bred or domesticated mammals. Data are presented in the same way as in Fig. 2. (A) Two-humped camel (*Camelus bactrianus*), a pregnant female: body mass, 650 kg; 6.1% fat. (B) Domesticated horse (*Equus caballus*): Data are means of measurements from two geldings; body mass, 444 and 570 kg; 5.2 and 4.0% fat. (C) Pregnant female lion (*Panthera leo*): body mass, 167 kg; 13.3% fat (data from Pond, 1998).

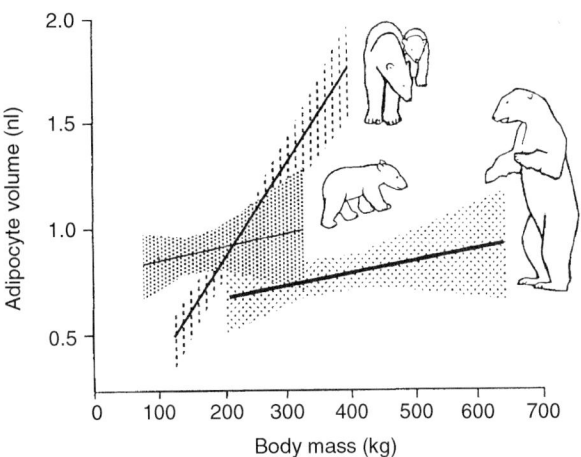

FIGURE 4 Means and 95% confidence intervals of the average volume of adipocytes in a sample of adipose tissue from wild polar bears in northern Canada measured in cubs accompanied by their mothers, weaned females (including pregnant and lactating mothers), and solitary males (data from Ramsay et al., 1992).

least change, so they are relatively massive (compared with those elsewhere in the body) in lean specimens and relatively small in obese ones.

V. ADIPOSE TISSUE FUNCTION

A. Adipocyte Metabolism

Molecular traffic into and out of adipocytes is regulated by a variety of receptors, carrier molecules, enzymes, and other substances, most of which are produced by the cells. As recently as a decade ago, adipocytes were known to produce only a few kinds of proteins. However, techniques to separate, identify, and synthesize proteins and other biological molecules are rapidly becoming cheaper and more efficient, so biochemists are constantly adding to the list of messenger molecules secreted from adipocytes, receptors that enable them to respond to such signals from each other and more remote tissues, and transport proteins and enzymes that regulate uptake of materials that at least some adipocytes can produce.

The genetic instructions for synthesizing such proteins are located in the nucleus, which is relatively small and is displaced from its usual position near the center. The lipid droplet (see Fig. 1) has no internal structure because it is almost pure liquid triacylglycerol. The thin rind of cytoplasm between the outer membrane and the lipid droplet contains the biochemical apparatus for building and storing receptors, messenger molecules, and enzymes plus a few mitochondria in which the fuel that powers their synthesis is produced.

Leptin (formerly known as Ob protein) is among the best known secretions of adipose tissue, identified in the early 1990s from the study of a spontaneous mutation that causes severe obesity in mice. This small protein passes from adipocytes (and, it has recently been discovered, certain other tissues, especially those involved in reproduction) into the blood, in which it is found in concentrations of approximately 10^{-8}–10^{-9} M, and then to the brain, in which it binds to specific receptors on many neurons in the hypothalamus, the region of the brain that controls appetite (and many other drives and emotions). This small protein seemed to have all the properties of the long-sought signal that regulates appetite in proportion to the level of fat reserves in adipose tissue. It has been identified in rats, humans, and *Sminthopsis macroura* (Marsupialia: Dasyuridae) and possibly in fish. In addition to regulating appetite, leptin adjusts energy expenditure and the physiological response to cold. Obesity could be caused by disruption of these chemical signals from the adipose tissue to the brain, leading to chronic overeating and/or very low rates of energy expenditure.

Excess lipids cannot be excreted; therefore, unless they can be oxidized to produce energy, they have to be deposited in the storage tissues. Most tissues oxidize lipids, but the largest consumers are cardiac and skeletal muscle, the liver, and, especially in neonatal mammals and hibernating species, brown adipose tissue. ATP production in mitochondria is never 100% efficient, a fraction of the energy used appears as heat. But heat production in brown adipose tissue can be greatly increased by "uncoupling" of fuel utilization from the formation of ATP in the mitochondria, implemented by a special protein. Until the late 1990s, uncoupling protein was believed to be unique to mammals and restricted to brown adipose tissue, but similar proteins have been found in a variety of tissues and organisms including green plants, although it is not clear that they all act to promote heat production and lipid utilization.

A large fraction of the volume of brown adipocytes is occupied by mitochondria so that there is not much space for storing lipid, which is present as numerous tiny droplets instead of a single large droplet in storage white adipose tissue. Active brown adipose tissue is only about 10% lipid by weight. Like white adipose tissue, brown adipose tissue is richly innervated by the sympathetic nervous system, and in both cases norepinephrine stimulates lipolysis. In brown adipose tissue, the fatty acids produced remain in the cell and are consumed for heat production, whereas those of white adipocytes are released into the blood, in which other tissues can take them up and use them.

When fully activated (e.g., during rapid warming of the body at emergence from hibernation), brown adipose tissue uses fuels and oxygen at up to 10 times the rate that muscles use them. More than 90% of the chemical energy it consumes is released as heat instead of being used to synthesize ATP, producing enough heat to raise the body temperature by more than 2°C per hour. In the newly born and in awakening hibernators, much of the brown adipocytes' fuel comes from white adipose tissue. The latter's stores therefore determine the neonate's ability to maintain its temperature until it can find food (mother's milk or a rich, digestible meal).

Under artificial conditions, brown adipose tissue can often be transformed into white and vice versa, and at least in certain species, it does so naturally. Many of the sites in which it actively generates heat in newborn mammals contain normal-looking white adipose tissue in the adult. In small mammals that live in cold climates, such as dwarf hamsters, at least some depots consist of a mixture of brown and white adipocytes throughout life. The proportion of space occupied by the brown ones increases when the animal is placed continuously in a cold environment, especially if the light–dark cycle is adjusted to resemble the long, dark nights of winter. In large mammals such as reindeer, brown adipocytes in newborn calves become white adipocytes within days, and cannot be reconverted back to brown.

B. Fatty Acid Composition of Storage Lipids

Animals can synthesize the common saturated (palmitic, stearic, and myristic) and monounsaturated (oleic) fatty acids and many can insert an additional double bond into long-chain fatty acids, although this capacity is minimal in higher vertebrates including mammals. Animal desaturases work slightly differently from the corresponding enzymes in plants and protoctists; therefore, fatty acids that originate in animals are distinct from those synthesized by primary producers. Except in ruminant animals (in which the rumen microbes alter the fatty acids before they can be absorbed), most fatty acids pass unaltered from plant to herbivore and from prey to predator. Therefore, fatty acids originally synthesized by microbes, protoctists, or plants often appear unaltered higher up the food chain.

The proportions of the various fatty acids, and their arrangement in the triacylglycerol molecules of most animal fats, are such that they are fluid at the higher body temperatures of birds and mammals. The lipids of cold-blooded animals generally contain a higher proportion of unsaturated fatty acids than those of warm-blooded animals, with species living in colder regions having the most polyunsaturates.

Recent studies suggest that the lipid composition of the diet may determine the habits and habitats of poikilothermic animals. The blue-tongued skink or shingle-backed lizard, *Tiliqua rugosa*, is quite common in the deserts of Western Australia. Its natural diet is mainly plants (at least when an adult) but also includes carrion and small prey; therefore, it can be maintained on an artificial diet in captivity. After a few weeks on experimental diets, those given food containing sunflower oil (rich in polyunsaturated fatty acids) chose to spend their days (and nights) in places that were up to 5°C cooler than those fed on mutton fat (containing mostly saturated fatty acids). These experiments show that diet can affect important aspects of their ecology, including the times of day at which they are active and where they forage.

The sea is always cold compared with the warm bodies of mammals and birds. Marine algae generally have more highly unsaturated fatty acids than terrestrial plants, and the animals that eat them both incorporate the algal fatty acids unchanged into their own tissues and elongate and/or desaturate them further. Thus, most plants cannot build fatty acids with more than 18 carbons, but many marine animals can elongate oleic acid to form gadoleic acid ($C20:1n-9$) and various 22- and 24-carbon fatty acids. Gadoleic acid is common in the lipids of many kinds of fish, especially those living at high latitudes. Other fatty acids are also readily elongated, but those of the $n-6$ series derived from linoleic acid are even more unstable (in the sense that they readily participate in chemical reactions, especially oxidation) than those of the $n-3$ series; therefore, long-chain fatty acids are in more common marine organisms.

Algae are the basis of nearly all marine food chains, so the storage and structural lipids of almost all animals that feed in or from the sea—ragworms, mussels, oysters, shrimps, crabs, fish, seals, whales, polar bears, gannets, albatrosses, and penguins—are rich in $n-3$ polyunsaturated fatty acids. Such mixtures of triacylglycerols rich in polyunsaturates are often called "fish oils," although they are not unique to fish, and the basic fatty acids from which they are derived are not even produced by fish. The lipids originate in the algae but may be modified by animals as they pass up the food chain, eventually reaching top predators such as large fish, seabirds, and marine mammals. Among the fish used as human food, cool-water species such as herring, sprats, capelin, mackerel, and halibut usually store more lipid than their tropical relatives because at high latitudes the supply of most kinds of food changes seasonally and many species routinely live on their fat reserves for weeks or months. The triacylglycerol fatty acids of large cold-water species include more monounsaturates than polyunsaturates because they are derived from waxes that are in the planktonic organisms that these species eat. Fish such as sardines, pilchards, and anchovies live in warmer waters in which the planktonic invertebrates rarely contain waxes, so their lipids are a rich source of $n-3$ polyunsaturated fatty acids for human consumers.

In addition to being chemically distinct, lipids of marine origin differ in the isotopic composition of the carbon atoms. Slightly different proportions of ^{13}C and ^{12}C are taken up by marine algae that obtain their carbon in solution as bicarbonate and terrestrial plants that take in carbon dioxide straight from the atmosphere; therefore, organic materials of marine origin contain about five to seven parts per thousand more of the heavier isotope (^{13}C) than similar molecules from animals that have eaten terrestrial foods.

This effect can be used to investigate the dietary habits of individual animals and populations. Storage fatty acids and cholesterol are particularly suitable because they pass intact up the food chain, are chemically quite stable, and are only slowly degraded by microbes; therefore, they can be detected in excreta and animal remains as well as in biopsy samples. The isotopic and chemical composition of triacylglycerol fatty acids can indicate the proportion of the diet that is derived from marine or terrestrial sources. This technique is particularly useful for widely ranging species with broad, locally variable diets, such as arctic foxes, that are not easily observed in the wild. Material from the marrow of buried human and other animal bones and in coral fragments has been used to study natural diets and food chains.

Some of the food of freshwater animals is derived from aquatic algae, but detritus from terrestrial plants and the animals that feed on them also makes a substantial contribution; therefore, fish and insects living in lakes and rivers usually have more fatty acids of the $n-6$ series than do oceanic animals. The fatty acid composition of predatory fish follows that of their diet: "Farmed" salmon fed on chow made from scraps of slaughtered livestock and other materials originating from the land acquire a correspondingly "terrestrial" fatty acid composition.

Mammalian adipocytes can selectively take up and

release certain fatty acids (there are no data on those of other vertebrates), making possible seasonal accumulation of triacylglycerols with physical properties appropriate to hibernation. Recently, site-specific differences have also been described. The adipocytes around lymph nodes, and to a lesser extent elsewhere, in node-containing depots sequester more polyunsaturates that are essential to the nutrition of the immune system than do large nodeless depots. Such variation must be considered when selecting tissue samples for the study of diet.

C. Adipose Tissue as Thermal Insulation

The notion that the superficial adipose tissue insulates the body against heat loss is one of the most firmly established of all dogmas in twentieth-century biology, but there is very little evidence that supports it. The hypothesis can be tested by determining whether its distribution and anatomical arrangement have evolved to carry out such a role more efficiently in animals that need more thermal insulation because they remain active during winter in cold climates or habitually swim in very cold water. The mammalian order Carnivora occur in a wide variety of habitats from hot tropics (e.g., desert foxes and jackals) to the Arctic (e.g., polar bears, wolverines, and arctic foxes) and almost all are predators, at least to some extent. Carnivora occur in an enormous range of sizes, from stoats and weasels to bears and tigers that can be several thousand times larger, and they are therefore suitable for investigating whether species differences in the relative abundance of superficial adipose tissue are determined by habits and habitat or by body size.

Figure 5 shows some measurements of the masses of all the intra-abdominal and superficial adipose tissue from various temperate, tropical, and arctic carnivores. To ensure comparability, the sample included only specimens that were of similar fatness to the bears. There is a fair amount of inter-individual variation in the partitioning of adipose tissue between internal and superficial depots, but overall the mass of superficial depots increases in simple proportion to body mass. In contrast, the intra-abdominal depots become proportionately smaller in larger mammals, no matter whether the animals are native to the Arctic or to warmer habitats.

The surface area over which this larger amount of superficial adipose tissue occurs is smaller in larger animals relative to their volume or mass. The body surface area scales as body mass$^{0.66}$, i.e., it becomes proportionately smaller with increasing body size. With

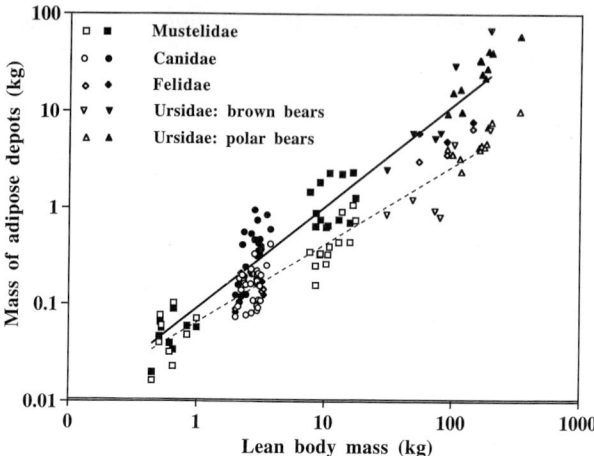

FIGURE 5 Allometric comparison of the masses of superficial (solid symbols) and intraabdominal (open symbols) adipose depots in various species of the order Carnivora. Each point represents a single individual. Squares, Mustelidae (mink, badgers, and wolverines); diamonds, Felidae (cats, jaguars, tigers, and a lion); circles, Canidae (arctic foxes); triangles, Ursidae (inverted triangles, brown bears). The lines are fitted to all the data except those from the polar bears (triangles) and fit the following: superficial depots α (lean body mass)$^{1.05}$, $r^2 = 0.867$; intraabdominal depots α (lean body mass)$^{0.81}$, $r^2 = 0.873$ [data from Pond and Ramsay (1992) and additional sources].

a smaller area to cover, the superficial depots are thicker in larger animals, even if the specimens are not fatter (i.e., if the ratio of the masses of lean and fat tissues is unchanged). The superficial adipose tissue is 10 times thicker in a 500-kg bear than in a 500-g rat or weasel from this effect alone, quite apart from the fact that in larger animals, a greater proportion of the adipose tissue is superficial and bears are often fatter than rat-sized animals.

The measurements from the wild polar bears are very near to the lines fitted to the data obtained from tropical and temperate-zone carnivores, indicating that the partitioning of adipose tissue between internal and superficial sites arises entirely from the fact that polar bears are very large and produce large quantities of fat, and it has nothing to do with adaptation to thermal insulation. There is no evidence for reorganization of adipose tissue in the smaller species of mammals that have evolved to become aquatic; for example, otters have many adaptations of the trunk, tail, and sense organs to their life as predators in lakes and rivers, but the arrangement of their adipose tissue is not different from that of their terrestrial relatives, stoats, weasels, wolverines, and badgers. Otter fur is thicker and less compressible than that of the terrestrial species, so it retains its insulating properties when in water.

The same arrangement is found in birds. Just before migration or breeding fasts, the superficial depots enlarge greatly and may become several millimeters thick. They also expand laterally so the depot around the wishbone extends over the anterior part of the flight muscles and along the neck. This arrangement is the only way that adipose tissue comprising 50% of the total body mass can be accommodated. In swimming animals, indefinite expansion of the abdomen is not compatible with streamlining while in water, but a sleek profile can be maintained if the additional fat is spread out over much of the body.

These effects may promote larger body size in polar species, but they cannot be strong compared to other determinants of size. Polar bears (*Ursus maritimus*) are obligate carnivores that need their bulk and strength for killing seals and swimming between ice floes, but other arctic mammals, such as the arctic fox and Svalbard reindeer (*Rangifer tarandus platyrhynchus*), are actually smaller than their relatives native to lower latitudes (the red fox, *Vulpes vulpes*, and the Canadian and European subspecies of *R. tarandus*).

This analysis highlights some pitfalls of using measurements of the thickness of superficial adipose tissue as a means of estimating fatness. Fatness is a ratio of masses—the mass of energy stores as a percentage of total body mass—but adipose tissue thickness is a linear measurement. With increasing body mass, the superficial adipose tissue becomes thicker for animals of the same total body composition because the ratio of their surface area to mass decreases. The superficial adipose tissue becomes thicker in larger animals, and depots that are adjacent but distinct in small species and the juveniles of large species merge in large adults to form a continuous layer, albeit of variable thickness. An almost continuous layer of superficial adipose tissue can create the impression that the specimen is obese, even when the layer represents only a small fraction of the total body mass.

The lactation habit, rather than thermoregulation, may be the main reason for the massive development of the superficial adipose depots in mammals. Reptiles are usually fattest just before maturation of the eggs begins, and the intra-abdominal fat bodies are much depleted by the time the clutch is ready to be laid. Heavily gravid females are often fasting, so the empty, inactive gut frees up more space inside the abdomen. In contrast to egg-laying vertebrates, the demands of lactation immediately follow those of gestation. Most mammals feed normally while they are pregnant and accumulate lipid in preparation for lactation. The reproductive organs and the gut cannot be excluded from the abdomen, but adipose tissue can be excluded. The subcutaneous depots may simply be the most convenient place to put a lot of adipose tissue.

In pinnipeds, cetaceans, and sirenians, the cellular and gross anatomy of adipose tissue is substantially altered to form blubber. Heat conducts through adipose tissue at only slightly less than half the rate of that through muscle or stagnant water and much faster than through fur or feathers. Therefore, blubber has to be thick to insulate thoroughly, and it is satisfactory as the sole form of insulation only for large animals with habits and habitats for which weight is not a major problem. Consequently, most small and medium-size seals retain fur as additional insulation, and extant sirenians and the smallest cetaceans, porpoises, and dolphins are restricted to temperate and tropical regions. Many of the great whales (e.g., humpbacks and fin whales) that feed in polar seas migrate to warmer waters to give birth to calves that are only one-twentieth of their size. Many such places provide little or no food for adult whales, so the mothers fast for several weeks while lactating, drawing on their lipid reserves. The whales do not return to cooler, food-rich waters until their calves have grown sufficiently large and fat for blubber insulation to be effective.

Seals are born on beaches or ice floes and most species do not go to sea until they are several weeks old. Neonates have thick coats of fluffy fur but little adipose tissue, although they fatten rapidly while suckling their mothers' rich, creamy milk. At about the time the seal pups are weaned, the neonatal fur is replaced by shorter, stiffer fur. However, fur of some sort continues to make an essential contribution to insulation in all seals except the adults of the largest species, the elephant seals (*Mirounga* spp.). In walruses (*Odobenus rosmarus*), the hair is very sparse throughout life, and at birth the single pup is very large compared to its mother, up to twice the size of neonatal elephant seals, although as adults the latter are much larger.

Both these very large pinnipeds occur in cold seas, *Odobenus* in the Arctic Ocean and one species of *Mirounga* in the northern Pacific including near Alaska and the other in the Southern Ocean. The skin of walruses hauled out on beaches basking in sunshine is noticeably pink due to the profuse flow of blood through it and the underlying blubber, which can be up to 15 cm thick. However, walruses are a ghostly pale gray when in the sea or exposed to cold weather because the blood is withdrawn from the skin and blubber to the muscles and other warm internal tissues. Blubber can work as an adjustable insulator in this way because blood flow through it and the skin can be almost completely shut down for long periods without ill effects.

If necessary, marine mammals can be very fat. As in the case of penguins, body bulk is only a slight impediment to swimming, and seals and whales can maintain near neutral buoyancy over a wide range of body compositions by adjusting the volume of air in the lungs. The maximum amount of blubber that whales or seals can have without risking overheating during vigorous swimming depends on body shape, the temperature of the water in which they are living, and how much heat they generate. Marine mammals dissipate excess heat by increasing blood flow through the blubber-free skin of the flippers and tail. On hot days, seals on land sometimes hold up a flipper to the breeze to accelerate cooling. A thickness of about 20 cm seems to be the upper limit. Blubber of that thickness apparently does not hold enough lipid to meet the energy storage needs of the very largest whales, fin whales, blue whales, and their relations. They have adipose tissue in the mesentery that supports the gut and around the kidneys and intermuscular adipose tissue in certain muscles, just as do terrestrial mammals.

VI. ENERGY STORAGE IN THE LIFE HISTORY

A. Optimizing Energy Stores

Healthy animals living wild usually have approximately 4–8% dissectible adipose tissue under ideal nutritional conditions but may transiently be much fatter just before some special event such as breeding, migration, or hibernation. Many animals that have less than 4% dissectible adipose tissue are probably having trouble finding enough food but not yet in danger of starvation; for others, however, having much less adipose tissue seems to be normal. Some common wild animals, such as rabbits, hares, moles, and foxes, seem to live and breed satisfactorily with less than 1% of the body mass being adipose tissue.

The concept of costs and benefits to accumulating fat stores has led to much theorizing but little experimentation about the ecological conditions that might define optimum fatness and how they change during the animal's life history. Some computer models incorporate data about the relationship between body composition and energy utilization and simulate real situations quite accurately, predicting how much adipose tissue is appropriate for birds engaged in certain activities under certain conditions. However, direct measurements of many species in the wild show that there is often much variation in fatness among members of the same species at the same times of year which cannot easily be linked to habits or the ecological conditions in which they live.

Polar bears are top carnivores living in the Arctic, where the climate is not only harsh but also extremely variable from time to time and place to place. These bears are among the very few large carnivores to have huge home ranges but no proper territories. They travel long distances between areas in which ice conditions make seal hunting feasible, eating large but highly irregular meals. Moderate levels of obesity seem to be essential to such habits. A sample of young adult bears studied in the Canadian Arctic in early November, when the sea ice was just beginning to freeze over and seal hunting was again becoming efficient, indicated that the bears comprised between 10 and 21% dissectible fat, which amounted to 73 kg of adipose tissue for the largest bear whose total body mass was more than 400 kg. Every year, a few individuals, adults as well as newly weaned juveniles, are found in a severely emaciated condition, sometimes attributable to an injury but sometimes apparently just following a run of bad luck in catching seals or finding carrion. Their superficial adipose tissue, which over the rump might be more than 10 cm thick in a successful hunter, has shrunk to a flimsy sheet of watery tissue. Their muscles are wasted, so the hips are narrow and the legs thin, and at first glance they look more like a dog than a bear.

Average fatness can be very different in different localities (and probably also from year to year, although there is less information on this topic), even for wild animals of the same species. For example, the common red fox is usually lean where food is available regularly but predators abound, even when, as now often happens, it frequents towns and eats the remains of people's fatty meals. Populations living in Scandinavia and parts of Canada, in which the food supply is very unpredictable and predators and potential competitors are scarce, are substantially fatter than those elsewhere in Europe.

Birds that are fat for only part of the year just before migration may have habits that reduce the risks associated with being heavy and sluggish. They may live and feed in flocks or herds in which the "safety in numbers" principle applies or spend time only in places where predators are few or absent. Small birds such as great tits that feed only by daylight become measurably heavier toward evening and lose weight during the hours of darkness, especially if the night is cold or prolonged. The amount of fat that they carry from day to day correlates with the likelihood of obtaining food, which depends on the presence of suitable prey, being senior enough in the "pecking" order to get sufficient access to it, and the prevalence of predators. Great tits in Britain became significantly heavier when their principal predator, the sparrowhawk, disappeared in the late 1950s and 1960s as a result of poisoning from agricultural insecticides. When improved farming practices allowed sparrowhawks to re-establish large populations in western Britain in the 1970s, the mean body mass of great tits declined, although that of wrens, a species rarely caught by hawks, remained unchanged.

B. Migration

In contrast to walking, running, and swimming, the muscle power required to stay airborne by active flapping flight increases disproportionately with increasing body mass. Other factors being equal, smaller birds can generate substantially more power than is needed to keep airborne, so they can carry proportionately more fuel, up to 50% of the body mass if necessary. Larger wings make flying energetically more efficient, so the longest nonstop journeys are undertaken by medium-sized birds, such as sandpipers, knots, turnstones, curlews, and godwits, with a body mass of approximately 100–800 g.

Small birds with a body mass of approximately 100 g that set out with adipose tissue triacylglycerols amounting to 100% of lean tissue mass (i.e., half the body mass is fuel) can fly for 3 or 4 days nonstop or 3000 to 4000 km (depending greatly on wind direction and other weather conditions). Flying time can be prolonged by slowing down, or it can be shortened by speeding up or by flying into a headwind. In general, larger birds can fly faster, and so arrive at their destination sooner, but because they can carry less fuel they often cannot fly as far as medium-sized birds. Birds with a lean body mass of approximately 100 g lose about 0.7% of the body mass per hour while migrating, but larger birds of approximately 500 g flying similarly loaded lose 1–1.5% per hour.

The rate of energy expenditure decreases as the birds become lighter through oxidization of fat. As they set out fully loaded, each gram of triacylglycerols takes them less than half as far as at the end of their journey, when their fuel is almost exhausted, because flying with half the body mass as fat uses three times as much muscle power as flying on "empty." A bird that sets out with half its total body mass as fuel and flies to exhaustion uses 40% more energy for the whole journey than one that sets out with a fuel load of 10% of the body mass, flies as far as it can, stops to feed, and continues on the next lap, repeating the cycle until it reaches its destination. Such a journey is slower because even with unlimited food, birds cannot fatten faster than about 10% of the body mass per day (usually closer to 3–6% per day). Therefore, increasing the body mass by 25% takes at least a week, and 3 or 4 weeks are needed to accumulate maximal fuel reserves.

The amount of fuel a large bird can carry is strictly limited, and most journeys begin with a full load and end with little to spare. Whooper swans (*Cygnus cygnus*) that fly between northwest Scotland and their breeding grounds in central Iceland are among the largest of all migratory birds. The swans set out with adipose tissue triacylglycerols amounting to about 20–25% of their lean body mass, the largest load they can take off with. If they encounter bad weather or adverse winds, they land on the sea and wait, sometimes for 30 h or more, until traveling conditions improve enabling them to continue their journey.

C. Hibernation

Hibernation and its hot weather equivalent, estivation, are periods of inactivity and seclusion, usually accompanied by cooler than normal body temperature, which enable animals to pass through seasons when food is scarce or inaccessible. Hibernation was among the first physiological states in which adipose tissue of wild mammals, reptiles, and amphibians was studied thoroughly. They depend on their fat reserves while their body temperature is too low to allow the collection and digestion of food, but reclaiming the stores is not entirely straightforward.

The enzymatic processes involved in fasting and starvation are essentially similar to those of slow exercise, but there is a critical difference. During exercise, the body is warm, often slightly warmer than when sedentary, but in hibernation the body temperature is low, sometimes close to 0°C or 35°C below normal. Enzymes do not work on solidified fats any more than they function in frozen water. Animals must still be able to metabolize the triacylglycerols in their adipose tissue, albeit much more slowly than when they are fully active. The fatty acid composition of triacylglycerols is largely irrelevant to their role as fuel when animals are warm, but it is crucial for their use during hibernation.

Experiments on captive chipmunks (*Eutamias amoenus*) and golden-mantle ground squirrels (*Spermophilus lateralis*) show that they enter hibernation more readily, remain cooler for longer, and are better able to survive long winters when plenty of polyunsaturated lipids are included in their diet during the weeks preceding hibernation than when they are fed saturated fats of similar calorific value. Unsaturated fatty acids may lower the melting point of the triacylglycerols and the membrane phospholipids, enabling them to remain more fluid, and hence retain their proper affinity for carrier molecules and enzymes, at cooler temperatures. The chemical composition of the storage lipids, together with other aspects of adipose tissue, is thus adapted to the physical conditions and its role in whole-body metabolism.

Squirrels obtain many such unsaturated fatty acids from the seeds and other plant parts that they eat. Hibernation is an active, physiologically controlled process, and metabolic preparations can be identified days or weeks before the animal actually allows its body to cool. As the weather becomes cooler and the days shorten in autumn, squirrels and other hibernatory rodents actively seek nuts and other foods that contain these lipids. The woodchuck or marmot (*Marmota flaviventris*) selectively retains linoleic acid (C18:2) before hibernation. The saturated fatty acids are released and oxidized by the muscles, liver, etc. while the animal is warm and active, but the polyunsaturates remain in the adipose tissue for use when the body is cold. Like other mammals, the squirrels cannot add more than one double bond to most kinds of long-chain fatty acids, so they depend on the increased availability in autumn of seeds rich in polyunsaturated lipids. Failure of a seed crop could prevent successful hibernation and thus lead to death from cold or starvation, even if plenty of other foods were available. Recent measurements from biopsies of gonadal and inguinal adipose tissue from alpine marmots (*Marmota marmota*) throughout the year indicate that selective release of certain fatty acids allows active regulation of the composition of storage triacylglycerols. The storage lipids remain in an appropriate state of fluidity both during deep hibernation and in the euthermic state by maintaining a high proportion of monunsaturates by retention of these fatty acids from the diet and, if necessary, by synthesis.

Sleeping undisturbed in a cool, secluded place uses storage materials only very slowly. Small tortoises may lose less than 5%, and rarely lose more than 15%, of the body mass during 5 months of hibernation at about 5°C, and much of this loss is water that is lost through evaporation from the lungs. Many reptiles and amphibians emerge from hibernation with a surprisingly large proportion of their lipid reserves still remaining and regain what they lost during the winter in a few weeks of feeding. These stores not only come in handy during spells of cold weather when food is scarce but also are often important for fueling mating and egg production. The less lipid a hibernator uses during the winter, the more it has left to fuel breeding the following spring.

D. Reproduction

The formation of large, yolky eggs that nourish the embryo until it is at an advanced stage of development is a very ancient means of reproduction among vertebrates. Most sharks, skates, and rays, whose ancestors were abundant in Devonian seas 400 million years ago, produce such eggs, which hatch into miniatures of the adults capable of feeding for themselves.

One-third of the fresh mass of the yolk of birds' eggs is storage fats, a proportion that may represent the maximum that can form a stable emulsion, holding the lipids in a state that the embryo or its mother can manage. The relative sizes of the yolk and the "white" part of eggs differ greatly between species, being lowest in species that feed their young in the nest for some time after hatching, such as pelicans (17% of the mass of the egg), gannets (18%), and crows (19%). Yolk content is highest in species whose relatively large eggs hatch to produce large, mature chicks that can walk well and feed themselves. The yolkiest eggs are those of kiwis (65%), megapodes (also known as brush turkeys or moundbuilders) (more than 51%), ducks (40%), and of course species of the order Galliformes, which includes turkeys, peafowl, pheasants, guineafowl, quail, and domestic poultry.

In developing domestic chickens, the cells lining the gut start "eating" droplets of yolk on approximately the 12th day of incubation and pass its lipids into the blood as lipoproteins. Simultaneously, the adipose tissue matures and starts to produce lipoprotein lipase in large quantities, enabling it to take up the yolk lipids. Both the number and the mean size of adipocytes increase until at least the 19th day of incubation and continue to increase after hatching on the 21st day. The adipose tissue serves as more than just a temporary store. As a cellular tissue rather than just a sack of yolk, its adipocytes and the lipoprotein lipase they produce discriminate between triacylglycerols of different fatty acid composition and thereby "ration" the embryo's irreplaceable lipid provisions to ensure that it develops properly.

Lipids are the main agents of transfer of energy from mother to offspring in most vertebrates and many other kinds of animals, including insects. Mammals are somewhat unusual in that their eggs are very small and contain almost no yolk because soon after fertilization they attach themselves to the lining of the uterus, forming a placenta through which the embryo obtains all the nutrients it needs. Lipids cross the placenta and enter the fetus's blood, but most seem to be deployed in structural roles, particularly the formation of membranes of the brain and nervous system, rather than being broken down for energy production. The situation changes abruptly at birth, when the neonate starts to suckle. The milk of most mammals is rich in triacylglycerols but contains only small quantities of carbohydrate as lactose. During suckling, carbohydrates and lipids almost swap the roles they had in the fetus. Suckling mammals convert lactose into glucose, most of which is incorporated into structural materials, often as "finishing touches" to proteins destined to move between cells or to attach to cell surfaces, and lipids become the main source of fuel.

As soon as the mother gives birth, an array of changes to hormones, their receptors, and enzymes gives the mammary gland preferential access to circulating lipids. The hormones that prompt the maturation of the mammary gland during the final stages of gestation and trigger milk secretion after birth also act on adipose tissue, disabling its uptake of triacylglycerols from the blood and prompting it to release its fatty acids. These changes direct fatty acids from the adipose tissue to the mammary gland, in which they are incorporated into milk triacylglycerols.

Although storage lipids contain only long-chain fatty acids, the enzymes of the mammary gland readily esterify medium-chain and even short-chain fatty acids into triacylglycerols. The precursors of milk of adequately fed animals come straight from the gut without passing via the adipose tissue so they may include fatty acids that never appear in storage triacylglycerols. However, during fasting, the adipose tissue becomes the major source of fatty acids, so the milk triacylglycerols from starving animals contain more long-chain fatty acids.

The composition of fatty acids in milk triacylglycerols is thus quite variable in many mammals, depending on what the mother has just eaten and how much is derived from her adipose tissue or obtained directly from the diet. The contribution of storage lipids is important for wild animals that come into contact with lipid-soluble pollutants such as DDT (dichlorodiphenyltrichloroethane, widely used as an insecticide from the 1940s to the 1970s) and PCBs (polychlorinated biphenyls, formerly used as components of large batteries and other electrical apparatus). Such contaminants eaten during lactation may go straight into the milk, and in fasting mothers such as bears, those that have accumulated in adipose tissue during the previous months are released with the fatty acids and thus transferred into the milk. These effects can produce alarmingly high concentrations of toxins in the offsprings' tissues, permanently impairing growth and development.

For some mammals, food supplies are too erratic, or obtaining them requires too many long absences from the nest, for lactation to be fueled simply by eating more. Milk production has to draw on the body's reserves of lipid from adipose tissue, calcium, and other minerals from the skeleton and protein from it, probably mainly muscle and liver. Therefore, many female mammals fatten during pregnancy and deplete their adipose tissue lipids during lactation.

The diet of most animals that are large as adults changes as they grow. Many lizards and snakes eat insects, slugs, and other small prey when young and, for obvious reasons, defer tackling larger prey until they are large and experienced. However, the composition of the diet of weanling mammals changes more radically and abruptly. The adjustment is just one of the many physiological adaptations of both mother and offspring to this uniquely mammalian means of nurturing the young. The lactation habit has the enormous advantage that it emancipates mammals from the limitation of breeding only where and when suitable food is available for the young and permits rapid growth of the neonate.

Lipid storage in adipose tissue is central to this breeding strategy, enabling mothers to store enough materials to synthesize milk at a high rate and the young to take full advantage of the nutrients their mother provides during the often brief period during which they are together. Its role is enhanced in large species that live in highly variable habitats. Polar bears (Figs. 4 and 5) can carry a large amount of storage tissue because they have no natural predators (other than humans) and hunt by stealth and skill rather than by speed and agility. The bears' massive adipose tissue enables females to separate feeding and mothering to a greater extent than can almost all other warm-blooded animals, thereby enabling them to breed successfully despite the vagaries of food supplies in the Arctic.

Marine mammals can afford to be fatter, at least for short periods, than can terrestrial species because they do not attempt to move far or fast on land and therefore are less constrained by the mechanical limitations of fatness. There are no large terrestrial predators in the Antarctic, so seals breeding on beaches or on sea ice can afford to suckle their young for several weeks. However, those living in and around the Arctic are obliged to share their habitat with polar bears, and they have evolved ways of transferring lipids from mother to young at a spectacular rate. Harp seals (*Phoca groenlandica*) that breed on pack ice throughout the Canadian and Russian Arctic are only 3% lipid by weight at birth but reach 47% lipid at weaning 13 days later. The single pup drinks an average of 3.7 kg of milk each day, gaining weight at an average rate of 2.3 kg per day. Transfer of lipid from mother to offspring increases as the young matures. The lipid content of the milk increases from 36% at the start of lactation to 57% just before weaning.

Lactation in the hooded seal (*Cystophora cristata*), which breeds on temporary ice floes in the North Atlantic, is even more compressed. The single pup weighs about 22 kg at birth, which is large relative to adult size but it is only slightly fatter than harp seal pups. It almost doubles its body mass to approximately 43 kg in only 3–5 days of suckling. More than 70% of the increase in body mass is blubber, which holds enough lipid to sustain the pup during a postweaning fast that can last several weeks. During the 4 days of lactation, the mother's body mass declines from 179 to 150 kg, with more than 80% of the loss from the blubber. She thus synthesizes milk at 2.5–6 times the rate of other seals that have been studied, which is probably a record for any mammal. Such haste may be necessary to minimize the risk that mother and pup become separated by storms and currents that move ice floes around and that of predators.

Most mammals, especially carnivores and other species for which skill and experience are essential for finding food, are weaned with storage tissues replete. Much of the triacylglycerols transferred to suckling seals are stored in their blubber and utilized during the long period between weaning and the time at which the pups become proficient at finding food for themselves.

See Also the Following Articles

CARNIVORES • DIAPAUSE AND DORMANCY • MIGRATION

Bibliography

Bard, S. M. (1999). Global transport of anthropogenic contaminants and the consequences for the Arctic marine ecosystem. *Mar. Pollution Bull.* **38**, 356–379.

Florant, G. L. (1998). Lipid metabolism in hibernators: The importance of essential fatty acids. *Am. Zool.* **38**, 331–340.

Frank, C. L., Dierenfeld, E. S., and Storey, K. B. (1998). The relationship between lipid peroxidation, hibernation, and food selection in mammals. *Am. Zool.* **38**, 341–349.

Lindström, Å., and Alerstam, T. (1992). Optimal fat loads in migrating birds: A test of the time-minimization hypothesis. *Am. Nat.* **140**, 477–491.

Murphy, D. J. (1994). Biogenesis, function and biotechnology of plant storage lipids. *Prog. Lipid Res.* **33**, 71–85.

Murphy, D. J., and Vance, J. (1999). Mechanism of lipid-body formation. *Trends Biochem. Sci.* **24**, 109–115.

Pond, C. M. (1996). Interactions between adipose tissue and the immune system. *Proc. Nutrition Soc.* **55**, 111–126.

Pond, C. M. (1998). *The Fats of Life.* Cambridge Univ. Press, Cambridge, UK.

Pond, C. M. (1999). Physiological specialisation of adipose tissue. *Prog. Lipid Res.* **38**, 225–248.

Pond, C. M., and Gilmour, I. (1997). Stable isotopes in adipose tissue fatty acids as indicators of diet in arctic foxes (*Alopex lagopus*). *Proc. Nutrition Soc.* **56**, 1067–1081.

Pond, C. M., and Ramsay, M. A. (1992). Allometry of the distribution of adipose tissue in Carnivora. *Can. J. Zool.* **70**, 342–347.

Ramsay, M. A., Mattacks, C. A., and Pond, C. M. (1992). Seasonal and sex differences in the cellular structure and chemical composition of adipose tissue in wild polar bears (*Ursus maritimus*). *J. Zool. London* **228**, 533–544.

Speake, B. K., Murray, A. M. B., and Noble, R. C. (1998). Transport and transformations of yolk lipids during development of the avian embryo. *Prog. Lipid Res.* **37**, 1–32.

Witter, M. S., and Cuthill, I. C. (1993). The ecological costs of avian fat storage. *Philos. Trans. R. Soc. London B* **340**, 73–92.

STRESS, ENVIRONMENTAL

John Cairns, Jr.
Virginia Polytechnic Institute and State University

I. What Is Environmental Stress?
II. Types of Environmental Stress
III. Stress Assessments
IV. Future Trends

GLOSSARY

acute An exposure to an environmental stress that is brief in relation to the temporal scale of the biological system exposed.
chronic An exposure to an environmental stress that is comparable in duration to the temporal scale of the biological system exposed.
ecosystem services The structures and functions of natural biological systems which directly or indirectly support human life.
environmental stress An action, agent, or condition that impairs the structure or function of a biological system.
function The performance of a biological system as a rate.
receptor A biological system that is exposed to the environmental stress.
response A particular structure or function of the receptor that is changed by exposure to the environmental stress.
structure The number, kinds, and arrangement of component parts at one point in time.
threshold The point at which a response begins to be produced.
uncertainty Imperfect knowledge concerning the current or future state of a system under consideration; a component of risk resulting from imperfect knowledge of the degree of hazard or of its spatial and temporal pattern of expression.

ENVIRONMENTAL STRESS is an action, agent, or condition that impairs the structure or function of a biological system. The terms "disturbance" and "perturbation" often are used to describe this concept. Examples of environmental stresses on biodiversity include floods, fire, drought, hurricanes, volcanic activity, climate change, land-use changes, introduction of exotic species, and chemical pollution.

I. WHAT IS ENVIRONMENTAL STRESS?

Three components are involved in the relationship that defines environmental stress. First, there is the environmental stress itself as defined previously. However, the environmental stress can only be defined in reference to its interaction with some biological system. Therefore, there must be a receptor—a biological system that is exposed to the environmental stress. Finally, there must be an adverse response—a particular structure or function of the receptor that is changed by exposure to the environmental stress to the detriment of that system. If the survival of that biological system (or another) is not threatened by the change, then there is no environmental stress.

II. TYPES OF ENVIRONMENTAL STRESS

A. Natural vs Anthropogenic

Environmental stress can be either natural or anthropogenic (i.e., resulting from human actions) in origin. Many environmental stresses, such as most hurricanes, droughts, floods, and fires, are a periodic feature of life on Earth. In contrast, environmental stresses such as the production and release of new chemical compounds and large-scale land-use changes result directly from human action. Ironically, the suppression of natural environmental stresses such as fires and floods can also be a source of stress to biological systems resulting from human actions. Some species have adapted to periodic disturbance and cannot continue without them. For example, some seeds will germinate only after exposure to the high temperatures of a fire. However, such adaptations take time to evolve. Other stresses, such as the meteor that probably wiped out 70% of Earth's species 65 million years ago, occur too quickly and intensely to result in adaptation.

Natural and anthropogenic stresses often have common components. For example, both hurricanes and wood harvesting result in downed trees. However, as a result of recent major increases in human population, technological capabilities, and standard of living globally, the amount of anthropogenic environmental stress has increased greatly. Anthropogenic environmental stress existed even 50,000 years ago, when fires set to aid hunters are thought to have altered the landscape in central Australia (Flannery, 1999). Since that time, human population increased slowly to 1 billion people in 1804 and then rapidly to 6 billion people in 1999, and the population is expected to increase by another 1 billion people every 12–15 years. The cumulative environmental stresses resulting from these exponential increases in human population are exacerbated by technologies that expand the character and scope of changes humans can make to their environments. Both the agricultural revolution (about 10,000 years ago) and the industrial revolution (about 200 years ago) expanded the types of anthropogenic stresses on the environment. In addition, increased affluence for people throughout the world increases the environmental stress on natural systems by increasing the per capita human use of natural systems. Humankind's collective ecological footprint (i.e., the amount of Earth's surface required to produce the resources used and to assimilate the wastes produced) is rapidly increasing (Rees, 1996) at the same time that productive land is decreasing through erosion, salinization, and unsustainable land-use changes. The ecological footprint for an average person can range from 0.4 ha required to provide for the lifestyle of one person in India, where the level of affluence is quite modest, to 5.1 ha for a person in the United States. If the lifestyle of every person living in 1996 was elevated to that of a typical North American, an additional two Earths would be needed to provide the surface area required (Rees, 1996).

B. Characterization of Environmental Stress

Environmental stresses of both natural and anthropogenic origin can be characterized on the basis of their spatial distribution, temporal distribution, intensity, and novelty (Kelly and Harwell, 1989). Spatial distribution of a stress describes its geographic extent and pattern. One basic concern is the size of the area affected by the stress, i.e., is the stress local, regional, or global in extent? Some kinds of environmental stress have an intrinsic spatial scale: A poorly dispersed chemical spill may be quite local in its effects, whereas air pollution can affect entire regions. Some stresses will be consistently spread over an area, whereas others will occur irregularly in patches. Cumulative environmental stresses, in which many individual small patches can join together to have larger impacts at a larger scale, have been documented. In one example at the local level, the cumulative loss of small wetland areas within a watershed had demonstrable adverse effects on water quality (Johnston et al., 1988). The magnitude of the biological responses to a stress will often be modified by the spatial distribution of environmental stress, for example, whether key features in the environment such as riverbanks or fencerows are affected or whether similar habitat patches nearby are left unaffected. The field of landscape ecology deals with these factors.

The temporal distribution of a stress describes its frequency and duration. Some stresses, such as chemical spills, are one-time occurrences. Others, such as the winter season or wildfires, can be expected to reoccur either at predictable or at unpredictable intervals. Some stresses rapidly ameliorate; others remain for long periods of time. The terms "acute" and "chronic" are used to describe the duration of a stress. Acute refers to stresses of short duration, whereas chronic refers to stresses that last longer in relationship to the duration of the biological system they affect. In some cases, the timing of the stress in relationship to other biological events may modify the magnitude of the biological response. For example, the spawning season for amphibians in northern North America in spring corresponds with the greatest thinning of the ozone layer and ultraviolet (UV) light penetrations. The developing eggs of some of these species have been shown to be affected adversely by UV light (Blaustein and Wake, 1995). This temporal factor may be contributing to observed declines in the numbers and kinds of some amphibian species present in those areas.

The intensity of an environmental stress describes its relative ability to evoke a response from the receptor. With increasing intensity, the impact may progress from a few, slightly affected, particularly sensitive components of the biological system to most components being grossly affected. Small changes in the histology or physiological state of individual species can be expected to occur at lower stress intensity and chronologically before changes in survival at a similar temporal and spatial scale. Similarly, at the community level, changes in species composition can be expected to occur at lower stress intensity and chronologically before changes in community functions. The intensity of hurricanes is routinely ranked from category 1 to category 5. In the case of chemical pollution, the intensity of the environmental stress can be described by the concentration of the chemical in the environment. Thus, copper is a natural, background constituent of water in a river or stream. Intake of copper is essential for both animal and plant life. However, concentrations higher than 10 μg/liter in river water can be expected to eliminate a few sensitive species and change the age class distributions in others by changing reproductive success. Higher concentrations can be expected to affect more components in more obvious ways. For example, in Shayler Run, Ohio, 120 μg/liter of copper caused fish kills at some times of the year, avoidance of the stream reach by other fish, declines in macroinvertebrate community species richness, and other gross responses (Geckler *et al.*, 1976).

The novelty of an environmental stress will determine whether or not biological systems will have mechanisms in place to deal with it. Environmental stresses that resemble naturally occurring stresses in their mode of action will be dealt with by the system in the same ways. Novel stresses may be more devastating because no mechanisms have evolved to cope with them. For example, when human harvests of wood products mimic relatively frequent natural events such as treefall or windthrow in their spatial extent, mechanisms are in place in the biological system to recover from this event. When human-created gaps are larger and more intense than historical disturbances, these same mechanisms may not help. Similarly, some natural systems can break down, render biologically unavailable, or disperse low levels of some chemical materials that are naturally occurring or that resemble naturally occurring substances without detectable disruption. The ability of a natural system to receive materials at some concentration, including anthropogenic wastes, without being degraded is its assimilative capacity. However, overloading a system with too much waste destroys both the structure and function of the ecosystem and its future assimilative capacity.

C. Receptors and Responses

An action, agent, or condition can be a stress to one biological system while simultaneously not affecting many others. Because environmental stress is defined by the observation of an impaired biological system, the probability of identifying an environmental stress is related to how thorough the search for impairment has been. Although there are an almost unlimited number of receptors and responses that could be affected by any particular environmental stress, it is generally impractical to monitor more than a small sample. Also, experience has shown that it is easy to overlook a response that may be important. For example, DDT caused eggshell thinning in some birds, although routine toxicity tests failed to identify this response.

The most useful responses to examine for studies that aim to influence decisions about environmental management tend to be those that are clearly related to stated environmental goals. These tend to be responses that are both biologically and socially relevant and that can be measured reliably. Sometimes, the responses that society cares most about cannot be measured directly. Other, presumably related, responses can be measured. In addition, responses that occur earlier in a chain of events, and which lead to an ecologically relevant event, may be useful as early warnings of conditions that have the potential to cause unacceptable damage. Table I summarizes some responses that have been used to evaluate environmental stress and guide environmental management.

As the goals in environmental protection have changed over time, so have the responses that are monitored. Most early tests of environmental stress were designed to protect one species—humans. This objective spurred tests that monitored the physiology of species used as human surrogates. Gradually this protection was extended, first to domesticated animals and

TABLE I

Examples of Receptors and Responses Used in Studies of Environmental Stress

Level of biological organization	Structural responses	Function responses
Individual	Condition	Growth
	Fat stores	Fecundity
	Histopathology	Physiological function
Population	Occurrence	Yield
	Abundance	Gross morbidity
	Age structure	
Community	Species richness	Production
	Trophic structure	Respiration
	Proportion of exotics	Extinction rate
Ecosystem	Nutrient pool size	Materials cycling
	Biomass	Materials export
Landscape	Habitat proportions	Regional production
	Patch size	Materials export
	Perimeter-to-area ratio	Resistance to stress

plants and then to commercially valuable wild species. These additional tests monitored the survival of populations of "important" species. Currently, goals extend beyond the protection of individual species and include the protection of biodiversity and ecosystem services (i.e., those structures and functions of natural biological systems that directly or indirectly support human life). Assessments should reflect these new goals because millions of species in the environment must be protected. Each species cannot be examined individually. In practice, a few species must serve as surrogates for many others, and a few systems serve as surrogates for many others.

Environmental responses may be characterized by type and scale. Responses are either structural (e.g., describing the number and kinds of components, such as the macroinvertebrate community structure) or functional (e.g., describing performance or flux, such as biological oxygen demand or primary production). Also, unique responses may occur at many distinct spatial and temporal scales and levels of biological organization (e.g., cells, tissues, organs, organisms, populations, communities, ecosystems, landscapes, biomes, and the world). Some attributes at higher levels of biological organization are not present at lower levels; for example, energy flow and nutrient spiraling are properties of ecosystems but not of organisms. Other attributes are present in some form at many levels; for example, one can measure the diversity of phenotypes at the population level and the diversity of species at the community level. Environmental goals can be stated on many of these levels, but tests of environmental stress are largely limited to those levels that are more accessible to human observation.

An awareness of scale provides two contrasting approaches to studying environmental stress. Top-down methods start with observed damage to a biological system of interest and investigations move down through hierarchical levels. Component structures and functions are examined in order to diagnose the causative agent and plan remedial actions. At the outset, the damage has already been done, so the relevance of the changes is known. However, the causative agent and the chain of events leading to unacceptable damage are not known. Bottom-up methods start with an environmental stress, and the effects of that stress on biological systems are determined through designed experiments. Because experiments on small and quick biological systems at lower scales are generally less expensive, these experiments are most common. In bottom-up assessments, the causative agent is known at the outset, but the importance of ultimate changes at any ecologically relevant higher scale is not known.

Microcosms and mesocosms are attempts to increase the spatial and temporal scales and level of complexity in biological systems that can be used in designed experiments of environmental stress. Microcosms and mesocosms simulate important attributes of natural systems in laboratory or outdoor conditions. As the names indicate, the main difference is in size. Microcosms are sometimes small enough to hold in one's hand; mesocosms may cover 1 acre. Neither is an exact reproduction of any real ecosystem, but they do enable studies of environmental stress in ways that avoid damaging natural systems. On rare occasions, environmental stress may be studied in designed experiments using entire ecosystems. For example, one whole system manipulation was carried out in the Hubbard Brook drainage basin in New York State (Bormann and Likens, 1979). Such efforts are of great value in calibrating models.

D. A General Environmental Stress Syndrome

A threshold is defined in Webster's *Third International Dictionary* as "the point at which a physiological or psychological effect begins to be produced." Moving upwards in biological systems, this effect can be generalized to the point at which a response begins to be produced. Woodwell (1974) asked the question, "Is it reasonable to assume that thresholds for effects of disturbance exist in natural ecosystems or are all disturbances effective, cumulative, and detrimental to the normal functioning of natural ecosystems?" Thresholds may be artifacts of testing procedures, reflecting the power of particular test designs rather than a feature of the system being studied. However, perhaps the more important question is "Can humans detect those environmental changes that are important to their own quality of life?" As is the case with human health, the gradient in environmental systems may be extensive between robust health and collapse in some cases, but an abrupt transition from health to collapse may occur in others. By reviewing information available about the behavior of ecosystems under stress, several researchers have tried to outline general ways in which ecosystems respond to various types of stress (Barrett et al., 1976; Odum, 1985; Rapport et al., 1985; Schindler, 1990). An environmental general stress syndrome at the ecosystem level may include the features listed in Table II;

however, experience is continually modifying this list. These efforts to derive an environmental stress syndrome are important because they define a progression of impact in which some minor changes precede other more serious ones. By recognizing changes early in the progression of impact, remediation could begin and crises could be averted. However, the challenge of finding one general description for widely varying systems, challenged by widely varying combinations of stress, is daunting.

Stressed ecosystems often recover once the stress has been removed. However, sometimes human assistance is required, and this process is called ecological restoration or rehabilitation. Restoration has as its goal the return of an ecosystem to a close approximation of its condition prior to the stress and the recreation of a functioning, self-regulating system that is integrated into the ecological landscape in which it occurs. The practice of ecological restoration often involves the reconstruction of physical conditions present prior to the stress, chemical cleanup, and biological manipulation, including revegetation and the reintroduction of native species.

TABLE II

Responses Expected in Stressed Ecosystems[a]

Energetics
   Community respiration increases
   Gross production/community respiration (P/R) becomes unbalanced
   Maintenance cost increase; gross production/standing crop biomass (P/B) and community respiration/standing crop biomass (R/B) ratios increase
   Importance of auxiliary energy increases
   Exported or unused primary production increases
Nutrient cycling
   Nutrient turnover increases
   Horizontal transport increases, vertical cycling of nutrients decreases
   Nutrient loss increases
Community structure
   Proportion of r-strategist increases
   Size of organisms decreases
   Life spans of organisms or parts decrease
   Food chains shorten
   Species diversity decreases, dominance increases, redundance declines
General system level trends
   Ecosystems become more open
   Autogenic successional trends reverse
   Efficiency of resource use decreases
   Parasitism increases, mutualism decreases
   Functional properties more robust than structural properties

[a] After Odum (1985).

III. STRESS ASSESSMENTS

Studies of environmental stress can have different purposes. In some studies, the purpose is accounting, i.e., what is the existing condition of this biological system? This question can be important for the purposes of disclosure, national environmental accounting, prioritization, and remediation. Studies of environmental stress can also be used for prediction, i.e., will this action cause a problem or which action is better? Predictive studies are used to register chemicals, rank risks, design processes, etc. Another distinct purpose for studies of environmental stress is to provide early warning of conditions that, if left unchecked, will result in damage significant to human quality of life. By detecting damage before it is of a magnitude that is unacceptable, crises can be averted.

A. Appraisal

Studies of environmental stress can assess the condition of biological systems that exist at a particular point in time. When repeated over time, trends in condition can be assessed. Appraising the condition of a biological system can confirm that environmental quality is adequate or can serve to define an existing environmental problem. Many countries are undertaking a national accounting of the health of their ecological systems. For example, the Canadian State of the Environment Reports and the Environmental Monitoring and Assessment Program in the United States measure the condition of rivers, lakes, forests, wetlands, arid lands, and agroecosystems. Although some of these programs measure common sources of environmental stress, as well as biological response, others focus solely on response. Once a problem is found, additional studies to diagnose the problem would include measures of environmental stress.

B. Prediction

Often, the easiest way to maintain environmental quality is to prevent damage before it occurs. This strategy requires prediction of the future. Such predictions can come from observations of the effects of similar stress on similar systems or from extrapolations or models from the effects of dissimilar stresses or the effects on dissimilar systems.

The ability to extrapolate the measured effects of environmental stress at one level to consequences at a higher hierarchical level depends on the use of mechanistic models that describe the interaction of component parts. The model is then calibrated, i.e., compared to observed behavior of a system under environmental stress and adjusted to maximize the accuracy of its predictions. Currently, there are a few calibrated models for large-scale predictions. Also, it is unlikely that calibrated models will be available for some processes because their spatial and temporal scales make testing impractical or unethical. All hypotheses and theories are more readily accepted if they have withstood rigorous testing. Predictions of environmental stress are no exception. As a general rule, multiple lines of evidence published in peer-reviewed professional journals whose contents have been reviewed by respected professionals result in acceptance both by the person carrying out research on environmental stress and by mainstream science. Generally, validation occurs in two primary ways: (i) a designed test of a hypothesis derived from a theory, especially by those having nothing to do with its development, and (ii) consilience with other well-accepted and tested theories. Predictions of environmental stress are particularly likely to be challenged outside the scientific community because taking precautionary measures to avoid conditions estimated to cause stress often requires changing societal and industrial practices and sometimes engenders costs. Environmental stress associated with global warming is a good illustrative example. Limiting greenhouse gases in the atmosphere will affect the lives of almost every person on the planet, as would significant global warming. In this case, the entire planet is the experimental unit, making designed tests of impact at the same hierarchical level as that of the environmental problem impossible since there is no "control" planet available. As a consequence, much uncertainty about the probable effect of increased greenhouse gases will persist. However, management decisions about such gases must be made and must be based on the best available information; managers must act even though there is uncertainty accompanying any estimate of risk. Fortunately, most forms of environmental stress, such as exposure to potential toxicants, are much more easily validated.

C. Early Warning

Monitoring is a systematic and orderly gathering of data to ensure that previously established quality control conditions are being met. Biomonitoring applies this activity to the detection of environmental stress. In this case, the goal is to provide an early warning that unacceptable levels of environmental stress have occurred. As is the case in an intensive care room in a hospital when heart and respiration rates are monitored and unacceptable conditions are detected, immediate action is mandatory. In any form of quality control, the more rapidly the information becomes available, the more quickly corrective action can be taken. Extensive use of information technology has made complete automation of some monitoring systems, including triggering the remedial action, possible. However, in the absence of carefully selected goals and objectives related to the decisions to be made, they can also generate huge amounts of unnecessary and inappropriate data.

Setting the corrective action threshold low to ensure early detection of deleterious change seems prudent, but it can also produce false-positive readings. A false positive is an indication that some deleterious effect has occurred when in fact none has occurred. Emergency team response to eliminate stress can be quite expensive and unpopular with management. False positives are usually most numerous when monitoring is in the early developmental stages and decrease substantially as experience is gained. Avoiding false positives by setting the action threshold well beyond the response threshold will probably result in false negatives. A false negative is information that no deleterious effects have occurred when in fact some have occurred. Thus, ecosystem damage occurs because no corrective action alert is produced. These false signals are clearly a matter of prime importance in the design of all monitoring systems. The dilemma is that one wishes to detect environmental stress at the earliest possible moment, but this may result in false signals since sensitive end points are often highly variable.

IV. FUTURE TRENDS

In view of unprecedented growth in human population and the desire to raise the standard of living above the subsistence level for most of the world's people, there are new challenges for those studying environmental stress. The level of environmental stress from food production and energy usage may increase. Simultaneously, there is great pressure to increase food production to feed the increasing human population, and the need to protect intact biological systems and ensure their robust functioning so that they can continue to provide necessary ecosystem services will become more pressing. Some approaches to these problems are of great interest.

The World Commission on Environment and Development (1987) of the United States published *Our Common Future*, which focuses attention on the future condition of the planet, arguably more so than any publication that preceded it. The commission defined sustainable development as "development that meets the needs of the present without compromising the ability of future generations to meet their own needs." This presents a curious combination of words because "sustain" means to continue and "development" is usually associated with growth. However, infinite growth on a finite planet is clearly not possible.

In December 1989, the General Assembly of the United Nations (UN) attempted to address the problems identified in *Our Common Future* (World Commission on Environment and Development, 1987) by organizing the UN Conference on Environment and Development (popularly known as the Earth Summit), which was held in Rio de Janeiro in June 1992. This conference resulted in a heightened awareness of the interrelatedness of environmental degradation and stress, population development, and depletion of natural resources. Previously, all had been viewed as separate problems, but now attempts are being made to address them as an interactive system. The resulting Rio Declaration on Environment and Development endorsed the following principles: (i) Nations should not cause damage to the environment of other states and areas beyond their borders; (ii) eradicating poverty and reducing disparities in worldwide standards of living are indispensable requirements for sustainable development; (iii) the polluter, in principle, should pay the cost of pollution; (iv) states should discourage or prevent transboundary movements of activities and substances that endanger health or environment; and (v) scientific uncertainty should not be a reason for postponing urgent measures to prevent environmental degradation.

Sustainable use of the planet will not be possible if excessive environmental stress impairs the ecological life support system that provides necessary ecosystem services. As a consequence, keeping environmental stress at tolerable levels has a direct bearing on the quality of life for humans. Events beyond human control, such as a large extraterrestrial object striking Earth, make it impossible to guarantee that reducing environmental stress will ensure sustainability.

The theory of "weak sustainability" asserts that human society is sustainable provided that the aggregate stock of manufactured and natural assets is not decreasing. Thus, the loss of the whaling industry would not impair sustainability if the proceeds of liquidation are invested in industries of comparable income-producing potential. Pearce and Atkinson (1993) dispute the assumption that natural and human-made capital are sustainable in this context. They assert that strong sustainability requires that natural capital stocks be held constant, independently of human-made capital. A weak sustainability scenario would permit considerably more environmental stress than a strong sustainability scenario. Conventional monetary analyses are biased against strong sustainability. For example, at a discount rate of 5%, the current value of ecological services for an American life span (about 76 years) from the present on is approximately 2.5¢. Using this approach, the farther into the future one projects, the less valuable natural systems appear, thus diminishing the significance of environmental stress.

Reducing environmental stress requires the cooperation of all of society. One of the most important components is the industrial system. Fortunately, the field of industrial ecology (IE) is developing worldwide (Hawken, 1993). The goal of IE is to reduce environmental stress at all stages: (i) extraction of raw materials, (ii) processing, (iii) disposal of manufacturing wastes, (iv) packaging, and (v) reincorporation into the environment at the end of the product's life in a nonstressful way, ideally in a way that enhances ecological integrity. Industrial ecology's primary goals are to (i) reuse materials as much as possible, (ii) reduce energy consumption per unit produced, and (iii) design both processes and products so that they can be reincorporated into the environment with minimal stress. Books such as *Engineering within Ecological Constraints* (Schulze,

1996), which was produced by the National Academy of Engineering, are directed toward achieving this goal.

See Also the Following Articles

DISTURBANCE, MECHANISMS OF • ECOLOGICAL FOOTPRINT, CONCEPT OF • ECOSYSTEM SERVICES, CONCEPT OF • HUMAN EFFECTS OF ECOSYSTEMS, OVERVIEW • STABILITY, CONCEPT OF • SUSTAINABILITY, CONCEPT AND PRACTICE OF

Bibliography

Barrett, G. W., Van Dyne, G. M., and Odum, E. P. (1976). Stress ecology. *Bio Science* **26**, 192–194.

Blaustein, A. R., and Wake, D. B. (1995). The puzzle of declining amphibian populations. *Sci. Am.* **272**(4), 52–57.

Bormann, R. H., and Likens, G. E. (1979). *Pattern and Process in a Forested Ecosystem*. Springer-Verlag, New York.

Flannery, T. F. (1999). Debating extinction. *Science* **283**, 182–183.

Geckler, J. R., Horning, W. B., Neiheisel, T. M., Pickering, Q. H., and Robinson, E. L. (1976). Validity of laboratory tests for predicting copper toxicity in streams, EPA 600/3-76-113. National Technical Information Service, Springfield, VA.

Hawken, P. (1993). *The Ecology of Commerce*. Harper Collins, New York.

Johnston, C. A., Detenbeck, N. E., Bonde, J. P., and Niemi, G. J. (1988). Geographic information systems for cumulative impact assessment. *Photogr. Eng. Remote Sensing* **54**, 1609–1615.

Kelly, J. R., and Harwell, M. A. (1989). Indicators of ecosystem response and recovery. In *Ecotoxicology: Problems and Approaches* (S. A. Levin, M. A. Harwell, J. R. Kelly, and K. D. Kimball, Eds.), pp. 9–35. Springer-Verlag, New York.

Odum, E. P. (1985). Trends expected in stressed ecosystems. *Bio Science* **35**, 419–422.

Pearce, D., and Atkinson, G. (1993). Capital theory and the measurement of sustainable development. *Ecol. Econ.* **8**(2), 103–108.

Rapport, D. L., Regier, H. A., and Hutchinson, T. C. (1985). Ecosystem behavior under stress. *Am. Nat.* **125**, 617–640.

Rees, W. E. (1996). Revisiting carrying capacity: Area-based indicators of sustainability. *Population Environ.* **17**, 195–214.

Schindler, D. W. (1990). Experimental perturbations of whole lakes as tests of hypotheses concerning ecosystem structure and function. *Oikos* **57**, 25–41.

Schulze, P. C. (1996). *Engineering Within Ecological Constraints*. National Academy Press, Washington, DC.

Woodwell, G. M. (1974). The threshold problem in ecosystems. In *Ecosystem Analysis and Predictions* (S. A. Levin, Ed.), pp. 9–21. SIAM Institute for Mathematical Society, Alta, VT.

World Commission on Environment and Development (1987). *Our Common Future*. Oxford Univ. Press, Oxford.

SUBSPECIES, SEMISPECIES, AND SUPERSPECIES

James Mallet
University College London

I. A Brief History of Subspecific Taxonomy
II. The Subspecies Today
III. Further Reading

DISTINCT POPULATIONS THAT REPLACE EACH OTHER GEOGRAPHICALLY were recognized either as full species or as lower level "varieties" or "forms" under the original Linnean taxonomy. A practical resolution of this ambiguity occurred in zoology largely between about 1880 and 1920 with the recognition of an additional taxon, the geographic subspecies. Since the 1980s, the fashion has changed once more. Some systematists are again recognizing geographical replacement forms as full species, even when they blend together at their boundaries.

I. A BRIEF HISTORY OF SUBSPECIFIC TAXONOMY

A. Variation below the Level of Species

Since the invention of binomial nomenclature by Linnaeus, there has been a conflict between "splitters," who named more or less well-defined local populations as separate species, and "lumpers," who ignored geographic variation and united local variants into a single species. The problem was compounded by early systematists' belief that species had an Aristotelian "essence," each fundamentally different from similar essences underlying other species. To Linnaeus' followers, it seemed important to decide which level of variation was fundamental. The terms "genus" and "species" both result from Aristotelian philosophy, and although Linnaeus is usually credited with establishing the species as the basal taxonomic unit, he confused matters, after recognizing that some plant species were of hybrid origin, by suggesting that genera were a more important taxonomic level (i.e., a separately created kind) than species.

Once evolution was accepted, it became clear that variation at all levels in the taxonomic hierarchy was due to more or less similar causes; the only difference between variation above the level of genus or species and below was one of degree. Darwin realized that species could evolve from intraspecific varieties. Darwin (1874) used the term species in a new and nonessentialist sense:

> The complete absence, in a well-investigated region, of varieties linking together two closely-allied forms, is probably the most important of all the criterions of their specific distinctness. . . . Geographical distribution is often brought into play unconsciously and sometimes consciously; so that forms living in two widely separated areas, in which most of the other inhabitants are specifically distinct, are themselves usually looked at as distinct; but in truth this affords no aid in distinguishing geographical races from so-called good or true species.

Darwin showed convincingly that there was no essential difference between species and "varieties"; species were simply varieties which had diverged more. However, with his term varieties, Darwin did not clearly distinguish between polymorphic variants within populations and the identifiable geographic populations that are today normally considered geographic "races" or "subspecies." To Darwin, the distinction was unimportant because polymorphic variants, clinal variation, geographic races or subspecies, and "good" species formed a continuum. Darwin demonstrated that this continuum was excellent evidence for an evolutionary origin of the taxa we call species.

B. The Trinomial Revolution

Many systematists wished to preserve the purity of the simple genus–species binomial nomenclature, but by the 1850s there were enormous stresses. It began to be realized that many clearly identifiable geographic replacement forms were an important intermediate stage between insignificant local variants and good species. Some lumped these replacement forms as varieties within species, whereas others continued describing these geographic forms as separate species: Practices varied widely, leading to considerable confusion. Although some Europeans had long advocated naming geographic forms as subspecies, the accumulation of major North American museum collections during the great push of colonization and railway construction westwards was probably the most important catalyst of a revolutionary new systematics. In this new approach, nomenclature consisted of a trinomial, genus–species–subspecies, which still is the dominant taxonomic practice today. The maxim was "intergradation [at the boundary between two geographic replacement forms] is the touchstone of trinomialism." Examples from commonly observed birds which intergrade are, in North America, the eastern rufous-sided towhee (*Pipilo erythrophthalmus erythrophthalmus*) replaced in the west by the spotted towhee (*Pipilo erythrophthalmus maculatus*), and in Europe the carrion crow (*Corvus corone corone*) found in the south and west replaced by the hooded crow (*Corvus corone cornix*) in the north and east. Among ornithologists responsible for this revolution in North America were Elliott Coues, who published a catalog of American birds in 1872 incorporating an early version of this trinomial nomenclature in which subspecies were prefixed by "var.", and Robert Ridgway, who dropped the "var." in his 1881 summary of North American bird nomenclature.

The American Ornithologists' Union soon adopted this policy, and the idea then spread to Europe, particularly England, where Walter Rothschild began amassing his vast collection of birds and butterflies and hired excellent and productive staff (the ornithologist Ernst Hartert and the entomologist Karl Jordan) to curate and describe the new material. Jordan was particularly important in spreading the idea of trinomial nomenclature to entomologists, and he was regarded by the Rothschilds as the "clever" member of the staff. He published important papers on the theory of systematics, justifying trinomial nomenclature and the recognition of the "subspecies" as a valid, identifiable taxon in its own right. Both Jordan and Hartert were Germans who contributed to and read German as well as English journals, and in Germany a similar revolution was taking place. Thus, these systematic ideas were able to spread to the rest of Europe during a time when science was often highly parochial. The standard trinomial nomenclature for subspecies soon became established in the International Code for Zoological Nomenclature and has remained there ever since.

C. Theories of Divergence

It is difficult to imagine the diversity of ideas by which the systematists of 100 years ago explained geographic variation. At that time, evolution by natural selection was far from generally accepted; in fact, many believed it had been disproved. One of the most influential ornithologists of the time was Otto Kleinschmidt, who believed that all species suddenly came into being long ago and thereafter remained completely separate. Replacement forms or subspecies developed via natural selection from the main species but, in Kleinschmidt's view, subspecies could never evolve into new species as the Darwinians supposed. To distinguish his new species concept from the older one in which geographical replacements might be named as separate species, Kleinschmidt called his theory of variation the *Formenkreis* (ring of forms) theory. The Formenkreis theory fitted neatly with, and indeed promoted, the new practices of naming subspecies and trinomial nomenclature.

In those times, there were many somewhat peculiar competing explanations for geographic variation and speciation, including Kleinschmidt's nonspeciation theory, J. P. Lotsy's hybridization theory, mutationism, inheritance of acquired characters, and natural selection. In Britain, Jordan and Rothschild argued eloquently and influentially against any new terminology (including Formenkreis) that had theoretical implications, and they proposed incorporating as little evolutionary theory into taxonomy as possible in view of the lack of agreement among scientists at the time. Although Rothschild and Jordan (1895, 1903), supported by Hartert, agreed both with the nomenclatural practice of naming subspecies and that subspecies were valid real taxa, they argued that the Linnean term species should be retained for the whole group of races and that the geographic forms were not true species—they were simply subspecies or incipient species.

However, others believed that the term species was too emotive to be used in the new, multiple-subspecies sense. Some scientists continued following the Formenkreis doctrine and had begun to name quite distinct taxa, which did not intergrade at their boundaries, as subspecies. This situation led in the 1920s and 1930s to the neo-Darwinian ornithologist Bernhard Rensch scrapping the term Formenkreis because of its theoretical limitations and instead substituting two new terms, *Rassenkreis* (circle of races) and *Artenkreis* (circle of species). Rassenkreise were again considered to be equivalent to species, composed of races or subspecies. However, now there was an additional layer in the taxonomy composed of groups of Rassenkreise that replaced another geographically—the Artenkreise. Thus, an Artenkreis could consist of multiple Rassenkreise. Rensch and many others believed that the subspecies was an incipient species, of which the geographic replacement species Artenkreise were a further development, until finally divergence was sufficient to allow complete geographic overlap, whereupon new Rassenkreise could again form.

These terms did not catch on, and most people came to the conclusion that the Rassenkreise were equivalent to the species referred to by Linnaeus and Darwin. Probably a major reason that we do not use these multiple taxonomic terms is due to the prolific work published in English by another German, Ernst Mayr. Mayr had worked for Walter Rothschild and knew Hartert. After Walter Rothschild was blackmailed by a lover, his enormous bird collection of 280,000 skins was sold in 1932 to the American Museum of Natural History, where Mayr was hired as curator. Mayr's experience of ornithology, contact with the European literature, and friendship with the geneticist Theodosius Dobzhansky (who helped to convince him of the lack of evidence for inheritance of acquired characters) resulted in a unique opportunity to influence the course of systematics and evolutionary biology. Mayr did not waste this opportunity. Mayr used the ideas underlying Rensch's new terms but renamed them in English. The Rassenkreis became simply the true species or "polytypic" species, with its geographic races being subspecies, whereas the Artenkreis became the "superspecies" and its component parts "semispecies," i.e., not very divergent true species. Mayr successfully blended the local species concept of Poulton and Dobzhansky based on interbreeding with the geographic Rassenkreis idea of species and renamed this combination of ideas "the biological species concept," a term which has remained strongly associated with Mayr's name. His many influential articles and books promoted a new program of species study, a science of the species still with us today. Central to Mayr's system was the belief that discrete taxa such as species or subspecies would normally diverge in "allopatry," i.e., in complete geographic isolation.

II. THE SUBSPECIES TODAY

A. Modern Views of Subspecies, Semispecies, and Superspecies

The view of Darwin, Wallace, Rensch, and Mayr that geographic replacement forms, subspecies and semispecies, which form a continuum with species, were in fact incipient species has few critics today. Most geographic replacement species or semispecies which do not intergrade when they meet must indeed have evolved from previously interbreeding subspecies. Modern genetic data have done nothing to cast doubt on this idea.

However, taxonomists were now required to describe subspecies, which has never been viewed as a particularly honorable or worthwhile activity in comparison with describing species, especially recently. A strong attack on the subspecies was mounted by Wilson and Brown (1953). Both were systematists working on ants, a group particularly riddled with poorly conceived trinomials at the time. Wilson and Brown argued that subspecies rarely if ever could be justified on the basis of multiple characters, and that therefore they were not "real taxa." The only real taxa were species, which in a sense were self-defining because interbreeding prevented divergent genes from flowing from one species to another. Subspecies which interbred at their boundaries, on the other hand, were not so endowed; therefore, genes and morphological characters could flow between them. Good examples were put forward of subspecies which undoubtedly would be difficult to justify on multiple-character grounds. This single article was enormously influential on systematics in the United States, and generations of systematists trained at Harvard and Cornell (where Wilson and Brown worked), their own many intellectual descendants, and their students' students in turn have eschewed the practice of naming subspecies.

Through genetic studies we now know, however, that many subspecies separated by hybrid zones differ at multiple morphological, behavioral, and genetic characters (Barton and Hewitt, 1985). For instance, the toad *Bombina bombina* meets its relative *Bombina variegata* across a broad front in Europe and differs strongly in call, morphology, skin thickness, the sizes of water bodies used, and egg size as well as in mitochondrial DNA and protein sequence. Their levels of differentiation suggest that the *Bombina* have evolved separately for many millions of years. (The two forms hybridize freely in the contact zone—although the hybrids can be shown to suffer some inviability—and so should be classified as members of the same polytypic species under the polytypic or biological species concept, but it has always seemed natural to place such well-defined forms in separate species despite the fact they have not truly "speciated.") This situation of multiple character changes has been shown to be true across many examples of hybrid zones, and gene flow can be shown to be almost completely blocked by hybrid zones such as these, despite abundant hybridization. Thus, although many named subspecies undoubtedly merited Wilson and Brown's (1953) scorn, genetic evidence shows that there are plenty of local replacement forms which hybridize at their boundaries but which do form "real" identifiable taxa and are valid subspecies under the Wilson and Brown criteria.

B. Subspecies, Species, and Conservation

Opposition among modern taxonomists to the subspecies can be traced as an influence on the recent "diagnostic" version of the phylogenetic species concept. The adherents of this view of species, led by the ornithologist Cracraft (1989), proposed a radical species concept so that even a single fixed character difference may define a geographic form as a separate species; multiple-character justification is not considered necessary by them even at the species level. The practical result of this new concept is that many local forms are again being recognized as species. In birds and butterflies, which often have many morphologically or genetically distinct subspecies, this could easily result in a 2- to 10-fold increase in the number of species, or even more in some groups.

It is probable that the revision of geographic forms upwards to the level of species is being driven not only by theoretical considerations but also by existing legislation, which proposes that "endangered species" are the valuable units to be conserved. If an area contains a taxon recognized as a species rather than just a local race, it may be viewed as more valuable for conservation purposes. The potential consequences for biodiversity and conservation of the continued instability of the term species are detailed elsewhere. Here, I only mention that today's conservationists are reducing emphasis on species conservation and are becoming increasingly aware of biodiversity at all the levels of the hierarchy of life, including well-marked subspecies. Thus, the legislative need for differentiating local races as species may ultimately become less of an impetus provided that future legislation falls more into line with prevailing biological thought.

III. FURTHER READING

Much of the historical overview in this article is discussed in the excellent reviews of Stresemann (1936, 1975), Mayr (1982), Rothschild (1983), and other sources cited previously.

See Also the Following Articles

NOMENCLATURE, SYSTEMS OF • SPECIATION, THEORIES OF • SPECIES, CONCEPTS OF • SYSTEMATIC, OVERVIEW • TAXONOMY, METHODS OF

Bibliography

Barton, N. H., and Hewitt, G. M. (1985). Analysis of hybrid zones. *Annu. Rev. Ecol. Syst.* **16**, 113–148.

Cracraft, J. (1989). Speciation and its ontology: The empirical consequences of alternative species concepts for understanding patterns and processes of differentiation. In *Speciation and Its Consequences* (D. Otte and J. A. Endler, Eds.), pp. 28–59. Sinauer, Sunderland, MA.

Darwin, C. (1874). *The Descent of Man and Selection in Relation to Sex*, 2nd ed. Murray, London.

Mayr, E. (1982). *The Growth of Biological Thought. Diversity, Evolution, and Inheritance*. Belknap, Cambridge, MA.

Rothschild, M. (1983). *Dear Lord Rothschild. Birds, Butterflies and History*. Hutchinson, London.

Rothschild, W., and Jordan, K. (1895). A revision of the Papilios of the eastern hemisphere, exclusive of Africa. *Novitat. Zool.* **2**, 167–463.

Rothschild, W., and Jordan, K. (1903). A revision of the lepidopterous family Sphingidae. *Novitat. Zool.* **9** (Suppl.).

Stresemann, E. (1936). The Formenkreis-theory. *Auk* **53**, 150–158.

Stresemann, E. (1975). *Ornithology. From Aristotle to the Present*. Harvard Univ. Press, Cambridge, MA. [Reprinted edition of Stresemann's 1951 "Entwicklung der Ornithologie" (H. J. Epstein and C. Epstein, Trans.; G. W. Cottrell, Ed.; Foreword and Epilogue by E. Mayr]

Wilson, E. O., and Brown, W. L. (1953). The subspecies concept and its taxonomic application. *Syst. Zool.* **2**, 97–111.

SUBTERRANEAN ECOSYSTEMS

David C. Culver
American University

I. The Subterranean Domain
II. Sources of Energy to Subsurface Environments
III. Adaptations to Subterranean Life
IV. Colonization and Speciation in Subterranean Environments
V. Taxonomic Survey of Subterranean Life
VI. Some Representative Subterranean Communities
VII. Geography of Cave Biodiversity
VIII. Conservation and Protection of Subterranean Habitats

GLOSSARY

amphibites Species that require both surface waters and subterranean waters in order to complete their life cycle.
anchialine (or anchihaline) Haline water, usually with restricted exposure to open air, and always with subterranean connections to the sea.
chemolithotrophy The process, utilized by some microbes, of obtaining energy from the breaking of inorganic chemical bonds.
epikarst Upper part of the percolation zone in karst.
hyporheic zone Interstitial space within the sediments of a streambed; a transition zone between surface water and permanent (phreatic) groundwater.
interstitial The spaces between sediment particles, especially in alluvial deposits.
karst A landscape in which the primary geomorphic agent is solution rather than erosion; typically formed in carbonates; landforms include caves, sinkholes, blind valleys, and large springs.
MSS The interstitial spaces deep in the soil and mantle/bedrock interface, as found in glacially fragmented zones. By origin, the *milieu souterrain superficiel*, or mesovoid shallow stratum, thus the acronym MSS.
stygobite Aquatic species that are obligate subterranean dwellers.
stygophile Aquatic species that can live and reproduce in both subterranean and surface environments.
troglobite Terrestrial species that are obligate subterranean dwellers; sometimes used for aquatic species as well.
troglomorphic Pertaining to the morphological, behavioral, and physiological characters that are convergent in subterranean species.
troglophile Terrestrial species that can live and reproduce in both subterranean and surface environments; sometimes used for aquatic species as well.

SUBTERRANEAN ECOSYSTEMS include a wide variety of aphotic, resource-poor sites, including caves, aquifers, and the underflow of rivers. This article discusses the kinds of ecosystems, the unique organisms that inhabit them, the evolution and adaptation of these organisms, and the energy flows through them.

I. THE SUBTERRANEAN DOMAIN

A. Caves

The subterranean domain, literally the domain below the earth, includes both air- and water-filled underground habitats. The most celebrated of these are caves. Most caves are formed by the dissolution of rock by acidic waters, but caves can also result from mechanical action, especially the flow of lava. Nearly all of the caves formed by the action of acidic waters are in carbonates (particularly limestone) and evaporites (particularly gypsum). Landscapes in which the primary agent molding the landscape is dissolution rather than erosion are called karst landscapes. Comprising approximately 15% of the earth's surface, karst represents 75% of the land area of Cuba, 40% of Slovenia, 25% of France and Italy, and 40% of the United States east of Tulsa, Oklahoma. Caves are present in the oldest Cambrian rocks to Holocene limestone less than 10,000 years old. Lava tube caves are common on volcanic islands such as Hawaii and in the northwestern United States. Although caves have a reputation as both exotic and rare habitats, they are actually quite common. More than 100,000 caves are known from Europe, and nearly 50,000 are known from the United States. Caves come in many sizes and varieties, ranging from a few meters long to Mammoth Cave in Kentucky with a surveyed length of more than 500 km. All caves share two important characteristics: complete darkness away from the entrance and reduced environmental variability relative to surface conditions.

Many organisms spend some or all of their life cycle in caves, particularly cave entrances. The moist rock of cave entrances provides habitat for many mosses and ferns, such as the Hart's tongue fern (*Phyllitis scolpendrium*). The entrance and twilight zone of a cave are refuges from temperature extremes of the surface. Some species, such as the spider *Meta menardi*, are specialized for the surface–subsurface ecotone at cave entrances. The entrance and twilight zones of caves are relatively predator free, at least for vertebrates. The neotropic oil bird *Steatornis caripensis* nests in caves; many other birds do so as well, albeit on a less regular basis. Many cave entrances in the eastern United States have one or more phoebe (*Sayornis phoebe*) nests. The best known visitors to caves are bats. Depending on the species, bats use caves as maternity colonies, as hibernacula, and as temporary roosts during the warmer months of the year. More than half of the species of bats found in the eastern United States use caves all or part of the year. About 95% of the more than 1 million gray bats, *Myotis grisescens*, hibernate in eight caves. Many bat

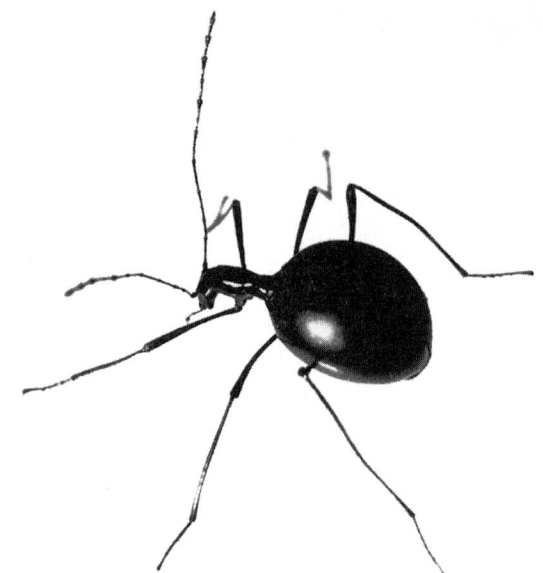

FIGURE 1 The first described troglobite, the beetle *Leptodirus hochenwartii* (family Carabidae) from a cave in Slovenia. The length of the beetle is 8 mm (photograph courtesy of Matija Gogala, Ljubljana, Slovenia).

species are at risk because of the vulnerability of their hibernating and maternity sites to disturbance.

Many species spend their entire life cycle in caves. In the case of terrestrial species, troglobites have an obligate dependence on caves and must complete their life cycle in caves. Troglophiles can complete their life cycle in caves, but they can also complete their life cycle in surface habitats. The equivalent terms for aquatic species are stygobites and stygophiles. Another useful term is troglomorphic, which denotes the morphological syndrome associated with life in caves—loss of eyes and pigment and the elaboration of extraoptic sensory structures. Most but not all troglobites are troglomorphic. Those that are not have likely not been isolated in caves for a long enough time for troglomorphic characteristics to evolve.

The remarkable cave amphibian, *Proteus anguinus*, was the first stygobite to be mentioned in scientific writing. Writing in 1689, Valvasor reported that local residents thought that this amphibian, common in caves in Slovenia, was the larvae of dragons. Its scientific description was accomplished by Laurenti in 1768. The first troglobite, the cave beetle *Leptodirus hochenwartii* (Fig. 1), was described by Schmidt from the same region in 1832. Currently, more than 4000 troglobites and 2000 stygobites have been formally named, and at least several times that number probably exist.

B. Other Shallow Subsurface Habitats

Terrestrial subsurface habitats also occur outside of caves. These include cracks and small voids in volcanic rocks and a superficial zone of rock fissures and debris slopes in schists, gneiss, and granite. Generally occurring at a depth of few meters, this habitat was called *milieu souterrain superficiel* (MSS) in French, or mesovoid shallow substratum in English. The MSS is generally found in mountains in temperate zones but apparently not in the tropics, where voids are usually filled with laterites and clay. Several dozen troglobites have been found in the MSS. For the most part, species in the MSS have been sampled by baits placed several meters deep in holes drilled for the purpose.

In karst, there is a similar habitat in the upper part of the rocks in the upper part of the percolation zone with numerous fissures and solution pockets—the epikarst. Either water or air filled, it also contains troglobites and stygobites. Although in many cases species found in the epikarst are also found in caves, there are species predominantly found in epikarst. Although the epikarst has occasionally been sampled directly by means of shallow wells and holes, most of the collections of the fauna of epikarst are from caves. Species collected in dripping water, in temporary pools, and in other habitats near the surface are denizens of the epikarst. Caves provide a window to this fauna.

A very different kind of subterranean aquatic habitat also contains an interesting fauna—the interstitial habitat. These habitats are aphotic and widespread. They include sands of marine and freshwater beaches—the psammon. The fauna in these habitats lives in the narrow voids between sand grains and belongs to the meiofauna, a size class of aquatic organisms too small to be consider macroinvertebrates and too large to be microorganisms. In addition to the absence of light, low oxygen levels are frequently encountered. The unconsolidated sediments of running water are also an important interstitial habitat. Because alluvial sediments are transported and deposited by currents, they have larger grain sizes and spaces than do sand beaches of lakes and oceans. Consequently, the fauna is larger and in many cases overlaps with the stygobitic fauna of caves. The hyporheic is the interstitial space within gravel deposits beneath stream channels. It is an ecotone between surface waters and permanent (phreatic) groundwaters. The hyporheic usually ranges from several centimeters beneath the channel to several meters away from the channel. The habitat is very complex, with areas of upwellings and downwellings and low oxygen and high oxygen often occurring within meters of each other.

Beneath the hyporheic and to the sides, the phreatic zone also contains an interesting fauna. In addition to stygobites and stygophiles, alluvial aquifers also have amphibites, species that require both surface waters and subterranean waters in order to complete their life cycle. Early sampling of alluvial aquifers was done by digging a hole near a stream and filtering the water that entered the hole through a plankton net. During the past 30 years, this technique has been supplemented by a small pump (Bou-Rouch Pump) attached to a pipe driven 1 or 2 m into the sediments. In addition to sampling in wells, sampling of the interstitial fauna of alluvial sediments has been done for only a few rivers, most of which are in Europe.

C. Deep Subsurface Habitats

Wells and artesian springs in karst areas often connect with water-filled underground voids tens to hundreds of meters deep that are part of the permanent water table—the phreatic zone. Few such sites have been sampled because of the difficulty in biological sampling in deep wells. However, the few sites that have been sampled have yielded a diverse array of macroscopic invertebrates. The artesian spring at San Marcos, Texas, which is connected to the Edwards Aquifer, has a diverse array of invertebrates, many of which are endemic, as well as several stygobitic fish and salamanders.

Perhaps the most exotic subsurface habitat is that of deep groundwater in porous aquifers. The water within both sedimentary and igneous rock contains an array of Eubacteria and Archaea. The presence of bacteria in the deep subsurface was first suggested by Edwin S. Bastin in the 1920s. He suggested that the curious presence of hydrogen sulfide and bicarbonate in water extracted from oil was the result of the activity of sulfate-reducing bacteria at depth. It was nearly 60 years before the technology was developed to sample bacteria at depth. To do so required the solution of several formidable technical problems. The first was to obtain samples uncontaminated by surface microbes, and the second was to characterize the microbes. Characterization of the samples is difficult because of the unusual environmental conditions in which the microbes live, making standard culturing techniques of limited value. Much of our knowledge of deep-groundwater microbes is derived from sampling wells near former nuclear weapons production sites in Washington State and in South Carolina. Bacteria and Archaea have been found at depths of more than 2 km. Their presence in sedimentary rock may be the result of the slow infiltration of water from the surface or they may have been present at the time of the deposition of the rock. The possibility that they were present at the time of deposition is plausible given the depth of the water and its age based on stable isotope ratios. Microbes present in igneous rock must have been deposited after it formed since it was too hot at the time of formation. In these habitats, chemolithotrophs, which are forms that use inorganic carbon as a carbon source and obtain energy from the oxidation of reduced inorganic chemicals, predominate. The possible great age of some of these microorganisms combined with their ability to survive in the absence of organic matter and sunlight makes it likely that they are more representative of some of the earliest forms of life on the planet than are surface-dwelling microorganisms. They may also be useful analogs for understanding what life on other planets, such as Mars, might be like. More than 9000 strains of organisms have been cataloged from subsurface habitats.

II. SOURCES OF ENERGY TO SUBSURFACE ENVIRONMENTS

All subsurface environments share an absence of light and hence an absence of photosynthesis. With the exception of a few caves and possibly most deep-groundwater habitats, there is no significant primary production. With these exceptions, all subsurface food webs rely on the import of surface organic matter in one form or another. For the most part, subterranean habitats are energy poor. The famous Romanian cave biologist Emil Racovitza stated that many cave organisms are "carnivores by predilection but saprophages by necessity" (translation by author). Also, the famous American cave biologist A. S. Packard noted more than a century ago that "cave animals, even the carnivorous species, take remarkably little food."

However, for most subsurface habitats, organic matter is brought in by the action of water. Dissolved organic matter (DOM) and fine particulate organic matter (FPOM) are important energy sources both in interstitial habitats and in many caves. In caves, water percolates through the soil into the cave, where it collects in pools or streams. In many caves, sinking streams (swallets) bring in additional organic matter—not only DOM and FPOM but also coarse particulate organic matter, including leaves and twigs. In a sense, many cave streams are not very different from a low-productivity stream on the surface. In many surface streams, allochthonous input greatly exceeds autochthonous production. The organic matter brought in by streams is also an important food source for terrestrial cave animals. The stranding of organic matter by flooding of streams is the primary source of food for many terrestrial cave communities.

For many subsurface terrestrial communities, the main source of energy is the result of the movement of animals in and out of the cave. In many North American caves, cave crickets in the subfamily Rhaphidophorinae periodically leave caves to forage and deposit their eggs in sandy substrates in the cave. These eggs are in turn the major part of the diet of some troglobitic carabid beetles. The fecal material of the cave crickets and the occasional dead cave cricket body are an important food supply for other troglobites. Fecal matter left by other cave visitors is also important in many caves. The most important source of fecal material is bats. In most temperate zone caves, only a few bats use the cave, and their guano is the basis for a group of species whose food base is transitory organic matter. In those temperate zone caves that serve as maternity sites or hibernacula and in many tropical caves, large bat colonies produce enough guano for there to be a community specialized on the guano. Most of these species are guano specialists rather than cave specialists.

There are other allochthonous sources of food. In a few caves, such as the lava tubes on the Canary Islands and some MSS. habitats, the major source of organic matter is wind-carried particles and debris, which penetrate into the subterranean habitats through cracks and fissures. In lava tubes and some shallow limestone caves, roots penetrating through the ceiling are an important food supply, and in some Hawaiian lava tubes there are species that specialize in this resource.

The possibility of a subterranean chemoautotrophic ecosystem has long intrigued biologists. Movile Cave, a small cave near the Black Sea in Romania, is such an ecosystem. This cave, discovered during geochemical prospecting in 1986, has no natural entrance. Its H_2S-, CH_4-, and NH_4^+-rich waters are slightly thermal. A series of air bells occur throughout the cave in which atmospheric composition is enriched in CO_2 and CH_4 and depleted in O_2. The water surface is covered with a floating mat of bacteria and fungi. There are nearly 50 species of stygobites and troglobites known from the cave, many of which are endemic to the cave. Chemoautotrophic production was demonstrated by incorporation of bicarbonate ion by the mat and by the presence of ribulose 1,5-biphosphate caroxylase-oxygenase, an enzyme characteristic of autotrophic organisms. A gram-negative rod-shaped bacteria was isolated from the mat that was an obligate autotroph using the energy of the oxidation of thiosulfate and sulfide. Furthermore, this autotrophic production is the basis not only for the aquatic food web but also for the terrestrial food web—the only documented case of a chemoautotrophic terrestrial ecosystem. Several other sulfiderich caves, including Frassati Caves in Italy, Villa Luz Cave in Mexico, and Cesspool Cave in Virginia, also exhibit many features of chemoautotrophic production. In Villa Luz Cave and Frassati Caves, a rich but nonstygobitic and nontroglobitic community occurs. The explanation for the rarity of troglobites and stygobites lies in their closer connection to surface waters, the penetration of light into many parts of the cave (Villa Luz and Cesspool Caves), and the relatively young age of the caves (Frassati and Cesspool Caves).

It is likely that many phreatic deep systems are chemoautotrophic (actually chemolithoautotrophic). A variety of chemical reactions are used by Bacteria and Archaea to provide the energy for biosynthesis. Among the reactions occurring in deep groundwater is methanogenesis:

$$4H_2 + CO_2 \to CH_4 + 2H_2O \quad \Delta G = -32.5 \text{ kcal}$$

The methanogenic reaction uses hydrogen gas as the energy source and inorganic CO_2 as the carbon source. Hydrogen derives from the reaction of the oxygen-poor water with iron-bearing minerals. Methanogens can persist indefinitely without any supply of carbon from the surface.

III. ADAPTATIONS TO SUBTERRANEAN LIFE

Compared to related surface-dwelling species, subterranean animals are characterized by the absence or reduction of a series of features, particularly eyes and pigment. Subterranean animals are also characterized by an elaboration of extraoptic sensory structures. This pattern of elaboration and reduction is of course extremely general when comparing two groups of organisms from different habitats. What makes subterranean animals particularly interesting in this regard are the obvious and unusual reductions, often called regressive evolution. Unlike some aspects of evolution, such as the evolution of sex and the origin of evolutionary novelty, eyelessness and regressive evolution in general have always appeared easy to explain. To an orthogeneticist such as Albert Vandel writing in the mid-twentieth century, eye loss in subterranean animals was a by-product of a senescent phyletic line. Animals were not blind because they were in caves; rather, they were in caves because they were blind. To a neo-Lamarckian such as A. S. Packard writing in the late nineteenth century, eye loss was a classic example of the role of disuse. To a neo-Darwinian, eye loss in subterranean animals was the result of an evolutionary tradeoff resulting from natural selection for elaborated nonoptic sensory structures. To neutral mutationists, eye loss in cave animals was easily explained by the accumulation of selectively neutral, structurally reducing mutations made possible by the lack of stabilizing selection. In a modern context, theories that account for regressive evolution in subterranean cave animals are classified in two groups: those that involve natural selection directly or indirectly and those that involve neutral mutation and genetic drift.

One of the best studied organisms in this regard is the amphipod *Gammarus minus* which occurs in springs and caves in much of the central Appalachians in West Virginia and Virginia. Populations independently isolated in caves show a strongly convergent morphology of appendage elongation and eye and pigment reduction. There was a strong correlation between the fitness components of mating and egg number with relative appendage length and eye size. Furthermore, this correlation was opposite for cave and spring populations. However, natural selection is not the whole explanation for eye and pigment reduction in *G. minus*. The rate of change of reduced structures is consistently greater by an order of magnitude than the rate of change of elaborated structures (Table I). Elaborated structures changed as a result of natural selection, whereas reduced structures changed as a result of both natural selection and the accumulation of structurally reducing, selectively neutral mutations. The increased variability of eyes in recently isolated cave populations is also most easily explained as a result of neutral mutation. The genetics of eye loss are best understood in the Mexican cave fish *Astyanax fasciatus*. Like *G. minus*, it has apparently been isolated in caves for less than 1 million years. Defects in developmental control genes are primarily responsible for eye reduction. Few if any of the sequences of opsin proteins and a structural lens protein α-A-crystallin are altered in cave populations. The master control gene of eye development *Pax-6* was not altered in cave populations. The defect causing eye degeneration is assumed to be in the regulation cascade responsible for eye development below the *Pax-6* and above the α-A-crystallin gene.

Among subterranean organisms, adaptations have been most thoroughly studied in cave animals. After eye and pigment reduction, the most characteristic trog-

TABLE I
Evolutionary Rates of Divergence for Different Morphological Characters of a Cave Population of the Amphipod *Gammarus minus* from Its Ancestral Spring Population[a]

Morphological character		Δ	Relative rate
Size	HL	4.0×10^{-7}	0.02
Eyes	EN	2.4×10^{-5}	1.00
	EA	1.5×10^{-5}	0.62
Antennae	P1	8.2×10^{-7}	0.03
	L1	4.1×10^{-7}	0.02
	N1	1.1×10^{-6}	0.05
	P2	1.1×10^{-6}	0.05
	L2	8.9×10^{-7}	0.04
	N2	1.4×10^{-6}	0.06

[a] The measure of divergence, Δ, is independent of the size of the character and depends only on its variability. The two populations compared are Organ Cave and Organ Cave spring, about 2 km apart in southern West Virginia. Time since divergence was set to 5×10^5 years. Traits were head length (HL), eye ommatidia number (EN), surface area of compound eye (EA), peduncle length of first antenna (P1), flagellum length of first antenna (L1), and number of flagellar segments of first antenna (N1) and the second antenna analogs (P2, L2, and N2). In the last column, all rates are compared to that of ommatidia number, which was set to 1. [Adapted from Culver et al. (1995) by permission of the publishers. © President and Fellows of Harvard College.]

lomorphic traits are the elaboration or elongation of extraoptic sensory structures. Among arthropods, the elongation of antennae is especially noticeable. A striking example is the first described troglobite, the Slovenian cave beetle *L. hochenwarti* (Fig. 1). Vertebrates such as the North American cave fish in the Amblyopsidae display a striking hypertrophy of the lateral line system. These changes are also reflected in the central nervous system. Areas associated with vision decrease in size and areas associated with the other senses increase (Fig. 2). The potential advantage of elaboration and increase in extraoptic sensory structures is to compensate for the absence of light by increasing tactile and chemical sensory sensitivity. In the resource-poor environment of caves, metabolic efficiency is often increased, particularly among species at or near the top of the food chain. Activity is often reduced, as is metabolic rate. The frequently observed decline in intraspecific aggressive behavior of troglobites and stygobites may also in part be an evolutionary response to food scarcity. Food scarcity is also apparently a strong selective force in life history patterns of cave animals. Cave animals live longer, begin reproduction at a later age, produce fewer offspring, and have larger eggs—all ad-

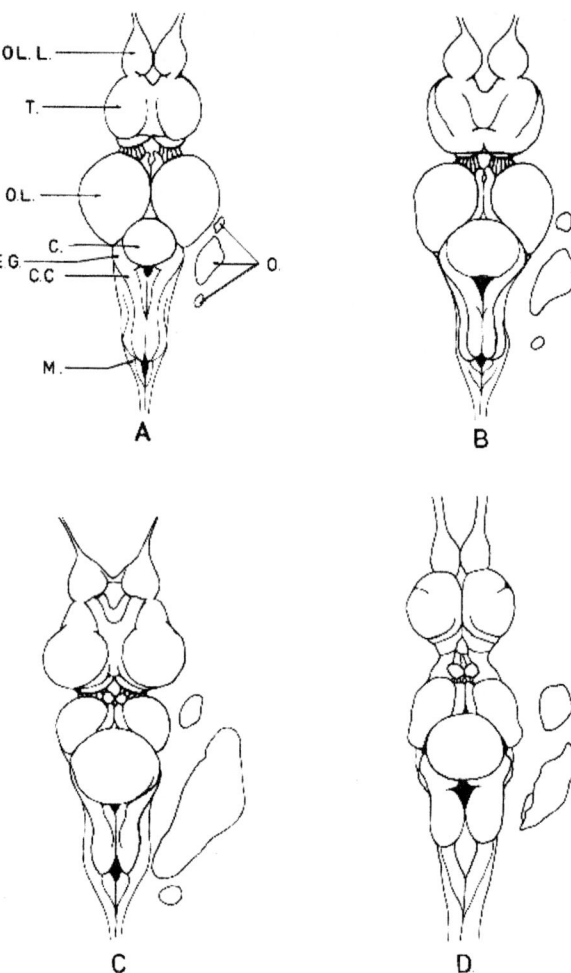

FIGURE 2 Brain morphologies of four fishes in the family Amblyopsidae: (A) *Chologaster cornuta*, a surface-dwelling species; (B) *Chologaster agassizi*, a stygophile found in both springs and caves; (C) *Amblyopsis rosae*, a stygobite; and (D) *Speoplatyrhinus poulsoni*, a stygobite. OLL, olfactory lobe; T, telencephalon; OL, optic lobe; C, cerebellum; EG, eminentia granularis; CC, cristae cerebelli; M, medulla oblongata; O, otoliths. Note the relative hypertrophy of extraoptic parts in stygobites. [Adapted from Poulson (1963) with permission.]

aptations to a relatively constant, food-poor environment. Perhaps the most striking example of life history modifications is that of a population of the crayfish *Orconectes australis australis* in Shelta Cave, Alabama. In an extensive mark-recapture study lasting more than 6 years, John E. Cooper found that the age at first reproduction was approximately 35 years, and that the crayfish lived to an age of more than 100 years.

Animals in other subterranean habitats also show many of the same adaptations since they share in common with cave animals an aphotic, resource-poor environment. However, there are two other environmental

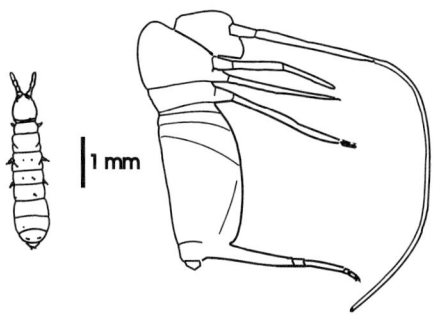

FIGURE 3 A typical soil collembolan (*Onychiurus armatus*, left) and a highly troglomorphic collembolan (*Pseudosinella christianseni*, right). (Drawing courtesy of K. A. Christiansen, Grinnell, Iowa.)

factors that exert strong selective pressures not present in cave environments: intragranular space size in interstitial habitats and oxygen levels in aquatic interstitial habitats. Interstitial animals tend to be both thin and small since they are constrained by interstice diameter. The morphological contrast between cave and interstitial animals can be quite striking. A terrestrial example comparing a cave with an epikarstic and deep-soil-dwelling Collembola is shown in Fig. 3. The small size of many aquatic interstitial species may occur as a result of progenesis, which is the precocious sexual maturation of an organism resulting in an adult descendent exhibiting the juvenile morphology of its ancestor. This process also results in reduced extremities, spines, and bristles (Fig. 4). More than cave dwellers, interstitial stygobites must be able to survive under conditions of low oxygen. Many interstitial crustaceans are able to regulate respiration rate, varying the ventilating activity of their pleopods according to oxygen availability.

Metabolic adaptations of deep-groundwater microorganisms are just beginning to be understood. Unlike the situation with eukaryotic organisms, the basic tenet of microbial ecology, Beijerinck's principle, holds in the deep subsurface: "Everything is everywhere, the environment selects." The environment in this case is an extremely food-poor one, with little or no organic matter. The major characteristic of phreatic microorganisms is the widespread occurrence of facultative or obligate chemoautotrophy. The Gibbs free energy available in chemoautotrophic responses is much less than that of photosynthetic organisms, reinforcing the concept of resource scarcity from a physical chemical standpoint.

IV. COLONIZATION AND SPECIATION IN SUBTERRANEAN ENVIRONMENTS

The terms troglobite, stygobite, troglophile, and stygophile are an ecological classification which attaches great importance to the isolation of species in caves. Troglophiles and stygophiles, species which can complete their life cycle in both surface and subterranean habitats, are often important components of cave communities. In many cases, especially in the tropics, they outnumber troglobites and stygobites. Troglobites and stygobites are typically but by no means always troglomorphic. The evolution of troglobitic (stygobitic) species that are troglomorphic (stygobitic) has received the most attention.

There are two components to evolution in subterranean habitats: colonization and subsequent isolation of the subterranean populations usually as a result of the extinction of the surface-dwelling populations. Colonization may be active or it may be passive, i.e., "colonization under constraint." An example of active colonization is the movement of surface-dwelling Homoptera in lava fields in Hawaii and elsewhere into lava tubes and cracks in the lava in which root sap is an abundant resource. In some cases, speciation has proceeded without the extinction of the surface populations. The mechanism of speciation is either parapatric or sympatric. Active colonization may be particularly important for interstitial organisms. The classic case of colonization under constraint is that of many troglobites in north temperate caves. Many of the surface-dwelling ancestors of these troglobites were leaf-litter-dwelling inverte-

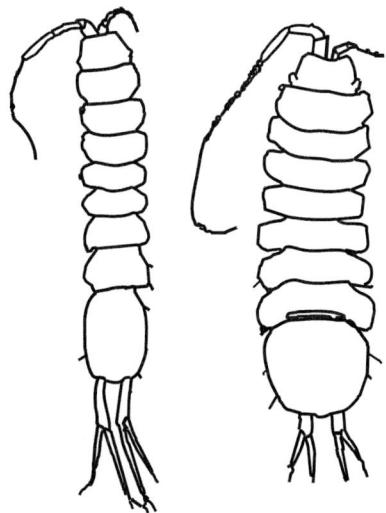

FIGURE 4 *Proasellus albigensis*, an isopod from interstitial habitats (left), is 5 mm long; *Proasellus vandeli*, an isopod from caves (right), is 4 mm long. (Drawing courtesy of G. Magniez, Dijon, France.)

brates. During Pleistocene interglacials, as warming occurred, many of these species were either forced into caves or up mountain slopes to cooler conditions. As warming continued, surface populations became extinct. For the beetle genus *Ptomaphagus*, various speciation events can be associated with particular interglacials (Fig. 5). For most cases of colonization under constraint, the factor that forced animals into caves, such as Pleistocene interglacials or the Messinian salinity crisis in the Mediterranean, was also the factor that resulted in the extinction of surface populations.

By whatever mechanisms organisms colonize subterranean environments and by whatever geological or climatological event surface populations become isolated in subterranean environments, subsequent dispersal is generally extremely limited. Most subterranean species have very restricted ranges. For example, more than 60% of the subterranean species described from the United States are known from a single county, and about half of these are known from a single locality. Species with large ranges, at least for subterranean species, have on closer examination been shown to be complexes of morphologically identical but genetically distinct species and semispecies. For example, the European cave salamander *Proteus anguinus* ranges more than 100 km from Italy to Croatia but actually represents several invasions of a common surface-dwelling ancestor that is extinct. For some species, there is little gene flow at distances of less than 1 km. This makes subterranean species excellent markers of biogeographic events. Once species are isolated in subsurface habitats, which are often a refuge from the vicissitudes of surface climatic change such as temperature and salinity, their distribution may leave a record of ancient events, especially continental drift. Examples abound in the Crustacea. The distributions of the order Syncarida and the amphipod family Bogidiellidae are Pangaean and predate the pre-Triassic breakup of Pangaea. The distributions of the amphipod family Crangonyctidae and the isopod family Asellidae are Laurasian; their isolation presumably dates back to the late Mesozoic. The order Remipedia and the ostracod genus *Danielopolina* have a Tethyan distribution. The distribution of the amphipod genus *Longipodacrangonyx* in Morocco closely matches the embayments of the Eocene sea, with little movement since then (Fig. 6). In most of these cases, extinction of surface populations was a result of marine regression. Of course, not all subterranean species are as old as the examples given here, but they do indicate the apparent great age of some subterranean groups combined with highly restricted dispersal.

A final example is the remarkable cave clam, *Congeria kusceri*, the only known stygobitic clam. Known from a series of caves in Bosnia and Herzegovina, Croatia, and Slovenia, *C. kusceri* is the only living representative of an otherwise extinct genus, known from the late Miocene. Marine ancestors were apparently victims of the Messinian salinity crisis 6 million years ago, and *C. kusceri* is likely a relic of Pliocene lakes and rivers in the region. As the lakes and rivers disappeared underground, *C. kusceri* went with them.

V. TAXONOMIC SURVEY OF SUBTERRANEAN LIFE

In all subterranean environments, a wide variety of heterotrophic bacteria and Protozoa occur, even in deep-groundwater sites. Early interest in autotrophs in caves centered on the potential role of nitrifiers such as *Nitrobacter* in the formation of saltpeter (KNO_3). There is a wide diversity of metabolic pathways among autotrophs, most important in the carbon fixing pathways: the Calvin cycle in bacteria, the acetyl-coenzyme A pathway in Archaea, and the reductive citric acid cycle found in Archaea and a few bacteria. There is also considerable diversity in the electron donor and acceptor among subsurface autotrophs. Electron donors include H_2, NH_4^+, S, and Fe^{2+}; acceptors include O_2, SO_4^{2-}, and NO_3^-.

Except for a few protozoans, the Eukarya is usually not found in deep groundwater. A wide taxonomic array of organisms are stygobitic or troglobitic, including the following phyla: Porifera, Hydrozoa, Platyhelminthes, Nemertina, Nematoda, Mollusca, Rotifera, Annelida, Onychophora, Arthropoda, and Chordata. However, for many of these groups, there are only a handful of subterranean species. Furthermore, many of these "oddball" species are found only in the Dinaric karst of Bosnia and Herzegovina, Croatia, and Slovenia.

Among stygobites, the Crustacea predominate: At least half of the described species are Crustacea. Among the Crustacea, the orders Remipedia, Spelaeographacea, and Thermosbaenacea are exclusively stygobitic. Numerically, crustacean stygobites are dominated by Copepoda (especially in the interstitial) and Amphipoda (especially in caves). Some Amphipoda genera are particularly speciose. For example, nearly 175 stygobitic species of the amphipod genus *Stygobromus* are known from North America. Other important groups include the planarians (Platyhelminthes: Tricladida), snails (Mollusca: Gastropoda), salamanders (Chordata: Cau-

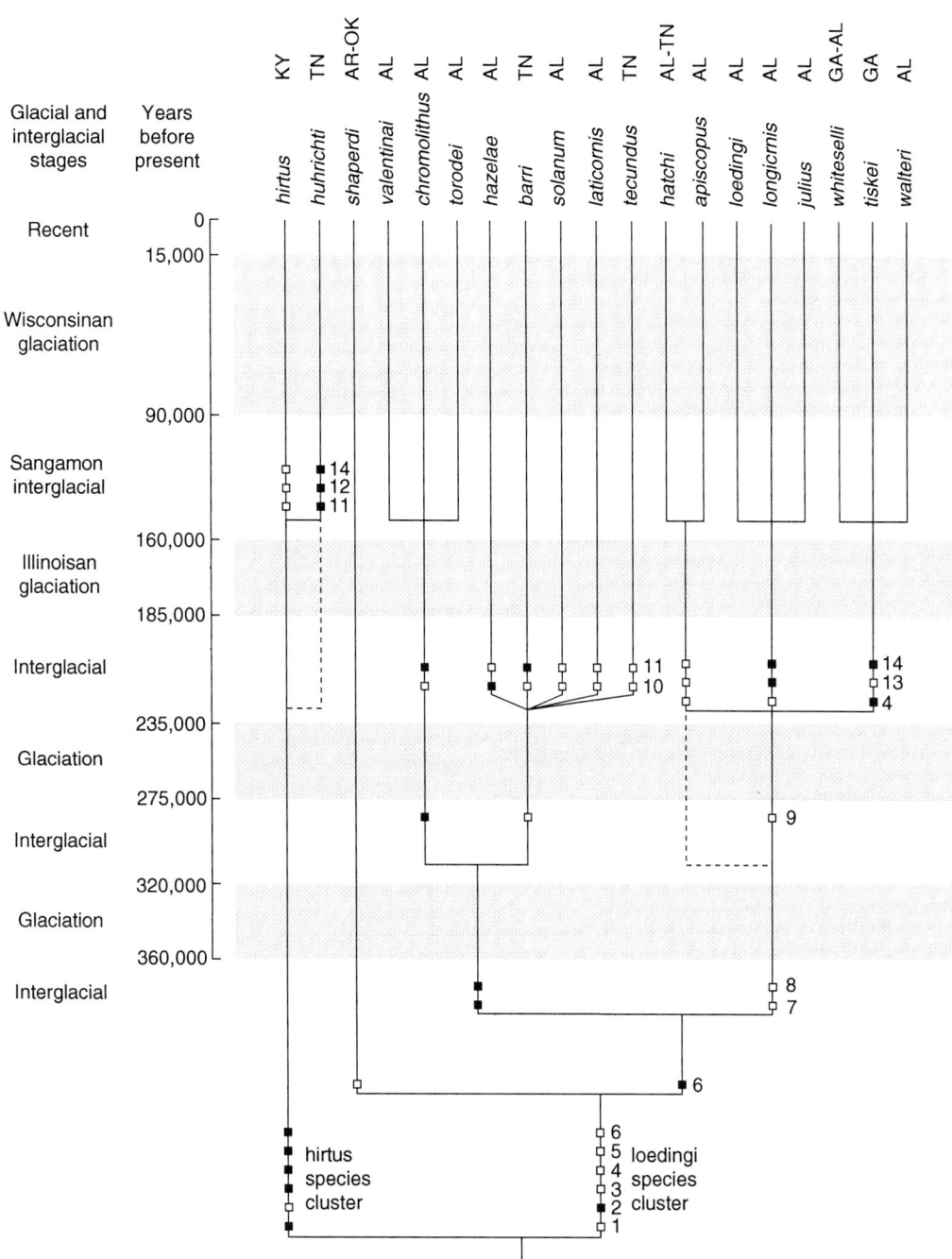

FIGURE 5 Phylogenetic hypothesis of evolution in the *Ptomaphagus hirtus* species group of beetles in the southeastern United States. Numbers refer to characters, transforming from the ancestral state (□) to the derived state (■). State abbreviations with species names indicate distribution. Lineage splits and character origin are shown for the latest likely times for the events (adapted with permission from Peck, 1984).

FIGURE 6 Localities of all known sites (stars) of *Longipodacrangonyx* amphipods plotted on a sketch map of Morocco showing the three marine embayments during the Eocene (shaded areas). [Adapted from *Hydrobiologia* 287, 49–64, Boutin (1994), Phylogeny and biogeography of Mectacrangenyctid amphipods in North Africa, Fig. 5, with kind permission from Kluwer Academic Publishers.]

data), and fish (Chordata: Pisces). The surface-dwelling ancestors of these species were largely benthic, and many species were adapted to environments with dim light and a detrital food base.

Among troglobites, the Arachnida, Diplopoda, and Insecta predominate: At least 90% of the species are in these groups. Among the Arachnida, spiders (Araneae), false scorpions (Pseudoscorpionida), and harvestmen (Opiliones) are common in temperate caves and in the tropics. Other groups, such as Schizomida, Ricinulei, and Amblypygi, are also important. Among the Insecta, springtails (Collembola) and bristletails (Diplura) are common, but the beetles (Coleoptera) predominate. More than 2000 troglobitic beetles are known. One genus in the United States (*Pseudanophthalmus*) accounts for nearly 250 species. The surface-dwelling ancestors of many of these species occurred in leaf litter in forests and are believed to have colonized caves during climatic vicissitudes of the Pleistocene.

Although bats are not troglobites since they spend part of their life cycle outside of caves foraging for food, their dependence on caves is every bit as profound. Among the nearly 1100 described species of bats, at least one-third utilize caves as day roosts, hibernacula, or maternity sites. For many species, especially those in the Vespertilionidae and Molossidae, there is an obligate dependence on caves. Some species form giant hibernating colonies: More than 90% of *Myotis grisescens* found east of the Mississippi hibernate in three caves in which numbers may reach more than 1 million. The Mexican free-tailed bat, *Tadarida brasiliensis*, forms large summer maternity colonies, with up to 20 million individuals in a single cave. This species declined in numbers due to the widespread use of DDT.

VI. SOME REPRESENTATIVE SUBTERRANEAN COMMUNITIES

A. Mammoth Cave, Kentucky

The longest cave in the world with more than 500 km of passage, Mammoth Cave harbors 28 troglobites and 15 stygobites. Most of the cave lies within the boundaries of Mammoth Cave National Park. The long length of the cave is due in part to a sandstone caprock that reduces erosion and loss of upper level passages. In the large upper level passages with sandy-bottomed floors (an unusual cave habitat), trogloxenic cave crickets (*Hadenoecus subterraneus*) deposit their eggs. Since *H. subterraneus* regularly leave the cave to feed, their eggs are an important energy source. Predaceous beetles, *Neaphaenops tellkampfi*, specialize on predation on cricket eggs. Such specialization is unusual because of the scarcity of resources in most caves. Also noteworthy is the co-occurrence of five species of trechine beetles in the closely related genera *Neaphaenops* and *Pseudanophthalmus*. The aquatic community is notable for its large population of the fish *Typhlichthys subterraneus* and the crayfish *Orconectes tellkampfi*.

B. Postojna–Planina Cave System, Slovenia

Consisting of 17 and 6 km of passages, respectively, connected by 2 km of flooded corridors, the Postojna–Planina Cave system has more known species of stygobites and troglobites than any other cave or other subterranean site. The sinking river in the main passages is inhabited by a rich assortment of stygobites, stygophiles, and accidental surface species. Hydrologically inactive parts of the system contain other aquatic and terrestrial habitats. Among the 34 troglobites is the first described troglobite—the beetle *L. hochenwartii* (Fig. 1). Beetles and Collembola dominate, with 5 species of the Collembolan genus *Onychiurus* co-occurring in the cave. Among the 48 stygobites is the European cave salamander *Proteus anguinus*. Both the snail (8 species) and crustacean (16 species) faunas are rich. The marine origin of some of the stygobitic species is evident in the hydrozoan *Velkovrhia enigmatica*. The Postojna–Planina Cave is one of the best studied caves in the world, and parts of it have been heavily visited by tourists since 1818.

C. Gua Salukkan Kallang–Towakkalak, Indonesia

This immense river cave system, with more than 20 km of passage and a large bat population, is unusual among lowland tropical caves in the richness of its troglobitic and stygobitic fauna. Food resources are abundant in the cave. There are large amounts of flood debris along the riverbanks, and bat and swiftlet guano is scattered along the underground galleries. In addition to troglobites and stygobites, there are many species specialized on guano as well. Most of the 7 stygobitic species are crustaceans, and arachnids and Collembola dominate the 21 troglobitic species.

D. Walsingham Caves, Bermuda

This complex of anchialine caves (ones with a direct connection to the sea) is approximately 1 km long, with most of the passages submerged. With a freshwater lens and a redox boundary between fresh and saltwater, Walsingham Cave likely has considerable productivity as the result of chemautotrophic and heterotrophic production. The fauna is entirely aquatic, with 37 species, 29 of which are crustaceans. Among the crustaceans is a representative of the exclusively subterranean order Mictacea and rich ostracod and copepod faunas with 6 species each.

E. Movile Cave, Romania

Movile is a small, mostly water-filled cave near the coast of the Black Sea with no natural connection to the surface. Extensive chemoautotrophic production in the form of sulfur oxidation occurs in the cave system. Nearly 50 stygobites and troglobites are known from the cave, and two-thirds of these are endemic to the cave or groundwater system that the cave intersects. Among the unusual components of the fauna are an endemic leech and an endemic water scorpion (*Nepa anophthalma*). Based on analysis of mitochondrial DNA sequences, *N. anophthalma* was isolated in the cave 2 million years ago. The high diversity is made possible by the high productivity, and the high endemism is a result of the lack of contact with the surface.

F. Edwards Aquifer, Texas

San Marcos Spring is an artesian spring that serves as an exit for the Edwards Aquifer, a large cavernous limestone reservoir that lies along the Balcones escarpment. The aquifer occupies an area of 10,000 km^2. Two-thirds of the 27 stygobites are crustaceans, and endemism is high as well. The subterranean amphipod fauna in this system may be the most taxonomically diverse of its kind in the world and includes 10 species of gammaridean amphipods, representing eight genera in four families. The vertebrate stygobitic fauna is rich with two catfishes and the Texas blind salamander *Typhlomolge rathbuni*. Oil and peat deposits above the aquifer as well as long-term and gradual input of organic material into the aquifer through numerous sinkholes on the adjacent Edwards Plateau are the primary inputs into the system. The aquifer is threatened by excessive draw down, both by agricultural interests and by the city of San Antonio.

G. Lobau Wetlands, Austria

Formed by a meander arm, the Lobau wetlands are part of the floodplain of the Danube River near Vienna. Extensive studies of the superficial gravel sediments, which are up to 20 m thick, using pumps and minivideo cameras in shallow wells revealed a complex habitat with areas of differing oxygen and permeability and a rich fauna. In a 1-km^2 area more than 100 species were found, one-third of which were stygobites. Stygobites were more common at depths of greater than 5 m and 100 m away from an surface water. The microcrustaceans—Copepoda and Ostracoda—were particularly diverse.

H. Rhone River Floodplain, France

After the Danube, the Rhone River is the most thoroughly studied river with respect to its interstitial fauna. Thirty-eight stygobitic species, dominated by 23 crustaceans and 10 oligochaetes, have been found in the interstitial waters of the Rhone. There are considerable faunal differences depending on the distance from the main channel of the river, depth, and upwelling and downwelling.

I. Flathead River Alluvial Aquifers, Montana

A tributary of the Columbia River, extensive aquifers occur in the Nyack and Kalispell Valleys. Utilizing an extensive network of shallow sampling wells, investigators discovered more than 200 species of invertebrates in the aquifer. Of these, 8 were stygobites and 6 were amphibitic stoneflies. The amphibites, species whose immature forms only occur in interstitial water, included 4 species in the genus *Isocapnia*. Stygobites were concentrated far from the river, whereas other interstitial species were near to or directly below the river.

VII. GEOGRAPHY OF CAVE BIODIVERSITY

Although other subsurface habitats have not been sufficiently studied to analyze biogeographical patterns, caves have been. On a worldwide basis, more than 10,000 caves have been biologically investigated. Except for in the United States, there is no thorough analysis of regional patterns. However, there is an extensive amount of information about the number of stygobites and troglobites at individual caves and karst wells. Because of the restricted possibilities for dispersal and the restricted abilities of stygobites and troglobites to disperse, species numbers in any one cave or karst well are low relative to the regional fauna. On a worldwide basis, only 18 caves and two wells are known to have 20 or more stygobites and troglobites (Table II). Several patterns emerge. First, there is a remarkable concentration of sites (six) in the Dinaric karst of Slovenia and Bosnia and Herzegovina. Diversity of stygobites in this region is especially high, and this reflects the opportunities for both colonization from marine embayments of the ancient Mediterranean Sea and isolation resulting from the periodic regression of the Mediterranean. Second, sites with high productivity, especially chemoautotrophy, compared to other subsurface sites are well represented; these include Movile Cave, Bayliss Cave, and Walsingham Caves. Gua Salukkang Kallang Towkkalak and the Postojna–Planina Cave system can also be considered as caves with high (secondary) productivity. This accords with the widely held view that caves are usually resource poor and food limited. Third, caves and wells that intersect the permanent saturated zone are also well represented. These include Movile Cave, Shelta Cave, Edwards Aquifer, all five sites in France,

TABLE II

Caves and Karst Wells with 20 or More Stygobites and Troglobites, Listed in Descending Order of the Number of Species.[a]

Locality	Stygobites	Troglobites	Total
Postojna–Planina system, Slovenia	48	36	84
Vjetrenica Jama, Bosnia and Herzegovina	39	21	60
Pestera Movile, Romania	18	29	47
Krizna Jama, Slovenia	29	16	45
Logarcek Jama, Slovenia	28	15	43
Mammoth Cave, Kentucky	15	26	41
Walsingham Caves, Bermuda	37	0	37
Sica–Krka system, Slovenia	27	7	34
Triadou Wells, France	34	0	34
Baget/Lachein–Peyrere–Sainte Catherine system, France	24	9	33
Salukkang Kallang–Towakkalak, Indonesia	7	21	28
Edwards Aquifer, Texas	27	0	27
Goueil di Her–Reseau Trombe-Henne Morte, France	14	13	27
Bayliss Cave, North Queensland, Australia	0	24	24
Shelta Cave, Alabama	12	12	24
Cent-fons, France	22	0	22
Grotta dell-Arena, Italy	6	14	20
Buso della Rana, Italy	15	5	20
Resurgence de Sauve, France	20	0	20
Grad (Osapska Jama), Slovenia	17	3	20

[a] Data assembled with assistance of Boris Sket, Ljubljana, Slovenia.

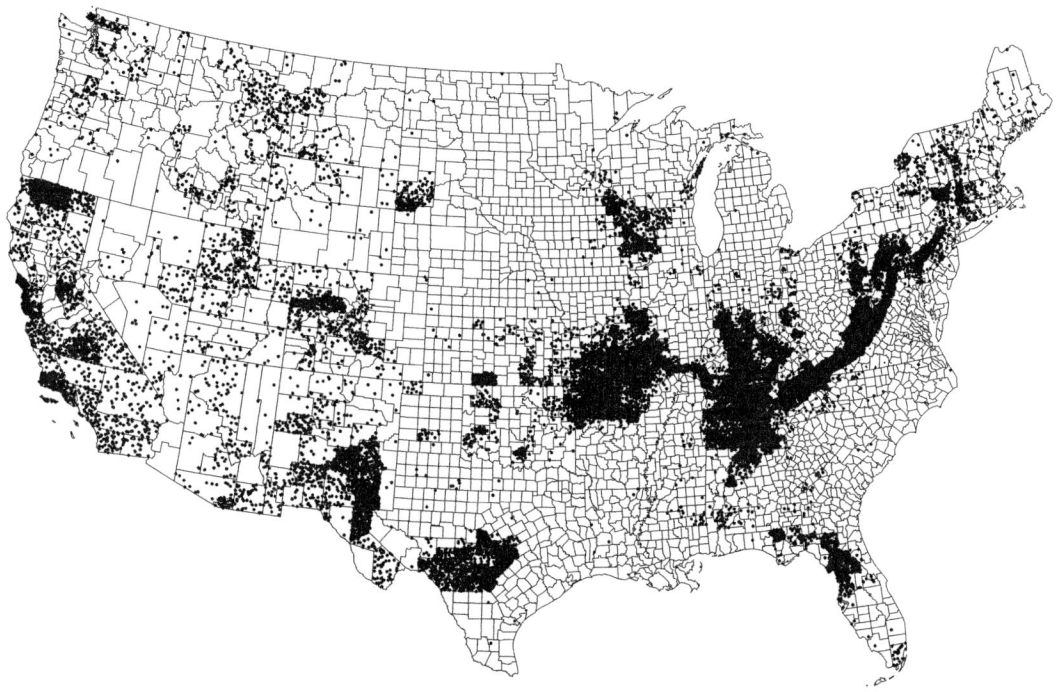

FIGURE 7 Dot map of the number of caves per U.S. county. Each dot represents one cave. [Adapted from Culver et al. (1999) with permission of National Speleological Society, © 1999.]

and all six caves in the Dinaric karst. Finally, many of the caves are long caves. Seven of the caves listed in Table II (Mammoth, Buso della Rana, Goueil di Her, Krizna, Postojna–Planina, Vjetrenica Cave, and Gua Sulukkang Kallang–Towakkalak) are more than 5 km long, but less than 0.1% of all caves are this long. A longer cave usually means a greater number of habitats.

More complete information on cave biodiversity patterns for the 48 contiguous states of the United States is available. Nearly 45,000 caves are known and their distribution by county is shown in Fig. 7. More than one-third of the counties (1144) have at least 1 cave. The major karst regions of the United States are apparent—the Appalachians (Pennsylvania to Alabama), the Interior Low Plateau (Kentucky, Tennessee, and Alabama) and associated areas immediately to the west, the Florida lime sinks, the Ozarks (Missouri, Arkansas, and Okalahoma), the Driftless Area (Illinois, Iowa, and Wisconsin), the Edwards Plateau (Texas), the Guadalupe Mountains (New Mexico), the Black Hills (South Dakota), and Mother Lode karst (California). Figure 8 shows the distribution of stygobites and troglobites by county. Some of the same regions are obvious: the Appalachians, the Interior Low Plateau, the Florida lime sinks, the Ozarks, and the Edwards Plateau. Cave areas to the north and west are not well represented by stygobites and troglobites. Diversity decreases to the north primarily as a result of covering of caves by the Pleistocene ice sheets, reducing the time available for colonization. Diversity decreases to the west primarily as a result of the fragmented nature of karst, the lack of suitable colonists due to the absence of forest litter, and progressive aridity in late Tertiary and Recent times.

VIII. CONSERVATION AND PROTECTION OF SUBTERRANEAN HABITATS

Subterranean sites present special conservation and protection problems. Literally out of sight, they are often neglected in decisions concerning conservation priorities. When they are considered, the high levels of endemism make it difficult to set priorities among endemic species. Once the decision has been made to protect a cave or other subsurface site, several principles apply. The protection of bat sites is most effectively done by gating the cave. However, there is considerable risk to the bats if the gate is not properly designed. Improperly

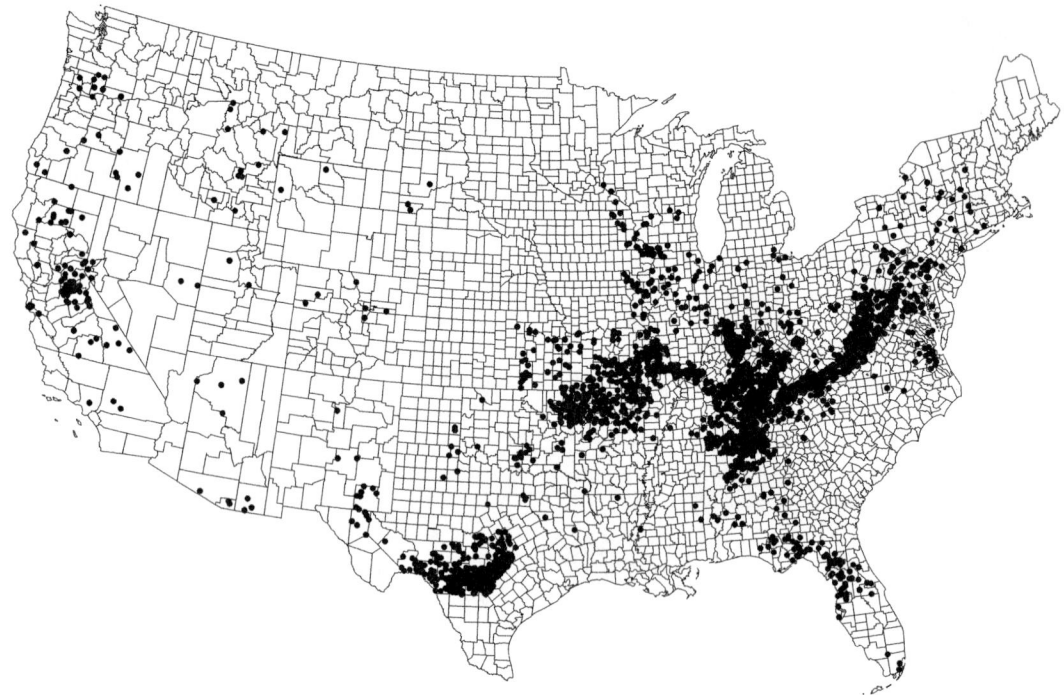

FIGURE 8 Dot map of the number of stygobites and troglobites per U.S. county. Each dot represents one county record of a stygobite or troglobite. [Adapted from Culver et al. (1999) with permission of National Speleological Society, © 1999.]

designed gates can change environmental conditions in the cave, making it unsuitable for bats, and if the bars on the gate are not properly spaced bats can suffer considerable predation at the gate. Second, protection of the subterranean aquatic fauna requires protection of the surface riparian habitat. In the case of cave streams, their upstream sources, whether they are sinking streams or sinkholes, must be protected. In the case of epikarstic species, it is the surface immediately above the cave that must be protected. Similar protection schemes are needed for the terrestrial riparian fauna. Third, protection of deep-groundwater species requires protection from both excessive draw down of the aquifer and contamination of the aquifer. Fourth, protection of troglobites that rely on the flow of organic matter into the cave entrance, such as those *Neaphaenops* beetles that are predators of cricket eggs, must include protection of the entrance area and any foraging area of the species responsible for bringing organic matter into the cave. Finally, conservation and protection of subsurface sites requires greater public awareness and appreciation.

See Also the Following Articles

SOIL BIOTA, SOIL SYSTEMS, AND PROCESSES • THERMOPHILES, ORIGIN OF • VENTS

Bibliography

Botosaneanu, L. (Ed.) (1986). *Stygofauna mundi*. Brill, Leiden, The Netherlands.

Boutin, C. (1994). Phylogeny and biogeography of metacrangonyctid amphipods in North Africa. *Hydrobiologia* 287, 49–64.

Chapelle, F. H. (1993). *Ground-Water Microbiology and Geochemistry*. Wiley, New York.

Chapman, P. (1993). *Caves and Cave Life*. Harper Collins, London.

Culver, D. C., Kane, T. C., and Fong, D. W. (1995). *Adaptation and Natural Selection in Caves. The Evolution of Gammarus minus*. Harvard Univ. Press, Cambridge, MA.

Culver, D. C., Hobbs III, H. H., Christman, M. C. and Master, L. L. (1999). Distribution map of caves and cave animals in the United States. *J. Cave Karst Stud.* 61, 139–140.

Gibert, J., Danielopol, D. L., and Stanford, J. (Eds.) (1994). *Groundwater Ecology*. Academic Press, San Diego.

Juberthie, C., and Decu, V. (1996). *Encyclopaedia Biospeologica. Tome I*. Societe de Biospeologie, Moulis, France.

Juberthie, C., and Decu, V. (1998). *Encyclopaedia Biospeologica. Tome II*. Societe de Biospeologie, Moulis, France.

Nowak, R. M. (1994). *Walker's Bats of the World*. Johns Hopkins Univ. Press, Baltimore.

Peck, S. B. (1984). The distribution and evolution of cavernicolous *Ptoma phagus* beetle in the southeastern United States (Coleoptera: Leiodidae: Cholevinae) with new species and records. *Can. J. Zool.* 62, 730–740.

Poulson, T. L. (1963). Cave adaptation in amblyopsid fishes. *Am. Midland Nat.*. 70, 257–290.

Wilkens, H., Culver, D. C., and Humphreys, W. F. (Eds.) (2000). *Subterranean Ecosystems*. Elsevier, Amsterdam.

SUCCESSION, PHENOMENON OF

H. H. Shugart
University of Virginia

I. Introduction
II. Organic Explanations of Ecological Succession
III. Mechanistic Explanations of Ecological Succession
IV. Succession and Biodiversity

GLOSSARY

autotrophic successions These generate energy from internal processes (photosynthesis).
climax community According to some theories of succession, the end result of succession in which successional change ends with a community that does not change and which is in equilibrium with the climate.
disturbance Major alternations of vegetation due to events such as wildfires, hurricanes, landslides, and human clearing.
heterotrophic successions Dependent on already fixed energy, such as the successional of communities associated with decomposition of dead logs.
individualistic view of succession Concept that succession is a consequence of species interacting with one another and their environment.
primary succession Succession on newly exposed substrates such as a sandbar or rubble at the foot of a receding glacier.
progressive succession Successions in which the dynamic changes are in the directions of increasing species diversity, structural complexity, greater biomass, and increased stability. **Retrogressive successions** are in the opposite directions.
secondary succession Succession on existing substrate (soil) following a disturbance.
succession The pattern of change expected in a community over time after a disturbance or after new substrate has been exposed.

ECOLOGICAL SUCCESSION is an ordered progression of structural and compositional changes in ecosystems toward an eventual stable condition. Descriptions of succession involve the nature of the changes and the factors that cause the changes. Ecologists have debated whether the succession is a community process or the summation of the consequences of individual species and their interactions with each other and the environment. Most ecologists doing research in this area currently favor the latter view. Investigations of the types of interactions among species have led to an increased interest in the mosaic nature of vegetation and to the application of computer models to project the expected patterns of change in vegetation over time. The biodiversity of landscapes may be highest when there is an intermediate level of disturbance. The appearance of species at different times in succession appears to be idiosyncratic to the particular vegetation; the rate of loss of species from communities seems to decrease logarithmically.

I. INTRODUCTION

Ecological succession is an ordered progression of structural and compositional changes in ecosystems toward an eventual stable condition. Primary successions are initiated on new substrates, such as a new volcanic island, a new sandbar in a river, or rubble fields at the foot of a receding glacier. Secondary successions involve the recovery of vegetation on established soils from land abandonment and from disturbances such as wildfires, hurricanes, or human alterations to vegetation. Other significant dichotomies used to categorize succession involve whether a given successional sequence is

1. Progressive (dynamically changing in the directions of increasing species diversity, structural complexity, greater biomass, and increased stability) or retrogressive (in the opposite directions)
2. Autotrophic (generating energy from internal processes) or heterotrophic (dependent on already fixed energy, such as the successional of communities associated with decomposition of dead logs)
3. Autogenic (changing due to interactions from inside the system) or allogenic (changing in response to changes in external variables)

Ecological succession is an early concept in ecology and was essential in the early definitions of ecological communities. Although the basic concepts of succession are easily understood, debates about succession (Box 1) have spawned considerable confusion and discussion. The mechanisms that drive ecological succession and its very existence as a natural phenomenon have been the subject of continual debate among ecologists. McIntosh (1985) identifies two contrasting views that typify traditional natural history and also can be found in current discussions of theoretical ecology. This dichotomy, which can be used to organize theories about ecological succession, contrasts mechanistic explanations and organic (holistic) explanations of the causes of succession. For mechanistic theories about succession, known laws explain the actions of the individual parts of a system and the whole system is the sum of these parts and their interactions. In the case of organic or holistic explanations, the whole system, its existence and design, explains the actions of the parts. Mechanistic explanations often are taken as the more modern interpretation of succession, but these "modern" views may have been the first developed and certainly developed as early as the organic views. Organic explanations of succession were most popular, at least in the United States, between the 1920s and early 1950s and have had a considerable impact on land use and conservation policies.

II. ORGANIC EXPLANATIONS OF ECOLOGICAL SUCCESSION

Organic explanations seek to understand succession in terms of principles that operate at the level of the whole system. In the 1920s and 1930s, F. E. Clements and particularly John Phillips were ascribing to ecological communities the attributes of a superorganism—a highly organized and coevolved assemblage of plants and animals interacting in a dynamic system. The ecosystem concept had its roots in debates regarding the organization and dynamics of natural systems. It was Tansley's negative view of Clements' and Phillips' interpretation of the community as a superorganism that in 1935 inspired his development of the ecosystem concept as an alternative to the organic term "community."

A. Clementsian Concepts of Ecological Succession

Between 1905 and 1935, F. E. Clements promoted a dynamic plant ecology built around a "supraorganismic" view of the ecological community. At the time of their conception, Clements' ideas represented a significant emphasis on the dynamics of vegetation and were an important early attempt to develop a formal theory predicting the pattern and expected change in ecological communities. The underlying conviction was that evolution and internal interactions would produce a homogeneous regional "climax" vegetation or community of regular species composition. The development of this "climatic climax" community was ecological succession, which was viewed as the community analog of the embryological interactions that produce an organism. In this view, succession was typified by a progressive sequence of seral stages or seres, communities that sequentially replaced one another over time until the climax stage was reached. Successional development was the result of a set of processes:

1. Nudation—the creation of a bare area (or partially bare area) to initiate succession

> **Box 1**
>
> **Complex Debates Arising in a Simple Definition**
>
> Is the pathway of successional change ordered because the mechanisms that cause succession need to occur in a proper sequence or can the steps in the order be skipped in some cases? Is there more than one ordered progression? How much variability can be tolerated before the apparent order in the progression becomes recognized as disordered or chaotic?
>
> ***Ecological succession is an <u>ordered progression</u> of structural and compositional changes in <u>ecosystems</u> toward an <u>eventual stable condition.</u>***
>
> In 1935, A.G. Tansley coined the term "ecosystem" to replace "community" because he felt the trend of thinking of the community as a direct analogue to a living organism — with succession an equivalent to embryological development — was scientifically and conceptually unwarranted.
>
> Classically, the eventual target of successional development is the climax community - a stable community associated with a given climate. Can this condition be reached before the dynamic climate changes? If so, is there more than one stable climax community?

2. Migration—the arrival of organisms at the location

3. Ecesis—the establishment of organisms at the location

4. Coaction—the interactions, particularly competition, among the organisms

5. Reaction (or facilitation)—the modification of the site by the organisms and the subsequent change in the relative abilities of the organisms to establish and survive

6. Stabilization—the development of a stable community called the climax community

Stabilization is less a process and more a consequence of the iterative reapplication of the migration, ecesis, coaction, and reaction processes until a stable community is reached. Clements' concept of the climax community was that there was only one stable vegetation type that was in equilibrium with the regional climate. Succession was the orderly, predictable, progressive, and linear development toward this climax community.

B. Application of Clementsian Concepts

The Clementsian succession paradigm had a pronounced effect on ecology in the United States in the first half of the twentieth century (much less so in Europe), and it shaped many of the laws and policies on the use of public lands. Earlier ecologists developed some of these ideas, and others evolved over Clements' and associates' scientific careers. Important aspects in the Clemetisian paradigm, many of which continue to be a part of natural land management today, are

Community-level management: The idea that one could use the state of ecological communities to evaluate their past and present conditions and to predict their future is a significant contribution of Clements and associates. Although it has been established by paleoecological studies that communities are not necessarily stable over century and millennial timescales, short-term changes and disturbances are often seen as changes in communities. Conservation groups often make considerable effort to preserve unique communities (as well as unusual or important species). Wildlife and endangered species management is often based on maintaining particular communities or habitat types appropriate to the survival of focal species of animal or plants.

Indicator species concept: Clements believed that the presence or absence of particular species could be used to assess the state of a community and its potential for agricultural conversion. For example, the presence of poisonous or distasteful plants on a range indicates overgrazing, the occurrence of certain species of Lupine (*Lupinus plattensis*) in a Nebraska prairie connotes a deep soil suitable for tilling and agriculture, and land with plants such as *Salicornia* might be so loaded with salt as to be unreclaimable. One could use indicator species to determine the past history of a landscape and to predict its future changes and potential uses.

Climax community concept: The climax community is the community that is stable in a given climate condition and is the ultimate product of successional processes. Current vegetation maps, particularly for large regions, display the expected climax community or potential natural vegetation expected in a region rather than the actual vegetation found there. Parks and wilderness areas are managed to maintain the natural climax community that is typical of the region. As an issue in conservation of diversity, often considerable effort is made to preserve unique communities (as well as species).

Progressive nature of succession: Clements viewed succession as being progressive (moving in a positive direction) toward the climax community. The succession progressed toward more diverse, more stable, and more desirable communities. This is a concept associated with eighteenth-century intellectuals and tied to other concepts such as the divine design of natural systems and ideas about the "balance of nature" and of the antiquity of certain ecological communities. Whether from Clements or earlier sources, these ideas have considerable influence on the aesthetics of conservation in the valuation of wilderness and the assessment of the importance of preserving certain species rather than others.

Perhaps the most enduring legacy from Clements is his pioneering attempt to synthesize quantitative observations about succession into a unified theory. Even strong detractors of the details of Clements' concepts are still involved in understanding succession as a general phenomenon.

C. Alternative Organic Theories on Succession

Clements' theory is the most frequently presented organic succession theory, so much so that textbook writers often term Clementsian theory to be "classic" succession theory. However, all American and British ecologists did not accept these ideas. This rejection was certainly the case for ecologists from continental Europe. Some of these ecologists emphasized a more individual species-oriented mechanistic description of succession. Others shared an interest in the holistic causes of successional change but differed from Clements in significant details. H. C. Cowles, who in 1899 was one of the first American ecologists to study ecological succession, characterized succession as "a variable approaching a variable": Succession was considered the change of a system perturbed away from—but moving toward—an equilibrium that is itself changing. Cowles believed that the climax community was never reached. In 1935, A. G. Tansley contrasted Clements' climatic climax or monoclimax theory with a polyclimax theory that had the climax vegetation of a region as a mosaic of local vegetation climaxes related to local conditions and disturbance history. In 1913, W. S. Cooper studied the forest succession on Isle Royale (Michigan) and found that the mature forest was a mosaic of patches of different ages and not the uniform climax community expected in Clementsian succession. He also believed that succession took multiple pathways and was not a linear progression of changes in seral stages. Similarly, in 1901, Cowles noted, "Succession is not a straight-line process. Its stages may be slow or rapid, direct or tortuous and often they are retrogressive" (*Bot. Gazette* **31**, 73–108, 145–182). Recognizing that there is and has been considerable difference in opinion among ecologists who use organic explanations of ecological succession, there is an even stronger contrasting of concepts between Clementsian succession and mechanistic explanations of ecological succession emphasized by some ecologists.

III. MECHANISTIC EXPLANATIONS OF ECOLOGICAL SUCCESSION

Some of the debate about the nature of succession concerns the degree to which the vegetation can be arranged into communities that are natural units of biological organization. Introductions of modern biology or ecology texts often have diagrams of biological organization that illustrate an organizational hierarchy from cells to tissues to individuals to populations to communities, etc. Such progressions convey the idea that the community is a unit of organization as demonstrable at a level such as the existence of the liver or spleen as an organizational unit is at some other level.

The American H. A. Gleason (in 1926) and the Russian L. G. Ramensky (in 1924) emphasized that vegetation was mostly the consequence of the chance arrival of species at a location and the subsequent interactions among the available species to produce the observed pattern of relative species abundance. These interactions did not produce distinct unit communities. The vegetation was believed to vary continuously with changes in the underlying environmental conditions. Under this "individualistic" view of ecological succession, the process of succession was a consequence of species interacting with one another in the context of the environment to produce vegetation dynamics or successional change. Today, most ecologists subscribe to this individualistic view of ecological succession.

A. Descriptive Mechanistic Models of Succession

When one compares modern descriptive models of succession with the Clementsian model, it is apparent that there is a substantial difference in the spatial scale considered by the two schools. Clements' climax community was considered by him to be a phenomenon that occurred over large areas. This is evident by the fact that Clementsian succession proceeded toward a regional-scale climax. Also, the union of the climax community (vegetation) and associated animals was a "biome" of which there were thought to be only 13 in nontropical North America. Clements believed that when the community in a given location was too small to have all of the species represented, succession might take different courses. However, he also believed that the succession processes he had described (nudation, ecesis, coaction, etc.) would still operate at these smaller scales.

In 1987, Pickett and colleagues produced a table of mechanisms and causes of succession that can be compared directly to earlier writing of Clements (Table I). In this comparison, one sees that many of the processes deemed important by Clements (notably, nudation, migration, ecesis, coaction, and reaction) are also represented by a mechanistic explanation of succession, albeit with different names and differing emphases. The differences are most pronounced with regard to the importance of the reaction (facilitation). Certainly in many primary successions, the changes produced by one set of species appear necessary for the success of the next. For example, the rapidly growing willows (*Salix* sp.) that stabilize sandbars in rivers seem to be a necessary preamble to the success of subsequent species; the ability of alder (*Alnus* sp.) to symbiotically fix nitrogen increases the fertility of sites uncovered by receding glaciers; and the organic acids produced by blue-green algae, lichens, and mosses speed the breakdown of granite and the development of a thin soil to support grasses, herbs, and even trees in primary successions on granite outcrops.

However, some species appear to block the success of others and hold sites against species that might otherwise succeed them. In some secondary successions, all the species involved in the succession are present as seeds or other propagules from the initiation. In these cases, the familiar successional sequence of grasses and herbs yielding to shrubs and then to trees may reflect a difference in rate of growth of individuals present from the start. Evolutionarily, it is difficult to explain why a species would evolve to help another take over a site it could otherwise occupy.

TABLE I
Clementsian Succession Processes Compared with a Mechanistic Explanation of the Causes of Succession[a]

General causes of succession	Contributing processes of succession	Modifying factors	Clementsian succession analog processes
Site availability	Coarse-scale disturbance	Size, severity and timing of disturbance, dispersion of species	Nudation and migration
Differential species availability	Dispersal	Landscape configuration, dispersal agents	Migration
	Propagule pool	Time since last disturbance, land use conditions	Migration and ecesis
	Availability of resources	Soil condition, topography, microclimate, site history	Ecesis
Differential species performance	Ecophysiology	Germination requirements, photosynthesis rates, growth rates, population differentiation	Coaction
	Life history	Photosynthesis allocation pattern, timing of reproduction, mode of reproduction	
	Competition	Hierarchy of competitive interactions, the presence of competitors, identity of competitors, within-community disturbances, predators, herbivores, resource base	
	Herbivory, predation, disease	Climate cycles, predator cycles, plant vigor, plant defenses, community composition, patchiness	
	Environmental stress	Climate cycles, site history, prior occupants	Reaction (in part)
	Allelopathy	Soil chemistry, soil structure, microbes, neighboring species	?

[a] The first three columns are from a review by Pickett *et al.* (1987).

Connel and Slatyer (1977) developed a descriptive model of the succession processes based on a mechanistic understanding of succession (Fig. 1). They used the reaction/facilitation issue to frame three models of succession based on mechanisms of interaction (the facilitation model, tolerance model, and inhibition model). In Fig. 1, the facilitation model is most like Clementsian succession as typically interpreted. The three models are different and have different implications for land management and particularly land reclamation. If one had the objective of restoring degraded landscapes, then one might speed the restoration by eliminating established species in the case of the inhibition model, but this would be ill advised in the facilitation model (Fig. 1).

B. The Landscape as a Mosaic

In addition to an expansion of the facilitation concept associated with Clementsian succession, there has also been an emphasis on understanding the mosaic nature of vegetation. Historically, this emphasis derives from earlier concepts of "cyclical succession" (exemplified by the cyclical replacement series involving a forest canopy gap) and the "polyclimax" (the mature forest as a mosaic trees, gaps, and recovering gaps). Theories about the dynamics of landscape mosaics have been developed in forests to a significant extent, probably because, regardless of other sources of spatial heterogeneity, a forest canopy is a mosaic of tree crowns (Fig. 2).

Starting with a small plot of land in a mature forest dominated by a single large tree, the large tree shades the ground and reduces the survival of smaller trees and seedlings below. There may be a few smaller shade-tolerating trees that survive under the large tree, but these are strongly suppressed in their rate of growth. The large tree dominates the resources (light, water, and nutrients) that are available at the site and blocks other trees from growing at the location. When this tree dies, the forest floor (where there had previously been little chance of a young tree's survival) becomes a nursery for small seedlings and saplings. There is adequate light and other resources, and hundreds of small trees survive and begin to grow toward the canopy. As these trees grow, they begin to compete with one another. Some of the trees lose to more vigorously growing competitors. Eventually one tree manages to win the race to be the local canopy tree and begins to

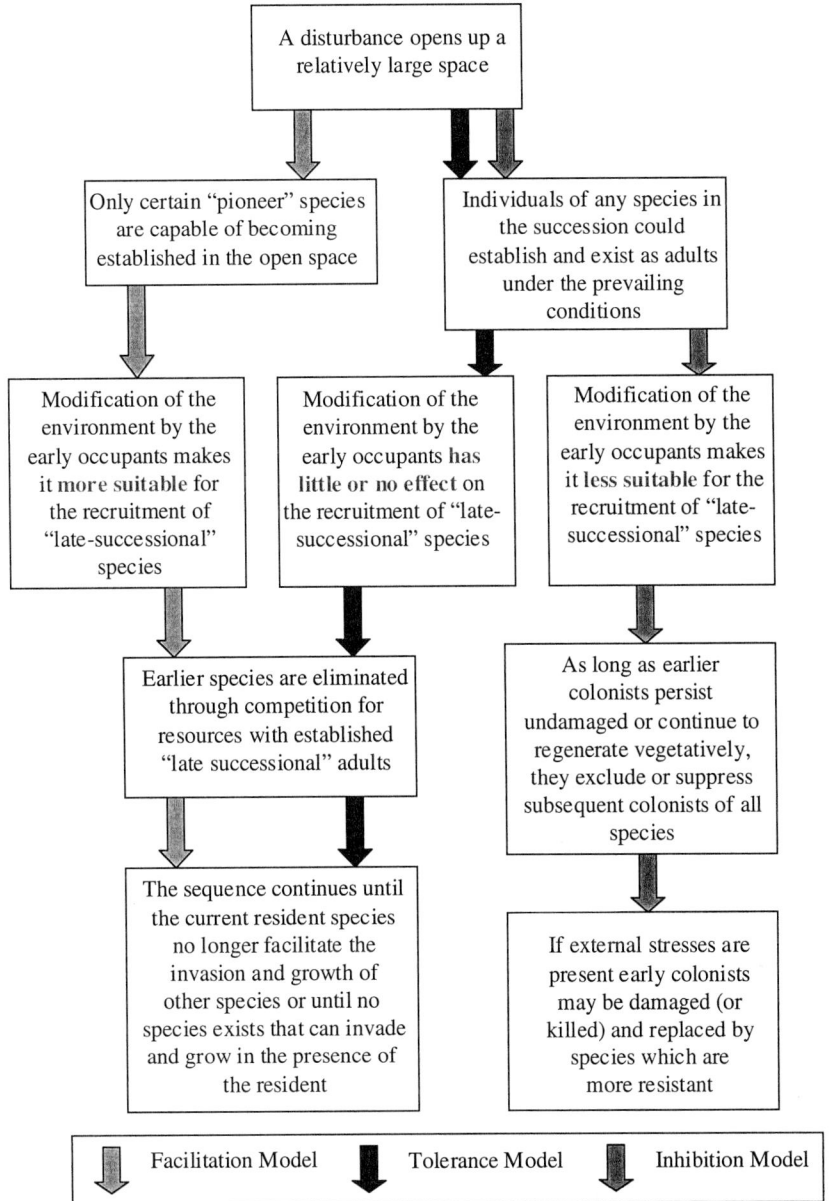

FIGURE 1 Mechanistic models of ecological succession (reproduced with permission from Connell and Slatyer, 1977).

eliminate the others. This represents the closure of the cycle with a large tree again dominating the site. Over the course of time this cycle is reinitiated by the death of the new dominant replacement tree. The implications of the cyclical nature of small-scale forest dynamics were clearly elucidated by Cooper in 1913 and by A. S. Watt in a classic paper in 1947.

The expected changes in the amount of living material (biomass) over multiple iterations of a gap generation and filling should create a "saw-toothed"-shaped curve. This curve drops abruptly with the death of a canopy dominant and then builds biomass as the regenerating trees grow, compete, and occupy the site (Fig. 3, top). The distances between the "teeth" in the saw-toothed, small-scale biomass curve are determined by how long a particular tree lives and how much time is required for a new tree to grow to dominate a canopy gap.

FIGURE 2 The forest canopy as a mosaic (photograph of South African rainforest from J. C. van Daalen).

and floristic richness, because processes of forest succession and many of the autecological properties of tree species, worked out long ago in the north temperate region, are cosmopolitan. There is a basic similarity of patterns in space and time because the same processes are at work.

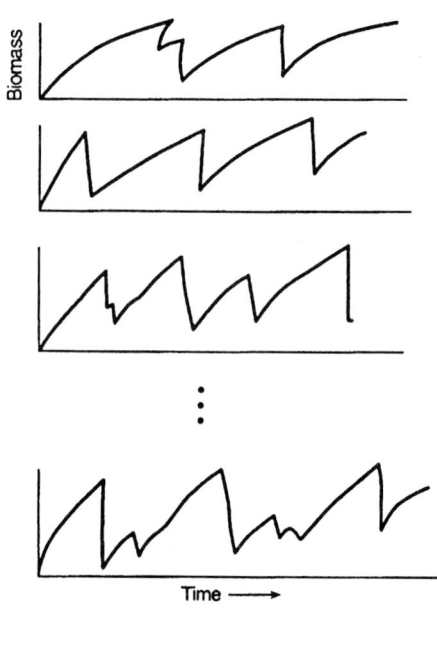

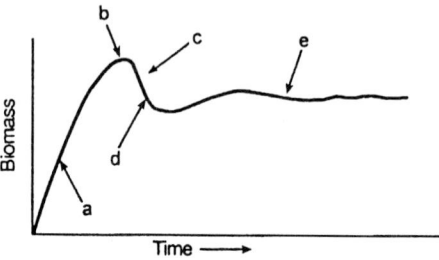

FIGURE 3 Biomass dynamics for an idealized landscape. The response is from a relatively large, homogeneous area composed of small patches with gap phase biomass dynamics. (Top) The individual dynamics of the patches that are summed to produce the landscape biomass dynamics. (Bottom) The landscape biomass curve rises (a) as all of the patches are simultaneously covered with growing trees. Next, local decreases in biomass are balanced by the continued growth of large trees at other locations and the curve levels at a maximum value (b). If the trees have relatively similar longevities, there is a period in which several (perhaps the majority) of the patches that comprise the forest mosaic all have deaths of the canopy-dominant trees and the curve decreases (c). Eventually, the local biomass dynamics become desynchronized (d) and the landscape biomass curve varies about some level (e) (reproduced with permission from Shugart, 1998).

The larger scale biomass dynamics (Fig. 3, bottom) is a simple statistical consequence of summing the dynamics of the parts of the landscape mosaic. If there has been a synchronizing event, such as a clear-cutting or other disturbance, one would expect the landscape mosaic biomass curve to increase as all of its parts are simultaneously covered with growing trees (Fig. 3,a). If the trees over the area have relatively similar longevities, there is also a subsequent period when the deaths (and biomass) on plots where a canopy tree happens to die are balanced by those on plots where large trees are still growing. During this period, the loss balances gain in biomass and the curve levels (Fig. 3,b). If there is sufficient synchronization in the sawtooth curves of the component plots, this is followed by a period during which many of the pieces that comprise the forest mosaic all have deaths of the canopy-dominant trees (Fig. 3,c) and landscape biomass decreases. Over time, the local biomass dynamics become desynchronized and the biomass curve varies about some level (Fig. 3,d) This mosaic of repairing gaps with all different stages of recovery represented on the landscape can be taken as the mature forest.

The mature forest should have patches with all stages of gap phase dynamics and the proportions of each should reflect the proportional duration of the different gap replacement stages. T. C. Whitmore believed that this was the expected pattern and process for all forests and asserted in 1982 that

> forests of the world are fundamentally similar, despite great differences in structural complexity

Whitmore is referring to mosaic forest canopies as a consequence of gap replacement processes. The occurrence of such patterns has been documented for several different kinds of forests. The presence of shade-intolerant trees occurring in patches in mature undisturbed forest is another observation consistent with the mosaic dynamics view of mature forests. The scale of the mosaics in many natural forests is somewhat larger than one would expect from gap filling of single tree gaps, indicating an importance of phenomena that cause multiple tree replacements. Also, relatively long records (approximately 40 years in most cases) of forest structure and composition indicate a tendency for the forest composition to fluctuate with species showing periods of relatively weak recruitment of individuals to replace large trees and strong recruitment in other periods.

C. Quantitative Mechanistic Models of Succession

A consequence of recognition of the mosaic nature of vegetation and an emphasis on more mechanistic representation of the succession process has been the development of quantitative models that can predict changes in vegetation structure (Shugart, 1998). Several of these models simulate the successional dynamics by accounting for the birth, growth, death, and interactions with the environment for the hundreds of individual plants living on a small plot of land. The predictions of hundreds of these plots are then combined to obtain a prediction of the change in ecological landscapes. Because they simulate the fates of each of the millions of plants involved in a landscape succession, these models are called "individual-based" model. These models require considerable computation to solve but they can be solved relatively easily as a consequence of the increased computational power of modern computers.

An advantage of individual-based models is that the following implicit simplifying assumptions associated with other modeling approaches (e.g., the Markov process or differential equation-based models) are not necessary: (i) The unique features of individuals are sufficiently unimportant to the degree that individuals are assumed to be identical, and (ii) the population is "perfectly mixed" so that there are no local spatial interactions of any important magnitude. Most ecologists are interested in variation in individuals (a basis for the theory of evolution and a frequently measured aspect of plants and animals) and appreciate spatial variation as being quite important. These assumptions seem particularly inappropriate for trees which are sessile and which vary greatly in size over their life span. This may be one of the reasons that tree-based forest models are among the earliest and most widely elaborated of this genre of models.

One group of individual-based models simulates the establishment, diameter growth, and mortality of each tree in an area the size of a gap left by the death of a canopy tree and is called gap models. Gap models can be used as an example of the more general individual-based modeling approach. In most gap models, calculations are on a weekly to annual time step. Early gap models were developed for a size unit (approximately 0.1 ha) approximately that of a forest canopy gap. Gap models feature relatively simple protocols for estimating the model parameters. For many of the more common temperate and boreal forest trees, there is a considerable amount of information on the performance of individual trees (growth rates, establishment requirements, and height/diameter relations) that can be used in estimating the parameters of such models. Gap models have simple, general rules for interactions among individuals (e.g., shading and competition for limiting resources) and equally simple rules for birth, death, and growth of individual trees (based on the natural history of each species).

Gap models differ in their inclusion of processes that may be important in the dynamics of particular sites being simulated (e.g., hurricane disturbance, flooding, and formation of permafrost) but share a common set of characteristics. These latter characteristics involve an emphasis on the demography and natural history of plant species, relatively general rules for physiological tradeoffs among species, and an emphasis on the understanding of successional processes at the whole plant level. Each individual plant is simulated as an independent entity with respect to the processes of establishment, growth, and mortality. This feature is common to most individual tree-based forest models and provides sufficient information to allow computation of species- and size-specific demographic effects. Gap model structure emphasizes two features important to a dynamic description of vegetation pattern: (i) the responses of the individual plant to the prevailing environmental conditions and (ii) how the individual modifies these environmental conditions. The models are hierarchical in that the higher level patterns observed (i.e., population, community, and ecosystem) are the integration of plant responses to the environmental constraints defined at the level of the individuals.

IV. SUCCESSION AND BIODIVERSITY

Since the initial formulation of a progressive, holistic concept of ecological succession, there has been a tendency to associate positive attributes to mature communities. Hence, there is an expectation for increasing successional age to be associated with increased biotic diversity. There are certainly ecological systems (notably many forest systems) that demonstrate this pattern, but there are other examples of ecological successions that show highest levels of species diversity (measured as the number of species or by standard indices of diversity) at intermediate successional ages (e.g. shortgrass prairie in Colorado) or even at initial successional stages (e.g., boreal forests in areas of Canada). Succession on sand dunes on the coast of Queensland, Australia, has the highest species richness in shrub-dominated communities at the beginning and end. The pattern of diversity in different successions seems to be an idiosyncratic consequence of the attributes of the participating species and the environmental conditions at a given location.

A. Biodiversity on Disturbed Landscapes

At the landscape level, the overall biodiversity can be related to the frequency and intensity of disturbance and there is evidence from both theoretical work and observations that an intermediate level of ecological disturbance can produce the most diverse landscapes (Huston and Smith, 1987). This occurs in part because disturbed landscapes have a mixture of species able to successfully occupy the differently aged patches created by the disturbance history of a given landscape. One would generally expect highly heterogeneous landscapes to be more diverse. Disturbances also prevent particularly well-adapted species from occupying the entire landscape.

In 1979, J. P. Grime developed a "triangle" based on three primary plant response strategies that can be used to develop rules that can be used to predict the proportions of each strategy (and associated life-forms) expected under a particular environmental regime. Grime recognized two types of external factors limiting the biomass of plants. The first was stress, involving the conditions that restrict plant productivity (shortage of light, H_2O, mineral nutrients, etc.). The second was disturbance, involving partial or total destruction of plant biomass (activities of herbivores, diseases, fire, frosts, etc.). These two external factors can operate independently so that there are four possible combinations of high or low stress and high or low disturbance.

Grime reasoned that the combined action of high stress and high disturbance created a condition from which the vegetation could not regenerate. Low-stress and low-disturbance environments would ultimately favor species that were able to compete effectively against other species (competitor strategy), high-stress and low-disturbance environments should be dominated by plants of species able to tolerate the particular stress (stress-tolerator strategy), whereas low-stress and high-disturbance environments should favor short-lived, fast-growing species (ruderal strategy). The general problem exemplified by Grime's work in development of these primary plant strategies is one of identifying plant functional types. An essential basis of this as well as other functional classifications of plants is that of "tradeoffs"—the idea that, due to underlying rules that derive from the species physiology and natural history, a single species cannot simultaneously be the best as a stress tolerator, a competitor, and a ruderal species.

B. The Gain and Loss of Species with Succession

One can view the richness of species on a parcel of land with the same disturbance history as a consequence of the gain and loss of species. The gain of species involves factors such as the migration and establishment of species (or the species gaining sufficient abundance or size so it can be sampled). There has been debate as to whether the gain of species at a site over succession is a consequence of the different species present at the site growing and developing at different rates or is actually due to species establishing themselves in an ordered sequence. The succession is a consequence of different growth rates of an initial innoculum of seeds of all the species involved germinating and growing at different rates and is sometimes called the initial composition model (Egler, 1954). The idea that one set of species is added to the community after a previous set has modified the site and lost the site has been termed the relay floristics model (Egler, 1954). To separate these two "models," a particular succession study has to be sampled intensively enough to actually detect all the species at a location, which rarely occurs. Considering a wide range of successions of plant communities, animals associated with succession, and heterotrophic successions (such as the progression of changes in a decaying log), one finds that the appearance of species with abundances sufficient to be counted varies with the particular collection of species and with the changes in the physical environment associated with the succession.

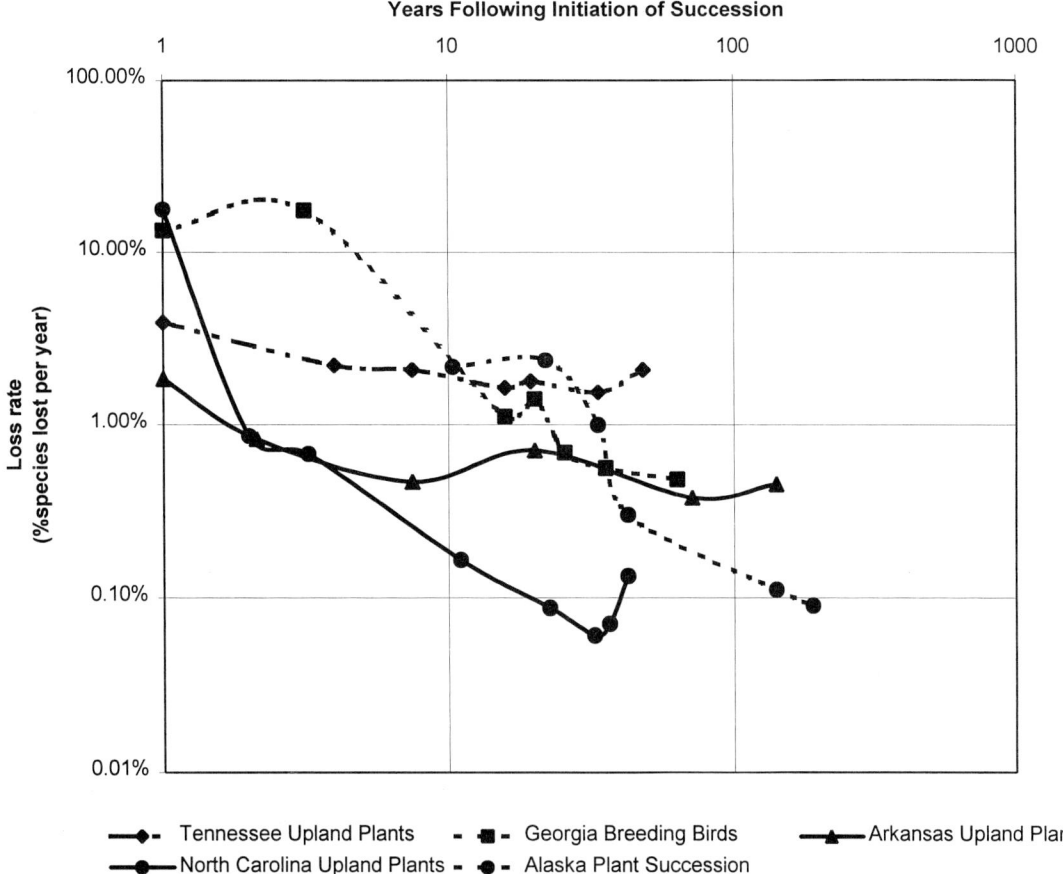

FIGURE 4 Loss rates of species from different successional sequences (reproduced with permission from Shugart and Hett, 1973).

The loss of species from successional communities tends to be somewhat more regular in its pattern of variation. Considering a wide range of communities (both autotrophic and heterotrophic successions), one finds that the rate of species local extinction through succession tends to decrease logarithmically over successional time. Successional sequences are often sampled and reported using a more or less logarithmic sampling regime (e.g., communities in a successional study might be sampled at 1, 2, 5, 10, 17, 35, 60, and 100 years). This reflects the pattern that one tends to sample successional sequences using designs where a proportion of the species in a site of a given age were found in the younger sites. Figure 4 illustrates this pattern for five different successional sequences in different locations. The initial loss rates of species are on the order of approximately 10% of the species found on a plot disappearing each year. In the later successional stages, this decreases to less that 0.1% species lost per year (Fig. 4). Because this pattern occurs across a wide range of successional sequences, this does not appear to be a consequence of the later successional species living longer (as is often the case in forest successions).

See Also the Following Articles

DISTURBANCE, MECHANISMS OF • ECOSYSTEM, CONCEPT OF • ECOSYSTEM FUNCTION, PRINCIPLES OF • INDICATOR SPECIES • LANDSCAPE DIVERSITY

Bibliography

Connell, J. H., and Slatyer, R. O. (1977). Mechanisms of succession in natural communities and their role in community stability and organization. *Am. Nat.* **111**, 1119–1144.

Egler, F. E. (1954). Vegetation science concepts. I. Initial floristic composition—A factor in old-field vegetation development. *Vegetatio* **4**, 412–417.

Grime, J. P. (1979). *Plant Strategies and Vegetation Processes*. Wiley, Chichester, UK.

Huston, M. A, and Smith, T. M. (1987). Plant succession: Life history and competition. *Am. Nat.* **130**, 168–198.

Pickett, S. T. A., Collins, S. L., and Armesto, J. J. (1987). Models, mechanisms and pathways of succession. *Bot. Rev.* **53**, 335–371.

Shugart, H. H. (1998). *Terrestrial Ecosystems in Changing Environments*. Cambridge Univ. Press, Cambridge, UK.

Shugart, H. H., and Hett, J. M. (1973). Succession: Similarities of species turnover rates. *Science* **180**, 1379–1381.

Whitmore, T. C. (1982). On pattern and process in forests. In *The Plant Community as a Working Mechanism* (E. I. Newman, Ed.), British Ecological Society Special Publ. No. 1, pp. 45–59. Blackwell, Oxford.

SUSTAINABILITY, CONCEPT AND PRACTICE OF

Kai N. Lee
Williams College

I. A Concept in Search of Practice
II. Trends and Transitions
III. Indicators

GLOSSARY

anthropogenic disturbances Disturbances of human origin, including modifications of ecosystem structure or function and displacement or removal of species from habitats.
demographic transition Transition of human populations from conditions of high birth rates and high death rates to low death rates, followed by low birth rates; in progress since the seventeenth century, with a large increase in the number of humans—projected to be completed in the twenty-first century.
ecosystem services Flows of services of natural origin which are valuable to human users and occupants of an ecosystem, e.g., water purification by flowing streams and crop production from fertile soils.
indicators Quantitative measurements of environmental and social variables that provide time series describing long-term trends; some of the trends may indicate a transition toward sustainability.
maximum sustainable yield Estimate of the size or proportion of standing stocks of a population that may be harvested without altering the long-term abundance of characteristics of the stocks.
sustainability transition A search for sustainable development, through action and research, pursued during the remaining decades of the demographic transition. A sustainability transition would be shaped by normative goals for human well-being and preservation of the life-support systems needed by human populations.
sustainable development A pattern and path of economic and social development compatible with the long-term stability of environmental systems, particularly those essential to human well-being.

BIODIVERSITY IS DEPENDENT ON SUSTAINABLE UTILIZATION OF THE NATURAL WORLD BY HUMANS, and the conservation of biodiversity may well be essential to the durability of the human species. An increase in human-caused impacts on the natural world during the past two centuries is the driving force behind a great extinction, a large-scale reduction of the diversity of biota on the planet that has occurred only five times in the geological record. This sixth great extinction is the first to be caused by a living species.

Is a sustainable economy possible? Recent scientific appraisals suggest that it is but that a transition toward sustainability will require significant social, political, and technological changes during the next two generations. This is also the time period in which human population seems likely to level off; hence, it is possible to think of a sustainability transition on the timescale of the demographic transition drawing to a close during the twenty-first century.

A sustainable economy is not a well-defined objective, so a transition toward sustainability must be a search rather than a march. Awareness of long-term trends and transitions together with indicators to inform our searches are important contributions that science can provide in addition to developing means for reconnecting human prosperity to the diverse and essential riches of the natural world.

I. A CONCEPT IN SEARCH OF PRACTICE

The idea of sustainability—that the fruits of nature, if harvested at moderate rates, may be reaped indefinitely—is ancient wisdom. However, translating that verity into workable policies is difficult and elusive. As economies and human population have grown rapidly during the past 200 years, exploitation of ecosystems for human gain has usually ignored sustainability and often depleted biodiversity. This may change in the next several decades as land transformation and human appropriation of ecosystem services surge toward natural limits and the growth rate of the human population declines toward zero. Still, more than 800 million people face hunger during at least a part of each year, and two-thirds of the human race live in developing countries, a label that implies priority for human welfare over the conservation of species and ecosystems. Is human well-being dependent on the survival of biodiversity, and if so how? Should attempts to improve the material conditions of human life be constrained by attempts to ensure the long-term survival of habitats and species? Species extinction is one of the only indications of irreversible environmental loss widely accepted by laypersons; what is its practical significance, both biologically and socially? The connection between sustainability and biodiversity is neither conceptually clear nor practically straightforward, but it is of fundamental significance.

A sustainable material economy may be defined as one in which anthropogenic disturbances to ecosystems—those of human origin—are smaller than or similar to those due to other causes. The magnitude and scope of anthropogenic change today are unprecedented and in many important dimensions have been accelerating during the past two centuries. Although disturbances of human origin raise obvious concerns about their sustainability, a practical operational definition of sustainable practices has been difficult to articulate or to implement.

In nature, unsustainable behavior is self-limiting, but material constraints on human welfare have been slow to appear. Humans live and work in virtually all parts of the planet, exchanging goods in a global economy. In some respects, richer societies can transfer the burdens of their unsustainable practices onto poorer ones; some of the best agricultural land in the tropics has been committed to the production of sugar, tea, coffee, chocolate, and other luxury goods for temperate-zone markets for centuries. However, reallocation of the costs of unsustainable activities is not the only process at work. Remarkably, indicators of well-being, such as life span, income, and education, have increased for nearly all human populations during the past several decades, when reliable data series have been collected. As a result, direct signals that the human situation cannot be maintained remain fragmentary and inconclusive, and they may have been obscured by technological innovation and changes in social organization. Although many species and ecosystems have been lost and transformed, there are few instances known to the general public in which changes have been forced on people by the destruction of biodiversity.

This article approaches sustainability as a long-term phenomenon reflected in local and national cultures as well as in a changing global sensibility about the place of humans in the natural world. From this perspective, the survival of today's biodiversity is likely to be determined by a larger historical process: whether the human economy can develop processes and governing mechanisms to achieve material sustainability. If not, our species will likely prove to be shorter lived than most in the geological record; however, we shall wreak havoc, as we are now doing, on a scale seen only a handful of times before in this record. Even if we are able to achieve material sustainability, this will not happen soon enough to save all that is endangered today. However, what would remain—under long-term human care—is likely to be a monument to our species worth contemplating seriously.

A. Science and Sustainability

The related ideas of sustainability and sustainable development have been influential throughout the rise of ecological science and environmental policy. Sustainability has in the course of this development acquired social significance in ways that were not intended by the scientific community but that matter nonetheless.

That ecosystems have a finite capacity to support any population is evident from the uneven production of crops throughout human history. As human numbers and consumption of natural resources increased in the nineteenth century, so did concern that the impact of our species on the natural world would have irreversible, ultimately self-destructive effects. That such fears had substance was clear from the extirpation of valued species such as the passenger pigeon; thoughtful writings such as George Perkins Marsh's *Man and Nature* (1864); the establishment of national parks beginning with Yellowstone (1872); and the founding of citizen groups as early as the British Commons, Open Spaces, and Footpaths Preservation Society, organized in 1863. However, it was also evident that farming, fishing, forestry, and manufactures had persisted for centuries in many places, often without noticeable decline in landscapes utilized and inhabited by humans.

The answer to this paradox seemed to lie in the idea of sustainable yield—the notion that most natural populations could be harvested, to a degree, without reducing their capacity to reproduce abundantly. Exceed the sustainable level and populations would dwindle even in the long run; harvest less, and part of the population would live out its life without benefiting humans. Therefore, there is an optimum (for humans) for each population, the maximum sustainable yield. This concept reinforced ideas of efficiency that were influential at the turn of the century in the United States when Gifford Pinchot, founder of the Forest Service, brought the concept of sustainable yield into public policy.

Sustainability held the promise of perpetual income, a flow of returns that humans might receive from nature in return for intelligent management. In this way, the stewardship ideals of an agricultural society might be extended to wild lands and sea. However, the promise was incomplete in significant ways. First, unlike the cultivation of annual crops, forests and fisheries were composed mainly of slower growing populations, in which the effects of habitat management or harvest on population size were delayed for years or decades. Moreover, the species of interest lived in ecosystems that humans could not manage as decisively as they did farmland or plantation. Third, mobile species such as fish were often exploited under open-access conditions, in which the short-term incentives facing harvesters ran counter to the interest of the community of harvesters or even themselves over the long term. Where these complications prevailed—as they often do even now—sustainable yield often proved to be a way to rationalize unsustainable behavior. Estimates, inevitably clouded by uncertainties and environmental fluctuations, would "inform" decisions that often led to overharvest or to irreversible habitat transformations such as the building of dams. Even the lessons of mismanagement are shaped by the analyst's understanding of ecology and sustainability. As Cronon (1992) has shown, the American Dust Bowl of the 1930s can be seen as a tragic ecological error or as a heroic struggle against nature; such contrasting narratives imply contrasting recommendations for current and future management.

That a scientific idea such as sustainable yield could be misused reinforces an implicit but important message: Science has normally been the servant of those who would transform natural ecosystems for human gain. Although the content and methods of science might be neutral, both the scientists who inferred practical implications and the way science was used often advanced or legitimated an extractive, exploitative agenda. Proponents of sustainability, like those advocating conservation and environmental protection, have found this imperialist mantle difficult to recognize and even more difficult to shed. Taking this burden into account is of instrumental significance in debates and activities in the future.

Science plays another role in sustainability as the wellspring of technology—the methods, tools, and means humans have used to extend their control over the natural world. Guided by the idea of carrying capacity, ecologists have estimated that humans now appropriate slightly less than half the net primary production on land and approximately the same fraction of fresh water. Such estimates indicate that further economic growth or improvements in human well-being—even if they occur—may not continue to increase the size of the material economy on which life depends. Indeed, it seems likely that achieving sustainable economies will require decreasing the burdens of wasted energy, discarded materials, and pollution that are now imposed on the environment. Technology plays a strategic role in this aspiration.

Surprisingly, the definition of a sustainable material economy remains problematic: Which stocks must be preserved? Which flows must be conserved and at what levels? Is there a single numeraire that can indicate sustainability or its opposite? There are no definitive answers to these questions. The following well-known statement was put forward in 1987 by the World Commission on Environment and Development chaired by Gro Harlem Brundtland:

Sustainable development is development that meets the needs of the present without compromising the ability of future generations to meet their own needs. It contains within it two key concepts:

- The concept of "needs," in particular the essential needs of the world's poor, to which overriding priority should be given; and
- The idea of limitations imposed by the state of technology and social organization on the environment's ability to meet present and future needs.

This definition of sustainable development describes environmental limitations in terms of technology and social organization and not the carrying capacity of nature in an objective sense. This language reflects an important conclusion: There is no consensus on which material conditions do or will limit activities that obviously cannot continue to expand in the long term. Instead, social constraints, in the form of decisions to follow some paths rather than others, may avoid some undesirable consequences even as new technologies enable *Homo sapiens* to continue to evade others. It should be noted that many people, including many natural scientists, doubt that constraints can emerge in a voluntary and humane fashion. If not, the needs of the poor may not be the driving force of a sustainable future, as proposed by the Brundtland Commission.

Since we lack a crisp operational definition of sustainability, the topic is best organized around questions. What is the evidence of recent trends on the magnitude and tendency of human-caused disturbances? Is there a way to define indicators of sustainability that could guide social constraints and the search for new technology? Is there a way to link such indicators to a process of social learning? In the following sections, the surprisingly mixed evidence of human impacts is summarized. Then the question of how to recognize sustainability and unsustainability is developed as a leading focus of social learning, the process (not yet clearly defined) of incorporating understanding of nature and its limits into human culture and governance. It is likely that social learning plays a critical role in searching for a long-term transition to sustainability: the fashioning of social constraint, the implementation of new technology, and the acceptance of limits to material endeavor that can jointly create a durable place for humans in the biosphere.

II. TRENDS AND TRANSITIONS

The prospects for sustainability should be judged against the situation of the present and the trends that have carried us to this point. Such a review demonstrates the grave risks faced by biodiversity in the next two or three human generations but also indicates important opportunities. Three trends—population, economic growth and consumption, and biodiversity—are reviewed in the following sections. The picture they provide is fleshed out with a wide-ranging review recently completed by the U.S. National Research Council. These set a context for the notion of a sustainability transition.

A. Population

Human population at the beginning of the twenty-first century is 6 billion people, more than triple the 1.65 billion alive in 1900. About four-fifths of today's population lives in the less developed areas of the world. With an annual growth of 1.3%, about 80 million people are being added to the planet each year, nearly all of the net increases occurring in the developing world. Global growth rates are declining and have been doing so since the peak rate in modern history, about 2.2% per year, occurred in the early 1960s. Because this slowly declining growth rate is applied to a large base, absolute population growth will remain high for the next few decades. Human population is expected to increase by 2 billion between 2000 and 2025, the same amount as in the last fourth of the twentieth century.

The deceleration of population growth, the closing phase of a process referred to as the demographic transition, first became apparent in studies of European demography. This scenario is now believed to be operating globally, although with considerable variations from the simple pattern described here. Within two centuries the European population went from conditions of high births and deaths to the current situation of low births and low deaths. Initially, deaths declined more rapidly than births, and population grew rapidly. Later, birth rates decreased to match or even exceed the decline in the death rates, and population stabilized and sometimes declined. We are now in the midst of a global demographic transition that is more rapid than its European prototype. Birth and death rates in developing countries have decreased at unexpectedly rapid rates. The average number of births for each woman of reproductive age has declined to three compared to six at the post-World War II peak of population growth. The mortality transition has also proceeded very rapidly, with life expectancy at birth having increased from 40 years in 1950 to about 64 years today. Whether these changes will lead to an increasing, decreasing, or stable population is unknown, although the rate of increase has clearly declined during the past three decades.

Today's population growth has immense momentum because large new generations of young people are reaching reproductive age. How much population will increase depends on their choices of family size and their ability to implement these choices. Policies designed to encourage such implementation may be able to slow growth considerably. Indeed, recent rates of decline in fertility have outpaced earlier projections, and the United Nations (UN) reduced its medium expectation of global population in 2050 from almost 9.8 billion in the 1994 projection to 9.4 billion in the 1996 projection. By the end of the twenty-first century, the world's human population is projected to reach 10.4 billion, a level that seems likely to be subject to significant change by policies or by inaction.

B. Economic Growth and Consumption

The growth of wealth has been as persistent as the growth of population. Moreover, barring severe disruptions to the global economy, income and consumption will continue to grow faster than human numbers. Supplying the energy and materials needed to support increasing consumption and addressing the environmental problems attendant on their extraction, consumption, and disposal may be the most significant challenge to sustainability, especially as more people adopt the materials-intensive, consumptive lifestyle now enjoyed by most people in industrialized nations.

There have been dramatic changes in human well-being since the early nineteenth century, when the modern pattern of industrial and information-intensive economic growth became clear in the historical record. Trends in gross domestic product (GDP)—a measure of the total economic activity in a nation's markets—reflect a nation's production and wealth per capita and hence give an indication of the well-being of that country's people.

There has been an average worldwide gain in GDP per person by a factor of 7.9 between 1820 and 1992; in the four "Western offshoots"—Australia, Canada, New Zealand, and the United States—economic growth has resulted in a gain of more than 17-fold over this span of approximately six generations, doubling economic output within each human life span. Even in Africa, the region with the weakest record, economic output per capita had tripled by 1980. Indeed, contemporary

TABLE I

Actual and Projected Changes in World Population, Food, Energy, and Economic Output[a]

	Actual—1950 compared to 1993	Scenario projection—1995 compared to 2050
Population	2.2×	1.6×
Food (grain)	2.7×	1.8×
Energy	4.4×	2.4×
Economy (GDP)	5.1×	4.3×

[a] Source: National Research Council (1999).

Africa is approximately at the level of the United States in the 1840s, when Henry David Thoreau undertook his famous sojourn to Walden Pond to escape his countrymen's materialism.

It is important to bear in mind that economic statistics such as GDP provide only a partial measure of human well-being. Pollution, which diminishes the value of ecosystem services or valued activities and assets, is excluded from conventional GDP accounts, whereas the costs of abating or repairing the damage caused by pollution are counted as contributions to economic output.

Averages also fail to account for disparities in the distribution of wealth. Disparities in incomes are widening and are likely to continue to do so in the absence of strong remedial actions. The gap is growing between rich and poor countries as a whole and between the rich and poor within many countries. On a global basis, the ratio of the income share of the richest 20% to the poorest 20% doubled during the past 30 years from 30:1 to 60:1.

Demand for energy and materials has approximately tracked growth in total economic output. As consumption has increased, however, use of energy and materials has become more efficient on average. Table I provides a description of consumption changes during the past 50 years and a scenario-based projection for the next 50 years. The scenario projection is intended as a reference case against which to test alternative sets of expectations; the reference scenario is not more likely than other forecasts, but it does provide one consistent set of assumptions as a benchmark.

C. Species and Ecosystems

In the geologic record, paleontologists have found five mass extinction events, each of which drastically reduced the number of species on Earth. Each time, enough life-forms survived to repopulate the waters and lands of Earth. The impact of human activities on the planet has accelerated the loss of species and ecosystems to a level comparable to a sixth mass extinction, the first driven by a living species.

Rates of species extinction have been estimated to be 100–1000 times higher than they were before large-scale human dominance of ecosystems. These rates of loss are driven primarily by the alteration of natural habitats. The extirpation can be much higher than the proportional loss of land area if habitats are fragmented by roads or other human clearings that reduce unconverted land to isolated patches. Depletion of species in these remaining patches can be delayed but will occur in the absence of human action to restore and reconnect habitat.

In addition to causing extinction, human activities also introduce species into ecosystems in which they have not been present. In some situations, exotic species proliferate, and the introduction of a new species can transform the ecological relationships of these habitats and further stress endangered species.

Accelerating rates of depletion and change cannot continue for long: Habitats are finite and the loss of species is irreversible. Efforts to slow the rate of depletion and change require some recognition of the value of the system, whether in terms of provision of resources and services or on aesthetic or ethical grounds. However, the value placed on species and ecosystems has been uneven historically. A small suite of edible and useful species have been valued, conserved, and propagated by humans. These number in the thousands out of tens of millions of species. However, for every species such as cacao, the far more complex ecosystems within which these valuable species live have been neither understood nor managed until very recently. In some cases, including cacao (which grows better in tropical forests than in plantations), the environing ecosystems play an important role in the economic return to human cultivation.

In a small number of cases, endangered species have recovered when protected by human efforts. For example, the bald eagle and some marine mammals in North America have rapidly rebuilt their numbers. These recoveries have taken place under conditions in which habitats required by the species were either intact or readily protected, the species had not been damaged beyond recovery, and the public will to preserve or conserve them was strong. Such circumstances do not currently apply to most regions of the world where species are threatened.

Land conversions and land degradation also degrade or destroy ecosystems and the services they provide to humans. Freshwater, coral reef, and forest ecosystems have suffered enormous assault from human activities. Covering less than 1% of the earth's surface, freshwater ecosystems have lost the largest proportion of species and habitat when compared with other ecosystems on land or with the oceans. Continuing overfishing, dam building and river development, and contamination will continue to place greater threats on freshwater ecosystems. Many estuaries and bays have deteriorated because of activities associated with land development and fishing pressure, undermining ecosystem services. For example, the oyster population of Chesapeake Bay once filtered a volume of water equal to that of the entire bay about once a week; overfished, they now filter that volume in about 1 year's time. This has adverse effects on the water quality of the estuary and on the many species that live in it. More than half of the world's coral reefs face changes in species composition, obliteration, and other major ecosystem effects. These losses in turn affect the livelihood of local communities that depend on the reef for food, tourism, and protection against damaging storms. Increasing need for fuelwood and land for agriculture, together with industrial logging, resulted in a net loss of approximately 180 million ha of forest between 1980 and 1995.

Losses of freshwater ecosystems, coral reefs, and forests inflict large losses of ecological services on local human communities, many of which are poor. These services include water purification, flood control, recycling of nutrients, mitigating climate and temperature extremes, and the production of crops and forest and marine species.

The scientific understanding of conservation biology has taken clear form only in the past 15 years, although humans motivated by aesthetic and cultural considerations have been seeking to preserve places for millennia. As these two currents of human activity converge, there has been increasing awareness of the need to conserve ecosystem processes, including evolution. This implies working at the scale of whole landscapes, with explicit attention both to the preservation of critical ecosystems and to the interactions between human activities and the managed and uninhabited ecosystems among them. It is an open question whether a reformulated conservation of this kind will prove workable in enough places to salvage the biological richness of the planet. Moreover, the abilities to assess and monitor the well-being of Earth's living resources and the services they provide are far from proven, but efforts in these areas will be essential to understand the role of species and ecosystems for a sustainable future.

D. An Overview of Trends

A broad review of long-term trends affecting sustainability was completed in 1999 by the Board on Sustainable Development of the U.S. National Research Council (NRC). The survey's findings included many that shape the social context in which a search for sustainability will take place:

• Although humans have modified their habitats for more than 10,000 years, during the past three centuries humankind has developed the capacity to change the environment on a scale that equals or exceeds natural rates of change. Most of this human-caused disturbance has occurred within the past two generations.

• The twentieth century was a time of notable transitions: The demographic transition was described previously; governance and politics have been changed by decolonization from 1945 to the dissolution of the Soviet Union, together with the rise of nongovernmental organizations; and several important biogeochemical cycles have moved from control by natural forces to anthropogenic dominance. These are historic changes in driving forces, with opportunities for sustainable governance as well as risks of catastrophe.

• The past generation has seen notable improvements in human well-being. Since 1960 life expectancy worldwide has increased by 17 years, infant mortality has been cut in half, access to safe drinking water has approximately doubled, primary school enrollments have increased by two-thirds, and per capita income has more than tripled. The developing world has gained as much ground in a single generation, in material terms, as the developed economies achieved in a century. These averages are misleading if reported without acknowledgment of continuing, worsening inequalities; the number of people living in absolute poverty (less than $1 a day using 1985 purchasing power as a standard) is more than 1 billion and increasing.

• The pattern of continued growth hides a sequence of technological transitions, each lasting two or three generations, that have been discerned by economic historians. Coal, the dominant energy source of the late nineteenth century, was displaced by oil and natural gas. Some national economies have shifted their centers of gravity from agriculture to manufacturing and then to services. Increasingly, products are transported throughout the global marketplace, whereas land and ecosystems remain fixed. The causes and details of these transitions are poorly understood, but their existence implies opportunities for sustainability, as suggested by the effective control of some pollutants in many developed countries. Technological transitions have been accompanied by dislocations of human communities, such as the displacement of rural African Americans into the industrial cities of the United States as mechanization transformed agriculture beginning in the 1920s. Such dislocations produce both permanent and transient social changes, exacerbate conflict, and affect attitudes toward the environment and sustainability.

• A majority of the human race live and work in urban centers, a shift that seems likely to persist. In the remainder of the demographic transition, approximately 100 million new urban dwellers will appear each year, almost all of them in developing countries. The challenges and opportunities of meeting their needs for ecosystem services such as clean water is likely to play a major role in the future of biodiversity, both in metropolitan regions and in agricultural areas, coastal zones, and forests.

• Human cultures are increasingly connected and increasingly aware of their diversity. Since 1950, trade between nations has grown more than twice as fast as their economies, with concomitant increases in the power of transnational firms and financial institutions. The diffusion of ideas through low-cost communications has spread Western images of high consumption and also environmental concern. A notable result of increasing connection has been the rise of the nongovernmental organization as an influential vehicle for articulating identity, demands, and pressures for institutional change and innovation. The notion of sustainability, and the expectation that a substantial degree of material equity must be part of a sustainable future, is likewise a product of this more connected global society. Diversity also fosters conflict under some conditions, bringing risks of terrorism, war, and civil disorder.

• The intensification of agriculture in the past 50 years has transformed the character of human dependence on the land. Increasing agricultural outputs have outpaced human population growth and dramatically reduced the incidence of famine. If increases in yield can be continued as the growth of human numbers slows, it may be possible to release a substantial fraction of land committed to crops today, permitting regrowth of forests similar to that observed in some developed countries. Intensification of agriculture also brings costs and risks to both nature and society, of course, and it remains an open question whether the advances of recent decades are sustainable.

- The scale and intensity of anthropogenic disturbances threatens to undermine the sustainability of human populations in some regions. Kasperson *et al.* (1995) performed a comparative study of nine regions that are under severe pressure using criteria that defined unsustainable damage in terms that encompassed the social capabilities of the human population. Six of these regions were found to be likely to be environmentally unsustainable in the future, but only one, the Aral Sea, was already unsustainable. More broadly, the global-scale analysis of Turner *et al.* (1990) concluded that the rates of anthropogenic disturbance are accelerating in some areas but decelerating in others; these judgments are summarized in Table II.

- Environmental pressures do not always increase with consumption and economic output. An interesting array of patterns has been found as shown in Fig. 1, which plots various indicators of environmental pressure against per capita GDP on a logarithmic scale. Economic growth in cities requires that people living in dense settlements avoid one another's pathogens; this seems to correspond to the availability of clean water, shown in the top two panels in Fig. 1. The two bottom panels indicate that material consumption, as measured by the production of trash and carbon dioxide, shows no sign of abating with income. However, the middle panels indicate that, in some cases, pollution increases with income until a demand for environmental quality appears and is met by governments. These patterns, which have also been found in other measures of environmental and resource exploitation, suggest the possibility of a long-term transition in environmental problems: from dirty water to industrial pollution to climate change and from public health to government regulation to technological transformation. Thus, the attention directed to managing municipal wastes and greenhouse gases may foreshadow effective action even in the lower panels of Fig. 1. The social dynamics driving these correlations have yet to be elucidated, however, and the policy implications are debated more than they are acted on. It seems imprudent, in any case, to rely solely on increasing income to produce effective environmental control before irreversible damage is done.

- The lower right panel of Fig. 1 shows a local effect, emission of carbon dioxide, whose impact is global. The climate changes due to this anthropogenic disturbance are already occurring, with time lags that are not well understood. Changes in the composition of the atmosphere from human activities have been small in percentage terms but exert a large effect because of the role of greenhouse gases as thermal gatekeepers in the heat balance of the planet. These changes in greenhouse gas concentrations suggest that the course on which the world economy is embarked is unsustainable at the global level, affecting all regions and ecosystems. The 1992 Framework Convention on Climate Change commits the nations of the world to a goal of stabilizing "atmospheric concentrations of greenhouse gases at a level which would prevent dangerous human interference in the climate system." However, agreements under this convention during the twentieth century aimed only at stabilizing emissions of greenhouse gases. Even if this goal were achieved, the inventory of climate-altering gases would continue to increase with continuing emissions. Climate change is likely to occur at rates far more rapid than vegetation, in particular, can migrate; therefore, the implications for biodiversity from continued anthropogenic disturbance to the atmosphere are large and perhaps grave for many ecosystems.

- In the marine environment of the United States, one-third of the species whose status is known are overfished. The prevalence of unsustainable exploitation of marine species used by humans is similar elsewhere. Forests continue to be lost in the tropics, where deforestation exacts much higher costs in biodiversity than it does in temperate forests. As with fisheries, the institutional means and political will to manage for multiple species and purposes on a sustainable basis remain elusive.

TABLE II

Magnitude, Recency, and Rate of Change in Human-Induced Transformation of Environmental Components[a]

	Magnitude of change since 10,000 years before present	
	50% total change reached in nineteenth century	50% total change reached in twentieth century
Rates decelerating since 1950	Terrestrial vertebrate diversity	Carbon tetrachloride, lead, sulfur releases
		Human population
		Marine mammals
Rates accelerating since 1950	Deforested area loss	Carbon, nitrogen, phosphorus releases
	Soil area loss	Floral diversity
		Sediment flows
		Water withdrawals

[a] Source: Turner *et al.* (1990).

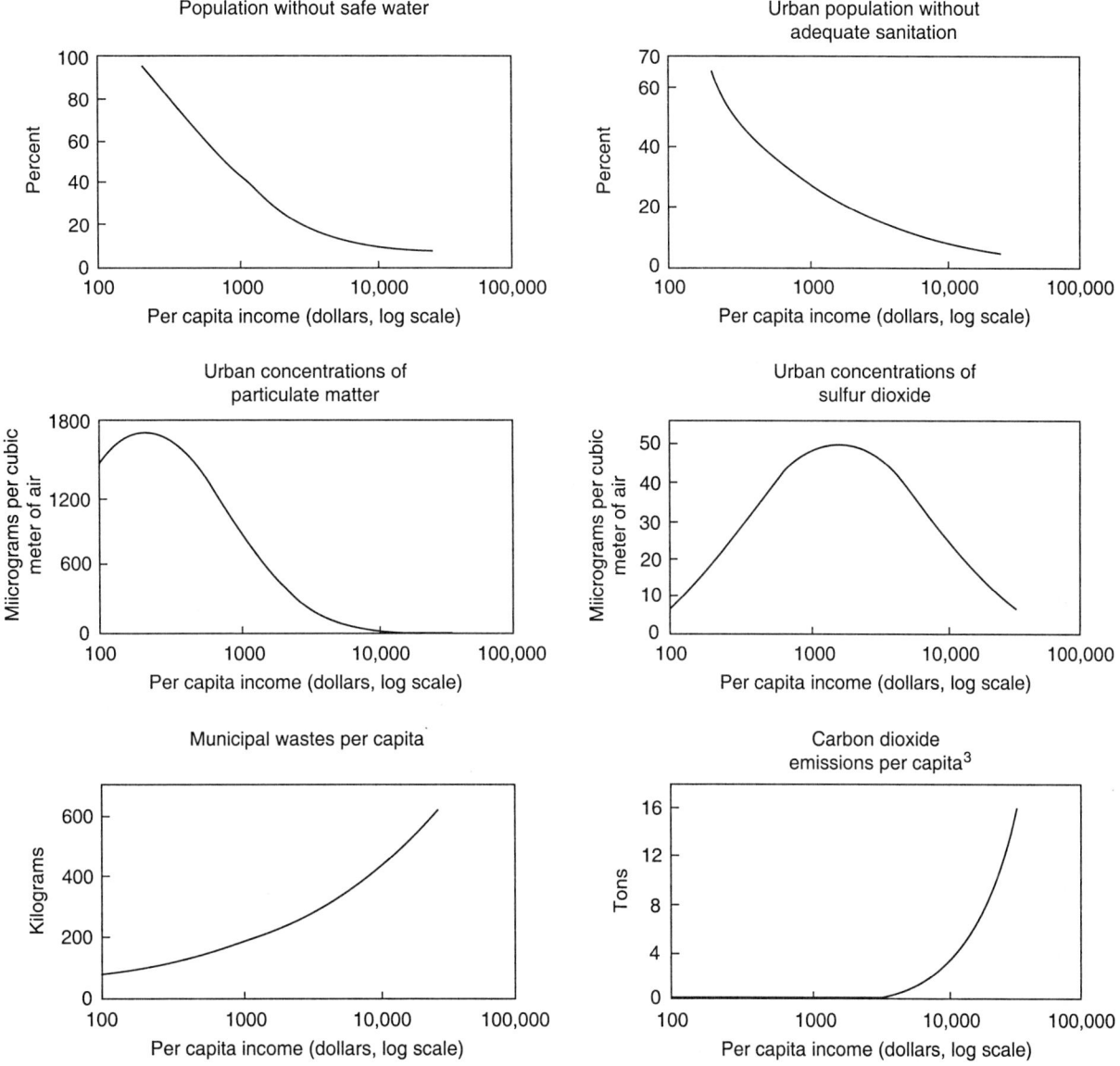

FIGURE 1 Environmental problems may worsen or improve with income growth. From World Development Report 1992 by World Bank, © 1992 by the International Bank for Reconstruction and Development/The World Bank. Used by permission of Oxford University Press.

• Fresh water, a basic necessity that cannot be economically transported over long distances, is in short supply or contaminated in many parts of the world. Many urbanizing areas are likely to face severe problems as their human populations grow.

• The NRC survey did not include trends in warfare, terrorism, or civil violence, but as events in the Congo River basin and the Andes demonstrate, civil disorder can affect biodiversity and the prospects for managing land substantially. The limited ability of governments to manage violent conflict must accordingly be borne in mind.

• Disease organisms and vectors may be expanding in response to anthropogenic disturbances. These disturbances take various forms—artificial selection of drug-resistant strains of pathogens, diseases jumping across species as humans unwittingly invade or enhance the habitats of pathogens and carriers of disease, and expansion of the ranges of some disease-bearing organisms with climate change and international trade.

Overall, sustainability is not a single or simple objective. Rather, sustainable human economies are likely to emerge, if they do at all, through social change and institutional adjustments that are only partially determined by rational processes and scientific knowledge. Rationality and science provide necessary guidance but are not likely to be sufficient in themselves.

In all, the NRC review demonstrated the abundant hazards and barriers to a sustainable future. However, the trends of the past have often elicited technological and social adaptations, turning a challenging hazard into a sometimes wrenching transition. The NRC study concluded that a determined search for a sustainability transition is worthwhile, as long as there is a commitment to recognizing inevitable surprises and to learning from experience and science.

E. A Sustainability Transition?

Using the demographic transition as a model, one can ask whether it may be possible, on scientific and institutional grounds, to achieve a sustainability transition. The demographic transition marks a long-term shift from high birth rates and high death rates to a society characterized by low birth rates and low death rates. Because death rates have been lowered first, there is a large net increase of population. A sustainability transition, by analogy, would be a change from today's patterns of highly dissipative use of energy and materials, together with severe and widening inequality, to a future in which use of energy and materials is far more conservative and material inequalities have been moderated, meeting the basic needs of all. Since the wasteful phase of growing consumption and high inequality came first, even a successful sustainability transition would entail significant losses of biological diversity and suffering in human populations; both are currently under way, of course. The NRC study summarized previously concluded that a sustainability transition is scientifically possible, although there remain many technological problems to be solved. It is unclear, however, whether the political will and leadership to assemble the needed social and institutional changes will be forthcoming.

A sustainability transition would pursue the following themes, inferred from the trends reviewed previously:

1. Completion of the demographic transition, with efforts to decrease the total population level by further facilitating choice of family size and additional moderating of population momentum
2. Conserving and rebuilding biodiversity using landscapes that reach beyond existing protected areas
3. Major improvements in the efficiency and end products of energy use, including alternatives to fossil fuels and mitigations and adaptations to climate change
4. Major improvements in industrial ecology: the ability to do more with less, in every material dimension, closing the loops of materials flows and increasing the efficiencies of production and use within these loops
5. Rethinking and changing the way cities are built, particularly in developing countries, so that the ecosystem services needed by urban populations can be affordable and accessible as cities continue to grow
6. Creating human and institutional capital to enlarge the existing capability to recognize, debate, and cross the hurdles that block the path to sustainability—to learn to negotiate a sustainability transition
7. Fostering social and political choices that can make sustainability transitions credible and feasible in democratic discourse

This is an agenda for scientific research, technological innovation, and social experimentation as well as activism. The first four items listed previously have long been staples of environmental studies and political action, whereas the last three raise new opportunities or reformulate existing ones. These themes entail social and behavioral changes that challenge widely held values. Like all social changes, these will be contested and outcomes will be shaped by struggle as well as technical and economic feasibility.

It is far from clear that a sustainability transition is socially feasible. However, the pace at which humans are disturbing natural processes and ecosystems leaves few options other that pursuing change as rapidly as practicable. The idea of a sustainability transition proposes a linked set of goals and strategies.

III. INDICATORS

Sustainability is clear in principle but obscure in practice. The search for a sustainability transition needs to take this difficulty seriously for several reasons. First, the histories of environmental science and environmental politics demonstrate the ubiquity of surprise. This is likely to be characteristic of a path toward sustainability as well; sustainability, like environmental quality, demands the reconciliation of multiple goals that are both partially incompatible and largely incommensurate. Second, the momentum of economic development and the social aspirations that propel it imply that those pursuing sustainability need to make good use of opportunities rather than await crystallization of a social consensus in favor of the drastic changes that seem likely to be necessary, at least in a cumulative sense.

For these reasons, it is important to think through the problem of indicators. That is, what quantitative variables and data sets should be used to assess progress toward or away from sustainability? Indicators matter, in business jargon, because "what gets measured gets managed." The significance of indicators, however, is larger than their use for managerial steering because some indicators such as the unemployment rate acquire a social definition, influencing broader choices such as elections or trade policy. That is, indicators have both social and managerial meanings. Over time, some indicators are likely to be widely accepted as approximate signals of movement toward sustainability, just as falling barometric pressure signals stormy weather. If sustainability acquires a social meaning, the public understanding needed to debate a sustainability transition will be strengthened.

Moreover, indicators are important to the conservation of biodiversity. The rising influence of geographic information systems has resulted in an important gain in public understanding. Biodiversity is no longer viewed only in terms of charismatic wildlife species but also as maps showing highly valued habitats and their condition. These maps provide a geographic template for considering indicators of sustainability. The maps also link biodiversity to human populations. The success or failure of these populations to meet their material needs will affect greatly the ability of conservationists to salvage the species and ecosystems of these landscapes. Indicators are an important, if incomplete, pathway to link biological knowledge and conservation to social change and economic development.

A. State of the Art

In day-to-day life we use prices, news and weather reports, and other routine methods of monitoring to guide our behavior and expectations. Indicators perform parallel functions for long-term changes and large-scale actions. As the members of the European Community prepared to institute a common currency, the nations agreed to meet numerical guidelines for their budget deficits as a fraction of GDP. This is a striking instance of the influence of indicators: For some time, these fractions superseded the electoral mandates of the national governments. Sustainability cannot be tracked by a single indicator; therefore, sets of parameters have been proposed to sense trends in social and environmental change.

These parameters have been selected by experts and citizens in quasi-governmental settings, usually using a pressure–state–response (PSR) framework. PSR posits linkages between human action and environmental consequences. Human activities exert pressures, such as burning gasoline in cars, that alter the state of environmental variables, such as the quality of air in a city. Impaired states in turn elicit responses, such as regulations governing pollution-control technology in new vehicles. These three classes of variables identified by PSR can be measured often using data already collected for administrative purposes. Combining these data with a simple but flexible scenario captures a central dynamic of sustainability: Humans can impair the life-support systems of the natural world, calling forth responses intended to protect environmental quality.

Using the PSR framework, governments and nongovernmental organizations have compiled numerous sets of indicators for sustainable development using various measurement regimes. However, because there is no widely accepted operational definition of the term "sustainable development," proposed indicators are often scrutinized more for their moral, economic, and political implications than for their scientific substance. To date, no single set of available indicators has gained wide acceptance.

Despite these difficulties, there are numerous efforts under way to assemble indicators of sustainable development. These efforts range on a governmental scale from municipal to international and on an ecological scale from watersheds to the planet as a whole. Hundreds of indicators and numerous schemes to collect, analyze, and aggregate the information needed to form sets of indicators have been proposed.

Two efforts by the UN Commission on Sustainable Development (UNCSD) and the World Bank demonstrate the way in which different approaches to indicators complement one another. The UN indicators were assembled using the PSR approach. Selected through a consensus process without an agreed operational definition of sustainable development, the 134 indicators are numerous, diverse in the methods used to measure development or sustainability, and include many indicators where reliable measurements do not exist.

The World Bank, in contrast, focused on estimating only three indicators, which it called capital accounts. Each attempts to capture the value to national economies of a vital aspect of the world. The most familiar account, "produced" capital, is normally called national wealth—physical capital and financial claims. A second account measures natural capital, the resources and capitalized value of services provided by the natural world. In principle, this would include standing timber, soil fertility, fish stocks, potable water, and the value of flood control by wetlands. Natural capital estimates are primitive in comparison to those for produced capital. The most recent World Bank study takes into account only the use values of natural resources, an approach that ignores unpriced damage to ecosystems as well as ecosystem services such as the flood-control capabilities of wetlands and aesthetic or moral dimensions of resource value. The third component of wealth, quantitatively the largest, is human resources—the economic value of labor, knowledge, and social institutions. The World Bank estimates this dimension of wealth as a residual by inferring the value of human resources needed to explain the generation of the actual flows observed in national income accounts. All three accounts, including the one measuring produced capital, are arguable in concept and subject to errors of estimation.

Although the World Bank's indicators are highly aggregated and estimated using drastic assumptions, they are conceptually clear. The wealth of nations should be considered in three parts. At least at the margins wealth can be transferred from one account to another in ways advantageous to people. In contrast, the UNCSD indicators do not warn unambiguously of imminent hazards in any ecosystem or society, nor do they provide guidance on how to pursue sustainable development.

The UNCSD indicator set includes variables that are not currently being measured, such wood-harvesting intensity or indices of local-level management of natural resources. This is a notable strength of the CSD process, reflecting the need to transfer reliable measurement methods to developing countries but also a realization that much of what needs to be sensed about sustainable development remains unclear. The UN has also sponsored a wide-ranging, if still diffuse, research effort under the aegis of the Scientific Committee on Problems of the Environment (SCOPE) of the International Council for Science (Moldan *et al.*, 1997). In evaluating measurements being taken in various categories, the SCOPE effort outlines a broad research agenda for indicators but does not provide a framework for monitoring.

In summary, the existing indicators of sustainable development are limited in their effectiveness by lack of agreement on the meaning of sustainable development. The projects carried out during the decade since the Brundltand Commission popularized the idea of sustainable development have drawn on the large bodies of work done in past decades on the measurement of human welfare and the condition of the environment. These efforts bring together many sources of illumination but have yet to produce a set of goals for social and natural conditions that can plausibly lead to prosperity for all while conserving the life-support systems on which human economies rest. Consequently, they have not provided indicators of sustainable development.

B. Navigating Transitions

That there is no consensus on the end point of sustainable development is different, however, from seeking indicators useful for a sustainability transition.

Like the slow loss of wetlands or the buildup of carbon dioxide in the atmosphere, a sustainability transition may well require quantitative indicators to be detected—that is, sensing the degree to which human needs are met and the degree to which life-support systems are put under stress. Human welfare, monitored since the depression of the 1930s, illuminates the role of governments in addressing hunger and poverty. Appropriate quantitative indicators have been developed, but at the turn of the twenty-first century they were not yet being reported regularly in a forum, such as the annual report of the UN Development Programme, in which crises could be highlighted and emergent transitions discerned.

Environmental indicators are numerous and need conceptual organization. One approach is geographic (NRC, 1999), focusing attention on four scales of significance to a long-term transition. These could be reported regularly by the UN Environment Programme, for example:

- Global circulatory systems, including the atmosphere, climate and ocean circulation, trade and travel, and the spread of exotic species and diseases, are affected by human activity. Trouble in the circulatory systems is important because the scale of circulation can implicate the entire planet more rapidly or persistently than governments can regulate. Although greenhouse gases and ozone-depleting substances had been carefully studied by the end of the twentieth century, biological and information-intensive flows were not assessed as components of a global circulatory system. In particular, practical ways to monitor the flow of invasive species and exotic pathogens still had to be developed, implemented, and reported together with the other circulatory systems.
- Regions vulnerable to critical environmental damage are being identified. As described previously, the Aral Sea has experienced severe, possibly irreversible damage; the Valley of Mexico appears to be headed in

the same direction (Kasperson *et al.*, 1995) as suburbanization continues apace. Because the governing factors, including a vulnerable natural setting, lack of feasible alternatives, and governing institutions unable to take effective action, vary among regions, there is not a single set of indicators that can monitor the combination of social and natural factors that lead to irreversible damage.

• Ecosystem services are only beginning to be monitored. The need is especially great in developing countries, where urbanization is continuing to accelerate, resulting in greater stress on geographically limited supplies, ecosystem services, and their associated infrastructures. Because many ecosystem services are not priced (e.g., flood control) or allocated by monopolies (e.g., water supply), market signals are absent or misleading. However, ecosystem services tend to be irreplaceable when damaged and inflexible because they are supplied by costly infrastructure.

• Parks and protected areas, created to enable their biota to persist indefinitely, have been identified on a place-by-place basis rather than through a consistent set of appraisals of their long-term sustainability. Attempts to conserve biodiversity on larger scales are just beginning. There are compilations at the species level of habitats with high levels of biodiversity but no regular comparative monitoring approach that could begin to build data series in which trends could be discerned.

In each of these settings, indicators form an indispensable but incomplete part of the intelligence needed to perceive and encourage a sustainability transition in the future.

A complementary function of indicators is to inform public policy and other decisions, large and small. Some navigational aids are available, although it seems likely that others may also come into wide use:

• National capital accounts, emerging from the World Bank's work, could provide a simple but crude measure of sustainability: Each of the three kinds of capital might be transformed by human activity, but as long as a nation's total capital increases over time its trajectory may lie in roughly the direction of a sustainability transition. The word "roughly" is important: A utilitarian metric is inadequate to assess sustainability. In addition, as the need of poor urban areas for ecosystem services demonstrates, deploying wealth in one account to meet needs in other accounts can raise difficult problems of politics and engineering. However, the capital accounts draw attention to transformations among forms of wealth—transformations that will continue through, and beyond, any long-term search for sustainability.

• Assessing policies using indicators assembled under the PSR framework can lead to adaptive management. Adaptive management treats policies as experiments, designing them so that lessons may be learned reliably from the implementation of policies, even those that fail. In the PSR context, this means using pressure and state indicators to test the effectiveness of responses.

• Monitoring ongoing transitions provides analysis of trends, particularly those that carry large future commitments, such as land conversion or population growth. Even if the population continues to stabilize, it will be essential to rapidly improve technology and energy efficiency so that humans can accomplish more, economically, with less impact on the natural world.

• Surprise diagnosis will be needed; a sustainability transition is an improbable development. Reflecting on environmental surprises during the past three decades, Kates and Clark (1996) concluded that we can learn from surprises so that we can better anticipate, avoid, or mitigate their consequences. However, knowing how to improve management is also a temptation to operate closer to the edge. Surprises could therefore become more frequent as humans gain better knowledge of the world; this possibility qualifies the conventional notion that science is valuable because it improves our ability to control or at least to predict danger. From this perspective, surprises are valuable indicators in themselves, although the meaning of surprises for sustainability will often be unclear. It is well accepted that surprise should produce humility. Surprise should also produce curiosity. On the timescale of the sustainability transition, curiosity and the learning it prompts are likely to be important, whether or not control can be extended in the short term.

C. Indicators and Social Learning

There is no agreement about which indicators to use to measure the movement of economies toward or away from sustainability. One should accordingly expect spirited debates over the value, biases, and meanings of indicators. In the related sphere of economic policy, one can observe that during the past 50 years there have been major disputes in the United States regarding economic growth, the incidence of poverty, unemployment, and inflation. All these are indicators that American politicians think will influence voter behavior. Remarkably, the independence of the data-gathering and analytical organizations has survived, despite their location within government agencies.

This is an important lesson: The independence of science is central to the social value of scientific information. Another lesson is that surprise is a valuable indicator. Governments and societies should anticipate unexpected things to happen. In a policy context, this idea is a kind of precautionary principle: Because surprise is likely, action should be undertaken with thought, humility, and caution.

Indicators used to report on a sustainability transition are likely to be biased, incorrect, inadequate, and indispensable. Obtaining the correct indicators is likely to be impossible in the short term. However, not trying to obtain the correct indicators will surely compound the difficulty of discovering, inventing, and achieving sustainable development.

1. Social Learning

As advisers to colonial powers and national governments, naturalists and ecologists have exercised influence on biodiversity and rare species through most of the history of ecological science. As the magnitude of anthropogenic disturbances increased during the past century, scientific understanding of the intricacy and vulnerability of ecosystems increased dramatically as well. The established means of preserving biological diversity—seed banks, zoos and captive breeding, and protected areas—do not suffice to safeguard the evolutionary heritage of the planet. A goal of environmentalists is to conserve biodiversity in more ambitious ways, working at the scale of landscapes that continue to be used and inhabited by people. These efforts should be viewed as part of a broader search for sustainable economies.

The search for sustainability, in turn, is a search rather than a march to a known destination. Salvaging the world's dwindling biodiversity is a race against anthropogenic disturbance. It will be lost in some places, as it has been in many already. The challenge is to learn from these experiences how better to provide a social environment in which people may meet the needs of the current generation while ensuring future generations the biological heritage with which to choose how better to be human in a natural world.

Acknowledgments

This article draws on work done by me and others for the Board on Sustainable Development (1999). I am indebted to Robert Kates, William Clark, Darby Jack, Jerry Mahlman, and Sherburne Abbott for guidance, assistance, and encouragement.

See Also the Following Articles

AGRICULTURE, SUSTAINABLE • ECONOMIC GROWTH AND THE ENVIRONMENT • ECOSYSTEM FUNCTION, PRINCIPLES OF • ECOSYSTEM SERVICES, CONCEPT OF • HUMAN EFFECTS ON ECOSYSTEMS, OVERVIEW • MASS EXTINCTIONS, NOTABLE EXAMPLES OF • POPULATION STABILIZATION, HUMAN

Bibliography

Adams, W. M. (1990). *Green Development. Environment and Sustainability in the Third World.* Routledge, London.

Bongaarts, J. (1994). Population policy options in the developing world. *Science* 263, 771–776.

Cohen, J. E. (1995). *How Many People Can the Earth Support?* Norton, New York.

Cronon, W. (1992, March). A place for stories: Nature, history, and narrative. *J. Am. History,* 1347–1376.

Fairhead, J., and Leach, M. (1996). Rethinking the forest–savanna mosaic: Colonial science and its relics in West Africa. In *West Africa, the Lie of the Land: Challenging Received Wisdom on the African Environment* (M. Leach and R. Mearns, Eds.). International African Institute, London.

Hammond, A. (1998). *Which World? Scenarios for the 21st Century.* Island Press, Washington, DC.

Hays, S. P. (1959). *Conservation and the Gospel of Efficiency. The Progressive Conservation Movement, 1890–1920.* Harvard Univ. Press, Cambridge, UK.

International Institute for Sustainable Development (1998). *Compendium of Sustainable Development Indicator Initiatives and Publications.* Web site: iisd1.iisd.ca/measure/compinfo.htm.

Kasperson, J. X., Kasperson, R. E., and Turner, B. L., II (Eds.) (1995). *Regions at Risk: Comparisons of Threatened Environments.* United Nations Univ. Press, Tokyo.

Lubchenco, J., et al. (1991). The Sustainable Biosphere Initiative: An ecological research agenda; a report from the Ecological Society of America. *Ecology* 72, 371–412.

Ludwig, D., Hilborn, R., and Walters, C. (1993). Uncertainty, resource exploitation, and conservation: Lessons from history. *Science* 260, 17, 36.

Maddison, A. (1995). *Monitoring the World Economy, 1820–1992,* Development Centre Studies. Organisation for Economic Co-operation and Development, Paris.

Marsh, G. P. (1864). *Man and Nature, or, Physical Geography as Modified by Human Action.* Scribner, New York.

Moldan, B., Billharz, S., and Matravers, R. (1997). *Sustainability Indicators. A Report on the Project on Indicators of Sustainable Development.* Wiley, New York. (Published on behalf of the Scientific Committee on Problems of the Environment).

National Research Council (1999). *Our Common Journey,* Report of the Board on Sustainable Development. National Academies Press, Washington, DC.

Ostrom, E. (1990). *Governing the Commons. The Evolution of Institutions for Collective Action.* Cambridge Univ. Press, Cambridge, UK.

Turner, B. L., Clark, W. C., Kates, R. W., Richards, J. F., Mathews,

J. T., and Meyer, W. B. (1990). *The Earth as Transformed by Human Action*. Cambridge Univ. Press, Cambridge, UK.

United Nations, Commission for Sustainable Development (1996, August). Indicators of sustainable development. Framework and methodologies, UN Publication No. E.96.II.A.16. United Nations, Commission for Sustainable Development, New York.

United Nations Development Programme (Various years). *Human Development Report*. Oxford Univ. Press, New York.

Vitousek, P. M., Mooney, H. A., Lubchenco, J., and Melillo, J. M. (1997). Human domination of Earth's ecosystems. *Science* 277, 494–499.

Wilson, E. O. (1993). *The Diversity of Life*. Harvard Univ. Press, Cambridge, MA.

World Bank (1997). *Expanding the Measure of Wealth: Indicators of Environmentally Sustainable Development*, Environmentally Sustainable Development Studies and Monographs Series No. 17. World Bank, Washington, DC.

SYSTEMATICS, OVERVIEW

Quentin D. Wheeler
Cornell University

I. What Is Systematics?
II. Elements of Biodiversity
III. Predictive Classifications
IV. The Missions of Systematics
V. Role of Taxonomy in Biodiversity Studies and Conservation

GLOSSARY

adjacency Relative position of alternative states of characters prior to any hypothesis of polarity.
allopatry Species or populations occupying separate areas.
anagenesis Origin of species (or characters) within a lineage without splitting (cladogenesis).
apomorphic Derived, relative to sister group.
area cladogram A branching diagram showing relative divergence of geographic areas.
autapomorphy A derived character unique to a single species or clade.
binomial (= binominal) nomenclature Naming species by combination of a specific epithet and the genus to which the species belongs.
biogeography, historical The study of spatial distribution of taxa through time.
character An attribute that is constantly distributed among all members of a species or a clade.
clade Branch on cladogram.
cladogenesis Origin of branch or clade.
cladogram A general graphic representation of a cladistic hypothesis; more general than a phylogenetic tree.
convergent characters Resemblance due to independent evolutionary events.
dendrogram Cladogram or tree; sometimes used specifically for phenogram.
dichotomous tree A cladogram in which all nodes are bifurcate.
dichotomy Two taxa arising from one node.
endemic taxa Species or clades unique to one geographic place.
holomorphology Totality of characters indicative of phylogeny (including molecular, behavioral, etc.).
homology Two structures hypothesized to share a common ancestry.
homoplasy Similar attributes not due to common ancestry but rather to convergent evolution, reversals, or mistaken interpretation.
macroevolution Evolution above the species level.
microevolution Evolution within species.
monophyletic group An ancestral species and all of its descendant species; evidenced by synapomorphy.
node Branching point on cladogram corresponding to hypothetical common ancestor.
ontogeny Sequence of embryonic and postembryonic changes in characters of an organism in the course of its life.
paraphyletic group A group including an ancestral species and some but not all of its descendants; evidenced by symplesiomorphy.

parsimony principle Seeks to minimize the number of character transformations (a special application of Occam's razor).

phenogram A branching diagram showing overall levels of phenetic resemblance rather than cladogenesis.

phylogenetic tree Common synonym for cladogram; or a more specific interpretation of phylogeny identifying ancestors and descendants.

phylogeny Evolutionary history of species and higher taxa.

phylogram cladogram; strictly, a cladogram interpreted to show terminal taxa and ancestral nodes (intermediate specificity between generalized cladogram and specific phylogenetic tree).

plesiomorphic Primitive, relative to sister group.

polarity Relative apomorphy/plesiomorphy of character states.

polyphyletic group A taxon for which all included species (or taxa) do not share a most recent common ancestor; based on homoplasies.

polytomy Many branches arising from a single node or an unresolved (portion or whole) cladogram.

Popperian A follower of philosopher of science Sir Karl Popper who accepts his falsificationist views.

semaphoront Individual organisms at a particular point in their developmental (ontogenetic) history.

stem Branch on cladogram between node and term.

sympatry Species or populations co-occurring in same area.

symplesiomorphy Shared, primitive similarity.

synapomorphy Shared, derived similarity.

taxon (plural taxa) Formally named species or clades.

taxonomy Classification of species and higher taxa, ultimately based on phylogeny.

term A terminal on a cladogram, either a taxon (in systematics) or a geographic area (in biogeography).

three-taxon statement Two species (or taxa) are more closely related to one another than either are to a third; the most basic cladistic hypothesis.

tokogeny Birth relationships responsible for reticulate patterns of relationships within and among populations.

tree Commonly a synonym for cladogram; or a cladogram interpreted in more evolutionary detail.

tritomy ("trichotomy," in error) Three undifferentiated branches or clades.

typification The practice of designating a single specimen (or, formerly, a series) as the name-bearer for a species; or a species for a higher taxon, especially genera.

THE TERMS TAXONOMY AND SYSTEMATICS are treated as synonyms in common usage with little loss of meaning. Two accounts of the relationships between these terms exist. According to one, championed by neo-Darwinists George Gaylord Simpson and Ernst Mayr, systematics encompasses broader studies of evolutionary relationships, whereas taxonomy concerns itself narrowly with classificatory and nomenclatural practices. Roy A. Crowson, on the other hand, believes that taxonomy is the broader field with the ultimate aim of predictive classifications; systematics, in this context, is limited to that part of taxonomy distinguishing species for study and analyzing phylogenetic relationships among them. Doing phylogenetic analyses to inform formal classifications fulfills a long tradition of taxonomic scholarship, predating the general acceptance of evolutionary thought. Precisely defined, *systematics* (or systematic biology) is that subset of taxonomy concerned with delimitation of species and analysis of phylogenetic relationships among them. *Taxonomy* is the broader field of science that explores species and phylogenetic diversity and applies that knowledge to the production of predictive classifications and formal sets of names.

I. WHAT IS SYSTEMATICS?

Because taxonomic knowledge subtends even the most fundamental observation and communication in biology, it is among those sciences essential to biodiversity study and conservation. Eventually, taxonomic classifications summarize and express in their evolutionary–historical context every source of comparative information from the molecular to the organismal, the fossil to the extant, and the species to the set of kingdoms. Taxonomy complements other biodiversity sciences by focusing on historical patterns rather than functional processes (as in ecology) and relationships among species and groups of species rather than among individuals and populations within single species (as in population genetics). Taxonomy depends on insights from population genetics and ecology to assist in the initial circumscription of species and sort out geographic and ecological distributions, and it provides a general conceptual framework within which data from other sciences may be synthesized and interpreted.

The boundary between population genetics and taxonomy was blurred in the early twentieth century by those who, like Ernst Mayr (in his 1942 *Systematics and the Origin of Species*), rejected studies of phylogeny and phylogenetic classification as speculative and embraced instead a wholesale focus on population biology. The widespread acceptance of this notion within biology assisted in the necessary growth of modern genetics but set the advance of taxonomy back by decades. Although many dedicated, ingenious taxonomists continued to make theoretical and practical progress against considerable odds, it was not until progress was made in phenetics ("numerical taxonomy" in the narrow, historic sense) and Hennig delivered the necessary theoretical justification that studies of phylogeny became unequivocally accepted again as solid science.

A predictive classification and the nomenclature that expresses it are invaluable for several reasons. First, a phylogenetic classification permits the retrieval of what is known about a taxon and predictions about attributes of taxa not yet observed. Second, formal nomenclature provides the vocabulary for biologists to communicate ideas and facts about biodiversity. With nearly 2 million species known and five or more times that number predicted, precision and accuracy in communication, retrieval of billions of facts about Earth's species, and the ability to predict where organisms with a particular property may be found will become immensely more important than they are already.

Taxonomy has been described as one of two fundamental divisions of modern biology. General biology is concerned with processes that are true for many kinds of living things or all of life. Experimental biologists fit within this division most of the familiar disciplines, i.e., ecology, ethology, physiology, and any others asking primarily "why?" and "how?" questions. A process is characterized in order to make predictions about the outcome of similar events in the same and as many other life-forms as possible, slicing through biodiversity in a functional or causal dimension. Taxonomists provide balance to this view through a broadly comparative and historically organized view of biological diversity. Given the immense variation in the ecology, behavior, physiology, and biochemistry of living things, there is in fact a single factor shared by all species: evolutionary history or phylogeny. Thus, a phylogenetic classification is the logical basis for the general reference system for biology.

Taxonomy is widely misunderstood, largely because of its historicity. Evolutionary history, not unlike the sequence of events in human society, is not replicable. The myriad of existing circumstances and preconditions cannot be enumerated, let alone recreated. Thus, the most fundamental assumptions underlying experimentation are violated by phylogeny. What is the universe of outcomes in the context of historical patterns among which one, at best, may be correct? Contributory processes and ancestral contingencies are so unpredictable *a priori* that efforts at randomization and modeling are little more than wishful thinking in most cases. Another effort has been made to force phylogenetic problems into a more familiar, population biology model.

Taxonomy is unique among the biological sciences for both theoretical and practical reasons. Most significant among these are the following:

Historical perspective: Taxonomy is not infrequently viewed with suspicion because of its nonconformity to standard practices of experimental biology. There is a good reason for this: Taxonomy is not experimental biology. Phylogenetic history's unrepeatable sequence of unique events cannot be seen by rolling back time or repeating evolutionary trajectories. Fortuitously, and unlike human history, evidence of phylogeny is preserved in the genetic makeup of organisms belonging to descendant species. Taxonomy is no less scientific than experimental biology. However, it does require a different ontology and epistemology for its justification.

Comparative method: Taxonomy is comparative at virtually every level. Whether assessing species boundaries, hypothesizing homologous characters, or choosing among several equally parsimonious cladograms, taxonomists are comparing the characters of as many species in the groups under study as possible. Few other areas of biology are comparative, and none to this same degree.

Worldwide collections: This comparative aspect of taxonomy means that taxonomists must have ready access to specimens for every species of a group being studied regardless of where on Earth these species might live. To make such work practical, taxonomists have assembled vast synoptic collections of certain taxa in major museums and herbaria. Whereas many biodiversity scientists need to study the same organisms or phenomena in the same study sites for extended periods of time, the same constraints do not exist for taxonomy. Just the opposite is true: Taxonomists require samples from the most disparate reaches of the geographic range of a species and representatives of as widely differing forms within a species or higher taxon as possible. In order to unambiguously sort out species, the taxonomist needs to see as much infraspecific variation as possible as well as to study specimens of closely related species whether or not they occur in the same places. In order to complete as comprehensive a study of an entire group as possible, the taxonomist needs to examine every species possible within that group regardless of where or when they live or lived. This is a very different requirement than that which faces the experimental biologist, and it demands access to collections from wherever in the world members of a specialist's taxon occur. Such collections are costly in time and money to be assembled and housed, laborious to maintain, and may only be accumulated with the combined efforts of many students and professionals over very long periods of time. The great natural history collections of the world have been built up over a period of one or two or more centuries and even the best of them contain but fractions of all the world's species.

Taxon inventories: Building taxonomic research collections demands that scientists with taxon-specific knowledge be encouraged and permitted to collect in as many locations within the range of their group as is possible. Long series of specimens are needed to address infraspecific and transgeographic variation. Two centuries of taxonomic practice has taught us that making species decisions in isolation (i.e., describing species based on material restricted to one locality, country, or in many cases even single continents) leads to redundant species descriptions, synonymy, and taxonomic confusion. Such collection requirements, and the minimum useful data associated with specimens, vary sometimes greatly from the requirements of a population geneticist or ecologist.

Taxon scholars: Taxonomists spend their entire careers working on one or a few higher taxa. Because there are genera of arthropods with as many as 2000 described species, it is not uncommon for a specialist in the megataxa (e.g., fungi and insects) to learn hundreds of species in the course of his or her career. This taxon focus creates a very different perspective than one centered on a phenomenon or ecosystem.

In the philosophy of science, the demarcation problem involves how we can distinguish science from nonscience. Experimental biologists sometimes confuse the "scientific (experimental) method" with the limits of science. The failure of taxonomic theories to fit the requirements of experimental biology to be quantifiable by one of a handful of statistics—or, more precisely, the failure of experimental biology to capture that which distinguishes science—does not make them unscientific. Rather, that which makes science is testability. Questions formulated in such a way that they may reasonably be shown false through observations (of which experiments in the narrow sense are but one example) are scientific. Questions not open to potential critical empirical testing and possible rejection, such as "Is there a God?," are simply outside the limits of science.

Stripped of the usual gauges of confidence limits in experimental biology, how can the historical patterns constructed by taxonomists be scientific? Phylogeneticists have found a powerful epistemic argument for their work within the hypotheticodeductive philosophy of science espoused by Sir Karl Popper. Knowledge does not exist with absolute certainty. Rather, "confidence" (in the common sense rather than statistical sense) increases in our ideas in proportion to the number of times that they are subjected to and survive critical testing. All-inclusive statements such as "all swans are white" are highly testable. Such law-like, all-or-nothing statements are in fact potentially falsified by a single observation. Aristotle took note of the hardened forewings of beetles called elytra. An implicit (and later explicit) hypothesis that "all beetles share elytra" has been corroborated approximately 400,000 times in the past 2000 years, but a single observation of a beetle without elytra tomorrow has the potential to refute it. Falsified hypotheses are discarded and new or modified hypotheses forwarded and tested in their place.

For the Popperian, the demarcation between science and nonscience is precisely that between the testable or potentially refutable assertion and the untestable claim. Unless it can be shown how a hypothesis is reasonably falsified, it is simply outside the bounds of science. This view of science is a general one, consistent with both experimental and comparative biological programs. Much of biology has become homogenized so that the protocols of experimental biology are viewed as more or less equivalent to the scientific method and not merely examples of it. As shown, Popper's hypotheticodeductive approach explains the significance of congruent character distributions and refutational role of incongruent ones. The inductive reasoning so popular in the eighteenth and nineteenth centuries does not suffice for taxonomy. The failure of phenetics demonstrated that large amounts of data alone does not suffice to demonstrate relationships.

A Popperian approach to science proceeds approximately as follows:

1. Hypothesis: An initial hypothesis can derive from anywhere—observations of the world or merely conjecture. Hypotheses not founded on prior observations or based on few or poor observations, of course, seem less likely to stand when subjected to critical tests. The point is that where a hypothesis comes from is secondary to its formulation in a falsifiable form.

2. Predictions: From the initial hypothesis logical predictions are made about the natural world. Cladograms predict the pattern which synapomorphies will follow as discovered.

3. The successful test: When these predictions are realized, the hypothesis is corroborated. Given many such tests and much accumulated corroboration, we feel increasing confidence in the accuracy of our assertion.

4. The unsuccessful test: When these predictions are not realized—even after many successful corroborative tests—then the hypothesis is called into question and the incongruent observations must be explained or the hypothesis rejected and modified or replaced with another.

5. Repeated tests: Regardless of whether it is corroborated (step 3) or replaced (step 4), the prevailing hypothesis is continually subjected to the rigors of further observations and continued critical tests.

Cladistics or phylogenetic systematics is the modern form of systematics that seeks to reconstruct patterns of evolutionary history. Darwin acknowledged the pioneering work on phylogenetic "trees" by Ernst Haeckel in Germany (Fig. 1), but a century would pass before a rigorous and cohesive theory of phylogenetic systematics emerged in Hennig's 1966 tome. Piecing together descent with modification is obviously complex, routinely encompassing worldwide patterns spanning millions of years. Despite the immense challenge of phylogenetic analysis, its fundamental tenets are surprisingly simple. Phylogenetic systematics was a logical extension and fully capitalized on centuries of steady growth in our knowledge of taxonomic diversity, merging as it did phylogeny and taxonomy.

Taxonomists study the patterns inherent among species due to relative recency of shared common ancestors. The distinction between pattern and process in evolutionary studies has evoked much controversy. Although different approaches and evidence might be required, all these varied ways of studying evolution are appropriate and ultimately necessary. Because taxonomists have sought to minimize the number of specific assumptions made about evolutionary processes, many microevolutionists have interpreted this to be a rejection or attack on their science. On the contrary, minimized process dependency of taxonomic hypotheses ensures that they are independent of any particular process, thus giving to microevolutionists a historical chronology within which any process of interest to them may be examined. The effort to purge phylogenetic analyses of unnecessarily specific assumptions does not mean that phylogenetic systematics is theory free or neutral. On the contrary, as most contemporary philosophers of science would agree, systematics, like all scientific assertions, is theory laden. Phylogeneticists have a detailed ontology and rich set of background

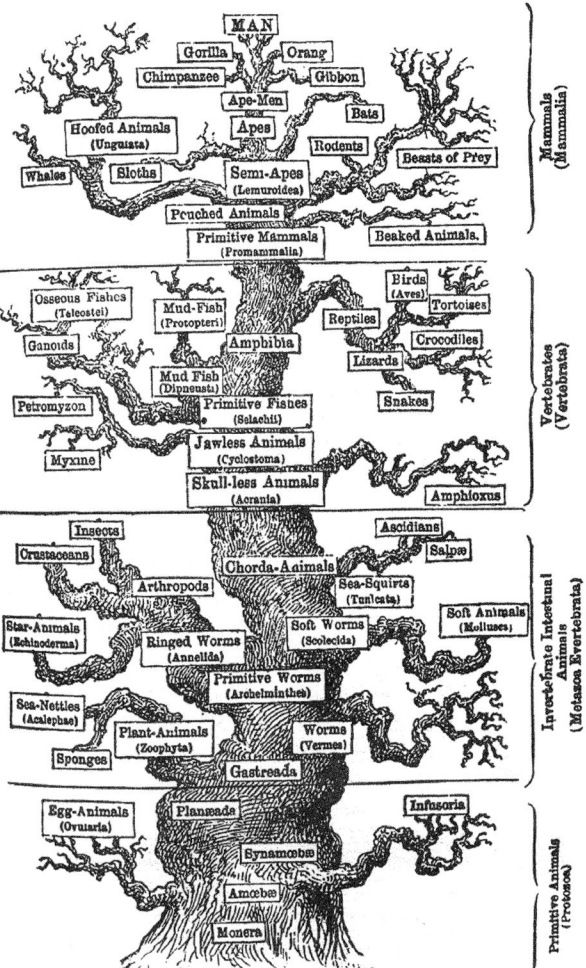

FIGURE 1 Treelike diagram depicting Ernst Haeckel's hypothesized phylogeny of life, cited for its genius by Charles Darwin (reproduced from Haeckel's *Generelle Morphologie der Organismen* published in 1866).

assumptions that subtend their science, as in their concept of characters discussed later.

The discovery of phylogenetic patterns not only can but also must precede insights into process. This is well established in respect to both how taxonomy is done and how it has progressed in the history of biology. Chapter 13 of the *Origin of Species* makes clear that Darwin was convinced of the fact of evolution from the patterns of character distribution documented by taxonomists. The existence of these character regularities led Darwin to seek an explanatory theory. Many taxa have been recognized for centuries, through changing ideas and social milieus. The same pattern has been visible to taxonomists whether they were creationists, microevolutionists, or cladists. A distinction exists between groups of species for which there are defining evolutionary novelties or synapomorphies (so-called "A groups") and those defined by lacking such synapomorphies ("not-A groups"). The history of taxonomy may be told in the search to eliminate not-A groups and recognize A groups in their place.

A. The Evidential Basis of Taxonomy

Taxonomy is comparative in nature, and those attributes of species and higher taxa that are observable and comparable are characters. For the taxonomist, character is a precise term that is central to hypothesis testing and at the heart of his or her epistemology. A *character* is an attribute of a species that is constantly distributed among all members of that species. In the context of phylogeny, a character is such an attribute present in an ancestral species and in all of its descendant species either in its original or in some subsequently modified state. Attributes shown to be inconstant in their distribution within terms on a cladogram (whether species or higher taxa), for example, are not characters in this strict sense but rather traits. Traits within a single species provide crucial evidence for the population geneticist but do not provide reliable indicia of common ancestry at the level of phylogeny. Traits at higher levels result in polymorphism in cladogram terms and the range of confusion associated with such. Considerable confusion persists in the literature that is actually nothing more than semantic misunderstandings spawned by not distinguishing these two concepts.

Any attribute that is heritable and that can differ constantly between species or clades is a potential character. Character sources are as diverse as might be expected, including anatomy, embryology, paleontology, molecular biology, biochemistry, ethology, and ecology. Hennig used the term *holomorphology* to express the totality of character sources, not limited in Hennig's time or since to morphology. Hennig emphasized that the source of character data is of less significance to taxonomy than how that evidence is analyzed. Simple observations with a hand lens can be rigorously analyzed, and the most expensive molecular techniques can generate data that are rendered useless by inappropriate analyses. Data that are not available for most taxa in an analysis or those of uncertain fidelity in their inheritance are frequently mapped onto a cladogram *a posteriori*. Recent arguments have been made in favor of "total evidence" analyses that combine all available sources of data. In practice, this has principally involved merging molecular and morphological data sets.

Mature sciences develop theories that are scientific (testable) yet not directly derivable from observations by the senses. Subatomic physics is based on the existence of particles never seen by anyone, but these ideas have been shown to map precisely onto predicted outcomes in power plants and atomic bombs. In taxonomy, characters represent a continuous intellectual tradition traceable to the notion of homology articulated by Richard Owen in 1842 and made consistent with evolutionary theory by Lankester in 1870, who differentiated homology from homoplasy. Today, homologies are theory-rich hypotheses of sameness that accommodate the "noise" of trait-level variation, incorporate existing knowledge from molecular and developmental genetics, and are more sophisticated than the simple phenetic idea that "they look the same." Even the choice regarding what specimens to compare to one another has relevance since some characters are only observable during particular periods in the life history of the organism. These character-bearing periods are *semaphoronts*; thus, it may be necessary to compare a third-instar larva with the same, a neonatal female with the same, and so forth.

Consider another Popperian statement that "all reptiles have scales" and assume that we have a study set with a few species each of lizard, snake, and birds. We wish to use scales and feathers to sort out their relationships. A simple phenetic view of these characters (using characters as mere adjectival words rather than as evolutionary hypotheses and claiming thereby to be "objective") would support the conclusion that lizards and snakes have scales and birds have feathers. This appears to be a matter of common sense and coding birds as lacking scales seems obvious. A more detailed consideration of the situation, taking into account phylogenetic insights as well as ontogenetic and genetic information, reveals that feathers are scales expressed in a highly modified form. The character "scale" in order to serve as a special similarity in the sense of Hennig must be interpreted to include its original state and all its subsequent modifications. Phylogeneticists would agree with pheneticists that scales are characteristic of a monophyletic group that we might call Reptilia. However, in order to be a monophyletic group Reptilia must be expanded to include birds. Snakes and lizards share nothing in common with one another not shared also by birds, either in an original or in a modified condition.

Hennig's evolutionary novelties or apomorphies may be hypothesized at the outset, as in Hennig's "search for the sister-group" hand method. Rationale used to determine the relative polarity of alternative conditions include outgroup comparisons, fossil precedence, and ontogenetic sequence. Alternatively, global parsimony among all available characters may be used to infer the relative status of a condition as apomorphic or plesiomorphic and the polarity read off the cladogram *post hoc*. As data sets increase in size (number of taxa and/or number of characters) or in complexity, the latter approach implemented by any of a number of computer algorithms will prove more satisfactory. For relatively small data sets, it is possible to seek exact solutions with some of these programs. For large, messy data sets, options are used that approximate the shortest trees through random movement of branches and other procedures.

The simplest phylogenetic hypothesis is the three-taxon statement, which states that two taxa share a more recent common ancestor with one another than either does with the third. Implicit in this simple statement is the prediction that all synapomorphies discovered to vary among these three taxa shall be distributed so as to be consistent with the three-taxon statement.

B. Historical Context

The history of taxonomy is an extremely long one, with its roots lost in prehistory. The ability to distinguish among species in nature is not unique to humans and is part of even simple forms of animal cognition. Our concern here, of course, is the more formal human efforts to classify and name the various kinds of living things in a disciplined way. People living close to the land know and have words to refer to a surprising number of species, and it is estimated by ethnobiologists that several tens of thousands of species are in daily use by people throughout the world. Early hominids, at least those who survived to become our ancestors, had to distinguish poisonous plants from edible ones and prey items from potential predators of humanoids. Not coincidentally, the rise of civilization during the past few thousand years has been correlated with the increase in our knowledge of "other" species.

In the third century B.C., Aristotle had a remarkable conspectus of biodiversity and distinguished many taxa, some of which are considered valid to this day (e.g., Coleoptera). After the subjugation of Greece by the Roman Empire, however, these early flowerings of formal taxonomy ceased and were not resumed for another 1500 years. During these dark centuries, Aristotle's insights were mongrelized with superstition into the essentialism attributed to him but not doing justice to his brilliance. An increasing interest in plants of medicinal value, published in ever-larger books known as herbals, helped fuel a renewed interest in classification. Cesal-

FIGURE 2 Prominent figures in history of taxonomy. (Left) Caroli Linnaeus (1707–1778), originator of modern classification and nomenclature, binomial nomenclature, and the hierarchic system of categorical ranks. He was also author of *Systema Naturae*, whose 10th edition in 1758 marks the beginning of credible names for most animal taxa. (Right) Willi Hennig (1913–1976), originator of phylogenetic systematic or cladistic theory that realized Haeckel's and Darwin's vision a century earlier for classifications that reflect phylogeny through differentiation of synapomorphy and monophyly and related concepts. He was the author of *Grundzuge einer Theorie der Phylogenetischen Systematik* in 1950 and the highly influential *Phylogenetic Systematics* in 1966.

pino's rudimentary species concept in *De Plantis Libri*, based on his observations that plant seeds give rise to like kinds, was a significant return to keen observation in taxonomy echoed in the animal world by John Ray in the middle of the seventeenth century.

The world of taxonomists rapidly expanded in the seventeenth century. Sailing ships were bringing many new kinds of plants and animals back to museums in Europe, and van Leeuwenhoek perfected simple microscopes revealing microbes to astonished biologists. Even as the known world increased in size, so too did efforts to learn more completely about the flora and fauna close to home. As naturalists sought to document species known to them, the absence of rules led to chaos, synonymy, and misunderstanding. Order came in the works of a Swedish taxonomist, Caroli Linnaeus (Fig. 2). In the writings of Linnaeus were deceptively simple criteria that, when applied with rigor, gave the foundations for all modern classifications. Among many contributions by Linnaeus, some deserve special mention. Linnaeus devised a stable system that avoided the rampant redundancy that had existed in names in the past. He adopted binomial nomenclature, the genius of which involved conveyance of ideas of relatedness in species names as well as maintenance of descriptive adjectives for use in multiple higher taxa. He proposed higher taxa that were ranked one within another, logically preadapted to meet needs for phylogenetic information content two centuries later. Also, he emphasized a character-based approach that made his ideas explicit and portable to the arrangement of specimens in collections and observations in the field.

The epistemic importance of a character-based system, and its logical justification, emerged from the comparative anatomical studies of Cuvier and the distinction by Richard Owen of analogy and homology. Darwin's *Origin* had little obvious direct impact on the practice of taxonomy, just as Darwin predicted. This was hardly surprising given the fact that Darwin derived his process ideas from the patterns so convincingly set out by taxonomists from Aristotle to Linnaeus and Darwin himself. On the other hand, the transformation of homology into evolutionary hypothesis by Lankester and his explicit recognition of homoplasy advanced the inevitable conceptual relationship between the classification of species and characters, the evidence for taxonomy.

Taxonomy entered a new period with the wide acceptance of evolutionary theory. Darwin boldly predicted that our classifications would come to be genealogies of species. Taxonomists almost immediately sought to realize this vision. John Henry Comstock, founder of the entomology department at Cornell University, wrote in 1893 that "the description of a species, genus, family, or order, will be considered incomplete until its phylogeny has been determined so far as is possible with the data at hand." Although the goal of phylogenetic systematics was soon clear to most post-Darwinian taxonomists, the method for achieving such phylogenetically informed classifications and names would not come together completely until the middle of the twentieth century. Comstock offered a method based on infusing functional considerations into the study of characters and taking on the study of character systems one at a time in sequence. Comstock's method foreshadowed the evolutionary taxonomy championed later by Mayr in the 1940s distinguished by (i) recognition of paraphyletic groups and (ii) desire to express both cladogenesis and the degree of divergence between branches. Because these neo-Darwinists wanted to express both degree of divergence and branching patterns in classifications, they ended up with many compromises and an excess of subjectivity from species limits to the highest taxa.

In part a reaction to this subjectivity—itself the product of combining unnecessary evolutionary process assumptions with the act of classification—pheneticists or numerical taxonomists developed explicitly objective alternatives. Early pheneticists argued that community of similarity equated to community of descent and went to excessive lengths to compile as much raw data as possible. This quasi-inductive approach proved incorrect: Overall similarity did not reveal phylogenetic relationships. Rather than abandon phenetics, its practitioners admitted their failure but added that they now believed evolution to be unknowable. Many of the massive phenetic data sets were subsequently shown to be suspect; morphometrics were used extensively to generate a large number of "characters." Unfortunately, independence among such evidence was never shown and in some studies the same ratios were being measured many times. More damning to the pheneticists, however, were two additional observations. First, their results were only repeatable given the same matrix (i.e., given no addition or subtraction of characters or taxa between analyses) and the same algorithm. By this time, there were scores of algorithms, any of which might give a different result. Although each was "objective," the choice among them was purely subjective. Second, J. S. Farris demonstrated in 1979 that the phylogenetic system was superior in its information content, the very criterion used to promote phenetics. Farris had delivered a death blow from which phenetics would never recover.

Hennig's 1950 *Grundzuge Einer Theorie der Phylogenetischen Systematik* had been read and cited by a handful of North American taxonomists, a few of which, such as Pedro Wygodzinski, had applied his principles. Almost immediately after the publication of Hennig's *Phylogenetic Systemtics* in 1966 (a new, English-language book), a theoretical revolution was afoot. Phylogenetic systematics developed rapidly along two parallel lines of scholarship that would converge years later. James S. Farris was transforming "numerical taxonomy" from the trivially objective phenetic methods that admitted to being uninformative about evolutionary history to sophisticated numerical approaches to phylogenetic analyses that took advantage of emerging computer technology to make possible the search for the shortest trees and the simultaneous analysis of large numbers of taxa and characters. Farris developed a rigorous quantitative taxonomy that was true to Hennig's theories and to the parsimony principle that he forged into the bedrock of phylogenetics. In so doing, Farris also laid the groundwork necessary to cope with massive numbers of characters that would later emerge from molecular sequence techniques. Simultaneously, Gareth Nelson, Norman Platnick, Eugene Gaffney, Donn Rosen, Joel Cracraft, E. O. Wiley, and other scientists associated primarily with the American Museum of Natural History were exploring the connection between the philosophy of science promulgated by Sir Karl Popper and Willi Hennig's theories and methods. The result was a powerful epistemic justification of Hennig and the transformation of cladistics to a rigorous science in its own right distanced from some of the evolutionary process assumptions common to taxonomic writings before Hennig and, to a lesser extent, inherent in Hennig's writings.

C. Linnaean Nomenclature

Modern taxonomy is generally taken to refer to classifications beginning with the work of Linnaeus in the eighteenth century. In zoology, for example, the earliest available names for all groups (except spiders) are those printed in the 10th edition of Linnaeus' *Systema Naturae*, arbitrarily assumed to have been published on January 1, 1758. Any names proposed prior to that date are set aside. All those contained therein or published thereafter meeting applicable rules are available and are evaluated with regard to their appropriateness following an international code of rules, the most recent edition of which was published in January 2000 by the International Commission on Zoological Nomenclature.

Efforts to reject and replace the Linnaeus system of names are proposed occasionally. To date, no such efforts have succeeded and the Linnaean system continues to serve as the backbone of information storage and communication about biological diversity. The most recent such proposition claims to be "phylogenetic" and would replace the familiar Linnaean categories with a rankless system in which names would be galvanized by reference to two common ancestral species. Proponents of rankless nomenclature emphasize putative stability of names in their arguments. This is ironic since phylogenetic theory demands names be sufficiently agile to reflect improved knowledge of monophyly, a necessary result of progress in cladistic hypothesis testing. Because this system cannot adapt to improvements in knowledge, from the perspective of cladistic information it is less stable than the existing Linnaean alternative.

The fact that the Linnaean system has been used continuously by taxonomists is not, of course, an accident nor a slavish adherence to tradition. Pre-Darwinians recognized patterns evident in character distributions that were used by Linnaeus and his students to group "related" animals and plants together. For Linnaeus, these relationships were the result of choices made by God in his Creation, but regardless of the causal process assumed, the evidence was expressible in hierarchic form. As the evolutionary implications of such patterns of resemblance became clear (as, for example, to Haeckel, Lankester, and other nineteenth-century biologists), improved explanations for the same undeniable patterns became available. Given a rigorous method for the study of phylogeny by Hennig, a classification system that was hierarchic in nature was a necessity to convey such relationships. The Linnaean system, due to its hierarchic logical structure and nested ranks, ideally met this need.

II. ELEMENTS OF BIODIVERSITY

We speak of biodiversity in quantitative terms, as in reference to one ecosystem having greater diversity than another, but sometimes fail to explicitly consider what the elements of biodiversity logically should be. In a general sense, of course, there are incalculable candidates: demes, ecosystems, populations, semaphoronts, genes, amino acids at particular sites, specific behavioral patterns, and so forth. Because species have traditionally served as the fundamental elements of nomenclature and of phylogeny, they clearly have special status in such attempts to quantify biodiversity. Ask any biologist about biological diversity of a place and his or her first response is likely to be some census or estimate of species numbers. Considering the necessary distinction between tokogeny and phylogeny, species are a given in such counts. However, species alone do not capture what most of us intuitively wish to express when comparing the diversity of areas.

Consider two islands of equal size with 10 species each. On one island are 10 closely related species of a single insect genus. On the other island are one each of species representing 10 higher taxa. In one trivial sense, the islands have equal diversity with 10 species each. In a deeper sense, the second island has more diversity, certainly in a qualitative sense but perhaps also in a quantitative sense. This is so if and when we take into account not only how many species live in a place but also what their relationships are one to another and in comparison with life on Earth elsewhere. Were all 20 of these hypothetical species unique endemics and we could save only one of the islands, I suspect that most of us would find this an easy decision.

In recent years, several numerical indices have been proposed to account for biodiversity at and above the species level. When biodiversity is seen through the lens of taxonomy, in terms of both species and phylogenetic diversity, needs and priorities for research and conservation can appear very different. Combined with research results from ecology, population biology, migration ethology, and assessments of bioeconomics and so-called biodiversity hot spots, these indices create a

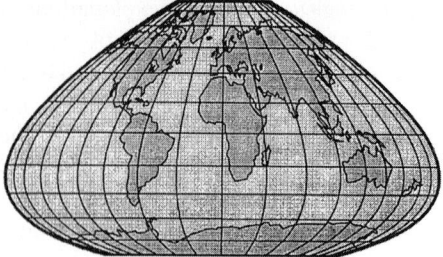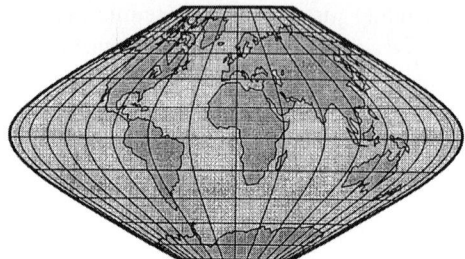

FIGURE 3 Visualization of two competing hypotheses of the global distribution of biodiversity. (Right) The tropical model, based on a simple count of numbers of species, argues that most biodiversity is equatorial with decreasing levels toward either pole. (Left) The Platnick model takes into account both phylogenetic diversity and species diversity and suggests that most biodiversity exists in both the tropical and the southern temperate regions, resulting in a pear-shaped globe.

powerful, science-centered basis for decision making in natural resource management and conservation biology.

Using the imagery of the world to represent broad scaled geographic models of biodiversity distribution on Earth, we can see three phases in the twentieth century. First was what Nelson and Platnick called the "Sherwin–Williams" model, with biodiversity originating in Northern Hemisphere centers of origin and spilling down the earth in decreasing abundance. This traditional view was biased by many factors, including a focus on certain vertebrate taxa but especially the location of major universities, museums, and herbaria in Europe and North America.

As tropical forests were finally explored in greater detail, it soon became obvious that there were far more species in these than in any of the supposed centers of origin to the north. From this species counting exercise, a tropical biology mythology arose that posited that the tropics are most diverse, with biodiversity levels diminishing as one moves either northward or southward away from the equator. There are many taxa with enormous numbers of tropical species, and based on existing species counts the tropics unequivocally house more species than any other region in the world. This tropical view morphed the world into one with a very exaggerated equatorial region. For an increasing number of taxa, however, the largest number of species are found in southern temperate regions (as, for example, with many spiders as shown by recent studies by N. I. Platnick). Whether the southern temperates could ever compete for raw numbers of species seems doubtful, but Platnick's work led to another major shift in our conception of the distribution of biodiversity on Earth (Fig. 3).

Taking into account clades and subclades, Platnick argued convincingly that the southern temperates are home to a disproportionate number of relictual early clades. Just as marsupial diversity in Australia challenges the traditional eurocentric view of mammal diversity, early arthropod clades make the southern end of the world diverse in amazing and unparalleled ways. According to Platnick's view, the world is most biodiverse in the tropics and in the southern temperates, resulting in a pear-shaped globe.

Moving beyond species and clades, individuals differ one from another in genetic and other respects and any effort to draw a line across levels of biological organization is arbitrary. In this view, biodiversity is to be found down to the molecular level, where individual amino acids ultimately are characterized by physical properties rather than by meaningful individuality. Such a reductionist argument suggests that it is impossible to accurately describe or meaningfully contrast levels of biodiversity. The experience and practice of biologists and laypersons, however, suggest otherwise.

Most conservation literature refers to an ecological "hierarchy" progressing from individual to deem to population to species to ecosystem. Such functional units of organization are necessary concepts for understanding the workings of living systems. They are not, strictly speaking, hierarchic. Because individuals may migrate into an area, die, or emigrate, the notion of a deem, population, or ecosystem is illusory. Actual combinations and numbers of organisms are in constant fluctuation and may change dramatically over geologic spans of time. Reference to key species is useful for this reason but does not negate the ephemeral nature of such assemblages. From the taxonomist's perspective, biodiversity is organized otherwise. Individuals belong to populations (at least in sexually reproductive forms) and populations to species. Subsequently, species belong to

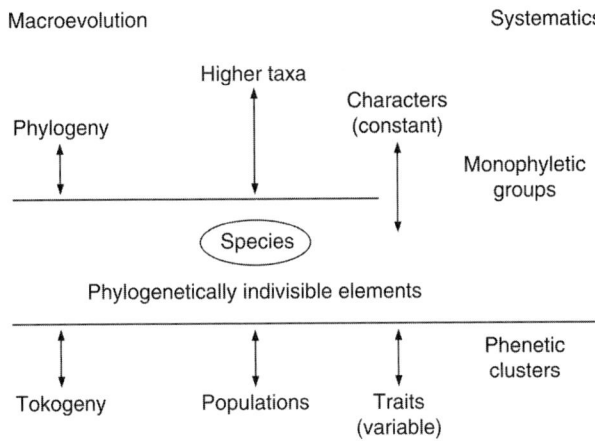

FIGURE 4 "Line of death" depicting the significance of species in Hennig's distinction between tokogeny or birth relationships (shared among individuals within and between populations of a single species) versus phylogeny or patterns of relative recency of common ancestry shared among species. Hennig's line—species—also corresponds with divisions between taxonomy and population biology, characters and traits, microevolution and macroevolution, and upward between species and higher taxa.

monophyletic clades of varied inclusivity. Because unambiguous indicia of common ancestry may only be retrieved in the realm of phylogeny rather than tokogeny, the elements of biodiversity are clearly species.

The hierarchic relationships that result from a shared evolutionary history may only be reconstructed among species. Higher taxa can be subdivided into less inclusive monophyla all the way down to pairs of sister species. Species, in contrast, cannot be divided into smaller units that retain the potential for unequivocal phylogenetic pattern recovery. Species, therefore, emerge as the logical elemental units for describing and discussing biodiversity. This also explains the intense interest in species concepts among biologists.

A great deal of confusion regarding systematic biology derives from frequent confounding of the kinds of genetic relationships described by Hennig as "tokogeny" and "phylogeny" (Fig. 4). Tokogeny refers to birth relationships, specifically the immediate genetic relations shared among actually or potentially interbreeding individuals. Tokogeny, then, is concerned with organisms within populations and the interchange of individuals (and thereby genes) among populations. In other words, tokogeny describes relationships below the species level. Phylogeny, in contrast, describes genetic relationships among species. Phylogeny refers to the pattern of descent with modification above the level of species.

To some, species concepts have played an inordinately important role in biology. Because species recognition is a prerequisite for research in either tokogeny or phylogeny, it is not surprising that so much emphasis has been given to the so-called "species problem."

Since the early twentieth century, geneticists have focused so intensely on population, then molecular, mechanisms that it is widely accepted that the definition of evolution is changes in gene frequencies within and among populations through time. This is true for that part of evolution called microevolution but not for macroevolution. Phylogeneticists seek to unravel the pattern of evolutionary history above the level of species. For clonal, asexual kinds of organisms, mutations mark new beginnings from which all (not subsequently mutated) offspring will be essentially exact copies. For sexually reproductive organisms, however, geneticists have demonstrated that patterns within species are reticulate (due to virtually random episodes of reproduction, even at very low frequencies) rather than hierarchic. Our ability to retrieve historical patterns depends on the availability of evidence reasonably interpretable as hierarchic. Certainly there are histories within and among populations, but this complex mix of parent–offspring relationships (called tokogenetic relationships by Hennig) obfuscates the history signal. What, then, are the elements of phylogeny? That is, the units about which such history is retrievable and below which it is not? The answer appears to be the traditional one—species—even though no single species concept is agreed on by all biologists.

III. PREDICTIVE CLASSIFICATIONS

Cladistic hypotheses are most succinctly expressed in graphic form as branching diagrams known as *cladograms* (also phylogenetic trees, trees, phylograms, or dedrograms, sometimes with special meanings). Species or the least inclusive higher taxa included in an analysis are plotted as *terms* (terminals) at the tips of the finest branches of the tree structure. They are grouped together by the observed pattern of distribution of shared-derived similarities or *synapomorphies*. Synapomorphies, or shared apomorphies, represent evolutionary novelties. Such novelties are heritable characters, and as such are constantly distributed within the terminals defined by them, whether species or higher taxa. Attributes that vary within a terminal, in contrast, are termed traits even at higher levels. Characters are shared by all members of a taxon, either in their original (ancestral) condition or in a subsequently modified form, the modern interpretation of character formalized by Norman Platnick in 1979. Synapomorphies may occur once on a cladogram or more often if the overall parsimonious distribution of characters suggests multiple origins (convergence), losses, or losses plus regains (reversals).

Hennig defined *monophyletic groups* as those including a common ancestral species and all of its descendant species. The goal of a phylogenetic classification is to make all groups monophyletic. Although monophyletic groups had long been recognized, Hennig's method was the first to require monophyly. Synapomorphies are taken as evidence of monophyly. *Symplesiomorphy* or shared primitive similarity erroneously leads to groups that include ancestral species and some but not all of its descendant species; these groups are known as *paraphyletic*. Errors are sometimes made while interpreting characters, and states are mistakenly grouped together as the same that have actually arisen from independent evolutionary events. Such instances of convergent evolution, parallelism, or reversal to ancestral character states are known as *homoplasy*. When groups are based on shared homoplasy, they include distantly related species that do not share a most recent common ancestor and are known as *polyphyletic*. Figure 5 shows examples of such groups referring to a primitively flightless silverfish, a damselfly, a beetle, and a fly. The best available insect classification suggests that fly + beetle constitutes one monophyletic group whose sister group is the damselfly. The silverfish is then sister to the other three combined. Were the silverfish and damselfly grouped together based on their comparatively simple metamorphosis (lacking a pupal stage), they would be paraphyletic because simple metamorphosis is a shared, primitive similarity in comparison to holometaboly. Were the small, aristate antennae of the damselfly and fly taken as similarity, the group might be described as polyphyletic since it is based on convergent similarity and not on common ancestry. This example was used for simplicity. In practice, paraphyletic and polyphyletic groups involve a much broader scale of grouping errors.

Hennig was concerned specifically with historical patterns above the level of species, and the concepts described previously pertain to how species are grouped together into higher taxa and clades. Such ideas, of course, presuppose that there is some agreement with regard to the definition of species.

Since the rise of modern genetic theory, biology in general has focused on the mechanistic aspects of species formation, basically asking "Why do species exist?" and "How do species originate and maintain their uniqueness?" Although the importance of such questions is obvious, another equally important and even more fundamental question was sometimes neglected: "What are species?" This "what" question differs fundamentally from the aforementioned questions in that it asks about a pattern rather than a process. This question, usually framed in terms of competing species concepts, was traditionally and is still logically within the purview of taxonomy. The answer to this question has direct and significant bearing on answers to related "why" and "how" questions.

The most elementary questions about biodiversity (How many species are there in a particular clade, place, or ecosystem?) require a scientific concept of species,

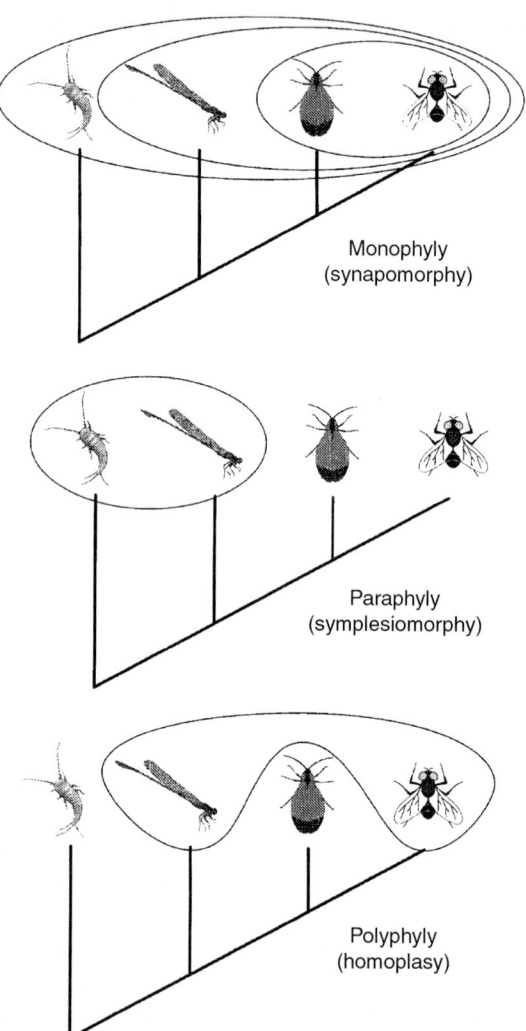

FIGURE 5 Example of Hennig's concepts of monophyly (based on synapomorphy), paraphyly (based on symplesiomorphy), and polyphyly (based on homoplasy) illustrated by four hexapods: silverfish, damselfly, beetle, and fly. See text for discussion.

as do successful strategies, policies, and laws to protect biological diversity. Because taxonomists are primarily concerned with the discovery, description, naming, and classification of species, they have traditionally provided the answer to the "what" question. More literature has probably been devoted to species and speciation than to any other single topic in biology, but species concepts remain highly contentious. Given so much attention, one might expect that the species "problem" is solved. On the contrary, there are more species concepts vying for adoption today than there were a century ago, and the debate among competing concepts rages on. Even if biologists could agree on one concept of species, it would remain true that we have only begun the process of exploring the species of Earth. Estimates of the total number of species living today vary tremendously from a few million to as many as 100 million. Although our ignorance about biodiversity is due largely to inadequate support for taxonomists to discover and describe the world's species, this work is impossible in the absence of an agreement regarding the concept of species to be applied.

Closely associated with the new synthesis was the biological species concept (BSC). The BSC was built on observations in the nineteenth century that sister species often lived in adjacent but separated areas, suggesting that allopatry was important in some way to species formation, particularly in animals. These ideas, combined with a shifting emphasis in biology to population genetic questions, made Ernst Mayr's persistent advocacy for the BSC extremely effective among zoologists. Currently, the majority of zoologists nominally accept the BSC, although the number of studies providing the kind of interbreeding information required by the concept are few. Botanists never accepted the BSC in large numbers, in part due to the incredibly diverse and complex genetic mechanisms in angiosperms and rampant polyploidy in pterydophytes.

With some level of discontent already in place, Hennig's writings forced a deeper consideration of species in the context of phylogenetic theory. Hennig pointed out that the BSC was at odds with evolutionary history since there were no clear breaks in the potential to interbreed among populations through geologic time. Projecting breeding patterns backwards through the geological record, there were no obvious places to demarcate one species from another. Hennig provided a fix for this dilemma and advocated a concept that in many respects resembled the BSC.

Donn Rosen noted that ancestral interbreeding, relative to extant populations, was plesiomorphic and therefore of little consequence to phylogeny; it was the loss of interbreeding that was of importance but which is indistinguishable from the absence of interbreeding. This, combined with a sincere desire to apply Hennigian theory to the species problem, led Rosen to develop an alternative species concept that attempted to apply cladistic analysis to populations. Rosen sought to ensure that species had novel status by making them equal to the smallest demonstrably autapomorphic units. Rosen's was the first concept explicitly couched in terms of phylogenetic theory, but it suffered from several problems. Ancestors were impossible to positively recognize given this concept, even though they clearly had existed. When two or more species arose from one polymorphic ancestral population, it was not clear that one or the other daughter was more or less apomorphic than the other. Also, this use of the idea of monophyly was at odds in logic and intent with Hennig's theories that dealt explicitly with supraspecific groupings. Despite these evident problems, some authors still advocate autapomorphic species.

George Gaylord Simpson developed an evolutionary species concept based on his view as a vertebrate paleontologist. His concept was revised and expanded by E. O. Wiley nearly 30 years later, bringing the arguments for evolutionary species explicitly in line with phylogenetic theory. This theory has no overt conflict with either phylogenetic theory or known evolutionary processes, but it is not clear how this concept is put into practice in an empirical sense.

About 20 years ago, a second generation of phylogenetic species concept emerged that was independent of cladistic analysis (so that it could provide the elements of phylogeny to be analyzed prior to such an analysis) and fully compatible with phylogenetic theory. Working simultaneously, two pairs of authors produced nearly identically worded versions of such a concept: Eldredge and Cracraft (1980) and Nelson and Plantick (1981). The phylogenetic species concept is very similar to the morphological concepts in broad use by taxonomists before they were set aside in favor of the BSC, but it is formulated specifically in a phylogentic framework. According to the phylogenetic species concept, a species is simply the smallest aggregation of (sexual) populations or (asexual) lineages diagnosable by a unique combination of character states.

Debate continues regarding the best alternative. This battle over species concepts is far from a mere academic exercise (Table I). Any effort to inventory or document biodiversity, to compare the biodiversity present in one habitat or geographic place versus another, and any wording in regulations aimed at natural resource management or biodiversity conservation will inevitably be

TABLE I
Alternative Species Concepts[a]

Biological species concept (Mayr, 2000, p. 17): "groups of interbreeding natural populations that are reproductively isolated from other such groups."

Hennigian species concept (Meier and Willmann, 2000, p. 31): " reproductively isolated natural populations or groups of natural populations. They originate via the dissolution of the stem species in a speciation event and cease to exist through either extinction or speciation."

Autapomorphic species concept (Mishler and Theriot, 2000, p. 44): "the least inclusive taxon recognized in a formal phylogenetic classfication . . . grouped into species because of evidence of monophyly" (see original for more extensive wording).

Evolutionary species concept (Wiley and Mayden, 2000, p. 73): "an entity composed of organisms that maintains its identity from other such entities through time and over space and that has its own independent evolutionary fate and historical tendencies."

Phylogenetic species concept (Wheeler and Platnick, 2000, p. 58): "the smallest aggregation of (sexual) populations or (asexual) lineages diagnosable by a unique combination of character states."

[a] Citations from Wheeler and Meier (2000).

described in terms of the numbers and kinds of species affected. Adopt the wrong concept, and we grossly under- or over-estimate biodiversity with potentially catastrophic effects. It is important that this debate proceed with all due haste, but it is no less important that we arrive at the correct answer and not simply an expeditious or democratically popular one. Every student of biodiversity has a responsibility to consider this question carefully and fully and has a stake in the outcome.

IV. THE MISSIONS OF SYSTEMATICS

Through a major initiative by the International Willi Hennig Society, the Society of Systematic Biologists, the American Society of Plant Taxonomists, and the Association of Systematics Collections, with funding from the U.S. National Science Foundation, the systematics community recently laid out a vision for the future of the field. Its visionary 1994 report, *Systematics Agenda 2000: Charting the Biosphere* (and parallel technical report), returns the focus of the systematics community to three traditional core missions that are energized by recent theoretical and technological advances and made urgent by the biodiversity crisis.

A. Mission 1: To Discover, Describe, and Inventory Earth's Species

This mission seeks to answer several of the most fundamental of all biodiversity questions, such as "What and how many species exist?" "How are they related?" "What properties do they share with other species?" and "Where do they occur?" In slightly more than two centuries, taxonomists have documented nearly 2 million species and today estimate that the total number of living species is between approximately 10 and 100 million species. The completion of a project begun in the eighteenth century by Linnaeus to make an inventory and description of every species on Earth is now complicated by projections of the extinction of thousands or millions of species. E. O. Wilson estimated that just the minimal description of a flora and fauna with 10 million species would require the life-time efforts of an army of taxonomists numbering 25,000. Although emerging technologies to manage this immense volume of data promise to facilitate the process, credible taxonomic work rests on rigorous theories and methods, worldwide collections of specimens, and the accumulated knowledge of specialists.

Deceptively simple in its questions, this taxonomic inventory is in itself "big science" on a scale that rivals the impressive ambitions of astronomers and physicists. Stated another way, the magnitude of this project becomes clear: the complete exploration, characterization, analysis, and documentation of every life-form on an entire planet. Taxonomy is rarely conceived on a planetary scale, but the evolutionary history that provides its conceptual framework requires approaching the task on a global basis. Earth is the only known planet on which biodiversity exists with certainty. This biodiversity, from a taxonomic perspective, is not partitioned by nor limited to particular geophysical boundaries or (necessarily) single ecosystems. The focus of taxonomic inventory is a taxon (clade) and includes as targets every species and subclade in that group that does or has lived no matter where on Earth living or fossil forms might be found.

Two models for taxonomic biodiversity inventories have been proposed. The All-Taxon Biodiversity Inventory (ATBI) seeks to enumerate every species living in a manageable-sized study site, e.g., 50–100 ha^2. The intent of the ATBI, conceived by Daniel Janzen, is to provide readily accessed identification information so that land and resource managers can make appropriate decisions and so that wild areas can be economically and socially valued possessions. This model is structured to facilitate long-term ecological, economic, and conservation work. These undeniably laudable goals, however, may be better achieved through an approach structured to take full advantage of taxonomic research, collections, and expertise.

A taxonomic alternative has been suggested: The All-Biota Taxon Inventory (ABTI) simply selects a taxon or monophyletic group as the object of study rather than a place and seeks to document all species of that group regardless of where they live. Advantages of the ABTI, proposed by Q. D. Wheeler, relate to both efficiency and science. The end product of an ABTI is monographic knowledge of all species of a taxon and a predictive (phylogenetic) classification for the group. Within this conceptual framework, ecologists and resource managers anywhere in the world can access what is known of these organisms, predicting what properties new or poorly known species are most likely to possess. Such predictivity, the hallmark of science, is lacking in the site-specific alternative. Furthermore, every aspect of taxonomy is made more difficult through the ATBI model. Species limits and nomenclatural decisions are equivocal when divorced from knowledge of species occurring outside the study area. Classifications are denied access to extralimital species that can make character coding and cladograms more reliable. Also, detailed knowledge and experience of the taxon specialist and synoptic collections are neither used nor built to their full capacity. Recall the unique comparative and historical requirements of taxonomy that cannot be met at single study sites. Taxonomic knowledge is complementary to ecological knowledge, but it is most efficiently and effectively derived from a very different approach to fieldwork and inventory. Historically, much of the time of taxonomists learning a particular group has been occupied with correcting mistakes that resulted from geographically isolated descriptions of species.

Ignorance about the nature of taxonomy and its attendant philosophy has contributed to an academic version of political correctness that pressures taxonomy to conform to the standards of other disciplines. Ecologists often do not understand the value of general collecting or trapping of specimens and insist that "modern" standards be applied that can be used to quantify data. In order to answer taxonomic questions, however, it is not necessary to know how many individuals occur per square meter or to gather data on relative humidity or feeding habits. Were taxonomists to uniformly adopt such data collection standards, they would cease to do taxonomy and spend much of their time doing secondrate ecology. Even extremely crude data can sometimes answer important questions. A primitive tribe of leiodid beetles (Neopelatopini) has been found in Chile, Argentina, and Australia. Does this clade show the often repeated southern temperate pattern of distribution? If so, we would expect one or more undescribed species to be found in South Africa. A neopelatopine with a label as vague as "South Africa" would be sufficient to answer this question. Additional information is desirable, of course, but must be weighed in the context of its cost to limited time and expertise.

If you are unconvinced, reverse the roles. Good ecology requires a great deal of work at the same study site. For the taxonomist, however, the goal is to see examples of a taxon from as many parts of its range as possible. Were taxonomists to insist that ecologists spend no more than 1 week at any particular site and then move on to another, I can only imagine the indignant outrage. If we are to understand biodiversity, it is critical to support all the biodiversity sciences. These fields are complementary to one another. Knowledge from all are needed for a complete understanding, but the best and most cost-effective data will come from each when its unique needs are met.

B. Mission 2: To Analyze Phylogeny and Predictively Classify Species

Despite the fact that Charles Darwin and taxonomists who came after him all recognized the importance of having classifications reflect phylogeny, it would be a century before Hennig gave us the means to do so in a rigorous and testable way. Confronted with the urgent need for taxonomic knowledge in the face of the biodiversity crisis, we are fortunate to have phylogenetic systematic theory, methods, and computer algorithms. As the century between *The Origin* and *Phylogenetic Systematics* attests, such theoretical advances could not be produced at will. With this necessity met, however, the taxonomic community is poised to meet the need for predictive classifications. Remaining obstacles are very practical ones related primarily to a shortage of taxonomic specialists, insufficient funds for their work, and inadequate resources for the maintenance and growth of natural history collections. In other words, we know what needs to be done and have the scientific foundation to do it; we need only place a sufficiently high priority on taxonomic revisions and monographs to get on with the analysis of phylogenetic relationships and the production of optimally useful and predictive classifications.

C. Mission 3: To Disseminate Data, Information, and Knowledge

Because good taxonomy is omnispective with regard to members of higher taxa, it produces a large volume of data, information, and, ultimately, knowledge. It can be critically important to know what labels are placed on a type specimen and the geographic location of every specimen studied in a monograph, not infrequently numbered in the thousands or even tens of thousands—raw data that can occupy many expensive pages in publications. The advent of computer technology and particularly the Internet has the potential to revolutionize the way in which and the speed with which these data are made accessible to the community.

In our rush to embrace this technology, many erroneous records have been typed into computer databases and foisted upon the world. The costs of identifying and correcting this disinformation will be immense, undoubtedly more than the cost of obtaining the correct information the first time. For relatively well-known taxa, it is appropriate to obtain existing museum data on the Internet in a searchable format. For poorly known taxa, such as most arthropods, fungi, microbes, and many plants, the blind typing of questionable data from museum specimens seems of questionable worth. As a minimum, collections need to capture data as new specimens are accessioned and, even more important, to capture taxonomic information as taxonomists return borrowed specimens following the completion of a monograph or revision. In large groups in which particular taxa are revised at most once per century, such data should be obtained when they are revised because these data are as good as they will likely be for the foreseeable future.

Credible taxonomic data are closely associated with specimens deposited in collections. As a result, there are considerable benefits to the maintenance of both collections and databases at a location where there are also scientists with taxon-specific knowledge and appropriate libraries. Because taxonomists ask geographically and temporally broad questions, reviewing an entire clade from its origin to the present anywhere on the earth that it has lived or currently exists, a typical study deals with thousands, often tens of thousands, of specimens and an enormous number of bits of data.

For a few taxa, it is possible for anyone sufficiently interested to quickly become adept at species identification. As a result, thousands of important records about the distribution, migration, and nesting of birds are accumulated in the absence of voucher specimens. For most taxa, and especially the largest ones (e.g., arthropods and fungi), such is not the case. Identifications are often only as believable as the credentials of the identifier are known. Major genera of insects are revised no more frequently than once per century, suggesting that at any given moment there may be no more than one or a few people who are fully competent to make an identification. Even such highly experienced specialists may require access to identified sets of specimens to verify their identifications.

The emergence of high-speed, high-capacity computers makes the dissemination and maintenance of biodiversity data at least feasible. It is possible to facilitate the exchange of information internationally on a time-scale never before dreamt of. Also, information can be shared in such a manner that each nation need not support an expert and a collection for every taxon that may be of importance to its agriculture or commerce. Traditional "products" of taxonomy can be done more rapidly and cost-effectively, and entire new kinds of output of data and knowledge are possible. The following are among the products of taxonomy:

Reference specimens: In many respects, the most important output of taxonomic research is in the form of authority-identified specimens. These specimens provide a permanent, three-dimensional, character-rich record of the concept of particular scientists with regard to the status of species and clades. These specimens in museums and herbaria are of daily value in accurate identification work and of long-term, historical importance to taxonomists who later revise the same taxon.

Monographs: Monography is the best example of taxonomic knowledge dissemination. In a single writing, one finds a review of essentially all that is known of a taxon (minimally an exhaustive review of every species ever thought to belong in it) and available new species and associated (geographic, ecological, etc.) information. A modern monograph includes a classification, expressed in carefully documented Linnaean names, based on the best corroborated existing cladistic hypothesis. Monographs have the decided advantage of taking into consideration every known member of a taxon at once, inevitably arriving at a better classification (or at least one that is better documented as such). Once a monograph is completed, it can be used for an incredibly diverse range of purposes and many more specific publications may be derived from it. For example, one might extract a checklist of the species of a particular country, generate a diagnostic key for their identification, or interpret some unusual biological attribute of an organism in its broadest evolutionary (historical) context. With an exiting monograph as a baseline, one can more confidently identify species from anywhere in the world and add new species descriptions or distribution records with great confidence. Even when a provincial study is done well (e.g., the species of a taxon for Mexico), there are equivocal species limits and nomenclatural uncertainties that can only be addressed when all included subgroups are included in the study regardless of where (or when) on Earth they occur.

Databases: Computer databases are an essential tool in dealing with billions of facts currently stored on labels attached to millions of specimens. These data, however, are only of great value when they are associated with a voucher specimen or specimens permanently deposited in a natural history collection. Also, the data rapidly degrade in importance and credibility if they are not routinely examined by taxonomists with knowledge of that taxon. As revisions are done, names changed, and so forth, it is essential that databases be brought up-to-date. This, however, is not always easy for the nonspecialist. Raw taxonomic data are of limited value. Far more important are the interpretive works of taxonomists that use such data to make both printed and electronic works which transmit instead information or knowledge.

Special-purpose publications: From a monographic or revisional study—a taxonomic study comprehensive enough to study all known and new members of a group—it is possible to extract less comprehensive works for special needs and purposes. For example, it will soon be possible to extract maps, checklists, or diagnostic keys for only those species of a particular country, park, or ecosystem from an electronic monograph.

Specialists: New generations of taxonomists are most easily and efficiently trained when students can have an established specialist as a mentor. An alarming number of taxa, some quite large and difficult to master, have few or no living authorities who can mentor a new generation of specialists.

Predictive classifications: As discussed previously, predictive classifications are perhaps the most useful of all taxonomic products.

V. ROLE OF TAXONOMY IN BIODIVERSITY STUDIES AND CONSERVATION

The significance of taxonomic knowledge in the study and conservation of biological diversity has not been sufficiently recognized nor supported. Recognition of the biodiversity crisis virtually requires a major shift in the way in which we think about biodiversity exploration and conservation. It is no longer (if it ever was) realistic to set out to maintain the status quo. Even the most optimistic among us no longer project a rosy future for all species. The prime question for conservationists is no longer "How can we save species threatened by extinction?" but rather "Which species shall we attempt to save?" Because every species, habitat, and ecosystem is unique in one or more valuable ways, it is difficult to prioritize conservation decisions based on ecological interactions or population genetic dynamics, even though success in any particular conservation effort depends heavily on the theory and practice of both ecology and genetics. One could crassly rank ecosystems according to their relative productivity or cash crop monetary value to current world economies, but this arbitrarily minimizes many valuable components of biodiversity and is based on an unwarranted arrogance that we know enough about biodiversity to predict which pieces are indispensable.

A more prudent approach might be to maximize the planet's biodiversity as measured in species and their relative contribution to overall biological diversity (i.e., an assessment of their unique contributions to clade diversity). Since choices among various clades and species seem increasingly inevitable, the obvious taxonomic question is whether we can succeed in conserving evidence of as many clades of the tree of life as possible along with all the benefits that would go with such a crosscut of biodiversity. As efforts proceed to understand and conserve biodiversity, regardless of specific goals, taxonomy will contribute in many ways.

A. Conceptual Framework

Taxonomy gives to biodiversity a historic framework within which every aspect of life on Earth may be understood and interpreted. Regardless of spatial, ecological, or genetic complexities, the only thing that every species on Earth shares in common is its organic history of diversification. Although it is essential to understand biotic and abiotic ecological parameters affecting spatial and seasonal distributions, it is no less important to understand the contribution of shared evolutionary history in the formation of observable current or past patterns of molecular, morphological, ecological, or geographic similarity.

B. Language of Biodiversity

Taxonomy's plant, animal, and microbial nomenclatures provide the language for communication about biodiversity. Because Linnaean names do not fully convey what we know about phylogeny, various conventions have been proposed to supplement its cladistic information content. Even before such augmentation, however, the Linnaean language of taxonomy is a powerful one. If I mention two staphylinoid families, for example, Leiodidae and Agyrtidae, you know with these three words several things. For example, the two families belong to the same superfamily (Staphylinoidea) and therefore have common remote ancestry, and they are two distinct groups (i.e., because each is ranked as a family, discernible by the "idae" ending, you may conclude that one is not subsumed within the other). Because taxonomists have worked diligently for a long time to make taxa monophyletic, you may assume also that the scientist who proposed this classification believed that all leiodids are more closely related to one another than any of them are to species of agyrtids.

C. Information Storage and Retrieval

Handling billions of pieces of data about millions of species is no small job. Although countless artificial, special-purpose classifications could be conceived (e.g., all fungi fruiting above 2725 m, all herbivores in forests, or all organisms that are green), the one thing shared in common by all forms of life on Earth is an evolutionary history or phylogeny. For this and more technical reasons, it may be stated that cladistic classifications offer the optimal available system in which to store or retrieve what is known of every aspect of biodiversity, the most logical general reference system for all biologists, and the best basis from which to predict what attributes species are likely to possess that have not yet been studied in sufficient detail.

D. Scale

Taxonomy provides a view of biodiversity on a scale not replicated by any other discipline. Many of taxonomy's questions are binary: Does species X occur in region A or not? Answers to such fundamental questions have far-reaching implications but require data that may not appear particularly sophisticated or detailed. A nineteenth-century survey performed by John LeConte for a railroad may record a particular species of beetle from "Lake Superior." Given some detective work, we might be able to reconstruct in greater detail where Dr. LeConte was working on a particular date, but the data would never pass for anything resembling standards for modern ecology. For a mostly tropical genus such as *Aglyptinus*, however, such data may be sufficient for these kinds of basic inquisitions. If we wish to determine whether the genus *Aglyptinus* occurs in North America or perhaps in Michigan, LeConte's data may be quite sufficient. For an ecologist asking about population density, it is obviously of no more use than a marker perhaps of a vague geographic area within which to begin to take appropriately quantified samples. There is a mistaken belief, however, that taxonomists' samples should conform to ecological standards. The argument goes more or less as follows: By carrying around a square-meter devise, the taxonomist could easily quantify his or her data and thereby make it useful to many scientists beyond taxonomy. Were this proposal without hidden costs, this argument would clearly prevail. In reality, any rigid sampling suitable to meet the constraints of serious ecological experimentation does incur substantial costs to taxonomy.

E. Evidential Record

The natural history museums (and botanical gardens and other institutions that care for collections) play a role in the documentation of biodiversity that can scarcely be overstated. Conservation organizations that have created databases for threatened species often include large numbers of observations not verifiable by reference to museum specimens. Such records cannot be critically questioned or truthed and are of far less value to science or society than those records that are evidenced in the form of permanently preserved museum specimens.

F. International Standards

Type specimens in museums and herbaria are the equivalent of an international bureau of standards for scientific names. The word "type" is sometimes naively used to vilify typificiation as a modern form of typology (in the Aristotelian sense). In fact, the type is used pragmatically to ground the name to observable features of one representative. Species are in taxonomy complex hypotheses and take into account ideas about a wide range of genetic, geographic, and morphological variation. When two or more species are hypothesized to exist where a single one did before, the type merely ties the name of the existing species to whichever of the two that the type specimen falls within. No one presumes that another representative will necessarily match the type in detail, except for those characters hypothesized to be diagnostic for the species. The insistence on types by the international codes of nomenclature has done much to establish and maintain stability in what is intended by species names.

G. Record Truthing

Information published in scientific journals or released in the context of a database is only credible if the identity of the species may be verified. In any taxon, especially in megadiverse taxa such as insects, a species that is believed to be well-known today may suddenly prove to be misidentified from some other geographic region, to be part of a complex of previously undistinguished sibling species, to be synonymic with a comparatively senior or junior name, or otherwise nominally changed in status. The only way to guard against such future uncertainties is to deposit a voucher specimen in a museum for permanent housing. In instances in which this has been done, it is possible to later revisit the question of identity and confirm or correct what was first reported in the literature. Literally thousands of scientific journal article are brought into question or must be summarily rejected due to the absence of a voucher. Despite this fact, few journals require such deposition of their authors.

H. Predictivity

Given a cladogram, it is possible to make predictions about the distribution of properties of organisms not yet studied. Even the simplest three-taxon statement makes a bold prediction. If A and B are hypothesized to share a more recent common ancestor than either one does with C, then the implicit prediction is that every apomorphy discovered in the future will be so distributed. The power of such predictivity when biologists are confronted with tens of millions of species can hardly be overstated. Where does a pharmacologist begin to look for the same or related chemical compounds found serendipitously in one species? The answer is obvious given a cladogram.

I. Authoritative Basis for Biodiversity Information

Although a taxonomic monograph may be too exhaustive or too technical for every user, it provides the authoritative comparative overview in which we may have confidence about species numbers and limits. As computer-based monography matures, it will soon be possible to computer generate user-friendly field guides, checklists, and other documents directly from the monographic database but limited to those species and that information necessary for a special purpose.

J. Efficient Inventories of Taxa

Taxon-specific knowledge and experience held by taxon specialists make the inventory of the species of a taxon more efficacious than alternative approaches to determine what and how many species live on Earth.

K. Epistemic Privilege

Taxonomists have a unique view of the world, necessitated by the comparative, historical, and parsimony-based nature of their science. Experimental biologists whose work does not depend on (and is not answerable by reference to) collections of specimens can hardly be expected to take the lead in building vast herbaria and museums. Nor can they be expected to place a major emphasis on the inventory of species and clades so close to the brink of extinction that only (not forthcoming) Herculean efforts could pull them back from the brink—those taxa referred to as the "living dead." This vision of the importance of phylogenetic hypotheses and taxonomic inventories places a strong burden on the taxonomic community to ensure that these essential aspects of biodiversity study are continued.

Acknowledgments

I thank Roberto Keller for a detailed reading of the completed manuscript, an illuminating argument over metaphysics, and assistance in the preparation of the figures. I also thank Nico Franz, Kelly Miller, and Kevin Nixon for reading and commenting on the manuscript.

See Also the Following Articles

BIODIVERSITY, EVOLUTION AND • CLADOGENESIS • DARWIN, CHARLES • NOMENCLATURE, SYSTEMS OF • PHYLOGENY • SPECIES, CONCEPTS OF • SPECIES DIVERSITY, OVERVIEW

Bibliography

Brooks, D. R., and McLennan, D. A. (1991). *Phylogeny, Ecology, and Behavior: A Research Program in Comparative Biology.* Univ. of Chicago Press, Chicago.

Crowson, R. A. (1970). *Classification and Biology.* Heinemann, London.

Eldredge, N., and Cracraft, J. (1980). *Phylogenetic Patterns and the Evolutionary Process.* Columbia Univ. Press, New York.

Farris, J. S. (1979). The information content of the phylogenetic system. *Syst. Zool.* **28**, 483–519.

Farris, J. S. (1983). The logical basis of phylogenetic analysis. *Adv. Cladistics* **2**, 7–36.

Forey, P. L., Humphries, C. J., Kitching, I. L., Scotland, R. W., Siebert, D. J., and Williams, D. M. (1992). *Cladistics. A Practical Course in Systematics*. Clarendon, Oxford.

Gaffney, E. S. (1979). An introduction to the logic of phylogeny reconstruction. In *Phylogenetic Analysis and Paleontology* (J. Cracraft and N. Eldredge, Eds.), pp. 79–111. Columbia Univ. Press, New York.

Hennig, W. (1966). *Phylogenetic Systematics*. Univ. of Illinois Press, Urbana.

Minelli, A. (1993). *Biological Systematics*. Chapman & Hall, London.

Nelson, G., and Platnick, N. I. (1981). *Systematics and Biogeography*. Columbia Univ. Press, New York.

Platnick, N. I. (1979). Philosophy and the transformation of cladistics. *Syst. Zool.* **28**, 537–546.

Popper, K. (1968). *Logic of Scientific Discovery*. Harper, New York.

Rieppel, O. (1988). *Fundamentals of Comparative Biology*. Birkhauser Verlag, Berlin.

Ross, H. H. (1970). *Biological Systematics*. Addison-Wesley, Reading, MA.

Schoch, R. M. (1986). *Phylogeny Reconstruction in Paleontology*. Van Nostrand–Reinhold, New York.

Schuh, R. T. (2000). *Systematic Biology*. Columbia Univ. Press, New York.

Wheeler, Q. D., and Meier, R. (Eds.) (2000). *Species Concepts and Phylogenetic Theory: A Debate*. Columbia Univ. Press, New York.

Wiley, E. O. (1981) *Phylogenetics*. Wiley, New York.

TAXONOMY, METHODS OF

R. I. Vane-Wright
The Natural History Museum London

I. The Tasks of Taxonomy
II. Building Blocks: Individuals and Characters
III. Special and General Classifications
IV. Differing Philosophies and Methods of Taxonomy
V. General Procedures
VI. From System to Classification
VII. Current Practice: Variations on a Cladistic Theme
VIII. Conclusions

GLOSSARY

clade A complete ancestor–descendant lineage (cf. monophyletic group).
cladistics Production of taxonomic system based on hierarchical patterns of homologous characters, expressed as cladograms.
cladogram A dendrogram expressing estimated cladistic relationships among taxa; a cladogram has no direct connotation of ancestry and the long axis does not connote time.
dendrogram A branching, nonreticulate diagram expressing nested hierarchical relationships or similarities (or both) between entities (e.g., taxa) such that the entities only appear at the tips of terminal branches.
diagnosis A set of attributes sufficient to define, characterize, or identify a given taxonomic group.
distance A measure of dissimilarity between two taxa based (normally) on the number of mismatches within a large set of characters or attributes compared for both taxa.
gap A large or relatively large difference in overall similarity between two taxa.
grade A group based on, for example, the functional level of organization or overall similarity rather than ancestor–descendant relationships.
ground plan (archetype, bauplan) A basic plan or general type, or a hypothetical ancestor.
jizz Characteristic, instantly recognizable appearance of an organism.
monophyletic group A group comprising a given (hypothetical) ancestor and all its descendants; in a restricted sense, within a given set of terminal taxa, all the members of a subset arising from a common ancestor that has given rise to no other member(s) of the whole set.
paraphyletic group A group comprising a given (hypothetical) ancestor and only some of its descendants.
phenetics Production of a taxonomic system based on overall similarity.
phenogram A dendrogram expressing overall similarity (long axis) between terminal taxa and sequentially linked groups of terminal taxa; the nodes do not connote specifiable characters and the long axis does not connote time.

polyphyletic group A group that does not include the most recent common ancestor of all of its members.

polythetic group A natural taxonomic group in which the terminal taxa are not known to share any universal unique character(s) but are nonetheless united by overall similarity (phenetics) or global parsimony (cladistics).

rank A specified categorical level in the taxonomic hierarchy (e.g., species, genus, family, and class) made coordinate in classification by definition but very frequently not coordinate in the taxonomic system.

similarity A measure of coincidence (matches) among a (large) set of characters compared for any two taxa.

sister groups Two terminal taxa or monophyletic groups that share a common ancestor that has not given rise to any other taxon under consideration.

taxon Any formally named or recognizable group in a taxonomic system (e.g., order, family, and genus), including all the particular terminal taxa (species).

Taxonomy is a highly controversial subject, and the issues are inextricably bound up with philosophical disputes which have endured for centuries. The problems are so important that no biologist can totally avoid facing them.—Michael T. Ghiselin (*The Triumph of the Darwinian Method*, 1969, p. 79)

No one would think much of a chemist who confused water and benzene "because they look alike."—Arthur J. Cain (*Animal Species and Their Evolution*, 1954, p. 11)

TAXONOMY plays the central role in the scientific discipline of systematics and, according to some authors, the two are largely or even entirely synonymous. Systematics makes an essential contribution to our understanding of biological diversity, including its origins, distribution, and maintenance. The primary task of taxonomy is systematization: to establish and give an account of biological order among the diversity of organisms. This involves enumerating the kinds of living things that exist and have existed in the past and determining the patterns of difference and connection among them. By giving expression to these patterns through naming inclusive sets, subordinate sets and least included entities (e.g., higher classes, genera, species, subspecies, and varieties), taxonomists have produced the general biological classification: a categorical arrangement of named, diagnosable groups and components (taxa) that can be used to refer to all known living and extinct organisms.

Although effective taxonomy predates Darwinism, systematics is now closely linked to the theory of evolution. Descent with modification gives justification to the general form of classification adopted: hierarchical, nonoverlapping sets rather than fuzzy sets or periodic tables. The fundamental methods of taxonomic systematization are few. Ultimately, there may prove to be only two—overall similarity methods (grouping by degree of genetic or phenetic similarity) and hierarchical methods (grouping by inclusive phylogenetic or cladistic relationship)—but the theoretical bases of taxonomy are various. This has led to a proliferation of approaches, often with variants. To understand alternative methods of taxonomy it is necessary to appreciate the philosophical differences between these approaches, and much of this article is devoted to exploring these differences rather than describing techniques. The methods of cladistics, currently the dominant approach, are the subject of a separate review.

I. THE TASKS OF TAXONOMY

Taxonomists perform five main functions: discrimination (discovery or primary recognition of taxa, also entailing formal description and diagnosis), comparison (assessing similarities, differences, and relationships among taxa), classification (production of summary schemes that encapsulate current knowledge of taxa and their main interrelationships), symbolization (application of names to taxa and classes: technical nomenclature), and identification (secondary recognition of taxa: matching unidentified material to the established system).

This article focuses on different approaches to primary discrimination and comparison of taxa (systematization) and the ways in which this knowledge can be expressed as summary schemes (classification). The procedures of taxonomy as an information retrieval system, which includes making checklists, catalogs, and databases, identification (naming specimens), and technical nomenclature (application and regulation of names: naming taxa), are not dealt with here. Nor, explicitly, are the means of gathering taxonomic data, e.g., from anatomy, karyology, biochemistry, and nucleic acid chemistry, and their corresponding subdisciplines (morphotaxonomy, cytotaxonomy, chemotaxonomy, molecular systematics, etc.). Although such subdisciplines are often described as taxonomic methods, in the wider context of systematics and taxonomy as a whole they are subsidiary techniques. Much the same methods and problems of interpretation apply, whatever the source of empirical data. This also applies to experimental taxonomy (the investigation of taxa based on predictions from alternative systems in an attempt to gather data bearing directly on particular taxonomic problems) and biometrics (quantitative comparison of related taxa).

Disagreements regarding taxonomic methods can lead to major differences in classification. Such discrepancies range from disputes over the validity of species, subspecies, and infrasubspecifics at one end of the scale to the extreme opposite where composition of even the most inclusive categories (domains and their component kingdoms) remains uncertain. By classifying the classifiers it may be possible to identify some of the fundamental reasons, but not all taxonomic disagreements are due to method alone because historical precedent and subjectivity still intervene. Moreover, in practice a great deal of constructive taxonomic work is done with little reference to philosophy or explicit method, being achieved by pragmatic intervention, notably the extension or modification of existing parts of the system (e.g., by description of new species and establishment of synonymy).

II. BUILDING BLOCKS: INDIVIDUALS AND CHARACTERS

Taxonomy is an empirical activity. On the basis of characters derived from sensory data, individual organisms or life cycles can be discriminated and then divided among or gathered into groups, and groups within groups, and so on to produce a taxonomic system. This can be done either top-down (divisive methods, as in Linnean divisions) or bottom-up (agglomerative methods, as in most numerical taxonomy). Although individual organisms and characters are the basic elements in this process, their definitions are not straightforward.

A. Individual Organisms

Organisms multiply, typically from spores or fertilized eggs, to reproduce the corresponding parental stage through a cycle of growth and differentiation. At any point along such an ontogenetic pathway, a particular organism can be referred to as an individual: fertilized egg, embryo, larva, subadult, adult, or postreproductive adult. However, some organisms multiply by splitting or propagation, rendering the parent–offspring distinction uncertain. Thus, land plants can spread by "runners" so that what appears to be a field of separate individuals can be identical clones of one original plant. In the case of social organisms (e.g., ants and termites) in which very large numbers of workers, guards, or other castes may be produced that never have an opportunity to reproduce, the entire colony has some of the properties of an individual. These distinctions are important because the reliability of the taxonomic process, other factors being equal, is dependent on the number of individual organisms sampled as well as the number of characters recognized and recorded. Multiple sampling from essentially the same individual organism can distort our views of natural variation or lack of it and lead to erroneous conclusions (Wiens, 1999).

B. Attributes and Characters

Taxonomists compare organisms by means of characters. Characters are abstractions derived from the detectable attributes of individual organisms or social groups (e.g., "large, two pointed prongs on head, color vision, always herd with tails erect"). To be informative, it is obvious that characters observed must not be universal throughout the organisms under investigation. The distribution of characters is the primary interest, and the degree to which they differentiate, coincide, and conflict will largely determine their usefulness for taxonomy.

In general, characters can be divided into two sorts: continuous (e.g., height, from tall to short) and discrete (e.g., paired horns versus no horns). However, this is not absolute but more a matter of scale. Thus, different individuals of a particular kind of fly might bear every possible combination of 1–20 spines on the thorax: Is this continuous or discrete variation? In addition, coding can be arbitrary (how tall is "tall"?), and subdivision can be ambiguous. If we observe different individuals with straight, curved, and forked horns, do we simply regard these as three separate characters, three states of one character (form of horn), or the intersection of two binary alternatives—horns straight/curved and horns simple/forked (with the expectation of being able to distinguish, in theory at least, two sorts of forked horns)? Moreover, there are other possibilities for coding such variables. The decision we make can affect analysis (e.g., by encoding spurious information derived from inapplicable characters). Thus, if an animal lacks horns, and this is coded as three separate pieces of data (not forked, not curved, and not straight), such redundancy can have undesirable effects on analysis (Strong and Lipscomb, 1999).

In theory, unit characters must not only be nonredundant but also homologous and independent (Pimentel and Riggins, 1987). If different parts of the body are functionally interdependent or developmentally linked, it will be misleading to count these attributes as separate characters. This can occur when a single gene has a pleiotropic effect, influencing, for example, the color of one organ and the form of another. At the extreme, it is clear that multiple characters should not be created by logical correlation (e.g., treating both the circumference and the diameter of a circular organ as two unit characters). Some organs (e.g., parts of the male genitalia of insects) may appear so complex that counting them as one character seems unreasonable, but any stopping rule for subdivision may be uncertain, whereas other characters that appear simple may prove to be complex (e.g., a single functional bone formed by fusion during development, representing several characters in comparison to related taxa in which the equivalent bones have not fused).

III. SPECIAL AND GENERAL CLASSIFICATIONS

On the basis of characters held in common, individual organisms can be grouped into a large number of classes, which are of two general kinds. On the one hand, individuals can be grouped in terms of a particular attribute (e.g., green, round, four-legged, marine, planktonic, nocturnal, and pollinating) or by small combinations of attributes to give prescriptions such as "plankton-feeding marine organisms" or "nonflowering, epiphytic semiparasitic plants." Alternatively, they can be placed into categories of species, genera, families, orders, and so on. The former are regarded as artificial or special classes and are generally defined by reference to given attributes, whereas the latter are viewed, ideally, as natural kinds or natural groups which are discovered but cannot be defined a priori (although they can be diagnosed a posteriori).

Special classes frequently overlap, such as the overlay of plankton feeders, not all of which are marine, and marine organisms, not all of which feed on plankton, to define the special class discussed previously; also, the same organism may recur in many different special classifications (e.g., eagle as flying organism, predator, or nest builder and in the conjunction of all three). In contrast, natural groups typically form nested, nonoverlapping sets in which each kind of organism or group only appears once. Thus, within the general classification of birds, all eagles are included as members of the family Accipitridae, which also includes vultures, buzzards, hawks, and kites.

A. Evolution, Genetic Relationships, and the Natural Hierarchy

Natural groups comprise individuals with many attributes in common, whereas individuals belonging to special classes have relatively few shared characters. In practice, the difference between, for example, "large marine animals" and Cetacea is simply that individuals of the latter natural class have far more in common than those of the former. Even so, the word "natural" has had many connotations in the context of taxonomy, including the essence of things classified (Aristotle), rationality (e.g., as in God's design), similarity, and explanatory power (Gilmour, 1940). Biology, however, has a unique theory that underpins its totality: the theory of organic evolution.

Ideas about evolution can be divided into a general theory of descent with modification and special theories about the processes affecting that descent (e.g., orthogenesis; natural, group, kin, sexual, and species selection; adaptive radiation; molecular drive; and molecular clocks). Most systematists agree that the general theory of evolution not only provides a compelling justification for seeking one natural, general classification of living organisms but also suggests the basis on which this classification is most securely founded. A hierarchical pattern of exclusive sets and inclusive subsets can reflect the primarily divergent sequence of ancestor–descendant relationships.

Hennig (1966) used the term "tokogenetic" relationships for the reticulate genealogical links that occur between parents and offspring within an evolving, sexually reproducing population. When a whole population splits to form two or more divergent subsystems within each of which tokogenetic relationships are maintained but between which processes of genetic recombination are largely or entirely discontinued, speciation (or phylogenesis) occurs. If we describe two extant taxa as phylogenetically more closely related to each other than to any third taxon (regarding them as sister taxa), we imply that the two do not share tokogenetic relations now but did so in the past within a common ancestor that they do not share with any other living taxon. Such ideas were basic to Hennig's concept of phylogenetic systematics. However, many organisms reproduce without genetic recombination (sex) and thus lack tokogenetic relationships. Moreover, the great variety of sexual processes suggests that tokogenetic relationships are not only nonuniversal but also may differ fundamentally in the many lineages where they do occur (Margulis and Sagan, 1984).

Species and other taxa grouped to reflect their phylogenetic relationships are expected, by virtue of their historical connections, to share far more characters in common than members of artificially formed groups. This expectation extends to unknown attributes, making phylogenetic classifications highly predictive. However, another type of genetical relationship is increasingly recognized as important in organic evolution: lateral gene transfer. The chimerical nature of lichens as fungal/algal symbionts has long been recognized. Following the work of Lynn Margulis, it is accepted that at least two of the cellular organelles found in all eukaryotes originated through symbiosis of fundamentally separate organisms. Evidence is accumulating from molecular phylogenetics that the genomes of many, perhaps all, major life-forms are chimeras formed from multiple original sources. The extent to which lateral gene transfer undermines current approaches to natural classification is uncertain, but at the domain level at least, a nonreticulate hierarchy based exclusively on divergence appears to be unrealistic (Doolittle, 1999).

IV. DIFFERING PHILOSOPHIES AND METHODS OF TAXONOMY

Taxonomic methods have developed over time. The massive edifice of taxonomic classification, involving millions of terminal and higher taxa, on which the study, scientific use, and conservation of biodiversity depends has been built up over centuries. This system has been produced by thousands of different minds using different methods, working with different knowledge, under different influences, and often seeking different goals. Some parts have been revised and reworked repeatedly, others hardly at all. To understand the strengths, weaknesses, and limitations of the taxonomic system, it is necessary to appreciate the ways in which systematists of contrasting persuasions have sought order in nature and tried to reflect that order in biological classification. In the sections that follows, the various approaches and methods are reviewed in a historical sequence.

A. Essentialism, Idealism, and Preevolutionary Taxonomy

In his early writings, it is clear that Linnaeus, the founding father of modern taxonomy, was trying to detect the pattern of Creation in the classes and species that he recognized and named. Linnaeus' system was, superficially at least, very simple. Having divided all life-forms into animals and plants, these kingdoms were then divided successively, on the basis of one or a few defining characters, into a series of smaller units down to the level of the genus. Within each genus a large number of terminal species were usually recognized, and each was given a short, diagnostic description. For each species, he also accepted that namable variations could occur but considered that these did not represent the fundamental plan of God's work.

It can be argued that Linnaeus started out as an essentialist, belonging to an intellectual tradition founded on Aristotle's methods of logical division (Linnaeus later modified his views considerably). Aristotle was perhaps more concerned with the classification of our knowledge about living things, or even the generation of knowledge through the process of classification, than he was with the classification of organisms as such. The young Linnaeus was trying to discover or reveal deep knowledge: the natural order of God's Creation. Through his search for an external criterion of verity, Linnaeus' concepts of species and higher taxa were fundamentally removed from those of the nominalists (see Section IV.C).

Essentialism and related but distinct ideas of the early Greek philosophers, notably Plato's idealism, continue to have influences on both the practice and understanding of taxonomy. According to the typological species concept, every species was thought to have its own idealized plan or design. The task of the taxonomist was then to recognize each of these theoretical designs, and describe, divine, or define the essential features of these "types" so that individual, real organisms could then be assigned to them. Thus, for a Platonist, "taxonomic names are the names, not of organisms, but of concepts" (Ghiselin, 1969). Even today, many taxonomists consider given species as concepts (e.g., fulfilling some ideal as gene pools, potential interbreeding units, or mate recognition systems) rather than empirical entities.

Idealism also had a great influence on concepts regarding major groups of organisms, especially animal phyla, which many morphologists believed could be formulated as a series of fundamentally different body plans (ground plans or Baupläne). The notion of the ground plan was strongly developed by the ideal morphologists, who evidently affected the developing ideas of Willi Hennig (see Section IV.F) in the 1930s and 1940s, even though he later rejected many of their notions. According to Hull (1965), by supposedly ridding taxonomy of its essentialist and idealist burdens, and moving on to an empirical and relativistic framework, Hennig and other modern systematists liberated the science of taxonomy from a philosophical bind that held back progress for more than 2000 years. Hull's evaluation, however, may have been premature (see Section VIII).

B. Empirical and Inductive Taxonomic Methods

According to Mayr (1969), empiricism should be included as an atheoretical approach embracing the plausible, commonsense view that once sufficient knowledge has been gathered about organisms, a natural system of classification will simply emerge or become self-evident. Although it seems difficult to accord such a process the status of a general method, particular empirical observations can and often do render theoretical disputes irrelevant.

When Vaughan Thompson discovered in 1829 that barnacles develop from a nauplius larva, they were readily transferred from the Mollusca (where they had often been placed) to the Crustacea. Although this change in classification could in retrospect be justified by appeal to arguments about evolution, overall similarity, bauplans, or synapomorphies, to the empiricist this would all appear unnecessary. To raise a brood of insects and find that all the males belong to one genus and all the females to another leads to an instant appreciation of generic synonymy, without any need for appeal to theory.

The most complete expression of empirical classification occurs outside science in the form of folk taxonomy. Because folk taxonomy is well developed even in illiterate cultures, instead of inquiring into method or philosophy, it can only be understood by description and comparison. Berlin (1992) identified seven major features of ethnobiological categorization: recognition is given to the most distinctive local species; classification is based on "affinities that humans observe" (not on cultural significance); systems have a limited hierarchic structure; recognized taxa are distributed among a few mutually exclusive universal ranks, approximately equivalent to kingdom, life-form (e.g., plant), family, genus, species, and variety; in different ethnic systems, taxa at each rank show striking similarities regarding their number of subordinate taxa; taxa at generic and specific levels have an internal structure in which some included members are viewed as "prototypical" and others as reminiscent of other taxa at the same rank, giving rise to a type of fuzzy set classification; and a large majority of the taxa differentiated, most notably at the generic level, correspond to groups recognized in formal taxonomy.

These regularities and correspondences are highly suggestive that there is a natural biological classification that is almost literally self-evident and independent of the observer, and that a significant number of natural elements and groupings can be recognized simply through extensive knowledge and contact with nature. Although the young Linnaeus thought he was discovering the handiwork of God, it seems possible he was also involved in formalizing preexisting European ethnobiological knowledge.

Thus, as empirical knowledge accumulates, a common sense or consensus view of certain issues arises, and in some cases such a view may seem irrefutable. Even so, there are many problems in taxonomy that cannot be decided (e.g., questions of relative rank or the status of paraphyletic groups) without an explicit theoretical framework.

C. Nominalism

One view of taxonomy is that, even though we may desire a general system, there are no independent means for assessment. The only objective realities are individual organisms, and all taxonomic groups are man-made abstractions ("categories of thought" according to Louis Aggasiz). Named groups are convenient for collective reference but have no independent basis separate from the human mind, simply being useful pigeonholes for dividing up or handling diversity. Many biologists have held similar views (apparently including Darwin, who once commented on species as a "term . . . arbitrarily given for the sake of convenience to a set of individuals closely resembling each other"; see Mayr, 1963, p. 14; also see Ghiselin, 1969).

By embracing the idea that taxonomic groups are established for convenience, nominalism has links with special classifications. If special classifications such as "trees" or "four-footed land vertebrates" are convenient, then there is no obvious reason why the nominalist should reject them, except insofar as it can be demonstrated that some comparable grouping has more heuristic or predictive power (e.g., trees extended to Tracheophyta and literal Tetrapoda extended to include snakes, birds, whales, and bats); in other words, they are in some sense more convenient. Convenient for who then becomes the question.

Berlin's (1992) review suggests that folk taxonomies are not nominalist, which perhaps might have been expected, but seminatural and based on extensive empirical knowledge. Folk systems often have terms for conspicuous natural groups such as Mammalia, but it is not clear to what extent paraphyletic or even polyphyletic taxa are also included. Because folk taxonomies have terms for life-forms, which are in effect grades, this suggests that humans may possess an innate mixed strategy for classification, sometimes grouping on palpable characters, whereas at other times forming groups on the basis of general resemblance, Gestalt, or jizz.

Although Panchen (1992) suggests that there are no extant practitioners of nominalism, the view that species are real while all higher taxonomic groups are artificial is widely expressed (even though quite misguided: contrast the difficulty of even "defining" many species of mice with the ease with which the class Mammalia can be recognized by many consistent features). There are major disagreements regarding the existence of fundamental differences between terminal taxonomic components (e.g., species) and higher groups (e.g., polytypic genera and families), as apparently accepted by Hennig, for example (and probably Mayr), and rejected by Gilmour (1940) and Nelson (1989).

D. Evolutionary Systematics: Grades and Clades

The Darwinian revolution provided a rationale "external" to the human mind for the basis of taxonomy: Hierarchical relationships among living things are explicable as the result of organic evolution, and taxonomists should strive to reflect these patterns as the logical basis of a general classification. However, if taxonomists looked to Darwin for a more detailed lead, what did they find? Unfortunately, at various places in the *Origin*, Darwin appeared to shift between general evolutionary statements ("the natural system is founded on descent with modification"), phylogenetic assertions ("the arrangement of groups within each class . . . must be strictly genealogical in order to be natural"), and nominalism ("I look at the term species, as one arbitrarily given for the sake of convenience"). Moreover, the pre-evolutionary taxonomic hierarchy had served the Darwinians well, presaging a continuing debate to this day: Is the theory of evolution relevant to the pursuit of taxonomy only as justification for seeking a single general system, or should we build into the taxonomic method theories about the way in which evolution has occurred?

Whatever the precise reason, the immediate impact of Darwinism on method was very limited. Taxonomy remained highly individualistic, and essentialist, nominalist, empiricist, and even creationist views all continued, although often cloaked in evolutionary language. The first move to formulate a general method post-Darwin did not occur until the 1920s, later epitomized by two books by Julian Huxley: *The New Systematics* (1940) and *Evolution: the Modern Synthesis* (1942). This emergent approach was developed and defended most notably by George Gaylord Simpson and Ernst Mayr, and it became evolutionary systematics or evolutionary taxonomy.

With respect to Darwin, the evolutionary systematists argued that the natural system (Fig. 1) should reflect descent with modification, to include the processes by which evolutionary change occurs, the measurable degree of modification (anagenesis), and the temporal sequence of divergence (cladogenesis). According to Simpson (1961, p. 107), evolutionary systematics requires that a natural classification should be "consistent with all that can be learned of the phylogeny of the group classified," but he was emphatic that this does not mean that natural classification be based on phylogenetic relationships alone [a view also clearly stated by Gilmour in *The New Systematics* (1940)]. This led Simpson to adopt a very broad notion of monophyly as "the derivation of a taxon through one or more lineages . . . from one immediately ancestral taxon of the same or lower rank."

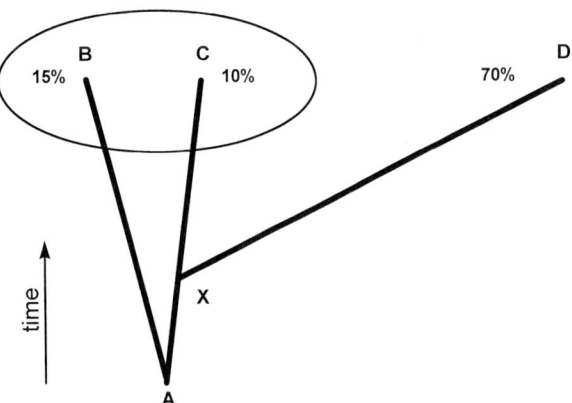

FIGURE 1 Evolutionary systematization. Horizontal axis, anagenesis; vertical axis, time. In this hypothetical case, the slope of each branch corresponds to the rate of anagenesis, with the percentage values representing the degrees of difference in the three lineages from the ancestral species (A). B is grouped with C because the two are more like their common ancestor than either is to D (even though D shares a common ancestor with C at X that it does not share with B). Note that this method requires a means of measuring similarity or distance (phenetics), estimating phylogenetic relationships (cladistics), and knowledge of ancestors (paleontology) [based on Mayr (1974) and Patterson (1982)].

To be natural, according to Mayr (1969), a classification must have explanatory, predictive, and practical values but also be emendable in the light of new evidence or understanding. Mayr then proposed that the fundamental basis for naturalness is the proportion or number of genes held in common by any two taxa: "If we knew the entire genotype of each organism, it would be possible to undertake a grouping of species that would accurately reflect their 'natural affinity'" (p. 81).

The empirical approach closest to this new ideal of evolutionary systematics is the use of pairwise distance data representing, in some way, the entire genome (e.g., immunological data and notably DNA–DNA hybridization data used by Sibley and coworkers for bird classification). In addition, complete nucleotide sequences for the entire genomes of various organisms are now becoming available for direct comparison. These approaches, however, founder on some of the fundamental problems of numerical taxonomy. Mayr's proposal that "genes-in-common" provides the ultimate arbiter of natural classification, however, is an important concept because it encapsulates the only well-articulated rival to the phylogenetic nexus idea first suggested by Darwin ("the arrangement of groups . . . must be strictly genealogical in order to be natural"). If lateral gene transfer undermines a strictly hierarchical approach, then the estimation of genes-in-common will certainly increase in importance as the basis of a natural system.

In practice, evolutionary systematics became a syncretistic, all-embracing method that included a regard for the absence of characters as informative and insisted on the primacy of paleontology for revealing phylogenetic sequences (Fig. 1). The method's most striking characteristic involved conflation of the two methods that soon sought to replace it: a desire to give expression to genetic distances (grades or anagenesis, as reflected in numerical taxonomy) and, at the same time, to ancestor–descendant relationships (clades, or cladogenesis, as reflected in cladistics). Because these two methods stem from fundamentally different philosophies, this led to an inevitable arbitrariness (Fig. 1). Thus, Simpson (1961, p. 107) was happy to write that although "taxonomy is a science . . . its application to classification involves a great deal of human contrivance and . . . there is a leeway for personal taste, even foibles." This lack of explicitness led to the demise of evolutionary systematics as the leading method in taxonomy. During its development, however, Ernst Mayr in particular made an enormous contribution, especially to ideas on the taxonomy of species, with which he was preoccupied as "the basic unit of classification" (Mayr, 1963, p. 11).

E. Numerical Taxonomy and Operationalism

Numerical taxonomy (or phenetics) emerged in the late 1950s, its origin associated with, among others, Charles Michener, Arthur Cain, and especially Robert Sokal and Peter Sneath. In Sokal and Sneath's original 1963 manifestation of the *Principles of Numerical Taxonomy*, any evolutionary approach is avoided in favor of an operational method based on direct comparison of phenotypes. As many characters as possible of the organisms to be compared, both continuous and discontinuous, are measured and counted from operational taxonomic units (OTUs), which can be individuals or samples from conventionally recognized taxa (typically species). On the basis of a matrix of variation in all features across all OTUs, the OTUs are then compared by overall similarity, affinity, or phenetic distance. Such measures can be obtained by transforming the raw matrix (Table I) to give the proportion of all character matches (affinity) or mismatches (distance: Table II) for every pairwise combination of OTUs. The results are displayed by means of a network (Fig. 2a), or OTUs are linked to each other by a clustering algorithm to produce a phenogram (Fig. 2b).

Although such a procedure appears objective at first, many different ways have been proposed to measure pairwise similarity or dissimilarity, and many different clustering methods have also been devised. Most clus-

TABLE I

Amino Acids Found at Eight Particular Positions in the Myoglobin Chain of Four Operational Taxonomic Units (OTUs): (A) Human, (B) Alligator, (C) Tuna fish, and (D) *Heterodontus* Shark[a]

	Characters							
OUT	0	1	2	3	4	5	10	11
A	—	G	L	S	D	G	V	L
B	M	E	L	S	D	Q	V	L
C	—	—	—	—	—	A	V	L
D	—	—	—	—	—	T	V	N

[a] Single letters stand for particular amino acids and dashes for positions not represented based on alignment of the four complete myoglobin sequences (from Patterson, 1980, p. 237).

TABLE II

Data of Table I Transformed Into a Distance Matrix, Based on the Proportion of Mismatches Summed across All Eight Positions, for All Six Pairwise Comparisons of the Four OTUs[a]

	A	B	C	D
A	0	.375	.625	.75
B		0	.75	.875
C			0	.25
D				0

[a] These distances are replicated in the branch lengths of an unrooted network (Fig. 2a). Taken as reciprocals to give measures of similarity, the data can also be used to produce a phenogram (Fig. 2b).

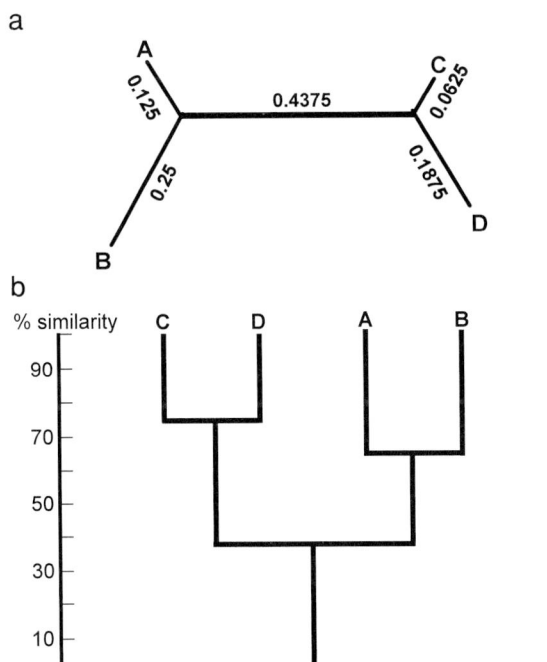

FIGURE 2 Numerical systematization. (a) The values in Table II between the four OTUs A–D represented by a distance network. The scaled lengths of the line segments are such that all the relative values in Table II are satisfied, but the angles of the four branches leading to the terminals have no special meaning. (b) Horizontal axis, linkages; vertical axis, overall similarity. By taking reciprocals of the values in Table II, the data give a similarity matrix—the basis of a phenogram. With a value of 75% similarity, (C + D) are the first OTUs clustered. Comparing A with B, and (C + D) with A and B separately, (A + B) form the next most similar cluster, linked at 62.5% similarity. The two clusters can then be linked together at 37.5% similarity based on three of the eight attributes occurring in one or both members of the two groups. Other linking procedures could be adopted.

tering algorithms employed in taxonomy are sequential, agglomerative, hierarchic, and nonoverlapping (SAHN). Among this class of methods there are subclasses (e.g., single linkage, complete linkage, and average linkage), which in turn have variations (Sneath and Sokal, 1973). The phenograms produced by SAHN algorithms are dendrograms and thus strictly hierarchical and lacking reticulations, appearing similar to a Linnean hierarchy or a cladogram.

Initially, it was expected that agreement would be reached on best if not ideal procedures, but this proved an illusion. As early as 1966, Hennig (p. 85) commented that "even the most recent authors present and recommend their own methods [for measuring morphological differences] but never explain why their method deserves preference over those of their predecessors." Despite these difficulties, the pheneticists pursued the ideas of numerical taxonomy with vigor based on four key hypotheses proposed by Sneath and Sokal that can be viewed as attempts to realize Mayr's goal of measuring genes-in-common (Panchen, 1992, p. 135). The nexus hypothesis stated that every character is likely to be affected by more than one gene, and every active gene is likely to affect more than one character. The nonspecificity hypothesis suggested that no major or distinct group of genes would exclusively affect one class of attributes (e.g., morphology and behavior); thus, classifications based on one character class or set should not differ from classifications based on another class or set, and there would be no a priori reason to prefer particular data sets to others. The hypothesis of the factor asymptote supposed that as data were added, the information content of the classification would reach an asymptote, whereas the hypothesis of the matches asymptote proposed that as the number of characters sampled increased, the similarity value for any pair of OTUs would tend toward a final, asymptotic value, thus conferring stability on the classificatory system.

Regrettably for phenetics, as discussed by Panchen (1992) and even acknowledged by Sneath and Sokal as early as 1973, these hypotheses, apparently basic to the validity of the methods, are now largely discredited (in particular the all-important nonspecificity hypothesis). More generally, phenetics offers no justification for choosing a hierarchical system unless it is accepted that the general theory of evolution dictates this as the most efficient representation. However, because numerical taxonomy conflates homologous and nonhomologous characters, and neither nodes nor branch lengths can be directly related to hypotheses about particular characters, the origin of characters, or even the degree of genetic change, the notion of overall similarity lacks analytical power.

Moreover, although it was originally expected that its procedures would lead to stability in classification, several factors preclude this (Eldredge and Cracraft, 1980, p. 176). In addition to the technical ambiguities already noted, every new taxon requires complete re-analysis, with the unique attributes and combinations of attributes usually affecting every branch length and often many branching points in the phenogram. Of course, other taxonomic methods, including cladistics, are not invulnerable to change due to new discoveries, but in cladistic analyses the effects are interesting (because they can be related to hypotheses about homology) and cladists never laid serious claim to the idea that stability in the face of new evidence was an important justification for the method.

Cladistics also suffers from its own algorithmic problems, but it is always possible to work out what different procedures are doing with respect to the data and the inferences drawn. With phenetic methods, the nature of the algorithms and the form of the results are not separable. Thus, a basic problem of numerical phenetics is the lack of a clear criterion of choice (such as parsimony, or even a model of evolution) by which one result can be judged against another. Even Mayr's concept of genes-in-common will not repair the difficulty because of innumerable problems related to the definition of a gene, duplication, alignment, position effects, and so on. The only other external arbiter is the indefinable goal that James Farris called Gilmour-naturalness ("a system of classification is the more natural the more propositions can be made regarding its constituent classes"), enthusiastically embraced by Sneath and Sokal (1973). This led Panchen (1992) to suggest that the original philosophy of numerical phenetics is closer to empiricism (not nominalism, as suggested by Mayr). Given enough observations and an appropriate technique, the numerical taxonomists expected that a single, stable, and predictive general system would simply emerge.

Although numerical taxonomy has been judged wanting, it has had a lasting and positive influence on current taxonomic methods and still flourishes in areas such as bacterial taxonomy, in which the general absence of tokogenetic relationships negates many advantages of a phylogenetic system. The most important contributions of numerical taxonomy have been technical, notably the use of data matrices, precise comparison across all taxa under scrutiny, and the adoption of algorithmic analytical methods. Advances in other methods, notably cladistics, have come not only from vigorous debate but also from the general adoption, post-Hennig, of data matrices and their exploration through numerical algorithms.

F. Phylogenetic Systematics and Cladistics

The fundamental methods of phylogenetic systematics were established by Willi Hennig, who wanted to base systematization and classification directly on the historical branching patterns of the phylogenetic nexus. He realized that it would never be possible to know the course of history precisely, and that the system would always be provisional. His goal was, by means of appropriate methods, to produce a "phylogenetic system" that would "approximate more closely than any other *the ideal system* [italics added] that reflects the phylogenetic relationships absolutely correctly" (Hennig, 1966, p. 29).

Hennig's method was "the search for the sister group," epitomized as the "three-taxon problem." For any group of three natural taxa, the expectation is that two have a common ancestor not shared by the third. Thus, for the trio shark, tuna, and human, of the three possible combinations of two from three, the one with tuna and human grouped to the exclusion of shark accords with what we know of ancestry (see Table III). According to Hennig's view, we should then recognize a taxon linking tuna and human as sister groups

TABLE III

Data from Table I Transformed to Show only Coincidences of Positive Attributes among the Four OTUs[a]

Taxa	Character			
	2	3	4	11
A	*	*	*	*
B	*	*	*	*
C				*
D				

[a] Characters 2–4 link (A + B), and character (11) links (A + B + C). These data are most efficiently represented by the nested-set pattern (D (C (A + B))); compare with the pattern ((D + C) (A + B)) implied by the phenetic analysis (Fig. 2). In this analysis, four characters (0, 1, 5, 10) in Table I are uninformative because there is no positive coincidence (5), only shared absence (0), both (1), or only shared coincidence (10).

(Osteichthyes) but not one linking shark and tuna: From a phylogenetic perspective, Pisces are paraphyletic, or a nongroup. Thus, Hennig defined phylogenetic relationships solely in terms of common ancestry (propinquity of descent) and did not include similarity, distance, grade, adaptive zone, or any comparable concept in his assessments.

This line of argument caused Mayr and other evolutionary systematists to object very strongly to Hennig's approach, accusing him of an unjustifiable restriction of the concept of relationship. For this reason, Mayr referred to Hennig and his followers as "cladists" (from clade or cladogenesis) to criticize them for their narrow approach—with the unexpected result that the term was happily adopted by the growing band he sought to reprove.

However, how could clades be recognized in the absence of prior knowledge or without acceptance of sequences simply "read" from the fossil record? Arguably Hennig's greatest methodological advance was the introduction of relativism to the concept of characters. He recognized three classes with respect to a set of taxa under study: autapomorphies (characters unique to individual terminal taxa), symplesiomorphies (characters present in all the taxa), and synapomorphies (characters shared by subsets of two or more taxa). According to Hennig, only characters of the third class provide evidence of grouping at the specified level of inclusiveness. Thus, a synapomorphy indicative of a natural group of species within the taxa under review (e.g., a genus) becomes a symplesiomorphy when addressing questions of relationship among species within that genus so delimited. Ideally, each and every character would be informative of relationships at just one level in the total hierarchy. Thus, virtually all the characters that might seem to unite shark and tuna to the exclusion of human (e.g., gills and fins) are actually characters that relate to more inclusive levels of the hierarchy (e.g., Chordata and Vertebrata). Observations such as "absence of mammary glands" or "absence of hair" are not characters linking "fish" but the counter conditions of autapomophies (of the Mammalia).

To see this in practice, the data in Table I have been transformed (Table III) to show only potential synapomorphies, i.e., the positive attributes shared by two or three of the four taxa, and thus able to provide evidence of grouping within (A–D). In this simple example there are no conflicting distributions, and there is just one maximally efficient solution. Efficiency means the hierarchical arrangement which captures all (or, where there is conflict, the largest number) informative characters (putative synapomorphies). In Fig. 3, all 15 possible fully resolved groupings of terminal taxa A–D are presented, with the characters that they can summarize marked at the relevant nodes. Arrangements 1–7, 9, 14, and 15 do not reflect any of the characters. Arrangements 10 and 12 capture one, and 8 and 13 capture three, but only arrangement 11 captures all four. The best cladistic arrangement, on the available data, is therefore (((A + B) C) D).

Colin Patterson (1980) summarized "the axioms of cladistics" as homologous characters having a hierarchical pattern in nature in which this pattern is efficiently expressed by cladograms, and the nodes connote the homologies shared by the organisms so grouped. The search for the sister group therefore reduces to finding the cladogram that summarizes the potentially homologous characters as parsimoniously as possible. Once this best fit cladogram has been found, the attributes at each node are hypothesized as homologous characters shared by the taxa subtended at that node.

One of the most challenging features of cladistics (or at least "transformed cladistics"—see Section VII) is the link between attributes, grouping, parsimony, characters, and homology. An attribute is accepted as a character if the weight of evidence from all other attributes under scrutiny suggests that it is a homologous feature peculiar to a particular group. In Table IV, two potential synapomorphies (x and y) have been added to the distribution of characters in Table III. Given these additional data, and presupposing independence of the attributes, the most efficient grouping of the four taxa based on positive shared features now changes and highlights a problem with the interpretation of attribute 11: Either it is not relevant as a character within the group and has undergone reversal in taxon D or it is not a character at all (it is nonhomologous). Information about this could be sought, for example, by studying the ontogeny or some more fundamental quality of attribute 11 in A–C, revealing a particular scientific strength of the cladistic method: Conflicts can be resolved by character analysis (Kitching *et al.*, 1998) involving recourse to additional data derived from previously unsampled or unused attributes or by investigation into the homology of the conflicting characters (they are not simply aggregated at face value, as in numerical taxonomy).

If characters x and y in Table IV were found to be due to pleiotropic effects of a single gene, then rejection of attribute 11 as a character relevant to resolving the relationships within the group would be premature. If, on the other hand, character 11 was a certain color produced, for example, by a pigment shared by taxa A and B, but the same hue was produced by a totally

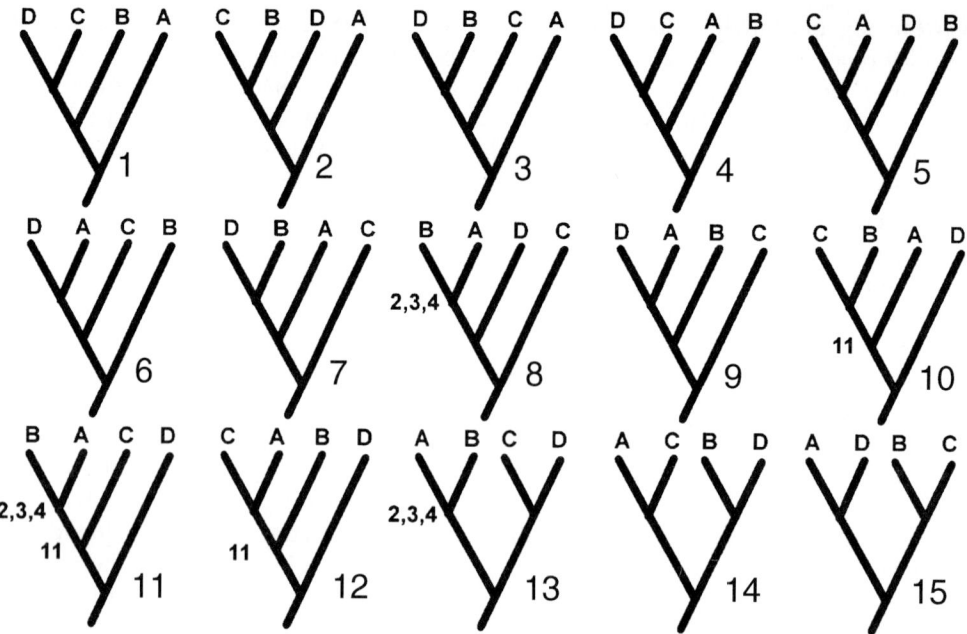

FIGURE 3 Cladistic systematization. Horizontal axis, convention; nodes, pattern of homologous characters. The putative synapomorphies in Table III plotted on all 15 possible arrangements (as fully resolved cladograms) for four terminal taxa. Numbers at given nodes indicate the shared attributes from Table III; arrangement 11 is the only one that reflects all four attributes, all of which can then be construed as homologous characters (based on Patterson, 1980).

different type of pigment in taxon C, this would be consistent with the idea that attribute 11 as originally formulated (color) was a noncharacter. The similarities and differences between taxa A–C in this regard should then be rescored as two new attributes (e.g., based

TABLE IV

Imaginary Character Matrix for Four Taxa and Six Characters[a]

Taxa	Character					
	2	3	4	11	x	y
A	*	*	*	*		
B	*	*	*	*		
C				*	*	*
D					*	*

[a] Characters 2–4 and 11 correspond to Table III. Additional characters x and y are in conflict with character 11. The most parsimonious solution, assuming that all characters are independent, is to group the four taxa as ((A + B) (C + D)). This implies that attribute 11 is no longer considered homologous [not "the same" in C as in (A + B)] or that it is symplesiomorphous with respect to the whole group (A + B + C + D) and does not appear in taxon D (e.g., in evolutionary terms, this would imply that character 11 had been secondarily lost in taxon D).

on chemistry), giving further support to the (A + B) grouping and removing conflict with the (C + D) grouping. The convergence in color between (A + B) and C could then be seen to be in the eye of the human observer. Such convergences are of intense interest for the study of evolution, but once detected they play no part in natural classification.

Thus, the relativistic nature of cladistics is emphasized. Based on the whole (attribute × taxon) matrix, the cladograms chosen and attributes thereby specified as homologous characters are probability statements: "Common ancestry versus convergence is tested by topographical correspondence [on the cladogram]. The resulting explanation is a statement of maximal likelihood rather than a denotion of lawful relations or processes" (Rieppel, 1988, p. 166; see Hennig, 1966, p. 29).

V. GENERAL PROCEDURES

The main operations involved in taxonomic research are listed as follows in an idealized order. In practice, this order is rarely followed precisely, and certain steps may be omitted in part or even altogether:

1. Individual specimens, or individuals representing selected taxa (ideally at least all immediately subordi-

nate taxa currently recognized within the group under study), including material representing suitable outgroups for comparison, are chosen.

2. A selection of attributes to be scored across all samples of taxa under study is made (in original work this will usually be a mixture of known and novel features).

3. The manifestations of these attributes are systematically recorded in an individual or taxon/attribute data matrix (in the process of doing this, it is normal for other potentially informative features to be recognized and these may be added to the list under stage 2; it is not unusual at this stage for additional study material to also be called for, when variation has become apparent or specimens prove to be incomplete or otherwise unsuitable).

4. Systematization of the data is carried out by cladistic analysis to produce a cladogram based on homologous characters defining each node (or for phenetic data, by an agglomerative numerical technique to produce a phenogram); in cladistic studies this may involve one or more rounds of Hennigian "reciprocal illumination," notably with respect to testing the coherence of putative homologies that seem to be in conflict (incongruent).

5. Inclusive monophyletic groups inferred from the parsimonious distribution of homologous characters are thus linked together and then named as taxa, to produce a hierarchical classification.

6. The homologous characters are used to help develop diagnoses for the various taxa recognized (as a potential aid for identification, information on characters that are absent may also be included, and keys, tables, or computerized identification programs may also be constructed at this point).

7. Publication of the results will ideally include the provision of necessary descriptions, indications (diagnoses), specifications of types, and formulation of names in accordance with the relevant code (e.g., zoological, botanical, and bacteriological codes) as well as a record of the complete character matrix and specification of the analytical procedures employed (e.g., computer programs and options and models in the case of maximum likelihood analyses).

Cladistic and numerical phenetic methods are essentially the same up to stage 3, but a crucial difference occurs at stage 4, the process of systematization. Stages 5 and 6 also differ, but both methods involve some arbitrariness with respect to where the phylogenetic nexus is to be cut or the phenogram to be divided (see Section VI). At stage 6 there is also some difference in that the very act of cladistic analysis gives the primary characters for diagnosis, whereas the numericist is faced with an additional searching process to enumerate a sufficient set (usually a polythetic set) of characters whereby given members of a specified portion of the phenogram can be recognized in isolation. Fundamental differences in approaches to systematization have already been discussed. The problems inherent in stage 5 are discussed next.

VI. FROM SYSTEM TO CLASSIFICATION

Classification involves translating a systematization scheme into words (or numbers). Based on the methods of logical division, Linnaeus (who established many of the conventions of formal classification) placed all known organisms within a descending series of fully nested hierarchical categories. The major ranks in the Linnean system were kingdom (e.g., plants and animals), class (e.g., Mammalia, Aves, Pisces, and Insecta), order (e.g., Hemiptera, Lepidoptera, Coleoptera, and Diptera), genus (e.g., *Culex*, *Tipula*, and *Musca*), and species [e.g., *Musca domestica* (housefly) and *Musca vomitoria* (bluebottle)]. For Linnaeus, systematization and classification were the same. Modern classifications attempt to summarize far more complex schemes. In this brief review, only the major differences between the major twentieth-century methods are considered.

According to the evolutionary method, classification should be viewed as a useful art, meaning that in addition to scientific analysis, human ingenuity is also needed to produce a practical classification. Simpson (1961) recognized three principles: A classification should reflect the most "biologically significant" relationships among the organisms, it should be consistent with the relationships on which it is based, and it should be as stable as possible without contravening the first two principles. These ideas were elaborated to embrace, most notably, grades and clades (anagenesis and cladogenesis), both of which were regarded as significant for classification. This led to the idea that, even if literally everything relevant were known about the connections among the members of a group, "innumerable different classifications could be made consistent with those interrelationships.... Selection among those alternatives is decidedly an art" (Simpson, 1961, p. 110).

To appreciate the difficulties resulting from such a view, consider Fig. 4. In cladistic terms, three paraphyletic groups (basal groups in A–C) are delimited, together with just 4 of the 34 subclades depicted. Why were these particular groupings chosen? Such a mysteri-

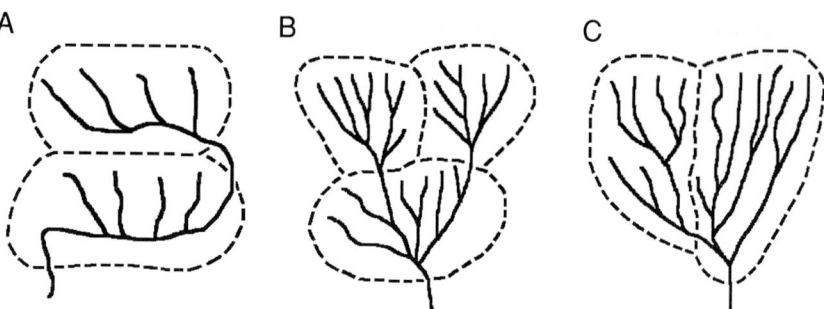

FIGURE 4 Evolutionary classification. The original caption reads: "Three hypothetical phylogenies . . . divided into taxa as shown by the broken lines. A is evidently less arbitrary . . . than B, and B somewhat less than C" (redrawn from Simpson, 1961, Fig. 7).

ous art proved difficult to follow (especially when the prescription for creating such trees in the first place is also imprecise). As discussed previously, Ernst Mayr urged that classifications be based on genes-in-common. Because it is a distance method, this transformation of evolutionary systematics encounters the same problems of classification faced by numerical taxonomists. Even so, the recent estimation that man and chimpanzee share 98.4% of their genome in common has led, for example, to the long-overdue abandonment of the paraphyletic or grade family Pongidae in favor of an expanded Hominidae, in contradistinction to the position adopted for decades by the evolutionary systematists, who argued repeatedly that the "large gap" between *Homo* and the rest of the great apes was reason enough to place them in separate families (Simpson, 1961; Mayr, 1963).

Sneath and Sokal (1973) proposed that good classifications have three desirable properties: naturalness, ease of manipulation, and practicality for information retrieval. Their concept of naturalness, as already discussed, was based on the ideal of incorporating all codified information about a very large numbers of characters. Manipulation related to the practical relationship between a classification and its degree of hierarchical structure (useful for memorizing) and its utility (e.g., for the construction of identification keys). Convenience for information retrieval was also viewed as important, but not if it conflicted with natural classification.

The typical product of systematization in numerical taxonomy is the phenogram. How is such a dendrogram to be turned into a summary classification of named groups? An early proposal was the use of phenon lines (Fig. 5). A phenon line cuts across the phenogram at a particular similarity level. The lines must be straight and are not allowed to "bend up and down according to . . . whim" (Sneath and Sokal, 1973), not only delimiting groups but also determining their rank. Thus, in Fig. 5, if OTUs 1–10 are species, the 80% phenon line could indicate seven subgenera, the 75% line four genera, the 65% line three subfamilies, and so on.

Such a procedure is not objective. The most fundamental problem is simply that the choices of percentage similarity levels and ranks are arbitrary, both within groups and between them. Other procedures, such as McNeill's method based on the assignment of rank levels to each node, also involve arbitrary limits and adjustments. The difficulty for phenetics is that "there is no necessary structural relationship between the pattern of clusters in hyperspace and the inclusive Linnean hierarchy into which they are converted" (Panchen, 1992, p. 151). If this is so, then there can be no nonarbitrary method for converting a phenogram (or any other distance-based method of systematization, including DNA–DNA hybridization trees) into a ranked summary classification.

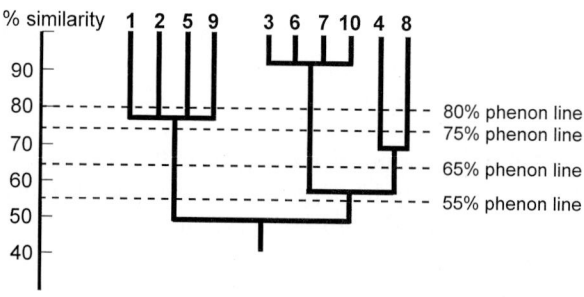

FIGURE 5 Phenetic classification. "The formation of phenons from a phenogram" (redrawn from Sneath and Sokal, 1973, Fig. 5.33).

Cladistic classification is based on the tenet that every putative monophyletic group can be named, but polyphyletic and paraphyletic groups should not. In a grand, top-down cladistic classification of all life, all coordinate monophyletic taxa (sister groups) could be placed at the same rank. Alternatively, taxa of the same geological age could be given the same rank (Hennig, 1966). Also, classifications could be formed bottom upwards by linking all terminal sister species starting with the lowest ranking pair or pairs and building up classes and ranks based on sister groups and ascending ranks held coordinate across the entire hierarchy. Such grand visions remain almost wholly impractical due to our very imperfect knowledge. Moreover, such systems would cause massive proliferation of named groups and ranks and be unstable in the face of most newly discovered taxa and every older geological find. In terms of producing a useful summary, cladistic classification thus faces potential problems of impracticality as well as arbitrariness.

In practice, when classifications are based on cladograms an attempt is usually made, within conventional ranks (e.g., order, family, and genus), to give expression to major monophyletic groups by naming inclusive taxa at intermediate levels. However, this usually means abandoning any equation of classificatory rank with clade level and any attempt to give all groups in the cladogram formal recognition. Thus, in a labeled cladogram of the main lineages of the order Lepidoptera (Fig. 6), not all putative monophyletic groups are named, and all terminals (despite being almost completely noncoordinate) are classified at the same superfamily rank except one of the cladistically most inferior groups, the crown group Ditrysia, which appears to retain its subordinal status accorded in earlier, precladistic classifications.

One partial solution to these difficulties is the "sequencing" convention, first proposed by Gary Nelson. This attempts to combine the practicalities of a limited Linnean hierarchy with a listing that enables the entire (cladistic) hierarchy to be recovered. A sequence of taxa named at the same rank indicates that the first is sister to all following at that rank, the next is sister to the remainder, and so on. Taxa of uncertain position can be annotated as *incertae sedis* and those comprising unresolved polytomies as *sedis mutabilis*. In presenting these sequences, comprehension is greatly facilitated by indentation. Thus, the cladogram for the major groups of Lepidoptera in Fig. 6 can be converted into a written classification (Table V).

In conclusion, the idea that classification can simply be equated with systematization is a vestige of preevolutionary taxonomy and should be abandoned (Minelli, 1993, p. 14). In practice, reflecting a Simpsonian view,

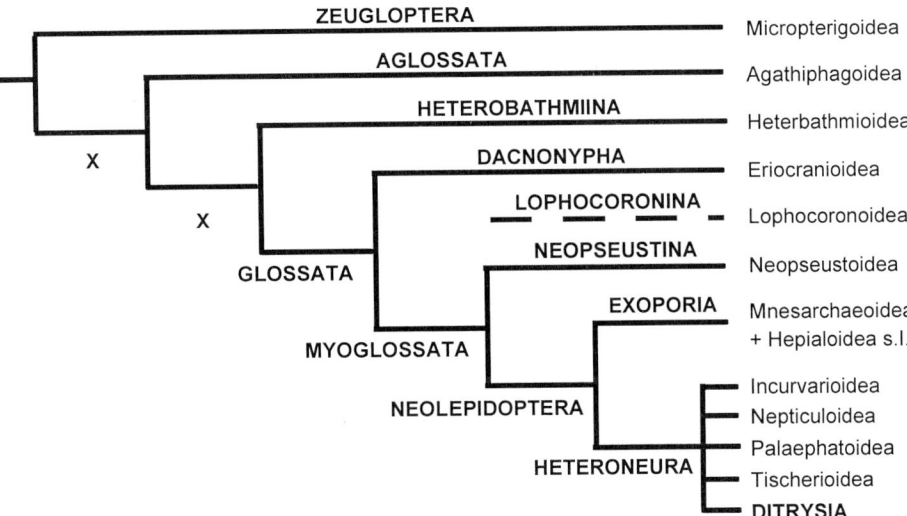

FIGURE 6 Cladistic classification. Labeled cladogram of main lineages of the order Lepidoptera (slightly modified from Scoble, 1992, Fig. 185). In this classification, two of the major monophyletic groups recognized are not named (X), and the ranks of the terminals are noncoordinate. The dashed line to the Lophocoronoidea indicates that the relationships of this group are uncertain: They may form the sister group of the Exoporia, the Neolepidoptera, or the Myoglossata (Scoble, p. 203).

TABLE V
Sequenced, Tabular Classification of the Lepidoptera, Reflecting Scoble's Cladogram (Fig. 4)

Order Lepidoptera
    Suborder Zeugloptera
    Suborder Aglossata
    Suborder Heterobathmiina
    Suborder Dacnonypha
    Suborder Lophocoronina *incertae sedis*
    Suborder Neopseustina
    Suborder Exoporia
        Infraorder "Mnesarchaeoidea"
        Infraorder "Hepialoidea"
    Suborder Heteroneura
        Infraorder "Incurvarioidea" *sedis mutabilis*
        Infraorder "Nepticuloidea" *sedis mutabilis*
        Infraorder "Palaephatoidea" *sedis mutabilis*
        Infraorder "Tischerioidea" *sedis mutabilis*
        Infraorder Ditrysia *sedis mutabilis*

classifications are made consistent, as far as possible, with current systematization. Named para- and polyphyletic assemblages are rejected whenever strong evidence of nonmonophyly becomes apparent, with such groups being very misleading for comparative biology, biogeography, ecology, and extinction studies (although "dicots," "fish," "reptiles," and "invertebrates" seem impossible to eliminate). Restraint is necessary to avoid unstable changes to the formal classification arising from partial or inconclusive evidence (careful use of consensus trees may be helpful here) and overelaborate nomenclature for the numerous hierarchical levels. With the real prospect of international consensus systems covering all major groups of organisms being developed on the Internet, a balance between stability and universally accessible periodic revision to reflect fundamental advances can be anticipated. Such IT-based bioinformatic systems should also solve most problems of information retrieval, including the ever-present difficulties of alternative classifications, synonymy, and misidentifications (the most pernicious of all classification problems affecting information retrieval).

VII. CURRENT PRACTICE: VARIATIONS ON A CLADISTIC THEME

Numerical taxonomy was afflicted by a proliferation of clustering algorithms, each tending to give different results for which no logical criterion of choice was available. The vaunted objectivity of phenetics dissolved, leaving the field of systematization open to the apparently more consistent and decisive methods of phylogenetic systematics. However, matters are rarely this simple.

In the 1970s, a partial schism opened between those phylogeneticists who followed directly in Hennig's footsteps, approaching character transformation from an evolutionary perspective, and the so-called "transformed cladists" (Patterson, 1982), who held that particular theories of evolution were unnecessary for cladistics, taking instead a "taxic approach" to cladistic analysis (Kitching *et al.*, 1998). The cladists embraced computer algorithms, character matrices, and global parsimony, leaving polarization of characters to the algorithms and outgroup choice. For Hennig, the nodes in a cladogram were hypothetical ancestors; for the transformed cladists the nodes indicate the hierarchical pattern of homologous characters. These differences have emerged in debate over congruence versus "total evidence" methods, now treated formally as partitioned versus simultaneous data analysis. Because many cladists claim to have demonstrated the superiority of simultaneous analysis for making cladograms, this debate links in turn to a seemingly fundamental disagreement over whether or not to employ Fisher's maximum likelihood statistics in phylogeny reconstruction, as first proposed by Edwards and Cavalli-Sforza and later implemented by Joseph Felsenstein as a general method (Huelsenbeck and Crandall, 1997).

The real source of these disagreements seems to be one of different scientific agendas: Those in favor of model-based likelihood methods and partitioned data sets are mainly seeking insights into evolutionary processes, most notably those affecting molecular evolution within different lineages or under different constraints. Those rejecting likelihood models in favor of global parsimony and simultaneous analysis of heterogeneous data sets are primarily concerned with tests of homology and the construction of cladograms (but see Wiens, 1999). Both approaches have different strengths and weaknesses, and both make simplifying assumptions in pursuing these different goals (Kitching *et al.*, 1998, p. 165). Because most work affecting taxonomy requires use of morphological data, or a combination of morphological and molecular data, for which plausible process models cannot be elaborated for simultaneous analysis using likelihood, the significance of likelihood methods for taxonomy per se (rather than understanding phylogenetic processes; Huelsenbeck and Crandall, 1997) remains to be demonstrated.

However, with respect to the reconstruction of phylogeny there are several other methods, global parsimony and maximum likelihood being only main contenders. These alternatives include James Lake's rate-invariant method for molecular data and pairwise distances (e.g., the DNA–DNA hybridization method noted previously). Even within the parsimony techniques generally adopted for cladogram construction, at least one fundamentally different challenge to global parsimony has emerged—the three-item statement analysis, first proposed by Gary Nelson and Norman Platnick. Three-item analysis discards any presupposition that characters undergo evolutionary transformation and, by decomposing a taxon × attributes matrix into an analysis of all possible three-taxon statements supportable by the data, focuses entirely on evidence for sister group relationships rather than character transformations. Although not widely acclaimed, this method is operational and often gives more or less different results from those of conventional parsimony methods, thus raising many fundamental issues. For a brief but informative discussion of the current debate, see Kitching et al. (1998, Chapter 9).

VIII. CONCLUSIONS

Taxonomy involves the search for a general pattern of order among living and extinct organisms, from which a universal reference system (or classification) is derived. As such, taxonomy is the primary discipline of biodiversity. Fundamental philosophical disagreements about the nature of human knowledge have given rise to several basic methods. All recent approaches agree that systematics is empirical and should be based largely (evolutionary systematics and maximum likelihood molecular systematics) or entirely (numerical taxonomy and cladistics) on the comparison of data in the form of characters observed and abstracted from specimens. The methods differ somewhat with respect to the type of data preferred, and fundamentally with regard to methods of analysis used to reveal pattern, and how the system, once obtained, is translated into a written classification.

Cladistics has become the dominant method of taxonomy (where a preferred method is made explicit). The research program of cladistics is based on the view that the ideal general system should reflect the phylogenetic nexus. However, the true nexus cannot be known. If the precise course of evolution, taking account of all the relevant details of every lineage, is real but unknowable, then the natural system cannot be discovered but only approximated by an indirect process of estimation. Because cladistic classification can only admit monophyletic groups (based on the selection of a particular cladogram), all admissible groups are necessarily defined because they must have at least one unique (presumptively) homologous character (or a unique loss within a more inclusive group that is specified by some other homologous feature). Thus, as noted by both Rieppel and Panchen, the cladistic method is in practice essentialist because the search for the sister group depends on discovery of locally unique and thus defining shared characteristics for each and every group. It has long been understood that although evolution is not bound by parsimony, scientific method admits nothing better than parsimony as the criterion of choice among competing hypotheses. Insofar as the divergent model of evolutionary history which justifies searching for a hierarchical pattern is flawed (Doolittle, 1999), the methods of cladistics may need further modification, notably to deal more adequately with hybrid origins.

In conclusion, we may have to accept the seeming paradox that although the principle to which cladistics aspires is natural (i.e., the groups we seek to recognize exist regardless of human perception; Rieppel, 1988, p. 163), its empirical methods are inescapably essentialist. The basic method of taxonomy may be equivalent to the construction of a universal, never to be experienced directly but derived from the sensory experience of particulars (samples of individuals and characters) by means of an unending iterative sequence of analysis, hypothesis formation, testing, and reanalysis. This is a manifestation of the "two ways of seeing" explored by Rieppel (1988): The world of being owes its form to the processes of becoming, but those evolutionary processes can only be inferred indirectly from the patterns we observe in nature.

Finally, it is notable that current approaches to the measurement of biodiversity that seek to maximize the number of expressible genes held in networks of germplasm banks or functional ecosystems depend on models incorporating information about branching sequences (cladistic relationships) and branch lengths (anagenesis). In practice, the taxonomic hierarchy and the raw data matrix (before removal of homoplasy and autapomorphies) from which it has been derived will often be the best or only available data for use in such models. For a review of this field, see Crozier (1997).

See Also the Following Articles

CLADISTICS • CLADOGENESIS • DIFFERENTIATION • EVOLUTION, THEORY OF • GENES, DESCRIPTION

OF • GENETIC DIVERSITY • NOMENCLATURE, SYSTEMS OF • NUCLEIC ACID BIODIVERSITY • PHENOTYPE, A HISTORICAL PERSPECTIVE • PHYLOGENETICS • SPECIATION, THEORIES OF • SPECIES, CONCEPTS OF

Bibliography

Berlin, B. (1992). *Ethnobiological Classification: Principles of Categorization of Plants and Animals in Traditional Societies.* Princeton Univ. Press, Princeton, NJ.

Crozier, R. H. (1997). Preserving the information content of species: Genetic diversity, phylogeny, and conservation worth. *Annu. Rev. Ecol. Syst.* **28**, 243–268.

Doolittle, W. F. (1999). Phylogenetic classification and the universal tree. *Science* **284**, 2124–2128.

Eldredge, N., and Cracraft, J. (1980). *Phylogenetic Patterns and the Evolutionary Process.* Columbia Univ. Press, New York.

Ghiselin, M. T. (1969). *The Triumph of the Darwinian Method.* Univ. of California Press, Berkeley. (1984 reprint by Univ. of Chicago Press, Chicago)

Gilmour, J. S. L. (1940). Taxonomy and philosophy. In *The New Systematics* (J. Huxley, Ed.), pp. 461–474. Oxford Univ. Press, Oxford.

Hennig, W. (1966). *Phylogenetic Systematics.* Univ. of Illinois Press, Urbana.

Huelsenbeck, J. P., and Crandall, K. A. (1997). Phylogeny estimation and hypothesis testing using maximum likelihood. *Annu. Rev. Ecol. Syst.* **28**, 437–466.

Hull, D. L. (1965). The effect of essentialism on taxonomy—Two thousand years of stasis. *Br. J. Philos. Sci.* **15**, 314–326; **16**, 1–18.

Kitching, I. J., Forey, P. L., Humphries, C. J., and Williams, D. M. (1998). *Cladistics,* 2nd ed. Oxford Univ. Press, Oxford.

Margulis, L., and Sagan, D. (1984). Evolutionary origins of sex. *Oxford Surv. Evol. Biol.* **1**, 16–47.

Mayr, E. (1963). *Animal Species and Evolution.* Harvard Univ. Press, Cambridge, MA.

Mayr, E. (1969). *Principles of Systematic Zoology.* McGraw-Hill, New York.

Mayr, E. (1974). Cladistic analysis or cladistic classification? *Z. Zool. Syst. Evol.* **12**, 94–128.

Minelli, A. (1993). *Biological Systematics. The State of the Art.* Chapman & Hall, London.

Nelson, G. (1989). Species and taxa: Systematics and evolution. In *Speciation and Its Consequences* (D. Otte and J. A. Endler, Eds.), pp. 60–81. Sinauer, Sunderland, MA.

Panchen, A. L. (1992). *Classification, Evolution, and the Nature of Biology.* Cambridge Univ. Press, Cambridge, UK.

Patterson, C. (1980). Cladistics. *Biologist* **27**, 234–240.

Patterson, C. (1982, April). Cladistics and classification. *New Scientist* **29**, 303–306.

Pimentel, R. A., and Riggins, R. (1987). The nature of cladistic data. *Cladistics* **3**, 201–209.

Rieppel, O. C. (1988). *Fundamentals of Comparative Biology.* Birkhäuser Verlag, Basel.

Scoble, M. (1992). *The Lepidoptera.* Oxford Univ. Press, Oxford.

Simpson, G. G. (1961). *Principles of Animal Taxonomy.* Columbia Univ. Press, New York.

Sneath, P. H. A., and Sokal, R. R. (1973). *Numerical Taxonomy. The Principles and Practice of Numerical Classification.* Freeman, San Francisco.

Strong, E. E., and Lipscomb, D. (1999). Character coding and inapplicable data. *Cladistics* **15**, 327–362.

Wiens, J. J. (1999). Polymorphism in systematics and comparative biology. *Annu. Rev. Ecol. Syst.* **30**, 327–362.

TEMPERATE FORESTS

John A. Silander, Jr.
University of Connecticut

I. Overview
II. Global Distribution Patterns of Temperate Forest Systems
III. General Characterization of Temperate Forests
IV. Trends in Biodiversity
V. Endemism and Range Size Rarity
VI. Explanations for Patterns of Diversity and Endemism
VII. Ecosystems Services
VIII. Conservation Status
IX. Conservation and Protection Strategies
X. Major Threats to Conservation
XI. Conservation Objectives and Research Needs

GLOSSARY

evapotranspiration The process of transferring moisture from the earth to the atmosphere by evaporation of water and transpiration from plants: actual evapotranspiration as observed at a locality or potential evapotranspiration, given unlimited water availability.
neogene The Miocene and Pliocene Epochs, accorded the status of a period when the Tertiary is considered an era.
physiognomy The outward appearance or morphology of a community as determined by the growth forms of the dominant plants present.
sclerophyll A plant with tough, leathery, evergreen leaves, usually associated with drought resistance.
sere The series of stages in an ecological succession sequence.
temperate region Any locality with at least 1 month of frost (for continental areas) or with one or more months with a mean temperature lower than 18°C (for maritime-influenced areas), and with at least 4 months with a mean temperature higher than 10°C.
tertiary The earlier part of the Cenozoic Era, occurring from about 65 to 2 million years ago.

A COMPELLING CASE CAN BE MADE that the temperate forests are afforded less protection and are more at conservation risk than all other forest systems of the world. They are more altered and reduced in extent globally than any other forest type. Temperate forests currently cover only about 30–35% of their current potential extent versus about 45 and 65% respectively for tropical and boreal forests. Moreover, temperate forests are afforded less conservation protection on average than are tropical forests. Very few if any remaining temperate forests have avoided human impact. Only about 1% of the remaining Northern Hemisphere temperate broad-leafed forests is substantially unaltered and old growth; the vast majority are either managed for wood production, are in plantations, or they reflect the pervasive, long-term effects of human land use practices. Because some of the highest human population densities are found in the temperate forest biome, some of the lowest forest areas per capita globally occur in

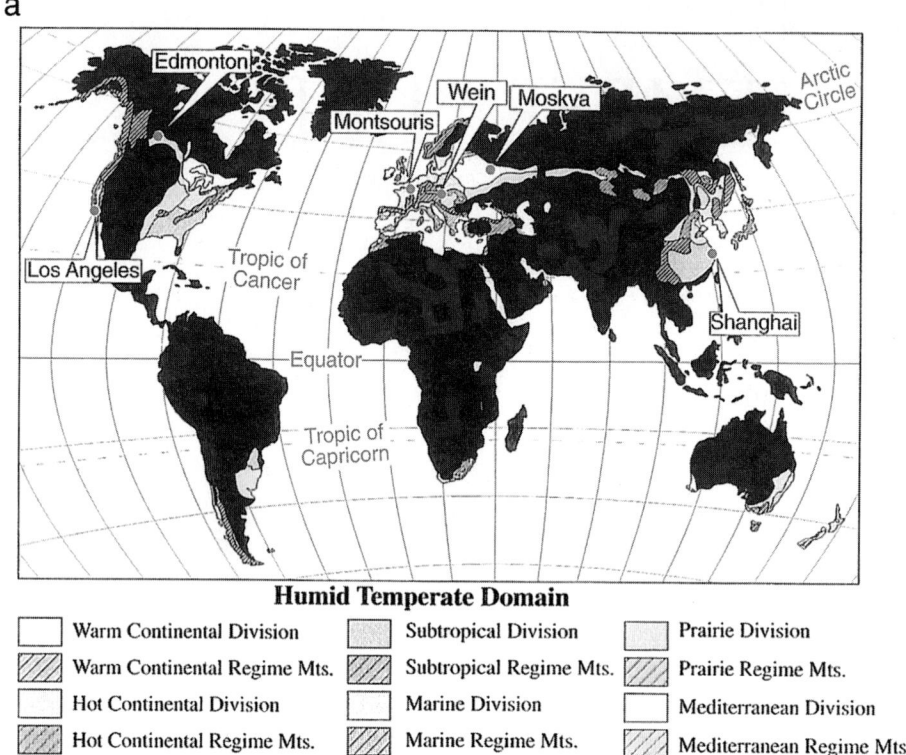

FIGURE 1 (a) Global distribution of the humid temperate ecoregions. These correspond to the potential distribution of temperate forest, grassland, and Mediterranean shrublands globally. (b) Representative climate diagrams for the various divisions of the humid temperate ecoregion domain [Reproduced with permission from R. G. Bailey, *Ecoregions: The Ecosystem Geography of the Oceans and Continents* (Figs. 6.2 and 6.3). © Springer-Verlag.]

this biome. Only the most isolated, inaccessible patches of forest remain unaltered by humans.

I. OVERVIEW

The temperate forests are globally important and unique. They host the largest and oldest organisms in the world. They serve as the world's major source of timber and wood products and are perhaps the only forests with some proven potential for sustainable management. The biomass of at least some temperate forests stands exceeds that of any tropical forest. The temperate forests of the world also provide critical ecosystem services locally and globally. Recent evidence indicates the global importance of carbon sinks in the temperate forest zone, especially in eastern North America. On a landscape level, temperate forests are critical to modulating hydrological, nitrogen, and carbon cycles. Although the biodiversity of temperate forests is typically much lower than that of tropical forests, some temperate forests approach the biodiversity observed at larger spatial scales in the tropics. Temperate forest biodiversity hot spots with high levels of endemism rival in importance those anywhere. They have a unique evolutionary history divergent from either the tropics or the boreal regions. Moreover, Northern and Southern Hemisphere forests are as different from each other as either is from tropical or boreal forests. This contrast reflects striking differences in climate, biogeography, evolutionary history, and the impact of humans.

The objectives here are to provide an overview of the distribution of temperate forests globally; their structure and composition; evolutionary history; diversity, endemism, and rarity; ecosystems services provided; current conservation status; and current threats.

II. GLOBAL DISTRIBUTION PATTERNS OF TEMPERATE FOREST SYSTEMS

A global view of the forests of the world reveals everything from broad continuous expanses of trees to mosaics of small forest patches in the landscape and from

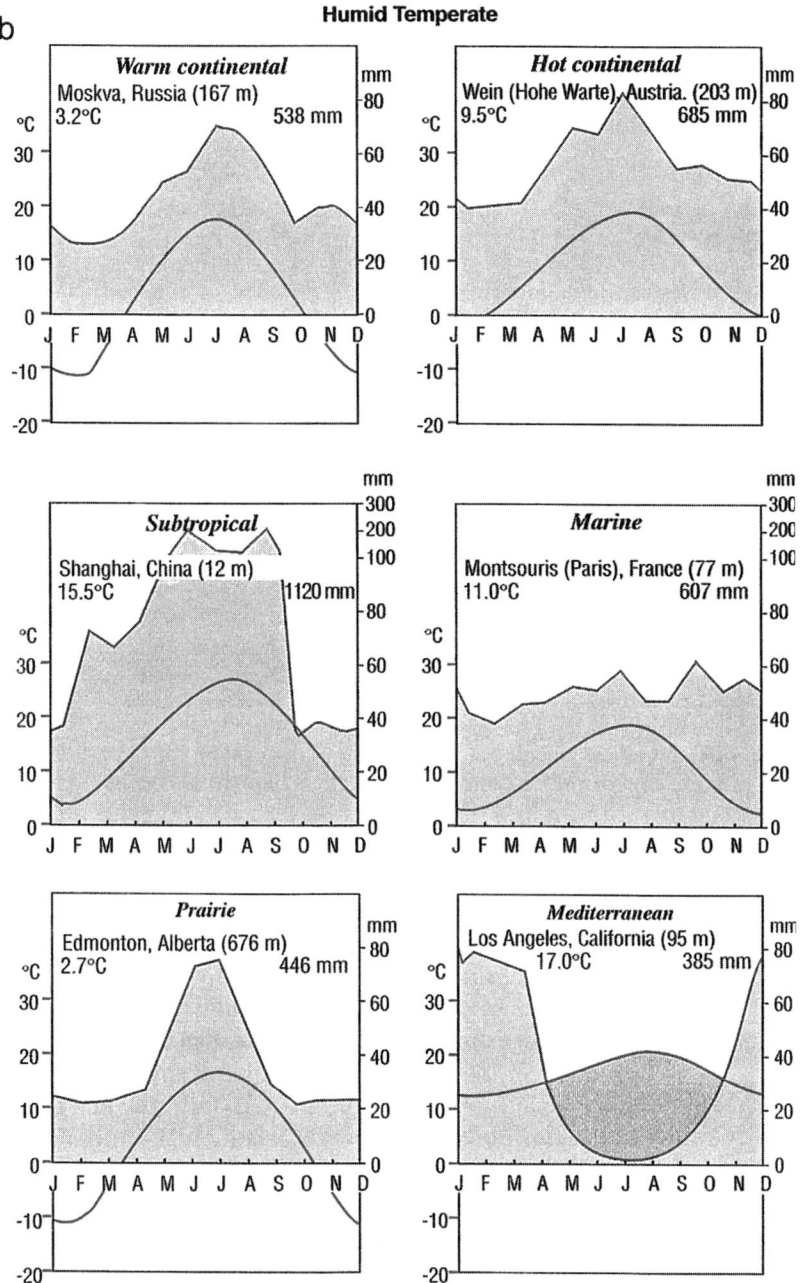

FIGURE 1 (continued)

dense, closed-canopy stands to open wooded parklands. As one progresses from the equator through the humid zones at midlatitude to tree line in the polar regions, changes in forest structure and composition typically occur gradually. Demarcating where the temperate forests begin and end along this continuum is difficult. Indeed, defining "temperate" is difficult. Of the many published maps portraying the extent of the temperate forest biome, few agree on boundaries. The broadest, most arbitrary definition of temperate forests includes all forested areas north or south of the tropics of Cancer and Capricorn, respectively. More common, general macroclimatic factors have been used to define boundaries. In such cases, the results are maps of the potential distribution of forests. Because of the pervasive effects of human and natural disturbances, such maps are considerably more extensive than those depicting extant forests (compare Figs. 1 and 2). Maps of existing forests

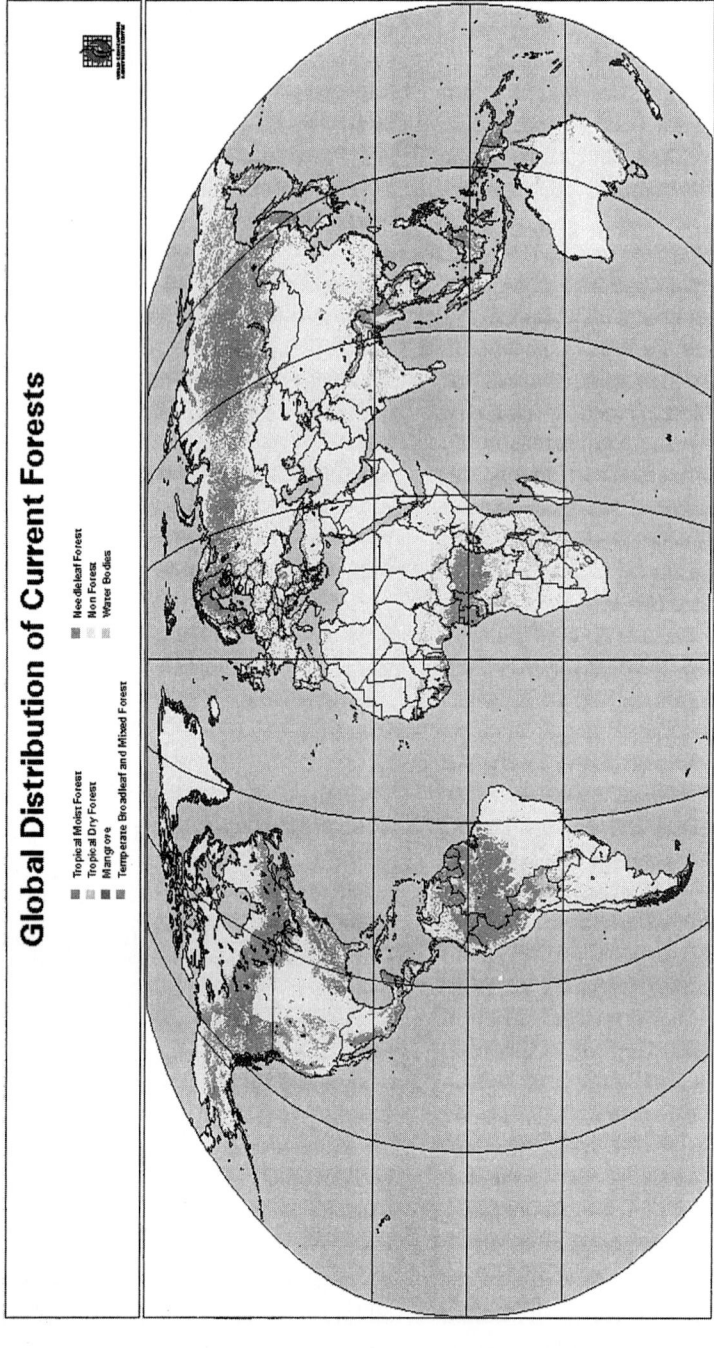

FIGURE 2 The current distribution of forests globally. The data on which this map is based were assembled by the World Conservation Monitoring Centre in collaboration with the World Wide Fund for Nature and published on the World Wide Web (http://www.wcmc.org.uk/forest/data/wfm.htm).

are generally derived from remotely sensed images, but even these vary in what is portrayed as forest. In consequence, Fig. 2 does contain acknowledged, inherent errors in misclassification and omission.

Figure 1a shows where globally temperate forested landscapes may potentially be found, delimited by the humid temperate domain or ecoregion. The boundaries are based on macroclimate, which distinguishes this zone from the polar ecoregion with boreal forests, the humid tropical region with tropical forests, and the dry ecoregions dominated by arid grassland/savanna or desert vegetation. The latitudinal boundaries are set by thermal regime as modified from Köppen and Trewartha, who developed the most commonly used climate classification scheme. These zones are similar to those of Holdridge and Walter but differ in detail. The temperate region defined in Fig. 1 includes any locality with at least 1 month with frost (for continental areas) or with 1 or more months with a mean temperature lower than 18°C (for maritime-influenced areas) and with at least 4 months with a mean temperature higher than 10°C. Moisture availability sets the remaining boundaries, with the humid temperate domain bounded by where precipitation equals or exceeds potential evapotranspiration. Most other maps of temperate forest biomes employ variations on this theme.

Biome or community boundaries are only approximate, reflecting the difficulty inherent in delineating features that are fuzzy spatially and temporally. Boundaries are often either broad ecotones or a mosaic of patches. Moreover, these patterns will change as the climate, biotic composition, or disturbance regimes inevitably change over time. Consequently, any map will be at best an abstract representation of reality.

Within any biome or ecoregion, subdivisions of convenience may be designated. In Fig. 1, the humid temperate domain can be subdivided into subclimatic zones: marine (areas with temperature fluctuation moderated by oceanic influences and with elevated moisture availability, in some cases producing rainforests) and continental (areas with comparatively greater temperature fluctuation and greater probability of drought). Temperature zones are delineated as well: subtropical (defined as having 8 or more months with mean temperatures higher than 10°C), continental hot temperate (4–7 months with temperatures higher than 10°C, warmest month higher than 22°C, and coldest month lower than 0°C), and continental warm temperate (as for hot temperate, but with warmest month lower than 22°C).

The excess of annual precipitation over evapotranspiration becomes less as one moves away from oceanic influences on continents at midlatitudes. Thus, temperate forests tend to be replaced by grasslands in central North America, central and eastern Europe, central eastern Asia, and eastern South America. This may be augmented by fire and grazers or browsers. The boundary between forest and grassland may be a broad transitional, open wooded parkland with scattered trees, or a broad mosaic of forest and grassland patches with forests restricted to favorable soils, sites with more moisture, or sites protected from fire.

Forests may also be replaced as aridity increases in Mediterranean climates with winter rainfall–summer drought regimes. As fire becomes a pervasive element in the landscape, forests tend to be replaced by sclerophyllous shrublands, thickets, or sometimes open woodlands. Again, the transition may be abrupt or a gradual mosaic, with forest patches restricted to favorable soils and/or moist sites protected from fires.

The boundaries between the temperate forested regions and the adjacent boreal or tropical regions are often imprecise. In both cases, there can be a very broad transition zone with considerable overlap in species composition. The subtropical–tropical boundary set by the 18°C mean monthly isocline is quite arbitrary. Some have attempted to show that this boundary approximately corresponds to the natural poleward limits of the distribution of palms.

Boreal forests are distinguished only for the Northern Hemisphere as typically conifer-dominated forests within specified climate regimes (i.e., monthly mean temperatures all lower than 22°C but 1–3 months with means higher than 10°C). This boundary approximately corresponds with the mean position of the summer polar front in the Northern Hemisphere.

Confusion remains regarding the classification of conifer-dominated forests. The boreal forests of the Northern Hemisphere and the conifer forests of the North American Pacific Northwest, the Asian Pacific northeast, and the Southern Hemisphere are treated differently by different authors. Using the previously mentioned temperature criteria, all of the Southern Hemisphere forests, except perhaps the southern tip of South America, would be classified as temperate, as would most of the Pacific forests of Asia and North America. The climatic definition for the temperate zone forests will be used here.

There are parallel inconsistencies in classifying certain subtropical forest regions. For example, northern Indian forests, which are climatically subtropical, are usually classed as tropical. The same is true for forests with some temperate affinities in subtropical regions of southern Brazil.

Dealing with mountainous areas is problematic since they typically contain multiple ecoregions (tropical or temperate to boreal and alpine depending on elevation and latitude). In the Americas, Africa, eastern Asia, and Australasia, one can find discontinuous bands of forest with temperate affinities extending from sea level in the temperate zone well into the tropics at higher elevations. The most common resolution is to classify all forest elements as part of unclassified mountainous regions, as part of the surrounding domain, or as separately classified biome units within the domain. One consequence of this inconsistent classification is to obfuscate biotic patterns in some of the most important global hot spots of biodiversity.

The temperate forest biomes, thus defined, comprise about 14,600,000 ha (estimates vary). By far the largest actual or potentially forested landscapes occur in the Northern Hemisphere (80% or more) (cf. Fig. 1 and Table II). The regional biomes include (i) eastern North America from the Atlantic coast west to about 95° latitude and from about 45° latitude south to about 28°; (ii) western North America from about 35° north to about 60° (and mainly from the Sierra–Cascade ranges, west); (iii) western and central Europe from the Atlantic coast north to about 60° and east through eastern Europe, but excluding the Mediterranean coastal zone and much of Spain, and then extending in a narrow strip around 55° east across Russia to west central Asia; (iv) a small, discontinuous temperate forest zone in the Middle East, especially along the south coast of the Black Sea, to the southern Caucasus and to the southern Caspian Sea; (v) eastern Asia from about 50° south to about 25° in southern China and from Japan and the Pacific coast northwest to about 120° and southwest to about 100°; and (vi) northern south Asia (India) and adjacent areas.

In the Southern Hemisphere the extent of temperate forests is much more restricted in extent: (i) eastern, coastal Australia from about 25° south to Tasmania, plus the southern tip of Western Australia; (ii) most of New Zealand; (iii) southern Chile and adjacent Argentina from about 40° south to about 55°; (iv) a small area of southern Brazil just below the tropic of Capricorn, plus adjacent Paraguay and Argentina; and (v) small patches of coastal and interior forest in south and southeastern South Africa.

Table I summarizes the potential and current extent of forested areas within each of the 11 temperate forest biomes, from World Conservation Monitoring Centre (WCMC, see Fig. 2), World Wide Fund for Nature (WWF), and the Food and Agriculture Organization data based on approximately 25 forest types. The more

TABLE I
Temperate Forest Cover by Regions

Forested region	Potential extent of temperate (mesic) forest cover (km² × 10⁶)	Maximum potential forest extent (including dry forests, woodlands, and thickets) (km² × 10⁶)	Current estimated extent of temperate forest cover (km² × 10⁶)	Conservation areas (IUCN classes I–VI) (km² × 10³)
Europe (including Mediterranean)	3.30	3.91	0.85–1.1	44.0–54.0
Russia	1.12	1.13	0.26–0.36	11.0–36.0
East Asia	3.21	3.79	0.62–0.75	24.0–26.0
North America	4.26	4.72	2.13–2.17	117.0–132.0
Eastern north	3.56	3.72	1.63–1.57	—
Western (Pacific)	0.7	1.0	0.5–0.6	—
Middle East	0.36	0.61	0.05–0.11	1.0–4.0
South Asia	0.87	1.34	0.20–0.31	20.0–34.0
South America	0.7	1.8	0.4–0.52	45.0–69.0
Southern (Chile and Argentina)	0.60	0.8	0.25–0.30	51.0–54.0
Southeastern (Southern Brazil and adjacent countries)	0.14	1.0	0.12–0.22	12.0–15.0
Australia	0.45	1.64	0.03–0.66	4.0–44.0
New Zealand	0.23	0.25	0.04–0.08	17.0–18.0
Southern Africa	0.1	0.4	0.01–0.1	4.0–6.0
Total	14.6	19.6	4.7–6.2	287.0–423.0

TABLE II
Worldwide Forest Cover

Forested biome	Potential extent of mesic forest cover (km² × 10⁶)	Maximum potential extent (including dry forests, woodlands, and thickets) (km² × 10⁶)	Current estimated extent of forest cover (km² × 10⁶)
Boreal forests	12.2	18.5	8.0–11.5
Tropical forests	29.0	40.0	14.2–15.6
Temperate forests	14.6	19.6	4.6–6.8
Total	55.8	78.1	26.8–33.9

conservative estimate of potential forest (Table I, column 2) is for mesic forest coverage only. Maximum potential extent (column 3) includes all types of dry forest, woodlands, and thickets but not savannas or other sparsely treed landscapes (i.e., forest cover <30%). The corresponding ranges under current forest extent are shown in the fourth column. These data are only approximate. The base data were obtained from different sources, with different categories or forest type classes reported for different regions. In some cases, conflicting information on forest cover is reported from different sources.

Almost 80–90% of the temperate forest biome types are found in the Northern Hemisphere. For current forest cover, the data are only slightly less for northern dominance. The major northern forest biomes of Europe, eastern North America, and east Asia all cover about the same potential area. However, eastern North America has by far the largest cover remaining (40–45%). The Middle East has the smallest percentage of potential forest cover (13–18%), followed by east Asia (<20%) and Europe (25–30%); western North America has the highest percentage of remaining forests (60–75+%). Estimates for western North America are only approximate (estimated from WCMC forest type information and other data), with the eastern and southern boundaries for this system fuzzy and disjunct. Similarly, the estimates for forest cover in south Asia are only approximate. For some regions these data can be deceptive. Much of the forest covering Europe, especially in western Europe, is intensively managed plantation or seminatural forest. The same is true for Japan, even though it retains more than one-third of its potential forest cover. China retains almost 12% of its original forest, but most of this is in the boreal to northernmost, mixed conifer forests.

Southern temperate forests are fragmented into six or more regional biomes, all much smaller in extent than their northern counterparts. The extreme ranges in forest cover tabulated for some of the southern biomes reflect the large contribution of dry forest systems within the regions. The largest of the southern biomes comprises the forests of southern Chile and adjacent Argentina. The smallest of all temperate forest biomes is that found in South Africa, perhaps rivaled in size only by the forests of Western Australia. Most significantly, the South African forests are the richest of all temperate forests in tree families, genera, and possibly species (certainly if the complete tree flora is included).

Compared to other forest system globally, the temperate forest biomes covers a slightly greater area than boreal forests. Boreal and temperate forests together approximate the tropical forests of the world (Table II).

III. GENERAL CHARACTERIZATION OF TEMPERATE FORESTS

A. Physiognomic Features

Each of the previously discussed 11 temperate forest regions may be characterized by general physiognomic features (i.e., the outward appearance and structure of the dominant vegetation):

1. Eastern North America: dominated by broadleaf deciduous forests, mixed with conifers to the north, locally dominated by conifers under drier, successional conditions or in fire-prone areas in the southeast and northwest, plus small patches of broadleaf evergreen forests in the south. Closed forest systems predominate.

2. Western North America: dominated by evergreen conifers with broadleaf trees contributing little to the forests; some northwestern coastal areas support rainforests (i.e., rainfall in excess of 2000–3000 mm per annum). The eastern edge of this biome is discontinuous and grades to open conifer woodlands and montane

boreal forests. To the south it grades to a mosaic of conifer or broad-leafed evergreen forests and woodlands or shrublands.

3. Europe: mainly broadleaf deciduous forests, mixed with conifers to the north and in mountainous areas, and to the south grading to a mosaic of broadleaf evergreen forests (many sclerophyllous leafed), conifer-dominated forests, and shrublands under Mediterranean climate influence.

4. Middle East: mainly broadleaf deciduous forests with a mosaic of broadleaf evergreen, sclerophyllous, and conifer forests and woodlands. Rainforests occur very locally near the southeastern Black Sea coast.

5. Eastern Asia: mainly broadleaf deciduous forests mixed with conifers to the north and broadleaf evergreen forests to the south; mostly mesic closed forests. Evergreen broad-leafed rainforests occur locally in southeastern Japan, one of the rarest forest types found in the Northern Hemisphere.

6. South Asia: dominated by broad-leafed evergreen to semievergreen monsoonal forests; temperate mixed forests (locally rainforests) occur in the foothills of the Himalayas.

7. Australia: dominated by broadleaf evergreen forests (mostly sclerophyllous); small patches of closed forest (<70% cover) and extensive areas of open forests grading to woodlands. Very restricted patches of rainforest are found in southeastern Australia and western Tasmania.

8. New Zealand: dominated by conifers and mixed with broadleaf evergreen forest patches, especially on the north island and locally on the south island. The western forests in New Zealand are rainforests.

9. Southern South America: dominated by broadleaf evergreen forests with some conifers. Broadleaf deciduous forests prevail in the southernmost areas and at higher elevation in the Andes; rainforests are restricted to the Pacific slope south of about 40° latitude.

10. Southern Brazil: characterized by the presence of southern conifer forests; elsewhere in this zone the forests are dominated by evergreen or semideciduous angiosperms with patches of open forest and thicket to the west.

11. South Africa: mixed broadleaf evergreen (many sclerophyllous leafed) with some conifers forming a very patchy mosaic in the landscape with thickets, shrublands, and savannas.

B. Dominant Floristic Features

The community dominance and floristic affinities for these regions as they exist today can be characterized very broadly. Detailed descriptions and characterizations may be found elsewhere. There are broad floristic affinities among the forested biomes in the Northern Hemisphere with many shared families and genera both now and in the fossil record. These include older lineages from the Tertiary flora of Asiamerica and recent lineages that evolved under cooler and/or drier climates. The Southern Hemisphere forests are very different floristically, with few important families or genera shared with the north. There are some Gondwanan lineages in common now or in the fossil record. However, the floras today reflect considerable divergence with many tropical affinities and many fewer common links than are seen in the north.

In eastern North America, in the northeast birch, maple, beech, and hemlock (*Betula, Acer, Fagus,* and *Tsuga*) dominate the landscape, with the latter two tending to form monodominant stands late in succession. Farther south and west the forests tend to be dominated by oaks (*Quercus* spp.) or hickories (*Carya* spp.). The central sections, especially in the central to southern Appalachians, tend to be the richest, with a diminished tendency toward dominance by one or a few canopy species, although the highest regional tree diversity is found in the southeast. In the extreme south, small patches of evergreen forest are found in protected areas dominated by evergreen oaks and *Magnolia*. Much of the southeast is now dominated by pine (*Pinus taeda*) plantations. Fire successional pines also dominate parts of the northwest. Drier sites throughout tend to be dominated by oaks or conifers, especially pines; wetter sites are dominated by conifers (e.g., *Tsuga* to the north and *Taxodium* to the south) or by locally adapted broad-leafed deciduous species (e.g., *Ulmus, Nyssa,* and *Acer*).

Western North American forests are dominated by a small number of large, long-lived conifer species; deciduous angiosperms are only minor components. The Pacific Coast rainforests are dominated by hemlocks (*Tsuga*), firs (*Abies*), spruce (*Picea*), and/or cedar (*Thuja*) from Alaska south to Washington State. Douglas fir (*Pseudotsuga*) becomes important in the central coast. Timber industries in these zones tend to actively manage the landscape for monospecific stands of native species (e.g., Douglas fir). Fragmented stands of redwoods (*Sequoia*), which globally are the tallest trees, occur more southerly extending to central California. On the drier eastern slopes of the coastal mountains (central sections), pines, Douglas fir, and sometimes poplars (*Populus*) dominate the landscape. Further south in the Sierras one finds local dominance by *Sequoiadendron*, the most massive tree globally, and *Pinus aristata*, the longest lived tree. Drier, lower elevation landscapes in the south (California) may be dominated by broad-leafed evergreen forests of tanoaks (*Lithocarpus*) and madrone (*Arbutus*) or open oak, pine, or mixed woodlands, which grade into Mediterranean shrublands or grasslands.

European forests are highly disturbed and fragmented following centuries of human habitation. Central European forests tend to be dominated by beech (*Fagus*) on many intermediate sites and by various oaks (*Quercus*) on drier and slightly wetter sites or very acidic sites. The wettest sites tend to be dominated by birches (*Betula*). Many of these forests have been converted to *Picea* or *Pinus* plantations. For example, German forests have gone from 90% deciduous broadleaf domination to 80% *Picea* plantations. To the south there is a transition to dominance by evergreen oaks and pines under Mediterranean influence. To the north and throughout much of the central uplands, conifers (*Pinus, Picea, Abies,* and *Larix*) can locally dominate the landscape or there are mixed forests with beech and birch. Many of the European forests not in plantation are still actively managed for timber products.

The Middle Eastern forests of northern Turkey to the Caucasus and the southern Caspian Sea are probably the poorest known and least studied of temperate forests. In the western Mediterranean-influenced zone, one finds sclerophyllous forests and woodlands dominated by pines and oaks. Along the coast of the Black and Caspian Seas and the southern Caucasus (Colchian and Hyrcanian regions) one finds highly diverse, mesic deciduous forests with oaks, maples, beech, chestnut, and many other species. At higher elevations the forests are dominated by beech with conifer-dominated or mixed stands above. Drier sites in the landscape are dominated by open oak forests or woodlands. Many of these forests have experienced a long history of human occupation and associated agriculture with overgrazing.

East Asia has the most diverse forests in the Northern Hemisphere. Along the Pacific coast in northern Japan, southern Russian, and China, one finds forests similar to those of the North American Pacific Northwest, with conifer dominance but also mixed with broad-leafed species of maples, birches, limes (*Tilia*), and elms. Central China has been extensively cultivated for centuries, and there is very little forest cover left. Remnant tracts here indicate a rich diversity of mixed deciduous trees mentioned previously plus many other genera, including oaks, elms, poplars, ash (*Fraxinus*), and rowan (*Sorbus*), with a rich understory. Locally, the understory may be dominated by bamboos. The same pattern is seen in Korea and Japan. In southern Japan, much of the forest is managed for native *Cryptomeria*, and in the north it is managed for *Abies* or *Picea*. South of the Yangtze River in China and in eastern and southern Japan the broad-leafed evergreen species increase in dominance as one approaches subtropics. These are the most diverse temperate forests in the Northern Hemisphere, with many of the same genera listed previously. However, these have become highly fragmented through human disturbance. In southern China there is a shift to forests with strong tropical affinities.

The temperate forests of south Asia are difficult to categorize. Most of the lowland and premontane remnant forests of northern India are climatically subtropical but the flora has strong tropical affinities. There is local strong dominance by Sal (*Shorea robusta*) and bamboos. In the northern hill forests there is a transition to strong temperate affinities with high species diversity. Oaks mixed with Lauraceae dominate the forest, but maples, *Castanopsis*, and *Magnolia* occur. In the montane zone diverse oak forests are mixed with conifer (*Abies, Picea,* and *Pinus*) forests and patches of *Rhododendron*. As is the case in much of Asia, these forests have been long affected by human disturbance. Throughout the lower elevations small stands of *Eucalyptus*, teak, pine, or *Populus* plantations are common.

The subtropical forests of the eastern mountains and coastal areas of Australia and the small areas of forest in Western Australia are dominated by the numerous sclerophyllous *Eucalyptus* species. Locally, stands tend to be dominated by one or only a few species. The eucalypt forests tend to form open-canopy stands grading to woodlands. The tallest angiosperm trees are found here. Only on Tasmania and in scattered pockets along the eastern mountain chain is there local dominance by closed-canopy or temperate rainforest species, including the southern beech (*Nothofagus*) and various southern conifers (*Dacrydium, Phyllocladus, Arthrotaxus,* or *Araucaria*). Throughout much of southeastern Australia plantations of Monterey pine (*P. radiata*) have become a pervasive component of the landscape.

The forests of New Zealand tend to be either multi-storied mixed conifer–broadleaf with species composition varying across the landscape or low-diversity southern beech (*Nothofagus*) forests. The mixed forests dominating in the lowlands have a scattered overstory of *Agathis* in the north or various podocarps (e.g., *Podocarpus, Dacrycarpus,* and *Phyllocladus*), with a sub-canopy of Lauraceae (*Beilschmeidia*), Myrtaceae (*Metrosideros*), Cunoniaceae (*Weinmannia*), and many other families and genera. Evergreen *Nothofagus* forests may form pure dense canopies in subalpine areas and may be a component in the lowland forests along with other broad-leafed species.

The forests of southern South America are confined to Chile and adjacent areas of Argentina. They vary from small remnants of sclerophyllous forests and woodlands in the Mediterranean zone to the species-rich Valdivian rainforests, the species-poor but still extensive north Patagonian and Magellanic forests, and depauperate deciduous *Nothofagus* forests at higher elevations and the interior south. The sclerophyll forests were dominated by *Acacia caven* and other species, with deciduous *Nothofagus* forests at high elevation. The Valdivian forests may be either broadleaf dominate, with *Nothofagus, Eucryphia* (Eucryphiaceae), *Laurelia* (Monimaceae), *Weinmannia*, and other species, or mixed with conifers [*Podocarpus, Araucaria*, or *Fitzroya* (Cupressaceae)]. The north Patagonian/Magellanic forests are dominated by evergreen *Nothofagus* mixed with *Podocarpus, Weinmannia*, and *Drimys* (Winteraceae). To the south, one of the deciduous *Nothofagus* species (*N. pumilo*) forms a pure stand or is mixed with *N. betuloides* at timberline. Many of the remaining Chilean forests are being clear-cut for chips and converted to *P. radiata* or *Eucalyptus* plantations.

The southeastern forests of Brazil have many south temperate affinities. These forests are characterized by the presence of *Araucaria*. However, other south temperate components include *Podocarpus, Weinmannia*, and *Drimys*, plus Sapindaceae, Proteaceae, and Myrtaceae. These forests have been largely cleared for agriculture. The subtropical forests to the west in adjacent Paraguay and Argentina have more tropical affinities.

The forests of South Africa are some of the richest in tree species of any in the temperate zone. However, this is also the smallest of all temperate forest biomes, and it is highly fragmented into many small forest patches. It is not clear how forest cover has changed during the Pleistocene. These forests do have south temperate affinities, with the presence of *Podocarpus*, Cunoniaceae, and Proteaceae, but the majority of the temperate forest flora have tropical affinities. Despite these tropical affinities, the level of tree endemism is high for a continental area contiguous with tropical forests. The "Afromontane" forest elements extend from southernmost South Africa at sea level to the mountains through northeastern Africa. The coastal, Indian Ocean (Maputaland/Pondoland) forests are quite different floristically, with high diversity and many local endemics. The vast majority of the forested landscape is now in *P. radiata* or *Eucalyptus* plantation.

IV. TRENDS IN BIODIVERSITY

Comparing trends in biodiversity within and among regions is fraught with difficulties. The results can differ depending on the scale of the sample unit compared (i.e., 0.001 vs 1, 1000, or 100,000,000 ha). For many regions of the world, data are available only for a limited scale range. For example, in east Asia there are very few accessible data records for small plots (0.1–100 ha). In other regions (e.g., the Middle East), species diversity numbers are either estimates for large areas or entirely lacking. For many records the number of tree species may be accurately reported but the number of herbaceous, especially ephemeral, plants may be significantly undercounted or not reported at all. Even simply listing tree species numbers can be misleading because authors vary widely in delimiting the threshold size for what constitutes a tree. Likewise, authors vary in classifying vegetation type with which tree taxa may be associated (closed forest to sparsely treed parkland). Nevertheless, some trends appear robust.

Table III summarizes regional tree taxon richness from a variety of different sources, and Fig. 3 plots tree richness tallies against area for 75 forest sample sites throughout the temperate zone (spanning 10^{-2} to 10^8 ha). Despite this large range in areas and the inherent variation in estimates and counts, there are significant differences in tree richness among biomes and between hemispheres. For these data, greater tree species richness occurs in the Southern Hemisphere across the full range of areas surveyed (Fig. 3). This is also reflected

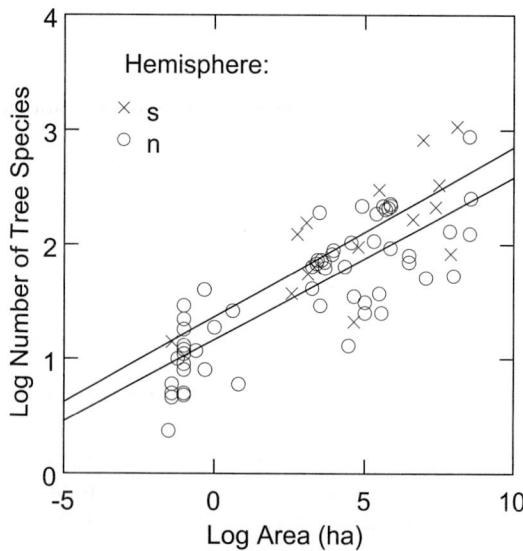

FIGURE 3 The relationship between tree species numbers and area sampled is shown, log transformed for both variables. The samples are from a wide range of different sources for most of the temperate zone forest biomes. These are simply classified here as Northern (n) and Southern (s) Hemisphere localities with best fit linear regression lines. There were significant differences between hemispheres ($p = 0.024, n = 75, r^2 = 0.695$) and among regions ($p < 0.001, r^2 = 0.916$).

TABLE III
Tree Species Diversity Patterns across the Temperate Forested Region

Region	Families	Genera	Species	Genus:family	Species:genus	Forest biome maximum extent (km² × 10³)
Europe	21	43	124	2.0	2.9	3300–3910
East Asia	67	177	876	2.6	4.9	3210–3790
Eastern North America	46	90	253	2.0	2.8	3560–3720
Western (Pacific) North America	24	47	131	2.0	2.8	700–1000
Chile	29	40	83	1.4	2.1	330–370
Southern Brazil	25	45	77	1.8	1.7	<100
Southeast Australia	37	78	331	2.1	4.2	300–700
New Zealand	47	74	212	1.6	2.9	230–250
Southeast South Africa	88	280	598	3.2	2.1	20–50

in the absolute number of families and genera. For species, the totals are similar, but the tabulated survey covers only a small part of the diverse tree flora for Australia and South Africa, respectively, and only part of southeastern Australia and Tasmania and Kwazulu-Natal province and adjacent Transkei. If all the tree species are included for South Africa alone (well over 1000 species in 370 genera and 97 families), the species numbers would be higher in the Southern Hemisphere despite the fact that southern forests only cover 10–20% of the area of northern forests. On an area basis, the taxon richness of the southern forests is at least an order of magnitude greater than that of the north.

This high southern diversity is contributed largely by the flora of South Africa, arguably one of the richest per unit area of any biome globally. Australia and New Zealand also contribute to this southern richness, each having a taxon richness per unit area of 4–10 or even 100 times that of northern forested regions. Just the rainforests of New South Wales and Victoria, covering less than 200,000 ha, have more than 250 tree species (not all included in Table III). Even the Chilean forests, perhaps the most depauperate in tree species of any temperate forest biome, have more taxa per unit area than any northern biome (except for total species in east Asia).

Among northern temperate forest biomes, east Asia has by far the richest tree flora. Europe and western North America are the most depauperate, with Europe having the lowest tree taxon diversity per unit area globally.

Figure 4 shows analogous species area data for all vascular plants tallied for about 200 plots or regions (spanning 10^{-6} to 10^8 ha) across all temperate forest biomes. Here, the hemisphere trends are reversed. The north has a slight but significantly higher total vascular flora than the south across the range of sample areas. East Asia and the Middle East (the latter, a smaller sample size) tend to have the highest vascular plant diversity, and New Zealand, South America, and western North American forests tend to have the lowest vascular plant diversities. Forested systems in eastern North America, Europe, Australia, and South Africa tend to be intermediate across areas sampled. There is a dramatic decrease in species diversity toward the pole, more so in the Southern Hemisphere. In pairwise comparisons at intermediate sample areas the trends tend to hold. The vascular plant flora for east Asia at 0.5–10

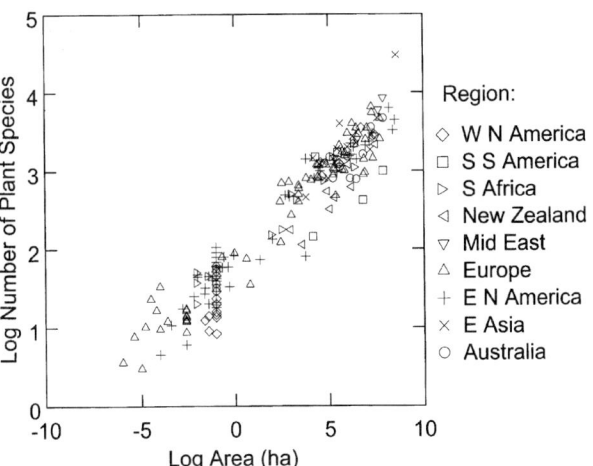

FIGURE 4 The relationship between total vascular plant diversity and area samples is shown, log transformed for both variables. The samples are from a variety of different sources covering most of the temperate zone forest biomes. The samples are classified by the biome region from which they were obtained. Significant differences are found between hemispheres ($p = 0.005$, $r^2 = 0.924$, $n = 198$) and among regions ($p < 0.001$, $r^2 = 0.938$).

ha is significantly richer than that of Europe or eastern North America. However, few differences can be seen among regions at the smallest plot sizes (<0.01 ha).

Trends in alpha, beta, and gamma diversity vary considerably within and among regions. Some of the large- to intermediate-scale patterns in species richness are undoubtedly related to spatial landscape heterogeneity. Regionally, high levels of species richness are associated with mountains [e.g., the Smoky Mountains (east North America), the Pyrenees, Alps, or Balkan Mountains (Europe), and the Sichuan Mountains or Mount Halla (east Asia)]. Complex mosaics of vegetation types [e.g., southeast Australia or Pondoland (South Africa)] or multiple successional states (e.g., Indiana Dunes National Seashore) all tend to support significantly higher diversity than adjoining areas. Even for the United Kingdom, with an exceptionally small flora overall, regional or county floras that encompass a diversity of habitats and seral stages may have as many species as are found in many similar-sized temperate regions elsewhere. Within Europe there many more tree species in the southern, Mediterranean-influenced region with complex spatial heterogeneity (almost 100 versus 12 common and 25 uncommon species in northern and central Europe). Perhaps the high tree and overall species diversity in South Africa is related to the extreme spatial heterogeneous and natural fragmentation of the vegetation (forest and otherwise). Locally or regionally, lower species diversity tends to be found under closed conifer forests and late successional stands dominated by trees that cast deep shade. Often, the highest diversity will be found in intermediate successional stage forests or in a landscape with a mosaic of seral stages. Locally, higher diversity will tend to be found on richer, fine-textured soils with circum neutral pH with higher cation exchange capacity and more humus. This contrasts with diversity trends elsewhere. For example, higher diversity tends to be found on poorer soils in Mediterranean shrubland biomes.

Compared to tropical forests, the species richness (of trees or all vascular plants) in the temperate zone will usually be smaller than that of comparable areas in the tropics, especially at the 0.05- to 10-ha scale. Most temperate forest stands tend to be dominated by one or a few species, with the other tree species being uncommon or rare. In the majority of tropical forest stands local dominance by one or a few species is rare; greater evenness is common. Perhaps only on a regional basis may floristic diversity of temperate forests (e.g., southern China and South Africa) begin to approach those observed in the tropics.

An alternative measure of global richness patterns may be obtained by focusing on clades or higher taxonomic richness. The data from Table III show more genera and families in the Southern Hemisphere forests sampled. Also, the ratios of species to genus and genus to family tend to be lower in the south. These trends indicate a potentially richer phylogenetic diversity (using higher taxa in lieu of cladistic information) in the south. A complementary, global perspective is provided by the British Natural History Museum (NHM) global mapping of plant family richness (395 in total) on 10° latitude/longitude grids. The trends reveal higher family diversity on average in the tropics, with the richest cell in Southeast Asia, and a dramatic decline poleward. However, family richness is as high in southern China as it is anywhere else in the world (other than Southeast Asia), and about as high in the central southern United States as in most of South America (including Amazonia) and southern Africa versus tropical Africa. Centers of plant family richness in the temperate zone at the 10° grid cell included southern east Asia, southern North America, and southern Africa, with lower diversity in Europe and temperate Australia and South America.

V. ENDEMISM AND RANGE SIZE RARITY

Endemism (species restricted to specific localities) and taxon rarity have received a similar amount of attention as has biodiversity. How do levels of plant endemism compare among temperate forested regions and with other biomes? Statistics on plant endemism tend to be available for only a limited range of spatial scales (often countywide, occasionally for states or provinces, and infrequently for localities) and are often incomplete or lacking for poorly known regions such as the tropics. Nevertheless, there are some trends in the data that are available. Boreal and cold temperate forested regions in the Northern Hemisphere tend to have very low levels of endemism based either on absolute numbers or the percentage expected on average per unit area [e.g., northern and western European countries (0–2%) and the eastern United States (0 to <1%)]. Regions that include warm temperate to subtropical regions with topographically heterogeneous landscapes (and therefore heterogeneous climates and vegetation types), and especially with isolated mountain ranges, tend to have higher than expected endemism per unit area [e.g., Bulgaria (9%), Turkey (31%), western North America, and countrywide for China (56%)]. Larger islands and peninsulas tend to have higher than expected species diversity [e.g., Korea (14%), Florida (12%), Japan (37%), and New Zealand (82%)]. The Southern Hemisphere forested regions tend to have some of the highest levels of floristic endemism globally (50–80% for Australia, New Zealand, Chile, and South Africa). In part this reflects their geographic isolation. The previous data are for species endemism for the entire flora. Tree species endemism is more difficult to determine and likely to be much lower. For example, in South Africa about 25% of the more than 1000 tree species are confined approximately to the borders (and enclaves) of the country and about 40% to southern Africa (versus 70–80% for the entire flora). Regional tree endemism is much lower (e.g., about 8% for Pondoland and 15% for Maputaland forest regions). However, very high levels of tree species continental endemism are found in southern South America: 85% of woody species and 34% of genera are endemic.

Compared to tropical forests, the absolute numbers of endemics are generally lower in the temperate zone, except perhaps for some of the Southern Hemisphere temperate zone. This reflects the larger floras in the tropics, which are still poorly known for many localities. However, on the basis of expected percentage endemism per unit area the trends are less clear. For example, although Venezuela (38%), Panama (13%), and the Congo (29%) have higher than expected levels of plant endemism, Columbia (4%), Nicaragua (<1%), and Nigeria (4%) have lower than expected endemism. On a subregion or local basis the tropics may well have substantially higher levels of endemism than equivalent localities in the temperate zone, but this pattern remains to be shown conclusively.

Species endemism is only one way to assess rarity or the geographical restrictions of taxa for conservation purposes. Range size rarity of taxa or clades evaluated in the absence of political or other arbitrary boundaries may be a more consistent and robust means of judging rarity. A global survey by the British NHM of range size rarity was done on a 10° grid for selected plant taxa. This showed that there were no clear trends between warm temperate and tropical regions in range size rarity. Many, perhaps the majority of the prime hot spots, are in temperate grids that include forested regions (e.g., southern South Africa, central China, southeastern Australia, and central to south-central Chile, with secondary centers in southern Chile, northern New Zealand, elsewhere in China, and the southeastern United States). Clade rarity and endemism need evaluation as an alternative to enumerating species endemism or range size rarity.

VI. EXPLANATIONS FOR PATTERNS OF DIVERSITY AND ENDEMISM

A variety of hypotheses have been put forward to explain local and global patterns of species distributions and therefore richness. One author lists as many as 120 named hypotheses for variation in species richness. Geological and biogeographic history provides one set of important related explanations for observed trends in richness and endemism. This may well explain some of the patterns within and between hemispheres.

A. Historical and Geological Explanations

In the Triassic period global plate tectonic activity had united major landmasses of the world to form a single supercontinent (Fig. 5) which created the potential for a common biota. However, by the beginning of the Cretaceous this landmass began to break up, forming northern and southern landmasses. This, together with the formation of a broad sea, effectively isolated the northern Asiamerica continent from the southern Gondwanaland. The angiosperm and gymnosperm floras on these separate landmasses thus began to evolve independently. This history is reflected in both the Tertiary floras of the fossil record and the floras we see today. Until approximately 40 million years into the Tertiary the climate was warm, moist, and relatively stable. In the Northern Hemisphere forests spanned North America and Eurasia into the present-day Arctic. In the mid-Cretaceous [100 million years ago (mya)] these forests were dominated by both conifers (Pinaceae, Ginkgoaceae, and Taxodiaceae) and angiosperm taxa. Early examples of widespread flowering plant genera include *Magnolia*, *Betula*, and *Platanus*. Many modern northern genera soon followed in the fossil record (e.g., *Quercus*, *Castanea*, *Carya*, *Ulmus*, *Juglans*, and *Acer*), also spanning the Northern Hemisphere.

By the beginning of the Neogene, approximately 25 mya, the climate began to change, becoming cooler and drier in certain regions as a consequence of mountain building and shifts in ocean currents. In the cold northern regions, conifers were favored, giving rise to the boreal forests. At midlatitude, summer drought and/or winter cold favored deciduous angiosperms. Midcontinent aridity also gave rise to the temperate steppe grasslands and deserts. This, together with Pleistocene glaciation, undoubtedly contributed to divergence in the north temperate forest floras and their biodiversity.

In Europe and central Asia, east–west trending mountain ranges and arid zones blocked the southward

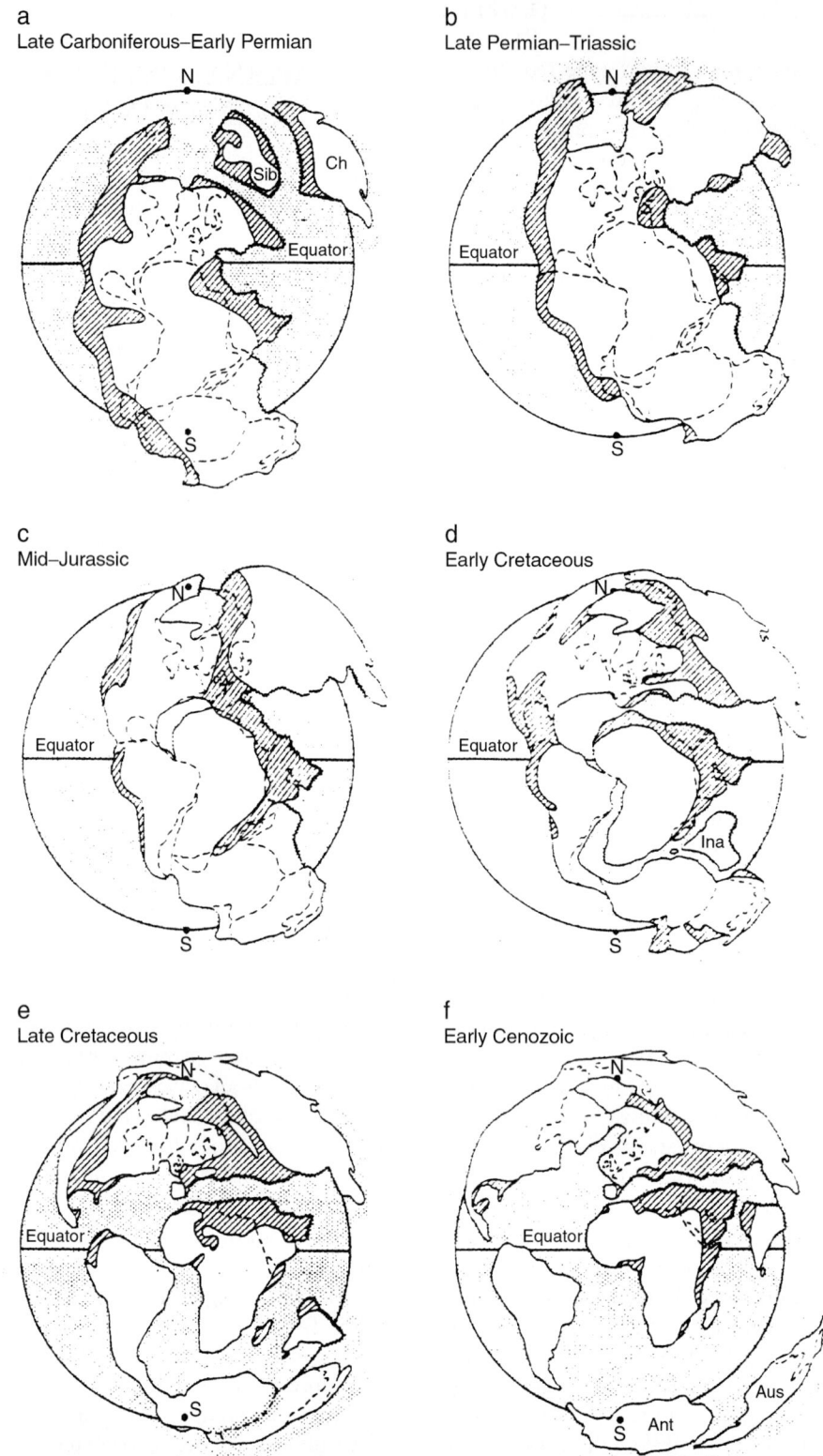

FIGURE 5 Changes in the locations of continental landmasses are shown from the Mesozoic to the early Tertiary in Lambert equal area projections. Areas on the back projection of the spheres are folded out for view. Dashed lines indicate shoreline of modern continents, hatched areas are epicontinental seas, and stippled areas are oceans (reproduced with permission from Cox and Moore, 1985).

retreat of the forests as the glaciers advanced. In consequence, there was apparently greater species extinction here than elsewhere in the Northern Hemisphere, and geographic isolation prevented recolonization from temperate or tropical locations elsewhere. In contrast, in east Asia, continental glaciation was less extensive and continuous connections with tropical wet forests permitted temperate forest species to retreat south and advance north with climate oscillations; there were many fewer extinctions and thus greater retention of an older phylogenetic diversity. This tropical connection undoubtedly accounts for the greater tropical affinities seen in the flora of east Asia today. In eastern North America the mountains trend north–south and thus migrations with the glacial oscillations were possible. However, the tropical connections were largely blocked by seas or arid zones. Compared to Asia, there were more extinctions with fewer opportunities for recolonizations by taxa with tropical affinities. In eastern North America the phylogenetic lineages have a more recent and more northerly balance. Western North America suffered the greatest extinctions. Extensive mountain building, accompanying aridity, and the development of a Mediterranean climate to the south favored conifers or sclerophyllous trees with divergent evolutionary lineages from those of the mesic Tertiary forests. To the north, cool, wet climates favored conifer forests (many with tertiary affinities) at the expense of angiosperms.

The forests of the Southern Hemisphere began similarly in the Cretaceous. Warm, mesic, mixed conifer and angiosperm forests spanned Gondwanaland. These were dominated by southern conifers, including Auracariaceae and Podocarpaceae, and southern angiosperms, including Cunonaceae, Proteaceae, Myrtaceae, and Sapindaceae. In addition to these, Casurinaceae and the genus *Nothofagus* were present throughout, except on what became Africa. Gondwanaland began to break up early in the Cretaceous (Fig. 5) well before Asiamerica. Southern Africa and India had become isolated by the early Cretaceous. This undoubtedly accounts for Africa's greater dissimilarity with the rest of the southern temperate flora. India eventually joined Asia and today it has primarily north temperate or Asian tropical floristic affinities. Not until early in the Tertiary did South America, Antarctica, and Australia finally separate. Climate changes in the Southern Hemisphere were less dramatic during the Neogene and Pleistocene than they were in the north. Aridity and glaciation were more localized; the climate remained less continental. In consequence, the forest flora retains more characteristics of the warm, mesic Tertiary flora than is the case in the north.

During the Neogene, temperate southern Africa experienced increased aridity. With climatic oscillations the forests expanded and contracted. Many of the Gondwanan elements became extinct, leaving only a couple of species of Podocarpaceae and a minor contribution of Cunonaceae and Proteaceae to the forest flora today. Connections along the east coast of southern Africa allowed connections with the tropical forest flora, and today the temperate South African forests have the strongest tropical affinities of any temperate region and the greatest tree diversity. Exceptionally high levels of spatial and temporal environmental heterogeneity may contribute to speciation here as well. With substantial tropical affinities also found in Australia and New Zealand, the southern forests as a group have a much stronger representation of tropical plant families than do the northern forests.

During the Pliocene the southern temperate forest region of Chile and adjacent Argentina became isolated from the rest of South America by the Andes and arid zones to the north and east. The isolation and relatively small extent of these forests undoubtedly account for the low diversity and high endemism found here. Only in the extreme south of Chile and at high elevations with lower mean and absolute temperatures are deciduous (*Nothofagus*) forests found in the Southern Hemisphere.

Australia and New Zealand retain much of their Tertiary flora. The maritime climate of New Zealand has changed relatively little during the Tertiary. Its long isolation from other landmasses is responsible for the high level of endemism. Increased aridity in Australia in the Neogene substantially affected the temperate forest structure and composition. Explosive radiation occurred in the genus *Eucalyptus* (Myrtaceae) and *Acacia* (Mimosaceae). Unique among temperate forests, there are more than 500 species of *Eucalyptus* in Australia (many restricted to temperate forests). These are sclerophyllous-, drought-, and fire-tolerant tree species that dominate most of the temperate forest and woodlands in the east and west. Only in moist, cool patches in the southeast or at higher elevations are remnants of the tertiary flora found. *Nothofagus*, Araucariaceae, Podocarpaceae, and other groups are part of the rich tree flora.

The previous account indicates that it is not simply history or geology but a complex process that incorporates the role of temporal and spatial heterogeneity, geographic isolation, recent Pleistocene effects, and the role of climate stability that jointly affect the patterns of taxon distributions, diversity, and endemism we see today. What about other explanatory variables?

B. Productivity

The differential effects of solar radiation figure prominently as an explanatory hypothesis, measured either directly or indirectly via actual or potential evapotranspiration (i.e., AET or PET) or productivity. There is conflicting evidence on the relative importance of this as an explanatory variable versus historical factors, at least at the regional scale. Examined at the plot level (~1 ha) across regions, correlations are found between diversity and AET. There are also correlations at the regional and latitudinal scales (boreal vs temperate vs tropical forests). However, other evidence shows that differences at regional and local scales can be well explained by historical factors after taking into account AET. Global joint correlations between species richness and productivity were examined by the British NHM for plant family diversity on a 10° grid scale. This study showed higher family richness than expected from productivity in southern Africa, southern North America, and southern east Asia and lower or expected diversity elsewhere in the temperate zone. Undoubtedly both sets of factors may be involved, but these remain only correlates; cause–effect relationships have not been demonstrated.

C. Spatiotemporal Heterogeneity and Other Explanations

The roles of disturbance regimes and environmental or habitat heterogeneity at local and regional levels have figured prominently as explanatory hypotheses for richness patterns. Higher diversity is predicted under intermediate levels of disturbance. On a local or landscape level there is considerable evidence to support this idea. Midsuccessional forests and successionally diverse landscape support a richer flora. For example, in northeastern North America many of the extinct or threatened species are those that occurred in open or early successional landscapes that were common 100 years ago but have since disappeared with the homogenizing reforestation of the landscape.

The predominant natural disturbances in temperate forest zones include fire and windstorm. Fires are most prevalent in the drier forests of western North America, Australia, South Africa, and the Mediterranean Basin. In many of these systems, periodic fires contribute to higher landscape biodiversity. The role of fires in other forest systems is less clear. Large cyclonic storms affect mainly eastern coastal, midlatitude regions in the Northern Hemisphere (eastern North America and Asia). Tornadic events are most prevalent in midcontinental North America. Such storms contribute to heterogeneous successional stages in the landscape and thus higher diversity. Occasionally, human activity can contribute to higher landscape diversity. Historically, Europeans husbanding the landscape for diverse forest products undoubtedly increased the local diversity over what it would have been naturally. Humans creating a moderate spatially and temporally heterogeneous landscape ironically may contribute slightly to increased regional diversity. The higher regional species richness in mountainous areas associated with the manifold habitats was discussed previously. Another example is the extremely rich South African forests (and other biomes). The high local and regional richness has been largely attributed to the environmental heterogeneity found at this scale.

In addition, there are many other explanatory variables for patterns in species richness, some of which were intimated or included in the topics discussed previously: environmental stability or predictability, abiotic rarifaction, land area, seasonality, aridity, range limits and geometric constraints, and many more. For many of these factors it is easy to establish correlation and much more difficult to establish cause–effect relationships.

VII. ECOSYSTEM SERVICES

Temperate forests provide important ecosystem services globally, regionally, and locally. Temperate forests contribute about 17% to global net primary productivity (versus about 49% for tropical forest systems and 8% for boreal forests). However, recent evidence from atmospheric and oceanic CO_2 data point to temperate forests, especially those in eastern North America, as globally important carbon sinks. The magnitude and causes of the net carbon uptake by temperate biomes are uncertain. However, it may be related to the reforestation that has occurred during the past century, especially in North America. CO_2 fertilization, anthropogenic N deposition, and global warming may contribute as well.

At local to landscape levels there are tight links among forest structure, composition, and species richness; soil attributes; mineral and hydrological cycles; and human disturbances. In the moist temperate zone species richness tends to be higher on soils that are better drained, warmer, and finer textured, with greater NO_3–N and P availability and higher cation saturation and lower Al^+ (toxic) levels, all associated with higher pH. These attributes are associated with many calcareous soils. Thinner calcareous soils, which apparently create greater spatial heterogeneity in forest canopies, are also associated with higher local richness. Hence, soil attributes are intimately related to pattern in species diversity. The extent to which these features are altered will directly affect biodiversity; the extent to which the community structure and composition are changed will feed back on these variables.

The strongest correlates of productivity tend to be soil texture and N cycling, which in turn are correlated with moisture retention, cation exchange, and maximum levels of humus accumulation. These correlations are strongest in undisturbed forests. Also, ecosystem control by soil texture extends over long time frames. Human disturbances can profoundly affect these links. Soil compaction with intensive forest management will have a cascading and long-term effect on nutrient and hydrological cycles. As the studies at Hubbard Brook Experimental Forest have demonstrated, forest clearing will dramatically alter watershed hydrology (increased water loss and sediment loss), nutrient cycling (with elevated nutrient loss), microclimate, and species composition for years or decades. Changes in forest species composition such as with conifer plantations will have an equally large effect. Substituting homogeneous conifers stands for mixed broad-leafed forests will increase C/N and lignin/C ratios, which in turn reduces decomposition rates, N mineralization rates, and pH, with cascading effects on cation exchange capacity and diversity in the soil flora and fauna. For example, in German *Picea* plantations, N cycling between canopy and forest soils is reduced by 75% from that observed in native beech forests. Moreover, with greater stem interception of precipitation, the conifer forest soils also become drier. In consequence, it may not be possible to successively reintroduce beech forests on these sites without large-scale soil amendments.

Even small changes in the broadleaf forest composition can alter decomposition rates and nutrient status. For example, sugar maple (*Acer saccharum*) and ash (*Fraxinus*) promote higher N mineralization, and sugar maple accumulates calcium. In contrast, oak-dominated forests have lower N mineralization. Substituting exotic species in plantations can dramatically alter ecosystem processes. *Eucalyptus* and *Melaleuca* tend to significantly lower water tables where they have been introduced. In the northeastern United States, Japanese barberry (*Berberis thunbergii*) has become a seriously invasive exotic, forming a continuous shrub layer. Under barberry canopies soil ammonium N levels are elevated, the soil flora and fauna are altered, and native species richness is depressed.

VIII. CONSERVATION STATUS

Table I shows the approximate extent of protected forests in each of the regional biomes based on WCMC global data for International Union for the Conservation of Nature (IUCN) conservation protection categories I–VI. Globally for temperate forests, 6 or 7% of the remaining forests receive some level of protection. This represents about 1 or 2% of the total temperate forest biome extent. Compared to other forest systems, temperate forests are slightly better protected than boreal forests (6 or 7% versus 5 or 6%) but apparently less well protected than tropical forests as a whole (10–12%).

There is considerable variation both within and among regions in the level of forest protection as well as among forest types. For example, temperate freshwater swamp forests are afforded the least protection (2.7% globally), whereas Southern Hemisphere evergreen broadleaf forests receive the highest level of protection (22.6% globally). Overall, Southern Hemisphere forests are afforded significantly better protection as a group than are northern forests.

The level of protection varies considerably among regions and countries within the temperate zone: Less than 3% of east Asian temperate forests are protected, but this varies with forest type from about 5% for cool to cold temperate and subtropical forests to 1% or less for most warm to hot temperate forests types (except for warm temperate rainforests, which only cover a tiny area and are about 15% protected). There is considerable variation among the countries of east Asia in the amount of forest afforded protection: almost 10% for Japan to less than 5% of China and less than 1% of North Korea. For Europe (excluding Russia) about 8% for the various forest types are given some protection. This also varies considerably among countries: As much as 25% of the broad-leafed deciduous forests in Germany are protected, whereas <1% in Bosnia and 2% in Russia are protected. For North America, the extent of forest protection varies from about 15% for conifer-dominated temperate rainforests to 7% for the northern cool or cold temperate forests and about 2% for the warm to hot temperate and subtropical forests versus more than 9% for North American boreal forests. The forests of the Middle East, which are the most reduced in extent, are afforded the least protection over all. Less than 3% of the forests are protected in this zone. This varies from <1% in Georgia and Azerbaijan to 1% in Turkey and perhaps as much as 12% in Iran. About 8% of India's temperate forests have some level of conservation protection. For temperate Australia about 9% of the forests are protected; for New Zealand the figure is about 43%. For South Africa, about 24% of its southernmost temperate forests are protected. For South America, about 23% of the temperate forests are afforded some protection, but the vast majority (more than 90%) of these are the wet evergreen forests of the south, mostly located in Chile. Less than 2% of the sclerophyllous and dry temperate forest are protected. An estimated 8% of Brazil's temperate mixed conifer forests are protected.

Overall, the most well protected forests, on a relative scale, are the Southern Hemisphere temperate rainforests, the Pacific wet forests of western North America and east Asia, the northernmost temperate forest (mixed with boreal elements), and mixed temperate forests in mountainous areas. The most poorly protected are dry and sclerophyll leafed forests of the Northern Hemisphere, wet and rainforest broadleaf deciduous and evergreen forests in the Northern Hemisphere (very limited in extent), and moist temperate deciduous and evergreen broad-leafed forests of the Southern Hemisphere. East Asia has the highest percentage of temperate forest types receiving no protection, followed by the Middle East and South America.

In some ways these data are misleading. Many of the forests classified as receiving protection are plantation or otherwise highly managed forests. These are largely native species forests in Europe, North America, and Asia, but they tend to be managed as homogeneous, even-aged, monocultures with consequent reduced diversity and ecosystem services. Natural, undisturbed, or old-growth forests comprise only a small fraction of the remaining forests: western Europe, 1%; eastern North America, 1%; eastern Asia, 1%; Australia, about 5% (but probably <1% is unlogged); South Africa, <1%. Somewhat better off are New Zealand with about 25%, northwest (Pacific) North America with about 13%, and southern Chile with about 45%.

Some temperate forests are unique in that significant reforestation has occurred during the past century (about 1 or 2% increase on average per annum), primarily in Europe and eastern North America. However, most of these net gains in forest cover are intensively managed (in Europe and the southeastern United States). In northeastern North America, during the past century the landscape has reverted naturally from 50–90% agricultural to mostly forested, following the extensive abandonment of agriculture. Most of this forest is highly fragmented and fairly homogeneous, being of similar age. Consequently, the ecological functionality of these forests is limited. Elsewhere, temperate forest cover continues to decline at rates that vary from 1 to 10% pa.

IX. CONSERVATION AND PROTECTION STRATEGIES

How does one develop a strategy for conserving or protecting the remaining temperate forests of the world? Some authors focus on "hot spots" of diversity or endemism. Globally, most attention has been focused on tropical systems and recently Mediterranean systems. Temperate forest-dominated systems have received less attention. For example, of Conservation International's (CI's) 24 hot spots, only New Zealand is a temperate zone forest-dominated locality. However, in his most recent iteration of "megadiversity countries," Mittermeyer highlights 4 of 17 countries that contain important temperate forest biomes: China, the United States, Australia, and South Africa. Three of these (Australia, China, and South Africa) rank among the top 12 countries worldwide in species richness across phyla and species endemism. The four largely temperate countries also rank among the top globally in the number of IUCN Red Data Book (RDB) "threatened" plant species. Although these predominantly temperate floras are better known, as a group they have more RDB species than all of the tropical countries together. The WWF and the IUCN have also identified approximately 240 centers of plant diversity globally. Of these, about 15% represent temperate forest-dominated systems and another 10% represent other temperate biomes (e.g., shrublands, grasslands, and arid lands). Only 1 or 2% are boreal or polar, and the balance (~74%) are tropical.

There have been similar attempts to identify hot spots regionally. For example, WWF identified more than 100 forest hot spots across Europe, the Mediterranean Basin, and the Middle East. Regional centers of diversity can also be detected from taxon turnover. Data from the Atlas Florae Europaeae project, which compared the joint diversity of Pinaceae and Fagaceae across Europe at a 50-km grid scale, show highest joint species diversity in these two families in the Balkans, the southern Alps, the Carpathians, and the southern Pyrenees. This obviously reflects the high spatial heterogeneity in these localities.

Alternatively, conservation assessments may include such factors as taxon or clade irreplaceability, minimum area sets, minimum viable niche space, or ecosystem integrity. A study was done by the British NHM to select conservation priority areas for selected plant groups. This exercise was done on a 10° grid globally and regionally for all of Europe on a 50-km grid. The global analysis revealed top-priority sites in eastern China, southeastern Australia, and central Chile along with six tropical areas, plus secondary centers in southern Chile, the southeastern and western United States, central and southern China, northern Japan, southwestern Europe, eastern and southwestern Australia, and southern Africa along with nine secondary centers in the tropics. Within Europe, many priority sites were identified in the Balkan Peninsula and margins of the Mediterranean (which include forested lands). However, there are prioritized areas elsewhere throughout Europe at lower densities.

World Resources Institute (WRI) provides another perspective on priority sites for forest conservation. They recognize "frontier forests"—large tracts of intact forested ecosystems sufficient to maintain viable populations of all indigenous species. This perspective includes large, wide-ranging predators and migratory species and takes into account the prevailing natural disturbance patterns. These are thus considered intact, fully functional forested ecosystems or landscapes. Very few of these are in the temperate zone, and most are under medium or high threat. This contrasts with the tropical and boreal zones, in which there are many more large tracts identified as frontier forest, even if there are substantial areas at risk. No intact forested landscapes are found in temperate Europe or Africa. In North America only one small patch of transitional boreal forest in central Ontario and discontinuous tracts of conifer forests in coastal British Columbia and Alaska (all under threat) can be considered intact landscapes. In Asia, only a few small forest patches in the inaccessible mountains of central and south-central China, boreal transition forest patches along the border between China and Russia, and the Primorski Krai region of Pacific Russia are considered intact temperate forest landscapes. All but one small patch are under threat. In South America the only frontier temperate forest is in southern Chile and adjacent Argentina. Most of this region is considered to be at risk. In Australia, only a small rainforest patch in Tasmania is classified as intact. Small patches of frontier forest occur on the west coast of the south island of New Zealand and one patch in the central north island. All are at risk except for the Tasmania rainforest.

X. MAJOR THREATS TO CONSERVATION

Many major threats exist for conserving the biodiversity and ecosystems services provided by temperate forests locally, regionally, and globally. Certainly a major threat is homogenization of the landscape. This is a result of intensively managing forests as near monocultures of either native or exotic species, managing landscape for similar forest age and size classes, introducing invasive exotic species, and relying globally on only a few taxa for forest plantations (e.g., *P. radiata* or *Eucalyptus*). The consequences of these landscape management strategies are reductions in local or regional diversity and alternation of many ecosystem processes (e.g., nutrient and hydrological cycles) and soil attributes. Large-scale increases in timber harvesting may well have a negative impact on carbon source–sink relationships and atmospheric CO_2 levels.

Increasing urbanization of the landscape and continued harvesting of ever more remote forests will lead to a more fragmented landscape. The likelihood of these fragments maintaining ecosystem functionality is ever diminished. Within the next few decades, it is likely that no fully intact, functional forested landscapes will remain in the temperate zone, given current trends.

Atmospheric pollution from acid precipitation, and the associated ecosystem N and S loadings, remain prevalent in the north temperate zone and will have long-term effects. There is evidence that calcium is being rapidly depleted in acidic soils as a consequence of long-term acid precipitation. This will have cascading effects on forest soils and on major tree species performance, particularly those most sensitive to calcium levels such as sugar maple.

Projected global warming will have a major effect on northern temperate forests. The main direct effect of global warming will likely be seen in the boreal forest zone, with decreased periods of snow cover and hence changes in surface albedo. This will likely favor the expansion of temperate forests well north into the boreal zone. The community response patterns will be determined by dispersal characteristics of the biota and the availability of propagules in source populations. This in turn will be dictated by the configuration of forest fragments in the landscape.

In the Southern Hemisphere, the effects of global warming will be much smaller due to the moderating maritime influences. Atmospheric pollution effects are very negligible or very localized. The major threats to temperate forests here are deforestation, especially in

South America, and the pervasive effects of introduced exotic species in plantation or as escapes. *Eucalyptus* and northern pines (especially *P. radiata*) are planted in monocultures across the Southern Hemisphere, with negative effects on biodiversity, soil attributes, and hydrology.

XI. CONSERVATION OBJECTIVES AND RESEARCH NEEDS

Where do we need to focus attention to improve our understanding of temperate forests and develop a more effective conservation plan? Clearly we need better and more accurate means of mapping temperate forests and assessing their conservation status. There are large discrepancies and errors in assessing the current and potential forest cover for all forest types. We need more information on patterns of species occurrences (simply the presence/absence is sufficient) at a variety of scales throughout the temperate zone. Because distribution (and hence richness) patterns vary across spatial scales, one cannot simply rely on particular, arbitrary sample sizes (e.g., 0.04-, 0.1-, 1-, or 50-ha plots). Effort should be placed on inventorying poorly known areas, such as the Middle East. We need to have a better, predictive understanding of the links between forest structure and composition, ecosystem functions, and the effects of human disturbances.

There needs to be a concerted, manifold effort at developing a variety of different conservation strategies for temperate forests. These include targeting more protection for forested biomes with high ratios of people to forest area and forest biomes or forest types that are poorly protected. Examples include east Asian forests in general, north temperate broad-leafed rainforests, and subtropical dry forests. Moreover, concerted effort should be placed on protecting the few remaining intact frontier forests, especially those at high risk, and old-aged forest stands particularly where these comprise miniscule components of the landscape. However, the strategy needs to be inclusive, focusing on conserving the complete spatial and temporal heterogeneity of the landscape, even if this necessitates some human husbandry of the landscape.

Strategies for identifying priority conservation sites should not simply target sites with high species diversity and/or endemism. Alternative means of evaluation need to be incorporated as well, such as phylogenetic richness and irreplaceability, range rarity, ecosystem or landscape integrity, landscape heterogeneity that incorporates migration in the face of natural and human-made disturbances, and minimum niche space for all biotic components of the landscape. Only with a manifold creative approach can we hope to come to grips with the conservation of this critical suite of temperate forest biomes.

See Also the Following Articles

BOREAL FOREST ECOSYSTEMS • DEFORESTATION AND LAND CLEARING • ENDEMISM • FOREST ECOLOGY • HOTSPOTS • TEMPERATE GRASSLAND AND SHRUBLAND ECOSYSTEMS • TROPICAL FOREST ECOSYSTEMS

Bibliography

Armesto, J. J., Rozzi, R., and Caspersen, J. (2000). Past, present, and future scenarios for biological diversity in South American temperate forests. In *Future Scenarios for Biological Diversity* (T. Chapin and O. Sala, Eds.), Springer-Verlag, New York.

Bailey, R. G. (1998). *Ecoregions: The Ecosystem Geography of the Oceans and Continents.* Springer-Verlag, New York.

Cox, C. B., and Moore, P. D. (1985). *Biogeography.* Blackwell, Oxford.

Dudley, N. (1992). *Forests in Trouble: A Review of the Status of Temperate Forests Worldwide.* Earth Resources Research, London.

Ehrlich, P. R. (1996). Conservation in temperate forests: What do we need to know and do? *Forest Ecol. Management* 85, 9–19.

Francis, A. P., and Currie, D. J. (1998). Global patterns of tree species richness in moist forests: Another look. *Oikos* 31, 598–602.

Latham, R. E., and Ricklefs, R. E. (1993). Continental comparisons of temperate-zone tree species diversity. In *Species Diversity in Ecological Communities. Historical and Geographical Perspectives* (R. E. Ricklefs and D. Schluter, Eds.). Univ. of Chicago Press, Chicago.

Ovington, J. D. (1983). *Ecosystems of the World 10: Temperate Broad-Leafed Evergreen Forests.* Elsevier, Amsterdam.

Qian, H., and Ricklefs, R. E. (1999). A comparison of the taxonomic richness of vascular plants in China and the United States. *Am. Nat.* 154, 160–181.

Rhode, K. (1992). Latitudinal gradients in species diversity: The search for the primary cause. *Oikos* 65, 514–527.

Röhrig, E., and Ulrich, B. (Eds.) (1991). *Ecosystems of the World 7: Temperate Deciduous Forests.* Elsevier, Amsterdam.

Schultze, E.-D., Bazzaz, F. A., Nedelhoffer, K. J., Koike, T., and Takatsuki, S. (1996). Biodiversity and ecosystem function of temperate deciduous broad-leafed forests. In *Scope 55: Functional Roles of Biodiversity: A Global Perspective* (H. A. Mooney, J. H. Cushman, E. Medina, O. E. Sala, and E.-D. Schultze, Eds.). Wiley, Chichester, UK.

TEMPERATE GRASSLAND AND SHRUBLAND ECOSYSTEMS

Osvaldo E. Sala, Amy T. Austin, and Lucía Vivanco
University of Buenos Aires and IFEVA-CONICET

I. Grassland Distribution
II. Extent of Biodiversity in Temperate Grasslands
III. Biodiversity and Ecosystem Functioning in Grasslands
IV. The Future of Biodiversity in Grasslands

GLOSSARY

convention on biological diversity The convention was first enacted in June 1992, and it has been signed by many countries. Its objectives are the conservation of biological diversity, the sustainable use of its components, and the fair and equitable sharing of the benefits arising from the utilization of genetic resources, including by appropriate access to genetic resources and by appropriate transfer of relevant technologies, taking into account all rights over those resources and technologies, and by appropriate funding.

functional type A group of species that share morphological and physiological characteristics that result in a common ecological role.

global biodiversity assessment The Global Biodiversity Assessment is an independent peer-reviewed analysis of the biological and social aspects of biodiversity commissioned by the United Nations Environment Programme.

international geosphere biosphere programme A scientific program that is part of International Council of Science and provides an international and interdisciplinary framework for the conduct of global change science.

niche complementarity Refers to how the ecological niches of species may not fully overlap and complement each other. Consequently, an increase in the number of species that complement each other may result in a larger volume of total resources utilized and in a higher rate of ecosystem processes.

sampling effect Refers to the phenomenon in which increases in the number of species increase the probability of including in the community a species with a strong ecosystem effect. This phenomenon yields an increase in ecosystem processes with increases in diversity without invoking niche complementarity.

scientific committee on problems of the environment An international organization that is part of the International Council for Science and is charged with synthesizing current scientific understanding associated with environmental issues.

GRASSLANDS are water-limited ecosystems, and water availability defines the distribution of grasslands. Grassland ecosystems occur in areas of the world that have an annual precipitation between 150 and 1200 mm and mean annual temperature between 0 and 25°C. Temperature controls the distribution of grasslands mainly indirectly by modulating water demand and consequently water availability. Increases in temperature result in increases in soil evaporation and plant transpiration; consequently, for a similar precipitation regime, the water balance becomes more negative as temperature increases. In contrast with most biological phenomena, primary production of North American grasslands decreases with increasing temperature, highlighting the indirect mechanism of the temperature control on the distribution of grasslands.

I. GRASSLAND DISTRIBUTION

How does the distribution in temperature and precipitation space translate into the distribution of grasslands in geographical space? Along precipitation gradients grasslands are located between forests and deserts. Several of the International Geosphere Biosphere Programme terrestrial transects (Koch et al., 1995) are located along precipitation gradients and they intersect forests, grassland–forest ecotones, grasslands, and deserts. This pattern is repeated in the North Eastern China transect, in the Great Plains transect in North America, and in the Patagonian transect in South America (Fig. 1).

Although temperature and precipitation are the major determinants of the distribution of grasslands, fire also may play an important role. Fire becomes particularly important in the grass–forest ecotones where the dominance of grasses or woody plants in many cases is determined by the frequency and intensity of fires. For example, in North American tallgrass prairie, the area covered with woody plants has increased dramatically in the past 100 years and the human intervention in reducing fire frequency is largely responsible for the change (Briggs et al., 1998). Similarly, data from pollen profiles, tree ring analysis, and photographic sources documented a shift in the grassland–forest ecotone in northern Patagonia with woody vegetation invading the grassland (Veblen and Markgraf, 1988). Again, changes in the human-induced fire regime were responsible for the forest expansion.

This article focuses exclusively on climatically determined grasslands, in contrast with grasslands resulting from human intervention. Anthropogenically determined grasslands are located in areas where potential natural vegetation is forest. Humans, in an attempt to produce forage for domestic animals, have logged forests and have maintained these plots as grasslands by mowing them periodically.

Worldwide, temperate grasslands are represented in all continents and potentially cover a vast area of 49×10^6 km^2 that represents 36% of the earth's surface

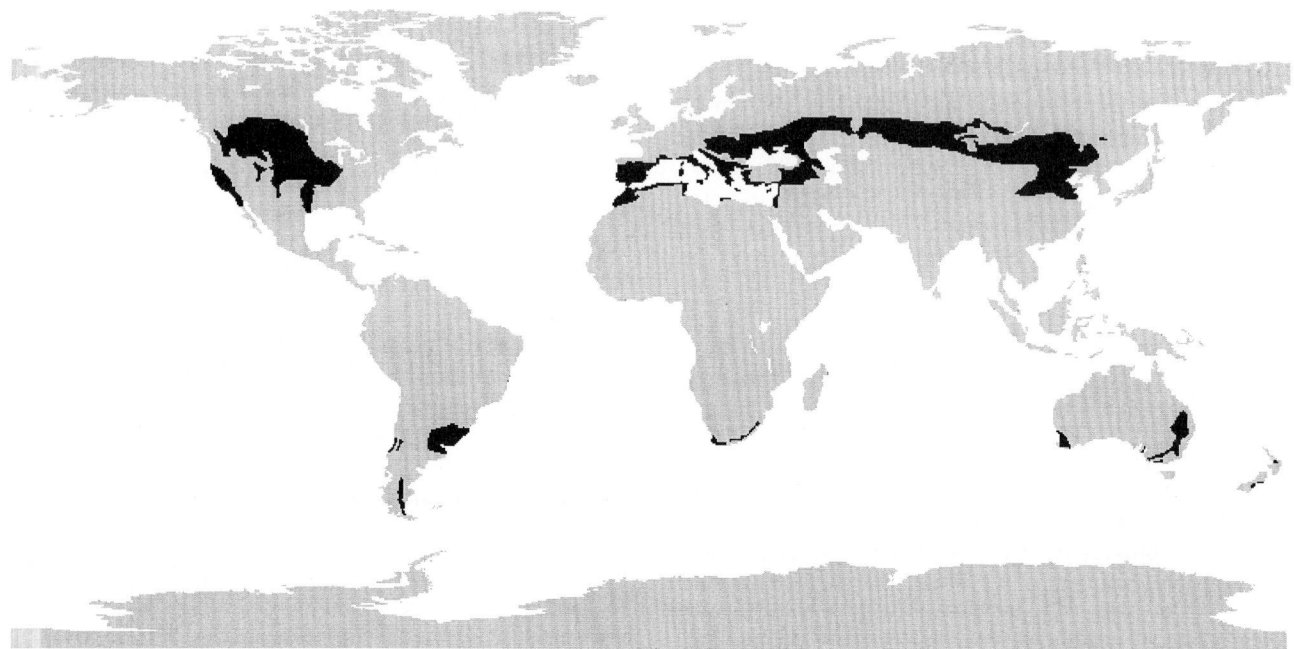

FIGURE 1 Map of the global distribution of temperate grasslands (adapted from Bailey, 1998).

(Shantz, 1954). Our definition of temperate grasslands excludes tropical and subtropical savannas but includes grass and shrub deserts. In North America, grassland is the potential natural vegetation of most of the Great Plains and it reaches from the Chihuahuan desert in the south to the deciduous forests of Canada in the north and from the Rocky Mountains in the west to the deciduous forest of the eastern United States (Fig. 1). In South America, grassland is the potential vegetation of the vast pampas and most of the Patagonian steppe. Finally, in Asia, grassland ecosystems cover a huge area from the Ukraine to China.

II. EXTENT OF BIODIVERSITY IN TEMPERATE GRASSLANDS

Biodiversity can be examined in many different ways, and multiple definitions exist for what constitutes "biodiversity." Nevertheless, the authors of the Global Biodiversity Assessment, using definitions originally proposed by the Convention on Biological Diversity, defined biological diversity as "variability among living organisms from all sources" (Heywood and Baste, 1995). Here, we focus on biodiversity in terms of taxonomically defined species, and the vast majority of studies quantifying ecosystem variation have used this measure. However, genetic biodiversity (genetic variation among a single species) and ecological diversity (including landscape diversity and functional group diversity) are also important components of biological diversity. The definition of biodiversity, therefore, will depend to a certain extent on one's objective and scale of interest, ranging from the gene to the ecosystem.

A. Diversity of Plants

Floristic diversity in grasslands varies broadly, with many natural grasslands having a very high level of plant diversity, at times approaching that seen in mainland tropical forests (Groombridge, 1992). For example, the Pampa region in Argentina represents some of the highest diversity grassland, with more than 400 species of grasses (Cabrera, 1970). In North America, more than 250 native species are found in tallgrass prairie (Freeman, 1998), the vast majority of which are perennial grasses.

Plant species of grasslands can be categorized into four functional types: grasses, shrubs, succulents, and herbs (Sala et al., 1997). The classification of plant species into functional types only has epistemological value and serves the purpose of facilitating the study of grasslands. This classification can be divided into many new subcategories or aggregated into fewer units depending on the needs of the analysis. The relative contribution of the four functional types depends on the seasonality of precipitation and the soil texture, which are the factors controlling the distribution of water availability in the soil profile. Water penetrates deeper into the soil profile in coarse-textured soils than in fine-textured soils. Similarly, water penetrates deeper into the soil in regions in which most of the precipitation occurs during the cold time of the year when evaporative demand is low. In general, grasses and shrubs have contrasting rooting patterns, with grasses having predominantly shallow roots and shrubs having deep roots (Jackson et al., 1996). Consequently, grasses dominate in regions in which the wet and warm seasons are in synchrony and in areas with predominantly fine-textured soils (Sala et al., 1997; Fig. 2).

Grassland plant species can also be classified according to their photosynthetic pathway into C_3 and C_4 species. The two groups of species have differences in the physiology of photosynthesis and in the morphology of leaves that result in different ecological characteristics that separate them in time and space. Regional analyses of the distribution of these two types of grass species showed that C_3 species decrease southward in North America and northward in South America and C_4 species show the opposite pattern (Paruelo et al.,

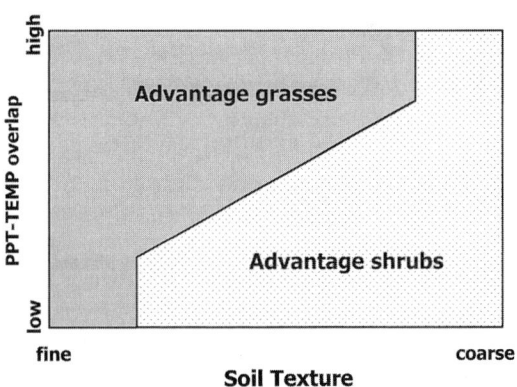

FIGURE 2 Conceptual model explaining the distribution of the functional groups grasses and shrubs with changes in seasonality of temperature and precipitation (congruence of warm and wet seasons) and soil texture. The gray shaded area represents conditions that are likely to favor the persistence of grasses, whereas the speckled area represents conditions that favor the shrub functional group. The intersection of the two areas represents points at which biotic influences are likely to be most pronounced (adapted from Sala et al., 1997).

1998). Similarly, the abundance of C_3 species increases whereas that of C_4 decreases along an altitudinal gradient (Cavagnaro, 1988). These biogeographical analyses correlate with ecophysiological studies showing that C_4 species have photosynthesis optima at higher temperature, have higher water use efficiency, and are better adapted to low water availability conditions (Kemp and Williams, 1980).

B. Aboveground Diversity of Animals

All major taxonomic groups are represented in grasslands, but despite their large areal extent (36% of the area of terrestrial ecosystems), overall faunal diversity is lower than in many other biomes. The number of bird and mammalian species that are found primarily in grasslands are estimated to be 477 and 245, respectively, representing only 5% of the world's species for each taxonomic group (Groombridge, 1992). Local diversity can be high in specific areas (e.g., there are an estimated 208 avian species for tallgrass prairie; Kauffman et al., 1998), but general patterns show lower diversity for most taxonomic groups compared with other ecosystems.

One of the striking features of grasslands in terms of animal diversity is the presence of large herbivores as a prominent component of secondary production. These large grazing mammals have an important impact on the functioning of grasslands, altering patterns of nutrient cycling, primary production, and plant species composition (McNaughton, 1993), although their presence and diversity varies across different continents. For example, in the Great Plains of North America, nearly all the large grazing mammals went extinct during the glaciation of the Pleistocene, but the proliferation of a very few species, particularly *Bison bison* (plains bison), dominated the plant–herbivore interactions until the introduction of domestic cattle at the beginning of the twentieth century (Lauenroth and Milchunas, 1992). In contrast, African grasslands contain a very high level of mammalian diversity of grazers, with up to 20 species coexisting in a single reserve (Cumming, 1982). Finally, many South American grasslands evolved without the presence of large grazers, and their primary herbivory prior to the introduction of sheep and cattle was due to insect species (Bucher, 1982). Thus, although there is variation in the diversity of the large herbivores, their presence and importance is a distinctive characteristic of grassland ecosystems.

Small mammals, birds, reptiles, amphibians, and insects also play an important role in the functioning of grasslands. Species richness of small mammals is actually higher than that for large mammals (168 vs 77 species overall), and they are mostly granivores or omnivores (Groombridge, 1992). In contrast, in Australian deserts small mammals are mostly insectivorous. Fluctuations in seed supply caused by unpredictable environmental conditions and the infertile soils could be an explanation for these differences (Morton, 1993).

Avian diversity in grasslands represents 5% of the total species of the world species diversity, and again the fluctuating climate has an important control on this distribution. In this case, birds can migrate to remote areas outside of the grassland biome to seek alternative resources in periods of unfavorable conditions. In North American grasslands, which demonstrate a strong seasonality, there are large annual variations of passerines in response to climatic conditions. Additionally, within the grassland ecosystems, there exists a gradient of avian biomass that decreases with precipitation and primary production from the tallgrass prairie to the mixed prairie and shortgrass steppe (Lauenroth and Milchunas, 1992).

Reptiles in grasslands are less diverse than mammals and birds, and amphibians are less diverse than reptiles in the tallgrass prairie of North America (Kauffman et al., 1998). Latitude has an effect on the biodiversity of reptiles and amphibians because they are ectotermic organisms, with an increase in the number of species from north to south of the tallgrass prairie (Kucera, 1992).

Insects are a diverse element of the terrestrial macrofauna of tallgrass prairie (Kauffman et al., 1998), reflecting general patterns of diversity for terrestrial ecosystems in which insects represent more than 50% of the species (Strong et al., 1984). They have a very important role as herbivores, pollinators, predators, parasitoids, and decomposers. Herbivorous insects are probably the most conspicuous functional group in tallgrass prairie (Kauffman et al., 1998) and may replace large grazing mammals as the primary consumer in some South American grasslands (McNaughton et al., 1993).

C. Diversity of Soil Organisms

The diversity of belowground organisms (bacteria, fungi, and micro-, meso-, and macrofauna) is known in much less detail than that of plants and animals aboveground, but it may constitute a very important component of the biota, often equaling or exceeding the aboveground biomass in grasslands (Paul et al., 1979). The large biomass of roots and other underground organs in grasslands and the high concentration of organic matter provide substrate for a large variety of bacteria, fungal, and nematode groups, and all are represented in grassland systems. Additionally, the arthropods, constituting the largest proportion of invertebrates in the shortgrass steppe and primarily herbivores, take advantage of the large amount of belowground primary production (Lauenroth and Milchunas, 1992). The studies that have been done show a very diverse group of organisms in grasslands soils in a variety of functional roles. For the shortgrass steppe, the relative importance in terms of biomass of the different functional groups is bacteria > fungi > nematodes > protozoa > macroarthropods > microarthropods (Lauenroth and Milchunas, 1992). In terms of species numbers, a soil invertebrate study in tallgrass prairie showed more than 200 species of nematodes, with fungivores constituting 40% of the nematode species (Ransom et al., 1998) and the nematode biomass was exceeded only by that of bacterial and fungal groups. Another study in the shortgrass steppe found soils to contain more than 100 species of fungi during the summer season (Christensen and Scarborough, 1969).

III. BIODIVERSITY AND ECOSYSTEM FUNCTIONING IN GRASSLANDS

The relationship between biological diversity and the functioning of ecosystems has been central in the development of ecological ideas, and grassland ecosystems have been crucial in testing these ideas. McNaughton (1993) traced the idea of the relationship between biodiversity and ecosystem functioning to statements by Charles Darwin about how increasing species diversity in a plot might result in higher productivity as a result of niche complementarity. Most species in Darwin's plot supposedly exploited different resources, and consequently the larger the number of species, the larger the volume of resources exploited. For example, plots containing only shallow-rooted or deep-rooted species should have lower productivity than plots containing both groups of species that jointly have access to water and nutrients stored in both upper and lower layers of the soil.

Species not only differ in their ability to exploit resources but also in their response to the environment. Species show a variety of responses to abiotic factors such as temperature or water availability as well as biotic factors such as predation or competition. These relationships led to the diversity–stability hypothesis or portfolio hypothesis that stated that higher species diversity results in greater ecosystem stability (McNaughton, 1993). Consequently, ecosystem processes will vary more in space or time in less diverse than in more diverse communities.

The basic theory of the relationship between biodiversity and ecosystem functioning has been established for decades, but its empirical support has been scarce and fragmentary. Pimm's (1984) review of experimental results and later the Scientific Committee on Problems of the Environment (SCOPE) project led by Hal Mooney (Schulze and Mooney, 1993) reinvigorated the field and highlighted the gaps in our understanding. The value of possessing a quantitative understanding of the relationship between biodiversity and ecosystem functioning is underscored by the fact that we are in the midst of the sixth major extinction event in the history of life (Chapin et al., 1998). The major difference between this and previous extinction episodes is that the current wave of extinction is the result of human activity. Society is urged to identify the consequences of the human-driven changes in biodiversity for the functioning of ecosystems that provide so many goods and services, including grasslands, that are at the basis of human well-being (Sala and Paruelo, 1997).

A large SCOPE project had the responsibility of assessing and evaluating the vast number of observations and experiments that addressed in different ways the role of biological diversity on the functioning of ecosystems. An initial meeting and corresponding volume synthesized the theory and formulated a plan of action (Schulze and Mooney, 1993). The second step was to address the same questions on a biome by biome basis. Finally, a large synthesis assessed the differences and commonalties among biomes in the relationship between biodiversity and ecosystem functioning (Mooney et al., 1995a, b, 1996). Results for the grassland biome emphasized the effects of changes in biodiversity on primary productivity, decomposition, water balance, atmospheric properties, landscape structure, and species interactions (Sala et al., 1995, 1996). This effort synthesized our knowledge and it also highlighted the gaps in our understanding and the need for further experimentation.

Grassland ecosystems played a key role in the experimentation regarding the functional roles of biodiversity mostly because of the small size and short life span of grasses, which made manipulative experiments feasible with few resources and in short periods of time. The most widely accepted hypothesis indicates that the rate of ecosystem processes, such as primary productivity or nutrient cycling, might increase linearly as species richness increases and that this relationship eventually saturates as ecological niche overlap increases (Fig. 3) (Vitousek and Hooper, 1993).

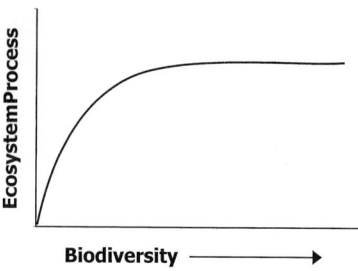

FIGURE 3 Theoretical relationship between biodiversity and ecosystem functioning. Species richness (number of species) is the most common indicator of biodiversity, although this axis term could also include genetic diversity (diversity within species) to landscape diversity. Ecosystem process could signify processes occurring at the ecosystem level, including primary production and decomposition, and components of nutrient cycling (adapted from Vitousek and Hooper, 1993).

The first experiment specifically designed to test this hypothesis was carried out under controlled conditions at the Imperial College Ecotron facility (Naeem et al., 1994). The experiment used synthetic communities with three trophic levels and three levels of species diversity, from the simplest with 1 species of a secondary consumer, 3 primary consumers, 2 producers, and 3 decomposers to the most diverse that had 31 species. The results of this experiment supported the hypothesis described in Fig. 3 in that carbon fixation (an indirect measure of primary productivity) increased with diversity. The first large-scale field experiment specifically designed to test this hypothesis was located in the North American tallgrass prairie (Tilman et al., 1996). The experiment consisted of sowed plots with seven levels of plant diversity and 20 replicates per level. Each replicate was a random draw from a pool of 24 native species. Consequently, replicates had the same diversity level but could have a different or equivalent species composition. Results follow a pattern similar to that of Fig. 3, with total plant cover (a nondestructive way of estimating light interception and production) and nutrient uptake increasing with species richness up to a level of approximately 10 species. Beyond 10 species, the different estimates of ecosystem functioning did not change. Recently, a large consortium of scientists organized a large-scale field experiment across Europe with the purpose of testing the same hypothesis (Hector et al., 1999). The same experimental design was repeated in eight locations in Europe from Sweden to Greece and it consisted of five diversity treatments with richness ranging from 1 to 32 species. Species identity varied among the sites since species for each treatment were always drawn from a pool of species adapted to local conditions. A single model represented the loglinear increase of aboveground biomass with increasing species richness for all sites.

The different experiments designed to test the effect of diversity on ecosystem functioning yielded results that are similar to those predicted from the hypothesis described in Fig. 3. These results can be interpreted as evidence of niche complementarity; that is, the higher the number of species with niches that do not overlap, the larger the total volume of resources exploited (Tilman et al., 1997). Alternatively, the same results can be interpreted as resulting from the sampling effect (Huston, 1997; Tilman et al., 1997). It indicates that as the number of species increases, the probability of including in the mix a species with strong ecosystem effects increases. This species may be a nitrogen fixer, a deep-rooted species, or simply a species with a combination of characters that maximizes production in these circumstances.

IV. THE FUTURE OF BIODIVERSITY IN GRASSLANDS

Biodiversity in grassland ecosystems is seriously threatened by human activity. Grassland ecosystems, in two of three possible scenarios of biodiversity change for the Year 2100, appear to be the most threatened biome; in the third scenario, grasslands appear behind only tropical forests, arctic ecosystems, and southern temperate forests (Sala et al., 2000).

What makes biodiversity in grassland ecosystems so vulnerable to human impact? Are grassland ecosystems particularly sensitive? Are they located in areas that will be affected the most? A recent study identified the most important drivers of biodiversity change in grasslands as changes in land use, climate, nitrogen deposition, biotic exchange (accidental or deliberate introduction of plant or animal species to an ecosystem), and atmospheric CO_2 and studied their expected change and the sensitivity of each biome (Sala et al., 2000). The study concludes that grasslands are both quite sensitive ecosystems and are located in parts of the world where ecosystems are going to be affected most by human activity (Fig. 4).

The expected change of each of the drivers was assessed using other scenarios, such as climate change or land use change, that were developed independently (Alcamo, 1994; Haxeltine and Prentice, 1996). The dif-

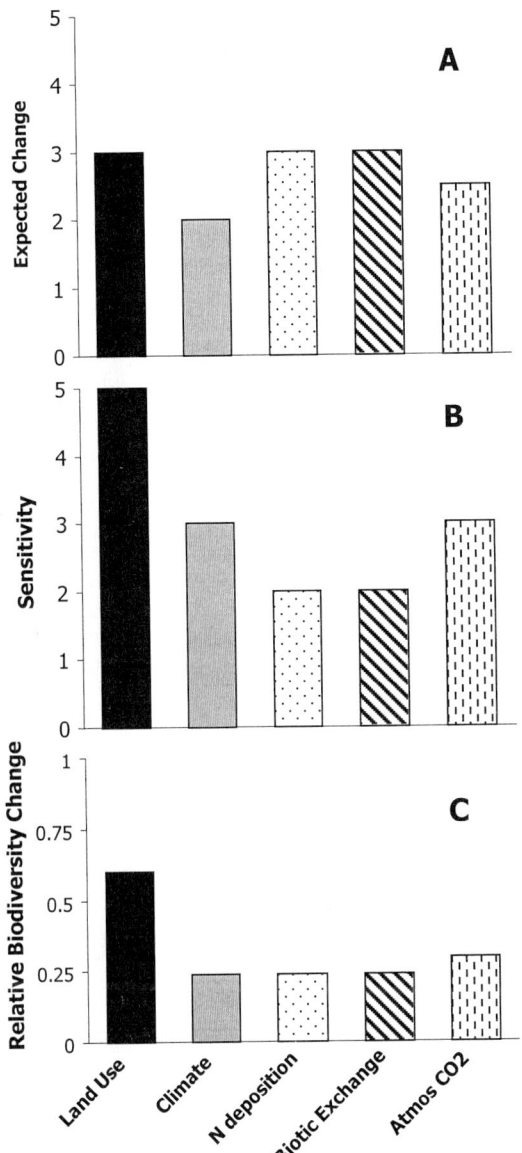

FIGURE 4 Scenarios of biodiversity change in grasslands for the Year 2100. (A) Expected change of the major drivers of biodiversity change in a relative scale from 1 to 5. (B) Sensitivity of grasslands to unit changes in each driver using the same relative scale. (C) Grassland biodiversity change resulting from changes in each driver and calculated as the product of the expected change of each driver multiplied by the sensitivity of grasslands to each driver (adapted from Sala et al., 2000).

ferent drivers of biodiversity change were originally expressed in different units from parts per million to hectares and degrees Celsius. In order to compare the rate of change among drivers, they were all converted into a 1 to 5 arbitrary scale. Land use, nitrogen deposi-

tion, and biotic exchange are expected to be the drivers that are going to change the most in grassland ecosystems (Sala et al., 2000) (Fig. 4A). The most dramatic changes in land use in grasslands are those that result in changes in land cover, such as the conversion into croplands. Grasslands are among the biomes that are going to experience the largest conversion in land use because of their mild climate and favorable soil conditions that made them quite suitable for agriculture. The conversion into agricultural land is not expected to be even across the world but driven by the patterns of food demand and population growth that indeed are quite idiosyncratic. For example, the IMAGE2 model (Alcamo, 1994) predicts for the Year 2100 a large increase in agricultural area in Africa and a reduction in North America resulting from an increase in demand and an increase in intensification, respectively. Biodiversity losses resulting from conversion to agriculture in one part of the world are not offset by a similar area that will be abandoned and reverts to grassland but that is located in a different part of the world. Therefore, total change in grassland area grossly underestimates the impact of land use change on biodiversity. Finally, the expected change of nitrogen deposition and biotic exchange in grasslands are among the highest of all biomes. Densely human-populated regions are predominantly located in temperate regions where the potential native vegetation is that of grasslands. Nitrogen deposition is associated with industrialization, whereas biotic exchange is associated with trade patterns and ultimately with human density.

Sensitivity to changes in each driver is the magnitude of change in biodiversity due to a change in a unit of driver. Sensitivity varies among biomes, and grasslands are quite sensitive to changes in land use, climate, and CO_2 concentration in the atmosphere (Fig. 4B). Grasslands are most sensitive to land use change, which means the conversion into croplands that implies plowing of native grasslands and sowing of a monospecific crop. This activity clearly results in the local extinction of all the native plant species that, in turn, determine the major characteristics of the habitat of animals and microorganisms. Consequently, land use change drives all plant species to local extinction and drastically affects the diversity of other organisms (Anderson, 1995). Biodiversity may be quite sensitive to changes in CO_2 concentration in ecosystems that are limited by water availability such as grasslands. There are well-known differences in the species response in water use efficiency due to changes in CO_2 (Jackson et al., 1994). Consequently, changes in CO_2 will first affect the competitive balance and then the relative abundance, and they may result in the local extinction of species.

The biodiversity change resulting from each driver (Fig. 4C) can be calculated as the product of the expected change (Fig. 4A) multiplied by the biome sensitivity to that driver (Fig. 4B). The largest biodiversity change for the Year 2100 in grasslands is expected to occur due to changes in land use. The second largest effect will result from the expected increase in CO_2 because although the atmospheric concentration of this trace gas will increase uniformly throughout the world (Fung *et al.*, 1987), grassland biodiversity may be particularly sensitive to it.

Finally, the total biodiversity change in grasslands expected for the Year 2100 can be calculated as the sum of the individual effects of each driver (Fig. 4C). This calculation assumes no interaction among drivers. Comparison of the total biodiversity change in grasslands against that of all the other terrestrial biomes of the world indicates that grasslands and Mediterranean ecosystems appear to be the biomes that will experience the largest change and will be affected the most (Sala *et al.*, 2000).

Other scenarios consider the assumption that there are synergistic interactions among the drivers of biodiversity change. We can envision several examples of the synergistic interactions among drivers. For example, the effect of biotic exchange will be amplified if combined with a land use change that results in fragmentation. Similarly, the effects of elevated CO_2 on biodiversity when occurring in conjunction with increased nitrogen deposition will be much larger than the sum of the individual effects. The scenario developed using the assumption of a synergistic interaction among drivers also indicated that grasslands and Mediterranean ecosystems are the most vulnerable (Sala *et al.*, 2000). The only scenario in which grasslands are not the most vulnerable ecosystem is when an antagonistic interaction was assumed and the total biodiversity change was equated with the change resulting from the driver with the maximum value. This scenario is plausible only under extreme conditions, such as those of slash-and-burn that destroy most of the plant and animal species, and further change due to the other drivers is not possible. Even in the antagonistic scenario, grasslands are among the most vulnerable biomes, following tropical, arctic, and southern temperate ecosystems.

Acknowledgments

We thank J. P. Guerschman for assistance with the figures. Research by OES reported in this chapter was supported by IAI, CONICET, FONCyT and the University of Buenos Aires. ATA was supported by Grant CRN-012 from the Inter-American Institute for Global Change Research. LV was supported by a fellowship from the University of Buenos Aires.

See Also the Following Articles

CARBON CYCLE • PLANT BIODIVERSITY, OVERVIEW • SLASH AND BURN FARMING, EFFECTS OF • TEMPERATE FORESTS

Bibliography

Alcamo, J. (1994). *Image 2: Integrated Modeling of Global Climate Change.* Kluwer, Dordrecht.

Anderson, J. M. (1995). The soil system. In *Global Biodiversity Assessment: Section 6* (H. A. Mooney, J. Lubchenco, R. Dirzo, and O. E. Sala, Eds.), pp. 406–412. Cambridge Univ. Press, Cambridge, UK.

Bailey, R. G. (1998). *Ecoregions: The Ecosystems Geography of the Oceans and Continents.* Springer, New York.

Briggs, J. M., Nellis, M. D., Turner, C. L., Henebry, G. M., and Su, H. (1998). A landscape perspective of patterns of processes in tallgrass prairie. In *Grassland Dynamics: Long-Term Ecological Research in Tallgrass Prairie* (A. K. Knapp, J. M. Briggs, D. C. Hartnett, and S. L. Collins, Eds.), pp. 265–279. Oxford Univ. Press, New York.

Bucher, E. H. (1982). Chaco and Caatinga—South American arid savannas, woodlands and thickets. In *Ecology of Tropical Savannas* (B. J. Huntley and B. H. Walker, Eds.), pp. 48–79. Springer-Verlag, New York.

Cabrera, A. (1970). *Flora de la Provincia de Buenos Aires: Gramineas.* Insituto Nacional de Tecnología Agropecuara, Buenos Aires.

Chapin, F. S., Sala, O. E., Burke, I. C., Grime, J. P., Hooper, D. U., Lauenroth, W. K., Lombard, A., Mooney, H. A., Mosier, A. R., Naeem, S., Pacala, S. W., Roy, J., Steffen, W., and Tilman, D. (1998). Ecosystems consequences of changing biodiversity. *BioScience* 48, 45–52.

Christensen, M., and Scarborough, A. M. (1969). Soil microfaunal investigations, Pawnee site, Colorado State University Report No. 23, U.S. IBP Grassland Biome, Colorado State University, Fort Collins.

Cumming, D. H. M. (1982). The influence of large herbivores on savanna structure in Africa. In *Ecology of Tropical Savannas* (B. J. Huntley and B. H. Walker, Eds.), pp. 217–245. Springer-Verlag, New York.

Epstein, H., Laurenroth, W., Burke, I., and Coffin, D. (1996). Ecological responses of dominant grasses along two climatic gradients in the Great Plains of the United States. *J. Vegetation Sci.* 7, 777–788.

Freeman, C. C. (1998). The flora of Konza prairie. A historical review and contemporary patterns. In *Grassland Dynamics: Long-Term Ecological Research in Tallgrass Prairie* (A. Knapp, J. Briggs, D. Hartnett, and S. Collins, Eds.), pp. 69–80. Oxford Univ. Press, New York.

Fung, I. Y., Tucker, C. J., and Prentice, K. C. (1987). Application of advanced very high resolution radiometer vegetation index to study atmosphere–biosphere exchange of of CO_2. *J. Geophys. Res.* 92D, 2999–3015.

Groombridge, B. (1992). *Global Biodiversity: Status of the Earth's Living Resources.* Chapman & Hall, London.

Haxeltine, A., and Prentice, I. C. (1996). BIOME3: An equilibrium terrestrial biosphere model based on ecophysiological constraints, resource availability, and competition among plant functional types. *Global Biogeochem. Cycles* **10**, 693–709.

Hector, A., Schmid, B., Beierkuhnlein, C., Caldeira, M. C., Diemer, M., Dimitrakopoulos, P. G., Finn, J. A., Freitas, H., Giller, P. S., Good, J., Harris, R., Högberg, P., Huss-Danell, K., Joshi, J., Jumpponen, A., Körner, C., Leadley, P. W., Loreau, M., Minns, A., Mulder, C. P. H., O'Donovan, G., Otway, S. J., Pereira, J. S., Prinz, A., Read, D. J., Scherer-Lorenzen, M., Schulze, E. D., Siamantziouras, A. S. D., Spehn, E. M., Terry, A. C., Troumbis, A. Y., Woodward, F. I., Yachi, S., and Lawton, J. H. (1999). Plant diversity and productivity experiments in European grasslands. *Science* **286**, 1123–1127.

Heywood, V. H., and Baste, I. (1995). Introduction. In *Global Biodiversity Assessment*, pp. 5–19. Cambridge Univ. Press, Cambridge, UK.

Huston, M. A. (1997). Hidden treatments in ecological experiments: Re-evaluating the ecosystem function of biodiversity. *Oecologia* **110**, 449–460.

Jackson, R. B., Sala, O. E., Field, C. B., and Mooney, H. A. (1994). CO_2 alters water use, carbon gain, and yield for the dominant species in a natural grassland. *Oecologia* **98**, 257–262.

Jackson, R. B., Candell, J., Ehleringer, J. R., Mooney, H. A., Sala, O. E., and Schulze, E. D. (1996). A global analysis of root distributions of terrestrial biomes. *Oecologia* **108**, 389–411.

Kauffman, D. W., Fay, P. A., Kaufman, G., and Zimmerman, J. L. (1998). Diversity of terrestrial macrofauna. In *Grassland Dynamics: Long-Term Ecological Research in Tallgrass Prairie* (A. Knapp, J. Briggs, D. Hartnett, and S. Collins, Eds.), pp. 101–112. Oxford Univ. Press, New York.

Kemp, P. R., and Williams, G. J. (1980). A physiological basis for niche separation between *Agropyron smithii* (C4) and *Bouteloua gracilia* (C4). *Ecology* **61**, 846–858.

Koch, G. W., Scholes, R. J., Steffen, W. L., Vitousek, P. M., and Walker, B. H. (1995). The IGBP Terrestrial Transects: Science Plan, Report No. 36. International Geosphere–Biosphere Programme, Stockholm.

Kucera, C. L. (1992). Tallgrass prairie. In *Natural Grasslands: Introduction and Western Hemisphere* (R. T. Coupland, Ed.), pp. 227–268. Elsevier, Amsterdam.

Lauenroth, W. K., and Milchunas, D. G. (1992). Short-grass steppe. In *Natural Grasslands: Introduction and Western Hemisphere* (R. T. Coupland, Ed.), pp. 183–226. Elsevier, Amsterdam.

Lieth, H., and Whittaker, R. (1975). *Primary Productivity of the Biosphere*. Springer-Verlag, New York.

McNaughton, S. J. (1993). *Biodiversity and function of grazing ecosystems*. In *Biodiversity and Ecosystem Function* (E. D. Schulze and H. A. Mooney, Eds.), pp. 361–383. Springer-Verlag, Berlin.

McNaughton, S. J., Sala, O. E., and Oesterheld, M. (1993). Comparative ecology of African and South American arid to subhumid ecosystems. In *Biological Relationships between Africa and South America* (P. Goldblatt, Ed.), pp. 548–567. Yale Univ. Press, New Haven, CT.

Mooney, H. A., Lubchenco, J., Dirzo, R., and Sala, O. E. (1995a). *Biodiversity and Ecosystem Functioning: Basic Principles*. Cambridge Univ. Press, Cambridge, UK.

Mooney, H. A., Lubchenco, J., Dirzo, R., and Sala, O. E. (1995b). *Biodiversity and Ecosystem Functioning: Ecosystem Analyses*. Cambridge Univ. Press, Cambridge, UK.

Mooney, H. A., Cushman, J. H., Medina, E., Sala, O. E., and Schulze, E. D. (1996). *Functional Roles of Biodiversity: A Global Perspective*. Wiley, Chichester, UK.

Morton, S. R. (1993). Determinants of diversity in animal communities. In *Species Diversity in Ecological Communities* (R. E. Ricklefs, Ed.), pp. 159–169. Univ. of Chicago Press, Chicago.

Paruelo, J., Jobbágy, E., Sala, O., Lauenroth, W., and Burke, I. (1998). Functional and structural convergence of temperate grassland and shrubland ecosystems. *Ecol. Appl.* **8**, 194–206.

Paul, E. A., Clark, F. E., and Biederbeck, V. O. (1979). Microorganisms. In *Grassland Ecosystems of the World: Analysis of Grasslands and Their Uses* (R. T. Coupland, Ed.), pp. 87–96. Cambridge Univ. Press, Cambridge, UK.

Pimm, S. L. (1984). The complexity and stability of ecosystems. *Nature* **307**, 321–326.

Ransom, M. D., Rice, C. W., Todd, T. C., and Wehmueller, W. A. (1998). Soils and soil biota. In *Grassland Dynamics: Long-Term Ecological Research in Tallgrass Prairie* (A. Knapp, J. Briggs, D. Hartnett, and S. Collins, Eds.), pp. 48–66. Oxford Univ. Press, New York.

Sala, O. E., and Paruelo, J. M. (1997). Ecosystem services in grasslands. In *Nature's Services: Societal Dependence on Natural Ecosystems* (G. C. Daily, Ed.), pp. 237–252. Island Press, Washington, DC.

Sala, O. E., Lauenroth, W. K., McNaughton, S. J., Rusch, G., and Zhang, X. (1995). Temperate grasslands. In *Global Biodiversity Assessment*, pp. 361–366. Cambridge Univ. Press, Cambridge, UK.

Sala, O. E., Lauenroth, W. K., McNaughton, S. J., Rusch, G., and Zhang, X. (1996). Biodiversity and ecosystem function in grasslands. In *Functional Role of Biodiversity: A Global Perspective* (H. A. Mooney, J. H. Cushman, E. Medina, O. E. Sala, and E. D. Schulze, Eds.), pp. 129–149. Wiley, New York.

Sala, O. E., Lauenroth, W. K., and Golluscio, R. A. (1997). Plant functional types in temperate semi-arid regions. In *Plant Functional Types* (T. M. Smith, H. H. Shugart, and F. I. Woodward, Eds.), pp. 217–233. Cambridge Univ. Press, Cambridge, UK.

Sala, O. E., Chapin, F. S., Armesto, J. J., Berlow, E., Bloomfield, J., Dirzo, R., Huber-Sanwald, E., Huenneke, L. F., Jackson, R. B., Kinzig, A., Leemans, R., Lodge, D. M., Mooney, H. A., Oesterheld, M., Poff, N. L., Sykes M. T., Walker, B. H., Walker, M., and Wall, D. H. (2000). Global biodiversity scenarios for the Year 2100. *Science* **287**, 1770–1774.

Schulze, E. D., and Mooney, H. A. (1993). *Biodiversity and Ecosystem Function*. Springer-Verlag, Berlin.

Shantz, H. (1954). The place of grasslands in the earth's of vegetation. *Ecology* **35**, 142–145.

Smith, T., Shugart, H. H., and Woodward, F. I. (1996). *Plant Functional Types*. Cambridge Univ. Press, Cambridge, UK.

Strong, D. R., Lawton, J. H., and Southwood, R. R. S. (1984). *Insects on Plants: Community Patterns and Mechansims*. Harvard Univ. Press, Cambridge, MA.

Tilman, D., Wedin, D., and Knops, J. (1996). Productivity and sustainability influenced by biodiversity in grassland ecosystems. *Nature* **379**, 718–720.

Tilman, D., Lehman, C., and Thomson, K. (1997). Plant diversity and ecosystem productivity: Theoretical considerations. *Proc. Natl. Acad. Science USA* **94**, 1857–1861.

Veblen, T. T., and Markgraf, V. (1988). Steppe expansion in Patagonia? *Quaternary Res.* **30**, 331–338.

Vitousek, P. M., and Hooper, D. U. (1993). Biological diversity and terrestrial ecosystem biogeochemistry. In *Biodiversity and Ecosystem Function* (E. D. Schulze and H. A. Mooney, Eds.), pp. 3–14. Springer-Verlag, Berlin.

TERRESTRIAL ECOSYSTEMS

Stephen Roxburgh and Ian Noble
Australian National University

I. The Earth's Terrestrial Ecosystems
II. Global Patterns of Biodiversity across Terrestrial Ecosystems
III. Summary

GLOSSARY

biome A group of ecosystems which are subject to similar climatic conditions and which share a similar range of vegetation structures, animal diversity, and soil types. Biomes represent major regional groupings of ecological systems and are therefore discernible at the global scale.
ecosystem All of the organisms living within or utilizing a particular area, together with their surrounding nonliving environment.
ecosystem classification A procedure for grouping together similar ecosystems based on a predefined set of criteria, e.g., vegetation structure, climate, land-use history, and soil type.
latitudinal biodiversity gradient The well-documented trend for there to be more species in tropical ecosystems compared with temperate or polar ecosystems.

THE TERRESTRIAL SURFACE OF THE EARTH comprises an enormous variety of environments and ecosystems. To reduce this complexity to a manageable level, ecosystems have been grouped together or "classified" using a range of different methods. There are consistent patterns in biodiversity across major ecosystem groups, with those from moist, tropical regions having significantly higher diversity than ecosystems from other regions.

I. THE EARTH'S TERRESTRIAL ECOSYSTEMS

A. Ecosystem Classifications: Overview

Terrestrial ecosystems cover approximately 148 million km^2, corresponding to 29% of the total surface area of the earth. They include such diverse habitats as the frigid regions around the poles, the searing heat of tropical deserts, and lush temperate and tropical rainforests. This enormous variety poses great difficulties for the study of ecosystems at regional to global scales, and it has led to the development of many classification systems which seek to reduce this complexity to a manageable level. Indeed, the development of such classification systems, and the identification of the spatial extent of the Earth's major ecosystem types, is a fundamental first step in managing the world's biological systems.

Despite the importance of this problem, the classification of the world's ecosystems into a consistent and useable framework is not without its problems. First, ecosystems usually grade into one another, hence making it difficult to mark ecosystem boundaries on a map. Second, there is no single agreed upon definition of an "ecosystem." To some researchers it represents an area within which the same broad mix of species can be found. To others it is merely an arbitrary area delimited solely for practical purposes. Third, different criteria can be used to generate different kinds of classifications. For example, human-induced land-use change is included in some classifications but not in others. Ecosystem classification systems are therefore approximations of reality and should be viewed as a set of tools which provide a simplified overview of the relative extents and spatial arrangements of the world's major ecosystem types.

Ecosystems are usually classified according to the dominant vegetation type for a region, which is in turn determined largely by geographical and environmental factors. Vegetation is considered the most appropriate criteria for classifying ecosystems because it constitutes the majority of biomass and also because of its role, through the process of photosynthesis, as the primary fixer of energy on which all other life depends.

B. The Biome Classification

The broadest units of ecosystem classification are called biomes. Biomes are complex mixtures of a large number of similar ecosystem types which share a similar physiognomy (vegetation structure) and climate but do not necessarily share the same species. Indeed, one of the strengths of the biome approach is that ecosystems with different species compositions, but with similar vegetation structures and ecosystem processes, can be compared. For example, the tropical rainforests of Central America and northern Australia are both recognized as belonging to the same biome "tropical rainforest," even though there would be little overlap in species composition between the two regions. The major climatic variables on which the biome classification is based are average annual temperature and average annual precipitation.

Many biome classifications have been suggested. These have classified the world into as little as 8 and into as many as 18 separate biomes. Figure 1 shows the classification of Cox and Moore, which maps 10 major terrestrial biomes. One feature of the spatial distribution of the world's biomes is that many are fragmented over the surface of the earth. For example, the "Mediterranean" biome is found in the areas surrounding the Mediterranean sea and also in Australia, South Africa, South America, and North America.

A brief description of the major terrestrial biomes is given in the following sections. More complete accounts of each can be found in the appropriate sections of this encyclopedia.

1. Tundra

Tundra occurs around the Arctic Circle at latitudes above where trees can survive and also at high altitudes above the tree line in other regions. The vegetation is of a very low stature, typically comprising low shrubs, sedges, grasses, lichens, and mosses. The climate is characterized by long, cold winters, during which average monthly temperatures seldom exceed 0°C and, at the highest latitudes, there is constant darkness for weeks at a time. Soils are permanently frozen (permafrost), making water unavailable to plant growth. Also, liquid water is available for only short periods at a time, thus making these ecosystems effectively "polar deserts." Due to these harsh climatic conditions and brief growing season (usually less than 60 days), plant growth and decomposition is slow and ecosystems are very slow to recover from disturbance. Animal diversity is similarly low, with a small number of permanent bird and mammal species and a number of seasonal insects. Many bird species and some mammals migrate to the tundra during the brief summer.

2. Taiga (Boreal Forest and Northern Coniferous Forest)

The taiga forms a broad band adjacent to the tundra biome, extending across North America, northern Europe, and northern Asia. The dominant vegetation consists of evergreen conifers, with some representation of deciduous broadleaved species in some environments. Within this biome it is common to find large areas of land dominated by only one or a few coniferous tree species. Winters are typically long and cold, contrasting with summers that, although brief, can be both warm and wet. The soils are characterized as being nutrient poor and acidic. Characteristic animals include insect- and seed-eating birds, small mammal herbivores such as the snowshoe hare, larger browsers such elk and moose, and carnivorous mammals such as the lynx, wolf, and grizzly bear. Many bird species migrate to the taiga during the summer to feed on swarms of seasonally breeding insects.

3. Temperate Forests

The temperate forest biome includes a wide range of forest types, including the temperate rainforests of the Pacific Coast of North America, southern Chile, and New Zealand; the temperate deciduous forests of eastern North America, eastern Asia, and Europe; and temperate evergreen forests such as the sclerophyllous eu-

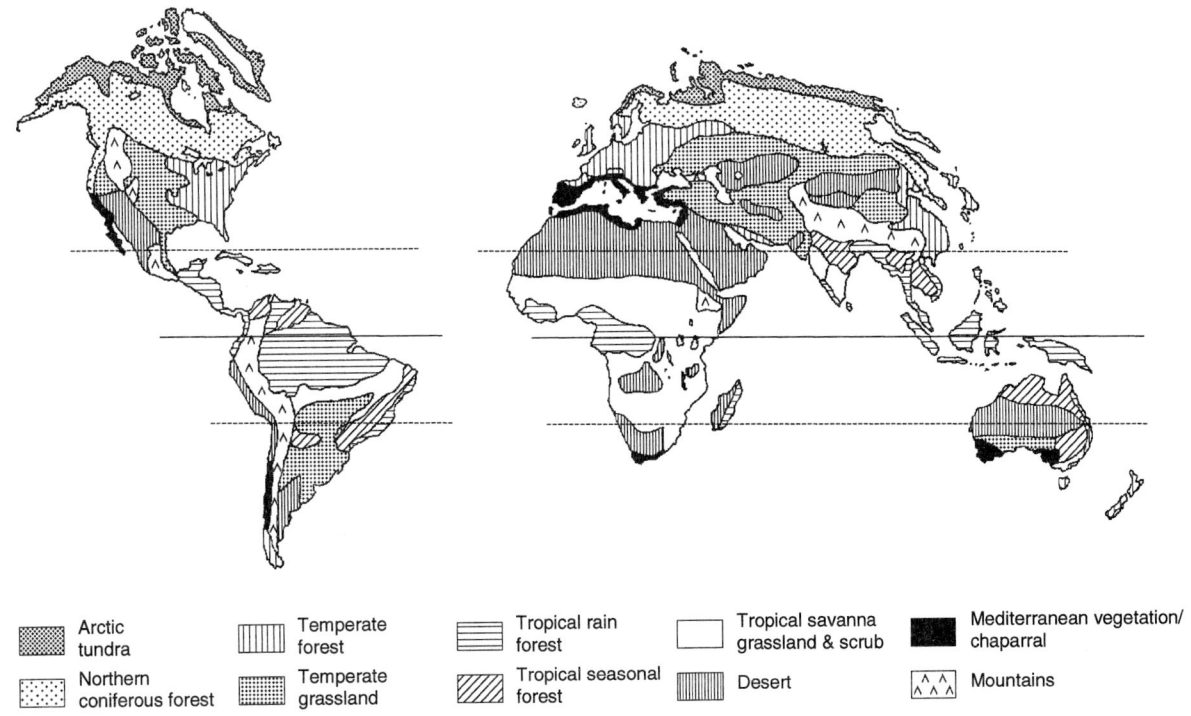

FIGURE 1 The world's major biomes, based on the classification of Cox and Moore (1985). Reproduced with the permission of Blackwell Scientific Publications.

calyptus forests of Australia and the southern beech forests of Chile and New Zealand. The forest structure is generally well developed, with various strata including emergent trees, canopy trees, small trees, shrubs, and a herbaceous layer. Along with an enhanced diversity of plant species and habitat structure, there is an associated general increase in the diversity of the fauna. The growing season is approximately 6 months. Because of the wide range of forest types and environments represented in this biome, soil nutrition, precipitation, and water availability are also highly variable.

4. Tropical and Temperate Grasslands

The grassland biome includes ecosystems from a wide range of geographical regions and environments. In the tropics these include the savanna, which can also be associated with an open cover of shrubs or trees. Examples of temperate grasslands can be found in North America (prairies), Europe and Asia (steppe), South America (pampas), and South Africa (veld). A common feature is the importance of herbivory, particular in the savanna, which has a high diversity of ungulate browsers. In comparison with tropical grasslands, temperate grasslands support a much lower faunal diversity. Fires are also an important component in many grassland ecosystems, in which repeated fires can prevent the establishment of woody species, thus preventing the grasslands from developing into shrublands or forests. Climate and soils vary with grassland type and region, with the tropical savanna experiencing distinct wet and dry seasons and temperate grasslands experiencing cold winters and hot summers.

5. Mediterranean Vegetation (Chaparral)

Examples of the Mediterranean biome can be found between approximately 30° and 40° latitude in both hemispheres and are associated with wet winters and dry summers. Mediterranean vegetation occurs most extensively on the coast surrounding the Mediterranean Sea, but it is also found as the "chaparral" in California and northwest Mexico, the "matorral" in Chile, the "fynbos" around the Cape of South Africa, and also in the southern regions of Australia. Because precipitation in this biome is generally less than in grasslands, and there are regular periods of drought, the vegetation is characterized by shrubs and small tress which have small, leathery "sclerophyllous" leaves adapted to drought stress. Fire is also a common feature of Mediterranean vegetation, and many plant species share adaptations to persist under a regime of repeated fire, e.g., resprouting from belowground and seeds that require fire to germinate. In general, soils are nutrient poor.

One striking feature of this biome is that high levels of endemism are common, but adaptations to the Mediterranean environment in geographically separate regions are very similar. This has been seen as evidence for convergent evolution, in which plant and animal species from different taxonomic groups respond independently, but similarly, in response to the same environmental pressures.

6. Deserts

Deserts occur in climates characterized by extreme aridity. Water limitation in these areas can arise due to many factors, for example, being located in hot tropical regions in which evaporation exceeds precipitation such as the Sahara desert; rainshadow regions in which the presence of a mountain range influences precipitation patterns, resulting in dry conditions on the leeward side of the range, such as in the Patagonian desert in Argentina; and coastal deserts in which prevailing winds prevent precipitation occurring, such as in the Namib desert in southern Africa. Continental-scale effects can also be important, where the interiors of major continents are too distant from moist airflow to receive sufficient rainfall. Vegetation in deserts is sparse, with shrubs being the most common life-form. Deserts are also characterized by succulents (e.g., cacti) and many ephemeral annual and perennial species which have dormancy as some part of their life cycle, enabling them to actively grow and reproduce when rain does occur. The animals of desert regions have also evolved in response to the harsh conditions. Examples include nocturnal activity to avoid the heat of day and dormancy during the periods between rainfall events.

7. Tropical Forests

This biome includes tropical seasonal forests, which occur in humid tropical climates with a pronounced dry season, as well as tropical rainforests. Tropical seasonal forests are characterized by a mix of evergreen and deciduous species, with some forest types dominated solely by deciduous species. In contrast, tropical rainforest thrives in regions in which rainfall is abundant and evenly distributed throughout the year and temperatures are uniformly warm. Tropical rainforests are noted for supporting the highest recorded diversities of both plants and animals. The soils of tropical forests are typically nutrient poor, with decomposition being rapid and with the nutrients stored mostly within the plant biomass.

C. Other Classification Systems

Although large-scale "biome" classification systems are useful in gaining an appreciation of the spatial extent of the world's major ecosystem types, for many applications a finer resolution is often required. Two approaches which aim to increase the detail but still retain a manageable number of categories are considered.

First, Olson's classification is an attempt to map the world's ecosystems as they were in 1980. Included in the classification are natural or seminatural ecosystems (e.g., "Tundra" and "Tropical dry forests") as well as those ecosystems that have undergone human modification (e.g., "Warm crops" and "Paddyland"). Olson's classification recognizes 44 different ecosystem complexes compared with the 10 major biomes in Fig. 1.

A different approach was adopted by Holdridge in his "life zone" classification. This classification aims to predict the potential vegetation that an area could support rather than what ecosystems might actually be occurring there. The classification is based on two main variables, labeled "Biotemperature" and "Annual precipitation." Annual precipitation is the average annual amount of rainfall a region receives, without taking into account its seasonal pattern. Biotemperature is the average annual temperature, excluding temperatures below 0°C and above 30°C. Note that these two climatic variables are essentially the same as those used to define the biomes discussed previously. A third variable in the Holdridge classification, "Potential evapotranspiration ratio," combines precipitation and biotemperature into an index which summarizes the water availability for a particular region and quantifies the balance between the amount of water entering the system through precipitation and that lost through the combined processes of evaporation and transpiration. Evaporation is the physical process whereby water is returned directly to the atmosphere from plant and soil surfaces, and its rate is directly influenced by the amount of incoming solar radiation. Transpiration is the biological process whereby soil water is taken up by plant roots, transported through the plant's vascular systems, and evaporated from the leaf surfaces through pores called stomata. Potential evapotranspiration ratios less than 1.0 indicate an excess of water in the ecosystem, and ratios greater than 1.0 indicate a deficit.

The classification produces 38 major life zone categories, ranging from hot-wet tropical forests and hot-dry temperate and tropical deserts to cold-dry polar deserts (Fig. 4).

The Holdridge classification has been used in global vegetation models to predict major changes in ecosystems types in response to global climate change, and Olson's classification has been similarly used to model the effects of global change, and in particular the dynamics of the global carbon cycle and carbon storage.

II. GLOBAL PATTERNS OF BIODIVERSITY ACROSS TERRESTRIAL ECOSYSTEMS

A. Description of Known Patterns

One of the best documented patterns of biodiversity is the tendency for tropical ecosystems to harbor more species than temperate or polar ecosystems, i.e., there is a gradient in diversity stretching from lower latitudes (high diversity) to higher latitudes (low diversity). Some examples of different terrestrial organisms which show this pattern are shown in Fig. 2. The trend is also seen in a wide range of marine organisms.

It is worth noting that these examples include only a subset of the total biodiversity for any given region, e.g., land-breeding birds or swallowtail butterflies. Furthermore, none of these examples are global in extent, with two of the four (landbirds and mammalian quadrupeds) being based on data from only a single continent.

The reason for the focus on small groups of species, at subglobal scales, is simply because the data required to perform global-scale, multispecies studies of biodiversity are still incomplete. Nevertheless, despite some minor exceptions, the majority of studies that have investigated latitudinal trends in diversity have shown that tropical regions are indeed characterized by high levels of species diversity.

Although distributional information for groups of individual species at the global level is in general poorly known, that for the higher taxonomic groupings (e.g., families) is considered more complete. One example of this approach, compiled by Williams and coworkers, is shown in Fig. 3. It shows the global distribution of the combined number of families of flowering plants, amphibians, reptiles, and mammals. Together, these four groups of families are considered to be the most completely known. It has been estimated that for the chordates 90% of all living species have been identified and for plants 84.4%. These values are far in excess of any other taxonomic group. Figure 3 shows that the trends at the family taxonomic level mirror those of individual species, with a concentration of families in the tropics, particularly around Central America and Southeast Asia, decreasing toward the higher latitudes in each hemisphere.

Figure 4 shows the relationship between the distribution of the family richnesses in Fig. 3 relative to the life zone classification of Holdridge. There is a consistent pattern, with warm/wet tropical ecosystems containing many more families than either the hot-dry (desert) or cold-wet (polar) ecosystems. Incorporating the spatial information of Fig. 3 into the Holdridge classification allows the trends in family richness across different ecosystem types to be displayed visually. The advantage of this approach, over traditional richness vs latitude relationships, is that different ecosystem types occurring at the same latitude but with different richnesses are able to be separated. For example, tropical rainforests and the Sahara desert both occur at tropical latitudes, but the latter is conspicuously lower in diversity.

A second major trend in terrestrial biodiversity is a decrease in diversity with increasing altitude. As altitude increases there is an associated change in the environment that is analogous to the changes that occur when moving from tropical to temperate latitudes. For example, temperature decreases with increasing altitude, and the total land surface area also tends to decrease with altitude (simply because mountains tend to be broader at the base than at the peaks). Because these changes appear to mirror those observed in the tropical to temperate gradient, many researchers have suggested that the factors controlling the pattern of lower diversity at higher altitudes are likely to be the same as those controlling the latitudinal gradient.

B. Possible Explanations

Although many plausible hypotheses have been suggested to explain the observed latitudinal trend in diversity, many of them have proved to be controversial and there is no consensus on which of the explanations, if any, are of primary importance in determining actual diversity gradients. A major source of the difficulty is that most of the explanations are based on correlative evidence, and it is well-known that correlation does not necessarily imply causation. Correlative evidence predominates because appropriate experimental tests are extremely difficult to devise at the spatial and temporal scales required.

For any given area on the earth, the number of species present (S) is due to many processes acting over a range of spatial and temporal scales. These processes can be summarized into three broad categories and combined into the following simple formula:

S = species which have immigrated from outside the area (dispersal)
  + species which have arisen through speciation
  − species which have been lost due to extinction.

At large spatial scales (e.g., regional or between biome), dispersal can be considered of limited importance compared with speciation or extinction. Therefore, the

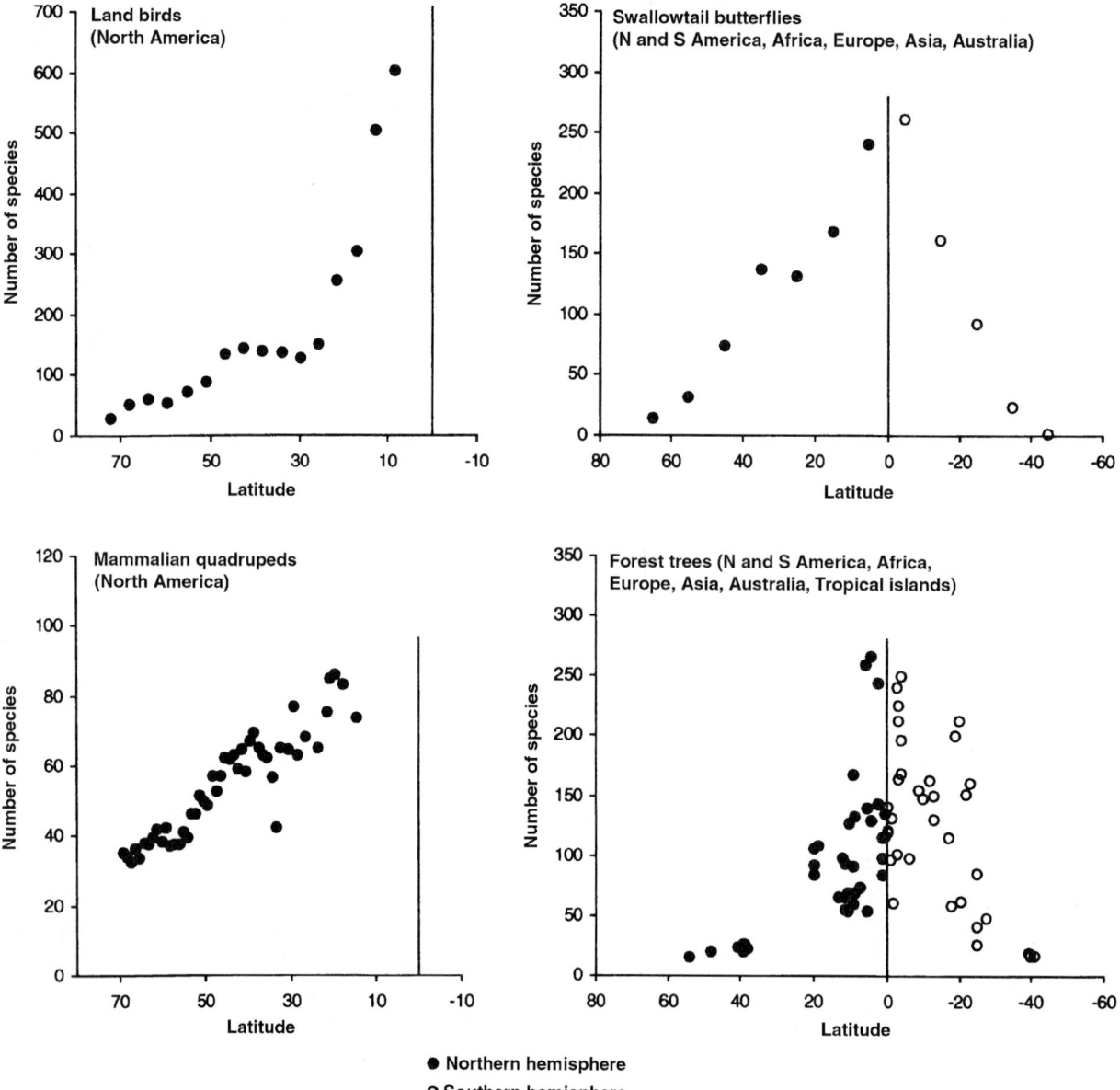

FIGURE 2 Examples of the latitudinal gradient in diversity, showing higher species richness near the equator (0° latitude) for all species groups. Swallowtail butterfly data redrawn from Sutton and Collins (1991). Land bird data-points represent diversity within areas approximately 230,000 km^2 (redrawn from MacArthur and Wilson, 1967). Mammalian quadraped data-points represent average diversity within areas approximately 62,500 km^2 (Rosenzweig, 1995, © Cambridge University Press. Reprinted with the permission of Cambridge University Press). Forest tree data-points represent diversity within 0.1ha plots (redrawn from Gentry, 1988).

hypotheses which aim to explain the latitudinal gradient in diversity must have, as a basis, processes which enhance speciation rates and/or reduce extinction rates in the tropics or, alternatively, which reduce speciation rates at higher latitudes and/or increase extinction at those latitudes. The hypotheses that have been suggested to explain the enhanced diversity of the tropics are described within this broad framework.

1. Time

One of the earliest explanations for the latitudinal diversity gradient is that tropical regions are more ancient,

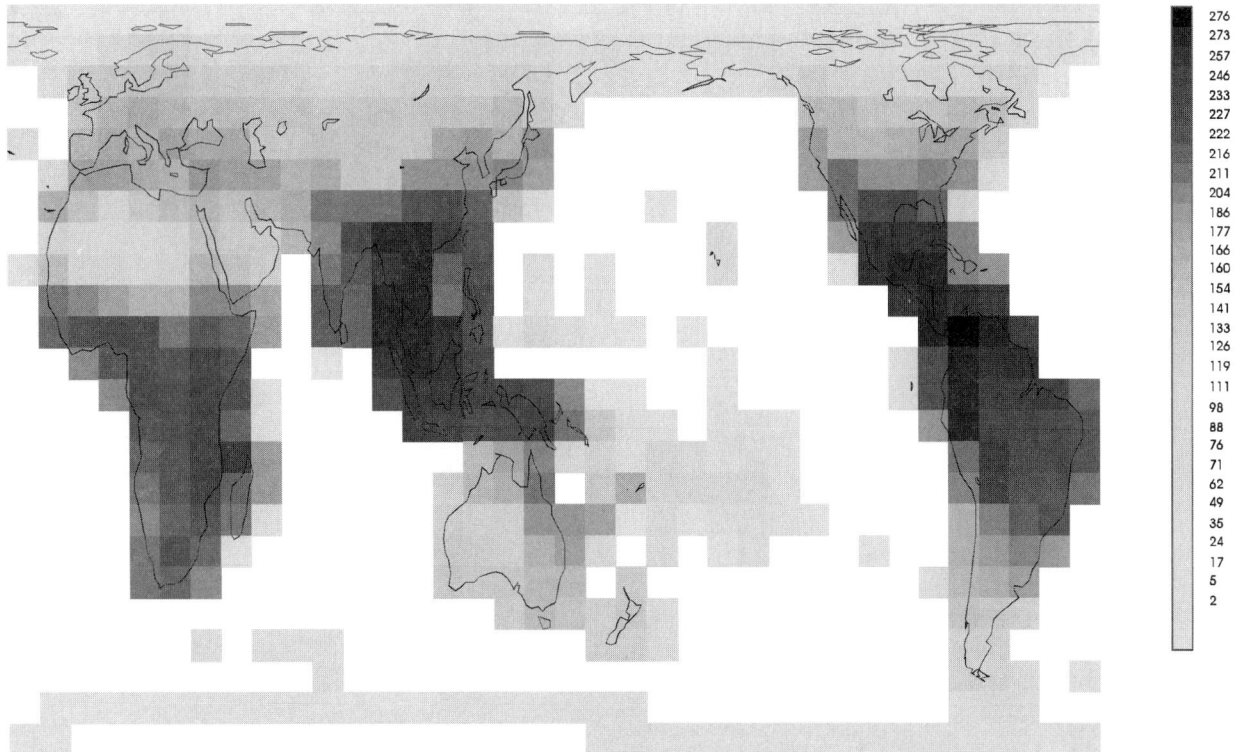

FIGURE 3 The global distribution of the combined family richness of flowering plants, amphibians, reptiles, and mammals (Williams *et al.*, 1997, reproduction kindly supplied by Dr. Williams). Grid cells are approximately 611,000 km^2.

and therefore less disturbed, than their temperate counterparts. The greater age of the tropics implies that there has been more time for speciation to occur, hence a longer time for species to accumulate. Particular emphasis was placed on glaciations because it was assumed that temperate regions are impacted upon by glacial activity more so than the tropics. However, the fossil record does not support this view. For example, there is evidence that the latitudinal gradient for angiosperms has existed for at least 110 million years. There is also evidence that during the most recent glaciation both temperate and tropical forests decreased substantially, with tropical forests being replaced by grassland and savanna. This suggests that tropical regions may not be as ancient or stable as once thought.

However, if the tropical rainforest remnants were more numerous, and their destruction less complete than in the temperate zone, then this continual isolation and reforming of tropical communities might promote diversity through the mechanism of allopatric speciation. Allopatric speciation is considered by many to be the most important of the speciation-promoting mechanisms. It occurs when a geographical barrier subdivides a population, such that dispersal and genetic exchange between the subpopulations is prevented. These subpopulations continue to evolve in isolation, eventually diverging enough so that two new species are produced. In this scenario, higher diversity in the tropics might be promoted due to enhanced allopatric speciation due to historical geological events, possibly combined with an increased extinction rate in the temperate regions.

2. Area

Calculated on a land area basis, tropical and subtropical ecosystems are approximately five times more extensive than temperate ecosystems and approximately twice as extensive as boreal and tundra ecosystems. The reason is due to the earth being approximately spherical so that a zone of a constant width running around the equator (the tropics) will contain much more area than a zone of equivalent width running around the earth's surface closer to the poles. Therefore, tropical ecosystems, which contain the highest diversities, are also those which cover the largest land area.

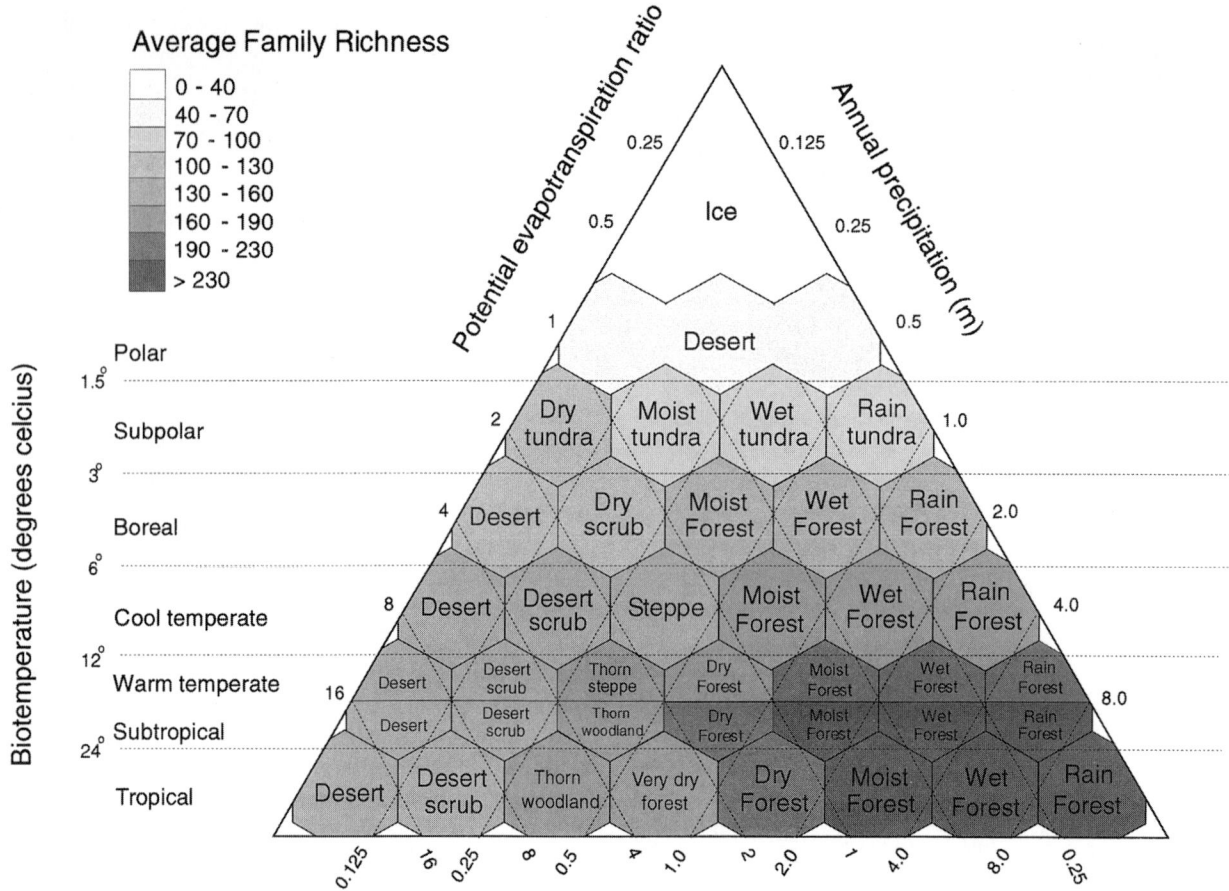

FIGURE 4 The global distributions of combined family richness incorporated into the Holdridge classification, which predicts 38 vegetation types based on the climatic variables 'biotemperature' and 'annual precipitation', as explained in the text. The scale is average combined family richness (flowering plants, amphibians, reptiles, and mammals) per 611,000 km² grid cell calculated for each of the major life zones.

Rosenzweig suggested three reasons why a greater land area in the tropics might result in greater diversity. They are all based on the assumption that the geographical ranges of tropical species are, or were in the past, larger than the ranges of species in other latitudes. First, a larger geographical range leads to a larger total population size, and a larger population size would be expected to buffer the species against accidental extinction. Second, a larger geographical range contains a greater number of potential refuges where the species might survive following catastrophic disturbances or climatic events, again buffering the species against accidental extinction. Third, a larger range has more chance of being subdivided by a geographical barrier, hence increasing the probability of allopatric speciation.

3. Habitat Heterogeneity

All other things being equal, larger areas also contain a greater variety of habitats than smaller areas, and hence they are capable of supporting more species, i.e., there is opportunity for the evolution of species to fill these available habitats. Because tropical ecosystems tend to be larger in extent than those in other regions, it follows that they should contain a greater number of species.

It has also been suggested that the increase in the diversity and complexity of vegetation as one approaches the tropics can enhance diversity by providing animals with a greater variety of arboreal habitats. This increased faunal diversity can enhance diversity because many of these species can become a resource on which species higher in the food chain can be supported. Although the increased vegetation complexity in the tropics potentially explains the increased diversity in animal species, it cannot explain the high plant diversity because there is no evidence to suggest that plant habitats are more diverse in tropical regions.

Despite many studies which illustrate a positive relationship between habitat heterogeneity and diversity, care must be taken with the interpretation of these results because habitats are often defined by the species that occur within them, hence leading to circular reasoning. To be a valid hypothesis, habitat heterogeneity must be defined independently from the species occupying those habitats.

4. Solar Energy

Tropical regions receive, per unit area and per unit time, greater amounts of solar radiation than any other ecosystems. This is again due to a spherical Earth, whereby light energy at higher latitudes intercepts the earth's surface at a more oblique angle compared with the tropics. Although the amount of incoming solar radiation directly affects productivity, and relatedly, the water balance of a region, it also has the potential to accelerate the rate at which new species evolve.

An acceleration of the evolutionary process in tropical regions can arise due to two factors. First, the greater amount of available energy might explain the tendency for tropical species to have shorter generation times. A shorter generation time means more offspring per unit time, hence faster evolution. Second, increased solar radiation would be expected to increase mutation rates, thereby increasing the genetic variability on which natural selection acts, leading to an increase in the rate of speciation. These two factors combine so that over evolutionary time more species would be expected to have evolved in tropical regions rather than those at higher latitudes.

5. Species Interactions

Species interactions, and in particular competition, are central to many hypotheses addressing the latitudinal diversity gradient. A common ingredient to these hypotheses is that there are factors in the tropics which allow competitive exclusion to be avoided or which allow shifts in the "ecological niche" of a species so that competition is no longer intense, enabling the coexistence of many species. The ecological niche is a multidimensional summary of the environmental ranges and resource tolerances of a species. Species with broad niches are considered generalists, and species with narrow niches are considered specialists. Most important, the amount of niche overlap between species is a measure of the amount of competition experienced. For example, two species with overlapping niches require the same range of resources and environmental conditions; hence, they would be expected to compete strongly, and in the absence of any other factors the weaker competitor of the pair would be expected to be eliminated by the stronger. This is known as the "competitive exclusion principle."

It has been suggested that competition between species in the tropics is of greater importance, and is potentially more intense, than that in temperate regions. The reason given is that the tropical environment is more predictable and less prone to large disturbance events which cause indiscriminate mortality; therefore, population sizes are greater so that species interactions are more likely. High competitive pressure and a stable environment allow natural selection to narrow the niche requirements of the species, allowing the environment and resources for which they are competing to be partitioned to a finer degree than otherwise possible. This would result in more species coexisting per unit area of habitat in the tropics than in temperate areas. This hypothesis assumes that competition is not as important in temperate and higher latitudinal regions, and that in these regions the physical environment plays a greater role in forcing natural selection and evolution. The validity of this assumption remains an open question.

A second hypothesis assumes there are more predators (including herbivores) in the tropics. Predators remove prey biomass, thereby reducing competition among the prey species by reducing the demand on available resources. Under this hypothesis, competition is also less intense in the tropics, not because the species have evolved to avoid it through partitioning the environment more efficiently but because prey biomass is reduced to such an extent that competition is significantly reduced. The assumption that predation is more important in tropical ecosystems also remains an open question.

6. Productivity

Productivity is defined as the rate at which biomass is accumulated per unit area. Many hypotheses have attempted to relate the productivity of different ecosystems to diversity; however, they are not always easy to disentangle from one other and from other hypotheses (e.g., the solar energy hypothesis).

One line of reasoning suggests that low-productivity environments (e.g., those at higher latitudes) can support only a limited resource base (e.g., a low-productivity environment might be able to support a herbivore species but at a population density too low to support a predator higher in the food chain). In contrast, in a high-productivity tropical environment a herbivore population might be large enough to support several predatory species. Also, in high-productivity environments food might be more abundant, allowing the opportunity for predators and herbivores to specialize more finely, resulting in a greater number of species (see the discussion in the previous section on the ecological niche).

Another suggestion is that although total productivity over the course of a year might be higher in the tropics than that in other regions, instantaneous rates at other latitudes are much higher (e.g., the flush of plant growth during spring at more temperate latitudes). This effect, combined with low nutrient availability in the tropics, would result in overall lower population growth rates in tropical species, which would tend to slow rates of competitive exclusion and hence extinction. Although slow rates of competitive exclusion cannot explain enhanced diversity on their own, when population sizes are continually impacted upon by external factors such as disturbances, under certain conditions the slow competitive exclusion can be converted into species coexistence, thus promoting higher diversity.

III. SUMMARY

There are consistent trends in global biodiversity across the various ecosystem types which broadly correlate with latitude. Many hypotheses have been proposed to explain this gradient; however, the spatial and temporal scales over which the processes of speciation and extinction take place make rigorous testing of these hypotheses extremely difficult. As a result, there is still no consensus on which of the hypotheses might be the most important in determining actual gradients. A further difficulty in identifying the primary causes of the latitudinal diversity gradient is that many of the hypotheses are intimately related, with the potential for many of them to be acting in unison to determine the diversity of a particular ecosystem. It is also likely that the relative importance of different hypotheses varies both across different ecosystems and through time.

Because the latitudinal diversity gradient appears to be a universal pattern across a wide range of different kinds of organisms, it is likely that there must be some common explanation. The challenge for the future is to identify which factors, particularly which combinations of factors, are the primary causes of this pattern.

See Also the Following Articles

ECOSYSTEM FUNCTION MEASUREMENT, TERRESTRIAL COMMUNITIES • ENDANGERED TERRESTRIAL INVERTEBRATES • INVERTEBRATES, TERRESTRIAL, OVERVIEW • MARINE ECOSYSTEMS

Bibliography

Cox, C. B., and Moore, P. D. (1985). *Biogeography. An Ecological and Evolutionary Approach.* Blackwell, Oxford.

Gentry, A. H. (1988). Changes in plant community diversity and floristic composition on environmental and geographical gradients. *Ann. Missouri Bot. Gard.* 75, 1–34.

Holdridge, L. R. (1967). *Life Zone Ecology.* Tropical Science Centre, San Jose, CA.

Leemans, R. (1990). Global data sets collected and compiled by the Biosphere Project, Working Paper. IIASA-Laxenburg, Austria.

MacArthur, R. H., and Wilson, E. O. (1967). *The Theory of Island Biogeography.* Princeton Univ. Press, Princeton, NJ.

Olson, J. S., Watts, J. A., and Allison, L. J. (1983). Carbon in Live Vegetation of Major World Ecosystems, Report No. ORNL-5862. Oak Ridge National Laboratory, Oak Ridge, TN.

Pianka, E. (1974). *Evolutionary Ecology.* Harper & Row, New York.

Rosenzweig, M. L. (1995). *Species Diversity in Space and Time.* Cambridge Univ. Press, Cambridge, UK.

Sutton, S. L., and Collins, N. M. (1991). Insects and tropical forest conservation. In *The Conservation of Insects and Their Habitats* (N. M. Collins and J. A. Thomas, Eds.), pp. 405–424. Academic Press, London.

Williams, P. H., Gaston, K. J., and Humphries, C. J. (1997). Mapping biodiversity value worldwide: Combining higher-taxon richness from different groups. *Proc. R. Soc. London B* **264**, 141–148.

World Conservation Monitoring Centre (1992). *Global Biodiversity: Status of the Earth's Living Resources.* Chapman & Hall, London.

THERMOPHILES, ORIGIN OF

Anna-Louise Reysenbach and Margaret L. Rising
Portland State University

I. Thermophiles and Hyperthermophiles Defined
II. Ecological Niche and Metabolic Diversity of Hyperthermophiles
III. Molecular Mechanisms for Thermophily
IV. Origin of Thermophiles: Fossil Record and Universal Phylogenetic Tree
V. Theories of the Origin of Life
VI. Conclusion

GLOSSARY

Archaea One of three domains of life; from the Greek *archaios* (ancient, primitive); prokaryotic cells; membrane lipids predominantly isoprenoid glycerol diethers or diglycerol tetraethers; formerly called archaebacteria.
Bacteria One of three domains of life; from the Greek *bacterion* (staff, rod); prokaryotic cells; membrane lipids predominantly diacyl glycerol diesters; formerly called eubacteria.
biomarker A macromolecule unique to a particular organisms or group of organisms such that its detection alone would suggest the presence of the organism or group of organisms.
chemolithotroph Organism deriving its energy from the oxidation of inorganic compounds.
Eukarya One of three domains of life; from the Greek *eu-* (good, true) and *karion* (nut; refers to the nucleus); eukaryotic cells; cell membrane lipids predominantly glycerol fatty acyl diesters.
heterotroph Organism deriving its energy from the oxidation of organic compounds.
hyperthermophiles From the Greek *hyper-* (over), *therme-* (heat), and *philos* (loving); includes organisms that grow best at temperatures warmer than 80°C.
photoautotroph Organism that uses light as its source of energy (photosynthesis) and inorganic carbon (CO_2) as its sole carbon source.
planetesimals Small, solid bodies similar to meteors in composition but revolving in orbit around a central gaseous nucleus, as do planets around the sun.
small subunit ribosomal RNA RNA (about 1500 bases in prokaryotes) that functions as part of the ribosome and the sequence permits the inference of evolutionary relationships among organisms.
thermophiles From the Greek *therme-* (heat) and *philos* (loving); includes organisms that grow best at temperatures between 50° and 80°C.

AT THE TIME when liquid water—a prerequisite for life as we know it—appears in the geological record (3.8 billion years ago), Earth was a hot, anoxic environment and under constant bombardment by meteors, many of which could have virtually vaporized the oceans. Early Earth, therefore, would have been an attractive home to heat-loving thermophiles and their extreme cousins, the hyperthermophiles, where thermophily would have offered a great selective advantage.

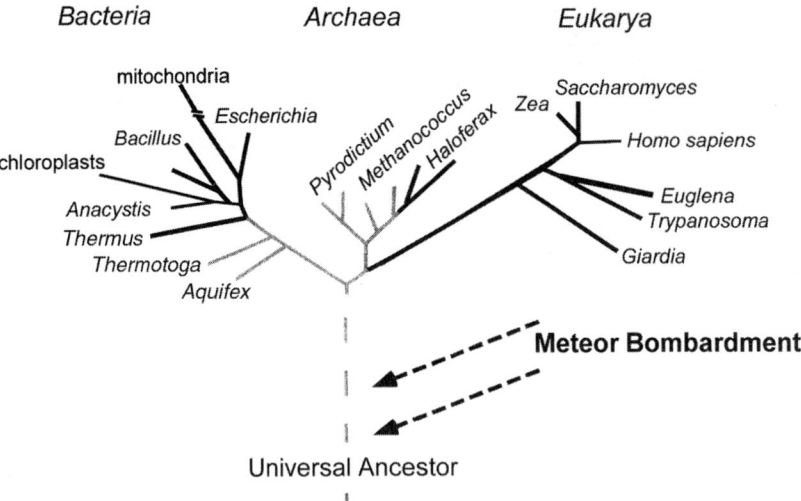

FIGURE 1 Rooted phylogenetic tree based on the small subunit rRNA molecule. The tree is not drawn to scale.

The geochemical and thermal characteristics of deep-sea hydrothermal vents and terrestrial hot springs are thought to approximate conditions on Earth at the earliest possible time that it could have supported life, providing modern analogs for testing early evolution of life hypotheses. In addition, the relatively recent use of the evolutionarily conserved molecules such as the small subunit ribosomal RNA (ss rRNA) sequences in phylogenetic analyses of all life has placed the thermophiles closest to the root of the universal tree of life (Fig. 1). On this tree, hyperthermophiles decidedly monopolize the lowest and shortest branches, suggesting that they may be related to the earliest microbes to inhabit Earth. However, the proposal that thermophiles may have originated early in Earth's history is much debated. Therefore, this article also considers alternative scenarios for the origins of thermophiles.

I. THERMOPHILES AND HYPERTHERMOPHILES DEFINED

All life can be placed in one of the three domains of life: the Archaea, Bacteria, or Eukarya. Archaea and Bacteria are prokaryotes, and the Eukarya are eukaryotes. However, phylogenetically, the Archaea are as different from the Bacteria as the Bacteria are from the Eukarya. Thermophiles are found in all domains as multicellular and unicellular organisms, such as fungi, algae, cyanobacteria, and protozoa, and they grow best at temperatures higher than 45°C. In contrast, the extreme thermophiles, or hyperthermophiles, grow best at temperatures higher than 80°C and are almost exclusively restricted to the Archaea, with only two hyperthermophilic orders in the Bacteria, namely, the Thermotogales and Aquificales. Commonly, thermophiles and hyperthermophiles are found associated with deep-sea and terrestrial hydrothermal vents. Thermophiles have also been obtained from deep (up to 3500 m), hot, subterranean areas, Jurassic oil-bearing sandstone and limestone formations, and suitable manmade environments such as smoldering coal refuse piles, coal-containing uranium mines, and boiling wastewaters from geothermal power plants.

II. ECOLOGICAL NICHE AND METABOLIC DIVERSITY OF HYPERTHERMOPHILES

Deep-sea hydrothermal and terrestrial hydrothermal vents form primarily as the result of plate tectonics, either as a result of seafloor spreading or as a tectonic plate moves across a hot spot. In both scenarios, fissures and faulting occur in the earth's crust, permitting water (seawater, groundwater, or rainwater) to percolate downward. As the fluid moves through the earth's crust and approaches the magma chamber, the fluid heats and reacts with the surrounding rocks, adding some minerals and gases (e.g., iron, manganese, carbon dioxide, hydrogen, and hydrogen sulfide) and removing others (e.g., sulfate and magnesium in seawater). The

FIGURE 2 Sampling a deep-sea hydrothermal vent chimney from the eastern Pacific Ocean. The porous sulfide chimney structures can reach several meters high and are ideal habitats for hyperthermophiles.

fluid is finally forced back to the surface as a highly altered hot fluid, its chemistry representing a history of its travels through the earth's crust. At deep-sea hydrothermal vents, this fluid can reach temperatures of 400°C, remaining in a liquid state due to the hydrostatic pressures at these depths (>2000 m). It rapidly mixes with cold, oxygenated water, and the minerals rapidly precipitate, giving the fluid a smoky appearance (Fig. 2). These vents are aptly named "black smokers." As the minerals precipitate they create porous sulfide–mineral structures called "chimneys." At terrestrial hot springs, the fluid may be ejected forcefully as a geyser or simply bubble into a thermal spring or mudpot (Fig. 3).

The high temperatures and unusual geochemical characteristics of deep-sea and terrestrial hydrothermal vents create ecological niches that are exploited by hyperthermophilic Bacteria and Archaea. The geochemical milieu provides different energy sources (electron donors) and carbon dioxide for a chemolithotrophic existence. The porous chimney structures provide temperature gradients along which thermophiles can situate themselves, and thermal springs create temperature gradients as the fluid moves away from the source. Comparable to photoautotrophs that harvest light energy to fix inorganic carbon, these chemolithotrophs harvest inorganic chemical energy to fix inorganic carbon. It has been calculated that the mixing of superheated thermal fluids at deep-sea vents with cold, oxygenated seawater causes a significant geochemical disequilibrium, significant enough to lower the Gibbs free energy and help drive biologically mediated redox reactions (Shock, 1996). The food chain does not stop there, though. Many thermophiles are heterotrophs, consuming organic carbon produced from biological activity. Additionally, at terrestrial vents, due to the available light energy, the springs are often colonized by a rich diversity of thermophilic phototrophs, including the cyanobacteria, which are oxygenic phototrophs that use water as the reductant and evolving oxygen in the process. Anoxygenic phototrophs are also found at terrestrial vents, primarily using the readily available hydrogen sulfide as the reductant in photosynthesis, releasing elemental sulfur in the process. Both of these types of phototrophs are part of a complex community with heterotrophs, and together they form thick green, purple, and orange microbial mats typical of many terrestrial thermal areas throughout the world.

The majority of hyperthermophiles are anaerobes, with a few exceptions that are microaerophilic, requiring low oxygen concentrations for growth. In contrast, aerobiosis is more common among the thermophiles because oxygen dissolves more readily in the lower temperatures of the cooler thermal springs. The thermophiles that have been isolated from deep-sea vents generally grow at pH values near neutrality. However, thermophiles from terrestrial hot springs have been grown at pH values from 1 to 10.

Although the highest temperature for life has not been determined, it has been suggested that 150°C may represent that threshold. In contrast, the lower temperature limit for most hyperthermophiles is 60°C, a temper-

FIGURE 3 Moose Pool, Yellowstone National Park, Wyoming. This acidic thermal spring contains elemental sulfur that accumulates on the gas bubbles (see inset). The archeon *Sulfolobus acidocaldarius* was isolated from this spring in the late 1960s by Thomas Brock.

ature fatally hot for most other organisms. Even 90°C is too low for one of the more extreme hyperthermophiles, *Pyrolobus fumarii*, which has a maximum temperature for growth of 113°C. However, for these thermophiles, the inability to grow does not result in death; they are capable of entering a dormant state when exposed to cold conditions and will thrive again when returned to favorable temperatures. Thermophiles are therefore very well suited to the deep-sea hydrothermal vent environment, in which the hot fluids are continually being mixed with cold, oxygenated seawater.

III. MOLECULAR MECHANISMS FOR THERMOPHILY

On a molecular level, thermophily is accomplished in many ways, contributing together to overall stability and success of life at high temperatures. In general, the amino acid composition of hyperthermophilic enzymes is surprisingly similar to that of homologous mesophilic enzymes. Because of the amino acid sequence similarity, it has been proposed that heat resistance is explained, instead, by the manner in which the polypeptide chains are folded on themselves, forming their tertiary conformational stability. It is a polypeptide's three-dimensional or tertiary shape in critical locations on the polymer that permits an enzyme to function. Proteins do not spontaneously fold into their active form after synthesis. Instead, many require the assistance of molecular chaperones for proper folding. A specialized form of molecular chaperone, called a thermosome, is of particular importance to hyperthermophiles because it assists proteins in refolding correctly after denaturation due to heat exposure. Also called heat shock proteins, thermosomes effectively increase heat tolerance of organisms so they can function at higher temperatures. For instance, at 108°C, approximately 80% of the protein of *Pyrodictium occultum* consists of a heat-induced thermosome, allowing it to survive 1 h of autoclaving at 121°C (Stetter, 1998).

Additional thermal resistance of DNA is conferred by the presence of DNA topoisomerases and histones and, to a lesser extent, by the base composition of the genome. All hyperthermophiles known to date have been shown to possess reverse gyrase, a unique type I DNA topoisomerase that causes stabilizing, positive supertwists in the DNA helix. Additionally, archaeal (but not bacterial) hyperthermophiles possess histones which significantly increase the temperature at which DNA denatures.

Because an increased guanine + cytosine (G + C) nucleic acid content increases the melting temperature of DNA, one would predict that hyperthermophiles have a higher G + C content in their genomes. In many cases, the opposite is true. However, analysis of the genome reveals that certain genes essential to survival, such as the ss rRNA gene, have higher G + C contents.

The presence of ethers instead of esters in cellular membranes also contributes to the stability of hyperthermophiles at high temperatures. Cell membranes of Bacteria and Archaea are generally distinguished by the presence of glycerol-linked ether lipid moieties in Archaea and ester lipid moieties in Bacteria. An exception to this rule is the presence of a glycerol ether lipid in *Thermotoga maritima*, one of the few bacterial hyperthermophiles. The presence of the ether lipids probably increases stability of cellular membranes against hydrolysis at high temperatures. Membranes of the archaeal hyperthermophiles contain lipids derived from diethers or tetraethers and are highly resistant against hydrolysis at high temperatures.

IV. ORIGIN OF THERMOPHILES: FOSSIL RECORD AND UNIVERSAL PHYLOGENETIC TREE

Were thermophiles the original ancestors to all life? Much of the debate about origins of life and early evolution of life on Earth centers on this question. Is thermophily an acquired feature derived from adaptation? As more information accumulates, the resolution of these questions becomes less clear. However, many different scenarios are emerging based on fossil record analyses, RNA and protein phylogenies, and our understanding of the conditions that prevailed on early Earth.

A. Early Earth and Its Hyperthermophilic Niche

Perhaps the most compelling evidence supporting the proposal that hyperthermophiles are living fossils of the first life that arose on Earth is that modern ecological niches of hyperthermophiles fit our current view of the conditions on primitive Earth, which was anoxic, hot, and volcanically active. During its first half billion years of existence (4.6–4.0 billion years ago), the temperature of the surface of Earth probably exceeded 100°C. From recent studies of ancient sedimentary rocks of the Isua formation in Greenland, traces of biologically produced carbon were detected, suggesting that the invention of life had already taken place 3.8 billion years ago. Thus, sometime prior to 3.8 billion years ago, it is likely that the first living organisms appeared at a time when Earth, although cooling, was much hotter than it is now, and all the essential geochemical energy sources were present for a chemolithotrophic and thermophilic existence. Heat-tolerant organisms such as hyperthermophiles would have been at a distinct advantage over heat-sensitive organisms.

In addition to the generally hot and anoxic conditions on early Earth, the bombardment of early Earth by meteors may have caused a bottleneck event that favored survival of hyperthermophilic organisms. An examination of the impact craters on Earth's geologically inactive moon strongly suggests that the earth was heavily bombarded by meteors from its formation 4.6 billion years ago until about 3.8 billion years ago. A large meteor impact, of which there were many, would have caused intense local heat and sent sufficient debris into the atmosphere to cause global cooling and a reduction in light intensity, detrimentally impacting photosynthesizing organisms. There is also evidence of extremely large meteor impacts that would have generated enough energy to heat the earth's surface to 2000 K and virtually vaporize the world's oceans. In either scenario, hyperthermophiles would have been uniquely adapted to survive. They could have tolerated lower light levels because many are chemolithotrophs, making a living independent of photosynthesis. Any life adapted to live near ocean-floor volcanic centers would have had the greatest chance of surviving one of these ocean-evaporating events.

B. The Fossil Record

Additional evidence for early ancestors of microbial life can be found in the rock record. Rocks older than 3.5 billion years are highly metamorphosed and deformed, precluding the preservation of morphological fossils. However, the 3.8-billion-year-old Isua rocks from Greenland, mentioned previously, offer indirect evidence that life existed earlier than 3.5 billion years ago. These sedimentary rock formations are associated with liquid water, a prerequisite for life as we know it. The rocks contain geochemical evidence of past biotic activity preserved within the minerals. Grains of apatite (calcium phosphate) contain a significant portion of carbonaceous inclusions that are isotopically "light," suggesting biological activity. Microorganisms preferentially incorporate the lighter carbon-12 isotope over the carbon-13 isotope when inorganic carbon is fixed into organic carbon using the enzyme ribulose bisphosphate carboxylase/oxygenase (Rubisco). This isotopic evidence suggests these putative life-forms were carbon-fixing chemolithotrophs or phototrophs and not heterotrophs.

Three hundred million years later, the first microfossils appear in rocks (about 3.5 billion years old) from Western Australia and South Africa. The fossils are simple rod-shaped and filamentous bacteria, indistinguishable morphologically from any similar-shaped bacteria today. Many are reminiscent of photosynthetic bacteria, such as modern cyanobacteria. However, there is some doubt that much oxygen evolved from photosynthesis into early Earth's atmosphere. Isotopic signatures associated with these fossils do suggest a lighter carbon preference, similar to that observed in the 3.8-billion-year-old rocks. Were these first fossils perhaps thermophiles? Unless we are able to identify specific thermophilic biomarkers in these fossils, we may never know the answer.

C. The Molecular Fossils

Although they are not a substitute for the existence of microfossils, evolutionarily conserved macromolecules within living organisms function as "molecular fossils," permitting an inference of relatedness to other organisms and of evolutionary distance from a hypothetical common ancestor. Carl Woese's pioneering work with the ss rRNA molecule (16S rRNA in prokaryotes) led to the generation of the universal phylogenetic tree (Fig. 1), in which life falls within three domains: the Archaea, Bacteria, and Eukarya. The suitability of ss rRNA as the macromolecule for evolutionary comparisons is based on several critical reasons. The ss rRNA exists in every living organism, its function in every living organism is constant, it is an ancient molecule (assuming protein synthesis was necessary in the earliest of life), it has undergone only moderate changes in nucleic acid sequence when compared between diverse biological domains, and the size of the molecule is large enough to contain considerable information but small enough to be manageable.

1. Molecular Phylogenies

The relatedness of organisms to each other as suggested by the universal phylogenetic tree is markedly different than the previously proposed phylogenies. Most notably, the three domains of biological life (the Bacteria, Archaea, and Eukarya) replace the five kingdoms (animals, plants, fungi, protista, and monera). The Eukarya domain includes the multicellular animals, plants, fungi, and protista. Distributed throughout the Bacteria and Archaea are the unicellular prokaryotes. Using alternative markers, such as 23S rRNA, RNA polymerase, elongation factor Tu, F_1F_0 ATPase β-subunit, RecA protein, and HSP60 heat shock protein, reveals general agreement with the major lineages in the ss rRNA tree. However, some discrepancies are evident.

One can root the ss rRNA tree by comparing it with a paralogous marker that arose from gene duplications (such as EF-Tu or ATPase) prior to the diversification of the three primary domains. This rooted tree reveals that all hyperthermophiles, whether bacterial or archaeal, comprise the deepest and earliest lineages within the tree (Fig. 1). In other words, they may be the closest living relatives to a common universal ancestor. However, mutational rates of ss rRNA and their associated base sequence changes are not consistent enough through time nor within lineages to assign a specific date to branching events, and so one cannot assign a clock to this tree. Nonetheless, the shorter branches leading to the hyperthermophiles indicate a slower evolutionary rate, whereas the longer branches leading to their mesophilic relatives suggest a faster evolutionary rate. It is possible that the hyperthermophiles have short lineages because they are so highly adapted to their ecological niche that they have no need for further adaptation. It is also possible that certain evolutionary constraints are imposed on thermophiles due to the extreme selectivity of the high-temperature environment. Nevertheless, the placement of thermophiles at the base of the universal phylogenetic tree strongly suggests that they are most closely related to the common ancestor of all life. In this sense, they are not unlike other "living fossils," or organisms whose morphology has changed very little based on comparisons of modern and ancient fossilized specimens, such as horseshoe crabs, club mosses, *Welwitschia*, and *Gingko bilboa*.

Although hyperthermophiles may be most closely related to the universal ancestor of all life, their metabolic machinery is anything but primitive. Viewed independently of the rooted ss rRNA universal phylogenetic tree, their heat-tolerant adaptations appear sophisticated or even highly evolved. If an analysis of another universal and highly conserved macromolecule placed hyperthermophiles away from the root of its phylogenetic tree, it would severely shake the topology of the phylogenetic tree discussed previously. A phylogenetic analysis of the conserved DNA-directed RNA polymerase has done just that (Klenk *et al.*, 1999). Although overall the RNA polymerase-based phylogeny corresponds very well with the ss rRNA-based phylogeny, the positions of the hyperthermophilic bacteria *A. pyrophilus* and *Thermotoga maritima* are in the middle of the RNA polymerase tree instead of in the most deeply rooted branch in the Bacteria as suggested by ss rRNA phylogeny. Other examples also exist in which the ss rRNA tree does not quite hold true. Additional analyses of universal and highly conserved macromolecules such as DNA-directed RNA polymerases or whole genomes may one day provide a comprehensive universal phylogenetic tree that may or may not place hyperthermophiles near its root.

The debate continues, and evidence against an early origin of thermophily accumulates. If mesophily preceded thermophily, then versions of heat-tolerant mechanisms that characterize hyperthermophiles would have evolved first in mesophiles and been exploited by hyperthermophiles to open up new ecological niches. Structural analysis of reverse gyrase and *Taq* polymerase support such a proposal, suggesting they evolved from mesophilic ancestors (Forterre, 1996). Additionally, the lipid moieties in hyperthermophilic Bacteria are not homologs of the lipid moieties in hyperthermophilic Archaea. Instead, they are analogs with opposite stereochemistry, suggesting the ether-based lipid moiety feature evolved independently in the Bacteria and Archaea instead of from a common ancestor (Forterre, 1996).

2. Lateral Gene Transfer and the Genetic Annealing Model

Lateral (horizontal) gene transfer and variable rates of evolution and mutation may have played a significant role in the evolution of life and may offer an explanation for the confusing results of the analyses of non-ss rRNA molecules (Woese, 1998; Doolittle, 1999). Lateral gene transfer refers to the exchange of genetic material from one organism to another. It is contrasted against vertical gene transfer, in which genetic information is passed vertically from parent to offspring. Bacteria and Archaea, which as a general rule have simpler cell designs, exhibit considerable horizontal gene transfer. In contrast, the Eukarya, which contain highly evolved cell designs, generally do not engage in horizontal gene transfer. Woese proposed in his "genetic annealing model" that the universal ancestor was not a discrete entity but rather a diverse community of primitive cells that evolved as a unit, engaging in horizontal gene transfer on a scale even greater than that occurring in Bacteria and Archaea today and developing into three different communities, which in turn gave rise to the three primary lines of descent as defined by the ss rRNA tree (Woese, 1998).

The ss rRNA universal tree, then, is not a conventional organismal phylogenetic tree but rather a history of the evolution of central components of the ribosome, with the deeply rooted branches of the universal tree representing a "gene tree" and not an "organismal tree." In other words, by the time the three primary lines of descent emerged, and the tree started to take form, self-replicating organisms had not yet taken form. Instead, "life," with its associated exchange of genetic information, existed in communal entities by way of lateral gene transfer. Additional cell complexity and function needed to evolve before there was life as we envision it today. Nonetheless, these communal entities could have evolved in the conditions on early Earth.

V. THEORIES OF THE ORIGIN OF LIFE

Theories of the origin of life are highly debated. Did life originate here on Earth, or did it develop on another planet, such as Mars, which shared similar planetary conditions 4.0 billion years ago? Did life originate in environments analogous to deep-sea hydrothermal vents, or did the first biological molecules form as a result of the reaction of electrical discharges within a prebiotic soup of chemicals? Closely tied to much of this debate are models pertaining to early Earth's atmosphere and the possible energy carbon sources (organic and inorganic). Given that this article deals with the origins of thermophiles, we focus our discussion on a possible high-temperature origin of life, highlighting some arguments that do not support this hypothesis.

In light of Earth's hot, reduced, and anoxic origins, it has been proposed that life may have arisen in environments very similar to present-day deep-sea hydrothermal vents. Here, life could take refuge and survive the planetesimal bombardment of early Earth in an environment rich in redox energy and inorganic carbon. The first biological entities could evolve rapidly as anaerobic, hyperthermophilic chemolithotrophs. The rapid mixing of superheated hydrothermal fluid, rich in reduced minerals, with cold oxygenated sulfate-rich seawater at deep-sea hydrothermal vents creates thermodynamic disequilibrium conditions that favor the production of organic molecules (Shock, 1996). The abundance of charged minerals associated with deep-sea vents led to the proposal by Gunter Wächtershäuser that the original source of reducing power for carbon fixation (and therefore a chemolithotrophic origin of life) may have come from exergonic "pyrite-pulled reactions"—the oxidative formation of pyrite (FeS_2) from ferrous sulfide (FeS) and hydrogen sulfide (H_2S or SH^-). Charged surfaces such as pyrite would attract and bind any negatively charged molecule in solution, such as carbonate, phosphate, and sulfide. These molecules would be maintained in sufficient proximity for subsequent metabolic interactions by "surface bonding," or anionic bonding to the positively charged pyrite surface, resulting in the formation of the first biomolecules on a charged surface. Many are skeptical of these high-temperature scenarios for the origin of life, offering evidence that many of the biomolecules of life, such as RNA, are not stable at high temperature and therefore would not support the proposal that RNA arose prior to DNA in an early RNA world (Miller and Bada, 1988). Furthermore, others have generated models to show that the ancestral rRNA would have a moderate G + C content, contrary to all thermophilic rRNAs which are generally rich in G + C contents (Galtier et al., 1999).

The "panspermia" hypothesis offers another scenario for the origin of life. It holds that life originated elsewhere in the galaxy and that microorganisms were propelled through space to Earth or, alternatively, that exogenous organic carbon arriving on planetesimals fueled a heterotrophic origin of life on Earth. Chyba and Sagan (1992) estimate that approximately 4 billion years ago about 100,000,000 kg/year of organic carbon was delivered by interplanetary dust particles, and about 10,000,000,000 kg/year of organic carbon was produced by postimpact plumes caused by meteor, asteroid, and comet bombardment. From what is known about the hostile environment on early Earth and the timing of the appearance of the first microfossils, life became considerably complex within a relatively short period of time. Therefore, life either evolved very rapidly after its inception or, consistent with the panspermia hypothesis, it was raining down on Earth from elsewhere in the galaxy. Because the panspermia hypothesis involves high temperatures as particles enter the earth's galaxy, it embraces, at the very least, a thermotolerant origin of life.

VI. CONCLUSION

Morphological features of early microfossils generally offer little information about their relatedness to extant microbes. Consequently, microbiologists have come to rely on the molecular record to define phylogenetic relationships and infer the history of microorganisms. The absence of a good microbial fossil record prevents the assignment of a time line on the molecular record. The ss rRNA universal phylogenetic tree places hyperthermophiles in the deepest and shortest branches of the tree, implying that they are the closest living relative to the common universal ancestor of life. However, as we sequence more microbial genomes and create detailed phylogenies from other molecules, the situation becomes increasingly confusing, with lateral gene transfer perhaps "muddying" the phylogenetic record. However, high-temperature conditions that prevailed on early Earth and its subsequent cooling favor the likelihood that at some time during Earth's early history thermophiles took advantage of the geochemical energy supplied by the hydrothermal fluid and evolved into the highly adapted chemolithotrophic and heterotrophic life that is found at high temperatures.

See Also the Following Articles

ARCHAEA, ORIGIN OF • BACTERIAL BIODIVERSITY • EUKARYOTES, ORIGIN OF • HIGH-TEMPERATURE ECOSYSTEMS • ORIGIN OF LIFE, THEORIES OF • PSYCHROPHILES • VENTS

Bibliography

Chyba, C., and Sagan, C. (1992). Endogenous production, exogenous delivery and impact-shock synthesis of organic molecules: An inventory for the origins of life. *Nature* 355, 125.

Corliss, J. (1990). Hot springs and the origin of life. *Nature* 347, 624.

Doolittle, W. (1999). Phylogenetic classification and the universal tree. *Science* 284, 2124.

Forterre, P. (1996). A hot topic: The origin of hyperthermophiles. *Cell* 85, 789.

Galtier, N., Tourasse, N., and Gouy, M. (1999). A nonhyperthermophilic common ancestor to extant life forms. *Science* 283, 220.

Jakosky, B. (1998). *The Search for Life on Other Planets*. Cambridge Univ. Press, New York.

Klenk, H., Meier, T., Durovic, P., Schwass, V., Lottspeich, F., Dennis, P., and Zillig, W. (1999). RNA polymerase of *Aquifex pyrophilus*: Implications for the evolution of the bacterial *rpoBC* operon and extremely thermophilic bacteria. *J. Mol. Evol.* 48, 528.

Madigan, M., Martinko, J., and Parker, J. (1997). *Brock Biology of Microorganisms*, 8th ed. Prentice Hall, New York.

Miller, S., and Bada, J. (1988). Submarine hot springs and the origin of life. *Nature* 334, 609.

Mojzsis, S., Arrhenius, G., McKeegan, K., Harrison, T., Nutman, A., and Friend, C. (1996). Evidence for life on earth before 3800 million years ago. *Nature* 384, 55.

Shock, E. (1996). Hydrothermal systems as environments for the emergence of life. In *Evolution of Hydrothermal Ecosystems on Earth (and Mars?)* (G. Bock and J. Goode, Eds.). Wiley, Chichester, UK.

Stetter, K. (1998). Hyperthermophiles: Isolation, classification, and properties. In *Extremophiles: Microbial Life in Extreme Environments* (K. Horikoshi and W. Grant, Eds.). Wiley-Liss, New York.

Wächtershäuser, G. (1988a). Pyrite formation, the first energy source for life: A hypothesis. *Syst. Appl. Microbiol.* 10, 207.

Wächtershäuser, G. (1988b). Before enzymes and templates: Theory of surface metabolism. *Microbiol. Rev.* 52, 452.

Woese, C. (1998). The universal ancestor. *Proc. Natl. Acad. Sci. USA* 95, 6854.

TIMBER INDUSTRY

Seppo Kellomäki, Jari Kouki, Pekka Niemelä, and Heli Peltola
University of Joensuu

I. Main Features of Boreal Forests, with Special Reference to Finland
II. Dynamics of Boreal Forest Ecosystems and Management Implications
III. Management of Forests for Optimal Productivity and Timber Production
IV. Impacts of Timber-Oriented Management on Biodiversity: Comparison between Managed and Natural Forest Areas
V. Conclusions

GLOSSARY

biodiversity of forests Variability in genetic structures, species composition, and/or habitat properties in forest ecosystems.
boreal forests A forest zone dominated by coniferous species, covering the northern latitudes of North America, the Nordic countries, and Russia.
forest management Manipulation of the properties of tree populations and/or modification of the properties of the habitats occupied by these populations.
sustainable forestry The management and utilization of forest resources in a way that balances human needs with undisturbed functioning of the forest ecosystem, and a long-term maintenance and sustainability of the resources and biodiversity of forest ecosystems.
timber industry Any kind of forest-based production using timber as a raw material.

THE MANAGEMENT OF FOREST RESOURCES can substantially change the structure and functioning of forest ecosystems at both the local and the regional level, with potential impacts on forest biodiversity. This article exemplifies these impacts in the context of boreal conditions with the aim of outlining how to maintain the biodiversity of boreal forest ecosystems. The management of forest resources and the impacts of this management on biodiversity are contrasted with the successional dynamics of boreal forests, which are also disturbance driven under natural conditions. The findings of this discussion are related to the structure of forests on opposite sides of the Finnish–Russian border, and the management of boreal forests in Finland and Russia is addressed against a background of the natural succession in such forests. Furthermore, these two forest areas offer a unique opportunity to study the effects that intensive management may have on the biodiversity of boreal forests. In this discussion, biodiversity is illustrated by reference to selected beetle populations. Outlines for the management of boreal forests with a view to preserving biodiversity are discussed.

I. MAIN FEATURES OF BOREAL FORESTS, WITH SPECIAL REFERENCE TO FINLAND

Forests currently cover about 3,440 million ha, or 27% of the world's land surface, and represent the largest land-based ecosystem types in the world. Forest species, along with related species communities and ecosystems, represent a large proportion of the global biological diversity and are of great importance in maintaining the functioning and structure of the biosphere. The boreal forests alone represent an area of 600 million ha and account for about 20% of the world's industrial timber supplies. Commercially, the most important coniferous species are pine (*Pinus*), spruce (*Picea*), fir (*Abies*), larch (*Larix*), juniper (*Juniperus*), thuja (*Thuja*), and hemlock (*Tsuga*), whereas the most common deciduous species in these forests are aspen (*Populus*), birch (*Betula*), willow (*Salix*), and alder (*Alnus*).

The boreal forests in northern Europe or Fennoscandia (including Norway, Sweden, Finland, and northwestern Russia) are probably the most intensively utilized of all; that is, they form 85% of the total forest area in Europe (956 Mha) and provide about 40% of the timber used in Europe (Table I). The stocking (68 m^3 ha^{-1}) and growth rate (1.1 m^3 ha^{-1} year^{-1}) in these forests are substantially lower than in more southerly regions, however, and consequently the timescale in forest production is long, with rotations of 40–160 years according to species and region. Growth and felling are currently well balanced, and the growth may even exceed the needs of the timber industry.

Forestry in northern Europe is mainly based on native tree species which invaded this region postglacially. In Finland, for example, Scots pine (*Pinus sylvestris*) is the dominant species throughout the country (about 45% of total stem volume) wherever nutrient and water supplies are limited at the site. Scots pine can also dominate at more fertile sites, but both Norway spruce (*Picea abies*) (about 37% of total stem volume) and birch (*Betula pendula* and *Betula pubescens*) can win out over it under such conditions. Birch species (about 18% of total stem volume) are most common on the most fertile soils.

In contrast with many developing countries, forest coverage in Fennoscandia has remained at the current level for decades or even increased slightly. Forest utilization in Finland has been occurring for centuries. Toward the late 1800s Finnish forests were intensively used for slash-and-burn cultivation and tar production without any actions to promote new tree generation. As a result, large areas, especially in southeastern Finland, became dominated by relatively young forests 100 years ago. Effective prevention of slash-and-burn cultivation and forest fires together with active timber-oriented management have led to an increase in the areas of closed-canopy forest during the past century.

In general, the number of species per unit area is low at northern latitudes. This also holds for Finland, where the estimated total number of species is about 50,000, of which about 43,000 are known. The reason for the relatively low total number of species is the short time that has elapsed since the last glaciation (10,000 years), with the consequence that immigration is still occurring. These species and the subsequent biodiversity involve a large contribution from natives of the eastern taiga (e.g., flying squirrel, Ural owl, and Siberian jay). Most of these taiga species are connected with spruce forests. Furthermore, the amount of dead or decaying wood is generally high in boreal forests,

TABLE I
Main Features of the Boreal Forest Zone in Europe[a]

Countries representing boreal forests in Europe	Examples of tree species of importance to forestry	Main parameters for forests	
Finland, Sweden, Norway excluding southwest part, northwestern Russia	*Pinus sylvestris, Picea abies, Betula pendula, Betula pubescens, Populus tremula, Pinus contorta, Larix sibirica*	Area (Mha)	808
		Total volume (Mm^3)	55,031
		Total growth (Mm^3)	878
		Total cut (Mm^3)	643
		Balance (Mm^3)	+235

[a] Source: UN-ECE/FAO (1992).

especially in the early phase of the succession after a fire or windstorm. Consequently, a high proportion of the forest species (20–25%) are dependent on dead wood (800 coleopterans, 1000 dipterans, 1000 fungi, 200 lichens, etc.). Many of these species are also specialized in living on recently burned tree material, whereas a high proportion live in peatland forests or on mires. Most of these species need a moist and/or shady habitat.

II. DYNAMICS OF BOREAL FOREST ECOSYSTEMS AND MANAGEMENT IMPLICATIONS

A. Management: Control of the Structure and Functioning of the Forest Ecosystem

Forests are ecosystems in which trees and other green plants occupy the site and intercept solar energy under the control of climatic and edaphic factors (Fig. 1). The interaction of populations with their environments forms complex food webs, in which the solar energy flows from producers to consumers. The links between the organisms are the key to the management of a forest ecosystem. Proper manipulation of forest dynamics will allow the production of timber or other items, with the management needs being related to the management objectives and to the changes in the structural properties of the forest ecosystem occurring with time. The management history and past structure of a forest ecosystem may also have a crucial impact on future management. Production in a forest can be maintained and/or increased by manipulation of the genetic properties of the tree populations and/or modification of the properties of the habitats occupied by these populations. This takes places by controlling the long-term functional and structural development of the forest ecosystems (the forest succession) in order to induce them to produce the items defined in the management goals. In other words, any kind of forest-based production is a result of environment, genotype, and the interaction between these.

B. Successional Dynamics of Forest Ecosystems and Management Implications

The long-term growth and development of tree populations are subject to disturbances due to wind, fire, and attacks by pests and pathogens. Under boreal

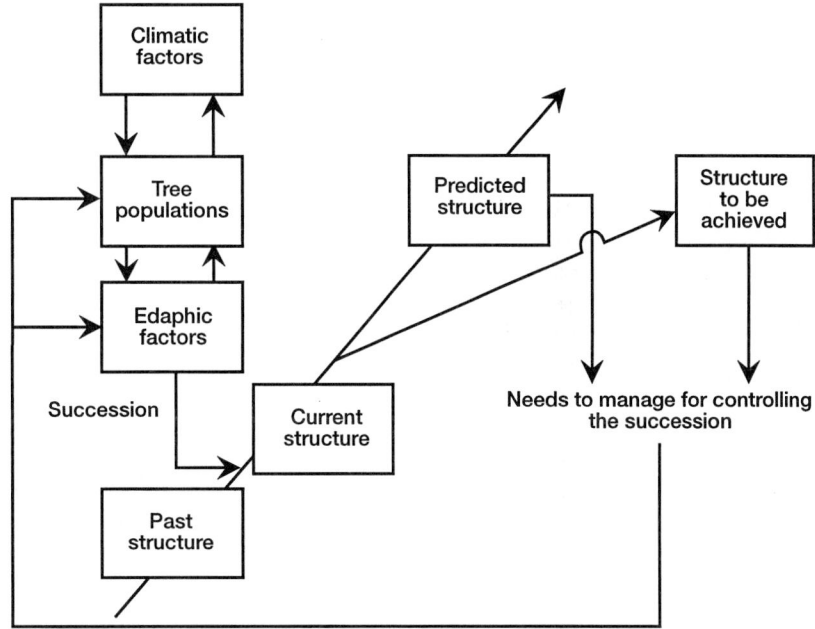

FIGURE 1 Schematic presentation of the structure and functioning of a forested ecosystem as an interaction between climatic and edaphic factors and the populations of organisms, with implications for management.

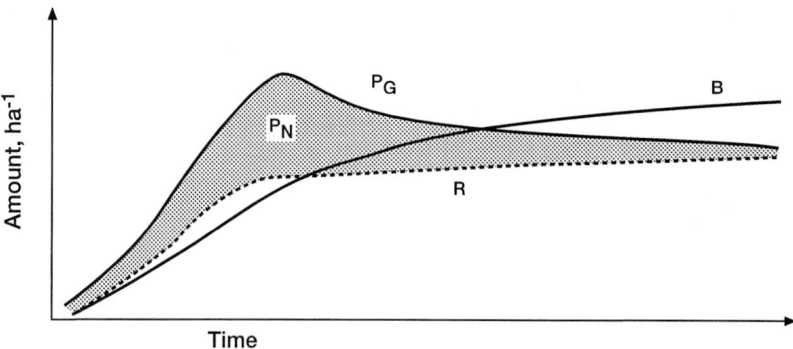

FIGURE 2 Schematic presentation of the course of total growth (P_G), net growth (P_N), respiration loss (R), and the accumulation of mass (B) in trees during the succession.

conditions in particular, such disturbances cause cycling in the dynamics of the forest ecosystem by returning the succession to an earlier phase. For example, a fire that kills old trees provides space for regeneration. For larger forest areas, this implies that forest landscapes contain a mosaic of tree stands of varying species composition and age, from open areas and saplings to mature trees.

During the succession, each stand or part of the mosaic undergoes a gradual change involving alterations in species composition and the accumulation of organic matter in trees and soils (Fig. 2). Many broad-leaved tree species can make better use of the abundant supply of resources at fertile sites than can coniferous trees. Broad-leaved species typically dominate the initial phase of the succession after a fire, and wherever site fertility is sufficiently high the changing conditions will later start to support shade-tolerant conifers more than the broad-leaved trees. This brings an invasion of conifers such as Norway spruce to the site, which finally replace the broad-leaved species unless a disturbance causes the succession to return to an earlier phase. These disturbances and the dynamics of the tree species result in substantial variability in the mass of organic matter in forest ecosystems. This cycling determines the management of the forest ecosystem and the timber yield obtained.

In terms of the energetics of the ecosystem, the energy intercepted by the trees and other green plants and the energy used at different trophic levels in the ecosystem's food webs are balanced by means of enhanced formation of dead and decaying wood and other organic matter. In this context, the food webs provide complex feedbacks through which the populations of the various species interact, with consequent control of the ecosystem dynamics. The food webs tend to increase in complexity during the succession.

C. Disturbance Dynamics of Forest Ecosystems and Management

The disturbance dynamics of forest ecosystems provide the framework in which management occurs. Two categories of disturbance may be distinguished which drive the succession: autogenic and allogenic (hence, we may speak of an autogenic and/or allogenic succession). Allogenic and autogenic successions are typical processes operating simultaneously in boreal forests.

In the autogenic succession, the structural dynamics of the forest ecosystem are driven by the regeneration, growth, and death of single trees, i.e., minor disturbances related to their life cycle (gap-phase dynamics). The deaths of single trees and the consequent gaps in the canopy release resources and enhance regeneration and growth. The concept of autogenic succession is applied in the form of selection forestry or uneven-aged management, in which single trees are the basic object of management. This system is widespread in the temperate zone, in which autogenic succession predominates in the dynamics of forest ecosystems.

Allogenic succession refers to dynamics driven mainly by major disturbances induced by fires, gales, and excessive snowfall. Under these conditions many trees may die simultaneously, releasing space for new trees to regenerate and grow to form a population with individuals of more or less the same age (cohort-phase dynamics). In allogenic succession the stand is the basic unit driving the structural dynamics of the forest ecosystem. The concept is applied to management in the form

TABLE II
Schematic Presentation of the Effects of Given Management Measures on Site Properties and Tree Populations[a]

Measure	Effects on site properties and populations				
	Nutrients	Moisture	Temperature	Soil physics	Species
Allogenic measures					
Site preparation	+++	++	+++	+++	+++
Ditching	++	+++	++	++	+++
Prescribed burning	+++	+	++	++	+++
Fertilization	+++				+++
Regenerative cutting	++	++	++	+	+++
Autogenic measures					
Selection for regeneration	+	+	+	+	+
Precommercial thinning	+	+	+	+	+++
Commercial thinning	++	++	+	+	++
Pruning	+	+	+		+

[a] +, slight effect; ++, moderate effect; +++, substantial effect.

of standwise forestry, or even-aged management, in which single stands form the basic object of management. This system is commonly employed in the boreal zone, in which allogenic succession predominates in the dynamics of forest ecosystems.

Management involves the use of a variety of measures to control the supply of resources and the properties of tree populations. The measures used to control the supply of resources may be classified into two categories in relation to the disturbances driving the succession methods liberating resources to a large extent (allogenic measures related to the allogenic succession) and methods liberating resources to a lesser extent (autogenic measures related to the autogenic succession) (Table II).

Allogenic measures mainly control the properties of the soil system, and subsequently the nutrient cycle and soil moisture, with the greatest enhancement of growth and regeneration of pioneer species being achieved through disturbance of the soil surface. Site preparation refers to mechanical measures used to reduce the effects of the ground vegetation on saplings and to adapt the physical and chemical properties of the soil. In this context, ditching is aimed only at lowering the groundwater level and reducing excess soil moisture. Prescribed burning affects the chemical properties of the soil and eliminates ground vegetation interference with sapling growth. Fertilizing implies the addition of nutrients at the site, leading to a substantial increase in the availability of nutrients for tree growth. Regenerative felling refers to terminal felling aimed at promoting reforestation of the site through natural regeneration by means of natural seeding or artificial regeneration by means of the sowing of seed or planting.

Autogenic measures are mainly represented by spacing, leading to enhancement of growth and regeneration among both pioneer and climax species. This is achieved through the thinning of sapling stands (precommercial thinning) and more mature stands (commercial thinning). The tending of sapling stands is aimed at proper spacing, as is also true of thinnings, and it also implies the elimination of unwanted tree species. Pruning involves the removal of dead or living branches from the lower crown in order to increase the amount of knot-free wood obtainable from the stem. Selection for regeneration refers to the removal of single mature trees or small groups of mature trees in order to create canopy gaps in which seedlings can become established and grow.

III. MANAGEMENT OF FORESTS FOR OPTIMAL PRODUCTIVITY AND TIMBER PRODUCTION

Among the key issues to be addressed in management is the optimal distribution of tree species and stand age (or tree size) over the forest area in order to maximize long-term growth. Assume a forest area of 100 ha to

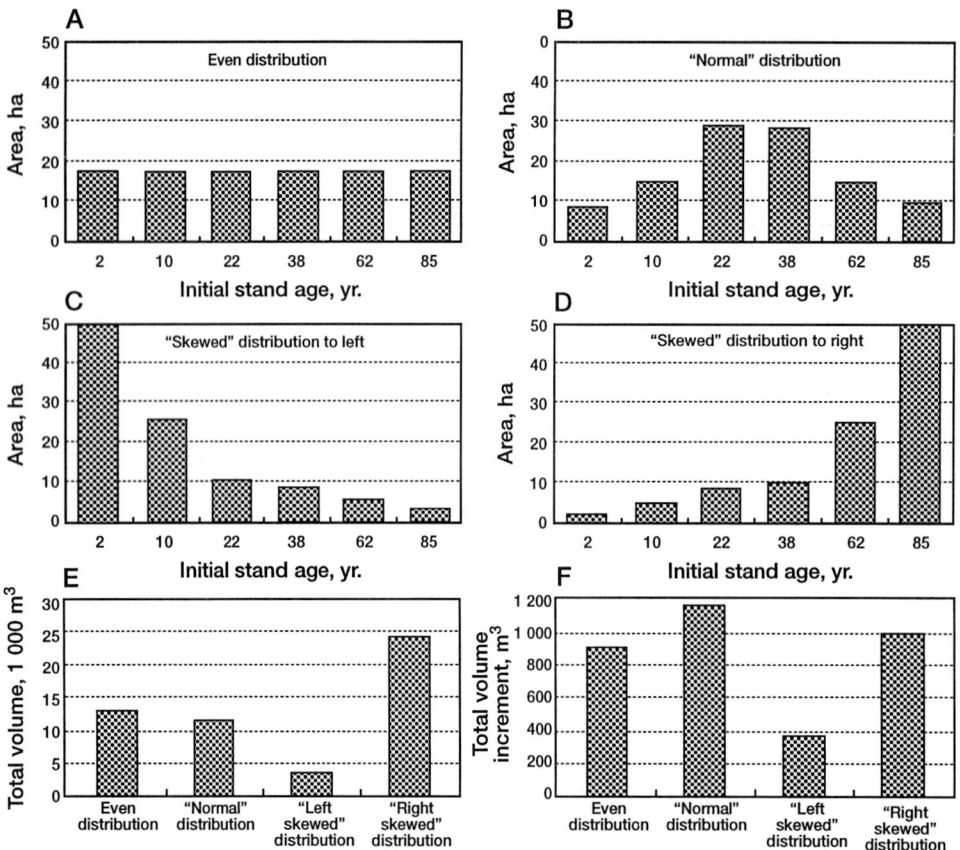

FIGURE 3 Computational example of how the age class distribution of tree stands can influence the productivity of a forest area. Distribution of tree stands with an overall area of 100 ha: (A) no single age class dominant (even distribution), (B) dominated by intermediate age classes (normal distribution), (C) dominated by young age classes (distribution skewed to the left), and (D) dominated by old age classes (distribution skewed to the right). (E and F) Stocking and growth, respectively, over the forest area for the various distributions.

be divided into six compartments (stands) allocated to initial age classes (i.e., 2, 10, 22, 38, 82, and +82 years) in different ways. Calculations by Kellomäki (1998, pp. 225–226) show growth during the next 10 years to be highest in the stand with an initial age of 38 years, with this growth being five times that in the stand with an initial age of 85 years. Consequently, a distribution skewed toward dominance by old stands represents the highest level of stocking but about 15% less growth than that for a normal distribution (Fig. 3). Where young stands are dominant, the total growth is only one-third of that for a normal distribution, whereas an even distribution gives 70–80% of the total growth achieved with a normal distribution.

In the short term, the maximization of timber production would thus require an age structure representing a normal distribution, i.e., a forest area that is fragmented into single stands of variable age and stocking with the inevitable impact that this will have on species with a habitat corresponding to mature or old-growth forests. In the long term, this age distribution will gradually lead to dominance by old stands, with increasing stocking and decreasing growth—a situation that is less than optimal for timber production. Terminal felling of stands previously of intermediate age but now mature will lead to a shift in the age distribution toward dominance by young stands with low stocking and absolute growth but high relative growth. The maximization of timber production over several rotations in a sustainable manner thus requires that the forest area should be divided into sections representing a mix of sapling, pole, and mature stands in such a way that the overall age distribution implies dominance by pole stands with high stocking and high growth. The forest area will still be fragmented in this case, which will mean major impacts on species with a preference for mature or old-growth forests.

IV. IMPACTS OF TIMBER-ORIENTED MANAGEMENT ON BIODIVERSITY: COMPARISON BETWEEN MANAGED AND NATURAL FOREST AREAS

A. Impact Mechanisms

Among the key issues in managing biodiversity in commercial forests is the maintenance of the functional and structural properties of the forest ecosystem, which will create and maintain heterogeneity among the habitats available for species to occupy. Two levels of heterogeneity may be distinguished: allogenic heterogeneity between stands and autogenic heterogeneity within stands. These concepts are related to allogenic and autogenic succession. Allogenic heterogeneity creates a landscape mosaic, whereas autogenic succession creates further heterogeneity within each element of the mosaic in terms of an even-aged stand structure or multilayered tree canopies (Table III). The contribution of allogenic and autogenic heterogeneity to the total heterogeneity of the landscape is probably multiplicative rather than additive.

The total impact of forestry on the availability of suitable habitats is difficult to predict. The optimization of age distribution for timber production may increase allogenic heterogeneity and fragmentation of the landscape. In particular, the remaining patches of mature or old-growth forests may be too small to sustain viable populations or too isolated for recolonization of areas where a population has temporally become extinct. Simultaneously, management may reduce autogenic heterogeneity in two ways. First, it may avoid or even curb successional processes that give rise to within-stand structural components and spatial patterns that are important for forest biodiversity (Table II). Second, regeneration practices and precommercial and commercial thinnings can reduce or even remove autogenic heterogeneity. This tendency is quite obvious in standwise management, especially if pure single tree populations in even-aged stands are preferred for reasons connected with timber production.

The critical question in the previous context is to what extent it is possible to maintain in commercial forests the processes, structural components, and patterns typical of mature or natural forests as listed in

TABLE III
Structural Components, Spatial Patterns, and Processes Important for Biodiversity in Natural Forests But Lacking or Altered in Managed Stands[a]

Processes
  Postfire succession
  Succession with tree species replacement
  Self-thinning
  Gap formation
  Snag and log formation
  Decomposition of coarse woody debris
Structural components
  Very old pine and spruce trees
  Old broad-leaved trees, particularly aspen and sallow
  Trees with abundant growth of epiphytic lichens
  Broken, stag-headed, and leaning trees
  Trees with holes and cavities
  Dead standing trees (snags)
  Fire-scarred trees, snags, and stumps
  Large fallen logs in various stages of decomposition
Spatial patterns
  A developed understory of tree saplings and shrubs
  Mixed stands with both conifers and broad-leaved trees
  Uneven-aged stand structure
  Multilayered tree canopies
  Patchy distribution of trees

[a] Source: Esseen et al., (1997).

Table III. Many of these elements are interrelated, and practically all are somehow related to different types of disturbances and the subsequent succession. The amount of coarse woody debris is highest during the early successional phases in natural forests, whereas in managed forests no dead or dying trees are left at this stage. The scarcity of such debris (dead and decaying wood) is generally accepted as the main reason for the disappearance of many species from the boreal forests that are being managed for timber production. In Finland, for example, more than 20% of the species with a preference for forest habitats are dependent on dead wood (e.g., there are about 1000 fungi and about 800 beetles whose occurrence is related to the presence of such material). This implies that 36% of the total number of species threatened by forestry practices are ones whose occurrence is related to the presence of dead wood.

It was not known for certain until recently whether species depending on dead wood require closed-canopy old-growth habitats or whether they could make use

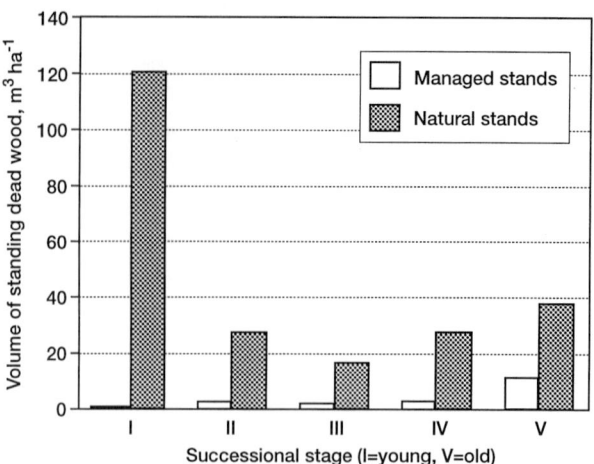

FIGURE 4 Volume of standing dead wood in different successional stages of managed and natural Norway spruce stands (data provided by Anneli Uotila, University of Joensuu).

of younger successional phases if a sufficient amount of dead wood was available. Recent evidence suggests (Fig. 4) that the latter may be possible and that many threatened species may survive in managed forests if certain properties typical of natural forests are maintained or even increased. On the other hand, many threatened species have highly specialized habitats, e.g., sand ridges, riverbanks, or large, overmature aspen trees (*Populus tremula*). The majority of these key habitats are of marginal importance for timber-oriented forestry since they represent sites of low productivity or are quite small in extent. Given careful planning and management, these key habitats could be set aside, which could have a major impact on the conservation of threatened species which prefer forest habitats. Along with measures to increase dead wood in commercial forests, this would provide new possibilities for reducing the impact of timber-oriented management on threatened species.

B. Properties of Forests under Intensive and Extensive Management: An Example

The impacts of timber-oriented management are exemplified in the following by referring to a comparison of selected beetle populations in an intensively managed forest area and an extensively managed one with a quite different management history (Siitonen and Martikainen, 1994; Siitonen et al., 1995). This material also provides a unique opportunity to contrast our current understanding with experimental data representing the long-term effects of forestry on the properties and biodiversity of boreal forests.

The two areas are located on either side of the border between Finland and Russia (62 or 63°N, 30–32°E) (Fig. 5). Before World War II, both areas belonged to Finland and were utilized and managed in similar ways. The border, created in 1944, split them in a biogeographically random manner.

The forests on the Finnish and Russian sides of the border have been used and managed quite differently for the past 50 years. Those in Finland have been subject to intensive use and a form of management that has included balancing growth with felling, systematic regeneration in the form of plantations established in clear-felled areas, or natural regeneration by means of seed trees or shelter wood fellings in favor of coniferous species. Precommercial thinnings and intermediate thinnings have been used to reduce the role of deciduous trees, and the latter have also been used to harvest the growth and yield that would otherwise have been lost in the natural self-thinning associated with the succession. Furthermore, the length of the rotation has been determined with the aim of maximizing the timber yield (80–100 years), which has further reduced the formation of dead and decaying wood. Additional reductions in dead wood have been achieved by effective fire fighting and the logging of trees blown down by storms or broken by heavy snowfall. The small-scale ownership of forest plots has further enhanced the fragmentation of the forests as each owner has aimed to maximize her or his timber yield and income.

On the Russian side of the border, clear-felled areas

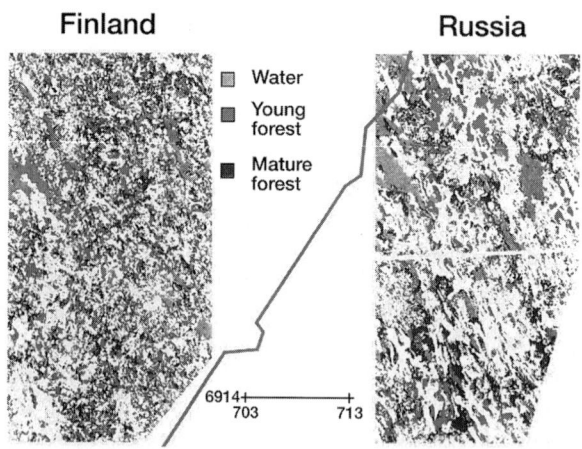

FIGURE 5 Landsat image (May 27, 1992) of part of the forest area on both sides of the border between Finland and Russia. Water is indicated in blue, young forest in red, and mature forest in green.

were not systematically regenerated, and thus fertile sites were occupied by mixtures of coniferous and deciduous trees. Since no precommercial thinnings or intermediate thinnings took place, the dominance of deciduous trees throughout the rotation was substantially higher than on the Finnish side. Later, dying trees were not logged in connection with thinnings, with the consequence that there is now a large amount of dead and decaying wood present. Furthermore, a low utilization rate of stems and the leaving of deciduous trees of low commercial value in the clear-felled areas were characteristic of logging practices. Simultaneously, uncontrolled forest fires were common and timber destroyed in other natural disturbances (e.g., storms and heavy snowfalls) remained unharvested. In other words, human intervention in the regeneration and growth of the forests has been substantially less than that on the Finnish side of the border.

C. Landscape and Local Patterns in Relation to Management

Systematic field samples from both sides of the border and analyses of satellite images indicated that the mean area of individual tree stands was quite similar but old-growth stands were substantially larger on the Russian side. Furthermore, the largest areas covered by homogeneous tree stands were larger on the Russian side. Most of the large stands on the Russian side were the result of extensive forest fires. On the other hand, the number of separate stands relative to the total area was larger on the Finnish side, i.e. the Finnish forest landscape is more fine grained and fragmented. This difference is quite evident even on the satellite image, which indicates the dominance of a coarse-grained landscape on the Russian side (Fig. 5).

Closer analysis of the properties of the forests shows that deciduous tree species are more common in the Russian forests than in the Finnish ones throughout the area (Fig. 6). Mature aspen in particular has been effectively eradicated from the Finnish forests because of its tendency to host a serious fungus disease that affects sapling stands dominated by pine. The most striking differences, however, concern sapling stands on fertile sites, which are dominated by Scots pine and Norway spruce in Finland but both Pendula and Pubescent birch in Russia. Coniferous tree species naturally dominated the sapling stands on poor sites in Finland, but they also did so on fertile sites, whereas in Russia deciduous species exceeded 20% of the total number of stems per hectare even on poor sites in many cases. In other words, the forest cover more frequently repre-

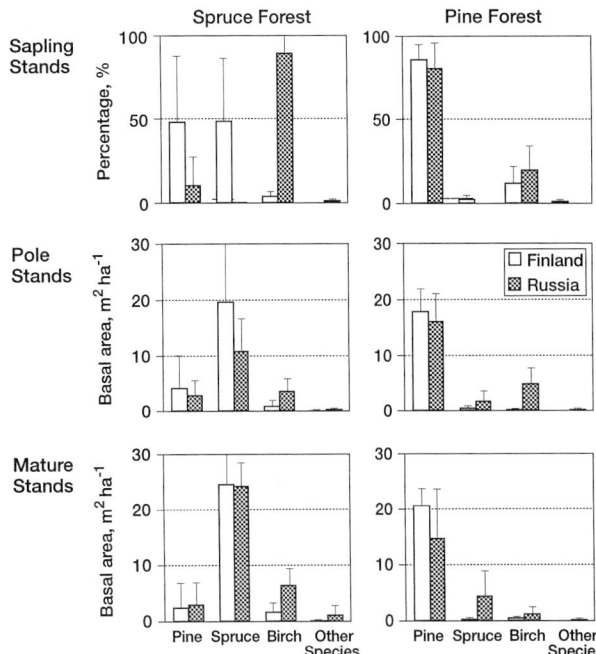

FIGURE 6 Tree species composition of sapling stands, pole stands, and mature stands in areas on the Finnish and Russian sides of the border.

sented a mixture of tree species on the Russian side than on the Finnish side.

In terms of dead and decaying wood, the difference between the two areas was clear; that is, the mean amount of dead wood in the Finnish forests was 4 m^3 ha^{-1} (range 0.1–26 m^3 ha^{-1}) and that in the Russian forests was 29 m^3 ha^{-1} (range 0.1–213 m^3 ha^{-1}) (Fig. 7). It is worth noting that the amount of coarse woody debris is high even in sapling stands in the Russian forests (up to 79 m^3 ha^{-1}), mainly as a result of forest fires. Here, the dead wood represents mainly dead trees (standing and lying), whereas in the Finnish forests it mainly comprises logging residues (stumps, branches, and tops of stems). These patterns held good over the whole range of variability in fertility and the age of sites and stands.

D. Beetle Fauna in Forests under Intensive and Extensive Management

Siitonen *et al.* (1995) assessed the impact of forest management on biodiversity in terms of the number of beetles species caught with window traps operating over one growing season. The total yield was about 29,000 individual beetles representing 623 species. There were

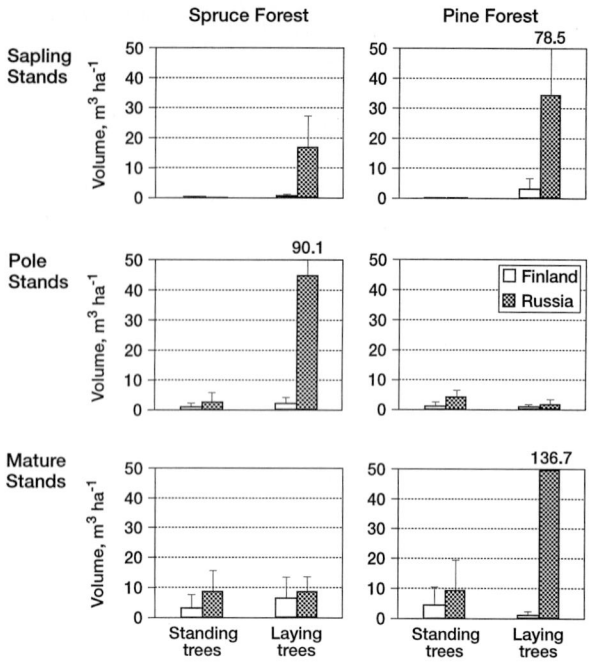

FIGURE 7 Volume of dead wood as a function of tree species and maturity of tree stands on the Finnish and Russian sides of the border. Spruce forest represents higher fertility than pine forest.

no major differences in the total number of species or in the main groups of species, except for the species preferring dead wood, with the consequence that the total number of saproxylic species was 179 in the Finnish forests and 213 in the Russian ones (Table IV).

It was found that 17 of the 20 most frequent species with a preference for dead wood were more frequent on the Russian side of the border. These represented mainly species with a preference for decaying birches (e.g., *Trichius fasciatus*, *Hylocoetus dermestoides*, *Leptura quadrifasciata*, and *Anaspis arctica*) or for the final phases of decay in coniferous trunks (e.g., *Anoplodera virens*, *Ampedus balteatus*, *Ampedus tristis*, and *Hadrobregmus pertinax*). The numbers of common and fairly common species in both countries (found consistently less than 25 times in Finland) were very similar, but the number of the rarest species (found consistently less than 12 times in Finland) was substantially greater on the Russian side (Table V).

The occurrence of species with a preference for dead wood was related to the amount of dead wood present in the tree stands, as shown in Fig. 8. On the other

TABLE IV
Total Numbers of Beetle Individuals and Species as a Function of Their Dependence on the Occurrence of Dead Wood

Parameter	Finnish forests	Russian forests	Total
No. of individuals			
Obligatory saproxylic	3,645	4,964	8,609
Facultative saproxylic	1,893	2,202	4,095
Other species	8,657	7,663	16,320
Total	14,195	14,829	29,024
No. of species			
Obligatory saproxylic	179	217	253
Facultative saproxylic	55	58	64
Other species	245	232	307
Total	479	507	623

TABLE V
Rarest Beetle Species with a Preference for Dead Wood Presented in Order of Abundance Based on the Total Number of Individuals

Species	Finnish forests	Russian forests
Found ≤12 times in Finland		
Pseudeuglenes pentatomus	—	1
Phymatura brevicollis	—	1
Hylis foveicollis	—	1
Enicmus planipennis	—	1
E. apicalis	—	1
Epuraea deubeli	—	1
E. longula	—	1
Cryptophagus confusus	—	1
Corticaria obsoleta	—	1
Found ≤25 times in Finland		
Stagetes borealis	1	30
Hylis procerulus	—	19
Euplectus fauveli	—	8
Lacon fasciatus	—	5
Triplax rufipes	—	4
Biploporus minutus	—	4
Mycetophagus quadripustulatus	—	3
Atrecus longiceps	—	3
Episernus angulicollis	—	2
Cryptophagus subdepressus	—	2
Corticaria crenicollis	—	2
Cis dentatus	—	1
Callidium aeneum	—	1
Epuraea muehlii	—	1
Orthocis linearis	1	—
Total	2	94

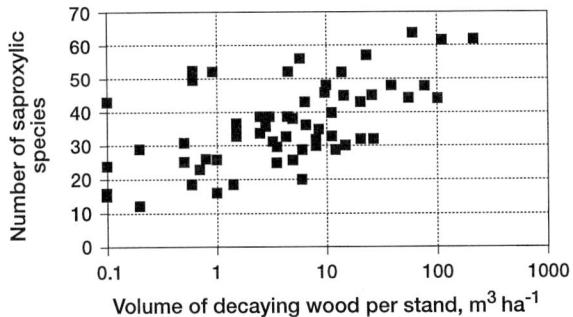

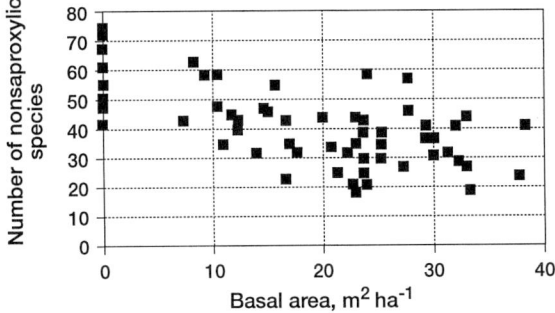

FIGURE 8 Correlation between the number of saproxylic species and the volume of decaying wood ($r = 0.548$, $d.f. = 58$, linear scale after logarithmic transformation of the x-axis) and correlation between the number of nonsaproxylic species and the basal area of the tree population ($r = -0596$, $d.f. = 58$, linear scale).

hand, there was no correlation between the occurrence of other species and the amount of dead wood. The numbers of specimens of other species were inversely correlated with the basal area of trees in the stand, however, which implies that there are more beetle species in sapling stands and pole stands than in mature or old-growth stands, regardless of tree species and site fertility.

V. CONCLUSIONS

Forests are a natural resource of prime importance, and they provide human communities with tangible items such as timber and wildlife and intangible benefits such as scenic beauty, wind reduction, and urban noise abatement. A unique feature of forest production is that its control requires only a small external input. Natural processes exert the greatest control, in the form of the biological diversity that controls the functioning and structure of forested ecosystems. The management and utilization of forest resources in a way that balances human needs with the undisturbed functioning of the forest ecosystem is the key to a form of sustainable forestry which provides ample space for maintaining biodiversity in the world's forests.

The current biodiversity of the boreal forests is a result of long-term evolution and changes in the environment and is thus related to the successional dynamics of the forest ecosystem. The long-term growth and development of tree populations is subject to disturbances due to wind, fire, and attacks by insects and pests, and under boreal conditions in particular, these disturbances cause a cyclic effect in the dynamics of the forest ecosystem by returning it to earlier phases in the succession. For example, a fire that kills old trees provides space for regeneration, enhancement of growth, and the accumulation of mass. Considered over large areas, this means that under natural conditions the boreal forests represent a mosaic of tree stands of varying age from saplings to mature trees and of varying tree species composition. The maximization of growth and timber yield implies that the area of mature or old-growth forests may be smaller than might be desirable in order to conserve the most specialized species.

The forest resources of northern Europe have been utilized intensively during the past 100 years, and this has inevitably had an impact on the current diversity of the forests. The management goal of maximizing the timber yield will inevitably lead to fragmentation of the forests and structural properties which deviate substantially from those arising under the influence of major disturbances. An increasing harvesting rate in terms of regular thinnings, together with a preference for coniferous species and a shorter rotation, will reduce the proportion of deciduous trees and the occurrence of dead and decaying wood. Furthermore, effective fire fighting and measures aimed at increasing the capacity of forests to resist the force of winds and heavy snowfalls will reduce the formation of dead wood and the occurrence of burnt wood, for which many species are specialized.

The previously mentioned tendencies will detract from the conditions which many rare and endangered species need in order to survive, as indicated by the positive correlation between the occurrence of these species and the amount of dead wood in the boreal forests. The increasing amount of dead wood probably indicates that there are more specialized microhabitats available under these conditions than in forests with a small amount of dead wood. These highly specialized species seem to be those affected most by the utilization and management of forest resources. On the other hand, the occurrence of species other than those confined specifically to dead wood seems to be quite similar regardless of the management history of the forests.

The protection of threatened biota may be based solely on a network of strictly protected areas. Currently, only 3.6% of the productive forestland in Finland is strictly protected, and opportunities to increase this within a short time span seem to be limited. Most of the valuable old-growth areas are already protected. Perhaps a better potential for protecting forest biodiversity and threatened species can be found in managed forests, in which the application of silvicultural practices that aim at restoring the natural properties of younger forests can be developed. Natural characteristics can be promoted quite rapidly in young managed forests since such forests are created continuously along with timber harvesting. On the other hand, most of the threatened species have highly specialized habitat requirements. By preserving these habitats (key habitats) in forest management, viable populations of these species can be maintained. Consequently, strictly protected areas and biodiversity-oriented management should be regarded as important complementary elements rather than alternatives when planning and developing new forestry practices.

See Also the Following Articles

BOREAL FOREST ECOSYSTEMS • FOREST ECOLOGY • LOGGED FORESTS

Bibliography

Aplet, G. H., Johnson, N., Olson, J. T., and Sample, V. A. (Eds.) (1993). *Defining Sustainable Forestry*. Island Press, Washington, DC.

DeGraaf, R. M., and Miller, R. I. (Eds.) (1996). *Conservation of Faunal Diversity in Forested Landscapes*. Chapman & Hall, London.

Esseen, P. A., Ehnström, B., Ericson, L., and Sjöberg, K. (1997). Boreal forests. *Ecol. Bull.* 46, 16–47.

Fiedler, P. L., and Jain, S. K. (Eds.) (1992). *Conservation Biology. The Theory and Practice on Nature Conservation, Preservation and Management*. Chapman & Hall, London.

Hansson, L. (Ed.) (1997). *Boreal Ecosystems and Landscapes: Structures, Processes and Conservation of Biodiversity*, Ecological Bulletins No. 46. Munksgaard International, Denmark.

Kellomäki, S. (1998). *Forest Resources and Sustainable Management. Papermaking Science and Technology*, Book 2. Gummerus Oy, Jyväskylä, Finland.

Kouki, J., Löfman, S., Martikainen, P., Rouvinen, S., and Uotila, A. (2000). Spatio-temporal patterns in the Fennoscandian forest fragmentation and habitat requirements of wood-associated threated species. *Scand J. Forest Res.* 15 (in press).

Mooney, H., Cushman, J. H., Medina, E., Sala, O. E., and Schulze, E. D. (1995). *Functional Roles of Biodiversity: A Global Perspective*. Wiley, Chichester, UK.

Siitonen, J., and Martikainen, P. (1994). Occurrence of rare and threatened insects living on decaying *Populus tremula*: A comparison between Finnish and Russian Karelia. *Scand. J. Forest Res.* 9(2), 185–191.

Siitonen, J., Martikainen, P., Kaila, L., Nikula, A., and Punttila, P. (1995). Kovakuoriaslajiston monimuotoisuus eri tavoin käsitellyillä metsäalueilla Suomessa ja Karjalan Tasavallassa. *Metsäntutkimuslaitoksen Tiedonantoja* 564, 43–63 in (Finnish).

UN-ECE/FAO (1992). *The Forest Resources of the Temperate Zones—The Forest Resource Assessments 1990*. United Nations, New York.

TOURISM, ROLE OF

Richard W. Braithwaite
CSIRO Tourism Research Program

I. Introduction
II. Negative Impacts of Tourism
III. Potential Benefits of Tourism
IV. The Big Issues

GLOSSARY

attraction A physical or cultural feature of a particular place that tourists feel meets an aspect of their leisure needs. Attractions are the main motivators for tourism trips and are the core of the tourism product. Without attractions there would be no need for other tourism services.

carrying capacity The level of usage of a site or area (constructed or natural) for specified activities and under a given level of management to provide for the recreational needs of visitors.

ecotourism Nature-based tourism that involves education and interpretation of the natural environment, improves the welfare of local people, and is managed to be ecologically sustainable.

recreation An activity voluntarily undertaken, primarily for leisure and satisfaction, during leisure time.

tourist There are many operational definitions. A reasonable example is a person who travels 40 km or more from home for any reason and who stays away for one or more nights.

TOURISM'S ROLE IN BIODIVERSITY might be to both build public support for biodiversity and help fund its conservation. However, the partnership has yet to be forged and current emphasis is on impacts.

I. INTRODUCTION

A. What Is Tourism?

Tourism can be defined as the sum of the phenomena and relationships arising from the interaction of tourists, business suppliers, host governments, and host communities in the process of attracting and hosting of these tourists and other visitors. Industries are typically thought of in terms of their output (for example, bales of wool), and measuring and assessing them involves taking into account activities that produce similar products. However, tourism does not conform to this model. It is not possible to reduce tourism to a simple relationship between inputs and the supply of a particular product. Tourism supply is defined in terms of the demand (of the buyer) rather than, as usual in other industries, in terms of the properties of the product and its production. The industry is dominated by marketing, and research has largely focused on estimating demand. It is part of the large and growing service sector of industry. While tourism does involve selling normal commodities, the core product that people are ultimately seeking is an experience. Experience is intangible, consists of activities rather than things, is produced and consumed simultaneously, and the customer has to be present and participate in the production process.

A region must have an attraction or attractions to be a tourist destination. Attractions are a necessary condition and they are arguably the most important component in any tourism product. Without attractions there would be no need for other tourism services. They not only entice, lure, and stimulate interest in travel, but they provide visitor satisfaction, the rewards from travel—the pure travel product. These attractions can be natural, manufactured, or of sociocultural character.

In one view, the "tourism system" consists of four components:

- Market. The internal and external factors affecting travel, the market inputs and the process by which purchasers of tourism products select a destination.
- Travel. The market segments, the passenger flows and forms of transportation.
- Destination. Procedures that a destination should follow to develop and service tourism activity.
- Marketing. The processes by which suppliers and destinations market products and services.

In essence, marketing sells the destination to the market while travel allows the market to get to the destination.

Another way of conceptualizing tourism is to look at the factors that contribute to a long-term profitable industry in a local area. Obviously not all of these conditions need be met to develop a successful local industry, but the more that are met the more likely the tourism destination is to be sustainable. Many of the insights represented in Figure 1 are derived from a large number of interviews done with tourism industry experts from regions across eastern Australia in 1997.

Different parts of this system are given different emphasis by different interests. Industry people tend to overemphasize the importance of marketing. Local government tends to focus on community support and local revenue from tourism. Tourism academics and management experts emphasize the chain of value represented by the series of experience points for the customer, of which the immediate experience of the attraction that the customers thinks they are purchasing is but one. Town planners focus on infrastructure and transport. Little, of course, would happen unless investors made money. Conservationists, in turn, focus on protection of the environment. Few people would have an overview of all the factors.

In the context of this chapter, it is important to recognize that protection of the environment is seen as only one of a range of important issues for tourism. When biologists talk of conducting research on the impact of tourism on biodiversity, many industry people feel threatened and hostile. As important as the impact issues are, a more effective approach is to package impact research in projects that deal with a wider range of issues. In particular, the quality of the tourism experience and using tourism as a vehicle for conservation education are two areas that naturally fit. However, they require social research and may need multidisciplinary collaboration with social scientists, a step many biologists are still unwilling to make.

It is also worth pointing out that the push for environmental care is mainly external. Governmental responsibility rests with sections of government outside the tourism system. This tends to be a recipe for tension and ineffectiveness. Educating the public in conservation, in turn, is unlikely to be high on the tourism industry agenda. It is high on the conservation agenda but will only be high on the tourism agenda if tourists are willing to pay for it.

B. Growth in Tourism and Its Contribution to Economic Development

Although people have always traveled to new places, the motivation of recreation rather than work is only a century or two old. The widespread phenomenon of today, in which most of the population of economically developed countries are tourists for part of the year, is a post–World War II development. Long-haul international destinations, such as are many of the "megadiverse" countries, were greatly enhanced as destinations by the reduction in airfares due to the widespread intro-

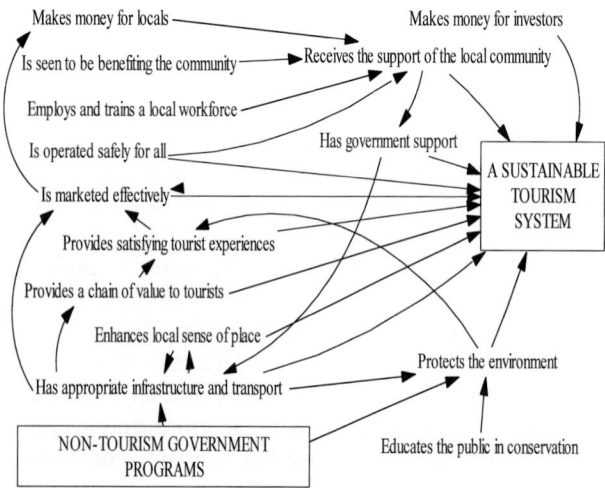

FIGURE 1

duction of the long-haul Boeing 747 in the 1970s. It was largely this technological development that made tourism the global economic force it is today.

According to the World Tourism Organization, the big three exports of world trade are crude petroleum products, motor vehicles parts and accessories, and tourism, each between 8 and 9% of the total. Tourism is the fastest growing sector of world trade, with an average annual growth rate of 9.6% in the 1980s. This compares with growth rates of 5.5% in merchandise exports and 7.5% in commercial services. From 1950 to 1991, international visitor arrivals worldwide increased 22 fold, from 25 million to 550 million, and this number was expected to increase by an additional 200 million, or 36%, in the 1990s. It is estimated that nature-based tourism might account for somewhere between 10 and 50% of all tourism, depending on whether direct or more indirect criteria are applied. This would indicate a contribution of between US$600 billion and US$3 trillion in 1996.

In countries with a low population density like Australia, the tourism industry is primarily based on the country's natural attractions, particularly its flora and fauna. While large city and other forms of urban tourism are important, tourism regions tend to be less developed and more peripheral regions. Indeed, one of their attractions is often an area's relatively unspoilt nature. Most communities, however, can only hope for small-scale economic advantages from tourism, and in such areas tourism can be no more than one element in a wider development strategy.

Critics of nature tourism relative to other forms of tourism believe that nature tourists are young and on low incomes, and therefore spend little and contribute little to the local economy. Recent World Wide Fund for Nature surveys in Caribbean countries on the importance attached by visitors to parks and protected areas show that when people classified their motivations as "main reason," "very important," or "not important," total expenditure per head was $2588, $1638, and $1531, respectively. The average lengths of stay were 13.0, 13.8, and 14.7 days.

Some major patterns in the growth of tourism are important for the future of biodiversity. First, the proportion of tourists traveling to the undeveloped world is increasing. It was 18.7% in 1980 but 21.8% by 1989. However, the percentage of receipts (i.e., income) over that period declined from 26.7% to 23% (Jenner and Smith, 1992). Second, tourists are becoming more active. The beach is still popular, but it is now much more of an outdoor gymnasium with sports and nature-based activities like scuba diving. Third, World Tourism Organization (WTO) figures show that tourist arrivals in the Asia/Pacific region have grown at double the world average rate (15%, compared with 7.3%) in the 1950–1991 period. And this is expected to be the highest growth area as we enter the twenty-first century.

Seventy-five percent of inbound tourists said they hoped to see a koala when making the decision to come to Australia, and 70% of departing tourists reported that they had actually seen one. In fact, 11% said they would not come to Australia but for the unique wildlife. On this basis, Hundloe and Hamilton (1997) estimated that the contribution of koalas to the Australian tourism industry might be as high as Aus$1.8 billion. They also estimated the amounts spent on viewing koalas and buying "koalabilia" to obtain a lower bound on the estimate of the economic contribution of koalas of Aus$336 million per annum. Similarly, Kenyan estimates of the value of an elephant from tourism revenue are about US$15,000 or $900,000 over the course of its life. It is also estimated that gross worth is $27,000 pa. for lions or $610,000 for a herd of elephants. These are social returns from tourism. Yet a poacher is not interested in the wider social benefits and will kill an animal to earn a few hundred dollars (Whelan, 1991). Obviously the economic feedback to the local community is an important issue.

C. What Do Tourists Want?

In the past, consumptive use of nature for recreation such as hunting and fishing were more important than they are now. A report for Alberta Tourism suggested that people involved in consumptive wildlife use were mainly male (90%) and few held degrees (5.6%), while among nonconsumptive users the genders were evenly balanced and 60% held degrees. Consumptive use is likely to be declining relative to nonconsumptive use of nature. In Canada, nonconsumptive use of wildlife is growing more rapidly than in the United States.

Concern about environmental issues has grown rapidly since the mid-1960s. In the United States, it peaked in 1970 with Earth Day and then declined until the 1980s when spectacular mishaps like the Exxon *Valdez* oil spill and the Chernobyl nuclear meltdown focused media attention on environmental matters. Concern continues to grow. In Europe, survey research indicates that public concern was more constant in the 1970s, rose steadily in the 1980s, and peaked in 1989 about the time of the European elections (Macnaghten and Urry, 1998). In Australia, support grew steadily from the mid-1960s and was boosted by successive Australian conservation issues, including the proposed mining of the Great Barrier Reef, Fraser Island, and in Kakadu

National Park, hydro power schemes drowning Lake Pedder, and clearing of native forests in New South Wales. Some development proposals were successfully stopped and others were not. However, the public debate generated further support for conservation issues generally. Support probably peaked around the 1990 federal election, which was fought primarily over green issues. Since then neither of the major parties has chosen to make it an issue and both promote a strong pro-conservation image.

U.S. polls have shown an increase in the number of people who feel that "we must sacrifice economic growth in order to preserve and conserve the environment" from 38% to 64% between 1976 and 1990. The "Canada in the World" survey in 1996 shows that only Germany, Japan, Australia, and the United Kingdom have a majority of people saying that economics should give way to environment when there is serious conflict. Generally the surveys show there is most concern about pollution issues that affect everyday quality of life (Macnaghten and Urry, 1998). At a local level, people are concerned about pieces of remnant vegetation. Obviously campaigns to save the tiger and other high-profile species also attract strong support. The origin of national parks lay in the recognition of the importance to people of encounters with nature. Philosophers tell us that what is actually valued is not wilderness, but the experiences it produces: feelings of awe, experiences of beauty. The issue is complex.

There are many surveys about why people choose the holiday experiences they do. They tend to show the growing importance of environmental issues in such decisions. The environmental factors influencing the choice of holiday destination of tourists in Spain were beautiful landscape (51%), unspoilt nature (43%), water quality (27%), air quality (22%), old customs (16%), and architecture (13%). The purpose of Japanese overseas trips were found to be to enjoy nature (72%), see famous tourist attractions, (56%), taste local food (48%), enjoy shopping (43%), rest and relax (38%), experience a different culture (36%), visit museums (31%), and stay in famous hotels (22%) (Jenner and Smith, 1992). Foreign holiday makers in Germany showed increasing environmental concerns, with the percentage of people noting serious environmental problems increasing from 21.9% in 1985 to 46.8% in 1988. The percentage of people noting problems at least doubled, whether seas, lakes, streams, rubbish, dirty beaches, dead trees, degraded landscape, or noise were under consideration.

In Europe, biodiversity issues are relatively few in the main communication media, but there is great interest in local birds among bird watchers. In all countries, there are nature-based tourism experiences available. Examples of the diversity may be found in compilations such as Geffen and Berglie (1993). However, in some countries, nature-based tourism is promoted more heavily than it is in other countries. For example, in Australia any hotel lobby has brochures promoting a range of nature-based tours and attractions, whereas one does not find this in the United Kingdom.

While tourists are increasingly seeking good quality nature-based experiences, there is also growing concern about the impact of tourism on host communities and environments. Activist groups, such as Tourism Concern in the United Kingdom, publicize examples of negative practices by their country's outbound operators in Third World countries. Such groups exist in many countries and primarily focus on the displacement of Third World peoples by tourism developments for people from the more developed countries. They frequently also attack conservation initiatives for the same reason. A range of somewhat anecdotal and non-quantitative assessments of the impacts of tourism are available for different regions (e.g., Croall, 1995; Pattullo, 1996).

Most countries conduct surveys to determine where tourists come from, what they do, and where they go. They also ask why the tourists made the decisions they did. The interpretation of these results is complex (e.g., Ryan, 1995). In this section, I have tried to establish that concern about the environment and the desire or need for good-quality, nature-based experiences are strong drivers for tourism. In the next section I examine the attitude to the environment and biodiversity by the tourism industry.

D. Engagement of Tourism Industry Sectors with the Environment

1. General Industry Leaders

Clearly, this concern reflects a much broader societal alarm about the degradation of our physical environment. Tourism, however, is perhaps even more sensitive and more dependant on a high-quality environment for its long-term success than are many other sectors. There are many quotes by industry leaders recognizing the importance of the environment. The leaders of the industry recognize that tourism is much more than recreational sites or facilities for accommodation and entertainment. Tourists also consider scenery, weather, transport, the hosts, and many intangible factors, any of which—singly or in combination—can be limiting factors.

A number of special features of the tourism industry offer peculiar challenges to our society and government. It is often said that the industry is fragmented and does not have a strong sense of itself. This fragmentation likely results because many of the major industry sectors are only partly in the tourism industry. For example, a restaurateur for whom 20% of customers are tourists and 50% are business people is likely to focus on the business trade. Many hoteliers, wine makers, and restaurateurs do not regard themselves as part of the tourism industry for the purposes of devising their business strategy. Economically, of course, a not inconsequential part of their income is derived from tourism. This diversity of viewpoints perhaps inhibits a strong common stand on issues.

The Pacific Asia Travel Association claims that the tourism income derived from what international visitors spend on sightseeing is only 3.3% in Hong Kong and 21.4% in the Philippines. In Australia, about 8% of the tourism industry income from international tourists is directly from tours, entertainments, and the like—that is, the sorts of things that people actually came to see. Perhaps half of that is strongly related to the natural environment and biodiversity. Put conversely, the tourists certainly did not come to experience the marvellous flight from another continent, to stay in a hotel, and to eat at a restaurant that could be in their home country. All industry sectors are dependent on the core product and are part of the "chain of value." If the core product fails, all sectors fail. Developing a sense of responsibility for the core product from all sectors may well prove to be the key to the development of a sustainable industry.

The results of a Tourism Council Australia survey of 208 Australian operators (accommodation, attractions, resorts, tours, and transport) asking about the attributes of the Australian tourism product are encouraging. The respondents gave 910 votes out of 1638 to natural environment categories ahead of hospitality/dining, service, price, entertainment, cultural/heritage attractions, special events, indigenous culture, and shopping.

2. Developers

Of the various components of the tourism industry, the people who build the resorts, hotels, shops, attractions, and so on seem least concerned about the natural environment. They build facilities that are usually sold and then managed by someone else as a tourism operation. Their main aim is to make money on their building activities. Architects generally try to produce an aesthetic building. Those that capture the "sense of place" are less common. That is, the buildings and environs become part of the environment in a harmonious and appropriate way. Such buildings are different and memorable. The landscaping around the buildings can not only enhance the appearance of the building but also greatly add to the amenity value of the development. Obviously, the use of local native plants or the re-creation of local ecosystems can be a strong contribution to generating a sense of place in a local community. This can make the place special, interesting, and thereby more attractive to tourists.

3. Attraction Operators

The people who run the attractions that tourists come to see are generally closest to the natural environment. In my experience, the majority regard themselves as conservationists. However, they may find themselves excluded from natural areas by conservation agencies. Sometimes this is due to overzealous application of the precautionary principle, and operators then call for research to provide more impartial information.

Wildlife can be an enormous long-term tourist drawcard, far outstripping the short-term profits to be made from transitory hallmark events. However, we need to get a lot better at managing wildlife tourism in a sustainable and holistic way.

Tourism has the ability to encourage vast numbers of people to take a more caring approach to the natural environment. It also has the ability, like any industry, to simply see the natural environment as a resource to be appropriated for corporate profit irrespective of environmental impacts. It is important to generate a succouring social environment that facilitates the former option. Very often this falls to protected area managers. They need to encourage appropriate behavior in operators rather than show an antagonistic attitude through officious behavior.

Ecotourism was one of the buzzwords of the 1990s (e.g., Lindberg et al., 1998). It is an idealized form of nature-based tourism encompassing holistic education of the tourist, maximizing benefits to the local community, and encouraging dedication to the pursuit of sustainability. The substantial literature on ecotourism seems to focus on the important issue of benefits to the local community. The environmental side is still poorly researched and discussions of it tend to be vague and nonquantitative. The word has been co-opted as an advertising gimmick by a range of operators and seems to have a confused image with many consumers.

For the tourism industry as a whole, ecotourism appears to have been a distraction from the pursuit of sustainability. As noble as the aims might be, the use of the term does seem to have had the effect of allowing other segments of the tourism industry to be less vigorous in pursuing sustainability than might otherwise have been the case. Sustainability should be everybody's business.

4. Accommodation

There has been considerable attention given to environmentally sound behavior by hotel chains. Some have attempted to develop a reputation in this area for marketing purposes. For example, in the early 1990s, the Canadian Pacific and Hotels Corporation conducted a thorough environmental audit of all its operations and set goals in waste reduction, reuse and recycling, increasing energy efficiency, as well as purchase of "nature friendly" hotel supplies. Various governments and industry groups have also produced publications to encourage best practice. However, while operator interest is substantial, application is patchy due to both economic and regulatory restrictions. In general, policies of this sector of the tourism industry appear to have little impact on biodiversity.

5. Restaurants

In the quest for tourism, regions are encouraged to express their regional identity in a variety of ways including the presentation of local foods. In Australia, some restaurants now prepare dishes incorporating indigenous species. These locally based foods are generally not foods prepared or presented in a traditional way ("bush tucker" in Aboriginal vernacular), but are usually the creative integration of Australian biodiversity products with more traditional European and Asian foods to produce new foods. Native fish and waterfowl species have long been used, but kangaroo, emu, and crocodile meat, insects, and a range of fruit, seeds, and leaves of Australian plants are now also used. To a fair extent they are a tourist curiosity, and most larger centers of population have at least one restaurant that specializes to some extent on modifications of "bush tucker."

6. Transport

"The business of tourism is, in reality, the renting out of the environment. It is therefore imperative for the industry to ensure that its 'product' is kept safe, unsullied and fresh not just for the next day but for every tomorrow" (Sir Colin Marshall, British Airways). Most of the biggest tourism operators are airlines. Because of the long lead times with purchase of aircraft, there is a greater interest in the long term. While airlines like Qantas and Air New Zealand promote themselves as environmentally aware, their interest has been focused on their own airline operations. Cathay Pacific, however, has focused its conservation effort on funding ecological studies of local high-profile species near the major destinations it serves.

Perhaps because of the media stories on the impact of affluent European tourists on the society and environment of Third World countries, British Airways has funded environmental and social audit studies in the Seychelles and more recently St. Lucia. The studies include an assessment of a range of environmental and social issues. They take a matrix approach in examining how water pollution, air pollution, noise impacts, aesthetic impacts, habitat loss (land), natural resource exploitation, and disruption of natural cycles are affected by tourism infrastructure, infrastructure operations, transport, use and consumption, and waste at a location.

II. NEGATIVE IMPACTS OF TOURISM

Tourists go to the Caribbean for its climate, sea and beaches . . . and this has put its coastlines under enormous pressure. . . . The large concrete hotels have been built close to the high water mark, groynes and piers erected, marinas for yachts and deep–water harbours for cruise ships constructed. . . . The great wetlands of the Caribbean have been grubbed out by developers eyeing their proximity to some of the regions best beaches. . . . In a generation the land and seascape have been transformed. (Pattullo, 1996)

There are a number of places around the world where tourism has seemingly run out of control and has come to generate a lot of problems and very few benefits for the local community.

The negative impacts of tourism may be divided into the direct impacts of tourists experiencing nature and the indirect ones resulting from the impact of providing facilities for tourists. While there are reviews of the negative impacts of particular plants and animals on tourists (e.g., Edington and Edington, 1986; Mieczkowski, 1995), the emphasis of this brief review is on the impact of tourists on biodiversity.

A number of frameworks have been used in the literature. For example, Wall and Wright (1977) describe the main ecosystem characteristics of impacts on water, soil, vegetation, and wildlife, while Edwards (1987) provides a matrix of habitats and direct tourism activities. There are many reviews of the negative impacts, but perhaps the most comprehensive are the 600-page tomes of Mieczkowski (1995) and Liddle (1997). There are also comprehensive treatments of specific ecosystem types (e.g. seacoasts, German Federal Agency for Nature Conservation, 1997).

A. Harvesting

The collecting, shooting, and trapping of wild food plants, wildlife, and fish, has the potential to be detrimental to populations of plants and animals. These consumptive activities can either be a tourist activity themselves or can arise from the goal of feeding or providing souvenirs for tourists.

There are many examples of overexploitation of resources by hunters and fishermen, some of this in a tourism context. For example, the Great Barrier Reef Marine Park Authority found that surveys conducted 10 years apart revealed that anglers' perceived catches had declined in numbers of fish and sizes of fish caught in four regions of the park.

Hunting and fishing, which are well regulated, are generally preferable to prohibition accompanied by illegal poaching (see Edington and Edington, 1986). The conservation of game species is enhanced in South Africa where well-managed game ranches have been established for decades and allow regulated hunting by residents and tourists. However, areas like Kenya have found nonconsumptive tourism more profitable (Whelan, 1991).

In Australia, the food used by indigenous Australians is called "bush tucker." Bush tucker tours are usually based on a range of bush products, thereby increasing the risk of nonsustainable use of particular resources. There are publications on bush tucker, which may encourage usage by a wider range of people.

B. Small-Scale Physical Impacts

The impacts of tourists and their vehicles crushing and breaking plants has been well studied (Liddle, 1997). In addition to physical abrasion and breaking of the plants, trampling compacts the soil, particularly with some soil types, and renders it unsuitable for plant growth primarily due to lack of water capacity of the soil. Unbordered paths tend to widen laterally over time as people walk out to the expanding limit of ground vegetation. People sometimes create their own paths where there is crowding, where the established pathway is overly circuitous, or where there are vantage points unserviced by paths. Poorly sited and designed paths may have the effects of trampling exacerbated by water-driven erosion during rainy periods. The number of passages by walkers to reduce the vegetation by 50% can vary between 12 and 1412 for different ecosystems (Liddle, 1997). The type of usage is also important. For example, per unit passage, horses have a greater impact on trail width than motorcycles, which in turn have a greater impact than hikers.

Campsites experience the same sort of trampling impacts as paths with a gradient of impact extending out from the most intensely used center (Cole, 1992). However, campsites are also affected by the collection of dead and living wood for campfires. Snow sports fragment habitat and destroy lichen cover (Jenner and Smith, 1992).

Boats can have a range of impacts through direct contact, wash, and propeller action, which can cause bank erosion, washing out of roots, and turbulence. These impacts cause redistribution of nutrients and increased turbidity with consequent effects on aquatic plant and animal life (Liddle and Scorgie, 1980).

There have also been studies of the impact of boats and divers on coral. Research done by the Great Barrier Reef Marine Park Management Authority indicates that coral breakages occur, but at levels that are statistically indistinguishable from the natural background of breakages from wave-action and storms. Floating hotels and day-trip pontoons are larger and have a wider range of impacts. The largest day-use pontoons may cater to in excess of 50,000 snorkellers per year with up to 700 at one time. Impacts are greater at low tide when visitors are able to stand on the coral. In Malaysia, beach closures occur at such times to protect the coral. In the Red Sea, it has been shown that careful briefing of divers decreases coral breakages.

C. Spreading Exotic Plants and Animals

In many parts of the world, local species have existed in isolation from the major land masses. Their biotas have developed high endemicity and consequential vulnerability to exotic invasion. Modern communication allows goods and people to move around the world in vast quantities. With the large volume of traffic moving between countries, there are numerous opportunities for organisms to be accidentally or illegally moved from one part of the world to another. For islands generally, with their long isolation, many potential pests have not arrived, but conversely some of those that have survived the journey have had a devastating impact. This is best documented for Hawaii and Australia.

It is likely that tourists have inadvertently moved

small plants and animals between countries. Certainly the spread of weeds into natural areas is well documented. For example, Lonsdale and Lane (1994) collected 1852 seeds of 76 species from 304 tourist vehicles entering Kakadu National Park; they found that 15 species had not previously been recorded in the park and 9 species were known tropical weed species. While these authors found most of their seeds trapped in the vehicles' radiator, in an African area with poorer roads many more were in mud stuck to the undersides of the vehicles where the seeds were also more likely to detach in a new location.

The transmission of diseases by tourists is a less common problem but is of great importance in a few situations. For example, direct transmission of disease is a long-standing concern for mountain gorillas, which are highly susceptible to human viruses and bacteria. These include tuberculosis, measles, and pneumonia, all of which could potentially wipe out a population of this highly endangered species (Roe *et al.*, 1997). There are similar concerns that the introduction of Newcastle Disease to Antarctica through infected poultry products could wipe out much of the bird life there.

D. Disturbance of Wildlife

There is a substantial North American literature on disturbance to wildlife by recreationists (Knight and Gutzwiller, 1995), and most of this is relevant to tourism. There are four broad causes of impact: harvest, habitat modification, pollution, and disturbance. Separately or in tandem they may cause death or modify the behavior of individuals. In the longer term, behavior, vigor, and productivity may be altered. Again death may result. At a population level, abundance, distribution, and demographic structure may be altered. Finally at a community level, interrelations between species (e.g., predation, competition) may be altered and species composition changed.

Sometimes tourism-derived impacts on a species can be diverse. For example, boats, sandcastle diggers, cars, motorbikes, beach umbrellas, and plastic bags are all hazards to loggerhead turtles at different stages of their life cycle on the Greek island of Zakynthos (Ryan, 1995). Conversely, some activities can have a range of impacts. For example, snowmobiles seem to have a range of impacts, but disturbance of wildlife during winter from this source can be particularly severe (Jenner and Smith, 1992). The Black Grouse (*Tetrao tetrix*) has been the subject of various studies in relation to the impact of skiing in the French Alps and is in some danger as a species. Another impact has been the increase in litter and waste due to tourist presence, which has attracted predators and severely reduced the success of egg laying (Jenner and Smith, 1992).

In northern Australia's Kakadu National Park, a study showed a range of species' responses to tourists on tour boats (Braithwaite *et al.*, 1996). Some species habituate quickly while others remain very sensitive to approach by humans. Each year as the waterbody contracts through the dry season the animals rehabituate to close contact with humans. Thirteen species were particularly sensitive to disturbance by the boats of tourists, and it was suggested that these be used as indicators of disturbance.

Currently, much of this wildlife is accustomed to the boats. Some neither move away nor move toward the boat, but allow the boat to come close. They behave this way only if they do not feel too threatened. If they feel threatened, some "freeze" and appear to hope not to be seen, but most move away. In moving away they experience stress, stop feeding, and expend considerable energy. If they do this often enough they are unable to breed. If they do it more they may not obtain enough energy to survive. Before these things happen, animals usually abandon the area. The relationship between the visitor and the fauna is often largely determined by a guide or operator.

One particularly difficult issue is that a crocodile leaping into the water or a large flock of magpie geese flying off is spectacular. Many tourists enjoy that spectacle. This means that the tourist may be more satisfied with their experience if the boat operator/guide regularly and severely disturbs the wildlife. We suspect that the right atmosphere is not being created in such cases. Some tourists will go to extraordinary lengths to get best photograph possible. Is there a compromise, a level of disturbance that satisfies the tourists, but is also acceptable to the animals?

A similar situation is found with wildlife tours in Nepal. Sloth bears and tigers are most easily disturbed by visitors and represent the most sensitive indicator species. In this situation, it is thought the availability of small-scale refuges from visitor disturbance is a key to sustainability.

A study for the Great Barrier Reef Marine Park Management Authority showed that large numbers of predatory fishes aggregated when the boats came to pontoons on the Great Barrier Reef. Regular fish feeding seemed to encourage such a buildup. However, there was no evidence that the feeding aggregations depleted fishes from other areas of the reef or that they affected local populations of fishes and invertebrates.

Another disturbance increased by tourists is wildfire.

Accidental ignition of wilderness areas is increased by visitation. Such changes in the frequency and time of year of these fires can have important impacts on both fauna and flora. Such problems are likely to be most common in fire-prone regions of the world such as those with Mediterranean type and wet-dry tropical climates.

E. Land Clearing as Part of Development

The primary cause of the loss of biodiversity is the habitat destruction that inevitably results from the expansions of human populations and activities. Obviously tourism developments and any facilities that even a few tourists use have often caused removal of natural vegetation. A substantial part of world tourism results from urban people going to less urban places in warmer climates. Many tourism developments occur close to attractive natural areas that are rich in biodiversity.

Even in Europe, land use for tourism has been growing significantly, particularly in the Alps, the Mediterranean coast, and the North Sea coast. The European Union (EU) expects this use and the more extensive forms of tourism such as hiking, cycling, fishing, and hunting to increase at the expense of land formerly dedicated to nature or agricultural purposes.

In some cases, tourism developments are less destructive of biodiversity than the extractive industries such as forestry and agriculture that they replace. For example, in Bali and other parts of Asia, the gardens created around hotels, resorts, and golf courses offer some improvement for biodiversity over the previous intensive rice agriculture.

Major landscape modification is an extreme form of land clearing. Two particular forms associated with tourism are building marinas and golf courses. The direct physical impacts of building marines include loss of marine habitat from land clearance, reclamation and dredging, changes in water flow within marina basins, and reduced water quality associated with waste disposal, refuelling, antifoulants, and effluent discharge (MacMahon, 1989).

With more than 50 million active golfers worldwide, golf is possibly the fastest growing sport, and golf course development has become the fastest growing property sector in the world. Improved technologies have enabled turf-sward treatment and large-scale removal of soil to change radically the landscape for golf courses. A scorched earth school of golf-course design has emerged whereby the land is flattened and recontoured from scratch. With a water demand of 3000 cubic meters a day, pumping up large amounts of underground water is depleting water in springs in natural areas. The fairways are increasingly being made broader to accommodate more people. Large amounts of fertilizers, pesticides, herbicides, fungicides, and other chemicals are needed to maintain the semblance of the Scottish coastal landscape at St Andrews. Japanese golf courses use 8.5 times as much pesticides as rice paddies, two tons per golf course annually. Much of it makes its way into surrounding systems (Pleumaron, 1992). The biodiversity impacts of these golf course developments are likely to be substantial but do not appear to be well documented. How can golf courses be made more environmentally benign? Clearly this is an important issue for biodiversity.

F. Pollution and Resource Use

If a country has a positive balance of payments for tourism, then the excess of inbound over outbound tourism effectively represents an increase in population size for such places. Such increase is a particularly heavy user of transport and all motorized transport contributes to pollution of the air and often water as well. The full range of human impacts, usually at the extravagant end of the per capita level of impact, are added to that of the resident population increasing the ecological footprint disproportionately.

In Hawaii, the discharge of partially treated sewerage effluents into Kaneohe Bay stimulated growth of the alga *Dictyosphaeria caverosa* such that it overgrew and killed large sections of the reef (Johannes, 1975).

The use of water by tourism ventures is in competition with other users, potentially domestic, industrial, and agricultural. In areas of the world where there is heavy dependence on groundwater, tourism activity will contribute to lowering the level of aquifers. This affects dependent fauna and flora as water no longer flows out on the surface. In drier parts of the world, such resources support major components of the biota, including many specialized species.

G. Perception of Impact by Tourists

Visitors to natural areas are most observant of, and regard as important, the direct impacts of other users (trail use for more than one activity, litter) but are becoming more aware of other impacts on the environment such as soil erosion (Hammitt *et al.*, 1996). In central Australia, M. Hillery and colleagues demonstrated a positive relationship between annual visitor numbers and level of measurable environmental impact, despite a relatively small level of impact. Fifteen percent of visitors were able to rate two or more sites for impact.

Over half of the visitors in the study identified concerns relating to tourism and introduced species. Only 8% identified other environmental issues. Surprisingly, some visitors were able to rate the area they were in against other parts of Australia. These results suggest that environmental quality and relative lack of impact are going to be commercially important in that consumers not only regard them as significant but also are able to discriminate with increasing accuracy. This bodes well for an improving relationship between tourism and conservation.

III. POTENTIAL BENEFITS OF TOURISM

The ugly face of tourism is often evident to travelers. This sometimes leads people to ignore the positive impacts of tourism and thinking about how the positive impacts might be maximized. Much of the writing on the positive effects is by social scientists and thus is different from the style of information accumulated by biologists on the negative impacts.

A. Building Public Support for Conservation

Many tourism destinations owe their popularity to conservation controversies. In Australia, Kakadu, South-West Tasmania, Fraser Island, and the Daintree River owe much of their tourism activity to all the free publicity on national media during conservation battles. Night after night, images of these areas are shown on prime-time television. They become very fashionable places to visit. A whitewater rafting operator in north Queensland suggested to me that announcement of a government intention to build a dam on the river he uses would be great for his business.

Conversely, travel to a beautiful place generally builds some bond with that place. If tourists have a positive experience they will care about it not being destroyed. Even the most sensitive new development destroys the place as people have it in their memories and is resented at least to some extent. The biological essence of a place needs to be emphasized as part of the tourism experience. It is not just identifying plants and animals, but natural history stories and, most important, big ideas about ecology and biogeography. Generally most tour guides in most places cover a small part of the range of information available for interpretation.

Obviously the more the quality of the interpretation improves, the better. Recent surveys of tourists visiting Kakadu National Park reveal that the "perceived needs for additional facilities and services" are largely to do with the desire for more information. The rankings of percentages were, in descending order, more information on native plants and animals, more opportunities to have direct contact with Aboriginal people, more information on Aboriginal culture, more information on the geology of the area, more information signs, places to buy Aboriginal arts and crafts, better maps of the park, more toilet facilities, more guided tours to places of interest, more places to buy food and drinks, and finally more picnic facilities. Clearly the demand is there. A huge amount of information is available but relatively little of it is accessible to tourists. The problem is getting it to the operators in contact with the tourists. KNP provide tour operators courses, a Tour Operators Handbook, information pamphlets, and an information center, but still many people want more.

B. Funding for Conservation Management

In the era of user pays, protected areas are already under great pressure to extract money directly from the public for the use of resources, both artificial and natural. Currently, the level of self-funding of protected areas ranges from zero to greater than 100%, with most less than 10%. The highest levels of self-funding are seen in Africa.

In the world of international trade where different traditions of subsidies have arisen in different countries, countries with large estates of publicly funded protected areas are already being accused of unfairly subsidizing their tourism industry. If world trade liberalization continues, international diplomatic pressure will add to the internal economic rationalist push. Others argue that public good requires ongoing subsidization of natural areas.

A key issue is how to do this in the most functional way. Some observers argue that resources are managed better in agencies that are funded out of fees rather than tax dollars and that full market prices should be charged. There is the issue of whether private industry, governments, or a parastatal organizations (i.e., corporate bodies within government) would work best for different aspects (entrance fees, accommodation, equipment rental, food sales, merchandise sales, etc.). Further, a range of financial mechanisms is available including user fees (either for general admission or for a specific activity), concession fees (charges to provider per visitor or as a general license fee on the operator), royalties on sales, special taxes, fines, and donations.

There seems to be much greater public acceptance of fees if it is clear that all the money goes to fund conservation rather than into "consolidated revenue." Multitiered systems with some groups paying more than others are also an important consideration. The most efficient and most acceptable method will vary with circumstances within and between countries. For example, some government agencies are reluctant to allow private concessions because the mechanisms for enforcement of standards of environmental care are inadequate or too prone to political interference.

C. Sustainable Regional Development

Across the world there have been many government schemes for redevelopment of rural areas in population and economic decline. The main problem has been the decline in value of agricultural commodities. However, in some cases the depletion of resources through use has also been a factor. Tourism is widely seen as an important option for economic recovery. The European Union increasingly views tourism as a better option than agricultural subsidy for adjusting for prosperity between regions. The EU's regional policy is increasingly focused on the role of tourism in socioeconomic development and the need to encourage responsible use of local resources for tourism. Economically disadvantaged rural areas are encouraged to develop ecotourism activities. Approximately 40% of the total budget of the ECU of $1755 million (Community contribution) for the community initiative called Leader II will be made available to develop rural tourism. Many Leader projects combine rural tourism activities with the protection of indigenous species and marketing based on protected areas of diversity.

In the United States, the Forestry Service has policies for aiding rural economic diversification through promoting the assessment of amenity resources. These refer to the aspects of the rural environment in which residents and visitors may find beauty, pleasure, and experiences that are unique to their locales. Such a process promotes a feeling of local pride and is more likely to result in a successful local tourism industry based on the local assets than might result from asking how the region can make money out of tourism.

D. Tourism in the Third World

Increasingly the nature-based tourism of people from more affluent countries is located in poor Third World countries where substantial natural areas are still available. In many of these less-developed countries, tourism has proved to be more valuable than traditional industries like agriculture. However, the key to the long-term success of such ventures is the level of returns to the local people. For example, in the Galapogos, limited returns to local people from international tourism forces them into unsustainable exploitation of sea cucumber and other natural resources, which may in time have large impacts on the dependent species in the ecosystems that are the basis of the tourism industry there (Southgate, 1998).

Many international aid projects now attempt to develop tourism sustainably in the poorer countries of the world. In fact, tourism is probably the only long-term prospect for funding conservation in much of the world. However, unless the local economic and social benefits are substantial, there is no prospect of success. Part of this problem is the lack of understanding by governments of the scale of economic benefits to be had from the tourism activity. For example, a study in the Virgin Islands National Park shows that the contribution by tourists to the island's economy was 10 times that of the annual management expenditure (Heywood and Watson, 1996). It is also essential that the solutions to the biodiversity-tourism nexus are harmonious with the local culture as well as consistent with local environmental and social conditions. The models for success are likely to be various.

Former WTO Secretary-General Antonio Enriquez Savignac attended the Rio Earth Summit in 1992 and was instrumental in getting tourism included in Agenda 21 as one of the only industries capable of providing an economic incentive for preservation of the environment. Secretary-General Francesco Frangialli renewed WTO's commitment to the goals of sustainable development in 1997 at the United Nations Earth Summit II in New York.

IV. THE BIG ISSUES

There are some difficult issues concerning biodiversity and tourism, which will require research and careful negotiation to resolve.

A. Engagement in Conservation Issues

While the tourism industry often has a bigger stake in environmental issues than traditional industries, it has yet to involve itself as a political protagonist in this area. The parts of the industry closest to the natural environment like tour operators are often ardent conservationists. Many are in it for the lifestyle rather than the money. In Australia, 85% of the industry is in small businesses and must focus strongly on the bottom line to survive financially. The larger businesses tend to be more remote from environmental issues and are more interested in other issues. As the industry matures, it is likely to take a higher profile on conservation issues in which it has a substantial stake.

B. Protected Areas and Tourism

The commercial world of the tourism industry and the governmental approach by park managers do not mesh easily. Control tends to rest with the park managers and to build a synergistic relationship with tourism they need to do the following:

- Develop empathy for the difficulties most small businesses have to survive financially.
- Respect third party legal obligations to tourists (i.e., not changing the rules at short notice).
- Sell the value of enlightened regulation to the tourism industry.
- Pursue policies and provide assistance, which will help to improve the quality of the tourism experience.

C. Carrying Capacity

The idea that beyond some point the resource degrades with greater usage, and the quality of the experience for visitors declines, seems simple enough. Many in the industry accept this in theory, but they do not like the idea that there are limits to the growth of their region or their business. The marketing paradigm suggests that the demand for a product is only limited by the quality of the salesmanship. The dominant view seems to be that the supply or carrying capacity can always be increased by appropriate "site-hardening." Supply can, however, exceed demand leading to cost cutting and a decline in quality and sometimes economic nonviability. On the other hand, low profitability may be offset at a community level by substantial employment. Research on the environmental, economic, and social tradeoffs in different situations is needed to develop a more sophisticated approach to this difficult issue.

D. Wildlife Tourism

Tourism based on the exposure to wildlife (animals) possibly offers the most satisfying of nature-based experiences but presents some special problems. For example, the experience may be intensified by severely disturbing the wildlife and the impacts may be more widespread and significant for conservation than with other nature-based experiences. On the other hand, the challenge is to increase the probability of the marvelous experiences that field biologists occasionally have in the course of their work.

E. Interpretation and Education

Visitors can gain from experiences physically, emotionally, and intellectually. However, satisfaction with a tourist experience, vacation, or holiday is the fulfillment of motivations. The motivations become goals, goals determine behavior, and in the search for a satisfying holiday, holiday makers engage in adaptive behaviors to secure the success of a satisfying vacation (Ryan, 1995). How does one meet expectations? How can people be helped to find the right places for them to do the things that they will most enjoy? What interpretative/educational material should be provided that will both meet the visitors' needs and also raise awareness and support for conservation issues? An example of this sort of research was done on the Skyrail chairlift over rainforest in north Queensland. The research showed visitors valued the interpretive computer interactives over all other components of the experience, including the interpretative center as a whole (Moscardo and Woods, 1998).

Much of the world of biodiversity is inaccessible to most tourists. Innovative ideas can allow a wider range of experiences to be shared with a wider range of people. For example, at the Naracoorte Caves Conservation Park in South Australia a new interpretative center substantially enriches the normal cave visits. Remote control cameras, infrared lighting, and image enhancement technology are used to enable visitors to see a range of bat behavior, including roosting, flocking, birthing, and feeding. Such visitor-driven systems could be applied to a wide range of sensitive habitats.

Part of the problem is that people often have a limited image of nature-based tourism. For example, the Penguin Parade at Phillip Island in southern Australia caters to a maximum of 4000 visitors per night. This well-managed attraction allows a large number of people to see hundreds of penguins come from hunting in the sea to return to their burrows in the sand hills. The facility illuminates the area, allows close proximity, and provides good quality interpretation, but it also monitors the welfare of the penguins closely. Run by a local trust, the facility provides an intimate nature experience to a large number of people with minimum impact on the animals. The admission fees, in turn, fund a major research program on penguins and the facility is a tourism icon, drawing visitors to a range of lesser known attractions in the region. It is a fine example of what can be done.

Tourism should be seen as entertainment, striving to satisfy tourist needs. To be successful, and therefore commercially viable, the tourism product must be packaged in a way that is attractive to the consumer. To assume tourists wish to be educated about biodiversity—or anything else—would be a mistake. Thus the tourism educational agenda for conservation must be creatively developed if it is to be successful.

See Also the Following Articles

CONSERVATION EFFORTS, CONTEMPORARY • ECONOMIC VALUE OF BIODIVERSITY, OVERVIEW • EDUCATION AND BIODIVERSITY • EDUCATION, BIODIVERSITY AND • ENVIRONMENTAL MOVEMENT • HUMAN IMPACT ON BIODIVERSITY • NATURAL RESERVES AND PRESERVES • RECREATIONAL USES AND ISSUES • ZOOS AND ZOOLOGICAL PARKS

Bibliography

Braithwaite, R. W., Reynolds, P. C., and Pongracz, G. B. (1996). *Wildlife Tourism at Yellow Waters. An Analysis of the Environmental, Social and Economic Compromise Options for Sustainable Operation of a Tour Boat Venture in Kakadu National Park.* CSIRO, Darwin. NT, Australia.

Cole, D. N. (1992). Modeling wilderness campsites: Factors that influence amount of impact. *Environmental Management* 16, 255–64.

Croall, J. (1995). *Preserve or Destroy: Tourism and the Environment.* Calouste Gulbenkian Foundation, London.

Edington, J. M., and Edington, M. A. (1986). *Ecology, Recreation and Tourism.* Cambridge University Press, Cambridge.

Edwards, J. (1987). The UK Heritage Coast: An assessment of the ecological impacts of tourism. *Annals of Tourism Research* 14(1), 71–87.

Geffen, A. M., and Berglie, C. (1993). *Eco Tours and Nature Getaways: A Guide to Environmental Vacations around the World.* Clarkson N. Potter, New York.

German Federal Agency for Nature Conservation. (1997). *Biodiversity and Tourism: Conflicts on the World's Seacoasts and Strategies for Their Solution.* Springer-Verlag, Berlin.

Hammitt, W. E., Bixler, R. D., and Noe, F. P. (1996). Going beyond importance—performance analysis to analyze the observance-influence of park impacts. *Journal of Park and Recreation Administration* 14(1):45–62.

Heywood, V. H., and Watson, R. (1996). *Global Biodiversity Assessment.* United Nations Environmental Program. Cambridge University Press, Cambridge.

Hundloe, T., and Hamilton, C. (1997). *Koalas and Tourism: an Economic Evaluation.* Discussion Paper No. 13. The Australia Institute, Canberra.

Jenner, P., and Smith, C. (1992). *The Tourism Industry and the Environment.* Economist Intelligence Unit Special Report No. 2453, London.

Johannes, R. E. (1975). Pollution and Degradation of Coral Reef Communities. In *Tropical Marine Pollution* (E. J. Wood and R. E. Johannes, Eds.), pp. 13–51. Elsevier Scientific Publishing, Amsterdam, Netherlands.

Knight, R. L., and Gutzwiller, K. J. (1995). *Wildlife and Recreation: Coexistence through Management and Research.* Island Press, Washington, DC.

Liddle, M. J. (1997). *Recreation Ecology: The Ecological Impact of Outdoor Recreation and Ecotourism.* Chapman and Hall, London.

Liddle, M. J., and Scorgie, H. R. A. (1980). The effects of recreation on fresh-water plants and animals: a review. *Biological Conservation* 17, 183–206.

Lindberg, K., Wood, M. E., and Engeldrum, D. (1998). *Ecotourism: A Guide for Planners and Managers,* Volume 2. The Ecotourism Society, North Bennington, VT.

Lonsdale, W. M., and Lane, A. M. (1994). Tourist vehicles as vectors of weed seeds in Kakadu National Park. *Biological Conservation* 69, 277–283.

Macnaghten, P., and Urry, J. (1998). *Contested Natures.* Sage Publications, London.

McMahon, P. J. T. (1989). The impact of marinas on water quality. *Water Science Technology* 21, 39–43.

Mieczkowski, Z. (1995). *Environmental Issues of Tourism and Recreation.* University Press of America, Lanham, MD.

Moscardo, G., and Woods, B. (1998). Managing tourism in the Wet Tropics World Heritage Area: Interpretation and the experience of visitors on Skyrail. In *Embracing and Managing Change* (E. Laws, B. Faulkner, and G. Moscardo, Eds.), pp. 307–323. Routledge, London.

Pattullo, P. (1996). *Last Resorts: The Cost of Tourism in the Caribbean.* Cassell, London.

Pleumarom, A. (1992). Course and effect: Golf tourism in Thailand. *The Ecologist* 22(3), 104–110.

Roe, D., Leader-Williams, N., and Dalal-Clayton, B. (1997). *Take only photographs, leave only footprints: The environmental aspects of wildlife tourism.* IIED Wildlife & Development Series No. 10, London.

Ryan, C. (1995). *Researching Tourist Satisfaction: Issues, Concepts, and Problems.* Routledge, London.

Southgate, D. (1998). *Tropical Forest Conservation: An Economic Assessment of Alternatives in Latin America.* Oxford University Press, Oxford.

Wall, G., and Wright, C. (1977). *The Environmental Impact of Outdoor Recreation.* Department of Geography Publication Series No. 11, University of Waterloo, Ontario, Canada.

Whelan, T. (1991). *Nature Tourism: Managing for the Environment.* Island Press, Washington, DC.

TRADITIONAL CONSERVATION PRACTICES

Carl Folke and Johan Colding
Stockholm University and Beijer International Institute of Ecological Economics

I. Introduction
II. Ecological Monitoring and Practices Framing Access to and Use of Species and Habitats
III. Practices of Multiple-Species Management, Resource Rotation, and Succession
IV. Practices Related to the Dynamics of Complex Ecosystems
V. Concluding Remarks

GLOSSARY

institutions Humanly devised constraints that shape human interaction and the way societies evolve through time; made up of formal constraints (rules, laws, constitutions), informal constraints (norms of behavior, conventions, self-imposed codes of conduct), and their enforcement characteristics.

natural disturbance Any relatively discrete event in time that disrupts ecosystem community or population structure and changes resources, substrate availability, or the physical environment; key for structuring biological communities and for maintaining resilience in ecological systems.

resilience The system's capacity to absorb disturbance and conserve opportunity for self-organization and evolution. Resilience has to do with how resistant the system is to fundamental reorganization such as a phase shift into another stability domain.

social taboo A prohibition imposed by social custom or as a protective measure.

traditional ecological knowledge A cumulative body of knowledge, practice, and belief, evolving by adaptive processes and handed down through generations by cultural transmission, about the relationship of living beings (including humans) with one another and with their environment.

traditional peoples Variously referred to as indigenous peoples, native peoples, or tribal peoples; peoples with cultures that differ from those of the mainstream of a national population, having distinct ethnic languages with locally evolved resource management systems; currently some 5000 distinct indigenous peoples exist in the world.

PEOPLE HAVE INHABITED terrestrial ecosystems of the world for thousands of years. Both resource management systems and cosmological belief systems have evolved and continue to develop. In fact, most, if not all, ecosystems and biodiversity have been altered by humans to various degrees (Nelson and Serafin, 1992). The human imprint has in many cases wiped out species and caused substantial land use change (e.g., Turner *et al.*, 1990; Wilson, 1992). However, there are practices of local peoples of both traditional and contemporary society that contribute to biodiversity conservation, practices that are more common than generally recognized (Berkes and Folke, 1998). For example, throughout the Amazonian tropics, scientists have found remnants of past agricultural management systems in landscapes previously believed to be free from human imprint, suggesting that such management systems were highly adapted to natural cycles of forest regeneration (Balée, 1992; Posey, 1992).

I. INTRODUCTION

Even traditional peoples with a relatively high human population number were able to utilize rain forest areas without destroying these environments and surrounding biological communities (Primack, 1993). However, the specific objective of these local practices is not necessarily directed toward the conservation of species, their habitats, and ecosystems. Rather, such practices are often geared for sustainable use of local resources and ecosystems, with biological conservation resulting as an indirect outcome. They are often tied to cultural belief systems, which makes it difficult to separate the belief component from actual management practices and the ecological knowledge system on which they build. Knowledge, practices, and beliefs tend to intermingle in most traditional management systems (Gadgil *et al.*, 1993). This constitutes the basis behind traditional ecological knowledge (TEK), denoting that resource management patterns are the products not only of a people's physical environment and its resources but also of their cultural perceptions of the environment and its resources (Ruddle, 1994).

Cultural belief systems of various peoples have protected species, their habitats, and even smaller ecosystems. For example, several verses in the Vedas and Upanishads mention conservation and protection of plants and animals, indicating that traditional conservation practices of many rural and indigenous groups of India go as far back as the Vedic period (5000 B.C.). In fact, a great deal of social mechanisms, such as social taboos, may be highly adaptive from an ecological perspective and contribute to biodiversity conservation (Colding and Folke, 1997). The term *conservation* is here used in the sense of sustainable use of natural resources for human benefit, without compromising the interests of future generations (WCED, 1987).

This article will provide examples of a diverse set of traditional management practices and institutions that exist among local resource users. We will illustrate that such practices play an important role for *in situ* conservation of biological resources. We deal with traditional management practices that lead to the sustainable use of biological resources and local institutions and belief systems that impose regulations on the use of species and access to ecosystems.

The word *traditional* refers here to the historical and cultural continuity of resource management, recognizing that societies are constantly redefining what is considered "traditional." The term "local resource users" refers to both traditional peoples and small-scale societies of Western countries with locally evolved management systems. Local resource users generally depend on a rather limited resource procurement base to provide them with a wide diversity of resources (Gadgil *et al.*, 1993). Many do not have access to fossil fuel dependent technology or capital markets. Their day-to-day survival depends on proper interpretations and knowledge of the dynamics of their local resources and ecosystems. Thus, they often have a stake in managing their resources for long-term endurance.

Many of the examples of management practices and local institutions presented in this article derive from the case study anthology edited by Berkes and Folke (1998). This anthology predominantly deals with local resource users that display success of long-term environmental management by using practices that contribute to building resilience in local ecosystems. A diverse set of management practices and social mechanisms from temperate and tropical regions and traditional and contemporary society were identified and analyzed in recognition of their importance for building resilience in combined social–ecological systems (Levin *et al.*, 1998). The focus of the analysis is presented in Fig. 1.

The examples stem from a wide range of local resource users, including hunters and gatherers, herders, fishers, agriculturists, and small-scale communities of industrial nations. This article describes these practices in relation to their nature conservation functions. Most of the practices described are interrelated and have multiple ecological functions. Nonetheless, we have categorized them in the manner outlined in Table I.

As indicated in Table I, the first category of practices includes ecological monitoring and regulation of the use of biological resources and ecosystems. They are dealt with in Section II of this article. The second category mainly concerns traditional agroforestry practices, with an emphasis on their effect for biological conservation. They are dealt with in Section III. The third category describes management practices that are related to the dynamics of complex ecosystems, in particular the processes that structure ecosystems at different temporal and spatial scales. These practices are dealt with in Section IV. Such conservation practices, often at the watershed and landscape levels, are important in secur-

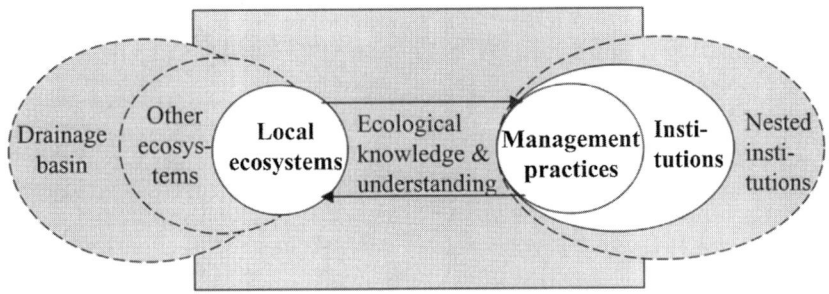

FIGURE 1 Focus of analysis of linked social–ecological systems. Ecological knowledge and understanding is a critical link between complex and dynamic ecosystems, adaptive management practices, and institutions.

ing a flow of natural resources and ecosystem services on which the local community depends.

In the concluding section, we stress lessons that can be learned from traditional conservation practices in building resilience through improved management of biodiversity. Resilience conserves options and opportunity for ecosystem renewal and evolutionary change, and its dynamic maintenance is a prerequisite for all forms of biological conservation.

TABLE I

Management Practices of Local Resource Users[a]

1. Ecological monitoring, and practices, framing access to and use of species and habitats
    Monitoring the status of the resource
    Total protection of certain species
    Protection of vulnerable life histories of species
    Protection of specific habitats
    Temporal restrictions of harvest
2. Practices of multiple-species management, resource rotation, and succession
    Multiple-species and integrated management
    Resource rotation
    Management of succession
3. Practices related to the dynamics of complex ecosystems
    Management of landscape patchiness
    Watershed management
    Managing ecological processes at multiple scales
    Responding to and managing pulses and surprises
    Nurturing sources of ecosystem renewal

[a] Source: Folke et al. (1998).

II. ECOLOGICAL MONITORING AND PRACTICES FRAMING ACCESS TO AND USE OF SPECIES AND HABITATS

The practices described below are often determined by local social institutions. These are often informal, based on traditional norms and conventions. Some appear to be more directly associated with resource management activities, while others appear to be more closely associated with cultural belief systems.

A. Monitoring the Status of the Resource

Monitoring the status of the resource is very common among local resource users (Folke et al., 1998). Monitoring leads to the acquisition of local ecological knowledge and helps local users respond to resource and ecosystem dynamics. Both qualitative and quantitative indicators develop as the result of experience and ecological knowledge based on monitoring (Johannes, 1998). Monitoring may provide information about location and timing of target resources (i.e., resource procurement) and also the condition and dynamics of individual populations and particular ecosystems for generating target resources. Ecological monitoring also provides the basis for the spatial and temporal regulations of resource use in local communities. Thus, various social response mechanisms evolve as a result of ecological monitoring. These are often in the form of local institutions, such as social taboos, area rotation regulations, and seasonal closures. The list of the importance of ecological monitoring for local resource users can be made very long. We will provide a few examples.

For resource procurement reasons, the Cree of James

Bay, Canada, monitor geographic distribution, migration patterns, and individual behavior of caribou, as well as the sex and age composition of the caribou herd, including the presence or absence of predators. They also predict herd size of caribou by examining the fat content of caribou (Berkes, 1998). The Inuit of Quebec and the Inuit and Innu of Labrador also use fat content as an indicator of health of both individual animals and the herd. Such monitoring may provide useful information for resource procurement and in the enforcement of hunting regulations when prey is low in abundance.

Similarly, the Cree of James Bay monitor the interaction between beaver and vegetation (Berkes, 1998). Such monitoring leads to knowledge about beaver populations and their ecological interaction with other biological resources. For example, when aspen is low in a particular area, trappers know that the beaver population is low in abundance.

Knowledge about the effect of natural disturbance regimes on target resources and other species is also based on ecological monitoring. Such knowledge requires monitoring over extensively long time periods and may be passed on from one generation to the next. For example, Cree beaver trappers know that 3–4 years after a fire occurs in an area, beavers will start to inhabit the area again (Berkes, 1998).

Also, monitoring combined with ecological knowledge may lead to resource procurement methods that do not deteriorate the habitats of target resources and secure a sustainable use of target resources. For example, in Tokelau in the South Pacific, islanders use, based on detailed knowledge of octopus behavior, a special octopus stick to extract the animal, which obviates the need for the destructive crushing of the coral or the use of poison (Ruddle, 1994). Another example is the use of the clam rake in the Maine soft-shell clam fishery, which is the only tool allowed for clam harvesting (Hanna, 1998). Such methods reduce the risk that a resource is overexploited, as compared to more advanced forms of technological methods.

Local institutions for management, based on ecological monitoring, may either be permanent in time such as in the two examples above, or "kick in" occasionally in times of resource scarcity. For example, ecological monitoring underlies the Cree's decision when to rotate beaver trapping grounds and fishing areas. Cree fishers rotate fishing grounds based on a declining catch per unit of effort and rest these sites when needed (Berkes, 1998). In this way, overfishing is limited. Coastal communities in Maine monitor clam populations to help determine the areas needing enhancement (Hanna, 1998). Such response mechanisms represent fine-tuned ways of responding to ecological feedback and are employed in order to avoid overexploitation of subsistence resources.

The widespread uses of closed seasons in Oceania are based on observations founded on local ecological knowledge about the spawning periods of key fish species, and fishing is prohibited during such periods. Pacific island groups often know when and where fish aggregate to spawn. At such aggregation sites, fishers monitor yearly changes of fish stock size and composition and reduce their fishing effort when stocks seem to be low (Johannes, 1978; Hviding, 1988).

Similarly, the use of fishing taboos among Pacific islanders at particular sea areas is based on local knowledge about the fluctuations of reef fish populations. On Satawal Island in the Central Caroline Island of Micronesia, fishing off one reef section is prohibited by taboo in order to conserve a breeding ground to supply the rest of the reef with resources (Ruddle, 1994). On Tokelau, the *lafu* (fishing taboo) is the most explicit conservation measure. *Lafu* is invoked by the Council of Elders. It bans all fishing in specific areas on the main reef and is announced to permit stock recovery.

Such examples allowing for ecosystem renewal is evident among a number of local resource users. For example, pastoral groups in arid and semiarid Africa monitor ecosystem change by way of tracking to determine daily movements of herds. This includes tracking ecological processes, such as climate, soil content, groundwater availability, forage availability, temporal environmental variability, and environmental degradation. Various forms of indicators are used for this, such as the behavior of fauna, specific plant indicators, soil color and texture, forage quantity and quality, and vegetation/plant composition (Niamir-Fuller, 1998).

Due to the proximity of local resource users to their resource base, ecological monitoring is often feasible on a daily basis. This facilitates detection and response to ecological change. Quite often, monitoring activities may be ascribed to particular individuals in local communities, such as resource stewards, elders, or shamans (Berkes and Folke, 1998).

B. Total Protection of Certain Species

Total protection of certain species is another practice that is particularly common among traditional peoples. For example, flora and fauna in India have been protected by indigenous belief systems for millennia and have been revered as the vehicles of Gods and Godesses.

Of particular interest are the different kinds of social taboos imposed on species by traditional societies

TABLE II
Threatened Species That Are Avoided as a Result of Taboos[a]

Species	Popular name	Local resource users/locality	IUCN status[b]
Kinosternon oaxacae	Oaxaca mud turtle	Pima Bajo, Papago, Yuman, U.S.A./Mexico	I
Chelonia mydas	Green sea turtle	Inhabitants of Buzios Island, Brazil	E
Naja oxiana	Oxus cobra	Local protection in vicinity of temples in India	K
Melanosuchus niger	Black caiman	Piro of Peru	V
Heloderma suspectum	Gila monster	Riverine Pima, Papago, U.S.A./Mexico	V
Pavo muticus	Green peafowl	Tamil Nadu, Rajasthan, Gujarath, India	V
Gorillas gorilla	Gorilla	Edo state, Nigeria	V
Colobus polykomos	Black and white colobus	Villages of Boabeng and Fiema, Ghana	V
Pan troglodytes	Chimpanzee	Edo state, Nigeria	V
Thomomys umbrinus emotus	Southern pocket gopher	Riverine Pima, Papago, Maricopa, U.S.A./Mexico	V
Perognathus alticola	White-eared pocket mouse	Riverine Pima, Papago, Maricopa, U.S.A./Mexico	V
Dipodomys gravipes	San Quintin kangaroo rat	Riverine Pima, Papago, U.S.A./Mexico	E
Dipodomys microps leucotis	Houserock chiseltoothed kangaroo rat	Riverine Pima, Papago, U.S.A./Mexico	K
Canis lupus	Grey Wolf	Bishnois of Thar desert, Rajasthan, India	V
Tremarctos ornatus	Spectacled bear	Achuar of Ecuador/Peru	V
Panthera tigris	Tiger	Local protection in vicinity of temples in India	E
Felis concolor	Puma	Maricopa, Yuman speakers, U.S.A./Mexico	E
Tapirus bairidi	Central American tapir	Coshiro-wa-teri of Brazil/Venezuela; Achuar of Ecuador	V
Myrmecophaga tridactyla	Giant anteater	Coshiro-wa-teri of Brazil/Venezuela; Achuar of Ecuador/Peru	V
Pridontes maximus	Giant armadillo	Achuar of Ecuador/Peru	V
Antilope cervicapra	Blackbuck	Bishnois of Thar desert, Rajasthan, India	V

[a] Source: Colding and Folke (1997).
[b] IUCN status: E, endangered; I, indeterminate; V, vulnerable; K, insufficiently known; R, rare.

throughout the World. Colding and Folke (1997) found that specific-species taboos protect threatened species as well as species considered keystone and/or endemic by ecologists. It was estimated that about 30% of the identified taboos protect species listed as threatened by IUCN. Table II displays a number of ecologically important and threatened species protected through taboos. This indicates the role that traditional local institutions, such as taboos, may have in biological conservation.

C. Protection of Vulnerable Life Histories of Species

This practice reduces the danger of overharvesting and the depletion of a population of target resources. For example, in the Maine fisheries, it is prohibited to gather lobsters with eggs (Acheson et al., 1998). Traditional fishing castes in the Bhandara district of Maharastra, India, never disturb the spawning aggregations of freshwater fish in hill streams, and the Phasepardhis of Maharastra traditionally let loose pregnant does caught in their snares (Gadgil, 1987). The Cree of James Bay never kill or disturb nesting geese (Berkes et al., 1995). The Tukano Indians of Colombia impose taboos on the collection of bird eggs and avoid the collection of reptiles during their breeding season as well as protect fish spawning aggregation sites in rivers (Reichel-Dolmatoff, 1976). Size restrictions are sometimes employed among South Pacific islanders on slow-moving or sessile marine species that are particularly susceptible to overharvesting (Johannes, 1978).

D. Protection of Specific Habitats

This practice is commonly found among local resource users. For example, pastoralists of arid and semiarid Africa use buffer zone areas of Sahelian rangelands which are protected from grazing except in the case of emergencies (Niamir-Fuller, 1998).

Also, whole forests, forest patches, coast stretches, rivers, or ponds may to various degrees be protected for human resource use. Usually, such areas are set aside by religious taboos and considered sacred to community members. In India, sacred groves were once extremely common. A sacred grove is a small part of a forest set aside for spiritual or religious purposes. The sizes of such protected areas vary from a clump of 5–10 trees to as much as 50 ha or more (Gadgil and Vartak, 1976). Sacred groves still exist in many parts of contemporary India—for example, in the Khasi Hills in Assam, in the Arvalli ranges of Rajasthan, all along the Western Ghats in the southern peninsula, in the districts of Bastar and Sarguja in Madya Pradesh, and in the Chanda district in Maharastra. Gadgil and Chandran (1992) indicate that the Indian traditional shifting cultivation system, *jhum* (described in Section III.C), is associated with sacred groves.

Kenya has sacred groves all along its coast, known as *kayas* (homesteads), used for ceremonies and burials (Wilson, 1993). So do the Yoruba of Ara in southwestern Nigeria (Warren and Pinkston, 1998). In South America, the Kuna Indians of Panama have spirit sanctuaries, places where spiritual animals, plants, or demons are believed to reside (Chapin, 1991). According to cosmological beliefs, the Kuna must respect these areas often located on choice agricultural land. The Cocnucos and Yanaconas of Colombia have similar sanctuaries (Redford and Maclean Stearman, 1993). The Tukano of the Uaupés basin on the Brazil–Colombia border reserve the forested river margin for fish and fishing. Fishing may be restricted to as little as 38% of the total river margin available (Chernella, 1987). The result is a management system that allows for, yet distinguishes, human use areas and animal refuge areas. Any deforestation of the river edge is prohibited.

The importance of such cultural beliefs for preserving patches of ecosystems should not be underestimated. New species of plants, and species that have disappeared from other areas, are still being discovered within sacred groves (Mohannan and Nair, 1981). A botanical survey in a Nigerian sacred grove yielded 330 plant species as compared to only 23 in surrounding nonprotected areas (Warren and Pinkston, 1998). Sacred groves are also important for maintaining ecological services, such as preserving local hydrological cycles, preventing soil erosion, serving as firebreaks, and serving as areas of recruitment of species, allowing for ecosystem renewal in face of various disturbances.

E. Temporal Restrictions of Harvest

This practice is adopted among some local resource users. The idea is that certain biological resources are protected from exploitation for certain periods of time. Among local resource users, the imposition of temporal taboos regulates access to resource(s) on either a sporadic, daily, weekly, or monthly basis (Colding and Folke, 2000).

For example, clans of Tikopia in the Solomon Islands impose sporadic taboos on particular foodstuffs they are associated with (Chapman, 1985). Several different durations of closed seasons exist among Vanuatu fishing villages, ranging from 1 month to 5 years (Johannes, 1998).

At the Sakumo and Djange Lagoons in Ghana, taboos are imposed on fishing during a particular day every week (Ntiamoa-Baidu, 1991). In India, taboos imposed on a monthly basis appear quite widespread. For example, many castes abstain totally from consumption of fish, poultry, and meat, and suspend all hunting as well, in the Hindu month of Sravana (roughly August), which is the peak of the main rainy season over most of India (Gadgil, 1987). Similarly, taboos on hunting certain animals from July to October exist in many Indian villages. The Oraons, the fourth largest tribal group in India, observe a taboo of hunting any wild animal or bird during the months of June and July (Xaxa, 1992).

In Papua New Guinea, the Maopa people in the Marshall Lagoon impose a seasonal ban on hunting (Kwapena, 1984). Every 3–4 years at the end of their cultivation work, a bush area is set on fire to chase and hunt any animals found. This traditional hunting ritual lasts for a week. Once a particular animal-rich area is selected and hunted on, it will be left alone for another 3–4 years. This ritualistic way of hunting allows for hunting areas to regenerate. It may also mimic natural disturbance regimes. Groups of Canadian Amerindian hunters use a similar system for rotating hunting, fishing, and trapping areas by their periodic resting (Berkes, 1998).

In the Maluku Islands of eastern Indonesia, temporal prohibitions on gathering are imposed on terrestrial plant and animal resources, such as coconut, nutmeg, areca, and cuscus, as well as on marine species, including fish, trochus, shell, sea cucumber, and seaweed. This traditional management institution, known as *sasi*, is continually changing, depending on changing social and ecological conditions (Soselisa, 1998).

Locally decided and implemented closed periods is one of the measures used for managing kombu kelp (*Laminaria augusta*) at Hokkaido Island, Japan. Regulations initiated for kelp harvesting as early as 1881 stated that villagers could not harvest kombu in the evening or on a rainy day, when good quality kombu cannot be produced (Iida, 1998).

III. PRACTICES OF MULTIPLE-SPECIES MANAGEMENT, RESOURCE ROTATION, AND SUCCESSION

This category constitutes management practices that largely can be described as agroforestry practices, although hunter and gatherers may employ methods that draw on the same principles. Agroforestry is the generic name used to describe "an old and widely practiced land use system in which trees are combined spatially and/or temporally with agricultural crops and/or animals" (Farrell, 1987). They are common in traditional agricultural systems throughout the world, especially in the tropics, but were once common in temperate regions as well (Beets, 1990).

The three types of practices described in the following are often related to the efficient use of local ecosystems in terms of space, nutrient availability, climatic conditions, and soil conditions. These practices serve to actively manage species diversity and intensity of use of subsistence resources for ecosystem renewal and species recruitment. Such practices often create habitat heterogeneity on the landscape scale, affecting forest structure and species composition by creating a mosaic of forest patches of different ages (Primack, 1993). Thus, the outcome of these practices often enhances diversity of biological resources at the local level (Orejuela, 1992).

A. Multiple-Species and Integrated Management

The cultivation of several species of crops on the same piece of land is often referred to as polycultures or intercrops. There is an enormous variety of this type of practice, and they may involve the mix of annual crops with other annuals, annuals with perennials, or perennials with perennials. Crops may be sown in different spatial arrangements and may range from the growing of two crops in alternate rows to complex assemblies of a dozen or more species (Altieri, 1987). Integration of crop cultivation with animal species is also common in traditional agroforestry systems.

Polyculture cropping systems constitute at least 80% of the cultivated area of West Africa and predominate in other parts of Africa as well as in the Latin American tropics and many Asian countries (Liebman, 1987). Frequently, more yield can be harvested from an area sown in polyculture than from an equivalent area of monoculture without the application of synthetic fertilizers, pesticides, and field machinery. Polyculture cropping systems may provide several social–ecological benefits as compared to monocultures (Liebman, 1987)—for example, by reducing the risk of total crop failure and by contributing to an increase of nutritional returns to local resource users.

Polycultures may also promote soil water conservation and nutrient recycling as well as reduce soil erosion and insect pests. Some polyculture systems may also reduce crop diseases and increase weed control. Polycultures may thus provide for many ecological services, including the maintenance and enhancement of biological diversity. In fact, diversity is used to enhance productivity and resilience (Berkes *et al.*, 1995)

The two most common polyculture systems in Java, Indonesia, are *Talun-kebun* (rotation between mixed gardens and tree plantation) and *pekarangan* (home garden intercropping system including animals) (Altieri, 1987). A *Talun-kebun* is an indigenous Sudanese agricultural system, consisting of three stages that each serves a different function. In the first stage, *kebun*, a mixture of annual crops is usually planted, mainly for cash income. After 2 years, tree seedlings begin to grow in the field, and the *kebun* gradually evolves into a *kebun-campuran*, where annual crops are mixed with half-grown perennials. This stage promotes soil and water conservation. After the annuals are harvested, the field is usually abandoned for 2 or 3 years and becomes dominated by perennials, the *talun* stage (perennial crop garden). The *talun* can be turned back into a *kebun* after the forest is cleared or be planted to rice paddy, depending on whether irrigation water is available. A *Talun-kebun* may be composed of up to 112 species of plants, of which about 40% provide for building materials and fuel wood, 20% are fruit trees, 15% are vegetables, and the remainder is used for ornamentals, medicinal plants, spices, and cash crops (Altieri, 1987).

The integration of animals in polyculture systems is a common practice in diverse regions of the world. In Southeast Asia, rice ecosystems include diverse animal species. For example, domestic ducks may be integrated in paddy rice cultivation to allow for insect and weed control, and fish species, such as common carp, may be integrated and harvested at the end of the rice-growing season. The rice/fish culture differs considerably from country to country and from region to region (Altieri, 1987).

In China, as in Southeast Asia, crops, chicken, duck, and fish are often integrated to provide for a very high overall production through waste recycling and use of residues (Yan and Yao, 1989). Also, natural control of insects and weeds is provided for in these systems (Beets, 1990). In Indonesia, traditional systems combined rice and fish culture, and wastes from this system often flowed downstream into brackish water aquaculture systems (*tambak*). The *tambaks* were polyculture ponds, often combining fish, vegetation, and tree crops (Gadgil et al., 1993).

In Rajasthan, India, local farmers grow crops and rear livestock under and among trees. The Bishnoi, inhabiting one of the driest climatic zones on Earth, integrate the cultivation of food crops and animal rearing among islands of khejri trees, *Prosopis cineraria*, which are strongly protected by religious taboo (Sankhala, 1993). The Bishnoi keep as many khejri trees as possible on their farms, providing them with material for fencing, firewood, and fodder and pods for cattle as well as providing excellent microclimatic conditions for crop cultivation. Even threatened species, such as the blackbuck and gray wolf, benefit from these farming systems (Colding and Folke, 1997).

In many parts of Africa, "farm trees" can also be found within and adjacent to farm fields. The trees are actively protected, harvested, and managed by farmers to yield construction poles, fuel wood, fodder, edible fruits, nuts and leaves, medicines, and other products without unduly competing with associated annual crops (Beets, 1990).

B. Resource Rotation

Resource rotation exists among several communities of local resource users (Folke et al., 1998). Chisasibi Cree hunters rotate trapping areas (ideally) of beavers on a 4-year cycle to allow for the recovery of beavers to these areas. They use a similar rotational practice for resting fishing areas, using a traditional pattern of fishing in remote lakes on a 4- or 5-year cycle. Also, Cree goose hunters rotate hunting areas on a 7-day cycle in order to reduce disturbance to feeding and resting geese and to harvest for subsistence needs with a minimum disruption of the large population that passes through an area (Gadgil et al., 1993).

The Awa Indians of southwestern Colombia and neighboring Ecuador hunt echimid rodents on a cyclical basis, which allows sufficient time for the recovery of wild populations (Orejuela, 1992).

Arid and semiarid African pastoralists migrate seasonally in accordance with plant availability, determined by precipitation and lunar cycles (Niamir-Fuller, 1998). The yearly cycle of nomads and their cattle provides a rotational management system that enables the recovery of heavily grazed rangelands. For example, the Wodaabe Fulani follow the lunar cycle when moving to new pastures, which means that the camp is moved every 2–3 days. By contrast, the Rufa'a al Hoi of Sudan move to a new pasture every 204 days. The Fulani of northern Sierra Leone once practiced "shifting pasturage," whereby they heavily stocked an area for 2–3 years and then moved elsewhere and rested the first area for 15–20 years. These are all examples of small-scale movement, or micromobility, practiced by many Sahel pastoralsist tribes (Niamir-Fuller, 1998). Local resource users enforce grazing rotation in parts of the Hindu–Kush Himalayas in a similar fashion (Jodha, 1998).

C. Management of Succession

Polycultures and crop rotation often involve management of succession, as exemplified by the different systems of shifting cultivation that exist in the world. Shifting cultivation is defined by FAO (1982) as "a farming system in which relatively short periods of cultivation are followed by relatively long periods of fallow." Although part of polycultures, it can be distinguished by fallow periods that are ideally very long.

Shifting cultivation, or "slash-and-burn" cultivation or "swiddens," involves the clearing of a plot of land, usually a forest area, its use for a few years, and, as soil fertility declines, its abandonment in favor of another plot of land to be cleared in the same fashion. It is one of the oldest forms of agriculture and most present-day agricultural systems have evolved from it. In shifting cultivation, agriculture becomes a sequential cropping of crops and noncrops. Presently, about 3–500 million people, or about 40% of the total agricultural population of developing nations, depend on shifting cultivation for their daily livelihood. In total, shifting cultivation covers about 30% of the world's exploitable soils (Beets, 1990).

Shifting cultivation is common in all tropical areas and was once common in temperate regions as well. It has received much attention as one of the major degrading processes in tropical forest areas due to human population increase that greatly has led to shortages in fallow periods with subsequent loss of soil fertility. Generally, if fallow is less than 5–7 years, land degradation occurs and species diversity may be greatly reduced (Berkes et al., 1995). However, if adequate long fallow is allowed for, shifting cultivation may be highly sustainable (e.g., Posey, 1985).

Extremely high crop diversity is a characteristic of many traditional shifting cultivation systems. For example, Philippine swidden cultivators can distinguish over 600 plant species (Beets, 1990). In general, swiddens do not compare in complexity to the surrounding forest. However, when shifting cultivation is analyzed as an agroforestry system, i.e., the use of trees is also taken into account, then the overall result of managing forest patches can lead to an enhancement of biodiversity (Berkes *et al.*, 1995). For example, the Runa Indian managed swiddens in the Ecuador Amazon increase species diversity in 5-year-old fallows as compared to unmanaged fallows. Between 14 and 35% of this enhanced species diversity was attributed to direct planting and protection of secondary species.

Different forms of swiddens and related fallow systems exist in the world. In tropical Mexico, as well as in other Mesoamerican countries, *milpa* (maize fields) is widely practiced (Alcorn and Toledo, 1998). *Milpa* involves the clearing of new fields in high forests, or secondary regrowth forests, for maize cultivation. If more than one successive crop of maize is taken in a short fallow *milpa*, weedy species come to dominate the plot and forest regeneration may not occur. *Milpa* is governed by strong informal institutions, which reinforces reciprocity and community-based control of natural resources. Farmers who mismanage their *milpa* are labeled as witches and punished by social pressure. Each stage of the *milpa* cycle is also named and marked by ritual activities, rendering *milpa* a strong sociocultural practice.

The Awa Indians of southwestern Colombia and neighboring Ecuador practice a shifting agriculture known as slash–mulch (Orejuela, 1992). It involves the cutting of natural vegetation and the mulching of this material for a temporary agricultural field. Maize and/or short-cycle varieties of red and white beans are planted by the slash–mulch method. Once yields decline after several cycles of harvest, the field is left fallow and permitted to regrow as an enriched secondary forest. A similar system involving mulching is used to clear lots to raise cattle.

In northeastern India, detailed studies on shifting agriculture (*jhum*) have described multispecies systems involving 4 to over 35 crop types based on locally adapted native strains (Berkes *et al.*, 1995). This practice requires sophisticated local ecological knowledge, including the use of soil nutrients by adequate changes in the crop mixture depending on the length of *jhum* cycles and the consequent availability of soil nutrients.

On hill slopes, farmers combine r-strategist species (cereals and legumes) with K-strategists, with emphasis on vegetative growth, such as leafy vegetables (Berkes *et al.*, 1995).

An interesting form of shifting cultivation by the Kayapó Indians of Brazil constitutes the creation of forest islands known as *apete* (Posey, 1985). This begins as small mounds of vegetation about 2 m in diameter (*apete-nu*). As planted crop and tree seedlings grow and the planted area expands, the taller vegetation in the center of the mounds is cut to allow light to enter. A full-grown *apete* has an architecture that creates zones that vary in shade and moisture (Fig. 2). The species mix includes medicinal plants, palms, and vines that produce drinking water. Of a total of 120 species found in 10 *apete*, Posey (1985) estimated that 75% may have been planted. *Apete* constitutes the manipulation of semidomesticated plants on which the Kayapó could survive during times of warfare. These old forest islands have been scattered for millennia in known spots throughout the forest and savanna.

Practices of multiple species, resource rotation, and succession management contribute to biodiversity conservation relative to more modern and technology-dependent ways of exploiting resources and ecosystems. By actively using biodiversity in production, they contribute to biodiversity conservation in areas outside protected areas and reserves.

IV. PRACTICES RELATED TO THE DYNAMICS OF COMPLEX ECOSYSTEMS

Apete is also a good example of traditional practices that actively promote patchiness and heterogeneity at the landscape level. Many agroforestry systems make use of patchiness for multifunctional benefits, although patchiness may also be used in smaller cultivation systems to provide for crop protection and natural pest management (Altieri, 1987).

A. Management of Landscape Patchiness

Such management is employed by Sahelian pastoralists in order to mimic the variability and unpredictability of the landscape. For example, an appropriate mix of herding animals is used to utilize different vegetation types and patches in a dynamic fashion (Niamir-Fuller, 1998). The progressive widening of grazing radius around wells as the wet season advances is also an example of active management at the landscape level.

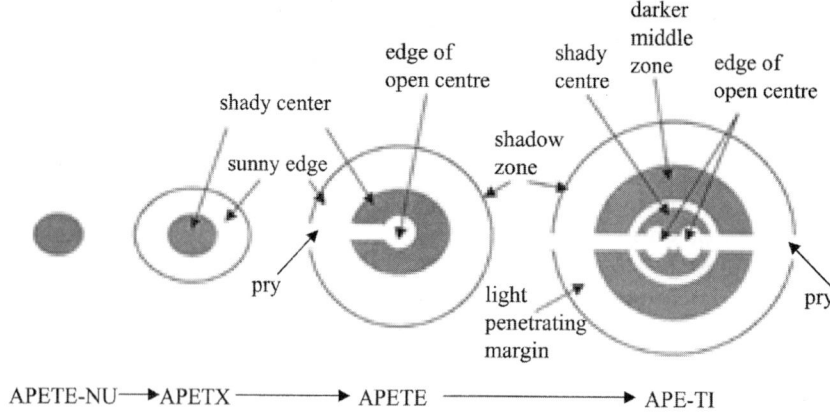

FIGURE 2 Among the Kayapo Indians of Brazil, enhancement of biodiversity is facilitated for by the creation of forest islands, *apete*. Through a number of management methods, this behavior promotes patchiness and heterogeneity in the landscape in time and space. Source: Gadgil et al., 1993.

This practice is employed by the Maasai of Kenya to leave enough forage around wells for the dry season (Niamir-Fuller, 1998).

Fire management among Australian aborigines and northern Canadian Amerindians was practiced widely to open up clearings (meadows and swales), corridors (trails, traplines, ridges, grass fringes of streams and lakes), and windfall forests. These clearings provided improved habitats for ungulates and waterfowl, thus increasing hunting success, and the corridors and windfall areas improved accessibility (Gadgil et al., 1993).

The use of different elevation zones for cultivators in the eastern Himalayas and other parts of the world represents practices that make use of habitat heterogeneity and create patchiness at the landscape level. The ancient *ahupua'a* system of Hawaii utilizes different elevation zones, determined by precipitation, that are used for various integrated farming practices (Costa-Pierce, 1987). In the ancient *ahupua'a*, both freshwater and seawater fish ponds were integrated with agriculture, and river valleys were managed as integrated systems, from the upland forest (left uncut by taboo) all the way down to the reef (Costa-Pierce, 1987).

B. Watershed Management

The *ahupua'a* also constitutes an example of traditional watershed management systems. Southeast Asia and Oceania had, and to lesser extent still have, a number of these prescientific ecosystem management practices. Examples include the Yap *tabinau*, the Fijian *vanua*, and the Solomon Islands *puava* (Ruddle et al., 1992). These all refer to generically similar watershed-based management systems. For example, the *vanua* concept is an integrated human–nature component that regards the land, water, and human environment as one unit, one and indivisible (Ruddle et al., 1992). Similarly, the *puava* includes all resources and land in a watershed, from the top of the mainland mountains to the open sea outside the barrier reef (Hviding, 1990). In all these cases, the social group inhabiting the ecosystem unit is considered to be part of the system.

C. Managing Ecological Processes at Multiple Scales

There is some evidence that traditional management systems may be useful in managing ecological processes at multiple scales. In *milpa*, described earlier in Section III.C, food crops are managed on a 1- to 3-year scale, and some tree crops and products on a 30-year scale (Alcorn and Toledo, 1998).

James Bay Cree hunters seem to be simultaneously managing beaver populations on a 4- to 6-year scale, lake fish on a 5- to 10-year scale, and caribou on an 80- to 100-year scale (Berkes, 1998). The holistic forestry of the Gitskan people of northern British Columbia simultaneously manages the production of fiber over several square kilometers with ecological processes involving soil bacteria at the spatial scale of a few square meters (Pinkerton, 1998).

D. Responding to and Managing Pulses and Surprises

Natural disturbances, such as fire, hurricanes, pest outbreaks, and heavy grazing, are inherent to the internal dynamics of ecosystems and often set the timing of ecosystem renewal processes (Holling et al., 1995).

An example of a practice that responds to disturbance and manages pulses and surprises is the establishment of range reserves within the annual grazing areas of African herders. These reserves provide an emergency supply of forage that serves as buffer when disturbance, such as drought, challenges the process and function of the dryland ecosystem (Niamir-Fuller, 1998). At the landscape level, this functions to maintain the resilience of both the ecosystem and the social system of the herders. Such practices may be considered ecological adaptations to unpredictable, low-rainfall environments and provide "ecological insurance" to the local communities.

Sacred groves in India absorb disturbance by serving as firebreaks for cultivated areas and villages (Gadgil et al., 1998). At the same time, these sacred habitats function as recruitment centers for species regeneration and ecosystem renewal. In times of natural disturbances and other emergencies, such areas of refugia are essential for providing social–ecological resilience.

Many of the traditional forms of polyculture previously described may provide for pest and insect protection that contributes to building resilience (Altieri, 1987). The Warlis of India control pests by placing certain kinds of tree branches in their paddy fields. This practice serves to attract birds for insect control and buffers against outbreaks of various pest populations (Pereira, 1992).

E. Nurturing Sources of Ecosystem Renewal

Many traditional societies seem to actively nurture sources of ecosystem renewal by creating small-scale disturbances. Traditional agroforestry practices such as shifting cultivation create forest gaps and enable people to produce crops or enhance wild foods without disrupting natural renewal processes (Berkes et al., 1995). Such practices may even enhance genetic and ecological diversity in the benign mosaic landscape of forest, fallow, and gardens (Orejuela, 1992). Also, traditional rotational cycles employed by hunters and gatherers, described in Section III, allow for species recovery of wild populations.

Aboriginal use of fire in as geographically diverse areas as Canada, Australia, and California had many elements and principles in common (Berkes et al., 1995). Until the late 1940s, Amerindians of northern Alberta, Canada, regularly used fire to create clearings (meadows and swales), corridors (trails, traplines, ridges, grass fringes of streams and lakes), and windfall forests (Lewis and Ferguson, 1988). These clearings provided improved habitat for species such as ungulates and waterfowl. Australian aborigines possessed detailed technical knowledge of fire and used it to improve feeding habitat for game and to assist in the hunt itself (Lewis, 1989).

African herders behave like a disturbance by following the migratory cycles of the herbivores from one area to another (Niamir-Fuller, 1998). Pulses of herbivore grazing contribute to the capacity of the semiarid grasslands of Africa to function under a wide range of climatic conditions. If this capacity of the ecosystem to deal with pulses is reduced, an event that previously could be absorbed can flip the grassland ecosystem into a relatively unproductive state, dominated and controlled by woody plants for several decades (Walker, 1993).

V. CONCLUDING REMARKS

One of the 10 principles of the *Global Biodiversity Strategy* concerns the linkage of biodiversity with cultural diversity and the conservation of the two together. At a broader level, traditional conservation practices as a part of cultural diversity may offer benefits in terms of local biological and ecological understanding, sustainable resource management systems, implementation of protected areas, development planning, and environmental impact assessment. Each of these potential benefits is related to biodiversity conservation, either directly or indirectly. Biodiversity is recognized as an important component of sustainable use. It is in this sense it is being conserved through traditional conservation practices as illustrated in this article through practices framing access to and use of species and habitats, practices of multiple-species management, and practices related to biodiversity conservation in complex dynamic ecosystems. Such practices seem to contribute to maintaining and building ecological resilience *sensu* Holling (1986).

Traditional conservation practices rely on the accumulation of ecological knowledge and understanding over many generations, and knowledge is embedded in local institutions and transmitted culturally. Practices are generally site and context specific although the ecological knowledge embedded in them may be spatially and temporally transmitted.

The locally generated knowledge and associated conservation practices are not necessarily complete in the sense of understanding of all aspects of an ecosystem and its dynamics. However, many traditional conservation practices are in line with the shifting scientific view on the nature of ecosystems as nonlinear, multiequilibrium, and full of surprises, threshold effects, and system flips. Predictability and controllability are not limited by the scientific data available but by the very nature of ecological systems.

Traditional ecological knowledge systems, based on detailed observations of the dynamics of the natural environment, feedback learning, social system–ecological system linkages, and resilience-enhancing mechanisms, seem akin to adaptive management (Berkes and Folke, 1998). Traditional conservation practices parallel adaptive management in their reliance on learning-by-doing and the use of feedback from the environment to provide corrections for management practice.

The parallels between adaptive management and indigenous management systems are probably not accidental. Flexible social systems that proceed along by learning-by-doing are better adapted for long-term survival than are rigid social systems that have set prescriptions for resource use. In light of this, adaptive management in modern society could be seen as a replication of traditional ecological knowledge systems in the framework of contemporary science. It is a sort of rediscovery of principles applied in traditional social–ecological systems. It is a search for sustainable use of ecosystems and biological diversity as a means of survival, a social and institutional response to resource scarcity and management failure. Even though there are no doubt major differences between the two, adaptive management may be viewed as the scientific analogue of traditional ecological knowledge and practices because of its integration of uncertainty into management strategies and its emphasis on practices that confer ecological resilience.

See Also the Following Articles

AGRICULTURE, TRADITIONAL • CONSERVATION MOVEMENT, HISTORICAL • HISTORIC AWARENESS OF BIODIVERSITY • INDIGENOUS PEOPLES, BIODIVERSITY AND • RELIGIOUS TRADITIONS AND BIODIVERSITY • SOCIAL AND CULTURAL FACTORS • SUSTAINABILITY, CONCEPT AND PRACTICE OF

Bibliography

Acheson, J. M., Wilson, J. A., and Steneck, R. S. (1998). Managing chaotic fisheries. In *Linking Social and Ecological Systems. Management Practices and Social Mechanisms for Building Resilience* (F. Berkes and C. Folke, Eds.). Cambridge Univ. Press, Cambridge, UK.

Alcorn, J. B., and Toledo, V. M. (1998). Resilient resource management in Mexico's forest ecosystems: The contribution of property rights. In *Linking Social and Ecological Systems. Management Practices and Social Mechanisms for Building Resilience* (F. Berkes and C. Folke, Eds.). Cambridge Univ. Press, Cambridge, UK.

Altieri, M. A. (1987). *Agroecology: The Scientific Basis of Alternative Agriculture.* Westview, Boulder, CO.

Balée, W. (1992). People of the fallow: A historical ecology of foraging in lowland South America. In *Conservation of Neotropical Forests. Working from Traditional Resource Use* (K. H. Redford and C. Padoch, Eds.). Columbia Univ. Press, New York.

Beets, W. C. (1990). *Raising and Sustaining Productivity of Smallholder Farming Systems in the Tropics.* AgBé Publishing, Alkmaar, Holland.

Berkes, F. (1998). Indigenous knowledge and resource management systems in the Canadian subarctic. In *Linking Social and Ecological Systems. Management Practices and Social Mechanisms for Building Resilience* (F. Berkes and C. Folke, Eds.). Cambridge Univ. Press, Cambridge, UK.

Berkes, F., and Folke, C. (1998). *Linking Social and Ecological Systems. Management Practices and Social Mechanisms for Building Resilience.* Cambridge Univ. Press, Cambridge, UK.

Berkes, F., Folke, C., and Gadgil, M. (1995). Traditional ecological knowledge, biodiversity, resilience and sustainability. In *Biodiversity Conservation* (C. A. Perrings, K.-G. Mäler, C. Folke, C. S. Holling, and B.-O. Jansson, Eds.). Kluwer Academic, Dordrecht/Norwell, MA.

Chapin, M. (1991). Losing the way of the great father. *New Scientist* Aug. 10.

Chapman, M. (1985). Environmental influences on the development of traditional conservation in the South Pacific region. *Environ. Conserv.* **12**(3).

Chernella, J. (1987). Endangered ideologies: Tukano fishing taboos. *Cult. Surv.* **11**(2).

Colding, J., and Folke, C. (1997). The relations among threatened species, their protection, and taboos. *Conserv. Ecol.* (online) **1**(1), 6. Available from the Internet: http://www.consecol.org/vol1/iss1/art6.

Colding, J., and Folke, C. (2000). The taboo system: Lessons about informal institutions for nature management. *The Georgetown Int'l Envtl. Law Review* **12**(2).

Costa-Pierce, B. A. (1987). Aquaculture in ancient Hawaii. Integrated farming systems included massive freshwater and seawater fish ponds. *BioScience* **37**(5).

FAO. (1982). *Improved Production Systems as an Alternative to Shifting Cultivation.* FAO Soils Bulletin No. 52, FAO, Rome.

Farrell, J. G. (1987). Agroforestry Systems. In *Agroecology: The Scientific Basis of Alternative Agriculture* (M. A. Alatieri, Ed.). Westview, Boulder, CO.

Folke, C., and Berkes, F. (1998). *Understanding Dynamics of Ecosystem–Institution Linkages for Building Resilience* (Beijer Discussion Paper Series No. 112). Beijer International Institute of Ecological Economics, The Royal Swedish Academy of Sciences, Stockholm, Sweden.

Folke, C., Berkes, F., and Colding, J. (1998). Ecological practices and social mechanisms for building resilience and sustainability. In *Linking Social and Ecological Systems. Management Practices and Social Mechanisms for Building Resilience* (F. Berkes and C. Folke, Eds.). Cambridge Univ. Press, Cambridge, UK.

Gadgil, M. (1987). Social restraints on exploiting nature: The Indian experience. *Development: Seeds of Change 1987, 1*.

Gadgil, M., and Chandran, M. D. S. (1992). Sacred groves. In *Indigenous Vision: Peoples of India Attitudes to the Environment* (G. Sen, Ed.). Sage Publications, New Delhi.

Gadgil, M., and Vartak, V. D. (1976). The sacred groves of Western Ghats in India. *Econ. Bot.* **30**.

Gadgil, M., Berkes, F., and Folke, C. (1993). Indigenous knowledge for biodiversity conservation. *Ambio* **22**(2–3).

Gadgil, M., Hemam, N. S., and Reddy, B. M. (1998). People, refugia, and resilience. In *Linking Social and Ecological Systems. Management Practices and Social Mechanisms for Building Resilience* (F. Berkes and C. Folke, Eds.). Cambridge Univ. Press, Cambridge, UK.

Hanna, S. S. (1998). Managing for human and ecological context in the Maine soft shell clam fishery. In *Linking Social and Ecological Systems. Management Practices and Social Mechanisms for Building Resilience* (F. Berkes and C. Folke, Eds.). Cambridge Univ. Press, Cambridge, UK.

Holling, C. S. (1986). Resilience of ecosystems; local surprise and global change. In *Sustainable Development of the Biosphere* (W. C. Clark and R. E. Munn, Eds.). Cambridge Univ. Press, Cambridge, UK.

Holling, C. S. (1996). Cross-scale morphology, geometry, and dynamics of ecosystems. In *Ecosystem Management. Selected Readings* (F. B. Samson and F. L. Knopf, Eds.). Springer-Verlag, New York.

Holling, C. S., Schindler, D. W., Walker, B. W., and Roughgarden, J. (1995). Biodiversity in the functioning of ecosystems: An ecological synthesis. In *Biodiversity Loss. Economic and Ecological Issues* (C. Perrings, K.-G. Mäler, C. Folke, C. S. Holling, and B.-O. Jansson, Eds.). Cambridge Univ. Press, Cambridge, UK.

Hviding, E. (1988). *Marine Tenure and Resource Development in Marovo Lagoon, Solomon Islands: Traditional Knowledge, Use, and Management of Marine Resources, with Implications for Contemporary Development* (FFA Report No. 88/35). South Pacific Forum Fisheries Agency, Honiaro, Solomon Islands.

Hviding, E. (1990). Keeping the sea: Aspects of marine tenure in Marovo Lagoon, Solomon Islands. In *Traditional Marine Resource Management in the Pacific Basin: An Anthology* (K. Ruddle and R. E. Johannes, Eds.). UNESCO/ROSTSEA, Jln. M. H. Thamrin No. 14, Jakarta, Indonesia.

Iida, T. (1998). Competition and communal regulations in the Kombu Kelp (*Laminaria angustata*) harvest. *Hum. Ecol.* **26**(3).

Jodha, N. (1998). Reviving the social system–ecosystem links in the Himalayas. In *Linking Social and Ecological Systems. Management Practices and Social Mechanisms for Building Resilience* (F. Berkes and C. Folke, Eds.). Cambridge Univ. Press, Cambridge, UK.

Johannes, R. E. (1978). Traditional marine conservation methods in Oceania and their demise. *Annu. Rev. Ecol. Syst.* **9**.

Johannes, R. E. (1998). The case for data-less marine resource management: Examples from tropical nearshore finfisheries. *Trends Ecol. Evol.* **13**.

Kwapena, N. (1984). Traditional conservation and utilization of wildlife in Papua New Guinea. *Environmentalist* **4** (Suppl. 7).

Levin, S. A., Barrett, S., Aniyar, S., Baumol, W., Bliss, C., Bolin, B., Dasgupta, P., Ehrlich, P., Folke, C., Gren, I. M., Holling, C. S., Jansson, A. M., Jansson, B.-O., Mäler, K.-G., Martin, D., Perrings, C., and Sheshinski, E. (1998). Resilience in natural and socioeconomic systems. *Environ. Dev. Econ.* **3**.

Lewis, H. T. (1989). Ecological and technical knowledge of fire: Aborigines versus park managers in Northern Australia. *Am. Anthropol.* **91**.

Lewis, H. T., and Ferguson, T. A. (1988). Yards, corridors and mosaics: How to burn a boreal forest. *Hum. Ecol.* **16**.

Liebman, M. (1987). Polyculture cropping systems. In *Agroecology: The Scientific Basis of Alternative Agriculture* (M. A. Altieri, Ed.). Westview, Boulder, CO.

Monahan, C. N., and Nair, N. C. (1981). Kunstlaria Prain: A new genus record for India and a new species in the genus. *Proc. Indian Acad. Sci.* **90**.

Nelson, J. G., and Serafin, R. (1992). Assessing biodiversity: A human ecological approach. *Ambio* **21**(3).

Niamir-Fuller, M. (1998). The resilience of pastoral herding in Sahelian Africa. In *Linking Social and Ecological Systems. Management Practices and Social Mechanisms for Building Resilience* (F. Berkes and C. Folke, Eds.). Cambridge Univ. Press, Cambridge, UK.

Ntiamoa-Baidu, Y. (1991). Conservation of coastal lagoons in Ghana: The traditional approach. *Landscape Urban Plan.* **20**.

Orejuela, J. E. (1992). Traditional productive systems of the Awa (Cuaiquer) Indians of Southwestern Colombia and neighboring Ecuador. In *Conservation of Neotropical Forests. Working from Traditional Resource Use* (K. H. Redford and C. Padoch, Eds.). Columbia Univ. Press, New York.

Pereira, W. (1992). The sustainable lifestyle of the Warlis. In *Indigenous Vision: Peoples of India Attitudes to the Environment* (G. Sen, Ed.). Sage Publications, New Delhi.

Pinkerton, E. (1998). Integrated management of a temperate montane forest ecosystem through wholistic forestry: A British Columbia example. In *Linking Social and Ecological Systems. Management Practices and Social Mechanisms for Building Resilience* (F. Berkes and C. Folke, Eds.). Cambridge Univ. Press, Cambridge, UK.

Posey, D. A. (1985). Indigenous management of tropical forest ecosystems: The case of the Kayapo Indians of the Brazilian Amazon. *Agrofor. Syst.* **3**(2).

Posey, D. A. (1992). Interpreting and applying the "reality" of indigenous concepts: What is necessary to learn from the natives? In *Conservation of Neotropical Forests. Working from Traditional Resource Use* (K. H. Redford and C. Padoch, Eds.). Columbia Univ. Press, New York.

Primack, R. B. (1993). *Essentials of Conservation Biology*. Sinauer, Sunderland, MA.

Redford, K. H., and MacLean Stearman, A. (1993). Forest-dwelling native Amazonians and the conservation of biodiversity: Interests in common or in collision? *Conserv. Biol.* **7**(2).

Reichel-Dolmatoff, G. (1976). Cosmology as ecological analysis: A view from the rainforest. *Man* **11**.

Ruddle, K. (1994). Local knowledge in the folk management of fisheries and coastal marine environments. In *Folk Management in the World's Fisheries: Lessons for Modern Fisheries Management* (C. L. Dyer and R. McGoodwin, Eds.). Univ. Press of Colorado, Niwot.

Ruddle, K., Hviding, E., and Johannes, R. E. (1992). Marine resources management in the context of customary tenure. *Mar. Res. Econ.* **7**.

Soselisa, H. L. (1998). Marine Sasi in Maluku, Indonesia. *Out of the Shell. Coastal Resources Res. Netw. Newsl.* **6**(3).

Turner, B. L., Clark, W. C., and Kates, W. C. (1990). *The Earth As Transformed by Human Action: Global and Regional Changes in the Biosphere over the Past 300 Years.* Cambridge Univ. Press, Cambridge, UK.

Walker, B. H. (1993). Rangeland ecology: Understanding and managing change. *Ambio* **22**(2–3).

Warren, D. M., and Pinkston, J. (1998). Indigenous African resource management of a tropical rainforest ecosystem: A case study of the Yoruba of Ara, Nigeria. In *Linking Social and Ecological Systems. Management Practices and Social Mechanisms for Building Resilience* (F. Berkes and C. Folke, Eds.). Cambridge Univ. Press, Cambridge, UK.

Wilson, A. (1993). Sacred forests and the elders. In *Indigenous Peoples and Protected Areas* (E. Kemf, Ed.). Earthscan Publications Ltd., London.

Wilson, E. (1992). *The Diversity of Life*. The Belknap Press of Harvard Univ. Press, Cambridge, MA.

World Commission on Environment and Development (WCED). (1987). *Our Common Future*. Oxford Univ. Press, Oxford.

Xaxa, V. (1992). Oraons: Religion, customs and environment. In *Indigenous Vision: Peoples of India Attitudes to the Environment* (G. Sen, Ed.). Sage Publications, New Delhi.

Yan, J., and Yao, H. (1989). Integrated fish culture management in China. In *Ecological Engineering: An Introduction to Ecotechnology* (W. J. Mitsch and S. E. Jörgensen, Eds.). Wiley, New York.

TROPHIC LEVELS

Peter Yodzis
University of Guelph

I. Food Chains and Trophic Levels
II. The Utility of Trophic Levels
III. Food Webs and Trophic Levels

GLOSSARY

basal species A species that eats no other species.
ecological transfer efficiency The ratio of energy ingested by a population's predators to energy ingested by the population.
food chain A sequential relationship of the form x_1 is eaten by x_2 which is eaten by x_3 which is eaten by ... which is eaten by x_n.
food web A specification of which species eat which in an ecosystem.
trophic level of a species 1 + a weighted average of the lengths of all food chains linking that species to basal species. Different weightings may be appropriate for addressing different questions.

THE TROPHIC LEVEL OF A SPECIES in an ecosystem is a measure of the length of food chains linking that species to basal species (autotrophs + detritus). This article will present the several different ways in which this concept can be construed, will indicate the appropriate context for each definition, and will discuss the significance of trophic levels both as categories of description universally applicable to all ecosystems and as expressions of ecosystem bioenergetics.

I. FOOD CHAINS AND TROPHIC LEVELS

Trophic ecology has to do with feeding relations—for instance, weasels eat mice, mice eat herbs—and is among the most basic organizing principles underlying biodiversity in natural ecosystems. One of the earliest attempts to identify ecosystem structure was based on trophic ecology, as follows. Some species in nature (mostly plants) do not eat anything; instead they utilize solar energy through photosynthesis and are called *primary producers* or *autotrophs*. Other species eat autotrophs and are called *herbivores* (Fig. 1). *Carnivores* eat herbivores, *secondary carnivores* eat carnivores, *tertiary carnivores* eat secondary carnivores, and so on (Fig. 1). These are, of course, highly aggregated entities: each of the categories just mentioned contains many species in any given ecosystem.

A *food chain* is a sequential relationship of the form x_1 is eaten by x_2 which is eaten by x_3 which is eaten by ... which is eaten by x_n, where the entities x_i might be individual species, or they might be aggregations such as those just defined.

Early trophic ecology (approximately the 1930s

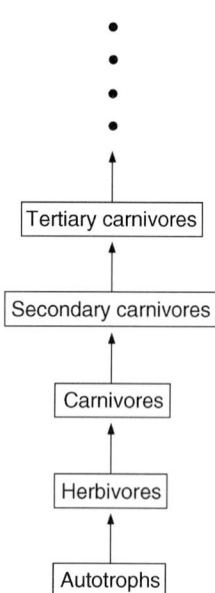

FIGURE 1 The IBP conception of an ecosystem as a linear sequence of trophic levels.

through 1970s) viewed an ecosystem as a single food chain involving the aggregated entities just defined, as depicted in Fig. 1. In this view of an ecosystem, autotrophs constitute "trophic level" 1, carnivores constitute "trophic level" 2, secondary carnivores are "trophic level" 3, and so on. [I am using "scare quotes" for the term "trophic level" because there are difficulties with this particular definition, which I will discuss in Section III.] The International Biological Program (IBP) of 1964–1974 was very much organized around this concept of ecosystem structure.

The concepts of food chain and trophic level are very closely related. Thus, in the formulation sketched in Fig. 1, species in the nth trophic level are linked to primary production (autotrophs) through a food chain of length $n - 1$.

There is another food chain that is very important in a great many ecosystems, namely, the one that is based on detritus. Detritus is certainly eaten by a wide variety of organisms, but it is not itself a living organism, so it cannot exactly be said to "eat" anything. However, biomass does move from living organisms into detritus as they decay after dying. If one is concerned to follow the recycling of specific nutrients, one needs to keep track of this entire dynamic. For purely trophic studies it is generally adequate (and far simpler) to treat detritus as though it were another organism that, like an autotroph, does not eat anything. Then autotrophs and detritus together are called *basal species*, and the detritivores are lumped together with the herbivores in trophic level number 1.

II. THE UTILITY OF TROPHIC LEVELS

The trophic level concept has been exceptionally durable: it has been one of the basic concepts of ecology for six decades and is one of the few ecological concepts contained in the vocabulary of most educated people. The reason for this distinguished place in the scheme of things is that the concept is both simple and useful. Furthermore, it is universal: it applies to all ecosystems.

Because of this universality, trophic levels enable us to compare the role of vastly different species in vastly different systems. For instance, we can discuss and understand a lake and the surrounding forest with a common language: the forest has its vegetation and its leaf litter; the lake has its phytoplankton and its dissolved organic matter (basal species). The forest has herbivorous insects, birds, and mammals; the lake has zooplankton (herbivores). And so on. We can use the same language to compare these two systems with any other ecosystem anywhere in the world.

This categorical and conceptual role can be made more quantitative and detailed, revealing important similarities and important differences among systems, by adopting a *bioenergetic* viewpoint, as follows.

Biological organisms contain caloric energy, which is transferred to organisms in the next step up a food chain: herbivores gain energy from consuming basal species, carnivores gain energy by consuming herbivores, and so forth. Each organism, or set of organisms such as a trophic level, *produces* energy at a certain rate. This is the maximum rate at which the next trophic level up the food chain could in principle ingest energy.

The rate of energy production by a trophic level must necessarily be less than the rate of energy ingestion by that trophic level. First, not all energy ingested by an organism is available to be metabolized by that organism. Some of it will be lost to excretion. The ratio of metabolizable energy to ingested energy is called *assimilation efficiency* and is typically about 0.45 for herbivores and 0.85 for carnivores. Of the metabolizable energy, some is lost to respiration, being used up by the organism to carry out its various activities and also simply to live, and the remainder is available for the production of new tissue, which can in principle be consumed by the next trophic level. The ratio of energy production to metabolizable energy is called *production efficiency* and ranges from about 0.1 to 0.4 for invertebrate ectotherms to about 0.01–0.03 for endotherms.

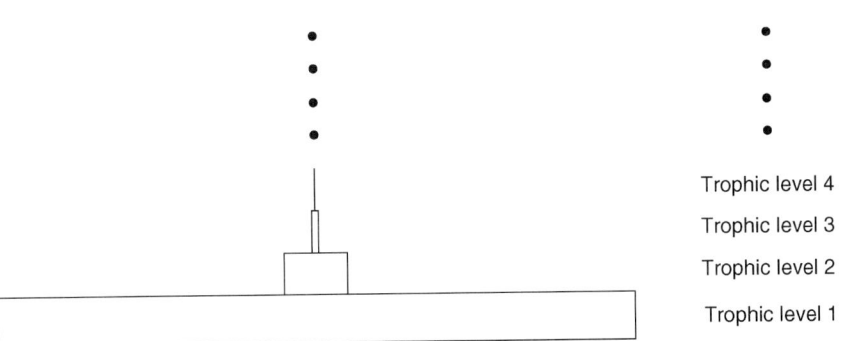

FIGURE 2 The pyramid of production in an ecosystem. The width of each layer is proportional to the rate of production of biomass in the corresponding trophic level. Because energy is dissipated in each transfer from one trophic level to the next, production decreases at an approximately geometric rate as trophic level increases.

Furthermore, not all of the energy produced at a trophic level will actually be ingested by the next higher trophic level; much of it will be missed and end up as detritus. There is no generally agreed term for this form of energy loss, nor is there a great deal of quantitative data for it.

The *ecological transfer efficiency* of a trophic level is the ratio of (energy ingested from that trophic level by the next highest trophic level) to (energy ingested by that trophic level). It is the product of the three efficiencies explicated in the preceding paragraph. Ecological transfer efficiency ranges from 0.001 or smaller (depending upon losses to detritus) up to a maximum of about 0.5.

It quickly becomes apparent that energy is dissipated quite rapidly as we ascend a food chain. Suppose, for example, that each ecological transfer efficiency in a food chain is 0.1, which is rather high. Then of the energy produced by basal species (primary production), one-tenth is produced by herbivores. One-tenth of that, or one-hundredth of primary production, is produced by carnivores. One-tenth of that, or one-thousandth of primary production, is produced by secondary carnivores, and so on up the food chain. Less and less energy is available to higher trophic levels as we move up a food chain. One can visualize this phenomenon as a "pyramid of production" for a food chain (Fig. 2).

This geometrically decreasing production as we move up a food chain, hence the rapidly decreasing available energy flow for the next higher trophic level, has been offered as one possible explanation for the apparently limited number of trophic levels in natural systems. Eventually, as we move up a food chain, the small fraction of primary production available to a putative next highest trophic level simply will not be enough to support a viable biological population.

The pyramid of production is an inescapable consequence of the dissipative processes, sketched above, that lead to ecological transfer efficiencies less than 1. There are a couple of similar "pyramids" that, while not universal in this way, are fairly typical of trophic levels. They follow from the circumstance that, for the most part, predators tend to be larger than their prey. For a predator to be larger (hence also faster and stronger) than its prey greatly facilitates the capture and consumption of prey.

Thus, as we move to higher trophic levels, we will, generally speaking, see larger animals. And yet, moving to higher trophic levels, these larger animals need to live on smaller energy production from the next trophic level down. As a result, there will usually be fewer animals at higher trophic levels. This "pyramid of numbers" is frequently, though not necessarily always, observed.

An obvious exception to the pyramid of numbers emerges if we treat parasites and parasitoids as "predators"; they are almost always smaller than their "prey." Even though parasitism is tremendously widespread in nature, these are not really trophic relationships, and so most trophic studies do not include parasites or parasitoids.

What about total biomass [= (number of animals) × (weight of each animal)] at each trophic level? The number of animals tends to decrease as trophic level increases, while the weight of each animal tends to increase. The result is equivocal. Particularly in aquatic systems, where very small organisms at low trophic levels have very rapid rates of biomass turnover and can be grazed to quite low levels, one frequently (but not always) sees "inverted pyramids" of biomass, with more biomass at higher trophic levels. But terrestrial systems typically (though by no means always) display pyramids of biomass, with less biomass at higher trophic levels.

There are exceptions to this scheme, but they prove the rule; that is, they make sense in terms of the ideas underlying the scheme. For instance, some of the very largest animals, such as elephants and big ungulates, are herbivores. These animals are so large that they could not possibly range far enough to live by eating, say, lions. The only way to get a high enough energy density to support such large animals is by feeding directly on plants.

Just as energy propagates upward through food chains, so may chemical substances contained in organisms. This becomes particularly interesting when toxic contaminants are present. If those toxic substances are absorbed and/or ingested by animals at some trophic level, then, depending upon the rate at which they are excreted, there may be residues in the tissue consumed by higher trophic levels. Under some circumstances, the concentration of toxins may increase as trophic level increases, which is called *biomagnification*.

III. FOOD WEBS AND TROPHIC LEVELS

The conception of an ecosystem as a linear chain of "trophic levels" (Fig. 1) is a useful starting point, but if we examine trophic relations with a higher degree of taxonomic resolution—that is, not lumping so many biological species together as we did in motivating Fig. 1—we find quite a different trophic structure. A *food web* is a specification of which species eat which in an ecosystem. For instance, Fig. 3 is a food web for Wytham Wood, a forest near Oxford in England. An arrow from one kind of organism to another indicates that the organisms at the head of the arrow eat the organisms at the other end.

One can detect something like "trophic levels" here (partly because of the way I chose to draw the picture), but there is certainly not a simple flow of energy through a linear sequence of levels as in Fig. 1; this picture is more "webby." For instance, we might put weasels at "trophic level" 5, because they eat titmice, which eat spiders, which eat insects, which eat herbs. But weasels also eat voles and mice, which eat herbs: this would put weasels at "trophic level" 3.

We do not want to throw away the trophic level concept altogether—it is too useful for that—but in the light of more refined data such as Fig. 3, we need to refine our concept of trophic level. In fact, there are a number of different ways that we may define the term "trophic level," and it seems imprudent to insist that any one definition is "The Right" one. Rather, different trophic level concepts may be appropriate for different purposes.

The constant theme linking all trophic level concepts together is the idea that trophic level has to do with the lengths of food chains linking a species to basal species. We just have to bear in mind that a species will generally be linked to basals through several food chains, which might be of different lengths. For instance, there are 13 food chains that link weasels to basal species in the Wytham Wood food web of Fig. 3. Three of these have length 2, 9 of them have length 3, and 1 of them has length 4. The relative importance assigned to these 13 food chains distinguishes several different definitions of "trophic level." Five commonly used definitions are listed here; following discussion of these, a sixth, in a considerably different spirit, will be addressed:

1. 1 + the length of the shortest food chain linking a species to some basal species.
2. 1 + the length of the longest food chain linking a species to some basal species.
3. 1 + the mean length of food chains linking a species to some basal species.
4. 1 + the weighted mean length of food chains linking a species to some basal species, where the weighting reflects energy flow through each food chain.
5. $\lambda + (\delta X_{\text{organism}} - \delta X_{\text{reference level}})/E$, where E is the average enrichment of a heavy isotope and $\delta X = [(R_{\text{sample}}/R_{\text{standard}}) - 1] \times 10^3$. Here X denotes the heavy isotope (for instance, ^{13}C, ^{15}N, or ^{34}S) and R denotes the heavy/light ratio (for instance, $^{13}C/^{12}C$, $^{15}N/^{14}N$, or $^{34}S/^{32}S$). The trophic level of the reference level organisms is λ: this might be basal species ($\lambda = 1$), or perhaps herbivores ($\lambda = 2$).

Definition 5 requires explanation. Certain heavy isotopes are enriched relative to the light isotope in biochemical reactions. As a result, the ratio of heavy to light isotope in an organism's tissue may bear a systematic relation to the ratio in that organism's diet. For instance, the ratio $^{15}N/^{14}N$ appears to be enriched by $E = 3.4\lambda‰$ ($\pm 1\lambda‰$) in a wide variety of organisms. Therefore, if we know the isotope ratio for organisms at some reference trophic level, we can use Definition 5 to calculate a trophic level from measurements of the isotope ratio in other organisms. The resulting numbers, which are a weighted average over all food chains from the reference level to the organisms in question, are probably fairly close to what we would get from Definition 4. This method requires far less effort than a direct calculation

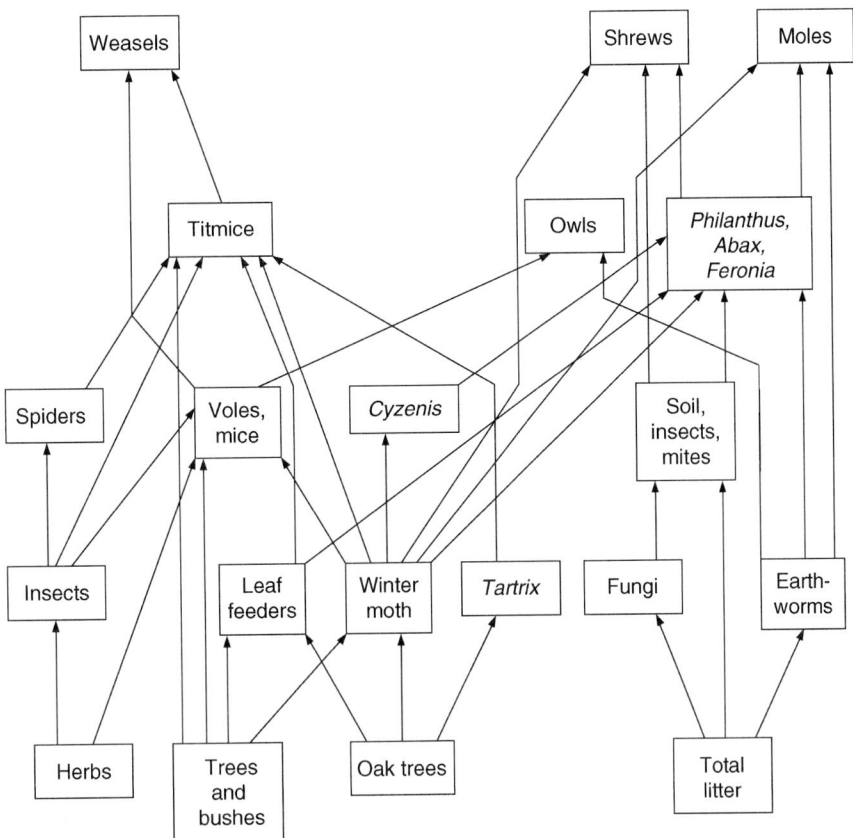

FIGURE 3 A food web for Wytham Wood, England (based on data in Varley, 1970).

of Definition 4 would; for instance, there is no need to measure dietary proportions. It does require careful calibration of isotope ratios at the reference level, including allowance for possible differences among species at that level. In practice, use of the $^{13}C/^{12}C$ ratio has not been fruitful, in part because so little ^{13}C is enriched at each trophic transfer, but as of this writing, the $^{15}N/^{14}N$ ratio appears to hold promise as a tool for trophic studies.

Table I shows, for each nonbasal species in the Wytham Wood food web, the frequency distribution of food chains linking that species to basals and the consequent trophic level according to Definitions 1–3. (Definitions 4 and 5 require more data than are available for the Wytham Wood food web.) For instance, depending upon the relative importance attached to the 13 food chains linking weasels to basal species, we may put weasels at trophic level 3 (shortest chain; Definition 1), 5 (longest chain; Definition 2), or 3.8 (mean length; Definition 3).

Generally speaking, one would like to use something like Definition 4, even though it means replacing the notion of discrete trophic levels with a trophic continuum. However, calculating trophic level in this way requires a tremendous amount of data. Definition 5 may be a good surrogate, but it still requires data beyond the food web itself. One is tempted to regard Definition 3 as a reasonably good substitute, but if we are thinking energetically, then because of the dissipation of energy as we move up a food chain, the shorter chains may well be more important energetically than the longer ones, so the equal weighting of Definition 3 may be deceptive. Definition 2 is what one has in mind implicitly when one draws tidy pictures such as Fig. 3, but the existence of very long food chains can be deceptive. Animals that have a food chain to basals with 8, 9, or even 10 links exist, but they invariably also have chains no longer than 3 links, and these shorter chains are likely more important energetically. However, very long food chains may be particularly significant if one is concerned with biomagnification of toxin concentrations.

Another viewpoint, which ought to produce a sixth trophic level definition if it could be articulated pre-

TABLE I

Food Chains and Trophic Levels in the Wytham Wood Food Web of Fig. 3

Species	Number of chains to basal species of length				Trophic level using		
	1	2	3	4	Shortest chain	Longest chain	Mean length
Insects	1	0	0	0	2	2	2.0
Winter moth	2	0	0	0	2	2	2.0
Tartrix	1	0	0	0	2	2	2.0
Leaf feeders	2	0	0	0	2	2	2.0
Earthworms	1	0	0	0	2	2	2.0
Fungi	1	0	0	0	2	2	2.0
Voles, mice	2	3	0	0	2	3	2.6
Spiders	0	1	0	0	3	3	3.0
Titmice	1	6	1	0	2	4	2.9
Cyzenis	0	2	0	0	3	3	3.0
Philanthus, Abax, Feronia	0	6	3	0	3	4	3.3
Soil insects, mites	1	1	0	0	2	3	2.5
Owls	0	3	3	0	3	4	3.5
Weasels	0	3	9	1	3	5	3.8
Shrews	0	3	7	3	3	5	4.0
Moles	0	3	6	3	3	5	4.0

cisely enough, emphasizes the *top-down* aspect of trophic relationships. This viewpoint, which has been put forward by S. Fretwell and L. Oksanen, counts a predator as one trophic level higher than its prey only if it significantly controls the biomass or dynamics of the prey species. Oksanen suggests that on this basis the distinction among carnivores, secondary carnivores, tertiary carnivores, and so on largely evaporates, leaving only three true trophic levels: basal species, herbivores, and carnivores—except in pelagic systems, where, due to the very small size of the primary producers and the consequent small size of zooplankton, planktivory and piscivory emerge as truly distinct trophic roles, permitting four trophic levels. This notion that there are actually only a few trophic levels despite the existence of some very long food chains is consonant with the implications of energetics noted in the preceding paragraph.

Thus, the term "trophic level" needs to be used with caution, and if we are to speak quantitatively of trophic levels, we need to specify exactly which definition we are using and to choose a definition that sheds the most light on the particular issues of concern.

See Also the Following Articles

ECOTOXICOLOGY • ENERGY FLOW AND ECOSYSTEMS • FOOD WEBS • PREDATORS, ECOLOGICAL ROLE OF • SPECIES INTERACTIONS

Bibliography

Polis, G. A., and Winemiller, K. O. (1996). *Food Webs: Integration of Patterns and Dynamics*. Chapman & Hall, New York.

Rundel, P. W., Ehleringer, J. R., and Nagy, K. A. (Eds.) (1989). *Stable Isotopes in Ecological Research*. Springer-Verlag, New York.

Varley, G. C. (1970). The concept of energy applied to a woodland community. In *Animal Populations in Relation to Their Food Resources* (A. Watson, Ed.). Blackwell Sci., Oxford.

Yodzis, P. (1989). *Introduction to Theoretical Ecology*. Harper-Collins, New York.

TROPICAL FOREST ECOSYSTEMS

Gary S. Hartshorn
Organization for Tropical Studies

I. Introduction
II. Physical Factors
III. Major Forest Ecosystems

GLOSSARY

andepts Soils derived from andesitic volcanic ash.
australasia Geographic region that encompasses Australia, eastern Indonesia, and nearby islands of the tropical southwest Pacific.
great faunal interchange Development of the Central American isthmus facilitated the interchange of fauna from north to south and vice versa.
indo-Malaya Geographic region from India to Malaysia and western Indonesia.
intertropical convergence zone Low-latitude region characterized by warm air masses and convectional rainfall; also known as the heat or thermal equator.
neotropics Tropical portion of Western Hemisphere ranging from southern Mexico to northeastern Argentina, including the Caribbean and the tip of Florida; New World Tropics is synonymous.
paleotropics Tropical regions of the Eastern Hemisphere extending from Africa through south and southeast Asia into the southwest Pacific; also known as Old World Tropics.
rain shadow Less rainy sector on leeward side of mountains and protected valleys caused by descending, warming air masses.
terra firme Nonflooded landscape in tropical lowlands, usually characterized by highly weathered, nutrient-poor soils.

TROPICAL FOREST ECOSYSTEMS are of global importance and interest for several reasons. They are home to the great majority of species on this planet Earth. They house an astounding complexity of ecological interactions among species as well as with their environment. They protect watersheds, modulate stream flow, and minimize soil erosion. Their protection in national parks and equivalent reserves is a high priority for international organizations, national agencies, and local communities. Yet the destruction of tropical forests depletes natural resources, destroys environmental services, and is a significant source of carbon dioxide and other greenhouse gases accumulating in the Earth's atmosphere that contribute to global climate change. If given the opportunity, tropical forests can reestablish surprisingly quickly to create forest habitat and to extract appreciable quantities of atmospheric carbon and store it as wood and organic matter.

I. INTRODUCTION

Though covering only 9% of the Earth's land, tropical forest ecosystems are famous for their exuberant greenness, impressive structure, and striking diversity of fauna and flora—and how poorly studied they are. Despite their attraction to nineteenth-century explorers such as Charles Darwin, Alexander von Humboldt, and Alfred Russell Wallace and the development of permanent field stations in the twentieth century, we still know very little about the biodiversity in tropical forests and how species function and interact in these ecosystems. For example, the huge uncertainty about the total number of species on our planet is largely due to how little is known about the invertebrates living in tropical forest ecosystems. Even though the phrase "tropical rain forest" still conjures up vivid images and myths about this signature ecosystem, there is a great amount of variety and heterogeneity in the types and extent of tropical forest ecosystems. This essay first provides an overview of the principal physical factors influencing the distribution and variation in tropical forest ecosystems and then describes the major types and features of tropical forest ecosystems.

II. PHYSICAL FACTORS

A. Geography

Tropical forest ecosystems are usually differentiated into neotropical (New World Tropics) and paleotropical (Old World Tropics). In contrast to the Amazon's dominance of the Neotropics, the Paleotropics include two major regions—Africa and Asia. African tropical forest ecosystems occur primarily in the Congo basin and the higher lands that border the basin. Outliers occur in a narrow coastal band of West Africa and in the Eastern Arc Mountains of East Africa. A narrow strip of tropical forests occurs along the eastern coast of Madagascar as well as in patches on other islands of the Indian Ocean. Because of striking biogeographic differences (e.g., Wallace's Line), Asia is further differentiated into the Indo-Malayan and Australasian regions.

Tropical biogeography is profoundly influenced by the degree of insularity of major landmasses. For example, the lengthy isolation and connection of Australia, New Guinea, and other islands of Australasia facilitated the striking evolution and diversification of marsupials (e.g., tree kangaroos) in the absence of Indo-Malayan carnivores. The tree family Dipterocarpaceae dominates tropical forest ecosystems in the Indo-Malayan region, but this family has far less importance in Australasian tropical forests. Similarly, the long geological isolation of South America led to a distinctive fauna (e.g., anteaters, sloths, and tapirs) but many of these species were wiped out during the Great Faunal Interchange via the Central American isthmus. In contrast, the continued isolation of oceanic islands (e.g., Galapagos and Madagascar) has enabled some of the unique fauna and flora to survive.

The latitudinal extent of tropical forest ecosystems is not defined by the astronomical limits of the tropic of Capricorn (23°30′ S) and the tropic of Cancer (23°30′ N). The ecological limits are determined by winter cold (which may not mean freezing temperatures) and inadequate rainfall to support forest vegetation. Generally, tropical forest ecosystems extend farther poleward where moisture is adequate, e.g., in the moist lowlands of Australia, Madagascar, and Mexico. Where aridity precludes the presence of lowland forest (e.g., Argentina and India), tropical forest ecosystems may extend to higher latitudes on the flanks of cooler, moister mountains.

Though approximately 85 countries contain tropical forest ecosystems covering 14,076,490 km^2, equivalent to 58% of North America, the top 10 countries comprise 75% of this total (see Table I). The Amazon basin, shared by eight countries, has by far the largest area of tropical forest ecosystems, with 5 of the top 10 countries by extent of forest. Vast tropical forest ecosystems also occur in the Congo basin and on the mega-islands of the tropical Far East (Borneo, New Guinea, Sumatra, and Sulawesi). Smaller, but significant tropical forest ecosystems occur in the Caribbean, Mesoamerica, trans-Andean South America, coastal Brazil, West Africa, East Africa, Madagascar, South Asia, Southeast Asia, southern China, the Philippines, and many tropical Pacific islands.

B. Physiography

Regional features such as mountains and water bodies play an important role, influencing climate, soils, and vegetation in addition to ecosystem functions. Because major mountain chains (above 2500 m) block the normal flow of air masses, they are the most prominent physiographic feature influencing tropical forest ecosystems. The best example is the Andes, running along the western coast of South America. On the eastern (windward) side, copious rains fall usually throughout the year, whereas on the leeward (i.e., rain shadow) side, coastal deserts dominate—especially in Peru and Chile with the oceanic upwelling of the cold Humboldt Current. Many other mountain ranges (e.g., New Guinea, western India, East Africa, and Mesoamerica) have similar windward–leeward effects on climate and vegetation. Though lesser ranges may not create such striking differences, they provide impressive altitudinal

TABLE I
Top 10 Countries with Tropical Forests

Rank	Country	Forests (km²)	Cover(%)	Annual loss (km²)	% loss/year
1	Brazil	4,427,200	52.4	25,544	0.5
2	Democratic Republic of the Congo	1,352,650	60.0	7,402	0.7
3	Indonesia	911,340	50.3	10,844	1.0
4	Peru	782,960	61.2	2,168	0.3
5	Bolivia	686,380	63.3	5,814	1.2
6	Venezuela	602,330	68.3	5,034	1.1
7	India	540,140	18.2	2,708	0.5
8	Colombia	535,540	51.6	2,622	0.5
9	Mexico	457,650	24.0	5,080	0.9
10	Angola	375,640	30.1	2,368	1.0
Subtotal top 10		10,671,830	46.5	69,584	0.65
Total 85 countries		14,076,490	10.8	154,000	0.8

Source: World Resources Report, 1998.

gradients in physical features and they may also be biological barriers. Volcanic massifs (e.g., Mount Kilamanjaro), isolated mesas (e.g., tepuís of southern Venezuela), ancient geologic shields (e.g., Guianan and Brazilian), and uplifted limestone platforms (e.g., Yucatán, northern Borneo, Palawan, and Papua New Guinea) all contribute to the physiographic differentiation of tropical forest ecosystems.

Mountainous regions and high rainfall generate impressive river systems such as the Amazon, Brahmaputra, Congo, Ganges, Mekong, Niger, and Orinoco. These complex hydrographic systems both facilitate and impede ecological connections. The Amazon basin has three major types of streams: white water (originating in the Andes and carrying relatively fertile sediments); clear water (originating from terra firme in the lower basin), and black water (originating from white sand in the basin). Tributaries cut deeply into mountain flanks and carry huge sediment loads. Their natural floodplains are vast, creating unique ecosystems such as the famous flooded forests in the Amazon basin. Major rivers also meander over surprisingly large areas (~25%), reworking the sediments into new landforms and creating oxbow lakes.

Continental islands (Australia), marine islands (e.g., Antilles, Galapagos, Hawaii, Madagascar, and south Pacific), islands on the continental shelf (e.g., Borneo, Hainan, Java, New Guinea, Philippines, Sri Lanka, Sulawesi, and Sumatra), and innumerable smaller islands are important physiographic features. Due to degree of isolation, proximity to the "mainland," and colonization/extinction events, islands may develop distinctive fauna and flora that make significant contributions to tropical biodiversity.

C. Climate

Tropical climates are characterized by predictable temperature patterns with modest seasonal variation and precipitation regimes that are much more unpredictable and highly variable from year to year. Daily temperature range (on a sunny day) is far greater than the seasonal variation between the warmest and coolest months. In mountainous terrain, the daily temperature variation may be as large as 20°C. Seasonal differences are usually determined by rainfall patterns, such that the drier season is called "summer" and the wet season is "winter." The rainy season(s) tracks the movement of the thermal equator, also known as the intertropical convergence zone (ITCZ), which is about 1 month behind the passage of the sun. The ITCZ brings a daily pattern of afternoon convectional rains, often via thunderstorms or the monsoon. When the ITCZ is seasonally away from the region, it is usually the dry season. For example, when the sun is in the Southern Hemisphere, the northeast tradewinds dominate the Northern Hemisphere subtropics, e.g., the Caribbean and Africa. In contrast to the persistent tradewinds, the Asian monsoon involves a reversal of winds, with a summer inflow of warm, moist air and a winter outflow of cool, dry air.

In the Caribbean, the tradewinds pass over warm seawater, bringing considerable moisture and clouds to mountainous land near the sea. The cloud cover and moisture create the renowned cloud forests of the windward Caribbean region such as Monteverde (Costa Rica), Blue Mountains (Jamaica), El Verde (Puerto Rico), and Rancho Grande (Venezuela). On the opposite side of the mountains, descending tradewinds produce much more arid forest ecosystems on the leeward slopes. The same windward–leeward phenomenon occurs on the slopes and valleys of major mountain ranges such as the Andes and Himalayas. The front ranges are exposed to moisture-laden airmasses and orographic rainfall, while interior valleys are much more arid.

In tropical regions, considerable between-year variation in total rainfall and its seasonality is the normal pattern, regardless of the average annual rainfall. Long-term rainfall records indicate that the annual maximum is usually more than twice the amount of the driest year. Near the equator, there are usually two alternating rainy and dry seasons, each about 3 months in duration. As one moves away from the equator, one dry season (when the sun is in the other hemisphere) lengthens and the other shortens or disappears. Rainfall regimes tend not to be simple or predictable, because short rainless periods can occur during the rainy season or isolated rains can occur during the dry season. The El Niño Southern Oscillation (ENSO) phenomenon occasionally eliminates or greatly delays one of the seasons; i.e., the rains may be very late in arriving or they may override a normal dry season.

D. Soils

Tropical soils have been influenced by major geologic phenomena such as plate tectonics, weathering of ancient landscapes, uplifting of mountain ranges, volcanic outputs, and oceanic submergence or emergence. Erosion of the ancient Guianan shield contributed the extensive white sand soils of the Río Negro basin, leaving the impressive flat-topped mesas ("tepuís") that dominate southern Venezuela. Many of the infertile, terra firme soils of the Amazon basin were eroded from the geologically ancient Brazilian shield. The active floodplains of several major western Amazonian rivers receive sediments eroded from the geologically young Andes, thus juxtaposing annually renewed, fertile floodplain soils with the infertile terra firme soils that are so typical of much of the Amazon basin.

Much volcanic activity has characterized the Quaternary, especially in mountainous regions of the Tropics. Volcanic ash deposits and mudflows tend to be more prevalent than lava flows (Hawaii is an exception). There are two major types of volcanic ash—andesitic and rhyolitic; the former is rich in certain nutrients, whereas the latter is high in silica. Soils developed from andesitic ash are termed "andepts" and are famous as the best soils for coffee or tea (e.g., Assam, Colombia, Central America, Hawaii, Jamaica, Java, and Kenya). The natural fertility of andepts is the foundation of the highly productive paddy rice cultures of Southeast Asia. In contrast, extensive soils derived from rhyolite in Honduras and Nicaragua have low agricultural potential and are dominated by open pine forests.

Tropical soils vary greatly in their principal chemical and physical properties, such as parent material, soil structure, drainage regime, and age. Variations in soil properties play an important role in the different types of tropical forest ecosystems; however, we do not know much about their controlling influence on the distribution of most tropical tree species. The more restrictive a site is, the more distinctive the vegetation may be. For example, the combination of nutrient-poor, white sand soils and a perched water table in southern Venezuela gives rise to a dwarf shrubby vegetation ("bana" or "caatinga") in a tropical rain forest region.

III. MAJOR FOREST ECOSYSTEMS

Tropical forest ecosystems are extremely rich in species of both fauna and flora. The highest tree species richness numbers—approximately 300 species per hectare—have been documented from the western Amazon and northern Borneo. Lowland forests of the tropical Far East and Amazonia are far richer in species than is the Congo basin. Even more striking is the fact that if one excludes tree species from the total floristic richness, tropical forest ecosystems are still the richest ecosystems in the world, due to the species diversity of non-trees such as epiphytes and climbers. Given similar rainfall regimes, central Amazonian forests tend to have fewer epiphytes than Mesoamerican forests, whereas tree species richness may be three times greater in the former.

There are many classification systems for differentiating tropical forest ecosystems that use bioclimatic, structural, or floristic criteria. Each has its purposes and merits, but they are too detailed for general overview purposes such as this chapter. Rather, a simple artificial system is used that first recognizes three elevational zones—lowlands (<500 m), low mountains (500–2000 m), and high mountains (>2000 m)—and then differentiates three major moisture regimes—perhumid (wet), humid (intermediate), and subhumid (seasonally dry). Brief overviews are also included for some distinctive tropical forest ecosystems such as mangroves, flooded forests, savannas, and páramo. The interested reader must keep in mind that these are arbitrary groupings along continuous gradients, analogous to differentiating the colors of the rainbow, and that it is impossible to assign regional altitudinal or rainfall limits for major types of tropical forest ecosystems.

A. Lowlands (<500 m)

Classic lowland tropical forest ecosystems extend onto the foothills and lower slopes, but it is difficult to define an upper ecological limit for the lowlands. Some authors prefer an elevation (e.g., 1500 m), whereas others use the upper limit of typical lowland crops (such as the "coffee line") or more conventional bioclimatic, structural, or floristic criteria. In order to include the very representative lowlands of the Amazon and Congo basins (with base elevation of ~300 m) as well as extensive foothills, the arbitrary upper limit of 500 m for lowlands is used here.

1. Mangroves

Probably the best known tropical forest ecosystem, mangrove forest ecosystems are characteristic of most tropical coastal zones due to the ability of several unrelated tree species to grow in saline conditions. Mangroves are typical of estuaries, deltas, and low-energy coasts lacking strong wave action. Due to greater tidal fluctuation and lower base flow of rivers associated with longer and more severe dry season, mangrove forests extend substantially farther inland on the western side of landmasses than on eastern coasts. For example, mangroves extend 190 km inland along the Gambia river in West Africa. Neotropical and West African mangrove forests have far fewer tree species than the mangroves of the Indian Ocean and the tropical Far East. Most mangrove species have evolved effective ways of dealing not only with their salty environment (e.g., leaves that exude salt through specialized glands) but also with the anaerobic sediments in which they grow. Mangrove ecosystems are important nurseries for shellfish as well as finfish and offer frontline buffering against tropical storms.

2. Flooded Forests

In the wet lowlands, flooded forests occur behind the mangroves (where salinity does not influence the floristic composition of the forests), along many major rivers such as the lower Amazon, and in low-lying areas with poor drainage. The frequency, depth, and length of flooding appear to be important physical factors determining the composition and dominance of flooded forests. Many flooded forests are pure stands of a single tree species, such as *Campnosperma brevipetiolatum* (New Guinea), *C. panamensis* (Panama and Costa Rica), *Mauritia flexuosa* (Amazon basin), *Manicaria saccifera* (Neotropics), *Mora oleifera* (circum-Caribbean), *Nypa fruticans* (New Guinea), *Prioria copaifera* (Central America), *Pterocarpus officinalis* (Costa Rica), *Raphia taedigera* (Neotropics and West Africa), and *Shorea albida* (Borneo). We do not have a good understanding of why any one of these species is able to dominate a particular ecosystem.

In contrast to flooded forests dominated by a single species, more heterogeneous flooded forests occur farther from the sea or adjoining the pure stands. The dominant tree species of pure stands also occur in the more mixed forests, usually with far less dominance. Swamp forests may have very large trees, but tree density tends to be lower in swamp forests than in adjoining terra firme forests. Most tree species are not restricted to swamp or flooded forests, but may occur on terra firme, where rainfall is adequate and especially where there is no effective dry season. Even with a mixture of tree species, these flooded forests are not nearly as species rich as nearby terra firme forests.

Many tree species that occur in flooded forests have large seeds that float, as can be seen in the flotsam on tropical beaches. One of the most characteristic trees of tropical beaches, the coconut (*Cocos nucifera*, Palmae), has such a wide pantropical distribution that it is difficult to determine its geographic origin. Several tree genera with large, buoyant seeds have the same or a sibling species across the Atlantic Ocean, such as *Elaeis* (Palmae), *Pentaclethra* (Mimosaceae), *Raphia taedigera* (Palmae), and *Sacoglottis* (Humiriaceae).

3. Perhumid Forests

Lowland perhumid forests are well known as classic tropical rain forest. Curiously, these forest ecosystems are more typical of the foothills and lower slopes of front ranges than of the extensive Amazon or Congo lowlands. The absence of major mountain ranges in tropical Africa may by a factor in the comparative scarcity of perhumid forest ecosystems there, especially in the Congo basin. Coastal West Africa from Guinea to Liberia is one of the few regions on that continent with average annual rainfall exceeding 3000 mm. Other lowland perhumid forests (e.g., southeastern Mesoamerica, Colombian Chocó, Gulf Province of Papua New Guinea, north Borneo, and Chittagong in Bangladesh) may have annual rainfall of 6000–10,000 mm.

Lowland perhumid forests on adequately draining

soils have high tree diversity, consistently exceeding 100 species per hectare for trees 10 cm or more in diameter (dbh (diameter at breast height)). Some centers of diversity usually exceed 200 tree species per hectare and a few exceptional areas (e.g., eastern Ecuador, northeast Peru, central Guyana, and northern Borneo) report more than 300 tree species per hectare. Though forest stature may be lower and tree density higher on poor soils, these forests typically have 500–600 trees per hectare (>10 cm dbh). Thus, the great majority of tree species are represented by only one individual per hectare. Lowland perhumid forests on terra firme soils with adequate drainage show considerable variation in the distribution of tree species over a mosaic of forest types.

Palms are very rich in species in lowland perhumid forest ecosystems. Because of their distinctive leaf form, palms are conspicuous in these forests, ranging from dwarf species (<1.5 m tall), to understory colonial species, to elegant single stems that reach the subcanopy, to the famous climbing rattans. A few palm genera have radiated into several hundred species of climbing palms in the Indo-Malayan region; commercial rattan comes from robust climbing palms. In neotropical forests the climbing palm, *Desmoncus*, has just a few dozen species and they are not nearly as abundant as the rattans of the tropical Far East. In some neotropical forests (e.g., La Selva and Costa Rica), palms comprise up to 25% of the stems (>10 cm dbh) in heterogeneous old-growth forests. Palms tend to be less abundant and less diverse in secondary forests.

Plants dependent on other plants for support, such as epiphytes, climbers, and stranglers, are abundant in lowland perhumid forests, especially on tree branches in the forest canopy. Many plant families have epiphytic species, but most typical of lowland perhumid forests are orchids, ferns, gesneriads, and bromeliads—the latter is restricted to the Neotropics. There are even epiphytic cacti (*Epiphyllum* and *Rhipsalis*) in lowland perhumid forests. The latter genus (with sticky, bird-dispersed seeds) is the only member of this neotropical plant family present in Africa. Lianas (woody climbers) are one of the most striking components of tropical forests because of their growth form, stem structure, and impressive size; they may be 35 cm in diameter and over 100 m long as they drape over the crowns of canopy trees. Many plant families are represented by lianas, but they are especially common in the Apocynaceae, Bignoniaceae, Dilleniaceae, Fabaceae, and Marcgraviaceae. In contrast to robust lianas, slender herbaceous climbers are typical of the Araceae, with hundreds of species in several genera.

Although some stranglers belong to the genus *Clusia* (Clusiaceae), most are figs (*Ficus*, Moraceae). A strangler begins life as an epiphyte, sending down aerial roots, eventually forming an interconnected network of roots that embrace the host tree. A strangler may survive as an independent tree for many years after the host tree dies and decomposes, leaving a "hollow cylinder" as its testimony. Interestingly, the strangling part of this relationship has not been confirmed; death may be associated with the inability of the host tree to tolerate shading by the strangler. Each fig species is pollinated by a unique species of tiny fig wasps, representing one of the best examples of coevolution. Figs are important keystone food resources for vertebrate fruit eaters in many tropical forests.

Lowland perhumid forests are extremely dynamic because tree falls are frequent events. The fall of a large canopy tree creates a sizeable opening in the forest canopy and prime regeneration sites on the forest floor. Up to 50% of the tree species in heterogeneous old-growth tropical forests may be shade intolerant and dependent on canopy gaps for successful regeneration. In addition to the well-known "fugitive" pioneer species (e.g., *Cecropia, Macaranga, Musanga, Ochroma,* and *Trema*) that characterize the rapid recolonization of disturbed areas, many valuable timber species—such as mahoganies, Spanish cedar, and dipterocarps—require gaps to reach the canopy. Numerous studies of many tropical forests over the past few decades indicate that turnover rates on the order of 100 years are the norm in tropical forests. Gap-phase dynamics appears to be an important ecological contributor to the maintenance of species-rich tropical forests.

4. Humid Forests

Lowland humid forests are the most widespread of tropical ecosystems, covering much of the Amazon and Congo basins, Mesoamerica, and the tropical Far East. Mean annual rainfall typically exceeds 1500 mm, but the upper limit is more related to the seasonality of rainfall. Whereas lowland perhumid forests have only a short or ineffective dry season, lowland humid forest ecosystems are characterized by a significant dry season. These latter forests are almost as equally impressive as the former (e.g., the tallest trees in Ghana reach 60 m); however, the number of deciduous trees is much greater in the seasonal forests. Epiphytic orchids and gesneriads are less abundant in the canopy of humid forests than in perhumid forests. Smaller, thin, supple woody vines are more common in humid forests, perhaps related to the greater proportion of deciduous species in the forest canopy. Large palms are common subcanopy components in humid forest ecosystems. Because of fire tolerance and the adaptive strategy of building the trunk belowground, some of these palms (e.g., *Attalea speciosa, Copernicia* spp., and *Scheelea rostrata*) are aggressive colonizers of pastures created from humid forests and maintained by fire.

Seasonal rains and drought not only determine the ecological functions and processes but also have strong influence on the populations of animals and plants. The synchronization of flowering with the dry season is truly impressive. Flowering during the dry season when tree crowns and lianas are deciduous facilitates visitation by pollinators such as large bees. The alternation of rainy and dry seasons and the use of seasonal changes to trigger physiological responses (i.e., flowering, insect emergence, nesting, and rearing of young) make these ecosystems unusually susceptible to climatic fluctuations such as ENSO events. The best example is the ENSO-caused rains during the dry season that occasionally limit flowering and cause famine among fruit-eating animals on Panama's Barro Colorado Island (BCI).

The pioneering studies of the spatial distribution of trees by Hubbell and associates on BCI indicate that very few tree species are hyperdispersed as suggested in the classic literature on tropical forests. Many species have the adults randomly distributed, while some species show clumping. For example, several caesalpinioid legumes (*Brachystegia laurentii, Cynometra alexandri, Gilbertiodendron dewevrei,* and *Julbernardia seretii*) form single-species stands in the Congo basin. A long-term study on the 50-ha plot at BCI has elucidated evidence for a strong density dependence within species, suggesting an important role for interactions between plants and their predators and pathogens. The spatial distribution of plants has important consequences for pollination, breeding systems, herbivory, and successful regeneration.

One of the most famous tropical trees is the Brazil nut (*Bertholletia excelsa*, Lecythidaceae), which has a patchy distribution in Amazonian humid lowland forests. Because adult trees occur in clumps, there has been considerable conjecture about indigenous influence on the abundance of this valuable species. Detailed studies document that the clumping is natural, probably due to the scatter-hoarding behavior of forest agoutis (*Dasyprocta leporina*, Heteromyidae) that cache seeds for later use. Another neotropical tree species with an unusual distribution is the true mahogany (*Swietenia macrophylla*, Meliaceae), which is reasonably abundant in the subtropical humid forests of northern Mesoamerica and the southern Amazon basin but quite scarce in the intervening tropical forests. Because of its high-value timber, there have been numerous attempts to promote or increase its abundance in natural forests, but they have mostly failed. Recently, it has been suggested that the mahogany requires large-scale disturbance such as by fire or river meanders, creating the high light regime needed for establishment and growth of young trees.

One of the more fascinating historical aspects of this ecosystem is the development of truly impressive human civilizations such as the Maya in northern Mesoamerica and the Khmer in Kampuchea (Cambodia). Both cultures developed intensive agriculture based on sophisticated engineering and management of water during the dry season through the use of raised fields, irrigation canals, and paddies. The Maya are believed to have also favored particularly useful plants such as mahogany, *Manilkara zapota* (Sapotaceae), and *Brosimum alicastrum* (Moraceae). About A.D. 900 the classic Maya civilization collapsed, perhaps due to overpopulation and/or a prolonged, severe drought. Given our current understanding of tropical forest dynamics, it is quite unlikely that the humid lowland forests now occupying the classic Maya region are successional forests.

5. Subhumid Forests

Lowland subhumid tropical forest ecosystems are much more extensive in the Paleotropics (e.g., Madagascar and India) than in the Neotropics. The high-value teak (*Tectona grandis*, Verbenaceae) forests of Burma, Thailand, and Java are representative of this ecosystem, as are the vast sal forests (*Shorea robusta*, Dipterocarpaceae) of northern South Asia. These ecosystems also occupy significant areas in the Caribbean, along the Pacific coast of Mesoamerica and northeast Brazil. Mean annual rainfall seldom exceeds 1500 mm, but the determining ecological factor is the long and severe dry season. Because of the historic pastoralist tradition in this ecosystem and the prevalence of nearly annual fires, most lowland subhumid forests have been destroyed or severely degraded by centuries of use. One of the most notable exceptions is the well-studied dry forest of Guánica on the south coast of Puerto Rico. This island is unusual in that the economic pressures to use land have not been nearly as intense this century compared to most developing countries.

Lowland subhumid forest ecosystems have far lower tree species richness than humid and perhumid forests. They are also much shorter in height and simpler in structure, but virtually all woody species are deciduous. In low-stature forests, the shrub layer is often dense and very spiny; thin lianas are abundant (including the *Vanilla* orchid), but epiphytes are less rich in species and abundance. Terrestrial cacti and bromeliads are more common, especially on extreme sites such as rock outcrops. At the latitudinal limits of the Tropics, lowland subhumid forest ecosystems include chaparral, open pine forests, and mixed oak–pine forests—depending primarily on moisture availability. Evergreen forests occur in riparian situations and in other wet sites such as in karst terrain, where underground moisture is available during the long dry season.

In contrast to the pure stands in some freshwater swamps mentioned earlier, pure stands in lowland subhumid forest ecosystems usually can be attributed to edaphic factors, such as rhyolitic ash (tuff or pumice) or montmorillonite clay (black cotton soils). One of the more distinctive forests is the pure stands of the only lowland oak species, *Quercus oleoides*, that has a patchy distribution on rhyolitic ash deposits from southern Mexico to northwestern Costa Rica. Appreciable areas of lowland subhumid forest ecosystems in Mesoamerica exhibit single-species dominance, such as *Celaenodendron mexicanum* (Euphorbiaceae), *Crescentia alata* (Bignoniaceae), *Erythrina fusca* (Fabaceae), and *Parkinsonia aculeata* (Caesalpiniaceae). Despite the attribution of these single-dominant stands to restrictive soil factors, there is precious little clarity as to why a species is able to attain such dominance.

6. Savannas

Savannas are defined as having a continuous cover of grass, but trees may be conspicuous components of savanna landscapes. Though vast areas in the major tropical regions meet the criteria for savannas, it is much more difficult to ascertain if it is natural or derived through human activities (e.g., burning). Tropical Africa epitomizes the image and grandeur of savannas with vast expanses of grass, sparse trees, migratory herds of herbivores, and their predators. Recently, the role of elephant herds in transforming woodland to savanna has been well documented. Extensive savannas also occur in the tropical Far East and the Neotropics (the llanos of Colombia and Venezuela, Rupununi in Guyana, and Beni in Bolivia).

The extensive pastures of Mesoamerica are considered to be derived savannas created by almost annual burning to "freshen" pastures and suppress woody invaders. Neotropical savannas have some very characteristic, fire-resistant tree species, such as *Acrocomia vinifera* (Palmae), *Byrsonima crassifolia* (Malpighiaceae), and *Curatella americana* (Dilleniaceae). Lowland pine (*Pinus caribaea* var. *hondurensis*, Pinaceae) savannas occur in northeastern Nicaragua, much of lowland Honduras, part of the Guatemalan Petén, and east central Belize. In contrast to the derived savannas of Mesoamerica, the pine savannas appear to be natural ecosystems maintained by frequent fire. The flat pine savannas of Belize occur on sandy soils that flood during the rainy season and undergo severe drought in the dry season.

B. Low Mountains

For purposes of this article, low-mountain ecosystems occur between 500- and 2000-m elevation. The rather arbitrary upper limit represents the maximum extent of tropical lowland flora near the equator. In many regions, the transition to montane flora more typically occurs by 1500 m, while nearer the latitudinal limits of the Tropics, this transition can be as low as 500 m. This elevational limit for tropical plants is more noticeable with crops such as highland coffee (*Coffea arabica*, Rubiaceae) or tea (*Camellia sinensis*, Theaceae). Low-mountain ecosystems typically have a mixture of tropical and temperate fauna and flora; for example, pines and oaks are conspicuous trees in northern Mesoamerican forests. Some tree genera have pairs of closely related species, with one occurring at lower elevations and the other in higher elevations in Mesoamerica, e.g., *Billia* (Hippocastanaceae), *Brunellia* (Brunelliaceae), *Hedyosmum* (Chloranthaceae), and *Symplocos* (Symplocaceae). In some ways, low-mountain ecosystems between 500 and 2000 m are transitional between lowland tropical and tropical montane floras. These ecosystems are typical of the major mountain ranges as well as the many smaller chains and high plateaus (e.g., Brazilian Shield, Serengeti Plain, and Great African Plateau).

1. Subhumid Forests

Low-mountain subhumid forest ecosystems are common on leeward slopes and in valleys and plateaus protected from orographic rain. Because of favorable climate, this ecosystem has had a long history of human activities, leaving little natural vegetation. Subhumid forests are relatively open and low in stature, with an abundance of spiny trees and shrubs, especially Mimosaceae. In northern Mesoamerica and continental Asia, conifers (*Pinus* spp.) are common in subhumid forest ecosystems. *P. kesiya* and *P. merkusii* occur in the Malay archipelago, while *P. wallichiana* and *P. roxburghii* are common on the lower flanks of the Himalayas.

This ecosystem is extensive on the plateaus that border both the Amazon and Congo basins. The Great African Plateau is dominated by miombo woodland (dominated by *Brachystegia*, *Julbernardia*, and *Isoberlinia*, Caesalpiniaceae), while the Brazilian shield is characterized by similar open forest ("cerrado"). The Brazilian cerrado are low, open forests or woodlands with characteristically crooked trees that occur on soils high in aluminum and low in pH. Cerrado vegetation exhibits considerable variation, ranging from grass-dominated savanna to fairly dense forest; however, most of the plant species are of tropical origin. Where soils are less restrictive, more typical tropical forest exists that is surprisingly tall (to 40 m), fairly open canopy, with a high proportion of deciduous species.

2. Humid Forests

Humid forest ecosystems often ring lower, more arid regions on low mountains, because of more moisture, cooler temperatures, and/or less severe drought. Low-mountain humid forests show enormous variation of forest type, structure, and floristic composition due to their widespread distribution and great heterogeneity of physiography, soils, and rainfall patterns. An excellent example of this variation occurs in northern Mesoamerica, where pine forests dominate the upper slopes and broad-leaved forests occur in the valleys. Pines are characteristic of low-fertility soils and frequently burned sites, with *P. oocarpa* often dominant. The broad-leaved forests usually have a mixture of tropical (e.g., *Alchornea*, Euphorbiaceae) and temperate tree genera. *Parinari excelsa* (Chrysobalanaceae) has an unusually broad distribution in Africa, dominating humid forests in both the Congo lowlands and the Guinea highlands.

Impressive tropical forests occur on fertile soils (i.e., andepts) in Mesoamerica, with a high density of canopy trees (typically in the Lauraceae and Sapotaceae) plus occasional tall (>50 m), emergent trees of *Ulmus mexicana* (Ulmaceae). Also occurring in these mid-elevation humid forests are canopy trees of Podocarpaceae (Southern Hemisphere conifers), Juglandaceae, and Proteaceae. The latter tree family (that includes *Macadamia*) is especially diverse and abundant in the Queensland Wet Tropics of northeastern Australia.

3. Perhumid Forests

These high-rainfall ecosystems are common on the eastern flanks of low mountains and ranges that receive moisture-laden clouds from the east. When the mountains are oriented more or less perpendicular to the prevailing air currents, they nurture the famous cloud forests of the Caribbean, Mesoamerica, northern and eastern Andes ("ceja de selva" and "yungas"), and even on isolated volcanic massifs. Perhumid forests on low mountains may have tree species richness comparable to lowland perhumid forests. They also exhibit complex vertical structure and fairly tall trees; the latter may be related to the better drainage and anchorage typical of dissected topography.

Canopy trees are usually evergreen (with abundant epiphylls) and with heavy epiphyte loads, including bryophytes, shrubs, and even small trees. The abundant mosses in the canopy give these perhumid forests a distinctive brownish hue visible from small planes. Epiphytic Ericaceae shrubs are usually very abundant in cloud forests, providing year-round nutrition to nectar-feeding birds. The sizeable and heavy (when wet) loads of organic material in tree crowns cause frequent branch falls as well as stimulating the host tree to produce feeding roots on the branches to take up nutrients. In perhumid forests on nutrient-poor soils, numerous species of insect-trapping, epiphytic pitcher plants (*Utricularia*, Lentibulariaceae) are quite abundant.

C. High Mountains (>2000 m)

Because several tropical lowland species and crops may grow on protected sites up to 2000-m elevation, we use this as the lower limit of high-mountain forest ecosystems. High mountains in the Tropics have a temperate climate characterized by warm diurnal temperatures (on a sunny day) and cold nocturnal temperatures, including occasional freezes. In Africa, the fairly isolated mountains are defined as the Afromontane zone. Treeline varies considerably between about 3300 and 3800 m, depending on moisture, disturbance by fire, and proximity to the equator. Despite the temperate climate, most forest ecosystems have distinctive tropical features such as epiphytes and lianas. Nevertheless, the high-mountain tropical flora has much stronger affinities with temperate zone flora.

1. Subhumid Forests

In the Afromontane region of East Africa, *Juniperus procera* (Cupressaceae) forms pure stands; however, it is unclear if human disturbance is a contributing factor to this species' success. It is possible that high-mountain subhumid forests once existed on the Andean Altiplano, but centuries of human activities have destroyed most natural vegetation. The only remnants may be the small stands of *Polylepis* (Rosaceae) that occur up to 4000-m elevation in the central Andes—the highest forests in the world. The forest patches persist on rocky slopes that may provide natural protection from frequent fires and soil temperatures may be slightly higher. *Polylepis* trees are small, twisted, and gnarled, with considerable trunk and branch dieback, especially on stand edges.

2. Humid and Perhumid Forests

The high elevations and cool temperatures tend to produce a preponderance of humid and perhumid forest ecosystems because of frequent cloud cover and low evapotranspiration. These ecosystems usually have the canopy dominated by holarctic genera (e.g., *Pinus*, *Quercus*, and Lauraceae), whereas the understory is more of a mixture of holarctic and tropical genera. Northern Mesoamerican forests are the center of diversity for oaks (*Quercus*, Fagaceae) and conifers (*Abies*, *Cupressus*, *Juniperus*, *Pinus*, *Podocarpus*, *Taxodium*, and *Taxus*). Except for *Podocarpus*, these are all holarctic genera. On poor soils, the coniferous forests tend to be open and frequently degraded by grazing and fire. More fertile and/or wetter sites also have conifers, but the forests are more of a mixture with oaks. Northern Hemisphere conifers do not occur naturally south of the Nicaraguan lowlands; thus oaks dominate the high-mountain forests of southern Mesoamerica and Colombia. The bamboos *Chusquea* in the Neotropics and *Arundinaria alpina* (both Poaceae) in East Africa are typically very abundant in the understory, where they may be impeding natural regeneration of trees. As with lower elevation perhumid forests, the large canopy trees carry huge epiphyte loads of mosses, tank bromeliads, and ericaceous shrubs.

3. Páramo

Although not strictly a forest ecosystem, high-mountain páramo merits brief mention as a unique high-mountain tropical ecosystem. Subalpine páramo occurs above treeline on high mountains in the Neotropics (Ecuador to Costa Rica) and East African volcanic massifs. Neotropical páramo is characterized by a high density of shrubs, including *Chusquea*, *Hypericum* (Hypericaceae) and *Vaccinium* (Ericaceae) and occasional emergent trees of *Clusia* (Clusiaceae). Where drainage is poor, bogs and a distinctive flora are frequent. But the most striking feature in Andean páramo is the abundant, postlike *Espeletia* (Compositae) that have a single, thick stem topped by a roseate of furry leaves. Afroalpine páramo has giant *Lobelia* (Lobeliaceae) and giant *Senecio* (Compositae) that are very similar in growth form to *Espeletia*. The thick trunk is fire-resistant and the roseate of leaves may offer some protection from nightly freezing temperatures.

See Also the Following Articles

AMAZON ECOSYSTEMS • BIODIVERSITY-RICH COUNTRIES • HOTSPOTS • MANGROVE ECOSYSTEMS • RAINFOREST ECOSYSTEMS, ANIMAL DIVERSITY • RAINFOREST ECOSYSTEMS, PLANT DIVERSITY • RAINFOREST LOSS AND CHANGE

Bibliography

Bullock, S., Mooney, H., and Medina, E. (Eds.) (1995). *Seasonally Dry Tropical Forests*. Cambridge Univ. Press, New York.

Churchill, S., Balslev, H., Forero, E., and Luteyn, J. (Eds.) (1995). *Biodiversity and Conservation of Neotropical Montane Forests*. New York Botanical Garden, Bronx, NY.

Gentry, A. (Ed.) (1990). *Four Neotropical Forests*. Yale Univ. Press, New Haven, CT.

Hamilton, L., Juvik, J., and Scatena, F. (Eds.) (1995). *Tropical Montane Cloud Forests* (Ecological Studies 110). Springer-Verlag, New York.

Hartshorn, G. (2000). Tropical and subtropical vegetation of Meso-America. In *North American Terrestrial Vegetation* (M. Barbour and D. Billings, Eds.), 2nd ed. Columbia Univ. Press, New York.

Holdridge, L., Grenke, W., Hatheway, W., Liang, T., and Tosi, J. Jr. (1971). *Forest Environments in Tropical Life Zones: A Pilot Study*. Pergamon, San Francisco.

Kellman, M., and Tackaberry, R. (1997). *Tropical Environments: The Functioning and Management of Tropical Ecosystems*. Routledge, New York.

Leigh, E., Jr., Rand, S., and Windsor, D. (Eds.) (1982). *The Ecology of a Tropical Forest: Seasonal Rhythms and Long-Term Changes*. Smithsonian Institution Press, Washington, DC.

Lieth, H., and Werger, M. (Eds.) (1989). *Tropical Rain Forest Ecosystems* (Ecosystems of the World 14B). Elsevier, Amsterdam.

Lugo, A., and Lowe, C. (Eds.) (1995). *Tropical Forests: Management and Ecology*. Springer-Verlag, New York.

McDade, L., Bawa, K., Hespenheide, H., and Hartshorn, G. (Eds.) (1994). *La Selva: Ecology and Natural History of a Neotropical Rainforest*. Univ. of Chicago Press, Chicago.

Richards, P. (1996). *The Tropical Rainforest: An Ecological Study*, 2nd ed. Cambridge Univ. Press, New York.

White, F. (1983). *The Vegetation of Africa*. UNESCO, Paris.

Whitmore, T. (1984). *Tropical Rain Forests of the Far East*, 2nd ed. Oxford Univ. Press, New York.

TRUE BUGS AND THEIR RELATIVES, DIVERSITY OF

Carl W. Schaefer
University of Connecticut

I. Suborders of Hemiptera
II. Heteroptera: General
III. Heteroptera: Specific
IV. Conclusions

GLOSSARY

beak (or rostrum) The thin elongate mouthparts, adapted for piercing the organisms upon which the bugs feed (the mouthparts of bugs are similar in function to, but quite different in structure from, those of mosquitoes).
food pump The sclerotized pump used by hemipterans to suck up the juices (plant or animal) upon which they feed.
hemelytron The heteropteran's forewing; so called because the anterior part of the wing is opaque and the posterior part is clear (membranous).
salivary pump The sclerotized pump used by hemipterans to inject saliva into the organisms upon which they feed.
scent glands Glands which in heteropterans secrete various compounds for various purposes, including defense, attraction of mates, and attraction of both sexes. Adult heteropterans have scent glands on their metathoraces, and nymphal heteropterans as well as some adults have scent glands on the dorsa of their abdomens.
scutellum The dorsal portion of the mesothorax, in most heteropterans triangular or shield-shaped (hence "little shield").

HETEROPTERA IS ONE of the most diverse groups of insects. Heteropterans ("true bugs") range in length from a single millimeter to more than 8 cm and in width from a millimeter to several centimeters. They occur in all colors and combinations of colors, including jewel-like combinations of gold or green on an iridescent purple background. They live in habitats as varied as spider webs, anthills, termite mounds, grain, the bodies of bats, under the bark of trees (feeding on fungi), on the bark of trees (feeding on other arthropods), high in the canopies of tropical trees, deep in the soil, on the surface of water (without getting wet), and in the homes and beds of people. Some use tools (to capture termites), the males of some carry the eggs about to keep them safe, others squirt foul liquids (to repel enemies), yet others eat foul liquids (milkweed sap) and store the bitterness. Heteropterans feed on everything from fungi and crop plants to other insects and vertebrates, including people. One group (triatomine bugs) is the vector of one of the most serious diseases in Latin America (Chagas' disease). Heteroptera is a suborder in the insect order Hemiptera. Like other hemipterans, heteropterans have narrow elongated mouthparts that pierce plants or animals and suck up the sap or the body fluids. The insects have incomplete metamorphosis (i.e., there is no pupal stage and the immatures look very much like the adults). The order is an old one: Hemipterans occurred in the Lower Permian, and heteropterans assignable to an extant infraorder have been found from the Upper Permian.

I. SUBORDERS OF HEMIPTERA

The number of actual suborders has recently come into question, because of the uncertain validity of Homoptera. However, three suborders are frequently accepted in most entomology texts: Homoptera, Heteroptera, and Coleorrhyncha. Nevertheless, many homopterists recognize two orders, Homoptera and Hemiptera (=Heteroptera). The reasons for this confusion are themselves confusing and have to do with the importance that heteropterists and homopterists give to the groups they study; this importance in turn is based on the extent to which the students of one group understand the work of the other group's students. Those who have undertaken the most thorough study of all hemipteran groups conclude that they share too many characteristics to be treated as separate orders.

A. "Homoptera"

This group includes the cicadas, leafhoppers, aphids, scale insects, whiteflies, and their relatives. Nearly all feed on plants (a few on fungi), and many are important pests of crops and ornamentals. Recent work on the origins and (especially) the evolutionary relationships of groups within "Homoptera" has led to the conclusion that the group did not have a unique ancestor common to all homopterans and to no other hemipterans. As a result, the higher classification of "Homoptera," and the systematic status of the group itself, are deeply in question. That is why I place the term "Homoptera" in quotation marks: The term is useful in grouping together a number of insects, but it is not useful in suggesting they compose a single evolutionary unit.

B. Coleorrhyncha

Members of the single family (Peloridiidae) have a classic Gondwana distribution: Australia, New Zealand, southern South America. They are associated with moss in *Nothofagus* (southern beech) forests and probably feed on the moss. Small, somewhat flattened, often with lovely semitransparent expansions of the head and thorax, these insects are poorly known and deserve much more study. Some believe Coleorrhyncha (formerly included in Homoptera) to be the sister group of Heteroptera—that is, that these two groups had a common ancestor. The evidence for this conclusion is stronger than that for the earlier inclusion of coleorrhynchans in Homoptera, where they were thought to be a sort of "nonmissing link" between Heteroptera and Homoptera.

C. Suborder Heteroptera

I estimate that there are about 37,000 described species of Heteroptera and (a much rougher estimate) perhaps 25,000 species remaining to be described: a total of about 62,000 species of Heteroptera altogether. This estimate may understate the number of undescribed species. Like all hemipterans, they feed by piercing other organisms and sucking their juices. Unlike Homoptera and Coleorrhyncha, however, Heteroptera includes many predacious insects, and many aquatic ones. Heteroptera are more closely allied (evolutionarily) with Coleorrhyncha than with Homoptera, although some have recently argued that the closest relatives of Heteroptera are certain homopteran groups.

II. HETEROPTERA: GENERAL

A. Why "True" Bugs?

Heteropterans are the only group of insects that entomologists agree may legitimately be called "bugs"; hence the name entomologists themselves use, "true bugs." Nonentomologists call all insects (and other arthropods) "bugs," and, although entomologists disapprove, they are resigned to this use.

Why are heteropterans called "bugs"? The reason almost certainly lies in the original Middle English meaning of the word: "bug" (as *bugge*) meant originally a *wraith or specter,* something unseen that emerges from the shadows and does harm. This is exactly what the best known heteropteran does: the bedbug (*Cimex lectularius* L.) sucks blood from the sleeping victim and disappears; in the morning only the irritating welt from the bite remains, its cause and origin unknown. Clearly, an unseen nocturnal shadow has caused this damage: a bug. (This original meaning of *bugge* is retained in Modern English in such words as *bugbear, bugaboo,* and *bogeyman;* one may also wonder if it is retained in the verb "to bug," meaning (1) to irritate and (2) to use a small and secret device for spying—a form of damage.)

B. Characteristics

Heteropterans share many structural features with other insects, of course; other features with other insects with incomplete metamorphosis; and of course some characteristics with other hemipterans. Still more aspects of their morphology are unique.

Like all insects, heteropterans have three basic body regions: head, thorax, abdomen; and, like those of nearly all insects, these regions are specialized: the head for grasping and preparing food and for exploring the environment, the thorax (with legs and wings) for locomotion, and the abdomen for processing food and for reproduction. Like most insects, heteropterans have three pairs of five-jointed legs and two pairs of wings, a pair of compound eyes (usually large, because true bugs are mostly diurnal and therefore sight is an important sense), a pair of antennae, and an aedeagus (or penis) for the introduction of sperm into the female, who in turn has an ovipositor for laying eggs.

Like other insects with incomplete metamorphosis, immature bugs look like their adults, except for the lack of functional wings and genitalia and for different body proportions. As the immatures develop, wings appear as small pads, which enlarge at each molt; later, the genitalia too may be discerned, and often the last immature stage can be sexed.

Heteropterans share more specific features with other hemipterans: The mouthparts are modified into a long sharp tube, with two internal openings. Through one of these, salivary fluid is pumped into the plant or animal (plant only, in the case of nonheteropteran hemipterans), and through the other opening the fluids are sucked up. A set of pumps within the head accomplishes these activities. This mechanism and method of feeding are perhaps the best evidence that all hemipterans are closely related, and this in turn is why most entomologists consider all hemipterans to compose a single order.

Heteropterans, like other hemipterans, have certain characteristic features of the digestive system and the excretory system (although heteropterans lack the "filter chamber" independently evolved in several homopteran groups). The primitively 12–13 ganglia of the ventral nervous system are fused to a single one (or to four in a few homopteran groups). The wing venation is reduced (less so in many homopterans); the chromosomes are holocentric (diffuse centromere). Many hemipterans can produce sounds, usually by scraping a "pick" over a corrugated surface. Often these structures appear the same and occur on the same parts of the body. But in most cases, these sound-producing structures have evolved independently; the appearance is constrained by their function, and their occurrence on the same body parts is constrained by the fact one body part must move and be in contact with an unmoving part.

Heteropterans themselves are more diverse in habit and habitat than the other hemipteran groups. Heteroptera contains the only hemipterans that live in, and sometimes on, the water; the only hemipterans that are partly marine; the only carnivorous hemipterans; and the only hemipterans that feed on vertebrate (including human) blood. It is not surprising, then, that within the suborder Heteroptera is a wide variety of sizes and shapes, nor is it surprising that structures of heteropterans (legs, etc.) are highly modified in some groups.

The beak, or rostrum, characteristic of Hemiptera in general, in some plant-feeding heteropterans may be longer than the body, and in some predacious heteropterans be so short as not to reach the thorax. It may be slender or stout, and the number of segments may be reduced from the original five. The beak arises closer to the back of the head in heteropterans than in homopterans, and thus a region of the head lies just in front of the rostrum's base; the presence of this region in coleorrhynchans is one reason why Coleorrhyncha and Heteroptera are thought to have had a common ancestor.

Highly characteristic of Heteroptera alone is the division of the forewing into two regions, the posterior one, clear and membranous (like the hindwing), and the anterior portion, harder, often leathery, and opaque: hence the name "heteroptera" ("different wings"). The two wings on each side are coupled together with a device on the forewing; this is found only in Heteroptera and Coleorrhyncha. Heteropteran sperm, too, has some unique features; coleorrhynchan sperm has not been examined.

Characteristic of nearly all heteropterans is a triangular (pointing backward) part of the top of the thorax, the *scutellum* ("little shield"); in a few heteropterans, especially in the group Pentatomoidea, the scutellum covers all or most of the dorsal surface of the insect, except the head (but including the wings).

Unique to Heteroptera are scent glands. These are found on the abdomen of immature heteropterans (from one to four pairs on either side of the midline). These glands are usually nonfunctional in adults, although in many species they do produce scent; the known number of such species is increasing with increasing study, and the function of the scent from these dorsal abdominal glands is becoming more clear. Adult heteropterans have a pair of glands (regardless of their possession of functional dorsal glands) on the last (third) thoracic segment. These lateral metathoracic scent glands are thought to repel predators (and have been shown to repel ants), but they certainly have other functions as well, including the attraction of conspecifics of the opposite (and sometimes of both) sex(es).

Heteroptera do not differ biologically from other insects and other hemipterans. They do the three basic things that all organisms do—feed, reproduce, and disperse—and, because these things are basic to survival as a species, bugs do them much as do other winged insects. However, bugs are remarkably more diverse than other hemipterans (and indeed than other insects with incomplete metamorphosis) in their choices of food and of habitats and in their interactions with their environments (for the latter, see below).

Heteropterans feed like other hemipterans. Their sharp elongated mouthparts are adapted for piercing; one internal pump pushes saliva into the food source, and another sucks back the food. In most cases, this food has been all or partly digested by salivary enzymes. Few bugs feed on the watery contents of plants' xylem, and so bugs lack the filter chamber that in many homopterans removes and excretes excess water (and sugars). (Diuresis does occur in blood-sucking bugs, however: see below.)

The salivary enzymes and substances of plant-feeding bugs serve other functions than digestion alone. Many phytophagous bugs form a small feeding cone on the surface of the plant where the mouthparts are inserted. This cone, made from salivary secretions, helps stabilize the mouthparts as they penetrate to the source of food. Other salivary products help break down cell walls and release cells' contents; other products mimic plant auxins and "fool" the plant into mobilizing nitrogen-rich materials at the point of feeding.

As a general rule, with many exceptions, plant-feeding heteropterans feed on the highly nutritional reproductive parts of plants—flowers, developing fruits and seeds, ripened fruits and seeds, and fallen fruits and seeds. This may be a matter of size: Many plant-feeding bugs are relatively large (i.e., larger than the majority of homopterans); larger hemipterans are better able to penetrate plant parts to the seeds within and to disperse to find flowers, fruits, and seeds—which are transitory resources. Smaller heteropterans, like many homopterans, feed on leaves, a resource more permanent, easier to penetrate, but of less nutritional value.

The flat bugs (Aradomorpha—see below) have very long, very slender mouthparts with which they feed on the linear array of fungal mycelia below the bark of dead or dying trees. A few of these bugs live in termite nests, where they probably feed also on fungi. With very few exceptions, these are the only bugs that feed on fungi.

Indeed, the great majority of phytophagous heteropterans feeds on flowering plants, angiosperms. Several small groups feed on gymnosperms, and a very few bugs feed on mosses or ferns. It seems likely that the great radiation of heteropterans—like the radiations of so many other insect groups—occurred at the same as time as, and perhaps in response to, the radiation of flowering plants.

Within the angiosperms, some groups are less preferred than others: grasses, for example, and legumes. Each of these groups has developed defenses against being eaten, and in each case one or more groups of bugs has overcome the defense and radiated on the plant group. It is significant that these two groups (grasses and legumes) have few *individual* species feeding on them but, rather, a few groups of species (subfamilies, families). This suggests that an early ancestor broached the defense (of grass or legume) and its descendants radiated on this newly available food source.

Many bugs are predacious. Indeed, it has been argued that this was the original way of life of Heteroptera (an argument that some dispute). The prey is other insects and often their eggs or other immature stages. Some of the smallest heteropterans are predacious, on arthropod eggs, mites, and scale insects and aphids. It was long thought that predacious bugs inserted their mouthparts and sucked up fluids inside the prey, much as plant feeders suck sap. Recently, it has been demonstrated how much more complex, and interesting, the situation really is (Cohen, 1998). A predacious bug actively pumps digestive enzymes and fluid into the prey. The enzymes digest the soft internal organs of the prey, and the fluid and digested food are sucked up. This "solid-to-liquid feeding" process is repeated over and over, the mouthparts within the prey moving throughout its body, probing even into the legs and antennae to digest the muscles there. There remains but a husk, a shell, the exoskeleton of the prey, all internal organs having been digested, liquefied, and sucked into the predator. As a consequence, a predacious bug need feed on only a few prey, because all of the prey except its exoskeleton is eaten. The implications of this for biological control using heteropterans are considerable (see below).

The predacious habit has arisen often in the Heteroptera, secondarily, probably more often than phytophagy. This suggests that the search for more concentrated and usable nitrogen is as important in this group as it is in others.

A few bugs feed on vertebrate blood. Heteropterans' mouthparts, like those of mosquitoes, are long and sharp, and so adapted for sticking into liquid-filled organisms. It seems inevitable that some would take vertebrate blood, although few feed on birds and none on cold-blooded vertebrates (with one exception). With few exceptions, the vertebrates fed upon are small mammals and bats—and man. Members of two closely related families (Polyctenidae and Cimicidae) feed on bats, and a few members of the latter family feed on humans: the bedbugs, some of whose relatives also feed on poultry. Members of the assassin bug subfamily Triatominae feed on small mammals, and some of them on humans; these spread the protozoan trypanosome which causes Chagas' disease.

The problem of feeding on vertebrate blood includes the problem of ingesting red blood cells through a narrow tube. The physics of this type of feeding have been studied, and maximal feeding is related to the inside diameter of the feeding tube, the concentration of red blood cells, and the duration of feeding. Such feeding also results in the accumulation of blood fluids. These are excreted (diuresis) in a manner roughly analogous to the excretion of plant juices by homopterans ("honeydew").

Two other heteropterans feed on vertebrates: a birdnest-living lygaeid (the "seed bug" family) apparently feeds only on the blood of the host bird; another group of lygaeids apparently feeds on small mammals in Africa (and may there occupy the niche occupied in Latin America by Triatominae). Giant water bugs (Belostomatidae) have been reported to attack baby ducklings (this may be a legend), and some species specialize in taking the fluids of small fish.

No heteropterans feed exclusively on dead or decaying organic matter (i.e., none is a saprophage). However, there are many anecdotal accounts of various bugs found feeding on bird droppings, carrion, and dead insects. It is likely these bugs sought water or a source of concentrated soluble nitrogen.

The great majority of heteropterans reproduces like other insects: Sexes find one another, there may be a brief period of courtship, mating occurs (fertilization is internal), cigarettes are smoked and promises made, and the fertilized eggs are laid near or on the appropriate food for the young.

The sexes find one another probably via sex attractants given off by one or the other sex, or perhaps by both (depending on the species; however, very few species have been studied). Courtship, if it occurs at all, is brief. In a few species, males fight for females; the enlarged legs of certain males are used to knock one another off a leaf to which sex attractants are drawing females. However, there seems to be very little intrasexual rivalry in Heteroptera. Mating is often end-to-end, or in many species the male lies at an angle across the female; in all cases both sexes are ventralside down. In some end-to-end species, the female, larger than the male, may be spotted moving around dragging her mate with her; or she may feed casually on a leaf, with the smaller male dangling helplessly in the air, attached only by the genitalia.

Soon after mating occurs, eggs are laid singly, or in characteristic batches, sometimes in several layers, depending on the species and, often, on the family. A female lays several batches of eggs in her lifetime, each one usually requiring a separate mating. The eggs of many species have tiny openings (micropyles) through which the sperm enter; other small openings (aeropyles) provide gas exchange; and some species' eggs have a round or oval line of weakening, through which the insect will hatch.

Eggs may be round, oval(!), cylindrical, or barrel-shaped. Many are brightly colored. The egg of Rhopalidae ("scentless plantbugs") is mounted on an elongate pedicel (like eggs of some neuropterans), presumably to protect it from predators.

Parental care of the eggs occurs in a few species. As is well known, the female of some giant water bugs (Belostomatidae) glues her eggs onto the back of the male, who carries them until hatch. By sweeping water across them, he keeps them aerated and prevents the attack of fungi. The females of some stinkbugs (Acanthosomatidae) and some lace bugs (Tingidae) protect their eggs and early immature stages, guarding them from attack from parasites and such predators as ants. Experiments have shown that in all these instances, the protection is effective.

I stress, however, that with respect to reproduction, as with respect to feeding and all other aspects of heteropterans' lives, we know very little about the lives of some bugs, and nothing about the lives of most.

One strange exception, unique to a few heteropterans, is "traumatic insemination." One of the male's claspers is sharp and sometimes scimitar-shaped. With it he slashes the female's abdomen and sperm is deposited via a duct running along the clasper. The sperm finds its way to the storage receptacle of the female. In the evolutionarily more advanced members of this group, a special pad of tissue on one side of the female's abdomen receives the sperm, which then moves to the storage site, passing on its way *through* the cells. This is the situation in the common bedbug. In evolutionarily less advanced species (certain members of the bedbug's family and some Anthocoridae), there is no such tissue, and the sperm are simply deposited in the abdomen. Counting the resulting mating scars allows one to learn how often a female has been mated.

Like those of most insects, the eggs of heteropterans hatch within the year; often they wait through cold seasons (in temperate regions) or dry ones (in tropical regions). Hatching often occurs through a weakened area of the egg and is aided by an "egg burster" on the hatchling, which is soon lost.

After hatching, the immature bug molts several times, the last time to adult; five immature stages occur in nearly all bugs, and four in the rest. The first immature of several plant-feeding bugs does not feed. Blood-feeding bugs require a bloodmeal to molt to the next stage; molting may require a full meal in other bugs (especially predacious ones), but in fact the stimulus to molt in nonbloodfeeders is unknown.

Each successive stage is larger than its predecessor, of course, and in each the wings are more fully developed. Development of the wings is so consistent in Heteroptera, that the immature stage can often be determined by the relative development of the wings. The sex can often be detected in fourth-stage, and always in fifth-stage, immatures, as the outlines of the external genitalia become discernible.

In general, the immatures feed on the same organisms as the adults. Predacious immatures may take smaller prey, or they may collaborate in taking larger prey. The young of some plant-feeding bugs feed on different plants than the adults, perhaps in response to differing seasonal availabilities of different plants; perhaps also some plants provide better food for growth and others better food for reproduction.

In temperate regions, most bugs go through a single life cycle, overwintering as egg or adult (there being no pupal stage). Eggs overwinter embedded in twigs or in leaf litter; diapausing adults overwinter in litter or just below the soil surface, near their host plants. In spring, overwintered adults mate, lay eggs, and the hatchlings actively seek food (those of some species do not feed, however). Overwintered eggs hatch usually in midspring and become adult in midsummer or late summer, mating and laying eggs that will overwinter. In warmer regions (and, for some species, in temperate regions), there may be more than one generation in a year, and these may overlap. In regions with dry seasons, adults or eggs may estivate. Aquatic heteropterans may overwinter as eggs in or near the water; several species are active below the ice in temperate-region winter; and adults of water striders and their relatives often overwinter on land, rather far from water.

Heteropterans vary widely in their ecological adaptations, both to the biotic and the abiotic aspects of their habitats. Their relationship with their sources of food is of particular significance, both ecologically and because it is this relationship that determines their impact on humans.

Because heteropterans have sucking mouthparts, they must take in liquid or semiliquid food and can accommodate only the smallest of particulate matter. This constraint lies at the base of their relationships to their food.

Plant-feeding heteropterans vary in their degree of host specificity. Some species feed on only a single species of plant, others on members of the same genus, others on many members of the same plant family; and many bugs feed very widely indeed. One coreid is listed from plants in 90 genera representing 35 plant families! It is of interest that a single genus may contain both very highly host-specific species and species feeding on a great variety of plants; this phenomenon is not restricted to Heteroptera.

Not surprisingly, the more host specific a bug is, the more closely its life cycle is tied to the host's. Eggs hatch when the preferred parts of the preferred host plant are available: Hatch occurs later in the plant's life cycle if the bug prefers reproductive parts, earlier if the bug prefers somatic parts. The triggers for hatch are not known in nearly all cases. The eggs of some mirids are laid in woody tissue of the host and prepare to hatch when the spring uptake of moisture causes the wood to swell and exert pressure. Several other species of Miridae specialize on grasses, and as these different grasses succeed one another in a field over several weeks, so do the mirids. Again, the dynamics and phenology of this succession remain to be studied.

Mirids in particular, many of which feed on short-lived annuals, are closely tied to their hosts' life cycles. The adults of several species of heteropterans feed on different plants than do the immatures. This may reflect differing needs of the stages: immatures need nutrients for growth, adults need nutrients for reproduction; or, the switch from one host to another may merely reflect that fact that these bugs live longer than the plants preferred by the immatures.

The newly hatched immatures of many phytophagous bugs do not feed at all, or take only water; some species take up symbionts from the egg-shells. This period of time when feeding does not occur may be a "hedge" against a delayed appearance of the host plant. It would be interesting to seek a correlation between species with nonfeeding hatchlings and the uncertainty of their host plants' appearance.

Many heteropterans prey on other insects; a few feed on vertebrates; and only a very few feed, occasionally, on noninsect arthropods. In general, predacious bugs feed on prey of appropriate size: very small predators feed on small prey, such as insect (and mite) eggs, mites, scales, and aphids. Large predators feed on larger prey. However, many larger prey organisms (such as caterpillars) are slow-moving and have defenses against predators on their body surfaces; such organisms are available to predators that avoid the body's surface (by penetrating it with their beaks). And so many small predacious bugs will attack large prey, often collectively; such bugs include small adults and the small early immatures of predators whose adults are large. There is some evidence that the collective effect of the predators' injected digestive enzymes benefits all the predators feeding on the single prey. These collective associations are all of the same species.

Aquatic and semi-aquatic bugs are nearly all predacious. Those bugs that live beneath the water's surface prey upon other small aquatic invertebrates (although one group of water bugs feeds on snails, and another on small fish). Bugs that live on the water's surface (semi-aquatic) prey upon land invertebrates that fall into the water and become trapped on its surface.

Predacious bugs—whether land, aquatic, or semi-aquatic—are either sit-and-wait predators or stalk-and-pounce predators. The great majority attacks its prey on foot, not in flight.

Bugs that feed upon vertebrate blood live associated with their hosts, not on them. The hosts are vertebrates that live in enclosed habitats: burrows, nests, caves—and houses. The bugs spend the day in cracks and crevices in these habitats and, at night, venture forth to feed. When feeding, the bug injects an anti-irritant, so that only later is the fact of feeding known. The host list is rather limited: a few birds that live communally (swallows, domestic poultry, and a few others), humans (which also live communally), and especially bats—two families of heteropterans (Cimicidae and Polyctenidae) appear to have arisen as specialists on bats. One may speculate that the bedbug itself (family Cimicidae) became acquainted with humans when the latter took shelter from the cold in caves, during early Ice Ages; indeed, the only wild bedbug collected came from a bat cave in the northern Middle East.

Surprisingly little is known about the influence of abiotic factors on the individual lives of heteropterans or on bugs' population or community ecologies.

The effects of, and the reactions of bugs to, light and humidity are very poorly known. A few species disperse (i.e., large numbers move from one place to another, although not at regular intervals), and this movement in a few species is influenced by amount of moonlight. In other cases, dispersal is a response to deteriorating environment, whether physical (drying, for example) or biological (dearth of food). Several such bugs (e.g., species of cotton stainers, *Dysdercus*) lose their flight muscles after such dispersal, presumably to divert resources to reproduction.

The existence of circadian rhythms has been documented as influencing a very few physiological phenomena of a very few bugs.

The greatest enemy of heteropterans is other arthropods—this is of course true of all groups of insects. Small heteropterans are fed upon by small predacious arthropods, and large heteropterans by large ones. Large heteropterans are also eaten by insectivorous vertebrates—fish, reptiles, birds, and small mammals. All bugs are attacked by ants.

It has been thought that the scent glands, whose possession is a unique characteristic of heteropterans, are defensive against predators. This is probably so, but it is certainly not their only function. Nevertheless, the few actual experiments and observations that have published support the idea that predators—especially ants and vertebrates—are repelled by the glands' secretions. More work on this is needed.

Many true bugs are brightly colored, arrayed in contrasting patterns of black with red, orange, yellow, or white. These bugs, said to be warningly colored, are assumed to be distasteful to predators (at least, to visually orienting predators, like vertebrates) and to warn such predators away. Experiments with the large milkweed bug (family Lygaeidae) confirm that this is the case. The bug feeds on the intensely bitter sap of milkweed and sequesters the cardenolides (which are also strong ATPase inhibitors); these have no effect on the bug itself.

There are many other bugs with conspicuous patterns like the milkweed bug's. Many of these bugs are also bad-tasting, but certainly many are not. Batesian mimicry (where a palatable organism looks like an unpalatable one, and thereby gains protection) is common in Heteroptera. So is Müllerian mimicry, where two unpalatable species resemble one another and presumably each gains from the protection afforded by the other. The difference between Batesian and Müllerian mimicry is graded, not absolute. Also, we need experimental evidence to show what type of mimicry two phylogenetically different but similar bugs represent.

In making these assessments, one should also realize that the types of patterns, and colors, "available" (in the ecological-evolutionary sense) to these bugs are limited. The dark parts of the pattern must be black or near black, and the colored parts red to yellow, because of pigment limitations. The types of patterns are also limited by the smallness of the canvas (the insect's body) on which they are to be painted. There may also be some genetic constraint, for many of the patterns are much the same: white spots or a line across the wings, for example (although this may not be for contrast but for pattern disruption; see below). For this reason, similar warning patterns may have arisen by chance, and not from a selective pressure for mimicry.

A few predator–prey mimicry complexes are known. In these, a brightly colored predacious species closely resembles its usual prey. Some have dubbed this "aggressive mimicry," thinking the predator "fools" the prey into allowing approach. This supposes that the prey sees as do human beings, something that is most unlikely. More probably, these are Batesian (or Müllerian) mimicry complexes, and what the prey loses in associating with the predator, it more than gains by being protectively—warningly—colored like the predator. This suggestion needs experimental testing.

Such complexes often involve large populations. Other (nonheteropteran) warningly colored species also live in large populations. Perhaps the mass of bright color warns motile predators (like birds) away from a distance.

Other bugs protect themselves by mimicking inedible bits of their environment. At least seven distantly related families and superfamilies contain species that mimic ants; and within several of these families such mimicry has evolved several times independently. Ants are unattractive as food, because of their high content of formic acid, and many other insects mimic ants. In the Heteroptera there are species that look so much like ants that even experts may be fooled. When mimics look like particular species of ants (rather than like ants in general), they look like *local* species. The immatures of some Alydidae look like small species of ants when young and like larger species of ants as they grow.

Three facts constrain such mimicry: First, large bugs cannot look like ants, because ants are small. Second, the bug must be at least somewhat ant-shaped: slender and elongate. This is probably why ant mimicry has not developed in the very large superfamily Pentatomoidea, whose members are large and squat. Third, ants are characterized by a constriction between thorax and abdomen (the "wasp waist"). Large bugs can do nothing about the first constraint. Small bugs over evolutionary time may overcome the third constraint, by "developing" a similar "waist." However, many true bugs "fake" the constriction: Where on an ant the "waist" would be, the bug (otherwise dark, like an ant) has white markings. Viewed from a slight distance, against the dark background of a branch or the ground, these markings give the appearance of an *absence* of body—hence, a "waist." A sequence of such antlike appearances may be seen in the Araphinae, a neotropical group of Largidae, and in the Micrelytrini, a tropical group of Alydidae only distantly related to Largidae. Many unrelated heteropteran ant mimics have one or more dorsal spines, presumably to harm the tender mouth tissues of a vertebrate not deterred by the antlike appearance.

Less common is the mimicry of wasps, which occurs in some adult Alydidae, and doubtless in a few other bugs.

Other heteropterans look like twigs, or leaves, or other things of no interest to a predator. Many run or fly rapidly and erratically.

Many bugs can defend themselves more actively, as anyone who has been "bitten" by a backswimmer or giant water bug knows. Predacious bugs especially, with their stout beaks designed to be driven through the hard exoskeleton of another insect, can defend themselves effectively against vertebrate predators—or against insect collectors.

Aquatic bugs would seem to be delightful prey for fish. Yet they are not so frequently eaten; indeed, even though trout live in the same streams as water striders, stomach contents indicate the striders are rarely eaten. Many water bugs "bite" fiercely (as most aquatic biologists know), and it seems likely that their scent glands help protect them, although the evidence for this is very slim.

Like many other animals, some true bugs break up their body outline by using spots or lines of white against a dark background (the "zebra effect"); indeed, the antlike appearance of some bugs may have arisen this way. Many Coreidae have a white line, straight or zigzagged, across the dark wings; many Pyrrhocoroidea have white spots in the same place, as do some members of other groups. Like the zebra's white stripes, these bits of white break up the bug's outline and make it more difficult to be seen—especially when moving in an erratic way.

Heteropterans seem not to be heavily parasitized. Some groups of parasites are specialized for parasitizing certain bugs (the subfamily Phasiinae of the dipteran family Tachinidae, for example); the eggs of heteropterans may at times be parasitized by hymenopterans; and certainly some species of heteropterans are quite heavily parasitized. But overall, the group seems less afflicted with parasites than (for example) the Lepidoptera.

The females of some lace bugs (Tingidae) ward off predators and parasites by protecting their eggs and young, and this behavior occurs in a few other heteropteran groups. Some heteropterans bury their eggs partly or completely in the substrate, probably to prevent their being parasitized: Many lace bugs, for example, embed their eggs on the lower surface (less conspicuous) of leaves and cover them with frass, thus concealing them from sight and (perhaps) from scent.

Finally, like all arthropods, heteropterans wear their skeletons on the outside. This makes their bodies hard and often smooth. As a result, several kinds of potential predators (especially ones with jaws) simply cannot grasp or crush heteropterans (or many other arthropods). For example, there are few records of predacious beetles feeding on bugs.

C. Economic Importance

Many heteropterans are serious pests of crops, and one group transmits the trypanosome protozoan that causes Chagas' disease. This group is the Triatominae, a subfamily of assassin bugs (Reduviidae) that lives in the New World Tropics (with one exception) and in which *Trypanosoma cruzi* develops. These bugs feed on vertebrate blood. Several live in association with humans and can transmit *Trypanosoma cruzi* to them. The result is Chagas' disease, which afflicts millions of people a year in Latin America, often killing its victims and costing several billions of dollars in lost lives, lost productivity, and disease prevention and treatment (see Schofield, 1994). A genus of Triatominae lives in India, but so far Chagas' disease does not occur there. Very rarely, the disease occurs in the southern United States, where the bugs, the trypanosomes, and the wild rodent hosts all may be found.

More widespread is the damage caused to plant crops by heteropterans. Several species of stinkbugs, called collectively Sunn pests, are the chief destroyers of wheat and barley in a broad swathe from eastern Europe through the Middle East into Pakistan. In parts of this range, these bugs may destroy up to 70% of the crop (see Javahery, 1995). The southern green stinkbug is a major pest of many crops throughout the world. In the United States, the chinch bug *Blissus leucopterus* (Lygaeoidea: Blissidae) is sometimes a serious pest of wheat. Cotton stainers (Pyrrhocoridae) damage cotton and cotton seed in many semitropical parts of the world (but not the United States). Lygus bugs (Miridae) are very serious pests of a wide variety of crops, especially in the temperate parts of North America. Many other heteropterans are general or specific pests, locally or worldwide, on many crops: A detailed account of these pest insects may be found in Schaefer and Panizzi (2000).

A few heteropterans are important in biological control. A very few feed on weeds and provide some control. Many are predacious and feed on other insects. However, like most predators, most heteropterans are not prey-specific and cannot usually be depended upon to control completely pests on a crop; moreover, they do not feed on many prey items. These insects would serve better as components of an Integrated Control Program. Among the families that have been studied are Reduviidae, Berytidae, Nabidae, Anthocoridae, (some) Miridae, Geocoridae, and Pentatomidae: Asopinae; members of several aquatic families have been tried in controlling mosquitoes, without much success (see Schaefer and Panizzi, 2000).

As noted above, predacious heteropterans feed on only a few prey. When it was believed that these bugs removed from their prey only the latter's liquids, it was also thought that each predator must feed on many prey in order to get enough nourishment. Now we know this is not so. As a result, those using these bugs in biocontrol programs must rethink the bugs' efficiency, because each bug actually feeds on far fewer prey items than had been thought. Many prey–predator models, and considerable biocontrol planning, have been rendered invalid by the recent discovery of this mode of feeding (Cohen, 1998).

III. HETEROPTERA: SPECIFIC

The suborder Heteroptera contains eight infraorders: Enicocephalomorpha, Dipsocoromorpha, Gerromorpha, Leptopodomorpha, Nepomorpha, Cimicomorpha, Pentatomomorpha, and Aradomorpha; "-morpha" means "in the form of," and each infraorder is derived from the generic name of a member: Hence the often pen- and tongue-shattering concatenation of syllables. These groups, and the families that compose them, are discussed very well in Schuh and Slater (1995).

Of these, the first two contain relatively few species (130 and 210, respectively), of poorly known insects; these are all predacious and many live in or on ground debris; it is certain that many more species will be discovered when this type of habitat is explored in the Tropics. They are of interest because they are phylogenetically primitive.

The Gerromorpha contain about 1500 species, most of them in the Gerridae, the water striders, or water skaters, or "Jesus bugs," which are common everywhere skimming the surface of still waters or the edges of moving waters. Here they capture insects that fall into the water; they rarely if ever capture aquatic organisms. One group of water striders is among the very few insects that live on or in saltwater: the marine water striders live on the ocean's surface, often far from land. Because they live on the water, not in it, gerromorphans are called "semi-aquatic" bugs, or, in an earlier terminology, Amphibicorisae.

The Leptopodomorpha too are semi-aquatic, most of the 300 or so species living near water but neither on it nor in it; a few species live in the splash zone of the sea. All are predacious. Most belong to the family Saldidae, the shore bugs, which are fairly common on rocks in or near water, where their ability to jump and conceal themselves in crevices frustrates the best of collectors.

The 2000 species of Nepomorpha, aquatic bugs (Hydrocorisae), live below the water's surface. Here are the familiar backswimmers (Notonectidae), water boatmen (Corixidae), giant water bugs (Belostomatidae; some Indian species may reach 4 in. in length), and water scorpions (Nepidae); there are several other nepomorphan families that have more tropical species than temperate. All of these are predacious, although water boatmen feed also on freshwater algae. Some giant water bugs specialize on freshwater snails and may be useful in controlling those harboring the schistosomiasis platyhelminth. Others sometimes cause some damage in fisheries, where they may attack young fish. From time to time, someone suggests using some of these aquatic bugs to control mosquitoes, but the attempts have yet to succeed.

The Cimicomorpha, with more than 19,000 species, is the largest of the heteropteran infraorders. This is because it contains the Miridae, or plant bugs, which, with almost 10,000 species, is the largest heteropteran family. Cimicomorpha also hold the Reduviidae, whose 6700 species make it the second-largest heteropteran family. Cimicomorphans are basically (primitively) predacious, but several groups have secondarily become phytophagous: the small family Thaumastocoridae feeds on palms; the Tingidae (1800 species), or lace bugs, feeds on leaves of many plants and may occasionally become serious pests; and most members of the Miridae feed on plants, although many mirids are predacious. It is possible that the evolutionary success of Miridae (as reflected in the family's many species) derives from mirids' association with annual weedy plants, plants that grow rapidly and yield to other species; mirids, too, develop rapidly, and many are quite host specific. Many of the predacious Miridae are site-specific, living on a single species of plant. If the plant should be of economic significance, these mirids can be useful in biological control.

Of the wholly predacious groups, the assassin bugs, Reduviidae, are the most important. Some are specialists (unusual for predators), one group feeding on millipedes, another living in spider webs and stealing food from the spider; yet another feeds on termites and sometimes lures them from the mound. The group's biocontrol potential has been little studied, except recently in southern India (see Schaefer and Panizzi, 2000). As mentioned above, members of the subfamily Triatominae feed on vertebrate blood, and the neotropical members transmit the Chagas' disease trypanosome to humans.

All 80 species of another cimicomorphan group, Cimicidae, feed on vertebrate blood, especially on that of cave-dwelling vertebrates such as bats, communally nesting birds—and man. The bedbugs (*Cimex lectularius* L. and *C. hemipterus* (Fabricius)) are parasites of man. *C. lectularius* feeds only on man and is one of a very few insects that occur *only* with man (the housefly and the head and body louse are among the others). *Cimex hemipterus* is at times a pest of poultry. *C. lectularius* is primarily a temperate-zone species, although it extends into the Tropics; *C. hemipterus* is tropical.

Other important cimicomorphan groups include the damsel bugs (Nabidae), a worldwide family of nearly 400 species, common in crop fields in North America and often studied for their biocontrol potential; and Anthocoridae, minute flower bugs, a group (600 species), also worldwide, of very small and often strikingly patterned insects much studied for their control of pests, especially in greenhouses.

The seventh infraorder, Pentatomomorpha, is also large, with at least 11,600 species. Two of its families are tied for third largest in Heteroptera: Pentatomidae and Lygaeidae, each with about 4100 described species (Note: Recent work has shown that many of the subfamilies of Lygaeidae are worthy of family rank). Pentatomomorpha is one of only two infraorders that appear to have arisen as plant feeders, although one important subgroup, the subfamily Asopinae (family Pentatomidae), is secondarily predacious and important in biocontrol programs. Most pentatomomorphs feed on the nitrogen-rich reproductive parts of plants, especially on their ripe and ripening seeds. Pentatomomorpha is also the most difficult infraorder to spell correctly.

A major pentatomomorph group is the Lygaeoidea, or seed bugs and milkweed bugs; most of these feed on seeds, although a few groups feed on grasses and members of one of these, Blissidae or chinch bugs, sometimes become pests on wheat. Another major group is the Pentatomidae (stinkbugs), which also feed on reproductive parts, although many others feed on somatic tissue. Included here is the southern green stinkbug, *Nezara viridula* (L.), a major pest of many crops across the world. Here also are the Sunn pests, a group of pentatomids and some Scutelleridae (a family closely related to Pentatomidae) that ravage wheat and barley throughout the Middle East and surrounding areas. Several small families are related to Pentatomidae, including the Cydnidae, many of which live in the soil and suck from the roots of plants. Members of Scutelleridae are often brilliantly and iridescently colored (although Sunn pests are not). Females of many Acanthosomatidae guard their eggs from parasites and their young from predators.

The Coreidae (leaf-footed bugs) (1300 species) and Pyrrhocoridae (cotton stainers) (400 species) are related to Lygaeoidea. The former includes the squash bug (*Anasa tristis* (De Geer)) and a group of bugs with expansions on their legs ("leaf-footed" bugs); some of the latter (in the Neotropics) are brilliantly colored, and others (in North America) come to houses for warmth when winter sets in. The largest genus in Pyrrhocoridae is the tropical *Dysdercus*, many of whose species are serious pests of cotton, which they damage in part by direct feeding on the seeds. Greater damage is caused by their feeding punctures' providing entry into the cotton boll of boll rot disease organisms; these organisms, and the bugs' excreta, destroy and stain the cotton fibers (hence "cotton stainers"). One other species, the European *Pyrrhocoris apterus* L., is famous for its role in the discovery of "paper factor," a discovery that led to the development of natural and artificial juvenile hormone analogs, useful in insect control.

The last (eighth) infraorder is Aradomorpha, until recently included as a member (superfamily Aradoidea) of the Pentatomomorpha, from whose other members it differs in several significant respects. Among these are the long mouthparts and the lack of long sensory hairs arranged in characteristic patterns on the underside of the abdomen; pentatomomorphans have shorter mouthparts (like those of other heteropterans) and have these abdominal sensory hairs (unlike most other heteropterans).

The Aradomorpha are a small group of brownish to grayish flattened bugs, nearly all of which live under the bark of dead or dying trees, where they feed on the long mycelia of fungi; adapted for this are the very long very slender mouthparts, which at rest are kept coiled in a special pouch within the head. These bugs range from about 3 to 10 mm long and are often wingless; the upper surface of the body is frequently "bumpy" or pebbly and the legs and antennae are short. One species, *Aradus cinnamomeus* Panzer, differs from the others and feeds on the sap of several pine species; from time to time it becomes a serious pest of commercial pines in northern Europe. It has a 2-year life cycle, and the discovery that some populations reproduce in odd-numbered years, and others in even-numbered years, has led to interesting biogeographical and population studies. Two other species (family Termitaphididae), distantly related to all the others (family Aradidae), live in termite nests, where they lay their eggs among the termites'. Only a few millimeters long, these bugs lack an ovipositor, eyes, and wings and probably feed on fungi within the nests. Very little indeed is known about them.

IV. CONCLUSIONS

Heteropterans share fundamental properties of structure and biology, especially of feeding: their elongated mouthparts so well adapted for the sucking up of the fluids of other organisms. Yet the diversity of heteropterans is remarkable, in size, form, habitat, and structure. Groups now considered rare are probably in fact common, as collecting in habitats hitherto disregarded has shown: Dipsocoromorpha are probably far more common than we believe, but live in soil and debris habitats in the Tropics, habitats no one studies. Recent work on tree-canopy habitats in the Tropics has revealed assemblages of unknown or poorly known heteropterans found (apparently) nowhere else. Too many entomologists have for too long scorned insects that cannot be collected with nets. Once such other collecting methods as searching the ground's surface became common, so did groups (like Lygaeoidea) once thought relatively scarce.

It follows that the diversity of Heteroptera, like the diversity of all other groups, is constrained not by evolution but by man's knowledge and by man's willingness to seek diversity in different habitats: For the diversity is there.

What I have sketched above is based in truth on a remarkably small number of observations and experiments on a remarkably small number of heteropterans. One hopes those observations are accurate and those heteropterans are representative. Much more work, on many more heteropterans, as well of course as much more collecting, will make even more clear just how diverse and fascinating true bugs are.

See Also the Following Article

INSECTS, OVERVIEW

Bibliography

Blatchley, W. S. (1926). *Heteroptera or True Bugs of Eastern North America, with Especial Reference to the Faunas of Indiana and Florida.* Nature Publishing Co., Indianapolis, IN.

Butler, E. A. (1923). *Biology of the British Hemiptera–Heteroptera.* H. F. & G. Witherby, London.

Cohen, A. C. (1998). Solid-to-liquid feeding: The inside(s) story on extra-oral digestion in predaceous Arthropods. *Am. Entomol.* 44, 103–116.

Dolling, W. R. (1991). *The Hemiptera.* Oxford Univ. Press, Oxford.

Javahery, M. (1995). *A Technical Review of Sunn Pests.* FAO, United Nations, Regional Office of the Near East, Cairo, Egypt.

Schaefer, C. W., and Panizzi, A. R. (2000). *Heteroptera of Economic Importance.* CRC Press, Boca Raton, FL.

Schofield, C. J. (1994). *Triatominae: Biology and Control.* Eurocommunica Publications, West Sussex, England.

Schuh, R. T., and Slater, J. A. (1995). *True Bugs of the World (Hemiptera: Heteroptera).* Cornell Univ. Press, Ithaca, NY.

Slater, J. A., and Baranowski, R. M. (1978). *How to Know the True Bugs (Hemiptera–Heteroptera).* Brown, Dubuque, IA.

Weber, H. (1930). *Biologie der Hemipteren: Eine Naturgeschichte der Schnabelkerfe.* Verlag-Springer, Berlin.

ULTRAVIOLET RADIATION

Andrew R. Blaustein and Nareny Sengsavanh
Oregon State University

I. Properties of Light
II. Ultraviolet Radiation
III. Measurement of Environmental UV Radiation
IV. Effects of UV Radiation on Biological Systems
V. Effects of UV Radiation on Biogeochemical Cycles
VI. Effects of UV Radiation on Humans

GLOSSARY

chlorofluorocarbon (CFC) A type of hydrocarbon that is composed of nonreactive, nonflammable organic molecules in which both chlorine and fluorine atoms replace some of the hydrogen atoms. Use of CFCs is one of the primary causes of stratospheric ozone depletion.
Montreal protocol A treaty signed by a number of nations to curtail CFC production
ozone Form of oxygen in which three atoms of oxygen occur together. Forms a natural layer in the stratosphere that blocks living organisms from harmful ultraviolet radiation from the sun
ultraviolet radiation Radiation with wavelengths shorter than violet light and with more energy

I. PROPERTIES OF LIGHT

Light is a natural phenomenon that allows us to see objects, shapes, and colors. It radiates from the sun, the stars, a flame, or a lightbulb. Light encompasses a broad spectrum of radiation that includes gamma rays, ultraviolet rays, infrared rays, microwaves, and radiowaves (Fig. 1). These types of radiation are all classified as nonvisible light and consist of the longest and shortest wavelengths and frequencies. A small section of the spectrum is the visible light spectrum. The wavelengths of this region range from 380 to 750 nanometers (nm) and are the only part of the electromagnetic spectrum (Fig. 1) that can be detected by the human eye. When light shines on an object, some of the light is absorbed and converted to heat; another percentage is scattered or dispersed in various directions; some is transmitted, and the rest may be reflected depending on the material on which the light strikes.

II. ULTRAVIOLET RADIATION

The electromagnetic spectrum is continuous but the types of electromagnetic radiation do not begin or end at precise points along the spectrum. For example, red light shades into invisible infrared (below red) radiation

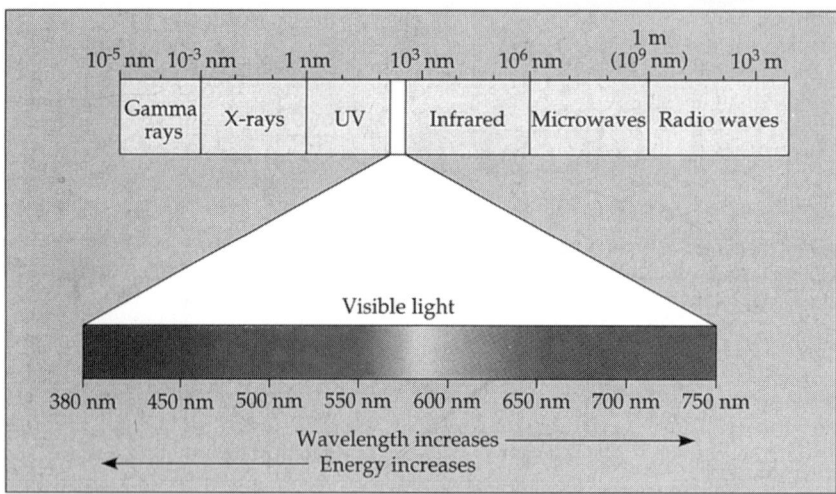

FIGURE 1 The electromagnetic spectrum. Visible light and other forms of electromagnetic energy radiate through space as waves of various lengths. From Campbell (1996). *Biology*, 4th ed., Figure 10.5, p. 187. Copyright © 1996 by Benjamin Cummings Publishing Company. Reprinted by permission.

and violet light shades into invisible ultraviolet (beyond violet) radiation. The major source of ultraviolet (UV) radiation for the earth is the sun. UV is responsible for producing suntans and vitamin D in the human body. In humans, overexposure to UV can lead to serious and sometimes irreparable harm. It can cause mutations in cellular DNA, which can ultimately cause significant alterations in cells, the main cause of cancer. Other damaging effects of UV include premature aging, blindness, and sterilization. Ultraviolet radiation also has the ability to kill microorganisms, plants, and animals.

UV can be divided into four wave bands. These are Vacuum UV (<200 nm), UV-C (200–280 nm), UV-B (280–315 nm), and UV-A (315–400 nm). At the earth's surface, vacuum UV and UV-C are not present because of their absorption by various gases such as oxygen and ozone. The formation of atmospheric oxygen and a stratospheric ozone layer was essential for the evolution of life on earth. The ozone layer shields the terrestrial surface from harmful UV radiation. Unfortunately, through anthropogenic emissions of chlorofluorocarbons (CFCs) and other gases, the ozone layer has been adversely affected. It has thinned and has developed "holes" in polar regions. Thus, there is potential for increased UV radiation hitting the earth's surface.

The ozone hole over Antarctica has dramatically increased since its discovery in the 1970s. At midlatitudes, ozone levels have also continued to decrease. Stratospheric ozone levels are at their lowest point since measurements began, so current UV-B radiation levels are thought to be close to the maximum. Global ozone measurements from satellites from 1979 to 1993 show increases in UV-B radiation at high and midlatitudes of both hemispheres, but only small changes in the tropics. These estimates assume that cloud cover and pollution have remained constant over the years. Increases in surface erythemal (sun-burning) UV radiation relative to values obtained in the 1970s are estimated to be about 7% at Northern Hemisphere midlatitudes in winter and spring and 4% in summer and fall, 6% at Southern Hemisphere midlatitudes throughout the year, 130% in the Antarctic in spring, 22% in the Arctic in spring. The resulting increases in UV radiation threaten humans, animals, plants, and microorganisms in both terrestrial and aquatic ecosystems.

The pioneering work of Sherwood Rowland and Mario Molina showing that CFCs were responsible for the decrease of stratospheric ozone caused great concern among atmospheric and environmental scientists. They reasoned that CFCs in the stratosphere would be subjected to intense UV radiation, which would break them apart, releasing free chlorine atoms.

All the chlorine of a CFC molecule would eventually be released due to further photochemical breakdown. The free chlorine atoms could then damage stratospheric ozone forming chlorine monoxide (ClO) and molecular oxygen. Molecules of ClO can react to release more chlorine and an oxygen molecule. These reactions are part of the chlorine cycle because of the continuous regeneration of chlorine as it reacts with ozone. Chlorine is the catalyst in the reaction because it promotes the chemical reaction without being used up. Because chlorine can last for 40 to 100 years, every chlorine atom in the atmosphere can potentially breakdown 100,000 molecules of ozone. CFCs are especially damaging because they transport agents that constantly move chlorine atoms into the stratosphere and because chlorine atoms are removed from the stratosphere very slowly.

By 1978 most Scandinavian countries and the United States banned the use of CFCs in spray cans. In 1983, most European countries proposed a voluntary reduction in the use of CFCs. As a result of increasing concern over the effects of ozone depletion, an international agreement was developed to reduce and eventually eliminate certain anthropogenic ozone destroying substances. A treaty, known as "The Montreal Protocol" was signed by a number of nations on September 16, 1987, in Montreal, Quebec. The protocol, which consisted of a plan that would dramatically cut CFC production, was recognized as the first worldwide effort to solve a massive environmental problem. Since 1987, more than 150 countries have signed the agreement to phase out all use of CFCs by 2000. Despite this agreement, the worsening environmental news about CFCs and ozone depletion in the 1990s led certain nations to adopt stricter measures limiting CFC production.

Several industrial companies have developed CFC substitutes. CFCs and several other chemicals that contribute to ozone depletion have been phased out in the United States and several other countries. Nevertheless, existing stockpiles of these chemicals can be used until the deadline. Moreover, developing countries are on a different time course and will phase out CFCs by 2006. Unfortunately, CFCs are very stable and estimates suggest that those in use today will continue to deplete stratospheric ozone for 50 to 150 years. Atmospheric scientists believe that the Antarctic ozone hole will reappear each year until about 2050. However, because of the Montreal Protocol and other measurements to limit ozone depleting substances, there is some optimism about ozone depletion being curtailed.

Bromine (Br) is another important source of ozone depletion. Bromine is present in smaller quantities than chlorine but is more destructive on an atom to atom basis. The largest source of Br is methyl bromide that naturally comes from the oceans and wildfires. However, a large portion of methyl bromide is human made and is used as a fumigant. Another important source of bromine is in fire extinguishers. Like CFCs, methyl bromide has a long atmospheric lifetime (approximately 72 years). Regulations on methyl bromide use are being debated.

III. MEASUREMENT OF ENVIRONMENTAL UV RADIATION

Although significant advances have been made, measuring UV radiation is difficult. There are significant differences in how various devices measure UV radiation. Moreover, a number of factors affect UV radiation. For example, there are geographical and seasonal variations in UV. Thus, recent studies have confirmed that there are generally higher UV-B levels at lower latitudes in the United States. Other measurements confirm latitudinal differences in Europe, Asia, and New Zealand.

Spectral measurements show higher summer values of both UV-A and UV-B radiation in New Zealand and Australia compared with Germany due to the yearly cycle of the sun-earth distance and to lower stratospheric ozone levels in the Southern Hemisphere and higher air pollution levels in Germany. Additional UV data are being accumulated for midlatitudes and eventually we will gain a more complete picture of the UV situation at these latitudes.

Cloud cover also affects UV-B radiation levels at the surface. Data from several locations in the United States suggest that monthly average UV levels are reduced by 10 to 50% by cloud cover, depending on the location and season. Aerosols (small particles suspended in the air) may also reduce UV levels in polluted areas. The magnitude of this effect is highly variable and may depend on the number of particles and their chemical and physical composition.

UV levels are expected to increase with increasing surface elevation above sea level due to a thinner atmosphere overhead. Measurements in a remote area of Chile, for example, showed increases of 4 to 10% per kilometer. Other locations showed larger vertical gradients of up to 40% per kilometer near Santiago, Chile, and 9 to 23% per kilometer in the Swiss Alps.

Because of the high spatial and temporal variability of surface UV radiation and the difficulty of maintaining calibration within instruments, it will be very difficult to provide global UV climatology and to represent long-term UV trends based on ground-monitoring stations alone. On the other hand, satellite-based measurements can provide global coverage and continuous long-term monitoring. Yet it is difficult to use remote satellite data in an attempt to estimate UV levels from specific microhabitats on earth. The derivation of surface UV irradiance from satellites is indirect because satellites detect radiation reflected by the atmosphere and the surface of the earth. The use of radiative transfer models is necessary to relate transmission, reflection, and atmospheric absorption. These models have been useful in showing general changes in UV radiation reaching the surface, computed for clear skies using satellite ozone measurements.

IV. EFFECTS OF UV RADIATION ON BIOLOGICAL SYSTEMS

A. DNA Repair

At the terrestrial surface, most UV radiation of biological concern is in the 280 to 315 nm range (UV-B band). UV light induces a variety of photoproducts in DNA. UV photoproducts impede gene expression by blocking transcription and they can cause cell death or mutagenesis. There are a number of responses to DNA damage by living cells. The two fundamental processes for repairing UV-induced DNA damage are direct reversal of DNA damage or its excision. These processes can explained by analogy. If we consider a rope with a knot, we can either undo the knot (direct reversal of damage) or we can cut the knot out and replace the rope with a new piece (excision of the knot). This descriptive analogy roughly represents how photoreactivating enzyme, photolyase, and excision repair, respectively, work to undo DNA damage.

Thus, harmful photoproducts can be removed by photoreactivating enzyme, photolyase. In some species, this is the most important mechanism for DNA repair. In excision repair, when DNA molecules are damaged, a segment of the strand of DNA containing the damage is cut out by one repair enzyme, and the resulting gap is filled in with nucleotides (building blocks of DNA) properly paired with nucleotides in the undamaged strand.

B. Effects of Increased Solar Ultraviolet Radiation on Terrestrial Plants

Obviously, it is extremely important to understand the effects increasing solar ultraviolet radiation has on terrestrial plants. Plants comprise most of the living matter in terrestrial ecosystems. Damage to plants within natural ecosystems and in agricultural systems can have far-reaching consequences to other organisms, including humans.

Both physiological and developmental processes of plants are affected by UV-B radiation. Several studies have shown that some plant species in greenhouses, in growth chambers, and in the field have reduced growth with reduced leaf area under close to ambient UV-B conditions compared with plants grown under reduced UV-B levels. Increased UV-B radiation may greatly affect photosynthesis. In certain species, such as soybean, sunflower, and corn seedlings, solar UV-B radiation reduced photosynthesis by about 15% when a 12% ozone depletion was simulated. UV-B radiation can also alter the time of flowering and the number of flowers in certain species. Alteration of flowering time can have a severe impact on plants because pollinator availability may be subsequently affected. Anther walls can absorb more than 98% of incident UV-B radiation and pollen walls contain UV-B-absorbing compounds. However, after the transfer to the stigma, pollen may be susceptible to UV-B radiation. Several experimental studies using Mylar or glass filters that shielded plants from UV-B radiation, showed that flowering was enhanced under these regimes. Different studies have shown different effects of UV-B on flowering in different species. For example, flowering is inhibited in such plants as *Melilotus* and *Trifolium* but *Zea mays* and *Sorghum* were not affected. In sexually reproducing populations of a desert plant, the effects of UV-B radiation on growth and biomass seem to accumulate in subsequent generations that were exposed to UV-B radiation. Thus, the effects of exposure to UV-B radiation may be amplified.

The yields of certain crop plants may be greatly affected by increases in UV-B radiation. The available data on the effects of UV-B radiation on yield illustrate significant interspecific variability and variability among cultivars complicated by differences in how the experiments were performed. Nevertheless, some species and certain cultivars seem to be more tolerant of UV-B effects than others. For example, results from greenhouse and field tests suggest that the soybean cultivar "Forrest" was more tolerant to UV-B radiation than the cultivar "Shore." A study of 10 crop species in Florida showed that under UV-B radiation, yields were reduced by 5 and 90% in half of them, including wheat (5% reduction), potato (21% reduction), and squash (90% reduction). In this study, rice, peanut, and corn were not affected.

Several studies have shown that plants may be more susceptible to pathogens and may affect insect pests when exposed to UV-B radiation. Thus, a number of studies have shown that ambient UV-B radiation can reduce insect herbivory of agricultural pests and native plants. Supplementation of solar UV-B radiation in field studies can reduce the population of herbivorous insects in certain systems. It is unclear why changes in herbivory occur. It is possible that the secondary compound plant defenses may be altered. Most of these studies suggest that the changes in insect herbivory are due to changes in host plant tissues.

Experiments conducted in greenhouses and in the laboratory suggest that viral and fungal pathogens react differently to UV-B radiation. In some studies UV-B radiation promotes the severity of disease, whereas in other studies, it seems to prevent the severity of the disease. For example, cucumber plants first exposed to UV-B radiation were more susceptible to subsequent infection by fungal pathogens. However, if exposed to UV-B radiation after infection, there was no effect on the severity of the disease. Other studies have shown that when UV-B radiation is removed, there is increased incidence of fungal infection.

Even roots of plants whose shoots were exposed to elevated UV-B radiation can be affected. For example, the microorganisms associated with the roots of sugar maple trees were altered when the shoots of the trees were exposed to UV-B radiation.

Because there are differences in tolerance to UV-B radiation in different species, it is suggested that a reduction of primary productivity in one plant species may lead to an increase in primary productivity in another more UV-B-tolerant species. Thus, it is possible that the overall productivity within an ecosystem may change, but the species composition of the system may not change. Even if the plant species composition does not change, individual plant form may change, which could affect how these plants compete for sunlight, moisture, and nutrients. This could lead to significant changes in the overall characteristics of ecosystems.

C. Effects of Radiation on Aquatic Systems

1. Plankton

The attenuation of solar radiation on the water column is dependent on a variety of factors. The transparency of the water to ultraviolet radiation is dependent on the type of water. For example, there can be highly turbid coastal waters that do not allow much UV-B to penetrate very deeply. In comparison, some ocean waters are clear enough so that UV-B penetration can be dozens of meters. In the Antarctic, 1% of the solar UV-B hitting the surface has been measured at a depth of 65m. Solar UV-B has been shown to degrade dissolved organic carbon (DOC). Increased breakdown of DOC and subsequent consumption by bacteria increases the UV-B penetration in the water column.

Globally, phytoplankton is the most important producer in aquatic ecosystems. Thus, damage to phytoplankton populations will affect higher trophic levels. A number of recent studies in a variety of aquatic ecosystems have shown that UV-B radiation affects the growth, survival, and distribution of phytoplankton. Phytoplankton exist on the top layers of the oceans and freshwater aquatic systems. The Antarctic is especially productive in phytoplankton and the region is significantly impacted by UV-B because of the Antarctic ozone hole. Therefore, a number of studies on the effects of UV-B on phytoplankton have been conducted in that region. In certain experiments, productivity was two to four times higher in tanks where UV-A and UV-B were excluded. Pigmentation was also affected. *In situ* incubations of natural phytoplankton assemblages in Antarctic waters indicated that photosynthesis was impaired by about 5% under the ozone hole. A similar result was found in the tropics. Screening of most UV <378 nm resulted in a 10 to 20% increase in photosynthesis. However, no significant decreases in stratospheric ozone have been observed in the tropics.

Many phytoplankton can actively move to different positions within a particular habitat. They may do this via flagella, cilia, or by utilizing buoyancy to adjust their position in the water column. Various chemical, magnetic, light, and gravity cues influence movement so that plankton can maintain specific positions within the water column. To cope with constantly changing environmental conditions, these organisms must constantly adjust their positions. If UV radiation affects motility, or the ability of phytoplankton to respond to external cues, this may negatively affect their growth and survival. There is growing evidence that many phytoplankton species are under stress from ambient levels of UV radiation.

2. Macroalgae and Seagrasses

In contrast to the motile phytoplankton, macroalgae and seagrasses are attached to their growing sites. Thus, they are restricted to certain depth zones. It is thought that this zonation is caused, at least in part, to limits of visible light penetration at various depths. Some species may be more tolerant to solar radiation than others. Thus, increased levels of UV-B radiation may expose algae and seagrasses to levels that they have not encountered, perhaps affecting growth, or photosynthesis. Several studies have shown that UV-B radiation inhibits photosynthesis in many red, brown, and green algae. Deep-water algae were most affected whereas intertidal algae were the least sensitive.

The DNA of algae appears to be poorly shielded when compared to higher plants. For example, doses of UV-C radiation (at 254 nm) necessary to kill leaves of higher plants appear to be about four orders of magnitude greater than that necessary to kill highly resistant algae. Flavenoids, highly effective UV screening compounds found in higher plants, have not been found in algae. In higher plants, flavenoids exist in high enough concentrations in the epidermis of leaves so that in combination with cuticle waxes and other cell wall components, the incidence of UV radiation is reduced by one or two orders of magnitude. However, algae do produce other UV-absorbing substances, including a yellow protein-carotenoid complex in some species and other substances that afford some protection form UV radiation.

A recent experimental study conducted in Canada illustrates how ecosystems may have complex responses to increased solar UV-B radiation. Solar UV radiation can reduce photosynthesis and growth in bottom-dwelling algal communities in shallow freshwater. However, in this study, greater amounts of algae accumulated in UV-exposed habitats than in UV-protected environments. UV-A and UV-B radiation inhibited insects that feed on algae. Because larval algal consumers are more sensitive to UV radiation than algae, algal abundance increased.

3. Invertebrates

Sunlight can be lethal to a wide variety of marine and freshwater plankton following exposure for just a few hours or a few days. Mortality rates are usually lower in zooplankton that contain photoprotective compounds including pigments derived from dietary plant carotenoids, melanin, and substances known as mycosporine-like amino acids that absorb in the UV-A and UV-B range. The presence of these photoprotective compounds suggests that there is significant selection pressure associated with the harmful effects of UV radiation.

A number of experimental studies have shown that natural levels of UV radiation are lethal to zooplankton. One recent study showed that zooplankton communities exposed to ambient levels of UV-B for three days in Pennsylvania experienced significant mortality when they were not shielded from UV-B radiation. However, mortality was significant only in an oligotrophic lake, not in a eutrophic lake where light penetration is not as great.

Marine invertebrates differ greatly in their sensitivity to UV-B radiation. For example, while one species of crustacean may suffer about 50% mortality at current levels of ambient UV-B radiation at the sea surface, some shrimp can tolerate irradiances higher than those predicted for a 16% ozone depletion. Bottom-dwelling invertebrates of the ocean may also be affected by UV-B radiation. For example, cleavage in sea urchins is impaired by UV radiation. Marine organisms associated with coral reefs, such as sponges, bryozoans, and tunicates, are also adversely affected by UV-B radiation.

Corals are affected by UV radiation in a number of ways. Depending on the species and the particular ecosystem, a number of studies illustrate that UV radiation can cause death, inhibit growth, and contribute to coral bleaching. Bleaching, a well-recognized phenomenon, occurs principally via the loss or expulsion of symbiotic algae from the coral host tissue. When the algae are lost, the coral loses its characteristic color and the remaining white skeleton is most noticeable. Although a number of events, including elevated seawater temperature and heavy rains, can contribute to bleaching, there is some evidence that UV radiation may also play a role in certain bleaching events.

UV radiation can inhibit photosynthesis in symbiotic algae. UV radiation affects respiration among corals and their symbiotic algae, but not in a consistent way. In some species, respiration increases under UV radiation whereas in other species it may decrease and in others respiration may be unaffected.

UV radiation inhibits the growth of symbiotic algae in culture. Reproduction in corals may be affected by UV-B radiation. Broadcast spawning at night, a nearly universal phenomenon among many reef invertebrates and corals, might be related to avoiding UV-induced DNA damage as well as to reduce predation. In at least one study the larvae of reef corals from shallow water were more resistant to UV-B radiation than those from deeper water.

In the Antarctic, a number of marine invertebrates sustain UV-induced DNA damage. These include worms and crustaceans. Krill, copepods, and gelatinous zooplankton generally have transparent eggs, larvae, and or adults stages that are pelagic, planktonic, and are often found in surface water for several months. Thus, these species are especially vulnerable to DNA damage from elevated UV-B exposure. These species are important components of the ecosystem and damage to them could have significant ecosystem consequences.

4. Fishes and Amphibians

A recent study of Antarctic zooplankton, including larval fish, showed that they sustain DNA damage during periods of increased UV-B flux. Fish larvae in Antarctic marine ecosystems sustained UV-B-induced DNA damage greater than the lethal limit determined for Antarctic diatoms and comparable to the lethal limit of damage for cultured goldfish cells. DNA damage has been shown to be especially correlated with daily UV-B flux in icefish eggs. Icefish larvae, however, showed patterns of DNA damage that correlated less well with daily UV-B flux. Antarctic fish appear primarily to use photolyase to remove harmful UV-B-induced photoproducts.

Other fish species may also be affected by UV-B radiation. For example, in the Arctic ecosystem, many economically important fish species, including cod, pollock, herring, and salmon, spawn in open shallow water and are subjected to increased solar UV-B radiation. Because many of their eggs are found near the surface, it is possible that marine fish productivity could decline in this region due to increased UV-B radiation. At this time, however, it is difficult to assess the impact of UV-B radiation on Arctic fish productivity. However, one recent laboratory has shown that salmon exposed to UV-B radiation are more prone to fungal infections and skin lesions.

Several laboratory studies have shown that UV radiation affects the growth, development, and hatching success of certain amphibian species. These studies have shown that under simulated UV light of various intensities, amphibians may develop skin lesions, edema, eye damage, curvature of the body, and behavioral abnormalities. Under relatively low-level but prolonged doses of UV radiation, the mortality of embryos increases compared with controls that were shielded from UV radiation.

a. Amphibian Declines and Ultraviolet Radiation

Amphibian populations are in serious decline in various areas of the world. Unfortunately, the causes for amphibian population declines have been difficult to assess. Much of the information on amphibian declines comes from observational or anecdotal accounts. Hypothesized causes for the declines include habitat destruction (the most obvious cause), pathogens, introduced exotic species, pollution, and increased ultraviolet radiation. These agents may act alone or in combination to contribute to the decline of amphibian populations.

The diversity of locations where amphibian populations have declined prompted consideration of atmospheric factors such as increased ultraviolet irradiance associated with depletion of stratospheric ozone. Several investigators have used field experiments to examine the potential role of ultraviolet-B radiation in amphibian population declines by measuring the mortality of embryos that were shielded from UV radiation compared to embryos that were unshielded. Continuous high mortality in early life stages may ultimately contribute to a decline at the population level.

Recent field experiments from North America, Europe, and Australia show that ambient UV-B damages the embryos of certain amphibian species but not others. Results of these experiments by several different investigators strongly indicate that the hatching success of at least nine species of amphibians, from widely separated locales, is reduced under ambient UV-B radiation. This includes a diverse group taxonomically of two frog species, one toad species, two salamander species, and a newt from North America, two frog species from Australia, and a species of toad from Europe. Some of these species are found in montane areas, others at sea level. A key characteristic shared by these species is that they often lay their eggs in shallow water, where they are exposed to solar radiation.

Hatching success of several other amphibian species in North America, Australia, and Europe were not affected by UV-B radiation. This is not surprising because many studies have demonstrated differential sensitivity of amphibians to various abiotic factors. There may be variation in response to UV-B radiation, perhaps even within a species at different locations. For example, embryos of western toads in Oregon are sensitive to ambient levels of UV-B while those of a different subspecies in Colorado are unaffected.

Based on a limited sample, there is a correlation between resistance to UV-B and the activity of photoreactivating enzyme, photolyase. Species with the highest photolyase activities seem to be more resistant to UV-B radiation. Furthermore, of the species examined, frogs and toads generally have more photolyase activity than salamanders. It is also possible that nuclear excision repair is also being used to counter UV-induced DNA damage, but this has not been measured in amphibian species taken from the wild.

In nature, more than one environmental agent may affect an animal as it develops. This seems also true for amphibians developing at their natural field sites. Field experiments have been used to examine at least three factors that seem to interact synergistically with UV-B: a pathogenic fungus, low pH, and fluoranthene, a polycyclic aromatic hydrocarbon that may pollute aquatic environments impacted by petroleum contamination. In certain combinations with UV radiation, these three agents increase mortality to levels greater than that contributed by UV radiation alone.

Obviously, UV-B cannot be invoked as contributing to all amphibian population declines. For example, UV-B radiation is unlikely to contribute to mortality in amphibians that are primarily nocturnal, live under dense forest canopies, lay eggs hidden from solar radiation, or have high photolyase activities.

b. Amphibian Deformities and UV Radiation

Reports of deformed amphibians have been given wide media attention and have been the subject of several recent scientific workshops. The most common reports of deformities are frogs and toads with extra or missing limbs. Three major agents are being examined; pesticides, UV radiation, and parasitic trematode infection. However, there are very few available data on this subject based on experimental work. One recent laboratory study of northern leopard frogs showed that upon exposure to 24 hr of simulated ambient UV light, frogs developed hindlimb malformations. It remains unclear as to whether UV radiation can cause extra limbs in wild amphibian species. However, other deformities in wild species have been observed. These include edema, curvature of the spine, and lesions in larvae and newly metamorphosed amphibians. A recent report showed severe retinal abnormalities consistent with UV damage in a basking frog species.

V. EFFECTS OF UV RADIATION ON BIOGEOCHEMICAL CYCLES

The effects of increased UV-B radiation on biogeochemical processes may be complex. For example, increased UV-B radiation may affect the magnitude and direction of trace gas on emissions and mineral nutrient cycling in terrestrial systems. Moreover, the effects on these processes may be species specific and may vary in different ecosystems. Increased UV-B radiation may alter the chemical composition of plant tissue, the photodegredation (breakdown) of dead plant matter, the release of carbon monoxide, the microbial decomposers, and nitrogen fixing organisms.

In aquatic ecosystems, increased UV-B radiation may affect the processes that produce organic matter and the processes that degrade organic matter. There may be photodegradation of dissolved organic matter, which may lead to the production of organic acids and ammonium. Photoinhibition of surface aquatic organisms may also affect biogeochemical processes.

The potential effects of increased UV-B radiation on terrestrial and aquatic carbon, nitrogen, sulfur, oxygen, and metal cycles have been explored in a number of studies. These effects and similar effects on biogeochemical cycles in the atmosphere may aid or help impede the buildup of greenhouse gases and aerosols in the atmosphere.

VI. EFFECTS OF UV RADIATION ON HUMANS

UV-B radiation does not penetrate far into the body because most of it is absorbed in the superficial tissue layers. Therefore, much of the UV damage affects the skin and eyes. However, there are also systemic effects. UV-B is the main cause of sunburn and tanning and the formation of vitamin D_3 in the skin. UV-B also affects the immune system. UV-B can cause snow blindness and is a significant factor causing cataracts. It also contributes significantly to the aging of the skin and the eyes and is effective in causing skin cancer.

A. Eyes

Photokeratitis is the effect most attributable to exposure to UV radiation. It is similar in effect to a sunburn and often occurs after short-term exposure to UV radiation. The eyeball becomes inflamed and reddens, often accompanied pain and photophobia (fear of light). This is frequently diagnosed in skiers as snow blindness.

Exposure of the eye to UV radiation can also have effects on the cornea. Thus, exposure to UV radiation can contribute to degeneration of the fibrous layer of the cornea. Under certain conditions, exposure to UV radiation can cause an outgrowth of the outermost mucous layer over the cornea, which results in the loss of transparency. Squamous cell carcinoma of the cornea, a malignant neoplasm, can also be found after exposure to solar radiation. These diseases are associated with outdoor living or working near areas of high reflectance (e.g., near water, concrete, or sand).

Cataracts are the leading cause of blindness in the world. They are characterized by a gradual loss in the transparency of the lens of the eye due to oxidized lens proteins. This can lead to blindness unless the affected lens is removed. There is a correlation between certain types of cataracts and exposure to UV-B radiation. Several studies have suggested that the relative risk associated with increased exposure to sunlight and cortical cataracts (those that develop in the outer layer of lens protein) is between about one- and three-fold. One model suggests that a sustained 10% loss of ozone worldwide would lead to an additional 30,000 blind people per year.

B. Sunburn

Sunburn is the most common effect of exposure to UV radiation. This results in the reddening of the skin and possibly blistering. Sensitivity to sunburn and tanning varies with pigmentation. Heavily pigmented individuals are less sensitive to sunburning than less heavily pigmented individuals. There are various categories with regard to sunburn and tanning sensitivity. The most sensitive individuals (skin type I) develop a moderate to severe burn and usually do not tan within an hour of exposure to summer sun. These individuals usually have very fair, freckled skin, red or blond hair, and blue eyes. The most resistant individuals (skin type VI) are darkly pigmented and become more pigmented after exposure to the sun.

C. Photoaging

Exposure to sunlight ages the skin. This is known as photoaging and is characterized by wrinkles, altered skin pigmentation, and an overgrowth of abnormal elastic fibers in the dermis.

D. Skin Cancer

There are various types of skin cancers. These are basal cell carcinoma (BCC), squamous cell carcinoma (SSC), and cutaneous melanoma (CM). The carcinomas of the skin are often referred to as the "nonmelanoma skin cancers" (NMSC). The nonmelanoma skin cancers are clearly correlated with sunlight. They occur primarily in light-skinned people and then usually on the areas of the body most exposed to sunlight.

BCC is the predominant form of NMSC in light-skinned people. The most susceptible people to BCC are those with the lightest skin and poor tanning ability. The incidence of BCC in light-skinned populations has recently been increasing in certain regions. Although it was originally thought that cumulative lifetime exposure to sunlight was directly related to developing BCC, recent information suggests that this may not be the case.

SCC is much less common than BCC but is much more common than CM in the United States. More than the other skin cancers, epidemiological data suggest that cumulative lifetime exposure to UV is a critical risk factor for developing SCC. Several studies found an increase in the risk of developing SCC with incidence of childhood sunburns. However, this may be related to a high level of childhood exposure to sunlight rather than the number of sunburns per se.

CM is relatively rare compared with BCC or SCC. It accounts for only about 2 to 3% of the skin cancers associated with solar radiation, but is also accounts for most of the mortality. Like BCC, there does not seem to be a clear relationship between developing CM and cumulative lifetime exposure to UV radiation. Furthermore, CM often appears on areas of the body that are not the most heavily sun-exposed. Several epidemiological studies have shown that exposure to sun during childhood increases the risk of developing CM. An additional risk factor is the appearance of freckles or moles.

E. Effects on the Immune System

In humans, the skin is the first line of defense against foreign bodies that may threaten an individual's health. Thus, the immune system helps maintain health against infectious diseases, cancers, and parasites. The skin incorporates a number of cells from the immune system that can mount or influence immune responses to foreign substances. substances entering the body, such as viruses, have to be "recognized" by the immune system as either "self" or "nonself" (foreign) entities. UV radiation can induce photochemical changes in the skin and potentially alter cell surface proteins that are used to determine "self" from "nonself" entities. Thus, UV radiation can act as an immunosuppressive. The immunosuppressive effects of UV-B radiation can influence the outcome of melanoma and nonmelanoma skin cancers, certain infectious diseases, some forms of autoimmunity, and allergy. For example, implants of UV-induced tumors between genetically identical mice are rejected in unexposed hosts but fail to be rejected by UV-exposed mice. UV-B radiation can inhibit local inflammatory responses within UV-irradiated skin. Thus, the response elicited by injection of an antigen into the skin of sensitized individuals may be diminished in UV-irradiated skin.

Cellular immune responses are of great importance in the defense of a variety of infectious agents. Some infectious agents can be harmed by exposure to UV radiation, whereas others are unaffected. Immunosuppression may also decrease an individual's resistance to certain infectious diseases. In animal models, human infectious diseases have been shown to be influenced by exposure to UV-B radiation. These diseases include herpes, tuberculosis, trichinella, candidiasis, leshmaniasis, listeriosis, and Lyme disease. Effects include suppression of immune responses to the organisms or their antigens, reactivation of latent infections, increased body loads of infectious organisms, decreased resistance to reinfection, and reduced survival. The impact of exposure on antigen presenting cells that are in the skin and act essentially as the skin's defense system suggests the possibility that UV-B radiation may exacerbate or ameliorate autoimmune diseases such as Lupus or HIV. At the very least, UV-induced immune suppression could affect the course of certain diseases within human populations.

See Also the Following Articles

BIOGEOCHEMICAL CYCLES • ENDANGERED REPTILES AND AMPHIBIANS • GREENHOUSE EFFECT • PLANKTON, STATUS AND ROLE OF • SEAGRASSES

Bibliography

AMBIO (1995, May). Environmental Effects of Ozone Depletion: 1998 Assessment. *AMBIO* **24**.

Blaustein, A. R., Kiesecker, J. M., Chivers, D. P., Hokit, D. G., Marco, A., Belden, L. K., and Hatch. A. (1998). Effects of ultraviolet radiation on amphibians: Field experiments. *American Zoologist* **38**, 799–812.

Häder, D. P. (Ed.) (1997). *The Effects of Ozone Depletion on Aquatic Ecosystems*. R. G. Landes, Georgetown, TX.

Nebel, B. J., and Wright, R. T. (1998). *Environmental Science*. Prentice Hall, Upper Saddle River, NJ.

Tevini, M. (Ed.) (1993). UV-B Radiation and Ozone Depletion. Lewis Publishers, Boca Raton, FL.

UNEP. (1998). Environmental Effects of Ozone Depletion: 1998 Assessment. *United Nations Environment Programme*. Nairobi, Kenya.

URBAN–SUBURBAN ECOLOGY

Ann P. Kinzig* and J. Morgan Grove†
*Arizona State University and †United States Department of Agriculture Forest Service

I. Introduction
II. Definitions
III. The Founding of Cities and Affected Biomes
IV. The Biogeophysical Determinants of Biodiversity
V. Human Ecology and Social Differentiation
VI. Human Impacts on Biodiversity in and around Urban Areas
VII. Cities in the Future

GLOSSARY

human ecology A type of ecology focused on a specific species, *Homo sapiens*. Human ecology may be thought of as a subset of social ecology, which is a life science focusing on the ecological study of various social species such as ants, bees, wolves, dolphins, or orangutans.

social differentiation A term taken from biology to describe the specialization of functions in a society and to characterize societies over time. Theories of social change propose that increased social differentiation emerges as societies increase in size and complexity. Differentiation is frequently accompanied by a need for increased coordination and increased interdependence in larger and more complex societies.

urban An area characterized by high human population densities, or significant commercial or industrial infrastructure. The boundaries between urban, suburban, and rural are not sharp and can be difficult to characterize.

urban ecology The study of urban systems from an ecological perspective; an emerging field within ecology that strives to understand human interactions in ecological systems in and around urban areas and to develop theories and analyses that include human communities as fundamental components of ecological systems.

ALTHOUGH CITIES MAY be the glory of humanity, *Homo sapiens* has lived, through most of its history, as relatively isolated bands of hunter–gatherers, migratory herders, or agriculturists in scattered farming villages, farmsteads, or small trading centers. It is only recently that large cities have become commonplace, and only recently that a large fraction of the world's people has found its home in them. This urbanization trend has significant—and global—ecological consequences. The land use and land cover changes that accompany urbanization can substantially alter biogeochemical cycles, biodiversity levels, and disturbance patterns, among other things.

I. INTRODUCTION

In spite of the evidence that humans are an increasingly global force in determining the structure and functioning of ecosystems, ecologists have in general been reluctant to study those areas in which a human presence—

and thus potentially their impacts on their surroundings—is most intense. Instead, ecologists have been drawn more to the "pristine" ecosystems where humans, if considered at all, are treated as an exogenous perturbing force. Nonetheless, some threads in twentieth-century scholarship offer a basis for studying the ecology of urban systems, beginning with human ecology in the 1920s and advancing to more recent ecological research programs that have begun to fully integrate humans into their ecological observations and explanations (such as the new Urban Long-Term Ecological Research sites in Baltimore and Phoenix, funded by the National Science Foundation).

Our purpose in this article is to examine the impacts of urban settlements on biodiversity. In Section II, we offer definitions that will serve as guides in subsequent sections. Section III gives a general history of urban impacts on biodiversity—the preferential settlement of some environments over others, and thus the differential impacts on biomes and species. In Section IV, we discuss determinants of biodiversity in the absence of human influences, and in Section V we examine those aspects of human social organization and activities needed to understand the patterns of biodiversity in urban systems. In Section VI we examine how human activities in urban settings have altered distributions and abundances of species. Finally, in Section VII we briefly discuss what some future trajectories of those alterations might be.

II. DEFINITIONS

There are many possible definitions of what might constitute an urban environment. We might begin with an examination of human population density and define urban as those areas with high, suburban as those areas with moderate, and rural as those areas with low human population densities. We might forego counting humans in favor of counting structures—commercial, manufacturing, and residential—that could define urban and suburban boundaries. As with any definition, the "dividing lines" between urban, suburban, and rural categories would be somewhat arbitrary. We could always resort to using the political boundaries that define cities but many political boundaries do not derive from the contours or features of the landscape, and therefore they do not tend to capture the boundaries one would want to use in an ecological assessment.

We will not attempt or employ an exact definition of urban, suburban, rural, and "pristine" ecosystems here. We will allow some measure of intuition to guide the reader with respect to where those boundaries might lie. We propose that urban ecology in general—and the study of patterns of biodiversity in urban settings in particular—cannot stop at the city boundary. The high human population densities that characterize cities impact their less populated surroundings through flows of resources and wastes, through increased temperatures and other climatic influences, and through the extension of city infrastructure. Thus, assessments of patterns of "urban biodiversity" must extend to what would be considered rural or even pristine environments.

We have and will use "pristine" ecosystems to define those areas that appear untouched or least affected by humans and human activities. The advance of global warming and the global dispersal of such pollutants as DDT mean that such systems no longer exist. Nonetheless, they exist in our minds—for many ecologists and environmentalists, they represent a "desirable" state prior to human interference, one to which we should strive to return ecosystems. The notion that one could—through conservation or restoration—re-create a pristine ecosystem itself contains a fundamental irony or philosophical absurdity; nonetheless, it is the idea of pristine, the idea of a wilderness, that in many ways defines our relationship to Nature. But, again, the boundary line between impacted and pristine ecosystems is fuzzy, and the latter category may no longer exist.

Similarly, defining "natural" ecosystems presents a special challenge to urban and human ecologists. Are humans to be seen as an intrinsic part of urban or other ecosystems—as natural in their impacts as beaver dams or termite mounds? If so, what becomes of the notion of a "natural" system or Nature apart from humans? If not, what is "unnatural" about humans? And how can the constant exclusion of humans as something outside of the "natural" system ever permit an integrated analysis of human ecological dynamics?

Philosophical considerations aside, we will sometimes—for want of a better vocabulary—use the word "natural" in a manner similar to "pristine"—relatively untouched by human impacts. Perhaps the most precise definition would be an ecosystem whose structure and functioning can be understood largely without reference to humans or human activities (except that it is humans who must examine and write about them). Nonetheless, we do believe that humans can, to a degree, be studied as other organisms are studied, in terms of their resource needs, interactions with other organisms, and impacts on their surroundings. From our perspective, there may be quantitative and even qualitative differences in these interactions and impacts, but there is nothing "unnatural" about the ecosystems humans have constructed for themselves.

Finally, we use the term "biodiversity" somewhat loosely in the coming pages. We recognize that biodiversity encompasses biological variability across many scales—from the level of the gene to the individual to the population, species, community, habitat, and landscape. We will frequently use species number as our measure of biodiversity, not because we believe that unit is necessarily the "proper" one for discussing distributions of biological resources or human impacts on those resources. It is, however, the measure frequently used in discussions of biological resources, and the measure frequently used for analysis in the literature.

III. THE FOUNDING OF CITIES AND AFFECTED BIOMES

Humans, through their creation of urban and near-urban environments, can profoundly affect the distribution and abundance of populations, species, communities, and ecosystems. They do so in many ways—through fragmentation, the mobilization of formerly recalcitrant or scarce resources, preferential cultivation, or destruction. We will examine these on-site or near-site disturbances more systematically below. But before turning to that discussion, it is worth noting that not all biomes are equally susceptible to the disruptions caused by urban settlements. There are certain characteristics that make sites more or less desirable for dense human settlements. These characteristics have changed over time as our technologies and societies have advanced, allowing acquisition of resources from afar or the defense of even physically vulnerable locations. Nonetheless, we are more apt to find cities nestled in valleys than perched on mountaintops; more likely to find ancient cities on fertile rather than barren soils; more likely to find cities situated on ecotones or "transition zones"—for instance, between the piedmont and coastal plain—because those locations allow access to more diverse resources.

Consider the appearance of the very first urban centers. They would have begun, in many cases, as villages—and later fortresses—around which agricultural fields were established. There would have to be a favorable climate and soil and a ready supply of water. The labor required for clearing would have to be minimal. There would also have to be locally available productive grains and legumes that were a part of the first "packages" of cultivatable materials. Frequently (but not always), the emergence of agriculture as a dominant means of food procurement over hunting and gathering was aided by the availability of domesticatable animals, to be used both as draft animals and protein supplements. It is only in sites sharing these characteristics that enough food could have been produced to allow population densities to become high enough to form settlements somewhat larger than villages, on their way to being urban centers. We know, in fact, many of the places where agriculture first arose—in southwest Asia by 8000 B.C.; Greece, Cyprus, and the Indian subcontinent by 6500 B.C.; Egypt by 6000 B.C.; Spain by 5200 B.C.; and Britain by 3500 B.C. The thick forests of the wet tropical regions and the unfavorable climate of the boreal regions made these areas less favorable for the early emergence of intensive agriculture, and thus these biomes were not initially impacted by urban centers.

Later, as civilizations advanced, the need for local agricultural self-sufficiency may have been offset to some extent by the emergence of trade routes. Some urban centers would have been established in areas of particularly favorable mineral or gemstone deposits. Similarly, circumstances of chance or talent would have allowed the emergence of an artisan class within an urban center; both types of sites would have served to anchor the locations of trade routes. Other urban centers may have emerged merely because they were particularly favorable stopping-off places—just a day's journey up or down the road from a final destination—or because they were at the crossroads of two routes. Nonetheless, there would have been certain characteristics of these sites as well that meant some biomes were more susceptible to disruption from urban centers than were other biomes—those along waterways or coastways or valley bottoms, for instance.

In current times, technology has allowed establishment of urban centers in new locations and biomes. Near-global transportation of food products, improvements in our ability to clear land, and our ability to control indoor climate have extended urban centers into deserts, wet tropical forests, and boreal regions. Nonetheless, many of the large urban centers remain in areas of the highest terrestrial productivity. A high percentage continue to be located along waterways or coasts. Much of the growth in urban centers in the coming century will come in those countries with the fastest growing populations; these countries are largely located in the subtropical or tropical regions of both the Northern and Southern Hemispheres. Thus, when we think about the types of biomes—and thus the types of flora and fauna—likely to be most significantly impacted by urban centers, we should concentrate on temperate forests and grasslands, coastal and riparian zones, savanna/woodland systems, and wet and dry tropical forests. Boreal forests and montane or alpine ecosystems are less likely to be impacted. Desert and arid systems will become increasingly vulnerable as populations grow in the northern regions of sub-Saharan Africa and the North American Southwest, among other locations.

In addition to significant spatial differentiation in the distribution of cities and consequent impacted biomes, there is temporal differentiation as well. Technology not only allows the founding of urban centers in previously unsettled areas but also changes the interactions of a city with its surroundings and the geographic region over which resources are acquired or wastes dispersed. Advances in transportation allow housing expansion in areas further removed from industrial or commercial centers; advances in architecture allow vertical rather than horizontal expansion of the city center. Figure 1 shows the development of Baltimore over the years and illustrates the impacts of technological advancement on that development.

Finally, the situation with respect to regional and global levels of development and trade can also profoundly influence the growth and structure of cities. Older cities can display the residual characteristics of their structures of a century ago—residential areas close to the city center, the robust industrial base—or skeleton of that industrial base—that fueled its urban growth. Newer cities, or newer sections of cities, can

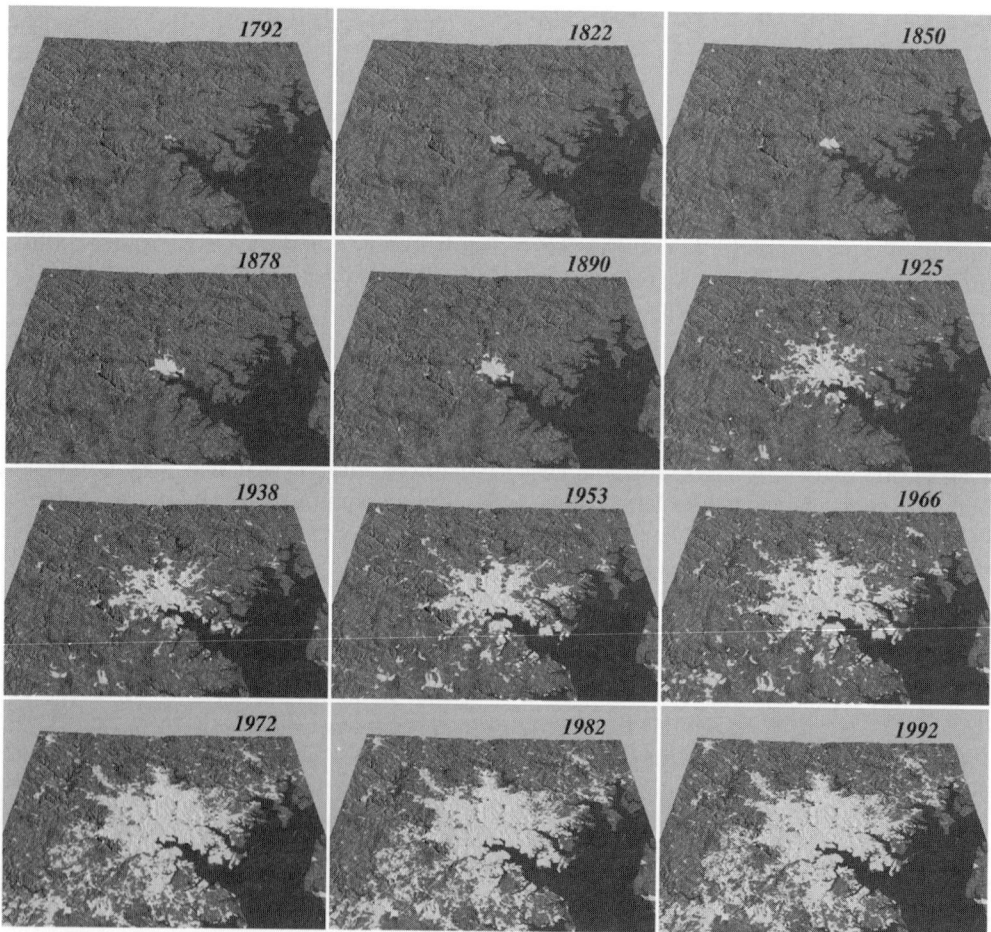

FIGURE 1 Perspective view of urban growth in Baltimore, MD over 200 years (1792–1992). Yellow polygons (light shading) are built-up areas as determined from historical maps and satellite imagery, green areas (medium shading) are forests and blue (dark) areas are water (Chesapeake Bay). Source: Penny Masuoka, UMBC, NASA Goddard Space Flight Center and William Acevedo, USGS, NASA Ames Research Center. See also color insert, Volume 1.

reflect the dramatic increase in commercial activity fueled by increasingly global and service-oriented industries and the desire for larger homes and more land that comes with economic development.

Once a human settlement is established in space and time, however, what are the impacts on biodiversity? What characteristics of both the prehuman ecological systems and the human social systems determine those impacts? In the next section we examine those mechanisms that allow multiple species to coexist and that determine those patterns of coexistence across the landscape before turning in Section V to an examination of the human social systems and their influences on human impacts on those patterns of biodiversity.

IV. THE BIOGEOPHYSICAL DETERMINANTS OF BIODIVERSITY

Before looking at the impacts of urban systems on biodiversity, it helps to first have at least a rudimentary understanding of what allows the coexistence and persistence of different types of organisms. Why are there so many species? Why aren't there more? Why do we see the types of species we do in a particular location?

Ecologists do not have a "unified theory" that explains patterns of biodiversity in all systems. Thus it is difficult to offer a complete set of causal mechanisms underlying those patterns. Nevertheless, we review some basic principles that influence the distribution and abundance of organisms in terrestrial systems, and then discuss in later sections the ways in which the organization and structure of urban systems—and the types of human activities taking place within them—influence these determinants of biodiversity and thus biodiversity itself. The following pages discuss in more detail six key concepts: (1) performance trade-offs; (2) environmental and resource heterogeneity; (3) disturbance regimes; (4) invasion dynamics; (5) interspecific interactions; and (6) spatial relationships.

Organisms face several different life "tasks"—to gather resources, to survive, to reproduce. Performing these tasks means *trade-offs*—resources allocated to seed cannot be allocated to leaves, for instance, and time spent hiding from predators cannot be devoted to gathering food. The ways in which organisms balance these trade-offs leads to a variety of different strategies for survival and reproduction, and thus different types—or species—of organisms. Consider the serpentine grasslands of Jasper Ridge, California. These hilltop systems are found in a Mediterranean climate, characterized by summer droughts and winter rains. Unlike most of the ecosystems of temperate North America, which exhibit growth in the summer and dormancy in the winter, the Mediterranean grasslands grow, flower, and set seed from late autumn to late spring and are dormant during the dry summer months. Serpentine soils are thin, and many of the plants rise a mere 10 cm from the grassland floor. Most of the 100 or more species found there are annual plants—growing, reproducing, and dying within a year.

Lepidium nitidum is an early bloomer. As soon as the autumn rains come, it uses the resources derived from photosynthesis to send out shallow roots, extracting water and nutrients needed for further growth from the near-surface layers of the soil. But it soon diverts those acquired resources to seed production, dispersing progeny, and dying a mere 3–4 months after the rains begin.

Contrast this evolved strategy to that of some of the longer-lived annuals. These send out deeper roots, waiting for the drier periods when most of its annual competitors—after drawing resources from shallow and intermediate layers of the soil to use for growth and reproduction—have already died. The deeper roots allow acquisition of moisture and nutrients in locations not reached by the more shallow-rooted plants and continue to garner resources for reproduction even during dry periods. These deep-rooted individuals set seed and die nearly a year after the rains begin.

The long-lived annuals may acquire more resources over their longer lifetimes than do the short-lived annuals, but they also must allocate more resources to rooting structures. Early senescence may be more favorable in a particularly dry year, while late senescence could flourish under generous late summer rains, but both strategies—and all the strategies in between—allow for successful competition for the resources required for reproduction, and thus coexistence.

These "performance trade-offs" and the resulting diversity of types are a feature of all ecosystems. Performance trade-offs—and the multitude of strategies that can evolve to balance these trade-offs—allow many different species to coexist in one ecosystem.

But not all strategies are equally viable under all conditions. The needlelike leaves of the temperate and boreal evergreen forests would not necessarily serve their wearers well in the Sonoran Desert. The water-loving plant that establishes in riparian zones would not do well under the more arid conditions that exist slightly uphill. Thus, performance trade-offs may allow many different species to emerge and coexist, but different conditions—from place to place around the globe, or from place to place within an ecosystem or landscape—impose constraints on the types of strategies and species that can persist there. Thus, *heterogeneities in resources or environments* influence patterns of biodiversity by creating conditions under which different types of organisms—employing different resource acquisition and reproduction tactics—are best suited to thrive.

We can see variations in resources and environments over both broad and small scales. Macroscopic features like mountains can affect climate by forcing air upward, causing the air to get colder and denser and thus "wringing out" the water vapor held within. This makes the upwind sides of mountains moist; the downwind side dry. But topology can play a significant role on smaller scales as well—the footprint that serves as a temporary wetland and home for water-loving bacteria, for instance. Similarly, soils and the nutrients they hold can vary from tropical to temperate to boreal regions, but can also vary as we move from tree trunk to root to root-free soil.

In addition to variability in resources and environments, *disturbance regimes* influence patterns of biodiversity as well. Frequent fires in the prairie, for instance, favor grasses that have allocated significant resources and reproductive structures belowground, where they can remain relatively unscathed by the fire. In the absence of fire, however, those grasses that have allocated more to aboveground reproductive or photosynthetic capacity might thrive. Disturbances also operate over a variety of spatial and temporal scales—from the 100-year floods or storms that devastate huge areas to the tree-fall or worm track that opens up new habitat on much smaller scales. (These examples illustrate another important principle—spatial and temporal scales are frequently related. Thus, large-scale disturbances occur relatively infrequently, small-scale disturbances more frequently.) In addition, disturbance—particularly large-scale disturbance—is frequently patchy. We can see this in large forest fires, which consume some patches while leaving others untouched. Thus, disturbance itself can create patterns of environmental heterogeneity that will influence patterns of biodiversity.

Invasion dynamics also shape communities and ecosystems. Ecosystems are continually "tested" by invading organisms—can an invader find a "niche" either by coexisting within the original complement of species or by outperforming another species, thus driving it to local extinction and creating a niche? The communities we see today are a result of these continuous invasions, a testing and retesting to find or create situations in which a species can survive and reproduce. Those systems subjected to longer or more continuous invasions may have filled more niches—"squeezing in" more types of organisms and strategies. Isolated systems, on the other hand, may contain organisms maladapted to the current environment but able to persist because they have never been tested against a more efficient, invading organism. In addition, the tests invading organisms face may change over time as community composition changes or as climate or other aspects of the disturbance regime change. Thus, a species that originally could not find a foothold may appear later, and one that thrived in an initial invasion may find itself the victim of changing circumstances. The impact of an invasive species on biodiversity levels will depend, in large part, on the interactions the invader encounters within its new home. Impacts can frequently be large, because invasive species often arrive without the diseases, pests, or competitors that keep them in check in their native environments.

Interspecific interactions can affect the complement of species in a given ecosystem. "Keystone" species are those that have a disproportionate influence on community composition and structure—their removal bodes dramatic consequences for biodiversity and the functioning of ecosystems. Less dramatic, but equally important for patterns of biodiversity, are those species that create niches for others—the beetles that burrow into bark, the mites that live on the feet of grasshoppers, the ferns freed from the surface when they colonize towering trees. Antagonistic interactions can influence biodiversity as well—the presence of a predator engenders new strategies for predator evasion; the evolution of a toxin-producing bacterium promotes evolution of strategies of defense.

Finally, *the area available* for species can influence levels of biodiversity. Ecologists have long observed that larger areas have more species and have been able to quantify this areal influence on species numbers through the species–area relationship. (The species–area relationship basically states that species number is proportional to the area available raised to the power z, where z usually falls between 0.1 and 0.35 and depends on the habitat and the taxonomic group of interest.) Thus, small islands in an island chain tend to have fewer species than large islands in the same chain, and small lakes in a region tend to have fewer species than large lakes in the same region. Connectivity among habitat patches can be important as well, allowing, for instance, replenishment of existing populations from other, separated populations that may serve as an important source of progeny or genetic diversity.

Humans have altered all of these biodiversity-determining conditions in urban, suburban, and rural systems, with consequent impacts on the patterns of biodiversity we see within and surrounding urban environments. A catalog or "snapshot" of those patterns of biodiversity is certainly possible—we could map the existing patterns of species from the city center to more pristine surroundings, and even compare the species lists and communities with those we surmise existed prior to the founding of the human settlement. Understanding those patterns, however, and developing predictions regarding past and future dynamics—and a comprehension of the types of policies or institutions capable of influencing those patterns—require much more than this mapping or snapshot approach. Instead, we must understand how the presettlement landscape influenced human settlement patterns, how human activities altered the landscape, and how these alterations further influenced the decisions and activities of the human inhabitants. We would also need to understand what changes in species distributions might have occurred even in the absence of human settlements—due to natural extinction or speciation events, for instance, or climatic variations. In other words, we need an *integrated* understanding of ecological and human-social dynamics.

Moreover, just as heterogeneity in climate, disturbance, soils, and community composition can affect patterns and dynamics of biodiversity, heterogeneity in human culture and community influences the evolving patterns of urban biodiversity. There can be significant differences in the ways in which city inhabitants and urban communities purposefully manipulate biodiversity or respond to changing conditions in ways that indirectly influence biodiversity. Understanding these differences requires that we further integrate knowledge of diverse socioeconomic and cultural conditions in our human ecological system, just as we must account for the diversity of strategies, species, and conditions inherent in the nonhuman ecological system.

V. HUMAN ECOLOGY AND SOCIAL DIFFERENTIATION

The study of cities from an ecological perspective in the United States began with Park *et al.*'s landmark publication, *The City*, in 1925, which formally introduced human ecology as a new research agenda for sociology and the study of cities in America. Their research focused on many of the social changes that had resulted at that time from the rapid expansion of America's urban areas due to the mass immigration of people from Europe and rural America. The explosive growth of the city, the confluence of people from diverse backgrounds, the breakdown of old ways, and the changes that were necessary for a viable new urban life caught the imagination of the authors.

Although Park *et al.* drew upon the work of European social and biological scientists such as Malthus, Darwin, and Spencer, the initial development of human ecology in America was influenced significantly by and contemporaneous with the emerging fields of plant and animal ecology in America. The Chicago School—as it came to be known—conceived of human ecology as an extension of the developing fields of plant and animal ecology.

The Chicago School articulated and developed an approach to human ecology that drew upon and paralleled earlier ecological work. First, Park and his colleagues applied a community ecology approach to the complexities of urban society in order to uncover a set of regular social patterns and processes in the apparent confusion of the urban melting pot. For instance, Park *et al.* employed ecological concepts such as succession, competition, and metabolism to describe stages of human community structure (organization) and function (processes): specifically, indicators of social disorganization such as disease, crime, vice, insanity, and suicide. Second, the Chicago School conceived of the city as a closed and functional system (community) that could be treated as an organism or "superorganism." Park and his colleagues also focused on the spatial and temporal dimensions of the city.

A significant product of this work was Burgess' ideal model of the city which the Chicago School used to describe and measure the city's spatial differentiation and development into zones and areas-within-zones through processes of concentration, centralization, segregation, invasion, and succession.

The Chicago School's conception of human ecology was criticized strongly by social scientists for several reasons, including opposition to the use of biological factors to explain individual human behavior and social structures; a singular reliance upon competition as the mechanism for explaining the organization of economic functions and the spatial distribution of human populations and services; and the use of macroscale processes and functional approaches to explain individual behavior from both a conceptual and a statistical point of view. Many of these criticisms were addressed in later studies in human ecology.

It is not enough, however, to treat human social systems as monolithic blocks. Just as performance trade-offs and the emergence of different strategies for resource acquisition and reproduction can influence the patterns of biodiversity on the landscape, differences in the structure and organization of human social systems need to be considered in order to understand how humans impact and manipulate the biodiversity surrounding them.

All social species are characterized to varying degrees by patterns and processes of social differentiation. In the case of *Homo sapiens*, social differentiation or social morphology has been a central focus of sociology since its inception. In particular, social scientists have used concepts of social identity (age, gender, class, caste, and clan) and social hierarchies (wealth, power, status, knowledge, and territory) to study how and why human societies become differentiated.

Social hierarchies—or social differentiation—is an important concept for understanding human ecological systems because it affects the *allocation of critical resources* (natural, socioeconomic, and cultural) and hence the patterns of and impacts on biodiversity. In essence, social differentiation determines "who gets what, when, how, and why." This allocation of critical resources is rarely equitable. Unequal access to and control over critical resources is a consistent fact within and between households, communities, regions, nations, and societies.

Wealth is access to and control over material resources in the form of natural resources, capital (money), or credit. The unequal distribution of wealth is a central feature of human ecological systems, and one that must be accounted for when examining the driving forces behind past changes in the environment and predicting the impacts of future policies and conditions. *Power* is the ability to alter others' behavior through explicit or implicit coercion. The powerful (often elites with political or economic power) typically have access to resources that are denied the powerless. *Status* is access to honor and prestige and the relative position of an individual (or group) in an informal hierarchy of social worth. Status is distributed unequally, even within small communities, and high-status individuals may or may not have access to either wealth or power. For instance, a minister or an imam may be respected and influential in a community even though he or she is neither wealthy nor has the ability to alter coercively other people's behavior. *Knowledge* is access to or control over specialized types of information (technical, scientific, religious, and so forth). Not everyone within a social system has equal access to different types of information. Knowledge often provides advantages in terms of access to and control over the critical resources and services of social institutions. Finally, *territory* is access to and control over critical resources through formal and informal property rights.

These types of hierarchies are important constraints that affect the allocation or supply of critical resources. It is also important to examine and understand variations in demand. For instance, additional social characteristics that may affect the demand for critical resources are often related to demographics or ethnic or religious backgrounds, since different age groups and various cultures may have different needs from or attitudes toward their environment and each other.

Processes of social differentiation of human ecological systems also have a spatial dimension that is usually characterized by patterns of territoriality that lead to spatial heterogeneity on many scales. This spatial understanding of social differentiation in an ecological context enables researchers to ask "who gets what, when, how, why, and *where*?" and, subsequently, to ask about the reciprocal relationships between spatial patterns and sociocultural and biophysical patterns and processes of a given area (Grove and Burch, 1997).

Today, it is increasingly difficult to determine where traditional biogeophysical ecology ends and human ecology begins (Golley, 1993). The articulation of a biosocial approach to human ecological systems has occurred over the past 20 years and continues from the development of and discourse between plant, nonhuman animal, and human ecologies and social sciences. Indeed this integration among ecological and social sciences is absolutely essential if we are to understand the functioning and dynamics of cities in an ecological landscape, and if we are to be able to both understand and predict the impacts of humans in urban centers on biodiversity.

VI. HUMAN IMPACTS ON BIODIVERSITY IN AND AROUND URBAN AREAS

Once an urban center has been established, human population densities and activities can affect biodiversity in many ways—both increasing and decreasing the variety of flora and fauna. These impacts span a variety of spatial scales. Thus, we can discuss the alteration of biodiversity in the urban centers or surrounding suburban regions, or examine the impacts at the landscape level due to the import and procurement of resources from surrounding rural regions to urban areas or the export of pollution or heat. Moreover, the effects on biodiversity span both ecological and evolutionary time scales. Finally, the intensity of impacts can vary—from the complete loss of species or the creation of entirely new ecosystem complexes to less obvious alterations in the genetic makeup of populations.

Recall those characteristics of the environment that could vary within and across ecosystems and influence community composition—resource availability, environmental heterogeneity, disturbance regimes, invasion dynamics, interspecific interactions, and habitat area and connectivity. Humans alter all of these determinants of biodiversity, and thus alter patterns of biodiversity within and around urban systems. In addition, humans create, alter, or destroy habitats in urban centers. Thus human influences on biodiversity fall into at least four categories:

1. Alteration of habitat
2. Alteration of resource flows
3. Alteration of disturbance regimes
4. Alteration of species composition (invasion dynamics and interspecific interactions)

Space precludes an exhaustive analysis of the extent and patterns of each of these impacts. Instead we briefly discuss each category below; a summary of these various categories is presented in Table I.

TABLE I
A Classification of Human Impacts That Affect Biodiversity in Urban Settings

Alteration of habitat
   Destruction of habitat
   Creation of habitat
   Fragmentation of habitat
Alteration of resource flows
   Reduction in net primary production
   Increase in regional temperature
   Mobilization and concentration of nutrients
   Dispersion of toxins
   Diversions of water and changes in timing of availability
   Degradation of water quality
Alteration of disturbance regimes
   Increase in small-scale disturbance
   Attempts to decrease large-scale disturbance
   Unintentional shifts in disturbance
Alteration of species composition
   Preferential cultivation or destruction of native species
   Introduction of nonnative species

A. Alteration of Habitat

Humans alter habitats in several different ways in and around urban settings. These alterations can roughly be grouped into the three categories of

1. Destruction of habitats
2. Creation of habitats
3. Fragmentation of habitats

1. Destruction of Habitats

Perhaps the most obvious destruction of habitat comes from the nearly complete elimination of the ecological community that existed prior to human settlement as the result of the paving of land or clearing for building or agriculture. Any species endemic to the converted areas would be lost and, by the species–area relationship, a reduction in the overall area of habitat would lead to a reduction in the number of species the remaining habitat could support. Habitat destruction could also lead to the loss of different populations within a species, and these lost populations may carry unique genetic information.

But human activities in and around urban settings destroy habitats in other ways. The vegetation structure in urban areas, for instance, can differ from the ecosystem the city replaced. Thus orchards can be depauperate in understory vegetation; similarly, many urban gardens lack primary producers that rise more than a few feet in height. This change in structure can affect the habitats available to small animals and birds, among other organisms. Similarly, the loss of many large predators from urban systems means the loss of any companion species—the mites, insects, or bacteria that may preferentially or obligatorily colonize the host.

The extent and pattern of habitat destruction will depend on the socioeconomic characteristics of the community in which land conversion is taking place and the power relationships between that community and nearby communities. Wealthy communities—frequently located at the urban fringes or suburban areas—are more likely to have the resources required to own and manage large residential lots and, in some cases, therefore allow natural vegetation to remain somewhat intact. These same resources, however, may be associated with increased applications of fertilizers and pesticides. Poorer communities frequently find themselves—in contrast—on the most infertile or poorly drained soils (the latter making these communities historically more susceptible to diseases such as cholera, malaria, and dysentery).

There have been few studies on how the rate of destruction of habitat depends on such characteristics as the age of the city, the stage of development of regional or global markets, the institutional arrangements for managing natural resources or urban expansion, or the biogeophysical characteristics of the region. Some ecologists working in urban settings, or along urban to rural gradients, have examined the differences in ecosystem structure that urban development can bring, but we do not yet have a comprehensive picture of the general patterns of those changes in structure or what those changes might mean for preferential selection or suppression of certain types of organisms.

2. Creation of Habitats

The transformations that accompany the creation of urban centers not only destroy habitat, they create habitat. Some of these new habitats are not necessarily unique to the urban environments—they are merely the result of any human settlement or presence. These types of habitats would include agricultural fields, gardens, barns, and houses. The impacts of these kinds of transformations and structures on the biological resources of the planet have been around for millennia and have, in some cases, led to the creation of new species or subspecies that are obligates for a human presence. Other urban habitats—such as sewage lines, landfills, and factories—have emerged in more recent times. Humans also create new ecological communities by mixing native and nonnative species in residential or commercial lots or parks and botanical gardens.

Some created "urban habitats" disappear as urban development progresses. Thus, raw surface sewage was not uncommon in English cities of the early nineteenth century; today, this habitat would be found primarily in the cities or urban fringes of those portions of the world lacking the resources required to ameliorate this problem. Other habitats, previously absent, appear in ever increasing abundance—witness the rise of golf courses over the latter half of this century. Finally, the economic development of a city and its surroundings can allow the return of previously destroyed habitat. Thus, we see regrowth of forests in the northeastern United States and in Europe, in part driven by more efficient agriculture and in part driven by a shift from agricultural to industrial to service economies in some regions.

3. Fragmentation of Habitats

Urban development fragments the landscape into patches of varying sizes, qualities, and land uses. The size and configuration of the patches will have important implications for organisms in the urban environment. Perhaps the most direct impact of that fragmentation is the elimination of species that require large contiguous ranges in which to obtain the resources required for survival. Thus, we see the elimination of much of the large-to-medium fauna in city centers and urban environments (except for humans and their associated pets). Further, some species must rely on dispersal to maintain populations in smaller or more widely separated patch types, as some patches may be too small to maintain indigenous populations. Thus, fragmentation may lead to shifts in species abundance that are related to dispersal characteristics. Migration routes may be adversely influenced by fragmentation of habitats as well.

Fragmentation also introduces the urban equivalent of "ecotones"—transition zones between pavement and yard, or a lawn and a garden, or a park and a commercial mall. The preponderance of edges in a highly fragmented landscape increases the abundance of edge-dwelling species and decreases the abundance of those species particularly sensitive to edge effects. Finally, fragmentation can affect the exchange of water, nutrients, and energy both within and across patches, leading to changes in the competitive outcomes among resident species.

Again, there have been few studies of the patterns of patch size, distribution, type, and quality in urban environments, how those patterns change over time and across urban to rural gradients, or how those patterns depend on the biogeophysical or socioeconomic characteristics of the region. Fragmentation may in general be more intense within the city proper and diminish as we move toward rural settings, but how does the city center compare to an urban–suburban transition zone with respect to distribution and quality of patches? Does fragmentation increase over time as more activities and structures are accommodated within the city boundaries, or does it decrease over time as economic growth allows greater landholdings and more preservation? And how will communities reassemble as patterns of fragmentation change—particularly if we have lost many species in response to fragmentation, climatic change, or other urban disturbances?

B. Alteration of Resource Flows

The availability of resources such as energy, temperature, nutrients, and water has profound influences on the distribution and abundance of species. Humans can and do alter all of these resource flows in and around the urban center. Biodiversity-impacting activities generally fall into six categories, as follows:

1. Reduction in net primary production
2. Increase in regional temperature
3. Mobilization and concentration of nutrients
4. Dispersion of toxins
5. Diversions of water and changes in the timing of water availability
6. Degradation of water quality

With respect to (1), conversion of habitat in urban regions frequently leads to a reduction in net primary productivity as vegetation is eliminated or thinned. Quite apart from the impacts of lost habitat area, the reduction in net primary productivity will impact the complexity of the plant and animal community and the number of species that can be supported given the available energy. Since all food transfers involve some loss of energy that cannot thus be put to productive purposes, the net primary productivity—or the amount of energy fixed by photosynthesizers and available to support nonphotosynthetic life forms—will determine the extent to which multiple trophic layers can be sustained within a food web. On the other hand, in some environments—such as the Sonoran desert of central Arizona—net primary production in irrigated and managed yards may exceed the primary production of the desert that used to be there.

The original capacity of the land to fix energy through photosynthetic activity is instead replaced or supplemented with the energy plants and combustion engines required to power residential, commercial, and industrial activity. There is also "waste heat" associated with these energy conversions, and this, in conjunction with the preponderance of heat-absorbing dark surfaces, leads to the well-known urban heat island effect, which can elevate city temperatures as much as 3°F above the surrounding countryside. Such an increase in average temperatures could prove detrimental to those nonsessile organisms living at the very edge of their optimum temperature range.

Human activities also mobilize and concentrate nutrients. The disruption of soils that accompanies land conversion and building can stimulate nutrient cycles and promote soil erosion—both can increase the nutrient loading of the river or groundwater draining the watershed in which a city is located. Further, there is a substantial import of nutrients from regions of agricultural productivity to the mouths of the city inhabitants. The waste generated by these inhabitants—and the nutrient loads they contain—can have significant impacts on the nutrient dynamics and balances of surrounding biomes. Similarly, the burning of fossil fuels for energy can lead to a "fixing" of nitrogen—conversion of atmospheric gas in the atmosphere (N_2) to nitrogen oxides, which are then deposited on surrounding ecosystems, increasing the nitrogen availability in those systems. In addition, through these wastes and through industrial processes, areas surrounding cities may experience elevated levels of elements usually present only in trace amounts—such as mercury—or that are completely foreign—such as plastics or manufactured pesticides.

Significant portions of the available flow of surface water can be diverted to support activities in urban regions. In the western portions of the United States, for instance, there are protracted and complicated political and social negotiations to determine allocation of water to cities versus agriculture versus "the environment"—basically all other flows that had previously occurred in the absence of human occupation. (Indeed, the water politics of the American West is a prime example of how power and wealth can determine ecological impacts within the human habitat.) The decisions made about water allocation may affect not only the quantity of water delivered to systems but the seasonal or diurnal timing of water availability as well.

Finally, water quality can also be degraded, through increased loading of sediments, nutrients, or toxics or through elevated temperatures (the result of first using the water for cooling in power plants or factories before returning it to downstream flow).

The impacts of these resource diversions or perturbations will depend on how they are distributed across the landscape. Biological communities simultaneously experiencing elevated levels of toxins in the soils, a reduction in net primary productivity, elevated temperatures, and increased nitrogen deposition will likely suffer more profound consequences than one in which, perhaps, only net primary productivity has been altered. Multiple stresses can have impacts that are more severe than one might expect from a consideration of the impacts of single stresses.

C. Alteration of Disturbance Regimes

On small scales, human activities in urban environments lead to a general increase in the frequency of disturbance. Lawns are clipped weekly, vegetation in parks is plucked or trampled daily, and buildings are razed and reraised over the years. This increased disturbance can lead to selection of "weedier," colonizing species at the expense of late-successional, slower-growing organisms. Increased frequency of disturbance can also favor species with shorter generation times. (Humans do also, however, tend longer-lived trees and perennials, thus potentially reducing the presence of weedier species—this influence is discussed below.)

On larger scales, in contrast, humans attempt to reduce the frequency and intensity of disturbances in and around urban settings. Thus, urban and suburban wooded or forested parks are frequently managed so as to minimize large fire events. Similarly, there are attempts to shelter coastal cities against the worst ravages of storms by building seawalls or planning the orientation of buildings and other structures. Floods are controlled, with varying success. This change in the disturbance regime can shift community composition relative to that which would emerge in the absence of the urban center—leading, for instance, to establishment of less fire-tolerant trees in urban forests.

Urban design can also change disturbance regimes in unintended ways. In Baltimore, the local climate has been changing—the precipitation from thunderstorms is occurring further from the city boundary, since storm fronts coming from the northwest are encountering warm air earlier—a result of the urban heat island. Similarly, the buildup of particulate matter in the lower atmosphere as a result of five continuous days of commuting has led to a greater frequency of weekend storms along the Northeastern seaboard—the particulate matter peaks by the end of the commute week and serves as a condensation medium for clouds.

D. Alteration of Species Composition

Humans intentionally and unintentionally manipulate the species compositions in urban settings. They do so indirectly through all of the other activities listed above, which can promote some species at the expense of others. In addition, however, there is

1. Preferential destruction or promotion of native species
2. Introduction of nonnative species

Humans have generally preferentially cultivated or destroyed species in urban settings based on aesthetics, values, or comfort. In hot environments, for instance, city residents might preserve and cultivate shade trees. There is also a clear aesthetic preference for flowering plants in most residential or commercial gardens. Residents may value preservation of native plants or feel strongly about providing habitat for songbirds. In contrast, most urban environments are depauperate in poisonous or prickly plants (the yards of cactus lovers aside), and there is in addition a general attempt to control "unappealing" organisms through the use of exterminators and traps. People's childhood homes, socioeconomic class, racial identification, and recreational preferences can all have a profound impact on those species that are deemed "desirable" versus "undesirable."

To understand these patterns of species portfolio management, ecologists will have to join with sociologists to examine how and why urban residents manipulate the neighborhood species complement the way they do. Such studies have been conducted in subsistence villages, where species manipulation is frequently a matter of survival, but fewer studies have examined the dynamics of aesthetic or cultural choices in high-population-density urban and suburban landscapes.

In addition to manipulating the indigenous suite of species, humans alter "invasion dynamics" through the intentional or unintentional introduction of nonnative species. On small, residential scales, aesthetic choices about landscaping can lead to significant introduction of nonnatives; similarly, the affinity for family pets can expose the small prey of a region to previously unknown predators. These aesthetics and affinities are very much influenced by culture and socioeconomic considerations. Immigrant communities may prefer vegetation reminiscent of their native homeland; families with children would prefer lawns to a more prickly groundcover; small lots are less likely to support large and complex vegetation structures than are large lots.

Finally, the ways in which humans manipulate species composition—or destroy or preserve habitat—depend on the value they place on biodiversity in general, or particular types of biodiversity in particular. Are habitats and species that reflect the prehuman landscape of a region prized most highly? Those that evoke memories of the childhood home? Are birds more highly valued than arthropods? Are rare species to be most highly valued, or certain combinations of species? These values will vary from person to person, town to town, country to country. They will play out differently at different levels of organization—from decisions made about species or habitat preservation at the federal level, to protection of native diversity at county levels, to recreation of preferred landscapes at the individual level.

VII. CITIES IN THE FUTURE

We cannot predict with certainty what cities will look like in the future. How will the economic structures of cities change in the context of regional and global markets? To what extent will changes in economy affect our stewardship of the biosphere? Will development bring with it not only the means for further destruction of biodiversity but also the means for more effective protection? How will information flow and people move on a daily basis with the advent of new technologies and changes in the availability of renewable and nonrenewable resources? How will ideologies change to either promote or hinder the protection of the other species with whom we share the planet? Will the political and social forces that determine the dynamics of a city and the impacts on the environment evolve to encompass a more integrated understanding of the city as part of a broader ecosystem?

It is difficult to predict how such changes will affect human settlement patterns; but it is evident that urbanization is a clear and dominant trend that will continue to characterize our species and the biodiversity of Earth. Ecologists have traditionally treated human social systems as separable from ecological systems and have traditionally treated anthropogenic impacts on biodiversity as an exogenous, rather than endogenous, perturbing force. But as an increasing number of the Earth's ecosystems come under the influence of these highly populated human habitats, and as an increasing number of species are influenced by the manipulations (either intentional or unintentional) of humans as they shape and respond to their surroundings, the scientific community will—we predict—discover that understanding the future of biodiversity on the planet—not just in cities—will require treating *Homo sapiens* as an interacting member of ecological communities, and not as a species that stands apart from them. Cities are an ideal environment for examining humans as integral parts of ecological systems, and therefore serve as an ideal proving ground for the theories of how human activities will influence biodiversity patterns in the future.

Acknowledgments

The authors thank James Collins, Simon Levin, and Gary Wade for their comments on the manuscript. For Morgan Grove, research support on the Baltimore Ecosystem Study was provided by the Burlington Laboratory (4454) and Global Change Program, Northeastern Forest Research Station, USDA Forest Service, the National Science Foundation (NSF Grant #DEB-9714835), and the Environmental Protection Agency (EPA Grant #R-825792-01-0).

See Also the Following Articles

ECOLOGICAL FOOTPRINT, CONCEPT OF • ENGERY USE, HUMAN • HUMAN EFFECTS ON ECOSYSTEMS, OVERVIEW • KEYSTONE SPECIES • LAND-USE PATTERNS, HISTORIC

Bibliography

Golley, F. B. (1993). *A History of the Ecosystem Concept in Ecology: More Than the Sum of the Parts*. Yale Univ. Press, New Haven, CT.
Grove, J. M., and Burch, W. R. Jr. (1997). A social ecology approach to urban ecosystem and landscape analyses. *J. Urban Ecosyst.* 1(4), 259–275.
Hough, M. (1984). *City Form and Natural Processes: Towards a New Urban Vernacular*. Van Nostrand–Reinhold, New York.
Park, R. E., Burgess, E. W., and McKenzie, R. D. (Eds.). (1925). *The City*. Univ. of Chicago Press, Chicago.
Pickett, S. T. A., Burch, W. R. Jr., Dalton, S., Foresman, T., Grove, J. M., and Rowntree, R. (1997). A conceptual framework for the study of human ecosystems in urban areas. *J. Urban Ecosyst.* 1(4), 185–199.
Pickett, S. T. A., and Ostfeld, R. S. (1995). The shifting paradigm in ecology. In *A New Century for Natural Resources Management* (R. L. Knight and S. F. Bates, Eds.), pp. 261–278. Island Press, Washington, DC.
Spirn, A. W. (1984). *The Granite Garden: Urban Nature and Human Design*. Basic Books, New York.

VENTS

Cindy Lee Van Dover
College of William & Mary

I. Midocean Ridges and the Hydrothermal Setting
II. Microbial Diversity
III. Invertebrate Diversity

GLOSSARY

autogenic species A species that creates habitat for other organisms; this habitat would not exist if the autogenic species was absent.

chemoautotrophic Primary production of organic matter using energy derived from chemical reactions rather than from photons.

dyke intrusion Upwelling magma; may reach the surface of the crust to flow out as lava and solidify as basalt.

end-member fluid In black smokers, the hot, chemically modified fluid that develops at depth in the crust and exits undiluted by seawater.

euphotic zone Sunlit upper layer of the ocean that can support photosynthesis.

heterotroph Organism that relies on organic material (other organisms, pieces of organisms, dissolved organic material) for its nutrition.

hydrothermal plumes Created by rising black smoker effluents that mix turbulently with the surrounding seawater, becoming dilute and eventually reaching neutral buoyancy (typically ~200 m above the black smoker source); once neutral, they spread laterally and are detectable by chemical and physical tracers for kilometers.

mid-ocean ridge Crustal accretion boundaries between major tectonic plates that make up the ocean crust.

phylotypes Molecularly differentiated taxa.

plume prospecting Use of chemical and physical anomalies in hydrothermal plumes to locate hydrothermal vents.

spreading rate The rate at which tectonic plates move apart from each other.

symbiont One of a pair of coexisting, mutually supportive organisms; in vent invertebrate–bacteria symbioses, the "symbiont" colloquially refers to the bacterium associated with the host invertebrate tissue; endosymbiont—within the host invertebrate (intra- or extracellularly); episymbiont—on the outer surface of the host invertebrate.

tectonism Forces that disturb and dislocate the Earth's crust (earthquakes).

vicariant event A historical event that isolates species' populations and leads to speciation (e.g., the closing of the Isthmus of Panama); usually considered in a geological context, but human activities also generate vicariant events.

IN THE CATALOG OF HABITATS ON EARTH, chemosynthetic ecosystems at deep-sea hydrothermal vents are singularly decoupled from climatic variations and from anthropogenic activities. Furthermore, chemosynthetic ecosystems are tightly coupled to the geophysical properties of tectonism and volcanism. As a consequence, one can look to tectonic history and volcanic periodicity as fundamental predictors of the boundaries of biogeographic provinces and of species richness within provinces. Chemosynthetic systems support endemic faunas, are globally distributed, and are insular habitats. In such putatively simple systems, hydrographic and topographic controls on biodiversity and biogeography of microbial and metazoan communities may be much more readily resolved than in systems where climate and human activities obscure their role. Systematic study of diversity in chemosynthetic systems within the context of planetary-scale processes will lead to an increased understanding of fundamental controls on biodiversity and biogeography within oceans (on Earth and other planetary bodies) where propagules are waterborne and subject to dispersal in an open system.

I. MIDOCEAN RIDGES AND THE HYDROTHERMAL SETTING

Although deep-sea hydrothermal vents occur in a variety of geological settings, including isolated back-arc spreading centers (such as the one that lies behind the Mariana Trench) and seamounts (e.g., Loihi, the incipient Hawaiian island), most hydrothermal systems lie along the network of volcanic midocean ridges that girdle the globe. Midocean ridges are the boundaries between the tectonic plates that make up the ocean crust; as such, they are the locus of episodic volcanism and earthquake activity associated with magmatic dyke intrusions and seafloor spreading. Most of the midocean ridge (and thus its hydrothermal systems) lies well below the euphotic zone, at depths greater than 2000 m. Where ridges intersect continents (e.g., at the head of the Gulf of California) or shoal to volcanic islands (e.g., Iceland and the Azores), hydrothermal systems may be found at shallower depths and even intertidally.

The geological pavement of deep-sea vents is typically hard basalt. Local relief of the basalt varies from flat, featureless sheets to smooth lobate surfaces a few to tens of centimeters high to rounded pillows of a meter or more relief. There are a small number of locales where sedimentation rate is in excess of volcanism, resulting in soft-sedimented bottoms with intruding basalt sills. The two best known sediment-hosted hydrothermal systems are in Guaymas Basin (Gulf of Mexico), where terrestrial runoff and pelagic production rates are high, and portions of the Gorda Ridge (off Oregon) and the northern Juan de Fuca Ridge (off Vancouver, Canada), where terrigenous inputs from the continental margin are significant.

Hot springs develop on midocean ridges where seawater can penetrate ~2 km into the porous basalt crust to reach hot rock. In this "reaction zone" at depth, seawater reacts with the basalt, losing all of its magnesium and oxygen and becoming enriched in metals and reduced compounds, including hydrogen sulfide. The resulting acidic, hot water is thermally buoyant and rises to the seafloor to exit as high-temperature (typically 350°C) fluids. As the vent fluid exits, it mixes with the cold, nearly neutral seawater and precipitates mineral sulfides of copper, iron, and zinc, forming substantial mineral chimneys known as black smokers. Where venting is persistent over time, chimneys coalesce to form larger, hydrodynamically and topographically complex sulfide edifices tens to hundreds of meters in diameter. Plumes emitted from black smokers rise ~200 m, until they become neutrally buoyant and spread out laterally. Hundredths of a degree deviations from ambient temperatures, elevated particulate concentrations, and modified chemistry allow water column geochemists to undertake "plume prospecting" by towing sensor packages at the depth of the neutrally buoyant plume. This prospecting method permits extensive mapping of ridge segments to yield integrated measures of hydrothermal output. Plume prospecting is also critical in locating new hydrothermal fields in unexplored ocean basins.

Not all of the vent fluid escapes as high-temperature fluid. Large volumes of warm, sulfide-laden water are emitted as "diffuse flow" vents where ambient seawater has mixed with the end-member fluid. It is this warm-water flow that sustains productive populations of free-living chemoautotrophic, thermophilic microorganisms (up to 115°C) and dense invertebrate populations (typically <40°C). These diffuse flows may issue from porous surfaces of sulfide-mineral structures or directly from cracks and fissures in basalt lavas and can contribute an order of magnitude more flow than the focused output of black smokers at some sites. Because there are large gradients in chemistry and temperature within vent ecosystems, there is a large potential for diverse microbial and invertebrate types.

Hydrothermal vent fields, typically consisting of multiple discrete zones of focused and diffuse flows, are the basic unit of hydrothermal activity on a ridge axis. Vent fields vary in size from hundreds to thousands of square meters and represent islands of productive habitat within an otherwise relatively barren, hard-substrate environment.

The tectonic plates that make up the crust do not all move apart at the same rate. Spreading rate is a function of magma supply, with fastest spreading centers (100–170 mm year^{-1}) associated with well-developed magma chambers and supporting a higher spatial frequency of hydrothermal systems than slow-spreading systems (moving apart at 10–50 mm year^{-1}). As will be seen below, spreading rate may have a profound influence on diversity and biogeography of vent invertebrate communities because of its control on the spacing between vents. Other features vary with spreading rate, including the relative longevity of venting sites and bathymetric isolation of near-bottom waters. The two end members of the spreading rate continuum include the slow-spreading Mid-Atlantic Ridge, characterized by long-lived (thousands of years) vent systems and deep, basinlike valleys within which vents are located, and the fast-spreading East Pacific Rise, where vents are relatively short-lived (months to years to decades) and occur on topographic highs.

II. MICROBIAL DIVERSITY

A. Free-Living Microorganisms

Deep-sea hydrothermal vent ecosystems are celebrated as sites where chemosynthesis by microorganisms sustains primary production and a huge biomass of primary and secondary invertebrate consumers. The generation of biomass in the system is driven primarily by the microbial oxidation of sulfide, which yields biochemical energy for the fixation of CO_2. Inasmuch as the oxygen used in sulfide oxidation comes from seawater and ultimately from photosynthesis, the larger part of the vent ecosystem cannot be said to be totally independent of sunlight.

The potential metabolic diversity of vent microorganisms is large, with both aerobic and anaerobic poises. Reduced sulfur substrates, iron, manganese, ammonia, methane, and hydrogen can all serve as electron donors, leading to sulfide, sulfur, and thiosulfate oxidation, iron and manganese oxidation, nitrification, methane and hydrogen oxidation, denitrification, sulfur and sulfate reduction, and methanogenesis. While microbial isolates may be induced to be strictly autotrophic in the laboratory, mixotrophy may be the rule in the natural environment, with assimilation of simple organic compounds (heterotrophy) whenever they are available. Physiological diversity of microorganisms may also be great, with microbial types optimized for life in extreme temperatures (perhaps to >115°C; superthermophiles), high temperatures (50–115°C; hyperthermophiles and thermophiles), intermediate temperatures (10–50°C; mesophiles), and cold temperatures (<10°C; psychrophiles). Microorganisms can be pressure tolerant or barophilic (unable to survive low pressure). Microbial habitats include low-temperature diffuse-flow environments, buoyant plumes, interstices of sulfide structures, and surfaces of rocks and organisms. There is also growing evidence for an expansive, deep subsurface biosphere within the porous upper layer of the ocean crust, wherever temperatures are compatible with life.

Diversity of microbial communities at the level of species or phylotype is one of the outstanding issues in microbial ecology. Isolates cultured from deep-sea samples may only deliver a few percent of the actual diversity within the system. Current research combines traditional culture methods with molecular characterization of natural communities. Within vent systems, microbial communities may be dominated by a single phylotype, as in the case of bacteria associated with mineral sulfides at black smoker chimneys on the Mid-Atlantic Ridge. Dominance by a small number of phylotypes is also observed at the Loihi Seamount hydrothermal vents. The dominance of microbial communities by a small number of taxa is reminiscent of species-abundance characteristics observed in studies of diversity in marine invertebrate communities from "extreme" environments. It is, however, too soon to conclude that there is any single pattern of microbial diversity in the free-living microbial populations at hydrothermal vents. An even greater unknown is the degree to which vent microorganisms express biogeographic patterns. The current dogma is that free-living bacteria have unlimited dispersal capabilities in the open ocean because they can drift indefinitely in an inactive state. Hydrothermal vents may prove to be an ideal system in which to test this hypothesis with rigor at a global scale.

Biotechnology industries have embraced the vent ecosystem as a source of novel microbial organisms and enzymes that operate under extreme conditions. The prospect of thermostable or barophilic enzymes in particular has led to systematic assays of deep-sea microbial communities for industrial applications. Where microorganisms live at meso- and thermophilic temperatures in organic-rich sediments, they are implicated in transformation and decomposition of freshly cracked organic material. Microbial hydrocarbon processors are obvious targets for extraction and design of enzyme systems by petroleum and environmental waste management industries. Microorganisms from vents are also screened for exceptional therapeutic properties. Biotechnology interests thus place an added emphasis on discovery and isolation of as many microorganisms as possible from vent ecosystems.

B. Bacterial–Invertebrate Symbioses

Deep-sea hydrothermal vents are distinguished from their shallow-water and terrestrial hot-spring counterparts by the hugely successful associations between chemoautotrophic, endosymbiotic microorganisms and their macroinvertebrate hosts. Endosymbiotic bacteria are now known from a number of deep-sea invertebrate groups, including the bivalve and gastropod mollusks, vestimentiferan and pogonophoran annelids, and sponges. It is the vestimentiferan tubeworms that exhibit the greatest anatomical and physiological accommodation of the symbionts. As adults, tubeworms lack a mouth and gut, having instead a special organ—the trophosome—derived from gut tissues and made up of host cells filled with bacteria, all bathed in hemoglobin-rich fluids. Symbiont acquisition in tubeworms is by ingestion from the environment when the worm is in its larval and/or postlarval stage and still has a digestive system. Is there more than one tubeworm symbiont species in the environment? Based on DNA–DNA hybridization techniques, the sulfide-oxidizing chemoautotrophic endosymbionts of two co-occurring species of vestimentiferans belonging to different genera (*Riftia pachyptila* and *Tevnia jerichonana*) were found to be identical. Geographically distant populations of *R. pachyptila* have been shown to host this same symbiont, although DNA fingerprinting can discriminate symbionts among populations. A geographically disjunct population of the vestimentiferan worm, *Ridgeia piscesae*, from within the same ocean basin also hosted this same symbiont. A second symbiont was discovered in the vestimentiferan *Lamellibrachia* sp., but, because this species occurs in the Gulf of Mexico, this may be a basin-scale difference in the distribution of infective symbiont species.

Vesicomyid clams found at hydrothermal vents also host chemoautotrophic bacteria and have nonfunctional guts. In contrast to the vestimentiferan tubeworms, symbionts are transmitted maternally in the clams and there is, accordingly, strict fidelity between a symbiont species and its host species. A map of phylogenetic relationships among species of vesicomyid clams is remarkably congruent with the phylogenetic map of their respective symbiont species, and is evidence of cospeciation of host and symbiont.

There is physiological diversity in microbial endosymbionts of vent taxa. Sulfide oxidizers predominate in vestimentiferan tubeworms, but there is at least one instance of a tubeworm that hosts a methanotrophic symbiont. Within bivalves, some symbionts use sulfide; others use the more oxidized and less toxic forms of sulfur, including thiosulfate or elemental sulfur. Several vent mussel species exhibit a dual symbiosis, harboring both methanotrophs and sulfur oxidizers. This diversity of symbiont capabilities within individuals increases metabolic flexibility and thus the range of environmental conditions under which the mussels may thrive.

Episymbiotic bacteria also occur in well-known associations with certain invertebrates. One of the best examples is the bacterial "fur coat" of the pompeii worm, *Alvinella pompejana*. This large polychaete worm lives in galleries on the sides of black smoker chimneys and experiences one of the steepest gradients of temperature along its body length of any invertebrate (40–50°C). The dorsal surface of *A. pompejana* is colonized by a morphologically diverse consortium of microorganisms, including large filamentous bacteria visible to the naked eye. The anatomical relationship between bacteria and worm is exquisite in its complexity, but the functional role of the members of the consortium remains unclear. Both trophic and detoxifying roles have been suggested. Episymbiotic, chemoautotrophic bacteria are also known from the chitinous cuticle of vent shrimp, *Rimicaris exoculata*. In this instance, the bacteria are thought to contribute to the diet of the shrimp; they are also likely to play some role in sulfide detoxification. Molecular characterization of the shrimp episymbionts shows them to be a single phylotype that matches the phylotype of the dominant free-living microbial species associated with the sulfide minerals on which the shrimp lives.

While there is much known and far more to be learned about microbial diversity at the interface between vent fluid and seawater, recent speculation about a vast, deep subsurface biosphere has fueled efforts to tap into the fluid-filled voids of the upper ocean crust. Efforts to understand microbial abundance, diversity, and function within such a biosphere may yield ancient lineages or otherwise novel species adapted to unusual conditions.

III. INVERTEBRATE DIVERSITY

It is the lush, gardenlike accumulations of invertebrate biota that make hydrothermal vent ecosystems a prized subject for television documentaries. Tubeworms often star in the title role, with clams and mussels as supporting actors, but this view of a simple ecosystem is little more than a Hollywood gimmick for audiences with short attention spans. The idea that vent systems have low invertebrate diversity is both pervasive and potentially misleading and is in large part due to the paucity of quantitative studies of diversity at vents. There is also a widespread failure to scale diversity measures at vents to the truly minute global acreage of vent habitat compared to extensive areas of soft-sediment deep sea or terrestrial rain forest, where diversity is so celebrated. These cautionary notes aside, there is an *a priori* reason to expect that diversity at vents might be low: Sulfide is a potent toxin, poisoning a major enzyme system of cellular respiration (cytochrome *c* oxidase) and bringing to a halt the aerobic production of ATP. Multicellular organisms that live at vents all must have some means of avoiding sulfide toxicity. This requirement may be a fundamental determinant of higher order invertebrate diversity at vents—only those groups that have efficient sulfide detoxification systems can occur where sulfide is present.

If we consider diversity at the phyletic level, vents are depauperate in several marine phyla normally found in the deep-sea benthos, including Porifera, Echinodermata, and the lophophorates (Phoronida, Bryozoa, Brachiopoda). The scarcity of sponges (Porifera) and echinoderms is attributed to the lack of excretory and blood vascular systems in these groups and a consequent inability to tolerate elevated levels of sulfide or other toxic compounds associated with hydrothermal vents; where these phyla are represented at vents, the animals are associated with the periphery of a vent field or other circumstances where sulfide levels are likely to be low (as in a waning vent site). The colonial motif and asexual reproduction are also apparently unsuccessful at vents. Colonial coelenterates such as gorgonians, antipatharians, and hydroids are common animals of the deep sea but so far are virtually unknown at vents. An exception is a colonial siphonophore (distantly related to the Portuguese Man o' War) that is often found in large numbers at dying vents.

A number of hypotheses have been put forth to explain the absence of colonial organisms at vents, including intense competition and the ephemerality of vents, which may preclude any advantage of vegetative reproduction. Colonial organisms would also be at a disadvantage by virtue of the connectivity between individuals, since exposure of one individual within the colony to sulfide jeopardizes the aerobic metabolism of the entire colony. Encrusting colonial species are difficult to posit, because the vent effluent is buoyant and would tend to lift off recruiting polyps. Further, if established, an encrusting species could cap the flow, resulting in either diversion of the flow to less resistant paths or pooling of the flow with concomitant, life-threatening elevation of sulfide concentrations and temperature. While strategies to cope with these kinds of physical and chemical challenges to colonial organisms are not inconceivable, they have so far not been observed.

A. Origins of Vent Invertebrate Taxa

The majority of species that occur at vents (~90%) are known only from vents. This high degree of endemicity is difficult to prove, given the lack of effective means of (and interest in) collecting organisms from nonvent, deep-sea basalts. Nevertheless, it is a distinctive fauna. Where does it come from? There is no single source. Some species belong to nonvent deep-sea genera (e.g., the squat lobster, *Munidopsis lentigo*); others have closest relatives among the shallow-water invertebrates (e.g., the polychaete *Ophryotrocha*). Several species, notably symbiont-containing invertebrates (tubeworms, clams, and mussels), have closest alliances to species known from other chemosynthetic ecosystems, such as brine and hydrocarbon seeps and whale skeletons. Some vent families (e.g., scale worms in the polychaete family Polynoidae) are broadly represented in the marine environment. Still others appear to be specialized taxa, known only from hydrothermal systems—the polychaete family Alvinellidae is a good example. Finally, some species appear to be most closely related to ancient Paleozoic or Mesozoic taxa, relict lineages that have found refuge in hydrothermal systems. These include several stalked barnacle species and an archaeogastropod limpet. Some taxa routinely found at vents have been hugely successful, undergoing substantial radiation. These include the siphonostome copepods, the archaeogastropod limpets, and the polynoid polychaetes; within each of these groups there are now dozens of described species.

B. Biogeography

When vents were first discovered in the eastern Pacific, there was much speculation about how many vents there might be, where they were, and whether the fauna would be cosmopolitan. We know now that there are likely to be hundreds, if not thousands, of hydrothermal vent fields along the midocean ridge system and that they can be found in every ocean basin. Furthermore, we know that the fauna is not cosmopolitan. Distinct biogeographic provinces occur, defined in part by the tectonic fabric and history of the ocean plates. Along a continuous ridge system such as the East Pacific Rise, which stretches from the head of the Gulf of California down to high latitudes in the Southern Hemisphere, the fauna of the hydrothermal vents undergoes subtle changes. The same species of tubeworms, mussels, and clams generally dominate the biomass (mussels are absent from the northernmost locales), but the details of their relative abundance change from site to site. These dominant taxa (especially the tubeworms and mussels) are autogenic bioengineers, creating complex three-dimensional habitat that becomes occupied by countless individuals of crustacean, mollusk, and polychaete species. We know little about how the ranges of the hundreds of smaller invertebrate species that live in association with these autogenic taxa vary along the length of the East Pacific Rise nor of how these ranges relate to the life history characteristics of each species.

At its northern limit in the Gulf of California, the East Pacific Rise goes terrestrial as a strike-slip fault (the San Andreas), reemerging in the submarine environment off northern California as the Mendocino fracture zone. From there, the spreading ridge axis between the Pacific and North American plates heads northward as the Gorda, Juan de Fuca, and Explorer Ridge systems. The vent fauna of these northern ridges is distinct from that of the East Pacific Rise at the species level, but there are alliances at the generic and familial levels. Verena Tunnicliffe (University of Victoria) notes that at one point in the geological history of these ocean plates, the East Pacific Rise and northern ridges were one continuous system. Overriding by the North American Plate was a vicariant event, bisecting the ridge system and resulting in subsequent development of the vent faunas in isolation and of two distinct, formerly related, biogeographic provinces.

Other biogeographic provinces are known: Deep-sea hydrothermal vents at back-arc spreading centers associated with western Pacific microplates support a distinctive invertebrate fauna at the species level, although at higher taxonomic levels, affiliation to the two eastern Pacific biogeographic provinces is patent. This affiliation may be related to the direct link that existed between the eastern and western Pacific via the now extinct Kula Ridge more than 40 million years ago. Deep-sea vents along the Mid-Atlantic Ridge support an even more disparate group of species and genera. Based on biogeographic characterization of these four major geographic regions, we expect that ridge systems in unexplored ocean basins (the Arctic and the Indian Oceans in particular) will yield hundreds of new species. Exploration and discovery remain hallmarks of the study of diversity and biogeography at vents.

C. Gene Flow and Genetic Diversity

Deep-sea hydrothermal vents on midocean ridges are distributed as linear, insular, and ephemeral arrays of endemic species. To persist in the face of certain local extinction, survival of a species in any region is only possible where propagules can disperse to other vents. An understanding of gene flow along these arrays is essential if we are to understand species distributions and speciation processes at vents. Two basic models of gene flow, borrowed from island biogeography, have been applied to genetic data from vent populations. In the "stepping-stone" model, gene exchange occurs locally, between neighboring populations, and genetic exchange decreases with increasing distance. In the "island" model, there is thorough mixing of gene pools over a regional scale. Vent taxa conform to no single model of gene flow. Tubeworms on the East Pacific Rise, for instance, appear to follow a stepping-stone model, while mussels on the same ridge follow the island model. Still other species display a ridge-based isolation model, where topography of the ridge system (e.g., deep east–west fracture zones that offset the shallower north–south-trending ridge) creates a genetic filter. Studies of genetic diversity in vent systems by Robert Vrijenhoek (Monterey Bay Aquarium Research Institute) and his students highlight the significance of the number of populations in sustaining genetic diversity: vent species with abundant populations have higher genetic diversity than species with fewer populations. Genetic diversity in vent organisms (measured as percentage of polymorphic loci within a population) is also correlated with the order of establishment of species at nascent vents, with early colonists having double the genetic diversity of later colonizing species.

D. Biodiversity

For two decades, basic community descriptors of species richness and diversity eluded vent ecologists. Sampling was qualitative, with uneven efforts, and typically resulted in lists of species for vent sites without regard to distributions within specific microhabitats. An exception is the work of Fred Grassle (Rutgers University), who compared diversity of soft-sediment vent infauna with infaunal diversity in the surrounding soft-sediment, nonvent environment. His quantitative data demonstrated that diversity in vent muds is low compared to the relatively high species diversity of the "normal" deep sea. Further, at vents, one or two species account for 70–90% of the infaunal individuals, whereas in the nonvent deep sea, abundances are more evenly distributed among species, with the most common species making up less than 20% of the total. Similar results were found in studies by others of soft-sediment meiofauna. But soft sediments are the exception rather than the rule for deep-sea hydrothermal vents.

What about hard-substrate biodiversity at vents? Species richness measured at a site on the Mid-Atlantic Ridge based on replicate samples within a single vent habitat (mussel beds) and species–effort curves has been compared to values obtained using a comparable sampling effort at an intertidal mussel bed in south-central Alaska. The intertidal, photosynthetic site supported more than twice as many species than the deep-sea chemosynthetic site, consistent with the idea that diversity at vents is low. But diversity in intertidal mussel beds is well-known to be geographically variable; some intertidal mussel beds (e.g., along the coast of Japan) are reported to have species richness values comparable to what is found at the Mid-Atlantic Ridge vent site.

Vents allow us to study patterns of regional diversity over different scales of spatial frequency of habitat and in the absence of confounding climatic and anthropogenic effects. The Mid-Atlantic Ridge is a slow-spreading ridge system, with vents spaced relatively far apart (hundreds of kilometers). Diversity is predicted to be higher on faster spreading centers where vents are more closely spaced. With the same sampling methods as used at the Mid-Atlantic Ridge vent site, diversity has been measured at two vent fields on the ultrafast spreading southern East Pacific Rise. Diversity at these vents is comparable to that of the Alaskan intertidal mussel bed. Thus, whether one considers diversity at vents to be high or low depends only on which end members one chooses to compare.

Of theoretical importance is the fact that patterns in diversity at midocean ridge hydrothermal vents may be controlled by fundamental properties associated with plate tectonics and volcanism that determine the spatial frequency of vent habitat. Where vents are closely spaced, the likelihood of extinction is relatively low, and species tend to accumulate. Where vents are far apart, a premium is placed on effective dispersal, extinction becomes much more likely, and allopatric speciation is enhanced. This hypothesis leads to the prediction that slow-spreading systems, with distant vents, support a greater number of biogeographic provinces, each with lower species diversity than fast-spreading systems with closely spaced vents.

Area–spreading rate relationships have also been proposed as controls on species diversity at vents. In general, the well-known area–species relationship is driven by increasing numbers of habitats with increasing area. While presumably relatively unimportant when comparing diversity within habitat types (e.g., mussel beds), the increasing vent area (and, by inference, habitat diversity) with spreading rate should contribute to increasing species diversity within vent fields. Quantitative assessment of species diversity within all representative microhabitats of vent fields and measures of areal extent of fields along ridges of different spreading rates is not a tractable approach at this time.

See Also the Following Articles

ARCHAEA, ORIGIN OF • HIGH-TEMPERATURE ECOSYSTEMS • MARINE ECOSYSTEMS • MICROBIAL BIODIVERSITY • THERMOPHILES, ORIGIN OF

Bibliography

Jollivet, D. (1996). Specific and genetic diversity at deep-sea hydrothermal vents: An overview. *Biodivers. Conserv.* **5**, 1619–1653.

Juniper, S. K., and Tunnicliffe, V. (1997). Crustal accretion and the hot vent ecosystem. *Philos. Trans. R. Soc. London, A* **355**, 459–474.

Karl, D. M. (Ed.). (1995). *The Microbiology of Deep-Sea Hydrothermal Vents.* CRC Press, Boca Raton, FL. Includes review articles on the geological and geochemical setting, free-living microbial communities, symbioses, thermophiles, microbe–metal interactions, plume microbiology, and biogeochemical cycles.

Gebruk, A. V., Galkin, S. V., Vereschaka, A. L., Moskalev, L. I., and Southward, A. J. (1997). Ecology and biogeography of the hydrothermal vent fauna of the Mid-Atlantic Ridge. *Adv. Mar. Biol.* **32**, 93–144.

Humphris, S. E., Zierenberg, R. A., Mullineaux, L. S., and Thomson, R. E. (Eds.). (1995). *Seafloor Hydrothermal Systems: Physical, Chemical, Biological and Geological Interactions.* American Geophysical Union, Washington, DC. Heavily oriented toward the physical, chemical, and geological contexts, but with some outstanding reviews of vent biology/ecology.

Peek, A. S., Feldman, R. A., Lutz, R. A., and Vrijenhoek, R. C. (1998). Cospeciation of chemoautotrophic bacteria and deep-sea clams. *Proc. Natl. Acad. Sci. USA* **92**, 7232–7236.

Tunnicliffe, V., McArthur, A. G., and McHugh, D. (1998). A biogeographical perspective of the deep-sea hydrothermal vent fauna. *Adv. Mar. Biol.* **34**, 353–442.

Van Dover, C. L. (1995). Ecology of Mid-Atlantic Ridge hydrothermal vents. In *Hydrothermal Vents and Processes* (L. M. Parson, C. L. Walker, and D. R. Dixon, Eds.), pp. 257–294. Geological Society, London. Special Publication 87.

Van Dover, C. L. (2000). *The Ecology of Deep Sea Hydrothermal Vents.* Princeton Univ. Press, Princeton, NJ. A comprehensive overview of geology, chemistry, microbiology, trophic and reproductive ecologies, evolution and biogeography, and community dynamics of vent communities, with chapters on cognate communities and the relevance of vent ecosystems to the origin of life and astrobiology.

Vrijenhoek, R. C. (1997). Gene flow and genetic diversity in naturally fragmented meta-populations of deep-sea hydrothermal vent animals. *J. Hered.* **88**, 285–293.

VERTEBRATES, OVERVIEW

Carl Gans* and Christopher J. Bell†
*Department of Integrative Biology, University of Texas at Austin and †Department of Geological Sciences, University of Texas at Austin

I. Introduction
II. General Vertebrate Characteristics
III. Early Chordate and Vertebrate History
IV. Vertebrate Classification
V. Definitions and Diagnoses of Major Chordate Groups

GLOSSARY

chordate A member of the group Chordata. The Chordata includes the most recent common ancestor of tunicates and cephalochordates and all of that ancestor's descendants. Tunicates, lancelets, hagfishes, and vertebrates are all chordates.

ectoderm An embryonic tissue that provides the future outside layer of the animal.

ectothermy A method of body temperature control in which the animal utilizes external sources for gaining and giving up heat, thus achieving temperature control without affecting metabolic rate.

endothermy A method of body temperature control in which the animal modifies its metabolic rate to achieve the desired body temperature.

neural crest An embryonic tissue intermediate between neurectoderm and ectoderm, with cells migrating widely to their final destination. This tissue gives rise to anterior skeletal elements, many portions of the future head and pharynx, and all pigment cells. Sometimes also referred to as mesectoderm.

neurectoderm An embryonic tissue that gives rise to the central tube of the nervous system.

notochord A stiff, flexible, longitudinal rod running along the middorsal portion of the chordate body. It is situated dorsal to the coelom and ventral to the central tube of the nervous system.

pharynx The anterior portion of the alimentary canal, characterized by lateral buds that provide skeletal support for the gill region.

tuberculum interglenoideum An anterior projection of the first (cervical) vertebra in salamanders. The tuberculum interglenoideum bears articular facets that insert into the foramen magnum of the skull and provide additional articulation points between the skull and the vertebral column.

VERTEBRATES INCLUDE ALL the fishes, amphibians, reptiles, birds, and mammals. These animals are united in a more inclusive group, the Chordata, that includes the closest living relatives of vertebrates, the hagfishes, lancelets, and tunicates. There are approximately 54,450 known species of chordates, over 51,000 of which are classified as members of Vertebrata. Nearly half (approximately 24,000) of the known species of vertebrates are members of a single group of ray-finned fishes, the actinopterygians. The vertebrates are found on all major land masses and in all major oceans and seas on Earth.

I. INTRODUCTION

The vertebrates include most of the major groups of animals that humans encounter and interact with on Earth, and in common parlance, the term "vertebrate" is often equivalent to "animal." Vertebrates include all the fishes, amphibians, reptiles, birds, and mammals; they are found on every major landmass and in all of the world's oceans and major seas. The fossil record of the group is excellent and extends back to the late Cambrian period of the early Paleozoic, more than 500 million years ago. The vertebrates are members of a nested series of more inclusive groups, extending all the way down the tree of life to a common ancestor they share with bacteria, the most primitive life forms known. In order to properly understand the vertebrates, it is necessary not only to briefly discuss the characteristic features unique to the group but also to evaluate the features they share with their closest living relatives, the hagfishes, lancelets, and tunicates. These organisms plus vertebrates constitute the group Chordata. In the pages that follow, we summarize some generalized characteristics of chordates and vertebrates, examine the early history of these groups as it is revealed from the study of the fossil record, and provide explicit definitions and diagnoses for all of the major chordate groups.

II. GENERAL VERTEBRATE CHARACTERISTICS

The diversity among vertebrate groups makes it possible to utilize functional analyses and paleontological data to reconstruct a phylogenetic sequence of vertebrates and to place the currently surviving (extant) vertebrate groups into a historical sequence. This sequence permits the development of explanatory schemes for some conditions observed in humans. Consequently, the study of the comparative morphology and physiology of vertebrates has proved important for the understanding of the human condition. These studies in turn led to the development of curricular variants for students training themselves for careers in human medicine. Examples are approaches to applied physiology and morphology of the precursors' conditions. Such instructional variants may be presented as the sequential development of the history of a topic such as gas exchange, digestion, or embryology. Alternatively, the study may involve a series of steps, each explaining the condition in one of the subgroups of animals representing putative precursors for the vertebrate condition (i.e., basal chordates or echinoderms), or historical stages in the development of groups of vertebrates (i.e., fishes, reptiles, and mammals).

Although a few arthropods and mollusks achieve large, indeed giant, sizes, most of the largest species, whether terrestrial, aquatic, or aerial, are vertebrates, and this latter group includes the largest animals known. However, some vertebrates are small, almost tiny, and this poses interesting questions of how they perform the basic functions of reproduction, locomotion, respiration, food processing, and waste discharge. Vertebrate size reduction involves two strategies: (1) maintenance of the adult pattern at a reduced body size and (2) the acquisition of metamorphosis, meaning that the life processes, and the structures that facilitate them, change markedly at one or more stages and sizes of development.

Vertebrates may be visualized as elongate, tubular creatures possessed of a central tubule, this being a gut or alimentary canal. The anterior end of the gut forms a mouth and buccal structures, whereas the posterior end is modified for the storage and ultimate elimination of waste. The central tubule is placed in a lined cavity, the coelom, that also houses various diverticulae of the gut that support aspects of such visceral functions as gas exchange (i.e., gills and lungs, heart, and vessels), digestion (pancreas and liver), fluid balance (kidneys), and reproduction (gonads). The entire animal is coated by a multilayered epidermis that protects the internal tissues from mechanical stresses and can serve as a physiological barrier. The space between coelom and epidermal layer is occupied by the somatic musculoskeletal system that facilitates locomotion and supports and maintains the shape of the coelom and hence of the entire animal.

The musculoskeletal system involves the cartilages and bones of the internal skeleton and the striated somatic musculature. Externally, the system underlies the epidermis, this portion forming a dermis having skeletogenic properties. Two additional median and tubular structures complete this architectural scheme. The first tube, the notochord, lies just dorsal to the gut and is filled with vacuolated cells, their turgor pressure turning the notochord into a stiffened skeletal rod. The second, still more dorsal, tube consists of neuronal cells that involve sensory function and transmission of signals to the musculature.

The entire animal may have various appendages such as fins or limbs, each having characteristic internal skeletal supports and motile capacity developed by the contraction of the attaching striated musculature, which is subdivided into myotomes, paired blocks that parallel the notochord. The notochord provides a capacity for resisting longitudinal stresses, while the appendages involve support of the animal above the substrate and the transmission of laterally directed force patterns.

Whereas this general pattern seems to characterize vertebrate structure, the group shows major curiosities. The various groups of nonvertebrate creatures can generally be diagnosed by a relatively few characteristics. In contrast, there are no fewer than 25 new characteristics shared by vertebrates (Table I). These involve multiple and major physiological systems, for instance the multiple senses, the head skeleton, the vertebral column, and the digestive system. This clear diversity of vertebrate specializations has long raised questions about the adaptive reasons to explain why these comparable characteristics are modified in vertebrates and why and how a multiplicity of such characteristics might be associated and consequently subject to possibly simultaneous selection. Actually, the answer to this conundrum has been available for some 30 years. Rather than looking at the functions of the many "vertebrate characteristics," namely, aspects seen in the adult condition, it is more informative to base comparisons on the stages of vertebrate embryology or development. The traditional rules that posited firm conditions, such as that the ectoderm formed all the epidermis and the mesoderm all of the striated musculature, soon broke down. In particular, the neural crest, an embryonic tissue intermediate between the ectoderm and the neurectoderm, proved to be the source of many diverse tissues. Similarly, the neurogenic epidermal placodes formed a set of embryonic tissues that provided a source for many shared vertebrate characteristics (Table I). The most interesting aspect was the discovery that many otherwise indistinguishable adult tissues could be derived from different embryonic sources, depending on their position in the embryo. Thus cephalic striated muscle has a histological pattern equivalent to that of the striated muscle of the trunk but is of neural crest rather than mesodermal origin. Similarly, the bones of the skull are derived from neural crest and the more posterior and serially equivalent vertebral bones are of (sclerodermal) mesodermal origin.

Consequently, the vertebrate characteristics either are derived from neural crest of neurogenic epidermal placodes or are induced by these tissues. In contrast to searching for 25 functional issues, the developmental approach makes it possible to recognize two or three associated aspects and note that these provided the cause of the vertebrate condition.

TABLE I
Derived Characters of Vertebrata and Their Embryonic Origin[a]

Vertebrate characters	Embryonic origin
Nervous system	
Sensory nerves with ganglia, cranial	NC, P
Sensory nerves with ganglia, trunk	NC
Peripheral motor ganglia	NC
Second- and higher-order motor neurons	NC
Forebrain	NC?
Chromatophores	NC
Paired special sense organs	
Nose	P
Eyes (accessory organs)	NC?, (P)
Ears	P
Lateral-line mechanoreceptors	P
Lateral-line electroreceptors	P
Gustatory organs	NC?, P
Pharyngeal and alimentary modifications	
Cartilaginous bars	NC
Branchiomeric muscle	ME
Smooth muscle of gut	MH
Calcitonin cells	NC
Chromaffin cells, adrenal cortex	NC
Circulatory system	
Gill capillaries, endothelium	ME
Major vessels, trunk	M
Wall of aortic arches	NC
Muscular heart	MH
Skeletal system	
Anterior neurocranium and sensory capsules	NC
Cephalic armor and derivatives	NC
Armor of trunk	NC??, ME?

[a] M, mesoderm; ME, mesoderm, epimere; MH, mesoderm, hypomere; NC, neural crest; NC?, intermediate between neurectoderm and neural crest; P, placodes; (P), placodes provide lens and peripheral component, but not sensory tissues. Origin of the various trunk armors is uncertain. Table modified from Gans (1993).

III. EARLY CHORDATE AND VERTEBRATE HISTORY

Vertebrates are members of a more inclusive group of animals, the chordates. The history of both chordates and vertebrates extends back into the Cambrian period in the early Paleozoic. Although the fossil record of early chordates is generally quite poor, several fossil localities have produced fossils that are likely to represent the remains of early chordate animals. Controversy over the interpretation of anatomical features of these animals is in part due to the fact that the important characteristics by which we recognize chordates are soft anatomical features that are not easily fossilized. Exquisite preservation of soft-bodied organisms in the Burgess Shale in Canada (middle Cambrian, approximately 530 million years ago) resulted in the discovery of several specimens of a chordate-like animal named *Pikaia gracilans*. These important fossils appear to display major chordate structural features, including myotomes (segmented muscle blocks) and a stiff, longitudinal rod running along the middorsal portion of the body, the notochord. Some controversy regarding the precise phylogenetic position of *Pikaia* persists due to a lack of adequate description of these fossils in the primary literature. Recent discoveries in early Cambrian deposits in China stimulated renewed interest in the origin and diversification of chordates. *Yunnanozoon lividum* was described as an early representative of the cephalochordates (Chen et al., 1995; Chen and Li, 1997), and *Haikouela lanceolata* was recently identified as a potential craniate from the early Cambrian (Chen et al., 1999) but these interpretations are controversial (Shu et al., 1996). The earliest fossils attributed to the Vertebrata are enigmatic bits of bone from the late Cambrian described under the name *Anatolepis heintzi*. The vertebrate affinity of these fossils was challenged on various grounds, but recent analyses showed they contain dentine, a form of the calcium phosphate mineral apatite (the major mineral component of bone) that is found only in vertebrate skeletal tissues (Smith et al., 1996).

Homoiosteles and stylophorans, Paleozoic animals with a highly asymmetrical body form and a calcite skeleton resembling that of echinoderms, are another problematic extinct group. Fossils range in age from middle Cambrian to middle Devonian. The body can be divided into two distinct regions: a compact, asymmetrical region and a long segmented appendage. Interpretation of these fossils as "calcichordates," a distinct assemblage of stem-group chordates (see Jefferies, 1986, 1997) is controversial. Most authors treat them as aberrant echinoderms, and the controversy over their affinity is based in part on interpretations of orientation and locomotion (see Parsley, 1997).

IV. VERTEBRATE CLASSIFICATION

The classification of chordates has a long and complicated history. Traditionally, the vertebrates were divided into fishes, which occupied aquatic environments, and the tetrapods, which mainly occupied terrestrial habitats. At present, there are two major classification systems in use. The Linnaean classification system was first published by Linnaeus about 1758 and organizes major divisions of life into kingdoms (Plantae, Animalia, Fungi, etc.), with major body plans organized into phyla (singular: phylum). The basic ranks descend from phylum to class, order, family, genus, and species. This classification system is familiar to most teachers and students and remains the dominant classification taught to elementary and high school students in the United States (Fig. 1). In the past several decades, a nonranked classification system was described by Hennig, a European entomologist. This classification system, termed "phylogenetic systematics" or "cladistics," organizes taxa into nested sets of monophyletic groups based on common ancestry (a monophyletic group contains an ancestor and all of that ancestor's descendants). A classification system based on natural groupings of taxa (based on ancestor–descendant relationships) is thus achieved. Phylogenetic systematics is rapidly becoming an important alternative classification

Kingdom Animalia
    Phylum Chordata (chordates)
        Subphylum Urochordata (tunicates)
        Subphylum Cephalochordata (lancelets)
        Subphylum Vertebrata (hagfish + vertebrates)
            Class Myxini (hagfishes)
            Class Cephalaspidomorphi (lampreys)
            Class Chondrichthyes (chimaeras, sharks, skates, rays)
            Class Osteichthyes (bony fishes)
                Subclass Actinopterygii (teleosts, etc.)
                    (many orders)
                Subclass Sarcopterygii (lobe-finned fishes)
                    Order Crossopterygii (includes coelacanths)
                    Order Dipnoi (lungfishes)
            Class Amphibia (amphibians)
                Order Gymnophiona (caecilians)
                Order Anura (frogs and toads)
                Order Caudata (salamanders)
            Class Reptilia (reptiles)
                Order Testudines (turtles)
                Order Rhynchocephalia (tuatara)
                Order Lacertilia (lizards)
                Order Serpentes (snakes)
                Order Amphisbaenia (amphisbaenians)
                Order Crocodylia (alligators and crocodiles)
            Class Aves (birds)
                (many orders)
            Class Mammalia (mammals)
                Order Monotremata (echidnas and platypus)
                Order Marsupialia (marsupials)
                (18 placental orders)

FIGURE 1 A traditional, Linnean classification of the major chordate groups. The major vertebrate groups are ranked into eight categories, all given the equivalent rank of class.

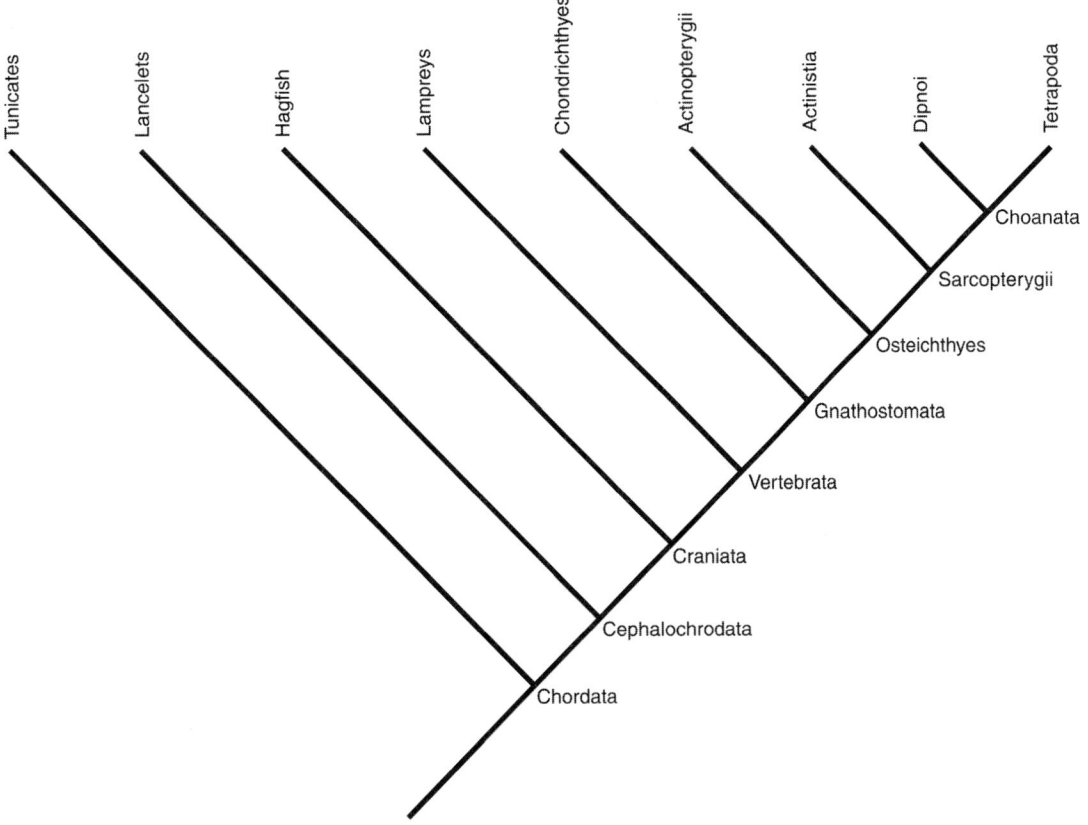

FIGURE 2 A graphic representation of a cladisitic classification of chordates. The major groups are arranged in a nested hierarchy, and taxon names are defined based on shared ancestry. The groups nested within Tetrapoda are depicted in Fig. 3. See text for definitions and diagnoses.

system and is increasingly being utilized in biological sciences (including paleontology). Although we provide ancestry-based definitions for all the major chordate groups discussed below, we include three figures to illustrate the differences in classification systems now in use. Figure 1 illustrates a version of a Linnaean classification system, in which taxa are placed in equivalent ranks. Figures 2 and 3 are graphic representations of cladistic classifications, showing the hierarchical arrangement of nested taxa (Fig. 2 shows the relationships of the major chordate groups, and Fig. 3 shows the relationships of the major groups nested within the taxon Tetrapoda).

V. DEFINITIONS AND DIAGNOSES OF MAJOR CHORDATE GROUPS

The major chordate groups discussed below are *defined* in terms of their ancestry; in other words, membership in a given taxon is dependent upon shared ancestry. Taxa are *diagnosed* by a list of characteristics hypothesized to have first appeared in the common ancestor of the members of the group; this assemblage of characters can be used to identify a taxon (see Rowe, 1987, for further discussion).

A. Chordata

Chordata is defined as the most recent common ancestor of tunicates and cephalochordates, and all of that ancestor's descendants. Chordates share a common, generalized body plan and can be diagnosed by four features shared by members of all the major groups: a notochord, pharyngeal slits, a hollow dorsal nerve tube, and a postanal tail (an extension of the body posterior to the anus). The notochord is a stiff, flexible rod running dorsal to the coelom along the length of the body beneath the central nervous system. In many vertebrates, the notochord is almost completely replaced by blocks of bone forming vertebrae. Pharyngeal slits are lateral openings from the pharynx (an organ situated behind

FIGURE 3 A graphic representation of a cladisitic classification of tetrapods. The major groups are arranged in a nested hierarchy, and taxon names are defined based on shared ancestry. See text for definitions and diagnoses.

the mouth that constitutes part of the digestive tract). These features are found in all chordates at some point during their lifetime, but they are not retained as fully functional units throughout life in all chordates.

The tunicates (urochordates) are a group of basal chordates represented by approximately 3000 species. They are entirely marine. Some tunicates spend their entire life as pelagic organisms, floating in the water column. One group of tunicates, the sea squirts, undergo a dramatic metamorphosis from a planktonic larval form to a sessile adult (attached to a substrate); a notochord is present in the tail of the larval stage but is resorbed in the adult during metamorphosis. The pharynx enlarges in the adult and expands to form a barrel-shaped branchial basket through which a constant flow of water passes as the animal filter-feeds.

B. Cephalochordata

Cephalochordata is defined as the most recent common ancestor of lancelets and Craniata, and all of that ancestor's descendants. Cephalochordates are diagnosed by the presence of segmented muscle blocks. The lancelets are marine organisms that live in shallow, tropical and temperate seas surrounding all the continental land masses except Antarctica. Two major groups are recognized and are classified under the names *Branchiostoma* (23 species) and *Epigonichthyes* (7 species; Poss and Boschung, 1996). Lancelets lack a well-differentiated head region. A rudimentary brain is present, but significant regions of the brain found in vertebrates appear to be absent. A recent volume addressing many aspects of lancelet biology (Gans et al., 1996) provides an important foundation for future research on this group.

C. Craniata

Craniata is defined as the most recent common ancestor of hagfishes and Vertebrata, and all of that ancestor's descendants. Craniates are diagnosed by the presence of a differentiated head, with well-developed paired sense organs designed to act as distance receptors. There are 43 described species of hagfishes; all are marine organisms, and the group has a worldwide distribution. Hagfishes are scavengers, feeding on dead and moribund fishes, worms, mollusks, and crustaceans. They lack recognizable vertebral structures but have weakly developed, irregularly shaped plates that are positioned almost randomly along the nerve chord in the tail region. Their eyesight is poor, but they have well-developed senses of smell and touch. Reproductive biology is poorly understood; only a few embryos are known. Hagfishes were not known as fossils until 1991. The first reported fossil was found in Pennsylvanian age rocks in Illinois (Bardack, 1991). It differs from living forms in at least a few features (better developed eyes, different position of the gills, and some aspects of the feeding apparatus).

D. Vertebrata

Vertebrata is defined as the most recent common ancestor of lampreys and Gnathostomata, and all of that ancestor's descendants. As noted above, Vertebrata is diagnosed by a suite of characters (see Table I). In all vertebrates there is at least some development of vertebral elements. The basal members of Vertebrata, the lampreys, have only rudimentary development of these structures. Lamprey vertebral elements consist of tiny distinct cartilages, situated above the dorsal nerve cord. The homologies of these elements with vertebral units in other vertebrates are not known. Many adult lampreys are parasites on other fish, and sometimes act as scavengers. They have a circular mouth with a rasping tongue; the mouth is used like a suction cup to attach the lamprey to a fish, and the tongue is used to open a hole in the body cavity of the fish, providing access to the body fluids on which the lampreys feed. Some lampreys spend significant portions of their adult lives in marine habitats, but all lampreys return to freshwater to breed. There are approximately 32 living species of lampreys.

Many earlier classifications placed lampreys, hagfishes, and a diverse suite of extinct jawless fishes into the class Agnatha ("without jaws"). A great diversity of jawless marine vertebrate groups populated Ordovician, Silurian, and Devonian seas. One enigmatic group, the conodonts, were first discovered in 1856 and subsequently were found to be remarkably abundant in marine rocks dating from the late Cambrian through the Triassic. For over a century, conodonts were known only from abundant, tiny, phosphatic, toothlike fossils that showed a remarkable diversity in form. These fossils were, and remain, important tools for establishing correlation and relative chronology of Paleozoic and early Mesozoic rocks. The systematic position of conodonts was debated for decades, and various authorities suggested affinities with many different animal groups. In 1983, a conodont fossil with soft-body preservation was reported for the first time (Briggs *et al.*, 1983). Subsequent finds that preserve greater anatomical detail revealed several features that can be interpreted as a notochord, a dorsal nerve cord, myotomes, and a distinct head region with very large eyes. These features suggest that conodonts may well have been chordates, or possibly vertebrates, but the relationships of conodonts and other jawless fishes are still quite controversial (for reviews, see Benton, 1997; Zimmer, 2000).

E. Gnathostomata

Gnathostomata is defined as the most recent common ancestor of Chondrichthyes and Osteichthyes, and all of that ancestor's descendants. Gnathostomata is diagnosed by the presence of jaws, an enlarged forebrain, and a skeleton in the paired pectoral and pelvic appendages. The evolution of jaws provided new opportunities for vertebrates by opening the door to truly predatory habits, with new food-handling capabilities and a broader diversity in potential diet. Among living vertebrates the chondrichthyans constitute the basal members of Gnathostomata. There are approximately 850 species of living chondrichthyans, including the sharks, skates, rays, and chimaeras. The majority are marine, but a few species inhabit freshwater environments. There is no bony skeleton; the skeleton is predominantly cartilaginous, with only minor calcification occurring throughout. Males have special structures called claspers that help facilitate internal fertilization. The sharks first appear in the fossil record in the late Devonian, chimaeras appear in the Carboniferous, and rays appear in the Jurassic. The whale shark (*Rhincodon typus*) is the largest living fish, achieving lengths of up to 15 meters.

F. Osteichthyes

Osteichthyes is defined as the most recent common ancestor of Actinopterygii and Sarcopterygii, and all of that ancestor's descendants. Osteichthyes is diagnosed by the presence of lungs (see below) and by most of the bony skeleton being formed from cartilaginous precursors. Many bony fishes have an elongated sac, the gas bladder, situated dorsal to the digestive tract. Gas bladders can serve either a buoyancy control function (swim bladder) or a gas exchange function (respiratory gas bladder). In the latter condition, they are highly vascularized and serve as a supplementary reserve for oxygen to be used in respiration. Although gas bladders and lungs differ in their position relative to the digestive tract (lungs are ventral to the digestive tract) and lungs tend to be paired while gas bladders are single, their developmental origins and gross anatomical features are quite similar. These similarities provide some evidence that the two structures are homologous. Actinopterygii includes a remarkably diverse assemblage of fishes, including more than half of the living species of known vertebrates; over 24,000 valid species are known (see Eschmeyer, 1998). The skeleton of actinopterygians is extensively ossified and the internal pectoral skeleton has broad bony plates that facilitate support of fin rays.

G. Sarcopterygii

Sarcopterygii is defined as the most recent common ancestor of Actinistia and Choanata, and all of that ancestor's descendants. Sarcopterygii is diagnosed by the presence of fleshy, lobed appendages, with differentiated proximal limb bones in the pectoral and pelvic fins, and by the presence of a ring of four or more bony plates in the eye (the sclerotic, or scleral, ring). Actinistia includes at least one, and possibly two, species of living coelacanths. These fishes fist appear in the fossil record in the Devonian and make their last appearance in the fossil record near the end of the Cretaceous, approximately 70 million years ago. The discovery of a extant species of coelacanth off the coast of South Africa in 1938 was therefore quite surprising. Waters near the Comoro Islands (situated between the northern tip of Madagascar and the African mainland) yielded a second specimen in 1952, and subsequently approximately 200 additional specimens were recovered. An additional population of coelacanths was recently discovered in Indonesia (Erdmann *et al.*, 1998; Forey, 1998). Analysis of the DNA of the Indonesian coelacanth suggests that it might belong to a new species, but this interpretation is controversial (Holden, 1999).

H. Choanata

Choanata is defined as the most recent common ancestor of Dipnoi and Tetrapoda, and all of that ancestor's descendants. Choanata is diagnosed by the presence of folded tooth enamel and choanae (internal nostrils that open directly into the palatal portion of the roof of the mouth). The lungfishes (Dipnoi) first appear in the fossil record in the Devonian and are represented by six living species. Their modern distribution is restricted to the southern continents of Africa, Australia, and South America. The Australian lungfish is the least derived of the three; it cannot live out of water but can survive in stagnant pools by coming to the surface to gulp air. The South American and African lungfishes can survive out of water for relatively long periods of time.

I. Tetrapoda

Tetrapoda is defined as the most recent common ancestor of Amphibia and Amniota, and all of that ancestor's descendants. Tetrapoda is diagnosed by the presence of differentiated fingers and toes and by a bony joint formed between the occipital condyles of the skull and the anterior vertebral elements.

Extant amphibians share three morphological characteristics: (1) loss of several skull bones found in extinct amphibian groups; (2) short or absent ribs; (3) pedicellate dentition (an upper tooth crown resting on a lower pedicel). The relationships among the three major extant amphibian groups are not well supported, but combined analyses of morphological and molecular data favor a sister taxon relationship between Anura (frogs and toads) and Caudata (salamanders). Batrachia is defined as the most recent common ancestor of Anura and Caudata, and all of its descendants. In this arrangement, Batrachia and Gymnophiona (the caecilians) are sister taxa (Fig. 1; see also Cannatella and Hillis, 1993).

There are approximately 165 species of living caecilians. All extant caecilians are limbless, and their distribution encompasses tropical and subtropical regions of Central and South America, Africa, India, and Southeast Asia. Most caecilians are fossorial, spending the majority of their life underground. One South American group (*Typhlonectes*) is secondarily aquatic. The otic capsules, exoccipital, basioccipital, and parasphenoid ossifications of the braincase are fused into a solid unit (the os basale). Caecilians possess a pair of sensory organs called tentacles that are unique among vertebrates. The tentacles are protrusible organs, usually situated between the external naris and the orbit, and probably function in chemoreception. Although the orbits of caecilians are often covered with skin (and sometimes bone), the eyes are photoreceptive in most species that have been examined. In some caecilians, the eye is closely associated with the tentacle and in at least one species, they eye is protruded from the skull during protrusion of the tentacle (Nussbaum, 1992). The derived groups of caecilians fuse many elements of the skull and lack a tail. The fossil record of caecilians is notably poor but extends back at least as far as the Cretaceous. The Jurassic *Eocaecilia* shares many characteristics with extant caecilians (including the reported presence of a tentacular foramen) but retains limbs and other anatomical features not seen in extant forms (Jenkins and Walsh, 1993).

There are approximately 415 described species of extant salamanders, the majority of which are found in the Northern Hemisphere. A significant radiation of tropical South American salamanders extends as far south as southern Brazil and Bolivia. Caudates share a number of morphological features of the skeleton, including the loss of the quadratojugal ossification and the presence of intravertebral spinal foramina in the atlas vertebra. Most salamanders also have a reduced number of bones in the pectoral girdle relative to extinct amphibian groups and frogs (caecilians lack a pectoral girdle), and they have an anteriorly projecting process on the atlas (the first vertebra), the tuberculum interglenoideum, which forms accessory articulations between the skull and the vertebral column. Members of one large assemblage of caudates (the Plethodontidae) lack lungs, and this condition is also found in one gymnophionan. The early history of salamanders is poorly known, but fossils belonging to Caudata (as defined above) are known from the Mesozoic. Stem-group salamanders (outside of Caudata as defined above, but on the evolutionary stem leading to modern salamanders) are reported from the Jurassic. A recent review of Mesozoic taxa was provided by Evans and Milner (1996).

There are more than 4100 described species of extant anurans, distributed on all the major continental landmasses except Antarctica. Skeletal features shared by frogs include the fusion of forearm bones (radius and ulna), fusion of the lower leg bones (tibia and fibula), elongation of the ankle bones, reduced number of trunk vertebrae (10 or fewer among extant frogs), and presence of a urostyle (a rodlike, posterior bony extension of the vertebral column). Many of these skeletal features are associated with jumping ability and evolved as early as the early Jurassic (Shubin and Jenkins, 1995). The fossil record of the group is extensive (see recent review by Sanchiz, 1998) and at least two stem-group frogs are now know from the early Triassic (Evans and Borsuk-Bialynicka, 1998).

Many amphibians have a larval stage followed by metamorphosis into the adult form. Metamorphosis was defined by Duellman and Trueb (1986, p. 173) as "a series of abrupt postembryonic changes involving structural, physiological, biochemical, and behavioral transformations." The transformations that take place in frogs are the most dramatic of any within the Tetrapoda. The majority of frog species have a larval stage in the life cycle (following hatching from an egg and preceding development of the mature adult body form) known as a tadpole. The tadpole has a more or less oval-shaped body with a tail; there is no distinct head region, but the mouth includes a beak and a number of rows of chitinous toothlike structures. During metamorphosis many morphological and physiological changes take place that completely transform the animal's appearance and the way it interacts with its environment (e.g., the tail is resorbed, limbs grow, the axial and appendicular skeleton is dramatically transformed, the digestive tract shortens, gills disappear, and respiration function is transferred to lungs). Many frogs show variants of direct development.

J. Amniota

Amniota is defined as the most recent common ancestor of Reptilia and Mammalia, and all of that ancestor's descendants. Amniota is diagnosed by a suite of characters, including presence of an amniote egg (see below), caniniform teeth, two or more vertebrae in contact with the pelvic girdle, internal fertilization, and keratin (a protein that acts as the building block for scales, nails, hooves, hair, and feathers). The amniote egg includes a series of extraembryonic membranes that surround the developing embryo and provide all the nutritional, waste disposal, and gas exchange requirements during development. In reptiles (including birds) and the Monotremata (basal mammals), the embryo and extraembryonic membranes are encapsulated in an egg with either a leathery exterior or a hard shell. Because this is a self-contained fluid-filled system, eggs can be laid away from water. Although fishes, amphibians, and most reptiles are often referred to as "cold-blooded," many species in these groups manage to maintain elevated body temperature at a constant level. This is achieved by shuttling between heat sources and heat sinks (ectothermy). However, birds and mammals generally maintain a constant temperature by changing the basal metabolic rate (endothermy). There is substantial evidence suggesting that some of the fossil reptiles did maintain body temperatures above the level of the environment. There has been much argument about the question of whether the mechanism by which this was achieved was ectothermy or endothermy. The earliest known members of the Amniota appear in the fossil record during the Pennsylvanian period.

K. Reptilia

Reptilia is defined as the most recent common ancestor of Chelonia and Sauria, and all of that ancestor's descendants. Reptilia is diagnosed by several features of the skull, including the presence of suborbital fenestrae (paired openings in the skull that are situated ventral to the orbital region).

There are approximately 260 species of extant Chelonia (turtles). They are found on all continental landmasses except Antarctica and also occupy all of the world's tropical and temperate oceans and seas. All turtles lack marginal teeth, but the earliest known stem-group turtles (*Proganochelys*) from the Triassic have a few small teeth on the palatal portion of the skull; all extant chelonians lack palatal teeth. The palate is firmly attached to the braincase. The presence of a hard, bony shell is one of the most distinguishing features of turtles. The upper portion of the shell (the carapace) is formed in part from the vertebral column and ribs. During development these elements fuse with dermal ossifications to form a generally continuous, hard bony shell. The generally flat, ventral plate (the plastron) of the shell is formed anteriorly by dermal bones and skeletal elements that are probably homologous with elements of the pectoral girdle in other vertebrates. The bony shell of most turtles is covered by a series of keratinous scutes; in some forms, these scutes are lacking and a leathery skin covers the carapace and plastron. Turtles are unique among vertebrates in having the pectoral girdle enclosed by the rib cage. Turtles first appear in the fossil record during the Triassic.

L. Sauria

Sauria is the most recent common ancestor of Lepidosauria and Archosauria, and all of that ancestor's descendants. Sauria is diagnosed by the presence of a specialized ankle joint, long and generally slender hind limbs, and relatively larger leg muscles.

Lepidosauria can be defined as the most recent common ancestor of Squamata and the extant rhynchocephalian (*Sphenodon*; see below), and all of that ancestor's descendants. Lepidosauria is diagnosed by various features of the skull and the presence of a transverse cloacal slit, a distally notched tongue, and a modified middorsal scale row. Rhynchocephalians were a very diverse and widespread group of lepidosaurs during the Mesozoic era, but by approximately 60 million years ago, they were extinct everywhere except in New Zealand. The last surviving rhynchocephalians are restricted to a few islands in New Zealand. These relict populations are classified in the genus *Sphenodon*, and two species are currently recognized.

Squamata includes all the lizards and snakes as well as the fossorial amphisbaenians. A number of anatomical features are shared by these three groups, including aspects of the skull, postcranial skeleton, musculature, and tongue. All squamates have paired, evertible hemipenes. Many squamate species lack, or have reduced, limbs and girdles.

There are approximately 4150 species of extant lizards, widely distributed on all the continents except Antarctica. A number of extinct groups were fully marine, but today only one species utilizes a marine habitat (the marine iguana of the Galapagos Islands feeds on marine algae). Many groups of lizards have lost or reduced limbs. The two North American species of *Heloderma* are venomous. Lizards first appear in the fossil record in the Jurassic.

Both morphological and molecular data indicate that snakes evolved from within lizards, but their precise phylogenetic position among other squamates is unresolved. There are approximately 2700 species of extant snakes, with a distribution encompassing tropical oceans and seas and all continents except Antarctica. Several species spend most or all of their lives in open ocean habitats. Most snakes are limbless, but some basal members retain vestigial pelvic girdle and limb elements. A few poorly known Mesozoic snakes had well-developed limbs. All snakes are carnivorous, and several lineages of advanced snakes evolved potent venom that lets them subdue prey. Most snake species have an extremely modified and flexible skull structure that permits them to eat animals much more bulky than themselves and to ingest prey that is of a greater diameter than their own head. Snakes first appear in the fossil record during the Cretaceous, but their fossil record as a whole is relatively poor and consists predominantly of isolated vertebrae.

Amphisbaenia is a small group of squamates with approximately 160 extant species. They are found in Africa, the Iberian Peninsula, the Middle East, Mexico, from Panama to Argentina in South America, and Georgia and Florida in the southeastern United States. As a result of their subterranean habits and the fact that almost one-third of the described species are known from only one specimen, many aspects of their biology and systematics are not well understood. Three species of *Bipes* retain anterior limbs, but all other species are limbless. The compact and well-ossified skull is the primary burrowing tool and is covered with relatively large scales. The scales of the skin are usually rectangular and are arranged in rings that encircle the body and allow rectilinear locomotion. Amphisbaenians first appear in the fossil record in the late Paleocene of North America.

Archosauria is defined as the most recent common ancestor of Crocodylia and Aves, and all of that ancestor's descendants. Archosaurs are diagnosed by a number of anatomical features and also by behavioral characteristics such as nest building and parental care of young. The two extant groups of archosaurs represent only a tiny fraction of the past diversity of this group, which includes the Mesozoic flying reptiles (pterosaurs), all the dinosaurs, and many extinct crocodylians.

There are 22 extant species of Crocodylia (alligators, crocodiles, and caimans), most of which are found in the Tropics. Two species of alligators extend into temperate regions in North America and China. The earliest known crocodylians were terrestrial, but all living species are found in aquatic habitats. Some extant species are found in nearshore marine habitats, but none are fully marine. Crocodylians have a complete secondary palate that forms the roof of the mouth (similar to the configuration seen in mammals). Stem-group crocodylians are known from the late Triassic, but the earliest true crocodylians are found in early Jurassic rocks.

Aves is defined as the most recent common ancestor of all extant birds, and all of that ancestor's descendants. There are approximately 9700 species of extant birds. In traditional classifications, Aves was considered to be a separate class of vertebrates, equivalent in rank to the Amphibia, Mammalia, and Reptilia. Cladistic classifications recognize birds as a distinctive group of vertebrates, but one that is nested within Reptilia, specifically within the Dinosauria. Although not accepted by all ornithologists, there is strong morphological support for this hypothesis. Birds and dinosaurs share such morphological features as an in-turned femoral head, opposable thumb, and a perforate acetabulum. An additional suite of characters indicate that birds are nested within theropod dinosaurs. Aves (as defined above) is diagnosed by the loss of teeth, the presence of a pneumatic foramen in the humerus (upper arm bone), and fusion of uncinate processes with the ribs. Most extant bird lineages appear in the fossil record at least by the early Cenozoic, and several are reported from the Cretaceous.

M. Mammalia

Mammalia is defined as the most recent common ancestor of Monotremata and Theria (marsupials plus placentals), and all of that ancestor's descendants. Mammalia is diagnosed by a large number of anatomical and physiological features, including the presence of mammary glands and a unique jaw joint. Mammary glands in females produce milk to nourish the rapidly growing young during postnatal development. In most vertebrates the jaw joint is formed between the quadrate bone of the skull and the articular bone of the lower jaw, but in mammals, the jaw joint is formed between the dentary bone (the only bone in the lower jaw of mammals) and the squamosal bone of the skull. Bones homologous with the quadrate and articular are still present in mammals, but their positional relationship and function has been transformed. The incus and malleus in the middle ear of mammals are homologous with the quadrate and articular (respectively) in other vertebrates. The 4810 species of extant mammals can be found on all continents and major bodies of water on Earth.

Monotremata is represented by three living species (duck-billed platypus and two species of echidna) that are found only in Australia and New Guinea, but the fossil record of this group extends back to the Cretaceous and reveals that monotremes were once found in South America. The monotremes are the only mammals that lay eggs. The marsupials are found today only in Australia and surrounding islands, North America, and South America. Their earliest appearance in the fossil record is in the Cretaceous of North and South America, and fossils are known from all continental landmasses. Marsupials differ from other mammals in a number of physiological and anatomical features, especially those related to reproduction. The gestation period in marsupials is very short compared with that of other mammals and the young are born in a nearly embryonic state. After birth, development takes place in a pouch (or sometimes under a fold of skin) on the mother's abdomen. The Placentalia includes a diverse assemblage of mammals that include many of the most familiar mammals and the only mammals to adopt a completely or predominantly marine lifestyle (whales, dolphins, porpoises, seals, sea lions, walruses) and to achieve powered flight (bats). Human beings are members of a placental mammal group, the Primates.

See Also the Following Articles

AMPHIBIANS, BIODIVERSITY OF • BIRDS, BIODIVERSITY OF • ENDANGERED BIRDS • ENDANGERED MAMMALS • ENDANGERED REPTILES AND AMPHIBIANS • FISH BIODIVERSITY OF • MAMMALS, BIODIVERSITY OF • REPTILES, BIODIVERSITY OF

Bibliography

Bardack, D. (1991). First fossil hagfish (Myxinoidea): A record from the Pennsylvanian of Illinois. *Science* 254, 701–703.

Benton, M. J. (1997). *Vertebrate Paleontology*, 2nd ed. Chapman & Hall, London.

Briggs, D. E. G., Clarkson, E. N. K., and Aldridge, R. J. (1983). The conodont animal. *Lethaia* 16, 1–14.

Cannatella, D. C., and Hillis, D. M. (1993). Amphibian relationships:

Phylogenetic analysis of morphology and molecules. *Herpetol. Monogr.* 7, 1–7.

Chen, J.-Y., and Li, C.-W. (1997). Early Cambrian chordate from Chengjiang, China. *Bull. Mus. Nat. Sci.* 10, 257–273.

Chen, J.-Y., Huang, D.-Y., and Li, C.-W. (1999). An early Cambrian craniate-like chordate. *Nature* 402, 518–522.

Chen, J.-Y., Dzik, J., Edgecombe, G. D., Ramsköld, L., and Zhou, G.-Q. (1995). A possible Early Cambrian chordate. *Nature* 377, 720–722.

Duellman, W. E., and Trueb, L. (1986). *Biology of Amphibians.* McGraw-Hill, New York.

Erdmann, M. V., Caldwell, R. L., and Moosa, M. K. (1998). Indonesian 'king of the sea' discovered. *Nature* 395, 335.

Eschmeyer, W. N. (1998). *Catalog of Fishes.* California Academy of Sciences Center for Biodiversity Research and Information Special Publication 1.

Evans, S. E., and Borsuk-Bialynicka, M. (1998). A stem-group frog from the Early Triassic of Poland. *Acta Palaeontol. Pol.* 43, 573–580.

Evans, S. E., and Milner, A. R. (1996). A metamorphosed salamander from the Early Cretaceous of Las Hoyas, Spain. *Philos. Trans. R. Soc. London, B* 351, 627–646.

Forey, P. (1998). A home from home for coelacanths. *Nature* 395, 319–320.

Gans, C. (1993). Evolutionary origin of the vertebrate skull. In *The Skull: Patterns of Structural and Systematic Diversity* (J. Hanken and B. K. Hall, Eds.), pp. 1–35. Univ. of Chicago Press, Chicago.

Gans, C., Kemp, N., and Poss, S. (Eds.) (1996). The lancelets (Cephalochordata): A new look at some old beasts. *Isr. J. Zool.* 42 (Suppl.).

Holden, C. (1999). Dispute over a legendary fish. *Science* 284, 22–23.

Jefferies, R. P. S. (1986). *The Ancestry of the Vertebrates.* Cambridge Univ. Press, Cambridge, UK.

Jefferies, R. P. S. (1997). How chordates and echinoderms separated from each other and the problem of dorso-ventral inversion. In *Geobiology of Echinoderms* (J. A. Waters and C. G. Maples, Eds.), pp. 249–266. *Paleontological Society Papers 3.*

Jenkins, F. A. Jr., and Walsh, D. M. (1993). An Early Jurassic caecilian with limbs. *Nature* 365, 246–250.

Nussbaum, R. A. (1992). Caecilians. In *Reptiles and Amphibians* (H. G. Cogger and R. G. Zweifel, Eds.), pp. 52–59. Smithmark Publishers, New York.

Parsley, R. L. (1997). The echinoderm classes Stylophora and Homoiostelea: Non Calcichordata. In *Geobiology of Echinoderms* (J. A. Waters and C. G. Maples, Eds.), pp. 225–248. *Paleontological Society Papers 3.*

Poss, S. G., and Boschung, H. T. (1996). Lancelets (Cephalochordata: Branchiostomatidae): How many species are valid? *Isr. J. Zool.* 42, S13–S66.

Rowe, T. (1987). Definition and diagnosis in the phylogenetic system. *Syst. Zool.* 36, 208–211.

Sanchiz, B. (1998). *Encyclopedia of Paleoherpetology: Salientia,* Part 4. Verlag Dr. Friedrich Pfeil, München.

Shu, D., Zhang, X., and Chen, L. (1996). Reinterpretation of *Yunnanozoon* as the earliest known hemichordate. *Nature* 380, 428–430.

Shubin, N. H., and Jenkins, F. A. Jr. (1995). An Early Jurassic jumping frog. *Nature* 377, 49–52.

Smith, M. P., Sansom, I. J., and Repetski, J. E. (1996). Histology of the first fish. *Nature* 380, 702–704.

Zimmer, C. (2000). In search of vertebrate origins: Beyond brain and bone. *Science* 287, 1576–1579.

VICARIANCE BIOGEOGRAPHY

Christopher John Humphries
The Natural History Museum London

I. Historical Biogeography
II. Vicariance, Dispersal, and Extinction
III. Examples of Vicariance and Dispersal
IV. Plate Tectonics
V. Cladistic Methods in Biogeography
VI. The Progression Rule
VII. Vicariance Biogeography
VIII. Poeciliid Fish in Middle America
IX. Component Analysis
X. *Heterandria* and *Xiphophorus*
XI. Conclusions

GLOSSARY

area of endemism An area recognized by congruent distributions of two or more groups of organisms.
biogeography The study of the distributions of organisms on Earth and why the patterns exist.
cladistic biogeography The combination of cladistics (phylogenetic systematics) with vicariance biogeography. A method that searches for patterns of relationship among areas of endemism.
cladistics (phylogenetic systematics) A method of phylogeny reconstruction concerned with branching patterns. Sister group relationships are hypothesized on the basis of shared derived, or synapomorphic, characters.
components Elements of groups of areas, or taxa, as determined by the branching pattern of a cladogram. For example, in a group comprising three taxa A, B, and C, when B is more closely related to C, there are two components, an ABC component and a BC component.
dispersal The movement, or spread, of an organism from one area to another independent from other organisms and of historical events that change the natural distribution of the organism (see Fig. 1).
ecological biogeography Spatial patterns of organisms considered in terms of their interactions with other organisms and the environment.
endemic A taxon, or group of taxa, restricted to a proscribed area and found nowhere else on Earth.
historical biogeography Spatial patterns of organisms interpreted in terms of their concordance to patterns of Earth history.
panbiogeography The examination of distributions and relationships on a worldwide scale; term introduced by Croizat (1964).
plate tectonics (including continental drift) The theory that the Earth's crust is composed of plates that move relative to each other through seafloor spreading and subduction. As continents are found on plates, continental drift is one result of this movement.
vicariance The existence of closely related taxa or biota in separate, disjunct areas, which have been separated by the formation of a natural barrier (vicariance event).
vicariance event The splitting of a taxon or biota into two or more geographical subdivisions by the formation of natural barriers, e.g., mountain building, glaciation, or stream capture (see Fig. 1).

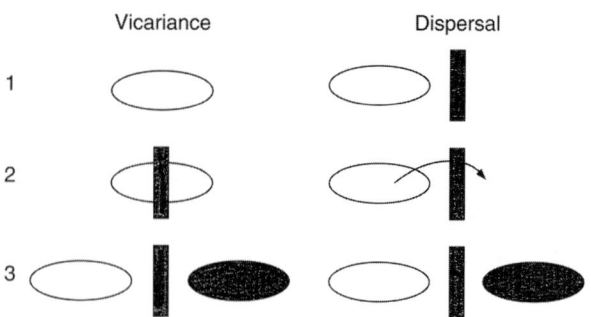

FIGURE 1 Vicariance and dispersal explanations: (a) vicariance; (b) dispersal. See text for explanation.

VICARIANCE BIOGEOGRAPHY, in the strict sense, is the study of repeated patterns of disjunct distributions within many members of a biota that may be explained by vicariance (or splitting) and other historical events. However, in the past 20 years or more, it has become a metaphor for historical biogeography, involving the study of life and Earth history using cladistic methods. The subject has developed in fits and starts but has today come to represent a method to classify areas of endemism on a global, regional, and local scale in terms of their historical relationships.

I. HISTORICAL BIOGEOGRAPHY

Biogeography means many different things to biologists and geographers of different persuasions depending on their outlooks and purposes of enquiry. It has been clearly noticed and well documented through accumulation of systematic distribution data for most groups of organisms over the past two centuries of global exploration that there are many organisms that exhibit particular distribution patterns on Earth. In efforts to interpret these patterns, biogeographers pose the question: "What lives where and why?" The usual answers invoke either ecological or historical explanations, or a combination of both (Nelson and Platnick, 1981; Craw et al., 1999; Humphries and Parenti, 1999).

For the most part ecological biogeographers classify distribution patterns as communities in a general hierarchy of organization. Populations and species live in habitats that belong to different land classes in ecosystems within biomes. Few would dispute that many similarities of climate, topography, habitat, and landscape exist between, for example, the rain forests of South America, Africa, and Southeast Asia, but, when examined closely, the resident organisms in each rain forest are specifically and generically quite different. Similar species in similar ecological niches belong to phylogenetically disparate taxa that have particular phylogenetic and distribution histories that have occurred through geological time. Historical biogeographers focus on the older developments of global biotas in an effort to study the history of the Earth rather than relatively more recent events such as short-term species interactions and recent perturbations in distribution due to climatic changes.

Data for historical biogeography come from comparative biologists and systematists who include distribution information in their monographs and revisions. A major synthesis is found in the works of Croizat, especially in *Space, Time, Form: The Biological Synthesis* (1964; see Craw et al., 1999). Croizat considered geological events and changes in Earth history and the "form-making" of species as two parts of a continuous historical process. To him, evolution of the Earth and the evolution of organisms are part and parcel of one series of historical events. Recently, greater attention has been given to these ideas, kindled by the general acceptance of plate tectonics and the notion that the continents are continually reshaped and changing in their spatial geometry (see Craw et al., 1999). At the same time, there has been a widening interest in the use of cladistics for systematic investigation of organisms (Hennig, 1966). Today, cladistics is considered as the primary means of reconstructing the relationships of organisms. Comparing different groups of organisms sharing similar distribution patterns and occupying similar areas of endemism with historical patterns in geology and Earth history allows biogeographers to provide retrodictive hypotheses of divergence and explanations of past and present diversity of organisms (Humphries and Parenti, 1999).

II. VICARIANCE, DISPERSAL, AND EXTINCTION

Particular species are endemic to certain areas today either because their ancestors originated there and their descendents have survived to the present day or because the ancestors occurred elsewhere and later some of the descendents dispersed into new areas. Such a difference invokes two, very different, historical explanations which in a shorthand way are described as vicariance or dispersal modes of diversification. In a vicariance explanation, an ancestral species (Fig. 1a, 1) divides into two separate populations when a physical barrier is formed across the population such that neither of the two new populations can cross over the barrier (Fig. 1a, 2). With time, the two separated populations evolve into two closely related species, shown here as black-and-white ovals (Fig. 1a, 3). In a dispersal explanation, the range of one species is already bounded at least in one direction by a barrier (Fig. 1b, 1), which is later crossed by some members of the population (Fig. 1b, 2). If the organisms of the new population survive, successfully interbreed, and yet remain isolated from the "ancestral" population, they eventually evolve into a different species (Fig. 1b, 3). Thus, in the vicariance explanation the appearance of the barrier is seen as the cause of the disjunction of the two species, whereas in the dispersal explanation the barrier is considered to be older than the disjunction.

III. EXAMPLES OF VICARIANCE AND DISPERSAL

By Victorian times and the emergence of evolutionary theory, it was clear that the distinction of dispersal and vicariance explanations was emerging. As described by Nelson and Platnick (1984), the English botanist Joseph Hooker (1817–1911) provided perhaps the earliest coherent vicariance explanations of plant distributions. Hooker studied the plants of the Southern Hemisphere in South Africa, Australia, southern South America, and New Zealand and concluded that their pattern of distribution suggested a former, more extensive flora, which had broken up by geological and climatic causes. Hooker compared the southern floras with those of the Northern Hemisphere and also concluded that the various southern floras were more closely related to each other than any were to the floras of the Northern Hemisphere. As a consequence, he also boldly concluded that divergence of the northern and southern floras must have occurred before the fragmentation of the southern floras. His vicariance hypothesis considered that initially a worldwide flora was first subdivided by a barrier into the northern and southern floras and then the latter were further subdivided by expanses of ocean—barriers severing once continuous floras that no longer exist today.

Nelson and Platnick (1984) compared Hooker's views with fellow Englishman Alfred Russell Wallace, another great biologist of the late nineteenth century. Wallace disagreed entirely with Hooker's hypothesis. In 1876, he said that the north and south division of biotas was not a vicariance pattern but represented the fact that

". . . the northern continents are the seat and birthplace of all higher forms of life, while the southern continents have derived for the greater part, if not the whole, from the north; . . . it implies the erroneous conclusion that chief southern lands—Australia and South America—are more closely related to each other than to the northern continent."

At that time the prevailing geological theory was one of fixed continents and slow, gradual changes in both Earth history and biogeographical developments. Hooker saw a close connection between cause and effect and said that the basic pattern of relationships divided the northern from the southern taxa, and the pattern was explained by vicariance events. Wallace, heavily influenced by Darwinian selection theory on the other hand, did not see a necessary connection between distribution and Earth history. For him, the basic pattern was east–west and the pattern was explained by dispersal from northern centers of origin.

IV. PLATE TECTONICS

It has become clear that the nineteenth century was characterized by the dominance of British Empire and European colonialism throughout the world. It was clear that the "higher forms of [human] life" inhabiting the north were invading and dominating elsewhere, particularly the countries of the Southern Hemisphere (Nelson and Platnick, 1984). It was no accident that the "favored races" were of northern origin as indicated in the title of Darwin's (1859) famous book: *On the Origin of Species by Means of Natural Selection, or the Preservation of Favoured Races in the Struggles for Life*. Although throughout his life his ideas changed, Wallace's earlier views on biogeography were part of Darwinian orthodoxy (Nelson and Platnick, 1984). Many groups of animals and plants were seen created in "centers of origin" in the Northern Hemisphere to then disperse and colonize other parts of the globe. Following the central dogma of Darwinism, the colonizing organisms were seen as the surviving fittest and superior competitors to all others that lay before them. The less fit southern forms became thrust aside and eventually extinct. This rather fanciful scenario has been repeatedly described in textbooks, even within the 1990s, and is sometimes known as the "monoboreal relict hypothesis" (Nelson and Platnick, 1984).

The origin of taxa from "centers of origin" and by dispersal to other areas is heavily entrenched in biogeographic literature, an idea that goes back to the Age of Reason in the eighteenth century. For example, Linnaeus believed that if the Earth was changing, with continents growing in size, and that the process had operated in the past

". . . the continent in the first ages of the world lay immersed under the sea, except a single island in the midst of this immense ocean; where all animals lived commodiously, and all vegetables were produced in the greatest luxuriance . . ." (translated from Linnaeus, 1781, p. 77; see Nelson and Platnick, 1980)

Linnaeus knew that different organisms had particular lifestyles and he suggested that the "center of origin" must have been a primordial island in the Tropics and "bore a very lofty mountain" inhabited at different altitudes by species with different ecological requirements. As new land emerged beyond this island and as the seas receded, the resident organisms migrated by various means to colonize the new parts with suitably similar ecological conditions (Nelson and Platnick, 1980). Embedded deeply within these comments are two important ideas: that species originate and disperse from a "center of origin" and that regularities in distribution are controlled by ecological conditions.

During the nineteenth century adherence to dispersal could be attributed to contemporary geological and geophysical theory. Such a belief allowed only stable continents separated by permanently wide oceans. Biogeographers, such as Hooker, who considered former land connections and favored vicariance explanations, were scorned as being out of line with the prevailing view (Nelson and Platnick, 1984). Even though the late nineteenth century and the first half of the twentieth century produced occasional theories that suggested continents were mobile and forever changing in shape and position through continental drift, mountain building, and erosion, dispersal theories still dominated the literature. It was not until the 1960s that geophysical studies eventually showed that continents and ocean bottoms were not permanent geographical features and that a general acceptance, one of the great paradigm shifts in science, came about. Early theories of continental drift were based on a model of the world that had a fixed diameter and that all of the present-day continents were joined up into one supercontinent, Pangea. During the Jurassic, some 180 million years ago, the supercontinent broke up into two major landmasses, the southern supercontinent, Gondwanaland, and the northern supercontinent, Laurasia. Laurasia contains those areas that have subsequently crystallized out into Europe, North America, and Asia, but exclude the Indian subcontinent. From the Cretaceous onward, the southern end of the world, Gondwanaland, broke up into those areas we now know as Antarctica, Australia, India, Africa, and South America.

The breakup of Gondwanaland lends credence to vicariance biogeography and hence Hooker's theory. The similarities among floras of the Southern Hemisphere suggest former land connections. It also revived the theory that Hooker's hypotheses of north–south connections exist for plants and contradicted Wallace's theory of east–west connections based on animals. As Nelson and Platnick (1984) asked, how can we determine which hypothesis is correct? Are both theories correct to some degree, or is there another hypothesis that should be erected which subsumes both into one explanation? The interesting point is that if the explanation is a general one, what process caused it and what methods are required to uncover the historical pattern? Paradoxically, through attempts to resolve the relationship between space, form, and time in terms of vicariance and dispersal, including application of track analysis in panbiogeography (Croizat, 1964; Craw et al., 1999) and cladistics in vicariance biogeography (Nelson and Platnick, 1981), it was clear that a more general issue was at stake. This was the question of how to interpret patterns of plant and animal distributions in a systematic way to provide independent theories of biogeography, rather than rely on geophysical and geological theories as prior explanations for present-day distribution of biotas (see Humphries and Parenti, 1999). To give some flavor of the rather fitful debate this idea created, the stages of application of cladistic methods that grew out of vicariance biogeography are briefly described.

V. CLADISTIC METHODS IN BIOGEOGRAPHY

The analysis of patterns in cladistic biogeography starts with finding the species or higher taxa that exist in the different biogeographic areas and to discover their relationships. Cladograms, or branching diagrams, are used to express hypotheses of relationship between organisms. An example is shown in Fig. 2a, where among the four species, C and D are more closely related to each other than either are to B, and also species B, C, and D are more closely related to each other relative to A. The cladogram specifies two inclusive informative

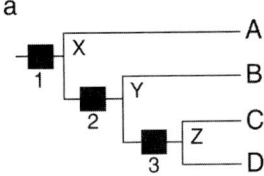

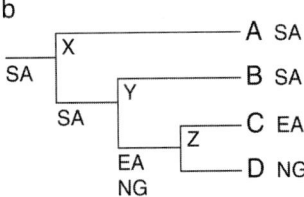

FIGURE 2 (a) Cladogram depicting the relationships between four species. A–D are four recent species, X–Z are hypothetical ancestors, and 1–3 are group-defining characters. (b) Area cladogram showing application of the progression rule to species A and B in South America (SA), C in eastern Australia (EA), and D in New Guinea (NG).

groups of species, BCD and CD. Cladograms are produced by the distribution of characters or features shared by the different species or taxa. Thus, in Fig. 2a, character 3 specifies the group CD, character 2 specifies group BCD, and character 1 specifies ABCD as the overall group. Characters 2 and 3 then specify unique groups that have characters not shared by any other organism. If such features are taken to be present in the ancestors of each group, then the cladogram can be taken as specifying the pattern of evolution or common descent of that group. Seen in this way, the cladogram of Fig. 2a may be taken to specify that taxa C and D have a unique ancestor Z that is not the ancestor of A or B and also that taxa B, C, and D have an ancestor Y that is not the ancestor of A. CD and BCD are monophyletic groups, compared, for example, with A and D, which is a polyphyletic group, and A and B, which comprise a paraphyletic group.

Cladistics was applied to biogeography in the 1960s at about the same time when theories of plate tectonics and shifting continents were becoming respectable. However, dispersal hypotheses to explain distribution patterns were still in vogue and thus the earliest applications in biogeography used cladograms to determine "centers of origin" of monophyletic groups. The German dipterist Willi Hennig (1966) reasoned that there is a recognizable, close relationship between a species and the space each occupies. He assumed that dispersal patterns are unique for each taxonomic group and that each had an independent history.

VI. THE PROGRESSION RULE

Central to Hennig's (1966) method to find a center of origin for a group of taxa from a particular cladogram was the idea that phylogenetically primitive members of a monophyletic group will be found near the center. That is, they are occupied by the species nearest to the basal nodes of cladograms. Within a continuous range of species of a monophyletic group, it was plausible that a series of characters transform and run parallel with distribution in space, such that the youngest members would be on the geographic periphery of the group.

To apply the "progression rule," one works backward considering the areas in the terminal positions of the cladogram and then gradually working down to the base of the cladogram, assigning distributions to the internal nodes or the common ancestors. If the areas occupied by the descendents do not overlap, they are combined together in assigning the probable distribution of the common ancestor. If they do overlap, the unshared element is eliminated. Consider the four species in Fig. 2b. Species A and B occur in South America, C occurs in eastern Australia, and species D occurs in New Guinea. Working from the most derived species, C and D, the distributions clearly do not overlap; hence they are added together to give the area for ancestor Z as eastern Australia plus New Guinea. For ancestor Y the descendent taxa do not overlap, with B occurring in South America and ancestor Z in eastern Australia and New Guinea. The two distributions do not overlap so the area for ancestor Y is South America, eastern Australia, and New Guinea. Moving down the cladogram to ancestor X, descendent species A occurs in South America and ancestor Y in South America, eastern Australia, and New Guinea. The two distributions thus overlap in South America, and eastern Australia and New Guinea are eliminated as the unshared area. Working back up the cladogram, the simplest hypothesis is to consider that South America is common to all three ancestral species, X, Y, and Z, and that species C and D dispersed to eastern Australia and New Guinea from South America. The conclusion from analysis of this biogeographic pattern is that the center of origin for the group is South America.

To demonstrate this early use of cladograms in biogeography, a simple example can be seen in the caddis fly distribution of *Wormaldia* (Ross, 1974; see Humphries and Parenti, 1999). *Wormaldia* and other caddis flies have been a lifelong study of Ross because they occur in almost all freshwater systems throughout the globe, show a high degree of endemicity in all areas, and are very speciose. Ross considered the geographical

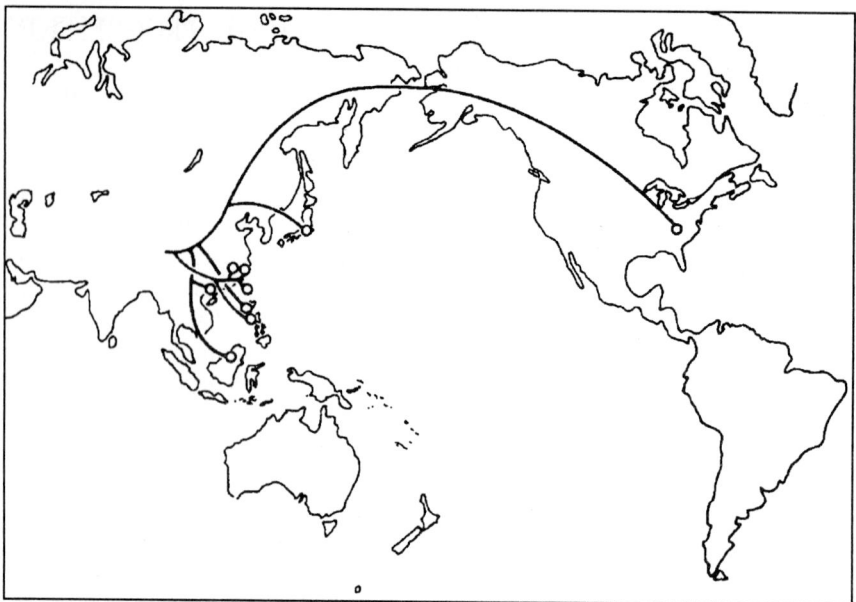

FIGURE 3 Distribution of the *Wormaldia kisoensis* group of caddis flies. The circle in Japan is *Wormaldia kissoensis*, and that in North America is *Wormaldia mohri*. (Redrawn from Ross, 1974)

distribution of the nine species of the *Wormaldia kisoensis* complex of caddis flies. Eight species are present in the western Pacific from northwestern Borneo in the south and Japan in the north and one species occurs in the Smoky Mountains of eastern North America (Fig. 3). The cladogram of phylogenetic relationships is shown diagrammatically in Fig. 4. Like Hennig, Ross (1974) assumed that the species nearest to the base of the stem of the cladogram denotes the ancestor of the group. He applied also the progression rule to the analysis to determine the center of origin of the group. In *Wormaldia* the most derived species pair occurs in Japan and eastern North America, which led Ross to propose the dispersal hypothesis of origin of the group in the western Pacific and a single dispersal of one species across the Bering straits to North America (Figs. 3 and 4).

Brundin's classic studies of chironomid midges (Brundin, 1966) showed that the southern temperate areas of South America, Southern Africa, Tasmania, southeast Australia, and New Zealand are inhabited by 600–700 species. Trans-Antarctic relationships are a recurring phenomenon throughout the group, so by way of an example consider the midges of subfamily Diamesinae, which show a double distribution: two major groups present in both Northern and Southern Hemisphere temperate areas but absent from the Tropics. The largest and most widespread is the relatively generalized tribe Heptagymi, represented by eleven species in Andean South America, two species in southeastern Australia, and five species in New Zealand (Fig. 5). Its sister group is the relatively more apomorphic and monotypic tribe Lobodiamesini of New Zealand.

The cladogram in Fig. 5 shows that there are a total of 25 terminal taxa in the Southern Hemisphere areas of South America, New Zealand, Australia, and South Africa and three groups in Laurasia. The monotypic genus *Heptagyia* occurs in South America and *Paraheptgyia* has five South American species. According to

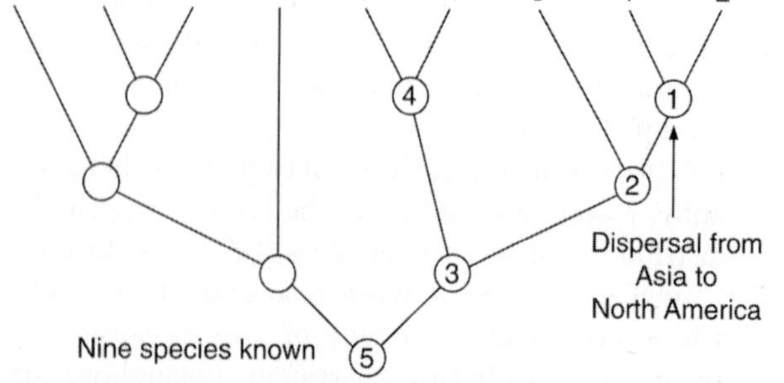

FIGURE 4 Cladogram for the *Wormaldia kisoensis* group of caddis flies. (Adapted from Ross, 1974)

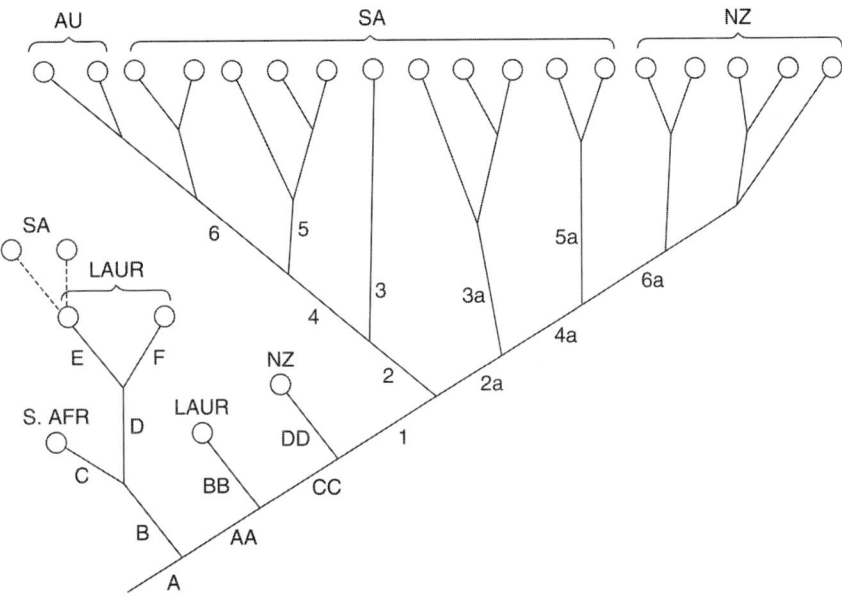

FIGURE 5 Partial area cladogram of the subfamily Diamesinae (Diptera; Chironomidae). Some named lineages are as follows: 1, Heptagyini; 3, *Heptagyia;* 4, *Paraheptagyia;* 3a, *Reissia;* 5a, *Limaya;* 6a, *Maoridiamesa;* A, Diamesinae; C, Harrisonini; E, Diamesini; F, Protanypodini; BB, Boreoheptagyini; DD, Lobodiamesini. (From Brundin, 1981, Fig. 3.7, p. 119)

Brundin, the southeastern Australian subgroup of two species is a younger evolved offshoot of the older South American group including *Heptagyia*. Brundin considered the Australian taxa to have dispersed from Patagonia or east Antarctica at stem 6 by the end of the Paleocene because they have derived characters and because the stem species (indicated by 1, 2, and 4) never occurred in Australia. The other stem (2a) includes the genera *Reissia*, with three species in South America, *Limaya*, with two species in South America, and *Maoridiamesa*, with five species in New Zealand. Brundin (1981) considered that *Maoridiamesa*, on a different stem from the Australian *Paraheptagyia* group, agreed well with plate tectonic theory for an early separation of New Zealand from western Antarctica in the Upper Cretaceous. The fact that *Maoridiamesa* is a comparatively younger, derived offshoot of an older group in South America was, according to Brundin, evidence of long-distance dispersal from South America via west Antarctica to New Zealand of stem species 4a rather than a vicariance event. In other words, Brundin thinks that the *Maoridiamesa* group is younger than the areas in which it occurs.

The progression rule can be interpreted as one particular optimization procedure applied to cladograms. It has problems in the sense that it can specify a different center of origin when applied to different data. Changes in data occur for a host of different reasons, but largely through the discovery of known taxa in new places, the discovery of new taxa, and reinterpretation of relationships. Patterson (1981) showed that the progression rule was very sensitive to the discovery of new taxa, and particularly fossils. The discovery of fossils simply added more areas to the problem and thus increased the number of possibilities for finding new centers of origin. Thus, for example, if we added two new fossils, V and W, to the example described earlier (see Figs. 2 and 6), then the possibility of a center of origin in North America and in turn Europe would alter the interpretation significantly. Patterson considered that for groups of organisms with a significant fossil record, the progression rule is heavily influenced by fossils and this would be especially true in the areas where the best fossils occurred. The progression rule is dependent on the notion that organisms occur in a center of origin. The problem is that when the progression rule is applied to a cladogram, it can always specify a center whether it may, or may not, have occurred. Suppose, for the sake of argument, the species of Fig. 6b in Europe, North America, South America, eastern Australia, and New Guinea were caused by the breakup of continents, a vicariance explanation; then the progression rule would provide a spurious hypothesis impossible to test.

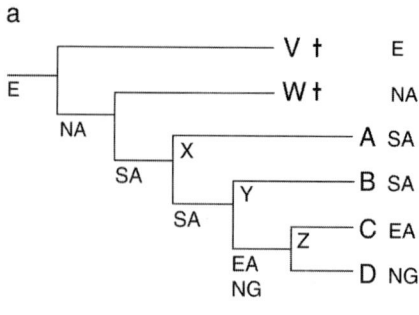

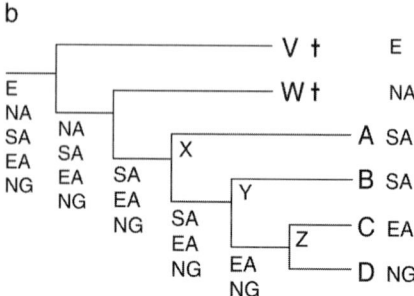

FIGURE 6 (a) Fossils and centers of origin. A–D represent four Recent species, A and B in South America (SA), and C and D in eastern Australia (EA) and New Guinea (NG), respectively. V and W are fossils, as indicated by the crosses distributed in North America (NA) and Europe (E), respectively. Areas on the internal nodes are ancestral areas assigned using the progression rule. (b). The same cladogram but applying a vicariance hypothesis. See text for explanation.

VII. VICARIANCE BIOGEOGRAPHY

The problem of interpreting cladograms as dispersal scenarios is the same as treating them as phylogenetic trees rather than as hierarchical classifications to show groups with relatively more inclusive groups of set membership. Such a procedure requires making ad hoc assumptions not based on the information on which they are based. Furthermore, interpreting individual cladograms as having individual histories leads to certain conceptual difficulties. A crucial one is how do we interpret repetitious distribution patterns? If we have many unrelated taxonomic groups repeating a pattern of distribution between major continents, such as South America, eastern Australia, and New Zealand (Fig. 7), it is improbable that the pattern was caused by separate unique dispersal events. This implies that species in each different group separately makes its own way from one continent to the other. A logical and more robust conclusion would be to suggest that at one time, the continents were in contact and that the present-day pattern was due to the breakup of a formerly continuous biota.

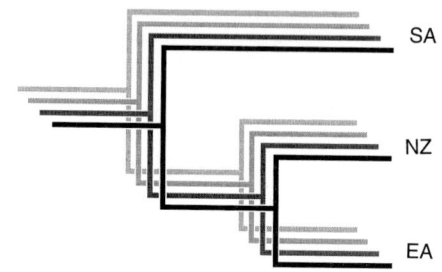

FIGURE 7 Four hypothetical groups of organisms showing the same area relationships.

The breakthrough in application of cladistic reasoning to biogeography came with the realization that disjunct distributions could occur because of such vicariance events. This is based on the idea that present-day endemic taxa occur in the same areas as their ancestors and the Recent taxa evolved *in situ*. Where dispersal models explain disjunctions by dispersal across pre-existing barriers, vicariance models explain them by the appearance of barriers fragmenting ancestral species ranges (Fig. 1). To historical biogeographers what became particularly clear was the important idea that distribution data are insufficient to resolve decisively either dispersal or vicariance as the cause of disjunct distribution patterns.

Platnick and Nelson (1978) argued that one should not worry about the cause of a disjunct distribution pattern between different related areas of endemism but whether or not it conforms to a general pattern of relationships shown by other groups of taxa endemic to the areas occupied. Thus, there is an analogy to the three-taxon statements in systematics as the most basic units for expressing relationships. In biogeography three-area statements are the most basic units for expressing relationships between areas of endemism. The generality of the area cladograms can be examined by comparison with other unrelated taxonomic groups endemic to the relevant areas and corroboration of a particular pattern is equivalent to a general statement for the relative recent ancestry of the biotas under scrutiny. However, there has been a long, drawn out saga of trying to find ways for applying the principles of cladistic biogeography because initial applications of the method (e.g., Rosen, 1976) encountered problems of incongruence between different groups of organisms. Theoretically, it should be possible to connect every area of endemism of the world into one huge general statement of interrelationships. However, our perception of the world is less than perfect for a variety of reasons—extinction, dispersal of widespread taxa, and restricted distributions of taxonomic groups. The significance of

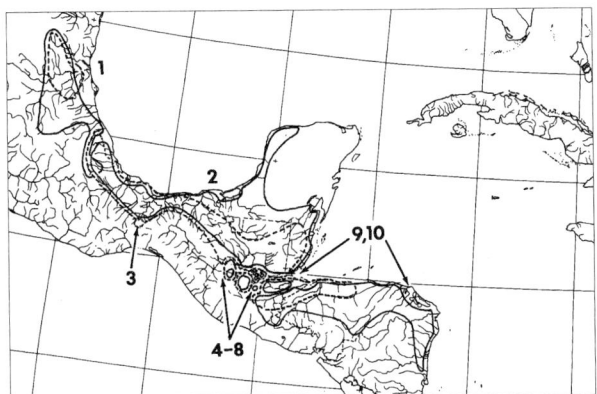

FIGURE 8 Map to compare the distributions of the species of *Heterandria* (full line) and *Xiphophorus* (broken line). Numbers refer to areas defined by the occurrence of taxa. (From Rosen, 1979, Fig. 45)

these problems can be demonstrated with poeciliid fish in Middle America.

VIII. POECILIID FISH IN MIDDLE AMERICA

To discover relationships of areas, at least two groups of taxa must be available for the same or similar set of areas. Rosen (1978, 1979) examined two groups of fishes, *Heterandria* and *Xiphophorus* from the Mesoamerican region. Both genera have close relatives elsewhere but each has a monophyletic subgroup inhabiting 11 areas in southern tropical Mexico, south to eastern Honduras in *Xiphophorus*, and further south to eastern Nicaragua in *Heterandria* (Fig. 8). Area cladograms for the two genera, which indicate areas occupied by particular species, are shown in Figs. 9 and Fig. 10.

IX. COMPONENT ANALYSIS

Cladistic biogeography would be uncomplicated if all groups of organisms were each represented by one taxon in each of the smallest identifiable areas of endemism. This is not the case. Unique patterns may be meaningful and cannot at the same time be incongruent with patterns determined from other taxa. Nevertheless, incongruence between two or more cladograms can occur for a variety of historical reasons. Particular groups of organisms can exhibit older or younger patterns than the groups to which they are being compared. Also, some taxa have dispersed and become widespread and by so doing have obscured the historical signal in

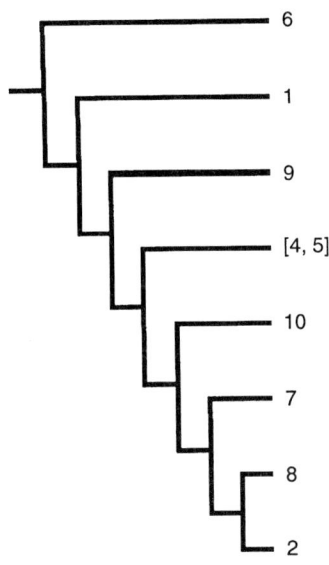

FIGURE 9 Area cladogram for *Heterandria*. The numbers refer to areas defined by the species as shown on the map in Fig. 8. (Redrawn from Rosen, 1979, Fig. 48)

the area patterns. Comparing several different groups of organisms is redundant in the sense that the same pattern is repeated over and over or areas are represented by two or more taxa. With extinction the observed pattern creates a spurious historical signal. All of these problems effectively cause sampling errors and lead to wrong predictions of the general patterns of area interrelationship. Nelson and Platnick (1981) used cladistic logic and component analysis to make comparisons between area cladograms to yield the maximum resolution in the general-area cladograms (see Box 1).

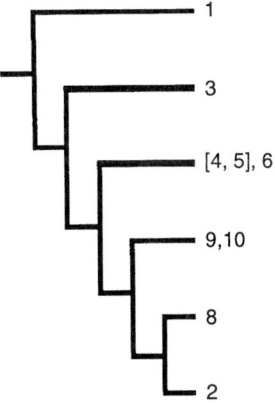

FIGURE 10 Area cladogram for *Xiphophorus*. The numbers refer to areas defined by the species as shown on the map in Fig. 8. (Redrawn from Rosen, 1979, Fig. 49)

Box 1

Component Analysis

To determine the degree of congruence between different groups of organisms occupying similar areas, the problems of "missing areas," widespread taxa, and redundancy are taken into consideration using component analysis and consensus.

Components are the elements of a group of areas, or group of taxa, as determined by the branching pattern of a cladogram. For example, in a group comprising three taxa (or areas) A, B, and C, when B is more closely related to C, there are two components, a general uninformative ABC component and an informative BC component.

Components of cladograms in vicariance biogeography can be manipulated to extract historical signal from even the most recalcitrant cases (Fig. 11). For example, consider the four areas of endemism, South Africa (SA), Pacific South America (PSA), New Zealand (NZ), and eastern Australia (EA). Now consider three groups of organisms distributed in these four areas; three species of lizards (L1–L3) in SA, PSA, and NZ, and three species each of birds (B1–B3) and fish (F1–F3) in PSA, NZ, and EA. The three numbered informative nodes (1–3) on the cladograms express the relationships for each species in the lizards, birds, and fishes. Notice that each group of organisms occurs in three areas, and in each case there is one area not present (EA for the lizards and SA for the birds and fish). Also, for each group there is one species endemic to an area, i.e., L2 in PSA, B2 in NZ, and F2 in EA. The remaining species are more widespread, occurring in either two or three areas (respectively L1, B1, and F1 in two areas and L3, B3, and F3 in three areas).

Because of widespread species and the lack of representation of one area in each of the area cladograms, at first sight relationships between the four areas seem difficult to resolve (Fig. 11). However, it can be assumed that each species originated in one area with extinction, dispersal, and failures to vicariate, causing the more complex patterns. Taking this view, we can examine the informative three-area relationships in the area cladograms to see if they tell us anything about the general-area relationships. Thus, in the lizard cladogram, L1 in SA, L2 in NZ, and L3 in PSA is the only informative statement that can be

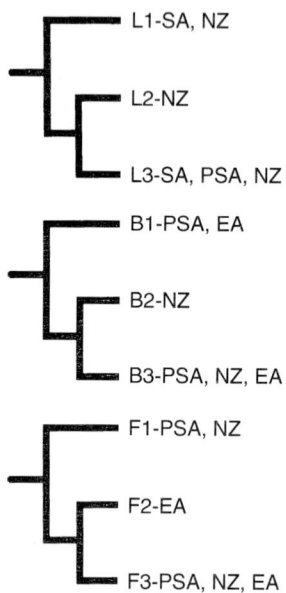

FIGURE 11 Three cladograms for three groups of lizards (L1–L3), birds (B1–B3), and fishes (F1–F3). The areas in which species occur include South Africa (SA), New Zealand (NZ), Pacific South America (PSA), and eastern Australia (EA). L2, B2, and F2 are endemics to NZ and EA, respectively. L1, B1, and F1 are widespread in two areas and L3, B3, and F3 are each widespread in three of the four areas. Lizards are not present in EA, and birds and fishes are not present in SA.

represented as SA(PSA, NZ) and scored as 0(1, 1) in a matrix (Table I, column 1). The "missing" area (EA) is scored as a dash. Informative area statements for birds are PSA(NZ, EA) and EA(NZ, PSA), and for fishes PSA(EA, NZ) and NZ(EA, PSA) similarly scored in the matrix (Table I, columns 2–5). Consequently, for the three cladograms there are five informative three-area statements (Table I). By using a standard parsimony

TABLE I
Informative Components (Three-Item Statements) of the Three Cladograms in Fig. 12 (Scored as 0(1,1).)

Area	Components expressed as three-item statements				
	1	2	3	4	5
Root	0	0	0	0	0
South Africa (SA)	0	–	–	–	–
New Zealand (NZ)	1	1	1	1	0
Pacific South America (PSA)	1	0	1	0	1
Eastern Australia (EA)	–	1	0	1	1

VICARIANCE BIOGEOGRAPHY

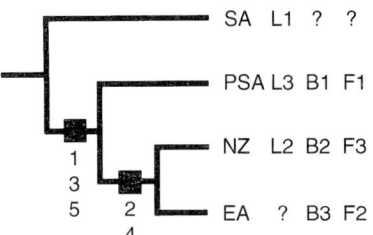

FIGURE 12 General-area cladogram for lizards (species of lizards (L1–L3), birds (B1–B3), and fishes (F1–F3). The acronyms for areas are as in Fig. 12. The five informative components (1–5) identify area relationships.

analysis, the best fitting tree for the five statements is shown in Fig. 12. Thus, despite the complexity of the patterns in the original three groups (Fig. 11), if there is one history that best explains the data, then it might be a vicariance hypothesis for the four areas as offered in Fig. 12—(SA, PSA, NZ, EA) (SA(PSA, NZ, EA)) (SA(PSA(NZ, EA)))—and shown as an area cladogram in Fig. 13, indicating the relationships of the four areas of endemism.

X. HETERANDRIA AND XIPHOPHORUS

To illustrate just how the methods to extract the best possible patterns of relationship work, consider the two groups of poeciliid fish genera *Heterandria* and *Xiphophorus* (Figs. 8–Fig. 10) based on the work of the American ichthyologist Donn Rosen (1978, 1979). The 11 identifiable areas of endemism, each determined by substituting species of *Heterandria* or *Xiphophorus* for the areas in which each occurs, include areas 1, 2, 4, 5, 8, 9, and 10 common to both groups. Areas 4 and 5 are occupied by one species from each group and thus treated as one area, [4–5]. An area 11 was described by Rosen as a putative hybrid area between areas 4–5 and 2, but to simplify matters, it is excluded here. A comparison of the two cladograms shows that *Xiphophorus* is less informative than *Heterandria* because it has two widespread species in areas [4–5], 6, 9, and 10 and is absent from area 7. In *Heterandria* areas [4–5], 6, 9, and 10 are all occupied by recognizable endemic species. However, the pattern is not quite complete in that there is no species present in area 3.

By using the assumption that one sequence of historical events can explain the patterns observed in both *Heterandria* and *Xiphophorus* and then using the logic applied for the hypothetical example used for the breakup of Gondwana (Box 1), a single result can be obtained. Platnick (1981) noted that if widespread taxa are uninformative, they cannot at the same time be incongruent. We can assume that the information on areas 6 and 9 in *Heterandria* is correct. Thus the incongruent information in the same areas for *Xiphophorus* (areas 6 or 4–5, but not both, and 9 or 10, but not both) is due either to dispersal or a failure to speciate in response to vicariance events that gave the pattern

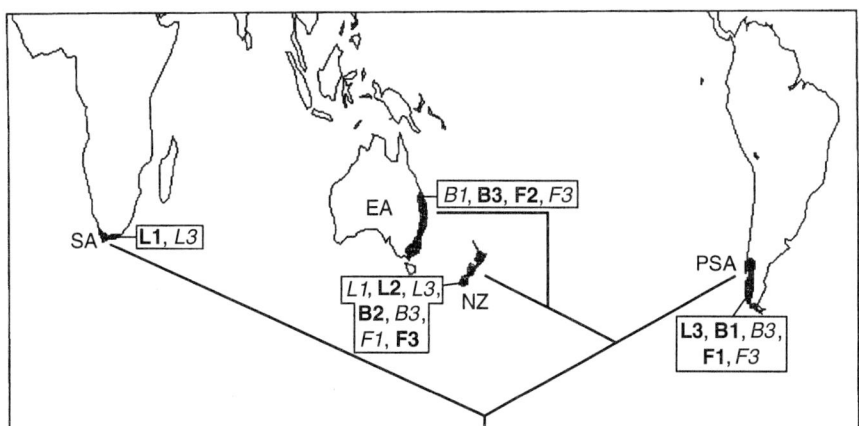

FIGURE 13 Sequence of vicariance events (1–4) for some areas of Gondwanaland hypothesized from the general-area cladogram in Fig. 12. The area acronyms are as for Fig. 11. This is just one scenario out of several where an ancestral area gradually divides into four separate areas by successive vicariance events. The taxa are given the same abbreviations as in Fig. 11 and those providing the relational signals are written in bold and roman type.

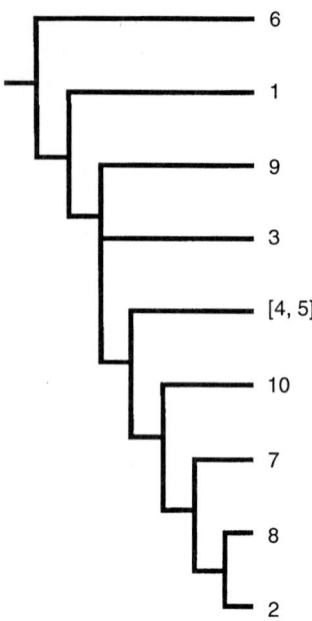

FIGURE 14 Consensus tree of the three possible intersecting cladograms of *Heterandria* and *Xiphophorus*. See text for explanation.

in *Heterandria*. Taken on their own, widespread taxa are uninformative. However, when considered with other cladograms, informative results are possible. Similarly, "absence" data (e.g., area 3 in *Heterandria* and area 7 in *Xiphophorus*) can never be incongruent with information at hand so unique areas should be placed in terms of whatever pattern exists (i.e., area 7 by *Heterandria* and area 3 by *Xiphophorus*). By using these assumptions, the *Xiphophorus* cladogram allows the populations in areas 9 and 6 to occur in any of 12 positions and the *Heterandria* cladogram allows area 3 to float onto any of the branches of the cladogram. Thus, of the 425 allowable trees for *Xiphophorus* and the 27 for *Heterandria*, the analysis yields three possible intersections. These are summarized as a consensus tree (Fig. 14).

By using this method, we have an area hypothesis that accounts for all 10 areas of endemism (i.e., excluding the hybrid area 11) that can be recognized from the two genera of fishes *Heterandria* and *Xiphophorus*. If the pattern has been created due to changes in Earth history, the question that we could ask now is: "What might have been the historical factors in Mesoamerica to cause this pattern and how might these be compared with the given biological distribution?" Ideally, we would require that geological information be assembled into cladograms, in the same way as biological cladograms, so that biotic and historical patterns can be compared. Until such time as geological data can be ordered for a more informative comparison, one thing we can say is that the observed patterns in species relationships of *Heterandria* and *Xiphophorus* in Mesoamerica have been formed over a period of at least the past 80 million years and are most likely to have been brought about by vicariance events where life and Earth evolved together (Humphries and Parenti, 1999; Rosen, 1978).

XI. CONCLUSIONS

At this time, methods of vicariance or cladistic biogeography have proved useful to analyze and compare biotic patterns at the highest resolution and organized in such a way as to compare them to independent sources of data such as geological patterns. Cladistics is a general method of determining class and subclass relations whatever the source of data without recourse to evolutionary narrative. A cladistic approach to Earth history makes it possible to express biogeographic interrelationships as hierarchical relations from biotic information. The development of cladistic methods that cater for the possibilities to consider complicating events but without ever using such information prior to analysis provides a system that allows the possibility of generating general biogeographic hypotheses even from seemingly ambiguous patterns. In other words, vicariance biogeography is a general empirical procedure without any prior assumptions in the analysis about dispersal, vicariance, redundancy, or extinction events but which at the same time never denies that they occur

See Also the Following Articles

BIOGEOGRAPHY, OVERVIEW • CLADISTICS • DARWIN, CHARLES • DIVERSITY, COMMUNITY/REGIONAL LEVEL • ENDEMISM • ISLAND BIOGEOGRAPHY

Bibliography

Brundin, L. (1966). Transantarctic relationships and their significance as evidenced by midges. *K. Sven. Vetenskapsakad. Handl. (Ser. 4)* 11, 1–472.

Brundin, L. (1981). Croizat's panbiogeography versus phylogenetic biogeography. In *Vicariance Biogeography: A Critique* (G. Nelson and D. E. Rosen, Eds.), pp. 94–158. Columbia Univ. Press, New York.

Craw, R. C., Grehan, J. R., and Heads, M. J. (1999). *Panbiogeography: Tracking the History of Life* (Oxford Biogeography Series No. 11). Oxford Univ. Press, Oxford.

Croizat, L. (1964). *Space, Time, Form, the Biological Synthesis*. Published by the author, Caracas.

Hennig, W. (1966). *Phylogenetic Systematics.* Univ. of Illinois Press, Urbana, IL.

Humphries, C. J., and Parenti, L. R. (1999). *Cladistic Biogeography: Analysing the Patterns of Animal and Plant Distributions,* 2nd ed. (Oxford Biogeography Series No. 12). Clarendon, Oxford.

Nelson, G., and Platnick, N. I. (1980). A vicariance approach to historical biogeography. *Bioscience* **30**, 339–343.

Nelson, G., and Platnick, N. I. (1981). *Systematics and Biogeography: Cladistics and Vicariance.* Columbia Univ. Press, New York.

Nelson, G., and Platnick, N. I. (1984). *Biogeography* (Carolina Biology Readers No. 119). Carolina Biological Supply Co., Burlington, NC.

Patterson, C. (1981). Methods of palaeobiogeography. In *Vicariance Biogeography: A Critique* (G. Nelson and D. E. Rosen, Eds.), pp. 446–489. Columbia Univ. Press, New York.

Platnick, N. I. (1981). Widespread taxa and biogeographic congruence. In *Advances in Cladistics: Proceedings of the First Meeting of the Willi Hennig Society* (V. A. Funk and D. R. Brooks, Eds.), pp. 223–227. The New York Botanical Garden, Bronx, NY.

Rosen, D. E. (1976). A vicariance model of Caribbean biogeography. *Syst. Zool.* **24**, 431–464.

Rosen, D. E. (1978). Vicariant patterns and historical explanation in biogeography. *Syst. Zool.* **27**, 159–188.

Rosen, D. E. (1979). Fishes from the uplands and intermontane basins of Guatemala: Revisionary studies and comparative geography. *Bull. Am. Mus. Nat. Hist.* **162**, 267–376.

Ross, H. H. (1974). *Biological Systematics.* Addison-Wesley, Reading, PA.

WETLANDS ECOSYSTEMS

Barbara L. Bedford,* Donald J. Leopold,† and James P. Gibbs†
*Cornell University and †State University of New York, Syracuse

I. Diversity of Wetland Ecosystems
II. Diversity within Wetland Ecosystems
III. Functional Diversity
IV. Threats to Wetland Diversity
V. Restoring and Maintaining Wetland Diversity
VI. Conclusions

GLOSSARY

benthic Located on the bottom of a water body, or pertaining to bottom-dwelling organisms.

biological diversity The variety of different forms of life on earth, usually considered to include variety at all levels of biological organization from genetic variation to variety seen in species, populations, communities, and ecosystems.

bog An acidic peatland that receives all water inputs from the atmosphere and hence has low pH (usually 3.4–4.6), low alkalinity, and negligible concentrations of base cations in surface waters; typically dominated by mosses of the genus *Sphagnum* and other plants able to tolerate low pH and low nutrient availability—that is, members of the heath family (Ericaceae).

calcareous fen A term variously used for wetlands with high calcium carbonate in water and soils. In the strict sense, the term refers to peat-accumulating wetlands with surficial deposits of calcium carbonate and a distinctive flora of calciphilic species. These wetlands typically develop where substantial amounts of cold, calcium-carbonate rich groundwater discharge to the surface.

detritivores Organisms that consume dead plant or animal material.

emergent vegetation Rooted plants that tolerate saturated or flooded soil but not extended periods of submersion.

floodplain Broad, flat areas adjacent to rivers and streams that become inundated during floods.

hydrologic regime The aggregate set of variables describing the behavior of water in wetlands, including water depth, magnitude and frequency of water level fluctuations, seasonal pattern of water fluctuations, direction and velocity of water flows, and water residence time.

hydrophytes Plants capable of growing and persisting in saturated or flooded soils.

littoral The shoreline zone where sunlight penetrates to a wetland's substrate and supports rooted plant growth.

macroinvertebrates Organisms lacking backbones and generally visible to the human eye.

marsh A wetland dominated by herbaceous species of plants (e.g., cattails, sedges, grasses, floating ferns) and typically occurring on mineral soils. Floating-leaved and submerged species of plants dominate the parts of marshes that grade into open water. Although the surface layer of some marsh soils may consist of shallow accumulations of organic matter over mineral soils, the depth of this layer is usually less than the minimum required for the site to be classified as peatland.

peat A type of soil formed from accumulation of the remains of dead plants, with high organic matter content (minimally 20% and as high as 90% by dry weight) and depth greater than 30 to 40 cm. Peat forms in some wetlands because anoxic conditions, low pH, low temperatures and short growing seasons, or plant tissues highly resistant to decay prevent rates of decomposition from keeping up with rates of organic matter production.

peatlands Wetlands in which dead organic matter accumulates to form peat more than 30 to 40 cm deep; typically includes bogs, poor fens, and rich fens but some swamps also occur on peat.

poor fen An acidic peatland similar to a bog but having a somewhat higher surface water pH (4.3–5.8) because it receives some groundwater or surface water in addition to precipitation; typically dominated by *Sphagnum* mosses or sedges. But for the presence of distinctive poor-fen indicators, vegetation may superficially resemble bog vegetation or sedge-dominated rich fens that lack *Sphagnum*.

rich fen An alkaline peatland that receives significant inputs from groundwater or surface water; typically having surface water pHs >6.0, higher alkalinity, and distinctly higher concentrations of base cations than bogs and poor fens; dominated by brown mosses (true mosses, largely of the Amblystegiaceae) and small sedges.

species richness The number of different species in a specified area.

submerged vegetation Plants that grow and reproduce while completely submerged.

swamp Wetlands dominated by woody species. Although the term has been variously used, it is used here in the restricted sense of forested wetland, either on mineral soils or peat. However, while many peatland ecologists classify forested fens on deep peat as swamps, they would not call a bog forest a swamp.

THE BIOLOGICAL DIVERSITY OF WETLANDS, or wetland biodiversity, encompasses the immense variety of life supported by the high number of different types of wetlands occurring worldwide. This chapter describes the diversity of wetland ecosystems occurring across major geographic regions, and the diversity of plant and animal species found within particular types of wetlands, including factors believed to control observed patterns in species diversity. The functional diversity of wetlands is discussed at the organismal level (i.e., in terms of various adaptations of plants and animals to the wetland environment), because these adaptations often allow a relatively large number of species to coexist within a wetland. The chapter also identifies threats to wetland diversity and discusses methods of managing to reverse trends in the loss of wetland diversity. Examples and data are drawn primarily from North America, with some reference to wetlands on other continents.

I. DIVERSITY OF WETLAND ECOSYSTEMS

Wetlands, and the organisms adapted to these unique environments, are as diverse as the myriad combinations of climate, topography, and geology that cause different types of wetlands to form. By definition, wetlands are ecosystems whose physical, chemical, and biological characteristics are determined by the constant or recurrent presence of water at or near the soil surface. Beyond this commonality, however, lies a seemingly infinite number of ways in which climate and landscape features combine to create wetlands with different: magnitudes and seasonal patterns of water level fluctuation, paths and rates of water movement, soils, water chemistry, nutrient status, and species composition. Out of this landscape diversity rises the large number of different types of wetland ecosystems seen worldwide.

Wetlands occur on every continent except Antarctica and in every climatic regime. The best available estimates indicate that they cover only between 4 and 6% of the Earth's land surface. Within this relatively small area, however, an astonishingly high number of different wetland ecosystem types occur. A recent global summary (Mitsch, 1994) hints at this diversity when it describes the world's wetlands in terms of coarse-scaled differences in climatic setting, geomorphologic setting, and major wetland type. According to that summary, wetlands "are found (a) in humid, cool regions as bogs, fens, and tundra; (b) along rivers and streams as riparian wetlands, seasonally flooded forests, and back swamps; (c) in the delta's of the world's great rivers; (d) along temperate, subtropical, and tropical coastlines as salt marshes, mudflats, and mangrove swamps; and (e) in arid regions as inland salt flats, seasonal playas, and vernal pools." They also are found

FIGURE 1 Southern deepwater swamp, Congaree National Monument, South Carolina, USA. Photo by Donald J. Leopold.

FIGURE 2 Forested wetland in upstate New York, USA. Photo by Donald J. Leopold.

as (f) marshes, wet meadows, and swamps in depressions that occur in a variety of landscapes and climatic settings; (g) marshes, wet meadows, and swamps on extensive flat areas where the land surface slope is minimal and climate is humid; and (h) marshes, bogs, and fens along the shores of lakes in many climatic regions. Wetlands even occur on (i) slopes as blanket bogs, seepage fens, and vegetated springs.

Finer-scaled differences in climate, geomorphologic setting, and biogeographic region yield an even greater diversity of wetland ecosystem types. Mangrove swamps, for instance, do not exhibit uniform species composition worldwide. The mangrove ecosystems of the Indo-western Pacific have a far greater number of plant species than do the mangrove swamps of Central America, and they certainly should be considered different ecosystems for conservation purposes. The nutrient-poor sawgrass (*Cladium jamaicense*) marshes of the Everglades differ profoundly from the nutrient-rich giant reed grass (*Phragmites australis*) marshes of the Camargue in France's Rhone delta in land form, water chemistry, vegetation, and animal life. Neither of these marsh ecosystems bears much similarity to the thousands of small marshes that dot the prairie pothole region of Canada and the north central United States. The montane cushion-plant herbaceous bogs of Chile contain flora and fauna quite distinct from that of the *Sphagnum* moss-dominated bogs of the Northern Hemisphere, and from the domed bogs of Sarawak. The shore pine (*Pinus contorta*) bogs of Vancouver Island's Pacific coast contain many plant species not seen in Canada's continental bogs. Seasonally flooded forests along rivers include ecosystems as dissimilar as the sometimes monotypic palm swamps of South America, the moderately species-rich hardwood swamps of the southeastern United States, and the highly species-rich *varzea* forests of the Amazon basin. Forested wetlands also include those occurring on steep windward slopes of Puerto Rico's Luquillo Mountains where annual precipitation as high as 5000 mm keeps the clayey soils continuously saturated. In Finland, where about one-third of the land is peatland, at least 30 different mire types are recognized. Some of the bogs and fens of Ireland and Scotland, Scandinavia, Finland, Alaska and Canada, the former Soviet Union, and Maine form intricately patterned peatlands, some of the most unique land forms on earth.

A. Classification of the Major Wetland Ecosystems of the World

Because wetlands take so many forms, occur in many different physiographic and climatic settings, and form part of the history and culture of many regions of the world, their diversity is not easily cataloged. Different regional and colloquial names for them abound, as do formal classification systems. Different common names may refer to the same type of wetland with similar species, while the same name may be applied in different parts of the world to wetlands with quite different species. For example, the terms mire, moor, muskeg, and peatland all refer to similar types of wetlands with similar plant growth forms and many common species. On the other hand, the term swamp is used for a variety of forested wetlands in the United States but for areas dominated by giant reed grass (*Phragmites*) and other herbaceous species in parts of Europe and Africa. Fens dominated by woody species may be called swamps, forested fens, or forested wetlands. In the strictest sense, bogs are peat-accumulating wetlands that receive water only from precipitation. However, the term also is applied to poor fens, which superficially resemble true bogs, and more loosely to any wet vegetated environment that offers less than firm footing.

Numerous formal classifications of the different types of wetlands have been developed. They give some indication of the vast diversity of wetland ecosystems but cannot be used to enumerate that diversity. Because they have been developed for different purposes and from different scientific perspectives, they vary in the criteria used for classification and in the level of resolution to which they classify wetlands. Nonetheless, the existence of literally dozens of wetland classifications worldwide implies the existence of such a high diversity of types that they must be grouped into similar classes for practical reasons of inventory, mapping, management, and scientific communication.

The high diversity of wetland ecosystems also is revealed by considering the many different bases for wetland classification: geomorphologic setting in the landscape, genesis (how they formed), shape or form (e.g., surface morphology, basin morphometry, morphology of the underlying mineral terrain), hydrology (e.g., flooding frequency and duration), sources of water supply, physiognomy of the vegetation, species composition of the vegetation, soil type, water chemistry, and various combinations of these factors. Given that each of these variables represents not a single factor but rather a continuum, the extremely large number of possible combinations becomes apparent. Thus, most classifications in wide use today are broadly based, interdisciplinary, and hierarchical (i.e., consisting of several nested levels with increasing detail only at lower levels in the hierarchy). Species composition, the basis for diversity within wetlands and the focus of many older classifications, is usually not a defining characteristic except at the lowest levels in current classifications. Rather, similarities in hydrology and geomorphology are more likely to form the highest levels of classification, reflecting the importance of the interaction of water with the landscape in determining the biotic characteristics of wetlands.

Nonetheless, the terms bog, fen, marsh, and swamp persist in both scientific and popular literature because of their long history of use and the powerful images they evoke. These terms are not distinguished primarily on the basis of geomorphology or hydrology; nor are they distinguished on any single criterion, nor used consistently. Perhaps the key distinction among them in terms of species diversity is their source or sources of water, which determine hydrologic regime and water chemistry. Greater surface water inputs generally lead to greater water level fluctuations and higher concentrations of base cations and nutrients in marshes and swamps on mineral soils than in bogs and fens. In this chapter, these terms follow current American usage (see glossary).

The current classification systems used in the United States (Cowardin et al., 1979) and Canada (National Wetlands Working Group, 1997) reveal much about the diversity of wetlands in these two countries. For example, the U.S. classification is hierarchical, recognizing 5 major wetland *systems* and 10 *subsystems*, each of which is divided further into 4 to 8 *classes*. A *system* is defined as "a complex of wetlands and deepwater habitats that share the influence of similar hydrologic, geomorphologic, chemical, or biological factors" (Cowardin et al., 1979, p. 4). The five systems are the marine, estuarine, riverine, lacustrine, and palustrine. The term *class* "describes the general appearance of the habitat in terms of either the dominant life forms of the vegetation or the physiography and composition of the substrate—features that can be recognized without the aid of detailed environmental measurements." Classes can be further divided into *dominance types* based on dominant plant species for vegetated sites or on the dominant sedentary or sessile animal species where vegetation is not the dominant cover. Special *modifiers* are used for water regime (type and duration of flooding), water chemistry, soils, and human alterations (e.g., excavated, impounded, diked). Fifty-five classes emerge before one ever gets to the level of dominant species, water regime, or water chemistry.

Most of Canada's wetlands fall within just one *system* (palustrine) of the U.S. classification. Canada's classification differentiates many types of palustrine wetland, especially peatlands. This classification also is hierarchical but with just three levels: *class, form*, and *type*. Five *classes* are recognized: bog, fen, marsh, swamp, and shallow water. *Forms* are defined on the basis of surface morphology of the wetland, position in the landscape, morphology of the underlying terrain, tidal effects, and proximity to surface water bodies. The Canadian system recognizes 16 bog forms and 4 subforms, 12 fen forms and 7 subforms, 8 marsh forms with 18 subforms, 8 swamp forms with 19 subforms, and 5 shallow water forms with 24 subforms. The 18 *types* based on the general physiognomy of the vegetation (e.g., hardwood treed, coniferous treed, low shrub) modify these 121 forms and subforms. Differences in species composition are left to individual users.

Only the wetlands of western Europe are as well documented as those of the United States and Canada. Specific sites on other continents have received consid-

FIGURE 3 Salt marsh in southern New Jersey, USA. Photo by Donald J. Leopold.

erable attention but, with few exceptions, comparative regional treatments began to emerge in the English language only 15 years ago, with most appearing since 1990 (see Bibliography). The exceptions include Chapman's books—for example, *Salt Marshes and Salt Deserts of the World* (1960), *Mangrove Vegetation* (1976), and *Wet Coastal Ecosystems* (1977). However, despite the general lack of scientific studies in many areas of the world, as of 1991, 62 countries from all continents had signed the Convention on Wetlands of International Importance Especially as Waterfowl Habitat (Ramsar Convention) and had designated 527 sites covering more than 30 million ha as being of international importance. Five international wetland symposia have been organized since 1980 by INTECOL (The International Society of Ecology), with a sixth planned for the year 2000 in Quebec City.

B. Major Regional Wetland Ecosystems

Efforts to classify wetlands into unique entities often obscure, either spatially or temporally, critical connections among parts of what might be viewed as integrated ecological systems. For example, various classification systems segment the great delta of the Mississippi and Atchafalaya rivers into many types of floodplain forest, freshwater marsh, salt marsh, streams, and lakes. They assign each prairie pothole in the prairie region of the United States and Canada to one of about 20 classes of prairie wetlands. The mangrove forests of Central America are made to seem distinct from the adjacent beds of sea grasses. Functionally, however, these diverse wetlands are linked and might be better viewed as major regional wetland ecosystems. Water, sediments, plant propagules, fish, birds, other animals, genetic material, and pollutants move among them. One type changes through time to become another. Diversity emerges in both space and time from the components as well as their interconnectedness.

In fact, prior to human modification of them, most wetlands existed as complexes of several different types functioning as integrated systems linked by movements of water, sediment, nutrients, plants, and animals. The vast rain-fed and riverine floodplains that form the Sudd in southern Sudan, for example, consist of a complex and dynamic mix of different types of grassland (e.g., *Echinochloa pyramidalis* or *Hyparrhenia rufa* dominated), wild rice marshes (*Oryza longistaminata*), woodlands (*Acacia seyal* or *Balanites aegyptiaca* dominated), and open water that cover more than 45,000 km^2. The great floodplain of the Orinoco River and its tributaries form a nearly 10-million ha wetland complex of streams, ponds, shallow oxbow lakes, riverine forest, marshes, and seasonally-flooded palm (*Copernicia tectorum*) savannas and grassland in Venezuela and eastern Columbia. The eastern shore of Lake Ontario, New York, is bordered by a 17-mile long interconnected system of streams, shallow ponds, marshes, fens, and shrub swamps. Most temperate zone estuaries include freshwater, brackish, and saltwater marshes along with tidal creeks and mudflats. Mangrove swamps, saline lagoons, sea grass beds, and coral reefs form interconnected systems along shallow coasts in subtropical and tropical regions.

These complexes, or assemblages of wetlands within regions, represent another level of wetland diversity that only partially remains today. Current classification systems do not capture such regional ecosystems, though clearly they exist. Many people recognize wetland complexes such as the Everglades and the Prairie Pothole Region as regional ecosystems. Classification systems, however, only recognize hierarchy in types, not in assemblages of wetlands within regions. Current efforts by conservation organizations and government agencies to move to ecoregional mapping attempt to overcome this limitation.

C. Continental and Regional Diversity

Climate and geomorphology, along with biogeographic factors, control the diversity of wetlands on different continents and in different regions. Wetland development is promoted where precipitation annually exceeds evapotranspiration on a long-term basis. In such climates, wetlands develop in a number of landscape settings, including depressions in the land surface, areas where minimal land surface slope slows surface water runoff, along breaks in the slope of the land surface, and even on slopes if precipitation greatly exceeds evapotranspiration. In semiarid, arid, and desert climates, wetlands occur far less frequently, but they are found (a) in areas with depressions or lowlying land along rivers or coasts where runoff converges from extensive uplands or mountains, (b) along coasts regularly inundated by tides, (c) in areas where permafrost or other impermeable layers below ground prevent or slow infiltration of surface water, and (d) in spring-fed oases.

In addition to topographic variation, variation in the composition and stratigraphy of surficial geological deposits promotes diversity in the types of wetland that occur within a region. The composition of subsurface materials plays a major role in determining the chemistry of water entering wetlands, and thereby the plant species composition of wetlands. For example, wetlands occurring in regions dominated by limestone deposits tend to have high-pH waters rich in base cations that promote the growth of distinctive rich fen vegetation. In contrast, waters draining areas with granitic bedrock tend to be low in pH and base cations; such areas more frequently contain bog-like vegetation. Areas with thick deposits of glacial till can contain a mix of rock types that, in turn, may support a variety of wetland types. A diversity of types in a given region also is promoted where surface topography and subsurface stratigraphy of geological deposits produce groundwater flow paths of varying lengths. Wetland water chemistry is strongly influenced not only by water source but also by the length of groundwater flow paths that enter the wetland and the number of different groundwater flow systems that feed the wetland. Different flow systems may contribute water with different chemistry because the water originates from different geological deposits. If the water originates from the same geological materials, longer flow paths may yield higher concentrations of solutes because longer flow paths generally correlate with longer groundwater residence time (i.e., longer contact time with geological materials).

Biogeographic factors also influence the number of wetland types occurring in different regions and continents. Aquatic plants generally are known for their remarkably wide geographic range, but many species have smaller ranges determined in part by biogeographic factors (i.e., the climatological, geological, glacial, and floristic history of a region). The distributions of these more restricted species help define distinctive biogeographic regions that contain wetlands with species compositions unlike those from other biogeographic regions. For example, shallow ponds that support many species of the Atlantic Coastal Plain flora might be classed broadly as emergent marshes, but this would miss their unique contribution to the diversity of North American wetlands. The flora of these coastal plain ponds, which has an unusual abundance of rare and uncommon species, bears little resemblance to the emergent marshes of other biogeographic regions.

While no systematic comparison has been made of wetland ecosystem diversity in different regions, the factors controlling regional wetland diversity are understood. All other things being equal, regions with the greatest climatic and geomorphologic variation are likely also to contain the highest diversity of wetland types. If wetland types are defined to the level of plant species composition, then biogeography enters into the equation. Thus, continental diversity will vary as a function of the number and combination of different climatic, geomorphologic, and biogeographic regions occurring on that continent.

II. DIVERSITY WITHIN WETLAND ECOSYSTEMS

A. Plant Species Richness within Wetlands

At the regional and global scale, wetlands clearly exhibit high diversity of types and an associated high diversity of plant species. Within-habitat (or *alpha*) diversity, and between-habitat (or *beta*) diversity, however, is highly variable among wetlands. Wetlands include some of the most species-rich plant communities in the world, as well as numerous examples of extensive stands dominated by a single species. Familiar examples of single-species wetlands include mangrove swamps, tidal marshes of salt marsh cordgrass (*Spartina alterniflora* or *Spartina patens*), cattail (*Typha* spp.) marshes, and papyrus (*Cyperus papyrus*) swamps. Species-rich herbaceous communities include calcareous fens with as many as 20 to 30 species of vascular plants and bryophytes per square meter. Intermediate examples include abandoned beaver ponds with 3 to 12 species per square meter and a range of coastal wetlands (bottomland hardwoods, wet prairies, fresh marsh, and salt marsh) from the southern United States with 1 to 17 species per square meter.

Change in plant species composition across gradients within wetland sites depends on the amount of change that occurs within a site in the physical and chemical variables controlling species distributions. A single small depression may encompass a wide range of water depths and the associated changes in species composition. Although adjacent zones may share some species, the deep-water end of the gradient, which may be dominated by floating or submerged species, shares no species with the sedge- and forb-dominated wet meadow zone at the other end of the gradient. Depending on the slope and width of the floodplain, wetlands along major rivers may include areas that experience very different frequencies of flooding and consist of very different plant communities, from the open water and marshes of backwater sloughs to hardwood-dominated forests along natural levees. Erosion and deposition of sediments within the floodplain create surfaces of varying ages that support plant communities of different successional status and composition. Much of the species diversity in the Peruvian Amazon has been attributed to the heterogeneity created by long-term patterns of erosion, deposition, and plant succession.

In other wetlands, environment changes little over relatively large distances, for example, where the accumulation of peat has filled in basins and minimized topographic changes. Though the presence of hummocks and hollows introduces small-scale heterogeneity, mean water depth, pH, and nutrient availability change little as one moves across the bog surface. Plant species composition in plots spaced at great distances, but at equivalent microtopographical elevation, may be quite similar.

Although many studies have examined plant species richness in small plots, systematic comparisons are difficult to make because plot size and sample number vary among studies. Given that species richness is known to increase with increasing sample area, conclusions cannot be drawn about relative richness among samples if sample areas are not equal. Nonetheless, some general statements can be made. First, the number of plant species present generally decreases as water depth and frequency of saturation or flooding increase. For example, in his study of Wisconsin plant communities, Curtis (1959) reported that the average number of species per stand ranged as follows: 7 in stands of submerged aquatics, 11 for emergent aquatics, 28 to 29 in sedge meadows, 44 in wet prairie, and 62 in wet-mesic prairie.

Second, the effect of moving water on diversity depends on flood velocity, magnitude, and frequency. Some degree of disturbance from flooding increases diversity by removing live biomass and litter, thereby creating openings for germination of new species. Extreme or very frequent flooding, however, lowers diversity. Few species can withstand both the high stress of extended periods of anoxia and the frequent disturbance of biomass removal.

Third, the vegetation of saline wetlands tends to be less species-rich than freshwater wetlands. For example, coastal salt marshes of the eastern United States rarely contain more than two species in 0.25-m^2 plots. Plots of comparable size in freshwater herbaceous communities frequently contain 10 or more species. Increasing the plot size in salt marshes adds few species but is likely to significantly increase species number in many types of freshwater wetlands. For example, the number of vascular and bryophyte species in rich fens in New York increases from about 20 in 1-m^2 plots to 40 in 25-m^2 plots to 65 in 100-m^2 plots.

Fourth, in the temperate and boreal zones, *alpha* diversity generally increases across the range of peatland types from bog to rich fen for vascular species. Continental raised bogs in eastern North America may have only 20 to 25 vascular species per site, one of the most impoverished vascular floras in North America. Rich fens may contain as many as 140 vascular species at a single site. The bryophyte diversity also tends to increase from bogs to rich fens in eastern North America. Continental raised bogs usually have fewer than 20 species of bryophytes in 100 m^2 while rich fens usually have between 20 and 60 bryophytes in a similar-sized plot. However, Vitt and others (1995) reported that bryophyte diversity in peatlands of continental western Canada could be similar for bogs and fens. Species richness in larger plots (2500 m^2) ranged from about 5 to 24 in bogs and poor fens and from about 5 to 28 in rich fens. Thus, even though the highest species richness was found in extreme rich fens, any individual bog could contain as many species of bryophytes as a rich fen. At the landscape scale, rich fens are distinctly more species rich than bogs and poor fens. Because their species composition varies more among sites, rich fens have greater diversity when an entire region is considered. In both Britain and continental western Canada, bogs have only about half the number of bryophyte species as fens.

Fifth, relative to the landscapes in which they occur, many types of wetlands appear to be islands of diversity. Fens in northeastern Iowa support 28% of the regional and 18% of the total state flora even though they currently cover only 0.01% of the land surface. Boreal old-growth swamp forests constitute only 5% of Sweden's forest land area but harbor more than half of all vascular forest plant species found in Sweden and about a third of the total Swedish boreal flora of bryophytes. Almost a third of the federally listed threatened and endangered plant species in the United States are wetland dependent. Furthermore, for some types of wetlands, rare and uncommon plant species tend to be associated with the most species-rich sites.

Finally, latitudinal patterns in plant species diversity of wetlands do not appear to parallel those for terrestrial environments. In sharp contrast to terrestrial habitats, wetlands in the tropics are not necessarily more species rich than those in the temperate and boreal zones. According to Ellison (in press), the mangrove swamps, palustrine forested wetlands, palm swamps, and hardwood swamp forests of Central America all exhibit low species diversity. On the basis of a comparison of representative aquatic families and aquatic habitats, Crow (1993) also concluded that aquatic plant diversity was higher in both warm and cool temperate latitudes than in the tropics. On the other hand, some tropical wetlands can be highly diverse. Ellison (in press) reported high plant species richness (292 species of 0.5 m in height or greater) in riparian forest fragments in Belize. Additional systematic and standardized studies are needed to address this issue.

FIGURE 4 Frog and duckweed in freshwater marsh in upstate New York, USA. Photo by Donald J. Leopold.

B. Major Groups of Wetland Animals

In terms of abundance, biomass, and species richness, macroinvertebrates are the most important wetland-dependent animals in freshwater wetlands. Four groups—the insects, mollusks, crustaceans, and annelids—make up the majority. The aquatic insects are the most functionally and taxonomically diverse, and include forms that occur throughout all habitats (the bottom, deep waters, and shallow, vegetated areas) in most wetlands. The midges (Chironomidae, Diptera) constitute perhaps the most widespread group of aquatic insects, and larval forms of midges typically represent the most abundant macroinvertebrates in wetlands. Other ecologically important, widespread groups include the mosquitoes, dragonflies and damselflies (Odonata), mayflies (Ephemeroptera), caddisflies (Trichoptera), stoneflies (Plecoptera), craneflies (Tipulidae), and water beetles (e.g., Corixidae, Belostomatidae, and Notonectidae). Among these groups, only the water beetles are entirely aquatic; members of other groups typically have aquatic larval stages that undergo a synchronized transformation ("hatch") into adult forms, which may live for periods ranging from 1 to 2 days (some mayflies) to weeks or months (many dragonflies).

The mollusks include the many species of filter-feeding clams and herbivorous snails. The most widespread and common are fingernail clams. Larger clams and mussels are equally widespread but less common, although they are sometimes abundant along large rivers and lake fringes. Most large clams dwell on the bottom of wetlands, whereas the fingernail clams often occur in submerged vegetation. Snails also occur in many wetlands and live upon stems, stalks, and leaves of aquatic vegetation, and graze on epiphytes.

The crustaceans, like the aquatic insects, are commonly encountered throughout most freshwater wetlands. Especially common in open-water and upon aquatic vegetation are isopods and amphipods. In protected open water areas, small zooplankters, including cladocerans and copepods are abundant. Crayfish are common bottom-dwelling crustaceans in many wetland ecosystems. The last major group of aquatic invertebrates, the annelids (mainly oligochaetes), are not particularly diverse taxonomically. However, they can be extremely abundant in some wetlands, where they burrow in the wetland bottom or attach to aquatic vegetation.

Vertebrate animals represent another important component of the wetland fauna and include the amphibians, reptiles, fishes, birds, and mammals. The amphibians are composed of two main groups, the frogs and toads, and the salamanders. Most species of frogs and toads have complex life cycles that include an aquatic, usually herbivorous, larval stage adapted for rapid growth (the tadpole stage), and a terrestrial, carnivorous adult stage adapted for dispersal among wetlands. Their relative use of wetlands versus uplands varies among groups. For example, the major group of frogs in North America, the Ranidae, includes (a) members that may be mostly terrestrial but return to wetlands to breed and sometimes to overwinter (e.g., members of the leopard frog complex, e.g. *Rana pipiens*) and (b) other species (e.g., bullfrogs, *Rana catesbeiana*) that are primarily aquatic and typically leave wetlands only to disperse to new areas. Among North American salamanders, all are wetland dependent except for members of the Plethodontidae, although even this family includes the Desmognathinae, a large group of stream-associated species. In salamanders, wetland dependency ranges from strictly aquatic (e.g., mudpuppies [*Necturus*], hellbenders [*Cryptobranchus*], Conger eels [*Amphiuma*], and sirens [*Siren*]), to semiaquatic in groups that spend much the growing season in wetlands but overwinter on land (e.g., newts [Salamandridae]), to primarily terrestrial species that return to wetlands only to breed (e.g., the vernal pool-breeding mole salamanders [Ambystomidae]).

In terms of numbers, species richness, and biomass, the majority of wetland fishes are small forage fishes. In North America, the primary groups are the killifishes (*Fundulus*), shiners (*Notropis*), sunfishes (*Lepomis*), and mosquito fishes (*Gambusia*). In deeper-water wetlands or wetlands connected to permanent water bodies, larger, bottom-feeding species occur, such as Ictalurid catfish (e.g., bullheads) and carp (*Cyprinus carpio*), as well as carnivorous species such as pickerel (*Esox*), perch (*Perca*), and bass (*Micropterus, Morone*).

Turtles are the most diverse group of wetland reptiles. Most turtles are highly aquatic and leave water only to lay their eggs and to disperse to new habitats. Commonly encountered, highly aquatic turtles in North America include the snapping turtle (*Chelydra serpentina*), mud and musk turtles (Kinosternidae), and softshells (Trionychidae). The Family Emydidae includes many other highly aquatic turtles, such as cooters, sliders, painted turtles (*Pseudemys* and *Chrysemys*), map turtles and sawbacks (*Graptemys*), and terrapins (*Malaclemys*), as well as other partially aquatic species that travel regularly among disjunct wetlands (e.g., spotted turtle [*Clemmys guttata*]) or that use wetlands only for hibernation (e.g., wood turtles [*Clemmys insculpta*]). Other wetland-dependent reptiles include the water snakes (e.g., *Nerodia*) that use wetlands primarily for feeding on fish, frogs, and crayfish, and rest above water on hydrophytes or on land at other times. Most wetland-dependent snakes are viviparous, that is, bear live young, and therefore do not undertake migrations to terrestrial habitats to lay eggs, as do turtles. Many other snakes, however, and some lizards, live primarily in uplands but feed opportunistically in wetlands (e.g., king snakes [*Lampropeltis*], rat snakes [*Elaphe*], and garter snakes [*Thamnophis*]).

Several large groups of birds occur nearly exclusively in wetlands. These include many carnivorous (mainly fish-eating) species, such as the loons (Gaviidae), herons and bitterns (Ardeidae), several waterfowl species (Anatidae), the kingfishers (Alcedinidae), terns (Sternidae), and several raptors (Accipitridae, e.g., ospreys *Pandion haliaetus*]). Omnivorous groups include the grebes (Podicipedidae), many species of diving and dabbling ducks (Anatidae), cranes (Gruidae), rails (Rallidae), and gulls (Laridae). Primarily insectivorous species include the shorebirds (Charadriidae) and many songbirds. Unlike other wetland inhabitants, birds are highly mobile and often use disjunct wetlands seasonally or even daily.

FIGURE 5 Foraging assemblage of scarlet ibis and barefaced ibis in the seasonally flooded grasslands of central Venezuela (Apure state). Photo by Mark Gregory.

The last group of wetland-associated vertebrate animals, the mammals, include few, wholly wetland-dependent species. Most prominent are rodents such as the muskrat (*Ondatra zibethicus*) and beaver (*Castor canadensis*). Other obligate wetland mammals include several primitive, carnivorous shrews ("water shrews" in the genus *Sorex*), some lagomorphs such as the swamp rabbit (*Sylvilagus aquaticus*) and marsh rabbit (*S. palustris*), and mustelids such as the river otter (*Lutra canadensis*). Wetlands are a critical resource, however, for many otherwise terrestrial species, which use wetlands for feeding and cover, such as moose (*Alces alces*), raccoon (*Procyon lotor*), and mink (*Mustela vison*).

Animal communities of saltwater wetlands can differ substantially from those in freshwater systems because of the effects of salinity on animal metabolism. Generally speaking, analogous taxa occur in freshwater and saltwater wetlands, such that similar niches are occupied, but species richness of animals generally declines along a freshwater-brackish-saltwater gradient. Reptiles from the Gulf of Mexico region provide an example: about 24 species occur in fresh marsh, 16 in brackish marshes, but just 4 in salt marshes. Of the salt marsh species, few are salt marsh specialists. Some terrapins are largely restricted to salt marsh habitats, but other regularly encountered salt marsh inhabitants, such as cooters and alligators, actually prefer freshwater situations. Amphibians show a particularly sharp trend along the wetland salinity gradient. Owing to their highly permeable skins and inability to counteract the drying conditions produced by high osmotic pressures of salt water, very few species can survive in saline wetlands and even then occur only infrequently (e.g., some toads).

Relatively few breeding birds specialize on salt marsh habitats, although sparrows and rails are well represented. However, salt marshes are noted for their importance as stopover sites for the vast populations of migrating species that breed in freshwater wetlands that freeze in winter. Waterfowl, shorebirds, songbirds, and owls can occur in brackish and salt marsh habitats in large numbers during the colder months resulting in very diverse, if unstable, assemblages during the winter period. Thus, though occupied for relatively short periods, and not by a large number of breeding species, coastal marshes nevertheless are a critical habitat in the annual cycle of many bird species.

Mammal communities in saltwater wetlands are generally quite species poor. However, the occasional effects of high densities of aquatic rodents (e.g., muskrats and nutria) can be of ecological significance. As in freshwater systems, these species can radically alter the wetland environment through so-called "eat-outs" of the vegetation. Otherwise, saline wetlands are infrequently visited by species more dependent on freshwater wetlands, such as otter, mink, or raccoon.

Overlap of freshwater and marine species, in conjunction with resident species that can tolerate the substantial fluctuations in salinity, produces highly complex, species-rich assemblages of fish and crustacean species in saltwater wetlands. Though characteristically species rich, these assemblages are temporally quite unstable because of daily tidal cycles, with freshwater-associated species descending into salt marsh systems on outgoing tides and marine species arriving on incoming tides. In particular, many marine fishes and crustaceans, especially shrimp, use salt marshes as nursery areas, particularly for post-larval and juvenile forms. Here they take advantage of the lower numbers of predators in salt marshes relative to the open ocean and have access to the high productivity of salt marshes.

C. Controls of Species Diversity

Within particular wetland ecosystems, plant and animal diversity reflects the interacting influence of a number of abiotic and biotic variables that operate across a range of spatial and temporal scales. Diversity in wetlands cannot be understood without attention to these gradients, the scales at which they operate, and their interactions. Of the physical and chemical variables, hydrologic regime, mineral ion concentrations, pH, nutrient availability, salinity, and disturbance have been shown to play particularly strong roles. They largely determine the number of species that are physiologically capable of living in a site. Important biotic variables include primary productivity and interspecific competition. Along with abiotic factors, they influence which plant and animal species actually occur within the site. Because the composition, structure, and density of vegetation form key elements of habitat for animal species, these gradients also strongly influence animal diversity.

Depending on the type of wetland, different abiotic and biotic variables assume greater or lesser importance in determining species distributions. Hydrologic regime is important in all wetlands for both plant and animal species. Saturation with water severely restricts diffusion of oxygen into the soil while microbes rapidly deplete available soil oxygen. Hydrologic regime, therefore, largely determines the distribution and frequency of the most universal stress to which wetland plants and animals must adapt—anoxia in the water column, shallow soils, and rooting zone of plants. Salinity generally assumes importance only in coastal wetlands and in the saline wetlands of semiarid and arid regions. However, inland salt marshes occur in the Great Lakes region and some of the wetlands of the prairie pothole region of North America are saline. Mineral ion concentrations and pH affect species composition in all types of wetlands but have particular significance in peatlands that exhibit both extremes of the gradient in these factors. The combination of low nutrient availability, low pH, and low concentrations of calcium and magnesium makes bogs inhospitable environments for most plant and animal species. The extremely high pH and calcium concentrations of some rich fens are associated with high species richness and a large number of rare and uncommon species of plants and snails. Nutrient availability, which is lower in all undisturbed peatlands than in mineral soil wetlands, has become an issue for many types of wetlands as humans have increased nutrient inputs to wetlands.

Flowing water, especially at faster rates and larger volumes, exerts an additional mechanical stress on wetland species in some environments. Plants growing in river floodplains, tidal marshes, and other wetlands subject to flooding with rapidly flowing water must be able to withstand the force of moving water or must have life histories that allow them to germinate and reproduce in periods between floods. Well-developed root systems and flexible stems are essential for perennial plant species. Animals in these environments must be able to move out of the reach of flood waters, either into sediments or the tops of trees or out of the flooded area.

Many types of disturbance—factors that remove plant biomass—affect wetland diversity. Extended periods of inundation with deep water kill most emergent plants, which allows new species to germinate when water levels drop. Muskrats and nutria eat marsh vegetation. Moose consume wetland shrubs and aquatic vegetation. Beavers fell trees, alter hydrologic regimes with their dams, and destroy existing vegetation in digging for mud to strengthen the dam. The flooding created by beaver ponds destroys flood-intolerant vegetation. Tides and storm surges deposit patches of debris that bury vegetation in coastal wetlands. Hurricanes rip trees out of swamps and deposit sediment and other debris in wetlands. Humans harvest timber from swamps in many parts of the world. In Africa, southeastern Asia, and parts of South America, humans harvest other wetland plants for various uses, including firewood, food, and shelter. Reeds (*Phragmites communis*) are still harvested in The Netherlands and England as thatching for roofs.

Change over time as well as space is the rule in wetlands. Because climate drives hydrology, seasonal, annual, and longer-term changes occur in water depth and all the chemical and physical factors it regulates. Extended periods of wet or dry years dramatically change the distribution of both plant and animal species. Some wetlands systems, such as the prairie potholes of North America, have 10- to 20-year cycles of wet to dry climate. In other regional wetlands, such as the deep-water cypress swamps of the southeastern United States, wet to dry cycles occur over much longer time periods. The effect of this temporal variation at many scales is to increase overall species diversity by promoting a dynamic community composition and spatial distribution among years. In wetlands where organic matter accumulates as peat, even more dramatic changes occur with time; the landscape itself is transformed. Over hundreds to thousands of years, a mineral-rich, deep-water pond with high pH and relatively high nutrient availability can change to a raised bog with low pH, low nutrient availability, and low mineral ion concentrations. The plants and animals of a pond are replaced with those of the bog. Long-term deposition of mineral sediments in estuaries can create similarly dramatic changes in topography, landform, and the hydrologic regime for wetland organisms.

1. Controls of Plant Species Richness

The question of what controls patterns in plant species richness is an old one that has received renewed attention with observed declines in species diversity over much of the globe. In wetlands, the question has been resolved only in part. The effects of many single factors are relatively well understood because of the over-riding importance of some physical and chemical variables in determining the pool of species physiologically capable of living in a site. Spatial and temporal heterogeneity, long recognized as critical determinants of species richness in other systems, are also positively correlated with species richness in wetlands. However, the relative importance of interacting sets of variables, operating in different types of wetlands and over different spatial and temporal scales, is only beginning to be understood. Because humans have increased nutrient inputs to many wetlands, the question of the relationship of plant species richness to nutrient availability, community productivity or biomass, and competition continues to motivate many wetland studies. Controls operating at regional and global scales have been little studied.

FIGURE 6 Raised bog, coastal Maine, USA. Photo by Donald J. Leopold.

Gradients in several physical and chemical factors show strong correlation with species richness in wetlands, especially at the ends of the gradients where stress on plant growth is high. While more than 6700 species, approximately one-third of the total United States vascular flora, grow in wetlands, few species have evolved to persist in wetlands with low pH, high salinity, continuous inundation with shallow water, or high flooding frequency. Hence, bogs (low pH), the low marsh zone of salt marshes (high salinity), deep-water marshes (continuous inundation), and wetlands in the active channels of rivers (high flooding frequency) are all species-poor. In general, species richness increases with decreasing stress due to abiotic factors.

Numerous studies have shown that water level fluctuations are critical to maintaining species diversity in a wide variety of wetlands. Alternating periods of inundation and exposure of the soil create temporal and spatial gradients of water depth that provide conditions for germination of the largest number of species. Species diversity always declines when water levels are artificially stabilized. Several studies have shown that coastal plain ponds, Great Lakes shoreline vegetation, prairie wetlands, and Carolina bay wetlands all require fluctuations in water depth to maintain overall species richness.

Spatial heterogeneity increases the number of species co-occurrences in wetlands through a number of mechanisms. The presence of hummocks and hollows in peatlands creates small-scale gradients in pH and moisture that allow species with a wider range of tolerances to coexist in relatively small areas. The presence of mounds and tree bases in floodplain swamps provides sites for germination of a number of species that do not germinate or persist in standing water. Old muskrat mounds serve the same function in cattail marshes. Microtopographic variation in riparian wetlands causes spatial variation in flood frequencies that results in higher species richness. Deposition of wrack (dead plant debris) by tides and storms increases species richness in salt marshes by producing openings in dense stands of *Spartina*; these openings have higher salinity than the surrounding vegetation and are colonized by more salt-tolerant species (e.g., *Distichlis spicata*). High variation in substrate pH, type, texture, and age creates microhabitat variation along arctic, boreal, and tropical rivers and increases the number of plant species occurring in riparian vegetation.

Some of the spatial heterogeneity associated with higher species richness in wetlands results from various types of disturbance. In general, richness first increases and then decreases along gradients of increasing disturbance so that richness is highest at intermediate points along the gradient. In many environments, lack of disturbance allows a few highly competitive species to displace other species through the process of competitive exclusion. Disturbance disrupts this process and allows other species to persist. However, richness decreases if disturbance is too frequent because few species are adapted to very high disturbance frequencies.

Fertility, or nutrient availability, appears to be another strong control on plant species richness. However, few studies have actually measured nutrient availability because of the difficulty of doing so. Rather, nutrient availability has been inferred from surrogates such as community biomass, percent soil organic matter, or substrate texture. Use of these surrogates generally suggests that species richness is highest at moderate levels of community biomass and presumably, therefore, at moderate levels of nutrient availability. Close inspection of the data shows that richness can be low or high at low to intermediate levels of community biomass but almost always decreases above some threshold of high biomass. The amount of biomass at which that threshold occurs, however, varies greatly among wetland types. Furthermore, the relationship between richness and biomass appears to be scale-dependent. When data from wetlands with different vegetation types are examined, richness follows a unimodal relationship with community biomass. Within small plots within types, the relationship breaks down.

The complex interplay of abiotic factors with biological variables such as competition and productivity has been examined in only a few freshwater wetlands and seldom through experimental manipulation of variables. Salt marshes have received more attention. Results from these studies emphasize that the relative importance and direction of the effect of biotic variables on species richness change along gradients of stress. Abiotic variables play the strongest role where stress is high but the presence of some species can ameliorate the stress for other species (e.g., by lowering salinity through shading, which decreases concentration of salts due to evapotranspiration). Species richness is thereby increased above what abiotic conditions might otherwise allow through positive interactions among species. Nonetheless, the number of species tends to decline monotonically with increasing stress from abiotic factors. Where stress in minimal, biotic control of species richness is greater. Here, variables related to density effects (e.g., productivity, biomass, light) are more likely to control variation in species richness, probably through competitive effects of dominant species on subordinate ones. In such environmentally benign wetlands, productivity is likely to be high and a few species are likely to dominate unless the process of competitive displacement is disrupted by disturbances that remove plant biomass and allow other species to establish.

The interplay of these local biotic and abiotic factors is played out against the background of regional and historical controls on the species pool (i.e., the number of species available to colonize a site). While local and current conditions determine which plant species do colonize a site, the regional species pool sets the upper limit to the number of species that could colonize the site. The geological, evolutionary, and ecological history of a region, along with its area, controls this pool. Geological age, processes of speciation and extinction, geographic dispersal, and chance historical events all leave their imprint on the species pools of particular regions. Few attempts have been made to relate the species pool to the number of species actually co-occurring in a site because of the difficulty of obtaining adequate data sets.

2. Controls of Animal Species Diversity

Patterns of diversity and productivity in wetland animals follow gradients strongly related to hydrologic regime, primarily because of its effects on the composition, structure, and productivity of wetland vegetation and on oxygen availability within the water column and wetland sediments. Oxygen levels most influence the distributions of wholly aquatic, nonmigratory wetland animals (i.e., the macroinvertebrates and fishes). The productivity and diversity of wetland invertebrates and fish, and, consequently, many water birds and mammals, are highest in semipermanent wetlands (e.g., those flooded during most of the growing season but dry otherwise). Most amphibians, however, occur in transient pools. Trends in reptile diversity and productivity along this inundation gradient are not clear.

Semipermanent and permanent water bodies are the primary habitats of fish, which generally lack the ability of many other wetland inhabitants to disperse overland or enter dormancy when wetlands dry. Thus, areas of sufficiently oxygenated water must occur in wetlands throughout the year to support fish populations, especially those of large-bodied species. Large, seasonally flooded wetlands with pools or channels that remain permanently flooded, and lakes with well-developed littoral vegetation tend to support the most diverse fish communities. Similarly, fish-eating birds and mammals frequent the habitats occupied by their fish prey. In contrast, temporary, unvegetated pools are the primary breeding habitats of amphibians, offering two main advantages. First, amphibian larvae are able to exploit a pulse of primary productivity that occurs shortly after pools fill with water. Second, temporary pools will not support fish, major predators on amphibian eggs and larvae.

Many other wetland animals exploit intermittently flooded wetlands. Waterfowl, particularly egg-laying females, frequently travel among shallow pools, ditches, and puddles to feed on the pulse of invertebrate production that occurs following flooding. These feeding pools often are remote from nesting areas at more permanent wetlands, where deep waters and dense vegetation protect nesting birds and their eggs from predators. Similarly, snakes and turtles often can be found in temporarily flooded wetlands, and seek refuge from subsequent low-water periods by migrating or entering dormancy in a wetland's sediments.

Much of the variation in cover and food for wetland fauna, and hence in animal diversity, is associated with variation in the species diversity, density, and structural characteristics of wetland plants. Many wetland animals show affinities for particular life-forms of hydrophytes (e.g., submerged plants, floating-leaved plants, or emergents), and some are associated with specific plant taxa. For example, most aquatic invertebrates seek protection from predation and safe oviposition sites in macrophytes. In general, plants with dissected leaves support more invertebrates than plants with broad leaves. Macroinvertebrate diversity and density generally are greater in vegetated than open water areas, and, within vegetated areas, are greatest in areas of mixed emergent and submerged vegetation. Shading by erect hydrophytes also influences macroinvertebrate abundance; unshaded areas of submerged vegetation often support dense populations owing, in part, to elevated water temperatures in unshaded areas. Similarly, shallow, vegetated wetland habitats fringing deep-water areas are critical nursery habitats for fishes, many species of which use stands of particular wetland plant species as spawning habitat. These areas provide protective cover for eggs and larval fishes and food resources for young fishes. Like aquatic insects, many fishes prefer unshaded, shallow areas where warm water temperatures hasten the development of eggs and larvae.

Two key water chemistry parameters that influence distributions and diversity of wetland animals are acidity and salinity. The relationship of acidity to the distribution of wetland animals can be complex. For example, many aquatic, fish-eating birds frequent wetlands with moderately acidic waters (pH 5.5–6.0). Although fish populations often are reduced in wetlands with waters of pH <6.0, suspended material may precipitate from the water column at low pH. The resulting increase in water transparency may improve encounter rates with prey and capture success for birds that pursue their prey underwater, and thereby compensate for decreased fish density. The reduction or loss of fish from moderately and strongly acidified waters also reduces competition between fish and insectivorous birds for macroinvertebrate prey. This may explain why dabbling ducks frequent waters of relatively high acidity that fish-eating ducks avoid. Highly acidic waters (i.e., pH <5.0) generally support few fish or amphibians and have low diversity of aquatic invertebrates. Water bird and mammal use of such wetlands is limited.

Salinity strongly limits animal diversity in wetlands, primarily through effects on the composition of plant and invertebrate communities and through its direct physiological effects on wetland animals. Difficulty in maintaining water balance leads many dabbling ducks, particularly those brooding young, to avoid wetlands with highly saline waters. Owing to their highly permeable skin, amphibians are intolerant of even moderately saline conditions. For similar reasons, highly saline conditions often limit the survival and hatching success of fish eggs and survival of larval fish. Many mammals, such as muskrats, are tolerant of highly saline conditions, but populations often are limited by the low quality of forage plants that grow in saline wetlands.

Three spatial characteristics of wetlands and their distribution in the landscape—size, connectivity, and isolation—also influence the diversity of wetland animals within and among sites. For example, small wetlands often support fewer species of birds than large wetlands, and isolated wetlands support fewer species than wetlands in complexes. Aquatic, fish-eating birds are particularly sensitive to wetland area and often shun small wetlands (<10 ha). Small wetlands may not provide a sufficient quantity of habitat or prey for many large-bodied species, and short inter-wetland distances may provide certain species with alternate foraging sites while minimizing their time in flight. Larger wetlands generally support a wider diversity of vegetative lifeforms and, therefore, may provide a greater variety of habitats for wetland species than smaller wetlands. This may account for the frequently observed pattern of increased fish species diversity in larger than smaller wetlands.

The proximity of diverse habitat types within a wetland, and connectivity among wetland environments and with upland habitats, is critical to maintaining high diversity of wetland animals, particularly amphibians, some birds, and fish. Aquatic breeding and terrestrial nonbreeding habitats must be contiguous for amphibian populations to persist. Local breeding populations undergo frequent extinction and recolonization events, and individuals often are exchanged among populations. The importance of migration in amphibian population ecology is indicated by the presence in many species of a juvenile stage adapted strictly for dispersal. Destruction of larval or adult habitats, or the connections between them (e.g., riparian corridors) often drives local amphibian populations to extinction.

Proximity of diverse habitat types within a wetland is critical to many breeding water birds and fish. Some water birds, such as rails, dwell within dense stands of emergent vegetation, while others, such as bitterns, herons, and shorebirds, forage mainly along the interface of wetland vegetation and open water. Still others, such as grebes and many dabbling ducks, are associated with floating and submerged vegetation, while loons and cormorants forage in deep, open waters. Most of these birds, however, build their nests within dense stands of emergents, on small islands, or in trees or shrubs.

Fish assemblages provide another example of the importance to wetland animals of spatial heterogeneity and connectivity in wetland habitats. Most large species of bottom-feeding and predatory species undergo small-scale migrations between feeding and spawning areas in shallow wetlands and resting areas in deeper, open-water habitats. These migrations often occur nightly, with fishes returning to deep waters to rest during the day. Critical to these migratory movements are the links established between shallow, vegetated areas and deep-water zones by seasonal regimes of wetland flooding. Finally, many fishes undergo seasonal migrations over hundreds of kilometers into river floodplains and return to the river channels as waters recede.

Some wetland animals create some of the spatial and temporal heterogeneity that promotes animal diversity in wetlands. Beavers alter stream channels and transform riparian forests into freshwater marshes and ponds with stands of dead timber. Beaver activities also permanently modify soils beneath dam sites and even change local topographies; beaver dams reduce current velocities and cause massive silt depositions, which remain long after the dams disintegrate. High levels of herbivory by muskrats and geese can result in "eat outs," whereby essentially all emergent plant material in a marsh is consumed over a few years. Such drastic, animal-caused modification of the wetland environment not only affects habitat availability for other wetland animals but can alter patterns of energy flow and biotic productivity within wetland ecosystems. Other, less drastic examples of habitat modification include the activities of alligators, which construct small basins that serve as critical refuges for many fishes, and important feeding areas for fish-eating birds, during low water periods. Crayfish burrows serve in many areas as moist refuges for water snakes, salamanders, and frogs.

III. FUNCTIONAL DIVERSITY

Wetland plants exhibit a diverse array of adaptations that allow many species to tolerate stresses associated with soil saturation, alternating periods of flooding and drought, high salinity, low pH, and low nutrient availability. Other adaptations assist wetland plants in dispersal and establishment. Animals also exhibit an array of adaptations to wetland environments, although unlike plants, many simply can move to escape stresses. Collectively, these adaptations explain much of the high species richness and diversity of many wetlands, especially those with strong hydrological and chemical gradients.

A. Adaptations to Anoxic Environments

Many wetland plants have one or more morphological and anatomical adaptations that allow them to tolerate soil saturation and anoxia for short to long time periods, primarily by allowing more oxygen to reach the plant root system. Key morphological adaptations include (a) aerenchyma, air spaces in roots and stems that allow oxygen diffusion from stems above water to roots; (b) hypertrophied lenticels, enlarged openings in stems and roots that allow gas exchange between internal plant tissue and the atmosphere; (c) adventitious or stem roots developed above the water line; and (d) the ability to grow new roots under anoxic conditions. In black mangrove (*Avicennia*), pneumatophores, vertically growing air roots, absorb oxygen that is transported to the connected, submerged, lateral growing roots. Prop roots of red mangrove (*Rhizophora*) function in much the same way. In both mangrove species, oxygen enters the plant through lenticels exposed above water and diffuses to roots in anoxic sediments. Many woody species of alluvial floodplains have extensive, shallow root systems placed where sediments are least likely to experience oxygen deficits.

Plant physiological adaptations generally involve tolerance to low soil oxygen and specialized chemical reactions. Specialized reactions include an accumulation of malate instead of ethanol, the production of high levels of nitrate reductase, and a reduction in ethanol production by reducing alcohol dehydrogenase activity. Additionally, rhizosphere oxygenation is an important mechanism for many species and can lessen toxic concentrations of some anaerobic soil compounds.

Wetland animals, with their characteristically high metabolic rates, have developed a variety of adaptations to low levels of oxygen and carbon dioxide. Perhaps most obvious is development of specialized regions of the body for gas exchange. Examples include gills in fish and crustacea, parapodia in polychaetes, and highly vascularized tissues on the lower lips of some tropical fishes or in the cloacas (uro-genital openings) of turtles. Other physical adaptations include modification of respiratory pigments to improve oxygen-carrying capacity in invertebrates. For example, midge larvae (*Chironomus*) are often colored brightly red, indicating the unusually high concentrations of hemoglobin present in these organisms, permitting them to survive long periods of hypoxia. Some vertebrates, particularly fishes, also increase densities of circulating red blood cells and thereby their oxygen-holding capacity.

Physiological adaptations of animals primarily involve shifts in metabolic pathways. For example, bivalves use alternative biochemical pathways, primarily a switch to glycolytic fermentation, to increase energy production under anoxic conditions. Some invertebrates also diversify the by-products of glycolysis to avoid toxic accumulation of any single compound, particularly ethanol. Turtles are remarkable for the ability of these lung-breathers to remain under water submerged in sediments for months during the winter season. Biochemical adaptation for natural anoxia tolerance in turtles includes well-developed antioxidant defenses that minimize or prevent damage by reactive oxygen species during the reoxygenation of organs after anoxic submergence. Last, many invertebrates store large quantities of respirable carbohydrate, usually glycogen, for breakdown and oxygen-liberation during periods of anoxia.

Behavioral adaptations also are critical and widespread, including dormancy or low locomotor activity during periods of oxygen stress, and migration from hypoxic to oxygen-rich environments. Many animals in low-oxygen situations have developed means of moving water more rapidly across respiratory surfaces. For example, benthic animals often use a variety of behavioral means (fanning, retreating into and out of burrows) to ventilate their burrows and increase the water flow across membranes during times of hypoxia.

These adaptations combine in interesting ways in particular taxa. Some fishes highly adapted to mud and shallow-water swamps cope by aestivating in mucous cocoons or by migrating overland while air breathing through modified swim bladders (e.g., lungfish). Some aquatic microinvertebrates, such as Collembola (springtails) and mites (Acari), have developed a "physical gill" that traps air gathered at the water surface in body surface hairs. Breathing of the trapped air, while underwater, occurs via a tracheal system, which opens to the body surface. Some aquatic insects, including mosquito larvae and chrysomelid beetles, tap the air within the aerenchyma of plant roots using a highly specialized, spinelike siphon attached to their abdomens. Last, some fly larvae use snorkel-like devices that extend above the surface of liquid mud or anoxic water and that permit the animal to air-breathe while remaining submerged in the anoxic substrate.

B. Adaptations to Salinity and Drought

Although halophytes (i.e., plants capable of persisting in saline environments) are not restricted to wetlands, these species dominate saline wetlands, such as inland and coastal salt marshes, and coastal fringe forests. Some species, such as various mangroves (*Avicennia*, *Laguncularia*, and *Rhizophora* spp.), are facultative halophytes and have mechanisms to inhibit salt absorption by their roots and to excrete salt from their leaves. Because nonhalophytes ("glycophytes") exhibit a range of tolerance to a salinity gradient, salinity and flooding regime determine plant species composition and richness along a gradient from salt marsh to tidal freshwater marsh. Plant species that use the C_4 pathway of photosynthesis are well adapted to habitats subjected to drought stress. Saline wetlands, therefore, often are dominated by C_4 plant species because the salt content makes water uptake more difficult. Plant species with C_4 photosynthesis are much more efficient in fixing carbon and in water use and have higher rates of net photosynthesis, growth, and dry matter production than C_3 plants.

Little is known about adaptations of wetland animals to salt stress. The fundamental problem of salt stress is its effect on water movement at the cellular level, which interferes with metabolic processes. Most simple microinvertebrates are osmo-conformers whose internal salt concentration tracks that of the external environment. More complex wetland animals, particularly vertebrates, typically osmoregulate—that is, retain internal salt concentrations independent of those external to the body. Osmoregulation primarily involves moving ions across body membranes against the concentration gradient and is thus an energy intensive process. This process is usually associated with specialized renal organs (kidneys and antennal glands) and salt-secreting nasal or rectal glands. Behavioral adaptations, particularly short-term migrations, are common in mobile species. In such a manner, both freshwater and marine species can occur in salt marshes, retreating during unfavorable periods either upstream (freshwater species) or downstream (marine species) during the course of the daily tidal cycle.

C. Adaptations to Low Nutrient Availability and High Acidity

Nitrogen fixation is relatively uncommon in wetland vascular plants but the few examples are noteworthy. Nitrogen fixation by sweet gale (*Myrica gale*) can contribute 3 to 4 g N m^{-2} yr^{-1} to *Sphagnum* bogs. Nitrogen

FIGURE 7 Acidic peatland, coastal Maine, USA. Photo by Donald J. Leopold.

fixation rates of speckled alder (*Alnus rugosa* = *A. incana* ssp. *rugosa*) have been measured at 85 to 167 kg N_2 ha^{-1} yr^{-1}. Some nonsymbiotic, aerobic and anaerobic bacteria, and blue-green algae are also significant nitrogen fixers in wetlands.

Most of the carnivorous species in the world occur in wetlands and other oligotrophic habitats. The relationships between insect capture, mineral nutrition (particularly nitrogen and phosphorus), and plant growth and reproduction have been investigated for numerous species. The relatively large diversity of organisms that can survive in the "pitchers" of some carnivorous plants also has been studied. Although benefits to carnivorous plants have been demonstrated, there is limited evidence that carnivory is necessary for most species to survive.

Evergreen, sclerophyllous shrubs that generally belong to the Ericaceae (heath family) are characteristic of nutrient-poor peatlands. The role of this foliage type in wetlands is uncertain, but it has been suggested that plant species with evergreen leaves may be more efficient in using nutrients than species with deciduous leaves. These leaves, which persist for up to four years, extend the period that the plant can photosynthesize, which conserves energy in the plant because all of the leaves do not have to be renewed each year. Furthermore, evergreen species tend to resorb more nutrients from senescing leaves than other species, thus increasing nutrient use efficiency. High nutrient resorption and long retention of nutrients in evergreen sclerophyllous leaves may be especially important in phosphorus-poor habitats such as bogs.

One of the most important groups of plants in colder, generally acidic wetlands of the world are *Sphagnum* mosses, a large genus consisting of more than 135 species. These mosses can thrive in conditions highly stressful to most plants (e.g., low pH, low nutrient availability, and saturated soils). They are the primary peat producers worldwide and have the capacity to modify the environment in ways that favor their own growth. Once established, *Sphagnum* mosses can further acidify their environment through cation exchange processes and by excretion of polygalacturonic acids. They also paludify their environment through their high water-holding capacity and their role in peat accumulation which can block water drainages. *Sphagnum* species also can tolerate great desiccation. Three major gradients exert the greatest influence on the bryophyte flora of peatlands: acidity–alkalinity, dry–wet, low light-high light. Numerous *Sphagnum* species can occur even within a relatively small peatland because of the affinities of particular species to very specific pHs, other water chemistry variables, height above water table, and light.

D. Adaptations to Reproducing in Wetlands

Seeds of many wetland species have features that enable their dispersal by water (hydrochory). Red mangrove (*Rhizophora mangle*) produces seeds that germinate while on the parent tree, resulting in viviparous seedlings. The germinated seed can root if it lands on sediment. Otherwise, the seedling floats until a suitable substrate is reached.

Seed banks, the store of seeds that accumulate in soils, play a critical role in the long-term vegetation dynamics of many types of wetlands worldwide. Seed banks allow wetland vegetation to adjust to rising and falling water levels and help maintain compositional diversity across water depth gradients and from year to year and over longer climatic cycles. Most wetland species do not germinate under water and only a few survive more than a year or two of continuously deep water. A series of wet years can eliminate most vegetation, but when water levels fall, many species that existed only as seeds in the seed bank germinate in the wet exposed mud. Thus, fluctuating water levels and seed banks drive wetland succession. Seed banks are particularly important in maintaining the species composition and zonation of wetlands that have a large number of annual species (e.g., tidal freshwater marshes). An understanding of the seed bank is essential in wetland management, especially in restoration efforts.

Seed density in wetland soils, usually determined by germination tests, typically ranges from 1000 to 14,000 seeds m^{-2} with density and viability decreasing with increasing depth in the substrate. Graminoids (grasses and sedges) are often the most abundant group in wetland seed banks, with seeds of broad-leaved herbaceous and woody species far less abundant. Although the composition of the seed bank and plant species present in the wetland can be highly correlated, in many cases the seed bank harbors a much greater species richness than found in the extant vegetation. Seeds of some species are transient, while others can persist for decades.

The greatly enlarged, rounded bases of water tupelo (*Nyssa aquatica*) and buttressed bases of bald- and pond cypress (*Taxodium spp.*) develop especially in deeply flooded conditions. These swollen bases likely aid in anchoring the tree. The "knees" of *Taxodium* also may function in this manner. *Taxodium* buttresses appear to provide a refugium above water for numerous species of herbaceous and woody plants that become established in the crevices where organic materials have accumulated over time. In some deep-water swamps, alternative colonization sites are limited for many plant species.

IV. THREATS TO WETLAND DIVERSITY

Worldwide, the loss of wetlands is estimated at about 50%. Functionally, the losses are much greater because extensive areas of boreal wetlands have been little altered while many different types of regional wetlands have been reduced by 80 to 90% elsewhere (e.g., California, southern Ontario, southwestern France, New Zealand). Wetland diversity and numerous plant and animal species have been impacted by a host of anthropogenic activities and natural disturbances for centuries. While single and cumulative effects of these activities have been assessed for many wetland species (primarily wildlife) and functions, few studies have examined the effects of these activities on species diversity. Because human activities often modify wetland hydrology and water chemistry, one should expect significant changes in diversity given the well-established relationships between wetland hydrology, water chemistry, and species distributions.

Despite the relatively recent appreciation of wetland functions and values, wetlands continue to be destroyed. A notable example is the Pantanal of Brazil, the world's largest wetland and home to many endangered or threatened plant and animal species. Various development plans, including dam and highway construction, threaten the diversity of this vast wetland. Although some human activities may be able to be carried out in a sustainable manner in wetland habitats, habitat preservation is the key to preserving wetland diversity around the world.

A. Direct Habitat Destruction

By far, the majority of palustrine wetland loss in North America has been due to agricultural activities, especially direct conversion of wetlands to agricultural land. Agricultural uses have converted over 80% of the Mississippi River bottomland hardwood forest. About 90% of the wetlands in California, Ohio, and Iowa have been destroyed, primarily for agriculture. Wetlands adjacent to agricultural lands also can be adversely affected because drainage effects can extend a significant distance into the wetland. Coastal wetlands, most first altered by agricultural activities, including grazing and salt production, are now impacted by draining and filling for urban and industrial development. Drainage of wetlands on organic substrates leads to accelerated decomposition, erosion of organic matter, and release of carbon dioxide (a "greenhouse" gas) into the atmosphere.

Riparian wetlands in the western United States can be seriously degraded by cattle grazing, which adversely affects the adjacent aquatic ecosystem. Wildlife species composition and diversity are modified because heavy grazing reduces the structural complexity of riparian wetlands.

Timber harvest, when done without extensive drainage and total land clearing, generally converts forested wetland to earlier successional stages, or changes species composition and structure. Logging, specifically the extraction of cut trees through mechanical skidding and transport of heavy log loads, adversely modifies soil physical properties. Early logging of southern forested wetlands was typically done with no regard for sustainability. Consequently, forest composition and structure were dramatically altered. Logging roads can modify wetland hydroperiods such that productivity is raised or lowered. Occasionally, especially in northern forested wetlands, a heavy timber harvest can raise the water table, flooding already established species and decreasing the regeneration sites for new plants. Whole-tree timber harvest of oligotrophic peatlands can greatly reduce a site's nutrient pool.

Tropical intertidal zones are being increasingly altered for mariculture, especially mangrove wetlands for shrimp farming. Many salt marshes have been ditched and diked, and often converted to impoundments, to favor waterfowl.

Peat extraction, primarily for fuel or horticultural uses, has destroyed millions of hectares of peatlands in North America, Great Britain, and northern Eurasia. Where the mined peatlands have been abandoned or actively managed for conservation values, the vegetation can slowly recover. Which species come to occupy the site depends on the hydrology of the site, availability of propagules, and time since abandonment.

Gravel and sand mining can adversely affect wetlands directly by habitat destruction, and indirectly by changing the hydrology and water chemistry of an adjacent wetland. These mining activities can be especially damaging to ombrotrophic and minerotrophic peatlands (e.g., bogs and fens) because many of the characteristic species occur at very specific water levels and chemistry parameters.

B. Direct Alteration of Wetland Hydrology

Humans alter wetland hydrology primarily by building dams at wetland outlets to regulate the depth and duration of inundation, and by constructing channels and ditches to drain wetlands and prevent prolonged flooding. Artificially regulated water levels often result in impoverished animal communities because few species of wetland animals have life histories adapted to the stable water regimes that result. This decrease in diversity occurs even where water control structures were originally built with the intention of benefiting wetland animals (e.g., at wildlife refuges). Management strategies for many water impoundments now seek to emulate the "natural" flooding regimes that were present before impoundments were built. Wetland drainage has obvious and drastic effects on wetland animal communities and has resulted in a loss of about half of the original habitat available to the wetland fauna of the United States. Among the most imperiled wetland habitats are the easily drained, intermittently flooded basins and pools that are critical to a large proportion of the wetland fauna.

Dam construction and water diversion projects also have significant effects on downstream riparian vegetation by altering peak and minimum flows, decreasing erosion and deposition of sediment, and lowering floodplain water tables. Altering the natural variability of river flow can adversely affect the species composition and structure of riparian wetland vegetation. Dredging, channelization, and levee construction have greatly impacted extensive wetland areas in the Mississippi River Delta.

Groundwater extraction is a primary cause of wetland modification because it lowers the regional water table and alters recharge and discharge patterns within the wetland and surrounding landscape. Irrigation with water supplied by aquifers that are recharged in part by wetlands has led to declines in the area of major regional wetlands in North America (e.g., the prairie pothole and Nebraska sandhills regions).

Although plant species of coastal wetlands may be well adapted to flooding, increases in the level or duration of flooding above that at which species have long persisted can cause serious deterioration of these wetlands. Sea-level rise and saltwater intrusion are important factors in the degradation and loss of wetland forests and coastal marshes in the southeastern United States. Increased flooding and salinity have been particularly damaging to forests dominated by baldcypress.

C. Landscape Fragmentation

Other human activities impoverish the wetland fauna in more subtle ways, even where wetlands receive nominal legal protection. Destruction of the uplands adjacent to wetlands destroys the habitat interface critical for amphibians and reptiles that migrate between wetland and upland habitats. Reductions in the connectivity among wetlands (e.g., by construction of roads and levees) blocks the migration routes of amphibians, fish, and birds, and can lead to local population extinctions. In contrast, dredging of canals that link major river systems causes sudden mixing of fish and reptile faunas that may have evolved in isolation for thousands of years.

D. Other Threats

Anthropogenic inputs of nutrients and toxins drastically alter the chemical environment of wetlands and render many sites unsuitable for wetland plants and animals. Atmospheric deposition of nitrogen and sulfur

FIGURE 8 Alligator in pool at Everglades National Park, southern Florida, USA. Photo by Donald J. Leopold.

is particularly high in parts of western Europe but is also high in the northeastern United States. Atmospheric deposition may be especially damaging to peatlands dominated by bryophytes (mosses and lichens), because these species are very sensitive to changes in nutrients and acidity-alkalinity. Wetlands in agricultural landscapes receive excess nitrogen via ground water draining from agricultural fields and excess phosphorus primarily via surface water runoff. Residential development also contributes excess nitrogen and phosphorus to wetlands. Coastal oil spills can adversely affect the vegetation of salt marshes and tidal freshwater wetlands.

Invasive, exotic wetland species typically cause a substantial loss of native wetland species and greatly alter wildlife habitat. In otherwise arid regions, invasive species use substantial amounts of water and can desiccate watercourses. Invasive species generally respond favorably to altered hydrologic regimes, substrate disturbance, or changes in water quality, especially eutrophication. Introduction of exotic wetland animals, for example, carp, nutria (*Myocaster coypu*), and bullfrogs, outside their native range have greatly altered wetland environments throughout much of North America, mostly to the detriment of the native wetland fauna.

Purple loosestrife (*Lythrum salicaria*), an emergent, herbaceous species of Eurasian origin, has become a serious invasive species of open wetlands in eastern North America. A single individual of this species can produce an average of 2,700,000 seeds. Another very serious invasive species in wetlands of the eastern United States is the common reed (*Phragmites communis*), which can produce 200 to 300 culms m^{-2} in fresh and brackish wetlands through an extensive network of rhizomes. Recent biological control methods (i.e.,

FIGURE 9 Yellow pitcher plants in pocosin, South Carolina, USA. Photo by Donald J. Leopold.

FIGURE 11 Water tupelo in southern deepwater swamp, South Carolina, USA. Photo by Donald J. Leopold.

leaf- and root-feeding beetles) appear promising for *L. salicaria*.

Salt cedars (*Tamarix* spp.) are rapid, arborescent invaders of riparian wetlands in the southwestern United States, especially downstream from dams where flooding is minimal. Another invasive tree species of riparian wetlands in the western United States is Russian-olive (*Elaeagnus angustifolia*). In southern Florida, the invasive tree, melaleuca (*Melaleuca quinquenervia*), has been aggressively colonizing shallow wetlands. Two other tree species that seriously threaten southern Florida wetlands are Australian pine (*Casuarina*) and Brazilian pepper (*Schinus terebinthifolius*). Glossy buckthorn (*Rhamnus frangula*) is becoming a serious problem in wetlands around the eastern Great Lakes basin.

Some native species become invasive following some hydrologic changes or an increase in nutrients and dominate wetlands previously occupied by other, more desirable species. Cattail is regarded as a threat to southern Florida marshes that have been dominated by sawgrass (*Cladium jamaicense*). The spread of cattail is believed due mostly to increased phosphorus input from agricultural lands and anthropogenic changes to the regional hydrology. Cattail is also considered a weed in many wetlands and aquatic habitats that are used for agriculture or to provide water resources (e.g., rice fields, irrigation canals, recreational lakes, and reservoirs). Mechanical (e.g., cutting, water level modification, fire, shading), chemical (selective versus nonselective herbicides), and biological control methods vary in their effectiveness, which varies widely among species. Maintaining water levels over 0.5 m is effective in controlling spread of cattail.

FIGURE 10 Acidic peatland in the Adirondacks, New York, USA. Photo by Donald J. Leopold.

E. Natural Disturbances

Severe flooding, hurricanes, and fire are the major natural disturbances that greatly affect wetlands. Although wetland plant and animal communities are adapted to each of these disturbances, events of exceptional magnitude and duration, or that occur during unusual times of the year, can have profound effects. A serious concern following any of these severe events is the spread of invasive species.

During the 1993 flooding of the Mississippi River, submerged aquatic, emergent wetland, and floodplain species were adversely affected by extreme sediment deposition, uprooting due to wave action, and probably increased inputs of agricultural chemicals. Recolonization of open spaces left by plant mortality can happen quickly by germination of local or upstream plant propagules.

FIGURE 12 Freshwater marsh in the Adirondacks, New York, USA. Photo by Donald J. Leopold.

Hurricanes especially affect the mangrove forests of the Gulf of Mexico and Caribbean, and the deep-water and alluvial swamps of the southern United States. These coastal forests, where many species have shallow root systems, are often subjected to the highest winds and succumb to windthrow. Trees that are especially resistant to windthrow (e.g., *Taxodium distichum*) may lose most of their crown but are able to reestablish a new canopy quickly by extensive sprouting along the bole. Seedling regeneration is generally higher after such disturbances because of increased light and microsite availability. Storm surges of saltwater occasionally cause greater damage than high winds to species intolerant of high salinity.

Forested wetlands on substantial peat deposits, such as the black spruce (*Picea mariana*) and tamarack (*Larix laricina*) swamps of the boreal forest and baldcypress swamps of the southeastern United States, have naturally burned during periods of extreme drought. In northern climates, fires can convert the forest to a more open peatland. In the southeastern swamps (e.g., Okeefenokee Swamp), fires can convert forested wetlands to prairie-like, tree-less wetlands. Fire can favor the regeneration of some tree species, like *Chamaecyparis thyoides* (Atlantic white-cedar), which regenerate best under full sunlight conditions on saturated peat. In the Everglades of southern Florida, open sawgrass marshes naturally burned every 1 to 5 years, ignited by lightning strikes from May to October when water levels are highest and substrates are saturated. However, human-set fires are most common from November to May when soils can be dry enough to ignite. These fires, to which the vegetation is not well adapted, can cause substantial ecological damage.

V. RESTORING AND MAINTAINING WETLAND DIVERSITY

Because of extensive loss and degradation of wetlands throughout the world, and continuing loss and degradation in much of the world, maintaining wetland diversity will require a strategy that includes restoration. The ultimate goal in any specific wetland restoration project is to have a self-sustaining, functioning ecosystem. The following discussion focuses on techniques to achieve this goal. However, in terms of restoring wetland diversity, the goal must be a broader one—best addressed at the landscape, regional, and continental scale—to restore the structure, function, and self-sustaining properties of a wide diversity of wetland types. Without attention to restoration of the full complement of wetland types, wetland diversity cannot be restored at any but the smallest scales.

Restoring self-sustaining wetland ecosystems initially, and at later times, may require various engineering, horticultural, and wildlife management activities (National Research Council, 1992). The intensity of these activities depends on whether the site was initially an upland (thus requiring wetland "creation"), a wetland that has been drastically disturbed (e.g., through conversion to agricultural crop production), or disturbed to lesser degrees. In some cases, a wetland may have reached a successional stage that does not support a species or group of species of concern, and manipulation is required to maintain these species. Nontidal emergent wetlands are generally easiest to restore. Long-term monitoring of site hydrology and vegetation is required to determine if and when project objectives are not being met.

Because wetlands serve many functions, and seldom can all functions be restored in a single site, restoration objectives must be clearly defined for every project. Restoring wetlands as habitat for rare and uncommon species, or for overall biological diversity, especially requires clear statements of objectives, as well as thorough knowledge of the ecology of the species of concern. Wetlands that could be correctly described as functioning wetlands have failed to attract the species for which they were restored. Consider the restoration of a bottomland hardwood forest. The restoration could be viewed as successful if a certain percentage of planted oak (*Quercus*) species and individuals were to become established after a given period of time. However, such a planting would be a failure in terms of restoring the diversity of neo-tropical migratory birds, which would fail to use the site because they require a more complex forest structure. In general, a higher overall plant and animal diversity is achieved by planting an array of fast- and slow-growing plant species, and where appropriate a mix of woody and herbaceous species tolerant of the range of hydrological and chemical conditions at a site.

The first step to restoring wetland plant composition, structure, and function is to ensure a range of hydrological conditions by implementing various techniques. Depending on the existing condition of the site and restoration goals, these techniques may include reestablishing stream flow or tidal fluctuations, restoring flood regimes, halting drainage, and reestablishing topography and microtopography. These conditions can be attained by excavating basins or channels, constructing dikes, grading an undulating topography, installing water control structures, plugging ditches, and removing drainage tile lines. If the site is to be used for the reintroduction of rare species, there are additional considerations, especially genetic, political, and legal issues.

Once a range of hydrological conditions is established, plant species best adapted to each specific hydrological condition can be selected. Many wetland restorations fail because the relatively few plant species selected grow poorly or die because conditions are too wet or dry for too long. If the goal of a particular restoration is to restore functions such as flood control or nutrient retention, then relatively few plant species need to be planted. If the goal of a project is to restore the natural heritage or educational values of a wetland, or to reestablish the compositional and structural complexity of highly diverse wetlands, then many plant species should be established.

Sources of native trees, shrubs, sedges, grasses, and herbaceous species include plant nurseries and other wetlands, especially those that are planned for future, permitted alteration. Seeds of many wetland species can be collected and grown for future transplanting or direct seeding. Some woody species can be established from stem cuttings. Many herbaceous species can be propagated from divisions off the parent. Spreading soil collected from a donor wetland, especially those planned for future alteration, will provide the greatest diversity of plant and animal organisms. Numerous woody and herbaceous plant species can arise from transplanted soil, via whole plants, seeds, roots, rhizomes, and stolons. Including leaf litter, detritus, and, for forested wetlands, large amounts of coarse woody debris are beneficial.

Herbivores cause many wetland restoration projects to fail, directly (by feeding) and indirectly (by burrowing, by dam building). Invasive species, such as *Lythrum salicaria*, *Phragmites australis*, and *Tamarix* spp., also must be controlled if plant and animal diversity are desired.

Understanding the ecological factors that control the diversity of animals in wetlands is key to altering these factors to restore, enhance, or create wetlands for the benefit of wetland fauna. Because hydrology underpins so much of wetland ecology, manipulation of wetland hydrology through water-level regulation is the most common and cost-effective way to manage wetland habitats for wetland fauna. More specifically, water-control structures can be used to flood or drain wetlands and thereby alter plant communities and the habitats available for wildlife. By drawing down water levels, a germination phase can be produced, which can be followed by gradual reflooding once plant communities become established. Timing, duration, and degree of drawdown (for example, shallow, growing season drawdowns versus complete overwinter drawdowns) influence subsequent plant species composition and the types of wildlife later attracted. Several years of stable water levels of moderate depth following drawdown are often required to establish the submerged plant communities that are important to many wetland animals (invertebrates, ducks, fish). Manipulation of wetlands to produce early stages of plant succession can create a diversity of habitat niches for wetland animals, particularly if practiced within a wetland complex in which management of different units is staggered for different successional stages. This practice permits the wetland fauna, elements of which can be quite mobile, to track habitat conditions over time and persist over the long-term within the same local area. Manipulation of wetland hydrology also has been used to manipulate salinity in coastal areas by blocking natural drainages, thereby permitting the fresh water that accumulates to leach out salt. The resultant shift in plant communities to more salt-intolerant species also changes wildlife communities.

In addition to water-level manipulation, a number of artificial procedures can be used to modify wetlands to enhance wildlife habitats. These include creation of openings in dense emergent vegetation through cutting, application of herbicides or, more permanently, by blasting. Planting of select forage species is sometimes attempted but is usually expensive and rarely successful. Provision of artificial nest sites for select species, especially ducks, has been one of the most common and visible forms of wetland management to benefit wildlife. Artificial nesting and loafing sites can be constructed for birds. Large, whole tree boles, including partially hollowed-out trees, can be placed within or adjacent to the restoration site. While perhaps aesthetically pleasing, the overall contribution of such efforts to boost local populations is at best modest given the small fraction of regional populations that can be supported by such structures. These artificial procedures, in general, are expensive and time-consuming and should be considered a minor complement to a larger strategy of natural management that uses hydrological manipulation to produce habitat changes over large areas in a cost-effective manner.

The tradeoffs between community types must be articulated and considered in management and planning. For example, flooding bogs and meadows to produce marshes, a not uncommon occurrence in North America, entails losses of certain species, many rare or unusual, and gains of others, many common and sought after by recreational users. In general, maintaining wetlands in a productive and natural state that meets the needs of the entire wetland fauna is more likely to meet diverse public needs than is species-specific management (e.g., for waterfowl only).

VI. CONCLUSIONS

Despite lack of systematic comparisons, the existing scientific literature on wetland ecosystems is unambiguous in conveying the high level of diversity supported by wetlands. The source of this high diversity lies in the exceptionally large number of different types of wetlands that occur worldwide. Because wetlands occur on every continent except Antarctica, in every climatic region, in most biogeographic regions of the globe, and in a wide array of geological and topographical settings, the number of different types of wetlands created by the myriad combinations of these factors is very large and has yet to be catalogued. The extraordinary richness of plant and animal species that depend on wetlands arises from this diversity of types. Relative to the area they occupy, wetlands support a disproportionately large fraction of the world's rare, endangered, and threatened plant and animal species. Maintaining the high diversity of wetland species means maintaining a high diversity of wetland types.

See Also the Following Articles

ESTAURINE ECOSYSTEMS • LAKE AND POND ECOSYSTEMS • RIVER ECOSYSTEMS • ECOSYSTEMS WETLANDS RESTORATION

Bibliography

Chapman, V. J. (1960). *Salt Marshes and Salt Deserts of the World.* Interscience, New York.
Chapman, V. J. (1976). *Coastal Vegetation*, 2nd ed. Pergamon Press, Oxford.
Chapman, V. J. (1977). *Wet Coastal Ecosystems.* Elsevier, Amsterdam.
Cowardin, L. M., Carter, V., Golet, F. C., and LaRoe, E. T. (1979). *Classification of Wetland and Deepwater Habitats of the United States.* U. S. Fish and Wildlife Service Publication FWS/OBS-79/31, Washington, DC.
Crow, G. E. (1993). Species diversity in aquatic angiosperms: Latitudinal patterns. *Aquatic Botany* 44, 229–258.
Crum, H. (1988). *A Focus on Peatlands and Peat Mosses.* The University of Michigan Press, Ann Arbor.
Curtis, J. T. (1959). *The Vegetation of Wisconsin.* University of Wisconsin Press, Madison, WI.
Ellison, A. M. (in press). Wetlands of Central America. In *Wetlands*

of the World. D. F. Whigham, D. Dykyjova, and S. Hejny, Eds., Volume 2. Kluwer, Dordrecht, The Netherlands.

Finlayson, M., and Moser, M. (Eds.) (1991). *Wetlands.* Facts on File, New York.

Good, R. E., Whigham, D. F., and Simpson, R. L. (Eds.) (1978). *Freshwater Wetlands: Ecological Processes and Management Potential.* Academic Press, New York.

Gore, A. J. P. (Ed.) (1983). *Mires: Swamp, Bog, Fen and Moor.* Elsevier, New York.

Lugo, A. E., Brinson, M., and Brown, S. (Eds.) (1990). *Forested Wetlands.* Elsevier, New York.

Mitsch, W. J. (Ed.) (1994). *Global Wetlands: Old World and New.* Elsevier, New York.

Mitsch, W. J., and Gosselink, J. G. (2000). *Wetlands.* (Third Edition.) John Wiley and Sons, Inc., New York.

National Research Council. (1992). *Restoration of Aquatic Ecosystems: Science, Technology, and Public Policy.* National Academy Press, Washington, DC.

National Wetlands Working Group. (1997). *The Canadian Wetland Classification System.* Wetlands Research Centre, University of Waterloo, Waterloo, Ontario, Canada.

Vitt, D. H., Li, Y., and Belland, R. J. (1995). Patterns of bryophyte diversity in peatlands of continental western Canada. *The Bryologist* 98(2): 218–227.

van der Valk, A. (Ed.) (1989). *Northern Prairie Wetlands.* Iowa State University Press, Ames, IA.

WETLANDS RESTORATION

Phil Benstead and Paul José
Royal Society for the Protection of Birds

I. What Is a Wetland?
II. The Value of Wetlands—Functions and Biodiversity
III. Wetlands—Threats and Losses
IV. What Is Wetland Restoration?
V. Restoration Planning and Implementation
VI. Wetland Restoration Case Studies
VII. Conclusion and the Way Forward

GLOSSARY

bund An artificial embankment keyed into the substrate and used to retain water.
groundwater Underground water that is held in the soil and in pervious rocks.
hydraulic conductivity A measure of the potential water flow through a soil.
ox-bow A crescent-shaped waterbody occurring on a river floodplain having once been part of a river meander that has been cut through and abandoned.
rehabilitation The partial return of structure and functions found in a predisturbance state to a disturbed habitat.
restoration The process of reinstating some or all preexisting functions to a lost wetland.
riparian Of or inhabiting a riverbank or margin.
stakeholder A person who has an interest or concern with a particular issue or business.
watershed The total area from which a single river collects water.
water table The upper surface of the zone of saturation in a soil or rock formation.

EFFECTIVE WETLAND RESTORATION is not only vital for the maintenance of biodiversity, but also for a number of other valuable functions. The chapter defines the term *wetland* and describes the reasons and extent of habitat loss. The nature and value of natural wetland functions is described. After briefly discussing project planning and implementation, a series of restoration case studies representative of a broad cross-section of wetland types from different continents are presented. The chapter concludes with an overview of current progress on wetland restoration and indications for future direction.

I. WHAT IS A WETLAND?

The Ramsar Convention on Wetlands[1] defines wetlands as "areas of marsh, fen, peat, and or water, whether

---

[1] The Ramsar Convention is an intergovernmental treaty that provides the framework for international cooperation for the conservation and wise use of wetlands and their resources.

natural or artificial, permanent or temporary, with water that is static or flowing, fresh, brackish or salt, including areas of marine water the depth of which at low tide does not exceed six meters." This Ramsar definition is perhaps the broadest of the many definitions of what constitutes a wetland and it is the one used throughout this chapter. Wetlands therefore include the following:

- marine—coastal wetlands such as coastal lagoons, rocky shores, and coral reefs
- estuarine—for example, deltas, tidal marshes and mangrove swamps
- lacustrine—wetlands associated with lakes
- riverine—wetlands along rivers and streams
- palustrine—marshes, swamps, and bogs
- human-made wetlands such as reservoirs, fish ponds, flooded mineral workings, saltpans, sewage farms, and canals

Because wetlands are dynamic systems and occupy the transitional zone between aquatic and terrestrial habitats there is often controversy over dictating boundaries and precise definitions. More restrictive definitions than the one presented earlier and methods for boundary delineation certainly exist and are the subject of several manuals published in North America where this is an important and politically charged issue (Mitsch et al., 1994).

II. THE VALUE OF WETLANDS—FUNCTIONS AND BIODIVERSITY

Aside from their importance for wildlife, wetlands are important, and sometimes essential, for the well-being of the people who live in or near them. In recent years there has been increasing awareness of the value of the hydrological and chemical functions of wetlands. To summarize, wetlands perform the following functions:

- flood alleviation—wetlands naturally regulate water flows within a watershed
- shoreline stabilization and erosion control
- storm protection—many wetlands, especially mangroves, act as windbreaks during storms, protecting coastal habitat and property
- groundwater recharge—wetlands retain water within a watershed enabling groundwater recharge
- water quality improvement—wetlands can retain nutrients, toxic substances and sediment
- climate—wetlands can influence local climatic conditions
- carbon storage—some wetlands such as peatlands store carbon
- biomass export
- habitat for wildlife—wetlands are crucial for the maintenance of biodiversity.

Economic benefits accrue from the functions just described, therefore when they are lost through the destruction of wetlands they often have to be replaced at enormous financial cost to society. These benefits include the following:

- water supply—wetlands can influence both water quantity and quality
- fisheries—nearly three-quarters of the world's fish harvest is linked to the health of wetland areas
- agriculture—floodplains provide some of the most fertile agricultural land in the world
- timber production
- energy resources, such as peat and plant material
- wildlife resources
- transport
- recreation and tourism opportunities

Additionally many wetlands have special cultural attributes; being part of religious, cosmological and folk lore beliefs, as well as providing aesthetic inspiration and wildlife sanctuaries. From a biodiversity perspective, wetlands are very important; most wetland ecosystems have high biodiversity and support suites of specialist species (Fig. 1).

III. WETLANDS—THREATS AND LOSSES

Wetlands have been lost at a phenomenal rate, especially in the developed countries of the Northern Hemisphere. Reasons for the destruction of wetlands center on the fact that their functions and values are often not directly "harvestable" by their owners for profit. Flood alleviation functions benefit those living downstream, fish and wildlife produced may migrate outside the system and be harvested by others, and so on. Other valuable functions such as groundwater recharge and sediment and nutrient removal are incredibly important but also cannot be exploited commercially.

Threats faced by wetlands are summarized in Table I. These threats have led to catastrophic declines in the

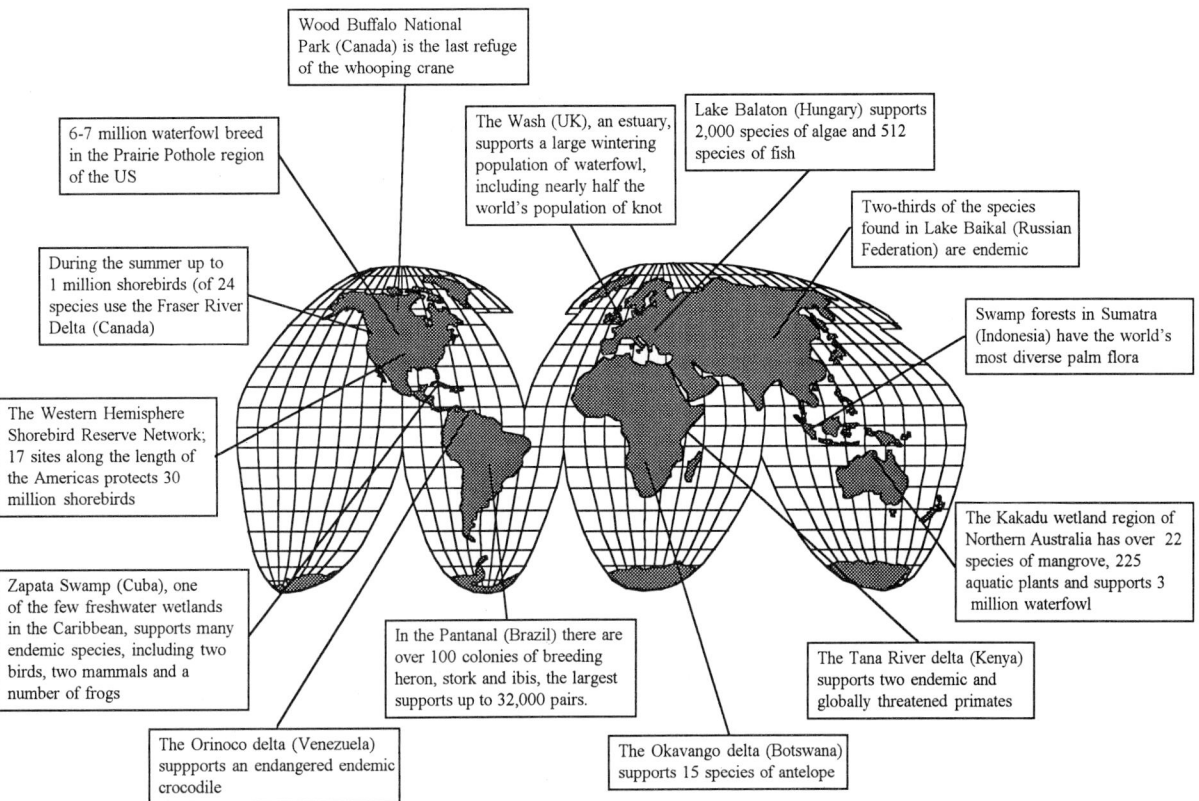

FIGURE 1 Examples from around the world of the importance of wetlands for biodiversity and the support specialist species.

coverage of wetlands around the world (Fig. 2). Any such statistics belie the real situation, however, as the remaining wetlands in the areas that have suffered serious loss are usually far from pristine. The extent of the loss of wetlands and the resultant financial cost incurred due to the loss of important wetland functions have created the need for restoration in some areas. To date restoration attempts have largely taken place in Europe and North America, regions which have seen the most devastating declines in wetland coverage.

IV. WHAT IS WETLAND RESTORATION?

Strictly speaking restoration is the process of reestablishing a naturally functional, self-sustaining system, implying a return to an original predisturbance state. In practice this is not possible as restored habitats rarely mimic all the properties of the original. Replacing or restoring preexisting biological resources is likewise not possible. Restoration is more realistically defined as the reinstatement of some or all preexisting functions to "lost" wetlands (Hollis, 1993). At this point it is also worth defining rehabilitation, which is the partial return of structure and functions to degraded wetlands. The process often involves the selection of desirable features and functions only. For the purposes of this chapter, restoration and rehabilitation will be treated as synonymous. Restoration and rehabilitation, together with management, form a continuum. Management can be applied to existing wetlands to develop certain functions artificially (Hollis, 1993).

Restored wetlands are unlikely to support diverse assemblages or rare species. The process of restoration can, however, rapidly enhance the biodiversity of an impoverished area. Restoration of wetland ecosystems is becoming increasingly important for the maintenance of biodiversity, particularly in developed countries because (a) remaining wetland sites are becoming increasingly fragmented and isolated making it difficult or impossible for biota to move from one area to another and (b) loss of natural processes (e.g., flooding) means that natural habitats are no longer dynami-

TABLE I
Threats Faced by Wetlands

- Hydrological management for flood control, irrigation, public water supply, and drainage
- Drainage for agriculture, forestry, urbanisation, mosquito control, and tourism development
- Dredging and stream channelisation for navigation
- Infilling for land reclamation and waste disposal
- Infrastructure developments such as embankments, coastal protection, and port facilities
- Dam construction for water supply and hydroelectricity generation
- Salinity changes
- Aquaculture development
- Wastershed development, which leads to watershed damage and erosion
- Surface mineral extraction, especially for gravel, peat, and clay
- Pollution and nutrient inputs from urban, agricultural, industrial, and mining sources
- Damage to fisheries due to overfishing, dynamite fishing, or conflict between artisanal and commercial fishermen
- Damage to forest resources caused by overexploitation, wastage, military activity, and illegal logging
- Damage to biodiversity through introduced species, hunting pressure, disturbance, cessation of traditional management practices, and livestock pressure
- Social factors such as poverty of wetland users, inadequate control by users of marketing of wetland products, low economic return of products, users not aware of sustainable alternatives, malnutrition, and outside vested interests

These factors are compounded by institutional and legal factors such as lack of legislation or its enforcement, destructive policies, land ownership, official attitudes toward wetlands, lack of adequately trained staff, and lack of appropriate planning.

cally destroyed and re-created, leading to the aging and eventual loss of wetland ecosystems.

Restoration (and creation) of wetland habitats is seen as vital to preserving biodiversity in the United Kingdom (Her Majesty's Stationery Office, 1995). Setting restoration (and creation) targets for priority habitats has been put forward as a key method for conserving populations of species of conservation concern such as the Eurasian bittern *Botaurus stellaris*.

The following case studies are biased geographically toward developed countries where the most damage has been inflicted on wetland ecosystems. The emphasis in the northern hemisphere tends to be toward restoring specific functions, usually for aesthetic or wildlife conservation reasons. There is a marked tendency to adopt quick technological fixes (engineering solutions), and it is still unusual for whole ecosystem restoration to be considered. Wetlands have a much broader significance in developing countries because greater importance is attached to their socioeconomic value; people depend on them directly because wetlands are very productive, providing crops and grazing; tidal wetlands support fish nurseries vital for offshore fisheries, as well as providing 10% of the total global catch; and they are an essential part of the hydrological system, buffering against drought and serious floods. The realization of the socioeconomic importance of wetlands in developing countries has promoted an interest, and an investment, in the preservation and restoration of natural wetlands around the world.

The extent of loss of wetlands in developing countries has, so far, not been as severe as in developed countries. The extent of remaining wetlands is such that only restoration of unique sites and habitats is usually put forward as a priority—for example, Tasek Bera, Malaysia (see the case study that follows). Relatively simple, single function restoration is becoming more common (e.g., conversion of fish ponds in the Mai Po reserve in Hong Kong to improve habitat for waterbirds and the replanting of mangroves after storm damage in tropical Asia).

The way forward for countries that can neither afford to make mistakes or rectify them has been recognized to be a close integration of the disciplines of wetland conservation and rural development. This is typified by the attempts in the Senegal River case study to restore functions with a wider socioeconomic value, as well as those which benefit biodiversity conservation within the National Park. Without action, losses on a similar scale to those experienced in developed countries are likely. Loss of wetland functions in developing countries will not only result in a considerable loss of biodiversity but will also have severe impacts on the human populations of affected regions (Dugan, 1988).

The progress of wetland restoration project is rarely assessed scientifically, although many pronouncements of success and failure have been made. For example, Zentner (1988) gave his opinion on projects undertaken in coastal California (during 1984–1987) judging 65% to be completely successful—that is, functioning or beginning to function like similar wetlands. Zentner considered that an additional 25% had created functioning wetland habitat but had failed to achieve stated aims (often the target habitat had not been restored). Zentner's conclusions were subjective, however; the study found only 36% of projects reviewed in coastal California had any form of monitoring, and that this was often only rudimentary. Restoration can never return natural wetland functions perfectly, it is nevertheless a

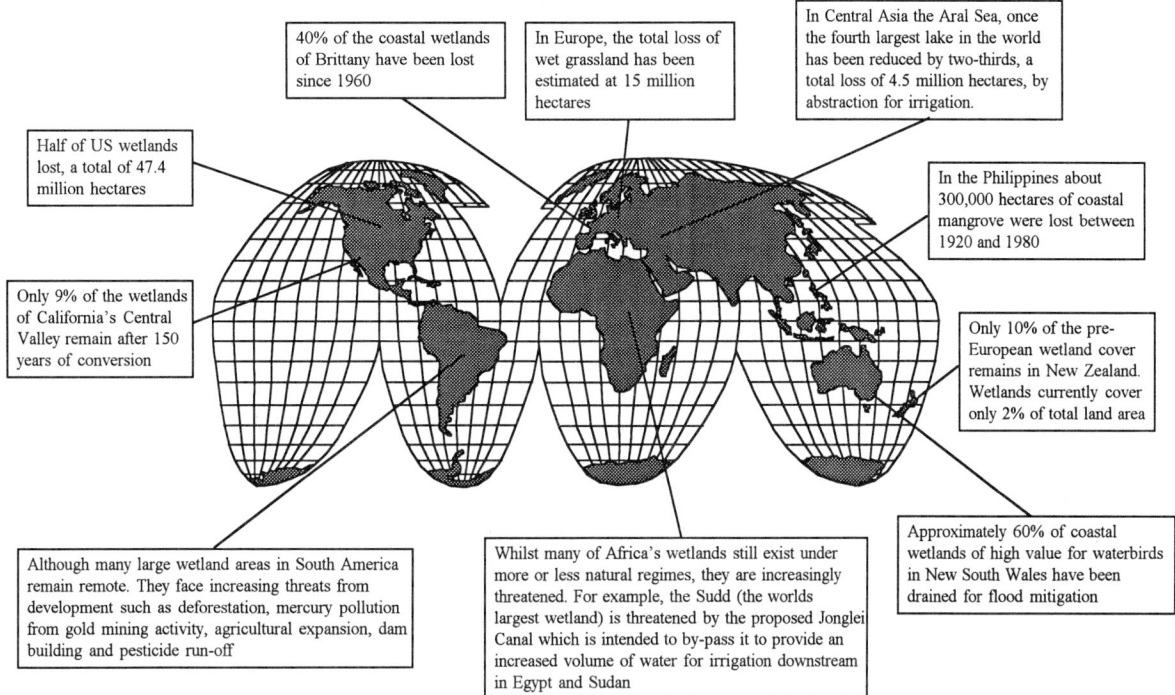

FIGURE 2 Examples from around the world of the scale of loss of wetlands to anthropomorphic factors. From Dugan (1993) and Finlayson and Moser (1991).

valuable wetland conservation tool alongside effective legislation to protect existing natural wetlands.

Creation of new wetland ecosystems on sites where they previously did not exist is increasingly common in developed countries. This is often undertaken as a requirement following development, either as part of a mitigation process or to return land to a beneficial use after mineral extraction. Creation is a distinct activity and outside the scope of this chapter, but many of the techniques involved in restoration are shared with this discipline.

V. RESTORATION PLANNING AND IMPLEMENTATION

The framework that follows describes the essential areas that should be addressed when developing restoration proposals. Suggested contents for a restoration plan are outlined in Table II. Restoration planning is required to do the following:

- Clearly state the aim and objectives of the project.
- Enable the safeguarding of any existing conservation interest.
- Identify appropriate target habitat and communities.
- Identify appropriate techniques for restoration.
- Identify project constraints (e.g., legal problems, access difficulties).
- Ensure that restoration operations are well coordinated.

TABLE II
Essential Components of a Restoration Plan

- An outline and details of the scope of project
- Site survey and appraisal—existing site conditions should be assessed in detail to identify topographical features, land use and classification, climate, water levels and movements, ownership boundaries, right of way, existing vegetation, and special factors (e.g., prevailing winds, views, and existing nature conservation interest)
- Formulation of restoration objectives, which should identify final land form and water levels, new positions of rights of way, planting details and management requirements of new vegetation, and treatment of unstable slopes
- Design of operational details and working plans to give precise details for implementation phase
- Afteruse management prescriptions

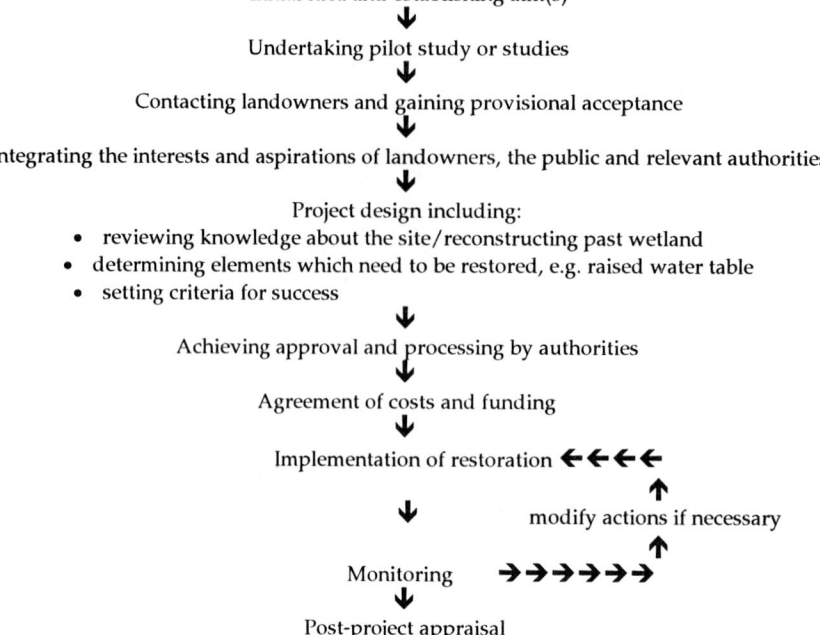

FIGURE 3 Essential stages of any restoration project. Adapted from Benstead et al. (1999).

- Detail monitoring/aftercare programs/post-project appraisal.

The stages which typically make up a restoration project are shown in Fig. 3. The timescale for achieving restoration will depend on a number of factors, such as the following:

- how degraded the site is (e.g., high levels of nutrients can take a long time to deplete)
- how isolated the site is (e.g., a site in an active river corridor is more likely to recolonize rapidly)
- the ability to manage the site for commercial/artisanal purposes, such as management for pastoral agriculture or regular cutting for roofing material, where such management is desirable
- climatic factors (e.g., low rainfall will prolong restoration and dry conditions will encourage invasion of ruderal, weedy plant species)
- the extent and knowledge of any previously installed land drainage system (e.g., location of underdrainage)
- nature of adjoining land use (e.g., risk of flooding of neighboring residential and agricultural land)

Achieving project objectives depends on a sound understanding of hydrological and ecological processes.

The following section describes the key areas and techniques that need to be addressed when implementing a restoration plan: hydrology, soils, and vegetation establishment/management.

A. Hydrology

Complete understanding of site hydrology and the feasibility of restoring a target hydrological regime is key to achieving restoration goals. Failure to provide suitable hydrological conditions is the likeliest cause of project failure. Restoration of a hydrological regime should aim to mimic natural conditions wherever possible. Planning and site assessment should include careful evaluation of the target water regime (levels, seasonality, etc.) and whether these are likely to be achievable and will meet the requirements of target habitats and species. Calculating the water budget for the site by assessing likely inputs and outputs is essential. It is possible to calculate water budgets using just climatic information to give a rough estimate of site water demand. However, in reality surface wetness is influenced by a range of other factors such as the relative hydraulic conductivity of different soil types and topography (Benstead et al., 1997). Hydrological restoration may include a range of options from reinstatement of natural flooding by setting back flood embankments, using existing drainage infrastructure in reverse, installing water retention features (see the Hornborgasjön case study, presented later), and removing or destroying existing drainage infrastructure (see the Abernethy case, presented later).

B. Soils

In all forms of restoration ecology particular attention should be paid to soils. Creating a functional soil profile with balanced aerobic, anaerobic, organic, and inorganic components is essential for the maintenance of wetland functions that are related to nutrient transformations, food chain support, and fish habitat.

Mineral soils that have been dewatered are likely to respond to rewetting. However, organic soils that have dried out are likely to have suffered irreversible physical and chemical alteration. Detailed investigation of soil properties (such as nutrient status, structure, moisture retention, and porosity) should be the subject of early site assessment (Allen, 1989). Soil invertebrate fauna of areas that have been drained are usually terrestrial in nature and are unlikely to survive "rewetting." Complete loss of earthworm fauna is possible and this can drastically lengthen the time it takes soil to recover from rewetting where structure has been lost. The loss of soil invertebrates also has implications for species that depend on them for prey. Recolonization of soil fauna can, in some cases, take decades and "inoculation" with cultures of suitable invertebrates may have to be considered (Butt and Frederickson, 1995).

C. Vegetation Reestablishment

Where sites are adjacent to existing wetlands, restoration can sometimes be left to natural regeneration. This occurs through seed drift from nearby areas, germination from the seed bank and the arrival of flood borne diaspores, and rhizome fragments. A number of techniques can be used to increase the rate at which species recolonize a wetland site where natural sources of material are not present (see the Upper Thames case study, presented later); these involve the introduction of biological material using rhizomes, seed, seedlings, mature plants, and cuttings.

VI. WETLAND RESTORATION CASE STUDIES

The following 13 case studies (Fig. 4) illustrate how restoration has been successfully undertaken on a range of habitats with a broad geographical spread. Initial case studies highlight the importance of project planning. Subsequent case studies work up from simple small-scale restoration projects to larger, more complex projects.

A. Advocacy and Planning Solutions—Creating the Framework for Successful Restoration at Tasek Bera

Tasek Bera, Peninsular Malaysia, is a 6150 ha freshwater lake system with open water and fringing vegetation surrounded by freshwater swamp forest. This habitat diversity coupled with relative isolation has combined to produce a diverse ecosystem with an unusually high degree of endemism. The biological community supported is unique within Malaysia, and in recognition of this Tasek Bera was recently designated as the country's first Ramsar site.

In the past 30 years Tasek Bera has come under threat from a number of directions:

- the forested watershed has been cleared for agriculture, increasing the sediment and nutrient loading of water entering the lake and also reducing the volume of water entering the system especially during dry periods
- pollution in the form of sewage from nearby rivers during high river levels (when flow reversal occurs)
- increased access due to new logging tracks and highways
- damage to hydrology by logging infrastructure, especially roads, which leads to altered drainage patterns and increased siltation

Because of the importance of the site and the problems it faces, the area is the subject of a restoration project. This project is in the early stages and before the required physical restoration action can be taken it is necessary to ensure that policy, planning and institutional arrangements are in place to ensure its ultimate success. This work includes the following:

- defining the site boundaries
- identifying an appropriate buffer zone and developing guidelines for neighboring land managers within this area
- baseline surveys on all aspects of the site
- writing the management plan for the site
- providing training for local government officers

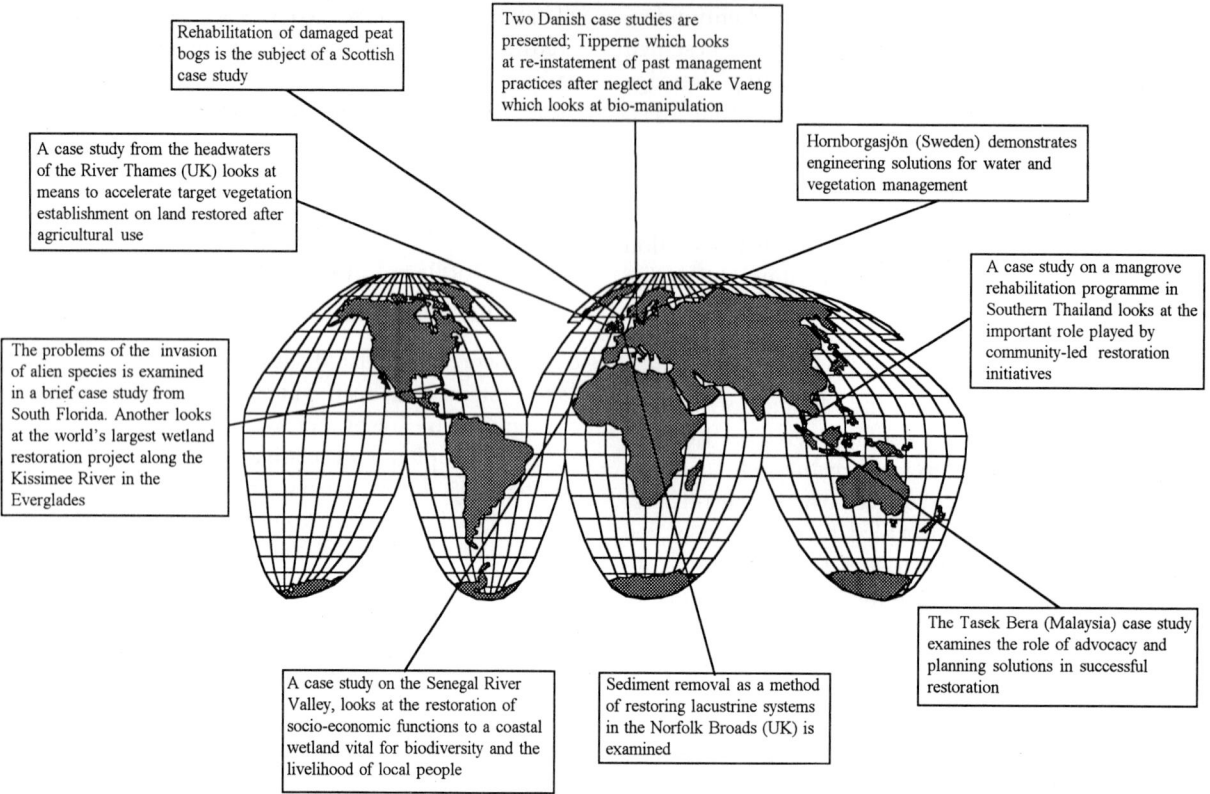

FIGURE 4 Location of case study sites.

- education work with local schools
- writing and implementing a community development strategic plan
- undertaking community pilot projects focusing on handicraft production and ecotourism involving the local aboriginal people
- planning ecotourism use of the site

This level of planning, site assessment, communication, and community involvement is essential if large projects are to achieve their objectives.

B. Simple Solutions—Reinstating Agricultural Management at Tipperne (Thorup, 1998)

Tipperne, Denmark, is a 700 ha nature reserve consisting mainly of brackish meadows, reedswamps, and dunes. A wealth of information on its bird populations is available for this century.

During the period 1928–1945 agricultural utilization of the site was fairly constant; the entire area was mown once or twice a year for hay in August/September, followed by aftermath grazing until November. After 1946 agricultural use declined and by the early 1960s the area had been abandoned. Invasion by rank vegetation and common reed *Phragmites australis* followed over the next 10 years. The loss of wet grassland habitat impacted on the numbers of breeding shorebirds using the site until only a handful remained.

In 1973 a management plan was initiated with the aim of restoring the wet grassland areas. During the first 10 years part of the reserve was grazed and quite dense populations of breeding waders and wildfowl reestablished in these areas. From 1984 management was further improved by mowing the remaining areas of the site; this increased the area of restored wet grassland dramatically. Breeding wet grassland birds increased again in response (see Fig. 5).

C. Reinstating Management Beneficial to the Maintenance of Biodiversity after Intensive Agricultural Use

Damage to wetlands, through intensification or neglect, may be reversible, although it can take many years for a

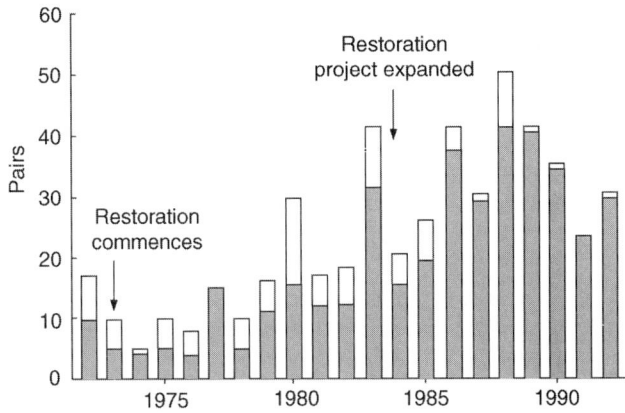

FIGURE 5 Graph showing the increase in pairs of breeding Northern Shoveler *Anas clypeata* at Tipperne between 1972 and 1992. Hatched bars are certain breeding pairs, while unhatched are possible pairs. Adapted with permission from Thorup (1998).

characteristic wetland community to recover. Dramatic reversals can be achieved after relatively short periods of time, however. Five years after reinstating cutting management to a wet grassland that had been neglected for 20 years in the Luznice river floodplain (Czech Republic), plant species diversity and composition were restored to a quality comparable with adjacent grasslands that had received uninterrupted management (Benstead *et al.*, 1999).

A similar time span was found for the restoration in 1991 of 141 ha of wet grazing pasture on a nature reserve in the North Kent Marshes, United Kingdom (Benstead *et al.*, 1997). The project involved manipulating the sward either by reseeding or through allowing natural regeneration and by raising water levels in the drainage channels using sluices, the creation of bunds, and blocking off piped drainage. The site was grazed by sheep owned by grazing tenants (April-December) followed by mowing to control persistent weed species. This simple process resulted in an increase in the number of shorebirds breeding on the site within 5 years (Table III).

In the Drenthe Aa wet grasslands of the Netherlands, the mean number of plant species increased from 18 to 26 (per field of 500 m^2) within 16 years after the cessation of fertilizer application (Bakker, 1989). These fields were adjacent to diverse unimproved meadow communities so that colonization from these populations could occur.

Reversion of agricultural land to wetland can be difficult because the sites may have any of the following:

- high nutrient availability (cropping or topsoil removal may have to be considered to lower nutrient levels)
- impoverished seed banks, repeated cropping, and the use of herbicides will have removed most of the naturally occurring plant material, and this will have been replaced by many undesirable ruderal species
- altered hydrological regimes (soil structures may have altered irreversibly)

Reversion of arable land to wetland habitats is therefore more easily undertaken on sites that have been under cultivation for a short time, that are adjacent to existing wetlands which can act as a seed source, and where appropriate hydrological conditions can be reestablished. Where livestock are used to manage sites, moving stock from an existing wetland to a site which is being restored can help. The livestock will transport both diaspores and invertebrates between the sites. Sheep are considered to be better at transporting biological material than other livestock: in a study in Germany, examinations of one sheep yielded 8500 diaspores of 85 plant species, and the sheep was also found to transport 13 species of Orthoptera (Fischer *et al.*, 1996). This transport process should be borne in mind when reintroducing grazing stock to restored sites as it may have a negative effect by introducing undesirable species.

D. Reintroduction—Accelerating Target Vegetation Community Development in the Floodplain Wet Grasslands of the Upper Thames Tributaries

Restoration of agricultural land to floodplain wet grassland has been investigated, along the River Ray (a tributary of the River Thames in the United Kingdom). Drainage improvements in the 1970s led to the conversion of traditional hay meadows to arable land. However, the combination of periodic flooding and a heavy clay soil make cultivation problematic and the recent introduction of agri-environment mechanisms (Countryside Stewardship and Environmentally Sensitive Areas schemes) aimed at restoring hay meadows has been well received by farmers. Restoration best-practice techniques have been the subject of experimentation.

The investigation looked at four arable reversion techniques: natural regeneration, use of cut plant material transported from another site, sowing seed mixes,

TABLE III
Number of Pairs of Breeding Waders at Northward Hill, North Kent Marshes, UK, after Restoration in 1991

Breeding population pairs	1991	1992	1993	1994	1995	1996	1997	1998
Vanellus vanellus	3	2	16	35	56	26	15	26
Gallinago gallinago	0	0	2	0	0	0	0	1
Tringa totanus	1	1	5	27	35	5	10	32
Avosetta recurvirostra	0	0	0	5	33	26	14	20

and the use of nurse crops (Manchester et al., 1998). The study revealed the following results:

- On ex-arable sites, natural regeneration would take decades to restore a seminatural vegetation type. Analysis of the available seed bank found that small quantities of desirable species were represented, but not in great enough abundance or reasonable diversity to restore target communities. "Seed rain" or floodwater seed dispersal from neighboring herb-rich areas appeared to be limited. However, several species did colonize the sward by natural regeneration.
- Scattering hay bales has several compelling advantages; it ensures that the supply of seeds are of known provenance and is very cheap. This technique was, however, unpredictable and less effective than sowing seed.
- Sowing seed mixtures resulted in significantly greater numbers of species in the resultant sward than either natural regeneration or hay bales. The most comprehensive seed mix produced the sward with the highest species richness and also the highest number of desired/potential species. Sowing seed in this way is far cheaper than transplanting mature plants or turfing areas, neither of which were considered in the experiment.
- Nurse cropping was ineffective in boosting the establishment of other species.

E. Rehabilitation of Damaged Peat Bogs at Abernethy by Removing Drainage Systems (Brooks and Stoneman, 1997)

Peat bogs have suffered around the world from damage caused by peat extraction, grazing and conversion to improved grassland, and afforestation. Peat extraction results in drainage of peat bogs and often leads to the invasion of scrub. Functional restoration of peat bogs following extraction therefore tends to revolve around hydrological control and vegetation management. Hydrological control techniques employed include blocking off drainage systems using dams and sluices, retaining water using bunds, and infilling drainage channels.

Vegetation control often involves the physical removal or chemical treatment of invasive and unwanted plant species, the re-ntroduction or transplanting of desirable plant species, and subsequent appropriate management of the desired vegetation community.

At Abernethy Forest (a reserve managed by The Royal Society for the Protection of Birds), areas of valley mire were threatened by an extensive forestry drainage network installed prior to purchase. Mire vegetation was slowly being replaced by drier heath communities. To counter this the drainage channels were initially blocked with peat dams, using hydraulic diggers. Peat was removed from borrow pits, either within or adjacent to the drainage channels. All vegetation and humified peat was removed, and as each successive layer of the dam was built up, the structure was compacted to ensure a watertight barrier. This method worked well where water levels were high but if they were low then the exposed peat dried out, cracked, and the dams leaked. The ploughlines were not touched and continue to have a limited drainage function.

An alternative method was tried later in another intensively drained mire on the site. This involved completely destroying the drainage network. The plough ridges were bulldozed into the drainage channels, filling them in, and then compacted down to reduce water flow. This method is useful in that it is simple and effectively removes all functioning components of the drainage system.

F. Biomanipulation—Fish Removal in Lake Vaeng, Denmark (Moss et al., 1996)

Eutrophication of shallow freshwater lakes initially results in a flush of plant growth and an increase in phytoplankton density. But eventually aquatic macrophytes disappear and algae take over. There is a net loss of biodiversity. This can be a continuing problem even when the original polluting element has been dealt with. The core of the problem appears to be that once open waters are created, fish find it easy to remove zooplankton, which are important grazers of phytoplankton. Algal dominance is further ensured by factors such as shading (algae begin growing earlier in the season and quickly create unsuitable light conditions for colonizing plants) and sediment changes (algae lay down a fine sediment, which contrasts markedly with the more fibrous deposits laid down by plant beds; this increases turbidity and decreases light values further disadvantaging plants).

One way to reverse this problem after preventing the source of nutrient pollution is by biomanipulation. This usually involves the removal of zooplanktivorous fish or the addition of piscivorous ones. Following on from this it may be necessary to protect reestablishing plants against physical disturbance and grazing. Upon successful reestablishment of aquatic plant communities, fish and grazing animal populations can be reinstated.

This technique was successfully employed at Lake Vaeng in Denmark. The 15 ha, shallow lake was polluted by sewage effluent and devoid of plants. Diversion of the sewage effluent in 1981 reduced nutrient levels but failed to restore water clarity and aquatic plant growth. During 1986-1987 roughly half the zooplanktivorous fish were removed. Successive removal in the following two seasons took the fish density down to 8 g live weight per square meter (Fig. 6). The results were rapid, within two years water clarity improved, the zooplankton community acquired *Daphnia* species, which are important algae grazers, blue-green algae disappeared, and nitrogen and phosphorous levels halved. Plants, however, were slower to respond but 80 to 90% plant coverage of the lake bed was achieved by 1991. This delay was attributed to grazing by birds.

G. Sediment Removal and Isolation as a Method for Restoring Lacustrine Systems in the Norfolk Broads (Moss *et al.*, 1996; Tickner *et al.*, 1991)

The Norfolk Broads (United Kingdom) are a system of shallow-flooded, medieval peat diggings connected by a set of rivers flowing through low-lying ground in East Anglia. This area supports high biodiversity but this has declined in recent years (George, 1992). Drainage improvements, the declines in traditional wetland man-

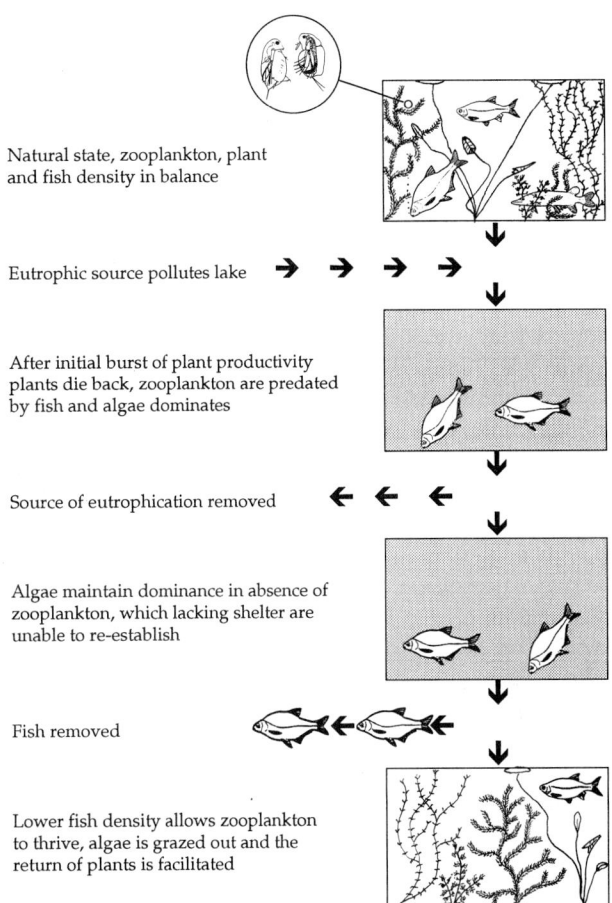

FIGURE 6 Biomanipulation is one way to deal with the effects of algal dominance in lake ecosystems resulting from eutrophication.

agement practices, and the pollution of rivers have led to widespread degradation of habitat.

Many of the broads were open to the river and subject to tidal flooding with eutrophic freshwater. Siltation gradually left few open water areas remaining (e.g., at Strumpshaw Broad) or greatly reduced the depth of the waterbodies (e.g., Cockshoot Broad). Aquatic macrophytes were lost and the remaining wetland plants were all indicators of eutrophic conditions. River drainage improvements also reduced the extent, depth, and duration of regular flood events, leading to the encroachment of undesirable trees species (*Salix* and *Alnus* spp).

In some cases isolation was achieved by sealing the channel connecting the broad to the river and building up the river bank to prevent flooding in all but the highest flood events. With no other inputs, rainwater alone was quickly found to be insufficient for the wetland requirements at Strumpshaw and a borehole was sunk into the underlying aquifer and up to 185,000

TABLE IV

Drainage Channel Flora Recovery: The Presence of Pollution-Tolerant and Pollution-Intolerant Aquatic Plants in Drainage Channels at Strumpshaw, 1976–1990

Species	1976	1979	1980	1986	1990
Potamogeton pectinatus	●	—	—	●	—
Callitriche stagnalis	●	●	●	●	—
Nuphar lutea	●	●	●	●	●
Filamentous algae	●	—	—	●	●
Elodea canadensis		□	□	□	—
Ranunculus aquatilis		○	○	○	—
Potamogeton crispus		□	□	□	—
Ceratophylum demersum		●	●	●	●
Hydrocharis morsus-ranae		□	□	□	
Chara spp.		○	○	○	—
Zannichellia palustris		□	□	□	—
Myriophyllum spicatum			□	□	□
Lemna minor			●	●	●
Utricularia vulgaris			○	○	○
Lemna trisulca			○	○	○
Polygonum amphibium			□	□	□
Cicuta virosa				□	□
Sparganium emersum				○	—
Acorus calamus				□	□
Potamogeton berchtoldii				○	○
Callitriche palustris					○
Callitriche platycarpa					○
Nymphaea alba					○
Rumex hydrolapathum					○
Ranunculus lingua					○
Typha angustifolia					○

● = pollution-tolerant species, □ = moderately pollution-tolerant species, ○ = pollution-intolerant species, — = not recorded. Adapted from Tickner *et al.* (1991).

cubic meters of water were abstracted each year. The largest management operation involves the removal of the underlying, nutrient-rich sediment from the former basins of the Broads (Moss *et al.*, 1996; Tickner *et al.*, 1991).

This trial work has shown considerable benefits to biodiversity. At Strumpshaw Broad, aquatic plants have made a steady recovery and pollution intolerant plants are making a comeback (Table IV). Invertebrate diversity has likewise increased, with a rise in the number of dragonfly species using the site. The increase in the area of open water has benefited wintering wildfowl. The restoration has certainly succeeded in improving water quality on the reserve and made the site more attractive to target bird species, but the current management regime is not sustainable, relying as it does on pumping water. Clearly isolation is not a long-term option either. Connectivity with the rivers in the system is ecologically desirable and therefore sources of pollution in main rivers must eventually be addressed. At Cockshoot Broad, simply removing the sediment and associated nutrients was not enough and biomanipulation techniques (such as those described in the Lake Vaeng case study) may have to be used in the future.

H. Removing Alien Species—The *Melaleuca* Problem in Southern Florida (Maltby, 1997)

Introductions of the tree *Melaleuca quinquenervia* from Australasia to the United States has been an ecological disaster. Having wide ecological tolerances the tree is able to survive in swamp environments and today the trees cover nearly 10% of the southern half of Florida. *Melaleuca* has much higher water requirements than the natural sawgrass vegetation it quickly replaces. The subsequent loss of swamp vegetation and drying effect threatens wildlife and may even lower water tables and affect public water supplies. The only solution in cases such as this is the elimination of the problem species. Mechanical removal of *Melaleuca* will cost $370–2000 ha^{-1}. Authorities in Florida are pinning their hopes on a biological control in the form of a weevil *Oxyops vitiosa* imported from Australia.

I. Engineering Solutions—Water Level and Vegetation Management at Hornborgasjön (Hertzman and Larsson, 1999)

Hornborgasjön, Sweden, is a large post-glacial lake, some 10,000 years old. Successive agricultural drainage projects between 1805 and 1935 lowered the lake water level and reduced the extent of wet grassland around the lake. The remaining shallow, open water was steadily invaded by *Phragmites australis*, sedge *Carex*, and willow *Salix* scrub until little open water remained by the mid-1960s. Drainage and land-use changes reduced the value of the lake for wildlife. Ambitious plans to restore the lake were proposed in 1965 and have recently been completed.

The restoration plan called for the raising of the lake water level by 0.85 m. In order to achieve this, a 3-km

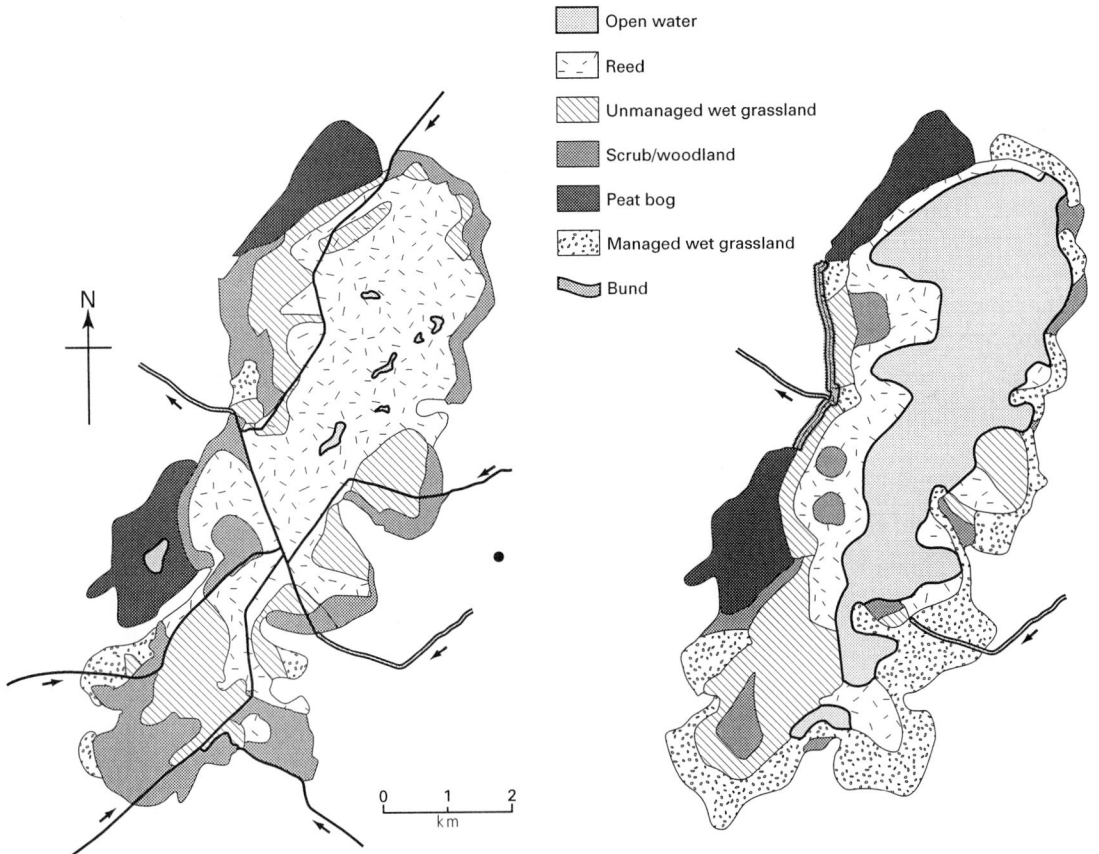

FIGURE 7 Extent of major vegetation communities at Hornborgasjön before and after restoration. Adapted from Benstead *et al.* (1999).

bund and sluice was built around the western boundary to protect neighboring arable farmland. Water levels were slowly raised between 1992 and 1995, restoring some 4100 ha to wetland. Approximately 800 ha of invasive birch *Betula* and alder *Alnus* woodland and *Salix* scrub were cut and 1200 ha of reedbed was burnt or chopped using specially designed amphibious machinery.

The open water character of the lake has been restored (Fig. 7). The reserve supports the only remaining Swedish breeding colony of black-necked grebe *Podiceps nigricollis* (110 pairs in 1997). Additionally, although maximum water levels have only just been achieved, black tern *Chlidonias niger* have returned to breed on the lake (65 pairs in 1997). The remaining reedbeds provide improved feeding conditions for many birds and support increased numbers of many species such as bittern *Botaurus stellaris*, marsh harrier *Circus aeruginosus*, and reed warbler *Acrocephalus arundinaceus*, despite a marked reduction in areal cover of this habitat.

Total costs exceeded $10 million (in 1992), including $2.7 million in compensation payments, $2.0 million in bund and sluice construction, and a similar sum spent on vegetation clearance.

J. Restoring Socioeconomic Functions—The Senegal River Valley (Wit, 1997)

In developing countries alteration of natural hydrological regimes can have drastic and far-reaching effects on the human populations relying on wetland resources. Construction of dams along the River Senegal, at Manatali in Mali and Diama at the river mouth, was originally undertaken without appreciation of the reliance of the entire socioeconomic system of a large region on natural river processes. These developments resulted in alteration of 75% of floodplain wetlands; 95% loss of the seasonal grazing land; 90% reduction in fish production in the river and estuary; salinization of half of the newly created, irrigated land; stagnation, which has led to a high incidence of disease and pollution; and threats to the biodiversity of two National Parks in the delta area.

To counter the threat to local communities and biodiversity, a trial restoration of 60,000 ha was undertaken. New sluices and embankments were installed to bypass the Diama dam and to allow fresh water to flood former estuarine wetlands connected to the sea. New floodgates allowed migratory shrimp and fish access between the former lakes and sloughs, allowing them to breed. Fish harvests have increased dramatically as a result. Results are positive and there are plans to increase the scope of the restoration.

K. Local Community Involvement in the Restoration of Mangrove Forest in Pattani Bay, Thailand (Erftemeijer and Bualuang, in press)

Development driven processes have resulted in the loss of over half of Thailand's mangrove resource since 1970. Fish catches, coastal biodiversity and incomes of coastal-dwelling communities have declined as a result and now restoration has a high priority. While the techniques for mangrove restoration are well documented, securing the long-term future of a restored area through the wise use of natural resources must involve the local community.

A 3-year project in Pattani Bay, an internationally important wetland for migratory shorebirds, aimed to demonstrate the ability of mangroves to strengthen the regional prosperity of those managing the resource. This was achieved by promoting community organization and partnership between all stakeholders, physical rehabilitation of degraded mangrove forest, supporting alternative livelihood initiatives, and effective communication.

Progress in any project of this nature is slow and incremental. Organization of village committees and generating support from key community leaders and the local administration was a lengthy process fraught by internal "political difficulties." Project staff were trained in conflict resolution. Early planting trials failed due to overgrazing and drought, but then the villagers drew up a restoration plan and microtopographical improvements were made to ensure survival of future plantings. To date 30 ha of forest have been replanted.

Lessons learned from this project highlight the need to ensure that community-led restoration projects:

Do not shift objectives away from broader long-term solutions to activities with immediate benefits.

Ensure involvement of all stakeholders in the process is long-term as progress is slow and incremental.
Promote and maintain community ownership and commitment. This is essential for long-term sustainability.

A similar approach has been used by the Yadfon Association in nearby Trang Province, where, over 13 years, 512 ha of mangrove forest have been restored and are being managed as a community forest.

L. Effects of Stream Restoration in Denmark (Iversen et al., 1995)

European rivers are often highly eutrophic being affected by excessive discharges of organic matter and nutrients. Recently, improved sewage treatment has reduced the transport of organic matter and phosphates leading to improved ecological conditions. However, nitrogen levels remain high where leaching from agricultural land is a problem. Several channelized Danish streams have been physically restored since 1989. Restoration involves remeandering, channel bed raising, replacement of riffle-pool sequences, replacement of gravel fish spawning beds, and removal of obstacles to fish migration.

Before restoration the straight channels provided little physical variety and low biodiversity. After remeandering, the new river course varied in respect to stream velocity, water depth, and substrate, improving habitat for wildlife (Fig. 8). Submerged and emergent macrophytes have been shown to increase following restoration (Iversen et al., 1995). Macroinvertebrate density and diversity also increased. Specialist species favoring stone and gravel beds were able to return. Fish populations benefited from the provision of suitable spawning habitat, increased shelter opportunities provided by the more diverse riverine topography, and the increase in prey density.

Additional benefits occur. Restoration produced higher water table levels resulting in wetter riparian corridors and a higher frequency of flooding. Wet riparian meadows facilitate de-nitrification and can play a valuable role in treating agricultural runoff before it enters streams and rivers. Typical nitrate removal rates of 400 kg ha^{-1} yr^{-1} have been recorded (Rebsdorf et al., 1994). Flood events allow sedimentation of particulate matter and associated nutrients over riparian corridors. Sediment can sequester phosphorous transported by rivers and streams. Restoring this nutrient removal process is important in countries like Denmark, where nonpoint source pollution is leading to excessive eutrophication of shallow coastal waters.

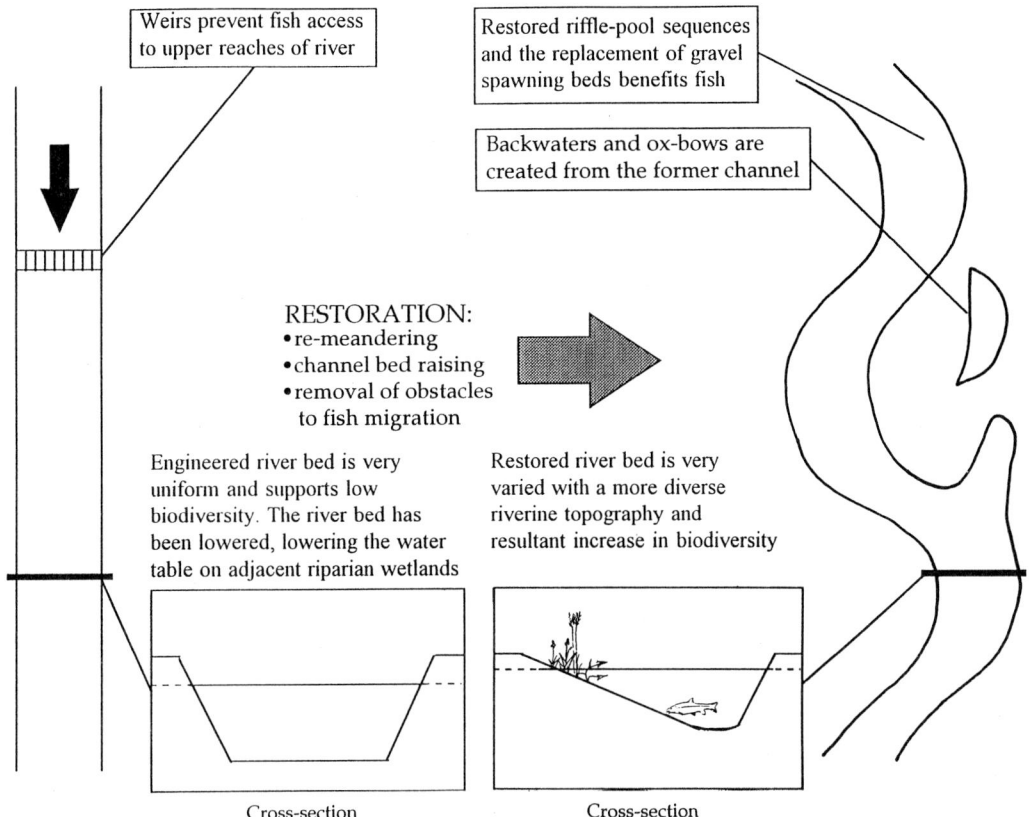

FIGURE 8 Restoration of Danish streams involved the remeandering of channelized stretches and greatly improves value for biodiversity. Other benefits include improved nutrient retention by newly created riparian wetlands, which reduces pollution of watercourses.

The restoration of streams benefits both the individual landowner and society as a whole. Experience from Denmark has shown that conflicts between landowners and environmental interest can be reduced. Landowners involved have indicated that it is particularly important that land redistribution is considered in such schemes. With remeandering, land on either side of a previously straightened river may have to be swapped so that the river remains the property boundary. If a land holding is primarily within the previously drained floodplain, land swaps to include some higher ground may be needed to ensure farming viability. Once again communication with, and involvement of, all user groups is essential for success.

M. Thinking Big on the Kissimmee River, Florida—The Ecosystem Management Approach (Dahm et al., 1995)

The Kissimmee River project in Florida's Everglades is one of the largest wetland restoration project conceived to date. The project aims to remeander 70 km of river and restore natural flooding conditions to 11,000 ha of floodplain wetland. This project, at a cost of $700 million, will reverse the drainage work and river channelization undertaken in the 1950s and 1960s. The detailed project aims are to recover natural ecological functions of the river system, including reflooding drained wetlands; maintain the physical, chemical, and biological integrity of the river; remeander the river and block all drainage channels; and expand the area of the Everglades National Park by 44,600 ha.

So far a 550-ha area of drained floodplain wetland has been used as a pilot study area where the following has been undertaken:

- The flow of water through remnant sections of the old river has been increased.
- New flow fluctuation schedules have been adopted.
- Hydrological and hydraulic modeling studies have been conducted.
- Monitoring has been established to discover the impact of the hydrological changes on wetland vegeta-

tion, fish, secondary productivity, benthic invertebrates, and river channel habitat characteristics.

The results are promising and show that reflooded areas still have viable seedbanks, even decades after drainage; riverine ecosystems responded favorably to the reinstatement of natural flow regimes; restoration of water flow to old river sections reestablished channel morphology, natural substrate characteristics, and benthic invertebrate species diversity; and increased flow through remnant channels cleared organic debris.

VII. CONCLUSION AND THE WAY FORWARD

The growing realization that wetland degradation has led to the loss of important economic and ecological functions has provoked many governments and organizations to consider the restoration of wetlands in order to reinstate beneficial functions. This chapter has presented a broad overview of the value of wetlands and the threats they face, as well as examples of a range of wetland restoration projects from different regions of the world.

Wetland restoration is costly and not guaranteed to succeed. The costs are so great in fact that currently only the developed countries can consider restoration. Globally this is where the bulk of the damage to wetlands has taken place. Where successful, wetland restoration can undoubtedly offset wetland losses, improve degraded sites, reestablish valuable functions, and increase biological capital.

Wetland restoration projects that achieve their objectives often have the following attributes:

- aims that address causes of damage and not symptoms
- good planning and legislative frameworks, which includes watershed planning and the provision for sympathetic land use in and around wetlands
- good planning of project implementation, aftercare, and subsequent management
- identification and involvement of all stakeholders from an early stage; successful projects depend on the cooperation between landowners, managers, user groups, public authorities and politicians at different levels, technical and scientific consultants, and nongovernmental organizations
- communications networks that keep people informed throughout all the phases of the project
- successful pilot projects that engender interest in wider schemes and clarify methodologies
- an appropriate and sustainable hydrological regime
- close proximity to, or surviving sources of, biological material suitable for recolonization
- compliance monitoring during construction phase
- monitoring systems that allow management prescriptions to be modified beneficially in the light of information received
- management systems installed that ensure long-term survival of the restored habitat—ensuring low revenue costs is important
- large areas; this is especially important for biodiversity conservation as large areas hold more species and contained populations are less vulnerable to external influences

Failure of wetland restoration projects often results from inadequate planning and typically arises for a variety of reasons:

- lack of scientific knowledge and the dearth of information in the field
- inadequate consideration of the site's hydrology
- inadequate consideration of the site's topography (actual or designed)
- lack of practitioner expertise in the two key fields of hydrology and soil science
- inadequate supervision of plant and site personnel during project implementation
- failure to adequately implement project design
- invasion of "weed" species
- inappropriately high grazing pressure on newly developed vegetation

A growing concern is that the increasing availability of restoration technology and improved best-practice must not be used by developers to justify the further damage of natural wetlands around the world. There is no replacement for *in situ* conservation of natural and seminatural habitats. Conservation of wetland biodiversity, especially in the Northern Hemisphere, can only be assured with a continued investment in restoration, however. Potential restoration sites must therefore be identified and safeguarded by the planning process to ensure that they remain undeveloped and available for restoration in the future.

It must be stressed that wetland restoration is more than the provision of habitat for waterfowl. Restoration of wetlands should be seen as a vital component of integrated watershed management planning. In the past many restoration projects have relied on quick technological fixes not the creation of self-maintaining, sustainable wetland ecosystems. In the future the scale of projects will increase to take in entire floodplains and the emphasis will be on restoring near-natural functions. Natural processes and ecosystem management will be embraced more widely. This is already apparent in the bold restoration plans for the Kissimmee River in southern Florida and other projects in North America, and this approach should become more commonplace in the future.

Restoration ecologists must seize every opportunity. Sea-level rise is ongoing, and while coastal wetlands will be threatened by marine incursion and the improvement of sea defenses, there is also the opportunity for wetland restoration that will provide flood defense functions and create habitat. The task for environmentalists is to lobby and work with coastal defense bodies to ensure that flood protection and managed retreat schemes result in no net loss of wetland habitat.

Finally, it must be stressed that setting clear objectives, monitoring results, and undertaking research must be undertaken to accurately appraise restoration projects and better the current understanding of the processes involved. The publication of both positive and negative results will ensure that future projects succeed in restoring lost functions and maximizing biodiversity conservation opportunities.

Acknowledgments

The authors thank the following for providing information for this chapter: Dr Paul Erftemeijer (Wetlands International), Tomas Hertzman (Swedish Environmental Protection Agency), Dr. Chris Joyce (Loughborough University), Hans Skotte Møer (Ministry of Environment and Energy, Denmark), Mogens Bjørn Nielsen (County of Sønderjylland), Crawford Prentice (Wetlands International—Asia Programme), Ole Thorup, and Professor Max Wade (University of Hertfordshire). Finally, we thank the anonymous reviewers of this chapter for their input.

See Also the Following Articles

RESTORATION, CHARACTERISTICS AND REQUIREMENTS • RESTORATION OF BIODIVERSITY, OVERVIEW • WETLANDS ECOSYSTEMS

Bibliography

Allen, S. E. (1989). *Chemical Analysis of Ecological Materials*, 2nd ed. Blackwell Scientific, Oxford.
Bakker, J. P. (1989). *Nature Management by Grazing and Cutting*. Kluwer Academic, Dordrecht.
Benstead, P., Drake, M., José, P., Mountford, O., Newbold, C., and Treweek, J. (1997). *The Wet Grassland Guide: Managing Floodplain and Coastal Wet Grassland for Wildlife*. Royal Society for the Protection of Birds, Sandy.
Benstead, P. J., José, P. V., Joyce, C. B., and Wade, P. M. (1999). *European Wet Grassland: Guidelines for Management and Restoration*. Royal Society for the Protection of Birds, Sandy.
Brooks, S., and Stoneman, R. (1997). *Conserving Bogs: The Management Handbook*. The Stationery Office, Edinburgh.
Butt, K. R., and Frederickson, J. (1995). Earthworm cultivation and soil inoculation: A practical technique for land restoration. *Journal of Practical Ecology and Conservation* 1(1), 14–19.
Dahm, C. N., Cummins, K. W., Valett, H. M., and Coleman, R. L. (1995). The ecosystem view of the restoration of the Kissimmee River. *Restoration Ecology* 3(3), 225–238.
Dugan, P. (1988). Wetlands restoration and creation—is it relevant to the developing world? In *Increasing Our Wetland Resource* (J. Zelazny and J. S. Feirabend, Eds.), pp. 259–263. National Wildlife Federation, Washington, D.C.
Dugan, P. (Ed.) (1993). *Wetlands in Danger*. Mitchell Beazley, London.
Erftemeijer, P. L. A., and Bualuang, A. (in press). Participation of local communities in mangrove forest rehabilitation in Pattani Bay, Thailand: Learning form successes and failures. *Proceedings of the Second International Conference on Wetlands and Development*. Dakar, Senegal, 8–14 November 1998.
Finlayson, M., and Moser, M. (1991). *Wetlands*. Facts on File, Oxford.
Fischer, S. F., Poschlod, P., and Beinlich, B. (1996). Experimental studies on the dispersal of plants and animals on sheep in calcareous grasslands. *Journal of Applied Ecology* 23, 1206–1222.
George, M. (1992). *The Land Use, Ecology and Conservation of Broadland*. Packard Publishing, Chichester.
Hertzman, T., and Larsson, T. (1999). Lake Hornborga, Sweden—The return of a bird lake. *Wetland International Publication 50*. Wetlands International, Wageningen.
Her Majesty's Stationery Office (HMSO). (1995). *Biodiversity*. (2 Vols.). Her Majesty's Stationery Office, London.
Hollis, G. E. (1993). Goals and objectives of wetland restoration and rehabilitation. In *Waterfowl and Wetland Conservation in the 1990s—A global perspective* (M. Moser, R. C. Prentice, and J. Van Vessem, Eds.), pp. 187–194. IWRB Special Publication No. 26. IWRB, Slimbridge.
Iversen, T. M., Kronvang, B., Hoffmann, C. C., Søndergaard, M., and Hansen, H. O. (1995). Restoration of aquatic ecosystems and water quality. In *Nature Restoration in the European Union*. (H. S. Møller,. Ed.), pp. 63–69. Proceedings of a seminar, Denmark, 29–31 1995, Ministry of Environment and Energy and The National Forest and Nature Agency, Denmark.
Maltby, E. (1997). A tale of two *Melaleucas*. *World Conservation* 3/97, 8–10.
Manchester, S., Treweek, J., Mountford, O., Pywell, R., and Sparks, T. (1998). Restoration of a target wet grassland community on ex-arable land. In *European Wet Grasslands: Biodiversity, Management and Restoration* (C. B. Joyce and P. M. Wade, Eds.), pp. 277–294. John Wiley, Chichester.
Mitsch, W. J., Mitsch, R. H., and Turner, R. E. (1994). Wetlands of the Old and New Worlds: ecology and management. In *Global Wetlands: Old World and New* (W. J. Mitsch, Ed.), pp 3–56. Elsevier, Amsterdam.

Moss, B., Madgwick, J., and Phillips, G. (1996). *A Guide to the Restoration of Nutrient-Enriched Shallow Lakes.* Broads Authority, Norwich.

Rebsdorf, A., Friberg, N., Hoffmann, C. C., and Kronvang, B. (1994). Interactions between rivers and riparian areas. *Miljøprojekt 275,* Ministry of Environment and Energy, Danish Environmental Protection Agency.

Thorup, O. (1998). Ynglefuglene på Tipperne 1928–1992. Dansk Orn. *Foren, Tidsskr.* **92,** 1–192.

Tickner, M., Evans, C., and Blackburn, M. (1991). Restoration of a Norfolk Broad: A case study at Strumpshaw Fen. *RSPB Conservation Review* **5,** 72–77.

Wit, P. (1997). Buffer zones, wetlands and community management. *World Conservation* **3/97,** 11–14.

Zentner, J. J. (1988). Wetland restoration success in coastal California. In *Increasing Our Wetland Resource* (J. Zelazny and J. S. Feirabend, Eds.), pp. 216–219. National Wildlife Federation, Washington, D.C.

WILDLIFE MANAGEMENT

David Saltz
Blaustein Institute for Desert Research, Ben Gurion University of the Negev

I. Introduction
II. Harvest and the Concept of Sustained Yield
III. Endangered Species and Reducing the Risk of Extinction
IV. Managing Wildlife for Conserving the Integrity of Ecosystems and Landscapes
V. Control of Wildlife
VI. Wildlife Management Techniques

GLOSSARY

density dependence Change in the birth and death rate of a population as a result of changes in population density; usually in the form of increased mortality rate and decreased fecundity as density increases.
focal species A species that can be monitored and managed to maintain the integrity of the ecological system of which they are a part.
intermediate disturbance hypothesis A hypothesis claiming that some level of disturbance is necessary to maintain biodiversity.
minimum viable population (MVP) A method for assessing the minimum size a population must be to bring its extinction over a given time span to below a predetermined level.
population viability analysis (PVA) A method of assessing the probability of extinction for a given population within a specified time span.
stochastic processes Processes that affect the dynamics of a population through random chance alone.
sustained yield The removal of a constant number of animals from a wild population that can be maintained through time while keeping the population at a predetermined size.

WILDLIFE MANAGEMENT is the science of manipulating wild populations to achieve a specific goal.

I. INTRODUCTION

Although the term wildlife can be expanded to include all living organisms, the science of wildlife management, as an academic program offered and taught in universities and as written in textbooks, generally refers only to mammals and birds.

Four major goals of wildlife management can be recognized:

1. Maximizing harvest/yield over time.
2. Preventing extinction and increasing population survival probability.
3. Maintaining and managing the integrity of ecosystems and landscapes of which wildlife populations are a part.
4. Controlling wildlife to minimize damage to human crops and assets caused by wild populations or to return the ecosystem to some predetermined state.

As a science, wildlife management relies heavily on the understanding of ecological theory and processes, but differs from the science of ecology in its research objectives. Ecological research is targeted at formulating new hypotheses and providing the data to support or reject them. Thus, the species or setting selected to perform the research is that best suited to test a specific ecological theory. In contrast, research in wildlife is often dictated by real needs of specific problem-species or specific environments. The management of wild populations requires four processes: assessing the problem and defining the management objectives for the target population, monitoring and analyzing the ecology of the target population, formulating and implementing management actions based on the identified problems and ecology of the target population, and evaluating the implemented methodology. These processes can be carried out on a local or global scale. In addition, because wildlife management is an applied science involving what is often a national asset, public opinion strongly influences its goals. Thus, education, law enforcement, and public relations are also legitimate aspects of wildlife management.

II. HARVEST AND THE CONCEPT OF SUSTAINED YIELD

The science and profession of wildlife management have evolved over time based on perceived needs and ecological understanding. Originally, the field of wildlife management addressed only game species and harvest issues, recognizing wildlife as a crop that is a product of the land. The realizations that populations may go extinct due to over-harvesting and that past harvest rates affect the future performance of wild populations were the main forces driving the development of the science (Leopold, 1933). The recognition that populations can be over-harvested is not new (a review of past examples of game management can be found in Aldo Leopold's book "Game Management"). As far back as biblical times, harvesting wildlife intelligently was considered important (Deuteronomy 22:6). According to Marco Polo's writings, Kublai Khan imposed hunting regulations throughout his Mongol Empire. The first refuge in the modern western world was established by the British Parliament in 1869. However, the formulation of wildlife management as a quantitative science occurred only during the early part of the 20th century, driven mostly by the concept of "sustained yield" and its maximization (e.g., maximum sustained yield, MSY). Leopold's "Game Management" was a keystone to this process, as was president Roosevelt's doctrine of conservation.

Sustained yield refers to a constant level of harvest that can be applied that will allow the harvested population to stabilize at a given size. This can be achieved by removing an amount that is equal to the population's rate of increase. In the absence of limiting factors, the rate of increase of a population should remain unchanged, and, therefore, absolute growth will increase with population size (more animals have more offspring). However, populations do not grow indefinitely, which suggests that at some point the population rate of increase begins to decline as the population increases. This "density-dependent" response is due to a decline in the amount of resources available per individual, which causes a subsequent reduction in reproductive success and increased mortality. The classic "logistic growth equation," introduced by P. F. Verhulst in 1838, describes the growth curve of such a population, depicting an S-shaped curve of population size against time (Fig. 1a). The rate of increase of the population at time t is the first derivative (dN/dt) and is given by

$$\frac{dN}{dt} = r_{max} N \left(1 - \frac{N}{K}\right), \quad (1)$$

where N is population size at time t, r_{max} is the maximum growth rate per individual when resources are not limiting, and K is the maximum size of the population that can be supported over time in a particular environment (K is often termed "carrying capacity"). As can be clearly seen, there are two cases when population growth is 0: when $N = 0$ and when $K = N$. Thus a sustained yield requires that the population be kept below K. Between $N = 0$ and $N = K$, growth rate initially increases with population size and then decreases (Fig. 1b). Each point on this curve represents the sustained yield for a given population size, and the highest point is the MSY.

We calculate MSY by finding the maximum value of dN/dt. This can be done analytically by taking the second derivative of the equation depicting the growth curve of the population and equating it to 0. In the case of the logistic equation,

$$\frac{d(dN/dt)}{dN} = r_{max}\left(1 - 2\frac{N}{K}\right) = 0. \quad (2)$$

This equation equals 0 when $r_{max} = 0$ (which is meaningless) or when $2\frac{N}{K} = 1$. Rearranging the latter, we find that $N = \frac{K}{2}$; that is, the population grows the fastest when it is $\frac{1}{2}K$. To find the rate of increase at that point, we simply substitute $\frac{1}{2}K$ for N in Eq. (1) and find

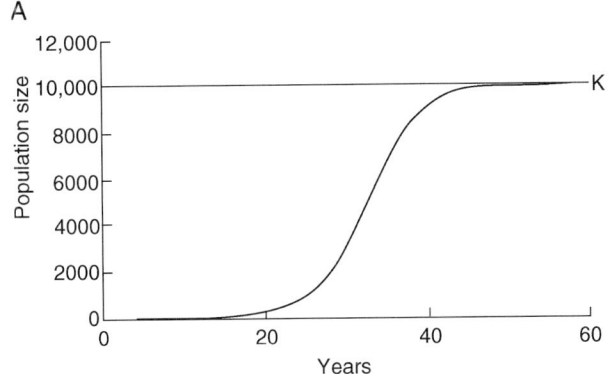

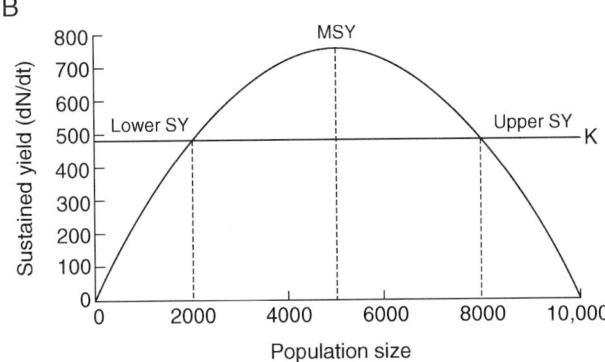

FIGURE 1 The logistic population growth curve as a function of time (a) and sustained yield as a function of population size (b).

responses may vary depending on the species and environment skewing the dN/dt curve; the age structure of the population must be considered; the environment (carrying capacity) constantly changes; other species (predators, competitors, etc.) interact with the population; and hunters may select for specific animal traits (sex, age group, size, etc.). Thus, much of the research on harvested populations has been directed at developing more sophisticated models with better predictive capabilities and estimating the parameters needed for them (Getz and Haight, 1989).

Although many advances have been made in this field, to date, harvest strategies in many places around the world are still implemented on a trial-and-error basis. However, even under such protocol, harvest theory has made an important contribution in demonstrating that uncontrolled harvest can very quickly lead to the demise of populations. Furthermore, while the logic of a sustained yield set somewhat above MSY is obvious, nonecological factors may have a strong influence on harvest policy. In many species, the long-term monetary income from a sustained yield may be less than the yield offered by the economic market on regular investments (Clark, 1976). In other words, economically, a greater profit can be realized by selling licenses to hunt the entire population and investing the profits than by limiting the number of licenses to a sustained yield that will ensure the continued existence of the population. Thus, in recent years effort has been directed at assessing the economic value of wildlife for nonconsumptive uses such as tourism and conservation.

III. ENDANGERED SPECIES AND REDUCING THE RISK OF EXTINCTION

With the biodiversity crisis coming into focus during the second half of the 20th century, the science of conservation began to evolve. Initially, conservation and wildlife management were treated as separate sciences. Eventually, the sciences merged in the area where they overlap (wildlife), and managing wildlife populations for conservation became an integral part of wildlife management. Topics such as management for recovery and sustainability, the dynamics of small population (especially as impacted by demographic, genetic, and environmental stochastic processes), fragmentation, and the importance of populations to the integrity of the ecosystem and biodiversity became a central focus of wildlife management. In terms of management for conservation, Caughley and Gunn (1996) recognize two central paradigms: the small population paradigm and the declining population paradigm.

that the maximum growth rate and, therefore, MSY, is $r_{\max}\frac{K}{4}$.

Note that except for the point of MSY, a given yield (growth rate) can be achieved at two different population sizes, one on either side of MSY ($\frac{K}{2}$)—the upper and lower sustained yields. Thus it appears that one may achieve the same level of sustained yield with a small population as with a large one. However, the two points differ considerably in terms of stability. The points to the right of MSY are stable because moderate over-harvesting will increase the growth rate of the population, thus compensating for the excessive harvest (and vice versa). In contrast, the points to the left of MSY are unstable because over-harvesting will cause a decrease in the rate of increase. As a result, the next harvest will further decrease the population and rate of increase and so on. Initially the impact may appear minor, but this "snowballing" effect can very quickly lead to the demise of the population.

Clearly, real populations do not behave deterministically as in as the logistic model. Density-dependent

The sensitivity of small wildlife populations to extinction was first recognized and expressed in 1938 in a book written by the famous American ecologist W. C. Allee. In his book, Allee writes: "The general conclusion seems to be that different species have different minimum populations below which the species can not go with safety, and that in some instances this is considerably above the theoretical minimum of one pair." In 1981, Shaffer defines four stochastic elements that play an important role in the dynamics of small populations and contribute to their increased risk of extinction: demographic stochasticity, genetic stochasticity (including loss of genetic variance and inbreeding depression), environmental stochasticity, and catastrophes.

The accelerated loss and fragmentation of wildlife habitat during the second half of the 20th century restricted many wildlife populations and required that management steps be taken to minimize these stochastic effects. The size of refuges and other areas set aside for conserving wild populations must be large enough to support a population equal to or larger than the minimum viable population (MVP) (Belovsky, 1987). Corridors connecting small populations are currently considered important (Bennett, 1999), but their effectiveness has not yet been fully demonstrated. Where the establishment of corridors is not feasible, translocation has been implemented. For instance, in South Africa, mountain zebras are translocated by airlift between isolated populations. Habitat improvement and resource supply during limiting periods of harsh environmental conditions are also being used to minimize population declines and reduce the impact of stochastic processes and risk of extinction.

The Management of declining species requires that the cause of decline be identified and managed directly or indirectly. Human-related decline in wildlife populations in the 20th century can be attributed to three main factors: over-harvesting, loss and fragmentation of habitat, and the introduction of alien species. With the improved control of harvest by science and legislation, loss of habitat and fragmentation became the major causes of wildlife population declines. This process has highlighted the importance of considering space in wildlife management. Management efforts have been concentrated on land allocation for conserving wildlife and securing refuges and linkages (corridors) between patches (fragments) of adequate habitat. Roads and fencing are some of the major causes of fragmentation, and special overpasses, underpasses, and gates have been developed to ensure wildlife movement.

IV. MANAGING WILDLIFE FOR CONSERVING THE INTEGRITY OF ECOSYSTEMS AND LANDSCAPES

Wildlife populations are important for maintaining biodiversity in the system of which they are a part. Many wild populations are critical for the functioning of the entire community. Beavers (*Castor canadensis*) are a classic example. Thus, in many instances, populations are managed in order to sustain species richness of the ecosystem of which they are a part. For example, research has shown that both overgrazing and undergrazing will reduce plant species richness. This is known as the intermediate disturbance hypothesis (Connell, 1978). Thus, in certain cases, herbivore populations are managed to maintain the diversity of the plant community. Elk (*Cervus elaphus*) populations were regularly controlled to maintain the existing landscape in Yellowstone National Park. Wolves (*Canis lupus*) were reintroduced in 1995 (Fritts and Carbyn) into Yellowstone with the hope that they would control the elk population. However, this is yet to be seen.

The use of focal species as a tool to manage ecosystems has gained popularity in recent years. The underlying assumption is that by managing the focal species, the entire system can be secured. Indicator, keystone, and umbrella species are the main types that have been suggested as focal species through which systems can be monitored, conserved, or managed.

Indicator species are species that testify to the well-being (health) of the entire ecosystem. These species are the first to respond to deterioration of the ecosystem, since they are more sensitive. By monitoring and managing the indicator species population only, we can estimate and protect the welfare of the entire ecosystem. We assume that as long as the status of the indicator species is satisfactory, the entire system is operating adequately.

Keystone species are species that play an important role in the ecosystem. These may be animals that hold most of the biomass in the ecosystem or that influence many of the other species or functions of the ecosystem. In this way, if the population of a keystone species is removed from the ecosystem, the ecosystem changes dramatically. Therefore, by managing only the population of the keystone species, we are managing most of the ecosystem.

Umbrella species are species that require a large area containing many types of habitats to sustain a viable population. Often these are large bodied homeotherms with large home ranges. By securing a tract of land large enough to sustain a viable population of these species, many others will come under the same protection. The focus here is on leaving enough area for the umbrella species and other members of the ecosystem, but there are no direct management implications for this approach.

V. CONTROL OF WILDLIFE

The ever-expanding human population ensures that human–wildlife conflicts will continue. Most of these conflicts are within the realm of damage to crops, equipment, and other assets. Thus, a major component of wildlife management is wildlife control. A good example is managing habitats around airports to reduce gull activity near runways that endangers aircraft during takeoff. Often, damage caused by wildlife is the result of human manipulation of the environment, such as outbreaks of herbivores following the eradication of predators and high concentration of canids around human waste sites leading to outbreak of rabies. Wildlife management strives to implement control measures in a manner enabling the other goals of wildlife management (sustained yield or maintaining biodiversity) to be met.

The four processes necessary for wildlife management generally apply for wildlife damage control, with some differences. These processes are the following: (1) defining the problem. The species causing the damage must be identified, the type of damage determined, and extent of the damage evaluated. (2) knowledge of the general ecology, behavior, and dynamics of the species causing the damage must be obtained and evaluated with special regard to its response to various control techniques. (3) The combination of (1) and (2) defines the methods that are feasible and *economical* to control the damage. (4) The methodology used must be evaluated. Unless the cost of control is less than the losses due to the damage, management should not be implemented.

VI. WILDLIFE MANAGEMENT TECHNIQUES

Because wildlife management is an applied science, methods and techniques for both research and implementation are important. As a result, a considerable amount of research has been directed at developing, improving, and assessing techniques. Techniques for managing wildlife are aimed at studying, reducing, increasing, or maintaining the population at its current level while securing its integrity (i.e., preventing loss of genetic diversity, enhancing long-term survival, preventing epizootics, etc.) and the integrity of its ecosystem. The techniques can be classified generally into two categories:

1. Applied techniques for manipulating populations by impacting survival, reproductive success, or distribution.
2. Techniques for studying, analyzing, and assessing population.

Many of the more common techniques are published in a series of "Wildlife Techniques" published by the Wildlife Society. The first of this series, edited by H. S. Mosby, was published in 1960. The most recent issue (the fifth edition) was published in 1994 and edited by T. A. Bookhout. The book covers four main topics: experimental design, laboratory techniques, population analysis and management, and habitat analysis and management.

A. Applied Techniques for Managing Wildlife

Applied techniques may be resource (habitat) manipulation, mechanical, behavioral, immunization and immunocontraception, biological, and direct population control.

Habitat and community manipulation. This refers to managing target species by impacting important resources and other species that interact with them (i.e., habitat manipulation, or impacting predator–prey relationships and competing species). By controlling key elements that affect the species' abundance (water, shelter, predators, etc.), it is possible to manipulate its densities and distribution. The increase in computer power enabled the use of spatially explicit models and Geographic Information Systems techniques to project the impacts of various habitat manipulations (Verner *et al.*, 1986).

Mechanical. The simplest and most successful method of controlling wildlife damage is exclusion by fencing. The advantages of fencing are that it is relatively fail-proof and nonlethal. However, it is expensive and labor intensive, demanding constant upkeep. Furthermore, fencing over large tracts of land is a major cause of fragmentation.

Behavioral. These are modern forms of scarecrows that operate on the animal's senses and are, therefore, acoustic, olfactory, or visual. A good example of visual repellents, other than the classic scarecrow, is the silhouettes of birds of prey that are placed on large glass windows to prevent other birds from slamming into them. Examples of acoustic deterrents are gas cannons that produce explosions at irregular time lapses or recordings of alarm calls of other animals. Ultrasonic deterrents work by using sounds at a high decibel level (above 120 dB), outside the range of human hearing, to cause a painful stimulus to animals with sensitive hearing, such as canids. Olfactory repellents are chemical compounds. Two types of chemical compounds can repel animals: (1) compounds that link a food source with induced sickness through a behavioral process called conditioned taste aversion and (2) compounds that repel through unpleasant stimuli of the nervous system. The first type is properly referred to as aversive agents, while the second type are true sensory repellents.

Immunization and immunocontraception. These techniques require the administration of a vaccine either directly, by capturing and injecting, or indirectly, by oral administration with bait. A major advancement in wildlife management in the past decade is the control of rabies by oral immunization. Immunocontraception is a very attractive method of control because it is potentially highly species specific, but it so far has had limited success.

Biological. Biological control has been mostly effective against insects. There are only a few cases in which biological control was effectively applied to wildlife. Of these, the introduction of the myxoma virus to control rabbits introduced into Australia is the most famous. Although complete eradication was not achieved, the virus now holds the rabbit population at 20% of its original size.

Direct population control. Direct population control refers to the removal (harvesting) or the addition (reintroduction and translocation) of animals. Removal is usually used when the animal is a game species and can also be controlled by hunting license quotas. Its main advantages are that it is species specific, the number of animals removed can be closely controlled, and it has limited impact on the environment. Reintroductions and translocations are important methods for enhancing the viability of existing populations and reducing the species risk of extinction.

B. Studying, Analyzing, and Population-Assessing Techniques

A major prerequisite of sound management is the accurate estimate of population size and dynamics, and the determination of how these are influenced by other factors. Consequently, a significant part of the science today is directed at the development and improvement of estimation techniques for assessing population density, survival, reproduction success, well-being, age structure, density-dependent responses, and interspecific interactions. Such data are the basic requirements for estimating sustainable harvest quotas in different and variable environments and assessing extinction probability and other risks.

Research and management of wildlife are carried out over large geographic areas, often with limited access and visibility. Considerable effort has been directed toward developing remote data collection techniques, such as radio telemetry and other marking methods, as well as indirect methods of assessing population condition by using individual animal physiological indicators (Harder and Kirkpatrick, 1994). Rigorous statistical methods, based on maximum likelihood estimation techniques, are continuously being refined. Computer software for assessing density and survival, such as DISTANCE, SURVIV, ESTIMATE, JOLLY, BROWNIE (Lancia *et al.*, 1994), and most recently MARK, were developed specifically for wildlife research and management.

Methods for assessing the viability of small populations (population and habitat viability analysis, PHVA) and for estimating the minimum population size necessary to ensure its long-term existence (MVP) have been developed based on stochastic theory (Burgman *et al.*, 1993). Commercial software such as RAMAS and Vortex are available for PHVA and MVP analysis; however, in most cases and due to the detailed knowledge required, the data are insufficient for a reliable analysis.

See Also the Following Articles

ENDANGERED BIRDS • ENDANGERED ECOSYSTEMS • ENDANGERED MAMMALS • INDICATOR SPECIES • KEYSTONE SPECIES • SUSTAINABILITY, CONCEPT AND PRACTICE OF

Bibliography

Ballou, J. D., Gilpin, M., and Foose, T. J. (1994). "Population Management for Survival and Recovery." Columbia University Press, Cichester, NY.

Belovsky, G. (1987). Extinction models and mammalian persistence. In *"Viable Populations for Conservation"* (M. E. Soule, ed.), pp. 35–58. Cambridge University Press, Cambridge, UK.

Bennett, A. F. (1999). "Linkages in Landscapes." IUCN, Gland, Switzerland.

Bookhout, T. A. (1994). "Research and Management Techniques for Wildlife and Habitats, Fifth Edition." The Wildlife Society, Bethesda, MD.

Burgman, M. A., Ferson, S., and Akcakaya, H. R. (1993). "Risk Assessment in Conservation Biology." Chapman & Hall, London, UK.

Caughley, G., and Gunn, A. (1996). Conservation Biology in Theory and Practice. Blackwell Science, Cambridge, MA.

Clark, C. W. (1976). "Mathematical Bioeconomics: The Optimal Management of Renewable Resources." John Wiley & Sons, New York, NY.

Fritts, S. H., and Carbyn, L. N. (1995). Population viability, nature reserves, and the outlook for gray wolf conservation in North America. *Restoration Ecol.* **3**, 26–38.

Getz, W. M., and Haight, R. G. (1989). "Population Harvesting." Princeton University Press, Princeton, NJ.

Harder, J. D., and Kirkpatrick R. L. (1994). Physiological methods in wildlife research. *In "Research and Management Techniques for Wildlife and Habitats, Fifth Edition"* (T. A. Bookhout, ed.), pp. 275–306. The Wildlife Society, Bethesda, MD.

Lancia, R. A., Nichols, J. D., and Pollock, K. H. (1994). Estimating the number of animals in wildlife populations. *In "Research and Management Techniques for Wildlife and Habitats, Fifth Edition"* (T. A. Bookhout, ed.), pp. 215–253. The Wildlife Society, Bethesda, MD.

Leopold, A. (1993). "Game Management." Charles Scribner's Sons, New York, NY.

Olney, P. J. S., Mace, G. M., and Feistner, A. T. C. (1994). "Creative Conservation." Champman & Hall, London, UK.

Starfield, A. M., and Bleloch, A. L. (1991). "Building Models for Conservation and Wildlife Management." Burgess International Group, Inc, Edina, MN.

Verner, J., Morrison, M. L., and Ralph, C. J. (1986). "Wildlife 2000: Modeling Habitat Relationships of Terrestrial Vertebrates" The University of Wisconsin Press, Madison, WI.

WORMS, ANNELIDA

Kristian Fauchald
Smithsonian Institution

I. Morphology
II. Reproductive Biology
III. Systematics, Diversity, and Phylogeny

GLOSSARY

chaetae Bristles composed of slender chitin cylinders glued together by scleroprotein, emerging from parapodia or body wall in most polychaetes.

nuchal organ Sensory organ with nerves coming from the posterior part of the brain, or the nerve ring, usually present as paired ciliated pits or short ridges at the posterior end of the prostomium; present in most polychaete species.

parapodium Segmentally arranged outpocket from the body wall; usually divided into two branches (rami), the dorsal notopodium and the ventral neuropodium; supports the chaetae; extremely variably developed and may be absent.

peristomium Region around the mouth of the larvae; if distinct in adults may carry peristomial cirri.

prostomium Morphologically anterior-most part of the body; may carry antennae and eyes and contain at least part of the brain.

segment In annelids a section of body set off from the rest of the body by septae at both ends containing a separate part of the secondary body cavity (coelom), with paired ventral ganglia, and usually nephridia, gonads, transverse muscles, and segmental blood vessels; carrying externally parapodia and gills.

trochophore Larva characteristic especially of annelids and molluscs consisting of an episphere, ciliated girdle around the middle called the prototroch, and the hyposphere. In feeding trochophores the mouth is located behind the prototroch and the anus is terminal, usually surrounded by a second ciliated band, the telotroch. Segmentation takes place between the prototroch and the telotroch and is initiated immediately in front of the telotroch.

POLYCHAETA is a commonly encountered group of annelids (segmented worms) best represented in the marine environment, although several species are present in freshwater and a few are known from moist soil on land. In addition to the polychaetes, the Annelida contains one additional species-rich group, the Clitellata, which includes the earthworm and their aquatic relatives and the leeches. The clitellates evolved from the polychaetes, but the exact relationship has not yet been diagnosed. The Polychaeta, the many-bristled worms, includes about 15,000 described species. Estimates of the total numbers of species varies, but judging from the rates of description of new species it is likely that more than 20,000 species will eventually be recognized, based on morphological characters in addition to the study of all genetically fixed properties of the worms. The shape of these worms varies a great deal.

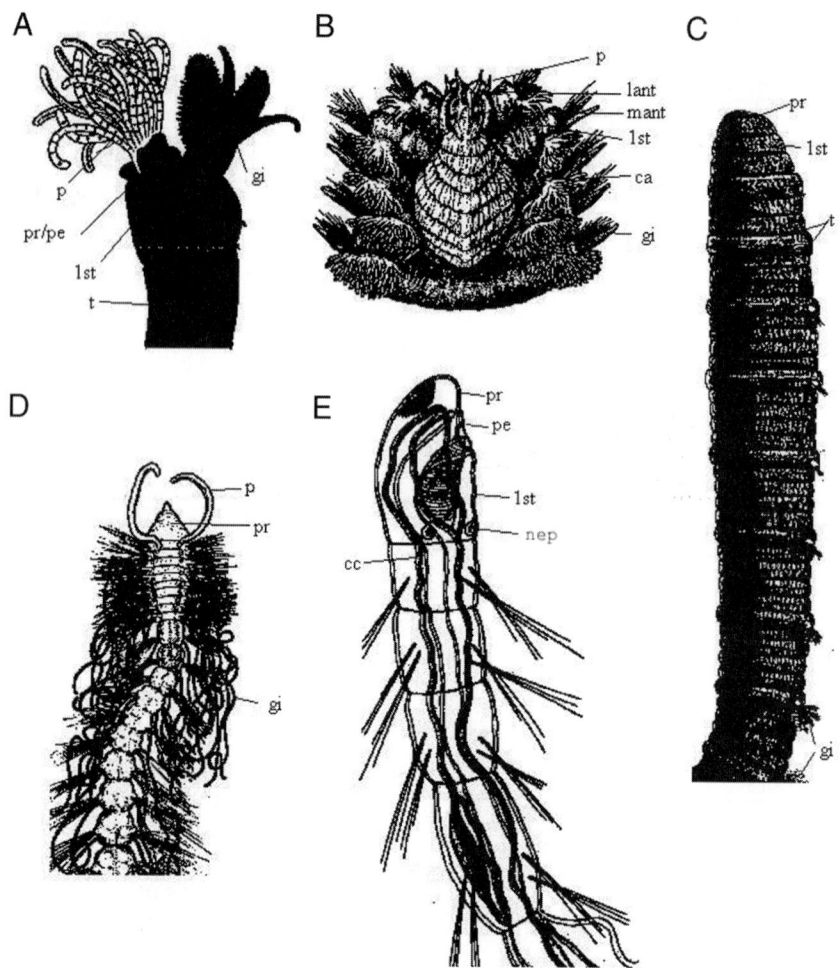

FIGURE 1 Polychaete anterior ends. (A) Ampharetid, lateral view. (B) Amphinomid, dorsal view. (C) Arenicolid, lateral view. (D) Cirratulid, dorsal view. (E) Ctenodrilid, lateral view. (Modified from Rouse and Fauchald, 1997.)

This is reflected in the description given later. No single feature can be used to diagnose the taxon uniquely; until recently the presence of the nuchal organ was considered such a feature, but even this structure is missing in several polychaetes and present in several nonpolychaete taxa.

I. MORPHOLOGY

The polychaetes are composed of three body-regions derived from structures present in the larvae. The presegmental prostomium and peristomium are followed by a segmented body and a postsegmental pygidium that carries the anus. Most polychaetes are less than 100 mm in length, with 100 or fewer segments; many are very small, with as few as 10 segments, and are adapted to life between sand grains in sandy beaches. Some species may become very large, with 2 to 3000 segments and lengths between 1 and 3 m; the longest recorded specimen was 6 m long when collected in Port Jackson, Australia. These very large worms may be as old as 50–100 years, but the documentation for this is rather poor. In general, small species are short-lived compared to larger relatives. Perhaps most polychaetes live to 12–18 months, while others go through three or four generations in a year. The general body shape may be earthworm-like with cylindrical bodies and very little in the way of external appendages, but more usually the protruding parapodia and chaetae makes the worms appear ragged. The ventral side may be flattened with two longitudinal ridges on either side of the mid-

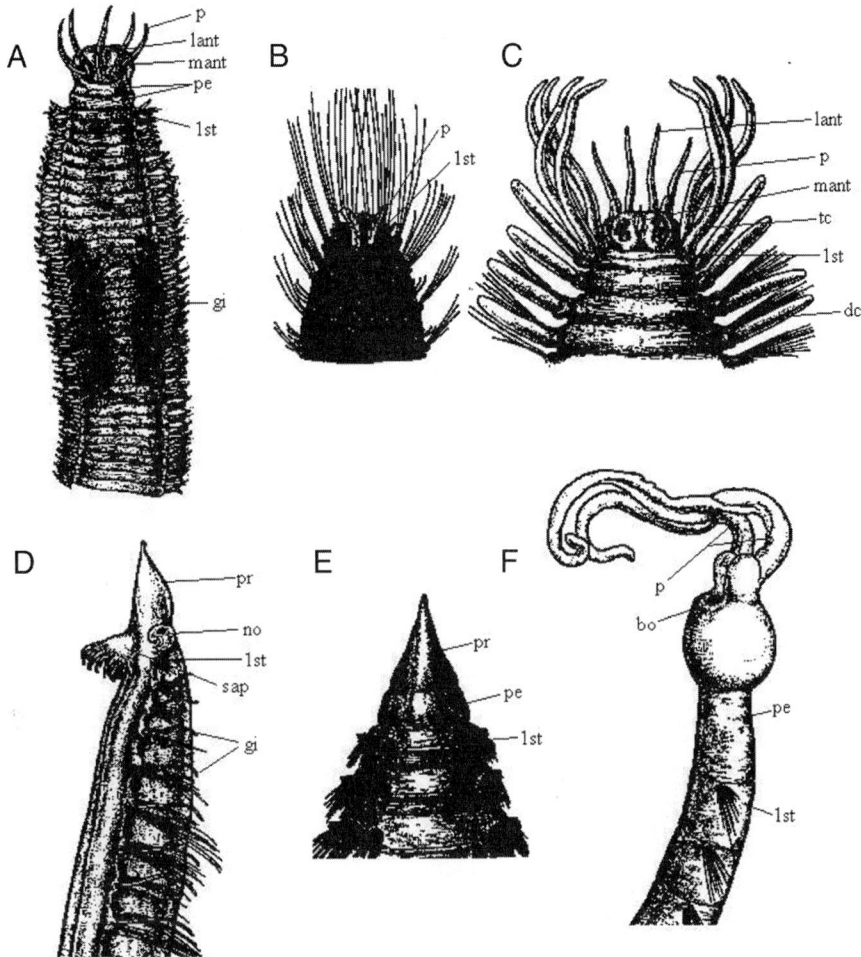

FIGURE 2 Polychaete anterior ends. (A) Eunicid, dorsal view. (B) Flabelligerid, dorsal view. (C) Hesionid, dorsal view. (D) Opheliid, lateral view. (E) Orbiniid, dorsal view. (F) Oweniid, lateral view. (Modified from Rouse and Fauchald, 1997.)

ventral nerve chord; the dorsal side is usually convex. This description fits best for free-living, medium-sized worms; the tubicolous worms are often cylindrical with inconspicuous parapodia, the very small species are often very slender, and the parasitic taxa may be flattened and disc-shaped.

A. The Head

The head consists of the prostomium and peristomium in adults, but one or more of the first segments may be fused to this structure (see Figs. 1–3). The prostomial part may carry antennae and eyes, and the fused anterior segments may have one pair of tentacular cirri per fused segment; these cirri are the dorsal and ventral cirri of each segment participating in the head structure.

In Canalipalpata, the head end can be complex and the numbers of segments included in the head may be difficult to count. Some of these worms have a tentacular crown (e.g., the sabellids and serpulids) or a very large number of antennae (e.g., terebellids). Certain polychaetes lack anterior appendages entirely (e.g., capitellids and maldanids).

B. Segments

Segments are always added just in front of the pygidium; that is, from the posterior end. They are often seen first as a low ridge in front of the pygidium; parapodia, chaetae, and other externally obvious structures are differentiated as more segments are added and, as a consequence, each segment no longer has the position

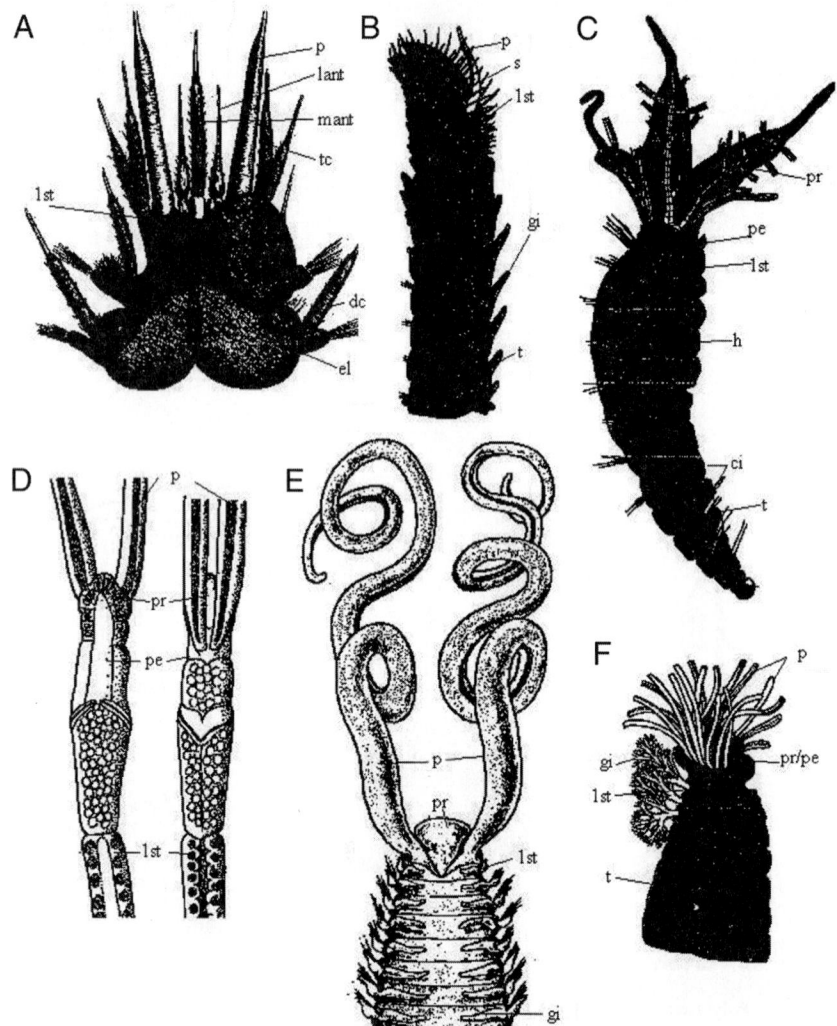

FIGURE 3 Polychaete anterior ends. (A) Polynoid, dorsal view. (B) Sabellariid, lateral view. (C) Sabellid, lateral view, whole body. (D) Siboglinid, dorsal and ventral view. (E) Spionid, dorsal view. (F) Terebellid, lateral view. (Modified from Rouse and Fauchald, 1997.)

of the last segment. The total number of segments may be fixed and limited (e.g., 20, 25, etc., in certain polynoids, maldanids, and opheliids). In many species the number of segments varies with relatively narrow ranges; for example, many nereidids species have 90–120 segments. Some species have an unlimited number of segments and the worm keeps adding segments throughout life. Normally each segment has a pair of parapodia (see Fig. 4). In many species the parapodia are poorly developed and marked only by the position of the chaetae; in other taxa the parapodia are complex structures with a variety of lobes and lamellae. In well-developed parapodia, the two rami often differ. The notopodia are smaller (or shorter) than the neuropodia in many species; when they are completely missing, the parapodia are called uniramous. In the identificatory literature a number of terms (e.g., sesquiramous) are used to characterize the level of reduction of the notopodia. These terms are used inconsistently, and it is better to specify the structures observed than to apply a poorly understood term. When the parapodia are best developed each ramus has a chaetal lobe supporting one or more bundles (or fascicles) of chaetae and pre- and postchaetal lamellae, often covering the bases of the emerging chaetae. The chaetal lobe is supported by one or more aciculae in many species. A dorsal cirrus is often found on the dorsal edge of the notopodium or on the adjacent body wall and a ventral cirri is along

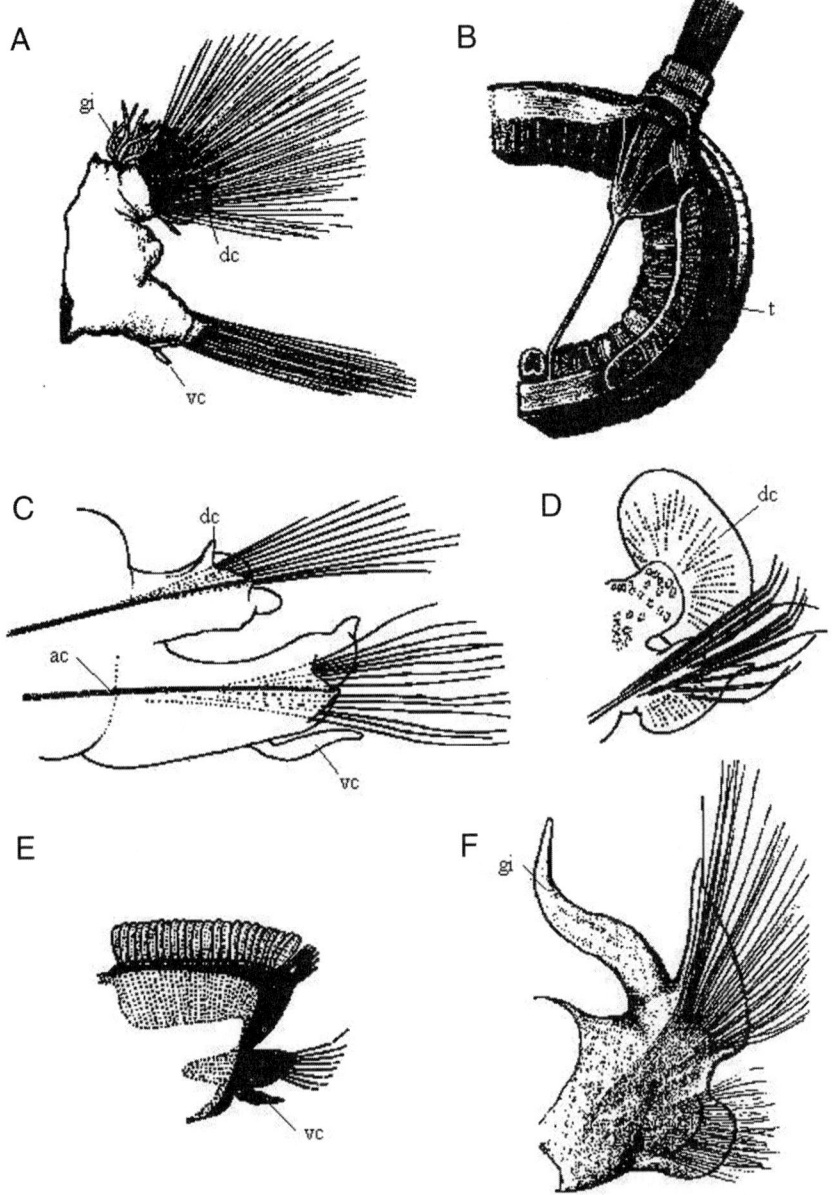

FIGURE 4 Polychaete parapodia. (A) Amphinomid; note similar rami. (B) Arenicold; note neuropodia modified to a torus and the musculature linking the base of the chaetae to the ventral body wall. (C) Paralacydoniid; note unequal rami and the notable presence of aciculae. (D) Phyllodocid; note large, foliose dorsal cirrus. (E) Chrysopetalid; note that some of the notopodial chaetae are large, flattened paleae covering the dorsum. (F) Spionid; note the dorsal flattened gill. (Modified from Rouse and Fauchald, 1997.)

the ventral edge of the neuropodium; both cirri are mainly sensory structures. In addition to the pre- and postchaetal lamellae, various other lamellar or finger-shaped structures may be associated with each ramus, and a complex terminology is associated with the position and shape of each structure.

C. The Pygidium

The Pygidium may be a rounded or tapering simple structure with the anus opening either dorsally or terminally, but frequently it carries anal (or pygidial) cirri. Commonly one or two pairs of cirri are present, but in some taxa the pygidium may be ornamented with numerous anal cirri or flattened lamellae (e.g., maldanids), and in the opheliids it may have an open hood with many papillae attached both inside the hood and along the margin.

D. Chaetae

Chaetae, the bristles, may be in bundles; however, in most cases the "bundles" consist of two or more short, oblique rows of a few chaetae each. Other polychaetes have chaetae in flattened fascicles usually arranged dorsoventrally. Each chaeta is produced from an invaginated epidermal cell (chaetoblast). The outer cell membrane of each chaetoblast is covered with microvilli, each of which produces a slender cylinder of chitin. These cylinders are glued together by scleroproteins produced by cells lining the invagination. All chaetae are formed the same way, through complex interactions between production of cylinders and glue secreted to keep the chaeta together and give the external shape. The chaetae may be tapering capillaries, but most chaetae are much more complex (Fig. 5). A species may have a single kind of chaetae or two or more kinds; in some parapodia each chaeta may be different, but more usually groups of chaetae with a similar structure are present. New chaetae are formed throughout the life of the worms, and distribution of different kinds of chaetae along the body is carefully controlled and changes as the worm grows. In rows of identical chaetae, new chaetae are produced at one end of the fascicle and migrate through the fascicle and drop out at the other end (e.g., hooks in terebellids). In syllids, which often have few chaetae, and in which each chaeta may differ from all other chaetae in a parapodium, replacement takes places in a one-for-one fashion. Special kinds of chaetae, such as aciculae, which are rods supporting long parapodia, and various large spines often show signs of wear, an indication that they are not replaced or at least not replaced as often as other chaetae. The base of the chaetae (chaetal sac) is anchored with muscles running across the body cavity, making it possible to protrude and retract the whole fascicle or bundle since in most cases only a single set of muscles is present. Aciculae and large spines often have their own investment of muscles and are independently movable.

E. The Nervous System

The nervous system consists of a dorsal brain connected by a double nerve ring running around the esophagus to double ventral nerve chords. The anterior part of the brain is located in the prostomium, but penetrates well into the peristomium or even into the next several segments in many species. In most groups the two parts of the ventral nerve chord are wholly or partially fused into a single entity; each segment always contains a pair of ganglia from which the segmental nervous system runs to the parapodial muscles, nephridia, gonads, and circulatory system. The gut is invested from a separate set of nerves from the lower part of the brain (stomatogastric nerves). Antennal nerves run from the anterior part of the brain, and a pair of nerves associated with the posterior part of the brain runs to the palps. Indeed, Orrhage has found a pattern in the emergence of nerves and internal connections in polychaete brains, showing that the limited number of nerves can be identified from one group to the next based on position and function.

The ventral nerve chord may contain two or more giant fibers associated with rapid retraction into burrows or tubes consisting of very thick nerves running from the anterior end innervation the longitudinal muscles directly with only a minimum of synapses. Other motor fibers are segment-to-segment loops passing along undulatory locomotory waves.

F. Sense Organs

Eyes are mostly eye-spots located on the prostomium or even directed on the upper surface of the brain. In some polychaetes the eyes become much larger when the species are sexually mature (e.g., nereididids), and the epidermis over the eye becomes a translucent lens. In a group of pelagic polychaetes (e.g., alciopin phyllodocids) the eyes are very large camera constructions with a distinct lens and with focal capabilities of considerable complexity. In the polychaete eyes, the sensory cells penetrate through a pigment layer (retina) into the space in which the lens is supported, and the optical nerve is collected behind the pigment layer; thus the retina is reversed compared to in vertebrate eyes.

The nuchal organs are usually located in pits, in short grooves, or on short ridges near the junction between the prostomium and the first segment. The nuchal organs may be very large, complexly folded structures (caruncles) in the euphrosinids and amphinomids, and the epaulettes present in some syllids are also nuchal organs. In some spionids the nuchal organs stretch along most of the anterior region of the worm as narrow, ciliated attached ridges. The presence of nuchal organs has been considered diagnostic for the polychaetes in relation to the clitellates, which lack these structures. However, some polychaetes lack nu-

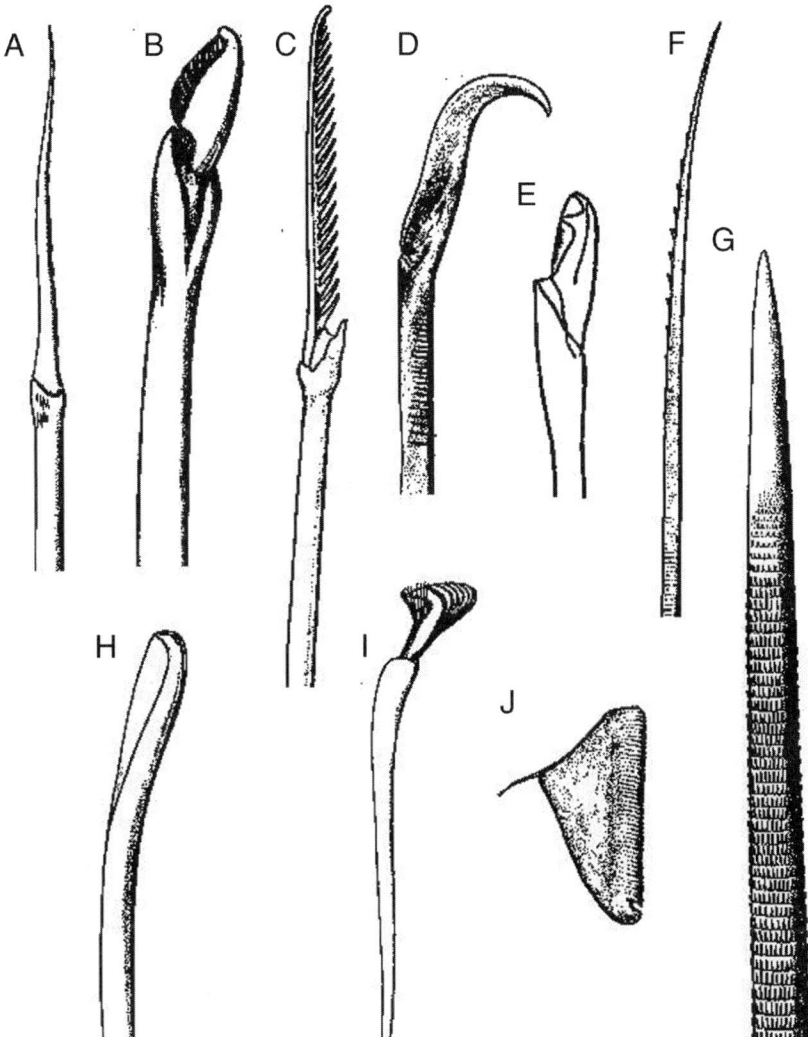

FIGURE 5 Polychaetae chaetae. (A–E) Compound chaetae showing a variety both of hinges and of the distal ends of the chaetae (A, sigalionid; B, nereidid; C, syllid; D, flabelligerid; E, eunicid). (F) Capillary chaeta of hesionid. (G) Spine of a polynoid. (H) Hook of a spionid; note the two teeth at the end. (I) Hook of a maldanid; note the many teeth at the end and the spray of fibers under the main tooth. (J) Uncinus of a chaetopterid; note the straight edge with the many small teeth along it. (Modified from Rouse and Fauchald, 1997.)

chal organs, and sensory structures closely similar to the nuchal organs are present in more distantly related worms (e.g., sipunculans, the peanut worms), so the presence of these sensory organs may be a shared feature of a large group of invertebrates and the loss among certain annelids may be a secondary phenomenon.

Ciliated lateral organs are located between the parapodial rami in many polychaetes; they are usually tufts of cilia in a shallow pit, but may form eversible papillae in some taxa. Similarly constructed patches of cilia have been reported from a variety of polychaetes and may form patterns of sensory organs characteristic of each of the major groups. All of these ciliated structures are assumed to be at least partially chemosensory, but the experimental evidence is usually missing.

G. Musculature

Musculature includes two pairs of longitudinal muscle bundles; these bundles are usually ovate in cross-section, but especially in species with massive longitudinal muscles, they are folded over presumably to avoid buckling during strong contraction. The ventral pair is usually heavier than the dorsal pair, which in some worms (e.g., terebellids) forms a thin layer between the upper edge of the notopodia and the dorsal mid-line. Circular muscles are usually considered to be present; however, at best they are present near the septal boundaries. Most of the other segmental muscles are associated with the parapodia, the chaetae, or are dorsoventral transverse muscles. Whether these antagonists to the longitudinal muscles represent the remnants of a former complete layer of circular muscles has yet to be determined.

H. Septa

Septa separating the segments are rarely complete, but in many cases they are sufficiently musclarized, with radiating and circular muscles to allow each septum to clamp down around the digestive tract and at least temporarily isolate each segment. In certain families one or more of the anterior septa may be complete (gular membranes) and may isolate an anterior part of the body cavity from the rest of the body.

I. Digestive Tract

The digestive tract is usually a more-or-less straight tube running from mouth to anus. The larval stomodaeal region (infolded ectodermal region in the mouth) may be elaborated into an eversible pharynx (proboscis). This pharynx may be axial or may include only the ventral side; it may be sac-like or muscular. Jaws are present in some taxa; they are thickened regions of the cuticle and as such consist primarily of sclerotinized protein in which a variety of metal ions may be imbedded, or they may be calcified. Paired salivary glands are often present. The middle of the gut may be invested with a large blood-sinus or is at least well supplied with segmental blood vessels. In postlarval siboglinids (Pogonophora + Vestimentifera *auctores*) the gut is blind-ending at both ends and the lumen of the gut is largely obliterated, but a very inflated specialized gut wall is present. The cells of this gut wall contain numerous commensal bacteriae. The scale worms have numerous segmentally arranged blind sacs attached to the mid-gut; at least some part of the intermediary metabolism takes place in these sacs, which also in part supplements the circulatory system in bringing nutrients to the tissues in which they are needed. Some polychaetes have a looped gut; in these taxa the septation is missing in the region that contains the loop. Looping allows for an increased length of gut in relation to the size of the worms.

J. Gills

Gills are present in a variety of taxa. They are usually associated with the notopodia but may be present anywhere along the body; they are recognizable as thin-walled extensions from the body wall containing a vascular loop in which the two limbs of the loop are linked through capillaries. This definition excludes the structures called gills in the capitellids and glycerids, both of which lack a circulatory system; these structures cannot at this time be considered homologous with the gills present in taxa with closed circulatory systems. Functionally, any thin-walled extension of the body wall will have a gas-exchange function whether it structurally belongs to a given category of structure or not. This issue is of considerable phylogenetic interest in that gills must have evolved independently several times, while in other clades, morphologically different structures, including the notopodial cirri, have taken on respiratory functions.

K. Circulatory Systems

Circulatory systems may be complete, consisting of a dorsal, often contractile vessel, in which the blood flows anteriorly, one or two major ventral vessels, and capillaries connecting these vessels to the gut and to the rest of the organs. Some polychaetes lack the capillary beds and thus have an open circulatory system, and small species often lack a circulatory system entirely, presumably relying on diffusion for transportation of oxygen, carbon dioxide, and so forth. Shared absence of a circulatory system is not considered a feature of phylogenetic importance; it seems to be linked to a reduced body size.

L. Nephridia

Nephridia may be present already in the larvae as protonephridia; the postlarval nephridia may be either protonephridia or metanephridia. Protonephridia open into the body cavity with intracellular slits; metanephridia open with a funnel. In all postlarval polychaetes the inner end of the nephridia is located in the segment preceding the one in which the duct opens to the exterior. About 100 years ago, Goodrich identified a series of possible fusions between the nephridia and the exit ducts for the gonads (e.g., protonephromixia, metanephromixia, mixonephridia). New information suggests that this classification may not represent the situation accurately, and that it is more complex than previously understood. Nephridia are often present in many segments, especially when the segments are similar in structure; however, in many polychaetes the number of pairs of nephridia may be limited to four to five. Excretory nephridia are always located anterior to nephridia, which functionally become gonoducts when the worms are sexually mature.

M. Gonads

Gonads may be present in many segments, or they may be limited to several or even just a few segments in the middle of the body. Sexes are most frequently separate, but hermaphroditic taxa are scattered throughout the group. The gonads are usually very simple, part of the peritoneal lining of each segment, sometimes attached to the transverse or oblique muscles. In many polychaetes only the first few cell cleavages leading to eggs and sperm take place in the gonad proper; the rest of the development takes place in the coelomic fluid in which the eggs and sperm float. The developing spermatocytes may have incomplete early mitotic cleavages, leading to flattened sheets or rounded balls of cells in which the head end of the sperm is embedded in the shared structure; these structures are called sperm morulae. The yolk of the developing eggs may be furnished by nurse cells, either in a layer or as two strings. However, in many species no nurse cells are present, but the coelomocytes function at least in part to furnish the building blocks of yolk for the eggs and the general nutrients to the developing sperm.

II. REPRODUCTIVE BIOLOGY

Most species spawn in the water, the site of fertilization as well as development up to settlement; indeed, several species swarm on the surface, such as the palolo-worms of tropical waters, sometimes while emitting luminescence (e.g., *Odontosyllis*). A few species have modified some of the parapodia for copulation, but this is rare and limited to certain small species. However, we are increasingly finding that females may store sperm in sperm receptacula, which may be segmentally arranged or at or near the anterior end. Sperm morphology has been related to the modes of fertilization. The round-headed sperm with a short acrosome and a long tail has been considered primitive since it is present in taxa that spawn into the water column. When fertilization takes place elsewhere, the sperm usually has another structure, often with greatly elongated heads and short tails.

In taxa with both feeding and nonfeeding (or brooded) larvae, eggs are much larger in the taxa in which the larvae do not feed. Nonfeeding larvae are present in all major groups of polychaetes and in some families are the only kind of larvae present. The shift between feeding and nonfeeding larvae must have evolved more than once; the direction in which this shift has taken place cannot be considered settled in many cases.

Early development may take place in open water, in deposited egg masses, in brood-chambers of various sorts, or internally, in the body of the females. After the early cleavages, which follow the pattern called spiral cleavage as in many related phyla, and gastrulation, a trochophora larva develops. The trochophore consists of an upper episphere with a sensory tuft of cilia on the top, and often a pair of small eyes. The episphere is separated from the hyposphere by an encircling prototroch, a ciliary band which in the polychaetes may consist of anything from a single row of ciliated cells to a broad densely ciliated girdle. The mouth is located on the ventral side, immediately posterior to the prototroch and is often associated with a longitudinal mid-ventral ciliary band, the gastrotroch. At the posterior end of the larva, a third ciliary band, the telotroch, is often also present. In feeding larvae the gut is fully developed and a pair of larval nephridia are present. When metamorphosis starts, segmentation starts posteriorly, just in front of the telotroch; internally the segments are first visible as paired mesoderm blocks on either side of the gut. These blocks eventually become hollow, and the mesoderm forms the peritoneal lining and muscles and most other structures in the developing segment. The space inside each block becomes the secondary body cavity, the coelom, which eventually contacts the exterior through nephridial ducts. Metamorphosis in most polychaetes is a gradual process of adding more segments with a progressive reduction in the prototroch and other larval structures; the body proportions of a newly settled juvenile may reflect what the adult worm will eventually look like, but in many cases considerable change in relative proportions of various body regions takes place during the later development. The body of the polychaetes may appear to consist of a series of repeated units, but as we become more familiar with postmetamorphic development it is becoming increasingly obvious that the polychaetes maintain integrated bodies with distinct proportionality between different regions.

III. SYSTEMATICS, DIVERSITY, AND PHYLOGENY

Traditionally the polychaetes were separated into errants and sedentaries (two equivalent groups); since

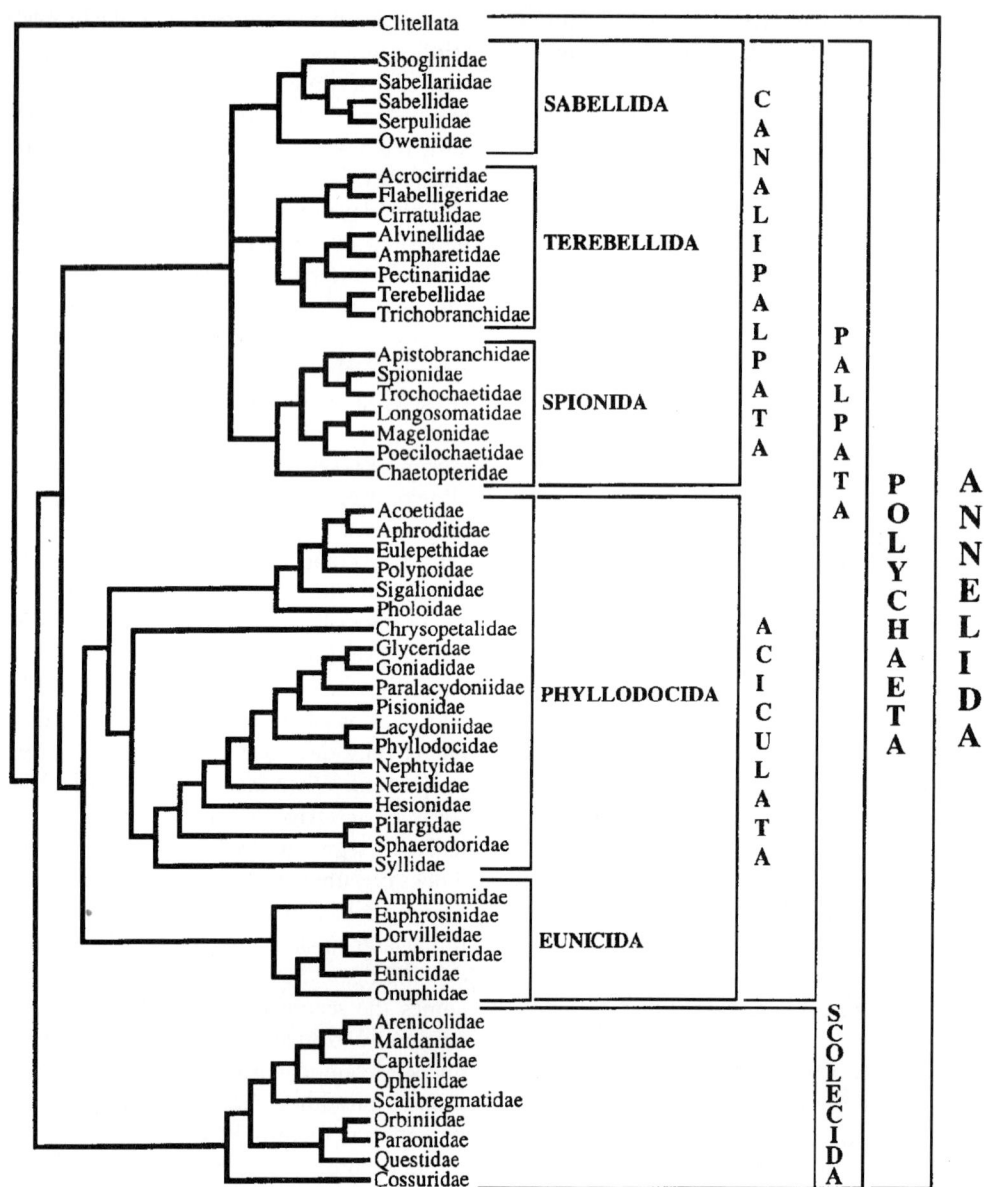

FIGURE 6 Diagram showing the current systematic subdivision of the Polychaeta. Only 54 of the 80 families have been included; most of the remaining families can be readily associated with the taxa where they fit but are sufficiently poorly known or lack so many features that including them would have distorted the current analysis. Eventually we expect to be able to incorporate all families. (Modified from Rouse and Fauchald, 1997.)

the 1950s this system has been replaced by one in which the group is separated into many orders, but without a clearly stated phylogenetic scheme to support the groups. The classification presented here was proposed by Rouse and Fauchald (1997) (Fig. 6); it will certainly be improved when more information becomes available. Especially important will be a clearly recognized linkage between the clitellates and the polychaetes.

In addition to the approximately 80 families traditionally included among the polychaetes, a major change was proposed by Rouse and Fauchald. The pogonophorans (including the vestimentiferans) have been considered a separate phylum (phyla). Their major features match the polychaetes well, and they are included here as a single family of polychaetes, Siboglinidae. Some of the unusual features, such as the "absence" of a gut is incorrect: the siboglinids have most tissues associated with the gut, but in the adults the gut is closed both anteriorly and posteriorly, and the gut lumen is nearly obliterated. Evidence from molecular systematics furnishes good criteria for including the siboglinids among the polychaetes, even if the exact placement among the canalipalpates is still uncertain.

Rouse and Fauchald included the myzostomids among the polychaetes; this group has been considered distinct for more than 100 years, and recently it has been shown that the apparent similarity to the polychaetes is incidental (see especially papers by Eeckhaut and collaborators). The myzostomids are associated with echinoderms and have a fossil record stretching back to the paleozoic, so they certainly have a long separate evolutionary history.

The fossil record of the polychaetes is, not surprisingly, uneven, considering that they are soft-bodied with very few readily fossilizable structures. They were present in the middle Cambrian (in the Burgess Shale) in taxa resembling, in certain respects, recent canalipalpates. In the same material is also found *Wiwaxia*, which may or may not have been a polychaete, but which appears to share certain details with members of a recent aciculate polychaete family. Jaws of various polychaetes have been reported present from Ordovician on; these jaws were originally considered a separate group, Scolecodonta, but are now considered disarticulated jaws of aciculate polychaetes, especially the euniceans, which are still common in all kinds of environments. Whole body fossils have also been reported from other paleozoic strata, and the presence of recent families has been suggested. Many records lack the diagnostic characters of recent taxa so the presence is often difficult to document. Calcified tubes similar to those present among the serpulids are present from early paleozoic strata on; some of the tubes may have belonged to nonannelid taxa, but certainly some of them were serpulids. The upshot of the scattered records shows that most major taxa of polychaetes must have been present by the Ordovician and some clades may be even older than that. This suggests that each group has a long separate evolutionary history. Most of the groups were thus present before Pangaea was formed, which agrees with the very wide dispersal of all major polychaete taxa: they have been rafting on continents. On the other hand, there is much evidence to show that individual species are capable of very rapid dispersal as invasive species, probably mostly through transportation of adults.

Polychaetes are important components of the bottom fauna at all depths of the oceans from beach sands to the soft muds of the deepest trenches. Recently a rich fauna of very unusual polychaetes have been found in both hot vents and cold seeps. Various polychaetes are also present in the pelagic environment throughout their life; more commonly the polychaetes have benthic adults but are present in the plankton as larvae for a varying period of time. In certain areas the polychaetes may make up as much as 90% of the standing crop of benthic macrofauna, so any account of life in and on the sea bottom must taken into account the activities of these worms.

Polychaetes often appear to be very widely dispersed geographically, and some species are considered cosmopolitan. Careful studies have in many cases shown that the records on which these distribution maps were based are erroneous. More than one species have been called by the same name: later investigations have shown that what was originally thought to be a single taxon often represented several distinct taxa. This correction of the geographical records is important for our understanding of the biodiversity of many areas, and is in part the result of increased quality of analysis based on the development of statistical techniques and the emergence of molecular systematics or has been demonstrated in experimental studies. Many polychaetes are genuinely widely dispersed, however, possibly because they can be transported on ship bottoms and in ballast water. Other forms of accidental introductions also have been demonstrated. Importation of a species of abalone from South Africa to California for use in aquaculture brought along a small sabellid polychaete that burrows into the shell of the South African abalone; the sabellid escaped and is now doing damage to native abalones not only in aquaculture but also in the wild. Other polychaetes, especially members of the family Spionidae, have been transported with importation of oysters; they burrow into the shells of the oysters, forming mud-blisters, which may have serious consequences for the oyster industry in certain areas.

Polychaetes may move around at the surface of the substrate on which they live, but even the most active species are usually cryptic, nestling in cracks and crevices, under the cover of sea weed or debris or along the edge of rocks where these meet sand or mud. Many polychaetes from a variety of higher taxa form tubes or burrows in which they live more or less permanently. Some, such as the onuphid *Hyalinoecia* are capable of moving their quill-like tubes along, but others, such as the related *Diopatra* are sessile; in this genus the tube consists of a flaky grey matrix which, where it reaches above the substrate, is decorated with shell fragments, wood or sea grass debris, and sea weed. The sabellids, especially the larger species, form tough but pliable tubes, but the closely related serpulids have their tubes impregnated with calcium carbonate. Some species, such as certain glycerids, live in complex galleries with several openings on the surface; the worm moves around in the galleries so that they can intercept small prey-organisms moving around on the surface of the mud or sand.

The presence of polychaete tubes creates a microenvironment in many areas so that the sheer presence of these worms modifies sedimentation patterns and influences the relationship between the surface of the sediment and the overlying water.

In general, polychaetes feed on anything organic. Some taxa are obligate carnivores, but most species will feed basically on any kind of plant or animal fragments present, live or dead. Several species take food particles out of the water column, mostly by creating small back eddies in flowing water so that particles in transport in the water will drop out of the flow. Only the chaetopterids are truly filtering the water for contained food and are specialists of that mode of feeding. The nereidid polychaetes have been shown to be capable of filtering as well, but they are not morphologically structured for this activity: This may also be true for several other polychaetes that are capable of building temporary tubes and driving water through the tube with undulatory motions. Incomplete mucus plugs will catch particles in the stream. Many polychaetes feed by capturing small, light-weight particles in mucus strands and transporting them to the mouth along the palps, feeding mainly on the light fraction of particles in bottomload transport at the water–sediment interface, often creating or taking advantage of turbulence developing around irregularities in the bottom. Some polychaetes feed at depth at the bottom of their tubes (maldanids) and are thus capable of turning over sediments. The arenicolids are especially well known for their feeding in the intertidal sands and mud flats; these worms are large and as a consequence have considerable effect on the life conditions for all other inhabitants in the bottom. They are buried sufficiently deeply to avoid being the prey of plovers and other birds.

Certain polychaetes, such as the capitellids, include some of the most pollution-resistant species in the sea, capable of surviving in inner harbor waters where the organic pollution load is very heavy. Other polychaetes such as some glycerids are extremely sensitive to change in sediment composition and oxygen levels in the water. This range of reactions to disturbance has made polychaetes much used in environmental research. Most of the taxa used have short generation spans and at most very short larval life, making them ideal subjects for laboratory studies of heavy metal ion toxicity. Most of the species used in this kind of research can be maintained under very simple conditions and have contributed to giving the polychaetes a reputation for being generally pollution resistant, but it is likely that this is a sampling error. These species used in research easily go into culture. The more sensitive taxa are much more difficult to keep in culture, but may be more generally characteristic of the group as a whole in terms of physiological requirements.

See Also the Following Articles

INVERTEBRATES, MARINE, OVERVIEW • WORMS, NEMATODA • WORMS, PLATYHELMINTHES

Bibliography

Fauchald, Kristian. (1977). The polychaete worms. Definitions and keys to the orders, families, and genera. *Natural History Museum of Los Angeles County, Science Series* **28**, 1–188.

Fauchald, Kristian, and Jumars, Peter A. (1979). The diet of worms: a study of polychaete feeding guilds. *Oceanogr. Marine Biol. Ann. Rev.* **17**, 193–284.

Fauchald, Kristian, and Rouse, Gregory W. (1997). Polychaete systematics: Past and present. *Zoologica Scripta* **26**(2), 71–138.

Garrison, Fredrick W., and Gardiner, Stephen L. (eds.) (1992). "Microscopic Anatomy of Invertebrates, Volume 7." Wiley-Liss Inc., New York, NY.

Rouse, Gregory W., and Fauchald, Kristian. (1997). Cladistics and polychaetes. *Zoologica Scripta* **26**(2), 139–204.

WORMS, NEMATODA

Scott L. Gardner
The University of Nebraska, Lincoln

I. What Is a Nematode? Diversity in Morphology
II. The Ubiquitous Nature of Nematodes
III. Diversity of Habitats and Distribution
IV. How Do Nematodes Affect the Biosphere?
V. How Many Species of Nemata?
VI. Molecular Diversity in the Nemata
VII. Relationships to Other Animal Groups
VIII. Future Knowledge of Nematodes

GLOSSARY

anhydrobiosis A state of dormancy in various invertebrates due to low humidity or desiccation.
cuticle The noncellular external layer of the body wall of various invertebrates.
gubernaculum A sclerotized trough-shaped structure of the dorsal wall of the spicular pouch, near the distal portion of the spicules; functions in guiding the spicules.
hypodermis The cellular, subcuticular layer that secretes the cuticle of annelids, nematodes, arthropods (see epidermis), and various other invertebrates.
pseudocoelom A body cavity not lined with a mesodermal epithelium.
spicule Blade-like, sclerotized male copulatory organ, usually paired, located immediately dorsal to the cloaca.
stichosome A longitudinal series of cells (stichocytes) that form the anterior esophageal glands in *Trichuris*.
stoma The buccal cavity, just posterior to the oval opening or mouth; usually includes the anterior end of the esophagus (pharynx).
synlophe In numerous Trichostrongylidae, an enlarged longitudinal or oblique cuticular ridge on the body surface that serves to hold the nematodes in place on the gut wall.
vermiform Worm-shaped with tapering form both posteriorly and anteriorly.

NEMATODES are the most speciose phylum of metazoa on earth. Not only do they occur in huge numbers as parasites of all known animal groups, but also they are found in the soils, as parasites of plants, and in large numbers in the most extreme environments, from the antarctic dry valleys to the benthos of the ocean. They are extremely variable in their morphological characteristics, with each group showing morphological adaptations to the environment that they inhabit. Soil-dwelling forms are extremely small; many marine species have long and complex setae; and parasitic species manifest amazingly great reproductive potential and large body size. Nematodes are one of the major synanthropic metazoans, with some species such as pinworms having coevolved with humans and their relatives since the beginning of the lineage of the primates. While estimates of the numbers of known species hover around 20,000 actual numbers of taxonomists/systematists with expertise in this group are decreasing yearly. This is despite the fact that the Nemata are probably the last great group of Metazoa to be well-documented and described. Estimates of the actual number of species of nematodes that remain to be described include several thousand from insects and millipedes, several thousand from vertebrates, and perhaps millions from marine habitats.

I. WHAT IS A NEMATODE? DIVERSITY IN MORPHOLOGY

A. General Characteristics and Synapomorphies

Despite numerous assertions that appear in the literature stating that nematodes are morphologically conservative, the contrary is actually true: species of the phylum Nemata are extremely variable in their morphological characteristics. Because of this diversity, almost any broad statement regarding their anatomy probably should be tempered or qualified. Nevertheless, nematodes are nonsegmented worms that generally lack external appendages. Most are vermiform, with tapering anterior (Fig. 1a) and posterior (Fig. 1b) ends, cylindrical in cross-section, and covered with a usually translucent, flexible, acellular cuticle (Fig. 1c) secreted by an underlying cellular hypodermis. The cuticle may be smooth (Fig. 2) or ornamented with rings (Figs. 3a–3c), longitudinal striations, spines (Fig. 3b), or spikes, or it may have well-developed wing-like structures called lateral alae that are very common on the externo-lateral surfaces (Fig. 4). Some marine species have long setae (genus *Draconema*) (Figs. 5a, 5b, and 6), modified sensory papillae that are probably used in movement and in detecting their environment. In contrast to all other nematodes, some marine species have eye spots that enable them to detect light in their environment (e.g., genus *Thoracostoma*) (Fig. 7).

The fine structure of the external body-cuticle is complex, acting to protect the animal from the external environment and allowing it to remain homeostatic inside. The cuticle can be extremely resistant, and depending on the nematode species and its life-history attributes. The cuticle of the nematode may be able to resist digestion in the most inhospitable stomachs in the vertebrate world. On the other hand, nematodes may be extremely delicate, able to exist intact only within the osmotically balanced tissues of other animals (e.g., members of the order Filaroidea), and if moved from the isosmotic solution to one with fewer salts, they may explode and die in a most amazing display.

Nematodes are called "pseudocoelomates" because in most forms, their coelom is not completely lined with mesodermally derived cells and they are triploblastic. Possessing no circular body muscles, movement is accomplished by contraction and relaxation of longitudinal muscles in apposition to a hydrostatic skeleton. The body wall is flexible and very strong. All nematodes maintain their form because their body fluids (in the hydrocoel) are under a positive pressure relative to their environment, analogous to the way a water balloon maintains its shape. The cross-section in Fig. 8 shows *Vexillata* and the round shape of the body under the cuticle that supports spines or arêtes (the whole structure is called the synlophe). Nematodes have a complete digestive system with an anterior stoma just behind the mouth and usually a tri-radiate or triple muscle-pumping type of esophagus (Figs. 1a, and 1c) that can be muscular or glandular in structure (or both). The intestine, tubular in form (Fig. 1c), is usually a single cell in thickness lined on the peritoneal side with a thin collagen-like material, internally lined with micro villi (Grasse, 1965; Maggenti, 1981, 1991a). The tube extends from the esophagus straight to the anus or cloaca, sometimes showing out pouched cecae or diverticulae near the esophageal end.

Most nematodes are sexually dimorphic with separate sexes (diecious or amphigonus) and are oviparous; however, some are ovoviparous and some are viviparous. Males usually have a cuticularized spicule or pair of spicules that are used to assist in the transfer of sperm to the females. Many species also have a gubernaculum (seen extended from the cloaca of a species of an aspidoderid in Fig. 23) that guides the spicule during copulation and spicular eversion. Some are hermaphrodytic; in these cases the nematode produces both sperm and ova from the ovotestis of the same individual at different times during ontogeny (Maggenti, 1991a; Malakhov, 1994).

A synapomorphy of the Nemata is the presence of noncontractile axon-like myoneural processes or extensions that run from the contractile or body portion of muscle cells to the neural junctions of the nerve cords. The width of nematodes is usually less than 2 mm, even when extremely long. However, even here there are exceptions, with the giant kidney worm, *Dioctophyme renale*, of canids attaining a size of 15 mm × 1000 mm. The eggs of nematodes are also very similar in size, with most ranging from 50 to 100 μm long by 20 to 50 μm wide, with an exception being the antarctic marine nematode *Deontostoma timmerchioi* (40 mm in length of body) that has eggs that range from 870 to 1100 μm long by 240 to 350 μm wide (huge by nematode standards).

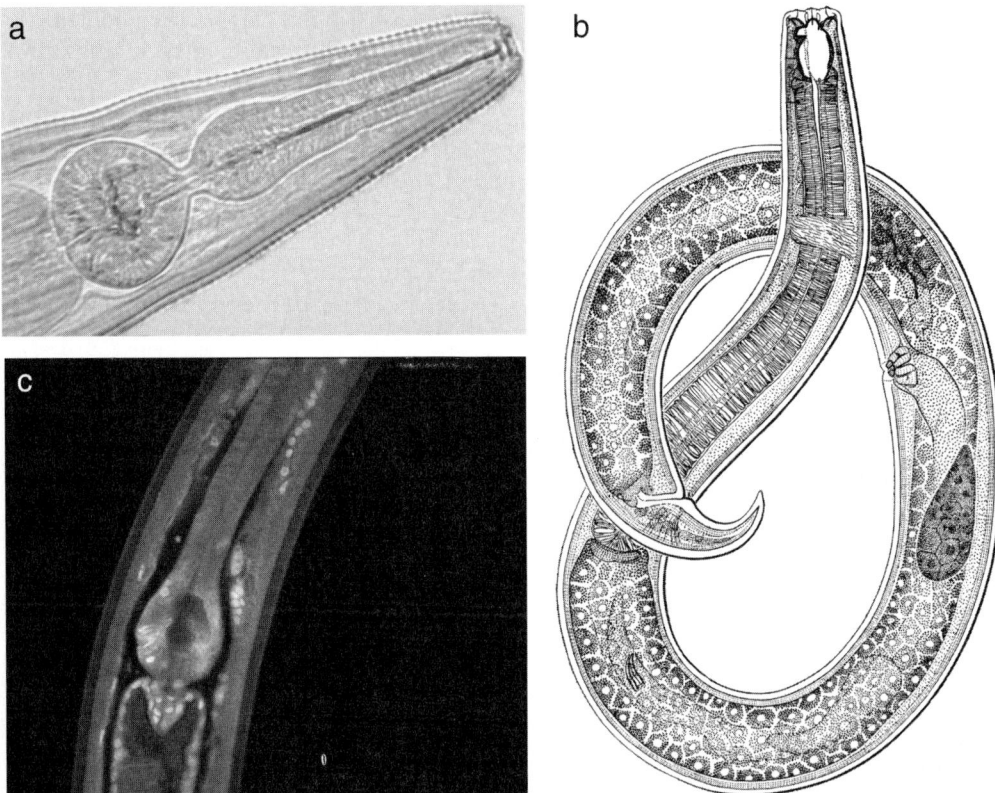

FIGURE 1 (a) Anterior end of *Didelphoxyuris*, the pinworm of South American Marsupials, showing the small anterior stoma at the narrow end, followed by the well-developed muscular esophagus with a large posterior bulb. The esophagus is tripartite and serves as a muscular pumping organ, pumping food into the intestine. (b) Whole drawing of a predatory nematode of the genus *Mononchus* showing the general structures of a nematode including the tapering anterior and posterior ends and the ventral vulval opening just posterior to midbody. The stoma is heavily cuticularized and bears a spike or spine that is used to puncture the cuticle of the prey (modified from Cobb, 1914). (c) Confocal microscopic image of the esophageal bulb and posterior part of the esophagus of a species of *Pratylenchus*. The cuticle can be seen as the outer light outline of the nematode with an underlying brighter layer showing the cellular hypodermis. The bulb shaped posterior part of the esophagus attaches via a valve to the single cell thick-tubular- intestine, shown here as an optical section just posterior to the esophageal bulb.

As mentioned previously, nematodes are extremely diverse in size, shape, and structure. The following will only briefly touch on the expansive subject of nematode morphological diversity. For additional reading and exploration see the essential works and references therein of Nickle (1991), Maggenti (1981), and Grassé (1965).

B. External Covering: The Cuticle

An example of diversity in shape is that of the complex cuticular arêtes found in species of the O. Strongylida: Trichostrongyloidea (parasitic in vertebrates) in which the exocuticle is modified into a series of cuticular aretes called the synlophe. In these forms, it is thought that the ridges running down the length of the body of the nematode are used in maintaining their position in the intestine of their hosts (Fig. 8). Other forms have an exocuticle composed of serrated ridges (Figs. 3a, 3b, 9, 10, 11), bumps, or may be very smooth (Fig. 2). Some groups (Heterakoidea) possess large cuticularized suckers just anterior to the cloaca that are surrounded by sensory papillae that evidently allow the male to find the female in the intestinal tract of the host as in *Paraspidodera* (Figs. 2, 12, 13) and other aspidoderids (Fig. 23). Marine nematodes of the genus *Draconema* (Figs. 5a, and 6) and *Desmoscolex* (Fig. 3c) have a wildly modified cuticle relative to most nematodes and very large hair-like and segmented cuticular setae that the nematodes use for both movement and detecting its environment (Figs. 5a and 6). One curious structure

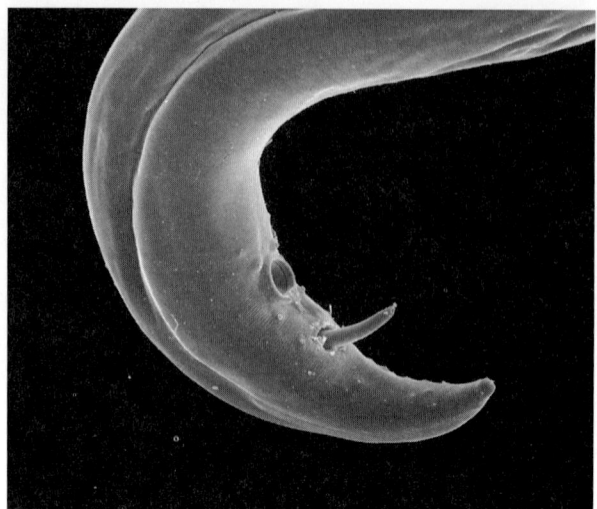

FIGURE 2 Posterior end (tail) of a male *Paraspidodera* showing the relatively smooth cuticle, a sucker just anterior to the opening of the cloaca, and one spicule protruding from the cloaca.

that occurs in all Nemata is the amphid, a highly variable sensory organ that can be very obvious as in *Desmodera* (Fig. 10) or very inconspicuous as in *Chambersiella* (Fig. 18).

C. The Alimentary Canal

The mouth of the nematode (Fig. 14a) provides the anterior opening to the external environment that connects to the stoma leading posteriad to the muscular esophagus (Figs. 15, 16, 17, 18). The stoma may be quite reduced or absent as in the case of members of the Trichostrongyloidea (O. Strongylida) or they may be well developed such as that found in *Clarkus* (Fig. 16) and *Mononchus* (Fig. 1b) and capable of inflicting damage on their prey, or it may be modified into horny tooth-like structures that are used to attach to the intestinal villi of the host animal (e.g., *Ancylostoma*) (Fig. 19). In mammalian gut parasites of the genus *Trichuris* (Adenophorea: Trichuridae), the stoma is lacking and the esophagus is formed of stichocytes comprising a stichosome that is glandular in function, but many other plant parasites such as the Adenophorean ectoparasitic, below-ground root-feeders (capable of transmitting viruses between and among plants *Xiphinema*, *Longidorus*, or *Dorylaimus*: O. Dorylamida: Longidoridae) have a tubular stoma with the posterior parts of the stoma modified into a spear that is used to penetrate plant cell walls as in *Dorylaimus stagnalis* (Fig. 20). The Secernentean plant endoparasitic nematode of the genus *Pratylenchus* (Fig. 21a) also has a modified stomatal spear (much more delicate than those found in *Dorylaimus* sp.) with which it penetrates plant cells as the nematode moves through the tissues and cells of the plant (usually below-ground).

D. The Reproductive System

The reproductive system of most animal parasitic nematodes is adapted to produce extremely large numbers

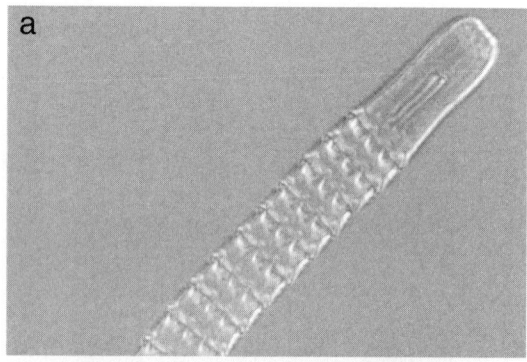

FIGURE 3 (a and b) Anterior ends of marine nematodes showing rings and spines on the cuticle. (c) Whole drawing of a marine nematode of the genus *Desmoscolex* showing heavily cuticularized rings and long setae (modified from Cobb, 1914).

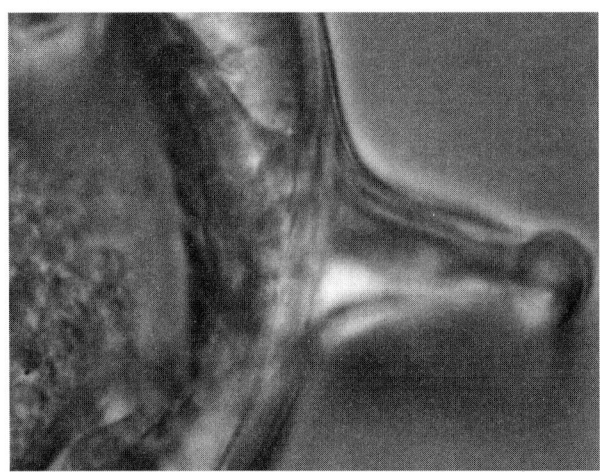

FIGURE 4 Lateral alae of *Aspidodera*, a parasite of South American xenarthrans, marsupials, and rodents. The ala, or wing, runs the length of the body, possibly providing support for the nematode in the gut of its host. The cellular hypodermis of the nematode can be seen just inside the translucent cuticular layer.

of eggs (e.g., *Ascaris*). Some characteristics of larger animal parasitic nematodes include very large body, very two large ovaries, and an equally large uterus. Free-living nematodes generally have much smaller bodies and are therefore individually less prolific. However, the reproductive systems of most nematodes have the following basic structural similarities: Most females have two ovaries, one anterior (as in *Pratylenchus*, Fig. 21b) and one posterior, each connected to an oviduct and uterus with the uteri connecting to the vagina, terminating in the vulva. Males usually have one testes as in the Secernentea or two as in the Adenophorea, a seminal vesicle, and a vas deferens (see Maggenti, 1991a) connecting to the outside via the cloaca (Fig. 13) which is a joining of the reproductive system and the rectum. Secondary sexual organs in male nematodes are usually much more pronounced and variable than that of the female (Figs. 2, 6, 11, 12, 13, 22, 23, 24, 25) and most males have one or two spicules (Fig. 22), a spicular pouch, sometimes with a spicular sheath (Fig. 24), and a gubernaculum (Fig. 23). Some males of the O. Strongylida have a well-developed copulatory bursa (Fig. 22).

Some plant parasitic nemas produce eggs directly into the plant where the nematode lives (in the case of species of the genus *Pratylenchus*, the female produces eggs that hatch within the tissues of the plant and the juveniles begin feeding), whereas in others such as *Heterodera* (which are ectoparasites) the body of the female fills with eggs, forming a sac that eventually dries and transforms into a resistant cyst, with the juveniles within capable of resisting environmental extremes.

II. THE UBIQUITOUS NATURE OF NEMATODES

There are few habitats on earth unoccupied by nematodes. More than a century of both biological surveys and informal collecting has led many biologists to believe that the phylum Nemata is probably the most ubiquitous of all animal groups. Early in the history of scientifically based biological investigations, pioneers of microscopy opened a new window into the previously unseen microscopic world of the soils.

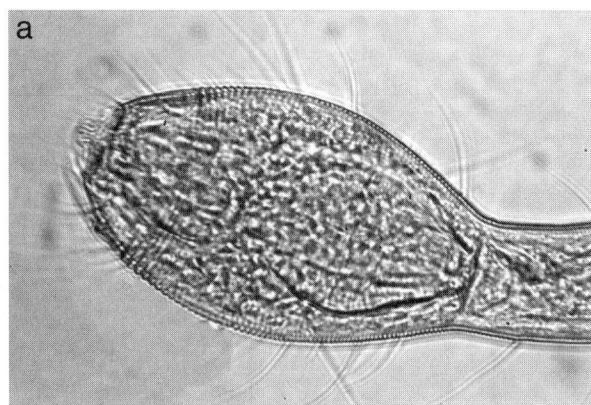

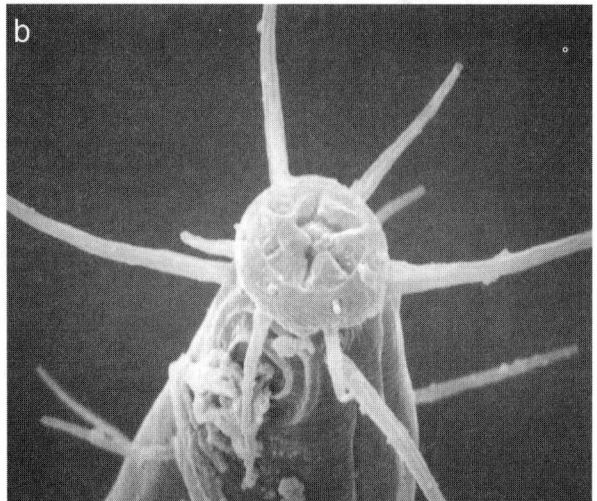

FIGURE 5 (a) The expanded head with visible cuticularized stoma of a species of *Draconema*, a marine nematode. Long setae that are probably used to sense the environment are visible in this photograph. (b) Scanning electron micrograph of the anterior end of a marine nematode showing the long setae and a well-developed amphid just ventral and posterior to the lip rings.

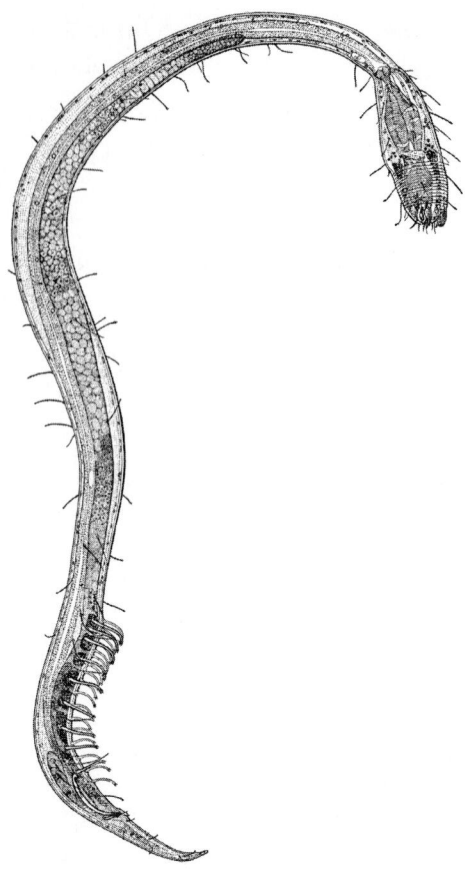

FIGURE 6 Drawing of a male of *Draconema* sp. illustrating the large setae, spicules, testis, and large expanded stoma and head (after Cobb, 1914).

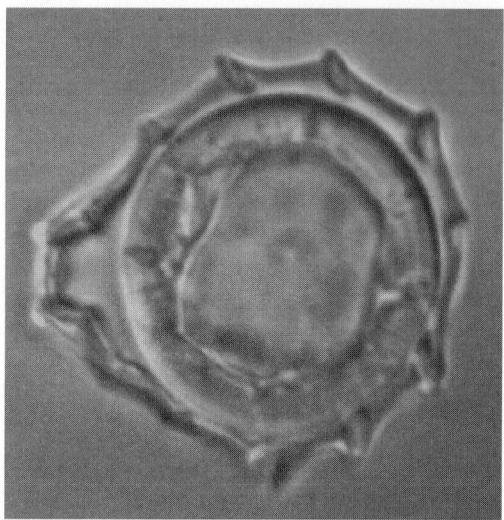

FIGURE 8 Cross-section at midbody of *Vexillata armandae*, a species of nematode parasitic in rodents of the genus *Perognathus* in New Mexico. This photograph shows the spines (arêtes) in the cuticle, called the synlophe in Trichostrongyloidea. Inside is the hypodermis lining the body of the nematode. The reproductive tract and the inestine are not visible in this photograph.

Descriptions of the "invisible" life seen by pioneers of microscopy such as Antony van Leeuwenhoek evoked images of a wonderfully diverse and dynamic community of worms and other organisms. Subsequent investigations by other early researchers began to open up the unseen world of the nematodes.

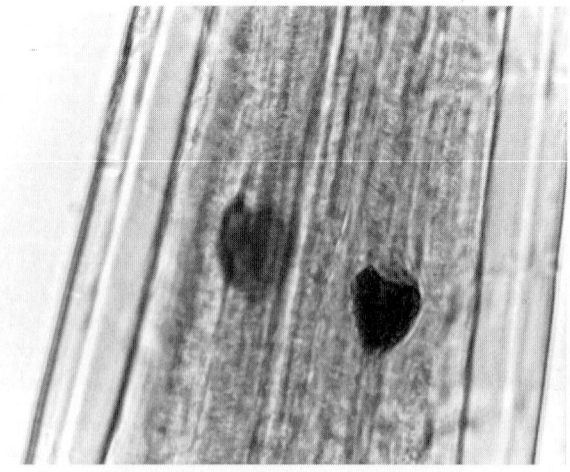

FIGURE 7 Cuticularized "eye spots" in the marine nematode *Thoracostoma*. Located at about the mid-part of the esophagus, this nematode can detect light in its environment.

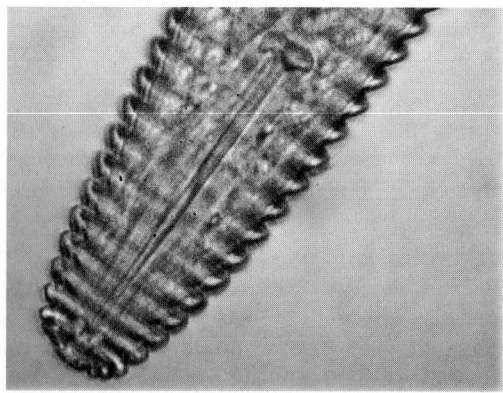

FIGURE 9 Anterior end of a species of *Criconemoides*, an external root feeding plant parasite. The rings of the cuticle can be seen from the head end posteriad. The strong cuticularized stomatal spear is clearly visible in this photograph.

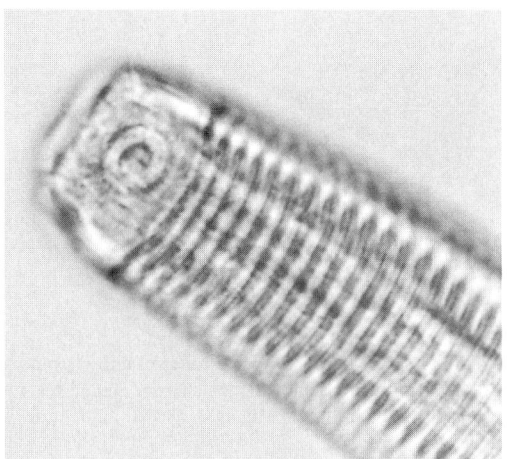

FIGURE 10 Anterior end of a marine nematode of the genus *Desmodera* showing the rings of the cuticle and a well-developed circular amphid.

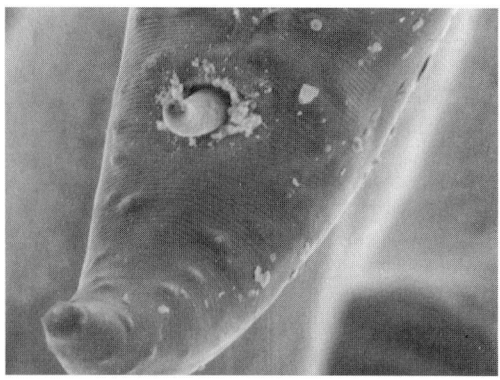

FIGURE 12 Scanning electron micrograph showing the posterior end of a species of *Paraspidodera* (a parasite of the cecum of rodents of the genus *Ctenomys* in Bolivia) with a spicule everted.

A. Early Views of Nematode Diversity: The Human Perspective

Humans have been parasitized by nematodes from the earliest times. Eggs of the pin-worm *Enterobius vermicularis* and the whip-worm *Trichuris trichuria* occur in coprolites dated to about 7000 years old (y.o.) from dry areas of Peru, and eggs of the hookworm *Ancylostoma duodenale* have been reported from coprolites dated to around 7230 y.o. In eastern Brazil. *Ascaris lumbricoides* has been positively identified from human coprolites dated to about 28,000 y.o. from caves in France, but this is the only occurrence of a record this old. The seeming dearth of other nematodes from human remains older than about 7000 y.o. appears to be due to the fact that organic material comprising the coprolites themselves do not preserve well enough to last that long (Karl J. Reinhard, personal communication).

In what is thought to be the oldest surviving written account of *Ascaris* in humans (dated to approximately 4700 y.o.) (in China), foods to avoid and a description of the symptoms of humans infected with these worms was accurately given (Maggenti, 1981). In the area of the Nile River Valley, early Egyptian physicians recorded the presence of both *Ascaris* and *Dracunculus* (the "Guinea worm") in an ancient papyrus manuscript (written by Egyptian physicians around 3552–3550 years ago) that was obtained and translated by the

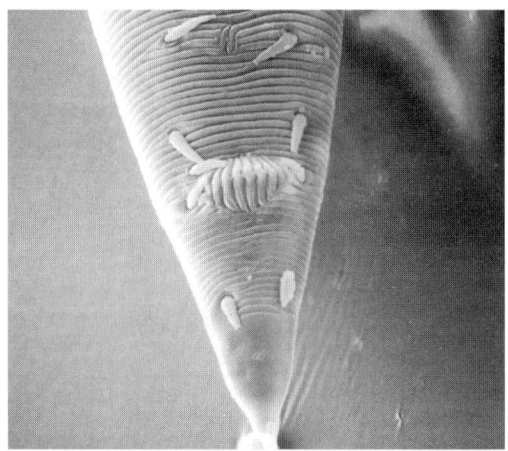

FIGURE 11 Scanning electron micrograph of the posterior end of a male marine nematode showing rings of the cuticle and papillae lateral to the cloaca.

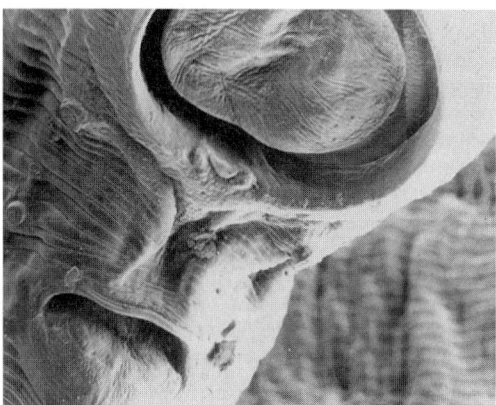

FIGURE 13 Close-up scanning electron micrograph of the sucker and cloacal opening of a species of *Paraspidodera* (a parasite of the cecum of rodents of the genus *Ctenomys* in Bolivia). Numerous sensory papillae can be seen on the cuticle around the cloaca.

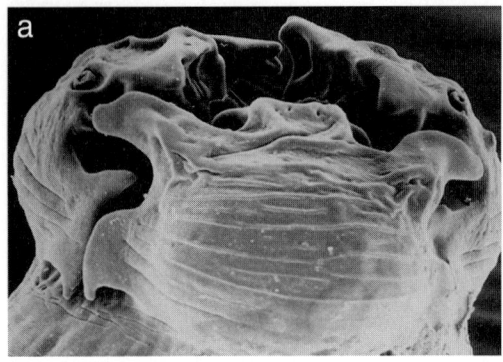

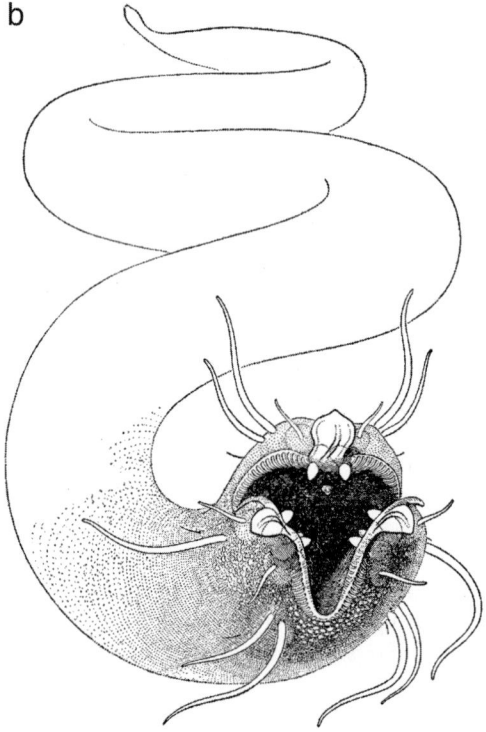

FIGURE 14 (a) Anterior end of a species of *Paraspidodera* (a parasite of the cecum of rodents of the genus *Ctenomys* in Bolivia) showing three huge lips, the stomatal opening (mouth) in the middle of the lips, and large anteriorly directed sensory papillae on lips 2 and 3 near the outer part of the photograph. (b) Head-on view of a predaceous nematode as might be seen from the perspective of a mild-mannered bacterial feeding form such as *Caenorhabditis* just before it is devoured (after Cobb, 1914).

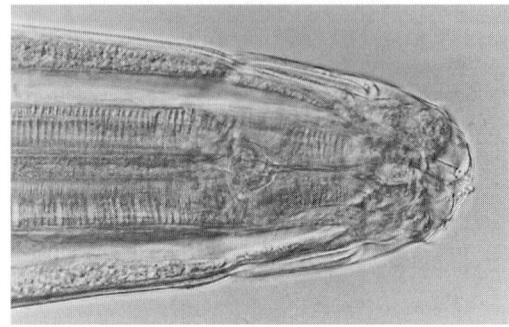

FIGURE 15 Anterior end of a species of the family Aspidoderidae from a rodent from Bolivia showing the anterior end of the muscular esophagus attaching to the partially muscular and cuticularized stoma.

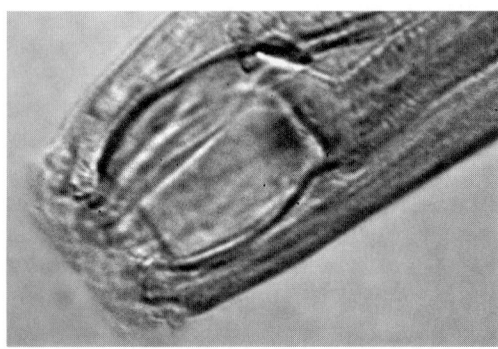

FIGURE 16 Heavily cuticularized stoma of a predaceous nematode of the genus *Clarkus*. This species is a soil predator living in rhizosoil of grapes in Northern California.

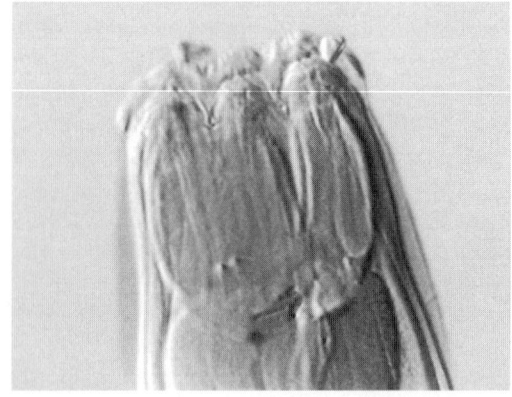

FIGURE 17 Anterior end of *Miconchus* sp. showing well-developed lips with sensory neurons running posteriad from the anteriorly directed sensory papillae on each lip.

Egyptologist "Ebers" in 1872 (see Chitwood and Chitwood, 1977; Maggenti, 1981). In the extant literature, the first mention of a nematode from a nonhuman animal was by Hippocrates about 2430 years ago; he described the occurrence of pinworm nematodes of horses and human females. From that time, little more was discovered until Albertus Magnus and Demetrios Pepa-

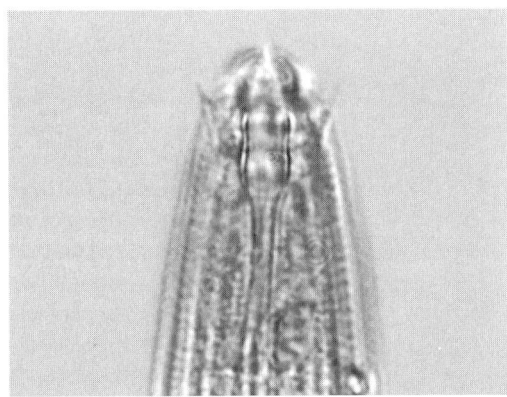

FIGURE 18 Delicate but well-cuticularized stoma of *Chambersiella*.

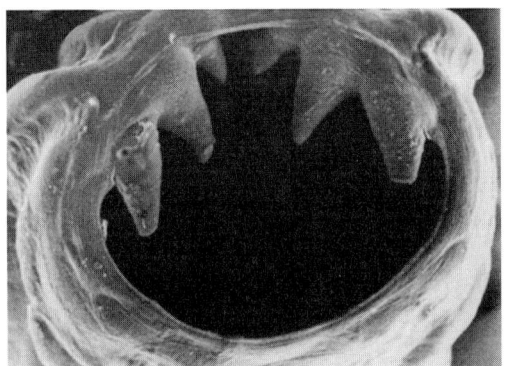

FIGURE 19 Scanning electron micrograph looking into the mouth of a species of mammalian parasite of the genus *Ancylostoma*. The teeth are used to attach to the villi of the host mammal.

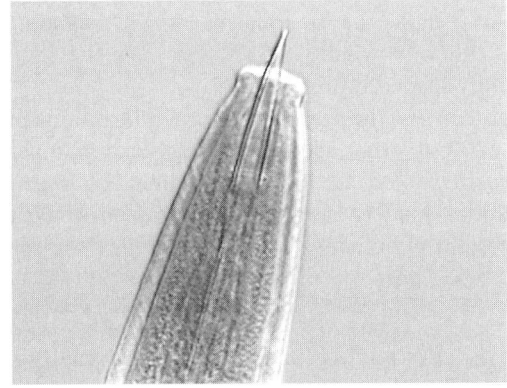

FIGURE 20 Anterior end of *Dorylaimus stagnalis*, a plant root ectoparasite showing the well-developed stomatal spear that is used to penetrate plant root cells.

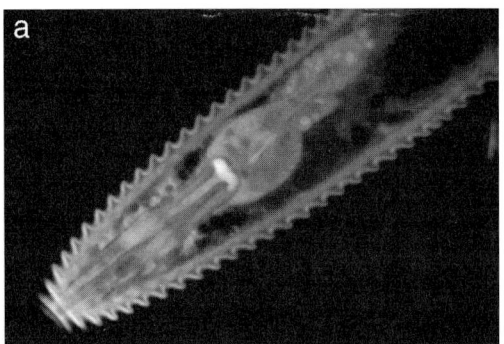

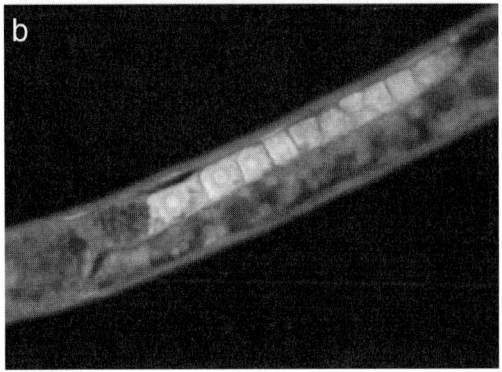

FIGURE 21 (a) Confocal image of *Pratylenchus* sp. showing the well-developed spear that the nematode uses to penetrate plant cells while moving through the roots of the plant. The muscles and glandular part of the esophagus are also visible. (b) Confocal image of *Pratylenchus* showing the anteriorly directed ovary with individual ovocytes and their nuclei visible.

gomenos (in the 13th century, cited from Rausch, 1983) recorded nematodes from falcons (also see Chitwood and Chitwood 1977). With the development of the microscope and the emergence of Europe from the dark ages, knowledge of nematodes as parasites of plants and animals and of free-living forms expanded rapidly. It

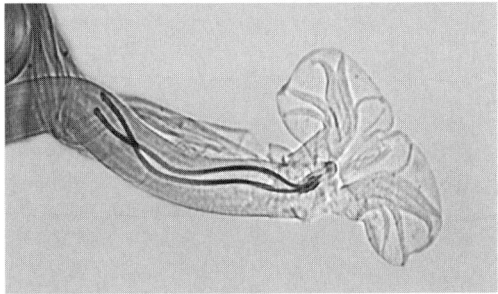

FIGURE 22 Posterior end of a male trichostrongyloid showing the well-developed copulatory bursa, bursal rays, and the long thin paired spicules.

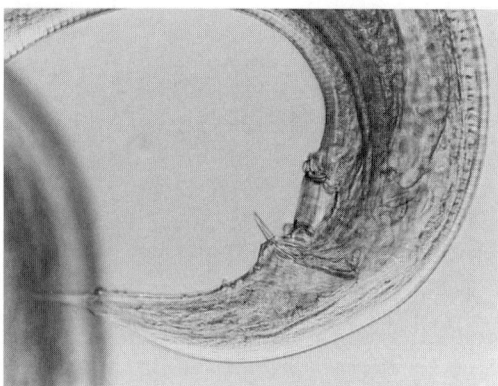

FIGURE 23 Posterior end of an aspidoderid from a rodent of the genus *Oxymycterus* from Bolivia. The gubernaculum can be seen protruding from the cloaca of this specimen. The cuticularized sucker is also easily visible surrounded by sensory papillae.

was found that nematodes occurred everywhere people looked; in fact, Anton van Leeuwenhoek first recorded the presence of vinegar eels (*Anguillula aceti*) in his vinegar stored for personal use in a letter dated 21 April 1676, although he was not aware that others had reported finding nematodes some time earlier (Dobell, 1932).

Estimates of infections of people with common human parasitic nematodes give the following numbers (from Crompton, 1999): of a total human population of about 6 billion individuals (in the year 2000) the strongylid hookworms *Ancylostoma duodenale* and *Necator americanus* infect about 1,298,000,000 (22%), and the large intestinal nematode *Ascaris lumbricoides* occurs in about 1,472,000,000 (25%) people at any one time in the world. Obviously, many people harbor more

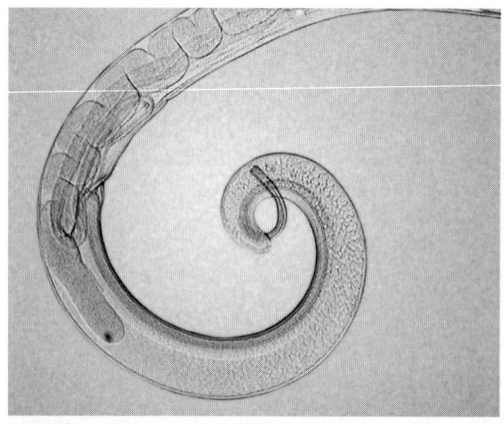

FIGURE 24 Posterior end of *Trichuris* from a Bolivian species of *Ctenomys*. The long spicule and everted spicular sheath can be seen.

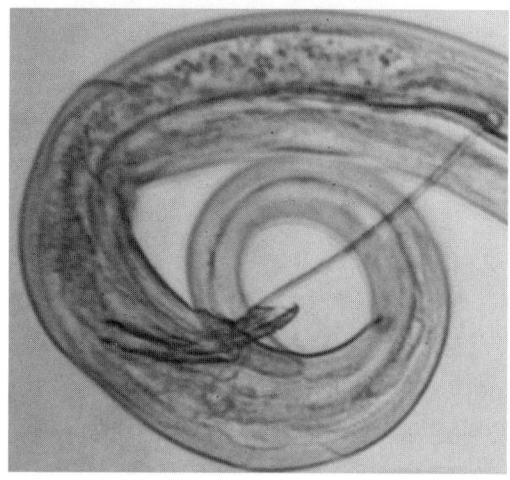

FIGURE 25 Posterior end of the filaroid nematode *Litomosoides* from *Ctenomys opimus* from high altitude western Bolivia. Note the dimorphic nature of the spucules in this species: one being long and filamentous, the other short and stubby.

than one species of nematode at a time, and it is common for people to sport *Ascaris, Necator, Trichuris*, and *Enterobius* simultaneously. I provide the following estimate to indicate just how important these organisms are in the web of life on Earth. To put the number of infections of humans in perspective, I made the following extrapolations: An adult female *Ascaris* produces approximately 200,000 eggs per day at an average rate of about 5 grams of eggs per year. Actual data for *Ascaris* in humans that are infected show an average of 18 worms per infected person. Given that half of these are females, I calculate that 9 worms/person will produce about 45 grams of eggs in the feces of the host per year. In one year the total population of *Ascaris* in humans worldwide is conservatively estimated to produce 66,240,000 kg or 66,240 metric tons (72,864 tons [English]) of eggs; this is equal in weight to about 348 large adult blue whales, 8,832 adult male elephants, or 364 fully loaded railroad coal cars.

Estimates of the number of human infections in the year 2000 by other species of parasitic nematodes are shown in Table I. At the present time, it is estimated that approximately 138 species of nematodes have been reported from humans (Crompton, 1999) with from 32 to 36 being host-specific.

B. The Science of Nematode Diversity

Nathan A. Cobb, often considered the father of nematology in North America, was a student of the renowned German zoologist Ernst Haeckel. After just a few years

TABLE I
Numbers of Common Nematode Infections in Humans Worldwide[a]

Species of nematode	Numbers infected	Distribution
Ancylostoma duodenale and *Necator americanus*	1,298,000,000	Worldwide
Ascaris lumbricoides	1,472,000,000	Worldwide
Brugia maylayi and *B. timori*	13,000,000	South Pacific, SE Asia, India
Dracunculus medinensis	80,000	Sub-Saharan Africa and Yemen
Loa loa	13,000,000	West and Central Sub-Saharan Africa and Yemen
Onchocerca volvulus	17,660,000	Central and South America and Sub-Saharan Africa
Strongyloides stercoralis	70,000,000	Temperate regions
Trichuris trichiura	1,049,000,000	Worldwide
Enterobius vermicularis	400,000,000	Temperate regions

[a] Data from Crompton (1999).

of research spanning the globe from Europe, Australia, and North America, Cobb amassed a huge amount of knowledge and came to have a deep appreciation for the immense number of species that existed. With scientific knowledge based on keen observational skills, he understood the nature of both the great numerical density and species diversity of nematodes in all habitats of the globe that he examined. Thus armed, he wrote the following:

> In short, if all the matter in the universe except the nematodes were swept away, our world would still be dimly recognizable, and if, as disembodied spirits, we could then investigate it, we should find its mountains, hills, vales, rivers, lakes, and oceans represented by a film of nematodes. The location of towns would be decipherable, since for every massing of human beings there would be a corresponding massing of certain nematodes. Trees would still stand in ghostly rows representing our streets and highways. The location of the various plants and animals would still be decipherable, and, had we sufficient knowledge, in many cases even their species could be determined by an examination of their erstwhile nematode parasites.
>
> We must therefore conceive of nematodes and their eggs as almost omnipresent, as being carried by the wind and by flying birds and running animals; as floating from place to place in nearly all the waters of the earth; and as shipped from point to point throughout the civilized world in vehicles of traffic.
>
> COBB (1914)

As if challenged by this assertion, scientists have tested Cobb's hypotheses by examining the extremes of the biosphere on Earth to evaluate the limits of nematode life. Through these investigations, biologists have now shown that nematodes are living and reproducing everywhere on Earth that water exists in a liquid state even for short periods of time annually.

III. DIVERSITY OF HABITATS AND DISTRIBUTION

A. General Distribution

The most obvious ecological characteristic that defines habitats for members of the phylum Nemata is that they all are aquatic animals—to move, live, eat, and reproduce, nematodes must exist in an aqueous environment. This environment includes soils, muds, sands, plants, and animals. They can be found living in soils with moisture contents as low as 5 to 10%, but in the majority of these cases, the nemas are associated with the roots of plants. It is evident that the environment in which nematodes live constrains their ultimate size. Soil-dwelling species live in the water film of the interstices of soil particles. Nematodes must live and carry out all functions of life in these spaces, thus free-living or plant-parasitic soil-dwelling forms are usually extremely small. Morphological diversification of nematodes can operate only within the constraints of their life history parameters, and evolutionary pathways for nematodes living in an interstitial–soil or sediment/sand environment are limited. Thus these forms have limited abilities to diversify into the nonaquatic regions of the Earth, and their size is constrained by this fact. A lingering question remains, why do we not see extremely large marine nematodes? Perhaps it is because we have not yet looked carefully in the marine environment.

In contrast to the small sizes of plant-parasitic, free-living marine, freshwater, or soil nematodes (some adults can be as small as *Criconemoides* with an adult length of around 250 μm), species that occur as parasites of animals are free from many of the physical constraints on their size, so the bodies of species in mammals are, therefore, relatively large. In fact, the largest nematode thus far recorded is *Placentonema gigantissima* from the blue whale with a length of more than 8 m (Maggenti, 1981). The ratio between the smallest adult nematode to the largest can be calculated as 0.250 mm (*Criconemoides*)/8,000 mm (*Placentonema*) = 0.000031, compared to the ratio between a shrew and a blue whale (50 mm/24,4000 mm = 0.12), indicating that the size differences in nematodes are three orders of magnitude greater than the mammals.

In the marine-benthic environment, Lambshead (1993) has estimated (based on transect data from deep sea benthic samples) that there may be as many as 100 million species of marine infaunal nematodes. As exorbitant as that estimate seems, deep-sea nematodes have been shown to be extremely rich in species diversity. For example, one marine sediment sample from the east Pacific benthos was reported to contain 148 species from a total of only 216 individual specimens examined (Lambshead, 1993). At the present time the true extent of species diversity in the marine benthos can only be imagined, because fewer than 20 studies of nematode community structure from marine benthic habitats have been reported (Boucher and Lambshead, 1995). Data from Boucher and Lambshead (1995) shows that in marine environments the highest diversity in nematodes occurs in abyssal benthic sediments.

We know that great numbers of individuals and species of nematodes live in sediments on the ocean floor distributed from the intertidal continental margins to the benthos of the abyssal zones, even though these marine habitats remain mostly unexplored. Nematodes occur in tissues and organs of all species of vertebrates that have been studied, and some, such as *Physaloptera spp.* live, feed, and reproduce in the strongest stomach acids of mammals, birds, and reptiles. Desiccated specimens from both the Arctic and Antarctic have been rehydrated to form viable colonies, and living nematodes have been found in the limited meltwater in some of the dry valleys of Antarctica, an area that is probably one of the most extreme biotopes on the earth.

B. Extreme Biotopes

Some of the most extreme soil habitats on the Earth exist in the dry valleys of the Antarctic, where the annual mean air temperature is −20°C, and soil temperatures at a 5-cm depth for the two "summer months" range from −2.7 to 15.9°C. In this area, no vascular plants grow and mosses and lichens are rare; this is the only terrestrial soil system known where nematodes are the final consumers and are at the apex of the food chain (in this case it seems more of a chain than a web). Three nematode species exist in these dry soils: *Scottnema lindsayae*, a microbivore (feeding on bacteria and yeast), *Plectus antarcticus* (a bacterial feeder), and *Eudorylaimus antarcticus*, an omnivorous predator that presumably feeds on individuals of the other two species (Powers *et al.*, 1998). Another biological extreme, Death Valley in California, has recorded some of the highest temperatures in North America, and the soils of the valley are teeming with nematodes, many of which have been discovered to possess similar adaptive traits to those found in the cold deserts of Antarctica.

Individuals of some species of Nemata are capable of resisting extended periods of dessication, for example, 3rd stage juveniles of *Anguina tritici* have been dried for more than 20 years in a state of anhydrobiosis in which all metabolic activities are shut down (Maggenti, 1981). These nematodes have been shown to have specialized proteins that fold into stable/preserved structures as the organism dries. In this state of crypto- or anhydrobiosis, individual nematodes can remain viable through incredible extremes of temperature, desiccation, hypoxia, and even synthetic nematicides designed to kill nematodes (of course the regular biochemical processes that nematicides interfere with are not operational, so the nematode does not notice this particular assault). When water again becomes available, the animal comes back to life when the molecular structures rehydrate and the proteins and enzymes spring back into normal operation.

C. Habitat Diversity

Nematodes occupy every conceivable life history niche. There are benthic deepwater marine forms that appear to consume mostly diatoms and others such as species of the genus *Draconema* (Figs. 5a and 6) that are mostly associated with marine algae, but it is still unclear what they actually eat. Nematodes of the genus *Dirofilaria* live in the aorta and left ventricles of canids and are transmitted by mosquitoes from dog to dog. Species of nematodes live in the hearts of sharks, and other species can occur by the thousands in the stomachs of pilot whales (S.L.G., personal observation). As the human consumption of raw marine fish (ceviche and sashimi) has increased, transfer of juvenile *Anasakis, Terranova,* and other anasakines from the fish intermediate host to humans is occurring more commonly, and these nematodes are turning up as parasites in the stomach of humans.

Some species feed on fungi in the soil while others are trapped and are themselves consumed by different species of fungi. Still other nematodes such as *Mononchus* (Fig. 1b) and *Clarkus* (Fig. 16) are predatory and hunt and eat other nematodes in the soil environment. Free-living bacterial feeding nematodes have been shown to be integral parts in the carbon and nitrogen cycles in healthy soils (Ferris *et al.*, 1998), and recent survey work has shown that undisturbed or natural soils in noncultivated habitats can have as many as 20 times more species than soil from similar areas that have been under cultivation (Al Banna and Gardner, 1996).

D. Abundance: Estimates and Facts

In addition to the large numbers of species that may occur in any given habitat, nematodes also occur in very great densities. For instance, in sheer numerical density of individuals in any given environment, nematodes exceed even the mites and beetles combined. More than 90,000 nematodes were recorded from a single decomposing apple, and one report showed that 1 cc of marine mud contained 45 nematodes representing 19 species. Nematodes in marine estuaries occur at high numerical densities with reports of $4,420,000/m^2$ in surface mud and 527,000,000/acre in the top 3 inches of sand on the Massachusetts coast. Counts and extrapolations for relatively moist soils (10 to 70% moisture content) worldwide show that in the uppermost levels, nematodes occur in mind-boggling abundance: 7 to 9 billion/acre in undisturbed sod in North China; from 800,000,000 to >1 billion/acre (representing just 35 species) in Utah and Idaho, and around 3 billion/acre in low-lying alluvial soils of Europe and other areas of North America.

IV. HOW DO NEMATODES AFFECT THE BIOSPHERE?

Because of the huge numbers of nematodes that have been shown to occur in plants, animals, soils, and the benthos, there has been much speculation about the role of nematodes in basic biological processes occurring in the soils of the Earth. Cobb (1914) speculated that some nematodes "are beneficial"; however, he also noted that this area of study was still in its infancy. In fact, in the year 2000, this area of study is still just developing, and recent work has shown that nematodes can be good indicators of biodiversity (Bongers and Ferris, 1999). Gardner and Campbell (1989) showed that mammalian parasites with complex life cycles may serve as excellent indicators of areas of high biological diversity. Because of the multifarious nature of parasitic nematodes in mammals, it is expected that these kinds of species may provide biologists with additional tools for identification of areas of high biological diversity.

The fact that parasitic nematodes occur in such high prevalences and numerical densities in mammals should give us pause. There is obviously a huge energy drain on any population of mammal that we should care to analyze, and this energy drain probably causes significant decreases in the number of offspring in any given population over long periods of time.

A. Soils and Plants

Studies have indicated that nematodes play a substantial role in the cycling of carbon and nitrogen in the soil environment (Bongers and Ferris, 1999), and it has been shown that the number of bacterial-feeding nematodes increases as the bacteria increase with annual warming of the soils. In the rainforest of Cameroon, average nematode abundances of $2.04 \times 10^6/m^2$ of rainforest soil were found, indicating that these nematodes play a significant role in carbon flux (CO_2 and CH_4) in this rain forest site (Lawton *et al.*, 1996).

Nematodes occur in, on, and around the roots, bulbs, rhizomes, stems, and leaves of plants. They can cause galls in both the somatic and germinal tissues of the plants, where the nematodes can encyst and dry, waiting for the next stage in their life cycle. There are species that are almost fully endoparasitic, exiting the plant root or stem only as juveniles. Others are fully ectoparasitic, such as species of *Xiphinema* (the vector of grape fanleaf virus). Some species spend part of their life cycle in a plant and part of it in the soil. As mentioned earlier, seed parasites of the genus *Anguina* may spend most of their existence as juveniles in a dried state, with the nematodes rehydrating, molting, and then as adults crawling up the outside of the plant during periods of high relative humidity, then laying eggs in the seed head. The seeds are consumed by the developing juveniles, and when they dry, the seed coats protect the also dried nematodes within and are distributed through the environment as are normal seeds (Maggenti, 1981). Nematodes of the genus *Pratylenchus* use their spear to move through the roots of the plants that they infect, penetrating the plant cells with repeated jabs of their stomatal armature.

B. Predators, Entomopathogenic Forms, Fungal and Bacterial Feeders

Many nematodes that are found in the soil are either predaceous forms eating other nematodes or forms that prey on mites or other soil macroorganisms. The more spectacular predators such as *Mononchus* and *Clarkus* have specialized buccal structures with which to puncture the cuticles of other nematodes (Figs. 1b and 16). Microbivorous fungal and bacterial feeding nematodes are also extremely abundant, with species specializing in their feeding habits on bacteria, fungi, diatoms, and other microscopic organisms. Some, such as juveniles of species of the genera *Heterorhabditis* and *Steinernema*, carry bacteria of the genus *Photorhabdus* in their digestive system. The 3rd stage juveniles wait in the soil until an unwary insect passes nearby. The nematode then homes in on and penetrates the hapless insect, making its way into the hemocoel where it releases the bacteria, which proliferate, killing the host. The nematode then feeds on the bacterial colony and reproduces in the insect, eventually again producing 3rd stage juveniles that leave the carcass of the insect and disperse into the soil, waiting there for another insect to invade (for more specific details of the life-cycle of these entomopathogenic nematodes, see Gaugler and Kaya, 1990).

Bacterial-feeding forms occur in the soil in extremely high numbers. In soils that have not been disturbed and that have a good layer of organic matter, large numbers of all kinds of nematodes occur. One of the most well-known groups is the Rhabditida, or the rhabditid nematodes. These forms feed on bacteria and yeasts growing in the soil and are typically found in high numbers in moist soils with high organic content. The most famous of these forms are members of the genus *Caenorhabditis*, of which the complete genome of *C. elegans* has been sequenced.

C. Aquatic and Marine Nematodes

This is an area that is wide open for future biologists. How do the trillions upon trillions of individuals and the millions of species that occur in the oceans really affect the biosphere? Nothing is known on the subject at the present time.

V. HOW MANY SPECIES OF NEMATA?

A. Estimates of Number Described

Estimates of the numbers of species of nematodes that are known (i.e., described species) vary widely. However, in 1819, Rudolphi summarized what was known of the nematodes, recording 11 genera and about 350 species. Just 115 years later, in 1934, Filipjev reported that 4601 species of nematodes had been described, with about half free-living and the other half parasitic. By 1950, Libbie Hyman estimated that approximately 9000 species were described (based on her analysis of the zoological record with descriptions being recorded at a rate of about 200 new species described per year). In 1981, Maggenti's summary showed around 15,000 described species. My analysis from counting additions to the zoological record shows that in the 5-year period from 1992 through 1996, numbers of descriptions were relatively stable, with approximately 776 new species described (average of approximately 155 descriptions per year). From 1996 through 1998 the numbers of descriptions decreased to 118 per year, most likely due to the continued retirement and expiration of knowledgeable taxonomists.

B. How Many Species of Vertebrate Parasitic Nematodes Exist?

All species of vertebrates examined thus far serve as hosts for at least one species of parasitic nematode. Some mammalian hosts harbor many species of nematodes that are distributed through several orders and families. Some of these nematodes are highly host-specific, surviving and reproducing successfully only in host individuals comprising a single species or perhaps a closely related group of species. Other nematodes show less specificity, being much more likely to jump from one suitable vertebrate host to another during opportune times during their life history (Brant and Gardner, 2000).

Within a host, many different types of habitats may be occupied by nematodes. As a species, *Homo sapiens* harbors approximately 35 species of host-specific parasitic nematodes (Chitwood and Chitwood, 1977). To illustrate the diversity of habitats in a single animal host, humans will be used as an example. In a human, nematodes can occur as juveniles in muscle tissues (usually smooth muscle such as the diaphragm or the tongue [*Trichinella*]) and in the mucosa of the intestine (*Strongyloides*); as migrating forms in blood and lungs (*Ascaris*); as microfilariae in blood or lymph (filarioids of humans); and as adults in the small and large intestines (*Ancylostoma, Necator, Ascaris*), in mesentaries and subcutaneous tissues (*Onchocerca, Loa, Wuchereria*), and in the large intestine and cecum (*Enterobius*).

Cobb was well acquainted with animal parasitic nematodes, and his familiarity with host-specificity in the Nemata led him to estimate as early as 1914 that well over 80,000 species of nematodes would eventually be found parasitizing vertebrates alone. Similar estimates could be derived for parasites of invertebrates and plants, resulting in totals of free-living and parasitic species numbers far in excess of the approximately 20,000 species of nematodes known today. Cobb (1914) stated "There must be hundreds of thousands of species of nematodes. Of this vast number only a very few thousand have been investigated, and of these, comparatively few with any degree of thoroughness."

Present estimates conclude that only about 14,000 species of parasitic nematodes have been described from all taxa of slightly more than 48,000 currently recognized species of vertebrates. If each of the approximately 4450 known species of mammals were infected with only two species of host-specific nematodes, we would expect to find a minimum of 8900 species of parasitic nematodes only in the class Mammalia. In addition, many species of wild mammals each harbor more than two host-specific species of nematode, so the above estimate of 8900 species of nematodes only from mammals can be considered very low.

As of this writing, the natural history, development, and transmission parameters of only around 561 species are known (Anderson, 1992). For parasites of vertebrates, Anderson (1992) estimated that there were about 2300 described genera distributed among 256 families comprising about 33% of all nematode genera known. This percentage is about equal to the percentage of genera of Nemata presently known in marine and freshwater habitats. It is agreed by most workers that this bias is due to the larger number of parasitologists working on selected groups of Nemata relative to the number of specialists working on the free-living marine and freshwater forms.

Pinworms, nematodes of the order Rhabditida, superfamily Oxyuroidea (Fig. 1), show high levels of host specificity, and it is well-known that almost all species of rodents and primates have one or more species-specific pinworms (for instance, *Homo sapiens* harbors both *Enterobius vermicularis* and *E. gregorii*). Both recent and historical studies have shown that pinworm nematodes exhibit high levels of coevolution, i.e., concomitant host–parasite speciation (and host specificity) with their primate and rodent hosts (Hugot, 1999). At the present time, slightly more than 716 species of Oxyuroidea have been described, with the vertebrates hosting about 496 species and invertebrates about 217 species. The greatest diversity in the Oxyuroidea to be found in the future is expected to come from examination of the Arthropoda, especially beetles, cockroaches (4000 species described and more than 20–30,000 expected to be found), and millipedes (17,000 species described and more than 60,000 species expected to be found). This gives huge numbers of pinworms occurring just in the cockroaches and the millipedes if each species harbors its own species of pinworm. At least two species of pinworms (genus *Thylastoma*) occur in laboratory colonies of *Periplaneta americana* and more are expected to be found in free-living populations (J. P. Hugot, personal communication).

When adequate surveys are completed, large numbers of species of Oxyuroidea are also expected to be described from Neotropical rodents of the family Muridae. Up to the present time, only around seven species of pinworms have been described from Neotropical murids, and these rodents potentially host from 400 to 800 undescribed species of Oxyuroidea (given that only one to two new species of oxyuroid nematodes are found in each species of rodent examined) (J. P. Hugot, personal communication).

Recent comparative studies in the Oxyuroidea show that the larger the body size of the host, the larger the body size of oxyurid nematodes that it harbors (Morand *et al.*, 1996).

C. Comparative Nematode Diversity of New World Subterranean Rodents: Geomyidae

Papers on nematode parasites from rodents of the Nearctic family Geomyidae covering the dates from 1857 to 1999 were reviewed. Combined with field-collected specimens from the early 1970s up to the present time, I discovered that six of the approximately 11 nematode parasites reported from pocket gophers in North America are host specific to only the Geomyidae (Table II).

Of members of the vertebrate class Mammalia, one of the most complete sets of nematode parasite data exists for rodents of the family Geomyidae. Of the approximately 35 known species of pocket gophers (Wilson and Reeder, 1993), only 15 species have been surveyed for parasitic nematodes. From those 15 species, six species of nematodes are known to be host-specific only to geomyids. Some nematodes such as the stron-

TABLE II

Nematode species	Classification and location in host
	Sub-Phylum Adenophorea
Trichuris fossor	Cecum and large intestine
	Sub-Phylum Secernentea
Ransomus rodentorum	Cecum/Small intestine
Vexillata vexillata	Duodenum
Vexillata convoluta	Duodenum
Heligmosomoides thomomyos	Duodenum
Litomosoides thomomydis	Mesentaries
Litomosoides westi	Mesentaries

gylid Ransomus rodentorum, the heligmosomid Heligmosomoides thomomyos, and filarioids of the genus Litomosoides have been reported from more than one species of gopher; other nematode species such as Vexillata vexillata occur only in gophers of the tribe Thomomyini (genus Thomomys) but do not appear to be host species specific.

These mammals occur in an extremely wide and ecologically variable geographic area (from southern Manitoba and British Columbia south to extreme northern Colombia); therefore, there may be many more undescribed or undetected species of nematodes in these hosts than this analysis provides. In addition, no studies on genetic diversity of nematodes (or any endoparasites) in these rodents have been published; therefore, levels of genetic variation in these nematodes are unknown and the true genetic diversity that exists will probably result in an increase in the number of nematode species that are recognized.

There is little if any evidence of phylogenetic coevolution of the nematode parasites and their pocket gopher hosts. However, all species listed in Table II are specific to species of the family Geomyidae, and both Litomosoides and Vexillata appear to exhibit some level of phylogenetic host specificity with two closely related North American species of Litomosoides being found only in geomyids (Brant and Gardner, 2000) and species of Vexillata occurring more generally in members of the Geomyoidea.

D. Nematodes of Tuco Tucos (Ctenomyidae)

A review of the nematode parasites occurring in Neotropical rodents of the genus Ctenomys (Table III) indicates a considerably more depauperate fauna of nematodes as compared with the nearctic Geomyidae. Data collected from 1984 on indicate that nematodes of the genera Trichuris and Paraspidodera have cospeciated with their hosts and exhibit different levels of phylogenetic congruence relative to their hosts. In addition, nematodes of the trichostrongyloid (O. Strongylida) genus Pudica were encountered only two times from the same species of Ctenomys in one locality (from a sample of more than 500 individuals and more than six species of hosts examined). The occurrence of A. caninum in Ctenomys appears to be a capture, as it only occurred in areas where dogs, humans, and ctenomyids lived in relatively close proximity (banana fields in lowland Santa Cruz, Bolivia).

Whereas most of the pocket gophers examined carefully generally harbor from one to several species of nematodes in the small intestine, very few nematodes are found in samples of tuco tucos. Even though the genus Ctenomys contains almost 40 species, comparatively few species of nematodes from them have been described or reported. This lack of parasites in a wide-open group of mammals might be a result of rapid speciation in the mammal group, with parasites failing to keep up with the speciation rate of the mammals themselves and actually losing parasites through time; there is some evidence that the ctenomyids have speciated rapidly in the recent past. The lack of a diverse fauna of nematodes in these mammals could also be due to historical accident, whereby the ancestor of the ctenomyids had a low diversity of nematode parasites (for whatever reason) thus giving rise to a phylogenetic lineage of mammals lacking a diverse fauna of parasites. But why have they not picked up more parasitic nematodes from other syntopic species of mammals?

There is some evidence of host-switching in nema-

TABLE III

Nematode species	Classification
	Sub-Phylum Adenophorea
Trichuris (>3 spp.)	Cecum and large intestine
	Sub-Phylum Secernentea
Ancylostoma caninum (host capture in synanthropic species of rodents?)	Duodenum
Pudica sp. (host capture from Muroid rodents)	Duodenum
Litomosoides (2 spp.)	Peritoneal cavity and mesenteries
Paraspidodera (>6 spp.)	Cecum and large intestine

todes of the genus *Litomosoides*, in that two species occur in *Ctenomys opimus* in high altitude western Bolivia, but these nematodes have not been reported from any other species of *Ctenomys* from throughout the Neotropics. Superficially this indicates a host-capture event from some other lineage of mammals (Brant and Gardner, 2000). Another example is the fact that nematodes of the genus *Pudica* found in *Ctenomys* have diverse relatives in other species of muroid rodents in South America but none in ctenomyids, leading to the conclusion that most are now found in the tucos because of host-switching events and not phylogenetic coevolution.

Nematodes of the genus *Paraspidodera* are found only in Hystricognath rodents in the Neotropical region and these nematodes appear to have had a long historical, coevolutionary association with ctenomyids (Gardner, 1991), showing varying levels of both cospeciation and host switching. The adenophorean whip worm genus *Trichuris* occurs in many diverse groups of rodents in the neotropics, but no analyses have yet been done to examine the levels of coevolution with ctenomyids.

The multifarious nature of nematode diversity in subterranean mammals in the Nearctic and Neotropical regions requires at the minimum that phylogenetic hypotheses for each group of mammals and their nematodes be developed so each host group can be compared with each parasite group. Much more detailed work must be paid to collecting parasites from some of the unknown species of *Ctenomys* throughout the Neotropics. The same can be said about the level of knowledge of parasitic nematodes in the Geomyidae in the northern Neotropics and southern Nearctic regions.

VI. MOLECULAR DIVERSITY IN THE NEMATA

Several molecules have been used to begin to assess phylogenetic and genetic diversity within the Nemata, and the number of investigations using molecular methods to try to quantify the diversity of the nematodes is rapidly increasing (see references in Dorris *et al.*, 1999; Blaxter *et al.*, 1998; Adams, 1998; Al-Banna *et al.*, 1997; Nielsen, 1996). However, because of the extremely large number of species that may exist, examination of levels of molecular or genetic diversity in representatives of the group as a whole is just beginning even though massive amounts of molecular data on nematodes are now pouring into the literature stream. The summary papers by Blaxter *et al.* (1998) and Dorris *et al.* (1999) indicate the utility and power of estimating the molecular–phylogenetic relationships among the Nemata using ribosomal DNA and other molecular sequence data. Initial studies of the molecular diversity within and among several lineages of the Strongylida show that the genetic diversity among these taxa is relatively low (Chilton *et al.*, 1997). From these works, it is clear that molecular phylogenies will provide robust tests of the hypotheses of morphological relationships among the Nemata. As more regions of DNA are used to examine the relationships among the nematodes, we expect a clarification of both the deep phylogenetic branches that are relatively obscure in the molecular phylogeny of the Nemata and the more rapidly evolving branch tips that represent extant species with valuable genetic information.

VII. RELATIONSHIPS TO OTHER ANIMAL GROUPS

An analysis grouped the nematodes, gastrotrichs, priapulids, kinorhynchs, and the loriciferans into a group (superphylum) called the Cycloneuralia (Fig. 26) based on the circular shape of the brains in these groups (Nielsen *et al.*, 1996). The aforementioned study and at least one other (Zrzavy *et al.*, 1998) using 18s rDNA sequences showed that the Nemata share a common ancestor with members of the phylum Nematomorpha,

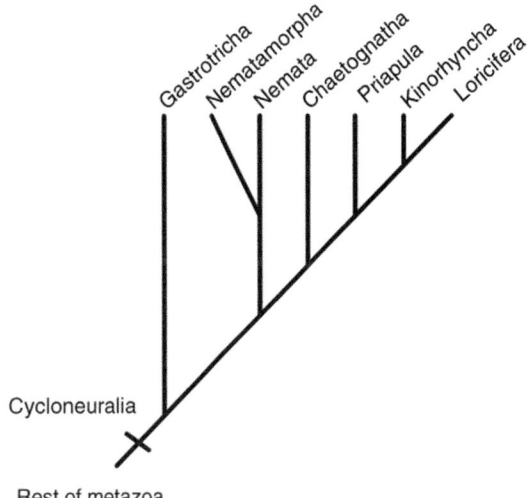

FIGURE 26 Phylogenetic tree showing the relationships of the Nemata to the rest of the Cycloneuralia (after Nielsen *et al.*, 1996).

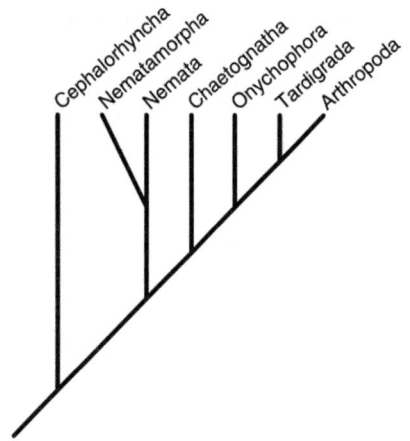

FIGURE 27 Phylogenetic tree showing the relationship of the Nemata to the rest of the Animalia (after Zrzavy et al., 1998).

but there was a shuffling of the other groups out of the "Cycloneuralia" (Fig. 27).

As mentioned previously, nematodes are soft-bodied and mostly very small organisms. Any larger forms that existed were probably parasites of vertebrates; however, these left no fossil traces. The only fossil nematodes that are known are insect parasitic or plant parasitic forms that occur very rarely in amber inclusions. Because there are no fossil records of nematodes of Cambrian or Precambrian ages, estimates of the age of the Nemata have been only speculative, and without fossils, it is difficult to calibrate molecular clocks for the nematodes. However, through application of various models of molecular evolution and molecular clock theory, estimates of the time of divergence of the nematodes from the rest of the animal groups appears to be about 1177 ± 79 million years (Wang et al., 1999) (this study showed a basal origin of the nematodes on a phylogenetic tree, in contrast to the relatively derived placement in the analyses shown in Figs. 26 and 27).

Most authors consider the ultimate origin of nematodes to be from a marine ancestor (Maggenti, 1981; Malhakov, 1994). The tri-radiate esophagus and the tubular body indicate a possible primarily sedentary existence with the posterior end attached to the substrate and the anterior end freely encountering the marine environment from all sides, thus the secondarily derived, somewhat radial symmetry.

A. Groups of the Nemata

Above the level of the order, confusion reigns relative to the classification and systematic arrangement of the nematodes. Maggenti (1991a) is usually followed in this regard, and his analyses followed corroborated historical analyses in recognizing the two main subphyla: The Secernentea and the Adenophorea. Recent work shows that these groups are substantiated both in morphological and in molecular analyses, although competing phylogenetic hypotheses and associated classifications have also been proposed (Dorris et al., 1999; Blaxter et al., 1998; Brooks and McLennan, 1993; Adamson, 1989).

B. Classification

Since 1949, at least eight authors have provided classifications for members of the phylum Nemata (see references in Malakhov, 1994; Brooks and McLennan, 1993; Maggenti, 1991). Of these, the classifications of Maggenti (1981, 1991a) have proven to be the most useful summary of all nematodes (free-living and parasitic); however, phylogenetic hypotheses have been proposed based on both molecular and morphological characteristics. This does not necessarily mean that a classification will be developed from a proposed phylogenetic tree (see Maggenti, 1991a; Brooks and McLennan, 1993).

VIII. FUTURE KNOWLEDGE OF NEMATODES

I hope that this summary treatment provides readers with sufficient knowledge to allow more in-depth research on nematodes. The group is so large and so ecologically, morphologically, and phylogenetically diverse that to attempt to discuss the diversity of the group in such an abbreviated way is practically futile at best. N. A. Cobb (1914) stated this clearly on the last page of his famous "Nematodes and Their Relationships" (p. 490):

> The foregoing fragmentary sketch may indicate to the student, as well as to the general reader, the vast number of nematodes that exist, the enormous variety of their forms, and the intricate and important relationships they bear to mankind and the rest of creation.

As more data on distribution and function of nematodes throughout the biosphere are obtained, the importance of this group of worms will surely be realized. We are just beginning to explore the oceans, and we are probably losing more species of nematodes from rainforest clearing than will ever be ultimately found, described, and classified.

I hope that mankind will generate more interest in the microscopic world of the Nemata, and I hope that this article will provoke the reader into action and disprove what Van Leeuwenhoek stated about his fellow man on 28 September 1715: "And over and above all, most men are not curious to know: nay, some even make no bones about saying, What does it matter whether we know this or not?" (Dobell, 1932, pp. 324–325).

Acknowledgments

Special thanks to Tom Powers for encouragement during this writing blitz and to the staff and students of the Manter Laboratory. Also thanks to Sue Ann, Teal, and Hudson for their help with reference materials and support.

See Also the Following Articles

PARASITISM • WORMS, ANNELIDA • WORMS, PLATYHELMINTHES

Bibliography

Adamson, M. (1989). Constraints in the evolution of life histories in zooparasitic nematoda. In "Current Concepts in Parasitology" (R. C. Ko, ed.), pp. 221–253. Hong Kong University Press, Hong Kong.

Al-Banna, L., Williamson, V. M., and Gardner, S. L. (1997). Phylogenetic analysis of nematodes of the *genus Pratylenchus* using nuclear 26s rDNA. *Mole Phylogen Evol.* 7, 94–102.

Al Banna, L., and Gardner, S. L. (1996). Nematode diversity of native species of *Vitis* in California. *Can. J. Zool.* 74, 971–982.

Anderson, R. C. (1992). "Nematode Parasites of Vertebrates: Their Development and Transmission." CAB International, Wallingford, UK.

Blaxter, M. L., De Ley, P., Garey, J. R, Liu, L. X., Scheldeman, P., Vierstraete, A., Vanfleteren, J. R., Mackey, L. Y., Dorris, M., Frisse, L. M., Vida, J. T., and Thomas, W. K. (1998). A molecular evolutionary framework for the phylum Nematoda. *Nature* 392, 71–75.

Bongers, T., and Ferris, H. (1999). Nematode community structure as a bioindicator in environmental monitoring. *Trends Ecol. Evol.* 14, 224–228.

Boucher, G., and Lambshead, P. J. D. (1995). Ecological biodiversity of marine nematodes in samples from temperate, tropical, and deep-sea regions. *Conserv. Bio.* 9, 1594–1604.

Brant, S. V., and Gardner, S. L. (2000). Phylogeny of species of the genus *Litomosoides* (Nemata: Onchocercidae): Evidence of rampant host-switching. *J. Parasitol.* 86, 545–554.

Brooks, D. R., and McLennan, D. A. (1993). "Parascript: Parasites and the Language of Evolution." Smithsonian Institution Press, Washington, DC.

Chilton, N. B., Gasser, R. B., and Beveridge, I. (1997). Phylogenetic relationships of Australian strongyloid nematodes inferred from ribosomal DNA sequence data. *Internatl. J. Parasitol.* 27, 1481–1494.

Chitwood, B. G., and Chitwood, M. B. (1977). "Introduction to Nematology." University Park Press, Baltimore, MD.

Cobb, N. A. (1914). "Nematodes and their relationships." In "Yearbook of Department of Agriculture for 1914," pp. 457–490. Government Printing Office, Washington, DC.

Crompton, D. W. T. (1999). How much human helminthiasis is there is the world? *J. Parasitol.* 85, 397–403.

Dobell, C. (1932). "Antony Van Leeuwenhoek and His 'Little Animals.'" Harcourt, Brace and Co., NY.

Dorris, M., De Ley, P., and Blaxter, M. L. (1999). Molecular analysis of nematode diversity and the evolution of parasitism. *Parasitol. Today* 15, 188–193.

Ferris, H., Venette, R. C., van der Meulen, H. R., and Lau, S. S. (1998). Nitrogen mineralization by bacterial feeding nematodes: Verification and measurement. *Plant Soil* 203, 159–171.

Gardner, S. L. (1991). Phyletic coevolution between subterranean rodents of the genus *ctenomys* (Rodentia: Hystricognathi) and nematodes of the genus *Paraspidodera* (Hetevakoidea: Aspidoderidae) in the neotropics: temporal and evolutionary implications. *Zool. J. Linnean Soc.* 102, 169–201.

Gaugler, R., and Kaya, H. K. (1990). "Entomopathogenic Nematodes in Biological Control." CRC Press; Boca Raton, FL.

Grasse, P. P. (1965). "Traité de Zoologie. Anatomie, Systématique, Biologie. Némathelminthes (Nématodes) (Nématodes–Gordiacées–Rotiféres–Gastrotriches–Kinorhynques.)" Tome IV. Fascicule II, III.

Hugot, J. P. (1999). Primates and their pinworm parasites: The Cameron hypothesis revisited. *System. Biol.* 48, 523–546.

Lambshead, P. J. D. (1993). Recent developments in marine benthic biodiversity research. *Océanis* 19, 5–24.

Lawton, J. H., Bigtnell, D. E., Bloemers, G. F., Eggleton, P., and Hodda, M. E. (1996). Carbon flux and diversity of nematodes and termites in Cameroon forest soils. *Biodiversity Conservation* 5, 261–273.

Maggenti, A. R. (1981). "General Nematology." Springer-Verlag, New York, NY.

Maggenti, A. R. (1991a). Nemata: Higher classification. In "Manual of Agricultural Nematology," pp. 147–187. Marcel Dekker, Inc., New York, NY.

Maggenti, A. R. (1991b). General nematode morphology. In "Manual of Agricultural Nematology," pp. 3–46. Marcel Dekker, Inc., New York, NY.

Malakhov, V. V. (1994). "Nematodes: Structure Development, Classification and Phylogeny." Smithsonian Institution Press, Washington, DC.

Morand, S., Legendre, P., Gardner, S. L., and Hugot, J. P. (1996). Body size evolution of Oxyurid (Nematoda) parasites—The role of hosts. *Oecologia.* 107, 274–282.

Musser, G. G., and Carleton, M. D. (1993). Family Muridae. In "Mammal Species of the World: A Taxonomic and Geographic Reference," (D. E. Wilson and D. M. Reeder, eds.), pp. 501–755. Smithsonian Institution Press, Washington, DC.

Nickle, W. R. (1991). "Manual of Agricultural Nematology." Marcel Dekker, New York.

Nielsen, C., Scharff, N., and Eibye, J. D. (1996). Cladistic analyses of the animal kingdom. *Biol. J. Linnean Soc.* 57, 385–410.

Powers, L., Mengchi, H., Freckman-Wall, D., and Virginia, R. A. (1998). Distribution, community structure, and microhabitats of soil invertebrates along an elevational gradient in Taylor Valley, Antarctica. *Arctic Alpine Res.* **30**, 133–141.

Rausch, R. L. (1983). The biology of avian parasites: Helminths. In *"Avian Biology, Vol. VII"* (D. S. Farner, J. R. King, and K. C. Parkes, eds), pp. 367–442. Academic Press, NY.

Wang, D., Kumar, S., and Hedges, B. (1999). Divergence time estimates for the early history of animal phyla and the origin of plants, animals and fungi. *Proc. Royal Soc. London Ser. B* **266**, 163–171.

Zrzavy, J., Mihulka, S., Kepka, P., Bezdek, A., and Tietz, D. (1998). Phylogeny of the Metazoa based on morphological and 18S ribosomal DNA evidence. *Cladistics* **14**, 249–285.

WORMS, PLATYHELMINTHES

Janine N. Caira* and D. Timothy J. Littlewood†
*University of Connecticut and †The Natural History Museum London

---

I. General Features
II. Major Groups
III. Nervous System
IV. Feeding and Digestion
V. Excretion and Osmoregulation
VI. Reproduction
VII. Ontogeny
VIII. Phylogenetic Relationships
IX. Host Associations
X. Medical Importance

GLOSSARY

acetabulum Ventral sucker used for attachment and/or locomotion in digeneans.
acoelomate Lacking a body cavity; see parenchyma.
cercomer Posterior parenchymous and/or muscular extension of body; homology among groups is controversial; some view both the hook-bearing regions of monogeneans and some larval cestodes to be modifications of the cercomer.
cirrus Male intromittent copulatory organ consisting of distal, invaginable portion of ejaculatory duct that is evaginated (by turning inside out) through the genital pore during copulation; may or may not be armed with spines, microtriches, etc.; see penis.
cortex The outermost region of parenchyma in taxa in which two distinct regions are present; inner boundary often marked by conspicuous circular muscle fibers; see medulla.
duogland adhesive system Combination of a cement, a viscid gland producing adhesive material, and a releasing gland; usually associated with cells of ventral epidermis; provides temporary adhesion to the substrate for many interstitial flatworm species.
haptor The posterior attachment organ of monogeneans.
medulla The innermost region of parenchyma in taxa in which two distinct regions are present; outer boundary often marked by conspicuous circular muscle fibers; see cortex.
monozoic Possessing only a single set of reproductive organs; see polyzoic.
microthrix A specialized surface extension of the neodermis of cestodes, characterized by an electron-dense cap composed of numerous microtubules separated from distal cytoplasm by a base-plate (plural microtriches).
neodermis A syncytial, nonciliated epidermis with cell bodies and nuclei (often referred to as cytons or pericaryon) laying below other body wall elements; unique to the Neodermata; completely replaces epidermis, usually at metamorphosis when larva encounters the first host.
parenchyma The connective tissue between cells of ectoderm and endoderm layers that fills spaces among organ systems; as platyhelminths are acoelomates, the space among internal organs is completely filled with parenchyma.

penis Male intromittent copulatory organ consisting of a muscular, often papilliform, retractable structure that is protruded (without turning inside out) through the genital pore during copulation; may or may not be armed with spines, etc.; see cirrus.

polyzoic Possessing multiple serial sets of reproductive organs; condition found in most cestode orders.

proglottid Compartmentalized regions of cestode body; each proglottid contains one or more sets of reproductive organs; controversial as to whether proglottids are homologous with segments of coelomates.

protonephridial system A system of branched canals, each branch with one or more specialized flame bulbs bearing one to many cilia; specialized for excretion and osmoregulation; found in many phyla; in platyhelminths gaps within or among cytoplasmic extensions of one or two canal cells and/or the terminal cell allow fluids to enter the canal system at a rate determined by the pressure difference created by the beating of the cilia of the flame bulb.

rhabdite A rod-shaped, proteinaceous secretory product found in the epidermis of most free-living platyhelminths; known functions include mucus production for ciliary gliding, cocoon formation, prey capture, and predator repulsion.

scolex Attachment organ in adult cestodes; often armed with suckers and/or hooks.

statocyst Static sense organ in which the movement of a statolith is detected, thereby indicating an animal's orientation.

strobila The portion of the body of cestodes exhibiting serial repetition of proglottids.

THE PHYLUM PLATYHELMINTHES is presently considered to contain approximately 20,000 species, although the number is likely to be a gross underestimate. These animals generally are bilaterally symmetrical, are dorsoventrally flattened, and lack a body cavity. None of these features, however, is unique to this phylum, and thus, none can be counted on for the definitive recognition of membership in the group. That is not to say the phylum does not consist of a natural assemblage of taxa. The platyhelminths appear to be unique in their possession of, for example, multiciliated gastrodermal cells and epidermal and protonephridial cilia that lack accessory centrioles. Unfortunately, these distinguishing features are most appropriately observed with the use of transmission electron microscopy and thus are not especially convenient for the casual identification of platyhelminths. In many cases it is more useful to turn to the rather distinctive features that characterize the platyhelminth subgroups.

Their lack of a protective outer body covering means that water figures prominently in the lives of the platyhelminths, and indeed, the majority of species are found in moist or aquatic environments; marine, freshwater, and terrestrial environments have all been colonized. The phylum includes species spanning a wide range of adult lifestyles, from those that are free-living interstitial dwellers to those that are obligate (internal or external) parasites of vertebrates. Consequently, the group includes species ranging enormously in size (400 μm to 80 m), shape, form, and occasionally color. Species that are obligate parasites of vertebrates belong to three distinctive, potentially monophyletic groups: Digenea, Monogenea, and Cestoda. A number of these are of significant medical or veterinary importance. There exists little evidence to suggest that the remaining platyhelminths, some of which are free-living and some of which exist as either commensals or parasites of invertebrates, form a single cohesive group, although collectively they are often referred to as "turbellarians." Present knowledge suggests that the parasitic groups are more diverse than the nonparasitic groups. However, there is no question that numerous free-living and parasitic taxa await discovery and formal description.

I. GENERAL FEATURES

A glance at the defining features of each of the major platyhelminth groups reveals that the characteristics that vary most among, and even within, these groups are primarily those associated with the digestive and reproductive systems. The digestive system generally lacks an anus. Thus, there is no unidirectional flow of food materials through the system; in many species food enters and solid wastes exit through the oral opening. Several major features of the female reproductive system are of particular importance. The female gonad of platyhelminths is either homocellular, in that it consists of only a single type of cell, or it is heterocellular, consisting of cells that produce either ova or vitelline cells. The former configuration results in the production of endolecithal eggs because vitelline (yolk) material is incorporated directly into the cytoplasm of the ova. Heterocellular platyhelminths usually conduct these functions in separate organs, possessing an ovary (or occasionally several ovaries) that produces ova and a vitellarium that produces vitelline cells containing the nutritive yolk. This configuration results in the production of ectolecithal eggs, as vitelline cells are packaged along with, but independent of, the zygote in the egg.

In addition, most platyhelminths possess a formalized excretory system consisting of generally two (rarely one or three) sets of protonephridia. The nervous system and sense organs are present to a greater or lesser extent among different taxa. Specialized organs functioning in respiration and circulation are entirely lacking in many groups of platyhelminths. Thus, respiration generally occurs by means of diffusion of gases directly through the surfaces of the body, a process that is greatly facilitated and made effective by the relatively flattened nature of the platyhelminth body. The "circulation" of materials throughout the various parts of the platyhelminth body is also accomplished by diffusion. However, a number of digeneans are known to possess a "lymphatic system" with contractile vessels and cellular inclusions that may have a circulatory function. The details of most organ systems are difficult to observe without the use of specialized staining and microscopic techniques.

One final, rather unusual feature of the platyhelminths is that the differentiated nonreproductive (somatic) cells of the body are unable to divide. Thus, mitosis does not occur in the differentiated somatic cells of platyhelminths; rather, these cells are replaced as necessary by undifferentiated stem cells located below the outer layers of the body, which undergo mitosis. This phenomenon helps to explain the remarkable regenerative abilities of certain platyhelminth groups.

II. MAJOR GROUPS

Diversity in the group is generally most conveniently represented at the level of the taxonomic rank of order. Within the platyhelminths this rank is relatively informative and much more numerically manageable than lower ranks such as family or genus (for example, there are 42 orders versus 401 families and 4241 genera). However, it is important to recognize that traditionally, taxonomists working on each of the three major groups exhibiting obligate parasitism (Digenea, Monogenea, and Cestoda) have generally worked independently of one another, and also independently from the taxonomists studying the platyhelminth groups exhibiting more free-living life history strategies. Thus, the concepts of higher ranks, such as order, are not necessarily comparable among the major groups. For example, the 243 genera of monogeneans are currently distributed among 12 orders, whereas the 2549 genera of digeneans are distributed among only three orders. Diagnoses are presented later for the orders currently recognized in all groups except the digeneans, which are treated at the level of superfamily. Numbers of families and genera for each of the orders (or superfamilies in the case of the digeneans) are given in Table I.

For more detailed information on specific groups, readers are referred to the works of Cannon (1986) and Rieger (1998) for the non-neodermatan groups; Yamaguti (1971) and Schell (1985) for the digeneans; Yamaguti (1963), Schell (1985), and Boeger and Kritsky (1993) for the monogeneans; and Schmidt (1986) and Khalil et al. (1994) for the cestodes. However, the treatments of these groups by Hyman (1951) and the five sections covering the platyhelminths in the Traité de Zoologie (Grassé, 1961) are classic and remain among the more comprehensive sources of detailed information on the platyhelminths. The many works of Klaus Rohde (for example, 1997), Ulrich Ehlers (for example, 1985), and Peter Ax (for example, 1996), among others, have done much to further our understanding of the morphology and phylogenetic relationships among these groups.

Discussions of diversity are most effectively organized around hypotheses of the phylogenetic relationships among the groups involved. Although investigators have not yet come to a consensus about the phylogenetic relationships among all of the major platyhelminth groups, the relationships among some of these groups are fairly well accepted. The tree in Fig. 1 illustrates some of the aspects of these relationships for which supporting morphological and molecular data are starting to accumulate. The presentation of diversity that follows is generally organized around the topology of this tree; the non-neodermatan orders are treated in the sequence in which they appear from left to right on this tree. A more detailed treatment of the phylogenetic relationships among the various groups is provided in Section VIII.

An illustration is included of a representative of each platyhelminth order (or superfamily for the digeneans). A consistent stippling pattern has been used throughout these figures for each different organ system so that the conditions of the major organs are readily visible and comparable among figures. The organs represented by each stippling pattern are labeled in Figs. 12 and 18.

1. Acoela

See Fig. 2. Glandular frontal organ present; lamellated rhabdites absent, some with unlamellated rhabdoids; duogland adhesive system absent; epidermis ciliated throughout; statocyst present, with single statolith and two parietal cells; some with paired ocelli; mouth usually ventral, occasionally subterminal, opens directly into parenchyma or into simple or occasionally strongly muscular pharynx; gut lumen absent; digestion in tem-

TABLE I
Platyhelminth Diversity

Text box no.	Figure no.	Order or Superfamily, etc.	No. families	No. genera
		Platyhelminthes	404	4,252
1	2	Acoela	13	87
2	3	Nemertodermatida	1	3
3	4	Catenulida	5	12
4	5	Haplopharyngida	1	1
5	6	Lecithoepitheliata	2	5
6	7	Macrostomida	3	21
7	8	Polycladida	37	146
8	9	Proseriata	7	68
9	10	Tricladida	10	101
10	11	Prolecithophora	11	26
11	12	Kalyptorhynchia	16	124
12	13	Typhloplanida	8	104
13	14	Dalyelliida	8	65
14	15	Temnocephalida	3	14
		Neodermata	279	3,475
15	17	Aspidogastrea	3	19
		Digenea	156	2,553
16	18	Allocreadioidea	3	179
17	19	Clinostomoidea	1	13
18	20	Cyclocoeloidea	4	45
19	21	Diplostomoidea	10	158
20	22	Echinostomatoidea	21	213
21	23	Hemiuroidea	16	358
22	24	Gymnophalloidea	12	176
23	25	Lepocreadioidea	10	162
24	26	Microphalloidea	3	85
25	27	Notocotyloidea	7	76
26	28	Opisthorchioidea	8	236
27	29	Paramphistomatoidea	15	162
28	30	Plagiorchioidea	29	462
29	31	Schistosomatoidea	3	84
30	32	Transversotrematoidea	1	5
31	33	Troglotrematoidea	7	29
32	34	Zoogonoidea	6	110
		Monogenea	52	254
33	36	Dactylogyridea	8	74
34	37	Gyrodactylidea	4	9
35	38	Udonellidea	1	7
36	39	Montchadskyellidea	1	1
37	40	Capsalidea	3	26
38	41	Monocotylidea	5	24
39	42	Polystomatidea	2	14
40	43	Chimaericolidea	1	2
41	44	Diclybothriidea	2	16
42	45	Mazocraeidea	27	89
		Cestoda	68	649
43	47	Amphilinidea	2	6
44	48	Gyrocotylidea	1	1
45	49	Caryophyllidea	4	41
46	50	Spathebothriidea	2	5
47	51	Pseudophyllidea	6	56
48	52	Haplobothriidea	1	1
49	53	Diphyllidea	2	3
50	54	Trypanorhyncha	19	45
51	55	Tetraphyllidea	8	53
52	56	Proteocephalidea	2	58
53	57	Lecanicephalidea	4	5
54	58	Nippotaeniidea	1	2
55	59	Tetrabothriidea	1	6
56	60	Cyclophyllidea	15	364

continues

porary vacuoles within syncytial endoderm; protonephridial system absent; germinal tissues discrete or diffuse; testes and ovary separate or combined in hermaphroditic gonad; if separate, testes paired or not; sperm filiform, biflagellated, flagellar axonemes incorporated into sperm cytoplasm, axoneme arrangement 9 + 2; seminal vesicle present or absent; penis present or absent; if present penis either inserted into seminal vesicle or not; some with well-developed hard accessory organs such as penis stylet; male pore variable in position; ovaries paired or not, diffuse or follicular; vitellarium absent; vagina and seminal receptacle present or absent; female pore probably absent (eggs released through rupture of epidermis); eggs endolecithal (yolk produced within oocytes).

Habitat: Predominantly marine; predominantly free-living, but some symbiotic in digestive system or body cavity of echinoderms (primarily holothuroids and echinoids); a number of species of Convolutidae and Sagittiferidae symbiotic with green algae.

2. Nemertodermatida

See Fig. 3. Glandular frontal organ present; lamellated rhabdites absent, some with unlamellated rhabdoids; duogland adhesive system generally absent; epidermis ciliated throughout; statocyst usually with two (occasionally one, three, or four) statoliths and several parietal cells; usually with, occasionally without, ventral mouth; usually without, occasionally with, simple pharynx; but generally with some form of lumen, with intes-

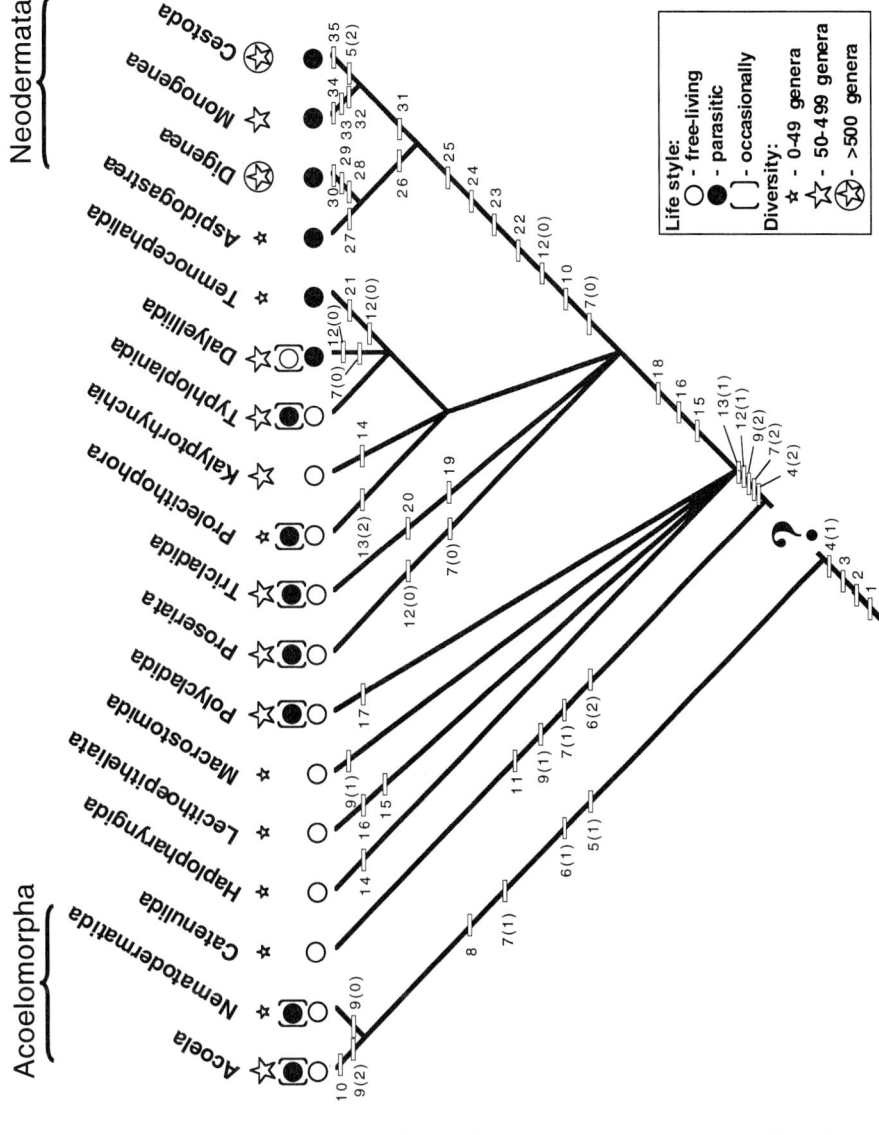

FIGURE 1 Hypothesized phylogenetic relationships among major platyhelminth groups. State "0" should be assumed prior to point at which character is marked on tree; state "1" should be assumed if character is marked on tree but no state is given. Topology is modified from Littlewood et al. (1999).

Character list:
1. protonephridial cilia accessory centriole: 0=present; 1=absent
2. epidermal cilia accessory centriole: 0= present; 1=absent
3. gastrodermal cell ciliation: 0=uniciliated; 1=multiciliated
4. number of cilia in protonephridial flame bulbs: 0= 1; 1= 2; 2=at least 4
5. gut: 0= well defined; 1=poorly defined (syncytial); 2= absent
6. protonephridial system: 0=paired; 1=absent; 2=unpaired
7. rhabdites: 0=absent; 1=non-lamellated present; 2=lamellated present
8. glandular frontal organ: 0=absent; 1=present
9. sperm ciliation: 0=monociliated; 1=aciliated; 2=biciliated
10. cilia of sperm cell: 0=not incorporated into cytoplasm of sperm cell; 1=incorporated into cytoplasm of sperm cell
11. position male and female pores: 0=ventral; 1=dorsal
12. duogland adhesive system: 0=absent; 1=present
13. arrangement of microtubules in sperm cilia: 0=9 + 2; 1=9 + "1"; 2=with elaborate intracellular membrane
14. proboscis (not associated with gut): 0=absent; 1=present
15. female gonad: 0=homocellular; 1=heterocellular
16. eggs: 0=endolecithal; 1=ectolecithal
17. intestinal branching: 0=absent; 1=extensive
18. relationship between ovarian and vitelline tissue: 0=in same organ; 1= in separate organs
19. intestine: 0=bifurcating; 1= triradiate
20. extra, embryonic intestine: 0=absent; 1=present
21. 2-12 anterior tentacles: 0=absent; 1=usually present
22. protonephridial filter: 0= formed from terminal cell only; 1= 2-celled weir formed from canal cell and terminal cell
23. number of rootlets in cilia of larval epidermal cells: 0=2 (rostral & caudal); 1=1 (rostral only)
24. form of outer layer of adult body: 0=cellular epidermis; 1= syncytial neodermis
25. epidermal cells: 0=not shed by larva; 1=shed by larva
26. epidermal cells of larva: 0=not separated from one another by neodermis; 1=separated from one another by neodermis
27. oviducts: 0=not divided into chambers; 1=divided into chambers
28. miracidium: 0=absent; 1=present
29. sporocyst/redia: 0=absent; 1=present
30. cercaria: 0=absent; 1=present
31. cercomer (with sickle-shaped hooks) in larva: 0=absent; 1=present
32. 4 rhabdomeric eyespots: 0=absent; 1=present
33. oncomiracidium: 0=absent; 1=present
34. cercomer in adult (=opisthaptor): 0=absent; 1=present
35. microtriches: 0=absent; 1=present

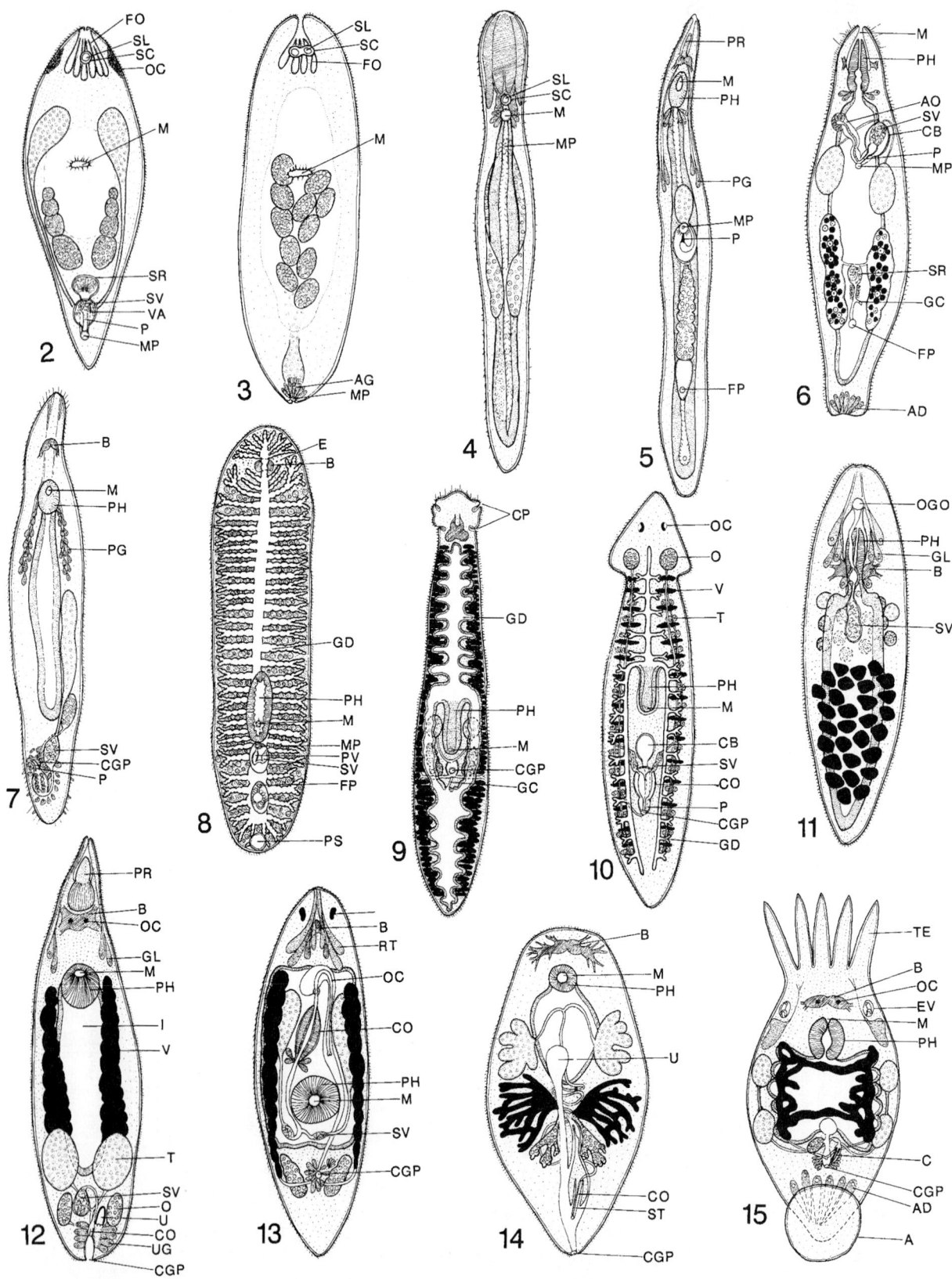

tinal epithelium and glandular club cells; protonephridial system absent; gonad hermaphroditic; sperm filiform, unflagellated, flagellum not incorporated into body of sperm, 9 + 2 microtubule arrangement; male antrum shaped as simple ciliated invagination of epidermis, eversible penis or hard parts absent; male pore posterior; vitellarium absent; oviducts absent, female genital pore usually absent, occasionally dorsal female pore present, eggs possibly released into lumen of gut; eggs endolecithal.

Habitat: Entirely marine; predominantly free-living, some symbiotic in body cavity or intestine of holothuroidean echinoderms.

3. Catenulida

See Fig. 4. Some with tail; lamellated rhabdites absent, unlamellated rhabdoids present; duogland adhesive system absent; epidermal ciliation sparse; some with ventrolateral ciliated furrow; brain lobed or not; statocyst often present, usually with one but occasionally with up to six statoliths and several parietal cells; mouth ventral, opening into simple pharynx; intestine in form of simple elongate sac; protonephridial system unpaired; single hermaphroditic gonad present; sperm aflagellated; penis present or absent, male and female genital pores dorsal; vitellarium absent; eggs endolecithal; some undergo asexual multiplication via paratomy or parthenogenesis.

Habitat: All species are free-living, primarily in freshwater, but some in marine environments.

4. Haplopharyngida

See Fig. 5. With protrusible anterior proboscis not connected to digestive system; lamellated rhabdites present; duogland adhesive system present; brain enclosed in connective tissue capsule (encapsulated), bilobed; statocyst absent; mouth opens into simple muscular pharynx; intestine in form of simple elongate sac; protonephridial system paired; single testis present; sperm aflagellated; penis present, with circle of hard, straight stylets; single ovary; vitellarium absent; eggs endolecithal; male pore anterior to female pore.

Habitat: All species are free-living in marine environments.

5. Lecithoepitheliata

See Fig. 6. Some with weak posterior adhesive disc; some with anterior pigmented girdle; lamellated rhabdites present; duogland adhesive system present; statocyst absent; mouth anterior; pharynx usually muscular, anterior; intestine in form of simple elongate sac; protonephridial system paired; testes compact or follicular, one or many; sperm biflagellated, flagella not incorporated into body of sperm, 9 + "1" microtubule arrangement; seminal vesicle present or absent; penis present, with or without stylet; muscular copulatory bulb present or absent; accessory male organ present or absent, if present opening into male atrium, prostatic vesicle present or absent; male pore opening to outside or into pharyngeal cavity; female organ single or tetrapartite, in form of heterocellular germovitellarium (vitelline-producing cells surrounding ovum-producing cells); seminal receptacle present or absent; genitointestinal canal sometimes present; female pore ventral; eggs ectolecithal (yolk not produced in oocytes).

Habitat: All species are free-living in freshwater, marine, or terrestrial environments.

6. Macrostomida

See Fig. 7. Frontal glands sometimes present; some with spatulate posterior region bearing adhesive papillae; lamellated rhabdites present; duogland adhesive system present; brain not encapsulated; statocyst absent; simple pharynx present, sometimes tubular; intestine in form of simple elongate sac, sometimes with pre-oral blind sacs; with protonephridial system paired; testes compact, single or paired; sperm aflagellated; seminal vesicle present or absent; prostatic vesicle present or absent; penis present or absent; penis stylet present or absent; ovary single or paired; vitellarium absent; eggs endolecithal; genital pores separate or combined; ability to asexually produce zooids present or absent.

Habitat: All species are free-living in marine, euryhaline, or freshwater environments.

FIGURES 2–15 **2** Acoela: adult of *Otocelis* sp. (modified from Cannon, 1986); **3** Nemertodermatida: adult of *Nemertoderma* sp. (modified from Cannon, 1986); **4** Catenulida: adult of *Retronects* sp. (modified from Cannon, 1986); **5** Haplopharyngida: adult of *Haplopharynx* sp. (modified from Cannon, 1986); **6** Lecithoepitheliata: adult of *Gnosenesima* sp. (modified from Cannon, 1986); **7** Macrostomida: adult of *Bradynectes* sp. **8** Polycladida: adult of *Cestoplana* sp. (redrawn from Beauchamp in Grasse, 1961); **9** Proseriata: adult of *Bothrioplanus* sp. (modified from Beauchamp in Grasse, 1961); **10** Tricladida: adult of *Dagesia* sp. (modified from Cannon, 1986); **11** Prolecithophora: adult of *Prolecithoplana* sp. (from several literature sources); **12** Kalyptorhynchia: adult of *Cystiplana* sp. (modified from Cannon, 1986); **13** Typhloplanida: adult of *Promesostoma* sp. (redrawn from Ehlers, 1974); **14** Dalyelliida: adult of *Syndesmis* sp. (modified from Cannon, 1986); **15** Temnocephalida: adult of *Temnocephala* sp. (modified from Baer in Grasse, 1961). (See abbreviations on pages 897–898.)

7. Polycladdida

See Fig. 8. Body oval or elongate, or thick and broad with folded margins; with (Cotylea) or without (Acotylea) pseudosucker posterior to female genital pore; anterior or nuchal (neck) tentacles present or absent; ventral prostatoid-like organs present or absent; dorsal surface smooth or papillate; body margins of some with hard spines; may be multicolored with complex patterns; lamellated rhabdites present; duogland adhesive system present; statocyst absent; numerous ocelli usually present, marginal or scattered anteriorly, some arranged in one or two pairs of clusters (cerebral and tentacular); pharynx variable in position, ruffled and vertically oriented or sometimes consisting of anteriorly directed tube, rarely consisting of multiple tubes arranged ventrally; intestine usually with numerous radiating branches, branches sometimes anastomosing protonephridial system paired; testes numerous, scattered between branches of gut; sperm biflagellated, flagella not incorporated into body of sperm, 9 + "1" microtubule arrangement; true seminal vesicle present or absent; intromittent organ absent or in form of a penis or cirrus; penis stylet present or absent; cirrus armed or not; prostatic vesicle present or absent; bursa copulatrix present or absent; some with multiple copulatory complexes; ovaries numerous, scattered between branches of gut; seminal bursa (or Lang's vesicle) opening into vagina present or absent; uterus present or absent (in form of expanded oviducts in some); vitellarium absent: eggs endolecithal; male and female gonopores usually separate (male anterior to female) but sometimes combined; some with hypodermic impregnation.

Habitat: Predominantly marine, a few freshwater or euryhaline; mostly free-living, some symbiotic with other invertebrates, for example, wrapped around the abdomen of hermit crabs, associated with gorgonian corals in mantle cavities of gastropod or lamellibranch molluscs, in genital bursa of ophiuroid or echinoid echinoderms; some live in association with their food, for example, in association with tunicate colonies feeding on zooids, or in association with oyster colonies feeding on oyster tissue; some seek shelter in empty shells.

8. Proseriata

See Fig. 9. Epithelial nuclei insunk or not; body of some with feeble or prominent tactile bristles; body uniformly ciliated (sometimes with exception of caudal tip) or cilia restricted to ventral surfaces and head; head with or without ciliated pits; lamellated rhabdites absent; duogland adhesive system present; brain encapsulated or not; statocyst usually present; pigmented ocelli present or absent; pharynx tube horizontal or vertical; intestine linear or tripartite, dorsal to pharynx; protonephridial system paired or tripled, occasionally single; testes paired or not, usually compact, rarely follicular; sperm biflagellated, flagella not incorporated into body of sperm, 9 + "1" microtubule arrangement; usually with single male copulatory organ but occasionally with many; occasionally with accessory (prostatoid) organ; bursa present or absent; with one pair of compact ovaries, ovary subdivided into smaller follicles consisting of oocytes surrounded by accessory cells in some; vitellarium follicular, follicles in paired lateral rows; vagina externa present or absent; genitointestinal canal usually present; eggs ectolecithal; genital pores common or separate.

Habitat: Predominantly free-living in marine environments; a few occur in freshwater; occasionally ectocommensal on marine crustaceans.

9. Tricladida

See Fig. 10. Usually large, flattened, elongate; body sometimes pigmented, terrestrial forms often colorful; some with triangular, semi-lunate, or pointed anterior region of body; caudal adhesive disc present or absent; lamellated rhabdites present in some; duogland adhesive system present; cilia uniform throughout all surfaces of body or restricted to ventral surface; statocyst absent; with many ocelli or one pair of ocelli (cave-dwelling species lack ocelli), ocelli sometimes with distinct lens; mouth in middle or posterior half of body; pharynx plicate or tubular, posteriorly directed, usually one, occasionally more pharynges; intestine tripartite with one anterior and two posterior branches, posterior branches may or may not fuse, diverticulae sometimes present; anterior branch sometimes anterior to brain; protonephridial system paired; testes, two or many; sperm biflagellated, flagella not incorporated into body of sperm, 9 + "1" microtubule arrangement; seminal vesicle present or absent; male copulatory structures complex; penis usually pyriform, usually unarmed, but sometimes armed with spines or stylet; prostate organ usually absent; one pair of small ovaries usually at anterior of body; female copulatory structures variable, with or without copulatory bursa; accessory genital organs may be present in form of extra bursae each with separate pore; vitellarium follicular, extensive; genitointestinal canal sometimes present; eggs ectolecithal; gonopore usually single, rarely two or three; temporary embryonic pharynx and intestine replaced by definitive pharynx during development to adult stage.

Habitat: Predominantly free-living in marine, euryhaline, freshwater, or moist terrestrial environments (suborders associated with habitats); one family known only from freshwater caves; some symbiotic with other invertebrates, for example, in mantle cavity of gastropods, or gill lamellae of horseshoe crabs, or on dorsal surface of skates.

10. Prolecithophora

See Fig. 11. With or without two ciliated grooves around anterior of body; anterior adhesive disc usually absent (present in Hypotrichinidae); extensive epidermal spicules present or absent; cilia uniform throughout all surfaces of body or restricted to ventral surface; brain encapsulated or not; statocyst lacking; one to three pairs of ocelli present or absent; mouth anterior, midventral or posterior; with simple, plicate or weakly bulbous pharynx; intestine in form of simple elongate sac, well defined (surrounded by tunic of connective tissue) in most; paired protonephridial system present; testes, one or two, compact or diffuse; sperm biflagellated, flagella not incorporated into body of sperm, mitochondria with elaborate outer cell membrane foldings; seminal vesicle in some; male copulatory stylet present or absent; ovary and vitellarium not fully separated in some, but separated in others; ovary compact or diffuse, paired or not; vitellarium diffuse; accessory female openings present or not; eggs ectolecithal; male pore opening anterior to female opening but into common atrium; combined mouth and gonopore (Combinata) or gonopore opening separate from mouth (Separata).

Habitat: Predominantly marine, some inhabiting freshwater; free-living, or symbiotic (commensal or parasitic) on gills, on body surfaces, in mantle cavity of gastropod and lamellibranch molluscs, or on surface of small crustaceans; rarely parasitic on anal and branchial regions of fish; some of these associations may be phoretic, with the worm using the host merely for transport.

11. Kalyptorhynchia

See Fig. 12. Body usually ciliated throughout; rarely with epidermal spicules of crystalline calcium carbonate; rarely with anterior girdle of highly vacuolated epidermal cells; anterior sheathed proboscis present or absent, divided or not, with or without hooks, gland cells, or specialized apex; lamellated rhabdites present in some; duogland adhesive system present; statocyst absent; ocelli present or absent; pharynx variable in position (anterior or posterior), bulbous, usually oriented ventrally, with or without sphincter or hard knobs, not associated with proboscis; gut in form of simple elongate sac; protonephridial system paired; testes single or paired, or in form of median row of follicles; sperm biflagellated, flagella not incorporated into body of sperm, 9 + "1" microtubule arrangement; internal seminal vesicle in copulatory bulb or not; copulatory organ with stylet or not; or three separately opening prostatic organs with their own spiny cirri, or one cirrus, or copulatory organ with no penial or cirrus structures; prostate vesicle incorporated in testis or not; prostate with spiny structures or not; ovary and vitellarium separated or not (ovo-vitellarium), paired or single; uterus generally present, with or without accessory piece; seminal receptacle present or absent, paired or not; bursa present or absent; eggs ectolecithal; usually 1 but sometimes 2 gonopores.

Habitat: All species free-living, primarily in marine, a few in euryhaline, several in freshwater environments.

12. Typhloplanida

See Fig. 13. Anterior usually modified with prominent rhabdoid tracts or rarely with an impermanent sheathed proboscis, or with a permanent terminal invagination with proboscis glands; dermal and adenal rhabdites present; duogland adhesive system present; cilia uniform throughout all surfaces of body or restricted to ventral surface; brain undivided, encapsulated; statocyst absent; paired ocelli present or absent; tubular spinous, buccal region occasionally present; mouth usually midventral or posterior; pharynx usually ventrally oriented, in form of simple tube with annular fold or bulbous; intestine in form of simple elongate sac; paired protonephridial system present; excretory pore opens into mouth or genital atrium; testes single or paired; sperm biflagellated, flagella not incorporated into body of sperm, 9 + "1" microtubule arrangement; single or paired seminal vesicles sometimes present; male system often with funnel-shaped stylet; sheathed, unarmed cirrus present; penis papilla sometimes present; ovary and vitellarium usually separate; ovary single, paired, or unpaired; vitellarium paired or a column of serially arranged follicles; usually with bursa copulatrix and seminal receptacle; vagina absent; eggs ectolecithal; male and female pores separate or combined, male anterior to female; common oro-genital pore present or absent.

Habitat: Species known from freshwater, terrestrial, and marine environments; mostly free-living; a few known from the body surfaces of polychaete annelids.

13. Dalyelliida

See Fig. 14. Lamellated rhabdites present in some; duogland adhesive system absent; cilia completely covering body, or just ventral; statocyst usually lacking; paired ocelli present or absent; mouth, pharynx, and intestine present or occasionally absent; pharynx bulbous, oriented anteriorly if present; intestine in form of simple elongate sac if present; paired protonephridial system present or occasionally absent; testes paired, compact or lobed; sperm biflagellated, flagella not incorporated into body of sperm, 9 + "1" microtubule arrangement; with or without male copulatory organ; if present male organ with or without stylet; ovary and vitellarium separate or occasionally combined; ovary single or paired, lobed; vitellarium single or paired, branched or unbranched; seminal receptacle present or absent; uterus present; copulatory bursa present or absent; single genital pore present or absent; at least one taxon is dioecious and sexually dimorphic (*Kronborgia*).

Habitat: Some species free-living in marine, freshwater, or terrestrial environments; many species commensals or parasitic: found, for example, in the connective tissue of tube feet of asteroid echinoderms, in the hemocoel of crabs, shrimp, amphipods, and isopods, in the mesenchyme of myzostomid annelids, in kidneys and gonoducts or mantle cavity and gut of gastropod molluscs, in stomach of lamellibranch molluscs, in mesenchyme of the turbellarian *Plagiostomum*, in either the coelom or the digestive tract of holothurioid, echinoid, or crinoid echinoderms; some also in sipunculans.

14. Temnocephalida

See Fig. 15. Usually with two to 12 anterior tentacles, tentacles usually adhesive; with posterior adhesive region, usually circular but occasionally crescentic or divided into two or more smaller regions, often pedunculate; some with lamellated rhabdites, if present restricted to anterior region of body; duogland adhesive system absent; locomotory cilia usually absent; outer body layer divided into a series of syncytial epidermal plates; statocyst absent; paired ocelli usually present; mouth terminal or ventral; pharynx tubular or bulbous, usually oriented anteriorly, sometimes reduced; intestine in form of simple sac or with weak lateral diverticulae; paired protonephridial system present, usually with two lateral excretory vesicles; with one to 10 pairs of testes; sperm biflagellated, flagella not incorporated into body of sperm, 9 + "1" microtubule arrangement; male organ usually armed, usually with eversible cirrus mounted on tubular or conical shaft; ovary single; vitellarium lateral or scattered in form of large follicles over intestine; eggs ectolecithal.

Habitat: Entirely freshwater; ectocommensal, primarily on external body surfaces, gills, or lining of branchial chambers of decapod crustaceans, occasionally found on amphipods, molluscs, turtles, and hemipterans.

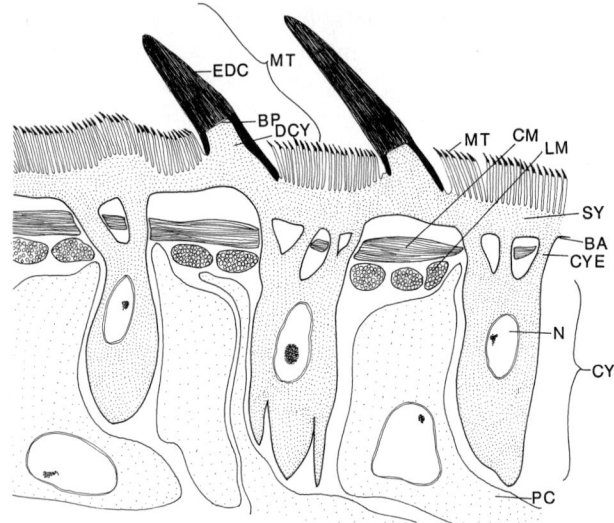

FIGURE 16 Neodermis of a cestode. (See abbreviations on pages 897–898.)

A. Neodermata

Ehlers championed the recognition of the remaining platyhelminth groups as the neodermatans because they share an outer body covering in the form of a neodermis, which is sometimes also referred to as a tegument (Fig. 16). Neodermatans generally begin life with a single-layered, cellular, ciliated epidermis, like most of the nonneodermatan platyhelminth groups, but as they mature they shed this outer epidermal layer and replace it with the neodermis. This transformation usually occurs when the first larval stage encounters its host. The cytoplasm of the neodermal cells is continuous throughout the surface of the animal, as there are no cell membranes separating adjacent cells; this layer is a syncytium. The nuclei of these cells are generally located below the basal lamina and outer muscle layers of the body, in structures called cytons or pericarya. These are, in fact, the nuclei of the stem cells, the cytoplasm of which grew outward to form the syncytial layer of the neodermis. The cytons are connected to the outer syncytial layer by thin cytoplasmic extensions.

The neodermatans share several other rather unique features. In these platyhelminths, the filters of the flame bulbs of the protonephridial system are in the form of a two-celled weir. Each filter is constructed from longitudinal cytoplasmic rods of the terminal cell that alternate with longitudinal cytoplasmic rods of the first canal cell. In addition, rather than possessing two ciliary rootlets, one directed anteriorly (the rostral rootlet) and one directed posteriorly (the caudal rootlet), the cilia of neodermatans appear to have lost the caudal rootlet, and possess only the anterior rootlet. Finally, the flagella of the sperm in neodermatans, rather than remaining separate from the cytoplasm of the main sperm body, are incorporated into the main sperm body, arranged in a proximal—distal direction along its lateral margins.

There are four major neodermatan groups: the Aspidogastrea and the Digenea, collectively referred to as the Trematoda, and the Monogenea and the Cestoda, collectively referred to as the Cercomeromorpha. The orders or superfamilies of each of these 4 groups are treated separately below.

15. Aspidogastrea

See Fig. 17. With ventral holdfast organ in form of linear series of 20-100 suckers, numerous transverse rugae, or extensive adhesive disc subdivided by septa into numerous loculi; disc with or without papillae; neodermis with microvilli in form of hemispherical microtubercles; buccal funnel at anterior end of body; oral sucker usually absent; oral lobes occasionally present; with one or two blind intestinal ceca; excretory vesicle present, V-shaped, excretory pore near posterior of body; with one, two, or many testes; sperm biflagellated, flagella incorporated into body of sperm, 9 + "1" microtubule arrangement; cirrus present; cirrus-sac present or absent; ovary single; vitellarium follicular or tubular, if follicular usually with multiple lateral follicles, rarely single median tube flanked by follicles, if tubular, two tubes; uterus generally long; Laurer's canal present or absent; Mehlis' gland present; eggs ectolecithal; single common genital pore.

Habitat: Adults endoparasites in kidneys, pericardial chamber, gill lamellae, or intestine of freshwater and occasionally marine lamellibranch molluscs, kidneys (renal cavities) and pericardial chamber of marine and freshwater gastropod molluscs, occasionally guts of turtles or freshwater and marine teleosts, gall bladder and bile ducts of elasmobranchs, or in lumen of rectal glands of Holocephali.

B. Class Digenea

Almost certainly monophyletic, the digeneans are most easily characterized by their possession of a unique series of larval stages including miracidia, sporocyst, redia, cercaria, and metacercaria. Variation in life cycles seen among digenean species suggests that perhaps only a subset of these ontogenetic stages should be considered to distinguish the class. Most digeneans possess an anterior sucker (or oral sucker), which is usually (Fig. 19), but not always (Fig. 24), associated with the mouth. Most also possess a second sucker (or acetabulum) that is usually midventral (Fig. 30), but occasionally posterior in position (Fig. 29). The body of some species is divided into two parts (Fig. 21). Other major variations in adult body form include a posterior telescoping ecsoma (Fig. 23), a sucker-like ventral adhesive tribocytic organ (Fig. 21), one or more anterior collars of large spines (Fig. 22), and paired anterior retractile proboscides. Most species possess an anterior mouth that opens into a bulbous pharynx and an intestine consisting of two or occasionally one blind ceca, with or without diverticulae. However, the ceca in some species are connected to one another posteriorly (Fig. 20) or open into the excretory bladder (Fig. 21). The excretory system generally consists of two sets of protonephridia, both of which empty into a common excretory vesicle. This vesicle is variable in shape among taxa and usually opens at the posterior end of the body. With the exception of the schistosomes (Fig. 31) and some didymozoids, all digeneans are hermaphroditic. The male intromittent organ is usually a cirrus, which may or may not be armed with spines, and may or may not be surrounded by a cirrus-sac. Testes are usually oval, but occasionally lobed or tubular; testes can be single, paired, or numerous. Sperm may be stored inside or outside the cirrus-sac in an internal or external seminal vesicle respectively; some taxa possess both vesicles. Most digeneans possess a single ovary that may be round, lobed, or tubular in form. Digeneans lack a vagina; instead, sperm enters the female system through the uterus, which opens, generally in combination with the male system, into a common genital pore. Some species possess a seminal receptacle in the form of a basal expansion of the uterus or as diverticulae of the Laurer's canal. A Laurer's canal, which may or may not open to the outside, may be present or absent. The function of the Laurer's canal is unknown, but it is considered by some to represent the remnant of a vagina. The vitellarium consists of columns of numerous follicles along both sides of the body in most taxa, but is occasionally present as one or a pair of compact follicles or vitelline chords. Almost all digeneans are endoparasites as adults in various organ systems of vertebrates.

With over 2500 nominal genera, the digeneans are by far the most diverse of the platyhelminth groups. Unlike all of the other major platyhelminth groups, family rather than order is the highest taxonomic category of general discussion and relative stability within the class. Unfortunately, the relationships among the

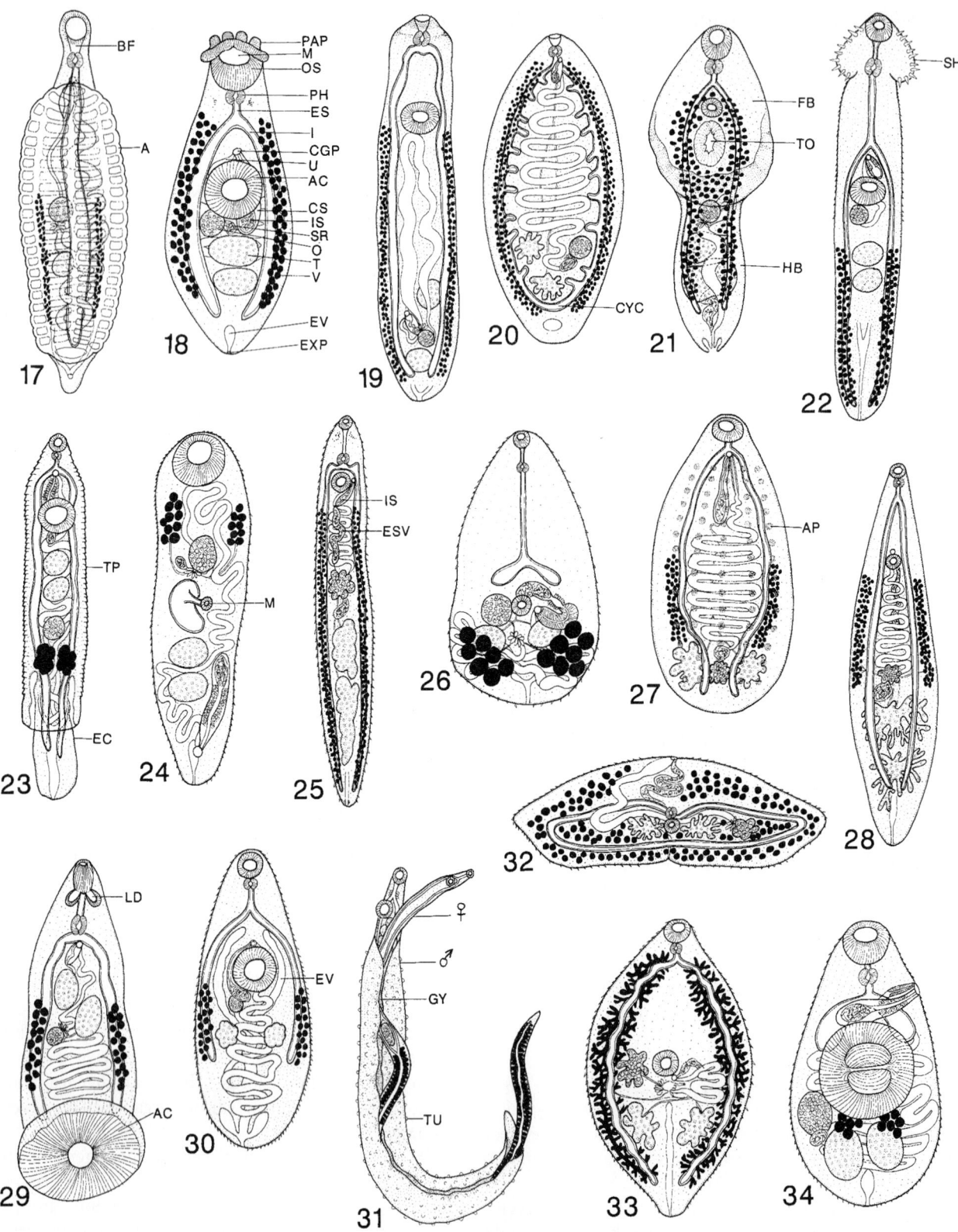

approximately 150 families remain poorly understood, and as a consequence, there is currently no generally accepted higher level classification scheme for the digeneans (for example, see Brooks et al., 1985; Pearson, 1992). As a consequence, this group is treated here following the superfamilies recently recognized by David Gibson and Rod Bray of The Natural History Museum in London. No tree of the relationships among these groups is presented; the superfamilies are treated below in alphabetical order. This scheme was selected primarily to demonstrate digenean diversity; the monophyly of many of these groups remains uncertain. It is important to note that the concepts of the superfamilies presented here are based on a combination of life history, adult morphology, and host association characters; thus as memberships in the groups change, so will the generalized characteristics and host generalizations for the groups. The fact that details of the life histories of many species remain unknown presents obvious problems with a classification scheme based to some extent on larval features. Not to be ignored is the diversity of hosts and sites within the hosts that these animals parasitize.

To make the following diagnoses more concise, unless mentioned it should be assumed that the superfamily possesses no specialized body regions or structures, both suckers are present, the mouth opens into the oral sucker, a muscular pharynx is present the gut consists of two blind intestinal ceca, the organisms are hermaphroditic, the testes are paired, the sperm are biflagellated and the flagella are incorporated into the body of the sperm, the flagella of the sperm possess a 9 + "1" arrangement of microtubules, a cirrus and cirrus sac are present, the ovary is single, the vitellarium is follicular and arranged in two lateral fields, there is a common genital pore, the eggs are ectolecithal and the epidermal cells of the miracidium are covered with cilia.

16. Allocreadioidea

See Fig. 18. Oral sucker with or without one or three pairs of muscular papillae; body surfaces usually inspined; cirrus sac usually present; genital pore opens near acetabulum. Larvae: Cotylocercous xiphidiocercaria, ophthalmoxiphidiocercaria, trichocercous or homalometrine cercaria; metacercaria usually encysts in invertebrate hosts, rarely on substrate.

Habitat: Adults parasitic in digestive tract and occasionally gall bladder of marine and freshwater teleosts, or occasionally gall bladder of freshwater turtles, or intestine of chameleons; some species progenetic in insects.

17. Clinostomoidea

See Fig. 19. Oral sucker weak, enveloped in collar-like anterior end of body, pharynx rudimentary or small; genital pore opens in posterior third of body. Larvae: Brevifurcate cercaria; metacercaria encysts on vertebrate hosts.

Habitat: Adults parasitic in mouth and esophagus of alligators and piscivorous birds.

18. Cyclocoeloidea

See Fig. 20. Oral sucker usually absent, occasionally present; vertral sucker absent or, if present, weakly developed; intestinal ceca usually fused posteriorly to form cyclocoel; genital pore opens near pharynx. Larvae: Cercaria usually tailless or with rudimentary tail; metacercaria encysts within sporocyst or redia in molluscan host.

Habitat: Adults parasitic in nasal sinuses, nasolacrimal sinus, infraorbital sinuses, respiratory tract, and occasionally coelom or intestine of birds.

19. Diplostomoidea

See Fig. 21. Body generally divided into anterior forebody and posterior hindbody; tribocytic organ present posterior to acetabulum; reproductive organs generally concentrated in hindbody; genital pore usually opens in posterior of body. Larvae: Cercaria variable; metacercaria generally encysts on vertebrate hosts, very rarely invertebrate hosts.

FIGURES 17–34 17 Aspidogastrea: adult of *Aspidogaster* sp.; 18 Allocreadioidea: adult of *Crepidostomum* sp.; 19 Clinostomoidea: adult of *Odhneriotrema* sp. (redrawn from Schell, 1985); 20 Cyclocoeloidea: adult of *Typhlocoelum* sp. (redrawn from Schell, 1985); 21 Diplostomoidea: adult of *Neodiplostomum* sp. (modified from Schell, 1985); 22 Echinostomatoidea: adult of *Echinochasmus* sp. (modified from Schell, 1985); 23 Hemiuroidea: adult of *Parahemiurus* sp. (modified from Schell, 1985); 24 Gymnophalloidea: adult of *Bucephalopsis* sp. (modified from Schell, 1985); 25 Lepocreadioidea: adult of *Echeneidocoelium* sp. (redrawn from Yamaguti, 1971); 26 Microphalloidea: adult of *Microphallus* sp. (modified from Schell, 1985); 27 Notocotyloidea: adult of *Quinqueserialis* sp. (redrawn from Schell, 1985); 28 Opisthorchioidea: adult of *Clonorchis* sp. (redrawn from Schell, 1985); 29 Paramphistomotoidea: adult of *Megalodiscus* sp. (modified from Schell, 1985); 30 Plagiorchioidea: adult of *Styphlotrema* sp. (redrawn from Schell, 1985); 31 Schistosomatoidea: male and female adults of *Schistosoma* sp.; 32 Transversotrematoidea: adult of *Transversotrema* sp. (redrawn from Yamaguti, 1971); 33 Troglotrematoidea: adult of *Paragonimus* sp. (modified from Schell, 1985); 34 Zoogonoidea: adult of *Brevicreadium* sp. (redrawn from Yamaguti, 1971). (See abbreviations on pages 897–898.)

Habitat: Adults parasitic in digestive system of birds and reptiles, especially crocodilians and snakes, sometimes amphibians, sometimes mammals including dogs, cats, dolphins, and some monotremes.

20. Echinostomatoidea

See Fig. 22. Spiny head collar often present; two anterior, retractile, spiny proboscides rarely present; body usually spined; ceca sometimes with lateral diverticula; hermaphroditic sac sometimes present, often with cirrus sac, cirrus occasionally absent; genital pore variable in position. Larvae: Cercaria echinostome-like, megalurous, gymnocephalous megaperid or haplosplanchnid; metacercaria usually encysts on aquatic vegetation, sometimes on vertebrates, rarely in invertebrates.

Habitat: Adults parasitic in the intestinal tract, gall bladder, bile duct, liver, lungs, nasal cavity, conjunctival sac, bursa fabrici, and orbit of vertebrates including a variety of marine and freshwater teleosts, birds, crocodiles, turtles, snakes, lizards, marine mammals including pinnepeds and cetaceans, and a variety of terrestrial herbivorous mammals.

21. Hemiuroidea

See Fig. 23. Body generally elongate; sometimes with posterior telescoping ecsoma; body surface aspinose but often with transverse plications; ceca sometimes fused to one another posteriorly, or fused posteriorly with excretory vesicle; usually hermaphroditic (monoecious) but one group (didymozoids) occasionally dioecious (if dioecious adult sexually dimorphic, male and female usually encysted in pairs); hermaphroditic duct almost always present; cirrus sac often absent; vitellarium often in form of rosette or tubular, or in two compact masses. Larvae: Miracidium spinous, lacking cilia; cercaria cystophorous; metacercaria usually unencysted in invertebrate or vertebrate hosts, occasionally within redia in molluscan host.

Habitat: Adults parasitic in digestive tract or swim bladder of freshwater and marine teleosts and rarely elasmobranchs; some associated with branchial cavities of marine teleosts and elasmobranchs; occasionally in mouth of frogs or stomach of sea snakes.

22. Gymnophalloidea

See Fig. 24. Mouth opening into oral sucker or ventral sucker (bucephalids); cirrus sac present or absent; cirrus sometimes absent; genital pore variable in position. Larvae: Cercaria often produced in branched sporocysts; cercaria often with rudimentary tail or tailless, often gasterostomous; metacercaria encysts on vertebrate hosts or within sporocyst in molluscan host; very rarely encysts in invertebrates.

Habitat: Adults parasitic in liver, bile duct, gall bladder, digestive tract of freshwater and marine teleosts, or large intestine, cloaca, bursa fabricius of birds, occasionally mammals (such as, for example, hedgehogs, rabbits, ruminants).

23. Lepocreadioidea

See Fig. 25. Spined or not; with or without circumoral spines; oral sucker occasionally with papilliform extensions; acetabulum ventral or rarely at posterior end of body; ceca open into excretory vesicle or not; one or usually two intestinal ceca present; with two to many testes; cirrus sac usually present, sometimes absent; hermaphroditic duct in some; external seminal vesicle in some; internal seminal vesicle bipartite in some. Larvae: Gymnophalous cercaria, produced in redia; metacercaria encysts on invertebrates, or on substrate, rarely in sporocyst or redia in molluscan host.

Habitat: Adults parasitic almost exclusively in digestive tract, rarely gall bladder or urinary bladder of marine teleosts; one record from lungs of reptile.

24. Microphalloidea

See Fig. 26. Body surfaces usually spined; oral sucker may be surrounded by semicircular collar of spines; gut lacking or poorly developed (with short ceca) in some; vitellarium usually in restricted symmetrical clusters of large follicles; genital atrium often with papilla. Larvae: Xiphiodiocercaria of ubiquita, microphallous, or virgulate type; metacercaria encysts on invertebrates or vertebrates, or occasionally in sporocyst in the molluscan host.

Habitat: Adults parasitic in intestine or ceca of birds most commonly, but also in digestive system of bats, snakes, varanid lizards, marine and freshwater teleost fishes, and mammals; occasionally bile ducts of mammals.

25. Notocotyloidea

See Fig. 27. Ventral sucker generally lacking (monostomes); often with adhesive papillae, ridges, or spines throughout ventral surface of body; head collar present or absent; male and female pores occasionally separate, pore(s) variable in position. Larvae: Cercaria monostomate, bi- or trioculate, with posterior adhesive glands at posterior of body; metacercaria encysts on substrate or on invertebrate hosts, rarely in sporocyst or redia within molluscan host.

Habitat: Adults parasitic in digestive system of marine turtles, marine iguanids, birds, mammals (including dugongs, cetaceans, bats, rodents, and insectivores), occasionally in oviducts of marine turtles or bursa fabrici of birds.

26. Opisthorchiodea

See Fig. 28. Body sometimes spined; oral sucker sometimes with ring of spines; cirrus sac usually lacking; hermaphroditic duct sometimes present opening into genital atrium; ventral sucker sometimes enclosed in ventrogenital sac, with or without gonotyl; seminal vesicle when present often bipartite. Larvae: Cercaria generally with ocelli, pleurolophocercous, parapleurolophocercous or gymnocephalus cercaria; metacercaria encysts on vertebrates, rarely invertebrates.

Habitat: Adults parasitic in stomach, intestine, occasionally bile ducts of marine teleost fishes, snakes, lizards, birds, and mammals; occasionally ovary of freshwater teleosts; occasionally progenetic encysted in musculature of frogs.

27. Paramphistomoidea

See Fig. 29. Usually unspined; oral sucker often with paired lateral diverticula, occasionally one unpaired diverticulum; acetabulum posterior or subterminal, sometimes with muscular lip, occasionally lacking; body occasionally with caudal appendages; usually with lymphatic system (usually with one to three longitudinal trunks); genital sucker sometimes present; ventral pouch sometimes present; cirrus sac present or absent; hermaphroditic sac in some. Larvae: Cercaria lacking acetabulum or with posterior acetabulum, develops from redia or branched sporocyst; metacercaria usually encysts on vegetation, but occasionally on frogs, metacercarial stage occasionally lacking.

Habitat: Adults parasitic in digestive tract of marine and freshwater teleost fishes, snakes, marine turtles, amphibians, birds and a wide diversity of marine and terrestrial mammals (including humans), or in lungs and trachea of turtles and tortoises.

28. Plagiorchioidea

See Fig. 30. (Perhaps most problematic of digenean superfamilies, clearly paraphyletic as currently circumscribed.) Body usually spined; oral sucker occasionally with pair of muscular lateral papillae; acetabulum rarely vestigial; fields of vitellarium usually restricted anteriorly and posteriorly; excretory vesicle Y-shaped. Larvae: Cercaria often xiphidioceraria of leptocercous, microcercous, cystocercus, armatao, ornatae, virgulate, ubiquita, or corcacriaeum type; often developing in daughter sporocysts in gastropod or lamellibranch molluscs; metacercaria encysts in terrestrial arthropods or gastropods.

Habitat: Adults parasitic in digestive system of mammals (especially bats), turtles, amphibians, reptiles, and occasionally fishes; liver and bile ducts of mammals, birds, and reptiles; cloaca and oviducts of birds and mammals; bursa fabricus of birds; urinary bladder of fishes and amphibians; coelom or pericardial chamber of elasmobranchs; occasionally progenetic in leeches and crustaceans.

29. Schistosomatoidea

See Fig. 31. Dioecious or monoecious, if dioecious male with ventral gynecophoral canal in which female resides for much of her life; surface of male body often with tubercles, with or without spines; pharynx usually lacking; both oral sucker and acetabulum present, or one or both lacking; ceca of gut often anastomosing at several points along length; one, two or many testes; genital pores separate or combined. Larvae: Cercaria apharynegeate, brevifurcate, sometimes also lophophorate; metacercarial stage lacking; cercaria develops into schistosomula; first host usually a mollusc but rarely an annelid.

Habitat: Adults parasitic in blood vessels and/or heart of freshwater and marine teleosts, elasmobranchs, holocephalans, usually freshwater (but sometimes marine) birds and mammals (including humans), crocodiles, and marine and freshwater turtles.

30. Transversotrematoidea

See Fig. 32. Body transversely elongated; spined; ceca united posteriorly to form transverse cyclocoel; excretory vesicle curved to right, receiving collecting vessels from both sides of body; cirrus absent; genital pore anterior. Larvae: Cercaria brevifurcate with pair of anterior lateral appendages; metacercarial stage lacking; cercaria infects definitive host cutaneously and develops directly into adult stage.

Habitat: Adults parasitic under scales of freshwater, euryhaline, and marine teleosts.

31. Troglotrematoidea

See Fig. 33. Body spined or not; oral sucker and pharynx sometimes weak; cirrus sac usually lacking; vitellarium follicular or often in two lateral dendritic clumps; genital pore anterior or posterior. Larvae: Microcercous xiphidiocercaria; metacercaria encysts on invertebrates or vertebrates.

Habitat: Adult parasitic in lungs of mammals (including humans) and frontal sinus, intestine and kidneys of some insectivores; also known from nasal fossa of foxes; abscesses in skin of cats; trachea and esophagus, or bursa fabric of passeriform birds.

32. Zoogonoidea

See Fig. 34. Body usually spined; 1–11 testes; cirrus sac usually present; cirrus armed with spines in some; vitelline follicles usually few, compact or often arranged in lateral bunches. Larvae: Cercaria chaetomicrocercous or tailless; metacercaria encysts in marine lamellibranchs, freshwater gastropods, turbellarians, aquatic annelids, or echinoderms; occasionally second intermediate host lacking.

Habitat: Adults parasitic in stomach or intestine of marine or freshwater fishes, or ovary of marine fishes, rarely coelom, gall bladder or urinary bladder of teleost fishes.

C. Class Monogenea

The Monogenea (or Monogenoidea) generally possess both anterior and posterior attachment structures. The form of the anterior attachment structure varies among groups, consisting of one or two suckers, pseudosuckers, or grooves. In some cases this region is equipped with gland cells. In the past this anterior structure has been referred to as a prohaptor. The posterior attachment structure also varies widely in morphology among taxa. This structure is generally now referred to as a haptor, although in the past the term opisthaptor has also been used. The haptor may be symmetrical or asymmetrical, may bear glands, suckers, and/or may be subdivided into numerous loculi. In addition, the haptor of most monogeneans is armed with a series of hardened structures greatly facilitating attachment to the host. The terminology of these structures is based to some extent on their morphology, but also to a large extent on when they appear in the course of development. For example, the small hooked structures that are usually found around the periphery of the haptor of the oncomiracidium and that persist into the adult stage of some monopisthocotyleans are termed hooklets or uncinuli. The larger, more centrally located hooked structures are generally termed hamuli or anchors, depending on whether or not they are present in the oncomiracidium. The more complex structures consisting of multiple sclerites, and which are found only in polyopisthocotyleans, are termed clamps. Adults of some, but not all, monogeneans possess one or more pairs of anterior ocelli, but monogeneans generally possess four rhabdomeric ocelli in at least one stage of their development. The mouth is almost always located in the anterior region of the body. Most monogeneans possess a con-

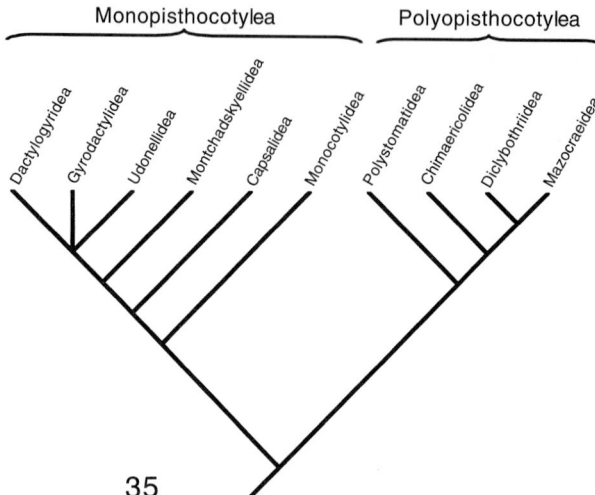

FIGURE 35 Hypothesized phylogenetic relationships among monogenean orders (after Boeger and Kritsky, 1993).

spicuous muscular pharynx and an intestine that is usually divided into two, rarely one, blind ceca, which are occasionally branched or anastomose posteriorly. All monogeneans are hermaphroditic. The male and female genital pores are usually combined. The number of testes ranges from one to many among different groups. Most groups possess a single ovary that varies somewhat in form from compact to double inverted U-shaped. The male intromittent organ can be a penis or a cirrus, but the exact form of this feature is not well known in some groups. The terminal genitalia are often complex. The excretory system consists of a pair of protonephridial systems that open separately to the outside, usually in the anterior of the body. Adult monogeneans are generally external parasites of cold-blooded vertebrates, but a few occur internally in the body cavity or very rarely the digestive system of their vertebrate hosts. The majority of the monogenean groups possess a ciliated larval stage known as an oncomiracidium, although the larvae of several groups are unciliated. The oncomiracidium bears a posterior haptor that develops into the haptor of the adult form.

The monogenean orders are treated below in sequence according to their position from left to right on the tree in Fig. 35. Membership in the families and even orders is quite stable although the monophyly of some orders remains uncertain.

To make the following diagnoses more concise, unless otherwise mentioned in the diagnosis of a group, it should be assumed that the order possesses a posterior, discoidal, undivided, symmetrical haptor, a conspicuous muscular pharynx, a gut consisting of two blind intestinal ceca, paired testes, biflagellated sperm, flagella incorporated into body of sperm, 9 + "1" microtubule arrangement, a single ovary; a vitellarium arranged in two extensive lateral fields of follicles, ectolecithal eggs, and a single, ventral genital pore.

33. Dactylogyridea

See Fig. 36. With two or more pairs of head organs, with or without cephalic glands; posterior half of body covered with anteriorly directed cuticular spines or not; haptor with one or two pairs of anchors supported by one to four transverse bars, with or without accessory plaques, occasionally with two suckers and median terminal anchor complex, usually 14 or 16 marginal hooklets; mouth subterminal, ceca fused posteriorly or not, with lateral diverticulae or not; testes, one to many; sperm with one axoneme; vas deferens encircling left caecum; seminal vesicle usually present; prostatic complex present; male copulatory organ sclerotized, elongated, muscular, ovary compact, lobed or tubular; vitellarium follicular or occasionally divided into tubular lobules in form of frond-like sprays spreading laterally in two or three groups (Protogyrodactylidae), coextensive with intestinal ceca; seminal receptacle and vagina present or absent; genito-intestinal canal absent; oviparous; eggs oval or tetrahedral; genital aperture median or marginal.

Habitat: Adults parasitic on gills and skin of marine and freshwater, occasionally euryhaline teleosts; rarely on skin of elasmobranchs.

34. Gyrodactylidea

See Fig. 37. With two head organs; haptor usually with one pair of anchors often supported by dorsal and ventral transverse bars with 16 marginal hooklets; mouth ventral, intestinal ceca, one or two, usually not fused posteriorly; testis single; sperm microtubules absent; vas deferens encircling left caecum; male copulatory organ sclerotized or muscular, spines present or absent; accessory piece present; vitellarium symmetrical, near posterior part of intestinal cecae; vagina absent or present, opening ventrally near left margin of body; genital pore marginal; many species viviparous; eggs oval or tetrahedral.

Habitat: Adults parasitic on gills of freshwater and marine teleosts, in stomach and intestine of amphibians, and gills of squid and marine crustaceans.

35. Udonellidea

See Fig. 38. Pair of head organs or pseudosuckers usually present; haptor unarmed in all developmental stages, often glandular, mouth terminal, intestine sacciform; with single testis; prostatic complex present; ovary compact; vagina absent; ciliated larval stage completely lacking. It should be noted that recent evidence suggests that the udonellids may represent an unusual group of gyrodactylideans.

Habitat: Adults parasitic on marine copepods that are themselves parasitic on marine teleosts or elasmobranchs.

36. Montchadskyellidea

See Fig. 39. Haptor with one pair of anchors lacking supporting bars, with 14 marginal hooklets; mouth subterminal, gut diverticulae present; testis single; male copulatory organ sclerotized; vas deferens encircling left caecum; accessory piece present; ovary encircling right caecum; one midventral vagina present; genital pores separate, opening on separate ventral papillae.

Habitat: Adults parasitic in gut of marine teleosts.

37. Capsalidea

See Fig. 40. Anterior end with paired petaloid head organs or paired glandular pseudosuckers; haptor sometimes subdivided by septa into central area and multiple peripheral loculi, with 14 marginal hooklets and two or four central anchors and two anterior sclerites; mouth ventral; intestinal ceca fused posteriorly, with lateral and axial diverticulae; testes, two to many; sperm microtubules absent; vas deferns forming bipartite external seminal vesicle and simple internal vesicle, or seminal vesicle absent; penis sac present or absent; penis present or absent; penis elongate, muscular, spines absent; prostatic complex present; vagina present or absent; seminal receptacle present or absent; genital apertures usually combined, marginal.

Habitat: Adults parasitic on skin or gills of marine teleosts (including remoras), or on skin or gills of elasmobranchs.

38. Monocotylidea

See Fig. 41. With single anterior sucker or several sucker-like depressions or paired head organs (lobes) with cephalic glands; haptor aseptate or divided by septa into loculi, often with one pair of anchors, with or without marginal papillae, usually with 14 marginal hooklets; ocelli present or absent; mouth ventral, intestinal ceca occasionally fused posteriorly, occasionally with lateral diverticulae; with 1 to several testes; sperm microtubules lying along 1/4 of cell periphery, with 2 axonemes 1 of which is reduced; male copulatory organ sclerotized; ovary tubular, encircling right intestinal

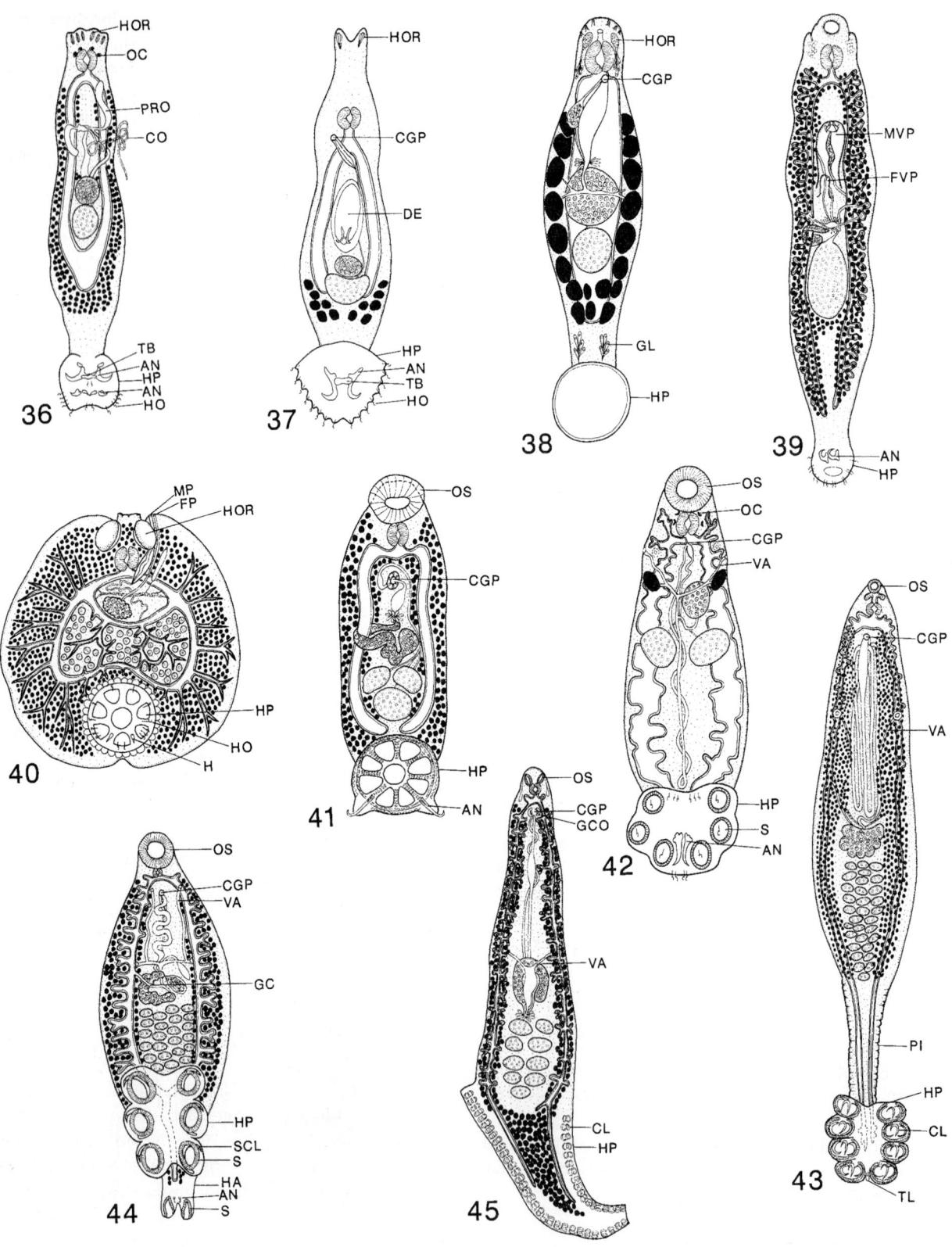

caecum; ventral vaginal pore sometimes present; genital pore usually median.

Habitat: Adults parasitic on skin, gills and sometimes nasal bulbs of elasmobranchs; occasionally in coelom, oviducts and cloaca of elasmobranchs; occasionally on skin of holocephalans.

39. Polystomatidea

See Fig. 42. With anterior muscular oral sucker; haptor with three but occasionally one pair of suckers, one to two pairs of hamuli usually present, accessory spines or spurs sometimes present, marginal hooklets often present, medial appendix-like prolongation sometimes present; ocelli usually absent; mouth terminal or subterminal; ceca occasionally fused posteriorly, often with lateral diverticulae; testes, one to many; seminal vesicle present; male intromittent organ usually armed with hooks or spines; genito-intestinal canal present; ovary compact; vitellarium rarely compact; uterus anterior or opposite ovary or reaching posterior of body; with two lateral vaginae.

Habitat: Adults parasitic in mouth, pharynx, esophagus, or urinary bladder of batrachian and anuran frogs; mouth, esophagus, nasal passages, or urinary bladder of turtles; urinary bladder of urodeles; eyes of hippopotamus; or skin and gills of batrachians.

40. Chimaericolidea

See Fig. 43. Single, weak circumoral sucker usually present; body with highly contractile, sometimes with very long, posterior peduncle; haptor with four pairs or four alternating rows of two short-stalked symmetrical clamps, and terminal lappet-bearing hooks; ceca with lateral diverticulae, extending posteriorly into haptor; with numerous testes; seminal vesicle present or absent; cirrus muscular, armed or not; prostatic complex present or absent; ovary lobed or branched; with two lateral vaginae opening into vitelline duct; seminal receptacle present or absent; uterus with numerous longitudinal loops, with or without lateral branches; genito-intestinal canal present.

Habitat: Adults parasitic on gills or in gill chamber of holocephalans.

41. Diclybothriidea

See Fig. 44. With anterior muscular, paired ventral bothria or oral sucker; haptor with three pairs of sessile suckers, each with single, large, curved, hooked sclerite, with medial, haptoral appendix bearing three pairs of hamuli or one pair of terminal suckers and two small anchors; mouth terminal or nearly so; ceca united posteriorly, with lateral diverticulae; testes numerous; male intromittent organ usually armed; ovary tubular, convoluted; with two lateral vaginae; uterus preóvarian; genito-intestinal canal present; eggs operculate, with or without polar filaments, occasionally forming chains.

Habitat: Adults parasitic on gills of sturgeon, paddlefish (chondrosteans) and elasmobranchs.

42. Mazocraeidea

See Fig. 45. With two anterior oral suckers, occasionally enlarged in form of discoid adhesive organ with conical papillae; haptor symmetrical or asymmetrical, set off from body by distinct peduncle or not, distinctly bilobed or not, usually with four to numerous pairs of clamps, without marginal hooklets, terminal lappet usually present; terminal lappet with one to three pairs of similar or dissimilar hook-like sclerites, bifid or not; testes numerous; copulatory apparatus consisting of sheaf of long spines or longitudinally striated sheath; male intromittent organ present or absent; armed or unarmed; ovary tubular, winding, convoluted, rarely compact or inverted U-shaped; vitellarium sometimes extending posteriorly into haptor; seminal receptacle present or absent; vagina present or absent, paired or not; vaginal aperture spined or not, conical sclerite in front of vaginal pore or not; eggs with two filaments; genito-intestinal canal present; genital pores common or separate, genital pore sometimes sucker-like; in rare cases (Diplozoidae) two adults permanently fused in form of letter X.

Habitat: Adults parasitic on gills and skin, or occasionally mouth cavity of marine, or sometimes freshwater teleosts; sometimes on crustaceans in mouth cavity of teleosts.

FIGURES 36–45 36 Dactylogyridea: adult of *Ancylodiscoides* sp. (modified from Yamaguti, 1963); 37 Gyrodactylidea: adult of *Gyrodactylus* sp. (modified from Yamaguti, 1963); 38 Udonellidea: adult of *Udonella* sp. (modified from Yamaguti, 1963); 39 Montchadskyellidea: adult of *Montchadskyella* sp. (modified from Bychowsky et al., 1970); 40 Capsalidea: adult of *Tristoma* sp. (modified from Yamaguti, 1963); 41 Monocotylidea: adult of *Monocotyle* sp. (modified from Yamaguti, 1963); 42 Polystomatidea: adult of *Neodiplorchis* sp. (modified from Yamaguti, 1963); 43 Chimaericolidea: adult of *Chimaericola* sp. (modified from Yamaguti, 1963); 44 Diclybothriidea: adult of *Erpocotyle* sp. (modified from Yamaguti, 1963); 45 Mazocraeidea: adult of *Heteraxinoides* sp. (modified from Yamaguti, 1963). (See abbreviations on pages 897–898.)

D. Class Cestoda

The cestodes, or tapeworms, conspicuously differ from almost all other platyhelminth groups in their lack of all organs of a digestive system, the only exceptions perhaps being the amphilinideans and gyrocotylideans, the anterior attachment organ of which is considered by some to represent a vestige of a digestive system. Thus, in general, the cestodes have no mouth, pharynx, esophagus, or intestine at any stage in their development. Rather, they absorb nutrients, in the form of very small molecules, directly through the outer layer of their body. There is evidence to suggest that neodermal extensions known as microtriches assist in this process. Microtriches appear to be unique to tapeworms. These structures typically consist of an outer electron dense cap composed of multiple microtubules, a baseplate, and a base consisting of an extension of the cytoplasm of the tegument (Fig. 1). The entire structure is surrounded by a plasma membrane. The form and height of the electron dense cap varies rather significantly among different tapeworm groups. These structures may also assist with attachment.

The majority of tapeworms have a body consisting of an attachment region called the scolex, followed by a series of proglottids collectively referred to as the strobila. Scolex morphology varies significantly among the various tapeworm groups; indeed, recognition of the major subgroups of tapeworms is based to a large extent on scolex form. The scolex is simple in form in some groups, but it is more commonly equipped with attachment structures such as hooks, suckers, or hooked tentacles. It is also common for the scolex to be subdivided into two flexible bothria or four muscular bothridia. Bothridia can be distinguished from bothria in their possession of discrete inner and outer membranes bounding their musculature, visible most readily in histological section.

The proglottids of most tapeworms are hermaphroditic, each containing one or more sets of male and female reproductive organs. A few groups of tapeworms (specifically the amphilinideans, gyrocotylideans, and caryophyllideans) lack proglottids and possess only a single set of male and female reproductive organs rather than multiple sets (i.e., they are monozoic rather than polyzoic). One group of tapeworms, the spathebothriideans, lack distinct proglottids but possess multiple sets of reproductive organs arranged in a linear series (i.e., they are nonproglottized but polyzoic). The majority of tapeworms, however, are proglottized, and each proglottid bears one or more sets of reproductive organs (i.e., they are proglottized and polyzoic). In proglottized

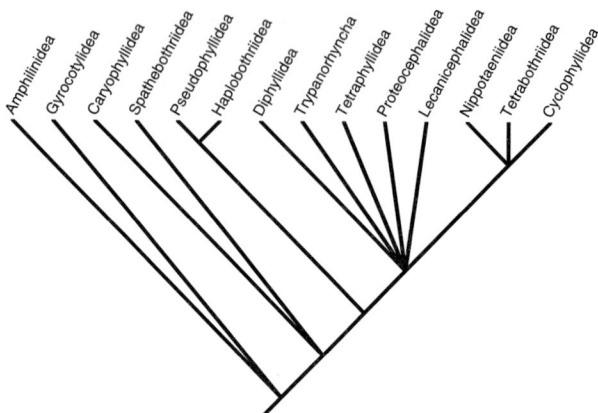

FIGURE 46 Hypothesized phylogenetic relationships among cestode orders; polytomies indicate uncertain relationships.

tapeworms, a germinitive region located directly behind the scolex consecutively produces the proglottids of the strobila. Thus, the proglottid nearest the scolex is the youngest, and that furthest from the scolex is the oldest. In some groups of tapeworms, the posterior-most proglottids, upon reaching a certain level of maturity, drop from the strobila (apolytic) and may continue their development independent of the main body of the worm. In other groups, the proglottids are retained on the strobila essentially throughout the life of the animal (anapolytic). In the former instance the proglottids may complete the production of eggs independent of the strobila; in the latter case proglottids on the strobila produce eggs that are released through uterine pores or tears. Proglottids filled with eggs are termed gravid. Consecutive proglottids either overlap (craspedote) or not (acraspedote). A very few species exhibit proglottids that are either male or female. Because these proglottids are found in the same strobila, individuals of these species are considered to be hermaphroditic, but the proglottids themselves are not. There is, however, evidence of dioecy in individuals of a few tapeworm species in which all of the proglottids of the strobila are either entirely male or entirely female.

With almost 650 recognized genera, the cestodes are the second most diverse group of neodermatans. The most recent comprehensive systematic treatment of the group is that of Khalil et al. (1994), in which 14 orders were recognized. The tree in Fig. 46 summarizes some of the ideas of cestode interordinal relationships that are more widely accepted. Membership in some orders remains, to a large extent, based on the vertebrate group parasitized by the adult worms, complicating ordinal circumscription.

The terms Cestoidea, Cestodaria, and Eucestoda are also sometimes used in reference to tapeworms. The term Cestoidea is synonymous with the term Cestoda; the terminology of Hyman (1951) and Grassé (1961) is followed here. In the past, the amphilinids and gyrocotylideans have been collectively referred to as the Cestodaria. Evidence against the monophyly of this group is mounting, and, thus, this term will not be used here. The term Eucestoda is used to collectively refer to the cestodes other than the amphilinids and gyrocotylideans. The eucestoda are distinguished from these groups, for example, by their possession of sperm that lack mitochondria.

To make the following diagnoses more concise, unless otherwise indicated in the diagnosis of a group, it should be assumed that the members of an order possess an anterior scolex of some form, lack all vestiges of a digestive system, and possess a strobila consisting of numerous proglottids that each contain one full set of both male and female reproductive organs (polyzoic). These proglottids contain numerous, compact testes filling much of the region anterior to the ovary, sperm lacking a mitochondrion, flagella that are incorporated into the main body of sperm, 9 + "1" microtubule arrangement in the flagella of the sperm, a cirrus armed with microtriches, a cirrus sac, a single posterior ovary, a vitellarium consisting of numerous follicles that extend in two lateral bands, one down each side of the proglottid, a Mehlis' gland located posterior to the ovarian bridge, a uterus that is sacciform, and male and female systems that open through a combined lateral genital pore, with pores of consecutive proglottids alternating irregularly down the length of the strobila.

43. Amphilinidea

See Fig. 47. Distinct scolex absent but with anterior coniform "sucker"; one set of male and female reproductive organs (monozoic) present; testes in two lateral bands; sperm with mitochondrion; cirrus sac absent, ovary varying in shape; uterine pore anterior, uterus often N-shaped and/or sinuous; seminal receptacle present; genital pores usually separate, posterior.

Habitat: Adults parasitic in body cavity of chondrosteans, some freshwater and marine teleosts, and Australian freshwater turtles.

44. Gyrocotylidea

See Fig. 48. Distinct scolex absent but with anterior "sucker," rosette usually present at posterior end of body; body usually "spined" (spines are likely to be microtriches), "spines" limited in distribution or not; excretory system reticulate; one set of male and female reproductive organs (monozoic) present; sperm with mitochondrion; cirrus sac absent; ovary follicular, in V- or U-shaped band around seminal receptacle; seminal receptacle present; uterus sinuous, in median region of anterior part of body, terminal portion expanded into sac, opening into ventral uterine pore; genital pores separate, both anterior.

Habitat: Adults parasitic in spiral intestine of holocephalans.

45. Caryophyllidea

See Fig. 49. Scolex usually with one to three pairs of weakly developed shallow depressions or grooves (considered to be loculi if temporary, bothria if more permanent), occasionally with terminal introvert or disc; with one set of male and female reproductive organs (monozoic); ovary H- or dumbbell-shaped; vitellarium usually preovarian, occasionally postovarian follicles also present; seminal receptacle present or absent; uterus often sinuous; female pore a combined utero-vaginal pore; genital pores separate or combined, median, ventral.

Habitat: Adults parasitic in intestine of siluriform and cypriniform freshwater fishes.

46. Spathebothriidea

See Fig. 50. Scolex usually bearing one or two sucker-like attachment organs, but occasionally absent; osmoregulatory system in form of canals on each side of strobila, with irregular commissures and lateral pores; body proglottization absent; with multiple sets of male and female reproductive organs serially repeated down body (polyzoic); testes in two irregular lateral bands that extend throughout the length of the body; ovary rosette-shaped or bilobed, median, posterior to genital pores; small seminal receptacle occasionally present; uterus often sinuous; male and female pores separate, opening close to one another along median line, usually irregularly alternating in adjacent proglottids between opening dorsally and ventrally.

Habitat: Adults parasitic in intestine of freshwater, euryhaline and marine chondrosteans and teleosts.

47. Pseudophyllidea

See Fig. 51. Scolex usually with two bothria (one dorsal and one ventral), margins of bothria occasionally crenulated or rolled, each bothrium occasionally armed with one pair of hooks or apical series of hooks, occasionally with posterior appendages, occasionally divided into two or several vertical loculi, with or without muscular apical disc, scolex occasionally replaced by scolex deformatus or pseudoscolex, scolex occasionally lacking; polyzoic, proglottization usually complete, proglottids

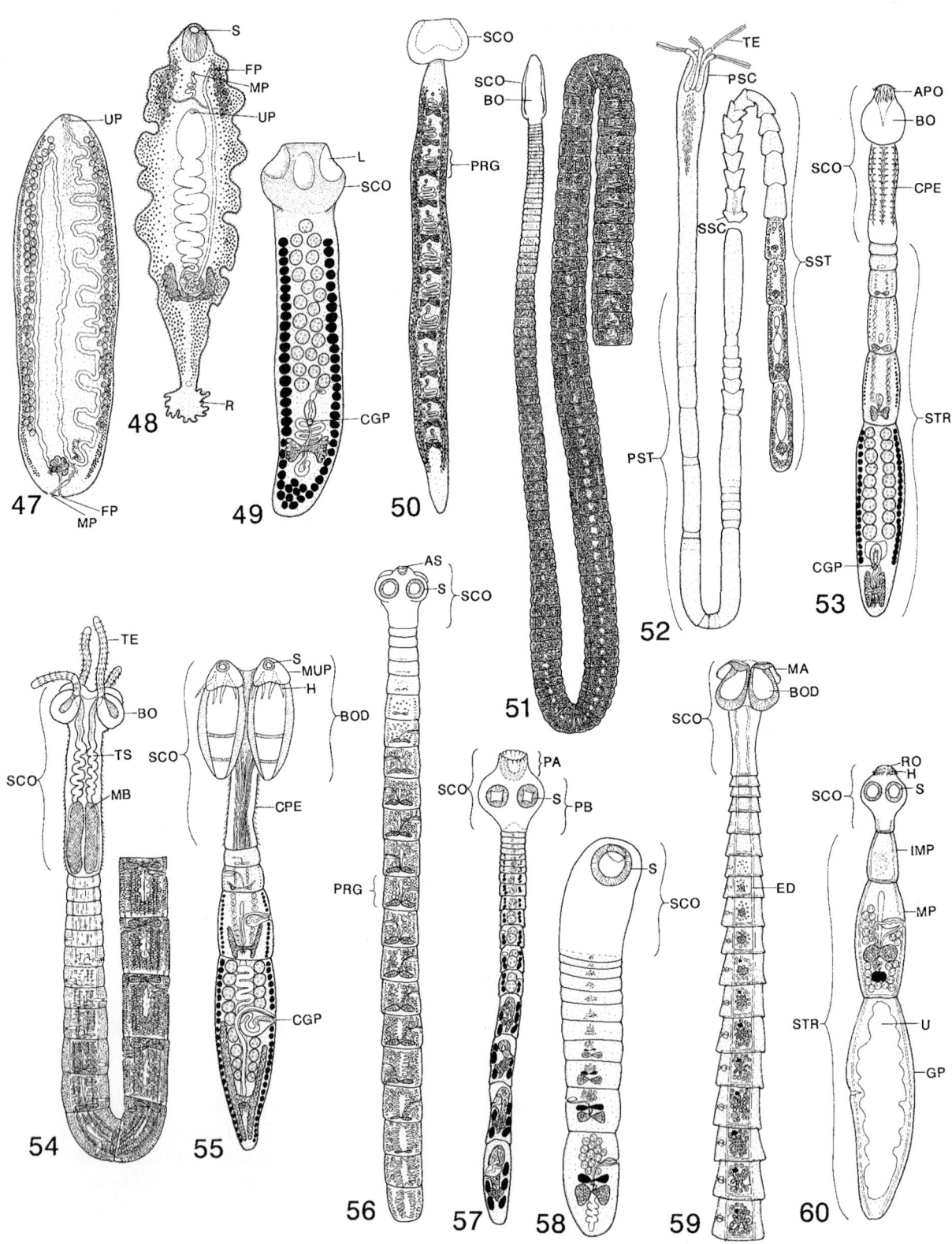

craspedote or acraspedote, generally anapolytic; usually with one set of male and female reproductive organs per proglottid, but occasionally with up to eight sets per proglottid; testes usually arranged in two lateral bands throughout length of proglottid; cirrus-sac occasionally absent; cirrus with or without microtriches; ovary compact or dendritic, vitellarium rarely in single compact postovarian mass, vitelline follicles lateral or encircling proglottid; uterus tubular or sacciform with lateral diverticulae, coiled or not; preformed ventral uterine pore present; genital pores usually common (one for each set of reproductive organs per proglottid), usually median, occasionally marginal.

Habitat: Adults parasitic in intestines of freshwater and marine fishes, sometimes piscivorous mammals (such as for example, bears, seals, cetaceans, felids), piscivorous birds, rarely varanid lizards, boid snakes, anurans and caudates.

48. Haplobothriidea

See Fig. 52. Body bearing primary and secondary scoleces and strobila; primary scolex with four tentacles with spiniform microtriches at bases, each tentacle with muscular sac into which it can be withdrawn, bothria, bothridia and suckers absent; primarily strobila proglottized, but primary proglottids never develop reproductive organs on primary strobila; proglottized regions of primary strobila separate from remainder of primary strobila and develop into secondary strobilae that produce proglottids, each proglottid bearing one set of reproductive organs; anterior-most proglottid of each secondary strobila modified as secondary scolex; secondary scolex with four shallow anterior indentations surrounding central, apical dome; cirrus-sac median, anterior, cirrus armed with microtriches; testes numerous, arranged in two lateral fields; ovary inverted U-shaped; vagina posterior to cirrus-sac; seminal receptacle present; follicles of vitellarium arranged in two lateral fields, confluent in midline at anterior and posterior regions of proglottid; uterus with dilated uterine sac; preformed uterine pore present or absent; genital pores separate, median, anterior; eggs operculate.

Habitat: Adults parasitic in intestine of the bowfin *Amia calva*.

49. Diphyllidea

See Fig. 53. Scolex with two bothria, proximal surfaces of bothria occasionally with spines; scolex with apical organ; apical organ usually armed with one row of large hooks on dorsal and ventral surfaces, often with row of smaller hooks on lateral margins, occasionally with small spines; bothria supported by cephalic peduncle; cephalic peduncle often armed with multiple columns of straight hooks with trifid bases; proglottids acraspedote; generally apolytic; testes usually filling proglottid anterior to ovary; ovary bilobed; lateral bands of follicles of vitellarium occasionally extending medially; vagina posterior to cirrus-sac; uterine pore lacking; common genital pore median, ventral.

Habitat: Adults parasitic in spiral intestine of rays, skates, and sharks of families Hemiscyllidae, Scyliorhinidae, and Triakidae (all marine).

50. Trypanorhyncha

See Fig. 54. Scolex with four retractable, hollow tentacles each armed with numerous spiral rows of hooks (hook patterns of enormous taxonomic importance); each tentacle everted and retracted by rhyncheal apparatus consisting of tentacle sheath, and retractor muscle originating within posterior muscular bulb (tentacles and rhyncheal apparatus rarely absent); scolex with two bothria and elongated peduncle housing rhyncheal apparatus; proglottids craspedote or acraspedote, apolytic or anapolytic; usually with one set of male and female reproductive organs per segment, but occasionally with two sets; cirrus-sac or hermaphroditic-sac usually present; external, internal and accessory seminal vesicles present or absent; ovary bi-, tetra-, or multi-lobed in cross-section; follicles of vitellarium lateral or completely encircling proglottid; vagina usually posterior to cirrus-sac, occasionally dorsal or ventral to cirrus-sac; uterus tubular or branched, with or without preformed uterine pore; common genital pore often associated with muscular lips.

FIGURES 47–60 47 Amphilinidea: adult of *Amphilinia* sp. (redrawn from Gibson in Khalil *et al.*, 1994); 48 Gyrocotylidea: adult of *Gyrocotyle* sp. (modified from Gibson in Khalil *et al.*, 1994); 49 Caryophyllidea: adult of *Paraglaridacris* sp. (after Mackiewicz in Khalil *et al.*, 1994); 50 Spathebothriidea: adult of *Spathebothrium* sp.; 51 Pseudophyllidea: adult of *Diphyllobothrium* sp.; 52 Haplobothriidea: adult of *Haplobothrium* sp.; 53 Diphyllidea: adult of *Echinobothrium* sp.; 54 Trypanorhyncha: adult of *Pterobothrium* sp.; 55 Tetraphyllidea: adult of *Acanthobothrium* sp.; 56 Proteocephalidea: adult of *Proteocaphalus* sp.; 57 Lecanicephalidea: adult of *Corrugatocephalium* sp.; 58 Nippotaeniidea: adult of *Nippotaenia* sp. (modified from Bray in Khalil *et al.*, 1994); 59 Tetrabothriidea: adult of *Tetrabothrius* sp.; 60 Cyclophyllidea: adult of *Echinococcus* sp. (See abbreviations on pages 897–898.)

Habitat: Adults parasitic in spiral intestine, stomach, rarely gall bladder, of elasmobranchs or very rarely holocephalans.

51. Tetraphyllidea

See Fig. 55. Scolex usually divided into four muscular sessile or pedunculated bothridia of a variety of forms, bothridia occasionally absent; bothridia with or without apical muscular pad with or without one or three accessory suckers, with or without one or two pairs of uni-, bi-, or tripronged hooks; bothridia may or may not be further subdivided into loculi, metascolex occasionally present; proglottids craspedote or acraspedote, generally apolytic; rarely with proglottids that are either male or female (or female proglottids retain male copulatory apparatus); usually with one strobila per scolex, rarely with multiple strobilae per scolex; ovary H-, dumbbell-, A-, or inverted U-shaped; vagina anterior to cirrus-sac, crossing male duct (vas deferens); follicles of vitellarium sometimes completely encircling proglottid, rarely condensed around ovary; uterine pores sometimes present or uterus opening through median slits in uterine wall; genital pores rarely separate; genital pores rarely unilateral along length of strobila.

Habitat: Adults parasitic in spiral intestine of marine and freshwater elasmobranchs.

52. Proteocephalidea

See Fig. 56. Scolex usually with four round suckers or bothridia; suckers or bothridia uni-, bi-, tri-, or tetraloculate; apex of scolex sometimes with armed rostellum, or large sucker, or occasionally with apical organ in form of two lappets; metascolex sometimes present as wrinkled region posterior to suckers that may or may not cover suckers of scolex; scolex with or without gland cells; proglottids craspedote or acraspedote, usually anapolytic; longitudinal muscle bundles usually conspicuous, occasionally inconspicuous; ovary usually dumbbell-shaped in dorso-ventral view; uterus usually with lateral branches (diverticulae) when gravid, opening through one or more longitudinal apertures; vagina usually anterior to, but occasionally posterior to, cirrus-sac; genital pores rarely separate; distinction between cortex and medulla usually marked.

Habitat: Adults parasitic in intestine of freshwater teleosts, occasionally in amphibians, terrestrial lizards (such as, for example, *Varanus*, monitor lizards) and snakes.

53. Lecanicephalidea

See Fig. 57. Scolex usually divided into two parts: an anterior region (pars apicalis) quite variable in form, elongate (myzorhynchus), pad-like (metaporhynchus), sucker-like, or consisting of a crown of nonhooked tentacles; posterior region (pars basalis) usually cushion-like, often bearing four circular suckers or bothridia; proglottids craspedote or acraspedote, usually apolytic; testes, three to many, arranged in a single or multiple columns; external seminal vesicle usually present, often conspicuous, extending length of proglottid; ovary bi-lobed, multi-lobed or irregular in dorso-ventral view; follicles of vitellarium sometimes encircling proglottid; vagina almost always posterior to cirrus-sac.

Habitat: Adults parasitic in spiral intestine of marine elasmobranchs, primarily rays.

54. Nippotaeniidea

See Fig. 58. Scolex in form of single apical sucker; proglottids acraspedote, anapolytic or apolytic; testes restricted to previtelline medulla of proglottid; cirrus-sac thin-walled; convoluted ejaculatory duct present; seminal vesicle present or absent; ovary dumbbell-shaped in dorso-ventral view, vitellarium consisting of two symmetrical compact masses, anterior to ovary; uterus in form of transverse coils in medulla; vagina posterior to cirrus-sac.

Habitat: Adults parasitic in digestive system of freshwater teleosts.

55. Tetrabothriidea

See Fig. 59. Scolex usually with four muscular bothridia; bothridia round, rectangular or occasionally triangular in form, often with auriculate muscular appendages; auricles sometimes fused to form apical "organ"; apical rostellum present or absent; proglottids craspedote; anapolytic; testes, few to many; ovary bi-lobed or highly lobed in dorso-ventral view; vitellarium in form of single compact mass, usually anterior to ovarian bridge; uterus transverse, with one or more dorsal pores; seminal receptacle present or absent; vagina occasionally armed with "spines"; genital pores combined or not; genital atrium often muscular, sometimes with distinct male atrial canal, or genital papilla, occasionally with "spinest"; genital pores usually unilateral along length of strobila; vagina usually posterior to cirrus-sac.

Habitat: Adults parasitic in intestine of seabirds, pinnipeds or cetaceans.

56. Cyclophyllidea

See Fig. 60. Scolex usually with four round suckers; suckers armed with spines or not; often with apical organ in form of single sucker (rarely four apical suckers), some with apical, often protrusible rostellum; rostellum armed with one or two rings of hooks or not, with or without rostellar pouch; scolex divided into scolex and metascolex, or with pseudoscolex in some; proglottids craspedote or acraspedote; apolytic or anapolytic; usually with one but rarely two strobilae per scolex; usually with one but occasionally two sets of male and female reproductive organs per proglottid; strobila rarely with proglottids that are entirely either male or female, or male and female proglottids that alternate regularly along strobila; testes, one to many, usually anterior to ovary, occasionally posterior to ovary; cirrus-sac usually present; cirrus usually armed with spiniform microtriches, occasionally with stylet, occasionally with accessory sac; external and internal seminal vesicles present or absent; ovary dumbbell-shaped, or asymmetrical in dorso-ventral view; vitellarium usually single compact mass, occasionally bilobed mass, usually posterior to ovary; vagina present or absent, replaced functionally by cirrus-sac and modified male duct, or replaced by supplementary ducts running between seminal receptacles in consecutive proglottids in some taxa; uterus ring-shaped, or saccate, with or without lateral branches (diverticulae), or reticular, transverse in some, replaced by parauterine organ(s) in some, replaced by uterine capsules or parenchymatous egg capsules in some, uterus single or double, usually with preformed uterine pore; common genital pore rarely midventral; genital pores occasionally regularly alternating or unilateral along strobila; genital atrium conspicuous or not, with or without prominent radial musculature; vagina usually posterior to cirrus-sac, occasionally dorsal to ventral to cirrus-sac.

Habitat: Adults parasitic in intestine of amphibians, snakes, lizards, birds or mammals.

III. NERVOUS SYSTEM

A bewildering variety of neural organization and complexity characterizes the platyhelminths (for example, see Halton and Gustafsson, 1996). The basic plan involves an apical subepidermal brain consisting of multiple neurons, which can be uni-, bi- (especially in Neodermata), or multilobed. This is a true brain because it controls reflexes in the peripheral nerve net. The number of nerve cells (neurons) in the brain varies between 50 and 550 in free-living species, but this number is less well known for the neodermatans. The brain is generally connected to two, or sometimes more, main longitudinal nerve cords, each consisting of axons with cell bodies distributed at irregular intervals along the length of the cords. These main nerve cords are usually connected to one another by numerous transverse commissures in a ladder-like arrangement. In addition, a peripheral array of nerve plexuses (or nerve-nets) is found throughout the body. The attachment organs and the components of the female reproductive system involved in egg formation are especially well supplied with nerve plexuses, but plexuses are also found associated with the pharynx and intestine and below the surface and below the muscle layers of the body. There are separate plexuses for sensory, integrative and motor activities. The central nervous system (brain plus main cords) is more emphasized in neodermatan groups than in non-neodermatan groups. Cestodes may exhibit regional differences in the complexity of the subtegumental plexuses along the length of the body. Transmission electron microscopy suggests that platyhelminths exhibit an extensive diversity of types of sensory receptors.

Several different neuronal cell types are present, including unipolar, bipolar, and multipolar cells. Whereas unipolar neurons are generally confined to the ganglia of the brain and/or longitudinal nerve cords, cells of plexuses are dominated by bipolar cells. Individual neurons are highly secretory and contain a variety of types of vesicles in their cytoplasm. Platyhelminths utilize a diversity of signaling molecules and at least two different groups of neuropeptides. The nervous systems of platyhelminths appear to possess compounds similar to all of the major neurotransmitters known in taxa as complex as vertebrates. In addition, there are two groups of neuropeptides that appear to be unique to flatworms: neuropeptide F and FMRPamide-related peptides (FaRPs) both of which are fairly extensively distributed among the various platyhelminth groups. The existence of glial or glial-like cells in platyhelminths remains somewhat controversial.

IV. FEEDING AND DIGESTION

Other than acoels, a few dalyelloids, and the cestodes, flatworms generally have a digestive system divisible into a foregut and a cecum. The foregut leads from the mouth to the blind-ended sac-like caecum that is lined with a thin gastrodermis. Free-living species tend to be microphagous or predatory and utilize bacteria, unicellular alga, protists, and/or almost all types of invertebrates as sources of nutrition. Characteristically, the free-living species have a highly muscular, bulbous, suctorial pharynx. Depending on the species, the pharynx can be protruded, everted, inserted, applied to, extended over, or used to envelop and swallow whole the various foodstuffs they actively seek or encounter. Food passes from the mouth via the pharynx to the esophagus and then to the caecum. The gastrodermis has glandular and phagocytic cells that digest and incorporate the nutrients.

Symbiotic and parasitic platyhelminths vary their feeding habits according to the hosts on or in which they live. Groups that possess suckers may have numerous secretory cells and sensory structures surrounding the mouth and sucker. Although similar in basic design to the foregut of free-living groups, the parasitic groups tend to exhibit distinct cell type configurations in the gastrodermis. In these groups, the junction between the foregut and caecum is invariably abrupt. Aspidogastreans have a single cell type in their gastrodermis that alternates cyclically between a secretory-absorptive phase and an autophagic-exocytotic phase. Digeneans have a syncytial or cellular gastrodermis with only one cell type, and digestion takes place extracellularly in the lumen of the caecum. There is evidence, however, that some digeneans appear capable of acquiring nutrients directly through microvillar extensions of their body surfaces. In contrast, digestion in the caecum of monogeneans is intracellular. The polyopisthocotylean monogeneans tend to be blood-feeders. The nonpigmented cells in these taxa are thought to support and protect the pigmented digestive cells in the gastrodermis. In addition, at least some of these taxa possess a bucco-esophageal canal between the buccal cavity and the esophagus that permits regurgitation of intestinal contents. The monopisthocotyleans tend to feed on the epidermal cells and mucus of their host and have only a single gastrodermal cell type involved with endocytosis and digestion.

In the Cestoda and some parasitic dalyelliids (for example, *Fecampia* and *Kronborgia*) all vestiges of a digestive system have been lost. These species rely solely on acquiring nutrients through their outer body wall. Among the Cestoda, and indeed all other neodermatans, the neodermis plays an active role in the biochemical transport of nutrients. In cestodes the surface area of the neodermis is markedly increased by the presence of microtriches, which enhance the availability of readily available nutrients to be actively absorbed into the parasite's body. Acoels have no gut lumen and digestion takes place in temporary vacuoles within the syncytial endoderm. Some acoels harbor Zooxanthellae and subsequently derive nutrients from the products of photosynthesis of these symbionts.

V. EXCRETION AND OSMOREGULATION

The protonephridial system is thought to function primarily in osmoregulation, whereas the removal of metabolic wastes generally occurs by way of diffusion through the outer body layer. However, removal of excess water can also be achieved through the gut, and in free-living species it is not uncommon to find excretory products stored in various tissues. Each protonephridial system is composed of a series of collecting ducts and ciliated flame bulbs. The beating of the cilia in the flame bulbs and collecting ducts directs and drives body fluids through filters formed by the terminal cells and/or the distal canal cells (of various forms depending on the taxon) into minor collecting ducts or canals, which empty into one or more major canals (nephridioducts) that either open to the outside through a nephridiopore or empty into an excretory vesicle that subsequently opens to the outside through a nephridiopore. This system is present in all platyhelminths except the acoelomorphs and a few catenulids in which excretion is thought to take place solely through the digestive syncytium. The protonephridial system is most developed in freshwater species. Indeed, this system is considered to have played a major role in allowing platyhelminths to invade freshwater habitats.

The protonephridial system of most taxa is paired, consisting of two nephridioducts, one running along the lateral margin of each side of the body; these nephridioducts connect to many flame bulbs throughout their length. Some of the catenulids are exceptional in their possession of a single dorsal nephridioduct that opens through a single posterior nephridiopore; some proseriatans also possess a single system, but some possess three protonephridial systems. In most of the non-neodermatan groups, the two nephridioducts open to the outside of the body through separate nephridiopores that are found in the anterior half of the body, generally on the ventral surface. In most temnocephalideans, the two nephridioducts open into separate saccate excretory vesicles, located in the anterior of the body on either side of the pharynx, and the left and right protonephridial systems are generally connected by two transverse ducts, one at the anterior and one at the posterior aspect of the system.

The aspidogastreans possess a protonephridial system much like that of the temnocephalans but without the transverse ducts. The excretory system of digeneans differs somewhat from that of the aspidogastreans in that the two nephridioducts generally open into a common posterior excretory vesicle. This vesicle takes on a number of different forms in different taxa; for example, it can be V-, Y-, or I-shaped. Unlike other platyhelminth taxa, the arrangement of the flame bulbs relative to the nephridioducts is regular and consistent in the cercarial stage of digeneans among individuals of the same species, but differs among individuals of different taxa, and thus this "flame cell formula" is often of significant taxonomic value. The larger number of flame bulbs and the difficulty of seeing most of the elements of the protonephridial system in the more robust bodies of the adult digeneans reduce the taxonomic utility of the flame bulb arrangement in adults.

The protonephridial system of monogeneans is also similar to that of temnocephalans, but the anterior transverse duct is absent in the polyopisthocotyleans. The amphilinideans possess an excretory system consisting of a network of collecting ducts, rather than just two lateral nephridioducts. This network empties into an elongated excretory vesicle that opens to the outside through a posterior nephridiopore. The flame bulbs are usually found in pairs, grouped in clusters throughout the length of the body. The gyrocotylidean system also includes two longitudinal nephridioducts that are connected to one another by a number of secondary canals. The main nephridioducts open into a nephridiopore on the dorsal surface in the anterior half of the body. In most of the other cestode groups the protonephridial system consists of two dorsal and two ventral longitudinal nephridioducts, the ventral ducts generally being larger in diameter than the dorsal ducts; the nephridioducts are connected to multiple flame bulbs throughout the length of the body. Some cestodes do, however, have up to 12 longitudinal ducts. The pairs of dorsal and ventral nephridioducts are connected to one another in the scolex, where they may undergo extensive winding. The left and right nephridioducts are often connected to one another by way of transverse excretory ducts located in the posterior end of each proglottid; this condition is most commonly found in anapolytic eucestode species. In some groups, the longitudinal vessels are connected to one another with numerous lateral nephridioducts, resulting in a network-like arrangement of vessels. In the first proglottid formed (located at the posterior end of the strobila) the ventral pair of nephridioducts often empty into a small vesicle at the posterior end of this proglottid. However, once this proglottid drops from the strobila, the ventral vessels open to the outside separately in the posterior end of the terminal proglottid of the strobila. Rohde (1991) provides a useful overview of the evolution of the protonephridial system in platyhelminths.

VI. REPRODUCTION

All platyhelminth species undergo sexual reproduction. Given the predominance of hermaphroditism in this phylum, there is some question about the relative frequency and importance of self- versus cross-fertilization in most species. In the few taxa in which this issue has been investigated in detail, cross- rather than self-fertilization appears to be the dominant form of sexual reproduction. Dioecy does, however, occur in some acoels, in some triclads, in some lecithoepiteliatans, in a few dalyelloids (for example, *Kronborgia*), within the digenean family Schistosomatidae, and in a few tetraphyllidean and cyclophyllidean cestodes. To date, no instance of dioecy in aspidogastreans or monogeneans are known. Most platyhelminths are simultaneous hermaphrodites, each individual possessing a full complement of male and female reproductive systems. There is a general tendency for the male system to mature before that of the female (protandry). Sperm transfer usually occurs during copulation. But, in a few species of triclads, polyclads, rhabdocoels, and cestodes, sperm is injected directly through the body wall and then migrates through the body to the female reproductive organs. Fertilization is always internal.

Asexual multiplication in one form or another is also found in a number of platyhelminths, either in the adult or, most commonly, in one or more of the larval stages in neodermatans. Adults of many of the catenulids, acoels, and macrostomids undergo forms of asexual multiplication either as paratomy (regeneration before fission) or as architomy (regeneration after fission). These phenomena result in a linear chain of individuals that eventually separate and each develop into a mature worm.

Some adult triclads possess the ability to regenerate complete individuals from, in some cases, a very small portion of the body of the original individual. For example, it is possible to divide an adult freshwater triclad posterior to the pharynx and produce two complete adult individuals. More dramatic is the ability of adults of species belonging to the triclad genus *Phagocata* to divide into a number of separate fragments, each of which subsequently encysts and develops into a complete adult worm. Their ability to regenerate has made freshwater triclads favored animals for studies in wound healing and developmental genetics.

With the exception of the aspidogastreans, asexual multiplication is found in at least some members of all of the neodermatan groups. Digeneans typically undergo asexual multiplication in the larval stages that parasitize the first, usually molluscan, intermediate host. All digeneans possess one or more generations of sporocysts and/or one or more generations of redia. Asexual multiplication occurs in the generation of all larval stages developing within any generation (mother or daughter) of both sporocysts and redia. This form of asexual multiplication, termed polyembryony, results in multiple genetically identical individuals of the subsequent larval stage that develop within the saccate body of the parent generation. Thus, first generation sporocyts (mother sporocyts) produce either multiple second generation sporocysts (daughter sporocysts) or multiple first generation redia (mother redia), which in turn produce multiple first generation redia (mother redia) or multiple second generation redia (daughter redia), respectively. Second generation sporocysts produce either multiple cercariae or multiple redia that then subsequently produce multiple cercariae.

Asexual multiplication is uncommon among the monogeneans. It is, however, considered to be present in the form of sequential polyembryony (hyperviviparity) in some members of the monogenean order Gyrodactylidea. Rather than releasing eggs containing oncomiracidia, an adult can produce a second generation juvenile *in utero* (without sexual reproduction) that in turn can produce a third generation juvenile *in utero,* until one single individual may yield in excess of 20 individuals of the subsequent generation from within its own body, apparently without the intervention of a second individual or of gametes. As a result, the body of such asexually multiplying individuals in effect resembles a series of Russian dolls. There is evidence, however, of sperm in the seminal receptacle of some of these embryos, which has led some to question this as a truly asexual form of multiplication.

In tapeworms, asexual multiplication occurs in the terrestrial members of the order Cyclophyllidea that develop through larval stages known as coeneri and hydatid cysts. The fluid-filled cavity of these larvae is lined with a layer of germinative tissue that produces multiple protoscoleces, in the case of coeneri, and daughter hydatids that in turn are lined with germinative tissue that produces multiple protoscoleces, in the case of hydatid cysts. Each protoscolex has the potential to become an adult worm. The reproductive potential in these groups is enormous, with the hydatid cysts of some species producing hundreds of thousands, perhaps even millions, of protoscoleces.

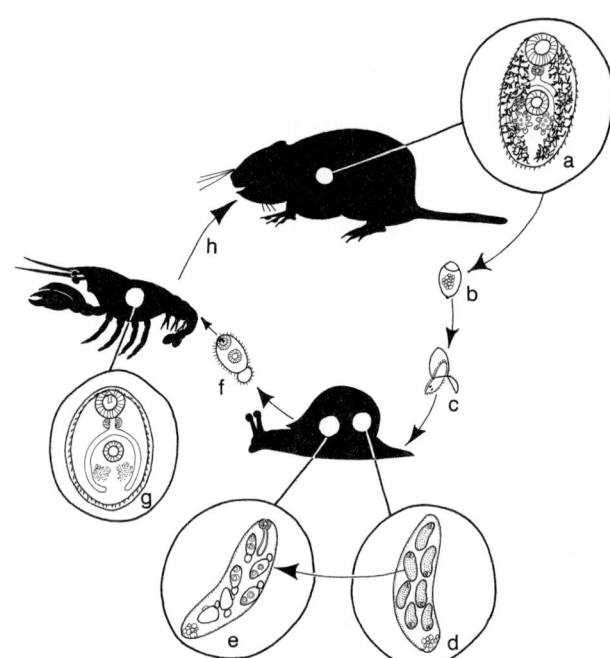

FIGURE 61 Generalized digenean life cycle. (a) Adult in definitive host, undergoes sexual reproduction and produces eggs; (b) eggs released with faeces of host; (c) miracidium hatches from egg and penetrates first intermediate host (usually a mollusc); (d) miracidium develops into sporocyst containing numerous redia; (e) redia leave sporocyst and produce numerous cercariae; (f) cercaria escapes from first host and finds second intermediate host; (g) cercaria sheds tail, encysts, and develops into metacercaria; (h) second intermediate host containing metacercaria is eaten by definitive host.

VII. ONTOGENY

There are a remarkable variety of larval forms found among the platyhelminths, especially within and among the classes of neodermatans. In general, the life cycle of each class of neodermatan is characterized by one or more unique larval stage (see Figs. 61–63). This diversity reflects the great range of habitats invaded and life history strategies employed. Most non-neodermatans undergo direct development such that the zygote develops directly into a form resembling a young adult. As discussed by Ruppert (1978), however, there are several notable exceptions to this pattern. One species of catenulid develops from a zygote into a pelagic larval stage known as a Luther's larva (Fig. 64); the larva is vermiform in shape, ciliated throughout, but in addition possesses several anterior rings of elongate cilia that aid in swimming. The zygotes of some polyclads develop into free-swimming, ciliated larval stages bearing arms. These larvae are generally referred to as Götte's

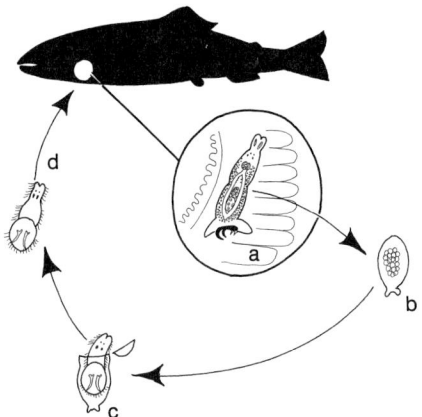

FIGURE 62 Generalized monogenean life cycle. (a) adult on gills of definitive host (usually an aquatic vertebrate), undergoes sexual reproduction, and produces eggs; (b) eggs released into water; (c) oncomiracidium hatches from egg; (d) oncomiracidium finds definitive host.

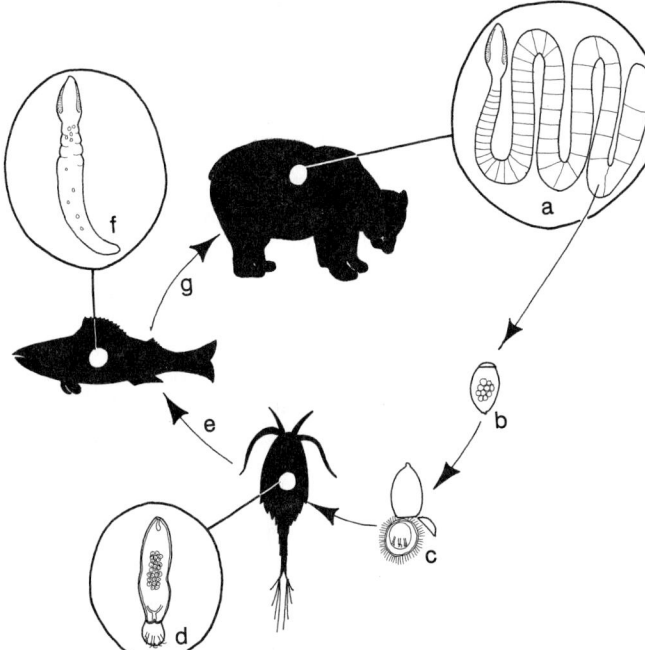

FIGURE 63 Generalized life cycle of aquatic cestode. (a) adult in definitive host, undergoes sexual reproduction, and produces eggs; (b) eggs released with faeces of host; (c) coracidium larva hatches from egg and penetrates first intermediate host (usually a copepod); (d) coracidium develops into procercoid; (e) first host eaten by second intermediate host; (f) procercoid develops into plerocercoid; (g) second host eaten by definitive host (usually a piscivorous mammal).

larvae (Fig. 65) or Müller's larvae (Fig. 66) depending on whether they possess four or seven to ten arms, respectively. Götte's larvae are found only in members of the polyclad suborder Acotylea; Müller's larvae are known from members of both suborders of polyclads. There does not seem to be a phylogenetic pattern to the distribution of these larval stages among the polyclads. For example, the polyclad genus *Stylochus* includes species that develop into Götte's larvae, species that develop into Müller's larvae, and species that undergo direct development.

Prior to developing into the adult form, aspidogastreans pass through a larval stage called a cotylocidium (Fig. 67), which is often but not always ciliated. This larval stage is free-swimming in at least some species. However, the life cycles of many species are not yet fully understood. In some cases it appears that the cotylocidium hatches from the egg and parasitizes the same host as the parent individual from which it was released. In other cases it appears that the cotylocidium hatches from the egg and moves into the water column to find a new host. In yet other cases the cotylocidium never hatches from the egg; rather, the egg is ingested by the molluscan host. In some species the cycle is direct, involving only a single host. In other species, the cycle may involve a second host. This host may be infected by ingesting infected molluscs.

Digenea undergo complex development in which certain intermediate life cycle stages are able to produce multiple copies of new generations before becoming sexually reproductive adults. Consequently the term "larva" is used loosely when discussing intermediate life cycle stages in the Digenea. The digeneans pass through a series of larval stages that are either parasitic or free-living, depending on their association with this sequence of hosts. A generalized digenean life cycle is illustrated in Fig. 61. As can be seen, the basic sequence of larvae is as follows: miracidium, sporocyst, redia, cercaria, and metacercaria. The miracidium (Fig. 68) possesses cilia on epidermal plates, or occasionally on tufts or bars. The neodermis is formed when the miracidium sheds its epidermal cell plates, thus all stages following the miracidium have an outer layer that is a neodermis. In most taxa the miracidium develops into a sporocyst, but occasionally may develop into a redia. Sporocysts and redia are both sacciform. The main difference between them is that the sporocyst has none of the components of a digestive system (Fig. 69), whereas the redia usually possesses at least a mouth, a pharynx, and often a single intestinal caecum (Fig. 70). These stages vary morphologically among species in their possession of lateral lobes, and in some cases birth pores;

the sporocysts of some species are branched. It is quite common for digeneans to possess more than a single sporocyt and/or redial generation in their life cycle. A first generation sporocyst (primary sporocyst) is essentially a miracidium that has lost its epidermal plates, whereas the second generation sporocysts (secondary sporocysts) are produced through asexual multiplication within the primary sporocyst. A first generation redia (primary redia) is most commonly produced through asexual multiplication within a primary or secondary sporocyst, but exceptionally develops directly from a miracidium that has shed its epidermal plates. In most taxa, the final result of several generations of sporocysts and/or redia is the development of a number of cercariae. Each cercaria generally possesses an anterior body and a posterior tail. The cercaria is by far the most morphologically diverse of the digenean larval forms. Indeed, classification of the digenean families is based to a large extent on cercarial morphology. An elaborate nomenclature based on morphology exists to describe these cercarial types (for example, see Cable, 1956). Examples of some of the more distinctive types include ophthalmoxiphidiocercaria (Fig. 71), furcocercous (Fig. 72), microcercous (Fig. 73), and cystocercous cercariae (Fig. 74). The metacercarial stage is usually achieved when the cercaria drops its tail and secretes a protective cyst around itself. The metacercaria is essentially a juvenile version of the adult worm (Fig. 75). However, variations on this general developmental pattern occur: the metacercarial stage may be lacking or it may be preceded by an unencysted mesocercarial stage.

Associations of these larval forms with their hosts are quite predictable. In general, the miracidium and cercaria are free-living forms. The miracidium hatches from the egg and seeks the first (molluscan) host; the cercaria escapes from the mollusc and seeks either the second intermediate or the definitive host. The sporocyst and redial stages generally parasitize the first (molluscan) host. The metacercarial stage encysts either on aquatic vegetation or in or on the tissue of a second intermediate host, which can be either an invertebrate or vertebrate, depending on the digenean species. Exceptions to this general pattern do occur. In some species, the miracidium is not free-swimming; rather, it remains within the egg, which is subsequently eaten by the first host. In some species, the cercaria does not leave the mollusc to swim freely; rather, it continues its development into a metacercaria, encysting within the body of the sporocyst or redia within the molluscan host. In species that lack metacercaria, such as the schistosomes, the cercaria penetrates the tissues of the definitive host directly. This stage is referred to as a schistosomula once it drops its tail and enters the host.

In general, the monogenean life cycle consists of two stages, the adult and a free-living, usually ciliated, larval stage known as the oncomiracidium. This larva possesses one or two pairs of ocelli, and is conspicuous in its possession of a posterior muscular attachment structure known as a haptor, which is usually armed with hooklets (uncinuli) and/or hamuli. The oncomiracidium differs from the ciliated larvae of other neodermatans in that its cilia are often restricted to three distinct regions. When the oncomiracidium hatches from an egg, which is sometimes anchored to the substrate, it swims or creeps to find its vertebrate host. An oncomiracidial stage is lacking from monogeneans such as the gyrodactylideans, that undergo sequential polyembryony. A generalized monogenean life cycle is shown in Fig. 62.

Both amphilinideans and gyrocotylideans possess a ciliated larval stage known as a lycophore larva (Figs. 77, 78 respectively). This larva possesses 10 hooks at the posterior end of its body; these hooks are of similar size and shape in gyrocotylideans but consist of six large and four small hooks, which also differ in shape, in the amphilinideans. This larval stage is also sometimes referred to as a decacanth. The life histories of many species in these two cestode groups are not well known. It appears that, in at least some amphilinideans, the lycophore does not escape from the egg until it is eaten by a small crustacean. In this intermediate host, the lycophore burrows through the gut and comes to rest in the hemocoel, where it develops into a procercoid by shedding its ciliated epidermal cells, elongating, and developing a posterior tail. It subsequently develops into a plerocercoid by shedding its tail and undergoing further elongation. In another case, the lycophore larva hatches and actively penetrates through the cuticle of the intermediate crayfish host. Development into the adult stage occurs only when the intermediate host is eaten by an appropriate vertebrate, at which point the plerocercoid sheds its tail and becomes sexually mature. It is possible that gyrocotylideans do not require an intermediate host to complete their development.

Most eucestodes require at least two hosts to complete their life cycles, but it is also common for additional paratenic hosts to occur in the life cycles of these platyhelminths. Paratenic hosts are not required, but serve to bridge food chain gaps, thereby facilitating infection of appropriate vertebrate hosts. A generalized aquatic eucestode life cycle is illustrated in Fig. 63. In general, eucestode zygotes initially develop into a larval stage known as a hexacanth embryo (Fig. 79). This embryo characteristically possesses three pairs of hooks. The hexacanth may or may not be surrounded by a layer termed an embryophore. A hexacanth surrounded by an embryophore is referred to as an oncosphere. An oncosphere in which the outer region of the embryophore is ciliated is referred to as a coracidium and is usually free-swimming. Some cestode hexacanths or oncospheres are surrounded by an egg shell; this shell possesses a cap-like operculum in some groups. A hexacanth enters the first intermediate host either when it is eaten or by actively swimming to and then penetrating this first host. The first intermediate host is an invertebrate in most eucestode groups, but some terrestrial cyclophyllideans utilize vertebrate first intermediate hosts. The developmental fate of the hexacanth remains to be discovered in many cestode species; this is especially true of the trypanorhynchs and tetraphyllideans. However, existing data suggest that, in general, the larval stages that follow the hexacanth vary somewhat in form among the different cestode orders. These forms are generally given different names depending on their morphology. The key features for distinguishing among larval types are the presence or absence of a scolex, the number of scoleces, whether the scolex (or scoleces) are invaginated (turned inside out) or retracted (withdrawn) into the main body of the larva, and whether a cavity known as a primary lacuna is present. Although current terminology does not accommodate all known combinations of these features, some common larval forms are given (see following). For example, procercoids are solid-bodied, lack a scolex, and usually possess a posterior extention of the body considered a cercomer by some (Fig. 80). Plerocercoids possess a scolex that is neither invaginated nor retracted (Fig. 84). Cysticercoids possess a conspicuous scolex that is retracted into the tissue of the neck alone, or, in some cases, the neck is then retracted into the main body of the larva (Fig. 81). Cysticerci possess one, or more invaginated scoleces. Cysticerci with multiple (Fig. 82) scoleces are referred to as either hydatid cysts (Fig. 83) or coeneri, depending on whether they have the ability to generate daughter cysts or not, respectively. The reproductive potential of hydatid cysts is enormous as each scolex (termed a protoscolex in hydatids) can go on to become either an adult worm or another protoscolex, if released from the original cyst within the first host. The numerous marine eucestode larvae referred to as "*Scolex plero-nectes*" or "*Scolex polymorphus*" are likely to be the plerocercoid larvae of tetraphyllideans.

VIII. PHYLOGENETIC RELATIONSHIPS

The tree shown in Fig. 1 provides an overview of the general understanding of the phylogenetic relationships among the major platyhelminth groups. The topology of this tree has been modified from several sources (e.g., Littlewood et al., 1999). The characters that have been mapped onto the tree include some of the more conspicuous characters that support these relationships, as well as some of the useful distinguishing characteristics of major groups. As this tree includes only a mapping of some useful characters it should not be interpreted as a cladogram.

Few unique features (synapomorphies) have been identified for the phylum Platyhelminthes. None of the characters normally used to describe the group (i.e., dorsoventral flattening, acoelomate, bilateral symmetry) is unique to this group. Multiciliated epidermal and gastrodermal cells are candidates as diagnostic features for the phylum. However, multiciliated epidermal cells are also found in the phylum Nemertini, suggesting that this character may be more appropriately considered as diagnostic at a higher taxonomic level. The fact that all of the cestodes and some parasitic dalyelliids lack a gastrodermis (and in fact a digestive system) means that gastrodermal ciliation is not universally diagnostic for the platyhelminths either. Nonetheless, transmission electron microscopy indicates that the lack of accessory centrioles from the base of the cilia of the protonephridial flame bulbs and epidermis may be truly synapomorphic for the phylum as a whole, as may be the presence of more than one cilium in the flame bulbs of the protonephridial system.

Some morphological (Smith et al., 1986) and recent molecular data tend to confirm the monophyly of many of the major platyhelminth groups but have highlighted the need for further systematic research. For instance, there is increasing evidence that the Acoela and Nemertodermatida (collectively known as the Acoelomorpha) may not belong in the Platyhelminthes. The nemertodermatidans are unique among the platyhelminths in their possession of sperm that is monoflagellated rather than biflagellated. The acoels show unique neural organization and developmental patterns. Typically appearing as basal bilaterians and not platyhelminths in molecular phylogenetic analyses, the acoelomorphs lack a protonephridial system and are the only flatworms with glandular frontal organs. In addition, these two groups lack a well-defined gut and possess nonla-

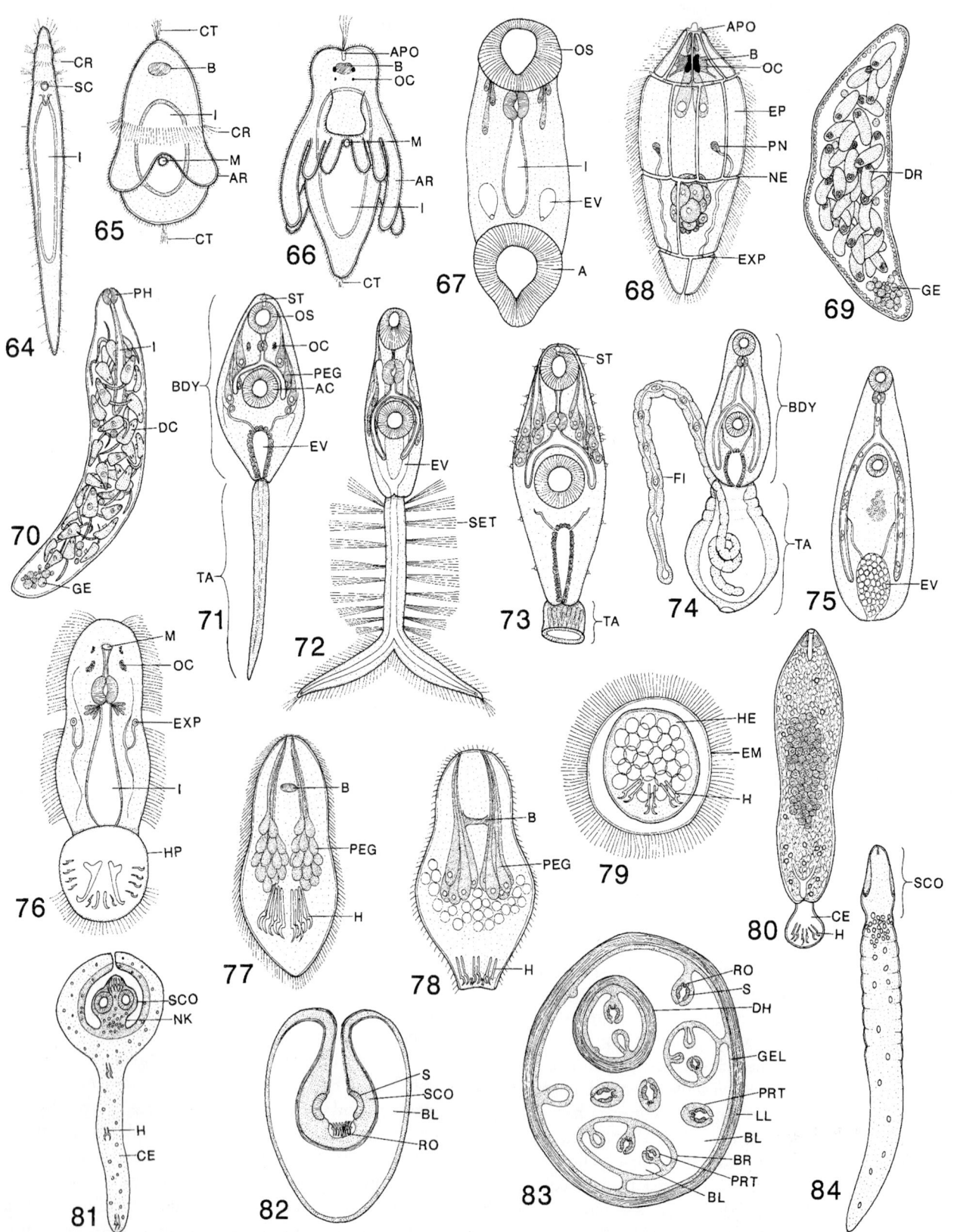

mellated rhabdoids rather than lamellated rhabdites. However, they have been included here until their phylogenetic position is more formally addressed. Relationships among the remaining nonneodermatan platyhelminth taxa are even less well understood. There is some agreement that the catenulids are among the more basal platyhelminths. The members of this order are exceptional among the platyhelminths in their possession of aflagellated sperm and one rather than two protonephridial systems as well as genital pores that are dorsal rather than ventral in position. With the exception of the acoelomorphs, catenulids, and proseriatans (and at least some dalyelliidans and temnocephalideans) all of the nonneodermatan groups possess a unique duogland adhesive system consisting of two different types of gland cells, one that produces an adhesive substance and one that produces a substance that reverses (releases) the action of the adhesive substance. In addition, all of the nonneodermatan taxa, with the exception of the acoelomorphs, catenulids, proseriatans and some dalyelliidans, possess lamellated rhabdites. Yet there is little evidence to suggest that these orders belong together in a distinct, monophyletic group. Rather, additional characters unite a subset of these orders with the neodermatans. Unfortunately, the affinities suggested by these characters are inconsistent. For example, all platyhelminths except the acoelomorphs and catenulids possess flame bulbs with more than three cilia. All platyhelminths except the acoelomorphs, catenulids, macrostomids, and prolecithophorans possess biflagellated sperm. All platyhelminths except the acoelomorphs, catenulids, prolecithophorans, macrostomids, and haplopharyngids share the unique 9 + "1" arrangement of microtubules in the flagella of their sperm. Three major, but somewhat related, characteristics unite the proseriatans, triclads, prolecithophorans, kalyptorhynchs, typhloplanids, dalyelliidans, and temnocephalidans with the neodermatans: separate ovary and vitelline organs, a heterocellular female gonad, and production of ectolecithal eggs. The latter two, however, are also found in the lecithoepitheliatans.

Unique features support the individual monophyly of many of the remaining nonneodermatan orders: the polyclads possess intestinal cecae that are highly branched, the triclads possess a triradiate intestine and an unusual embryonic intestine that does not appear to be homologous to that of the adult, all but one of the temnocephalans possess two to 12 anterior tentacles, and the haplopharyngids and kalyptorhynchs both possess an anterior proboscis independent of their digestive systems. The probosci of the two groups are not thought to be homologous.

The monophyly of the Neodermata, consisting of the aspidogastreans and digeneans (collectively the Trematoda) along with the monogeneans and cestodes (collectively the Cercomeromorpha), is now one of the most well-established aspects of platyhelminth relationships. The members of this group are characterized by their unique syncytial outer body layer known as the neodermis, which generally replaces the cellular epidermis when the first larval stage encounters the first host. In addition, the neodermatans possess protonephridial filters in the form of a two-celled weir. The sperm of neodermatans is unusual in that the two flagella are incorporated into the cytoplasm of the sperm cell and are arranged in a proximal–distal direction, rather than remaining separate. All neodermatans also lack lamellated rhabdites and the duogland adhesive system and possess cilia with only one rather than two rootlets. All neodermatans are obligate parasites at some time during their lives.

The Trematoda are characterized by their possession of epidermal cells in their first larval stage (miracidia and cotylocidia) that are separated from one another by portions of the neodermis. In addition, the first or primary host in the life cycle of the trematodes is a mollusc. The Aspidogastrea are unique in their possession of oviducts separated into chambers by septa and neodermal microvilli in the form of hemispherical microtubercles. The Digenea are distinctive in their possession of several unique larval stages, the miracidium, sporocyst and/or redia, and cercaria. Perhaps in part because they are by far the most diverse of the major

---

FIGURES 64–84 64 Catenulida: Luther's larva (modified from Ruppert, 1977); 65 Polycladida: Götte's larva (modified from Ruppert, 1977); 66 Polycladida: Müller's larva (modified from Ruppert, 1977); 67 Aspidogastridea: cotylociduum; 68 Digenea: miracidium; 69 Digenea: sporocyst; 70 Digenea: redia; 71 Digenea: ophthalmoxiphidiocercaria; 72 Digenea: furcocercous cercaria; 73 Digenea: microcercous cercaria; 74 Digenea: cystocercous cercaria; 75 Digenea: unencysted metacercaria; 76 Monogenea: oncomiracidium; 77 Cestoda: gyrocotylidean lycophore larva (redrawn from Beauchamp in Grasse, 1961); 78 Cestoda: amphilinidean lycophore larva (redrawn from Beauchamp in Grasse, 1961); 79 Cestoda: coracidium; 80 Cestoda: procercoid (modified from Joyeux and Baer in Grasse, 1961); 81 Cestoda: cysticercoid (scolex retracted); 82 Cestoda: cysticercus (scolex retracted and invaginated); 83 Cestoda: hydatid cyst (modified from Hyman, 1951); 84 Cestoda: plerocercoid. (See abbreviations on pages 897–898.)

platyhelminth groups, there is currently no generally accepted, higher level classification scheme for the digeneans. A preliminary attempt at the generation of a tree for the families in this subclass was presented by Brooks et al. (1985). However, it is clear from the work of others (for example, Pearson, 1992) that a more detailed analysis of characters is required before these relationships can be resolved with any confidence.

The Cercomeromorpha are unique in their possession of, at least at some time in their development, a posterior extension of the body that is equipped with a number of sickle-shaped hooklets. The name for this group comes from the notion that this posterior extension of monogeneans and cestodes is homologous, an idea that is not universally accepted. The monophyly of both the monogeneans and the cestodes is fairly well established. The Monogenea are unique among the platyhelminths in their general possession of four rhabdomeric ocelli at some point in their lives, a ciliated larval stage known as the oncomiracidium (with the exception of the few viviparous forms), and the retention of the posterior haptor in the adult stage. The results of the most recent extensive cladistic treatment of the group (see Fig. 35 and Boeger and Kritsky, 1993) support recognition of two major subgroups within the class: the Polyopisthocotylea, which possess a haptor that is subdivided into multiple suckers and/or bears distinctive attachment structures known as clamps, and the Monopisthocotylea, which possess a haptor that, although it may bear loculi, is undivided and, although it may bear anchors or hamuli, never bears clamps. Although the monophyly of the Monogenea has been called into question by some molecular phylogenetic analyses, it seems likely that this reflects a rapid divergence of the two major subgroups rather than lack of support for monogenean monophyly. The interrelationships among the monogenean orders within these two major subgroups are not universally agreed upon. The phylogenetic position of the udonellids is particularly uncertain.

There is little doubt that the cestodes are monophyletic; all cestodes lack a gut and thus also a ciliated gastrodermis. All cestodes possess unique extensions of the neodermis called microtriches (Fig. 16). The relationships among the cestode orders remain controversial despite the fact that this issue has received much recent attention (for example, Hoberg et al., 1997; Mariaux, 1998; Olson and Caira, 1999). The tree in Fig. 46 summarizes some of the more generally accepted ideas on these relationships. There is evidence to suggest that many, but not necessarily all, of the individual cestode orders are monophyletic.

IX. HOST ASSOCIATIONS

It is useful to consider host associations in general terms. The tendencies of platyhelminths to live in association with other organisms are summarized on the tree in Fig. 1. One of the most striking aspects of this mapping is that it becomes apparent that all but five of the major platyhelminth groups (catenulids, haplopharyngids, lecithoepitheliatans, macrostomids, and kalyptorhynchs) include at least some species that live in association with other organisms. Thus, symbiosis, in its broadest sense, is a widely distributed feature among the platyhelminths. Within the major platyhelminth groups, however, symbiosis varies in its occurrence. Only the four neodermatan groups and the temnocephalids consist entirely of obligate symbionts. The dalyelliidans are predominantly symbiotic, although a few free-living dallyelliidan species are known. The remaining groups (acoelans, nemertodermatidans, polyclads, proseriatans, triclads, prolecithophorans, and typhloplanidans) are predominantly free-living, but all of these groups also include at least some species that are symbiotic.

The hosts of almost all of the symbiotic nonneodermatans are invertebrates; molluscs, arthropods, and echinoderms figure prominently in this role. The neodermatans are unlike almost all of the other symbiotic platyhelminths in that they are generally associated with at least one vertebrate host group at some point in their lives. The aspidogastreans are somewhat exceptional in that some species are known only from freshwater clams and snails; however, records of aspidogastreans from the digestive system of turtles, elasmobranchs, and teleosts are not uncommon. Although uncharacteristic of the group as a whole, some monogeneans are found associated with cephalopods, and others are found associated with crustaceans that parasitize teleosts or elasmobranchs.

The neodermatan groups are also unique in that their life cycles generally involve more than a single host species. All digeneans possess at least two different hosts in their life cycles. In general, the first (intermediate) host is a mollusc, while the final (definitive) host is a vertebrate. In species possessing life cycles involving more than two hosts, there is one (or sometimes more) additional intermediate (or paratenic) host that is either an invertebrate or a vertebrate depending on the digenean group. In such taxa the host that precedes the vertebrate is generally an invertebrate, often some sort of mollusc or arthropod. The monogeneans are exceptional in this respect as their life cycles involve only a single host.

Platyhelminths are found either in or on their hosts. The nonneodermatan groups are often found on, rather than in, their hosts. The exceptions are the symbiotic acoels, nemertodermatidans, and dalyelliidans, which are most commonly found inhabiting the coelom or digestive tract of their hosts. Among the neodermatans, the monogeneans are unique in that they are generally found on, rather than in, their hosts; the remaining neodermatan groups are most commonly found as endoparasites of their respective hosts.

X. MEDICAL IMPORTANCE

The medical and economical importance of the various platyhelminth groups is closely tied to the type and intimacy of the host associations of the various groups. Species that use vertebrates as hosts are obviously of much greater concern than those that do not. Thus, the neodermatans are by far the most medically and economically important platyhelminth groups. Space limitations do not permit discussion of the economic consequences of infections with neodermatans, but some of the medically important species are briefly discussed below.

Within the neodermatans only the digeneans and cestodes include species that infect humans. Among the 17 superfamilies of digeneans, human parasites are essentially restricted to four. In the majority of these species it is the adult stage that parasitizes humans. Many of these species are zoonotic; that is, their normal hosts are usually vertebrates other than humans. Humans acquire infections with such species when they unwittingly ingest the metacercarial stage. These zoonotic groups infect a variety of different organs in their human hosts. They include a total of eight species of lung digeneans in the superfamily Troglotrematoidea, five species of liver digeneans in the superfamily Opisthorchioidea, and seven species in the superfamilies Echinostomatoidea, Opisthorchioidea, and Troglotrematoidea that inhabit the digestive tract of their hosts. Finally, human infection with cercariae of species of the superfamily Schistosomatoidea that normally infect birds and aquatic mammals is responsible for an irritating condition in humans known as cercarial dermatitis or swimmer's itch.

Only six species of digeneans parasitize humans as their normal final (definitive) hosts. These include one species of liver fluke in the superfamily Opisthorchioidea and five species of blood flukes in the superfamily Schistosomatoidea. The unusual life-cycle of the schistosomes requires that humans come into contact with the free-swimming cercariae of these species to acquire an infection. Schistosomes, which cause a disease known as schistosomiasis in humans, are without question the most deadly of the digeneans. Hundreds of millions of people globally are infected with these species. Schistosomes are responsible for the deaths of tens of thousands of people each year.

Species that parasitize humans belong to only two of the 14 orders of tapeworms. The adults of at least four species of pseudophyllideans (all of which are members of the genus *Diphyllobothrium*) have been reported from the digestive system of humans; evidence suggests that all of these species are zoonotic in humans, as they normally parasitize other fish-eating vertebrates as adults. Plerocercoids of some species of pseudophyllideans of the genus *Spirometra* are known to parasitize the musculature and subcutaneous tissues of humans, causing a condition known as sparganosis. By far the most pathogenic of the human tapeworms, however, are the cyclophyllideans. Nine species in five genera parasitize humans. These species are known from humans as either larval or adult stages. Larval stages are generally much more pathogenic than adult stages. The hydatid cysts, coeneri, and cysticerci of *Echinococcus*, *Multiceps*, and *Taenia* species, respectively, are arguably the most life-threatening of the human-infecting cestodes. These larval stages often develop in difficult-to-treat extraintestinal sites, such as the brain, or grow extremely large and/or invasively into one or more organs of the body, making surgical removal of the parasite difficult. Thus, infection with any of these larval stages can be fatal. Infection with an adult tapeworm is generally debilitating rather than life-threatening, because these stages remain in the digestive system and thus are relatively easy to treat. With the exception of *Taenia solium*, it appears that tapeworm infections in humans are also all zoonotic. Infection with adult tapeworms can be prevented by properly cooking food. Infection with larval stages usually involves contact with fecal matter from humans or other animals.

Abbreviations: A, adhesive disc; AC, acetabulum; AD, adhesive glands; AG, accessory gland cells; AN, anchor; AO, accessory male organ; AP, adhesive papillae; APO, apical organ; AR, arm; AS, apical sucker; B, brain; BA, basal lamina; BDY, body; BF, buccal funnel;

BL, bladder; BO, bothrium; BOD, bothridium; BP, base plate; BR, brood sac; C, cirrus; CB, copulatory bulb; CE, cercomer; CGP, common genital pore; CL, clamp; CM, circular muscle; CO, copulatory organ; CP, ciliated pits; CPE, cephalic peduncle; CR, ciliary ring; CS, cirrus-sac; CT, ciliary tuft; CY, cyton; CYC, cyclocoel; CYE, cytoplasmic extension; DC, developing cercaria: DCY, distal cytoplasm; DE, developing embryo; DH, daughter hydatid cyst; DR, developing redia; EC, ecsoma; ED, excretory duct; EDC, electron dense cap; EM, embryophore; EP, epidermal cell; ES, esophagus; ESV, external seminal vesicle; EV, excretory vesicle; EXP, excretory pore; FB, forebody; FI, filament; FO, frontal organ; FP, female genital pore; FVP, female ventral papilla; GC, genitointestinal canal; GCO, genital corona; GD, gut diverticula; GE, germ cells; GEL, germinal layer; GL, gland cells; GP, gravid proglottid; GV, germinovitellarium; GY, gynecophoral canal; H, hook; HA, haptoral appendix; HB, hindbody; HE, hexacanth embryo; HG, hermaphroditic gland; HO, hooklet; HOR, head organ; HP, haptor; I, intestine; IMP, immature proglottid; IE, inner envelope; IS, internal seminal vesicle; L, loculus; LD, lateral diverticula of oral sucker; LL, laminated layer; LM, longitudinal muscle; M, mouth; MA, muscular appendage; MB, muscular bulb; MP, male genital pore; MP, mature proglottid; MT, microthrix; MUP, muscular pad; MVP, male ventral papilla; N, nucleus; NE, neodermis; NK, neck; O, ovary; OC, ocellus; OE, outer envelope; OGO, orogenital opening; OS, oral sucker; P, penis; PA, pars apicalis; PAP, papilla; PB, pars basalis; PC, parenchymal cell; PEG, penetration glands; PG, proboscis glands; PH, pharynx; PI, posterior isthmus; PN, protonephridium; PR, proboscis; PRG, proglottid; PRO, prostatic reservoir; PRT, protoscolex; PS, pseudosucker; PSC, primary scolex; PST, primary strobila; PV, prostatic vesicle; R, rossette; RO, rostellum; RT, rhabdoid tracts; S, sucker; SC, statocyst; SCL, sclerite; SCO, scolex; SET, setae; SH, spiny head collar; SL, statolith; SR, seminal receptacle; SSC, secondary scolex; SST, secondary strobila; ST, stylet; STR, strobila; SV, seminal vesicle; SY, syncytium, T, testis; TA, tail; TB, transverse bar; TE, tentacle; TL, terminal lappet; TO, tribocytic organ; TP, transverse plications; TS, tentacle sheath; TU, tubercles; U, uterus; UG, uterine gland; UP, uterine pore; V, vitellarium; VA, vagina.

See Also the Following Articles

PARASITISM • WORMS, ANNELIDA • WORMS, NEMATODA

Bibliography

Ax, P. (1996). "Multicellular Animals: A New Approach to the Phylogenetic Order in Nature, Vol. I." Springer-Verlag, New York.

Boeger, W. A., and Kritsky, D. C. (1993). Phylogeny and a revised classification of the Monogenoidea Bychowsky, 1937 (Platyhelminthes). *System. Parasitol.* **26**, 1–32.

Brooks, D. R., O'Grady, R. T., and Glen, D. R. (1985). Phylogenetic analysis of Digenea (Platyhelminthes: Cercomeria) with comments on their adaptive radiation. *Can. J. Zool.* **63**, 411–443.

Bychowsky, B. E., Korotajeva, V. D. and Nagibina, L. F., (1970). *Montchadskyella intestinale* gen. et sp.n., a new member of endoparasitic monogeneans. *Akad. Nauk SSR, Parazitoiogiia* **4**, 451–457. (In Russian.)

Cable, R. M. (1956). Scientific survey of Porto Rico and the Virgin Islands. In "Marine Cercariae of Puerto Rico," XVI, Part 4, pp. 491–577. New York Academy of Sciences, NY.

Cannon, L. R. G. (1986). "Turbellaria of the World." Queensland Museum, Brisbane, Australia.

Ehlers, U. (1974). Interstitielle Typhloplanoida (Turbellaria) aus dem Litoral der Nordseeinsel Sylt. *Mikrof. Meeves.* **49**, 427–526.

Ehlers, U. (1985). "Das Phylogenetische System der Plathelminthes." Gustav Fischer Verlag, Stuttgart, Germany.

Grassé, P. P. (1961). "Traité de Zoologie: Plathelminthes, Mésozaires, Acanthocéphales, Némertiens," Volume IV, (Masson and Cie, eds.), Paris, France.

Halton, D. W., and Gustafsson, M. K. S. (1996). Functional morphology of the platyhelminth nervous system. *Parasitology.* **113**, 47–72.

Harrison, F. W., and Bogitsh, B. J. (eds). (1991). "Microscopic Anatomy of Invertebrates. Volume 3. Platyhelminthes and Nemertinea.", Wiley-Liss, New York.

Hoberg, E. P., Mariaux, J., Justine, J. L., Brooks, D. R., and Weekes, P. J. (1997). Phylogeny of the orders of the Eucestoda (Crcomeromorphae) based on comparative morphology: Historical perspectives and a new working hypothesis. *J. Parasitol.* **83**, 1128–1147.

Hyman, L. H. (1951). "The Invertebrates. II. Platyhelminthes and Rhynchocoela. The Acoelomate Bilateria." McGraw-Hill, New York.

Khalil, L. F., Jones, A., and Bray, R. A. (1994). "Keys to the Cestode Parasites of Vertebrates." CAB International, Wallingford.

Littlewood, D. T. J., Rohde, K., Bray, R. A., and Herniou, E. A. (1999). Phylogeny of the platyhelminths and evolution of parasitism. *Biol. J. Linnean Soc.* **68**, 257–287.

Mariaux, J. (1998). A molecular phylogeny of the Eucestoda. *J. Parasitol.* **84**, 114–124.

Olson, P. D., and Caira, J. N. (1999). Evolution of the major lineages of tapeworms (Platyhelminthes: Cestoidea) inferred from 18S ribosomal DNA and Elongation Factor-1α. *J. Parasitol.*

Pearson, J. C. (1992). On the position of the digenean family Heronimidae: An inquiry into a cladistic classification of the Digenea. *System. Parasitol.* **21**, 81–166.

Rieger, R. M. (1998). 100 years of research on 'Turbellaria.' *Hydrobiologia* **383**, 1–27.

Rohde, K. (1997). The origins of parasitism in the Platyhelminthes: A summary interpreted on the basis of recent literature. *Internat. J. Parasitol.* **27**, 739–746.

Rohde, K. (1991). The evolution of protonephridia of the Platyhelminthes. *Hydrobiologia* **227**, 315–321.

Ruppert, E. E. (1978). A review of metamorphosis of turbellarian larvae. *In* "Settlement and Metamorphosis of Marine Invertebrate Larvae," pp. 65–81. Elsevier, New York.

Schell, S. C. (1985). "Handbook of Trematodes of North America North of Mexico." University Press of Idaho, Moscow, ID.

Schmidt, G. D. (1986). "CRC Handbook of Tapeworm Identification." CRC Press, Inc., Boca Raton, FL.

Smith, J. P. S. III, Tyler, S., and Rieger, R. M. (1986). Is the Turbellaria polyphyletic? *Hydrobiologia* **132**, 13–21.

Yamaguti, S. (1963). "Systema Helminthum. Monogenea and Aspidocotylea. Vol. IV.," Interscience Publishing (John Wiley & Sons), New York, NY.

Yamaguti, S. (1971). "Synopsis of Digenetic Trematodes of Vertebrates," Vols. I and II. Keigaku Publishing Co., Ltd., Tokyo.

ZOOS AND ZOOLOGICAL PARKS

Anna Marie Lyles
The Wildlife Conservation Society

I. The Changing Role and Façade of Zoos
II. Choosing Animals to Promote Biodiversity Conservation
III. Keeping Biodiversity
IV. Supporting Diversity in Nature

GLOSSARY

exotic animals Animals that have not been domesticated for use by humans in agriculture or as pets; also connotes that the wild animals are from another region of the world.

ex situ Outside the natural range. Hence, a captive breeding program in Europe for a South American species would be an *ex situ* program. (Conversely, a Chinese program to breed native panda bears in captivity is an *in situ* breeding program because it is within the natural range.)

founders Animals that were collected from their natural range and that have produced progeny in captivity (and are assumed to be unrelated). Thus, these animals are the founding stocks of the captive population. Also generalized to include animals with wild parents that are collected in order to supplement the genetic diversity of an established captive population.

International Species Information System (ISIS) A central repository for zoo animal records founded in 1973; has also developed the computer programs that most zoos use for keeping their animal records, as well as software used by zoos for studbooks.

studbooks Also called genealogies, registries, or pedigrees, a database of historic and living animals intended to go beyond passively recording bloodlines and records; typically kept in computer databases that are also useful for actively managing the genetics, demography, and husbandry of captive populations.

ZOO is a handy three-letter word, and thus in this article "zoos" shall be used in a very general way to include any public or private institutions that house live exotic animals, largely for the purposes of exhibition to the public. Zoos are permanent establishments that are open to the public to provide education, recreation, and cultural enjoyment. Zoos, for purposes of this discussion, are not transient shows such as circuses or petting zoos. Nor are zoos private collections of exotic animals. "Zoos," defined thus, include the facilities that are traditionally considered zoos, such as zoological parks and aquariums, and also many wildlife centers and "bioparks," as well as some insect houses and *ex situ* breeding centers. There are currently over 1,200 well-established zoos worldwide.

Zoos reach a vast audience; for example, North American zoos attract more attendees than the major professional sports combined. The best modern zoo exhibits are sublime, invoking in visitors a sense of awe and respect for nature. Zoos are increasingly operated by specialized, professional staff in partnership with government and private interests, with the express goal of conserving wildlife. The main focus of this article is on how zoos are evolving into institutions for wildlife conservation, and also on the art and science of managing biodiversity in zoos.

I. THE CHANGING ROLE AND FAÇADE OF ZOOS

A. Zoos from Ancient to Recent Times

Zoos probably arose as places of spectacle and entertainment shortly after civilization became urbanized. The first recorded zoological collection dates to 2,500 B.C., when several thousand wild animals are documented in Saqqara, Egypt. Ancient Egyptians kept many wild species for use in religious ceremonies, or perhaps because they hoped to tame them. These ancient animal collections were acquired as tribute gifts or via expeditions that were mounted specifically to collect wildlife.

Zoo collections have sprung up independently on many continents and in many civilizations. For example, Chinese emperors as far back as 1150 B.C. had walled animal collections they called "parks of knowledge." When Cortés reached Mexico City in 1519, he witnessed, and then destroyed, the Aztec ruler Montezuma's immense zoo and gardens. Montezuma's zoo was so large that 300 keepers were employed for the aviaries alone, and it even housed malformed humans as part of the collection. Most of the zoological collections of antiquity were imperial and exclusive—symbols of wealth and power. They were menageries in the sense that they were collections of living trophies to be managed.

Collections during Roman times were also for the amusement of citizens. Large numbers of wild animals were used for blood sport and spectacles; holding areas that also allowed public viewing, called *viveria*, were associated with arenas. After the fall of Rome, collections of wildlife were not so common in Europe. During Europe's Middle Ages, imperial menageries once again became more significant. Wild animals were used as diplomatic gestures—as tributes and ceremonial gifts.

The roots of public zoos are somewhat contemporaneous with the emergence of Europe's great cities. London gained a zoo, of sorts, in 1252 when Henry III transferred his menagerie to the Tower of London; London's citizens were pressed to support the menagerie, and in turn the public could view some of the animals, such as a polar bear and an elephant. Urban menageries, under the sponsorship of merchants and other citizens, began to emerge, first in Frankfurt in the late 1300s, then in the 1500s at The Hague and in Augsburg.

The rising profile of zoos can be traced to the Renaissance, when trade and exploration brought more animals into Europe. Colonial officers in far-flung empires served as conduits for animal requests. Public interest was most piqued by the wildlife of Africa and the Far East. The first zoological garden, which combined displays of plants and animals, was established in Versailles by Louis XIV in 1665; in 1794 this collection became a division of Paris' natural history museum, creating a formal link to zoological research that was later emulated by many other zoos and museums. The first "modern" zoo that survives to this day is probably the Schönbrunn Zoo in Vienna, originally established as a private collection in the 1770s, but to which the public were intermittently admitted.

The 19th century has been hailed as the century of science and the golden age of museums. This century witnessed an explosive increase in the numbers of zoos and museums. Scientific societies sprang up, including zoological societies. Zoological research was nominally based in 19th century zoos, but the inquiry was mostly limited to collecting and classifying. Local governments tended to operate zoos of this era with the aim of providing recreation and edification for the increasingly better educated citizenry that arose with the Industrial Revolution. Zoos were sources of civic pride, and they competed to exhibit as many kinds of animals as possible. The rise of print media made it possible for civic leaders to compare collections more readily with rival zoos. Natural social groupings were largely ignored, and frequently solitary animals were displayed. Generally species were grouped to correspond with scientific classification, for instance, into bird, reptile, or carnivore houses. These animal collections were viewed as analogous to collections of art, and even the zoo architecture of the period came to be considered art.

An innovation at the beginning of the 20th century was habitat exhibits, as pioneered by Carl Hagenbeck at Tierpark, outside Hamburg, Germany. In these zoos of living panoramas, it became increasing common to display animals along geographic lines rather than strictly taxonomic ones. Habitat exhibits were well suited to zoo designs that intended to teach ecological principles, and thus led to the display of animals in more natural social groupings. Eventually mixed-species exhibits became more common, and these conglomerations might even be grouped by ecological biomes such as "polar zone" or "tropical zone," rather than along strictly geographical or taxonomic lines.

Later in the 20st century, world culture became increasingly electronic and graphic. Zoos found themselves in competition with wildlife film documentaries. No longer would the animal displayed in a sterile cage suffice for an increasingly sophisticated audience. "Immersion" exhibits became the ideal for most new zoo development. Immersion exhibits are ones that seek to make visitors feel that they are in the habitat with the animals. This may be accomplished by incorporating the same details of rocks, logs, or plants into both the animal and the public space, while using these same details to hide any cage elements, and also by special effects such as mist machines and animal call background soundtracks that help create an illusion. Barriers to separate animals and people may be relatively invisible (e.g., glass) or carefully placed so they are unobtrusive. Special feeders and techniques of animal management may be used to encourage animals to be active and visible, for example, hidden heating pads for animals to lounge upon or feeders that intermittently spray out small amounts of food near a viewing station to encourage visible foraging behavior. Zoos often try to craft the visitor space so that visitors do not see other visitors while observing the animals. In many instances, the designers' intent is to invoke an almost religious experience of awe and respect for nature, much as the grand cathedrals did in early eras.

B. Rise of the Zoo Conservation Ethic

Zoos, as we have seen, arose as places of spectacle, not as conservation organizations. But many modern zoos have chosen wildlife conservation as their primary mission. The conservation ethic has developed in zoos over the last century in parallel with the non-zoo world, and in many cases one can point to zoos as the places that incubated the people and ideas that have led to important conservation developments. The National Zoo in the United States, founded in 1887, is considered to be the first zoo created to preserve endangered species.

Conservation of wildlife and wildlife habitat has become the major goal of most of the prominent zoos and zoo professional organizations. Although altruism is part of the reason for the shift, compelling utilitarian reasons also drive these lofty goals. The supply of wild animals to stock zoos is drying up. As humanity expands and wild places contract, exotic animals are getting scarcer. As regulators seek to simultaneously protect their natural resources and their agricultural sectors, governments make it ever more difficult to import and export wildlife. Wild animals may carry diseases that will harm livestock and perhaps even spread to the human populace; increasing understanding of disease has led to import restrictions on quarantines. For example, the bird-importing era, at least in the United States, slowed markedly in the early 1970s, when quarantine laws were enacted to guard against Newcastle's disease. In the 1990s the Wild Bird Trade Act was passed in the United States to restrict importation of any birds protected by CITES, the Convention on International Trade in Endangered Species of Wild Fauna and Flora. Most professional animal dealers have gone out of business due to the increasing restrictions, and zoos cannot easily collect wild animals.

Simultaneously, zoos have needed to import fewer animals because they became more proficient at keeping and breeding wildlife. Zoo keeping is a complicated art and science, and zoo keepers and managers have found it useful to pool their knowledge. Furthermore, single zoos are rather like islands from a population biology point of view, in that they are not big enough to sustain viable populations. So cooperative breeding programs for animals have become key, while cheap, fast transport, usually by air, makes it feasible for distant zoos to exchange breeding stocks. In many regions of the world, the leading zoos and their staff now belong to professional organizations. There are also world zoo organizations that help link the regions. These organizations have codes of ethics and accreditation (certification) programs. For example, North America's American Zoo and Aquarium Association was founded in 1924 and now includes nearly 200 institutional members (this is about one-tenth of the number of animal exhibitors that are licensed with the U.S. government).

Public support is yet another motive for many zoos to become conservation organizations. Zoos are having difficulties in the competition for public funding. Modern zoos are costly, with expensive professional staffs, sophisticated veterinary medicine facilities, and a public that demands complex, educational, and artistic exhibits. Municipal governments are increasingly burdened by social programs and other competing demands for funds and now only fund a fraction of zoo operations compared to earlier times (on average 21% of the zoo budget in North America). But while the funding situation is tough in the richer nations, it is often quite desperate in poorer regions of the world. Most zoo-based conservation is being done by the "haves" while much of the need for conservation and access to endangered species is in the "have-not," largely tropically located zoos of the world.

There is enormous public support for environmental causes and grants available for conservation programs. However, zoos face some challenges in getting or generating environmental funds, in part because of public ambivalence about the ethics of keeping animals in zoos.

C. Animal Welfare in Zoos

Advocates for animals are becoming increasingly strident, even to the extent of using terrorist tactics. Their ire is more typically directed against laboratories that use animals, but zoos also have come under fire. Some animal rights extremists have expressed a goal of putting zoos out of existence. Extreme though these views may be, they signal a shift in public attitudes towards animal welfare. Many people now consider animals as fellow creatures with needs and feelings that should be respected. Zoos that have not become attuned to perceptions of animal welfare have often experienced an erosion of public support. Some of the poorer, old-fashioned zoos have closed, and even some prominent zoos have almost succumbed.

These animal advocates often confuse animal rights with animal welfare, and their anthropomorphic ideas about what animals need are often at odds with what is best for animals. For example, well-meaning protestors tried to prevent the transfer of a gorilla named Timmy, who had never procreated, to a zoo in New York that had considerable success in breeding gorillas as well as many potential fertile mates for Timmy. The protestors did not want Timmy to be separated from his current, infertile, female partner. However, in nature gorillas do not form monogamous bonds and they do change partners. Eventually, after court battles, Timmy was moved, formed bonds with new females, and has begat numerous offspring.

A related problem that we shall return to in Section III is the problem of limited space in zoos. Thus, breeding unwanted, "surplus" animals is a problem for zoos because surplus animals inhabit valuable space. Some zoos took the apparently logical stance, from a utilitarian animal management vantage, of culling their surplus stocks or selling them to the private collectors. In private hands, some animals have ended up on game ranches where they are shot for sport. This has caused major public relations imbroglios, as a result of which zoos have had to examine their ethical policies and restrict their animal trade to within accredited zoos. Zoos have developed contraceptive and husbandry techniques to prevent unplanned breeding, and they have developed scientific breeding plans to keep population demographics aligned with available space. Furthermore, there has been a trend of exchanging animals without fees as they become increasingly priceless.

Improved contraception for animals is just one example of how health care for exotic animals is becoming increasingly sophisticated. Improved veterinary care deserves much of the credit for allowing zoos to shift from relatively sterile concrete and steel cages to more natural housing, which is often, literally, dirty (as the substrate is frequently dirt or other organic matter). Most zoos now have their own veterinarian and veterinary hospital. Some even have pathologists, endocrinologists, and nutritionists providing preventative care, better drugs, and diagnostics. The understanding of wild animals' nutritional needs is advancing rapidly, and there are now numerous commercial brands of food for exotic animals, such as "Flamingo Fare." Nutritionists are even improving on fish diets (which are expensive and become depleted in certain vitamins during shipping and storage) for animals such as penguins and sea lions by devising a gelled, artificial fish in a fish shape. Most zoos have "browse" programs to grow and harvest fresh plant foods for their denizens.

There are numerous technical periodicals, books, and other publications related to zoo husbandry, including "Zoo Biology," "International Zoo News," the "International Zoo Yearbook," "The Shape of Enrichment," and "The Dodo." As a testament to the improving welfare of zoo animals, zoos are now recording many longevity records, and zoo animals frequently senesce and die of old age—an almost unheard of occurrence in nature. As zoos have become more successful at keeping exotic animals alive and meeting their physical needs, the attention of zoo biologists has begun to shift toward behavior.

Zoos have also added applied psychology to their toolkits. They now must reassure their public that animals are not distressed but are potentially "happy." New techniques are being developed to stimulate zoo animals, reduce psychological stress, and elicit cooperation rather than coerce desired behaviors. Some zoos train animals to perform in shows, which is not a new phenomenon. But modern shows demonstrate natural behaviors to the public and keep the performing animals occupied, rather than emphasizing cute tricks or an animal's fierce nature. These new techniques range from behavior management and operant conditioning to environmental enrichment. Behavioral techniques aim to banish boredom, eliminate stereotypic tics, and otherwise encourage behavioral patterns that approximate those of animals in nature.

D. Emerging Trends

Zoos, as we have seen, have evolved from imperial menageries to the modern habitat and immersion exhibits that featured animals in more natural settings and social groupings. In the 21st century, we expect this trend toward more ecological themes to continue, with more zoos featuring animals in diverse, mixed-species exhibits. Zoos that predominantly have these diversity-designed exhibits have been given the moniker "biopark," or "biodome." This sort of zoo may feature botanical collections housed with terrestrial and aquatic vertebrates and invertebrates. There is a considerable art in developing these mixed species exhibits, because zoos must ensure that co-housed animals do not decimate the foliage or each other. Animals that do wreak havoc with plants or other animals are either excluded from the biopark or tend to be isolated, although their cage boundaries may be crafted carefully so that this is not obvious. Some zoos are extending this idea to focus on the ecology of a certain region using both zoo exhibits and more museum-like exhibits; typically the focus may be the region in which they are located, for example, the "living desert museum" in the American southwest near Tucson, Arizona.

Also emerging at the turn of the 21st century seems to be a new trend, which we shall call "conservation edu-tainment." These thematic exhibits have a storyline to which visitors may be introduced, sometimes through videos or costumed actors, as they wait to enter the exhibit. This visit may even include an amusement-style ride; for example, Sea World visitors to the polar region exhibit experienced a "tundra buggy" simulated ride before viewing polar bears and other creatures. Some call this the "Disneyfication" of zoos, after a new zoo that the Walt Disney Company opened in Florida in the late 1990s; one feature is a drive-through, safari-style bus ride, where the storyline is that visitors are joining a patrol to catch poachers.

The actual educational value of these new-style zoo exhibits is debatable, and a still greater unknown is whether the lessons learned by visitors will translate into conservation actions. Conservation education, zoos have argued, is one of their major contributions, and one that is becoming increasingly important as citizens become further distanced from nature. Zoos are trying to justify these claims, and also to do a better job of conservation education. Thus, zoos are beginning to scrutinize their exhibits and programs using evaluation techniques. Evaluation methods, including surveys, interviews, or observation of visitors, are sometimes employed during the design phase to improve the message before expensive construction begins. They may also be used at other times to help fine-tune exhibits or otherwise inform about how the zoo is working or what the audience is like. The science of evaluation is rapidly developing in zoos and museums, and it will probably become an increasingly important part of zoo planning.

As the world becomes more wired to the Internet, zoos are following suit. Many zoos now have Web sites where visitors can find information about the zoo. Electronic mail is becoming a valuable communication tool for sharing the specialized knowledge that zoo keeping requires. Some zoo databases, including the vast ISIS (International Species Information Service) registry of zoo animals, are now available at Internet Web sites. Some zoos are incorporating electronic media into their educational efforts, either integrated with exhibit graphics or separately. Cooperative breeding plans, and the data needed to manage them, will increasingly be shared via electronic means. And some Web site programs, for example, one for trumpeter swans, allow school children and others to dynamically follow the movements of reintroduced animals via satellite tracking. Virtual zoos will not keep real animals, which is the primary business of zoos, so zoos will probably continue to find ways to use the Internet that enhance rather than radically change what they do.

As the supply of wild-born founders dwindles, and more amenities are expected for the animals in hand, zoos are abandoning the one-of-each-kind collecting approach. Within-zoo species diversity is decreasing in order to make room for zoos to house a breeding nucleus of at least several pairs and their young offspring. Many zoos are trying to allocate a proportion of their animal space for off-exhibit breeding efforts, and some are operating special facilities devoted to breeding of endangered species—mostly *ex situ* but also some *in situ*. Many zoos are developing specializations, and concentrating their efforts on the specific groups of animals that they can keep best and exhibit most dramatically. Zoo workers are also becoming more specialized, both along taxonomic lines and also professionally into fields such as educators, veterinarians, nutritionists, population biologists, behaviorists, life-support system operators, pathologists, and horticulturists. With the increasing need for cooperation, more formal committees and other structures are developing through the professional organizations.

II. CHOOSING ANIMALS TO PROMOTE BIODIVERSITY CONSERVATION

A. Space, the Final Limit

Modern zoos are expensive to build and upgrade. In North America alone, zoos spend around two million dollars per year on capital or physical improvements, with much larger expenditures for major new exhibits or zoo facilities. A major exhibit can cost in the tens of millions of dollars and typically takes at least three years to plan and build. Maintaining a zoo and its animals is also very expensive. Therefore, there are practical limits to how many zoos can be supported and thus to how many animals can be maintained by zoos. For example, the number of animals that are presently kept in the 181 accredited zoos of North America is approximately 800,000.

Thus, conservation-oriented zoos choose their inhabitants with care in order to maximize the biodiversity that they preserve in their finite space. A biodiversity trade-off must be made between the number of species that are maintained by zoos versus the genetic diversity or size of the captive populations that are maintained. Zoo planners have to think long-term because they are trying to maintain viable populations for decades or centuries. In many cases, zoo populations are, essentially, closed gene pools because it is difficult or impossible to acquire fresh bloodlines or founders. Extinction in nature is the most obvious cause for a closed captive population, and zoos work with several "extinct" species such as the Prezwalski's horse. But species that are extant in nature can also be effectively isolated in zoos by government import or export regulations, concerns about disease transmission, rarity in nature, political issues, or the extreme difficulty and expense of collecting some species. In most cases zoo planners also try to prevent domestication of their wild animals, or, in other words, they seek to freeze evolution or at least retard inbreeding. Arguably, this is a prudent and responsible strategy should the animals be needed someday for restocking wild habitats.

Hence, zoo population plans usually strive to breed animals better than would be done randomly to retard natural selection, and they try to do this using as few animals as possible. This usually requires a captive population of several hundred animals. In this effort, attention must often be paid to picking the right unit for conservation, or the evolutionarily significant unit. Zoos have funded a considerable amount of applied molecular genetic research in order to determine subspecies and species characteristics, so that they are breeding animals that should interbreed, and so that they are not maintaining subviable numbers of several subspecies or species rather than a viable population of one.

B. Metamanagement of Zoo Biodiversity

Each zoo is an island too small to do much conservation breeding on its own. To facilitate cooperation, organized conservation programs involving captive animals are beginning to develop around the world through regional zoo professional associations. Most of the labor that goes into these programs is pro bono or volunteer. Much of the work is accomplished through workshop meetings and via committee, at considerable personal and institutional expense.

The basic unit of these programs is the studbook, which is the database of historic and living animals in a particular captive population. Studbooks are usually maintained through the efforts of one "studbook keeper." Populations that are monitored with studbooks can be managed through regional or even global captive breeding plans, usually overseen by a committee, which specifies where each animal should be and whether it should be bred.

In order to prioritize how much captive space various populations should be allocated, zoo specialist groups for each kind of animals have formed, and are known as "taxon (for taxonomic) advisory groups." Taxon advisory groups typically work closely with the appropriate taxonomic and scientific specialist groups of the World Conservation Union (IUCN), as well as with scientific advisors. Regional taxon advisory groups tend to develop priorities that are complementary to those in other areas; for example, Europe may choose to allocate space to breed a different species of penguin than North America. Taxon advisory group priorities are formalized into "regional collection plans," which ideally include a justification for including or excluding each species or subspecies in the taxonomic group, as well as setting a target population size for each managed population. These plans may also specify areas where research is needed, or where field conservation should be a priority. Tools are being developed to facilitate regional planning, including meta-collection software that has been developed by Australian zoos.

The representation of biodiversity in zoos is biased toward animals that are furry, large, brightly colored, or otherwise captivating for people. In general, the more closely related to humans, the greater the representation. So, most kinds of primates are kept in zoos and managed through cooperative programs, whereas only a fraction of rodents or bats are represented. Similarly, most kinds of flamingos can be found in captivity, whereas only a fraction of the small perching birds are kept. A smaller proportion of birds are maintained in organized programs than mammals, but only a fraction of the scaly creatures (fish, reptile, and amphibian) are kept. Very few invertebrates are represented in captivity, but this seems to be changing as butterfly exhibits and insect displays are gaining in popularity. Invertebrate husbandry is relatively primitive for both aquatic and terrestrial species; the public generally considers the collection of wild invertebrates acceptable so there has been less pressure to develop husbandry techniques. The situation for fish, and most other nonmammalian aquarium denizens, is similar to that of invertebrates. Much work is needed to develop breeding programs for aquariums and invertebrate collections.

C. Institutional Planning

Zoos as individual institutions are also becoming very mission-oriented. Many zoos have developed mission statements and master plans for growth and development. These plans provide the overall organization of the zoo exhibits and the visitor traffic patterns as well as locations of facilities for merchandise, refreshments, and bathrooms. (Bathroom location is amazingly important for a positive visitor experience!) Master plans usually designate the kind of educational messages the zoo will try to convey and the themes of the exhibits. They may be informed by practical considerations of what animals and plants will do best in the zoo, and these may be based on local climate or the experience and expertise of the zoo's staff.

Within the context of the master plan, zoo planners, typically curators, will develop an animal collection plan. Ideally, curators will develop these plans after studying the regional collection plans that have been developed for metamanagement of zoo collections. This collection plan will usually list each species that the zoo keeps or wishes to obtain, along with a justification for what the species will contribute to the zoo and to conservation of the species. The plan should say how many individuals there are of each sex, and how many are desired. The plan will specify where these animals will be housed. If the species is managed by a cooperative breeding plan, then the institutional collection plan should include the recommendations for which animals are to be bred and which not; if not part of a cooperative plan, the curator should have a breeding and possibly a research plan of their own. Institutional collection plans usually need to be updated annually.

III. KEEPING BIODIVERSITY

A. Zoo Animal Records Keeping

Basic to managing animals and improving husbandry is keeping track of the inventory and gathering data. The memory of zoo staff is no longer sufficient to keep these records, as animals live longer and breed better, the government expects more information, and both animals and keepers are more mobile. Zoos are tracking an ever-increasing amount of information about their animals. Each animal in the inventory, current and historic, is accessioned, which means it is assigned a unique identification number. The animal also must be physically identifiable, so most are tagged or marked; this can be done with traditional bands and tags or with more high-tech injectable microchips. Records are kept for diverse variables, including characteristics that allow an animal to be individually identified, dates of birth and death and other life-stage transitions, sex, parents, offspring, current and previous locations, health notes, cage mates, previous owners, physical parameters such as measurements, species, and subspecies, and behavioral observations. Zoos are standardizing their data collection and records so that all this information can be pooled.

Keeping these records has turned into a specialty of its own, and the keepers of the data are usually called "registrars." Registrars may also be responsible for some of the government permits and paperwork associated with modern zoo keeping. These data are generally organized into a computer database. The most widely used program is ARKS, the Animal Records Keeping System that was developed by ISIS in 1985. ARKS generates a variety of standard reports. In addition, ARKS is useful for looking up simple questions about individual animals, for example, potential exposure to disease, as well as for mining the data to seek trends or clues to larger questions.

Most zoos send their ARKS data to ISIS at regular intervals so that it is backed-up and also so that parts of the central database can be combined to create an international inventory. This international data is useful for various other uses such as cooperative breeding programs, scientific research, or for governmental authorities. A weakness of ARKS is its limited ability to manage data for group-living or colonial species. Thus ARKS is of little use for managing animals such as schooling fish, many invertebrates, colony-nesting birds, or other animals that are kept in groups with multiple adult males and females. Similarly, animals that breed in relatively large enclosures (such as ducks in large ponds or fish in large tanks) are difficult to keep track of with individual records. Because of this weakness, animal records for aquaria and invertebrate collections are not standardized.

B. Zoo Population Management Strategies

Zoos have historically made much of their "babies," as births used to be relatively rare and exciting events. But as zoos have become more cooperative, an unfortunate cycle has developed. Rare breeding successes for a particular species tended to be replicated and replicated again. Then, zoo curators would notice that there was a surplus of the species and they could no longer find good homes for the babies. So all zoos would stop breeding the species. Some years later, as the stocks of the animal were aging and starting to die off, there would again be room and demand for babies, and curators would try to resume breeding. But all too often, it would be too late for the aging stocks and the captive population would be lost.

To prevent these boom-and-bust breeding cycles, and also to avoid inbreeding and subsequent loss of vigor, zoo biologists have developed population management strategies and tools. Key to zoo population management is gathering information about all the animals in the managed, captive population (the managed population often is not identical to the captive population because some zoos may opt out of the cooperative program and some individual private breeders may participate). The population data are kept in studbooks, which contain genealogical information as well as demographic information. As in zoo records keeping, ISIS distributes the most popular database for studbooks, a program called SPARKS, which stands for single population analysis and records keeping system. Many European studbook keepers use a similar program called "Zoobook." These studbook programs integrate with computer software that performs much of the demographic and genetic analyses that are essential to zoo population management.

Two principle kinds of conservation breeding programs are the current standards. The first is a program that participants essentially sign a contract to join, and agree to abide by the decisions of an elected committee. This type of program was first introduced in the United States, and was named the species survival plan or SSP; nearly identical programs are known by other names in other regions of the world. The second kind of program is a less intensive one, in which recommendations are issued but individual zoos elect whether or not to cooperate. In North America this second kind of program is known as a population management plan or PMP. These two kinds of programs follow similar population management strategies but differ more in the intensity and politics of the actual management. The SSP also has conservation, research, and education components, whereas the PMP is strictly for population management.

Planning for conservation breeding programs generally begins by determining the purpose of the program. Will animals from the population eventually be returned to the wild, or will the primary purpose of the population be for exhibition in zoos? The managers should decide what the appropriate population is for management purposes, for example, whether to limit the management to a subset of the animals that are known to derive from founders of the same subspecies, while excluding those that are of more uncertain ancestry. Once these decisions are made, demographic and genetic goals are set. Genetic and demographic analyses are done using the studbook data and specially designed software. The analyses are used to estimate parameters, which are needed for the long-range population forecasts that are used to guide the goals. Goal setting is done with the help of analytical tools that consider the number of founders (amount of genetic diversity that was likely to be captured), the effective population size N_e, the current population size, the rate of population growth, the number of years since the program was founded, and the generation time. Typically, conservation-breeding programs aim to preserve greater than 90% of the original diversity for at least 100 years. Sometimes it becomes apparent that demographic intervention is needed, perhaps to increase the population growth rate, lengthen the generation time, or increase the number of spaces available so that the population can grow. At other times managers realize that more founders must be imported, if possible from zoos in another region of the world, but sometimes from nature.

As an aside, zoos have sustained much criticism from conservationists for taking animals from the wild for exhibition, including charges that this has led to the depletion of wild stocks. Zoos rebut that the numbers of animals that they take are trivial compared with animals that die through habitat loss, hunting, or other human activities. Good captive management reduces the number of founders, or wild-born breeding stock, needed for long-term population viability. Theoretically, the required number of founders is 25 individuals picked at random from a wild population, but many captive populations start with only a fraction of this number. Even with good management, new founders may need to be added, since not all founders will produce the ideal number of offspring. Founders should be collected with minimal impact on the wild population. The way that this is best accomplished depends upon the natural history of the species and the specifics of the program. For example, one low-impact approach to collecting founders is to salvage orphaned or injured nonreleasable wild animals; this opportunistic approach has the disadvantage that it may be slow and that disabled wildlife can constitute poor breeding stock.

Once the goals of the population are determined, and plans are made to redress problems such as too few founders, a detailed conservation-breeding plan is developed. These plans give a recommendation for each animal in the managed population, which will include whether the animal should be moved, if it should be bred, which animal it should be paired with, and whether any special actions are needed, such as fertility testing.

Scientific principles are given a heavy weight in the planning. Zoo biologists tend to start with demography, using a Leslie matrix approach that incorporates both current and historic data. The demographic software tools were developed at the National Zoological Park by Jon Ballou and Laurie Bingaman-Lackey. Demographic data sets for zoos are often small, but contain highly detailed information compared to wild populations. Therefore zoo demographic estimates are often very crude and may require considerable tweaking. Early in the analysis, zoo biologists look at the population's age structure and sex ratio. If these are skewed, efforts may be needed to create a better demographic balance. The next consideration is rates of fertility, infant and adult mortality; these estimates are used to project the number of breeding pairs that will be required in order to end up with the desired number of births that will balance deaths and also generate the recommended population growth or shrinkage.

Genetics usually are the primary variables considered in deciding the animal-by-animal breeding priorities, except that aging animals with only a few breeding years left may be given special breeding priority. Genetic algorithms, largely developed by Brookfield Zoo's Robert Lacy, use pedigree-based genetic simulations to assess the genetics of each animal and potential pairing in the population. Managers attempt to simultaneously maximize three parameters when recommending breeding matches: (1) good matches will improve the genetic diversity of the population (i.e., by increasing underrepresented lineages), (2) good matches will attempt to balance the genetic value of male and female, and (3) good matches will also avoid inbreeding with close relatives. Balancing the genetic value, which is done by looking at a parameter called mean kinship, or the effective number of close kin, is important because it breeds big families with other big ones and little families with other little ones. It is detrimental to mix overrepresented and underrepresented lineages because pedigree webs become increasingly hard to untangle through time, and it is easier to balance representation over time when the lines are kept separate.

Although the science of conservation breeding has advanced considerably, practical considerations also must enter into planning. Wild animal keeping in practice is a complex art, and it would be foolish to ignore details such as the distance of a proposed transfer and behavioral and physiological observations. For example, there is no point in recommending a pairing of two gorillas that are incompatible. Similarly, if an animal has not bred after many attempts at several zoos, there may be a physiological problem or perhaps managers have not yet tried the right trick.

Group breeding species are a special challenge to zoo population biologists as they are for zoo animal records keepers. The methods outlined above work poorly for populations in which a significant proportion of the pedigree is unknown, and also when managers cannot control which individuals mate with which, as is generally the case with group breeding animals. An approach that is sometimes used is to completely ignore the pedigree and manage the species as a subdivided population, with occasional transfers of animals between demes or subpopulations. But the pedigree-free approach often throws out large amounts of information about a population and results in overly conservative management that will require larger populations than might otherwise be needed. Similarly, species whose life history is more size-dependent than age-dependent are not well served by current methods, and may be better managed using stage-based population management. As zoos move beyond management of animals with relatively simple breeding biology, such as tigers, and tackle species with more complex life histories, new population approaches will be needed. Similarly, as the "brave new world" of reproductive technology develops, frozen zoos will need to become integrated into population planning.

C. Frozen Zoos and Other High-Tech Solutions

An alluring prospect is that someday we may be able to preserve biodiversity in bottles on a warehouse shelf. Another enticing prospect is the use of reproductive technology to propagate animals that have not birthed through conventional techniques. New technologies are making these prospects possible in some cases.

"Assisted reproduction" is a catchall phrase for a variety of veterinary or endocrine interventions used to promote breeding. Artificial insemination, using fresh or frozen sperm, has been in use for many years in the domestic livestock industry and also for some zoo animals. The techniques can be readily adapted to zoo animals if there is a closely related domestic model in which the technique has been developed; for example, rare species of cows can benefit from research with domestic cows. Zoos have also employed other techniques, such as *in vitro* fertilization with harvested eggs, or embryo transfer both within and between species. The idea of growing a rare species in the womb of a common one is appealing, and has been achieved in some instances.

"Frozen zoos" are being developed by several zoos. In these super-chilled repositories, sperm, eggs, and embryos are committed to long-term storage. Now that cloning is also a real possibility, frozen zoos of nonreproductive tissue are also being developed. Saving biodiversity in freezers is tantalizing, but there are many limitations to this approach and hurdles to be overcome.

Assisted reproductive techniques are difficult and costly to adapt to animals that do not have a closely related model, and also for those with complex reproductive physiology, for example, animals with induced ovulation. By and large, the techniques have not been cracked for animals with shelled eggs. We cannot, as of yet, grow embryos in a test tube, so those derived from a frozen zoos must be grown in a real womb (or egg). This means a real population of surrogates will be needed to mother the frozen zoo, although in some cases these surrogate mothers can be of a different species. Surrogate mothers may produce animals with species-inappropriate behaviors and possibly even a hybrid phenotype.

Frozen zoos will not preserve learned animal behaviors (culture), and they will not eliminate the need to keep real populations. Frozen zoos, cloning techniques and the like, as San Diego's Oliver Ryder predicts, will probably become powerful adjuncts to the management of populations. So far, genetic and demographic models for managing zoo populations have not factored in frozen resources, but someday they probably will. This will allow managers to keep smaller numbers of living animals. In addition, relative to the transfer of whole animals, the transfer of sperm, egg, and embryos between wild and captive populations can be done with far less risk of cotransferring disease-causing pathogens. Thus, these techniques may be highly useful in importing new gene lines into small populations, both wild and captive.

IV. SUPPORTING DIVERSITY IN NATURE

A. The Ark Paradigm

Enamored, perhaps, with their newfound ability to breed and manage exotic animals, zoos and their commentators were for a time fixated upon the "ark" paradigm of conservation. That is, like the biblical Noah, they talked about saving species in captivity until some time in the future when the flood of human population shall recede and there will again be the space, and also the will, to restore Nature. Zoos have actually had some success saving species this way. The list of species that have been rescued from extinction by taking refuge for some time in zoos is ever-growing and includes black-footed ferret, California condor, European bison, Mongolian wild horse, Arabian oryx, Partula snails, and Guam rails. However, some species have become extinct during their captive sojourn and no opportunity has arisen to restore some others, for example, the birds of Guam. We are seeing that restoration of a fauna may be only a remote possibility and the utility of reintroduction overstated in some ways.

B. The Post-Ark Paradigm

Maintenance of captive stocks has several functions, but the primary purpose is not, as it is popularly misperceived, for future reintroduction efforts. Zoos may indeed rescue individual animals and sequester them in the safety of captivity. As zoo professionals look beyond the ark, they are developing many other ways to contribute to the conservation of exotic animals.

Exciting, educational exhibits are potentially zoos' preeminent contribution to conserving animals and their habitats. Worldwide, zoos are estimated to attract over 600 million visitors annually. The majority of these visitors live in urban areas and otherwise have little contact with wildlife. Zoos are beginning to study and develop approaches for mobilizing visitors for conservation through innovative exhibit and educational programs. North America's accredited zoos report that each year 9 million students visit and enjoy onsite education programs, and in addition 85,000 teachers profit from teacher training; Some zoos are even developing "museum schools" in which a zoo or museum actually serves as an alternative school for a small number of students who typically combine rigorous college preparatory curricula with hands-on or applied learning in the zoo setting (more information about museum schools can be found at http://www.astc.org). Zoo-based conservation education programs need not end at the exit gates. Zoo educators and graphic artists can also develop conservation curricula that can be used in other parts of the world, including the native range of some endangered species. There is a remarkable dearth of educational materials about most endangered species within their native ranges, and any effort that zoos can make to redress this problem will be most helpful. Zoos may also need to develop curricula that more directly address the biodiversity crisis, and they may need to reach out more effectively to decision makers (politicians).

Zoos can serve as refuges for displaced, injured, or confiscated wildlife. Thus, zoos help wildlife authorities enforce regulations by providing homes for seized wildlife. In addition, many zoos maintain patches of artificial or natural habitats for native wildlife and can provide significant breeding refuges for some birds and butterflies, and for fish and small animals that need relatively little space for a viable population.

Zoos sponsor significant basic research, often studies that would be difficult or impossible to accomplish with animals in the wild. Furthermore, zoos invest a considerable amount of money in research; for example, accredited zoos in North America spent 51 million dollars on scientific research in 1998. While developing technology and methodology for captive animals under relatively controlled conditions, zoos create expertise that is useful *in situ*, for example, guidelines for medication dosages.

Direct financial support of field conservation is another way that zoos are helping preserve biodiversity. In 1998, accredited North American zoos were involved in more than 700 conservation projects in 80 countries, including providing grants to non-zoo researchers. Another form of *in situ* conservation funding is for zoos in the wealthier nations to help improve the conservation impact of zoos in lesser developed nations. Many zoos also sponsor member tours to see wildlife in nature, and they try to make arrangements that will have a positive impact on conservation.

C. The Ark Revisited

Human populations are still increasing and wildlife will continue to disappear. More direct conservation measures will be needed from all zoos. Zoos will be called upon to provide refuge for more species as the wild populations dwindle. Shrinking wild populations in postage-stamp parks may need to be managed more like zoo populations, with infusions of unrelated blood, potentially from zoo stocks. Intensive, joint, meta-population management of *in situ* and *ex situ* wild animals will require both scientific and political innovation. Zoos may indeed help to save some wildlife from the flood of humanity, but they probably will not be acting as closed vessels. Rather, expect to see zoo-based conservation continue to become increasingly dynamic and interactive.

See Also the Following Articles

BREEDING OF ANIMALS • CAPTIVE BREEDING AND REINTRODUCTION • CONSERVATION EFFORTS, CONTEMPORARY • EDUCATION AND BIODIVERSITY • ETHICAL ISSUES IN BIODIVERSITY PROTECTION • EX SITU, IN SITU CONSERVATION • NATURAL RESERVES AND PRESERVES

Bibliography

Conway, W. G. (1986). The practical difficulties and financial implications of endangered species breeding programmes. *Internatl. Zoo Yearbook* 24/25, 210–219.

Gibbons, E. F. Jr., Durrant, B. S., and Demarest, J. (eds.) (1995). "Conservation of Endangered Species in Captivity." State University of New York Press, Albany, NY.

Hoage, R. J., and Deiss, William A. (eds.) (1996). "New Worlds, New Animals: From Menagerie to Zoological Park in the Nineteenth Century." Johns Hopkins University Press, Baltimore, MD.

IUDZG/CBSG (IUCN/SSC). (1993). "The World Zoo Conservation Strategy; the Role of the Zoos and Aquaria of the World in Global Conservation." Chicago Zoological Society, Brookfield, Illinois.

Norton, B. G., Hutchings, M., Stevens, E. F., and Maple, T. L. (eds.) (1995). "Ethics on the Ark: Zoos, Animal Welfare and Wildlife Conservation." Smithsonian Institution Press, Washington, DC.

Ryder, O. A., and Benirschke, K. (1997). The potential use of "cloning" in the conservation effort. *Zoo Biol.* **16**, 295–300.

Shepperdson, D., Mellen, J., and Hutchins, M. (eds.) (1998). "Second Nature: Environmental Enrichment for Captive Animals." Smithsonian Institution Press, Washington, DC.

Web site resources:

American Zoo and Aquarium Association: http://www.aza.org
ISIS (International Species Inventory System): http://www.isis.org
IUCN Captive Breeding Specialist Group: http://www.cbsg.org
ZooLex (animal exhibit design): http://www.zoolex.org

Contributors

TAKUYA ABE (deceased)
ISOPTERA
   Center for Ecological Research
   Kyoto University
   Kyoto, Japan

JOACHIM ADIS
ARTHROPODS (TERRESTRIAL), AMAZONIAN
   Max-Planck-Institute for Limnology
   Plön, Germany
   in cooperation with the National Institute for
   Amazonian Research, Brazil

GREGORY H. ADLER
RAINFOREST ECOSYSTEMS, ANIMAL DIVERSITY
   University of Wisconsin, Oshkosh
   Oshkosh, Wisconsin, USA

ROSS A. ALFORD
AMPHIBIANS, BIODIVERSITY OF
   James Cook University
   Townsville, Australia

EDITH B. ALLEN
RESTORATION OF ANIMAL, PLANT, AND
MICROBIAL DIVERSITY
   University of California, Riverside
   Riverside, California, USA

MICHAEL F. ALLEN
RESTORATION OF ANIMAL, PLANT, AND
MICROBIAL DIVERSITY
   Center for Conservation Biology
   University of California, Riverside
   Riverside, California, USA

ALFONSO ALONSO
FRAMEWORK FOR ASSESSMENT AND MONITORING
OF BIODIVERSITY
   Smithsonian Institution
   Monitoring and Assessment of Biodiversity Program
   (SI/MAB)
   Washington, DC, USA

PETER ALPERT
STEWARDSHIP, CONCEPT OF
   University of Massachusetts
   Amherst, Massachusetts, USA

MIGUEL A. ALTIERI
AGRICULTURE, TRADITIONAL
   University of California, Berkeley
   Berkeley, California, USA

SONIA ALTIZER
DISEASES, CONSERVATION AND
   Princeton University
   Princeton, New Jersey, USA

OLOF ANDRÉN
INVERTEBRATES, TERRESTRIAL, OVERVIEW
   Swedish University of Agricultural Sciences
   Uppsala, Sweden

MARCOS A. ANTEZANA
DIVERSITY, MOLECULAR LEVEL
   University of Chicago
   Chicago, Illinois, USA

MICHAEL F. ANTOLIN
GENES, DESCRIPTION OF
   Colorado State University
   Fort Collins, Colorado, USA

J. DAVID ARCHIBALD
DINOSAURS, EXTINCTION THEORIES FOR
San Diego State University
San Diego, California, USA

MIKE ASHMORE
AIR POLLUTION
University of Bradford
West Yorkshire, England, UK

AMY T. AUSTIN
TEMPERATE GRASSLAND AND SHRUBLAND ECOSYSTEMS
Universidad de Buenos Aires
Buenos Aires, Argentina

JONATHON BALLOU
INBREEDING AND OUTBREEDING
Smithsonian Institution
Washington, DC, USA

ROBERT BARBAULT
LOSS OF BIODIVERSITY, OVERVIEW
Institut Fédératif d'Ecologie Fondamentale et Appliqúee (CNRS-FR3)
Paris, France

RICHARD T. BARBER
OCEAN ECOSYSTEMS
Duke University
Beaufort, North Carolina, USA

TODD J. BARKMAN
HEMIPARASITISM
Pennsylvania State University
College Park, Pennsylvania, USA

JOHN A. BARONE
DEFENSES, ECOLOGY OF
Mississippi State University
Mississippi State, Mississippi, USA

NICHOLAS H. BARTON
DIFFERENTIATION
Institute of Cell, Animal and Population Biology
University of Edinburgh
Edinburgh, Scotland, UK

SONIA D. BATTEN
PELAGIC ECOSYSTEMS
Sir Alister Hardy Foundation for Ocean Science
Plymouth, England, UK

KAMALJIT S. BAWA
LOGGED FORESTS
Ashoka Trust for Research in Ecology and the Environment
Bangalore, India

F. A. BAZZAZ
RESOURCE PARTITIONING
Harvard University
Cambridge, Massachusetts, USA

KAREN H. BEARD
CONSERVATION EFFORTS, CONTEMPORARY
Yale University
New Haven, Connecticut, USA

JOHN BEDDINGTON
RESOURCE EXPLOITATION, FISHERIES
Imperial College of Science, Technology and Medicine
London, England, UK

BARBARA L. BEDFORD
WETLANDS ECOSYSTEMS
Cornell University
Ithaca, New York, USA

ANDREA BELGRANO
PELAGIC ECOSYSTEMS
The Royal Swedish Academy of Sciences
Stockholm, Sweden

CHRISTOPHER A. BELL
VERTEBRATES, OVERVIEW
University of Texas
Austin, Texas, USA

PHILIP BENSTEAD
WETLANDS RESTORATION
Royal Society for the Protection of Birds
Surrey, England, UK

FIKRET BERKES
RELIGIOUS TRADITIONS AND BIODIVERSITY
University of Manitoba
Winnipeg, Manitoba, Canada

R. J. BERRY
PHENOTYPE, A HISTORICAL PERSPECTIVE ON
University College London
London, England, UK

MALCOLM BEVERIDGE
AQUACULTURE
University of Stirling
Stirling, Scotland, UK

WILLIAM C. BLACK, IV
GENES, DESCRIPTION OF
Colorado State University
Fort Collins, Colorado, USA

ANDREW R. BLAUSTEIN
ULTRAVIOLET RADIATION
Oregon State University
Corvallis, Oregon, USA

WILLIAM BOND
FIRES, ECOLOGICAL EFFECTS OF
University of Cape Town
Rondebosch, South Africa

CRISTINA BONILLA-WARFORD
RESTORATION OF BIODIVERSITY
University of Wisconsin
Madison, Wisconsin, USA

MICHAEL G. BOOTH
CONSERVATION EFFORTS, CONTEMPORARY
   Yale University
   New Haven, Connecticut, USA

ELGENE O. BOX
ASIA, ECOSYSTEMS OF
   University of Georgia
   Athens, Georgia, USA

RICHARD W. BRAITHWAITE
TOURISM, ROLE OF
   CSIRO Tourism Research Program
   Lyneham, Australia

SUSANNE BRAKMANN
ORIGIN OF LIFE, THEORIES OF
   Max-Planck Institut für Biophysikalische Chemie
   Göttingen-Nikolausberg, Germany

ROB BROOKER
ARCTIC ECOSYSTEMS
   Institute of Terrestrial Ecology
   Banchory Kinkardineshire, Scotland, UK

DANIEL R. BROOKS
DIVERSITY, ORGANISM LEVEL
   University of Toronto
   Toronto, Ontario, Canada

BRIAN V. BROWN
FLIES, GNATS, AND MOSQUITOES
INSECTS, OVERVIEW
   Natural History Museum of Los Angeles County
   Los Angeles, California, USA

JOEL S. BROWN
RESTORATION OF ANIMAL, PLANT, AND MICROBIAL DIVERSITY
   University of Illinois
   Urbana-Champaign, Illinois, USA

ROBERT BROWNELL
MARINE MAMMALS, EXTINCTIONS OF
   National Marine Fisheries Service
   Seattle, Washington, USA

JOANN M. BURKHOLDER
EUTROPHICATION/OLIGOTROPHICATION
   North Carolina State University
   Raleigh, North Carolina, USA

GUY L. BUSH
SPECIATION, PROCESS OF
   Michigan State University
   East Lansing, Michigan, USA

PHILIP J. CAFARO
ENVIRONMENTAL ETHICS
ETHICAL ISSUES IN BIODIVERSITY PROTECTION
   Colorado State University
   Fort Collins, Colorado, USA

JANINE N. CAIRA
WORMS, PLATYHELMINTHES
   University of Connecticut
   Storrs, Connecticut, USA

JOHN CAIRNS, JR.
STRESS, ENVIRONMENTAL
   Virginia Polytechnic Institute and
   State University
   Blacksburg, Virginia, USA

TERRY V. CALLAGHAN
ARCTIC ECOSYSTEMS
   University of Sheffield
   Abisko Scientific Research Station
   Sheffield, England, UK

JAMES T. CARLTON
ENDANGERED MARINE INVERTEBRATES
   Williams College—Mystic Seaport
   Mystic, Connecticut, USA

JUAN CARLOS CASTILLA
MARINE ECOSYSTEMS, HUMAN IMPACT ON
   Pontificia Universidad Católica de Chile
   Santiago, Chile

S. CATOVSKY
RESOURCE PARTITIONING
   Harvard University
   Cambridge, Massachusetts, USA

NEIL CHALMERS
MUSEUMS AND INSTITUTIONS
   The Natural History Museum
   London, England, UK

M. G. CHAPMAN
INTERTIDAL ECOSYSTEMS
   Centre for Research on Ecological Impacts of
   Coastal Cities
   University of Sydney
   Sydney, Australia

BRIAN CHARLESWORTH
POPULATION GENETICS
   University of Edinburgh
   Edinburgh, Scotland, UK

YURI CHERNOV
ARCTIC ECOSYSTEMS
   Severtzov Institute of Evolution
   Moscow, Russia

PETER CHESSON
METAPOPULATIONS
   University of California, Davis
   Davis, California, USA

ELLEN W. CHU
ENVIRONMENTAL IMPACT, CONCEPT AND
MEASUREMENT OF
   Northwest Environment Watch
   Seattle, Washington, USA

COLIN W. CLARK
COMMONS, CONCEPT AND THEORY OF
   University of British Columbia
   Richmond, British Columbia, Canada

W. A. CLEMENS
MAMMALS (PRE-QUATERNARY), EXTINCTIONS OF
University of California, Berkeley
Berkeley, California, USA

JONATHAN A. CODDINGTON
ARACHNIDS
Smithsonian Institution
Washington, DC, USA

DEBRA P. C. COFFIN
LANDSCAPE DIVERSITY
United States Department of Agriculture
Agricultural Research Service
Jornada Experimental Range
Las Cruces, New Mexico, USA

JOHAN COLDING
TRADITIONAL CONSERVATION PRACTICES
Stockholm University and Beijer International
Institute of Ecological Economics
Stockholm, Sweden

DAVID C. COLEMAN
SOIL BIOTA, SYSTEMS AND PROCESSES
University of Georgia
Athens, Georgia, USA

PHYLLIS D. COLEY
DEFENSES, ECOLOGY OF
University of Utah
Salt Lake City, Utah, USA

N. J. COLLAR
ENDANGERED BIRDS
BirdLife International
Cambridge, England, UK

ROBERT K. COLWELL
ARACHNIDS
University of Connecticut
Storrs, Connecticut, USA

JAMES A. COMISKEY
FRAMEWORK FOR ASSESSMENT AND MONITORING OF BIODIVERSITY
Smithsonian Institution
Monitoring and Assessment of Biodiversity Program
(SI/MAB)
Washington, DC, USA

SEAN R. CONNOLLY
FOSSIL RECORD
University of Arizona
Tucson, Arizona, USA

EDWARD F. CONNOR
SPECIES-AREA RELATIONSHIPS
San Francisco State University
San Francisco, California, USA

PETER CONVEY
ANTARCTIC ECOSYSTEMS
British Antarctic Survey
National Environment Research Council
Cambridge, England, UK

HOWARD V. CORNELL
DIVERSITY, COMMUNITY/REGIONAL LEVEL
University of Delaware
Newark, Delaware, USA

ALAN P. COVICH
ENERGY FLOW AND ECOSYSTEMS
Colorado State University
Fort Collins, Colorado, USA

R. M. COWLING
ENDEMISM
University of Cape Town
Rondebosch, South Africa

PAUL ALAN COX
PHARMACOLOGY, BIODIVERSITY AND
National Tropical Botanical Garden
Lawai, Hawaii, USA

CATHERINE L. CRAIG
EVOLUTION, THEORY OF
Harvard University
Cambridge, Massachusetts, USA
Tufts University
Medford, Massachusetts, USA

KEVIN CROOKS
PREDATORS, ECOLOGICAL ROLE OF
University of California, San Diego
San Diego, California, USA

DAVID C. CULVER
SUBTERRANEAN ECOSYSTEMS
American University
Washington, DC, USA

GRETCHEN DAILY
OSYSTEM SERVICES, CONCEPT OF
Stanford University
Stanford, California, USA

FRANCISCO DALLMEIER
FRAMEWORK FOR ASSESSMENT AND MONITORING OF BIODIVERSITY
Smithsonian Institution
Monitoring and Assessment of Biodiversity Program
(SI/MAB)
Washington, DC, USA

AVINOAM DANIN
NEAR EAST ECOSYSTEMS, PLANT DIVERSITY
The Hebrew University of Jerusalem
Jerusalem, Israel

PARTHA DASGUPTA
ECONOMIC VALUE OF BIODIVERSITY, OVERVIEW
University of Cambridge
Cambridge, England, UK

SHAMIK DASGUPTA
ECOSYSTEM SERVICES, CONCEPT OF
University College London
London, England, UK

FRANK W. DAVIS
DISTURBANCE, MECHANISMS OF
    University of California, Santa Barbara
    Santa Barbara, California, USA

F. A. A. M. DE LEIJ
ECOTOXICOLOGY
    University of Surrey
    Surrey, England, UK

IAN DENHOLM
INSECTICIDE RESISTANCE
    IACR—Rothamsted
    Hertfordshire, England, UK

CLAUDE W. DEPAMPHILIS
HEMIPARASITISM
    Pennsylvania State University
    University Park, Pennsylvania, USA

GREG DEVINE
INSECTICIDE RESISTANCE
    IACR—Rothamsted
    Hertsfordshire, England, UK

PHILIP J. DEVRIES
BUTTERFLIES
    Center for Biodiversity Studies
    Milwaukee Public Museum
    Milwaukee, Wisconsin, USA

MARK DEYRUP
ENDANGERED TERRESTRIAL INVERTEBRATES
    Archbold Biological Station
    Lake Placid, Florida, USA

SANDRA DÍAZ
ECOSYSTEM FUNCTION MEASUREMENT, TERRESTRIAL COMMUNITIES
    Universidad Nacional de Córdoba—CONICET
    Córdoba, Argentina

RODOLFO DIRZO
BIODIVERSITY-RICH COUNTRIES
CENTRAL AMERICA, ECOSYSTEMS OF
    Instituto de Ecología
    Universidad Nacional Autonoma de México
    Mexico City, Mexico

ANDREW P. DOBSON
CONSERVATION BIOLOGY, DISCIPLINE OF
    Princeton University
    Princeton, New Jersey, USA

BARBARA S. DRAUSAL
EDIBLE PLANTS
    Universidad Nacional de Comahue and CONICET
    Neuguen, Argentina

CARLOS M. DUARTE
SEAGRASSES
    Instituto Mediterráneo de Estudios Avanzados (CSIC-UIB)
    Esporles, Islas Baleares, Spain

PAUL V. DUNLAP
MICROBIAL BIODIVERSITY
    Center of Marine Biotechnology
    Biotechnology Institute
    University of Maryland
    Baltimore, Maryland, USA

JENNIFER A. DUNNE
GREENHOUSE EFFECT
    University of California, Berkeley
    Berkeley, California, USA

DONALD N. DUVICK
BREEDING OF PLANTS
    Iowa State University
    Johnston, Iowa, USA

G. J. EBLE
BIODIVERSITY, ORIGIN OF
    Sante Fe Institute
    Sante Fe, New Mexico, USA

JOAN G. EHRENFELD
PLANT-SOIL INTERACTIONS
    Rutgers University
    New Brunswick, New Jersey, USA

PAUL R. EHRLICH
HUMAN EFFECTS ON ECOSYSTEMS, OVERVIEW
    Stanford University
    Stanford, California, USA

BRIAN ENQUIST
PLANT COMMUNITIES, EVOLUTION OF
    Cornell University
    Ithaca, New York, USA

DOUGLAS H. ERWIN
MASS EXTINCTIONS, NOTABLE EXAMPLES OF
    National Museum of Natural History
    Washington, DC, USA

TERRY L. ERWIN
FOREST CANOPIES, ANIMAL DIVERSITY
    Smithsonian Institution
    Washington, DC, USA

JAMES ESTES
PREDATORS, ECOLOGICAL ROLE OF
    University of California, Santa Cruz
    Santa Cruz, California, USA

JULIAN EVANS
DEFORESTATION AND LAND CLEARING
    Imperial College of Science, Technology and Medicine
    University of London
    London, England, UK

TIMOTHY J. FAHEY
FOREST ECOLOGY
    Cornell University
    Ithaca, New York, USA

PAUL FALKOWSKI
BIOGEOCHEMICAL CYCLES
  Rutgers University
  New Brunswick, New Jersey, USA

KRISTIAN FAUCHALD
WORMS, ANNELIDA
  Smithsonian Institution
  Washington, DC, USA

PHILIP M. FEARNSIDE
SOUTH AMERICAN NATURAL ECOSYSTEMS, STATUS OF
  National Institute for Research in the Amazon (INPA)
  Manaus, Brazil

TOM FENCHEL
MICROORGANISMS, ROLE OF
  University of Copenhagen, Denmark
  Helsingør, Denmark

BLAND J. FINLAY
PROTOZOA
  Centre for Ecology and Hydrology Windermere
  Cumbria, England, UK

CARL FOLKE
AQUACULTURE
TRADITIONAL CONSERVATION PRACTICES
  Stockholm University and Beijer International Institute of Ecological Economics
  Stockholm, Sweden

PETER L. FOREY
CLADISTICS
  The Natural History Museum
  London, England, UK

JOHANNES FOUFOPOULOS
DISEASES, CONSERVATION AND
  Princeton University
  Princeton, New Jersey, USA

RICHARD FRANKHAM
INBREEDING AND OUTBREEDING
  Macquarie University
  North Ryde, Australia

TESS L. FREIDENBURG
KEYSTONE SPECIES
  Oregon State University
  Corvallis, Oregon, USA

ROBERT S. FRITZ
PLANT HYBRIDS
  Vassar College
  Poughkeepsie, New York, USA

RAINER FROESE
FISH STOCKS
  International Center for Living Aquatic Resources Management
  Laguna, Philippines

KAZUE FUJIWARA
ASIA, ECOSYSTEMS OF
  Yokohama National University
  Yokohama, Japan

DOUGLAS J. FUTUYMA
COEVOLUTION
  State University of New York, Stony Brook
  Stony Brook, New York, USA

MADHAV GADGIL
POVERTY AND BIODIVERSITY
  Indian Institute of Science
  Bangalore, India

ANDREA GAGER
DISEASES, CONSERVATION AND
  Princeton University
  Princeton, New Jersey, USA

KATHLEEN A. GALVIN
HUNTER-GATHERER SOCIETIES, ECOLOGICAL IMPACT OF
  Colorado State University
  Fort Collins, Colorado, USA

CARL GANS
VERTEBRATES, OVERVIEW
  University of Texas
  Austin, Texas, USA

SCOTT L. GARDNER
WORMS, NEMATODA
  University of Nebraska
  Lincoln, Nebraska, USA

KEVIN J. GASTON
INDICATOR SPECIES
  University of Sheffield
  Sheffield, England, UK

ALENE GELBARD
POPULATION STABILIZATION, HUMAN
  Population Reference Bureau, Inc.
  Washington, DC, USA

LEAH GERBER
MARINE MAMMALS, EXTINCTIONS OF
  University of California, Santa Barbara
  Santa Barbara, California, USA

WILLIAM H. GERWICK
PLANT SOURCES OF DRUGS AND CHEMICALS
  Oregon State University
  Corvallis, Oregon, USA

JABOURY GHAZOUL
DEFORESTATION AND LAND CLEARING
  Imperial College of Science, Technology and Medicine
  University of London
  London, England, UK

MICHAEL J. GHISELIN
DARWIN, CHARLES
  California Academy of Sciences
  San Francisco, California, USA

JAMES P. GIBBS
WETLANDS ECOSYSTEMS
State University of New York, Syracuse
Syracuse, New York, USA

ROSEMARY G. GILLESPIE
ADAPTIVE RADIATION
University of California, Berkeley
Berkeley, California, USA

JOSHUA R. GINSBERG
MAMMALS, BIODIVERSITY OF
Wildlife Conservation Society
Bronx, New York, USA

ALEXANDER N. GLAZER
NATURAL RESERVES AND PRESERVES, RESEARCH AND TRAINING
University of California Natural Reserve System
Oakland, California, USA

H. C. J. GODFRAY
PARASITOIDS
Imperial College at Silwood Park
Berkshire, England, UK

SERGEI I. GOLOVATCH
MYRIAPODS
Russian Academy of Sciences
Moscow, Russia

PATRICK GONZALEZ
ENERGY USE, HUMAN
U.S. Agency for International Development
Washington, DC, USA

SARAH C. GOSLEE
LANDSCAPE DIVERSITY
United States Department of Agriculture
Agricultural Research Service
Jornada Experimental Range
Las Cruces, New Mexico, USA

JOHN GRACE
CARBON CYCLE
University of Edinburgh
Edinburgh, Scotland, UK

J. FREDERICK GRASSLE
MARINE ECOSYSTEMS
Institute of Marine and Coastal Sciences
Rutgers University
New Brunswick, New Jersey, USA

JEREMY J. D. GREENWOOD
BIRDS, BIODIVERSITY OF
British Trust for Ornithology
Norfolk, England, UK

J. P. GRIME
HERBACEOUS VEGETATION, SPECIES RICHNESS IN
The University of Sheffield
Sheffield, England, UK

LOUIS J. GROSS
COMPUTER SYSTEMS AND MODELS, USE OF
The Institute for Environmental Modeling
University of Tennessee
Knoxville, Tennessee, USA

J. MORGAN GROVE
URBAN/SUBURBAN ECOLOGY
United States Department of Agriculture Forest Service
Washington, DC, USA

MACE A. HACK
MIGRATION
Princeton University
Princeton, New Jersey, USA

NELSON G. HAIRSTON, JR.
DIAPAUSE AND DORMANCY
Cornell University
Ithaca, New York, USA

TIM HALLIDAY
ENDANGERED REPTILES AND AMPHIBIANS
The Open University
Milton Keynes, England, UK

NATHALIE HAMEL
LAND-USE ISSUES
University of Washington
Seattle, Washington, USA

SUSAN S. HANNA
PROPERTY RIGHTS AND BIODIVERSITY
Oregon State University
Corvallis, Oregon, USA

JOHN HARTE
GREENHOUSE EFFECT
University of California, Berkeley
Berkeley, California, USA

GARY S. HARTSHORN
TROPICAL FOREST ECOSYSTEMS
Organization for Tropical Studies
Durham, North Carolina, USA

GREGG HARTVIGSEN
BIODIVERSITY, EVOLUTION AND
CARRYING CAPACITY, CONCEPT OF
State University of New York, Geneseo
Geneseo, New York, USA

PAUL H. HARVEY
BIODIVERSITY GENERATION, OVERVIEW
University of Oxford
Oxford, England, UK

RICHARD R. HARWOOD
AGRICULTURE, SUSTAINABLE
Michigan State University
East Lansing, Michigan, USA

JOHN HASKELL
PLANT COMMUNITIES, EVOLUTION OF
   Cornell University
   Ithaca, New York, USA

ALAN HASTINGS
POPULATION DYNAMICS
   University of California, Davis
   Davis, California, USA

STEFAN HAUSER
SLASH AND BURN AGRICULTURE, EFFECTS OF
   International Institute of Tropical Agriculture
   Croydon, England, UK

DAVID L. HAWKSWORTH
NOMENCLATURE, SYSTEMS OF
   MycoNova
   London, England, UK

MARK HAY
GRAZING, EFFECTS OF
   Georgia Institute of Technology
   Atlanta, Georgia, USA

GEOFFREY HEAL
BIODIVERSITY AS A COMMODITY
   Columbia University
   New York, New York, USA
   and Stanford University
   Stanford, California, USA

GENE S. HELFMAN
FISH, BIODIVERSITY OF
   University of Georgia
   Athens, Georgia, USA

JOSEPH HELLER
NEAR EAST ECOSYSTEMS, ANIMAL DIVERSITY
   The Hebrew University of Jerusalem
   Jerusalem, Israel

JESSICA J. HELLMANN
SPECIES INTERACTIONS
   Stanford University
   Stanford, California, USA

PAUL HENDERSON
MUSEUMS AND INSTITUTIONS
   The Natural History Museum
   London, England, UK

GEORGE R. HENDREY
ACID RAIN AND DEPOSITIONS
   Brookhaven National Laboratory
   Upton, New York, USA

HENRY A. HESPENHEIDE
BEETLES
   University of California, Los Angeles
   Los Angeles, California, USA

MASAHIKO HIGASHI (deceased)
ISOPTERA
   Center for Ecological Research
   Kyoto University
   Kyoto, Japan

ROBERT S. HILL
SOUTHERN (AUSTRAL) ECOSYSTEMS
   University of Adelaide
   North Terrace, Australia

PETER J. HOGARTH
MANGROVE ECOSYSTEMS
   University of York
   York, England, UK

HOPE HOLLOCHER
SPECIATION, THEORIES OF
   Princeton University
   Princeton, New Jersey, USA

JODIE S. HOLT
HERBICIDES
   University of California, Riverside
   Riverside, California, USA

ROBERT D. HOLT
PREDATORS, ECOLOGICAL ROLE OF
SPECIES COEXISTENCE
   Museum of Natural History and Center for
   Biodiversity Research
   University of Kansas
   Lawrence, Kansas, USA

TAMARA L. HORTON
NUCLEIC ACID BIODIVERSITY
   Princeton University
   Princeton, New Jersey, USA

FRANCIS G. HOWARTH
ADAPTIVE RADIATION
   Bishop Museum
   Honolulu, Hawaii, USA

ROBERT W. HOWARTH
NITROGEN, NITROGEN CYCLE
   Cornell University
   Ithaca, New York, USA

WILLIAM HOWARTH
LITERARY PERSPECTIVES ON BIODIVERSITY
   Princeton University
   Princeton, New Jersey, USA

JENNIFER B. HUGHES
POPULATION DIVERSITY, OVERVIEW
   Stanford University
   Stanford, California, USA

CHRISTOPHER J. HUMPHRIES
CLADOGENESIS
VICARIANCE BIOGEOGRAPHY
   The Natural History Museum
   London, England, UK

GARY R. HUXEL
FOOD WEBS
   University of California, Davis
   Davis, California, USA

DAVID W. INOUYE
POLLINATORS, ROLE OF
    University of Maryland
    College Park, Maryland, USA

PAMELA JAGGER
MARKET ECONOMY AND BIODIVERSITY
    Resources for the Future
    Washington, DC, USA

DANIEL H. JANZEN
LATENT EXTINCTION—THE LIVING DEAD
    University of Pennsylvania
    Philadelphia, Pennsylvania, USA

PAUL C. JEPSON
PESTICIDES, USE AND EFFECTS OF
    Oregon State University
    Corvallis, Oregon, USA

CHRISTOPHER N. JOHNSON
NATURAL EXTINCTION (NOT HUMAN-INFLUENCED)
    James Cook University
    Townsville, Australia

MARA J. JOHNSON
MICROBIAL BIODIVERSITY, MEASUREMENT OF
    University of California, Davis
    Davis, California, USA

NORMAN F. JOHNSON
HYMENOPTERA
    Ohio State University
    Columbus, Ohio, USA

PAUL JOSÉ
WETLANDS RESTORATION
    Royal Society for the Protection of Birds
    Surrey, England, UK

RONEN KADMON
REMOTE SENSING AND IMAGE PROCESSING
    The Hebrew University of Jerusalem
    Jerusalem, Israel

PETER KAREIVA
ECOLOGY, CONCEPT AND THEORIES IN
SALMON
    National Marine Fisheries Service
    Seattle, Washington, USA

JAMES R. KARR
ENVIRONMENTAL IMPACT, CONCEPT AND MEASUREMENT OF
    University of Washington
    Seattle, Washington, USA

J. BOONE KAUFFMAN
RANGE ECOLOGY, GLOBAL LIVESTOCK INFLUENCES
    Oregon State University
    Corvallis, Oregon, USA

NILS KAUTSKY
AQUACULTURE
    Stockholm University and Royal Swedish Academy of Sciences
    Stockholm, Sweden

SEPPO KELLOMÄKI
TIMBER INDUSTRY
    University of Joensuu
    Joensuu, Finland

CYNTHIA KICKLIGHTER
GRAZING, EFFECTS OF
    Georgia Institute of Technology
    Atlanta, Georgia, USA

A. ROSS KIESTER
SPECIES DIVERSITY, OVERVIEW
    United States Department of Agriculture Forest Service
    Corvallis, Oregon, USA

ANN P. KINZIG
URBAN/SUBURBAN ECOLOGY
    Arizona State University
    Tempe, Arizona, USA

IAN J. KITCHING
CLADISTICS
    The Natural History Museum
    London, England, UK

WILLIAM R. KONSTANT
PRIMATE POPULATIONS, CONSERVATION OF
    Conservation International
    Washington, DC, USA

CHRISTIAN KÖRNER
ALPINE ECOSYSTEMS
    University of Basel
    Basel, Switzerland

ABRAHAM B. KOROL
RECOMBINATION
    University of Haifa
    Haifa, Israel

JARI KOUKI
TIMBER INDUSTRY
    University of Joensuu
    Joensuu, Finland

CLAIRE KREMEN
HUMAN EFFECTS ON ECOSYSTEMS, OVERVIEW
    Stanford University
    Stanford, California, USA

HANS KRUUK
CARNIVORES
    Institute of Terrestrial Ecology
    Banchory Kinkardineshire, Scotland, UK

P. SAM LAKE
INVERTEBRATES, FRESHWATER, OVERVIEW
    Monash University
    Melbourne, Australia

DAVID LAMB
REFORESTATION
    University of Queensland
    Brisbane, Australia

P. J. D. LAMBSHEAD
INVERTEBRATES, MARINE, OVERVIEW
The Natural History Museum
London, England, UK

LAURA F. LANDWEBER
NUCLEIC ACID BIODIVERSITY
Princeton University
Princeton, New Jersey, USA

JOHN H. LAWTON
INDICATOR SPECIES
Imperial College of Science, Technology and Medicine
London, England, UK

MARTIN J. LECHOWICZ
NORTH AMERICA, PATTERNS OF BIODIVERSITY IN
McGill University
Montréal, Québec, Canada

KAI N. LEE
SUSTAINABILITY, CONCEPT AND PRACTICE OF
Williams College
Williamstown, Massachusetts, USA

WILLIAM G. LEE
INTRODUCED PLANTS, NEGATIVE EFFECTS OF
Landcare Research
Dunedin, New Zealand

CLARENCE L. LEHMAN
STABILITY, CONCEPT OF
University of Minnesota
St. Paul, Minnesota, USA

JOHN T. LEHMAN
ECOSYSTEM FUNCTION MEASUREMENT, AQUATIC AND MARINE COMMUNITIES
University of Michigan
Ann Arbor, Michigan, USA

EGBERT GILES LEIGH, JR.
MUTUALISM, EVOLUTION OF
Smithsonian Tropical Research Institute
Washington, DC, USA

WADE LEITNER
MEASUREMENT AND ANALYSIS OF BIODIVERSITY
University of Arizona
Tucson, Arizona, USA

DONALD J. LEOPOLD
WETLANDS ECOSYSTEMS
State University of New York, Syracuse
Syracuse, New York, USA

CHRISTIAN LÉVÊQUE
LAKE AND POND ECOSYSTEMS
Institut de Recherche pour la Developpement (IRD)
Paris, France

JEFFREY S. LEVINTON
EXTINCTION, RATES OF
State University of New York, Stony Brook
Stony Brook, New York, USA

ANDRÉ LEVY
COEVOLUTION
State University of New York, Stony Brook
Stony Brook, New York, USA

K. E. LIMBURG
RIVER ECOSYSTEMS
State University of New York
College of Environmental Science and Forestry
Syracuse, New York, USA

DAVID R. LINDBERG
MOLLUSCS
University of California, Berkeley
Berkeley, California, USA

DAVID B. LINDENMAYER
POPULATION VIABILITY ANALYSIS
Australian National University
Canberra, Australia

ROBERTO LINDIG-CISNEROS
RESTORATION OF BIODIVERSITY
University of Wisconsin
Madison, Wisconsin, USA

SIMON H. LININGTON
GENE BANKS
Royal Botanic Gardens
Surrey, England, UK

TIMOTHY J. LITTLEWOOD
WORMS, PLATYHELMINTHES
The Natural History Museum
London, England, UK

J. M. LOCK
AFRICA, ECOSYSTEMS OF
Royal Botanic Gardens
Surrey, England, UK

MARK V. LOMOLINO
BIOGEOGRAPHY, OVERVIEW
Oklahoma Biological Survey
Oklahoma Natural Heritage Inventory and University of Oklahoma
Norman, Oklahoma, USA

GÁBOR L. LÖVEI
EXTINCTIONS, MODERN EXAMPLES OF
Danish Institute of Agricultural Sciences
Slagelse, Denmark

ANNA MARIE LYLES
ZOOS AND ZOOLOGICAL PARKS
Medarex, Inc.
Princeton, New Jersey, USA

J. M. LYNCH
ECOTOXICOLOGY
University of Surrey
Surrey, England, UK

JENNIFER L. MACALADY
MICROBIAL BIODIVERSITY, MEASUREMENT OF
University of California, Davis
Davis, California, USA

CARLOS A. MACHADO
DIVERSITY, MOLECULAR LEVEL
   Rutgers University
   New Brunswick, New Jersey, USA

JAMES A. MACMAHON
DESERT ECOSYSTEMS
   Utah State University
   Logan, Utah, USA

SHIRLEY M. MALCOM
EDUCATION, BIODIVERSITY AND
   The American Association for the Advancement of Science
   Washington, DC, USA

KARL-GÖRAN MÄLER
ECONOMIC GROWTH AND THE ENVIRONMENT
   Beijer International Institute of Ecological Economics
   Stockholm, Sweden

JAMES MALLET
SPECIES, CONCEPT(S) OF
SUBSPECIES, SEMISPECIES, AND SUPERSPECIES
   University College London
   London, England, UK

LYNN MARGULIS
EUKARYOTES, ORIGIN OF
   University of Massachusetts
   Amherst, Massachusetts, USA

BRIAN MARQUEZ
PLANT SOURCES OF DRUGS AND CHEMICALS
   Oregon State University
   Corvallis, Oregon, USA

GARY J. MARTIN
ETHNOBIOLOGY AND ETHNOECOLOGY
   The Global Diversity Foundation
   Marrakech-Medina, Morocco

PAUL S. MARTIN
MAMMALS (LATE QUATERNARY), EXTINCTIONS OF
   University of Arizona
   Tucson, Arizona, USA

MICHELLE MARVIER
ECOLOGY, CONCEPT AND THEORIES IN
   Santa Clara University
   Santa Clara, California, USA

JOHN M. MARZLUFF
LAND-USE ISSUES
   University of Washington
   Seattle, Washington, USA

NICOLÁS MATEO
BIOPROSPECTING
   Instituto Nacional de Biodiversidad (INBio)
   Santo Domingo de Heredia, Costa Rica

NADYA MATVEYEVA
ARCTIC ECOSYSTEMS
   Komarov Botanical Institute
   St. Petersburg, Russia

MIKE MAUNDER
PLANT CONSERVATION, OVERVIEW
   Royal Botanic Gardens
   Surrey, England, UK

NIGEL MAXTED
EX SITU, IN SITU CONSERVATION
   University of Birmingham
   Birmingham, England, UK

BRIAN H. McARDLE
POPULATION DENSITY
   University of Auckland
   Auckland, New Zealand

MICHAEL A. McCARTHY
POPULATION VIABILITY ANALYSIS
   Australian National University
   Canberra, Australia
   and The University of Adelaide
   Adelaide, Australia

EARL D. McCOY
SPECIES-AREA RELATIONSHIPS
   University of South Florida
   Tampa, Florida, USA

KEITH R. McDONALD
AMPHIBIANS, BIODIVERSITY OF
   Queensland Parks and Wildlife Service
   Townsville, Australia

ANTON McLACHLAN
COASTAL BEACH ECOSYSTEMS
   College of Science
   Sultan Qaboos University
   Al-Khodh, Oman

JEFFREY A. McNEELY
SOCIAL AND CULTURAL FACTORS
   IUCN—The World Conservation Union
   Gland, Switzerland

ROBIN MEADOWS
CAPTIVE BREEDING AND REINTRODUCTION
   Freelance Science Writer
   Greenlawn, New York, USA

CURT MEINE
CONSERVATION MOVEMENT, HISTORICAL
   International Crane Foundation
   Baraboo, Wisconsin, USA

ROBERT MENDELSOHN
ECONOMIC VALUE OF BIODIVERSITY, MEASUREMENTS OF
   Yale University
   New Haven, Connecticut, USA

BRUCE A. MENGE
KEYSTONE SPECIES
   Oregon State University
   Corvallis, Oregon, USA

MARK C. MERWIN
FOREST CANOPIES, PLANT DIVERSITY
   The Evergreen State College
   Olympia, Washington, USA

KEN MILLIGAN
PLANT SOURCES OF DRUGS AND CHEMICALS
 Oregon State University
 Corvallis, Oregon, USA

E. J. MILNER-GULLAND
MAMMALS, CONSERVATION EFFORTS FOR
 Imperial College London
 London, England, UK

ALESSANDRO MINELLI
MYRIAPODS
 University of Padova
 Padova, Italy

RUSSELL A. MITTERMEIER
PRIMATE POPULATIONS, CONSERVATION OF
 Conservation International
 Washington, DC, USA

JOHN C. MOORE
DIVERSITY, TAXONOMIC VERSUS FUNCTIONAL
 University of Northern Colorado
 Greeley, Colorado, USA

RICHARD Y. MORITA
PSYCHROPHILES
 Oregon State University
 Corvallis, Oregon, USA

MAX MORITZ
DISTURBANCE, MECHANISMS OF
 California Polytechnic University
 San Luis Obispo, California, USA

CRAIG L. MOYER
PSYCHROPHILES
 Western Washington University
 Bellingham, Washington, USA

LADISLAV MUCINA
EUROPE, ECOSYSTEMS OF
 University of Stellenbosch
 Stellenbosch, South Africa

DIETER MUELLER-DOMBOIS
ISLAND BIOGEOGRAPHY
 University of Hawaii
 Honolulu, Hawaii, USA

NORMAN MYERS
HOTSPOTS
 Oxford University
 Oxford, England, UK

WERNER NADER
BIOPROSPECTING
 Instituto Nacional de Biodiversidad (INBio)
 Santo Domingo de Heredia, Costa Rica

NALINI M. NADKARNI
FOREST CANOPIES, PLANT DIVERSITY
 The Evergreen State College
 Olympia, Washington, USA

SHAHID NAEEM
COMPLEXITY VERSUS DIVERSITY
 University of Washington
 Seattle, Washington, USA

PIOTR NASKRECKI
GRASSHOPPERS AND THEIR RELATIVES
 University of Connecticut
 Storrs, Connecticut, USA

RAN NATHAN
DISPERSAL BIOGEOGRAPHY
 Princeton University
 Princeton, New Jersey, USA

EVIATAR NEVO
GENETIC DIVERSITY
 University of Haifa
 Haifa, Israel

M. E. J. NEWMAN
BIODIVERSITY, ORIGIN OF
 Santa Fe Institute
 Sante Fe, New Mexico, USA

RAYMOND C. NIAS
ENDANGERED ECOSYSTEMS
 World Wide Fund for Nature
 Sydney, Australia

JURGEN NIEDER
FOREST CANOPIES, PLANT DIVERSITY
 Botanisches Institut der Universität Bonn
 Bonn, Germany

PEKKA NIEMELÄ
TIMBER INDUSTRY
 University of Joensuu
 Joensuu, Finland

KARL J. NIKLAS
PLANT COMMUNITIES, EVOLUTION OF
 Cornell University
 Ithaca, New York, USA

KEVIN C. NIXON
PHYLOGENY
 Cornell University
 Ithaca, New York, USA

IAN NOBLE
TERRESTRIAL ECOSYSTEMS
 Australian National University
 Canberra, Australia

LINDSEY NORGROVE
SLASH AND BURN AGRICULTURE, EFFECTS OF
 King's College London
 London, England, UK

DAVID R. NOTTER
BREEDING OF ANIMALS
 Virginia Polytechnic Institute and State University
 Blacksburg, Virginia, USA

BEATE NÜRNBERGER
ECOLOGICAL GENETICS
   Ludwig-Maximilians-Universität München
   Munich, Germany

EUGENE P. ODUM
ECOSYSTEM, CONCEPT OF
   University of Georgia
   Athens, Georgia, USA

JENNIFER L. O'HARA
CONSERVATION EFFORTS, CONTEMPORARY
   Yale University
   New Haven, Connecticut, USA

GORDON H. ORIANS
AESTHETIC FACTORS
   University of Washington
   Seattle, Washington, USA

ELINOR OSTROM
COMMONS, INSTITUTIONAL DIVERSITY OF
   Center for the Study of Institutions, Population, and Environmental Change and Workshop in Political Theory and Policy Analysis
   Indiana University
   Bloomington, Indiana, USA

MARGARET A. PALMER
INVERTEBRATES, FRESHWATER, OVERVIEW
   University of Maryland
   College Park, Maryland, USA

DANIEL PAULY
FISH STOCKS
   Fisheries Centre
   University of British Columbia
   Vancouver, British Columbia, Canada

HELI PELTOLA
TIMBER INDUSTRY
   University of Joensuu
   Joensuu, Finland

KENNETH PETREN
HABITAT AND NICHE, CONCEPT OF
   University of Cincinnati
   Cincinnati, Ohio, USA

DAVID PIMENTEL
AGRICULTURAL INVASIONS
   Cornell University
   Ithaca, New York, USA

PRABHU PINGALI
AGRICULTURE, INDUSTRIALIZED
   International Maize and Wheat Improvement Center
   Lisboa, Mexico

GARY A. POLIS (deceased)
FOOD WEBS
   University of California, Davis
   Davis, California, USA

CAROLINE M. POND
STORAGE, ECOLOGY OF
   The Open University
   Milton Keynes, England, UK

DOROTA L. PORAZINSKA
SOIL CONSERVATION
   Colorado State University
   Fort Collins, Colorado, USA

JAMES W. PORTER
REEF ECOSYSTEMS
   University of Georgia
   Athens, Georgia, USA

HUGH P. POSSINGHAM
POPULATION VIABILITY ANALYSIS
   National Center for Ecological Analysis and Synthesis
   The University of Adelaide
   Adelaide, Australia

F. HARVEY POUGH
REPTILES, BIODIVERSITY OF
   Arizona State University West
   Glendale, Arizona, USA

ALISON G. POWER
ECOLOGY OF AGRICULTURE
   Cornell University
   Ithaca, New York, USA

THOMAS O. POWERS
WORMS, NEMATODA
   University of Nebraska
   Lincoln, Nebraska, USA

GHILLEAN PRANCE
AMAZON ECOSYSTEMS
   Royal Botanic Gardens
   Surrey, England, UK

RICHARD B. PRIMACK
ENVIRONMENTAL ETHICS
ETHICAL ISSUES IN BIODIVERSITY PROTECTION
EXTINCTION, CAUSES OF
   Boston University
   Boston, Massachusetts, USA

JURGENNE PRIMAVERA
AQUACULTURE
   Southeast Asian Fisheries Development Center
   Bangkok, Thailand

HUGH W. PRITCHARD
GENE BANKS
   Royal Botanic Gardens
   Surrey, England, UK

DAVID PYKE
RANGE ECOLOGY, GLOBAL LIVESTOCK INFLUENCES
   Oregon State University
   Corvallis, Oregon, USA

OLIVER RACKHAM
LAND-USE PATTERNS, HISTORIC
   University of Cambridge
   Cambridge, England, UK

NISHANTA RAJAKARUNA
PLANT BIODIVERSITY, OVERVIEW
   The University of British Columbia
   Vancouver, British Columbia, Canada

KATHERINE RALLS
CAPTIVE BREEDING AND REINTRODUCTION
INBREEDING AND OUTBREEDING
   Smithsonian Institution
   Washington, DC, USA

EDUARDO H. RAPOPORT
EDIBLE PLANTS
   Universidad Nacional de Comahue and CONICET
   Bariloche, Argentina

JOHN A. RAVEN
PHOTOSYNTHESIS, MECHANISMS OF
   University of Dundee
   Dundee, Scotland, UK

G. CARLETON RAY
ESTUARINE ECOSYSTEMS
   University of Virginia
   Charlottesville, Virginia, USA

MAJORIE L. REAKA-KUDLA
CRUSTACEANS
   University of Maryland
   College Park, Maryland, USA

WILLIAM E. REES
ECOLOGICAL FOOTPRINT, CONCEPT OF
   School of Community and Regional Planning
   University of British Columbia
   Vancouver, British Columbia, Canada

PHILIP C. REID
PELAGIC ECOSYSTEMS
   Sir Alister Hardy Foundation for Ocean Science
   Plymouth, England, UK

C. S. REYNOLDS
PLANKTON, STATUS AND ROLE OF
   NERC Freshwater Biological Association
   Institute of Freshwater Ecology
   Cumbria, England, UK

JAMES F. REYNOLDS
DESERTIFICATION
   Duke University
   Durham, North Carolina, USA

ANNA-LOUISE REYSENBACH
THERMOPHILES, ORIGIN OF
   Portland State University
   Portland, Oregon, USA

STEPHEN J. RICHARDS
AMPHIBIANS, BIODIVERSITY OF
   Queensland Parks and Wildlife Service
   Brisbane, Australia

DAVID M. RICHARDSON
PLANT INVASIONS
   University of Cape Town
   Rondebosch, South Africa

MARGARET L. RISING
THERMOPHILES, ORIGIN OF
   Portland State University
   Portland, Oregon, USA

G. PHILLIP ROBERTSON
AGRICULTURE, SUSTAINABLE
   Michigan State University
   Hickory Corners, Michigan, USA

SCOTT K. ROBINSON
NEST PARASITISM
   University of Illinois at Urbana-Champaign
   Urbana-Champaign, Illinois, USA

GEORGE RODERICK
ADAPTIVE RADIATION
   University of California, Berkeley
   Berkeley, California, USA

JON PAUL RODRIGUEZ
CONSERVATION BIOLOGY, DISCIPLINE OF
   Centro de Ecología
   Instituto Venezolano de Investigaciones Cientificas
   Caracas, Venezuela

DEREK ROFF
LIFE HISTORY, EVOLUTION OF
   McGill University
   Montréal, Québec, Canada

KLAUS ROHDE
PARASITISM
   University of New England
   Armidale, Australia

PATRIK RÖNNBÄCK
AQUACULTURE
   Stockholm University and Royal Swedish Academy of Sciences
   Stockholm, Sweden

RICHARD B. ROOT
GUILDS
   Cornell University
   Ithaca, New York, USA

TERRY L. ROOT
CLIMATE CHANGE AND ECOLOGY, SYNERGISM OF
   University of Michigan
   Ann Arbor, Michigan, USA

MICHAEL R. ROSE
ADAPTATION
   University of California, Irvine
   Irvine, California, USA

STEPHEN I. ROTHSTEIN
NEST PARASITISM
   University of California, Santa Barbara
   Santa Barbara, California, USA

STEPHEN ROXBURGH
TERRESTRIAL ECOSYSTEMS
    Australian National University
    Canberra, Australia

DANIEL I. RUBENSTEIN
MIGRATION
SOCIAL BEHAVIOR
    Princeton University
    Princeton, New Jersey, USA

PHILIP W. RUNDEL
MEDITERRANEAN-CLIMATE ECOSYSTEMS
    University of California, Los Angeles
    Los Angeles, California, USA

CARL SAFINA
FISH CONSERVATION
    National Audubon Society
    Islip, New York, USA

DORION SAGAN
EUKARYOTES, ORIGIN OF
    Sciencewriters, Inc.
    Amherst, Massachusetts, USA

ROWAN F. SAGE
C_4 PLANTS
    University of Toronto
    Toronto, Ontario, Canada

OSVALDO E. SALA
TEMPERATE GRASSLAND AND SHRUBLAND ECOSYSTEMS
    Universidad de Buenos Aires
    Buenos Aires, Argentina

DAVID SALTZ
WILDLIFE MANAGEMENT
    Blaustein Institute for Desert Research
    Ben-Gurion University of the Negev
    Beersheva, Israel

KAJ SAND-JENSEN
FRESHWATER ECOSYSTEMS, HUMAN IMPACT ON
    University of Copenhagen
    Hillerod, Denmark

JOSÉ A. SARUKHÁN
BIODIVERSITY-RICH COUNTRIES
    Instituto de Ecología
    Universidad Nacional Autonoma de México
    Mexico City, Mexico

KATHRYN A. SATERSON
GOVERNMENT LEGISLATION AND REGULATION
    Brandywine Conservancy
    Chadds Ford, Pennsylvania, USA

CARL W. SCHAEFER
TRUE BUGS AND THEIR RELATIVES
    University of Connecticut
    Storrs, Connecticut, USA

P. H. SCHALK
INVERTEBRATES, MARINE, OVERVIEW
    ETI
    Universiteit van Amsterdam
    Amsterdam, The Netherlands

MICHAEL H. SCHIEWE
SALMON
    National Marine Fisheries Service
    Seattle, Washington, USA

OSWALD J. SCHMITZ
CONSERVATION EFFORTS, CONTEMPORARY
    Yale University
    New Haven, Connecticut, USA

DAVID CLAYTON SCHNEIDER
SCALE, CONCEPT, AND EFFECT OF
    Memorial University of Newfoundland
    St. John's, Newfoundland, Canada

STEPHEN H. SCHNEIDER
CLIMATE CHANGE AND ECOLOGY, SYNERGISM OF
    Stanford University
    Stanford, California, USA

EGBERT SCHWARTZ
MICROBIAL BIODIVERSITY, MEASUREMENT OF
    University of California, Davis
    Davis, California, USA

KATE M. SCOW
MICROBIAL BIODIVERSITY, MEASUREMENT OF
    University of California, Davis
    Davis, California, USA

REINMAR SEIDLER
LOGGED FORESTS
    University of Massachusetts
    Boston, Massachusetts, USA

NARENY SENGSAVANH
ULTRAVIOLET RADIATION
    Oregon State University
    Corvallis, Oregon, USA

J. JOHN SEPKOSKI, JR. (deceased)
MASS EXTINCTIONS, CONCEPT OF
    University of Chicago
    Warwick, Rhode Island, USA

JONATHAN H. SHARP
MARINE AND AQUATIC COMMUNITIES, STRESS FROM EUTROPHICATION
    University of Delaware
    Lewes, Delaware, USA

BRYAN SHORROCKS
COMPETITION, INTERSPECIFIC
    Centre for Biodiversity and Conservation
    University of Leeds
    Leeds, England, UK

H. H. SHUGART
SUCCESSION, PHENOMENON OF
    University of Virginia
    Charlottesville, Virginia, USA

JOHN A. SILANDER, JR.
TEMPERATE FORESTS
    University of Connecticut
    Storrs, Connecticut, USA

DANIEL SIMBERLOFF
INTRODUCED SPECIES, EFFECT AND DISTRIBUTION
 University of Tennessee
 Knoxville, Tennessee, USA

ELLEN L. SIMMS
PLANT-ANIMAL INTERACTIONS
 University of California Botanical Garden
 Berkeley, California, USA

R. DAVID SIMPSON
MARKET ECONOMY AND BIODIVERSITY
 Resources for the Future
 Washington, DC, USA

K. D. SINGH
RAINFOREST LOSS AND CHANGE
 Harvard University
 Cambridge, Massachusetts, USA

L. B. SLOBODKIN
LIMITS TO BIODIVERSITY (SPECIES PACKING)
 State University of New York, Stony Brook
 Stony Brook, New York, USA

MELINDA SMALE
AGRICULTURE, INDUSTRIALIZED
 International Maize and Wheat Improvement Center
 Mexico City, Mexico

DAVID SMITH
HEMIPARASITISM
 University of Maryland
 College Park, Maryland, USA

WILLIAM H. SMITH
POLLUTION, OVERVIEW
 Yale University
 New Haven, Connecticut, USA

PAUL V. R. SNELGROVE
MARINE SEDIMENTS
 Memorial University of Newfoundland
 St. John's, Newfoundland, Canada

LUIS A. SOLORZANO C.
SOUTH AMERICA, ECOSYSTEMS OF
 The Woods Hole Research Center
 Woods Hole, Massachusetts, USA

ALISON SPECHT
AUSTRALIA, ECOSYSTEMS OF
 School of Resource Science and Management
 Southern Cross University
 Lismore, Australia

RAYMOND L. SPECHT
AUSTRALIA, ECOSYSTEMS OF
 The University of Queensland
 St. Lucia, Australia

LESLIE E. SPONSEL
HUMAN IMPACT ON BIODIVERSITY, OVERVIEW
 University of Hawaii
 Honolulu, Hawaii, USA

ERKO STACKEBRANDT
BACTERIAL BIODIVERSITY
 DSMZ—Deutsche Sammlung von Mikroorganismen
 und Zellkulturen GmbH
 Braunschweig, Germany

ROBERT S. STENECK
FUNCTIONAL GROUPS
 University of Maine
 Walpole, Maine, USA

THOMAS J. STOHLGREN
ENDANGERED PLANTS
 BRD/USGS Colorado State University
 Fort Collins, Colorado, USA

DAVID L. STRAYER
ENDANGERED FRESHWATER INVERTEBRATES
RIVER ECOSYSTEMS
 Institute of Ecosystem Studies
 Millbrook, New York, USA

D. P. SWANEY
RIVER ECOSYSTEMS
 Cornell University
 Ithaca, New York, USA

IAN R. SWINGLAND
BIODIVERSITY, DEFINITION OF TERM
 Durrell Institute of Conservation and Ecology
 Kent, England, UK

DAVID TAKACS
HISTORICAL AWARENESS OF BIODIVERSITY
 California State University, Monterey Bay
 Seaside, California, USA

GISELLE TAMAYO
BIOPROSPECTING
 Instituto Nacional de Biodiversidad (INBio)
 Santo Domingo de Heredia, Costa Rica

SANDY L. TARTOWSKI
NITROGEN, NITROGEN CYCLE
 Cornell University
 Ithaca, New York, USA

BRUCE TIFFNEY
PLANT COMMUNITIES, EVOLUTION OF
 Cornell University
 Ithaca, New York, USA

DAVID TILMAN
FUNCTIONAL DIVERSITY
 University of Minnesota
 St. Paul, Minnesota, USA

VICTOR M. TOLEDO
INDIGENOUS PEOPLES, BIODIVERSITY AND
 Institute of Ecology
 National University of Mexico
 Morelia, Michoacán, Mexico

LIK TONG TAN
PLANT SOURCES OF DRUGS AND CHEMICALS
 Oregon State University
 Corvallis, Oregon, USA

JENNIFER I. TOUGAS
REEF ECOSYSTEMS
University of Georgia
Athens, Georgia, USA

MICHAEL TRAVISANO
BACTERIAL GENETICS
University of Houston
Houston, Texas, USA

MAX TROELL
AQUACULTURE
Stockholm University and Royal Swedish Academy of Sciences
Stockholm, Sweden

ANDREAS TROUMBIS
CATTLE, SHEEP, AND GOATS, ECOLOGICAL ROLE OF
University of the Aegean
Lesbos, Greece

ROY TURKINGTON
BOREAL FOREST ECOSYSTEMS
University of British Columbia
Vancouver, British Columbia, Canada

I. M. TURNER
RAINFOREST ECOSYSTEMS, PLANT DIVERSITY
Singapore Botanic Gardens
Singapore

WILL TURNER
MEASUREMENT AND ANALYSIS OF BIODIVERSITY
University of Arizona
Tucson, Arizona, USA

A. J. UNDERWOOD
INTERTIDAL ECOSYSTEMS
Centre for Research on Ecological Impacts of Coastal Cities
University of Sydney
Sydney, Australia

JAMES W. VALENTINE
GEOLOGIC TIME, HISTORY OF BIODIVERSITY IN
University of California, Berkeley
Berkeley, California, USA

CINDY LEE VAN DOVER
VENTS
College of William and Mary
Williamsburg, Virginia, USA

GLENN VANBLARICOM
MARINE MAMMALS, EXTINCTIONS OF
University of Washington
Seattle, Washington, USA

R. I. VANE-WRIGHT
TAXONOMY, METHODS OF
The Natural History Museum
London, England, UK

GIORGINI AUGUSTO VENTURIERI
CROP IMPROVEMENT AND BIODIVERSITY
Federal University of Santa Catarina
Santa Catarina, Brazil

CONSTANTINO VETRIANI
ARCHAEA, ORIGIN OF
Rutgers University
New Brunswick, New Jersey, USA

ROSS A. VIRGINIA
ECOSYSTEM FUNCTION, PRINCIPLES OF
Dartmouth College
Hanover, New Hampshire, USA

LUCÍA VIVANCO
TEMPERATE GRASSLAND AND SHRUBLAND ECOSYSTEMS
Universidad de Buenos Aires
Buenos Aires, Argentina

KRISTIINA A. VOGT
CONSERVATION EFFORTS, CONTEMPORARY
Yale University
New Haven, Connecticut, USA

THOMAS J. VOLK
FUNGI
University of Wisconsin, LaCrosse
LaCrosse, Wisconsin, USA

DAVID L. WAGNER
MOTHS
University of Connecticut
Storrs, Connecticut, USA

DIANA H. WALL
ECOSYSTEM FUNCTION, PRINCIPLES OF
SOIL CONSERVATION
Colorado State University
Fort Collins, Colorado, USA

THOMPSON WEBB III
PALEOECOLOGY
Brown University
Providence, Rhode Island, USA

PETER WESTON
SOUTHERN (AUSTRAL) ECOSYSTEMS
Royal Botanic Gardens
Surrey, England, UK

ROBERT G. WETZEL
FRESHWATER ECOSYSTEMS
University of Alabama
Tuscaloosa, AL, USA

QUENTIN D. WHEELER
SYSTEMATICS, OVERVIEW
Cornell University
Ithaca, New York, USA

JEANNETTE WHITTON
PLANT BIODIVERSITY, OVERVIEW
The University of British Columbia
Vancouver, British Columbia, Canada

RICHARD G. WIEGERT
HIGH-TEMPERATURE ECOSYSTEMS
University of Georgia
Athens, Georgia, USA

DAVID M. WILLIAMS
CLADISTICS
  The Natural History Museum
  London, England, UK

PAUL WILLIAMS
COMPLEMENTARITY
  The Natural History Museum
  London, England, UK

THOMAS WILLIAMSON
PLANT SOURCES OF DRUGS AND CHEMICALS
  Oregon State University
  Corvallis, Oregon, USA

MICHAEL R. WILLIG
LATITUDE, COMMON TRENDS WITHIN
  Texas Tech University
  Lubbock, Texas, USA

A. WISEMAN
ECOTOXICOLOGY
  University of Surrey
  Surrey, England, UK

ISA WOO
RESTORATION OF BIODIVERSITY
  University of Wisconsin
  Madison, Wisconsin, USA

R. WOODROFFE
MAMMALS, CONSERVATION EFFORTS FOR
  University of Warwick
  Warwick, England, UK

DAVID S. WOODRUFF
POPULATIONS, SPECIES AND CONSERVATION GENETICS
  University of California, San Diego
  La Jolla, California, USA

F. I. WOODWARD
CLIMATE, EFFECTS OF
  University of Sheffield
  Sheffield, England, UK

DONALD J. WUEBBLES
ATMOSPHERIC GASES
  University of Illinois at Urbana-Champaign
  Urbana-Champaign, Illinois, USA

PETER YODZIS
TROPHIC LEVELS
  University of Guelph
  Guelph, Ontario, Canada

PETER ZAHLER
ENDANGERED MAMMALS
  Wildlife Conservation Society
  New York, New York, USA
  University of Massachusetts
  Amherst, Massachusetts, USA

JOY B. ZEDLER
RESTORATION OF BIODIVERSITY
  University of Wisconsin
  Madison, Wisconsin, USA

DANIEL ZOHARY
DOMESTICATION OF CROP PLANTS
  The Hebrew University of Jerusalem
  Jerusalem, Israel

Glossary

A

ablation Direct transfer of water molecules from ice crystals to the vapor phase, without transition through the liquid phase.

abundance The number of individuals of a given species in a region.

abyssal plains Relatively flat areas of the ocean bottom below ~4000 m depth.

acceptance A behavior pattern in which hosts treat brood parasitic eggs as if they were their own eggs.

accession A collection, usually a sample (e.g., seed lot) but may be a set of genetically related samples.

accidental epiphyte Plant that normally grows terrestrially but that occasionally grows to maturity in a tree crown, usually in terrestriallike microsites such as the crotches of branches.

accounting price In this context, a measure of the economic value of a resource in social terms, described as the increase in social well-being that would be enjoyed if one unit more of the resource were made available without cost, and expressed as the difference between the market price of the resource and the tax or subsidy that ought to be imposed on it.

accretionary Accumulating sand.

acetabulum Ventral sucker used for attachment and/or locomotion in digeneans.

acetylcholinesterase (AChE) The enzyme responsible for breaking down the neurotransmitter acetylcholine (ACh) at nerve synapses, thereby preventing hyperexcitation of cholinergic pathways in the nervous system.

acid deposition The combination of acid rain plus dry deposition; a term preferred over "acid rain." It is the anthropogenic acidification of terrestrial and freshwater ecosystems by (primarily) sulfuric acid, derived from sulfur dioxide produced by burning oil and coal and deposited in rain and snow (acid rain), directly as particles (dry deposition) and as cloud droplets. Typically defined as rain, fog, snow, sleet, or hail with a pH less than 5.6.

acidophiles From the Latin *acidus* (sour) and the Greek *philos* (loving). Includes organisms that grow optimally at low pH.

acidophilic Organisms preferring or requiring a low pH environment.

acid rain The popular term for acid deposition (see above).

acoelomate Lacking a body cavity.

actinopterygia Ray-finned bony fishes.

active restoration Requiring manipulation by humans for successful colonization and/or establishment of organisms and ecosystem functioning.

acute An exposure to an environmental stress that is brief in relation to the temporal scale of the biological system exposed.

acute toxicity A damaging effect caused by a single short period of exposure to high concentrations of a pollutant.

adaptation Any genetically controlled characteristic that increases an organism's fitness, usually by ensuring the organism to survive and reproduce in the environment it inhabits.

adaptationism The doctrine that all important evolutionary processes are dominated by natural selection, and that all significant biological characters increase an organism's fitness.

adaptive landscape A graph of average fitness against the state of the population, represented by allele frequencies or trait means.

adaptive management A systematic, cyclical process for continually improving management policies and practices based on lessons learned from operational programs.

adaptive radiation The evolution of one or a few forms into many different species that occupy different habitats within new geographical areas or habitats.

adaptive shift A change in the nature of a trait (morphology, ecology, or behavior) that enhances survival and/or reproduction in an ecological environment different from that originally occupied.

adaptive syndrome A coordinated set of adaptations.

adipocytes Large cells unique to vertebrates that take up and release fatty acids, produce receptors and respond to circulating and locally produced agonists, and secrete a variety of protein and lipid informational molecules.

adipose fin A small, fleshy fin without supporting spines or rays, set far back on the dorsal surface of many catfishes, characins, salmons, and other groups.

adipose tissue, brown Thermogenic tissue unique to mammals that consists of small adipocytes containing numerous mitochondria and many small lipid droplets plus vascular and neural tissue. Thermogenesis is controlled by the degree of uncoupling of ATP (adenosine triphosphate) synthesis in the mitochondria and can be very high in neonates and adults emerging from hibernation.

adipose tissue, white Storage tissue unique to vertebrates and best developed in tetrapods that consists almost entirely of expandable adipocytes (q.v.) that contain a large droplet of lipid plus vascular and neural tissue.

adjacency Relative position of alternative states of characters prior to any hypothesis of polarity.

adult parasite A parasite associated with a host during part or the whole of its mature phase.

aerenchyma Porous root tissue, especially well developed in wetland plants, that allows diffusive flux of oxygen from above-ground tissues to root tips. This tissue both supports the respiratory demand of the root tissues and allows oxygen to leak into the surrounding soil.

aerial roots In mangrove species such as *Rhizophora*, roots branch out from the stem some distance above the soil surface. Lenticels (pores) in the aerial portion of these roots enable gas exchange to take place, through aerenchyma tissue, with the respiring underground portions of the root.

aerobe An organism that utilizes or requires the presence of oxygen for growth.

aerobic In the presence of oxygen.

aerosols Microscopic airborne particles; Fine particulate matter suspended in the atmosphere, with diameters less than 5.5 μm.

aesthetics The field of investigation, from philosophical and psychological perspectives, that attempts to discover the rules and principles that govern the sense of beauty and ugliness.

affiliative behavior Behavior that is supportive and brings individuals together.

afforestation The process of planting trees in areas currently devoid of them. It is done in desert regions and is increasingly common in Europe, where grasslands and pastures are being converted into woodlands and forests.

aggregation The process of combining several potentially separate components of a system to simplify analysis.

agonistic behavior Behavior that is aggressive, threatening, combative, and submissive.

agricultural intensification Management of agricultural land to increase yield principally through the use of high-yielding crop varieties, chemical fertilizers and pesticides, irrigation, shorter rotations, larger fields, and mechanization.

agroecosystem A simplified natural ecosystem subjected to exploitation for purposes of food and fiber production. (e.g., crop field and grazing pasture).

agroforestry The practice of combining agricultural crops or animal husbandry with the maintenance and cultivation of trees on the same patch of land.

agroscape The agricultural, ranching, and plantation countryside, with its roads, irrigation ditches, buildings, and so on. The agroscape stands in contrast to the wildland countryside that is not directly managed by humanity (though it is strongly impacted by it). The agroscape intergrades with wildlands in the form of woodlots, abandoned fields, poor soil sites, hedgerows, and edges of wildlands.

air pollution Lowering of air quality due to release of toxic materials by factories, automobiles, fires, and other human activities.

alate An imago (adult insect) still possessing its wings.

albedo The fraction of light hitting a surface that is reflected.

algae Primitive plantlike organisms that photosynthesize. These organisms can be unicellular, filamentous, or colonial microscopic forms (microalgae), or they can be macroscopic (macroalgae), consisting of a primitive plant body (thallus) that lacks vascular tissue in most species.

alien (introduced, exotic, nonindigenous, nonnative, invasive) Describing a species that has been transported by human activity (i.e., mariculture), intentionally or accidentally, to a site at which it does not naturally occur.

alien invertebrates Invertebrates intentionally or accidentally imported by humans into new geographic areas.

alien plants Plant taxa whose occurrence in a region is due to their introduction (intentionally or accidentally) as a result of human activity. Synonyms include "exotic plants" (not recommended), "nonindigenous plants," and "nonnative plants."

alien species Species that has been moved and established outside of its native range as a result of human activities; also called exotic species, introduced species, nonindigenous species.

alkalinity The acid-neutralizing capacity (ANC) of water: $ANC = [HCO_3^-] + [CO_3^{2-}] + [OH^-] - [H^+]$.

Allee effect Biological phenomenon whereby the survival of the individuals of a population increases by spatial aggregation. For population size to be regulated, it must exhibit a negative density dependence. That is, the population growth rate must decline as the population gets larger. However, under certain circumstances, some populations exhibit positive density dependence. This phenomenon, in which population growth rate increases as the population gets larger, is called an Allee effect. An Allee effect may generate a critical minimum population size, below which extinction will occur.

allele frequency The frequency of a variant form of a genetic locus within a population.

alleles Variants of nucleotides within the genome that cause changes in a given protein; one of two or more alternative forms of a gene. A particular variant of a gene.

allochthonous Referring to the production of organic matter outside of the ecosystem; in streams and rivers, this would be inputs from the upstream watershed.

allopatric Describing two taxonomic entities whose geographic ranges do not intersect with each other; populations, species, or taxa occupying different and disjunct geographical areas.

allopatric speciation The evolutionary development of new species in the presence of a geographical barrier which reduces gene flow and promotes genetic divergence.

allopatry Species or populations occupying separate areas.

allopolyploid (amphiploid) Species with chromosome sets derived from interspecific hybridization and doubling of chromosomes of the sterile hybrid, which restores fertility.

allozyme Enzyme locus that produces more than one electrophoretic variant.

all taxa biodiversity inventory (ATBI) The idea, first suggested by D. H. Janzen, that it might be feasible to produce a complete species list for all the organisms living in one place, a hectare of tropical forest, for example. The goal has so far proved elusive.

alluvial fan Fine sediments deposited by water that form a conical fan when laid down by streams or rivers

alpha diversity The diversity of species, often estimated as species richness, within a local community or site.

alpha landscape diversity Number and dominance of patch types within a landscape.

alpine Refers to the life zone above the climatic high-elevation treeline, irrespective of latitude. Though originating in the Alps (a pre-Roman word for high mountains), the term is applied globally. The reader should be aware that this term is often used in a much wider sense in common language, and is also applied to regions with mountains in general, including settlements and resorts, which is not the meaning here.

alternation of generations In plants, a reproductive cycle in which a multicellular gametophytic phase alternates with a multicellular sporophytic phase to complete the life cycle.

altruism Behavior that enhances the recipient's reproduction while reducing the actor's.

alvin A term for newly hatched salmon with unabsorbed yolk sac.

amniota Taxon of vertebrates including mammals and reptiles (including birds).

amphibian Member of a class of vertebrates (the Amphibia), comprising frogs and toads (order Anura), newts and salamanders (order Caudata), and caecilians (order Gymnophiona), which typically return to water to breed and pass through an aquatic larval stage with gills. Amphibians have a moist skin without scales, which is permeable to water and gases.

amphibites Species that require both surface waters and subterranean waters in order to complete their life cycle.

AMS dates Radiocarbon dates obtained by directly measuring the amount of carbon 14 in a sample by using an accelerator mass spectrometer.

anadromous Describing aquatic organisms that spawn and undergo early development in freshwater, migrate to sea to grow and mature, and return to freshwater to reproduce.

anaerobe An organism able to grow in the absence of oxygen.

anagenesis A pattern of evolutionary change involving the transformation of an entire population, sometimes to a state different enough from the ancestral population to justify renaming it as a separate species; also called phyletic transformation of an unbranched lineage of organisms which changes to such an extent as to be justifiably called new species or taxa. This is considered "biological improvement" (sensu Huxley, 1958), covering all types of change from detailed adaptation to generalized organizational advance.

analysis, emic and etic Concepts derived from the linguistic terms "phonetics" (representing speech sounds by precise and unique symbols and by technical descriptions of articulation, as practiced by trained linguists) and "phonemics" (characterization of speech through a minimal number of symbols, typically recognized by the speakers of a language). By extension, etic refers to the external explanation of cultural knowledge and practice (such as the use of Linnean taxonomy or scientific nomenclature to describe local useful plants), whereas emic denotes the internal perspective of local people (e.g., ethnobiological categories and nomenclature).

anamniotic Eggs are not surrounded by the complex membranes that distinguish the amniotic eggs of reptiles, birds, and mammals.

ancestor The taxon from which descendant species are derived, often synonymized as primitive or generalized. Ancestral traits or conditions are those which appear in an ancestor.

ancestral Describes a character or character state of the organism being considered that retains the primitive condition for its evolutionary lineage.

anchialine (or anchihaline) Haline water, usually with restricted exposure to open air, and always with subterranean connections to the sea.

ancient lakes Lakes with a persistence of more than 100,000 years are called long-lived or ancient lakes.

andepts Soils derived from andesitic volcanic ash.

anemophily Pollination by wind.

angiosperms Flowering vascular plants in which the seeds, which are produced from the ovules of the flowers, are enclosed in fruits developed from the matured ovaries; the largest class of vascular plants, as opposed to cryptogams (mosses, etc., which do not produce seeds) and conifers (in which the seeds are not enclosed in fruits).

anhydrobiosis Ability of certain organisms to lose all detectable water from their bodies under certain conditions, yet remain viable when subsequently rehydrated. A state of dormancy in various invertebrates due to low humidity or desiccation.

animal unit month (AUM) Quantity of forage needed to sustain one 1000-lb cow or five sheep for one month.

animism Belief in spiritual beings. The term is associated with the anthropology of E. B. Tylor, who described the origin of religion and primitive beliefs in terms of animism in *Primitive Culture* (1871). Tylor considered animism a minimum definition of religion and asserted that all religions, from the simplest to the most complex, involve some form of animism. (From Latin *anima*, "breath" or "soul.")

anion Negatively charged atom or molecule.

Annelida Segmented worms, a relatively species-poor phylum.

annual A plant, sometimes called an ephemeral, that survives year to year as a seed rather than as a "plant."

anoxia The absence of molecular oxygen; a condition in which there is almost no dissolved oxygen in the water (<0.1 mg of dissolved oxygen [DO]/L).

antennules (*pl.* antennae) The first pair of sensory appendages on the head, and antennae are the second pair of sensory appendages on the head of crustaceans.

antheridium A unicellular or multicellular structure that produces male gametes (sperm).

anthophile A flower-visiting animal, which may or may not be a pollinator (transfer pollen).

anthropocentric Human-centered in perspective, especially with regard to the value of the natural world.

anthropocentrism The philosophical, ethical, or political position that only human beings have moral worth or intrinsic value.

anthropochoric Geographic distribution due to human agency.

anthropogenic Caused by humans; related to the activities of human beings.

anthropogenic disturbances Disturbances of human origin, including modifications of ecosystem structure or function and displacement or removal of species from habitats.

anthropoids "Higher" primates such as marmosets, tamarins, monkeys, and apes.

anticancer agent A chemical substance that is capable of bringing about a remission or cure of the family of diseases known as cancer.

apex predator An organism that occupies a food web's highest trophic level.

apomixis A very common but "hidden" mode of clonal propagation by seeds, the embryos of which are 100% genetic copies of the source plant. Seeds are produced without fertilization, but often pollination is required to induce apomictic seed production. Apomictic plants also reproduce sexually, but these are very rare events.

apomorphic Derived; relative to sister group.

apomorphy A derived character or character state; if

two or more taxa share apomorphies, these are referred to as synapomorphies.

aposematic 1. Describing an organism that is rendered less susceptible to predation by advertising its obvious unpalatability. 2. Specifically, warningly colored; boldly colored, usually involving reds, oranges, or yellows, as well as black and white.

aquaculture Commercial farming of aquatic organisms, including seaweeds, raising captive-bred fish, and raising wild-born fish in captivity. Mariculture refers specifically to marine (saltwater) aquaculture. Farming implies some sort of intervention in the rearing process to enhance production, such as regular stocking, feeding, or protection from predators. Farming also implies individual or corporate ownership of the stock being cultivated. For statistical purposes, aquatic organisms which are harvested by an individual or corporate body which has owned them throughout their rearing period contribute to aquaculture, whereas aquatic organisms which are exploitable by the public as a common property resource, with or without appropriate licenses, are the harvest of fisheries.

aquifer Geological formation that contains and allows movement of groundwater.

arbicolous Living on the trees, or at least off the ground in shrubs and/or on tree trunks.

Archaea One of three domains of life. From the Greek *archaios* (ancient, primitive). Prokaryotic cells; formerly called Archaeobacteria.

archaebacteria (Archaea) One of two major groups of prokaryotes, including methane-producing bacteria, extreme halophiles, and extreme thermophiles.

archegonium A multicellular structure that produces the female gamete (egg).

Arctic Geographically, the region lying to the north of latitude 66.7°N, but environmentally, and in the context of this article, the region to the north of the climatically controlled northern latitudinal treeline which corresponds approximately to the mean July isotherm of 10°C.

area cladogram A branching diagram showing relative divergence of geographic areas.

area of endemism An area recognized by congruent distributions of two or more groups of organisms.

ark paradigm The concept that threatened species can be preserved in or at a special facility; a term derived from the Biblical account of Noah's Ark saving every extant species on earth during a great flood.

arrhenotoky Reproductive mode in which unfertilized eggs develop into haploid males and fertilized eggs develop into diploid females.

Arthropoda Joint-legged animals, the most species-rich phylum existing today.

artificial ecosystems Ecosystems dominated by human artifacts and human induced material and energy flows.

assemblage 1. Collective occurrence of several individuals representing the same species or several species without interspecific relationships. 2. The collection of animals and plants found together in a patch of habitat.

assembly rules Constraints that one community imposes on subsequent configurations of that community.

associational refuge When a potential prey escapes or deters a consumer by associating with another organism that interferes with the ability of the consumer to locate or attack the prey.

assortative mating The tendency of like to mate with like whether it is based on similarities in genotypes or similarities in phenotypes.

atmospheric lifetime Timescale characterizing the rate of removal of a gas from the atmosphere, generally defined as the time to remove 63.2%, the e-folding time.

ATP synthetase Membrane-associated protein complex that couples exergonic fluxes of H^+ across the membrane to ADP phosphorylation, or vice versa.

attraction A physical or cultural feature of a particular place that tourists feel meets an aspect of their leisure needs. Attractions are the main motivators for tourism trips and are the core of the tourism product. Without attractions there would be no need for other tourism services.

Australasia Geographic region that encompasses Australia, eastern Indonesia, and nearby islands of the tropical southwest Pacific.

autapomorphy A derived character unique to a single species or clade.

autocatalysis Self-acceleration of certain chemical reactions that yield catalytically active products. The reaction rate of autocatalytic processes increases exponentially.

autochthonous Referring to the production of organic matter within the ecosystem, for example, primary production by aquatic plants.

autocorrelation Densities from adjacent samples are likely to be more similar than are ones far apart.

autogenic species A species that creates habitat for other organisms; this habitat would not exist if the autogenic species were absent.

autotroph Literally, a self-feeder; an organism that is able to utilize inorganic carbon (carbon dioxide) as the sole carbon source for growth; for example, green plants and certain bacteria.

autotrophic Describing an organism able to utilize carbon dioxide as its source of carbon; i.e., An organism that obtains its nutrition by synthesizing organic substances from inorganic substances acquired from its environment.

autotrophic respiration The respiration of photosynthetic organisms.

autotrophic successions Succession processes that generate energy from internal processes (photosynthesis).

autotrophy The ability of organisms to grow and reproduce independently of external sources of organic carbon compounds.

available Of a scientific name of an organism, one that must be taken into account when deciding the correct (q.v.) name to be used for it.

available (or bioavailable) concentration The concentration of a contaminant in soils or surface waters which can be taken up by the target organism.

B

background concentration The concentration of a contaminant in an environment which has not been measurably influenced by anthropogenic sources.

background extinction A distinctly lower rate of extinction, more typical of most of the fossil record. These rates of extinction that characterized the major part of the evolution of life.

background selection Selection that takes place at loci other than that of interest. Such events can strongly increase the intensity of genetic drift and hinder the action of selection at a locus of interest. These effects are normally modeled by reducing the effective population size applicable to the locus of interest.

Bacteria One of three domains of life. From the Greek *bacterion* (staff, rod). Prokaryotic cells; formerly called Eubacteria.

bacterial Relating to or involving the Bacteria.

balancing selection Selection that acts to maintain two or more alleles in a stable equilibrium.

ballast Water or other material carried by ships to balance them after they have unloaded their cargo.

bands The basic economic, social, and political unit of hunter-gatherer societies.

barophile Pressure-loving bacteria.

barophilic Able to tolerate high pressures and growing better under high pressure.

barotolerant Pressure-loving; barophilic.

barreais Soils naturally enriched with organic matter and clay carried by the rain and/or a river.

basal rate of metabolism The minimum amount of energy spent by an adult animal that has not eaten recently, at normal body temperature during rest (usually sleep).

basal species A species that eats no other species.

Batesian mimicry A form of mimicry in which the target organism is rendered less susceptible to predation by its resemblance in morphology or coloration to a different species that is unpalatable.

batha A biblical term for semishrub communities, mainly of seral vegetation of old field succession in the Mediterranean part of the Near East.

beak (or rostrum) The thin elongate mouthparts, adapted for piercing the organisms upon which the bugs feed (the mouthparts of bugs are similar in function to, but quite different in structure from, those of mosquitoes).

beauty The qualities of a perceived or imagined object whereby it evokes feelings of admiring pleasure.

benthic Living or located at or on the bottom of the sea, streams, lakes, or riverbeds, or another aquatic environment.

benthos 1. Bottom-living organisms, including those that reside on hard and soft bottom surfaces and others that reside between sediment grains. 2. The substrate at the bottom of the sea or fresh waters.

Beringia A region at the northwestern corner of North America separated from Asia only by the shallow waters of the Bering Strait. When ocean levels drop during glaciation, Asia and North America are connected by land in this region.

beta diversity The degree of turnover in species (and changes in their abundances) among communities or sites along a gradient or within a larger area.

bet-hedging strategy In this context, a trait of an organism, living in a variable environment, that leads to low variation in fitness. In general, such a trait provides an organism greater net fitness over a range of environmental conditions than would a trait specialized for any single environment. A bet-hedging trait is expected to evolve when the environment in which a species lives fluctuates over a fixed range of conditions that is sufficiently broad that fitness varies significantly, and when precisely which state the environment will take in the immediate future is unpredictable.

between-habitat diversity Degree of dissimilarity in species composition between two habitats. The greater the number of species shared by two habitats, the lower the between-habitat diversity.

bifurcation point A transitional boundary in the state of a dynamic system (such as an ecosystem), at which the stability or basic nature of the system changes due to alterations in its fundamental characteristics.

binomen, binomial The name of a species, consisting of the name of the genus in which it is placed followed by a word peculiar to that species.

binomial (= bionominal) nomenclature The practice or principle of naming species by combination of a specific epithet and the genus to which the species belongs.

bio-accumulation The accumulation of toxicants or other chemicals by successive organisms in the food chain.

bioassay (biological assay) A laboratory test for evaluating the response of organisms to a toxin and for diagnosing the presence or absence of resistance.

bioassay 1. The use of living cells or organisms to make quantitative or qualitative measurements of the amounts or activity of substances. 2. Specifically a test for pharmaceutical activity in substance conducted either in living organisms (*in vivo* assays) or in the test tube (*in vitro* assays).

biocentric Valuing the existence and diversity of all biological species.

biocentrism Position that all living beings have moral worth or intrinsic value.

biocontrol Biological control; the control of agricultural pests with the use of predators and other beneficial organisms (e.g., control of turf grass crickets with parasitic nematodes).

bio-cultural axiom Recognition that biological and cultural diversity are mutually dependent and geographically coterminous.

biodiversity Biological diversity; a term variously defined by different authors; e.g.: 1. "The variability among living organisms from all sources and the ecological systems of which they are a part" (Wilson, 1988). 2. The diversity of life, including genetic biodiversity (diversity within a species), species biodiversity (diversity among species), and ecosystem biodiversity (diversity among ecosystems). 3. A measure of the relative abundance of species or higher taxonomic units found in a certain area at a particular time. In ancient communities this is often simply referred to as the number of species or taxa found at a site, usually referred to as "species richness" in the ecological literature. 4. The number of species on the planet. Popularly supposed to refer to the spectrum of all species on Earth, the concept should also include species' subunits (genetic diversity) and the diversity of ecosystems and ecological processes. 5. The variety of living organisms in an area, including the variety in genes, species, functional types of organisms, and ecosystems. 6. Species, genetic, and ecosystem diversity in an area, sometimes including associated abiotic components such as landscape features, drainage systems, and climate. 7. Hereditary variation in life-forms at all levels of organization. Examples are diversity among wild species, among individuals within those species, among crop species, and among and within varieties of a given crop. 8. Number and composition of genotypes, species, functional types, and/or landscape units present in a given system. 9. Total inventory of genes, organisms, species, populations, communities, and their habitats. 10. Diversity of microbial, animal, and plant species in an ecosystem that performs distinct ecological functions and services.

biodiversity assessment A biodiversity assessment entails the identification and classification of species, habitats, and communities within a given area or region. The overall purpose is to provide information needed to evaluate whether management is necessary to conserve biological diversity. Assessments also provide data and information that can be applied to monitoring programs or for providing basic information for scientific inquiry.

biodiversity of forests Variability in genetic structures, species composition, and/or habitat properties in forest ecosystems.

biogeochemical Consisting of biological and abiotic transformations.

biogeochemical cycles Pathways and storage of chemical elements through the biota and geologic resources, including the atmosphere, hydrosphere, and lithosphere.

biogeochemistry The discipline which studies biotic controls on the chemistry of the environment and geochemical controls on the structure and function of ecosystems.

biogeographic regions Major areas of the world where large assemblages of species originated and that differ considerably from other regions.

biogeography The study of the distribution of plant and animal life in the earth's environment and of the biological and historical factors that produced this distribution.

biological altruism Behavior of an organism such that the fitness of another organism is increased while its own fitness is decreased.

biological control Introduction of a natural enemy (parasite, predator, herbivore, or pathogen) of an undesirable species, usually itself introduced.

biological diversity Biodiversity; the variety of different forms of life on earth, usually considered to include variety at all levels of biological organization from genetic variation to variety seen in species, populations, communities, and ecosystems.

biological indicators Species or communities that enable an evaluation of environmental conditions and detect changes. Indicator species are normally surrogates of other species in the area of interest and are usually sensitive to environmental change. Environmental variables may also be used as indicators.

biological integration Organization of life functions in distinct, self-regulating units (cells, tissues, organs, organisms, populations, communities, and ecosystems).

biological integrity The wholeness of a living system, including the capacity to sustain the full range of organisms and processes having evolved in a region.

biological soil crusts Collective term referring to the combination of nonvascular and non-seed-forming plants that commonly occur on the soil surface of many arid and semiarid ecosystems. These crusts may consist of mosses, lichens, liverworts, green algae, and cyanobacteria. Synonyms: cryptogams, microfloral, cryptobiotic, cryptogamic, microbiotic, organogenic, biogenic, biotic, and microphytic soil crusts.

biological species concept Concept of a species as a population or series of populations that are reproductively isolated from other groups, as well as the degree of morphological similarity.

bioluminescence Light production by living organisms.

biomagnification The increase in body burden of persistent pollutants in organisms at higher trophic levels in food webs.

biomarker A macromolecule unique to a particular organisms or group of organisms such that its detection alone would suggest the presence of the organism or group of organisms.

biomass 1. The total dry mass of a population or community of animals or plants; collective weight or mass of all the members of a given population or stock at a given time, or, on the average, over a certain time period. 2. Specifically, the mass of biological material after removal of water by oven-drying at 70–100°C, often expressed as mass per area of ground surface.

biome A group of ecosystems which are subject to similar climatic conditions and which share a similar range of vegetation structures, animal diversity, and soil types. Biomes represent major regional groupings of ecological systems and are therefore discernible at the global scale.

bionics Science of systems, which function in the same way as or similar to living systems.

bionomic equilibrium In a biological common property resource, an equilibrium between resource production rate and exploitation rate, characterized by the dissipation of economic rents.

bionomics Mode of life of a species.

biopiracy A term for the illegal appropriation or exploitation of genetic and biochemical resources.

bioprospecting The search for commercially valuable biochemical and genetic resources in plants, animals, and microorganisms.

bioquads Occurrence record of organisms, serving as key units for biodiversity research and consisting of four elements (species names, location, time, and source).

bioregion 1. A landscape subunit at the scale of the continent that is set apart by its coherent geological and biotic history. 2. Specifically, one of six biogeographic divisions of South America consisting of contiguous ecoregions. Bioregions are delimited to better address the biogeographic distinctiveness of ecoregions.

biosphere The region of the earth where life exists, including parts of the lithosphere, hydrosphere, and atmosphere, the thin layer of life at the surface of the earth.

biosphere people A term for people enjoying access to resources garnered from the entire biosphere and made available through markets.

biosphere reserve An international conservation program aiming to protect representative ecosystems throughout the world in ways compatible with development efforts.

biosynthesis The biochemical process by which a plant produces molecules of utility or adaptation.

biota Living things, in particular, the flora and fauna of a region.

biotic impoverishment Systematic reduction in the earth's capacity to support life.

biotic interaction When the growth or behavior of one species affects those of another species; such interactions can be antagonistic (e.g., competition for limiting resources, predation, herbivory, or parasitism) or facilitory (e.g., pollination or other forms of mutualism).

biotope Region that is distinguished by particular environmental conditions (climate, soil, altitude, etc.) and therefore a characteristic assemblage of organisms.

bivoltine Describing species with a reproductive pattern of two generations per year.

black box In this context, an entity that can be examined at the system level without specifying its internal contents.

bleaching The loss of symbiotic zooxanthellae from corals. Bleaching is usually caused by elevated sea surface temperatures, but it can also be caused by sedimentation, salinity variation, or bacterial infection.

bloom Proliferation of algae in river, lake, estuarine, or marine waters. Older literature referred to a bloom as 5000 or more algal cells per liter (L) (or 5 cells per ml), although this density generally is too low to discolor the water. Algal "blooms" range in cell density from eukaryote blooms (e.g., dinoflagellates at 10^3 to 10^4 cells/ml, to cyanobacteria at 10^8 to 10^9 cells/ml).

blubber Specialized superficial adipose tissue found only in pinnipeds, cetaceans, and sirenians that serves as thermal insulation and provides energy storage.

bog An acidic peatland that receives all water inputs from the atmosphere and hence has low pH (usually 3.4–4.6), low alkalinity, and negligible concentrations of base cations in surface waters; typically dominated by mosses of the genus *Sphagnum* and other plants able to tolerate low pH and low nutrient availability—that is, members of the heath family (Ericaceae).

booklungs In certain insects, one to four pairs of abdominal respiratory organs consisting of a thin, multifolded membrane (the book's "pages") over which blood circulates and that is open to an air-filled cavity on the outside, itself open to the exterior via a spiracle. Gases passively diffuse back and forth across the membrane.

boreal Pertaining to the northern high latitudes (but not polar), regions which contain large continental landmasses and thus continental climates with moderate summers but long, severely cold winters (From Greek *boreas*, the north wind).

boreal forest A forest zone dominated by coniferous species, covering the northern latitudes of North America, the Nordic countries, and Russia.

bottom-up effects A term for a condition in which physical parameters (such as nutrients) allow increased primary productivity and the effects of this cascade up through higher trophic levels. A simple example involves nutrient additions to a lake resulting in increased phytoplankton growth, this increasing resource of food allowing increased numbers of herbivorous zooplankton, and the abundance of zooplankton leading to increased densities of zooplankton-eating fishes.

bottom-up forces Population-regulating processes based on the availability of food, nutrients, and energy.

BP Before the present, usually measured in billions of years (Ga) or millions of years (Ma).

brachyodont Low-crowned cheek tooth commonly found in omnivorous or burrowing animals.

brachypterous Having tegmina and wings shorter than the abdomen but overlapping or touching each other on the dorsum.

brackish Less salty than sea water; somewhat salty.

breed "A homogenous, subspecific group of domestic livestock with definable external characteristics that enable it to be separated by visual appraisal from other similarly defined groups within the same species, or . . . for which geographical separation from phenotypically similar groups has led to general acceptance of its separate identity" (Food and Agriculture Organization, 1995).

breeding line Any distinctive livestock population. The term may be synonymous with "breed" but is also often applied to somewhat distinct subpopulations within a breed or to breeding populations, either purebred or crossbred, that are of fairly recent origin. Breeding lines may exist within a breed, and newly created lines may become recognized breeds.

broodstock Fish or shellfish from which a first or subsequent generation may be produced in captivity, whether for growing as aquaculture or for release to the wild for stock enhancement.

browse 1. That part of leaf, twig, and reproductive growth of shrubs, woody vines, and trees available for animal consumption. 2. The act of consuming leaves, twigs, or reproductive parts of shrubs, woody vines, and trees.

bryophyte Nonvascular plant of the division Bryophyta (a moss, liverwort, or hornwort).

bulk density The mass of the undisturbed dry soil per unit volume.

bund An artificial embankment keyed into the substrate and used to retain water.

bundle sheath Cell layer at the periphery of the vascular bundles in leaves. In C_4 plants, the reactions of the photosynthetic carbon reduction cycle are localized in this layer, in contrast to C_3 plants, where they occur in the mesophyll tissue. In comparison to C_4 species, the bundle sheath cells of C_3 species are small, with few chloroplasts.

bush-fallow cropping system An agronomic system in which soil fertility is maintained by allowing native vegetation to regrow following several years of cropping.

bycatch Any living thing caught unintentionally in fishing gear; sometimes called **bykill** because so many such creatures are discarded dead. About one-fourth of the total world catch is bykill.

C

caatinga Within the Amazon region this term is applied to open forest on white sand, which occurs in the basin of the Rio Negro. The term is an indigenous word from the Tupi language meaning an open place.

caboclos Cultural and/or racial crossing between European and indigenous populations.

calanoid copepods These belong to the Crustacea and are very abundant in the zooplankton. They play a major ecological role in the food web of the pelagic ecosystem as grazers of phytoplankton and as a food source for, e.g., larval and adult fish.

calcareous fen A term variously used for wetlands with high calcium carbonate in water and soils. In the strict sense, the term refers to peat-accumulating wetlands with surficial deposits of calcium carbonate and a distinctive flora of calciphilic species. These wetlands typically develop where substantial amounts of cold, calcium-carbonate rich groundwater discharge to the surface.

calcification The deposition of calcium carbonate skeletons by aquatic plants or animals. In reef-building corals, calcium is deposited in its aragonitic mineral form.

Cambrian explosion A term for a period of about 25 million years during the Cambrian, when the diversity of multicellular life greatly increased in a relatively short time.

canopy Usually the highest tree layer in the vertical stratification of forests. The herbaceous foods for canopy-dwelling animals are found among the branches and leaves of the canopy.

cantharophily Pollination by beetles.

capoeira Brazilian term for secondary forest on cleared ground.

carbon cycle Photosynthetic organisms preferentially use the carbon-12 isotope instead of carbon-13, enriching living organisms in C-12. This differentiates the organic from inorganic carbon reservoirs. Quantifying shifts between the two isotopes of carbon reveals shifts between the two reservoirs.

carboxylase Enzyme that catalyzes the formation of a $C-C$ bond between inorganic C (CO_2 or HCO_3^-) and some organic molecule.

carnivore An organism that consumes other animals; literally, a meat-eater.

carrying capacity The maximum population that can be sustained indefinitely in a given area without changing the ecosystem in ways that will eventually reduce the sustainable population. This balance between population and resources is a dynamic one that involves changes in technology and other factors.

case law Interpretations of the U.S. Constitution, statutes, and regulations provided by the judicial branches of U.S. federal or state government. Such interpretations are provided when two or more parties disagree as to the meaning of a law in a specific context and bring it to the courts to decide.

caste A group of individuals in a colony that are both morphologically distinct and specialized in behavior.

catches The fish (or other aquatic organisms) of a given stock killed during a certain period by the operation of fishing gear(s). This definition implies that fish not landed, that is, discarded at sea, or killed by lost gear ("ghost fishing"), should be counted as part of the catch of a fishery.

cation Positively charged atom or molecule.

cation exchange capacity The capacity of a soil to adsorb cations such as calcium, magnesium, and potassium on the surfaces of soil particles and organic matter.

cerci Paired, usually not segmented structures at the end of the abdomen, sometimes used by males during mating to grasp the female's abdomen.

cercomer Posterior parenchymous and/or muscular extension of body; homology among groups is controversial; some view both the hook-bearing regions of monogeneans and some larval cestodes to be modifications of the cercomer.

cerotegument Hydrophobic secretion layer covering the body of several arthropods adapted to temporary submersion.

cerrado The Brazilian term for the large area of savanna and savanna forests that are the dominant vegetation of the Planalto of Central Brazil. Cerrado vegetation is further divided in four categories: (a) campo limpio (i.e., clean field), (b) campo sujo (i.e., dirty field or grasslands with scattered shrubs), (c) campo cerrado (i.e., closed fields or grasslands with numerous trees and shrubs), and (d) cerradao (i.e., when the vegetation is dominated by a closed canopy of trees).

certification Procedure of determining whether a particular process is conducted in an ecologically benign fashion (also referred to as eco-labeling).

chaco Region located between the Paraná basin, the central Brazilian shield, and the Andes. The ecosystems present in the Chaco region include temperate grasslands, savannas, and arid and semiarid environments.

chaetae Bristles composed of slender chitin cylinders glued together by scleroprotein, emerging from parapodia or body wall in most polychaetes.

champsosauridae Extinct taxon (Choristodera) of reptiles superficially resembling crocodilians

change matrices A way to relate area of land use classes at the beginning with that at the end of a period. From such matrices, it is possible to follow the exact path of change viz. how much land was transferred from a given use to other uses, how much came into that use, and what remained unchanged during the period.

chaparral Evergreen sclerophyllous-leaved shrublands that cover large areas of California.

character An attribute that is constantly distributed among all members of a species or a clade.

character displacement The displacement of characters caused by competition between species that live in the same place; divergence in a morphological character between two species when their distributions coincide in the same ecological environment compared to overlap of the character in question in the two species when they are geographically separated.

chelate, subchelate, chela (s.), chelae (pl.) Having a pincer-like claw (usually due to an extension of the second from terminal segment beside the terminal segment, forming the claw). In subchelate forms, the terminal segment merely folds back on the second from terminal segment. The chela is the clawlike appendage.

chelicerae (chelate) The first pair of preoral appendages. They are at most three segmented, usually two, and usually the distal segment acts against the penultimate to grab or hold prey or objects. If the basal segment has a finger-shaped outgrowth against which the distal segment operates, the chelicerae are chelate (as in scorpions or harvestmen). If not, the chelicerae are subchelate, as in spiders and tailless whip scorpions. In parasitic mites the chelicerae are modified into piercing stylets.

chemical defenses Compounds used by plants to deter or poison herbivores and pathogens.

chemical prospecting The process of searching for natural compounds in wild living plants, animals, and microorganisms with a potential for the development of chemical products like pharmaceuticals, pesticides, cosmetics, or food additives.

chemiosmotic energy coupling Exergonic (photo)-chemical redox reaction, or an exergonic hydrolysis of a phosphate anhydride or thioester, coupled to endergonic ion (normally H^+) transfer across a membrane, or vice versa.

chemoautotrophic Describing organisms (microbes) that use inorganic compounds as a source of carbon and energy and function as primary producers.

chemoautotrophs Organisms able to synthesize organic compounds by the oxidation of energy-rich inorganic sources. Light is not used.

chemoautotrophy The use of energy-yielding chemical reactions as an energy source for synthesis of organic matter from inorganic precursors; metabolism exclusively based on the oxidation of inorganic compounds.

chemolithotrophs Organisms deriving their energy from the oxidation of inorganic compounds.

chemolithotrophy The process, utilized by some microbes, of obtaining energy from the breaking of inorganic chemical bonds.

chemostat Apparatus for growing microorganisms in a continually replenished medium.

chemotrophs Organisms that utilize chemicals as sources of energy.

chiasmata χ-like configurations.

chiropterophily Pollination by bats.

chlorofluorocarbon (CFC) A type of hydrocarbon that is composed of nonreactive, nonflammable organic molecules in which both chlorine and fluorine atoms replace some of the hydrogen atoms. Use of CFCs is one of the primary causes of stratospheric ozone depletion.

chordate A member of the group Chordata. The Chordata includes the most recent common ancestor of tunicates and cephalochordates and all of that ancestor's descendants. Tunicates, lancelets, hagfishes, and vertebrates are all chordates.

chromatin Complex of DNA and associated proteins that make up chromosomes of eukaryotes.

chromophore Organic molecule that absorbs photosynthetically active radiation and either carries out photochemistry or transfers excitation energy to another chromophore that carries out photochemistry. Always bound to a polypeptide in the photosynthetic apparatus.

chromosome Threadlike structure that includes DNA and proteins (containing genes arranged in a linear sequence along the thread), which can be visualized when condensed during cell division.

chronic exposure An exposure to an environmental stress that is comparable in duration to the temporal scale of the biological system exposed.

chronic toxicity A damaging effect caused by a long period of exposure to low or moderate concentrations of a pollutant.

chronostratigraphic correlation Matching strata in different locations according to the points in time to which their boundaries correspond.

circannual rhythms Endogenous, or internal, rhythmic cycles of one year in duration that govern the onset and cessation of migratory behaviors.

cirrus In this context, a male intromittent copulatory organ consisting of distal, invaginable portion of ejaculatory duct that is evaginated (by turning inside out) through the genital pore during copulation; may or may not be armed with spines, microtriches, etc.

clade The branch between the nodes on a cladogram or phylogenetic tree; the unit of cladogenesis; i.e., a group of organisms including its common ancestor and all its descendants.

cladistic Describing a classification based entirely on monophyletic taxonomic groupings within a phylogeny; taxonomic units that are paraphyletic or polyphyletic are rejected. A **cladist** is one who practices cladistics, usually in the sense of using parsimony to adjudicate between data from multiple characters in the construction of a cladogram, which is an estimate of the true phylogeny.

cladistic biogeography The combination of cladistics (phylogenetic systematics) with vicariance biogeography. A method that searches for patterns of relationship among areas of endemism.

cladistics (phylogenetic systematics) A method of phylogeny reconstruction concerned with branching patterns. Sister group relationships are hypothesized on the basis of shared derived, or synapomorphic, characters.

cladogenesis branching evolution; the origin of diverging new forms from an ancestral lineage; the form of divergence and phyletic splitting, including speciation; adaptive radiation of species to major divergence of families and phyla. Evidence for cladogenesis is from those fossil groups that increase in the number of recognizable taxa over time.

cladogram A general graphic representation of a cladistic hypothesis; more general than a phylogenetic tree. A cladogram has no direct connotation of ancestry and the long axis does not connote time.

cladogram support Tests that permit some evaluation of how well data fit a cladogram.

classification In this context, the process of assigning the pixels of an image to discrete categories.

climate change Variability of weather over time.

climate sensitivity Long-term change in global mean surface temperature following a doubling of equivalent carbon dioxide (CO_2) concentration in the atmosphere.

climatic models Mathematical descriptions of the flows of energy, momentum, and materials around the atmosphere and oceans, typically run on large computers to stimulate the earth's climate naturally and when disturbed by greenhouse gases or other so-called forcings or disturbances.

climax According to some theories of succession, the end result of succession in which successional change ends with a community that does not change and which is in equilibrium with the climate.

climax community The stable ecosystem that is supposed to result from a sufficiently long period of unchanging environment.

clique A group of organisms in a food web in which any pair of species in the clique shares at least one prey species. Thus, it is a group of species that have similar diets and are different from other such groups.

cloaca The common chamber in which the reproductive, excretory, and digestive tracts of amphibians unite before exiting the body.

clonal growth A vegetative mode of propagation and expansion of plants by runners, tillers, or plant fragment dispersal, which is very important in the alpine life zone. (See also apomixis.) Clonal plants also produce sexual offspring by seed, but their clonal propagules often show higher survival. All clonal offspring of one source plant have the same genome and hence belong to the same "genet."

clone 1. A group of individuals formed by vegetative reproduction from a parent. This term applies particularly to those trees and plants that reproduce by creeping stems or by suckers from roots, forming circular patches of genetically identical individuals. 2. A population of cells all descended from a single cell; a number of copies of a DNA fragment obtained by allowing the DNA fragment to be replicated by a phage or plasmid.

clupeid fish Pelagic fish, including the anchovy, sardine, and herring, which feed on plankton.

clutch size The number of eggs a bird lays in its nest at one time.

Cnidaria The marine invertebrate phylum containing the reef-building corals.

coadapted gene complexes Genes that are adapted to function well together; genes that have been selected to work in a coordinated fashion to confer high fitness.

coalescent process The merging of lineages traced backward in time from a sample of genes.

coastal Describing estuaries, semi-enclosed seas, and shallower regions of the ocean, including areas influenced by rivers and runoff from land.

coastal sage scrub Semiwoody shrublands dominated by drought-deciduous species and occurring in semiarid areas along the coast of Southern California and the interior transition from chaparral to desert.

coastal zone Zone whose terrestrial boundary is defined by (a) the inland extent of astronomical tidal influence or (b) the inland limit of penetration of marine aerosols within the atmospheric boundary layer and including both salts and suspended liquids, whichever is greater; the seaward limit is defined by (a) the outer extent of the continental shelf (approximately 200 m depth) or (b) the limits of territorial waters, whichever is greater (Hayden *et al.*, 1984).

Code In nomenclature, one of the sets of internationally agreed rules governing the scientific names of organisms.

coevolution Process of reciprocal genetic changes in interacting species that result from the mutual selection pressures that strongly interacting species can exert on one another; long-term evolutionary adaptation of species to each other (e.g., mutually beneficial relationships between bees and flowering plants).

coevolutionary arms race. The continuing bouts of co-evolving defenses and counterdefenses that occur between hosts and parasites.

coexistence 1. The ability of species to persist together in an ecological community indefinitely in the absence of any major environmental change. 2. More generally, the state of two or more species being found in the same place at the same time.

cognition Act or process of knowing and understanding.

cohesion The sum total of forces or systems that hold a species together. The term is used especially in the interbreeding and cohesion species concepts. Cohesion mechanisms include isolating mechanisms in sexual species as well as stabilizing ecological selection, which may cause cohesion even within asexual lineages.

cohort A group of individuals of the same age that can be identified within a population.

combating desertification Activities specifically aimed at prevention and/or reduction of land degradation, rehabilitation of partly degraded land, and reclamation of desertified land.

combinatorial chemistry Laboratory methods to produce all possible combinations from various sets of chemical building blocks in a short period of time and to generate molecular diversity for the screening of new bioactive compounds.

commensalism An association of animals in which one uses food supplied in the internal or external environment of a host without affecting the host in any way.

commensal, symbiotic Living in association with (e.g., on or in) another organism.

commensal 1. Living in association with (e.g., on or in) another organism. 2. Specifically, describing a protozoon that is loosely but not obligately associated with another organism (e.g., the ciliates attached to the external surfaces of crustacean zooplankton).

common-pool resources Resources that include all ecosystems that are large enough such that excluding potential beneficiaries from their use is a nontrivial task and each individual's consumptive use (e.g., harvesting of a boatload of fish or a truckload of forest products) reduces what is available to others.

commons (or common property resource) Any resource asset that is not privately owned and controlled.

community 1. A group of species co-occurring in an area and interacting through trophic and spatial relationships; all forms of life that coexist and interact with each other in a particular habitat. Alternatively, some designated subset, such as the avian community or the vascular plant community. 2. A social unit consisting of members who are in direct interaction with each other and who have a collective identity, however defined. Relationships in such a group are principally primary rather than secondary in nature, and conformity to group norms is achieved mainly by peer pressure.

community-based conservation Conservation-oriented management of communal or public property by local residents.

community diversity Number of ecosystems and associated plant and animal species per 1° latitude by 1° longitude grid-cell.

community structure The web of potential biological interactions among members of a community that may be characterized in terms of diversity, complexity, hierarchy, and stability.

compass orientation Navigation in a particular direction without reference to landmarks or sites of origin or destination. Migrants are known to use compass information from magnetic fields, chemical gradients, and visual features such as the stars, sun, and planes of light polarization.

compensation depth Depth where photosynthesis and respiration are in balance.

competition 1. The use or defense of a resource by an individual which reduces resource availability to other individuals. 2. A reduction in one species' growth rate because of the effects of another species.

competitionism The view that competition regulates populations.

competitive Describing the displacement of one species from its habitat or ecological niche by another. When humans appropriate other species' "ecological space," it often leads to the local or even global extinction of the nonhuman organism.

competitive exclusion Extirpation or extinction of one species by another in a given area through competition for resources.

competitive release The expansion of a species-realized niche that is associated with the absence or removal of competition from other species.

complementary conservation Application of a range of conservation techniques (including *ex situ* and *in situ*) to conserve the target taxon, one technique acting as a backup to another. The degree of emphasis placed on each technique depends on the conservation aims, the type of species being conserved, the resources available, and whether the species has utilization potential.

complex adaptive systems A group of individuals or types that exhibit variability in which large-scale patterns emerge from small-scale interactions. The system is adaptive in the sense that it is subject to extrinsic selection that leads to a change in structure and/or dynamics over time.

components In this context, elements of groups of areas, or taxa, as determined by the branching pattern of a cladogram. For example, in a group comprising three taxa A, B, and C, when B is more closely related to C, there are two components, an ABC component and a BC component.

compound Poisson distribution A family of probability distributions of numbers per area where some areas have a higher expected density than others.

CONABIO The Mexican National Commission on Biodiversity.

conditions In this context, physical features of the environment, such as substrate type, ambient temperature, or salinity, that affect the ability of organisms to survive, grow, and reproduce.

congener A species that belongs to the same genus as the one under discussion.

conjugation Direct cell-to-cell transfer of DNA.

connectivity web This type of food web illustrates only feeding links without reference to strength of interaction or energy flow.

consensus cladogram (tree) Branching diagram that summarizes the common branching patterns from two or more cladograms.

conservation Maintenance of the diversity of living organisms, their habitats, and the interrelationships among organisms and their environment.

conservation biology An integrative approach to the protection and management of biological diversity that uses appropriate principles and experiences from the natural sciences, the social sciences, and various resource management fields.

conservation efforts Any action that aims to reduce the probability of extinction of a taxon over a specified time period.

conservation status Relative likelihood of extinction of a species or community.

conservation tillage Tillage that reduces soil disruption in order to conserve organic matter and water and reduce erosion.

conserved Of a scientific name, one that the appropriate international body has decided should continue to be used for an organism in cases where a strict application of the *Code* (q.v.) would mean it had to be replaced.

conspecific Of the same species.

conspecific brood parasites Brood parasites that lay their eggs in the nests of other individuals of the same species.

constitutive defenses Defenses that are manufactured and maintained by a plant, regardless of whether it has been attacked by an herbivore or pathogen.

contamination The presence of elevated concentrations of a toxic substance, compared with normal ambient concentrations.

continental drift Model first proposed by Alfred Wegener that states that the continents were once united and then were displaced over the surface of the globe; the movement of continents as a result of plate tectonic processes. This was especially important for Southern Hemisphere continents.

continental rise An area at the base of the continental slope between 3000 and 4000 m where the bottom slope is slight and sediments often accumulate.

continental shelf A region of ocean bottom extending from the low water mark at the edge of continents to a depth (~200 m) at which the incline increases markedly and the continental slope begins.

continental slope Ocean bottom extending from the edge of the continental shelf at an ~4° incline to a depth (3000–10,000 m) at which the slope decreases and the continental rise begins.

contingency Possibility or uncertainty; an event that may occur but is not likely.

contracted vegetation Vegetation restricted to wadis that receive additional water supply.

convective mixing Vertical mixing produced by the increasing density of a fluid in the upper layer, especially during winter in temperate and polar regions.

Convention on Biodiversity (CBD) An international convention signed by 175 countries, which originated at the Earth Summit of Rio de Janeiro. The convention was first enacted in June 1992, and it has been signed by many countries. Its objectives are the conservation of biological diversity, the sustainable use of its components, and the fair and equitable sharing of the benefits arising from the utilization of genetic resources, including by appropriate access to genetic resources and by appropriate transfer of relevant technologies, taking into account all rights over those resources and technologies, and by appropriate funding.

convergence The evolution of similar characters in genetically unrelated or distantly related species, often as the result of selection in response to similar environmental pressures.

convergent Describing species that develop similar behavioral, physiological, or morphological characteristics in response to living in similar environments in different places.

convergent characters Resemblance due to independent evolutionary events.

convergent evolution Distant or unrelated organisms evolving the same anatomical, morphological body plan characteristics, or ecological function.

coppice Regrowth of a felled tree from the stump.

coral reef Benthic environments characterized by reef-building corals with symbiotic dinoflagellates.

correct Of a scientific name of an organism, one that conforms to the appropriate international *Code* (q.v.) for the position in which it is placed.

corridor Strip of land or water that differs from the adjacent landscape on both sides.

cortex The outermost region of parenchyma in taxa in which two distinct regions are present; inner boundary often marked by conspicuous circular muscle fibers; see medulla.

cowbirds American blackbirds in the genus *Molothrus*, which contains important brood parasites.

craniomandibular Where the head, or cranium, and the jaw, or mandible, meet.

Crenarchaeota One of two kingdoms of organisms of the domain Archaea. From the Greek *crene* (spring, fountain), for the resemblance of these organisms to the ancestor of the Archaea, and *archaios* (ancient). Include sulfur-metabolizing, extreme thermophiles.

crepuscular Describing organisms that are active in dim light conditions, such as dusk or dawn.

criteria and indicators A monitoring system for the assessment of the economic, social, or ecological data about land management practices as an aid to improving sustainability.

critical habitat Habitat of a threatened or endangered species that is itself threatened by destruction, disturbance, modification, or human activity, potentially resulting in a reduction in the numbers, distribution, or reasonable expansion or recovery of that species.

critically endangered A species facing an extremely high risk of extinction in the wild in the immediate future.

crochets The minute hooklets on the fleshy abdominal prolegs of a caterpillar.

crop genetic diversity The diversity of the sets of genes carried by different individuals within a crop species. It occurs in the form of nucleotide variation within the genome.

crop varieties Named populations of crop plants that possess recognizable features and known utility for food, feed, or fiber.

crossbreeding/crossbred The mating of animals of different breeds to produce commercial market animals or breeding animals. The resulting crossbred animals often exhibit improved fitness and performance compared to their purebred parents.

crossing-over Complex interaction of homologs within bivalents at pachytene resulting in reciprocal exchange of genetic material between non-sister chromatids.

cross-resistance The ability of a single gene or mechanism to confer resistance to more than one toxin.

crown group A clade including a group of modern species, the common ancestor, and all its descendants, including extinct lineages.

cryobiosis Anabiosis (latent life) due to freezing.

cryopreservation Long-term storage of gametes or embryos in liquid nitrogen.

cryoturbation Process by which soil particles and stones are moved and mixed by the frequent formation and subsequent melting of ice crystals in the soil column.

cryptic Describing an organism that is concealed or obscured by the similarity of its appearance to the surrounding environment.

cryptic species Species that are virtually indistinguishable by normal morphological analysis and are, instead, defined by a combination of genetic, behavioral, and other characters.

cryptoendolithic Living within the surface of rocks.

cryptogam Plant without apparent reproductive organs; plant that reproduces by spores or gametes rather than seeds; includes bryophytes and lichens.

ctenidia (sing. ctenidium) The molluscan gill, comprising a characteristic muscular axis from which multiple filaments arise. The relative placement of ciliary bands, nerves, muscles, and blood spaces is similar in all extant taxa.

cuckoos A family of birds of which approximately half the species (61 of 125) are obligate interspecific brood parasites.

cultivar Cultivated variety or genetic strain of a domesticated food plant.

cultural diversity Variety of human groups distinguished through beliefs, lifeways, dress, food, languages, sexual behavior, forms of productive organization, art, and conceptions of nature.

cultural ecology Analysis of how culture influences the interactions between a human population and the ecosystems in which they reside; also called ecological anthropology.

cultural ecosystem Ecosystem produced by the long-term interaction of wild plants and animals with human activities.

cultural eutrophication The acceleration of nutrient overenrichment caused by humans; can be caused by direct, point-source pollution (sewers) or by diffuse, non-point-source pollution (such as fertilizer runoff from farm fields).

culture 1. A system of socially learned, shared, and patterned ideas, institutions, behaviors, and their material products that distinguishes a particular society. 2. All capabilities and habits acquired by people as members of society.

cursorial Adapted for running.

cuticle The noncellular external layer of the body wall of various invertebrates.

cyanobacteria A large group of eubacteria with chlorophyll *a* and oxygenic photosynthesis; formerly known as "blue-green algae."

Cyperaceae Sedge family.

cytochrome P_{450} monooxygenases A ubiquitous group of enzymes involved in the NAPDH-mediated oxidation and metabolism of a broad range of endogenous and exogenous substrates.

cytotoxicity The ability of a chemical compound to kill cells in an experimental system, such as in a petri dish.

D

database Assemblage of information organized in tables with a logical structure. It is a fundamental tool used to gather, organize, and analyze biodiversity information.

decomposition The biotic breakdown of dead organic matter (detritus) by bacteria and fungi that releases carbon dioxide and nutrients for recycling.

deep ecology Activist philosophy that advocates radical personal and political change to protect wild nature.

deep sea The seabed and immediately overlying water covered by seas at least 200 m deep.

deforestation The complete or almost complete removal of tree cover and conversion of forested land to other uses as a result of human activities; the change from forest to other land uses such as agriculture or ranching; or depletion of forest crown cover to less than 10% density. Because shifting cultivation involves a change of land use, it is considered deforestation.

degraded Describing an area with a loss of forest structure, productivity, and native species diversity. A degraded site might still contain trees (i.e., a degraded site is not necessarily deforested) but it has lost at least some of its former ecological integrity.

degradation A decrease in ecosystem productivity or structure, and/or declines in native species diversity (sometimes with concomitant increases in exotic species) due to land use practices. Directly related to declines in biodiversity, ecosystem degradation encompasses soil impoverishment (e.g., compaction, erosion, salinization, loss of biological soil crusts) and hydrological alterations that diminish water availability. Similar to desertification.

deme Smallest population unit in population genetic models.

demographic transition Transition of human populations from conditions of high birth rates and high death rates to low death rates, followed by low birth rates; in progress since the seventeenth century, with a large increase in the number of humans—projected to be completed in the twenty-first century.

dendrogram Cladogram or tree; a branching, nonreticulate diagram expressing nested hierarchical relationships or similarities (or both) between entities (e.g., taxa) such that the entities only appear at the tips of terminal branches.

denitrification Bacterial nitrate respiration of organic matter to elemental nitrogen under oxygen-free conditions in soils and aquatic sediments; the reduction of nitrate or nitrite to gaseous nitrogen products, mainly N_2 and N_2O, by bacteria.

density dependence Effects on the growth rate of a population due to interactions among the members of that population; the condition that environmental factors influence population growth rate in relation to population size. Density dependence usually is seen as an linear, inverse relationship between population growth rate and population density (i.e., population growth decreases as density increases) and may occur if individuals compete or predators are more effective as a prey population increases.

density independence The absence of environmental factors that influence population growth as a function of density. This may occur if mortality removes a fixed percentage of a population, independent of population size.

depauperate Of low diversity, lacking in species; opposite of speciose.

dependent community The group of species that requires a particular host plant to complete some or all of their life cycle.

depletion In fisheries, reduction to population levels low enough to reduce or threaten future productivity.

deposit feeders Organisms that feed primarily by ingesting organic material occurring on or between sediment grains.

deposition The rate of influx of a substance from the atmosphere, usually expressed as mass per unit area of ground; the delivery of material inputs to the earth's surface from the atmosphere.

derived Describes a character or character state of the organism being considered that has changed from the primitive condition for its evolutionary lineage.

descriptive taxonomy The description of new species usually involving written accounts, with illustrations, of the characteristics of a specimen.

desertification Process of changing a nondesert area into what appears to be a desert, regardless of the climate. This process often has a negative connotation relating it to overgrazing by domestic animals, or general human misuse; however, factors, not related to humans, may cause desertification.

deterministic system A dynamical system whose detailed future behavior can be predicted, in principle, for all time, assuming perfect knowledge of the system at the present.

detoxification Reduction in the toxic effect of a pollutant by its chemical or biological transformation to another less toxic chemical.

detrital shunts Energy and nutrients from the saprovore web reenter the plant herbivore predator food web when detritivores are eaten by predators that also eat plants, herbivores, or other predators.

detritivores Organisms that consume dead plant or animal material.

detritivory The process of feeding on fallen and generally dead organic debris; although the term covers both plant or animal matter, in moths the term is especially apt to apply to leaf litter feeders.

detritus Nonliving organic matter in both soluble and particulate forms; dead organic matter.

diadromy Migrations that take individuals between fresh and salt-water habitats, a common phenomenon for many migratory fish species.

diagnosis A set of attributes sufficient to define, characterize, or identify a given taxonomic group.

diagnostic species Species that emerge from semiquantitative analysis as being the key species associated with a floristic group.

diapause A state of dormancy in some animals that is induced by a "token" environmental cue, such as day length. The token cue serves as a reliable indicator of a coming onset of harsh environmental conditions, but is not by itself harsh.

diaspore A plant part distributed by dispersal, regardless of its developmental and morphological origins. A diaspore may be a naked seed, a seed enclosed in a fruit, or many seeds enclosed in a fruit. It may also mean bulbs or lengths of rhizomes. A good synonym is propagule.

diastema A large gap between incisors and premolars that is thought to permit an animal space to manipulate food with the tongue.

dichotomous tree A cladogram in which all nodes are bifurcate.

dichotomy Two taxa arising from one node.

diffused vegetation Vegetation occupying all slopes and most habitats.

digitigrade Literally, walking on the digits; a posture where the majority of the weight is borne by the metacarpal and metatarsal bones and the heel/palm is raised off the ground.

dimension Set of measurement units that are completely similar. Commonly used dimensions are length (inches, meters, etc.), time (seconds, years, etc.), and mass (grams, pound-mass, etc.).

dimensions of desertification The interactions and feedbacks of meteorological, ecological, and human components of land degradation.

dinosauria Taxon of reptiles including Ornithischia (bird-hipped dinosaurs) and Saurischia (reptile-hipped dinosaurs, including birds).

diploid Individual (i.e., one who has two sets of chromosomes, usually received from different parents), in contrast to a haploid organism such as most microorganisms; the gametophyte stage of higher plants or some sexual forms (e.g., male ants and honeybees) may have the same form or allele of a gene on both members of a chromosome pair (in which case it is heterozygous) in different forms (alleles) (in which case it is homozygous).

diploid hybrids Species formed from hybridization where chromosomes from species of equal chromosome numbers become stabilized through recombination.

Diptera Group of insects to which the flies, including gnats and mosquitoes, belong.

dipterocarp A member of the Dipterocarpaceae family (comprising 22 genera) of South Asian and African timber trees.

discount rate Rate (usually annual) at which future revenues are discounted to calculate a present value. Personal discount rates often exceed market rates by a wide margin.

disease Any impairment of the normal physiological functions of an organism. While disease normally refers to infection by bacterial fungal, protozoan, or viral pathogens, technically bleaching could also be classified as a disease based on its physiological effect.

disharmonic fauna Describing a fauna in which a strongly reduced number of major lineages is represented, as in oceanic islands, due to their improbable colonization from distant sources.

disparity Range of diverse morphological architectures present in higher taxa such as classes, phyla, and kingdoms; it is used to explain the origins and maintenance of this diversity of life-forms and body plans.

dispersal The movement, or spread, of an organism from one area to another independent from other organisms and of historical events that change the natural distribution of the organism. Specifically, the movement of an individual organism away from its place of origin to the place where it breeds (also, movement by an adult from one breeding location to another).

dispersal barrier The species-specific physical and biological restrictions a propagule must overcome to accomplish successful colonization.

dispersal biogeography Usually refers specifically to "a branch of historical biogeography that attempts to account for present-day distributions based on the assumption that they resulted from differences in the dispersal abilities of individual lineages" (Brown and Lomolino, 1998; p. 628). More generally, it is the study of the relationships between dispersal and the geographical distribution of organisms. I adopt the latter definition for this article.

dispersal route The particular set of physical and biological conditions that allow organisms to cross dispersal barriers.

disruptive selection Selection acting to preserve extreme phenotypes in a population. Speciation usually involves disruptive selection, because intermediates (hybrids between incipient species) are disfavored (see also stabilizing selection).

disseminule The unit of dispersal; any part or stage in the life cycle of an organism that is used for dispersal.

dissipative beaches Wide, flat beaches.

distance In this context, a measure of dissimilarity between two taxa based (normally) on the number of mismatches within a large set of characters or attributes compared for both taxa.

disturbance Any relatively discrete event in time that disrupts ecosystem, community, or population structure and changes resources, substrate availability, or the physical environment; e.g., major alternations of vegetation due to events such as wildfires, hurricanes, landslides, and human clearing.

disturbance regime The collective spatial, temporal, physical, and ecological characteristics of a disturbance process operating in an area.

divergence A fixed genetic difference between two species or populations, or the process of evolving such a difference.

diversity The variety of species within a given taxonomic group; a quantitative measure of the taxonomic or physiognomic composition of coexisting organisms at one of three scales: alpha (local diversity), beta (diversity change between adjacent local sites), and gamma (regional diversity).

diversity distribution A frequency distribution showing the number of classes (usually species) that have one, two, three, etc. individuals per class (per species).

diversity index A numerical measure combining information about the number of species present in a sample or habitat and information about their relative abundances.

diversity principle General geographical coincidence between high concentrations of both biological and cultural diversity, usually in the tropics.

diversity-productivity hypothesis The proposal that greater diversity would lead, on average, to greater total biomass or productivity.

diversity-stability hypothesis The proposal that ecosystems containing more species would be more stable.

DNA Deoxyribonucleic acid, the molecule of inheritance that stores genetic information that is passed from one cell to another and from one generation to succeeding generations; it is composed of four nucleotides: adenine (A), cytosine (C), guanine (G), and thymine (T).

dodder Viney parasites of the genus *Cuscuta* that form

haustoria on the stems of their hosts. The unrelated parasitic vine, *Cassytha*, is often referred to as Laurel Dodder.

domains The highest taxonomic rank defined to classify organisms into Archaea, Bacteria, and Eucarya, which differ from each other in fundamental genomic and phenetic properties; i.e., the major divisions of the biota on earth.

domesticate Plant that has been selected by humans and adapted for use as a food crop, nutrient, fiber, or other purpose.

dominant species Species that have the greatest influence on ecosystem structure and function by virtue of their abundance, biomass, or coverage. Such organisms account for most of the biomass in a community, and thus are the primary components of community structure. Trees in forests, mussels in rocky intertidal habitats, grasses in grasslands, and kelps or corals in nearshore subtidal habitats are all dominant species.

donor control Consumer population growth is affected by their resources but consumers do not affect the renewal rate of these resources and hence cannot depress their resources.

dormancy Any state of reduced metabolic activity of an organism. Typically, dormant organisms have associated characteristics such as cessation of development, the absence of reproduction, and enhanced resistance to harsh environmental conditions. Some disciplines have distinct meanings for dormancy that add further constraints to its meaning.

dormant Describing a condition under which a seed will not germinate, even if all environmental requirements for germination are met.

dorsal ventral The dorsal part of the organism is the "back," on the opposite side of the body from the mouth; ventral is the same side of the body as the mouth and legs for a crustacean.

drought-deciduous Descriptive of plant species that lose their leaves during the dry season as soil moisture becomes limited.

dry deposition Deposition of dry pollutants from the atmosphere including gases and aerosols.

drylands Arid and semiarid croplands, pastures, rangelands, and subhumid woodlands in which the index of aridity is less than 0.65; drylands cover about two-fifths of the land surface of the earth and are home to more than 20% of the human population.

duogland adhesive system Combination of a cement, a viscid gland producing adhesive material, and a releasing gland; usually associated with cells of ventral epidermis; provides temporary adhesion to the substrate for many interstitial flatworm species.

durable resistance Inherited resistance to a disease or insect pest, with relatively long effective lifetime. It is usually imparted by several genes of individually small effect.

dyke intrusion Upwelling magma; may reach the surface of the crust to flow out as lava and solidify as basalt.

dynamical system A set of rules defining how certain variables change with time. Ecological models are dynamical systems representing what are believed to be (vastly more complex) dynamical systems in nature.

dynamical variable A quantity in a dynamical system that changes with time according to the rules of the system (contrast with parameter).

dynamic model Mathematical description of a system that has components that vary in time.

E

Earth energy balance Average balancing of incoming solar energy by outgoing terrestrial radiation for Earth as a whole.

ecocriticism Interdisciplinary study of literature, history, religion, and philosophy with an emphasis on places, evolutionary biology, and environmental problems.

ecofeminism Liberation philosophy that draws connections between preserving nature and promoting women's rights.

ecogeography Analysis of a species' ecological, geographical, and taxonomic characteristics to assist in the formulation of collection and conservation priorities.

ecological biodiversity Variety of biotic communities (plant, animal, and microbial communities) and their complexes (ecosystems, landscapes, and biomes).

ecological biogeography Spatial patterns of organisms considered in terms of their interactions with other organisms and the environment.

ecological community An assemblage of species occurring together in a given space and time.

ecological deficit An ecological deficit exists when the "load" (see *human load*) imposed by a given human population on its own territory or habitat (e.g., region, country) exceeds the productive capacity of that habitat. In these circumstances, if it wishes to avoid permanent damage to its local ecosystems, the population must use some biophysical goods and services imported from elsewhere (or, alternatively, lower its material standards).

ecological footprints Spatial coverage of ecological impacts consequent on the activities of a given group of people.

ecological functions Attributes or properties of assemblages or habitats that are dependent on the biodiversity of animals and plants present. Examples are production of nitrogen, sequestering of heavy metals, sustained production of harvested resources, maintenance of a diverse food web, and so on.

ecological integrity The balanced state of an ecosystem under normal environmental conditions, including the capacity of an ecosystem to absorb disruption and to recover from the disruption.

ecological model A vastly simplified mathematical representation of an ecological system intended to capture the full system's essence. Qualitatively, ecological models are to natural ecological systems as line drawings are to full-color photographs.

ecological niche The set of requirements that must be met if a particular species is to survive. It is sometimes used to mean the place in which those requirements are met.

ecological processes Direct or indirect interactions between species—such as grazing, predation, competition, responses to disturbance, and so on—that cause spatial or temporal patterns in distributions and abundances of species.

ecological refugees A term for people who have lost access to their traditional base of natural resources yet have very limited access to resources through markets.

ecological release Expansion of habitat, or ecological environment, often resulting from release of species from competition.

ecological transfer efficiency The ratio of energy ingested by a population's predators to energy ingested by the population.

ecological transition Tendency for societies to be in growing disequilibrium with their biophysical environment as they increasingly deplete natural resources and degrade their habitat, thereby reaching new thresholds of environmental impact.

economic botany As originally conceived, a branch of applied botany that arose during the colonial period to identify and characterize economically important plants and the products derived from them. Currently, it is a scientific endeavor that seeks to document the properties of useful plants through agronomic, archaeological, ecological, ethnobotanical, genetic, historical, phytochemical, and other empirical approaches. It overlaps broadly with ethnobiology because both fields have witnessed a similar development in theory and methodology in recent years.

economic efficiency State of affairs in which all goods in an economy are allocated in such a way that no one can be made better off without making someone else worse off; also known as Pareto optimality.

economic impact Capacity of weeds to limit productive use of terrestrial environments and the costs associated with their control in both managed and natural areas.

economic rent The flow of net economic benefits (revenue minus costs) derived from the exploitation of a resource asset.

ecoregion A geographically distinct assemblage of natural communities that share a large majority of their species and ecological dynamics, share similar environmental conditions, and interact ecologically in ways that are critical for their long-term persistence.

ecosystem A community of organisms (plants, animals, and microbes) and their physical environment interacting as an ecological unit, such as a lake, a wetland, a forest, or an agricultural landscape. An ecosystem is a conceptual view of an assemblage of organisms and of physical and chemical components in their immediate environment, and the flow of materials and energy between them. The set of species in an ecosystem interact in characteristic fashion, and generate among them biomass flows that are stronger than those linking that area to adjacent ones.

ecosystem classification A procedure for grouping together similar ecosystems based on a predefined set of criteria, e.g., vegetation structure, climate, land-use history, and soil type.

ecosystem composition The list of species or functional groups that are present in a given ecosystem.

ecosystem diversity Diversity of habitats, ecosystems, and the accompanying ecological processes that maintain them.

ecosystem engineer A term for animals (e.g., beavers) that modify an ecosystem by physically changing it (e.g., building dams) or by being present in such a great abundance (e.g., large grazing animals) that they have a disproportionately large impact on ecosystem functions.

ecosystem engineering The concept whereby members of the macrofauna (e.g., termites and earthworms) are actually moving parts of the soil volume for their own uses (e.g., making macropores, which permit flow of large amounts of water rapidly through the soil).

ecosystem engineer species Species that directly or indirectly modulate the availability of resources (other than themselves to other species) by causing physical state changes in biotic or abiotic material and in so doing modify, maintain, and/or create habitats.

ecosystem function(ing) The sum total of processes such as the cycling of matter, energy, and nutrients operating at the ecosystem level.

ecosystem goods Commodities, such as timber and seafood, supplied free to humanity by natural ecosystems.

ecosystem health Condition of an ecosystem, whether natural, managed, or human-dominated, that is free from influences that would damage or destroy its characteristic structures and functions.

ecosystem integrity Condition of an ecosystem that is largely free from human interference and possesses a species composition and functional organization comparable to those of natural ecosystems in the region.

ecosystem management Approach to management that attempts to maintain high ecosystem integrity while providing the services, uses, values, and products from that system for the long term. It explicitly integrates social, economic, and natural system sustainability.

ecosystem people People meeting the bulk of their resource requirements from a limited area near their habitation through gathering or low input agriculture and animal husbandry.

ecosystem services The wide array of conditions and processes through which ecosystems, and their biodiversity, confer benefits on humanity; these include the production of goods, life-support functions, life-fulfilling conditions, and preservation of options.

ecosystems, anthropogenic and natural Communities of organisms and their environment formed either through human action or through natural processes. In practice, it is difficult to establish the extent to which an ecosystem is anthropogenic or natural, reflecting the current and historical impact of people on the environment.

ecosystem stability Capacity of an ecosystem to persist in the same state. It has two components. Ecosystem resistance is the ability to stay in the same state in the face of perturbation. Ecosystem resilience is the ability to return to its former state following a perturbation.

ecosystem structure The organisms, their communities, biodiversity, and habitats that comprise an ecosystem.

ecotone A zone of transition between two different

habitats that may contain a community of organisms distinct from either habitat.

ecotourism Nature-based tourism that involves education and interpretation of the natural environment, improves the welfare of local people, and is managed to be ecologically sustainable.

ecotoxicology The use of test organisms (e.g., the water flea, *Daphnia*) to study the toxicity, pathways of accumulation, and breakdown of chemicals, particularly those manufactured by humans (e.g., pesticides).

ecotype or ectotypic Refers to genetic (evolutionary) differentiation within a given species (a specific "race") that reflects an obvious advantage in a given environment. Ecotypic traits are retained when individuals are transplanted into a different environment where these traits have no advantage.

ectoderm An embryonic tissue that provides the future outside layer of the animal.

ectomycorrhiza (pl. ectomycorrhizae; adj., ectomycorrhizal) Part of a mutualistic relationship between a fungus (usually a basidiomycete, but sometimes an ascomycete) and a host plant in which hyphae aggregate as an extra surface around roots of the plant, aiding nutrient transfer between the plant and fungus.

ectoparasite A parasite living on the surface of a host.

ectothermic Deriving the energy needed to raise body temperature from sources outside the body.

ectothermy A method of body temperature control in which the animal utilizes external sources for gaining and giving up heat, thus achieving temperature control without affecting metabolic rate.

effective population size (N_e) Size of the ideal population used in population genetics theory that would have the same rate of increase in inbreeding or decrease in genetic diversity as the actual population under study. The size can vary across genes due to changes in background selection across the genome.

efficiency The degree of success in reaching a conservation goal relative to the cost (or to some surrogate for cost, such as the number of areas). For goals such as maximizing the number of species represented for a particular cost, a distinction should be made between merely selecting areas on the basis of any records of species and choosing areas that maximize persistence of species in the long term.

effluent line Level on the beach face where the water table intersects the surface.

egg bank An accumulation of long-lived diapausing eggs (i.e., those eggs that persist in diapause for longer than a single growing season) of aquatic invertebrates in the sediments of marine or freshwater habitats.

egg rejection Host responses to parasitism that include ejection of parasitic eggs, abandonment of parasitized nests (usually with renesting), or a new cup built over a parasitized clutch.

El Niño Southern Oscillation (ENSO) Regarded as quasi-periodic fluctuations occurring in the equatorial region of the Pacific Ocean. The ENSO is associated with changes in the sea surface temperature, sea surface levels, and rainfall patterns in the tropics and with a possible influence on the weather at higher latitudes. ENSO events have also been associated with changes in fisheries and ecosystem processes.

elasmobranchii Sharks and their relatives, such as skates and rays.

elective culture The provision of appropriate physical and chemical conditions that elicit the growth of specific metabolic types of microbes.

electromagnetic radiation The energy transmitted through space in the form of electric and magnetic waves. Can be detected by various types of sensors.

electrophoresis Separating charged molecules (such as polypeptides or polynucleotides) between the two poles of an electric field.

elfin woodland Small, stunted tree growth characteristic of forests at higher elevations in warm, moist regions.

Eltonian niche The role, or occupation, of a species in a community.

emergent A very tall tree that emerges above the general level of the forest canopy.

emergent vegetation 1. Plants that are attached to the ocean bottom but extend up through the water column above the ocean surface. 2. Rooted plants that tolerate saturated or flooded soil but not extended periods of submersion.

emission The release of a contaminant to the environment.

enchytraeids Segmented worms that are related to but smaller than earthworms.

endangered Describing a species that is at substantial risk of extinction as a result of human activities.

endangered species Species that are likely to become extinct in the near future because of normal human activities. Examples of such activities are land clearing for agriculture or housing and accidental importation of invasive species through commerce.

endangerment Condition in which a species is at risk of extinction.

endemic Restricted or native to a geographically defined area; describing a plant or animal species limited to a small area of the world, for example, one island or mountain range; i.e., not introduced.

endemicity State of a species or other taxon being restricted to a given area, such as a specific habitat, region, or continent.

endemic languages Languages that are restricted to a single country and, like their species counterparts, hold a high percentage of the unique traits in human language.

endemics Those species that are limited to relatively small areas, being found nowhere else on Earth.

endemic species Species confined in their distribution to a particular geographic region, especially the area where they evolved. The size of the region is arbitrary (a species can be endemic to North America or to a tiny island).

endemic taxa Species or clades unique to one geographic place.

endemism The fact of being found in a specific location of limited size, rather than being widely distributed.

end-member fluid In black smokers, the hot, chemically modified fluid that develops at depth in the crust and exits undiluted by seawater.

endocrine disruptor Chemical that interferes with the chemical communication (mainly performed by hormones) within an organism.

endogenous forces Forces within a population that affect the dynamics of the population.

endolithotrophic Living inside rocks, usually sandstone.

endomycorrhiza (pl. endomycorrhizae; adj. endomycorrhizal) Part of a mutualistic relationship between a fungus (a zygomycete) and a host plant in which no sheath is formed. Nutrient transfer occurs with the aid of highly branched arbuscules.

endoparasite A parasite living inside a host.

endophyte 1. Fungus or other organisms residing or growing within plant tissues. 2. The portion of a parasitic plant that is embedded inside host tissue.

endophytic parasite Parasites with vegetative bodies that are entirely endophytic and unobservable unless flowering.

endosymbiont An oganism living in a specialized form of symbiosis in which one partner thrives within cells, lumen, or tissues of the host organism; most obligate endosymbionts belong to the group of uncultured organisms.

endosymbiosis Condition of one organism living inside another; includes intracellular symbiosis (endocytobiosis) and extracellular symbiosis.

endosymbiotic Describing a organism living in a long-term association inside a host organism to their mutual benefit (e.g., the endosymbiotic methanogenic bacteria that live inside anaerobic protozoa and utilize waste H_2 produced by the host).

endotherm An animal whose body temperature is largely determined by its own metabolism rather than by the temperature of the environment.

endothermic Deriving the energy needed to raise body temperature from within the body—i.e., from metabolic heat production.

endothermy A method of body temperature control in which the animal modifies its metabolic rate to achieve the desired body temperature.

energetic web This type of food web quantifies the amount of energy (or material) that flows across links joining species.

energy The capacity to perform work. Potential energy is this capacity stored as position (e.g., in a gravitational or electromagnetic field) or as structure (e.g., chemical or nuclear bonds). Kinetic energy is this capacity as manifested by the motion of matter. The joule (J) is the common SI unit of energy, where 1 J equals the amount of energy required to increase by one Kelvin the temperature of one gram of water. Other units include kilocalories (kcal), kilowatt-hours (kWh), and British thermal units (BTU).

energy efficiency A measure of the performance of an energy system. First law efficiency, the most commonly used measure, equals the ratio of desired energy output to the energy input. Second law efficiency equals the ratio of the heat or work usefully transferred by a system to the maximum possible heat or work usefully transferable by any system using the same energy input.

energy, industrial Forms of energy generally transformed in bulk at centralized facilities by means of complex technology. The major forms of industrial energy are oil, coal, natural gas, nuclear, and hydroelectric. In addition to hydroelectric, industrial energy also includes other technologically complex methods of harnessing renewable energy, including photovoltaics, electricity-generating wind turbines, and geothermal turbines.

energy, nonrenewable Forms of energy whose transformation consumes the energy source. The major forms include oil, coal, natural gas, and nuclear.

energy, renewable Forms of energy whose transformation does not consume the ultimate source of the energy, harnessing instead solar radiation, wind, the motion of water, or geologic heat. The major forms of renewable energy are solar, biomass, wind, hydropower, and geothermal. The forms of renewable energy that depend on complex technology are forms of industrial energy. The simpler renewable systems are forms of traditional energy.

energy, traditional Forms of energy generally dispersed in nature, renewable, utilized in small quantities by rural populations, and often not counted in government statistics. The principal forms of traditional energy are firewood, charcoal, crop residues, dung, and small wind and water mills.

entropy An expression or measurement of the energy available for use by a system, including living systems; often described as the extent to which the system tends toward a state of disorder or randomness. The entropy of a completely ordered system (e.g., a system at a temperature of absolute zero) is zero.

environment Surroundings; the complex of physical, chemical, and biotic factors acting upon a living system and influencing its form and survival.

environmental ethics Philosophical discipline that specifies proper human relationships to the natural world.

environmental impact Ability of weeds to displace native fauna and flora, alter key ecosystem functions, and change disturbance regimes.

environmental stress An action, agent, or condition that impairs the structure or function of a biological system.

environmental weeds Invasive alien plants that impact on natural or seminatural ecosystems, for example, by eliminating native organisms or altering ecosystem functioning (also known as "wildland weeds"). Native species can be environmental weeds, especially when the disturbance regime or resource levels have been altered.

eolian Wind blown; transported by the wind.

epidemic A sudden increase in parasite prevalence or intensity beyond what is normally present.

epigenetic level Study of molecules which are the product of gene expression.

epikarst Upper part of the percolation zone in karst.

epiphyll (folicolous) Plant that grows on the leaf surface of another plant.

epiphyte A plant that uses another plant, usually a tree, for support but not for nourishment.

epiphytic Describing a nonparasitic plant that uses another plant as mechanical support but does not derive nutrients or water from its host.

epiphytic material Live and dead canopy vascular and nonvascular plants, associated detritus, microbes, invertebrates, fungi, and crown humus.

epistasis Interactions between genes at different chromosomal locations in the determination of phenotypic character values.

epizootic Describing disease outbreaks among animal

populations (as distinguished from an epidemic in human populations).

equilibrium A condition of stasis in some dynamical variable.

ERIN The Environmental Resources Information Network of Australia.

estimator A statistic calculated from data to estimate the value of a parameter.

estuary Semi-enclosed coastal body of water that has a free connection with the open sea and within which seawater is measurably diluted with freshwater derived from land drainage.

ethical holism Position that complex aggregates such as species, ecosystems, or human societies have intrinsic value.

ethics Codes that exert a palpable influence on human behavior. Embedded in worldviews, ethics provide models to emulate, goals to strive for, and norms by which to evaluate actual behavior.

ethnobiology A term coined in 1935 that has been defined as the study of the reciprocal interactions between people and the biological organisms in their local environment and, recently, as the study of biological sciences as practiced in the present and the past by local people throughout the world. Many researchers consider that ethnobiology comprises numerous subfields, such as ethnobotany, ethnoecology, ethnoscience, and ethnozoology, but there is no consensus on this point.

ethnobotany Study of the variety, natural history, and characteristics of the plants used by human cultures.

ethnoecology Typically defined as the study of local knowledge and management of ecological interactions. Recently, some researchers have proposed an alternate definition, considering ethnoecology as an emerging field that focuses on local peoples' perception and management of complex and coevolved relationships between the cultural, ecological, and economic components of anthropogenic and natural ecosystems. It is concerned with the interaction between knowledge, practice, and production, and it is oriented toward applied research on conservation and community development.

ethnoscience A discipline that arose as a minor subfield of ethnography concerned with recording in great detail local peoples' knowledge of biological organisms and the physical environment. Later, the term came to be used in a more restricted sense by cognitive and linguistic anthropologists to refer to local classificatory systems (as an object of study) and their semantic analysis (as a methodological approach). In France, the term is used to refer to ethnobiological studies in general.

ethnozoology The study of the reciprocal interactions between people and the animals in their local environment.

ethology The study of the behavior and social relations of animals.

etiology The study of causes or origins; stories that explain the origins of phenomena.

Eucalyptus The dominant tree genus in Australia today.

eucaryote Another spelling of eukaryote.

Eukarya One of three domains of life. From the Greek *eu* (good, true) and *karion* (nut; refers to the nucleus).

eukaryotes Organisms with genetic material organized into chromosomes that are contained within a membrane-bound nucleus in the cell; eukaryotic cells undergo mitosis and meiosis during cell division, ensuring the equal division of chromosomes among daughter cells. Eukaryotes include animals, fungi, and plants as well as a large number of microbial groups collectively referred to as protists.

eukaryotic Describing an organizational state of cellular organisms in which the genome of the cell is stored in chromosomes enclosed in a membrane-bound nucleus; all protists (algae and protozoa), fungi, plants, and animals have eukaryotic cells.

euphotic The top layer of a water body through which sufficient light penetrates to support net photosynthetic gain and the growth of photosynthesizing organisms. Rarely more than 100 m in depth, the euphotic layer can be as little as 1 m in turbid waters.

euphotic zone Sunlit upper layer of the ocean that can support photosynthesis.

Euryarchaeota One of two kingdoms within the domain Archaea. From the Greek *eurys* (broad, wide), for the relatively broad patterns of metabolism of these organisms, and *archaios* (ancient). Includes halophiles, methanogens, and some anaerobic, sulfur-metabolizing, extreme thermophiles.

euryhaline Able to live over a wide range of salinities, from brackish to fully marine waters.

eusociality Cooperative behavior among individuals of the same species characterized by reproductive division of labor, overlap of generations, and cooperative nesting.

eutherian Mammal in which the embryo is attached to the mother by a placenta.

eutrophic Trophic status of an aquatic ecosystem that is characterized by relatively low phytoplankton species diversity but high phytoplankton production (biomass as mean chlorophyll *a* ca. 15–40 μg/L), with the phytoplankton often dominated by cyanobacteria in lakes, and by dinoflagellates or other flagellates in estuaries); high nutrient concentrations and loadings (for example, in lakes, mean inorganic N ca. 1900 μg/L, mean total P ca. 80 μg/L), high decomposition in the bottom water and surface sediments (with abundant organic materials available for this process); and bottom-water dissolved oxygen deficits, sometimes with occasional to frequent fish kills. Eutrophic lakes typically are shallow with well-developed littoral zones (area where light penetration is sufficient to support growth of rooted plants), sometimes extending across most of the bottom area.

eutrophication The process by which a body of water becomes overenriched in dissolved nutrients (nitrogen, phosphorous) that stimulate growth of aquatic plant life (e.g., algae), usually resulting in the depletion of oxygen from water. This frequently creates unfavorable conditions for fish and other biota.

evaporative aerodynamics Study of the influence of aerodynamics on the development of both leaf structure and foliage distribution throughout a plant community.

evapotranspiration The process of transferring moisture from the earth to the atmosphere by evaporation of water and transpiration from plants: actual evapotranspiration as observed at a locality or potential evapotranspiration, given unlimited water availability.

evolution The morphological or genetic change in species over time. Small changes that do not lead to reproductive isolation among members of a group are referred to as "microevolution." Speciation, or the generation of new species, is generally referred to as "macroevolution."

evolutionary Describing biological and physical processes leading to heritable change in characteristics of populations or species over time. Key agents of evolutionary change are mutations, genetic drift, gene flow, and natural selection. The ability of a population to adapt to environmental changes is controlled genetically and is termed evolvability.

evolutionary equilibrium hypothesis The hypothesis that frequencies of acceptance of brood parasitism reflect an equilibrium between costs and benefits of host defenses against parasitism.

evolutionary lag hypothesis The hypothesis that hosts lack defenses against brood parasites because the defenses have not yet had time to evolve.

evolutionary theory A body of statements about the general laws, principles, or causes of evolution

executive order A directive from the U.S. president, or a state governor, specifying actions by government officials and agencies. State and federal executive orders often describe how to administer a provision of a statute, treaty, or the constitution and usually have the force of law.

exoenzymes Digestive enzymes excreted by fungi into the environment to digest materials externally.

exogamy The practice of a person seeking a mate outside of his or her group.

exogenous forces Forces outside a population that affect the dynamics of the population.

exons Set of segments of interrupted genes that remain after cutting and splicing of messenger RNA, and that include the parts of the gene that are translated into proteins.

exotic animals Animals that have not been domesticated for use by humans in agriculture or as pets; also connotes that the wild animals are from another region of the world.

exotic species A species that occurs in an area outside

of its historically known range and that has been introduced into a new habitat or ecosystem, either intentionally or accidentally.

exploitative competition Interaction among two or more species that use a common resource that is limited, in which one species benefits more than the other.

ex situ Outside the natural range. Hence, a captive breeding program in Europe for a South American species would be an *ex situ* program. (Conversely, a Chinese program to breed native panda bears in captivity is an *in situ* breeding program because it is within the natural range.)

***ex situ* conservation** Conservation activities that involve individuals held outside their native habitats (e.g., in zoos or seed banks). *In situ conservation:* activities which aim to conserve wild populations in their native habitats.

***ex situ* preservation** Maintenance of a breed outside the environment and agricultural production system(s) in which it was developed and used. *Ex situ* preservation often involves cryopreserved gametes and/or embryos but may also involve live animals.

extant Having some living representatives (i.e., not extinct) at a particular time.

externality Cost or benefit imposed on others as the result of some economic activity.

extinct Describing the disappearance of the last living individual of a species. A species can be locally extinct if it concerns a definite population or location; we speak of "extinct in the wild" when the only individuals alive of a species are in captivity and of "globally extinct" when no living individual remains of a species.

extinction The final disappearance of a species unable to evolve, to adapt to changes in its local physical or biological environment, or to shift its geographic range to avoid such changes. Extirpation is the local extinction of a population on a habitat patch that may subsequently be recolonized by dispersal from other populations of the same species. Risk of extinction is codified, and vulnerable, endangered, and critical imply a significant probability of extinction within years or decades rather than centuries; threatened is used as a general term.

extinction cascade A chain of extinctions triggered by the extinction of a particular species on which many others depend. Species affected by other species are directly (parasites that live only on that species) or indirectly (predators that rely heavily on the species for food) linked with the extinct species through ecological links. Most species support other ones: a number of specialist herbivores can depend on a plant species for food or many parasites are host specific (can only parasitize one species). When these supporting species die out, the dependent species also go extinct. This can trigger a chain of extinctions, termed an "extinction cascade." For example, when the passenger pigeon died, at least two feather lice parasites followed them into extinction. When more than one species dies out at the same time (this by definition must happen when the host of an obligate parasite dies out), the term "coextinction" is also used.

extinction rate The number or proportion of taxa becoming extinct per unit time or after an important geological temporal boundary.

extinction spasm or pulse A catastrophic burst of extinctions, peaking in less than a millennium.

extinction vortex Positive feedback loops that increase the risk of extinction of a population as it declines in size.

extirpated Locally extinct.

extremophile An organism that requires, or grows optimally in, extreme environmental conditions; e.g., extremes of temperature, pressure, or acidity.

extremophilic Describing an organism that lives and thrives in a habitat characterized by extremes of temperature, acidity, alkalinity, salinity, light, or pressure that would prove lethal to most other organisms.

F

facultative epiphyte Plant or lichen that commonly grows epiphytically and terrestrially, usually exhibiting preference for one or the other habit in a particular habitat.

facultative parasite A parasite that can also live without a host.

Fagus grandifolia American beech trees.

fallow A phase in which no crop is on the land and the volunteer vegetation regrows.

family A category, in the classification of evolutionary lineages, of related organisms that ranks above a genus and below an order. A family usually contains many genera.

family planning The conscious effort of couples to regulate the number and spacing of births through artificial and natural methods of contraception. Family planning connotes conception prevention to avoid pregnancy and abortion, but it can also refer to efforts of couples to induce pregnancy.

farming intensity In a broad continuum, extensive systems are those which are closest to natural fisheries, requiring minimal inputs and offering relatively low yields, whereas intensive systems require a large amount of inputs to maintain an artificial culture environment, with high yields. Between these extremes are the varying degrees of semi-intensive aquaculture, where definitions are less distinct: (i) extensive aquaculture does not involve feeding of the organism, (ii) semi-intensive aquaculture involves supplementation of natural food by fertilization and/or the use of feeds, and (iii) intensive aquaculture is when the culture species is maintained entirely by feeding with nutritionally complete diets.

fat body Storage tissue of arthropods that stores lipids and also has endocrine and immunological functions.

fatty acid Any of hundreds of different aliphatic hydrocarbons with an acid group. The main differences are the number and position of one or more double bonds and the presence of substitutions.

fauna The animal species living in a defined area.

feedback Change in a system component that triggers effects that eventually change the original component again. Feedbacks can be positive (self-reinforcing) or negative (self-dampening).

feed conversion The efficiency of farmed animals to incorporate given feed into biomass. Feed conversion is usually expressed in terms of the feed conversion ratio of weight of diet used to fish/shellfish flesh biomass produced. The ratio is affected by the relative moisture content of both feed and aquaculture product as well as the metabolic characteristics of the farmed species, farming techniques, and husbandry.

fellfield Dry, windswept habitat of cold regions comprising mineral soil, gravels, stones, and rock, dominated by cryptogams (especially mosses and lichens) and, outside the Antarctic, by compact cushion-forming phanerogams and short grasses.

felsic rock Continental crustal rocks relatively rich in silicon and aluminum.

fertility The actual reproductive performance of an individual, a couple, a group, or a population measured by a variety of rates, including the crude birth rate (number of births per 1000 population) and total fertility rate.

final (definitive) host A host that harbors sexually mature stages of a parasite.

fingerprint In this context, a pattern produced by DNA fragments or lipids that represents a community or species.

fire regime The type of fire, mean and variance in fire frequency, intensity, severity, season, and areal extent of a burn in an ecosystem.

first contact Initial human arrival on a landmass followed by human colonization.

first-contact extinctions A wave of extinction of species native to a continent or island, following the first arrival of humans to that area.

first theorem of welfare economics The statement that if all goods are private and all private and social costs are equal, then an economy with a complete set of competitive markets operates in a way that is Pareto efficient.

fishery A collective effort to gather, collect, or catch wild aquatic wildlife or plants for recreational or commercial purposes. Fisheries extract large numbers of wild fish, sea urchins, corals, seaweeds, shrimp, snails, clams, scallops, squids, turtles, whales, and other creatures.

fishing effort The level of fishing activity quantified in terms of the number and power of vessels and duration of fishing.

fissipedia A suborder of the Carnivora with divided toes.

fitness Lifetime reproductive success of an individual with a particular genotype relative to another individual with a different genotype. Natural selection typically favors "survival of the fittest" and thus facilitates the future evolvability of a population.

fixation The process of forming organic molecules from inorganic gases.

flagship species Charismatic or well-known species that is associated with a given habitat or ecosystem and that may increase awareness of the need for conservation action.

flexibility The degree to which alternatives exist for one or more selected areas in the context of reaching a particular conservation goal. When seeking a set of areas to represent maximum diversity, flexibility for a selected area may be absent, incomplete (replacing the selected area while still reaching the conservation goal would require substitution of two or more areas or one or more areas of greater cost), or complete (other areas could be substituted, one-for-one by number or by cost, with the current choice).

floodplain Broad, flat areas adjacent to rivers and streams that become inundated during floods.

flood resistance Submersion ability of weeks or months.

flood tolerance Submersion ability of few hours up to several days.

floristic group A plant community that contains a "homogeneous" suite of species sorted by some classificatory program.

focal species A species that can be monitored and managed to maintain the integrity of the ecological system of which they are a part.

foliage projective cover (FPC) The horizontal cover of leaves (in crowns of varying foliage density) in overstory and understory strata of an open-structured plant community, measured using vertical cross-wire sighting tubes.

food chain A representation of the links between consumers and their resources, for example nutrients → plant → herbivore → carnivore. In these representations, energy or material flows up the chain in a linear fashion. In addition, a food chain can be a linear set of species within a food web.

food pump The sclerotized pump used by hemipterans to suck up the juices (plant or animal) upon which they feed.

food pyramid A graphic representation of the energy or biomass relationships of a community, in which the total amount of biomass, or total amount of energy available, at each successive trophic level is proportional to the width of the pyramid at the appropriate height.

food web A set of trophic (feeding) relationships among species in a community; schematically, the pattern often resembles a web of species each connected by trophic interactions with other species.

foraminifera An order of animal-like protists, many of which secrete calcareous skeletons ("tests").

forbs Non-graminoid herbaceous plants.

forest change processes A collective term for all activities on the forest land which affect the stand or site and, in particular, the biological diversity. These processes include forest exploitation, forest fragmentation, and establishment of new plantations.

forest degradation Damage to forest ecosystems through human activities that does not result in the total elimination of forest cover.

forest management Broadly, a pattern of human activities to derive economic or other utility from a forest. In its narrower sense—as a concept or guiding principle for the multiple and sustainable use of a forest—it is a complex ecological and sociological concept in which exceptional skill has to be exercised on the part of forest manager. Forests could be managed for the extraction of timber or nontimber forest products, forage for animals, watershed protection, or recreational use.

forest stand A relative homogeneous forest landscape unit that can be distinguished from neighboring units by forest age or composition.

forest structure The arrangement of all the parts of

the forest stand—stems, branches, leaves, roots, and so on.

formation Large-scale (subcontinental or continental) vegetation complex defined primarily on the basis of a combination of dominating life-forms (hence, vegetation structure).

fossil Anything found in strata of rocks or sediments that is recognized as the remains of an organism from a former geological time.

fossil fuels Forms of stored energy produced by the action of pressure and temperature on organic matter buried over geologic time. The major types of fossil fuels are oil, natural gas, and coal.

fossorial 1. Adapted for living or digging underground. 2. A burrowing animal.

founder effect Random genetic sampling in which only a few "founders" derived from a large population initiate a new population. Since these founders carry only a small fraction of the parental population's genetic variability, radically different gene frequencies can become established in the new colony.

founder genome equivalent The number of equally contributing founders that would have produced the same genetic diversity found in an existing captive population if there had been no random loss of founder alleles.

founders Animals that were collected from their natural range and that have produced progeny in captivity (and are assumed to be unrelated). Thus, these animals are the founding stocks of the captive population. The term is also generalized to include animals with wild parents that are collected in order to supplement the genetic diversity of an established captive population.

fractal Length dimension with exponent other than L^1 (length), or L^2 (area), or L^3 (volume).

fractionation A method of obtaining a pure compound by extracting components of a mixture with solvents of different solubility.

free rider In this context, a term for one who enjoys the benefit of a good or service without paying the cost.

frequency dependence Phenomenon that the dynamics of a given ecological or genetic type depend on its frequency in a population.

frequency-dependent selection Selection in which fitnesses vary with the frequency of different genotypes.

fruiting body Sexual reproductive structure of a fungus.

frustule The hard, silica-containing skeleton of diatoms (green algae).

fry Juvenile salmon that have absorbed their yolk sac and emerged from the gravel.

function 1. A process carried out by an organism or group of organisms. 2. The performance of a biological system as a rate.

functional constraint A limit to the kinds of nucleotides or amino acids which can appear in a gene or protein without compromising function; often called "selective" constraints.

functional dependency Dependency of one species on another to complete a particular ecosystem process or function; an example would be the dependency of plants on decomposers and decomposers on plants to recycle nutrients.

functional diversity The range and value of those species and organismal traits that influence ecosystem functioning.

functional groups Aggregations of species that perform similar ecosystem processes, such as grazers, suspension or filter feeders, leaf shredders, predators, and decomposers.

functional redundancy, functional similarity The fact or condition of having species that may be substituted because their contributions to ecosystem processes are similar or overlapping.

functional type A group of species that share morphological and physiological characteristics that result in a common ecological role.

functional web This type of food web quantifies the strength of interaction between species linked using data from manipulative experiments.

fungus A member of the kingdom Fungi. Members of this kingdom are heterotrophic (requiring a preformed organic source, i.e., not able to make their own food), eukaryotic, have walls of chitin, and reproduce by means of spores.

furca (pl. furcae) Forked structure, usually associated with the telson in crustaceans.

fynbos Habitat type in southern Africa that is characterized by thickets and low shrubs, in which fire plays a dominant role in ecosystem maintenance. Plant endemism is particularly high in these areas.

G

GABA receptor Part of the inhibitory ion channel complex gated by GABA (γ-aminobutyric acid) in postsynaptic nerve membranes.

gadoid fish Includes the cod *Gadus morhua* L.; the life cycle of cod can be summarized as four stages: spawning, larvae, juveniles, and adults. During the larval and early juvenile phases they feed largely on copepods on the nursery grounds; after this phase they become largely stationary and feed on benthic or epibenthic animals.

gametogenesis The developmental process for the formation of functional male and female haploid reproductive cells that combine at fertilization to produce the zygote.

gametophyte In an organism with an alternation of generations, the haploid, gamete-producing phase.

gamma diversity The diversity of species, often estimated as species richness, in a larger area as a consequence of both alpha and beta diversity.

gamma landscape diversity Total number of patch (homogenous) landscape types contained within a geographic region.

gap 1. An opening in the vegetation created from disturbances such as clearing, logging, fires, diseases, storms or the natural death of a tree. 2. A large or relatively large difference in overall similarity between two taxa.

gap analysis A biodiversity policy tool which compares the actual distribution of species and vegetation classes to areas managed for the long-term protection of biodiversity and other classes of land management. Species and classes that are poorly represented in protected areas are gaps in the protection of all biodiversity. Such species and classes are candidates for proactive protection.

garrigue Low-growing secondary evergreen shrublands that dominate extensive areas of the Mediterranean Basin.

gene Historically, the inherited factor which determines a trait. Tends to be used somewhat loosely; more strictly represents a place or locus on the chromosomes which codes for a particular function.

gene (allele) frequency The frequency of an allele in a population.

gene bank Facility where genetic material is stored in the form of seeds, pollen, embryos, or semen, or, in some cases, as live plants or animals living in a field, a greenhouse or other installation.

gene conversion A process of nonreciprocal (unidirectional) transfer of genetic information.

gene flow Movement of genes between populations, usually via immigration and mating of whole genotypes, but sometimes single genes may undergo horizontal gene transfer via transfection by microorganisms.

gene pool The genetic diversity contained within a population, species, or crop. The primary gene pool of a crop represents the biological species, the secondary gene pool includes species that can be crossed with it, allowing at least some transfer of genes, and the tertiary gene pool includes related wild species where such gene transfer involves specialized techniques.

gene prospecting Search for genes in wild living plants, animals, and microorganisms for the breeding or genetic engineering of plants, animals, and microorganisms in agriculture, fermentation, and cell culture for agricultural and industrial production.

general circulation model (GCM) Computer models developed to simulate global climate and widely used for global climate change predictions.

generalist brood parasites Species that parasitize many (up to more than 200) host species.

genetic architecture Characterization of the number

and types of genes and their interactions that underlie a particular trait.

genetic code Language that specifies how DNA will be translated into protein sequences by means of three-nucleotide "words" (codons) that specify the 20 amino acids and regulators of transcription (start and stop codons).

genetic diversity The diversity of the sets of genes carried by different organisms. It occurs in the form of nucleotide variation within the genome. When this variation causes a change in a given protein, the variants are called alleles. Allelic variation occurs at various genetic loci or gene positions within a chromosome. This diversity allows the population or species to adapt and evolve in response to changing environments and natural selection pressures.

genetic drift The variation of allele frequency from one generation to the next that occurs due to chance. Genetic drift leads to the loss of genetic variation in small populations due to the random loss of founder alleles during reproduction.

genetic enhancement Management actions taken to increase the genetic variability and viability of a population; includes translocations and reintroductions.

genetic erosion The loss of genes from a gene pool attributed to the elimination of populations caused by factors such as the adoption of high-yielding varieties, farmers' increased integration into the market, land clearing, urbanization, and cultural change.

genetic marker Genetically variable locus that produces distinct variants when analyzed by standard methods such as electrophoresis. It may be based on protein or DNA and may or may not be selectively neutral.

genetic recombination The physical exchange of genetic material between a pair of chromosomes during meiosis.

genetic variation The allelic variation that occurs at various genetic loci or gene positions within a chromosome. Genetically variable loci are termed polymorphic or are said to show polymorphism.

genome A complete single set of genes of an organism or organelle; also the basic haploid chromosome set.

genomics The study of the molecular organization of genomic DNA and physical mapping.

genotype The particular combination of genes carried by an organism; the genetic makeup of an individual.

genotypic Relating to the genetic composition of an organism.

gentes (sing. gens) Lineages of cuckoos in which individual females specialize on a single host and lay mimetic eggs.

geographic information systems (GIS) Computer systems that store, enhance, analyze, and display layers of geographic data and connect these to alphanumeric databases.

geomorphology The study of the surface configuration of the earth, especially the nature and evolution of current land forms, their relationships to underlying structures, and the history of geological activity as represented by such surface features.

geothermal heating Water heated at depth in the earth and released to the surface as thermal outflows.

germplasm Sources of hereditary material. In animal breeding the term is commonly applied to breeding animals and to fresh or frozen sperm cells, ova, and embryos.

Global Biodiversity Assessment An independent peer-reviewed analysis of the biological and social aspects of biodiversity commissioned by the United Nations Environment Programme.

global change disturbances Alterations or disturbances to the natural conditions of the global atmosphere (e.g., greenhouse gases produced by human activities) or to enough regions (e.g., largescale deforestation or the wide-spread introduction of exotic, invasive species in many places throughout the world) to have global impacts.

global circulation The regular global circulation pattern of the earth's atmosphere, which generates the world's basic climate types; the system involves an Intertropical Convergence zone of low pressure near the equator, subtropical high-pressure belts near the Tropics of Cancer and Capricorn, trade winds flowing from these high-pressure belts toward the equatorial low, and westerly winds in the midlatitudes.

global circulation models (GCMs) Computer models

of atmospheric circulation patterns and surface energy balance used for weather forecasting and to predict climates at regional and global scales. GCMs are so large that they must be run on supercomputers.

global climate change Current and predicted changes in global temperature, rainfall, and other aspects of weather due to increased human production of carbon dioxide and other greenhouse gases, including phenomena such as global warming, severe storm frequency and intensity, and glacial melting.

global warming The additional heating of the earth's climate from the incremental injection of greenhouse gases to the atmosphere from human activities such as deforestation or fossil fuel consumption.

glutathione S-transferases (GSTs) Enzymes that catalyze the metabolism of a range of substrates following their conjugation with the endogenous tripeptide glutathione.

glycogen An insoluble polymer of glucose; the main storage carbohydrate in animals.

goals An explicit and precise statement of conservation aims. Goals should express the values of those people who provide the mandate for conservation. Making the goal explicit allows efficiency to be measured as an aid to accountability. A statement should include which attributes are valued (such as genetic diversity), which surrogates for this value are actually surveyed (such as higher taxa, species, and threatened species), which areas are to be considered (such as land management units or grid cells), and how constraints of viability, threat, and cost are to be measured and accommodated. Goals are not universal but depend on people's values and their situations. Consequently, different goals may conflict, and areas necessary to meet one goal may be insufficient to meet broader goals.

Gondwana (Gondwanaland) A large supercontinent in what is now the Southern Hemisphere that separated during the Mesozoic, forming the modern continents of Australia, Antarctica, South America, and Africa. Named after an ancient kingdom in India. Gondwana rifted apart over many millions of years, giving rise to the current Southern Hemisphere landmasses.

gonochoric A mode of reproduction in which individuals of the species are either male or female and produce either eggs or sperm within a single colony.

gonochoristic Describing an individual that is either a male or a female throughout life.

governance systems Sets of rule configurations used to govern human–ecosystem relationships at operational, collective-choice, and constitutional levels of analysis.

grade A group based on, for example, the functional level of organization or overall similarity rather than ancestor–descendant relationships.

graminoids Grasses (family Poaceae) and grass-like plants, mostly sedges (family Cyperaceae) and rushes (family Juncaceae).

granivores Animals that eat seeds or achenes (grass fruits).

grassland Land that has a vegetation cover dominated by grasses.

graze The act of consuming herbaceous plants, including grass, grass-like plants, and forbs.

grazer A consumer that removes only a part of each prey it attacks and thus rarely kills a prey in the short term. The term is commonly applied to animals that eat plants (herbivores) but can also be argued to apply to nonlethal microbial pathogens and to animals that cause tissue loss, but not death, when they feed from colonial animals such as sponges.

grazing intensity Frequency and closeness of grazing.

grazing pressure Stocking rate, the units of grazing animals per land area.

great faunal interchange Development of the Central American isthmus facilitated the interchange of fauna from north to south and vice versa.

greenhouse gases Gases such as water vapor and carbon dioxide that can affect climate through their absorption and reemission of terrestrially emitted infrared radiation. They selectively let more of the solar energy into and out of the atmosphere than they permit the transmission of long-wave infrared radiation. They contribute to the greenhouse effect, which traps radiant energy in the lower atmosphere and makes the earth warmer.

grilse Atlantic salmon that spends only one year at sea before returning to spawn.

Grinnellian niche The requirements and behaviors expressed by a species wherever it normally occurs.

ground plan (archetype, bauplan) A basic plan or general type, or a hypothetical ancestor.

groundwater A "reservoir" of water residing below ground in saturated soils and beneath geologic formations.

group selection Selection between different populations or sub-populations based on attributes of the entire group, where these attributes usually are either selected against or not favored at the level of individual selection.

gubernaculum A sclerotized trough-shaped structure of the dorsal wall of the spicular pouch, near the distal portion of the spicules; functions for reinforcement of the dorsal wall.

guild A group of species having similar functional roles in the community (i.e., herbivores).

H

habitat The type of environment in which an organism or species lives, grows, and reproduces.

habitat corridors or connections Strips of habitat that connect isolated habitat patches in a landscape transformed by human land use. Connections can be achieved through the conservation of existing habitat or by ecological restoration.

habitat creation Construction of one habitat type from another, often a disturbed upland excavated to make a wetland.

habitat fragmentation Process by which a continuous area of habitat is divided into two or more fragments by roads, farms, fences, logging, and other human activities; splintering of once contiguous land cover into isolated pieces. Fragmentation happens when habitat is lost from the interior, rather than the edge, of a large block of cover. The resulting habitat patches are sometimes referred to as "habitat islands" and the intervening, converted land is called the "matrix."

habitat restoration Active modification of the current state of a degraded habitat in order to return it to a former, preferred state.

habitat specialists Species found only in a specific habitat (e.g., species found only in forests of Sequoia trees).

habitat structure Analogous to community structure, but limited to the physical structural aspects of a habitat. The structure of habitats may be characterized by such measures as complexity, heterogeneity, regularity, stratification, and fractal dimensionality.

Hadley cell A pattern of air circulation that causes deserts at about 30° N and 30° S of the equator. Sun heats the equator, air rises, pressure decreases, moist air moves north and south, cooling as it goes and falls onto the spinning earth, drying as it falls. The falling dry air increases pressure preventing the incursion of moist air.

halophiles From the Greek *halos* (salt) and *philos* (loving). Organisms that grow optimally at high salt concentrations.

halophilic Describing an organism requiring high levels of salts for growth.

haptor The posterior attachment organ of monogeneans.

harmful algae Algae that are undesirable to humans because (a) they become too abundant in response to nutrient overenrichment and then, at night, use most or all of the oxygen in the water for their respiration, so that fish and other organisms suffocate or become seriously physiologically stressed; (b) they become too abundant in response to nutrient enrichment, and overgrow beds of desirable rooted vegetation so that the beneficial plants cannot receive enough light to survive; (c) they cause or promote disease in other plants or animals; or (d) they produce toxins that hurt or kill finfish, shellfish, or other higher trophic levels including humans. "Harmful algae" include prokaryotic cyanobacteria or blue-green algae. More recently, the term has been used to include organisms that are not photosynthetic, primitive plantlike organisms—for example, certain nontoxic animal-like dinoflagellates, which cause fish disease (e.g., *Amyloodinium ocellatum*), and toxic animal-like dinoflagellates (e.g., the toxic *Pfiesteria* complex), which do not have their own chloroplasts for photosynthesis, but which resemble plantlike dinoflagellates in appearance and other general features.

harvest To gather a crop. "Harvest" is an appropriate word for farming operations, including fish farming, but not for catching or collecting wild animals or plants. This term is widely misused in industry public relations to make the extraction of wild fish, natural stands of trees, and other wild organisms seem like agriculture, though nothing is planted or nurtured and these things are merely taken for profit. For wildlife, including wild fish, appropriate words include, among others: catch, fish for, take, extract, land, gather, and collect.

haustorium (pl. haustoria) The organ found in all parasitic plants that penetrates vascular tissue of the host plant and forms a functional bridge for uptake of water and nutrients from the host.

health A flourishing condition, well-being; capacity for self-renewal.

heath Open vegetation dominated by dwarf undershrubs usually of the family Ericaceae.

hectare A metric measure of surface equal to 10,000 square meters or 2.471 acres (U.S.).

hedonic price Price of a good expressed as a function of the particular attributes it embodies.

helminth One of several classes of parasitic worms: nematodes, cestodes, trematodes (monogeneans and digeneans), and acanthocephalans.

hemelytron The heteropteran's forewing; so called because the anterior part of the wing is opaque and the posterior part is clear (membranous).

hemimetabolous Having incomplete metamorphosis, that is, showing gradual change from molt to molt, with externally developing wing pads, and lacking any larval and pupal stages.

hemiparasite A parasite that photosynthesizes.

herbal A compilation of medicinal plants and their properties often used in earlier times as a reference for physicians prescribing medical treatments based on plant therapies.

herbicide Chemical used to suppress or kill plants, or to severely interrupt their normal growth processes.

herbicide resistance Inherited ability of a plant to survive and reproduce following exposure to a dose of herbicide normally lethal to the wild type. In a plant, resistance may be naturally occurring or induced by techniques such as genetic engineering or selection of variants produced by tissue culture or mutagenesis.

herbicide tolerance Inherent ability of a species to survive and reproduce after herbicide treatment. This implies that there was no selection or genetic manipulation to make the plant tolerant; it is naturally tolerant.

herbivores Animals that eat plants. Usually excludes instances when a single animal eats an entire plant, which is categorized as predation.

herbivory 1. The process in which animals feed on plants. 2. Damage to plant tissues by herbivores or pathogens.

heritability The proportion of the variance in a trait that is due to additive genetic effects.

hermaphroditism A mode of reproduction in which individuals of the species produce both eggs and sperm within a single colony, sometimes within the same polyp.

hermatypic Reef-building; more recently, this term has been replaced by the term zooxanthellate to refer to those coral species with symbiotic algae.

heterochromatin Regions of chromosomes that do not include coding DNA, generally make up the structure of chromosomes, and always remain condensed during a cell's life cycle.

heterogamety Having two different sex chromosomes.

heterogeneity Variation in environmental conditions through space (spatial) or time (temporal).

heterosis Hybrid vigor, exhibited when offspring are larger, more vigorous, and more productive than the parents.

heterostyly Having two or more kinds of flowers with stamens and pistils of different lengths.

heterotroph Literally, a feeder on others; an organism that is dependent on organic material from an external source to provide carbon for growth; for example, vertebrates.

heterotrophic Describing a mode of nutrition in which carbon is obtained from the organic compounds made by autotrophic organisms.

heterotrophic respiration The respiration of all those life forms that feed on photosynthetic organisms, including bacteria, fungi, and animals.

heterotrophic successions Dependent on already fixed energy, such as the successional of communities associated with decomposition of dead logs.

heterotrophy The ability of organisms to grow and reproduce through the consumption of organic carbon sources, taken in particle form (i.e., obtained by feeding).

heterozygosity The average number of different heterozygotes across loci divided by all loci studied.

heterozygote An individual having two different alleles at a genetic locus.

hexapod Group including insects and their primitive relatives.

high seas Parts of the ocean outside national boundaries, usually beyond 200 miles of any nation's coast.

historical biogeography Spatial patterns of organisms interpreted in terms of their concordance to patterns of Earth history.

historical ecology Transdisciplinary and diachronic analysis of how human societies and ecosystems change and in turn transform one another through time in local and regional landscapes. Data are drawn from geology, archaeology, history, and other sources.

historic rainforest All land areas with a potential to support the tropical rainforest as determined by the climatic and physiographic conditions whether currently forested or not. The concept is interesting because it provides a baseline to calculate the rainforest loss in a country or region over a historic period.

Holarctic A biotic realm in the temperate, boreal, and arctic regions of the Northern Hemisphere with strong affinities among the regional floras and faunas.

Holometabola Insects characterized by complete metamorphosis, a wingless larval stage, and an intermediate pupal stage.

holomorphology Totality of characters indicative of phylogeny (including molecular, behavioral, etc.).

holoparasite A parasitic plant that lacks chlorophyll and cannot photosynthesize.

homeophasic adaption The adaptation of the membrane to maintain the bilayer phase.

hominid A great ape from the lineages most closely related to humans, where this may be a lineage ancestral to humans.

homoiohydry Ability to maintain a constant internal water balance independent of fluctuating environmental conditions.

homologous Denoting common ancestry. Structures, processes, sequences, behaviors, etc. are said to be homologous if there is evidence that they are derivations from a common ancestral structure. In molecular biology the term indicates a significant degree of similarity between DNA or proteins.

homology A condition in which two characters or structures are hypothesized to share a common ancestry.

homoplasy A character or character state acquired by parallel or convergent evolution that bears resemblance to a character in a different group; similar attributes not due to common ancestry.

homozygosity In a diploid organism, the presence of the same alleles at a given locus (e.g., A_1A_1).

homozygous Possessing the same gene form (allele) on both chromosomes.

host An individual that is parasitized by a parasite.

hotspots Those areas that (a) feature exceptional concentrations of endemic species and (b) face imminent threat of habitat destruction.

human ecology A type of ecology focused on a specific species, *Homo sapiens*. Human ecology may be thought of as a subset of social ecology, which is a life science focusing on the ecological study of various social species such as ants, wolves, dolphins, or orangutans. It includes issues such as human interaction with the environment and with other species.

human load The total "human load" imposed on the "environment" by a specified population is the product of population size times average per capita resource consumption and waste production. The concept of 'load' recognizes that human carrying capacity is a function not only of population size but also of aggregate material and energy throughput. Thus, the human carrying capacity of a defined habitat is its maximum sustainability supportable load.

Hutchinsonian niche The set of environmental conditions, or opportunities, that will permit a species to exist indefinitely. The set of opportunities that are available to a guild can be referred to as a "nook."

hybrid An individual that is heterozygous (intermediate) for one or more heritable characters that distinguish two or more populations, including individuals that are F_1s, F_2s, and the set of all backcrosses. Hybrids are initially formed when gametes from two species, subspecies, or races combine to form F_1 plants.

hybrid inviability The phenomenon whereby one or both sexes of progeny produced in crosses between two different species are unable to survive.

hybridization Interbreeding of individuals from two or more populations of species, subspecies, or races, which are distinguishable by one or more heritable characters.

hybrid sterility The phenomenon whereby one or both sexes of progeny produced in crosses between two different species are able to survive, yet are unable to reproduce.

hybrid swarm Mixture of hybrid genotypes, including F_1s, F_2s, and backcrosses, due to hybridization between two or more species that co-occur in a certain locality.

hybrid zone Areas of overlap or points of contact between two populations that are distinguishable based on one or more heritable characters where viable or partially fertile hybrids are formed.

hydraulic conductivity A measure of the potential water flow through a soil.

hydrologic cycle The movement of water between the surface of the earth and the atmosphere.

hydrologic regime The aggregate set of variables describing the behavior of water in wetlands, including water depth, magnitude and frequency of water level fluctuations, seasonal pattern of water fluctuations, direction and velocity of water flows, and water residence time.

hydrophily Underwater pollination developed by most seagrasses and many freshwater angiosperms.

hydrophytes Plants capable of growing and persisting in saturated or flooded soils.

hydrothermal plumes Created by rising black smoker effluents that mix turbulently with the surrounding seawater, becoming dilute and eventually reaching neutral buoyancy (typically ~200 m above the black smoker source); once neutral, they spread laterally and are detectable by chemical and physical tracers for kilometers.

hypercycle Cyclic sequence of self-supporting reactions between primitive prebiotic biomolecules, nucleic acids, and proteins. Hypercyclic organization of reacting systems was postulated to explain the spontaneous emergence of replicating systems. As a result of the hypercyclic reaction principle, functional characteristics of replicating molecules (i.e., phenotypes) are connected to their hereditary characteristics or their genotypes. Consequently, both phenotypes and appertaining genotypes are evaluated and selected in feedback loops.

hypermetamorphic A life cycle that includes two or more larval forms, with each often specialized for a different feeding function.

hyperparasite (of first, second, etc. degree) A parasite living on or in another parasite.

hypersaline More salty than seawater.

hyperthermophile From the Greek *hyper* (over), *therme* (heat), and *philos* (loving). An organism that grows optimally at temperatures higher than 80°C.

hyperthermophilic Describing organisms that grow best at temperatures warmer than 80°C.

hypodermis The cellular, subcuticular layer that secretes the cuticle of annelids, nematodes, arthropods (see epidermis), and various other invertebrates.

hypognathous Position of the head when the mouth is directed toward the ventral side of the body.

hyporheic zone The area of saturated soils beneath a stream or river channel; interacts with the stream through the processes of hydraulic upwelling and downwelling.

hyporheic zone Interstitial space within the sediments of a streambed; a transition zone between surface water and permanent (phreatic) groundwater.

hypoxia Condition in which the water has depressed levels of oxygen that are too low to sustain healthy fish populations (usually considered as ≥0.1 to <2 [sometimes <4 or <4.5] mg DO/L). Note that hypoxic levels of 3–4 mg DO/L can stress or kill sensitive egg and larval stages of some finfish and shellfish species, and that many motile fish actively avoid hypoxic areas.

hypsodont High-crowned cheek tooth commonly found in grazing mammals.

I

ideal population A hypothetical population widely used in population genetics theory. In this ideal population, the breeding sex ratio is equal, mating is random, generations do not overlap, selection and mutation do not occur, and the lifetime number of offspring produced by individual parents has a Poisson distribution.

idiobiont A parasitoid that develops on a paralyzed, incapacitated host.

igapó Vegetation periodically flooded by acidic black water or clear water rivers. Specifically, the floodplain of the blackwater rivers of the Amazon basin.

image processing The manipulation of digital image data, including operations such as image display, restoration, enhancement, and classification.

imago The adult insects; the final developmental stage when insects possess wings.

immobilization The process wherein nutrients are taken up or immobilized in litter and other organic detritus until later (usually weeks to months) in the decomposition process.

impact A forceful contact; a major effect of one thing on another.

INBio The Instituto Nacional de Biodiversidad (National Biodiversity Institute), a Costa Rican association created in 1989 to generate and disseminate knowledge and promote sustainable uses of biodiversity.

inbreeding The mating of close biological relatives; mating between closely related individuals more often than would be expected by chance.

incidence The number of samples containing at least one of a given species in a census.

index of aridity Ratio of mean annual precipitation to mean annual potential evapotranspiration.

indicators Quantitative measurements of environmental and social variables that provide time series describing long-term trends; some of the trends may indicate a transition toward sustainability.

indigenous, local, and traditional Adjectives used by anthropologists, ethnobiologists, and other academics to describe people, practices, and knowledge. Indigenous denotes people (and their cultural practices and knowledge) who claim to be the original or long-term inhabitants of a particular place, in contrast to more recent colonizers. Traditional refers to established lifestyles, practices, and beliefs that guide cultural continuity and innovation—a definition that recognizes that traditions are always in a process of adaptation and change. Local, preferred by many researchers because it is the broadest and least value-laden term, indicates cultures that are found in a specific part of the world. It is commonly used to refer to people, whether long-term residents or recent arrivals (rural or urban), who make a living from the land and are knowledgeable about the biological resources in their environment.

indigenous intellectual property rights Rights to intellectual properties belonging to indigenous peoples, including but not limited to iconongraphical representations, terminologies, names, phrases, legends, methods, and techniques of traditional cultivation, healing, identification, preparation, and use of biodiversity.

indigenous peoples Peoples who have resided in the same geographical area for many generations and who possess legends, proverbs, geneologies, languages, and other unique cultural features linking them to the land.

individualism of species Refers to the observation that the distributions and abilities of most species do not exactly coincide because each species is the product of a unique evolutionary history. Coevolution and convergence may occur, but rarely are these processes so complete that groups of species are distributed as a unit or superorganism.

individualistic view of succession Concept that succession is a consequence of species interacting with one another and their environment.

individual selection Selection driven by differences in the net reproduction of individual organisms.

indo-Malaya Geographic region from India to Malaysia and western Indonesia.

induced defenses Plant defenses, including both chemical and physical defenses, that are produced, at least in their final form, only after the plant has been damaged by herbivores or pathogens.

industrialized agriculture Modern form of agriculture that differs from traditional agriculture in the use of elaborate and expensive machinery, the control of pests with toxic chemicals rather than biocontrols, fertilization by synthetic rather than organic products, excessive consumption of water, and farm ownership and management by corporations rather than individuals.

industrial melanism Selection for darker pigmentation as a result of industrial pollution, particularly in moths and butterflies.

infauna Organisms living below the sediment–water interface and between sediment grains.

input environment Collective term for all energy and materials moving into a given system.

inselberg Granitic domes that rise above the forest in the older geological formations to the north and south of the alluvial plains of Amazonia and elsewhere.

in situ Literally, in or at the site or location; in this context referring to conservation efforts within the natural habitat.

in situ **conservation** Conservation of ecosystems and natural habitats and the maintenance and recovery of viable populations of species in their natural surroundings and, in the case of domesticates or cultivated species, in the surroundings where they have developed their distinctive properties.

in situ **preservation** Maintenance of a breed in the environment and agricultural production system(s) in which it was developed and used.

instar A larval stage; the first instar hatches from the egg and upon molting enters the second instar. Most moths undergo five or six instars prior to pupation.

institutions Humanly devised constraints that shape human interaction and the way societies evolve through time; made up of formal constraints (rules, laws, constitutions), informal constraints (norms of behavior, conventions, self-imposed codes of conduct), and their enforcement characteristics.

instrumental value The value of something relative to human interests or desires.

integrated conservation and development project Project that links biological conservation with human development in a local area or set of areas.

integrated weed management Approach for suppressing weeds that combines information on the biology and ecology of the weed with all available control technologies so that no one method is used exclusively.

intellectual property rights The rights of developers of ideas and techniques to require payment for their use by others and to prevent their use by others unless such payment has been made.

intensity In this context, either the mean number of parasites within the subset of infected hosts or the average parasite load of the entire population.

interaction webs (functional webs) Subset of species that through their interactions and responses to abiotic factors make up the dynamic core of food webs or communities. These webs include keystone species, dominants, and other strong interactors.

interference competition Interaction among two or more species that use a common resource that is not limiting, in which one species is harmed by having its access to the resource restricted.

intermediate disturbance hypothesis A hypothesis claiming that some level of disturbance is necessary to maintain biodiversity.

intermediate host A host that harbors sexually immature, developing stages of a parasite.

international geosphere biosphere programme A scientific program that is part of International Council of Science and provides an international and interdisciplinary framework for the conduct of global change science.

International Species Information System (ISIS) A central repository for zoo animal records founded in 1973; has also developed the computer programs that most zoos use for keeping their animal records, as well as software used by zoos for studbooks.

interspecific brood parasites Brood parasites that lay their eggs in the nests of other species.

interstices The spaces between sediment particles, especially in alluvial deposits.

interstitial Living among sand or silt grains; usually refers to an aquatic benthic environment, the intertidal, or in groundwater.

intertropical convergence zone Low-latitude region characterized by warm air masses and convectional rainfall; also known as the heat or thermal equator.

intraspecific parasitism A parasitic association of members of the same species.

intrinsic rate of growth A parameter, often designated as the variable r, reflecting the rate of change in population size due to births and deaths.

intrinsic value A measurement or description of the values inherent in a resource by virtue of its existence (e.g., a living organism), as opposed to its value for present or future human use.

introduced plants Species transported by humans to regions outside their natural geographic range.

introgression Permanent transfer of genes from one or more species, subspecies, or race into another species, subspecies, or race via hybridization and backcrossing.

introns Segments of interrupted genes that are removed after transcription and before translation of messenger RNA to proteins.

invasibility The properties of a community or ecosystem that render it susceptible (or resistant) to invasion by alien plants.

invasive Describing native and exotic species that are capable of displacing indigenous species or spreading into habitats where they were not common previously.

invasiveness The delimitation of features of an organism (e.g., life-history traits) that enable it to invade (i.e., to overcome various barriers to invasion).

invasive plants Alien plants that frequently recruit offspring, often in very large numbers, in natural or seminatural ecosystems at considerable distances from parent plants (100 m; often much farther).

invasive species Species that have recently colonized some geographic regions different from the one in which they were initially described.

invasive weed Introduced plant species that establish self-maintaining populations and spread, with and without human assistance, into new areas where they frustrate human intentions in production and natural landscapes.

invertebrates Animals without a backbone (vertebrae), ranging from unicellular Protozoa to multicellular, complex organisms such as insects, i.e., most of the animal kingdom.

in vitro Literally, in glass; i.e., in a test tube; more

broadly, in a laboratory or other artificial setting rather than in nature.

invisible hand The term used by Adam Smith to describe the capacity of a decentralized market system to attain efficiency.

irreplaceability A property of areas that include species (or other valued attributes) restricted to so few areas that all such areas would be needed in order to meet a conservation goal.

island biogeography, theory of A theory that the number of species in the biota of each island (or island-like habitat) is in dynamic equilibrium, with species frequently going extinct on the island and new species frequently arriving.

isocline A line along which some property remains constant.

isolating mechanisms The sum total of all types of factors that prevent gene flow between species, including premating mechanisms (mate choice) and postmating mechanisms (hybrid sterility and inviability). Modern authors deny that these "mechanisms" have necessarily evolved to preserve the species' integrity as originally assumed, though this may sometimes be the case in reinforcement of premating isolation. Isolating mechanisms are a subset of the factors that cause cohesion of species under the interbreeding and cohesion species concepts.

isotope Many elements exist in several forms, called isotopes. Carbon has eight isotopes, of which ^{12}C forms 98.9% of naturally occurring CO_2 and ^{13}C forms 1.1%. Other carbon isotopes are unstable, although ^{14}C has a half-life as long as 5730 years.

isotopic discrimination The tendency of chemical and physical reactions to "prefer" one isotope against another. The enzyme ribulose bisphosphate carboxylase oxygenase (Rubisco), which is responsible for capturing CO_2 in photosynthesis, shows strong discrimination for $^{12}CO_2$ against $^{13}CO_2$.

iteroparity Repeat breeding (see *semelparity*).

iteroparous Describing aquatic species that can spawn more than one time.

J

jack A sexually precocious male salmon that spends one winter or less at sea before returning to its natal stream to spawn.

jizz An informal term for the characteristic, instantly recognizable appearance of an organism.

K

karst A landscape in which the primary geomorphic agent is solution rather than erosion; typically formed in carbonates; landforms include caves, sinkholes, blind valleys, and large springs.

karyogamy Fusion of nuclei.

kelt A spawned out Atlantic salmon or steelhead that is returning to the ocean to recuperate, sexually mature, and spawn again.

key innovation A trait that increases the efficiency with which a resource is used and can thus allow entry into a new ecological zone; a property of a species that promotes the process of speciation.

key-industry species Prey of intermediate trophic status that support a large group of consumers.

keystone species Consumers that have a large effect, and one that is disproportionately large relative to their abundance, on communities and ecosystems. Uniquely, the strong effects of keystone species on their interacting species exert extensive influence, often indirectly, on the structure and dynamics of communities and ecosystems. They are a distinct subset of a more broadly defined set of "strong interactors" that also include species having strong effects on interacting populations but not necessarily on communities or ecosystems. Keystone species can include predators, parasites, pathogens, herbivores, pollinators, and mutualists of higher trophic status, but generally are not plants, sessile animals, or "resources." The sea otter (*Enhydra lutris*) has been described as a keystone species in certain Pacific Coast habitats of North America.

kHz A unit of frequency equal to 1000 hertz; the number of hertz (abbreviated Hz) equals the number of cycles per second (e.g., sound waves).

koinobiont A parasitoid that develops on a mobile, active host.

Korarchaeota Proposed third kingdom within the domain Archaea. From the Greek *koros* (young man), for the early divergence of this group during the evolution of the Archaea, and *archaios* (ancient). Includes a small group of ribosomal RNA sequences retrieved from geothermally heated sediments.

Kranz anatomy Specialized anatomy of the C_4 leaf in which the bundle sheath tissue is enlarged and enriched with chloroplasts, whereas the mesophyll is reduced in size and often forms a lighter green halo around the bundle sheath. This produces a wreath-like ("Kranz" in the original German) appearance.

kwongan Evergreen heathlands that cover extensive areas of southwestern Australia.

K/T Abbreviation for Cretaceous/Tertiary, usually in reference to the K/T boundary.

K-T event A notable large-scale extinction event occurring about 65 million years ago, at the boundary of the Cretaceous and the Tertiary, involving the extinction of about 70% of extant species, including the dinosaurs and many other land-dwelling vertebrates.

L

land The terrestrial ecosystem that encompasses soils, vegetation, other biota, and the ecological, biogeochemical, and hydrological processes that operate therein.

land cover Physical and biotic character of the earth's terrestrial surface. Land cover is typically the vegetation (e.g., tropical forest, marsh, desert, cornfield, shrubland) or human construct (e.g., road, dwelling, industrial area) that covers the surface. It may be grossly defined as in the preceding examples, or defined at finer scales (e.g., moist deciduous tropical forest, moist evergreen tropical forest, dry tropical forest).

land degradation Reduction or loss of the biological and economic productivity and complexity of terrestrial ecosystems, including soils, vegetation, other biota, and the ecological, biogeochemical, and hydrological processes that operate therein.

land ethic Valuing land as part of a biotic community, not merely as property; philosophy of proper treatment of lands and the natural communities on them.

landraces The originally adapted but variable crop populations on which farmers based their selections. They are geographically or ecologically distinctive populations which are conspicuously diverse in their genetic composition both between populations and within them. They have certain genetic integrity and are recognizable morphologically; farmers have names for them and different landraces are understood to differ in adaptation to soil type, time of seeding, date of maturity, height, nutritive value, use, and other properties.

landscape Spatially heterogeneous area composed of a mosaic of interacting components (patches, corridors, and area of matrix).

land use The use that humans make of the earth's terrestrial surface, usually to obtain goods or benefits. Examples include agriculture, mineral exploration, settlement, natural reserve, and timber production. Land use refers to the dominant human *activity* that occurs on the earth's surface, and it often modifies natural characters to change land cover (e.g., replacing native grassland with a cornfield or a road).

land use change The changes in the way land is used at a given location, for example when a forested land is converted to agricultural fields or when a forestry plantation is changed to other types of use.

language A system of conventional spoken or written symbols that enable human beings to communicate as members of a social group and participants in its culture.

larva (pl. larvae) Early stages in the development of an organism; for invertebrates, often morphologically quite distinct from the adult.

larval parasite An organism that is parasitic only at a larval stage.

Late Quaternary The past 21,000 years since the last time of maximum glaciation.

Late Quaternary extinction event Selective prehistoric extinctions, typically catastrophic, eliminating within the past 40,000 years two-thirds or more of large land mammals of America, Australia, and Madagascar and at least half the species of land birds on remote islands of the Pacific. Humans are present or suspected to be present in virtually all cases.

latent parasitism Parasitism without obvious symptoms.

lateral transfer Genetic information passed between organisms through means other than inter-breeding.

latitudinal biodiversity gradient The well-documented trend for there to be more species in tropical ecosystems compared with temperate or polar ecosystems.

latitudinal gradient A gradual change in a characteristic of interest (e.g., species richness) with a gradual change in latitude; a gradient is well defined if it adheres to a particular mathematical relationship.

laurisilva Warm-temperate woodlands characterized by evergreen broadleaves such as laurel (*Laurus*) trees.

law of thermodynamics, first Physical principle that energy is neither created nor destroyed, only converted between different forms. Energy is therefore conserved. In thermodynamic terms, the change in energy of a system equals the difference of the heat absorbed by the system and the work performed by the system on its surroundings.

law of thermodynamics, second Physical principle that any system will tend to change toward a condition of increasing disorder and randomness. In thermodynamic terms, entropy must increase for spontaneous change to occur in an isolated system.

leaf area index Single-sided total leaf area of the vegetation per unit area of land.

leaf beetles Members of the family Chrysomelidae.

legacy ecosystems Ecosystems that carry an intact "memory" of a change in some physical, chemical, or structural attribute or a change in species composition that modifies the resistance and resilience characteristics of the ecosystem.

legumes Plants belonging to the family Leguminosae, many of whose members can form symbiotic associations with nitrogen-fixing bacteria.

lethal equivalent Group of mutant alleles that would cause an average of one death if homozygous; for example, one lethal equivalent might represent two mutant alleles, each with a 50% probability of causing death, or any other combination of mutant alleles that would produce an average of one death.

lichen Composite organism consisting of a fungus (the mycobiont) and an alga and/or a cyanobacteria (the phycobiont) that live in a symbiotic relationship.

life expectancy The average number of additional years a person could expect to live if current mortality trends were to continue for the rest of that person's life; most commonly cited as life expectancy at birth.

life-form A system of classification of the plants in communities based on the position of their perennating structures relative to the ground; i.e., the size and stature of a plant under natural life conditions. Environmental constraints can cause the life-form to differ substantially from that of a plant that develops under more favorable conditions, where genotype morphology, the "growth form," finds full expression. For instance, the growth form "tree" may be modified to the life-form "shrub" in the alpine zone.

lignin Heterogeneous carbon polymer associated with cellulose to form wood; resistant to decay by bacteria, and broken down by some fungi.

linkage disequilibrium Nonrandom association of alleles on chromosomes.

Linnean system/hierarchy The hierarchical arrangement of inclusive categorical ranks in which, apart from the lowest rank, subordinates members of all lower ranks; it contrasts with exclusive classifications (e.g., *scala naturae*).

literature Imaginative and crafted writings, in the form of poetry, prose, fiction, or drama.

littoral Relating to or located in the shoreline zone where sunlight penetrates to a wetland's substrate and supports rooted plant growth.

littoral zone Region of a lake or river between the land and the open water (pelagic zone) that is colonized by emergent, floating-leaved, and submersed aquatic plants and their attendant sessile microbiota (periphyton).

living dead A term coined by the American tropical biologist Daniel Janzen, denoting the last living individuals of a species destined to extinction. By definition, extinction happens when the last individual of a species dies. In reality, however, extinction of a species can be certain even earlier. Most species need both male and female to reproduce. if there are no fertile individuals of one sex, the species is doomed even if several individuals are still alive. Similarly, below a certain population size, a species cannot form a self-sustaining population, and its numbers dwindle. The decline may take many years but its course cannot be easily altered.

llanos Region of tropical South America, in the east side of the Andean range of Colombia and Venezuela. The llanos landscapes occupy lower elevations (2–300 m), and the landforms vary from flat to rolling terrain. The dominant vegetation is savanna with areas of dry forest, gallery forest, and morichales.

local diversity Number of species or ecosystems in a small geographic area.

local population That part of the population of a species found in a particular habitat patch.

local species richness Number of species found at a local site; distinguished from regional or global species richness that "sums" the number of species across a number of individual sites.

loci Gene positions within a chromosome. Allelic variation occurs at various genetic loci.

locus A precise location in the genome, whether a gene is found there or not; formerly this term was used interchangeably with gene, but the definition has become more specific in the era of molecular genetics.

logging The operation of harvesting trees, sawing them into appropriate lengths (bucking), and transporting them (skidding) to a sawmill. In modern day, this method is mostly mechanized.

logistic growth Regulated population growth that follows the logistic equation $dN/dt = rN(1 - N/K)$. Populations growing according to this equation increase rapidly at low densities and the growth rate decreases as they approach carrying capacity (K).

lognormal distribution A statistical distribution which is normal or Gaussian ("bell curve") when the logarithm of the original data is used. Species abundance data from many communities fit this distribution well. The lognormal distribution often occurs as the result of the law of large numbers in which many small factors interact multiplicatively to produce a result such as species diversity.

Lotka–Volterra competition model A model describing the growth rates of two competing species as linear, declining functions of the abundances of each species.

M

macaxeira Group of varieties of manioc (*Manihot esculenta* Crantz) with low levels of cyanidric acid. These varieties may be consumed without prior treatment to lower their toxicity, as must be done with other varieties of wild manioc. Synonyms: sweet manioc and aipim.

macrobiota Organisms large enough to see, usually greater than 1 mm.

macroevolution Evolution above the species level.

macrofauna Animals large enough to be retained on a 300- or 500-μm sieve.

macroinvertebrates Organisms lacking backbones and generally visible to the human eye.

macrophytes Vascular plants, mosses, liverworts, and macro-algae.

macropod A marsupial in the family Macropodidae, the kangaroos and wallabies.

macropterous Having wings that are fully developed, reaching or exceeding the end of the abdomen.

macrotidal Describing a spring tide range of more than 4 m.

mafia effect A term for a behavior pattern in certain birds in which interspecific brood parasites that may destroy clutches from which parasitic eggs have been ejected by the host.

mafic rock Oceanic crustal rocks relatively rich in magnesium and iron.

major ecosystem type Groups of ecoregions that share minimum area requirements for conservation, response characteristics to major disturbance, and similar levels of β diversity (i.e., the rate of species turnover with distance).

major habitat type Groups of ecoregions that have similar general structure, climatic regimes, major ecological processes, β diversity, and flora and fauna with similar guild structures and life histories.

mammal Member of the order Mammalia. A species that provides milk for its young and has fur or hair. They typically (but not exclusively) bear live young.

management Management in relation to biodiversity conservation involves decisions that have consequences for biological resources. Management can be designed to protect or restore biological resources, especially under conditions where the lack of intervention would lead to an irreversible or undesired change.

mangal A term sometimes used to specify the mangrove habitat as a whole as opposed to "mangrove" applying specifically to the trees themselves. For the most part, however, mangrove is considered to apply to both trees and habitat.

mangrove A collective term for environments characterized by mangrove trees.

mantle An ectodermal tissue layer that secretes calcium carbonate in the form of spicules or shell.

maquis Tall vegetation cover of the Mediterranean Basin dominated by evergreen sclerophyllous shrubs and trees.

marginal benefits and costs Additional benefit or cost incurred in response to a small incremental change in the quantity of some variable (in the limit, the first derivative of the benefit or cost function).

marginal value Economic value of the next incremental unit of something. In this context, marginal values are those associated with managing the next small unit of an ecosystem in a particular way (e.g., preserving, rather than clearing, the next unit of forest).

marine reserves Designated areas where no fishing, mining, or other consumptive use is allowed, usually for purposes of replenishing nearby fishing grounds or maintaining normal evolution, growth, and fecundity.

market-based incentives Policy designed to change behavior by altering the price paid, or cost borne, by a person engaged in an activity that affects biodiversity.

market economy Organization of economic activity in which private individuals buy and sell goods at prices that balance supply and demand.

market value Benefit of wildlife or species as an input into the economy.

maromba Elevated flooring made of wood, used in flooded areas for placing cattle or plants during river flood periods.

marsh A wetland dominated by herbaceous species of plants (e.g., cattails, sedges, grasses, floating ferns) and typically occurring on mineral soils. Floating-leaved and submerged species of plants dominate the parts of marshes that grade into open water. Although the surface layer of some marsh soils may consist of shallow accumulations of organic matter over mineral soils, the depth of this layer is usually less than the minimum required for the site to be classified as peatland.

marsupial Mammal without placenta and with a pouch to carry the young.

marsupium Folds of skin that envelope the mammary glands and provide protection to infants: found in metatherian and protherian mammals.

mass extinction An extinction occurring over a short period of time that is of large magnitude, wide biogeographic impact, and involving the extinction of many taxonomically and ecologically distant groups. Can be defined as, a rapid loss of a large fraction of biodiversity on timescales of 10^0–10^6 years, generally involving a variety of unrelated groups.

mass ratio hypothesis Hypothesis stating that ecosystem processes are largely determined by the dominant contributors of the overall plant biomass, that is, dominant species will exert greater influence on processes than will subordinate species.

mast fruiting A period or year in which a heavy crop of fruits/seeds is synchronously produced by trees and shrubs. This phenomenon, uniquely characterized by synchronicity, high variability, and periodicity of heavy fruit production, distinguishes it from nonmasting plants.

matorral Evergreen sclerophyllous-leaved shrublands that dominate large areas of central Chile.

matrix Background landform, habitat, or ecosystem in a landscape, characterized by extensive area, high connectivity, and major control over landscape dynamics.

matrix species Dominant plant with broad coverage.

maxillae The second pair of mouthparts, located between the mandibles and the labium (the third pair of mouthparts).

maximum sustainable yield 1. Estimate of the size or proportion of standing stocks of a population that may be harvested without altering the long-term abundance of characteristics of the stocks. 2. Specifically, the maximum level of catch that can be removed continually from a fish population without depleting the population.

meadow Grassland maintained by mowing.

mean kinship This value, calculated for every living member of a captive population, is the average kinship between that individual and all members of the population (including itself). Typically, living founders are excluded in the calculation of mean kinships. A population's average mean kinship is the average of the mean kinships of all the individuals in the population.

mechanistic models of population dynamics Models which are explicit about resource consumption, the relationship of consumption to demography, and mortality factors.

median, medial Along or toward the midline of the body.

Mediterranean-type ecosystem (MTE) Habitat characterized by mild wet winters and warm dry summers. MTEs occur in California, central Chile, the Mediterranean Basin, the Cape Region of South Africa, and southwestern and South Australia, all at N and S latitudes of 30 to 35°.

medulla The innermost region of parenchyma in taxa in which two distinct regions are present; outer boundary often marked by conspicuous circular muscle fibers; see cortex.

mega-city Modern city with a large, expanding population, characterized by high consumption levels of energy, water, and food from sources outside the city.

megadiverse An area or a country possessing a much larger proportion of species than would be expected by its extent, latitudinal position, and other factors.

megadiversity The fact of being megadiverse; specifically, the phenomenon of at least 70% of all species being confined to 17 "megadiversity" countries.

megafauna Large terrestrial mammals that are wolf-sized, deer-sized, and larger. Commonly used in reference to the many species of extinct "Pleistocene megafauna" that 9000 years ago populated the New World. The elimination of this megafauna by hunting (of the herbivores) and starvation (of the herbivore-deprived carnivores) was probably the first, and certainly the most dramatically irreversible, of the anthropogenic macroalterations of New World ecosystems. Today, of the extinct Pleistocene megafauna, only the horse remains—evolutionarily developed in the New World but surviving in the Old World until brought back as a gift from the Pleistocene by Spanish soldiers.

meiofauna Animals small enough to pass through a 500-μm sieve but large enough to be retained on a 63-μm sieve.

meiotic drive Preferential segregation of a parasitic gene during gamete production.

melittophily Pollination by bees.

meltwater Ice or snow melted by radiant energy in the polar regions.

mesh size The size of the holes in a fishing net.

mesic Moist, but not wet, environment.

mesophiles From the Greek *mesos* (middle) and *philos* (loving). Includes organisms that grow optimally at temperatures between 20 and 50°C.

mesophyll tissue Photosynthetic cells that are located between the arrays of vascular bundles and bundle sheath cells of a leaf. In C_4 plants, PEP carboxylase and pyruvate-phosphate dikinase are localized in the mesophyll tissue, while Rubisco is absent.

mesopredator A small to mid-sized predator.

mesotrophic Intermediate in production or species typical of intermediate-production environments.

mesotrophy Trophic status of an aquatic ecosystem that is characterized by moderate phytoplankton production and moderate nutrient concentrations and loadings.

metacommunity An assemblage of species in a patchy landscape, each species comprising a metapopulation in that its distribution shifts among habitat patches because the distribution is determined by colonization and extinction dynamics.

metapopulation A group of local populations among which individuals migrate relatively frequently; however, the rate of migration is slow enough that the populations fluctuate independently. Each subpopulation is subject to periodic extinction and subsequent recolonization from other occupied patches. A key feature is the distinction between local (transient) and regional (persistent) dynamics in one or several interacting species. An important recent realization is that most species exist as metapopulations and that this is probably the original, "natural" state of all species.

metatherian A member of the infraclass of mammals that includes the marsupials.

metazoa Organisms whose individuals have more than one cell (multicellular).

metazoan Literally, a multicelled animal.

methanogens Strictly anaerobic Archaea that produce (Greek *gen*: to produce) methane.

microbes Single-celled organisms, such as bacteria, archaea, protists, and unicellular fungi.

microbial response Expected responses of microscopic algae to nutrient enrichment is excess production beyond what can be consumed by grazers and species shift to noxious species; this expected response is not necessarily what happens.

microclimate 1. The climate that plants, small animals, and soil microbes experience, and that differs substantially from the "macroclimate" reported by weather stations. This difference is related to surface warming by the sun or cooling at night, as well as wind shelter effects, and is largely driven by relief, exposure, ground cover, and plant stature. 2. In popular use, any localized climate in a relatively limited area; e.g., a particular valley.

microcosm studies Laboratory experiments involving population dynamics of small organisms with short generation lengths.

microevolution Evolution within species.

microfauna Animals that have high turnover rates, and live in water films in soils. Small mesofauna, such as nematodes, are also water-film dwellers.

microhabitat A small-scale, self-contained environmental unit occupied by a specific subset of interacting species.

micropterous Having wings that are greatly shortened, not overlapping or touching on the dorsum.

microthrix A specialized surface extension of the neodermis of cestodes, characterized by an electron-dense cap composed of numerous microtubules separated from distal cytoplasm by a base-plate (plural microtriches).

microtidal Describing a spring tide range of less than 2 m.

microtubule Hollow cellular structure that is 24–25 nm wide and made of the protein tubulin; it is the main component of centrioles, kinetosomes, the mitotic spindle, and the undulipodium or eukaryotic flagellum.

middomain effect A gradient wherein species richness increases symmetrically from latitudinal extremes to the middle of a region as a consequence of the random placement of species ranges within a geographic domain (also known as Perinet effect).

mid-ocean ridge Crustal accretion boundaries between major tectonic plates that make up the ocean crust.

mimicry 1. A defensive mechanism in which organisms of two species resemble each other, ordinarily because at least one of them is distasteful. 2. Specifically, brood parasitic eggs or nestlings that closely match those of the hosts.

mineralization The conversion of a substance from an organically bound form to an inorganic form; the availability of inorganic nutrients in the decomposition of organic detritus, occurring after the immobilization phase; see above.

minimum viable population (MVP) A method for assessing the minimum size a population must be to bring its extinction over a given time span to below a predetermined level.

miombo Woodland habitat type widespread in south-central Africa, characterized by numerous species of the tree genus *Brachystegia* and *Isoberlinia*, which form nearly closed canopies. Fire is an annual event in this habitat, which supports relatively low populations of large mammals.

mistletoe Parasites of the closely related families Loranthaceae or Viscaceae with haustoria that originate as root tissue and penetrate stems or roots (in Loranthaceae) of the host plants.

mitochondrion (pl. mitochondria) A eukaryotic cellular organelle which is used for cellular respiration and energy production.

mixed farming system An agronomic system that incorporates different combinations of herbaceous crops, trees, and animals within a single farm unit.

mixing ratio Ratio of the concentration of a gas to the concentration of air. It can be defined in terms of volume (e.g., molecules per m^3) or mass (e.g., g per m^3).

mixotrophic Describing form of nutrition involving both autotrophic (photosynthetic) and heterotrophic carbon acquisition.

mixotrophy The ability of a normally autotrophic organism to switch, circumstantially, to phagotrophy, or to support an otherwise meager food supply by resorting to the ingestion and assimilation of bacteria or their presents.

molecular phylogenetics The study of the history of evolutionary relationships according to similarities and differences among (usually extant) organisms in molecular characters (such as DNA).

monitoring In this context, monitoring involves the repeated collection and analysis of observations and measurements to evaluate changes in populations of species and environmental conditions. Monitoring also helps in assessing progress toward meeting a management objective. Monitoring can serve as a warning system, alerting managers that changes in biodiversity may require changes in biodiversity management regimes to ensure protection of biological resources.

monoculture Usually refers to growing a single uniform plant variety over a large area.

monogamy A reproductive pattern in which one male mates with one female exclusively.

monograph effect A term for the fact that the apparent existence of intensive evolutionary activity in a given interval actually reflects the fact that researchers have thoroughly studied a particular section of the fossil record which reveals this activity, while other such investigations elsewhere might just as well have produced the same number of previously unknown species.

monophyletic group Ideally a group comprising a given (hypothetical) ancestor and all its descendants; in a restricted sense, within a given set of terminal taxa, all the members of a subset arising from a common ancestor that has given rise to no other member(s) of the whole set.

monophyletic Describing a group in which all species are descended from a single common ancestor and all descendants of the ancestor are classified in the group; characterized by shared derived characters.

monophyly A true, historical, evolutionary lineage consisting of an ancestor and all of its descendants; defined by shared, derived characters.

monotheism Belief in the unity of the Godhead, or in one God, as opposed to pantheism and polytheism. Monotheism is a firm tenet of Islam and Judaism. Christianity, with its concept of Trinity, alone among the three monotheistic religions, dilutes monotheism. (From Greek *mono*, "one," and Greek *theos*, "god.")

monotreme A mammal in the family Monotremata, which includes spiny anteaters and the platypus.

monozoic Possessing only a single set of reproductive organs; see polyzoic.

monsoon system A wind system covering the eastern half of Asia in which winter cooling of the large landmass produces strong, stable high pressure, with clear skies and outward flow of cold, dry air, and summer warming produces low pressure, drawing wet air masses, with clouds and rain, inward from the adjacent oceans.

Montreal protocol A treaty signed by a number of nations to curtail CFC production

moor Open vegetation, dominated especially by dwarf Ericaceae and *Sphagnum* mosses, on peaty soil.

morbidity Host weakness or lethargy caused by disease.

morichal Plant community characteristic of tropical savannas, it is seasonally flooded and the presence of the palm *Mauritia flexuosa* is conspicuous.

morphodynamics Interactions between the physical structure and the water and sediment movement in a beach and surf zone environment.

morphology 1. The study of form, especially organismal form. 2. The particular form of a given organism.

morphometrics A statistical procedure by which morphological traits are quantified and analyzed.

mortality Deaths as a component of population change. Measures include the crude death rate (number of deaths per 1000 population).

mss The interstitial spaces deep in the soil and mantle/bedrock interface, as found in glacially fragmented zones. By origin, the *milieu souterrain superficiel*, or mesovoid shallow stratum, thus the acronym MSS.

mulch Retained slash covering soil surface; this can be *in situ* slashed or imported biomass.

Mullerian mimicry When two or more distasteful species come to resemble one another. In Batesian mimicry there is a model (unpalatable species) and one or more mimics (palatable species).

multidimensional niche A tempero-spatial region defined by meeting a set of different requirements for viability of a particular kind of organism.

multilines Planned mixtures of different selections of the same variety that differ only in genes for disease or insect resistance.

multimodel Single integrated model that links together models taking different approaches.

multiple resistance The occurrence of more than one resistance mechanism in the same individual or pest population.

multituberculata Extinct taxon of mammals superficially resembling rodents.

mushroom Vernacular word for a large, fleshy fruiting body consisting of a cap with gills or pores on the undersurface (or sometimes flat or with teeth or folds), usually on a stalk, and producing sexual spores. The term is usually reserved for members of the Basidiomycota.

musth Annual reproductive period in male elephants characterized by extreme aggressiveness and secretions from temporal glands.

mutation Change in an allele, producing a different allele; rate of occurrence affected both physically (especially by joining radiation) and by many chemicals: It may also refer to changes in a chromosome (involving duplication or inversion of a segment).

mutation rate The frequency with which new mutations arise per generation.

mutualism Interactions between organisms of different species that increase the fitness of both participants; two organisms living together for the mutual benefit of both.

mutualistic Describing a mutually beneficial association or interaction, temporary or permanent, among organisms of the same or different species.

Müllerian mimicry A form of mimicry in which two or more unpalatable species resemble each other, with the effect that predators are more likely to avoid any species with this appearance.

mycelium (pl. mycelia) Vegetative filamentous body of a fungus; mass of hyphae from a single individual.

mycorrhizae A root tip that is infected with fungi in a mutually beneficial partnership. The fungal hyphae explore large volumes of bulk soil, absorbing nutrients and transferring them to the plant; the plant supplies the organic carbon necessary for growth and energy production to the fungus. Different groups of fungi form vesicular-arbuscular mycorrhizae (fungal hyphae invaginate into the plant root cells) and ectomycorrhizae (fungal hyphae grow between plant root cells and form a thick sheath over the root tip, but they do not invaginate). Several other forms are specific to particular plant families (Ericaceae, Orchidaceae).

mycotrophic (mycoparasitic) plant A plant that obtains energy through mycorrhizal associations with saprophytic or other biotrophic fungi.

myiophily Pollination by flies.

myrmecophily The ability to form symbiotic associations with ants.

myth A story, usually of unspecified origin and at least partially traditional, that ostensibly relates actual events to explain some practice, belief, institution, or natural phenomenon and that is frequently associated with religious beliefs; myths provide models for human behavior and often emphasize behaviors that support conservation of biodiversity.

myths of desertification Controversy stemming from failure to consider all dimensions (meteorological, ecological, and human) of the problem, and alarmist tone connoted by the word desertification, which incorrectly suggests the action of deserts "moving" across the landscape, engulfing fertile lands and leaving starving people in their wake.

N

native (indigenous) species A species that occurs naturally in a given area.

natural classification Systems of classification portraying as accurately as possible the entire pattern of life and its relationships.

natural disturbance Any relatively discrete event in time that disrupts ecosystem community or population structure and changes resources, substrate availability, or the physical environment; key for structuring biological communities and for maintaining resilience in ecological systems.

natural enemy Any species which consumes or parasitizes another species; a general term that includes predators, herbivores, parasites, pathogens, and parasitoids.

natural forest An area of land which supports a minimum of 20% tree cover that has arisen as a result of natural processes of establishment and succession.

naturalized plants Alien plants that reproduce and sustain populations without the input of resources or direct intervention by humans; they often recruit offspring freely, but mainly very near adult plants, and do not invade natural ecosystems (*cf.* invasive alien plants).

natural regeneration The process of replacement by natural seedlings in gaps created by the selective cutting of marketable timber.

natural resources management Management of natural systems to maintain or provide yields of wood, water, and other resources for people.

natural selection Preferential reproduction and survival of certain species under given environmental conditions. Growth advantages of selected individuals reflect the efficiency of their reproduction pathways.

nature The material world and the physical forces or processes that control it.

N_e Symbol for the genetic effective size of a population, an important indicator of future evolvabilty or extinction risk; usually much less than the observed census size (N).

nearly neutral mutation A mutation whose population genetical dynamics are influenced by both genetic drift and selection.

nectar robber An animal that bites holes in flowers to obtain nectar, typically from a flower with a long corolla.

nekton Actively swimming pelagic organisms.

nematocysts Harpoon-like stinging cells found in the tentacles of all cnidarians. They are used to pierce, immobilize, and capture zooplankton food.

neodermis A syncytial, nonciliated epidermis with cell bodies and nuclei (often referred to as cytons or pericaryon) laying below other body wall elements; unique to the Neodermata; completely replaces epidermis, usually at metamorphosis when larva encounters the first host.

neogene The Miocene and Pliocene Epochs, accorded the status of a period when the Tertiary is considered an era.

neotenic A secondary reproductive with juvenile morphological characters. Neotenics derive from larvae, nymphs, pseudergates, or workers through at least one special moult.

Neotropics A biogeographic region that includes the New World tropics, extending from southern Mexico south through the Southern Cone of South America to Tierra del Fuego, including the Caribbean and the tip of Florida; New World Tropics is synonymous. Many different ecosystems are found here, including tropical rainforest.

net mineralization/immobilization The net result of microbial decomposition of organic matter is either the incorporation of nutrient elements (particularly nitrogen) into the microbial biomass, rendering it unavailable for plant uptake *(immobilization)*, or their release into the soil solution (mineralization) after microbial demand for each element has been satisfied.

net primary production Plant photosynthesis less plant respiration.

net social benefit A measure of the social benefit of a project, obtained by identifying the projects positives (favorable outputs) and its negatives (inputs, unfavorable outputs), then multiplying these by the accounting price of each commondity involved. The net result of comparing these positive and negative values will indicate the social benefit of the project.

neural crest An embryonic tissue intermediate between neurectoderm and ectoderm, with cells migrating widely to their final destination. This tissue gives rise to anterior skeletal elements, many portions of the future head and pharynx, and all pigment cells. Sometimes also referred to as mesectoderm.

neurectoderm An embryonic tissue that gives rise to the central tube of the nervous system.

neutral Having no effect on fitness.

neutral mutations Mutations whose effects on fitness are either nonexistent or so small that their fate is controlled by genetic drift rather than selection.

niche complementarity Refers to how the ecological niches of species may not fully overlap and complement each other. Consequently, an increase in the number of species that complement each other may result in a larger volume of total resources utilized and in a higher rate of ecosystem processes.

niche differentiation Differences in the morphology, physiology, or behavior of species that can influence their abundances, dynamics, and interactions with other species, including the ability of various competing species to coexist.

niche dimensions Ranges of values of environmental measurements in an ecological niche. A range of temperatures, salinities, or oxygen concentrations may be niche dimensions for a population of fish.

niche 1. An organism's biotic and abiotic resource requirements, collectively enabling the organism's trophic and behavioral role in the community. 2. Pattern of response of an individual, a population, or a species to the physical and biological gradients of its environment.

niche overlap The proportion of available resources that are shared by two species. Usually used in the context of a single resource that limits population growth.

nitrification The oxidation of ammonia to nitrite and nitrate by bacteria.

nitrogen fixation The transformation of atmospheric nitrogen into a form usable by plants.

nitrogen saturation Nitrogen levels in excess of the biological nitrogen needs of an ecosystem, potentially causing adverse ecological effects.

node Branching point on cladogram corresponding to hypothetical common ancestor.

nonavian dinosaurs A collective term for dinosauria excluding birds.

nongovernmental organization An organization that is not an agency or part of a government, typically one that is not-for-profit and not owned by individuals.

nonindigenous species A species occurring beyond its natural range or potential natural dispersal range. Synonyms used here are exotic and alien species.

nonsynonymous mutation A nucleotide change in a coding region that changes an encoded amino acid.

North Atlantic Oscillation (NAO) The difference in atmospheric pressure between Ponta Delgada, Azores, and Stykkisholmur, Iceland. The variations in the NAO are usually associated with a positive or negative phase related to changes in the direction and strength of the westerly wind as well as in sea surface temperature. During a positive phase, winters over Scandinavia are warmer and vice versa. Recently, the fluctuations observed in the NAO have been related to changes in ecosystem processes.

Nothofagus A genus of Southern Hemisphere trees with a classic "Gondwanic" distribution in southern South America, southeastern Australia, New Zealand, New Caledonia, and New Guinea.

notochord A stiff, flexible, longitudinal rod running along the middorsal portion of the chordate body. It is situated dorsal to the coelom and ventral to the central tube of the nervous system.

nuchal organ Sensory organ with nerves coming from the posterior part of the brain, or the nerve ring, usually present as paired ciliated pits or short ridges at the posterior end of the prostomium; present in most polychaete species.

nucleotide Subunit of DNA and RNA composed of a ringed five-carbon sugar, a ringed nitrogen-rich base, and phosphates; nucleotides are often referred to as base pairs because individual types form complementary hydrogen bonds (G-C and A-T in DNA, G-C and A-U in RNA) to make double-stranded DNA and RNA molecules.

nucleus Compartment of a cell in which DNA is stored on chromosomes.

NUE Nitrogen-use efficiency, a measure of photosynthesis or growth relative to the nitrogen content of the leaves or plant.

nunatak Isolated ice-free summit of a mountain whose bulk is permanently below the surface level of an ice sheet.

nurse plant A plant that shelters and facilitates the growth of others.

nutrient cycle (biogeochemical cycle) The repeated pathway of mineral elements, such as carbon, nitrogen, phosphorus, and water, from the environment through organisms and back into the environment.

nutrient enrichment Usually excess of nitrogen and phosphorus nutrients to aquatic systems.

nutrient limitation The addition of a nutrient or nutrients to an ecosystem causes an increase in net primary production.

nutrients Dissolved mineral salts necessary for primary productivity and phytoplankton growth: Macronutrients are phosphate, nitrate, and silicate; micronutrients are iron, zinc, manganese, and other trace metals.

nutrient uptake capacity The instantaneous rate of nutrient acquisition, usually measured in brief (1–2 hr) incubations. Uptake capacity reflects the abundance of transport sites on the root cell membranes and their affinity for nutrient ions.

nutritionalism The view that bottom-up forces regulate populations.

nymph An immature individual on the developmental pathway to the imago and which possesses external wing buds.

O

obligate epiphyte Plant that always grows on another plant for structural support, but derives no nutrients from the host.

obligatory parasite A parasite that cannot survive without a host.

oligonucleotide A short, single-stranded nucleic acid molecule either obtained from an organism or synthesized chemically.

oligotrophic Term applied to water and soils that are particularly poor in nutrients, such as the weathered white sand soils of Amazonia.

oligotrophy Trophic status of an aquatic ecosystem that is characterized by relatively high phytoplankton species diversity but low phytoplankton production (chlorophyll $a <$ ca. 10 [g/L]); low nutrient concentrations and loadings (for example, mean inorganic N $<$ ca. 700 μg/L, mean total P $<$ 10 μg/L); low decomposition (with little organic material available to decompose); and plentiful oxygen throughout the water column.

ontogeny The developmental history of an organism from egg to adult. Includes embryogenesis that describes the generative phase and the allometric phase of growth and maturity.

ontology The branch of philosophy concerned with beliefs.

open access resource Resource that can be exploited by any person, regardless of the consequences of that exploitation for other would-be users; a resource for which *property rights* (see below) are not well defined.

open communities Communities which receive immigrants from external sources and may export emigrants as well.

operon The basic unit of bacterial gene organization, containing transcription start and stop sites.

opportunistic species Species typical of transient, unstable, unpredictable, frequently disturbed, or periodically extreme environments, usually having strong dispersal abilities and faster growth, smaller size, and shorter life spans than potential competitors.

optimization Procedure for reconstructing the most parsimonious sequence of character change on a cladogram.

option value A variation on the concept of USE VALUE (see below); a measurement or description of the potential value of a resource for possible future use.

organelles Membrane-bound structures within eukaryotic cells that carry out specific functions and that may contain their own DNA; typical organelles are mitochondria and chloroplasts.

organism The smallest entity of life that can function as a whole, distinct from others of the same type; a single living being.

organismal diversity Number and relative abundance of all species living in a given area.

organizing centers Soils are centers of history and activity in terrestrial ecosystems. See the legacy concept in forest ecology.

ornithophily Pollination by birds.

orobiome Mountain range characterized by particular climatic pattern and characteristic sequence of vegetation zones.

orogenesis Formation of mountains.

out of school experiences A term for things that individuals experience that may support or reinforce learning. These might include participation in youth-serving organizations or hobbies such as bird watching.

output environment Collective term for all energy and materials moving out of a given system.

overcompensatory growth In grazed plants, the reaction to tissue removal by enhanced primary production compared to in undefoliated controls.

overexploitation Harvesting of a natural resource, such as fish or timber, at a rate more rapidly than it can be naturally replenished.

overfishing Extracting marine organisms faster than they can reproduce.

overgraze Consumption of forage to the extent that declines in productivity or desirable species composition are probable (i.e., degradation). Overgrazing is excessive use in terms of the amount, duration, and frequency of plant utilization.

overshoot A population is in overshoot when it exceeds available carrying capacity. It may survive temporarily but will eventually crash as it depletes vital natural capital (resource) stocks. A population in overshoot may permanently impair the long-term productive potential of its habitat, reducing future carrying capacity.

ovipositor Modified appendages of the seventh and eighth abdominal segments used for egg-laying paraphyletic: A group in which only some of the species descended from an ancestor are classified together; characterized by shared ancestral characters.

ovoviviparous Young are born alive, but the mother simply retains eggs within her body until they hatch.

ox-bow A crescent-shaped waterbody occurring on a river floodplain having once been part of a river meander that has been cut through and abandoned.

oxic and anoxic conditions Environmental conditions with and without oxygen, respectively.

oxidizing capacity Self-cleansing ability of the atmosphere through oxidative reactions.

ozone Molecule composed of three oxygen atoms that is extremely important to life on Earth because of its absorption of solar ultraviolet radiation; ozone is also a greenhouse gas.

ozone layer A natural layer in the stratosphere that blocks living organisms from harmful ultraviolet radiation from the sun.

P

packing In this context, the presence of a relatively large number of competitive species in a confined area.

paedomorphosis Reproduction while retaining at least some larval characteristics.

paleoclimatology The study of past climates from fossils and other traces left in the geological record.

paleotropics Tropical regions of the Eastern Hemisphere extending from Africa through south and southeast Asia into the southwest Pacific; also known as Old World Tropics.

Palynology The study of pollen and spores.

pampa(s): Region of temperate South America, west of the Andes in Argentina. The Pampa region contains several ecosystem types, including temperate grasslands, dry forest, and xeric shrub lands.

panbiogeography The examination of distributions and relationships on a worldwide scale; term introduced by Croizat (1964).

pantanal Largest wetland of South America drained by the Paraguay River and its tributaries.

pantheism The doctrine that identifies the universe with God. In Western thought, the term is associated with the Dutch philosopher Baruch Spinoza. His view represents an important criticism of the "orthodox" view of a god whose reality is somehow external to the reality of the world. (From Greek *pan*, "all," and Greek *theos*, "god.")

parameter A quantity in a dynamical system that is fixed as part of the rules of the system (contrast with dynamical variable).

parametric statistics Statistical methods that assume data have certain characteristics, such as a particular kind of distribution (Poisson, normal, binomial, etc.).

parapatric Describing two taxonomic entities whose geographic ranges are distinct but adjacent to each other.

paraphyletic Describing a group of taxa that includes some, but not all, of the descendants of the most recent common ancestor for the group.

paraphyletic group A group comprising a given (hypothetical) ancestor and only some of its descendants; a taxonomic grouping of animals that does not meet the cladistic criterion of including the most recent common ancestor and all its descendants.

paraphyly The status of a group that includes a most recent common ancestor and only some, not all, of its descendants.

parapodium Segmentally arranged outpocket from the body wall; usually divided into two branches (rami), the dorsal notopodium and the ventral neuropodium; supports the chaetae; extremely variably developed and may be absent.

parasite An organism that obtains its nutrition from a living host, and in so doing (typically) harms the host.

parasitism A close association of two organisms in which one, the parasite, depends on the other, the host, deriving some benefit from it without necessarily (though usually) damaging it.

parasitoid An insect whose larval stage feeds on a second (host) species, killing the host in the process; a parasitoid attack is equivalent to delayed predation and therefore the phenomenon is distinguished from true parasitism.

parataxonomist A layperson who has received training in practical basic biology, ecology, and taxonomy, as well as in the collection and preparation of biological specimens, so that he or she can undertake a specific part of a biodiversity inventory.

parenchyma The connective tissue between cells of ectoderm and endoderm layers that fills spaces among organ systems; as platyhelminths are acoelomates, the space among internal organs is completely filled with parenchyma.

Pareto efficient A pattern of operation of an economy is said to be Pareto efficient if it is impossible to change its operation so that everyone gains or at least someone gains and no one loses.

parr Older juvenile salmon that are distinguished by parr marks—prominent oval spots on their sides.

parsimony General scientific principle that given alternative explanations or hypotheses for a set of observations or data, the most corroborated is that requiring the fewest ad hoc (ancillary or additional) hypotheses (a special application of Occam's razor).

parthenogenesis Reproduction in which eggs are not fertilized by males.

partial migration The case where intrapopulational variation in migratory behavior leads some individuals to migrate while others within the same population may only migrate locally or remain sedentary.

passive restoration Relying on natural successional processes for restoration after the stresses that caused the disturbance have been removed.

pastoralism Agrarian life and work; also literary accounts of rural life, often simplified or idealized.

pasture Lands in which the native vegetation has been removed in favor of cultivated grasses or other forage plants; grassland maintained by grazing domestic animals.

patagium The skin that stretches across the arms and legs of flying and gliding mammals.

patch An area of suitable habitat for a particular species or particular collection of species, ideally bounded by unsuitable habitat or habitat with different physical properties. Normally, it is one of many such areas in a region.

patch disturbance The measurable habitat and ecosystem modification caused by large animals, including humans, as they forage for food or other resources. Patch disturbance is most pronounced near the den site, temporary camp, or other central place within the overall home range of the individual or group.

patchiness Variation from place to place (or time to time) in the abundances of animals or plants, caused by the interaction of numerous processes.

path dependence The influence of previous actions on the present state.

pathogen A microbial agent that causes disease.

patrilocal residence The practice of married couple's living in the husband's community.

páramo Ecosystem of the high Neotropical mountains, found in South America above tree line in the northern Andean mountains of Colombia, Venezuela, Ecuador, Peru, and Bolivia. The dominant vegetation is similar to an alpine meadow or grassland; cacti and giant rosette plants are conspicuous.

peat A type of soil formed from accumulation of the remains of dead plants, with high organic matter content (minimally 20% and as high as 90% by dry weight) and depth greater than 30 to 40 cm. Peat forms in some wetlands because anoxic conditions, low pH, low temperatures and short growing seasons, or plant tissues highly resistant to decay prevent rates of decomposition from keeping up with rates of organic matter production.

peatlands Wetlands in which dead organic matter accumulates to form peat more than 30 to 40 cm deep; typically includes bogs, poor fens, and rich fens but some swamps also occur on peat.

pedipalps The second pair of preoral appendages. They are multisegmented and primitively leg-like. They may be raptorial or sensory (like antennae) or used as walking legs.

pegging The penetration of the fertilized groundnut flower into the soil.

pelagic Pertaining to the water column in aquatic environments; i.e., the (open-water) part of the aquatic environment that is far from the shore and the bottom bed.

pelagic organism A free-swimming (nekton) or floating (plankton) organism that lives exclusively in the water column.

pelagic zone Open-water portion of a lake or reservoir beyond the littoral zone.

penetrometer resistance Energy required to push a probe of defined size and shape into the soil.

penis Male intromittent copulatory organ consisting of a muscular, often papilliform, retractable structure that is protruded (without turning inside out) through the genital pore during copulation; may or may not be armed with spines, etc.; see cirrus.

perennating structures The tissue that carries a plant into a new growing season. For example, buds of trees, shrubs, etc., bulbs of some plants (e.g., tulips) or seeds of annuals where the "plant" dies each year.

perfectly competitive economy Ideal organization in which all goods are owned by perfectly rational, perfectly informed individuals, all goods can be bought and sold in markets, and no one individual can influence the prices at which transactions can take place. A perfectly competitive economy is efficient.

periodic parasite A parasite visiting a host at intervals.

periphyton Bacteria, fungi, algae, and sessile microfauna growing attached to substrata (sediments, rock, plants, animals, sand).

peristomium Region around the mouth of the larvae; if distinct in adults may carry peristomial cirri.

permafrost The phenomenon of water which is permanently at or below 0°C. Usually, but not always, the water is in the solid state. The Arctic is characterized by the presence of large, continuous areas of permafrost that have the form of lower soil layers that are permanently frozen with a shallow (usually <1 m) "active layer" which freezes and thaws each year and accommodates belowground biological activity.

permanence A property of communities that ensures long-term coexistence of species because community trajectories (variation in numbers through time) have no species approaching very low numbers.

permanent parasite A parasite associated with a host for long periods.

persecution Deliberate killing of animals perceived to be a nuisance.

perturbation A temporary change in one or more dynamical variables or parameters of a dynamical system due to external factors.

pesticide A chemical substance used for controlling, preventing, destroying, or mitigating a pest organism.

phagotroph An organism that ingests solid food particles (e.g., bacteria).

phagotrophic Feeding on particulate matter.

phagotrophy A type of heterotrophy that involves the consumption of protists, plants, or animals as food.

phanerogam Plants with conspicuous reproductive organs (flowers); plants that reproduce by seeds.

phanerozoic The geological interval of abundant animal fossils, beginning approximately 545 years ago.

pharate A "cloaked" or hidden stage, for example, the adult moth just prior to its emergence from the pupa.

pharynx In certain organisms, the anterior portion of the alimentary canal, characterized by lateral buds that provide skeletal support for the gill region.

phenetic A classification or grouping based purely on overall similarity. Pheneticists use matrices of overall similarity rather than parsimony to construct a phenogram as an estimate of the phylogeny. Examples of phenetic methods of estimation include unweighted pair group analysis (UPGMA) and neighbor joining.

Cladists reject phenetic classifications on the grounds that they may result in paraphyletic or polyphyletic groupings.

phenetics The study of phenotypes, usually describing the grouping of organisms into taxa on the basis of estimates of similarity.

phenogram A dendrogram expressing overall similarity (long axis) between terminal taxa and sequentially linked groups of terminal taxa; the nodes do not connote specifiable characters and the long axis does not connote time.

phenotype The appearance (function and behavior) of an organism; the state of an individual with respect to a specific trait.

phenotypic Relating to observable characteristics of an organism.

phenotypic plasticity Property of a given genotype to express different phenotypes as a function of ecological conditions.

pheromone A chemical released by one individual that elicits a response in a second individual of the same species.

phobia Obsessive fear or dread of an object or situation.

phoresis An association in which one organism uses another as a means of transport and/or protection.

phoresy A method of long-range dispersal in which the dispersing animal attaches itself to another animal (e.g., beetle, wasp, or bird) that carries the disperser along with it until the disperser drops off or disembarks.

phosphoenolpyruvate carboxylase (PEPCase) PEP carboxylase, the initial carboxylation enzyme in C_4 photosynthesis.

photoautotroph An organism that uses inorganic material as a source of carbon for growth, and light as an energy source; e.g., plants.

photoautotrophic Describing organisms able to synthesize organic compounds from water and inorganic nutrients, using the energy in photons of light.

photoautotrophy A type of autotrophy in which organisms gather light energy in order to reduce carbon dioxide to organic carbon; characteristic of green plants, most algae, and some prokaryotes.

photochemical Relating to chemical reactions with rates which are increased by radiation of particular wavelengths.

photochemical source or sink Production or loss of an atmospheric gas involving photodissociation and/or chemical reactions.

photodissociation Destruction of a molecule through absorption of solar radiation and subsequent breaking of one or more chemical bonds.

photorespiration Biochemical process in which O_2 is assimilated by the oxygenation of RuBP, and CO_2 is given off in the metabolism of the products of RuBP oxygenation.

phototroph An organism utilizing the energy of light, as in sunlight, for growth.

phototrophy Energy metabolism based on light energy.

phrygana Dwarf shrub land of the eastern Mediterranean Basin characterized by a dominance of species with seasonal leaf dimorphism.

phylesis Line(s) of descent between ancestors and descendants.

phyllopod A leaflike appendage.

phylogenetic Pertaining to the true (i.e., evolutionary) pattern of relationship, usually expressed in the form of a binary branching tree, or phylogeny. If hybridization produces new lineages, as is common in many plants and some animals, the phylogeny is said to be "reticulate." Phylogenies may be estimated using phenetics, parsimony (cladistics), or methods based on statistical likelihood.

phylogenetic species concept Concept of a species in which species-level identity is determined by members sharing distinct characteristics.

phylogenetic tree A hypothesis for describing the history and relationships among living species.

phylogeny The evolutionary relationships among taxa (groups of related organisms), often portrayed with some kind of branching diagram, with branches representing speciation events. Typically, the true phylogeny of a group is hidden deep in the past and evolutionary biologists must infer relationships. Various types of data and methods of analysis are used in this effort and there is considerable contention among groups of scientists as to which are most likely to estimate the true phylogeny.

phylogeography The study of the relationship between and among genetics, morphology, paleontology, and ecology to better understand the spatial aspects of evolution.

phylogram cladogram; strictly, a cladogram interpreted to show terminal taxa and ancestral nodes (intermediate specificity between generalized cladogram and specific phylogenetic tree).

phylotypes Molecularly differentiated taxa.

phylum (pl. phyla) Highest level of taxonomic division in the animal world, below kingdom, and followed in descending order by class, order, family, genus, and species.

physiognomy The outward appearance or morphology of a community as determined by the growth forms of the dominant plants present.

physiography The landforms that give shape and character to the continental landscape, for example, the configuration of mountains and drainage basins.

phytochorion (pl. phytochoria) Region within which a substantial proportion of the flora is endemic.

phytophagous Feeding on plants; also, herbivorous.

phytophagy Plant feeders, herbivores.

phytoplankton Unicellular and chain-forming algal plants that are suspended and transported in the water above the bottom. They are photoautotrophic plant cells ranging in size from 1 μm to 1 mm. They are divided into 12 taxonomically defined divisions, including diatoms, dinoflagellates, and coccolithophorids.

phytotoxic Damaging to plants.

picophytoplankton The smallest (<2 μm) size class of photoautotrophic plankton.

pinnipedia Suborder of the Carnivora with fin-like limbs (seals, sea lions, and walrus).

pioneer guild The first groups of species to colonize a newly formed or denuded habitat.

piscivores Predators that consume fish in aquatic habitats.

pixel Abbreviated from "picture element"; the smallest part of a picture (image).

placental mammal See *eutherian*.

planetesimals Small, solid bodies similar to meteors in composition but revolving in orbit around a central gaseous nucleus, as do planets around the sun.

planktivores Predators that consume zooplankton in aquatic habitats.

plankton Organisms that float freely in the water column and do not maintain their position independent of water movements. Phytoplankton (literally plant plankton) are plankton with photosynthetic pigments and zooplankton are animals of the plankton.

planktonic Living in the water column of lakes and/or the sea; describing small organisms with no or limited powers of locomotion that are suspended in the water and largely dispersed by turbulence and other water movements.

planktotrophic Larval forms of invertebrates that feed on plankton to survive and grow.

plantation An area of land that supports planted forest, usually for commercial exploitation.

plantigrade Walking on the soles of the feet.

planula A coral larva. This ciliated planktonic stage rarely lasts for more than 1 or 2 weeks prior to settlement.

plasmids DNA molecules that replicate independently of the major bacterial chromosome(s).

plasmogamy Fusion of the cytoplasm of two cells.

plasticity Ability of a plant to respond to temporal changes or spatial variation in environmental conditions by altering the size or the distribution of plant parts. These are phenotypic, rather than genetic changes.

plastron Thin layer of air, held by specific body structures, into which oxygen from the surrounding water is added by means of diffusion to the same extent as oxygen is withdrawn by breathing.

plate tectonics A modern geological theory of tectonic activity according to which the earth's crust is divided into a small number of large, rigid plates whose independent movements relative to one another cause deformation, volcanism, and seismic activity along their margins.

pleiotropic gene A gene that has more than one independent phenotypic effect.

pleiotropy Phenomenon of a gene affecting more than one trait.

pleiotropy One gene effects more than one trait.

Pleistocene The geological time that ends with the last glacial period and the appearance of humans. It started 2 million years ago and finished 10,000 years ago with the end of the last Ice Age.

Pleistocene glaciation The expansion of ice sheets over the large northern continents, especially northern Europe and North America (excluding the northwest) during the "Ice Ages."

pleopod Abdominal appendage in crustaceans.

plesiomorphic Primitive, relative to sister group.

plesiomorphy An apomorphic character or character state that specifies a more inclusive group than that under consideration.

Pliocene The period of geological time from 3.5 to 1.8 million years ago.

plume prospecting Use of chemical and physical anomalies in hydrothermal plumes to locate hydrothermal vents.

plurivoltine Species with several generations per year (also "multivoltine").

pneumatophores In some species of mangrove, such as *Avicennia* and *Sonneratia*, underground roots spread laterally from the main stem. Pneumatophores grow vertically from these, typically standing 10–20 cm above the soil surface, enabling gas exchange to take place with the underground roots.

Poaceae Grass family.

podzol A soil type highly characteristic of boreal forests and developed as a consequence of podzolization.

podzolization The process of acid leaching whereby clay, organic particles, and mineral ions (primarily iron and aluminum) are carried downwards and deposited in the B soil horizon, leaving an impoverished and leached A horizon. This occurs as a consequence of low temperatures and precipitation in excess of the needs of evapotranspiration.

poikilohydry Condition of internal water balance varying with changes in ambient humidity.

Poisson distribution The probability distribution of numbers per area if the organisms are distributed completely at random in space.

polarity Relative apomorphy/plesiomorphy of character states.

pollard A tree repeatedly cut at about 6–10 feet above ground to produce successive crops of wood.

pollination The transfer of a pollen gran (male gametophyte) to a flower's stigma.

pollinium A packet of pollen, characteristic of flowers in the Orchidaceae and Asclepiadaceae.

pollution 1. Introduction of substances in quantities that are threatening to living resources, biological processes, and human health and activities. 2. Move generally, any materials or chemicals in excess of natural levels caused by the activities of humans.

polyandrous A female that has more than one (male) mate.

polyandry Mating system in which a female mates with more than one male.

polycentric governance arrangements Complex, multitiered governance systems in which there is no single center of authority.

polyculture Intensive growing of two or more crops either simultaneously or in sequence on the same piece of land.

polygamy Males and females both have several different mates.

polygynous A male that has more than one (female) mate.

polygyny Mating system in which a male mates with more than one female.

polymerase chain reaction (PCR) A method for amplifying DNA *in vitro* involving the use of oligonucleotide primers complementary to nucleotide sequences in a target gene and replication of the target sequences by the action of DNA polymerases.

polymorphic Relating to the existence of more than one discrete intraspecific type such as distinct phenotypes or genotypes at one locus or several loci.

polymorphism A genetic difference between two or more individuals of the same species; to exclude new mutations from the definition, the underlying mutations are required to segregate above an arbitrarily set minimal frequency.

polyphagous Eating plants from more than two unrelated plant families.

polyphyletic Describing a group of species or lineages that have independent origins instead of being descended from a single common ancestral group.

polyphyletic group A group of animals that, although grouped together in a phylogeny or evolutionary tree, do not share a single common ancestor. "Winged vertebrates" (including birds and bats) give an example of a polyphyletic group.

polyphyly The status of a group in which the most recent common ancestor of the included taxa is excluded from the group. Defined by convergent, nonhomologous characters.

polyploid Having more than one set of homologous chromosomes.

polythetic group A natural taxonomic group in which the terminal taxa are not known to share any universal unique character(s) but are nonetheless united by overall similarity (phenetics) or global parsimony (cladistics).

polytomy Many branches arising from a single node or an unresolved (portion or whole) cladogram.

polyunsaturated fatty acid Fatty acid with two or more double bonds in the chain of carbon atoms. They occur in distinct families defined by the position of the double bonds, of which the most common are *n*-3 and *n*-6.

polyzoic Possessing multiple serial sets of reproductive organs; condition found in most cestode orders.

poor fen An acidic peatland similar to a bog but having a somewhat higher surface water pH (4.3–5.8) because it receives some groundwater or surface water in addition to precipitation; typically dominated by *Sphagnum* mosses or sedges. But for the presence of distinctive poor-fen indicators, vegetation may superficially resemble bog vegetation or sedge-dominated rich fens that lack *Sphagnum*.

Popperian A follower of philosopher of science Sir Karl Popper who accepts his falsificationist views.

population A group of individuals with common ancestry that are much more likely to mate with one another than with individuals from another such group; members of a single species that can potentially interact with each other.

population change An increase or decrease in the size, composition, or distribution of a population resulting from the interaction of births, deaths, and migration in a population in a given period of time.

population composition The distribution of a population, usually by age and sex, measured by the number and proportion of males and females in different age groups.

population dynamics The change in the number of individuals in a population over time.

population momentum The tendency for population growth to continue beyond the time that replacement-level fertility has been achieved because of the relatively young age structure of the population.

population regulation The constraint of positive population growth. The study of population regulation deals with the factors that cause this constraint, such as competition for food or predation.

population stability The tendency for populations to return to a previous size after a disturbance, such as reductions due to hunting or disease or increases due to immigration. Stable populations may be locally stable (return after small disturbances) or globally stable (return after severe or catastrophic disturbances).

population structure Subdivision of a given population according to features such as location, age, size, or social status.

population viability analysis (PVA) A method of assessing the probability of extinction for a given population within a specified time span.

postemergence herbicide Herbicide applied after the emergence of the specified weed or crop.

post-transcriptional modification Cutting and splicing of mRNA in eukaryotic cells, in some cases to produce alternative proteins with different structural or regulatory properties (e.g., sex determination in *Drosophila*).

potential natural vegetation The vegetation cover which would develop naturally in an area and become stable (not replaced by a subsequent stage) if all outside disturbances were eliminated.

power The rate of energy transformation over time. The watt (W) is the common SI unit of power, where 1 W equals the power expended by the transformation of one joule in one second.

P/R The ratio between photosynthetic and respiratory rates of the combined coral host and zooxanthellate symbiont. A ratio greater than 1 ($P/R > 1$) indicates a net gain of energy that is then available for growth and reproduction.

predation An interaction between two species in which one species consumes the other.

preemergence herbicide Herbicide applied to the soil prior to the emergence of the specified weed or crop.

preservation The school of conservation philosophy that emphasizes the protection of natural features and landscapes from human exploitation.

presoldier A transitional morph that always precedes the soldiers; an unsclerotized individual whose head shows signs of soldier differentiation.

prevalence The proportion of hosts in a population that are infected or diseased.

primary forest Forests that appear to be undisturbed by human influence.

primary hemiepiphyte Plant that begins its life cycle anchored in a tree crown and ultimately becomes rooted in the ground (e.g., strangler fig).

primary producer An organism that is able to synthesize organic matter from carbon dioxide.

primary production 1. The synthesis of organic matter from carbon dioxide by photosynthesis. 2. Specifically, an estimate of the rate of carbon fixation by phytoplankton; this can account for up to 75% of the photosynthetic production on Earth.

primary reproductives Dealate reproductives that founded a new colony after nuptial flight.

primary succession Succession on newly exposed substrates such as a sandbar or rubble at the foot of a receding glacier.

priority 1. Areas that need the most urgent management action to avert threat in order to meet a conservation goal (such as to increase the probability of persistence of valued species or other attributes). Ranking of areas by priority may differ from their ranking by value. 2. Of a scientific name, its date of valid publication; the first published name is generally the one to be used, subject to the provisions of the pertinent *Code* (q.v.).

prisoner's dilemma A paradigm in which uncertainty about others' behavior leading to choices that are individually rational but collectively irrational.

proboscis (pl. proboscides) Tongue of an insect.

procaryote Another spelling of prokaryotes.

process intricacy Complexity of temporal or spatial patterns of ecological processes such as population dynamics or production.

production Amount of new organic biomass formed over a period of time, including any losses from respiration, excretion, secretion, injury, death, and predation; the increase in biomass of an individual, population or community as it grows by converting energy-food into biomass.

proglottid Compartmentalized regions of cestode body; each proglottid contains one or more sets of reproductive organs; controversial as to whether proglottids are homologous with segments of coelomates.

prognathous Position of the head when the mouth opening is directed forward.

Progressive Era The period in American political history, especially coinciding with the presidency of Theodore Roosevelt (1901–1909), in which the conservation movement gained definition.

progressive succession Successions in which the dynamic changes are in the directions of increasing species diversity, structural complexity, greater biomass, and increased stability. **Retrogressive successions** are in the opposite directions.

prokaryotes Life forms that are members of the domains Archaea and Bacteria as opposed to Eukarya, comprising organisms with a cell nucleus; single-celled organisms defined by having their DNA arranged as a circular molecule not contained within a nucleus, and which reproduce by simple fission.

prokaryotic Describing an organizational state of cells lacking a membrane-bound nucleus and certain other organelles. Bacteria, including the Cyanobacteria, are typically prokaryotic.

propagules Reproductive units (spore, seed, bulb, cyst, egg, bud, larva, etc.) that give rise to new individuals; any part of an organism capable of developing into a new individual.

property right Legal and social construct under which the owner of any good cannot be deprived of it without the payment of compensation that is acceptable to her.

property rights systems Bundles of property rights with their associated rules and obligations.

prosimians "Lower" primates such as lemurs, lorises, and galagos.

prostomium Morphologically anterior-most part of the body; may carry antennae and eyes and contain at least part of the brain.

protandry Breeding system in which individuals change sex from male to female.

protein A three-dimensional macromolecule constructed of amino acids which is formed based on an RNA sequence.

protist A member of a diverse collection of eukaryotes, defined only by their exclusion from the groups plants, animals, and fungi; includes all microbial eukaryotes.

protogynous, protandrous Describing different types of hermaphroditism (in contrast to gonochoristic). Individuals in protogynous species develop into a reproductive female early in life, then change sex and become a reproductive male later in life. Individuals in protandrous species are functional males first, then females.

protogyny A breeding system in which individuals change sex from female to male.

protonephridial system A system of branched canals, each branch with one or more specialized flame bulbs bearing one to many cilia; specialized for excretion and osmoregulation; found in many phyla; in platyhelminths gaps within or among cytoplasmic extensions of one or two canal cells and/or the terminal cell allow fluids to enter the canal system at a rate determined by the pressure difference created by the beating of the cilia of the flame bulb.

protophytes A diverse assemblage of partly unrelated groups of eukaryotic, phototrophic microorganisms.

protozoa A diverse assemblage of partly unrelated groups of eukaryotic, mostly motile and phagotrophic microorganisms; some phototrophic flagellates and fungi-like protists are traditionally included.

proximate cause(s) of extinction The actual immediate agent(s) that cause(s) a species to become extinct.

psammophilic Sand-loving.

pseudergate (false worker) A temporarily nonreproductive individual serving the colony in nutrition, construction, or brood care, which results from a late, reversible deviation from the pathways to the imago and is characterized by reduced wing buds compared to nymphs of the same stage.

pseudocoelom A body cavity not lined with a mesodermal epithelium.

pseudocopulation Pollination by a male insect attempting to copulate with a flower mimicking a female insect.

pseudoextinction The disappearance of a species from the fossil record due not to the death of all its members but to an evolutionary change that results in it being classified as a new species.

pseudofecal pellet Fiddler crabs and their relatives collect soil with their mouthparts, separate organic particles from mineral components by a complex flotation process, ingest the former, and discard the latter in the form of compact pellets. These are known as pseudofecal because, although extraction has taken place, the waste material has not passed through the gut.

psychrophile An organism that grows more optimally at low temperature or that requires low temperature for growth.

public good A good whose consumption is nonrival and nonexcludable. Nonrival means that one person's consuming it does not preclude another from doing likewise. Nonexcludable means that the provider of the good cannot ensure that only those who have paid can benefit from its provision. Knowledge is a public good: My knowing something does not conflict with your knowing the same fact, and those who develop knowledge cannot ensure that only people and institutions that have contributed to the costs can benefit. An apple, in contrast, is a private good: If I eat it, you cannot. Also, apple producers can ensure that only those who pay for them can eat them.

Pull of the Recent A term reflecting the fact that recently formed rock is more accessible and offers better preserved fossils; this may produce a bias toward the conclusion that diversity has increased toward the present time.

puna Ecosystem of the high Andean mountains found above tree line from Bolivia to southern Peru. The Puna comprises four distinct ecological regions: wet, dry, thorny, and desert Puna. The dominant vegetation types vary from grassland/shrub lands to desert.

punctuated equilibrium The morphological stasis of fossil species over long periods of time punctuated by seemingly instantaneous speciation and morphological change.

pupoid Earliest postembryonic stage with incompletely developed appendages, as in insect pupae, hence motionless.

purebred An animal that is a member of a recognized livestock breed. Breed membership may be determined by pedigree records, geographical location, or knowledge of the breeding structure of the herd or flock. Animals that possess the full range of characteristics commonly associated with a specific breed are sometimes also designated as purebreds, but the designation must be recognized as subjective in these cases.

pure culture An organism growing in the absence of all other organisms.

pycnocline The layer in which density changes most rapidly with depth and separates the surface mixed layer from the deep ocean waters.

Q

Q_{10} Mathematical index expressing the effect of temperature on respiration or decomposition: $Q_{10} = 2$ means that the rate doubles for a 10°C increase in temperature.

quality-of-life value Benefit of wildlife or species to people's well-being or happiness.

quasispecies Hierarchically ordered population of mutants that results from erroneous copying of a genotypic ancestor. Whole mutant distributions behave like single species. They are subjected to natural selection. The average sequence is defined as a consensus sequence that represents the "center of gravity" of the distribution. The consensus sequence is synonymous to a wild type.

Quaternary The past 2 million years (approximately) of Earth history, including the Pleistocene and Holocene (or Recent) epochs, and characterized by the extreme fluctuations in global temperature that produce the ice ages.

queens and kings Females and males actively reproduce in a colony.

quiescence Dormancy in which inhibition of development depends directly on environmental factors.

R

R_0 The basic reproductive ratio of disease; a parameter that describes the number of new infections generated by a single infected host entering an entirely susceptible population.

radiation All the species descended from a single common ancestor; tends to be a speciose group in comparison with others of a similar age.

radiative forcing Measure used to express and compare the potential of climate change factors to perturb the Earth energy balance, reported in watts per square meter ($W\ m^{-2}$). A positive radiative forcing tends to warm the Earth's surface and a negative radiative forcing tends to cool the surface.

radiocarbon dates Dates of organic matter in geological samples using the radioactive decay of carbon 14, which has a half-life of 5750 years.

radiocarbon dating An isotopic or nuclear decay method for inferring age of organic materials. Carbon 14 is produced in the upper atmosphere by cosmic ray bombardment and oxidized to form $C^{14}O_2$. Distributed through the earth's atmosphere and oceans, a small percentage is incorporated into a variety of organic materials to decay with a half-life of 5700 years. By dating tree rings of known age, and other methods, radiocarbon determinations can be calibrated. Although routine application of radiocarbon dating is usually limited to dates of less than 40,000 years, ages up to 75,000 years have been measured.

radula A ribbon-like structure bearing transverse rows of small "teeth," typically used to collect food and move it into the mouth.

rainforest The tropical forest formation which occurs approximately within a latitudinal belt of 10° on either side of the equator. It may be noted that temperate and sub-tropical zones have forest formations structurally similar, but biologically not as diverse as the tropical rainforest.

rain shadow Less rainy sector on leeward side of mountains and protected valleys caused by descending, warming air masses.

rami (sing. ramus) A cuticular extension (usually relatively long and thin, can be leaflike) from the body (e.g., caudal rami = extention from the posterior segment of the body, usually paired in crustaceans).

random genetic drift The random fluctuation in allele frequency caused by random variation in fitness between individuals.

range condition Current productivity and composition of plants on a rangeland relative to its ecological potential. Often range condition is classified into arbitrary classes (i.e., excellent, good, fair, and poor).

range expansion The spatiotemporal process in which successive colonization events increase the size of a species' geographical range. The terms *invasion*, *migration* (as used in the paleobotanical literature), and *spatial spread* are roughly synonyms for range expansion. *Invasion*, in particular, is often used to describe colonization outside the species range.

rangeland Land on which the potential native vegetation is predominately grasses, grass-like plants, forbs, or shrubs. Rangelands include prairies, marshes, tundra, wet meadows, savannas, shrubland steppe, chaparral, desert grasslands, and woodlands.

rank A specified categorical level in the taxonomic hierarchy (e.g., species, genus, family, and class) made coordinate in classification by definition but very frequently not coordinate in the taxonomic system.

Rapoport effect A latitudinal gradient wherein the sizes of the distributional ranges of species decrease with decreasing latitude.

rare species Species with small world populations that are not presently listed as endangered or vulnerable, but are at risk because of their small population size.

rDNA Genes coding for rRNA that play a fundamental role in the translation process; the most thoroughly studied molecule in prokaryotic cells; used in comparative phylogenetic studies.

reaction center (Bacterio)chlorophyll–protein complex that performs the primary photochemical reactions of photosynthesis, the earliest products being the oxidized reaction center (bacterio)chlorophyll at the outside/thylakoid lumen side and a reduced (bacterio)chlorophyll (RC1) or reduced (bacterio)-phaeophytin (RC2) on the cytosol/stroma side of the photosynthetic membrane.

receptor A biological system that is exposed to the environmental stress.

recipient control Consumers substantially depress populations of their resources.

reciprocal selection Evolutionary change as a result of species acting as selective agents on each other through interspecific interactions.

reclaim To recover productivity at a degraded site using mostly exotic tree species. The original biodiversity is not recovered although the protective function and many of the original ecological services may be reestablished.

reclamation A revegetation or land management goal that includes a lower diversity of species and may include substitutions by introduced species.

recolonization The reappearance of a species in an area where it has earlier been present, then went extinct.

recombination The shuffling of gene combinations in the production of gametes, possibly by the physical breaking and rejoining of pieces of chromosomes.

reconstructed evolutionary trees Phylogenies linking contemporary species, therefore containing no information about lineages that have gone extinct.

recreation An activity voluntarily undertaken, primarily for leisure and satisfaction, during leisure time.

recruitment Arrival of new juvenile animals or plants into a habitat or an older stage of the population. For many marine animals, recruitment occurs some time after the settlement of planktonic larvae in the adults' habitat.

recruits The young fish that are first caught by a fishery.

redfield ratio From large-area and time averaging, ratio of carbon, nitrogen, phosphorus, and oxygen for normal ocean plankton and deep-ocean nitrate and phosphate pools.

Red Lists Compilations of recently extinct, threatened, and vulnerable species for a country or a larger region.

reference area An undisturbed or natural area chosen to compare with a restored site to determine the success of restoration.

reflective beaches Narrow, steep beaches.

reforestation The reestablishment of trees and understory plants at a site previously occupied by forest cover.

refugia In the context of the Arctic, land areas which were not covered by ice sheets or glaciers during the last glaciation. Consequently, some biota could survive there and recolonize adjacent areas when the ice retreated. Therefore, these areas are associated with high biodiversity and endemism. Refugia usually occurred in coastal areas (now continental shelves) and ice-free mountain tops or "nunataks."

region Large geographic area that contains more than one landscape.

regulations Rules and administrative codes required by statutes and issued by local, state, and federal government agencies. Regulations have the force of law since they are adopted under the authority granted by statutes.

regulon A group of operons jointly regulated by one factor.

rehabilitate To reestablish the productivity and some, but not necessarily all, of the plant and animal species thought to be originally present at a site. For ecological or economic reasons, the new forest might also include species not originally present at the site. The protective function and many of the ecological services of the original forest may be reestablished.

rehabilitation Creation of an alternative ecosystem following a disturbance, different from the original and having utilitarian rather than conservation values.

reification The assignment of empirical reality to the referenda of a word or theory, regardless of the existence of any such referenda.

reintroduction Releasing individuals of a species into an area where that species no longer occurs in an effort to reestablish a wild population. Reintroduced individuals may be captured from a healthy wild population in another area or may be derived from a captive population if there are no healthy wild populations remaining.

relative abundance distribution The frequency distribution depicting the number of species in a community as a function of the number of individuals comprising each species.

religion Human recognition of superhuman controlling power, and especially of a personal God or gods entitled to obedience and worship; the relationship between people and that which they regard as holy, often in supernatural terms. Religions include a wide range of ceremonies and ritual practices aimed at supporting the moral and ethical values of a society and maintaining natural or built sacred spaces where such ceremonies and practices can take place.

remediation Removal of toxicants from a contaminated environment using chemical, physical or biological means.

remote sensing The measurement or acquisition of information of landscape pattern by a recording device that is not in physical or intimate contact with the landscape under study.

renosterfeld Evergreen needle-leaved shrubland on richer soils of the Cape Floristic Region of South Africa and dominated by the resinous shrub *Elytropappus rhinocerotis* (Asteraceae); the name literally means rhinoceros bush.

rent seeking The attempt to gain advantage through claims on resource surpluses.

repetitive DNA Regions of DNA that include the same DNA sequence repeated up to several hundred or thousand times; regions with repeated segments that involve only 2–5 base pairs of DNA are called microsatellites.

replacement Divergence due to a change of an amino acid in a protein, or the process to evolve such a change.

replacement-level fertility The level of fertility at which a couple replaces itself in the population. In low mortality countries, a total fertility rate of 2.1 is considered replacement because some women will die before the end of their childbearing years.

representation The occurrence of species (or other attributes) within a set of selected areas. A distinction must be made between records of a species and areas with high probability of persistence for the species in the long term.

reproductive isolation A condition in which a derived population is not able, for some reason, to breed and obtain fertile progeny with the original population; intrinsic barriers to the production of offspring.

reptile Member of a class of vertebrates (the Reptilia), comprising turtles and tortoises (order Testudinata), lizards, snakes, and worm-lizards (order Squamata), the tuatara (order Rhynchocephalia), and crocodiles and alligators (order Crocodylia), which typically lay eggs with a leathery, impermeable shell. Reptiles have a dry, horny skin with scales, plates, or scutes.

reserve selection algorithms Mathematical techniques used to maximize efficiency in the selection of protected areas for conservation. The efficiency criteria vary with circumstances but may, for example, be the minimum number of reserves with every species represented, or minimum cost.

resilience Rate at which an ecosystem is able to recover to conditions similar to those that existed prior to the imposition of a disturbance; the rate at which an ecosystem returns to original conditions after a perturbation.

resistance 1. An individual's capacity to reduce the damage inflicted upon it by an enemy (predator, pathogen, etc.) 2. The degree to which an ecosystem moves away from or is able to maintain current structure and function when faced with a disturbance.

resolution The magnitude of a biological response due to a chemical insult.

resource Any necessity that is used or consumed for organismal growth, reproduction, or survival, including food, space, shelter, and nutrients.

resource catchment Locality from which the resources consumed by a group of people are derived.

resource conservation The protection and enhancement of resources on which sustainability depends.

resource dynamics Inputs, outputs, and internal cycling of key resources, such as carbon, water, and mineral nutrients, in an ecosystem.

resource partitioning Ecological arrangement in which two or more species use different, nonoverlapping resources in a given habitat, such as warblers foraging for insects in different locations within a tree or canopy.

resources Physical and biotic features of the environment, such as shelter sites or foods, that are required by organisms and are consumed such that use by one individual reduces their availability to others.

response A particular structure or function of the receptor that is changed by exposure to the environmental stress.

restoration The reestablishment of similar structures and functions of an ecosystem or parts of an ecosystem that are no longer present because of past land uses or disturbances.

restore To reestablish the presumed structure, productivity, and species diversity of an ecosystem originally present at a site. The ecological processes and functions of the restored system will closely match those of the original.

reticulate evolution Pattern of speciation where new species arise from interspecific hybridization between two species coupled with establishment of reproductive isolation.

revisionary taxonomy The reevaluation of entire groups of organisms based on new and old evidence.

rhabdite A rod-shaped, proteinaceous secretory product found in the epidermis of most free-living platyhelminths; known functions include mucus production for ciliary gliding, cocoon formation, prey capture, and predator repulsion.

rhizodeposition The mixture of sloughed cells, mucilages, and small-molecular-weight sugars, amino acids, and other compounds leaked from root cells, which are deposited in the soil adjacent to the surface of fine roots. Exudation takes place from the root tip back to the zone of suberization. The chemical quality and quantity of the exudate is altered by the presence of mycorrhizae.

rhizosphere Volume of soil adjacent to, and strongly influenced by, a plant root. The rhizosphere is usually considered to extend about 2 mm from the root surface, and includes the "rhizoplane," or soil directly in contact with the root surface.

ribosomal rRNA Universally distributed molecule among cellular life forms. Widely used to infer the evolutionary relationships among organisms.

rich fen An alkaline peatland that receives significant inputs from groundwater or surface water; typically having surface water pHs >6.0, higher alkalinity, and distinctly higher concentrations of base cations than bogs and poor fens; dominated by brown mosses (true mosses, largely of the Amblystegiaceae) and small sedges.

richness Simplest measure of diversity, represented by a count of the total number of species present in some ecologically meaningful unit (such as a community, metacommunity, region) without regard to variation in the number of individuals present per species.

ricochetal Bouncing movement that involves the release of stored energy in stretched tendons, movement that involves quick changes of direction, as in *to ricochet*.

rights In this context, justified claims that others respect and protect one's important interests.

riparian Relating to or located on the band or edge of a river; describing the interface between streams and terrestrial uplands. These zones are three-dimensional areas of direct interaction between the terrestrial and aquatic ecosystems. Riparian zones often contain unique species and high levels of biological diversity.

riparian zone The vegetated areas on either side of a stream or river, with saturated soils usually underlying aerated soils.

river continuum concept The first of a series of conceptual, unifying models to explain and predict the structure and function of river ecosystems.

rivers in Amazonia The three main types of rivers in the Amazon region are known as white, black, and clear water. White water rivers are muddy with much suspended sediment and are neutral or only slightly acidic. Black water rivers are dark because of dissolved tannic matter and are acidic (pH about 4). Clear water has neither mud nor humic matter and is usually slightly acidic.

RNA Ribonucleic acid, composed of nucleotides like DNA, but differing from DNA in that the base uracil (U) in RNA replaces thymine (T) and single-stranded

RNA molecules form important structural and regulatory parts of cells.

root parasite Parasites that form haustoria on roots of the host plant.

rotation The sequence of crops grown in a single field.

Rubisco Ribulose-1,5-bisphosphate carboxylase/oxygenase, the enzyme in all plants that catalyzes the formation of two phosphoglycerate molecules from RuBP and CO_2, and catalyzes the competing reaction of RuBP oxygenation.

rules Commonly understood, normative statements that specify who must, must not, or may take some action or affect some outcome at a particular node in a decision tree.

rumen The first, and usually largest, chamber of the four-chambered stomach found in ruminant ungulates; the place where cellulose digestion occurs.

ruminant Grazing mammal with a complex four-chambered stomach (including a rumen). Common ruminants include cattle, sheep, goats, deer, antelopes, and giraffes.

Russell cycle Named after F. Russell, the Russell cycle describes a major change in the biota of the English Channel from 1925 to 1972.

S

SABONET The South African Botanical Diversity Network of Angola, Botswana, Lesotho, Malawi, Mozambique, Namibia, South Africa, Swaziland, Zambia, and Zimbabwe.

salivary pump The sclerotized pump used by hemipterans to inject saliva into the organisms upon which they feed.

salt wedge Water from the ocean, with higher salt content, that moves into an estuary along the bottom of the water column, beneath less dense fresh water that has moved into the same area from a river. The salt content makes the ocean water heavier than the fresh (riverine) water, so that under calm conditions, the estuarine water becomes density-stratified. This "salt wedge" of bottom water can become somewhat isolated from the overlying fresh or less brackish water. The longer the period in which the total water column is not mixed by winds or storms, the more distinct the two water layers or strata become. Salt wedges most often develop in warm seasons when plantlike phytoplankton production is high in the surface waters and respiration by heterotrophs (bacteria, fungi, animals) is also high, especially in the lower water column and sediments. Nutrients typically are higher within the salt wedge than in the overlying water, because of decomposition processes. At the same time, the bottom-water salt wedge can become hypoxic or anoxic, underlying waters can be saturated or supersaturated with oxygen from phytoplankton photosynthesis.

saltation Sand particles jumping across the surface due to wind movement.

sampling effect The phenomenon in which increases in the number of species increase the probability of including in the community a species with a strong ecosystem effect. This phenomenon yields an increase in ecosystem processes with increases in diversity without invoking niche complementarity.

sampling scope The area, interval of time, and taxonomic grouping over which sampling takes place.

sapromyiophily Pollination by carrion and dung flies.

saturated fatty acid A fatty acid with the maximum complement of hydrogen atoms and no double bonds.

saturation Upper limit to species richness within a community, set by species interactions, and independent of the pool of colonists to which that community is accessible.

savanna Ecosystem with a more or less continuous herbaceous layer dominated by graminoids and broad-leaved herbs with an overstory of trees or shrubs that covers less than 10% of the area. In contrast, grasslands are typified by a pure graminoid or herbaceous layer with no (or very few) tree or shrub elements that rise above the grass layer.

Scala naturae (Great Chain of Being) One of the most pervasive ideas in western thought derived from two concepts Plato's principle of plenitude and the linear series of Aristotle and Plotinus (Panchen, 1992); in plain English, the belief in a linear progression from the simplest forms of life to the most perfect.

scale Unit measure (resolution, inner scale) relative to largest multiple (range, outer scale).

scale dependence A condition in which either the form or the parameters of a relationship between two variables (e.g., richness and latitude) is contingent on spatial or temporal attributes.

scansorial Using both the forest floor and canopy for movement and seeking resources.

scent glands Glands which in heteropterans secrete various compounds for various purposes, including defense, attraction of mates, and attraction of both sexes. Adult heteropterans have scent glands on their metathoraces, and nymphal heteropterans as well as some adults have scent glands on the dorsa of their abdomens.

Scientific Committee on Problems of the Environment An international organization that is part of the International Council for Science and is charged with synthesizing current scientific understanding associated with environmental issues.

Scleractinia The taxonomic order of cnidarians that includes the reef-building corals.

sclerophyllous Descriptive of leaves with a leathery texture due to the presence of sclerenchyma with large amounts of lignin and cellulose in their tissues.

sclerophyll A plant with tough, leathery, evergreen leaves, usually associated with drought resistance.

scolex Attachment organ in adult cestodes; often armed with suckers and/or hooks.

scope Ratio of largest multiple to unit measure.

scutellum The dorsal portion of the mesothorax, in most heteropterans triangular of shield-shaped (hence "little shield").

seagrasses Marine angiosperms, all monocots, able to grow submersed in marine waters, to which they are restricted.

seasonal leaf dimorphism A phenological trait in which plant species change the morphology of leaves through the year to better adapt them to prevailing temperature and water stress.

Secchi-depth Measurement of the transparency of water by determining the depth at which a white Secchi-disc (20-cm diameter) lowered from the surface disappears out of the sight of the observer.

secondary compounds A synonym for chemical defenses in plants; contrasts with chemical compounds used in primary metabolism, such as photosynthesis and cellular respiration.

secondary extinction Extinction of a species resulting from the extinction of another species on which it relies.

secondary forest An area of previously forested land that was subsequently degraded or deforested through human or natural action but which now supports regenerating or mature natural forest.

secondary hemiepiphyte Plant that begins its life cycle as a terrestrial seedling, ascends a tree, and can later lose root connections with the ground, including (a) lianas, woody climbing plants with relatively thick stems that generally grow in mature habitats, and (b) vines, herbaceous climbing plants that regularly grow in disturbed habitats or forest edges.

secondary metabolite An unusual compound that does not play a role in primary metabolism (e.g., production and storage of energy). These metabolites were initially argued to be waste products but have often been shown to defend the producer from consumers, pathogens, or competitors.

secondary productivity The rate of assimilation and growth by animals (biomass per unit area of habitat per unit time or, in some cases, biomass per unit volume per unit time).

secondary reproductives Reproductives that differentiated in an established colony, whatever their origin and morphology. They may be supplementary reproductives if older reproductives are still present or replacement reproductives if not.

secondary succession Succession on existing substrate (soil) following a disturbance.

sedimentation Particulate material falling out of the water column onto the seafloor.

seed A term used to describe eggs, larvae, postlarvae,

or juveniles (fry and fingerlings) stocked into aquaculture production systems.

seed bank An accumulation of dormant plant seeds that persist in dormancy for longer than a single growing season.

seed lot A sample of seeds with a common harvest and post-harvest history.

segment In annelids a section of body set off from the rest of the body by septae at both ends containing a separate part of the secondary body cavity (coelom), with paired ventral ganglia, and usually nephridia, gonads, transverse muscles, and segmental blood vessels; carrying externally parapodia and gills.

segregation Allocation of genetic variants ("alleles") to different gametes during sexual reproduction.

selection Differential reproduction within a set of genes, individuals, groups, etc. caused by differential performance in fitness determining tasks. The appropriate unit of selection (gene, individual, or group) is determined by the level at which the causation of the fitness differences must be described (interactions or performances at the level of genes, individuals, groups, etc.).

selective Describing the differential survival or reproductive success of individuals, associated with differences in phenotype or genotype.

selectivity Phenomenon in which some plants are killed with doses of herbicides that have little or no effect on other plants.

self-organization Spontaneous formation of complex structural aggregates from biological macromolecules such as proteins, nucleic acids, or lipids. The assembly pathways are usually determined by physicochemical properties of the constituent molecules.

semaphoront Individual organisms at a particular point in their developmental (ontogenetic) history.

semelparity The reproductive pattern of breeding once and dying; sometimes called "big bang" reproduction.

semelparous Describing species that breed once and die.

seminatural Describing vegetation that owes its character to human activity, but in which the plants are wild, not sown or planted. Examples: coppice woods; most heathland; some savannas.

semisteppe batha Semishrub communities developing on most soil types at the boundary of the Mediterranean zone of the Near East.

sensor A device that gathers energy, converts it to a digital value, and presents it in a form suitable for obtaining information about the environment.

separatrix Locus of points on the state space of a dynamical system separating different basins of attraction.

sequestration Term used to describe the uptake of carbon from a dilute source (the atmosphere) to a concentrated form (biomass).

sere The series of stages in an ecological succession sequence.

serotiny Seeds stored on the plant with dispersal triggered by fire.

setae (pl.) Bristle- or hairlike extensions of the cuticle that are characteristic of arthropods and related phyla.

settlement Land used for human habitation, including for the construction of cities, towns, villages, rural settlements, and roads. In a gross sense, settlement is also a land cover when used to refer to the land occupied by the variety of trappings accompanying human living space, including homes, gardens, roads, transportation centers, and industrial areas.

sexual selection Selection that acts directly on mating success through direct competition between members of one sex for mates or through choices made between the two sexes or through a combination of both modes.

shifting agriculture The practice of clearing a plot of land for cultivation for a short period of time, then abandoning it and allowing it to revert to its natural vegetation when the cultivation moves to another plot.

shifting cultivation Any temporally or spatially cyclical agricultural system that involves the clearing of land followed by cultivation and fallow periods.

shoot growth Biomass production (per hectare) of foliage shoots produced annually in the overstory of a plant community.

sibling species A pair of closely related, morphologically similar species (usually sister species).

siderophore Chemicals secreted by roots (primarily non-protein-forming amino acids), which complex with insoluble metal ions bringing them into solution and permitting their transport to and uptake into the root.

silviculture An applied science and branch of forestry concerned with the theory and practice of controlling forest establishment, composition, and growth.

similarity A measure of coincidence (matches) among a (large) set of characters compared for any two taxa.

Simpson diversity index An index of diversity based on species abundance data. If p_i is the frequency of species i in a community of n different species, then

$$\text{Simpson diversity} = \left(1 - \sum_{i=1}^{n} p_i^2\right)$$

and is equal to the probability that two randomly chosen individuals from a given community are of the same species. This index of diversity is statistically unbiased, making it especially useful in practice.

sister group The evolutionary lineage most closely related to the one being discussed; one of two terminal taxa or monophyletic groups that share a common ancestor that has not given rise to any other taxon under consideration.

sister species A pair of species that have arisen from a single speciation event; each is the other's closest relative.

sister taxa 1. From a species perspective, the sister taxon of a particular species is the one with which it had the most recent common ancestor. 2. From the perspective of a node in a bifurcating phylogenetic tree, each daughter lineage gives rise to a set of contemporary species. The two sets are sister taxa with reference to the node.

small subunit ribosomal RNA RNA (about 1500 bases in prokaryotes) that functions as part of the ribosome and the sequence permits the inference of evolutionary relationships among organisms.

smolt A downstream migrating salmon that has began the physiological transition to seawater.

social costs The total costs to society of an action. These may exceed (or in some cases be less than) the private costs, which are the costs of that action to the individual or institution executing it.

social differentiation A term taken from biology to describe the specialization of functions in a society and to characterize societies over time. Theories of social change propose that increased social differentiation emerges as societies increase in size and complexity. Differentiation is frequently accompanied by a need for increased coordination and increased interdependence in larger and more complex societies.

social taboo A prohibition imposed by social custom or as a protective measure.

soil aggregate A crumb-sized unit of soil, composed of aggregated soil minerals, microbes, and soil microfauna, which are cemented together by a combination of biological materials such as polysaccharide secretions, fungal hyphae, and chemical substances such as precipitated carbonates or silicates. Aggregates are classified by size and stability in water (disintegrating versus retaining their structure and integrity).

soil food web Representation of all feeding interactions among organisms in the soil (who eats whom).

soil organic matter Material in soil containing organic carbon derived from the decomposition of mainly plant residue.

soil organic matter Organic substances, including a wide variety of carbohydrates, proteins, lipids, waxes, phenolic, and humic compounds, which accumulate in soil as a result of both plant and microbial growth. These compounds include small-molecular weight materials, which are rapidly decomposed to carbon dioxide; larger compounds, which may be slowly decomposed over years to decades; and large, complex, aromatic substances, which may be stable within the soil for millennia. Soil organic matter affects all aspects of the soil's biology, chemistry, and physics.

soil structure Spatial arrangement of soil particles.

soil texture The relative abundance of sand (50 μm $< \phi <$ 2 mm), silt (2 μm $< \phi <$ 50 μm), and clay

($\phi < 2$ μm) particles in the soil (USDA criteria). Particle size distribution determines the distribution of pore sizes, which in turn strongly affects the behavior of water in the soil.

solar-heated Describing water raised to temperatures significantly exceeding the regional temperatures of lakes and streams. Many of the thermally tolerant and thermal-opportunist groups are thought to have evolved in shallow solar-heated ponds and water margins.

soldier In this context, an individual with a strongly sclerotized head showing defensive adaptations, such as enlarged mandibles, a stopperlike shape, or a frontal gland able to produce a defensive secretion.

Southern Ocean The circumpolar ocean in the Southern Hemisphere between the Subtropical Front and the continent of Antarctica.

spallation Thermally induced neutron and proton ejection following high-energy proton collision.

spatially structured population A population whose reproductive and survival rates vary over the region that it inhabits, and whose members stay long enough in a locality to experience the local reproductive and survival rates.

spatial and/or temporal patterns of genetic variation The systematic changes in the alleles occurring at specific loci along spatial and/or temporal dimensions.

spatial heterogeneity In community sampling, variation in species capture probabilities across space; may be due to habitat variation, intraspecific clumping, interspecific association, or other factors.

spatial subsidies Input from other habitats of organic carbon, nutrients, and prey or the movement of consumers. These resources can influence greatly the energy, carbon, and nutrient budget of recipient habitats. In general, nutrient inputs (nitrogen, phosphorus, and trace elements) increase primary productivity; detrital and prey inputs produce numerical responses in their consumers.

spawner Mature individual of a stock responsible for reproduction.

specialist brood parasites Species that parasitize only one or a few host species.

specialization Evolutionary adaptation in a particular mode of life or habitat.

speciation The evolutionary process of the origin of a new species.

species A group composed of individuals that interbreed or potentially interbreed and are separated reproductively from other such groups.

species abundance distribution The number of species found in each interval of abundance in a community.

species accumulation curve A plot of the total number of species observed in a census against some measure of cumulative sampling effort.

species–area curve A graphical depiction of the dependence of species richness on area.

species–area model A function used to describe species–area curves.

species–area relationship The dependence of the number of species in a sample region on the area or size of the region.

species composition Types of species that constitute a given community or sample.

species density The number of species within a sampling unit of fixed size.

species diversity A feature of biological communities or assemblages that reflects the variety of organisms in an area and that includes two components, species richness and species evenness (the degree to which all species have the same proportional abundance). Global species diversity refers to all species. Local species diversity refers to some geographic region such as Hawaii or New York City.

species evenness A measure of the relative abundance of species in an area.

species flock An aggregate of closely related species that share a common ancestor and are endemic to a geographically circumscribed area.

species life span The time between the first record of a species in the fossil record to its disappearance. This time span is typically in the range of millions of years.

species packing The study of how species on the same trophic level coexist in a limited region or container.

species pool Set of species that are able to colonize a habitat of interest.

species richness The number of species occupying a particular area (such as an island) or biological entity (such as a branch of a tree) without regard to any other properties of the species. Species richness may also be expressed as the list of species that generates that number.

species turnover The change in the composition of species in an area due to the extinction of some species and the replacement by a new species by colonization.

specific mate recognition systems (SMRS) Fertilization and mate recognition systems in the recognition concept of species, the factors leading to premating compatibility within a species. See also cohesion, which is similar to SMRS, but includes postmating compatibility as well.

specific resistance Narrowly targeted resistance to a specific genotype of disease or insect pest. It is usually imparted by a single gene of large effect, and typically with a relatively short effective lifetime.

specificity The ability to relate a chemical effect to a distinct biological function or organism.

speciose Of high diversity, having many species in a group or area.

spermatophore A chitinous container produced by the male to hold sperm. It may be attached to the substrate for the female to find or passed to the female from the male during mating.

spicule Blade-like, sclerotized male copulatory organ, usually paired, located immediately dorsad to the cloaca.

spinnerets Usually three, rarely four pairs of modified terminal abdominal appendages in spiders bearing one to hundreds of hollow spigots from which silk is drawn.

spirochetes Helically shaped gram-negative bacteria with flagella in the periplasmic space between the two cell membranes.

spore One- to several-celled propagule of totipotent cytoplasm with cell walls, produced by cell division with concomitant meiosis or mitosis, that may serve for dispersal or overseasoning, but that does not contain an embryo.

sporophyte In an organism with an alternation of generations, the diploid, spore-producing phase.

spreading rate The rate at which tectonic plates move apart from each other.

squamata Lizards and snakes.

stabilizing selection Selection on a continuously varying trait toward an intermediate optimum; selection that favors intermediate phenotypes.

stable population A population with an unchanging rate of growth and an unchanging age composition as a result of age-specific birth and death rates that have remained constant over a sufficient period of time. Zero population growth is a special case of population stabilization in which no growth occurs because the number of births and deaths are the same.

stakeholder A person who has an interest or concern with a particular issue or business.

stasis (stasigenesis) The persistence of organisms through long geological periods of time that greatly resemble their fossil forebears. The "process" is used to recognize delimitable anagenetic units, or "grades."

statocyst Static sense organ in which the movement of a statolith is detected, thereby indicating an animal's orientation.

statutes or legislation Laws passed by the legislative branch of U.S. federal or state governments. The U.S. Congress passes federal laws and state legislatures pass state laws. Local laws are usually called municipal ordinances.

stem Branch on cladogram between node and term.

stem parasite Parasitic plants that form haustoria on stems of the host.

stenohaline Able to live only in fully marine waters.

stenotopic Referring to taxa with restricted habitat requirements (i.e., confined to a single biotope) and hence restricted distributions.

stenotopy The status of a species (or higher taxon) with very restricted geographic range.

stewardship 1. Management of a thing for someone or something else. 2. In the context of biodiversity, responsibility to protect biological diversity, as given by God or accepted from society.

stichosome A longitudinal series of cells (stichocytes) that form the posterior esophageal glands in *Trichuris*.

stochastic Relating to or being the result of chance and random processes.

stochasticity A stochastic process is one in which the state of the system cannot be precisely predicted given its current state and even with a full knowledge of all the factors affecting that process. In a population context, we may know the detailed life history parameters of a species. However, various unpredictable (stochastic) processes, such as the chance nature of birth and death (demographic stochasticity), year-to-year variation in climate (environmental stochasticity), and catastrophes, means that accurately predicting the precise size of a population in the future is not possible. Despite this, we can use probability theory and simulation models to make probabilistic predictions. A stochastic population model is one in which each possible future population size has an associated probability. This approach to prediction is the same as stating that the chance of getting a head with the next toss of a fair coin is 50%. Our prediction is accurate but we cannot say if the outcome will be a head or a tail.

stochastic event An event that occurs by chance; a random event.

stochastic processes Processes that affect the dynamics of a population through random chance alone.

stochastic system A dynamical system whose detailed future behavior cannot be predicted, even in principle, due to random forces inherent in the system.

stock recruitment relationship The relationship between the adult breeding stock and the number of recruits produced by that stock.

stoma The mouth or buccal cavity, from the oral opening; usually includes the anterior end of the esophagus (pharynx).

storage effect A general mechanism for the maintenance of biodiversity within a single habitat based on differences between competing species in their responses to environmental conditions. A resistant life-history stage can allow coexistence by the storage effect if each species reproduces successfully under the conditions favorable for that species and can survive through unfavorable periods (e.g., when a competing species dominates) in the resistant stage. Often, the resistant stage has prolonged dormancy. This mechanism can also serve to promote the maintenance of genetic diversity within a single population.

storage organ Any structure that sequesters storage materials (usually the carbohydrates, starch or glycogen, or triacylglycerol) for export to other tissues that utilize it.

strain A population of cells all descended from a single cell.

stratification The formation of distinct layers with different densities; stratification inhibits mixing.

stratigraphic interval A stratum with identifiable upper and lower boundaries.

stratigraphic range The strata through which a taxon is known to have been extant.

stratigraphic section An outcrop of rock (with fossils in this case) or a drilled core. The term can also refer to a composite for a region in which fossil ranges and stratigraphic events have been summarized into a synthesized rock column.

stratum (pl. strata) An interval of sedimentary rock that is distinguishable from previous and subsequent intervals.

stream order A numbering system to denote stream size within a network of streams. There are several stream order systems. Stream order is also dependent on the scale of observation.

strict metapopulation (Also called a classical metapopulation.) A metapopulation satisfying the following conditions: (i) Local populations are partially isolated from one another and are frequently capable of sustaining themselves for several to many generations in the absence of immigration from other local populations, (ii) local population extinction occurs on a timescale of several to many generations, and

(iii) migration between local populations leads to reestablishment of local populations following local extinction.

stridulatory apparatus An organ of sound production based on the mechanism of rubbing one part of the body against another.

strobila The portion of the body of cestodes exhibiting serial repetition of proglottids.

strong interactors (foundation species) Species that have a large effect on the species (one or a few) with which they interact. Communities and ecosystems may have many strong interactors, and such species may occur at all trophic levels. "Strong interactors" is a more general term that can include keystone species, but not all strong interactors are keystone species. A similar idea is "foundation species," defined as the group of critical species whose effects and interactions define much of the structure of a community.

structural intricacy Complexity of patterns of links or connections among species in a community that are created by biotic interactions, shared pathways of nutrient and energy flow, or phylogenetic relationships.

structure The number, kinds, and arrangement of component parts at one point in time.

studbooks Also called genealogies, registries, or pedigrees, a database of historic and living animals intended to go beyond passively recording bloodlines and records; typically kept in computer databases that are also useful for actively managing the genetics, demography, and husbandry of captive populations.

stygobite Aquatic species that are obligate subterranean dwellers.

stygophile Aquatic species that can live and reproduce in both subterranean and surface environments.

subarctic The ecotone (ecological boundary zone) connecting the treeless tundras in the north with the taiga or coniferous boreal forests in the south. The area is characterized by the presence of scattered, deciduous or coniferous trees of low stature and is sometimes termed "forest tundra." (Other Arctic vegetation zones are described in the text.)

submerged vegetation Plants that grow and reproduce while completely submerged.

sub-nivean Pertaining to the environment beneath a temporary, seasonal, or permanent covering of snow.

subordinate species Species that have a minor influence on ecosystem structure and function, presumably because of their lesser abundance and biomass compared to dominant species.

subpolar Pertaining to the regions between the polar and temperate zones, but for the oceans the boundaries are the Subtropical Front and the Polar Front.

substitution Divergence due to a change of base in a DNA sequence, or the process to evolve such a change.

substitution possibility The possibility that the depletion of a given resource can be compensated for by one or more of various forms of substitution; e.g., by substituting a more plentiful natural resource for a scarce one, or by substituting a manufactured product for a natural one.

substrate Material or host from which a fungus derives its nutrition.

subsurface oxygen depletion Due to isolation of waters below the surface, metabolic breakdown of organic matter from the surface waters can cause depletion of dissolved oxygen concentration, leading to very low oxygen (hypoxia) or no oxygen (anoxia).

subtropical Pertaining to the regions which, under the influence of the trade winds, are permanently stratified.

succession 1. The predictable change in species that occupy an area over time caused by a change in biotic or abiotic factors benefiting some species but at the expense of others. 2. The pattern of change expected in a community over time after a disturbance or after new substrate has been exposed.

sulfur bacteria Bacteria that depend on the phototrophic or chemotroph oxidation of reduced sulfur compounds.

supercontinent The amalgamation of many continental masses into a single mass through continental drift. The supercontinent of Pangea (from 300 to approximately 190 million years ago) included most continental areas other than those that comprise east Asia.

supralittoral Immediately above the intertidal zone.

suspension feeders Organisms that feed primarily on organic material suspended in the water above the bottom.

sustainability Ability of a society or a particular human activity to continue indefinitely without depleting resources or damaging the environment. State that defines the biogeophysical, economic, social, cultural, and political thresholds in between which it is acceptable to continue to use and obtain services or products from a given piece of land. This definition requires all five factors to be considered when determining whether future generations will be able to acquire the same natural resources as the current generation.

sustainability gap The global ecological deficit—that is, the difference between any excessive human load on the ecosphere and the long-term carrying (or load-bearing) capacity of the planet.

sustainability transition A search for sustainable development, through action and research, pursued during the remaining decades of the demographic transition. A sustainability transition would be shaped by normative goals for human well-being and preservation of the life-support systems needed by human populations.

sustainable In general, capable of meeting economic goals in a manner that does not degrade the quality of the underlying ecosystem.

sustainable agriculture Form of agriculture that is environmentally sound, culturally sensitive, socially acceptable, and economically viable.

sustainable development Meeting human needs without damaging the ecosystems that produces these resources and at the same time providing equitable distribution and access to these resources around the world. A pattern and path of economic and social development compatible with the long-term stability of environmental systems, particularly those essential to human well-being.

sustainable forestry The management and utilization of forest resources in a way that balances human needs with undisturbed functioning of the forest ecosystem, and a long-term maintenance and sustainability of the resources and biodiversity of forest ecosystems.

sustainable use Exploitation of wildlife in a manner that avoids depletion of the resource (e.g., limited hunting, ecotourism).

sustained yield The management and harvesting of renewable resources in a manner that does not exceed their rate of replacement or reproduction.

sustained yield The removal of a constant number of animals from a wild population that can be maintained through time while keeping the population at a predetermined size.

swamp Wetlands dominated by woody species. Although the term has been variously used, it is used here in the restricted sense of forested wetland, either on mineral soils or peat. However, while many peatland ecologists classify forested fens on deep peat as swamps, they would not call a bog forest a swamp.

swiddening Umbrella term including diverse types of horticulture in which a small section of forest is cut and burned to plant crops in a temporary garden; shifting cultivation is used as a synonym, but slash-and-burn cultivation is now considered to be a pejorative term.

symbiogenesis Production of new organelles, tissues, organs, or species by the symbiotic integration of two different organisms.

symbiont One of a pair of coexisting, mutually supportive organisms; in vent invertebrate–bacteria symbioses, the "symbiont" colloquially refers to the bacterium associated with the host invertebrate tissue; endosymbiont—within the host invertebrate (intra- or extracellularly); episymbiont—on the outer surface of the host invertebrate.

symbiosis A mutualism among members of different species in which members of one species (the symbiont) live inside or on a body of the other (the host).

symbiotic Describing species interaction in which two organisms live in close proximity. Symbiosis can be antagonistic or mutualistic. Often, the larger individual is called the *host*, and its inhabitant is called the *guest*.

symmetry The correspondence in size, form, and arrangement of parts on opposite sides of a two- or three-dimensional object.

sympatric Describing populations, species, or taxa that occur together in the same geographical area within the dispersal range of one another.

sympatric speciation The process of genetic divergence between populations occupying the same geographic range leading to distinct species; the separation of a single population living in one place into two species.

sympatry Species or populations co-occurring in same area.

symplesiomorphy Shared, primitive similarity.

synapomorphic Shared derived characters; homologies of monophyletic groups.

synapomorphy A character state that is derived relative to the ancestral state (plesiomorphy). Synapomorphies are special cases of homology that are evidence of monophyly among the taxa that bear them relative to taxa that bear the plesiomorphic state(s).

synaptonemal complex A three-part proteinaceous structure that affects the number and distribution of crossovers and converts crossovers into functional chiasmata.

synergisms Interactions among many factors, which collectively may have a much larger or smaller effect than the sum of the effects of each factor acting independently.

synergist A chemical used, at sublethal concentrations, to inhibit particular groups of detoxifying enzymes and therefore to implicate the involvement of these enzymes in resistance.

synlophe In numerous Trichostrongylidae, an enlarged longitudinal or oblique cuticular ridge on the body surface that serves to hold the nematodes in place on the gut wall.

synonym One of two or more scientific names applied to the same organism.

synonymous mutation A nucleotide change in a coding region that does not change the encoded amino acids.

syntaxon (pl. syntaxa) A category of vegetation typology based on the floristic–sociological approach (known also as the Braun-Blanquet approach); the basic syntaxon rank is "association," which further groups into "alliances," which group into "orders," and orders group into "classes."

system functions Processes that occur in an ecosystem that are typically measured as changes in the carbon or nutrient cycles regulated by decomposers, consumers, or primary producers.

T

taiga Mostly used as a synonym for boreal forest, but more precisely it is a Russian word applied to Eurasian conifer forests described as damp and almost impenetrable. It is also defined as a coniferous forest with no admixture of nonconiferous tree species except *Betula* and *Populus*.

target taxon Species or species group that the conservation action is focused upon.

taungya The Burmese term for an agricultural system in which crops are interplanted with plantations of trees. As the trees grow and shade the areas, cultivation of crops is abandoned.

taxon (pl. taxa) A general term used to indicate groups of related organisms at any level in a taxonomic hierarchy. In the classification of evolutionary lineages, the term can refer to species, genera, families, orders, etc.

taxon cycle The repetitive pattern by which widespread dispersive stage I populations or species give rise to more restricted and specialized stage II populations or species; subsequent divergence leads to stage III local endemics.

taxonomic Relating to or being a group of individuals representing a classification of organisms, such as genus, family, or order. Higher taxa are those above the species level.

taxonomic diversity The number and the relative abundance of species in a community.

taxonomic rank The level in the Linnean hierarchy to which a particular taxon or group of species is assigned. For example, mammals are at the level of the class (Mammalia) and hippopotamuses are at the level of family (Hippopotamidae).

taxonomic richness Number of species, genera, and families from a given time period or fossil excavation.

taxon sorting The argument that in certain taxa, species undergo relatively rapid speciation and extinction, while in others the species speciate and die out relatively slowly; over time this difference will mean that the highly volatile families become extinct more quickly than the less volatile ones.

taxonomy Classification of species and higher taxa, ultimately based on phylogeny.

techno-ecosystem Technology-based ecosystem in the contemporary world that is fundamentally distinct from natural ecosystems in the use of energy sources other than sunlight (fossil fuels, nuclear power), an urbanized concentration of human population, and the generation of substantial amounts of air and water pollutants and waste materials.

tectonism Forces that disturb and dislocate the Earth's crust (earthquakes).

tegmina (sing. tegmen) Thickened forewings.

teleology The imputation of goal-directed behavior or structures.

telson Most posterior segment of the body in crustaceans, usually flattened and often armored.

temperate Pertaining to the climates and landscapes of the midlatitude regions, which are seasonally warmer and cooler, with winter frost even in most coastal areas and at least partial dormancy or collapse of the vegetation.

temperate region Any locality with at least 1 month of frost (for continental areas) or with one or more months with a mean temperature lower than 18°C (for maritime-influenced areas), and with at least 4 months with a mean temperature higher than 10°C.

temporal dispersal The emergence of individuals from dormancy over a range of years (or other time interval), when those individuals entered dormancy in a single year. Often, the years (or other time intervals) have different environmental qualities for growth and reproduction.

temporal heterogeneity In community sampling, variation in species capture probabilities over time; may arise from temporal environmental variation, migration, speciation, or other factors.

temporal migration The avoidance of harsh environmental conditions in an environment by an individual organism that enters dormancy before conditions become harsh and emerges from dormancy when favorable conditions return.

temporary parasite A parasite found in or on a host only for short periods.

tepui Venezuelan term for the sandstone table mountains of the Guayana Highland or Lost World region.

term In this context, a terminal on a cladogram, either a taxon (in systematics) or a geographic area (in biogeography).

terra firme Brazilian term for areas that are above the level of periodic inundation by the rivers; nonflooded landscape in tropical lowlands, usually characterized by highly weathered, nutrient-poor soils.

terra firme forest Continuous hardwood forest of the nonflooded or upland parts of the Amazon rain forest.

terrestrial A species that lives on land (as opposed to marine or fresh water) for the majority of its life cycle.

terrestrial animals Animals that live on land for their entire lives (e.g., spiders) as opposed to animals that live in water for their entire lives (e.g., lobsters). There are also many amphibious animals that spend part of their lives in both places (e.g., dragonflies).

terrestrial invertebrates Animals that are not vertebrates (such as fish, reptiles, amphibians, birds, and mammals) that live on land for their entire lives. Examples are all insects except for those with aquatic larvae; nonaquatic mites and nematodes; and all spiders, millipedes, centipedes, and scorpions.

terricolous Soil inhabitants.

Tertiary The earlier part of the Cenozoic Era, occurring from about 65 to 2 million years ago.

tetrapods The terrestrial vertebrate classes amphibians, reptiles, birds, and mammals, so named because they primitively possess four legs.

tetrapod vertebrates Vertebrate animals with four limbs (or vertebrates that have evolved from such animals, such as snakes).

thelytoky Reproductive mode in which unfertilized eggs develop into diploid females.

therapsids An order of reptiles that existed during the late Paleozoic and early Mesozoic and from which mammals are believed to have evolved.

thermally tolerant Describing organisms able to live at temperatures significantly higher than the regional norm. They can also exist at cooler temperatures, but may not compete well in the latter situation.

thermal opportunist Organisms that have developed life history characteristics that permit the exploitation of temporary cooler spots in thermal systems.

thermal systems Outflows of geothermally heated water, usually thought of in terms of those at the surface of the earth (see *thermal vent*) where light is present.

thermal vent Geothermally heated water issuing from cracks in the ocean floor. Very high temperatures and pressures and the complete absence of light are characteristic.

thermocline In the stratification of warm surface water over cold, deeper water, the transition zone of rapid temperature decline between the two layers.

thermophiles From the Greek *therme* (heat) and *philos* (loving). Organisms that grow optimally at temperatures between 50 and 80°C.

thermophilic Literally, heat-loving; denoting organisms that grow best at temperatures between 50° and 80°C.

thermophily 1. Those organisms adapted to living at high temperatures and unable to survive at lower temperatures. 2. More generally, any type of adaptation to thermal systems.

thoracopod Appendage on the thorax (paired in crustaceans).

threatened Vulnerable, endangered, and critically endangered species.

threatened species Those species that are likely to become endangered in the foreseeable future throughout all or a significant portion of their range.

three-taxon statement Two species (or taxa) are more closely related to one another than either are to a third; the most basic cladistic hypothesis.

threshold The point at which a response begins to be produced.

threshold theorem of epidemiology If the density of susceptible individuals in a population falls below some critical threshold, then a disease may not be able to maintain itself in the population.

time horizon Time period over which future benefits and costs are considered.

toadstool vernacular name for a poisonous mushroom.

tokogeny Birth relationships responsible for reticulate patterns of relationships within and among populations.

tolerance An individual's capacity to sustain damage by an enemy with limited decrease in fitness.

top-down effects When feeding by higher trophic-level organisms has cascading effects on lower trophic levels. A simple example is when fishes (predators) consume zooplankton (herbivores) and, by lowering the numbers of these herbivores, allow increased densities of photosynthetic phytoplankton (primary producers).

top-down forces Population-regulating processes that originate from consumer limitation.

torper A condition, usually in birds and mammals, when the body temperature falls a few degrees below ambient for a short period of time. Differs from hibernation in (a) the amount the body temperatures fall and (b) the length of the resting state.

torsion The 180° counterclockwise rotation of the visceral mass and mantle cavity during the larval stage of gastropods.

total allowable catch The level of permitted catch established by fishery regulations.

total fertility rate The average number of children that would be born to a woman during her lifetime if she were to pass through her childbearing years conforming to the age-specific fertility rates of a given year.

tourist Generally, any person who travels 40 km or more from home for any reason and who stays away for one or more nights.

tracheae A system of hollow, branched or unbranched air-conducting tubes used for respiration, opening via abdominal spiracles. They may or may not extend into the cephalothorax or legs.

Tracheophytes Vascular plants.

trade-off Negative correlation between traits, such that a benefit due to changes in the value of one trait is associated with a cost produced by changes in the value in another trait.

traditional agriculture Indigenous form of ecologically based agriculture resulting from the coevolution of local cultural and environmental systems.

traditional ecological knowledge A cumulative body of knowledge, practice, and belief, evolving by adaptive processes and handed down through generations by cultural transmission, about the relationship of living beings (including humans) with one another and with their environment. A subset of indigenous knowledge, which is local knowledge held by indigenous peoples or local knowledge unique to a given culture or society.

traditional or local environmental knowledge (TEK) Detailed and accurate knowledge about the environment, including biotic species and ecological processes, that many indigenous and other peoples have developed, accumulated, and apply in their daily and intimate interactions with their habitats and in their system of natural resource use and management. The study of TEK is included within ethnoecology.

traditional peoples Variously referred to as indigenous peoples, native peoples, or tribal peoples; peoples with cultures that differ from those of the mainstream of a national population, having distinct ethnic languages with locally evolved resource management systems; currently some 5000 distinct indigenous peoples exist in the world.

traditional societies Groups in which knowledge, practice, and belief are handed down through generations largely by cultural transmission. Tradition itself evolves by adaptive processes, but not all tradition is necessarily adaptive.

tragedy of the commons Process whereby a commons is overexploited because individual users cannot expect to realize the potential benefits of resource conservation; overuse of resources resulting from open access and the incomplete accounting for costs.

trait 1. A quality possessed by an individual organism that affects its ability to survive and reproduce. 2. Specifically, a recognizable characteristic of a crop variety, such as plant height, grain color, specific disease resistance, yield potential, or tolerance to heat and drought.

transactions costs Costs other than price associated with the trade of environmental goods and services: information, negotiation, decisions, and enforcement.

transduction Viral-mediated genetic recombination.

transect A linear series of samples, for example, across the intertidal zone.

transformation Uptake and integration into the host genome of free DNA from the environment.

transgenes Genes imparted to an organism by means of biotechnology rather than sexual hybridization.

transgenic crop Crop whose genome contains a gene(s) from a distinct species that has been inserted by genetic engineering.

transient species Species that are present as scattered seedlings or small immature individuals; many of these species occur as dominant or subordinate species in neighboring vegetation associated with different environmental conditions or management regimes.

transpiration Amount of water taken up by the vegetation and released through the leaves to the atmosphere.

transport host A host that harbors sexually immature stages of a parasite that do not develop.

transposable elements Fragments of DNA containing genes that provide the ability for the DNA fragment to change its location in the genome.

transposable element A DNA sequence that can move from one place in the genome to another.

treaties Multilateral agreements between or among nations. International treaties, often called "conventions," are entered into in the United States by the president but must be ratified by the Senate. Treaties supersede federal, state, and local laws that might have contradictory goals.

tree Commonly a synonym for cladogram; or a cladogram interpreted in more evolutionary detail.

treeline Also known as the forest line, this describes the high-elevation limit of (usually fragmented) forest. Most often there is no "line" visible in the landscape, so treeline position represents a convention. "Outpost" tree individuals may occur at higher elevations (the tree species line), and the boundary of closed, tall forest with timber-size stems (the timberline) commonly occurs at an elevation that is 50–150 m lower. The whole transition from timberline to the tree species line is called the "treeline-ecotone."

tree of porphyry A tree classification constructed as a dichotomous key that at any rank divides on the basis of differential characters to give two taxa at the rank below. The net result is a comblike branching diagram intended to classify individuals within a general scheme.

trench A steep-sided depression in the ocean floor ranging from 6000 to 10,000 m depth and associated with areas of tectonic plate subduction.

triacylglycerol Ester of glycerol and three fatty acids; the most common lipid storage molecule in animals and green plants.

trichobothria Long, delicate, slender setae set in broad, shallow innervated sockets in the cuticle. Trichobothria are sensitive to vibration or near-field air movement and are a major sense organ of arachnids.

tritomy ("**trichotomy**," in error) Three undifferentiated branches or clades.

trochophore Larva characteristic especially of annelids and molluscs consisting of an episphere, ciliated girdle around the middle called the prototroch, and the hyposphere. In feeding trochophores the mouth is located behind the prototroch and the anus is terminal, usually surrounded by a second ciliated band, the telotroch. Segmentation takes place between the prototroch and the telotroch and is initiated immediately in front of the telotroch.

troglobite Terrestrial species that are obligate subterranean dwellers; sometimes used for aquatic species as well.

troglomorphic Pertaining to the morphological, behavioral, and physiological characters that are convergent in subterranean species.

troglophile Terrestrial species that can live and reproduce in both subterranean and surface environments; sometimes used for aquatic species as well.

trophic Relating to feeding; describing feeding habits or the kind of nutrition used by a group of organisms.

trophic cascades A chain reaction of top-down interactions across multiple trophic levels. These occur when changes in the presence or absence (or shifts in abundance) of a top predator alter the production at several lower trophic levels; primary and secondary production at lower levels are alternately constrained or unconstrained by the feeding activities of consumers at upper levels.

trophic complexity The number and types of organisms that feed at different trophic levels within a community; also known as food web complexity.

trophic efficiency The percentage of material or energy that moves, without loss, from one trophic level to the next. Most food chains have trophic efficiencies around 10%. Through tight internal recycling, corals routinely achieve trophic efficiencies in excess of 90%.

trophic interaction Feeding relationship between two species; these include predation, herbivory, parasitism, bacterivory, frugivory, or any other interaction that involves individuals of one species consuming individuals or parts of individuals from another species; trophic interactions represent a subset of all biotic interactions in a community.

trophic level Feeding level in a food chain or pyramid (e.g., carnivores); Position of a species within a food web. Plants—autotrophs that convert solar energy to chemical energy and utilize mineral nutrients—constitute the first trophic level. Primary consumers—animals that feed on living plants—constitute the second trophic level. The third trophic level is composed of secondary consumers—animals that feed on primary consumers. Predation, parasitism, grazing, and herbivory are intertrophic level interactions; competition occurs among species in the same trophic level.

trophic status Ranking system for aquatic ecosystems, based on the amount of organic production and nutrient (N,P) levels. The major component that is usually considered in assigning trophic status is phytoplankton production, but this can be misleading. For example, some lakes are classified as oligotrophic because water-column nutrients and phytoplankton production are low, despite the fact that benthic plant production (e.g., of rooted angiosperms) is high.

trophic transfer The amount of biomass and/or energy which is transferred from the primary producers to the herbivore-based food chain rather than directly from plants to detritivores.

trophogenic region Region where net production of organic matter occurs by photoautotrophy or chemoautotrophy.

tropholytic region Region where respiration and decomposition of organic matter proceed in the absence of primary production.

tropical Astronomically, the region lying between the Tropic of Cancer (23½°N) and Tropic of Capricorn (23½°S); more generally, involving the climates and landscapes characteristic of this region, which are essentially frost free in the lowlands and permit biological activity as long as water is available.

tuberculum interglenoideum An anterior projection of the first (cervical) vertebra in salamanders. The tuberculum interglenoideum bears articular facets that insert into the foramen magnum of the skull and provide additional articulation points between the skull and the vertebral column.

tundra This is a type of vegetation characteristically occupying the Arctic. However, the term is used in many ways, from characterizing individual plant communities of the Arctic which consist of dwarf shrubs and sedges to characterizing all vegetation above the altitudinal treeline and between the latitudinal treelines and the poles in both hemispheres. In this article, the term is used in the Russian sense to characterize Arctic vegetation lying between the taiga and the region of the polar deserts.

turbidity Particulate material suspended in the water column that reduces water clarity, light penetration, and hence photosynthesis.

turnover Changes in the taxonomic and possibly physiognomic composition of a community over evolutionary timescales in response to environmental or evolutionary change.

type Of an organism, the specimen, culture, or other element on which a scientific name of an organism is based and that fixes its application; the name-bearing type.

type I community An assemblage of species where the richness is a constant proportion of the number of species occurring in a larger geographic unit in which the community is embedded.

type II community An assemblage of species where the richness is independent of the number of species in a larger geographic unit in which the community is embedded.

typification The practice of designating a single specimen (or, formerly, a series) as the name-bearer for a species; or a species for a higher taxon, especially genera.

U

ultimate cause of extinction Being rare (few in numbers) and of limited distribution are precursors to extinction. The causes leading to rarity are the ultimate causes of extinction.

ultraviolet radiation Radiation with wavelengths shorter than violet light and with more energy.

umbrella species Species that have either large habitat needs or have other requirements whose conservation results in many other species being conserved at the ecosystem or landscape level.

uncertainty Imperfect knowledge concerning the current or future state of a system under consideration; a component of risk resulting from imperfect knowledge of the degree of hazard or of its spatial and temporal pattern of expression.

uncultured prokaryotes Organisms, the presence of which has been detected by molecular methods in the environment but they have not been cultured under artificial laboratory conditions.

understory The broad spectrum of plants at the ground level of forests that often provide forage for grazing animals. These transitory plants are often relatively sparse in dense forests with closed canopies.

undulipodium Cilium or eukaryotic flagellum composed of a [9(2) + 2] microtubular axoneme.

uniramous, biramous, triramous, polyramous An extension of the body (usually an appendage) having one, two, three, or many branches (e.g., triramous antennae have three main branches or flagellae).

Univoltine (Bivoltine) Species with one (two) generations per year.

upwelling Upward vertical movement of water through the bottom of the surface mixed layer produced by a divergence at the surface; transport of water from the deep ocean to the surface.

urban Designating an area characterized by high human population densities, or significant commercial or industrial infrastructure. The boundaries between urban, suburban, and rural are not sharp and can be difficult to characterize.

urban ecology The study of urban systems from an ecological perspective; an emerging field within ecology that strives to understand human interactions in ecological systems in and around urban areas and to develop theories and analyses that include human communities as fundamental components of ecological systems.

use values Values that are obtained by using a natural resource, such as timber, fuelwood, water, and landscapes. These include direct, indirect, option, and nonuse values.

utilitarian The school of conservation philosophy that emphasizes the value of the natural world in contributing to human well-being, especially in terms of economic standard of living.

V

valid Of a scientific name, one that conforms to the conditions of valid publication proscribed in the relevant *Code* (q.v.).

values Values are interpreted here in the broadest sense to include monetary and nonmonetary values. Biodiversity attracts many different values, which are not universal and which depend on social and economic situations. Consequently, different values may conflict.

variation The abundant normal genetic differences between individuals in a population. Such variation is ultimately due to changes at the DNA base pair sequence level but is typically monitored at higher levels of expression. Overall genetic variation is associated with population viability and future evolvability. Conservation genetics is focused mainly on the protection and maintenance of genetic variation.

varzea Floodplain of the whitewater rivers of the Amazon basin; Brazilian term for vegetation periodically flooded by whitewater rivers.

vector An animal that transmits parasites among definitive hosts; for example, mosquitos are vectors of malaria.

vegetation The total plant cover of an area.

vegetation megazone Large-scale vegetation complex characteristic for a zonobiome.

vegetation survey Product of research activity aimed at the description and classification of vegetation cover on various levels of complexity in a certain geographic area using various field and data-evaluation methods.

vermiform Worm-shaped with tapering form both posteriorly and anteriorly.

vertical transmission The process of infection from parents to offspring (as opposed to horizontal transmission, in which parasites are transferred by direct contact or vectors).

vestigial Describing an organ or structure that is present only in rudimentary form (probably lost from a previous condition).

viability The status of a population is one deemed to have a reasonable chance of long-term persistence. It has been stated that a viable population is one that has a 99% chance of persisting 1000 years. Different managers, researchers, and policymakers use different measures that range from a 99% chance of persisting 1000 years to a 95% chance of persisting 100 years. The population size that just meets the definition of being viable is called a minimum viable population (MVP). One of the key questions of reserve design is how big does a reserve, or network of reserves, need to be to contain an MVP. This size is called a minimum viable habitat area for a particular species.

vibratory papillae Mobile, grooved, rod-like appendages arising from the distal edge of the first thoracic segment, used for communicating.

vicariance The existence of closely related taxa or biota in separate, disjunct areas, which have been separated by the formation of a natural barrier (vicariance event).

vicariant event A historical event that isolates species' populations and leads to speciation (e.g., the closing of the Isthmus of Panama); usually considered in a geological context, but human activities also generate vicariant events.

virulence The degree to which a parasite reduces the probability of survival or the reproductive capacity of an individual host. A relatively avirulent (benign) parasite has little impact on its host's fitness.

voltage-gated sodium channel A large transmembrane protein that regulates the flow of sodium ions across axonal membranes and mediates the rising phase of action potentials.

voucher A term for a representative specimen of a plant or animal that is properly collected, prepared, and preserved in a herbarium or museum to facilitate expert identification of the species.

vulnerable A species not critically endangered or endangered but facing a high risk of extinction in the wild in the medium-term future.

W

wadi Dry water course.

water-holding capacity Capacity of soil to hold water (e.g., sandy soils have very low water-holding capacity).

water pollution Lowering of water quality due to input of sewage, pesticides, agricultural run-off, and industrial wastes that can result in harm to aquatic plants and animals.

water table The upper surface of the zone of saturation in a soil or rock formation.

watershed (catchment) A unit of landscape in which precipitation drains to a common stream or lake; alternatively, a watershed is the divide between catchments.

weak interactors Species that have little effect on other species, at least under average conditions. Under some circumstances, weak interactors may occupy important roles in ecological communities as a result of changes that lead to temporary increases in their abundance, size, or biomass.

weed Plant that interferes with the growth of desirable plants and is unusually persistent and pernicious. Weeds negatively affect human activities and as a result are undesirable.

weed control Reducing or suppressing weeds in a defined area to an economically acceptable level without necessarily eliminating them.

weevils Members of the beetle family Curculionidae.

wet deposition The influx of contaminants from the atmosphere in precipitation (i.e., rain, snow, mist, fog, and clouds).

wilderness Large area that remains essentially unmanaged and unmodified by human beings.

wild manioc Maniocs with a high level of cyanidric acid.

wild or nondomesticated species Those species that have not been brought into regular cultivation. Many may, however, have well-known uses.

wildwood Wholly natural forest not affected by sedentary human activities such as agriculture and pasturage, widespread in prehistoric times (including earlier interglacial periods) and still surviving in remote places.

willingness-to-pay Expression by people of strength of preference in dollar terms.

woodland Forest forming part of a cultural landscape; ecosystem in which the mature overstory tree or tall shrub cover makes up 10 to 60% of the area. Typically, a well-developed herbaceous understory is present. Note: forests typically have a mature tree cover >60%.

wood-pasture Cultural ecosystems, often savanna-like, combining trees and domestic animals.

worker An individual resulting from an early, irreversible deviation from the pathway to the imago, and performing helper tasks. Workers are primarily characterized by the loss of the ability to proceed to the winged imago, but they need not be permanently sterile.

worldview The larger conceptual complex in which ethics are embedded. A. N. Whitehead called it the conceptual order, or one's general way of conceiving the universe, which supplies the concepts by which one's observations of nature are invariably interpreted. In general, world-views limit and inspire human behavior, shape observations, and perceptions. A. Toynbee's *Weltanschauung*.

WUE Water-use efficiency, a measure of the amount of photosynthesis or growth per unit of water lost in transpiration.

X

xenobiotics Toxic substances not naturally produced within organisms.

xeric A dry environment.

xeromorphic Describing a form that has developed in response to highly arid conditions.

Z

zero-growth isocline In a graph with axes describing factors important to population growth (e.g., resource abundance), a line along which a population's growth rate is zero.

zonation The tendency of climate, soil, and natural landscape types to occur in distinct latitudinal zones (e.g., tropical, subtropical, and temperate), as generated by the global atmospheric circulation system.

zonobiome Broad ecological topographical unit characterized by a certain climatic pattern.

zooplankton From the Greek *zoon* meaning animal, zooplankton comprises those animals that are found passively drifting or weakly swimming in the water column. Zooplankton can be divided into two major categories: holoplankton, which are organisms that spend their entire lives as plankton, and meroplankton, which are organisms that spend part of their life cycle as plankton and part on the seafloor as benthic invertebrate larvae or as nekton (e.g., fish larvae).

zooxanthellae Symbiotic dinoflagellate algae in corals and other tropical marine invertebrates.

zugenruhe A term for a behavior pattern of restlessness exhibited by some migratory species, especially birds, if not allowed to migrate during their usual migratory period. It reflects an underlying physiological transition to a migratory state.

Index

*Volume numbers are boldfaced, separated from the first page reference with a colon.
Subsequent references to the same material are separated by commas.*

A

Abernethy peat bogs, UK, restoration of, 5:814
Ablation, definition, 1:171
Abundance, definition, 4:123
Abyssal plains, definition, 4:1
Acari (mites, ticks), 1:213
   ecology, 1:214–215
   phylogeny/taxonomy, 1:215–217
   reproduction/dispersal, 1:215
   soil, 5:312–313
Acceptance, definition, 4:365
Accession, definition, 3:165
Accidental epiphyte, definition, 3:27
Accounting price, definition, 2:291
Accretionary, definition, 1:741
Acetabulum, definition, 5:863
Acetylcholinesterase
   definition, 3:465
   inhibiting drugs, 1:475
Acid deposition, 1:2; *see also* Acid rain
   definition, 1:1; 3:437; 4:731
   effects on
      aquatic microbial communities, 1:10
      crops, 1:5
      fish, 1:13
      forests, 1:5–8
      groundwater, 1:9
      macroinvertebrates, 1:12–13

      macrophytes, 1:12
      marine waters, 1:10
      periphyton, 1:11–12
      phytoplankton, 1:11
      soils, 1:8–9
      surface waters, 1:9–10
      zooplankton, 1:12
   indicator species for, 3:437–438
   regulation of, 1:15
   as threat to freshwater fauna, 3:541
Acidification
   freshwater ecosystems, 3:97–98
      biological consequences, 3:105–106
   by nitrogen, effects on biodiversity, 4:384–385
Acidophiles, definition, 1:219
Acidophilic, definition, 4:349
Acid rain, 1:1–2; 4:737; *see also* Acid deposition
   causes of, 1:2–4
   definition, 1:1
Acoela, 5:865–866
Acoelomate, definition, 5:863
Acrididae (true grasshoppers/locusts), 3:263
Acridoidea, 3:261–262
Across Trophic Level System Simulation (ATLSS) Project, 1:852
Actinopterygia, definition, 2:95
Acylalanine fungicides, 4:514

Adaptation, 1:693
   after adaptationism, 1:22–23
   Darwinian theory of, 1:18
   definition, 1:17; 5:285
   evidence for, 1:20–21
   historical concept, 1:18
   homeophasic, definition, 4:917
   as process, 1:18
   as product of evolution, 1:18–20
   rates of, and coevolution, 1:757
   role of mutations, 2:254
   thermal, of Archaea, 1:227–228
Adaptationism
   critique of, 1:21–22
   definition, 1:17
   Lamarckian, 2:192
Adaptative syndrome, definition, 3:295
Adaptive landscape, 2:89
   definition, 2:85
Adaptive management, 3:64–65
   assessment and monitoring, 3:69
      baseline information and pilot projects, 3:71
      data collection, 3:71–72
      data management and analysis, 3:71–72
      monitoring programs, 3:69–70
      qualitative and quantitative monitoring, 3:71
      sampling and data gathering designs, 3:70–71

Adaptive management (*continued*)
    sampling objectives, defining, 3:70
    site-specific sampling, 3:71
    definition, 3:63
    evaluation and decision making, 3:72
    implementation, 3:69
    keys to success, 3:72
    plan and design, 3:65
        biological inventory assessment, 3:65–67
        design management, 3:69
        identification of resource needs, 3:67
        objective setting, 3:67–69
        spatial/temporal scales, definition of, 3:67
        stakeholder participation, 3:65
Adaptive radiation
    after extinction events, 1:395–396
    definition, 1:693
    factors underlying, 1:26–27
        environmental change, 1:29–31
        isolated landmass colonization, 1:31
        key innovations, 1:27–29
    initiation
        ecological changes, 1:31–32
        founder events, 1:31–32
        rapid proliferation/hybridization, 1:32
    Jurassic–Cretaceous, 1:416
    origins of, 1:406–407
    predisposition of species for, 1:31
    process, case studies
        African cichlids, 1:36
        Canadian sticklebacks, 1:37, 39
        Caribbean lizards, 1:36–37, 38
        Galapagos finches, 1:33–34
        Hawaiian *Drosophila*, 1:35
        Hawaiian honeycreepers, 1:35–36
        Hawaiian silverswords, 1:36
        Hawaiian swordtail crickets, 1:39–41
        Hawaiian *Tetragnatha* spiders, 1:37–39
        partula land snails, 1:41
        Pseudomonas in culture, 1:41
        research potential, 1:41
Adaptive shift, definition, 1:25
Adaptive systems, complex, definition, 1:393
Adipocytes
    definition, 5:495
    size and numbers, 5:500–501

Adipose fin, definition, 2:755
Adipose tissue, 5:499–500
    anatomical arrangement of, 5:501–503
    definition, 5:495
    as thermal insulation, 5:508–510
Adjacency, definition, 5:569
Adult parasite, definition, 4:463
Aerenchyma, 3:855
    definition, 3:853; 4:689
Aerial photography, 5:125–126
    in estimation of land use/land cover, 3:671–672
Aerial roots, definition, 3:853
Aerobe, definition, 4:191
Aerobic, definition, 2:311
Aerodynamics, evaporative, definition, 1:307
Aerosols, definition, 1:1; 3:277
Aesthetics
    and biodiversity, 1:53
    and cognition, 1:53–54
    definition, 1:45
    environmental, evolutionary approaches, 1:46–47
    and stress reduction, 1:53
Affiliative behavior, definition, 5:295
Afforestation, definition, 2:23
AFLP, *see* Amplified fragment length polymorphism
Africa
    climate, 1:56
        past fluctuations in, 1:58
    deforestation in, 2:24
    desertification in, 2:64–65, 67
    ecosystems
        coastal, 1:67–68
        deserts, 1:65; 2:45–46
        fire, 1:58–59
        forest, 1:59–60
        future for, 1:69–70
        lakes, 1:66–67
        Madagascar, 1:69–70
        major phytogeographic/ecoclimatic zones
            Cape Region, 1:59
            Mediterranean North Africa, 1:58
            montane/afroalpine, 1:65–66
            tropical, 1:60
        rain forests
            deforestation, 5:30–31
            historic distribution, 5:27
        freshwater fishes, diversity, 2:779
        geology, 1:56
        late Quaternary mammalian extinctions, 3:835–837

        Madagascar, 2:735–736; 3:834–835
        past climatic/environmental fluctuations, 1:59
        postglacial extinctions, 2:734–735
        seasonal tropical vegetation, 1:60–61
            bushland/thicket, 1:63
            grasslands, 1:63–64
            shrublands, 1:64–65
            woodland, 1:61–63
        soils, 1:56–57
    sub-Saharan
        characteristic crops, 2:225–226
        remote sensing in measurement of interannual vegetation changes, 5:132–133
        traditional agricultural areas, 2:220
        traditional energy use, biodiversity impacts of, 2:539–542
        Usambara Mountains, Tanzania, diversity management, 4:655
    wetlands, 1:67
Agaricus bisporus (white button mushroom), 3:158–159
Aggregation, 1:849
    definition, 1:845
Agnathans, 2:757
Agonistic behavior, definition, 5:295
Agricultural intensification, 1:87
    biodiversity impacts, 3:798–799
    and crop commercialization, implications for genetic diversity, 1:94
    definition, 3:659
    in Europe, 3:670
    implication for genetic diversity, 1:88–92
Agriculture; *see also* Agroecosystems
    in Africa, 1:69
    applications of biodiversity
        genetic diversity in reserve, 1:550–551
        spatial diversity, 1:549
        temporal diversity, 1:549–550
    and biodiversity, economic issues, 4:88–89
    as biodiversity-eroding factor, 3:774
    commercial, and deforestation, 2:30
    costs of nonindigenous species
        in Australia, 1:78, 79
        in Brazil, 1:81, 82
        in South Africa, 1:79, 80
        in South India, 1:80, 81
        in U.K., 1:77–78
        in U.S., 1:72–73, 76

emergence of, 2:218–219
   geographic centers, 2:219–220
land use issues, 3:660, 662–663
microbe introductions in, in U.S., 1:76
modernization of, 1:86–87
pollinators in, 4:728–729
   introduced species, 4:728–729
   as vectors/victims of engineered genes, 4:728–729
and reduced biodiversity, 1:547
regions, characteristic crops, 2:224–226
slash and burn, see Swiddening
soil conservation in, methods, 5:319–320
   diversified cropping, 5:320–321
   manipulation of biotic community, 5:321–322
   pollutant bioremediation, 5:321
   reduced tillage, 5:320
sustainable, definition, 1:109
traditional, 2:218
   biodiversity impacts, 3:794–795
   definition, 1:109
   natural pesticides in, 1:476
   and resource management, 5:290–292
Agroecosystems
   definition, 1:109; 5:315
   impact of planned diversity, 2:270
      on pest regulation, 2:272
      on productivity/stability, 2:270–272
      on soil processes, 2:272–273
   planned and unplanned diversity, 2:269–270
   traditional
      and biodiversity-based strategies, 1:115–116
      biodiversity of, 1:112–113
         ecological advantages, 1:113–114
         preservation of, 1:114–115
         farmer's knowledge of, 1:112–113
         structural/functional elements, 1:113
   unplanned diversity and ecological services, 2:273
      and pest regulation, 2:273–274
      and soil processes, 2:273–274
Agroforestry, definition, 2:23
Agromyzidae (leaf-miner flies), 2:821
Agroscape, definition, 3:689
Air pollution/pollutants; see also Acid deposition
   definition, 2:698
   dose–response relationships, classification, 1:122
   effects
      on forest ecosystems, 1:128–129
      heathland communities, 1:128
      on interaction between organisms, 1:123–125
      on lichen biodiversity, 1:126–128
   evolutionary responses to, 1:123
   and extinction, 2:708
   and extinctions, 2:708
   global distribution, 1:121
   historical overview, 1:120
   interactions with environmental factors, 1:123
   levels, and gross domestic product, 2:298
   in lower atmosphere, 1:304
   sources and impacts, 1:120
Alate, definition, 3:581
Albedo
   definition, 3:277
   snow/ice, as feedback, 3:281
Algae
   brown, 2:197
   definition, 2:649
   green, temperature limits for, 3:354
   harmful, definition, 2:649–650
   lineages, 2:196
   red, neurotoxic anthelminthics from, 4:716–717
Algal blooms
   and anthropogenic nutrient enrichment, 2:655
   definition, 2:649
   harmful, definition, 4:1
   and oxygen depletion, 2:652
Alien plants, definition, 4:677
Alien species; see also Exotic species
   definition, 2:425; 3:729
Aliphatic acid herbicides, 4:515
Alkalinity, definition, 1:1
Alkaloids
   bioreactive, 4:719
   as plant chemical defenses, 2:12–13
Allee effects, 3:766
   and coexistence, 5:425
   definition, 1:855; 4:601, 769
Allele frequency, definition, 4:777
Alleles, 3:183
   definition, 1:85; 2:85; 4:759
   and protein polymorphisms, 3:197
Alligatorids, 5:146
   conservation issues, 5:157–158
Allochthonous, definition, 2:311; 5:213

Allopatric, definition, 4:329; 5:371
Allopatric speciation, 5:389–390
   definition, 1:25; 5:1
   dichopatric speciation, 5:374–376
   peripatric speciation and founder effect principle, 5:376
Allopatry, definition, 5:569
Allopolyploid, definition, 4:659
Allopolyploid speciation, 5:380
Allozyme
   definition, 2:245
   diversity, nature of, 3:209
All taxa biodiversity inventory, definition, 3:437
Alluvial fans, 2:40
Alpha diversity, 3:705; 5:248–249
   definition, 3:215, 701, 747
Alpha landscape density, definition, 3:645
Alpine, definition, 1:133
Alpine ecosystems
   animal diversity, 1:141
   biodiversity
      functions of, 1:142
      and global change, 1:142–143
   climatic microhabitats, 1:135–136
   life zone, 1:134
   microbial diversity, 1:141–142
   morphotype diversity, 1:136–138
   physiotype diversity, 1:138–139
   reproductive diversity, 1:139
   substrate diversity, 1:135–136
   taxonomic diversity, 1:139–141
Alternation of generations, definition, 4:1
Altman, Sidney, 4:444
Altruism, definition, 5:295
Alvin, definition, 5:233
Amazonia
   arthropods
      biotope/habitat specificity, 1:256–257
      genetical species, 1:259
      morphological species, 1:257–259
      origin/distribution, 1:250–253
      species richness, 1:253–256
      taxa and species numbers, 1:250
   crop domestication in, 1:899–911
   floodplain ecosystems
      arthropod species, 1:252–253
      igapò, 1:151–152
      lakes, 1:152
      mangrove forest, 1:153
      pirizal, 1:153
      swamps/buritizal, 1:152–153
   nonflooded terre firme ecosystems

Amazonia (*continued*)
  dry semideciduous southern fringe forest/cerradão, 1:149–150
  rain forests, 1:146–148
  white sand oligotrophic formations, 1:148–149
  rainforests
    deforestation, 2:24–25; 3:799
      biodiversity implications, 5:32
    fate of, 5:22
    historic distribution, 5:26–27
    plant diversity, 5:14, 15, 18, 19
  river types, definition, 1:145
  savanna
    flooded, of eastern Amazon, 1:154
    nonflooded/cerrado, 1:153–154
  secondary forests, 1:154–155
  transition ecosystems
    babassu palm forest, 1:150
    bamboo forest of western Amazon, 1:151
    liana forest, 1:150–151
Ambiregnal organisms, nomenclature systems, 4:397
Amblypygi (whip spiders), 1:210–211
American Camping Association, 2:771
American Museum of Natural History, New York, 4:277
Americas; *see also* Central America; North America; South America
  characteristic crops, 2:226
  pre-Columbian agriculture, 2:220
Amino acids
  nonprotein, as plant chemical defenses, 2:13
  prebiotic formation, experiments, 4:443
  substitutions, rate of, 2:186
g-Aminobuteric acid receptors, *see* GABA receptors
2-Aminopyrimidine fungicides, 4:514
Amniota, 5:763
  definition, 2:95
Amoebae, 4:906–908
Amphibians
  biodiversity in Cape Region of South Africa, 4:157
  biodiversity in Chile, 4:152–153
  biogeographic history, 1:160
  conservation status of, 2:480
  declining populations, 2:485–486
    and UV radiation, 5:729–730
  definition, 2:479
  deformities in, 2:486
  diversity, 1:160, 169
  as ecosystem components, 1:168–169
  evolutionary history, 1:159–160
  human uses of, 1:168
  as indicator species, 3:439
  larval, ecology and functional morphology
    anurans, 1:165–167
    caecilians, 1:165
    salamanders, 1:165
  of mangrove ecosystems, 3:858
  migration patterns of, 4:229
  morphology and functional anatomy, 1:160–162
  nonindigenous, in U.S., 1:75
  North American, diversity, 4:410
  postlarval, behavior and ecology, 1:167–168
  reproductive biology and life histories, 1:162–163
    anurans, 1:163–165
    caecilians, 1:163
    salamanders, 1:163
  response to climate change, 1:719
  threats to
    climate change, 2:481–482
    commercial exploitation, 2:483–484
    exotic species introduction, 2:484
    habitat destruction, 2:481
    pollution, 2:482–483
    species at high risk, 2:484–485
    ultraviolet radiation, 2:482
Amphibites, definition, 5:527
Amplified fragment length polymorphism (AFLP), 3:197
AMS dates, definition, 4:451
Anadromous, definition, 5:233
Anaerobe, definition, 4:191
Anagenesis, 1:694
  definition, 1:693; 5:569
  mechanisms for, 1:695–696
Anamniotic, definition, 1:159
Ancestral, definition, 2:755; 5:145
Anchialine (anchihaline), definition, 5:527
Andepts, definition, 5:701
Anemophily
  definition, 4:723
Angiosperms
  adaptations to parasitism, 4:471
  definition, 1:261; 4:601
  evolution of, 2:198
  parasitic, 4:484
  radiation, and insect pollination, 4:612–613
  role in austral biota, 5:368
Anhydrobiosis, definition, 1:171; 5:843
Animal rights, and conservation ethics, 2:600–601
Animal unit month, definition, 5:33
Animism, 5:286–287
  definition, 5:109
Anion, definition, 1:437
Annelida (segmented worms), 3:563; 5:831; *see also* Polychaeta
  definition, 3:561
Anoxia, definition, 2:649; 4:377
Antarctica
  atmospheric/meteorological circulation patterns, 1:181
  biodiversity, 1:178
    bipolar comparison, 1:178–179
  biogeographic zones, 1:172–173
    continental, 1:174
    maritime, 1:174
    sub-Antarctic, 1:173
  colonization, 1:180
    dispersing propagules, 1:181
    mechanisms of, 1:180–181
  human influences, 1:181–182
    alien species introduction, 1:182
    environmental change, 1:182–183
      biological effects, 1:183–184
  microflora, 4:918
  tectonic/glacial/biological history, 1:172
  terrestrial biota
    fauna, 1:175–176
    flora, 1:176–177
    freshwater, 1:177–178
    habitats, 1:175
    microbial, 1:177
    origins/antiquity of, 1:179–180
Antennules/antennae, definition, 1:915
Anthelminthics, from red algae, 4:716–717
Antheridium, definition, 4:1
Anthophile
  definition, 4:723
Anthropocentric/anthropocentrism
  definition, 1:883; 2:545
  and views on human impacts, 3:793
Anthropochoric, definition, 4:291
Anthropogenic; *see also* Human impacts
  definition, 2:441; 3:277; 4:731
Anthropogenic disturbances, definition, 5:553
Anthropogenic ecosystems, definition, 2:609
Anthropoids, definition, 4:879

Antibiotic pesticides, 4:512
Antibiotics, from fungi, 3:159
Anticancer agents
    from cyanobacteria, 4:712–713
    definition, 4:711
    from higher plants
        camptothecin, 4:718–719
        taxol, 4:717–718
Ants, see Hymenoptera
Anurans (frogs)
    larval, ecology and functional
        morphology, 1:165–167
    morphologic features, 1:162
    rainforest, diversity of, 5:5
    reproductive biology and life
        histories, 1:163–165
Apex predator, definition, 4:857
Apicomplexans, 4:915
Aplacophora, 4:239
Apocrita, 3:418
Apomixis, definition, 1:133
Apomorphic characters, 1:678
Apomorphy, definition, 1:677; 5:569
Aposematic, definition, 1:559; 4:249
Aquaculture, 3:799
    definition, 1:185; 2:783
    development and practices,
        1:190–191
    ecological footprint as tool in,
        1:195–196
    feed, 1:187
    and fish conservation, 2:791
    habitat modification and
        biodiversity, 1:192–193
    impacts on biodiversity, 1:186–187
    impacts on marine ecosystems, 4:34
    integrated, 1:197
    land use, 1:, 187–188, 190
    seed, 1:191
    waste impacts, 1:188–189, 193–195
    water use, 1:188
    wild capture of larvae/spawners for,
        1:193
Aquaria, role in conservation biology,
    1:863
Aquatic biota, acid deposition effects,
    1:10–14
Aquatic ecosystems; see also
        Freshwater ecosystems; Lakes/lake
        ecosystems; Marine ecosystems;
        Oceans; Riverine ecosystems;
        Wetlands/wetland ecosystems
    eutrophication stress, 4:1–11
    and nitrogen cycle perturbations,
        4:737–738
    pesticide impacts, 4:521–522

Aquatic fauna, diversity in logged
    forests, 3:757
Aquifer, definition, 2:425
Arachnida
    ecology, 1:201–202
    order Acari, 1:213
        ecology, 1:214–215
        phylogeny/taxonomy, 1:215–217
        reproduction/dispersal, 1:215
    order Amblypygi, 1:210–211
    order Araneae, 3:562
        ecology, 1:203–204
        phylogeny/taxonomy, 1:204–206
        reproduction/growth, 1:204
    order Opiliones, 1:208
        ecology, 1:209
        phylogeny/taxonomy, 1:209–210
        reproduction/growth, 1:209
    order Palpigradi, 1:211–212
    order Pseudoscorpiones, 1:212–213
    order Ricinulei, 1:212
    order Schizomida, 1:211
    order Scorpiones
        ecology, 1:206–207
        phylogeny/taxonomy, 1:207–208
        reproduction/growth, 1:207
    order Solifugae, 1:213
    order Uropygi, 1:211
    overview, 1:200–201
    palentology, 1:202–203
    phylogeny/taxonomy, 1:202
    reproduction/growth, 1:201
Aradomorpha, 5:721
Aral Sea, 3:541
Araneae (spiders), 3:562
    ecology, 1:203–204
    phylogeny/taxonomy, 1:204
    reproduction/growth, 1:204
Arbicolous, definition, 3:19
Arboricolous, definition, 1:249
Archaea, 2:195–196; 4:200–202
    definition, 1:219; 4:201; 5:647
    halophilic, 1:224–225
    historical background, 1:220
    methanogenic, 1:222–224
    nonextremophilic, 1:225
    origin and evolution, 1:226–227
        early respiratory processes, 1:227
        temperature adaptation,
            1:227–228
    photosynthesis in, 4:550
    physiological types, 1:220–221
    separation from Bacteria/Eucarya,
        1:328
    taxonomy/phylogeny, 1:220
    thermophilic, 1:221–222

uncultured, in marine environment,
    1:334–335
Archaean eon, microbial diversity in,
    3:220–221
Archaeobacteria, see Archaea
Archean eon, 1:411
Archegonium, definition, 4:1
Archetype, definition, 5:589
Arctic
    biodiversity
        bipolar comparison, 1:178–179
        circumpolar patterns, 1:240–241
        consequences of, 1:243–245
        controls on, 1:242–243
        latitudinal patterns, 1:237–240
        threats to, 1:246–247
        topography effects, 1:241–242
    definition, 1:232
    environmental characteristics,
        1:233–236
    subzones, 1:234–235
    microflora, 4:918
    natural resources/exploitation,
        1:245–246
Area cladogram, definition, 5:569
Area per se hypothesis, of species–area
    relationship, 5:398
Area selection
    complementarity in, 1:814–816
    prioritizing, 1:823–824
Area–species relationship, in
    estimation of future extinction
    rates, 3:764–765
Aridity, index of, 2:63
    definition, 2:61
Aristotle, 4:713; 5:575
Ark paradigm, 4:646; 5:910
    definition, 4:645
Arrhenius, Svante, 4:441
Arrhenotoky, definition, 3:417
Arroyos, 2:41
Arthropods, 2:202; see also Arachnida;
    Insecta
    Amazonian
        biotope/habitat specificity,
            1:256–257
        genetical species, 1:259
        morphological species, 1:257–259
        origin and distribution
            Andes/Guyanian shield, 1:252
            neotropics, 1:250
            non-flooded/floodplain forests,
                1:252
            treeless floodplain, 1:252–253
        species richness, 1:253–256
        taxa and number of species, 1:250

Arthropods (continued)
  arbicolous, 3:19, 22
  nonindigenous
    in Australia, 1:79
    in U.K., 1:77
    in U.S., 1:75–76
  radiation, ecology of, 1:919–921
  terrestrial, 3:561, 562
Artificial ecosystems, definition, 4:845
Artiodactyla, 3:796–797
Ascaris lumbricoides, 5:849–852
Ascension Island, habitat restoration, 4:651
Ascomycota, 2:200; 3:146–147
Asia
  artificial ecosystems, 1:289
  biodiversity, 1:265–266
  boreal forest, 1:283–284
  characteristic crops, 2:224–225
  conservation status, 1:289–291
  deforestation in, 2:24, 411–412
  deserts of, 2:46, 48–49
  freshwater fishes, diversity, 2:779–780
  humid monsoon (extra-tropical), 1:280
    laurel forest, 1:280
    mountains, 1:282–283
    pine forest, 1:281–282
    temperate deciduous forest, 1:280–281
      Japan/Korea, 1:281
      Manchuria/eastern China, 1:281
  humid tropical forest, 1:272–275
  humid tropical/subtropical mountains, 1:276–277
  late Quaternary mammalian extinctions, 3:835–837
  monsoon forest, 1:275
  Neolithic agriculture, 2:219–220
  physiographic features, 1:263–266
  polar/mountain tundra, 1:284–285
  rain forests
    deforestation, 5:30
    historic distribution, 5:27
  regionalization and biomes of, 1:267, 269, 272
  southeast, deforestation in, 3:667–668
  tropical dry forest, 1:276
  western/interior dry regions, 1:277
    Mediterranean ecosystems, 1:278–279
    subtropical deserts, 1:277–278
    temperate grasslands, 1:279
    temperate semideserts/deserts, 1:279–280
    Dzungaria, Gobi, and Tarim Basin, 1:280
    middle Asian, 1:280
    Tibetan Plateau, 1:280–281
  wetlands/coastal ecosystems, 1:286–290
Asilidae (robber flies), 2:821
Aspergillus spp., 3:143, 160
Assemblages
  climate change impacts on, 3:292–293
  definition, 1:249; 3:485
  ecological, structure of, 5:451
  functional changes over evolutionary time, 3:124–125
    origin of functionally distinct organism and functional stasis, 3:125–126
Assembly rules, definition, 5:185
Asses, Asiatic wild, 5:302–303
Associational refuge, definition, 3:265
Assortative mating, definition, 5:383
Asynchrony, from global warming, 3:290
ATLSS Project, 1:852
Atmosphere
  as common property resource, 1:773–774
  human impacts on, 2:558–559
  lower, air pollutants in, 1:304
  nitrogen deposition from, 4:380–381
  oxidizing capacity, 1:295
  transparency/visibility, acid deposition effects, 1:14–15
  upper, microflora, 4:918
Atmospheric gases, concentrations, 1:294
  changes in, 1:295
    carbon dioxide, 1:295–296
    carbon monoxide, 1:299–300
    halogens, 1:298–299
    methane, 1:297–298
    nitrogen oxides, 1:301
    nitrous oxide, 1:298
    non-methane hydrocarbons, 1:300–301
    sulfur gases, 1:301
  CO_2, analysis, 1:611–612
  following accretion, 1:439
Atmospheric lifetime, definition, 1:293
ATP synthase
  in CO_2 fixation, 4:555–557
  definition, 4:549
Attraction (tourist), definition, 5:667
Australasia, definition, 5:701

Austral ecosystems, 5:361–362
  biogeography, history of, 5:362–363
  patterns, 5:363
    distributional, and dispersal means/climate/area, 5:363–364
    fossil record, 5:365–366
    general-area cladograms, 5:364
    molecular clocks, 5:364–365
  processes, 5:366
    climate, 5:367–368
    continental drift, 5:366
    disturbance, 5:368
    high latitude plant growth, 5:368–369
    long-distance dispersal, 5:366–367
Australia
  biogeographic regions, 1:316
  bioprospecting initiatives, 1:481–482
  desertification in, 2:67–68
  deserts of, 2:49
  diagnostic floristic groups, 1:316–319
  diagnostic species
    Acacia vegetation
      Australian arid zone, 1:314–315
      subtropical eastern Australia, 1:314
    aquatic vegetation
      temperate southern Australia, 1:315
      tropical/subtropical northern Australia, 1:315
    chenopod low shrublands, southern Australian arid zone, 1:315
    closed-forests, semideciduous monsoonal northern Australia, 1:311
    subtropical eastern Australia, 1:311–312
    coastal dune vegetation, 1:315
    coastal wetland, 1:315–316
    eucalypt open-forests/woodlands
      monsoonal northern Australia, 1:312
      montane southeastern Australia, 1:313
      subtropical eastern Australia, 1:312
      temperate southwestern Australia, 1:313
      wetlands, 1:313
    heathlands/shrublands

monsoonal northern Australia, 1:313
montane southeastern Australia, 1:314
subtropical eastern Australia, 1:313–314
temperate southeastern Australia, 1:314
temperate southwestern Australia, 1:314
hummock grasslands, Australian arid zone, 1:315
mallee eucalypt open-scrubs
monsoonal northern Australia, 1:313
subtropical eastern Australia, 1:313
temperate southeastern Australia, 1:313
temperate southwestern Australia, 1:313
tussock grasslands, 1:314
ecosystems, 1:308
community diversity of, 1:321–323
Environmental Resources Information Network (ERIN), 1:435
evaporative aerodynamics, 1:308
exotic species
invertebrates, 1:79
plants, 1:78
vertebrates, 1:78–79
floristic groups, diagnostic species, closed-forests (rain forest)
subtropical eastern Australia, 1:310–311
temperate southeastern Australia, 1:311
tropical northeastern Australia, 1:310
freshwater fishes, diversity, 2:779–780
late Quaternary mammalian extinctions, 3:833–834
major plant communities, 1:308–310
mammalian endemism, 2:448
mound springs of Great Artesian Basin, 2:436
postglacial extinctions, 2:733–734
species richness, 1:319–321
Autapomorphy, definition, 5:569
Autocatalysis, definition, 4:439
Autochthonous, definition, 5:213
Autocorrelation, definition, 4:745

Autogenic species, definition, 5:747
Autopolyploid speciation, 5:379
Autotroph, definition, 2:305; 4:191
Autotrophic respiration, definition, 1:609
Autotrophic successions, definition, 5:541
Autotrophs, definition, 2:305; 4:1
Autotrophy, definition, 4:569
Availability concentration, definition, 1:119
Available, definition, 4:389
Ayurveda, 2:613

B

Babassu palm forests, 1:150
Bacillus thuringiensis, 1:76, 477; 4:269–270, 513
potential resistance to, 3:527
resistance to, 3:469
as threat to pollinators, 4:729
transgenic plants, 3:476
Background concentration, definition, 1:119
Background extinction, definition, 3:841
Background selection, definition, 2:179
Bacteria, 4:198–200; *see also* Prokaryotes; Proteobacteria
biomass and species numbers, 5:309
codes of nomenclature, 4:393
definition, 1:219; 5:647
estimated number of species, 4:199
evolution of, 1:349; 2:195–196
genetic exchange
evolutionary effects of, 1:348–349
mechanisms, 1:346
conjugation, 1:347
transduction, 1:346–347
transformation, 1:347–348
genomic structure
informational
codon usage, 1:343
GC bias, 1:343
gene organization/expression
operon, 1:344–345
regulation, 1:345–346
physical
chromosomes, 1:341–342
plasmids, 1:342–343
size, 1:341
lineages (phyla), 1:340
metabolic diversity of, 4:203–205
phylogenetic diversity, 1:328–330

principle properties of, 4:202–203
psychrophilic, *see* Psychrophiles
role for element cycling, 4:205
aquatic sediments, 4:206–207
extreme environments, 4:207
terrestrial soils, 4:207
water column of oceans/lakes, 4:205–206
symbiotic
evolutionary origin, 1:331–332
identification, 1:331
phylogenetic affiliation, 1:332–333
Bacterioplankton, 4:586–587
Baikal, Lake, 3:641
invertebrate species, 2:438
Balancing selection, definition, 2:85
Bald Eagle Protection Act (1940), 3:238
Baldwin effect, 4:545
Ballast water, definition, 2:783; 4:27
Baltica, 1:924
Bamboo forests, Amazonian, 1:151
Bands, definition, 3:411
Barophile, definition, 4:917
Barophilic, definition, 4:191
Barotolerant, definition, 4:191
Barreais, definition, 1:897
Basal species, definition, 5:695
Basidiomycota, 2:200; 3:146, 147
Batesian mimicry, 1:759–760; 2:92
in butterflies, 1:567; 2:256
definition, 1:559
Bateson, William, 4:538
Batha, definition, 4:353
Battani Bay, Thailand, restoration of, 5:818
Bauplan, definition, 5:589
Beaches, sandy
biodiversity of, 4:21
and beach type, 1:745–747
biological factors, 1:749
latitudinal effects, 1:747
paradigms for, 1:749–750
physical factors, 1:747–749
defining variables, 1:742
microfauna composition and zonation, 1:744–745
swash climate, 1:748
transect surveys, 1:743–744
types of, 1:742
Bears, polar
evolutionary history, 4:44
features and habitat boundaries, 4:42
Beauty, definition, 1:45

Beetle, flour (*Tribolium* spp.), in interspecific competition studies, 1:802–804
Beetles, *see* Coleoptera
Behavioral evolution, 2:677
Beluga whale (*Delphinapterus leucas*), endangered status of, 4:54–55
Benthic, definition, 5:781
Benthic organisms, definition, 4:27
Benthic zone
  biodiversity of, 4:23
  definition, 1:915; 3:531, 543; 4:1, 13
  environment of, 4:2–4
Benthos
  and benthic environment, 4:2–4
  definition, 3:75; 4:1, 13, 901
Benzimidazole fungicides, 4:514
Benzonitrile herbicides, 4:516
Beringia, definition, 4:403
Beta diversity, 3:713; 5:248–249
  definition, 1:419; 3:215, 701, 747
Bet-hedging strategy, definition, 2:79
Between-habitat diversity, definition, 2:161
The Bible, pastoralism in, 3:741–742
Bichirs, 2:763
Biebricher, Christof, 4:444
Bifurcation point, definition, 2:291
Binomen/binomial, definition, 4:389
Binomial nomenclature, definition, 5:569
Bioaccumulation, 2:367
  definition, 2:363
Bioassay, definition, 3:437, 465; 4:523
Bioavailable concentration, definition, 1:119
Biocentric/biocentrism, definition, 1:883; 2:545
Biocontrol; *see also* Biological control
  definition, 5:315
Bio-cultural axiom, definition, 3:452
Biodiversity
  aesthetic factors, 1:53
  alpine ecosystems
    functions of, 1:142
    and global change, 1:142–143
  Antarctic, 1:178
    bipolar comparison, 1:178–179
  approaches to studying
    community approach, 2:206
    ecosystem approach, 2:206–207
    individual/population approach, 2:206
  Arctic
    circumpolar patterns, 1:240–241
    consequences of, 1:243–245
    controls on, 1:242–243
    latitudinal patterns, 1:237–240
    threats to, 1:246–247
    topography effects, 1:241–242
  Asia, 1:265–266
  assessment
    definition, 3:63
    and functional significance, 5:174–175
  biogeochemical determinants of, 5:737–739
  biologists and promotion of, 3:366–367
  and centralized regimes, 1:779–780
  and climate change, 3:283–284
    study methods, 3:284–286
  concepts and definitions, 1:835
  conservation of
    and economic development, 3:414
    and empowerment of indigenous peoples, 3:462–463
    goals of, 2:685–686
    market-based incentives, 4:94–95
    methods
      biodiversity utilization, 2:692–693
      community-based conservation, 2:692
      conservation objectives, 2:688–689
      conservation projects, 2:692
      conservation strategies, 2:689–690
      conserved product dissemination, 2:692
      ecogeographic survey and preliminary survey mission, 2:688
      ex situ techniques, 2:690–691
      field exploration, 2:689
      project commission, 2:687–688
      selection of target taxa, 2:687
      in situ techniques, 2:691–692
    and property rights, 4:897–898
    sustainable and integrated conservation, 2:693–694
  contribution to crop varieties
    broad adaptation, 1:553
    hybrid vigor, 1:552
    new cultural systems, 1:555–557
    new traits, 1:554
    pest/pathogen resistance, 1:552–553
    transgenes and key biological functions, 1:555
    yield, 1:552
  contribution to productivity, 1:360–362
  creation of term, 3:364–365
  crisis in, and human ecological footprint, 2:238–240
  and cultural diversity, relationship, 3:454–456
  current patterns of, 1:397–398
  definition, 1:109, 378–380; 3:365–366
    contextual variations, 1:386–387
    attributes of, 1:387–388
    classifying, 1:387
    implications, 1:389–390
    precautionary principle, 1:389
    resource asset/management objectives, 1:388
  deforestation consequences, 2:32
  diffuse benefits of, 4:92–93
  and disease
    in maintenance of, 2:114–116
    as threat to, 2:117–118
  and disturbance, 2:155
  versus ecological complexity, 1:831–832
  economic value of, 2:291–292
    institutional failures and poverty, 2:294–295
    market failure in determining, 2:292–294
    measurement, 2:286
      diversity, 2:286–287
      local diversity, 2:287
      populations, 2:287–288
    resources as luxury, 2:297–298
    substitution possibilities, 2:298–300
    valuing resources and evaluating projects, 2:295–297
  and ecosystem services, 1:364–366
  element of, 5:577–579
  estimators of
    based on extrapolation, 4:140
      birth chains, 4:141
      Michalis–Menton equation, 4:141
      species–area curves, 4:140–141
    choosing, 4:141–142, 143
    evaluation, 4:142–143
  in existing markets, 4:88
    agriculture, 4:88
    harvest of wild organisms, 4:89–90
  ex situ conservation, 1:386
  fish
    information management

bioquads as key data sets, 2:812
conceptual challenges, 2:810–811
preservation
  marine protected areas, 2:813–814
  traditional approaches, 2:812–813
in fossil record, 3:215–216
future of, 3:801–803
gene diversity distributions, 5:247–258
generation of, components, 1:403
and genetic knowledge, 1:364
and habitat fragmentation, 3:313–314
herbicide effects on, 3:344
historical awareness of, before 1986, 3:363–364
and human flourishing, 2:551
  aesthetic/recreational enjoyment, 2:552
  artistic expression and scientific knowledge, 2:552–553
  health and wealth, 2:551–552
  historical understanding and religious inspiration, 2:553
human impacts on
  in cities, 3:796–797
  colonialism, 3:797–798
  modern transportation/commerce, 3:798
importance of carrying capacity, 1:647
and insurance, 1:362–364
intertidal
  as indicator of ecological function, 3:495–497
  measurement, 3:494–495
and language, 5:289–290
loss of, causes, 1:384–385
management of
  applied dispersal biogeography in, 2:150–151
  ethical issues, 2:598–599
marine, processes controlling, 3:545
and markets, 1:366–368; 4:86–87
  ecosystem services, 1:371–373
  insurance, 1:368–369
  knowledge, 1:369–371
  preservation problems in, 4:87
  productivity, 1:368
measurement
  community diversity, 1:382–383
  reasons for, 4:123–124
  species diversity, 1:381–382

symbols used in, 4:125
taxonomic diversity, 1:382–383
measures of, 4:124
microbial, 4:178
  measurement of, 4:178–179
    counting methods, 4:180–181
    enrichment/isolation versus direct measurement of communities, 4:179
    environmental heterogeneity, 4:179–180
    level of taxonomic resolution, 4:179–180
over geologic time
  Cambrian explosion, 3:221–222
  disparity, 3:216–217
  factors
    climate change, 3:229–230
    extinction events, 3:230–231
    plate tectonics/global heterogeneity, 3:228–229
  fossil record, 3:215–216
    sampling problems, 3:217–219
  microbial diversity of Archaean/Proterozoic eons, 3:220–221
  Phanerozoic diversity in marine environment, 3:222–226
  Phanerozoic marine life, 3:221–222
  Phanerozoic terrestrial diversity, 3:226–228
  species richness, 3:216
quality-of-life aspects
  diversity, 2:288
  local diversity, 2:288–289
  populations, 2:289
of recombination systems, 5:61–63
and religion, 5:286
  animism, 5:286–287
  Buddhism, 5:287
  Christianity, 5:287
  Hinduism, 5:287
  Islam, 5:287–288
  Jainism, 5:288
  Judaism, 5:288
  Shinto, 5:288
  Sikhism, 5:288–289
  Taoism, 5:289
restoration of, 5:185–187
  abiotic/biotic restraints on, case studies identifying, 5:206–208
  multispecies approaches, 5:187–188
  novel opportunities for, 5:192
    tree plantations, 5:192

utility line rights-of-way, 5:193
single-species approaches, 5:190–192, 191
sampling considerations
  abundance and incidence, 4:126–127
sampling bias influence, limitation of, 4:125–126
sampling scope, 4:126
surrogates for quadrats and species, 4:126–127
sampling theory, estimators based on, 4:127
  estimation on heterogeneous samples, 4:139–140
  models of sampling process, 4:127–128
  moment estimators, 4:132
    Chao's abundance-based estimator, 4:134–135
    Chao's incidence-based estimator, 4:135–137
    Fisher's a, 4:132–134
  resampling methods, 4:130
    bootstrap estimator, 4:131–132
    jackknife estimator, 4:130–131
  sample convergence-derived estimates, 4:137
    Chao's abundance-based coverage estimator, 4:137–138
    Chao's incidence-based coverage estimator, 4:137–139
  species observed, 4:128–130
scalable quantities, 5:245–246
scaling relations for, 5:253
in situ conservation, 1:385–386
spatial changes, 1:384
taxonomic diversity distributions, 5:246–247
temporal changes, 1:383–384
terrestrial invertebrates, 3:564
theories of, 2:264–265
of thermal ecosystems, 3:359–360
threats to, 2:684–685
tropical versus temperate gradients, 3:447
understanding by science-literate adult, 2:766–767
urban versus rural perceptions, 5:293
values of, 3:367–368, 497
and property prices, 4:91
Biogeochemical, definition, 2:311
Biogeochemical reactions/cycles, 1:442–443

Biogeochemical reactions/cycles (*continued*)
  anthropogenic alterations in, 2:563–564
  biomineralization, 1:444
  of carbon, 1:443
  definition, 2:345; 4:731
  evolution of, 1:452–453
  hydrolysis, 1:442
  nitrogen, 1:448–449
  oceanic solubility/biological pumps, 1:448
  oxidation–reduction, 1:442–443
  oxygen, 1:444–445
  phosphorus, 1:451
  primary production, 1:447–448
  sequestration/burial, 1:445–447
  sulfur, 1:450–451
  trace metals, 1:451–452
Biogeochemistry, definition, 4:377
Biogeographic regions, definition, 1:307, 419
Biogeography, 3:565
  applied dispersal, 2:150–151
  and biodiversity conservation, 1:469–470
  cladistic methods in, 5:770–771
  continental drift/plate tectonics, 1:462–465
  definition, 1:456, 665; 5:767
  dispersal, definition, 2:127
  ecological, definition, 5:767
  fundamentals of, 1:456–457
  geographic template dynamics, 1:461–462
  glacial cycles of Pleistocene, 1:465–466
  historical, methods, 2:146
  history of, 1:457–461
  island, *see* Island biogeography
  species diversity/composition gradients, 1:466–469
Biological altruism, definition, 1:17
Biological control
  definition, 3:517
  in disease intervention, 2:125
  pathogens in, 2:116
  in traditional agriculture, 1:113–114
  use of herbivorous insects in, 3:522
Biological diversity, *see* Biodiversity
Biological field stations, 4:322–323
Biological indicators, definition, 3:63
Biological integration, definition, 2:363
Biological integrity, definition, 2:557
Biological soil crusts, 5:45–46
  definition, 5:33

Biological species concept, 2:86; 5:372, 373
  definition, 2:395
Bioluminescence, definition, 4:191
Biomagnification, definition, 4:731
Biomarkers, definition, 5:647
Biomass
  conversion systems, 2:538–539
  definition, 1:609; 2:801
  patterns of, and food web energy, 3:9
    community structure, 3:10
    complex food webs, 3:10
    exploitation ecosystem hypothesis, 3:9–10
    green world hypothesis, 3:9
    multichannel omnivory, 3:10
    trophic cascades, 3:10
    trophic levels, 3:9
  and termite abundance, in tropical regions, 3:610
Biomass appropriation, role of indigenous peoples, 3:456
Biomass pyramid, definition, 3:1
Biomes
  classification, 5:638–640
    chaparral, 5:639–640
    deserts, 5:640
    temperate forests, 5:638–639
    temperate grasslands, 5:639
    tropical forests, 5:640
    tropical grasslands, 5:639
    tundra, 5:638
  definition, 1:261; 2:407; 5:637
  major threats to, 2:421
  net productivity, and carbon cycle, 1:613
Biome theory, in island biogeography, 3:567–568
Biomineralization, 1:444
Biomolecules, 4:440
Biomonitoring, 2:367–369
Bionic prospecting, 1:478–479
Bionics, definition, 1:471
Bionomic equilibrium
  definition, 1:769
  theory of, 1:770
Bionomics, definition, 1:249
Biopiracy, 2:471
  definition, 1:471
Bioprospecting
  benefit sharing models/experience, 1:480–483
  concepts/practices, in Costa Rica
    agreement criteria, 1:485–486
    lessons learned, 1:486–487

    mechanisms/legal framework, 1:484–485
    nature-based initiatives, 1:483–484
    partnerships/management issues, 1:484
  definition, 4:271
  economic value, 4:90–91
    biodiversitiy-related markets, 1:479–480
    biopiracy, 1:480
    conservation, 1:479–480
  modern, 1:472–473
    bionic, 1:478–479
    chemical, 1:473–476
    gene, 1:476–478
  traditional, 1:471–472
Biopulping, 3:161–162
Bioquads
  definition, 2:801
  as key biodiversity data sets, 2:812
Bioregions, definition, 4:403; 5:345
Bioremediation, use of fungi in, 3:161–162
Biosphere
  definition, 2:363, 557
  human impacts on, 2:558–559
Biosphere 2, 2:354–355
Biosphere people, definition, 4:845
Biosphere reserve; *see also* Nature reserves; Protected areas
  definition, 4:845
Biosphere Reserve program (UNESCO), 4:854
Biosynthesis, definition, 4:711
Biota, definition, 2:1, 557
Biotechnology, in plant breeding, 1:555
Biotic impoverishment, 2:560–561
  definition, 2:557
  direct depletion of human life, 2:566
    cultural diversity loss, 2:567
    cumulative effects, 2:569
    environmental injustice, 2:567–568
    epidemics/emerging diseases, 2:566–567
    political instability, 2:568–569
    reduced quality of life, 2:567
  direct depletion of nonhuman life, 2:564
    biotic homogenization, 2:565
    genetic engineering, 2:565–566
    habitat fragmentation/loss, 2:565
    overharvest of renewables, 2:564–565

indirect biotic depletion, 2:561
  altered biogeochemical cycles, 2:563–564
  chemical contamination, 2:562–563
  global climate change, 2:564
  soil degradation, 2:562
  water degradation, 2:561–562
Biotic interaction, definition, 1:831
Biotope, definition, 1:249; 2:497
Bipyridilium herbicides, 4:515
Biramous, definition, 1:916
Birds
  aquatic, eutrophication effects, 2:667
  biodiversity in Cape Region of South Africa, 4:156–157
  biodiversity in Chile, 4:152
  biogeographic regions, 1:509–510
  of boreal forests, 1:541
  brood parasites
    cowbird
      adaptations, 4:370–371
      –host systems, 4:369
      management, 4:373
    cuckoo
      adaptations, 4:368–369
      host defenses against, 4:369
      –host systems, 4:368
    impacts on host population dynamics, 4:372–373
    mating systems, 4:374
    obligate, adaptations, 4:366
    spacing behavior, 4:374
    survey of, 4:366–367
    vocal behavior, 4:374–375
  canopy, 3:23–24
  current threats to, 1:516–517
    reduction of, 1:517–518
  as diaspore dispersers, 4:615
  diets of, 1:494
  ecological diversity, 1:492–494
  endangered
    causes of endangerment, 2:399–401
    conservation approaches
      research and synthesis, 2:401–402
      site/habitat conservation, 2:402
      trade controls and international legislation, 2:402–403
      critically, management techniques
        captive breeding, 2:403–404
        control/restriction of aliens and natives, 2:403
        cross-fostering and crossbreeding, 2:405
        habitat restoration, 2:403
        nest-site provision/enhancement, 2:405
        reintroduction and translocation, 2:404
        role of concerned citizen, 2:405
        supplementary feeding, 2:404–405
    numbers, criteria and extinction rate predictions, 2:398
    regions, countries, habitats, 2:398–399
    target units, identification of criteria for, 2:397–398
      and scale, 2:396–397
      and taxonomy, 2:396
  endemism, centers of, 1:508
  evolutionary convergence, 1:510
  extinction
    fragmentation as cause, 1:513–514
    history of, 1:514–515
    recent prehistory of, 1:515–516
  field observations of interspecific competition
    New England warblers, 1:807–808
    Pacific island birds, 1:808–809
  flight, constraints/opportunities, 1:490
  flightless, 1:496–497
  guilds, 3:296–297
  as indicator species for large scale environmental change, 3:443–444
    farmland bird declines in Northwest Europe, 3:445
    migratory songbird declines in North America, 3:444–445
    peregrine falcons and DDT, 3:444
  introduced species, 1:514
  island, evolution of, 1:511–512
  island colonizations by, 1:510–511
  legislation protecting, 3:238
  in logged forests, 3:755–756
  of mangrove ecosystems, 3:858
  migration, 1:494–496
  migration patterns of, 4:223–225
  nonindigenous
    in Australia, 1:78–79
    in Brazil, 1:81–82
    in India, 1:81
    in South Africa, 1:80
    in U.K., 1:77
    in U.S., 1:75
  North American, diversity, 4:410
  pesticide impacts, 4:521
  as pollinators, 4:726–727
  rainforest, diversity of, 5:5–7
  response to climate change, 1:719, 723
  size, constraints on, 1:490–491
  species abundance, 1:491–492
  species–area relationship, 1:512–513
  species diversity
    and altitude, 1:505–506
    continental differences, 1:500
    geographical, losses of, 1:518–519
    geographical variation, 1:503–505
    and habitat, 1:499–500
    and habitat complexity, 1:500–502
    and habitat diversity, 1:505
    and latitude, 1:506–507
      gradients in, causes, 1:507–508
    and other animals, comparison, 1:507
    and productivity, 1:505
    and succession, 1:503–505
  species replacements, 1:508
  taxonomic diversity, 1:497–499
Bivalvia, 4:239–240
Bivoltine, definition, 1:249
Biwa, Lake, 3:641
Black box, 2:307–308
  definition, 2:305
Bleaching (coral)
  definition, 5:73
  and elevated sea surface temperature, 5:93–94
Blubber, definition, 5:495
Blue whale (*Balaenoptera musculus*), endangered status of, 2:443; 4:56, 59
Bog, definition, 5:781
Bombyliidae (bee flies), 2:821
Booklungs, definition, 1:199
Boreal, definition, 1:261
Botanical gardens, 2:476, 690
  in biodiversity education, 2:772
  canopy plants, 3:38
  as gene banks, 3:180
  role in conservation biology, 1:863
Botany, economic, as discipline, 2:615
Bottom-up effects, definition, 3:265; 4:857
Boulder fields, intertidal, 3:490–491
Bowfin (*Amia calva*), 2:763
Bowhead whale (*Balaena mysticetus*), 4:62

Boys and Girls Clubs of America, 2:771
Boy Scouts of America, 2:771
Brachyodont, definition, 3:777
Brachypterous, definition, 3:247
Brackish, definition, 1:915
Brazil
  deforestation and economic policy, 2:294
  exotic species
    invertebrates, 1:82
    plants, 1:81
    vertebrates, 1:81–82
Breed, definition, 1:533
Breeding line, definition, 1:533
Bridgeoporus nobilissimus, 3:161, 162
Bromus tectorum, 3:504–505
Brood parasitism, 4:365–366
  conspecific, 4:375
  impacts on host population dynamics, 4:372–373
  modeling host–parasite coevolution, 4:371–372
  natural history, 4:366
    black-headed duck (*Heteronetta atricapilla*), 4:367
    conspecific brood parasitism, 4:367–368
    cuckoo-finch, 4:367
    Cuculinae, 4:366–367
    Icterinae, 4:367
    Indicatoridae, 4:367
    nonavian, 4:368
    obligate avian parasites, adaptations, 4:366
    viduine finches, 4:367
  research needs, 4:375–376
Broodstock, definition, 1:185
Brown tree snake (*Boiga irregularis*), 2:712; 3:518, 772
Browse, definition, 5:33
Bryophytes
  canopy, 3:29
    taxa diversity, 3:32–33
  definition, 3:27; 4:451
Buddhism, 5:115, 287
Buffon, Comte de, 1:458
Bulk density, definition, 5:269
Bund, definition, 5:805
Bundle sheath, definition, 1:575
Bush-fallow cropping system, 1:104
  definition, 1:99
Bushland, African, 1:63
Buzzards (*Buteo buteo*), 5:137
Bycatch, 2:787–788; 3:387; 5:166
  definition, 2:783

C

Caatinga, definition, 1:145; 5:327
Caboclos, definition, 1:897
Caecilians
  larval, ecology and functional morphology, 1:165
  morphologic features, 1:162
  reproductive biology and life histories, 1:163
Calanoid copepods, definition, 4:497
Calcareous fen, definition, 5:781
Calcification (coral), 5:80–81
  definition, 5:73
  and elevated CO_2, 5:94
California, biodiversity in, 4:148–149
  vascular plants, 4:149
  vertebrates, 4:149
Cambrian explosion, 1:411–412, 416; 2:93; 3:124, 222, 223
  definition, 1:411
Cambrian period, extinctions of, 2:720
Camp Fire Boys and Girls, 2:771
CAM photosynthesis, 1:576; 4:557
Camptothecin, 4:718–719
Canaima National Park (Venezuela), 4:318
Candida albicans, 3:155
Cane toad, 1:75, 79
Canidae (dogs, foxes), 1:630
  African wild dog (*Lycaon pictus*), 3:815–817
  feral dogs, 1:74–75, 81
  food caching by, 1:634
  foraging behavior, 1:634
  interspecific competition, 1:809–810
  mating systems, 5:298
  social systems, 1:638
Canopy
  Amazonian rain forest, 1:146
  architecture and animal substrate, 3:20–21
  definition, 3:20, 28, 747
  exploration of, 3:21
  gaps, remote sensing in study of, 5:139–140
  historical roots, 3:28–29
  information sources, 3:29
  invertebrates, 3:21–23
  logged forests, 3:751
  plants of
    areas for further study, 3:38–39
    conservation, 3:37
    effects of forest fragmentation/habitat conversion, 3:37
    effects of global environmental change, 3:37
    growth habit diversity, 3:36
    physiological diversity, 3:36
    resource acquisition/retention, diversity of modes of, 3:36–37
    habitat diversity
      host tree specificity, 3:36
      microclimate, 3:35
      spatial scales of diversity, 3:35–36
    hemiepiphytes, 3:30–31
      taxa diversity, 3:33–35
    nonvascular epiphytes, 3:29–30
      taxa diversity, 3:32–33
    parasitic, 3:31
      taxa diversity, 3:35
    vascular epiphytes, 3:29
      ant-associated, 3:37
      atmospheric epiphytes, 3:37
      bark epiphytes, 3:37
      characteristics, 3:30
      human epiphytes, 3:36
      tank and trash basket epiphytes, 3:36
      taxa diversity, 3:31–32
    vulnerability to extinction and invasion, 3:37
  vertebrates, 3:23–24
Canopy biology, as discipline, 3:19
Cantharophily
  definition, 4:723
Capoeira, definition, 1:145
Captive breeding
  critically endangered birds, 2:403–404
  elements of success, 1:600
  mammal, 3:815
Captive breeding programs; *see also* Reintroduction programs
  difficulties
    adaptation to captivity, 1:601
    breeding failure in captivity, 1:601
  disease in, 2:119–120
  general aspects of, 1:600–601
  genetic/demographic management of population
    during capacity phase, 1:603–604
    during founding phase, 1:602–603
    during growth phase, 1:603–604
    starting program, 1:602
  need for, 1:600
Carbamate
  herbicides, 4:515
  insecticides, resistance to, 3:469
Carbamate pesticides, 4:512

Carbon
  biogeochemical processes, 1:443
  in oceans, 1:448
Carbon cycle
  atmospheric analysis, 1:611–612
  definition, 4:111
  effects of deforestation, 1:609–611
  and fossil fuel reserves, 1:619
  general features, 1:609–611
  geological process, 1:618–619
  history of, 1:622–624
  management of, 1:625
    forest sequestration, 1:625–626
    geological sequestration, 1:627
    ocean sequestration, 1:626–627
  ocean analysis, 1:612–613
  perturbation of, 4:732
    climate change, 4:732–734
    ecosystems and, 4:732–734
  role of rivers, 1:617–618
  role of soil, 1:614–615, 617
Carbon dioxide
  atmospheric, 1:295, 611–612
    elevation, effects on
      coral reefs, 5:93, 94
      photosynthesis and growth, 1:620–621
    sources of, 1:296
  emissions
    and climate prediction, 1:714
    from fossil fuels, 1:619
  equivalent CO2 concentration
    definition, 3:277
    in prediction of climate change, 3:279
  fixation, NAD(P)H and ATP in, 4:555–557
  in greenhouse effect, 3:278
  sinks, 1:610, 612
    ocean, 1:612–613
      effects of climate change, 1:620–621
    terrestrial, 1:613–617
      effects of climate change, 1:621–622
Carboniferous period, 1:925
Carbon monoxide, atmospheric, 1:299–309, 611
Carbon pump, biological, 3:283
Carboxylase, definition, 4:549
Carnivora, 3:793–794
  definition, 4:857
  diets of, 1:634
    specialization in, 1:635–636
  diversity, changes in, 1:639–640
  foraging behavior, 1:632–634

  and phylogeny, 1:634–636
  guilds, in ecosystems, 1:636–638
  size and ecology of, 1:631–632
  social systems of, 1:638–639
  species diversity, 1:629–631
  taxonomy, 1:630–631
Carnot efficiency, 2:530
b-Carotene, in photosynthesis, 4:552
Carrying capacity
  definition, 1:643–644; 2:229; 5:285
  human, 1:647
    and ecofootprint analysis, 2:233–234
    importance to biodiversity, 1:647
  origin of concept, 1:642–643
  of populations, 1:644–645
    determination, 1:645–646, 647
Carson, Rachel, 3:364
Case law, definition, 3:233
Castes
  definition, 3:581
  differentiation in termites, 3:595
Catastrophes, and demographic population fluctuation, 4:834
Catches, definition, 2:801
Catenulida, 5:869
Caterpillars, 1:564–565
  ant associations, 1:567
  call production, 1:567
  semiochemical production, 1:566
Cation, definition, 1:437
Cation exchange capacity, definition, 1:1; 5:269
Catlin, George, 1:886
Cats, see Felidae
Caulerpa taxifolia (alga), 3:519–520
Caves, 5:528; see also Subterranean ecosystems
  biodiversity, geography of, 5:538–539
  microflora, 4:918
Cech, Thomas, 4:444
Cecidomyiidae (gall midges), 2:821
Cells, information transfer in, 3:186
Cenozoic era, mass extinction of, 4:121
Center for Plant Conservation (U.S.), 4:656
Central America
  as biogeographic region, 1:665–669
  high-elevation ecosystems, 1:674
  historic rain forest distribution, 5:26–27
  life zones and ecosystems, 1:669–670
  seasonally dry tropical forests, 1:671–672

  temperate forests, 1:673–674
  tropical cloud forests, 1:672–673
  tropical rain forests, 1:670–671
Cephalochordata, 2:757; 5:760
Cephalopoda, 4:243–244
Ceratopogonidae (biting midges), 2:821
Cerci, definition, 3:247
Cercomer, definition, 5:863
Cerotegument, definition, 4:291
Cerrado, 1:149–150
Cerrado, 1:153–154
  definition, 5:327
Certification, definition, 4:85
Cetacea, 3:794; 4:37
  diving capabilities, 4:39
  evolutionary history, 4:42–43
  features and habitat boundaries, 4:39–40
CFCs, see Chlorofluorocarbons
Chaco, definition, 5:327
Chaetae, definition, 5:831
Champsosauridae, definition, 2:95
Change matrices, definition, 5:25
Chaparral, definition, 4:145
Character, definition, 5:569
Character displacement, 1:757
  definition, 1:25, 753; 2:85
Characters, taxonomic
  in cladogram evaluation, 1:687–689
  coding, for cladistic analysis, 1:684
  homology between, 1:682
  as phylogenetic evidence, 1:683–684
  plesiomorphic versus apomorphic, 1:678
  polarization, and cladogram rooting, 1:685–686
  quantitative versus qualitative, 1:682–683
  recognition, 1:682
Cheetah, in African grasslands, 1:65
Chelate, definition, 1:915
Chelicerae, definition, 1:199
Chemical contamination, 2:562–563
Chemical defenses, definition, 2:11
Chemical prospecting, 1:473–476
  definition, 1:471
Chemical sensitivity, 1:48
Chemiosmotic energy coupling, definition, 4:549
Chemoautotroph, 1:437
  definition, 1:437; 5:747
Chemoautotrophs, 1:452
  definition, 2:509; 4:349
Chemoautotrophy, definition, 4:201
Chemolithotroph, definition, 5:647

Chemolithotrophy, definition, 5:527
Chemostat, definition, 3:729
Chemotroph, definition, 4:191
Chesapeake Bay, 2:587–589
Chestnut blight fungus (*Endothis parasitica*), 2:118; 3:523, 524, 527
Chiasmata, definition, 5:53
Chile
   biodiversity and traditional agriculture, 1:116
   biodiversity in, 4:149–150
      birds, 4:152
      reptiles and amphibians, 4:152–153
      vascular plants, 4:150–151
      vertebrates, 4:151–152
Chilopoda (centipedes), 4:292
Chimaeras (Chondrichthyes), 2:760
China
   desertification in, 2:68
   eastern, temperate deciduous forests, 1:281
   Neolithic agriculture, 2:219–220
   in situ livestock preservation, 1:542
Chironomidae (midges), 2:821
Chiroptera, 3:791
Chiropterophily
   definition, 4:723
Chlorofluorocarbons, 2:368, 563
   atmospheric, 1:298–299
   definition, 5:723
Chlorophylls, 4:550
Chloropidae (fruit flies), 2:822
Chloroplast DNA, 3:191
   diversity in, 3:201
Choanata, 5:762
Chordata, 5:759–760
Chordate, definition, 5:755
Chordates
   early history, 5:757–758
   major groups, 5:759
      Amniota, 5:763
      Cephalochordata, 5:760
      Choanata, 5:762
      Chordata, 5:759–760
      Craniata, 5:760–761
      Gnasthostomata, 5:761
      Mammalia, 5:765
      Osteichthes, 5:761–762
      Reptilia, 5:763–764
      Sarcopterygii, 5:762
      Sauria, 5:764–765
      Tetrapoda, 5:762–763
      Vertebrata, 5:761
Christianity, 5:287
   spread of, in Gangte community, 4:848–849

Chromatin, 3:183, 189
Chromophore, definition, 4:549
Chromosomes, 3:183
   bacterial, 1:341–342
   definition, 4:759
   mutations in, 3:191–192
Chronic toxicity, definition, 1:119
Chronostratigraphic correlation, 3:56–57
   definition, 3:53
Chytridiomycota, 3:145
Cichlids, African, 1:67
   adaptive radiation, 1:30, 406
      case studies, 1:36
   sympatric speciation, 5:378
Ciguatoxin, 4:714
Ciliates, 4:912–914
   and nuclear dualism, 4:416
Cimicidae, 5:720
Circannial rhythms, definition, 4:221
Cirrus, definition, 5:863
Cities
   functional footprints of, 2:236–237
   and human-dominated techno-ecosystem, 2:308–309
   versus rural perceptions of biodiversity, 5:293
Clades
   definition, 1:677, 693; 3:841; 5:569, 589
   in evolutionary systematics, 1:698
Cladistic biogeography
   and component analysis, 5:775–776
   definition, 5:767
Cladistics, 5:573
   definition, 1:677; 5:589, 767
   progression rule, 5:771–773
   relationships in, 1:677–678
Cladogenesis, 1:694
   definition, 1:693; 5:569
   mechanisms for, 1:695–696
Cladograms
   conditional data combination, 1:692
   consensus, 1:689–690
   construction, 1:684–685
   definition, 1:677; 5:569, 589
   evaluation
      character fit, 1:687–688
      character weighting, 1:688–689
   Fitch optimization, 1:687
   partitioned analysis, 1:691–692
   versus phylenogenetic trees, 1:681–682
   rooting, and character polarization, 1:685–686
   simultaneous analysis, 1:691

   Wagner optimization, 1:686–687
Cladogram support
   definition, 1:677
   group, 1:690–691
Classifications, systematic, natural classification, 1:697–698
Claudistics, monophyly and paraphyly in, 1:698, 700–701
Clements, Frederic, 5:174
Climate
   African, 1:56, 58
   austral ecosystems, 5:367–368
   boreal zone, 1:534
   and forest ecology, 3:42–43
   and species richness
      experiments, 1:739–740
      global-scale observations, 1:730–733
      observations by area, 1:730–731
      theories, 1:733–734
   stability, and megadiversity, 1:425
   tropical forest ecosystems, 5:703–704
Climate change, 2:564; 3:386
   abrupt, 1:724
   ancient trends, 3:278
   and Antarctic ecosystems, 1:182–183
      biological consequences, 1:183–184
   anthropogenic, 3:676
      biodiversity threats from, 3:283–284
   and biodiversity over geologic time, 3:229–230
   biota likely to benefit from, 3:289
   biota most at risk from, 3:288
   biotic responses to
      adjustment, 3:286–287
      evolution, 3:287
      extinction, 3:291–292
      migration, 3:287, 289–291
   and Central American ecosystems, 1:674–675
   and coral reefs, 5:93–94
   definition, 4:731; 5:73
   and disease, 2:121–122
   ecological responses to, predictions
      animals, 1:719–721
      approaches to, 1:721–723
      vegetation, 1:718
   ecosystems and, 4:734
      coastal-marine, 4:735
      forests, 4:734–735
      freshwater, 4:735–736
   effects on biota

current impacts, evidence for, 3:291
  integrated research, 3:286
  manipulation studies, 3:285–286
  models for, 3:284–285
  and natural climate variability, 3:285
  paleobiological studies, 3:285
effects on fire ecology, 2:752–753
endangerment of freshwater invertebrate from, 2:432–433
external versus internal forces of, 1:712–713
and extinctions, 2:709–710
and fish conservation, 2:792
and forest biodiversity, 3:50
general circulation models for, 3:279
impacts of desertification, 2:74–75
impacts on species' social systems, 5:300–301
and indicator species
  fossil record, 3:442–443
  species distribution, contemporary changes in, 3:443
prediction of future trends, 2:709; 3:279–280
projections, 1:713–714
recent trends, 3:279
response of CO_2 sinks
  ocean, 1:620–621
  terrestrial, 1:621–622
salmon threat from, 5:240–241
synergisms with habitat fragmentation, 1:710
threats to amphibians from, 2:481–482
Climate modeling, relevance to regional changes, 1:716–717
  downscaling predictions to, 1:717–718
Climate sensitivity, definition, 3:277
Climate–vegetation classification systems, 3:284
Climatic models, definition, 1:709
Climatic stress, evolution across phylogeny caused by, 3:206–209
Climax, definition, 3:675
Climax community, definition, 5:541
Clique, definition, 5:441
Cloaca, definition, 1:159; 3:777
Clonal growth, definition, 1:133
Clone, definition, 3:675; 4:177
Cloning, in preservation of livestock biodiversity, 1:544–545
Clupeid fish, definition, 4:497
Clutch size, definition, 1:17

Cnidaria, definition, 5:73
Coadapted gene complexes, definition, 3:427; 5:383
Coal, biodiversity impacts of, 2:533–535
Coalescent process, definition, 2:85
Coastal, definition, 4:13
Coastal ecosystems
  African, 1:67–68
  Asian, 1:286–290
  and climate change, 4:734–735
  endangered, 2:418
  eutrophication effects, 4:383–384
  major threats to, 2:418–419
  and megadiversity, 1:424–425
  scrub habitats, predators, 4:866–868
  seagrasses
    biogeography/species richness, 5:258–259
    genetic diversity, 5:262–263
    origins and evolution, 5:255–257
  seagrass habitats, biodiversity in, 5:263–266
Coastal sage shrub, definition, 4:145
Coastal zone, definition, 2:579
Code, definition, 4:389
Codon usage, in bacteria, 1:343
Coelacanth, 2:761
Coelescleritophora, 4:237
Coevolution
  brood parasites–host, modeling, 4:371–372
  definition, 2:1; 3:265; 4:365
  evidence for, 1:755–757
  herbivore–plants
    resistance, tolerance, and overcompensation, 1:763–764
    specialist and generalist herbivores, 1:764–765
  host–parasite, 2:116–117
  and intraspecies competition, 1:757–759
  major consequences, 1:765
    community structure, 1:766–767
    phenotypic diversity, 1:765–766
    species diversity, 1:766
  meaning and varieties of, 1:753–755
  mutualism, 1:765
  of plants and herbivores, 4:606–607
  role in adaptive radiation, 1:28–29
Coevolutionary arms race, definition, 4:365
Coexistence, see Species coexistence
Cognition, definition, 1:45
Cohesion, species, definition, 5:427

Cohort, definition, 2:311
Cold seeps, 3:558–559
Coleoptera (beetles), 1:351–352; 3:482–483
  competition experiments, 5:416–417
  fauna, in forests under intensive/extensive management, 5:664–665
  global diversity, estimation of, 1:356–357
  local faunas, 1:355–356
  major families
    Buprestidae, 1:352, 354
    Carabidae, 1:352
    Cerambycidae, 1:354
    Chrysomelidae, 1:354
    Curculionidae, 1:354
    Scarabaeidae, 1:352
    Staphylinidae, 1:352
    Tenebrionidae, 1:354
  as pollinators, 4:724
  rainforest, diversity of, 5:4
  species richness, explanation, 1:354–355
  taxonomic and ecological patterns, 1:355
Coleorrhyncha, 5:712
Coleus forskohlil (Willdenow), 2:471
Collections, biological, 4:271–272
  in biodiversity-rich versus poor nations, 1:432–434
  content, 4:272
  definition, 3:165
  and taxonomy, 5:571
  uses and roles of
    biological research, 4:274
    environmental conditions, 4:274
    inventories and conservation, 4:274
    legislation implementation, 4:274
    resource management, 4:274–275
    taxonomy studies, 4:272–274
Collembolans (springtails), soil, 5:311–312
Colonization
  determinants of success, 2:133
    ecological opportunity, 2:134–135
    physical access, 2:134
    preadaptations, 2:133–134
  of virgin habitats, observation of, 2:137–139
Combinatorial chemistry, definition, 1:471
Comisiòn Nacional Para el Estudio y Uso de la Biodiversidad (CONABIO), 1:434–435

Comisión Nacional Para el Estudio y
  Uso de la Biodiversidad
  (CONABIO) (continued)
  definition, 1:419
Commensal, definition, 1:915; 4:901
Commensalism, 4:464
  definition, 4:463
Common-pool resources
  definition, 1:777
  incentives, using rules to change,
    1:780–781
    authority rules, 1:782–783
    boundary rules, 1:781–782
    information, scope, and
      aggregation rules, 1:783–784
    payoff and position rules,
      1:782–783, 783
    rule changing, as adaptive
      process, 1:786–788
    self-organized resource government
      systems, 1:784–785
Commons (common property
  resource)
  and biological reserves, 1:775
  community control of, 1:774–775
  definition, 1:769, 777
  economic concepts, 1:775–776
  examples of, 1:770
  management
    allocated quotas, 1:773
    of incentives versus resources,
      1:773
    international, 1:773–774
    risk/uncertainty in, 1:775
  policies, focus of, 1:778
  privatization of, 1:774
  and regulated bionomic equilibrium,
    1:772–773
  theory of bionomic equilibrium,
    1:770
  tragedy of, 1:769; 4:89–90
    definition, 4:891
    polycentric systems in coping
      with, 1:789–790
Community-based conservation, 3:414
Community/communities
  climax, definition, 5:541
  and conservation genetics,
    4:805–806
  definition, 1:249; 3:1; 4:13; 5:285
  diversity
    definition, 1:307
    local control of
      diversity–productivity
        relationship, 2:164–166
      prediction of, 2:167

  species–energy theory,
    2:163–164
  Volterra–Gause perspective,
    2:162–163
  regional control of
    metacommunity processes,
      2:169–171
    regional processes, 2:167–169
  ecological, latitudinal variation in,
    3:705–708
  and fossil record, 1:396–397
  and regional perspectives, synthesis,
    2:174–175
    comparative approach, 2:175–176
    manipulation of productivity and
      species pool, 2:174–175
  saturated, testing for, 2:171–173
  stability of, 1:398
    as complex adaptive systems,
      1:398–399
    effects of disturbance on
      biodiversity, 1:399–400
  structure
    and coevolution, 1:766–767
    definition, 3:303
  structure and functioning of,
    3:131–133
  taxa saturation, 1:398
  type I and II, definition, 2:161
Compass orientation, definition, 4:221
Compensation depth, definition, 2:311
Competition
  as constraint on species coexistence,
    5:415
  definition, 1:71; 3:729; 4:769; 5:413
  and evolution, 5:460–461
  interspecific
    field experiments
      Anolis lizards, 1:806–807
      ants and mice, 1:806
      barnacles, 1:805–806
    field observations
      African wild dogs and hyenas,
        1:809–810
      New England warblers,
        1:807–808
      Pacific island birds, 1:808–809
    laboratory experiments
      Drosophila, 1:804–805
      flour beetles, 1:802–804
      Paramecium, 1:801–802
      yeast, 1:800–801
      yeast and Paramecium,
        1:800–802
    models
      aggregation model, 1:798–799

      Lotka–Volterra, 1:794–798, 800
      resource utilization curves,
        1:798–799
    and niche descriptions, 3:311
    intraspecific, and coevolution,
      1:757–759
    and pairwise interactions,
      5:454–457
Competitionism, definition, 4:857
Competitive exclusion, definition,
  1:71; 2:229
Competitive release, 3:265
Complementarity
  and area selection, 1:814
  and biodiversity indicators,
    1:815–816
  biological basis of, 1:814
  history of, 1:815
  need for, 1:814–815
  use of, 1:815–816
  assessment of
    comparing area sets, 1:827
    efficiency, 1:824–825
    flexibility, 1:825
    irreplaceability, 1:825–827
  definition, 1:814
  techniques based on, 1:819
    branch-and-bound algorithms,
      1:820
    exhaustive search, 1:819–820
    heuristic algorithms, 1:820–823
    prioritizing areas, 1:823–824
Complementary conservation,
  definition, 2:683
Complex adaptive systems, 1:784–785
  definition, 1:393
  properties and mechanisms, 1:785
    building blocks, 1:786
    internal models, 1:786
    tag use, 1:785–786
Component analysis, 5:775–776
Components (cladistic), definition,
  5:767
Comstock, John Henry, 5:576
CONABIO, see Comisión Nacional
  Para el Estudio y Uso de la
  Biodiversidad
Condor, California, restoration of,
  5:194–195
Confucius, 2:612, 613
Conjugation
  bacterial, 1:347
  definition, 1:339
Connectivity web, definition, 3:1
Connectivity webs, 3:3–5
Consensus cladogram, definition,
  1:677

Conservation
  biodiversity
    duties to humans, 2:549–551
    and economic development, 3:414
    environmental philosophies
      deep ecology, 2:553–554
      ecofeminism, 2:554–555
      humanism, 2:555
    intrinsic value arguments
      anthropocentric denials of
        intrinsic value, 2:547–548
      from economic to ethics, 2:546
      ethical holism, 2:548–549
      extensionist arguments for
        intrinsic value, 2:546–547
      nonextensionist arguments for
        intrinsic value, 2:547
    religious stewardship, 2:551
  community-based, 3:414
  definition, 5:481
  consideration of social behaviors in,
    5:302–303
  definition, 2:683; 3:811
  emerging themes in, 1:895–896
  evolving thinking in, analysis
    human influences on
      and invasive species, 1:973–975
      and sustainable development,
        1:972–973
    species to ecosystem shift,
      1:968–971
  future directions and challenges in,
    1:979–980
  goals of, 2:594
    health, 2:594–595
    integrity, 2:595–597
    sustainability, 2:597–598
  historical
    background of, 1:883–884
    historic precedents, 1:884–886
    modern origins, 1:886–887
      gaining definition of,
        1:887–888
      Progressive Era and utilitarian/
        preservationist split,
        1:888–889
    prehistoric precedents, 1:884
  implications of keystone species,
    3:629–630
  and the living dead, 3:698–699
  mammals
    captive breeding/reintroduction,
      2:450–451
    economic incentives, 2:451–452
    legislation, 2:449–450
    protected areas and preserve sizes,
      2:449–450
    species survival commissions,
      2:449–450
    trends, 2:452–453
  new concepts in, 1:966–968
  non-biodiversity foci of, problems
    with, 3:365
  planning, effect of scale on,
    5:253–254
  of plant biodiversity, 4:9–10
  representing biodiversity for
    constraints, 1:819
    goal identification, 1:816–817
    maximum-coverage sets,
      1:817–819
    minimum-cost sets, 1:817
  role of biogeography, 1:469–470
  role of collections, 4:274
  species, tools for, 1:975, 977
    economic evaluations, 1:979
    models, 1:977
    reserve designs, 1:978–979
    species and spatial relationships in
      landscape, 1:977–978
  and subspecific taxonomy, 5:526
  of subterranean ecosystems,
    5:539–540
  support for, and tourism, 5:676
  traditional practices, 5:682–683
    ecological process management,
      5:690–691
    habitat protection, 5:685–686
    management of landscape
      patchiness, 5:689–690
    management of natural
      disturbances, 5:691
    multiple-species/integrated
      management, 5:687–688
    nurturing of ecosystem renewal
      sources, 5:691
    protection vulnerable life
      histories, 5:685
    resource rotation, 5:688
    resource-status monitoring,
      5:683–684
    species protection, 5:684–685
    succession management,
      5:688–689
    temporal harvest restriction,
      5:686–687
    watershed management, 5:690
  twentieth century, evolution of,
    1:889
    from conservation to
      environmentalism, 1:892–894
    during Great Depression/World
      War II, 1:891–892
    institutionalization of
      conservation, 1:890–891
    reintegration of conservation,
      1:894–895
  types of, 3:814–815
  umbrella principle, 2:452
Conservation biologists
  ethical consensus of, 2:593–594
  goals of, 2:594
    health, 2:594–595
    integrity, 2:595–597
    sustainability, 2:597–598
  ideal of, 2:605–606
Conservation biology
  definition, 1:883
  ecosystems and communities,
    1:856–857
    corridors and connectivity, 1:860
    GAP analysis, 1:858–859
    reserve selection algorithms,
      1:859–860
    SLOSS debate, 1:857–858
    spatial considerations, 1:858
  future challenges, 1:863–864
  individuals and genes, 1:862
    measuring genetic variability,
      1:862–863
    role of zoos, aquaria, botanical
      gardens, and gene banks,
      1:863
  and pollination, 4:729–730
  species and populations, 1:860–861
    population viability analysis,
      1:861
    small/declining populations,
      demography, 1:861–862
Conservation genetics
  evolutionary processes
    gene flow, 4:808–809
    genetic drift, 4:809
    genetic erosion, 4:810–811
    hybridization, 4:808
    inbreeding, 4:807
    mating system, 4:806
    mutation, 4:806
    natural selection and adaptation,
      4:811
    outbreeding, 4:807–808
    population bottlenecks,
      4:809–810
  evolvability and future of
    biodiversity, 4:815–816
  genetic management, examples,
    4:811–812
    conservation management,
      4:814–815

Conservation genetics (*continued*)
  defining species, 4:813–814
  genetic consusing, 4:812–813
  mating system, 4:812
  pedigree management in very small populations, 4:812
  phylogeography, gene flow, and population structure, 4:813
  problems of hybrids, 4:812
  reintroductions, translocations, and genetic enhancement, 4:814
 levels of interest
  communities, 4:805–806
  genes, 4:802–803
  populations, 4:803–804
  species, 4:805
  subspecies, 4:804–805
Conservation status, definition, 2:407
Conservation strategies
  high-value ecosystems, 3:378–379
  wilderness areas, 3:378
Conservation tillage, definition, 1:99
Conserved (scientific name), definition, 4:389
Conspecific, definition, 4:759
Conspecific brood parasites, definition, 4:365
Conspecific brood parasitism, 4:375
Constitutive defenses, definition, 2:11
Contamination, definition, 1:119
Continental drift, 1:462–465
 and austral ecosystems, 5:366
  Cenozoic period, 1:928
  definition, 1:456; 5:361
  Jurassic period, 1:926–927
  of Ordovician, 1:925
Continental rise, definition, 4:1
Continental shelf
  definition, 4:1, 22, 403
Continental slope, definition, 4:1
Continents, early, 1:924
Contingency, definition, 3:739
Contracted vegetation, definition, 4:353
Convective mixing, definition, 4:427
Convention on Biological Diversity (1993), 1:433; 2:471–473, 765; 3:236–237; 4:647, 722
  definition, 1:419; 5:627
Convention on International Trade in Endangered Species of Wild Fauna and Flora (1973), 1:326; 3:236
Convention on the Law of the Sea (United Nations), 5:164

Convergence, definition, 1:25
Convergent characters, definition, 5:569
Convergent evolution, 3:126
  definition, 3:121
  groupings among mobile organisms, 3:126–128
  groupings among sessile organisms, 3:126–128
  mammalian, 3:801–802, 803
Cooloolidae (cooloola monsters), 3:258
Cooperative Holocene Mapping Project, 1:711, 712
Copepods, calanoid, definition, 4:497
Cope's Rule, 3:799–800
Coppice, definition, 3:675
Coral reefs, 2:807; 5:74
  benefits from, 5:81–82
  biodiversity, 4:21–22
  control of scleractinian diversity, 5:77–78
  coral biology, 5:80
    anatomy, 5:80
    calcification, 5:80–81
    photosynthesis, 5:81
    physiological limitations, 5:81
    reproduction and recruitment, 5:80
  coral disease
    diversity effects, 5:90–92
    human influence, 5:93
    identification, 5:87–90
    model of, 5:92
  crustacean biodiversity, 1:918
  definition, 4:13
  diversity
    and ecosystem function, 3:134–135
    and stability of, 3:133–134
  ecosystem insurance, 3:135–137
  endangered, 2:419–420
  equivalent species and ecosystem redundancy, 3:135–137
  eutrophication effects, 4:384
  fish diversity, 2:781
  and global climate change, 5:93–94
  human impacts, 4:34; 5:84
    altered salinity, 5:86–87
    altered temperatures, 5:86
    coastal urbanization, 5:82–83
    eutrophication, 5:85–86
    heavy metals/toxins, 5:86
    sedimentation, 5:83, 85
  loss of, 2:704
  modern, age of, 1:929

  phyletic diversity, 5:75
  plant–herbivore interactions, 3:266, 267, 269
  restoration, use of grazers in, 3:275
  species diversity, 5:75–77
  species loss, 5:78–80
  species richness, 1:423
Corridor, definition, 3:645
Cortex, definition, 5:863
Cospeciation, 1:754
Costa Rica, Instituto Nacional de Biodiversidad (INBio), 1:435, 480–481, 482, 483–487; 4:275–276
Cowbirds, *see* Icterinae
Craniata, 5:760–761
Craniomandibular, definition, 3:777
Crassulacean acid metabolism photosynthesis, 1:576; 4:557
Crenarchaeota, 1:328
  definition, 1:219
Crepuscular, definition, 1:915
Cretaceous period, 1:927
  end, mass extinction of, 4:118–120
  mammalian evolution during, 3:846–847
Cretaceous–Tertiary event, 1:414; 3:223, 230; 4:98–99, 100
  consequences, 3:847–848
  definition, 1:411; 2:95
  extinctions, 2:101, 719, 723–724
    vertebrate, and survival patterns, 2:99–100
  extinction theories, testing with vertebrate fossil record, 2:100
    impact theory, 2:100–102
    marine regression theory, 2:103–106
    volcanism theory, 2:102–103
Crick, Francis, 4:442, 444
Crickets, Hawaiian swordtail, adaptive radiation, case studies, 1:39–41
Criteria and indicators, definition, 2:23
Critically endangered, definition, 2:455
Crochets, definition, 4:249
Crocodilians, 5:146–147
  conservation issues, 5:157–158
  reproduction and parental care, 5:152
Crop genetic diversity, definition, 1:85
Cropping systems
  multiple, in traditional agriculture, 1:110
  sustainable
    bush-fallow rotation, 1:104
    definition, 1:99

mixed farming, 1:104–106
  definition, 1:99
Crop production
  commercialization of, 1:92–93
    implications for genetic diversity, 1:94–96
    seed industries, 1:93–94
  intensification of, 1:87
Crops
  acid deposition effects, 1:8
  domesticated
    conscious versus unconscious selection, 2:222
    maintenance methods, 2:221
    reproductive systems, 2:220–221
      and plant purposes, 2:222–223
    sowing/reaping impacts, 2:223–224
  domestication, 2:217
    in Africa, 1:69
    early centers of, 2:219–220
  fungal pathologies, 3:153
  hybridization of, 4:674–675
  and introduced pests, 3:522
  multiline, 1:557
  productivity and stability, effects of planned diversity, 2:270–271
  properties of, 3:677
  rotation, and biodiversity, 1:556
  transgenic, see Transgenic crops
Crop varieties, definition, 1:547
Crossbreeding, definition, 1:533
Crossing-over, definition, 5:53
Cross-resistance, 3:470
  definition, 3:465
  versus multiple resistance, determination, 3:471
Crown group, definition, 3:841
Crustacea
  endangered species, 2:460, 461
  fossil record, 1:924–929
  geographic distributions, 1:937–938
  habitats, 1:930–931
  intertidal, 1:744–745
  life history patterns
    development types, 1:936
    reproduction, 1:932–936
  of mangrove ecosystems, 3:860–862
  morphological diversity, 1:921–924
  newly discovered group, 1:918
  role in biological communities, 1:931–932
  species flocks in ancient lakes, 3:641
  summary and biodiversity in, 1:916–917
  taxonomic diversity, centers of, 1:921
  vulnerability to extinctions
    fossil record, 1:938–940
    in modern era, 1:940–942
Cryobiosis, definition, 4:917
Cryopreservation
  definition, 1:533
  in preservation of livestock biodiversity, 1:544
Cryoturbation, definition, 1:171
Crypic species, definition, 3:777
Cryptic, definition, 1:559
Cryptoendolithic, definition, 4:191
Cryptogam, definition, 1:171
Cryptosporidium spp., 4:915
Ctenidia, definition, 4:235
Ctenomys spp., nematode diversity in, 5:858–859
Cuckoos, see Cuculinae
Cuculinae (cuckoos)
  adaptations of, 4:368–369
  definition, 4:365
  host defenses against, 4:369
  –host systems, 4:368
Culicidae (mosquitoes), 2:821
Cultivar, definition, 2:375
Cultural diversity
  and biological diversity, relationship, 3:454–456
  correlation with biodiversity, 3:802
  definition, 1:419; 3:452
  in megadiverse countries, 1:421
Cultural ecology, definition, 3:790
Cultural ecosystems, definition, 3:675
Cultural eutrophication, definition, 5:213
Cultural evolution, human, trends, and biodiversity impacts, 3:791–792
Culture, definition, 3:790; 5:285
Curacin A, 4:713
Curlew, Eskimo (*Numenius borealis*), 2:399
Cursorial, definition, 3:777
Cuscuta spp., see Dodder
Cuticle, definition, 5:843
Cyanobacteria, 2:195
  anticancer agents from, 4:712–713
    cryptophycin, 4:713
    curacin A and chemotype concept, 4:713
  definition, 4:201, 349
Cyperaceae, definition, 4:451
Cytochrome b gene, mitochondrial, and genetic distances in vertebrates, 3:201
Cytochrome oxidase, 1:227
Cytochrome P450 monooxygenases, 3:467
  definition, 3:465
Cytochromes P-450, 2:364
  as biomarkers, 2:369
Cytotoxicity, definition, 4:711

D

Dams, 2:536–538; 3:385
  salmon threat from, 5:240
  as threat to freshwater fauna, 3:541–542
  threat to riverine ecosystems, 5:226–228
Dartmoor National Park (UK), 4:319
Darwin, Charles, 1:460, 463, 694; 2:672; 3:267, 363, 742–743; 4:440
  Beagle voyage, 2:2–3
  childhood and education, 2:2
  on coevolution, 1:755
  and concept of niche, 3:305
  contribution to biodiversity studies
    classification, 2:7–8
    competitive natural economy, 2:5–7
  on convergent evolution, 3:126
  Darwinism following, 2:8
  discovery of natural selection, 2:3
  on ecological complexity, 1:832
  later publications, 2:4–5
  on nature of the organism, 2:191–192
  Origin of Species, 2:4
  on species dispersal, 2:128
Darwinism
  and adaptation, 1:18–19
  and adaptationism, 1:22
  alternatives to, 2:8
  contemporary developments, 2:9
  and literature, 3:742–743
  reconciliation with genetics, 2:8–9
Dasyuridae (marsupials), 1:631
Database, definition, 1:419
Databases, canopy plants, 3:38
DDT, 2:366–367, 368, 370–371, 563; 4:742
  and peregrine falcons, 3:444
Declarations of Assisi
  Buddhist declaration, 5:114
  Christian declaration, 5:111
  Hindu declaration, 5:114
  Jewish declaration, 5:112
  Moslem declaration, 5:112
Decomposition, definition, 2:509; 3:531

Deep ecology, 2:553–554
  definition, 2:545
Deep sea, definition, 3:543; 4:13
Defenses, plant
  and animal self-medication, 2:19–20
  chemical
    definition, 2:11
    diversity and function of, 2:11–12
      alkaloids, 2:12–13
      nonprotein amino acids, 2:13
      phenolic compounds, 2:12
      saponins, 2:13
      terpenes, 2:13
      toxic proteins, 2:13
  human uses of, 2:18
    hunting and fishing, 2:18–19
    spices, 2:20
  mutualisms, 2:14
    ants, 2:14–15
    domatia, mites, and other predators, 2:15
    endophytic fungi, 2:15
    predators/parasitoids, 2:15–16
  phenological strategies, 2:16
  physical/structural, 2:14
  plant investments in
    assumptions, 2:16
    costs/benefits, 2:16–17
    induced defenses, 2:17
  plasticity in, 2:18
  theories of, 2:17–18
Definitive host, definition, 4:463
Deforestation, 3:384, 663, 799
  and carbon flux, 1:614
  in Central America, 1:675–676
  and C4 grasses, 1:594–595
  and climate change, 3:283–284
  consequences of
    biodiversity, 2:32
    economic losses, 2:33
    ecosystem services, 2:32–33
    global climate change, 2:33
    social, 2:33
  contemporary, 2:24–25
  definition, 2:23, 407; 5:25
  direct causes of
    cattle ranching/livestock grazing, 2:30
    commercial agriculture, 2:30
    fire, 2:32
    fuelwood collection/charcoal production, 2:31
    infrastructural development/industrial projects, 2:30–31
    logging, 2:31–32
    plantations, 2:31
    shifting cultivation, 2:30
  and economic policy, in Brazil, 2:294
  global, quantification
    conflicting estimates, 2:26
    improvement of assessment accuracy, 2:25–26
    remote sensing developments, 2:26
  historical, 2:24
  indirect causes of, 2:26–27
    developmental policies/tax incentives, 2:27
    forest/forest product undervaluation, 2:29
    market demands, 2:28–29
    population growth/poverty, 2:27
    resettlement programs, 2:27
    tenurial policies, 2:27–28
    weak government institutions, 2:29–30
  and the living dead, 3:691–693
  patterns of, 3:771
  rain forests, assessment techniques, 5:25–26
  in Southeast Asia, 3:667–668
  sustainable development alternatives
    improving productivity of subsistence agriculture/ranching, 2:34–35
    joint forest management, 2:34
    policy and institutional reforms, 2:35
    protected area systems, 2:33–34
    sustainable timber harvest procedures, 2:34
    tree plantations, 2:35
  tropical, extrapolation of population extinctions from rates, 4:766
Degradation, definition, 5:33, 97
De Materia Medica (Dioscorides), 4:713
Deme, definition, 2:245
Demographic stochasticity, 4:832
Demographic transition, definition, 5:553
Dendrogram, definition, 5:569, 589
Denitrification, definition, 3:89; 4:377
Density dependence, definition, 1:599; 4:769; 5:823
Density independence, definition, 1:641
Deoxyribonucleic acid, *see* DNA
Deoxyribonucleic acid (DNA), 4:446
Depauperate, definition, 2:441, 755
Dependent community, definition, 4:659
Depletion, definition, 2:783
Deposit feeders, definition, 4:1
Deposition, definition, 1:119; 4:377
Derived (characters), definition, 5:145
Derived (taxa/traits), definition, 2:755
Dermoptera, 3:790–791
Descriptive taxonomy, definition, 4:271
Desertification
  in Africa, 2:64–65, 67
  in Australia, 2:67–68
  in China, 2:68
  and climate change, 3:283
  conceptual model of
    ecological dimensions of land degradation, 2:73–74
    human dimensions, 2:75–76
    meteorological aspects, 2:74–75
    stepwise degradation model, 2:76–77
    threshold of response, 2:73
  defining, 2:70–71
  definition, 2:23, 61; 3:659; 5:33
  dimensions of problem, 2:62–63
  and extinctions, 2:704–705
  global extent of
    limitations of estimates, 2:71–73
    UNEP estimates, 2:71
  impact of livestock, 1:654–655
  in Mediterranean region, 2:67
  native rangelands, and livestock grazing, 5:43–44
  reality versus myths of, 2:62
  UN efforts to combat, 2:68–70
Deserts, 5:38–39; 5:640
  African, 1:65
  animals, 2:51–52
    adaptation to
      food scarcity, 2:52
      high radiation, 2:53
      loose substrates, 2:52–53
      temperature, 2:52
      water scarcity, 2:52; 4:338–339
    role in ecosystem processes, 2:54–55
  anthropogenic disturbances, 2:57
  Asian
    subtropical, 1:276–277
    temperate, 1:278–280
  causes of, 2:37–38
  climates, 2:38
    continental deserts, 2:39–40
    cool coastal deserts, 2:38–39
    rain shadow deserts, 2:38
    subtropical deserts, 2:40
  conservation issues, 2:57–58

definition, 2:37
distribution and characteristics, 2:43
    Africa, 2:45–46
    Asia, 2:46, 48–49
    Australia, 2:49
    Middle East, 2:46
    North America, 2:43–44
    South America, 2:44–45
ecosystem processes, 2:53–54
    nutrients, 2:54
    primary productivity, 2:54
    role of animals in, 2:54–55
    succession, 2:54
geomorphology of, 2:40
    alluvial fans, 2:40
    arroyos, 2:41
    dunes, 2:41
    playas, 2:40–41
human use of, 2:55–56
natural disturbances, 2:56–57
Near East, vegetation, 4:363
    savannoid, 4:363–364
plants, 2:49–50
    morphology and behavior, 2:50–51
    physiological processes, 2:51
soils
    characteristics, 2:41–42
    crusts, 2:42–43
    desert varnish, 2:42
    pavements, 2:42
Desmodus Plasminogen Activator, 1:476
Desmostylia, 4:41
    evolutionary history, 4:44
    extinctions over evolutionary time, 4:50
Deterministic system, definition, 5:467
Detoxification, definition, 1:119
Detrital shunts, 3:12–13
    definition, 3:1
Detritivores, definition, 4:249; 5:781
Detritus, definition, 3:75
Deuteromycetes, 3:147
Developing countries
    population growth in, 4:822–823
        impact of education, 4:825–826
    population policies, 4:828–829
    wilderness protection in, 2:604–605
Developmental biology, evolutionary, 2:678–679
Devonian period, 1:925
    late, mass extinction of, 1:414; 4:112–115
Diadromy, definition, 4:221
Diagnosis (taxonomic), definition, 5:589

Diagnostic species, definition, 1:307
Diapause, 2:80–81
    as adaptive trait, 2:81
    and biodiversity maintenance, 2:83
    definition, 2:79; 4:249
    as migrations from past, 2:82–83
    in moths, 4:259
    variations in, 2:81
Diaspores
    definition, 4:601
    dispersers
        plant adaptation to, 4:616–618
        types of, 4:615–616
Diastema, definition, 3:777
Diatoms, 2:196–197
Dicarboxymide fungicides, 4:514
Dichopatric speciation, 5:374–376
Dichotomous tree, definition, 5:569
Dichotomy, definition, 5:569
Dictyosteliomycota, 3:151
Differentiation
    definition, 2:86
    ecological, 2:92–93
    and gene flow, 2:90–92
Diffused vegetation, definition, 4:353
Digital image interpretation, 5:127
Digitalis, 4:719–720
Digitigrade, definition, 3:777
Dinitroaniline herbicides, 4:515
Dinitrophenol fungicides, 4:513
Dinitrophenol pesticides, 4:512
Dinoflagellates, 2:656, 657; 4:912
    toxins from, 4:714
        ciguatoxin, 4:714
        Gymnodinium breve, 4:714
        maitotoxin, 4:714
        Pfisteria piscicida, 4:714, 716
Dinosauria, definition, 2:95
Dinosaurs
    ecological and evolutionary diversity, 2:97–98
    extinction
        current theories, 2:98–99
        impact theory, 2:100–102
        marine regression theory, 2:103–106
        misconceptions about, 2:96
        scenario for, 2:106–108
        volcanism theory, 2:102–103
    nonavian, definition, 2:95
Dioscorides, Padanius, 2:611; 4:713
Diphenyl ether herbicides, 4:515
Diploid, definition, 4:537
Diploid hybrids, definition, 4:659
Diplopoda (millipedes), 4:292–295
Diptera (flies), 3:483–484

adult, 2:823
biogeographical distributions, 2:815–816
conservation biology of, 2:825–826
definition, 2:815
larvae
    herbivorous, 2:822–823
    non-free living, 2:823
    parasitic, 2:823
    parasitoid, 2:823
    predatory, 2:823
    scavenger, 2:822
major subdivisions, 2:816–822
    Brachycera, 2:822
    nematocerous families, 2:822
multispecies restoration of, 5:188
as pollinators, 4:724–725
special associations
    aquatic Diptera, 2:823
    kleptoparasitism and phoresy, 2:824–825
    with phytotelmata, 2:823–824
    with social insects, 2:824
swarming behavior, 2:825
Dipteroaceae, 5:15
Dipterocarp, definition, 3:747
Disaster species, 4:108
Discount rate, definition, 1:769; 4:891
Diseases
    in aquaculture, 1:193–195
    assessment of threats, 2:123–124
    and biodiversity maintenance, 2:114–116
    as biodiversity threat, 2:117–118
    in captive breeding programs, 2:119–120
    endemic diseases, 2:118–119
    definition, 2:109, 698; 5:73
    emerging, 2:566–567
    and extinctions, 2:713
    interactions with wildlife/human activities
        commerce/travel, 2:122
        global climate change, 2:121–122
        habitat fragmentation, 2:120–121
        pollution, 2:121
    intervention methods, 2:124–125
    prevention of outbreaks, 2:122–123
    requirements of, 2:110
    threat to mammals from, 2:444–445
    threat to reptiles/amphibians from, 2:483
Disharmonic fauna, definition, 4:291
Disparity, definition, 3:215
Dispersal
    biogeographic and evolutionary consequences, 2:149

Dispersal (*continued*)
    extinction, 2:150
    speciation, 2:149–150
  biogeographic perspective, 2:130
    gradual range expansion, 2:130
    jump range expansion, 2:130–133
  definition, 2:127; 4:305; 5:767
  determinants of success, 2:133
  ecological perspective, 2:129–130
  evolutionary and population genetics
    perspectives, 2:130
  examples of, 5:769
  geography of, 2:136
  mechanisms and adaptations for,
    2:131–133
  methodological approaches
    barrier removal, 2:139–140
    inference from current
      distributions, 2:144–147
    modeling
      dispersal models, 2:148–149
      reproduction-dispersal models,
        2:148–149
    natural range expansion
      Holocene postglacial spread,
        2:140–141
      recent, 2:140
      spread of early hominids,
        2:141, 143
    observation, 2:136–137
    spread of alien species, 2:143–144
    virgin habitat colonization,
      2:137–139
Dispersal barrier, definition, 2:127
Dispersal biogeography, definition,
  2:127
Dispersal route, definition, 2:127
Dispersal–vicariance analysis, 2:147
Disseminule, definition, 2:127
Dissipative beaches, definition, 1:741
Distance (taxonomic), definition,
  5:589
Disturbance
  and biodiversity, 2:155
  boreal forests, dynamics and
    management implications,
    5:658–659
  definition, 2:153; 3:41; 4:27; 5:361
  ecology of, 2:153–155
  by herbivores and predators, 2:158
  humans as agents of, 2:159
  interaction among mechanisms,
    2:158–159
  mechanical, 2:155–156
  natural
    definition, 5:681

    and human, 2:557–558
    traditional management of, 5:691
  physico-chemical, 2:156–158
  and species richness, 1:730
  wildlife, tourism impacts, 5:675
Disturbance regimes
  alterations in urban areas, 5:744
  definition, 2:153
Dithiocarbamate fungicides, 4:513
Divergence, definition, 2:179
Divergence of character, 1:757
Diversity
  versus composition, 3:111
  definition, 4:879
  estimation of
    species evenness, 2:208
    species richness, 2:208
  taxonomic versus functional, 2:205
    linkage of, 2:213
      diversity of C3 and C4 grasses,
        2:213
      exotic species invasion,
        2:213–214
      island biogeography theory, test
        of, 2:213–214
      natural selection, 2:214–215
Diversity indices
  definition, 1:377; 3:485; 5:245
  Simpson, 5:441
Diversity principle, 3:802–803
  definition, 3:790
Diversity-productivity hypothesis,
  definition, 3:109
Diversity-stability hypothesis,
  definition, 3:109
DNA banks, 3:179
DNA (Deoxyribonucleic acid)
  comparative analysis across
    genomes, 3:198
  conservation of, 2:691
  definition, 3:183; 4:415
  diversity within/among species,
    3:200–202
  evolution of, 1:704–706
  mitochondrial (mtDNA), 3:200–201
    in animals, 3:190–191
    in plants, 3:191
  polymorphisms, 3:197–198
  repetitive sequences, 3:190
  ribosomal (rDNA)
    definition, 1:325
    16S, in bacterial phylogenetics,
      1:327
  sequence variation, measurement
    and interpretation, 4:779
  structure of, 3:185–186

    UV radiation effects, 5:726
DNA fingerprinting, 3:190
  definition, 4:177
  in measurement of microbial
    biodiversity, 4:179–180
Dodder (*Cuscuta* spp.), definition,
  3:317, 327
Dodo bird (*Raphus cucullatus*), 2:400
Dogs, *see* Canidae
Dolichopodidae (long-legged flies),
  2:821
Domains, 5:309
  definition, 1:325; 5:305
  tripartition of, 1:328
Domesticate, definition, 2:375
Domesticated animals, in logged
  forests, 3:756–757
Domestication
  genetic recombination and, 5:62–63
  livestock, 1:651–653; 5:291
  plant
    Amazonian, selection, fixation and
      status of
      environment, 1:905–906
      management systems,
        1:906–911
      reproductive isolation, 1:905
    definition, 1:897–898
    process of, 1:898–899
      discovery of useful plants,
        1:901–905
      genetic diversity, 1:899–901
Dominant species, 3:332–334, 622
  definition, 3:329, 613
Domingo, Esteban, 4:446
Domoic acid, 4:716–717
Doñana National and Natural Parks
  (Spain), 4:321
Donor control, definition, 3:1
Dormancy, 2:80–81
  as adaptive trait, 2:81
  and biodiversity maintenance, 2:83
  definition, 2:79
  as migrations from past, 2:82–83
  seed, definition, 5:269
  variations in, 2:81
Dorsal, definition, 1:915
Double-stranded DNA, 3:185
*Draft BioCode: The Prospective
  International Rules for the Scientific
  Names of Organisms*, 4:396–397
Drosophila spp.
  genetic studies in, 2:87, 88, 181,
    182
  Hawaiian, adaptive radiation, case
    studies, 1:35

in interspecific competition,
1:804–805
Drosophilidae (vinegar flies), 2:821
Drugs, see Herbals; Pharmaceuticals
Dry deposition, 1:2
definition, 1:1
Drylands
definition, 2:61
global
distribution, 2:66
extent and classification, 2:63
land degradation, 2:61–62
index of aridity, 2:63
Dunes, 2:41
Dune vegetation, Australia, diagnostic species, 1:315
Duogland adhesive system, definition, 5:863
Durable resistance, definition, 1:547
Dust Bowl, 1:891
Dyke intrusion, definition, 5:747
Dynamical system, definition, 5:467
Dynamical variable, definition, 5:467
Dynamic model, 1:850
definition, 1:845
Dzungaria, 1:289

E

E. coli, see *Escherichia coli*
Earth, distribution of elements on, 1:438
Earth energy balance, definition, 3:277
Earth Summit, Rio de Janeiro, 2:68–69
Earthwatch Institute, 2:773
Earthworms, 5:313
Ebola virus, RNA editing, 4:426
Ecdysozoa, 2:202
Echinacea, 4:721
Ecocriticism, definition, 3:739
Ecofeminism, 2:554–555
definition, 2:545
Ecogeography, definition, 2:683
Ecological biodiversity, definition, 2:635
Ecological biogeography, definition, 5:767
Ecological community, definition, 5:173
Ecological complexity
versus biological diversity, 1:831–832
community. phylogenetic, and function complexity, 1:835–836
entangled bank metaphor of, 1:832
measures of, 1:836

community complexity, 1:836–838
functional complexity, 1:838–839
phylogenetic complexity, 1:839–840
utility of, 1:840
significance of, 1:840
community complexity, 1:840–841
ecosystem functioning, 1:841
phylogenetic complexity, 1:841–842
Ecological deficit, definition, 2:229
Ecological footprint
analysis, 2:230
basic concepts, 2:234–235
methodology, 2:235–236
conceptual and methodological strengths, 2:240–241
definition, 4:845
human
and biodiversity crisis, 2:238–240
implications for global sustainability, 2:236
functional footprints of cities, 2:236–237
global development and social equity, 2:237–238
methodological limits, 2:241–243
Ecological functions
definition, 3:485
intertidal biodiversity as indicator of, 3:495–497
Ecological genetics
research objectives, 2:245
of species interactions, mutualism, 2:257–258
Ecological integrity, definition, 1:965; 2:783
Ecological models, definition, 5:467
Ecological niche, definition, 3:729
Ecological processes, definition, 3:485
Ecological refugees, definition, 4:845
Ecological release, definition, 1:25
Ecological speciation, see Nonallopatric speciation
Ecological transition, definition, 3:790
Ecological variants
adaptivity of, 2:249
evolution, limits to, 2:256–257
genetics of, 2:253–254
in spatially structured populations
scale, 2:249
selection versus gene flow, 2:249–251
selection versus genetic drift, 2:251–252

Ecology, central concepts and theories in
biodiversity, 2:264–265
development, 2:259–260
ecosystem theories, 2:266–267
evolutionary theory, 2:267–268
population dynamics, 2:260–261
scaling laws, 2:265–266
spatial patterns, 2:265
species interactions, 2:261–264
synthesis, 2:268
Economic botany
definition, 2:609
as discipline, creation of, 2:614–615
Economic efficiency, definition, 4:85
Economic growth, 2:277–278
accounting, and technological change, 2:283–284
and environmental Kuznets curves, 2:282–283
Economic impact, definition, 3:501
Economic rent, definition, 1:769; 4:891
Economics
of conservation efforts, evaluations, 1:979
and fishery subsidies, 2:788–789
losses, from deforestation, 2:33
resource versus environmental, 2:292
value of
diversity, 2:286–287
local diversity, 2:287
populations, 2:287–288
welfare, first theorem of, definition, 1:359
Ecoregions, 2:410–411
definition, 2:407; 4:403; 5:345
Ecosystem classification, definition, 5:637
Ecosystem composition
definition, 3:109
versus diversity, 3:111
Ecosystem conservation, 2:421–423
Ecosystem diversity, 1:381
definition, 1:377
Ecosystem engineers, definition, 1:965; 4:27; 5:305
Ecosystem functions
and acidification of freshwater ecosystems, 3:107
and biodiversity, 2:338
in lakes and pond ecosystems, 3:638
and biomass production, 2:325–326
decomposition, 2:332–334

Ecosystem functions (*continued*)
  definition, 2:321, 345; 3:109, 121
  under disturbance, 2:335–337
  and diversity, in coral reefs, 3:134–135
  ecosystem productivity, limits to, 2:347–348
  and ecosystem services, 2:346
  ecosystem stability, 2:350–351
  keystone species, 2:348–349
  measurement
    reasons for, 2:321–323
    short-term resource dynamics and long-term, large-scale ecosystem process approaches, 2:323–325
    whole ecosystem and community structure-/composition-based approaches, 2:323
  monitoring, field manipulations, and synthesized ecosystems, 2:339
    experimental manipulations, 2:340–343
    hard- and soft-trait approaches, 2:339–340
  nutrient cycling, 2:349
    controls over, 2:330–331
    open versus tight cycles, 2:331
    by plants, 2:329–330
    rainfall effects, 2:331
    soil type, land use, and vegetation structure effects, 2:331–332
  succession, 2:349–350
  and plant functional traits, 2:338
  positive-feedback switches, 2:339
  and taxonomic/functional diversity, 2:337–338
  and trophic transfer
    food chain controls, 2:329
    herbivore effects on nutrient cycling/primary productivity, 2:329
    herbivore performance factors, 2:327–328
  water dynamics, 2:334–335
Ecosystem goods/services, 3:383
  definitions, 3:383
  overexploitation of, 3:387
Ecosystem health, definition, 2:593
Ecosystem integrity, definition, 2:593
Ecosystem management, definition, 1:965
Ecosystem people; *see also* Indigenous peoples
  characteristics, 4:846
  definition, 4:845
  local knowledge and sacred respect, 4:854–855
Ecosystem processes, 3:110
  and functional diversity, early work on, 3:111–112
Ecosystems
  bifurcations, and discontinuous value functions, 2:300–302
  categories, and peoples, relationships, 4:846–847
  concept, 2:306–308
    development, 2:345–346
    history of, 4:427–428
    utility of, 4:428–429
  definition, 1:261; 2:345; 3:1; 5:345
  designer, 5:202
  endangered
    in Australia, 2:409–410
    ecoregion approach, 2:410–411
    general reviews of, 2:409
    in U.S., 2:409
  energy flow, 2:510
    biodiversity effects on, 2:519–520
      regulation of, 2:520–521
      stressful environments, 2:521–522
    boundaries, 2:510–512
      ecosystem analysis, 2:514–515
      food webs, 2:513
        controls of energy flow, 2:514
        trophic dynamic concept, 2:512–513
      trophic levels, 2:513
  high-value, 3:378
  internal/external nutrient cycling, 2:517–518
    loops, spirals, and chains, 2:518–519
  living dead, 3:697–698
  multiple energy pathways, 2:515
    depth/spatial heterogeneity, importance of, 2:517
    lake ecosystems, 2:515, 517
  net productivity, and carbon cycle, 1:613
  signs of deterioration, 3:388
    epidemiological changes, 3:389
    faltering food production, 3:388–389
    fisheries decline, 3:389
    water quality decline, 3:390
  structure and functioning of, 3:131–133
  trophic structure, guilds as framework for, 3:300
Ecosystem scientists, areas of study, 2:346–347
Ecosystem services, 1:103–104
  classification of, 2:354
  conceptual framework, 2:311–312
  definition, 4:891; 5:553
  definitions, 1:99; 2:353; 3:383
  deforestation consequences, 2:32–33
  in food production, 2:356–357
    climate stability, 2:357
    natural pest control, 2:358
    pollination, 2:358
    soil-supplied services, 2:357–358
  history of concern for, 2:359–360
  marine sediments, 4:7–8
    filtration, 4:10
    linkage with diversity, 4:10–11
    nutrient cycling, 4:9
    pollutant cycling, 4:9–10
    secondary production, 4:10
    sediment and shoreline stability, 4:10
  safeguarding of, 2:312–313
  temperate forests, 5:622–623
  value of population diversity, 4:763–764
Ecosystem stability, definition, 2:321
Ecosystem structure
  definition, 3:121
  and diversity, in coral reefs, 3:134
Ecotone
  definition, 3:303; 5:371
  and swiddening, 3:794–795
Ecotourism, 2:452; 4:91–92
  definition, 5:667
Ecotoxicology
  biological processes, 2:369–370
  biomonitoring, 2:367–369
  definition, 2:363; 3:437; 4:731
  early warning systems, 2:370–371
  environmental fates of pollutants, 2:367
  related disciplines
    biochemistry, 2:365
    biotechnology, 2:366
    chemistry, 2:365
    ecology, 2:366
    microbiology, 2:365–366
    toxicology, 2:366–367
  specificity and resolution, 2:371–373
Ecotype/ectotypic, definition, 1:133
Ectoderm, definition, 5:755
Ectomycorrhiza, definition, 3:141
Ectoparasite, definition, 4:463
Ectoparasites, 4:464

Ectothermy, definition, 5:145; 5:755
Edge effects
 and extinctions, 2:705
 in species–area relationship, 5:415
Education
 biodiversity
  colleges and universities, 2:774–775
  in informal sector, 2:770–771
   zoos, botanical gardens, museums, 2:772–773
   K–12, 2:766
   3–5, 2:767
   6–8, 2:767–768
   9–12, 2:768
   K–2, 2:766–767
   and school science, 2:768–769
  role of museums, 4:278
  systematics research and training, 2:775
 environmental, 2:769–770
Edwards Aquifer, Texas, 5:537
Effective population size, definition, 1:599; 2:179; 3:427
Efficiency, definition, 1:813
Egg bank, definition, 2:79
Egg rejection, definition, 4:365
Eigen, Manfred, 4:442
Elasmobranchii, 2:757–760
 definition, 2:95
Elective culture, definition, 4:191
Electricity transmission, 2:535
Electromagnetic radiation, 5:123–124
 definition, 5:121
 remote sensors, 5:124–125
Electrophoresis, definition, 3:195
Elements, origin of, 1:438–439
 biological assimilation, 1:442
 biological transport, 1:441–442
 hydrological cycle, 1:439, 441
 phase state transitions, 1:439
 vulcanism/orogenesis, 1:441
 weathering, 1:439, 443
Elephant seal, northern (*Mirounga angustirostris*), 4:64–65
Elephants (Proboscidea), 3:794–795
 in African forests, 1:60
 in African grasslands, 1:64
El Niño–southern oscillation, 1:620, 712–713; 2:119; 4:498
 definition, 1:609; 4:497
 and ocean ecosystems, 4:434, 437
 and pelagic fishes, 4:504–505
 and South American ecosystems, 5:333
Elton, Charles, 3:111, 364

Eltonian niche, definition, 3:295
Embryos, cryopreservation of, 1:543; 2:690
Emergent, definition, 3:19
Emergent vegetation, definition, 4:1; 5:781
Emic analysis, definition, 2:609
Emission, definition, 1:119
Emotional responses
 to environmental stimuli
  negative, 1:49
  positive, 1:50–53
 in food selection, 1:51–53
 in habitat selection, 1:50–51
 to symmetry/beauty, 1:50
Empididae (dance flies), 2:821
Encephalartos longiforlius, 2:472–473, 475
Enchytraeids, definition, 3:561
Endangered, definition, 2:455
Endangered species
 of Asia, 1:290
 definition, 2:425, 487
 fungal, 3:162
 livestock breeds, 1:538–541
Endangered Species Act (1973), 3:239–240, 365
 application to fungi, 3:162
Endangered Species Act, U.S., 1:861; 2:423, 471
Endangerment
 definition, 2:395
 priority levels, 2:395
Endemic, definition, 1:261; 2:109; 3:675; 4:317; 5:767
Endemicity, definition, 1:377
Endemic languages, 3:454
 definition, 3:452
Endemics, definition, 3:371
Endemic species, definition, 2:425, 487; 3:437, 531
Endemic taxa, definition, 5:569
Endemism
 Amazonian arthropods, 1:253–256
 area of, definition, 5:767
 Asia
  higher animals, 1:269
  plant species, 1:267
 birds, centers of, 1:509
 categories
  evolutionary age and affinity, 2:497–498
  local abundance, 2:498
  spatial distribution, 2:497
 concept of, 2:498–499
 conservation of, 2:506

correlates and causes of, 2:501–502
 abiotic environmental factors, 2:503
 area, 2:503
 biology, 2:503–505
 biotope, 2:503
 endemism and speciation, 2:506
 regional species richness, 2:502–503
 taxonomy and phylogeny, 2:505
definition, 1:419; 4:329
freshwater invertebrates, 2:427–428, 427–429
measurement, 2:499
 biases, 2:499–500
 percentage versus counts, 2:499
 units of, 2:499
in megadiverse countries, 1:421
 geological factors, 1:423
myriapods, 4:299–300
patterns
 centers, 2:501
 congruence, 2:501
 latitudinal gradients, 2:500–501
End-member fluid, definition, 5:747
Endocrine disruptors, 2:364
 definition, 2:363
Endogenous forces, definition, 4:769
Endolithotrophic, definition, 4:917
Endomycorrhiza, definition, 3:141
Endoparasites, 4:464
 definition, 4:463
Endophyte, definition, 2:11; 3:317
Endophytic parasite, definition, 3:317
Endosymbiont, definition, 1:325; 4:901
Endosymbiosis
 definition, 2:623
 and eukaryote origins, 2:628
  serial endosymbiosis theory, 2:629–632
 in species creation, 2:627
Endothermy
 active reptiles and evolution of, 5:150–151
 definition, 5:145
Endothermy, definition, 3:777; 5:755
Energetic web, definition, 3:1
Energy, definitions, 2:525
Energy efficiency, definition, 2:525
Energy flow
 biodiversity effects on, 2:519–520
  regulation of, 2:520–521
  stressful environments, 2:521–522
 food web controls of, 2:514
 trophic dynamic concept, 2:512–513

Energy use, human
  biodiversity impacts of, 2:542
    coal, 2:533–535
    hydroelectric, 2:536–538
    natural gas, 2:533
    nuclear fission, 2:535–536
    oil, 2:530–533
    renewable technologies,
      2:538–539
  future energy paths, 2:542–544
  and laws of thermodynamics,
    implications of, 2:529–530
  patterns and scale, 2:526–529
Entropy, definition, 2:191, 525
Environmental Education Act, U.S.,
  2:769
Environmental ethics, definition, 2:545
Environmental impact
  definition, 3:501
  measuring, 2:570–574
  recognizing and managing,
    2:574–576
Environmentalism, and land ethics, in
  literature, 3:744
Environmental Resources Information
  Network (ERIN), 1:435
  definition, 1:419
Environmental stochasticity,
  3:767–768; 4:832–834
Environmental toxicology, and
  indicator species, 3:439–440
Environments, cyclical, stabilizing
  selection in, 3:211
Enzymes
  gene prospecting for, 1:476–477
  genetic coding of, 3:186
  kinetic diversity, adaptive evolution,
    3:199–200
Eocene–Oligocene boundary,
  extinction during, 3:850
Eolian, definition, 1:437
Epibenthic, definition, 1:915
Epidemics, 2:566–567
  definition, 2:109
Epidemiology, threshold theorem of,
  definition, 2:259
Epigenetic level, definition, 1:325
Epikarst, definition, 5:527
Epiphyll, definition, 3:27
Epiphytes
  definition, 3:27
  nonvascular, 3:29–30
  vascular, 3:29
    characteristics, 3:30
    taxa diversity, 3:31–32
Epiphytic material, definition, 3:19

Epistasis, definition, 1:17; 5:371, 383
Epizootic, definition, 5:73
Equilibrium, definition, 5:467
Equilibrium theory, of island
  biogeography, 1:469
Eremic, definition, 4:329
Ergosterol biosynthesis inhibitors,
  4:514
ERIN, see Environmental Resources
  Information Network
Erosion, 5:317–318
  impact of livestock, 1:654–655
Escherichia coli, 1:341
  divergence in, 1:349
  gene expression in, 3:188–189
Essentialism, 5:593
Estimator, definition, 4:123
Estuarine ecosystems, 2:579–581
  biodiversity, 2:582–584
    ecological functions of, factors,
      2:584–587
      altered delivery of freshwater,
        2:585
      biogeochemical processes, 2:585
      biological properties,
        2:584–585
      changes in/fate of suspended
        matter, 2:585
      chemical modification, 2:585
      ecosystem modification, 2:585
      extreme events, 2:585
      material exchange, 2:584
      natural climate variations,
        2:585
      physiochemical properties,
        2:584
      sea level changes, 2:585
      temperature, 2:585
      wind, 2:585
    Chesapeake Bay case study,
      2:587–589
    classification systems, 2:581–582
    cultural eutrophication of
      macroalgae, 2:658–660
      macrophytes, 2:658–661
      microfauna, 2:664–665
    future challenges, 2:590
    high nutrient, low growth concept
      in, in estuaries, 4:6–7
Estuary, definition, 2:579
Ethical holism, definition, 2:545
Ethics, definition, 2:545; 5:109
Ethnobiology
  current trends, 2:616
    cognitive mapping, 2:616–617
    ethnobiological classification,
      2:618–619

    internationalization, 2:620
    knowledge variation, 2:619
    mechanisms of change, 2:619–620
    resource management and
      valuation, 2:617–618
    ritual, religion, and symbolism,
      2:620
    scientific covalidation, 2:618
  definition, 2:609
  as discipline, creation of, 2:615–616
  historical development, 2:611
    in Arab world, 2:612
    in China, 2:612–613
    creation of new fields in,
      2:614–616
    early European, 2:611
    in India, 2:613
    in Middle Ages, 2:611–612
    in New World, 2:613
    in Renaissance, 2:613–614
  synthesis of disciplines in,
    2:620–621
Ethnobotanical screens, in drug
  discovery, 4:718–719
Ethnobotanies, 1:112
Ethnobotany
  of canopy plants, 3:38
  definition, 2:375, 609; 4:523
  as discipline, creation of, 2:614–615
Ethnoecology, 1:112
  and biodiversity, 3:457–462
  definition, 1:109; 2:610; 3:452
  as discipline, creation of, 2:615–616
Ethnoscience
  definition, 2:610
  as discipline, creation of, 2:615
Ethnozoology, definition, 2:609
Ethology, definition, 4:281
Etic analysis, definition, 2:609
Etiology, definition, 3:739
Eucalyptus spp., 5:365–366
Eukaryotes, 4:202
  comparison with prokaryotes,
    2:624–626
  definition, 1:219; 3:183; 4:201;
    5:647
  endosymbiosis in origins of, 2:628
    serial endosymbiosis theory,
      2:629–632
  evolution of, 2:196
  fungi, 4:202–204
  genome of, 3:184, 189–190
  Protista, 4:202
  soil, numbers and biodiversity
    collembola, 5:311–312
    earthworms, 5:313

fungi, 5:310–311
microfauna, 5:311
mites, 5:312–313
nematodes, 5:311
termites, 5:313
species number, versus prokaryotic, 1:326
temperature limits for, 3:354
unicellular, 4:209
phagotrophic, 4:210
photoautotrophs, 4:209–210
symbiotic, 4:210
Eumantacidae (monkey grasshoppers), 3:260
Eumastacoidea, 3:260
Eunicella verrucosa (sea fan), 2:460
Euphotic zone, definition, 2:311; 4:427, 569; 5:747
Europe
characteristic crops, 2:224
ecosystems of, 2:635–636
biomes, 2:638
classification of vegetation, 2:636
dominance approach, 2:637
floristic–sociologic approach, 2:636–637
habitat classification, 2:637
diversity of vegetation types, 2:638–644
future challenges, 2:646
habitat biodiversity and vegetation, sources, 2:637–638
habitat types, diversity, 2:644–646
late Quaternary mammalian extinctions, 3:835–837
Mediterranean region, desertification, 2:67
river management and indicator species, 3:438–439
Euryarchaeota, 1:328
definition, 1:219
Euryhaline, definition, 2:455
Eurypterids (sea scorpions), 1:920
Eusociality, 3:424
definition, 3:417
Eutherian, definition, 1:629; 3:777
Eutheria (placental mammals), 3:780–781, 789
Artiodactyla, 3:796–797
Carnivora, 3:793–794
Cetacea, 3:794
Chiroptera, 3:791
Dermoptera, 3:790–791
Hyracoidea, 3:796
Insectivora, 3:790
Lagomorpha, 3:798

Macroscelidea, 3:798–799
Perissodactyla, 3:795–796
Pholidota, 3:797–798
primates, 3:791–793
Proboscidea, 3:794–795
Rodentia, 3:798
Scandentia, 3:790
Sirenia, 3:794
Tubulidentata, 3:796
Xenarthra, 3:789–790
Eutrophic, definition, 2:649; 4:377
Eutrophication, 2:650; 3:274
cultural, 2:650
definition, 5:213
freshwater ecosystems, 3:96–97
biological consequences, 3:102
oligotrophication in reversal of, 2:667–669
definition, 2:425; 4:1; 5:73
effects on
aquatic macrophytes
estuarine/marine communities, 2:661–663
freshwater communities, 2:660–661
macroalgae
estuarine/marine assemblages, 2:658–660
freshwater assemblages, 2:658
microalgae
algal blooms and nutrient enrichment, 2:655, 657, 658
long-term human influence, 2:654–655
species shift across nutrient gradients, 2:653–654
effects on biodiversity, 4:382–384
effects on structural stability, 5:477–478
invertebrate macrofauna, 2:665–666
in lake and pond ecosystems, 3:639; 4:3
hardwater lakes, 3:102–103
in marine ecosystems, 4:24
and marine sediments, 4:12
microfauna
estuarine/marine communities, 2:664–665
freshwater communities, 2:663–664
stoichiometry
changing nutrient ratios, 4:8–10
Redfield ratios, 4:7
silicon, 4:7–8
threat to coral reefs, 5:85–86

vertebrate macrofauna, 2:666–667
Evaporative aerodynamics
Australia, 1:308
definition, 1:307
Evapotranspiration, definition, 5:607
Evolution
across phylogeny caused by climatic stress, 3:206–209
adaptation as product of, 1:18–20
and adaptationism, 1:21–22
Archaea, 1:226–229
Carnivora, 1:639–640
convergent, definition, 3:121
Darwinian
dual aspects of, 1:694
phyletic gradualism hypothesis, 1:696
quantitative treatment of, 4:444–446
and speciation, 1:695
definition, 1:393; 2:671; 4:799; 5:383
in environmental aesthetics, 1:46–47
influence on ecological systems, models, 1:400
of introduced species, 3:527–528
within lineages, 1:407–408
morphological, 1:703–704
and multispecies interactions, 5:465
mutualism and, 4:287–289
and pharmacology, 4:523–525
and phenotypes, 4:545–546
of psychrophiles, 4:921
rates of, 1:702–703
responses to air pollution, 1:123
reticulate, definition, 4:659
and species interactions, 5:460
competition, 5:460–461
mutualism, 5:461
predation, 5:461–462
of thermophily, 3:355–359
Evolutionary biology, structure and study of, 2:674–676
behavioral evolution, 2:677
evolutionary developmental biology, 2:678–679
evolutionary ecology, 2:676–677
evolutionary genetics, 2:680
evolutionary paleontology, 2:677–678
evolutionary physiology and morphology, 2:679
evolutionary systematic biology, 2:679
human evolution, 2:679–680
molecular evolution, 2:680–681

Evolutionary equilibrium hypothesis,
    definition, 4:365
Evolutionary lag hypothesis, definition,
    4:365
Evolutionary theory
    Darwinian, 2:672
    definition, 2:671
    development of, 2:671–672
    molecular revolution and current
        thought, 2:673–674
    protein/nucleotide diversity,
        discovery of, 2:673
    synthetic theory, 2:672–673
Evolutionary trees, reconstructed
    definition, 1:403
    mammalian, 3:783
    versus real, 1:404–406
Executive order, definition, 3:233
Exoenzymes
    definition, 3:141
    and heterotrophism, 3:144
Exogamy, definition, 3:411
Exogenous forces, definition, 4:769
Exons, definition, 3:183
Exotic animals, definition, 5:901
Exotic species
    agricultural pollinators, 4:728–729
    and animal absence from site/region,
        5:193
    bird, 1:514
    and bird endangerment, control of,
        2:403
    and conservation efforts, 1:973–975
    as constraint on restoration,
        5:207–208
    definition, 1:71; 3:531
    endangerment of freshwater
        invertebrate from, 2:432
    examples of threats from, 1:976
    extinction cascades due to,
        examples, 3:772
    and extinctions
        aquatic habitats, 2:712–713
        islands, 2:712
    and forest biodiversity, 3:49
    influences of livestock, 5:48
    invertebrate, introductions
        into Australia, 1:79
        into Brazil, 1:82
        into India, 1:81
        into South Africa, 1:80
        into U.K., 1:77–78
        into U.S., 1:75–76
    in island ecosystems, 3:386
    island predators, 4:869–870
    in logged forests, 3:754

    in marine ecosystems, 4:24; 4:33
    onshore waters, 3:386
    pathogens, 3:386
    and plant–soil interactions,
        4:706–707
    and taxonomic versus functional
        diversity, 2:213–214
    threats from
        to amphibians/reptiles, 2:484
        to freshwater fauna, 3:542
        to mammals from, 2:447
        to riverine ecosystems, 5:228–229
    tourism in spread of, 5:673–674
    vertebrate, introductions
        into Australia, 1:78–79
        into Brazil, 1:81–82
        into India, 1:80–81
        into South Africa, 1:79–80
        into U.K., 1:77
        into U.S., 1:73–75
Exploitation competition, definition,
    1:793
Exploitation ecosystem hypothesis,
    3:9–10
    and trophic cascades, 3:11
Exploitative competition, definition,
    3:295
Ex situ, definition, 5:901
Ex situ conservation/preservation,
    1:386; 3:165
    botanical/zoological gardens, 2:690
    definition, 2:683, 689; 3:811; 4:645
    field gene banks/livestock parks,
        2:690–691
    of livestock genetic resources,
        1:541–543
    of plants, 4:654
    seed/embryo storage, 2:690
    in vitro conservation, 2:690
Extant, definition, 1:915; 2:755; 3:53
Externality, definition, 1:769; 4:891
Extinction(s), 3:229–230
    amphibian, 1:169–170
    anthropogenic, 3:771–772,
        793–794, 839
        current rates of, 3:803
        prehistorical, 1:884; 2:232; 3:384
    avian, 2:399–400
        fragmentation as cause,
            1:513–514
        history of, 1:514–515
        recent prehistory of, 1:515–516
    background
        decline in, 2:728–729
        definition, 2:715; 3:841
        mammals, 3:843–844

        versus mass extinction, 2:718
        and turnover, 2:727–728
    canopy plant vulnerability to, 3:37
    carnivore, 1:640
    cascade
        definition, 2:731
        examples, 2:738–739
    causes of
        desertification, 2:704–705
        disease, 2:713
        exotic species
            aquatic habitats, 2:712–713
            islands, 2:712
        global climate change, 2:709–710
        habitat degradation/pollution,
            2:706–707
            air pollution, 2:708
            pesticides, 2:707
            water pollution, 2:707–708
        habitat fragmentation, 2:705
            dispersal barriers, 2:705
            edge effects, 2:705
            interspecific interactions,
                2:705–706
        habitat loss, 2:701–702
            coral reefs, 2:704
            grasslands, 2:702
            mangroves, 2:704
            rain forests, 2:702
            tropical dry forests, 2:702
            wetlands/aquatic habitats,
                2:702, 704
        human population expansion,
            2:699–700
        large-scale development
            projects, 2:700–701
        natural resources, unequal use
            of, 2:700
        multiple factors, 2:713–714
        overexploitation, 2:710–711
        rain forest loss, 2:702
    and climate change, 1:719–721
    –colonization dynamics, 5:423–424
    current crisis of, primary causes,
        3:773–775
    definition, 2:731; 3:761; 4:799; 5:73
    deterministic processes and
        extinction vortices, 4:835
    dinosaur
        current theories, 2:98–99
        impact theory, 2:100–102
        marine regression theory,
            2:103–106
        misconceptions about, 2:96
        scenario for, 2:106–108
        volcanism theory, 2:102–103

dispersal effects on, 2:150
Eocene–Oligocene, 3:850
evolutionary radiations following, 1:395–396
extrapolation from tropical deforestation rates, 4:766
first-contact, 2:735
   definition, 2:731
   Madagascar, 2:735–736
   New Zealand, 2:736–737
   Pacific islands, 2:736
and inbreeding/genetic diversity loss, 3:431
   in captivity, 3:432
   in wild populations, 3:432–433
and introduced pathogens, 2:117
latent, 3:689–690; *see also* Living dead
late Triassic, 1:926
mammal
   current crisis in, 3:806–807
      patterns, 3:808
   future, projection of, 3:808–809
   patterns and causes, 3:842–843
   Pleistocene, 2:442–443
   and pseudoextinctions, 3:843
   recent, 2:443
marine mammal
   modern anthropogenic, 4:50–52
      factors and processes, 4:66–68
      minimization of, 4:68–69
   natural versus unnatural, 4:50
   over evolutionary time, 4:50
   vulnerability to, factors, 4:44–47
mass, 1:937
   abrupt extinctions, 4:99
   causes of, 2:721–722
   Cenozoic, 4:121
   Cretaceous, 1:928
   of current era, documentation, 1:400
   definition, 2:715; 3:841; 4:97; 4:111
   definitions and identification of, 2:718–721
   end-Cretaceous, 4:118–120
   end-Triassic, 4:117–118
   fossil data, 4:99–100
   gradual extinctions, 4:99
   history of concept, 4:97–99
   kill curve, 4:103–104
   late Devonian, 4:112–115
   late Ordovician, 4:111–112
   late Permian, 4:115–117
   magnitudes of, 4:100
   mammals, 3:843–844
   temporal scale of, 3:843–844
   marine, 4:100
   Mesozoic, 4:121
   models of, 4:99
   Neoproterozoic and Paleozoic, 4:120
   pace of, 2:722–726
   periodicity, hypothesis of, 4:102–103
   Permian–Triassic, 1:415–416, 926; 3:125; 4:100
   Phanerozoic, 1:413–415
      taxon sorting effect, 1:415–416
   at present, 2:743
   recoveries from, 4:107
      delayed diversification, 4:107–108
      disaster species and Lazarus taxa, 4:108
      large-scale diversity rebound, 4:107
   selectivity of, 4:105–107
   self-organized criticality, 4:104–105
   since end of Pleistocene, 3:762–763
   stepwise extinctions, 4:99
mechanisms, 3:765–766
   isolated populations, 3:766–768
   metapopulations, 3:768–769
modern era, 2:737; 4:108–109
   Hawaiian Islands, 2:737–738
   New Zealand, 2:737
   paradoxes of, 2:738
natural, 4:305–306
   and biodiversity trends, 4:313–315
   causes of, 4:305–306
      environmental changes, 4:305–306
      interaction with other species, 4:306–307
   geographic patterns in, 4:313
   risk of, and species age, 4:308–309
   selectivity of, 4:308–309
      and body size, 4:311–312
      and dispersal ability, 4:311
      and reproductive/life history traits, 4:313
      and specialization, 4:312–313
      and species rarity, 4:309–311
   periodicity of, 2:726–727
plant, 2:465–466
   numbers and locations, 4:652–653
and population size, 1:399
postglacial
   African, 2:734–735
   Australian, 2:733–734
   North America, 2:733
problems in understanding, 2:739
   geographical bias, 2:740
   inherent biological problems, 2:741
   methodological obstacles, 2:740–741
   number of species, 2:739
   record keeping, 2:739–740
   taxonomic/habitat biases, 2:740
proximate causes, definition, 2:732
rates
   current, 3:763–764
   deduction from fossil record, 3:762
   definition, 2:715
   future, estimation from area–species relationship, 3:764–765
   measures and types of, 2:716
   problems in measuring
      accurate estimate of fossil ranges, 2:717
      biased preservation and convergence, 2:717
      comparison with origination rates, 2:717–718
      pseudoextintion, 2:718
      taxon-level bias, 2:716–717
and rescue effect, 3:710
risk of, IUCN categories and criteria of, 2:456–457
role of stochasticity in time to, 4:835
secondary, definition, 4:305
theoretical aspects of, 2:741
   density-related traits, 2:742
   insights from population dynamics, 2:742
   introduced species, 2:741–742
   time factors, 2:742–743
ultimate causes, definition, 2:732
Extinction spasm (pulse), definition, 3:825
Extinction vortex, definition, 1:855
Extirpated, definition, 2:755
Extremophiles, 2:195–196
   definition, 2:191; 4:191, 523
Exxon *Valdez*, 2:531–532

F

Facultative epiphyte, definition, 3:27
Facultative parasite, definition, 4:463
Fagus grandifolia (American beech tree), 4:451
Falcons, peregrine, and DDT, 3:444
Fallow
  definition, 5:269
  length, in slash and burn systems, 5:270–271
Family, definition, 3:295
Family planning, 4:822–823
  definition, 4:819
FAO Global Databank, 1:538–539
Farmers, traditional, 1:548
  knowledge of, 1:112–113
Farming intensity, definition, 1:185
Farris, James S., 5:576
Fat body, definition, 5:495
Fatty acids
  analysis, in measurement of microbial diversity, 4:186–187
  composition of storage lipids, 5:506–508
  definition, 5:495
Faulkner, William, 3:743–744
Fauna
  definition, 1:351
  disharmonic, definition, 4:291
Feedback, in greenhouse effect
  biogeochemical, 3:281
    marine, 3:281, 283
    terrestrial, 3:283
  definition, 3:277
  geophysical, 3:280–281
Feed conversion, definition, 1:185
Felidae (cats), 1:630, 631
  as bird threat, 2:400
  feral, economic costs from, 1:78
  food caching by, 1:634
  hunting behavior, 1:634
  as island exotic predators, 4:869
  mating systems, 5:298
  saper-tooth species, 1:639–640
Fellfield, definition, 1:171
Felsic rock, definition, 1:437
Fens, definition, 5:781, 782
Fertile crescent, and Neolithic agriculture, 2:219
Fertility
  definition, 4:819
  human, 4:821–822
Fig wasps, 3:422
Final (definitive) host, definition, 4:463

Finches, Galapagos, 1:19; 2:248
  adaptive radiation, case studies, 1:33–34
  rates of adaptation, 1:757
Finland, boreal forests, main features, 5:656–657
Fire
  in African ecosystems, 1:57–58
  and C4 plants, 1:594–595
  determinants of, 2:748
    fire regime, 2:748–749
  as disturbance mechanism, 2:157–158
  in earth history, 2:745–746
  ecological effects
    on ecosystem structure/function, 2:749
    on species and populations, 2:749–750
  ecology, effects of global change, 2:752–753
  elimination of, as constraint on restoration, 5:207, 208
  as greenhouse gas source, 2:753
  human use of, biodiversity impacts, 3:794
  interactions with herbivory, 2:750–751
  and invasive species, 2:751
  and landscape fragmentation, 2:751
  management of, 2:751
    policies, 2:751–752
  remote sensing studies, 5:139
  species response to burning
    animals, 2:747–748
    plants, 2:746–747
  as surrogate for grazing, 3:273
  and world biomes, incidence and influence, 2:746
Fire regimes, 2:748–749
  definition, 2:745
  effects of livestock grazing, 5:50
First contact, definition, 3:825
Fish conservation
  factors affecting, 2:789
    aquaculture, 2:791
    global atmospheric change, 2:792
    habitat issues, 2:790
    human population growth, 2:790
    introduced species, 2:790–791
    pollutants, 2:791–792
  marine
    biodiversity, 2:792
    genetic diversity, 2:793–794
    species diversity, 2:794
  marine protected areas, 2:797

  solution and reasons for optimism, 2:797–799
Fisher, R.A., 4:538–539
Fisheries
  bycatch, 2:788; 3:387; 5:166
  catching methods, 5:163
  catch processing, 5:163
  changes since 1945, 5:163
  decline in, 3:389
  definition, 2:783
  economic issues, 5:165–166, 170–171
  ecosystem impacts
    fishing down food webs, 2:809–810
    historical trends, 2:808–809
    sustainability concepts, 2:809
  ecosystem management, 5:171–172
  history of, 5:161–163
  intensity and limits of, 2:787
  international conflicts, 2:786
  and lake and pond ecosystems, 3:643
  management, 2:812–814; 5:163–164, 169–170
  overharvest, 2:564
  science of, 5:167
  social and economic concerns, 2:785
  stock monitoring, 5:170
  subsidies and economics, 2:788–789
  and theory of bionomic equilibrium, 1:770–772
  trophic level of catches, 5:166
Fishery Conservation and Management Act (1976), 3:240
Fishes
  acid deposition effects, 1:13
  agnathans, 2:757
  anatomy and physiology, 2:802–803
  bony
    bichirs, 2:760
    bowfin, 2:763
    coelacanth, 2:761
    gars, 2:763
    lungfishes, 2:761
    sturgeons/paddlefishes, 2:761–763
    teleosts, 2:763
      class Actinopterygii, 2:763–764
      subdivision Clupeomorpha, 2:764–765
      subdivision Elopomorpha, 2:764
      subdivision Euteleostei, 2:765–768
      superorder Acanthopterygii, 2:768, 770–777

brood parasitism among, 4:368
cartilaginous, 2:757–758
    chimaeras, 2:760
    sharks, 2:758–760
    skates/rays, 2:760
cephalochordates, 2:757
clupeid, definition, 4:497
as commodities versus wildlife, 2:786–787
as diaspore dispersers, 4:615
estuarine dependency, 2:583–584
eutrophication effects, 2:666–667
freshwater diversity, 2:778
    African (Ethiopian) region, 2:779
    Australian region, 2:779–780
    Nearctic region, 2:778
    Neotropical region, 2:778
    North America, 4:410–411
    Palearctic region, 2:779
gadoid, definition, 4:497
geographic diversity, 2:777–778
marine diversity, 2:780
    Antarctic region, 2:782
    Arctic region, 2:782
    deep sea, 2:783
    eastern Atlantic region, 2:781–782
    eastern Pacific region, 2:781
    and ecosystem diversity, 2:795–797
    extinction
        stages of, 2:794–795
        vulnerability to, 2:794–795
    Indo-West Pacific region, 2:780–781
    pelagic regions, 2:782–783
    temperate regions, 2:782
    western Atlantic region, 2:781
migration patterns of, 4:227–229
multispecies restoration of, 5:188
pelagic, 4:504–507
poeciliid fish, and vicariance biogeography, 5:775
    component analysis, 5:777–778
reproduction/recruitment, 2:803–804
respiratory constraints, 2:804
    adaptation to, 2:804
        growth–mortality relationships, 2:805
riverine ecosystems, 5:224–225
species flocks in ancient lakes, 3:640–641
taxonomic diversity, 2:755–757
UV radiation effects, 5:728–730
Fishing effort, definition, 5:161

Fish stocks
    definition, 2:801
    exploited, distribution of, 2:806–807
        adaptations to open-ocean habitats, 2:807
        shelf communities
            demersal stocks, 2:807–808
            neritic stocks, 2:807
            status of, 2:808
            pelagic stocks, 2:808
    traditional approaches to management, 2:812–813
Fitness
    definition, 1:17; 2:85; 4:601; 4:799; 5:383
    increased, evidence for, 1:21
    measurement of, 1:18–20
    reproductive, and inbreeding depression, 3:430
Fitzroya cupressoides (alerce tee), 2:471–472, 473
Fixation
    carbon, NAD(P)H and ATP in, 4:555–557
    definition, 1:437
    nitrogen, definition, 1:99
Flagship species, definition, 1:377
Flamingo (*Phoenicopterus* spp.), 2:399–400
Flathead River alluvial aquifers, Montana, 5:537–538
Flatworms, see Platyhelminthes
Fleming, Alexander, 4:719
Flexibility, definition, 1:813
Flies, see Diptera
Floodplain
    definition, 5:781
    hydrologic regime, 5:781
Floodplain ecosystems, Amazonian, 1:151–153
Flood pulse concept, of riverine ecosystems, 5:218–221
Flood resistance, definition, 1:249
Flood tolerance, definition, 1:249
Floristic group, definition, 1:307
Fluorescent *in situ* hybridization, in measurement of microbial diversity, 4:186
Focal species, definition, 5:823
Foliage projective cover, definition, 1:307
Folicolous, definition, 3:27
Folk taxonomies, 1:112
Food, Drug, and Cosmetic Act (FDCA), 3:344
Food and Agriculture Organization, U.N.
    deforestation data, 5:26, 27–28
    and plant conservation, 4:646
Food (biomass) pyramid, definition, 3:1
Food caching, in carnivores, 1:633–634
Food chains, definition, 2:509; 3:1; 5:695
Food pump, definition, 5:711
Foods, selection, aesthetic factors, 1:51–53
Food webs, 2:513
    aquatic, and nutrient enhancement, 4:3–4
    and community organization, in forests, 3:48–49
    controls of energy flow in, 2:514
    definition, 1:831; 2:509; 3:1; 4:27; 5:413
    detrital, composition of, 2:334
    generalized theory, and predators, 4:860
    in lake and pond ecosystems, 3:638
    and microbe restoration, 5:199
    and multispecies interactions, 5:462–463
    omnivory and structure of, 3:7–8
    studies, current topics and trends
        age structure effects, 3:13
        detritus, 3:12
        energetic webs and population/community dynamics, 3:14–15
        interaction strength, 3:14
        intermediate levels of complexity, 3:16–17
        modeling, 3:15–16
        as open systems, 3:11–12
        role of nutrients and stoichiometry, 3:13–14
    types of, 3:2–3
        connectivity webs, 3:3–5
        energetic webs, 3:5–6
        functional (interaction) webs, 3:6–7
Foraging, as disturbance mechanism, 2:158
Foraging behavior, carnivores, 1:632–636
Foraminifera, 4:908–909
    definition, 4:97
Forbs, definition, 5:33
Ford, E.B., 4:539
Foreign Assistance Act (1962), 3:242
Forest change processes, definition, 5:25

Forest management, definition, 5:655
Forestry
  and conservation ethics, 2:602
  conservation gains in, 1:890
  sustainable, definition, 5:655
  sustainable harvest procedures, 2:34
Forests, 3:41–44; *see also* Rain forests; Reforestation
  Amazonian
    babassu palm, 1:150
    bamboo, 1:151
    dry semideciduous southern fringe, 1:149–150
    liana, 1:150–151
    mangrove, 1:153
    secondary, 1:154–155
  Asian
    boreal, 1:283–284
    humid tropical, 1:272–275
    laurel forest, 1:281
    monsoon forest, 1:275
    pine forest, 1:282–283
    temperate deciduous, 1:282
    tropical dry forest, 1:276
  biodiversity
    ecosystem function and, 3:50–51
    human activity and, 3:49
      alien species introductions, 3:49
      forest conversion/fragmentation, 3:49
      harvest, 3:49
      pollution, 3:49–50
      rapid climate change, 3:50
  biodiversity in
    food webs and community organization, 3:48–49
    general patterns, 3:44–45
    structure/pattern, and disturbance regimes, 3:45–48
  biodiversity of, definition, 5:655
  boreal, 1:533
    animals of, 1:539–541
    Asian, 1:284–285
    climate, 1:534–535
    definition, 1:261, 533; 5:655
    disturbance dynamics, management implications, 5:657–658
    fire, 1:536
    forest–tundra ecotone, 1:539
    geography, 1:533–534
    under intensive/extensive management, properties of, 5:662–663
    main features, 5:656–657
    predators, 4:866
    soils, 1:535–536
    successional dynamics, management implications, 5:657–658
    succession in, 1:541–542
    vegetation, 1:536–538
      latitudinal classification, 1:536–538
      longitudinal classification, 1:539
  and climate change, 4:734–735
  coniferous, remote sensing in estimation of cover, 5:130–131
  degradation
    from acid deposition, 1:5–8
    definition, 2:23
    methods of overcoming, 5:98–99
  endangered ecosystems, 2:412–414
  fragmentation and selective harvesting of, 3:384–385
  fungal pathologies, 3:153–154
  and historic land-use patterns, 3:678–680
  logged
    abiotic changes
      hydrology, 3:752–753
      insolation, temperature, and wind patterns, 3:752–753
    biotic changes, 3:753
      aquatic fauna, 3:757
      domestic animals/nonforest species, 3:756–757
      fungi and mycorrhizae, 3:754
      invertebrates, 3:754–755
      plants, 3:753–754
      vertebrates, 3:755–756
    ecological interactions, 3:757–758
    genetic effects and evolutionary processes, 3:758
    management, 3:749
      definition, 3:747
      impacts on biodiversity, 5:661–662
      implications of plant–soil interactions, 4:706
      landscape and local patterns, 5:663–664
      nontimber forest products, 3:749–750
      for optimal production, 5:659–661
      timber, 3:671, 749
    secondary versus managed, 3:750
    structure, alterations in
      canopy, 3:751
      fragmentation, 3:752
      gaps and edges, 3:751–752
      soil effects, 3:752
      spatial mosaic of forest types, 3:752
      vertical structure, 3:751
    synergistic anthropogenic effects, 3:758–759
  major threats to, 2:411–412
  in management of carbon cycle, 1:625–626
  natural, definition, 2:23
  and nitrogen cycle perturbations, 4:738–739
  ozone effects, 1:128–129
  pine, multispecies restoration of, 5:191–192
  primary, definition, 3:747
  restoration, use of grazers in, 3:273–274
  secondary, definition, 2:23
  succession, in northeastern Minnesota, remote sensing in study of, 5:134
  sulfur/nitrogen disposition effects, 1:129–130
  temperate, 5:607, 638–639
    biodiversity trends, 5:616–618
    conservation/protection
      major threats to, 5:625–626
      objective and research needs, 5:625–626
      status, 5:623–624
      strategies, 5:624–625
    dominant floristic features, 5:614–616
    ecosystem services, 5:622–623
    endemism, and range size rarity, 5:618–619
    endemism/diversity patterns
      historical and geological explanations, 5:619–621
      productivity, 5:622
      spatiotemporal heterogeneity, 5:622
    global distribution patterns, 5:608–613
    physiognomic features, 5:613–614
    remote sensing in delineation of tree crowns, 5:131–132
  tropical, 5:640, 701–702
    climate, 5:703–704
    conversion to livestock pasture, 5:44, 46–47
    geography, 5:702
    and indigenous populations, 3:456–457

major ecosystems, 5:704–705, 705
  flooded forests, 5:705
  humid forests
    high mountains, 5:710
    lowland, 5:706–707
    low mountains, 5:709
  mangroves, 5:705
  pàramo, 5:710
  perhumid forests
    high mountains, 5:710
    lowland, 5:705–706
    low mountains, 5:709
  savannas, 5:708
  subhumid forests
    high mountains, 5:709–710
    lowland, 5:707–708
    low mountains, 5:708–709
  in megadiverse countries, 1:421
  physiography, 5:702–703
  predators, 4:868–869
  soils, 5:704
  South American, 5:334, 353, 354
    distribution and structure, 5:334
    functional aspects, 5:336–337
  tropical dry, loss of, 2:702
Forest stand, definition, 3:41
Forest structure, definition, 3:41
Formamidine pesticides, 4:512
Formation, vegetation, definition, 2:635
Fossil fuels
  and carbon cycle, 1:619–620
  consumption, CO2 emissions from, 1:296
  definition, 2:525–526
  and ecological footprint, 2:242–243
Fossil record, 1:411
  amphibian, 1:160
  austral ecosystems, 5:365–366
  biases in, 1:412
  biodiversity in, 1:394; 3:215–216
    taxonomic remedy for spotty record, 3:219–220
  and communities, 1:396–397
  and contemporary biodiversity, 1:404
  C4 plants, 1:590–591
  dinosaur, quality of, 2:96–97
  early chordate, 5:757–758
  estimating diversity trends
    capture-recapture estimates, 3:59
      closed population model, 3:60
      Jolly–Seber model, 3:59–60
    generalized inverse Gaussian–Poisson distribution, 3:60–61

nonparametric models, 3:61
rarefaction, 3:57–59
unexplored methods and future studies, 3:61–62
  of evolutionary processes, 2:93
  extinction measured in, 2:715–716; 3:762; 4:99–100
  hymenopteran, 3:418–419
  K–T event, 4:98, 100
  mammalian, 3:784
  microbial diversity in Archaean/Proterozoic eras, 3:220–221
  models for diversity increases in, 1:413
  of plant–herbivore interactions, 3:271–272
  sampling effects in
    extent of sampling, 3:55–56
    incomplete record, 3:54, 217–219
    preservation quality, 3:54–55
    time averaging, 3:56–57
  seagrasses, 5:255–257
  value of, 1:396
  vertebrate, in testing of K/T extinction theories, 2:100
Fossils
  definition, 1:393
  nomenclature systems, 4:397
Fossorial, definition, 2:441; 3:777
Foundation species, definition, 3:613
Founder, definition, 1:599
Founder effect, 5:390–391
  definition, 1:25
  and peripatric speciation, 5:376
  of transient species, 3:335–336
Founder genome equivalent, definition, 1:599
Founders, definition, 5:901
Fox, Sidney, 4:443
Fractal, definition, 5:245
Fractionation, definition, 4:523
Free rider, definition, 4:891
Frequency dependence, definition, 2:245
Frequency-dependent selection, 2:92
  definition, 2:85
Freshwater biota, origins and peculiarities, 3:634
Freshwater ecosystems
  acid deposition effects, 1:9–10
    on aquatic biota, 1:10–13
    on fisheries, 1:13
  Antarctic systems, 1:177–178
  biodiversity, 3:83–87
    factors influencing, 3:536–537
      habitat age and isolation, 3:538

habitat heterogeneity and disturbance, 3:538
invertebrate origins, 3:537–538
species interactions, 3:538–539
  issues, 2:784
  biogeochemistry and nutrient cycling, 2:317–318
    new and regenerated production, 2:318–319
  biological quality
    Red Lists and historical development, in Danish and U.S. streams, 3:99–101
    species decline in Europe and North America, 3:98–99
  and climate change, 4:735–736
  crustacean biodiversity, 1:917
  cultural eutrophication effects
    factors influencing, 2:652–653
    on fish populations, 2:666–667
    on macrophytes, 2:660–661
    on microfauna, 2:660–661
  dominant taxonomic groups, 3:535–536
  effects of livestock, 5:48
  fish diversity, 2:778
    African (Ethiopian) region, 2:779
    Nearctic region, 2:778
    Neotropical region, 2:778
    Palearctic region, 2:779
  global diversity patterns, 3:536–537
  habitat diversity, 3:80–82, 90–91
  human impacts on, 3:92–93
    acidification, 3:97–98
      biological consequences, 3:105–106
    agricultural/industrial chemicals, 3:96
      biological consequences, 3:101–102
    amelioration of, 3:94
    area/variability restrictions, 3:93–94
    cultural eutrophication, 3:96–97
      biological consequences, 3:102
    hydrological cycle changes, 3:94–95
    organic pollution, 3:95–96
      biological consequences, 3:101–102
  interactive regulatory mechanisms, 3:82–83
  isotopes, in ecosystem function management, 2:319–320
  loss of, 2:702
  major habitats, 3:531–532

Freshwater ecosystems (*continued*)
  major species groups and lifestyles, 3:532, 534
  and marine ecosystems, biological comparisons, 2:312–313
  physical structure
    biogeochemical cycles, 3:78–80
    light, 3:76–77
    water, 3:75–76
  phytoplankton of, 4:572
  primary production, measurement, 2:313
    carbon-based methods, 2:313
    C-14 methods, 2:314
    fluorescence methods, 2:314–315
    interpretation and analysis, 2:315
    oxygen-based methods, 2:313
  secondary production, measurement, 2:315
    bioenergetic analysis, 2:315–316
    biomass accrual, 2:316
    birth and death analysis, 2:316
    egg ratio method, 2:316–317
    instar analysis, 2:317
  sediments, species in, 3:92
  species, African, 1:67–68
  species evolution and richness, 3:91–92
  threats to fauna of
    altered water regimes, 3:541–542
    exotic species invasions, 3:542
    habitat loss, 3:540
    reduced water quality, 3:540–541
  zooplankton of, 4:573
Frogs, *see* Anurans
Fruiting body, definition, 3:141
Frustule, definition, 4:349
Fry, definition, 5:233
Fumigant pesticides, 4:512
Functional constraint, definition, 2:179; 4:177
Functional dependency, definition, 1:831
Functional diversity
  active, definition, 5:195
  definition, 2:205, 579; 3:109
  and ecosystem processes, early work on, 3:111–112
  effects of
    experimental studies, 3:114, 116–118
    theory and concepts, 3:112–114
  measurement of, 3:110
  and stability
    experimental/observational studies, 3:118–119
    theory and concepts, 3:118
Functional groupings
  among mobile organisms, 3:126–128
  among sessile organisms, 3:128–131
Functional groups, 3:110–111
  definition, 2:205, 345, 509; 3:109, 121
  evolution, versus phyletic evolution, 3:124
  versus keystone species, 3:137–138
  similarity versus relatedness, 3:122–124
  utility of, 3:122
Functional (interaction) web, definition, 3:1
Functional redundancy/similarity, active, definition, 5:195
Functional type, definition, 5:627
Functional webs, 3:623
  definition, 3:613
Fungi, 2:199–200; 4:202–204
  in animal pathology, 3:156
  chestnut blight, 3:523, 524, 527
  commercial applications
    biopulping/bioremediation, 3:161–162
    drug production, 3:159
    food use, 3:160
    mushroom cultivation, 3:158–159
    wine/beer making, 3:160
  in crop pathology, 3:153
  definition, 3:141
  deuteromycetes, 3:147
  distinguishing characteristics, 3:142
  endangered species, 3:162
  endophytic, in plant defenses, 2:15
  excluded taxa, 3:149–151
  in forest pathology, 3:153–154
  in human pathology, 3:154–156
  importance of studying, 3:143–144
  life cycles, 3:144–145
  major phyla
    Ascomycota, 3:146–147
    Basidiomycota, 3:146, 147
    Chytridiomycota, 3:145, 146
    Zygomycota, 3:145, 146
  and mycorrhizae, in logged forests, 3:754
  pleomorphic, nomenclature systems, 4:398
  roles in ecosystem, 3:143
  as saprophytes, 3:153
  soil, 5:310–311
  surface area and reproduction, 3:151–153
  symbiosis with termites, 3:594–595, 601–602
  taxonomy, 3:143
  temperature limits for, 3:354
Fungicides
  inorganic, 4:513
  organic, 4:513–514
  systemic compounds, 4:514
Furca/furcae, definition, 1:915
Fure, Lake, eutrophication of, 3:102–103
Fynbos, 1:59
  definition, 1:56; 4:145

G

GABA receptors, 3:469
  definition, 3:465
Gadoid fish, definition, 4:497
Gall makers, hymenopteran, 3:422
Gametogenesis, definition, 5:383
Gametophyte, definition, 4:1
Gamma diversity, 3:705; 5:248–249
  definition, 3:701
Gamma landscape density, definition, 3:645
Gap
  forest, definition, 3:747
  taxonomic, definition, 5:589
Gap analysis, 1:851, 858–859; 4:325–326
  definition, 5:441
Gardens, home, 1:110
Garrigue, definition, 4:145
Gars, 2:763
Gastropoda, 4:242–243
Genealogy, species concepts based on, 5:432
Gene banks; *see also* Seed banks
  animal germplasm, 3:178–179
  definition, 1:855
  DNA banks, 3:179
  field, 2:690–691; 3:179–180
  microorganisms, 3:179
  pollen, 3:177
  role in conservation biology, 1:863
  somatic/zygotic plant embryos, 3:177
  spores, 3:177
  vegetative tissues, 3:177–178
Gene conversion, definition, 5:53
Gene flow, 2:90–92
  and conservation genetics, 4:808–809
  definition, 2:85; 4:759; 5:427
  measures of, 2:252

versus selection, in spacially
structured populations,
2:249–251
Gene frequency, definition, 4:537
Gene mapping, in preservation of
livestock biodiversity, 1:544
Gene pool, definition, 2:683; 3:165;
4:759; 5:383, 427
Gene prospecting, 1:476–478
definition, 1:471
Generalist brood parasites, definition,
4:365
Genes, 3:186–187
coadapted complexes, definition,
3:427
and conservation genetics,
4:802–803
definition, 1:855
encyclopedia of, 3:198
insecticide resistance, selection of,
3:471–474
numbers, in organisms, 3:186
Genetic architecture, definition, 5:383
Genetic change
and founder events, 1:31–32; 5:376
in seeds, 1:88
Genetic code, definition, 3:183
Genetic diversity, 1:380; 3:195
of crops
in reserve, 1:550–551
spatial diversity, 1:549
temporal diversity, 1:549–550
definition, 1:85, 377
generation, 2:181
mechanisms, 3:210
genetic partition of, 3:200
and host–parasite coevolution,
2:116–117
implications of crop intensification
in landraces versus modern
varieties, 1:90
spatial patterns, 1:88–90
loss of, and extinction, 3:431
maintenance of, 2:181; 3:196
in nature, 3:210–211
in marine ecosystems, 4:19–20
methodologies, 3:196–197
as phylogenetic divergence
empirical observations, 2:186–187
molecular evolution models,
2:187–188
substitution process, 2:186–187
as polymorphism
cytological/ontogenetic
considerations, 2:182–183
empirical observations, 2:181–182

molecular biological
considerations, 2:182
molecular evolution models,
2:183–185
mutation fates, factors
influencing, 2:183
statistical tests, 2:185
significance of, 3:210
in single-species restoration,
5:191–193
Genetic drift, 5:388
and conservation genetics, 4:809
definition, 1:17; 2:1, 179; 3:761;
4:777; 5:371
versus natural selection, in small
isolated populations, 3:211–212
versus selection, in spacially
structured populations,
2:251–252
Genetic engineering, 2:565–566
Genetic enhancement, definition,
4:799
Genetic erosion, definition, 1:85;
2:683; 4:799
Genetic factors, in phobias, 1:49
Genetic loss, threat to mammals from,
2:444
Genetic markers
definition, 2:245
spatial distribution
F_{st} interpretation, 2:252–253
measures of gene flow, 2:252
Genetic recombination, definition,
5:383
Genetics, development of, 2:8
Genetic variation
definition, 1:85
significance, 4:800–802
spatial patterns
definition, 1:85
implications of crop
intensification, 1:88–90
temporal patterns
definition, 1:85
within/between populations,
2:246–247
Genomes
definition, 3:195; 4:537
eukaryotic, 3:189–190
mitochondrial, 3:200–201
of prokaryotic, 3:187–189
sizes of, 3:186
and critical climatic tolerances,
1:740
Genomic, definition, 3:195
Genomic balance, 4:541

Genotype, definition, 1:547, 693; 2:85;
3:165; 4:537; 5:383
Genotypic, definition, 4:177
Gens/gentes, definition, 4:365
Geobotany, definition, 3:566
Geographic Information Systems,
1:850
definition, 5:121
integration of digital images in,
5:129
Geographic range, and transspecific
patterns of population density,
4:757–758
Geographic speciation, see Allopatric
speciation
Geography, and allopatric speciation,
5:374–376
Geological process, and carbon cycle,
1:618–619
Geology, continental Africa, 1:57
Geomorphic–trophic hypothesis, of
riverine ecosystems, 5:222
Geomorphology, definition, 1:665
Geomyidae rodents, nematode
diversity in, 5:857
Geosphere, human impacts on,
2:558–559
Geothermal heating, definition, 4:349
Germplasm, definition, 1:533
Gerromorpha, 5:719
Ginseng, 4:720–721
Girl Scouts of the U.S.A., 2:771
Girls Incorporated, 2:771
Glaciation, 1:929
pleistocene, definition, 1:261–262
Glaciations
atmospheric data from, 1:711
Younger Dryas, 1:724
Glansdorff, Paul, 4:442
Gleason, Henry, 5:174
Global Biodiversity Assessment,
definition, 5:627
Global change disturbances, definition,
1:709
Global circulation
definition, 1:261
models, definition, 1:609
Global climate change, see Climate
change
Global Environment Facility, 4:855
Global Learning and Observation to
Benefit the Environment
(GLOBE), 2:770–771
Global warming, 3:800; see also
Climate change
and alpine biodiversity, 1:142–143

Global warming (*continued*)
and Antarctic ecosystems, 1:182–183
and Arctic greenness index, 1:246–247
critiques and responses, 3:282
definition, 1:709
and deforestation, 2:33
effects on canopy plants, 3:37
and fish conservation, 2:792
forecasts
elements of, 1:714–715
verification techniques, 1:715–716
past climate changes, 3:278
Glutathione *S*-transferases, 3:468
definition, 3:465
Glycogen, definition, 5:495
Gnasthostomata, 5:761
Goals, definition, 1:813
Gobi Desert, 1:279
Godwana(land), 1:924; 2:726; 5:362
definition, 2:755; 4:145; 5:361
Gonochoric, definition, 5:73
Gonochoristic, definition, 1:915
Gould, Steven Jay, 3:740
Governance systems, definition, 1:777
Grades
versus clade, 1:695
definition, 1:693; 5:589
in evolutionary systematics, 1:698
Gradient gel electrophoresis, in measurement of microbial diversity, 4:182–183
Graminoids, definition, 5:33
Grande Gallerie de l'Evolution (Museum National d'Histoire Naturelle, Paris), 4:276–277
Granivores, definition, 4:601
Grass carp (*Ctenopharyngodon idella*), 3:522, 525
Grasshoppers; *see also* Orthoptera
Melanoplus spp., 2:493
Grasslands, 5:35–38
Australian, diagnostic species
hummock, 1:315
tussock, 1:314
and C4 plants, 1:585
definition, 3:659
endangered ecosystems, 2:414–415
high Andean
distribution and structure, 5:338
functional aspects, 5:338
and historic land-use patterns, 3:682–683
restoration, use of grazers in, 3:272–273

temperate, 5:639
Asian, 1:278
biodiversity of, 5:629
aboveground animals, 5:630
and ecosystem functioning, 5:631–632
future concerns, 5:632–634
plants, 5:629–630
soil organisms, 5:631
distribution, 5:628–629
South American
distribution and structure, 5:338
functional aspects, 5:338–339
tropical, 5:639
Gray whale (*Eschrichtius robustus*)
eastern North Pacific, 4:64
northern Atlantic, 2:798
modern anthropogenic extinction, 4:51–52
western North Pacific, endangered status of, 4:54
Grazers/Grazing, 3:265
conversion of natural ecosystem to, 3:385
definition, 5:33
direct versus indirect effects, 3:267–268
as disturbance regimes, 3:267
diversity of, and host chemistry, 3:271
in ecosystem restoration
coral reefs, 3:275
forests, 3:273–274
grasslands, 3:272–273
lakes, 3:274–275
effects on nitrogen cycle, 4:383
exotic, plant endangerment by, 2:470
by introduced species, 3:520
livestock
biological impoverishment from, 5:39–40
and desertification of native rangelands, 5:43–44
and greenhouse effect, 5:47–48
remote sensing studies, 5:139
selectivity of, 2:375
spatial/temporal scales, roles of, 3:268–269
between-habitat patterns, 3:269
geographic patterns, 3:269
between habitat patterns, 3:269
microhabitat patterns, 3:270–271
within-habitat patterns, 3:270
Grazing intensity, definition, 1:651

Grazing pressure, definition, 1:651
Great apes, in African forests, 1:60
Great Depression, conservation efforts during, 1:891–892
Great faunal interchange, definition, 5:701
Green algae, metabolites, 4:717
Greenhouse effect; *see also* Climate change; Global warming
basic mechanism, 3:277–278
biogeochemical feedbacks
marine, 3:283
terrestrial, 3:283
geophysical feedback, 3:280–281
and livestock grazing, 5:47–48
natural, 1:710
Greenhouse gases, 3:278; 4:732–733
carbon dioxide, 1:295–296
chlorofluorocarbons/halocarbons, 1:298–299
definition, 1:293, 709; 3:277
fire as source of, 2:753
methane, 1:297–298
nitrogen oxides, 1:301
nitrous oxide, 1:298
non-methane hydrocarbons, 1:300–301
ozone, 1:301–304
regulation of, 1:304
sulfur gases, 1:301
Green Revolution, 3:389
biodiversity impacts, 3:798–799
limitations of, 1:115
Green world hypothesis, 3:9–10
and trophic cascades, 3:11
Grilse, definition, 5:233
Grinnellian niche, definition, 3:295
Gross domestic product, air pollutant levels and, 2:298
Ground plan (taxonomic), definition, 5:589
Groundwater
acid deposition effects, 1:9
definition, 2:425; 3:531; 5:805
Group selection, definition, 1:17
Growth rate, intrinsic, definition, 5:453
Gryllacrididae (raspy crickets, leaf-rolling crickets), 3:259
Gryllacridoidea, 3:258–260
Gryllidae (true crickets), 3:256–257
Grylloidea, 3:255–257
Gryllotalpidae (mole crickets), 3:256
Guadalquivir ecosystem (Spain), 4:321
Gua Salukkan Kallang, Indonesia, 5:537

Gubernaculum, definition, 5:843
Guilds
　carnivore, in ecosystems, 1:636–638
　definition, 1:629; 2:205; 3:121; 4:27
　definition and properties of, 3:296–297
　problems defining, 3:300–301
　utility in search for organization/patterns, 3:297–299
Guinea Savanna, 1:62–63
Gymnodinium breve, 4:714
Gymnosperms, 2:198

H

Habitat
　and bird species diversity, 1:499–500
　　and habitat complexity, 1:500–502
　　and habitat diversity, 1:505
　　structure and foraging niches, 1:502–503
　definition, 1:45, 249; 2:635
　diversity, in megadiverse countries, 1:421
　living dead, 3:697–698
　modification, by introduced species, 3:520–521
　versus niche, 3:303–305
　　multispecies comparisons, 3:307–310
　　origins of concept, 3:307
　selection, aesthetic factors, 1:50–51
　structure, definition, 3:303
Habitat conversion, land use issues, 3:667–668
Habitat corridors/connections, definition, 1:855
Habitat creation, definition, 5:185
Habitat degradation, land use issues, 3:670–671
Habitat destruction
　from aquaculture, 1:187–188, 190
　and butterfly diversity, 1:571–572
　endangerment of freshwater invertebrate from, 2:430
　and extinction
　　coral reefs, 2:704
　　from desertification, 2:704–705
　　grasslands, 2:702
　　habitat loss, 2:701–702
　　mangroves, 2:704
　　rain forest loss, 2:702
　　tropical dry forest, 2:702
　　wetlands/aquatic habitats, 2:702, 704
Habitat diversity hypothesis, of species–area relationship, 5:398
Habitat fragmentation, 2:565; 3:770–771
　and animal absence from site/region, 5:193
　and biodiversity, 3:313–314
　and bird extinctions, 1:513–514
　and climate change, synergisms, 1:710–711
　definition, 2:698; 3:659
　and disease, 2:120–121
　effects of fire, 2:751
　and extinction
　　dispersal barriers, 2:705
　　edge effects, 2:705
　　interspecific interactions, 2:705–706
　and forest biodiversity, 3:49
　land use issues, 3:668–670
　threats to reptiles/amphibians from, 2:481
　threat to mammals from, 2:444
　in urban areas, 5:742–743
Habitat heterogeneity, and latitudinal gradients, 5:644–645
Habitat loss, 3:769
　and bird endangerment, 2:399
　in logged forests, 3:755–756
Habitat restoration
　definition, 1:855
　and site assessment, 5:194–195
Habitat specialists
　definition, 2:487
　special perils of, 2:491–492
Habitat Suitability Index, 1:850–851
Haeckel, Ernst, 3:743; 5:573
Hagfishes, 2:757
Haldane, John Scott, 4:442
Haldane's Rule, 5:393
Halocarbons, atmospheric, 1:298–299
Halophiles, definition, 1:219; 4:191
Halophilic Archaea, 1:224–225
Haplopharyngida, 5:869
Haptor, definition, 5:863
Harvest, definition, 2:783
Harvey, William, 4:441
Haustorism, 3:323
Haustorium, definition, 3:317
Hawaiian Islands
　endangered invertebrates of, 2:488
　modern extinctions, 2:737–738
Heath, definition, 3:675
Heathland
　air pollution effects on, 1:128
　definition, 3:675
　and historic land-use patterns, 3:683
Heathlands
　Australia, diagnostic species, 1:313–314
　endangered ecosystems, 2:414–415
Hectare, definition, 3:19; 4:317
Hedonic price, definition, 4:85
Helianthus sp., hybrid speciation in, 5:379–380
Helminth, definition, 2:109
Hemelytron, definition, 5:711
Hemiepiphytes, 3:27
　canopy, 3:30–31
　taxa diversity, 3:33–35
Hemimetabolous, definition, 3:247
Hemiparasite, definition, 3:317
Hemiptera, suborders, 5:712
　Coleorrhyncha, 5:712
　Heteroptera, 5:712
　Homoptera, 5:712
Hemoglobin, evolution of, 1:704
Herbals, 4:713–714, 720–721
　definition, 4:523
Herbicide resistance
　definition, 3:339
　selection of, 3:346
Herbicides, 3:342
　classification of, 3:343
　definition, 3:339
　fate of, in environment, 3:342–344
　inorganic, 4:514
　organic, 4:514–516
　regulation of, 3:344
Herbicide tolerance, definition, 3:339
Herbivores
　definition, 4:601; 4:857
　diversity of, in hybrid plant communities, 4:669
　evolution of, 2:198–199
　functional groups, in coral reefs, 3:134–135
　generalist and specialist, hybrid plant resistance to, 4:667–668
　on hybrid plants
　　effects on plant fitness, 4:673
　　population dynamics, 4:672–673
　as keystone species, 3:617
Herbivory
　definition, 2:11; 5:315
　by introduced species, 3:522
　in marine communities, 3:266
　and pairwise interactions, 5:459
Heritability, 2:247
　definition, 4:777

Hermaphroditic, definition, 5:73
Hermaphroditism, as adaptation to parasitism, 4:468
Hermatypic, definition, 5:73
Herpestidae (mongooses), 1:630, 631
　as bird threat, 2:400
　Indian (*Herpestes javanicus*), 1:74; 3:520, 522, 527
　social systems, 1:638
Heterandria spp., component analysis, 5:777–778
Heterochromatin, definition, 3:183
Heterocyclic nitrogen herbicides, 4:515
Heterogamety, definition, 5:383
Heterogeneity, definition, 5:173
Heteronetta atricapilla (black-headed duck), 4:367
Heteroptera (true bugs), 5:711
　Aradomorpha, 5:721
　Cimicomorpha, 5:720
　economic importance, 5:719
　general characteristics, 5:712–719
　Gerromorpha, 5:719
　Leptopodomorpha, 5:720
　Lygaeoidea, 5:720–721
　Nepomorpha, 5:720
　Reduviidae (assassin bugs), 5:720
Heterosis, definition, 1:547
Heterostyly, definition, 2:1
Heterotrophic, definition, 4:901
Heterotrophic respiration, definition, 1:609
Heterotrophs, definition, 1:437; 2:191; 4:191; 5:647
Heterotrophy, definition, 4:569
Heterozygosity, definition, 3:195, 715
Heterozygote, definition, 1:17
Hexapoda
　definition, 3:479
　major subdivisions, 3:479–480
Hibernation, energy storage in, 5:511–512
Hieracium pilosella, 3:505–507
High seas, definition, 2:783
High-yield modern varieties (HYV), and genetic erosion, 1:114
Hinduism, 5:114–115, 287
Hippocrates, 4:713
Hirudin, 1:476
Historical biogeography, 5:767–768
　definition, 5:767
Historical ecology, definition, 3:790
Historic rainforest, definition, 5:25
HNLC (high nutrient, low chlorophyll) concept, and nearshore nutrient enrichment, 4:5–6

HNLG (high nutrient, low growth) concept, in estuaries, 4:5–6
Holarctic, definition, 4:403
Holocene period, 1:711
　dispersal and natural range expansion, 2:140–141
Holometabola, definition, 3:417
Holomorphology, definition, 5:569
Holoparasite, definition, 3:317
Homeobox (*Hox*) genes, 2:674
　definition, 2:671
Hominid, definition, 1:17
Hominids
　early, spread of, 2:141
Homoiohydry, definition, 3:27
Homology, definition, 1:325, 677; 4:559; 5:569
Homoplasy, definition, 1:677; 5:569
Homoptera, 5:712
Homozygosity
　definition, 3:715
　and transspecific patterns of population density, 4:757
Homozygous, definition, 3:761
Honeybee (*Apis mellifera*), 3:419; 4:726
　African subspecies, 3:425
　products from, 3:425–426
Honeycreepers, Hawaiian, adaptive radiation, case studies, 1:35–36
Hooker, Joseph, 5:362, 363
Hornborgasjön, Sweden, restoration of, 5:816–817
Horticulture, impacts on biodiversity, 3:794–795
Host–parasite interactions
　cleaning symbiosis, 4:472–473
　coevolution, 2:116–117; 4:472
　modeling, 4:371–372
　effects on host individuals/populations, 4:473–474
　immune/tissue reactions and resistance, 4:473
　models, microparasites, 2:111
Host–parasitoid interactions, and population dynamics, 4:774–775
Hosts
　definition, 4:365
　population dynamics, impacts of brood parasitism, 4:372–373
Host specificity
　as adaptation to parasitism, 4:468–469
　latitudinal gradients, 4:480–481
Hotspots
　conservation impact of, 3:379–380

　conservation strategies
　　high-value ecosystems, 3:378
　　megadiversity countries, 3:379
　　wilderness areas, 3:378
　criteria for, 3:373
　definition, 3:371, 437; 4:645
　indicator species, taxa number and indicator accuracy, 3:447–448
　as originally identified, 3:372
　revised analysis, 3:372–374
　　findings of, 3:374
　　　action responses, 3:377–378
　　　congruence among species categories, 3:375–376
　　　higher taxa assessment, 3:377
　　　"hotter" hotspots, 3:374–375
　　　"hottest" hotspots, 3:376–377
　　　species/area relationships, 3:375
　Usambara Mountains, Tanzania, diversity management, 4:655
Hot springs, bacterial diversity, 1:335
Hoyle, Fred, 4:441
Hubbell's unified theory, 1:383
Hudson–Kreitman–Aguade (HKA) test, 2:189
Human ecology
　definition, 5:733
　and social differentiation, 5:739–741
Human impacts
　on biodiversity
　　alien invasions, 3:386–387
　　in cities, 2:699–700; 3:796–797
　　colonialism, 3:797–798
　　commercial farming, 3:798–799
　　driving forces, 3:390
　　　economic arrangements, 3:391
　　　overconsumption, 3:391
　　　overpopulation, 3:390–391
　　　political arrangements, 3:391
　　　social arrangements, 3:391
　　　technology use, 3:391
　　forest, 3:49–50
　　habitat alteration, 3:384–386
　　modern transportation/commerce, 3:798
　　overexploitation, 3:387–388
　　proximate solutions, 3:392
　　types, 3:791–792
　　ultimate solutions, 3:392–393
　causes of, 2:569–570
　eutrophication, 4:2
　history of, 2:558–560
　on marine ecosystems
　　coastal and oceanic water linkages, 4:32
　　ecosystem engineer and invasive species, 4:33

mariculture, 4:33–34
  rocky intertidal communities, 4:28–30
  rocky subtidal communities, 4:30–32
 measuring, 2:570–574
 on molluscs, 4:245–246
 on nitrogen cycle, 4:379–382
 on species interactions, 5:465
 on species' social systems, 5:300–302
 on urban biodiversity
  disturbance regime alteration, 5:744
  habitat creation, 5:742
  habitat destruction, 5:741–742
  habitat fragmentation, 5:742–743
  resource flow alteration, 5:743–744
  species composition alteration, 5:744–745
Humanism, as environmental philosophy, 2:555
Human load, definition, 2:229
Human rights, and conservation ethics, 2:601–602
Humans
 acid deposition health effects, 1:14
 as agents of disturbance, 2:159, 557–558
 and carnivore extinctions, 1:640
 carrying capacity, 2:233–234
 cultural evolution, thresholds of impact, 3:791–792
 emotional response to landscape, 1:50–51, 53
 energy use patterns, 2:526–529
 evolutionary theory and, 2:679–680
 fungal pathogens, 3:154–155
 and global ecological change, ecofootprint analysis, 2:230–231
 and microbes, 4:212
 nematode parasites of, 5:849–852, 853
 origins of ecological knowledge, 2:610
 as part of ecosystems, 3:414–415
 as patch disturbance species, 2:232–233
 plant consumption by, 2:376–377
 population, carrying capacity, 1:648
 population growth, 3:390–391
  causes and effects, 4:824–825
   economic development and environment, 4:826–827
   education, 4:825–826
  history of, 4:820
   changing age profiles, 4:824
   demographic transition, 4:820–821
   mortality, fertility and natural increase, 4:821–822
   twentieth century, 4:821
  responses to, 4:828–829
 property rights and conservation ethics, 2:602–604
 rise as geophysical force, 3:383–384
 settlements, land use issues, 3:663
 techno-ecosystem dominated by, 2:308–310
 threats to terrestrial mammals from, 3:813
 use of deserts by, 2:55–56
 UV radiation effects, 5:730
  eyes, 5:730
  immune system, 5:731
  photoaging, 5:731
  skin cancer, 5:731
  sunburn, 5:730–731
Humpback whale (Megaptera novaeangliae), 4:62, 64
Humus, 1:615
Hunter-gatherer societies; see also Indigenous peoples
 conservation among, 3:412–413
 land tenure, institutions, and biodiversity, 3:413–414
 modernization processes and, 3:413
 and natural resource exploitation, 2:710; 3:412
Hunting, commercial, 2:445
 and animal absence from site/region, 5:193
Hutchinsonian niche, definition, 3:295
Huxley, Thomas, 4:442
Hyaenidae (hyenas), 1:631
 social systems, 1:638
Hybrid inviability, definition, 5:383
Hybridization
 and conservation genetics, 4:808
 of crops and native plants, 4:674–675
 definition, 3:517; 4:659
 in disease intervention, 2:125
 of introduced species, 3:527
 in speciation, 5:391–392
 threat to mammals from, 2:444
Hybrids
 definition, 4:659
 nomenclature systems, 4:397–398
 relative fitness of, 3:434
Hybrid speciation
 introgressive, 5:379
 recombinational, 5:379–380
Hybrid sterility, definition, 5:383
Hybrid swarm, definition, 4:659
Hybrid zone
 definition, 2:85; 4:659
 plant
  dynamics, models of, 4:663
  fitness in, 4:663
  types of, 4:662
Hydantoinases, 1:477
Hydraulic conductivity, definition, 5:805
Hydrocarbons, non-methane, atmospheric, 1:300–301
Hydroelectric power, biodiversity impacts of, 2:536–538
Hydrologic cycle, definition, 4:731
Hydrologic regime, definition, 5:781
Hydrology, modification by introduced species, 3:521
Hydrolytic reactions, 1:442
Hydrophily, definition, 5:255
Hydrophytes, definition, 5:781
Hydrosphere, human impacts on, 2:558–559
Hydrothermal plumes, definition, 5:747
Hydrothermal vents, 3:351, 352–353, 558–559; 4:23–24
 bacterial–invertebrate symbiosis, 5:750
 invertebrate diversity, 5:750–751
  biodiversity, 5:752–753
  biogeography, 5:751–752
  origins of taxa, 5:751
 marine ecosystems, invertebrate diversity
  gene flow and genetic diversity, 5:752
  origins of taxa, 5:751
 microbial diversity, 5:749–750
 midocean ridges and setting, 5:748–749
Hyena, spotted
 in African grasslands, 1:64
 and wild dogs, interspecific competition, 1:809–810
Hymenoptera, 3:484
 ants
  leaf-cutting, 3:422–423
  in mangrove ecosystems, 3:857–858
  and plant defenses, 2:14–15
  symbioses

Hymenoptera (continued)
    with butterflies, 1:566
    with canopy epiphytes, 3:37
    with caterpillars, 1:567
    with plants, 4:610–611
  classification, 3:418
  as introduced biocontrol agents, 3:523
  oviposition behavior, 3:420
  pest species, 3:424–425
  phylogeny and fossil record, 3:418–419
  phytophagy among, 3:422–423
  as pollinators, 3:425; 4:725–726
    introduced species, 4:728
  reproduction, 3:419
  social behavior, 3:423–424
Hyolitha, 4:237
Hypercycle
  characteristics of, 4:448
  definition, 4:439
  as ordering principle, 4:447–448
Hypermetamorphic, definition, 4:249
Hyperparasite, definition, 4:463
Hypersaline, definition, 1:915
Hyperthermophiles
  definition, 1:219; 5:647, 648
  ecological niche and metabolic diversity, 5:648–650
Hypodermis, definition, 5:843
Hypognathous, definition, 3:247
Hyporheic zone, definition, 5:213; 5:527
Hypoxia
  and algal blooms, 2:652
  definition, 2:650
Hypsodont, definition, 3:777
Hyracoidea, 3:796

I

Ibn Es-Sûri, Rachid-eddin, 2:612
Ice Age–Holocene transition, 1:711–712
Icterinae (cowbirds)
  adaptations, 4:370–371
  definition, 4:365
  host defenses against, 4:371
  –host systems, 4:370
  management, 4:373
Idealism, 5:593594
Ideal population, definition, 1:599
Idiobiont parasitoids, 3:420–421
  definition, 3:417
Igapò, 1:151–152
  definition, 1:145; 5:327

IKONOS satellite, 5:142
Image processing, definition, 5:121
Imago, definition, 3:581
Immobilization, definition, 5:305
Impact theory, of dinosaur extinction, 2:100–102
INBio, see Instituto Nacional de Biodiversidad
Inbreeding
  and conservation genetics, 4:807
  definition, 1:17; 3:747; 4:777
  in wild populations, preventive mechanisms, 3:430
Inbreeding coefficient, 3:427
Inbreeding depression, 3:427
  evidence for, 3:428
  factors affecting severity of, 3:428–429
  measuring, as lethal equivalents, 3:429–430
  reduction of, 3:431
  in small populations, 3:430–431
  in total reproductive fitness, 3:430
  variation in susceptibility to, 3:430
Incidence, definition, 4:123
India
  exotic species
    invertebrates, 1:81
    plants, 1:80
    vertebrates, 1:80–81
  Gangte tribe
    as ecological refugees, 4:851–853
    land use changes, 4:850–851
    and market forces, 4:849–850
    political/economic subjugation, 4:848–849
    value appropriation, 4:851
  in situ livestock preservation, 1:541–542
Indicatoridae (honey guides), 4:367
Indicators, definition, 5:553
Indicator species, 3:437–438
  for acid deposition, 3:437–438
  amphibian decline as example of, 3:439
  of biodiversity, 3:445–446
    difficulties with, 3:446–447
    suggested taxa, 3:445–446
    taxa number and indicator accuracy, 3:447–449
  for climate change
    fossil record, 3:442–443
    species distribution, contemporary changes in, 3:443
  for environmental change, 3:440
    large scale, birds as, 3:443–445

plants and carbon dioxide, 3:441–442
and environmental toxicology, 3:439–440
interpretation of, 3:439
lake acidification, 3:440–441
and management of European rivers, 3:438–439
widespread application, 3:439
Indices
  of diversity
    definition, 1:377; 3:485; 5:245
    Simpson, 5:441
Indices, of scientific names, 4:401
Indigenous, definition, 2:610
Indigenous intellectual property rights, definition, 4:523
Indigenous peoples; see also Ecosystem people; Hunter-gatherer societies
  agricultural systems, 1:111; 3:461–462
  Amazonian, crop domestication by, 1:901–905
  conserving biodiversity by empowerment of, 3:462–463
  criteria for, 3:453
  definition, 3:452; 4:523; 5:285
  ecological knowledge, 3:458–460
  and ethanobotanical research, 1:475–476
  ethnomedical traditions, 4:525–526
  and future of biodiversity, 3:802
  Huilliche Indians, 1:116
  and pantheistic tradition, 5:115–116
  populations, 3:452, 453
  P'urhepecha Indians, 1:111
  resource management among, 5:290–292
  Tarahumara Indians, 1:111
Indigenous species, definition, 1:71
Individualism of species, definition, 3:295
Individual selection, definition, 1:17
Indo-Malaya, definition, 5:701
Induced defenses, definition, 2:11
Industrial energy
  biodiversity impacts of
    coal, 2:533–535
    hydroelectric, 2:536–538
    natural gas, 2:533
    nuclear fission, 2:535–536
    oil, 2:530–533
    renewable technologies, 2:538–539
  definition, 2:525
  use, scale of, 2:527

Industrialism, ecology and, in literature, 3:743–744
Industrialized agriculture
  biodiversity impacts, 3:798–799
  definition, 2:305
Industrial melanism, definition, 1:17
Infauna, definition, 3:543; 4:1
Infection, monitoring populations for, 2:123
Iniquiline communities, 4:609–610
Input environment, definition, 2:305
Insecta, 3:561–562; *see also specific orders*
  brood parasitism among, 4:368
  major subdivisions, 3:479
    Endopterygota, 3:482–484
    Neoptera, 3:481
    Paleoptera, 3:480–481
    Pterygota, 3:480
  of mangrove ecosystems, 3:857–858
  migration patterns of, 4:229–230
Insect growth regulator pesticides, 4:512
Insecticide, Fungicide and Rodenticide Act, Federal (FIFRA), 3:344
Insecticide resistance, 3:465
  combating, 3:474–475
  cross-/multiple resistance, 3:470–471
  diagnosis of, 3:471
  extent of, 3:466–467
  fitness of resistance individuals, 3:474
  mechanisms, 3:467
    increased detoxification, 3:467–468
    target site alterations, 3:468–469
  in nonpest species, 3:475–476
  origins of, 3:467
  resistance genes
    homology, 3:469
    origins of, 3:469–470
    selection of, 3:471–472
      ecological influences, 3:472–473
      genetic influences, 3:472
      operational influences, 3:473–474
Insecticides
  inorganic, 4:511
  organic, 4:511–513
Insectivora, 3:790
Inselberg, 1:63
  definition, 1:145
Inselbergs, of Amazonia, 1:155
In situ conservation/preservation, 1:385–386, 856

definition, 1:533, 855, 861; 2:683; 4:645
  genetic reserve conservation, 2:691
  of livestock genetic resources, 1:543
  on-farm conservation, 2:691–692
Instar, definition, 3:479; 4:249
Institutions; *see also* Museums and institutions
  definition, 4:891; 5:681
Instituto Nacional de Biodiversidad (INBio), 1:435, 480–481, 482, 483–487; 4:275–276
  definition, 1:419, 471
Instrumental value, definition, 2:545, 593
Insurance, definition, 1:359
Integrated conservation and development program, definition, 5:481
Integrated weed management, definition, 3:339
Intellectual property rights, definition, 1:359
Intensity, definition, 2:109
Interaction webs, 3:623
  definition, 3:613
Interference competition, definition, 1:793
Intermediate disturbance hypothesis, definition, 5:823
Intermediate host, definition, 4:463
International Conference on Population and Development (1994), 4:829
International Convention for the Regulation of Whaling (1946), 3:236
International Cooperative Biodiversity Group (ICBG), 1:482
International Geosphere Biosphere Programme, definition, 5:627
International Long Term Ecological Research Network, 4:323–324
International Panel on Climate Change, 2:564
International Species Information System, definition, 5:901
International treaties, 3:234, 236
  Convention on Biological Diversity, 3:236–237
  Convention on International Trade in Endangered Species of Wild Fauna and Flora, 3:236
  International Convention for the Regulation of Whaling, 3:236
International Union for the Conservation of Nature, 4:318

Interspecific brood parasites, definition, 4:365
Interstitial, definition, 1:915; 5:527
Intertidal ecosystems, 3:485
  biodiversity
    as indicator of ecological function, 3:495–497
    measurement, 3:494–495
    values of, 3:497–498
  boulder fields, 3:490–491
  general features, 3:486
  mangrove forests, 3:491–493
  Pisaster ochraceus as keystone species, 3:614–615
  rocky shores, 4:21
    competition, 3:487–488
    niche descriptions, 3:311
    physical factors, 3:489
    predation, 3:488–489
    predators, 4:860–861
    recruitment, 3:487
  sandy beaches/mudflats, 3:493–494; 4:21
Intertropical convergence zone, definition, 5:701
Intraspecific parasitism, definition, 4:463
Intrinsic value, definition, 2:291, 545
Introduced species; *see also* Exotic species
  chain reaction effects, 3:524–525
  concept of, 3:517–518
  distribution among habitats, 3:519–520
  effects
    competition, 3:521
    habitat modification, 3:520–521
    herbivory, 3:522
    indirect, 3:524
    parasitism/disease, 3:522–523
    predation, 3:521–522
    quantification, 3:528
  geography and magnitude of invasions by, 3:518–519
  hybridization and introgression, 3:523–524
  invasional meltdown, 3:525
  island vulnerability, 3:520
  mutualism among, 3:526
  plant
    in Australia, 1:78
    in Brazil, 1:81
    Bromus tectorum, 3:504–505
    definition, 3:501
    Fallopia japonica, 3:509–510
    Hieracium pilosella, 3:505–507

Introduced species (*continued*)
  impacts, prediction and evaluation, 3:514–515
  in India, 1:80
  Lantana camara, 3:510–512
  Mimosa pigra, 3:512–513
  Opuntia spp., 3:508–509
  Pinus pinaster, 3:513–514
  Salvina molesta, 3:503–504
  in South Africa, 1:79
  Spartina alterniflora, 3:507–508
  in U.K., 1:76–77
  Undaria pinnatifida, 3:502–503
  in U.S., 1:72
    economic costs, 1:72–73
  plants
    habitat modification by, 3:525
    pollination/dispersal by introduced animals, 3:525
    and resource/interference competition, 3:521
  timing of introductions, 3:519
Introgression
  definition, 3:517; 4:659
  by introduced species, 3:523–524
Introns, definition, 3:183
Invasibility, plant, 4:684–685; *see also* Plant invasions
  definition, 4:677
Invasive alien plants, definition, 4:677
Invasiveness, plant, 4:684; *see also* Plant invasions
  definition, 4:677
Invasive species; *see also* Exotic species
  definition, 1:965; 3:729
Invertebrates
  alien, 2:494
    definition, 2:487
  aquatic
    acid deposition effects, 1:12
    UV radiation effects, 5:728
  of boreal forests, 1:541
  canopy, 3:21–23
  definition, 1:915; 3:543, 561
  as diaspore dispersers, 4:615–616
  Epibenthic, multispecies restoration of, 5:188
  freshwater
    endangered, 2:426
      number and distribution of, 2:433, 435
      protection of, 2:435–436, 438, 440
    endangerment, causes of, 2:426–427, 427
      alien species, 2:432
      direct harvest, 2:431–432
      global climate change, 2:432–433
      habitat destruction and degradation, 2:430
      pollution, 2:430–431
      small ranges, 2:427–429
      sparse populations, 2:429–430
  life cycles, 3:534
  major groups, 3:532–534
  number of species, 3:533, 535
  origin of, 3:537–538
  role in ecological processes, 3:539
    material uptake and transfer, 3:539
    nutrient recycling, 3:539
    plant production, 3:539
  hydrothermal vent, 5:750–751
    biodiversity, 5:752–753
    biogeography, 5:751–752
    gene flow and genetic diversity, 5:752
    origins of taxa, 5:751
  in logged forests, 3:754–755
  marine
    causes of endangerment, 2:459–460
    diversity
      global, 3:550–551
      over evolutionary time, 3:556–558
    endangered, difficulties in determination, 2:458–459
    examples of, 2:460–463
    pharmaceuticals derived from, 4:719–720
    potential scale of endangerment, 2:463–464
    sampling and assessment
      benthos, 3:545–547
      pelagos, 3:547–548
      taxonomy, 3:548–550
  nonindigenous, in U.S., 1:75–76
  riverine ecosystems, 5:224–225
  species flocks in ancient lakes, 3:641
  storage in, 5:496–497
  temperature limits for, 3:354–355
  terrestrial
    definition, 2:487
    endangered
      awareness and understanding of, 2:488
      logistical problems, 2:488–490
      strategic considerations, 2:490–491
      evidence for, 2:488
      significance of, 2:492–494
    environment, 3:563–564
    migration patterns of, 4:229–230
    taxonomic groups, 3:561–563
Invisible hand, definition, 1:359
In vitro, definition, 4:645
Irreplacability, definition, 1:813
Islam, 5:287–288
Islands
  alien invasions, 3:386
  biogeography
    equilibrium theory, 1:469; 2:128
    research, experimental sampling design, 3:578
      horizontal within ecosystem gradients, 3:578
      vertical between ecosystem gradients, 3:578–579
    spatial scales, 3:566
    theory of, 1:857; 3:525, 566–567, 567–568
      definition, 3:517
      in taxonomic versus functional diversity, 2:213–214
    timescales, 3:566
    vegetation analysis methods
      data processing/display methods, 3:574
      dendrograph technique, 3:576
      improved display methods, 3:576–578
      two-way table technique, 3:574, 576
    floristic checklist, 3:568–570
    sampling methods, 3:570
      quantitative methods, 3:571
      count-plot method, 3:572
      distance methods, 3:572–574
      relevè method, 3:570–571
  endangered ecosystems, 2:420
  exotic species on, 2:712; 3:520; 4:869–870
  mammalian endemism, 2:447–448
  megadiversity, 1:424
  myriapod fauna, 4:300
  oceanic, late Quaternary mammalian extinctions, 3:837–838
Isoclines
  definition, 3:729
  zero-growth, definition, 5:413
Isolating mechanisms, definition, 5:371, 427
Isoprene, atmospheric, 1:611

Isoptera (termites), 3:581–582
  caste differentiation, 3:594–595
  distribution, 3:585–586
  eusociality of, 3:586–587
  families and principle genera, 3:583
  food, 3:587–588
  foraging, organization of, 3:599–601
  fossil record, 3:584–585
  functional role, 3:607, 609
  fungus-growing, life history of, 3:601–602
  global diversification
    dispersal ability, 3:603
    evolutionary radiation, 3:603–604
    expansion and diversification, 3:604
    major events, 3:603
    radiation of higher termites, 3:604–605
  and human disturbances, 3:610–611
  nest and life types, 3:588–590
  phylogeny, 3:582, 583
  predators of, 3:588
  social life diversity, 3:597–598
  and soil, 3:609; 5:313
  symbiosis
    carbon/nitrogen balance, 3:593–594
    cellulose decomposition, 3:592–593
    digestive tube, 3:590–591
    with fungi, 3:594–595
    types of, 3:591–592
  in terrestrial ecosystems, 3:605–607
Isotopes
  in aquatic/marine ecosystem function management, 2:319–320
  definition, 1:609
Isotopic discrimination, definition, 1:609
Israel, *see* Near East ecosystems
Iteroparity, definition, 3:715
Iteroparous, definition, 5:233
IUCN Red Data Book, *see* Red Data Book

J

Jack, definition, 5:233
Jainism, 5:288
Japan, temperate deciduous forests, 1:281
Jharkhand movement, 4:854
Jizz, definition, 5:589
Johannsen, Wilhelm, 4:538

Jordan, *see* Near East ecosystems
Jurassic–Cretaceous radiation, 1:416
Jurassic period
  crustacean fossil record, 1:927
  mammalian evolution during, 3:846–847

K

Kainic acid, 4:716
Kalyptorhynchia, 5:871
Karst, definition, 5:527
Karyogamy, 3:144
  definition, 3:141
Kauffman, Stuart, 4:442
Kelp beds
  biodiversity, 4:21
  plant–herbivore interactions, 3:266
Kelt, definition, 5:233
Key-industry species, 3:622
  definition, 3:613
Key innovation, 1:403
  and adaptive radiations, 1:406–407
  definition, 1:25
Keystone species
  concept, critique and reevaluation, 3:620–621
    related concepts, 3:621–622
  conservation implications, 3:629–630
  definition, 1:377; 2:259; 3:41; 4:27
  and ecosystem functioning, 2:348–349
  in ecotoxicology, 2:372
  Enhydra lutris, 3:615–617
  versus functional groups, 3:137–138
  habitat modifiers, 3:620
  and habitat/niche concepts, 3:314
  herbivores, 3:617, 619
  hybrid plant communities, 4:670
  identification of, 3:623
    context dependency, 3:627–628
    experimental approaches, 3:623–625
    interaction strength, 3:628–629
    nonexperimental approaches, 3:626
    a priori, 3:626–627
  mutualists, 3:620
  Pisaster ochraceus, 3:614–615
  plants, 3:619–620
  predators, 4:859
  terminology, 3:622–623
Kheyr Al-Ichbili, Abû-l-, 2:612
kHz, definition, 3:247

Kinetoplastids, RNA editing, 4:419–421
King, definition, 3:581
Kissimmee River, Florida, restoration of, 5:819–820
Kittlewell, H.B.D., 4:539, 540
Kleptoparasitism, in Diptera, 2:824–825
Koinobiont parasitoids, 3:421
  definition, 3:417
Korarchaeota, 1:328
  definition, 1:219
Korea, temperate deciduous forests, 1:281
Kranz anatomy, definition, 1:575
K-T event, *see* Cretaceous–Tertiary event
Kuznets curves, environmental, 2:282–283
Kwongan, definition, 4:145
Kyoto Protocol, 1:626

L

Lacey Act (1900), 3:237
Lagomorpha, 3:798
Lakes/lake ecosystems
  acidification, indicator species, 3:440–441
  African, 1:66–67; 3:640
  ancient
    definition, 3:633
    and species flocks, 3:639–640
  bacterial element cycling, 4:205–206
  biodiversity, 3:636
    and biological productivity/eutrophication, 3:639
    and ecosystem function, 3:638
      food webs, 3:638
      top-down control, 3:638
    and ecosystem stability, 3:639
    in fish, 3:636–637
    in freshwater sediments, 3:636
    of plankton and microbial, 3:636
    threats to, 3:641
      fisheries practices, 3:643
      habitat alteration, 3:641–642
      pollution, 3:643
      species introductions, 3:642–643
      water competition, 3:641
  characteristics, 3:633–634
  energy pathways, 2:515
  eutrophication, 4:3
    arctic and tropical lakes, 3:104–105

Lakes/lake ecosystems (continued)
   hardwater lakes, 3:102
      resource and predatory control in food webs, 3:103–104
      reversals, 3:283–285
      and species richness, 3:105
      whole-lake experiments, 3:103
   freshwater biota, origins and peculiarities, 3:634
   geothermally heated, 3:351, 352
   hardwater, eutrophication of, 3:102–103
   latitudinal gradient, 3:634–635
   predators, 4:864
   role of intra-/interspecies communication systems, 3:639
   species richness
      and area, relationship, 3:635
      and morphometry, 3:636
   vertical distribution, 3:635
Lamarkian adaptationism, 2:1, 192
Lampreys, 2:757
Lanao, Lake, 3:641
Land, definition, 2:61
Land cover, definition, 3:659
Land degradation
   definition, 2:61
   impact of livestock, 1:654–655
Land ethic
   definition, 3:739; 5:481
   and environmentalism, in literature, 3:744
Landraces, 1:88
   definition, 1:85
   versus modern varieties, genetic diversity, 1:90
   survival, and farmer incentives, 1:94
   and Vavilov centers, 3:796
Landsat program, 3:671; 5:126–127, 140
Landscape(s)
   biodiversity planning in, 3:653–654
      case study: Sevilleta National Wildlife Refuge, 3:654–657
   characteristics, 3:645–646
   definition, 3:645
   description of, 3:648–649
   disturbed, succession and biodiversity, 5:550
   diversity, controls on, 3:649–650
   dynamics, 3:651–653
   function, 3:650–651
   human emotional response to, 1:50–51
   structure, description of, 3:646
      patch, 3:646–648

Land use/land use issues, 3:659–660
   agriculture, 3:660, 662–663
   and alpine biodiversity, 1:143
   biodiversity effects, 3:665–667
      habitat conversion, 3:667–668
      habitat degradation, 3:670–671
      habitat fragmentation, 3:668–670
   in boreal ecosystems, 1:542, 543
   changes, CO_2 emissions from, 1:296
   definition, 3:659
   human settlement, 3:663
   and land cover
      effects, 3:663, 665
      estimation, 3:671–672
   natural resource extraction, 3:663
   patterns
      and Central American ecosystems, 1:675–676
      crops and domestic animals, properties of, 3:677
      fields, hedges, and terraces, 3:677–678
      forest and woodland, 3:678–680
      historic, 3:675–676
         and climate change, 3:676
         forests, 3:678–680
         grassland, 3:682–683
         heathland, 3:683
         implications for biological conservation, 3:686–687
         moorland, 3:683–684
         rise and fall of, 3:684–686
         savannas, 3:680–682
         woodlands, 3:678–680
         wood pasture, 3:680–682
Language
   and biodiversity, 5:289–290
   definition, 5:285
   in literature and science, 3:740
Lantana camara, 3:510–512
Larger, definition, 5:305
Larvae
   definition, 2:815; 3:531, 581
   dipteran
      herbivorous, 2:822–823
      non-free living, 2:823
      parasitic, 2:823
      parasitoid, 2:823
      predatory, 2:823
      scavenger, 2:822
Larval parasite, definition, 4:463
Late Quaternary extinction event, definition, 3:825
Latin America
   deforestation in, 2:24, 411
   tropical forests and indigenous populations, 3:457

Latitudinal gradients
   and Arctic biodiversity patterns, 1:237–240
   in bird species diversity, 1:506–507, 506–508
      causes, 1:507–508
   and boreal forests vegetation, 1:536–538
   in butterfly species diversity, 1:568
   definition, 3:701; 5:637
   lake and pond ecosystems, 3:634–635
   and megadiversity, 1:421–422
   parasites
      in host ranges and specificity, 4:480
      in reproductive strategies, 4:478, 480
      in species richness/abundance, 4:478
   possible explanations, 5:641–642
      area, 5:643–644
      habitat heterogeneity, 5:644–645
      productivity, 5:645–646
      solar energy, 5:645
      species interactions, 5:645
      time, 5:642–643
   and species richness, 1:731; 3:702
      theories/hypotheses
         climate extremes, 1:737–738
         competition, 1:736
         disturbance, 1:736–737
         energy availability, 1:737
   synthetic model of, 3:712–713
   and transspecific patterns of population density, 4:757
Laurical, 1:477–478
Laurisilva, definition, 4:291
Laurussia, 1:924
Lawn and garden management, costs of agricultural invasions, 1:73, 76
Lazarus taxa, 4:108
Leaf area index, definition, 5:269
Leaf beetles (Chrysomelidae), 1:354
   definition, 1:351
Lecithoepitheliata, 5:869
Legacy ecosystems, definition, 1:965
Legislation; *see also* Regulations
   collections in implementation of, 4:274
   definition, 3:233
   habitat/ecosystem conservation, 3:240
      federal lands, 3:240–241
      private land, 3:241–242
   influencing biodiversity, 3:235

international treaties, 3:234, 235–237
local ordinances, 3:243–244
national, 3:237
   conservation of genetic resources, 3:237
   conservation of species, 3:237–240
   state legislation, 3:242–243
Legumes, definition, 1:99
Leopold, Aldo, 1:889, 892; 3:364, 744
Lepidoptera, 3:484
   butterflies
      diversity and habitat destruction, 1:571–572
      endangered, 2:492
      evidence of coevolution, 1:755
         rates of adaptation, 1:757
      family Lycaenidae, 1:561–562, 566
      family Nymphalidae, 1:562–564
      family Papilionidae, 1:560–561
      family Pieridae, 1:561
      geographical patterns of diversity
         latitudinal gradients, 1:568
         in neotropical sites, 1:569
         regional patterns, 1:569–570
         in space and time, 1:570–571
      host relationships, 1:566
      life cycle stages
         adult, 1:565–566
         caterpillar, 1:564–565
         egg, 1:564
         pupa, 1:565
      mimicry and diversity in, 1:568
   moths, 4:249–251
      coloration, 4:267
      conservation of, 4:269–270
      defensive chemistry, 4:263–264
      dispersal and migration, 4:264
      ditrysia, 4:253–254
      diurnality/nocturnality, 4:267–268
      feeding biology
         adults, 4:261–262
         larva, 4:259–261
      fossil history, 4:254
      Glossatan infraorders, 4:251–253
      heteroneurans, 4:253
      importance of, 4:268–269
      key adaptations, 4:264–265
      life cycle stages
         diapause, 4:258–259
         egg, 4:256
         larva, 4:256–258
         pupa, 4:258–259
      natural enemies, 4:262–263

North American, diversity, 4:410
pheromones, 4:264–265
as pollinators, 4:725
rainforest, diversity of, 5:4–5
species diversity and biogeography, 4:255–256
suborders, 4:251
taxonomic diversity, 1:559–560
Lepidosaurs
   conservation issues, 5:158
   Rhynchocephalia, 5:147
   Squamata, snakes, characteristics, 5:155–157
Leptopodomorpha, 5:720
Lerner, Michael, 4:541
Lethal equivalents
   definition, 3:427
   measuring inbreeding depression as, 3:429–430
Liana forests, Amazonian, 1:150–151
Lichens, 3:153
   air pollution effects on, 1:128; 2:708
   canopy, 3:29–30
   definition, 3:27
   nomenclature systems, 4:398
   woodlands, in boreal forests, 1:538–539
Life
   earliest record of, 1:394
   origins of, hypotheses and theories, 4:441–443; 5:553
   temperature limits to, 4:353–355
Life cycles, complex, origins of, 4:471–472
Life expectancy, definition, 4:819
Life form, definition, 1:133
Life Galleries (Natural History Museum, London), 4:277
Life history, 3:715–716
   analysis, tradeoffs in, 3:719–720
   evolution of, 3:720
      in environments with stochastic and predictable components, 3:727
      in predictable environments, 3:723–727
      in static environments, 3:720–723
      in stochastic environments, 3:720–723
   genetic models, 3:718
      quantitative, 3:718–719
      simple Mendelian, 3:718
   phenotypic models, 3:716–717
      game theory, 3:717–718
      optimality modeling, 3:717
Light, properties of, 5:723

Lignin, definition, 1:609
Linkage disequilibrium, definition, 1:17
Linnaeus, Carolus, 1:458, 696; 3:363, 742; 5:575–576
Linnean system, definition, 1:693, 697
Lipids
   plant, composition of, 5:498–499
   storage, fatty acid composition, 5:506–508
Li Shizhen, 2:613
Literature
   definition, 3:739
   and science relations, 3:740
Littoral zone, definition, 3:75; 5:781
Livestock
   breeds, definition, 1:534
   diversity, current trends, 1:535–536
   domestication, 5:291
   as food source for predators, 1:660–661
   genetic resources
      coordinating use/preservation, 1:536–538
      management strategies for conservation, 1:538
      ex situ strategies, 1:543
      inventory/characterization, 1:538–541
      in situ strategies, 1:541–543
      new technologies, impacts
         cloning, 1:544–545
         cryopreservation, 1:544
         gene mapping, 1:544
   grazing
      biological impoverishment from, 5:39–40
      and deforestation, 2:30
      and desertification of native rangelands, 5:43–44
      effects on
         aquatic ecosystems, 5:48
         biogeochemical cycling, 5:49
         exotic species/pathogens, 5:48
         fire regimes, 5:50
         and greenhouse effect, 5:47–48
      inbreeding depression in, 3:428
      integration in traditional agriculture, 1:114
      intensive production and nutrient cycling, 1:657
      and invertebrate fauna, 1:661
      microbe introductions affecting, 1:79, 80, 81
      in mixed farming systems, 1:104–105

Livestock (*continued*)
    as parasite/pathogen hosts,
        1:662–663
    populations, genetic structure,
        1:534–535
    properties of, 3:677
    as seed dispersers, 3:274
    terrestrial/aquatic ecosystem effects,
        5:40–41
        primary influences, 5:41–43
        secondary/tertiary influences, 5:43
    threat to mammals from, 2:444–445
    and vertebrate fauna, 1:661–662
Living dead
    animals, 3:695–696
    case study, 4:653
    definition, 2:731; 3:689; 4:645
    deforestation and, 3:691–693
    and predators, 3:696–697
    and restoration biology, 3:698–699
    small plants, 3:694–695
Lizards, 5:147–148
    Caribbean, adaptive radiation,
        1:36–37
    characteristics, 5:153–155
Llanos, definition, 5:327
Lobau wetlands, Austria, 5:537
Local, definition, 2:610
Local diversity, definition, 2:285
Local species richness, definition,
    3:531
Locus/loci
    definition, 1:85; 3:183, 761
    patterns of diversity among, 3:200
Loeb, Lawrence, 4:446
Logistic growth, definition, 1:641
Lognormal distribution, definition,
    5:441
Longhurst, Alan, 4:429, 430
Loranthaceae, *see* Mistletoe
Lotka, A.J., 2:306
Lotka–Volterra model, 1:794–798,
    800; 2:292; 3:730–731
    definition, 4:857; 5:413
Lowlands, climatic zonation, in Asia,
    1:266
Lungfishes, 2:761
Lycaenidae, 1:561–562
Lyell, Charles, 1:460, 463
Lygaeoidea, 5:720–721

M

Macaxeira, definition, 1:897
Macrobiota, definition, 1:915
Macroevolution, definition, 5:569

Macrofauna, definition, 4:1
Macroinvertebrates
    aquatic, acid deposition effects,
        1:12–13
    definition, 5:781
Macroparasites
    characteristics, 2:110
    host–parasite interactions, models,
        2:113–114
Macrophytes
    aquatic, acid deposition effects, 1:12
    definition, 1:1
Macropod, definition, 3:777
Macropterous, definition, 3:247
Macroscelidea, 3:798–799
Macrostomida, 5:869
Macrotidal, definition, 1:741
Madagascar
    ecosystems of, 1:68–69
    first-contact extinctions, 2:735–736
    late Quaternary mammalian
        extinctions, 3:834–835
Mafia effect, definition, 4:365
Mafic rock, definition, 1:437
Magnuson Act (1976), 3:240
Maimonides, 2:612
Maitotoxin, 4:714
Maize
    genetic narrowing, 1:91–92
    genetic variation, 1:89–90
    trends in genetic diversity, 1:92
Major ecosystem type, definition,
    5:345
Major habitat type, definition, 5:345
Major histocompatibility locus (MHC),
    2:116–117
Malthus, Thomas Robert, 2:3, 672;
    4:826
Mammalia, 5:765
    of boreal forests, 1:539–541
    conservation
        captive breeding/reintroduction,
            2:450–451
        economic incentives, 2:451–452
        legislation, 2:449–450
        protected areas and preserve sizes,
            2:449–450
        species survival commissions,
            2:449–450
        trends, 2:452–453
        umbrella principle, 2:452
    definition, 3:811
    diagnostic characteristics, 3:779
    as diaspore dispersers, 4:615
    diversity of, 2:441–442
    endangered

        geographic analysis, 2:447
        IUCN Red Data Book, 2:448–449
        by taxa, 2:449
    evolutionary history, 2:442;
        3:841–842
        during age of dinosaurs,
            3:846–847
        during Tertiary, 3:849–850
    evolutionary trends
        brain size, 3:799
        convergent evolution, 3:801–802
        Cope's Rule, 3:799–800
        Island Rule, 3:800–801
        locomotion, 3:801
    island endemism, 2:447–448
    of mangrove ecosystems, 3:859
    migration patterns of, 4:225–227
    new discoveries
        in Annamite mountain range,
            3:805–806
        cryptic species, phylogeography,
            and evolutionary significant
            units, 3:806
        patterns of, 3:804–805
    nonindigenous
        in Australia, 1:78
        in Brazil, 1:81
        in India, 1:80–81
        in U.K., 1:77
        in U.S., 1:73–75
    North American, diversity, 4:410
    overview of class, 3:779
        Eutheria, 3:780–781
        Metatheria, 3:780
        Monotremata, 3:779–780
        Theria, 3:780
    phylogeny
        early history, 3:782
        ordinal diversification
            patterns of, 3:784
                marsupials, 3:785–789
                monotremes, 3:784
                placentals, 3:789–799
            structure of evolutionary tree,
                3:782
            timing of, 3:782–784
    placental versus marsupial,
        differentiation
        cranial and skeletal differences,
            3:781
        reproduction, 3:781–782
    plant species consumed by, 2:376
    as pollinators, 4:726
    rainforest, diversity of, 5:7–8
    South Africa (Cape Region),
        biodiversity, 4:156

species richness
    gradients of, 3:802, 804
    hotspots of, and endemmicity,
        3:804
    terrestrial
        conservation efforts, case studies
            African wild dog, 3:815–817
            rhinoceroses, 3:817–819
            saiga antelopes, 3:819–821
            themes of, 3:821–822
        conservation needs, 3:811–813
        threats to, 3:813–814
    threats to, 2:443
        exotic species introductions,
            2:447
        exploitation, 2:445
        genetic loss, 2:444
        habitat loss, 2:444
        interference, 2:446
        livestock and disease, 2:444–445
        persecution, 2:445–446
        pollution, 2:446–447
        relationship with physiology,
            2:443
Mammalian extinctions
    background and mass, 3:843–844
    Cretaceous–Tertiary, 3:847–848
    current crisis in, 3:806–807
    distal and proximate causes,
        3:844–845
    early Paleocene recovery, 3:848–849
    late Quaternary, 3:825
        Australia, 3:833–834
        comparative approach, 3:826–827
        Eurasia and Africa, 3:835–837
        Madagascar, 3:834–835
        North America, 3:826–827
        oceanic islands, 3:837–838
        patterns and causes, 3:838–839
            anthropogenic models, 3:839
            climatic models, 3:839–840
        South America, 3:832–833
    mass, temporal scale of, 3:844
    Paleocene, 2:442–443
    patterns and causes, 3:842–843
    pseudoextinctions, 3:843
    recent, 2:443
Mammoth Cave, Kentucky, 5:536
Management; *see also* Adaptive
    management
    biodiversity, ethical issues,
        2:598–599
    definition, 3:63; 5:285
    ecosystem, definition, 5:481
    natural resources, definition, 5:481
Manatee, West Indian (*Trichechus*

manatus), endangered status of,
    4:60–61
Manchuria, temperate deciduous
    forests, 1:281
Mangal, definition, 3:853
Mangal/mangle, definition, 5:327
Mangrove forests, 5:705
    Amazonian, 1:152–153
    Asian, 1:286, 287
    Australian, diagnostic species,
        1:315–316
    biodiversity of, 3:491–493
    diversity of, 3:864
        global patterns, 3:864–865
        local variation in species
            distribution and diversity,
            3:867–868
        regional patterns, 3:866–867
    endangered, 2:419
    environmental adaptations
        reproduction, 3:855–856
        salinity, 3:854
        waterlogging, 3:855
    genetic diversity, 3:869
    habitat, 3:854
    in Israel, 4:364
    loss of, 2:704
    marine fauna, 3:859
        Crustacea, 3:860, 861–862
            as ecosystem engineers, 3:862
            grapsid crabs, 3:860–861
            ocypodid crabs, 3:861
        distribution, 3:868–869
        fish, 3:860
        meiofauna, 3:863–864
        mollusks, 3:863
        root communities, 3:859–860
    South American, 5:341–342
    terrestrial fauna, 3:856–857
        amphibia and reptiles, 3:858
        birds, 3:858
        insects, 3:857–858
        mammals, 3:859
    uses/abuses of, 3:869
Mangrove trees, 3:853
Manioc, wild, definition, 1:897
Mantle, definition, 4:235
Maquis, definition, 4:145
Marginal benefits/costs, definition, 4:85
Marginal value, definition, 2:353
Mariculture, *see* Aquaculture
Marine ecosystems; *see also* Pelagic
    ecosystems
    algal blooms, 2:655
    bacterial element cycling, 4:206–207
    benthic invertebrates

large-scale diversity patterns
    bathymetric gradients,
        3:554–555
    latitudinal gradients, 3:552–554
    sampling and assessment,
        3:545–547
    small-scale diversity patterns,
        3:555–556
biodiversity and ecological function,
    relationship, 12
biogeochemistry and nutrient
    cycling, 2:317–318
    new and regenerated production,
        2:318–319
Cambrian, 1:919
cold seeps, 3:558–559
crustacean biodiversity, 1:917–919
cultural eutrophication effects
    algal blooms, 2:656–657
    on fish populations, 2:666
    on macroalgae, 2:658–660
    on macrophytes, 2:658–661
    on microfauna, 2:664–665
ecosystem diversity
    anthropogenic changes,
        consequences, 4:24
    benthic, 4:23
    continental shelves, 4:22
    coral reefs, 4:21–22
    intertidal beaches, 4:21
    kelp beds, 4:21
        predators, 4:861–864
    mid-ocean ridges and
        hydrothermal vents, 4:23–24
    pelagic zone, 4:22–23
    rocky shores, 4:21
ecosystem function, 4:20
    measurement, conceptual
        framework, 2:311–312
ecosystem services and sedimentary
    diversity, 4:7–8
    filtration, 4:10
    nutrient cycling, 4:9
    pollutant cycling, 4:9–10
    secondary production, 4:10
    sediment and shoreline stability,
        4:10
ecosystem units, 4:14
endangered, 2:418, 420–421
as feedback mechanism, 3:281, 283
fish diversity
    Antarctic region, 2:782
    Arctic region, 2:782
    deep sea, 2:1535
    eastern Atlantic region,
        2:781–782

Marine ecosystems (*continued*)
  eastern Pacific region, 2:781
  Indo-West Pacific region, 2:780
  pelagic regions, 2:782–783
  temperate regions, 2:782
  western Atlantic region, 2:781
 and freshwater ecosystems, biological comparisons, 2:312–313
 genetic diversity, 4:19–20
 higher taxa of, 4:16–18
 human impacts on
  coastal and oceanic water linkages, 4:32
  ecosystem engineer and invasive species, 4:33
  mariculture, 4:33–34
  rocky intertidal communities, 4:28–30
  rocky subtidal communities, 4:30–32
 hydrothermal vents, 3:558–559
  bacterial–invertebrate symbiosis, 5:750
  invertebrate diversity, 5:750–751
   biodiversity, 5:752–753
   biogeography, 5:751–752
   gene flow and genetic diversity, 5:752
   origins of taxa, 5:751
  microbial diversity, 5:749–750
  midocean ridges and setting, 5:748–749
 introduced species in, 3:519–520
 isotopes, in ecosystem function management, 2:319–320
 versus land environments, comparison, 4:14–16
 in megadiverse countries, 1:421
 nearshore, nutrient response, 4:5–6
 nonanthropogenic changes and variability, 4:34–35
 pelagic invertebrates
  diversity patterns, 3:551–552
  sampling and assessment, 3:547–548
 Phanerozoic biodiversity, 3:221–222
 phytoplankton of, 4:572
 plant–herbivore interactions, 3:266, 268, 270
 predators, 4:865–866
 primary production, measurement, 2:313
  carbon-based methods, 2:313
  C-14 methods, 2:314
  fluorescence methods, 2:314–315

  interpretation and analysis, 2:315
  oxygen-based methods, 2:313
 role in carbon cycle, 1:612–613
 sandy beaches
  biodiversity of
   and beach type, 1:745–747
   biological factors, 1:749
   latitudinal effects, 1:747
   paradigms for, 1:749–750
   physical factors, 1:747–749
  defining variables, 1:742
  microfauna composition and zonation, 1:744–745
  swash climate, 1:748
  transect surveys, 1:743–744
  types of, 1:742
 secondary production, measurement, 2:315
  bioenergetic analysis, 2:315–316
  biomass accrual, 2:316
  birth and death analysis, 2:316
  egg ratio method, 2:316–317
  instar analysis, 2:317
 sedimentary habitats, 4:4
  biodiversity patterns, global estimates, 4:6–7
  deep-sea areas, 4:5–6
  intertidal areas, 4:4
  regulation of diversity
   deep sea, 4:7–8
   shallow water, 4:7–8
  species numbers, global estimates, 4:6
  subtidal areas, 4:4–5
 siltation of onshore waters, 3:385–386
 species diversity, 4:18–19
 threats to sedimentary diversity, 4:11
  exotic species introductions, 4:12–13
  fishing impacts, 4:11
  global climate change, 4:13
  habitat destruction, degradation, and shoreline modification, 4:11–12
  pollution and eutrophication, 4:12
 zooplankton of, 4:573
Marine Mammal Protection Act (1972), 3:238–239
Marine mammals
 endangered taxa/populations, 4:53, 57–59
 extinction
  identification and monitoring vulnerable populations, factors hindering, 4:47–49

  modern anthropogenic, 4:50–52
   approaches to minimize, 4:67–68
   Caribbean monk seal, 4:51
   factor/processes facilitating, 4:66–67
   Japanese sea lion, 4:51
   northern Atlantic gray whale, 4:51–52
   regions of greatest concern, 4:67
   Steller's sea cow, 4:51
  over evolutionary time, 4:50
  vulnerability to, factors, 4:44–47
 general features and habitat boundaries, 4:38–39
 migratory patterns, 4:226–227
 recovering taxa/populations, 4:63
Marine regression theory, of dinosaur extinction, 2:103–106
Marine reserves, definition, 2:783
Marine sediments, 4:1–2
 bacterial element cycling, 4:206–207
Market-based incentives, definition, 4:85
Market economy, 4:85–86
 and biodiversity, 4:86–87
  diffuse benefits of, 4:92–93
  preservation, and property rights, 4:93–94
 definition, 4:85
Market value, definition, 2:285
Maromba, definition, 1:897
Marsh, definition, 5:781–782
Marsh, George Perkins, 1:887; 3:363–364
Marsupials; *see also* Metatheria
 definition, 1:629
Marsupium, definition, 3:777
Mass ratio hypothesis, definition, 3:329
Mast fruiting, definition, 3:747
Mate recognition systems, evolution of, 5:373
Material corrosion, from acid deposition, 1:14
Mating, assortative, definition, 5:383
Mating systems
 and conservation genetics, 4:806
 impacts on social systems, 5:296–297
Matorral, definition, 4:145
Matrix, definition, 3:645
Matrix species, definition, 5:185
Maxillae, definition, 4:249
Maximum sustainable yield

definition, 5:553
fishery, 5:167–168
definition, 5:161
McClintock, Barbara, 3:184
McDonald–Kreitman test, 2:189
Meadow, definition, 3:675
Mean kinship, definition, 1:599
Mecoptera (scorpionflies), 3:483
Median/medial, definition, 1:915
Mediterranean ecosystems, 4:145–147;
see also Near East ecosystems
Asian, 1:277–278
California, biodiversity in,
4:148–149
vascular plants, 4:149
vertebrates, 4:149
Cape Region (South Africa),
biodiversity in, 4:154–155
birds, 4:156–157
invertebrates, 4:157
mammals, 4:156
reptiles and amphibians, 4:157
vascular plants, 4:155–156
chaparral, 5:639–640
characteristic crops, 2:224
Chile, biodiversity in, 4:149–150
birds, 4:152
reptiles and amphibians,
4:152–153
vascular plants, 4:150–151
vertebrates, 4:151–152
definition, 4:145
Mediterranean basin, biodiversity in,
4:153
vascular plant diversity,
4:153–154
vertebrates, 4:154
natural disturbance regimes,
4:147–148
north African, 1:58
Southwestern Australia, biodiversity
in, 4:157–158
vascular plants, 4:158
vertebrates, 4:158
speciation patterns, 4:148
Medulla, definition, 5:863
Mega-city, 2:310
definition, 2:305
Megadiverse countries, 3:379
biodiversity threats in
deforestation/habitat
fragmentation, 1:427–428
species overexploitation, 1:429
biological information management,
1:434
Comisión Nacional Para el

Estudio y Uso de la
Biodiversidad (CONABIO),
1:434–425
Environmental Resources
Information Network, 1:435
Instituto Nacional de
Biodiversidad (INBio), 1:435;
4:275–276
South African Botanical Diversity
Network, 1:435–436
scientific development levels, 1:431
sociodemographics and biodiversity,
1:429–431
Megadiversity
and cultural diversity, correlation,
3:802
definition, 1:419; 3:371
facets of, 1:420–421
underlying causes
ecological diversity, 1:425
geohistorical factors, 1:423–425
latitudinal position, 1:421–422
physical factors, 1:422–423
rain forest presence, 1:425–426
Megafauna
definition, 3:689, 825; 4:1
and latent extinction, 3:695–696
terrestrial, drug/chemicals from,
4:720
Megaloptera, 3:483
Megazone, vegetation, definition, 2:635
Meiofauna
definition, 4:1
of mangrove ecosystems, 3:863–864
Meiotic drive, definition, 1:17
Melaleuca spp., 5:816
Melittophily, 4:726
definition, 4:723
Mendel, Gregor, 4:538
Mendelism, 2:8
Mesh size, definition, 5:161
Mesophiles, definition, 1:219
Mesophyll tissue, definition, 1:575
Mesopredator, definition, 4:857
Mesothorax, definition, 3:479
Mesotrophic, definition, 2:650; 4:377
Mesozoic–Cenozoic event, see
Cretaceous–Tertiary event
Mesozoic era, mass extinctions of,
4:121
Messenger RNA (mRNA), 3:186
transcription in prokaryotes,
3:188–189
Metabolism
basal rate of
in carnivores, 1:631

definition, 1:629
and size of birds/mammals, 1:491
and Darwinian evolution, 4:444–445
types, in bacteria, 4:204
Metacommunity
definition, 2:161
processes, and regional control of
species richness, 2:169–171
Metals, toxic
deposition/environmental pathways,
1:121
effects near large smelters,
1:125–126
sources and impacts, 1:120
Metapopulations
definition, 1:159; 2:161; 3:761;
4:161
models of, 2:265
spatial variation, quantitative effects,
4:166–167
competitive interactions,
4:172–174
predators and parasitoids,
4:171–172
single species dynamics,
4:167–171
strict, 4:162
definition, 4:161
multispecies considerations
competition, 4:165–166
predators and parasitoids,
4:162–164
single species considerations,
4:162–164
theory, lessons from
conservation, 4:174–175
nonequilibrium dynamics, 4:175
population regulation, 4:175
Metatheria (marsupials), 3:780,
785–786
carnivorous, 1:631
social systems, 1:638
Dasyuromorphia, 3:786–787
definition, 3:777
Didelimorphia, 3:786
differentiation from placental
mammals, 3:781–782
Diprodontia, 3:787–789
Microbiotheria, 3:786
Notoryctemorphia, 3:787
Paucituberculata, 3:786
Peramelemorphia, 3:787
Metathorax, definition, 3:479
Metazoa
definition, 4:349, 569
evolution of, 2:200–201

Methane
  atmospheric, 1:297, 611
    sources of, 1:298
  from livestock, 5:47-48
Methanogenesis, 1:222-223
Methanogenic Archaea, 1:222-224
Methanogens, definition, 1:219
Mexico
  biodiversity and traditional agriculture, 1:116
  Comisiòn Nacional Para el Estudio y Uso de la Biodiversidad (CONABIO), 1:434-435
  sociodemographics and biodiversity, 1:431-432
Microalgae
  long-term human influence, 2:654-655
  species shifts across nutrient gradients, 2:653-654
Microarrays, 3:198
Microbes
  biodiversity, 4:178, 191-192
    biological significance, 4:194-195
    measurement of, 4:178-179
      counting methods, 4:180-181
      enrichment/isolation versus direct measurement of communities, 4:179
      environmental heterogeneity, 4:179-180
      level of taxonomic resolution, 4:179-180
      statistical approaches, 4:188-189
    scientific neglect of, 4:195
    scope of, 4:192-194
  cellular constituents, analysis
    fatty acids, 4:186-187
    nucleic acids
      PCR-dependent methods, 4:182-185
      PCR-independent methods, 4:185-186
  definition, 4:191, 201-202
  enzyme assays, 4:187-188
  major groups, 4:197-198
    domain Archaea, 4:200-202
    domain Bacteria, 4:198-200
    domain Eucarya, 4:202
      fungi, 4:202-204
      Protista, 4:202
  and man, 4:212
  pharmaceuticals derived from, 4:719
  roles over geological time and in contemporary biosphere, 4:210-212
    substrate utilization patterns, 4:187
Microbial pesticides, 4:512
Microbiology
  Delft school of, 4:195-196
  Woesean reformation of, 4:197
Microbiota
  diversity, in Archaean/Proterozoic eons, 3:220-221
  diversity and functional groups, 5:198
    free-living saprobes, 5:198-199
    soil animals and food webs, 5:199
    symbionts, 5:199-200
  diversity and functional redundancy, 5:196-197
  establishment, 5:200-201
  riverine ecosystems, 5:222
  spatial/temporal arrangement, 5:197-198
Microclimate, definition, 1:133, 171
Microcosm studies, definition, 5:413
Microevolution, definition, 5:569
Microfauna
  definition, 5:305
  soil, 5:311
Microhabitats
  climatic, in alpine ecosystems, 1:135-136
  definition, 1:171; 3:303
Microparasites
  characteristics, 2:110
  host-parasite interactions, models, 2:111-113
Micropterous, definition, 3:247
Microsatellites, 3:190, 197
Microthrix, definition, 5:863
Microtidal, definition, 1:741
Microtubule, definition, 2:623
Middle East, deserts of, 2:46
Middomain effect, definition, 3:701
Midocean ridge, definition, 5:747
Migration, 4:221-222
  definitions of, 4:222-223
  energy storage, 5:511
  partial, definition, 4:221
  patterns of, 4:223
    amphibians/reptiles, 4:229
    birds, 4:223-225
    fish/aquatic species, 4:227-229
    mammals, 4:225-227
    terrestrial invertebrates, 4:229-230
Migratory Bird Treaty Act (1918), 3:238
Migratory species, human disturbance attributes and susceptibility to, 4:230-231
    ecological consequences, 4:230-231
    evolutionary consequences, 4:232-233
Milankovitch cycle, 1:466, 467
Mildew, downy (Oomycota), 3:149
Miller, Stanley, 4:443
Mimicry, definition, 2:1; 4:365
Mimosa pigra, 3:512-513
Mimulus sp. (monkey flower), parapatric speciation in, 5:377-378
Mineralization
  definition, 4:377; 5:305, 315
  net, definition, 4:689
Minimum viable population, 1:861
  definition, 5:823
Mining, 3:385
  biodiversity impacts, 2:533-534
Minisatellites, 3:197
Miombo, definition, 1:55
Miombo woodland, 1:63
Missense mutations, 3:192
Mistletoe (Loranthaceae, Viscacaea), 3:326-327
  definition, 3:27, 317
Mites, *see* Acari
Mitochondria, definition, 4:415
Mitochondrial DNA (mtDNA)
  in animals, 3:190-191
  diversity in, 3:200-201
  in plants, 3:191
"Mitochondrial Eve," 2:86
Mixed farming system, 1:104-106
  definition, 1:99
Mixing ratio, definition, 1:293
Mixotrophy, definition, 2:650; 4:569
Mnemiopsis leidyi (ctenophore), 3:520
Models
  applications
    GAP analysis, 1:851
    habitat suitability indices, 1:850-851
    individual-based approaches, 1:851-852
    metapopulation models, 1:851
  competition
    aggregation model, 1:798-799
    Lotka-Volterra, 1:794-798
    resource utilization curves, 1:798-799
  definition, 1:845-846
  dynamic, 1:849
    definition, 1:845
  future trends
    multimodeling and regional assessment, 1:852

species behavioral dynamics,
   1:852–853
in life history analysis
   genetic models
      quantitative, 3:718–719
      simple Mendelian, 3:718
   phenotypic models, 3:716–717
      game theory, 3:717–718
      optimality, 3:717
limitations of, 1:848–849
purposes of, 1:846
   control, 1:846–847
   description, 1:846
   mechanism, 1:846
   prediction, 1:846
statistical approaches, 1:849–850
types of, 1:847–848
Molds, 3:149–151
Molecular biology, "central dogma of," 3:186
Molecular evolution
   major discoveries in, 3:195
   models
      divergence
         nearly neutral theory, 2:188
         neutral theory, 2:187–188
         selection models, 2:188
      polymorphism, 2:183–184
         nearly neutral theory, 2:184–185
         neutral theory, 2:184
         selection models, 2:184–185
      problems with, 2:189–190
   rates of, hemoglobin, 1:704
   selection versus neutrality theories, 3:195
Molecular phylogenies, 1:403
   definition, 3:53
Molecular toxicology, 2:364
Mollusca, 3:562
   body plan, 4:235–236
   classification, 4:246–247
   ecology of, 4:244–245
   endangered species, 2:460, 461
   human impacts, 4:245–246
   major groups, 4:237
      Aplacophora, 4:239
      Bivalvia, 4:239–240
      Cephalopoda, 4:243–244
      Coeloscleritophora, 4:237
      Gastropoda, 4:242–243
      Hyolitha, 4:237
      Monoplacophora, 4:240–241
      Polyplacophora, 4:238–239
      Rostroconchia, 4:237–238
      Scaphopoda, 4:241–242

of mangrove ecosystems, 3:863
species flocks in ancient lakes, 3:641
temporal/spatial distributions, 4:236–237
Mongoose, see Herpestidae
Monitoring, definition, 3:63
Monk seal, Mediterranean (*Monachus monachus*), endangered status of, 4:56
Monoculture, 1:551
   definition, 1:547
Monogamy, definition, 5:383
Monograph effect, 1:412
   definition, 1:411
Monophyletic, definition, 1:629; 3:417, 777; 4:249; 5:427
Monophyletic groups, 1:681
   definition, 5:569, 589
Monophyly
   in claudistics, 1:698, 700–701
   definition, 1:199, 677, 693; 4:559
   species concepts based on, 5:432
Monoplacophora, 4:240–241
Monotheism, definition, 5:109
Monotremata, 3:779–780, 784–785
Monotreme, definition, 3:777
Monozoic, definition, 5:863
Monsoon systems
   Asian, 1:265
   definition, 1:261
Montane
   fringing ecosystems, of Amazonia, 1:154–155
      Andes foothills, 1:155–156
      arthropod species, 1:252
      campo rupestre, 1:156–157
      granite inselbergs, 1:155
      low hills, 1:155
      tepuis, of Guayana highland, 1:155
   southeastern Australia, diagnostic species, 1:313, 314
Monte Carlo simulation, in population viability analysis, 4:838–839
Montreal Protocol on Substances That Deplete the Ozone Layer, 1:299, 304, 774; 5:490
   definition, 5:723
Moors
   definition, 3:675
   and historic land-use patterns, 3:683–684
Morbidity, definition, 2:109
Morichal, definition, 5:327
Morphodynamics, definition, 1:741
Morphology, definition, 2:1

Morphometrics, definition, 5:383
Mortality, definition, 4:819
Mosquitofish (*Gambusia affinis*), 3:522
Movile Cave, Romania, 5:537
Mss, definition, 5:527
Muir, John, 1:888, 889
Mulch, definition, 5:269
Müllerian mimicry, 1:760
   in butterflies, 1:567; 2:255
   definition, 1:559; 4:249
Multidimensional niche, definition, 3:729
Multilines, definition, 1:547
Multimodal, 1:852
   definition, 1:845
Multiple resistance, 3:470
   versus cross-resistance, determination, 3:471
   definition, 3:465
Multituberculata, definition, 2:95
Muscidae (house flies), 2:821
Museums and institutions
   advisory roles, 4:279
      consulting, 4:279
      in government policy, 4:279–280
      regional and international initiatives, 4:280
   exhibitions, 4:276–277
   publications and outreach, 4:278–279
   role in education and training, 2:772–773; 4:278
   role in research and biodiversity, 4:275–276
Mushrooms
   commercial cultivation, 3:158–159
   definition, 3:141
   poisonous, 3:155–156
Mustelidae (martin family), 1:630
Musth, definition, 5:295
Mutability, and Darwinian evolution, 4:445
Mutation rate, definition, 4:777
Mutations, 4:440, 539; 5:389
   and conservation genetics, 4:806
   definition, 2:179; 4:537
   and determinist population genetics, 4:779–780
   factors influencing fate of, 2:183
   favorable, survival of, 4:781
   fixation of, 2:187
   and gene differentiation, 2:86–87
   heritability of, 2:183
   major, role in adaptation, 2:254
   neutral, definition, 4:777
   and organism characteristics, 2:182

Mutations (*continued*)
  types of, 3:191–192
Mutualisms, 4:464
  among introduced species, 3:525
  and coevolution, 1:765
  definition, 1:753; 3:141; 4:281, 463
  ecological genetics of, 2:257–258
  and evolution, 5:461
  evolution and maintenance of
    gains from cooperation, 4:283
    kin selection, 4:283–284
    mutual enforcement, 4:285–286
    mutualism and the common good, 4:286–287
    selection among groups, 4:284–285
  in fungi, 3:153
    lichens, 3:157–158
    mycorrhizae, 3:157
  importance of, 4:281–283
  and niche concept, 3:311
  and pairwise interactions, 5:459–460
  plant–animal, 4:610
  in plant defenses, 2:14
    ants, 2:14–15
    domatia, mites, and other predators, 2:15
    endophytic fungi, 2:15
    predators/parasitoids, 2:15–16
  plant–pollinator, 4:727
    threats to, 4:729–730
  and progressive evolution, 4:287
    and footprints of natural selection, 4:288–289
    major transitions of evolution, 4:287
    modularity and evolvability, 4:287–288
Mutualists, as keystone species, 3:620
Mycelium, definition, 3:141
Mycetismus, 3:154, 155–156
Mycology, importance of, 3:143–144
Mycorrhizae, 3:153, 157
  definition, 4:689
  lack of, as constraint on restoration, 5:189
Mycoses, 3:154–155
Mycotoxicosis, 3:154, 156
Mycotrophic, definition, 3:317
Myiophily, 4:725
  definition, 4:723
Myriapoda, 4:291
  basic biology, 4:295–296
  basic morphology
    Chilopoda, 4:292

  Diplopoda, 4:292–295
  Pauropoda, 4:295
  Symphyla, 4:295
developmental diversity, 4:296–297
distribution by continent, 4:299–300
endemism and speciation, 4:299–300
geographical patterns, 4:298–299
habitats and adaptations, 4:301
  burrowing, 4:301–302
  open-country, 4:302
  special habitats, 4:302–303
number of species known/expected, 4:297–298
reproduction, 4:296–297
Myrmecophilidae (ant-loving crickets), 3:257–258
Myrmecophily, 1:566–567
  canopy epiphytes, 3:37
Myrmecotrophy, 4:611
Myths
  definition, 5:285–286
  of desertification, definition, 2:61
Myxoma virus, 1:757; 2:115–116; 3:523
Myxomycetes (slime molds), 3:149–151
  RNA editing, 4:421–422
Myxomycota, 3:149–151

N

NAD(P)H, in CO_2 fixation, 4:555–557
Namib Desert, 1:66
National Biological Information Infrastructure (NBII), 2:773
National Cancer Institute (U.S.), bioprospecting program, 1:480–481
National Environmental Policy Act (1969), 3:242
National Forest Management Act (1916), 3:241
National 4-H Clubs, 2:771
National Oceanic and Atmospheric Administration, U.S., satellite program, 5:127
National parks, research in, 4:321–322
National Park Service Act (1916), 3:241
Native species, definition, 1:71
Natural classification, definition, 1:693
Natural ecosystems, definition, 2:609
Natural enemies

  as constraint on species coexistence, 5:415
  definition, 5:413
Natural forest, definition, 2:23
Natural gas, biodiversity impacts of, 2:533
Naturalized plants, definition, 4:677
Natural regeneration, definition, 3:747
Natural resources management, definition, 5:481
Natural rights, and animal fables, 3:744–745
Natural selection, 2:87–90; 3:211; 5:387–388
  and conservation genetics, 4:811
  Darwin's discovery of, 2:3
  definition, 1:25; 4:439
  evidence for, 1:20–21
  versus evolution by common descent, 1:694
  versus genetic drift, in small isolated populations, 3:211–212
  in genetic variants in spacially structured populations
    versus gene flow, 2:249–251
    versus genetic drift, 2:251–252
  and limiting resource concept, 2:92–93
  synthetic theory of, 1:695
  in taxonomic versus functional diversity, 2:214–215
Nature, etiological tales of, 3:740–741
Nature reserves
  Gir (India), 4:853
  Keoladev (Ghana), 4:853–854
  origins of protected areas, 4:317–318
  selection of, 3:449
  sustained ecological research, 4:324–325
Ne, definition, 4:799
Near East ecosystems
  animal adaptations
    to dry conditions, 4:338–340
    to sand, 4:340–341
  biodiversity
    determining factors, 4:336
    rain, 4:336–338
    substratum, 4:339–340
    vegetation, 4:341–342
  extent of, 4:330
  human impact, 4:345
    agricultural practices, 4:350–351
    habitat destruction, 4:345–347
    hunting, 4:345

toxic pollution, 4:347–350
endemic versus widespread animals, 4:335–336
environmental conditions, 4:353
  climate, 4:355–356
  rock types, geomorphology, and edaphic conditions, 4:355
  topography, 4:353–355
flora, 4:356–357
freshwater diversity
  historic factors, 4:342–344
  human impacts, 4:346–347
historic zoogeography, 4:330–335
the Levant, 4:329
vegetation, 4:357
  Ceratonia siliqua and *Pistacia lentiscus* open forests, 4:359
  desert, 4:363
    savannoid vegetation, 4:363–364
  mangroves, 4:364
  maquis and forest, 4:357–358
  Mediterranean savanoid vegetation, 4:360
  montane forest of Mt. Hermon, 4:358–359
  oases with Sudanisn trees, 4:363
  patterns in dry areas of Israel, Jordon, and Sinai, 4:361
  Quercus calliprinos woodlands, 4:358
  Quercus ithaburensis open forests, 4:359
  sand, 4:363
  semisteppe batha, 4:360
  shrub–steppes, 4:361–362
    with trees, 4:362–363
  swamps and reed thickets, 4:364
  synanthropic, 4:364
  tragacanth vegetation of Mt. Hermon, 4:360–361
  wet salinas, 4:364
  Ziziphus lotus with herbaceous vegetation, 4:359–360
Nectar robber
  definition, 4:723
Negev Desert, *Acacia* trees, demography, remote sensing in study of, 5:136
Nekton, definition, 4:13
Nematocytes, definition, 5:73
Nematoda (roundworms), 3:563; 5:311
  abundance, 5:854–855
  alimentary canal, 5:846
  biospheric affects
    aquatic/marine nematodes, 5:856

    predators and entomopathogenic forms, 5:856
    soils and plants, 5:855–856
  classification, 5:859–860
  cuticle, 5:845–846
  diversity
    early views of, 5:849–852
    science of, 5:852–853
  extreme biotopes, 5:854
  general characteristics and synapomorphies, 5:844–845
  general distribution, 5:853–854
  habitat diversity, 5:854–855
  molecular diversity in, 5:859
  parasitic, 4:484
  relationships to other animal groups, 5:859–860
  reproductive system, 5:846–847
  species numbers, 5:856
    comparative diversity
      Ctenomys spp. rodents, 5:858–859
      Geomyidae rodents, 5:857–858
    parasitic forms, 5:856–857
  ubiquity of, 5:847–848
Nematostella vectensis (salt marsh sea anemone), 2:460
Nemertodermatida, 5:866, 869
Neo-Darwinism, 2:1
Neodermata, 5:872–873
  Aspidogastrea, 5:873
  class Cestoda, 5:882–883
    Amphilinidea, 5:883
    Caryophyllidea, 5:883
    Cyclophyllidea, 5:886–887
    Diphyllidea, 5:885
    Gyrocotylidea, 5:883
    Haplobothriidea, 5:885
    Lecanicephalidea, 5:886
    Nippotaeniidea, 5:886
    Proteocephalidea, 5:886
    Pseudophyllidea, 5:883, 885
    Spathebothriidea, 5:883
    Tetrabothriidea, 5:886
    Tetraphyllidea, 5:886
    Trypanorhyncha, 5:885–886
  class Digenea, 5:873, 875
    Allocreadioidea, 5:875
    Clinostomoidea, 5:875
    Cyclocoeloidea, 5:875
    Diplostomoidea, 5:875–876
    Echinostomatoidea, 5:876
    Gymnophalloidea, 5:876
    Hemiuroidea, 5:876
    Lepocreadioidea, 5:876
    Microphalloidea, 5:876

    Notocotyloidea, 5:876–877
    Opisthorchioidea, 5:877
    Paramphistomoidea, 5:877
    Plagiorchioidea, 5:877
    Schistosomatoidea, 5:877
    Traglotrematoidea, 5:877–878
    Transversotrematoidea, 5:877
  class Monogenea, 5:878–879
    Capsalidea, 5:879
    Chimaericolidea, 5:881
    Dactylogyridea, 5:879
    Diclybothriidea, 5:881
    Gyrodactylidea, 5:879
    Mazocraeidea, 5:881
    Monchadskyellida, 5:879
    Monocotylidea, 5:879, 881
    Polystomatidea, 5:881
    Udonellidea, 5:879
Neodermis, definition, 5:863
Neogene, definition, 5:607
Neoproterozoic, mass extinction of, 4:120
Neotenic, definition, 3:581
Neotropics
  definition, 4:403; 5:1; 5:701
  freshwater fishes, diversity, 2:778
Nest parasitism, *see* Brood parasitism
Net mineralization, definition, 4:689
Net national product, 2:278
  Hamilton as constant-equivalent utility, 2:281
  marginal cost–benefit rule, 2:280–281
  and population/welfare, 2:279–280
  and social well-being, 2:278–279
  concept of sustainability, 2:281
Net social benefit, definition, 2:291
Neural crest, definition, 5:755
Neuroectoderm, definition, 5:755
Neuroptera, 3:483
Neutral evolution, 2:86–87
  definition, 2:85
Neutral mutations, definition, 2:179
New Zealand
  first-contact extinctions, 2:736–737
  modern extinctions, 2:737
NFTPs, *see* Nontimber forest products
Ngerukewid Islands Wildlife Preserve, 4:318–319
Niche complementarity, definition, 5:627
Niche differentiation
  definition, 3:109; 5:173
  models, 3:112, 114
  resources and
    spatial heterogeneity, 5:177–178

Niche differentiation (continued)
    temporal heterogeneity,
        5:178–180
    theoretical considerations,
        5:175–177
Niche overlap, definition, 3:303
Niches
    concept of, origin, 3:305–307
    definition, 5:173
    dimensions, definition, 3:729
    Eltonian, definition, 3:295; 5:413
    evolution of, 3:312–313
    Grinnellian, definition, 3:295; 5:413
    versus habitat, 3:303–305
        indirect interactions, 3:311–312
        multispecies comparisons,
            3:307–310
    Hutchinsonian, definition, 3:295;
        5:413
    parasite, 4:474–475
    temporal, and stability, 5:475
    temporal partitioning, and species
        coexistence, 5:423
        nonlinear consumption, 5:423
        storage effect, 5:423
Niche theory, and taxonomic versus
    functional diversity, 2:207
    functional groupings of species,
        2:207–208
Nigeria, sacred groves and traditional
    forest conservation, 5:116
Nitrification, definition, 4:377
Nitrogen cycle
    ecological importance, 4:377–379
    effects on biodiversity, 4:382
        acidification by nitrogen,
            4:384–385
        eutrophication, 4:382–384
        nitrogen dynamics, 4:385–387
    in fresh versus sea water, 1:621
    future considerations, 4:387
    in grazed ecosystems, 1:656–657
    human impacts on, 4:379–382
    perturbation of, 4:736
        acid rain, 4:737
        aquatic ecosystems, 4:737–738
        forest ecosystems, 4:738–739
        nitrogen oxides as trophospheric
            ozone precursor, 4:739–740
        nitrogen saturation, 4:739
    in soils, 5:308
Nitrogen fixation
    definition, 1:99; 4:377
    symbiotic, 4:208–209
Nitrogen oxides
    in acid deposition, 1:2, 3–4
    atmospheric, 1:300–301
    deposition/environmental pathways,
        1:121
    effects on forest ecosystems,
        1:129–130
    sources and impacts, 1:120
    as trophospheric ozone precursor,
        4:739–740
Nitrogen saturation, definition, 4:377,
    731
Nitrogen-use efficiency (NUE)
    definition, 1:575
    enhancement in C4 photosynthesis,
        1:581–582
Nitrous oxide, atmospheric, 1:298
NOAA, see National Oceanic and
    Atmospheric Administration, U.S.
Node (cladogram), definition, 5:569
Nomenclature systems, 4:389–390
    alternative systems, 4:398–399
    codes of, 4:393
        ambiregnal organisms, 4:397
        bacteriological, 4:393
        botanical, 4:393–394
        cultivated plants, 4:394–395
        lichens, 4:398
        pleomorphic fungi, 4:398
        virological, 4:395
        zoological, 4:395–396
    current names, determination,
        4:400–401
    Draft BioCode, 4:396–397
    hierarchical, 4:390–392
    indices of names, 4:401
    Linnaean, 5:577
    newly discovered species, 4:400
    origins of, 4:390
    principles of, 4:392–393
    purpose of, 4:390
    reasons for name changes,
        4:399–400
Nominalism, 5:596–598
Nonallopatric speciation
    allopolyploid speciation, 5:380
    autopolyploid speciation, 5:379
    direct/reticulate speciation, 5:380
    introgressive hybrid speciation,
        5:379
    parapatric speciation, 5:377–378
    recombinational hybrid speciation,
        5:379–380
    speciation by interspecific
        hybridization, 5:379
    spontaneous thelytokous speciation,
        5:379
    sympatric speciation, 5:378

Nonavian dinosaurs, definition, 2:95
Nongovernmental organizations
    definition, 5:481
    and environmental stewardship,
        private companies/individuals,
        5:493–494
Nonindigenous Aquatic Nuisance
    Prevention and Control Act
    (1990), 3:237, 238
Nonindigenous species; *see also* Exotic
    species; Introduced species
    definition, 1:71; 5:185
Nonrenewable energy, definition,
    2:525
Nonsense mutations, 3:192
Nonsynonymous mutations, definition,
    2:179
Nontimber forest products, 3:749–750;
    4:90
Norfolk Broads, U.K., restoration of,
    5:815–816
North America
    changes in biodiversity, 4:411–412
    continental diversity, 4:404
    deserts of, 2:43–44
    diversity in major groups of
        organisms, 4:408
        amphibians, 4:410
        birds, 4:410
        butterflies, 4:410
        freshwater fishes, 4:410–411
        land mammals, 4:410
        plants, 4:409
        reptiles, 4:410
    late Quaternary extinctions,
        3:827–832
    midcontinent prairies, predators,
        4:868
    postglacial extinctions, 2:733
    terrestrial bioregions, 4:404–406
        Canadian shield, 4:406
        Caribbean, 4:408
        eastern North America,
            4:406–407
        southernmost North America,
            4:408
        Southwestern North America,
            4:407–408
        western North America, 4:407
North American Association for
    Environmental Education,
    2:769–770
North Atlantic oscillation, 4:498
    definition, 4:497
    and pelagic fishes, 4:505–506
    and zooplankton, 4:501–502

NorthEast Svalbard Nature Reserve, 4:318
Nothofagus spp., 5:361, 365
Notochord, definition, 5:755
Notocotyloidea, 5:876–877
Noxious Weed Act (1974), 3:237–238
Nuchal organ, definition, 5:831
Nuclear power industry, biodiversity impacts of, 2:535–536
Nucleic acids, 4:440
　analysis, in measurement of microbial diversity
　　fluorescent *in situ* hybridization, 4:186
　　gradient gel electrophoresis, 4:182–183
　　intergenetic transcribed spacer analysis, 4:183–184
　　membrane hybridization, 4:185–186
　　oligonucleotide assay, 4:186
　　restriction fragment length polymorphisms, 4:184
　　solution hybridization, 4:185
　　ssrDNA clone library sequencing, 4:184–185
　sequence modification, 4:415
　　gene scrambling, 4:415–416
　　　ciliates and nuclear dualism, 4:416
　　　DNA processing during macronuclear formation, 4:416
　　　unscrambling process, 4:417–419
　　RNA editing, 4:419
　　　Ebola virus, 4:426
　　　human Apo B and NF1, 4:424
　　　A → I deamination, 4:424–425
　　　kinetoplastids, 4:419–421
　　　mitochondrial tRNA editing, 4:423–424
　　　myxomycetes, 4:421–422
　　　paramyxoviruses, 4:425
　　　plant organelles, 4:422–423
Nucleotides, 3:185
　definition, 3:183
　diversity, and evolutionary theory, 2:673
　mutations in, 3:192
　polymorphisms, 3:197–198
　　single, 3:202
Nucleus, definition, 4:415
Nunatak, definition, 1:171
Nurse plants
　definition, 5:185

　lack of, 5:189
Nutrient cycle, definition, 2:345
Nutrient cycling
　controls over, 2:330–331
　and ecosystem functioning, 2:349
　effects of introduced species, 3:521
　in freshwater ecosystems, 3:539
　in grazed ecosystems, 1:655–657
　open versus tight cycles, 2:331
　by plants, 2:329–330
　rainfall effects, 2:331
　soil type, land use, and vegetation structure effects, 2:331–332
Nutrient enrichment
　definition, 4:1
　estuaries, 4:6–7
　and food web complexity, 4:3–4
　lakes, 4:3
　nearshore ocean, 4:5–6
　response of nearshore waters, 4:4–5
Nutrient limitation, definition, 4:377
Nutrients, definition, 4:427
Nutrient uptake capacity, definition, 4:689
Nutritionalism, definition, 4:857
Nymph, definition, 3:581
Nymphalidae, 1:562–564

O

Obligate epiphyte, definition, 3:27
Obligatory parasite, definition, 4:463
Oceans; *see also* Pelagic ecosystems
　and carbon cycle
　　in management of, 1:626–627
　　stability of, 1:627
　carbon gradient, 1:448
　ecosystems, 4:22–23
　　distinguishing characteristics, 4:434–436
　　and global change, 4:436–437
　　shared characteristics, 4:433–434
　elemental profile, 1:449–450
　fishing situation, 2:784–785
　　international conflicts, 2:786
　　management problems, 2:785–786
　　social and economic concerns, 2:785
　hydrothermal vents and cold seeps, 3:558–559
　natural functional units, partitioning, 4:429
　　biome concept, 4:429–430
　　central problem, 4:429
　　ecosystem concept, 4:430–433
　physical characteristics, 3:544–545

Odobenidae, 4:50
Odum, Howard W., 2:306–307
Oil
　biodiversity impacts of, 2:530–533
　environmental impacts, 3:800
Oligonucleotide microarrays, 3:198
　definition, 4:177
Oligonucleotides, assay of, in measurement of microbial diversity, 4:186
Oligotrophic, definition, 1:145; 2:650; 4:377; 5:73
Oligotrophication, 2:650
　in reversal of cultural eutrophication, 2:667–669
Omnivory
　multichannel, and food web energy, 3:10
　and structure of, 3:7–8
Ontogeny, definition, 1:693
Ontology, definition, 5:569
Onycophorans (velvet worms), 1:920
Oomycota, 3:149
Oparin, Alexander, 4:442
Open access resource, definition, 4:85
Open communities, definition, 5:413
Operationalism, 5:596–598
Operon, 1:344–345
　definition, 1:339
Opiliones (harvestman), 1:208
　ecology, 1:209
　phylogeny/taxonomy, 1:209–210
　reproduction/growth, 1:209
Opportunistic species, definition, 4:377
Optimization, definition, 1:677
Option value, definition, 2:291
Opuntia spp., 3:508–509
Ordovician period, 1:925
　late, mass extinction of, 4:111–112
　origination during, 1:416
Organelles
　chloroplast DNA, 3:191
　definition, 3:184
　mitochondrial DNA
　　in animals, 3:190–191
　　in plants, 3:191
Organic selection, principle of, 4:545
Organism
　definition, 2:191
　as energy flow systems, 2:192–193
　as information systems, 2:193–194
　selection processes of, interaction with
　　cohesive properties, 2:194
　　hierarchical organization, 2:194–195

Organism (*continued*)
  historical uniqueness, 2:194
Organismal diversity, definition, 1:377
Organizing centers, soils as, definition, 5:305, 307
Organochlorine pesticides, 4:511
Organometallic fungicides, 4:514
Organophosphate fungicides, 4:514
Organophosphate insecticides, resistance to, 3:468
Organophosphate pesticides, 4:512
Organotin pesticides, 4:512
Orgel, Leslie, 4:444
Origination
  and extinction rate, of nonavian dinosaurs, 3:845
  in Phanerozoic, 1:416–417
Origin of Species (Darwin), 2:4; 3:744; 5:573
Ornithophily, 4:726
  definition, 4:723
Orò, Juan, 4:443
Orobiome, definition, 2:635
Orogenesis, and elemental transport, 1:441
  definition, 1:437
Orthoptera, 3:247
  food and feeding, 3:251
  overview, 3:248–249
  palentology, 3:251
  phylogeny/taxonomy, 3:251
  reproduction and growth, 3:250
  sound production, 3:249–250
  superfamily Acridoidea, 3:261–262
    Acrididae (true grasshoppers/locusts), 3:263
    Pamphagidae (earth hoppers), 3:262
    Paulinidae (aquatic grasshoppers), 3:263
    Pyrgomorphidae (bush hoppers), 3:262–263
    Romaleidae (lubber grasshoppers), 3:263
  superfamily Eumastacoidea, 3:260
    Eumantacidae (monkey grasshoppers), 3:260
    Proscopidae (false walking sticks), 3:260–261
  superfamily Gryllacridoidea
    Cooloolidae (cooloola monsters), 3:258
    Gryllacrididae (raspy crickets, leaf-rolling crickets), 3:259
    Rhaphidophoridae (cave crickets, camel crickets), 3:259–260

Stenopelmatidae (Jerusalem crickets, wetas), 3:259
superfamily Grylloidea, 3:255
  Gryllidae (true crickets), 3:256
    major subfamilies, 3:256–257
  Gryllotalpidae (mole crickets), 3:256
  Myrmecophilidae (ant-loving crickets), 3:257–258
superfamily Pneumoroidea (bladder hoppers), 3:261
superfamily Tetrigoidea (pigmy grasshoppers, grouse locust), 3:263–264
superfamily Tettigonioidea, 3:252
  Prophalangopsidae (haglids), 3:252
  Tettigoniidae (katydids), 3:252–253
    major subfamilies, 3:253–255
superfamily Tridactyloidea (false mole crickets, sandhoppers), 3:263–264
Osteichthes, 5:761–762
Otters, marine
  California (*Enhydra lutris*)
    as keystone species, 3:615–617
    status of, 4:61–62
  evolutionary history, 4:44
  features and habitat boundaries, 4:41–42
  –kelp forest system, 4:861–862
  northern and Russian subspecies (*Enhydra lutris*), 4:65–66
Outbreeding, and conservation genetics, 4:807–808
Outbreeding depression, 3:427
  causes of, 3:433
  evidence for, 3:433, 435
  variation in susceptibility to, 3:435
Output environment, definition, 2:305
Overcompensatory growth, definition, 1:651
Overexploitation, definition, 2:698
Overfishing, 2:421, 711, 784–785
  definition, 2:783
  impacts on marine ecosystems, 4:34
  marine, 4:24
  types of, 2:787–788
Overgraze, definition, 5:33
Overshoot, definition, 2:229
Oviposition, parasitoids, strategies, 4:488–490
Ovipositor, definition, 3:417
Ovoviviparous, definition, 1:199
Oxathiin fungicides, 4:514

Ox-bow, definition, 5:805
Oxic/anoxic conditions, definition, 3:89
Oxidation–reduction reaction, 1:442–443
Oxidizing capacity, definition, 1:293
Oxygen cycle, 1:444–445
Oxygen depletion, subsurface, definition, 4:1
Ozone
  definition, 1:293; 4:731; 5:723
  depletion, 3:386
    in Antarctic, 1:183, 303
  deposition/environmental pathways, 1:121
  effects on forest ecosystems, 1:128–129
  sources and impacts, 1:120
  stratospheric
    causes/consequences of depletion, 1:303–304
    importance of, 1:301–302
    trends in, 1:302–303
      projected, 1:304
  trophospheric, nitrogen oxides as precursor, 4:739–740

P

PABITRA (Pacific-Asia Biodiversity Transect) initiative, 3:565, 579–580
Pacific islands, first-contact extinctions, 2:736
Packing, definition, 3:729
Paddlefishes, 2:761, 763
Paedomorphosis, definition, 1:159
Paleocene epoch, early, mammalian recovery during, 3:848–849
Paleoclimatology, definition, 3:437
Paleoecology, 4:451–452
  age models and mapping intervals, 4:460
  data and sensing systems, 4:452
  dating uncertainties, 4:459–460
  scaling factors, 4:452–453
  spatial characteristics, 4:456–457
  taxonomic and numerical characteristics, 4:453–456
  temporal characteristics, 4:457–459
  temporal resolution, 4:459
  zoom lens perspective, 4:460–461
Paleotropics, definition, 5:701
Paleozoic era, mass extinction of, 4:120

Palpigradi (micro-whip scorpions), 1:211–212
Palynology, definition, 4:451
Pampas, definition, 5:327
Pamphagidae (earth hoppers), 3:262
Panbiogeography, definition, 5:767
Panda, giant (*Ailuropoda melanoleuca*), 2:444; 3:813
Panspermia, 4:441
Pantanal, definition, 5:327
Pantheism, definition, 5:109
Pantheistic traditions, and indigenous knowledge, 5:115–116
  biodiversity conservation, 5:117–118
  context of, 5:116–117
Papilionidae, 1:560–561
Paramecium spp., in interspecific competition studies, 1:801–802
Parameter, definition, 1:845; 5:467
Parametric statistics, definition, 3:53
Pàramo, 1:674
  definition, 1:665; 5:327
Paramyxoviruses, RNA editing, 4:425
Parapatric speciation, 5:391–392
Paraphyletic, definition, 5:428
Paraphyletic groups, 1:681
  definition, 3:841; 5:145; 5:569, 589
Paraphyly
  in claudistics, 1:698, 700–701
  definition, 1:199, 693
Parapinnixa affinis (Northeastern Pacific crab), 2:460
Parapodium, definition, 5:831
Parasite–host relationships, and coevolution, 1:760–762
  special features, 1:762
    evolution of virulence and avirulence, 1:762–763
    gene-for-gene systems, 1:762
    sex selection, 1:762
Parasites
  brood, ecology and social behavior, 4:373
    foraging ecology, 4:373
    mating systems, 4:373
    spacing behavior, 4:373
    vocal behavior, 4:373–375
  communities as general ecological models, 4:476–477
  community structure
    community ecology concept, 4:475–476
    empirical evidence, 4:476–477
  definition, 3:141
  distribution in animal and plant kingdoms, 4:478, 479
  domestic ruminants as hosts of, 1:662–663
  ecological niches of, 4:474–475
  economic and hygienic importance, 4:484
    in fisheries, 4:484
    in humans, 4:482–483
    in livestock, 4:483–484
    in plants, 4:484
  exotic, 2:117–118
  micro- versus macro-, 2:110
  population dynamics
    concepts of population growth, 4:477
    ecological strategies, 4:477–478
  types of, 4:464–465
  zoogeography of
    as biological markers, 4:481–482
    latitudinal gradients
      in host ranges and specificity, 4:480
      in reproductive strategies, 4:478, 480
      in species richness/abundance, 4:478
    von Ihering method, 4:482
Parasitism
  adaptations to
    aggression, 4:468
    asexual reproduction, 4:468
    complexity, 4:465–466
    dispersal mechanisms, 4:467
    of flowering plants, 4:471
    hermaphroditism, 4:468
    host specificity, 4:468–469
    infection mechanisms, 4:467–468
    life cycles, 4:469–470
    parthenogesis, 4:468
    physiological, 4:470–471
    reproductive capacity, 4:466–467
    site specificity, 4:469
    size, 4:465
  in crustaceans, 1:932
  definition, 4:281, 463
  evolution of, 2:201
  by introduced species, 3:522–523
  origins of, 4:471–472
  and pairwise interactions, 5:459
  virulence, evolution of, 4:472
Parasitoids, 3:420, 484; 4:485–486
  community ecology, 4:493–495
  definition, 3:417; 4:249
  host acceptance and oviposition strategy, 4:488–490
  –host interactions, and population dynamics, 4:774–775
  host location, 4:487–488
  hybrid plant communities, 4:670–672
  importance of, 4:493–495
  life history patterns, 3:421; 4:486–487
  and metapopulation structure, 4:164
  in plant defenses, 2:15–16
  population dynamics, 4:491–493
  resistance and virulence, 4:490–491
Parataxonomist, definition, 1:419; 4:271
Parenchyma, definition, 5:863
Pareto efficiency, 1:776
  definition, 1:769
Parr, definition, 5:233
Parsimony, 1:679–681
  definition, 1:677
Parsimony principle, definition, 5:570
Parthenogesis, as adaptation to parasitism, 4:468
Parthogenesis
  arrhenotokous, 3:419
  definition, 3:417
Partial migration, definition, 4:221
Passenger pigeon (*Ectopistes migratorius*), 2:399; 3:387
Passive sampling hypothesis, of species–area relationship, 5:398–399
Pasteur, Louis, 4:440, 441
Pastoralism
  definition, 3:739
  in poetry and scripture, 3:741–742
Pasture
  definition, 3:675; 5:33
  tropical forest conversion to, 5:46–47
Patagium, definition, 3:777
Patch, definition, 3:645; 4:161
Patch disturbance, definition, 2:229
Patchiness
  definition, 3:485
  on rocky shores, processes, 3:487
    competition, 3:487–488
    physical factors, 3:489
    predation, 3:488–489
    recruitment, 3:487
    traditional management of, 5:689–690
Path dependence, definition, 4:891
Pathogens
  definition, 2:109
  domestic ruminants as hosts of, 1:662–663
  influences of livestock, 5:48

Pathogens (*continued*)
  introduced species, 3:522–523
  and niche descriptions, 3:311
  plant, hybrid plant communities, 4:670–672
  vectoring by introduced species, 3:525
Patrilocal residence, definition, 3:411
Paulinidae (aquatic grasshoppers), 3:263
Pauropoda, 4:295
PCBs, *see* Polychlorinated biphenyls
PCR, *see* Polymerase chain reaction
Peat, 1:619
  definition, 5:781–782
Peatlands, definition, 5:781–782
Pedipalps, definition, 1:199
Pegging, definition, 5:269
Pelagic, definition, 4:13, 569
Pelagic ecosystems, 4:22–23
  fishes of, 4:504–507
  outlook for, 4:507
  phytoplankton and primary production, 4:498–500
  zooplankton, 4:500–504
Pelagic organisms, definition, 4:27
Pelagic zone
  biodiversity of, 4:22–23
  definition, 3:75, 543
  provinces of, 4:14, 15
Penetrometer resistance, definition, 5:269
Penicillium spp., 2:200; 3:143, 159, 160
Penis, definition, 5:863
Pentatomomorpha, 5:720
Peregrine falcons, and DDT, 3:444
Perfectly competitive economy, definition, 4:85
Periatric speciation, 5:390–391
Periodic parasite, definition, 4:463, 464
Periphyton
  acid deposition effects, 1:11–12
  definition, 1:1
Perissodactyla, 3:795–796
Peristomium, definition, 5:831
Permafrost
  definition, 1:232, 533; 4:917
  microflora, 4:918
Permanence, definition, 5:413
Permian period, 1:926
  mass extinction of, 1:414; 2:719, 720; 3:230, 845–846; 4:100; 4:115–117
Persecution, definition, 3:811

Persistent organic pollutants
  deposition/environmental pathways, 1:121
  sources and impacts, 1:120
Perturbation, definition, 5:467
Peru, biodiversity and traditional agriculture, 1:116
Pest control
  and planned diversity, 2:272
  and unplanned diversity, 2:273–274
Pesticide resistance, in introduced species, 3:527
Pesticides; *see also* Fungicides; Herbicides; Insecticides
  definition, 4:509
  delivery and bioavailability, 4:516
  deposition obstacles, 4:518
  ecotoxicological impacts
    aquatic systems, 4:521–522
    avian, 4:521
    farms-scale observations, 4:520–521
  ecotoxicology and management, 4:518–520
  efficiency, 4:516
  and extinctions, 2:707
  formulations, 4:510
  management tools, 4:522
  soil application, 4:517
  spray application, 4:517–518
  threat to repitles/amphibians from, 2:482–483
  types, 4:510–511
  uses of, 4:510
Pests, pressure from, and megadiversity, 1:425–426
Petroleum industry
  biodiversity impacts of, 2:530–533
  environmental impacts, 3:800
Petroleum oil pesticides, 4:512
Pfisteria piscicida, 4:714, 716
Phage reproduction, 1:346
Phagotroph, definition, 4:901
Phagotrophy, definition, 4:201, 569
Phanerochaete spp., 3:161–162
Phanerogam, definition, 1:171
Phanerozoic eon, 1:394, 412
  definition, 4:97
  marine biodiversity in, 3:221–222
    chordates, 3:224–226
    invertebrates, 3:222–224
    taxonomic turnover, 3:226
  mass extinctions of, 1:413–415
  terrestrial biodiversity in
    chordates, 3:228
    invertebrates, 3:226, 228

  plant, 3:226
Phanerozoic revolutions, 1:30
Pharate, definition, 4:249
Pharmaceuticals
  biodiversity and ethical issues, 4:534–535
  and indigenous intellectual property rights, 4:535–536
  research, bilateral agreements, 4:535
  biological, 1:474–476
  bioreactive alkaloids, 4:719
  from cyanobacteria, 4:712–713
  from fungi, 3:159–160
  from higher plants
    camptothecin, 4:718–719
    taxol, 4:717–718
  from marine organisms, 4:719–720
  microbial products, 4:719
  from red algae, 4:716–717
  from terrestrial animals, 4:720
  from terrestrial plants, 4:720–722
Pharmacognosy, 2:614
Pharynx, definition, 5:755
Phenetics, definition, 4:537; 5:428; 5:589
Phenograms, definition, 5:570, 589
Phenol-derived herbicides, 4:515–516
Phenols, as plant chemical defenses, 2:12
Phenotypes
  constraints on social behavior, 5:297–298
  definition, 1:17, 694; 4:537; 5:383
  and evolution, 4:545–546
  formal analysis, 4:543–544
  making of, 4:540–543
  simple versus complex traits, 2:253–254
  variation in, 4:544–545
    as adaptation, 2:249
    heritability of, 2:247–248
Phenotypic, definition, 4:177
Phenotypic diversity, and coevolution, 1:765–766
Phenotypic plasticity, 2:248–249
  definition, 2:245
Phenoxyaliphatic acid herbicides, 4:514
Phenylamide fungicides, 4:514
Pheromones
  definition, 4:249
  in Hymenoptera, 3:424
  moth, 4:265–267
Philippines, deforestation in, remote sensing in study of, 5:134–136

Phobia, definition, 1:45
Pholidota, 3:797–798
Phoresis/phoresy, 4:464
　definition, 1:199; 4:463
　in Diptera, 2:824–825
Phoridae (humpbacked flies, scuttle flies), 2:821
Phosphoenolpyruvate carboxylase, definition, 1:575
Phosphorus cycle, 1:451
Photoautotroph, definition, 1:437; 2:191; 5:647
Photoautotrophs, 4:209–210
　definition, 2:509; 4:349
　temperature limits for, 3:354
Photoautotrophy, definition, 2:311; 4:569
Photochemical, definition, 1:119
Photochemical source/sink, definition, 1:293
Photodissociation, definition, 1:293
Photorespiration, definition, 1:575
Photosensitivity, 1:48
Photosynthesis
　C3, 1:575
　　reaction centers
　　　light harvesting/excitation energy transfer to, 4:550–552
　　　and primary photochemistry, 4:552–553
　C4, 4:557
　　benefits of, 1:577
　　climate change and future of, 1:593–594
　　costs of, 1:581
　　membrane-associated reactions leading to ATP and NAD(P)H, 4:553–555
　　O2 and CO2 requirements, 1:578–580
　　temperature requirements, 1:580
　CAM, 1:576; 4:557
　and carbon cycle, 1:610
　and climate change adaptation, 3:286–287
　effects of CO2 elevation, 1:621–622
　evolution of, 1:452, 453
　net, saturation of with global warming, 1:279
Phototroph, definition, 4:191
Phototrophy, definition, 4:201
Photovoltaic cells, 2:538
Phrygana, definition, 4:145
pH scale, and acid rain, 1:2
Phthalamide fungicides, 4:514

Phylesis, definition, 1:694
Phyllopod, definition, 1:915
Phylogenetic, definition, 1:17; 5:428
Phylogenetic congruence, 1:753
Phylogenetic divergence, genetic diversity as
　empirical observations, 2:186–187
　molecular evolution models, 2:187–188
　substitution process, 2:186–187
Phylogenetic species concept, 2:86
　definition, 2:395
Phylogenetic trees
　versus cladograms, 1:681–682
　definition, 2:671; 5:570
　small subunit RNA-based, 2:674
Phylogeny, 4:559–560
　analytical methods
　　phenetic models, 4:560–561
　　phylogenetic models, 4:561
　　　maximum likelihood, 4:562
　　　neighbor-joining, 4:562
　　　parsimony, 4:561
　　　parsimony analysis, 4:561
　applications
　　adaptation, 4:566
　　general evolutionary context, 4:566
　　vicariance biogeography, 4:566
　Arachnida, 1:202, 207–208, 215–217
　Archaea, 1:220
　and classification, 4:564–565
　　ranks and priority, 4:565–566
　data sources, 4:563–564
　definition, 1:219; 2:1; 3:417; 4:177; 5:570
　historical overview, 4:559–560
　molecular, definition, 1:403
　and organismal diversity, 2:195
　phylogenetic diversity, 4:566–567
　phylogenetic interpretation, 4:564–565
　support methods
　　bootstrap, 4:562–563
　　Bremer support, 4:563
　　jackknife, 4:563
Phylogeography, definition, 3:777
Phylogram, definition, 5:570
Phylotypes, definition, 5:747
Phylum, definition, 1:419; 3:561
Physiognomy, definition, 5:607
Physiography, definition, 4:403
Phytases, 1:477
Phytochorion, definition, 1:55
Phytogeographic regions, Amazonia, 1:147–148

Phytophagous, definition, 1:351
Phytophagy
　definition, 3:417
　in Hymenoptera, 3:422–423
Phytoplankton
　acid deposition effects, 1:11
　definition, 1:1; 4:1, 427, 497, 857
　eutrophication effects, 4:384
　and primary production in pelagic ecosystems, 4:498–500
Phytotelmata, and Diptera, associations, 2:823–824
Phytotoxic, definition, 1:119
Picophytoplankton, definition, 4:569
Pieridae, 1:561
Pigeon, domestic (*Columba livia*), economic impact, 1:75, 77
Pigs, feral
　in Australia, 1:78
　in South Africa, 1:79
　in U.S., 1:74
Pinatubo volcanic eruption, 1:298, 303
Pinchot, Gifford, 1:888, 889
Pinnipedia, 4:37
　definition, 1:629
　diving capabilities, 4:39
　evolutionary history, 4:43
　features and habitat boundaries, 4:40–41
Pinus pinaster, 3:513–514
Pioneer guild, definition, 3:747
Pirizal, 1:153
Pisaster ochraceus (sea stars), identification as keystone species, 3:614–615
Piscivores, definition, 4:857
Pixel, definition, 5:121
Placental mammals, *see* Eutheria
Planetesimals, definition, 5:647
Planktivores, definition, 4:857
Plankton, 4:569–570
　community structure, 4:570–571
　　assembly and autopoiesis of, 4:588–591
　　explanation of species diversity, 4:575–576
　　freshwater zooplankton, 4:574–575
　　marine zooplankton, 4:571, 574
　　phytoplankton, 4:571
　　quantification of, 4:587–588
　　quantification of diversity in, 4:588
　　resource limitation and diversity, 4:592
　　system exergy and disturbance, 4:591–592

Plankton (continued)
    definition, 2:311; 3:75; 4:13
    diversity, mechanisms promoting/
            maintaining
        within habitats, 4:592–594
        species richness among habitats,
            4:594–596
    habitat constraints, 4:576–579
    phagotrophic
        basic adaptations, 4:584–585
        life history strategies, 4:585–586
    photoautotrophic
        basic adaptations, 4:579–580
        marine, traits in, 4:584
        seasonality of, 4:582–584
        specializations among, 4:580–582
    UV radiation effects, 5:727
Planktonic, definition, 3:531, 543;
    4:901
Planktotrophic, definition, 2:455
Plant–animal interactions
    antagonistic, 4:602
    carnivorous plants, 4:608
        benefits of carnivory, 4:608–609
        costs of carnivory, 4:608–609
        nonprey guests, 4:609–610
        prey capture mechanisms, 4:608
    community and ecosystem effects,
        4:607–608
    mutualistic interactions, 4:610
        ants, 4:610–611
        fruit and seed dispersers
            adaptations to diaspore
                dispersers, 4:616–618
            types of, 4:615–616
        pollinators, 4:611–612
            competition for, 4:614
            pollination syndromes,
                4:613–614
            role in plant diversification,
                4:612–613
    plant consumers
        evolutionary responses to, 4:604
            chemical defenses, 4:604–605
            escape, 4:605–606
            mechanical defenses, 4:605
            tolerance and compensatory
                growth, 4:605–606
        types of, 4:602–604
    types of, 4:602
Plantations
    definition, 2:23
    and deforestation, 2:31
    in restoration of biodiversity, 5:192
Plant communities, 4:631
    abundance, 4:641–643
    allocation and reproductive
        strategies, 4:635–636
    body size and plant form,
        4:634–635
    diversity equilibria, 4:639–640
    latitudinal gradients, 4:640–641
    niche diversification and
        competition, 4:634
    stability and resilience, 4:638–639
    succession, 4:638
Plant–herbivore interactions, in fossil
    record, 3:271–272
Plantigrade, definition, 1:629
Plant invasions
    concepts and terminology,
        4:678–679
    current extent of, 4:680–681
    history of, 4:679–680
    invasiveness and invasibility,
        4:684–685
        plant traits and environmental
            features, linkage, 4:685–687
    managing, 4:687
        confounding factors, 4:688
        ecological management, 4:687
        prevention, 4:687
        removal, 4:687
    modeling, 4:687
    processes, 4:681
        dispersal to new area, 4:681
        establishment/naturalization,
            4:682
        spread, 4:682–684
Plant pathogens, introduced
    in Australia, 1:79
    in Brazil, 1:82
    in India, 1:81
    in South Africa, 1:80
    in U.K., 1:77–78
    in U.S., 1:76
Plants; see also Introduced species,
    plants; Vegetation
    aquatic
        acid deposition effects, 1:11–12
        Australia, diagnostic species,
            1:315
    biodiversity
        angiosperms, 4:6–7
        conservation of, 4:9–10
        defining, 4:1–2
        endemism, patterns of, 4:7
        gymnosperms, 4:5–6
        measures of, 4:7
        nonvascular plants, 4:3–4
        seedless vascular plants, 4:4–5
        threats to
            affecting susceptibility, 4:7–8
            anthropogenic, 4:8–9
            natural, 4:9
    C4
        biogeography
            primary controls, 1:582–583
            secondary controls, 1:583–586
        common features, 1:577
        dominant ecosystems, human
            impacts on, 1:595–596
        evolution of, 1:590
            carbon isotope discrimination,
                1:591–592
            driving forces, 1:592–593
            fossil evidence, 1:590–591
            phylogenetic approaches, 1:592
        photosynthetic enhancements,
            1:577–581
        polyphyletic origins, 1:587–590
        species numbers, 1:576, 586
        taxonomic diversity, 1:586–
        water/nitrogen use efficiency,
            1:581–582
    centers of diversity, 1:420
    codes of nomenclature, 4:393–394
        cultivars, 4:394–395
        hybrids, 4:397–398
    conservation
        ex situ, 4:654
        facilities and skills for, 4:655–657
        history of, 4:645–648
        integrated approaches to,
            4:654–655
    conservation approaches, 2:473–477
    defenses, see Defenses, plant
    desert, 2:49–50
        morphology and behavior,
            2:50–51
        physiological processes, 2:51
    diversity
        distribution and loss of,
            4:648–649
        human influences, 4:649–652
    drug/chemicals from, modern
        approaches to discovery,
        4:714–715
        bioassays, 4:716
        ecological screens, 4:718
        ethnobotanical screens,
            4:718–719
        fractionation, 4:716–717
        phylogenetic screens, 4:717–718
        random screens, 4:717
        structural determination, 4:717
        vouchering, 4:715–716
    edible, 2:375

diversity of, 2:377
geographic patterns, 2:380
most prolific taxa, 2:377–378
parts, 2:378–380
endangered
examples, 2:471–473
legal mandates protecting, 2:471
endangerment
causes of, 2:469–470
patterns, 2:467, 469
evolution of, 2:197–199
extinctions, 2:465–466
numbers and locations, 4:652–653
recoveries from, 2:476
growth
effects of CO2 elevation, 1:621–622
in high latitude austral ecosystems, 5:368–369
hybrid
communities
diversity of herbivore species, 4:669
structure, 4:669–672
conservation of, 4:673–674
fitness, phytophage effects on, 4:673
genetic basis of resistance in
evidence
generalist and specialist herbivores, 4:667–668
genetic and environmental effects, 4:667
resistance traits, 4:665–666
theory, 4:663–665
phytophages on, population dynamics, 4:672–673
hybridization of, 4:659–660
diversity threats from, 4:674
hybrid complexes, 4:660–661
hybrid types, 4:661–662
occurrence of, 4:660
as keystone species, 3:619–620
medicinal, history of, 4:713–714
North American, diversity, 4:409
parasitic
advantages, 3:324
clades, relationship to other plants, 3:319–321
degree of, evolution, 3:321–322
haustorium, 3:323
host specificity, 3:324–325
natural history, 3:318
parasite–host interaction, physiology of, 3:324

as pests, 3:325
prehaustorial events, 3:323
and related phenomena, 3:322–323
pollination by deception, 4:727–728
and pollinators
in agriculture, 4:728–729
coevolutionary history, 4:727
management of, 4:729–730
in natural systems, 4:727–728
threats to mutualisms, 4:729
rare species, distribution, 2:468
reactions to grazing, 1:657–659
responses to burning, 2:746–747
riverine ecosystems, 5:222–223
small, and latent extinction, 3:693–694
terrestrial
and climate, 4:636–637
coevolution, 4:637–638
drug/chemicals from, 4:720–722
camptothecin, 4:718–719
digitalis, 4:719–720
Echinacea, 4:720–721
ginseng, 4:720–721
taxol, 4:717–718
tropane alkaloids, 4:719
evolution of, 4:632–633
UV radiation effects, 5:726–727
traits of, effects on ecosystem properties, 3:330
transgenic, and insecticide resistance, 3:476
tropical rainforests
classification, 5:16
numbers, 5:14
as units of biodiversity, 4:2
vascular
biodiversity in California, 4:149
biodiversity in Cape Region of South Africa, 4:155–156
biodiversity in Chile, 4:150–151
biodiversity in Mediterranean basin, 4:153–154
biodiversity in Southwestern Australia, 4:158
temperature limits for, 3:355
tropical rainforests, diversity, overview, 5:14–15
Plant–soil interactions, 4:690–692
large scale interactions
effects of species composition, 4:703–705
soil geographic patterns and vegetation, 4:703
management/conservation implications

atmospheric CO2 increase, 4:707–708
biodiversity and ecosystem function, 4:708
ecosystem restoration, 4:706
exotic species invasions, 4:706–707
nutrient pollution, 4:707
mesoscale processes, 4:696
nutrient acquisition and root growth
individual plants, 4:696–700
litter inputs and plant tissue chemistry, 4:701–703
plant–microbe interactions, 4:701
stand scale, 4:700–701
rhizosphere, 4:692–694
root–soil interface, 4:694–696
Planula, definition, 5:73–74
Plasmid DNA, 3:187
definition, 1:339
Plasmodiophoromycota, 3:151
Plasmodium spp., 2:117, 122
Plasmogamy, 3:144
definition, 3:141
Plasticity (plant), definition, 4:689
Plastron, definition, 1:249
Plate tectonics, 1:462–465; 5:769–770
in biodiversity over geologic time, 3:228–229
definition, 1:456, 665; 5:767
and species dispersal, 2:128
Platyhelminthes (flatworms), 3:563
excretion and osmoregulation, 5:888–889
feeding and digestion, 5:887–888
general features, 5:864–865
host associations, 5:896–897
major groups, 5:865
Acoela, 5:865–866
Catenulida, 5:869
Dalyelliida, 5:871–872
Haplopharyngida, 5:869
Kalyptorhynchia, 5:871
Lecithoepitheliata, 5:869
Macrostomida, 5:869
Nemertodermatida, 5:866, 869
Neodermata, 5:872–873
Aspidogastrea, 5:873
class Cestoda, 5:882–883
Amphilinidea, 5:883
Caryophyllidea, 5:883
Cyclophyllidea, 5:886–887
Proteocephalidea, 5:886
Pseudophyllidea, 5:883

Platyhelminthes (flatworms) (*continued*)
    class Digenea, Opisthorchiodea, 5:877
    class Monogenea, 5:878–879
        Chimaericolidea, 5:881
        Dactylogyridea, 5:879
        Diclybothriidea, 5:881
        Gyrodactylidea, 5:879
        Mazocraeidea, 5:881
        Monchadskyellida, 5:879
        Monocotylidea, 5:879, 881
        Polystomatidea, 5:881
        Udonellidea, 5:879
    Polycladdida, 5:870
    Prolecithophora, 5:871
    Proseriata, 5:870
    Temnocephalida, 5:872
    Tricladida, 5:870–871
    Typhloplanida, 5:871
  medical importance, 5:897
  nervous system, 5:887
  ontogeny, 5:890–893
  phylogenetic relationships, 5:893, 895–896
  reproduction, 5:889–890
Playas, 2:40–41
Pleiocene, definition, 4:451
Pleiotropic gene, definition, 5:371
Pleiotropy, definition, 2:1; 5:383
Pleistocene
  biogeographical dynamics, 1:465–466
  definition, 1:456; 3:761
  mammalian extinctions, 2:442–443
  mass extinction since end of, 3:762–763
Pleistocene glaciation, definition, 1:260–261
Pleopod, definition, 1:915
Plesiomorphic characters, 1:678
Plesiomorphy, definition, 1:677; 5:570
Plume prospecting, definition, 5:747
Plurivoltine, definition, 1:249
Pneumatophores, definition, 3:853
Pneumocystis spp., 4:915
Pneumoroidea (bladder hoppers), 3:261
Poaceae, definition, 4:451
Podzol, definition, 1:533
Podzolization, definition, 1:533
Poetry, pastoralism in, 3:741–742
Poikilohydry, definition, 3:27
Poisson distribution, definition, 4:745
Polarity, definition, 5:570
Polaroid, definition, 3:675

Policy, environmental, experimentation in, 1:788–789
Pollen
  conservation of, 2:691; 3:177
  fossil, 1:396, 397
    and climate change, 1:711–712
    Quaternary, in paleoecology studies, 4:451–452
      age models and mapping intervals, 4:460
      data and sensing systems, 4:452
      dating uncertainties, 4:459–460
      scaling factors, 4:452–453
      spatial characteristics, 4:456–457
      taxonomic and numerical characteristics, 4:453–456
      temporal characteristics, 4:457–459
      temporal resolution, 4:459
Pollination, 4:723–724
  and conservation biology, 4:729–730
  insect, and angiosperm radiation, 4:612–613
  self, in domesticated crops, 2:221
  and self-fertilization, preventive mechanisms, 3:430
  syndromes, 4:613–614
Pollinators
  in agriculture
    introduced species, 4:728
    as vectors/victims of engineered genes, 4:728–729
  beetles, 4:724
  birds, 4:726–727
  flies, 4:724–725
  hymenopteran, 3:425; 4:725–726
  lepidopteran, 4:725
  mammals, 4:726
  and plants
    coevolutionary history, 4:727
    obligate/facultative mutualisms, 4:727
    pollen/nectar robbers, 4:727
    pollination by deception, 4:727–728
Pollutants
  entry, movement, and fate in ecosystems, 2:367
  trace, of global importance, 4:742
Pollution; *see also* Air pollution; Water pollution
  Arctic, 1:246
  biodiversity impacts, 3:799–800
  definition, 1:119; 2:783; 4:731
  and disease outbreaks, 2:121

  and extinction, 2:706–707
  pesticides, 2:707
  and forest biodiversity, 3:49–50
  soil, bioremediation, 5:321
  threat to mammals from, 2:446–447
  tourism impacts, 5:675
Polyandry, definition, 5:295
Polycentric governance arrangements
  in coping with tragedies of the commons, 1:789–790
  definition, 1:777
Polychaeta
  morphology, 5:832–833
    chaetae, 5:836
    circulatory systems, 5:838
    digestive tract, 5:838
    gills, 5:838
    gonads, 5:839
    head, 5:833
    musculature, 5:837–838
    nephridia, 5:838–839
    nervous system, 5:836
    pygidium, 5:835–836
    segments, 5:833–835
    sense organs, 5:836–837
    septa, 5:838
  reproductive biology, 5:839
  systematics, diversity, and phylogeny, 5:839–842
Polychlorinated biphenyls, 2:563; 4:742
  threat to reptiles/amphibians from, 2:482
Polycladdida, 5:870
Polyculture, 1:551
  and biodiversity, 1:556–557
  definition, 1:109, 547; 2:269
Polydnavirus, 4:490–491
Polyembryony, in Hymenoptera, 3:421
Polygamy, definition, 5:383
Polygyny, definition, 5:295
Polymerase chain reaction (PCR), 3:197; 4:648
  definition, 4:177
  in measurement of microbial biodiversity, 4:179
    gradient gel electrophoresis, 4:182–183
    intergenetic transcribed spacer analysis, 4:183–184
    restriction fragment length polymorphisms, 4:184
    ssrDNA clone library sequencing, 4:184–185
Polymers, prebiotic formation, experiments, 4:443–444

Polymorphism, genetic
  allozyme diversity across phylogeny, analysis of
    global, 3:202–204
    regional, 3:204–205
  definition, 2:179, 245; 3:195; 4:777
  DNA, 3:197–198
  evolutionary significance, 3:209–210
  protein, 3:197
  statistical tests, 2:185
Polyphagous, definition, 4:249
Polyphyletic, definition, 1:915; 3:777; 5:428
Polyphyletic groups, definition, 5:569, 590
Polyphyly, definition, 1:199, 694
Polyplacophora, 4:238–239
Polyploid, definition, 5:371
Polyramous, definition, 1:916
Polythetic groups, definition, 5:590
Polytomy, definition, 5:569
Polyunsaturated fatty acid, definition, 5:495
Polyzoic, definition, 5:863
Poor fen, definition, 5:781–782
Popper, Karl, 5:572
Popperian, definition, 5:569
Population change, definition, 4:819
Population composition, definition, 4:819
Population density
  definition, 1:641
  measuring, 4:745
    as causal agent, 4:746
    as response variable, 4:745–746
  spatial density variation
    determinants of, 4:751
      external forcing, 4:751
      lifestyle characteristics, 4:751–752
      population dynamics, 4:752
    genetic/evolutionary consequences, 4:753
    population dynamic consequences, 4:752–753
    spatial patterns, 4:746–747
      density–biomass thinning laws, 4:751
      statistical patterns in spatial variation, 4:747–749
      Taylor's power law, 4:749–751
  temporal density variation, 4:754
    consequences, 4:755
    external forces, 4:754
    life history characteristics, 4:754
    population dynamics, 4:754–755

temporal patterns, 4:753
  distribution shape, 4:753
  red-shifted variance, 4:753
  Taylor's power law, 4:753–754
transspecific patterns
  density comparison across species, 4:755
  distribution of density over species, 4:755–756
  relationship with other traits, 4:756–758
  transspecies, 3/2 thinning law, 4:756
Population differentiation, versus speciation, 2:86
Population diversity, 4:759–760
  changes in
    amelioration of, 4:766–767
    impacts of, 4:766–767
  divergence detection, 4:760–761
    chromosomal variation, 4:761
    DNA variation, 4:761
    migration rates and population size, 4:761
    morphological variation, 4:761
    protein variation, 4:761
  divergence indices, 4:761–762
  divergence processes, 4:760
    factors influencing, 4:760
    genetic drift, 4:760
    mutation, 4:760
    natural selection, 4:760
  extent of, 4:764
    extrapolation of population extinctions from tropical deforestation rates, 4:766
    populations per area, 4:764–765
  importance, 4:762
    aesthetic value, 4:762
    direct economic value, 4:762–763
    ecosystem service value, 4:763
    genetic value, 4:762
    species conservation value, 4:762
  variability within species, 4:760
Population dynamics, 2:260–261
  density dependence
    continuous time, 4:772–773
    discrete time, 4:773
  density-independent
    age structure, 4:771–772
    of single species without age structure, 4:770–771
  role of stochasticity, 4:773
  two-species interactions, 4:773–774
    competition, 4:775–776
    disease dynamics, 4:775–776

    host–parasitoid dynamics, 4:774–775
    predation, 4:774
Population ecology, in resource economics, 2:292
Population genetics
  deterministic
    allele and genotype frequency, 4:779
    genetic load, 4:784–786
    maintenance of variation by selection, 4:783–784
    multiple loci, 4:786–787
    mutation, 4:779–780
    mutation–selection balance, 4:784
    selection at single locus, 4:780–782
    selection on quantitative traits, 4:782–783
  interaction of drift with deterministic forces
    diffusion equations, 4:793–794
    directional selection at linked loci, 4:793
    group selection, 4:795
    kin selection, 4:795–796
    linkage to balanced polymorphisms, 4:793
    Muller's ratchet, 4:794–795
    population subdivision, 4:791–793
    shifting balance theory, 4:796
  random genetic drift, 4:787
    coalescent process, 4:790–791
    increase homozygosity, 4:787–788
    molecular evolution and variation, 4:788–790
    population differentiation, 4:788
Population genetic studies, across sharp ecological contrasts
  evolution across phylogeny caused by climatic stress, 3:206–209
  past microgeographic studies, 3:205–206
Population momentum, definition, 4:819
Population regulation/control
  definition, 1:641
  and metapopulation theory, 4:175
  of zoo populations, 5:888–890
Population(s); see also Metapopulations
  carrying capacity of, 1:644–645
    current research, 1:646–647
    determination, 1:645–646, 647
  and conservation genetics, 4:803–804

Population(s) (continued)
  definition, 1:641, 855; 4:769; 5:467
  effective size, definition, 1:599;
    3:427
  genetic variation among, 2:246–247
  human
    in Africa, 1:79
    and fish conservation, 2:790
    growth in, 3:390–391, 796
      causes and effects, 4:824–825
        economic development and
          environment, 4:826–827
        education, 4:825–826
      history of, 4:820
        changing age profiles, 4:824
        demographic transition,
          4:820–821
        mortality, fertility and natural
          increase, 4:821–822
        twentieth century, 4:821
      prospects, 2000–2050,
        4:827–828
      responses to, 4:828–829
  ideal, definition, 1:599
  livestock, genetic structure,
    1:534–535
  minimum viable, definition, 5:823
  molecular variation, protein
    electrophoresis, 4:778
  phenotypic variations, 4:777
    concealed variability, 4:778
    discrete variation, 4:777–778
    interpretation, 4:778
    quantitative variation, 4:778
  regulation, definition, 1:641
  spacially structured, ecological
    variants in
      scale, 2:249
      selection versus gene flow, 2:249
  stochastic processes and extinction
    deterministic extinction processes
      and vortices, 4:835–836
    genetic stochastic processes,
      4:834–835
    role of stochasticity in time to
      extinction, 4:835
    types of demographic population
      fluctuation, 4:832–834
Population size
  effects on species coexistence, 5:423
  and extinction, 3:768
Population structure, definition, 2:245
Population viability analysis, 1:861
  applications, 4:836
  criticisms of, 4:842
  definition, 4:831–832; 5:823

demographic models, 4:837–838
  analytic methods and
    approximations, 4:838
  matrix population models, 4:838
  Monte Carlo simulation models,
    4:838–839
  numerical methods, 4:838
  problematic issues, 4:840
    density dependence, 4:840
    genetics, 4:840
    habitat dynamics, 4:840–841
    space, 4:840
  sensitivity of, 4:839–840
  time series analysis methods,
    4:838
in management decisions,
  4:836–837
other viability measures, 4:837
  extrapolation from similar species,
    4:841
  historical data, 4:841
  testing, 4:841–842
Porphyry, tree of, definition, 1:694
Porpoise, see Cetacea; Vaquita
Postemergence herbicides, definition,
  3:339, 342
Postojna–Planina cave system, Slovenia,
  5:536–537
Post-transcriptional modification,
  3:189
  definition, 3:184
Potential natural vegetation, definition,
  1:261
Pouchet, Fèlix-Archiméde, 4:441
Pound, Ezra, 3:740
Poverty
  and ecosystems, 4:845–848
    conservation costs, 4:853–854
    ecodevelopment, 4:854–855
    ecological refugees, 4:851–853
    land use changes, 4:850–851
    market forces, 4:849–850
    role of political/economic
      subjugation, 4:848–849
    value appropriation, 4:851
  and resource degradation,
    2:294–295
Powell, John Wesley, 1:887
Precipitation
  acid, see Acid rain
  Antarctic, changing patterns in,
    1:183
  boreal zone, 1:534
  chemistry, 1:4–5
  and C4 plants, 1:585
  effects on nutrient cycling, 2:331

and species richness, 1:732
Predation, 4:464
  and biodiversity, theoretical studies,
    4:871
  as density-independent mortality
    factor, 4:871–872
  generalist predators
    and biodiversity, 4:873–875
    and community stability, 4:875
  numerical and functional
    responses, 4:873
  specialist predators and
    biodiversity, 4:873
  definition, 4:769
  and evolution, 5:461–462
  by introduced species, 3:521–522
  and pairwise interactions,
    5:457–459
  and patchiness of rocky shores,
    3:488–489
  and population dynamics, 4:774
  and species coexistence, 5:417
Predator–prey relationships
  in carnivores, 1:637–638
  and coevolution, 1:760–762
  and mammalian evolution, 3:801
  and metapopulation structure, 4:164
  and structural stability, 5:476–477
Predators, 4:857–858
  apex, definition, 4:857
  case studies
    boreal/temperate forests, 4:866
    exotic predators on islands,
      4:869–870
    fragmented coastal scrub habitats,
      4:866–868
    kelp forests, 4:861–864
    lakes, 4:864
    midcontinent North American
      prairies, 4:868
    oceanic systems, 4:865–866
    rivers and streams, 4:865
    rocky shores, 4:860–861
    summary and synthesis of,
      4:870–871
    tropical forests, 4:868–869
  definition, 4:858
  future concerns, 4:876–878
  generalized food web theory, 4:860
  hymenopteran, 3:421–422
    versus parasitoids, 3:420
  indirect effects, 4:859
  influences, generality of, 4:875–876
  keystone species, 4:859
  and latent extinction, 3:696–697
  in plant defenses, 2:15–16

responses to, and coevolution
escape space and divergent
defenses, 1:759
mimicry, 1:759–760
species interactions, 4:859
top-down forces, 4:859
trophic cascades, 4:859–860
Preemergence herbicides, definition, 3:339, 342
Preservation; see also Conservation
definition, 1:883
Presoldier, definition, 3:581
Prevalence, definition, 2:109
Prigogine, Ilja, 4:442
Prigogine–Glansdorff principle, 4:445–446
Primary forest, definition, 3:747
Primary producers, and ecosystem diversity and function, in coral reefs, 3:135
Primary production/productivity
in aquatic/marine ecosystems, measurement, 2:313
carbon-based methods, 2:313
C-14 methods, 2:314
fluorescence methods, 2:314–315
interpretation and analysis, 2:315
oxygen-based methods, 2:313
definition, 1:419; 2:509; 4:377, 497
desert, 2:54
in freshwater ecosystems, 3:539
hydrothermal vents, 3:558
increased, in megadiverse countries, 1:422
in mangrove ecosystems, 3:864
net
and carbon cycle, 1:613
in contemporary world, 1:447–448
definition, 1:727
in diversity studies, 1:732–733
effects of elevated CO2, 1:621
in ocean ecosystems, 4:436
in pelagic ecosystems, 4:498–500
Primary reproductives, definition, 3:581
Primary succession, definition, 5:541
Primates, 3:791–793
conservation efforts, 4:885, 887–888
conservation status, 4:885
diversity and conservation, 4:880–882
endangered species, 4:886
future outlook, 4:888
threats to, 4:882
habitat destruction, 4:882–883

hunting, 4:883–884
live capture, 4:884–885
Priority (scientific name), definition, 4:389
Prisoner's dilemma, definition, 4:891
Procaryotes
definition, 4:349
thermophilic, 3:353
Process intricacy, definition, 1:831
Procyonidae (raccoon family), 1:630
Production, definition, 2:311; 3:75, 543
Productivity
contribution of biodiversity to, 1:360–362; 3:112
in environmental structuring, 3:131
and latitudinal gradients, 5:645–646
Proglottid, definition, 5:863
Prognathous, definition, 3:247
Progressive Era
conservation efforts during, 1:888–889
definition, 1:883
Progressive succession, definition, 5:541
Prokaryotes, 2:195
comparison with eukaryotes, 2:624–626
definition, 1:325; 3:184; 4:201, 569
versus eukaryotes, species numbers, 1:326
genome of, 3:184, 187–189
phylogenetic classification, 1:327–328
soil, species numbers, 5:309
symbiotic, 1:330
evolutionary origin, 1:331–332
identification, 1:331
phylogenetic affiliation, 1:332–333
uncultured free-living, diversity, 1:333–334
hot spring environments, 1:335
marine environments, 1:335
soils, 1:335–336
Prolecithophora, 5:871
Propagules
in Antarctic colonization, 1:181
definition, 1:171; 2:127; 5:185
Property rights, 4:891
and biodiversity protection, 4:897
alignment of private and social goals, 4:898
distribution of benefits, 4:897–898
exclusivity, 4:897

uncertainty, 4:897
context of
ownership, 4:892–893
people and biodiversity, 4:892
scope, 4:892
values, 4:892
definition, 4:85
evolution of, 4:896
ecosystem use, 4:896–897
expectation about use, 4:897
single-species use, 4:896
forms of, 4:893
common property, 4:893
open access, 4:893
private property, 4:893–894
state property, 4:893
functions of, 4:895
externalities, 4:895–896
scale, 4:896
transaction costs, 4:896
uncertainty, 4:895
optimal forms, 4:894–895
over goods/services, 4:894
Property rights systems, definition, 4:891
Prophalangopsidae (haglids), 3:252
Proscopidae (false walking sticks), 3:260–261
Proseriata, 5:870
Prosimians, definition, 4:879
Prostomium, definition, 5:831
Protandrous, definition, 1:915
Protandry, definition, 5:295
Protected areas
coexistence with humanity, 4:319–321
IUCN classification of, 4:318–319
origins of, 4:317–318
Proteins, 4:440
definition, 4:415
diversity
and evolutionary theory, 2:673
within/among species, 3:198–200
genetic coding of, 3:186
polymorphisms, 3:197
toxic, as plant chemical defenses, 2:13
Proteobacteria, 1:329
subclasses, symbionts of, 1:332–333
Proterozoic eon, 1:411
microbial diversity in, 3:220–221
Prothorax, definition, 3:479
Protista, 3:149; 4:202
evolution of, 2:196–197
Protists, definition, 4:201, 415
Protoctists, definition, 2:623

Protogynous, definition, 1:915
Protogyny, definition, 5:295
Protonephridial system, definition, 5:863
Protophytes, definition, 4:201
Protozoa, 4:901–902
  competition experiments, 5:416
  definition, 4:201, 349
  diversity and global species richness, 4:905
    amoeboid protozoa, 4:906–910
    apicomplexans, 4:915
    ciliated protozoa, 4:913–914
    flagellated protozoa, 4:910–912
    pathogenic, 4:915
    saphrophytic protozoa, 4:913–914
  and ecosystem function, 4:904–906
  in eutrophication, 2:663
  functional roles, 4:902–903
  species, nature of, 4:903–904
  temperature limits for, 3:354
  terrestrial, 3:563
P/R ratio, definition, 5:74
Psammophilic, definition, 1:741
Pseudocoelom, definition, 5:843
Pseudoextinctions, 2:718
  definition, 2:732; 4:305
  mammalian, 3:843
Pseudofecal pellet, definition, 3:853
Pseudomonas spp., adaptive radiation in culture, case studies, 1:41
Pseudoscorpiones, 1:212–213
Psudergate, definition, 3:581
Psychodidae (moth flies), 2:821
Psychrophiles, 4:917–918
  biodiversity of, 4:920–921
  definition, 4:191
  environments, 4:918
    Antarctic, 4:919
    Arctic, 4:918–919
    caves, 4:918
    oceans, 4:919–920
    permafrost, 4:919
    sea ice, 4:920
    upper atmosphere, 4:918
  evolution of, 4:921
  metabolic activity, 4:921–922
    enzymes, 4:922–923
    membranes, 4:923
Pterostylis truncata (brittle greenwood), 2:473, 476
Pterygota, 3:480
Public awareness, of environmental issues, 2:773–774
Public good, definition, 1:359; 4:85
Public goods, definition, 4:891

"Pull of the Recent," definition, 1:411
Puna, definition, 5:327
Punctuated equilibrium, definition, 1:694
Pupoid, definition, 4:291
Purebred, definition, 1:533
Pure culture, definition, 4:177
Pycnocline, definition, 4:427
Pycnogonids (sea spiders), 1:920
Pyrethroid pesticides, 4:512
  resistance to, 3:469
Pyrgomorphidae (bush hoppers), 3:262–263
Pyridine herbicides, 4:515
Pyrimidine fungicides, 4:514
Pyrosere, 3:794

Q

Q10, definition, 1:609
QTL, definition, 2:85
QTL mapping, 3:527
Quality-of-life values
  definition, 2:285
  diversity, 2:288
  local diversity, 2:287
  populations, 2:287–288
Quantitative genetics, 2:246
Quantitative trait locus (QTL), definition, 2:85
Quasispecies, definition, 4:439
Quaternary period
  definition, 3:825; 4:305; 4:451
  late
    definition, 4:451
    extinction event
      Australia, 3:833–834
      comparative approach, 3:826–827
      definition, 3:825
      Eurasia and Africa, 3:835–837
      Madagascar, 3:834–835
      oceanic islands, 3:837–838
      patterns and causes, 3:838–839
        anthropogenic models, 3:839
        climatic models, 3:839–840
      South America, 3:832–833
  and South American biogeography, 5:333
Queen, definition, 3:581
Quiescence, definition, 1:249
Quinone fungicides, 4:514

R

R0, 2:111–112
  definition, 2:109

Rabbit, European (*Oryctolagus cuniculus*), 1:757; 2:115–116; 3:522, 523, 524
  economic impacts, 1:77, 78
Radiation; *see also* Adaptive radiation; Electromagnetic radiation; Ultraviolet radiation
  species
    definition, 1:403
    nonadaptive, 1:26
Radiative forcing, definition, 3:277
Radiocarbon dating
  definition, 3:825; 4:451
  uncertainties in, 4:459–460
Radula, definition, 4:235
Rainforests
  Amazonian, on terre firme, 1:146–148
  animal diversity
    future of, 5:10–11
    geographical patterns of, 5:8–9
    overview, 5:3–4
  Asian, 1:271–274
  Australian, floristic groups, diagnostic species, 1:310–311
  beetle diversity, 5:4
  biodiversity implications of change, 5:28–32
  bird diversity, 5:5–7
  butterfly diversity, 5:4–5
  changes within, types, 5:28
  current area and rate of loss, 5:27–28
  definition, 5:25
  and geographical context, 5:2–3
  deforestation, assessment techniques, 5:25–26
  as factor in megadiversity, 1:425–426
  frog diversity, 5:5
  high species richness, hypotheses, 5:9–10
  historic area, 5:26–27
  loss of, 2:702
  mammal diversity, 5:7–8
  tropical, 5:13
    biodiversity overview, 5:14
    communities, 5:15–19
    fate of, 5:21–22
    plant diversity
      reasons for, 5:19–21
      value of, 5:21–22
  types of, 5:3
Rain shadow, definition, 5:701
Rami, definition, 1:915–916

Random genetic drift, 2:86, 89, 91
    definition, 2:85
Range condition, definition, 5:33
Range expansion
    definition, 2:127
    gradual, 2:130
    jump, 2:130–133
    observed rates of, 2:142–143
Rangelands, 5:34
    conversion to croplands/pasture, 5:48
    definition, 5:33–34
    desertification, and livestock grazing, 5:43–44
    grasslands and savannas, 5:35–38
    hot deserts, 5:38–39
    temperate/cold-desert shrublands, 5:38
    tundra, 5:34–35
Range management, conservation gains in, 1:890
Rank (taxonomic), definition, 5:590
Rapoport effect, 3:709–710
    definition, 3:701
Rare species
    definition, 2:465
    kinds of rarity, 2:466–467
Rats
    as bird threat, 2:400
    economic costs from, 1:74, 78, 79, 80–81
    as island exotic predators, 4:869
Rays, 2:760
Reaction centers
    definition, 4:549
    light harvesting/excitation energy transfer to, 4:550–552
    and primary photochemistry, 4:552–553
Recipient control, definition, 3:1
Reciprocal selection, definition, 5:173
Reclamation, definition, 5:97, 195
Recolonization, definition, 2:732
Recombination
    artificial sequence optimization, 5:69
    and cancer, 5:68–69
    definition, 1:17–18
    evolutionary effects of, 5:63
        environmental adaptation, 5:63
        horizontal gene transfer, 5:64–65
        novelties, 5:63–64
        sequence polymorphism, 5:64
    evolution of
        experimental, 5:67–68
        theoretical models, 5:65–66
    genetic control and mechanisms of, 5:58–61

as genetic mapping tool, 5:68
as source of genetic variation in breeding, 5:68
systems
    biodiversity of, 5:61–63
    and life history traits, 5:66
    and species ecology, 5:66–67
targeted gene replacement, 5:69
types of, 5:54
    crossing-over and chiasmata, 5:54
        crossover distribution in eukaryotic chromosomes, 5:57
        environmental conditions, effects, 5:58
        genetic interference and mapping functions, 5:56–57
        recombination rate and map distance, 5:56
        sex effects, 5:57–58
    ectopic recombination, 5:55
    extranuclear recombination, 5:55
    illegitimate recombination and horizontal transfer, 5:55
    mitotic recombination, 5:54–55
    nonhomologous chromosome segregation, 5:54
    nonreciprocal exchange, 5:55
Recreation
    conservation gains in, 1:890
    definition, 5:667
    ecosystem impacts, 3:386
    influence on wildlife communities, 3:670–671
Recruitment
    definition, 2:801; 3:485
    and patchiness of rocky shores, 3:487
    competition, 3:487–488
Recruits, definition, 5:161
Red algae, neurotoxic anthelminthics from, 4:716–717
Red Data Book (World Conservation Union)
    definition, 3:89
    future extinction rates deduced from, 3:766
    for threatened animals, 1:600; 2:443, 448–449
    birds, 2:397
    categories and criteria of extinction risk, 2:456–457
    marine invertebrate species, 2:460
    terrestrial mammals, 3:813–814
    for threatened plants, 4:646

Redfield ratio, 1:620
    definition, 4:1
Redi, Francesco, 4:441
Reduviidae (assassin bugs), 5:720
Reefs; *see also* Coral reefs
    artificial, 4:34
Reference area, definition, 5:195
Reflective beaches, definition, 1:741
Reforestation
    case studies
        degraded lands in Nepal, 5:107
        endangered biota habitats in New Zealand, 5:106–107
        tropical forest lands in Brazil, 5:106
    choice of areas, 5:103–104
    definition, 5:97
    ecosystem recovery following, 5:102–103
    methods of, 5:99
        direct seeding, 5:100
        planting densities and species richness, 5:101–102
        seedling planting, 5:100
        species choice, 5:100–101
    need for, 5:97–98
    socioeconomic issues, 5:104–106
    specialized conditions, 5:104
Refugia, definition, 1:232
Region, definition, 3:645
Regulations; *see also* Legislation
    acid deposition, 1:15
    air quality/pollutant emission, 1:130
        defining thresholds for, 1:131
    command and control versus market-based incentives, 4:94–95
    definition, 3:233
    of greenhouse gases, 1:304
    of herbicide use, 3:344
    post-World War II environmental, 1:893–894
    and soil conservation, 5:325
Regulon, definition, 1:339
Rehabilitation, definition, 5:97, 195; 5:805
Reification, definition, 3:729
Reintroduction
    definition, 1:599
    programs, 1:604–605; *see also* Captive breeding programs
    criteria for success, 1:606
Relative abundance distribution, definition, 5:413
Relevé method, of vegetative sampling, 3:570–571

Religion
  Asian traditions, 5:113–115
  and biodiversity, 5:286
    animism, 5:286–287
    Buddhism, 5:287
    Christianity, 5:287
    Hinduism, 5:287
    Islam, 5:287–288
    Jainism, 5:288
    Judaism, 5:288
    Shinto, 5:288
    Sikhism, 5:288–289
    Taoism, 5:289
  definition, 5:109, 286
  as means of encoding ethics, 5:110–111
  monotheistic traditions
    as base for environmental ethics, 5:112–113
    Judeo-Christian tradition, debates over, 5:111–112
  pantheistic traditions, 5:115–116
Remediation, definition, 2:363
Remote sensing
  aerial photography, 5:125–126
  applications, 5:129
    species distribution/richness, modeling, 5:137
      case studies, 5:137–139
    vegetation changes, measurement, 5:132
      case studies, 5:132–136
    vegetation mapping and land cover patterns, 5:129–130
      case studies, 5:130–132
  data analysis, 5:127–129
  definition, 2:23
  electromagnetic radiation, 5:123–125
  fundamentals, 5:122–123
  satellite imagery, 5:126–127, 140–142
Renewable energy
  biodiversity impacts of, 2:538–539
  definition, 2:525
Renosterfeld, definition, 4:145
Rent seeking, definition, 4:891
Repetitive DNA, definition, 3:184
Replacement, definition, 2:179
Replacement-level fertility, definition, 4:819
Representation, definition, 1:813
Reproduction
  amphibian, 1:162–165
  crustacean, 1:932–936
  haplodiploid, in Hymenoptera, 3:419

  in metazoans, 2:201
  Nematoda, 5:846–847
  Polychaeta, 5:839
  storage energy in, 5:512–514
  systems, of domesticated plants, 2:220–221
Reproductive behavior, and trait evolution, 1:407
Reproductive fitness, and inbreeding depression, 3:430
Reproductive isolation, definition, 1:897; 5:383
Reproductive Revolution, 4:822–823
Reptilia, 5:145; 5:763–764
  ancestral and derived characters, 5:148
    metabolism and energetics, 5:149–150
    temperature and water relations, 5:148–149
  biodiversity in Cape Region of South Africa, 4:157
  biodiversity in Chile, 4:152–153
  conservation status of, 2:480
  crocodilians
    conservation issues, 5:157–158
    extant evolutionary lineages, 5:146–147
  definition, 2:479
  as diaspore dispersers, 4:615
  extant evolutionary lineages, 5:145–146
  lepidosaurs
    conservation issues, 5:158
    Rhynchocephalia, 5:147
    Squamata, 5:147
      characteristics, 5:152–153
      lizards, 5:147–148
        characteristics, 5:153–155
      snakes, 5:148
        characteristics, 5:155–157
  of mangrove ecosystems, 3:858
  migration patterns of, 4:229
  North American, diversity, 4:410
  response to climate change, 1:719
  threats to
    commercial exploitation, 2:483–484
    exotic species introduction, 2:484
    habitat destruction, 2:481
    pollution, 2:482–483
    species at high risk, 2:484–485
  turtles
    conservation issues, 5:157
    extant evolutionary lineages, 5:146

Requisite Variety, Law of, 1:778
Reserves and preserves, see Nature reserves; Protected areas
Reserve selection algorithms, 3:449
  definition, 3:437
Resilience
  definition, 1:965; 2:153, 269; 4:27; 5:681
  effects of natural resource degradation, 2:299–300
Resistance
  definition, 1:753, 965; 2:153, 269
  pesticide
    in introduced species, 3:527
    and selection pressure, 3:526
Resolution, in ecotoxicology, 2:371–373
  definition, 2:363
Resource, definition, 4:305; 5:173; 5:453
Resource catchment, definition, 4:845
Resource concentration hypothesis, of species–area relationship, 5:399
Resource conservation, 1:102–103; see also Conservation
  definition, 1:99
Resource dynamics
  biomass production, 2:325–326
  definition, 2:321
  measurement, whole ecosystem and community structure-/composition-based approaches, 2:323
  nutrient cycling
    controls over, 2:330–331
    open versus tight cycles, 2:331
    by plants, 2:329–330
    rainfall effects, 2:331
    soil type, land use, and vegetation structure effects, 2:331–332
  trophic transfer, 2:325–326
Resource government systems, self-organized, as complex adaptive systems, 1:784–785
Resource management
  among indigenous peoples, 5:290–292
  role of collections, 4:274–275
  and traditional belief systems, 5:292–293
Resource partitioning, 3:265
  definition, 1:793
  and niche differentiation
    empirical evidence, 5:177–181
    theoretical considerations, 5:175–177

and species diversity, 5:180–181
Restoration
  active, definition, 5:195
  of animal diversity, 5:202–203
    reasons for animal absences, 5:203–204
    sources of animals for, 5:205–206
  assessment of success, 5:201–202
  definition, 1:965; 5:195; 5:805
  of diverse plant communities, 5:197–199
    improving plant species diversity, 5:190–191
    limits of, 5:199–200
  of eutrophic systems, 2:667–669
  forest
    definition, 5:97
    implications of plant–soil interactions, 4:706
    possibility of, 5:99
  goals of, 5:195–197
  microbes
    diversity and functional groups, 5:208
      free-living saprobes, 5:208–209
      soil animals and food webs, 5:209
      symbionts, 5:209–210
    diversity and functional redundancy, 5:196–197
    establishment, 5:210–211
    spatial and temporal arrangement, 5:207–208
  passive, definition, 5:195
  use of grazers in
    coral reefs, 3:275
    forests, 3:273–274
    grasslands, 3:272–273
    lakes, 3:274–275
Restriction fragment length polymorphism (RFLP), 3:196, 197
  in measurement of microbial diversity, 4:184
Reticulate speciation, 5:380
  definition, 4:659
Reverse gyrase, 1:228
Revisionary taxonomy, definition, 4:271
RFLP, see Restriction fragment length polymorphism
Rhabdite, definition, 5:863
Rhagoletis sp. (fruit fly), sympatric host race formation and speciation in, 5:378
Rhaphidophoridae (cave crickets, camel crickets), 3:259–260

Rhinoceroses, 3:817–819
  black (*Diceros bicornis*), 5:302
Rhizodeposition, definition, 4:689
Rhizosphere, 4:692–694
  definition, 4:689
Rhone River floodplain, France, 5:537
Ribonucleic acid, see RNA
Ribosomal DNA (rDNA)
  definition, 1:325
  16S, in bacterial phylogenetics, 1:327
Ribozymes, discovery of, 4:444
Ribulose-1,5-bisphosphate, 1:622; 4:555, 556, 557
  kinetics, temperature effects, 1:580
Rice
  genetic narrowing, 1:91
  genetic variation, 1:89–90
  trends in genetic diversity, 1:92
Rice paddies, 3:795–796
Rich fen, definition, 5:781–782
Ricinulei (ricinuleids), 1:212
Ricochetal, definition, 3:778
Rights, definition, 2:545
Rinderpest virus, 1:810; 2:117–118; 3:523
Riparian zone, definition, 5:213; 5:805
Riparian zones, 5:42
  definition, 5:34
River continuum concept, 5:216–217
  definition, 5:213
Riverine ecosystems
  assessment and management, 5:229–230
  biodiversity threats, 5:226
    dams, 5:226–228
    introduced species, 5:228–229
    nutrient loading, 5:228
    overharvesting, 5:228
    stream channelization, 5:228
    toxic substances, 5:228
  conceptual models, 5:216
    flood pulse concept, 5:218–221
    geomorphic–trophic hypothesis, 5:222
    resource spiraling, 5:217–218
    river continuum concept, 5:216–217
    serial discontinuity, 5:218
    telescoping ecosystem model, 5:221–222
  fish, 5:224–225
  invertebrates, 5:224–225
  large animals, 5:225–226
  microbiota, 5:222

  plants, 5:222–223
  predators, 4:865
Rivers
  anatomy of, 5:214–215
  biogeochemistry, 5:215–216
  black versus white water, of Amazonia, 1:151–152
  modification of, 3:93–94; see also Dams
  Southeast U.S., endangered invertebrate fauna, 2:433
RNA editing, 4:419
  Ebola virus, 4:426
  human Apo B and NFI, 4:424
  A (r) I deamination, 4:424–425
  kinetoplastids, 4:419–421
  mitochondrial tRNA editing, 4:423–424
  myxomycetes, 4:421–422
  paramyxoviruses, 4:425
  plant organelles, 4:422–423
RNA (ribonucleic acid), 4:446
  definition, 3:184; 4:415
  messenger, 3:186
    transcription in prokaryotes, 3:188–189
  ribosomal (rRNA), 3:186
    definition, 1:219
  small subunit ribosomal, definition, 5:647
  structure of, 3:185–186
  transfer (tRNA), 3:186
Rodentia, 3:798
Rodents, in U.S., 1:74
Romaleidae (lubber grasshoppers), 3:263
Roosevelt, Theodore, 1:888, 889
Root parasite, definition, 3:317, 325–326
Root–soil interface, 4:694–696
Rostroconchia, 4:237–238
Rostrum, definition, 5:711
Rotation, definition, 1:99
Roundworms, see Nematoda
rRNA, see RNA, ribosomal
Rubisco, see Ribulose-1,5-bisphosphate
RuBP, see Ribulose-1,5-bisphosphate
Rules, definition, 1:777
Rumen, definition, 3:778
Ruminants; see also Livestock
  definition, 5:34
  domestication of, 1:651–653
  ecosystem effects, 1:653
    on vegetation, 1:653–654
  and symbiotic bacteria, 4:208
Rumphius, 2:614

Russell cycle, 4:507
    definition, 4:497

S

SABONET, see South African Botanical Diversity Network
Saccaharomyces cerevisiae, 3:144
    commercial applications, 3:160
Sacred groves
    and traditional forest conservation in Nigeria, 5:116
Sagan, Carl, 4:443
Sahara Desert, 1:65
Saiga antelopes (Saiga tatarica), 3:819–821
Salamanders
    Ensatina spp., dichopatric race formation and speciation in, 5:374–376
    larval, ecology and functional morphology, 1:165
    morphologic features, 1:162
    reproductive biology and life histories, 1:163
Salinity, and C4 plants, 1:585–586
Salinization, as threat to freshwater fauna, 3:541
Salivary pump, definition, 5:711
Salmon
    biodiversity, value of, 5:241
    classification, 5:233–234
    historical status and overview of decline, 5:238
    life histories, 5:234
        Atlantic salmon, 5:236–237
        chinook (king) salmon, 5:235–236
        chum (dog) salmon, 5:235
        coho (silver) salmon, 5:235
        pink (humpback) salmon, 5:234–235
        sea-run cutthroat trout, 5:237
        sockeye (red) salmon, 5:235
        steelhead salmon, 5:237
    major threats to, 4:232; 5:238–239
        environment and climate change, 5:240–241
        habitat, 5:239
        harvest, 5:239
        hatcheries, 5:240
        hydropower, 5:240
    migratory patterns, 4:227–228
    recovery strategies, 5:242
    special adaptations
        anadromy, 5:237–238

homing, 5:238
Sal (Shorea robusta), 4:852, 853
Salt wedge, definition, 2:650
Salvina molesta, 3:503–504
Sampling effect
    definition, 3:109; 5:627
    model, 3:112
Sampling scope, definition, 4:123
The Sand County Almanac (Leopold), 3:744
Sand dunes, endangered, 2:419
Saponins, as plant chemical defenses, 2:13
Saprophytes, fungal, 3:153
Sarcophagidae (flesh flies), 2:821
Sarcopterygii, 5:762
Sarracenia rubra (pitcher plant), 2:472, 474
Saturated fatty acid, definition, 5:495
Saturation, definition, 2:161
Sauria, 5:764–765
Savannas, 5:35–38
    Amazonian
        flooded, of eastern Amazon, 1:154
        nonflooded/cerrado, 1:153–154
    Asian, 1:275
    definition, 3:675; 5:34
    and historic land-use patterns, 3:680–682
    tropical, South American, distribution and structure, 5:337–338
    tropical lowlands, 5:708
Scala naturae, definition, 1:694
Scale
    concepts of
        complete and incomplete similarity, 5:251–253
        multiscale analysis, 5:249–250
        scaling relations for biodiversity, 5:253
        scope, 5:250–251
    definition, 5:245
    effect of
        on conservation planning, 5:253–254
        on diversity estimates, 5:253–254
Scale dependence, 3:702
    definition, 3:701
Scandentia, 3:790
Scansorial, definition, 3:19
Scaphopoda, 4:241–242
Scent glands, definition, 5:711
Schizomida (schizomids), 1:211
Schrödinger, Erwin, 4:442
Scientific Committee on Problems of

the Environment, definition, 5:627
Scleractina, definition, 5:74
Sclerophyll, definition, 5:607
Sclerophyllous, definition, 4:145
Scolex, definition, 5:863
Scope, definition, 5:245
Scorpiones
    ecology, 1:206–207
    phylogeny/taxonomy, 1:207–208
    reproduction/growth, 1:207
Scrub, endangered ecosystems, 2:414–415
Scutellum, definition, 5:711
Seafloor spreading, and continental drift, 1:465
Seagrasses
    biogeography/species richness, 5:258–259
    and cultural eutrophication, 2:661–663
    definition, 5:255
    genetic diversity, 5:262–263
    habitat biodiversity
        function of, 5:266
        invertebrates, 5:264–265
        managing threats to, 5:266–267
        microalgae, 5:264
        microbial, 5:263–264
        vertebrates, 5:265–266
    origin/evolution of, 5:255–257
    species diversity
        distribution and controls of global, 5:259–260
        regional, 5:260
        local, maintenance of, 5:260–262
    taxonomy, 5:257–258
    UV radiation effects, 5:727–728
Sea ice, microflora, 4:920
Seal, Caribbean monk (Monachus tropicalis), 4:51
Sea lion, Japanese (Zalophus japonicus), 4:51
Seasonal leaf dimorphism, definition, 4:145
Secchi-depth, definition, 3:89
Secondary compounds, definition, 2:11
Secondary extinction, definition, 4:305
Secondary forests
    Amazonian, 1:154–155
    definition, 2:23; 3:747
Secondary metabolite, 3:265
    definition, 4:711
Secondary production, in aquatic/marine ecosystems, measurement, 2:315

bioenergetic analysis, 2:315–316
biomass accrual, 2:316
birth and death analysis, 2:316
egg ratio method, 2:316–317
instar analysis, 2:317
Secondary reproductives, definition, 3:581
Secondary succession, definition, 5:541
Sedimentation, definition, 5:74
Seed banks, 3:166
collection value, 3:169–171
community, 3:169
definition, 2:79
in food/agriculture, current status, 3:167–168
historical context, 3:166–167
management
characterization and evaluation, 3:176
collection, 3:173
distribution to users, 3:176
duplication, 3:176
packaging, 3:174–175
procedures, 3:172–173
regeneration, 3:176
seed bank design, 3:177
seed cleaning, 3:173
seed drying, 3:173–174
seed lot viability monitoring, 3:175–176
seed storage standards, 3:173–174
storage temperature, 3:175
scientific principles underlying, 3:171
genetic considerations, 3:172
storage conditions, 3:171–172
for wild species, 3:169
Seed lot, definition, 3:165
Seeds
in aquaculture, 1:188
definition, 1:185
commercial production, 1:93–94
dispersal
of Amazon terre firme forest, 1:148
and germination, effects of livestock, 1:659–660
by grazers, 3:273–274
vàrzea/igapò forests, 1:151–152
domesticated crops, and retention on stalk, 1:657–658
energy storage in, 5:497–498
protection from predators, 5:499
and plant evolution, 2:198
technical change, traditional/modern, 1:87–88

Segment (annelid), definition, 5:831
Segregation, definition, 1:18
Selection
definition, 2:179; 4:777
disruptive, definition, 5:427
stabilizing, definition, 5:428
Selectivity, herbicide, definition, 3:339
Self-organization, definition, 4:439
Self-replication, and origin of life, 4:444
Self-reproduction, and Darwinian evolution, 4:445
Selous Game Reserve (Tanzania), 4:319
Semaphoront, definition, 5:569
Semelparity, definition, 3:715
Semelparous, definition, 3:778; 5:233
Seminatural, definition, 3:675
Semiochemical pesticides, 4:512
Semispecies, modern views of, 5:525–526
Semisteppe batha
definition, 4:353
vegetation of, 4:360
Senegal River valley, restoration efforts, 5:817–818
Sensor, definition, 5:121
Sentinelese Island, 4:846–848
Sequestration, definition, 1:609
Sere, definition, 5:607
Serial endosymbiosis theory, in eukaryote origins, 2:629–632
Serotiny, definition, 2:745
Setae, definition, 1:916
Settlement, definition, 3:659
Sevilleta National Wildlife Refuge, 3:654–657
Sexual selection, 5:388–389
definition, 1:25; 2:1
Shadow price, definition, 2:291
Shannon index, of biodiversity, 1:382, 383
Sharks, 2:758–760
Shifting agriculture
definition, 3:748
and long-term intensive forest management, 3:750
Shifting cultivation, definition, 2:23
Shinto, 5:288
Shoot growth, definition, 1:307
Sibling species, definition, 5:428
Siderophore, definition, 4:690
Sikhism, 5:288–289
Silent Spring (Carson), 3:744, 799
Silurian period, 1:925
Silverswords, Hawaiian, adaptive radiation, case studies, 1:36

Silviculture, definition, 3:748
Similarity, definition, 5:590
Simple sequence repeats, 3:201–202
Simpson diversity index, definition, 5:441
Siphonaptera (fleas), 3:483
Sirenia, 3:794; 4:37
diving capabilities, 4:39
evolutionary history, 4:43–44
extinctions over evolutionary time, 4:50
features and habitat boundaries, 4:41
SIR models, 2:111
Sister groups, definition, 5:145; 5:590
Sister species, definition, 5:371
Sister taxa
and adaptive radiations, 1:406–407
definition, 1:403
Skates, 2:760
Slash-and-burn agriculture, *see* Swiddening
Slime molds, *see* Myxomycetes
Smog, photochemical, 1:295
Smolt, definition, 5:233
Snails, partula, adaptive radiation, case studies, 1:41
Snakes, 5:148, 155–157
Snares Islands Nature Reserve, 4:318
Snow, C.P., 3:740
Social behavior, 5:295–296
evolution of, 2:201–202
hymenopteran, 3:423–424
to social organization, formation patterns/mechanisms, 5:296–297
ecological challenges and phenotypic constraints, 5:297–298
kinship and demography, 5:298–299
Social benefit, net
definition, 2:291
evaluation of, 2:295–297
Social costs, definition, 1:359
Social differentiation
definition, 5:733
and human ecology, 5:739–741
Social organization, formation patterns/mechanisms, 5:296–297
Social taboo, definition, 5:681
Soil aggregate, definition, 4:690
Soil food web, definition, 5:315
Soils; *see also* Plant–soil interactions
acid deposition effects, 1:8–9
alpine ecosystems, 1:136

Soils (*continued*)
- microbial diversity, 1:141–142
- Amazonian, 1:905
- bacterial element cycling, 4:207
- biodiversity in, 5:309
  - bacterial, 1:335–336
  - collembola, 5:311–312
  - earthworms, 5:313
  - evolutionary history, 5:308
  - fungi, 5:310–311
  - microfauna, 5:311
  - mites, 5:312–313
  - nematodes, 5:311
  - prokaryotes, 5:309–310
  - termites, 5:313
- biological crusts, 5:45–46
  - definition, 5:33
- boreal zone, 1:535–536
- and carbon cycle, 1:266–267, 269, 279
- classification of, in traditional agroecosystems, 1:112
- consequences of burning, 5:274–276
- conservation
  - in agriculture, methods, 5:319–320
    - diversified cropping, 5:320–321
    - manipulation of biotic community, 5:321–322
    - organic amendments, 5:321
    - pollutant bioremediation, 5:321
    - reduced tillage, 5:320
  - gains in, 1:890–891
  - policies and regulations, 5:325
  - of urban soils, 5:321
- as constraint on restoration, 5:188, 189
- continental Africa, 1:56–57
- definition, 5:315–316
- depletion of, 2:562
- desert, 2:41–43
- erosion of, 5:317–318
- fate of herbicides in, 3:342–344
- fertility, and species richness, 1:739
- forest, 3:43
  - acidification, 1:129–130
- forming factors, 5:305–306
- high-quality, attributes of, 5:316–317
- logged forests, effects of management systems, 3:752
- mineralization and immobilization of nutrients in, 2:334
- organic matter, 1:100, 103
  - and cation exchange capacity, 5:317
- definition, 1:99; 4:690
- pesticide treatment of, 4:517
- physical structure of, 4:696–697
- and planned diversity, 2:272–273
- polyphasic nature of, biotic influences, 5:306–307
- processes
  - immobilization and mineralization, 5:307
  - nitrogen cycle, 5:308
- quality
  - assessment and monitoring, 5:322–323
  - effects of natural systems conservation, 5:318–319
  - indicators
    - biological attributes, 5:324–325
    - chemical attributes, 5:323–324
    - physical attributes, 5:323
  - role of, 5:316
  - in carbon cycle, 1:614–615, 617
  - termite effects on, 3:609
  - tropical forest, 5:704
- types, and nutrient cycling, 2:331–332
- and unplanned diversity, 2:274
- wetland, restoration, 5:811

Soil structure, definition, 5:315
Soil texture, definition, 4:690; 5:315
Solar energy systems, 2:538
Solar heated, definition, 4:349
Soldiers
- definition, 3:581
- termite, evolution of, 3:595–597

Solifugae (wind spiders), 1:213
South Africa
- Cape Region, biodiversity in, 4:154–155
  - birds, 4:156–157
  - invertebrates, 4:157
  - mammals, 4:156
  - reptiles and amphibians, 4:157
  - vascular plants, 4:155–156
- exotic species
  - invertebrates, 1:80
  - plants, 1:79
  - vertebrates, 1:79–80

South African Botanical Diversity Network (SABONET), 1:435–436
- definition, 1:419

South America; *see also* Amazonia
- aquatic ecosystems, threats to, 5:358
- biogeography, 5:332
  - early history and associations, 5:332
  - late Tertiary, 5:332–333
  - Quaternary, 5:333
- conservation priorities, 5:358–359
- deserts of, 2:44–45
- human use
  - of converted areas, 5:347
  - of remaining natural habitats, 5:347, 353
- land cover in, 5:356
- late Quaternary mammalian extinctions, 3:832–833
- major ecosystems of, 5:334
  - high Andean grasslands
    - distribution and structure, 5:338
    - functional aspects, 5:338
  - mangrove, 5:341–342
  - temperate grasslands
    - distribution and structure, 5:338
    - functional aspects, 5:339–340
  - tropical forest, 5:334, 353, 354
    - distribution and structure, 5:334–336
    - functional aspects, 5:336–337
  - tropical savannas, functional aspects, 5:338–339
  - wetlands, 5:342
  - xeric formations, 5:340
    - distribution and structure, 5:340–341
    - functional aspects, 5:341
- physical environment, 5:328
  - climate, 5:330–332
  - geology and geomorphology, 5:328–330
- protected areas, status of, 5:358
- terrestrial ecosystems, 5:348–352
  - original extent of, 5:345–346
  - present extent of, 5:346–347
  - threats to, 5:357

Southern oceans, definition, 4:427
Spallanzani, Lazzaro, 4:441
Spallation, definition, 1:437
Sparrow, English (*Passer domesticus*), 1:75
- hybridization of, 3:527

Spartina alterniflora, 3:507–508
Spartina anglica (salt marsh grass), allopolyploid speciation in, 5:380
Spatial heterogeneity, definition, 4:123
Spatially structured population, definition, 4:161
Spatial patterns of genetic variation, 1:88–90
- definition, 1:85

Spatial subsidies, definition, 3:1

Spawners, in aquaculture
  definition, 1:185
  wild capture of, 1:193
Specialist brood parasites, definition, 4:365
Specialization
  definition, 1:753
  and selectivity of extinction, 4:312
    conditions, 4:312
    interactions, 4:312–313
    resources, 4:312
Speciation, 3:216
  allopatric
    definition, 1:25
    dichopatric speciation, 5:374–376
    peripatric speciation and founder effect principle, 5:376
  coincident, 1:754
  definition, 5:428
  dispersal effects on, 2:149–150
  and endosymbiosis, 2:627
  and evolutionary ecology, 2:676–677
  evolutionary forces in, 5:387
    common modes of, 5:389
      classical allopatric, 5:389–390
      parapatric/hybridization, 5:391–392
      periatric/founder effect, 5:390–391
      sympatric, 5:392
    genetic drift, 5:388
    mutation, 5:389
    natural selection, 5:387–388
    sexual selection, 5:388–389
  and extinction rates, 2:717–718
  genetic patterns of, 5:392–393
    genetic divergence patterns, 5:393–394
    postzygotic reproductive isolation and Haldane's rule, 5:393
  modes of, 5:373
    and coexistence, 5:425
  myriapods, 4:299–300
  new directions for studying
    developmental models, 5:394–395
    genealogical models, 5:394
  nonallopatric, 5:377
    allopolyploid speciation, 5:380
    autopolyploid speciation, 5:379
    direct/reticulate speciation, 5:380
    introgressive hybrid speciation, 5:379
    parapatric speciation, 5:377–378
    recombinational hybrid speciation, 5:379–380
    speciation by interspecific hybridization, 5:379
    spontaneous thelytokous speciation, 5:379
    sympatric speciation, 5:378
  versus population differentiation, 2:86
Species
  average duration of, 1:400
  basal, definition, 5:695
  concepts of, 5:371–373
    cohesion, 5:386–387, 434
    diagnostic, 5:432–433
    ecological, 5:431
    evolutionary and lineage, 5:433–434
    genealogy, 5:386, 432
    importance for biodiversity/ conservation
      biodiversity in space and time, 5:439
      differences as ecological markers, 5:438–439
      genetic differences versus species status, 5:438
      as real entities, 5:438
    interbreeding, 5:430–431
    monophyly, 5:431–432
    phenetic, 5:385, 435
    populations as evolutionary units, 5:435
    purpose, 5:428–429
    recognition, 5:386, 431–432
    taxonomic practice, 5:434–435
  and conservation genetics, 4:805
  crypic, definition, 3:777
  Darwin's morphological criterion, 5:429–430
  definition, 1:393; 4:177; 5:453
  dominant, 3:622
    definition, 3:613
  Eurasian, invasiveness of, 3:518
  focal, definition, 5:823
  functional groupings of, and taxonomic versus functional diversity, 2:207–208
  genotypic cluster criterion, 5:435–437
  importance in ecosystem processes, 2:522
  individualistic nature of, 3:300–301
  interactions between
    theories of, 2:261–264
    types of, 1:793
  lifetimes of
    and extinction risk, 4:308–309
    variation among taxonomic groups, 4:307–308
    variation within taxonomic groups, 4:308
  loss of, causes, 3:769
    changes in biotic environment, 3:771–772
    habitat change and destruction, 3:769
    habitat fragmentation, 3:770–771
    population exploitation and persecution, 3:771
  matrix, definition, 5:185
  maximal information content of, 4:446–447
  migratory, human disturbance
    attributes and susceptibility to, 4:230–231
    ecological consequences, 4:230–231
  numbers of, 2:739
  patterns in fossil records, 1:394
  patterns of change over time, 1:397–398
  polytypic, 5:430
  rarity, and extinction selectivity, 4:309
    geographic range, 4:309
    local abundance, 4:309–310
    relationship between range and abundance, 4:310–311
    spatial structure of populations, 4:311
  representative, primary habitat and niche components, 3:306
  spatial patterns, theories, 2:265
  in traditional agroecosystems, 1:110
  unreality of, in space and time, 5:437–438
  wild/nondomesticated, definition, 3:165
Species abundance distribution, definition, 4:123
Species accumulation curve, definition, 4:123
Species–area curves
  in conservation biology, 5:406
  loss of species
    from area reduction, 5:406–408
    slowing, 5:408–410
    role of incidence functions and nestedness, 5:410
  definition, 5:397
Species–area model, definition, 5:397
Species–area relationship
  definition, 5:397

Species–area relationship (*continued*)
  functional form of, 5:401–403
    extreme value model and random placement, 5:404
    self-similarity and power-function model of, 5:403–404
  models, interpretation of parameters, 5:404
    slope values
      canonical, 5:404–405
      island–mainland differences in, 5:405
      isolation effects, 5:405
      other interpretations, 5:405–406
  multiple causes of, 5:399–400
  sampling practice, 5:400
    independent areas, 5:400–401
    nested areas, 5:401
  statistical practice, 5:401
  underlying mechanisms, 5:397–398
    area per se hypothesis, 5:398
    edge effects, 5:399
    habitat diversity hypothesis, 5:398
    passive sampling hypothesis, 5:398–399
    resource concentration hypothesis, 5:399
Species coexistence
  competition experiments, 5:416
    beetles, 5:416–417
    protozoa, 5:416
  competitive exclusion principle, 5:417–418
  constraints on, 5:415
    community and spatial contexts, 5:415–416
    competition, 5:415
    natural enemies, 5:415
  definition, 5:173; 5:413
  and exclusion, 5:416
  future directions, 5:425
  landscape heterogeneity, 5:424
  mechanisms, 5:418–419
    for local coexistence, 5:421
      food web effects, 5:421–422
      habitat heterogeneity, 5:422
      individual discreteness, consequences, 5:422–423
      models of multiple limiting factors, 5:421
  in open communities, 5:423
    autecology and population size effects, 5:423
    colonization–extinction dynamics, 5:423–424

and permanence, 5:414–415
and predation, 5:417
role of interactions, 5:414
role of scale, 5:414
temporal niche partitioning, 5:423
  nonlinear consumption, 5:423
  storage effect, 5:423
traditional approaches to, 5:419
  classical niche partitioning, 5:419
  limiting similarity, 5:420
  Lotka–Volterra model, 5:419–420
  manipulative field experiments, 5:420–421
Species composition
  alterations in urban areas, 5:744–745
  definition, 1:456
Species density, definition, 3:701
Species distribution, remote sensing in modeling of, 5:137–139
Species diversity, 1:380–381; 5:441
  and biodiversity, 1:381–382
  and biodiversity policy, 5:450
  and climate, experiments, 1:739–740
  and coevolution, 1:766
  definition, 1:377, 727; 3:701, 729; 4:27; 4:123; 5:1, 441
  forms of, 5:447–450
  maintenance mechanisms, 5:181
    chance, 5:181–182
    frequency-dependent effects, 5:181
    spatial structure, 5:182
  in marine ecosystems, 4:18–19
  and net primary productivity, 1:732–733
  and species packing, 3:735
Species evenness
  definition, 2:205
  in diversity estimation, 2:208
    linkage with species richness, 2:208–209
Species flocks
  ancient lakes, 3:639–640
  definition, 2:755; 3:633
Species interactions, 5:453–454
  asymmetrical, 5:460
  and evolution, 5:460
    competition, 5:460–461
    mutualism, 5:461
    predation, 5:461–462
  human impacts on, 5:465
  and latitudinal gradients, 5:645
  multispecies, 5:462
    and evolution, 5:465

food webs and trophic levels, 5:462–463
keystone species, 5:463
trophic cascades and indirect interaction, 5:463–465
pairwise, 5:454
  competition, 5:454–457
  mutualism and symbiosis, 5:459–460
  parasitism and herbivory, 5:459
  predation, 5:457–459
Species life span, 2:732–733
definition, 2:732
Species packing
  coexisting species, similarity of, 3:733–734
  definition, 3:729
  laboratory experiments, 3:731–733
  niche separation and anatomical difference, 3:734–735
  and species diversity, 3:735
  theories, 3:730
    ecological niche, 3:731
    expansion to multiple species, 3:733
    Lotka–Volterra equations, 3:730–731
Species pool, definition, 2:161
Species richness
  in African forests, 1:60
  Amazonian semideciduous forest, 1:149–150
  Amazonian várzea forest, 1:151
  and area, relationships, 1:727–729
  Australian ecosystems, 1:319–321
  computation and expression of, 5:442–446
  definition, 1:262; 2:161; 3:329; 4:123; 5:1
  determinants of, 5:447
  in diversity estimation, 2:208
    linkage with species evenness, 2:208–209
  diversity–productivity relationship, 2:164–167
  estimators, theoretical properties, 4:125
  freshwater fishes, 2:778
  herbaceous vegetation
    decline in, and ecosystem reassembly, 3:334–335
    dominant species, 3:332–334
    high levels of, effects, 3:332
    mass ratio hypothesis, 3:330
    subordinate species, 3:332–334
    filter effects, 3:335

transient species, 3:334
    founder effects, 3:335–336
  latitudinal gradients in, 3:702
    mechanisms, 3:708
      evolutionary speed, 3:709
      geographical area hypothesis, 3:708–709
      geometric constraint hypothesis, 3:710–712
      Rapoport–rescue hypothesis, 3:709–710
  local, definition, 3:531
  Mediterranean-type ecosystems, 4:146
  in megadiverse countries, 1:420
    geological factors, 1:423
  metacommunity processes, 2:169–171
  myriapods, 4:299–300
  patterns of, 5:446–447
  rainforest, hypotheses, 5:9–10
  regional
    and endemism, 2:502–503
    regional processes, 2:167–169
  remote sensing in modeling of, 5:137
    case studies, 5:137–139
    in restored habitats, case studies, 5:187–188
  spatial scale, 2:173–174
  species–energy theory, 2:163–164
  time dependency, 1:730
  Volterra–Gause perspective, 2:162–163
Species turnover, definition, 2:205
Specificity, in ecotoxicology, 2:371–373
  definition, 2:363
Specific mate recognition systems, definition, 5:428
Specific resistance, definition, 1:547
Speciose, definition, 2:755
Sperm, cryopreservation of, 1:543
Spermatophore, definition, 1:199; 3:247
Sperm competition, 1:407
Sphaeroceridae (lesser dung flies), 2:821
Sphagnum moss, and air pollution, 1:120
Spicule, definition, 5:843
Spiders, *see* Araneae
Spiegelman, Solomon, 4:444
Spinnerets, definition, 1:199
Spirochetes, definition, 2:623
Spontaneous thelytokous speciation, 5:379

Spore, definition, 3:141
Sporophyte, definition, 4:1
SPOT (Systéme Pour L'Observation de la Terre) program, 5:127
Spreading rate (tectonic plates), definition, 5:747
Squamata, 5:147
  characteristics, 5:152–153
  definition, 2:95
  lizards, 5:147–148
    characteristics, 5:153–155
  snakes, 5:148
    characteristics, 5:155–157
SSR, *see* Simple sequence repeats
Stability
  aspects of, 5:469
  basic theory
    algebraic methods, 5:471–472
      discrete-time systems, 5:472–473
      multiple species, 5:472
      single species, 5:472
      strength of return, 5:473
    equilibrium, 5:469–470
    geometric interpretation
      local versus global stability, 5:471
      multiple species, 5:471
  under changing conditions
    biodiversity and temporal stability, 5:473–475
    emerging deterministic stability, 5:475–476
    stability amid chaos, 5:476
    temporal niches, 5:475
  definition and meaning, 5:467–468
  diversity–stability question, 5:468–469
  evolutionary, 5:479
  spatial, 5:478–479
  structural, 5:476
    effects of eutrophication, 5:477–478
    predator–prey systems, 5:476–477
Stabilizing selection, definition, 2:85
Stable population, definition, 4:819
Stakeholder, definition, 5:805
Starlings, European, 1:75
Stasis, definition, 1:694
Statocyst, definition, 5:863
Statutes (legislation)
  definition, 3:233
  influencing biodiversity, 3:235
Steinbeck, John, 3:744
Steller's sea cow (*Hydrodamalis gigas*), 4:51

Steller's sea lion (*Eumetopius jubatus*), 4:59–60
Stem (cladogram), definition, 5:569
Stem parasite, definition, 3:317
Stenohaline, definition, 2:455
Stenopelmatidae (Jerusalem crickets, wetas), 3:259
Stenotopic, definition, 2:497; 4:291
Stewardship
  of biodiversity
    counter arguments, 5:485–487
    instrumental arguments, 5:483–484
    intrinsic arguments, 5:484–485
  contexts, 5:487
    cultural and political factors, 5:488–489
    ecological factors, 5:487–488
    economic factors, 5:488
  definition, 2:545; 5:481
  entymology of, 5:482
  rationales, 5:482–483
  types of stewards, 5:489
    governmental, 5:489–491
    nongovernmental organizations, 5:491–992
Stichosome, definition, 5:843
Sticklebacks, Canadian, adaptive radiation, case studies, 1:37
Stochastic, definition, 3:761
Stochastic event, definition, 2:441
Stochasticity
  definition, 4:831
  demographic, 3:766–767
  environmental, 3:767–768
  role in population dynamics, 4:773
  role in time to extinction, 4:835
Stochastic processes, definition, 5:823
Stochastic system, definition, 5:467
Stock–recruitment relationship, definition, 5:161
Stoma, definition, 5:843
Storage
  energy, in life history
    hibernation, 5:511–512
    migration, 5:511
    optimizing stores, 5:510–511
    reproduction, 5:512–514
  in invertebrates, 5:496–497
  plants, 5:497
    lipid composition in, 5:498–499
    in seeds, 5:497–498
      protection from predators, 5:499
  substances amenable to, 5:496
  vertebrates, adipose tissue, 5:499–500

Storage effect, definition, 2:79
Storage organ, definition, 5:495
Strain, definition, 4:177
Stramenopila, 3:149
Stratification, definition, 4:427
Stratigraphic interval/range, definition, 3:53
Stratigraphic section, definition, 4:97
Stratiomyidae (soldier flies), 2:821–822
Stratosphere, ozone, 1:301–302
  causes/consequences of depletion, 1:303–304
Stratum/strata, definition, 3:53
Stream order, definition, 5:213
Strepsiptera, 3:483
Stress
  abiotic, and latitude, 3:712–713
  climatic, evolution across phylogeny caused by, 3:206–209
  environmental
    assessments, 5:519
      appraisal, 5:519–520
      early warning, 5:520–521
      future trends, 5:521–522
      predation, 5:520
    characterization of, 5:516–517
    definition, 5:515–516
    general, syndrome, 5:518–519
    natural versus anthropogenic, 5:516
    receptors and responses, 5:517–518
    definition, 5:515
  and genetic ecological diversity, 3:212
  marine and aquatic ecosystems, from eutrophication, 4:1–11
Stridulatory apparatus, definition, 3:247
Strobila, definition, 5:863
Strong interactors, 3:622
  definition, 3:613
Structural intricacy, definition, 1:831
Studbook, definition, 1:599; 5:901
Sturgeons, 2:761
Stygobite, definition, 5:527
Stygophile, definition, 5:527
Subarctic, definition, 1:232
Subchelate, definition, 1:915
Submerged vegetation, definition, 5:782
Subnivean, definition, 1:171
Subordinate species, 3:332–334
  definition, 3:329
  filter effects, 3:335

Subpolar, definition, 4:427
Subspecies
  and conservation genetics, 4:804–805
  modern views of, 5:525–526
Substituted amide herbicides, 4:515
Substituted aromatic fungicides, 4:514
Substituted urea herbicides, 4:515
Substitution
  amino acids, rate of, 2:186
  definition, 2:180
  process of, 2:187
Substitution possibilities, 2:298–300
  definition, 2:291
Substrate, definition, 3:141
Subterranean ecosystems, 5:527
  adaptations to, 5:531–533
  caves, 5:528
    biodiversity, geography of, 5:538–539
  colonization and speciation in, 5:533–534
  conservation and protection of, 5:539–540
  deep subsurface habitats, 5:529–530
  Edwards Aquifer, Texas, 5:537
  energy sources, 5:530–531
  Flathead River alluvial aquifers, Montana, 5:537–538
  Gua Salukkan Kallang, Indonesia, 5:537
  Lobau wetlands, Austria, 5:537
  Mammoth Cave, Kentucky, 5:536
  Movile Cave, Romania, 5:537
  Postojna–Planina cave system, Slovenia, 5:536–537
  Rhone River floodplain, France, 5:537
  shallow subsurface habitats, 5:529
  taxonomic survey, 5:534–536
  Washington Caves, Bermuda, 5:537
Subtropical, definition, 4:427
Succession
  and biodiversity, 5:550
    on disturbed landscapes, 5:550
    gain/loss of species, 5:550–551
  and bird species diversity, 1:503–505
  in boreal forests, 1:541–542
  definition, 2:345; 3:41; 3:748
  heterotrophic, definition, 5:541
  individualistic view of, definition, 5:541
  logged forests, early phases, 3:753–754
  mechanistic explanations of, 5:545

  descriptive models, 5:545–546
  landscape as a mosaic, 5:546–549
  quantitative models, 5:549
  and nutrient cycling, 2:349–350
  organic explanations of, 5:542
    alternative theories, 5:544–545
    Clementsian concepts of, 5:542–544
    application, 5:544
  primary, definition, 5:541
  progressive, definition, 5:541
  on restored lands, 5:201–202
  secondary, definition, 5:541
  theory of, in island biogeography, 3:568
Sulfur bacteria, definition, 4:201
Sulfur cycle, 1:450–451
  perturbation of, 4:740–741
Sulfur oxides
  in acid deposition, 1:2–3
  atmospheric, 1:301
  deposition/environmental pathways, 1:121
  effects on forest ecosystems, 1:129–130
  sources and impacts, 1:120
Supercontinent, definition, 4:111
Superspecies, modern views of, 5:525–526
Supralittoral, definition, 1:741
Surface waters, acid deposition effects, 1:9–10
Surveys, vegetation, definition, 2:635
Survival, and environmental information/problem solving, 1:47
  fundamental concepts, 1:48–49
Suspension feeders, definition, 4:1
Sustainability
  concept of
    development, 1:100–102
    and practice, 5:554
  definition, 1:965
  and economic growth/social well-being, 2:281–282
  elements of, 1:99–100
  in fisheries management, 2:809
  global, and human ecological footprints, 2:236
  functional footprints of cities, 2:236–237
  global development and social equity, 2:236–237
  indicators of, 1:102; 5:562–563
    ecosystem services, 1:103–104
    resource conservation, 1:102–103
    and social learning, 5:565–566

state of the art, 5:563–564
science and, 5:554–556
transition to, 5:564–565
  social feasibility, 5:562
  trends and transitions, 5:556
    economic growth and consumption, 5:557
    overview, 5:559–562
    population, 5:556–557
    species and ecosystems, 5:557–559
Sustainability gap, definition, 2:230
Sustainability transition, definition, 5:553
Sustainable agriculture, definition, 1:109
Sustainable development
  in conservation formula, 1:972–973
  definition, 1:965–966; 5:553
Sustainable forestry, definition, 5:655
Sustainable use, definition, 3:811
Sustained yield, definition, 1:883; 5:823
Swamps
  Amazonian, 1:152–153
  definition, 5:782
  near East, 4:364
Swiddening, 5:269, 291
  alternatives to
    improved short fallow systems, 5:283–284
    pastures, 5:283
    perennial crop/tree plantations, 5:283
  biodiversity impacts, 3:794–795
  consequences of burning
    global effects, 5:276
    heat evolved during, 5:274
    on soil organisms, 5:276
    on soils, 5:274–276
    on vertebrates, 5:276
    on weeds, 5:276
  and C4 vegetation, 1:594–595
  definition, 3:790
  effects
    during cultivation phase
      crop/weed residue management, 5:279
      nutrient export during harvesting, 5:279
      pest and disease suppression, 5:278–279
      prolonged cropping phases, 5:279–280
      tillage, 5:277
      weeding, 5:277–278
    during fallow phase, 5:280–282
  and field establishment, 5:273–274
  geographic distribution, 5:272
  history of, 5:271–272
  labor requirements, 5:272
  reasons for, 5:270
  scale dependence of effects
    temporal and spatial scales, 5:282–283
    topographic preferences and fragmentation, 5:282
  typical fallow/crop sequences in, 5:272–273
Symbiogenesis, 2:626
  definition, 2:623
Symbiosis/symbionts, 4:464; *see also* Endosymbiosis
  in adaptive radiation, 1:28
  bacterial, 4:208–209
    evolutionary origin, 1:331–332
    identification, 1:331
    phylogenetic affiliation, 1:332–333
  bacterial–invertebrate, of hydrothermal vents, 5:750
  butterfly–ant, 1:566–567
  definition, 4:281, 463; 4:601; 5:747
  eukaryotic, 4:210
  and origins of eukaryotes, 2:626–628
  and pairwise interactions, 5:459–460
  termite, 3:590–595
Symbiotic, definition, 1:915
Symmetry, definition, 1:45
Sympatric, definition, 3:778; 5:371
Sympatric speciation, 5:378
  definition, 1:25; 2:85
Sympatry, definition, 5:569
Symphyla, 4:295
Symphyta, 3:418, 419
Symplesiomorphies, 1:678, 681; 5:580
  definition, 5:569
Synapomorphies, 1:678; 5:579–580
  definition, 1:694; 4:559; 5:569
Synapsida, 3:841–842
  origins and Permian–Triassic mass extinctions, 3:845–846
Synaptonemal complex, definition, 5:53
Synergisms, definition, 1:709
Synergist, definition, 3:465
Synlophe, definition, 5:843
Synonymous mutations, definition, 2:180
Synonym (scientific name), definition, 4:389

Syntaxa, definition, 2:635
Syrphidae (hover flies, flower flies), 2:821
Systema Naturae (Linnaeus), 3:742; 5:577
Systematics; *see also* Taxonomy
  definition, 5:570
  missions of, 5:582
    information dissemination, 5:584–585
    phylogenic analysis, 5:583
    species discovery/description, 5:582–583
  predictive classifications, 5:579–582
  versus taxonomy, 5:570–572
System functions, definition, 1:966
Systems theory, and biodiversity, 2:209
  species and process modeling, 2:212
  species interaction modeling, 2:212
  stability, 2:211–212
  steady state, 2:209–211

T

Tabanidae (horseflies, deer flies), 2:821
Taiga, 5:638; *see also* Forests, boreal
  definition, 1:533
Taoism, 5:115, 289
Taphonomy, 1:396
Tardigrades, 2:202
Tardigrades (water bears), 1:920
Target taxon
  definition, 2:683
  selection of, 2:687
Tarim Basin, 1:279
Tasek Bera, Malaysia, restoration efforts, 5:811–812
Taungya, definition, 3:748
Taxol, 4:717–718
Taxon cycle, definition, 1:25
Taxonomic diversity, definition, 2:205
Taxonomic rank, definition, 5:442
Taxonomic richness, definition, 3:215
Taxonomy; *see also* Systematics
  attributes and characters, 5:591–592
  and colonial expansion, in literature, 3:742
  current practice, 5:604–605
  definition, 5:569
  descriptive, definition, 4:271
  evidential basis of, 5:574–575
  evolution, genetic relationships and natural hierarchy, 5:592–593
  folk, 1:112
  general procedures, 5:600–601

Taxonomy (*continued*)
  and genetic diversity, relationship, 4:546–547
  historical context, 5:575–577
  individual organisms, 5:591
  philosophies/methods, 5:592–593
    empirical methods, 5:594
    evolutionary systematics, 5:595–596
    nominalism, 5:594–595
    numerical taxonomy and operationalism, 5:595–596
    preevolutionary, 5:593
  revisionary, definition, 4:271
  role in biodiversity studies/ conservation, 5:585
    as authoritative basis for information, 5:587
    conceptual framework, 5:585
    epistemic privilege, 5:587
    evidential record, 5:586
    information storage/retrieval, 5:586
    international standards, 5:586
    as language of biodiversity, 5:585–586
    predictivity, 5:587
    provision of scale, 5:586
    record truthing, 5:587
    taxa inventories, 5:587
  special and general classifications, 5:592
  subspecies, semispecies, and superspecies, modern views of, 5:525–526
  subspecific, history of
    divergence theories, 5:524–525
    trinomial revolution, 5:524
    variation, 5:523–524
  versus systematics, 5:570–572
  system to classification, 5:601–604
  tasks of, 5:590–591
Taxonomy studies, role of collections, 4:272–274
Taxon sorting, definition, 1:411
Taxon/taxa
  definition, 1:916; 2:1; 3:295; 5:569
  derived, definition, 2:755
  evolution, lineage effects, 2:187
  number of, in fossil record, 1:395
  sister, definition, 1:403
  structure of, 5:450
  target, definition, 2:683
Techno-ecosystem, 2:308–310
  definition, 2:305
Tectonism, definition, 5:747

Tegmina, definition, 3:247
Teleology, definition, 1:18
Teleosts, 2:763
  class Actinopterygii, 2:763–764
  subdivision Clupeomorpha, 2:764–765
  subdivision Elopomorpha, 2:764
  subdivision Euteleostei, 2:765
    superorder Cyclosquamata, 2:767
    superorder Lampridiomorpha, 2:768
    superorder Ostariophysi, 2:765–767
    superorder Paracanthopterygii, 2:768
    superorder Polymixiomorpha, 2:768
    superorder Protacanthopterygii, 2:767
    superorder Scopelomorpha, 2:767–768
    superorder Stenopterygii, 2:767
  superorder Acanthopterygii, 2:768
    series Atherinomorpha, 2:770
    series Mugilomorpha, 2:770
    series Percomorpha, 2:770–771
      order Beryciformes, 2:771
      order Gasterosteiformes, 2:771
      order Perciformes, 2:771–776
      order Pleuronectiformes, 2:776
      order Scorpaeniformes, 2:771
      order Synbranchiformes, 2:771
      order Tetraodontiformes, 2:777
Telescoping ecosystem model, of riverine ecosystems, 5:221–222
Telson, definition, 1:916
Temnocephalida, 5:872
Temperate, definition, 1:262
Temperate region, definition, 5:607
Temperature
  Antarctica, 1:172
  Arctic, 1:235
    changes in, 1:246–247
  boreal zone, 1:534
  and C4 photosynthesis, 1:580
  and decomposition rates, 1:617
  limits for vascular plants, 3:355
  limits to life, 4:353–355
  and species richness, 1:731–732
Temporal dispersal, definition, 2:79–80
Temporal heterogeneity, definition, 4:123
Temporal migration, definition, 2:80
Temporal patterns of genetic variation, definition, 1:85

Temporary parasite, definition, 4:463
Tephritidae (fruit flies), 2:821
Tepui, definition, 1:145
Tepuis, 1:155
Term (cladogram), definition, 5:569
Termites, *see* Isoptera
Terpenes, as plant chemical defenses, 2:13
Terra firme
  Amazonian
    definition, 1:145
    nonflooded ecosystems, 1:146–150
  definition, 5:701
Terra firme forest, definition, 3:19
Terrestrial, definition, 3:811
Terrestrial animals, definition, 2:487
Terrestrial ecosystems
  biome classification, 5:637–640
  classifications, 5:637
  global biodiversity patterns, known patterns, 5:641
  latitudinal gradients, possible explanations, 5:641–642
    area, 5:643–644
    habitat heterogeneity, 5:644–645
    possible explanations, solar energy, 5:645
    productivity, 5:645–646
    species interactions, 5:645
    time, 5:642–643
  other classification systems, 5:640
Terrestrial environments, 3:563–564
Terrestrial invertebrates, definition, 2:487
Terricolous, definition, 1:249
Tertiary period
  crustacean radiations, 1:929
  definition, 5:607
  late, and South American biogeography, 5:332–333
  mammalian evolution during, 3:849–850
Tethys Sea, 1:927
Tetragnatha spiders, Hawaiian, adaptive radiation, case studies, 1:37–39
Tetrapoda, 5:762–763
  definition, 1:159
Tetrapod vertebrates, definition, 4:97
Tetrigoidea (pigmy grasshoppers, grouse locust), 3:263–264
Tettigoniidae (katydids), 3:252–253
Tettigonioidea, 3:252–255
Thalicinidae, 1:631
Thelytokous speciation, spontaneous, 5:379

Thelytoky, 3:419
    definition, 3:417
Theophrastus, 2:611; 4:713
Therapsids, definition, 2:441
Theria, 3:780
Thermally tolerant, definition, 4:350
Thermal opportunist, definition, 4:349
Thermal systems
    biodiversity of, 4:359–360
    definition, 4:349
    general characteristics and
        definitions, 4:350–351
    geothermally heated, classification
        and description, 4:351–353
Thermal vents, definition, 4:349–350;
    see also Hydrothermal vents
Thermoadaptation, Archaea,
    1:227–228
Thermocline, definition, 4:917
Thermodynamics
    first law of, definition, 2:526
    laws of, implications for human
        energy use, 2:529–530
    and living organisms, 4:441
    second law of
        definition, 2:526
        and human ecology/economy,
            2:231–232
Thermophiles
    Archaea, 1:221–222
    Bacteria, 1:335, 349
    definition, 1:219; 5:647, 648
    origins of, 5:650
        early earth, 5:650–651
        fossil record, 5:651
        molecular fossils, 5:651–562
            lateral gene transfer/genetic
                annealing model,
                5:562–653
Thermophily
    definition, 4:350
    evolution of, 4:355–359
Thiocarbamate herbicides, 4:515
Thiophantes fungicides, 4:514
Thoracopod, definition, 1:916
Thoreau, Henry David, 3:740, 743
Threatened, definition, 2:455
Three-taxon statement, definition,
    5:569
Threshold theorem of epidemiology,
    definition, 2:259
Ticks, see Acari
Tillage
    conservation, definition, 1:99
    effects on soil quality, 5:318
Timber industry, definition, 5:655

Time horizon, definition, 4:891
Titicaca, Lake, 3:640
Toadstool, definition, 3:141
Tokogeny, definition, 5:569
Tolerance, definition, 1:753
Top-down effects, definition, 3:265;
    4:857
Torsion, definition, 4:235
Total allowable catch, definition, 5:161
Total fertility rate, definition, 4:819
Tourism
    future issues
        carrying capacity, 5:678
        interpretation/education,
            5:678–679
        protected areas, 5:678
        tourism industry engagement,
            5:677–678
        wildlife tourism, 5:678
    growth in, and economic
        development, 5:668–669
    industry sectors, engagement with
        environment, 5:670
        accommodation, 5:672
        attraction operators, 5:671–672
        developers, 5:671
        industry leaders, 5:670–671
        restaurants, 5:672
        transport, 5:672
    negative impacts, 5:672–673
        harvesting, 5:673
        land clearing/development, 5:675
        perception by tourists, 5:675
        pollution/resource use, 5:675
        small-scale physical, 5:673
        spread of exotic species,
            5:673–674
        wildlife disturbance, 5:674–675
    potential benefits
        conservation management
            funding, 5:676–677
        public conservation support,
            5:676
        sustainable regional development,
            5:677
        third world tourism, 5:677
Tourist
    definition, 5:667
    desires of, 5:660–680
    perception of negative impacts,
        5:675–676
Toxicity
    acute, definition, 1:119
    chronic, definition, 1:119
    mushroom, 3:155–156
    pesticide, 3:344

Toxicology, see Ecotoxicology
Tracheae, definition, 1:199
Tracheophytes, definition, 4:451
Trade-off, definition, 1:753; 5:173
Traditional agriculture, see Agriculture,
    traditional
Traditional energy
    biodiversity impacts of, 2:539–542
    definition, 2:525
    use, scale of, 2:527
Traditional environmental knowledge
    (TEK), 1:112–113; 3:458–460,
    802; 5:116–117
    and biodiversity conservation,
        5:117–118
    definition, 1:112–113; 3:790; 5:109;
        5:681
Traditional peoples; see also
        Indigenous peoples
    definition, 5:681
Traditional societies
    definition, 5:109
    and pantheistic tradition, 5:115–116
Tragedy of the commons, definition,
    4:891
Traits
    definition, 1:547; 5:453
    derived, definition, 2:755
Transduction
    bacterial, 1:346–347
    definition, 1:339
Transect, definition, 1:741
Transfer RNA (tRNA), 3:186
Transformation
    bacterial, 1:347–348
    definition, 1:339
Transgenes
    definition, 1:547
    and plant breeding, 1:555
Transgenic crops
    definition, 2:269
    ecological risks of, 2:274–275
Transient species, 3:334
    definition, 3:329
    founder effects, 3:335–336
Transpiration, definition, 5:269
Transport host, definition, 4:463
Transposable elements
    definition, 2:85; 3:184
    discovery of, 3:184
Treaties, definition, 3:233
Tree (cladogram), definition, 5:569
Treeline, definition, 1:133–134, 232
Trees
    and forest biodiversity, 3:44–45
    and the living dead, 3:693–694

Trench, definition, 4:1
Triacylglycerol, definition, 5:495
Triassic period
  end, mass extinction of, 4:117–118
  therapsid radiation during, 3:846
Trichobothria, definition, 1:199
Tricladida, 5:870–871
Tridacna derasa (giant clam), 2:460
Tridactyloidea (false mole crickets, sandhoppers), 3:263–264
Trilobites, 1:920
Triramous, definition, 1:916
Tritomy, definition, 5:569
Trochophore, definition, 5:831
Troglobite, definition, 5:527
Troglomorphic, definition, 5:527
Troglophile, definition, 5:527
Tropane alkaloids, bioactive, 4:719
Trophallaxis, 2:202; 3:424
Trophic, definition, 5:315
Trophic cascades, 4:857; *see also* Food webs
  and indirect multispecies interactions, 5:463–465
  and predators, 4:859–860
Trophic complexity, definition, 1:99
Trophic efficiency, definition, 5:74
Trophic interaction, definition, 1:831
Trophic levels, 2:513
  definition, 2:311; 3:1–2; 4:601; 5:74
  and food chains, 5:695–696
  and food webs, 5:698–700
  and multispecies interactions, 5:462–463
  utility of, 5:696–698
Trophic status, definition, 2:650
Trophic transfer, 2:326–327
  definition, 2:321
  food chain controls, 2:329
  herbivore performance factors, 2:327–328
Trophogenic region, definition, 2:311
Tropholytic region, definition, 2:311
Tropical, definition, 1:262
Tropical ecosystems; *see also* Forests, tropical; Rainforests, tropical
  biomass, and termite abundance, 3:610
  conversion to pasture, 5:46–47
  grasslands, 5:639
  lowland savannas, South Americann, 5:708
  savannas, South American
    distribution and structure, 5:337–338

  functional aspects, 5:338–339
  soils, 5:704
  versus temperate biodiversity gradients, 3:447
  termite abundance and biomass, 3:610
  woodlands, 1:62–64
True bugs, *see* Heteroptera
Tsetse fly (*Glosina* sp.), 1:61
Tuatara, 5:147
  characteristics, 5:152
Tuberculum interglenoideum, definition, 5:755
Tubulidentata, 3:796
Tundra, 1:235, 539; 5:34–35; 5:638
  Asian, 1:284–285
  definition, 1:232
Turbidity, definition, 5:74
Turner, Frederick Jackson, 3:743
Turtles (testudines/chelonia), 5:146
  characteristics, 5:151–152
  conservation issues, 5:157
Type (organism), definition, 4:389
Typhloplanida, 5:871
Typification, definition, 5:569

U

Ultraviolet radiation
  in Antarctic, 1:183
  biogeochemical cycle effects, 5:730
  biological system effects
    on aquatic systems
      amphibians, 5:729–730
      fishes, 5:728–729
      invertebrates, 5:728
      macroalgae and seagrasses, 5:727–728
      plankton, 5:727
    on DNA, 5:726
    on terrestrial plants, 5:726
  definition, 5:723
  environmental radiation, measurement, 5:725–726
  human effects, 5:730
    eyes, 5:730
    immune system, 5:731
    photoaging, 5:731
    skin cancer, 5:731
    sunburn, 5:730–731
Umbrella species, definition, 1:966
Uncertainty, definition, 3:63
Uncultured prokaryotes, definition, 1:325
Undaria pinnatifida, 3:502–503

Underground ecosystems, *see* Subterranean ecosystems
Understory, definition, 3:748
Undulipodium, definition, 2:623
Ungulates; *see also* Grazers/Grazing; Livestock
  in African forests, 1:60
  in African grasslands, 1:64
Uniramians, 1:920
Uniramous, definition, 1:916
United Kingdom, exotic species
  invertebrates, 1:77–78
  plants, 1:76–77
  vertebrates, 1:77
United Nations
  Charter for Nature (1982), 5:489–491
  efforts to combat desertification, 2:68–70
  Environmental Programme, estimates of global desertification, 2:71
  fisheries regulation and data collection, 5:164–165, 171
  Framework Convention on Climate Change, 1:610, 625
United States, exotic species
  invertebrates, 1:75–76
  plants, 1:72
  vertebrates, 1:73–75
Univoltine, definition, 1:249
Upwelling, definition, 4:427, 917
Urban, definition, 5:733
Urban ecology
  biogeochemical determinants of biodiversity, 5:737–739
  cities in the future, 5:745
  definitions, 5:733, 734–735
  founding of cities and affected biomes, 5:735–737
  human impacts on biodiversity in, 5:741
    disturbance regime alteration, 5:744
    habitat creation, 5:742
    habitat destruction, 5:741–742
    habitat fragmentation, 5:742–743
    resource flow alteration, 5:743–744
    species composition alteration, 5:744–745
  social differentiation and human ecology, 5:739–741
Urbanization
  and biodiversity commodification, 3:796–797

in western Washington state, 3:668
Urey, Harold Clayton, 4:443
Uropygi (whip scorpions), 1:211
Ursidae (bears, pandas), 1:630
Use values, definition, 1:377
Utilitarian, definition, 1:883

V

Vaeng, Lake (Denmark), restoration of, 5:814–815
Valid (scientific name), definition, 4:389
Values, definition, 1:813
Vaquita (*Phocoena sinus*), endangered status of, 4:55
Variation, definition, 4:799
Varzea, definition, 5:327
Vàrzea, definition, 1:145
Vavilov centers, 3:796
Vector, definition, 2:109
Vegetation; *see also* Plants
  analysis methods, in island biogeography
    data processing/display methods, 3:574
      dendrograph technique, 3:576
      improved display methods, 3:576–578
      two-way table technique, 3:574, 576
    floristic checklist, 3:568–570
    sampling methods, 3:570
      quantitative methods, 3:571
        count-plot method, 3:572
        distance methods, 3:572–574
      relevè method, 3:570–571
  changes in, remote sensing, 5:132
    case studies, 5:132–136
  contracted, definition, 4:353
  CO_2 uptake, 1:613–614
  definition, 1:262
  diffused, definition, 4:353
  emergent, definition, 5:781
  impacts of domestic ruminants, 1:653–654
  and land cover patterns, remote sensing, 5:129–130
    case studies, 5:130–132
  ozone effects, 4:739–740
Vegetation megazone, definition, 2:635
Vegetation surveys, definition, 2:635
Ventral, definition, 1:915
Vermiform, definition, 5:843
Vertebrata, 5:755–756, 761
  adipose tissue in, 5:499–500
  function, 5:505–506
  measuring in wild animals, 5:503–505
  as thermal insulation, 5:508–510
biodiversity
  in California, 4:149
  in Chile, 4:151–152
  in Mediterranean basin, 4:154
  in Southwestern Australia, 4:158
canopy, 3:23–24
classification, 5:758–759
early history, 5:757–758
general characteristics, 5:756–757
in logged forests, 3:755–756
species richness in Wyoming, remote sensing in modeling of, 5:137–139
temperature limits for, 3:355
Vertical transmission, definition, 2:109
Vestigial, definition, 1:916
Viability, definition, 4:831
Vicariance
  definition, 5:767
  examples of, 5:769
Vicariance biogeography, examples of, 5:774–775
Vicariance event, definition, 5:767
Vicariant event, definition, 5:747
Vicariant speciation, *see* Dichopatric speciation
Virchow, Rudolf, 4:440
Virulence, definition, 1:753; 2:109
Viruses
  codes of nomenclature, 4:395
  as quasi-organisms, 5:309–310
Viscacaea, *see* Mistletoe
Viverridae (genets, civits), 1:630–631
Volcanism theory, of dinosaur extinction, 2:102–103
Voltage-gated sodium channels, definition, 3:465
von Baer, Karl-Ernst, 4:441
Voucher, definition, 4:523
Vulcanism, and elemental transport, 1:441
Vulnerable, definition, 2:455

W

Wächtershäuser, Günther, 4:444
Wadi, definition, 4:353
Wallace, Alfred Russell, 1:460, 461; 4:440
Warfare, environmental impacts, 3:800–801
Washington Caves, Bermuda, 5:537
Water
  decline in quality of, 3:390
  distribution in biosphere, 3:76
  human impacts on, 2:561–562
  inland, salinity of, 3:77–78
  surface, acid deposition effects, 1:9–10
Water-holding capacity, definition, 5:315
Water hyacinth (*Eichhornia crassipes*), 3:520
Water pollution, 3:386
  agricultural/industrial chemicals, of freshwater ecosystems, 3:96
    biological consequences, 3:101–102
  and aquatic ecosystem stress, 4:10–11
  definition, 2:698
  effects of livestock grazing, 5:49–50
  endangerment of freshwater invertebrate from, 2:430–431
  and extinction, 2:707–708
  and fish conservation, 2:791–792
  impacts on marine ecosystems, 4:34
  and lake and pond ecosystems, 3:643
  and marine sediments, 4:12
  cycling, 4:9–10
  nutrient
    in eutrophication, 2:650
      factors influencing impacts of, 2:652–653
    of freshwater ecosystems, 3:95–96
      biological consequences, 3:101–102
    and harmful algal blooms, 2:655, 657, 658
  as threat to freshwater fauna, 3:540–541
  threat to riverine ecosystems, 5:228
Watershed, definition, 3:531; 5:213; 5:805
Water table, definition, 5:805
Water-use efficiency (WUE)
  definition, 1:575
  enhancement in C4 photosynthesis, 1:581–582
Weak interactors, 3:622
  definition, 3:613
Weed control
  chemical, 3:341–342
  definition, 3:339
  effects on biodiversity, 3:344–345

Weed control (*continued*)
    other organisms, 3:346–347
    selection of herbicide resistance, 3:346
    wild plants, 3:346–347
  integrated approaches, 3:341
  and slash and burn agriculture, 5:270, 276
  tools and methods, 3:341
  in wildlands, 3:347–348
Weeds
  C4, 1:594–595
  characteristics, 3:340
  definition, 3:339
  edible, 2:380–381
  environmental, definition, 4:677
  impacts of, 3:340
  invasive, definition, 3:501
  nonindigenous
    in Australia, 1:78
    in South Africa, 1:79
    in U.K., 1:77
    in U.S., 1:72–73
  root parasitic, 3:325–326
  in traditional agriculture
    benefits of, 1:111
    control of, 1:114
Weed science, 3:340–341
Weevils (Curculionidae), 1:354
  definition, 1:351
Wegener, Alfred L., 1:464
Weismann, August, 4:538
Weissmann, Charles, 4:446
Welfare economics, first theorem of, definition, 1:359
Wet deposition, definition, 1:119
Wetlands/wetland ecosystems
  adaptations
    to anoxic environments, 5:795
    to low nutrient availability/high acidity, 5:796–797
    plant reproduction, 5:796–797
    to salinity and drought, 5:796
  African, 1:66
  Asian, 1:286–289
    most threatened, 1:288
  Australian, diagnostic species, 1:315–316
  classification of, 5:783–785
  continental and regional diversity, 5:785–786
  diversity of, 5:782–783
    animals
      diversity controls, 5:793–794
      major groups, 5:788–790
      controls on, 5:790–791
    plant species richness, 5:786–788
      controls on, 5:791–793
    threats to, 5:797–798
      exotic species, 5:799–800
      habitat destruction, 5:798
      hydrological alterations, 5:798–799
      landscape fragmentation, 5:799
      natural disturbances, 5:801
      pollution, 5:799
  drainage of, 3:385
  endangered ecosystems
    Asian, 2:417
    European, 2:416–417
    neotropical, 2:417
    North American, 2:417
  functional diversity of, 5:794–795
  functions and biodiversity, 5:806
  loss of, 2:702
  major regional, 5:785
  restoration, 5:801–803
    attributes of successful projects, 5:820
    case studies, 5:811
      Abernethy peat bogs, UK, 5:814
      after extensive agricultural use, 5:812–813
      Battani Bay, Thailand, 5:818
      Hornborgasjön, Sweden, 5:816–817
      Kissimmee River, Florida, 5:819–820
      Lake Vaeng, Denmark, 5:814–815
      Melaleuca problem, southern Florida, 5:816
      Norfolk Broads, U.K., 5:815–816
      Senegal River valley, 5:817–818
      stream restoration, Denmark, 5:818–819
      Tasek Bera, Malaysia, 5:811–812
      Upper Thames tributaries, UK, 5:813–814
    defining, 5:807–809
    planning and implementation, 5:809–810
    hydrology, 5:810–811
    soils, 5:811
    vegetation reestablishment, 5:811
  South American, 5:342
  threats and losses, 5:806–807, 808
Whales
  beluga (*Delphinapterus leucas*), 4:54–55
  blue (*Balaenoptera musculus*), 4:56, 59
  bowhead (*Balaena mysticetus*), 4:62
  endangered status of, 2:798; 4:52
    modern anthropogenic extinction, 4:51–52
  gray (*Eschrichtius robustus*)
    eastern North Pacific, 4:64
    North Atlantic, 4:52
    western North Pacific, endangered status of, 4:54
  humpback (*Megaptera novaeangliae*), 4:62
  predation by, 4:866
  right (*Balaena glacialis*), 4:52, 54
  sperm (*Physeter macrocephalus*), social relationships and harvesting plans, 5:302
Wheat
  genetic narrowing, 1:91
  genetic variation, 1:89
  trends in genetic diversity, 1:92
Wild Bird Conservation Act (1992), 3:238
Wilderness, definition, 2:593
Wilderness areas, 3:378
Wildlands, weed management in, 3:347–348
Wildlife, trade in, 2:710–711
  Convention on International Trade in Endangered Species, 2:471–473
Wildlife management
  conservation gains in, 1:890
  control of wildlife, 5:827
  for ecosystem and landscape conservation, 5:826–827
  endangered species and extinction risk reduction, 5:825–826
  goal of, 5:823
  harvest and sustained yield concept, 5:824–825
  techniques, 5:827
    applied, 5:827–828
    population assessment, 5:828
Wildwood, definition, 3:675
Willingness-to-pay, definition, 2:285
Wilson, E.O., 3:367, 368
Woese, Carl, 4:444
Woodlands
  African, 1:61–63

definition, 3:675; 5:34
and historic land-use patterns, 3:678–680
lichen, in boreal forests, 1:538–539
Wood pasture
definition, 3:675
and historic land-use patterns, 3:680–682
Workers
definition, 3:581
termite, evolution of, 3:595–597
World Bank, 2:298
Worldview, definition, 5:109
Worms, *see* Annelida; Nematoda; Platyhelminthes; Polychaeta

X

Xenarthra, 3:789–790
Xenobiotics, definition, 2:363
Xeric formations, South American, 5:340–341
Xeromorphic, definition, 4:329
Xiphophorus spp., component analysis, 5:777–778
Xiphosurans (horseshoe crabs), 1:920

Y

Yeasts, 3:160
biology of, 3:144
in interspecific competition studies, 1:800–801
Younger Dryas miniglacial, 1:724

Z

Zacatonal, 1:674
Zen Buddhism, 5:114–115
0/00, definition, 5:74
Zero-growth isocline, definition, 5:412
Zonation, definition, 1:262
Zonobiome, definition, 2:635
Zoological Museum of Copenhagen, 4:277
Zooplankton, 2:803
acid deposition effects, 1:12
basic adaptations, 4:584–585
definition, 3:75; 4:427, 497
freshwater ecosystems, 4:573
life history strategies, 4:585–586
marine ecosystems, 4:571, 573, 574
pelagic ecosystems, 4:500–504
Zoos
animal records keeping, 5:907–908
animal welfare in, 5:904–905
in biodiversity education, 2:772
captive breeding/reintroduction programs, 1:600
choice of animals
institutional plannning, 5:907
metamanagement of biodiversity in, 5:906–907
space limits, 5:906
conservation ethic, rise of, 5:903–904
disease concerns, 2:120
emerging trends, 5:905–906
frozen zoos, 5:910
as gene banks, 3:180
history of, 5:902–903
population management strategies, 5:908–910
role in conservation biology, 1:863
supporting diversity in nature
ark paradigm, 5:910
post-ark paradigm, 5:910–911
Zooxanthellae, definition, 5:74
Zugenruhe, definition, 4:221
Zygomycota, 3:145

ISBN 0-12-226864-4